STRUCTURE REPORTS

for 1984

Volume 51B

Part 2

Structure Reports is prepared under the guidance of a Commission of the International Union of Crystallography. The members of the Commission sometime concerned with the preparation of this volume are listed below.

STRUCTURE REPORTS

for 1984

Volume 51B (Part 2)

ORGANIC SECTION

General editor

G. Ferguson

Section editor

G. Ferguson

Published for the

INTERNATIONAL UNION OF CRYSTALLOGRAPHY

KLUWER ACADEMIC PUBLISHERS

DORDRECHT / BOSTON / LONDON

First published in 1992

ISSN 0166–7033

Printed on acid-free paper

ISBN 0–7923–1758–0

Printed in the Netherlands

TABLE OF CONTENTS

INTRODUCTION

The present volume continues the aim of Structure Reports to present critical accounts of all crystallographic structure determinations. Details of the arrangement in the volumes, symbols used etc. are given in previous volumes (e.g. 41B or 42A, pages vi–viii).

University of Guelph,
Guelph, Ontario, Canada

G. FERGUSON

26 January 1992

(μ$_2$-1,4-DIOXANE-O,O')-TRICHLOROGALLIUM
$C_4H_8Cl_3GaO_2$ $GaCl_3.C_4H_8O_2$

A. BOARDMAN, S.E. JEFFS, R.W.H. SMALL and I.J. WORRALL, 1984. Inorg. Chim. Acta, 87, L27-L28.

Monoclinic, C2/c, a = 10.85, b = 19.62, c = 11.18 Å, β = 123.2°, Z = [4]. Mo radiation, R = 0.045 for 906 reflexions.

The crystal structure consists of distorted planar $GaCl_3$ units bridged by dioxane molecules in chair conformation, to form infinite chains parallel to the c-axis (Fig. 1). The geometry about Ga is distorted trigonal bipyramidal with the following parameters: Ga-Cl(1) 2.152, Ga-Cl(2) 2.159, Ga-O 2.210 Å; Cl(1)-Ga-Cl(2) 117.6, Cl(2)-Ga-Cl(2') 124.8°. The structure is isomorphous with that of $TlBr_3$.1,4-dioxane (1).

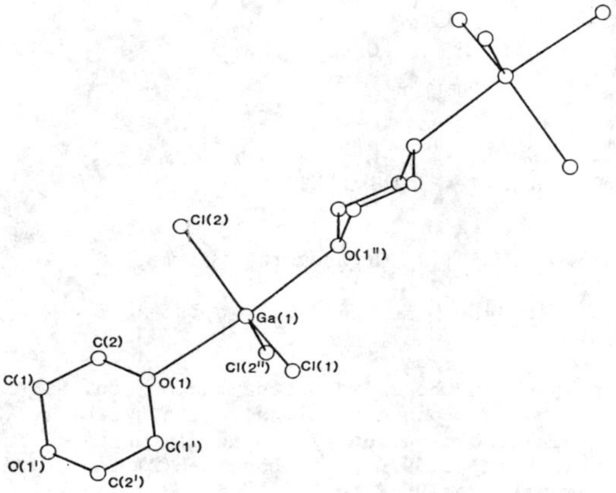

Fig. 1. Molecular structure of $C_4H_8Cl_3GaO_2$.

1. Structure Reports, 50B.

1,3,5,7-TETRAIODO-2,4,6,8-TETRAMETHYL-2,4,6,8,9,10-HEXATHIA-1,3,5,7-TETRAGALLATO-ADAMANTANE
$C_4H_{12}Ga_4I_4S_6$ $[Ga_4I_4(SCH_3)_4S_2]$

A. BOARDMAN, S.E. JEFFS, R.W.H. SMALL and I.J. WORRALL, 1984. Inorg. Chim. Acta, 83, L39-L40.

Monoclinic, $P2_1/c$, a = 10.85, b = 19.62, c = 11.18 Å, β = 94.1°, Z = 4. Mo radiation, R = 0.062 for 2322 reflexions.

The molecule has an adamantane-like cage structure (Fig. 1) consisting of four Ga-S six-membered rings, each of which has two SMe groups. The average bond distances are: Ga-I 2.516, Ga-S(C) 2.336, Ga-S 2.204 Å. Since there are two possible sites for the methyl group in the S-Me bridging group, four stereoisomers (plus one optical) are expected. The one observed here effectively has C_s symmetry.

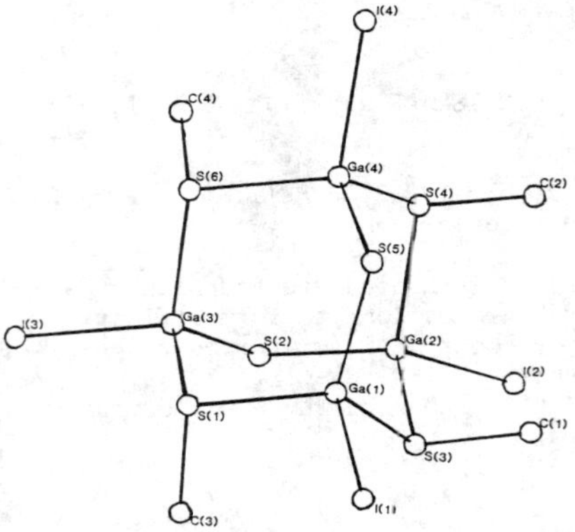

Fig. 1. Molecular structure of $C_4H_{12}Ga_4I_4S_6$.

2,2-DIMETHYL-4,4-DICHLOR-3-AZONIA-4λ^4-INDAPENTANE
$C_5H_{14}Cl_2InN$ $Cl_2(CH_3)InNH_2C(CH_3)_3$

M. VEITH and O. RECKTENWALD, 1984. J. Organometal. Chem., 264, 19-27.

Orthorhombic, $P2_12_12_1$, a = 15.446, b = 10.714, c = 6.286 Å, Z = 4. Mo radiation, R = 0.041 for 1587 reflexions.

The compound is a polymer in the solid state, the molecules being held together by N-H...Cl interactions forming a one-dimensional chain structure (Fig. 1). The In atom has fourfold tetrahedral coordination with bond distances of In-N 2.21 Å, In-C(1) 2.14 Å, In-Cl(1) 2.41 Å, In-Cl(2) 2.39 Å. The bond N-C(2) is 1.54 Å in length. The angles about the In atom range from 95.2 to 122.1°.

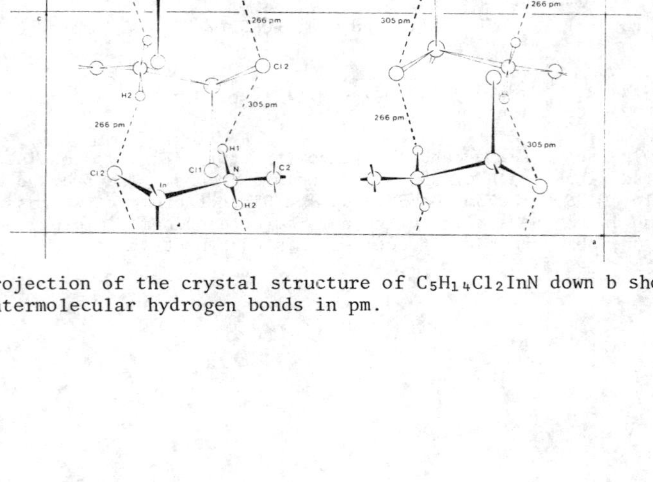

Fig. 1. Projection of the crystal structure of $C_5H_{14}Cl_2InN$ down b showing intermolecular hydrogen bonds in pm.

AMMONIUM TRIS(OXALATO)GALLATE(III) TRIHYDRATE
$C_6H_{12}GaN_3O_{12},3H_2O$ (I) $(NH_4)_3[Ga(C_2O_4)_3],3H_2O$

AMMONIUM TRIS(OXALATO)ALUMINATE(III) TRIHYDRATE
$C_6H_{12}AlN_3O_{12},3H_2O$ (II) $(NH_4)_3[Al(C_2O_4)_3],3H_2O$

N. BULC, L. GOLIČ and J. ŠIFTAR, 1984. Acta Cryst., C40, 1829-1831.

I. Triclinic, $P\bar{1}$, a = 7.901, b = 10.674, c = 10.710 Å, α = 83.14, β = 70.18, γ = 69.88°, D_m = 1.823, Z = 2. Mo radiation, R = 0.047 for 3326 reflexions.

II. Triclinic, $P\bar{1}$, a = 7.888, b = 10.667, c = 10.643 Å, α = 83.33, β = 70.44, γ = 69.10°, D_m = 1.664, Z = 2. Mo radiation, R = 0.047 for 3210 reflexions.

The Ga and Al atoms in these isomorphous compounds are coordinated by the O atoms from the bidentate oxalate ligands in a distorted octahedral arrangement with Ga...O (average) 1.97(1) and Al...O (average) 1.90(1) Å. The coordination octahedral are linked by hydrogen bonds. The oxalate groups deviate from planarity with maximum deviations of the O atoms from the least squares plane of 0.07 (I) and 0.08 Å (II). The maximum torsion angles along C-C are 7.3 (I) and 7.6° (II). The NH_4^+ ions are each surrounded by five O atoms with N...O 2.91-2.96 Å. One water molecule is disordered in the Ga compound while the Al compound has two disordered water molecules.

BIS(MORPHOLINE)ALUMINIUM TRICHLORIDE
$C_8H_{18}AlCl_3N_2O_2$ $[Al(C_4H_9NO)_2]Cl_3$

G. MÜLLER and C. KRÜGER, 1984. Acta Cryst., C40, 628-630.

Triclinic, $P\bar{1}$, a = 7.3110, b = 8.0598, c = 12.8353 Å, α = 80.583, β = 89.959, γ = 71.148°, Z = 2. Cu radiation, R = 0.059 for 2632 reflexions.

In the molecule (Fig. 1) the coordination geometry around aluminium is slightly distorted trigonal bipyramidal with chlorine in equatorial and nitrogen in axial positions. The Al-N bond lengths are 2.093(3) and 2.064(3) Å, Al-Cl 2.173(1), 2.197(1) and 2.181(1) Å. The crystal structure consists of discrete molecules, all intermolecular distances being greater than 2.8 Å.

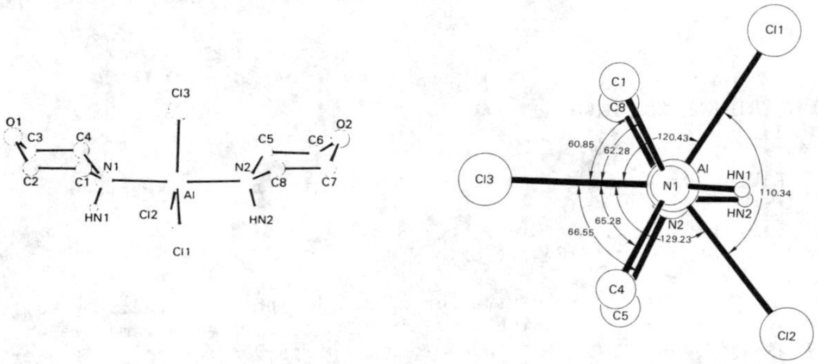

Fig. 1. $[Al(C_4H_9NO)_2]Cl_3$: perspective view of the molecule and the projection of the molecule down the N(1)-Al-N(2) axis showing selected dihedral angles.

TRIS(O-ETHYLXANTHATO)GALLIUM(III)
$C_9H_{15}GaO_3S_6$ (I) $Ga(S_2COC_2H_5)_3$

TRIS(O-ETHYLXANTHATO)INDIUM(III)
$C_9H_{15}InO_3S_6$ (II) $In(S_2COC_2H_5)_3$

B.F. HOSKINS, E.R.T. TIEKINK, R. VECCHIET and G. WINTER, 1984. Inorg. Chim. Acta, 90, 197-200.

I. Trigonal, $R\bar{3}$, a = 15.065, c = 13.421 Å, D_m = 1.63, Z = 6. Cu radiation, R = 0.044 for 668 reflexions.

II. Trigonal, $R\bar{3}$, a = 15.319, c = 13.522 Å, D_m = 1.75, Z = 6. Cu radiation, R = 0.040 for 1153 reflexions.

Crystals of compounds I and II are isomorphous and consist of enantiomorphous pairs of discrete molecules. Each molecule has a threefold axis of symmetry (Fig. 1), the S atoms forming two parallel equilateral triangles separated by 2.540 Å for I and 2.633 Å for II, with one triangle rotated by 40.27° (I), and 46.43° (II) relative to the other. The metal-sulphur distances in I and II are 2.465, 2.405 and 2.622 and 2.574 Å respectively, similarly the S-C distances 1.678, 1.704 and 1.693, 1.705 Å. The S(1)-Ga-S(2) and S(1)-In-S(2) angles are 73.50 and 69.77° respectively, and the S(1)-C(1)-S(2) angles are 118.9 and 121.9° for I and II respectively.

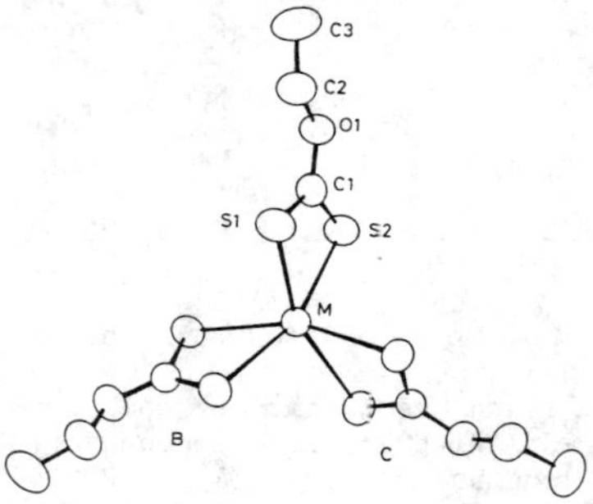

Fig. 1. Molecular structure of $C_9H_{15}MO_3S_6$ (where M = Ga, In). Ligands B and C are related by the symmetry operations $\bar{y}$,x-y,z and y-x,$\bar{x}$,z respectively.

TRIBROMOBIS(PYRIDINE)THALLIUM(III)
$C_{10}H_{10}Br_3N_2Tl$ $TlBr_3(C_5H_5N)_2$

S.E. JEFFS, R.W.H. SMALL and I.J. WORRALL, 1984. Acta Cryst., C40, 65-67.

Monoclinic, $P2_1$, a = 8.991, b = 21.59, c = 8.393 Å, β = 117.8°, D_m = 2.68, Z = 4. Mo radiation, R = 0.073 for 2184 reflexions.

The unit cell contains two crystallographically independent distorted trigonal-bipyramidal molecules (Fig. 1) which have similar geometrical parameters within the limits of experimental error. The two molecules differ in respect to the orientation of the pyridine ligands which occupy axial positions. A weak interaction between the two independent molecules accounts for the large distortions observed in the Br-Tl-Br equatorial bond angles (113.0°-131.1°). Main bond lengths are Tl-Br 2.518 to 2.557(9), Tl-N 2.41 to 2.47(2) Å.

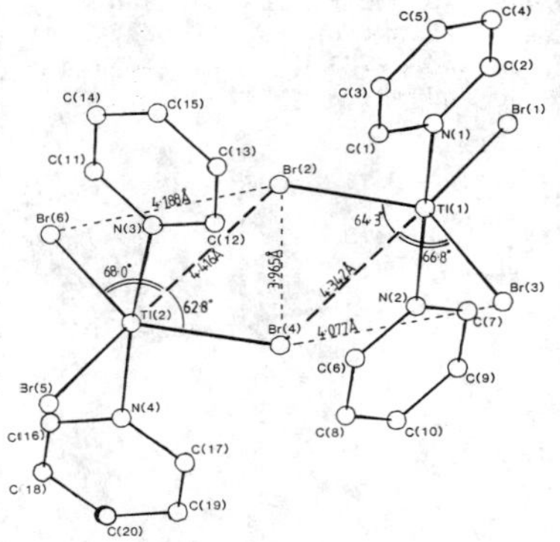

Fig. 1. The structure of $TlBr_3(py)_2$ showing intermolecular distances and angles.

DICHLORO-12-CROWN-4-ALUMINIUM TRICHLOROETHYLALUMINATE
$C_{10}H_{21}Al_2Cl_5O_4$ (I) $C_8H_{16}AlCl_2O_4^+ . C_2H_5AlCl_3^-$

DICHLORO-18-CROWN-6-ALUMINIUM TRICHLOROETHYLALUMINATE
$C_{14}H_{29}Al_2Cl_5O_6$ (II) $C_{12}H_{24}AlCl_2O_6^+ . C_2H_5AlCl_3^-$

J.L. ATWOOD, H. ELGAMAL, G.H. ROBINSON, S.G. BOTT, J.A. WEEKS and W.E. HUNTER, 1984.
J. Inclusion Phenomena, 2, 367-376.

I. Monoclinic, $P2_1/c$, a = 7.497, b = 22.121, c = 12.339 Å, β = 94.99°, Z = 4. Mo radiation, R = 0.085 for 1678 reflexions.

II. Triclinic, $P\bar{1}$, a = 8.414, b = 12.193, c = 12.394 Å, α = 73.14, β = 86.07, γ = 81.52°, Z = 2. Mo radiation, R = 0.063 for 2605 reflexions.

Views of the cations of I and II are in Fig. 1. In both cases the aluminum atom has octahedral coordination with Al-Cl 2.202(5), 2.200(8) Å (in I) and 2.148(3), 2.210(2) Å (in II), and Al-O 1.92-1.99(1) Å (in I) and 1.946(5)-2.065(4) Å (in II).

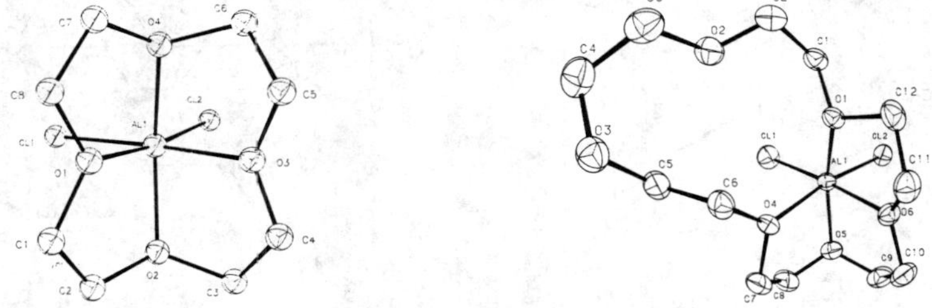

Fig. 1. Views of the cations in $C_{10}H_{21}Al_2Cl_5O_4$ (left) and $C_{14}H_{29}Al_2Cl_5O_6$ (right).

TRIIODOBIS(3-METHYLPYRIDINE-N-OXIDE)THALLIUM
$C_{12}H_{14}I_3N_2O_2Tl$ $TlI_3(MepyO)_2$

M.R. BERMEJO, A. CASTINEIRAS, M. GAYOSO, W. HILLER, U. ENGLERT and J. STRAHLE, 1984.
Z. Naturforsch., B39, 1159-1161.

Monoclinic, C2/c, a = 17.565, b = 7.807, c = 13.971 Å, β = 110.26°, Z = 4. Mo radiation, R = 0.047 for 2345 reflexions (at 190 K).

The molecule (Fig. 1) has twofold crystallographic symmetry with Tl and I1 lying on the twofold axis. The Tl coordination is slightly distorted trigonal bipyramidal, with Tl-O 2.383(3), Tl-I 2.700 and 2.703(1), O-N 1.334(4) Å, O-Tl-O' 169.1(2), Tl-O-N 117.9(2), I-Tl-I 108.30 and 125.85(1)°.

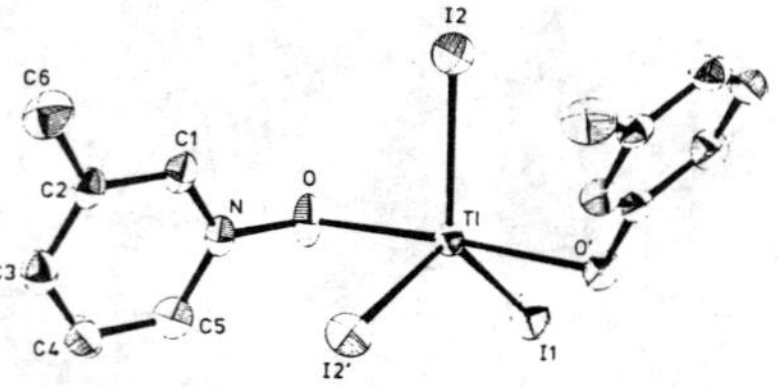

Fig. 1. A view of $TlI_3(MepyO)_2$.

(η^6-HEXAMETHYLBENZENE)GALLIUM(I) TETRACHLOROGALLATE(III)
$C_{12}H_{18}Cl_4Ga_2$ $[Me_6C_6Ga][GaCl_4]$

U. THEWALT, T. ZAFIROPOULOS and H. SCHMIDBAUR, 1984. Z. Naturforsch., B39, 1642-1646.

Rhombohedral, R3m, a = 12.449 Å, α = 91.01°, Z = 4. Mo radiation, R = 0.086 for 1057 reflexions (at -40°C).

In the structure (Fig. 1) tetrameric clusters are identified containing two crystallographically independent formula units, one having 3m and the other three having m symmetry (Fig. 1). In both types of (arene)Ga^+ cations, the Ga atoms are symmetrically positioned over the arene ring, 2.43(3) and 2.51(3) Å from the appropriate ring plane. Four (arene)Ga^+ cations and four $GaCl_4^-$ anions are arranged in such a way that Ga^I atoms are surrounded by six or seven chlorine atoms (at 3.16-3.67 Å). The four Me_6C_6 molecules form a protective cover around the inorganic $[Ga^+.GaCl_4^-]_4$ framework.

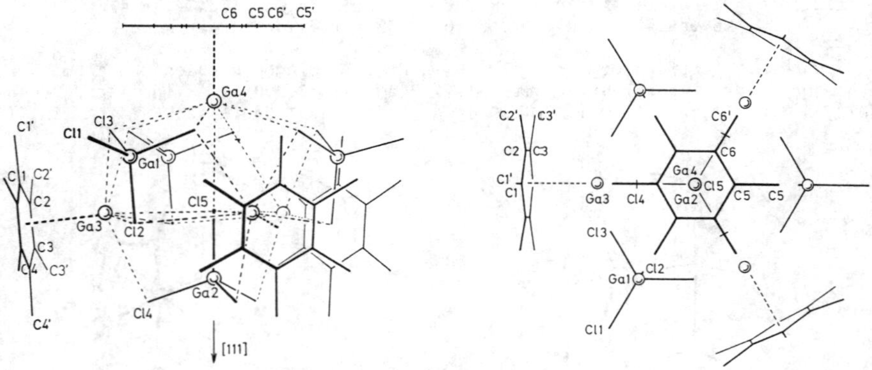

Fig. 1. Projections of the tetrameric $[C_6Me_6Ga^+GaCl_4^-]_4$ system onto the [111] axis (left) and along the [111] axis (right).

BIS(N,N,N',N'-TETRAMETHYLETHYLENEDIAMINE)TRIBROMOIODODIINDIUM
$C_{12}H_{32}Br_3IIn_2N_4$

M.A. KHAN, C. PEPPE and D.G. TUCK, 1984. Can. J. Chem., 62, 601-605.

Orthorhombic, Pbca, a = 22.795, b = 17.518, c = 12.396 Å, Z = 4. Mo radiation, R = 0.0409 for 1527 reflexions.

The structure (Fig. 1) is disordered, with each halogen site (X) occupied by 75% Br, 25% I. The molecule consists of two X_2(tmen)In units (tmen = N,N,N',N'-tetramethylethanediamine) with distorted trigonal bipyramidal geometry, joined by an In-In bond 2.775(2) Å in length.

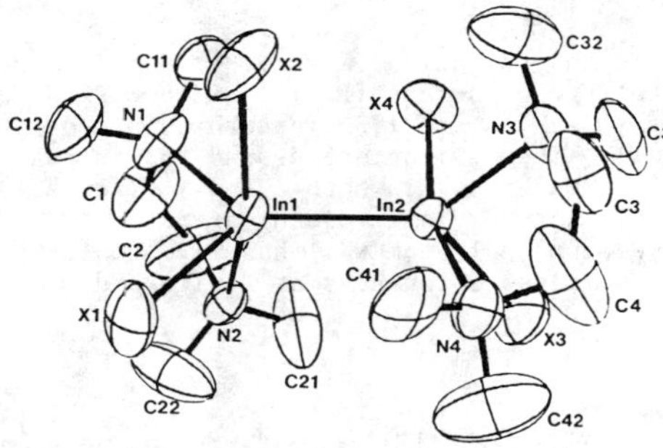

Fig. 1. The structure of $In_2Br_3I.2$(tmen) (X = 75% Br, 25% I).

DIPOTASSIUM TETRAKIS(TRIMETHYLALUMINIUM)SULPHATE

$C_{12}H_{36}Al_4K_2O_4S$ (I) $2K^+[Al_4(CH_3)_{12}SO_4]^{2-}$

DIPOTASSIUM TETRAKIS(TRIMETHYLALUMINIUM)SULPHATE p-XYLENE SOLVATE

$C_{12}H_{36}Al_4K_2O_4S,0.5C_8H_{10}$ (II) $2K^+[Al_4(CH_3)_{12}SO_4]^{2-},0.5p\text{-}(CH_3)_2C_6H_4$

R.D. ROGERS and J.L. ATWOOD, 1984. Organometallics, 3, 271-274.

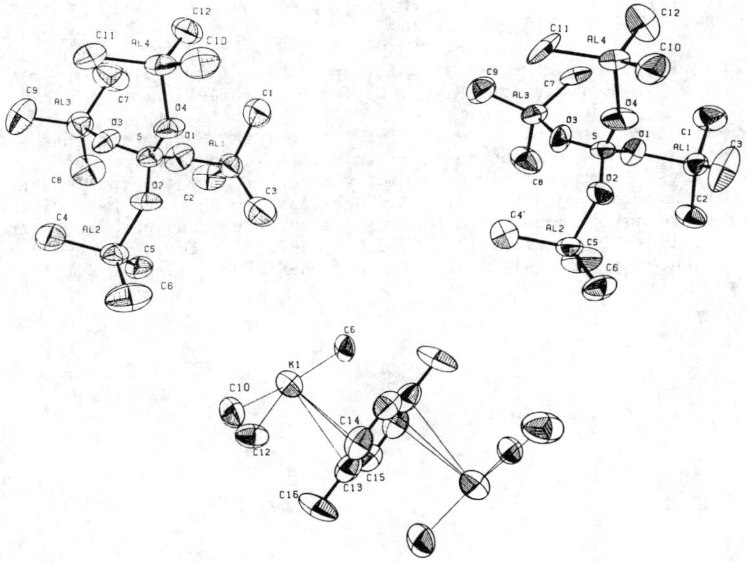

Fig. 1. Structure of the anion in $C_{12}H_{36}Al_4K_2O_4S$ (top left), and in $C_{12}H_{36}Al_4K_2O_4$-S,$0.5C_8H_{10}$ (top right), and packing of the p-xylene molecule in compound II (bottom).

I. Monoclinic, $P2_1/c$, a = 10.223, b = 20.225, c = 14.039 Å, β = 112.20°, Z = 4. Mo radiation, R = 0.095 for 1576 reflexions.

II. Monoclinic, $P2_1/c$, a = 9.773, b = 15.497, c = 20.442 Å, β = 92.51°, Z = 4. Mo radiation, R = 0.055 for 1300 reflexions.

The major difference in the two anions involves a rotation about the O(1)-Al(1) bond (Fig. 1). The Al-O bond lengths range from 1.88 to 1.92 Å for I and from 1.886 to 1.977 Å for II. The O-S-O angles are 109° in both compounds and the S-O lengths 1.47 Å in I and 1.45 Å in II. The environment around the sulphate ion appears crowded and the S-O-Al angles of 142° (average) in both I and II (range 137 to 156° and 132 to 150°) are quite large. In II the p-xylene molecules "bridge" or "insulate" K(1) and its symmetry related atom whilst in I atom K(1) has five contacts with methyl C atoms out to 3.7 Å and K(2) has two close O contacts at 2.91 Å and 3.01 Å.

[DIMETHYLBIS(1-PYRAZOLYL)GALLATO-N^2,$N^{2'}$](η^4-1,5-CYCLOOCTADIENE)RHODIUM(I)
$C_{16}H_{24}GaN_4Rh$ (I)

BIS(μ_2-3,5-DIMETHYLPYRAZOLYL-N,N')BIS[DICARBONYLRHODIUM(I)]
$C_{14}H_{14}N_4O_4Rh_2$ (II)

BIS(μ_2-3,5-DIMETHYLPYRAZOLYL-N,N')BIS[(η^4-1,5-CYCLOOCTADIENE)RHODIUM(I)]
$C_{26}H_{38}N_4Rh_2$ (III)

B.M. LOUIE, S.J. RETTIG, A. STORR and J. TROTTER, 1984. Can. J. Chem., 62, 1057-1067.

I. Orthorhombic, Pnma, a = 22.7447, b = 11.9776, c = 6.4635 Å, Z = 4. Mo radiation, R = 0.026 for 1824 reflexions.

II. Monoclinic, $P2_1/c$, a = 8.7403, b = 16.3594, c = 25.434 Å, β = 95.286°, Z = 8. Mo radiation, R = 0.024 for 5135 reflexions.

III. Triclinic, $P\bar{1}$, a = 10.350, b = 11.401, c = 12.355 Å, α = 112.485, β = 99.394, γ = 105.387°, Z = 2. Mo radiation, R = 0.037 for 6109 reflexions.

Views of the three complexes are in Fig. 1. In I the GaN_4Rh ring has a boat conformation and in II and III the Rh_2N_4 rings also have boat conformations. Molecule I has crystallographic mirror symmetry and there are two independent molecules in the asymmetric unit of II. In each case the Rh atoms have distorted square planar coordination geometry. Important mean structural parameters are Rh-N 2.087(2), Rh-C 2.129(2), Ga-N 1.991(2), Rh...Ga 3.5527(5) Å for (I); Rh-N 2.063(6), Rh-CO 1.852(9), Rh...Rh 3.262(4) Å for (II); and Rh-N 2.084(1), Rh-C 2.131(7), Rh...Rh 3.1538(3) Å for (III).

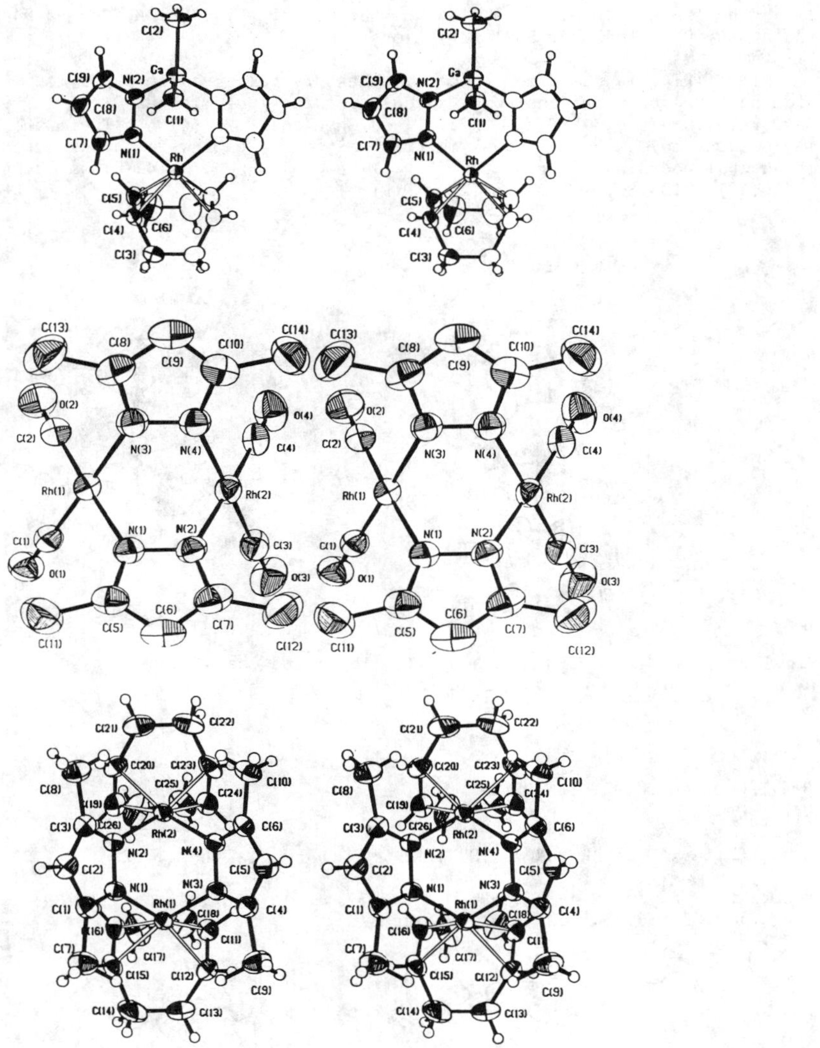

Fig. 1. Stereoviews of the pyrazolyl-rhodium complexes I (top), II (centre) and III (bottom).

BIS(μ_2-PYRIDINE-2-METHANOLATODIMETHYLGALLIUM)
$C_{16}H_{24}Ga_2N_2O_2$ (I)

(DIMETHYL(3,5-DIMETHYL-1-PYRAZOLYL)(2-PYRIDYLMETHOXY)GALLIUM)TRICARBONYLRHENIUM
$C_{16}H_{19}GaN_3O_4Re$ (II)

S.J. RETTIG, A. STORR, J. TROTTER and K. UHRICH, 1984. Can. J. Chem., 62, 2783-2792.

I. Monoclinic, C2/m, a = 16.716, b = 7.6513, c = 7.7591 Å, β = 108.349°, Z = 2. Mo radiation, R = 0.029 for 809 reflexions.

II. Monoclinic, $P2_1/n$, a = 9.7457, b = 36.991, c = 10.8821 Å, β = 92.944°, Z = 8. Mo radiation, R = 0.042 for 5321 reflexions.

In I (Fig. 1) the dimeric molecule has crystallographic C_{2h} symmetry and the Ga atom has irregular trigonal-bipyramidal coordination, with Ga-O(eq) 1.939(3), Ga-O(ax) 2.086(3), Ga-N(ax) 2.276(3) and Ga-C(eq) 1.981(4) Å. There are two independent molecules in the asymmetric unit of II each with the expected fac octahedral coordination geometry (Fig. 1) and with Re-O 2.142(6) and 2.143(8), Re-N 2.166(10)-2.194(8), Re-CO 1.89(2)-1.919(11) Å.

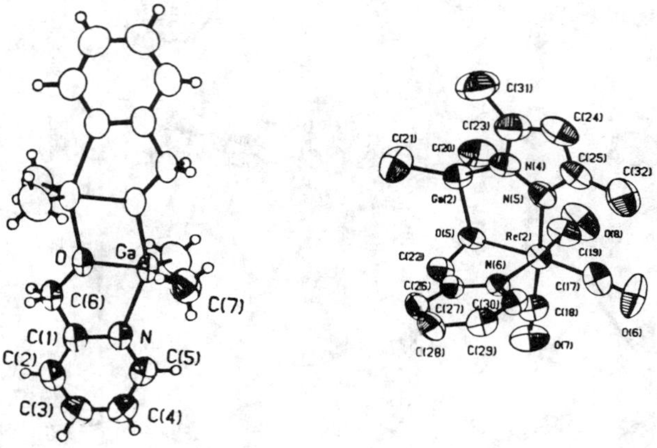

Fig. 1. Views of $C_{16}H_{24}Ga_2N_2O_2$ (I) (left) and one of the $C_{16}H_{19}GaN_3O_4Re$ molecules (II) (right).

BIS(TROPOLONATODIMETHYLGALLIUM)
$C_{18}H_{22}Ga_2O_4$

S.J. RETTIG, A. STORR and J. TROTTER, 1984. Can. J. Chem., 62, 1705-1708.

Monoclinic, $P2_1/n$, a = 10.2669, b = 9.3237, c = 10.8671 Å, β = 112.896°, Z = 2. Mo radiation, R = 0.034 for 1746 reflexions.

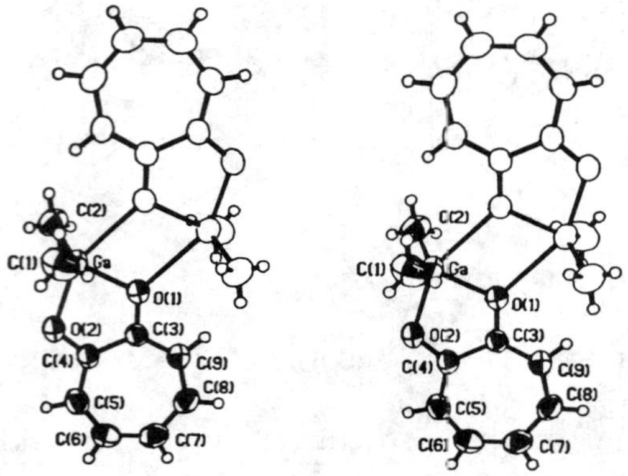

Fig. 1. A stereoview of $C_{18}H_{22}Ga_2O_4$.

The centrosymmetric binuclear dimer (Fig. 1) features five-coordinate gallium atoms having irregular trigonal bipyramidal coordination geometry and a planar four-membered Ga_2O_2 ring. Important bond lengths (corrected for libration) are Ga-O(equatorial) 1.972(2), Ga-O(axial) 2.025(2) and 2.551(2) Å, and Ga-C 1.951(4) and 1.954(4) Å.

GALLIUM(I) BIS(1,3,5-TRIMETHYLBENZENE) TETRACHLOROGALLATE(III)
$C_{18}H_{24}Cl_4Ga_2$ $2(C_9H_{12}).Ga^+.GaCl_4^-$

H. SCHMIDBAUR, U. THEWALT and T. ZAFIROPOULOS, 1984. Chem. Ber., 117, 3381-3387.

Monoclinic, Cc, a = 17.713, b = 10.856, c = 12.959 Å, β = 110.21°, Z = 4. Mo radiation, R = 0.036 for 1790 reflexions.

In the crystal (Fig. 1) the Ga(I) cation is situated above the centres of the two mesitylene rings (which are mutually inclined at 39.7°). The metal-ring centre distance is 2.673 Å. The $GaCl_4^-$ anions are linked via one chlorine atom each to the Ga(I) atoms of two cation units to form a chain-like polymer. Ga(1)-Cl distances are 2.156-2.178(3) Å. Ga(2)-Cl distances are longer (3.276 and 3.250(2) Å).

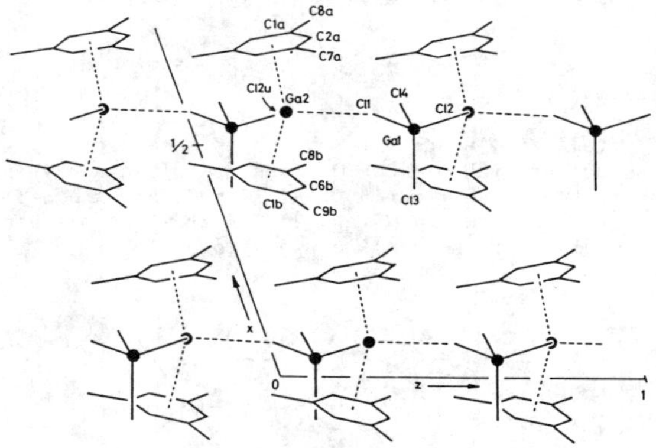

Fig. 1. The gallium coordination in and crystal structure of $C_{18}H_{24}Cl_4Ga_2$.

mer-TRICHLOROTRIS(PYRIDINE)INDIUM(III) PYRIDINE(1/1)
$C_{20}H_{20}Cl_3InN_4$ (I) $InCl_3(C_5H_5N)_3.C_5H_5N$

mer-TRICHLOROTRIS(PYRIDINE)THALLIUM(III) PYRIDINE(1/1)
$C_{20}H_{20}Cl_3N_4Tl$ (II) $TlCl_3(C_5H_5N)_3.C_5H_5N$

S.E. JEFFS, R.W.H. SMALL and I.J. WORRALL, 1984. Acta Cryst., C40, 1329-1331.

I. Orthorhombic, $C222_1$, a = 9.117, b = 16.83, c = 14.66, D_m = 1.59, Z = 4. Mo radiation, R = 0.050 for 1260 reflexions.

II. Orthorhombic, $C222_1$, a = 9.040, b = 16.88, c = 14.75, D_m = 1.94, Z = 4. Mo radiation, R = 0.061 for 1379 reflexions.

The crystal structures are isomorphous and contain neutral, octahedral $MCl_3(py)_3$ molecules with mer (trans) geometry and C_2 symmetry together with pyridine molecules of crystallisation (Fig. 1). The interplanar angle of the trans pyridines is 90(3)° and these planes are at 45(3)° to the M-Cl and M-N directions in the intermediate plane. Some mean bond lengths are In-Cl 2.474(4), In-N 2.33(2), Tl-Cl 2.505(5), Tl-N 2.41(1) Å.

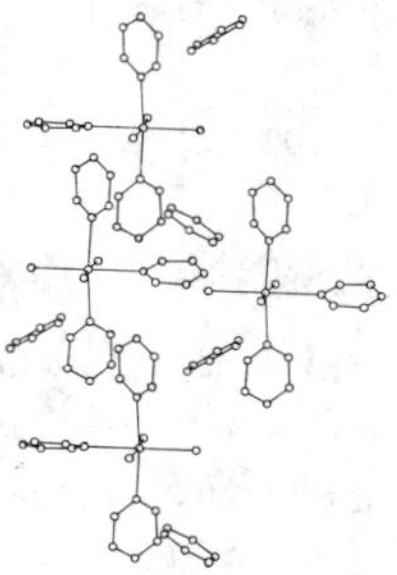

Fig. 1. Part of the crystal structure of In(or Tl)$Cl_3(C_5H_5N)_3.C_5H_5N$.

DICESIUM 18-CROWN-6 TRIS(TRIMETHYLALUMINIUM)SULPHATE

$C_{21}H_{51}Cs_2Al_3O_{10}S$ $[Cs_2.18\text{-crown-6}][Al_3Me_9SO_4]$

C.M. MEANS, N.C. MEANS, S.G. BOTT and J.L. ATWOOD, 1984. J. Am. Chem. Soc., 106, 7627-7628.

Monoclinic, $P2_1/n$, a = 13.552, b = 16.731, c = 17.029 Å, β = 94.79°, Z = 4. R = 0.064 for 2311 reflexions.

The compound crystallizes in infinite chains as shown in Fig. 1. Each Cs atom is 9-coordinate (interacting with 6 crown O atoms, 2 sulphate oxygens and the other Cs atom). The Cs...Cs contact is short at 3.92 Å. The Cs atoms are found asymmetrically disposed with respect to the crown, 1.79 and 2.37 Å from the plane of the O atoms. There is possibly some disorder in the 18-crown-6 molecule.

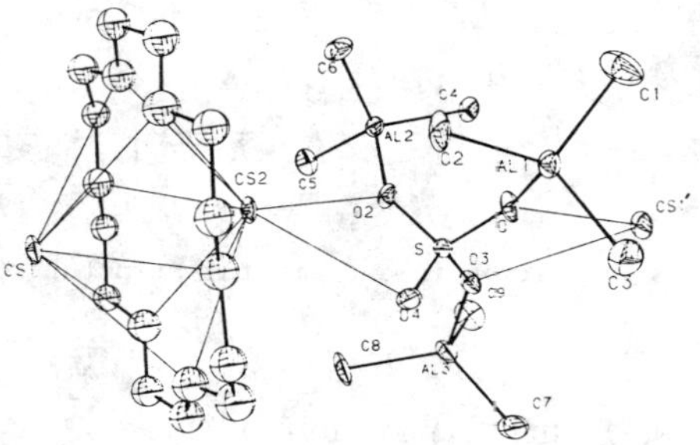

Fig. 1. View of the $[Cs_2.18\text{-crown-6}][Al_3Me_9SO_4]$ in the crystal lattice.

1,1-BIS(ACETYLACETONATO)-BIS(μ-TRIMETHYLSILOXY)-2,2-BIS(TRIMETHYLSILOXY)-DIALUMINIUM(III)

$C_{22}H_{50}Al_2O_8Si_4$ (I) $Al_2(C_5H_7O_2)_2(C_3H_9OSi)_4$

ACETYLACETONATOBIS(TRIPHENYLSILOXY)ALUMINIUM(III)

$C_{41}H_{37}AlO_4Si_2$ (II) $Al(C_5H_7O_2)(C_{18}H_{15}OSi)_2$

1,1,2-TRIS(ACETYLACETONATO)-1,2;1,2;2,3;2,3-TETRA-μ-ISOPROPOXY-3,3-DIISOPROPOXYTRI-ALUMINIUM(III)

$C_{33}H_{63}Al_3O_{12}$ (III) $Al_3(C_3H_7O)_6(C_5H_7O_2)_3$

M.F. GARBAUSKAS, J.H. WENGROVIUS, R.C. GOING and J.S. KASPER, 1984. Acta Cryst., C40, 1536-1540.

I. Trigonal, $P3_121$, a = 11.170, c = 25.87 Å, Z = 3. Mo radiation, R = 0.062 for 1320 reflexions.

II. Monoclinic, $P2_1/c$, a = 16.05, b = 8.961, c = 26.42 Å, β = 104.78°, Z = 4. Mo radiation, R = 0.06 for 2934 reflexions.

III. Monoclinic, $P2_1$, a = 9.77, b = 24.81, c = 9.98 Å, β = 115.7°, Z = 2. Mo radiation, R = 0.074 for 2335 reflexions.

In I (Fig. 1) the tetrahedrally and octahedrally coordinated Al atoms lie on the same twofold axis, in II (Fig. 1) the Al atom is tetrahedral and in III (Fig. 1) there is one tetrahedrally and two octahedrally coordinated Al atoms. The Al-O distances are 1.872-1.980 Å in bridging groups bound to octahedral Al; 1.766-1.794 Å in bridging groups bound to tetrahedral Al and 1.662-1.716 Å in terminal alkoxy and siloxy groups in tetrahedral positions.

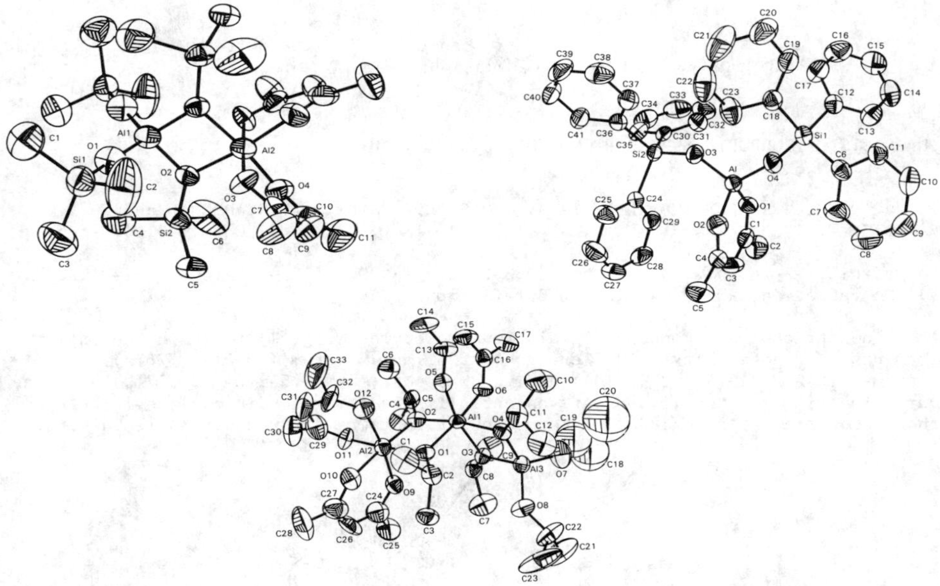

Fig. 1. Views of $Al_2(C_5H_7O_2)_2(C_3H_9OSi)_4$ (upper left), $Al(C_5H_7O_2)(C_{18}H_{15}OSi)_2$ (upper right), and $Al_3(C_3H_7O)_6(C_5H_7O_2)_3$ (lower).

TRIS(μ_2-CARBONYL)BIS(METHYLTRIS(1-PYRAZOLYL)GALLATO)DIRHODIUM BENZENE SOLVATE
$C_{23}H_{24}Ga_2N_{12}O_3Rh, C_6H_6$

B.M. LOUIE, S.J. RETTIG, A. STORR and J. TROTTER, 1984. Can. J. Chem., 62, 633-637.

Monoclinic, $P2_1$, a = 11.660, b = 15.2917, c = 11.750 Å, β = 119.09°, Z = 2. Mo radiation, R = 0.033 for 2604 reflexions.

The molecule (Fig. 1) contains a Rh-Rh single bond of 2.5818(8) Å bridged by three CO ligands, the overall molecular symmetry being close to D_{3h}. Other important mean structural parameters are: Rh-C 2.007(10), Rh-N 2.181(14) Å, Rh-C-Rh 80.1(5), and Rh-C-O 140.0(7)°.

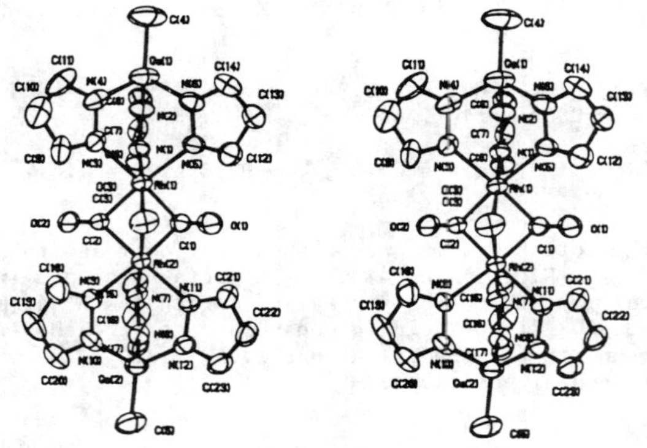

Fig. 1. A stereoview of $[MeGa(N_2C_3H_3)_3]_2Rh_2(CO)_3$.

CHLOROBIS(2-ISOPROPYL-8-MERCAPTOQUINOLINATO-N,S)INDIUM
$C_{24}H_{24}ClInN_2S_2$

L. Ya. PECH, V.S. FUNDAMENSKII, Ya.K. OZOLS, A.P. STURIS, V.N. PURMAL and Yu.A. BANKOVSKII, 1983. Koord. Khim., 9, 1126-1130 [Sov. J. Coord. Chem., 9, 639-643].

Monoclinic, $P2_1/n$, a = 14.713, b = 10.801, c = 15.295 Å, β = 104.66°, D_m = 1.58, Z = 4. Mo radiation, R = 0.042 for 2810 reflexions.

The structure is composed of discrete molecules (Fig. 1) in which the indium atom has a trigonal bipyramidal environment of one Cl atom (In-Cl 2.376(2)), two S atoms (In-S 2.419(5), 2.422(4)) and two N atoms (In-N 2.352(6), 2.377(5) Å). The quinoline rings of the ligands are essentially planar. The dihedral angle between their mean planes is 114.80°. Intermolecular contacts are van der Waals in nature.

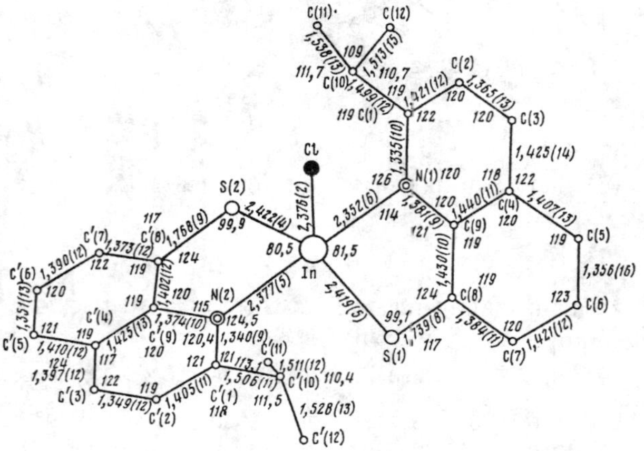

Fig. 1. $C_{24}H_{24}ClInN_2S_2$: the molecular geometry.

DIPHENYLETHYLTHIOGALLANE
$C_{28}H_{30}Ga_2S_2$ $[C_2H_5SGa(C_6H_5)_2]_2$

G.G. HOFFMANN and C. BURSCHKA, 1984. J. Organometal. Chem., 267, 229-236.

Monoclinic, $P2_1/c$, a = 10.873, b = 14.094, c = 8.556 Å, β = 97.39°, D_m = 1.44, Z = 2. Mo radiation, R = 0.031 for 1593 reflexions.

The molecule (Fig. 1) consists of a centrosymmetric cyclic dimer with a planar, four-membered Ga_2S_2 ring. The coordination about the Ga atoms is distorted tetrahedral, with angles of C(1)-Ga-C(7) 121.2, S-Ga-C(1) 108.1, S-Ga-C(7) 109.0 and S-Ga-S' 93.58°. Other angles are: Ga-S-Ga' 86.42, Ga-S-C(13) 106.1, and Ga'-S-C(13) 102.4°. Important distances are: Ga-S 2.373, Ga-S' 2.384, Ga-C(1) 1.964, Ga-C(7) 1.964, S-C(13) 1.838 and C(13)-C(14) 1.485 Å.

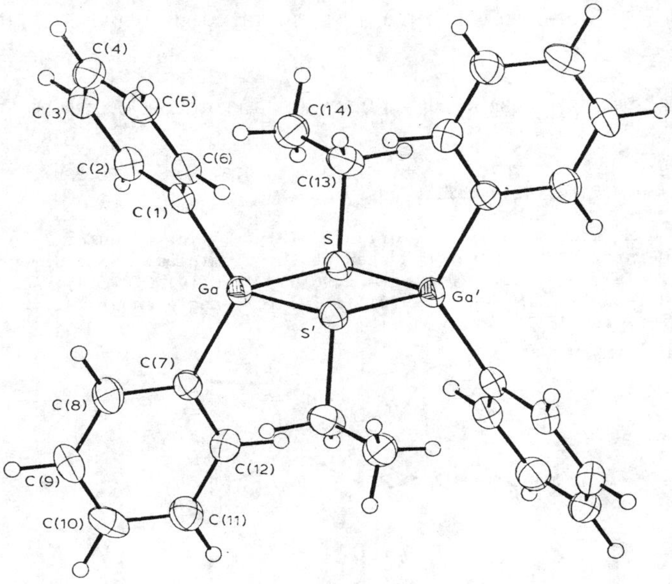

Fig. 1. Molecular structure of $C_{28}H_{30}Ga_2S_2$.

1,1,1-TRIS((5-METHOXYSALICYLALDIMINO)METHYL)ETHANE-N,O-GALLIUM
$C_{29}H_{30}GaN_3O_6$ Ga[(5-MeOsal)$_3$tame]

M.A. GREEN, M.J. WELCH and J.C. HUFFMAN, 1984. J. Am. Chem. Soc., 106, 3689-3691.

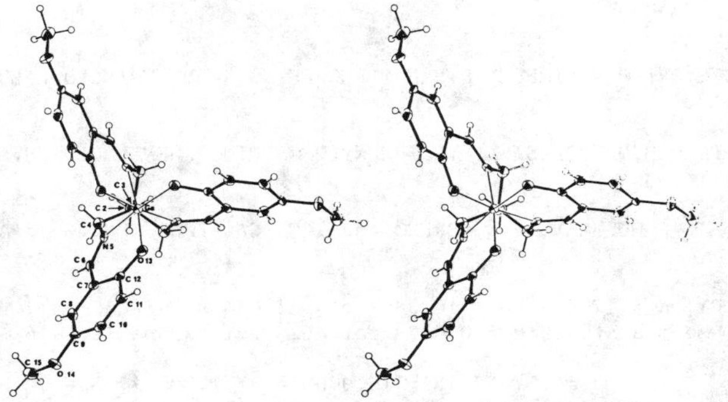

Fig. 1. A stereoview of Ga[(5-MeOsal)$_3$tame] viewed along the threefold axis.

Rhombohedral, R3c, a = 15.574, c = 17.584 Å, Z = 6. R = 0.0178 for 473 reflexions (at -160°C).

The molecule (Fig. 1) has threefold crystallographic symmetry, with Ga-N(5) 2.107(5), Ga-O(13) 1.941(5) Å, O(13)-Ga-O(13)' 92.63(16), N(5)-Ga-N(5)' 82.98(20)°. The twist angle in the structure is 52.4°.

(μ_2-HYDRIDO)(μ_2-9,10-DIHYDRO-9,10-ANTHRYLENE)(DIETHYLALUMINIUM)TRIS(TETRAHYDROFURAN)-MAGNESIUM
$C_{30}H_{45}AlMgO_3$

H. LEHMKUHL, K. MEHLER, R. BENN, A. RUFINSKA, G. SCHROTH and C. KRUGER, 1984. Chem. Ber., 117, 389-403.

Monoclinic, $P2_1/n$, a = 11.707, b = 18.687, c = 14.0164 Å, β = 94.299°, Z = 4. Cu radiation, R = 0.077 for 4240 reflexions.

The Al and Mg atoms are axially oriented (Fig. 1) in the anthrylene system; the magnesium atom interacts with the hydrogen on the aluminium. Some dimensions are: Mg-O 2.048-2.106(3), Mg-C 2.195(4), Mg-H 1.96(3), Al-C(10) 2.036(4), Al-C(Et) 1.980, 1.984(5), Al-H 1.62(3) Å, Mg-H-Al 139(2)°. The fold angle about C9...C10 in the anthrylene moiety is 148.7°.

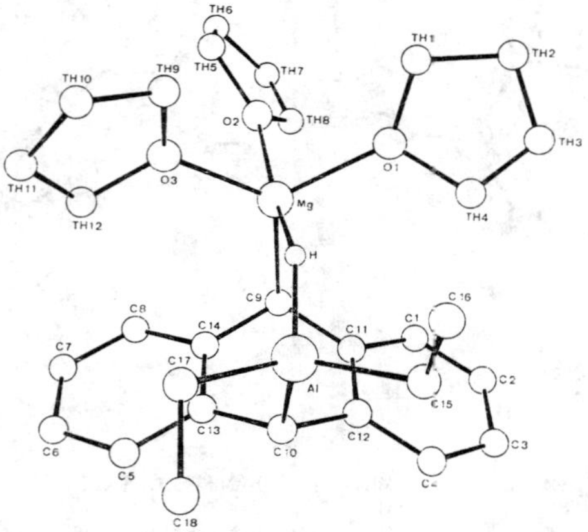

Fig. 1. A view of $C_{30}H_{45}AlMgO_3$.

TRIS(μ_2-HYDRIDO)HYDRIDOTRIS(METHYLDIPHENYLPHOSPHINO)RHENIUMDIMETHYLALUMINIUM
$C_{41}H_{49}AlP_3Re$ (I)

BIS(μ_2-HYDRIDO)TETRAHYDRIDOBIS(METHYLDIPHENYLPHOSPHINO)RHENIUMDIMETHYLALUMINIUM
$C_{28}H_{38}AlP_2Re$ (II)

W.A. SKUPINSKI, J.C. HUFFMAN, J.W. BRUNO and K.G. CAULTON, 1984. J. Am. Chem. Soc., 106, 8128-8136.

I. Triclinic, $P\bar{1}$, a = 17.815, b = 10.386, c = 11.094 Å, α = 111.47, β = 86.08, γ = 95.78°, Z = 2. Mo radiation, R = 0.0204 for 4627 reflexions (at -162°C).

II. Monoclinic, $P2_1/a$, a = 15.053, b = 15.900, c = 11.705 Å, β = 92.59°, Z = 4. Mo radiation, R = 0.0454 for 4014 reflexions (at -164°C).

Compound (I) is based on a ReH_6P_2 dodecahedron with $AlMe_2$ bound to two hydride ligands forming a η^2-H_2AlMe_2 unit (Fig. 1). There is a nearly planar ReP_2Al framework and the molecule closely approaches (non-crystallographic) C_2 symmetry, with Re-P 2.390, 2.391(2), Re-Al 2.508(2), Re-H 1.33(9)-1.70(8), Al-H 1.79(8) and 2.02(12) Å. Compound (II) (Fig. 1) has a ReH_4P_3 pentagonal bipyramid, with $AlMe_2$ attached through three hydride ligands (Re-P 2.358, 2.360, 2.382(1), Re-Al 2.501(2), Re-H 1.53(8)-1.73(5), Al-H 1.76(5)-2.10(8) Å.

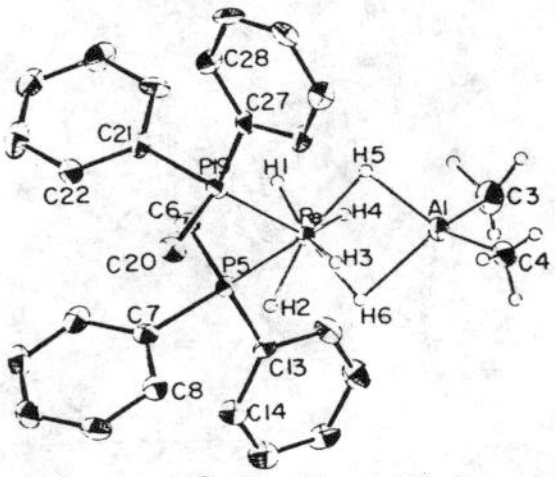

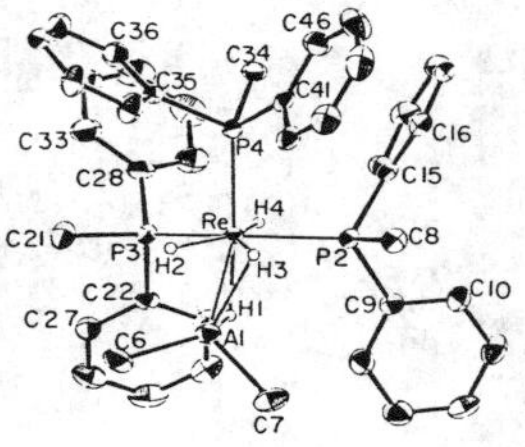

Fig. 1. Views of $C_{41}H_{49}AlP_3Re$ (I) (left) and $C_{28}H_{38}AlP_2Re$ (II) (right) showing the bridging H atoms.

2-HYDROXYETHANETHIOLATOLEAD(II) NITRATE

$C_2H_5NO_4PbS$ $Pb(SCH_2CH_2OH)(NO_3)$

W. CLEGG, I.L. ABRAHAMS and C.D. GARNER, 1984. Acta Cryst., C40, 1367-1369.

Orthorhombic, Pbca, a = 17.969, b = 11.860, c = 6.043 Å, Z = 8. Mo radiation, R = 0.038 for 935 reflexions.

In this polymeric sheet structure (Fig. 1) Pb has an irregular sevenfold coordination with nitrate and 2-hydroxyethanethiolato groups both acting as bidentate ligands and as bridges between Pb atoms. Bond length ranges are, Pb-S 2.786-3.113(2), Pb-O 2.539-2.688(7) Å.

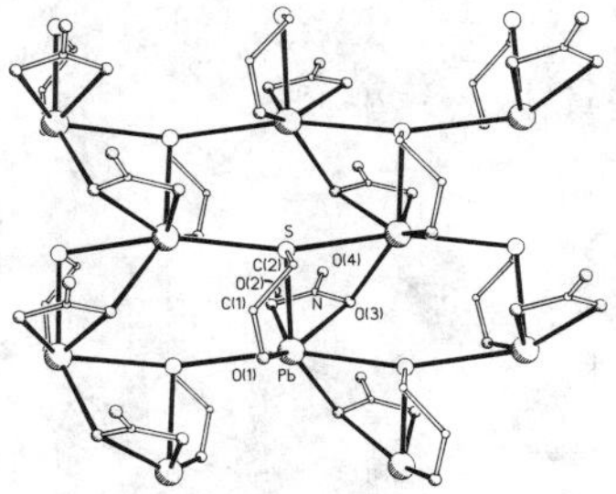

Fig. 1. Part of the polymeric $Pb(SCH_2CH_2OH)(NO_3)$ structure.

DIMETHYLTIN MOLYBDENUM TETRAOXIDE

$C_2H_6MoO_4Sn$ $(CH_3)_2SnMoO_4$

Y. SASAKI, H. IMOTO and O. NAGANO, 1984. Bull. Chem. Soc. Japan, 57, 1417-1418.

Tetragonal, $P4_2/mbc$, a = 13.705, c = 7.271 Å, Z = 8. Mo radiation, R = 0.031 for 879 reflexions.

The Sn atom has trans-octahedral coordination and the Mo atom is tetrahedrally-coordinated by four oxygen atoms (Fig. 1). All O atoms bridge the Sn and Mo atoms and the structure extends three-dimensionally in the form of zig zag chains of $(\text{-Sn-O-Mo-O-})_\infty$ along the a axis at Z = 0, and identical chains along the b axis at Z = 0.5. The structure consists of octahedra and tetrahedra in the ratio 1:1 and these polyhedra share only apices.

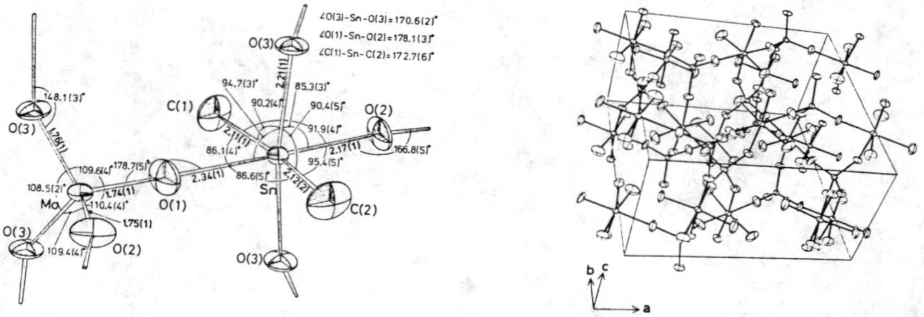

Fig. 1. Molecular structure of $C_2H_6MoO_4Sn$ showing bond lengths (Å) and angles (°), and a view of the unit cell.

3,3-BIS(TRIBROMOGERMYL)PROPIONIC ACID
$C_3H_4Br_6Ge_2O_2$

S.N. GURKOVA, A.I. GUSEV, N.V. ALEKSEEV, T.K. GAR and N.A. VIKTOROV, 1984. J. Struct. Chem., 25, 504-505 [Zh. Strukt. Khim., 25(3), 180-181].

Monoclinic, $P2_1/c$, a = 7.604, b = 20.574, c = 9.876 Å, β = 106.55°, Z = 4. Mo radiation, R = 0.097 for 830 reflexions.

The molecule is shown in Fig. 1. The coordination about the Ge atoms is slightly distorted tetrahedral with Ge-Br 2.249(7)-2.274(8) Å, Ge-C 1.94(4) and 1.98(4) Å, and bond angles ranging from 107.1(3) to 116(1)°.

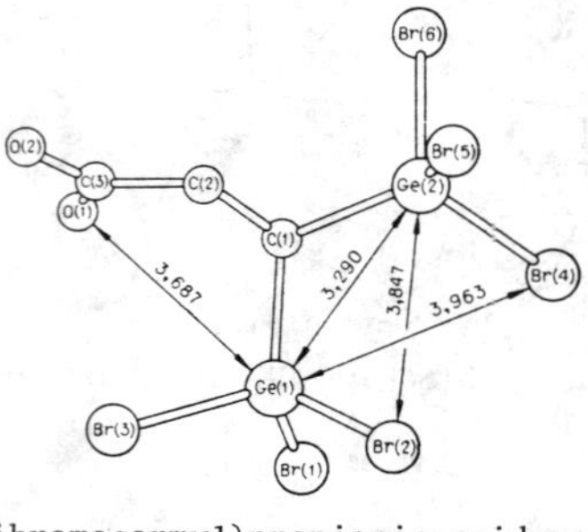

Fig. 1. The 3,3-bis(tribromogermyl)propionic acid molecule.

3-CHLOROPROPYLTRICHLOROSTANNANE
$C_3H_6SnCl_4$

V.I. SHIRYAEV, T.G. BASANINA, S.N. GURKOVA, A.I. GUSEV, G.V. DOLGUSHIN, V.P. FESHIN, V.P. ANOSOV, G.M. APAL'KOVA and V.S. NIKITIN, 1983. Koord Khim., 9, 780-786 [Sov. J. Coord. Chem., 9, 455-461].

Monoclinic, $P2_1/b$, a = 7.744, b = 12.654, c = 10.476 Å, γ = 115.42°, Z = 4. Mo radiation, R = 0.077 for 1161 reflexions.

In the molecule (Fig. 1) there is an intramolecular interaction between the Cl atom of the γ-chloropropyl group and the Sn atom (Sn...Cl(4) 3.279 Å). The conformation about Sn is distorted trigonal bipyramidal with the axial positions occupied by Cl(3) (Sn-Cl(3) 2.336(4) Å) and Cl(4). The Cl(3)-Sn-Cl(4) angle is 179°. The Sn-Cl(1) and Sn-Cl(2) distances are 2.314(4) and 2.320(4) Å respectively. The C(3)-Cl(4) distance is 1.85(2) and C(1)-C(2) and C(2)-C(3) 1.52(2) and 1.55(2) Å respectively. There is an intermolecular Sn...Cl contact (3.82 Å) involving the Cl(2) atom of a neighbouring molecule which is related to the original molecule by a centre of symmetry.

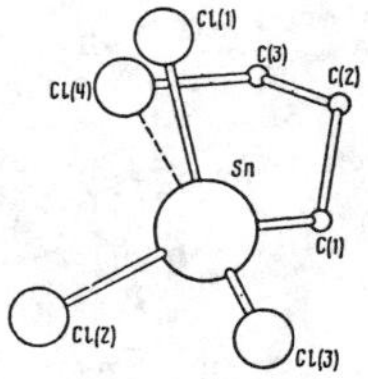

Fig. 1. $C_3H_6SnCl_4$: the molecular structure.

BIS(ACETATO)LEAD(II) TRIHYDRATE
$C_4H_{12}O_7Pb$ $Pb(O_2C_2H_3)_2.3H_2O$

R.G. BRYANT, V.P. CHACKO and M.C. ETTER, 1984. Inorg. Chem., 23, 3580-3584.

Monoclinic, C2/m, a = 15.803, b = 7.269, c = 9.049 Å, β = 109.55°, Z = 4. Mo radiation, R = 0.043 for 1124 reflexions.

The Pb atom is nine-coordinate with two bidentate acetate groups, three coordinated water molecules, and two bridging bonds to oxygens on neighboring acetate groups (Fig. 1), resulting in a polymeric structure. Intramolecular Pb-O distances range from 2.446(8) to 3.11(1) Å.

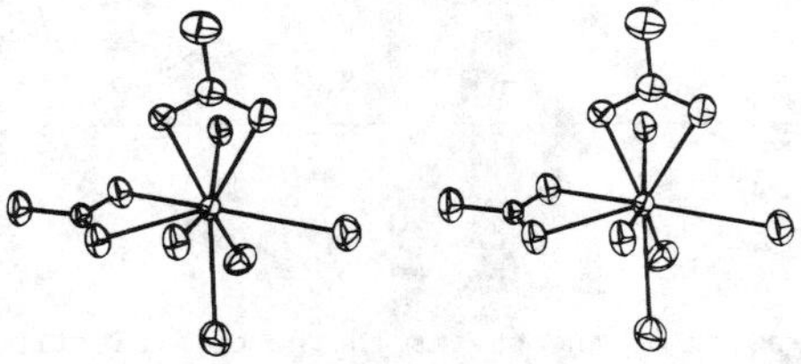

Fig. 1. Stereoview of the Pb environment in $Pb(O_2C_2H_3)_2.3H_2O$.

CHLORODIMETHYL(N,N-DIMETHYLDITHIOCARBAMATO)GERMANIUM(IV)
$C_5H_{12}ClGeNS_2$ $(CH_3)_2ClGe(S_2CN(CH_3)_2)$

R.K. CHADHA, J.E. DRAKE and A.B. SARKAR, 1984. Inorg. Chem., 23, 4769-4770.

Monoclinic, $P2_1/c$, a = 12.682, b = 7.532, c = 12.085 Å, β = 109.88°, D_m = 1.56, Z = 4. Mo radiation, R = 0.032 for 1299 reflexions.

The germanium atom is in a distorted trigonal bipyramidal environment with the Cl and one S atom of S_2CNMe_2 in axial positions (S(ax)-Ge-Cl 159.2(1)°). Bond lengths: Ge-C 1.929(2) and 1.925(2), Ge-Cl 2.251(1), Ge-S(eq) 2.254(1), and Ge-S(ax) 2.896(1) Å.

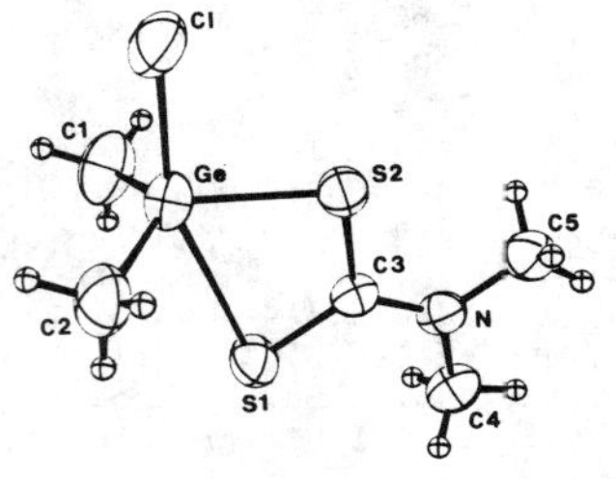

Fig. 1. The $Me_2ClGe(S_3CNMe_2)$ molecule.

(μ_2-DIMETHYLTHIOPHOSPHINATO-O,S)TRIMETHYLTIN
$C_5H_{15}OPSSn$ $(CH_3)_3SnOSP(CH_3)_2$

A.-F. SHIHADA, I.A.-A. JASSIM and F. WELLER, 1984. J. Organometal. Chem., 268, 125-133.

Monoclinic, $P2_1/n$, a = 9.453, b = 10.107, c = 11.4C2 Å, β = 102.85°, Z = 4. Mo radiation, R = 0.052 for 1697 reflexions.

The crystal structure consists of chains about the twofold screw axes of planar $Sn(CH_3)_3$ units and tetrahedral $OSP(CH_3)_2$ grcups which are linked by the O and S atoms of the thiophosphinate (Fig. 1). The coordination about the Sn atom is trigonal bipyramidal, with Sn-C distances of 2.132, 2.119 and 2.142 Å, and Sn-O and Sn-S distances of 2.267 and 2.737 Å respectively. The equatorial C-Sn-C angles are 118.4, 120.6 and 121.0° whilst the axial constituents with Sn make an O-Sn-S angle of 172.4°. The coordination angles about P range from 104.3 to 115,0°, with O-P-S being 115.0°. The P-C distances are 1.78 and 1.79 Å and the P-O and P-S bond lengths are 1.505 and 1.997 Å respectively. The P-O-Sn and P-S-Sn angles are respectively 145.0 and 104.3°.

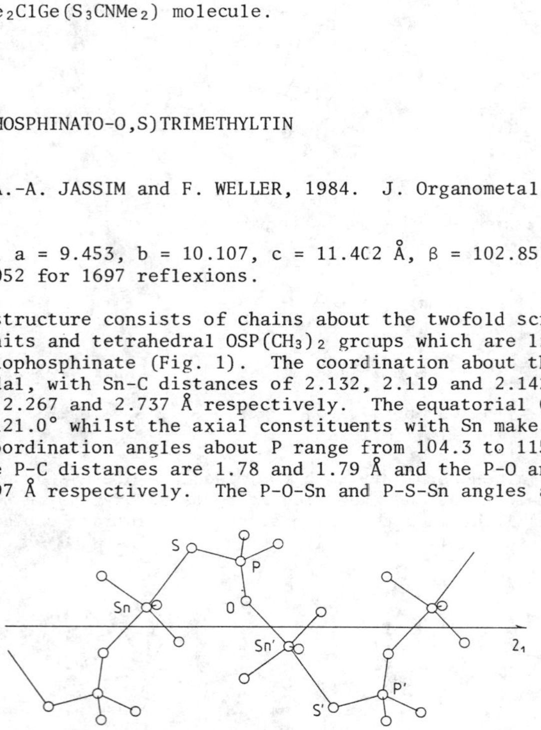

Fig. 1. Schematic drawing of the chains in the crystal structure of $C_5H_{15}OPSSn$.

LEAD SALT OF 2,4,6-TRINITRO-1,3-BENZENEDIOL MONOHYDRATE (α-LEAD STYPHNATE)
$C_6HN_3O_8Pb,H_2O$ $Pb^{2+}.C_6HN_3O_8{}^{2-},H_2O$

M.A. PIERCE-BUTLER, 1984. Acta Cryst., C40, 63-65.

Monoclinic, $P2_1/a$, a = 10.0299, b = 12.5418, c = 8.0331 Å, β = 97.782°, Z = 4. Mo radiation, R = 0.032 for 2399 reflexions.

For details of the β-polymorph see (1). The Pb atoms are seven-coordinate (Fig. 1) and are linked via oxygen bridges about an inversion centre. The average Pb-O distance is 2.580 Å (range 2.383 to 2.765 Å). The benzene ring (Fig. 1) shows considerable angle distortion and unequal C-O distances. The water molecule is coordinated to the metal and is also hydrogen bonded to the anion.

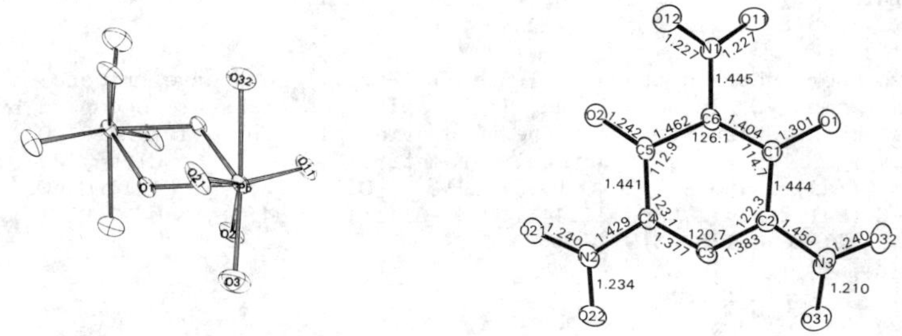

Fig. 1. The Pb coordination and a view of the styphnate ion in α-lead styphnate.

1. Structure Reports, 49B, 94.

MONOBASIC LEAD SALT OF 2,4,6-TRINITRO-1,3-BENZENEDIOL (β POLYMORPH) (β-MONOBASIC LEAD STYPHNATE)

$C_6H_3N_3O_{10}Pb_2$ $Pb_2(OH)_2^{2+}.C_6HN_3O_8^{2-}$

M.A. PIERCE-BUTLER, 1984. Acta Cryst., C40, 1364-1367.

Monoclinic, P2/c, a = 16.606, b = 16.635, c = 20.462 Å, β = 127.5°, Z = 16. Mo radiation, R = 0.076 for 4479 reflexions.

The two independent $Pb_4(OH)_4^{4+}$ units within the structure each have Pb atoms at the vertices of a tetrahedron with hydroxyl O atoms capping the faces at 1.0(4) Å from the planes of the nearest three Pb atoms. Each Pb atom is further coordinated to between one and five O atoms from the styphnate NO_2 groups (Fig. 1). Pb-O(H) distances are in the range 2.29(3) to 2.65(4) and Pb-O(NO) 2.53(4) to 2.77(6) with Pb-Pb 3.591(4) to 3.936(4) Å. All rings are planar within experimental error (maximum deviation 0.08(5) Å) with NO_2 groups rotated out of the ring plane by between 15 and 80° - the largest rotations being for NO_2 groups ortho to both OH groups. In addition in each ring one -O substituent is bent significantly out of the ring plane.

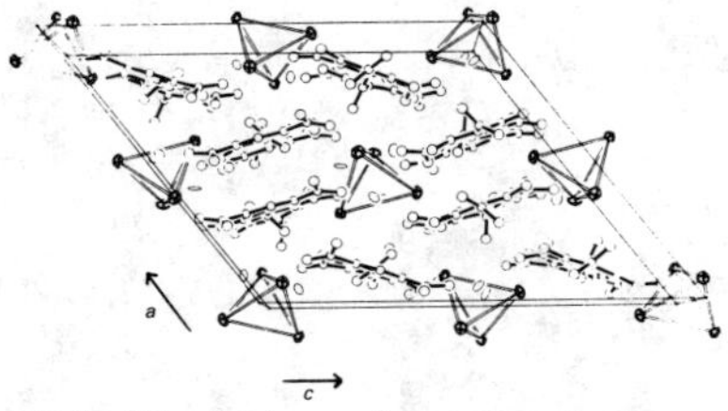

Fig. 1. The contents of half of the unit cell of β-monobasic lead styphnate.

1-(1-TRICHLOROGERMYLMETHYL)-2-PYRROLIDONE

$C_6H_{10}Cl_3GeNO$ (I)

N,N'-DIMETHYL-2-METHYL-3-(TRICHLOROGERMYL)PROPIONAMIDE

$C_6H_{12}Cl_3GeNO$ (II)

S.N. GURKOVA, A.I. GUSEV, N.V. ALEKSEEV, T.K. GAR and N.A. VIKTOROV, 1984. J. Struct. Chem., 25, 825-828 [Zh. Strukt. Khim., 25(5), 170-173].

I. Orthorhombic, $P2_12_12_1$, a = 9.061, b = 9.941, c = 12.297 Å, Z = 4. Mo radiation, R = 0.056 for 1425 reflexions.

II. Monoclinic, $P2_1/n$, a = 9.397, b = 10.984, c = 11.483 Å, β = 75.75°, Z = 4. Mo radiation, R = 0.028 for 1274 reflexions.

The five-coordinate Ge atoms in I and II (Fig. 1) both have trigonal bipyramidal coordination geometry. In both I and II the axial positions are occupied by O and Cl, and the Ge atoms are displaced from the equatorial plane by 0.19 and 0.20 Å towards Cl, respectively. Important structural parameters: (for I) Ge-Cl(eq) 2.134(3) and 2.138(4), Ge-Cl(ax) 2.253(3), Ge-C 1.967(10), Ge-O 2.140(7) Å, Cl(1)-Ge-O 178.3(2)°; (for II) Ge-Cl(eq) 2.134(1) and 2.144(1), Ge-Cl(ax) 2.264(1), Ge-C 1.931(5), Ge-O 2.123(3) Å, Cl(2)-Ge-O 175.63(9)°.

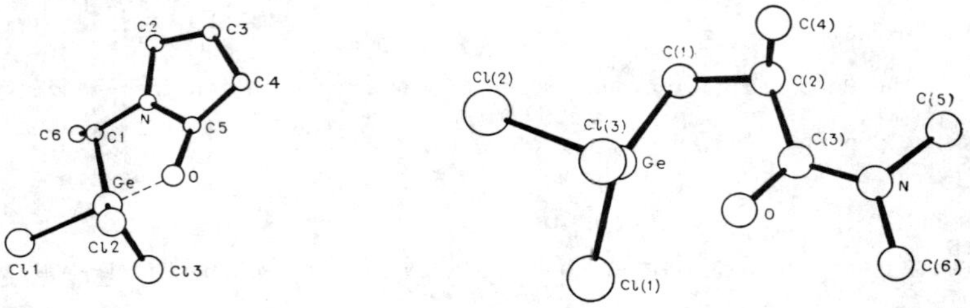

Fig. 1. The 1-(1-trichlorogermylmethyl)-2-pyrrolidone (left) and N,N'-dimethyl-2-methyl-3-(trichlorogermyl)propionamide (right) molecules.

DIACETATOBIS(THIOUREA)TIN(II)

$C_6H_{14}N_4O_4S_2Sn$ (I) $(CH_3COO)_2Sn(SC(NH_2)_2)_2$

DITIN(II)TETRABROMOPENTA(THIOUREA) DIHYDRATE

$C_5H_{20}Br_4N_{10}S_5Sn_2,2H_2O$ (II) $((NH_2)_2CS)_5Sn_2Br_4,2H_2O$

J.D. DONALDSON, S.M. GRIMES, S. CALOGERO, U. RUSSO, G. VALLE and P.J. SMITH, 1984. Inorg. Chim. Acta, 84, 173-174.

I. Monoclinic, Pc, a = 11.932, b = 10.937, c = 21.919 Å, β = 96.5°, D_m = 1.32, Z = 8. Mo radiation, R = 0.060.

II. Orthorhombic, Pnma, a = 27.83, b = 16.13, c = 6.11 Å, Z = 4. Mo radiation, R = 0.077.

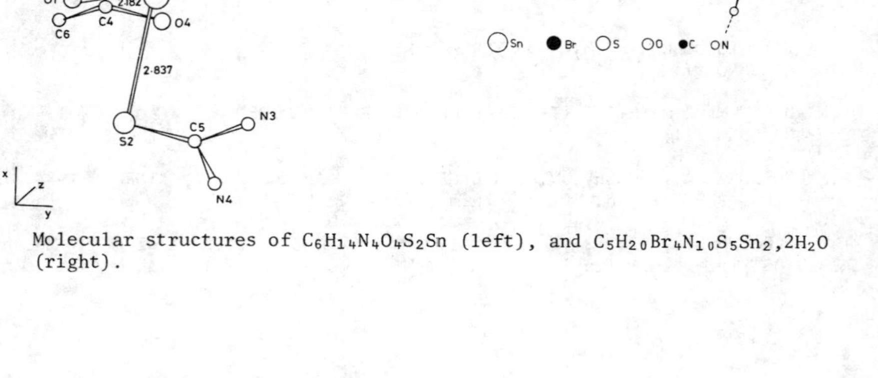

Fig. 1. Molecular structures of $C_6H_{14}N_4O_4S_2Sn$ (left), and $C_5H_{20}Br_4N_{10}S_5Sn_2,2H_2O$ (right).

In compound I the Sn atom has a square pyramidal environment (Fig. 1). It is bonded to two thiourea S atoms and to two carboxylate O atoms. Two longer Sn...O bonds (2.80, 3.09 Å) to O atoms from surrounding acetate groups, complete a distorted octahedral environment about the Sn atom. There are two Sn atoms present in compound II, both of which lie on mirror planes and have trigonal pyramidal coordination. The crystal packing of II is mainly determined by the water O atom in O-H...Br(3), O-H...S(3) hydrogen bonds and by the non-central thiourea N in N(1)-H...O and N(2)-H ...O hydrogen bonds. The O...Br, O...S and N...O distances are 3.31, 3.38, 2.92 and 3.03 Å respectively.

DIMETHYLTIN(IV) BIS(DIMETHYLDITHIOARSINATE)
$C_6H_{18}As_2S_4Sn$ $(CH_3)_2Sn[S_2As(CH_3)_2]_2$

L.S. DUMITRESCU, I. HAIDUC and J. WEISS, 1984. J. Organometal. Chem., 263, 159-165.

Orthorhombic, Ccmm, a = 6.310, b = 11.324, c = 23.160 Å, Z = 4. Mo radiation, R = 0.045 for 1121 reflexions.

The molecule has crystallographic mm symmetry and the coordination at Sn can be described as distorted tetrahedral (Fig. 1). The relevant angles are: C-Sn-S 111.8°, S-Sn-S 80.3° and C-Sn-C 121.8° with distances Sn-C 2.108 Å and Sn-S(1) 2.471 Å. The dimethyldithioarsinato ligand has distances of As-S(1) 2.171 Å, As-S(2) 2.089 Å, and As-C(2) 1.905 Å and the valency angles about As range from 104.6 to 114.0°. The angle Sn-S(1)-As is 101.7°.

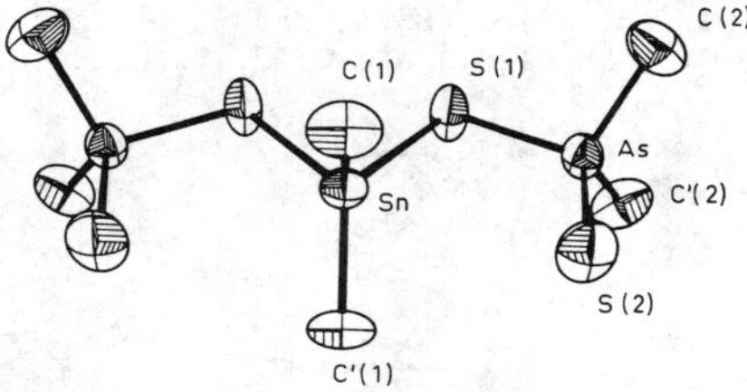

Fig. 1. Molecular structure of $C_6H_{18}As_2S_4Sn$.

DIMETHYLBIS(O-ETHYLXANTHATO)TIN(IV)
$C_8H_{16}O_2S_4Sn$ $(CH_3)_2Sn(S_2COC_2H_5)_2$

D. DAKTERNIEKS, B.F. HOSKINS, E.R.T. TIEKINK and G. WINTER, 1984. Inorg. Chim. Acta, 85, 215-218.

Orthorhombic, Pnma, a = 14.165, b = 7.6759, c = 13.977 Å, D_m = 1.71, Z = 4. Cu radiation, R = 0.041 for 1229 reflexions.

The structure consists of discrete molecular units with the Sn atom and the xanthate moieties lying on a crystallographic mirror plane (Fig. 1). The Sn atom is six-coordinate and the bidentate xanthate ligands display gross asymmetry in their mode of coordination to the metal. Sn-S distances are: Sn-S(1) 2.486, Sn-S(2) 3.088, Sn-S(3) 2.501 and Sn-S(4) 3.151 Å. The 'bite' angles are S(1)-Sn-S(2) 63.68° and S(3)-Sn-S(4) 62.49°, with angles S(1)-Sn-S(3) and S(2)-Sn-S(4) 83.52 and 150.31° respectively. The Sn-C(7) bond length is 2.111 Å and the angle C(7)-Sn-C(7') is 130.1°.

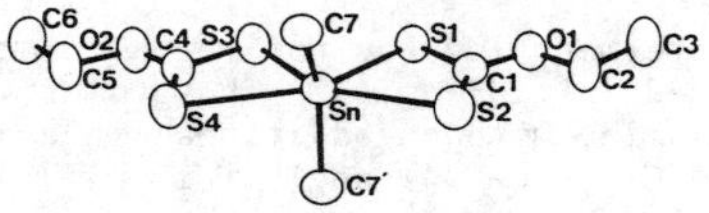

Fig. 1. The molecular structure of $C_8H_{16}O_2S_4Sn$. C(7) is related to C(7') by the crystallographic mirror plane.

DICHLOROBIS(N,N'-DIMETHYLTHIOUREA-S)DIMETHYLTIN
$C_8H_{22}Cl_2N_4S_2Sn$ (I) $(CH_3)_2SnCl_2.2(SC(NHCH_3)_2)$

DICHLOROBIS(N,N'-DIMETHYLTHIOUREA-S)DIPHENYLTIN
$C_{18}H_{26}Cl_2N_4S_2Sn$ (II) $(C_6H_5)_2SnCl_2.2(SC(NHCH_3)_2)$

CHLOROTRIPHENYL(TETRAMETHYLUREA-O)TIN
$C_{23}H_{27}ClON_2Sn$ (III) $(C_6H_5)_3SnCl.(OC(N(CH_3)_2)_2)$

S. CALOGERO, G. VALLE and U. RUSSO, 1984. Organometallics, 3, 1205-1210.

I. Monoclinic, $P2_1/n$, a = 9.843, b = 9.864, c = 9.089 Å, β = 100.1°, D_m = 1.62, Z = 2. Mo radiation, R = 0.026 for 1369 reflexions.

II. Monoclinic, $P2_1/n$, a = 10.791, b = 11.468, c = 10.147 Å, β = 93.9°, D_m = 1.45, Z = 2. Mo radiation, R = 0.045 for 1424 reflexions.

III. Orthorhombic, Pcab, a = 29.977, b = 14.235, c = 11.034 Å, D_m = 1.40, Z = 8. Mo radiation, R = 0.058 for 2125 reflexions.

In compounds I and II the Sn atoms have trans-octahedral coordination geometry (Fig. 1) with the following bond lengths: Sn-S 2.729 and 2.686, Sn-C 2.122 and 2.151, and Sn-Cl 2.616 and 2.589 Å respectively, for I and II. There are possible hydrogen bonds S(1)...HN(1) at distances of 2.76 and 2.40 Å respectively in I and II. The lattice of compound III contains discrete molecules with Sn having distorted trigonal bipyramidal geometry with the following bond lengths: Sn-C (equatorial) 2.12, 2.15 and 2.11 Å, Sn-Cl (apical) 2.485 Å, and Sn-O (apical) 2.383 Å. The angles O(1)-Sn-Cl, C(6)-Sn-C(12), C(6)-Sn-C(18), and C(12)-Sn-C(18) are 179.2, 122.6, 115.9 and 119.8° respectively.

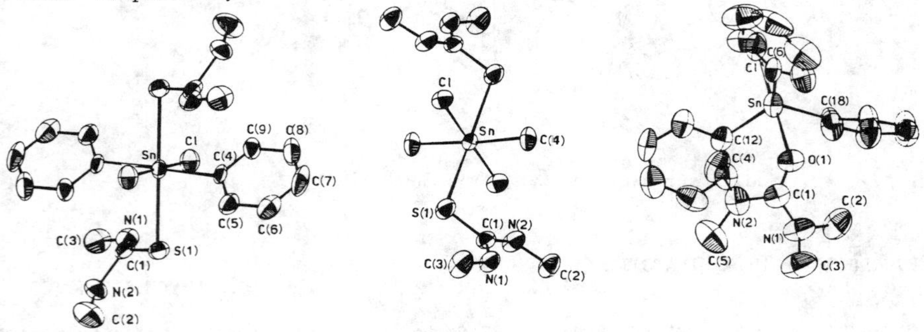

Fig. 1. Molecular structures of $C_8H_{22}Cl_2N_4S_2Sn$ (left), $C_{18}H_{26}Cl_2N_4S_2Sn$ (centre) and $C_{23}H_{27}ClON_2Sn$ (right).

DI-μ_3-OXO-BIS(μ-DICHLORO)BIS[μ-DIMETHYLTIN(IV)]BIS[CHLORODIMETHYLTIN(IV)]
$C_8H_{24}Cl_4O_2Sn_4$ $[(CH_3)_4Sn_2Cl_2O]_2$

D. DAKTERNIEKS, R.W. GABLE and B.F. HOSKINS, 1984. Inorg. Chim. Acta, 85, L43-L44.

Monoclinic, $P2_1/c$, a = 7.281, b = 18.532, c = 8.7218 Å, β = 109.82°, D_m = 2.27, Z = 4. Mo radiation, R = 0.032 for 2096 reflexions.

This structural re-determination has not revealed any additional structural details to those established initially (1). The geometry around Sn(1) is a distorted cis-R_2SnXYZ trigonal bipyramid, with an additional interaction with a Cl atom (Cl(2") 3.339 Å) which bisects the C(3)-Sn-C(4) angle (Fig. 1). This increases the coordination to 6 and results in a polymer in the solid state. The environment around Sn(2) is a grossly-distorted cis-R_2SnXYZ trigonal bipyramid with a Cl(I') interaction (3.409 Å) producing a distorted octahedral coordination. The four Cl atoms, four Sn atoms and two O atoms are planar, with Sn(2) 0.068 Å from the plane. Important distances are Sn-Cl(1) 2.445, Sn(1)-Cl(2) 2.789, Sn(1)-O 2.031, Sn(1)-C(3) 2.112, Sn(1)-C(4) 2.114, Sn(2)-Cl(2) 2.702, Sn(2)-O 2.038, Sn(2)-C(1) 2.107 and Sn(2)-C(2) 2.124 Å.

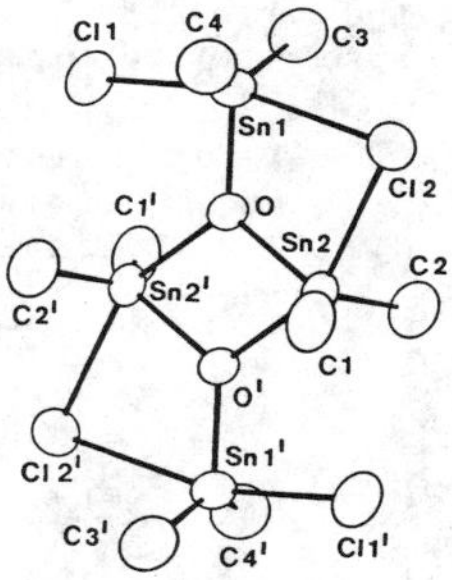

Fig. 1. Molecular structure of $C_4H_{12}Cl_2OSn_2$.

1. Structure Reports, 46B, 737.

BIS[BIS(DIMETHYLAMIDO)TIN(II)]
$C_8H_{24}N_4Sn_2$ $[Sn(N(CH_3)_2)_2]_2$

M.M. OLMSTEAD and P.P. POWER, 1984. Inorg. Chem., 23, 413-415.

Triclinic, P$\bar{1}$, a = 6.210, n = 7.094, c = 8.802 Å, α = 81.59, β = 72.46, γ = 76.97°, D_m = 1.8, Z = 1. Mo radiation, R = 0.035 for 1623 reflexions (at 85K).

The compound is a centrosymmetric dimer with the tin atoms bridged by two dimethylamido groups (Fig. 1). Each tin atom is three-coordinate, being bonded also to a terminal NMe_2 group. The geometry is consistent with the presence of a stereochemically active lone pair at each metal. Bond lengths: Sn-N(1) 2.067(4), Sn-N(2) 2.266(5) and 2.265(4) Å.

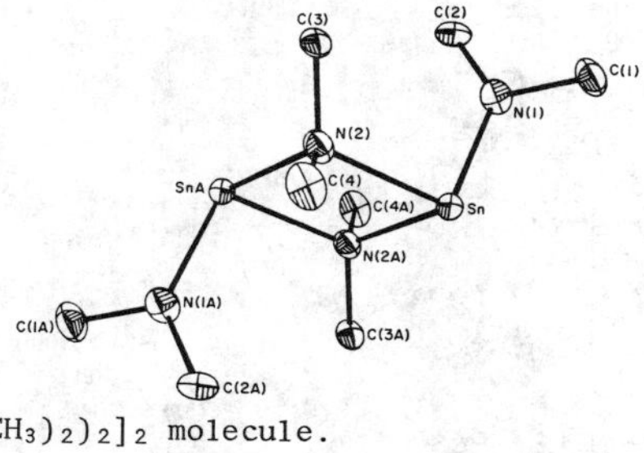

Fig. 1. The $[Sn(N(CH_3)_2)_2]_2$ molecule.

catena-[HEXAKIS(μ_2-THIOUREA)BIS(THIOUREA-S)OCTABROMOTETRALEAD(II)] (Form I)
$(C_8H_{32}Br_8N_{16}Pb_4S_8)_n$

F.H. HERBSTEIN and G.M. REISNER, 1984. Z. Kristallogr., 169, 83-93.

Monoclinic, $P2_1$, a = 13.608, b = 26.771, c = 6.016 Å, β = 95.61°, D_m = 3.14, Z = 2. Mo radiation, R = 0.063 for 3510 reflexions.

The asymmetric unit is a tetramer (Fig. 1) with two six-coordinate and two seven-coordinate Pb atoms linked through bridging S atoms of thioureas. These tetramers are linked in polymeric chains through bicoordinate Br ions. There is extensive N-H...Br hydrogen bonding and/or electrostatic interaction within and between these units. Pb-S distances are in the range 2.86-3.17(1) Å and Pb-Br distances are 2.926-3.169(4) Å. Crystal data are also reported for six other $PbBr_2$ thiourea polymorphs.

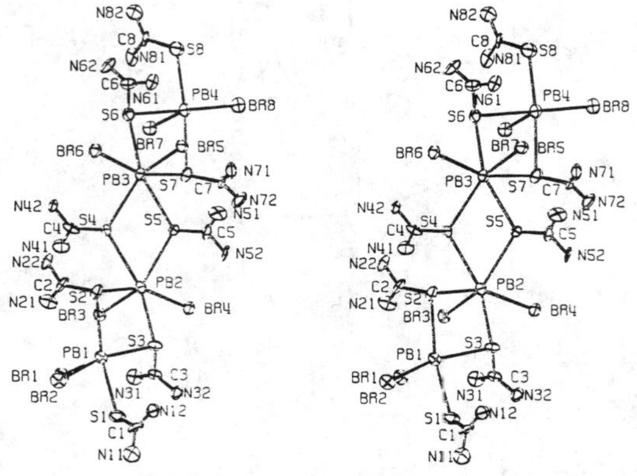

Fig. 1. A stereoview of the 4[$PbBr_2$.2(thiourea)] asymmetric unit.

DIPOTASSIUM TRIS(μ_2-MALONATO-O,O')DITIN(II) MONOHYDRATE
$(C_9H_6K_2O_{12}Sn_2)_n,(H_2O)_n$

Z. ARAFIN, E.J. FILMORE, J.D. DONALDSON and S.M. GRIMES, 1984. J. Chem. Soc. Dalton, 1965-1968.

Monoclinic, $P2_1/n$, a = 11.233, b = 18.776, c = 8.171 Å, β = 90.15°, Z = 4. Mo radiation, R = 0.027 for 5805 reflexions.

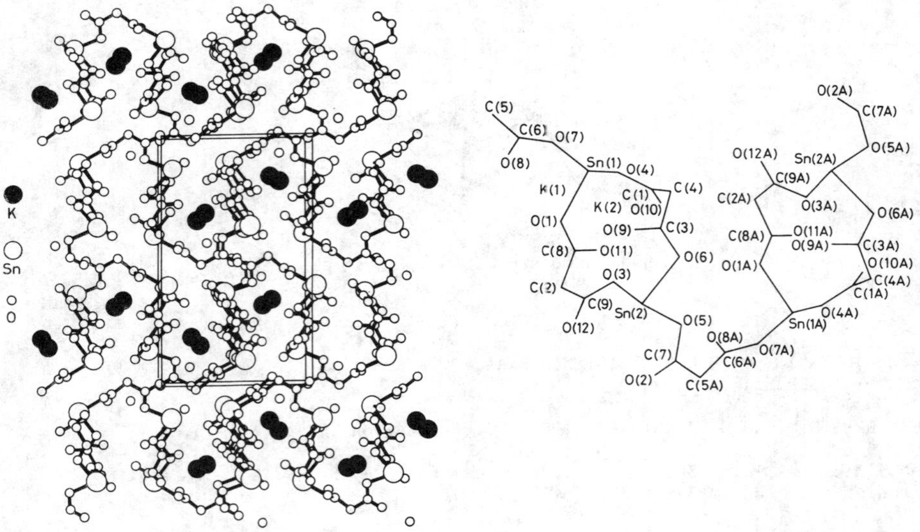

Fig. 1. $(C_9H_6K_2O_{12}Sn_2)_n,(H_2O)_n$: the asymmetric unit and a projection of the unit cell viewed down z.

The structure (Fig. 1) consists of polymeric anionic layers with discrete potassium ions and water molecules located between the layers. Both Sn atoms in the asymmetric unit are coordinated through single oxygen atoms of two of the three malonate groups forming a twelve-membered ring. The third malonate group is bonded to two tin atoms of neighbouring twelve-membered rings. The malonate groups have typical bond lengths and angles. The two K ions have different environments. One is surrounded by seven nearest-neighbour malonate O atoms (K...O 2.887-3.132 Å) and an O atom of water at 2.745 Å while the other is coordinated to eight malonate O atoms (K...O 2.688-3.229 Å) and an O atom of water at 2.971 Å. Coordination about each Sn atom is trigonal pyramidal. Sn(1) has two short (2.177 Å each) and one long bond (2.199 Å) to oxygen whilst Sn(2) has one short (2.161 Å) and two long bonds (2.206, 2.240 Å) to oxygen. The malonate groups form bridges between different Sn atoms which gives rise to the extended polymeric anionic system.

3-(1-GERMATRANYL)PROPIONAMIDE
$C_9H_{18}GeN_2O_4$ (I)

9-(2-HYDROXYETHYL)-1,6,12-TRIOXA-9-AZA-5-GERMA-SPIRO[4.7]DODECAN-2-ONE
$C_9H_{17}GeNO_5$ (II)

S.N. GURKOVA, A.I. GUSEV, N.V. ALEKSEEV, T.E. GAR, N.Yu. KHROMOVA and N.A. VIKTOROV, 1984. J. Struct. Chem., 25, 458-462 [Zh. Strukt. Khim. 25(3), 135-140].

I. Monoclinic, $P2_1/c$, a = 7.963, b = 14.689, c = 10.555 Å, β = 110.25°, Z = 4. Mo radiation, R = 0.035 for 1235 reflexions.

II. Monoclinic, Aa, a = 7.916, b = 19.041, c = 7.678 Å, β = 101.16°, Z = 4. Mo radiation, R = 0.029 for 857 reflexions.

The structures of the two molecules are shown in Fig. 1. In I the germanatrane skeleton has its usual structure. The Ge atom has trigonal bipyramidal coordination geometry with Ge-O(all equatorial) 1.799-1.817(3), Ge-C 1.941(6), and Ge-N 2.231(4) Å. In molecule II a bond is formed with a carboxylate oxygen atom while one of the branches of the triethanolamine ligand is displaced. The Ge-C bond becomes equatorial in this trigonal bipyramidal complex. Bond lengths for II: Ge-O(eq) 1.758(6) and 1.784(5), Ge-O(ax) 1.916(6), Ge-C(eq) 1.958(8), and Ge-N 2.143(6) Å.

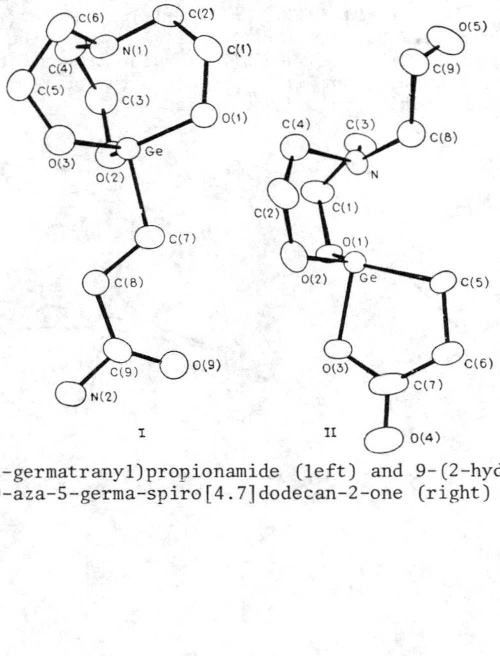

Fig. 1. The 3-(1-germatranyl)propionamide (left) and 9-(2-hydroxyethyl)-1,6,12-trioxa-9-aza-5-germa-spiro[4.7]dodecan-2-one (right) molecules.

3,3-BIS(CHLORODIMETHYLSTANNYL)PROPYL ETHYL SULPHOXIDE
$C_9H_{22}Cl_2OSSn_2$ (I) $[(ClMe_2Sn)_2CHCH_2CH_2S(O)Et]$

3,3-BIS(DICHLOROMETHYLSTANNYL)PROPYL ETHYL SULPHOXIDE
$C_7H_{16}Cl_4OSSn_2$ (II) $[(Cl_2MeSn)_2CHCH_2CH_2S(O)Et]$

3,3-BIS(BROMODIMETHYLSTANNYL)PROPYL ETHYL SULPHOXIDE
$C_9H_{22}Br_2OSSn_2$ (III) $[(BrMe_2Sn)_2CHCH_2CH_2S(O)Et]$

3,3-BIS(DIBROMOMETHYLSTANNYL)PROPYL ETHYL SULPHOXIDE
$C_7H_{16}Br_4OSSn_2$ (IV) $[(Br_2MeSn)_2CHCH_2CH_2S(O)Et]$

K. SWAMI, J.P. HUTCHINSON, H.G. KUIVILA and J.A. ZUBIETA, 1984. Organometallics, 3, 1687-1694.

I. Triclinic, $P\bar{1}$, a = 7.002, b = 9.383, c = 13.771 Å, α = 104.35, β = 98.67, γ = 90.37°, Z = 2. Mo radiation, R = 0.068 for 1306 reflexions.

II. Triclinic, $P\bar{1}$, a = 9.099, b = 9.477, c = 10.476 Å, α = 113.34, β = 99.68, γ = 91.87°, Z = 2. Mo radiation, R = 0.032 for 1842 reflexions.

III. Triclinic, $P\bar{1}$, a = 7.074, b = 9.325, c = 14.070 Å, α = 77.45, β = 81.03, γ = 88.23°, Z = 2. Mo radiation, R = 0.078 for 1739 reflexions.

IV. Monoclinic, $P2_1/n$, a = 8.965, b = 17.887, c = 11.166 Å, β = 102.29°, Z = 4. Mo radiation, R = 0.076 for 1128 reflexions.

Structures I and III consist of discrete binuclear units with non-equivalent Sn atoms bridged symmetrically by a C atom of the propyl ethyl sulphoxide group, and unsymmetrically by one or two halide atoms, depending upon the degree of halide substitution. On the other hand, molecules II and IV exhibit halide bridging between two adjacent binuclear units to produce a loosely associated bimolecular or tetranuclear unit (Fig. 1). In I and III bond lengths of interest are: Sn-Cl(1) 2.478, Sn(1)-Me(1) 2.13, Sn(1)-Me(2) 2.09, Sn(1)-C(1) 2.13, Sn(1)...Sn(2) 3.498, Sn(2)-Cl(2) 2.572, Sn(2)-Me(4) 2.12, Sn(2)-C(1) 2.15, Sn(2)-O 2.30 and Sn(1)...Cl(2) 2.925 Å; Sn(1)-Br(1) 2.626, Sn(2)-C(1) 2.11, Sn(1)...Sn(2) 3.527, Sn(1)-Br(2) 2.748, Sn(2)-O 2.17, Sn(1)...Br(2) 3.065 Å. In II and IV the following bond distances are of interest: Sn(1)-Cl(1) 2.366, Sn(1)-Cl(3) 2.374, Sn(1)-C(1) 2.148, Sn(1)...Sn(2) 3.440, Sn(1)...Cl(2) 3.118, Sn(2)-Cl(2) 2.537, Sn(2)-Cl(4) 2.371, Sn(2)-C(1) 2.14, Sn(2)...Cl(3) 3.632 Å; Sn(1)-Br(1) 2.520, Sn(1)-Br(3) 2.529, Sn(1)-C(1) 2.17, Sn(1)...Sn(2) 3.471, Sn(1)...Br(2) 3.335, Sn(2)-Br(2) 2.690, Sn(2)-Br(4) 2.534, Sn(2)-C(1) 2.11, Sn(2)-O 2.22, Sn(2)...Br(3) 3.584 Å. In I, the geometry about Sn(1) is distorted pyramidal, that about Sn(2) regular trigonal bipyramidal. In II we have distorted pyramidal about Sn(1) and distorted octahedral about Sn(2). The bromine derivatives exhibit similar geometries.

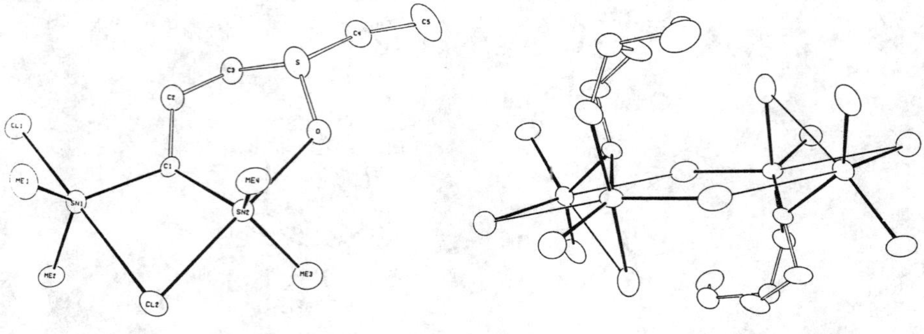

Fig. 1. Views of molecule I $C_9H_{22}Cl_2OSSn_2$ (left) and of the bimolecular unit formed by II $C_7H_{16}Cl_4OSSn_2$ (right).

2,2,6,6-TETRAMETHYL-4-THIA-8,8,9,9-TETRACHLORO-8,9-DIGERMABICYCLO[5.2.0]NON-1(7)-ENE
$C_{10}H_{16}Cl_4Ge_2S$

A.A. ESPENBETOV, Yu.T. STRUCHKOV, S.P. KOLESNIKOV and O.M. NEFEDOV, 1984. J. Organometal. Chem., 275, 33-37.

Monoclinic, $P2_1/n$, a = 9.301, b = 11.059, c = 16.031 Å, β = 99.30°, Z = 4. Mo radiation, R = 0.043 for 2485 reflexions, (-50°C).

The digermacyclobutene ring is planar, and the atoms C(1), C(2), C(3) and C(7) are also coplanar, the dihedral angle between these two planes being 2.8°. The conformation of the seven-membered 4-thiacycloheptene ring is an S(α)-chair (Fig. 1). The Ge-Ge distance is 2.380 Å and the Ge-Cl distances are 2.150, 2.146, 2.146 and 2.145 Å respectively for Cl(1),(2),(3) and (4). The Ge-C distances are 1.998 and 2.002 Å. The length of the double bond C(1)-C(2) is 1.319 Å and the S-C(4) and S-C(8) distances are 1.790 and 1.779 Å respectively. The angles C(1)-Ge(1)-Ge(2) and C(2)-Ge(2)-Ge(1) are equal at 74.7°.

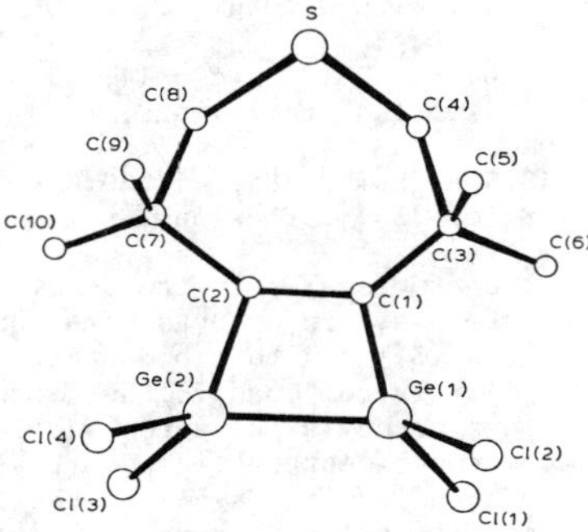

Fig. 1. Molecular structure of $C_{10}H_{16}Cl_4Ge_2S$.

DICHLOROBIS(N-METHYLIMIDAZOLE)DIMETHYLTIN(IV)
$C_{10}H_{18}Cl_2N_4Sn$ $(CH_3)_2SnCl_2(C_3H_3N_2CH_3)_2$

R. BARDI, A. PIAZZESI, R. ETTORRE and G. PLAZZOGNA, 1984. J. Organometal. Chem., 270, 171-176.

Triclinic, $P\bar{1}$, a = 7.218, b = 7.441, c = 7.972 Å, α = 114.1, β = 104.1, γ = 97.8°, D_m = 1.72, Z = 1. Mo radiation, R = 0.039 for 1300 reflexions.

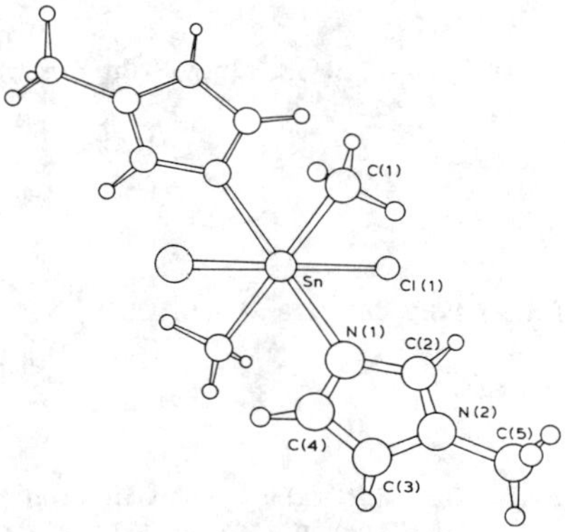

Fig. 1. Molecular structure of $C_{10}H_{18}Cl_2N_4Sn$.

The molecule has crystallographic inversion symmetry and the coordination geometry about the Sn atom is all trans (Fig. 1). The NMiz ligands are bonded to the metal through their pyridine-like N atom, and the rings are planar with no atom deviating from the least-squares plane through the five atoms by more than 0.002 Å. The Sn-C, Sn-Cl and Sn-N bond distances are 2.118, 2.571 and 2.329 Å respectively. Distances within the ring are: N(1)-C(2) 1.324, N(1)-C(4) 1.367, N(2)-C(2) 1.331, N(2)-C(3) 1.350 and C(3)-C(4) 1.359 Å. The cis angles about Sn are: N(1)-Sn-C(1) 89.8, Cl(1)-Sn-C(1) 88.6, Cl(1)-Sn-N(1) 89.2°.

TETRABROMO(N,N'-DI-n-BUTYLDITHIOOXAMIDE-S,S')TIN(IV) BENZENE SOLVATE
$C_{10}H_{20}Br_4N_2S_2Sn,0.33(C_6H_6)$ (I)

TRICHLOROBIS(N,N'-DIETHYLDITHIOOXAMIDE-S,S')BISMUTH(III) ACETONE SOLVATE
$C_{12}H_{24}BiCl_3N_4S_4,C_3H_6O$ (II)

M.G.B. DREW, J.M. KISENYI and G.R. WILEY, 1984. J. Chem. Soc. Dalton, 1723-1726.

I. Monoclinic, C2/c, a = 20.766, b = 21.245, c = 14.835 Å, β = 91.4°, D_m = 2.09, Z = 12. Mo radiation, R = 0.10 for 1247 reflexions.

II. Monoclinic, $P2_1/c$, a = 10.568, b = 8.780, c = 30.086 Å, β = 101.2°, D_m = 1.76, Z = 4. Mo radiation, R = 0.08 for 2341 reflexions.

Compound I has two equivalent molecules in the asymmetric unit, one with imposed C_2 symmetry (Fig. 1). The Sn atoms are in an octahedral environment and are bonded to four Br atoms (2.527(11)-2.625(13)) and to a bidentate dithio-oxamide ligand through sulphur (Sn-S 2.55(3)-2.61(3) Å). In compound II the geometry about the Bi atom is pentagonal bipyramidal. Cl atoms occupy axial sites (2.63(1), 2.72(1)) while the equatorial plane is occupied by two bidentate ligands (Bi-S 2.82(1)-3.041) and another Cl atom (Bi-Cl 2.65(1) Å). In both structures all of the ligand -NH groups are hydrogen-bonded for the most part to halogen atoms.

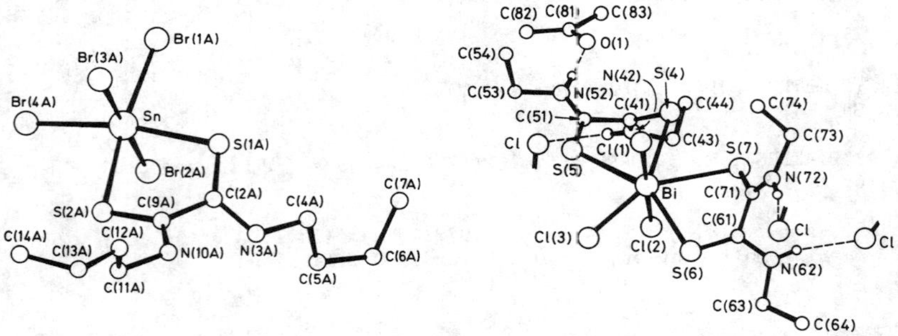

Fig. 1. The molecular structures: $C_{10}H_{20}Br_4N_2S_2Sn$ (left) and $C_{12}H_{24}BiCl_3N_4S_4$,-C_3H_6O (right).

2,2-DIBUTYL-1,3,2-DIOXASTANNOLANE
$C_{10}H_{22}O_2Sn$

A.G. DAVIES, A.J. PRICE, H.M. DAWES and M.B. HURSTHOUSE, 1984. J. Organometal. Chem., 270, C1-C3.

Orthorhombic, $P2_12_12_1$, a = 8.103, b = 7.409, c = 21.563 Å, Z = 4. Mo radiation, R = 0.069 for 1011 reflexions.

The compound exists as an infinite ribbon coordination polymer with six-coordinate tin (Fig. 1). The polyhedron about the Sn atom is a highly distorted octahedron with bond lengths Sn-C 2.13 and 2.14 Å to the butyl groups, Sn-O 1.98 and 2.097 Å to O(1) and O(2) respectively of the "chelating" diol, and Sn-O 2.50 and 2.52 Å to O(1a) and O(2b) of two neighbouring units to complete a polymeric chain along the 2_1 screw axis. The endocyclic angle O(1)-Sn-O(2) is 79.0°, and the exocyclic angle C(3)-Sn-C(7) is 138.6°.

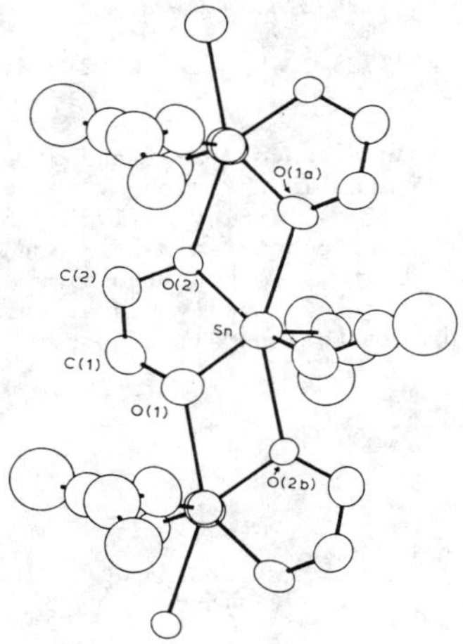

Fig. 1. The structure of $C_{10}H_{22}O_2Sn$.

TRIS(PYRAZOLYL)BORATE(TRIMETHYL)TIN
$C_{12}H_{19}BN_6Sn$ $HB(C_3H_3N_2)_3Sn(CH_3)_3$

B.K. NICHOLSON, 1984. J. Organometal. Chem., 265, 153-157.

Trigonal, P$\bar{3}$, a = 11.722, c = 8.211 Å, Z = 2. Mo radiation, R = 0.023 for 1386 reflexions.

The molecule (Fig. 1) has crystallographic threefold symmetry. The coordination about the Sn atom is distorted octahedral with three fac Sn-C(methyl) bonds and three Sn-N bonds to the tris(pyrazoyl)borate ligand. The C(4)-Sn-C(4') angles are opened out to 105.2° with a concomitant closing of the N(1)-Sn-N(1') angles to 74.8°. The Sn-C(4) and Sn-N(1) distances are respectively 2.165 and 2.388 Å which are within the accepted range for such bonds. The N(1)-N(2) and B-N(2) distances are 1.359 and 1.539 Å respectively, and the N(2)-B-N(2') angle is 108.4°. The Sn-N(1)-N(2) and N(1)-N(2)-B angles are 125.5 and 120.4° respectively.

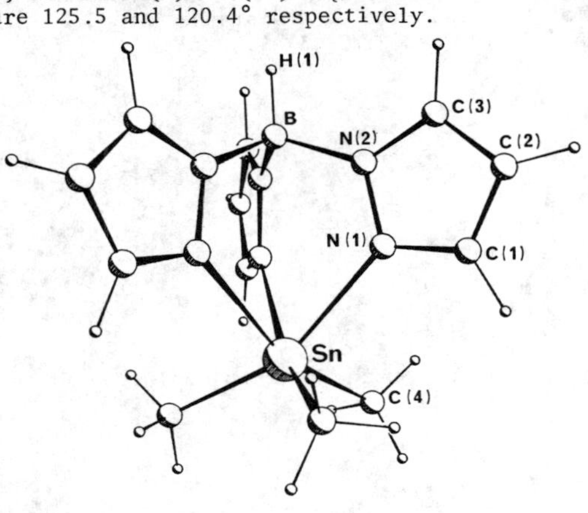

Fig. 1. Molecular structure of $C_{12}H_{19}BN_6Sn$.

LEAD(II) D-GLUCONATE

$C_{12}H_{22}O_{14}Pb$ $Pb[C_6H_{11}O_7]_2$

T. LIS, 1984. Acta Cryst., C40, 374-376.

Trigonal $P3_221$, a = 5.498, c = 47.28 Å, D_m = 2.40, Z = 3. Mo radiation, R = 0.037 for 838 reflexions.

The Pb atom (Fig. 1) lies on a twofold axis and is surrounded by six O atoms from four gluconate ligands; each Pb atom is chelated by the α-hydroxyl O atom at 2.48(2) Å and a carboxylate O atom at 2.43(2) Å forming a five-membered ring. The second carboxylate O atom (from another gluconate anion) is bonded at 2.71(2) Å. Thus the carboxylate groups bridge two Pb atoms forming two-dimensional nets parallel to (001). The gluconate ligands have the zig-zag, nearly planar carbon-chain conformation with their long molecular axis approximately parallel to c.

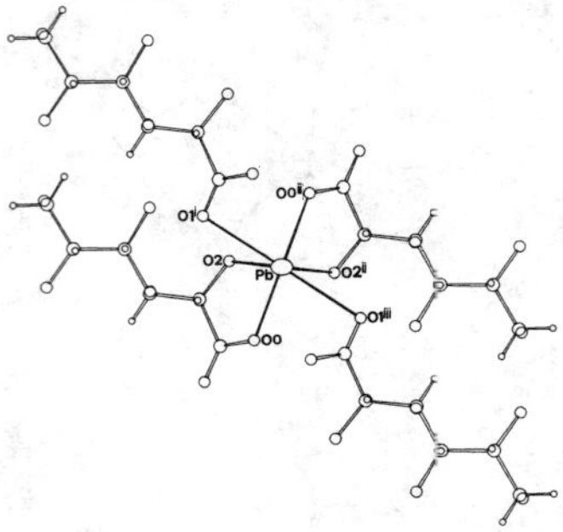

Fig. 1. A view of the $C_{12}H_{22}O_{14}Pb$ structure.

1,2-DI-t-BUTYL-3,3-DIETHYL-1,2,3-DIPHOSPHAGERMIRAN

$C_{12}H_{28}GeP_2$ $(t\text{-}BuP)_2GeEt_2$

K.-F. TEBBE and B. FRECKMANN, 1984. Acta Cryst., C40, 254-257.

Monoclinic, C2/c, a = 51.935, b = 9.786, c = 10.137 Å, β = 100.02°, Z = 12. Mo radiation, R = 0.064 for 2903 reflexions.

One molecule lies in a general position in the asymmetric unit and the other lies on a twofold axis (Fig. 1). The Ge-P distances are 2.296, 2.300 and 2.301(2) Å. The P-P distances are 2.232(4) and 2.226(3) Å.

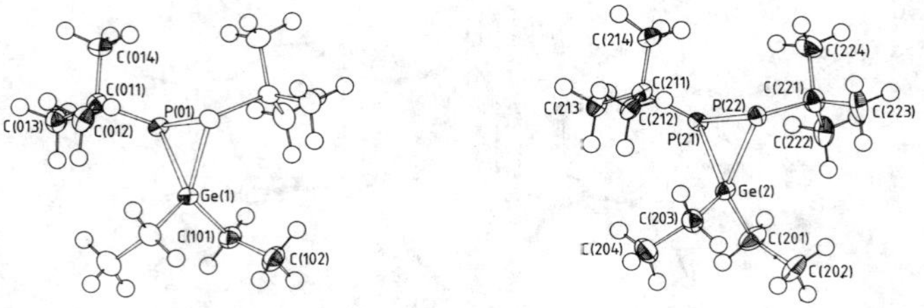

Fig. 1. The two independent $(t\text{-}BuP)_2GeEt_2$ molecules.

cis-1,1,2,2,3,4,4,5,5,6-DECAMETHYL-1,2,4,5-TETRASTANNACYCLOHEXANE
$C_{12}H_{32}Sn_4$

H. PREUT, P. BLECKMANN, T. MITCHELL and B. FABISCH, 1984. Acta Cryst., C40, 370-372.

Monoclinic, $P2_1/n$, a = 10.650, b = 12.313, c = 16.409 Å, β = 95.27°, Z = 4. Mo radiation, R = 0.030 for 2057 reflexions.

The molecule (Fig. 1) has a boat conformation with Sn-Sn 2.775 and 2.766(2) Å and Sn-C 2.14-2.17(1) Å.

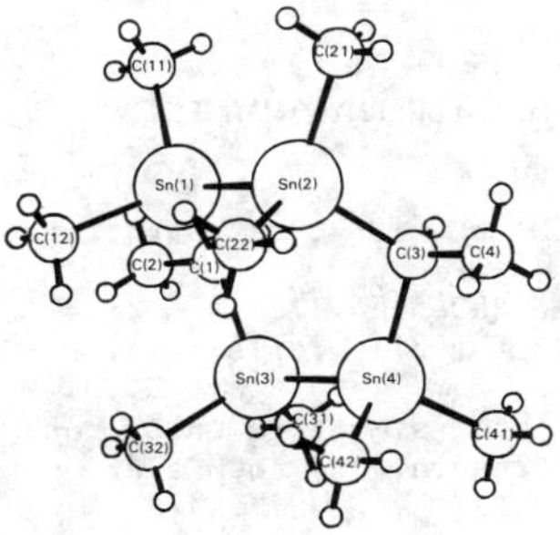

Fig. 1. A general view of $C_{12}H_{32}Sn_4$.

N-METHYL-5,5-DI-t-BUTYLDIPTYCHOXAZSTANNOLIDINE
$C_{13}H_{29}NO_2Sn$ (I) $(t\text{-}C_4H_9)_2Sn(OCH_2CH_2)_2NCH_3$

N-METHYL-5,5-DIMETHYLDIPTYCHTHIAAZSTANNOLIDINE
$C_7H_{17}NS_2Sn$ (II) $(CH_3)_2Sn(SCH_2CH_2)_2NCH_3$

R.G. SWISHER and R.R. HOLMES, 1984. Organometallics, 3, 365-369.

I. Monoclinic, $P2_1/c$, a = 22.655, b = 25.990, c = 9.036 Å, β = 106.97°, Z = 12. Mo radiation, R = 0.074 for 4121 reflexions.

II. Orthorhombic, $P2_12_12_1$, a = 8.844, b = 9.253, c = 14.309 Å, Z = 4. Mo radiation, R = 0.038 for 1415 reflexions.

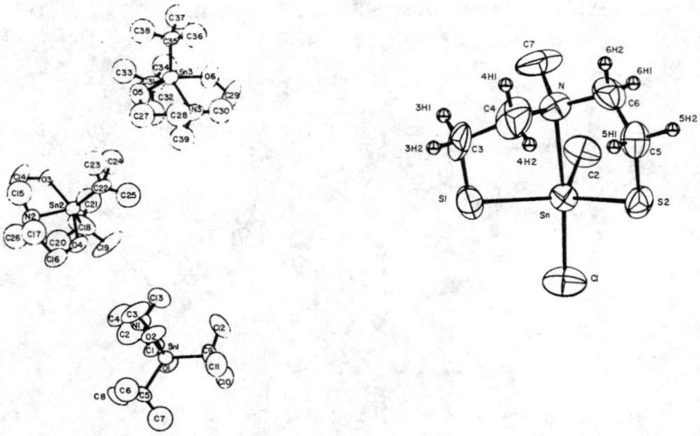

Fig. 1. Molecular structures of the three independent molecules in $C_{13}H_{29}NO_2Sn$ (left), and of $C_7H_{17}NS_2Sn$ (right).

Both compounds have trigonal bipyramidal structures (Fig. 1). In I, the O atoms are in apical positions: Sn(1)-O(1) 2.05, Sn(1)-O(2) 2.04, Sn(2)-O(3) 2.06, Sn(2)-O(4) 2.04, Sn(3)-O(5) 2.01 and Sn(3)-O(6) 2.02 Å (three independent molecules). The equatorial Sn-ligand bonds (Sn-N and Sn-C) range from 2.30 to 2.34 Å and from 2.17 to 2.37 Å respectively. In II, the methyl group (Sn-C(1) 2.16 Å) and the N atom (Sn-N 2.58 Å) are axial ligands and the S atoms (Sn-S(1) 2.433 and Sn-S(2) 2.429 Å) and the second methyl group (Sn-C(2) 2.15 Å) form the equatorial ligands. The Sn atom lies in the equatorial plane in I but is displaced some 0.37 Å from the plane towards C(1) in II. The angles O-Sn-O in the three independent molecules are 150.1, 150.2 and 151.3° whilst N-Sn-C(1) in II is 167.0°.

BIS(1,2-DIMETHOXYETHANE)BIS(TRIMETHYLGERMYL)MAGNESIUM
$C_{14}H_{38}Ge_2MgO_4$

L. ROSCH, C. KRUGER and A.-P. CHIANG, 1984. Z. Naturforsch., B39, 855-860.

Triclinic, P$\bar{1}$, a = 9.375, b = 16.809, c = 17.769 Å, α = 116.47, β = 99.85, γ = 98.25°, Z = 4. Mo radiation, R = 0.0907 for 3120 reflexions.

There are two molecules in the asymmetric unit, both with similar geometry (Fig. 1). The Mg atoms have octahedral and the Ge atoms tetrahedral coordination, with mean Mg-Ge 2.719(6), Mg-O 2.22(6), Ge-C 1.94(3) Å, Ge-Mg-Ge 99.3(2)°.

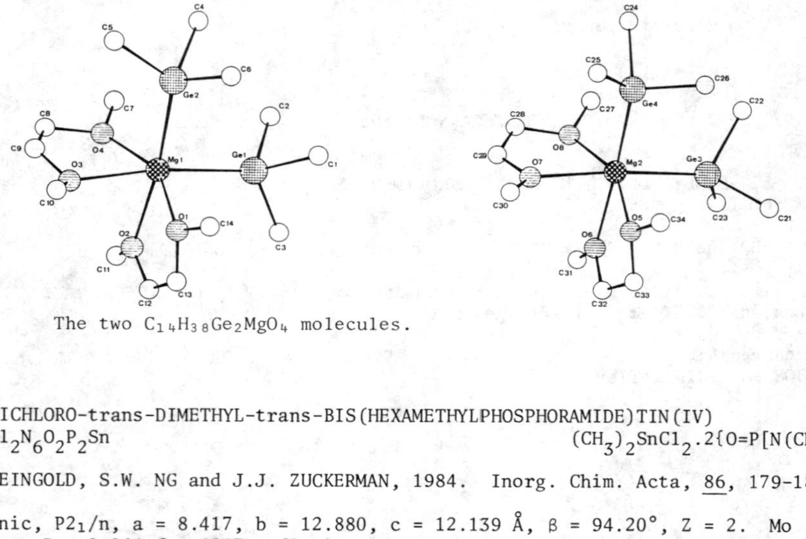

Fig. 1. The two $C_{14}H_{38}Ge_2MgO_4$ molecules.

trans-DICHLORO-trans-DIMETHYL-trans-BIS(HEXAMETHYLPHOSPHORAMIDE)TIN(IV)
$C_{14}H_{42}Cl_2N_6O_2P_2Sn$ $(CH_3)_2SnCl_2.2\{O=P[N(CH_3)_2]_3\}$

A.L. RHEINGOLD, S.W. NG and J.J. ZUCKERMAN, 1984. Inorg. Chim. Acta, 86, 179-183.

Monoclinic, $P2_1/n$, a = 8.417, b = 12.880, c = 12.139 Å, β = 94.20°, Z = 2. Mo radiation, R = 0.044 for 1897 reflexions.

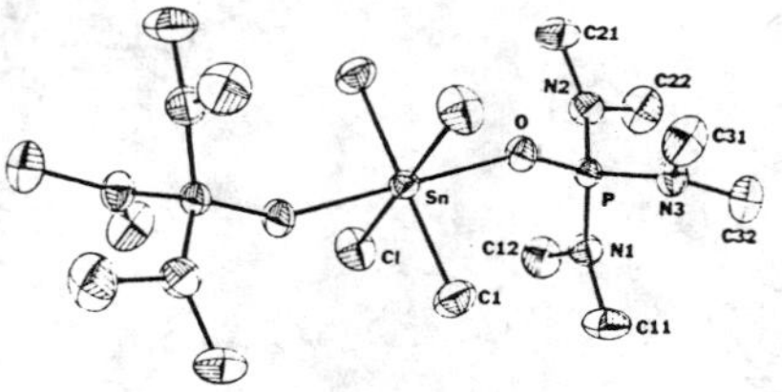

Fig. 1. Molecular structure of $C_{14}H_{42}Cl_2N_6O_2P_2Sn$.

Dimethyltin(IV) dichloride forms a discrete hexacoordinated, octahedral complex with two molecules of hexamethylphosphoramide in an all-trans structure (Fig. 1). The Sn atom is sited on an inversion centre, imposing 180° bond angles for the trans ligands. The Cl-Sn-O, Cl-Sn-C(1) and O-Sn-C(1) angles are 90.9, 90.6 and 91.7° respectively, and the Sn-O-P angle is 162.7°. Important distances are: Sn-Cl 2.575, Sn-O 2.231, Sn-C(1) 2.105 and O-P 1.496 Å. The molecular geometry does not differ markedly from that obtained in a previous study (1).

1. Structure Reports, 43B, 925.

1,3-DIPHENYL-3-METHYL-3-(TRICHLOROGERMYL)BUTAN-1-ONE
$C_{16}H_{15}Cl_3GeO$

S.N. GURKOVA, A.I. GUSEV, N.V. ALEKSEEV, T.K. GAR and N.A. VIKTOROV, 1984. J. Struct. Chem., 25, 829-831 [Zh. Strukt. Khim., 25(5), 174-176].

Monoclinic, Cc, a = 16.6774, b = 12.2078, c = 12.4884 Å, β = 138.072°, Z = 4. Mo radiation, R = 0.028 for 1640 reflexions.

The molecule is shown in Fig. 1. The five-coordinate Ge atom has trigonal bipyramidal coordination geometry, the Ge atom being displaced 0.42 Å from the equatorial plane towards Cl(1). The angle between axial substituents (Cl(1)-Ge-O) is 176.9(2)°. Bond lengths: Ge-C(eq) 2.138(1) and 2.144(3), Ge-Cl(ax) 2.181(2), Ge-C 1.994(6) and Ge-O 2.507(5) Å.

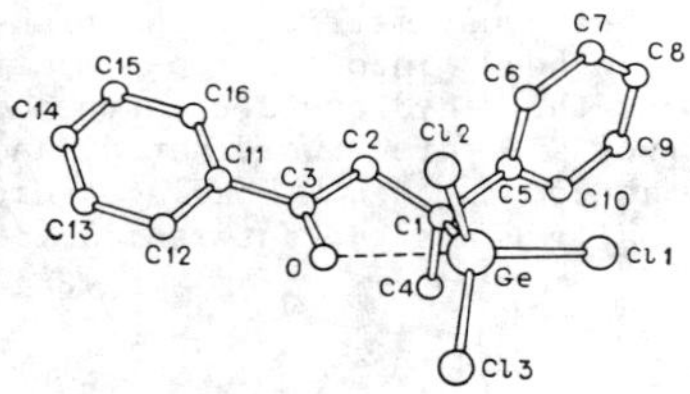

Fig. 1. The 1,3-diphenyl-3-methyl-3-(trichlorogermyl)butan-1-one molecule.

(η^5-PENTAMETHYLCYCLOPENTADIENYL)PYRIDINETIN TRIFLUOROMETHANESULPHONATE
$C_{16}H_{20}F_3NO_3SSn$ (I) $C_{15}H_{20}NSn^+.CF_3O_3S^-$

(2,2'-BIPYRIDYL)(PENTAMETHYLCYCLOPENTADIENYL)TIN TRIFLUOROMETHANESULPHONATE
$C_{21}H_{23}F_3N_2O_3SSn$ (II) $C_{20}H_{23}N_2Sn^+.CF_3O_3S^-$

F.X. KOHL, E. SCHLUTER, P. JUTZI, C. KRUGER, G. WOLMERSHAUSER, P. HOFMANN and P. STAUFFERT, 1984. Chem. Ber., 117, 1178-1193.

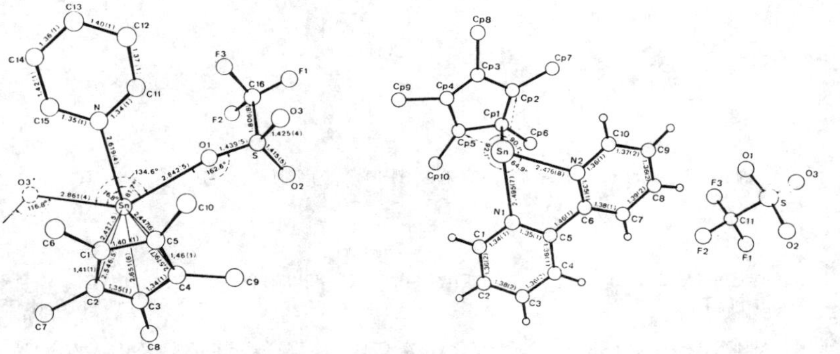

Fig. 1. Details of dimensions and structure for $C_{16}H_{20}F_3NO_3SSn$ (I) (left) and $C_{21}H_{23}F_3N_2O_3SSn$ (II) (right).

I. Monoclinic, $P2_1/c$, a = 8.627, b = 19.170, c = 12.024 Å, β = 93.90°, Z = 4. Mo radiation, R = 0.047 for 3518 reflexions.

II. Triclinic, $P\bar{1}$, a = 8.6372, b = 13.471, c = 11.844 Å, α = 85.32, β = 80.82, γ = 89.66°, Z = 2. Mo radiation, R = 0.065 for 4644 reflexions.

These studies of complexes of the $C_5Me_5Sn^+$ cation complexes with pyridine and bipyridine show that in I the η^5 bonding is distorted, with Sn-C(Cp) 2.427 to 2.651(6) Å (Fig. 1); in II there is even more distortion, with Sn-Cp1 2.41(1), Sn-Cp2 2.59(1), Sn-Cp3 2.80(1), Sn-Cp4 2.78(1) and Sn-Cp5 2.55(1). In I, polymeric chains are formed by Sn-O interactions.

BIS[2-(AMINOCARBONYL)ANILINIUM] DIMETHYLTETRACHLOROSTANNATE(IV) DIHYDRATE
$C_{16}H_{24}Cl_4N_4O_2Sn,2H_2O$ $[2\text{-}H_3NC_6H_4C(O)NH_2]_2^{+}[(CH_3)_2SnCl_4]^{2-},2H_2O$

F.A.K. NASSER, M.B. HOSSAIN, D. van der HELM and J.J. ZUCKERMAN, 1984. Inorg. Chem., 23, 606-609.

Triclinic, $P\bar{1}$, a = 7.389, b = 11.965, c = 6.995 Å, α = 102.29, β = 90.87, γ = 84.74°, Z = 1. Mo radiation, R = 0.019 for 1662 reflexions (at 138K).

The structure (Fig. 1) is centrosymmetric at the trans-Me_2Sn(IV) octahedral units which are held by a network of H-bonds into a three-dimensional, associated lattice. The equivalent H_2O molecules of the dihydrate are three-coordinate, the two different trans pairs of Cl atoms are two- and three-coordinated, and the amido N of the benzamido unit is three-coordinated through H-bonding, which symmetrically dimerizes the benzamide units and internally bonds the anthranilamide units.

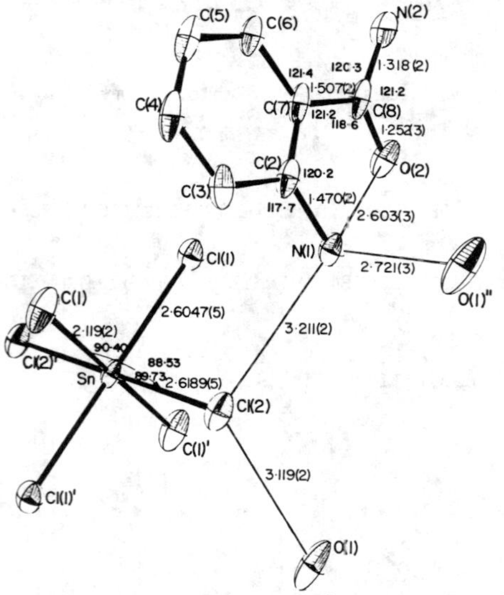

Fig. 1. The structure of $[2\text{-}H_3NC_6H_4C(O)NH_2]_2^{+}[(CH_3)_2SnCl_4]^{2-},2H_2O$.

TETRAETHYLAMMONIUM 2-FLUORO-2,2-DIMETHYL-1,3,2λ[6]-BENZOXATHIASTANNOLATE
$C_{16}H_{30}FNOSSn$ (I) $Et_4N^+[(C_6H_4OS)SnMe_2F]^-$

TETRAETHYLAMMONIUM 2-CHLORO-2,2-DIMETHYL-4-OXO-1,3,2λ[6]-BENZOXATHIASTANNINATE
$C_{17}H_{30}ClNO_2SSn$ (II) $Et_4N^+[(C_6H_4CO_2S)SnMe_2Cl]^-$

TETRAPHENYLPHOSPHONIUM 2-IODO-2,2-DIMETHYL-1,3,2λ^6-BENZOXATHIASTANNOLATE
$C_{32}H_{30}IOPSSn$ (III) $Ph_4P^+[(C_6H_4OS)SnMe_2I]^-$

J.F. VOLLANO, R.O. DAY and R.R. HOLMES, 1984. Organometallics, 3, 750-755.

I. Orthorhombic, Pbca, a = 17.794, b = 12.060, c = 18.095 Å, Z = 8. Mo radiation, R = 0.043 for 1642 reflexions.

II. Monoclinic, $P2_1$, a = 8.971, b = 10.631, c = 11.949 Å, β = 111.29°, Z = 2. Mo radiation, R = 0.028 for 1182 reflexions.

III. Monoclinic, $P2_1/n$, a = 11.951, b = 13.169, c = 20.125 Å, β = 106.03°, Z = 4. Mo radiation, R = 0.028 for 2952 reflexions.

Compounds I and II have pentacoordinated metals in their anions, and their geometries are not too distorted from trigonal bipyramidal (Fig. 1). The ring in each spans axial-equatorial positions with the more electronegative O atom located axially. In I, Sn-O, Sn-F, Sn-S, Sn-C(8) and Sn-C(7) are respectively 2.145, 2.041, 2.432, 2.159 and 2.120 Å, and the angle S-Sn-O is 81.6°. In II, Sn-Cl, Sn-C(8), Sn-C(9), Sn-O(1) and Sn-S are respectively 2.558, 2.142, 2.130, 2.191 and 2.422 Å, and the angle O(1)-Sn-S is 84.9°. Compound III can be viewed as having trigonal-bipyramidal Sn geometry distorted by the presence of an elongated Sn-I axial bond and a weak intermolecular Sn-O bond. The latter is formed via an inversion-related O atom (O'), producing a dimer in which the Sn atom is six-coordinated. Important bonds and their lengths are: Sn-I 3.225, Sn-C(7) 2.127, Sn-C(8) 2.114, Sn-O 2.117, Sn-S 2.452 and Sn-O' 2.646 Å. The angles Sn-O-Sn' and O-Sn-O' are respectively 113.3 and 67.1°.

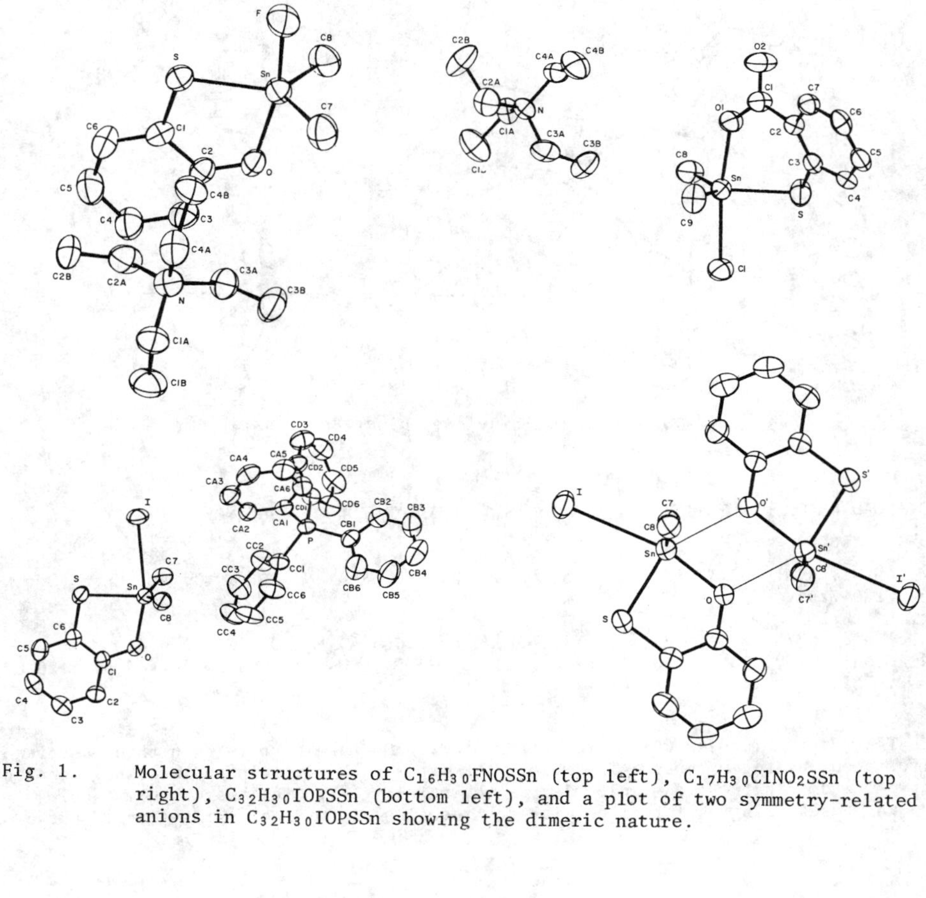

Fig. 1. Molecular structures of $C_{16}H_{30}FNOSSn$ (top left), $C_{17}H_{30}ClNO_2SSn$ (top right), $C_{32}H_{30}IOPSSn$ (bottom left), and a plot of two symmetry-related anions in $C_{32}H_{30}IOPSSn$ showing the dimeric nature.

DI-μ_3-OXO-BIS(μ-MONOCHLOROACETATO-O,O')BIS(MONOCHLOROACETATO)TETRAKIS[DIMETHYLTIN(IV)]
$C_{16}H_{32}Cl_4O_{10}Sn_4$ $\{[(CH_3)_2SnO_2CCH_2Cl]_2O\}_2$

G. VALLE, V. PERUZZO, G. TAGLIAVINI and P. GANIS, 1984. J. Organometal. Chem., 276, 325-329.

Triclinic, $P\bar{1}$, a = 11.426, b = 10.190, c = 7.667 Å, α = 112.17, β = 89.98, γ = 107.68°, D_m = 2.12, Z = 1. Cu radiation, R = 0.099 for 722 reflexions.

The compound consists of dimeric units located on centres of symmetry, and the dimers are weakly bonded to form rows oriented in the a direction (Fig. 1). The tin atoms have quite different environments; Sn(1) which has been considered pentacoordinate could be described as hexacoordinate if Sn(1)...O(4') (2.66 Å) is considered a bond. Sn(2) has distorted trigonal bipyramidal coordination but this changes to seven-coordination if Sn(2)...O(5) and Sn(2)...O(5') are taken into consideration (3.09 and 3.18 Å separations respectively). Other distances are: Sn(1)-O(1) 2.07, Sn(1)-O(1') 2.10, Sn(1)-O(3) 2.36, Sn(1)-C(1) 2.11, Sn(1)-C(2) 2.09, Sn(2)-O(1) 2.03, Sn(2)-O(4) 2.17, Sn(2)-O(2) 2.20, Sn(2)-C(3) 2.21, Sn(2)-C(4) 2.09 Å.

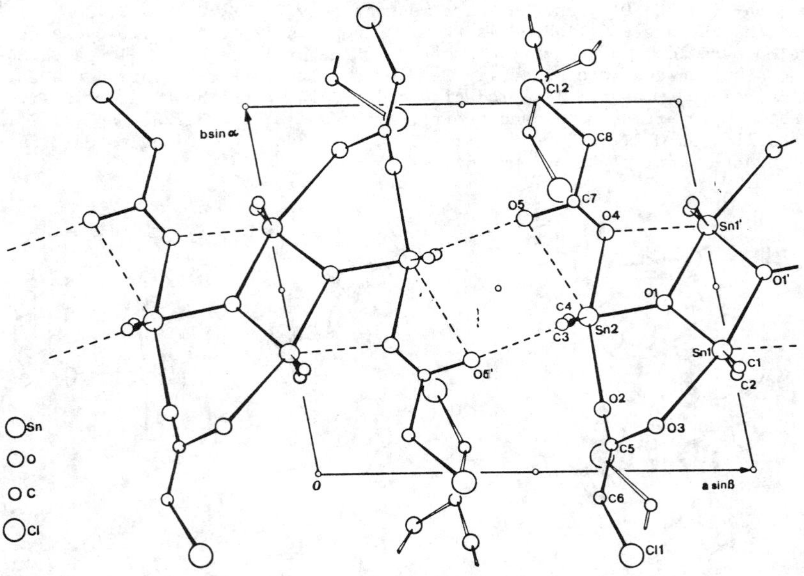

Fig. 1. Structure of $C_{16}H_{32}Cl_4O_{10}Sn_4$ viewed along the c axis.

BIS(AQUADICHLORODIMETHYLTIN) 1,4,7,10,13,16-HEXAOXACYCLOOCTADECANE
$C_{16}H_{40}Cl_4O_8Sn$ $2(C_2H_8Cl_2OSn),C_{12}H_{24}O_6$

M.M. AMINI, A.L. RHEINGOLD, R.W. TAYLOR and J.J. ZUCKERMAN, 1984. J. Am. Chem. Soc., 106, 7289-7291.

Monoclinic, C2/m, a = 16.348, b = 11.668, c = 15.063 Å, β = 93.22°, Z = 4. Mo radiation, R = 0.0331 for 3341 reflexions (at 23°C). Using data collected at -120°C R = 0.0282.

The water molecules which are coordinated to Sn take part in twin three-centre (bifurcated) hydrogen bonds. The relevant H atoms each participate in a short contact to two O atoms in two crown molecules as shown in Fig. 1. The two crown molecules have an unusual ring conformation with the sequence of torsion angles [(anti-gauche-anti)$_2$gauche-gauche-anti]$_2$. Each of the Sn atoms lies on a badly distorted O_h site with Sn-Cl in the range 2.393-3.311 (1) Å.

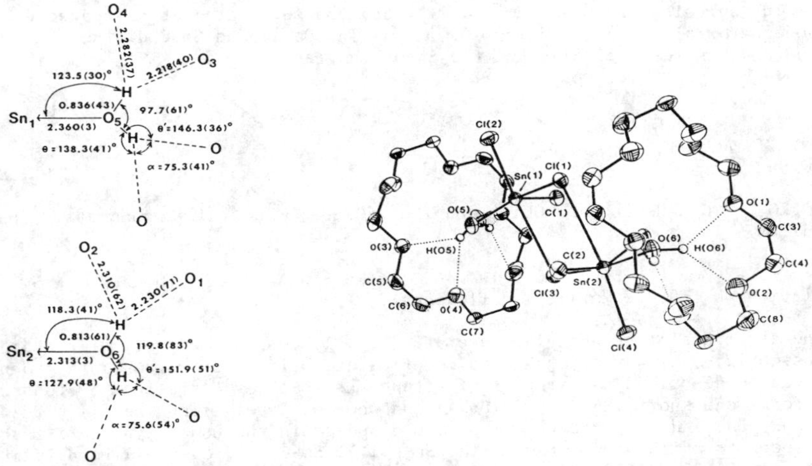

Fig. 1. $C_{16}H_{40}Cl_4O_8Sn$: details of the geometry about the two water molecules and a view of the complex.

DIPHENYLTIN(IV) DICHLORIDE 2,6-DIMETHYLPYRIDINE N-OXIDE
$C_{19}H_{19}Cl_2NOSn$ (I) $Ph_2SnCl_2.2,6\text{-}(CH_3)_2C_5H_3NO$

TRIMETHYLTIN(IV) CHLORIDE 2,6-DIMETHYLPYRIDINE N-OXIDE
$C_{10}H_{18}ClNOSn$ (II) $Me_3SnCl.2,6\text{-}(CH_3)_2C_5H_3NO$

A.L. RHEINGOLD, S.W. NG and J.J. ZUCKERMAN, 1984. Organometallics, 3, 233-237.

I. Triclinic, $P\bar{1}$, a = 9.359, b = 10.534, c = 10.996 Å, α = 97.46, β = 114.74, γ = 93.26°, Z = 2. Mo radiation, R = 0.025 for 3102 reflexions.

II. Triclinic, $P\bar{1}$, a = 8.174, b = 9.599, c = 9.895 Å, α = 115.33, β = 105.50, γ = 81.45°, Z = 2. Mo radiation, R = 0.028 for 2207 reflexions.

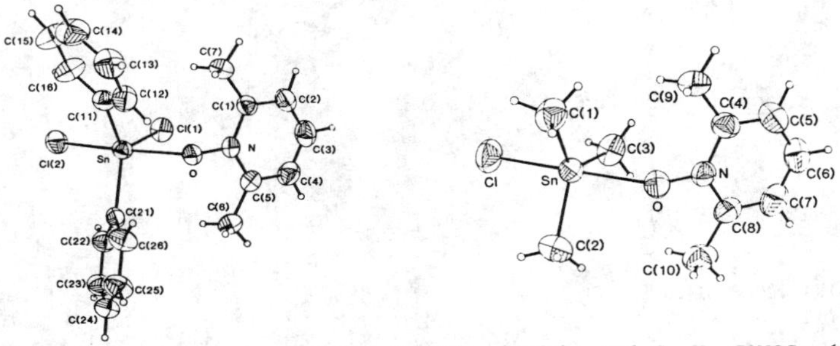

Fig. 1. Molecular structure of $C_{19}H_{19}Cl_2NOSn$ (left), and $C_{10}H_{18}ClNOSn$ (right).

In both complexes the Sn atom has trigonal bipyramidal coordination geometry (Fig. 1). In I the angles C(11)-Sn-C(21), C(11)-Sn-Cl(1) and C(21)-Sn-Cl(1) are 124.1, 120.2 and 114.7° respectively with the Sn atom displaced 0.128 Å from the equatorial plane towards the Cl(2) atom. The Sn-Cl(1) (equatorial) bond (2.366 Å) is shorter than the Sn-Cl(2) axial bond (2.471 Å). The Sn-O distance is 2.296 Å and the two Sn-C bond distances are approximately equal (2.119 and 2.128 Å), and the angle

O-Sn-Cl(2) is 172.8°. The Sn-O-N angle is 125.7°. In II, the Sn-C(methyl) bond distances are 2.119, 2.120 and 2.125 Å, and the C-Sn-C angles are 114.1, 121.8 and 123.1°. The apical Cl-Sn-O angle is 177.4° and the metal atom is displaced 0.236 Å from the equatorial plane towards the Cl atom. The Sn-Cl and Sn-O distances are 2.533 and 2.400 Å respectively, and the Sn-O-N angle is 130.6°.

TRIPHENYLTIN ACETATE
$C_{20}H_{18}O_2Sn$ $(C_6H_5)_3SnOOCCH_3$

K.C. MOLLOY, T.G. PURCELL, K. QUILL and I.W. NOWELL, 1984. J. Organometal. Chem., 267, 237-247.

Monoclinic, $P2_1/c$, a = 8.969, b = 10.146, c = 19.540 Å, β = 93.70°, D_m = 1.51, Z = 4. Mo radiation, R = 0.022 for 1841 reflexions.

The structure of triphenyltin acetate (Fig. 1) consists of polymeric chains in which essentially planar triorganotin moieties are bridged by carboxylate groups. The chains can be described as being of "flattened helical" or "stretched S" construction. The Sn forms one short (Sn-O(2) 2.185 Å) and one relatively long (Sn-O(1') 2.349 Å) bond to oxygen, and the acetate groups are disposed in the usual syn, anti-fashion with respect to pairs of Sn atoms. The Sn-C(11), Sn-C(21), Sn-C(31) bond lengths are respectively 2.120, 2.138 and 2.133 Å. There is a weak Sn-O interaction (Sn-O(1) 3.206 Å), suggesting this as the first crystallographically authenticated example of a six-coordinate triorgantin compound. Angles of interest are: C(11)-Sn-C(21) 113.0, C(11)-Sn-C(31) 111.2, C(21)-Sn-C(31) 135.2, O(1)-Sn-O(1') 139.8, O(1)-Sn-O(2) 43.4, O(1')-Sn-O(2) 173.6°.

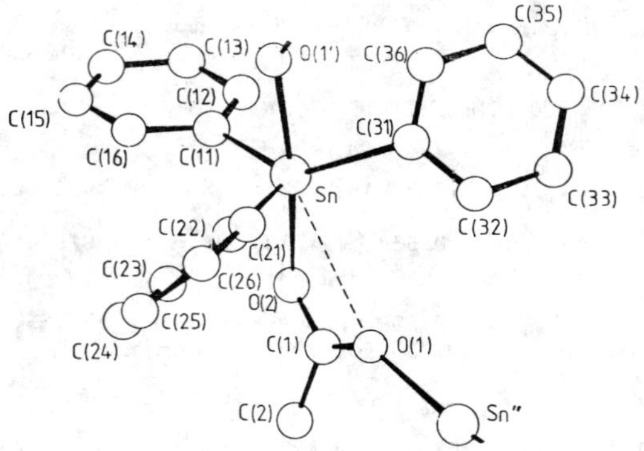

Fig. 1. Molecular structure of $C_{20}H_{18}O_2Sn$.

METHYLTHIOTRIPHENYLMETHANE
$C_{20}H_{18}S$ (I) $(C_6H_5)_3CSCH_3$

METHYLTHIOTRIPHENYLSILANE
$C_{19}H_{18}SSi$ (II) $(C_6H_5)_3SiSCH_3$

METHYLTHIOTRIPHENYLGERMANE
$C_{19}H_{18}GeS$ (III) $(C_6H_5)_3GeSCH_3$

METHYLTHIOTRIPHENYLSTANNANE
$C_{19}H_{18}SSn$ (IV) $(C_6H_5)_3SnSCH_3$

METHYLTHIOTRIPHENYLPLUMBANE
$C_{19}H_{18}PbS$ (V) $(C_6H_5)_3PbSCH_3$

G.D. ANDREETTI, G. BOCELLI, G. CALESTANI and P. SGARABOTTO, 1984. J. Organometal. Chem., 273, 31-45.

I. Orthorhombic, $P2_12_12_1$, a = 27.619, b = 7.562, c = 7.535 Å, Z = 4. Mo radiation, R = 0.053 for 1375 reflexions.

II. Triclinic, $P\bar{1}$, a = 10.406, b = 9.625, c = 9.107 Å, α = 81.3, β = 98.0, γ = 70.6°, Z = 2. Mo radiation, R = 0.036 for 3488 reflexions.

III. Triclinic, $P\bar{1}$, a = 10.422, b = 9.641, c = 9.118 Å, α = 81.3, β = 97.8, γ = 71.1°, Z = 2. Mo radiation, R = 0.030 for 3758 reflexions.

IV. Monoclinic, $P2_1/n$, a = 9.744, b = 18.866, c = 9.915 Å, β = 106.1°, Z = 4. Mo radiation, R = 0.041 for 2047 reflexions.

V. Monoclinic, $P2_1/n$, a = 9.651, b = 19.077, c = 9.895 Å, β = 106.8°, Z = 4. Mo radiation, R = 0.070 for 1779 reflexions.

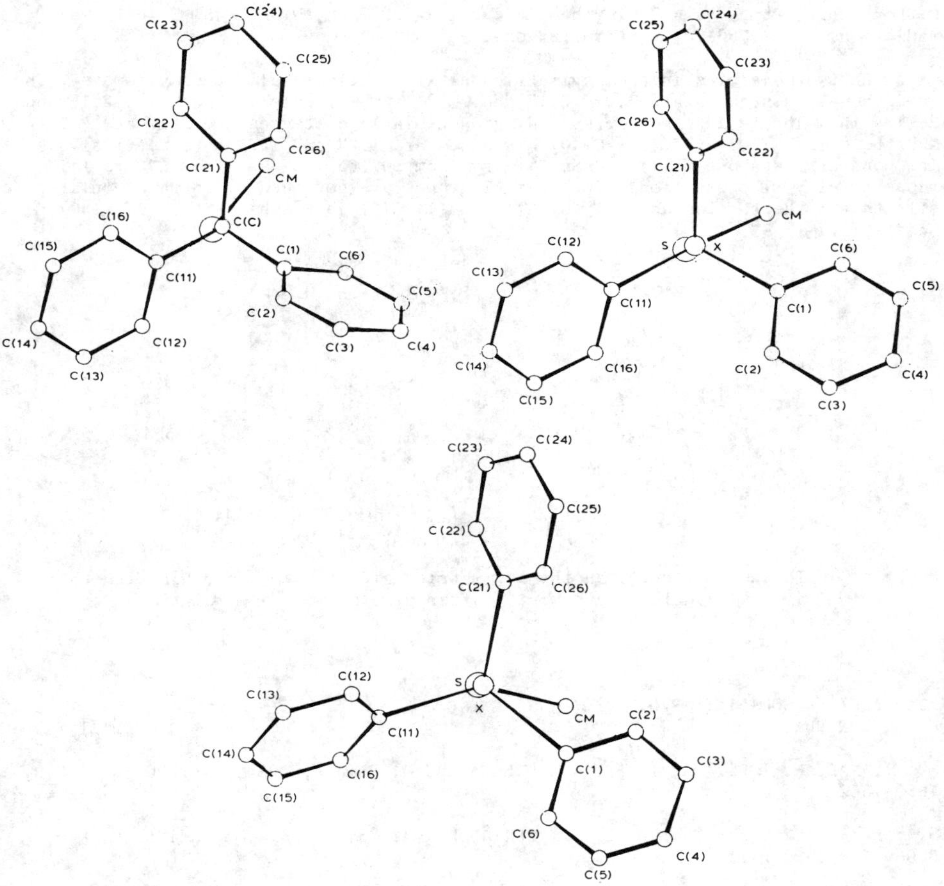

Fig. 1. Projections of the structures of $C_{20}H_{18}S$ (top left), $C_{19}H_{18}SSi$ and $C_{19}H_{18}GeS$ (top right), and $C_{19}H_{18}SSn$ and $C_{19}H_{18}PbS$ (bottom)

A distorted tetrahedral coordination around the central atom is a common feature in this series of derivatives (Fig. 1). Compounds II and III are isomorphous as are compounds IV and V. The greatest distortion from regular tetrahedral geometry involves the S-X-C(11) angle which is close to 102° for the first three compounds and to 104° for the other two. The other two S-X-C angles are closer to the ideal 109.5° angle. All phenyl rings are planar within experimental error and are oriented in propeller fashion. The X-S distances in I, II, III, IV and V are 1.852, 2.138, 2.224, 2.391 and 2.489 Å respectively and the S-CM distances range from 1.770(12) to 1.809(5) Å. The average X-C(phenyl) distances in I, II, III, IV and V are respectively 1.53, 1.861, 1.931, 2.130 and 2.19 Å.

TRIPHENYLVINYLTIN
$C_{20}H_{18}Sn$ $(C_6H_5)_3(CH=CH_2)Sn$

F. THEOBALD and B. TRIMAILLE, 1984. J. Organometal. Chem., 267, 143-149.

Orthorhombic, $Pbn2_1$, a = 7.016, b = 16.442, c = 29.75 Å, D_m = 1.48, Z = 8. Mo radiation, R = 0.080 for 2286 reflexions.

The two molecules in the asymmetric unit (Fig. 1) are almost enantiomers. The Sn-C(phenyl) distances are 2.07, 2.15 and 2.12 Å in molecule 1, and 2.15, 2.16 and 2.11 Å in molecule 2. The corresponding Sn-C(vinyl) distances are 2.17 and 2.15 Å and the C(19)=C(20) and C(39)=C(40) distances are 1.27 and 1.31 Å respectively. The bond angles about Sn(1) and Sn(2) vary from 107 to 112° and from 108 to 111° respectively. The positions of the vinyl groups are governed by intramolecular steric effects and the torsion angle C(20)-C(19)-Sn(1)-C(7) is 5° whilst that of C(40)-C(39)-Sn(2)-C(33) is -9°.

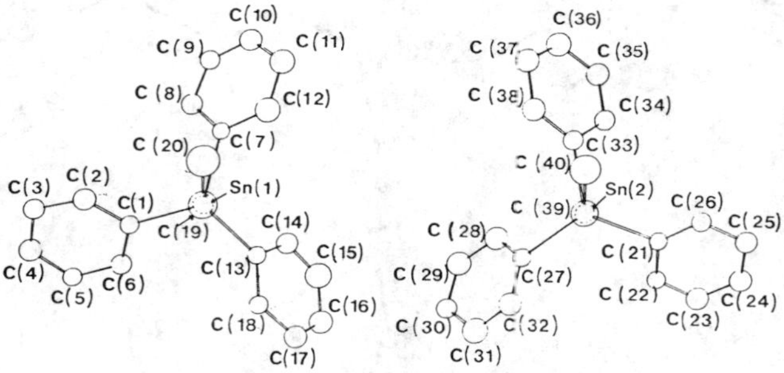

Fig. 1. The two molecules in the asymmetric unit of $C_{20}H_{18}Sn$ exhibiting their almost enantiomeric symmetry after displacement and rotation.

catena-DI-μ-IODO-BIS(PYRIDINE)LEAD(II)
$C_{20}H_{20}I_4N_4Pb_2$ $Pb_2I_4(C_5H_5N)_4$

H. MIYAMAE, H. TORIYAMA, T. ABE, G. HIHARA and M. NAGATA, 1984. Acta Cryst., C40, 1559-1562.

Trigonal, $P3_121$, a = 10.103, c = 24.584 Å, D_m = 2.78, Z = 3. Mo radiation, R = 0.069 for 1630 reflexions.

The two independent Pb atoms lie on twofold axes. One Pb has a regular octahedral coordination with the two pyridine molecules trans while the other has a very distorted octahedral coordination with the two pyridines cis (Fig. 1). Pb atoms are bridged through two I atoms (mean Pb-I 3.258(3) Å) to form helical chains along c.

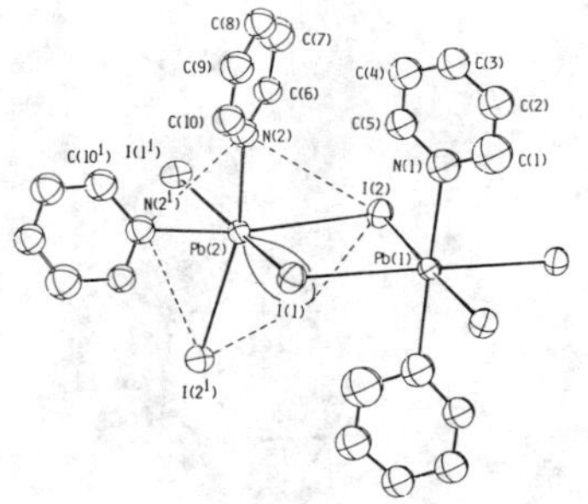

Fig. 1. The $Pb_2I_4(C_5H_5N)_4$ unit. (The lobe indicates a possible lone-pair site and broken lines show the pentagonal plane of the cis form.)

9,9-DI-n-BUTYL-9-STANNAFLUORENE
$C_{20}H_{26}Sn$

V.K. BEL'SKII, 1984. J. Struct. Chem., 25, 963-965 [Zh. Strukt. Khim., 25(6), 136-138].

Monoclinic, $P2_1/a$, a = 8.176, b = 9.668, c = 25.072 Å, γ = 109.35°, Z = 4. Mo radiation, R = 0.023 for 2400 reflexions.

The molecule is shown in Fig. 1. The tricyclic stannafluorene system is planar to within 0.012(4) Å. The Sn atom has distorted tetrahedral geometry with Sn-C distances ranging from 2.132(3)-2.142(4) Å (mean 2.138 Å) and an intracyclic C10-Sn-C12 bond angle of 83.0(1)°. Both n-butyl groups are planar and are oriented approximately perpendicular to one another (dihedral angle between the mean planes is 87.1°). The n-butyl groups have different configurations (cis-trans and trans-trans).

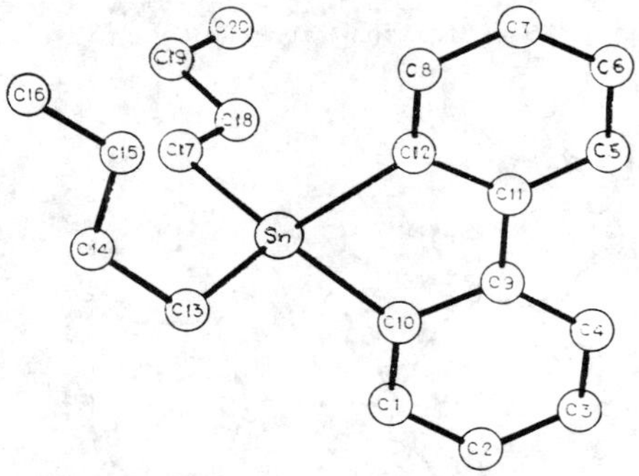

Fig. 1. The 9,9-di-n-butyl-9-stannafluorene molecule.

TETRACHLOROBIS(ETHYLCINNAMATE-O)TIN(IV)
$C_{22}H_{24}Cl_4O_4Sn$ $(C_{11}H_{12}O_4)_2SnCl_4$

F.D. LEWIS, J.D. OXMAN and J.C. HUFFMAN, 1984. J. Am. Chem. Soc., 106, 466-468.

Triclinic, $P\bar{1}$, a = 9.853, b = 8.557, c = 8.016 Å, α = 106.85, β = 103.01, γ = 99.05°, Z = 1. Mo radiation, R = 0.023 for 1557 reflexions (at -163°C).

The complex (Fig. 1) displays octahedral geometry about Sn with the metal at a centre of symmetry. Complexes show head-to-tail stacking of symmetry-related ester molecules with distances of 4.023(8) and 4.125(8) Å between reactive double bonds.

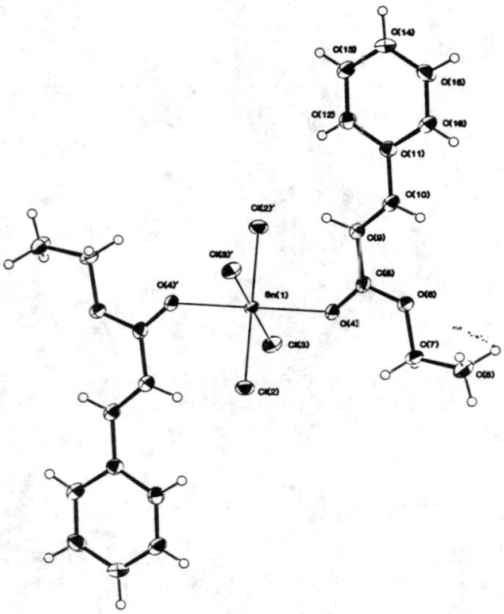

Fig. 1. The $(C_{11}H_{12}O_4)_2SnCl_4$ molecule.

BIS(ACETONITRILE)TRICYCLOHEXYLTIN HEXAFLUOROANTIMONATE
$C_{22}H_{39}F_6N_2SbSn$ $[(C_6H_{11})_3Sn(CH_3CN)_2]^+[SbF_6]^-$

W.A. NUGENT, R.J. McKINNEY and R.L. HARLOW, 1984. Organometallics, <u>3</u>, 1315-1317.

Monoclinic, $P2_1/n$, a = 13.923, b = 16.501, c = 11.891 Å, β = 93.85°, Z = 4. Mo radiation, R = 0.033 for 6896 reflexions (at -100°C).

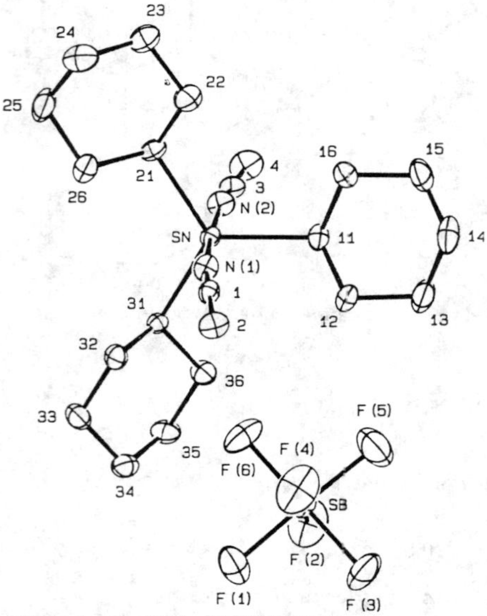

Fig. 1. Molecular structure of $C_{22}H_{39}F_6N_2SbSn$.

The coordination about the Sn atom is trigonal bipyramidal with the acetonitrile ligands axial (Fig. 1), and loosely coordinated with the Sn (Sn-N bonds are long at 2.374 and 2.472 Å). The Sn-C(11), Sn-C(21) and Sn-C(31) bond distances are 2.161, 2.159 and 2.158 Å respectively, and the equatorial angles C(11)-Sn-C(21), C(11)-Sn-C(31) and C(21)-Sn-C(31) are respectively 122.3, 120.5, 116.9°. The angle N(1)-Sn-N(2) is 177.2°. The cyclohexyl groups have slightly different conformations and these probably affect the Sn-N distances and the C-Sn-C angles. Additional angles are: Sn-N(1)-C(1) 168.3°, Sn-N(2)-C(3) 173.2°, N(1)-C(1)-C(2) 178.1° and N(2)-C(3)-C(4) 179.5°. In the anion the average Sb-F distance is 1.850 Å

(η^5-CYCLOPENTADIENYL)(TRICHLOROGERMYL)(TRIPHENYLPHOSPHINE)NICKEL(II) HEMIBENZENE SOLVATE

$C_{23}H_{20}Cl_3GeNiP,0.5(C_6H_6)$ η^5-$C_5H_5(GeCl_3)Ph_3PNi^{II},0.5(C_6H_6)$

N.A. BELL, F. GLOCKLING, A. McGREGOR, M.L. SCHNEIDER and H.M.M. SHEARER, 1984. Acta Cryst., C40, 623-625.

Monoclinic, C2/c, a = 15.080, b = 11.889, c = 29.113 Å, β = 93.38°, D_m = 1.52, Z = 8. Mo radiation, R = 0.051 for 2862 reflexions.

In the molecule (Fig. 1) there is a short Ni-Ge bond of 2.248(1) Å, suggesting appreciable d_π-d_π interaction between the two metal atoms. The Ni, Ge, P and one Cl atom are located in a plane approximately perpendicular to that of the distorted cyclopentadienyl ring. Ni is bonded to the C_5H_5 ring, to Ph_3P and to Ge while the latter is also bonded to three Cl atoms. The benzene ring lies on the twofold axis.

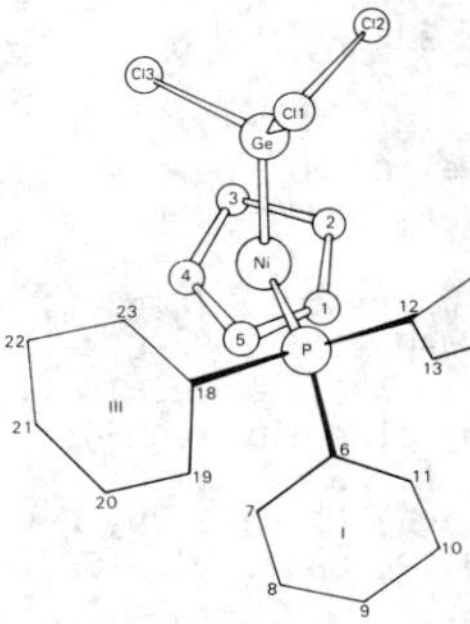

Fig. 1. A perspective view of the η^5-$C_5H_5(GeCl_3)Ph_3PNi^{II}$ molecule.

(μ_2-METHYLENE)BIS[(3-METHYL-3-AZAPENTANE-1,5-DITHIOLATO-N,S,S')PHENYLTIN]

$C_{23}H_{34}N_2S_4Sn_2$ $CH_2[(C_6H_5)Sn(SCH_2CH_2)_2NCH_3]_2$

R. WILLEM, M. GIELEN, J. MEUNIER-PIRET, M. VAN MEERSSCHE, K. JURKSCHAT and A. TZSCHACH, 1984. J. Organometal. Chem., 277, 335-350.

Monoclinic, $P2_1/c$, a = 8.279, b = 16.146, c = 21.331 Å, β = 102.72°, Z = 4. Mo radiation, R = 0.048 for 3457 reflexions.

Each Sn atom exhibits distorted trigonal bipyramidal geometry but the bridging methylene group is in an apical position in the Sn(1) environment, and an equatorial site in the Sn(2) coordination sphere (Fig. 1). The phenyl groups likewise occupy either an equatorial or an apical site whereas the N atoms are both apical. Distances involving Sn are: Sn(1)-S(1) 2.423, Sn(1)-S(2) 2.409, Sn(1)-N(1) 2.654, Sn(1)-C(1) 2.162, Sn(1)-C(2) 2.129, Sn(2)-S(3) 2.424, Sn(2)-S(4) 2.418, Sn(2)-N(2) 2.651, Sn(2)-C(1) 2.153 and Sn(2)-C(8) 2.179 Å. The angles C(1)-Sn(1)-N(1) and C(8)-Sn(2)-N(2) are 162.8 and 166.4° respectively, whilst the equatorial plane angles range from 110.9 to 126.2°.

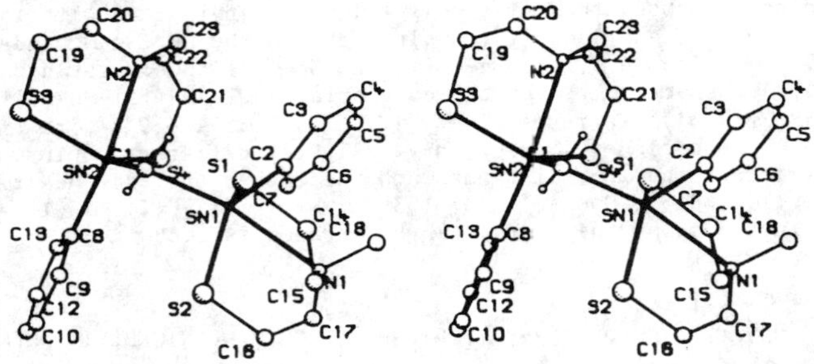

Fig. 1. Stereoview of $C_{23}H_{34}N_2S_4Sn_2$.

DI-t-BUTYLTIN OXIDE
$C_{24}H_{54}O_3Sn_3$ (I) $[(t\text{-}C_4H_9)_2SnO]_3$

DI-t-AMYLTIN OXIDE
$C_{30}H_{66}O_3Sn_3$ (II) $[(t\text{-}C_5H_{11})_2SnO]_3$

H. PUFF, W. SCHUH, R. SIEVERS, W. WALD and R. ZIMMER, 1984. J. Organometal. Chem., 260, 271-280

I. Trigonal, $R\bar{3}c$, a = 10.350, c = 51.444 Å, Z = 6. Mo radiation, R = 0.045 for 693 reflexions at 185K.

II. Monoclinic, C2/c, a = 21.390, b = 10.961, c = 19.000 Å, β = 121.75°, Z = 4. Mo radiation, R = 0.054 for 2875 reflexions at 185K.

Molecule I has $\bar{3}$ crystallographic symmetry (Fig. 1) with the Sn and O atoms forming a six-membered planar ring (Sn-O 1.965, Sn-C 2.194, C-C (average) 1.56 Å; O-Sn-O 106.9, Sn-O-Sn 133.1, C-Sn-C 119.5°). Molecule II has crystallographic twofold symmetry, with the axis of rotation passing through Sn(1) and O(1) (Fig. 1). Relevant distances and angles are: Sn(1)-O(2) 1.953, Sn(2)-O(1) 1.952, Sn(2)-O(2) 1.978, Sn(1)-C(10) 2.28, Sn(2)-C(20) 2.24, Sn(2)-C(30) 2.29, C-C (average) 1.59 Å; O(2)-Sn(1)-O(2') 106.3, O(1)-Sn(2)-O(2) 105.9, Sn(1)-O(2)-Sn(2) 133.7, Sn(2)-O(1)-Sn(2') 134.5°

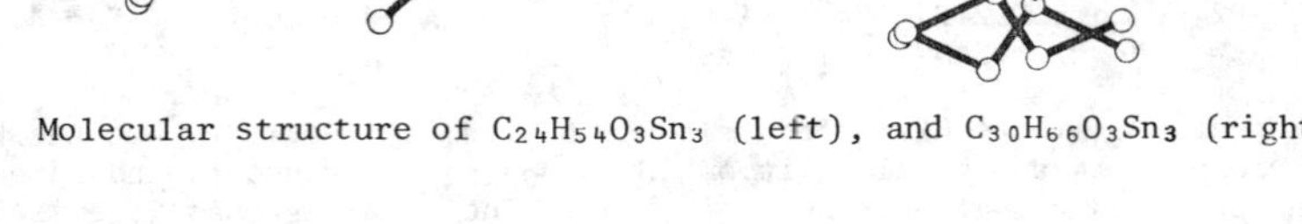

Fig. 1. Molecular structure of $C_{24}H_{54}O_3Sn_3$ (left), and $C_{30}H_{66}O_3Sn_3$ (right).

(SALICYLATO-O)TRIPHENYLTIN(IV)
$C_{25}H_{20}O_3Sn$ (I) $Ph_3Sn[o-(OH)C_6H_4CO_2]$

(o-ANISATO-O)TRIPHENYLTIN(IV)
$C_{26}H_{22}O_3Sn$ (II) $Ph_3Sn[o-(CH_3O)C_6H_4CO_2]$

(p-METHYLTHIOBENZOATO)TRIPHENYLTIN(IV)
$C_{26}H_{22}O_2SSn$ (III) $Ph_3Sn(p-(CH_3S)C_6H_4CO_2]$

J.F. VOLLANO, R.O. DAY, D.N. RAU, V. CHANDRASEKHAR and R.R. HOLMES, 1984. Inorg. Chem., 23, 3153-3160.

I. Monoclinic, $P2_1/c$, a = 13.318, b = 12.113, c = 14.590 Å, β = 116.47°, Z = 4. Mo radiation, R = 0.022 for 3218 reflexions.

II. Triclinic, $P\bar{1}$, a = 10.632, b = 11.452, c = 11.719 Å, α = 62.86, β = 71.95, γ = 62.31°, Z = 2. Mo radiation, R = 0.024 for 2364 reflexions.

III. Triclinic, $P\bar{1}$, a = 8.814, b = 9.816, c = 14.430 Å, α = 89.39, β = 80.19, γ = 69.89°, Z = 2. Mo radiation, R = 0.025 for 3712 reflexions.

The three structures are shown in Fig. 1. Compounds II and III are pentacoordinated structures containing Ph_3Sn (Sn-C 2.122(4)-2.135(3) Å) coordinated to the carboxylate group. The structural distortion for each is a displacement from the tetrahedron toward a trigonal bipyramid. Sn-O1 and Sn-O2 are 2.054(3) and 2.781(3) Å in II, 2.060(2) and 2.783(3) Å in III. For I, the structure contains an intermolecular O atom interaction (Sn-O3' 3.035(2) Å), in addition to H-bonding effects, with a resultant geometry close to tetrahedral. Bond lengths for I: Sn-C 2.117(3)-2.132(3), Sn-O1 2.083(2), Sn-O2 3.071(2) Å.

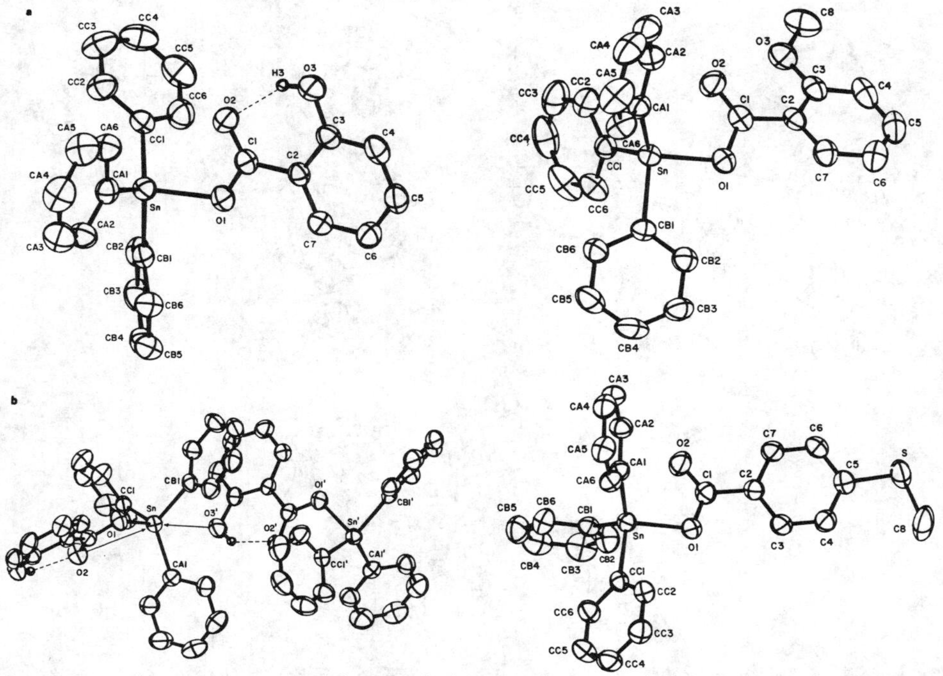

Fig. 1. The structures of $Ph_3Sn[o-(OH)C_6H_4CO_2]$ (left, intermolecular interaction below), $Ph_3Sn[o-(CH_3O)C_6H_4CO_2]$ (top right), and $Ph_3Sn[p-(CH_3S)C_6H_4CO_2]$ (bottom right).

(o-AMINOBENZOATO-O)TRIPHENYLTIN(IV)
$C_{25}H_{21}NO_2Sn$ (I)

(o-DIMETHYLAMINOBENZOATO-O)TRIPHENYLTIN(IV)
$C_{27}H_{25}NO_2Sn$ (II)

(p-AMINOBENZOATO-O)TRIPHENYLTIN(IV) BENZENE SOLVATE
$C_{25}H_{21}NO_2Sn,0.5(C_6H_6)$ (III)

R.G. SWISHER, J.F. VOLLANO, V. CHANDRASEKHAR, R.O. DAY and R.R. HOLMES, 1984. Inorg. Chem., 23, 3147-3152.

I. Monoclinic, $P2_1/c$, a = 12.476, b = 21.974, c = 7.962 Å, β = 98.64°, Z = 4. Mo radiation, R = 0.033 for 2964 reflexions.

II. Orthorhombic, Pccn, a = 37.111, b = 17.119, c = 7.479 Å, Z = 8. Mo radiation, R = 0.037 for 1799 reflexions.

III. Monoclinic, $P2_1/c$, a = 15.920, b = 9.111, c = 17.468 Å, β = 99.38°, Z = 4. Mo radiation, R = 0.029 for 3308 reflexions.

All three molecules (Fig. 1) have pseudo-pentacoordinated structures containing Ph_3Sn (Sn-C 2.122(4)-2.138(3) Å) coordinated to the carboxylate group. The structural distortion in each is a displacement from the tetrahedron toward the trigonal bipyramid. The Sn-O1 and Sn-O2 distances are 2.043(3) and 2.823(3) Å for I, 2.115(6) and 2.564(7) Å for II, 2.072(2) and 2.629(2) Å for III. Intramolecular N-H...O hydrogen bonding is present in I, while intermolecular N-H...O hydrogen bonds are present in III.

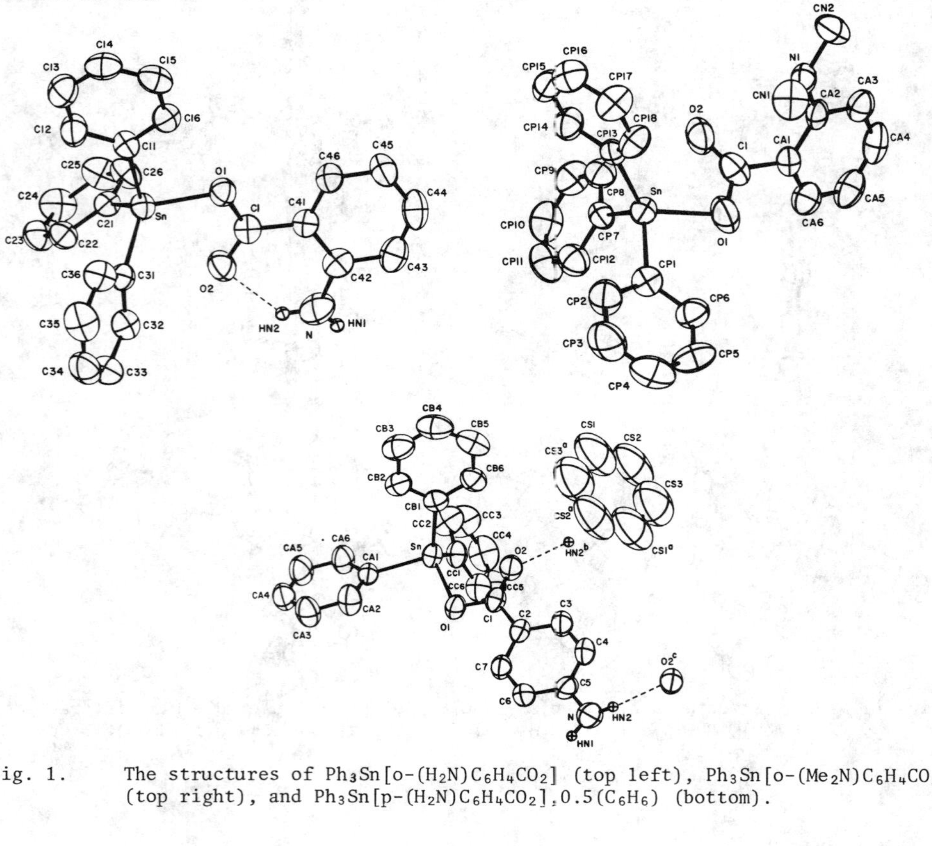

Fig. 1. The structures of $Ph_3Sn[o-(H_2N)C_6H_4CO_2]$ (top left), $Ph_3Sn[o-(Me_2N)C_6H_4CO_2]$ (top right), and $Ph_3Sn[p-(H_2N)C_6H_4CO_2].0.5(C_6H_6)$ (bottom).

6,7-DIHYDRO-6,6-DIPHENYL-5H-DIBENZO[c,e]STANNEPINE
$C_{26}H_{22}Sn$ (I)

6,7-DIHYDRO-6,6-DIPHENYL-5,7-BIS(TRIMETHYLSILYL)-5H-DIBENZO[c,e]STANNEPINE
$C_{32}H_{38}Si_2Sn$ (II)

L.M. ENGELHARDT, W.-P. LEUNG, C.L. RASTON, P. TWISS and A.H. WHITE, 1984. J. Chem. Soc. Dalton, 331-340.

I. Monoclinic, $P2_1/n$, a = 15.42, b = 13.334, c = 10.042 Å, β = 95.90°, Z = 4. Mo radiation, R = 0.037 for 1621 reflexions.

II. Monoclinic, C2/c, a = 9.937, b = 18.231, c = 17.316 Å, β = 99.20°, Z = 4. Mo radiation, R = 0.034 for 1974 reflexions.

Compound I consists of discrete molecules (Fig. 1) in which Sn-C distances range from 2.139(8) to 2.159(9) Å. The angle subtended at Sn by the bidentate ligand is 98.7(3)°. There is no symmetry whatever in the molecule and the mean torsion angle about the bond between the two halves of the bidentate ligand is 66.0°. Sn atom deviations from bidentate aromatic planes are 1.78 and 2.02 Å while the deviations from phenyl ring planes are 0.14 and 0.09 Å. In II a crystallographic twofold axis passes through the Sn atom and the midpoint of the bidentate ligand. The Sn-C(phenyl) distances average 2.136(4) Å and the Sn-C(alkyl) distances 2.184(4) Å. The angle subtended at Sn by the bidentate ligand is 99.4(2)° and the angle between the monodentate ligands is 116.1(2)°. The torsion angle about the central bond of the bidentate ligand is 70.8°. The Si-C distances range from 1.848(8) to 1.887(5) Å and the Si-C-Sn,C angles are 114.1(2) and 120.8(3)°. Sn deviates from the bidentate ligand plane by 2.04 Å and from the phenyl ligand plane by 0.08 Å.

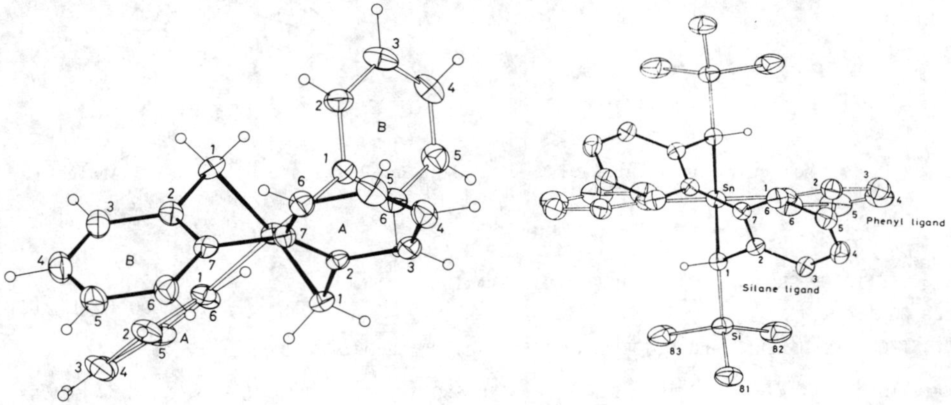

Fig. 1. A single molecule of $C_{26}H_{22}Sn$ projected down the bisector of the angle subtended at tin by two phenyl C atoms (left) and a single molecule of $C_{32}H_{38}Si_2Sn$ projected down the crystallographic twofold axis (right).

t-BUTYLCHLORO(2,6-DIACETYLPYRIDINE)BIS(2-AMINOBENZOYLHYDRAZONE)TIN(IV)
$C_{27}H_{30}ClN_7O_2Sn$ (I)

(2,6-DIACETYLPYRIDINE)BIS(2-AMINOBENZOYLHYDRAZONE)DIPHENYLTIN(IV)
$C_{35}H_{31}N_7O_2Sn$ (II)

C. PELIZZI, G. PELIZZI and G. PREDIERI, 1984. J. Organometal. Chem., 263, 9-20.

I. Orthorhombic, Pnma, a = 21.466, b = 17.286, c = 7.449 Å, Z = 4. Cu radiation, R = 0.042 for 1396 reflexions.

II. Monoclinic, $P2_1/a$, a = 23.389, b = 14.362, c = 9.388 Å, β = 87.14°, Z = 4. Cu radiation, R = 0.071 for 2653 reflexions.

Both complexes display distorted pentagonal bipyramidal structures with the deprotonated and quinquedentate hydrazone ligand spanning the pentagonal plane (Fig. 1). Molecule I has crystallographic m symmetry and the butyl ligand is disordered about the mirror plane. In I the Sn atom lies 0.18 Å out of the equatorial plane (towards C(13)) whilst in II it lies in the plane of the five donor atoms. The Sn-O distance in I is 2.144 Å and in II the distances are 2.207 and 2.187 Å. The Sn-N distances are almost identical in the two compounds: 2.278 and 2.318 Å in I, and 2.266, 2.308 and 2.324 Å in II. The Sn-Cl and Sn-C(13) distances in I are respectively 2.446 and 2.14 Å whilst the Sn-C(phenyl) distances in II are 2.182 and 2.170 Å. There are hydrogen bonds in both structures. In I, an N...N intramolecular hydrogen bond (N(1)...N(2), 2.666 Å) and an intermolecular one (N(1)...Cl, 3.310 Å), and in II two intramolecular bonds (N(1)...N(2) 2.72 Å, N(7)...O(2) 2.67 Å).

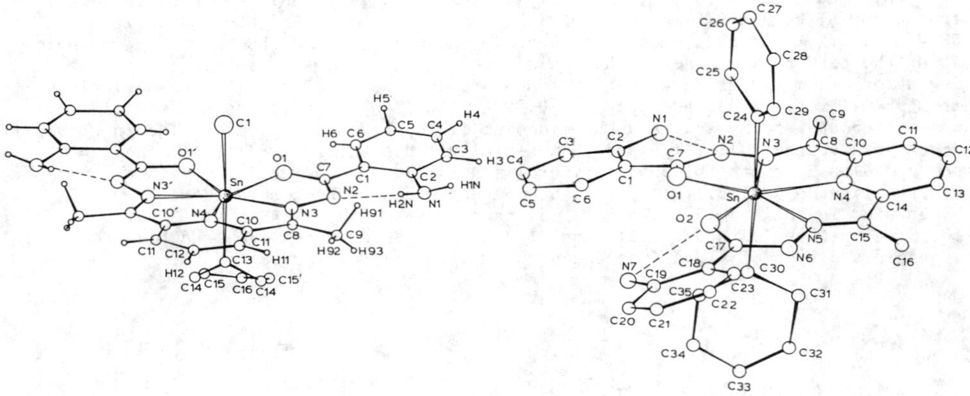

Fig. 1. Molecular structures of $C_{27}H_{30}ClN_7O_2Sn$ (left) and $C_{35}H_{31}N_7O_2Sn$ (right).

1,2-BIS(TRICHLOROACETATO)TETRAPHENYLDIGERMANIUM

$C_{28}H_{20}Cl_6Ge_2O_4$ $[(Cl_3CCOO)(C_6H_5)_2Ge]_2$

D. SIMON, K. HÄBERLE and M. DRÄGER, 1984. J. Organometal. Chem., 267, 133-142.

Triclinic, $P\bar{1}$, a = 9.062, b = 9.298, c = 10.580 Å, α = 107.48, β = 94.66, γ = 114.33°, D_m = 1.71, Z = 1. Mo radiation, R = 0.048 for 2942 reflexions.

The Ge-Ge bond straddles a crystallographic centre of symmetry (Fig. 1) and is bridged by both acetates. The Ge-Ge distance is 2.393 Å and the Ge-O bond lengths are 2.073 and 2.314 Å. Each Ge atom has trigonal bipyramidal coordination with equatorial angles of Ge(1')-Ge-C(11) 120.2, Ge(1')-Ge-C(21) 123.0 and C(11)-Ge(1)-C(21) 116.4°. The angle O(1)-Ge(1)-O(2') is 175.4° and that of Ge(1)-O(1)-C(1) is 119.6°.

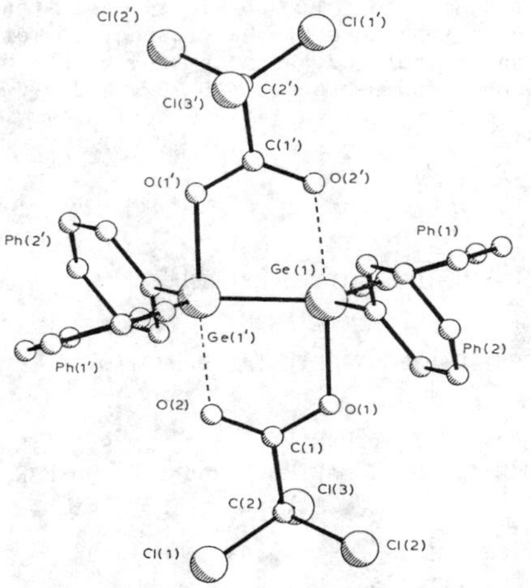

Fig. 1. Molecular structure of $C_{28}H_{20}Cl_6Ge_2O_4$.

TETRA-o-TOLYLGERMANIUM

$C_{28}H_{28}Ge$ $Ge(C_6H_4CH_3)_4$

V.K. BELSKY, A.A. SIMONENKO and V.O. REIKHSFELD, 1984. J. Organometal. Chem., 265, 141-143.

Triclinic, $P\bar{1}$, a = 8.245, b = 8.210, c = 17.072 Å, α = 79.48, β = 84.20, γ = 91.55°, Z = 2. Mo radiation, R = 0.026 for 2460 reflexions.

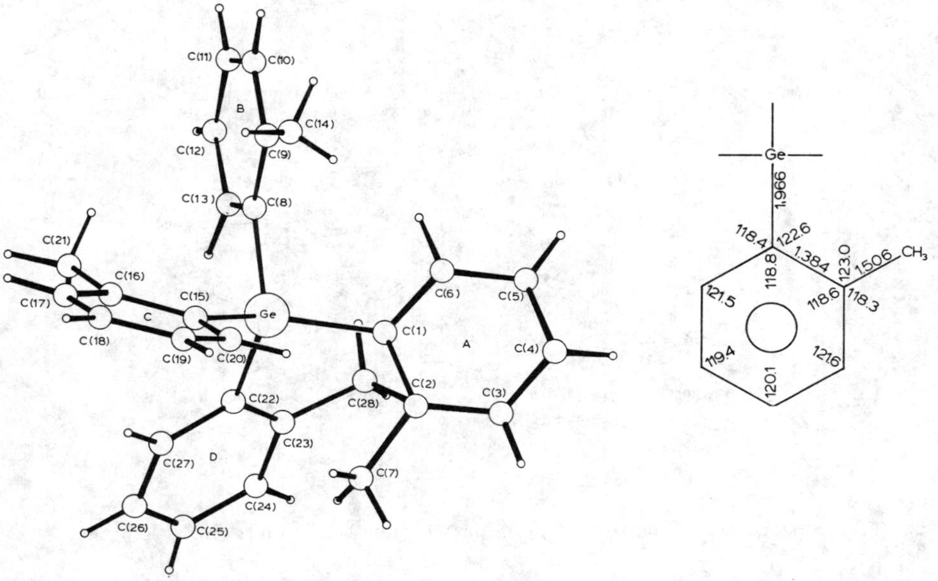

Fig. 1. Molecular conformation of $C_{28}H_{28}Ge$ (left), and mean interatomic distances and bond angles (right).

The mean value for the Ge-C bond (1.966 Å) is longer than that found in tetraphenylgermanium (1) presumably because of the presence of methyl groups in the ortho position (Fig. 1). The angles between the mean ring planes are equal in pairs: A-B 108.6 and C-D 117.1, A-C 66.6 and B-D 66.7, A-D 78.3 and B-C 77.1°. Intermolecular contacts correspond to normal van der Waals interactions.

1. Structure Reports, 38B, 722.

CARBONYLBIS(CYCLOPENTADIENYL)NIOBIUMTRIPHENYLTIN
$C_{29}H_{25}NbOSn$ (I) $(\eta^5\text{-}C_5H_5)_2Nb(CO)Sn(C_6H_5)_3$

CARBONYLBIS(METHYLCYCLOPENTADIENYL)NIOBIUMTRICHLOROTIN
$C_{13}H_{14}Cl_3NbOSn$ (II) $\eta^5(CH_3C_5H_4)_2Nb(CO)SnCl_3$

Yu.V. SKRIPKIN, O.G. VOLKOV, A.A. PASYNSKII, A.S. ANTSYSHKINA, L.M. DIKAREVA, V.N. OSTRIKOVA, M.A. PORAI-KOSHITS, S.L. DAVYDOVA and S.G. SAKHAROV, 1984. J. Organometal. Chem., 263, 345-357.

I. Monoclinic, $P2_1/n$, a = 10.102, b = 17.463, c = 14.247 Å, β = 95.58°, Z = 4. Mo radiation, R = 0.062 for 2407 reflexions.

II. Monoclinic, $P2_1/n$, a = 8.367, b = 13.359, c = 13.929 Å, β = 99.49°, Z = 4. Mo radiation, R = 0.056 for 4008 reflexions.

The Nb-Sn bond in II (2.764 Å) is shorter than that in I (2.825 Å) (Fig. 1, 2); in both cases the Nb-Sn bond is significantly shorter than the sum of the covalent radii (3.06 Å) indicating multiple bond character. The angle between the ring planes in I is 38.9° whilst that between the MeCp planes in II is 42.2°. The Sn-Cl distances in II are 2.391, 2.400 and 2.339 Å. The Sn-Nb-C(CO) angles in I and II are respectively 85.2 and 88.8°, and the angles about Sn in II are heavily distorted from tetrahedral (range: 95.3 to 121.61°).

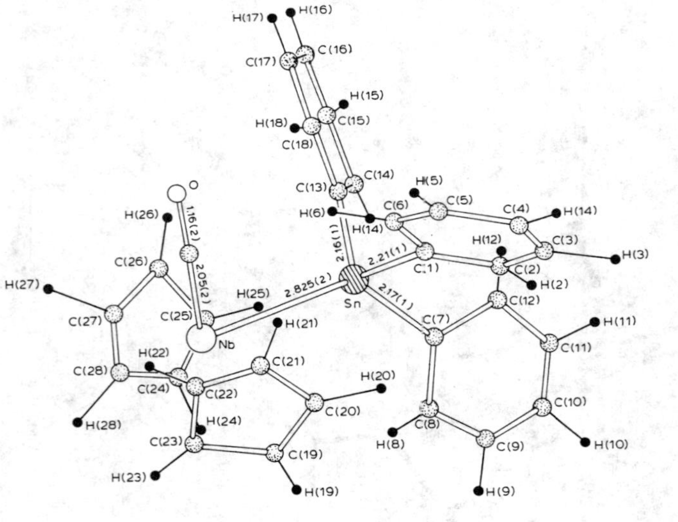

Fig. 1. Molecular structure of $C_{29}H_{25}NbOSn$ (I) with selected bond lengths.

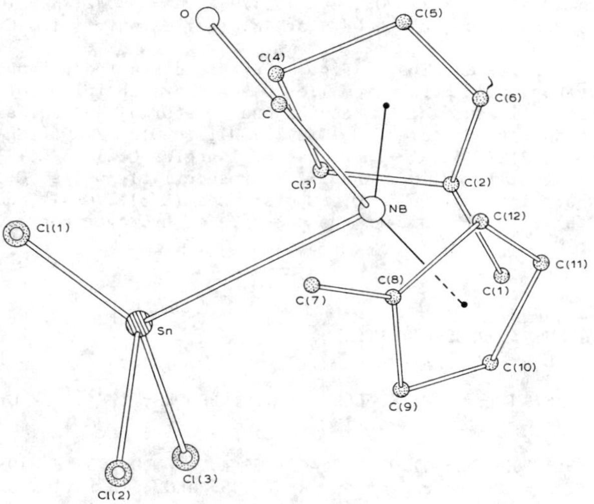

Fig. 2. The structure of $C_{13}H_{14}Cl_3NbOSn$ (II).

3,5-di-t-BUTYL-1,2-PHENYLENEDIOXY-BIS(o-TRIFLUOROMETHYLPHENYL)METHANOLTIN(IV) METHANOL SOLVATE

$C_{29}H_{32}F_6O_3Sn,2CH_4O$ $(C_6H_4CF_3)_2Sn(CH_3OH)(C_{14}H_{20}O_2),2CH_3OH$

T. BUTTERS, K. SCHEFFLER, H.B. STEGMANN, U. WEBER and W. WINTER, 1984. J. Organometal. Chem., 272, 351-361.

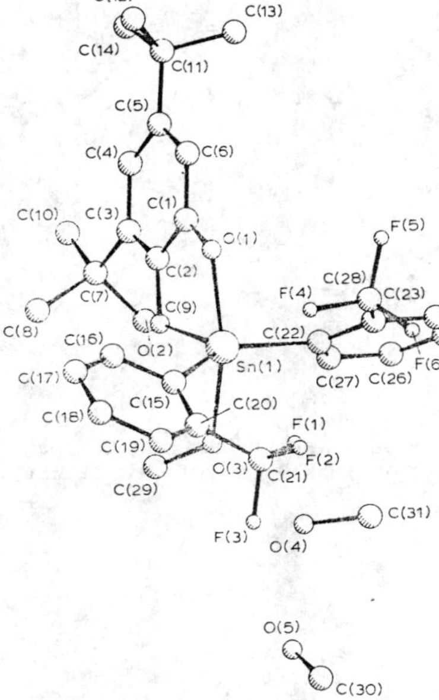

Fig. 1. Perspective drawing of $C_{29}H_{32}F_6O_3Sn,2CH_4O$.

Triclinic, $P\bar{1}$, a = 10.451, b = 11.905, c = 15.379 Å, α = 96.42, β = 104.17, γ = 111.43°, Z = 2. Mo radiation, R = 0.047 for 4810 reflexions, (-100°C).

The tin atom is five-coordinate (Fig. 1) with a distorted trigonal bipyramidal arrangement of ligands. The metal atom lies 0.5 Å above the plane of atoms C(1)-C(6) in the phenylenedioxy ligand, which occupies one equatorial and one axial coordination site. Relevant bond distances are: Sn-O(1) 2.091, Sn-O(2) 2.006, Sn-O(3) 2.251, Sn-C(15) 2.135 and Sn-C(22) 2.132 Å. The equatorial angles O(2)-Sn-C(15), O(2)-Sn-C(22) and C(15)-Sn-C(22) are 111.7, 110.2 and 137.4° respectively whilst O(1)-Sn-O(3) is 161.4°. The short intermolecular O...O distances (O(3)...O(4) 2.55, O(4)...O(5) 2.63 and O(1)...O(5) 2.70 Å) indicate the presence of hydrogen bonds.

BROMOTRITHIENYL(TRIPHENYL PHOSPHITE)TIN

$C_{30}H_{24}BrOPS_3Sn$ $(C_6H_5)_3PO.SnBr(C_4H_3S)_3$

D.W. ALLEN, D.J. DERBYSHIRE, I.W. NOWELL and J.S. BROOKS, 1984. J. Organometal. Chem., 260, 263-270.

Triclinic, $P\bar{1}$, a = 14.700, b = 11.002, c = 9.805 Å, α = 104.05, β = 95.04, γ = 101.71°, D_m = 1.58, Z = 2. Mo radiation, R = 0.055 and 4375 reflexions.

The crystal structure consists of discrete molecules in which the Sn atom is five-coordinate (Fig. 1). Three thienyl groups occupy equatorial sites in the trigonal bipyramidal arrangement (Sn-C(41) 2.121, Sn-C(51) 2.122, Sn-C(61) 2.125 Å), and the axial sites are occupied by Br (Sn-Br 2.651 Å) and O (Sn-O 2.335 Å). The metal atom is displaced by 0.182 Å out of the equatorial plane and towards the axial Br. The P-O distance is 1.510 Å and the Sn-O-P angle is 146.9°. Two of the thienyl ligands exhibit rotational disorder (C(41) and C(61)) whilst the third is free from disorder, presumably because of close proximity to a phenyl group.

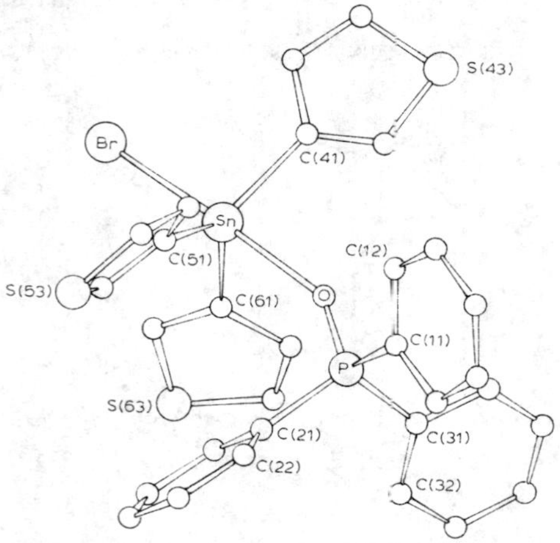

Fig. 1. Molecular structure of $C_{30}H_{24}BrOPS_3Sn$. Only one conformation of the two disordered thienyl ligands is shown, for clarity.

BIS[PENTA(METHOXYCARBONYL)CYCLOPENTADIENYL]TIN(II)
$C_{30}H_{30}O_{20}Sn$ $Sn[C_5(COOCH_3)_5]_2$

P. JUTZI, F.-X. KOHL, E. SCHLÜTER, M.B. HURSTHOUSE and N.P.C. WALKER, 1984. J. Organometal. Chem., 271, 393-402.

Orthorhombic, $P2_12_12$, a = 15.656, b = 15.424, c = 7.033 Å, Z = 2. Mo radiation, R = 0.037 for 1098 reflexions.

The Sn atom, which lies on the crystallographic twofold axis, has fourfold O coordination from carbonyl O atoms of two methoxy carbonyl groups on each of two ligands, with Sn-O distances of 2.240 and 2.271 Å (Fig. 1). The coordination is not "pyramidal" in the simplest sense, more based on a pseudo-trigonal bipyramid with the two ligands spanning two axial/equatorial edges and the third equatorial site occupied by a lone pair. The chelate "bite" gives an $O_{ax.}$-Sn-$O_{eq.}$ angle of 77.2°, the $O_{ax.}$-Sn-$O_{ax.}$ angle is 146.4°, and the $O_{eq.}$-Sn-$O_{eq.}$ angle 79.4°. The SnL_2 molecules stack along the twofold axis such that the metal forms two intermolecular Sn...O interactions (2.925 Å) to a carbonyl O on each of two symmetry-related ligands resulting in a very distorted octahedral SnO_6 coordination.

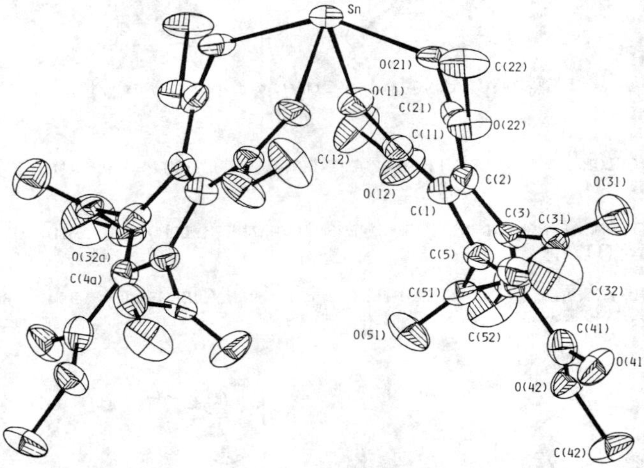

Fig. 1. Molecular structure of $C_{30}H_{30}O_{20}Sn$.

TRIS(1,3-DI-t-BUTYL-2,2-DIMETHYL-4-CHLORO-1,3-DIAZA-2-SILA-4-GERMA-4-CYCLOBUTYL)-PHOSPHINE
$C_{30}H_{72}Cl_3Ge_3N_6PSi_3$

M. VEITH, M. GROSSER and V. HUCH, 1984. Z. Anorg. Chem., 513, 89-102.

Triclinic, $P\bar{1}$, a = 19.552, b = 13.783, c = 10.743 Å, α = 90.4, β = 121.6, γ = 97.8°, Z = 2. Mo radiation, R = 0.07 for 6273 reflexions.

Although no crystallographic symmetry is required by the space group, the molecule (Fig. 1) has close to C_3 symmetry. The Ge, Cl, Si atoms lie in a plane, with the P atom disordered over two sites off this plane, such that the mean Ge-P-Ge angle is 115.0°. Other mean dimensions are: Ge-P 2.310(4), Ge-N 1.824(7), Ge-Cl 2.179(2), Si-N 1.736(7) Å.

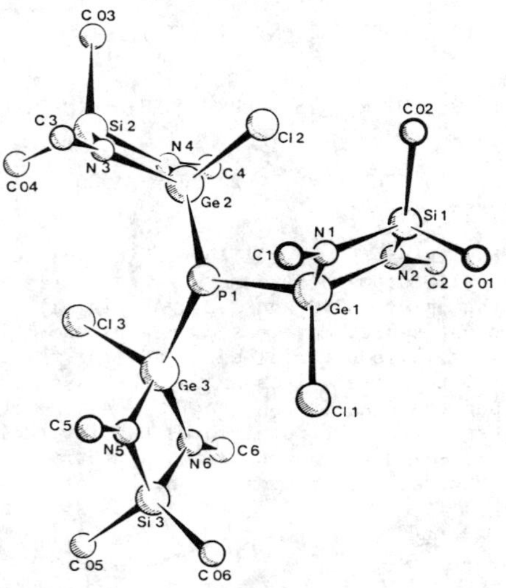

Fig. 1. A view of $C_{30}H_{72}Cl_3Ge_3N_6PSi_3$ with the t-butyl methyl groups omitted.

(μ_2-METHYLENE)(CHLORODIPHENYLTIN)(CHLORODIPHENYLHEXAMETHYLPHOSPHORAMIDE-O)TIN
$C_{31}H_{40}Cl_2N_3OPSn_2$ (I)

(μ_2-METHYLENE)(BROMODIPHENYLTIN)(BROMODIPHENYLHEXAMETHYLPHOSPHORAMIDE-O)TIN
$C_{31}H_{40}Br_2N_3OPSn_2$ (II)

M. GIELEN, K. JURKSCHAT, J. MEUNIER-PIRET and M. VAN MEERSSCHE, 1984. Bull. Soc. Chim. Belg., 93, 379-391.

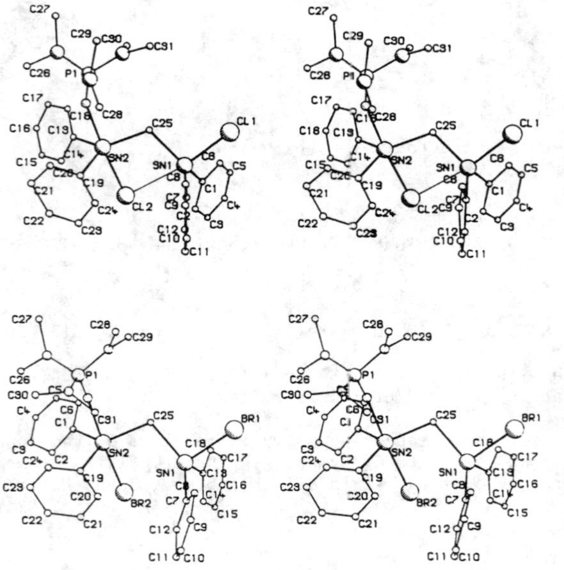

Fig. 1. Stereoviews of $C_{31}H_{40}Cl_2N_3OPSn_2$ (top) and one of the $C_{31}H_{40}Br_2N_3OPSn_2$ molecules (bottom).

I. Monoclinic, $P2_1/n$, a = 10.388, b = 21.484, c = 16.094 Å, β = 92.50°, Z = 4. Mo radiation, R = 0.056 for 4502 reflexions.

II. Monoclinic, $P2_1/n$, a = 18.092, b = 12.892, c = 31.998 Å, β = 94.77°, Z = 8. Mo radiation, R = 0.101 for 4696 reflexions.

There are two independent molecules in (II). In both (I) and (II), the Sn atoms have distorted trigonal bipyramidal geometry. The two Sn atoms in each molecule are non-equivalent. Carbon atoms are equatorial, and halogens and oxygens axial; in each case the neighbourhood around Sn(1) is more distorted than Sn(2). Some dimensions for (I) are: Sn-Cl(1) 2.431(2), Sn(1)...Cl(2) 3.019(2), Sn(1)-C 2.123-2.142(7), Sn(2)-Cl(2) 2.609(2), Sn(2)-O(1) 2.210(5), Sn(2)-C 2.136-2.160(7), Sn...Sn 3.573(1) Å.

TETRAETHYLAMMONIUM TRICARBONYL-η^5-CYCLOPENTADIENYLTRIPHENYLSTANNYLNIOBATE(I)
$C_{34}H_{40}NNbO_3Sn,3C_4H_8O$ $[(C_2H_5)_4N]^+[Nb(C_5H_5)\{Sn(C_6H_5)_3\}(CO)_3]^-,3C_4H_8O$

F. NÄUMANN, J. KOPF and D. REHDER, 1984. J. Organometal. Chem., 267, 249-255.

Monoclinic, $P2_1/n$, a = 13.833, b = 15.206, c = 17.918 Å, β = 97.89°, Z = 4. Mo radiation, R = 0.080 for 4189 reflexions.

In the complex anion (Fig. 1) the Nb-Sn bond length is 2.820 Å. Three CO groups and Sn span a tetragonal plane parallel to the cyclopentadienyl ring. The Nb-C(CO) distances are 2.13, 2.09 and 2.04 Å whilst the Nb-Cp(ring centroid) distance is 2.118 Å and the Nb-tetragonal plane centroid distance is 1.193 Å. The Sn atom is in a greatly distorted tetrahedral environment: the angles at Sn are 116.8 (Nb-Sn-Ph average) and 101.3° (Ph-Sn-Ph average). There are three occluded disordered solvent molecules.

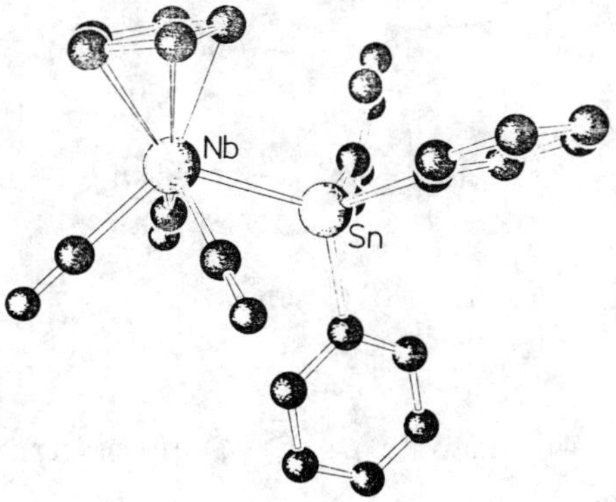

Fig. 1. Molecular structure of the anion in $C_{34}H_{40}NNbO_3Sn$.

TETRA-n-BUTYLAMMONIUM TRIS(PHENYLTHIOLATO-S)MERCURY
$C_{34}H_{51}HgNS_3$ (I)

TETRA-n-PROPYLAMMONIUM TRIS(PHENYLTHIOLATO-S)LEAD
$C_{30}H_{43}NPbS_3$ (II)

G. CHRISTOU, K. FOLTING and J.C. HUFFMAN, 1984. Polyhedron, 3, 1247-1253.

I. Monoclinic, $P2_1/a$, a = 20.663, b = 16.812, c = 9.757 Å, β = 95.52°, Z = 4. Mo radiation, R = 0.057 for 3437 reflexions (at -160°C).

II. Triclinic, $P\bar{1}$, a = 12.689, b = 11.255, c = 12.046 Å, α = 107.93, β = 109.64, γ = 86.01°, Z = 2. Mo radiation, R = 0.046 for 3487 reflexions (at -160°C).

The anion in I has a trigonal planar coordinated mercury (Fig. 1) with Hg-S 2.407-2.507(3), mean 2.448(43) Å, Hg-S-C 104.8-110.7(4)°. There are no weaker intermolecular Hg...S interactions. The anion in II has a trigonal pyramidal PbS_3 unit, with Pb-S 2.633-2.696(3) Å, mean 2.675(15) Å, mean S-Pb-S 95.4°, Pb-S-C 93.7-98.1(3)°.

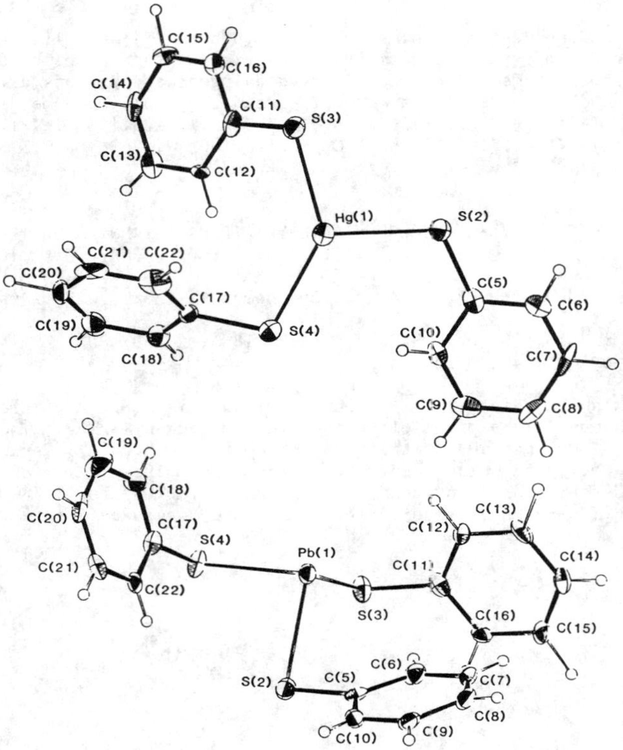

Fig. 1. The $[Hg(SPh)_3]^-$ and $[Pb(SPh)_3]^-$ anions.

BIS{μ-[4-CHLORO-1,3-BENZENEDITHIOLATO(2-)-S,S']}-BIS[DIPHENYLTIN(IV)]

$C_{36}H_{26}Cl_2S_4Sn_2$ $[Sn_2(C_6H_3ClS_2)_2(C_6H_5)_4]$

H. PREUT, K. GRÄTZ and F. HUBER, 1984. Acta Cryst., C40, 941-943.

Monoclinic, C2/c, a = 15.886, b = 7.560, c = 28.883 Å, β = 100.93°, Z = 4. Mo radiation, R = 0.062 for 1805 reflexions.

The molecule (Fig. 1) is a centrosymmetric dimer in which each of the two Sn atoms is tetrahedrally bonded by two C(phenyl) and two S atoms with Sn-C 2.16(1) and 2.07(1), Sn-S 2.419(4) and 2.420(4) Å. The Sn atoms are connected via two 4-chloro-1,3-benzenedithiolato moieties so that a twelve-membered ring, consisting of two Sn, four S and six C atoms, is formed. Bond lengths conform with those of similar compounds; there are some differences in bond angles which may be ascribed to packing effects and intramolecular strains.

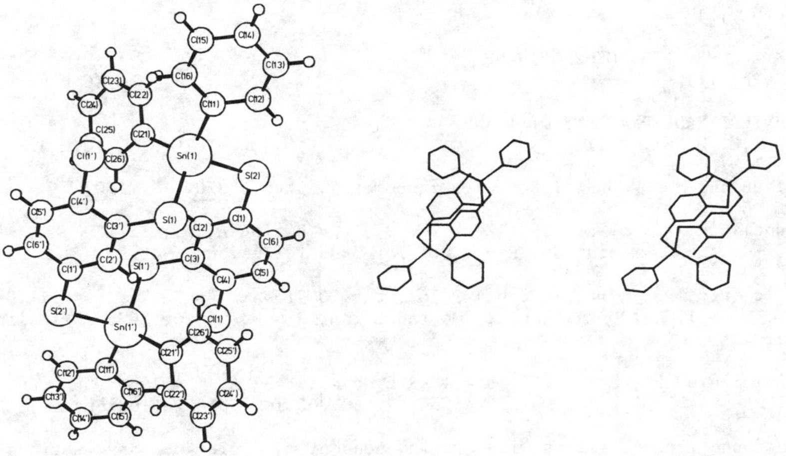

Fig. 1. General and stereoscopic views of the $Sn_2(C_6H_3ClS_2)_2(C_6H_5)_4$ molecule.

TETRAKIS(η^1-INDENYL)GERMANIUM
$C_{36}H_{28}Ge$ (I) $Ge(\eta^1-C_9H_7)_4$

TETRAKIS(η^1-INDENYL)TIN
$C_{36}H_{28}Sn$ (II) $Sn(\eta^1-C_9H_7)_4$

J.L. ATWOOD, A.D. McMASTER, R.D. ROGERS and S.R. STOBART, 1984. Organometallics, 3, 1500-1504.

I. Triclinic, $P\bar{1}$, a = 11.115, b = 11.259, c = 11.319 Å, α = 114.26, β = 95.11, γ = 90.65°, Z = 2. Mo radiation, R = 0.020 for 1537 reflexions.

II. Monoclinic, $P2_1/c$, a = 11.684, b = 22.254, c = 11.432 Å, β = 117.04°, Z = 4. Mo radiation, R = 0.032 for 1342 reflexions.

Although not isomorphous, both compounds are structurally identical (Fig. 1). Coordination about the metal centre is almost exactly tetrahedral in both molecules (range of angles is 108.3 to 110.6° in I and 108.3 to 111.3° in II) with mean Ge-C and Sn-C bond lengths of 2.00 and 2.19 Å respectively.

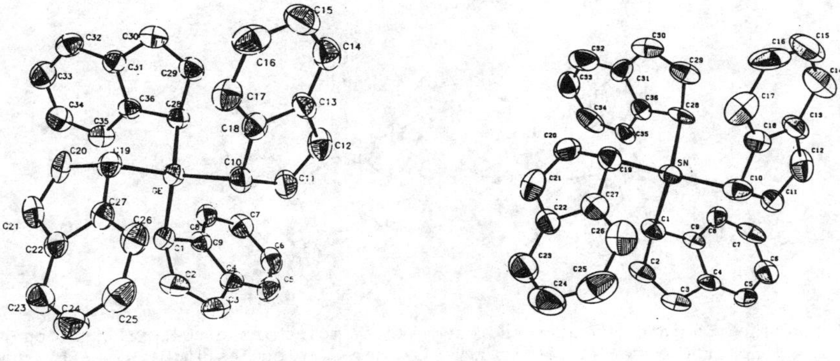

Fig. 1. Molecular structures of $C_{36}H_{28}Ge$ (left), $C_{36}H_{28}Sn$ (right).

TRIPHENYLPLUMBYL-TRIPHENYLSTANNANE
$C_{36}H_{30}PbSn$ (I) $(C_6H_5)_3Pb-Sn(C_6H_5)_3$

TRIPHENYLPLUMBYL-TRIPHENYLGERMANE
$C_{36}H_{30}GePb$ (II) $(C_6H_5)_3Pb-Ge(C_6H_5)_3$

HEXAPHENYLDILEAD HEXAPHENYLDISILANE
$C_{36}H_{30}Pb_2, C_{36}H_{30}Si_2$ (III) $(C_6H_5)_3Pb-Pb(C_6H_5)_3, (C_6H_5)_3Si-Si(C_6H_5)_3$

N. KLEINER and M. DRÄGER, 1984. J. Organometal. Chem., 270, 151-170.

I. Monoclinic, $P2_1/n$, a = 17.112, b = 9.37, c = 20.754 Å, β = 111.39°, D_m = 1.74 (20°C), Z = 4. Mo radiation, R = 0.061 for 3445 reflexions (-40°C).

II. Triclinic, $P\bar{1}$, a = 8.742, b = 9.762, c = 10.713 Å, α = 96.51, β = 111.03, γ = 112.72°, D_m = 1.62 (20°C), Z = 1. Mo radiation, R = 0.053 for 3633 reflexions (-50°C).

III. Triclinic, $P\bar{1}$, a = 9.079, b = 10.832, c = 15.657 Å, α = 79.91, β = 86.77, γ = 85.28°, D_m = 1.57, Z = 1. Mo radiation, R = 0.068 for 5318 reflexions.

Compound I crystallizes with two independent molecules per asymmetric unit (each molecule on a centre of symmetry). This results in statistical disorder of the metal atoms, each Pb and Sn having occupancy factors of 0.5 (Fig. 1). The Pb-Sn distances in the two molecules are 2.809 and 2.848 Å and the Pb/Sn-C distances average 2.21 Å. The C-Pb/Sn-C angles average 108.7°. Compound II has a similar arrangement but with one molecule per asymmetric unit (Fig. 1). A disordered (static alternation) molecule lies on the inversion centre at 0.5, 0.5, 0.5 and the distance between the heavy atom centres (Pb and Ge) on one atomic site is 0.33 Å. The Pb-Ge distance is 2.623 Å and the Ge-C and Pb-C distances are 2.00, 1.97, 2.11 and 2.17, 2.188 and 2.13 Å respectively. Compound III is a 'mixed phase' compound containing one half-molecule each of Pb_2Ph_6 and Si_2Ph_6 per asymmetric unit (Fig. 1). The molecules are centrosymmetric, lying on the inversion centres at 0, 0, 0 and 0.5, 0.5, 0.5. The Pb(1)-Pb(1') and Si(1)-Si(1') distances are respectively 2.846 and 2.519 Å. The average Pb-C distance is 2.222 Å and the average Si-C distance is 1.892 Å. The average angles C-Pb-C and C-Si-C are 107.7 and 111.8° respectively.

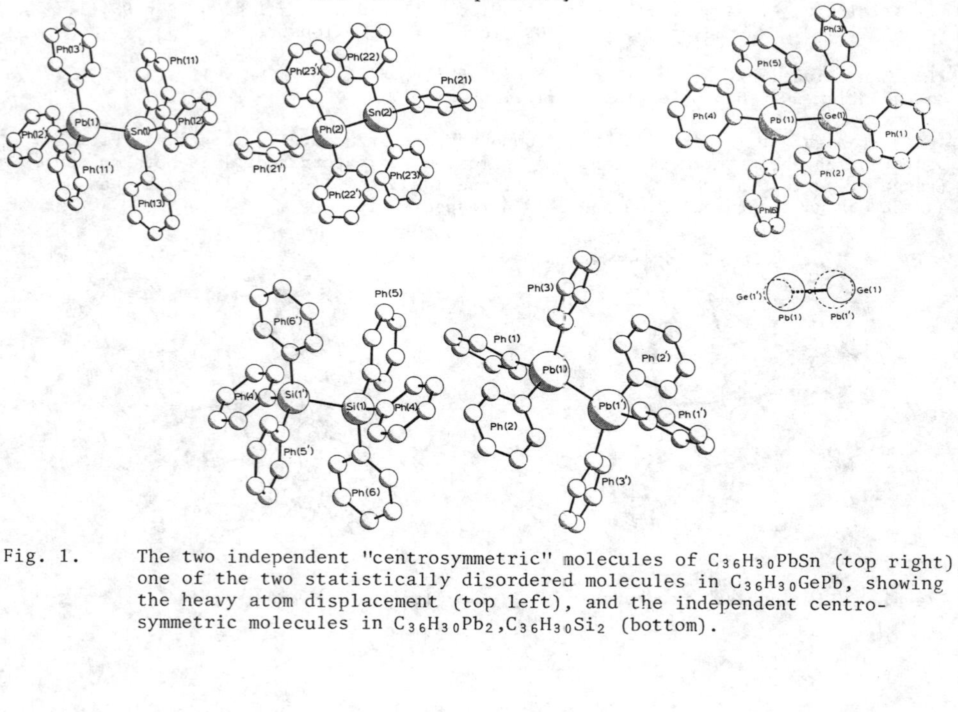

Fig. 1. The two independent "centrosymmetric" molecules of $C_{36}H_{30}PbSn$ (top right), one of the two statistically disordered molecules in $C_{36}H_{30}GePb$, showing the heavy atom displacement (top left), and the independent centrosymmetric molecules in $C_{36}H_{30}Pb_2, C_{36}H_{30}Si_2$ (bottom).

CARBONYLDIHYDRIDO(TRIMETHYLGERMYL)BIS(TRIPHENYLPHOSPHINE)IRIDIUM(III) HEMIBENZENE SOLVATE

$C_{40}H_{41}GeIrOP_2,0.5(C_6H_6)$ $H_2(Me_3Ge)CO(Ph_3P)_2Ir,0.5(C_6H_6)$

N.A. BELL, F. GLOCKLING, M.L. SCHNEIDER, H.M.M. SHEARER and M.D. WILBEY, 1984. Acta Cryst., C40, 625-628.

Monoclinic, $P2_1/c$, a = 13.63, b = 11.044, c = 26.07 Å, β = 92.52°, D_m = 1.54, Z = 4. Mo radiation, R = 0.052 for 3560 reflexions.

In the complex molecule Ir has distorted octahedral coordination (Fig. 1). The two phosphine groups are cis to each other, one being trans to a hydride hydrogen atom and the other trans to the germyl group. Some bond lengths are Ir-Ge 2.484(2), Ir-P(1) 2.353(4), Ir-P(2) 2.367(4), Ir-C(1) 1.85(2) and mean Ge-C 1.99(2) Å. The two benzene solvate molecules are located on centres of symmetry.

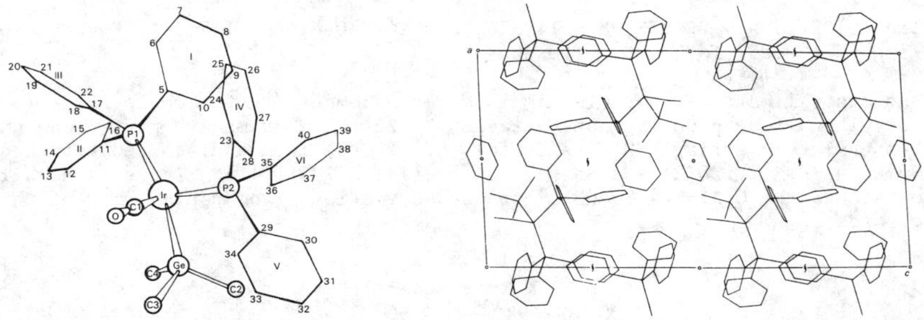

Fig. 1. Perspective view of the $H_2(Me_3Ge)CO(Ph_3P)_2Ir$ molecule and the projection of the crystal structure on the (010) plane.

TETRAPHENYLARSONIUM TRIS(PHENYLTHIOLATO)LEAD(II)

$C_{42}H_{35}AsPbS_3$ (I) $[(C_6H_5)_4As][Pb(SC_6H_5)_3]$

TETRAPHENYLARSONIUM TRIS(PHENYLSELENOLATO)LEAD(II)

$C_{42}H_{35}AsPdSe_3$ (II) $[(C_6H_5)_4As][Pb(SeC_6H_5)_3]$

P.A.W. DEAN, J.J. VITTAL and N.C. PAYNE, 1984. Inorg. Chem., 23, 4232-4236.

I. Monoclinic, $P2_1/c$, a = 10.758, b = 17.561, c = 20.273 Å, β = 105.11°, D_m = 1.654, Z = 4. Mo radiation, R = 0.026 for 4866 reflexions.

II. Monoclinic, $P2_1/c$, a = 10.899, b = 17.638, c = 20.617 Å, β = 105.84°, D_m = 1.856, Z = 4. Mo radiation, R = 0.034 for 3074 reflexions.

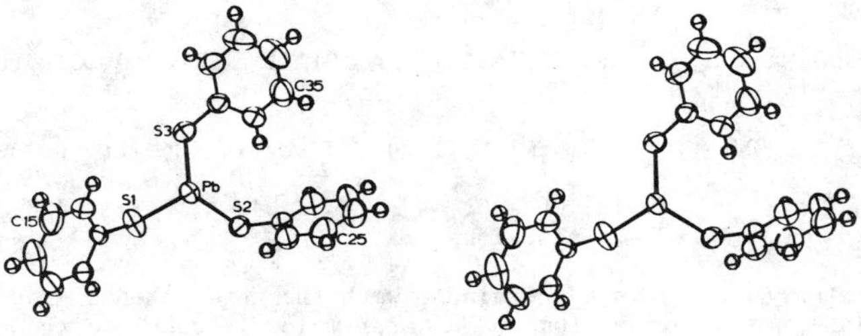

Fig. 1. Stereoview of the $[Pb(SC_6H_5)_3]^-$ anion.

The discrete $[Pb(EC_6H_5)_3]^-$ anions (Fig. 1) are isostructural; the three S or Se atoms form the base of a trigonal pyramid whose apex is occupied by the Pb atom. The Pb-S distances in I are 2.619(1), 2.647(1), and 2.623(1) Å, while in II the Pb-Se distances are 2.762(1), 2.727(1), and 2.733(1) Å. Angles at Pb are 90.32(4)-96.14(5)° in I and 88.82(4)-96.57(4)° in II. The phenyl groups adopt a propeller-like conformation and are disposed in equatorial positions above the basal E_3 planes.

3,3-DI-t-BUTYL-N,4,5,6,7-PENTAPHENYL-1,2-DIHYDRO-1,2,3-DIAZASTANNEPINE-1,2-DICARBOXIMIDE MONOHYDRATE
$C_{44}H_{43}N_3O_2Sn,H_2O$

H. PREUT, R. MARX and W.P. NEUMANN, 1984. Acta Cryst., C40, 1793-1795.

Monoclinic, C2/c, a = 22.838, b = 14.526, c = 26.270 Å, β = 115.32°, Z = 8. Mo radiation, R = 0.057 for 2500 reflexions.

The central part of the molecule is a seven-membered heterocycle (Fig. 1) with two t-butyl and four phenyl groups attached. The two N atoms are part of an annellated N-phenylurazole ring. N(1) and C(4), C(7), C(8) form a distorted tetrahedron with Sn with Sn-N 2.120(8) and Sn-C 2.164, 2.174, 2.185(11) Å. There is significant torsion around the C(3)=C(4) double bond and deviations from coplanarity of the two double bonds.

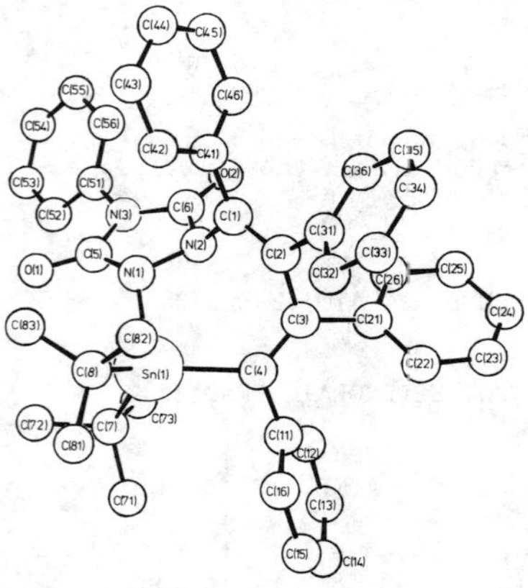

Fig. 1. The $C_{44}H_{43}N_3O_2Sn$ molecule.

3-(2-(1,10-PHENANTHROLYL))-5,6-DIPHENYL-1,2,4-TRIAZINE CHLOROAQUOTRIPHENYLTIN(IV)
$C_{45}H_{34}ClN_5OSn$

M.F.C. LADD, D.C. POVEY and F.E. SMITH, 1984. J. Cryst. Spectrosc. Res., 14, 249-259.

Orthorhombic, $Pca2_1$, a = 19.195, b = 9.144, c = 21.642 Å, D_m = 1.425, Z = 4. Cu radiation, R = 0.031 for 3287 reflexions.

The tin atom (Fig. 1) is 5-coordinate with the three phenyl groups forming the equatorial plane. A chlorine atom and a water molecule complete the coordination. The triazine moiety does not coordinate directly to the metal atom. The only interaction is due to two N...H-O hydrogen bonds formed between two nitrogen atoms from the ligand and the water molecule. Main bond lengths are Sn-O 2.359(4), Sn-Cl 2.503(2), Sn-C 2.133, 2.163, 2.161(6) Å.

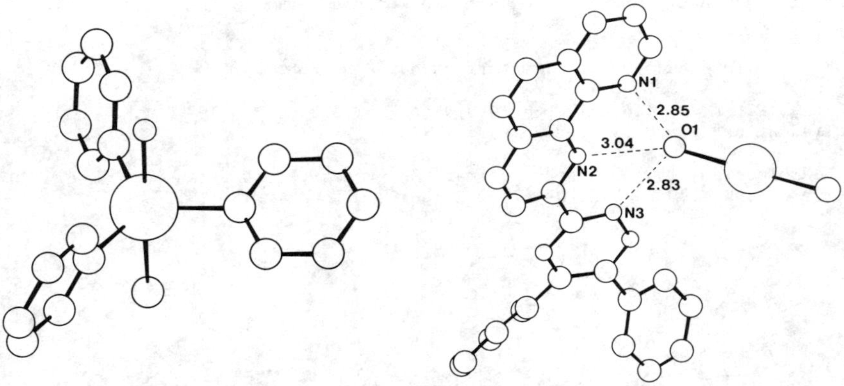

Fig. 1. The Ph_3SnClO moiety and a diagram showing the diphenyltriazine-phenanthroline linkage to the water molecule bonded to Sn.

DIPHENYLGERMANIUMDIPHENYLSILICON OXIDE
$C_{48}H_{40}Ge_2O_4Si_2$ (I) $[((C_6H_5)_2Ge)O((C_6H_5)_2Si)O]_2$

DIMETHYLGERMANIUMDIPHENYLSILICON OXIDE
$C_{28}H_{32}Ge_2O_4Si_2$ (II) $[((CH_3)_2Ge)O((C_6H_5)_2Si)O]_2$

H. PUFF, M.P. BÖCKMANN, T.R. KÖK and W. SCHUH, 1984. J. Organometal. Chem., 268, 197-206.

I. Monoclinic, $P2_1/c$, a = 14.764, b = 14.442, c = 21.911 Å, β = 111.39°, Z = 4. Mo radiation, R = 0.142 for 3288 reflexions.

II. Triclinic, $P\bar{1}$, a = 9.136, b = 9.379, c = 10.083 Å, α = 70.30, β = 79.15, γ = 67.84°, Z = 1. Mo radiation, R = 0.052 for 1600 reflexions.

Both compounds exist as non-planar, eight-membered Si-Ge-O rings in the solid state. In compound I the Ge and Si atoms are disordered (Fig. 1). Compound II has crystallographic inversion symmetry (Fig. 1) and the Si and four O atoms are virtually coplanar with the Ge atoms lying 0.886 Å above and below this plane. Important distances in II are: Ge(1)-O(1) 1.758, Ge(1)-O(2) 1.785, Si(1)-O(1) 1.615, Si(1)-O(2') 1.590 Å. The Ge-O-Si angles are 135.8 and 138.0°.

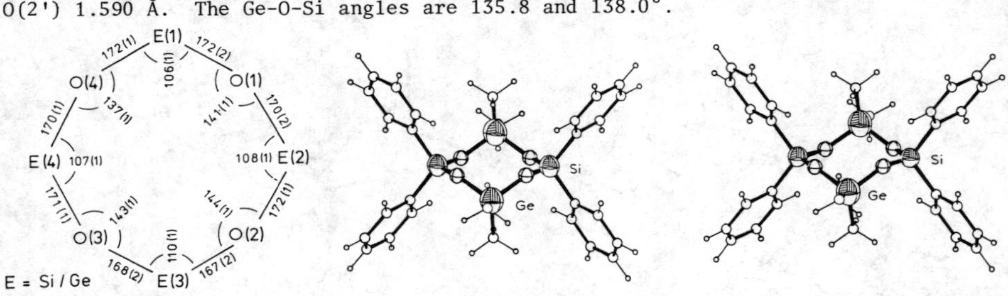

Fig. 1. Bond distances (pm) and angles (°) in $C_{48}H_{40}Ge_2O_4Si_2$, and stereoview of $C_{28}H_{32}Ge_2O_4Si_2$.

OCTAPHENYLTETRAOXATETRAGERMACYCLOOCTANE (Monoclinic form)

$C_{48}H_{40}Ge_4O_4$ $(Ph_2GeO)_4$

L. ROSS and M. DRAGER, 1984. Z. Naturforsch., B39, 868-875.

Monoclinic, $P2_1/n$, a = 14.881, b = 14.584, c = 21.372 Å, β = 108.59°, D_m = 1.45, Z = 4. Mo radiation, R = 0.051 for 2679 reflexions.

The molecule (Fig. 1) has approximate S_4 symmetry with dimensions Ge-O 1.747-1.762, Ge-C 1.900-1.946 Å, Ge-O-Ge 131.2-136.9, O-Ge-O 107.8-110.3°.

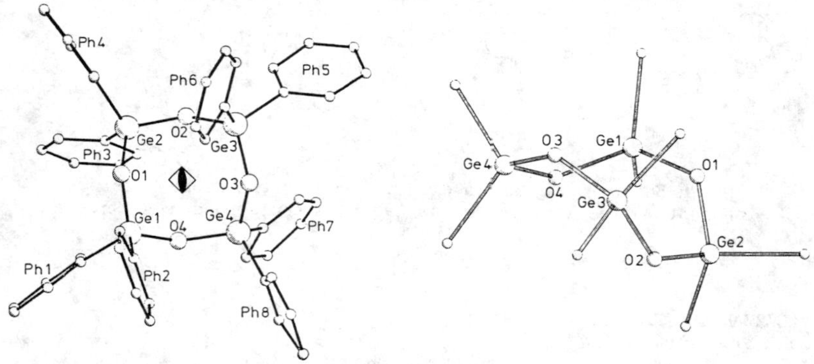

Fig. 1. A view of $[Ph_2GeO)_4$ and the Ge_4O_4 ring system.

1,3,3,5,7,7,10,10-OCTAPHENYL-2,4,6,8,9,11-HEXAOXA-1,3,5,7,11-PENTAGERMABICYCLO[3.3.3]-UNDECANE ACETONE SOLVATE

$C_{48}H_{40}Ge_5O_6,0.5(C_3H_6O)$ $Ph_8Ge_5O_6,0.5(C_3H_6O)$

K. HABERLE and M. DRAGER, 1984. Z. Naturforsch., B39, 1541-1547.

Monoclinic, Pc, a = 10.120, b = 22.161, c = 20.855 Å, β = 92.07°, D_m = 1.54, Z = 4. Mo radiation, R = 0.062 for 8177 reflexions.

The asymmetric unit contains two independent $Ph_8Ge_5O_6$ cages (Fig. 1) and one disordered acetone. Although no crystallographic symmetry is required, the molecular symmetry is close to D_3 and the symmetry of the $(GeO)_4$ rings is near C_2. Dimensions are Ge-O 1.72-1.80, Ge-C 1.91-1.95 Å, Ge-O-Ge 128.0-133.8, O-Ge-O 107.5-110.6°.

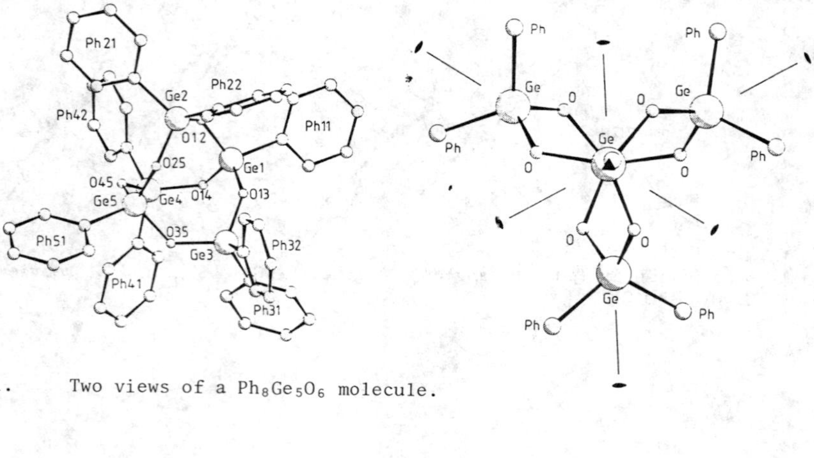

Fig. 1. Two views of a $Ph_8Ge_5O_6$ molecule.

BIS(μ_3-OXO)BIS(μ_2-HYDROXO)DICHLOROOCTAPHENYLTETRATIN ACETONE SOLVATE
$C_{48}H_{42}Cl_2O_4Sn_4,2C_3H_6O$ (I) $(Ph_4Sn_2ClO_2H)_2,2(CH_3)_2CO$

BIS(μ_3-OXO)BIS(μ_2-CHLORO)DICHLOROOCTAPHENYLTETRATIN
$C_{48}H_{40}Cl_2O_2Sn_4$ (II) $(Ph_4Sn_2ClO)_2$

J.F. VOLLANO, R.O. DAY and R.R. HOLMES, 1984. Organometallics, 3, 745-750.

I. Monoclinic, $P2_1/n$, a = 11.758, b = 19.327, c = 11.649 Å, β = 92.87°, Z = 2. Mo radiation, R = 0.028 for 3856 reflexions.

II. Monoclinic, $P2_1/c$, a = 10.284, b = 10.513, c = 22.143 Å, β = 100.75°, Z = 2. Mo radiation, R = 0.023 for 2387 reflexions.

Both compounds are centrosymmetric dimers and the two independent Sn atoms in each molecule are pentacoordinated with distorted trigonal bipyramidal geometry (Fig. 1). Both compounds have 'ladder' structures as evidenced by the near coplanarity of the atoms comprising the fused ring systems. In I the atoms Sn(1), Sn(2), O(1), O(2), Sn(1'), Sn(2'), O(1') and O(2') are coplanar to within ±0.07 Å while in II, the corresponding atoms Sn(1), Sn(2), O(1), Cl(1), Sn(1'), Sn(2'), O(1') and Cl(1') are coplanar to within ±0.03 Å. Important bond lengths in I are: Sn(1)-O(1) 2.061, Sn(1)-O(1') 2.127, Sn(1)-O(2) 2.160, Sn(2)-O(1) 2.024, Sn(2)-Cl(2) 2.465, Sn(2)-O(2) 2.212 Å. The corresponding lengths in II are: 2.050, 2.107, Sn(1)-Cl(1) 2.688, 2.040, 2.430 and Sn(2)-Cl(1) 2.697 Å. The angles O(1)-Sn(1)-O(1') and Sn(1)-O(1)-Sn(1') are 74.0 and 106.0° in I and 74.5 and 105.5° in II. In I, the H atom of the bridging OH forms a hydrogen bond to the O atom (O(3)) of the acetone of solvation. The O(2)...O(3) distance is 2.914 Å and the angle O(2)-H(2)...O(3) is 173.4°.

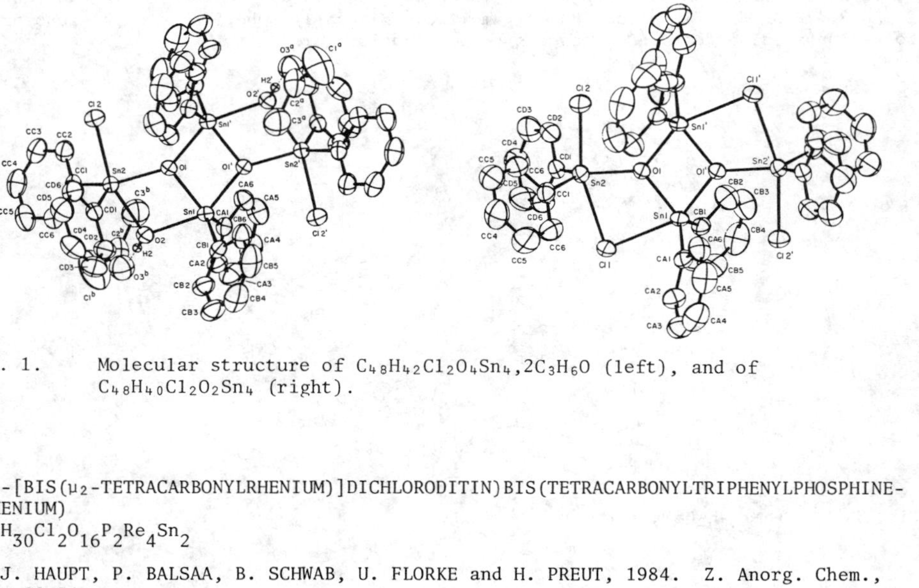

Fig. 1. Molecular structure of $C_{48}H_{42}Cl_2O_4Sn_4,2C_3H_6O$ (left), and of $C_{48}H_{40}Cl_2O_2Sn_4$ (right).

(μ_2-[BIS(μ_2-TETRACARBONYLRHENIUM)]DICHLORODITIN)BIS(TETRACARBONYLTRIPHENYLPHOSPHINE-RHENIUM)
$C_{52}H_{30}Cl_2O_{16}P_2Re_4Sn_2$

H.-J. HAUPT, P. BALSAA, B. SCHWAB, U. FLORKE and H. PREUT, 1984. Z. Anorg. Chem., 513, 22-34.

Orthorhombic, Pbcn, a = 19.189, b = 16.034, c = 19.042 Å, Z = 4. Mo radiation, R = 0.058 for 1885 reflexions.

The molecule (Fig. 1) has crystallographic inversion symmetry and the Re_2Sn_2 moiety is necessarily planar. Principal dimensions include Re1-Sn1 2.764(3), Re1-Sn1' 2.760(3), Sn1-Re2 2.695(3), Re...Re 3.160(2), Sn-Cl1 2.43(1), Re2-P1 2.424(9) Å, Re1-Sn1-Re1' 69.78(6)°.

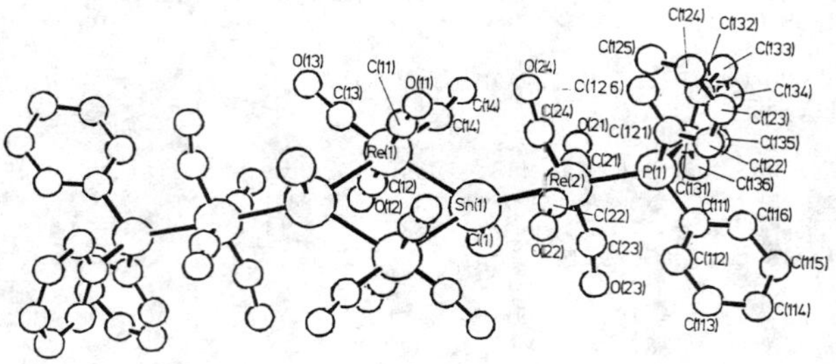

Fig. 1. A view of $C_{52}H_{30}Cl_2O_{16}P_2Re_4Sn_2$.

DECAPHENYLCYCLOPENTAGERMANE BENZENE SOLVATE
$C_{60}H_{50}Ge_5,C_6H_6$

L. ROSS and M. DRAGER, 1984. Z. Anorg. Chem., 519, 225-232.

Monoclinic, C2/c, a = 39.69, b = 12.834, c = 22.724 Å, β = 99.70°, Z = 8. Mo radiation, R = 0.085 for 5069 reflexions (at -25°C).

The conformation of the Ge_5 ring is between C_s and C_2, with Ge-Ge in the range 2.438-2.473 Å (Fig. 1). The $(Ph_2Ge)_5$ molecule is nearly spherical; the benzene of solvation is disordered and occupies trigonal prism holes.

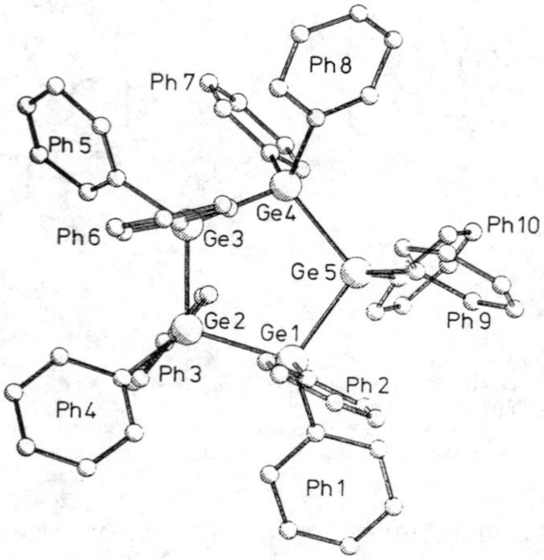

Fig. 1. The $(Ph_2Ge)_5$ molecule.

TETRAETHYLAMMONIUM TETRACARBONYLTRIS(TRIPHENYLTIN)CHROMIUM
$C_{66}H_{65}CrNO_4Sn_3$ $[Et_4N]^+[(Ph_3Sn)_3Cr(CO)_4]^-$

J.T. LINN, G.P. HAGEN and J.E. ELLIS, 1984. Organometallics, 3, 1288-1292.

Triclinic, P$\bar{1}$, a = 14.371, b = 20.560, c = 10.730 Å, α = 99.21, β = 105.07, γ = 83.29°, Z = 2. Mo radiation, R = 0.048 for 8042 reflexions.

The compound consists of discrete cations and anions with no unusually short interionic contacts. The anion (Fig. 1) contains a seven-coordinate Cr atom bound to four terminal CO groups and three well-separated triphenylstannyl units. The anion has approximate C_{3v} symmetry - a triangular face of tin atoms is capped by a CO unit. The three Cr-Sn distances are: 2.695, 2.728 and 2.719 Å, and the four Cr-C(CO) distances are: 1.855, 1.879, 1.865 and 1.848 Å. The Sn-C(phenyl) distances range from 2.155 to 2.178 Å, and the Cr-Sn-C(phenyl) angles range from 110.1 to 123.5°. The mean N-C and C-C distances in the cation are 1.52 and 1.59 Å respectively, and the mean N-C-C and C-N-C angles are 111 and 109°.

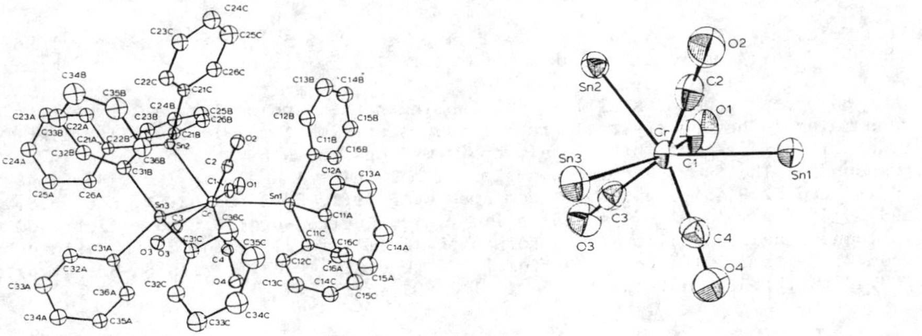

Fig. 1. Molecular structure of the anion in $C_{66}H_{65}CrNO_4Sn_3$, and its coordination core.

DECAPHENYLSTANNOCENE

$C_{70}H_{50}Sn$ $(Ph_5C_5)_2Sn$

M.J. HEEG, C. JANIAK and J.J. ZUCKERMAN, 1984. J. Am. Chem. Soc., 106, 4259-4261.

Monoclinic, $P2_1/n$, a = 10.458, b = 13.462, c = 18.130 Å, β = 97.81°, Z = 2. Mo radiation, R = 0.058 for 2560 reflexions (at 138 K).

The Sn atom sits on an inversion centre (Fig. 1) between related (equidistant) cyclopentadiene ring systems that are planar (and staggered and parallel). The Sn-(ring centre) distance is 2.401(6) Å and the mean Sn-C distance is 2.692(8) Å. The attached phenyl rings are canted in a double-opposed paddle-wheel fashion to give molecules of S_{10} symmetry.

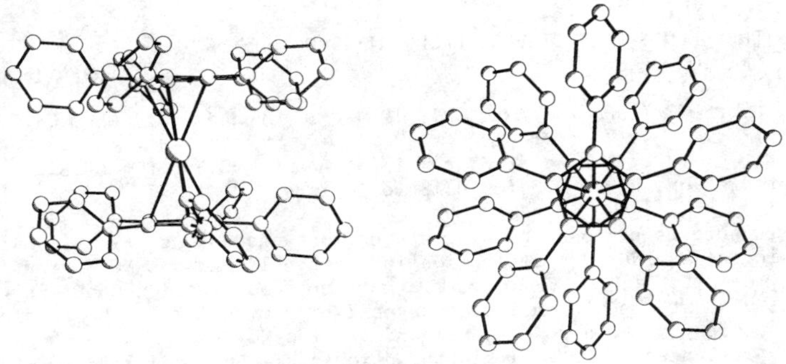

Fig. 1. Two views of decaphenylstannocene.

DODECAPHENYLCYCLOHEXASTANNANE

$C_{72}H_{60}Sn_6$ (I) $[(C_6H_5)_2Sn]_6$

DODECABENZYLCYCLOHEXASTANNANE DIMETHYLFORMAMIDE

$C_{84}H_{84}Sn_6, C_3H_7NO$ (II) $[(C_6H_5CH_2)_2Sn]_6, (CH_3)_2NCHO$

H. PUFF, C. BACH, H. REUTER and W. SCHUH, 1984. J. Organometal. Chem., 277, 17-28.

I. Monoclinic, $P2_1/c$, a = 19.363, b = 18.772, c = 17.615 Å, β = 96.71°, Z = 4. Mo radiation, R = 0.048 for 5459 reflexions.

II. Triclinic, $P\bar{1}$, a = 14.072, b = 15.726, c = 20.823 Å, α = 72.27, β = 77.66, γ = 64.39°, Z = 2. Mo radiation, R = 0.052 for 7152 reflexions.

Both compounds crystallize with two independent molecules per asymmetric unit, each molecule having crystallographic inversion symmetry (Fig. 1). There are considerable differences in the chair conformations shown by the molecules in the two compounds. The Sn ring torsion angles of molecules A and B in I are 43.5, 48.7 and 50.1°, and 52.9, 56.9 and 57.7° compared with 59.1, 59.8 and 62.8°, and 56.8, 59.2 and 59.5° for molecules A and B in II. These angles suggest a more flattened chair in I and a near-normal chair conformation in II (Fig. 1). The Sn-Sn distances range from 2.780 to 2.791 Å in I and from 2.792 to 2.811 Å in II, with Sn-Sn-Sn angles from 107.42 to 121.61° in I and 105.54 to 112.89° in II.

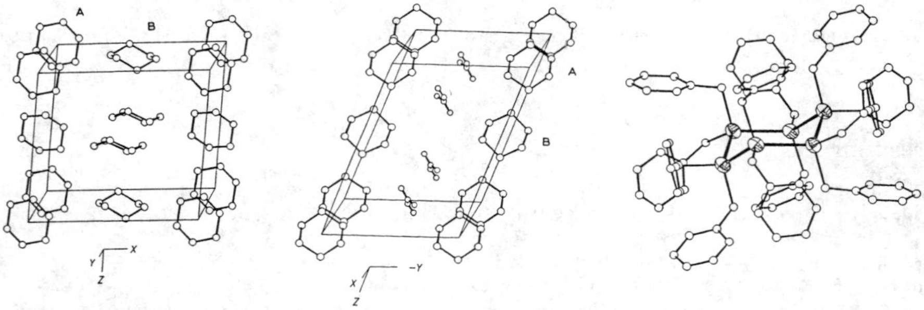

Fig. 1. Views of the unit cells of $C_{72}H_{60}Sn_6$ (without phenyl rings) (left), $C_{84}H_{84}Sn_6, C_3H_7NO$ (without benzyl rings) (centre) and of the molecule of $C_{84}H_{84}Sn_6$ (right).

TETRAKIS(TRIPHENYLPHOSPHINE)SILVER CHLOROTRINITRODIPHENYLTIN

$C_{84}H_{70}AgCl_{0.5}N_{2.5}O_{7.5}P_4Sn$ $[Ag(PPh_3)_4]^+[SnPh_2(NO_3)_2(Cl,NO_3)]^-$

C. PELIZZI, G. PELIZZI and P. TARASCONI, 1984. J. Organometal. Chem., 277, 29-35.

Triclinic, $P\bar{1}$, a = 22.43, b = 14.13, c = 13.96 Å, α = 90.59, β = 69.82, γ = 64.58°, Z = 2. Mo radiation, R = 0.073 for 4315 reflexions.

The structure is composed of discrete cations and anions (Fig. 1), with the latter showing a disordered distribution involving Cl^- and one NO_3^- group, each with an occupancy of 0.5. Taking disorder into account, coordination about Sn can be described as distorted pentagonal or hexagonal bipyramidal with the equatorial positions occupied by the disordered ligand and two NO_3 groups, the phenyl rings at the apices. Distances of note within the anion are: Sn-C(73) 2.14, Sn-C(79) 2.08, Sn-O(1) 2.33, Sn-O(2) 2.52, Sn-O(4) 2.56, Sn-O(5) 2.35, Sn-Cl 2.569, Sn-O(7) 2.24 and Sn-O(8) 2.43 Å. In the cation the Ag atom is bonded to four triphenylphosphine molecules in a slightly distorted tetrahedral environment. The Ag-P bond lengths are 2.631, 2.746, 2.645 and 2.615 Å with P-Ag-P angles ranging from 105.9 to 113.0°.

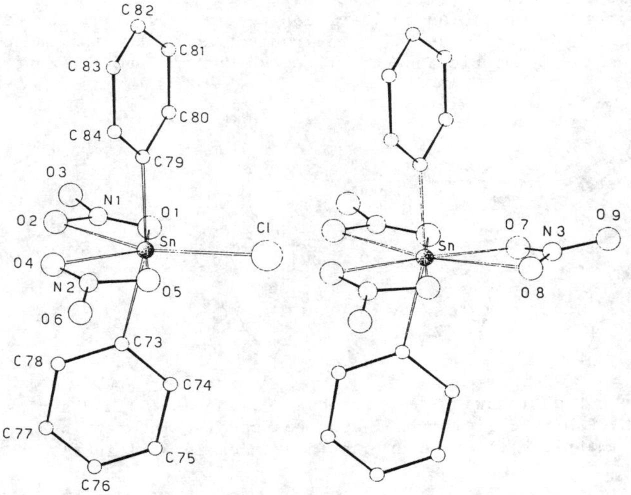

Fig. 1. A view of the disordered anion in $C_{84}H_{70}AgCl_{0.5}N_{2.5}O_{7.5}P_4Sn$.

1,2,5-TELLURADIAZOLE
$C_2H_2N_2Te$

V. BERTINI, P. DAPPORTO, F. LUCCHESINI, A. SEGA and A. DE MUNNO, 1984. Acta Cryst., C40, 653-655.

Orthorhombic, Pnma, a = 12.308, b = 7.840, c = 4.136 Å, D_m = 3.01, Z = 4. Mo radiation, R = 0.031 for 339 reflexions.

In the molecule the Te-N bond length is 2.023(6) Å. The molecules are linked in the crystal by weak Te-N bonds of length 2.764(6) Å into a ribbon polymer structure (Fig. 1). This accounts for the exceptionally high melting point and low solubility of the compound.

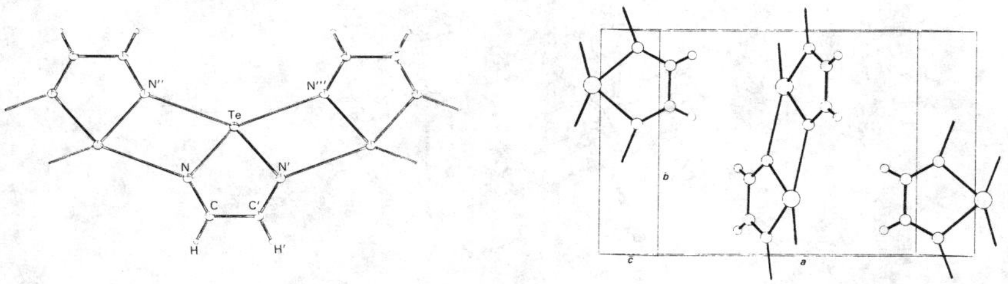

Fig. 1. $C_2H_2N_2Te$: views of the polymeric array of the structure and of the crystal packing.

TRITHIOUREATELLURIUM(II) DI(HYDROGEN DIFLUORIDE)
$C_3H_{14}F_4N_6S_3Te$ $Te[CS(NH_2)_2]_3{}^{2+}.(FHF)_2{}^{2-}$

M.A. BENO, R. SUNDELL and J.M. WILLIAMS, 1984. Croat. Chem. Acta, 57, 695-700.

Monoclinic, $P2_1/c$, a = 5.846, b = 20.46, c = 11.433 Å, β = 94.69°, Z = 4. Mo radiation, R = 0.038 for 5243 reflexions (at 133 K).

The trithiourea-tellurium(II) molecules crystallize as dimeric distorted square-planar cations located on an inversion centre (Fig. 1). The cations are linked by N-H...F hydrogen bonds. The environment about the two independent $(FHF)^-$ anions is decidedly asymmetric and therefore the H atoms are not centred between the fluorine atoms.

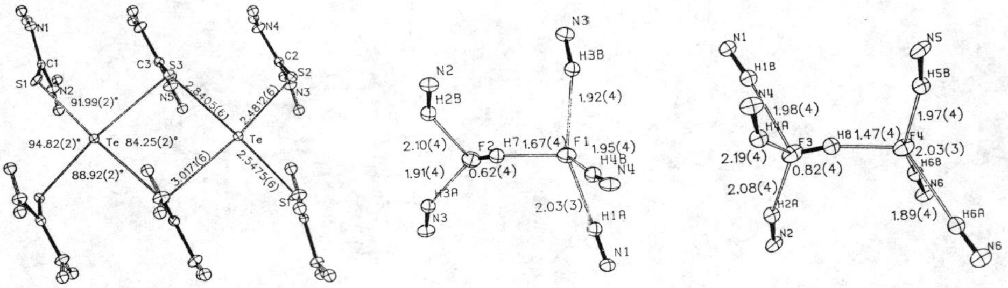

Fig. 1. The cation (showing selected bond distances (Å) and angles (°) in $C_3H_{14}F_4N_6S_3Te$ (left), and hydrogen bonding interactions to the two asymmetric bifluoride anions (distances in Å) (right).

TRIETHYLTELLURONIUM BROMIDE

$C_6H_{15}BrTe$ $(C_2H_5)_3TeBr$

R.K. CHADHA, J.E. DRAKE and M.A. KHAN, 1984. J. Organometal. Chem., 260, 73-80.

Cubic, $I\bar{4}3m$, a = 12.595 Å, D_m = 1.95, Z = 8. Mo radiation, R = 0.029 for 164 reflexions.

The Te and Br atoms are linked by weak secondary interactions to form isolated cubane-like tetramers (Fig. 1). The tetrameric unit $Te_4Br_4(Et)_{12}$ has T_d symmetry imposed by the space group. The Te...Br contact is 3.564 Å, the Te-C bond length 2.14 Å and the C-Te-C bond angle 89.3°. The Br'...Te...Br'' and Te...Br'...Te' angles are 95.9 and 83.8° respectively. Atoms C(1) and C(2) of the ethyl groups are disordered across the crystallographic mirror plane.

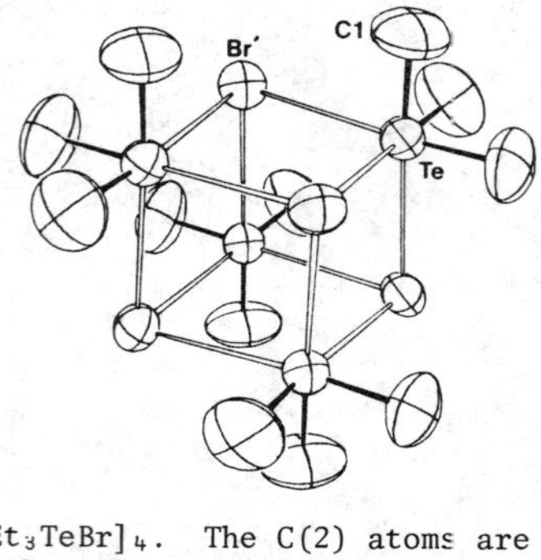

Fig. 1. The tetramer $[Et_3TeBr]_4$. The C(2) atoms are not drawn, and the C(1) atoms are plotted at the mean position of the disordered atoms.

4-NITROBENZYL TELLUROCYANATE
$C_8H_6N_2O_2Te$ (I) $NO_2C_6H_4CH_2TeCN$

4-NITROBENZYL SELENOCYANATE
$C_8H_6N_2O_2Se$ (II) $NO_2C_6H_4CH_2SeCN$

4-NITROBENZYL THIOCYANATE
$C_8H_6N_2O_2S$ (III) $NO_2C_6H_4CH_2SCN$

BENZYL SELENOCYANATE
C_8H_7NSe (IV) $C_6H_5CH_2SeCN$

K. MAARTMANN-MOE, K.A. SANDERUD and J. SONGSTAD, 1984. Acta Chem. Scand., A38, 187-200.

I. Monoclinic, $P2_1/c$, a = 4.749, b = 16.717, c = 11.753 Å, β = 93.26°, Z = 4. Mo radiation, R = 0.027 for 2083 reflexions.

II. Monoclinic, $P2_1/c$, a = 4.747, b = 16.293, c = 11.671 Å, β = 94.41°, Z = 4. Mo radiation, R = 0.036 for 1659 reflexions.

III. Monoclinic, $P2_1/c$, a = 8.342, b = 15.495, c = 7.044 Å, β = 99.78°, Z = 4. Mo radiation, R = 0.039 for 1439 reflexions.

IV. Monoclinic, $P2_1/c$, a = 6.029, b = 7.566, c = 17.578 Å, β = 95.25°, Z = 4. Mo radiation, R = 0.037 for 1075 reflexions.

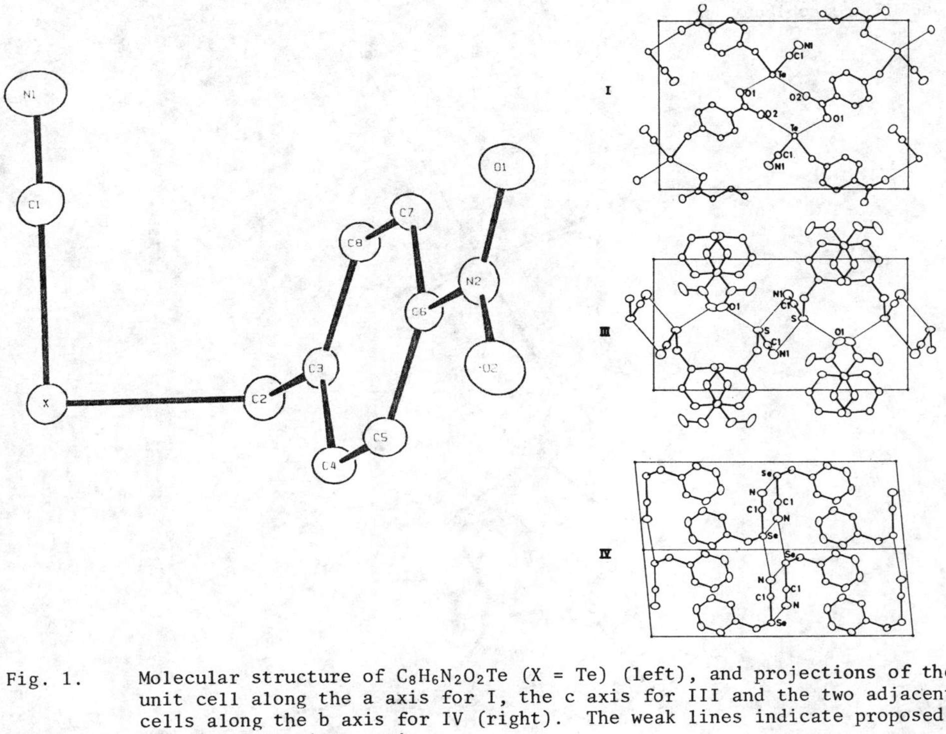

Fig. 1. Molecular structure of $C_8H_6N_2O_2Te$ (X = Te) (left), and projections of the unit cell along the a axis for I, the c axis for III and the two adjacent cells along the b axis for IV (right). The weak lines indicate proposed intermolecular interactions.

Crystals of I and II are isomorphous. The basic structure of the four compounds is shown in Fig. 1 and the C-X-C bond angles are 90.6° in I, 94.0° in II, 100.3° in III and 96.0° in IV. The X-C(CN) bond lengths, 2.060 Å in I, 1.844 Å in II, 1.680 Å in III and 1.829 Å in IV are distinctly shorter than the X-C(CH_2) bond lengths, 2.167 Å in I, 1.972 Å in II, 1.822 Å in III and 1.971 Å in IV. The pseudohalide group is syn-clinal (gauche) to the C(CH_2)-C(Ar) bond in all compounds with torsion angles of 77.4° (I), 75.3° (II), 58.3° (III) and 64.0° (IV). The Te atom in I forms two fairly strong intermolecular bonds to O atoms from neighbouring -NO_2 groups of 2.949 and 3.182 Å (Fig. 1). Compound II has similar, weaker Se-O bonds, 3.005 and 3.174 Å and atom X in compounds I and II can be considered as having distorted square planar coordination. In III, the S atom is in principle only three-coordinated, with a weak intermolecular S-O bond of 3.174 Å and a very weak S-N bond, 3.549 Å. The Se atom in IV is four-coordinated with two weak intermolecular Se-N bonds, 3.107 and 3.641 Å to neighbouring selenocyanate groups in a non-planar arrangement.

TETRACHLORO[(3-ALLYL-2-METHYL-OXAZOLIN-5-YL)METHYL]TELLURIUM
$C_8H_{13}Cl_4NOTe$

J. BERGMAN, J. SIDEN and K. MAARTMANN-MOE, 1984. Tetrahedron, 40, 1607-1610.

Tetragonal, $P4_12_12$, a = 10.542, c = 25.817 Å, Z = 8. Mo radiation, R = 0.049 for 2236 reflexions.

A view of the molecule is in Fig. 1. Atoms of the five-membered ring and C5 and C6 are coplanar.

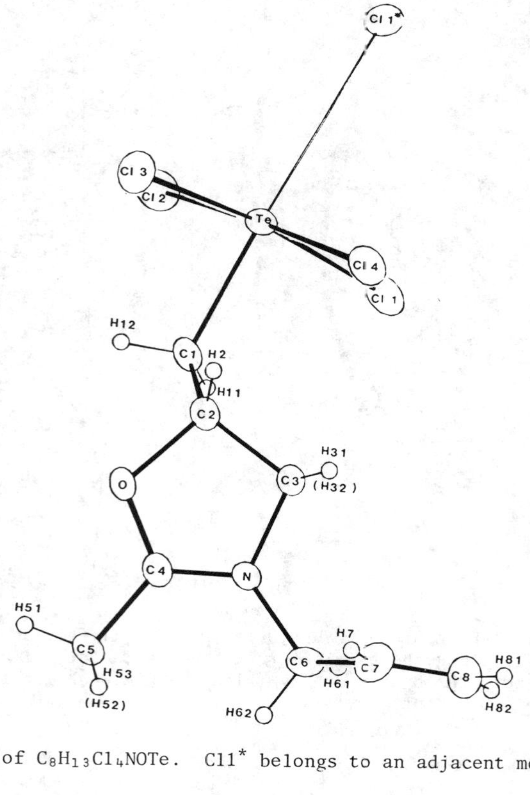

Fig. 1. A view of $C_8H_{13}Cl_4NOTe$. Cl1* belongs to an adjacent molecule.

BIS(DIETHYLDITHIOPHOSPHATO)TELLURIUM(II)
$C_8H_{20}O_4P_2S_4Te$ $(C_2H_5O)_2P(S)S-Te-S(S)P(OC_2H_5)_2$

L.S. REFAAT, K. MAARTMANN-MOE and S. HUSEBYE, 1984. Acta Chem. Scand., A38, 147-155.

Triclinic, $P\bar{1}$, a = 12.247, b = 12.656, c = 20.135 Å, α = 101.69, β = 103.29, γ = 104.72°, D_m = 1.680, Z = 6. Mo radiation, R = 0.029 for 5357 reflexions (at -135°C).

There are three crystallographically-independent molecules per asymmetric unit (Fig. 1) and each molecule has a P-S-Te-S-P chain in the trans form containing approximate twofold symmetry axes through the Te atoms. Average molecular dimensions are: Te-S 2.42, P-S 2.08, P=S 1.93 Å and S-Te-S 100.2, 96.6 and 95.2° in molecules I, II and III respectively. The molecules are joined together by weak intermolecular Te...S bonds with distances varying between 3.21 and 3.54 Å. Each Te atom participates in two such bonds at an S-Te-S angle of 70.3, 101.0, 91.7° in molecules I, II and III respectively. There is a tendency to square planar coordination around Te in all three molecules.

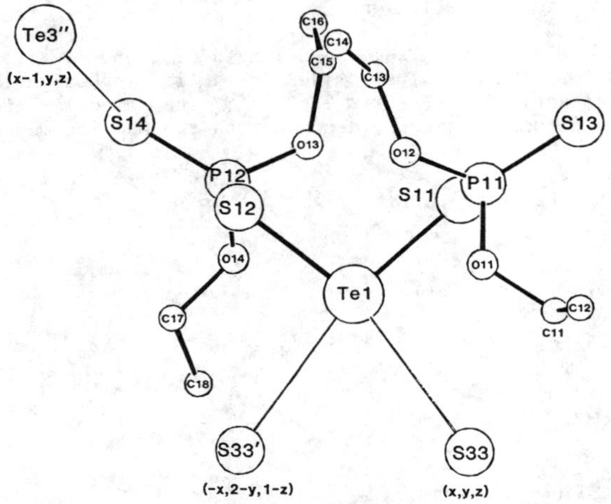

Fig. 1. Molecule I seen along the normal to the Te(1),S(11),S(12) plane in $C_8H_{20}O_4P_2S_4Te$.

3,4-QUINOXALINO-1-TELLURACYCLOPENTANE
$C_{10}H_8N_2Te$

A.E. McCARTHY and H.B. SINGH, 1984. J. Organometal. Chem., 275, 57-62.

Tetragonal, $I4_1/a$, a = 25.315, c = 6.010 Å, D_m = 1.91, Z = 16. Mo radiation, R = 0.041 for 744 reflexions.

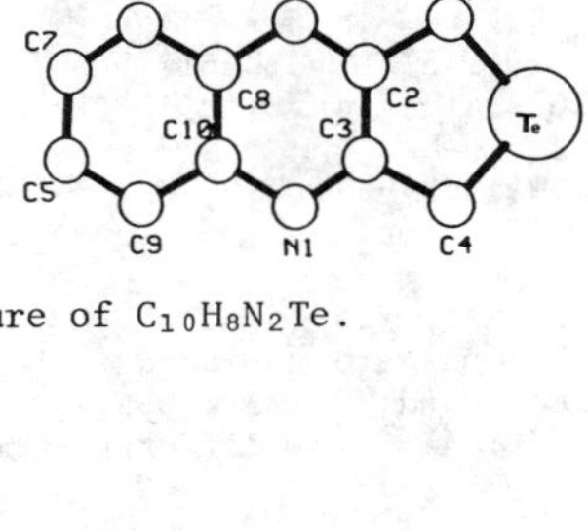

Fig. 1. Molecular structure of $C_{10}H_8N_2Te$.

The two aromatic moieties of the quinoxaline residue are almost planar but the Te-bonded residue is distorted from planarity and is tilted so that the Te atom is displaced 0.68 Å from the best plane through the molecule (Fig. 1). The Te-C bonds average 2.13 Å and the C-N bonds 1.34 Å. The C(1)-Te-C(4) bond angle is 80.7°. There are short intermolecular bonds between Te atoms of molecules related by the symmetry transformations x,y,z and 1-x, 0.5-y,z (3.791 Å) and x,y,z and 0.25+y, 0.75-x, 0.75-z (3.998 Å).

BIS(p-CHLOROPHENYL) DITELLURIDE
$C_{12}H_8Cl_2Te_2$

G. VAN DEN BOSSCHE, M.R. SPIRLET, O. DIDEBERG and L. DUPONT, 1984. Acta Cryst., C40, 1011-1012.

Orthorhombic, $P2_12_12_1$, a = 10.602, b = 14.543, c = 8.713 Å, D_m = 2.49, Z = 4. Mo radiation, R = 0.045 for 1959 reflexions.

In the molecule (Fig. 1) the Te-Te bond length of 2.680(1) Å is shorter than the values reported for diphenyl ditellurides in earlier investigations (1). The value of the dihedral angle C-Te-Te-C is 88.8(4)°. The molecular geometries observed in diphenyl ditellurides seem dependent on the packing modes alone.

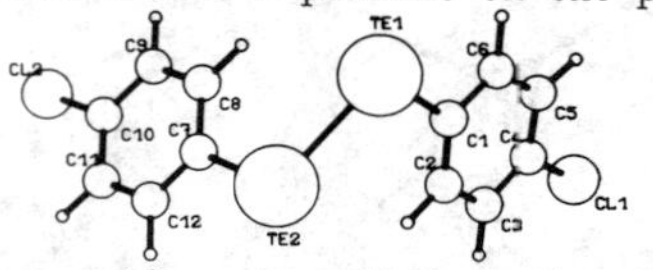

Fig. 1. $C_{12}H_8Cl_2Te_2$: perspective view of the molecule.

1. Structure Reports, 32B, 726; 45B, 840; 21, 595.

DIIODODIPHENYLTELLURIUM(IV) (α form)
$C_{12}H_{10}I_2Te$ (I)

DIIODODIPHENYLTELLURIUM(IV) (β form)
$C_{12}H_{10}I_2Te$ (II)

cis-BIS(μ_2-IODO)TETRAIODODIPHENYLDITELLURIUM(IV)
$C_{12}H_{10}I_6Te$ (III)

trans-BIS(μ_2-IODO)TETRAIODODIPHENYLDITELLURIUM(IV)
$C_{12}H_{10}I_6Te$ (IV)

N.W. ALCOCK and W.D. HARRISON, 1984. J. Chem. Soc. Dalton, 869-875.

I. Tetragonal, $I4_1$, a = 11.684, c = 10.196 Å, D_m = 2.61, Z = 4. Mo radiation, R = 0.030 for 575 reflexions.

II. Monoclinic, $P2_1/c$, a = 9.722, b = 15.627, c = 19.338 Å, β = 107.08°, D_m = 2.48, Z = 8. Mo radiation, R = 0.034 for 3303 reflexions.

III. Orthorhombic, $Cmc2_1$, a = 19.915, b = 12.617, c = 8.867 Å, D_m = ca. 3.3, Z = 4. Mo radiation, R = 0.044 for 1267 reflexions (at 173 K).

IV. Orthorhombic, Pnnm, a = 8.911, b = 11.559, c = 13.506 Å, D_m = 2.82, Z = 2. Mo radiation, R = 0.048 for 1201 reflexions.

In I and II, the primary geometry is pseudotrigonal bipyramidal with axial I atoms (Fig. 1). In II the Te-I bonds are variable with the two independent molecules having one longer and one shorter bond (2.883-2.959(1)) while in I the value falls in the middle of this range (2.928(1) Å). The difference between I and II lies in the packing. In I, two secondary bonds complete an irregular octahedron about each Te and link molecules into a three-dimensional network. In II, Te(2) also has two secondary bonds but in Te(1) an octahedron with one vacant position is produced. The resulting three dimensional network is extremely complicated. III and IV are dimers of $PhTeI_3$, having the phenyl groups on the same side or on opposite sides of the Te_2I_6 plane. The geometry at Te in each is square pyramidal with apical phenyl groups (III; Te-C(1) 2.16(2), Te-I(1-4) 3.152, 3.179, 2.775, 2.792(2) Å; IV; Te-C(1) 2.12(2), Te-I(1,2) 3.285, 2.780(1) Å. Both Te atoms lie slightly above the I_4 plane and both show I...I secondary bonding (3.848(2)-4.263(3) in III and 3.886(1) Å in IV).

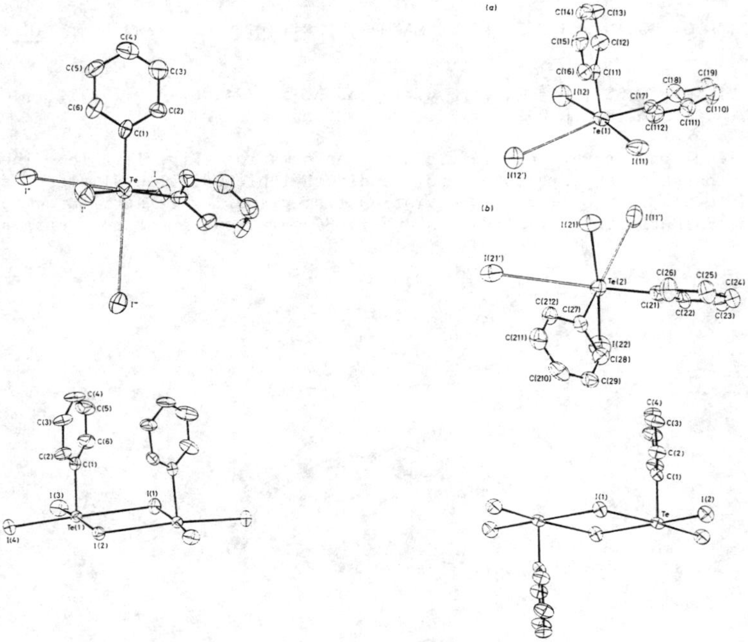

Fig. 1. The molecular structures: α-$C_{12}H_{10}I_2Te$ (top left), β-$C_{12}H_{10}I_2Te$ (top right), cis-$C_{12}H_{10}I_6Te$ (bottom left) and trans-$C_{12}H_{10}I_6Te$ (bottom right).

BIS(THIOBENZOATO-S)TELLURIUM(II)
$C_{14}H_{10}O_2S_2Te$

T. SUBRAHMANYAN, G. ARAVAMUDAN, G.C. ROUT and M. SESHASAYEE, 1984. J. Cryst. Spectrosc. Res., 14, 239-248.

Monoclinic, C2/c, a = 24.369, b = 4.333, c = 14.569 Å, β = 109.56°, Z = 4. Mo radiation, R = 0.027 for 772 reflexions.

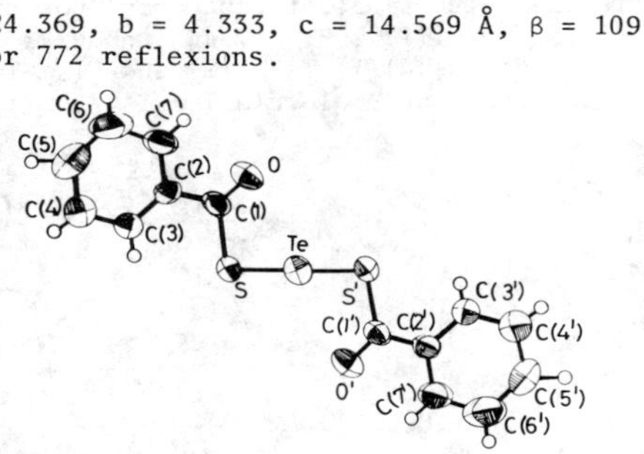

Fig. 1. A view of $C_{14}H_{10}O_2S_2Te$.

The tellurium atom lies on a twofold symmetry axis (Fig. 1), with Te-S 2.372(2), S-C 1.832(6) Å, S-Te-S 103.1(1)°, Te-S-C 102.5(2)°. There are also two weak Te...O interactions (3.242(2) Å).

BIS(DIMETHYLTHIENO)TETRATELLURAFULVALENE

$C_{14}H_{12}S_2Te_4$

K. LERSTRUP, D.O. COWAN and T.J. KISTENMACHER, 1984. J. Am. Chem. Soc., 106, 8303-8305.

Triclinic, P$\bar{1}$, a = 9.589, b = 11.232, c = 8.635 Å, α = 94.08, β = 102.12, γ = 89.41°, Z = 2. Mo radiation, R = 0.049 for 5914 reflexions.

The molecule has a very asymmetric boat conformation (Fig. 1). The dihedral angles between the C_2Te_4 core plane and the dimethylthienoditellural planes are 47.1(3) and 16.0(3)°. Some of this deviation from planarity is attributed to the formation of "dimers" in the solid as shown in Fig. 1 about inversion centres.

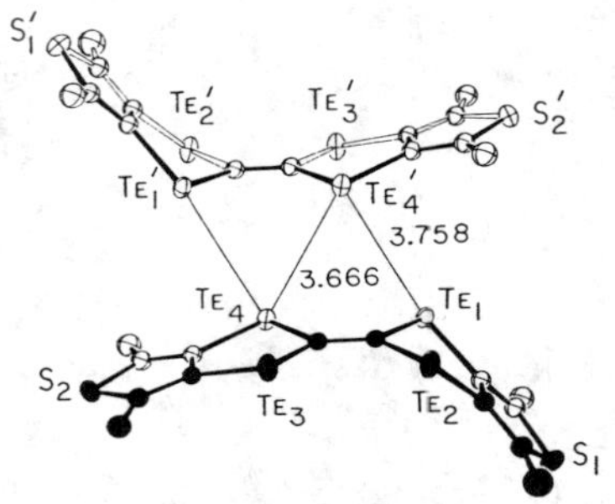

Fig. 1. The structure of $C_{14}H_{12}S_2Te_4$.

BIS(4-HYDROXY-3-METHYLPHENYL)TELLURIUM(IV)DICHLORIDE (β-ISOMER)

$C_{14}H_{14}Cl_2O_2Te$ $(C_7H_7O)_2TeCl_2$

R.H. JONES and T.A. HAMOR, 1984. J. Organometal. Chem., 262, 151-155.

Monoclinic, C2/c, a = 14.651, b = 12.787, c = 8.279 Å, β = 92.64°, Z = 4. Mo radiation, R = 0.030 for 1907 reflexions.

The molecule has twofold crystallographic symmetry and the coordination about the Te atom is trigonal bipyramidal with a vacant equatorial position (Fig. 1). The Cl atoms occupy the apical sites (Te-Cl(1) 2.518 Å) and the aryl ligands the equatorial sites (Te-C(1) 2.104 Å). The angles Cl(1)-Te-Cl(2) and C(1)-Te-C(21) are respectively 174.8 and 99.2°. A hydrogen bond (intermolecular) exists, O(1)-H(7)...Cl(1''), with O...Cl 3.222 Å.

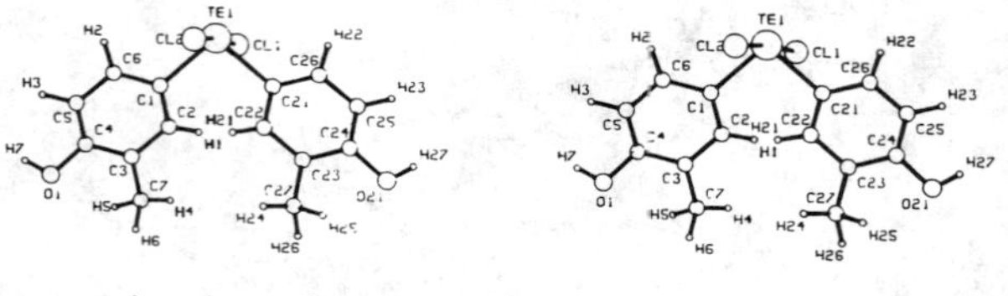

Fig. 1. Stereoscopic view of $C_{14}H_{14}Cl_2O_2Te$.

DICHLOROBIS(p-METHOXYPHENYL)TELLURIUM(IV)
$C_{14}H_{14}Cl_2O_2Te$ $TeCl_2(C_7H_7O)_2$

R.K. CHADHA and J.E. DRAKE, 1984. Acta Cryst., C40, 1349-1352.

Triclinic, $P\bar{1}$, a = 10.243, b = 12.480, c = 13.654 Å, α = 109.22, β = 80.42, γ = 77.30°, D_m = 1.74, Z = 4. Mo radiation, R = 0.028 for 3406 reflexions.

The structure contains isolated, "step-like" tetramers in which two independent molecules and their centrosymmetric equivalents are linked through Te...Cl interactions (Fig. 1). The equatorial Te-C bonds are of length 2.105 to 2.121(6) Å with C-Te-C 97.0(2) and 95.9(2)° in molecules 1 and 2. The axial bonds have lengths Te-Cl(1) 2.518(1), Te-Cl(2) 2.514(2), Te-Cl(3) 2.474(1), Te-Cl(4) 2.569(1) Å. The phenyl rings are in the propeller arrangement with dihedral angles between the TeCl plane and the rings of 40.2 and 37.3 (in 1) and 46.4 and 38.6° (in 2). The angles between the ring planes are 55.3 and 60.2° respectively.

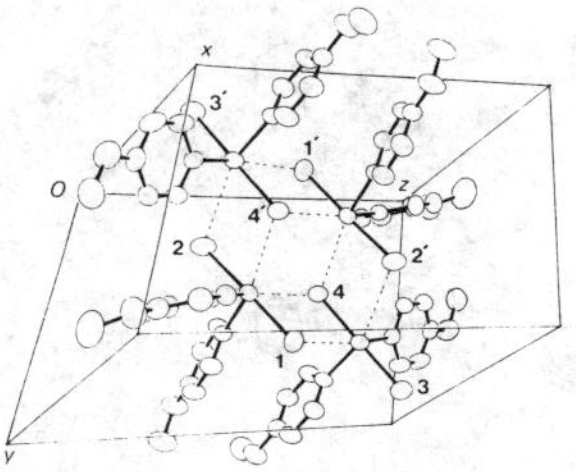

Fig. 1. The unit cell of $TeCl_2(C_7H_7O)_2$ showing Te...Cl interactions.

(μ-TELLURIUM)BIS[(η^5-CYCLOPENTADIENYL)TRICARBONYLCHROMIUM]
$C_{16}H_{10}Cr_2O_6Te$ $(Te[(\eta^5-C_5H_5)Cr(CO)_3]_2$

W.A. HERRMANN, J. ROHRMANN, M.L. ZIEGLER and T. ZAHN, 1984. J. Organometal. Chem., 273, 221-232.

Monoclinic, $P2_1/a$, a = 13.241, b = 7.310, c = 19.152 Å, β = 104.73°, Z = 4. Mo radiation, R = 0.034 for 2419 reflexions.

The molecule consists of two $CpCr(CO)_3$ units bridged by a Te atom (Fig. 1). The Te bridge acts as a two electron ligand which stabilises these two units. The Cr(1)-Te and Cr(2)-Te distances are 2.814 and 2.799 Å respectively, and the Cr(1)-Te-Cr(2) angle is 117.2°. The Cr-Cp(c) distances range from 2.158-2.207 Å for C(7)... C(11) and from 2.159 to 2.214 Å for C(12)...C(16), and the Cr-C(carbonyl) distances range from 1.831 to 1.881 Å.

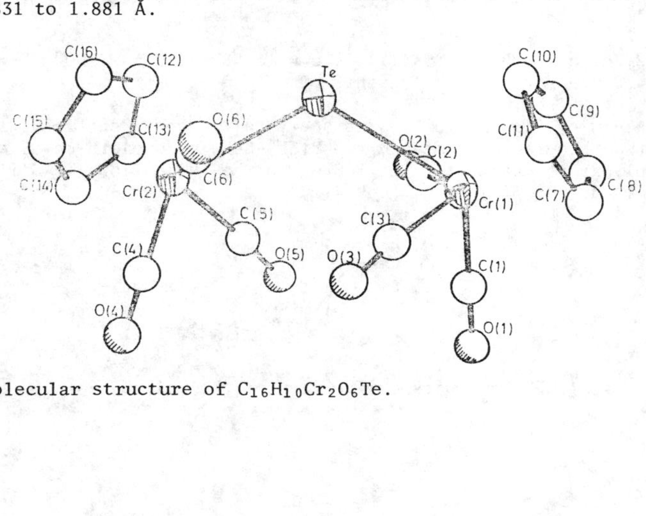

Fig. 1. Molecular structure of $C_{16}H_{10}Cr_2O_6Te$.

DIPHENYLTELLURIUM DI(TRIFLUOROACETATE)

$C_{16}H_{10}F_6O_4Te$ $(C_6H_5)_2Te(O_2CCF_3)_2$

S.B. BULGAREVICH, B.B. RIVKIN, N.G. FURMANOVA, O.O. EXNER, D.Ya. MOVSHOVICH, T.A. YUSMAN, I.D. SADEKOV and V.I. MINKIN, 1984. J. Struct. Chem., 25, 85-92 [Zh. Strukt. Khim. 25(1), 97-105].

Monoclinic, $P2_1/c$, a = 14.156, b = 13.689, c = 19.566 Å, β = 100.04°, Z = 8. Mo radiation, R = 0.064 for 4317 reflexions.

The structure contains two crystallographically independent centrosymmetric dimers (Fig. 1). Monomers are linked by weak Te...O interactions. Excluding the weak interaction, the coordination about Te is trigonal bipyramidal with the lone pair of electrons occupying an equatorial position. Mean bond lengths and angles are shown in Fig. 1.

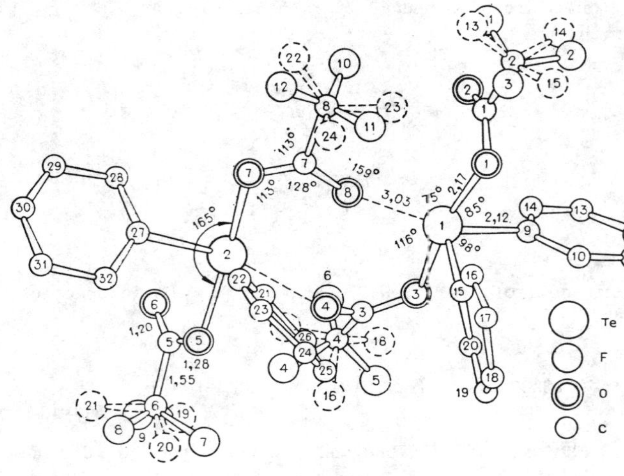

Fig. 1. The diphenyltellurium di(trifluoracetate) dimer.

trans-BIS[2(3H)-BENZIMIDAZOLETHIONE]BIS(THIOUREA)TELLURIUM(II) CHLORIDE

$C_{16}H_{20}Cl_2N_8S_4Te$ $[Te(CH_4N_2S)_2(C_7H_6N_2S)_2]Cl_2$

G.C. ROUT, M. SESHASAYEE, G. ARAVAMUDAN and S. SOWRIRAJAN, 1984. Acta Cryst., C40, 963-965.

Monoclinic, $P2_1/c$, a = 6.623, b = 11.387, c = 16.085 Å, β = 93.90°, Z = 2. Mo radiation, R = 0.029 for 2068 reflexions.

In the molecule (Fig. 1) Te lies at a crystallographic centre of symmetry and is bonded to two thiourea S atoms and two 2(3H)-benzimidazolethione S atoms in a trans-square-planar arrangement. The TeS_4 group has dimensions Te-S(1) 2.694(2), Te-S(2) 2.685(1) Å, and S(1)-Te-S(2) 89.4(1)°.

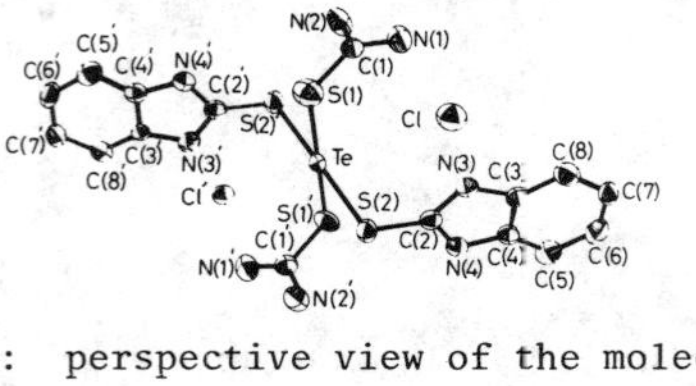

Fig. 1. $C_{16}H_{20}Cl_2N_8S_4Te$: perspective view of the molecule.

DIETHYLAMMONIUM TETRACHLORO(p-PHENOXYPHENYL)TELLURATE
$C_{16}H_{21}Cl_4NOTe$ $Et_2NH_2^+.C_6H_5OC_6H_4TeCl_4^-$

R.K. CHADHA, J.E. DRAKE and M.A. KHAN, 1984. Can. J. Chem., 62, 32-35.

Monoclinic, C2/c, a = 19.675, b = 7.634, c = 28.716 Å, β = 105.58°, Dm = 1.64, Z = 8. Mo radiation, R = 0.030 for 2636 reflexions.

The Te atom has approximate square-pyramidal geometry (Fig. 1) with the organic group apical. The vacant sixth site is available for a secondary-bonding interaction to an adjacent chlorine atom. Principal dimensions are Te-C(11) 2.127(5), Te-Cl(eq) 2.451-2.589(1), Te...Cl 3.619 Å.

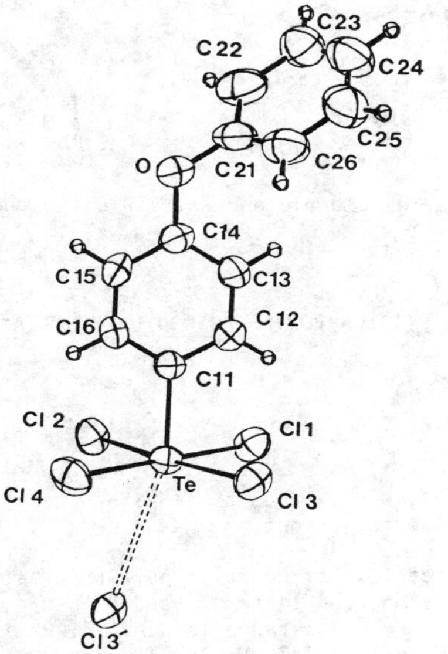

Fig. 1. A view of the $[C_6H_5OC_6H_4TeCl_4]^-$ anion.

THIOCYANATOTRIS(2,2'-IMINODIETHANOLDITHIOCARBAMATO)TELLURIUM(IV) MONOHYDRATE
$C_{16}H_{30}N_4O_4S_7Te,H_2O$

G.C. ROUT, M. SESHASAYEE, G. ARAVAMUDAN, K. RADHA, 1984. J. Cryst. Spectrosc. Res. 14, 193-204.

Monoclinic, $P2_1/n$, a = 13.291, b = 21.306, c = 10.450 Å, β = 104.25°, Z = 4. Mo radiation, R = 0.025 for 2894 reflexions.

The Te atom has pentagonal bipyramidal geometry (Fig. 1), with the thiocarbamate groups bonding to Te in an unequal bidentate manner. A network of hydrogen bonds links the complex molecules and water.

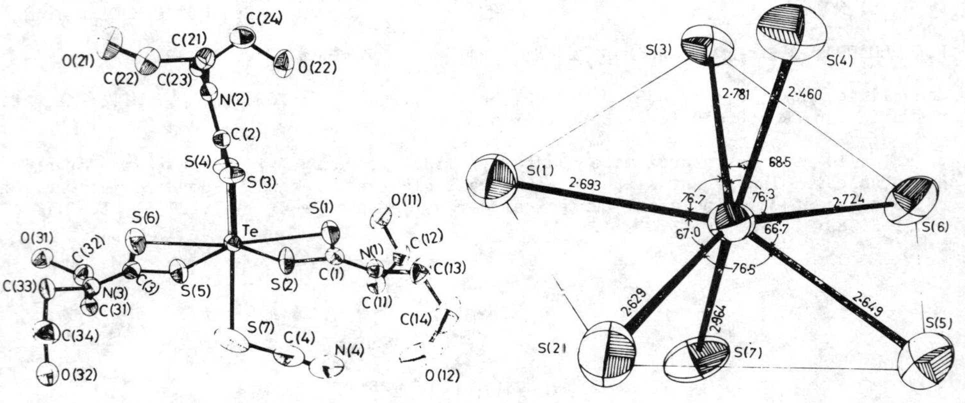

Fig. 1. The $C_{16}H_{30}N_4O_4S_7Te$ molecule and details of Te coordination.

(μ_3-TELLURIUM)-triangulo-BIS(TRICARBONYLFERRIO)DICARBONYL(η^5-PENTAMETHYLCYCLOPENTADIENYL)MANGANESE(Fe-Fe,2Fe-Mn)

$C_{18}H_{15}Fe_2MnO_8Te$ $(C_5Me_5)MnFe_2Te(CO)_8$

W.A. HERRMANN, C. HECHT, M.L. ZIEGLER and T. ZAHN, 1984. J. Organometal. Chem., 273, 323-331.

Monoclinic, $P2_1/n$, a = 9.824, b = 14.836, c = 15.355 Å, β = 97.22°, Z = 4. Mo radiation, R = 0.036 for 3456 reflexions.

The molecule is of a metallatetrahedral type with a tetranuclear cluster of two Fe atoms, and one Te and one Mn, the latter π-bonded to the pentamethylcyclopentadienyl ligand (Fig. 1). The Fe-Fe distance is 2.677 Å, whilst the remaining metal-metal bond lengths are: Mn-Fe(1) 2.743, Mn-Fe(2) 2.736, Mn-Te 2.518, Fe(1)-Te 2.472, and Fe(2)-Te 2.476 Å. The non-linear Mn-C(1)-O(1) and Mn-C(2)-O(2) bonds (angles 155.6 and 155.3°) suggest bridging carbonyls. The Fe(1)-C(2) and Fe(2)-C(1) distances are 2.315 and 2.320 Å respectively. The Mn-C(Cp) distances range from 2.138 to 2.208 Å and the Mn-Cp(centroid) distance is 1.806 Å.

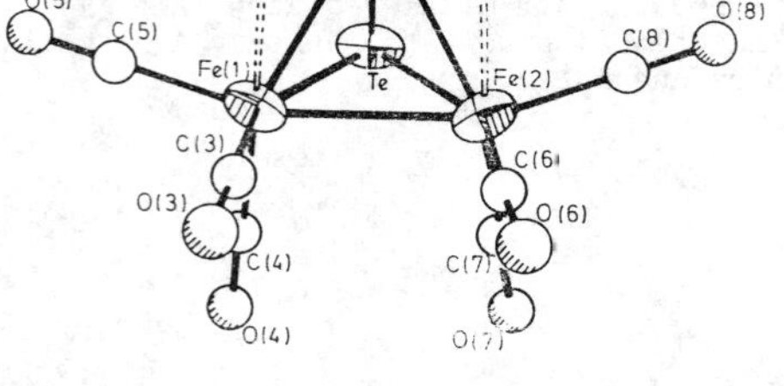

Fig. 1. Molecular structure of the cluster compound $C_{18}H_{15}Fe_2MnO_8Te$.

cis-1,1-DIFLUORO-1,1-DIHYDRO-3,3,3',3'-TETRAMETHYL-1,1'-SPIROBI[3H-2,1-BENZOOXA-TELLUROLE]
$C_{18}H_{20}F_2O_2Te$ (I)

3,3,3',3'-TETRAKIS(TRIFLUOROMETHYL)-1,1'-SPIROBI[3H-2,1-BENZOOXATELLUROLE]
$C_{18}H_8F_{12}O_2Te$ (II)

R.S. MICHALAK, S.R. WILSON and J.C. MARTIN, 1984. J. Am. Chem. Soc., 106, 7529-7539.

I. Monoclinic, C2/c, a = 12.170, b = 11.976, c = 25.596 Å, β = 111.26°, D_m = 1.67, Z = 8. Mo radiation, R = 0.026 for 2577 reflexions.

II. Monoclinic, $P2_1/c$, a = 9.154, b = 20.488, c = 10.829 Å, β = 106.08°, D_m = 2.08, Z = 8. Mo radiation, R = 0.023 for 3873 reflexions.

Views of (I) and (II) are in Fig. 1. Main bond lengths are for (I) Te-F 1.937, 1.942(2), Te-O 1.939, 1.943(2), Te-C 2.073, 2.078(3) Å; for (II) Te-O 2.077, 2.082(2), Te-C 2.103, 2.106(2) Å.

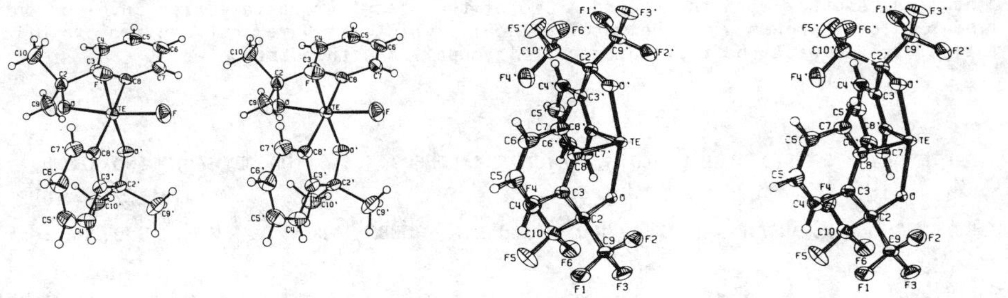

Fig. 1. Stereoviews of $C_{18}H_{20}F_2O_2Te$ and $C_{18}H_8F_{12}O_2Te$.

BIS[(ACETATO)(p-METHOXYPHENYL)]TELLURIUM(IV)
$C_{18}H_{20}O_6Te$ (I)

(μ_2-OXO)BIS(DIPHENYLTRIFLUOROACETATOTELLURIUM(IV)) HEMIHYDRATE
$C_{28}H_{20}F_6O_5Te_2$,0.5(H_2O) (II)

BIS(TRIFLUOROACETOXY)IODOBENZENE
$C_{10}H_5F_6IO_4$ (III)

N.W. ALCOCK, W.D. HARRISON AND C. HOWES, 1984. J. Chem. Soc. Dalton, 1709-1716.

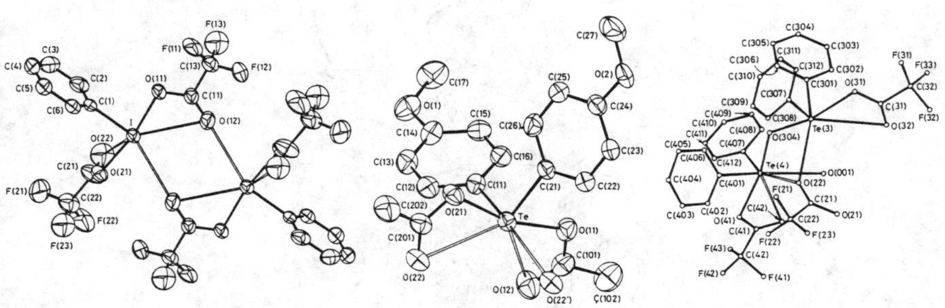

Fig. 1. $C_{10}H_5F_6IO_4$, centrosymmetric dimer (left); $C_{18}H_{20}O_6Te$ (centre) and one molecule and the shared water molecule, O(001), with one trifluoroacetate of the other molecule (right).

I. Monoclinic, $P2_1/c$, a = 9.529, b = 11.984, c = 17.035 Å, β = 101.70°, Z = 4. Mo radiation, R = 0.022 for 3033 reflexions.

II. Triclinic, $P\bar{1}$, a = 13.988, b = 14.287, c = 15.689 Å, α = 80.40, β = 81.89, γ = 86.65°, D_m = 1.62, Z = 2. Mo radiation, R = 0.042 for 5290 reflexions.

III. Triclinic, $P\bar{1}$, a = 9.787, b = 9.055, c = 7.674 Å, α = 91.45, β = 99.78, γ = 89.21°, D_m = 2.11, Z = 2. Mo radiation, R = 0.037 for 2104 reflexions.

All three compounds have pseudo-trigonal bipyramidal primary geometry (Te-C 2.098(3), 2.103(4), I-C 2.074(4); Te-O(μ-oxo) 1.981(7), Te-O and I-O (oxy-anion) 2.138-2.352 Å) and form secondary bonds to give pentagonal planar systems around Te and I (Fig. 1). There are also linear C-Te...O systems. In III there are two secondary bonds with I...O 3.00-3.04 Å. These complete a pentagonal plane around iodine along with the ClO_2 unit and link the molecules into centrosymmetric dimers. I has similar geometry with Te...O distances 2.95-3.02 Å, although both acetates link to the same Te to give monomers. In II two 'half'-pentagonal planes are formed from a pseudo-trigonal bipyramid containing two Te-C bonds. One plane is formed with one C atom and another with the second. Two of the phenyl groups are also involved in linear C-Te...O bonds with the water molecule, which is shared between two oxo-bridged molecules, and with the trifluoroacetate groups links them into dimers.

1,3-DIHYDRO-2,2-DIIODOTELLUROLO[3,4-b]QUINOXALINE 2,3-BIS(IODOMETHYL)QUINOXALINE

$C_{20}H_{16}I_4N_4Te$ $C_{10}H_8I_2N_2Te,C_{10}H_8I_2N_2$

H.B. SINGH, W.R. McWHINNIE, T.A. HAMOR and R.H. JONES, 1984. J. Chem. Soc. Dalton, 23-28.

Triclinic, $P\bar{1}$, a = 7.667, b = 11.826, c = 13.739 Å, α = 93.26, β = 98.37, γ = 83.29°, Z = 2. Mo radiation, R = 0.059 for 2998 reflexions.

The structure consists of discrete Te containing molecules and free 2,3-bis-(iodomethyl)quinoxaline molecules. Two molecules of each are linked by weak Te...I interactions into centrosymmetric dimers (Fig. 1). Coordination about Te is octahedral with two Te-C bonds (2.14(1)), two axial Te-I bonds (2.886(2), 2.911(2)) and two longer Te-I bonds (3.761(2), 3.788(2) Å). Large deviations of bond angles from ideal octahedral values occur (I(1')-Te-I(3) 125.4(1), I(1')-Te-C(1) 71.9(4), I(3)-Te-C(12) 76.8(4)°). A further very weak interaction links the dimers into chains along x. The organic moieties have normal bond lengths and angles.

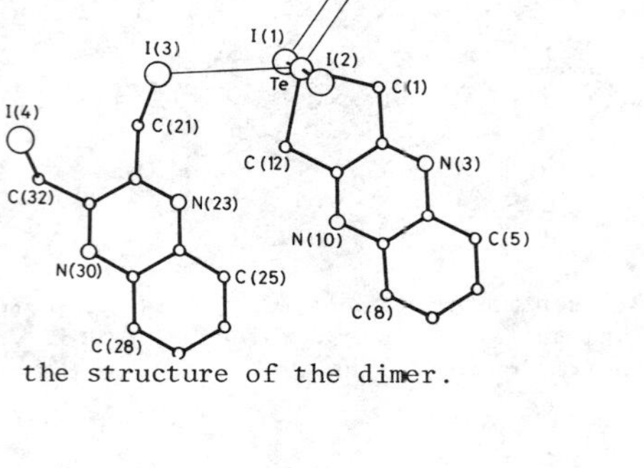

Fig. 1. $C_{20}H_{16}I_4N_4Te$: the structure of the dimer.

DICHLOROBIS(p-TRIMETHYLSILYLTOLYL)TELLURIUM
$C_{20}H_{30}Cl_2Si_2Te$ $(p-(CH_3)_3SiCH_2C_6H_4)_2TeCl_2$

R.K. CHADHA and J.E. DRAKE, 1984. J. Organometal. Chem., 268, 141-147.

Orthorhombic, Pbcn, a = 17.890, b = 17.047, c = 8.169 Å, D_m = 1.40, Z = 4. Mo radiation, R = 0.033 for 1490 reflexions.

The compound has twofold crystallographic symmetry and adopts four-coordinate geometry which may be considered as based on a distorted, trigonal bipyramid with an equatorial lone pair (Fig. 1). The Te atom is, however, involved in intermolecular interactions with two Cl atoms of the neighbouring molecules (Te...Cl'' 3.615 Å) resulting in distorted octahedral coordination and the formation of infinite chains parallel to the c axis (Fig. 1). The Te-Cl and Te-C distances are respectively 2.516 and 2.098 Å, and the angle C-Te-C is 96.1°. Other angles are: Cl''-Te-Cl''' 87.0, C(1')-Te-Cl''' 90.1 and C(1)-Te-Cl''' 165.7°. The Si-C distances range from 1.848 to 1.870 Å and the angles about Si from 107.3 to 111.4°.

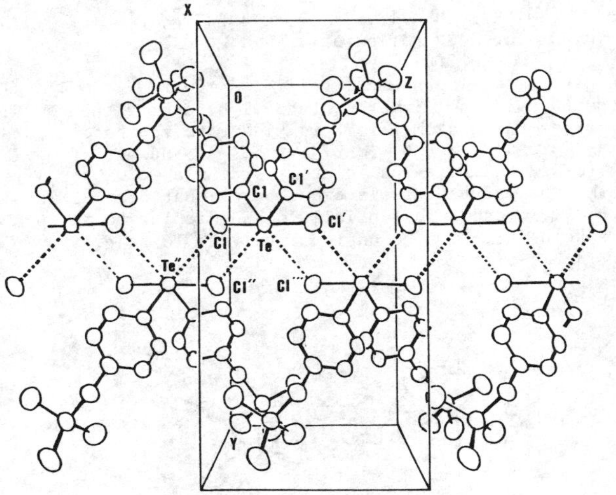

Fig. 1. Crystal structure of $C_{20}H_{30}Cl_2Si_2Te$ viewed down a.

TETRAKIS[BIS(2-HYDROXYETHYL)DITHIOCARBAMATO]TELLURIUM(IV)
$C_{20}H_{40}N_4O_8S_8Te$

G.C. ROUT, M. SESHASAYEE, G. ARAVAMUDAN and K. RADHA, 1984. Acta Cryst., C40, 1142-1145.

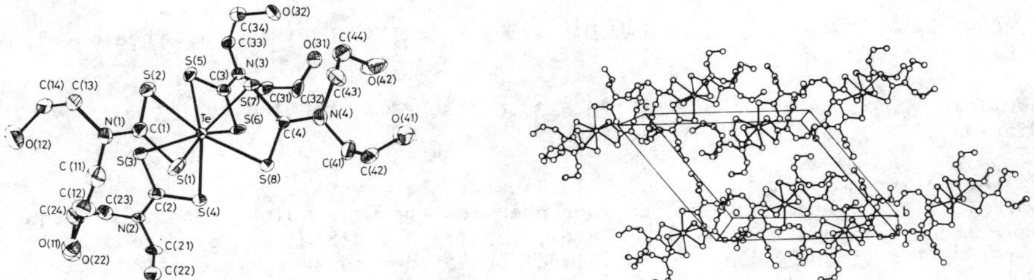

Fig. 1. $C_{20}H_{40}N_4O_8S_8Te$: perspective views of the molecule and of the molecular packing.

Triclinic, $P\bar{1}$, a = 9.008, b = 17.194, c = 14.501, α = 128.34, β = 95.60, γ = 97.13°, Z = 2. Mo radiation, R = 0.042 for 3482 reflexions.

In the molecule (Fig. 1) Te is bonded to all eight S atoms in a highly distorted dodecahedral configuration. The two planar trapezoids, each containing a TeS_4 group, make an angle of 86.50(1)° with each other. The Te-S bond lengths are in the range 2.637(4)-2.857(4), average 2.738(4) Å. The extensive hydrogen bonding of the terminal hydroxyl groups in the range 2.678(13)-2.763(14) Å hinders the auto redox reaction in the complex molecule and also provides stability for the crystal packing.

(μ_2-4,4'-BIPYRIDYL)BIS(O-ETHYLXANTHATOTELLURIUM(II))

$C_{22}H_{28}N_2O_4S_8Te_2$ $[Te(S_2COC_2H_5)_2]_2(C_{10}H_8N_2)$

B.F. HOSKINS, P.J. OLIVER and G. WINTER, 1984. Inorg. Chim. Acta, 86, L21-L23.

Monoclinic, $P2_1/c$, a = 9.0249, b = 13.211, c = 14.347 Å, β = 102.3°, D_m = 1.77, Z = 2. Cu radiation, R = 0.053 for 2487 reflexions.

The molecule has crystallographic inversion symmetry (Fig. 1). The outstanding feature of the structure is the planar pentagonal nature of the coordination around the Te atom which comprises the S atoms from two bidentate xanthate ligands (Te-S(1) 2.548, Te-S(2) 2.814, Te-S(3) 2.839, Te-S(4) 2.539 Å) and one of the N atoms from the bipyridyl ligand (Te-N 2.892 Å). The Te atom is displaced 0.013 Å from the best plane through atoms TeS_4N, the largest displacement being that of the N atom, 0.111 Å. The plane of the bipyridyl ligand is essentially perpendicular to the pentagonal plane, the dihedral angle being 88.1°. The angles involving Te range from 65.95° (S(3)-Te-S(4)) to 78.56° (S(4)-Te-S(1)).

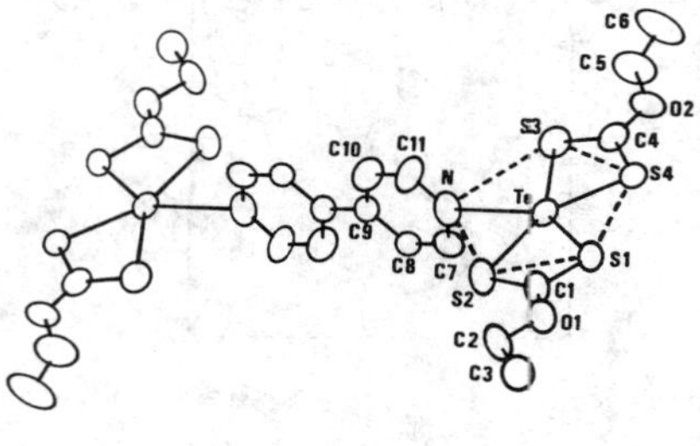

Fig. 1. Molecular structure of $C_{22}H_{28}N_2O_4S_8Te_2$.

TETRAKIS(PHENYLENETHIOUREA-S)TELLURIUM(II) DICHLORIDE DIHYDROCHLORIDE

$C_{28}H_{26}Cl_4N_8S_4Te$ $Te(ptu)_4Cl_2.2HCl$

G.C. ROUT, M. SESHASAYEE, G. ARAVAMUDAN and S. SOWRIRAJAN, 1984. Polyhedron, 3, 921-927.

Monoclinic, $P2_1/c$, a = 13.939, b = 26.523, c = 4.873 Å, β = 100.29°, Z = 2. Mo radiation, R = 0.055 for 918 reflexions.

The tellurium atom in the molecule lies at the crystallographic centre of symmetry (Fig. 1) and is bonded to four phenylenethiourea sulphur atoms in a square planar arrangement with Te-S(1) 2.678(6), Te-S(2) 2.674(5) Å and S(1)-Te-S(2) is 90.5(3)°. The ligand behaves as a thione. Chlorine atoms remain outside the coordination sphere of the Te and stabilise the packing arrangement in the unit cell through hydrogen bondings to nitrogen atoms.

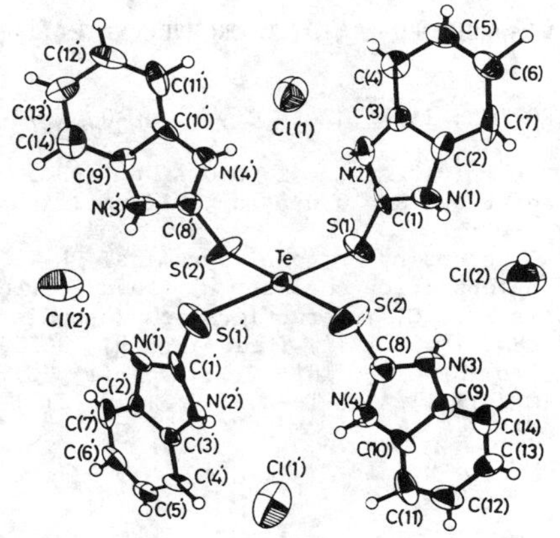

Fig. 1. A view of $Te(ptu)_4Cl_2.2HCl$.

1-METHYL-3,4-BENZO-1-TELLURACYCLOPENTANE TETRAPHENYLBORATE
$C_{33}H_{31}BTe$ (I) $C_8H_8TeCH_3{}^+B(C_6H_5)_4{}^-$

1-PHENYL-1-TELLURACYCLOPENTANE TETRAPHENYLBORATE
$C_{34}H_{33}BTe$ (II) $C_4H_8TeC_6H_5{}^+B(C_6H_5)_4{}^-$

R.H. JONES and T.A. HAMOR, 1984. J. Organometal. Chem., 269, 11-20.

I. Monoclinic, $P2_1/n$, a = 9.175, b = 17.402, c = 16.998 Å, β = 98.92°, Z = 4. Mo radiation, R = 0.051 for 1641 reflexions.

II. Triclinic, $P\bar{1}$, a = 9.635, b = 17.721, c = 16.858 Å, α = 89.77, β = 104.36, γ = 90.16°, Z = 4. Mo radiation, R = 0.090 for 6466 reflexions.

In both I and II, Te is three-coordinated in a pyramidal geometry with C-Te-C bond angles in the ranges 86.6-95.7° (I) and 84.9-99.5° (II), and C-Te bond distances in the range 2.07-2.14 Å (I) and 2.10-2.17 Å (II) (Fig. 1). In II one of the two independent cations in the asymmetric unit is disordered. The telluracyclopentane ring system of I is almost planar whereas that in II has a twist conformation with C(2) and C(3) on either side of the plane through Te, C(1) and C(4). In both structures there are short contacts of 3.4-3.5 Å between Te and C atoms of the tetraphenylborate anion.

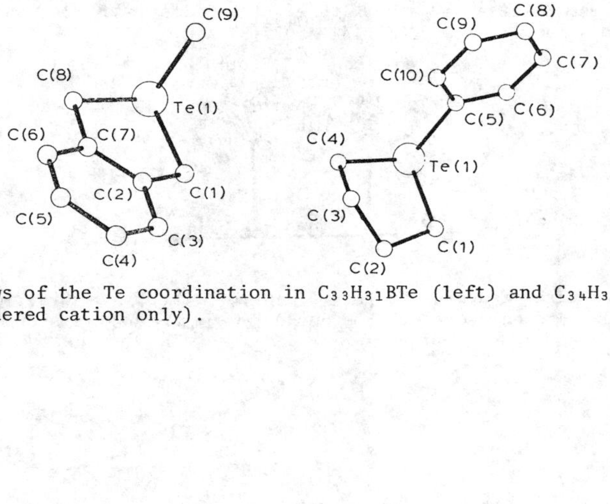

Fig. 1. Views of the Te coordination in $C_{33}H_{31}BTe$ (left) and $C_{34}H_{33}BTe$ (right) (ordered cation only).

BIS(TRIPHENYLTELLURIUM) DI-μ-CHLORO-BIS[DICHLOROMERCURATE(II)]
$C_{36}H_{30}Cl_6Hg_2Te_2$ $[Te(C_6H_5)_3^+]_2[Hg_2Cl_6^{2-}]$

M.N. PONNUSWAMY and J. TROTTER, 1984. Acta Cryst., C40, 1671-1673.

Triclinic, P$\bar{1}$, a = 9.090, b = 10.224, c = 10.820 Å, α = 95.54, β = 92.47, γ = 99.48°, D_m = 2.249, Z = 1. Mo radiation, R = 0.046 for 3334 reflexions.

The structure (Fig. 1) contains trigonal pyramidal Ph_3Te^+ cations and chloro-bridged dimeric $Hg_2Cl_6^{2-}$ anions which consist of two distorted $HgCl_4$ tetrahedra sharing an edge; interionic Te...Cl interactions complete distorted octahedral geometry at Te. Hg-Cl 2.684(2), 2.717(2) (bridging, b), 2.350(2) (terminal, t) Å, Cl(b)-Hg-Cl(b) 89.6(1), Cl(b)-Hg-Cl(t) 101.2(1), 106.6(1), Cl(t)-Hg-Cl(t) 141.3(1), Hg-Cl(b)-Hg 90.4(1)°; Te-C 2.11(1) Å, C-Te-C 94.4(2), 95.2(2), 101.3(2)°, Te...Cl 3.337(2), 3.489(2), 3.574(2) Å.

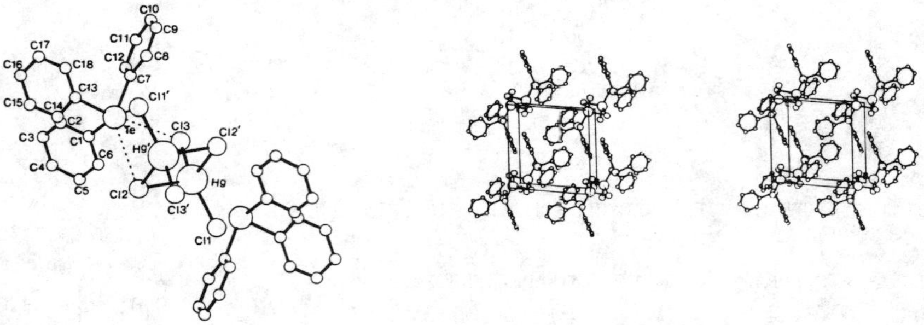

Fig. 1. A drawing of the $C_{36}H_{30}Cl_6Hg_2Te_2$ structure and a stereoview of the unit cell.

TETRAPOTASSIUM TETRAKIS(μ-DIPHOSPHITO-P,P')IODO(METHYL)DIPLATINUM(III) DIHYDRATE
$CH_{11}IK_4O_{20}P_8Pt_2,2H_2O$ $K_4[Pt_2(P_2O_5H_2)_4(CH_3)I].2H_2O$

C.-M. CHE, T.C.W. MAK and H.B. GRAY, 1984. Inorg. Chem., 23, 4386-4388.

Tetragonal, I4/mmm, a = 9.475, c = 15.786 Å, D_m = 2.97, Z = 2. Mo radiation, R = 0.037 for 961 reflexions.

The complex anion is disordered about a site having crystallographic 4/mmm symmetry resulting in a scrambling of the axial CH_3 and I ligands. Geometry of the unique part of the anion is shown in Fig. 1. The K^+ ions are each coordinated by eight oxygens, one K^+ at a $\bar{4}$m2 site with eight K-O distances of 2.819(5) Å and another at a mm site with K-O 2.837(14)-3.027(8) Å.

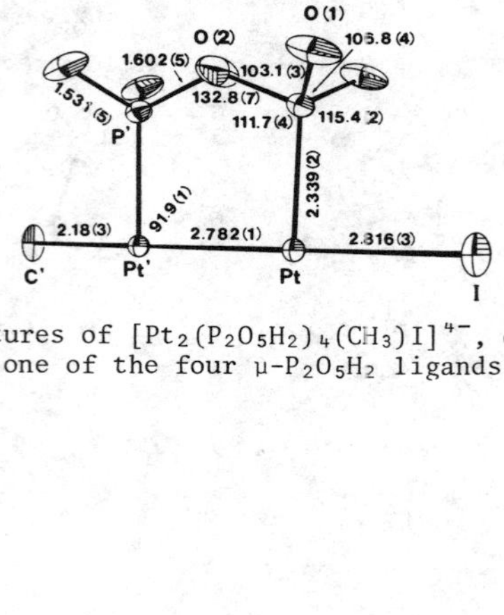

Fig. 1. Structural features of $[Pt_2(P_2O_5H_2)_4(CH_3)I]^{4-}$, disorder of Me and I not shown and only one of the four μ-$P_2O_5H_2$ ligands is shown.

TRIS(CHLOROMERCURI)ACETIC ACID DIMETHYLSULPHOXIDE SOLVATE
$C_2HCl_3Hg_3O_2,C_2H_6OS$ $(ClHg)_3CCOOH,(CH_3)_2SO$

D. GRDENIĆ, B. KORPAR-ČOLIG and M. SIKIRICA, 1984. J. Organometal. Chem., 276, 1-8.

Monoclinic, $P2_1/c$, a = 12.427, b = 10.351, c = 11.40 Å, β = 115.4°, D_m = 4.42, Z = 4. Mo radiation, R = 0.08 for 979 reflexions.

The main feature of the structure is the slightly deformed Hg_3CC tetrahedron (Fig. 1) in which the mean values of the Hg-C-Hg and Hg-C-C angles are 111° and 108° respectively. The deformation is due to the intermolecular interaction manifested by the Hg...O distances of 2.73 (the DMSO oxygen) and 2.85 Å (carboxyl oxygen) which are less than the sum of the van der Waal's radii. The Hg-C(1) distances are 2.06, 2.07 and 2.06 Å and the Hg-Cl distances are 2.33, 2.35 and 2.32 Å. The Hg...Hg intramolecular distances are 3.343, 3.482 and 3.352 Å.

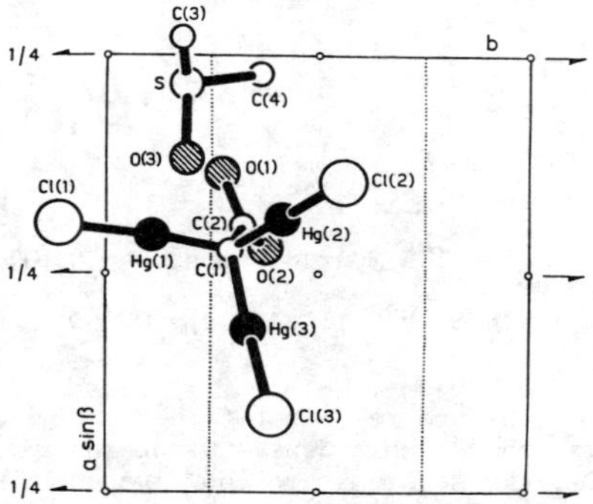

Fig. 1. The asymmetric unit in the structure of $C_2HCl_3Hg_3O_2,C_2H_6OS$ viewed along the c axis.

POTASSIUM TETRACARBONYLCOBALT
C_4CoKO_4 (I)

RUBIDIUM TETRACARBONYLCOBALT
C_4CoO_4Rb (II,III)

P. KLUFERS, 1984. Z. Kristallogr., 167, 253-260.

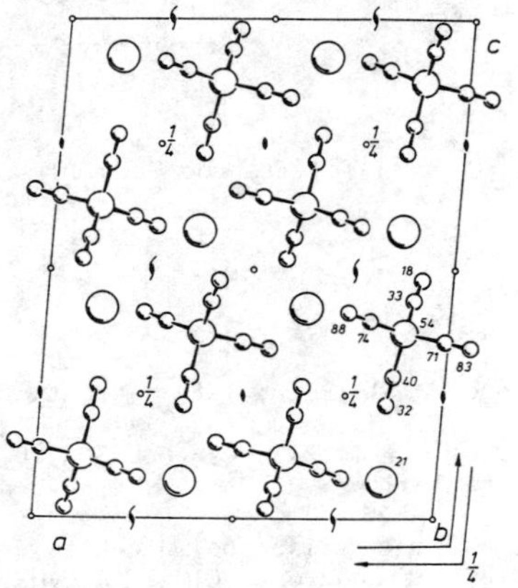

Fig. 1. A view of $RbCo(CO)_4$ at 117 K.

I. Monoclinic, I2/c, a = 14.271, b = 5.688, c = 17.632 Å, β = 94.15°, Z = 8. Mo radiation, R = 0.053 for 1561 reflexions.

II. Monoclinic, I2/c, a = 14.621, b = 5.746, c = 18.220 Å, β = 93.92°, Z = 8. Mo radiation, R = 0.061 for 1669 reflexions.

III. Monoclinic, I2/c, a = 14.228, b = 5.832, c = 18.010 Å, β = 94.31°, Z = 8. Mo radiation, R = 0.042 for 2165 reflexions (at 117 K).

The structures (Fig. 1) contain distorted tetrahedral $[Co(CO)_4]^-$ moieties which are coordinated to the alkali cations. In I, the K atom is surrounded by 9 CO groups; in II, the Rb atom is surrounded by 10 CO groups.

LITHIUM TETRACARBONYLCOBALT
C_4CoLiO_4 (I)

SODIUM TETRACARBONYLCOBALT
C_4CoNaO_4 (II)

P. KLUFERS, 1984. Z. Kristallogr., 167, 275-286.

I. Cubic, $P\bar{4}3m$, a = 5.542 Å, Z = 1. Mo radiation, R = 0.015 for 120 reflexions.

II. Orthorhombic, Pbcm, a = 5.528, b = 13.741, c = 9.209 Å, Z = 4. Mo radiation, R = 0.054 for 1072 reflexions.

In (I), the $[Co(CO)_4]^-$ ions are regular tetrahedral by space group symmetry (Fig. 1). The structure is discussed as cubic dense CO-packing with Li and Co atoms in tetrahedral voids. In (II), the Na atoms are coordinated by four O atoms forming a distorted square and by the $Co(CO)_3$ moiety of a $Co(CO)_4$ tetradehedron (Na...Co 3.182 Å). The structure (Fig. 1) can be considered as a distorted cubic dense CO-packing, with interstitial Na and Co atoms.

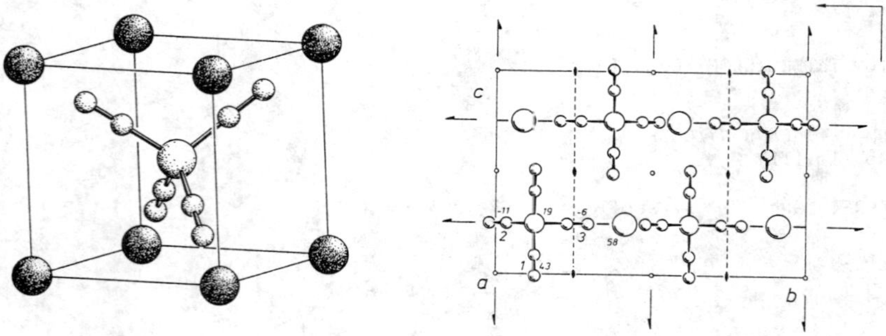

Fig. 1. Views of the $LiCo(CO)_4$ (I) (left) and $NaCo(CO)_4$ (II) (right) structures.

2,2-BIS(CHLOROMERCURI)BUTYRALDEHYDE
$C_4H_6Cl_2Hg_2O$ $C_2H_5C(HgCl)_2CHO$

B. KORPAR-COLIG, Z. POPOVIĆ and M. SIKIRICA, 1984. Croat. Chem. Acta, 57, 689-694.

Orthorhombic, $Pna2_1$, a = 11.146, b = 11.213, c = 6.573 Å, D_m = 4.51, Z = 4. Mo radiation, R = 0.049 for 866 reflexions.

The crystal structure consists of discrete molecules each of which can be described as a substituted C(2) tetrahedron, $(ClHg)_2.C(C_2H_5)CHO$ (Fig. 1). Both Hg atoms have effective octahedral coordination but the octahedron about Hg(1) differs from that round Hg(2). Hg(1)-C(2) (2.16 Å) is longer than Hg(2)-C(2) (2.09 Å) whilst the Hg(1)...O contact of 2.94 Å is significantly shorter than the Hg(2)...O distance of 3.12 Å. Other distances are: Hg(1)-Cl(1) 2.33, Hg(2)-Cl(2) 2.33 Å, and pertinent angles: C(2)-Hg(1)-Cl(1) 173, C(2)-Hg(2)-Cl(2) 172, Hg(1)-C(2)-Hg(2) 103°.

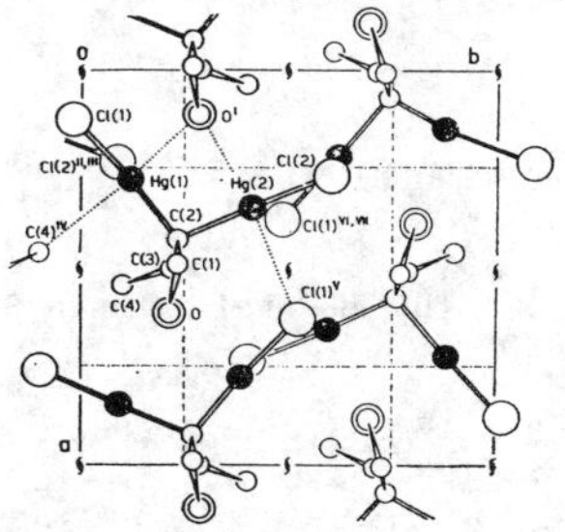

Fig. 1. The structure of $C_4H_6Cl_2Hg_2O$ viewed along the c axis.

HEXACARBONYLMOLYBDENUM

C_6MoO_6 $Mo(CO)_6$

T.C.W. MAK, 1984. Z. Kristallogr., <u>166</u>, 277-281.

Orthorhombic, Pnma, a = 12.019, b = 11.415, c = 6.488 Å, Z = 4. Mo radiation, R = 0.025 for 1316 reflexions.

The metal atom and two CO groups lie on a crystallographic mirror plane; the molecule conforms closely to idealized m3m symmetry. Mean bond lengths are Mo-C 2.059(3), C-O 1.125(5) Å.

PHENYLMERCURY(II) TRIFLUOROACETATE

$C_8H_5F_3HgO_2$ (I) $C_6H_5HgOCOCF_3$

PHENYLMERCURY(II) ACETATE

$C_8H_8HgO_2$ (II) $C_6H_5HgOCOCH_3$

B. KAMENAR, M. PENAVIĆ and A. HERGOLD-BRUNDIĆ, 1984. Croat. Chem. Acta <u>57</u>, 145-152.

I. Monoclinic, $P2_1/c$, a = 10.240, b = 5.159, c = 17.96 Å, β = 104.24°, D_m = 2.82, Z = 4. Cu radiation, R = 0.051 for 1014 reflexions.

II. Monoclinic, $P2_1/c$, a = 5.185, b = 24.10, c = 7.462 Å, β = 113.80°, D_m = 2.58, Z = 4. Mo radiation, R = 0.069 for 1508 reflexions.

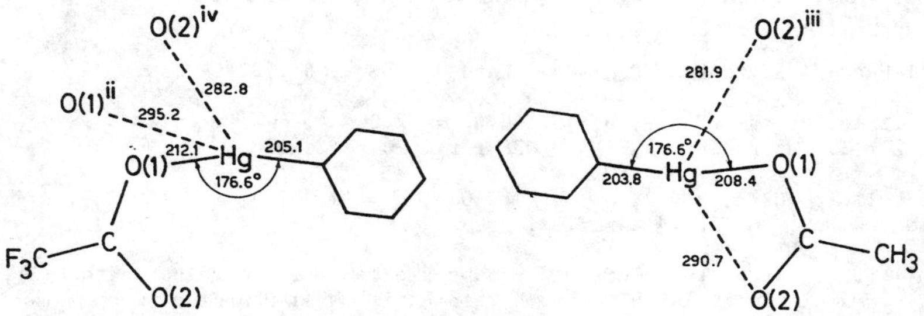

Fig. 1. Details of Hg coordination in $C_8H_5F_3HgO_2$ and in $C_8H_8HgO_2$ showing distances in pm.

Compounds I and II have essentially the same molecular structure (Fig. 1). The Hg atoms are diagonally coordinated with almost collinear C-Hg and O-Hg bonds. The effective coordination number of Hg in both structures is four as the metal atoms are additionally surrounded, in (I) by two trifluoroacetato-O atoms, and in (II) by two acetato-O atoms.

HYDROGEN cis-DIACETYLTETRACARBONYLRHENIUM
$C_8H_7O_6Re$

A.J. SCHULTZ, K. SRINIVASAN, R.G. TELLER, J.M. WILLIAMS and C.M. LUKEHART, 1984. J. Am. Chem. Soc., 106, 999-1003.

Monoclinic, $P2_1/n$, a = 6.258, b = 17.533, c = 10.088 Å, β = 96.02°, Z = 4. Neutron radiation, R = 0.128 for 1170 reflexions.

For previous X-ray work see (1). The positions of all hydrogen atoms were determined (Fig. 1). The enolic hydrogen participates in an intramolecular hydrogen bond O-H 1.15(4) and 1.26(4) Å; the O...O distance is 2.40(2) Å.

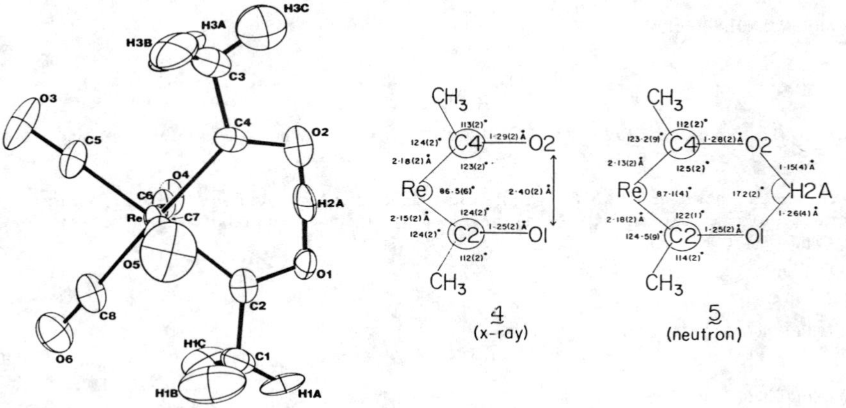

Fig. 1. The $C_8H_7O_6Re$ molecule and details of X-ray and neutron dimensions.

1. H. RATAJCZAK and W.J. ORVILLE-THOMAS, 1967. J. Mol. Struct., 1, 449-461.

[μ_2-XANTHINATO-N^3,N^7,N^9-TRIS(METHYLMERCURY(II)] NITRATE MONOHYDRATE
$C_8H_{11}Hg_3N_5O_5,H_2O$ (I)

[μ_2-XANTHINATO-N^1,N^3,N^7,N^9)TETRAKIS(METHYLMERCURY(II)] NITRATE
$C_9H_{13}Hg_4N_5O_5$ (II)

F. ALLAIRE and A.L. BEAUCHAMP, 1984. Can. J. Chem., 62, 2249-2259.

I. Triclinic, $P\bar{1}$, a = 7.319, b = 9.466, c = 12.004 Å, α = 96.06, β = 95.00, γ = 97.69°, Z = 2. Mo radiation, R = 0.038 for 1610 reflexions.

II. Monoclinic, C2/c, a = 24.55, b = 17.016, c = 19.108 Å, β = 110.26°, Z = 16. Mo radiation, R = 0.046 for 1418 reflexions.

In I (Fig. 1) the three CH_3-Hg^+ groups are linearly coordinated to N(3), N(7) and N(9) of a dianionic xanthine which retains its N(1)-H proton (mean Hg-N 2.07(3) Å). There are two independent formula units in II; each cation (Fig. 1) possesses methylmercury groups linearly coordinated to the four nitrogen atoms with mean Hg-N 2.07(5) Å.

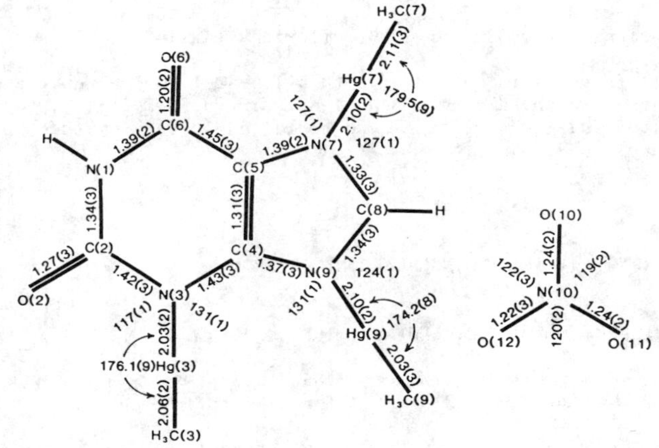

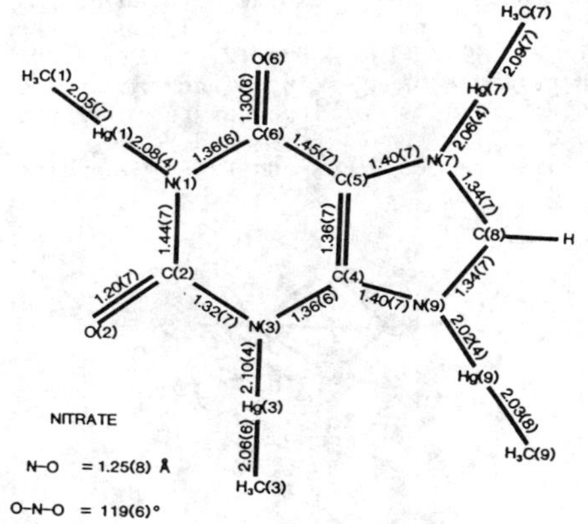

Fig. 1. Some dimensions for the xanthinato derivatives (I) (top) and (II) (averaged values for the two independent ions) (bottom).

DIIODOTETRAKIS(METHYL ISOCYANIDE)DIPALLADIUM(I) DICHLOROMETHANE

$C_8H_{12}I_2N_4Pd_2,CH_2Cl_2$ $(CH_3NC)_4Pd_2I_2.CH_2Cl_2$

N.M. RUTHERFORD, M.M. OLMSTEAD and A.L. BALCH, 1984. Inorg. Chem., 23, 2833-2837.

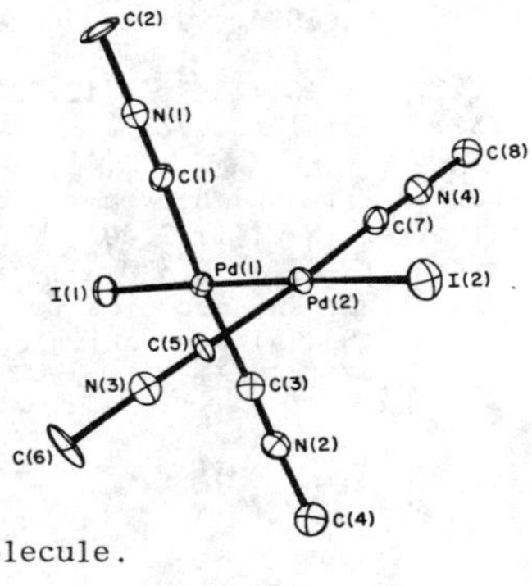

Fig. 1. The $(CH_3NC)_4Pd_2I_2$ molecule.

Triclinic, $P\bar{1}$, a = 7.188, b = 10.933, c = 12.803 Å, α = 80.58, β = 75.45, γ = 81.22°, Z = 2. Mo radiation, R = 0.039 for 2060 reflexions (at 140 K).

The molecule (Fig. 1) is unbridged with a linear I-Pd-Pd-I unit and two terminal isocyanide ligands on each Pd atom. Bond lengths: Pd-Pd 2.533(1), Pd-I 2.699(1) and 2.697(1), Pd-C 1.959(9)-1.974(9) Å. The dihedral angle between the two square coordination planes is 85.3°.

(μ_3-TETRAHYDROBORATO)(μ_2-HYDRIDO)NONACARBONYLTRIIRON

$C_9H_5BFe_3O_9$

J.C. VITES, C. EIGENBROT and T.P. FEHLNER, 1984. J. Am. Chem. Soc., 106, 4633-4634.

Triclinic, $P\bar{1}$, a = 8.823, b = 11.523, c = 7.850 Å, α = 98.89, β = 109.82, γ = 87.92°, D_m = 1.88, Z = 2. Mo radiation, R = 0.061 for 2157 reflexions.

The structure (Fig. 1) consists of a metal triangle of $Fe(CO)_3$ fragments capped with a boron. Each iron atom is six-coordinate with an approximately octahedral arrangement of nearest neighbors and has three CO ligands in an eclipsed arrangement. There are one long (2.674(2) Å) and two short (2.602(2), 2.591(2) Å) iron-iron distances in the triiron triangle, suggesting placement of the metal-metal bridging hydride on the Fe_1Fe_3 edge. This is confirmed by the displacement of the carbonyl groups away from that edge: e.g. the $Fe_1Fe_3C_9$ and $Fe_3Fe_1C_3$ angles are 108.6(3) and 108.9(3)° respectively, while the $Fe_2Fe_3C_8$ and $Fe_2Fe_1C_2$ angles are 92.7(3) and 90.8(3)°, respectively.

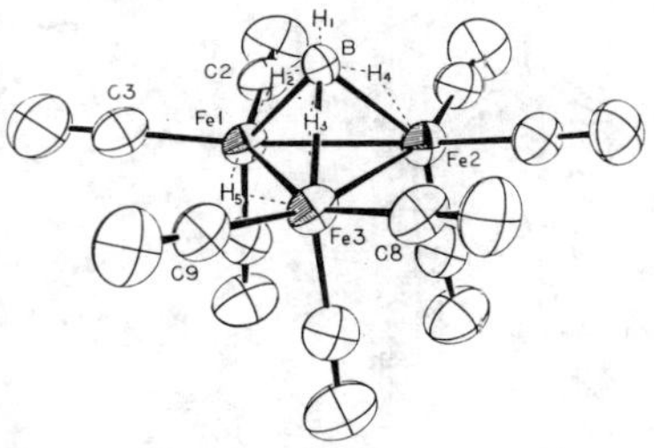

Fig. 1. A view of $C_9H_5BFe_3O_9$.

2-t-BUTYL-3,4-DIMETHYL-1,1,1-TRICHLORO-1-TUNGSTACYCLOBUTADIENE

$C_9H_{15}Cl_3W$ $W[C(C(CH_3)_3)C(CH_3)C(CH_3)]Cl_3$

M.R. CHURCHILL and H.J. WASSERMAN, 1984. J. Organometal. Chem., 270, 201-210.

Monoclinic, $P2_1/c$, a = 10.271, b = 10.113, c = 12.721 Å, β = 96.10°, Z = 4. Mo radiation, R = 0.046 for 2327 reflexions.

The W atom has a distorted trigonal bipyramidal coordination geometry (Fig. 1). Two Cl ligands are in axial sites (W-Cl(1) 2.404, W-Cl(2) 2.394 Å) while a third occupies an equatorial position (W-Cl(3) 2.330 Å). The tungstenacyclobutadiene system, W-C(1)-C(6)-C(7), is planar and occupies two equatorial sites. The W-C(α) linkages (W-C(1) 1.861, W-C(7) 1.864 Å) are equivalent, as are the C(α)-C(β) bond lengths (C(1)-C(6) 1.46, C(6)-C(7) 1.48 Å).

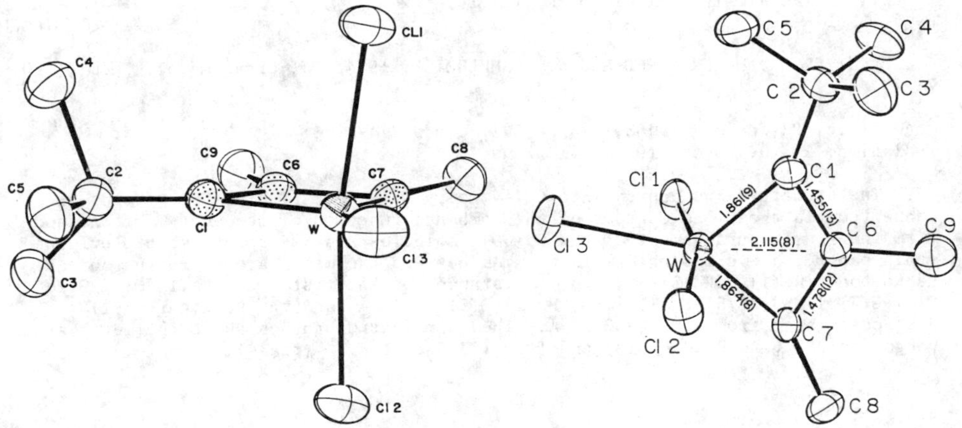

Fig. 1. Equatorial view (left) and axial view (showing selected bond lengths) (right) of $C_9H_{15}Cl_3W$.

NONACARBONYLTRIS(μ_2-HYDRIDO)(μ_3-METHYLIDYNE)TRIOSMIUM
$C_{10}H_4O_9Os_3$

A.G. ORPEN and T.F. KOETZLE, 1984. Acta Cryst., B40, 606-612.

Triclinic, $P\bar{1}$, a = 9.399 (9.295), b = 11.665 (11.457), c = 15.651 (15.622) Å, α = 112.32 (111.66), β = 90.11 (90.32), γ = 97.80 (96.93)°, Z = 4. Mo (neutron) radiation, R = 0.044 (0.022) for 4121 (4962) reflexions at 200 (11) K. [Data in parentheses for the neutron study at 11 K.]

The structure contains two crystallographically independent molecules (Fig. 1), each of which shows 3m (C_{3v}) symmetry to within experimental error. All carbonyl ligands are terminal and near linear, the hydride H atoms bridge the Os-Os vectors symmetrically, and the Os_3 triangle is bridged by a methylidyne group. Molecular geometric parameters averaged over 3m symmetry and both molecules, at 11 K, are: bond lengths Os-Os 2.893(2), Os-H 1.834(3), Os-C(H) 2.101(2), Os-C(trans to H) 1.911(2), Os-C(trans to C) 1.964(2), C-O 1.141(1), and C-H 1.086(2) Å; bond angles Os-H-Os 104.2(2), Os-C-Os 87.0(1), and Os-C-O 178.0(3)°. The neutron data provide a measure of the vibrational behaviour of the H atoms and allow estimation of the frequencies of the internal hydrogen vibrational modes.

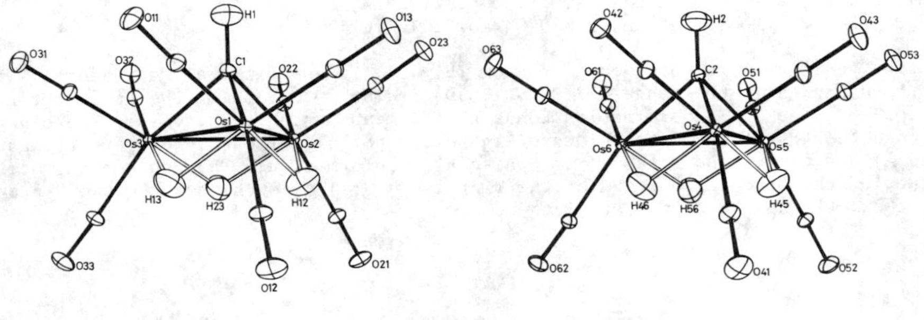

Fig. 1. Views of the two independent $C_{10}H_{14}O_9Os_3$ molecule at 11 K.

μ-ACETYL-OCTACARBONYL-μ-HYDRIDODIRHENIUM(IV)
$C_{10}H_4O_9Re_2$ $Re_2(CO)_8H(COCH_3)$

C.G. KREITER, K.-H. FRANZREB and W.S. SHELDRICK, 1984. J. Organometal. Chem., 270, 71-79.

Monoclinic, $P2_1/c$, a = 7.063, b = 22.503, c = 9.774 Å, β = 104.79°, Z = 4. Mo radiation, R = 0.073 for 1946 reflexions.

The structure has approximate C_s symmetry (Fig. 1). Two $Re(CO)_4$ groups are connected via the C and O atoms of the carbonyl function of the acetyl bridge and from the steric arrangement of the $Re(CO)_4$ moieties and NMR spectroscopic evidence (this paper), a hydrido bridge is assumed between the metal atoms, ruling out an Re-Re bond (Re(1)...Re(2) 3.208 Å). Distances of interest are: Re(1)-O(5) 2.204, O(5)-C(5) 1.254, C(5)-C(51) 1.56, Re(2)-C(5) 2.115 Å, and the Re-C(carbonyl) distances range from 1.87 to 2.08 Å. The coordination angles about the metal atoms range from 83.6 to 96.0° (cis) and from 173.5 to 178.1° (trans).

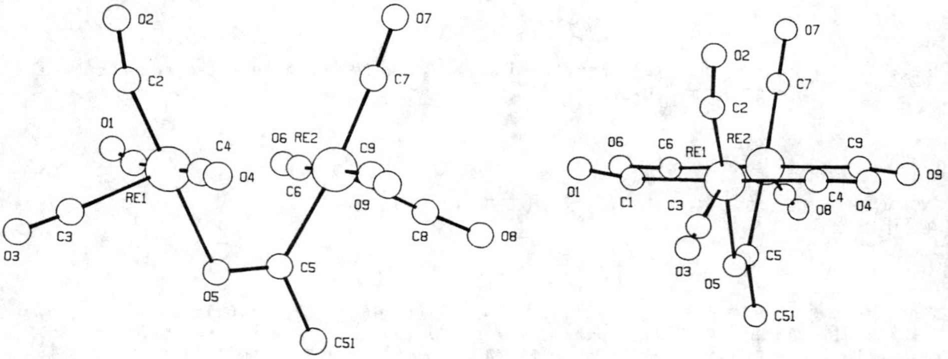

Fig. 1. Molecular structure of $C_{10}H_4O_9Re_2$, perspective views onto the Re(1)-O(5)-C(5)-Re(2) plane and along this plane.

BIS(μ_2-CHLORO)(μ_2-1,3,5,7-TETRATHIOCANE-S^1,S^3)BIS(TRIMETHYLPLATINUM(IV))
$C_{10}H_{26}Cl_2Pt_2S_4$

E.W. ABEL, G.D. KING, K.G. ORRELL, V. SIK, T.S. CAMERON and K. JOCHEM, 1984. J. Chem. Soc. Dalton, 2047-2058.

Triclinic, $P\bar{1}$, a = 10.344, b = 11.915, c = 12.758 Å, α = 132.42, β = 91.16, γ = 113.50°, Z = 2. Mo radiation, R = 0.04 for 2017 reflexions.

The configuration about each Pt atom (Fig. 1) is approximately octahedral with interbond angles in the range 177.3 to 178.9° and 86.2 to 93.5°. The Pt...Pt contact is 3.633 Å. The Pt-S coordination bonds have a mean length of 2.475 Å. The eight S-C bond lengths of the tetrathiocane ligand are not significantly different (mean 1.800 Å) and the ligand retains its boat-chair conformation. The non-bonding S...S contact of the bridging portion of the ring is 3.19(1) Å, such that the two S atoms do not exactly span the Pt...Pt distance.

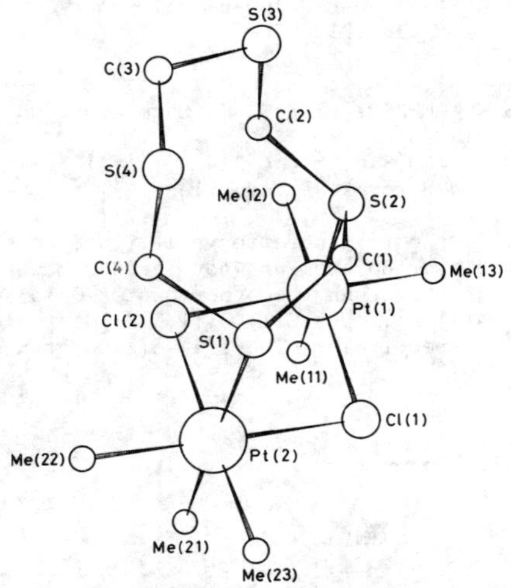

Fig. 1. $C_{10}H_{26}Cl_2Pt_2S_4$: the molecular structure.

(η^5-CYCLOPENTADIENYL)(t-BUTYL ISOCYANIDE)(TRITHIOCARBAMATO-S,S')COBALT(III)
$C_{11}H_{14}CoNS_3$

J. DOHERTY, J. FORTUNE, A.R. MANNING and F.S. STEPHENS, 1984. J. Chem. Soc. Dalton, 1111-1116.

Monoclinic, $P2_1/c$, a = 9.085, b = 12.386, c = 12.997 Å, β = 107.7°, Z = 4. Mo radiation, R = 0.074 for 1593 reflexions.

The Co atom (Fig. 1) lies in a pseudo-octahedral environment with a planar cyclopentadienyl ring occupying three coordination positions and two S and the isocyanide C atoms occupying the other three (Co-Cp 1.690, Co-C(1-5) 2.041(9)-2.08(1), Co-S(1) 2.235(2), Co-S(2) 2.252(5), Co-C(N) 1.827(8) Å). The t-butyl isocyanide group is linearly coordinated to the metal. The four-membered CoS_2C chelate ring is approximately planar.

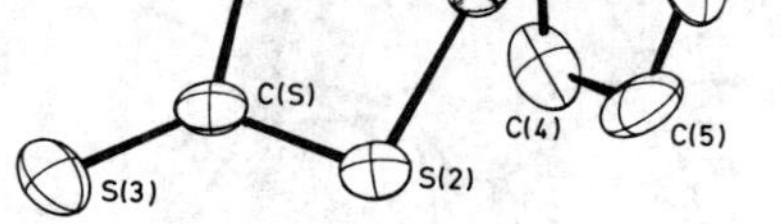

Fig. 1. $C_{11}H_{14}CoNS_3$: the molecular structure.

1,1,1,2,2,2,3,3,3-NONACARBONYL-1,2-μ-HYDRIDO-μ_3-σ:1'-2'-η-3',3',3'-TRIFLUOROPROPYNYL-$C^{1'}(Os^3)C^{1'-2'}(Os^1,Os^2)$-triangulo-TRIOSMIUM
$C_{12}HF_3O_9Os_3$

Z. DAWOODI, M.J. MAYS and K. HENRICK, 1984. J. Chem. Soc. Dalton, 1769-1772.

Triclinic, $P\bar{1}$, a = 14.354, b = 9.510, c = 6.714 Å, α = 103.21, β = 93.81, γ = 90.66°, Z = 2. Mo radiation, R = 0.028 for 2645 reflexions.

In the complex (Fig. 1) the trifluoropropynyl ligand acts as a five-electron donor. The position of the hydrido atom was not located but the H-bridged Os(1)-Os(3) distance 2.828(1) is much shorter than the other two Os-Os distances (Os(1)-Os(2) 2.872(1), Os(2)-Os(3) 2.874(1) Å). The C(1)-Os(1,2,3) distances are 2.262(10), 1.937(10) and 2.271(10) Å respectively. The C(1)-C(2) distance is 1.331(13) Å and the C(1)-C(2)-C(3) angle 133.9(9)°.

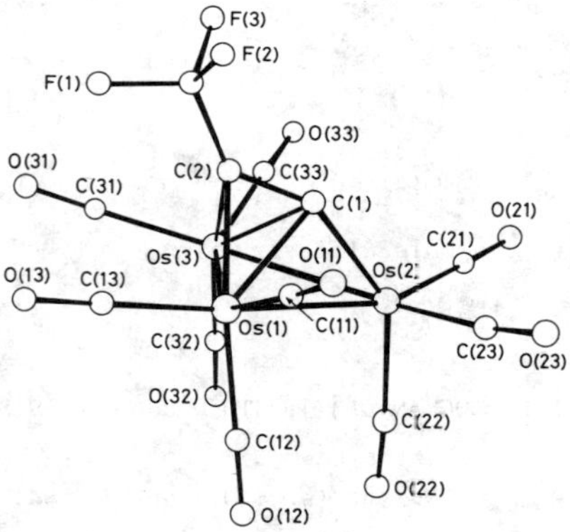

Fig. 1. $C_{12}HF_3O_9Os_3$: the molecular structure.

(μ_3-1,3-DITHIOLAN-1-YL-2-YLIDENE)(μ_3-SULPHIDO)TRIS(TRICARBONYLIRON)
$C_{12}H_4Fe_3O_9S_3$ $[Fe_3(CO)_9\{\dot{C}S(CH_2)_2\dot{S}\}S]$

G.J. KRUGER, L. LINFORD, H.G. RAUBENHEIMER and A.A. CHALMERS, 1984. J. Organometal. Chem., 262, 69-79.

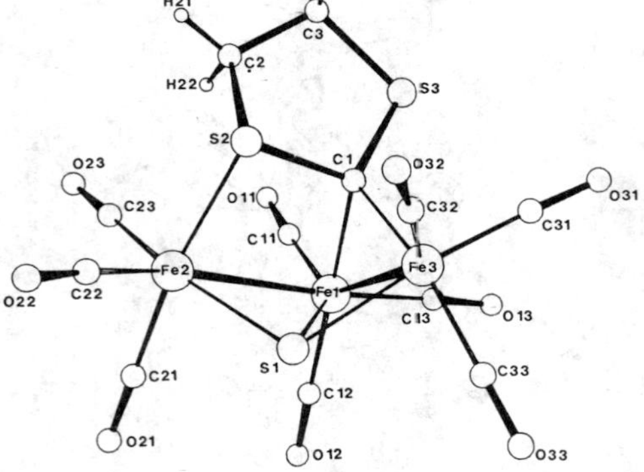

Fig. 1. Molecular structure of $C_{12}H_4Fe_3O_9S_3$.

Monoclinic, $P2_1/n$, a = 16.549, b = 11.301, c = 9.919 Å, β = 94.66°, Z = 4. Mo radiation, R = 0.039 for 2011 reflexions.

The carbene C(1) atom bridges two Fe atoms (Fe(1)-C(1) 2.080 Å, Fe(3)-C(1) 1.965 Å) while a sulphur atom of the ligand donates two electrons to Fe(2) (Fe(2)-S(2) 2.275 Å) (Fig. 1). Atom S(1) bridges all three Fe atoms with distances of Fe(1)-S(1) 2.228, Fe(2)-S(1) 2.244 and Fe(3)-S(1) 2.228 Å. The μ_3 bridge results in a wide Fe(3)-Fe(1)-Fe(2) angle of 91.4°. The Fe(1)-Fe(2) and Fe(1)-Fe(3) distances are 2.681 and 2.572 Å respectively.

2'-NITROBENZENESULPHONANILIDE
$C_{12}H_{10}N_2O_4S$ (I)

(2'-NITROBENZENESULPHONANILIDE-N)PHENYLMERCURY(II)
$C_{18}H_{14}HgN_2O_4S$ (II) $(C_{12}H_9N_2O_4S)Hg(C_6H_5)$

L.G. KUZ'MINA, Yu.T. STRUCHKOV, A.S. PEREGUDOV and D.N. KRAVTSOV, 1984. J. Struct. Chem., 25, 99-106 [Zh. Strukt. Khim., 25(1), 113-121].

I. Triclinic, $P\bar{1}$, a = 8.019, b = 8.572, c = 9.993 Å, α = 86.07, β = 67.07, γ = 77.81°, Z = 2. Mo radiation, R = 0.043 for 2286 reflexions.

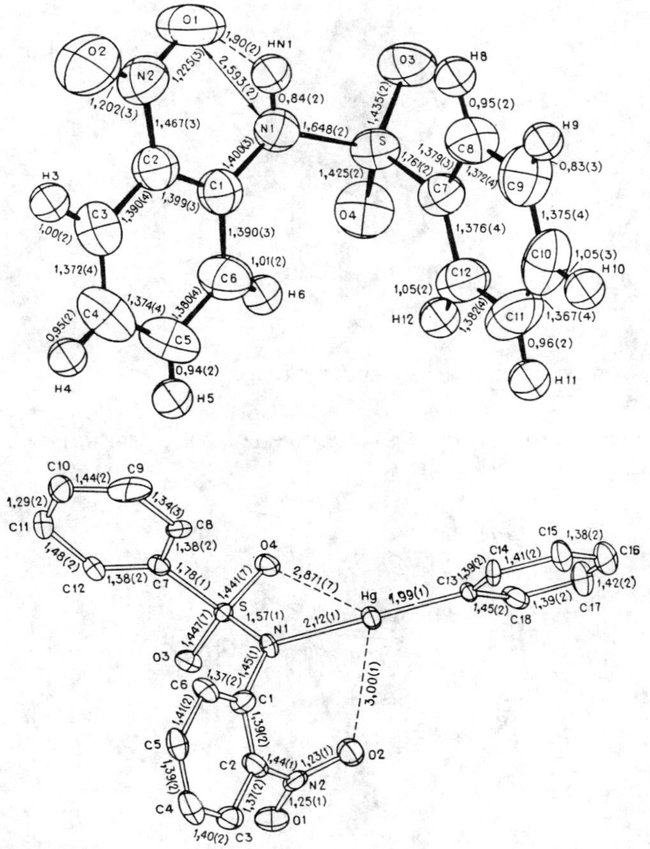

Fig. 1. Geometry of the 2'-nitrobenzenesulphonanilide (top) and (2'-nitrobenzene-sulphonanilide-N)phenylmercury(II) (bottom) molecules.

II. Orthorhombic, $Pn2_1a$, a = 8.191, b = 10.927, c = 20.069 Å, Z = 4. Mo radiation, R = 0.033 for 2069 reflexions (at -120°C).

Bond distances in the two molecules are shown in Fig. 1. In compound I an intramolecular N-H...O hydrogen bond involving a nitro group oxygen atom forms a six-membered ring. An analogous ring is formed in II by a weak N-Hg...O interaction. Geometrical data indicate that bond strengths in the six-membered chelate ring of I (involving H) are greater than in II (involving Hg).

CHLOROTRIMETHYL[HYDROTRIS(PYRAZOLYL)BORATO]TANTALUM(V)
$C_{12}H_{19}BClN_6Ta$ $[HB(N_2C_3H_3)_3]Ta(CH_3)_3Cl$

D.L. REGER, C.A. SWIFT and L. LEBIODA, 1984. Inorg. Chem., 23, 349-354.

Monoclinic, $P2_1/n$, a = 9.425, b = 12.567, c = 14.315 Å, β = 106.00°, Z = 4. Mo radiation, R = 0.045 for 1578 reflexions (at 123K).

The coordination geometry about Ta is nearly regular capped octahedral with a Me ligand capping the face defined by the other two Me ligands and the Cl^- ligand. Bond lengths involving Ta are: Ta-Cl 2.380(3), Ta-N 2.228(9)-2.294(8), Ta-C 2.20(1)-2.25(1) Å. The molecule is shown in Fig. 1.

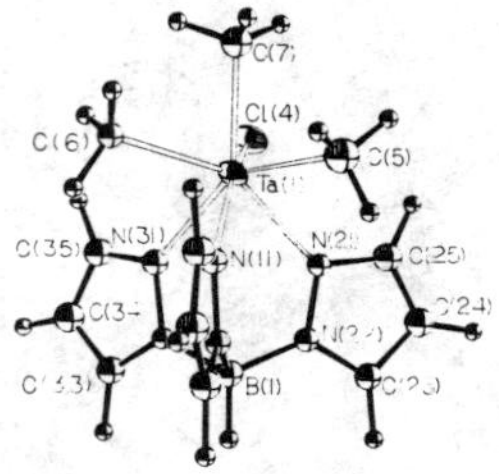

Fig. 1. The $[HB(N_2C_3H_3)_3]Ta(CH_3)_3Cl$ molecule.

UNDECACARBONYL(μ_2-TRIFLUOROMETHYL ISOCYANIDE)TRIIRON
$C_{13}F_3Fe_3NO_{11}$

I. BRUDGAM, H. HARTL and D. LENTZ, 1984. Z. Naturforsch., B39, 721-726.

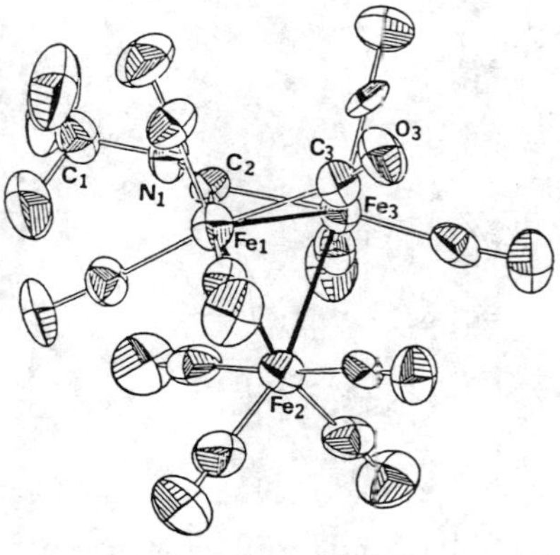

Fig. 1. A view of one of the $C_{13}F_3Fe_3NO_{11}$ molecules.

Triclinic, $P\bar{1}$, a = 14.607, b = 14.289, c = 8.967 Å, α = 92.40, β = 92.70, γ = 90.66°, Z = 4. Cu radiation, R = 0.066 for 2083 reflexions.

The isocyanide ligand occupies a bridging position between Fe1 and Fe3 (Fig. 1). There are two independent molecules in the asymmetric unit; mean dimensions are: Fe1-Fe2 2.693(3), Fe1-Fe3 2.531(2), Fe2-Fe3 2.685(3), Fe1-C3 2.01(1), Fe3-C3 1.97(1), Fe1-C2 1.94(1), Fe3-C2 1.99(1), Fe-C(terminal) 1.77-1.82(1) Å.

METHYLAMINOPHENYLCARBENEPENTACARBONYLTUNGSTEN(0)
$C_{13}H_9NO_5W$

J.T. GUY Jr., and D.W. BENNETT, 1984. Transition Met. Chem., 9, 43-46.

Monoclinic, P2/c, a = 10.892, b = 10.637, c = 12.581 Å, β = 97.51°, Z = 4. Mo radiation, R = 0.058 for 1417 reflexions.

The analysis shows that there is little metal-carbene π-bonding in this complex (Fig. 1), with W-C(6) 2.186(22), W-C(1) 1.940(28), W-C(eq) 2.00-2.11(4), N-C(6) 1.299(26) Å. The phenyl ring plane makes a dihedral angle of 78° with the N,C(6),W plane.

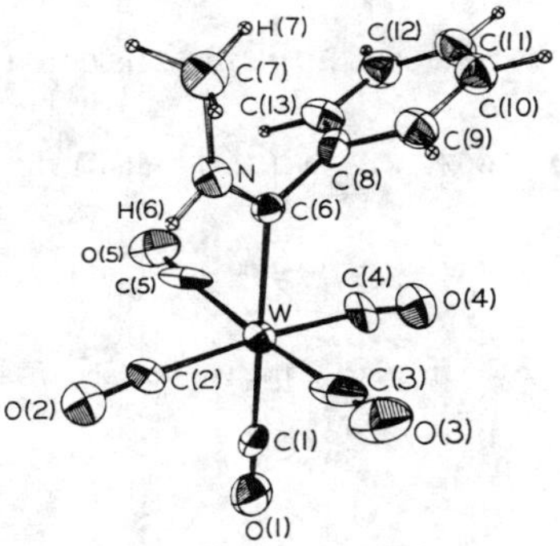

Fig. 1. A view of $C_{13}H_9NO_5W$.

cis-TETRACARBONYL[METHOXY(1-METHOXY-2-PHENYL)CARBENE]CHROMIUM
$C_{13}H_{10}CrO_6$ $(CO)_4Cr(C_6H_4(OCH_3)COCH_3)$

K.H. DÖTZ, W. STURM, M. POPALL and J. RIEDE, 1984. J. Organometal. Chem., 277, 267-275.

Triclinic, $P\bar{1}$, a = 8.376, b = 8.869, c = 9.350 Å, α = 81.15, β = 105.50, γ = 96.74°, Z = 2. Mo radiation, R = 0.037 for 1877 reflexions.

The metal atom has distorted octahedral coordination (Fig. 1) with cis-angles ranging from 77.6° (C(5)-Cr-O(5)) to 102.1° (C(4)-Cr-C(5)). The Cr atom is part of a "chromabenzofuran" system in which the five-membered metallacycle is coplanar with the six-membered aromatic ring. The Cr-C(carbonyl) distances are 1.894, 1.906, 1.906 and 1.821 Å, the shortest bond being opposite to the Cr-O(5) bond whose length is 2.183 Å. Other distances are: Cr-C(5) 2.003, C(5)-C(51) 1.470, C(51)-C(56) 1.394, C(56)-O(5) 1.376, O(5)-C(57) 1.447, C(5)-O(6) 1.316 and O(6)-C(6) 1.440 Å.

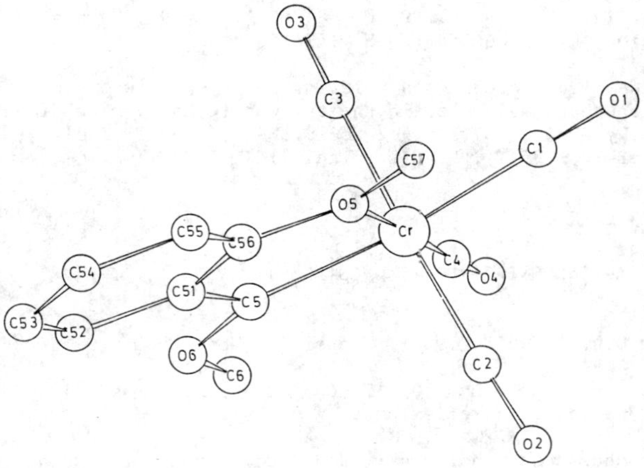

Fig. 1. Structure of $C_{13}H_{10}CrO_6$.

SODIUM BIS(η^5-CYCLOPENTADIENYL)(η^2-KETENE-C,O)METHYLZIRCONIUM TETRAHYDROFURAN SOLVATE
$C_{13}H_{15}NaOZr,2(C_4H_8O)$ $((Cp)_2Zr(OCCH_2)CH_3)^-.Na^+,2(THF)$

S.C.H. HO, D.A. STRAUS, J. ARMANTROUT, W.P. SCHAEFER and R.H. GRUBBS, 1984. J. Am. Chem. Soc., 106, 2210-2211.

Triclinic, $P\bar{1}$, a = 10.745, b = 14.449, c = 12.060 Å, α = 119.40, β = 91.26, γ = 114.83°, Z = 2.

The $Zr(COCH_2)CH_3$ moiety is planar with dimensions shown in Fig. 1.

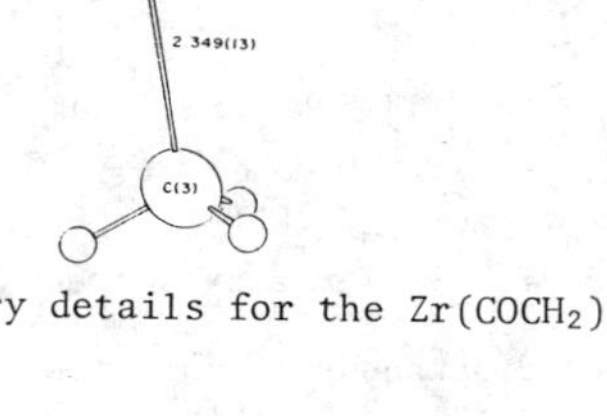

Fig. 1. Molecular geometry details for the $Zr(COCH_2)CH_3$ moiety in $Cp_2Zr(COCH_2)CH_3$.

TETRACARBONYL(DIISOPROPYLAMINOETHOXYCARBENE)IRON
$C_{13}H_{19}FeNO_5$ (I)

1,1,1-TRICARBONYLIRON-2-DIISOPROPYLAMINO-3,5-DIOXA-4-ETHYL-CYCLOPENTA-1,4-DIENE
$C_{13}H_{19}FeNO_5$ (II)

E.O. FISCHER, J. SCHNEIDER and K. ACKERMANN, 1984. Z. Naturforsch., B39, 468-473.

I. Monoclinic, $P2_1/n$, a = 9.840, b = 16.095, c = 10.439 Å, β = 104.57°, Z = 4. Mo radiation, R = 0.036 for 1988 reflexions (at -30°C).

II. Monoclinic, $P2_1/m$, a = 8.960, b = 10.285, c = 9.452 Å, β = 116.38°, Z = 2. Mo radiation, R = 0.051 for 1094 reflexions (at -30°C).

The coordination of the Fe atom in I is shown in Fig. 1; principal dimensions are Fe-C1 1.763(4), Fe-C2 1.785(4), Fe-C3 1.789(4), Fe-C4 1.784(3), Fe-C5 2.01(3) Å, C1-Fe-C5 171.6(2), C2-Fe-C3 107.9(2), C2-Fe-C4 122.6(2), C3-Fe-C4 129.5(2)°. Molecule (II) lies about a mirror plane. Principal dimensions include Fe-C1 1.755(4), Fe-C2 1.783(2), Fe-C3 1.904(6), Fe-O4 2.052(4) Å, C1-Fe-C1' 112.8(2), C2-Fe-C3 170.4(2), C1-Fe-O4 123.4(2)°.

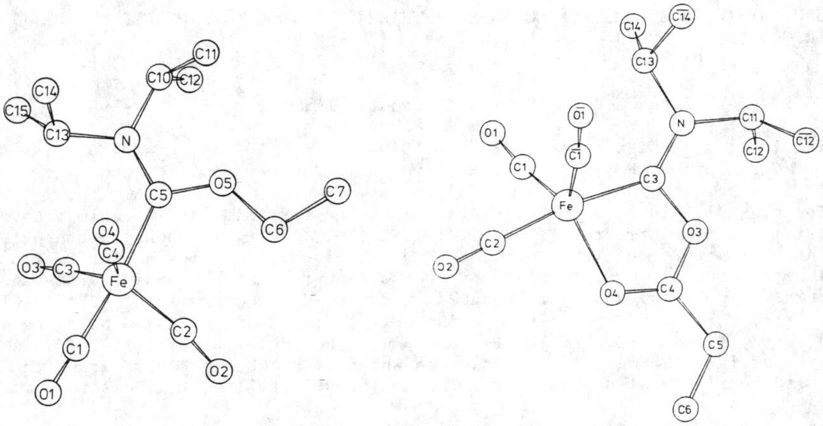

Fig. 1. Views of the $C_{13}H_{19}FeNO_5$ isomers (I) (left) and (II) (right).

FORMATO(2,6-BIS(DIMETHYLAMINOMETHYL)PHENYL-C,N,N')NICKEL(II)
$C_{13}H_{20}N_2NiO_2$

D.M. GROVE, G. VAN KOTEN, H.J.C. UBBELS, R. ZOET and A.L. SPEK, 1984. J. Organometal. Chem., 263, C10-C12. Organometallics, 3, 1003-1009.

Monoclinic, $P2_1/a$, a = 11.470, b = 11.007, c = 11.560 Å, β = 111.06°, Z = 4. Mo radiation, R = 0.026 for 2593 reflexions.

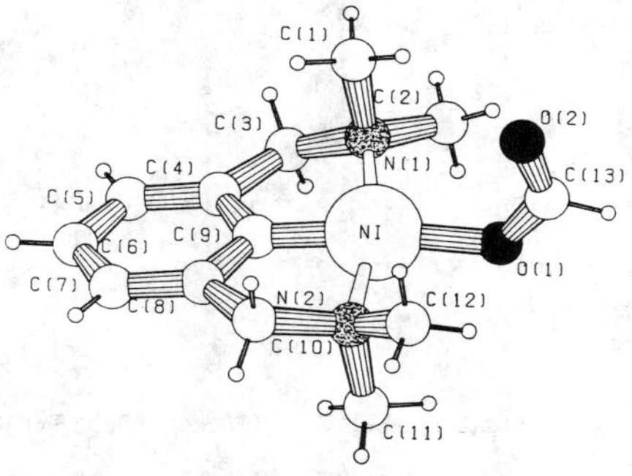

Fig. 1. Molecular structure of $C_{13}H_{20}N_2NiO_2$.

The Ni centre has approximate square planar coordination geometry, comprising two mutually trans N donor atoms, the C-bonded phenyl group of the N-C-N system and trans to this, a planar O-bonded OOCH group (Fig. 1). This formato ligand is oriented almost perpendicular (torsion angle N(1)-Ni-O(1)-C(13) 84.9°) to the coordination plane, with a non-bonding Ni-O(2) distance of 3.126 Å. The Ni-N distances are 1.976 and 1.973 Å, the Ni-C(9) and Ni-O(1) distances 1.814 and 1.937 Å, and the formato ligand bond distances O(1)-C(13) 1.270 Å and C(13)-O(2) 1.209 Å. The bond angles round Ni range from 83.2 to 98.7° and the angles Ni-O(1)-C(13) and O(1)-C(13)-O(2) are 122.9 and 127.3° respectively.

(μ_2-IODO)-BIS(μ_2-BIS(DIMETHYLPHOSPHINO)METHANE-P,P')IODOTRIMETHYLDIPLATINUM(II,IV) BIS(TRIODIDE)

$C_{13.5}H_{83.5}I_{4.5}P_4Pt_2$

L. MANOJLOVIĆ-MUIR and K.W. MUIR, 1984. Croat. Chem. Acta, 57, 587-595.

Triclinic, P$\bar{1}$, a = 9.240, b = 9.328, c = 18.044 Å, α = 102.39, β = 94.63, γ = 96.93°, Z = 2. Mo radiation, R = 0.049 for 5045 reflexions.

The crystal structure is built of cations and anions. The I_3^- anions have a symmetrical, linear structure (I(3)-I(4)-I(5) 179.2°) with I-I bond lengths 2.909 and 2.900 Å. The structure of the cations (Fig. 1) reveals the presence of Pt(II) and Pt(IV) metal centres, characterised by square planar and octahedral coordination geometries. The metal atoms are spanned by two bridging dmpm ligands (Pt(1)-P(1) 2.290, Pt(1)-P(3) 2.290, Pt(2)-P(2) 2.391, Pt(2)-P(4) 2.392 Å) and a bridging iodide ligand (Pt(1)-I(1) 2.652, Pt(2)-I(1) 2.756 Å). The angle Pt(1)-I(1)-Pt(2) is 91.2°. There is partial disorder in the cation in which I^- or CH_3^- ligands occupy the L sites with nearly equal probability. The non-bonded Pt(1)...Pt(2) distance is 3.863 Å.

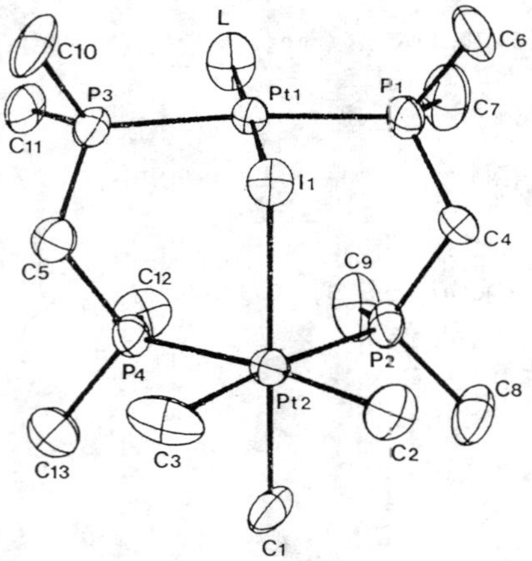

Fig. 1. A view of the structure of the cation in $C_{13.5}H_{83.5}I_{4.5}P_4Pt_2$. The disordered site L is occupied with almost equal probability by I(2) or the methyl carbon atom C(14).

1,1,1,1,2,2,2,2,3,3,3,3-DODECACARBONYL-1,2,3-TRISOSMACYCLOPENTAN-4-ONE

$C_{14}H_2O_{13}Os$

E.D. MORRISON, G.R. STEINMETZ, G.L. GEOFFROY, W.C. FULTZ and A.L. RHEINGOLD, 1984. J. Am. Chem. Soc., 106, 4783-4789.

Monoclinic, $P2_1/n$, a = 9.414, b = 15.369, c = 13.940 Å, β = 107.9°, Z = 4. Mo radiation, R = 0.033 for 3015 reflexions.

A view of the molecule and principal dimensions are in Fig. 1. The coordination about each Os atom is essentially octahedral and the Os-Os bond lengths are consistent with their being single bonds.

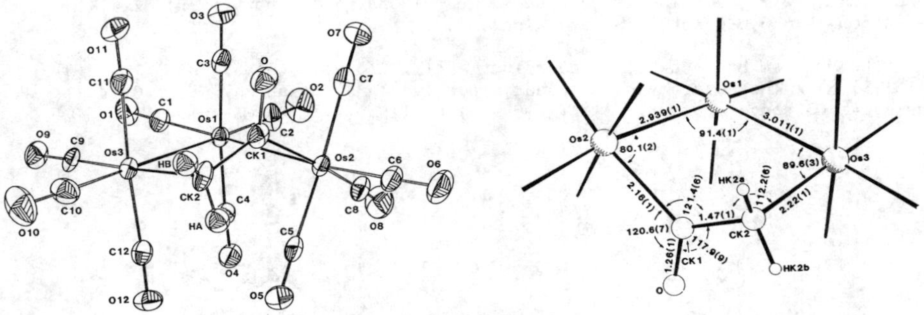

Fig. 1. A view of $C_{14}H_2O_{13}Os_3$ and principal dimensions.

(μ_3-ETHYLIDYNE)(μ_2-HYDRIDO)DODECACARBONYLTETRAIRON

$C_{14}H_4Fe_4O_{12}$ $HFe_4(CO)_{12}(CCH_3)$

P.L. BOGDAN, K.H. WHITMIRE, J.W. KOLIS, D.F. SHRIVER and E.M. HOLT, 1984. J. Organometal. Chem., 272, 169-177.

Trigonal, $P3_221$, a = 8.977, c = 41.49 Å, Z = 6. Mo radiation, R = 0.052 for 1502 reflexions.

The 60 electron cluster displays the expected tetrahedral arrangement of metal atoms (Fig. 1) with the ethylidyne group bridging one triangular face (Fe-C average 1.946(14) Å) in symmetric fashion (Fe-C-C angles 129.3(12)-131.6(9)°). The C(1)-C(2) distance is 1.513(20) Å. Fe-Fe distances in the basal plane (Fe(1),Fe(2),Fe(3)) average 2.562(3) Å and are shorter than apical-basal Fe-Fe distances (average 2.603(3) Å). Proton positions were not evident from a difference Fourier synthesis but the proton is assumed to bridge an apical-basal Fe-Fe bond - probably Fe(4)-Fe(3) because of its length (2.628(2) Å).

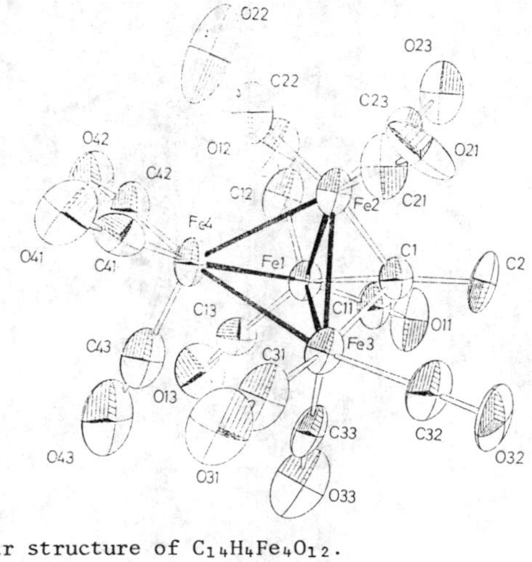

Fig. 1. Molecular structure of $C_{14}H_4Fe_4O_{12}$.

PENTACARBONYL(5H-DIPYRROLO[1,2-c.2',1'-e]IMIDAZOL-5-YL)MANGANESE
$C_{14}H_7MnN_2O_5$

U. BURGER, C. PERRET, G. BERNARDINELLI and E.P. KUNDIG, 1984. Helv. Chim. Acta, 67, 2063-2067.

Monoclinic, $P2_1/n$, a = 10.311, b = 9.294, c = 15.382 Å, [β not given], Z = 4. Mo radiation, R = 0.040 for 1635 reflexions.

The Mn atom has octahedral coordination (Mn-C(O) 1.854-1.876(6) Å, Mn-C(5) 2.181(5) Å). The heterocyclic ligand is nearly planar (mean deviation 0.04 Å). Bond angles at C5 range from 98.9(3)° (N4-C5-N6) to 115.9(3)° (N4-C5-Mn).

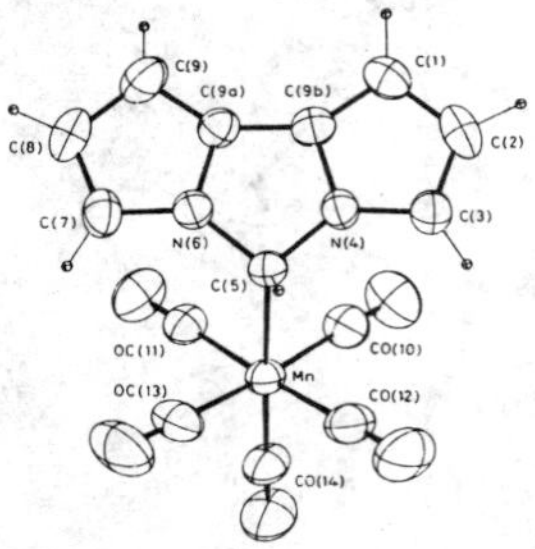

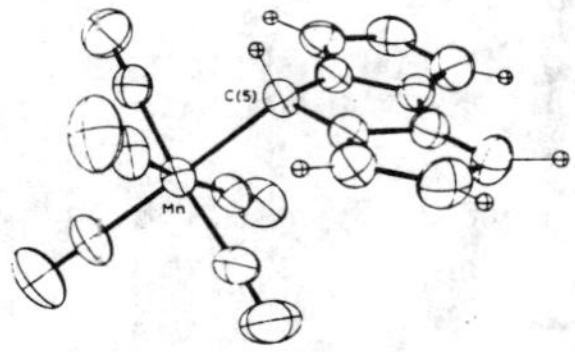

Fig. 1. Two views of $C_{14}H_7MnN_2O_5$.

DECACARBONYL(μ_2-HYDRIDO)(μ_2-METHYLCARBAMIC N-METHYLFORMIMIDIC ANHYDRIDATO)TRIOSMIUM
$C_{14}H_8N_2O_{12}Os_3$ $HOs_3(\mu\text{-}C(OC(O)N(H)CH_3)=NCH_3)(CO)_{10}$

Y.-C. LIN, A. MAYR, C.B. KNOBLER and H.D. KAESZ, 1984. J. Organometal. Chem., 272, 207-229.

Monoclinic, $P2_1/c$, a = 13.413, b = 16.712, c = 9.400 Å, β = 95.01°, Z = 4. Mo radiation, R = 0.031 for 2936 reflexions (-158°C).

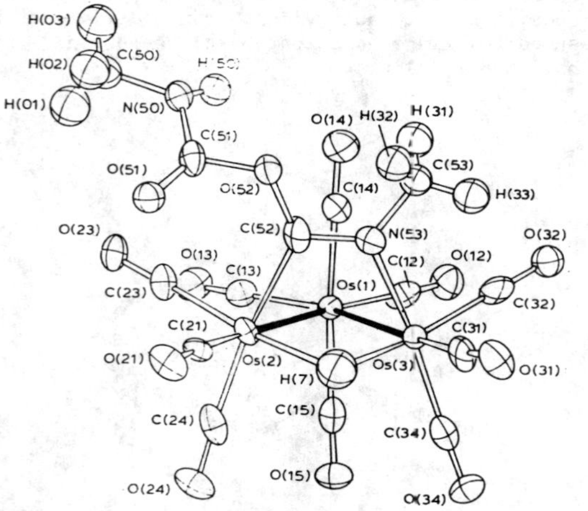

Fig. 1. Projection of $C_{14}H_8N_2O_{12}Os_3$.

The crystal consists of discrete molecules (Fig. 1). Three Os atoms form an approximate isosceles triangle (Os(1)-Os(2) 2.882, Os(1)-Os(3) 2.881, Os(2)-Os(3) 2.923 Å), and an N-methyl-iminyl group with a urethane derivative on the C atom bridges Os(2) and Os(3) through the C and N atoms respectively (Os(2)-C(52) 2.13, Os(3)-N(53) 2.122, C(52)-N(53) 1.27 Å). The two metal atoms Os(2) and Os(3) are also bridged by a hydrogen atom (H(7)) outside the trimetal plane and away from the bridging iminyl group. The distances Os(2)-H(7) and Os(3)-H(7) are respectively 1.96(12) and 1.72(12) Å.

(t-BUTYL ISOCYANIDE)NONADICARBONYLDIRHENIUM(0)
$C_{14}H_9NO_9Re_2$ $(CO)_9Re_2CNC(CH_3)_3$

M.O. ALBERS, J.C.A. BOEYENS, N.J. COLVILLE and G.W. HARRIS, 1984. J. Organometal. Chem., 260, 99-104.

Orthorhombic, Pbca, a = 17.160, b = 18.300, c = 12.370 Å, Z = 8. Mo radiation, R = 0.099 for 2043 reflexions.

The two halves of the molecule are staggered with respect to each other and the isonitrile ligand occupies an equatorial position (Fig. 1). The axial Re-CO bond distance (average 1.93 Å) is shorter than the average equatorial Re-CO bond distance (2.04 Å) excluding Re-C(1) in which the CO group is trans to the t-BuNC group (1.91 Å). The Re-Re bond length is 3.048 Å and the Re(1)-C(10) and C(10)-N distances are respectively 2.05 and 1.10 Å.

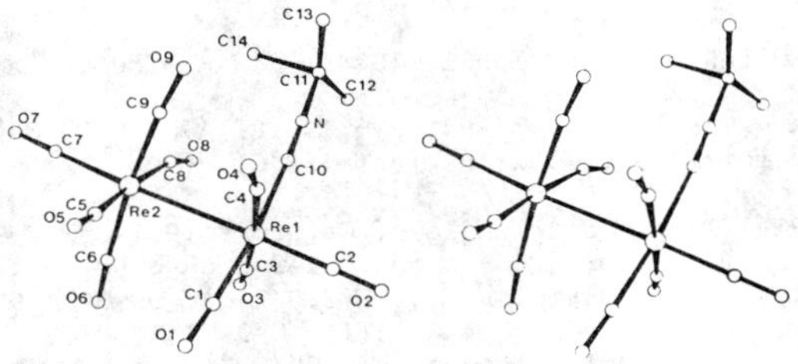

Fig. 1. Stereoscopic view of $C_{14}H_9NO_9Re_2$.

(μ_2-HYDRIDO)(μ_3-σ^2,η^2-3-DIMETHYLAMINOPROP-1-ENE-2,3-DIYL)NONACARBONYLTRIRUTHENIUM
$C_{14}H_9NO_9Ru_3$

S. AIME, D. OSELLA, A.J. ARCE, A.J. DEEMING, M.B. HURSTHOUSE and A.M.R. GALAS, 1984. J. Chem. Soc. Dalton, 1981-1986.

Monoclinic, $P2_1/n$, a = 13.067, b = 10.398, c = 15.544 Å, β = 112.78°, Z = 4. Mo radiation, R = 0.019 for 3138 reflexions.

In the molecule (Fig. 1) the μ_3 ligand is bound such that there is extensive π donation from the NMe_2 group to the metal. The molecule is therefore zwitterionic in character. The C(3)-N(4) distance (1.303(5) Å) is short implying a high degree of C-N multiple bonding. Both C(3) and N(4) have planar environments. Selected interatomic distances are: Ru(1)-Ru(2) 2.861(1), Ru(2)-Ru(3) 2.787(1), Ru(1)-Ru(3) 2.849(1), Ru(1)-C(carbonyl) 1.899(5)-1.935(6), Ru(2)-C(carbonyl) 1.892(6)-1.936(6), Ru(3)-C(carbonyl) 1.884(6)-1.926(6) Å.

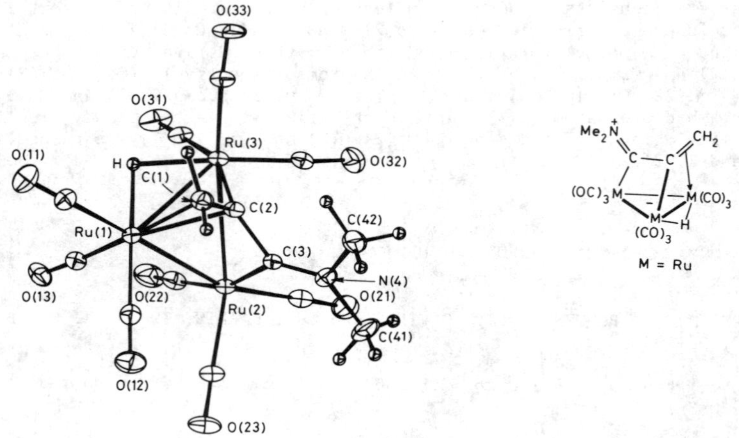

Fig. 1. $C_{14}H_9NO_9Ru_3$: the molecular skeleton and structure.

(μ_2-ACETYL-C,O)(μ_2-HYDRIDO)(METHOXYMETHYLCARBENE)NONACARBONYLTRIOSMIUM
$C_{14}H_{10}O_{11}Os_3$

C.M. JENSEN, C.B. KNOBLER and H.D. KAESZ, 1984. J. Am. Chem. Soc., 106, 5926-5933.

Monoclinic, $P2_1/n$, a = 9.578, b = 13.494, c = 15.187 Å, β = 96.30°, Z = 4. Mo radiation, R = 0.050 for 2728 reflexions (at -158°C).

The molecule (Fig. 1) has a triangle of Os atoms (Os1-Os2 2.934(2), Os1-Os3 2.852(1), Os2-Os3 2.892(1) Å) bridged on the Os1-Os2 edge by an acetyl group and an H atom on opposite sides of the Os_3 plane (Os-H 2.0(2) Å, Os-H-Os 94(9)°). Bond lengths at the acetyl group are Os1-O34 2.13(1), Os2-C34 2.05(2), O34-C34 1.29(2) Å. In the carbene group the C25-Os1 distance is 1.99(2) Å, with C25-O27 1.34(2) Å.

Fig. 1. A view of $C_{14}H_{10}O_{11}Os_3$.

(3,3,3,3-TETRACARBONYL-3-RHENAACETYLACETONEIMINATO)(CHLORO)PHENYLBORON
$C_{14}H_{12}BClNO_5Re$

P.G. LENHERT, C.M. LUKEHART and K. SRINIVASAN, 1984. Inorg. Chem., 23, 438-440.

Triclinic, $P\bar{1}$, a = 11.138, b = 12.348, c = 6.823 Å, α = 103.76, β = 103.71, γ = 103.99°, Z = 2. Mo radiation, R = 0.055 for 3865 reflexions.

A delocalized rhena β-keto iminato moiety is coordinated to boron as a bidentate chelating ligand (Fig. 1). Bond lengths: Re-CO 1.963(9)-2.00(2), Re-C 2.126(9) and 2.125(7), B-Cl 1.88(2), B-C 1.60(2), B-O 1.51(2), B-N 1.49(2), C-N 1.26(1), and C-O(ring) 1.289(9) Å.

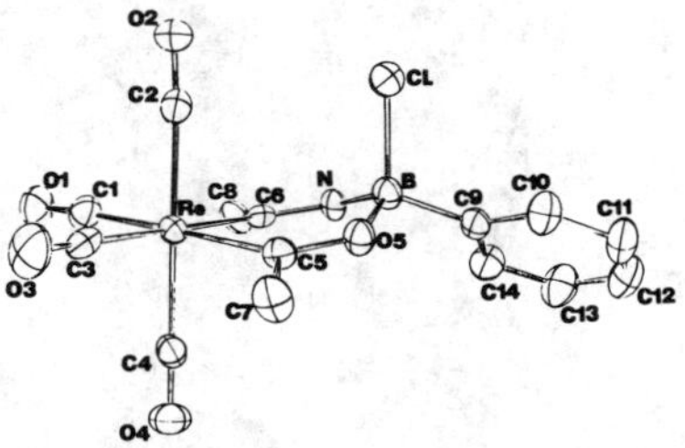

Fig. 1. The $C_{14}H_{12}BClNO_5Re$ molecule.

cis-μ_2-CARBONYL-μ_2-METHYLENEBIS[CARBONYL(η^5-CYCLOPENTADIENYL)RUTHENIUM]
$C_{14}H_{12}O_3Ru_2$

D.L. DAVIES, S.A.R. KNOX, K.A. MEAD, M.J. MORRIS and P. WOODWARD, 1984. J. Chem. Soc. Dalton, 2293-2299.

Triclinic, P$\bar{1}$, a = 6.771, b = 9.315, c = 11.864 Å, α = 103.12, β = 100.61, γ = 103.40°, Dm = 1.94, Z = 2. Mo radiation, R = 0.036 for 2725 reflexions.

The structure (Fig. 1) comprises two Ru atoms bridged symmetrically by a CO and a CH_2 ligand, with each Ru atom also carrying terminal carbonyl and η^5-cyclopentadienyl groups. The Ru(1)-Ru(2) distance is 2.707(1), Ru(1)-C(3) 2.077(6), Ru(1)-C(4) 2.019(4), Ru(2)-C(3) 2.079(5), and Ru(2)-C(4) 2.035(6) Å. The terminal CO groups are arranged cis to one another as are the cyclopentadienyl groups. The interplanar angle between the two bridge systems is 161°.

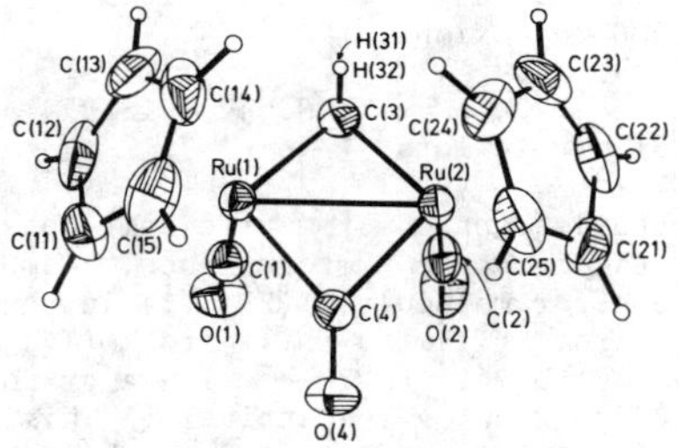

Fig. 1. $C_{14}H_{12}O_3Ru_2$: the molecular structure.

TRICARBONYL-(σ^1,η^3-1,4,6,6-TETRAMETHYL-2-OXOCYCLOHEPT-4-ENE-1,3-DIYL)IRON
$C_{14}H_{16}FeO_4$

P. EILBRACHT, R. JELITTE and L. WALZ, 1984. Chem. Ber., 117, 3473-3489.

Monoclinic, $P2_1/c$, a = 12.730, b = 8.863, c = 14.691 Å, β = 120.107°, D_m = 1.35, Z = 4. Mo radiation, R = 0.049 for 2329 reflexions.

The ligand conformation is shown in Fig. 1. The Fe-ring distances are Fe-C1 2.137(4), Fe-C3 2.160(5), Fe-C4 2.090(3), Fe-C5 2.137(4) Å. In the ring the C3-C4 and C4-C5 distances are 1.392(6) and 1.407(6) Å respectively.

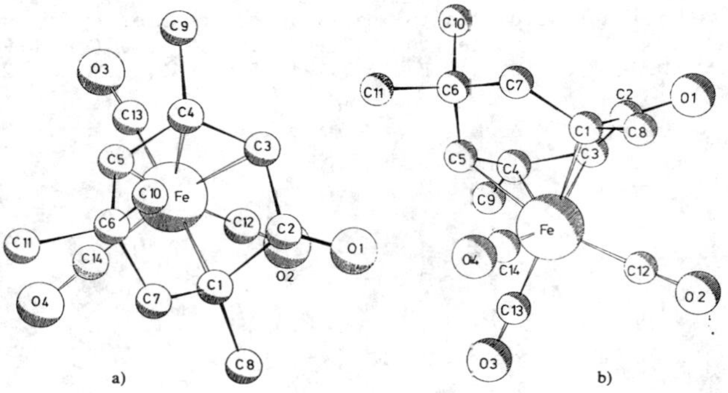

Fig. 1. Two views of $C_{14}H_{16}FeO_4$.

BIS(η^5-2,4-CYCLOPENTADIEN-1-YL)[(DIMETHYLSILYLENE)BIS(METHYLENE)]ZIRCONIUM
$C_{14}H_{20}SiZr$ (I) $(C_5H_5)_2Zr(CH_2Si(CH_3)_2CH_2)$

BIS(η^5-2,4-CYCLOPENTADIEN-1-YL)[(DIMETHYLSILYLENE)BIS(METHYLENE)]NIOBIUM
$C_{14}H_{20}NbSi$ (II)

BIS(η^5-2,4-CYCLOPENTADIEN-1-YL)[(DIMETHYLSILYLENE)BIS(METHYLENE)]MOLYBDENUM
$C_{14}H_{20}MoSi$

W.R. TIKKANEN, J.W. EGAN, Jr. and J.L. PETERSEN, 1984. Organometallics, 3, 1646-1650.

I. Monoclinic, $P2_1/m$, a = 8.075, b = 11.810, c = 15.524 Å, β = 102.48°, Z = 4. Mo radiation, R = 0.044 for 2702 reflexions.

II. Monoclinic, $P2_1/m$, a = 7.859, b = 11.578, c = 15.697 Å, β = 102.74°, Z = 4. Mo radiation, R = 0.034 for 2606 reflexions.

III. Monoclinic, $P2_1/m$, a = 7.627, b = 11.395, c = 15.874 Å, β = 102.40°, Z = 4. Mo radiation, R = 0.043 for 2515 reflexions.

In each case, the crystallographic asymmetric unit consists of two independent molecules which lie on crystallographic mirror planes. Comparison of the corresponding structural parameters for molecule 1 (Fig. 1) in each of the three structures reveals that the transition from Zr(IV) to Nb(IV) to Mo(IV) produces a decrease in the C-M-C bond angle from 81.0 to 76.1 to 72.4° with a gradual increase in the M-C distance from 2.240 to 2.275 to 2.301 Å respectively. This is further accompanied by a continual increase in the folding of the four-membered ring (MC_2Si) along the C→C vector, from 4.7 to 10.4 to 14.3° respectively.

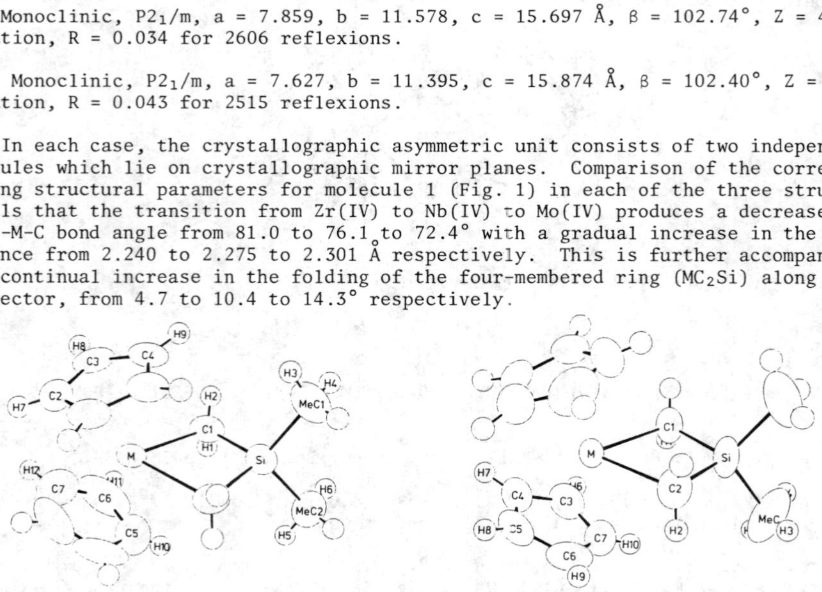

Fig. 1. Perspective views of the molecular configurations of the two molecules of $C_{14}H_{20}SiM$ (where M = Zr, Nb and Mo) in the asymmetric unit.

IODOTRIMETHYL[BIS(3,5-DIMETHYL-1-PYRAZOLYL)METHANE]PLATINUM(IV)
$C_{14}H_{25}IN_4Pt$ $Me_3PtI[(Me_2pz)_2CH_2]$

H.C. CLARK, G. FERGUSON, V.K. JAIN and M. PARVEZ, 1984. J. Organometal. Chem., 270, 365-373.

Orthorhombic, Pcmn, a = 11.936, b = 14.462, c = 10.138 Å, Z = 4. Mo radiation, R = 0.022 for 1719 reflexions.

The crystal structure consists of discrete monomeric molecules with crystallographic m symmetry (Fig. 1). The metal atom has octahedral geometry with one methyl group trans to iodine and two methyl groups trans to the N atoms of the bidentate ligand (Pt-I 2.843, Pt-N 2.236, Pt-C 2.077 (trans to I), and 2.032 Å (trans to N)). The unique pyrazolyl ring is planar to within 0.002 Å and the molecular dimensions are in accord with accepted values. The six-membered ring, PtN_4C, has a boat conformation with the Pt...C(6) fold angle 119.7°.

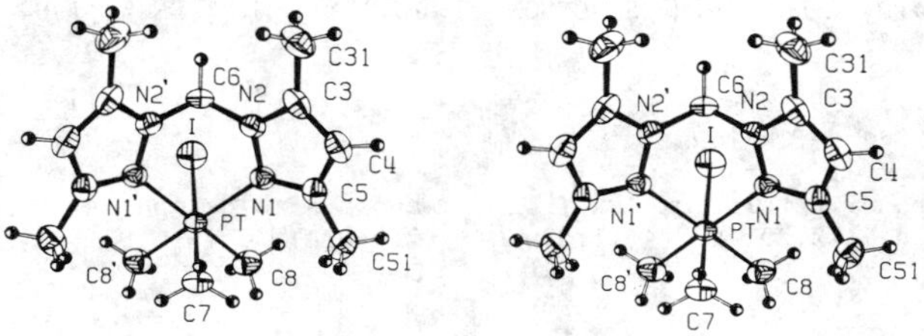

Fig. 1. Stereoview of $C_{14}H_{25}IN_4Pt$.

(σ-CYCLOPENTADIENYL)TRIISOPROPYLPHOSPHINEGOLD
$C_{14}H_{26}AuP$ $(\sigma-C_5H_5)AuP(C_3H_7)_3$

H. WERNER, H. OTTO, T. NGO-KHAC and Ch. BURSCHKA, 1984. J. Organometal. Chem., 262, 123-136.

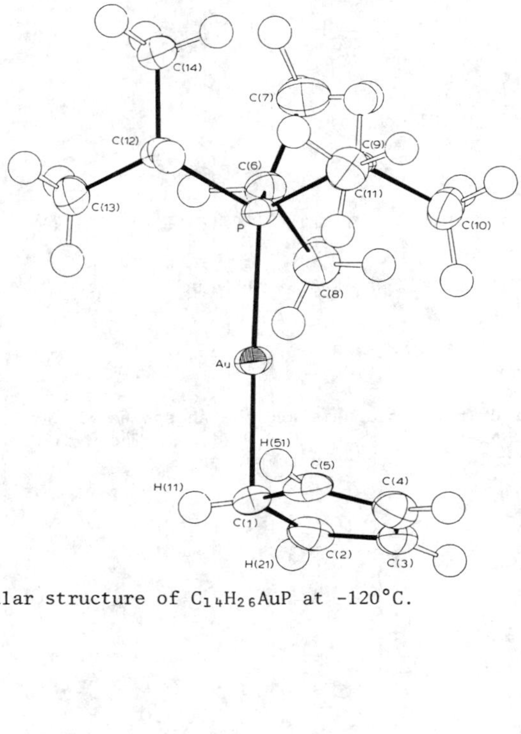

Fig. 1. Molecular structure of $C_{14}H_{26}AuP$ at -120°C.

Monoclinic, $P2_1/c$, a = 8.225, b = 15.623, c = 12.483 Å, β = 106.81°, Z = 4. Mo radiation, R = 0.037 for 2417 reflexions (at -120°C).

The Au atom is in a linear arrangement (Fig. 1) with P-Au-C(1) 177.0°, and Au-P and Au-C(1) distances of 2.267 and 2.175 Å respectively. The metal atom is primarily σ-bonded to the cyclopentadienyl ring which shows a weak "slip distortion" toward an η^3 mode of coordination (Au-C(2) 2.781, Au-C(5) 2.688 Å, Au-C(1)-C(2) 98.9, Au-C(1)-C(5) 94.0°).

1-(TRICHLOROZINC)-3,6,10,13,16,19-HEXAAZABICYCLO[6.6.6]ICOSANECOBALT(III) CHLORIDE DIHYDRATE

$C_{14}H_{31}Cl_4CoN_6Zn,2(H_2O)$

I.I. CREASER, J.D. LYDON, A.M. SARGESON, E. HORN and M.R. SNOW, 1984. J. Am. Chem. Soc., 106, 5729-5731.

Monoclinic, $P2_1/c$, a = 8.625, b = 16.273, c = 15.381 Å, β = 93.12°, D_m = 1.68, Z = 4. R = 0.032 for 2544 reflexions.

The Zn-C bond in this complex (Fig. 1) appears to be unusually stable. Some bond lengths are: Zn-Cl 2.291-2.374(1), Zn-C(1) 2.049(4), Co-N 1.971(4)-1.988(4).

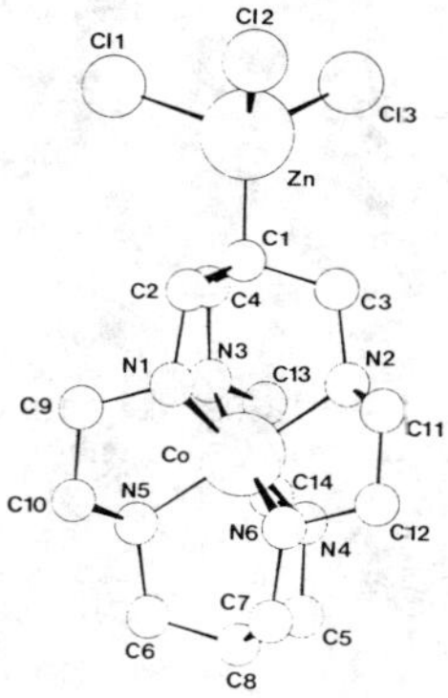

Fig. 1. A view of $C_{14}H_{31}Cl_4CoN_6Zn$.

trans-BIS(TRIMETHYLSILYLMETHYL)BIS(TRIMETHYLPHOSPHINE)NICKEL(II)

$C_{14}H_{40}NiP_2Si_2$

E. CARMONA, M. PANEQUE, M.L. POVEDA, R.D. ROGERS and J.L. ATWOOD, 1984. Polyhedron, 3, 317-323.

Orthorhombic, Pbcn, a = 14.345, b = 12.656, c = 12.815 Å, Z = 4. Mo radiation, R = 0.077 for 535 reflexions.

The molecule lies on a twofold axis and the phosphine ligands are trans (Fig. 1) with P-Ni-P 146.9(3)°, Ni-P 2.158(4), Ni-C 2.08(2) Å. The C-Ni-C angle is 170° and the nickel coordination is distorted-square-planar.

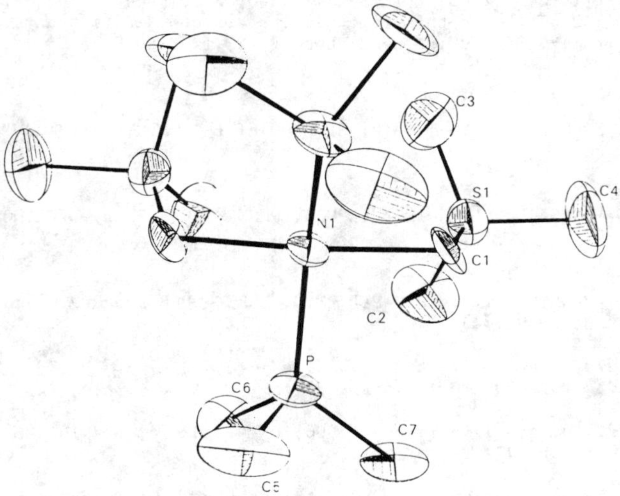

Fig. 1. A view of $C_{14}H_{40}NiP_2Si_2$.

TETRAMETHYLBIS[μ-[METHYLENEBIS[DIMETHYLPHOSPHINE]-P:P']]DIPLATINUM
$C_{14}H_{40}P_4Pt_2$ (I) $[Me_2Pt(Me_2PCH_2PMe_2)_2PtMe_2]$

BIS[μ-[METHYLENEBIS[DIPHENYLPHOSPHINE]-P:P']]TETRAMETHYLDIPLATINUM
$C_{54}H_{56}P_4Pt_2$ (II) $[Me_2Pt(Ph_2PCH_2PPh_2)_2PtMe_2]$

L. MANOJLOVIĆ-MUIR, K.W. MUIR, A.A. FREW, S.S.M. LING, M.A. THOMSON and R.J. PUDDEPHATT, 1984. Organometallics, 3, 1637-1645.

I. Monoclinic, $P2_1/a$, a = 16.634, b = 11.112, c = 12.469 Å, β = 103.40°, Z = 4. Mo radiation, R = 0.043 for 3646 reflexions.

II. Monoclinic, $P2_1/c$, a = 13.957, b = 17.218, c = 21.649 Å, β = 106.45°, D_m = 1.65, Z = 4. Mo radiation, R = 0.068 for 3818 reflexions.

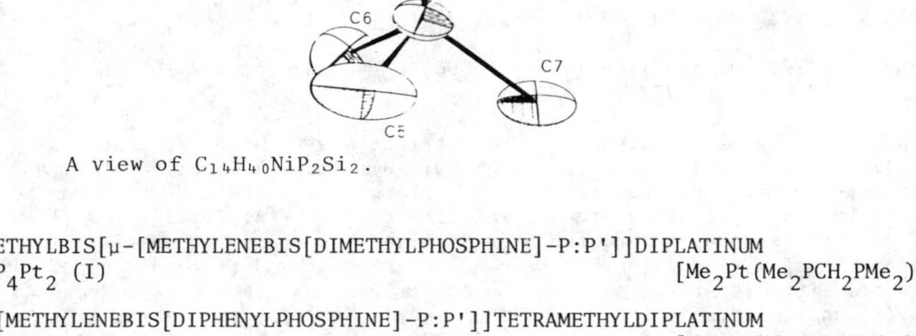

Fig. 1. Perspective views of $C_{14}H_{40}P_4Pt_2$ (left), and $C_{54}H_{56}P_4Pt_2$ (right).

Each molecule contains two cis-$PtMe_2$ fragments held together by two bridging $R_2PCH_2PR_2$ ligands (Fig. 1). Compound I has two independent half molecules per asymmetric unit, lying about crystallographic inversion centres. Compound II has approximate C_2 molecular symmetry. The eight-membered $Pt_2P_4C_2$ dimetallacycles adopt twist-chair (I) and twist-boat conformations (II) respectively. In I, the (average) bond distances of interest are: Pt-P(1) 2.272, Pt-P(2) 2.278, Pt-C(1) 2.13, Pt-C(2) 2.18, P(1)-C(3') 1.85 and P(2)-C(3) 1.84 Å. The cis angles about Pt range from 82.8 to 102.8°. In II, the bonds of interest are: Pt(1)-P(1) 2.299, Pt(1)-P(2) 2.294,

Pt(1)-C(1) 2.09, Pt(1)-C(2) 2.08, Pt(2)-P(3) 2.300, Pt(2)-P(4) 2.304, Pt(2)-C(3) 2.14 and Pt(2)-C(4) 2.05 Å. In this molecule the cis-coordination angles about the Pt(1) atom range from 75.0 to 97.5°, and about Pt(2) from 80.0 to 98.4°.

1,1,2,2,2,2,3,3,3,3-DECACARBONYL-1,2-(μ2-HYDRIDO)-1-(1'-METHYL-3'-OXOBUT-1'-EN-1'-YL-C',O)TRIOSMIUM
$C_{15}H_8O_{11}Os_3$ (I)

1,1,2,2,2,3,3,3,3-NONACARBONYL-1-DIMETHYLPHENYLPHCSPHINE-1,2-(μ2-HYDRIDO)-1,2-(μ2-HYDROXO)TRIOSMIUM
$C_{17}H_{13}O_{10}Os_3P$ (II)

A.J. DEEMING, P.J. MANNING, I.P. ROTHWELL, M.B. HURSTHOUSE and N.P.C. WALKER, 1984. J. Chem. Soc. Dalton, 2039-2045.

I. Monoclinic, C2/c, a = 17.377, b = 14.293, c = 16.476 Å, β = 91.79°, Z = 8. Mo radiation, R = 0.034 for 2906 reflexions.

II. Monoclinic, $P2_1/n$, a = 17.535, b = 9.500, c = 14.326 Å, β = 101.86°, Z = 4. Mo radiation, R = 0.041 for 3474 reflexions.

Compound I (Fig. 1) consists of an Os_3 triangle with Os-Os distances of 3.030(1) (Os(1)-Os(2)), 2.878(0) (Os(1)-Os(3)) and 2.916(0) Å (Os(2)-Os(3)). Os(2) and Os(3) are coordinated by terminal CO groups only while on Os(1) the chelating Me-C=CHCOMe ligand occupies an axial and an equatorial site (Os(1)-C(14) 2.048(9), Os(1)-O(11) 2.144(6) Å). It is thought that the hydride ligand bridges Os(1)-Os(2). In II the PMe_2Ph ligand is positioned in the Os_3 plane trans to the Os(1)-Os(2) bond. It produces very little distortion in the molecular framework. The Os-OH bond distances are 2.135(5) and 2.118(5) Å for Os(1)-O(10) and Os(3)-O(10) respectively. The Os(1)-Os(2) and Os(2)-Os(3) distances are 2.830 and 2.821 Å respectively. The hydride ligand was not located but probably bridges Os(1)-Os(3) (2.801 Å).

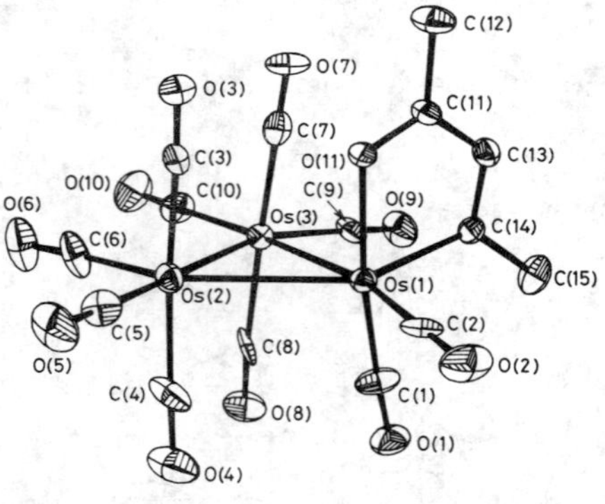

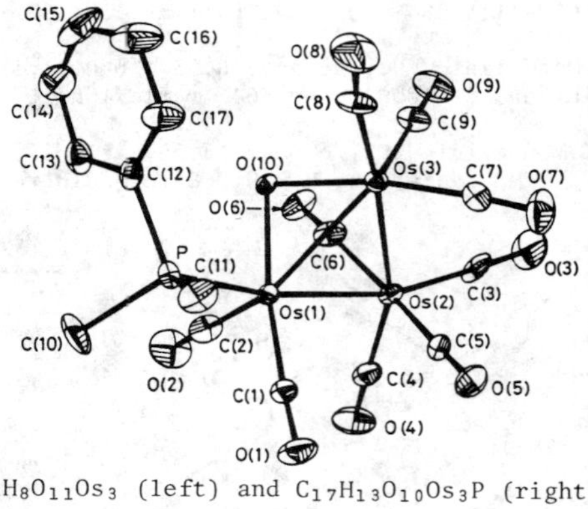

Fig. 1. The molecular structures: $C_{15}H_8O_{11}Os_3$ (left) and $C_{17}H_{13}O_{10}Os_3P$ (right).

(μ2-CARBONYL)(μ2-η5,η5-FULVALENE)(μ2-VINYLENE)DICARBONYLDIRUTHENIUM
$C_{15}H_{10}O_3Ru_2$ (I)

(μ2-η2,η2-DIMETHYLETHYNEDICARBOXYLATE)(μ2-η5,η5-FULVALENE)TRICARBONYL(η2-DIMETHYL-ETHYNEDICARBOXYLATE)DIMOLYBDENUM
$C_{25}H_{20}Mo_2O_{11}$ (II)

(μ2-η5,η5-FULVALENE)(η2-DIPHENYLACETYLENE)DICARBONYLOXOMOLYBDENUMRUTHENIUM
$C_{26}H_{18}MoO_3Ru$ (III)

J.S. DRAGE, M. TILSET, P.C. VOLLHARDT and T.W. WEIDMAN, 1984. Organometallics, 3, 812-814.

I. Orthorhombic, Pbca, a = 8.052, b = 15.011, c = 21.824 Å, Z = 8. Mo radiation, R = 0.017 for 1609 reflexions.

II. Monoclinic, $P2_1/c$, a = 9.6984, b = 15.438, c = 16.486 Å, β = 95.880°, Z = 4. Mo radiation, R = 0.022 for 2692 reflexions.

III. Monoclinic, $P2_1/c$, a = 10.1614, b = 16.776, Z = 4. Mo radiation, R = 0.016 for 2539 reflexions. [Values for c and β not quoted in the paper.]

Irradiation of the carbonyl(η^5=η^5-fulvalene)dimetal complexes of Mo and Ru in the presence of alkynes produces compounds I, II and III (Fig. 1) which are quite different. In I the alkyne (acetylene) is bound in the relatively rare μ-η^2 mode. In II, two alkyne molecules ($CH_3OOCC{\equiv}CCOOCH_3$) are bound to the metals, one as a two electron ligand and the other as a four electron ligand. The Mo-Mo distance is 2.906 Å. Compound III is a novel oxo complex in which the diphenylethyne ligand is bound in η^2-fashion to the Mo atom which is also coordinated to the oxo ligand O(3).

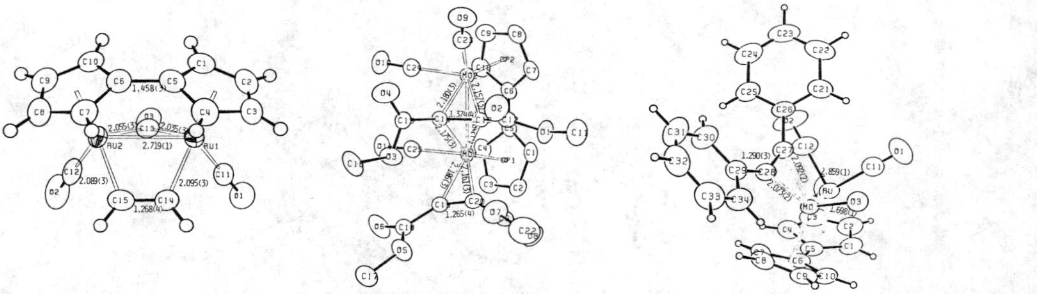

Fig. 1. Molecular structure with selected bond lengths (Å) of $C_{15}H_{10}O_3Ru_2$ (left), $C_{25}H_{20}Mo_2O_{11}$ (centre), and $C_{26}H_{18}MoO_3Ru$ (right).

BENZAMIDINIUM DIACETYLTETRACARBONYLRHENIUM

$C_{15}H_{15}N_2O_6Re$ $[C_6H_5C(NH_2)_2]^+[cis\text{-}(OC)_4Re(CH_3CO)_2]^-$

P.G. LENHERT, C.M. LUKEHART, P.D. SOTIROPOULOS and K. SRINIVASAN, 1984. Inorg. Chem., 23, 1807-1810.

Triclinic, $P\bar{1}$, a = 9.037, b = 13.155, c = 8.004 Å, α = 106.74, β = 103.16, γ = 89.91°, Z = 2. Mo radiation, R = 0.048 for 5174 reflexions.

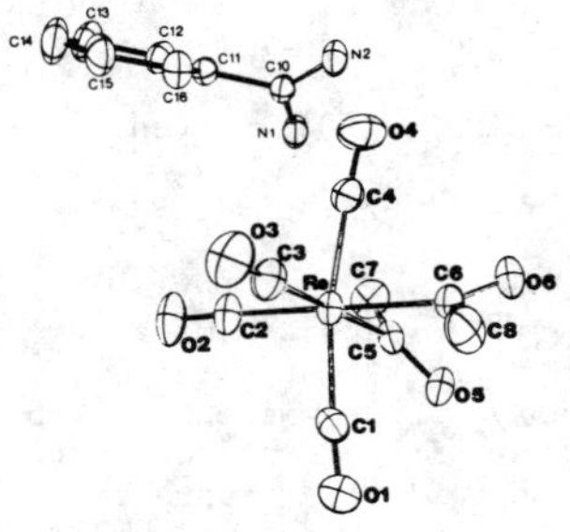

Fig. 1. The asymmetric unit of $C_{15}H_{15}N_2O_6Re$.

The structure consists of $[C_6H_5C(NH_2)_2]^+$ cations and $[cis\text{-}(OC)_4Re(CH_3CO)_2]^-$ anions (Fig. 1) linked by an extensive network of N-H...O hydrogen bonds. The coordination about Re is irregular octahedral with Re-CO 1.965(7)-1.974(6) and Re-C(O)CH_3 2.181(6) and 2.183(6) Å.

[BIS(N-METHYLIMIDAZOL-2-YL)(PYRIDIN-2-YL)METHANOL-N,N',N'']METHYLMERCURY(II) NITRATE
$C_{15}H_{16}HgN_6O_4$ $[CH_3Hg(C_{14}H_{13}N_5O)](NO_3)$

A.J. CANTY, J.M. PATRICK and A.H. WHITE, 1984. Inorg. Chem., 23, 3827-3830.

Monoclinic, $P2_1/n$, a = 15.786, b = 15.065, c = 7.715 Å, β = 91.24°, D_m = 1.95, Z = 4. Mo radiation, R = 0.051 for 1798 reflexions.

The cation (Fig. 1) contains a four-coordinate Hg^{2+} with two strong bonds (Hg-C 2.07(1), Hg-N(1') 2.12(1) Å) making an angle N-Hg-C 170.5(7)°. Opposite this angle are two weak Hg-N interactions (Hg-N 2.58(1) and 2.89(1) Å).

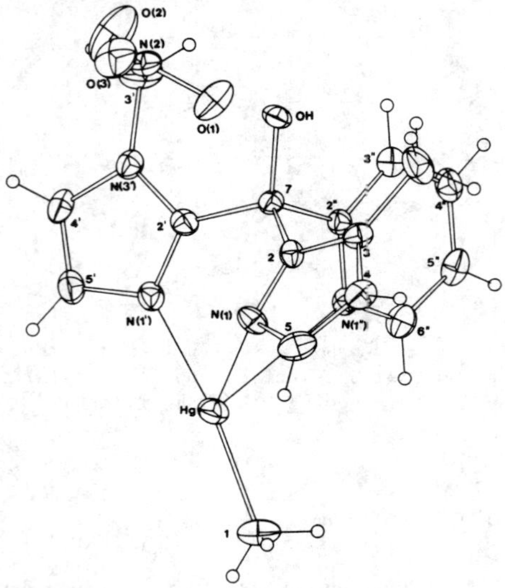

Fig. 1. The $[CH_3Hg(C_{14}H_{13}N_5O)]^+$ cation.

CARBONYL-(η^5-PENTAMETHYLCYCLOPENTADIENYL)BIS(METHOXYCARBONYL)RHODIUM
$C_{15}H_{21}O_5Rh$ $(\eta^5\text{-}Me_5Cp)Rh(CO)(COOCH_3)_2$

P.L. BURK, D. VAN ENGEN and K.S. CAMPO, 1984. Organometallics, 3, 493-495.

Orthorhombic, Pnma, a = 10.499, b = 14.323, c = 11.064 Å, Z = 4. Mo radiation, R = 0.037 for 843 reflexions.

The molecule (which has crystallographic mirror symmetry) has its planar Me_5Cp ligand symmetrically coordinated to the Rh atom (Fig. 1). The most striking feature is the presence of two carbomethoxy groups unidentately bound to the metal. The Rh-COO moieties are essentially planar with the planes of the COO groups intersecting at an angle of 79.5°. Atoms C(2) and C(2') are separated by 2.65 Å. The Rh-C(2) bond length is 2.041 Å and the ester C-O bonds of 1.185 and 1.349 Å are normal. Other bond distances are: Rh-C(1) 1.879, Rh-Cp(centroid) 1.897, C(1)-O(1) 1.136, and C(3)-O(3) 1.418 Å.

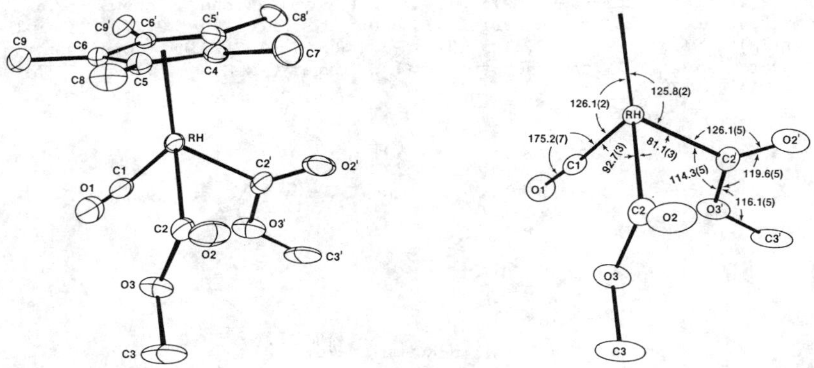

Fig. 1. Molecular structure of $C_{15}H_{21}O_5Rh$ (left), and schematic showing important bond angles (right).

6,6-BIS(η^5-CYCLOPENTADIENYL)-4,4-DIMETHYL-1-OXA-4-SILA-6-ZIRCONA-CYCLOHEXANE
$C_{15}H_{22}OSiZr$ $(C_5H_5)_2\dot{Z}r(OCH_2CH_2Si(CH_3)_2\dot{C}H_2)$

W.R. TIKKANEN and J.L. PETERSEN, 1984. Organometallics, 3, 1651-1655.

Monoclinic, $P2_1/c$, a = 12.422, b = 6.438, c = 19.593 Å, β = 95.16°, Z = 4. Mo radiation, R = 0.029 for 2767 reflexions.

The pseudotetrahedral ligand arrangement about the Zr(IV) centre is characterized by Cp(1)-Zr-Cp(2) and O-Zr-C(3) bond angles of 130.2 and 91.0° respectively (Fig. 1). The 1-oxa-4-sila-6-zirconacyclohexane ring adopts a distorted boat conformation. The insertion of an OCH_2 fragment into one of the Zr-C bonds is accompanied by the formation of a relatively short Zr-O bond of 1.941 Å and produces a rather obtuse Zr-O-C(1) bond angle of 143.1°. Other bond distances and angles of interest are: Zr-C(3) 2.297, Zr-Cp(1) (centroid) 2.242, Zr-Cp(2) (centroid) 2.251, O-C(1) 1.409, C(1)-C(2) 1.513, C(2)-Si 1.877, Si-MeC(1) 1.861, Si-MeC(2) 1.874, Si-C(3) 1.849 Å; Zr-C(3)-Si 109.8, O-C(1)-C(2) 111.5, C(1)-C(2)-Si 114.9, C(2)-Si-C(3) 112.0 and MeC(1)-Si-MeC(2) 108.8°.

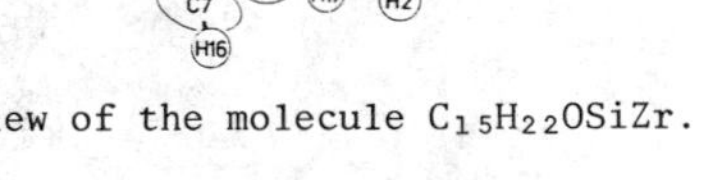

Fig. 1. Perspective view of the molecule $C_{15}H_{22}OSiZr$.

(η^5-CYCLOPENTADIENYL)IODOBIS(t-BUTYL ISONITRILE)RUTHENIUM
$C_{15}H_{23}IN_2Ru$ $[(\eta^5\text{-}C_5H_5)Ru(CNBu)_2I]$

J.C.A. BOEYENS, N.J. COVILLE and K. SOLDENHOFF, 1984. S. Afr. J. Chem., 37, 153-160.

Triclinic, $P\bar{1}$, a = 14.707, b = 11.751, c = 11.296 Å, α = 91.09, β = 98.68, γ = 105.23°, Z = 4. Mo radiation, R = 0.075 for three-dimensional data.

There are two independent molecules of the complex; one is shown in Fig. 1. The Ru atoms have octahedral coordination with mean distances Ru-I 2.717(3), Ru-C (isonitrile) 1.91(3), C-N 1.18(3) and 1.46(3), Ru-C(cp) 2.223 Å. One isonitrile is near linear (C-N-C 175°) and the other with a larger C-N-C bend (162°). The mean Ru-C-N angle is 175°.

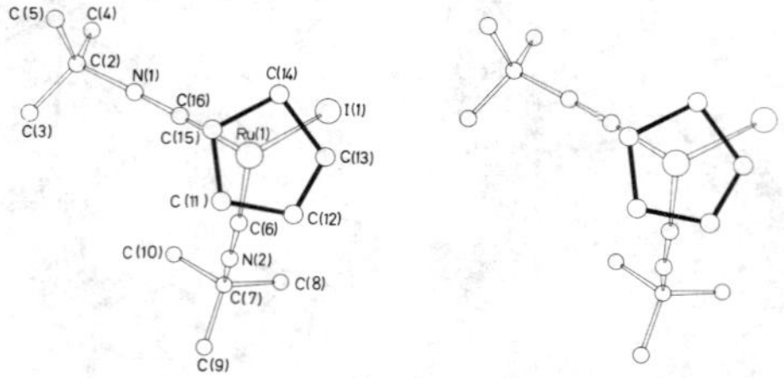

Fig. 1. A stereoview of a $C_{15}H_{23}IN_2Ru$ molecule.

CARBONYLCHLOROTRIS(TRIMETHYLPHOSPHINE)(η²-TRIMETHYLSILYL-C,O)MOLYBDENUM
$C_{15}H_{38}ClMoO_2P_3Si$ (I)

(σ^3-ACETYL-C,C,H)CARBONYL(N,N-DIMETHYLDITHIOCARBAMATE-S,S')BIS(TRIMETHYLPHOSPHINE)-MOLYBDENUM
$C_{12}H_{27}MoNO_2P_2S_2$ (II)

E. CARMONA, L. SANCHEZ, J.M. MARIN, M.L. POVEDA, J.L. ATWOOD, R.D. PRIESTER and R.D. ROGERS, 1984. J. Am. Chem. Soc., 106, 3214-3222.

I. Orthorhombic, $P2_12_12_1$, a = 10.747, b = 12.511, c = 18.897 Å, Z = 4. Mo radiation, R = 0.025 for 2141 reflexions.

II. Monoclinic, $P2_1/c$, a = 11.633, b = 9.962, c = 17.266 Å, β = 95.80°, Z = 4. Mo radiation, R = 0.029 for 1116 reflexions.

In (I) there is octahedral coordination at Mo and an η²-acyl ligand (Fig. 1), with Mo-C2 2.024(6), Mo-O2 2.324(4), Mo-P1 2.483(1), Mo-P2 2.504(2), Mo-P3 2.467(2), Mo-Cl 2.550(1), Mo-C1 1.929(6) Å. Compound (II) is prepared from (I) and has a strong Mo-C-H interaction (Fig. 1), with Mo-C3 2.60(1), Mo-H1 2.06(9), Mo-S1 2.547(2), Mo-S2 2.529(2), Mo-P1 2.435(2), Mo-P2 2.430(3), Mo-C1 1.83(1), Mo-C2 2.05(1) Å, Mo-C2-O2 149.2(8), Mo-C2-C3 90.9(8)°.

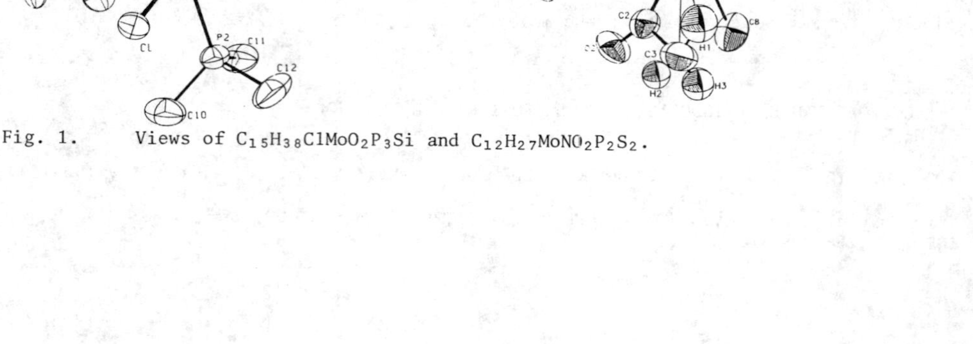

Fig. 1. Views of $C_{15}H_{38}ClMoO_2P_3Si$ and $C_{12}H_{27}MoNO_2P_2S_2$.

TETRAKIS(μ_2-SILVER)TETRACARBONYLCOBALT)
$C_{16}Ag_4Co_4O_{16}$ $[AgCo(CO)_4]_4$

P. KLUFERS, 1984. Z. Kristallogr., 166, 143-151.

Monoclinic, C2/m, a = 11.923, b = 17.837, c = 6.374 Å, β = 95.61°, Z = 2. Mo radiation, R = 0.042 for 832 reflexions.

The analysis reveals a WOF_4-type (NbF_5-type) structure with cubic dense packing in the $Ag(CO)_4$-partial structure and tetrameric $[AgCo(CO)_4]_4$ units (Fig. 1).

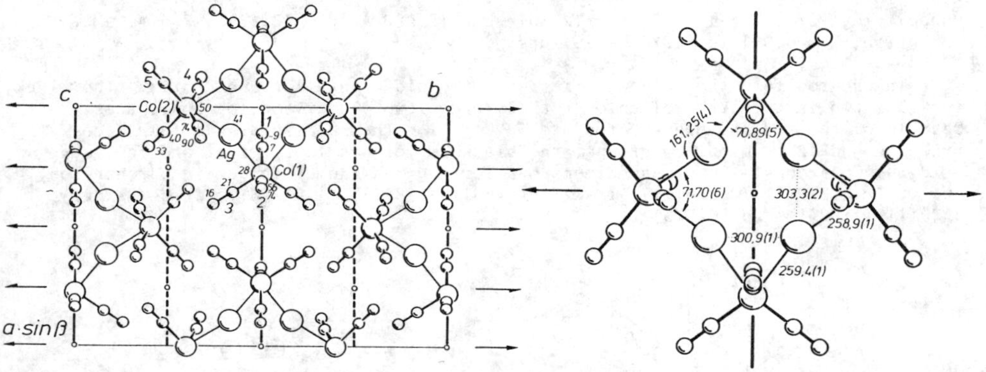

Fig. 1. Views of the $[AgCo(CO)_4]_4$ structure and some dimentions (pm and °).

(μ_6-DICARBIDO)HEXAKIS(μ-CARBONYL)OCTACARBONYL(μ_4-SULPHIDO)HEXACOBALT
$C_{16}Co_6O_{14}S$ $Co_6(\mu_6\text{-}C_2)(\mu\text{-}CO)_6(CO)_8(\mu_4\text{-}S)$

G. GERVASIO, R. ROSSETTI, P.L. STANGHELLINI and G. BOR, 1984. Inorg. Chem., 23, 2073-2079.

Orthorhombic, $P2_12_12_1$, a = 18.618, b = 13.528, c = 9.151 Å, Z = 4. Mo radiation, R = 0.045 for 1509 reflexions.

The molecule (Fig. 1) consists of a "boat" array of six Co atoms, of which the four basal ones form an essentially regular square, metal-metal bonded on all four sides (Co-Co 2.590(3)-2.619(3) Å). The two apical Co atoms are connected through a Co-C-C-Co array. The two C atoms of the "peripheral" dicarbido unit are bonded also to the four basal Co atoms. These four atoms, in turn, are capped also by the μ_4-S atom. Each of the six peripheral Co-Co bonds bears a μ-CO ligand, five of them showing different amounts of asymmetry. Other bond lengths: Co-Co 2.490(3)-2.562(3), Co-S 2.273(5)-2.304(5), Co-(μ_6-C_2) 1.81(2)-2.03(2), C-C 1.37(2) Å.

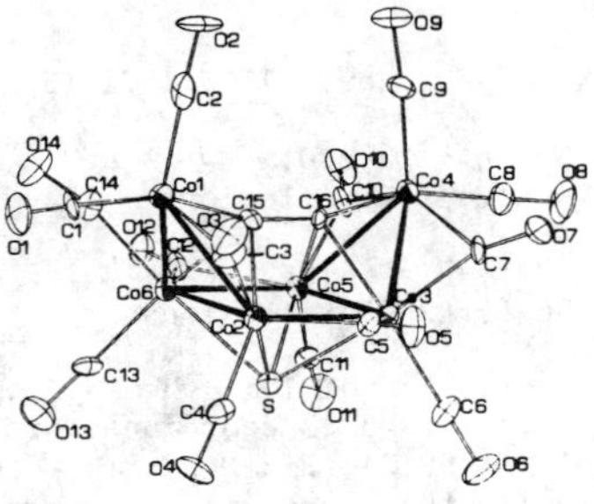

Fig. 1. The $Co_6(\mu_6\text{-}C_2)(\mu\text{-}CO)_6(CO)_8(\mu_4\text{-}S)$ molecule.

TETRAKIS(μ_3-CARBONYL)DODECACARBONYLHEXAIRIDIUM (Red and Black Isomers)
$C_{16}Ir_6O_{16}$

L. GARLASCHELLI, S. MARTINENGO, P.L. BELLON, F. DEMARTIN, M. MANASSERO, M.Y. CHIANG, CHIAU-YU WEI and R. BAU, 1984. J. Am. Chem. Soc., 106, 6664-6667.

Red form:
Monoclinic, C2/c, a = 16.656, b = 9.761, c = 16.776 Å, β = 118.09°, Z = 4. Mo radiation, R = 0.034 for 1508 reflexions.

Black form:
Monoclinic, P2/c, a = 13.622, b = 12.807, c = 18.589 Å, β = 132.65°, Z = 4. Mo radiation, R = 0.051 for 1973 reflexions.

In the red form the molecule lies on a twofold crystallographic axis. The black form has two independent molecules both lying on twofold crystallographic axes. Details of the red and black forms are in Fig. 1. The red isomer has four face-bridging and 12 terminal CO groups; the black isomer has 12 terminal and four edge-bridging CO groups. Some mean dimensions for the red isomer (black in parentheses) are Ir-Ir 2.779(1) (2.778(6)), Ir-C (terminal) 1.886(3) (1.882(8)), Ir-C (bridge) 2.20(1) (1.99(1) and 2.25(3)) Å.

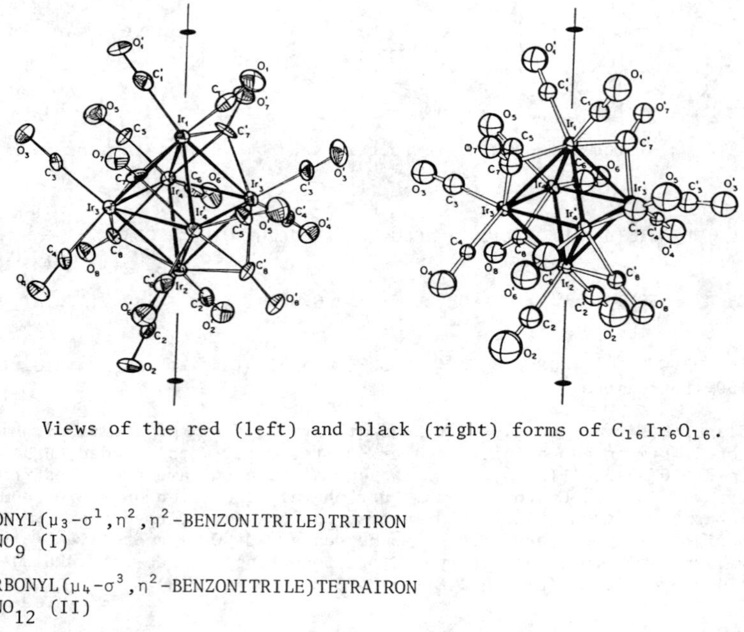

Fig. 1. Views of the red (left) and black (right) forms of $C_{16}Ir_6O_{16}$.

NONACARBONYL(μ_3-σ^1,η^2,η^2-BENZONITRILE)TRIIRON
$C_{16}H_5Fe_3NO_9$ (I)

DODECACARBONYL(μ_4-σ^3,η^2-BENZONITRILE)TETRAIRON
$C_{19}H_5Fe_4NO_{12}$ (II)

E. KELLER and D. WOLTERS, 1984. Chem. Ber., 117, 1572-1582.

I. Monoclinic, $P2_1/c$, a = 7.659, b = 35.270, c = 13.967 Å, β = 90.40°, D_m = 1.82, Z = 8. Mo radiation, R = 0.079 for 3647 reflexions.

II. Monoclinic, $P2_1/c$, a = 10.797, b = 18.292, c = 14.725 Å, β = 126.15°, D_m = 1.86, Z = 4. Mo radiation, R = 0.056 for 2693 reflexions.

There are two independent molecules of I in the asymmetric unit; one molecule is shown in Fig. 1. The benzonitrile ligand is coordinated to the Fe atoms formally via one σ-bond and two π-bonds (Fe1-N 1.99(2), Fe2-N 1.98(2), Fe3-N 1.80(2), Fe1-C1 2.12(2), Fe2-C1 2.08(3) Å). In II, the benzonitrile coordinates via three σ-bonds and one π-bond (Fe1-N 1.87(1), Fe2-N 1.97(1), Fe3-N 1.832(7), Fe2-C1 2.13(1), Fe4-C1

1.94(1) Å). The Fe-Fe distances are in the ranges 2.480(5)-2.661(5) Å (for I) and 2.498(2)-2.679(2) (for II).

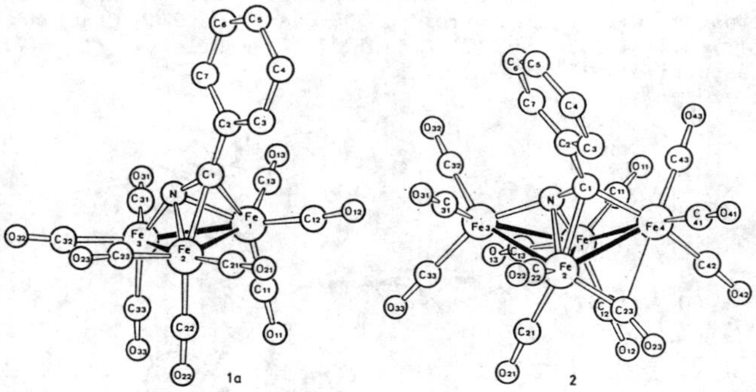

Fig. 1. Views of one of the $C_{16}H_5Fe_3NO_9$ molecules and of the $C_{19}H_5Fe_4NO_{12}$ molecule.

μ-CARBONYLDICARBONYLBIS(η^5-2,4-CYCLOPENTADIEN-1-YL)-μ-CYCLOPROPYLIDENEDIIRON(Fe-Fe)
$C_{16}H_{14}Fe_2O_3$

E.L. HOEL, G.B. ANSELL and S. LETA, 1984. Organometallics, 3, 1633-1637.

Orthorhombic, Pnma, a = 13.008, b = 14.126, c = 7.835 Å, Z = 4. Mo radiation, R = 0.032 for 446 reflexions.

The molecule has crystallographically-imposed mirror symmetry which includes the plane of the cyclopropylidene ligand and the atoms of the bridging CO ligand (Fig. 1). The two Fe atoms are related by this plane at an Fe-Fe bond distance of 2.503 Å. The Fe-Cp atom distances are: Fe(1)-C(4) 2.116, Fe(1)-C(5) 2.099, Fe(1)-C(6) 2.086, Fe(1)-C(7) 2.092 and Fe(1)-C(8) 2.091 Å. The Fe-cyclopropylidene ring distance Fe(1)-C(1) is 1.943 Å, and the intra-ring bond lengths are: C(1)-C(2) 1.52, C(1)-C(3) 1.51 and C(2)-C(3) 1.50 Å. The Fe-CO distances are: Fe(1)-C(9) 1.728 and Fe(1)-C(10) 1.914 Å. The angles at the bridging ligands are Fe(1)-C(1)-Fe(1') 80.2° and Fe(1)-C(10)-Fe(1') 81.7°.

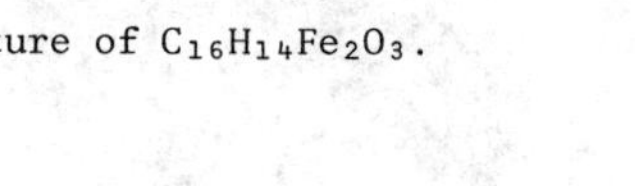

Fig. 1. Molecular structure of $C_{16}H_{14}Fe_2O_3$.

(3,3-DIMETHYLBUTYNIDE)DICYCLOPENTADIENYLVANADIUM
$C_{16}H_{19}V$ $(C_5H_5)_2VC\equiv CC(CH_3)_3$

W.J. EVANS, I. BLOOM and R.J. DOEDENS, 1984. J. Organometal. Chem., 265, 249-255.

Orthorhombic, Pnma, a = 9.075, b = 9.807, c = 16.444 Å, Z = 4. Mo radiation, R = 0.060 for 1500 reflexions.

The molecule (Fig. 1) has crystallographic m symmetry with the V atom and C atoms 1, 2, 3, 5, 8 and 9 lying on the mirror plane. The mean V-Cp(centroid) distance is 1.941 Å and the Cp(centroid)-V-Cp(centroid) angle is 146.6°. The distance V-C(1) is 2.075 Å and the angles C(1)-V-Cp(centroids) 107.4 and 106.0°. Other noteworthy distances are C(1)-C(2) 1.191 and C(2)-C(3) 1.474 Å. Angles V-C(1)-C(2) and C(1)-C(2)-C(3) are respectively 177.0 and 177.8°.

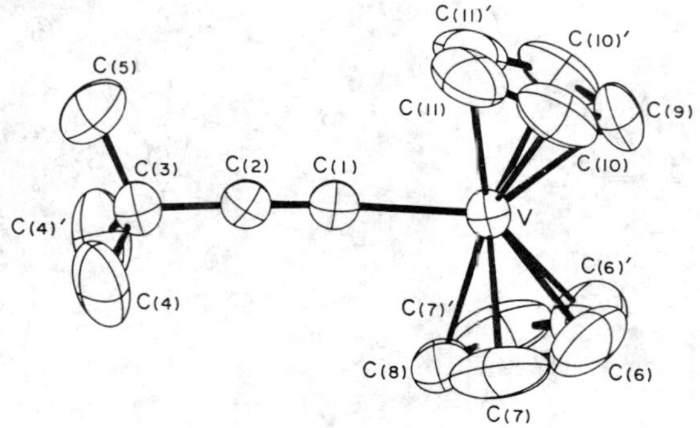

Fig. 1. Molecular structure of $C_{16}H_{19}V$.

HEXACARBONYL[3-(tert-BUTYLAMINO)-1-t-BUTYL-1-AZAALLYL]IRONMANGANESE
$C_{16}H_{21}FeMnN_2O_6$

J. KEIJSPER, J. MUL, G. van KOTEN, K. VRIEZE, H.C. UBBELS and C.H. STAM, 1984. Organometallics, 3, 1732-1739.

Monoclinic, $P2_1/c$, a = 16.044, b = 18.220, c = 17.330 Å, β = 129.69°, Z = 8. Cu radiation, R = 0.042 for 1287 reflexions.

The two independent molecules in the asymmetric unit have only slight differences in their molecular geometries (Fig. 1). The Fe-Mn distances of 2.616 and 2.640 for molecules A and B respectively are indicative of an Fe-Mn single bond. The mean bond lengths within the 3-amine-1-azaallylic N(1)∴C(7)∴C(8)N(2)H moiety are N(1)-C(7) 1.38, C(7)-C(8) 1.39 and C(8)-N(2) 1.45 Å. The ligand is an eight electron donor: (Mn-N(1) 2.02 Å (mean), Mn-N(2) 2.12 Å (mean)) (two σ bonds), Fe-C(7) 2.01 Å (mean), Fe-C(8) 2.08 (mean) and Fe-N(1) 2.04 Å (mean), (η^3-coordination). There are five terminal CO groups with normal M-C(carbonyl) bond lengths and M-C-O angles, and one semi-bridging CO group: Fe-C(4) 1.70 and 1.74 Å for molecules A and B respectively, with corresponding Mn-C(4) distances of 2.63 and 2.55 Å. The Fe-C(4)-O(4) angles are 169.3 and 165.3° for molecules A and B respectively.

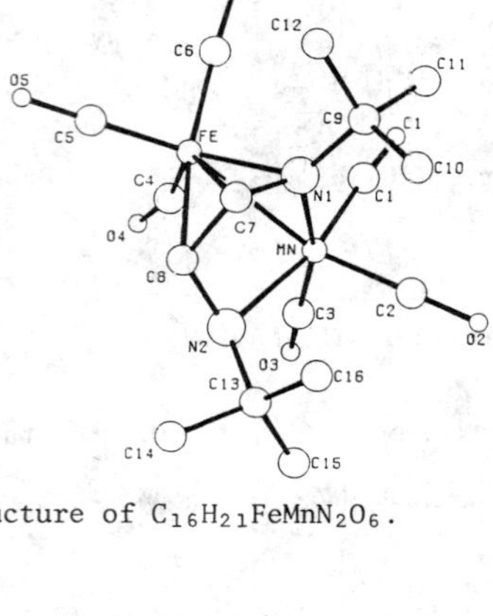

Fig. 1. Molecular structure of $C_{16}H_{21}FeMnN_2O_6$.

2-CARBONYL-5,6-BIS(METHOXYCARBONYL)-2,3-(S-METHYLEPITHIO)-2,2-BIS(TRIMETHYLPHOSPHINE)-1-THIA-2-FERRACYCLOHEX-5-EN-4-ONE
$C_{16}H_{28}FeO_6P_2S_2$

D. TOUCHARD, P.H. DIXNEUF, R.D. ADAMS and B.E. SEGMÜLLER, 1984. Organometallics, 3, 640-642.

Triclinic, $P\bar{1}$, a = 8.957, b = 10.440, c = 14.005 Å, α = 70.60, β = 83.40, γ = 67.96°, Z = 2. R = 0.033 for 3295 reflexions.

The equatorial sites of the distorted triangular-bipyramidal molecule are occupied by the PMe_3 groups and the C(13)-S(2) single bond which is roughly in the same plane as the Fe, P(1) and P(2) atoms (Fig. 1). The axial positions are occupied by the remaining CO group and the sulphido atom S(1) (C(1)-Fe-S(1) 176.50°) of the novel chelating ligand. Selected bond distances and angles are: Fe-S(1) 2.305, Fe-S(2) 2.272, Fe-P(1) 2.224, Fe-P(2) 2.252, Fe-C(1) 1.742, Fe-C(13) 2.067, C(13)-S(2) 1.783, C(14)-S(2) 1.820, S(1)-C(10) 1.718 Å; P(1)-Fe-P(2) 103.67, S(2)-Fe-C(13) 48.24, Fe-C(13)-S(2) 71.90, Fe-C(13)-C(12) 121.3, and S(2)-C(13)-C(12) 123.6°.

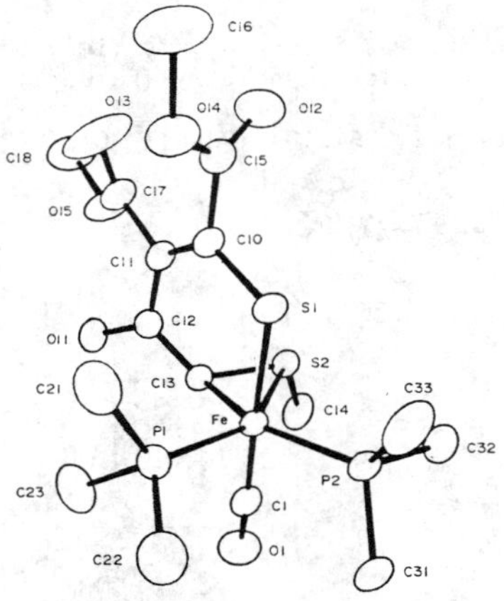

Fig. 1. Molecular structure of $C_{16}H_{28}FeO_6P_2S_2$.

1-(η^5-PENTAMETHYLCYCLOPENTADIENYL)-1-TRIMETHYLPHOSPHINE-1-RHODACYCLOBUTANE
$C_{16}H_{30}PRh$

R.A. PERIANA and R.G. BERGMAN, 1984. J. Am. Chem. Soc., 106, 7272-7273.

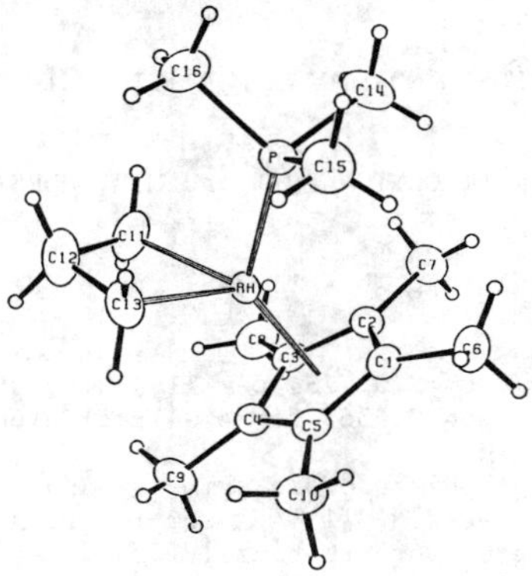

Fig. 1. A view of $C_{16}H_{30}PRh$.

Monoclinic, $P2_1/n$, a = 10.271, b = 12.964, c = 12.802 Å, β = 94.25°, Z = 4. Mo radiation, R = 0.0167 for 2206 reflexions.

The metallacyclobutane ring (Fig. 1) is essentially planar and symmetrical about the Rh-C12 axis, with Rh-C11, Rh-C13 2.085, C11-C12 1.512, C12-C13 1.527 Å, C11-Rh-C13 67.61, C11-C12-C13 99.55°.

BIS(μ_2-CHLORO)DICHLOROBIS[2-(N,N-DIISOPROPYLIMINIO)ETHYL-C]DIPALLADIUM(II)
$C_{16}H_{34}Cl_4N_2Pd_2$

R. McCRINDLE, G. FERGUSON, G.J. ARSENAULT, A.J. McALEES and D.K. STEPHENSON, 1984. J. Chem. Res., (S) 360-361, (M) 3301-3344.

Orthorhombic, Pbca, a = 13.914, b = 12.017, c = 14.010 Å, Z = 4. Mo radiation, R = 0.039 for 1141 reflexions.

The molecule (Fig. 1) occurs as a centrosymmetric dimer with the unique Pd atom having slightly distorted square-planar geometry (Pd-Cl(1) 2.444(2), Pd-Cl(1') 2.331(2), Pd-Cl(2) 2.299(3), Pd-C(1) 2.047(9) Å). There is also an interaction between the Pd atom and C(2) (Pd...C(2) 2.705(9) Å, Pd-C(1)-C(2) 101.0(6)°).

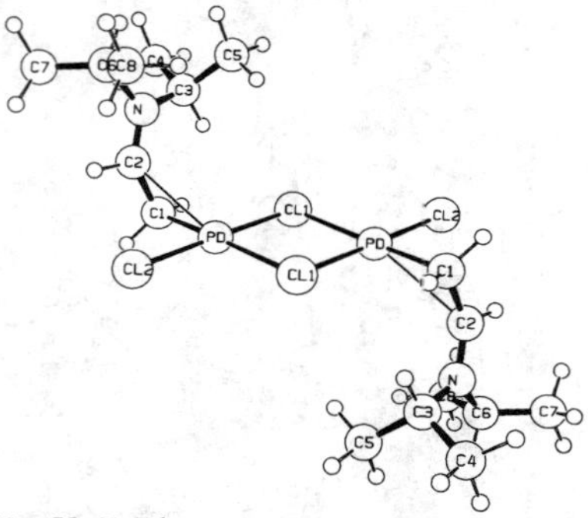

Fig. 1. A view of $C_{16}H_{34}Cl_4N_2Pd_2$.

BIS[1,2-BIS(DIMETHYLPHOSPHINO)ETHANE-P,P']TETRAMETHYLZIRCONIUM(IV)
$C_{16}H_{44}P_4Zr$ (I) $ZrMe_4(dmpe)_2$

TETRABENZYL[1,2-BIS(DIMETHYLPHOSPHINO)ETHANE-P,P']ZIRCONIUM(IV)
$C_{34}H_{44}P_2Zr$ (II) $Zr(CH_2Ph)_4(dmpe)$

G.S. GIROLAMI, G. WILKINSON, M. THORNTON-PETT and M.B. HURSTHOUSE, 1984. J. Chem. Soc. Dalton, 2789-2794.

I. Orthorhombic, $P2_12_12$, a = 12.490, b = 11.245, c = 8.882 Å, Z = 2. Mo radiation, R = 0.0598 for 1524 reflexions.

II. Triclinic, $P\bar{1}$, a = 11.307, b = 13.773, c = 11.452 Å, α = 91.713, β = 96.750, γ = 114.47°, Z = 2. Mo radiation, R = 0.0363 for 4694 reflexions.

Molecule I has crystallographic twofold symmetry and dodecahedral geometry (Fig. 1) with Zr-C 2.440(6), Zr-P 2.814(1) Å. Complex II has severely distorted octahedral geometry consisting of three σ-benzyl ligands and one partially π-bonded benzyl group, with Zr-C(σ) 2.264(7)-2.368(5), Zr-C_α(π) 2.359(6), Zr...C_β(π) 2.788(8), Zr-P 2.862(4), 2.885(4) Å.

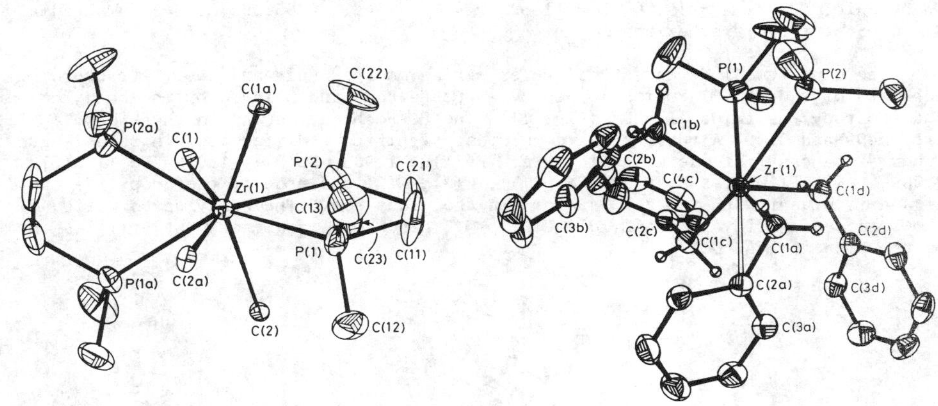

Fig. 1. Views of $ZrMe_4(dmpe)_2$ and $Zr(CH_2Ph)_4(dmpe)$.

(2,2-DIMETHYL-2-SILAPROPANE-1,3-DIYL)TETRAKIS(TRIMETHYLPHOSPHINE)OSMIUM(II)
$C_{16}H_{46}OsP_4Si$

T. BEHLING, G.S. GIROLAMI, G. WILKINSON, R.G. SOMERVILLE and M.B. HURSTHOUSE, 1984. J. Chem. Soc. Dalton, 877-881.

Monoclinic, $P2_1/n$, a = 19.690, b = 9.411, c = 27.584 Å, = 91.55°, Z = 8. Mo radiation, R = 0.039 for 6572 reflexions.

There are two crystallographically independent molecules in the asymmetric unit (Fig. 1) whose coordination geometries are equal and can be described as octahedral. The chelating ligand bonds symmetrically to Os with Os-C 2.236(9)-2.253(9) Å and the four-membered ring is essentially planar in each molecule. There are two distinct groups of Os-P bonds; those with P trans to P (2.339(9)-2.361(4)) and those with P trans to C (2.309(5)-2.324(4) Å).

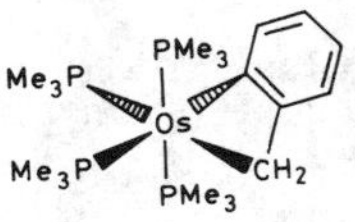

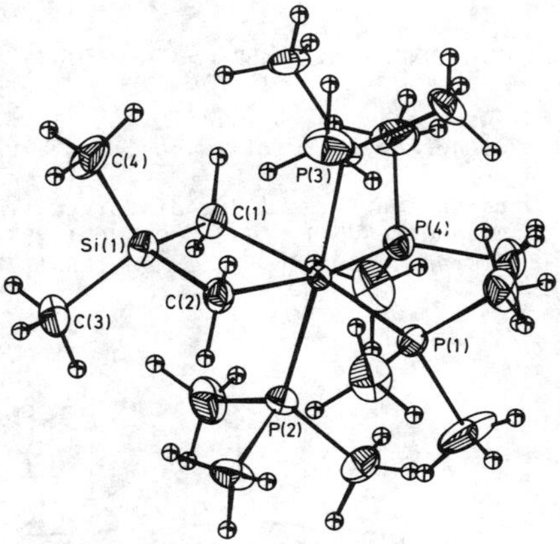

Fig. 1. $C_{16}H_{46}OsP_4Si$: the molecular skeleton and structure.

(μ-BENZYLCARBOXY)(μ-PROPYLIDENEAMIDO)BIS(TRICARBONYLIRON)(Fe-Fe)
$C_{17}H_{13}Fe_2NO_7$ $Fe_2(CO)_6(OCCH_2C_6H_5)(N{=}C(CH_3)_2)$

O. GAMBINO, G. GERVASIO, R. ROSSETTI and P.L. STANGHELLINI, 1984. Inorg. Chim. Acta, 84, 51-55.

Monoclinic, $P2_1/n$, a = 19.70, b = 11.62, c = 16.78 Å, β = 95.19°, Z = 8. Mo radiation, R = 0.056 for 2107 reflexions.

There are two independent molecules per asymmetric unit and their structure consists of two $Fe(CO)_3$ units joined by an Fe-Fe bond and bridged by the benzylcarboxy and isopropylene amido ligands (Fig. 1). The Fe-Fe bond lengths in the two molecules are 2.499 and 2.487 Å but there are no other significant differences in bond lengths. Other distances (for one molecule) are: Fe(1)-N 1.908, Fe(2)-N 1.945, Fe(1)-O(7) 2.009, Fe(2)-C(10) 1.943 Å. The N, C(7), C(8), C(9) skeleton of the propylideneamido ligand is planar with the two Fe atoms in the plane too. The benzylcarboxy ligand shows good planarity in the C(10), C(11), O(7) and Fe(2) framework but Fe(1) is out of the plane by ca. 0.1 Å.

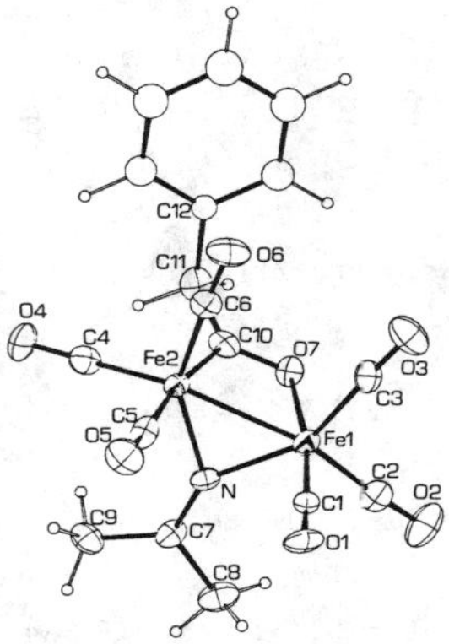

Fig. 1. Molecular structure of $C_{17}H_{13}Fe_2NO_7$.

TETRACARBONYL[(5,5-DI-t-BUTYL-2-OXO-1,3-DIOXOLAN-4-YLIDENE)ETHENYLIDENE]IRON
$C_{17}H_{18}FeO_7$

H. BERKE, U. GROSSMANN, G. HUTTNER and L. ZSOLNAI, 1984. Chem. Ber., 117, 3432-3442.

Monoclinic, $P2_1/c$, a = 11.36, b = 11.70, c = 14.82 Å, β = 108.5°, Z = 4. Mo radiation, R = 0.046 for 1933 reflexions.

The Fe coordination (Fig. 1) is distorted trigonal-bipyramidal with the carbene ligand equatorial and Fe-C5 1.803(5), C5-C6 1.255(5), C6-C7 1.315(6) Å. The Fe-C(carbonyl) distances are in the range 1.801(7)-1.821(6) Å.

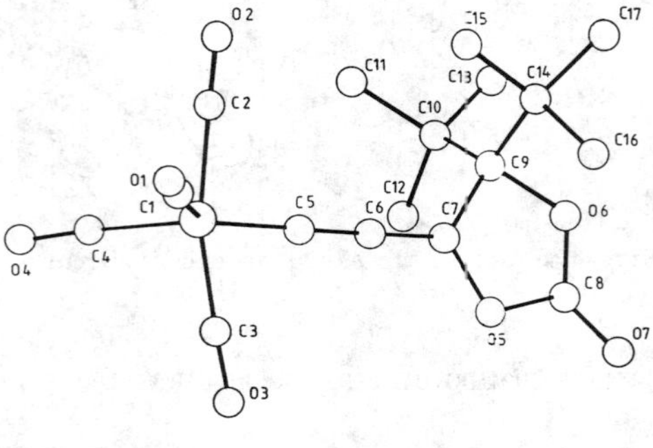

Fig. 1. The $C_{17}H_{18}FeO_7$ molecule.

[(R)-1-(METHOXYCARBONYL)ETHYL]PYRIDINEBIS(DIMETHYLGLYOXIMATO)COBALT(III) METHANOL SOLVATE
$C_{17}H_{26}CoN_5O_6,CH_4O$ (I)

[(R)-1-(METHOXYCARBONYL)ETHYL]PYRIDINEBIS(DIMETHYLGLYOXIMATO)COBALT(III)
$C_{17}H_{26}CoN_5O_6$ (II)

T. KURIHARA, A. UCHIDA, Y. OHASHI, Y. SASADA and Y. OHGO, 1984. J. Am. Chem. Soc., 106, 5718-5724.

I. Monoclinic, $P2_1$, a = 9.249, b = 20.82, c = 11.45 Å, β = 94.06°, Z = 4. Mo radiation, R = 0.053 for 3843 reflexions (at 223 K).
Monoclinic, $P2_1$, a = 9.289, b = 21.059, c = 11.499 Å, β = 94.16°, Z = 4. Mo radiation, R = 0.132 for 3422 reflexions (at 298 K).

II. Monoclinic, $P2_1$, a = 9.296, b = 20.09, c = 11.39 Å, β = 94.27°, Z = 4. Mo radiation, R = 0.104 for 3029 reflexions (at 298 K).

Crystals of (I) were studied at 223 and 298 K. The crystals lose methanol of solvation without loss of crystallinity to yield (II). When the methanol left (I), the conformation of the (methoxycarbonyl)ethyl group changed drastically to fill the void space due to loss of solvent. On warming (II) to 343 K the cell data change again, the space group becomes $P2_1/n$ and the material racemizes to give a structure which at 293 K is identical to the racemate described previously (1). Views of (I), (II) and the racemic form are in Fig. 1.

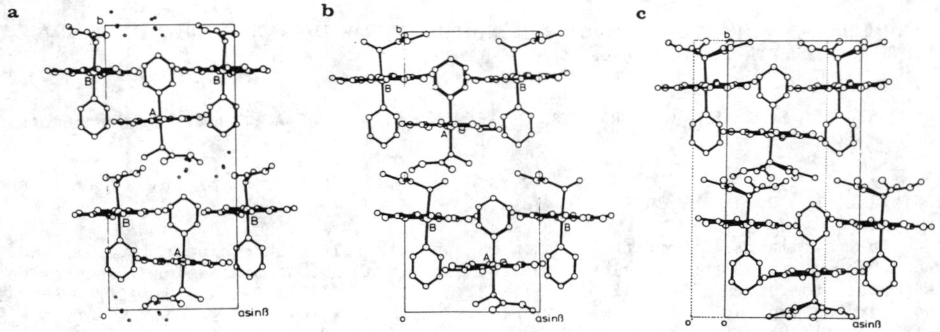

Fig. 1. Crystal structures of (a) $C_{17}H_{26}CoN_5O_6,CH_4O$ (I), (b) $C_{17}H_{26}CoN_5O_6$ (II) and (c) the racemate.

1. Structure Reports, 49B, 955.

trans-BIS(DIMETHYLGLYOXIMATO)ETHYL(METHYL-4-PYRIDINECARBOXIMIDATE)COBALT(III)
$C_{17}H_{27}CoN_6O_5$

L.G. MARZILLI, M.F. SUMMERS, J.H. RAMSDEN Jr., N. BRESCIANI-PAHOR and L. RANDACCIO, 1984. J. Chem. Soc. Dalton, 511-514.

Orthorhombic, $P2_12_12_1$, a = 12.776, b = 9.606, c = 16.884 Å, D_m = 1.47, Z = 4. Mo radiation, R = 0.039 for 1968 reflexions.

The four N atoms of the $Co(Hdmg)_2$ unit (Fig. 1) are coplanar, with the Co atom 0.05 Å above their mean plane towards N(5) of the pyridinecarboximidate ligand. The two Hdmg ligands make an interplanar angle of 9.1°. The pyridine ligand bends towards N(1), N(2) making an angle of 96.7° with the plane of the four equatorial N atoms. The ethyl group is oriented such that it lies parallel to one of the Co-N(equatorial) bonds and the $N-Co-CH_2CH_3$ fragment is characterised by N-Co-C and Co-C-C angles of

177.1(2) and 117.8(4)° respectively and by Co-N, Co-C and C-C distances of 2.081(3), 2.035(5) and 1.519(8) Å respectively. The Co-N(1-4) distances range from 1.879(4) to 1.887(4), Co-N(5) is 2.081(3) and Co-C(9) is 2.035(5) Å.

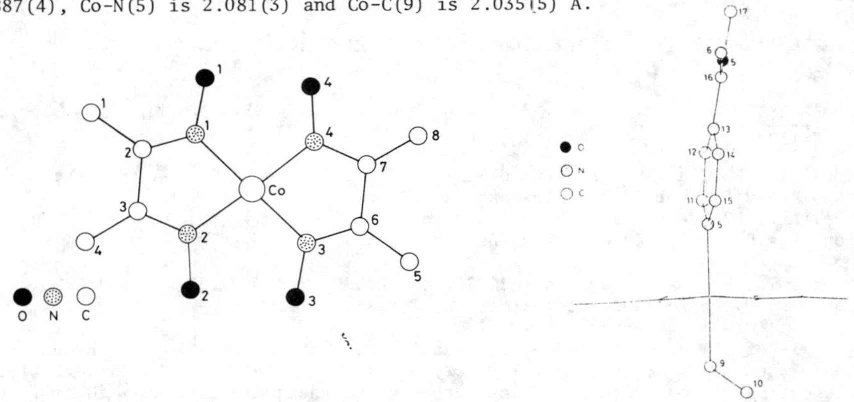

Fig. 1. $C_{17}H_{27}CoN_6O_5$: the Co(Hdmg)$_2$ unit and a side view of the molecule.

5,6-DIMETHYL-3,4-DIMETHOXYCARBONYL-2-(N,N,N',N'-TETRAMETHYLETHYLENEDIAMINE)-1-OXA-2-NICKELACYCLOHEPTA-3,5-DIEN-6-ONE
$C_{17}H_{28}N_2NiO_6$

H. HOBERG, D. SCHAEFER, G. BURKHART, C. KRÜGER and M.J. ROMÃO, 1984. J. Organometal. Chem., 266, 203-224.

Monoclinic, $P2_1/c$, a = 12.3175, b = 12.8536, c = 13.7439 Å, β = 103.715°, Z = 4. Cu radiation, R = 0.041 for 3709 reflexions.

The coordination about the Ni atom is approximately square planar (Fig. 1) and the conformation of the nickela-heptadiene ring is a tub. Coordination distances to the metal atom are: Ni-N(1) 2.015, Ni-N(2) 1.951, Ni-C(1) 1.903 and Ni-O(1) 1.867 Å. The coordination angles are: N(1)-Ni-N(2) 87.7, N(2)-Ni-C(1) 96.9, C(1)-Ni-O(1) 86.8 and O(1)-Ni-N(1) 88.7°. The distance 1.288 Å for C(5)-O(1) is typical of a carboxylate group (C(5)-O(2) is 1.239 Å). Bonds C(1)-C(2) and C(3)-C(4) have formal double bond lengths, 1.345 and 1.338 Å respectively.

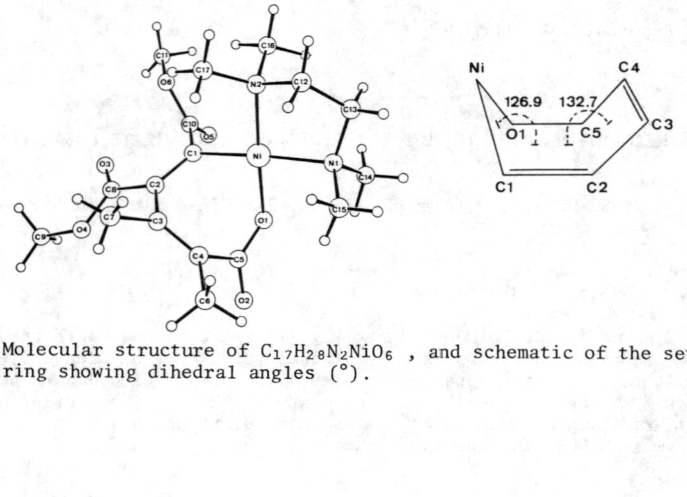

Fig. 1. Molecular structure of $C_{17}H_{28}N_2NiO_6$, and schematic of the seven-membered ring showing dihedral angles (°).

(μ_5-CARBIDO)-1,1,1,2,2,2,3,3,3,4,4,5,5,5-TETRADECACARBONYL-2,4-(μ_2-ETHOXYCARBONYL)-3,4-(μ_2-HYDRIDO)-cyclo-PENTAOSMIUM
$C_{18}H_6O_{16}Os_5$ (I)

(μ_5-CARBIDO)-1,1,1,2,2,2,3,3,3,4,4,5,5,5-TETRADECACARBONYL-3,4-(μ_2-IODO)-2,4-(μ_2-METHOXYCARBONYL)-cyclo-PENTAOSMIUM
$C_{17}H_3IO_{16}Os_5$ (II)

B.F.G. JOHNSON, J. LEWIS, W.J.H. NELSON, J.N. NICHOLLS, M.D. VARGAS, D. BRAGA, K. HENRICK and M. McPARTLIN, 1984. J. Chem. Soc. Dalton, 1809-1814.

I. Monoclinic, $P2_1/n$, a = 15.665, b = 16.761, c = 10.198 Å, β = 90.4°, Z = 4. Mo radiation, R = 0.038 for 3461 reflexions.

II. Orthorhombic, $Pna2_1$, a = 17.019, b = 10.063, c = 15.810 Å, Z = 4. Mo radiation, R = 0.049 for 2460 reflexions.

The Os core of compound I (Fig. 1) has a 'bridged-butterfly' geometry. All the CO ligands are terminal with Os-C-O angles in the range 174-179°. Os(4) is coordinated to two CO groups, the other Os atoms to three CO groups each. The equatorial Os(4) and Os(5) groups are bridged by a μ-CO_2Et group (Os(4)...Os(5) 3.604 Å). It is inferred that the H ligand bridges atoms Os(1) and Os(4). The Os geometry in II is best described as a distorted trigonal bipyramid with three very long, non-bonded Os...Os distances in the equatorial plane. Coordination of Os atoms to CO groups follows the same pattern as in I as does the coordination of the μ-Co_2Me group.

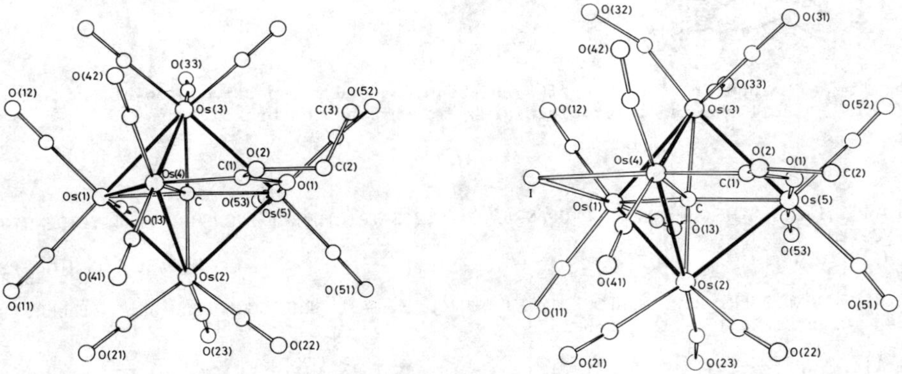

Fig. 1. The molecular structure of $C_{18}H_6O_{16}Os_5$ (left) and of $C_{17}H_3IO_{16}Os_5$ (right).

TETRACARBONYLBIS(η^5-2,4-CYCLOPENTADIEN-1-YL)[μ-[1,2-ETHANEDIYLBIS(IMINOCARBONYL)]]-DIIRON
$C_{18}H_{16}Fe_2N_2O_6$ (μ-$CONH(CH_2)_2NHCO)[CpFe(CO)_2]_2$

A.J. LINDSAY, S. KIM, R.A. JACOBSON and R.J. ANGELICI, 1984. Organometallics, 3, 1523-1530.

Triclinic, $P\bar{1}$, a = 9.504, b = 8.435, c = 12.989 Å, α = 101.20, β = 104.80, γ = 100.49°, D_m = 1.64, Z = 2. Mo radiation, R = 0.043 for 1485 reflexions.

The unit cell comprises two crystallographically-centrosymmetric independent molecules. The dimensions and conformations of the two independent half-molecules do not differ significantly from each other. Each Fe atom is coordinated to a planar cyclopentadienyl ring, two CO ligands and a planar carbamoyl moiety (Fig. 1). The overall geometry shows the metal centres to be in a transoid orientation about the bridging ligand as dictated by the inversion centre, and the central part of the structure adopts a gauche-trans-gauche conformation. The mean Fe-C(Cp) ring distance is 2.10 Å and the average ring C-C distance is 1.403 Å. The Fe-C (carbamoyl) distance (1.991 Å) is some 0.24 Å longer than the terminal Fe-CO distances. The bonds C(3)-O(3), C(3)-N, N-C(4) and C(4)-C(4') are respectively 1.228, 1.346, 1.466 and 1.522 Å in length. Weak N-H...O hydrogen bonds (N-H(1)...O(3) where N-H...O is 158° and N...O 2.894 Å) result in an extended sheet of molecules along the b axis.

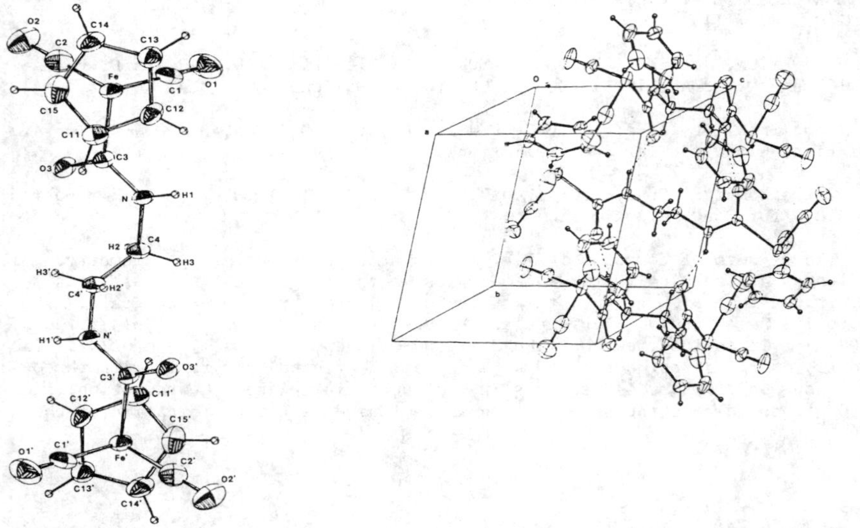

Fig. 1. Molecular structure of $C_{18}H_{16}Fe_2N_2O_6$, and perspective view of the unit cell showing intermolecular hydrogen bonds as broken lines.

TRIS(1,1,1,3,3,3-HEXAFLUORO-2-PROPANOLATO)(1,2,3-TRIETHYL-1-PROPEN-1-YL-3-YLIDENE)-TUNGSTEN

$C_{18}H_{18}F_{18}O_3W$ $W(C_3Et_3)[OCH(CF_3)_2]_3$

J.H. FREUDENBERGER, R.R. SCHROCK, M.R. CHURCHILL, A.L. RHEINGOLD and J.W. ZILLER, 1984. Organometallics, 3, 1563-1573.

Triclinic, P$\bar{1}$, a = 9.949, b = 16.419, c = 18.485 Å, α = 112.81, β = 93.53, γ = 98.60°, Z = 4. Mo radiation, R = 0.050 for 4540 reflexions.

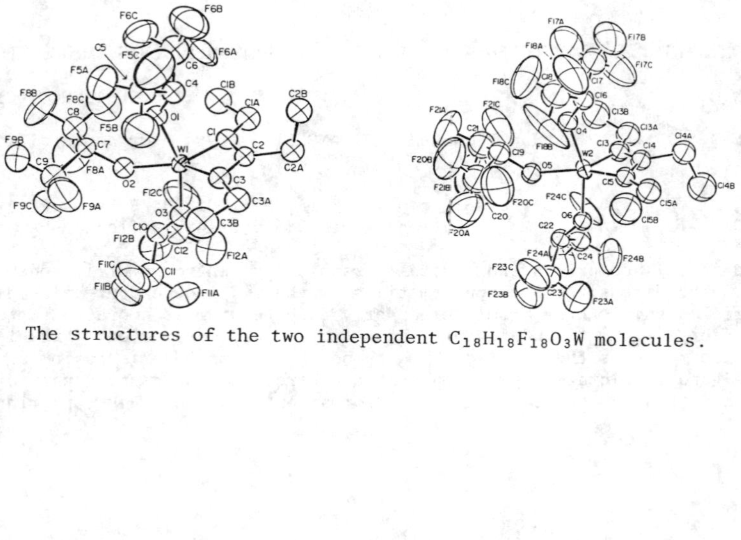

Fig. 1. The structures of the two independent $C_{18}H_{18}F_{18}O_3W$ molecules.

Two independent molecules define the crystallographic asymmetric unit. Each is a trigonal bipyramid containing a planar WC_3 ring system in the equatorial plane (Fig. 1). The two molecules differ primarily in the orientation of the β-ethyl group. In one molecule the W-C_α bond lengths are 1.90 and 1.86 Å with the W...C_β distance 2.09 Å and C_α-C_β 1.43 and 1.44 Å. Axial and equatorial alkoxide ligands show similar W-O-C angles of 129.4, 133.6 and 138.6° and W-O distances of 1.98, 1.96 and 1.93 Å, consistent with a poor π-electron-donating ability by the hexafluoro-isopropoxide ligand.

BIS(η^5-CYCLOPENTADIENYL)(o-PHENYLENEDIMETHYLENE-C,C)TITANIUM(IV)
$C_{18}H_{18}Ti$ (I)

BIS(η^5-CYCLOPENTADIENYL)(o-PHENYLENEDIMETHYLENE-C,C)HAFNIUM(IV)
$C_{18}H_{18}Hf$ (II)

BIS(η^5-CYCLOPENTADIENYL)(o-PHENYLENEDIMETHYLENE-C,C)ZIRCONIUM(IV)
$C_{18}H_{18}Zr$ (III)

BIS(η^5-TRIMETHYLSILYLCYCLOPENTADIENYL)(o-PHENYLENEDIMETHYLENE-C,C)NIOBIUM(IV)
$C_{24}H_{34}NbSi_2$ (IV)

G.S. BRISTOW, M.F. LAPPERT, T.R. MARTIN, J.L. ATWOOD and W.F. HUNTER, 1984. J. Chem. Soc. Dalton, 399-413.

I. Orthorhombic, $P2_12_12_1$, a = 8.158, b = 10.992, c = 15.713 Å, Z = 4. Mo radiation, R = 0.026 for 963 reflexions.

II. Orthorhombic, $P2_12_12_1$, a = 8.251, b = 11.070, c = 15.748 Å, Z = 4. Mo radiation, R = 0.024 for 1189 reflexions.

III. Orthorhombic, $P2_12_12_1$, a = 8.276, b = 11.075, c = 15.721 Å, Z = 4. Mo radiation, R = 0.019 for 970 reflexions.

IV. Triclinic, $P\bar{1}$, a = 12.371, b = 14.249, c = 14.514 Å, α = 78.51, β = 89.47, γ = 73.67°, Z = 4. Mo radiation, R = 0.027 for 4667 reflexions.

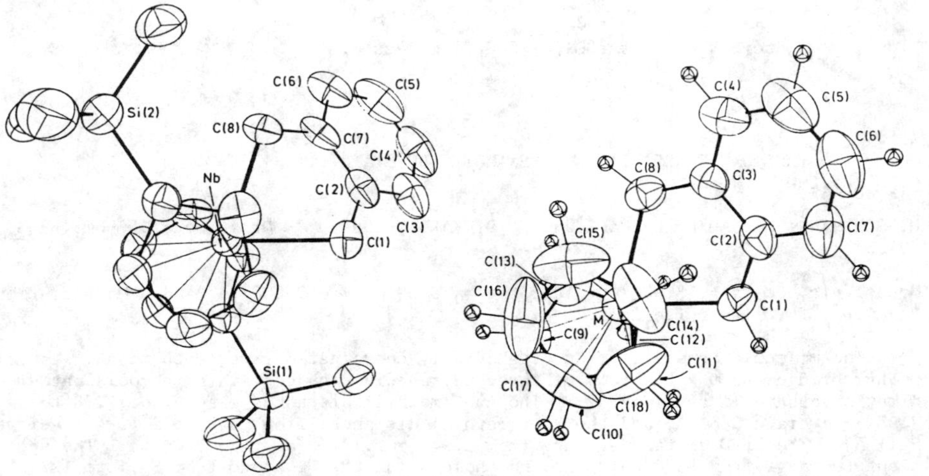

Fig. 1. The molecular structures: $C_{24}H_{34}NbSi_2$ (left) and $C_{18}H_{18}M$ (M = Ti, Hf, Zr) (right).

Compounds I, II and III are isostructural and do not possess crystallographically imposed symmetry (Fig. 1). The M-C(sp^3) distances are 2.203(6), 2.28(1) and 2.300(8) Å for Ti, Hf and Zr respectively. Likewise the average M-C(η) distances are 2.382(18), 2.51(1) and 2.520(15) Å. The 'bite angles' in the molecules, C(1)-M-C(8), are 76.7(2), 77.2(4) and 77.3(2)° for Ti, Hf and Zr respectively. The fold angle ϕ, the angle between the normal to the plane of M, C(1), C(8) and the normal to the four C atoms of the metallocyclopentene ring, C(1),C(2),C(3),C(8), is 41° in I, 53° in II and 53° in III. The C_6 rings are planar and the α-CH_2's lie respectively 0.10, 0.12 and 0.12 Å out of this plane on the same side as the M atom in I, II and III. In the Nb complex IV Nb-C(sp^3) is 2.286(9) Å, C(1)-Nb-C(8) is 72.4(7)° and ϕ is 44°.

BIS(DIMETHYLGLYOXIMATO)[(S)-1-METHOXYCARBONYLETHYL](4-PYRIDINECARBONITRILE)COBALT(III)
$C_{18}H_{25}CoN_6O_6$

T. KURIHARA, A. UCHIDA, Y. OHASHI, Y. SASADA and Y. OHGO, 1984. Acta Cryst., C40, 1557-1559.

Orthorhombic, $P2_12_12_1$, a = 14.575, b = 16.174, c = 9.268 Å, Z = 4. Mo radiation, R = 0.058 for 1938 reflexions.

The C(16) methyl group (Fig. 1) is cis with respect to O(5). The C(17)-C(15)-Co-N(4) and N(4)-Co-N(5)-C(9) torsion angles are -13.4(6) and 48.9(7)°. The Co-N distances are in the range 1.886-2.076(6) with Co-C 2.098(7) Å. The fact that the crystal is gradually decomposed by X-rays but only at 353K may be due to the mce groups being isolated from each other.

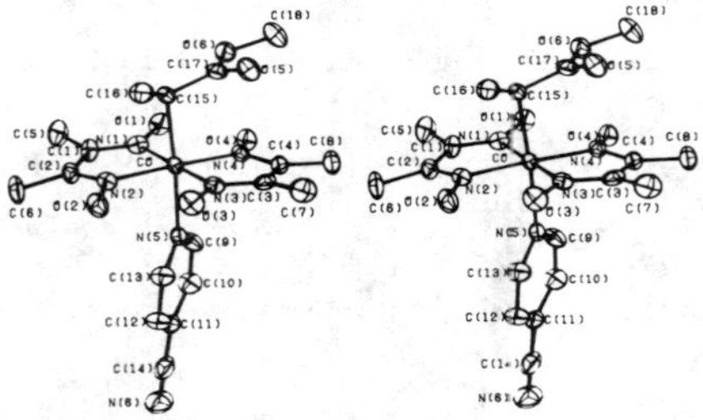

Fig. 1. Stereoview of the $C_{18}H_{25}CoN_6O_6$ molecule.

TRIS[(1,2-DIMETHOXYETHANE)LITHIUM] HEXAMETHYLLUTETATE(III)
$C_{18}H_{48}Li_3LuO_6$

H. SCHUMANN, H. LAUKE, E. HAHN and J. PICKARDT, 1984. J. Organometal. Chem., 263, 29-35.

Monoclinic, Cc, a = 13.90, b = 10.83, c = 18.39 Å, β = 94.08°, Z = 4. Mo radiation, R = 0.045 for 1124 reflexions.

The molecule consists of an octahedrally-coordinated rare earth metal, with methyl bridges to the Li atoms of three tetrahedra each containing a coordinated dimethoxyethane ligand (Fig. 1). The Lu-C(methyl) distances range from 2.48 to 2.57 Å (average 2.53 Å) and the Li-C(methyl) distances from 2.02 to 2.19 Å (average 2.11 Å). The Li-O distances range from 1.94 to 2.14 Å (average 2.04 Å). The angles about Lu range from 83.0 to 96.2° and those about the Li atoms from 81.1 to 122.9° (average 90.1 and 108.6° respectively).

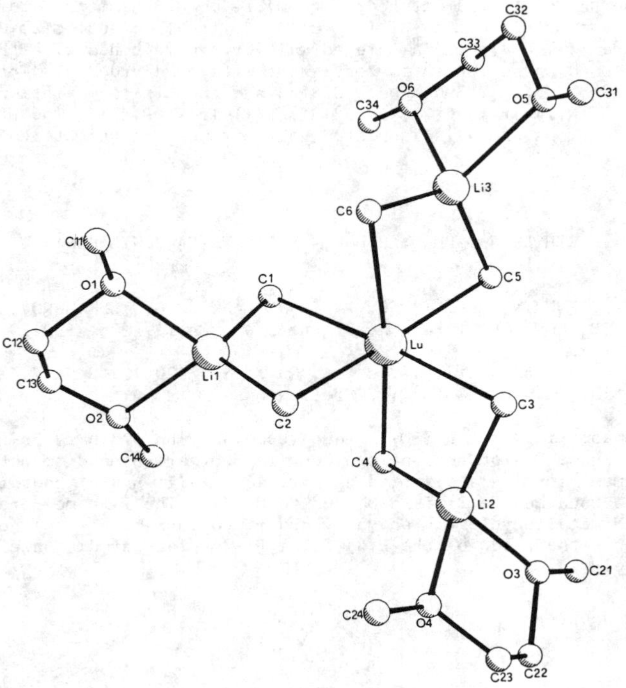

Fig. 1. Molecular structure of $C_{18}H_{48}Li_3LuO_6$.

2-METHYL-4,4,4,4-TETRAKIS(TRIMETHYLPHOSPHINE)-1-OXA-3-AZA-4-IRIDACYCLOPENT-2-ENETRIMETHYLPHOSPHONIUM HEXAFLUOROPHOSPHATE BIS(TETRAFLUOROBORATE)

$C_{18}H_{52}B_2F_{14}IrNOP_6$ $[Ir(CH_2OC(CH_3)=NH)(PMe_3)_4]^{2+}[HPMe_3]^+[PF_6^-]2[BF_4^-]$

D.L. THORN and J.C. CALABRESE, 1984. J. Organometal. Chem., 272, 283-293.

Monoclinic, $P2_1/c$, a = 13.772, b = 13.436, c = 19.506 Å, β = 90.02°, Z = 4. Mo radiation, R = 0.079 for 5312 reflexions, (-100°C).

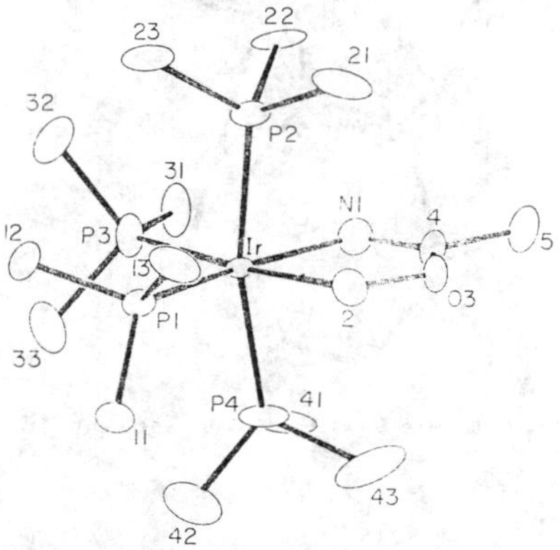

Fig. 1. Perspective drawing of the organometallic dication in $C_{18}H_{52}B_2F_{14}IrNOP_6$.

The dication has distorted octahedral geometry about the metal atom. Bond distances within the metallacycle, of Ir-C(2) 2.14, Ir-N(1) 2.08, C(2)-O(3) 1.44, C(4)-O(3) 1.33 and N(1)-C(4) 1.27 Å, are normal for single bonds with the exception of the N(1)-C(4) bond whose length is consistent with bond order 2. The Ir-P distances are 2.304, 2.384, 2.390, and 2.398 Å; and the cis coordination angles about Ir range from 76.3 (N(1)-Ir(1)-C(2)) to 98.3° (P(1)-Ir(1)-P(4)). The $HPMe_3^+$ cation, the PF_6^- and BF_4^- anions are all disordered at their respective lattice sites.

(μ_2-σ^3,η^2-PHENYLAZOMETHYLENE-o-PHENYLENE-C,N,N)BIS(TRICARBONYLIRON)

$C_{19}H_{11}Fe_2NO_6$ $Fe_2[\mu_2\text{-}\eta^2\text{-o-}C_6H_4CH_2NC_6H_5](CO)_6$

N.S. NAMETKIN, V.D. TYURIN, V.V. TRUSOV, A.I. NEKHAEV, G.G. ALEXANDROV, N.A. PARPIEV, M.T. TASHEV and H.B. DOUSTOV, 1984. J. Organometal. Chem., 276, 199-204.

Triclinic, $P\bar{1}$, a = 15.866, b = 7.614, c = 8.405 Å, α = 100.13, β = 97.42, γ = 106.55°, Z = 2. Mo radiation, R = 0.058 for 1970 reflexions.

The complex comprises two $Fe(CO)_3$ groups bonded by an ordinary Fe-Fe bond and a bridging ligand, whose N atom and one benzene fragment are bonded to both metal atoms, the benzene fragment forming a σ,π-bridge (Fig. 1). Fe(1) has η^2 coordination to the C(8)-C(13) double bond and Fe(2) is σ-bonded to C(13). The five-membered Fe(2), C(13),C(8),C(7),N metallocycle has an "easy chair" conformation with the N and C(7) atoms situated on either side of the C(8),C(13),Fe(2) plane at distances of 0.46 and 0.25 Å respectively.

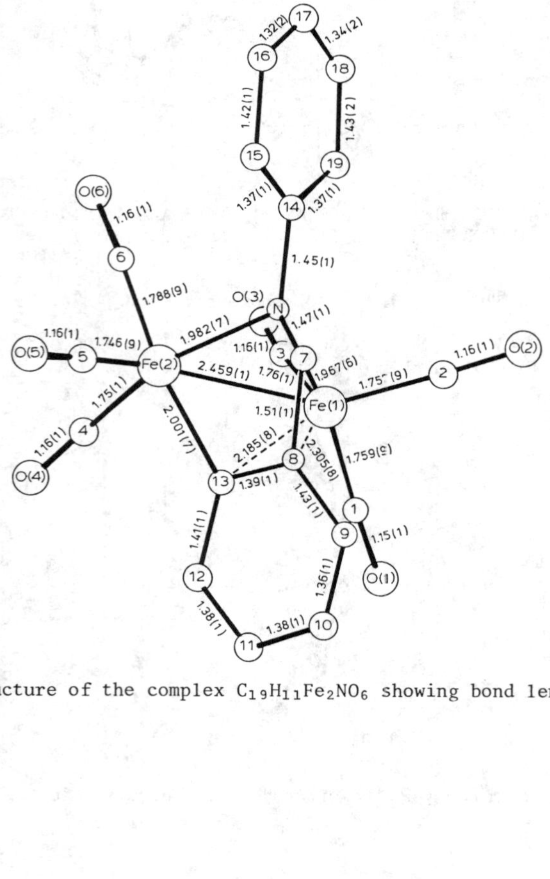

Fig. 1. Structure of the complex $C_{19}H_{11}Fe_2NO_6$ showing bond lengths (Å).

CARBONYL(η^5-CYCLOPENTADIENYL)(DIPHENYLCARBENE)NITROSYLCHROMIUM
$C_{19}H_{15}CrNO_2$ (I) $(\eta^5-C_5H_5)Cr(CO)(NO)[C(C_6H_5)_2]$

DICARBONYL(η^5-CYCLOPENTADIENYL)(DIPHENYLCARBENE)MANGANESE
$C_{20}H_{15}MnO_2$ (II) $(\eta^5-C_5H_5)Mn(CO)_2[C(C_6H_5)_2]$

W.A. HERRMANN, J.L. HUBBARD, I. BERNAL, J.D. KORP, B.L. HAYMORE and G.L. HILLHOUSE, 1984. Inorg. Chem., 23, 2978-2983.

I. Monoclinic, $P2_1/n$, a = 11.090, b = 10.579, c = 13.804 Å, β = 92.43°, D_m = 1.40, Z = 4. Mo radiation, R = 0.042 for 740 reflexions.

II. Monoclinic, $P2_1/c$, a = 12.882, b = 7.380, c = 17.243 Å, β = 106.95°, D_m = 1.41 (at 25°C), Z = 4. Mo radiation, R = 0.033 for 2937 reflexions (at -160°C).

The two molecules are shown in Fig. 1. In I the carbene plane is nearly coplanar with the N-Cr-C(carbene) fragment and nearly perpendicular to the C(carbonyl)-Cr-C(carbene) plane. The Cr-C(carbene) bond is short at 1.912(8) Å, other distances in I being Cr-N 1.672(8), Cr-C(Cp) 2.225(5)-2.255(6), Cr-CO 1.885(9) Å. The structure of II sharply contrasts with that of I in that the carbene plane is coincident with the mirror plane of $CpMn(CO)_2$. Bond lengths for II are: Mn-C(carbene) 1.885(2), Mn-C(Cp) 2.153(2)-2.182(2), and Mn-CO 1.793(2) and 1.783(2) Å.

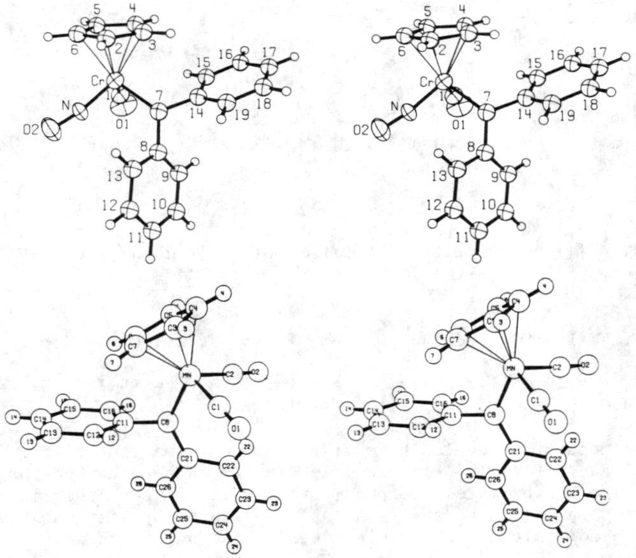

Fig. 1. Stereoviews of $(\eta^5-C_5H_5)Cr(CO)(NO)[C(C_6H_5)_2]$ (top) and $(\eta^5-C_5H_5)Mn(CO)_2[C(C_6H_5)_2]$ (bottom).

1,1,1,2,2-PENTACARBONYL-1,2-BIS(η^5-CYCLOPENTADIENYL)-2-(1'-OXACYCLOPENT-2'-YLIDENE)-MOLYBDENUMTUNGSTEN
$C_{19}H_{16}MoO_6W$ (I)

(μ_2-TETRAMETHYLENE)BIS(TRICARBONYL(η^5-CYCLOPENTADIENYL)MOLYBDENUM)
$C_{20}H_{18}Mo_2O_6$ (II)

H. ADAMS, N.A. BAILEY and M.J. WINTER, 1984. J. Chem. Soc. Dalton, 273-278.

I. Monoclinic, $P2_1$, a = 8.017, b = 15.026, c = 7.981 Å, β = 92.88°, D_m = 2.10, Z = 2. Mo radiation, R = 0.065 for 1256 reflexions.

II. Triclinic, $P\bar{1}$, a = 6.978, b = 7.005, c = 11.315 Å, α = 102.38, β = 95.81, γ = 105.91°, D_m = 1.76, Z = 1. Mo radiation, R = 0.026 for 1646 reflexions.

The structure of I (Fig. 1) comprises two fragments each involving a metal, cyclopentadienyl and three unidentate ligands joined by a long Mo-W bond. The coordination polyhedra which are square-based pyramidal, are mutually staggered. The molecule as a whole has approximate C_s symmetry. The Mo-W bond is 3.239(4) Å and the cyclic carbene ligand is bonded to W rather than Mo and is trans to the Mo-W bond. Molecular dimensions are of low precision. II possesses a crystallographically imposed centre of symmetry at the midpoint of the tetramethylene chain which links the Mo atoms. Each Mo coordination polyhedron is square-based pyramidal with the centroid of the Cp ligand occupying the axial site. Some bond distances are: Mo(1)-C(9) 2.356(4), Mo-C(CO) 1.977(5)-1.996(5), Mo-C(Cp) 2.314(7)-2.359(5), C(9)-C(10) 1.531(6) and C(10)-C(10') 1.533(6) Å.

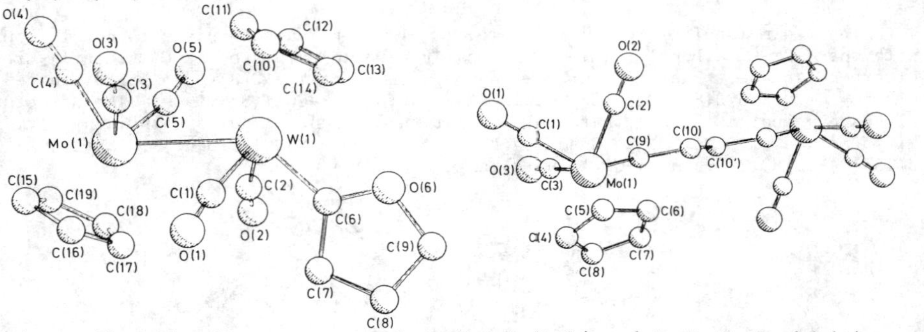

Fig. 1. The molecular structure: $C_{19}H_{16}MoO_6W$ (left) and $C_{20}H_{18}Mo_2O_6$ (right).

cis-TETRACARBONYL-{[((Z)-η^2-N,2,2-TRIMETHYL-3-BUTENYL)AMINO](p-TOLYL)CARBENE}-TUNGSTEN(0)
$C_{19}H_{21}NO_4W$

C.P. CASEY, N.W. VOLLENDORF and K.J. HALLER, 1984. J. Am. Chem. Soc., 106, 3754-3764.

Monoclinic, $P2_1$, a = 10.287, b = 11.651, c = 16.319 Å, β = 98.83°, Z = 4. Mo radiation, R = 0.039 for 4295 reflexions (at 183 K).

The two independent molecules have different conformations of the chelating carbene-alkene ligand (Fig. 1). In one molecule the carbene carbon-tungsten bond and alkene C=C bond are approximately perpendicular, while in the other molecule these bonds are almost parallel. The analysis was complicated by some substitution (30°) of the molecule A for molecule B and vice-versa and bond length data are as anticipated.

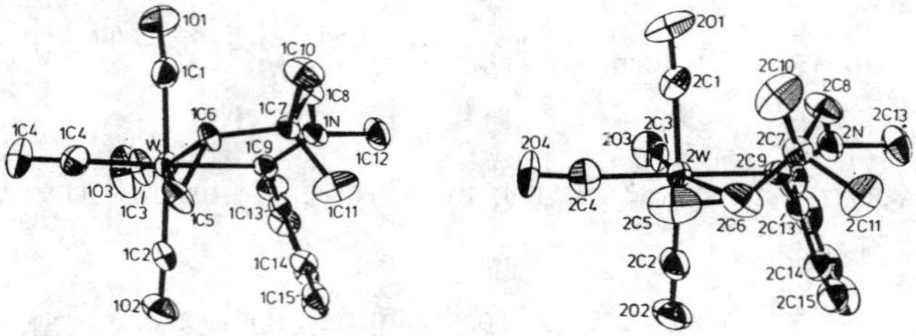

Fig. 1. Views of the two $C_{19}H_{21}NO_4W$ molecules.

BROMO(η[5]-PENTAMETHYLCYCLOPENTADIENYL)(PHENYL)(TRIMETHYLPHOSPHINE)RHODIUM
$C_{19}H_{29}BrPRh$ $(C_5Me_5)Rh(PMe_3)(C_6H_5)Br$

W.D. JONES and F.J. FEHER, 1984. Inorg. Chem., 23, 2376-2388.

Monoclinic, $P2_1/c$, a = 11.459, b = 9.008, c = 19.298 Å, β = 90.03°, Z = 4. Mo radiation, R = 0.047 for 2009 reflexions.

The molecule (Fig. 1) has a three-legged piano stool geometry with Rh-Br 2.533(2), Rh-P 2.458(3), and Rh-C(11) 2.054(9) Å.

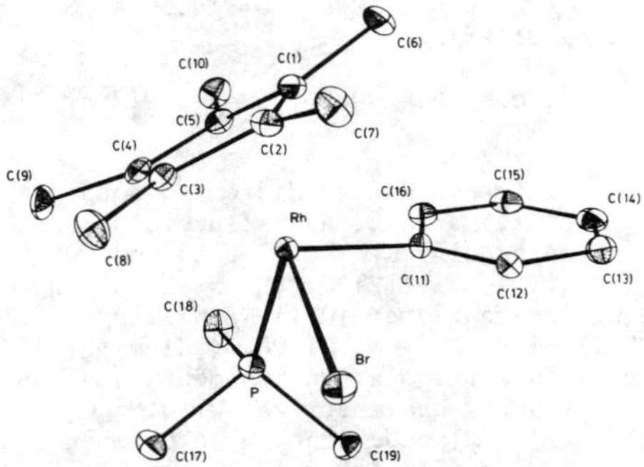

Fig. 1. The $(C_5Me_5)Rh(PMe_3)(C_6H_5)Br$ molecule.

TRI-t-BUTOXYBENZYLIDYNETUNGSTEN
$C_{19}H_{32}O_3W$ $(Me_3CO)_3WCPh$

F.A. COTTON, W. SCHWOTZER and E.S. SHAMSHOUM, 1984. Organometallics, 3, 1770-1771.

Monoclinic, $P2_1/n$, a = 10.011, b = 21.041, c = 10.064 Å, β = 94.66°, Z = 4. Mo radiation, R = 0.023 for 2917 reflexions, at -107° C.

The coordination geometry of the W atom is almost ideally tetrahedral with all ligand-W-ligand bond angles in the range 107-111° (Fig. 1). The average W-O distance is 1.865 Å, and the W-O-C bond angles range from 139 to 143°. The W≡C(1) bond length is 1.758 Å and the W≡C(1)-C(2) bond angle is 175.8°.

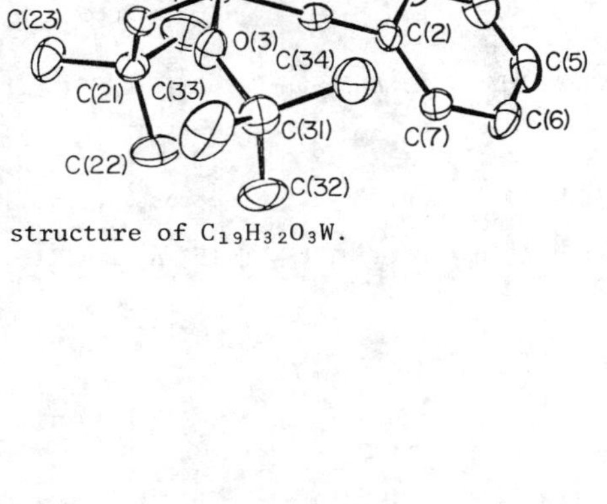

Fig. 1. Molecular structure of $C_{19}H_{32}O_3W$.

CHLORO[3-OXO-1-PHENYL-4-(TRIMETHYLSILYL)-1-BUTENYL-C,O]BIS(TRIMETHYLPHOSPHINE)NICKEL
$C_{19}H_{35}ClNiOP_2Si$ (I)

CHLORO[TRIMETHYL[(1,2-η)-5-METHYL-3-OXO-1,5-DIPHENYL-1-HEXENYL]PHOSPHONIUM](TRIMETHYL-PHOSPHINE)NICKEL
$C_{25}H_{37}ClNiOP_2$ (II)

E. CARMONA, E. GUTTIÉRREZ-PUEBLA, A. MONGE, J.M. MARÍN, M. PANEQUE and M.L. POVEDA, 1984. Organometallics, 3, 1438-1440.

I. Monoclinic, $P2_1/n$, a = 12.532, b = 14.099, c = 14.36 Å, β = 98.51°, Z = 4. Mo radiation, R = 0.056 for 1958 reflexions.

II. Orthorhombic, Pbca, a = 20.638, b = 20.09, c = 12.913 Å, Z = 8. Mo radiation, R = 0.056 for 1277 reflexions.

In I the Ni atom is in a five-coordinate distorted square pyramidal environment (Fig. 1) with the keto vinylic O(1) atom at a distance of 2.535 Å. This Ni...O interaction lengthens the corresponding C(1)-O(1) bond to 1.25 Å. Other bond lengths are: Ni-C(4) 1.896, Ni-P(1) 2.191, Ni-P(2) 2.189, Ni-Cl 2.299, C(1)-C(3) 1.46, C(3)-C(4) 1.34 Å. Bond angles of interest are: O(1)-Ni-Cl 105.6, O(1)-Ni-C(4) 76.9, Ni-C(4)-C(3) 118.8, C(4)-Ni-P(2) 89.5, P(2)-Ni-Cl 90.3, Cl-Ni-P(1) 89.5 and P(1)-Ni-C(4) 90.5°. In II, the Ni atom is in a trigonal environment, and the bonding in the nickelacyclopropane unit is slightly unsymmetrical with Ni-C(2) 1.95 Å and Ni-C(1) 1.97 Å. The α twist of the C(1)-C(2) bond (the angle between the C(1)-C(2) vector and the Ni,Cl,P(1) plane) is 8.5°. The ylidic P(2)-C(2) bond (1.77 Å) is shorter than the average P(2)-Me bond length (1.81 Å). Other bond distances are: Ni-Cl 2.266, Ni-P(1) 2.188, C(1)-C(2) 1.44 Å. Angles of interest are: C(1)-Ni-C(2) 43.2, C(1)-Ni-P(1) 117.4, P(1)-Ni-C(1) 93.8 and Cl-Ni-C(2) 105.7°.

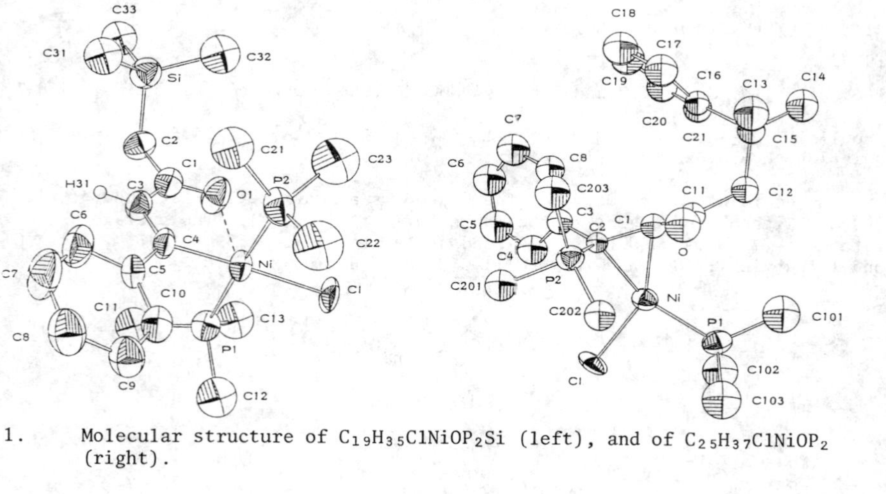

Fig. 1. Molecular structure of $C_{19}H_{35}ClNiOP_2Si$ (left), and of $C_{25}H_{37}ClNiOP_2$ (right).

μ_3-[1η:1-2η-BIS(TRIFLUOROMETHYL)VINYLENE]-μ_3-CARBONYL-TRIS(η-CYCLOPENTADIENYL)-triangulo-TRICOBALT
$C_{20}H_{15}Co_3F_6O$ $OC(CoC_5H_5)_3(CF_3CCCF_3)$

J.M. GROMEK and J. DONAHUE, 1984. J. Cryst. Spectrosc. Res., 14, 227-237.

Monoclinic, $P2_1/n$, a = 9.171, b = 25.176, c = 9.192 Å, β = 117.08°, Z = 4. Mo radiation, R = 0.057 for 1784 reflexions.

A view of the molecule (with average dimensions) is shown in Fig. 1. The bondings of the bis(trifluoromethyl)vinylene group to one side of the Co_3 triangle and of the carbonyl group to the other side are decidedly asymmetric.

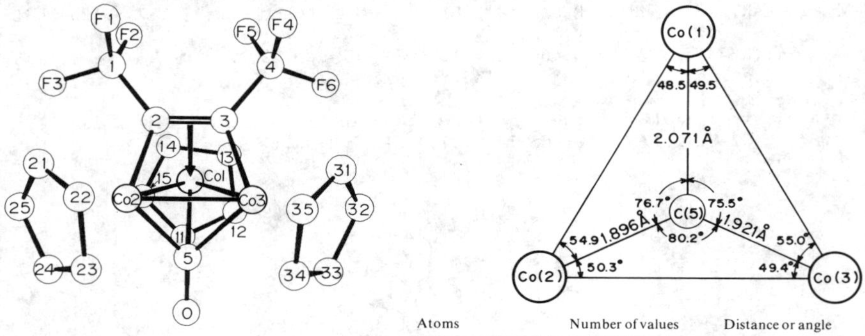

Atoms	Number of values	Distance or angle
Co–Co	3	2.456(9) Å
Co–C (of C_5H_5)	15	2.064(17)
C–C (in C_5H_5)	15	1.370(38)
C–F	6	1.318(18)
C–C–F	6	113.5(16)°
F–C–F	6	105.2(20)
C–C–C (in C_5H_5)	15	108.0(22)

Fig. 1. The $C_{20}H_{15}CO_3F_6O$ molecule and average dimensions.

2,2-DICARBONYL(η^5-CYCLOPENTADIENYL)-1,1-DIPHENYL-1-PHOSPHA-2-MOLYBDACYCLOPROPANE $C_{20}H_{17}MoO_2P$

E. LINDNER, E.U. KUSTER, W. HILLER and R. FAWZI, 1984. Chem. Ber., 117, 127-141.

Monoclinic, $P2_1/c$, a = 7.944, b = 16.388, c = 14.239 Å, β = 94.99°, Z = 4. Mo radiation, R = 0.032 for 3468 reflexions.

In the molecule (Fig. 1) the P-CH_2 distance is rather short (1.743(4) Å), and Mo-P and Mo-C are 2.364(1) and 2.311(4) Å respectively. The Mo atom is 2.03 Å from the cyclopentadienyl ring plane.

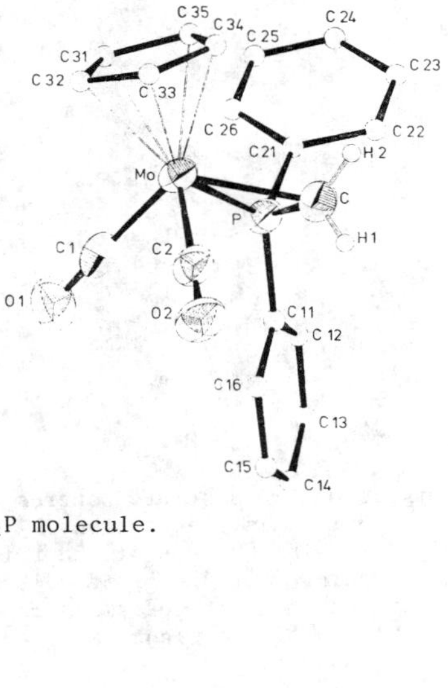

Fig. 1. The $C_{20}H_{17}MoO_2P$ molecule.

(μ_2-CARBONYL)(μ_2-DI-t-BUTYLALLENYLIDENE)BIS(TETRACARBONYLMANGANESE)
$C_{20}H_{18}Mn_2O_9$

H. BERKE, P. HARTER, G. HUTTNER and L. ZSOLNAI, 1984. Chem. Ber., 117, 3423-3431.

Triclinic, $P\bar{1}$, a = 8.181, b = 8.466, c = 19.40 Å, α = 89.29, β = 97.98, γ = 117.61°, Z = 2. Mo radiation, R = 0.0945 for 2148 reflexions (at 233 K).

The analysis establishes the structure shown in Fig. 1 with Mn-Mn 2.739(2), Mn-C1 2.06(1), C1-C2 1.27(2), C2-C3 1.39(2), Mn-C15 2.08(1) Å.

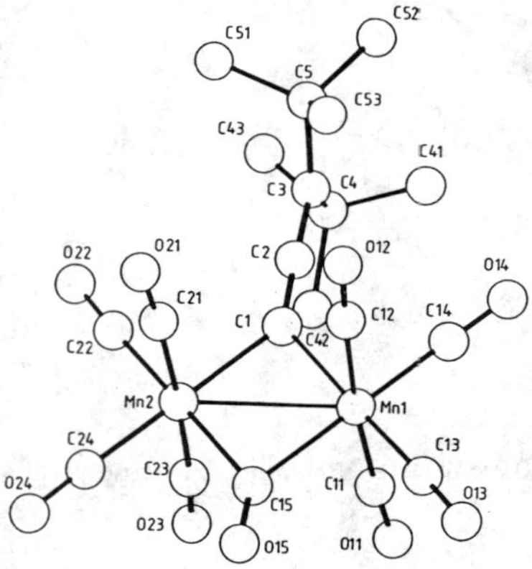

Fig. 1. A view of $C_{20}H_{18}Mn_2O_9$.

BIS(DIMETHYLPHENYLPHOSPHINE)BIS(TRICHLOROETHENYL)NICKEL
$C_{20}H_{22}Cl_6NiP_2$ $[Ni(C_2Cl_3)_2(PMe_2Ph)_2]$

J. CARVAJAL, G. MULLER, J. SALES, X. SOLANS and C. MIRAVITLLES, 1984. Organometallics, 3, 996-1002.

Monoclinic, $P2_1/c$, a = 9.203, b = 16.005, c = 9.604 Å, β = 114.79°, Z = 2. Mo radiation, R = 0.072 for 1535 reflexions.

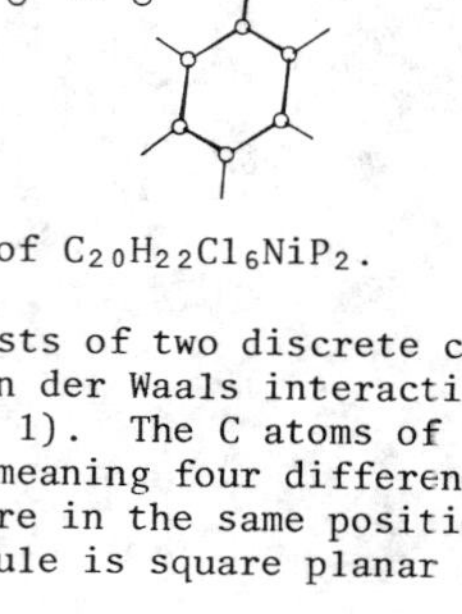

Fig. 1. Molecular structure of $C_{20}H_{22}Cl_6NiP_2$.

The crystal structure consists of two discrete centrosymmetric molecules per unit cell separated by normal van der Waals interactions. The syn and anti isomers are present in the lattice (Fig. 1). The C atoms of the trichlorovinyl group can be located in two different sites, meaning four different molecular distributions, two for each isomer. The Cl atoms are in the same positions for both isomers. The inner coordination sphere of the molecule is square planar in the anti isomer but is distorted square planar in the syn isomer. The Ni-P bond distance is 2.201 Å and the Ni-C(9) and Ni-C(9') distances are 1.91 and 1.93 Å respectively. Other distances are: P-C(1) 1.808, P-C(2) 1.821, P-C(3) 1.843, C(9)-C(10) 1.37, C(9')-C(10') 1.28, C(9)-Cl(12) 1.846, C(9')-Cl(11) 1.855, C(10)-Cl(11) 1.698, C(10)-Cl(13) 1.690, C(10')-Cl(12) 1.62 and C(10')-Cl(13) 1.85 Å

BIS[(μ_2-CHLORO)(9-O-NITROTRICYCLO[5.2.1.0^{2,6}]DECAN-3-EN-8-YL-C,N)PALLADIUM(II)] DICHLOROMETHANE SOLVATE

$C_{20}H_{24}Cl_2N_2O_4Pd_2,CH_2Cl_2$

M.A. ANDREWS, T.C.-T. CHANG, CHI-WEN F. CHENG, T.J. EMGE, K.P. KELLY and T.F. KOETZLE, 1984. J. Am. Chem. Soc., 106, 5913-5920.

Orthorhombic, Pbca, a = 24.717, b = 9.733, c = 20.655 Å, Dm = 1.95, Z = 8. Mo radiation, R = 0.062 for 1882 reflexions (at 120 K).

The structure is established as shown in Fig. 1. The molecule is nearly centrosymmetric although not required to be so by space group symmetry. The coordination geometry about the Pd atoms is nearly square planar. The Pd-Cl bonds trans to C (2.499 and 2.491(4) Å) are longer than those trans to N (2.331 and 2.328(4) Å). The Pd-N distances (1.94 and 1.95(1) Å) are very short. The crystal packing is characterised by a two-dimensional array of weakly interacting columns of metallacycles parallel to bc diagonals.

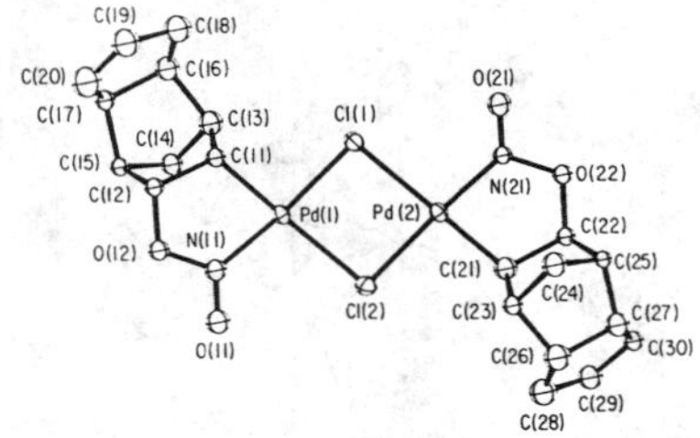

Fig. 1. The molecular structure of $C_{20}H_{24}Cl_2N_2O_4Pd_2$.

BIS[(μ_2-CHLORO)(3-(TRICHLOROTIN)HEXAFLUOROBUT-2-ENYL)TRIETHYLPHOSPHINEPLATINUM]

$C_{20}H_{30}Cl_8F_{12}P_2Pt_2Sn_2$ $[ClPt(PEt_3)C(CF_3)=C(CF_3)SnCl_3]_2$

H.C. CLARK, G. FERGUSON, A.B. GOEL and B.L. RUHL, 1984. Organometallics, 3, 15-17.

Triclinic, P$\bar{1}$, a = 9.501, b = 12.952, c = 9.263 Å, α = 108.77, β = 106.99, γ = 90.13°, Z = 1. Mo radiation, R = 0.037 for 3253 reflexions.

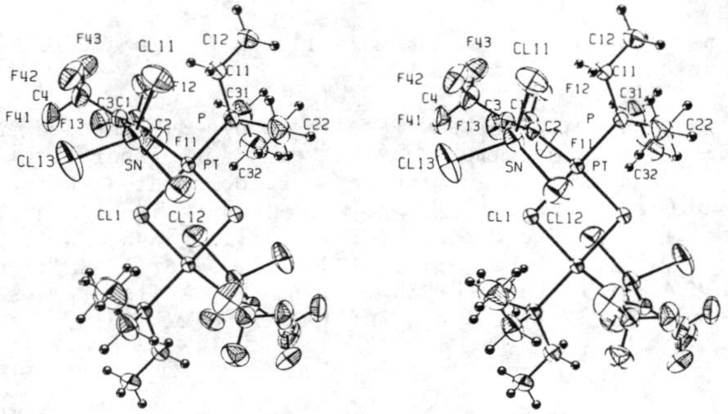

Fig. 1. Stereoview of the dimer $C_{20}H_{30}Cl_8F_{12}P_2Pt_2Sn_2$.

The unique Pt atom has close to square-planar geometry with the PtC_4Sn plane approximately normal (78.3°) to the $PtCl_2PC$ coordination plane (Fig. 1). Principal dimensions are as follows: Pt-Cl(1) (trans to PEt_3) 2.415, Pt-Cl(1') (trans to C) 2.396, Pt-C(2) 1.982, C(2)=C(3) 1.339, C(3)-Sn 2.127, Sn-Cl(11) 2.306, Sn-Cl(12) 2.288, Sn-Cl(13) 2.302, and Pt-P 2.230 Å. Atom Cl(12) of the $SnCl_3$ moiety is constrained to lie slightly off the normal to the Pt coordination plane with Pt...Cl(12) 3.430 Å. The C(3)-Sn-Cl(12) angle (129.02°) is considerably enlarged over the anticipated tetrahedral value as a result of Pt...Cl repulsion. The other C(3)-Sn-Cl angles are just less than tetrahedral (107.4 and 108.4°).

DIBROMOTETRAKIS(t-BUTYL ISOCYANIDE)OSMIUM DICHLOROMETHANE
$C_{20}H_{36}Br_2N_4Os,2(CH_2Cl_2)$

R.A. JONES, B.R. WHITTLESEY, J.L. ATWOOD and W.E. HUNTER, 1984. Polyhedron, 3, 385-387.

Monoclinic, $P2_1/c$, a = 9.702, b = 10.143, c = 18.497(5) Å, β = 104.28°, Z = 2. Mo radiation, R = 0.044 for 1105 reflexions.

The osmium atom lies on a crystallographic centre of inversion and all bond lengths and angles in the molecules are normal with Os-Br 2.558(3), Os-C(1) 1.99(2), Os-C(6) 2.00(3) Å. The angles about Os are very close to those expected for an octahedron indicating that there are no major distortions from this geometry (C(1)-Os-C(6) 88.9(8)° and C(1)-Os-Br 90.9(6)°).

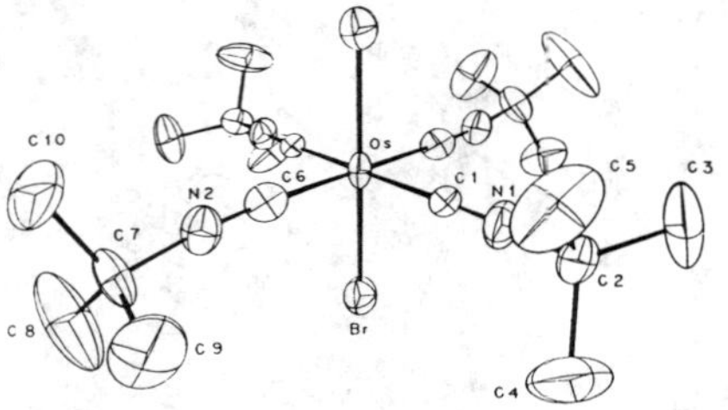

Fig. 1. A view of $C_{20}H_{36}Br_2N_4Os$.

BIS[(μ_2-O'-t-BUTYLACETATO-C,O)BROMO-TETRAHYDROFURANZINC]
$C_{20}H_{38}Br_2O_6Zn_2$ $(BrZnCH_2COO\text{-}t\text{-}Bu.THF)_2$

J. DEKKER, P.H.M. BUDZELAAR, J. BOERSMA, G.J.M. van der KERK and A.L. SPEK, 1984. Organometallics, 3, 1403-1407.

Monoclinic, $P2_1/n$, a = 10.322, b = 12.358, c = 11.655 Å, β = 112.65°, Z = 2. Mo radiation, R = 0.055 for 881 reflexions.

There are two dimeric units (with crystallographically-imposed centres of symmetry) in the unit cell. The dimer consists of a non-planar eight-membered ring $(ZnCCO)_2$ in which the Zn atoms are surrounded by a coordinated THF molecule, a Br atom, a C atom and an O atom, in a roughly tetrahedral fashion (Fig. 1). The bond distances and angles of most interest are: Zn(1)-Br(1) 2.346, Zn(1)-O(21) 2.02, Zn(1)-O(11) 2.05, Zn(1)-C(23) 1.98, O(21)-C(22) 1.31 Å; Br(1)-Zn(1)-O(11) 103.6, Br(1)-Zn(1)-O(21) 108.4, Br(1)-Zn(1)-C(23) 127.6, O(11)-Zn(1)-O(21) 93.2, O(11)-Zn(1)-C(23) 106.9, O(21)-Zn(1)-C(23) 111.2, and Zn(1)-O(21)-C(22) 123°.

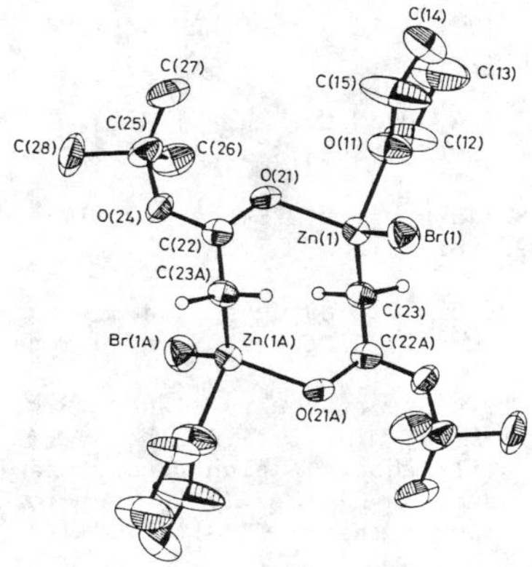

Fig. 1. Molecular structure of $C_{20}H_{38}Br_2O_6Zn_2$.

BIS((μ$_2$-CHLORO)CHLOROBIS(TRIMETHYLPHOSPHINE)TRIMETHYLPHOSPHINEMETHYLIDYNETUNGSTEN) BIS(TETRACHLOROALUMINATE)

$C_{20}H_{54}Al_2Cl_{12}P_6W_2$ $[W_2(CPMe_3)_2(PMe_3)_4Cl_4]^{2+}2AlCl_4^-$

S.J. HOLMES, R.R. SCHROCK, M.R. CHURCHILL and H.J. WASSERMAN, 1984. Organometallics, 3, 476-484.

Monoclinic, $P2_1/c$, a = 9.773, b = 24.80, c = 12.633 Å, β = 92.85°, Z = 2. Mo radiation, R = 0.087 for 2408 reflexions.

The results are of limited accuracy due to severe disorder of the $AlCl_4^-$ anions but fortunately the cation is well ordered (Fig. 1). The cation lies on a crystallographic inversion centre and the geometry about each W atom is roughly octahedral. The molecule contains a planar $P_2W(\mu\text{-}Cl)_2WP_2$ framework with mutually transoid $CPMe_3$ ligands. The W-W distance is 2.970 Å, the W-Cl(1), W-Cl(1'), and W-Cl(2) distances 2.458, 2.470 and 2.545 Å respectively, the W-P distances 2.543 and 2.535 Å, and the W-C(1) distance 1.83 Å, comparable to other W≡C 'triple' bonds. The angle W-C(1)-P(3) is 174°, and the angles W-Cl(1)-W' and Cl(1)-W-Cl(1') are respectively 74.1 and 105.9°.

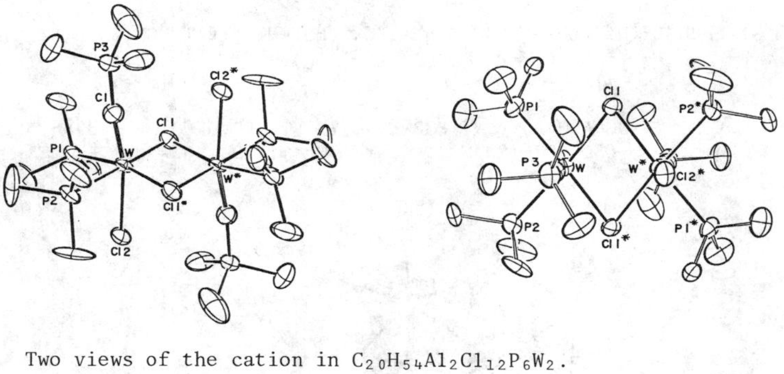

Fig. 1. Two views of the cation in $C_{20}H_{54}Al_2Cl_{12}P_6W_2$.

(η^2-1,1,3,3-TETRAKIS(TRIMETHYLSILYL)-2-PHOSPHAPROP-1-ENE-C,P)BIS(TRIMETHYLPHOSPHINE)-NICKEL

$C_{20}H_{55}NiP_3Si_4$ (I)

BIS(TRIMETHYLSILYL)METHYLENE-(2,4,6-TRI-t-BUTYLPHENYL)PHOSPHINE
$C_{25}H_{47}PSi_2$ (II)

A.H. COWLEY, R.A. JONES, J.G. LASCH, N.C. NORMAN, C.A. STEWART, A.L. STUART, J.L. ATWOOD, W.E. HUNTER and H.-M. ZHANG, 1984. J. Am. Chem. Soc., 106, 7015-7020.

I. Monoclinic, $P2_1/c$, a = 8.742, b = 19.225, c = 20.111 Å, β = 99.21°, Z = 4. Mo radiation, R = 0.031 for 1492 reflexions.

II. Triclinic, $P\bar{1}$, a = 9.118, b = 11.860, c = 14.403 Å, α = 69.41, β = 79.61, γ = 83.63°, Z = 2. Mo radiation, R = 0.061 for 3135 reflexions.

The Ni geometry in (I) is close to square planar (Fig. 1) with Ni-P(1) 2.239(2), Ni-C(8) 2.020(8), Ni-P(2) 2.195(3), Ni-P(3) 2.202(3) Å, P(1)-Ni-C(8) 48.9(2), P(2)-Ni-P(3) 102.1(1)°. Compound (II) adopts a planar conformation about the P=C bond (Fig. 1) with P=C 1.665(4) Å and C=P-C 110.7(2)°. Overcrowding causes the P-C7-Si2 angle (133.1(3)°) to be much larger than P-C7-Si1 (107.7(2)°). The dimensions are correlated with NMR data.

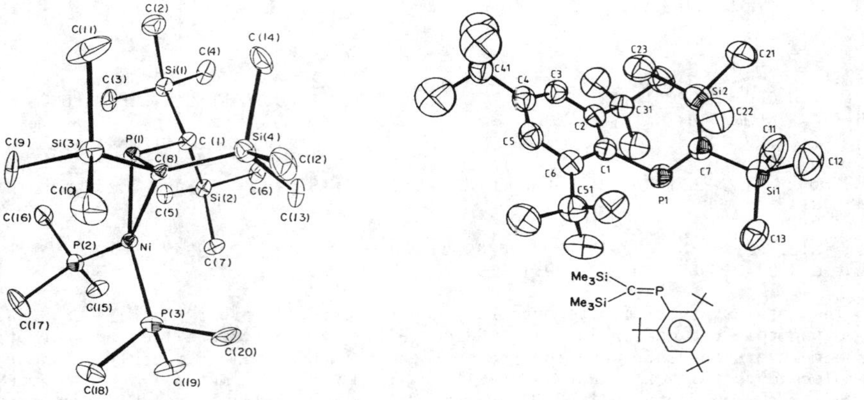

Fig. 1. Views of $C_{20}H_{55}NiP_3Si_4$ and $C_{25}H_{47}PSi_2$.

(μ_2-σ^2,η^2,η^2-2-PHENYLETHYNYLDICARBONYL(η^5-CYCLOPENTADIENYL)TUNGSTEN)BIS(TRICARBONYL)-IRON
$C_{21}H_{10}Fe_2O_8W$

N.A. USTYNYUK, V.N. VINOGRADOVA, V.N. KORNEVA, Yu.L. SLOVOKHOTOV and Yu.T. STRUCHKOV, 1983. Koord. Khim., 9, 631-637 [Sov. J. Coord. Chem., 9, 363-369].

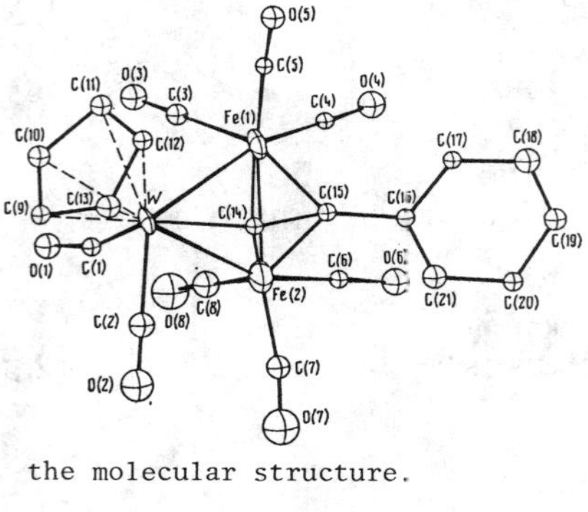

Fig. 1. $C_{21}H_{10}Fe_2O_8W$: the molecular structure.

Monoclinic, $P2_1/c$, a = 8.656, b = 14.261, c = 37.355 Å, β = 92.17°, Z = 8. Cu radiation, R = 0.12 for 4654 reflexions.

There are two crystallographically independent molecules (A,B) in the asymmetric unit. These differ with respect to the angle of rotation of the phenyl ring relative to the heteronuclear cluster. The molecules consist of an Fe_2W triangle (Fe-W 2.827(4)-2.887(5), Fe-Fe 2.480(7) (A), 2.467(7) Å (B)) which is bridged by a σ,π-C≡CPh ligand (Fig. 1). The W-C(14) distances are 1.91(4), 1.97(3), Fe-C(14) 2.06-2.19(3) and Fe-C(15) 2.02-2.15(3) Å. The Fe-C(CO) distances range from 1.72(3) to 1.86(5) and W-C(cp) from 2.28(5) to 2.45(5) Å. The dihedral angle between the Fe_2W triangle and the phenyl ring is 118° in molecule A and 102° in molecule B.

(μ_2-σ^3-3-(PHENYLCARBAMOYL)(1,1,1,4,4,4-HEXAFLUOROBUTAN-2-YLIDEN-3-YL))BIS(η^5-CYCLOPENTADIENYL)DIRHODIUM

$C_{21}H_{15}F_6NORh_2$ $(\eta^5\text{-}C_5H_5)_2Rh_2\{\mu_2(\eta^3\text{-}C(CF_3)C(CF_3)C(O)N(C_6H_5)\}$

R.S. DICKSON, R.J. NESBIT, H. PATERAS, J.M. PATRICK and A.H. WHITE, 1984. J. Organometal. Chem., 265, C25-C28.

Tetragonal, $I4_1/a$, a = 24.49, c = 14.298 Å, Z = 16. Mo radiation, R = 0.046 for 2559 reflexions.

The structure reveals that the cyanate C atom (C(3)) has condensed with a C atom of hexafluorobut-2-yne to generate a bridging amide ligand (Fig. 1). The metal to bridging N atom distances are: Rh(1)-N(3) 2.169 and Rh(2)-N(3) 2.049 Å. Other distances are: Rh(1)-Rh(2) 2.612, Rh(1)-C(1) 2.020, Rh(2)-C(1) 2.148, Rh(1)-C(2) 2.138, C(1)-C(2) 1.45, C(2)-C(3) 1.48, C(3)-O(3) 1.20 and C(3)-N(3) 1.41 Å.

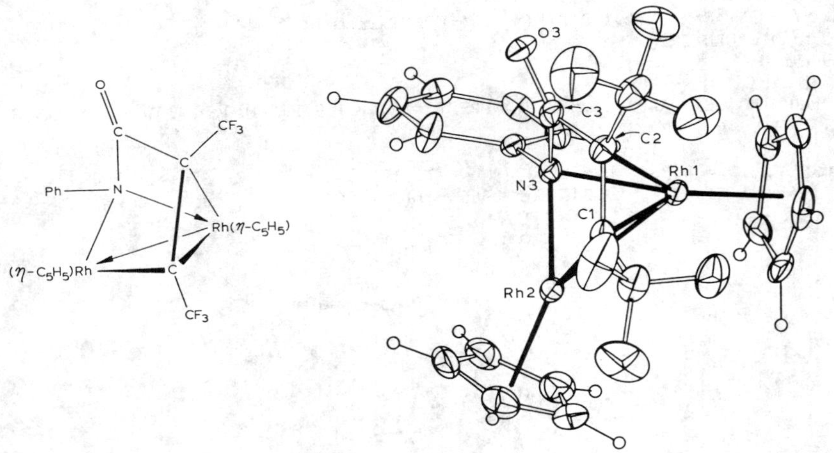

Fig. 1. Coordination mode of the bridging ligand in $C_{21}H_{15}F_5NORh_2$ (left), and molecular structure (right).

PENTACARBONYL[DIPHENYLMETHYLENEAMINO(ETHYLTHIO)CARBENE]TUNGSTEN

$C_{21}H_{15}NO_5SW$ $(CO)_5W{=}C(SC_2H_5)NC(C_6H_5)_2$

H. FISCHER, S. ZEUNER, K. ACKERMANN and U. SCHUBERT, 1984. J. Organometal. Chem., 263, 201-212.

Triclinic, $P\bar{1}$, a = 8.811, b = 10.082, c = 13.307 Å, α = 99.61, β = 97.61, γ = 109.15°, Z = 2. Mo radiation, R = 0.077 for 2664 reflexions.

The molecule consists of a relatively undistorted octahedral arrangement of five CO groups and a carbene ligand about a tungsten atom (Fig. 1). The carbene plane (C(11),W,S,N) is approximately coplanar with the best plane through C(11), C(12), C(13) and C(14). The angles involving W range from 87.0 to 91.7° and the W-C (equatorial) distances from 1.96 to 2.02 Å. The axial W-CO distance is 1.88 Å and the W-C(11) distance is 2.20 Å. Other distances and angles of note are: C(11)-S 1.76, C(11)-N 1.28, S-C(2) 1.87 Å; W-C(11)-S 118.0, W-C(11)-N 127.3, S-C(11)-N 114.7, C(11)-S-C(2) 106.3 and C(11)-N-C(1) 149.4°.

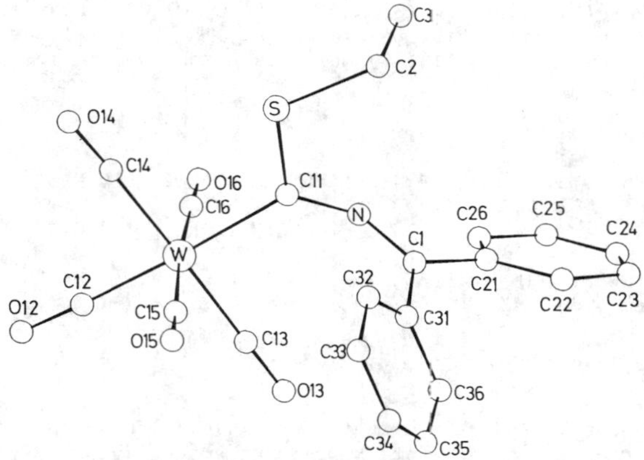

Fig. 1. Molecular structure of $C_{21}H_{15}NO_5SW$.

(μ2-CARBONYL)PENTACARBONYL-1,2-BIS(η^5-CYCLOPENTADIENYL)(μ4-1'-4'-η^3-METHYLBUTENYL-1'-YLIDYNE)DINICKELDIIRON
$C_{21}H_{16}Fe_2Ni_2O_6$

M.L.N. MARCHINO, E. SAPPA, A.M.M. LANFREDI and A. TIRIPICCHIO, 1984. J. Chem. Soc. Dalton, 1541-1546.

Triclinic, P$\bar{1}$, a = 8.881, b = 17.224, c = 7.912 Å, α = 78.87, β = 114.95, γ = 91.60°, Z = 2. Mo radiation, R = 0.046 for 2965 reflexions.

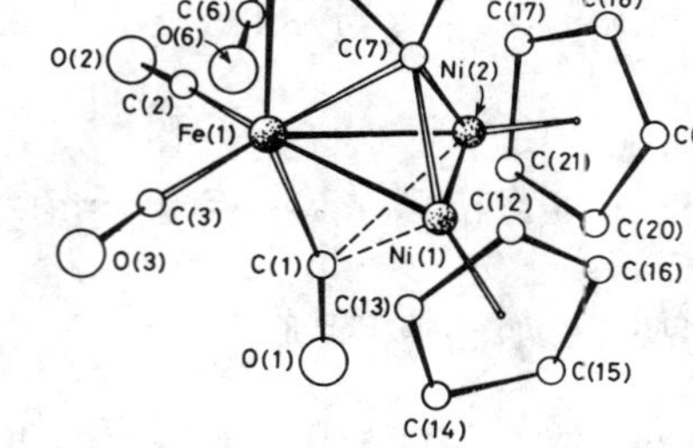

Fig. 1. $C_{21}H_{16}Fe_2Ni_2O_6$: view of the molecular structure of the complex.

The complex (Fig. 1) consists of a 'spiked-triangular' metal cluster in which two Ni atoms and one Fe atom form an isosceles triangle and a second Fe atom is bound axially to the first (Fe-Fe 2.755(3), Fe-Ni 2.384(3), Ni-Ni 2.337(3) Å. Each Ni atom is η-coordinated to a cyclopentadienyl ligand. Of the six CO groups, five are terminal, and the sixth triply bridges the triangle. The four adjacent C atoms C(7-10) of the alkyne ligand are involved in extensive η-bonding with Fe(2) (2.066-2.266 Å. C(7) also caps the triangle interacting with all three metal atoms (C(7)-Fe(1), Ni(1), Ni(2) 1.905(8), 1.868(8), 1.913(7) Å) through three sigma bonds.

(2,2'-BIPYRIDINE-N,N')(8-CARBOXYLATOTRICYCLO[5.2.1.0^{2,6}]DEC-4-EN-9-YL)NICKEL

$C_{21}H_{20}N_2NiO_2$

D. WALTHER, E. DINJUS, J. SIELER, L. ANDERSEN and O. LINDQVIST, 1984. J. Organometal. Chem., 276, 99-107.

Monoclinic, $P2_1/n$, a = 11.867, b = 11.426, c = 13.474 Å, β = 108.4°, Z = 4. Mo radiation, R = 0.053 for 2932 reflexions.

The metal coordination scheme is approximately square planar (Fig. 1) with the following distances: Ni-C(8) 1.929, Ni-N(1) 1.902, Ni-N(2) 1.961 and Ni-O(1) 1.845 Å. The angle between the planes through atoms Ni, N(1), N(2) and Ni, O(1) and C(8) is 5.2°, and the coordination angles about the metal are: O(1)-Ni-C(8) 87.6, O(1)-Ni-N(2) 92.0, N(1)-Ni-N(2) 83.2 and N(1)-Ni-C(8) 97.3°. Distances within the norbornene ligand are: C(8)-C(9) 1.557, C(7)-C(8) 1.540, C(7)-C(10) 1.549, C(10)-C(1) 1.530, C(1)-C(9) 1.544, C(7)-C(6) 1.551, C(1)-C(2) 1.543, C(2)-C(6) 1.565, C(2)-C(3) 1.498, C(3)-C(4) 1.390, C(4)-C(5) 1.410 and C(5)-C(6) 1.502 Å. The bridgehead angle C(7)-C(10)-C(1) is 93.9°, and additional distances are C(9)-C(11) 1.511, C(11)-O(2) 1.216 and C(11)-O(1) 1.313 Å.

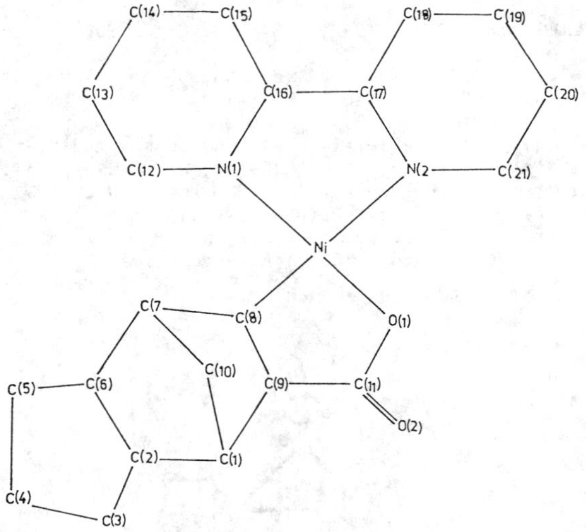

Fig. 1. Schematic diagram of the complex $C_{21}H_{20}N_2NiO_2$.

(μ_2-CARBONYL)-2-CARBONYL-2-(η^5-CYCLOPENTADIENYL)(μ_2-σ^1,η^3-p-TOLYLMETHYLENE)-1,1-BIS-(TRIMETHYLPHOSPHINE)PLATINUMTUNGSTEN TETRAFLUOROBORATE

$C_{21}H_{31}BF_4O_2P_2PtW$

J.C. JEFFERY, J.C.V. LAURIE, I. MOORE, H. RAZAY and F.G.A. STONE, 1984. J. Chem. Soc. Dalton, 1563-1659.

Monoclinic, $P2_1/c$, a = 13.903, b = 16.213, c = 11.859 Å, β = 96.89°, Z = 4. Mo radiation, R = 0.054 for 3430 reflexions.

In the cation (Fig. 1) the Pt-W bond (2.795(1) Å) is bridged by the tolylmethylene group in such a way that C, C(41), C(42) are η^3-cocrdinated to the tungsten atom (W-C 2.17, 2.41, 2.60(1) Å) while the C atom is also σ-bonded to the platinum atom (Pt-C 2.05(1) Å). The W atom carries a cyclopentadienyl ligand and a terminally bound CO group. The remaining carbonyl semi-bridges the metal-metal bond (W-C-O 155.1(14)°). The environment of the Pt atom is essentially square-planar.

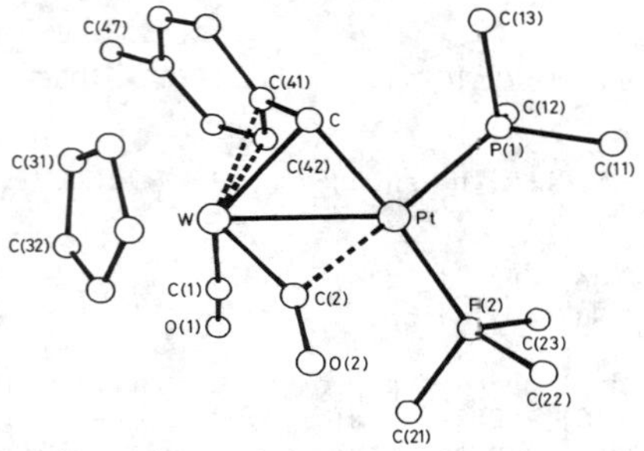

Fig. 1. $C_{21}H_{31}BF_4O_2P_2PtW$: the cation.

2,2-DICARBONYL-2-(η^5-CYCLOPENTADIENYL)(μ_2-HYDRIDO)(μ_2-p-TOLYLMETHYLENE)-1,1-BIS(TRIMETHYLPHOSPHINE)PLATINUMTUNGSTEN
$C_{21}H_{32}O_2P_2PtW$

J.C. JEFFERY, I. MOORE and F.G.A. STONE, 1984. J. Chem. Soc. Dalton, 1571-1580.

Monoclinic, $P2_1/c$, a = 10.554, b = 11.624, c = 19.393 Å, β = 97.64°, Z = 4. Mo radiation, R = 0.039 for 3083 reflexions.

In the molecule (Fig. 1) the metal-metal bond is spanned by the tolylmethylene and hydride ligands (Pt-W 2.895(1), Pt-C 2.109(9), W-C 2.259(9) Å). Coordination about the W atom is completed by two terminal CO groups (W-C 1.95 Å) and a η^5-C_5H_5 ligand. The Pt atom is part of a cis-$Pt(PMe_3)_2$ group. It has an essentially planar environment with an angle of 7° between the PtP_2 and PtCW planes. The Pt-P(1) and Pt-P(2) distances are 2.265(3) and 2.305(3) Å respectively.

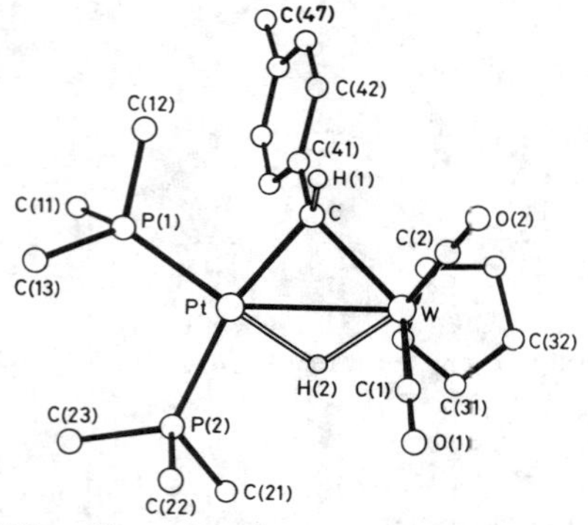

Fig. 1. $C_{21}H_{32}O_2P_2PtW$: the molecular structure.

[μ-[(1-η:2,3-η)-2,4-CYCLOPENTADIEN-1-YL]]DI-2,4-CYCLOPENTADIEN-1-YL[μ-[1,1,1-TRIMETHYL-N-(TRIMETHYLSILYL)-SILANAMINATO]]DIZINC

$C_{21}H_{33}NSi_2Zn_2$ $(\mu\text{-}C_5H_5)(\mu\text{-}N(SiMe_3)_2)(ZnC_5H_5)_2$

P.H.M. BUDZELAAR, J. BOERSMA, G.J.M. van der KERK and A.L. SPEK, 1984. Organometallics, 3, 1187-1190.

Monoclinic, $P2_1/c$, a = 15.447, b = 11.049, c = 14.503 Å, β = 108.35°, Z = 4. Mo radiation, R = 0.045 for 3424 reflexions.

The molecule (Fig. 1) contains two Zn atoms connected by a bridging cyclopentadienyl group and a bridging amide group. In addition, each Zn atom carries a terminal Cp group. The large Zn...Zn separation of 3.087 Å precludes any direct Zn-Zn bonding. The Zn(1)-N(1) and Zn(2)-N(1) distances are 2.005 and 2.024 Å respectively. For each Zn-Cp interaction there is one short Zn-C distance (Zn(1)-C(31) 2.146, Zn(1)-C(41) 2.059, Zn(2)-C(33) 2.140 and Zn(2)-C(51) 2.07 Å) representing a normal Zn-C single bond. There is also an intermediate distance of 2.4-2.6 Å to another C atom of the same ring (Zn(1)-C(35) 2.495, Zn(1)-C(42) 2.473, Zn(2)-C(32) 2.611 and Zn(2)-C(52) 2.55 Å) suggesting that both σ and π-type interactions contribute to the Cp-Zn bonds.

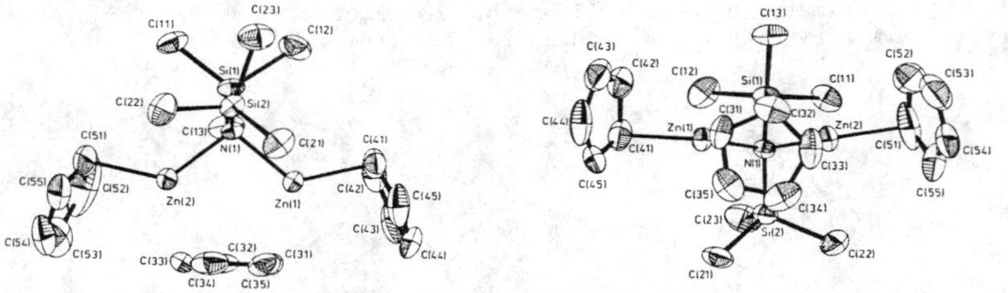

Fig. 1. Two orientations of the molecular structure of $C_{21}H_{33}NSi_2Zn_2$.

(DICARBONYL)(IODO)(IODOMETHYL)BIS(TRIISOPROPYL PHOSPHITE)IRON(II)

$C_{21}H_{44}FeI_2O_8P_2$

H. BERKE, R. BIRK, G. HUTTNER and L. ZSOLNAI, 1984. Z. Naturforsch., B39, 1380-1388.

Monoclinic, $P2_1/c$, a = 9.19, b = 12.86, c = 27.21 Å, β = 92.8°, Z = 4. Mo radiation, R = 0.095 for 2954 reflexions (at -60°C).

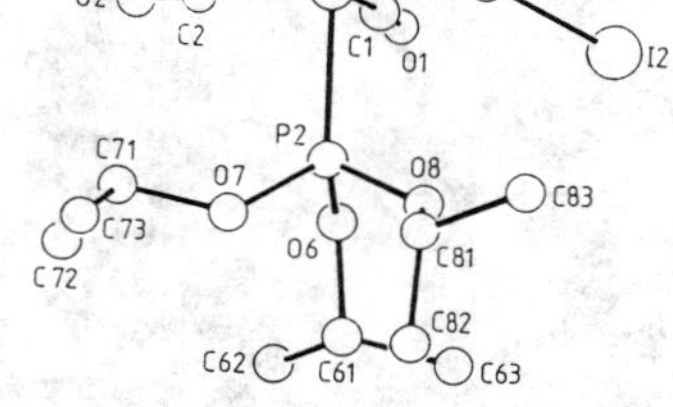

Fig. 1. A view of $C_{21}H_{44}FeI_2O_8P_2$.

The Fe atom has octahedral coordination (Fig. 1), with Fe-P 2.210 and 2.219(8), Fe-I(1) 2.618(6), Fe-C(1) 1.76(2), Fe-C(2) 1.86(3), Fe-C(3) 2.05(2), C(3)-I(2) 2.21(2), mean P-O 1.57 Å, P-Fe-P 174.6(3), I(1)-Fe-C(1) 174.3(8), C(2)-Fe-C(3) 172(1)°.

DICARBONYL(η^5-CYCLOPENTADIENYL)(2,2-DIPHENYLVINYLCARBYNE)MANGANESE TETRAFLUOROBORATE
$C_{22}H_{16}BF_4MnO_2$ $[\eta^5\text{-}C_5H_5(CO)_2Mn{\equiv}C\text{-}CH{=}CPh_2]^+BF_4^-$

N.E. KOLOBOVA, L.L. IVANOV, O.S. ZHVANKO, O.M. KHITROVA, A.S. BATSANOV and Yu.T. STRUCHKOV, 1984. J. Organometal. Chem., 262, 39-47.

Monoclinic, $P2_1/c$, a = 10.179, b = 15.286, c = 12.961 Å, β = 103.57°, Z = 4. Mo radiation, R = 0.054 for 2103 reflexions (at -120°C).

The Mn atom is bonded to the planar η^5-C_5H_5 ring (Mn-Cp mean plane distance is 1.772 Å), two linear carbonyls and the C(3) atom of the carbyne ligand, and has a 'piano stool' configuration (Fig. 1). The Mn-C(3) bond is very short at 1.665 Å and the angle Mn-C(3)-C(4) is 174.6°. With the exception of the C(5)Ph_2 moiety, the cation has approximate, non-crystallographic m symmetry. The BF_4^- anion has the usual tetrahedral symmetry with B-F distances ranging from 1.28 to 1.33 Å.

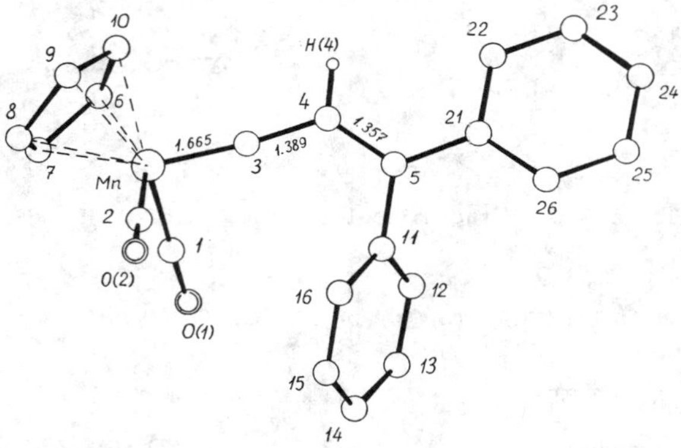

Fig. 1. Molecular structure of the cation in $C_{22}H_{16}BF_4MnO_2$ showing selected bond distances.

cis-BIS(2-PHENYLPYRIDINE-C,N)PLATINUM(II)
$C_{22}H_{16}N_2Pt$ $(C_6H_4C_5H_4N)_2Pt$

L. CHASSOT, E. MÜLLER and A. von ZELEWSKY, 1984. Inorg. Chem., 23, 4249-4253.

Monoclinic, $P2_1/n$, a = 12.339, b = 11.790, c = 11.788 Å, β = 106.02°, Z = 4. Mo radiation, R = 0.020 for 3012 reflexions.

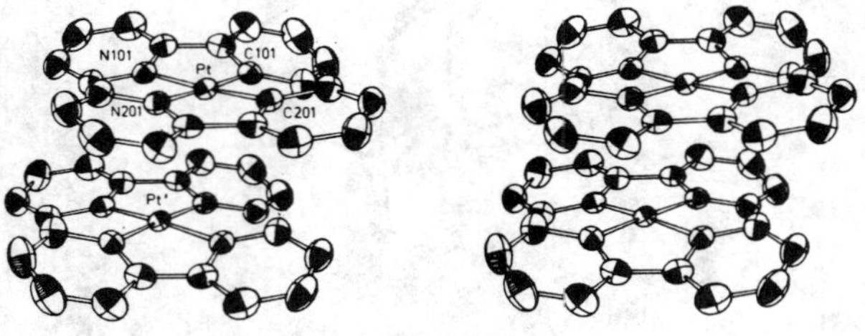

Fig. 1. Stereoview of the $(C_6H_4C_5H_4N)_2Pt$ dimer.

The molecule is a cis isomer with bowlike distortions of the Phpy$^-$ ligands and Pt-N 2.125(3) and 2.128(3), Pt-C 1.984(4) and 2.002(3) Å. The molecules are associated to dimeric units with essentially parallel planes at a distance of 3.37 Å (Fig. 1). The Pt...Pt distance in the dimer is 3.528(1) Å.

DICARBONYL(η^5-CYCLOPENTADIENYL)(10,11-DIHYDRO-5H-DIBENZO[a,b]CYCLOHEPTEN-5-YLIDENE)-MANGANESE

$C_{22}H_{17}MnO_2$ (I) $(\eta^5\text{-}C_5H_5)Mn(CO)_2(C_{15}H_{12})$

DICARBONYL(η^5-METHYLCYCLOPENTADIENYL)(η^2-10,11-DIHYDRO-5H-DIBENZO[a,d]CYCLOHEPTEN-YLIDEN-5-KETENE)MANGANESE

$C_{24}H_{19}MnO_3$ (II) $(\eta^5\text{-}C_5H_4CH_3)Mn(CO)_2[\eta^2\text{-}C(C_{14}H_{12})CO]$

W.A. HERRMANN, J. PLANK, G.W. KRIECHBAUM, M.L. ZIEGLER, H. PFISTERER, J.L. ATWOOD and R.D. ROGERS, 1984. J. Organometal. Chem., 264, 327-352.

I. Monoclinic, $P2_1/n$, a = 10.456, b = 7.899, c = 21.536 Å, β = 101.56°, D_m = 1.3, Z = 4. Mo radiation, R = 0.075 for 1272 reflexions.

II. Monoclinic, $P2_1/n$, a = 9.989, b = 13.023, c = 14.666 Å, β = 91.33°, D_m = 1.3, Z = 4. Mo radiation, R = 0.057 for 1893 reflexions.

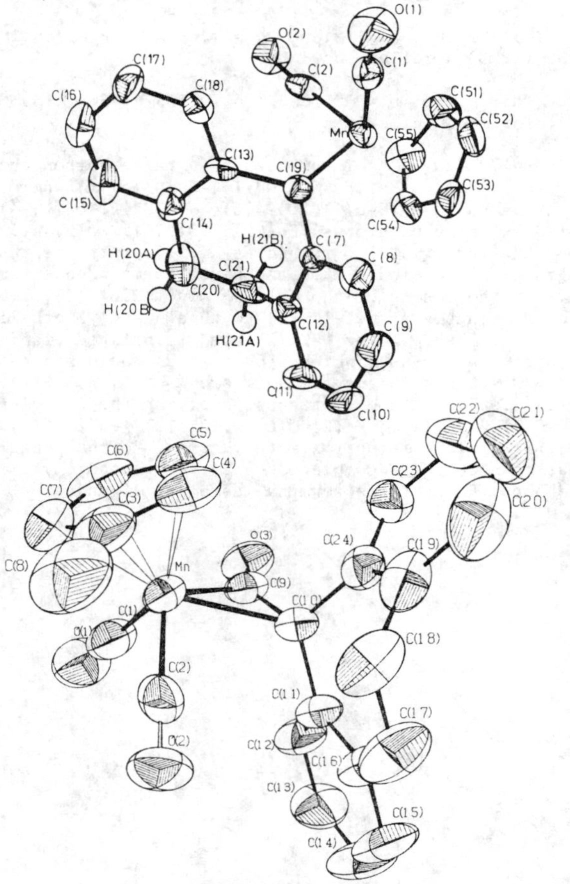

Fig. 1. Molecular structures of $C_{22}H_{17}MnO_2$ (top) and $C_{24}H_{19}MnO_3$ (bottom).

In I, the Mn-C(19) (carbene C atom) distance is quite short, being 1.853 Å and the angles around the carbene C atom sum to 360.1° indicating almost perfect planarity of the surrounding bonding atoms (Mn, C(7), C(13)). The Mn-C(Cp) distances range from 2.147 to 2.182 Å, and the distances C(19)-C(7) and C(19)-C(13) are respectively 1.520 and 1.500 Å (Fig. 1). The metal-carbonyl C distances are 1.778 and 1.789 Å. In II, the metal interaction with the ketene system >C(10)=C(9)=O(3) is assumed to consist of contributions from a metallacyclopropanone system and a π-olefin complex. Relevant distances are: Mn-C(9) 1.940, Mn-C(10) 2.212, C(9)-C(10) 1.401 and C(9)-O(3) 1.224 Å. The Mn-C(Cp) distances range from 2.109 to 2.190 Å and the Mn-CO distances are 1.78 and 1.811 Å (Fig. 1). The Mn,C(9),C(10),O(3) moiety is planar.

1,1,1,2,2-PENTACARBONYL-2-(η^5-CYCLOPENTADIENYL)-(μ_2-1',2'-DIMETHYL-3'-p-TOLYLPROPEN-1'-YL-3'-YLIDENE)IRONTUNGSTEN
$C_{22}H_{18}FeO_5W$ (I)

1,3-(μ_2-BIS(TRIMETHYLSILYL)ACETYLENE)-1,2-(μ_2-CARBONYL)-1,1,1,2,2,3-HEXACARBONYL-3-(η^5-CYCLOPENTADIENYL)(μ_3-p-TOLYL-METHYLIDYNE)DIIRONTUNGSTEN
$C_{28}H_{30}Fe_2O_7Si_2W$ (II)

J.C. JEFFERY, K.A. MEAD, H. RAZAY, F.G.A. STONE, M.J. WENT and P. WOODWARD, 1984. J. Chem. Soc. Dalton, 1383-1391.

I. Monoclinic, C2/c, a = 26.01, b = 12.880, c = 12.554 Å, β = 105.10°, Z = 8. Mo radiation, R = 0.039 for 3334 reflexions.

II. Monoclinic, $P2_1$, a = 8.922, b = 38.765, c = 10.055 Å, β = 114.30°, Z = 4. Mo radiation, R = 0.051 for 7443 reflexions.

In I (Fig. 1) Fe-W (2.720(1) Å) is bridged by a three-carbon chain. All the atoms of the chain are bonded to Fe (Fe-C 2.047(6)-2.103(6) Å) whereas only the two terminal atoms are attached to W (W-C(6) 2.194(6), W-C(8) 2.164(7) Å). Coordination about Fe is completed by three CO groups (Fe-C 1.770(7)-1.788(6) Å). In II the two molecules which comprise the crystallographic asymmetric unit are themselves related by a non-crystallographic inversion centre and hence form an enantiomorphous pair which differ mainly in the length of the Fe(2)-W bond (2.648(2), 2.682(2) Å). Each molecule consists of a triangle of metal atoms with a μ_3-tolylmethylidyne ligand capping one face of the triangle. The W-Fe(2) bond is bridged transversely by a bis(trimethylsilyl)acetylene ligand, which tips towards the opposite face of the metal atom triangle. Distances in the triangle are W-Fe(1) 2.775(2), W-Fe(2) 2.648(2), Fe(1)-Fe(2) 2.640(3) Å and distances involving the acetylene bridge are W-C(6) 2.053(14), W-C(7) 2.107(12), Fe(2)-C(6) 2.184(12), Fe(2)-C(7) 2.126(13) Å. The Si(1),C(6),C(7),Si(2) plane is approximately normal to the plane of the metal triangle. Coordination about W is completed by a cyclopentadienyl ligand and a carbonyl ligand and about Fe by three terminal carbonyl ligands.

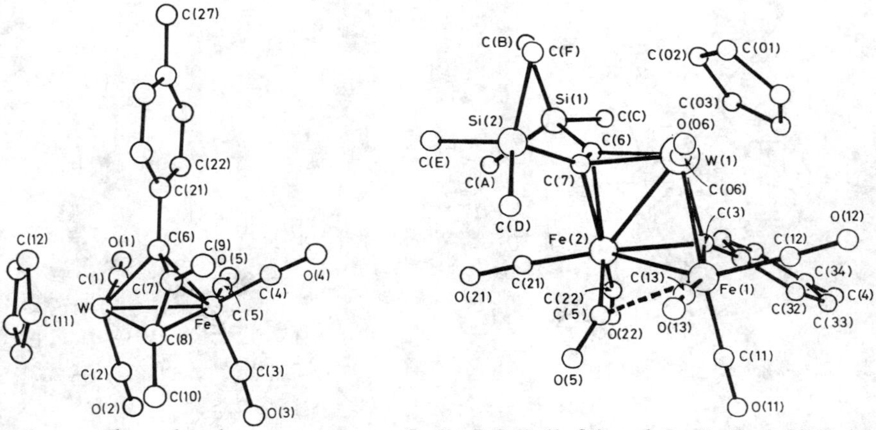

Fig. 1. The molecular structures: $C_{22}H_{18}FeO_5W$ (left) and $C_{28}H_{30}Fe_2O_7Si_2W$ (right).

trans-BIS(2,6-BIS(METHOXY)PHENYL)-BIS(TRIMETHYLPHOSPHINE)NICKEL(II)
$C_{22}H_{36}NiO_4P_2$ (I)

trans-BIS(2,6-BIS(METHOXY)PHENYL)-BIS(DIMETHYLPHENYLPHOSPHINE)NICKEL(II)
$C_{32}H_{40}NiO_4P_2$ (II)

trans-BIS(3,5-DIBROMO-2,6-BIS(METHOXY)PHENYL)-BIS(TRIMETHYLPHOSPHINE)NICKEL(II)
$C_{22}H_{32}Br_4NiO_4P_2$ (III)

trans-BIS(3,5-DIBROMO-2,6-BIS(METHOXY)PHENYL)-BIS(DIMETHYLPHENYLPHOSPHINE)NICKEL(II)
$C_{32}H_{36}Br_4NiO_4P_2$ (IV)

D. XU, Y. KAI, K. MIKI, N. KASAI, K. NISHIWAKI and M. WADA, 1984. Bull. Chem. Soc. Japan, 57, 1459-1468.

I. Monoclinic, $P2_1/c$, a = 8.737, b = 18.282, c = 16.319 Å, β = 105.06°, D_m = 1.284, Z = 4. Mo radiation, R = 0.062 for 4138 reflexions.

II. Triclinic, $P\bar{1}$, a = 16.993, b = 10.180, c = 17.177 Å, α = 112.44, β = 117.59, γ = 66.25°, D_m = 1.299, Z = 3. Mo radiation, R = 0.060 for 6006 reflexions.

III. Monoclinic, $P2_1/c$, a = 9.294, b = 14.091, c = 10.770 Å, β = 92.06°, D_m = 1.884, Z = 2. Mo radiation, R = 0.067 for 2284 reflexions.

IV. Triclinic, $P\bar{1}$, a = 9.370, b = 11.387, c = 9.013 Å, α = 94.52, β = 107.24, γ = 77.35°, D_m = 1.727, Z = 1. Mo radiation, R = 0.058 for 3206 reflexions.

There are three main structural features common to all four complexes (Fig. 1). The first is that the Ni atom assumes square-planar coordination with two phenyl and two phosphine ligands in trans-positions. The second is that the ring planes of the phenyl ligands are nearly perpendicular to the coordination plane of the Ni atom, and the third feature is that the phosphine ligands take a very characteristic conformation, with one methyl group located roughly on the coordination plane of the metal atom regardless of the type of phosphine ligand (PMe_3 or PMe_2Ph). The Ni-P distances in I, II, III and IV respectively (averaged where necessary) are: 2.173, 2.175, 2.220 and 2.224 Å, and the Ni-C distances 1.944, 1.937, 1.962 and 1.945 Å.

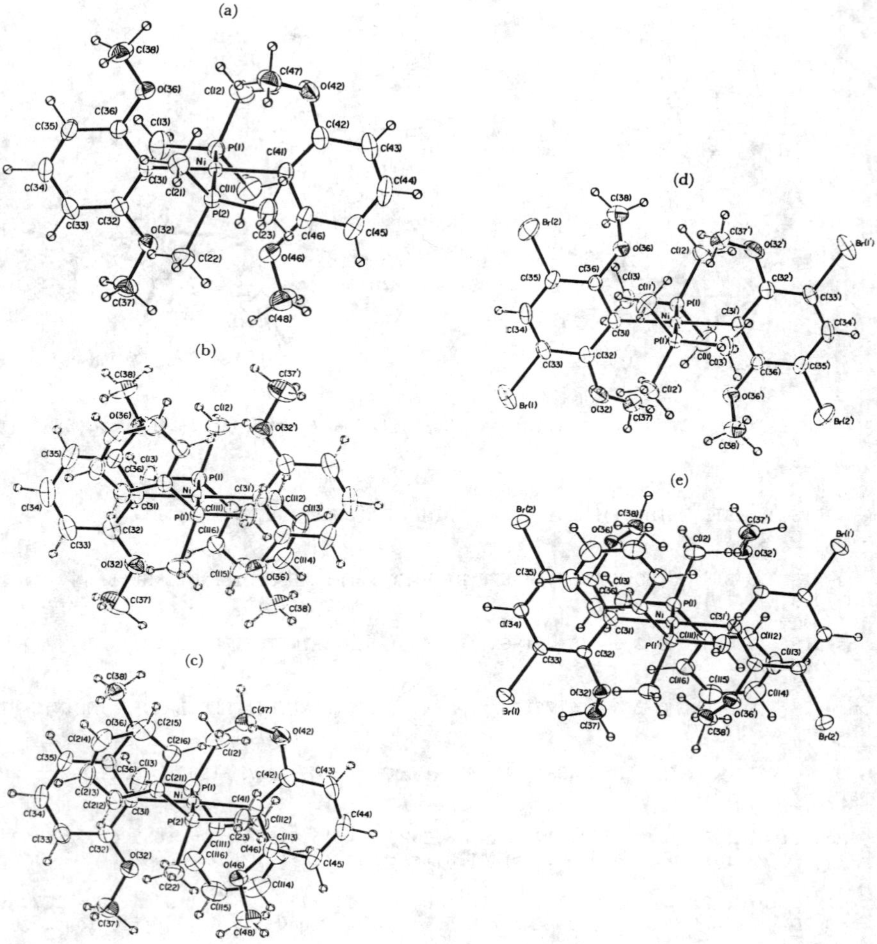

Fig. 1. Molecular structure of $C_{22}H_{36}NiO_4P_2$, (a), $C_{32}H_{40}NiO_4P_2$ (centrosymmetric molecule), (b), $C_{32}H_{40}NiO_4P_2$ (non-centrosymmetric molecule), (c), $C_{22}H_{32}Br_4NiO_4P_2$, (d), and $C_{32}H_{36}Br_4NiO_4P_2$ (e).

cis-TETRACARBONYLBIS(DIISOPROPYLAMINOETHOXYCARBENE)TUNGSTEN
$C_{22}H_{38}N_2O_6W$

E.O. FISCHER, R. REITMEIER and K. ACKERMANN, 1984. Z. Naturforsch., B39, 668-674.

Monoclinic, $P2_1/c$, a = 9.344, b = 15.102, c = 19.448 Å, β = 99.05°, Z = 4. Mo radiation, R = 0.044 for 3404 reflexions.

The complex is shown to be the cis-isomer with distorted octahedral coordination at tungsten (Fig. 1). Main dimensions are: W-C(carbonyl) 1.989-2.045(10), W-C5 2.300(9), W-C6 2.304(10), O5-C5 1.361(14), C5-N5 1.339(13), C6-O6 1.354(11), C6-N6 1.340(12) Å, C5-W-C6 90.1(3), C1-W-C3 168.3(4), C2-W-C4 85.9(4)°.

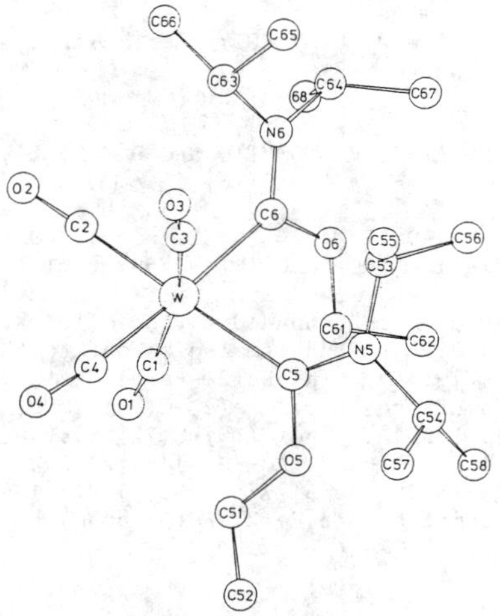

Fig. 1. $C_{22}H_{38}N_2O_6W$: a view of the structure.

BIS{[μ_2-TRIS(DIMETHYLPHOSPHINO)METHANIDE-C,P,P']TETRAHYDROFURANLITHIUM}
$C_{22}H_{52}Li_2O_2P_6$ (I)

TETRACARBONYL(1,1-BIS(DIMETHYLPHOSPHINO)ETHYL-1-DIMETHYLPHOSPHONIOMETHANIDE-C,P)-CHROMIUM(0)
$C_{13}H_{23}CrO_4P_3$ (II)

H.H. KARSCH, L. WEBER, D. WEWERS, R. BOESE and G. MULLER, 1984. Z. Naturforsch., B39, 1518-1528.

I. Monoclinic, $P2_1/n$, a = 9.285, b = 12.508, c = 14.079 Å, β = 100.07°, Z = 2. Mo radiation, R = 0.029 for 2158 reflexions (at -40°C).

II. Monoclinic, $P2_1/c$, a = 12.185, b = 8.817, c = 18.241 Å, β = 106.66°, Z = 4. Mo radiation, R = 0.063 for 1615 reflexions.

Molecule I is a dimer lying about an inversion centre (Fig. 1); the Li atom has pseudo-tetrahedral coordination with Li-O 1.923(3), Li-P(1) 2.588, Li-P(2) 2.684, Li-C(1)* 2.290, Li-P(1)* 2.945, Li-P(2)* 2.998, P(1)-C(1) 1.795, P(2)-C(1) 1.795, P(3)-C(1) 1.813(2) Å. Molecule II has a five-membered P_2C_2Cr ring in an approximate envelope conformation with P(1) at the flap (dihedral angle between C(5),Cr,P(3),C(6) and C(5),P(1),C(6) 138.3°). Main bond lengths are: Cr-P(3) 2.343(2), P(3)-C(6) 1.906, C(6)-P(1) 1.841, P(1)-C(5) 1.726, Cr-C(5) 2.200(9), Cr-C (carbonyl) 1.804-1.868(6) Å.

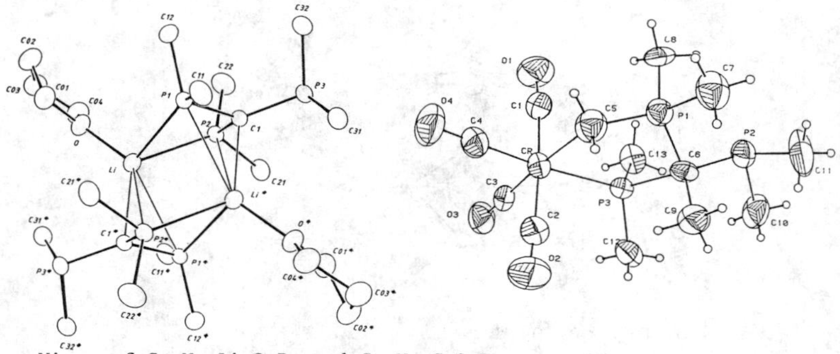

Fig. 1. Views of $C_{22}H_{52}Li_2O_2P_6$ and $C_{13}H_{23}CrO_4P_3$.

TRICARBONYL[μ_3-[1,1'-(η:η^2:η^2-1,2,-ETHYNEDIYL)BIS[BENZENE]]](HEXACARBONYLDIIRON)-RUTHENIUM(Fe-Fe)(2Fe-Ru)

$C_{23}H_{10}Fe_2O_9Ru$ $C_6H_5C_2C_6H_5RuFe_2(CO)_9$

V. BUSETTI, G. GRANOZZI, S. AIME, R. GOBETTO and D. OSELLA, 1984. Organometallics, 3, 1510-1515.

Triclinic, P$\bar{1}$, a = 9.146, b = 9.137, c = 16.321 Å, α = 80.81, β = 75.84, γ = 62.58°, D_m = 1.82, Z = 2. Mo radiation, R = 0.081 for 2696 reflexions.

In the complex the Ru atom is σ-bonded to the C(10) atom of the diphenylacetylene ligand (Ru-C(10) 2.20 Å) and is also bonded to the two Fe atoms (Ru-Fe(2) 2.564, Re-Fe(3) 2.582 Å) (Fig. 1). The Fe(2)-Fe(3) distance is 2.588 Å, and the Fe-diphenylacetylene ligand distances are: Fe(2)-C(10) 2.09, Fe(3)-C(10) 2.05, Fe(2)-C(11) 1.94 and Fe(3)-C(11) 1.96 Å. The C(10)-C(11) distance is 1.41 Å. Bond angles of interest are: Ru-Fe(2)-Fe(3) 60.2, Ru-Fe(3)-Fe(2) 59.5, Fe(2)-Ru-Fe(3) 60.4, Ru-C(10)-C(11) 126.6, Fe(2)-C(10)-Fe(3) 77.4, Fe(2)-C(11)-Fe(3) 83.1, and Fe(2)-M-Fe(3) (where M is the midpoint of the Fe(2)-Fe(3) bond) 86.8°.

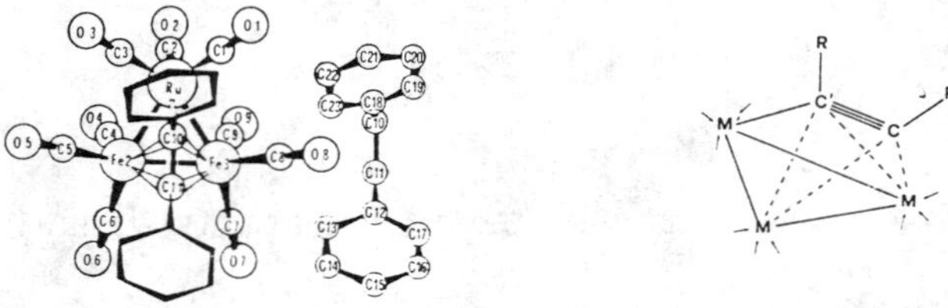

Fig. 1. Schematic diagram of the complex $C_{23}H_{10}Fe_2O_9Ru$ (M = Fe, M' = Ru, R = Ph) (left), and molecular structure (right).

(μ_3-2-(p-TOLYL)ETHYLIDYNE)(μ_2-OXO)(η^5-CYCLOPENTADIENYLTUNGSTEN)TRIS(TRICARBONYLOSMIUM)

$C_{23}H_{14}O_{10}Os_3W$

J.R. SHAPLEY, J.T. PARK, M.R. CHURCHILL, J.W. ZILLER and L.R. BEANAN, 1984. J. Am. Chem. Soc., 106, 1144-1145.

Triclinic, P$\bar{1}$, a = 9.704, b = 9.783, c = 14.599 Å, α = 108.11, β = 93.80, γ = 94.57°, Z = 2. Mo radiation, R = 0.039 for 3438 reflexions.

The molecular structure is shown in Fig. 1. The oxo ligand is best viewed as forming a double bond to the tungsten atom and acting as a donor to the adjacent osmium atom. Important distances: metal-metal, W-Os(1) 2.663(1), W-Os(2) 2.814(1), W-Os(3) 2.655(1), Os(1)-Os(2) 2.836(1), Os(1)-Os(3) 2.875(1), Os(2)-Os(3) 2.772(1) Å; metal-(μ_3-C), W-C(15) 2.030(12), Os(1)-C(15) 2.291(12), Os(2)-C(15) 2.100(10); metal-(μ-O), W-O(10) 1.812(7), Os(1)-O(10) 2.169(20) Å.

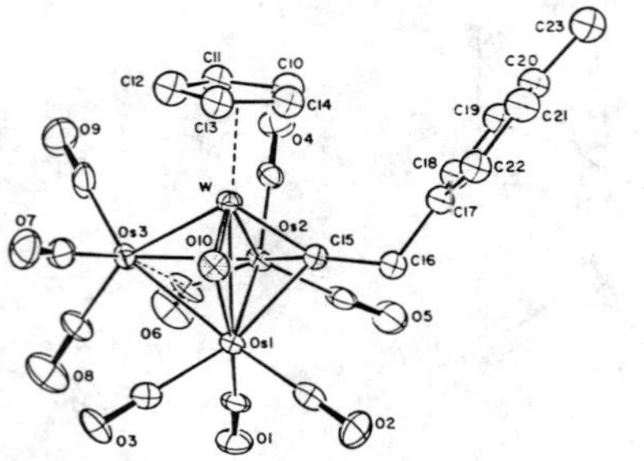

Fig. 1. A view of $C_{23}H_{14}O_{10}Os_3W$.

TETRACARBONYL(METHYLENETRIPHENYLPHOSPHORANE-C)IRON(II)
$C_{23}H_{17}FeO_4P$ $Ph_3PCH_2Fe(CO)_4$

L. TOUPET, B. WEINBERGER, H. DES ABBAYES and U. GROSS, 1984. Acta Cryst., C40, 2056-2058.

Triclinic, $P\bar{1}$, a = 9.535, b = 10.824, c = 11.099 Å, α = 105.40, β = 92.15, γ = 108.71°, D_m = 1.4, Z = 2. Mo radiation, R = 0.068 for 2960 reflexions.

In the molecule (Fig. 1), the Fe-C5-P angle is 118.6(2)° and Fe-C-O angles are 172.4(3) to 178.9(3)°. Some bond lengths are: Fe-C (equatorial) 1.763-1.783(4), Fe-C (axial) 1.744(4), Fe-C5 2.122(3), C5-P 1.755(3), P-C(Ph) 1.787-1.799(3) Å.

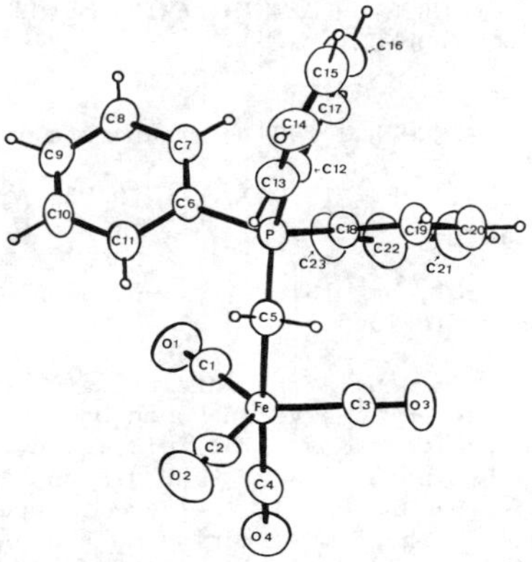

Fig. 1. The $C_{23}H_{17}FeO_4P$ molecule.

3-t-BUTYL-4-DIMETHYL-2-(2-PYRIDONO)PENT-2-EN-1-YLIDENEHEPTACARBONYLDIIRON(0)
$C_{23}H_{23}Fe_2NO_7$

H. BERKE, U. GROSSMANN, G. HUTTNER and O. ORAMA, 1984. Z. Naturforsch., B39, 1759-1766.

Monoclinic, $P2_1/c$, a = 10.599, b = 15.074, c = 20.830 Å, β = 136.55°, Z = 4. Mo radiation, R = 0.035 for 2567 reflexions (at -60°C).

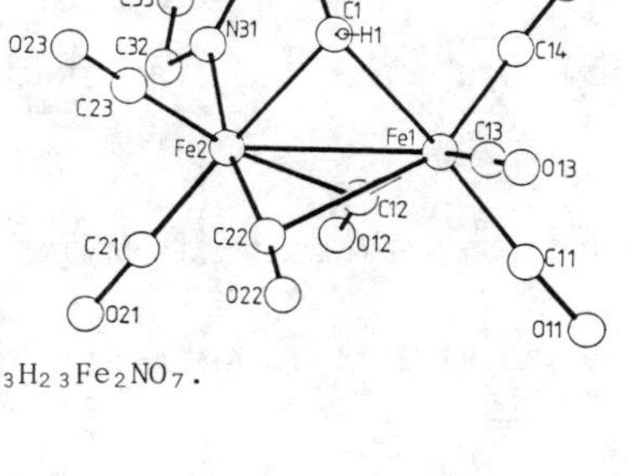

Fig. 1. A view of $C_{23}H_{23}Fe_2NO_7$.

In the molecule (Fig. 1) the bridging carbonyls are asymmetrically disposed (Fe(1)-C(12) 1.817(5), Fe(2)-C(12) 2.360(6), Fe(2)-C(22) 1.781(5), Fe(1)-C(22) 2.766(6). Other principal dimensions are Fe(1)-Fe(2) 2.595(2), Fe(1)-C(1) 2.071(7), Fe(2)-C(1) 2.014(5), C(1)-C(2) 1.467(8), C(2)=C(3) 1.36(1), Fe-C (terminal) 1.784-1.825(6), Fe(2)-N(31) 2.034(4) Å.

BIS(μ_2-CARBONYL)-(μ_2-CHLORO)-(DICARBONYLCOBALT)BIS(DIMETHYLBENZYLAMINE-C,N)-DIPALLADIUM
$C_{23}H_{26}ClCoN_2O_4Pd_2$ (I)

DI-μ-CARBONYL-μ_3-CARBONYL[μ-CHLOROBIS[2-[(DIMETHYLAMINO)METHYL]PHENYL-C,N]DIPALLADIUM]-(η^5-2,4-CYCLOPENTADIEN-1-YL)MOLYBDENUM
$C_{26}H_{29}ClMoN_2O_3Pd_2$ (II)

M. PFEFFER, J. FISCHER and A. MITSCHLER, 1984. Organometallics, 3, 1531-1539.

I. Orthorhombic, Pnma, a = 21.844, b = 11.947, c = 10.754 Å, Z = 4. Mo radiation, R = 0.024 for 1974 reflexions.

II. Monoclinic, $P2_1/n$, a = 20.501, b = 15.624, c = 8.403 Å, β = 99.83°, Z = 4. Mo radiation, R = 0.040 for 2330 reflexions.

Molecule I consists of two Pd(dmba) fragments bridged by a Cl atom and a $Co(CO)_4$ moiety; it may be considered to have an "open book" structure (Fig. 1). Two CO groups are semibridging to the Pd atoms. The Pd(1)...Pd(2) distance is 3.442 Å and the Co-Pd distances are 2.594 and 2.595 Å. The distances Pd(1)-Cl, Pd(2)-Cl, Pd(1)-N(1), Pd(2)-N(2), Pd(1)-C(5) and Pd(2)-C(14) are respectively 2.434, 2.434, 2.141, 2.146, 1.998 and 2.003 Å. The semi-bridging CO groups have Pd-C distances of 2.170 and 2.138 Å, and Co-C distances of 1.780 and 1.809 Å. The angles Pd(1)-Cl-Pd(2), Pd(1)-Co-Pd(2), Pd(1)-C(1)-Co and Pd(2)-C(2)-Co are 90.01, 83.13, 81.4 and 81.7° respectively. In II, two cyclopalladated units are held together by a Cl atom and an $Mo(CO)_3Cp$ moiety. The Pd-Mo distances are short (2.832 and 2.788 Å) and the Pd(1)...Pd(2) interaction is 3.241 Å. The Pd-Cl distances are 2.470 and 2.454 Å with a Pd(1)-Cl-Pd(2) angle of 82.32 Å. The C(3)-O(3) carbonyl group triply-bridges the metal atoms with Mo-C(3) 2.05, Pd(1)-C(3) 2.39 and Pd(2)-C(3) 2.37 Å. Other distances are: Pd(1)-C(1) 2.28, Pd(1)-N(1) 2.15, Pd(1)-C(9) 1.99, Pd(2)-C(2) 2.20, Pd(2)-N(2) 2.15, Pd(2)-C(18) 1.99, Mo-C(1) 1.98 and Mo-C(2) 1.97 Å.

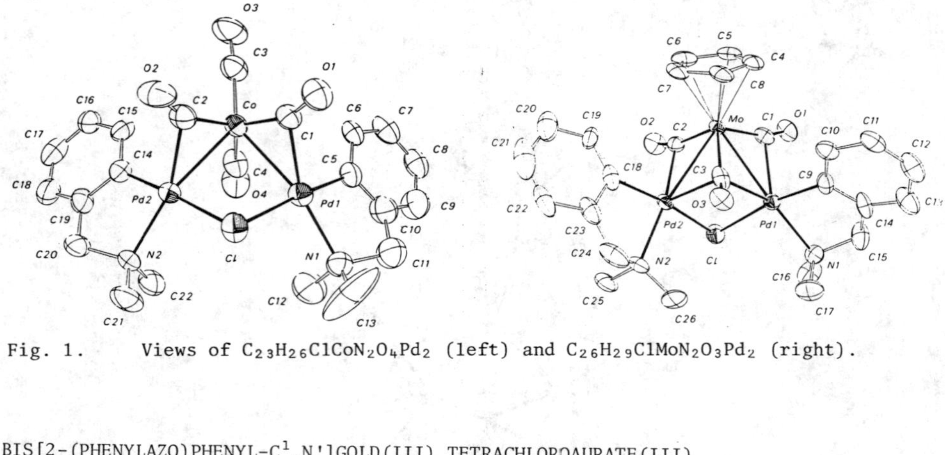

Fig. 1. Views of $C_{23}H_{26}ClCoN_2O_4Pd_2$ (left) and $C_{26}H_{29}ClMoN_2O_3Pd_2$ (right).

BIS[2-(PHENYLAZO)PHENYL-C[1],N']GOLD(III) TETRACHLOROAURATE(III)
$C_{24}H_{18}Au_2Cl_4N_4$

J. VICENTE, M.-T. CHICOTE, M.-D. BERMUDEZ, X. SOLANS and M. FONT-ALTABA, 1984. J. Chem. Soc. Dalton, 557-562.

Triclinic, $P\bar{1}$, a = 18.113, b = 9.752, c = 7.515 Å, α = 82.84, β = 101.90, γ = 99.17°, Z = 2. Mo radiation, R = 0.063 for 2753 reflexions.

The cation (Fig. 1) has square-planar geometry with two cis N atoms (Au-N 2.17(2) av.) and two cis-C atoms (Au-C 2.00(2) Å av.) bonded to gold. The Au atoms in $AuCl_4^-$ are located on inversion centres and display regular square-planar coordination (Au-Cl 2.284(9) Å, Cl-Au-Cl 90.0(4)°). The structure consists of discrete ions linked by van der Waals and ionic forces.

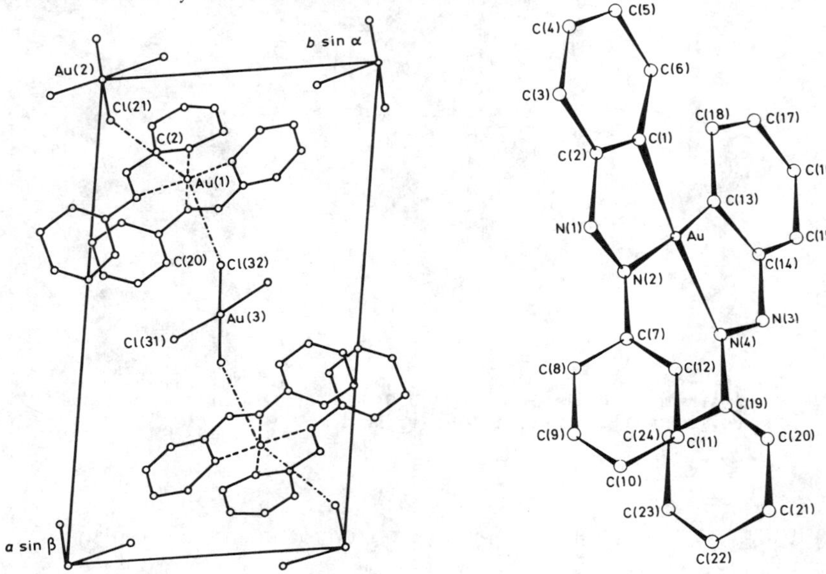

Fig. 1. $C_{24}H_{18}Au_2Cl_4N_4$: the structure of the cation and the unit-cell contents of the complex.

TRIS(o-PHENYLENEDIMETHYLENE-C(α),C(α'))TUNGSTEN(VI) BENZENE SOLVATE
$C_{24}H_{24}W,0.5(C_6H_6)$ (I)

BIS[OXOBIS(o-PHENYLENEDIMETHYLENE-C(α),C(α'))TUNGSTEN(V)]TETRAKIS(TETRAHYDROFURAN)-MAGNESIUM
$C_{48}H_{64}MgO_6W_2$ (II)

M.F. LAPPERT, C.L. RASTON, G.L. ROWBOTTOM, B.W. SKELTON and A.H. WHITE, 1984. J. Chem. Soc. Dalton, 883-891.

I. Trigonal, $R\bar{3}$, a = 10.125 Å, α = 91.65°, Z = 2. Mo radiation, R = 0.046 for 952 reflexions.

II. Orthorhombic, Pbca, a = 18.061, b = 16.759, c = 15.637 Å, Z = 4. Mo radiation, R = 0.047 for 1102 reflexions.

Compound I is composed of discrete complex molecules and solvent while the complex in II lies on a crystallographic inversion centre. The o-phenylenedimethylene ligands behave as chelating ligands in both I and II (Fig. 1) but in II the two ligands are bonded differently. One is typically metallacyclic in nature with W-C(α) 2.16(2) Å and W-C(α)-C (aromatic) 109(1)° and a fold angle of 42.4° whereas the other has a larger fold angle (66.1°), with W-C(α) 2.16(4) Å, and W-C(α)-C (aromatic) 98(2), 93(1)°. For the trischelate I all the ligands are equivalent and the tungsten symmetry is quasi 3/m. Here W-C is 2.21, W-C(aromatic) is 2.49 Å, W-C-C(aromatic) is 82.7 and the fold angle is 83.3°. In II the Mg atom is octahedrally coordinated (Mg-O 2.9 Å) with four thf ligands disposed equatorially. The thf ligands are disordered. The W-O distance is short (1.71(1) Å). In I the solvent molecule has imposed symmetry $\bar{3}$.

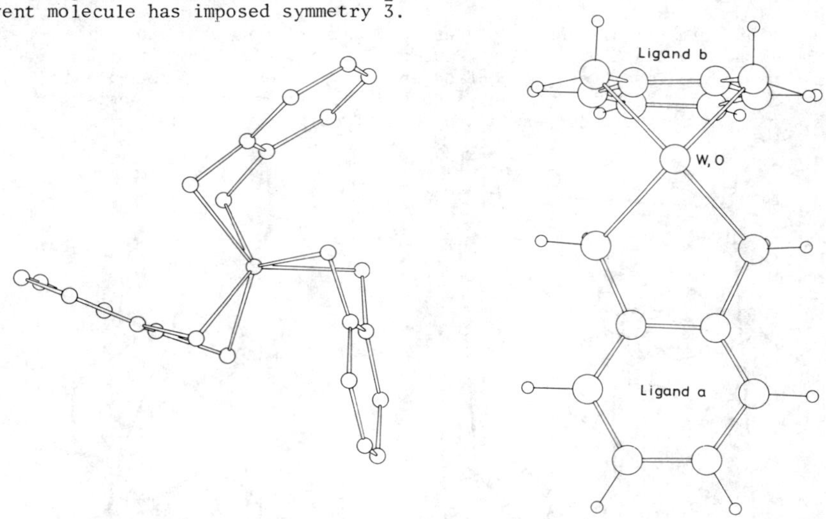

Fig. 1. A single molecule of $C_{24}H_{24}W,0.5(C_6H_6)$ (left) and the tungsten environment in $C_{48}H_{64}MgO_6W_2$ (right).

(μ_2-FORMYLMETHYLENE)BIS(CARBONYL(η^5-PENTAMETHYLCYCLOPENTADIENYL)RHODIUM)
$C_{24}H_{32}O_3Rh_2$

W. KALCHER, W.A. HERRMANN, C. PAHL and M.L. ZIEGLER, 1984. Chem. Ber., 117, 69-78.

Monoclinic, $P2_1/c$, a = 17.339, b = 16.584, c = 17.923 Å, β = 111.51°, Z = 8. Mo radiation, R = 0.078 for 3901 reflexions.

The analysis established the structure as shown in Fig. 1. Some dimensions are Rh-Rh 2.675(2), Rh-C1 1.99(3) and 2.00(2), Rh-C(Cp) 2.18(2)-2.35(2) Å.

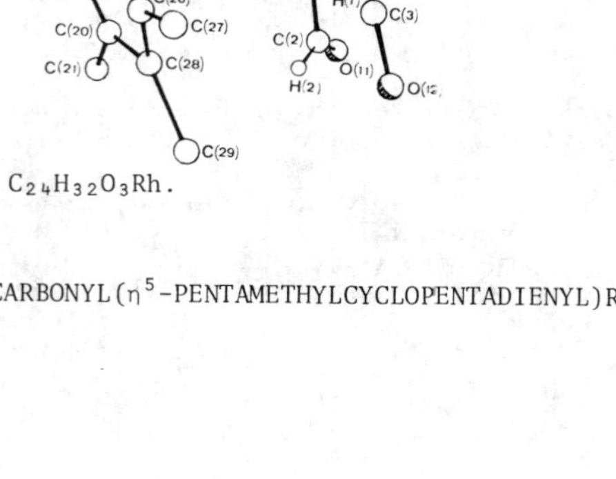

Fig. 1. A view of $C_{24}H_{32}O_3Rh$.

BIS(μ_2-METHYLENE)BIS(CARBONYL(η^5-PENTAMETHYLCYCLOPENTADIENYL)RHODIUM(IV)) BIS(TETRAFLUOROBORATE)
$C_{24}H_{34}B_2F_8O_2Rh_2$

K. ISOBE, S. OKEYA, N.J. MEANWELL, A.J. SMITH, H. ADAMS and P.M. MAITLIS, 1984. J. Chem. Soc. Dalton, 1215-1221.

Orthorhombic, Pbca, a = 13.717, b = 14.355, c = 14.482 Å, Z = 4. Mo radiation, R = 0.056 for 2321 reflexions.

The cation (Fig. 1) is centrosymmetric with the inversion centre in the middle of the planar di-μ-methylenedirhodium ring. Each Rh atom is η^5-bonded to one C_5Me_5 ring (average Rh-C(cp) 2.234 Å) and σ-bonded to two methylene C (2.060) and one carbonyl (1.908 Å). The Rh-Rh distance is 2.659 Å. The BF_4^- anion is disordered.

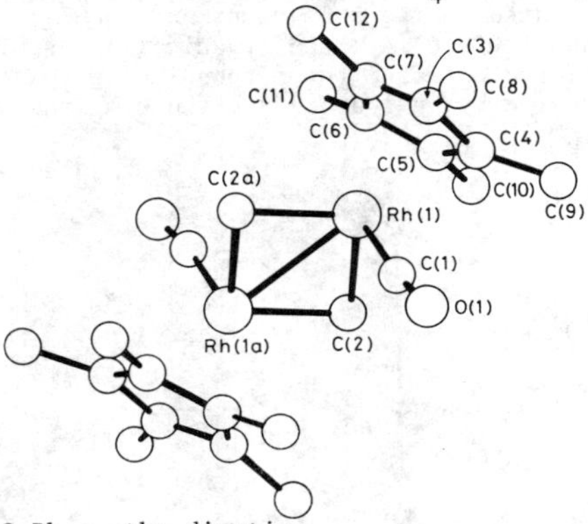

Fig. 1. $C_{24}H_{34}B_2F_8O_2Rh_2$: the dication.

(α,α'-BIS(TRIMETHYLSILYL)-o-PHENYLENEDIMETHYLENE-C(α),C(α'))BIS(η^5-CYCLOPENTADIENYL)-ZIRCONIUM(IV) BENZENE SOLVATE
$C_{24}H_{34}Si_2Zr,1.33(C_6H_6)$

M.F. LAPPERT, C.L. RASTON, B.W. SKELTON and A.H. WHITE, 1984. J. Chem. Soc. Dalton, 893-902.

Hexagonal, $P6_3/m$, a = 17.849, c = 16.923 Å, Z = 6. Mo radiation, R = 0.044 for 1645 reflexions.

The unit cell comprises discrete complex molecules (Fig. 1) lying on a mirror plane, plus a benzene molecule, also of m symmetry, the mirror plane passing through the midpoints of two C-C bonds, together with a disordered benzene molecule centred at the origin. The mirror plane bisects each cyclopentadienyl ligand and the metallacyclic framework. Bonding within the metallocene is the usual (bis-η^5) arrangement, Zr-C being in the range 2.490(5)-2.535(5) Å. The centroid-Zr-centroid angle is 125.0°. Pseudo-tetrahedral coordination of Zr is completed by the o-phenylenedimethylene chelating ligand (Zr-C(4) 2.305(4) Å). The C(4)-Zr-C(4') bite angle is 80.2(2)°. The fold angle of the ligand (dihedral angle between the ZrC(4)C(4') plane and the C_8 extension of the aromatic plane) is 66.7°.

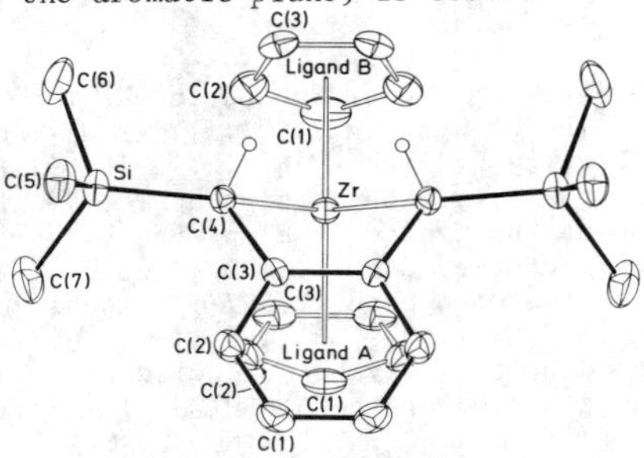

Fig. 1. $C_{24}H_{34}Si_2Zr,1.33(C_6H_6)$: the molecular structure.

BIS(μ-1,8-DIISOCYANO-p-MENTHANE-C,C')TETRAIODODIPALLADIUM(II)
$C_{24}H_{36}I_4N_4Pd_2$ $Pd_2(C_{12}H_{18}N_2)_2I_4$

C.-M. CHE, F.H. HERBSTEIN, W.P. SCHAEFER, R.E. MARSH and H.B. GRAY, 1984. Inorg. Chem., 23, 2572-2575.

Monoclinic, $P2_1/n$, a = 11.766, b = 15.123, c = 9.703 Å, β = 93.93°, Z = 2. Mo radiation, R = 0.080 for 3556 reflexions.

The centrosymmetric binuclear molecule is shown in Fig. 1. The Pd atoms are trans-square planar coordinated to two RNC groups and two I atoms (Pd-I 2.595(1) and 2.587(1), Pd-C 1.942(6) and 1.929(6) Å); the organic ligands are disordered. The coordination planes are tilted 35° from the perpendicular to the Pd-Pd vector, and a fifth (axial) site is occupied by an I atom from the other half of the dimer (Pd-I 3.852(1) Å).

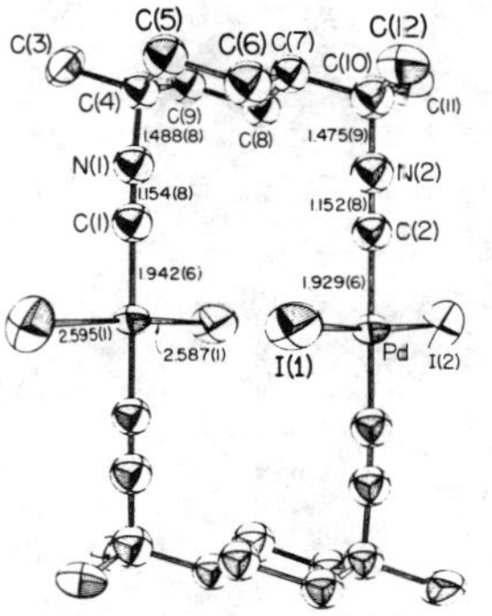

Fig. 1. The $Pd_2(C_{12}H_{18}N_2)_2I_4$ molecule.

2'-(2-3-η-BUT-2-YNE)(μ_2-CARBONYL)-2'-(η^5-CYCLOPENTADIENYL)(μ_2-p-TOLYLMETHYLENE)-1',1'-BIS(TRIMETHYLPHOSPHINE)PLATINUM TUNGSTEN TETRAFLUOROBORATE TETRAHYDROFURAN SOLVATE
$C_{24}H_{37}BF_4OP_2PtW,C_4H_8O$ (I)

(μ_2-CARBONYL)-2'-CARBONYL-2'-(η^5-CYCLOPENTADIENYL)(μ_2-σ.1-3-η-2-METHYL-1-p-TOLYLBUT-2-ENYLIDENE)-1'-(η^5-PENTAMETHYLCYCLOPENTADIENYL)COBALT TUNGSTEN TETRAFLUOROBORATE
$C_{29}H_{34}BCoF_4O_2W$ (II)

J.C. JEFFERY, I. MOORE, H. RAZAY and F.G.A. STONE, 1984. J. Chem. Soc. Dalton, 1581-1588.

I. Monoclinic, $P2_1/n$, a = 10.647, b = 10.214, c = 32.494 Å, β = 99.21°, Z = 4. Mo radiation, R = 0.063 for 3798 reflexions.

II. Monoclinic, $P2_1/c$, a = 10.793, b = 16.503, c = 15.957 Å, β = 93.52°, Z = 4. Mo radiation, R = 0.037 for 3678 reflexions.

In I (Fig. 1) the but-2-yne is η^2-coordinated to the tungsten centre (W-C(4) 2.05(2), W-C(5) 2.02(2) Å). The Pt-W bond (2.771(1)) is asymmetrically bridged by the CO ligand (W-C(2) 1.99(2), Pt-C(2) 2.17(2) Å, W-C(2)-O 151.9(15), Pt-C(2)-O 124.3(13)°) and symmetrically bridged by the tolylmethylene group (C(1)-W 2.10(2), C(1)-Pt 2.11(2) Å). Coordination about the W atom is completed by a η^5-C_5H_5 ligand while a planar environment about Pt is completed by two PMe_3 groups (Pt-P(1),P(2) 2.309(6), 2.316(5) Å). In compound II, the Co-W bond (2.552(1)) is asymmetrically bridged by a CO ligand (W-C(6) 2.014(7), Co-C(6) 1.99(2) Å, W-C(6)-O 150.7(6), Co-C(6)-O 130.1(6)°) and is also spanned by the tolylbut-2-enylidene ligand via an η^3-allylic attachment to the W atom and a σ bond of the $C(C_6H_4Me\text{-}4)$ end of the chain to the Co atom (W-C(1),C(2),C(3) 2.225(7), 2.352(7), 2.377(8), Co-C(1) 1.939(6) Å). The W atom carries a terminal CO ligand and the η^5-C_5H_5 group and the Co atom the η^5-C_5Me_5 ring.

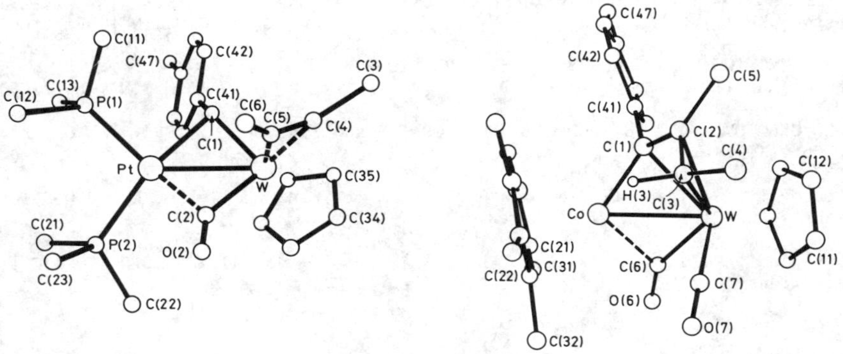

Fig. 1. The cations in $C_{24}H_{37}BF_4OP_2PtW,C_4H_8O$ (left) and $C_{29}H_{34}BCoF_4O_2W$ (right).

(μ_2-σ^2,η^4-BUTADIENE-1,4-DIYL)(μ_2-ISOPROPOXY)(η^2-ACETYLENE)PENTAKIS(ISOPROPOXY)-DITUNGSTEN

$C_{24}H_{48}O_6W_2$ (I)

(μ_2-σ^2,η^4-1,2,3,4-TETRAMETHYLBUTADIENE-1,4-DIYL)(μ_2-ISOPROPOXO)(η^2-DIMETHYLACETYLENE)-PENTAKIS(ISOPROPOXY)DITUNGSTEN

$C_{30}H_{60}O_6W_2$ (II)

M.H. CHISHOLM, D.M. HOFFMAN and J.C. HUFFMAN, 1984. J. Am. Chem. Soc., 106, 6806-6815.

I. Triclinic, $P\bar{1}$, a = 14.754, b = 12.135, c = 9.011 Å, α = 105.51, β = 88.66, γ = 70.53°, Z = 2. Mo radiation, R = 0.037 for 3188 reflexions (at -160°C).

II. Orthorhombic, Pcab, a = 16.352, b = 16.036, c = 26.335 Å, Z = 8. Mo radiation, R = 0.053 for 2720 reflexions (at -169°C).

Both molecules (I) and (II) have similar coordination geometry (Fig. 1). One W atom isin a distorted octahedral environment being coordinated to three terminal OR ligands, a bridging OR ligand, and forming two W-C σ bonds (mean W-C 2.13(2) Å) to the μ-C_4R_4' ligand. The other W atom is coordinated to one bridging and two terminal OR ligands, the η^2-alkyne (mean W-C 2.07(2) Å, mean C-C-C 136(3)°) and the four C atoms of the η^4-C_4R_4' ligand (mean W-C 2.40(2) Å). The η^2-C_2 and η^4-C_4 ligands are normal to one another. W-W distances are 2.877(1) (I) and 2.851(1) (II), indicative of single bonds.

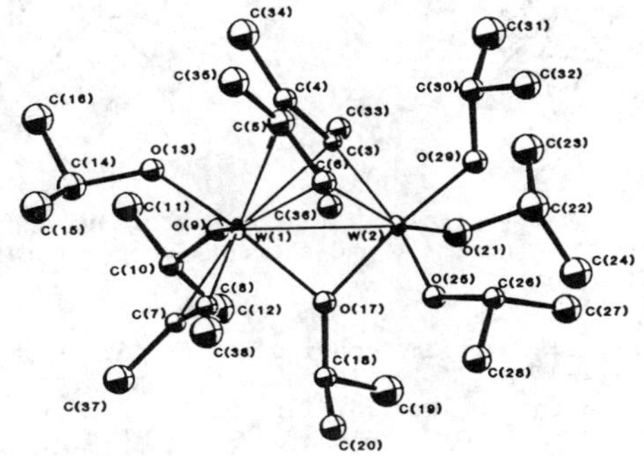

Fig. 1. A view of $C_{30}H_{60}O_6W_2$.

TETRAETHYLAMMONIUM (μ_5-CARBIDO)HEXADECACARBONYLPENTAIRONRHODIUM
$C_{25}H_{20}Fe_5NO_{16}Rh$ (I) $(C_2H_5)_4N^+[RhFe_5C(CO)_{16}]^-$

TETRAETHYLAMMONIUM HEXADECACARBONYLPENTAIRONRHODIUM
$C_{24}H_{20}FeNO_{16}Rh_5$ (II) $(C_2H_5)_4N^+[Rh_5Fe(CO)_{16}]^-$

Yu.L. SLOVOKHOTOV, Yu.T. STRUCHKOV, V.E. LOPATIN and S.P. GUBIN, 1984. J. Organometal. Chem., 266, 139-146.

I. Tetragonal, P4/n, a = 12.875, c = 9.755 Å, Z = 2. Mo radiation, R = 0.035 for 928 reflexions (-120°C).

II. Tetragonal, P4/n, a = 13.439, c = 9.465 Å, Z = 2. Mo radiation, R = 0.033 for 1598 reflexions (-120°C).

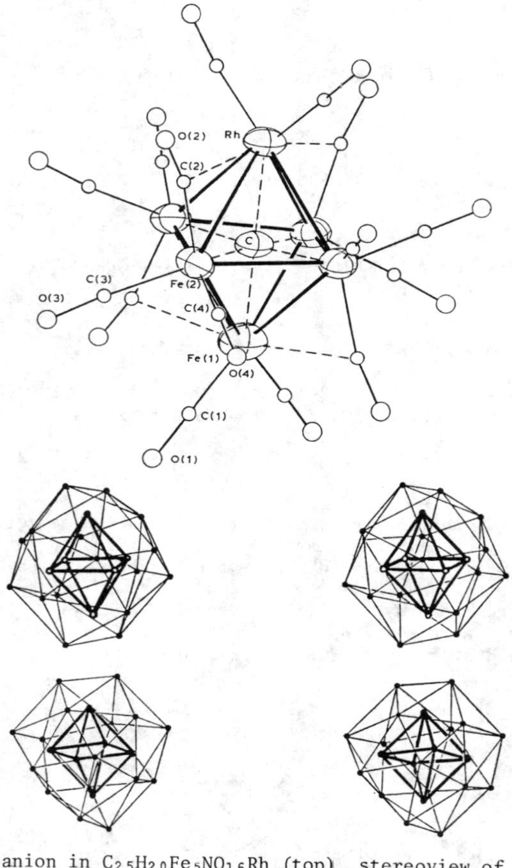

Fig. 1. Cluster anion in $C_{25}H_{20}Fe_5NO_{16}Rh$ (top), stereoview of cluster I within its ligand polyhedron (centre), and stereoview of cluster II within its ligand polyhedron (bottom).

In both anions identical CO-ligand configurations corresponding to the minimum ligand repulsion are realised (Fig. 1). The orientation of the disordered M_5M' cluster nucleus relative to the CO-shell differs in I (12 terminal and 4 semi-bridging CO) and II (12 terminal and 4 μ_3-CO). In the anion of I (Fig. 1) the Rh and Fe atoms occupy with equal probability the 'axial' vertices on the $\bar{4}$ axis. In anion II the Fe atom is statistically disordered at all the six vertices. In I the Rh-Fe(2), Fe(1)-Fe(2), Fe(2)-Fe(2') distances are 2.823, 2.723 and 2.666 Å respectively while in II the 'M(1)'-'M(2)' and 'M(2)'-'M(2')' distances are 2.742 and 2.748 Å respectively. The Rh-C(carbide) and Fe(1)-C(carbide), Fe(2)-C(carbide) distances are 2.037, 1.900 and 1.884 Å respectively, in I. The M...C(2) distances in I decrease to 2.574 Å (Rh) and 2.558 Å (Fe(1)) accompanied by an apparent distortion in the linear coordination Fe(2)-C(2)-O(2) (163.3°) and elongation of the Fe(2)-C(2) bond to 1.855 Å as compared to the other two bonds (Fe(2)-C(3) 1.771 and Fe(2)-C(4) 1.762 Å). These distortions indicate the presence of the four semi-bridging ligands.

(μ_2-p-TOLYLMETHYLIDYNE)DICARBONYLTRIS(η^5-CYCLOPENTADIENYL)VANADIUMTUNGSTEN
$C_{25}H_{22}O_2VW$

U. BEHRENS and F.G.A. STONE, 1984. J. Chem. Soc. Dalton, 1605-1607.

Monoclinic, $P2_1/c$, a = 8.118, b = 14.172, c = 18.457 Å, β = 90.71°, Z = 4. Mo radiation, R = 0.097 for 1699 reflexions.

The V-W bond (2.994(5) Å) is bridged (Fig. 1) by the p-tolylmethylidyne ligand (V-C 2.21(2), W-C 1.86(1) Å). One of the two CO groups on W semi-bridges the V-W bond (W-C-O 167(3)°, V-C(O) 2.63(3), W-C(O) 1.97(3) Å). The V and W atoms are also η^5-bonded to two and one cyclopentadienyl groups, respectively.

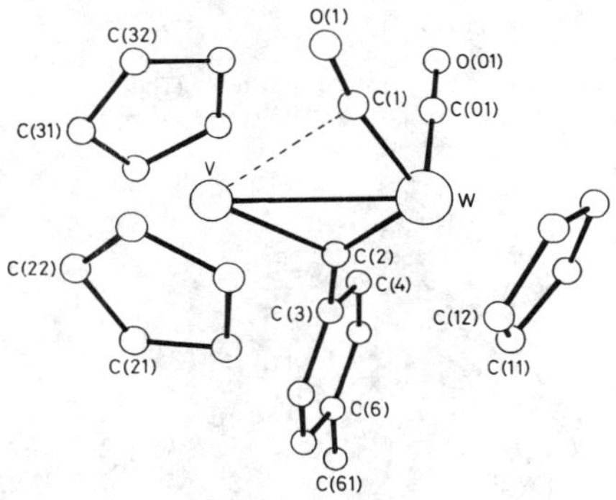

Fig. 1. $C_{25}H_{22}O_2VW$: the molecular structure.

(DICARBONYL-BIS(DIMETHYLPHENYLPHOSPHINE)(4,5-BIS(METHOXYCARBONYL)-1,3-DITHIOL-2-YLIDENE)IRON(0))DICHLOROMERCURY(II) DIETHYLETHER SOLVATE
$C_{25}H_{28}Cl_2FeHgO_6P_2S_2,0.25(C_4H_{10}O)$

H. LE BOZEC, P.H. DIXNEUF and R.D. ADAMS, 1984. Organometallics, 3, 1919-1921.

Monoclinic, C2/c, a = 35.059, b = 11.706, c = 17.600 Å, β = 101.96°, Z = 8. Mo radiation, R = 0.044 for 5250 reflexions.

The complex consists of a combination of the two $HgCl_2$ and $Fe(CO)_2(PMe_2Ph)_2[CS_2-C_2(COOMe)_2]$ moieties, with the Lewis acid ligand $HgCl_2$ in the position trans to the electron-releasing carbene ligand (Hg-Fe-C(9) 176.5°) (Fig. 1). The CO and phosphine ligand are trans to each other. The Hg-Fe bond length is 2.546 Å and the two Hg-Cl bonds are 2.486 and 2.514 Å. The two Cl-Hg-Fe angles are different, Cl(1)-Hg-Fe is 132.86° whilst Cl(2)-Hg-Fe is 123.05°. The Hg atom is 0.157 Å out of the plane Cl(1),Fe,Cl(2), and the dihedral angle between planes Cl(1),Cl(2),Hg and Fe,C(1),C(2), C(9) is 18.2°. Other bonds of interest are: Fe-C(9) (the carbene bond) 1.912 Å, Fe-P(1) 2.282, Fe-P(2) 2.260, C(9)-S(1) 1.688, C(9)-S(2) 1.704, S(1)-C(3) 1.743, S(2)-C(4) 1.739 and C(3)-C(4) 1.324 Å.

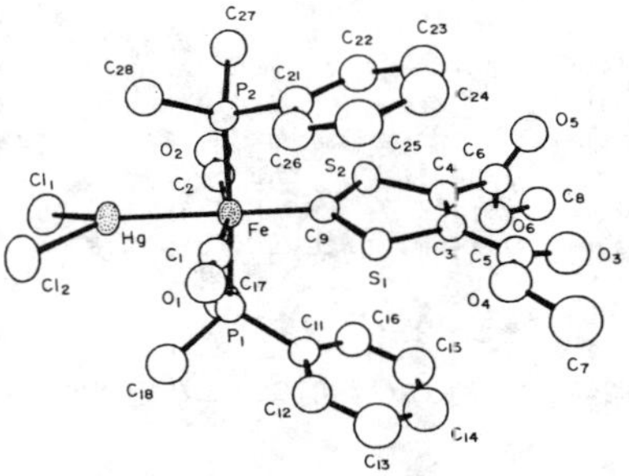

Fig. 1. View of the molecule $C_{25}H_{28}Cl_2FeHgO_6P_2S_2$.

trans-CHLOROMETHYLTRIETHYLPHOSPHINETRIPHENYLARSINEPLATINUM
$C_{25}H_{33}AsClPPt$

A.B. GOEL, S. GOEL and D. VANDERVEER, 1984. Polyhedron, 3, 373-375.

Orthorhombic, $P2_12_12_1$, a = 9.898, b = 10.719, c = 24.342 Å, D_m = 1.72, Z = 4. Mo radiation, R = 0.0521 for 2635 reflexions.

The Pt coordination is close to square-planar (Fig. 1) with Pt-P 2.271(6), Pt-As 2.423(2), Pt-Cl 2.397(5), Pt-C 2.076(16) Å, P-Pt-As 175.5(2), Cl-Pt-As 87.1(1), Cl-Pt-P 92.2(2), C-Pt-P 89.6(6), C-Pt-Cl 177.8(6)°.

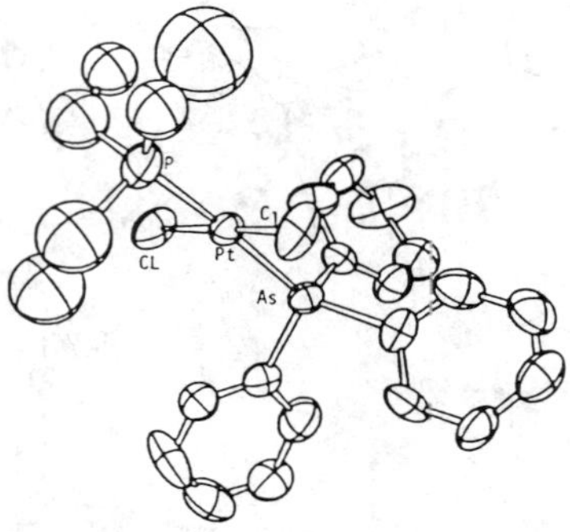

Fig. 1. A drawing of $C_{25}H_{33}AsClPPt$.

σ-PENTAFLUOROPHENYLBIS(TRIETHYLPHOSPHINE)(TRIETHYLPHOSPHONIODITHIOFORMATE-S)PALLADIUM-(II) PERCHLORATE CHLOROFORM SOLVATE
$C_{25}H_{45}ClF_5O_4P_3PdS_2,CHCl_3$

R. USON, J. FORNIES, R. NAVARRO, M.A. USON, M.P. GARCIA and A.J. WELCH, 1984. J. Chem. Soc. Dalton, 345-350.

Monoclinic, $P2_1/a$, a = 14.5965, b = 15.356, c = 19.324 Å, β = 109.257°, Z = 4. Mo radiation, R = 0.105 for 4093 reflexions. [Some crystal decay was detected during data collection.]

The X-ray analysis established the structure of the complex (Fig. 1) and the trans arrangement of the PR_3 ligands (Pd-P(1) 2.318(4), Pd-P(2) 2.351(4) Å). The $P(2)Et_3$ group is disordered. The triethylphosphoniumdithiocarboxylate ligand is bound unidentately to the Pd atom (Pd-S(1) 2.342(3) Å) and adopts a cisoidal conformation. The Pd...S(2) distance is 3.644(4) Å. The perchlorate ion and solvate molecule are unexceptional.

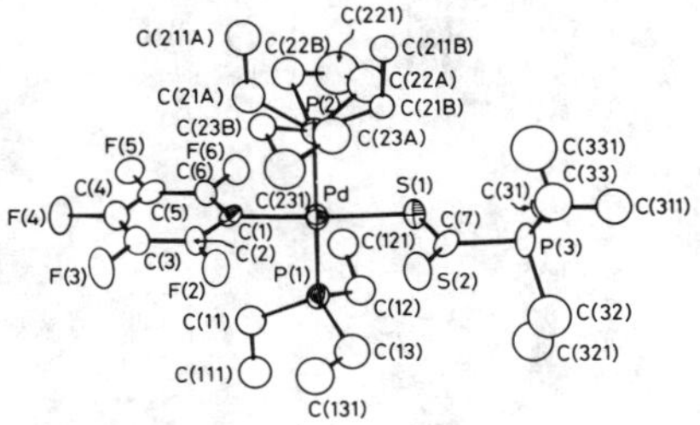

Fig. 1. $C_{25}H_{45}ClF_5O_4P_3PdS_2,CHCl_3$: the cation.

CARBONYLCHLOROHYDRIDO(DI-t-BUTYLCYCLOPROPYLMETHYLPHOSPHINE)(3-DI-t-BUTYLPHOSPHINO-2-METHYL-PROP-1-ENYL-PC^3)IRIDIUM

$C_{25}H_{50}ClIrOP_2$ $IrClH(CO)P(t\text{-}Bu)_2CH_2C(CH_3)\dot{C}H)(P(t\text{-}Bu)_2CH_2CHCH_2CH_2)$

W.J. YOUNGS, B.L. SIMMS and J.A. IBERS, 1984. J. Organometal. Chem., 272, 295-307.

Monoclinic, $P2_1/n$, a = 15.610, b = 15.763, c = 11.973 Å, β = 104.74°, Z = 4. Cu radiation, R = 0.089 for 2326 reflexions, (-151°C).

The Ir atom is in a distorted octahedral array of donor ligands with the hydride ligand at the vacant coordination site (Fig. 1). The coordination angles for Ir are: C(3)-Ir-P(1) 79.9, C(3)-Ir-P(2) 92.3, C(25)-Ir-P(1) 94.3, C(25)-Ir-P(2) 96.5, C(25)-Ir-C(3) 89.6, P(1)-Ir-Cl 93.9 and P(2)-Ir-Cl 93.3°. The angles C(3)-Ir-Cl (173.6°) and P(1)-Ir-P(2) (166.6°) are indicative of the expected folding of the molecule toward the smaller hydride ligand trans to atom C(25).

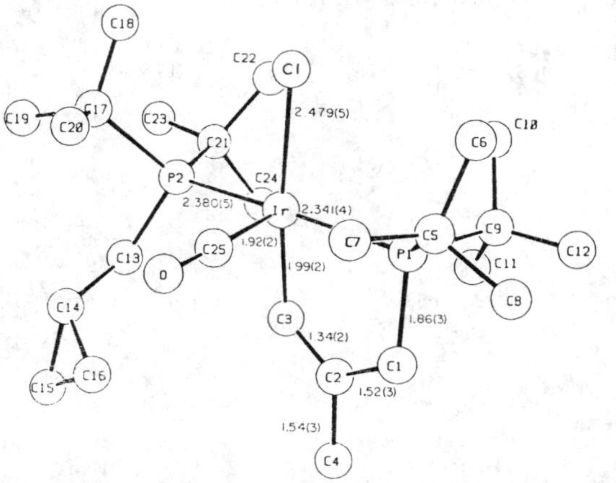

Fig. 1. Molecular structure of $C_{25}H_{50}ClIrOP_2$ showing selected bond distances (Å).

HEXAKIS(μ_2-METHYL)TRIS(TETRAMETHYLETHYLENEDIAMINELITHIUM)METHYLTHORIUM(IV) TETRA-METHYLETHYLENEDIAMINE

$C_{25}H_{69}Li_3N_6Th,C_6H_{16}N_2$

H. LAUKE, P.J. SWEPSTON and T.J. MARKS, 1984. J. Am. Chem. Soc., 106, 6841-6843.

Monoclinic, $P2_1/c$, a = 14.311, b = 12.458, c = 22.757 Å, β = 99.38°, Z = 4. Mo radiation, R = 0.040 for 4744 reflexions (at -110°C).

The Th atom has a distorted monocapped trigonal prismatic array of ligands (Fig. 1). The Th-C(7) distance (2.571(9) Å) is slightly shorter than the other Th-methyl distances (2.667(8)-2.765(9) Å). The Li geometry is as expected with mean Li-C 2.18(3) Å and mean Li-N 2.12(2) Å. The TMEDA molecule of solvation is disordered about an inversion centre.

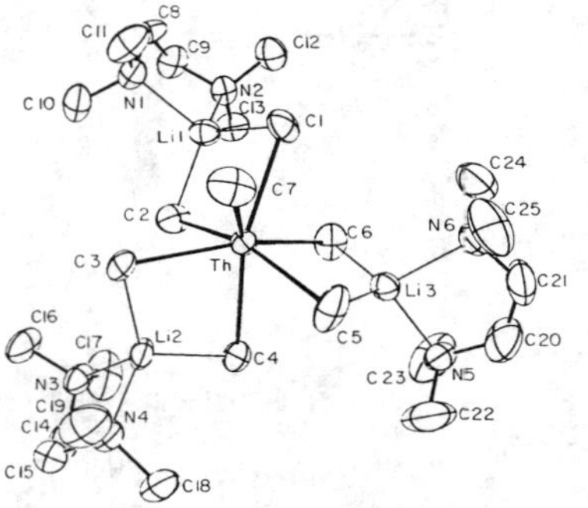

Fig. 1. A view of $C_{25}H_{69}Li_3N_6Th$.

(1-ETHYLYNE-2,3,4,5,6-PENTAFLUOROBENZENE)(TRIPHENYLPHOSPHINE)GOLD(I)
$C_{26}H_{15}AuF_5P$ $[Au(C_2C_6F_5)P(Ph)_3(C_{18}H_{15}P)]$

M.I. BRUCE, E. HORN, J.G. MATISONS and M.R. SNOW, 1984. Aust. J. Chem., 37, 1163-1170.

Monoclinic, $P2_1/c$, a = 16.104, b = 8.653, c = 17.560 Å, β = 112.98°, D_m = 1.92, Z = 4. Mo radiation, R = 0.033 for 1761 reflexions.

This compound crystallises as discrete molecules (Fig. 1) (Au-C 1.993(14), Au-P 2.274(3) Å, P-Au-C 177.9(3), Au-C≡C 175.4(10), C≡C-C 178.4(12)°).

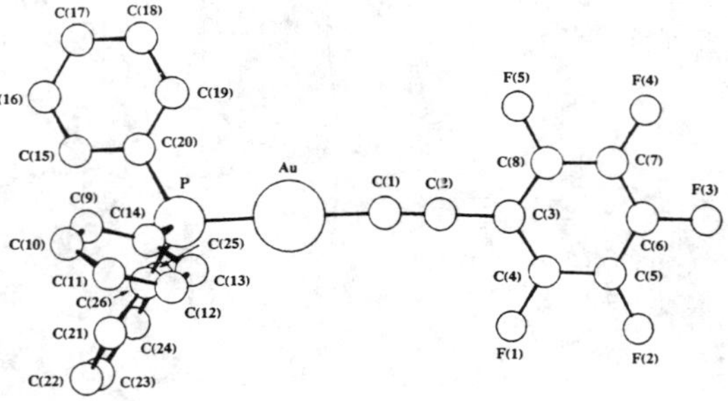

Fig. 1. The structure of $C_{26}H_{15}AuF_5P$.

BIS(μ-METHYLENEDIPHENYLTHIOPHOSPHINATE)DIGOLD(I)
$C_{26}H_{24}Au_2P_2S_2$ (I) $[AuCH_2P(S)(C_6H_5)_2]_2$

BIS(μ-METHYLENEDIPHENYLTHIOPHOSPHINATE)DIIODODIGOLD(II)
$C_{26}H_{24}Au_2I_2P_2S_2$ (II) $[Au(CH_2P(S)(C_6H_5)_2)I]_2$

BIS(μ-METHYLENEDIPHENYLTHIOPHOSPHINATE)DIIODODIGOLD(I,III) 1,2-DICHLOROETHANE SOLVATE
$C_{26}H_{24}Au_2I_2P_2S_2,C_2H_4Cl_2$ (III) $[Au(CH_2P(S)(C_6H_5)_2)_2AuI_2].C_2H_4Cl_2$

A.M. MAZANY and J.P. FACKLER, Jr., 1984. J. Am. Chem. Soc., 106, 801-802.

I. Monoclinic, C2/c, a = 23.874, b = 9.030, c = 12.172 Å, β = 105.61°, Z = 4. Mo radiation, R = 0.047 for 2161 reflexions.

II. Monoclinic, P2/n, a = 12.615, b = 12.804, c = 19.303 Å, β = 94.30°, Z = 4. Mo radiation, R = 0.052 for 3473 reflexions.

III. Monoclinic, C2/c, a = 25.187, b = 6.4465, c = 42.5443 Å, β = 92.14°, Z = 8. Mo radiation, R = 0.10 for 3747 reflexions.

The three structures are shown in Fig. 1. The Au(I)-Au(I) dimer I has a trans ligand geometry, which is retained upon oxidation to the isovalent Au(II) iodide dimer. The Au...Au distance is 3.040(1) Å in I and the Au-Au bond averages 2.609(1) Å in II. The Au(I)-Au(III) dimer in III exhibits a cis chelate ligand configuration with both a linear S-Au(I)-S geometry and a trans-$Au^{III}I_2C_2$ planar arrangement. Bond lengths: for I; Au-C 2.115(9), Au-S 2.323(3); for II; Au-C 2.092(15) and 2.101(15), Au-S 2.370(5) and 2.369(4), Au-I 2.693(2) and 2.681(1); for III; Au-C 2.12(4) and 2.13(5), Au-S 2.308(13) and 2.316(12), Au-I 2.615(4) and 2.611(4) Å.

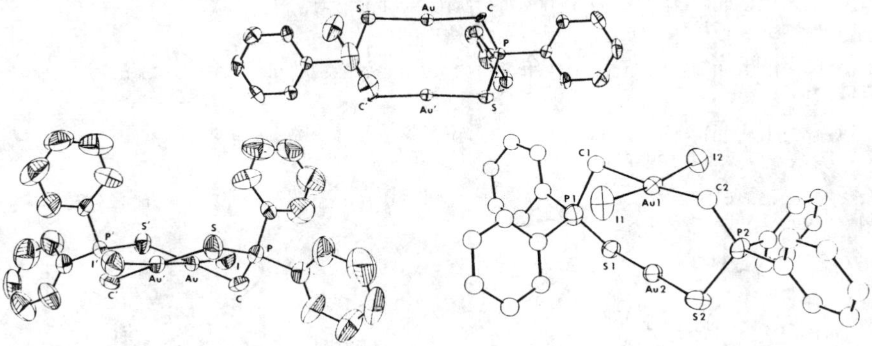

Fig. 1. The structures of $[AuCH_2P(S)(C_6H_5)_2]_2$ (top), $[Au(CH_2P(S)(C_6H_5)_2)I]_2$ (lower left), and $[Au(CH_2P(S)(C_6H_5)_2)_2AuI_2]$ (lower right).

BIS[(1,1'-DIMERCURY)-2-(DIMETHYLAMINOMETHYL)FERROCENOPHANE]
$C_{26}H_{30}Fe_2Hg_2N_2$

L.G. KUZ'MINA, Yu.T. STRUCHKOV, D.A. LEMONOVSKII, I.F. URAZOVSKII, I.E. NIFANT'EV and E.G. PEREVALOVA, 1983. Koord. Khim., 9, 1212-1219 [Sov. J. Coord. Chem., 9, 694-700].

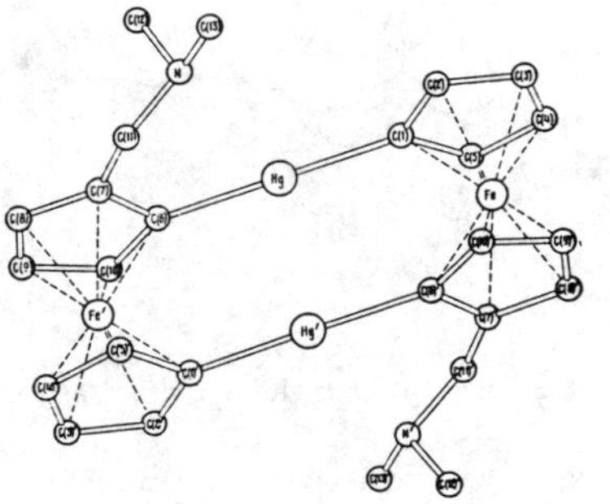

Fig. 1. $C_{26}H_{30}Fe_2Hg_2N_2$: the structure of one of the crystallographically independent molecules.

Triclinic, $P\bar{1}$, a = 10.062, b = 11.684, c = 12.517 Å, α = 114.46, β = 111.33, γ = 96.17°, Z = 2. Mo radiation, R = 0.030 for 4725 reflexions (at -120°C).

Each of the Hg atoms in the two independent centrosymmetric molecules (A,B) of the ferrocenophane is bonded to the C atoms in the Cp rings of two ferrocenyl moieties (Hg-C(1) 2.041(7), 2.064(6), Hg-C(6) 2.065(7), 2.055(6) Å) (Fig. 1). The Hg...N distances are 2.922(6) and 3.049(6) Å in A and B respectively. The ferrocenyl groups have the expected geometry with eclipsed conformations. The intramolecular Hg...Hg' distances are 3.323(2) and 3.362(2) Å in A and B respectively.

(μ_2-σ^2,η^2,η^6-3,4,5,6,7,8-HEXAMETHYLDECA-3,5,7-TRIENE-2,9-DIYLIDENE)BIS(η^5-CYCLOPENTADIENYL)DIMOLYBDENUM
$C_{26}H_{34}Mo_2$ (I)

(μ_2-σ^2,η^2,η^6-3,4,5,6,7,8-HEXAMETHYLDECA-3,5,7-TRIEN-2-YL-9-YLIDENE)BIS(η^5-CYCLOPENTADIENYL)DIMOLYBDENUM BIS(TRIFLUOROACETATO)HYDROGENATE
$C_{30}H_{36}F_6Mo_2O_4$ (II)

M. GREEN, N.C. NORMAN, A.G. ORPEN and C.J. SCHAVERIEN, 1984. J. Chem. Soc. Dalton, 2455-2465.

I. Triclinic, $P\bar{1}$, a = 8.449, b = 10.114, c = 14.872 Å, α = 86.66, β = 80.88, γ = 63.78°, Z = 2. Mo radiation, R = 0.020 for 4635 reflexions.

II. Monoclinic, $P2_1/c$, a = 10.648, b = 15.119, c = 19.655 Å, β = 105.30°, Z = 4. Mo radiation, R = 0.029 for 3781 reflexions.

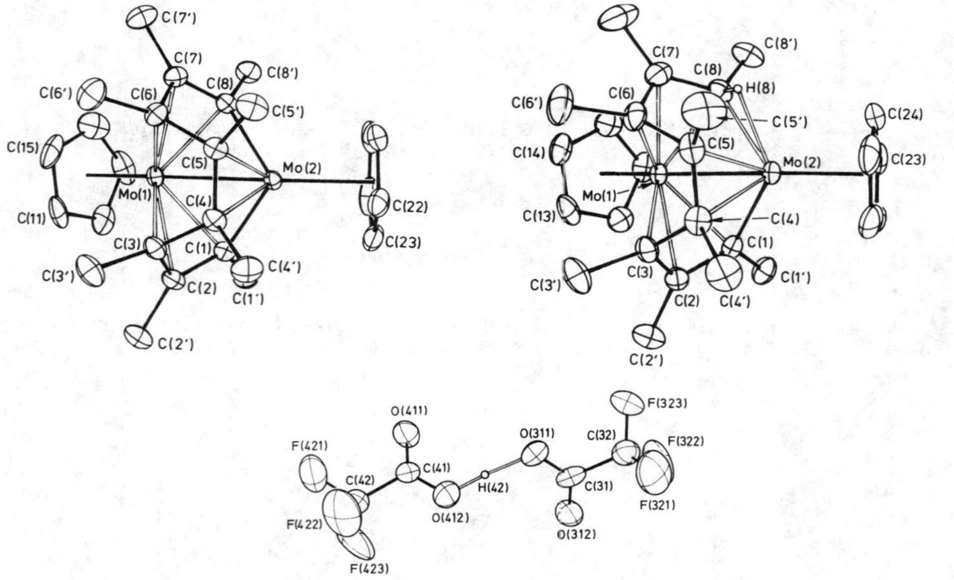

Fig. 1. The molecular structure of $C_{26}H_{34}Mo_2$ (left), the complex cation in $C_{30}H_{36}F_6Mo_2O_4$ (top right) and the complex anion (bottom right).

The molecule of complex I shows non-crystallographic C_s symmetry and there is a formal Mo=Mo double bond (2.595(1) Å) (Fig. 1). The C_8 carbon chain binds to the dimolybdenum unit such that it forms σ bonds to one Mo atom (Mo(2)) of mean length 2.092(2) Å. Each Mo atom carries an η^5-cyclopentadienyl group. Complex II comprises a complex cation and a bis(trifluoroacetato)hydrogenate anion. The non-hydrogen atom geometry of the complex is similar to that of I, with the exception that one of the terminal Mo-C σ bonds is markedly extended (2.196(5) Å). This bond is bridged by a hydrogen atom which shows Mo-H and C-H distances of 1.88(8) and 0.89(8) Å and an Mo-H-C angle of 99(5)°. The anion contains two CF_3CO_2 units linked by a strong, short hydrogen bond (O...O 2.429(8), O(311)-H(42) 1.35(7), O(412)-H(42) 1.08(7) Å).

DIBENZYLTETRAKIS(μ2-PROPIONATO)DITUNGSTEN
$C_{26}H_{34}O_8W_2$ (I) $W_2(CH_2Ph)_2(O_2CEt)_4$

DINEOPENTYLTETRAKIS(μ2-PROPIONATO)DITUNGSTEN
$C_{22}H_{42}O_8W_2$ (II) $W_2(CH_2\text{-t-Bu})_2(O_2CEt)_4$

M.H. CHISHOLM, D.M. HOFFMAN, J.C. HUFFMAN, W.G. VAN DER SLUYS and S. RUSSO, 1984. J. Am. Chem. Soc., 106, 5386-5388.

I. Monoclinic, $P2_1/c$, a = 10.877, b = 8.414, c = 15.153 Å, β = 94.65°, Z = 2. R = 0.035 for three-dimentional data (at -157°C).

II. Triclinic, $P\bar{1}$, a = 12.141, b = 9.133, c = 7.37 Å, α = 63.06, β = 101.85, γ = 80.47°, Z = 1. R = 0.059 for three-dimensional data (at -158°C).

Both molecules have crystallographic inversion symmetry and similar structures; that for I is in Fig. 1. Main bond lengths (averaged were appropriate) are for I (with data for II in parentheses) W-W 2.186(1) (2.187(2)), W-C 2.18(1) (2.21(2)), W-O 2.084(9) (2.08(1)) Å, W-W-C 178.7(4) (169.7(4))°.

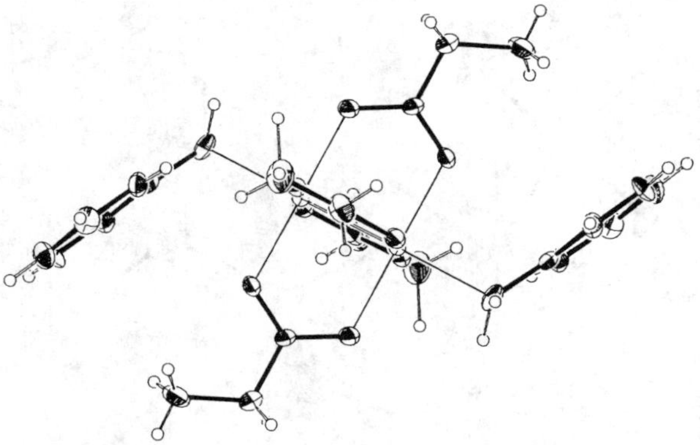

Fig. 1. A view of the centrosymmetric $W_2(CH_2Ph)_2(O_2CEt)_4$ molecule.

TETRACARBONYLBIS(1,1,3,3-TETRAMETHYL-2-THIOCARBONYLCYCLOHEXANE)DICOBALT(Co-Co)
$C_{26}H_{36}Co_2O_4S_2$ (I) $(C_{11}H_{18}S)_2Co_2(CO)_4$

DICYCLOPENTADIENYLBIS(1,1,3,3-TETRAMETHYL-2-THIOCARBONYLCYCLOHEXANE)DICOBALT(Co-Co)
$C_{32}H_{46}Co_2S_2$ (II) $(C_{11}H_{18}S)_2Co_2(C_5H_5)_2$

(2-t-BUTYL-3,3-DIMETHYL-1-METHYLTHIO-1-BUTENYLIUM)(CYCLOPENTADIENYL)(TRIMETHYL-PHOSPHINE)COBALT IODIDE
$C_{19}H_{35}CoIPS$ (III) $[(C_5H_5)(P(CH_3)_3)Co(SCH_3)CC(C_4H_9)_2]^+.I^-$

D. WORMSBÄCHER, R. DREWS, F. EDELMANN and U. BEHRENS, 1984. J. Organometal. Chem., 270, 93-108.

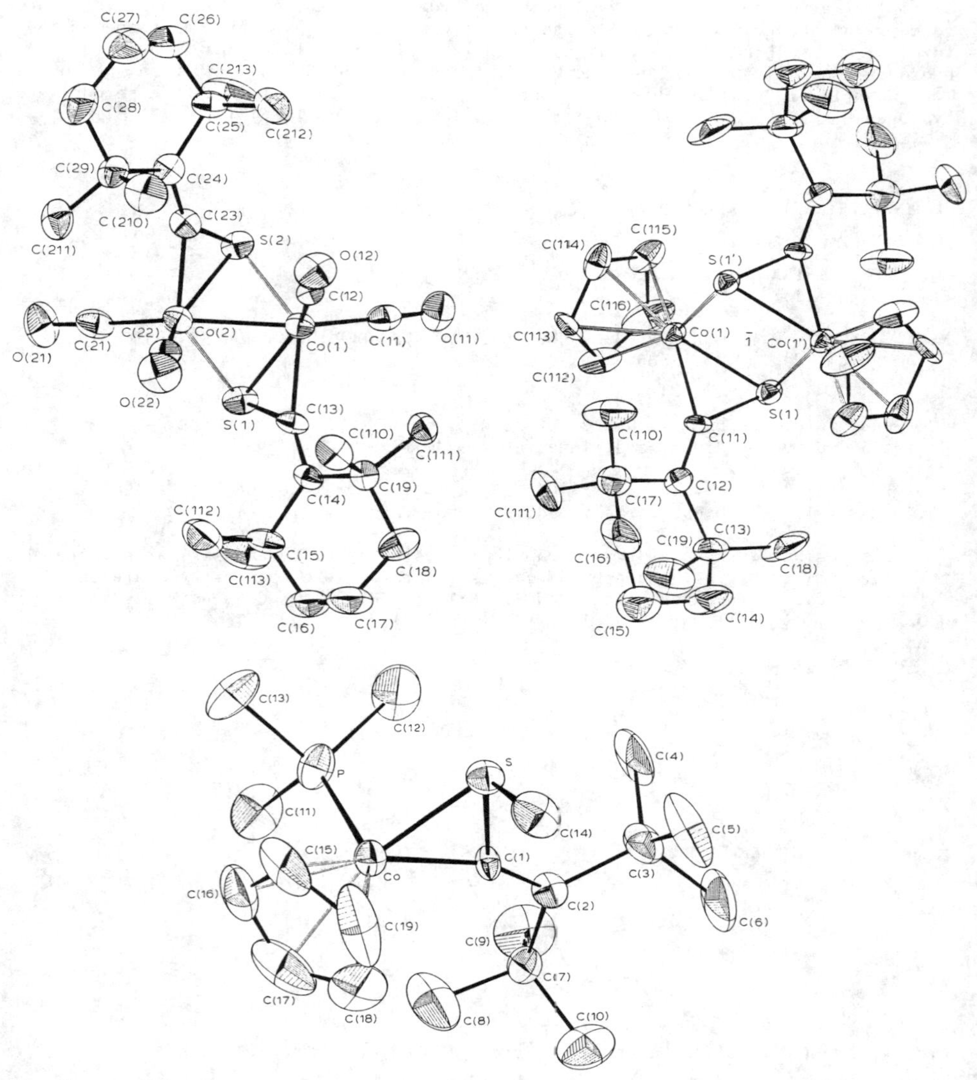

Fig. 1. Molecular structure of $C_{26}H_{36}Co_2O_4S_2$ (top left), one of two independent molecules of $C_{32}H_{46}Co_2S_2$ (top right), and the cation in $C_{19}H_{35}CoIPS$ (bottom).

I. Monoclinic, $P2_1/n$, a = 17.67, b = 8.862, c = 19.90 Å, β = 115.46°, Z = 4. Mo radiation, R = 0.058 for 2356 reflexions.

II. Triclinic, $P\bar{1}$, a = 7.501, b = 13.928, c = 15.736 Å, α = 101.77, β = 90.08, γ = 101.97°, Z = 2. Mo radiation, R = 0.097 for 2834 reflexions.

III. Tetragonal, $P\bar{4}2_1c$, a = 12.568, c = 28.399 Å, Z = 8. Mo radiation, R = 0.052 for 1337 reflexions.

In compounds I and II the central unit consists of a Co_2S_2 backbone (Fig. 1). In I, however, there is a Co(1)-Co(2) bond (2.405 Å) whilst in II there is no Co-Co bond, the metal atoms are related by a centre of symmetry and the average Co...Co separation for the two independent molecules is 3.226 Å. The Co-S distances are, in I: Co-S(1) 2.173, Co(2)-S(1) 2.286 Å; in II, Co(1)-S(1) 2.199 and Co(1)-S(1') 2.250 Å. Molecule I has approximate C_2 symmetry and the distances quoted are the average of 'symmetry-related' bonds. The Co(1)-C(13) distance in I is 1.984 Å and the S(1)-C(13) bond length 1.718 Å. The corresponding distances in II are: Co(1)-C(11) 1.955 Å and S(1)-C(11) 1.75 Å. Relevant angles are: in I, Co(1)-S(1)-Co(2) 65.2, S(1)-Co(1)-S(2) 90.0, S(1)-C(13)-C(14) 137.8°; in II, Co(1)-S(1)-C(1') 92.9, S(1)-Co(1)-S(1') 87.1 and S(1)-C(11)-C(12) 138.0°. Compound III exists as a complex cation (Fig. 1) and an iodide anion. The metal atom is π-bonded to a cyclopentadienyl ring (average Co-C(Cp) 2.10 Å) and to three other ligands (Co-P 2.208, Co-S 2.172, Co-C(1) 1.95 Å). The S-C(1) distance is 1.78 Å and the S-C(14) distance is 1.83 Å. Angles S-Co-C(1), Co-S-C(1) and S-C(1)-Co are respectively 50.7, 58.2 and 71.1°.

(μ_2-CYCLOBUTYLIDENE)BIS(CARBONYL(η^5-PENTAMETHYLCYCLOPENTADIENYL)RHODIUM)
$C_{26}H_{36}O_2Rh_2$

W.A. HERRMANN, C. WEBER, M.L. ZIEGLER and C. PAHL, 1984. Chem. Ber., 117, 875-894.

Triclinic, P$\bar{1}$. a = 9.633, b = 9.882, c = 15.449 Å, α = 98.88, β = 101.46, γ = 113.18°, Z = 2. Mo radiation, R = 0.040 for 4918 reflexions.

Main dimensions in the molecule (Fig. 1) are Rh-Rh 2.655(1), Rh-C(3) 2.050 and 2.062(6) Å. The four-membered ring is puckered as shown in Fig. 1, with C(3)-C(4) 1.530(7), C(3)-C(6) 1.548(10), C(4)-C(5) 1.600(17), C(5)-C(6) 1.473(11) Å.

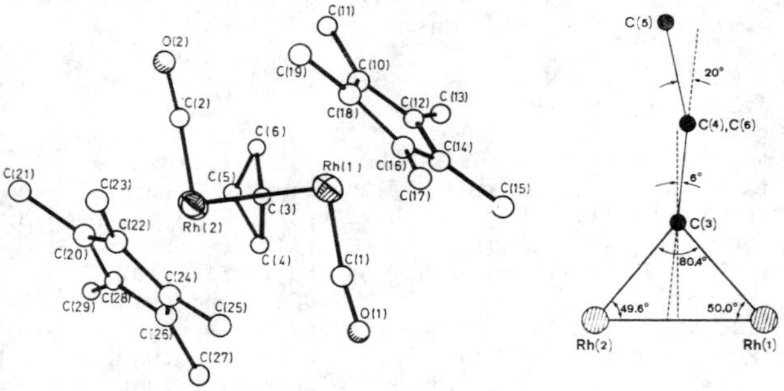

Fig. 1. The $C_{26}H_{36}O_2Rh_2$ molecule and a view of the four-membered ring coordination.

BENZYLCHLORO(QUINOLINE-8-ALDEHYDE-N,C)BIS(PYRIDINE)RHODIUM
$C_{27}H_{23}ClN_3ORh$

J.W. SUGGS and C.-H. JUN, 1984. J. Am. Chem. Soc., 106, 3054-3056.

Monoclinic, $P2_1/n$, a = 10.101, b = 16.476, c = 14.179 Å, β = 93.89°, Z = 4. Mo radiation, R = 0.032 for 2531 reflexions.

The Rh atom is six-coordinate (Fig. 1) with Rh-Cl 2.545(1), Rh-N(1) 2.042(3), Rh-N(2) 2.093(3), Rh-N(3) 2.221(3), Rh-C(10) 1.949(4), Rh-C(11) 2.103(5) Å. The strong trans influence of the benzyl group is apparent (Rh-N(3) is 0.13 Å longer than Rh-N(2)).

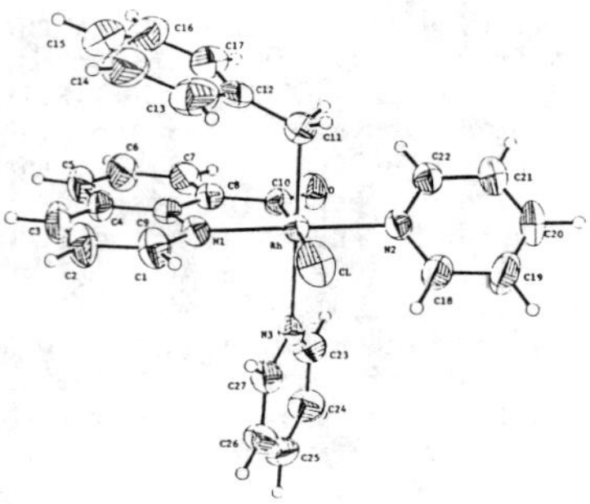

Fig. 1. A view of $C_{27}H_{23}ClN_3ORh$.

BIS(DIETHYLMALONATO-C)[2,2-BIS(2-PYRIDYL)-1,3-DIOXOLANE]PALLADIUM(II)
$C_{27}H_{32}N_2O_{10}Pd$ (I)

BROMO(DIETHYLMALONATO-C)[2,2-BIS(2-PYRIDYL)-1,3-DIOXOLANE]PALLADIUM(II)
$C_{20}H_{23}BrN_2O_6Pd$ (II)

G.R. NEWKOME, V.K. GUPTA, H.C.R. TAYLOR and F.R. FRONCZEK, 1984. Organometallics, 3, 1549-1554.

I. Monoclinic, $P2_1/c$, a = 9.837, b = 21.647, c = 14.089 Å, β = 95.45°, Z = 4. Mo radiation, R = 0.043 for 3507 reflexions.

II. Monoclinic, $P2_1/c$, a = 8.982, b = 14.318, c = 17.368 Å, β = 101.71°, Z = 4. Mo radiation, R = 0.046 for 2783 reflexions.

In both complexes the Pd atom has square-planar coordination (Fig. 1). In I the metal atom is bonded to two N atoms of the dipyridyl ketal ligand and σ-bonded to two malonate C atoms. In II, one of the bonding sites is occupied by Br and the other by the central malonate C atom. The Pd-N(1), Pd-N(2) and Pd-C(14) bond distances are, in I 2.103, 2.117 and 2.097 Å, and in II 2.107, 2.064 and 2.107 Å respectively. The Pd-C(17) distance in I is 2.092 Å and the Pd-Br distance is 2.404 Å in II. The six-membered chelate ring assumes a boat conformation in both compounds. The ketal rings both adopt a half-chair conformation but C(13) lies on the local twofold axis in I, whereas in II O(2) occupies that unique position. The ethoxycarbonyl groups of both complexes are disordered, exhibiting large thermal parameters. The cis angles about Pd range from 82.5 to 96.1° in I and from 85.7 to 95.7° in II.

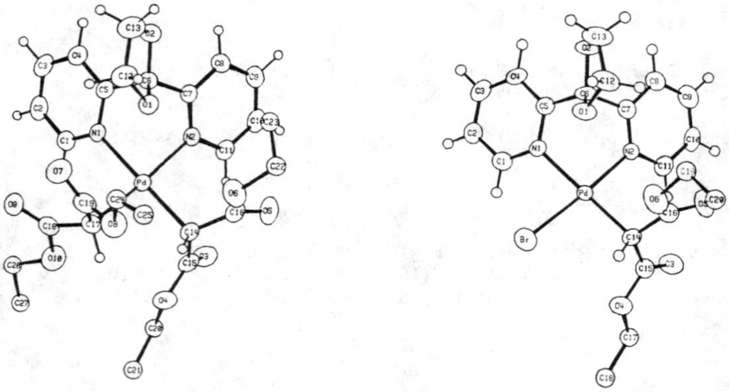

Fig. 1. Molecular structure of $C_{27}H_{34}N_2O_{10}Pd$ (left) and $C_{20}H_{23}BrN_2O_6Pd$ (right).

(η^2-PHENYLISOTHIOCYANATE-C,S)DECAMETHYLVANADOCENE
$C_{27}H_{35}NSV$ (I) $[(\eta^5\text{-}C_5Me_5)_2V(PhNCS)]$

(η^5-PENTAMETHYLCYCLOPENTADIENYL)BIS[(PENTAMETHYLCYCLOPENTADIENYL)(N-PHENYLTHIOAMINO-N,S)]VANADIUM(III)
$C_{44}H_{55}N_2S_2V$ (II) $[(\eta^5\text{-}C_5Me_5)V(PhNC(S)C_5Me_5)_2]$

S. GAMBAROTTA, M.L. FIALLO, C. FLORIANI, A. CHIESI-VILLA and C. GUASTINI, 1984. Inorg. Chem., 23, 3532-3537.

I. Monoclinic, $P2_1/n$, a = 9.876, b = 14.650, c = 16.784 Å, β = 94.40°, Z = 4. Mo radiation, R = 0.052 for 1686 reflexions.

II. Orthorhombic, Pbca, a = 29.791, b = 14.292, c = 18.698 Å, Z = 8. Cu radiation, R = 0.045 for 2644 reflexions.

The two complexes are shown in Fig. 1. In the CS fragment is accommodated in the cavity provided by the two bent C_5Me_5 ligands (V-C 2.042(7), V-S 2.444(2), C-S 1.745(7) Å). Complex II is a V(III) derivative containing a η^5-C_5Me_5 ring and two bidentate chelating thioamido groups (V-S 2.458(4), 2.449(3); V-N 2.154(4), 2.150(4) Å).

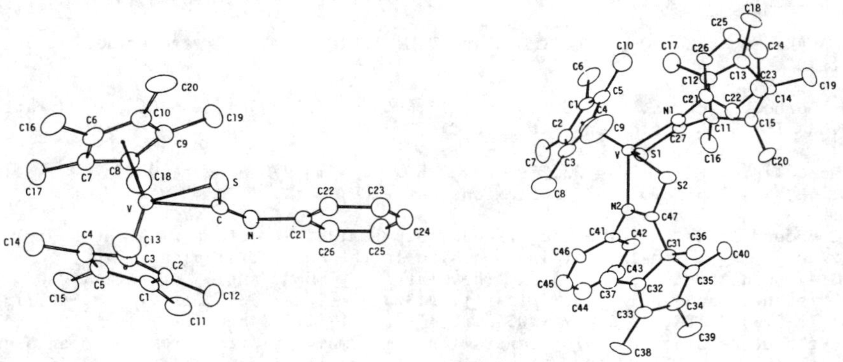

Fig. 1. The structures of $[(\eta^5\text{-}C_5Me_5)_2V(PhNCS)]$ (left) and $[(\eta^5\text{-}C_5Me_5)V(PhNC(S)\text{-}C_5Me_5)_2]$ (right).

TRIS[o-((DIMETHYLAMINO)METHYL)PHENYL]LUTETIUM
$C_{27}H_{36}LuN_3$ $Lu\text{-}[o\text{-}C_6H_4CH_2N(CH_3)_2]_3$

A.L. WAYDA, J.L. ATWOOD and W.E. HUNTER, 1984. Organometallics, 3, 939-941.

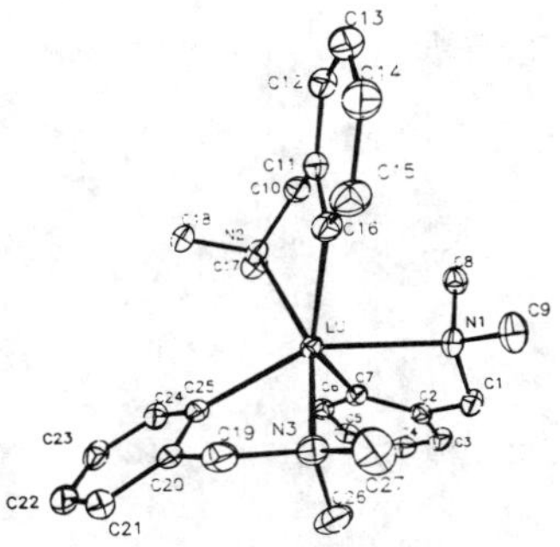

Fig. 1. Molecular structure of $C_{27}H_{36}LuN_3$.

Monoclinic, C2/c, a = 24.237, b = 9.388, c = 26.423 Å, β = 123.95°, Z = 8. Mo radiation, R = 0.023 for 2607 reflexions.

The Lu-C distances are 2.426 Å (C(7)), 2.455 Å (C(16)), and 2.425 Å (C(25)), and average 2.435 Å. The Lu-N distances fall into a two-short, one-long pattern (Fig. 1), with Lu-N(2) 2.478, Lu-N(3) 2.468 and Lu-N(1) 2.588 Å. In the five-membered metallocyclic rings, the three C atoms and the Lu atom are planar to 0.02 Å and the N atom resides 0.7 Å out of the plane. The torsion angles that involve the N and three C atoms of the rings are -25, 31 and -38°, and the bond angles N(1)-Lu-C(7), N(2)-Lu-C(16), N(3)-Lu-C(25), N(1)-Lu-N(2), N(1)-Lu-N(3) and N(2)-Lu-N(3) are respectively 68.6, 70.1, 71.3, 114.0, 91.2 and 149.8°.

[1,1'-(η^2-1,2-ETHYNEDIYL)BIS[4-METHYLBENZENE]](METHOXYMETHYL)TRIS(TRIMETHYLPHOSPHINE)-IRIDIUM(I)
$C_{27}H_{46}IrOP_3$ (I) $Ir(CH_2OCH_3)((p\text{-}tol)_2C_2)(PMe_3)_3$

cis-1,1,1-TRIS(TRIMETHYLPHOSPHINE)-1-BROMO-2,3-DI-p-TOLYL-1-IRIDACYCLOBUT-2-ENE
$C_{26}H_{43}BrIrP_3,CH_2Cl_2$ (II) $Ir(CH_2C(p\text{-}tol)=C(p\text{-}tol))Br(PMe_3)_3,CH_2Cl_2$

J.C. CALABRESE, D.C. ROE, D.L. THORN and T.H. TULIP, 1984. Organometallics, 3, 1223-1230.

I. Orthorhombic, Pbca, a = 18.310, b = 17.696, c = 18.408 Å, Z = 8. Mo radiation, R = 0.034 for 3283 reflexions, at -100°C.

II. Monoclinic, $P2_1/n$, a = 23.279, b = 8.579, c = 16.273 Å, β = 103.97°, Z = 4. Mo radiation, R = 0.031 for 4827 reflexions, at -100°C.

Compound I has the expected trigonal bipyramidal structure with the acetylene entity occupying a single coordination site in the equatorial plane (Fig. 1). The C(3)≡C(4) bond length is 1.30 Å and the C(aryl)-C≡C bond angles average 142.0°. Other distances and angles are: Ir-P(1) 2.323, Ir-P(2) 2.306, Ir-P(3) 2.340, Ir-C(1) 2.136, Ir-C(3) 2.128, Ir-C(4) 2.103 Å, P(3)-Ir-C(1) 170.3 and Ir-C(1)-O 116.4°. The iridacyclobutene ring in II is planar with a maximum deviation from the mean plane of 0.032 Å. The Ir-C(1) and Ir-C(3) distances are respectively 2.134 and 2.166 Å, C(1)-C(2) and C(2)-C(3) 1.344 and 1.525 Å respectively, and the angles C(1)-Ir-C(3), Ir-C(1)-C(2), Ir-C(3)-C(2) and C(1)-C(2)-C(3) 64.5, 98.0, 91.2 and 106.1° respectively. Other distances are: Ir-Br 2.6010, Ir-P(1) 2.246, Ir-P(2) 2.361 and Ir-P(3) 2.340 Å.

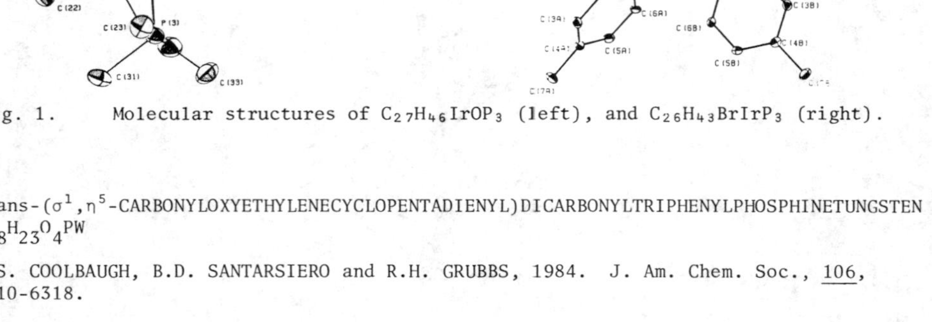

Fig. 1. Molecular structures of $C_{27}H_{46}IrOP_3$ (left), and $C_{26}H_{43}BrIrP_3$ (right).

trans-(σ^1,η^5-CARBONYLOXYETHYLENECYCLOPENTADIENYL)DICARBONYLTRIPHENYLPHOSPHINETUNGSTEN
$C_{28}H_{23}O_4PW$

T.S. COOLBAUGH, B.D. SANTARSIERO and R.H. GRUBBS, 1984. J. Am. Chem. Soc., 106, 6310-6318.

Monoclinic, $P2_1/n$, a = 8.127, b = 16.823, c = 17.623 Å, β = 101.98°, Z = 4. Mo radiation, R = 0.043 for 5335 reflexions.

The metalloester has a trans-configuration (Fig. 1) with W-P 2.495(3), W-C (Cp ring) 2.361, W-C (ring) 2.233(15), W-C(carbonyl) 1.95(2) Å.

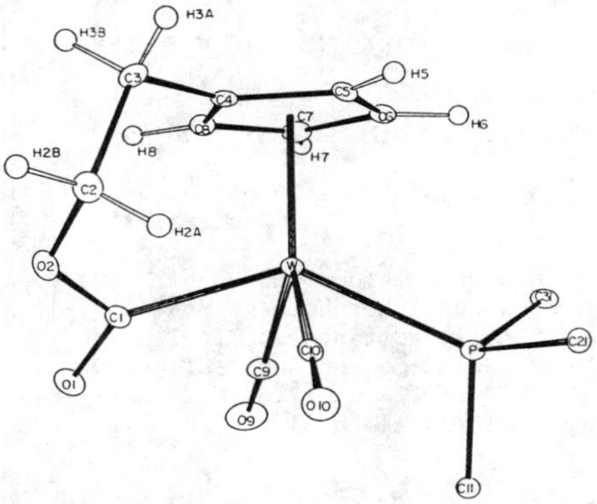

Fig. 1. A view of $C_{28}H_{23}O_4PW$ with only the first of each phenyl carbon atom shown.

(μ_2-CARBONYL)BIS(μ_4-σ^2,η^4-1-METHYL-2-PHENYLVINYLENE)DECACARBONYLTETRARUTHENIUM

$C_{29}H_{16}O_{11}Ru_4$ $Ru_4(CO)_{11}(CH_3C_2C_6H_5)_2$

S. AIME, G. NICOLA, D. OSELLA, A.M.M. LANFREDI and A. TIRIPICCHIO, 1984. Inorg. Chim. Acta, 85, 161-168.

Monoclinic, C2/c, a = 22.383, b = 9.048, c = 18.268 Å, β = 127.25°, Z = 4. Mo radiation, R = 0.034 for 1420 reflexions.

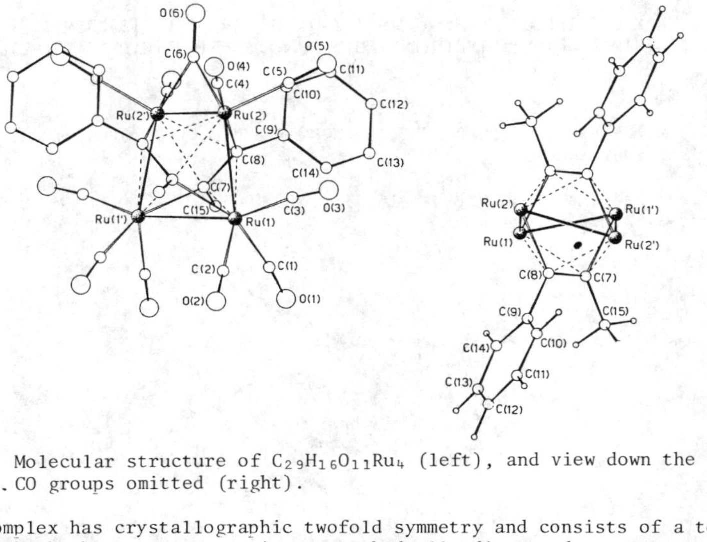

Fig. 1. Molecular structure of $C_{29}H_{16}O_{11}Ru_4$ (left), and view down the C_2 axis with CO groups omitted (right).

The complex has crystallographic twofold symmetry and consists of a tetranuclear cluster in which the Ru atoms are in a tetrahedrally-distorted square arrangement (Fig. 1). In the Ru_4C_4 core, the metal and C atoms occupy the vertices of a triangulated dodecahedron. Ten CO groups are terminal (Ru-C distances range from 1.86 to 1.90 Å) and one (C(6)-O(6)) symmetrically bridges an edge of the cluster (Ru(2)-C(6) 2.04 Å). Each of the two alkyne ligands is σ-bonded to two Ru atoms on the opposite vertices of the cluster (Ru(1)-C(7') 2.16, Ru(2)-C(8) 2.16 Å), and π-bonded to the other two (Ru(1)-C(7) 2.382, Ru(1)-C(8) 2.301; Ru(2)-C(7') 2.30, Ru(2)-C(8') 2.31 Å). The metal-metal distances are: Ru(1)-Ru(2) 2.829, Ru(1)-Ru(1') 2.739 and Ru(2)-Ru(2') 2.748 Å.

(CHLOROMETHYL)BIS[μ-[(DIPHENYLPHOSPHINIDENIO)BIS(METHYLENE)]]IODODIGOLD(Au-Au)
$C_{29}H_{30}Au_2ClIP_2$ $[Au(CH_2)_2PPh_2]_2(CH_2Cl)(I)$

H.H. MURRAY, J.P. FACKLER, Jr., and A.M. MAZANY, 1984. Organometallics, 3, 1310-1311.

Triclinic, $P\bar{1}$, a = 13.003, b = 13.370, c = 10.061 Å, α = 103.40, β = 100.95, γ = 90.26°, Z = 2. Mo radiation, R = 0.057 for 1246 reflexions.

The molecule consists of two Au atoms in roughly square-planar coordination environments bridged by two diphenylphosphine groups (Fig. 1) and also linked by a metal-metal bond. The Au(1)-Au(2) distance is 2.631 Å, and other bond distances are: Au(2)-I 2.858, Au(1)-C 2.00, C-Cl 1.75 and Au(1)-C(1) 2.03 Å. Selected bond angles are: Au(1)-Au(2)-I 179.1, Au(2)-Au(1)-C 177.6, C(2)-Au(2)-C(4) 177.3, C(1)-Au(1)-C(3) 174.5, Au(1)-C-Cl 115.6 and C(1)-Au(1)-Au(2) 93.0°.

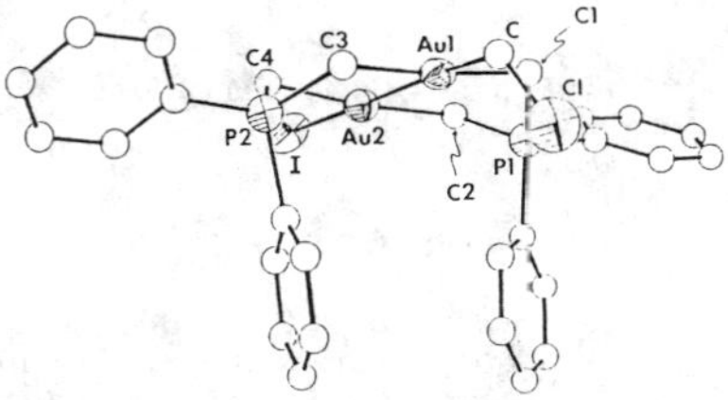

Fig. 1. Molecular structure of $C_{29}H_{30}Au_2ClIP_2$.

[μ_2-1,2,3,4-TETRAKIS(TRIFLUOROMETHYL)BUTA-1,3-DIENE-1,4-DIYL]{[2-(DIMETHYLAMINO-METHYL)PHENYL-C,N]PALLADIUM}[DIMETHYLAMINO(o-PHENYLENE)(TRIMETHYLSILYLMETHYL)-C,N]-PALLADIUM
$C_{29}H_{32}F_{12}N_2Pd_2Si$

F. MAASSARANI, M. PFEFFER, G. LE BORGNE, E. WEHMAN and G. VAN KOTEN, 1984. J. Am. Chem. Soc., 106, 8002-8004.

Orthorhombic, $P2_12_12_1$, a = 10.793, b = 16.037, c = 19.445 Å, Z = 4. Mo radiation, R = 0.0499 for 3432 reflexions.

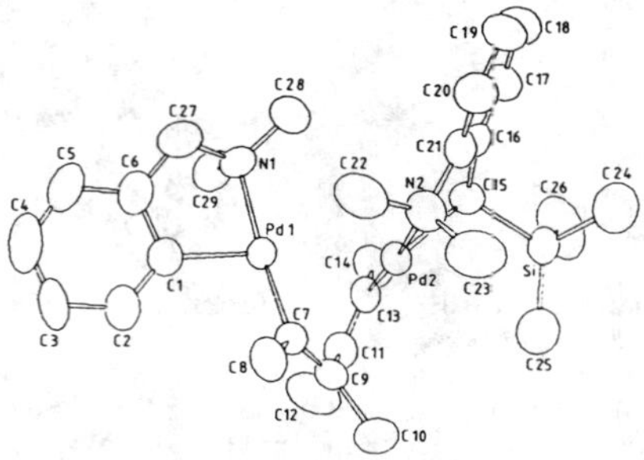

Fig. 1. The molecular structure of $C_{29}H_{32}F_{12}N_2Pd_2Si$.

A C_4 chain is σ-bonded to both Pd(1) and Pd(2) via C(7) and C(13) respectively (Fig. 1). The C(9)-C(11) distance is 1.55(1) Å and there is no delocalization along the C_4 chain. Some bond lengths are: Pd(1)-N(1) 2.121(6), Pd(1)-C(1) 1.979(8), Pd(1)-C(7) 2.031(8), Pd(2)-C(13) 2.023(7), Pd(2)-N(2) 2.177(6), Pd(2)-C(15) 2.042(8), Pd(1)-Pd(2) 2.756(1) Å.

CHLORO(η[5]-CYCLOPENTADIENYL)TETRAKIS(t-BUTYL ISOCYANIDE)NIOBIUM(III)OXOTETRACHLORO-(TETRAHYDROFURAN)NIOBIUM(V) TETRAHYDROFURAN SOLVATE

$C_{29}H_{49}Cl_5N_4Nb_2O_2,C_4H_8O$ $[(C_5H_5)Nb(CNC(CH_3)_3)_4Cl][NbCl_4O(C_4H_8O)],C_4H_8O$

H.C. ASPINALL, M.M. ROBERTS and S.J. LIPPARD, 1984. Inorg. Chem., 23, 1782-1784.

Orthorhombic, Pnma, a = 12.150, b = 13.852, c = 27.394 Å, D_m = 1.35, Z = 4. Mo radiation, R = 0.061 for 2169 reflexions.

The structure consists of discrete $[(C_5H_5)Nb(CNCMe_3)_4Cl]^+$ cations (Fig. 1), $[NbCl_4O(C_4H_8O)]^-$ anions (Fig. 1), and uncoordinated THF molecules, all of which lie on crystallographic mirror planes. The formally eight-coordinate cation has Nb-C(Cp) 2.37(1)-2.40(1), Nb-CNR 2.203(9) and 2.210(10), and Nb-Cl 2.505(4) Å. The anion has distorted C_{4v} stereochemistry with Nb displaced 0.33 Å toward the oxo ligand from the Cl_4 mean plane with Nb=O 1.665(11), Nb-O(THF) 2.394(12), and Nb-Cl 2.376(4)-2.387(4) Å.

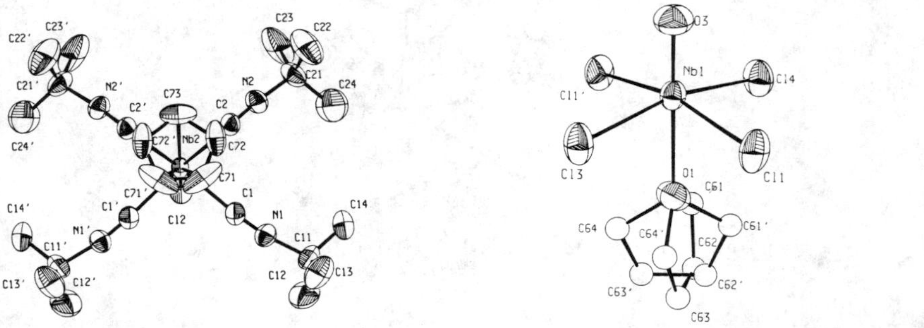

Fig. 1. The $[(C_5H_5)Nb(CNCMe_3)_4Cl]^+$ cation (left) and the $[NbCl_4O(C_4H_8O)]^-$ anion (right).

DIBENZYLTRIMETHYLPHOSPHINETETRAISOPROPOXYDIMOLYBDENUM

$C_{29}H_{51}Mo_2O_4P$

M.H. CHISHOLM, J.C. HUFFMAN and R.J. TATZ, 1984. J. Am. Chem. Soc., 106, 5385-5386.

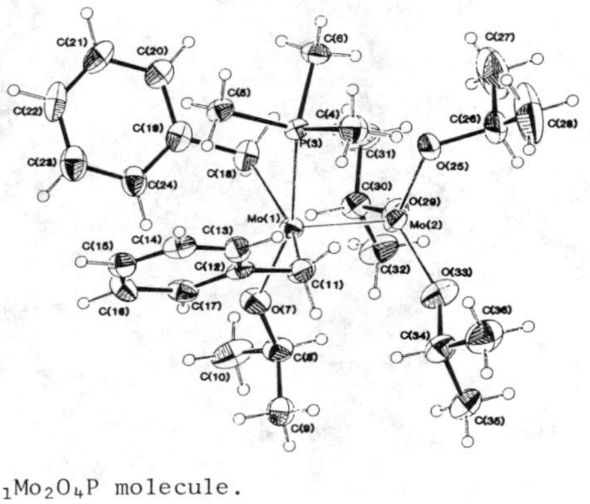

Fig. 1. The $C_{29}H_{51}Mo_2O_4P$ molecule.

Monoclinic, $P2_1/c$, a = 16.779, b = 10.104, c = 19.555 Å, β = 90.59°, Z = 4. Mo radiation, R = 0.037 for 3665 reflexions (at -159°C).

The complex is shown in Fig. 1. Some pertinent bond distances (Å) and angles (°) (averaged where appropriate): Mo-Mo 2.235(1), Mo-C 2.22(1), Mo(1)-O(7) 1.920(4), Mo(2)-O 1.89(2), Mo-P 2.581(2); Mo-Mo-C 101.3(6), Mo(2)-Mo(1)-O(7) 106.5(1), Mo-Mo-P 94.65(4), Mo(1)-Mo(2)-O 107.2(20), Mo-C-C 114.4(11).

(μ_3-ETHYLIDYNE)TRIS(μ-ISOPROPOXO)HEXAKIS(ISOPROPOXY)TRITUNGSTEN
$C_{29}H_{66}O_9W_3$ $W_3(\mu_3\text{-}CCH_3)(\mu\text{-}O\text{-}i\text{-}C_3H_7)_3(O\text{-}i\text{-}C_3H_7)_6$

M.H. CHISHOLM, D.M. HOFFMAN and J.C. HUFFMAN, 1984. Inorg. Chem., 23, 3683-3684.

Triclinic, $P\bar{1}$, a = 18.255, b = 11.396, c = 11.500 Å, α = 110.55, β = 72.56, γ = 75.16°, Z = 2. Mo radiation, R = 0.045 for 4755 reflexions (at -158°C).

The molecule (Fig. 1) contains three W atoms, each with a square-pyramidal O_4C donor set with the μ_3-CCH_3 apical. The three units are fused via μ_3-CCH_3 and μ-OR ligands as well as W-W bonds. Mean distances: W-W 2.75(1), W-O (terminal) 1.92(2), W-O (bridging) 2.04(3), and W-C 2.06(2) Å.

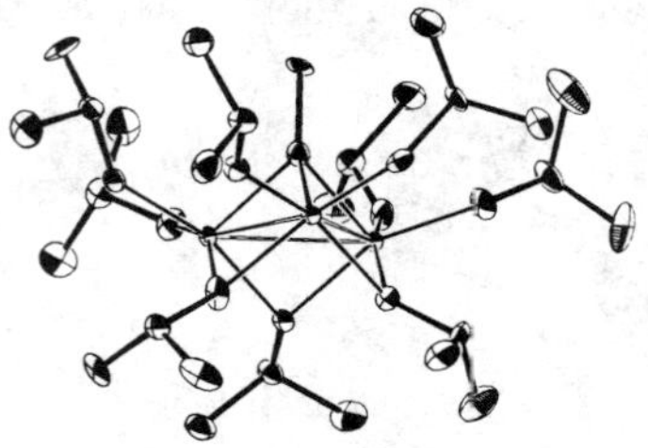

Fig. 1. The $W_3(\mu_3\text{-}CCH_3)(\mu\text{-}O\text{-}i\text{-}C_3H_7)_3(O\text{-}i\text{-}C_3H_7)_6$ molecule.

(η^5-CYCLOPENTADIENYL)(TETRACHLORO-2-OXIDOPHENOXYCARBONYL-C^2,O^2)TRIPHENYLPHOSPHINE-RHODIUM(III) DICHLOROMETHANE SOLVATE
$C_{30}H_{20}Cl_4O_3PRh,0.25(CH_2Cl_2)$

N.G. CONNELLY, M.J. FREEMAN, I. MANNERS and A.G. ORPEN, 1984. J. Chem. Soc. Dalton, 2703-2712.

Triclinic, $P\bar{1}$, a = 15.420, b = 16.677, c = 15.587 Å, α = 128.89, β = 90.83, γ = 99.83°, Z = 4. Mo radiation, R = 0.053 for 5296 reflexions.

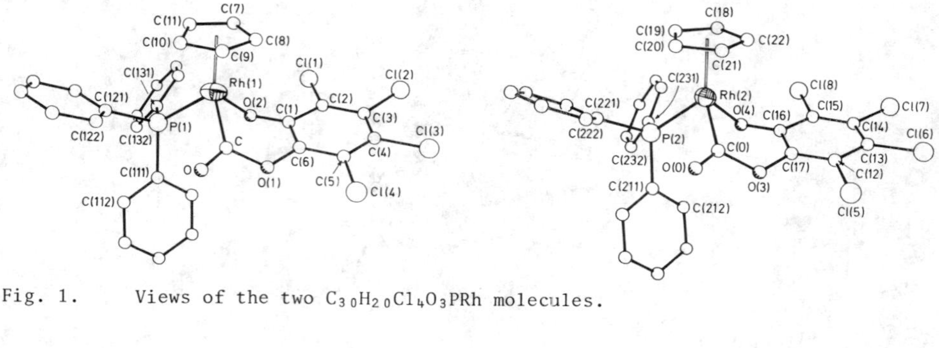

Fig. 1. Views of the two $C_{30}H_{20}Cl_4O_3PRh$ molecules.

The crystal structure contains two independent molecules of the complex (differing only in the orientation of phenyl rings) as well as molecules of dichloromethane disordered about inversion centres. The Rh coordination is shown in Fig. 1; mean bond lengths are Rh-P 2.298(3), Rh-C(metallacycle) 1.994(12), Rh-O 2.079(6) Å. The metallacycle has a distorted boat conformation.

(μ_2-2,3-DIPHENYLBUT-2-ENE-1,4-DIYLIDYNE)BIS((η^5-CYCLOPENTADIENYL)DICARBONYLCHROMIUM)
$C_{30}H_{20}Cr_2O_4$

N.A. USTYNYUK, V.N. VINOGRADOVA, V.G. ANDRIANOV and Yu.T. STRUCHKOV, 1984. J. Organometal. Chem., 268, 73-78.

Monoclinic, C2/c, a = 19.474, b = 15.282, c = 8.710 Å, β = 110.06°, Z = 4. Mo radiation, R = 0.027 for 2180 reflexions (-120°C).

The molecule consists of two "carbyne" moieties joined by a 1,2-diphenylethylene group which is sited on a centre of symmetry (Fig. 1). The Cr atom has a piano stool coordination, the coordination polyhedron being slightly distorted as a result of unequal bond angles between the "legs" of the stool and their deviation from an ideal value of 90°. The Cp(centroid)-Cr-C(3) angle is 132.5°. The bridging ethylene moiety C(3),C(4),C(5),C(4'),C(5'),C(3') is completely planar and the phenyl rings are turned by 68.5° with respect to this moiety. The short Cr-C(3) distance of 1.707 Å is consistent with a metal-carbon triple bond.

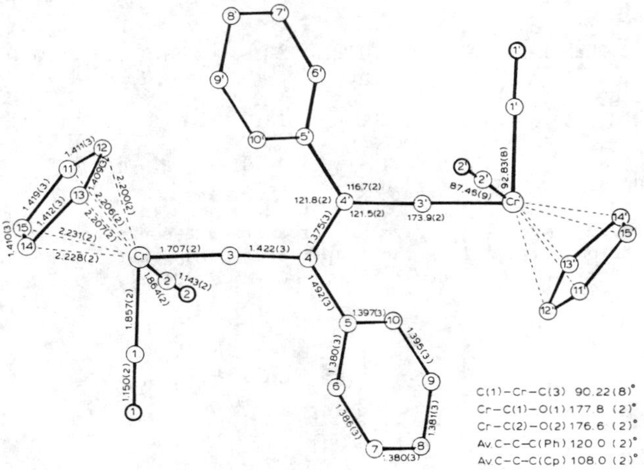

Fig. 1. Molecular structure of $C_{30}H_{20}Cr_2O_4$ showing bond lengths (Å) and angles (°).

(μ-HYDRIDO)(μ-N,N-DIMETHYLFORMAMIDO-C,O')NONACARBONYL(TRIPHENYL PHOSPHITE)TRIRUTHENIUM
$C_{30}H_{22}NO_{13}PRu_3$ $Ru_3\{\mu-H,\mu-OCNMe_2\}(CO)_9P(OPh)_3$

A. MAYR, Y.C. LIN, N.M. BOAG, C.E. KAMPE, C.B. KNOBLER and H.D. KAESZ, 1984. Inorg. Chem., 23, 4640-4645.

Monoclinic, Pc, a = 18.434, b = 10.166, c = 19.741 Å, β = 114.90°, Z = 4. Mo radiation, R = 0.030 for 4593 reflexions (at -158°C).

One of the two essentially equivalent independent molecules is shown in Fig. 1. The three Ru atoms of each molecule define a triangle of unequal sides: Ru(1)-Ru(2) 2.884(2) (x2), Ru(1)-Ru(3) 2.830(2) and 2.841(2), Ru(2)-Ru(3) 2.856(2) (x2) Å. Both the hydrido and carboxamido ligands bridge Ru(1)-Ru(2) on opposite sides of the Ru_3

plane (Ru-H 1.57(16)-1.98(16), Ru(1)-O 2.118(9) and 2.092(9), Ru(2)-C 2.102(13) and 2.084(13) Å). All three Ru atoms have approximately octahedral coordination geometry if the Ru(1)-Ru(2) bond is ignored. The double bond is delocalized in the μ-OCNMe$_2$ ligand: C-O 1.255(17) and 1.291(16), C-N 1.355(17) and 1.339(17) Å. Other bond lengths: Ru-P 2.270(4) and 2.281(4), Ru-CO 1.834-1.975(13) Å.

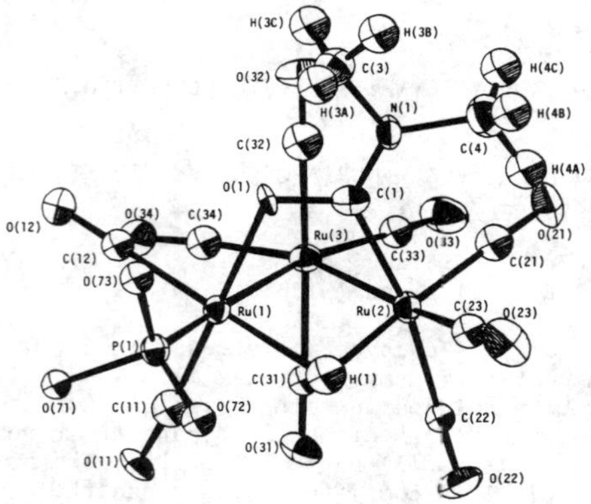

Fig. 1. The Ru$_3${μ-H,μ-OCNMe$_2$}(CO)$_9$P(OPh)$_3$ molecule.

TETRACARBONYL(((DIPHENYLARSINO)METHYL)DIPHENYLARSONIUMMETHYLIDE)CHROMIUM(0)
$C_{30}H_{24}As_2CrO_4$

L. WEBER, D. WEWERS, W. MEYER and R. BOESE, 1984. Chem. Ber., 117, 732-742.

Monoclinic, C2/c, a = 27.537, b = 14.185, c = 21.942 Å, β = 138.87°, Z = 8. Mo radiation, R = 0.035 for 2155 reflexions.

In the molecule (Fig. 1) the five-membered arsenic-containing ring has an envelope conformation, with As2 as the flap. Some bond lengths are: Cr-As1 2.450(1), Cr-C(0) 1.823(7) (trans to C(6)) to 1.885(7), Cr-C6 2.221(1), As1-C5 1.971(5), As2-C5 1.914(7) Å.

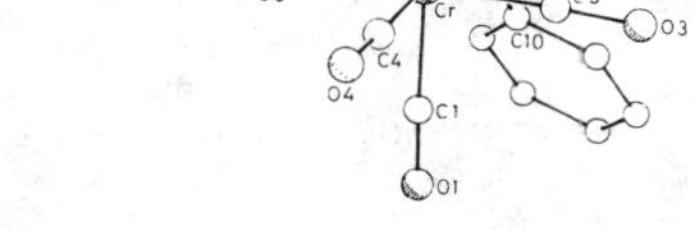

Fig. 1. A view of $C_{30}H_{24}As_2CrO_4$.

1,1,3,3-TETRACARBONYL-1,3-BIS(η^5-CYCLOPENTADIENYL)-1,2,2,3-BIS(μ_2-p-TOLYLMETHYLIDYNE)-GOLDDITUNGSTEN HEXAFLUOROPHOSPHATE
$C_{30}H_{24}AuF_6O_4PW_2$ (I)

1,2,2,3-BIS(μ_2-σ^1,η^2-CARBONYL)-1,2,2,3-TETRACARBONYL-1,3-BIS(η^5-CYCLOPENTADIENYL)-1,2,2,3-BIS(μ_2-p-TOLYLMETHYLIDYNE)MOLYBDENUMDITUNGSTEN
$C_{32}H_{24}MoO_6W_2$ (II)

G.A. CARRIEDO, J.A.K. HOWARD, K. MARSDEN, F.G.A. STONE and P. WOODWARD, 1984. J. Chem. Soc. Dalton, 1589-1595.

I. Orthorhombic, Pnna, a = 10.024, b = 22.247, c = 13.905 Å, Z = 4. Mo radiation, R = 0.050 for 1748 reflexions (at 200 K).

II. Triclinic, P$\bar{1}$, a = 7.829, b = 14.709, c = 15.579 Å, α = 118.27, β = 92.21, γ = 105.32°, Z = 2. Mo radiation, R = 0.038 for 3711 reflexions (at 210 K).

The cation in I (Fig. 1) has a bent W-Au-W spine (162.8(1)°) with Au-W bonds (2.752(1) Å) bridged by the p-tolylmethylidyne groups. The dihedral angle between the two dimetallacyclopropene rings is 62°. Coordination about each W atom is completed by a η^5-C_5H_5 ligand and a terminal CO group. Compound II comprises a bent W-Mo-W spine (148.2(2)°) with the Mo-W bonds (2.938(1) Å) bridged by CC_6H_4Me-4 groups. The Mo atom carries two terminal CO ligands. The bisector of the angle between these two ligands is an approximate twofold axis of symmetry for the molecule as a whole. The two Mo-W-carbene ring systems are not coplanar, but are twisted relative to one another. Of the two terminal carbonyl ligands on each tungsten atom, one is η^2-bonded to the central molybdenum atom.

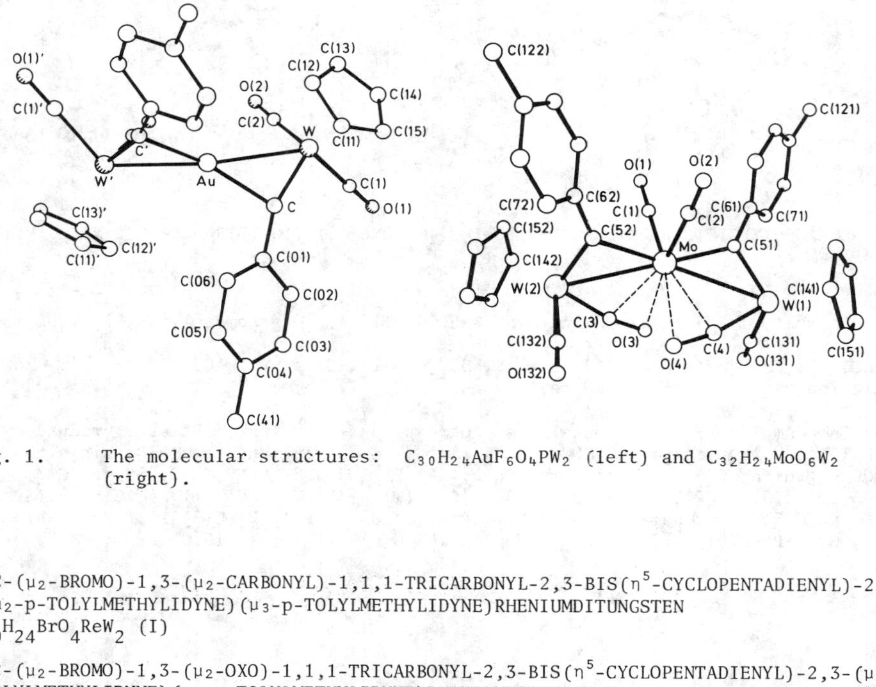

Fig. 1. The molecular structures: $C_{30}H_{24}AuF_6O_4PW_2$ (left) and $C_{32}H_{24}MoO_6W_2$ (right).

1,2-(μ_2-BROMO)-1,3-(μ_2-CARBONYL)-1,1,1-TRICARBONYL-2,3-BIS(η^5-CYCLOPENTADIENYL)-2,3-(μ_2-p-TOLYLMETHYLIDYNE)(μ_3-p-TOLYLMETHYLIDYNE)RHENIUMDITUNGSTEN
$C_{30}H_{24}BrO_4ReW_2$ (I)

1,2-(μ_2-BROMO)-1,3-(μ_2-OXO)-1,1,1-TRICARBONYL-2,3-BIS(η^5-CYCLOPENTADIENYL)-2,3-(μ_2-p-TOLYLMETHYLIDYNE)(μ_3-p-TOLYLMETHYLIDYNE)RHENIUMDITUNGSTEN
$C_{29}H_{24}BrO_4ReW_2$ (II)

G.A. CARRIEDO, J.C. JEFFERY and F.G.A. STONE, 1984. J. Chem. Soc. Dalton, 1597-1603.

I. Monoclinic, P2_1/c, a = 8.893, b = 26.587, c = 13.071 Å, β = 115.26°, Z = 4. Mo radiation, R = 0.042 for 3951 reflexions.

II. Orthorhombic, P2_1cn, a = 12.145, b = 12.767, c = 18.079 Å, Z = 4. Mo radiation, R = 0.043 for 2730 reflexions.

In I a ReW_2 triangle of metal atoms (Fig. 1) (W-W 2.620(1), Re-W 2.867(1), 2.912(1) Å) is asymmetrically capped by a CC_6H_4Me-4 ligand and is edge-bridged by CC_6H_4Me-4, Br and CO groups. The CO group is asymmetrically bound (W-C-O 155(1)°). The two W atoms carry cyclopentadienyl ligands and the Re atom three terminally bound CO groups. Compound II is similar to I but has the μ_2-CO group replaced by a μ_2-O group and the capping CC_6H_4Me-4 very asymmetrically bound to the metal triangle (C-Re 2.515(20), C-W 2.017(19), 2.114(18) Å).

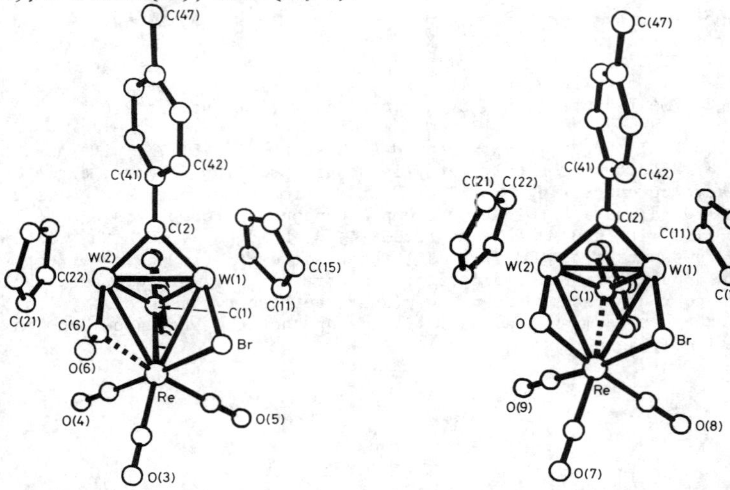

Fig. 1. The molecular structures: $C_{30}H_{24}BrO_4ReW_2$ (left) and $C_{29}H_{24}BrO_4ReW_2$ (right).

[1,2-ETHANEDIYLBIS(DIPHENYLPHOSPHINE)-P,P']METHYL(1,1,2,2-TETRAFLUORO-2-METHOXYETHYL)-PLATINUM

$C_{30}H_{30}F_4OP_2Pt$ (DPPE)$PtCH_3(CF_2CF_2OCH_3)$

H.E. BRYNDZA, J.C. CALABRESE and S.S. WREFORD, 1984. Organometallics, 3, 1603-1604.

Monoclinic, $P2_1/c$, a = 11.764, b = 15.503, c = 16.069 Å, β = 105.03°, Z = 4. Mo radiation, R = 0.045 for 4290 reflexions.

The complex (Fig. 1) contains a square planar-coordinated Pt atom bridged by a DPPE ligand with methyl and $CF_2CF_2OCH_3$ ligands occupying the remaining cis sites. Bond lengths and angles of interest are: Pt-P(1) 2.298, Pt-P(2) 2.258, Pt-C(3) 2.106, Pt-C(5) 2.074, C(5)-C(6) 1.54, C(6)-O(1) 1.35, O(1)-C(4) 1.46, C(6)-F (average) 1.38 and C(5)-F (average) 1.40 Å; P(1)-Pt-P(2) 84.9, P(1)-Pt-C(5) 95.2, P(2)-Pt-C(3) 90.4, C(3)-Pt-C(5) 89.7, Pt-C(5)-C(6) 114.5, C(6)-P(1)-C(4) 115.6°.

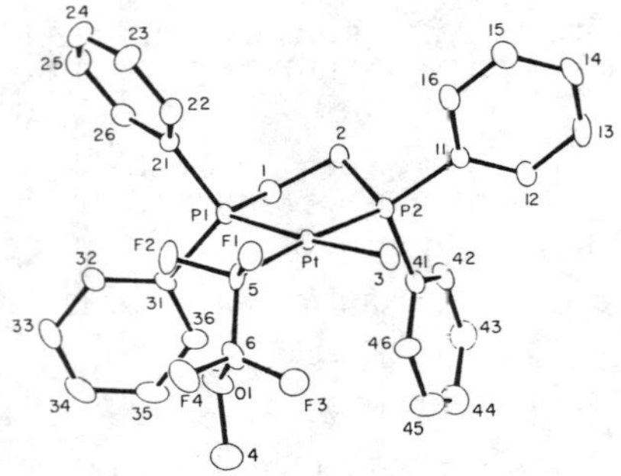

Fig. 1. Molecular structure of $C_{30}H_{30}F_4OP_2Pt$.

CARBONYLDICHLORO(CYCLOPROPYL)BIS(METHYLDIPHENYLPHOSPHINE)IRIDIUM(III)
$C_{30}H_{31}Cl_2IrOP_2$ $IrCl_2(C_3H_5)(CO)(PMePh_2)_2$

P.N. SWEPSTON, N.L. JONES, J.A. IBERS, SHANG MAOYU, HUANG JINLING and LU JIAXI, 1984. Acta Cryst., C40, 39-41.

Rhombohedral, $R\bar{3}$, a = 17.863, α = 108.97 Å, Z = 6. Mo radiation, R = 0.060 for 2866 reflexions.

The Ir atom is six-coordinate with the chloro ligands having a cis configuration and the phosphine ligands having a trans configuration. The cyclopropyl ligand is σ bonded to the Ir atom and has a mean C-C bond length of 1.56(2) Å. Other bond lengths are Ir-Cl(1) 2.397(4), Ir-Cl(2) 2.470(4), Ir-P(1) 2.363(4), Ir-P(2) 2.374(4), Ir-C(1) 2.11(2), Ir-C(4) 1.83(2) Å.

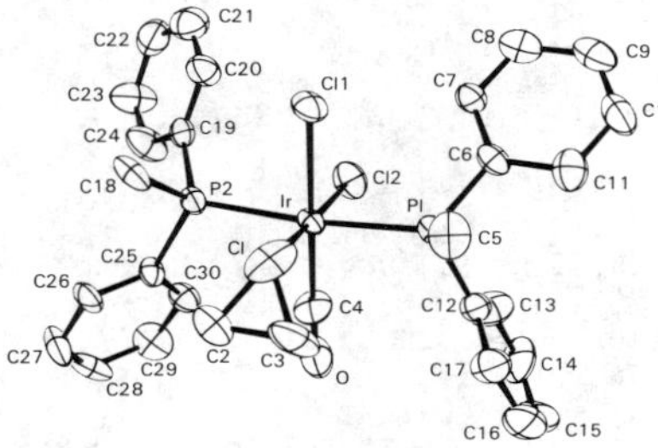

Fig. 1. A view of $C_{30}H_{31}Cl_2IrOP_2$.

trans,trans-BIS(BIS(p-TOLYLAMINO)CARBENE)GOLD(I) TETRAFLUOROBORATE
$C_{30}H_{32}AuBF_4N_4$ $[\{(CH_3C_6H_4NH)_2C\}_2Au]^+BF_4^-$

G. BANDITELLI, F. BONATI, S. CALOGERO and G. VALLE, 1984. J. Organometal. Chem., 275, 153-160.

Monoclinic, $P2_1/n$, a = 14.506, b = 20.753, c = 10.344 Å, β = 105.7°, Z = 4. Mo radiation, R = 0.068 for 2297 reflexions.

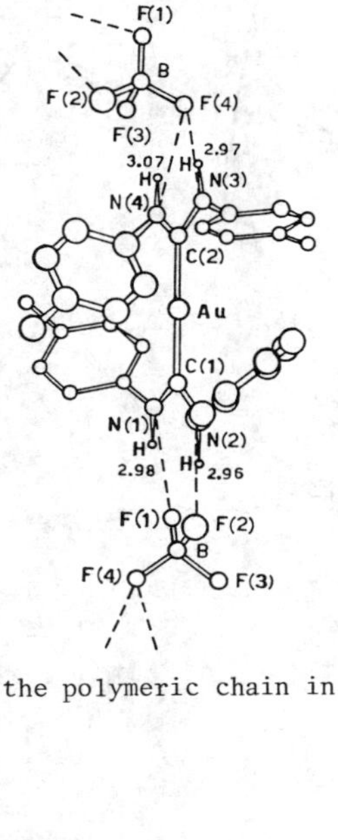

Fig. 1. Perspective view of the polymeric chain in $C_{30}H_{32}AuBF_4N_4$.

The ionic complex tends to exist as a linear polymer in the solid state (Fig. 1), with the cation linked via hydrogen bonds to two anions. In the cation the Au atom is linearly-coordinated with a C-Au-C angle of 177.4°. The Au-C distances are both 2.02 Å. In each ligand the two N atoms, the carbene C, and the Au atoms lie in a plane, and the two planes make a dihedral angle of 68.1°. In each ligand the carbene atom is sp^2-hybridized (the angles round it range from 114-126 and 114-123°) and the C•••N distances (1.32 to 1.37 Å) are intermediate between the values observed for single and double bonds in similar environments. In the anion, the B-F distances range from 1.24 to 1.45 Å and the F-B-F angles from 96 to 120°.

HEXACHLORO-(μ_2-1,4-DI-t-BUTYL-1,4-DIAZABUTADIENE-C,C',N,N')TETRAKIS(t-BUTYL ISO-CYANIDE)DINIOBIUM(III)
$C_{30}H_{54}Cl_6N_6Nb_2$ (I)

HEXACHLORO-(μ_2-1,4-DIISOPROPYL-1,4-DIAZABUTADIENE-C,C',N,N')TETRAKIS(ISOPROPYL ISOCYANIDE)DITANTALUM(III)
$C_{24}H_{42}Cl_6N_6Ta_2$ (II)

F.A. COTTON, S.A. DURAJ and W.J. ROTH, 1984. J. Am. Chem. Soc., 106, 6987-6993.

I. Monoclinic, Pn, a = 14.102, b = 13.486, c = 11.157 Å, β = 93.26°, Z = 2. Mo radiation, R = 0.054 for 2438 reflexions.

II. Monoclinic, $P2_1/c$, a = 18.014, b = 11.862, c = 17.948 Å, β = 98.23°, Z = 4. Mo radiation, R = 0.062 for 2706 reflexions.

Both (I) and (II) contain a tetradentate RNCCNR ligand formed by dimerization of isocyanide, linking the two metal atoms as shown in Fig. 1. In each case, one metal atom has roughly pentagonal bipyramidal geometry while the other has octahedral coordination. Some dimensions are: Nb-N 2.010 and 2.050(11), Nb-C11/21 2.093(13) and 2.106(11), C(11)-C(21) 1.42(2) Å. The corresponding data for the Ta complex are: 2.01(2), 2.01(2), 2.39(2), 2.35(2) and 1.40(2) Å respectively.

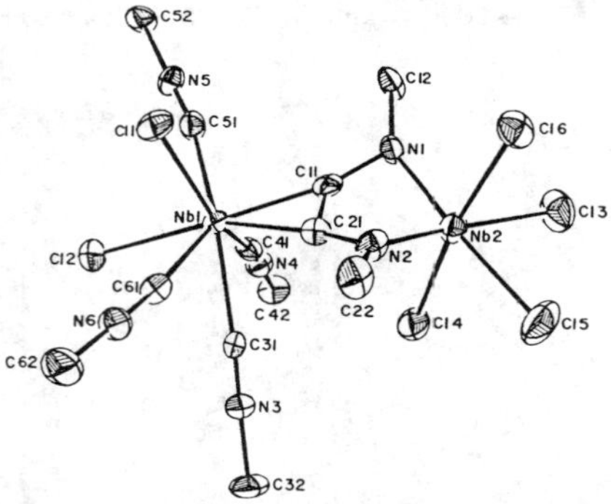

Fig. 1. A view of the Nb complex $C_{30}H_{54}Cl_6N_6Nb_2$; the Ta complex has similar coordination.

(μ_2-OXO)DIOXOHEXANEOPENTYLDITUNGSTEN
$C_{30}H_{66}O_3W_2$

I. FEINSTEIN-JAFFE, D. GIBSON, S.J. LIPPARD, R.R. SCHROCK and A. SPOOL, 1984. J. Am. Chem. Soc., 106, 6305-6310.

Rhombohedral, $R\bar{3}$, a = 10.232, c = 61.446 Å, D_m = 1.49, Z = 6. Mo radiation, R = 0.019 for 1290 reflexions.

The molecule has crystallographic C_3 symmetry and contains a linear O=W-O-W=O moiety (Fig. 1). Main dimensions include W=O 1.726(10) and 1.689(13), W-O (bridge) 1.923(10) and 1.977(10), W-C 2.141(9) and 2.127(14) Å. The two ends of the molecule are twisted by 31.9(5)° with respect to one another.

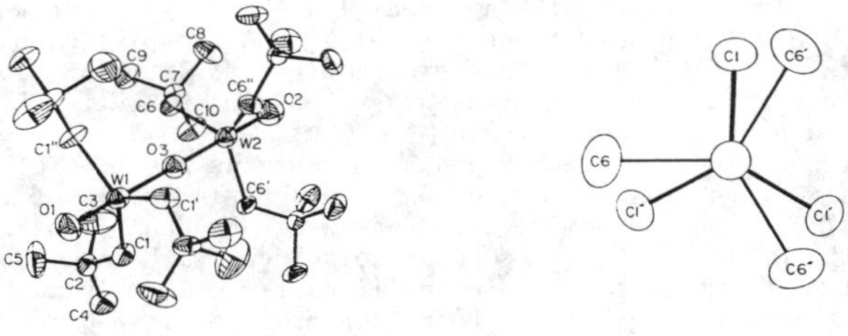

Fig. 1. $C_{30}H_{66}O_3W_2$: a view of the molecule and a view along the C_3 axis showing the ligand staggering.

BIS(μ_3-p-TOLYLMETHYLIDYNE)(μ-HYDRIDO)(μ-CARBONYL)NONACARBONYL(η^5-CYCLOPENTADIENYL)-TUNGSTENTRIOSMIUM
$C_{31}H_{20}O_{10}Os_3W$ (I) $(\eta^5\text{-}C_5H_5)WOs_3(CO)_{10}(\mu_3\text{-}CC_6H_4CH_3)_2H$

BIS(μ_3-p-TOLYLMETHYLIDYNE)(η^5-CYCLOPENTADIENYL)HYDRIDONONACARBONYLTUNGSTENTRIOSMIUM
$C_{30}H_{20}O_9Os_3W$ (II) $(\eta^5\text{-}C_5H_5)WOs_3(CO)_9(\mu_3\text{-}CC_6H_4CH_3)_2H$

J.T. PARK, J.R. SHAPLEY, M.R. CHURCHILL and C. BUENO, 1984. Inorg. Chem., 23, 4476-4486.

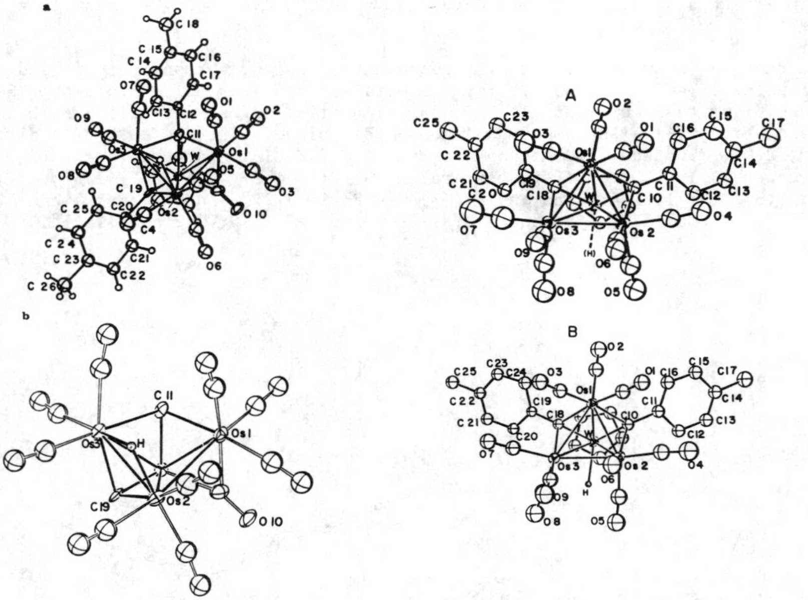

Fig. 1. The structure of $(\eta^5\text{-}C_5H_5)WOs_3(CO)_{10}(\mu_3\text{-}CC_6H_4CH_3)_2H$ (left, core at bottom) and the two independent molecules of $(\eta^5\text{-}C_5H_5)WOs_3(CO)_9(\mu_3\text{-}CC_6H_4CH_3)_2H$ (right).

I. Monoclinic, $P2_1/n$, a = 10.610, b = 16.000, c = 19.394 Å, β = 104.33°, Z = 4. Mo radiation, R = 0.071 for 3001 reflexions.

II. Monoclinic, $P2_1/n$, a = 19.067, b = 16.848, c = 19.947 Å, β = 98.62°, Z = 8. Mo radiation, R = 0.054 for 6169 reflexions.

The two molecular structures are shown in Fig. 1. In I the four metal atoms adopt a butterfly configuration with a dihedral angle of 98.7° between the Os(1)-Os(2)-W and Os(3)-Os(2)-W planes. One μ_3-CTol ligand occupies a normal position on the outside of the Os(2)-Os(3)-W triangle; the other takes up an unusual position, bridging the "hinge" W atom and the two "wing-tips" of the butterfly. The $WOs_3(\mu_3\text{-}C)_2$ fragment defines a capped square pyramid (base: Os(1-3), C(11) and principal apex W). II adopts a tetrahedral Os_3W geometry and also contains two μ_3-CTol ligands, which are located in almost equivalent environments on opposing sides of the pseudo plane of symmetry. Os(1), a unique osmium atom, and W are bonded to both alkylidyne fragments, and the Os(1)-W vector is almost coincident with the pseudo plane of symmetry. The two independent molecules of II differ primarily in the conformation of the CO ligands surrounding Os(3). Bond lengths: for I; W-Os 2.781(1)-2.921(1), Os-Os 2.883(1) and 2.913(1), W-μ-CO 2.075(26), Os-μ-CO 2.160(24), Os-CO 1.882(30)-1.974(30), W-μ_3-C (2x) 2.012(21), Os-μ_3-C 2.185(21)-2.270(21), W-C(Cp) 2.316(28)-2.374(27) Å; for II; W-Os 2.827(1)-2.914(1), Os-Os 2.791(1)-2.857(1), Os-CO 1.796(33)-1.926(24). W-μ_3-C 1.977(22)-2.013(2), Os-μ_3-C 2.135(20)-2.254(21), W-C(Cp) 2.308(24)-2.384(24) Å.

2-ETHOXY-5-(3-ETHOXY-3-DIETHYLIMINOPROPAN-2-YLIDENE)-3,4-DIPHENYL-1,1,1,1-TETRA-CARBONYL-1-TUNGSTACYCLOPENT-1-ENE
$C_{31}H_{35}NO_6W$

N.H.T. HUY, E.O. FISCHER, J. RIEDE, U. THEWALT and K.H. DÖTZ, 1984. J. Organometal. Chem., 273, C29-C32.

Orthorhombic, Pbca, a = 14.614, b = 25.577, c = 16.390 Å, Z = 8. Mo radiation, R = 0.058 for 2755 reflexions.

The tungsten atom is octahedrally-coordinated to four carbonyl groups and atoms of the complex ligand such that it forms a metallacycle (Fig. 1). Atoms W, C(5), C(7) and C(8) of this metallacycle lie approximately (largest deviation is <0.02 Å) in a plane while C(6) sits some 0.35 Å above this plane. The W-C(carbonyl) distances are 2.00, 2.02, 2.06 and 2.10 Å, and the W-C(5) and W-C(8) distances are 2.15 and 2.28 Å respectively. Other distances are: C(5)-C(6) 1.57, C(6)-C(7) 1.56, C(7)-C(8) 1.54 and C(5)-O(5) 1.31 Å. The angle C(5)-W-C(8) is 77.1° and the angles W-C(5)-C(6), C(5)-C(6)-C(7), C(6)-C(7)-C(8) and W-C(8)-C(7) are 119, 110, 112 and 115.7° respectively.

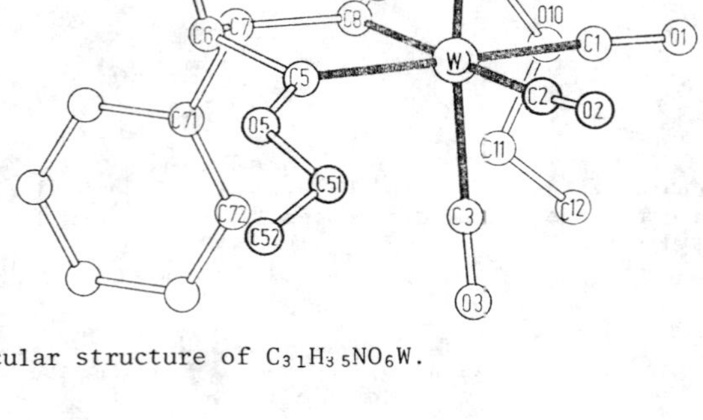

Fig. 1. Molecular structure of $C_{31}H_{35}NO_6W$.

2,3-(μ_2-CARBONYL)-2-CARBONYL-2-(η^5-CYCLOPENTADIENYL)-1-(η^5-PENTAMETHYLCYCLOPENTA-DIENYL)(μ_3-p-TOLYLMETHYLIDYNE)-3,3-BIS(TRIMETHYLPHOSPHINE)COPPERPLATINUMTUNGSTEN
$C_{31}H_{45}CuO_2P_2PtW$ (I)

BIS(μ_3-CARBONYL)TRIS(η^5-PENTAMETHYLCYCLOPENTADIENYL)-triangulo-COPPERDIRHODIUM
$C_{32}H_{45}CuO_2Rh_2$ (II)

G.A. CARRIEDO, J.A.K. HOWARD and F.G.A. STONE, 1984. J. Chem. Soc. Dalton, 1555-1561.

I. Monoclinic, $P2_1/c$, a = 10.516, b = 17.898, c = 18.134 Å, β = 94.25°, Z = 4. Mo radiation, R = 0.055 for 2238 reflexions.

II. Orthorhombic, $Pca2_1$, a = 15.50, b = 11.146, c = 18.07 Å, Z = 4. Mo radiation, R = 0.095 for 1022 reflexions.

In compound I the μ-CCuPtW core (Fig. 1) adopts a butterfly configuration with an angle of fold of 90.6° between the CWPt and CWCu planes. The Cu...Pt separation is somewhat long at 2.807(3) Å. The tungsten atom carries an η^5-cyclopentadienyl group and two CO ligands, one of which semi-bridges the Pt-W bond (Pt-W 2.779(2) Å, W-C-O 162(2)°). The C_5Me_5 ligand on copper is η^5 bonded to Cu with a Cu-W distance of 2.648(3) Å. The structure of II is based on a triangle of metal atoms which is positionally disordered. The mean metal-metal distance is 2.56(1) Å and the two CO groups are involved in semi-triply bridging behaviour.

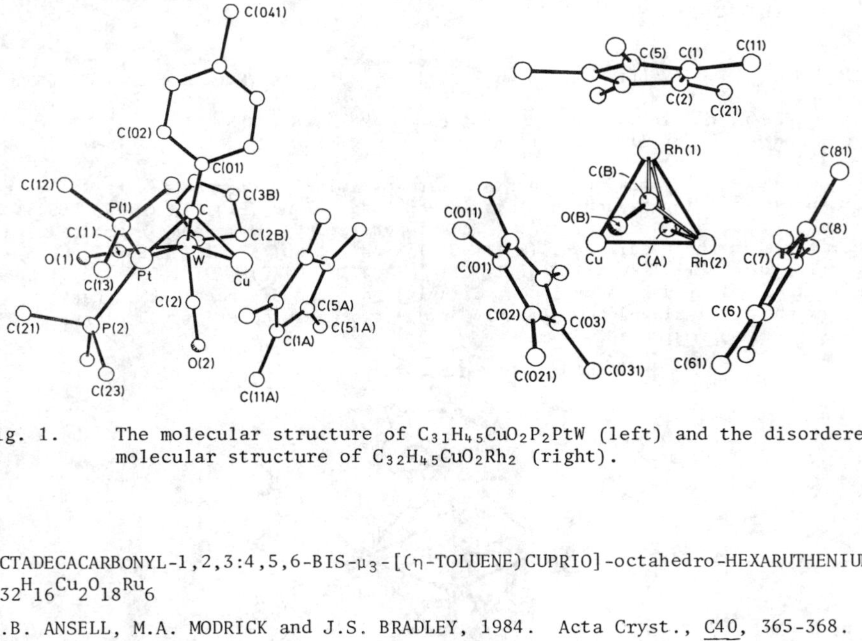

Fig. 1. The molecular structure of $C_{31}H_{45}CuO_2P_2PtW$ (left) and the disordered molecular structure of $C_{32}H_{45}CuO_2Rh_2$ (right).

OCTADECACARBONYL-1,2,3:4,5,6-BIS-μ_3-[(η-TOLUENE)CUPRIO]-octahedro-HEXARUTHENIUM
$C_{32}H_{16}Cu_2O_{18}Ru_6$

G.B. ANSELL, M.A. MODRICK and J.S. BRADLEY, 1984. Acta Cryst., C40, 365-368.

Triclinic, P1, a = 10.236, b = 10.395, c = 19.267 Å, α = 82.64, β = 75.27, γ = 82.44°, Z = 2. Mo radiation, R = 0.0377 for 4357 reflexions.

Two essentially similar (Fig. 1) but independent metal-carbonyl clusters contain Ru_6 octahedra capped by two Cu atoms on opposite trigonal faces. Three terminal carbonyl ligands are bound to each Ru atom, and a toluene molecule is η-bonded to each Cu atom. Cu-Ru distances average 2.66(6) Å. Ru-Ru distances range from 2.87 to 2.94 Å.

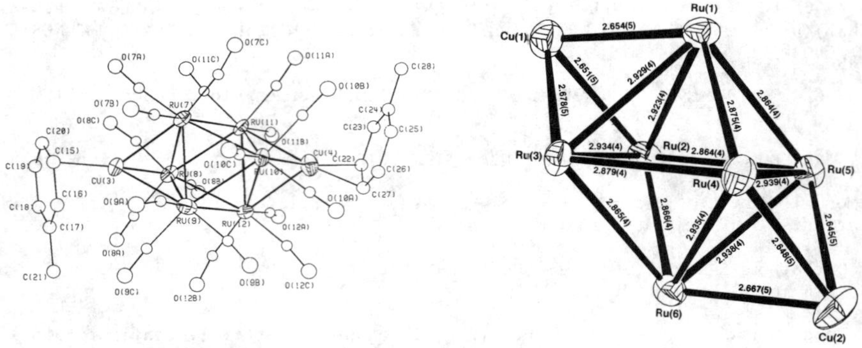

Fig. 1. One of the $C_{32}H_{16}Cu_2O_{18}Ru_6$ molecules and average metal-metal bonds within the Ru_6Cu_2 framework.

2,5,4,5-BIS(μ_2-CARBONYL)-2,2,4,4,5,5-HEXACARBONYL-1,3-BIS(η^5-CYCLOPENTADIENYL)(μ_4-η-DIPHENYLVINYLENE)CYCLODINICKELTRIRUTHENIUM

$C_{32}H_{20}Ni_2O_8Ru_3$

A. TIRIPICCHIO, M.T. CAMELLINI and E. SAPPA, 1984. J. Chem. Soc. Dalton, 627-632.

Monoclinic, $P2_1/n$, a = 16.459, b = 20.280, c = 9.423 Å, β = 105.67°, Z = 4. Mo radiation, R = 0.040 for 3114 reflexions.

The pentanuclear framework adopts a square pyramidal arrangement with Ru(2) at the apex and alternating Ni and Ru atoms in the basal plane (Fig. 1) (Ni-Ru 2.545(2)-2.673(3), Ru-Ru 2.724(2), 2.735(2) Å). The diphenylacetylene ligand interacts with the four metals in the basal plane (Ru(1)-C(9) 2.24(1), Ru(1)-C(10) 2.28(1), Ru(3)-C(9) 2.29(1), Ru(3)-C(10) 2.23(1), Ni(1)-C(10) 1.92(1), Ni(2)-C(9) 1.93(1) Å). Each Ni atom is η-coordinated by a cyclopentadienyl ligand and of the eight carbonyl groups, six are terminal and two slightly bridge two Ru-Ru edges of the cluster. The complex exhibits approximate C_2 symmetry.

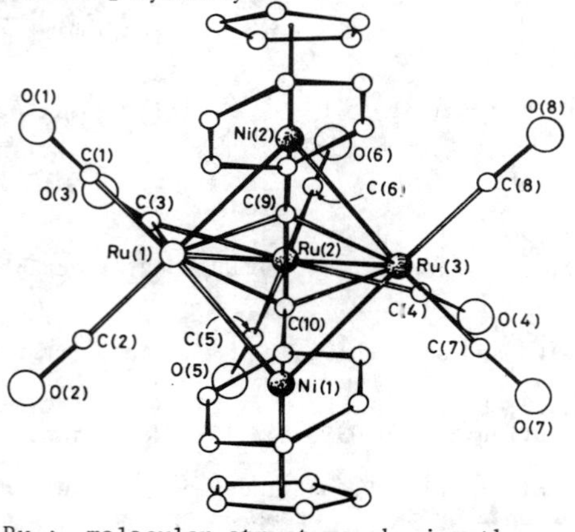

Fig. 1. $C_{32}H_{20}Ni_2O_8Ru_3$: molecular structure showing the approximate C_2 symmetry.

trans-BROMOBIS(METHYLDIPHENYLPHOSPHINE)(PENTABROMOPHENYL)NICKEL(II)
$C_{32}H_{26}Br_6NiP_2$

ZHANG ZHENGZHI, WANG SUKUN, YANG GUANGDI, SHEN CHENG and FAN YUGUO, 1984. J. Struct. Chem., 3, 119-123.

Orthorhombic, $P2_12_12_1$, a = 11.654, b = 15.812, c = 18.556 Å, Z = 4. Mo radiation, R = 0.066 for 2187 reflexions.

The Ni atom has a slightly distorted square planar coordination (Fig. 1) with Ni-Br 2.338(4), Ni-C 1.856(20), Ni-P 2.214 and 2.229(6) Å. The structure is compared with corresponding pentafluorophenyl derivative (1).

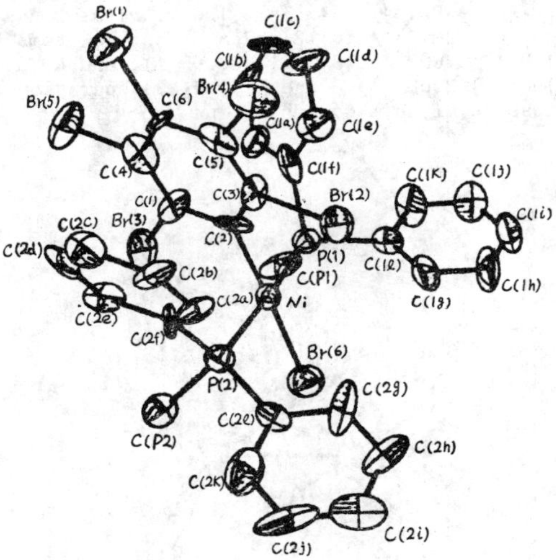

Fig. 1. A view of $C_{32}H_{26}Br_6NiP_2$.

1. Structure Reports, 39B, 574.

{1-[TRIS(η^5-CYCLOPENTADIENYL)URANIUMOXY]-2-(DIMETHYLPHENYLPHOSPHINE)VINYL}(η^5-CYCLOPENTADIENYL)DICARBONYLMANGANESE TOLUENE TETRAHYDROFURAN SOLVATE
$C_{32}H_{32}MnO_3PU,0.5(C_7H_8),0.25(C_4H_8O)$

R.E. CRAMER, K.T. HIGA and J.W. GILJE, 1984. J. Am. Chem. Soc., 106, 7245-7247.

Orthorhombic, Pba2, a = 22.423, b = 35.328, c = 8.648 Å, Z = 8.

There are two independent molecules of the complex in the asymmetric unit; one is shown in Fig. 1. The geometry at the U atom is normal with U-O 2.13, 2.14(2) Å. The Mn-C(23) distances are 1.98(3) and 2.02(4) Å, with C(23)-O(23) 1.36 and 1.38(4) Å and U-O-C 158 and 163(2)°.

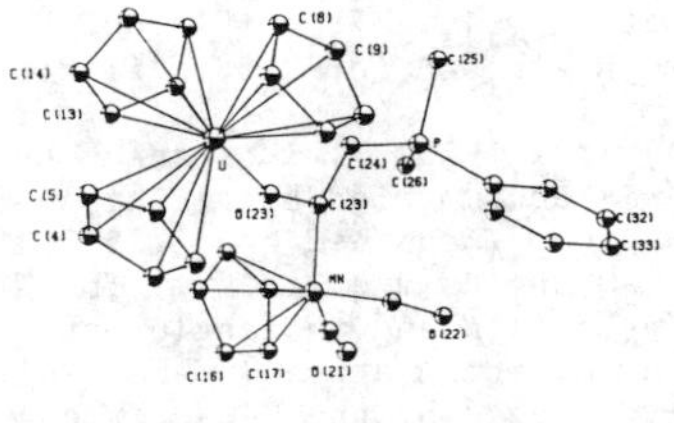

Fig. 1. A view of one molecule of $C_{32}H_{32}MnO_3PU$.

(μ_2-HYDRIDO)(μ_2-o-PHENYLDIAZO-C,N,N)(μ_3-(PHENYLARSINO)(DIPHENYLARSINO)METHANE-As,As,-As')OCTACARBONYLTRIRUTHENIUM

$C_{33}H_{22}As_2N_2O_8Ru_3$ $[HRu_3(CO)_8(\mu\text{-}C_6H_4N_2)(\mu_3\text{-}PhAsCH_2AsPh_2)]$

M.I. BRUCE, E. HORN, M.R. SNOW and M.L. WILLIAMS, 1984. J. Organometal. Chem., 276, C53-C57.

Triclinic, $P\bar{1}$, a = 10.846, b = 12.084, c = 15.884 Å, α = 72.63, β = 68.25, γ = 68.07°, D_m = 1.94, Z = 2. Mo radiation, R = 0.028 for 3676 reflexions.

The μ_3-$PhAsCH_2AsPh_2$ ligand sits on one side of a somewhat enlarged Ru_3 triangle (Fig. 1). Two of the edges are bonding (Ru(1)-Ru(2) 2.903, Ru(2)-Ru(3) 3.055 Å) while the third is not (Ru(1)-Ru(3) 3.547 Å). The metal-As bonds are: Ru(1)-As(1) 2.498, Ru(2)-As(2) 2.456 and Ru(3)-As(2) 2.437 Å. The two atoms Ru(1) and Ru(3) are bridged by N(1) (Ru(1)-N(1) 2.092, Ru(3)-N(1) 2.065 Å), and Ru(3) is bonded direct to the phenyl ring (Ru(3)-C(46) 2.083 Å). The N(1)-N(2) separation (1.267 Å) is consistent with the presence of an N=N double bond.

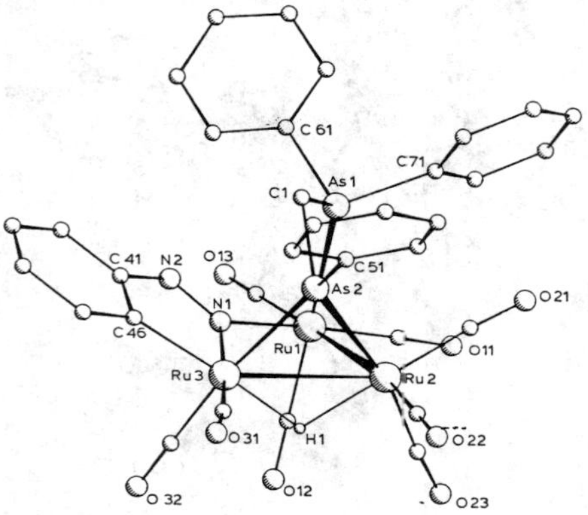

Fig. 1. Molecular structure of $C_{33}H_{22}As_2N_2O_8Ru_3$.

2,2-DICARBONYL-2-(η^5-CYCLOPENTADIENYL)(μ_2-p-TOLYLMETHYLIDENE)-1-TRIPHENYLPHOSPHINE-GOLDTUNGSTEN DICHLOROMETHANE SOLVATE

$C_{33}H_{28}AuO_2PW,CH_2Cl_2$ (I)

3,3-DICARBONYL-3-(η^5-CYCLOPENTADIENYL)(μ_3-p-TOLYLMETHYLIDYNE)-1,2,2-TRIS(TRIMETHYL-PHOSPHINE)GOLDPLATINUMTUNGSTEN HEXAFLUOROPHOSPHATE

$C_{24}H_{39}AuF_6O_2P_4PtW$ (II)

G.A. CARRIEDO, J.A.K. HOWARD, F.G.A. STONE and M.J. WENT, 1984. J. Chem. Soc. Dalton, 2545-2551.

I. Triclinic, $P\bar{1}$, a = 9.219, b = 9.845, c = 18.393 Å, α = 90.35, β = 104.77, γ = 87.27°, Z = 2. Mo radiation, R = 0.044 for 4983 reflexions (at 200 K).

II. Triclinic, $P\bar{1}$, a = 10.920, b = 11.237, c = 15.002 Å, α = 90.18, β = 113.63, γ = 91.79°, Z = 2. Mo radiation, R = 0.088 for 2168 reflexions.

The Au-W bond (2.729(1)) in I (Fig. 1) is spanned by the $CH(C_6H_4Me\text{-}4)$ group. A short W-C(1A) bond (2.06(1)) is accompanied by a long Au-C(1A) distance (2.27(1) Å). The W atom carries a η^5-C_5H_5 group (mean W-C(cp) 2.36(2)) and two CO ligands (W-CO 1.97(2), 1.94(2) Å). The W-C-O angles are linear. The $W(CO)_2$ plane makes an angle of 98° with the plane of Au(μ-C(1A))W and the angle between the latter and CC_6H_4Me is 78°. In II a triangular array of metal atoms (Au-W 2.801(2), Au-Pt 2.956(2), Pt-W 2.770(2) Å) is asymmetrically bridged by the C-C-C_6H_4Me-4 ligand (μ_3-C-Au 2.21(4), μ_3-C-Pt 1.97(4), μ_3-C-W 2.01(4) Å). The Au-Pt separation suggests there is little or no direct metal-metal bonding between these two metal atoms. The W atom carries the η^5-C_5H_5 ligand and two CO groups, but the latter are appreciably non-linear (W-C-O 162(3), 145(5)°). The Au and Pt atoms are ligated by one and two PMe_3 groups respectively (Au-P 2.27(1), Pt-P 2.26(1), 2.32(2) Å).

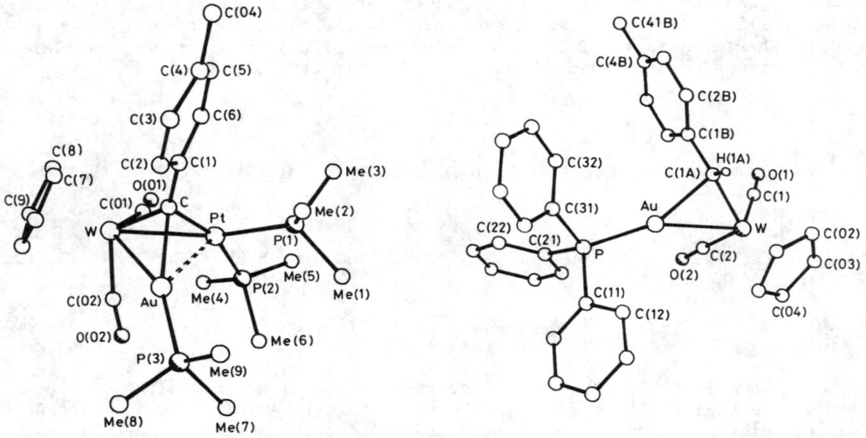

Fig. 1. The molecular structures: $C_{33}H_{28}AuO_2PW$ (left) and $C_{24}H_{39}AuF_6O_2P_4PtW$ (right).

MESITYLBIS(TRIMETHYL PHOSPHITE)TRIPHENYLPHOSPHINERHODIUM(I)

$C_{33}H_{44}O_6P_3Rh$ (I) $Rh((CH_3)_3C_6H_2)[P(OCH_3)_3]_2P(C_6H_5)_3]$

MESITYLTRIS(TRIMETHYL PHOSPHITE)RHODIUM(I)

$C_{18}H_{38}O_9P_3Rh$ (II) $Rh((CH_3)_3C_6H_2)[P(OCH_3)_3]_3]$

L. DAHLENBURG, A. YARDIMCIOGLU and N. HÖCK, 1984. Inorg. Chim. Acta, 89, 213-217.

I. Triclinic, $P\bar{1}$, a = 9.822, b = 10.269, c = 18.683 Å, α = 81.61, β = 78.62, γ = 73.37°, Z = 2. Mo radiation, R = 0.046 for 4930 reflexions.

II. Monoclinic, $P2_1/n$, a = 8.422, b = 17.590, c = 18.311 Å, β = 92.32°, Z = 4. Mo radiation, R = 0.072 for 4223 reflexions.

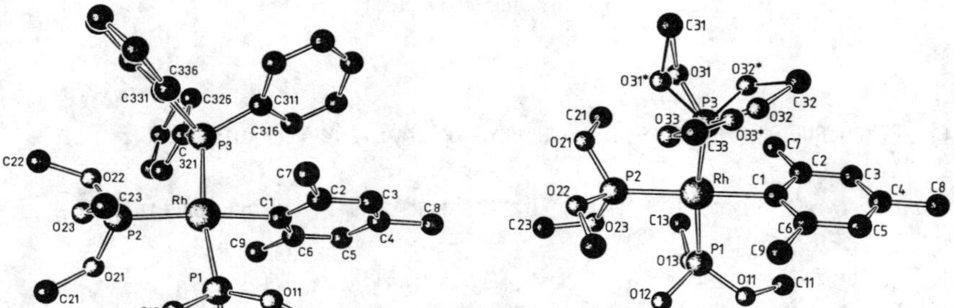

Fig. 1. Molecular structure of $C_{33}H_{44}O_6P_3Rh$ (left), and $C_{18}H_{38}O_9P_3Rh$ (right).

The geometry of both I and II can be described as planar with a slight distortion to tetrahedral (Fig. 1). The cis bond angles about Rh in I range from 82.8 to 95.1° and from 85.0 to 94.9° in II. The trans angles are reduced to 169.6 and 169.1° in I and II respectively. The Rh-P distances in I are 2.223, 2.221 and 2.329 Å, and in II are 2.265, 2.234 and 2.243 Å. The Rh-C(1) distances in I and II are respectively 2.096 and 2.115 Å. There is twofold orientational disorder of the P-O-C bonds around the Rh-P(3) axis in II (Fig. 1). Two sets of O atom positions, each with 0.5 occupancy factors were introduced and refined satisfactorily.

BIS(PERFLUOROPHENYL)(3-PHENYL-1,4-BIS(p-TOLUIDINO)-2,3-DIAZABUT-1-EN-1-YL-4-YLIDENE)-GOLD

$C_{34}H_{21}AuF_{10}N_4$

R. USON, A. LAGUNA, M.D. VILLACAMPA, P.G. JONES and G.M. SHELDRICK, 1984. J. Chem. Soc. Dalton, 2035-2038.

Monoclinic, $P2_1/c$, a = 12.230, b = 20.276, c = 13.123 Å, β = 101.21°, Z = 4. Mo radiation, R = 0.036 for 5822 reflexions.

The Au atom (Fig. 1) exhibits a distorted square-planar coordination probably because of the rigidity of the five-membered ring (C(1)-Au-C(2) 77.5°). The Au-C(1) and Au-C(2) distances are 2.053(5) and 2.053(6) respectively while Au-C(11) is 2.067(6) and Au-C(21) 2.097(5) Å. In the five-membered ring C(1)-N(2), N(2)-N(3) and N(3)-C(2) are 1.294(7), 1.431(6) and 1.312(7) Å respectively.

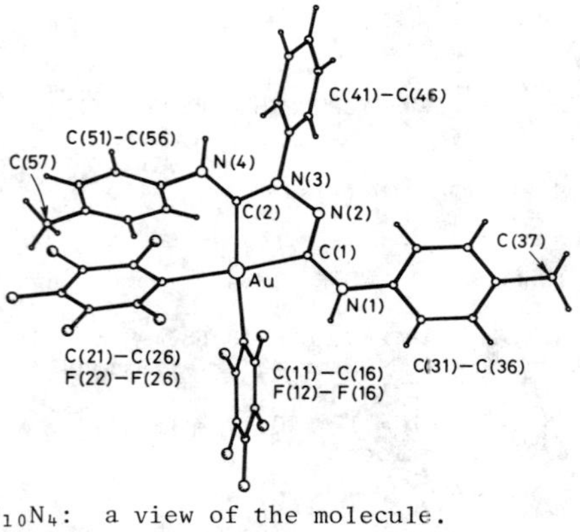

Fig. 1. $C_{34}H_{21}AuF_{10}N_4$: a view of the molecule.

(μ_2-1,1-FERROCENEDIYL)BIS(BIS(η^5-METHYLCYCLOPENTADIENYL)VANADIUM)

$C_{34}H_{36}FeV_2$

F.H. KOHLER, W.A. GEIKE, P. HOFMANN, U. SCHUBERT and P. STAUFFERT, 1984. Chem. Ber., 117, 904-914.

Monoclinic, C2/c, a = 25.75, b = 7.738, c = 17.634 Å, β = 129.89°, Z = 4. Mo radiation, R = 0.053 for 1662 reflexions.

The molecule has twofold crystallographic symmetry (Fig. 1) with the Fe atom on the twofold axis. The ferrocene bridge is distorted, with Fe-C1 (2.119(6) Å) much longer than the other Fe-C distances (2.031-2.041(6) Å). The V-C1 bond is rather long (2.171(7) Å); the V-C(ring) distances are 2.255(6)-2.330(5) Å.

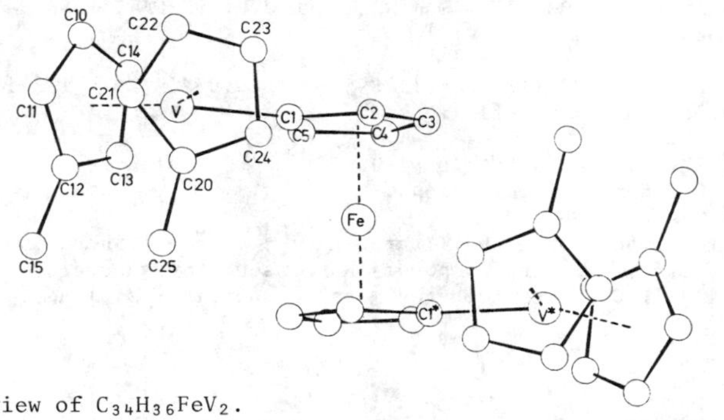

Fig. 1. A view of $C_{34}H_{36}FeV_2$.

TETRABENZYLTHORIUM 1,2-BIS(DIMETHYLPHOSPHINO)ETHANE
$C_{34}H_{44}P_2Th$ (I) $Th(CH_2Ph)_4(Me_2PCH_2CH_2PMe_2)$

TRIBENZYLMETHYLURANIUM 1,2-BIS(DIMETHYLPHOSPHINO)ETHANE
$C_{28}H_{40}P_2U$ (II) $U(CH_2Ph)_3Me(Me_2PCH_2CH_2PMe_2)$

P.G. EDWARDS, R.A. ANDERSEN and A. ZALKIN, 1984. Organometallics, 3, 293-298.

I. Triclinic, $P\bar{1}$, a = 11.463, b = 16.151, c = 21.527 Å, α = 106.28, β = 95.85, γ = 107.78°, Z = 4. Mo radiation, R = 0.047 for 7033 reflexions.

II. Monoclinic, $P2_1/c$, a = 13.035, b = 15.381, c = 14.540 Å, β = 98.06°, Z = 4. Mo radiation, R = 0.024 for 2891 reflexions.

In I there are two independent molecules in the asymmetric unit. The Th atoms are each bonded to four benzyl groups and to two P atoms from the DMPE ligand (Fig. 1). The average Th-C distance is 2.55 Å and the average Th-P distance is 3.17 Å. In complex II the U atom is bonded to a methyl group, three benzyl groups, and the two P atoms of the DMPE ligand. The U-C(methyl) distance is 2.41 Å and the average U-C(benzyl) distance is 2.50 Å.

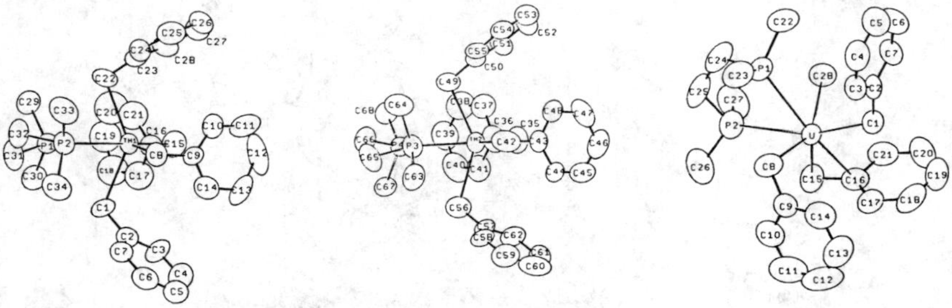

Fig. 1. Molecular structure of $C_{34}H_{44}P_2Th$ molecule 1 (left), molecule 2 (centre), and of $C_{28}H_{40}P_2U$ (right).

IODO(μ_2-OXYCARBENE-C,O)BIS(η^5-CYCLOPENTADIENYL)BIS(η^5-PENTAMETHYLCYCLOPENTADIENYL)-TRIMETHYLPHOSPHINEDIZIRCONIUM BENZENE SOLVATE
$C_{34}H_{50}IOPZr_2,C_6H_6$ (I)

4-[BIS(η^5-CYCLOPENTADIENYL)IODOZIRCONIUM]-2,2-BIS(η^5-PENTAMETHYLCYCLOPENTADIENYL)-1,3-DIOXA-2-ZIRCONACYCLOPENT-4-ENE BENZENE SOLVATE
$C_{32}H_{41}IO_2Zr_2,0.5(C_6H_6)$ (II)

P.T. BARGER, B.D. SANTARSIERO, J. ARMANTROUT and J.E. BERCAW, 1984. J. Am. Chem. Soc., 106, 5178-5186.

I. Monoclinic, C2/c, a = 27.318, b = 19.895, c = 19.932 Å, β = 132.188°, Z = 8. Mo radiation, R = 0.082 for 7179 reflexions.

II. Monoclinic, $P2_1/c$, a = 15.866, b = 10.673, c = 20.561 Å, β = 105.5°, Z = 4. Mo radiation, R = 0.096 for 3891 reflexions.

Molecular dimensions for I and II are in Fig. 1. In I there is a very short Zr-C bond length of 2.117(7) Å. Both compounds have pseudotetrahedral coordination geometry at Zr. In II the five-membered metallacene ring is planar.

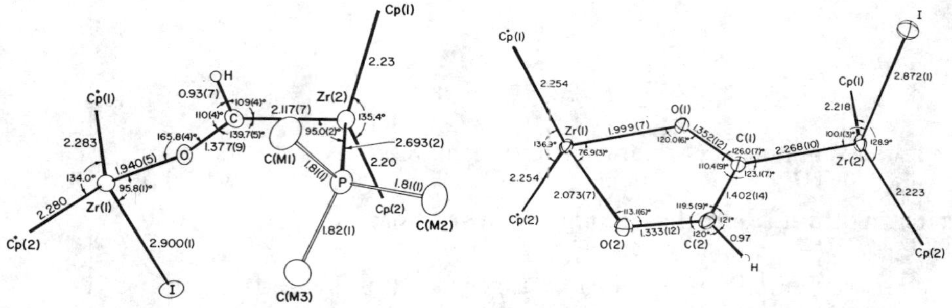

Fig. 1. Views of $C_{34}H_{50}IOPZr_2$ (I) (left) and $C_{32}H_{41}IO_2Zr_2$ (II) (right) with important dimensions.

(μ_2-σ^2,η^4-3-OXO-1,2,4,5-TETRAPHENYLPENTA-1,4-DIENE-1,5-DIYL)BIS(TRICARBONYLIRON)

$C_{35}H_{20}Fe_2O_7$ $Fe_2(CO)_6(PhC_2Ph)_2CO$

HUANG JINSHUN, LI SHIMEI, HE MEIYUN and WANG MANFANG, 1984. J. Struct. Chem., 3, 105-108.

Orthorhombic, Pcab, a = 10.434, b = 19.528, c = 28.305 Å, Z = 8. Mo radiation, R = 0.083 for 2138 reflexions.

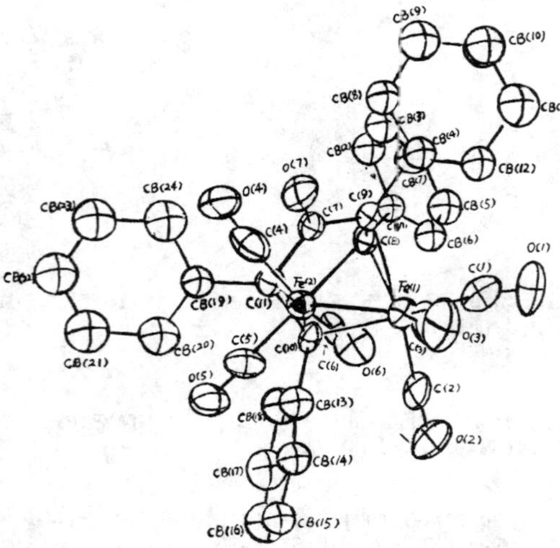

Fig. 1. $C_{35}H_{20}Fe_2O_7$.

The structure of the complex is shown in Fig. 1. There is an Fe-Fe bond (2.537(2) Å). The C-C bonds linking the phenyl rings are 1.396(14) and 1.408(15) Å. Fe-C (ligand) distances are 2.00-2.18 Å.

BIS(η^5-CYCLOPENTADIENYL)BIS(DIPHENYLPHOSPHINOMETHYL)TITANIUM
$C_{36}H_{34}P_2Ti$ $[Cp_2Ti(CH_2PPh_2)_2]$

M. ETIENNE, R. CHOUKROUN, M. BASSO-BERT, F. DAHAN and D. GERVAIS, 1984. Nouv. J. Chim., 8, 531-535.

Monoclinic, $P2_1/n$, a = 7.933, b = 37.493, c = 10.562 Å, β = 108.13°, Z = 4. Mo radiation, R = 0.036 for 2889 reflexions.

In the molecule (Fig. 1) the geometry around the Ti atom is normal with Ti-C(1) 2.183(4), Ti-C(2) 2.210(4), Ti-Cp1 2.085(4), Ti-Cp2 2.077(4), P(1)-C(1) 1.830(4), P(2)-C(2) 1.811(3) Å, C(1)-Ti-C(2) 95.7(1), Cp1-Ti-Cp2 131.9(2), Ti-C-P 122.3 and 122.9(2)°.

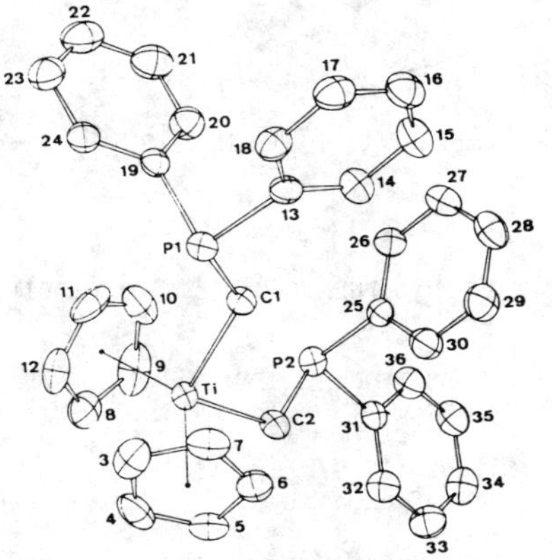

Fig. 1. A view of $C_{36}H_{34}P_2Ti$.

(μ_3-σ^1,η^4-PHENYLACETYLIDE)BIS(TRICARBONYLCOBALT)TETRACARBONYLTRICYCLOHEXYLPHOSPHINE-MANGANESE(I) ETHANOL SOLVATE
$C_{36}H_{38}Co_2MnO_{10}P,0.5(C_2H_6O)$ $[(C_6H_{11})_3PMn(CO)_4C{\equiv}CC_6H_5][Co_2(CO)_6],0.5C_2H_5OH$

G.A. CARRIEDO, V. RIERA, D. MIGUEL, A.M.M. LANFREDI and A. TIRIPICCHIO, 1984. J. Organometal. Chem., 272, C17-C20.

Monoclinic, $P2_1/c$, a = 13.545, b = 18.691, c = 16.944 Å, β = 110.89°, Z = 4. Mo radiation, R = 0.061 for 3402 reflexions.

The acetylide ligand is σ-bonded to the Mn atom and π-bonded to both Co atoms (Fig. 1) and the Mn-C(11) distance is 2.063 Å. The interaction with the Co atoms results in a lengthening of the acetylenic C(11)-C(12) bond (1.313 Å) which is close to that of the standard ethylenic double bond length (1.337 Å), and in a marked deviation from linearity (Mn-C(11)-C(12) 144.5°). The Co(1)-Co(2) distance is 2.469 Å, and other important distances and angles are: Co(1)-C(12) 2.002, Co(2)-C(12) 1.971, Co(1)-C(11) 2.053, Co(2)-C(11) 2.045, Mn-P 2.362 Å; P-Mn-C(11) 178.8 and C(11)-C(12)-C(13) 143.3°.

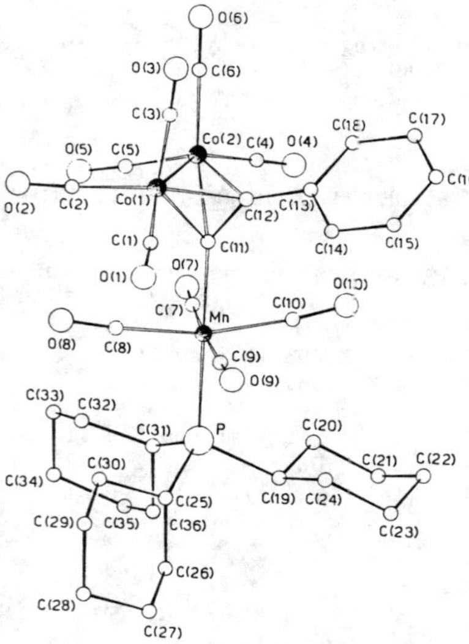

Fig. 1. A view of the compound $C_{36}H_{38}Co_2MnO_{10}P$.

TETRAKIS(μ_2-2-(DIMETHYLAMINOMETHYL)PHENYL-C^1,C^1,N)DIGOLDDILITHIUM
$C_{36}H_{48}Au_2Li_2N_4$

G. VAN KOTEN, J.T.B.H. JASTRZEBSKI, C.H. STAM and N.C. NIEMANN, 1984. J. Am. Chem. Soc., 106, 1880-1881.

Monoclinic, $P2_1/c$, a = 17.34, b = 13.91, c = 15.21 Å, β = 93.25°, Z = 4. Mo radiation, R = 0.034 for 2494 reflexions.

The structure consists of two ArAuAr anionic units and two cationic N_2Li sites (Fig. 1). The latter species form together a unique (virtually planar) $Au_2Li_2C(ipso)_4$ entity that has a square planar Au_2Li_2 arrangement with LiAuLi and AuLiAu angles close to 90° and AuLi distances of 2.87 Å (mean). The orientation of each of the aryl groups suggests that they are primarily bonded via C(ipso) to Au by a two-electron, two-centre bond. The mean Au-C bond length is 2.06 Å.

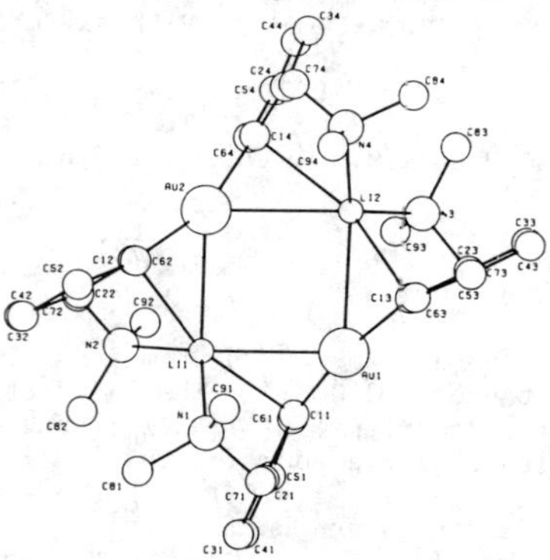

Fig. 1. A view of $C_{36}H_{48}Au_2Li_2N_4$.

BIS[DIMETHYL(PHENYL)PHOSPHINE]-3,3,6,6,9,9,13,13-OCTAMETHYL-anti,syn,anti,syn-11-PALLADAPENTACYCLO[$10.1.0.0^{2,4}.0^{5,7}.0^{8,10}$]TRIDECANE
$C_{36}H_{54}P_2Pd$

H.M. BUCH and C. KRUGER, 1984. Acta Cryst., C40, 28-30.

Triclinic, P$\bar{1}$, a = 11.236, b = 11.774, c = 14.682 Å, α = 74.01, β = 86.92, γ = 75.59°, Z = 2. Mo radiation, R = 0.031 for 6282 reflexions.

The nine-membered metallacycle contains a slightly distorted square-planar Pd atom (Fig. 1), with Pd-P 2.359 and 2.360(1), Pd-C 2.083 and 2.085(3) Å. The four cyclopropane rings impose steric rigidity on the ring system.

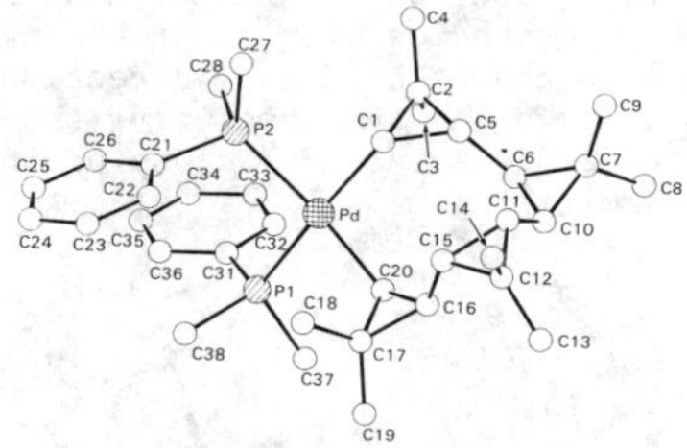

Fig. 1. A general view of $C_{36}H_{54}P_2Pd$.

(μ_2-HYDRAZIDO)BIS[(η^5-t-BUTYL-TETRAMETHYLCYCLOPENTADIENYL)IODONEOPENTYLIDYNE-TUNGSTEN(VI)] DICHLOROMETHANE
$C_{36}H_{62}I_2N_2W_2,CH_2Cl_2$

M.R. CHURCHILL, Y.-J. LI, L. BLUM and R.R. SCHROCK, 1984. Organometallics, 3, 109-113.

Monoclinic, C2/c, a = 17.831, b = 14.588, c = 17.169 Å, β = 99.95°, Z = 4. Mo radiation, R = 0.037 for 3869 reflexions.

The dinuclear tungsten(VI) complex has crystallographically imposed C_2 symmetry with a symmetrical trans geometry for the W-NH-NH-W skeleton (Fig. 1). The neopentylidyne ligands are essentially linearly coordinated to tungsten with a formal W-C triple bond. The W-C(1)-C(2) angle is 171.9°. The W-I distance is 2.714 Å and the metal-to-cyclopentadienyl ring C atom distances range from 2.350 to 2.525 Å. The W-N distance of 1.932 Å is rather short and indicative of partial double bond character, and the angles N-W-C(1), N-W-I, and I-W-C(1) are respectively 99.9, 101.6 and 94.0°.

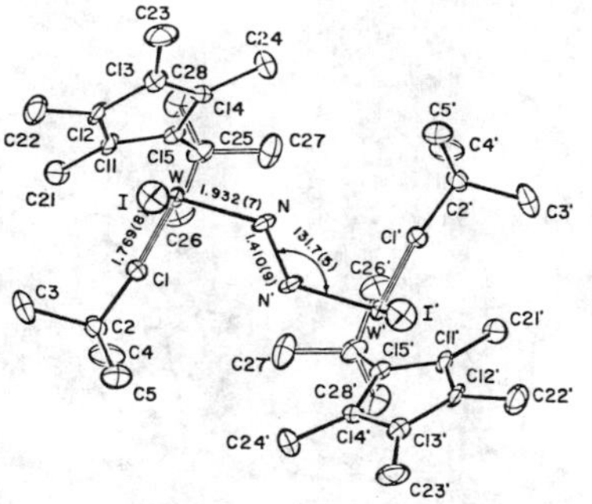

Fig. 1. Molecular structure of $C_{36}H_{62}I_2N_2W_2$ showing selected bond distances (Å) and angles (°).

TETRAKIS(TETRAHYDROFURAN)LITHIUM BIS(TRIS(TRIMETHYLSILYL)METHYL)COPPER

$C_{36}H_{86}CuLiO_4Si_6$ $[Li(THF)_4][Cu\{C(SiMe_3)_3\}_2]$

C. EABORN, P.B. HITCHCOCK, J.D. SMITH and A.C. SULLIVAN, 1984. J. Organometal. Chem., 263, C23-C25.

Triclinic, $P\bar{1}$, a = 9.404, b = 11.946, c = 23.262 Å, α = 94.37, β = 89.98, γ = 94.35°, Z = 2. Mo radiation, R = 0.144 for 2500 reflexions.

The crystal structure consists of one cation and two non-equivalent $[Cu\{C(Si(Me_3)_3\}_2]^-$ anions, each anion having crystallographic inversion symmetry (Fig. 1). Successful refinement of the structure required that some iso-structural $[Li\{C(SiMe_3)_3\}_2]^-$ anions were incorporated in a random manner at the Cu(2) site centred on 0,0,0, the Cu and Li occupancies of the site being in ca. 60/40 ratio. The $C(SiMe_3)_3$ groups are staggered about the C-Cu-C direction. Within experimental error the two independent anions have the same geometry (the Cu(1)-C(1) and Cu(2)-C(11) bond lengths are 2.027 and 2.070 Å though because of the Cu/Li disorder in the 2nd anion the latter value is unreliable).

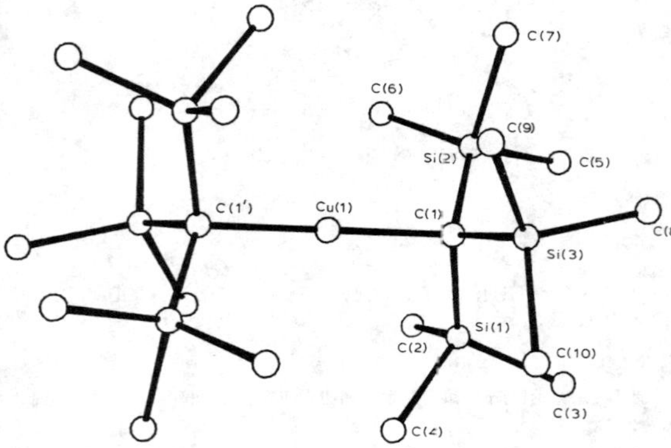

Fig. 1. Structure of one of the anions in $C_{36}H_{86}CuLiO_4Si_6$.

BIS(μ_2-4,4-DIMETHYLPENT-1-ENYL-1-OLATO-C,O)(μ_2-METHYL)(DIMETHYLALUMINIUM)BIS(ZIRCONOCENE)

$C_{37}H_{53}AlO_2Zr_2$

R.M. WAYMOUTH, B.D. SANTARSIERO and R.H. GRUBBS, 1984. J. Am. Chem. Soc., 106, 4050-4051.

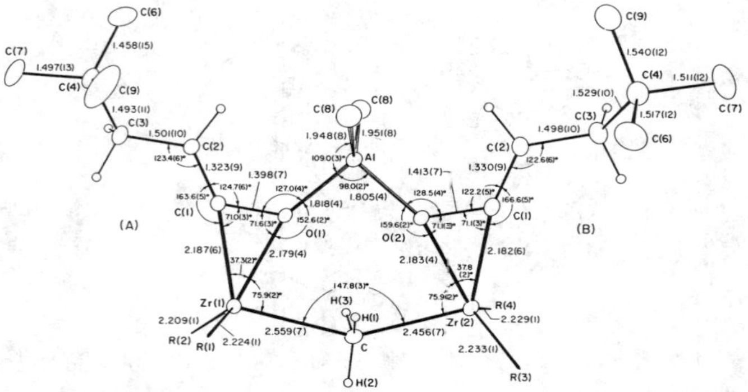

Fig. 1. The $C_{37}H_{53}AlO_2Zr_2$ molecule with dimensions; the cyclopentadienyl rings are represented by R(1)-R(4).

Monoclinic, $P2_1/c$, a = 10.2089, b = 20.214, c = 18.421 Å, β = 94.375°, Z = 4. Mo radiation, R = 0.042 for 3380 reflexions.

The bridging methyl group in this structure (Fig. 1) is very nearly planar and approaches a trigonal-bipyramidal geometry. The methyl bridge is not symmetrical, being 0.1 Å closer to Zr(2) than Zr(1). Details of dimensions are in Fig. 1.

(μ_2-σ^2,η^2,η^2-1,2-DIPHENYL-3-TRIMETHYLSILYLALLYL)(μ_2-TRIMETHYLSILYLMETHYLIDYNE)-TETRAKIS(TRIMETHYLSILYLMETHYL)DITUNGSTEN

$C_{38}H_{72}Si_6W_2$

M.H. CHISHOLM, J.A. HEPPERT and J.C. HUFFMAN, 1984. J. Am. Chem. Soc., 106, 1151-1153.

Orthorhombic, $P2_12_12_1$, a = 20.884, b = 13.734, c = 16.842 Å, Z = 4. Mo radiation, R = 0.052 for 5579 reflexions (at -160°C).

The analysis establishes the structure shown in Fig. 1. Main dimensions (Å and degrees) are W-W 2.548(1), W(1)-C(9) 2.24(1), W(1)-C(10) 2.45(1), W(1)-C(17) 2.29(1), W(2)-C(9) 2.20(1), W(2)-C(10) 2.68(1), W(2)-C(17) 2.17(1), W(1)-C(32) 1.96(1), W(2)-C(32) 2.00(1), C(9)-C(10) 1.41(2), C(10)-C(17) 1.41(2), W-C(alkyl) 2.13(2) (av), C(9)-C(10)-C(17) 106(1), W(1)-C(32)-W(2) 79.9(4), W(1)-C(9)-W(2) 69.9(3), W(1)-C(10)-W(2) 59.3(3), W(1)-C(17)-W(2) 69.8(3), C(22)-W(1)-C(27) 104.7(5), C(37)-W(2)-C(42) 98.7(6).

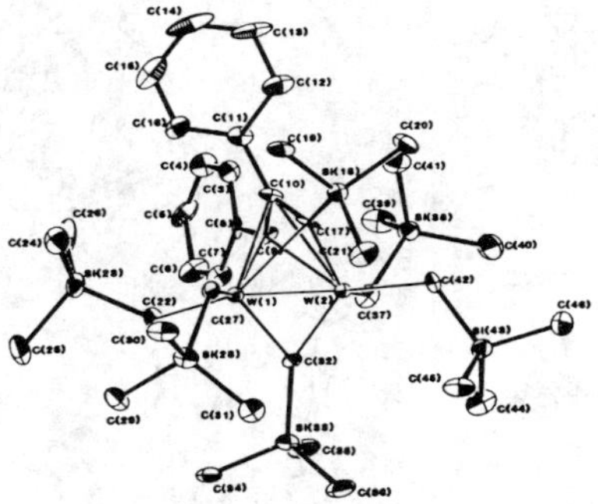

Fig. 1. A view of $C_{38}H_{72}Si_6W_2$.

[μ-[[(1,2,3,4-η)-3,4-BIS(1,1-DIMETHYLETHYL)-1,3-CYCLOBUTADIENE-1,2-DIYL]BIS[DIPHENYL-PHOSPHINE]-P,P']]TRICARBONYL(DICHLOROPLATINUM)IRON CHLOROFORM DICHLOROMETHANE SOLVATE

$C_{39}H_{38}Cl_2FeO_3P_2Pt,2CHCl_3,0.5(CH_2Cl_2)$ (I)

[μ-[(2,3,4,5-η)-3,4-BIS(DIPHENYLPHOSPHINO)-2,5-DIPHENYL-2,4-CYCLOPENTADIEN-1-ONE-P,P']]TRICARBONYL(DICHLOROPLATINUM)IRON CHLOROFORM BENZENE SOLVATE

$C_{44}H_{30}Cl_2FeO_4P_2Pt,3CHCl_3,0.5(C_6H_6)$ (II)

[3,4-BIS(DIPHENYLPHOSPHINO)-2,5-DIPHENYL-2,4-CYCLOPENTADIEN-1-ONE-P,P']BIS(PENTA-FLUOROPHENYL)PLATINUM CHLOROFORM SOLVATE

$C_{53}H_{30}F_{10}OP_2Pt,CHCl_3$ (III)

T. RUKACHAISIRIKUL, S. ARABI, F. HARTSTOCK, N.J. TAYLOR and A.J. CARTY, 1984. Organometallics, 3, 1587-1589.

I. Monoclinic, $P2_1/n$, a = 14.649, b = 18.517, c = 18.135 Å, β = 91.52°, Z = 4. R = 0.049 for 4736 reflexions.

II. Monoclinic, $P2_1/c$, a = 23.609, b = 13.192, c = 18.845 Å, β = 107.20°, Z = 4. R = 0.047 for 4262 reflexions.

III. Triclinic, P$\bar{1}$, a = 12.106, b = 14.161, c = 16.377 Å, α = 111.69, β = 107.52, γ = 92.05°, Z = 2. R = 0.036 for 5321 reflexions.

Molecule I consists of a planar cyclobutadiene ring with cis-di-tert-butyl and cis-diphenylphosphino groups, coordinated in η^4 fashion to an $Fe(CO)_3$ moiety (Fig. 1). The C_4 ring is slightly rectangular with the C(4)-C(5) and C(6)-C(7) bond lengths (average 1.492 Å) being slightly longer than the C(5)-C(6) and C(4)-C(7) bonds (average 1.447 Å). The Fe atom is symmetrically bound to the C_4 ring (Fe-C(4) 1.99, Fe-C(5) 2.11, Fe-C(6) 2.14 and Fe-C(7) 2.02 Å). The molecular structure of II reveals the presence of a bis(diphenylphosphino)diphenylcyclopentadienone ligand coordinated to an $Fe(CO)_3$ unit. Comparison of II and III shows that both molecules contain non-planar C_5O ring systems. In II the dihedral angle between the plane C(4), C(5), C(7) and C(8) and the plane C(5), C(7) and C(6) is 16.9°. In III the corresponding angle is 5.1°.

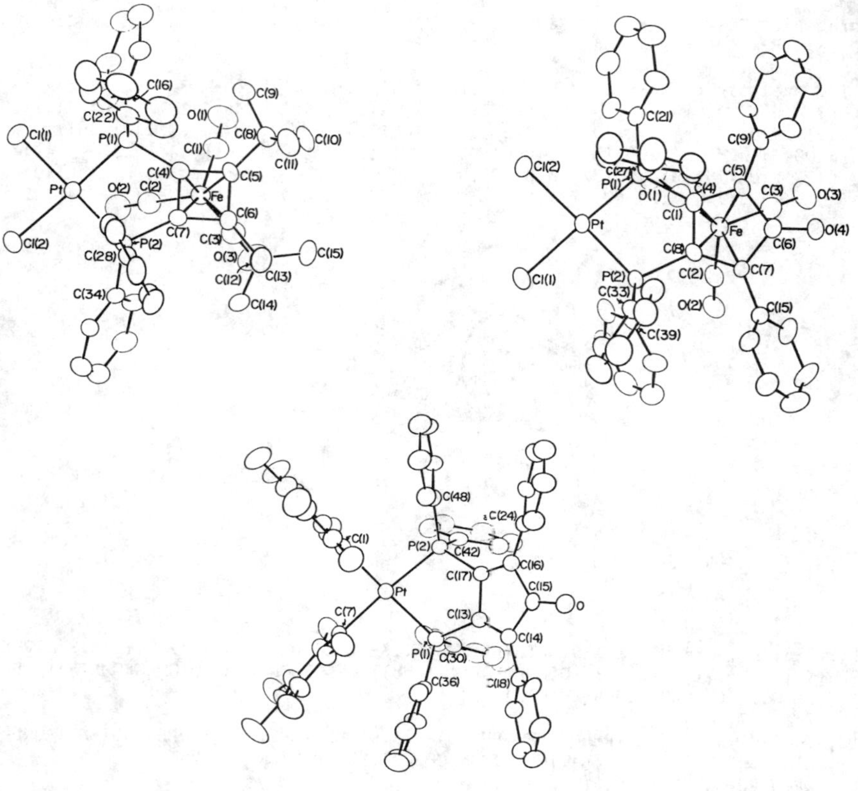

Fig. 1. Molecular structure of $C_{39}H_{38}Cl_2FeO_3P_2Pt$ (top left), $C_{44}H_{30}Cl_2FeO_4P_2Pt$ (top right), and $C_{53}H_{30}F_{10}OP_2Pt$ (bottom).

(μ_2-N,N-DIMETHYLAMINOCYANO-C,N)(μ_2-N,N-DIMETHYLAMINOCYANO-C,N,N)(N,N-DIMETHYLAMINO-CYANO-C,N)HEXAKIS(NEOPENTOXY)DITUNGSTEN

$C_{39}H_{84}N_6O_6W_2$ $W_2(OCH_2Bu)_6(NCNMe_2)_3$

M.H. CHISHOLM, K. FOLTING, J.C. HUFFMAN and N.S. MARCHANT, 1984. Polyhedron, 3, 1033-1035.

Triclinic, P$\bar{1}$, a = 16.784, b = 15.959, c = 9.839 Å, α = 76.58, β = 97.61, γ = 82.28°, Z = 2. Mo radiation, R = 0.057 for 4993 reflexions (at -160°C).

In the compound the W...W distance (3.85 Å) is a non-bonding one and based on Me_2NC-N or W-N distances (Fig. 1), each Me_2NCN molecule may be viewed as a 2- ligand.

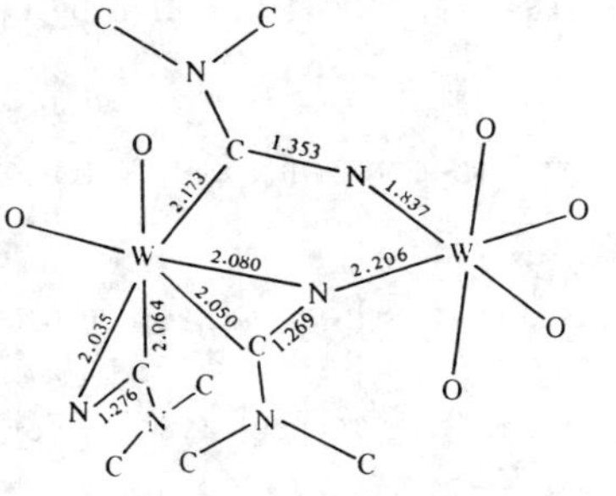

Fig. 1. Principal bond lengths in $W_2(OCH_2Bu)_6(NCNMe_2)_3$.

DICARBONYL(η^5-CYCLOPENTADIENYL)-(3,3-DIPHENYL-1-TRIPHENYLPHOSPHONIUM-ALLENYLIDENE(PENTACARBONYL)MANGANESE DICHLOROMETHANE SOLVATE

$C_{40}H_{30}MnO_2P,CH_2Cl_2$ $C_5H_5(CO)_2Mn^-C(P^+(C_6H_5)_3)$=C=C$(C_6H_5)_2$,$CH_2Cl_2$

N.E. KOLOBOVA, L.L. IVANOV, O.S. ZHVANKO, O.M. KHITROVA, A.S. BATSANOV and Yu.T. STRUCHKOV, 1984. J. Organometal. Chem., 265, 271-281.

Triclinic, P$\bar{1}$, a = 10.266, b = 10.603, c = 16.320 Å, α = 79.22, β = 84.30, γ = 74.70°, Z = 2. Mo radiation, R = 0.031 for 5126 reflexions (-120°C).

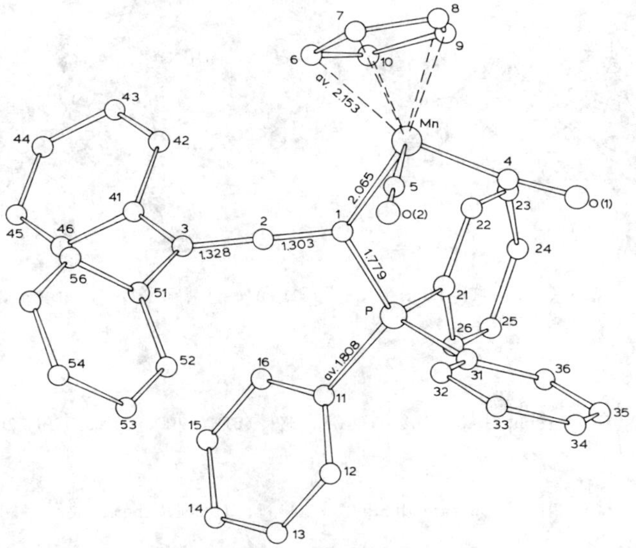

Fig. 1. Molecular structure of $C_{40}H_{30}MnO_2P$ showing selected bond distances in Å.

In the phosphonium ylide complex (Fig. 1) the Mn atom has a slightly distorted three-legged piano stool coordination. The cyclopentadienyl ring is planar with the Mn-C and C-C distances averaging 2.153 and 1.410 Å. The Mn-allene C(1) bond distance of 2.065 Å is characteristic of a single bond as is the P-C(1) distance of 1.779 Å. The double bond distances C(1)-C(2) and C(2)-C(3) of 1.303 and 1.328 Å respectively are comparable to those in a free allene molecule. The Mn...P distance is 3.369 Å, indicating no interaction and the coordination geometries of Mn (octahedral), P (tetrahedral) and C(1) (planar trigonal) exhibit no significant distortion. Within the crystal structure, infinite channels are occupied by CH_2Cl_2 molecules.

BIS(μ-ACETATO)BIS[μ-2-(DIPHENYLPHOSPHINO)PHENYL-C,P]DICHLORODIOSMIUM ACETIC ACID SOLVATE

$C_{40}H_{34}Cl_2O_4Os_2P_2,C_2H_4O_2$ (I)

BIS(μ-PROPIONATO)BIS[μ-2-(DIPHENYLPHOSPHINO)PHENYL-C,P]DICHLORODIOSMIUM

$C_{42}H_{38}Cl_2O_4Os_2P_2$ (II)

A.R. CHAKRAVARTY, F.A. COTTON and D.A. TOCHER, 1984. Inorg. Chem., 23, 4697-4700.

I. Monoclinic, C2/c, a = 16.315, b = 14.210, c = 19.271 Å, β = 109.34°, Z = 4. Mo radiation, R = 0.030 for 2747 reflexions.

II. Orthorhombic, Pbcn, a = 11.901, b = 18.473, c = 17.578 Å, Z = 4. Mo radiation, R = 0.027 for 2233 reflexions.

Both molecules (Fig. 1) have exact C_2 symmetry and consist of diosmium units bridged by two cisoid carboxylato ligands and by two 2-$Ph_2PC_6H_4$ ligands, each Os bearing an axial chloride ligand. The Os-Os triple bond distances (2.271(1) Å in I, 2.272(1) Å in II) are the shortest yet reported. The Cl-Os-Os-Cl chains are far from linear with Os-Os-Cl 161.08(5)° in I and 162.51(5)° in II. Other bond lengths are (respectively for I and II): Os-Cl 2.372(2) and 2.396(2), Os-P 2.345(2) and 2.353(2), Os-O (trans to C) 2.223(5) and 2.187(5), Os-O (trans to P) 2.105(5) and 2.105(5), Os-C 2.080(7) and 2.045(7) Å.

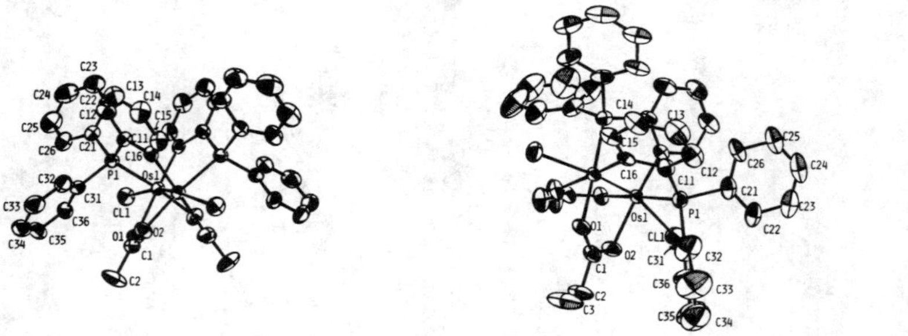

Fig. 1. The $Os_2(O_2CR)_2[Ph_2PC_6H_4]_2Cl_2$ molecules, R = CH_3 (left) and R = C_2H_5 (right).

(μ_2-PHENYLVINYLIDENE)PHENYLACETYLIDETETRAKIS(TRIETHYLPHOSPHINE)DIPLATINUM TETRAFLUOROBORATE

$C_{40}H_{71}BF_4P_4Pt_2$

D. AFZAL, P.G. LENHERT and C.M. LUKEHART, 1984. J. Am. Chem. Soc., 106, 3050-3052

Monoclinic, $P2_1/c$, a = 13.130, b = 15.792, c = 24.958 Å, β = 113.07°, Z = 4. Mo radiation, R = 0.069 for 4444 reflexions.

In the cation (Fig. 1) there is a Pt-Pt single bond (2.750(2) Å). The coordination at Pt1 is described as nearly square (excluding Pt2) and atoms Pt1,P1,P2,C50,C60 are nearly coplanar (maximum deviation 0.11 Å for Pt1). The principal coordination geometry at Pt2 (excluding Pt1) is described as distorted "Y" or "T" shaped. Some dimensions are: Pt1-P1 2.325(5), Pt1-P2 2.329(5), Pt2-P3 2.228(8), Pt2-P4 2.351(8), Pt1-C60 2.17(2), Pt1-C50 2.10(2), Pt2-C50 1.93(2) Å.

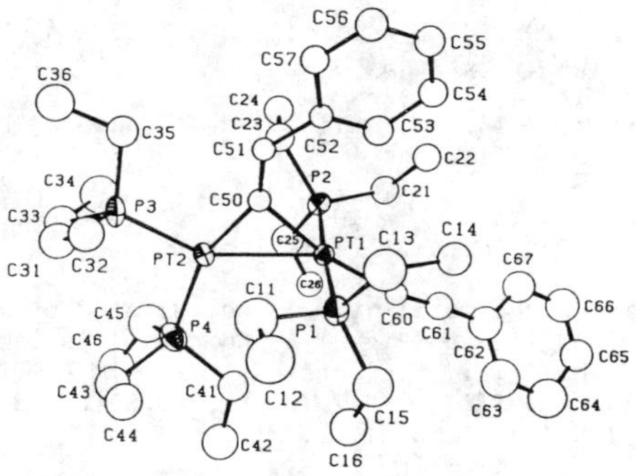

Fig. 1. The cation in $C_{40}H_{71}BF_4P_4Pt_2$.

trans-CHLORO(η'-PHENYL)BIS(TRIPHENYLPHOSPHINE)PLATINUM(II)
$C_{42}H_{35}ClP_2Pt$ $(C_6H_5)Pt(Cl)(PPh_3)_2$

W. CONZELMANN, J.D. KOOLA, U. KUNZE and J. STRÄHLE, 1984. Inorg. Chim. Acta, 89, 147-149.

Orthorhombic, Pbca, a = 11.800, b = 23.705, c = 25.549 Å, Z = 8. Cu radiation, R = 0.073 for 2575 reflexions.

The crystal structure is composed of discrete molecules exhibiting square-planar coordination about the Pt atom, with the tertiary phosphine ligands occupying trans positions (Fig. 1). The P-Pt-P angles range from 87.8 to 92.3° as a consequence of steric repulsion among the ligands. Important bond lengths and angles are: Pt-Cl 2.408, Pt-P(1) 2.303, Pt-P(2) 2.297 and Pt-C(1) 2.00 Å; P(1)-Pt-P(2) 176.0 and Cl-Pt-C(1) 179.9°.

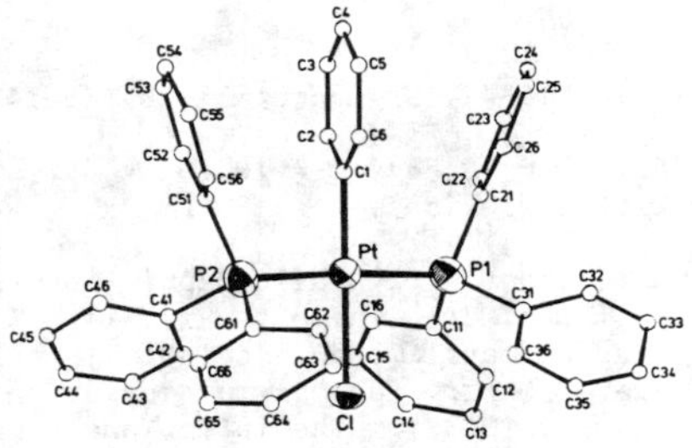

Fig. 1. Molecular structure of $C_{42}H_{35}ClP_2Pt$.

(η^2-CARBON DISULPHIDE-C,S)[1,1,1-TRIS((DIPHENYLPHOSPHINO)METHYL)ETHANE-P,P',P'']-NICKEL(0)
$C_{42}H_{39}NiP_3S_2$ (I) $[CH_3C(CH_2PPh_2)_3]Ni(\eta^2-CS_2)$

(η^2-N-PHENYLISOTHIOCYANATO-C,S)[1,1,1-TRIS((DIPHENYLPHOSPHINO)METHYL)ETHANE-P,P',P'']-NICKEL(0) DICHLOROMETHANE
$C_{48}H_{44}NNiP_3S,0.5(CH_2Cl_2)$ (II) $[CH_3C(CH_2PPh_2)_3]Ni(\eta^2-SCNPh).0.5CH_2Cl_2$

(η^2-N-PHENYLISOTHIOCYANATO-C,S)[1,1,1-TRIS(2-(DIPHENYLPHOSPHINO)ETHYL)AMINE-P,P',P'']-COBALT(0) n-BUTANOL
$C_{49}H_{47}CoN_2P_3S,1.33(C_4H_{10}O)$ (III) $[N(CH_2CH_2PPh_2)_3]Co(\eta^2-SCNPh).1.33C_4H_{10}O$

C. BIANCHINI, D. MASI, C. MEALLI and A. MELI, 1984. Inorg. Chem., 23, 2838-2844.

I. Orthorhombic, $Pn2_1a$, a = 20.924, b = 17.112, c = 10.283 Å, D_m = 1.36, Z = 4. Mo radiation, R = 0.048 for 2104 reflexions.

II. Orthorhombic, Pbac, a = 23.142, b = 21.429, c = 17.204 Å, D_m = 1.33, Z = 8. Mo radiation, R = 0.072 for 1499 reflexions.

III. Monoclinic, $P2_1/a$, a = 36.435, b = 10.801, c = 13.026 Å, β = 97.62°, D_m = 1.19, Z = 4. Mo radiation, R = 0.074 for 2123 reflexions.

In each of the complexes the metal atom is coordinated by an η^2-CS moiety and by the three P atoms of a tripod ligand (Fig. 1). Complex I is isomorphous with the Co analog (1). Bond lengths for I-III, respectively, are: M-P 2.235(2)-2.296(3), 2.156(6)-2.258(6), 2.269(4)-2.303(4); M-S 2.197(3), 2.259(6), 2.220(5); M-C 1.86(1), 1.88(2), 1.87(2); C-S(M) 1.63(1), 1.68(2), 1.72(2) Å.

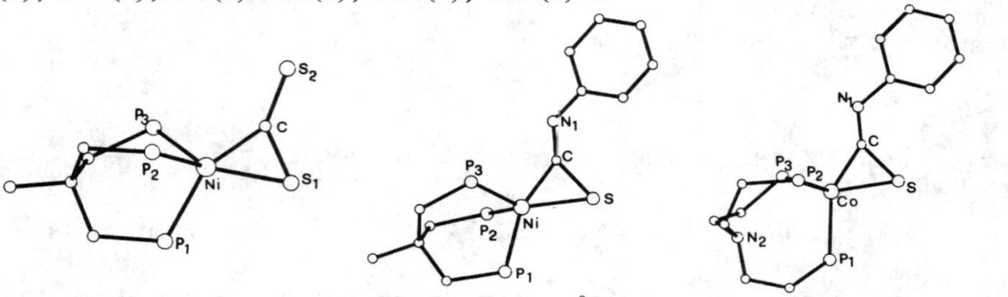

Fig. 1. The structures of $[CH_3C(CH_2PPh_2)_3]Ni(\eta^2-CS_2)$ (left), $[CH_3C(CH_2PPh_2)_3]$-$Ni(\eta^2-SCNPh)$ (middle), and $[N(CH_2CH_2PPh_2)_3]Co(\eta^2-SCNPh)$ (right).

1. Structure Reports, 46B, 857.

BIS(μ-IODO)BIS[o-(DI-o-TOLYLPHOSPHINO)BENZYL]DIPALLADIUM(II)

$C_{42}H_{40}I_2Pd_2$

A.L. RHEINGOLD and W.C. FULTZ, 1984. Organometallics, 3, 1414-1417.

Monoclinic, $P2_1/n$, a = 9.451, b = 10.864, c = 19.307 Å, β = 92.22°, Z = 2. Mo radiation, R = 0.033 for 2160 reflexions.

The molecule is situated about a crystallographic inversion centre thus requiring strict planarity of the Pd_2I_2 rhombus (Fig. 1). The Pd...Pd distance of 3.929 Å precludes any meaningful metal-metal bonding. The coordination geometry about Pd(II) is clearly square planar, the maximum deviation from perfect planarity (0.008 Å) is slight and occurs at C(37). The interplanar angle relating the central rhombus to the Pd coordination plane is 2.4°. The five-membered ring is folded along the P...C(37) vector at a dihedral angle of 40.6°.

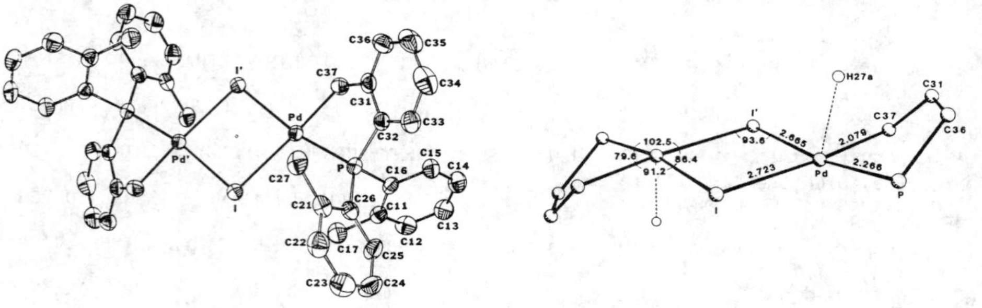

Fig. 1. Molecular structure of $C_{42}H_{40}I_2Pd_2$, and the central Pd_2I_2 rhombus and Pd coordination geometry.

TRIS(μ-CYCLOHEXYL ISOCYANIDE)TRIS(CYCLOHEXYL ISOCYANIDE)-triangulo-TRIPALLADIUM
$C_{42}H_{66}N_6Pd_3$ $Pd_3(CNC_6H_{11})_6$

C.G. FRANCIS, S.I. KHAN and P.R. MORTON, 1984. Inorg. Chem., 23, 3680-3681.

Monoclinic, $P2_1/n$, a = 14.002, b = 10.966, c = 29.067 Å, β = 92.30°, Z = 4. Mo radiation, R = 0.047 for 2624 reflexions.

The molecule (Fig. 1) consists of an equilateral triangle of Pd atoms, each side being bridged by a CNC_6H_{11} ligand with the three remaining CNC_6H_{11} ligands terminally bound to the three Pd atoms. Mean structural parameters: Pd-Pd 2.651(2), Pd-C_t 2.004(14), Pd-C_b 2.072(13) Å, Pd-C_b-Pt 79.5(5), Pd-C_t-N_t 169.1(13), $[C-N-C]_t$ 170.2(14), and $[C-N-C]_b$ 128.9(12)°.

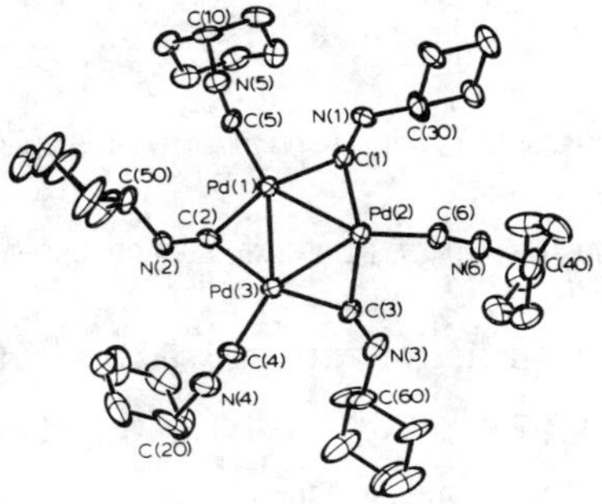

Fig. 1. The $Pd_3(CNC_6H_{11})_6$ molecule.

[μ_2-BIS(DIPHENYLPHOSPHINO)METHANIDO-P^1P^2(Au^1),C(Au^2)]-1,1,2-TRIS(PENTAFLUOROPHENYL)-DIGOLD(I,III)
$C_{43}H_{21}Au_2F_{15}P_2$ (I)

[μ_2-BIS(DIPHENYLPHOSPHINO)METHANIDO-P^1P^2(Au^1),C(Au^3)][μ_2-BIS(DIPHENYLPHOSPHINO)-METHANIDO-P^1P^2(Au^2),C'(Au^3)][1,1,2,2-TETRAKIS(PENTAFLUOROPHENYL)TRIGOLD(I,II,III)] PERCHLORATE CHLOROFORM SOLVATE
$C_{74}H_{42}Au_3ClF_{20}O_4P_4$,x($CHCl_3$) (II)

R. USON, A. LAGUNA, M. LAGUNA, B.R. MANZANO, P.G. JONES and G.M. SHELDRICK, 1984. J. Chem. Soc. Dalton, 839-843.

I. Monoclinic, $P2_1/m$, a = 10.003, b = 17.389, c = 11.754 Å, β = 103.53°, Z = 2. Mo radiation, R = 0.034 for 4016 reflexions.

II. Monoclinic, $P2_1/m$, a = 17.456, b = 17.704, c = 17.749 Å, β = 92.86°, Z = 2. Mo radiation, R = 0.096 for 5230 reflexions. [High R factor due to presence of disordered solvent molecules.]

The molecule of I (Fig. 1) possesses crystallographic m symmetry: both Au atoms, the methanide CH and the C_6F_5 groups at Au(I) lie in the mirror plane. In the Au-P-C-P ring the C atom lies 0.64 Å out of the Au(III) ligand plane and the P-C bond lengths are 1.815(6) Å. Some bond distances are: Au(1)-P 2.354(2), Au(2)-C(1) 2.094(7), Au(1)-C(11) 2.075(6) and Au(2)-C(21) 2.048(8) Å. The cation in II also possesses crystallographic m symmetry: all three Au atoms and the methanide C lie in the mirror plane. The chemically equivalent halves of the cation have very different geometries: Au(2)-P(1)-C(1)-P(1') is approximately planar (C(1) <0.1 Å out of the Au(2) ligand plane) while Au(3)-P(2)-C(2)-P(2) is not (C(2) is 0.8 Å out of the Au(3) ligand plane). Associated with this is a greater bending of the cation at C(2). Some bond distances in the cation are: Au(1)-C(1) 2.13(3), Au(2)-P(1) 2.347(7), Au(3)-P(2) 2.356(7), Au(1)-C(2) 2.09(3), Au(2)-C(51) 2.11(1), Au(3)-C(61) 2.11(1) Å.

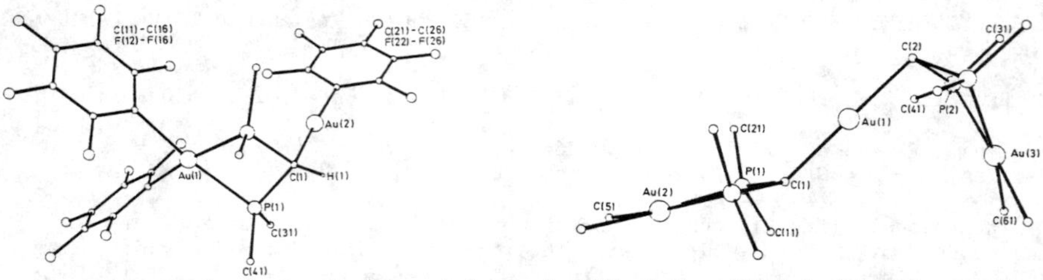

Fig. 1. $C_{43}H_{21}Au_2F_{15}P_2$ (left) and $C_{74}H_{42}Au_3ClF_{20}O_4P_4,x(CHCl_3)$ (right): a view of the molecule and cation.

HEXAKIS(PHENYL ISOCYANIDE)CHROMIUM(I) TRIFLUOROMETHYLSULPHONATE

$C_{43}H_{30}CrF_3N_6O_3S$ (I) $[Cr(CNC_6H_5)_6][CF_3SO_3]$

HEXAKIS(PHENYL ISOCYANIDE)CHROMIUM(II) BIS(HEXAFLUOROPHOSPHATE)

$C_{42}H_{30}CrF_{12}N_6P_2$ (II) $[Cr(CNC_6H_5)_6][PF_6]_2$

HEXAKIS(PHENYL ISOCYANIDE)CHROMIUM(III) TRIS(HEXACHLOROANTIMONATE) DICHLOROMETHANE SOLVATE

$C_{42}H_{30}Cl_{18}CrN_6Sb_3,CH_2Cl_2$ (III) $[Cr(CNC_6H_5)_6][SbCl_6]_3,CH_2Cl_2$

D.A. BOHLING and K.R. MANN, 1984. Inorg. Chem., 23, 1426-1432.

I. Monoclinic, $P2_1/n$, a = 10.757, b = 19.994, c = 10.242 Å, β = 110.69°, D_m = 1.314, Z = 2. Cu radiation, R = 0.042 for 3679 reflexions.

II. Monoclinic, $P2_1/n$, a = 13.888, b = 19.638, c = 16.009 Å, β = 91.20°, D_m = 1.423, Z = 4. Cu radiation, R = 0.060 for 2335 reflexions.

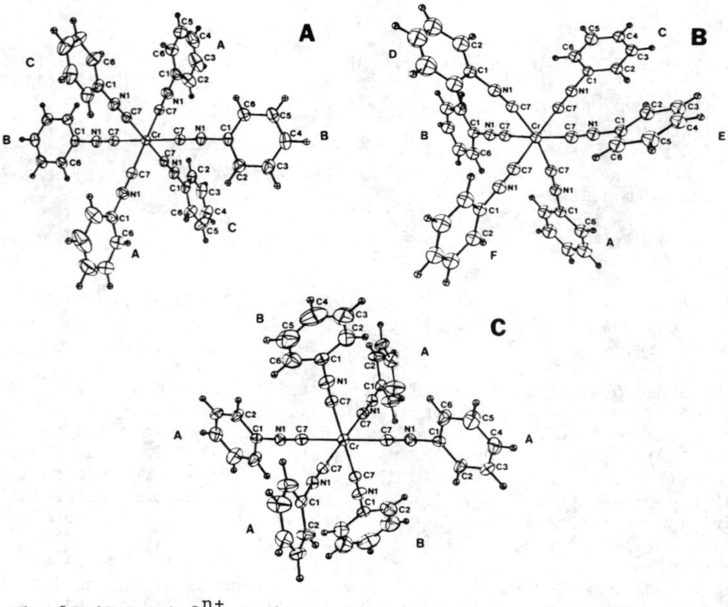

Fig. 1. The $[Cr(CNC_6H_5)_6]^{n+}$ cations: A, n=1; B, n=2; and C, n=3.

III. Orthorhombic, Cccm, a = 14.818, b = 18.450, c = 23.714 Å, D_m = 1.746, Z = 4. Mo radiation, R = 0.058 for 1445 reflexions.

The $[Cr(CNC_6H_5)_6]^{n+}$ (n = 1,2,3) cations are shown in Fig. 1. A minor Jahn-Teller distortion is observed in the Cr-C bond lengths (Cr-C 1.964(2), 1.971(2), 1.990(2) Å) of the centrosymmetric $[Cr(CNC_6H_5)_6]^+$ cation. For $[Cr(CNC_6H_5)_6]^{2+}$ the ligands show a pronounced C_{2v} angular distortion with C-Cr-C bond angles ranging from 83.7(3) to 97.0(2)°. The Cr-C distances range from 2.009(7) to 2.021(7) Å (mean 2.014 Å). The $[Cr(CNC_6H_5)_6]^{3+}$ cation has exact C_{2h} symmetry with equatorial and axial Cr-C bond lengths of 2.049(10) and 2.091(14) Å, respectively.

2,4-BIS(METHOXYCARBONYL)-1,1-BIS(TRIPHENYLPHOSPHINE)PLATINACYCLOBUTAN-3-ONE MONO-HYDRATE

$C_{43}H_{38}O_5P_2Pt,H_2O$

D.A. CLARKE, R.D.W. KEMMITT, M.A. MAZID, P. McKENNA, D.R. RUSSELL, M.D. SCHILLING and L.J.S. SHERRY, 1984. J. Chem. Soc. Dalton, 1993-2002.

Triclinic, P$\bar{1}$, a = 17.59, b = 12.736, c = 10.27 Å, α = 118.2, β = 94.7, γ = 74.0°, Z = 2. Mo radiation, R = 0.035 for 7263 reflexions.

The molecule (Fig. 1) can be considered as a highly puckered platinacyclobutan-3-one ring (fold angle 50.4(4)°) with a weak transannular Pt-C bond of 2.416(5) Å or a slipped η^3-oxodimethylenemethane compound. Coordination distances and angles are: Pt-P(1) 2.300(1), Pt-P(2) 2.272(1), Pt-C(1) 2.133(5), Pt-C(3) 2.155(6) Å, P(1)-Pt-P(2) 101.9(1), P(1)-Pt-C(3) 91.2(2), P(2)-Pt-C(1) 99.1(2) and C(1)-Pt-C(3) 68.4(2) Å. The C(1)-C(2) and C(2)-C(3) distances are 1.470(9) and 1.493(10) Å respectively. The torsion angles P(1)-Pt-C(3)-C(6) and P(2)-Pt-C(1)-C(4) are -96.5(4) and -43.0(5)° respectively, one methoxycarbonyl substituent occupying a pseudo-axial position and the other a pseudo-equatorial position.

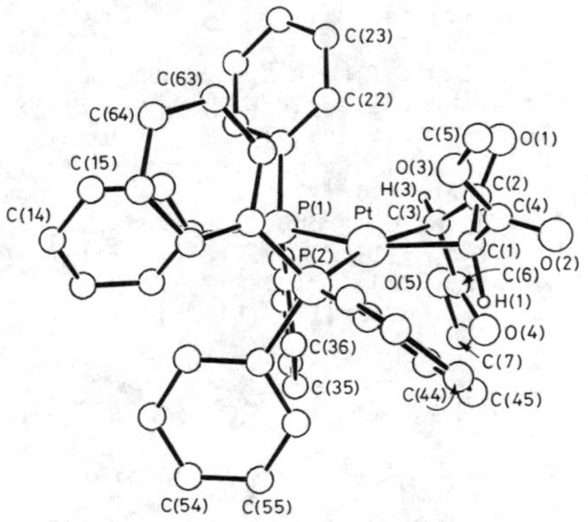

Fig. 1. $C_{43}H_{38}O_5P_2Pt,H_2O$: the molecular structure.

BIS(BENZOATO-O)BIS[μ-[(DIPHENYLPHOSPHINIDENIO)BIS(METHYLENE)]][μ-(NITROMETHYLENE)]-DIGOLD

$C_{43}H_{39}Au_2NO_6P_2$ $[Au((CH_2)_2PPh_2)(OC(O)Ph)]_2CHNO_2$

H.C. KNACHEL, D.S. DUDIS and J.P. FACKLER, Jr., 1984. Organometallics, 3, 1312-1313.

Monoclinic, P2_1/n, a = 10.781, b = 17.643, c = 21.739 Å, β = 101.35°, Z = 4. Mo radiation, R = 0.039 for 3158 reflexions.

The complex consists of an "A-frame" species with two Au atoms bridged by a nitromethylene group (at the apex) (Fig. 1). There is rotational disorder about the N-C(7) bond with roughly 1:1 occupancy for the two sets of O atoms. Bond lengths of interest are: Au(1)...Au(2) 3.073, Au(1)-O(1) 2.073, Au(1)-C(7) 2.03, Au(2)-C(7) 2.03, Au(1)-C(4) 2.11, Au(1)-C(1) 2.12, P(1)-C(1) 1.78, P(1)-C(11) 1.81, N-O (average) 1.2, and C(7)-N 1.48 Å.

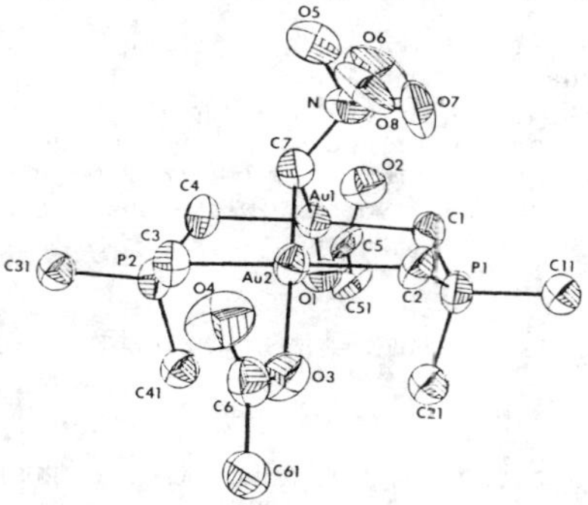

Fig. 1. Molecular structure of $C_{43}H_{39}Au_2NO_5P_2$ showing rotational disorder in the NO_2 group. Complete phenyl rings are not shown.

DI-μ-CARBONYLCARBONYLBIS(η^5-2,4-CYCLOPENTADIEN-1-YL)[μ_3-(PHENYLMETHYLIDYNE)](TRI-CARBONYLDIIRIDIUM)[μ_3-[(1-η:1,2,3-η:3-η)-1,2,3-TRIPHENYL-1-PROPEN-1-YL-3-YLIDENE]]-DITUNGSTEN

$C_{44}H_{30}Ir_2O_6W_2$ $(\eta^5-C_5H_5)_2W_2Ir_2(CO)_6(\mu_3-CPh)(\mu_3-\eta^3-C_3Ph)$

J.R. SHAPLEY, C.H. McATEER, M.R. CHURCHILL and L.V. BIONDI, 1984. Organometallics, 3, 1595-1596.

Monoclinic, $P2_1/c$, a = 14.253, b = 15.629, c = 18.731 Å, β = 96.86°, Z = 4. Mo radiation, R = 0.058 for 3395 reflexions.

The molecule contains an almost flat W_2Ir_2 core, consisting of two triangles fused at the W(1)-Ir(1) edge (dihedral angle W(1)-Ir(1)-W(2)/W(1)-Ir(1)-Ir(2) is 170.33°) (Fig. 1). A $\mu_3-\eta^3$-1,3-dimetalla-allyl (C_3Ph_3) moiety is π-bonded to Ir(2) and σ-bonded to Ir(1) and W(1). A μ_3-CPh moiety bridges Ir(1), W(1) and W(2). Ir(1) is linked to one terminal CO, whereas Ir(2) has two. Each W atom is attached to an $\eta^5-C_5H_5$ ring and to a semi-bridging CO group. Bond distances of interest are: W(1)-W(2) 3.080, W(1)-Ir(1) 2.723, W(1)-Ir(2) 2.852, Ir(1)-Ir(2) 2.720, W(2)-Ir(1) 2.665, W(1)-C(1) 2.23, Ir(2)-C(1) 2.23, Ir(2)-C(2) 2.30, Ir(2)-C(4) 2.30, Ir(1)-C(4) 2.07, W(1)-C(3) 2.01, W(2)-C(3) 2.08, Ir(1)-C(3) 2.18 Å.

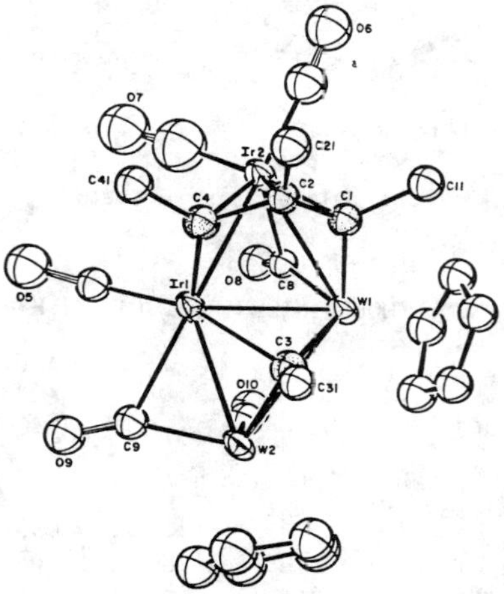

Fig. 1. Molecular structure (omitting phenyl rings) of $C_{44}H_{30}Ir_2O_6W_2$.

[BIS[3-(DIPHENYLPHOSPHINO)PROPYL]PHENYLPHOSPHINE-P,P',P'']PEROXOCARBONATO(p-TOLYL)-RHODIUM

$C_{44}H_{44}O_4P_3Rh$

L. DAHLENBURG and C. PRENGEL, 1984. Organometallics, 3, 934-936.

Monoclinic, $P2_1/n$, a = 12.217, b = 22.170, c = 14.964 Å, β = 102.07°, Z = 4. Mo radiation, R = 0.078 for 3206 reflexions.

The overall molecular geometry about the central Rh atom corresponds to distorted octahedral with the tris(phosphine) spanning meridional positions (Fig. 1). The coordinated peroxocarbonato group acts as an O,O' bidentate giving rise to a five-membered RhOOC(O)O chelate structure with the peroxo O(1) trans to the PhP fragment and the oxo O(2) trans to the tolyl ring. Important distances are: Rh-P(1) 2.269, Rh-P(2) 2.289, Rh-P(3) 2.348, Rh-O(1) 2.034, Rh-O(2) 2.127, Rh-C(1) 2.056, O(1)-O(3) 1.43, O(2)-C 1.27, O(3)-C 1.32 and O(4)-C 1.22 Å. Bond angles within the five-membered ring are: O(1)-Rh-O(2) 77.3, Rh-O(2)-C 113, O(2)-C-O(3) 121, C-O(3)-O(1) 115, O(3)-O(1)-Rh 112.3, O(2)-C-O(4) 126 and O(4)-C-O(3) 113°.

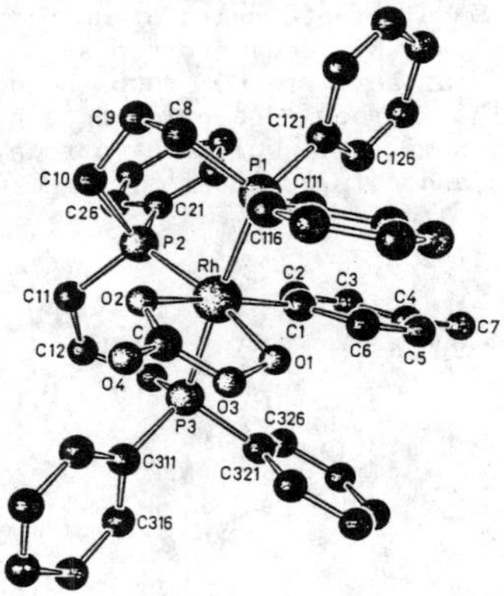

Fig. 1. Molecular structure of $C_{44}H_{44}O_4P_3Rh$.

(μ_2-DIPHENYLPHOSPHINO)(μ_2-DIPHENYLPHOSPHINOMETHOXYMETHYLENE-C,P)(METHYLDIPHENYLPHOSPHINODICARBONYLOSMIUM)TETRACARBONYLTUNGSTEN DICHLOROMETHANE SOLVATE

$C_{45}H_{37}O_7OsP_3W,1.5(CH_2Cl_2)$

S. ROSENBERG, R.R. WHITTLE and G.L. GEOFFROY, 1984. J. Am. Chem. Soc., 106, 5934-5940.

Monoclinic, $P2_1/n$, a = 13.907, b = 13.367, c = 24.344 Å, β = 90.60°, Z = 4. Mo radiation, R = 0.037 for 4967 reflexions.

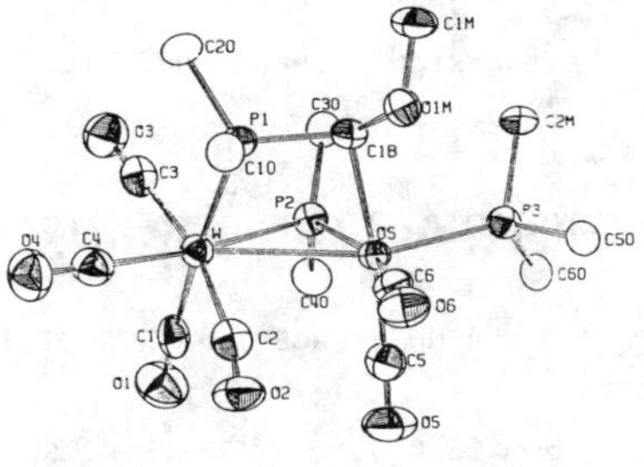

Fig. 1. A view of the inner core of $C_{45}H_{37}O_7OsP_3W$.

The molecular structure is shown in Fig. 1. Principal dimensions are: W-Os 3.022(1), W-P(1) 2.521(3), W-P(2) 2.538(3), W-C 1.99-2.05(1), Os-P(2) 2.357(7), Os-P(3) 2.314(3), Os-C(1B) 2.24(1), Os-C(carbonyl) 1.90, 1.94(1) Å. The dihedral angle between the W-P(2)-Os and W-P(1)-C(1B)-Os planes is 76.6°.

TRIS(TETRAETHYLAMMONIUM) CARBIDOHEXADECAKIS(μ_2-CARBONYL)TETRACARBONYLNONANICKELTRI-COBALT
$C_{45}H_{60}Co_3N_3Ni_9O_{20}$

A. CERIOTTI, R.D. PERGOLA, G. LONGONI, M. MANASSERO and M. SANSONI, 1984. J. Chem. Soc. Dalton, 1181-1186.

Monoclinic, $P2_1/c$, a = 12.543, b = 12.404, c = 38.249 Å, β = 98.53°, D_m = 1.91, Z = 4. Mo radiation, R = 0.055 for 4040 reflexions.

The metal framework of this compound is best described as a square antiprism of metal atoms tetra-capped on two alternate pairs of adjacent triangular faces (Fig. 1). The twenty carbonyl ligands may be divided into two sets consisting of sixteen edge-bridging carbonyl groups and four terminal CO groups bonded to the four capping atoms (Ni-C 1.684, C-O 1.169 Å). The sixteen edge-bridging ligands may be further divided into four sets of four stereochemically equivalent carbonyl groups. The crystal consists of discrete ammonium and complex ions.

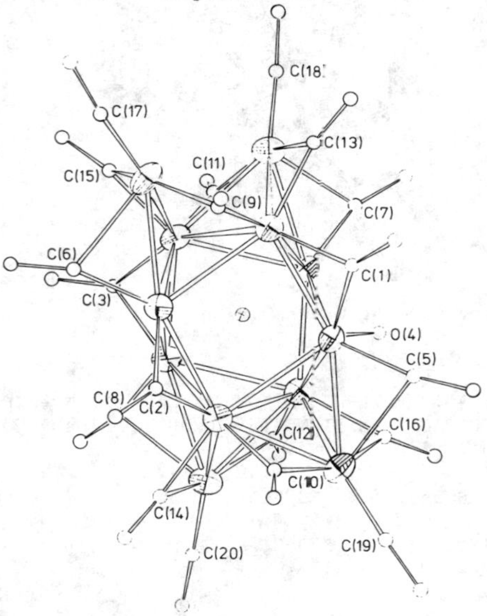

Fig. 1. $C_{45}H_{60}Co_3N_3Ni_9O_{20}$: the structure of the trianion.

TRIS[2,6-BIS(1-METHYLETHYL)PHENOLATO](1,2,3-TRIETHYL-1-PROPEN-1-YL-3-YLIDENE)TUNGSTEN
$C_{45}H_{66}O_3W$ $W(C_3Et_3)[O-2,6-C_6H_3(i-Pr)_2]_3$

M.R. CHURCHILL, J.W. ZILLER, J.H. FREUDENBERGER and R.R. SCHROCK, 1984. Organometallics, 3, 1554-1562.

Monoclinic, $P2_1/c$, a = 19.271, b = 13.000, c = 17.195 Å, β = 96.55°, Z = 4. Mo radiation, R = 0.042 for 3449 reflexions.

The sterically-crowded molecule is approximately a trigonal bipyramid in which the planar tungstenacyclobutadiene ring is located in the equatorial position with the two W-C_α bond lengths being 1.883 and 1.949 Å and the W...C_β bond distance being 2.159 Å (Fig. 1). (The C_β atom is C(2)). The axial W-O(2) and W-O(3) bonds are, at 2.008 and 1.979 Å respectively, considerably longer than the W-O(1) equatorial bond which is 1.885 Å in length. Some bond angles of interest are: O(1)-W-O(2) 81.8, O(1)-W-O(3) 84.3, O(1)-W-C(1) 150.0, O(1)-W-C(3) 127.7, O(2)-W-O(3) 163.5 and O(1)-W...C(3) 169.3°. The C(3)-W-C(1) angle is 82.3°, and the internal angle at the β-carbon atom C(1)-C(2)-C(3) is 120.8°.

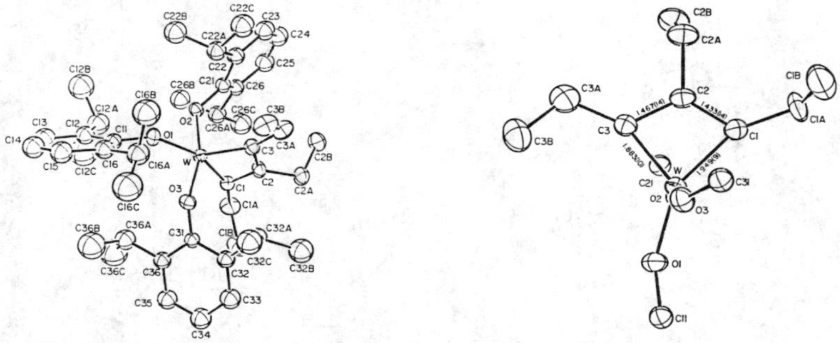

Fig. 1. Molecular structure of $C_{45}H_{66}O_3W$ (left) and a portion of the molecule projected onto the WC_3 plane (right) with selected bond lengths.

(μ_3-1-DIPHENYLPHOSPHINO-2-ETHOXY-2-OXOETHYL)(μ_3-HYDROXO)TRIS(8-QUINOLINYLMETHYL)-TRIPALLADIUM HEXAFLUOROPHOSPHATE
$C_{46}H_{41}F_6N_3O_3P_2Pd_3$ (I) $\{[Pd(C_{10}H_8N)]_3(\mu_3\text{-}Ph_2PCHCOOC_2H_5)(\mu_3\text{-}OH)\}PF_6$

TRIS(μ_3-CARBONYL)(μ_3-CHLORO)(η^5-CYCLOPENTADIENYLMOLYBDENUM)TRIS(8-QUINOLINYLMETHYL)-TRIPALLADIUM TETRAFLUOROBORATE
$C_{38}H_{29}BClF_4MoN_3O_3Pd_3$ (II) $\{[Pd(C_{10}H_8N)]_3[\mu_3\text{-}Mo(CO)_3(C_5H_5)](\mu_3\text{-}Cl)\}BF_4$

BROMO(ETHYL(DIPHENYLPHOSPHINO)ACETATE-P)(8-QUINOLINYLMETHYL)PALLADIUM MONOHYDRATE
$C_{26}H_{25}BrNO_2PPd,H_2O$ (III) $Pd(C_{10}H_8N)Br(Ph_2PCH_2COOC_2H_5).H_2O$

P. BRAUNSTEIN, J. FISCHER, D. MATT and M. PFEFFER, 1984. J. Am. Chem. Soc., 106, 410-421.

I. Monoclinic, $P2_1/n$, a = 23.573, b = 15.789, c = 12.381 Å, β = 100.21°, D_m = 1.73, Z = 4. Cu radiation, R = 0.073 for 3179 reflexions.

II. Monoclinic, C2/c, a = 23.877, b = 17.315, c = 19.142 Å, β = 111.80°, D_m = 1.90, Z = 8. Mo radiation, R = 0.047 for 5083 reflexions.

III. Monoclinic, C2/c, a = 18.691, b = 14.940, c = 17.722 Å, β = 97.27°, D_m = 1.65, Z = 8. Cu radiation, R = 0.041 for 3354 reflexions.

The structures of the complex cations in I and II and the molecule III are shown in Fig. 1. The cation in I consists of three $Pd(C_{10}H_8N_2)$ fragments held together by a μ_3-OH group and the $Ph_2PCHCOOC_2H_5^-$ ligand μ_3-bonded through the P, α-C, and carbonyl O atoms, respectively, to each of the Pd atoms. The Pd atoms have square-planar coordination geometry with Pd-(μ_3-O) 2.144(9)-2.28(1), Pd-O 2.17(1), Pd-N 1.99(1)-2.06(1), Pd-P 2.231(4), Pd-C 2.01(1)-2.06(1) Å. The structure of the cation in II shows three separated $Pd(C_{10}H_8N)$ units bridged by a μ_3-Cl atom and a μ_3-$Mo(C_5H_5)(CO)_3$ moiety. The four metal atoms, the three CO carbon atoms, and the μ_3-Cl atom define a distorted cubane structure for the inorganic core of II. The bonding mode of the $Mo(C_5H_5)(CO)_3$ involving the Mo and all three CO ligands provides the first example where such a unit is coordinated to three metal centres. Each CO caps a Pd_2Mo face in a semi triply bridging mode. The Pd atoms are six-coordinate and the Mo atom is coordinated to a η^5-C_5H_5, three CO, and three Pd atoms. Bond lengths: Pd-

Mo 2.791(1) (x2), 2.800(1), Pd-Cl 2.520(2)-2.525(2), Pd-N 2.083(6)-2.118(6), Pd-C 2.006(9)-2.023(9), Pd-CO 2.235(7)-2.295(8), Mo-CO 1.991(8)-2.020(8) Å. The Pd atom in III has square planar coordination geometry with Pd-Br 2.566(1), Pd-P 2.232(1), Pd-N 2.094(5), and Pd-C 2.042(6) Å.

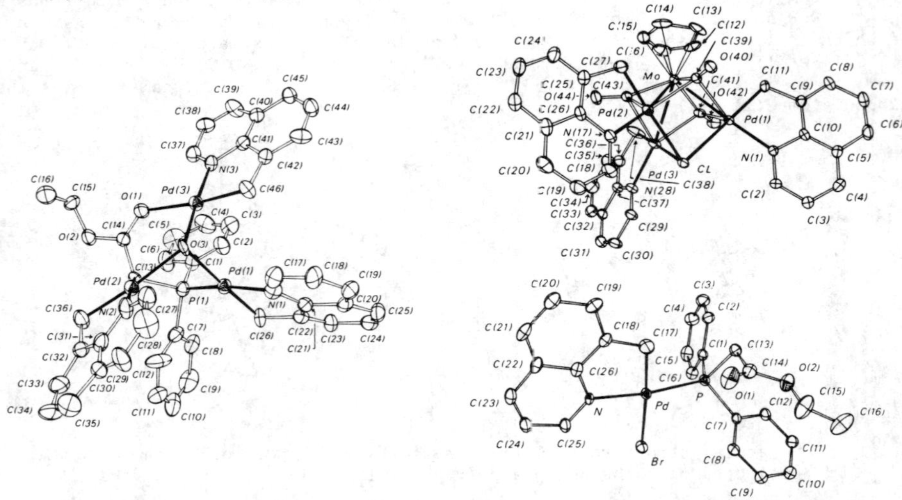

Fig. 1. The structures of $\{[Pd(C_{10}H_8N)]_3(\mu_3\text{-}Ph_2PCHCOOC_2H_5)(\mu_3\text{-}OH)\}^+$ (left), $\{[Pd(C_{10}H_8N)]_3[\mu_3\text{-}Mo(CO)_3(C_5H_5)](\mu_3\text{-}Cl)\}^+$ (top right), and $Pd(C_{10}H_8N)\text{-}Br(Ph_2PCH_2COOC_2H_5)$ (bottom right).

cis-BIS(3-HYDROXY-3-METHYLBUT-1-YNYL)BIS(TRIPHENYLPHOSPHINE)PLATINUM MONOHYDRATE
$C_{46}H_{44}O_2P_2Pt,H_2O$ (I)

trans-CHLORO(3-HYDROXY-3-METHYLBUT-1-YNYL)BIS(TRIPHENYLPHOSPHINE)PLATINUM TOLUENE SOLVATE
$C_{41}H_{37}ClOP_2Pt,0.5(C_7H_8)$ (II)

trans-CHLORO(3-HYDROXY-3-METHYLPENT-1-YNYL)BIS(TRIPHENYLPHOSPHINE)PLATINUM ACETONE SOLVATE
$C_{42}H_{39}ClOP_2Pt,C_3H_6O$ (III)

A. FURLANI, S. LICOCCIA, M.V. RUSSO, A.C. VILLA and C. GUASTINI, 1984. J. Chem. Soc. Dalton, 2197-2206.

I. Monoclinic, $P2_1/c$, a = 15.473, b = 21.725, c = 13.439 Å, β = 113.95°, Z = 4. Mo radiation, R = 0.052 for 3110 reflexions.

II. Triclinic, $P\bar{1}$, a = 11.695, b = 19.690, c = 9.214 Å, α = 101.88, β = 104.26, γ = 85.55°, Z = 2. Mo radiation, R = 0.053 for 3961 reflexions.

III. Triclinic, $P\bar{1}$, a = 15.544, b = 12.567, c = 11.719 Å, α = 91.12, β = 105.36, γ = 111.39°, Z = 2. Mo radiation, R = 0.046 for 4704 reflexions.

Complex I (Fig. 1) consists of monomeric units in which the Pt atoms are coordinated via σ bonds to two acetylide ligands and two triphenylphosphine molecules in a nearly square-planar cis arrangement (Pt-C(1) 2.02(1), Pt-C(6) 1.99(2), Pt-P(1) 2.307(5), Pt-P(2) 2.331(5) Å). The water O atom forms two strong hydrogen bonds with the two cis-oriented hydroxy groups acting as a donor atom. The presence of these hydrogen bonds affects the angles at Pt which range from 84.4(5) to 98.7(2)°. Packing is determined by an intermolecular hydrogen bond O(1)-H...O(2') involving the OH groups of adjacent molecules (O(1)...O(2') 2.81(2) Å). II and III both consist of monomeric units where the organic ligand, trans to the Cl atom is σ-bonded to Pt.

Coordination geometry is very similar in the two molecules apart from some difference in bond angles. The Pt atom has a square-planar environment in each complex with Pt-Cl 2.356(5), 2.336(4), Pt-C 1.95(2), 1.988(9), Pt-P 2.311(4), 2.321(4), Pt-P 2.320(4), 2.320(4) and C≡C 1.18(2), 1.15(2) Å for II, III respectively. The hydroxy groups are free in both complexes.

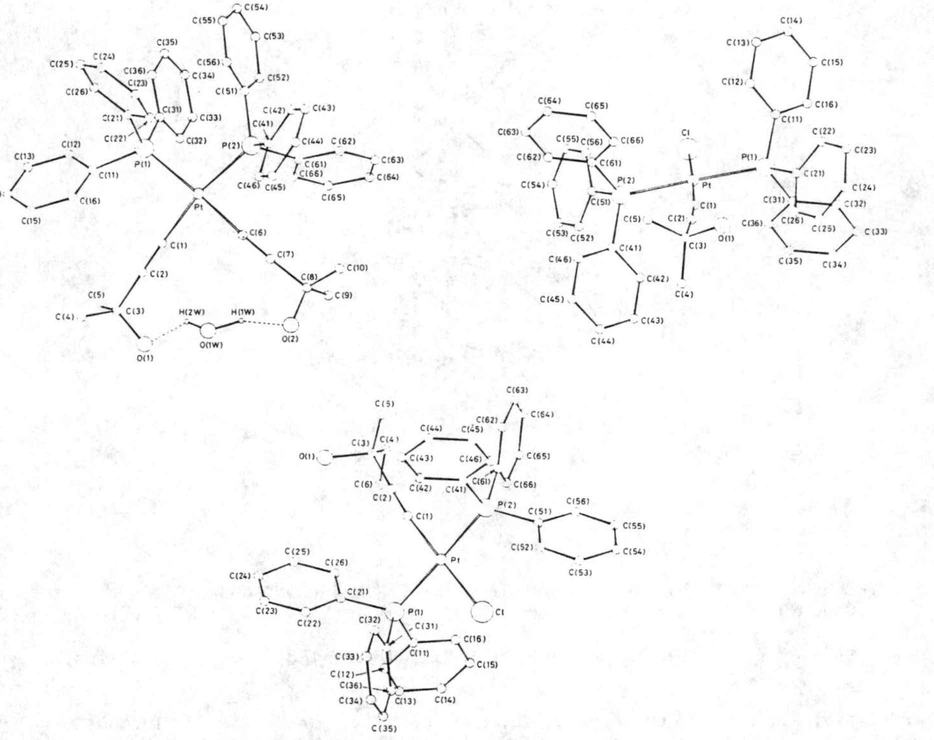

Fig. 1. The molecular structures: $C_{46}H_{44}O_2P_2Pt,H_2O$ (top left), $C_{41}H_{37}ClOP_2Pt$,-0.5(C_7H_8) (top right) and $C_{42}H_{39}ClOP_2Pt,C_3H_6O$ (bottom).

BIS[BIS(ACETONITRILE)DICARBONYLRHODIUM] TRIS(μ_2-CARBONYL)BIS[(μ_6-CARBIDO)NONAKIS(μ_2-CARBONYL)CYCLOHEXAKIS(CARBONYLRHODIO)RHODATE] DICHLOROMETHANE SOLVATE
$C_{47}H_{12}N_4O_{37}Rh_{16}$,0.3($CH_2Cl_2$)

S. MARTINENGO, D. STRUMOLO, P. CHINI, V.G. ALBANO and D. BRAGA, 1984. J. Chem. Soc. Dalton, 1837-1841.

Monoclinic, C2/m, a = 13.496, b = 37.175, c = 15.858 Å, β = 113.9°, D_m = 2.7, Z = 4. Mo radiation, R = 0.042 for 3460 reflexions.

The dianion (Fig. 1) has idealised D_{3h} symmetry and contains two Rh_7C moieties related by a crystallographic mirror plane. Each unit is a C_{3v} monocapped trigonal prism. The two halves are linked by a Rh-Rh bond between the capping atoms and by three bridging carbonyl ligands. The remaining thirty carbonyl groups are distributed fifteen on each Rh_6 prism. The Rh-Rh bond between the two Rh_7 units is remarkably short at 2.570(2) Å. Other average bond distances include: Rh-Rh 2.778, Rh-C (carbide) 2.14, Rh-C and C-O for terminal and bridging ligands 1.89 and 1.14 and 2.07 and 1.17 Å, respectively.

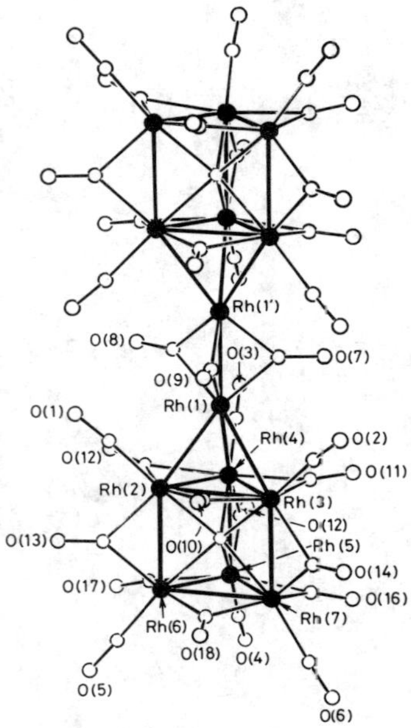

Fig. 1. $C_{47}H_{12}N_4O_{37}Rh_{16}$: the dianion.

BIS(TRIPHENYLPHOSPHINE)IMINIUM (μ_3-KETENYLIDENE)NONACARBONYLCOBALTDIIRON
$C_{47}H_{30}CoFe_2NO_{10}P_2$ $[Ph_3P]_2N^+.[Fe_2Co(CO)_9(CCO)]^-$

J.W. KOLIS, E.M. HOLT, J.A. HRILJAC and D.F. SHRIVER, 1984. Organometallics, 3, 496-498.

Triclinic, $P\bar{1}$, a = 18.360, b = 16.370, c = 20.567 Å, α = 76.16, β = 103.40, γ = 65.85°, Z = 4. Mo radiation, R = 0.086 for 4327 reflexions.

The two independent molecules present in the asymmetric unit are similar in structural detail. Both show the ketenylidene bound to the face of the metal triangle, and three terminal CO ligands bound to each of the metal atoms (Fig. 1). Metal-metal distances range from 2.527 to 2.545 Å and the CCO ligand (C(98)-C(99)-O(99)) is bound asymmetrically to the triangular face with a 24° tilt of the least-squares line through CCO, relative to the perpendicular to the metal triangle. It was not possible to differentiate Fe from Co in the refinement so the Co position is arbitrary.

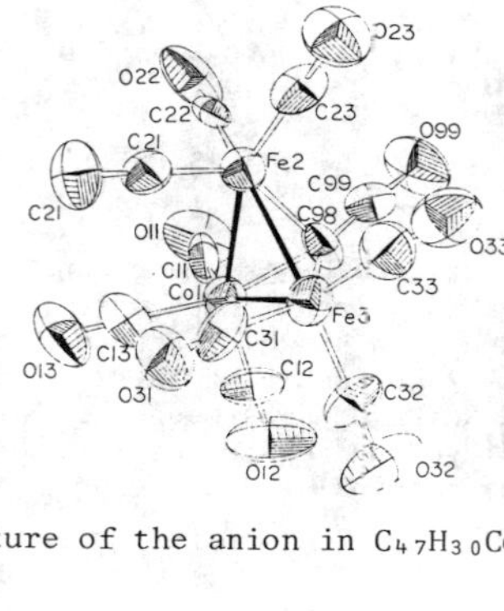

Fig. 1. Molecular structure of the anion in $C_{47}H_{30}CoFe_2NO_{10}P_2$.

[[2-[(DIPHENYLPHOSPHINO)METHYL]-2-METHYL-1,3-PROPANEDIYL]BIS[DIPHENYLPHOSPHINE]-P,P',P'']PHENYL-(T-4)-COPPER
$C_{47}H_{44}CuP_3$ [$MeC(CH_2PPh_2)_3Cu(Ph)$]

S. GAMBAROTTA, S. STROLOGO, C. FLORIANI, A. CHIESI-VILLA and C. GUASTINI, 1984. Organometallics, 3, 1444-1445.

Orthorhombic, $Pna2_1$, a = 18.382, b = 12.589, c = 17.154 Å, Z = 4. Cu radiation, R = 0.037 for 2535 reflexions.

The structure consists of a monomeric unit whose coordination geometry is that of a distorted tetrahedron defined by three P atoms from the triphos ligand and a C atom from the phenyl ring (Fig. 1). The angles at the Cu atom are far from the expected tetrahedral geometry, the C-Cu-P and P-Cu-P angles ranging from 119.4 to 126.9 and from 90.0 to 93.3° respectively, as a consequence of the strain of the chelating phosphine ligand. The Cu-P distances are: Cu-P(1) 2.342, Cu-P(2) 2.276, Cu-P(3) 2.295 Å and Cu-C(1) is 2.020 Å in length. The Cu atom lies some 0.033 Å out of the plane of the coordinated phenyl ring.

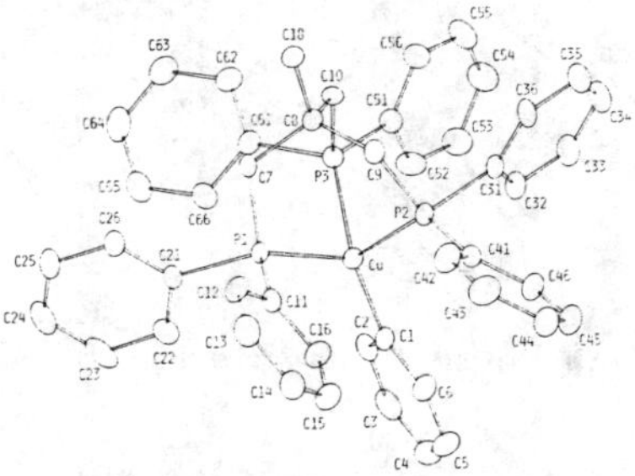

Fig. 1. Molecular structure of $C_{47}H_{44}CuP_3$.

(η^5-CYCLOPENTADIENYL)TRIS[(DIPHENYLPHOSPHINIDENO)BIS(METHYLENE)]URANIUM
$C_{47}H_{47}P_3U$

R.E. CRAMER, A.L. MORI, R.B. MAYNARD, J.W. GILJE, K. TATSUMI and A. NAKAMURA, 1984. J. Am. Chem. Soc., 106, 5920-5926.

Monoclinic, $P2_1/c$, a = 14.062, b = 26.752, c = 24.220 Å, β = 114.84°, Z = 8. Mo radiation, R = 0.050 for 6590 reflexions.

There are two independent crowded molecules in the crystal (Fig. 1). The overall structure of both is pentagonal bipyramidal with a Cp and a CH_2 group axial and five CH_2 moieties equatorial. The mean U-C σ-bond length is long at 2.66(3) Å. The Cp rings are 3.05 and 3.06 Å from the equatorial plane.

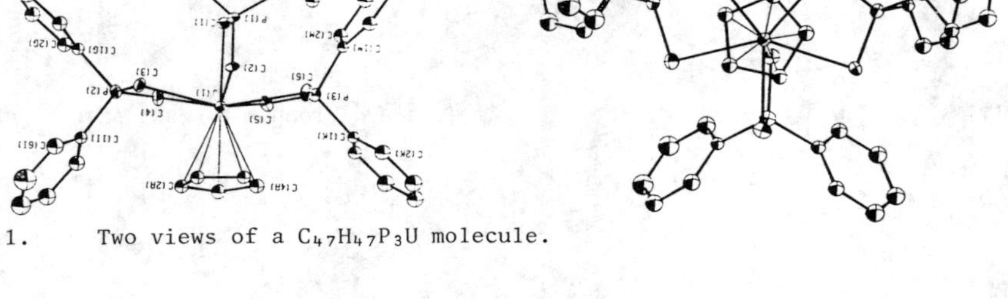

Fig. 1. Two views of a $C_{47}H_{47}P_3U$ molecule.

cis-BIS(2-NITROPHENYL)BIS(TRIPHENYLPHOSPHINE)PLATINUM(II)
$C_{48}H_{38}N_2O_4P_2Pt$

H.-A. BRUNE, M. WIEGE and T. DEBAERDEMAEKER, 1984. Z. Naturforsch., B39, 359-368.

Triclinic, $P\bar{1}$, a = 15.008, b = 12.701, c = 12.252 Å, α = 98.96, β = 92.87, γ = 100.95°, Z = 2. Mo radiation, R = 0.038 for 7302 reflexions.

The Pt atom has planar coordination geometry slightly distorted from square (Fig. 1) with Pt-P 2.313 and 2.332(1), Pt-C 2.060 and 2.061(6) Å, P-Pt-P 100.4(1), C-Pt-C 82.9(2)°.

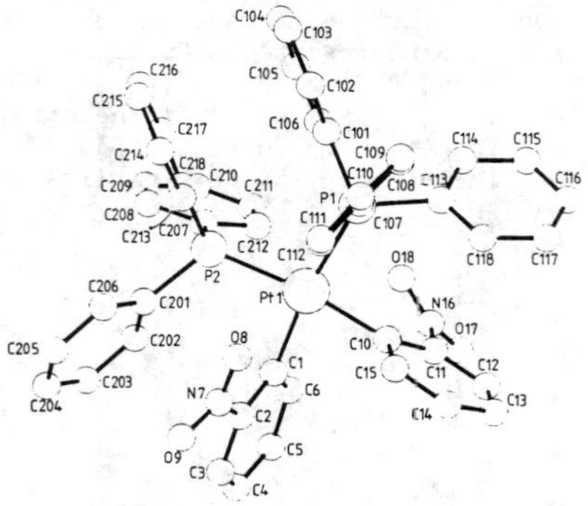

Fig. 1. A projection of $C_{48}H_{38}N_2O_4P_2Pt$ onto the PtP_2 plane.

BENZAMIDO(o-PHENYLENEDIPHENYLPHOSPHINO-P)(o-(DIPHENYLPHOSPHINO)-N-BENZOYLANILINE-C,N,P)PLATINUM(II) CHLOROFORM SOLVATE
$C_{50}H_{38}N_2O_2P_2Pt,0.2(CHCl_3)$

D. HEDDEN, D.M. ROUNDHILL, W.C. FULTZ and A.L. RHEINGOLD, 1984. J. Am. Chem. Soc., 106, 5014-5016.

Triclinic, $P\bar{1}$, a = 12.038, b = 12.576, c = 16.220 Å, α = 105.80, β = 104.84, γ = 95.75°, Z = 2. R = 0.0375 for 5115 reflexions.

Main dimensions in this molecule (Fig. 1) are: Pt-P(1) 2.298(2), Pt-P(2) 2.259(2), Pt-N(1) 2.063(7), Pt-C(3) 2.065(8) Å; P(1)-Pt-P(2) 100.6(1), P(1)-Pt-N(1) 83.7(2), P(1)-Pt-C(3) 164.8(3), P(2)-Pt-N(1) 175.6(2), P(2)-Pt-C(3) 94.5(3), N(1)-Pt-C(3) 81.2(3)°.

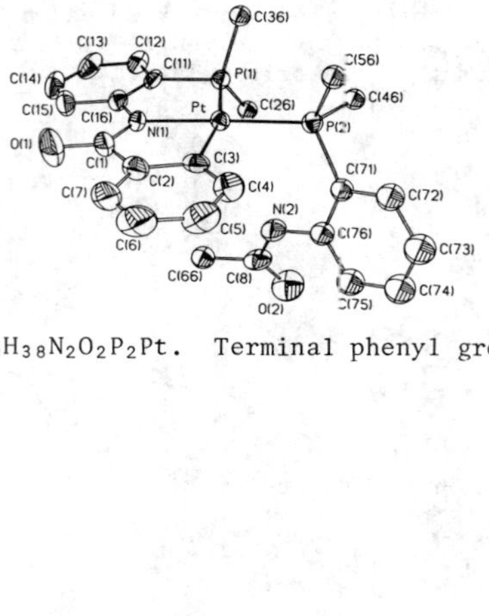

Fig. 1. A view of $C_{50}H_{38}N_2O_2P_2Pt$. Terminal phenyl groups shown as pivotal atom only.

TETRAPHENYLPHOSPHONIUM (μ_2-σ^2,η^2-1,2-BIS(METHOXYCARBONYL)-3-OXO-4,5-DIPHENYL-PENTA-1,4-DIEN-1-YL-C,O)BIS(TRICARBONYLIRON) DICHLOROMETHANE SOLVATE
$C_{51}H_{37}Fe_2O_{11}P,0.5CH_2Cl_2$ (I)

TETRAPHENYLPHOSPHONIUM (μ_2-σ^5-1,2-DIPHENYL-5,5,5-TRIFLUORO-3-TRIFLUOROMETHYL-PENTA-2,3,4-TRIYL-1-YLIDENE)BIS(TRICARBONYLIRON)
$C_{48}H_{31}F_6Fe_2O_6P$ (II)

J. ROS, X. SOLANS, M. FONT-ALTABA and R. MATHIEU, 1984. Organometallics, <u>3</u>, 1014-1020.

I. Monoclinic, $P2_1/n$, a = 17.244, b = 23.419, c = 12.218 Å, β = 94.31°, Z = 4. Mo radiation, R = 0.080 for 3196 reflexions.

II. Triclinic, $P\bar{1}$, a = 15.397, b = 14.495, c = 11.258 Å, α = 107.35, β = 96.92, γ = 110.85°, Z = 2. Mo radiation, R = 0.075 for 3475 reflexions.

Both compounds consist of discrete cations and anions, and I includes dichloromethane solvate (Fig. 1). In I, the organic group is σ-bonded to Fe(1) by C(4) (1.97 Å), π-bonded by the C(3)-C(4) bond to Fe(2) (Fe(2)-C(3) 2.16, Fe(2)-C(4) 1.95 Å), and bonded by the O(1) atom of the carbonyl group to Fe(1) (1.99 Å). The Fe(1),O(1),C(2),C(3),C(4) five-membered ring has an envelope conformation with Fe(1) tilting up -0.335 Å from the best plane through O(1),C(2),C(3),C(4). Fe(2) is some +1.890 Å from this plane. The Fe(1)-Fe(2) distance is 2.596 Å. The C(3)-C(4) bond distance is 1.42 Å. In II, the organic part of the molecule is σ-bonded to Fe(2) through atoms C(2) (2.09 Å) and C(5) (1.99 Å), and π-bonded by C(3), C(4) and C(5) to Fe(1) (2.09, 2.08, 2.07 Å respectively). The H atom bonded to C(2) occupies an exo position. The five-membered ring Fe(2), C(2), C(3), C(4) and C(5) has an envelope conformation too, with C(2) -0.82 Å from the plane through Fe(2), C(3), C(4) and C(5). Fe(1) is removed +1.740 Å from the same plane. The C(3)-C(4) and C(4)-C(5) bond lengths are respectively 1.49 and 1.43 Å.

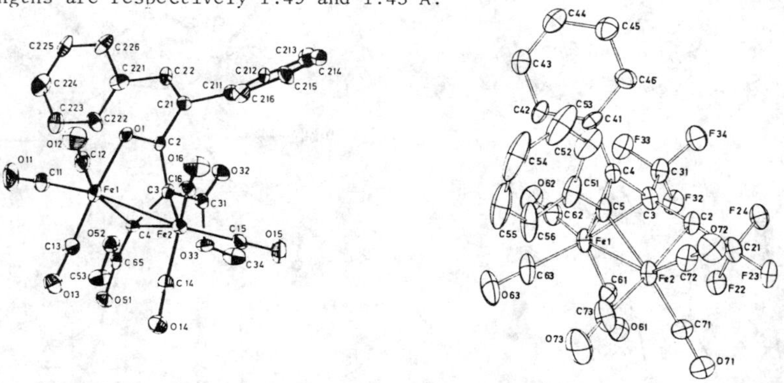

Fig. 1. Molecular diagrams of $C_{51}H_{37}Fe_2O_{11}P$ (left), and of $C_{48}H_{31}F_6Fe_2O_6P$ (right).

TRICHLORO(μ_2-2-(PHENYLPHOSPHINO-o-PHENYLENE)PYRIDYL-C,N,P)BIS(μ_2-2-(DIPHENYLPHOSPHINO)-PYRIDYL-N,P)DIRHENIUM
$C_{51}H_{41}Cl_3N_3P_3Re_2$ (I)

DICHLOROTETRAKIS(μ_2-2-(DIPHENYLPHOSPHINO)PYRIDYL-N,P)DIRHENIUM(II) BIS(HEXAFLUOROPHOSPHATE) ACETONE SOLVATE
$C_{68}H_{56}Cl_2F_{12}N_4P_6Re_2,2(C_3H_6O)$ (II)

TETRACHLOROBIS(μ_2-2-(DIPHENYLPHOSPHINO)PYRIDYL-N,P)TRIETHYLPHOSPHINEDIRHENIUM(II)
$C_{40}H_{43}Cl_4N_2P_3Re_2$ (III)

T.J. BARDER, F.A. COTTON, G.L. POWELL, S.M. TETRICK and R.A. WALTON, 1984. J. Am. Chem. Soc., <u>106</u>, 1323-1332

I. Monoclinic, $P2_1/n$, a = 12.160, b = 21.418, c = 18.464 Å, β = 99.04°, Z = 4. Mo radiation, R = 0.050 for 1956 reflexions.

II. Monoclinic, $P2_1/c$, a = 24.316, b = 12.101, c = 27.070 Å, β = 105.67°, Z = 4. Mo radiation, R = 0.0587 for 4667 reflexions.

III. Triclinic, $P\bar{1}$, a = 10.683, b = 20.033, c = 10.544 Å, α = 102.29°, β = 107.69, γ = 94.19°, Z = 2. Mo radiation,, R = 0.0311 for 4736 reflexions.

In I (Fig. 1) the two Re atoms have distorted octahedral coordination, with Re-Re 2.336(2), Re-Cl(ax) 2.578 and 2.568(9), Re-Cl(3) 2.429(9) Å. The four five-membered rings all have envelope conformations. The cation of II (Fig. 1) has no crystallographic symmetry but does have an approximate twofold axis coincident with the Re-Re bond. Each Re has approximate octahedral geometry, with Re-Re 2.300(1), Re-Cl(1) 2.575(5), Re-Cl(2) 2.628(6) Å (the difference in Re-Cl lengths is attributed to steric effects). Molecule III (Fig. 1) also has distorted octahedral coordination at Re, with Re-Re 2.270(1), Re-N(1) 2.076(7), Re-N(2) 2.206(7), Re-Cl(1) 2.616(2), Re-Cl(2) 2.406(2) Å. In all three structures there is rotation about the Re-Re bond away from an eclipsed conformation.

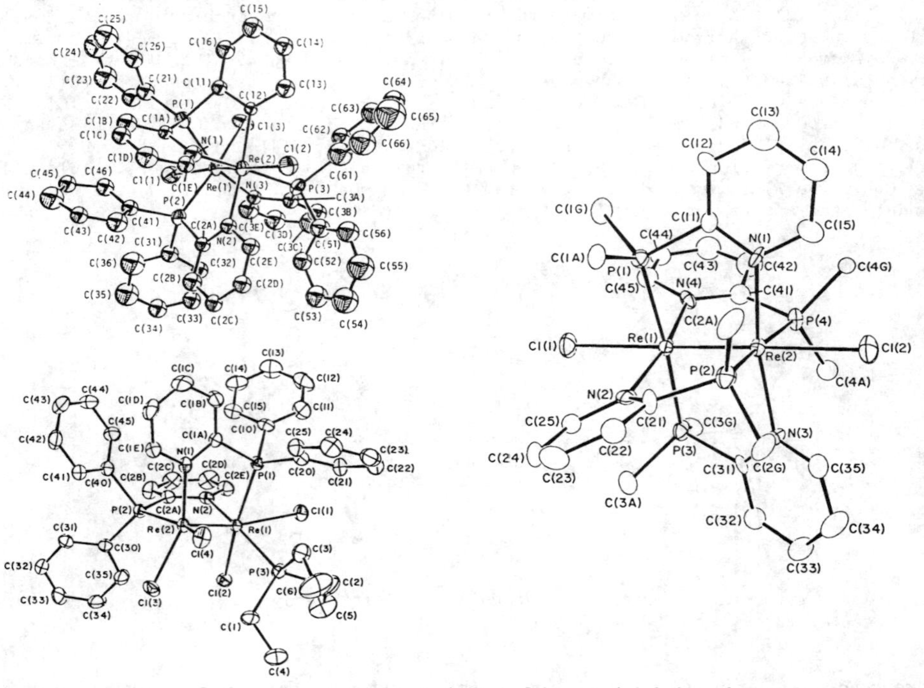

Fig. 1. Views of the Re_2 complexes I (top left), II (right) and III (lower left).

2',2'-DICARBONYL-1'-CHLORO-[μ_2-1-(p-METHOXYPHENYLIMINO)-2-(METHYLIMINO)PROPYL-C^1,N,N']-1',1'-BIS(TRIPHENYLPHOSPHINE)PLATINUMRHODIUM DICARBONYLDICHLORORHODATE

$C_{51}H_{43}Cl_3N_2O_5P_2PtRh_2$

A. MANTOVANI, M. PELLOSO, G. BANDOLI and B. CROCIANI, 1984. J. Chem. Soc. Dalton, 2223-2232.

Monoclinic, $P2_1/a$, a = 19.521, b = 17.212, c = 15.910 Å, β = 104.11°, D_m = 1.70, Z = 4. Mo radiation, R = 0.055 for 3892 reflexions.

The crystal consists of discrete well-separated cationic binuclear species and cis-dichlorodicarbonylrhodate anions (Fig. 1). Coordination about Pt is essentially square planar with mutually trans triphenylphosphine ligands (P(1)-Pt-P(2) 172.3°, Pt-P(1) 2.329(4), Pt-P(2) 2.332(4) Å). P(1) and P(2) lie below and Cl(1) and C(1) above the mean equatorial plane. Pt-Cl(1) is 2.375(3) and Pt-C(1) 1.97(1) Å. The planar N(1)-C(1)-C(11)-N(2) moiety forms a dihedral angle of 81.3° with the Pt coordination plane. Both Rh atoms have essentially square planar geometry. The average Rh-CO distance is 1.81(2) and Rh-N(1) and Rh-N(2) are 2.09(1) and 2.06(1) Å respectively.

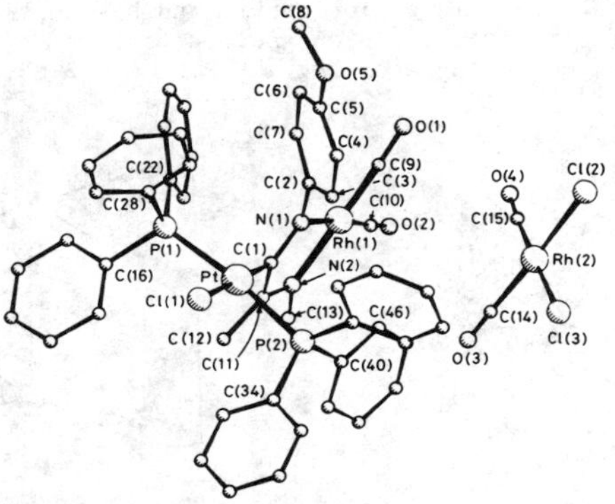

Fig. 1. $C_{51}H_{43}Cl_3N_2O_5P_2PtRh_2$: the molecular structure.

trans-BIS(2,4,6-TRIMETHOXYPHENYL)BIS(TRIPHENYLPHOSPHINE)PLATINUM(II)
$C_{54}H_{52}O_6P_2Pt$

H.A. BRUNE, M. WIEGE and T. DEBAERDEMAEKER, 1984. Z. Naturforsch., B39, 907-914.

Triclinic, $P\bar{1}$, a = 12.040, b = 10.613, c = 10.778 Å, α = 85.52, β = 116.90, γ = 109.40°, Z = 1. Mo radiation, R = 0.041 for 6623 reflexions.

The Pt atom lies on an inversion centre and the coordination is necessarily planar. Main dimensions are: Pt-P 2.295(1), Pt-C 2.104(5) Å, P-Pt-C 89.6(1)°. The plane of the phenyl ring bonded to Pt is inclined at 78.4° to the coordination plane.

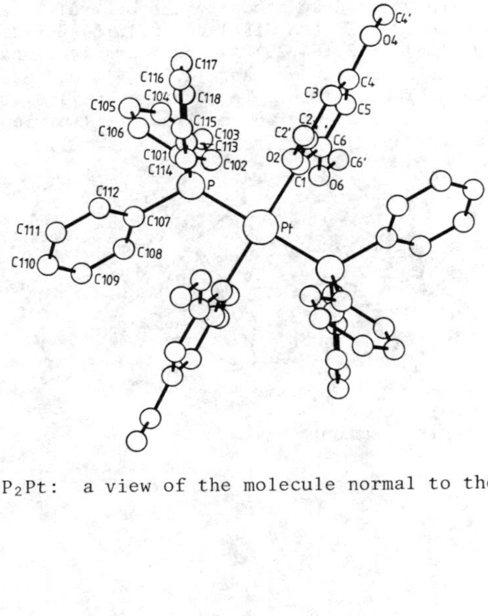

Fig. 1. $C_{54}H_{52}O_6P_2Pt$: a view of the molecule normal to the coordination plane.

BIS(TETRA-n-BUTYLAMMONIUM) TETRAKIS(μ$_2$-CHLORO)TETRAKIS(PENTAFLUOROPHENYL)DI-PLATINUM(II)DISILVER

$C_{56}H_{72}Ag_2Cl_4F_{20}N_2Pt_2$ $(NBu_4)_2[Pt_2Ag_2Cl_4(C_6D_5)_4]$

R. USON, J. FORNIES, M. TOMAS, F.A. COTTON and L.R. FALVELLO, 1984. J. Am. Chem. Soc., 106, 2482-2483.

Triclinic, P$\bar{1}$, a = 12.438, b = 13.160, c = 11.129 Å, α = 103.53, β = 101.27, γ = 74.75°, Z = 1. R = 0.079 for 1616 reflexions.

The anion lies about an inversion centre and has an essentially planar $Ag_2Pt_2Cl_4$ core. Dimensions are in Fig. 1. An unexpected feature is the close approach of two ortho-fluorine atoms to each Ag atom (Ag...F 2.60 and 2.69(1) Å).

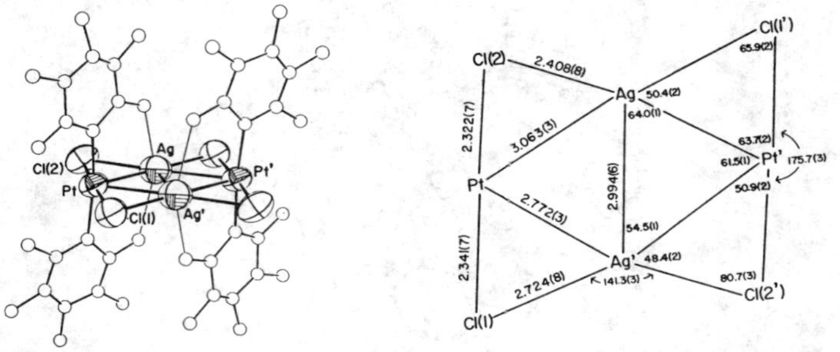

Fig. 1. A drawing of the $[Pt_2Ag_2Cl_4(C_6F_5)_4]^{2-}$ ion and dimensions of the core.

TETRAPHENYLARSONIUM BIS(μ$_6$-CARBIDO-2,3;4,5-DI-μ-CARBONYL-1,1,1,2,2,3,3,4,4,5,5,6,6,6-TETRADECACARBONYL-octahedro-HEXARUTHENIO-Ru^3,Ru^4)THALLATE(1-)(4Tl-Ru)

$C_{58}H_{20}AsO_{32}Ru_{12}Tl$ $As(C_6H_5)_4[Ru_6C(CO)_{16}]_2Tl$

G.B. ANSELL, M.A. MODRICK and J.S. BRADLEY, 1984. Acta Cryst., C40, 1315-1318.

Triclinic, P1, a = 15.325, b = 17.301, c = 17.356 Å, α = 115.25, β = 106.01, γ = 94.77°, Z = 1. Mo radiation, R = 0.078 for 3634 reflexions.

The $(Ru_6C(CO)_{16})_2Tl^-$ anion (Fig. 1) has pseudo 222 symmetry with distorted Ru_6 octahedra linked by two edge-related Ru atoms tc the central Tl atom. Each Ru_6 octahedron encapsulates a central C atom displaced 0.13(3) Å from the centroid of the octahedron (Ru-C 1.938(25) to 2.184(25) Å) and bears 14 linear and two semi-bridging CO ligands. The Ru-Ru distances range 2.813(4) to 2.938(4) Å except for those bridged by Tl which average 3.10 Å. The average Ru-Tl distances are 2.780(3) and 2.864(2) Å with the bonds deviating from coplanarity (torsion angle 35.9°) due to non-bonded O...O interactions.

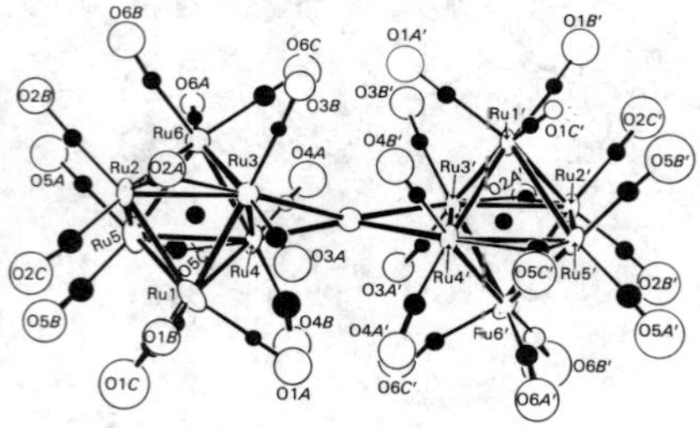

Fig. 1. The $(Ru_6C(CO)_{16})_2Tl^-$ anion.

BIS(μ_2-BIS(DIPHENYLPHOSPHINO-P,P')METHANE)(μ_2-HEXAFLUOROBUT-2-YNE)CHLOROBIS(METHYL ISOCYANIDE)DIRHODIUM TETRAFLUOROBORATE

$C_{58}H_{50}BClF_{10}N_2P_4Rh_2$

M. COWIE, R.S. DICKSON and B.W. HAMES, 1984. Organometallics, 3, 1879-1890.

Monoclinic, $P2_1/n$, a = 16.366, b = 18.685, c = 20.425 Å, β = 104.35°, D_m = 1.521, Z = 4. Cu radiation, R = 0.059 for 6800 reflexions.

The cation in the complex has two Rh atoms bridged by two mutually trans DPM ligands (Rh(1)-P(1) 2.301, Rh(1)-P(3) 2.302, Rh(2)-P(2) 2.371, Rh(2)-P(4) 2.384 Å) (Fig. 1). In the equatorial plane (perpendicular to the Rh-P vectors) the Rh atoms are bridged by a hexafluoro-2-butyne molecule coordinated as a cis-dimetalled olefin, while Rh(1) has a terminal Cl ligand and Rh(2) has two terminal linear methyl isocyanide ligands. The metals are also connected by a metal-metal bond yielding distorted octahedral coordination at both metals.

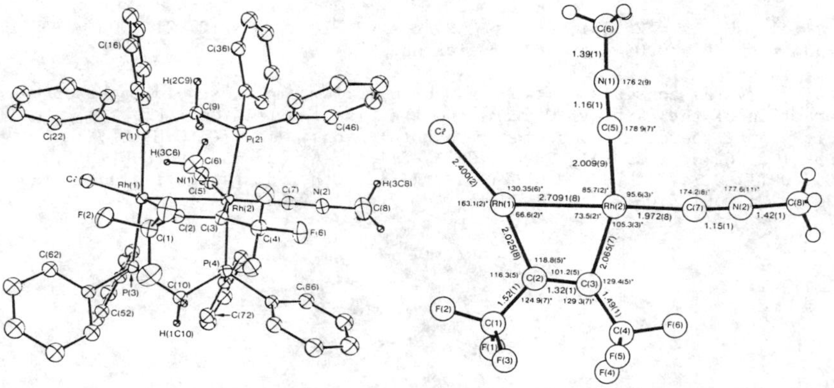

Fig. 1. Perspective view of the cation in $C_{58}H_{50}BClF_{10}N_2P_4Rh_2$, and view of the inner coordination sphere in the approximate plane of the metals and the hexafluoro-2-butyne group showing selected bond distances (Å) and angles (°).

(μ_2-DIMETHYLACETYLENEDICARBOXYLATE)BIS(BIS(DIPHENYLPHOSPHINO)METHANE-P,P')DICARBONYLDICHLORODIIRIDIUM DICHLOROMETHANE SOLVATE

$C_{58}H_{50}Cl_2Ir_2O_6P_4,2CH_2Cl_2$

B.R. SUTHERLAND and M. COWIE, 1984. Organometallics, 3, 1869-1878.

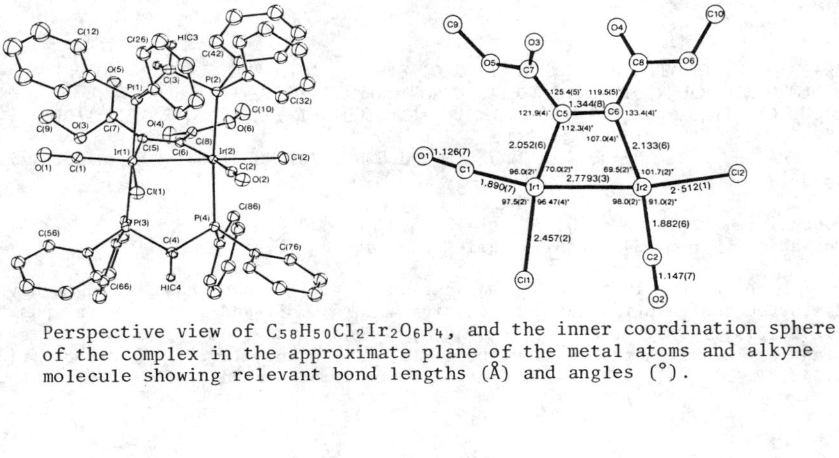

Fig. 1. Perspective view of $C_{58}H_{50}Cl_2Ir_2O_6P_4$, and the inner coordination sphere of the complex in the approximate plane of the metal atoms and alkyne molecule showing relevant bond lengths (Å) and angles (°).

Triclinic, $P\bar{1}$, a = 11.880, b = 23.514, c = 11.689 Å, α = 95.429, β = 110.389, γ = 77.999°, Z = 2. Cu radiation, R = 0.037 for 7796 reflexions.

Each metal has a slightly distorted octahedral geometry in which the coordination sites are occupied by the two trans P atoms (Ir(1)-P(1) 2.342, Ir(1)-P(3) 2.383, Ir(2)-P(2) 2.354, Ir(2)-P(4) 2.342 Å), a mutually cis terminal CO and Cl ligand, one end of the bridging acetylene group, bound as a cis-dimetallated olefin, and the Ir-Ir bond (Fig. 1).

DISODIUM [BIS(μ-CHLORO)BIS(BIS(DIETHYLMALONATO-C)PALLADATE(II))]DIPOTASSIUM [(μ-OXALATO)BIS(BIS(DIETHYLMALONATO-C)PALLADATE(II))] (1:1) TETRAHYDRATE
$C_{58}H_{88}Cl_2K_2Na_2O_{36}Pd_4,4H_2O$

T. KAWATO, T. UECHI, H. KOYAMA, H. KANATOMI and Y. KAWANAMI, 1984. Inorg. Chem., 23, 764-769.

Monoclinic, Cc, a = 28.331, b = 16.212, c = 22.510 Å, β = 122.21°, D_m = 1.571, Z = 4. Mo radiation, R = 0.056 for 4691 reflexions.

The crystal consists of two kinds of dimeric complexes held together by interaction of the carbonyl O atoms and the alkali-metal ions (Fig. 1). Pd(II) has square-planar coordination geometry in both anions: in $[(Pd(C_7H_{11}O_4)_2)_2Cl_2]^{2-}$, Pd-(μ-Cl) 2.362(5)-2.399(4), Pd-C 2.06(3)-2.12(1) Å, and in $[(Pd(C_7H_{11}O_4)_2)_2(C_2O_4)]^{2-}$, Pd-O 2.09(1)-2.14(1), Pd-C 2.05(2)-2.08(2) Å. Other distances include: Na-O 2.70(1)-2.82(2), Na...Cl 3.03(1) and 3.36(1), K-O 2.63(1)-2.85(2) Å.

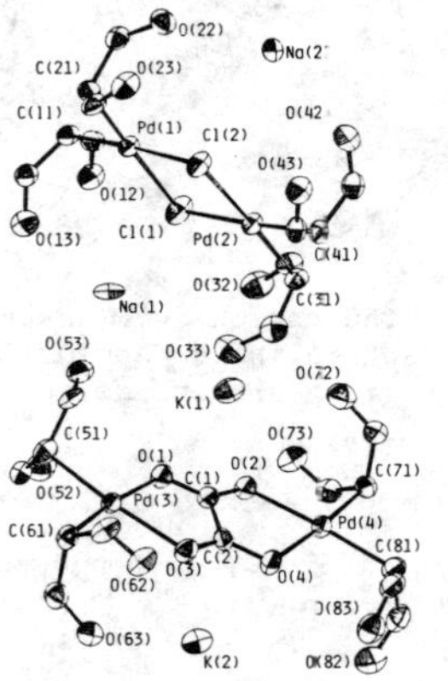

Fig. 1. The structure of $Na_2[(Pd(C_7H_{11}O_4)_2)_2Cl_2]K_2[(Pd(C_7H_{11}O_4)_2)_2(C_2O_4)],4H_2O$.

(μ_2-DIMETHYLETHYLENEDICARBOXYLATE-C,C')BIS(μ_2-BIS(DIPHENYLPHOSPHINO)METHANE-P,P')-CHLOROTRICARBONYLDIIRIDIUM HEXAFLUOROPHOSPHATE ACETONITRILE SOLVATE
$C_{59}H_{50}ClF_6Ir_2O_7P_5,C_2H_3N$ $[Ir_2(CO)_3Cl(\mu\text{-}H_3CO_2CC{=}CCO_2CH_3)(DPM)_2]^+PF_6^-,CH_3CN$

J.T. MAGUE, C.L. KLEIN, R.J. MAJESTE and E.D. STEVENS, 1984. Organometallics, 3, 1860-1869.

Monoclinic, $P2_1/n$, a = 10.531, b = 27.522, c = 20.233 Å, β = 92.40°, Z = 4. Mo radiation, R = 0.034 for 8129 reflexions (at 100 K).

Each Ir atom in the binuclear cation (Fig. 1) can be considered to have a distorted square-pyramidal coordination with a CO ligand in the apical position (Ir(1)-C(9) 1.935, Ir(2)-C(11) 1.910 Å). Phosphorus atoms of each DPM ligand occupy trans positions in the basal plane (Ir(1)-P(1) 2.366, Ir(1)-P(2) 2.368, Ir(2)-P(3) 2.342, Ir(2)-P(4) 2.339 Å), a C atom of the dimetallated olefin occupies the third basal site (Ir(1)-C(5) 2.077, Ir(2)-C(4) 2.055 Å), and a second CO ligand completes the basal plane of Ir(1) (Ir(1)-C(10) 1.962 Å) while for Ir(2) it is a chloride ligand (Ir(2)-Cl 2.461 Å). The Ir(1)-Ir(2) distance of 2.8933 Å indicates the presence of a metal-metal single bond.

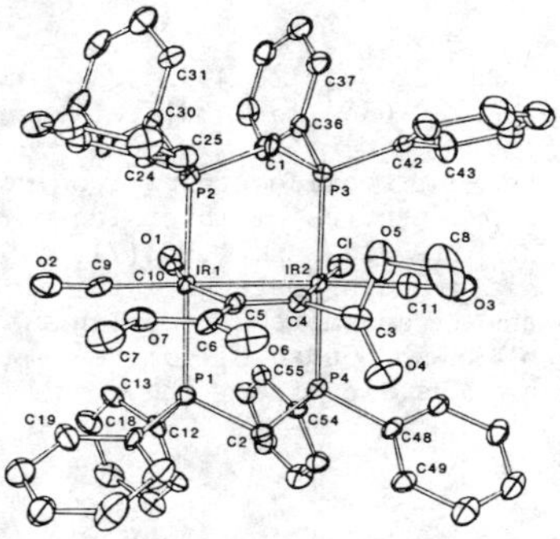

Fig. 1. Perspective view of the dimeric cation in $C_{59}H_{50}ClF_6Ir_2O_7P_5$.

CHLOROBIS(1,2-BIS(DIPHENYLPHOSPHINO)ETHANE-P,P')-2-PHENYLVINYLIDENERHENIUM TETRAHYDROFURAN SOLVATE

$C_{60}H_{54}ClP_4Re,C_4H_8O$ $[ReCl(C{=}CHPh)(Ph_2PCH_2CH_2PPh_2)_2],C_4H_8O$

A.J.L. POMBEIRO, J.C. JEFFERY, C.J. PICKETT and R.L. RICHARDS, 1984. J. Organometal. Chem., 277, C7-C10.

Triclinic, P1, a = 9.901, b = 13.569, c = 13.132 Å, α = 103.56, β = 107.05, γ = 114.86°, Z = 1. Mo radiation, R = 0.041 for 3887 reflexions.

The Re atom has distorted octahedral coordination (Fig. 1) with angles of P(1)-Re-P(2) 80.2, P(3)-Re-P(4) 78.7, P(1)-Re-P(4) 180.0, P(2)-Re-P(3) 177.7, and C(5)-Re-Cl 176.8°. The Re-C(carbene) distance of 2.046 Å is shorter than the expected Re-C single bond length of 2.297 Å, and the distance between the α and β-carbon atoms (C(5) and C(6)) of the vinylidene ligand (1.308 Å) is typical for a C=C double bond. The angle at the C_β atom C(5)-C(6)-C(91) is 125.5°, in agreement with sp^2 hybridization at this atom, and the angle Re-C(5)-C(6) is 166.6°. The Re-P distances are 2.389, 2.402, 2.440 and 2.441 Å, and the Re-Cl bond length is 2.454 Å.

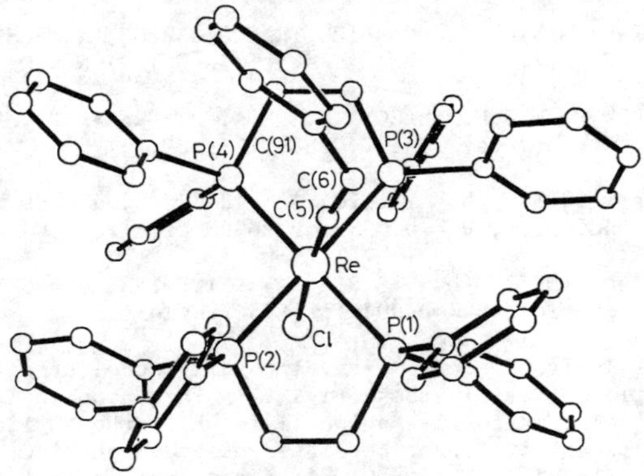

Fig. 1. Molecular structure of $C_{60}H_{54}ClP_4Re$.

TETRAKIS(2,6-DIMETHYLPHENYL ISOCYANIDE)RHODIUM(I) TETRAPHENYLBORATE ETHANOL SOLVATE
$C_{60}H_{56}BN_4Rh,C_2H_6O$ $[Rh(C_9H_9N)_4][B(C_6H_5)_4],C_2H_6O$

T.V. ASHWORTH, D.C. LILES, H.E. OOSTHUIZEN and E. SINGLETON, 1984. Acta Cryst., C40, 1169-1172.

Triclinic, P$\bar{1}$, a = 11.615, b = 23.582, c = 10.409 Å, α = 91.20, β = 104.19, γ = 102.36°, Z = 2. Mo radiation, R = 0.0554 for 4321 reflexions.

In the crystal (Fig. 1) the unit cell contains two independent monomeric square-planar cations lying across centres of inversion at 0,0,0 and 0.5,0.5,0 respectively; Rh-C distances are 1.957(6) and 1.963(7), 1.969(6) and 1.953(8) Å, respectively. Because of steric crowding between ortho methyl groups on adjacent ligands, the phenyl rings of the ligands are not coplanar with the plane of coordination. The formation of Rh-Rh bonded oligomers is thus sterically inhibited. The geometry of the BPh_4 anion is as expected. The ethanol molecule is disordered.

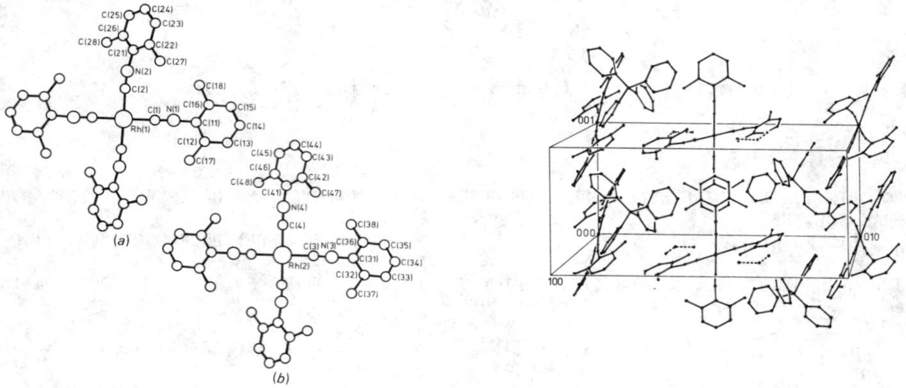

Fig. 1. $[Rh(C_9H_9N)_4][B(C_6H_5)_4],C_2H_6O$: perspective views of the two independent cations and the unit cell viewed perpendicular to the bc plane; broken lines indicate one orientation of the disordered ethanol molecule.

BIS(TRIPHENYLPHOSPHINE)IMINIUM (μ_8-CARBIDO)(μ_2-NITROSYL)TETRACOSACARBONYLDECAOSMIUM
$C_{61}H_{30}N_2O_{25}Os_{10}P_2$ (I) $[((C_6H_5)_3P)_2N^+][Os_{10}C(CO)_{24}(\mu_2\text{-}NO)]^-$

BIS(TRIPHENYLPHOSPHINE)IMINIUM (μ_6-CARBIDO)TRICOSACARBONYLNITROSYLDECAOSMIUM
$C_{60}H_{30}N_2O_{24}Os_{10}P_2,CH_2Cl_2$ (II) $[((C_6H_5)_3P)_2N^+][Os_{10}C(CO)_{23}(NO)]^-,CH_2Cl_2$

B.F.G. JOHNSON, J. LEWIS, W.J.H. NELSON, J. PUGA, D. BRAGA, K. HENRICK and M. McPARTLIN, 1984. J. Organometal. Chem., 266, 173-189.

I. Triclinic, P$\bar{1}$, a = 20.389, b = 14.670, c = 12.333 Å, α = 99.55, β = 94.43, γ = 103.03°, Z = 2. Mo radiation, R = 0.095 for 2699 reflexions.

II. Triclinic, P$\bar{1}$, a = 19.374, b = 16.813, c = 11.791 Å, α = 85.00, β = 101.81, γ = 99.43°, Z = 2. Mo radiation, R = 0.094 for 8736 reflexions.

In I the nitrosyl ligand adopts a novel bonding mode, bridging the wingtips of a "butterfly" indentation of metal atoms (Fig. 1). The anion in I undergoes rearrangement and CO loss in solution to give anion II in which the NO ligand is bonded in a terminal fashion (although it cannot be distinguished from a CO) to the tetrahedral Os_{10} metal skeleton (Fig. 1). The Os-Os distances in I range from 2.730 Å (Os(6)-Os(9)) to 2.984 Å (Os(3)-Os(4)), the Os-C (carbonyl) distances from 1.77 to 2.12 Å, the Os-C (non-carbonyl) distances from 2.00 to 2.18 Å whilst the "wingtip" Os...Os distance (Os(0)...Os(4)) is 3.517 Å and the Os(4)-N(1) and Os(0)-N(1) distances are 2.09 and 2.11 Å respectively. In II, the Os-Os distances range from 2.767 Å (Os(8)-

Os(9)) to 2.896 Å (Os(6)-Os(10)), the Os-C (carbonyl) distances from 1.80-1.93 Å, and the Os-C (non-carbonyl) distances from 2.01 to 2.07 Å.

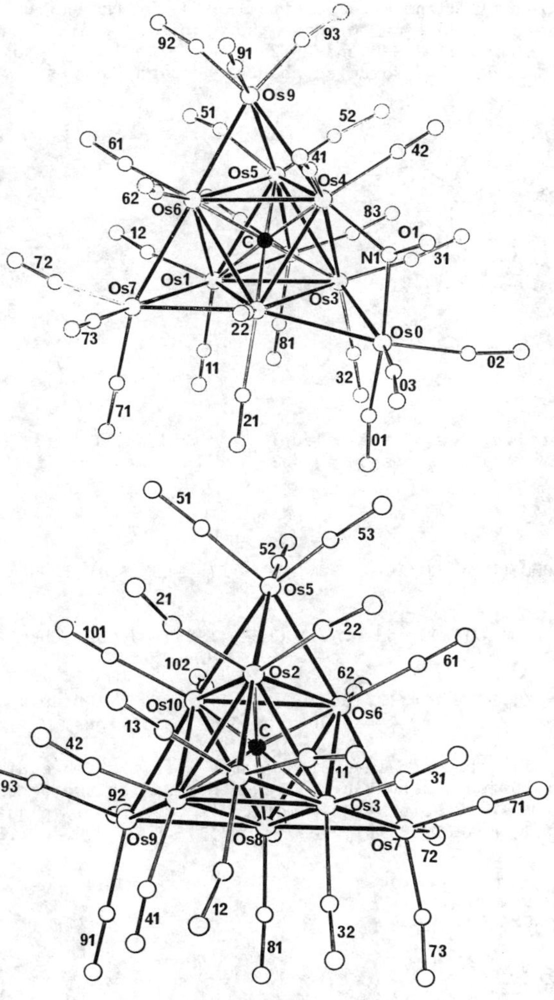

Fig. 1. The anion in $C_{61}H_{30}N_2O_{25}Os_{10}P_2$ (top), and that in $C_{60}H_{30}N_2O_{24}Os_{10}P_2$ (bottom).

CARBONYLDICHLORO[μ-[ETHYLPHENYL[THIOXO[(THIOXOMETHYL)AMINO]METHYL]CARBAMATO(2-)]]BIS-[μ-[METHYLENEBIS[DIPHENYLPHOSPHINE]-P:P']]DIRHODIUM(Rh-Rh)

$C_{62}H_{54}Cl_2N_2O_3P_4Rh_2S_2,0.5C_7H_8$ $[Rh_2Cl_2(CO)(\mu\text{-}SCNC(S)N(Ph)C(O)Et)(Ph_2PCH_2PPh_2)_2],0.5C_6H_5CH_3$

J.A.E. GIBSON and M. COWIE, 1984. Organometallics, 3, 984-996.

Monoclinic, $P2_1/c$, a = 17.964, b = 25.235, c = 15.861 Å, β = 97.02°, Z = 4. Mo radiation, R = 0.070 for 2112 reflexions.

The unusual tridentate ligand may be viewed as a new isothiocyanate group, RNCS, with R = C(S)N(Ph)C(O)OEt (Fig. 1). This group bridges the metals and is bound to one via the S atom and to the other via the C atom of the isothiocyanate group. It is additionally coordinated to one metal centre (Rh(1)) by the thiocarbonyl moiety. Other bond distances, in addition to those shown in Fig. 1 are: Rh(1)-P(1) 2.350, Rh(1)-P(3) 2.322, Rh(2)-P(2) 2.320 and Rh(2)-P(4) 2.329 Å. Two toluene molecules of crystallisation are severely disordered about inversion centres.

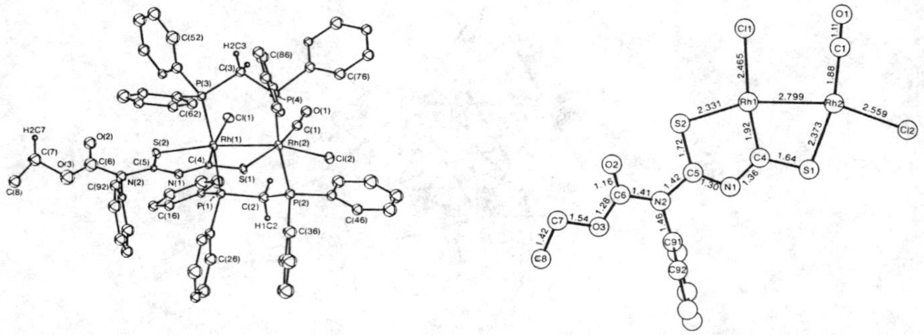

Fig. 1. Perspective view of the molecule $C_{62}H_{54}Cl_2N_2O_3P_4Rh_2S_2$ (left), and inner coordination sphere showing selected bond lengths (right).

BIS(μ_2-O-DIPHENYLPHOSPHINOBENZAMIDATO-N,N,P)BIS(μ_2-BENZAMIDATO-N,O)DIPHENYLDIRUTHENIUM

$C_{64}H_{52}N_4O_4P_2Ru_2$

A.R. CHAKRAVARTY, F.A. COTTON and D.A. TOCHER, 1984. J. Am. Chem. Soc., 106, 6409-6413.

Triclinic, P$\bar{1}$, a = 11.800, b = 12.164, c = 10.316 Å, α = 108.09, β = 97.89, γ = 101.33°, Z = 1. Mo radiation, R = 0.066 for 3157 reflexions.

The molecule lies about an inversion centre with the two equivalent Ru atoms surrounded by a distorted octahedron of ligand atoms; the two octahedra share an edge as shown in Fig. 1. The P-Ru(μ-N)$_2$Ru-P moiety is essentially planar. Mean dimensions are: Ru-Ru 2.566(1), Ru-P 2.305(2), Ru-O 2.093(5), Ru-N 2.024-2.127(4), Ru-C 2.104(6) Å.

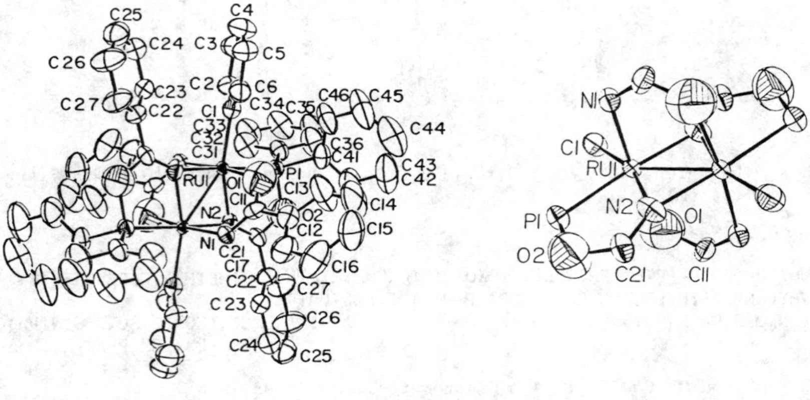

Fig. 1. The $C_{64}H_{52}N_4O_4P_2Ru_2$ molecule and a drawing showing only the metal and immediately adjacent ligand atoms.

BIS-DIMETHYLAMINOMETHYL[1,1-DIMERCURIO]FERROCENOPHANE TRIPHENYLCYANOBORONATE

$C_{64}H_{60}B_2Fe_2Hg_2N_4$ $[Me_2NCH_2C_5H_3HgFe]_2^{2+}(Ph_3BCN^-)_2$

L.G. KUZ'MINA, Yu.T. STRUCHKOV, D.A. LEMENOVSKY and I.F. URAZOWSKY, 1984. J. Organometal. Chem., 277, 147-151.

Monoclinic, $P2_1/c$, a = 9.788, b = 10.316, c = 27.040 Å, β = 98.29°, Z = 2. Mo radiation, R = 0.057 for 3107 reflexions (at -120°C).

The crystals consist of dications on inversion centres (Fig. 1) and anions in general positions. The dication is a centrosymmetric tetranuclear complex, the dimethylaminomethylferrocene residues being linked by two Hg bridges. The angle C(1)-Hg-C(6) is 179.5°. The chelate ring Hg,C(1),C(2),C(11),N is not ideally planar, the torsion angle C(1),C(2),C(11),N(1) being -20.7°. The cyclopentadienyl rings are in an eclipsed conformation but they are slightly non-parallel, the angle between their planes being 7.38°. The B atom of the anion has a distorted tetrahedral coordination (Fig. 1).

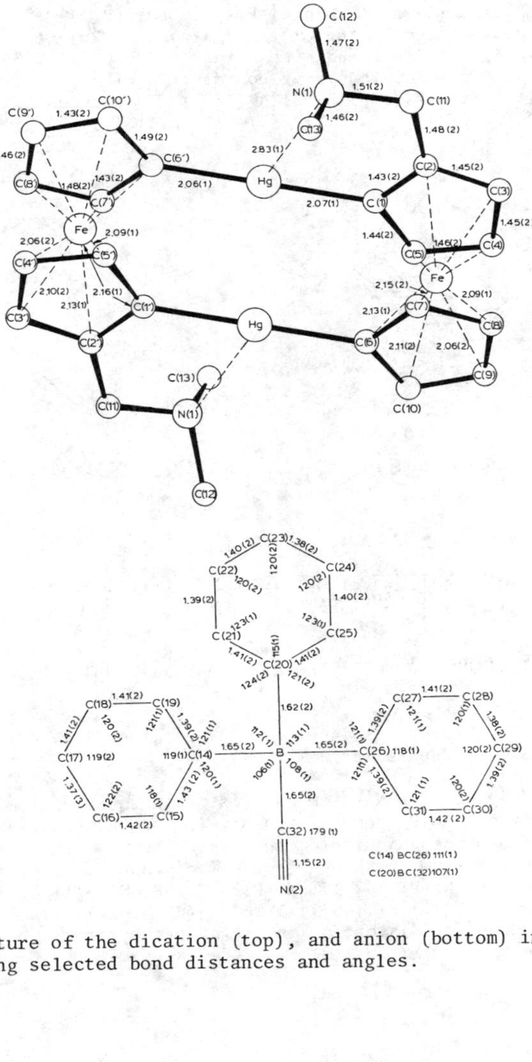

Fig. 1. Structure of the dication (top), and anion (bottom) in $C_{64}H_{60}B_2Fe_2Hg_2N_4$ showing selected bond distances and angles.

[μ_2-2,3-BIS(DIPHENYLPHOSPHINO)-N-METHYLSUCCINIMIDE-C^2,C^3,P,P']BIS(CHLOROTRIPHENYL-PHOSPHINEPALLADIUM) ACETONE SOLVATE
$C_{65}H_{53}Cl_2NO_2P_4Pd_2,3(C_3H_6O)$ (I)

[μ_2-2,3-BIS(DIPHENYLPHOSPHINO)-N-METHYLSUCCINIMIDE-C^2,C^3,P,P']BIS(ACETONITRILECHLORO-PALLADIUM) ACETONITRILE SOLVATE
$C_{33}H_{29}Cl_2N_3O_2P_2Pd_2,3(C_2H_3N)$ (II)

D. FENSKE and W. BENSMANN, 1984. Z. Naturforsch., B39, 1819-1822.

I. Monoclinic, $P2_1$, a = 14.897, b = 12.830, c = 18.163 Å, β = 103.43°, Z = 2. Mo radiation, R = 0.052 for 3526 reflexions (at 180 K).

II. Monoclinic, $P2_1/n$, a = 11.365, b = 22.189, c = 15.820 Å, β = 94.10°, Z = 4. Mo radiation, R = 0.043 for 6075 reflexions (at 180 K).

These binuclear Pd complexes are shown in Fig. 1. In (I) Pd(1) has square planar coordination with Pd(1)-Cl(1) 2.343, Pd(1)-P(1) 2.285, Pd(1)-P(2) 2.344, Pd(1)-P(3) 2.352 Å. Pd(2) is coordinated by the olefin (Pd(2)-C(1) 2.135, Pd(2)-C(2) 2.087(12)), a phosphine (Pd(2)-P(4) 2.337(4)) and a chlorine (Pd(2)-Cl(2) 2.390(3) Å). In complex (II) Pd(1) also has square-planar coordination (Pd(1)-Cl(1) 2.360, Pd(1)-Cl(2) 2.352(1), Pd(1)-P(1) 2.249, Pd(1)-P(2) 2.240(1) Å). Distances involving Pd(2) are: Pd(2)-C(1) 2.082, Pd(2)-C(2) 2.061(5), Pd(2)-N(1) 2.115(4), Pd(2)-N(2) 2.101(6) Å.

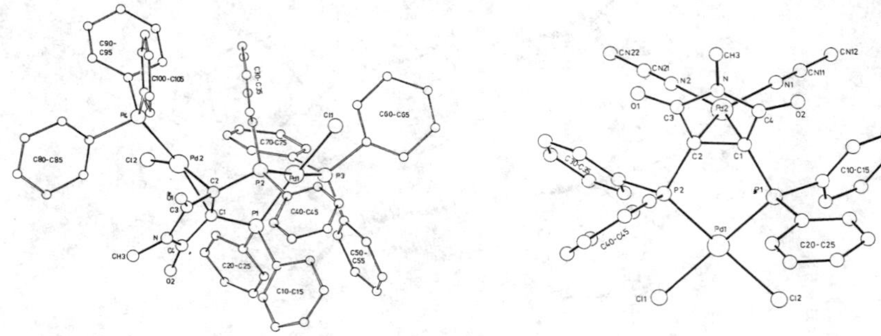

Fig. 1. Views of the N-methylsuccinimide complexes (I) and (II).

(BIS(2-DIPHENYLPHOSPHINOETHYL)(2-P-((METHYLTHIO)THIOCARBONYL)DIPHENYLPHOSPHINOETHYL)-AMINE-C,S,P,P')NICKEL TETRAPHENYLBORATE
$C_{68}H_{65}BNNiP_3S_2$

C. BIANCHINI, C.A. GHILARDI, A. MELI and A. ORLANDINI, 1984. J. Organometal. Chem., 270, 251-263.

Monoclinic, C2/c, a = 28.15, b = 14.448, c = 33.92 Å, β = 112.11°, Z = 8. Mo radiation, R = 0.095 for 1989 reflexions.

In the cation (Fig. 1) the Ni atom has distorted square-pyramidal geometry. The basal ligands consist of two P atoms of the np_3 ligand and the C atom and the un-methylated S atom of a phosphoniodithiomethyl ester group. The apex of the pyramid is occupied by the np_3 N atom, although the Ni-N distance is very long (2.59 Å). The basal plane distances are Ni-P(1) 2.18, Ni-P(2) 2.188, Ni-S(1) 2.117 and Ni-C(7) 2.03 Å. The basal angles P(1)-N(1)-S(1) 159.7, P(2)-Ni-C(7) 142.0 and S(1)-Ni-C(7) 49.0° indicate the amount of distortion in the square pyramid.

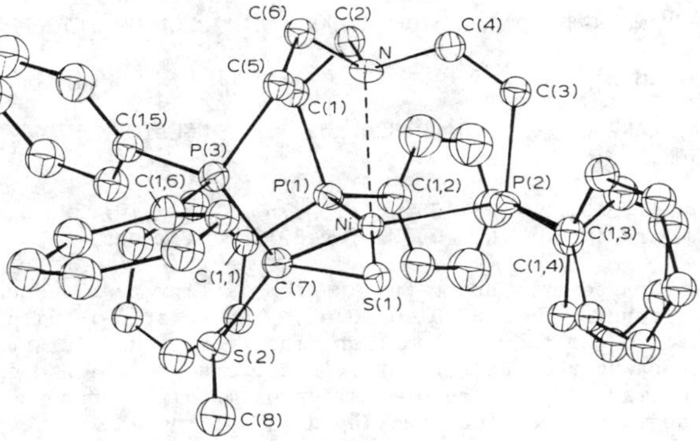

Fig. 1. Perspective view of the complex cation in $C_{68}H_{65}BNNiP_3S_2$.

BIS-μ_2-[BIS(DIPHENYLPHOSPHINO)METHANE-P,P']-2,2,2-TRICARBONYL-μ_2-p-TOLYLETHYNYL-1-p-TOLYLETHYNYLPLATINUMTUNGSTEN
$C_{71}H_{58}O_3P_4PtW$

A. BLAGG, A.T. HUTTON, P.G. PRINGLE and B.L. SHAW, 1984. J. Chem. Soc. Dalton, 1815-1822.

Monoclinic, $P2_1$, a = 11.847, b = 16.023, c = 18.351 Å, β = 116.23°, Z = 2. Mo radiation, R = 0.026 for 4043 reflexions.

The Pt and W atoms are linked by two diphenylphosphinomethane bridges to give an eight-membered PtP_4C_2W ring in a boat conformation (Fig. 1). The Pt...W distance is 3.037(1) Å. The two metal centres are asymmetrically bridged by one of the p-tolylethynyl groups (W...C(5) 2.398(9), Pt...C(5) 2.094(9) Å). The C(5)-C(6) distance is 1.19(1) Å.

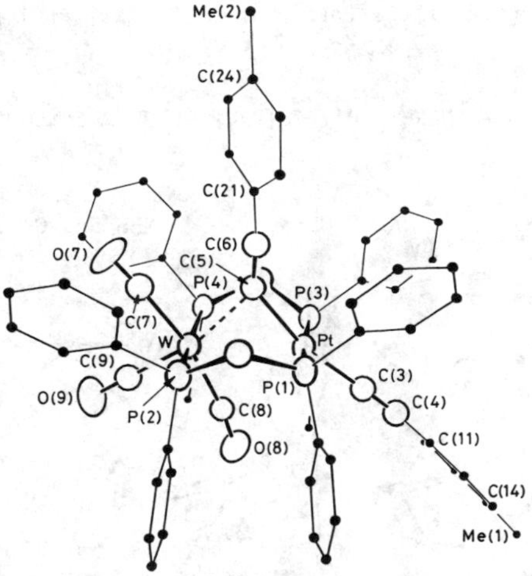

Fig. 1. $C_{71}H_{58}O_3P_4PtW$: the molecular structure.

BIS(1,2-BIS(2-ISOCYANO-4-t-BUTYLPHENOXY)ETHANE-C,C')RHODIUM TETRAPHENYLBORATE ACETONITRILE SOLVATE

$C_{72}H_{76}BN_4O_4Rh,1.5(C_2H_3N)$ $[Rh(C_4H_9C_6H_3OCH_2(NC))_4]^+BPh_4^-,1.5(CH_3CN)$

D.T. PLUMMER, B.A. KARCHER, R.A. JACOBSON and R.J. ANGELICI, 1984. J. Organometal. Chem., 260, 347-362.

Triclinic, P$\bar{1}$, a = 17.355, b = 21.135, c = 10.757 Å, α = 101.29, β = 98.36, γ = 113.92°, Z = 2. Mo radiation, R = 0.070 for 5912 reflexions.

The coordination geometry about the Rh atom is very nearly square planar (Fig. 1), with C-Rh-C angles ranging from 87.5 to 90.6°. The Rh atom is displaced 0.11 Å from the best plane through the four C atoms, and the average Rh-C distance is 1.965 Å. In the crystal structure the $[Rh(t\text{-}BuDiNC)_2]^+$ cations are aligned in a "slipped stack" face-to-face manner with the dimers approximately planar and Rh...Rh distances of 3.384 Å. The angle between the stacking axis and the Rh-Rh vector is 22.7°.

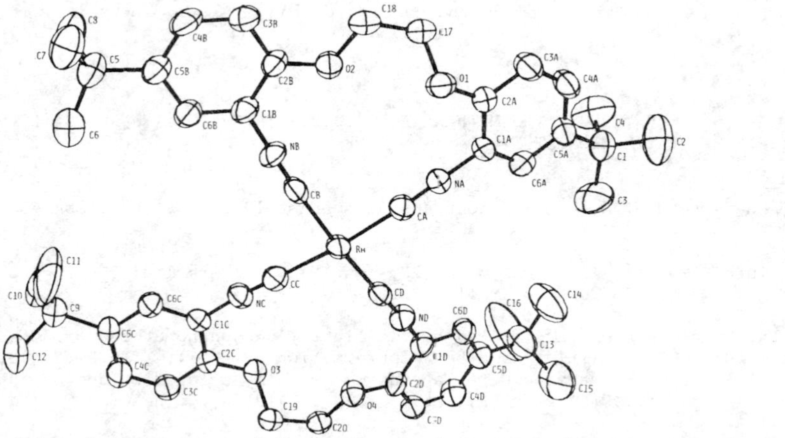

Fig. 1. Molecular structure of the cation in $C_{72}H_{76}BN_4O_4Rh,1.5(CH_3CN)$.

BIS(μ_2-BIS(DIPHENYLPHOSPHINO)METHANE-P,P')BIS(PENTAFLUOROPHENYLGOLD)BIS(PENTAFLUOROPHENYL)PALLADIUM

$C_{74}H_{22}Au_2F_{20}P_4Pd$

R. USÓN, A. LAGUNA, J. FORNIÉS, I. VALENZUELA, P.G. JONES and G.M. SHELDRICK, 1984. J. Organometal. Chem., 273, 129-139.

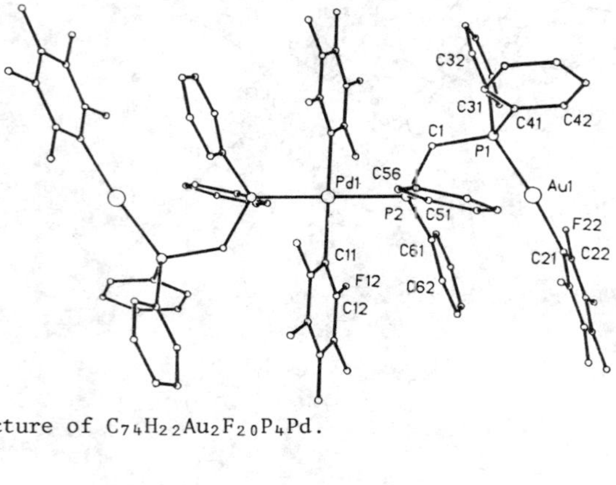

Fig. 1. Structure of $C_{74}H_{22}Au_2F_{20}P_4Pd$.

Monoclinic, $P2_1/c$, a = 17.091, b = 19.892, c = 10.254 Å, β = 90.85°, Z = 2. Mo radiation, R = 0.059 for 4324 reflexions.

The Pd atom lies on a centre of symmetry (Fig. 1), with almost perfect square-planar geometry. The Pd-C(11) and Pd-P(2) distances are 2.106 and 2.321 Å respectively and the angle P(2)-Pd-C(11) is 90.5°. The centrosymmetrically-related Au-containing moieties have Au-P(1) and Au-C(21) distances of 2.266 and 2.074 Å respectively, and an angle P(1)-Au-C(21) of 174.6°. The C-F distances range from 1.30 to 1.35 Å and the P-C distances are: P(1)-C(1) 1.83, P(2)-C(1) 1.84, P(1)-C(31) 1.807 and P(1)-C(41) 1.819 Å. There are no short Au...Au contacts.

TRIS(μ-HYDRIDO)-BIS[HYDRIDO(t-BUTYL ISOCYANIDE)BIS(TRIPHENYLPHOSPHINE)RHENIUM](Re-Re) HEXAFLUOROPHOSPHATE

$C_{82}H_{83}F_6N_2P_5Re_2$ $[Re_2H_5[P(C_6H_5)_3]_4(C_4H_9NC)_2]PF_6$

J.D. ALLISON, F.A. COTTON, G.L. POWELL and R.A. WALTON, 1984. Inorg. Chem., 23, 159-164.

Triclinic, $P\bar{1}$, a = 13.840, b = 23.951, c = 12.667 Å, α = 96.10, β = 111.80, γ = 76.51°, Z = 2. Mo radiation, R = 0.041 for 7225 reflexions.

The cation (Fig. 1) consists of two $ReH(PPh_3)_2$(t-BuNC) halves joined by three bridging H atoms on a Re-Re bond (2.604(1) Å). There is an approximate twofold axis passing through one bridging hydrogen atom and bisecting the Re-Re bond. Bond lengths: Re-P 2.366(3)-2.402(3), Re-C 1.978(11) and 1.976(11), Re-H_t 1.68(7) and 1.50(8), Re-H_b 1.65(9)-1.99(7) Å.

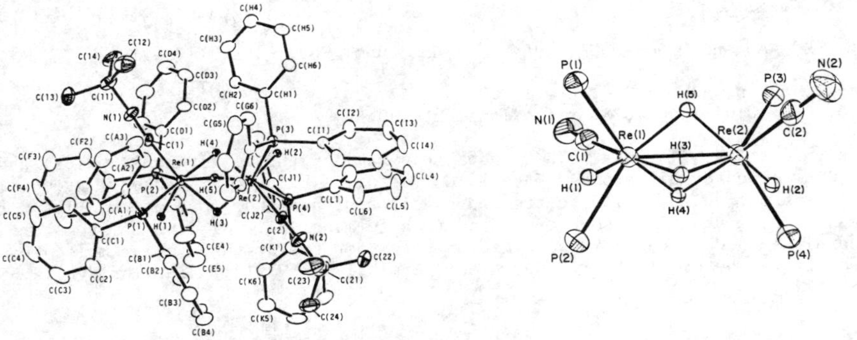

Fig. 1. The $[Re_2H_5(PPh_3)_4(t\text{-}BuNC)_2]^+$ cation, inner coordination group at right.

(μ_2-CARBONATO-O,O')BIS[BIS(TRIBENZYLPHOSPHINE)PHENYLPLATINUM] TOLUENE SOLVATE

$C_{97}H_{94}O_3P_4Pt_2,C_7H_8$

R.A. MICHELIN, G. STRUKUL, N. BRESCIANI-PAHOR, E. ZANGRANDO and L. RANDACCIO, 1984. Inorg. Chim. Acta, 84, 229-232.

Monoclinic, C2/c, a = 26.38, b = 13.576, c = 28.05 Å, β = 116.6°, D_m = 1.41, Z = 4. Mo radiation, R = 0.045 for 3511 reflexions.

The crystals consist of discrete dimers lying on crystallographic twofold axes. passing through atoms C(49) and O(2) of the bridging carbonate ligands (Fig. 1). The coordination about the Pt atoms is square planar with the metal atom displaced 0.06 Å from the mean plane through P(1),P(2),O(1),C(43). The coordination planes of the two Pt atoms have a dihedral angle of 39.0° while the interplanar angle between the phenyl ring and the coordination plane is 95.9°. The carbonate group is crystallographically constrained to be planar and its plane makes an angle of 97.3° with the Pt(1) coordination plane.

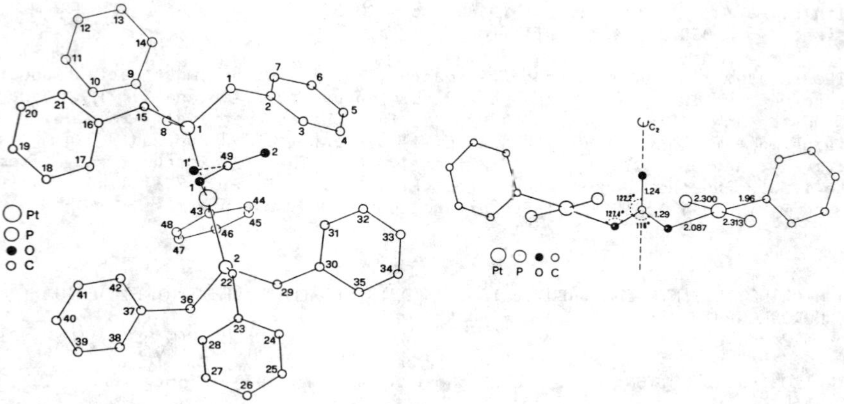

Fig. 1. The crystallographically-independent moiety of the dimer $C_{97}H_{94}O_3P_4Pt_2$ (left), and a sketch of the dimer showing selected bond lengths (Å) and angles (°) (right).

HEXAKIS(μ_2-PHENYL)TRICOPPERDILITHIUM BIS(μ_3-CHLORO)BIS(TRIS(DIETHYL ETHER-O)LITHIUM)-TETRAKIS(DIETHYL ETHER)DILITHIUM

$C_{112}H_{160}Cl_2Cu_3Li_6O_{10}$ $[Li_2Cu_3Ph_6]_2[Li_4Cl_2(Et_2O)_{10}]$

H. HOPE, D. ORAM and P.P. POWER, 1984. J. Am. Chem. Soc., 106, 1149-1150.

Monoclinic, $P2_1/n$, a = 13.214, b = 18.20, c = 24.42 Å, β = 97.17°, Z = 2. Mo radiation, R = 0.0632 for 5287 reflexions (at 140 K).

The anion (Fig. 1) contains a trigonal bipyramid with two lithium (apical) and three copper (equatorial) atoms. Each Li-Cu vector is bridged by a phenyl group. The cation lies about an inversion centre and consists of a planar core of two Li and two Cl atoms; each Cl atom is also coordinated to a terminal Li. All the Li atoms are four-coordinate due to further coordination to either two or three ether molecules. Some geometrical details: Li-O 1.96, Li-Cl(terminal) 2.697(12), Li-Cl(bridge) 2.537 and 2.505(12) Å.

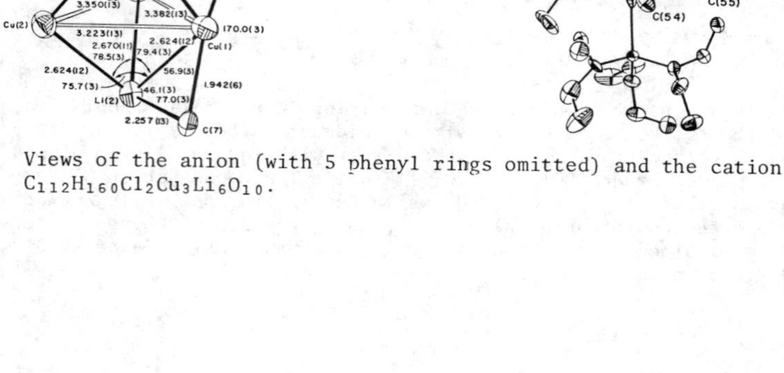

Fig. 1. Views of the anion (with 5 phenyl rings omitted) and the cation in $C_{112}H_{160}Cl_2Cu_3Li_6O_{10}$.

BIS(μ_2-CHLORO)(μ_2-σ^1,η^2-1-CYANOBUTA-1,3-DIENE)DICOPPER(I)

C_5H_5ClCuN $CuCl.CH_2CHCHCHCN$

V.V. OLIINIK, 1984. Kristallografiya, 29, 745-748 [Sov. Phys. Crystallogr., 29, 440-441].

Monoclinic, $P2_1/b$, a = 6.834, b = 13.717, c = 7.450 Å, γ = 111.62°, Z = 4. Mo radiation, R = 0.060 for 698 reflexions.

The Cu and Cl atoms form centrosymmetric fragments in the shape of deformed squares. There fragments are "cross-linked" by double chains of 1-cyanbutadiene-1,3 and form layers oriented along the [2$\bar{1}$0] direction (Fig. 1). The Cu-Cl bond distances are 2.443 and 2.291 Å and the angles Cl-Cu-Cl' and Cu-Cl-Cu' are 99.2 and 80.8° respectively. The Cu...Cu' distance is 3.071 Å. Each Cu atom is coordinated by an N atom and a C=C double bond remote from the cyano group in different ligand molecules resulting in a deformed tetrahedral environment of Cl, Cl' and N together with the centre of the double bond. The Cu-C(4) and Cu-C(5) distances are 2.15 and 2.12 Å respectively, and the Cu-N distance is 2.005 Å. The organic molecule is practically planar and exists as the trans isomer in this complex.

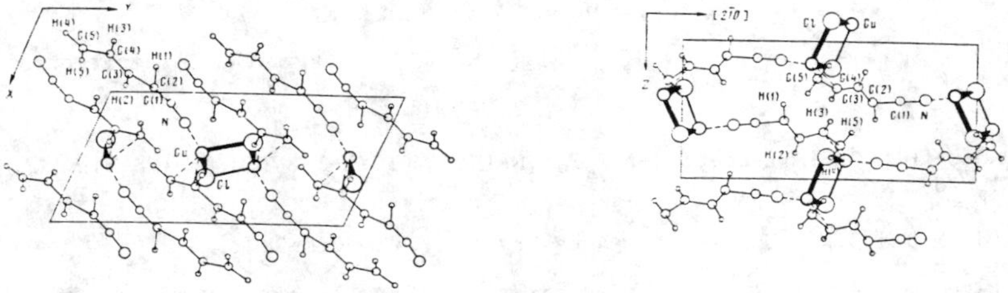

Fig. 1. Projection of the structure of C_5H_5ClCuN on the xy plane (left) and on the [2$\bar{1}$0]z plane (right).

BIS(η^2-BUT-2-YNE)OXOIODORHENIUM(III)

$C_8H_{12}IORe$ $Re(O)I(MeCCMe)_2$

J.M. MAYER and T.H. TULIP, 1984. J. Am. Chem. Soc., 106, 3878-3879.

Monoclinic, $P2_1/c$, a = 9.749, b = 7.307, c = 16.106 Å, β = 106.42°, Z = 4. Mo radiation, R = 0.027 for 2087 reflexions (at -100°C).

The structure contains isolated molecules which have virtual mirror symmetry (Fig. 1) with Re-I 2.691(1), Re-C 2.038-2.066(5), Re-O 1.697(3), C1-C2 1.278-1.288(7) Å, C-C-C 144.0-145.8(5)°.

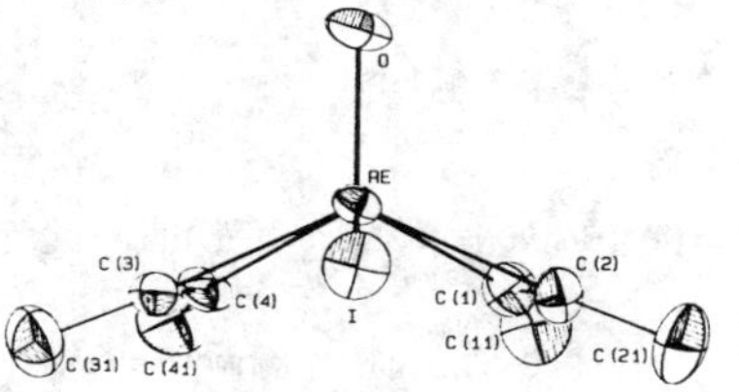

Fig. 1. A perspective drawing of $C_8H_{12}IORe$.

DICHLORO[1,2-ETHANEDIONE BIS(DIMETHYLHYDRAZONE)](η-ETHYLENE)PLATINUM(II)
$C_8H_{18}Cl_2N_4Pt$ $[PtCl_2(C_2H_4)(C_6H_{14}N_4)]$

A. BAVOSO, M. FUNICELLO, G. MORELLI and V. PAVONE, 1984. Acta Cryst., C40, 2035-2037.

Monoclinic, $P2_1$, a = 8.998, b = 8.133, c = 9.872 Å, β = 106.72°, D_m = 2.09, Z = 2. Mo radiation, R = 0.057 for 1404 reflexions.

The compound has trigonal-bipyramidal geometry (Fig. 1). The bidentate ligand is in a cis-conformation, with some distortion from planarity.

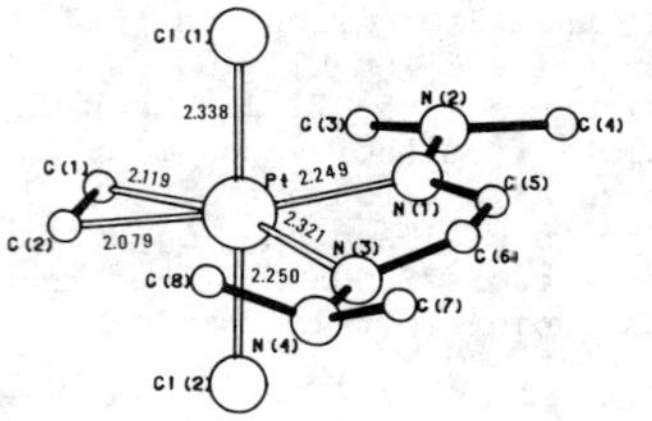

Fig. 1. A perspective view of $[PtCl_2(C_2H_4)(C_6H_{14}N_4)]$ with coordination distances.

a-CHLORO-b-CHLOROMETHYL-dc-(3-4-η-2,2,N,N-TETRAMETHYL-3-BUTENYLAMINE-N)PALLADIUM(II)
$C_9H_{19}Cl_2NPd$

G. FERGUSON and B.L. RUHL, 1984. Acta Cryst., C40, 2020-2022.

Monoclinic, $P2_1/n$, a = 6.574, b = 12.171, c = 16.158 Å, β = 101.42°, Z = 4. Mo radiation, R = 0.052 for 2471 reflexions.

The Pd atom has a slightly distorted square-planar coordination (Pd-Cl 2.334(2), Pd-N 2.198(6), Pd-C(olefin) 2.175(7) and 2.143(8), Pd-CH_2Cl 2.039(8) Å) with the CH_2Cl moiety trans to the N atom. The five-membered ring has an envelope conformation, with the quaternary C atom 0.61(1) Å above the plane of the other four atoms.

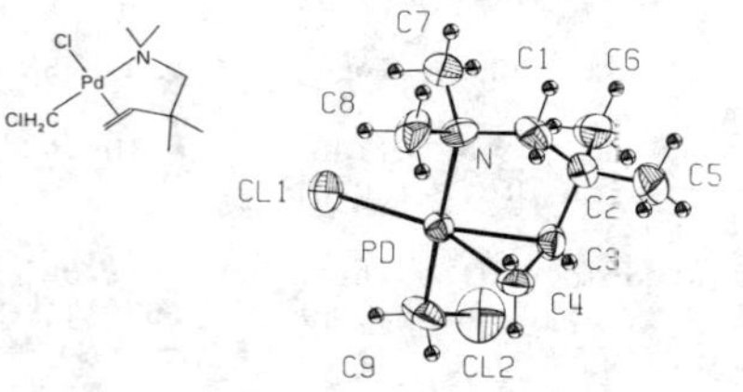

Fig. 1. A view of $C_9H_{19}Cl_2NPd$.

(μ_2-η^4-2-BUTYNEDIOIC ACID)BIS(TRICARBONYLCOBALT)DICARBOXYACETYLENEHEXACARBONYLDICOBALT
$C_{10}H_2Co_2O_{10}$ (I)

(μ_2-η^4-2-BUTYNE-1,4-DIOL)BIS(TRICARBONYLCOBALT) MONOHYDRATE
$C_{10}H_6Co_2O_8,H_2O$ (II)

(μ_2-η^4-HEXAFLUORO-2-BUTYNE)BIS(TRICARBONYLCOBALT)
$C_{10}Co_2F_6O_6$ (III)

F. BAERT, A. GUELZIM and P. COPPENS, 1984. Acta Cryst., B40, 590-595.

I. Orthorhombic, $P2_12_12_1$, a = 7.059, b = 11.107, c = 17.516 Å, D_m = 1.80-2.00, Z = 4. Mo radiation, R = 0.028 for 2286 reflexions.

II. Monoclinic, $P2_1/c$, a = 7.974, b = 12.591, c = 14.551 Å, β = 104.31°, D_m = 1.80-2.00, Z = 4. Mo radiation, R = 0.048 for 4427 reflexions.

III. Monoclinic, $P2_1/c$, a = 7.538, b = 30.804, c = 12.681 Å, β = 100.49°, D_m = 1.95-2.10, Z = 8. Mo radiation, R = 0.052 for 4914 reflexions at 100 K.

These complexes (and two others) are of the general form $Co_2(CO)_6(R\text{-}C{\equiv}C\text{-}R)$ and are analyzed in terms of variation in their geometry as a function of increasing electronegativity of the substituent R (CF_3 > COOH ≃ CH_2OH). The metal-acetylenic ligand distances decrease with increasing electronegativity of R, while the acetylenic C-C distance increases, indicating the strengthening of the Co-C(acetylene) bonds to be due to increased back donation. The effect is attributed to a stabilization of the ligand a_2 and b_1 orbitals by the electronegative substituents. Various correlations between geometric distortion parameters are discussed. Views of the complexes are in Fig. 1.

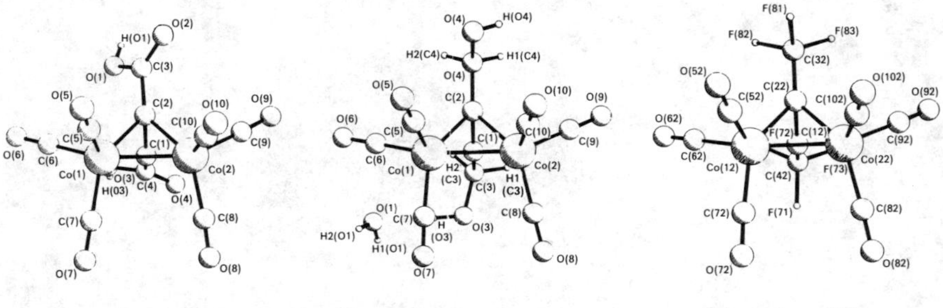

Fig. 1. Views (left to right) of the $Co_2(CO)_6R\text{-}C{\equiv}C\text{-}R$ complexes I - III.

(μ_2-σ^3,σ^2-3-(2-METHYL-4,4,6,6-TETRACHLOROCYCLOTRIPHOSPHAZEN-2-YL)PROPANE-1,2-DIYLIDENE)BIS(TRICARBONYLCOBALT)

$C_{10}H_6Cl_4Co_2N_3O_6P_3$ (I) $N_3P_3Cl_4(CH_3)[C_3H_3.Co_2(CO)_6]$

(μ_2-σ^2,σ^2-3-(2-ALLYL-4,4,6,6-TETRACHLOROCYCLOTRIPHOSPHAZEN-2-YL)PROPANE-1,2-DIYLIDENE)BIS(TRICARBONYLCOBALT)

$C_{12}H_8Cl_4Co_2N_3O_6P_3$ (II) $N_3P_3Cl_4(C_3H_5)[C_3H_3.Co_2(CO)_6]$

(μ_2-σ^2,σ^2-1-(2-METHYL-4,4,6,6-TETRACHLOROCYCLOTRIPHOSPHAZEN-2-YL)PROPANE-1,2-DIYLIDENE)BIS(TRICARBONYLCOBALT)

$C_{10}H_6Cl_4Co_2N_3O_6P_3$ (III) $N_3P_3Cl_4(CH_3)[C_3H_3.Co_2(CO)_6]$

H.R. ALLCOCK, R.A. NISSAN, P.J. HARRIS and R.R. WHITTLE, 1984. Organometallics, 3. 432-440.

I. Monoclinic, $P2_1/c$, a = 17.238, b = 7.470, c = 17.853 Å, β = 99.91°, Z = 4. Mo radiation, R = 0.048 for 2837 reflexions.

II. Monoclinic, $P2_1/n$, a = 8.038, b = 22.080, c = 13.064 Å, β = 91.51°, Z = 4. Mo radiation, R = 0.045 for 2911 reflexions.

III. Orthorhombic, Pnma, a = 8.592, b = 14.58, c = 17.787 Å, Z = 4. Mo radiation, R = 0.039 for 1965 reflexions.

The cyclotriphosphazene rings in I, II and III are planar but with the same distortion of ring symmetry. The P-N bonds closest to the organic units are long, the two bonds adjacent to these are short, and those furthest from the organic units are of intermediate length (Fig. 1). In each compound the $Co_2(CO)_6$ unit is π-bonded to the appropriate acetylenic site. The Co-Co distances in I, II and III are 2.454, 2.437 and 2.494 Å respectively. The short (0.889 Å) bond length in the allyl group of compound II is not real and is attributed to disorder of the terminal C(3B) atom.

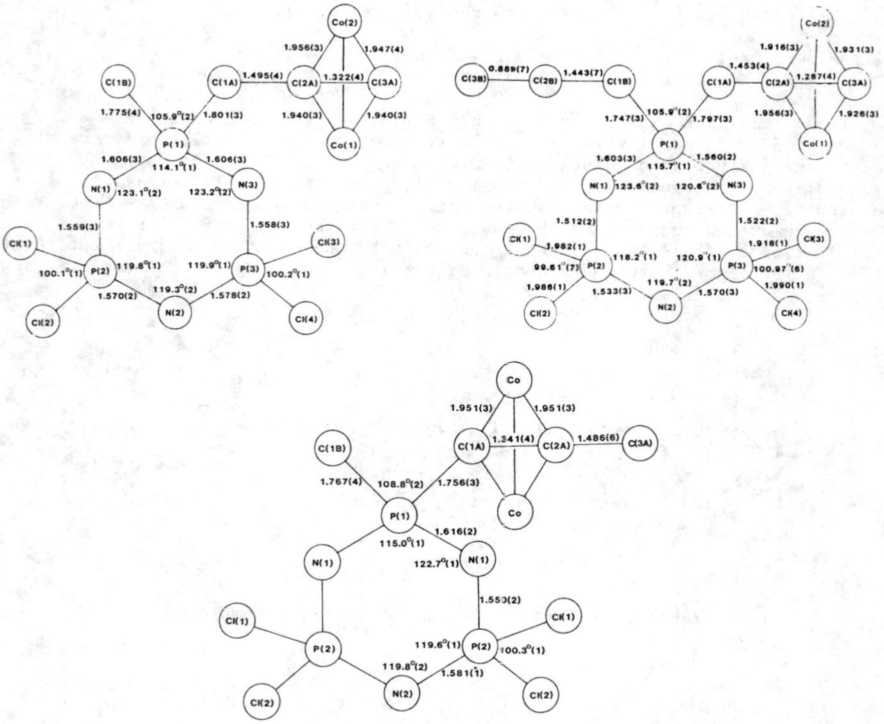

Fig. 1. Schematic figures showing selected bond lengths and angles in $C_{10}H_6Cl_4Co_2N_3O_6P_3$ (top left), $C_{12}H_8Cl_4Co_2N_3O_6P_3$ (top right), and $C_{10}H_6Cl_4Co_2N_3O_6P_3$ (bottom).

DICARBONYL(η^5-CYCLOPENTADIENYL)(η^3-2-METHYLALLYL)MOLYBDENUM(II)

$C_{11}H_{12}MoO_2$ $[Mo(C_4H_7)(C_5H_5)(CO)_2]$

N.W. MURRALL and A.J. WELCH, 1984. Acta Cryst., C40, 401-403.

Monoclinic, $P2_1/n$, a = 6.1105, b = 12.7885, c = 13.7247 Å, β = 98.787°, Z = 4. Mo radiation, R = 0.025 for 2867 reflexions.

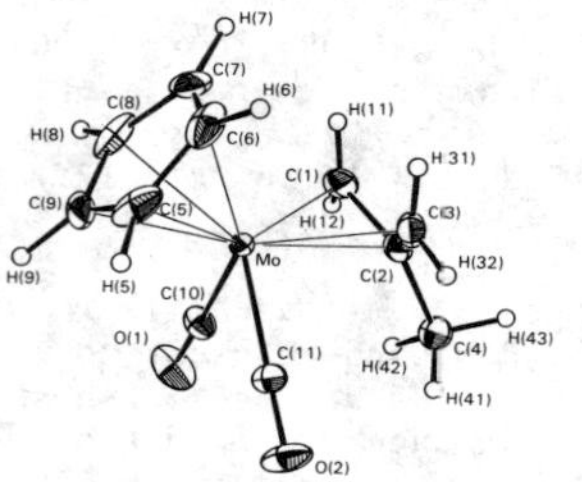

Fig. 1. A perspective view of $C_{11}H_{12}MoO_2$.

The η^3-2-methylallyl ligand (Fig. 1) is in the endo conformation and the three Mo-C(allyl) distances are equal within experimental error, average 2.3163(13) Å. The C_5H_5 ligand shows some degree of rotational disorder.

(η^5-2,4-CYCLOPENTADIEN-1-YL)BIS(η^3-2-PROPENYL)MOLYBDENUM
$C_{11}H_{15}Mo$ [$Mo(\eta^3-C_3H_5)_2(\eta^5-C_5H_5)$]

P.W. JOLLY, C. KRÜGER, C.C. ROMÃO and M.J. ROMÃO, 1984. Organometallics, 3, 936-937.

Monoclinic, $P2_1/a$, a = 11.241, b = 7.4232, c = 12.5643 Å, β = 102.715°, Z = 4. Mo radiation, R = 0.022 for 2493 reflexions.

The organic ligands adopt an essentially trigonal-planar arrangement around the central metal atom (Fig. 1). The two η^3-allyl groups adopt an endo, cis conformation with an angle of 25.6° between the two planes that brings the meso-H atoms H(2) and H(5) within 2.171 Å of each other. The Mo-Cp (centroid) distance is 2.025 Å and the Mo-D1 and Mo-D2 distances (D1 and D2 are the midpoints of the allyl groups C(1)-C(2)-C(3) and C(4)-C(5)-C(6) respectively) are 1.991 and 1.993 Å. Important distances are: C(1)-C(2) 1.407, C(2)-C(3) 1.407, C(4)-C(5) 1.396, C(5)-C(6) 1.414, Mo-C(1) 2.250, Mo-C(2) 2.248, Mo-C(3) 2.250, Mo-C(4) 2.262, Mo-C(5) 2.237, Mo-C(6) 2.252 Å. The angles Cp(centroid)-Mo-D1, Cp(centroid)-Mo-D2 and D1-Mo-D2 are respectively 128.4, 130.1 and 101.5°. The plane defined by atoms C(1), C(3), C(4) and C(6) is parallel to the plane of the Cp ring (dihedral angle 1.09°).

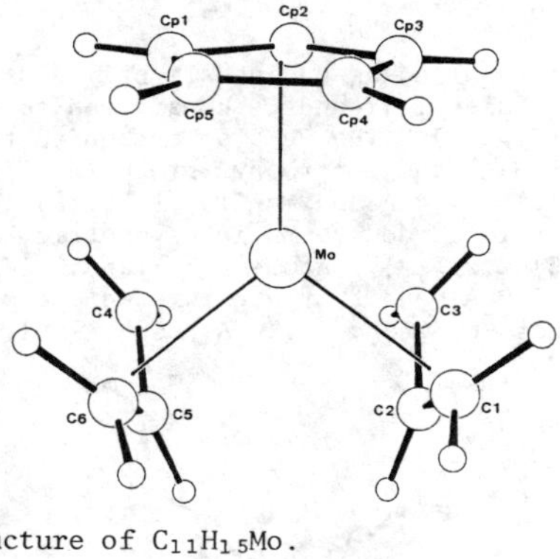

Fig. 1. Molecular structure of $C_{11}H_{15}Mo$.

DICHLORO-(η^2,η^2-4,4-DIACETYLHEPTA-1,6-DIENE)PLATINUM(II)
$C_{11}H_{16}Cl_2O_2Pt$

R. GRIGG, J.F. MALONE, T.R.B. MITCHELL, A. RAMASUBBU and R.M. SCOTT, 1984. J. Chem. Soc. Perkin I, 1745-1754.

Triclinic, $P\bar{1}$, a = 9.46, b = 11.08, c = 6.72 Å, α = 96.9, β = 88.5, γ = 110.1°, Dm = 2.24, Z = 2. Cu radiation, R = 0.101 for 1801 reflexions. [No absorption corrections applied.]

The molecule (Fig. 1) shows π-complexation of the olefinic groups with the 1,6-diene in the 'chair' conformation. Coordination distances to Pt are Pt-Cl(1) 2.322(7), Pt-Cl(2) 2.306(6), Pt-X(1) 2.06(1), Pt-X(2) 2.09(1) (X(1) and X(2) are the midpoints of C(1)-C(2) and C(6)-C(7) respectively). Elongated octahedral coordination around Pt is completed by a carbonyl oxygen atom (Pt...O 3.34) and the Cl atom of a centrosymmetrically related molecule (Pt...Cl 3.71 Å).

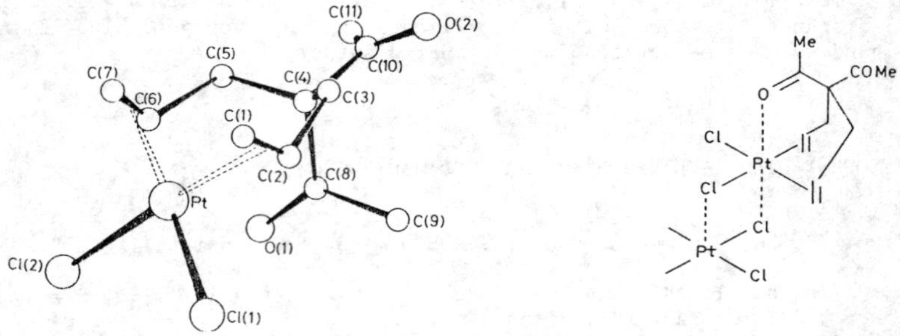

Fig. 1. $C_{11}H_{16}Cl_2O_2Pt$: the complex (left) and extended coordination scheme (right).

TRICARBONYL[(1,2,3-η)-2,4-PENTADIENYL](TRIMETHYLPHOSPHINE)MANGANESE

$C_{11}H_{16}MnO_3P$ $(C_3H_7)Mn(CO)_3PMe_3$

M.A. PAZ-SANDOVAL, P. POWELL, M.G.B. DREW and R.N. PERUTZ, 1984. Organometallics, 3, 1026-1033.

Monoclinic, $P2_1$, a = 7.170, b = 13.757, c = 7.660 Å, β = 113.2°, Z = 2. Mo radiation, R = 0.061 for 1041 reflexions.

The metal atom can be regarded as having slightly distorted octahedral coordination if the η^3-pentadienyl ligand is considered to span two coordination positions on the Mn atom (Fig. 1). The Mn atom is bonded to a mutually trans phosphorus group (Mn-P 2.300 Å) and a carbonyl group (Mn-C(1) 1.818 Å), two mutually cis carbonyl groups (Mn-C(2) 1.774 Å and Mn-C(3) 1.798 Å) and the pentadienyl ligand. The η^3-pentadienyl fragment is bonded to Mn in asymmetric fashion (Mn-C(11) 2.199, Mn-C(12) 2.124 and Mn-C(13) 2.255 Å). Atoms C(13) and C(14) are not bonded to the metal (Mn...C distances >3.0 Å) and they are twisted out of the plane of C(11), C(12), C(13) by distances of -0.86 and -1.22 Å respectively.

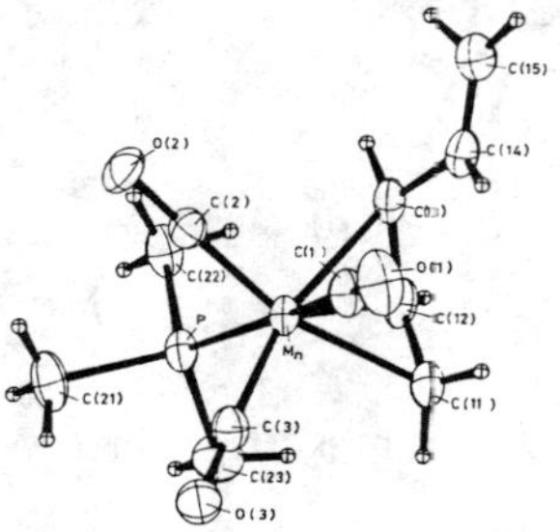

Fig. 1. Molecular structure of $C_{11}H_{16}MnO_3P$.

BIS(η^4-1,3-BUTADIENE)(TRIMETHYLPHOSPHINE)IRON

$C_{11}H_{21}FeP$ $Fe(\eta\text{-}C_4H_6)_2PMe_3$

J.M. McCALL, J.R. MORTON, Y. LE PAGE and K.F. PRESTON, 1984. Organometallics, 3, 1299-1303.

Monoclinic, $P2_1/c$, a = 7.729, b = 21.204, c = 8.096 Å, β = 115.18°, Z = 4. Mo radiation, R = 0.029 for 981 reflexions.

The molecular structure consists of an Fe atom square-pyramidally coordinated by one apical PMe_3 and two basal cis-butadiene ligands (Fig. 1). The Fe-P distance is 2.202 Å. Metal-butadiene distances are: Fe-C(1) 2.080, Fe-C(2) 2.017, Fe-C(3) 2.020, Fe-C(4) 2.083 Å; Fe-C(5) 2.090, Fe-C(6) 2.026, Fe-C(7) 2.019 and Fe-C(8) 2.089 Å. Distances within the butadiene ligands are: C(1)-C(2) 1.399, C(2)-C(3) 1.403, C(3)-C(4) 1.390 Å; C(5)-C(6) 1.392, C(6)-C(7) 1.391 and C(7)-C(8) 1.393 Å. The angles C(1)-C(2)-C(3), C(2)-C(3)-C(4), C(5)-C(6)-C(7) and C(6)-C(7)-C(8) are respectively 118.8, 117.6, 118.4 and 120.1°.

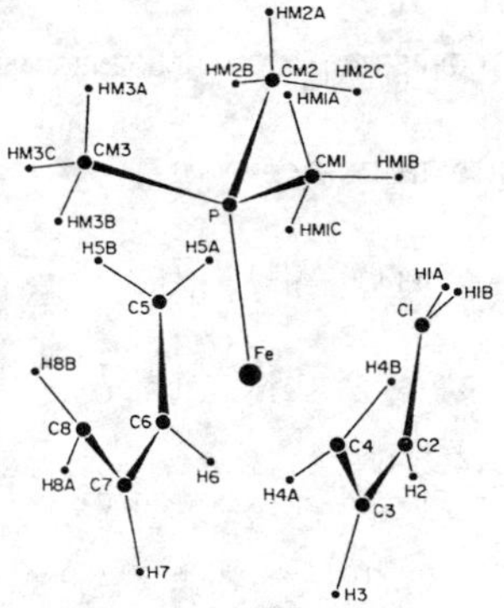

Fig. 1. Schematic molecular structure of $C_{11}H_{21}FeP$.

BIS(μ_2-O-ETHYLTHIOFORMATE-C,S)HEXACARBONYLDIIRON

$C_{12}H_{12}Fe_2O_8S_2$ $[HC(S)OC_2H_5]_2Fe_2(CO)_6$

D. SEYFERTH, G.B. WOMACK, M. COWIE and B.W. HAMES, 1984. Organometallics, 3, 1891-1897.

Triclinic, $P\bar{1}$, a = 8.237, b = 9.448, c = 6.392 Å, α = 107.31, β = 104.59, γ = 97.18°, Z = 1. Mo radiation, R = 0.029 for 1533 reflexions.

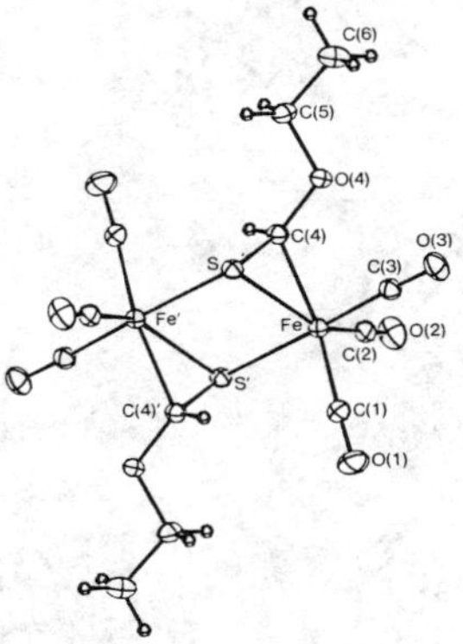

Fig. 1. Perspective view of $C_{12}H_{12}Fe_2O_8S_2$.

The two inversion-related Fe atoms are not bound directly to each other but are bridged by the two thioester groups (Fig. 1). Each group is π-bonded through its C=S moiety to a different metal and is then coordinated to the other metal via a dative bond from sulphur. The O atoms of these ligands are aimed away from the bridging positions. Principal bond lengths and non-bonded contacts are: Fe-S 2.2810, Fe-S' 2.3146, S-C(4) 1.774, Fe-C(1) 1.813, Fe-C(2) 1.785, Fe-C(3) 1.794, Fe-C(4) 1.987, Fe...Fe' 3.483, S...S' 2.999, Fe-C(4') 3.188, and H(C4)-C(1') 2.59 Å.

Distortions from idealized octahedral geometry arise mostly from the strain imposed by the bridging C-S group which results in an acute C(4)-Fe-S angle of 48.51°. The C(4)-O(4), C(5)-O(4), and C(5)-C(6) distances are 1.374, 1.429 and 1.479 Å respectively.

(ETHYLENE)(DI-2-PYRIDYLAMINE)COPPER(I) PERCHLORATE
$C_{12}H_{13}ClCuN_3O_4$ (I) $[Cu(NH(C_5H_4N)_2)(C_2H_4)]ClO_4$

(ACETYLENE)(DI-2-PYRIDYLAMINE)COPPER(I) TETRAFLUOROBORATE
$C_{12}H_{11}BCuF_4N_3$ (II) $[Cu(NH(C_5H_4N)_2)(C_2H_2)]BF_4$

(PERCHLORATO)CARBONYL(DI-2-PYRIDYLAMINE)COPPER(I)
$C_{11}H_9ClCuN_3O_5$ (III) $[Cu(NH(C_5H_4N)_2)(CO)ClO_4]$

J.S. THOMPSON and J.F. WHITNEY, 1984. Inorg. Chem., 23, 2813-2819.

I. Monoclinic, C2/c, a = 9.829, b = 17.105, c = 8.672 Å, β = 93.806°, Z = 4. Mo radiation, R = 0.050 for 1260 reflexions (at -100°C).

II. Monoclinic, C2/c, a = 10.140, b = 17.782, c = 7.852 Å, β = 105.34°, Z = 4. Mo radiation, R = 0.046 for 1259 reflexions (at -100°C).

III. Triclinic, P$\bar{1}$, a = 8.922, b = 10.755, c = 7.379 Å, α = 91.51, β = 99.24, γ = 81.25°, Z = 2. Mo radiation, R = 0.028 for 2774 reflexions (at -100°C).

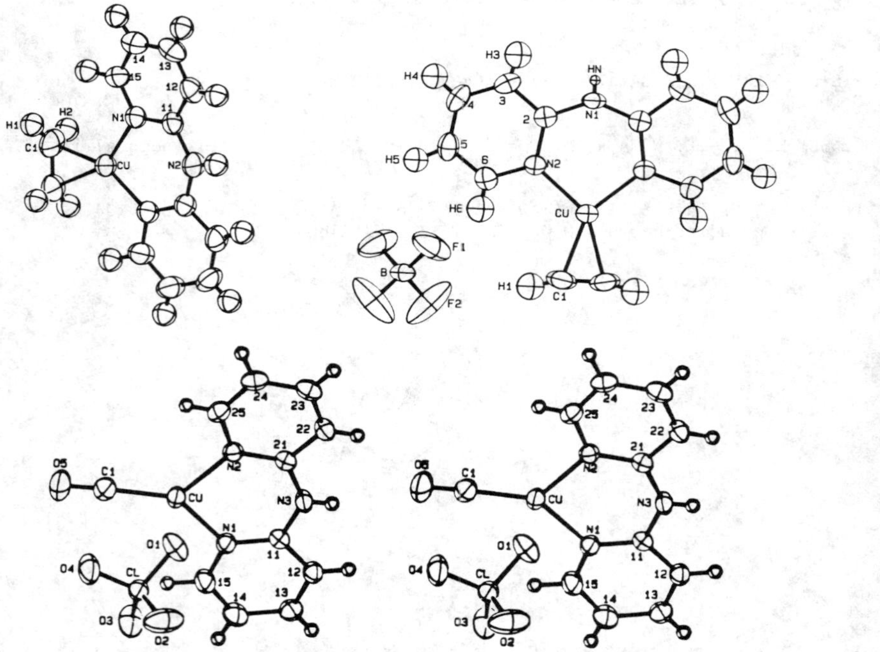

Fig. 1. The structures of $[Cu(NH(C_5H_4N)_2)(C_2H_4)]^+$ (top left), $[Cu(NH(C_5H_4N)_2)(C_2H_2)]^+$ (top right), and $[Cu(NH(C_5H_4N)_2)(CO)ClO_4]$ (bottom).

The three complexes are shown in Fig. 1. The geometry about Cu(I) in both I and II is trigonal planar, the Cu being coordinated to two pyridyl N atoms and an ethylene or acetylene molecule. The cations (and anions) in both I and II possess exact C_2 symmetry. Bond lengths: Cu-N 1.963(2), Cu-C 2.019(3), C1-C1' 1.359(7) Å in I; Cu-N

1.968(3), Cu-C 1.971(4), C1-C1' 1.188(11) Å in II. In III the Cu(I) ion is coordinated to two pyridyl N atoms (Cu-N 1.985(1) and 1.996(1) Å) and a CO group (Cu-C 1.808(2) Å) in a trigonal planar arrangement. There is also a ClO_4^- oxygen atom at a distance of 2.429(1) Å from the Cu atom, which suggests a weak interaction between cation and anion.

BIS(η^2-ETHYLENE)-METHYL-BIS(TRIMETHYLPHOSPHINE)-(CHLORODIMETHYLALUMINIUM-CHLORO)-TUNGSTEN(II)

$C_{12.3}H_{32.8}AlCl_{2.7}P_2W$ $WMe(PMe_3)_2(C_2H_4)_2(ClAlMe_xCl_{3-x})$

P.R. SHARP, 1984. Organometallics, 3, 1217-1223.

Monoclinic, $P2_1/n$, a = 9.922, b = 7.419, c = 29.241 Å, β = 92.79°, Z = 4. Mo radiation, R = 0.025 for 1362 reflexions.

The solid state structure consists of monomeric units with cis-ethylenes, trans-phosphines, and a weakly coordinated aluminate (Fig. 1), constituting overall, distorted octahedral geometry. The aluminate anion is coordinated through a long (2.861 Å) and weak W-Cl(1) bond. The other distances are: W-P(1) 2.501, W-P(2) 2.511, W-CW (methyl group) 2.13, W-C(1)(ethylene 1) 2.18, W-C(1) (ethylene 2) 2.169, W-C(2)(ethylene 1) 2.20 and W-C(2)(ethylene 2) 2.17 Å. The angle P(1)-W-P(2) is 160.1°, and the others P(1)-W-Cl(1), P(2)-W-Cl(1), P(1)-W-CW and P(2)-W-CW are 84.23, 83.57, 81.1 and 80.3° respectively. A complicating feature of the structure is disorder at two of the terminal aluminate Me/Cl positions. Cl* is a composite image of 92% Cl and 8% C whereas C* is a composite image of 75% C and 25% Cl. Crystals can be considered to be composed of an isomorphous mixture of 67% $[ClAlClMe_2]^-$, 25% $[ClAlCl_2Me]^-$ and 8% $[ClAlMe_3]^-$.

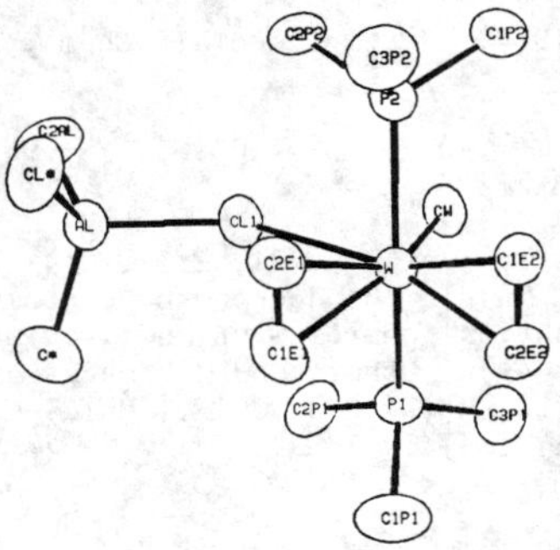

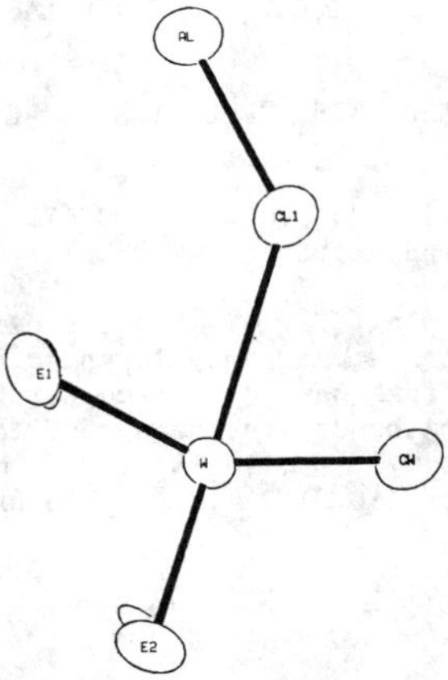

Fig. 1. Perspective views of $C_{12.3}H_{32.8}AlCl_{2.7}P_2W$ (left), and the molecule down the P-W-P axis (right).

DICARBONYL(η^5-CYCLOPENTADIENYL)(3-METHYLTHIO-3-OXO-1,2-BIS(TRIFLUOROMETHYL)PROPEN-2-YL-1-YLIDENE)TUNGSTEN

$C_{13}H_8F_6O_3SW$

J.L. DAVIDSON, M. SHIRALIAN, L. MANOJLOVIC-MUIR and K.W. MUIR, 1984. J. Chem. Soc. Dalton, 2167-2176.

Monoclinic, $P2_1/c$, a = 7.950, b = 14.416, c = 14.210 Å, β = 105.29°, Z = 4. Mo radiation, R = 0.040 for 3282 reflexions.

In the complex (Fig. 1) a vinyl group is bonded to a metal via both carbon atoms and therefore acts as a three-electron donor ligand. The structure consists of an approximately octahedrally coordinated tungsten atom bonded to an η^5-C_5H_5 ligand which occupies one face of the octahedron. Two CO ligands occupy two octahedral sites while the third is occupied by the η^2-$C(CF_3)C(CF_3)C(O)SMe$ ligand which adopts an orientation such that the vinylic C-C axis lies approximately parallel to one of the W-CO axes. Bond lengths include: W-C(1) 1.975(8), W-C(2) 1.994(8), W-C(3-7) 2.310(9)-2.373(10), W-C(8) 2.193(7) and W-C(9) 1.962(7) Å. The C(8)-C(9) distance is 1.443(10) and S-C(12) and S-C(13) are 1.773(7) and 1.834(10) Å respectively.

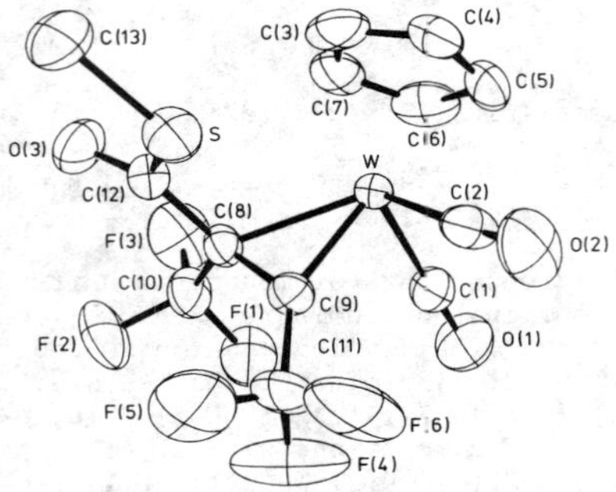

Fig. 1. $C_{13}H_8F_6O_3SW$: the molecular structure.

HEPTACARBONYL-μ-(2,5:2-5-η-2,4-HEXADIENE-2,5-DIYL)DIMANGANESE
$C_{13}H_8Mn_2O_7$ $Mn_2(CO)_7(CH_3CCHCHCCH_3)$

M. LEYENDECKER, W.S. SHELDRICK and C.G. KREITER, 1984. J. Organometal. Chem., 270, C37-C40.

Triclinic, $P\bar{1}$, a = 9.076, b = 9.172, c = 9.055 Å, α = 90.49, β = 103.63, γ = 96.10°, Z = 2. Mo radiation, R = 0.020 for 2388 reflexions.

The complex (Fig. 1) has approximate C_s symmetry. Atoms C(2), C(3), C(4), C(5) and Mn(1) form a manganacyclopentadienyl ring with Mn(2) π-bonded to the C atoms. Both metal atoms have distorted octahedral coordination spheres with Mn(1) actually forming seven bonds (including one to Mn(2) (2.654 Å)). Other bond distances are: Mn(1)-C(2) 2.071, Mn(1)-C(5) 2.074, Mn(2)-C(2) 2.126, Mn(2)-C(3) 2.131, Mn(2)-C(4) 2.134 and Mn(2)-C(5) 2.116 Å. The angle C(2)-Mn(1)-C(5) is 79.2°.

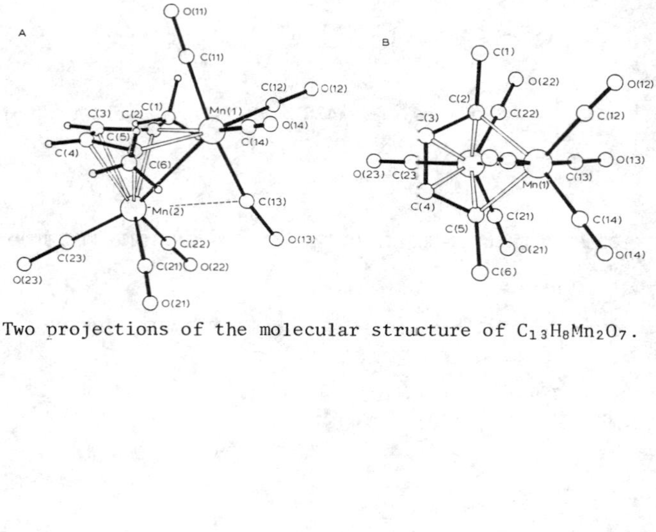

Fig. 1. Two projections of the molecular structure of $C_{13}H_8Mn_2O_7$.

(1,4-DIISOPROPYL-1,4-DIAZA-1,3-BUTADIENE)PENTACARBONYLDIRUTHENIUM
$C_{13}H_{16}N_2O_5Ru_2$ $Ru_2(CO)_5(C_8H_{16}N_2)$

J. KEIJSPER, L. POLM, G. van KOTEN, K. VRIEZE, G. ABBEL and C.H. STAM, 1984. Inorg. Chem., 23, 2142-2148.

Orthorhombic, $P2_12_12_1$, a = 12.775, b = 13.518, c = 9.636 Å, Z = 4. Cu radiation, R = 0.045 for 1480 reflexions.

The molecule is shown in Fig. 1. The single Ru-Ru bond (2.741(1) Å) is spanned by an asymmetric bridging carbonyl group with Ru(1)-C(3) (2.15(1) Å) significantly longer than Ru(2)-C(3) (2.03(1) Å), while the C(3)-O(3) vector is almost perpendicular (87°) to the Ru-Ru vector. The terminal Ru-CO distances are equal, 1.86(1) Å (mean). The $C_8H_{16}N_2$ ligand has a flat N=CC=N skeleton and is chelate bonded to Ru(2) with two equal bond lengths (2.14(1) Å) and η^2-C=N, η^2-C'=N' to Ru(1) with four equal bond lengths (2.27(1) Å).

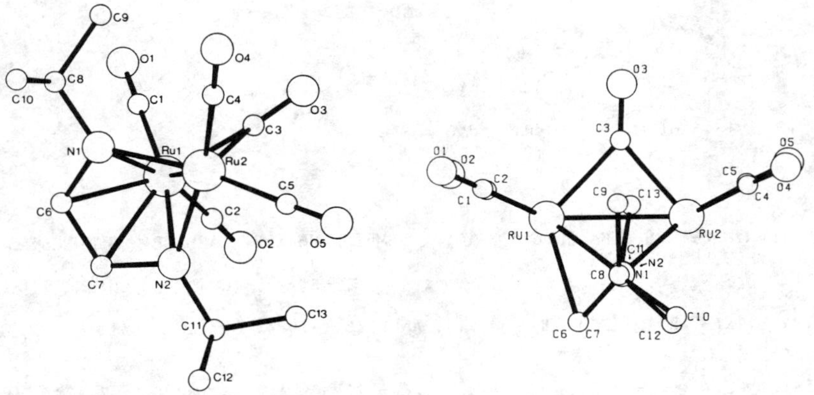

Fig. 1. Two views of $Ru_2(CO)_5(C_8H_{16}N_2)$.

DICARBONYL(η^4-trans-1,3-PENTADIENE)BIS(TRIMETHYLPHOSPHINE)CHROMIUM(0)
$C_{13}H_{26}CrO_2P_2$ (I)

DICARBONYL(η^4-trans,trans-2,4-HEXADIENE)BIS(TRIMETHYLPHOSPHINE)CHROMIUM(0)
$C_{14}H_{28}CrO_2P_2$ (II)

C.G. KREITER, M. KOTZIAN, U. SCHUBERT, R. BAU and M.A. BRUCK, 1984. Z. Naturforsch., B39, 1553-1564.

I. Hexagonal, $P6_3$, a = 16.251, c = 11.199 Å, Z = 3. Mo radiation, R = 0.131 for three-dimensional data (at 173 K).

II. Orthorhombic, Pbcn, a = 17.300, b = 31.310, c = 13.889 Å, Z = 8. Mo radiation, R = 0.073 for three-dimensional data.

Views of molecule I and one of the two independent molecules of II are in Fig. 1. The complexes have distorted octahedral geometry, with Cr-P 2.287-2.340, Cr-C(diene) 2.18-2.49 Å, P-Cr-P 151.6-159.7°.

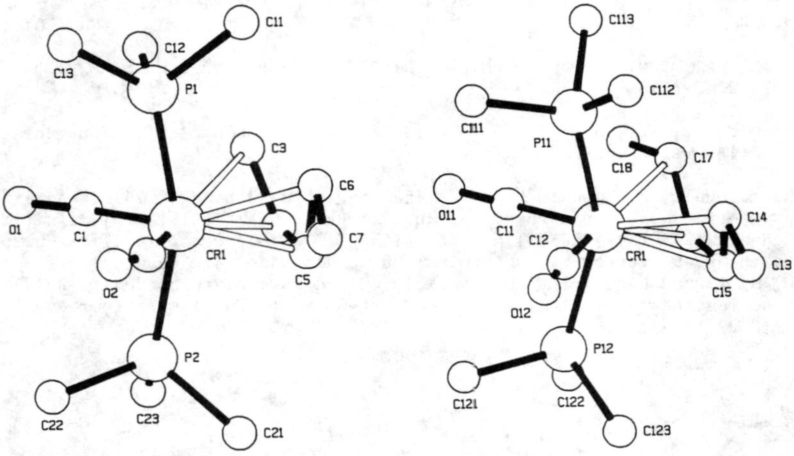

Fig. 1. Views of $C_{13}H_{26}CrO_2P_2$ and $C_{14}H_{28}CrO_2P_2$.

DICARBONYL-HYDRO-TRIS(PYRAZOLYL)BORATO(η^2-S-METHYLMERCAPTOCARBENE-C,S)TUNGSTEN TRIFLUOROMETHYLSULPHONATE

$C_{14}H_{14}BF_3N_6O_5S_2W$ $\{[HB(pz)_3](CO)_2W[\eta^2\text{-}CH(SMe)]\}^+.CF_3SO_3^-$

H.P. KIM, S. KIM, R.A. JACOBSON and R.J. ANGELICI, 1984. Organometallics, 3, 1124-1126.

Monoclinic, $P2_1/c$, a = 9.841, b = 11.374, c = 21.305 Å, β = 109.15°, Z = 4. Mo radiation, R = 0.072 for 2111 reflexions (at -40°C).

The crystal structure analysis reveals the η^2-carbene ligand bonded to the W atom through both the C and S atoms (Fig. 1). The W-C(3) distance of 1.93 Å indicates considerable double bond character and suggests a carbene-like interaction. The W-S(1) distance is 2.481 Å and the C(3)-S(1) distance 1.72 Å. The dihedral angle between the C(3)-S(1)-C(4) and W-C(3)-S(1) planes is 103.8°. Other bond distances and angles are: W-N (average) 2.20, W-CO (average) 2.06, S(1)-C(4) 1.85 Å, S(1)-W-C(3) 43.8, average N-W-N 82.4, C(1)-W-C(2) 89.3, W-S(1)-C(3) 51.0, W-S(1)-C(4) 110.5, W-C(3)-S(1) 85.3, and C(3)-S(1)-C(4) 105.9°.

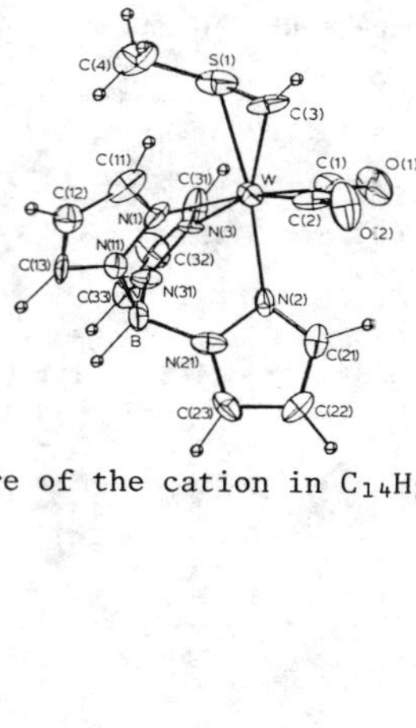

Fig. 1. Molecular structure of the cation in $C_{14}H_{14}BF_3N_6O_5S_2W$.

BIS(η^5-2,4-DIMETHYLPENTADIENYL)VANADIUM
$C_{14}H_{22}V$ (I) $V(\eta^5-C_7H_{11})_2$

BIS(η^5-2,4-DIMETHYLPENTADIENYL)CHROMIUM
$C_{14}H_{22}Cr$ (II) $Cr(\eta^5-C_7H_{11})_2$

C.F. CAMPANA, R.D. ERNST, D.R. WILSON and J.-Z. LIU, 1984. Inorg. Chem., 23, 2732-2734.

I. Monoclinic, C2/c, a = 13.719, b = 10.838, c = 18.173 Å, β = 77.74°, Z = 8. Mo radiation, R = 0.053 for 1102 reflexions.

II. Tetragonal, $P4_12_12$, a = 8.090, c = 20.847 Å, Z = 4. Mo radiation, R = 0.055 for 589 reflexions.

The two "open metallocene" compounds both exist in staggered conformations. The V-C distances in I range from 2.179(4) to 2.236(5) Å with a mean of 2.211(2) Å and the Cr-C distances in II range from 2.147(7) to 2.177(8) Å, averaging 2.165(4) Å.

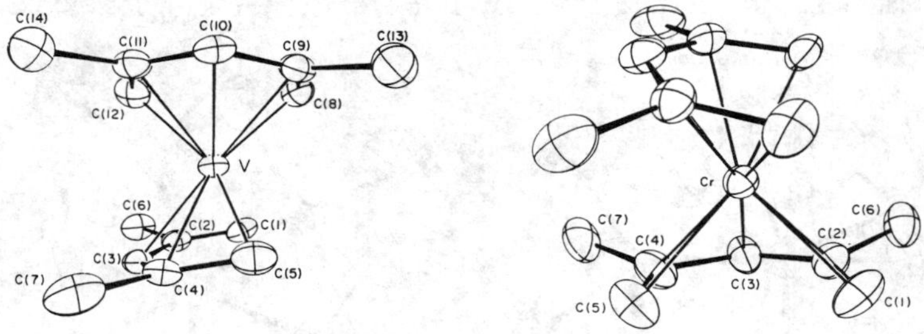

Fig. 1. The $M(\eta^5-C_7H_{11})_2$ molecules: M = V (left) and M = Cr (right).

(η^4-TETRAMETHYLDIVINYLDISILOXANE)BIS(TRIMETHYLPHOSPHINE)COBALT(0)
$C_{14}H_{36}CoOP_2Si_2$

H.-F. KLEIN, J. GROSS, H. WITTY and D. NEUGEBAUER, 1984. Z. Naturforsch., B39, 643-648.

Orthorhombic, Pbcm, a = 8.945, b = 17.685, c = 13,977 Å, Z = 4. Mo radiation, R = 0.045 for 1381 reflexions.

The molecule lies on a mirror plane (Fig. 1) passing through Co,P1,P2,C11,C21,O1. The Co coordination is tetrahedral with Co-P1 2.194(2), Co-P2 2.294(2), Co-C1 2.039(4), Co-C2 2.071(5), C2-Si 1.833(4), Si-O 1.649(2) Å, P-C-P 101.4(1), Si-O-Si 128.6(2)°.

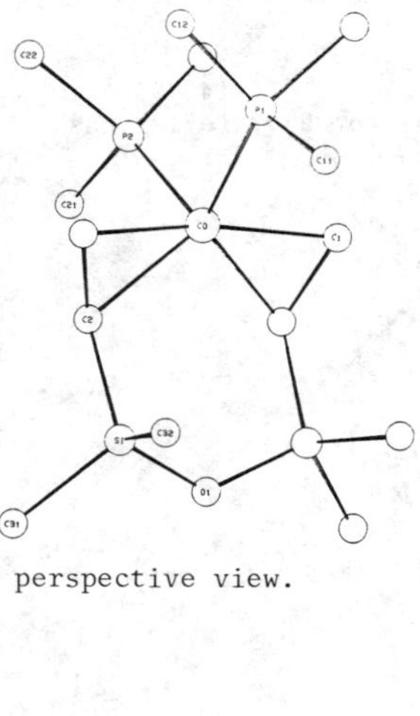

Fig. 1. $C_{14}H_{36}CoOP_2Si_2$: a perspective view.

(η^2-ETHENE)TETRAKIS(TRIMETHYLPHOSPHINE)RUTHENIUM(II)
$C_{14}H_{40}P_4Ru$ (I) $Ru(C_2H_4)(PMe_3)_4$

cis-HYDRIDOETHYLTETRAKIS(TRIMETHYLPHOSPHINE)RUTHENIUM(II)
$C_{14}H_{42}P_4Ru$ (II) $RuH(Et)(PMe_3)_4$

WAI-KWOK WONG, K.W. CHIU, J.A. STATLER, G. WILKINSON, M. MOTEVALLI and M.B. HURSTHOUSE, 1984. Polyhedron, 3, 1255-1265.

I. Monoclinic, $P2_1/n$, a = 13.744, b = 18.174, c = 9.379 Å, β = 108.03°, Z = 4. Mo radiation, R = 0.0336 for 3035 reflexions.

II. Triclinic, $P\bar{1}$, a = 9.237, b = 9.562, c = 15.133 Å, α = 103.50, β = 90.68, γ = 118.21°, Z = 2. Mo radiation, R = 0.0287 for 3300 reflexions.

The C_2H_4 group in I (Fig. 1) is coordinated in an almost symmetrical fashion (Ru-C 2.16. 2.18(1) Å); the ethylene C-C bond is long (1.439(9) Å). The P-Ru distances are in the range 2.297(3)-2.327(3) Å. In II, there is distorted octahedral geometry at Ru, with Ru-P 2.312-2.326(3) Å, Ru-C 2.214(6) and Ru-H 1.55(4) Å.

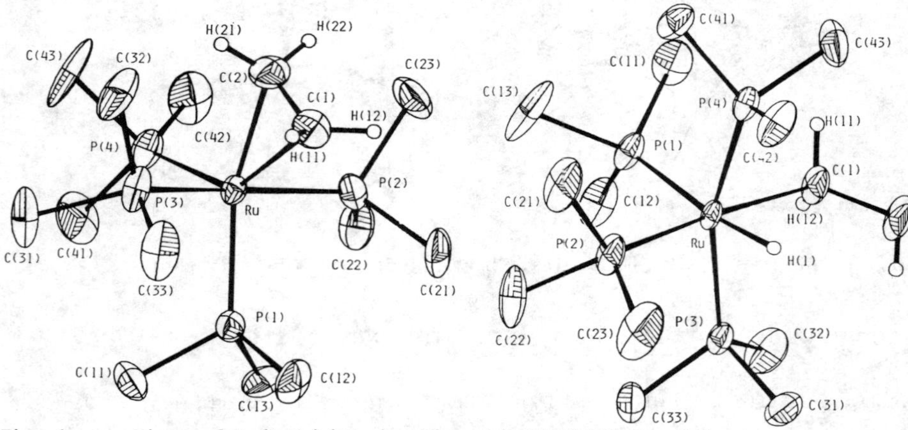

Fig. 1. Views of $Ru(C_2H_4)(PMe_3)_4$ (I) and $RuH(Et)(PMe_3)_4$ (II).

(μ-HYDRIDO)(μ-OXO)(μ_3-η^2-ACETYLENE)(η^5-CYCLOPENTADIENYL)OCTACARBONYLTUNGSTENTRIOSMIUM
$C_{15}H_8O_9Os_3W$ $(\eta^5-C_5H_5)WOs_3(CO)_8(\mu-O)(\mu_3-\eta^2-C_2H_2)(\mu-H)$

M.R. CHURCHILL, C. BUENO, J.T. PARK and J.R. SHAPLEY, 1984. Inorg. Chem., 23, 1017-1021.

Triclinic, $P\bar{1}$, a = 7.526, b = 9.372, c = 14.995 Å, α = 104.17, β = 92.06, γ = 109.74°, Z = 2. Mo radiation, R = 0.034 for 2507 reflexions.

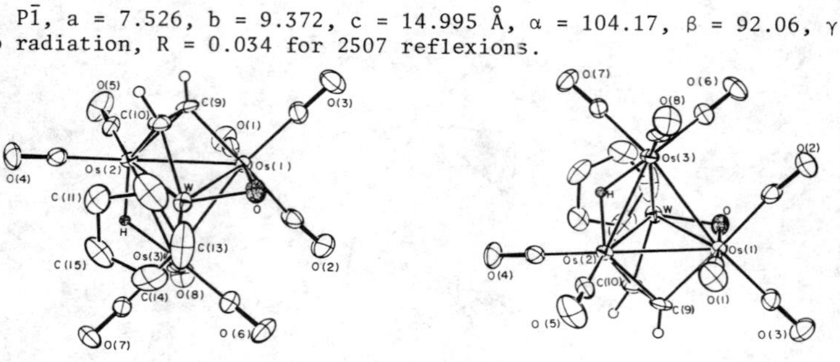

Fig. 1. Two views of $(\eta^5-C_5H_5)WOs_3(CO)_8(\mu-O)(\mu_3-\eta^2-C_2H_2)(\mu-H)$.

The molecule (Fig. 1) contains a tetrahedral WOs_3 core and possesses the following structural features. (1) The μ_3-η^2-C_2H_2 ligand bridges the W-Os(1)-Os(2) triangular face (Os(1)-C(9) 2.160(14), Os(2)-C(9) 2.095(15), Os(2)-C(10) 2.191(14), W-C(10) 2.050(13) Å). (2) The μ-O ligand bridges a W-Os edge, forming a triangular system in which W=O 1.761(8), Os(1)-O 2.200(8), W-Os(1) 2.885(1) Å and W=O→Os 92.8(4)°. (3) The μ-H ligand spans the Os(2)-Os(3) bond, Os-H 1.57(18) and 1.53(18) Å. Other distances: W-Os(2) 2.779(1), W-Os(3) 2.742(1), Os(1)-Os(2) 2.844(1), Os(1)-Os(3) 2.942(1), and Os(2)-Os(3) 2.726(1) Å.

cis-DICHLORO[(S)-α-METHYLBENZYLAMINE][(-)-2-VINYLTETRAHYDROPYRAN]PLATINUM(II)
$C_{15}H_{23}Cl_2NOPt$

R. LAZZARONI, G. UCCELLO-BARRETTA, S. BERTOZZI, C. BERTUCCI and F. MARCHETTI, 1984. J. Chem. Res., (S) 286-287, (M) 2570-2596.

Orthorhombic, $P2_12_12_1$, a = 19.390, b = 9.812, c = 9.371 Å, Z = 4. Mo radiation, R = 0.046 for 1159 reflexions.

The structure and absolute configuration are shown in Fig. 1. The Pt-coordination is undistorted square-planar. The coordination plane and the plane of Pt,C(9),C(10) are almost perpendicular (86°). The 2-vinyltetrahydropyran ring has a chair form with the vinylic group equatorial and torsion angle C9-C10-C11-O 11.6°. There is also an intramolecular N-H...O hydrogen bond (Fig. 1) (N...O 2.74 Å) and an intermolecular N-H...Cl hydrogen bond (N...Cl 3.39 Å).

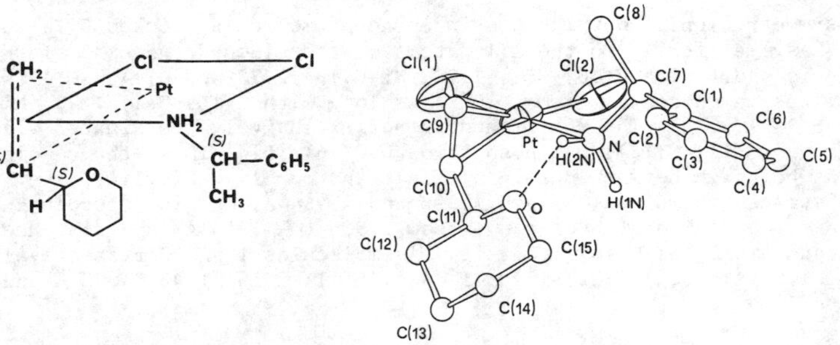

Fig. 1. The $C_{15}H_{23}Cl_2NOPt$ structure.

(η^5-PENTAMETHYLCYCLOPENTADIENYL)(η^3-1,3-DIMETHYLALLYL)CHLORORHODIUM
$C_{15}H_{24}ClRh$ (η^5-$C_5(CH_3)_5$)(η^3-$CH_3HC\underline{...}CH\underline{...}CHMe$)RhCl

A.S. BATSANOV and Yu.T. STRUCHKOV, 1984. J. Organometal. Chem., 265, 305-310.

Monoclinic, $P2_1/c$, a = 7.454, b = 14.655, c = 13.791 Å, β = 95.21°, Z = 4. Mo radiation, R = 0.041 for 1790 reflexions (-120°C).

The molecule has non-crystallographic m symmetry (Fig. 1). The cyclopentadienyl ring is bent by 6.8°, acquiring an envelope-like conformation (a C(3) envelope). The distances Rh-C(3) 2.151 Å and Rh-C(1), C(2), C(4), C(5) 2.217, 2.235, 2.215 and 2.206 Å combined with shorter C(1)-C(2), C(4)-C(5) bonds, one intermediate C(1)-C(5) and two longer bonds C(2)-C(3), C(3)-C(4) are consistent with a bonding scheme in which C(1), C(2), C(4) and C(5) form a π-coordinated pseudo-diolefin moiety and the C(3) atom is σ-bonded to Rh. The π-allyl ligand is symmetrically-coordinated to the Rh atom with Rh-C(6), C(7) and C(8) distances of 2.199, 2.117 and 2.191 Å respectively. The π-allylic moiety is tilted by its central atom C(7) towards the cyclopentadienyl ring and forms a dihedral angle of 13.7° with the mean plane of the latter. The Rh-Cl distance is 2.418 Å.

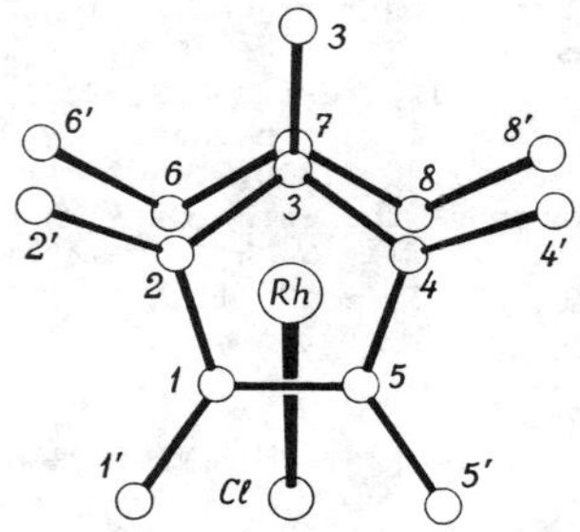

Fig. 1. Molecular structure of $C_{15}H_{24}ClRh$.

(2,2'-BIPYRIDINE)(η^3-2-METHYLPENT-2-ENAL)NICKEL HEMIHYDRATE

$C_{16}H_{19}N_2NiO,0.5H_2O$ $[(C_{10}H_8N_2)Ni(C_2H_5CH=C(CH_3)CHO],0.5H_2O$

E. DINJUS, H. GÖRLS, E. UHLIG, D. WALTHER, J. SIELER, O. LINDQVIST and L. ANDERSEN, 1984. J. Organometal. Chem., 276, 257-265.

Triclinic, P$\bar{1}$, a = 10.843, b = 9.650, c = 17.116 Å, α = 93.28, β = 101.28, γ = 120.98°, Z = 4. Mo radiation, R = 0.05 for 3654 reflexions.

The asymmetric unit contains two independent molecules linked by a water molecule which bonds asymmetrically to the O(1) atoms of the aldehyde groups. In molecule 1 O(1)...O(2) is short with C(16)-O(1) 1.28, Ni-C(16) 2.28 and O(1)...H(1)(water) 1.73 Å, whereas in molecule 2, O(1)...O(2) is long with C(16)-O(1) 1.25, Ni-C(16) 2.40 and O(1)...H(1) 1.92 Å. The MEA ligand is not exclusively coordinated via the olefinic group (C(11)=C(12)) but also through C(16) (Fig. 1) so the bonding can be described as being in between purely olefinic and purely π-allylic. The Ni-C(11) and Ni-C(12) distances in molecule 1 are 1.996 and 1.954 Å, and in molecule 2 are 1.976 and 1.945 Å. The Ni-N distances range from 1.935 to 1.943 Å, and distances within the MEA ligand are C(13)-C(14) 1.56, 1.44 Å (molecules 1 and 2 respectively); C(12)-C(13) 1.53, 1.55 Å; C(11)-C(12) 1.44, 1.44 Å; C(11)-C(16) 1.43, 1.42 Å and C(11)-C(15) 1.53, 1.52 Å.

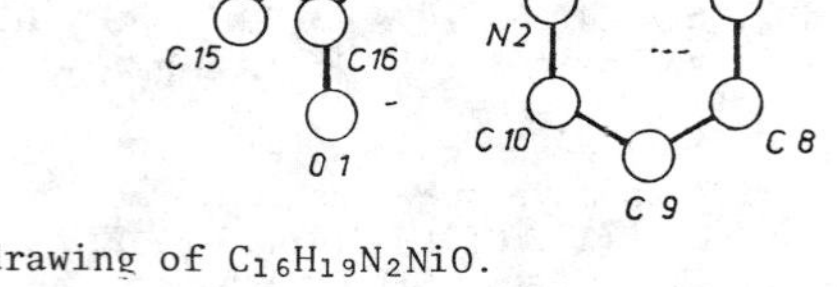

Fig. 1. Schematic drawing of $C_{16}H_{19}N_2NiO$.

DICARBONYL(η^5-PENTAMETHYLCYCLOPENTADIENYL)(η^4-TRIMETHYLENEMETHANE)MOLYBDENUM TETRA-FLUOROBORATE

$C_{16}H_{21}BF_4MoO_2$

S.R. ALLEN, S.G. BARNES, M. GREEN, G. MORAN, L. TROLLOPE, N.W. MURRALL, A.J. WELCH and D.M. SHARAIHA, 1984. J. Chem. Soc. Dalton, 1157-1169.

Orthorhombic, Pbca, a = 12.822, b = 12.311, c = 22.660 Å, D_m = 1.5, Z = 8. Mo radiation, R = 0.067 for 2747 reflexions.

The cation (Fig. 1) has effective C_s symmetry about the plane defined by the Mo, C(11), C(44), C(3) and C(8) atoms. The η^4-trimethylenemethane moiety is pyramidal with the CH_2 groups bent towards the Mo atom by an average of 12.4°. This group is oriented syn with respect to the OC-Mo-CO angle. There is intramolecular congestion involving the C_5Me_5 and $C(CH_2)_3$ groups that may contribute towards the observed asymmetric bonding of the former group and the Mo atom. Distances to Mo are: Mo-C(cyclopentadienyl) 2.296-2.374(9), Mo-C(11) 2.213(10), Mo-C(22) 2.324(9), Mo-C(33) 2.358(9), Mo-C(44) 2.392(11), Mo-C(102) 2.022(10) and Mo-C(102) 2.032(10) Å. The C(11)-C(22,33,44) distances are 1.46(2), 1.39(1), 1.40(2) Å respectively. The BF_4^- anion is disordered.

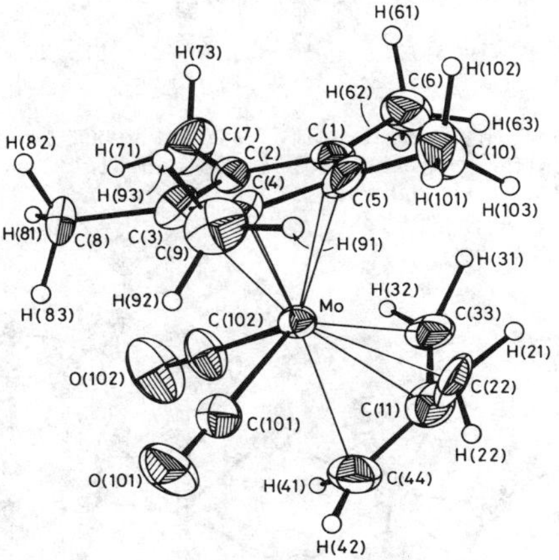

Fig. 1. $C_{16}H_{21}BF_4MoO_2$: the structure of the cation.

syn,syn-BIS(η^3-PENTADIENYL)BIS(TRIMETHYLPHOSPHINE)IRON
$C_{16}H_{32}FeP_2$ $(C_5H_7)_2Fe(P(CH_3)_3)_2$

J.R. BLEEKE and M.K. HAYS, 1984. Organometallics, 3, 506-508.

Monoclinic, C2/c, a = 14.746, b = 9.201, c = 14.021 Å, β = 92.66°, Z = 4. Mo radiation, R = 0.031 for 1501 reflexions.

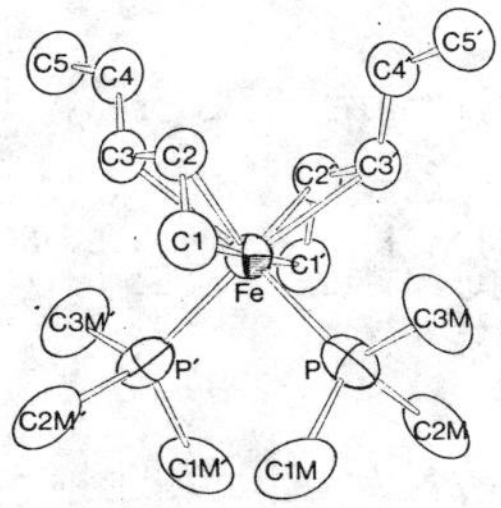

Fig. 1. Molecular structure of $C_{16}H_{32}FeP_2$.

The molecule resides on a twofold rotation axis and the molecular geometry is pseudooctahedral (Fig. 1). The six coordination sites around the Fe atom are occupied by the two C(1) atoms and two C(3) atoms of the pentadienyl ligands and the two P atoms of the mutually cis $P(CH_3)_3$ ligands. As a result of the trans influence exerted by the P ligands, Fe-C(3) is significantly longer (2.226 Å) than Fe-C(1) (2.115 Å).

Fe-C(2) at 2.034 Å is the shortest of the Fe-allylic C atom bonds. The η^3-ligands are W-shaped and C(1), C(2), C(3) and C(4) are coplanar while C(5) lies 0.26 Å out of the plane away from the Fe atom. Fe-P is 2.211 Å and the bond distances within the allylic ligand are: C(1)-C(2) 1.397, C(2)-C(3) 1.406, C(3)-C(4) 1.457 and C(4)-C(5) 1.302 Å. Angles C(1)-C(2)-C(3) and C(2)-C(3)-C(4) are 120.5 and 121.7° respectively whilst C(3)-C(4)-C(5) is 127.3°.

BIS[(μ_2-CHLORO)TRICHLORO(η^2-BIS(TRIMETHYLSILYL)ACETYLENE)TUNGSTEN]
$C_{16}H_{36}Cl_8Si_4W_2$

E. HEY, F. WELLER and K. DEHNICKE, 1984. Z. Anorg. Chem., 514, 18-24.

Triclinic, $P\bar{1}$, a = 9.282, b = 9.385, c = 10.805 Å, α = 115.29, β = 91.93, γ = 100.00°, Z = 1. Mo radiation, R = 0.046 for 2231 reflexions.

The structure is a centrosymmetric dimer linked by Cl bridges with W-Cl 2.444(3) and 2.724(3) Å. The tungsten atom achieves seven-coordination with three terminal Cl ligands (W-Cl 2.307-2.331(3) Å) and the side-on acetylene ligand (W-C 2.025, 2.028(11) Å).

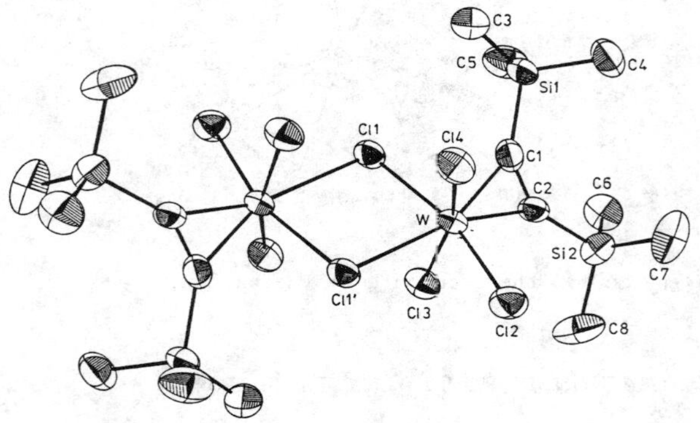

Fig. 1. A view of $C_{16}H_{36}Cl_8Si_4W_2$.

(η^2-BUT-2-YNE)(η^5-CYCLOPENTADIENYL)(p-NITROPHENYLTHIO)(TRIMETHYL PHOSPHITE)MOLYBDENUM
$C_{18}H_{24}MoNO_5PS$ (I)

(η^2-BUT-2-YNE)CARBONYL(η^5-CYCLOPENTADIENYL)[o-(PHENYLTHIO)PHENYLTHIO-S]MOLYBDENUM
$C_{22}H_{20}MoOS_2$ (II)

S.R. ALLEN, T.H. GLAUERT, M. GREEN, K.A. MEAD, N.C. NORMAN, A.G. ORPEN, C.J. SCHAVERIEN and P. WOODWARD, 1984. J. Chem. Soc. Dalton, 2747-2756.

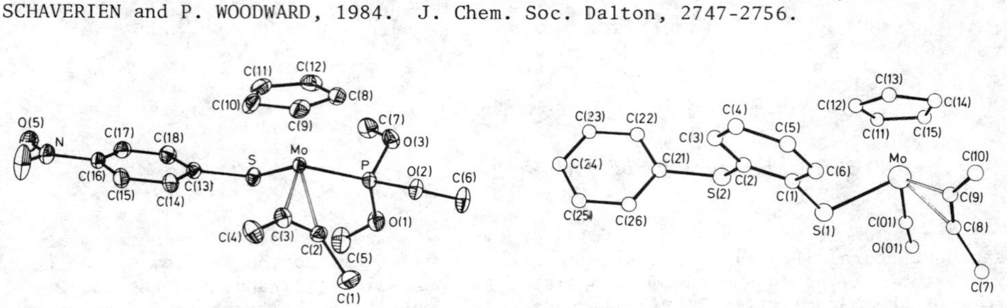

Fig. 1. Views of $C_{18}H_{24}MoNO_5PS$ (I) (left) and $C_{22}H_{20}MoOS_2$ (II) (right).

I. Monoclinic, $P2_1/n$, a = 8.968, b = 20.104, c = 11.452 Å, β = 95.77°, Z = 4. Mo radiation, R = 0.028 for 3601 reflexions (at 200 K).

II. Orthorhombic, $P2_12_12_1$, a = 9.852, b = 11.486, c = 18.476 Å, Z = 4. Mo radiation, R = 0.059 for 1714 reflexions.

Molecule (I) (Fig. 1) adopts a three-legged piano stool geometry, in which the but-2-yne ligand is oriented such that the central C-C bond lies approximately parallel to the Mo-P vector. Main dimensions are: Mo-P 2.332(1), Mo-C(2) 2.057(3), Mo-C(3) 2.020(3), Mo-S 2.406(1), C(2)-C(3) 1.305(4) Å, C(1)-C(2)-C(3) 137.2(3), C(2)-C(3)-C(4) 140.4(3)°. Molecule (II) (Fig. 1) also has piano stool geometry but with the but-2-yne ligand lying parallel to the Mo-CO vector, with Mo-S(1) 2.362(2), Mo-C(8) 2.088(8), Mo-C(9) 2.033(7), C(8)-C(9) 1.271(12), Mo-C(01) 1.929(8) Å, C(7)-C(8)-C(9) 142.0(8), C(8)-C(9)-C(10) 144.5(8)°.

(η^3-ALLYL)(METHYLMENTHYL-t-BUTYLPHOSPHINE)CHLORONICKEL(II)
$C_{18}H_{36}ClNiP$

H. BRANDES, R. GODDARD, P.W. JOLLY, C. KRUGER, R. MYNOTT and G. WILKE, 1984. Z. Naturforsch., B39, 1139-1150.

Tetragonal, $P4_3$, a = 13.835, c = 10.909 Å, Z = 4. Cu radiation, R = 0.0327 for 2191 reflexions.

The absolute configuration was determined (Fig. 1) and the Ni atom has slightly distorted square-planar coordination; the meso-carbon atom C21/C22 is disordered above and below the coordination plane. Main dimensions are: Ni-Cl 2.194(1), Ni-P 2.211(1), Ni-C1 1.992(6), Ni-C21 2.008(8), Ni-C22 2.002(16), Ni-C3 2.076(5) Å, P-Ni-Cl 100.1(1), P-Ni-C1 93.5(2), Cl-Ni-C3 94.1(2), C1-Ni-C3 72.2(2)°. The menthyl group bonded to P apparently blocks a coordination site at the Ni atom with its isopropyl group.

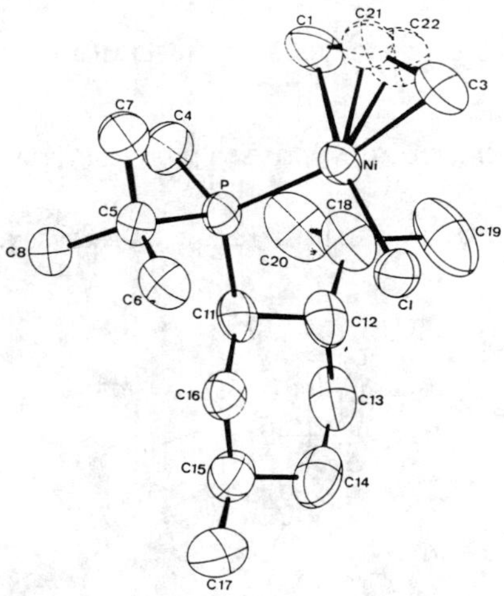

Fig. 1. The $C_{18}H_{36}ClNiP$ structure.

[(1,2,3,4,5-η)-2,4-DIMETHYL-2,4-PENTADIENYL]BIS(TRIETHYLPHOSPHINE)COBALT
$C_{19}H_{41}CoP_2$ (I) $(C_7H_{11})Co[P(C_2H_5)_3]_2$

[(1,2,3-η)-2,4-DIMETHYL-2,4-PENTADIENYL]TRIS(TRIMETHYLPHOSPHINE)COBALT
$C_{16}H_{38}CoP_3$ (II) $(C_7H_{11})Co[P(CH_3)_3]_3$

J.R. BLEEKE and W.-J. PENG, 1984. Organometallics, 3, 1422-1426.

I. Monoclinic, $P2_1/c$, a = 7.928, b = 18.041, c = 15.592 Å, β = 92.81°, Z = 4. Mo radiation, R = 0.038 for 3443 reflexions.

II. Monoclinic, $P2_1/c$, a = 8.927, b = 15.008, c = 16.694 Å, β = 104.33°, Z = 4. Mo radiation, R = 0.058 for 3144 reflexions.

The five coordinated C atoms of the U-shaped 2,4-dimethylpentadienyl ligand are essentially coplanar in I (Fig. 1) and the distance from the Co atom to the five C atoms ranges from 2.060 to 2.145 Å. The methyl C atoms are displaced out of the plane of the five C atoms toward the Co atom by 0.141 Å (C(6)) and 0.240 Å (C(7)). The angles C(1)-C(2)-C(3), C(2)-C(3)-C(4) and C(3)-C(4)-C(5) are respectively 122.9, 128.2, and 121.0° and the Co-P distances are 2.149 Å (P(1)) and 2.180 Å (P(2)). In II, the geometry about the Co atom approximates a square pyramid with two of the allylic C atoms (C(1) and C(3)) occupying adjacent basal sites, two of the $P(CH_3)_3$ P atoms (P(1) and P(2)) occupying the remaining basal sites and the P(3) atom residing in the axial site. The Co-P(3) distance is 2.197 Å, longer than the Co-P(1) and Co-P(2) distances 2.173 and 2.179 Å, and the four basal atoms (C(1), C(3), P(1) and P(2)) are coplanar to within 0.036 Å. The Co-C(1), Co-C(2) and Co-C(3) distances are respectively 2.053, 1.990 and 2.112 Å. The vinylic moiety (C(4), C(7) and C(5)) is bent out of the plane of the allylic moiety (C(1), C(2), C(6) and C(3)) away from the Co atom.

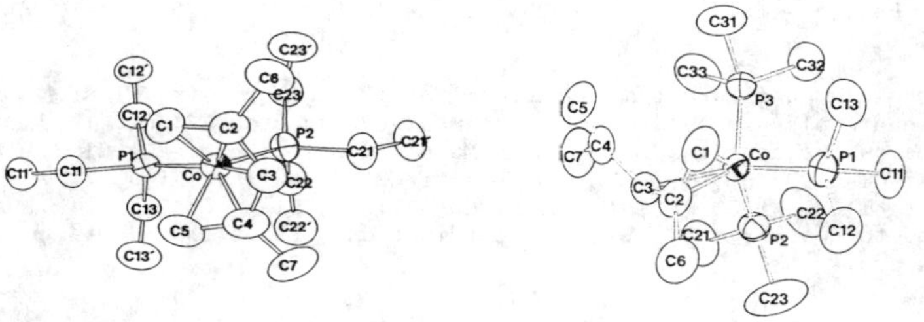

Fig. 1. Molecular structure of $C_{19}H_{41}CoP_2$ (left) and $C_{16}H_{38}CoP_3$ (right).

[σ^1,η^3-BUTADIENE(PENTACARBONYLCHROMIUM)CARBONYL]BIS(η^5-CYCLOPENTADIENYL)ZIRCONIUM
$C_{20}H_{16}CrO_6Zr$

G. ERKER, U. DORF, R. BENN, R.-D. REINHARDT and J.L. PETERSEN. 1984. J. Am. Chem. Soc., 106, 7649-7650.

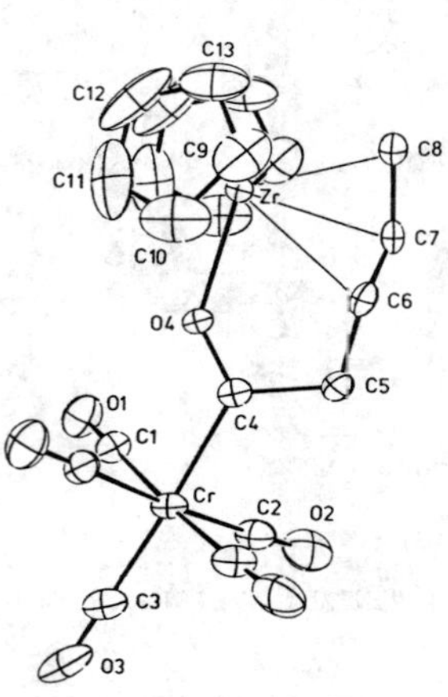

Fig. 1. A view of $C_{20}H_{16}CrO_6Zr$.

Orthorhombic, Pnam, a = 22.398, b = 7.283, c = 12.302 Å, Z = 4. R = 0.036 for 1857 reflexions.

The molecule (Fig. 1) lies on a mirror plane (through Zr,Cr,C3,O3,C4,O4). The four C atoms from butadiene occupy twofold disorder positions on either side of the plane. In the metallocene unit the mean Zr-C distance is 2.486(5) Å with Cp-Zr-Cp 129.2(4)°. The Cr atom has octahedral coordination, with Cr-C(4) 2.063(4), Cr-C(1) 1.887(3), Cr-C(2) 1.882(4), Cr-C(3) 1.879(5) Å. The C(4)-O(4) bond is remarkably short (1.254(5) Å) and the Zr-O(4) bond (2.188(3) Å) is rather long. The Zr-O(4)-Cr angle is 136.1(3)°.

(η^4-1,2,5,6-TETRAMETHYL-3,4-DIMETHYLENETRICYCLO[3.1.0.0^{2,6}]HEXANE)ZIRCONOCENE
$C_{22}H_{26}Zr$ (I) $(\eta^5-C_5H_5)_2Zr(C_{12}H_{16})$

(η^4-2,3-DIMETHYLENEBICYCLO[2.2.1]HEPTANE)ZIRCONOCENE
$C_{19}H_{22}Zr$ (II) $(\eta^5-C_5H_5)Zr(C_9H_{12})$

G. ERKER, K. ENGEL, C. KRÜGER and G. MÜLLER, 1984. Organometallics, 3, 128-133.

I. Triclinic, P$\bar{1}$, a = 8.578, b = 14.466, c = 15.717 Å, α = 75.00, β = 88.566, γ = 89.12°, Z = 4. Mo radiation, R = 0.035 for 7110 reflexions.

II. Triclinic, P$\bar{1}$, a = 7.4540, b = 8.8377, c = 11.8934 Å, α = 84.704, β = 88.922, γ = 79.765°, Z = 2. Mo radiation, R = 0.021 for 3165 reflexions.

In I, the two independent molecules have molecular geometries which are identical within experimental error. The framework of the diene ligand (C(1), C(2), C(3), C(4)) is planar and the metal-carbon distances are: Zr-C(1) 2.311, Zr-C(2) 2.542, Zr-C(3) 2.558 and Zr-C(4) 2.323 Å (Fig. 1). The C-C distances are: C(1)-C(2) 1.444, C(2)-C(3) 1.398, C(3)-C(4) 1.447 Å. There is a striking similarity in the essential structural features of (diene)zirconocene units in I and II. Again, (in II) rather long C(1)-C(2) (1.456 Å) and C(3)-C(4) (1.451 Å) bonds are opposed by a short C(2)-C(3) bond (1.391 Å). The metal forms σ bonds to the diene termini (2.307 Å (Zr-C(1)), 2.329 Å (Zr-C(4))) and exhibits substantial π-interaction with the internal C atoms of the diene unit (2.569 Å (Zr-C(2) , 2.578 Å (Zr-C(3))).

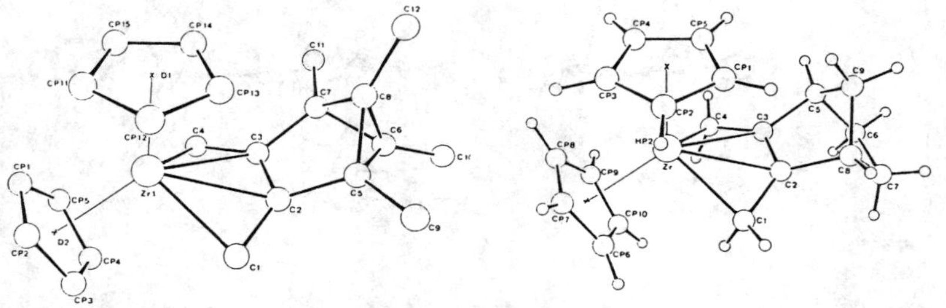

Fig. 1. Molecular structures of $C_{22}H_{26}Zr$ (left), and $C_{19}H_{22}Zr$ (right).

HEXACARBONYL-μ-(1-η^2-2-DIPHENYL-3-ETHOXY-η^3-η^1-ALLYL)DIIRON
$C_{23}H_{16}Fe_2O_7$ $Fe_2(CO)_6(\mu-C(OC_2H_5)C(C_6H_5)C(C_6H_5)H)$

J. ROS, R. MATHIEU, X. SOLANS and M. FONT-ALTABA, 1984. J. Organometal. Chem., 260, C40-C42.

Monoclinic, P2$_1$/n, a = 17.801, b = 10.006, c = 12.971 Å, β = 106.03°, Z = 4. Mo radiation, R_W = 0.054 for 2288 reflexions.

The molecule (Fig. 1) contains a Fe-Fe single bond (2.635 Å), and Fe(1) interacts with a π-allyl group while Fe(2) interacts with two C atoms of one of the phenyl groups. Important distances are: Fe(1)-C(6) 2.069, Fe(1)-C(5) 2.085, Fe(1)-C(4) 2.129, Fe(2)-C(4) 1.970, Fe(2)-C(61) 2.431 and Fe(2)-C(62) 2.433 Å. As a result of the Fe(2)-ring interaction, the C-C bonds within the ring show evidence of localization: C(61)-C(62) 1.393, C(62)-C(63) 1.443, C(63)-C(64) 1.370, C(64)-C(65) 1.405, C(65)-C(66) 1.381 and C(61)-C(6) 1.421 Å.

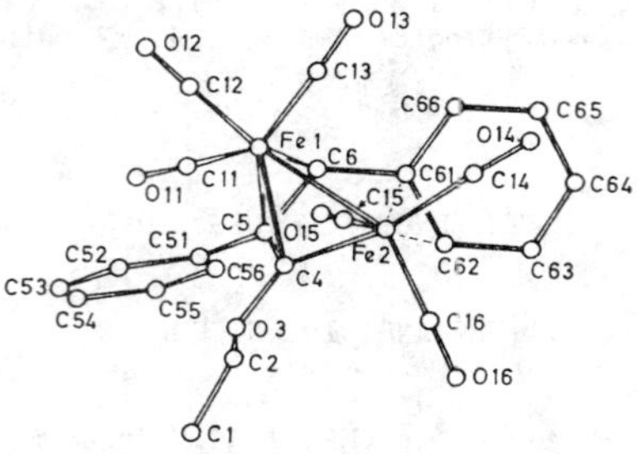

Fig. 1. Molecular structure of $C_{23}H_{16}Fe_2O_7$.

(S,S)-(η^3-ALLYL)CARBONYLNITROSYL(1-PHENYLETHYLAMINODIPHENYLPHOSPHINO)IRON
$C_{24}H_{25}FeN_2O_2P$ (I)

(RS)-(η^3-ALLYL)CARBONYLNITROSYL(1-PHENYLETHYLMETHYLAMINODIPHENYLPHOSPHINO)IRON
$C_{25}H_{27}FeN_2O_2P$ (II)

H. BRUNNER, H. WEBER, I. BERNAL and G.M. REISNER, 1984. Organometallics, 3, 163-170.

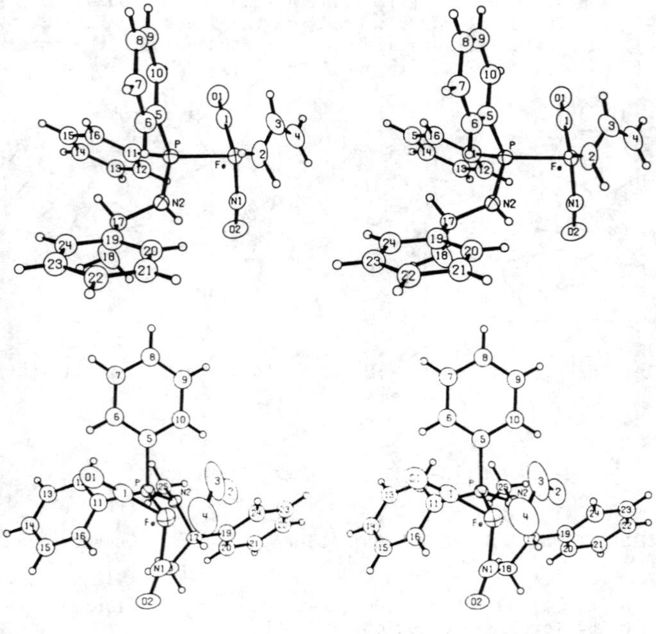

Fig. 1. Stereoviews of $C_{24}H_{25}FeN_2O_2P$ (top), and $C_{25}H_{27}FeN_2O_2P$ (bottom).

I. Tetragonal, $P4_1$, a = 9.605, c = 25.161 Å, Z = 4. Mo radiation, R = 0.035 for 1437 reflexions.

II. Orthorhombic, $P2_12_12_1$, a = 12.191, b = 13.747, c = 13.906 Å, Z = 4. Mo radiation, R = 0.055 for 2081 reflexions.

The coordination polyhedron for both compounds may be regarded as trigonal bipyramidal (Fig. 1) if allyl bonding to Fe is assumed via the two terminal C atoms and ignoring the central C atom. The trigonal plane is defined by CO, NO and C(2) (one of the allyl C atoms), and the axial positions are occupied by C(4) and the P atom. The π-allyl ligand is coordinated in η^3 fashion to the Fe atom. The central C atom (C(3)) in I and II is closer to the metal (2.084 and 2.112 Å) than the terminal atoms (2.142, 2.117 and 2.202, 2.151 Å). The Fe-N distances are 1.646 and 1.677 Å respectively in I and II and the Fe-C(carbonyl) 1.753 and 1.713 Å. The Fe-P distances in I and II are 2.222 and 2.245 Å respectively. The P-N(2) distances are identical at 1.671 Å.

(μ_2-CARBONYL)(μ_2-HYDRIDO)(μ_2-σ^1,η^2-2-PHENYLVINYL)(CARBONYL(η^5-CYCLOPENTADIENYL)-MANGANESE)DECACARBONYLTRIOSMIUM 1,4-DIOXANE SOLVATE

$C_{25}H_{13}MnO_{12}Os_3$,$1.5C_4H_8O$ $(\pi C_5H_5)MnOs_3(\mu\text{-}CH{=}CHC_6H_5)(\mu\text{-}H)(\mu\text{-}CO)(CO)_{11}$,$1.5C_4H_8O$

A.B. ANTONOVA, S.V. KOVALENKO, E.D. KORNIYETS, A.A. JOHANSSON, Yu.T. STRUCHKOV and A.I. YANOVSKY, 1984. J. Organometal. Chem., 267, 299-307.

Monoclinic, $P2_1/c$, a = 8.998, b = 12.491, c = 32.088 Å, β = 96.55°, Z = 4. Mo radiation, R = 0.046 for 2348 reflexions.

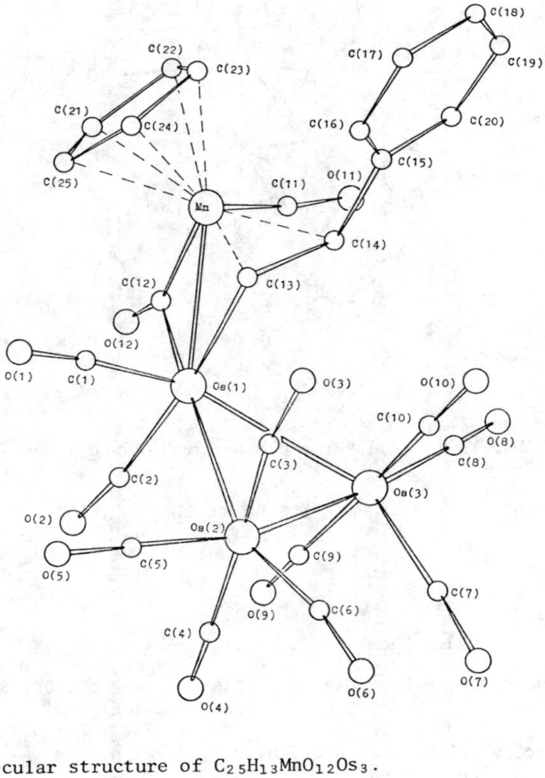

Fig. 1. Molecular structure of $C_{25}H_{13}MnO_{12}Os_3$.

The metal framework consists of an Os_3 triangle with the Mn atom linked to the Os(1) atom (Fig. 1). Two Os-Os edges, Os(1)-Os(2) 2.851 and Os(2)-Os(3) 2.889 Å are almost equal but the third, Os(1)-Os(3) is 3.069 Å implying the presence of a bridging hydride atom. The coordination polyhedron of the Os(1) atom can be regarded as a distorted octahedron if Mn, C(12) and C(13) are assumed to occupy two coordination sites. The Os(1)-Mn distance is 2.765 Å and is shortened by the presence of the bridging ligands. The Mn atom is π-bonded to the cyclopentadienyl ring and to the C(13)=C(14) double bond of the phenylvinyl group, with C(13)-C(14), Mn-C(13) and Mn-C(14) distances of 1.46, 2.08 and 2.28 Å respectively. The solvating dioxane molecules have chair conformations with one in a general position and one on an inversion centre.

BIS(DIMETHYLPHOSPHINO)ETHANE(ETHENE)(TRIPHENYLPHOSPHINE)NICKEL(0)
$C_{26}H_{35}NiP_3$

K.R. PORSCHKE, R. MYNOTT, C. KRUGER and M.J. ROMAC, 1984. Z. Naturforsch., B39, 1076-1081.

Triclinic, P$\bar{1}$, a = 9.942, b = 11.028, c = 12.327 Å, α = 89.503, β = 76.067, γ = 87.652°, Z = 2. Cu radiation, R = 0.044 for 4701 reflexions.

The Ni atom has distorted tetrahedral coordination (Fig. 1) with Ni-P1 2.180(1), Ni-P2 2.155(1), Ni-P3 2.179(1), Ni-C1 2.050(3), Ni-C2 2.045(3), Ni-(C=C mid point) 1.938, C1-C2 1.320(6) Å, P2-Ni-P3 90.1°.

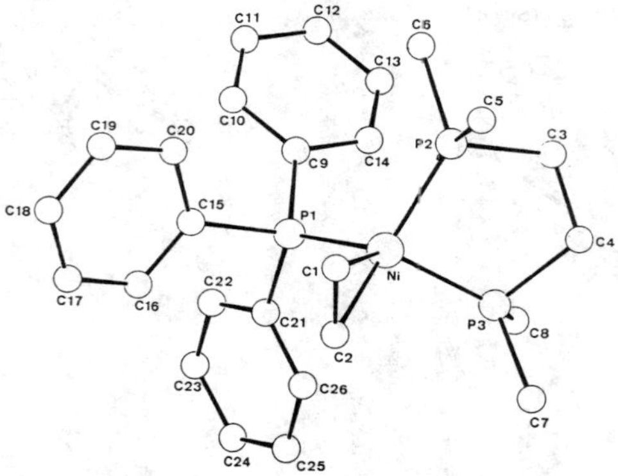

Fig. 1. A view of $C_{26}H_{35}NiP_3$.

CARBONYL(η^5-CYCLOPENTADIENYL)(η^4-2,3,4-TRIPHENYL-1,3-BUTADIEN-1-ONE)MANGANESE
$C_{28}H_{21}MnO_2$ (I)

(η^5-CYCLOPENTADIENYL)(η^4-2,3,4-TRIPHENYL-1,3-BUTADIEN-1-ONE)COBALT
$C_{27}H_{21}CoO$ (II)

2,3-DIPHENYLINDENONE
$C_{21}H_{14}O$ (III)

K.-J. JENS and E. WEISS, 1984. Chem. Ber., 117, 2469-2478.

I. Triclinic, P$\bar{1}$, a = 9.044, b = 13.642, c = 19.647 Å, α = 101.04, β = 99.09, γ = 111.46°, Z = 4. Mo radiation, R = 0.059 for 3736 reflexions.

II. Monoclinic, $P2_1/c$, a = 13.397, b = 8.766, c = 17.124 Å, β = 95.77°, Z = 4. Mo radiation, R = 0.055 for 2248 reflexions.

III. Monoclinic, $P2_1/n$, a = 9.456, b = 17.152, c = 9.874 Å, β = 111.0°, Z = 4. Mo radiation, R = 0.077 for 1818 reflexions.

There are two independent molecules of I in the asymmetric unit. Molecules I and II are very similar (Fig. 1) with Mn-C(1-4) 1.881(8)-2.217(8), Mn-C(cp) 2.116(11)-2.158(8) Å; the corresponding distances for II are 1.835(6)-2.090(6) and 2.045(8)-2.114(8) Å. The Mn-C(O) distances in I are 1.765(9) and 1.778(7) Å. The metal-ligand binding is asymmetric. Dimensions in III are as anticipated; the phenyl rings are rotated 38 and 129° out of the indenone plane.

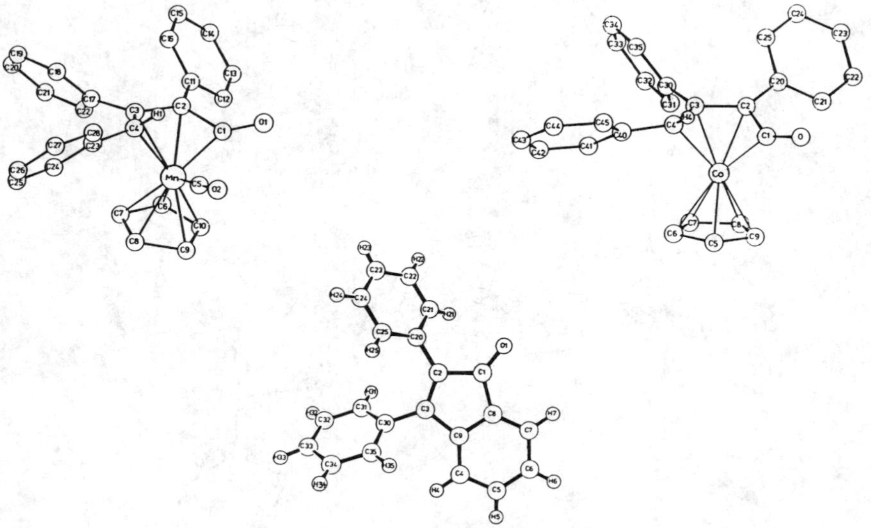

Fig. 1. Views of $C_{28}H_{21}MnO_2$ and $C_{27}H_{21}CoO$ (top) and of $C_{21}H_{14}O_1$ (bottom).

1-ACETYL-3-[4-PHENYLBUTADIENYLIRONTRICARBONYL]-5-FERROCENYLPYRAZOLINE (DIASTEREOMER A)
$C_{28}H_{24}FeN_2O_4$ (I)

1-ACETYL-3-[4-PHENYLBUTADIENYLIRONTRICARBONYL]-5-FERROCENYLPYRAZOLINE (DIASTEREOMER B)
$C_{28}H_{24}FeN_2O_4$ (II)

V.N. POSTNOV, V.G. ANDRIANOV, Yu.T. STRUCHKOV, A.M. BARAN and V.A. SAZONOVA, 1984.
J. Organometal. Chem., 262, 201-213.

I. Triclinic, $P\bar{1}$, a = 7.948, b = 12.960, c = 13.126 Å, α = 103.19, β = 103.62, γ = 101.95°, Z = 2. Mo radiation, R = 0.037 for 4065 reflexions.

II. Triclinic, $P\bar{1}$, a = 9.888, b = 10.850, c = 13.437 Å, α = 98.02, β = 105.25, γ = 111.38°, Z = 2. Mo radiation, R = 0.040 for 2594 reflexions.

Molecules I and II are geometric isomers (Fig. 1) with different orientations of the phenylbutadienyliron tricarbonyl substituents. The torsion angles C(18)C(17)-C(16)N(2) and C(18)C(17)C(16)C(15) in I are -158.3 and 18.0° respectively, and in II are -1.5 and 179.5° respectively. Thus in I we have a trans configuration of the double bonds C(17)=C(18) and C(16)=N(2), and in II a cis configuration.

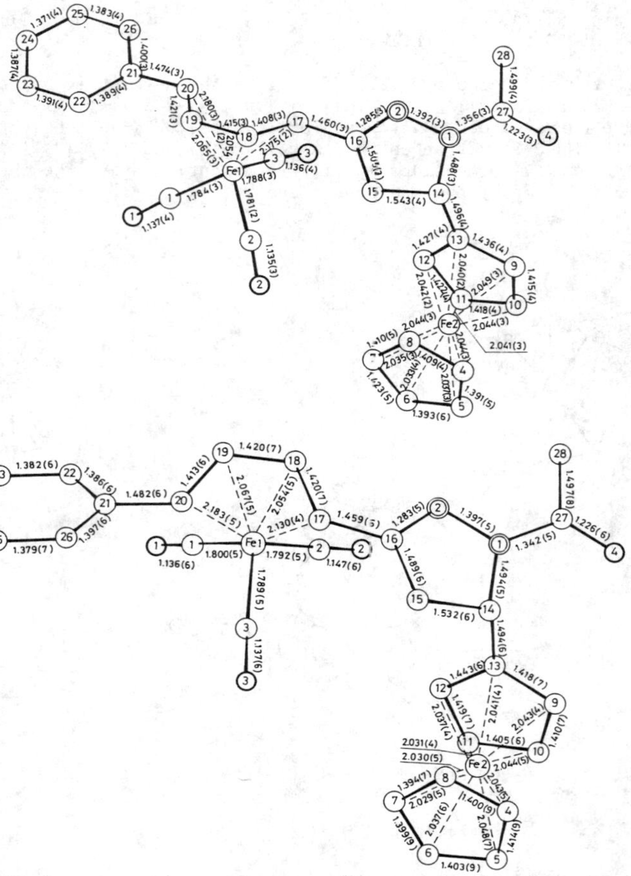

Fig. 1. Molecular structure of $C_{28}H_{24}FeN_2O_4$ (I, top) and II (bottom) showing bond lengths (Å).

[μ-[(2,3,4,5-η:6,7,8,9-η)-2,9-DIMETHYL-2,4,6,8-DECATETRAENE]]-BIS[(1,2,3,4,5-η)-2,4-DIMETHYL-2,4-PENTADIENYL]DICOBALT

$C_{28}H_{44}Co_2$ $[Co(C_7H_{11})_2]_2$

D.R. WILSON, R.D. ERNST and M.S. KRALIK, 1984. Organometallics, 3, 1442-1444.

Monoclinic, $P2_1/c$, a = 8.468, b = 21.71, c = 7.433 Å, β = 111.73°, Z = 2. R = 0.048 for 1154 reflexions.

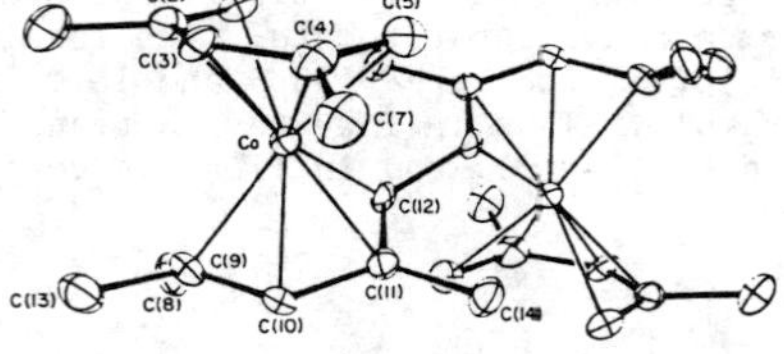

Fig. 1. Perspective view of the dimer $C_{28}H_{44}Co_2$.

The structure analysis confirms the dimeric nature of the (centrosymmetric) molecule (Fig. 1). The dimer has each Co atom coordinated by an η^5-2,4-dimethylpentadienyl ligand and a butadiene portion of half of the 2,4-dimethylpentadienyl dimer in η^4 fashion. The metal-ligand distances are: Co-C(1) 2.14, Co-C(2) 2.08, Co-C(3) 2.086, Co-C(4) 2.066, Co-C(5) 2.123, Co-C(9) 2.134, Co-C(10) 2.017, Co-C(11) 2.014 and Co-C(12) 2.087 Å. Selected C-C distances are: C(12)-C(12') 1.46, C(1)-C(2) 1.46, C(2)-C(3) 1.44, C(3)-C(4) 1.40, C(4)-C(5) 1.40, C(8)-C(9) 1.51, C(9)-C(10) 1.40, C(10)-C(11) 1.42 and C(11)-C(12) 1.43 Å. Bond angles of interest are: C(1)-C(2)-C(3) 123.8, C(2)-C(3)-C(4) 126.0, C(3)-C(4)-C(5) 122.5, C(8)-C(9)-C(10) 123.9, C(9)-C(10)-C(11) 123.5, C(10)-C(11)-C(12) 118.6 and C(11)-C(12)-C(12') 123.8°.

(η^2-BUT-2-YNE)(η^5-CYCLOPENTADIENYL)(α-β-η-o-DIPHENYLPHOSPHINOSTYRENE-P)MOLYBDENUM(II) TETRAFLUOROBORATE DICHLOROMETHANE SOLVATE
$C_{29}H_{28}BF_4MoP,CH_2Cl_2$ (I)

ACETONITRILE(η^5-CYCLOPENTADIENYL)[1'—4'-η-(o-(3'-METHYLPENTA-2',4'-DIENYL)PHENYL)-DIPHENYLPHOSPHINE-P]MOLYBDENUM(II) TETRAFLUOROBORATE
$C_{31}H_{31}BF_4MoNP$ (II)

S.R. ALLEN, M. GREEN, G. MORAN, A.G. ORPEN and G.E. TAYLOR, 1984. J. Chem. Soc. Dalton, 441-449.

I. Orthorhombic, $Pna2_1$, a = 22.324, b = 9.249, c = 14.566 Å, Z = 4. Mo radiation, R = 0.040 for 2228 reflexions.

II. Monoclinic, $P2_1/c$, a = 10.836, b = 8.551, c = 30.896 Å, β = 94.80°, Z = 4. Mo radiation, R = 0.029 for 3828 reflexions.

In I the Mo atom is bonded to an η-C_5H_5 group, a η^2-but-2-yne ligand and the P atom and o-vinyl substituent of the o-diphenylphosphinostyrene ligand (Fig. 1). But-2-yne shows the usual bending back of its methyl groups in a cisoid manner. The Mo-C(alkyne) distances show slight asymmetry (Mo-C(22) 2.037(7), Mo-C(21) 1.996(8) Å). Other bond distances are: Mo-P(1) 2.442(2), Mo-C(25-29) 2.303(7)-2.416(8), Mo-C(19) 2.266(7) and Mo-C(20) 2.265(8) Å. Compound II consists of isolated BF_4^- anions and complex cations. In the cation the Mo atom is bonded to an η-C_5H_5 group, the N of acetonitrile and a ligand derived from the condensation of the but-2-yne and phosphinostyrene ligands present in I. This ligand is η^4-bound to Mo via a 1,3-diene moiety which is the ortho substituent on a phenyl ring attached to a diphenylphosphine unit P bonded to Mo. The methyl groups of but-2-yne are mutually cis and the diene has an exo conformation relative to η-C_5H_5. Some bond distances are: Mo(1)-C(2) 2.425(3), Mo(1)-C(3) 2.281(3), Mo(1)-C(5) 2.216(3), Mo(1)-C(6) 2.263(3), C(2)-C(3) 1.400(4), C(3)-C(5) 1.415(4), C(5)-C(6) 1.430(4), Mo-P(1) 2.437(1), Mo-N(1) 2.138(2), Mo-C(25-29) 2.274(4)-2.393(4) Å.

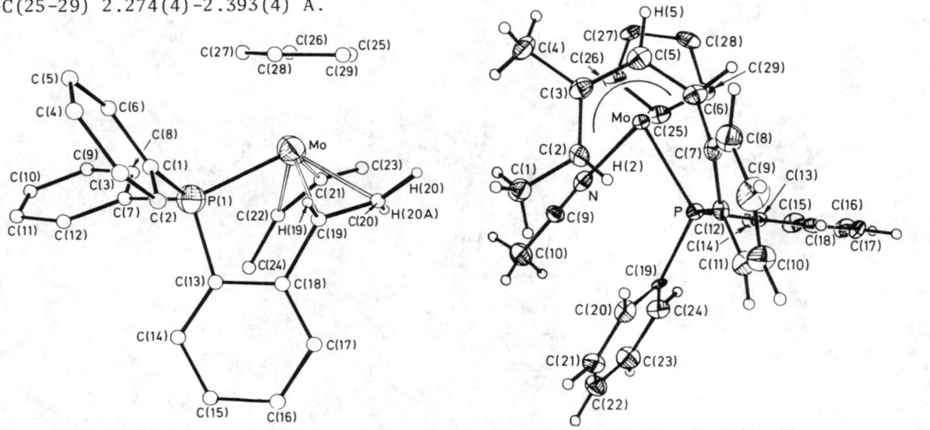

Fig. 1. The structures of the cations: $C_{29}H_{28}BF_4MoP,CH_2Cl_2$ (left) and $C_{31}H_{31}BF_4MoNP$ (right).

(η[5]-CYCLOPENTADIENYL)(TRIPHENYLGERMYL)(η[3]-HEXENYL)-NITROSYLMOLYBDENUM
$C_{29}H_{31}GeMoNO$

F. CARRÉ, E. COLOMER, R.J.P. CORRIU and A. VIOUX, 1984. Organometallics, 3, 970-977.

Monoclinic, $P2_1/n$, a = 9.257, b = 20.950, c = 13.375 Å, β = 98.03°, D_m = 1.49, Z = 4. Mo radiation, R = 0.043 for 1725 reflexions.

The cyclopentadienyl ring and the triphenylgermyl moiety show no unusual features (Fig. 1). The nitrosyl group is approximately linear with an Mo-N-O angle of 173.2° and an Mo-N distance of 1.78 Å. The Mo-Ge bond length is 2.604 Å which is considered short, and the average of the Mo-C distances for the 2-hexenyl ligand, 2.330 Å, is comparable with the average of the Mo-C(Cp) distances (2.366 Å). There is disorder in the 2-hexenyl ligand (Fig. 1) corresponding to a 57:42 ratio mixture of two isomers, one exo and one endo, in the solid state.

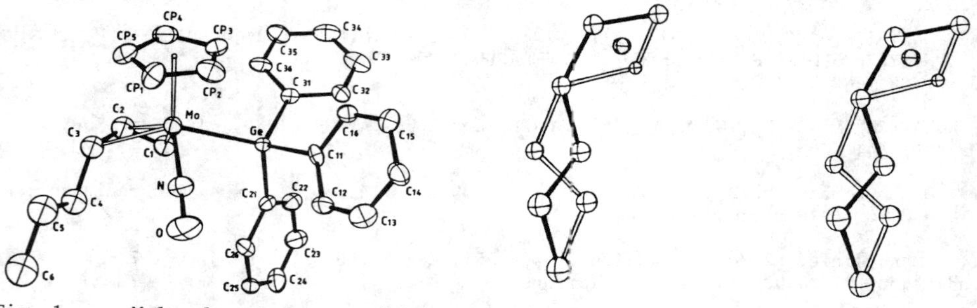

Fig. 1. Molecular structure of $C_{29}H_{31}GeMoNO$ (left), and stereoview of the two observed conformations in the 2-hexenyl ligand (right). The exo conformation is represented by the plain black bonds.

(μ-1,2-η-1,4-DIPHENYLBUTA-1,3-DIYNE)BIS((η[5]-CYCLOPENTADIENYL)DICARBONYLMOLYBDENUM)
$C_{30}H_{20}Mo_2O_4$ $[(C_5H_5)(CO)_2Mo]_2(C_6H_5C{\equiv}C{-}C{\equiv}CC_6H_5)$

N.A. USTYNYUK, V.N. VINOGRADOVA, V.N. KORNEVA, D.N. KRAVTSOV, V.G. ANDRIANOV and Yu.T. STRUCHKOV, 1984. J. Organometal. Chem., 277, 285-296.

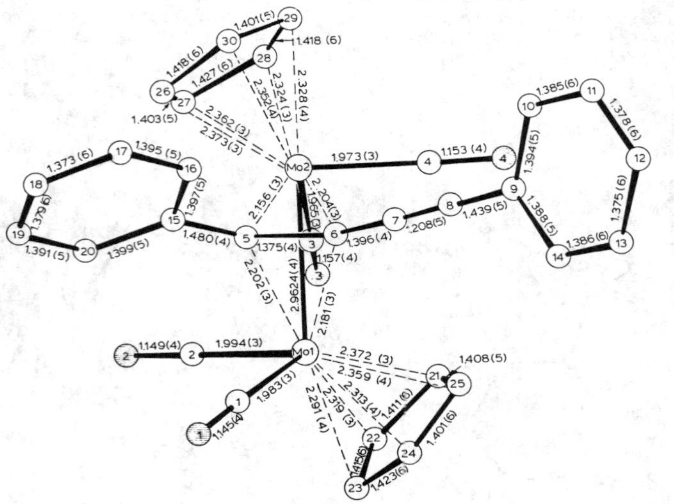

Fig. 1. Molecular geometry of the binuclear complex $C_{30}H_{20}Mo_2O_4$.

Monoclinic, $P2_1/n$, a = 8.545, b = 24.131, c = 11.969 Å, β = 92.51°, Z = 4. Mo radiation, R = 0.029 for 3951 reflexions (at -120°C).

In the dinuclear structure, the planar cyclopentadienyl rings are in the trans position with a pseudo-torsion angle Q(1)-Mo(1)-Mo(2)-Q(2) (Q(1) and Q(2) are Cp ring centroids) of 154.0° (Fig. 1). The coordinated C(5)≡C(6) triple bond is perpendicular to the Mo(1)-Mo(2) bond (92.1°) and is considerably elongated (to 1.375 Å) compared to acetylene (1.205 Å). The central fragment C(15), C(5), C(6), C(7), C(8) and C(9) is almost planar but the C(9)...C(14) and C(15)...C(20) phenyl rings are turned by 74.2 and 22.7°, respectively, relative to this plane.

μ-CHLORO-BIS[(η^3-ALLYL)-2,2'-BIPYRIDYLDICARBONYLMOLYBDENUM(II)] TETRAFLUOROBORATE DICHLOROMETHANE SOLVATE

$C_{30}H_{26}BClF_4Mo_2,CH_2Cl_2$ $[(bipy)(CO)_2(\eta^3-C_3H_5)Mo]_2(\mu-Cl)^+BF_4^-,CH_2Cl_2$

M.D. CURTIS and N.A. FOTINOS, 1984. J. Organometal. Chem., 272, 43-54.

Triclinic, $P\bar{1}$, a = 11.831, b = 10.142, c = 15.618 Å, α = 93.96, β = 104.33, γ = 91.41°, Z = 2. Mo radiation, R = 0.057 for 3742 reflexions.

In the cation (Fig. 1) the two Mo atoms are bridged by a single Cl atom. The Mo-Cl distances are 2.554 and 2.519 Å and the Mo-Cl-Mo angle is 134.0°. The η^3-allyl group is oriented so that its open face points towards the two cis-carbonyl groups. One allyl group has large thermal parameters but the Mo-C(allyl) distances in the other group are Mo(2)-C(5) 2.35, Mo(2)-C(6) 2.23 and Mo(2)-C(7) 2.33 Å. The angle C(5)-C(6)-C(7) is 115.0°. There of the four Mo-N distances cluster round 2.25 Å, and the other, Mo(1)-N(3), is somewhat shorter at 2.215 Å.

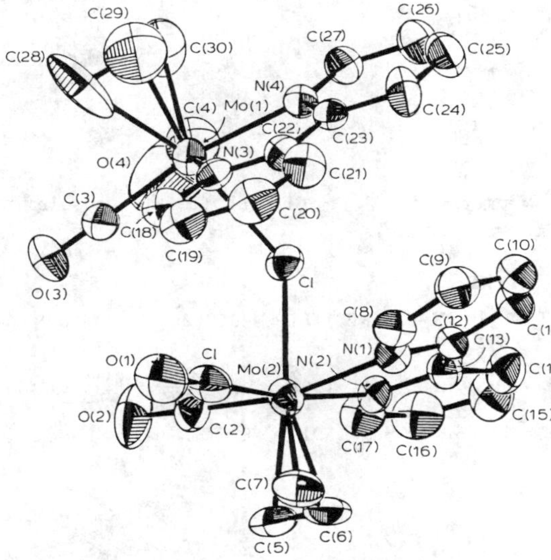

Fig. 1. Structure of the complex cation in $C_{30}H_{26}BClF_4Mo_2$.

TRICHLORO-1,6-BIS(DIPHENYLPHOSPHINO)-trans-HEX-3-ENERHODIUM(III)

$C_{30}H_{30}Cl_3P_2Rh$ (I)

DICHLOROCARBONYL-1,6-BIS(DIPHENYLPHOSPHINO)-trans-HEX-3-ENERUTHENIUM(II)

$C_{31}H_{30}Cl_2OP_2Ru$ (II)

G.R. CLARK and K. MARSDEN, 1984. J. Organometal. Chem., 265, 215-224.

I. Orthorhombic, Pbca, a = 16.109, b = 20.359, c = 17.194 Å, D_m = 1.55, Z = 8. Cu radiation, R = 0.030 for 1644 reflexions.

II. Orthorhombic, Pbca, a = 16.279, b = 20.160, c = 17.334 Å, D_m = 1.52, Z = 8. Cu radiation, R = 0.067 for 2089 reflexions.

The complexes are isostructural and have their central metal atoms bonded to the BDPH moieties through the olefin and two P atoms, with the P atoms occupying mutually trans positions (Fig. 1). In II, one Cl is trans to the olefin, and the remaining Cl and the CO group are statistically disordered. The M-Cl distances range from 2.346 to 2.366 Å in I and from 2.414 to 2.597 Å (disordered) in II. The M-P distances are 2.377, 2.402; 2.388, 2.416 Å in I and II respectively. The two metal-to-olefin C bonds are relatively long and the C=C bond relatively short in the two compounds: 2.315, 2.326, 1.376 Å in I; 2.30, 2.32, 1.37 Å in II, indicating weak metal-olefin bonding. Both complexes have a twisted olefin orientation (Fig. 1).

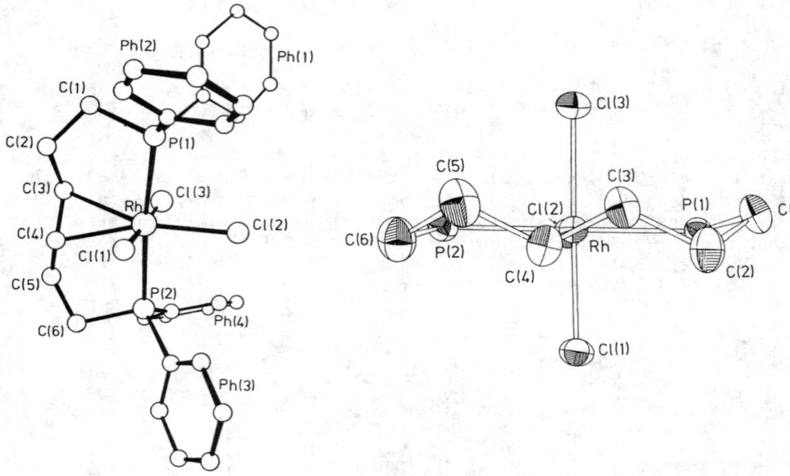

Fig. 1. Molecular geometry for $C_{30}H_{30}Cl_3P_2Rh$ (left), and olefin orientation (right).

(2-3-η-2-BUTYNE-1,4-DIOL)[N,N'-ETHANEDIYLIDENEBIS(2,6-DIISOPROPYLANILINE)]NICKEL(0)
$C_{30}H_{42}N_2NiO_2$ $Ni(C_{26}H_{36}N_2)(C_4H_6O_2)$

R. DIERCKS, J. KOPF and H. TOM DIECK, 1984. Acta Cryst., C40, 363-365.

Tetragonal, $I4_1/a$, a = 18.655, c = 17.591 Å, Z = 8. Mo radiation, R = 0.0525 for 2066 reflexions.

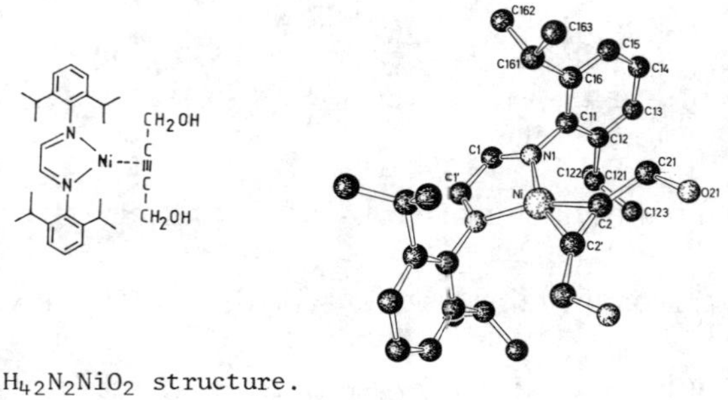

Fig. 1. The $C_{30}H_{42}N_2NiO_2$ structure.

In the complex which has crystallographic twofold symmetry (Fig. 1) the coordinated alkyne has a cis-geometry (C2'-C2-C21 152.4°); principal bond lengths are Ni-N1 1.886(3), Ni-C2 1.840(3), N1=C1 1.304(4), N;-C11 1.440(4) Å.

BIS[(μ-σ^3-CARBON DISULPHIDE-C,S,S')(DI-t-BUTYLPHENYLPHOSPHINE)PLATINUM]
$C_{30}H_{46}P_2Pt_2S_2$ $Pt_2(\mu-CS_2)_2[P(t-C_4H_9)_2C_6H_5]_2$

D.H. FARRAR, R.R. GUKATHASAN and S.A. MORRIS, 1984. Inorg. Chem., 23, 3258-3261.

Monoclinic, $P2_1/n$, a = 8.339, b = 14.034, c = 15.123 Å, β = 91.781°, Z = 2. Mo radiation, R = 0.027 for 2460 reflexions.

The centrosymmetric binuclear molecule (Fig. 1) features a nearly planar system of three fused rings. The coordination about Pt is essentially planar, the geometry being consistent with a distorted square planar coordination. Bond lengths: Pt-P 2.349(2), Pt-S(1) 2.337(2), Pt-S(2)' 2.383(2), Pt-C 1.970(6), C-S(1) 1.674(6), and C-S(2) 1.651(6) Å.

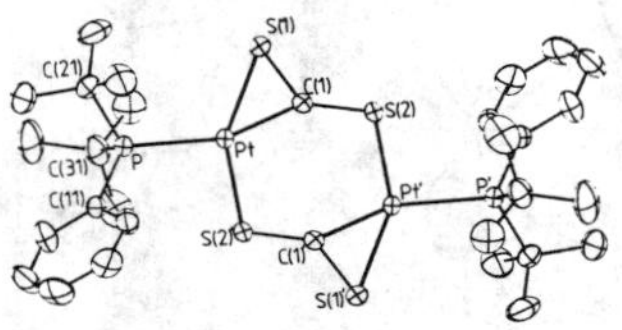

Fig. 1. The $Pt_2(\mu-CS_2)_2[P(t-C_4H_9)_2C_6H_5]_2$ molecule.

(μ_2-η^4-ETHYNE)BIS(μ_2-ISOPROPOXO)TETRAISOPROPOXYBIS(PYRIDINE-N)DITUNGSTEN
$C_{30}H_{54}N_2O_6W_2$ (I)

(μ_2-NEOPENTOXO)(μ_2-η^4-DIMETHYLETHYNE)PENTAKIS(NEOPENTOXY)BIS(PYRIDINE-N)DITUNGSTEN
$C_{44}H_{82}N_2O_6W_2$ (II)

(μ_2-t-BUTOXO)(μ_2-η^4-ETHYNE)(PENTA-t-BUTOXY)(PYRIDINE-N)DITUNGSTEN PYRIDINE SOLVATE
$C_{31}H_{61}NO_6W_2$,0.5(C_5H_5N) (III)

M.H. CHISHOLM, K. FOLTING, D.M. HOFFMAN and J.C. HUFFMAN, 1984. J. Am. Chem. Soc., 106, 6794-6805.

I. Monoclinic, $P2_1/a$, a = 19.061, b = 15.674, c = 12.234 Å, β = 108.08°, Z = 4. Mo radiation, R = 0.056 for 4709 reflexions (at -165°C).

II. Monoclinic, C2/c, a = 37.822, b = 12.207, c = 22.163 Å, β = 101.01°, Z = 8. Mo radiation, R = 0.052 for 5548 reflexions (at -160°C).

III. Triclinic, $P\bar{1}$, a = 17.399, b = 11.336, c = 11.216 Å, α = 71.53, β = 113.72, γ = 98.90°, Z = 2. Mo radiation, R = 0.040 for 4064 reflexions (at -160°C).

In each of the compounds (Fig. 1) there is a pseudotetrahedral $W_2(\mu-C_2)$ core with C-C distances (1.38 - 1.44 Å) and W-W distances (2.57 - 2.67 Å) that approach those of single bonds. These data are interpreted to imply a substantial rehybridization of triple bonds towards the idealized dimetallatetrahedrane. The t-butoxy compound (III) is based on two fused trigonal-bipyramidal moieties sharing a common equatorial (C_2R_2) and axial (OR) edge; the neopentoxide (II) has W atoms that share an edge of a trigonal bipyramid and an octahedron, and the isopropoxide (I) has two octahedral W atoms fused along a common face formed by two O-i-Pr ligands and the $\mu-C_2H_2$ ligand.

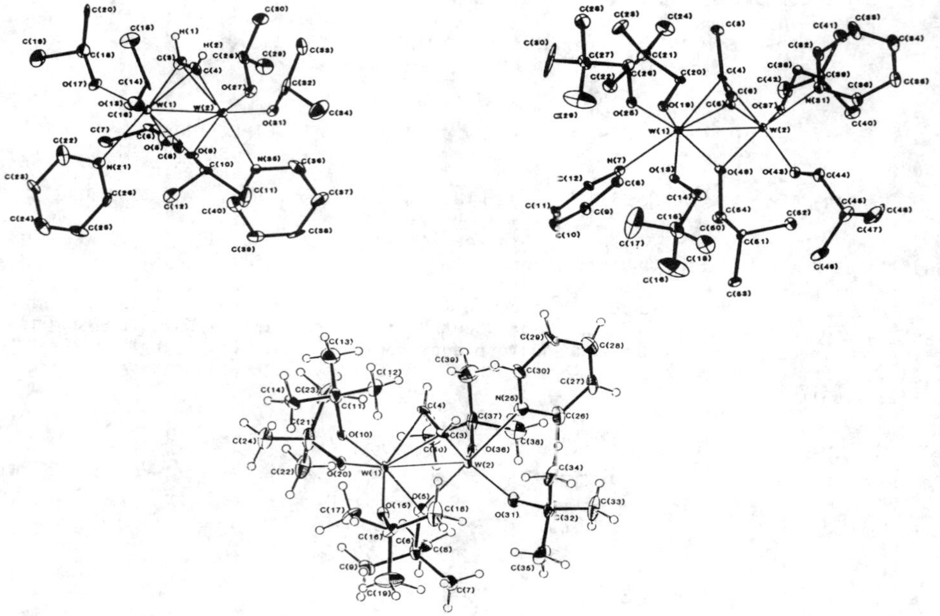

Fig. 1. Views of the alkoxides (I) - (III) clockwise from top left.

$(\mu_2$-HYDRIDO)-$(\mu_2$-σ^1-η^2-VINYL)BIS(1,2-BIS(DIISOPROPYLPHOSPHINO)ETHANE)DIRHODIUM
$C_{30}H_{69}P_4Rh_2$

M.D. FRYZUK, T. JONES and F.W.B. EINSTEIN, 1984. Organometallics, 3, 185-191.

Monoclinic, C2/c, a = 38.209, b = 14.596, c = 13.472 Å, β = 93.98°, Dm = 1.33, Z = 8.
Mo radiation, R = 0.025 for 4728 reflexions.

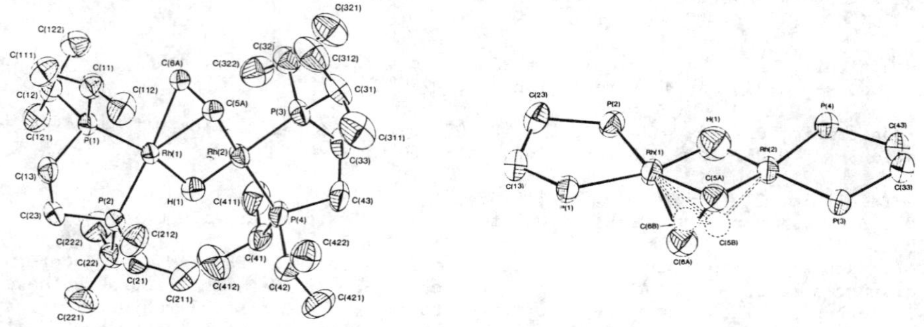

Fig. 1. Molecular structure of $C_{30}H_{69}P_4Rh_2$ (left) and view of the disordered vinyl group (right).

The binuclear Rh core of the molecule contains a disordered vinyl group and a bridging hydride ligand which is asymmetrically bound (Fig. 1). The Rh(1)-H(1) and Rh(2)-H(1) distances are respectively 1.85 and 1.66 Å. The Rh-C(vinyl) bond lengths are 2.210, 2.216, 2.295 and 2.314 Å (C(5A), C(5B), C(6A), C(6B) respectively). The Rh(1)-H(1)-Rh(2) angle is 109° and the metal-P distances range from 2.197 to 2.250 Å with P-Rh-P angles of 85.89 and 86.44°. Within the disordered vinyl ligand the fractional occupancies are 0.69 for C(5A)-C(6A) and 0.31 for C(5B)-C(6B). The C-C distances for the A and B sets are 1.362 and 1.33 Å respectively.

(μ_2-CARBONYL)(μ_2-σ^2,η^2-(2-DIPHENYLPHOSPHINOCARBONYL)-1,2-DIPHENYLVINYLENE)PENTA-CARBONYLRUTHENIUMCOBALT

$C_{33}H_{20}CoO_7PRu$ $RuCo(CO)_5(\mu\text{-}CO)[\mu\text{-}\eta^2\text{-}PPh_2C(O)C(Ph)C(Ph)]$

R. REGRAGUI, P.H. DIXNEUF, N.J. TAYLOR and A.J. CARTY, 1984. Organometallics, 3, 814-816.

Triclinic, P$\bar{1}$, a = 9.280, b = 12.419, c = 14.966 Å, α = 99.58, β = 99.55, γ = 90.10°, Dm = 1.60, Z = 2. Mo radiation, R = 0.028 for 4191 reflexions.

The molecule contains a bridging $Ph_2PCOC(Ph)C(Ph)$ ligand coordinated to Ru via P and atom C(8) of the original alkyne, and to Co via an η^2 interaction (atoms C(8) and C(9)) (Fig. 1). The Ru-P distance is 2.3335 Å and the P-C(7) distance of 1.865 Å is considerably elongated compared to the P-C(phenyl) bond lengths which average 1.820 Å. The Ru-Co distance is 2.6577 Å.

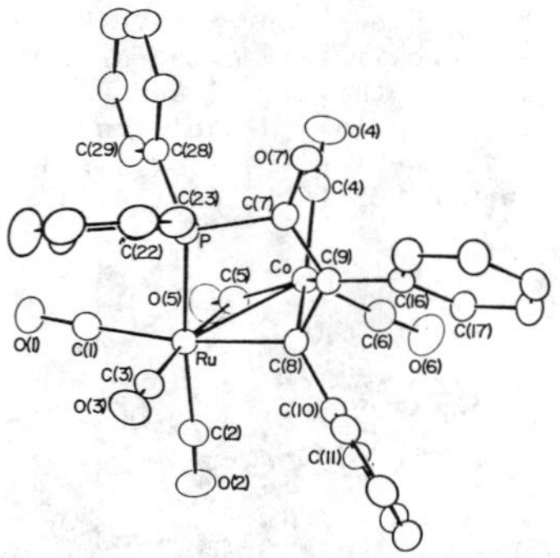

Fig. 1. Perspective view of the molecular structure of $C_{33}H_{20}CoO_7PRu$.

BIS[(μ_2-CHLORO)DICHLORO(η^2-DIPHENYLACETYLENE)TETRAHYDROTHIOPHENENIOBIUM(III)] TOLUENE SOLVATE

$C_{36}H_{36}Cl_6Nb_2S_2,C_7H_8$ $Nb_2Cl_6(C_4H_8S)_2(C_6H_5C{\equiv}CC_6H_5)_2.C_6H_5CH_3$

F.A. COTTON and W.J. ROTH, 1984. Inorg. Chim. Acta, 85, 17-21.

Triclinic, P$\bar{1}$, a = 10.516, b = 12.663, c = 9.068 Å, α = 100.63, β = 91.71, γ = 68.36°, Z = 1. Mo radiation, R = 0.058 for 2321 reflexions.

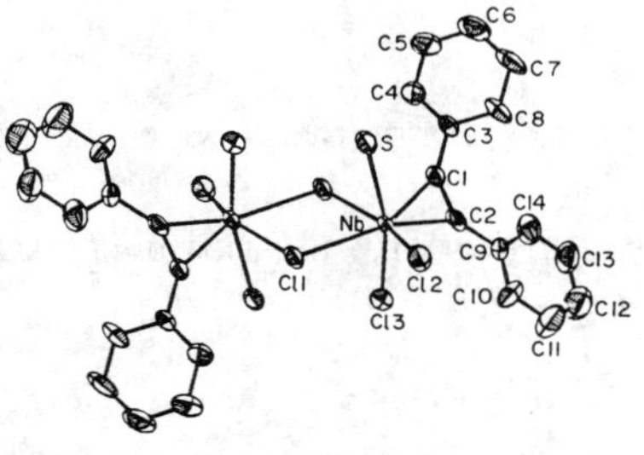

Fig. 1. Molecular structure of $C_{36}H_{36}Cl_6Nb_2S_2$ with tetrahydrothiophene C atoms omitted for clarity.

Both molecules in the crystal lattice, namely the Nb complex and the toluene (which is disordered), reside on crystallographic inversion centres (Fig. 1). The metal atom has two relatively short bonds (2.052 and 2.047 Å, respectively) to C(1) and C(2) and the existence of a π bond between the alkyne and Nb is suspected. Other bond distances involving Nb are: Nb-S(tetrahydrothiophene) 2.632, Nb-Cl(1) 2.758, Nb-Cl(1') 2.483, Nb-Cl(2) 2.354 and Nb-Cl(3) 2.372 Å. The Nb...Nb' distance is 4.137 Å. The angles C(1)-Nb-C(2) and Cl(1)-Nb-Cl(1') are 37.0 and 75.9° respectively, with Nb-Cl(1)-Nb' 104.1°.

(μ-DINITROGEN)BIS[DICHLORO(1,2-DIMETHOXYETHANE)(η^2-DIPHENYLACETYLENE)TUNGSTEN] 1,2-DIMETHOXYETHANE SOLVATE
$C_{36}H_{40}Cl_4N_2O_4W_2,0.5(C_4H_{10}O_2)$

M.R. CHURCHILL, Y.-J. LI, K.H. THEOPOLD and R.R. SCHROCK, 1984. Inorg. Chem., 23, 4472-4476.

Monoclinic, C2/c, a = 36.016, b = 14.131, c = 16.650 Å, β = 96.23°, Z = 8. Mo radiation, R = 0.048 for 3276 reflexions.

The dinuclear complex (Fig. 1) has approximate C_2 symmetry. The coordination geometry about each W atom is approximately octahedral with the acetylene in a position cis to the μ-N_2 ligand and the Cl^- ligands trans to one another. The W_2N_2 system is essentially linear, with a long N-N distance 1.292(16) Å consistent with a μ-$N_2{}^{4-}$ ligand. Other distances: W-Cl 2.414(2)-2.442(4), W-O 2.245(10)-2.299(10), W-N 1.776(11) and 1.735(11), W-C 2.032(15)-2.063(14) Å.

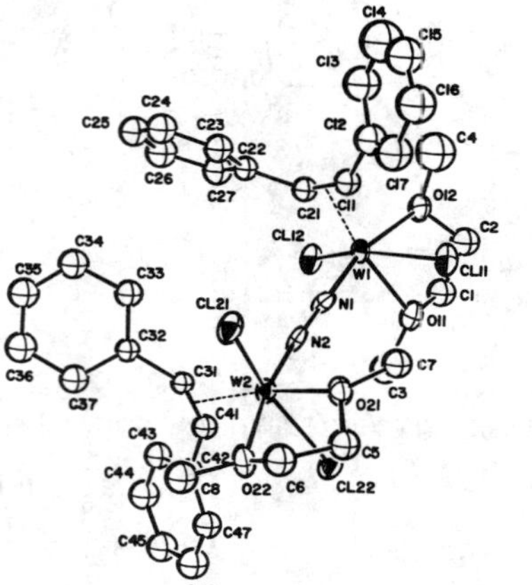

Fig. 1. The $[W(C_2Ph_2)(CH_3OCH_2CH_2OCH_3)Cl_2]_2(\mu\text{-}N_2)$ molecule.

TETRAPHENYLPHOSPHONIUM (η^2-DIPHENYLDIACETYLENE)PENTACHLOROTUNGSTEN CHLOROFORM SOLVATE
$C_{40}H_{30}Cl_5PW,CCl_4$ (I)

TETRAPHENYLPHOSPHONIUM (η^2-1,2-DIBROMO-1,4-DIPHENYLDIACETYLENE)PENTACHLOROTUNGSTEN CHLOROFORM SOLVATE
$C_{40}H_{30}Br_2Cl_5PW,CCl_4$ (II)

K. STAHL, F. WELLER and K. DEHNICKE, 1984. Z. Anorg. Chem., 518, 175-186.

I. Monoclinic, $P2_1/n$, a = 14.090, b = 13.852, c = 22.628 Å, β = 104.17°, Z = 4. Mo radiation, R = 0.054 for 2623 reflexions.

II. Monoclinic, $P2_1/c$, a = 9.283, b = 33.34, c = 14.919 Å, β = 106.92°, Z = 4. Mo radiation, R = 0.054 for 2537 reflexions.

Views of the anions of (I) and (II) are in Fig. 1; both complexes have the acetylene ligand bonding side-on in an approximately symmetrical way. There is a pronounced trans effect resulting in W-Cl(5) (2.463(5) (I), 2.467(5) Å (II)) being significantly longer than the other W-Cl bonds (2.356-2.364(4) (I), 2.346-2.397(4) Å (II)).

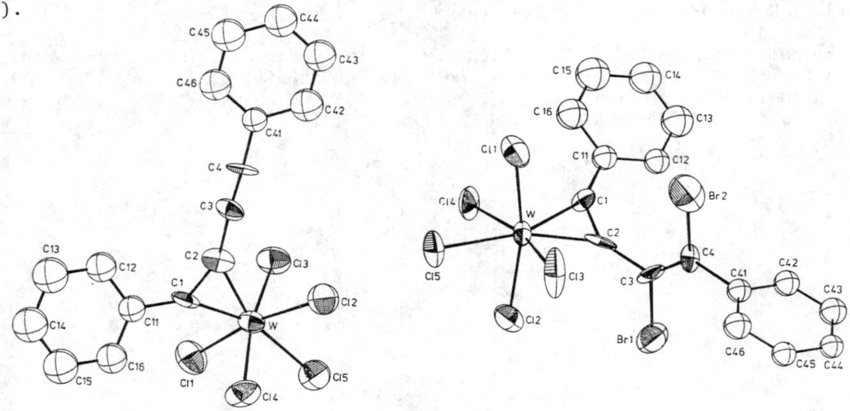

Fig. 1. Views of $[WCl_5PhC_4Ph]^-$ and $[WCl_5PhC_4Br_2Ph]^-$.

(η^3-2-METHYLENEBICYCLO[2.2.1]HEPTANE)BIS(TRIPHENYLPHOSPHINE)PALLADIUM HEXAFLUORO-PHOSPHATE

$C_{44}H_{41}F_6P_3Pd$ $[(PPh_3)_2Pd(\eta^3-C_8H_{11})]^+.PF_6^-$

S.A. GODLESKI, K.B. GUNDLACH, H.Y. HO, E. KEINAN and F. FROLOW, 1984. Organometallics, 3, 21-28.

Monoclinic, $P2_1/n$, a = 10.328, b = 17.767, c = 21.984 Å, β = 100.21°, Z = 4. Cu radiation, R = 0.053 for 5234 reflexions.

The structural study establishes unequivocally that the title compound exists as an exo-bis(phosphine) cationic complex. The metal atom is σ-bonded to the two phosphine groups (Pd-P(2) 2.331, Pd-P(3) 2.321 Å) and π-bonded to the allyl-norbornyl derivative (Fig. 1). The Pd-C distances are: Pd-C(40) 2.182, Pd-C(41) 2.197 and Pd-C(42) 2.231 Å and the bond lengths within the allyl grouping are: C(40)-C(41) 1.392 and C(41)-C(42) 1.422 Å indicating more "olefin" character in C(40)-C(41) despite the shorter Pd-C(40) distance.

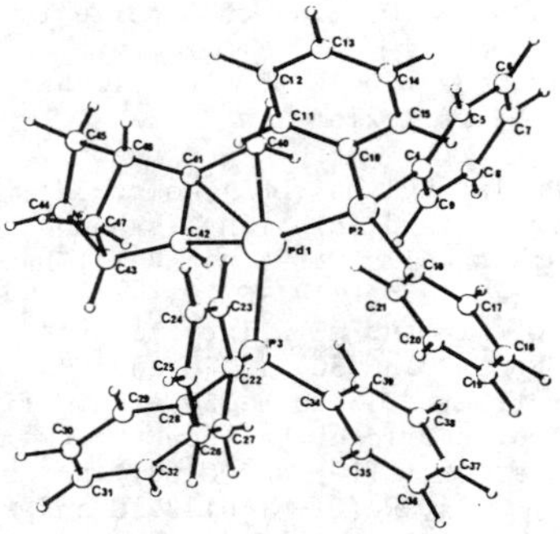

Fig. 1. Molecular structure of $C_{44}H_{41}F_6P_3Pd$.

BIS(TRIPHENYLPHOSPHINE)IMINIUM (η^3-1-ISOPROPYLALLYL-1,2-DIONATO-O,O'-DIFLUOROBORON)-TETRACARBONYLMANGANESE
$C_{46}H_{39}BF_2MnNO_6P_2$

P.G. LENHERT, C.M. LUKEHART and K. SRINIVASAN, 1984. J. Am. Chem. Soc., 106, 124-130.

Monoclinic, $P2_1/c$, a = 14.573, b = 17.472, c = 20.175 Å, β = 123.03°, Z = 4. Mo radiation, R = 0.084 for 3270 reflexions.

The molecule has the η^3-allyl structure shown in Fig. 1. The dihedral angle between the allyl carbon plane and the molecular equatorial plane (Mn,C(2),C(3)) is 115°. Bond lengths: Mn-CO(eq) 1.75(2) and 1.78(2), Mn-CO(ax) 1.83(2) and 1.82(2), Mn-C(5) 2.38(2), Mn-C(6) 2.20(2), and Mn-C(7) 2.16(2) Å.

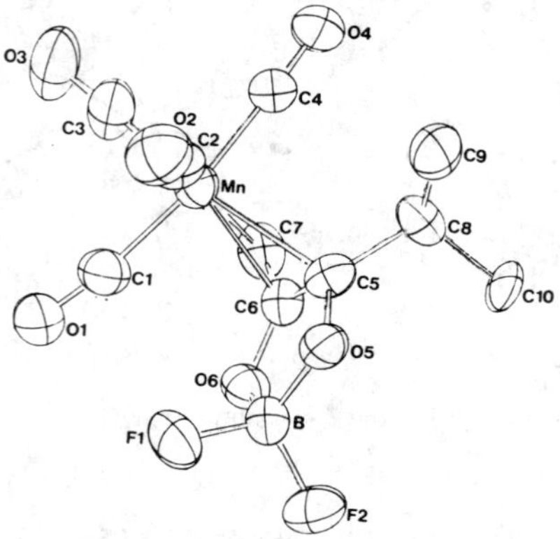

Fig. 1. The anion in $C_{46}H_{39}BF_2MnNO_6P_2$.

(μ_2-σ^2,η^4-TETRAPHENYLBUTA-1,3-DIENE-1,4-DIYL)(η^2-DIPHENYLACETYLENE)PENTACARBONYLDI-RHENIUM
$C_{47}H_{30}O_5Re_2$ (I) $Re_2(CO)_5(C_4Ph_4)(PhC{\equiv}CPh)$

(μ_2-σ^2,η^3,η^3-HEXAPHENYLHEXADIENE-1,6-DIYL)(η^2-DIPHENYLACETYLENE)TETRACARBONYLDI-RHENIUM HEXANE SOLVATE
$C_{60}H_{40}O_4Re_2,0.5C_6H_{12}$ (II) $Re_2(CO)_4(C_6Ph_6)(PhC{\equiv}CPh),0.5C_6H_{12}$

D.B. POURREAU, R.R. WHITTLE and G.L. GEOFFROY, 1984. J. Organometal. Chem., 273, 333-346.

I. Triclinic, $P\bar{1}$, a = 11.277, b = 12.301, c = 14.351 Å, α = 97.37, β = 104.68, γ = 104.49°, Z = 2. Mo radiation, R = 0.031 for 4548 reflexions.

II. Monoclinic, $P2_1/c$, a = 17.048, b = 16.676, c = 18.569 Å, β = 106.70°, Z = 4. Mo radiation, R = 0.078 for 3568 reflexions.

Compound I has two diphenylacetylene (DPA) molecules linked together in a rhenacyclopentadiene ring (Fig. 1) and this ring is bound in a π-fashion to the other Re atom (Re(2)). The Re-Re distance is 2.878 Å. Re(1) has an η^2-DPA ligand and the distances Re(1)-C(10), Re(1)-C(20) and C(10)-C(20) are 2.043, 2.006 and 1.329 Å respectively. The distances within the metallacycle are: Re(1)-C(30) 2.170, Re(1)-C(60) 2.099, C(30)-C(40) 1.388, C(40)-C(50) 1.446 and C(50)-C(60) 1.407 Å, and the Re(2)-C(ring) distances range from 2.255 to 2.359 Å. In II, three alkynes have linked together to form a "flyover bridge" between the two Re atoms. The Re-Re distance is 2.849 Å. The end atoms (C(10) and C(60)) are bonded to both Re atoms (Re(1)-C(10) 2.27, Re(2)-C(10) 2.04, Re(1)-C(60) 2.18 and Re(2)-C(60) 1.99 Å). Each alkyne C atom is involved in an η^3-allyl ligand π-bonded to one Re. The single DPA ligand has Re(2)-C(70), Re(2)-C(80) and C(70)-C(80) distances of 2.03, 1.95 and 1.22 Å respectively.

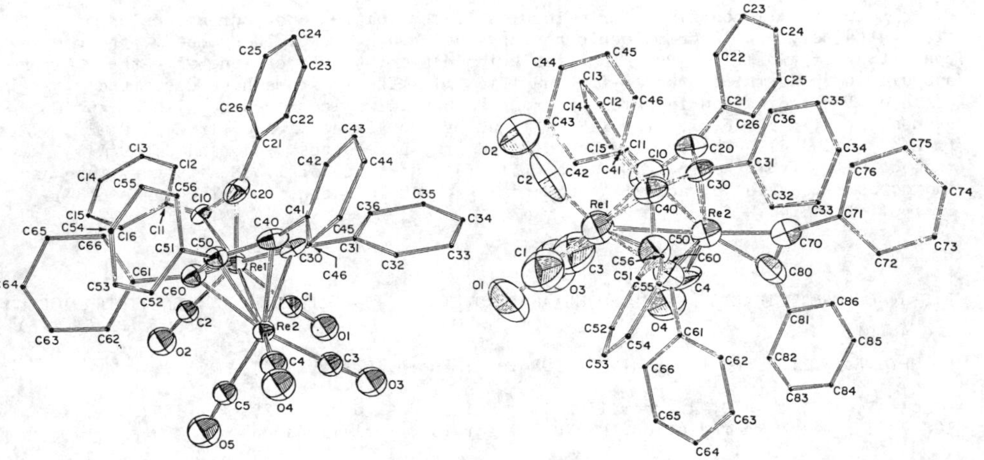

Fig. 1. Molecular structure of $C_{47}H_{30}O_5Re_2$ (right), and $C_{60}H_{40}O_4Re_2$ (left).

DIPHENYLACETYLENEBIS(TRIPHENYLPHOSPHINEDICARBONYLIRIDIUM) BENZENE SOLVATE

$C_{54}H_{40}Ir_2O_4P_2,C_6H_6$ $\{[Ir(CO)_2P(C_6H_5)_3]_2(CC_6H_5)_2\},C_6H_6$

M. ANGOLETTA, P.L. BELLON and F. DEMARTIN, 1984. J. Organometal. Chem., 267, 199-205.

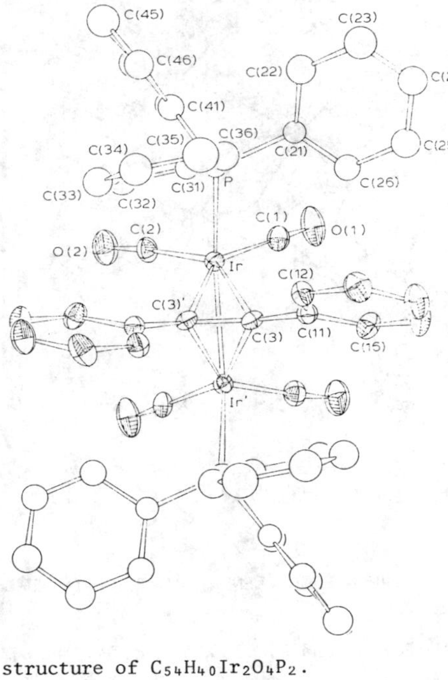

Fig. 1. Molecular structure of $C_{54}H_{40}Ir_2O_4P_2$.

Monoclinic, C2/c, a = 19.714, b = 14.257, c = 18.953 Å, β = 109.85°, D_m = 1.70, Z = 4. Mo radiation, R = 0.028 for 3411 reflexions.

The unit cell contains four molecules of the complex and four molecules of clathrated solvent. Each molecule has crystallographic twofold symmetry, the complex consisting of two $Ir(CO)_2PPh_3$ moieties bridged by diphenylacetylene with the C≡C bond approximately normal to the Ir-Ir bond (Fig. 1). The Ir atoms have distorted octahedral coordination in which the four equatorial apices are occupied by two CO groups (Ir-C(1) 1.906, Ir-C(2) 1.910 Å) and by the alkyne ligand (Ir-C(3) 2.127, Ir-C(3') 2.090 Å). The axial apices (P and Ir') have sites 2.346 and 2.689 Å respectively, from the Ir atom. The C(3)-C(3') distance of 1.398 Å is very long and indicates π-donation of acetylene towards the binuclear cluster. The angle C(3')-C(3)-C(11) is 142.2°.

BIS(μ_3-CHLORO)TETRAKIS(μ_2-CHLORO)HEXACHLOROTETRAKIS(μ_2-DIPHENYLACETYLENE)TETRANIOBIUM
$C_{56}H_{40}Cl_{12}Nb_4$ $[NbCl_3(Ph\text{-}C{\equiv}C\text{-}Ph)]_4$

E. HEY, F. WELLER and K. DEHNICKE, 1984. Z. Anorg. Chem., 514, 25-38.

Triclinic, P$\bar{1}$, a = 10.743, b = 13.896, c = 12.994 Å, α = 104.28, β = 107.99, γ = 108.72°, Z = 1. Mo radiation, R = 0.059 for 1548 reflexions.

The compound is a centrosymmetric tetramer (Fig. 1) which is described as a distorted double hexahedron with two corners missing. The units are linked by chlorine bridges. The diphenylacetylenes are bonded side-on with Nb-C 2.033(17)-2.068(19) Å. Nb-Cl distances are in the range 2.320-2.945(5) Å.

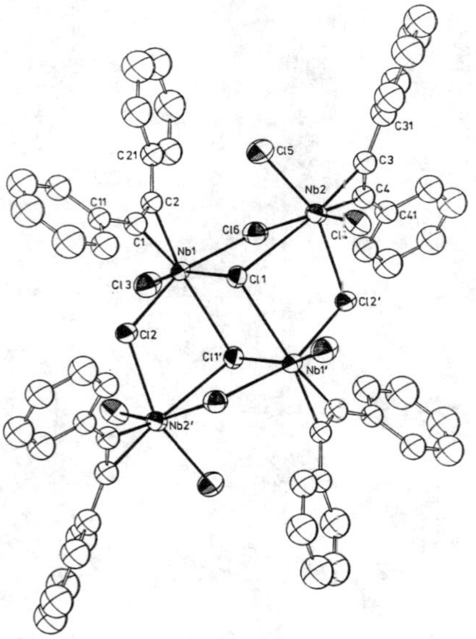

Fig. 1. A view of $[NbCl_3(Ph\text{-}C{\equiv}C\text{-}Ph)]_4$.

TETRAETHYLAMMONIUM TRIS(η^2-DIPHENYLACETYLENE)(TRIPHENYLTIN)TUNGSTEN
$C_{68}H_{65}NSnW$ $[NEt_4][W(PhC{\equiv}CPh)_3SnPh_3]$

J.M. MAHER, J.R. FOX, B.M. FOXMAN and N.J. COOPER, 1984. J. Am. Chem. Soc., 106, 2347-2353.

Monoclinic, $P2_1/n$, a = 19.655, b = 12.662, c = 24.084 Å, β = 111.07°, Z = 4. Mo radiation, R = 0.0498 for 7949 reflexions.

The immediate coordination sphere of tungsten approximates a tapered trigonal prism, defined by the C atoms of the acetylenes and which is capped on the larger trigonal face by the $SnPh_3$ ligand. Main dimensions are W-C 2.051-2.084(6), W-Sn 2.807(1), Sn-C 2.185-2.212(7), C-C (bonded to W) 1.299(8)-1.307(9) Å. The extent of tapering is marked and is quantified by the angles which the C-C bonds make with the W-Sn vector (11, 17 and 18°).

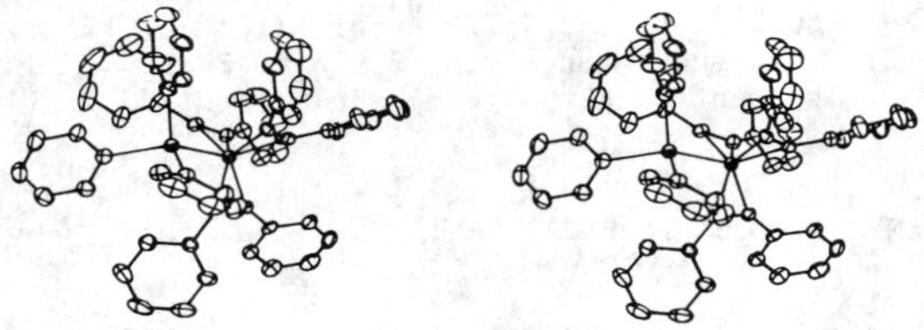

Fig. 1. A stereoview of $[W(PhC{\equiv}CPh)_3SnPh_3]^-$.

TRICHLORO(η^5-CYCLOPENTADIENYL)TITANIUM(IV)
$C_5H_5Cl_3Ti$ (I) $C_5H_5TiCl_3$

TRICHLORO(η^5-CYCLOPENTADIENYL)ZIRCONIUM(IV)
$C_5H_5Cl_3Zr$ (II) $C_5H_5ZrCl_3$

L.M. ENGELHARDT, R.I. PAPASERGIO, C.L. RASTON and A.H. WHITE, 1984. Organometallics, 3, 18-20.

I. Monoclinic, $P2_1$, a = 6.588, b = 10.419, c = 6.680 Å, β = 115.86°, Z = 2. Mo radiation, R = 0.031 for 872 reflexions.

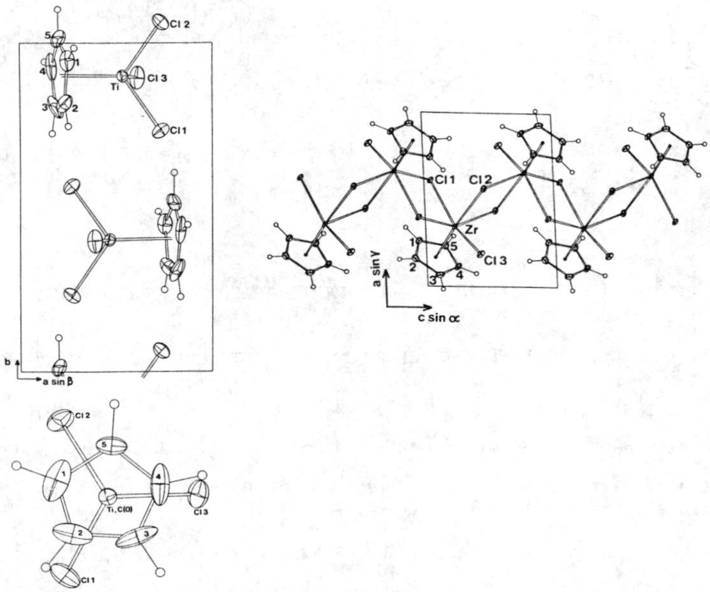

Fig. 1. Unit cell contents of $C_5H_5Cl_3Ti$ (left), and of $C_5H_5Cl_3Zr$ (right) projected down c and b respectively.

II. Triclinic, $P\bar{1}$, a = 8.629, b = 7.133, c = 6.326 Å, α = 100.68, β = 91.67, γ = 98.78°, Z = 2. Mo radiation, R = 0.042 for 1107 reflexions.

Compound I comprises a monomeric species with approximate tetrahedral geometry at the metal atom (Fig. 1). The Ti-Cl distances are 2.201, 2.248 and 2.221 Å and the Cl-Ti-Cl angles are 102.2, 102.3 and 104.1°. The ring centroid to metal distance is 2.01 Å whilst the Ti-C distances range from 2.26 to 2.36 Å. Compound II however, consists of a polymeric species (Fig. 1) in which the metal coordination is distorted octahedral. The Zr-Cl (terminal) distance is 2.419 Å and the Zr-Cl (bridging) distances are 2.518, 2.602, 2.663 and 2.728 Å for Cl(2''), Cl(1'), Cl(1) and Cl(2) respectively. The Zr-C distances range from 2.476 to 2.518 Å and the Zr-centroid distance is 2.196 Å. The centroid-Zr-Cl(cis) angles are 103.4, 106.5 and 100.8°.

1-(η^5-CYCLOPENTADIENYL)-1-COBALTAPENTABORANE
$C_5H_{13}B_4Co$

T.L. VENABLE, E. SINN and R.N. GRIMES, 1984. J. Chem. Soc. Dalton, 2275-2279.

Tetragonal, P4/n, a = 7.909, c = 6.985 Å, Z = 2. Mo radiation, R = 0.052 for 552 reflexions.

The cyclopentadienyl ring in the molecule is disordered. The molecule has a crystallographically imposed fourfold axis and the B_4 ring is therefore constrained to square-planar geometry parallel to the disordered $C_5H_5^-$ ligand (Fig. 1). The complex has a sandwich structure with bond lengths and angles in the expected ranges. The unusually short Co-B distance (1.977(2) Å) suggests a strong covalent metal-$B_4H_8^{2-}$ linkage. The terminal H atoms on the $B_4H_8^{2-}$ ligand are only slightly out of the B_4 plane.

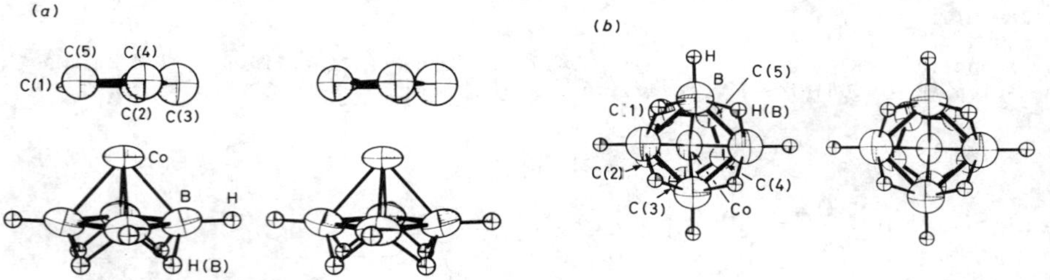

Fig. 1. $C_5H_{13}B_4Co$: stereoviews of the molecule (a) parallel to the C_5H_5 ring: (b) normal to the B_4 face. One position of the fourfold disordered C_5H_5 ring is depicted in each case.

TRICARBONYL(η^5-CYCLOPENTADIENYL)PENTACHLOROCYCLOTRIPHOSPHAZENECHROMIUM
$C_8H_5Cl_5CrN_3O_3P_3$ (I)

TRICARBONYL(η^5-CYCLOPENTADIENYL)(CYCLOPENTADIENYLTETRACHLOROCYCLOTRIPHOSPHAZENE)-MOLYBDENUM
$C_{13}H_{10}Cl_4MoN_3O_3P_3$ (II)

H.R. ALLCOCK, G.H. RIDING and R.R. WHITTLE, 1984. J. Am. Chem. Soc., 106, 5561-5567.

I. Monoclinic, $P2_1/m$, a = 8.334, b = 12.897, c = 8.783 Å, β = 105.44°, Z = 2. Mo radiation, R = 0.054 for 1214 reflexions.

II. Orthorhombic, $Pca2_1$, a = 16.075, b = 8.832, c = 14.780 Å, Z = 4. Mo radiation, R = 0.041 for 1864 reflexions.

Molecule (I) has crystallographic mirror symmetry; dimension details for (I) and (II) are in Fig. 1. In both compounds the N-P-N bond angle at the metal-bearing phosphorus is unusually narrow (112.3(3)° in (I) and 111.0(3)° in (II)). For (I) the phosphazene ring is significantly non-planar.

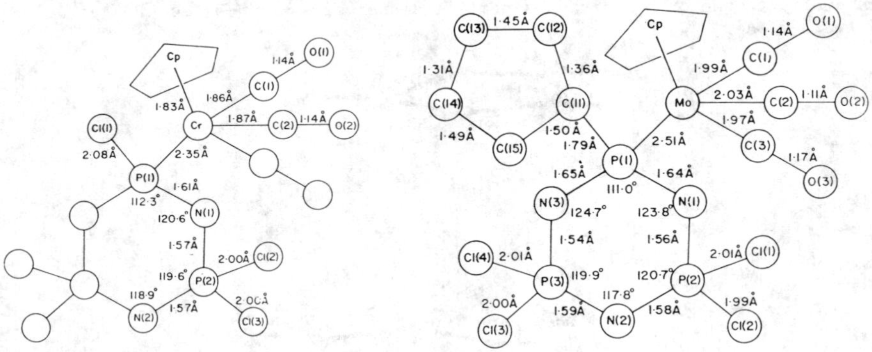

Fig. 1. Some dimensions for $C_8H_5Cl_5CrN_3O_3P_3$ (I) and $C_{13}H_{10}Cl_4MoN_3O_3P_3$ (II).

BROMO(η^4-CYCLOPENTA-2,4-DIEN-1-ONE)(η^5-CYCLOPENTADIENYL)RUTHENIUM(II)
$C_{10}H_9BrORu$ $Ru(C_5H_5)(C_5H_4O)Br$

T.P. SMITH, K.S. KWAN, H. TAUBE, A. BINO and S. COHEN, 1984. Inorg. Chem., 23, 1943-1945.

Orthorhombic, Pnma, a = 13.588, b = 8.876, c = 7.495 Å, Z = 4. Mo radiation, R = 0.0305 for 564 reflexions.

The complex (Fig. 1) has mirror symmetry, with the η^5-C_5H_5 and η^4-cyclopentadienone rings staggered and bent away from the halide. Bond lengths: Ru-Br 2.556(1), Ru-C 2.167(7)-2.263(7) Å.

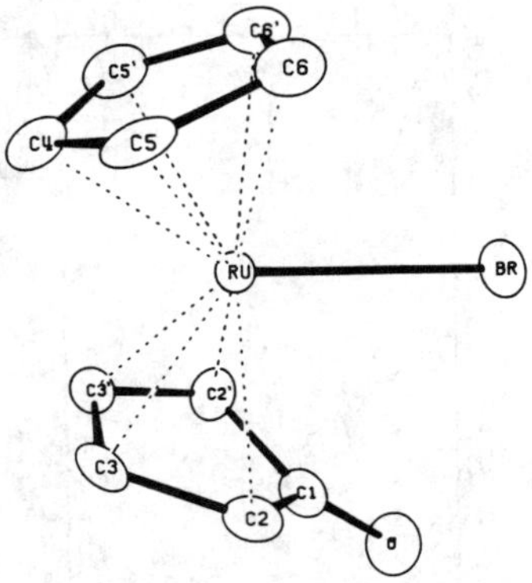

Fig. 1. The $Ru(C_5H_5)(C_5H_4O)Br$ molecule.

BIS(CYCLOPENTADIENYL)BERYLLIUM
$C_{10}H_{10}Be$

K.W. NUGENT, J.K. BEATTIE, T.W. HAMBLEY and M.R. SNOW, 1984. Aust. J. Chem., 37, 1601-1606.

Monoclinic, $P2_1/n$, a = 5.993, b = 7.478, c = 8.978 Å, β = 85.94°, Z = 2. Mo radiation, R = 0.035 for 357 reflexions, T = 128 K.

"Beryllocene", has a "slip-sandwich" structure with the beryllium disordered between two equivalent sites with each site linked centrally to one ring and peripherally to the other which has partial diene character (Fig. 1, Be-Be' 1.303(12), Be-C 1.925(6) Å).

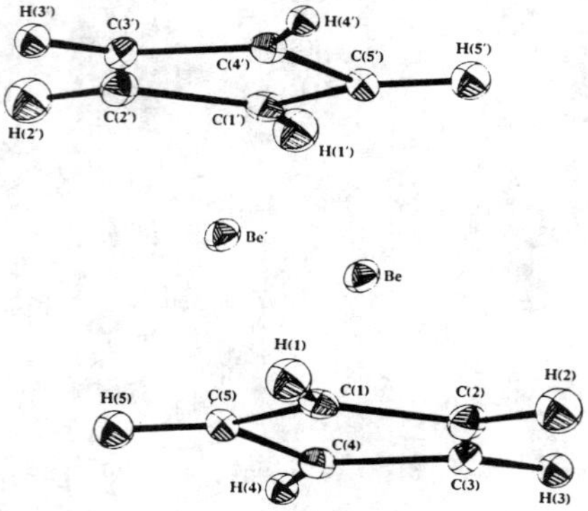

Fig. 1. Sideview of "beryllocene".

BIS(η^5-CYCLOPENTADIENYL)RHENIUM DICHLOROCUPRATE
$C_{10}H_{10}Cl_2CuRe$ $[Cp_2Re]^+.[CuCl_2]^-$

V.M. ISHCHENKO, B.M. BULYCHEV, G.L. SOLOVEICHIK, V.K. BEL'SKII and O.G. ELLERT, 1984. Polyhedron, 3, 771-774.

Orthorhombic, $Pm2_1n$, a = 13.696, b = 7.317, c = 5.969 Å, Z = 2. Mo radiation, R = 0.047 for 717 reflexions.

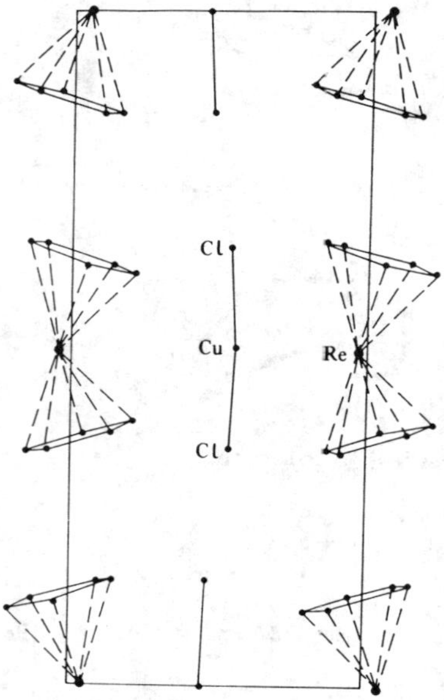

Fig. 1. A projection of the $[Cp_2Re]^+.[CuCl_2]^-$ structure on the (010) plane.

Both cation and anion (Fig. 1) have crystallographic mirror symmetry. The Cp rings form a bent-sandwich with an angle between the ring centres and Re of 150.1°. Main dimensions are Re...Cu 4.35(3), 4.38(3), Cu-Cl 2.097(5), Re-(ring centroid) 1.898, mean Re-C 2.275 Å, Cl-Cu-Cl 174.8(2).

BIS(η^5-CYCLOPENTADIENYL)(THIOSULPHATO-O,S)MOLYBDENUM(IV)
$C_{10}H_{10}MoO_3S_2$ $(C_5H_5)_2Mo(S_2O_3)$

G.J. KUBAS and R.R. RYAN, 1984. Inorg. Chem., 23, 3181-3183.

Monoclinic, $P2_1/n$, a = 13.600, b = 10.623, c = 7.610 Å, β = 100.08°, Z = 4. Mo radiation, R = 0.023 for 1722 reflexions (at -95°C).

The structure (Fig. 1) shows a bidentate coordination of thiosulphate to Mo (Mo-O 2.134(5), Mo-S 2.450(2) Å) with a dihedral angle of 46.8° between the Cp rings and a planar $MoOS_2$ ring. Mo-C distances range from 2.267(6) to 2.380(6) Å.

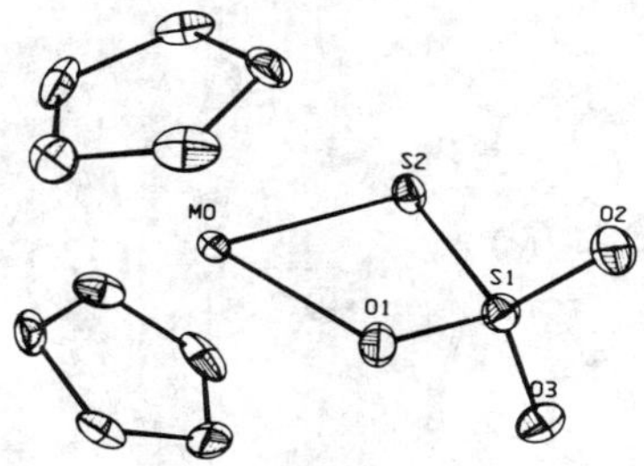

Fig. 1. The $(C_5H_5)_2Mo(S_2O_3)$ molecule.

BIS(η^5-CYCLOPENTADIENYL)HYDRIDO(TRICHLOROTIN)MOLYBDENUM
$C_{10}H_{11}Cl_3MoSn$ $Cp_2Mo(H)SnCl_3$

A.I. GUSEV, N.I. KIRILLOVA, A.N. PROTSKY, B.M. BULYCHEV and G.L. SOLOVEICHIK, 1984. Polyhedron, 3, 765-769.

Monoclinic, $P2_1/m$, a = 8.313, b = 13.109, c = 6.384 Å, β = 106.52°, Z = 2. Mo radiation, R = 0.028.

The molecule (Fig. 1) has crystallographic mirror symmetry (Mo, H, Sn and Cl(2) lie in the mirror plane) and the Cp rings are in eclipsed conformation. Some bond lengths are Mo-H 1.74(7), Mo-Sn 2.652(1), Sn...H 2.88, mean Mo-C 2.291(7), mean Sn-Cl 2.381(2) Å.

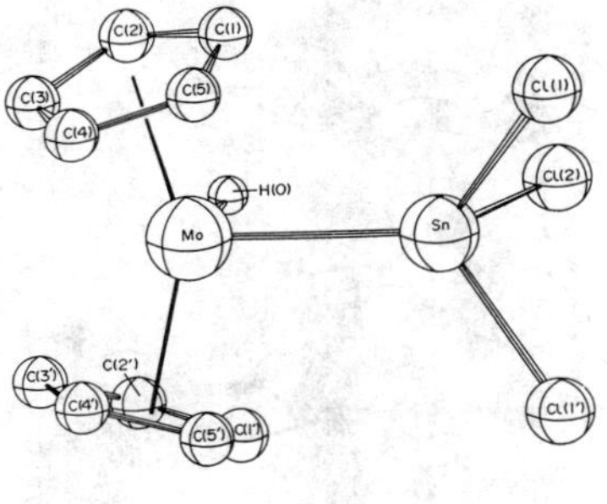

Fig. 1. The $Cp_2Mo(H)SnCl_3$ structure.

BIS-μ-HYDROXO-BIS(η-CYCLOPENTADIENYLDINITRATOZIRCONIUM(IV)) TETRAHYDROFURAN SOLVATE
$C_{10}H_{12}N_4O_{14}Zr_2,2C_4H_8O$ $[(\pi C_5H_5)Zr(NO_3)_2(\mu\text{-}OH)]_2,2C_4H_8O$

W. LASSER and U. THEWALT, 1984. J. Organometal. Chem., 275, 63-70.

Monoclinic, $P2_1/c$, a = 8.825, b = 12.137, c = 13.018 Å, β = 97.61°, D_m = 1.77, Z = 2. Mo radiation, R = 0.052 for 2063 reflexions.

The unit cell contains two dinuclear centrosymmetric metal complexes (Fig. 1) and four molecules of tetrahydrofuran of crystallization. The Zr atom has a pentagonal bipyramidal coordination geometry with the Cp group occupying an axial position, two chelating nitrato ligands and one OH bridging group in the equatorial plane, and one OH group in the other axial position. Important bond lengths are: Zr-O 2.167, Zr-O' 2.090, Zr-O(11) 2.320, Zr-O(12) 2.274, Zr-O(21) 2.314, Zr-O(22) 2.278 Å. The N-O(coordinated) distances range from 1.269 to 1.297 Å and the two uncoordinated O-N distances are 1.211 and 1.201 Å. The O-Zr-O' angle is 69.7° and that of Zr-O-Zr' is 110.3°. The distance of the metal atom to Cp ring centroid is 2.220 Å.

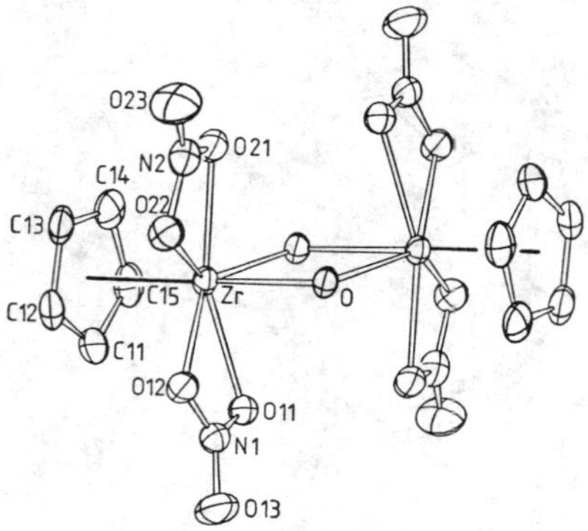

Fig. 1. Molecular structure of $C_{10}H_{12}N_4O_{14}Zr_2$.

(η[5]-CYCLOPENTADIENYL)(η[5]-FORMYLCYCLOPENTADIENYL)IRON (FORMYLFERROCENE)
$C_{11}H_{10}FeO$

K. SATO, M. IWAI, H. SANO and M. KONNO, 1984. Bull. Chem. Soc. Japan, 57, 634-638.

Orthorhombic, $P2_12_12_1$, a = 10.525, b = 11.294, c = 7.639 Å, Z = 4. Mo radiation, R = 0.0372 for 1001 reflexions.

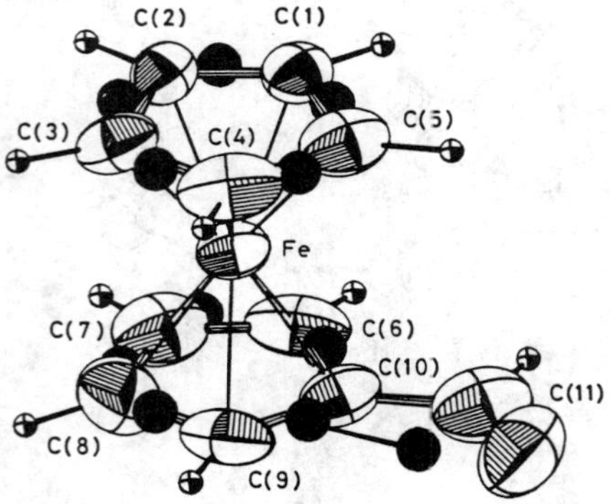

Fig. 1. A view of $C_{11}H_{10}FeO$; solid circles refer to the minor component of the disordered model.

The structure (Fig. 1) is disordered with two alternative ring orientations in the occupancy ratio 0.85/0.15. At 50°C there is a plastic phase with orthorhombic cell constants a = 11.9, b = 16.4, c = 7.0 Å. For previous work see 1.

1. M.F. DANIEL, A.J. LEADBETTER and M.A. MAZID, 1981. J. Chem. Soc. Faraday Trans. 2, 77, 1863.

TRICARBONYL(η^5-CYCLOPENTADIENYL)(η^1-DIMETHYLCARBAMODITHIOATO)TUNGSTEN
$C_{11}H_{11}NO_3S_2W$ $(\eta^5-C_5H_5)W(CO)_3(S_2CNMe_2)$

H.B. ABRAHAMSON, M.L. FREEMAN, M.B. HOSSAIN and D. van der HELM, 1984. Inorg. Chem., 23, 2286-2293.

Triclinic, P$\bar{1}$, a = 7.616, b = 9.962, c = 9.931 Å, α = 113.38, β = 94.68, γ = 95.83°, Z = 2. Mo radiation, R = 0.025 for 3714 reflexions (at 138 K).

The molecule (Fig. 1) has a four-legged piano stool conformation with a monodentate dithiocarbamate ligand. Bond lengths: W-C(Cp) 2.297(3)-2.383(4), W-CO 1.998(4)-2.007(4), W-S 2.496(1) Å. The non-bonded W...S distance is 3.905(1) Å.

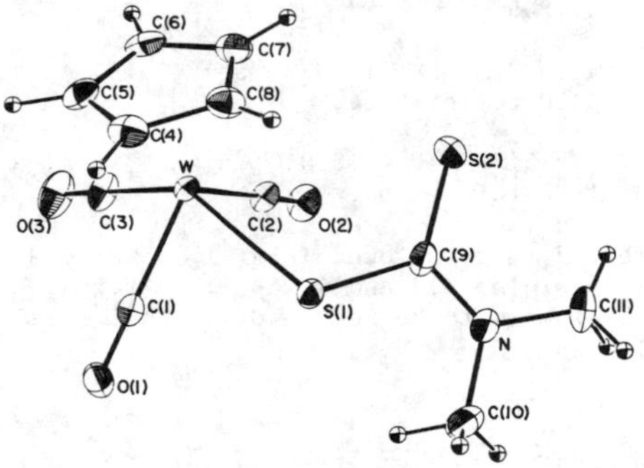

Fig. 1. The $(\eta^5-C_5H_5)W(CO)_3(S_2CNMe_2)$ molecule.

TRICHLORO(η^5-ETHYLTETRAMETHYLCYCLOPENTADIENYL)TITANIUM(IV)
$C_{11}H_{17}Cl_3Ti$ $Ti(C_{11}H_{17})Cl_3$

N.W. ALCOCK, G.E. TOOGOOD and M.G.H. WALLBRIDGE, 1984. Acta Cryst., C40, 598-600.

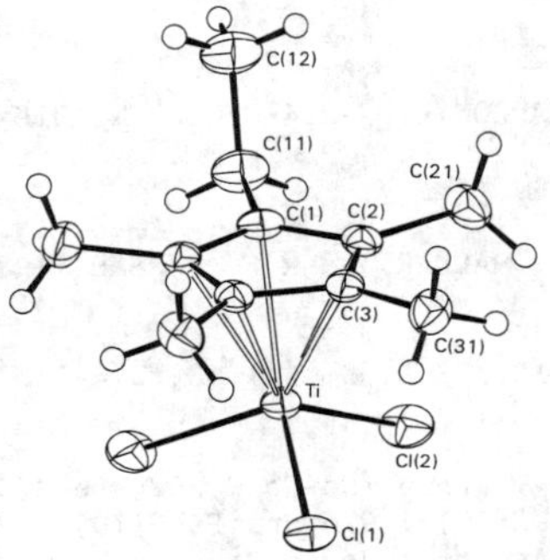

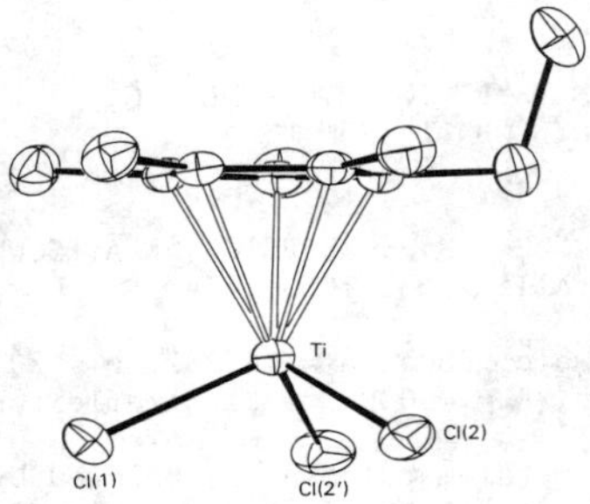

Fig. 1. $Ti(C_{11}H_{17})Cl_3$: perspective views of the molecule.

Monoclinic, $P2_1/m$, a = 6.618, b = 11.400, c = 9.262 Å, β = 95.81°, Z = 2. Mo radiation, R_w = 0.029 for 1368 reflexions (at 153K).

In the molecule (Fig. 1) the C_5 ring is slightly dished with mean Ti-C 2.352(15) Å. The mean Ti-Cl distance of 2.243(10) Å is significantly shorter than in the parent trichloro(η^5-cyclopentadienyl)titanium(IV) (1); this shortening may result from the inductive effect of the alkyl groups. There are no significantly short intermolecular distances and the crystal packing is dominated by $CH_3...CH_3$ and $CH_3...Cl$ contacts.

1. A.W. DUFF, 1982. D. Phil. Thesis, University of Sussex.

(μ_2-DISULPHIDO)(μ_2-TRISULPHIDO)BIS[CARBONYL-(η^5-CYCLOPENTADIENYL)RHENIUM]
$C_{12}H_{10}O_2Re_2S_5$ (I)

(μ_2-SULPHIDO)BIS[(η^5-CYCLOPENTADIENYL)DICARBONYLRHENIUM]
$C_{14}H_{10}O_4Re_2S$ (II)

M. HERBERHOLD, D. REINER, K. ACKERMANN, U. THEWALT and T. DEBAERDEMAEKER, 1984. Z. Naturforsch., B39, 1199-1205.

I. Monoclinic, $P2_1/n$, a = 10.775, b = 15.137, c = 11.152 Å, β = 114.21°, Z = 4. Mo radiation, R = 0.039 for 2180 reflexions.

II. Monoclinic, $P2_1/n$, a = 13.787, b = 12.010, c = 9.398 Å, β = 105.90°, Z = 4. Mo radiation, R = 0.050 for 1992 reflexions.

In (I) a seven-membered ring is found (Fig. 1) with Re-S 2.271-2.247(3), S-S 2.075-2.090(5), Re-Cp(centroid) 1.914 and 1.964 Å. A formal Re-Re single bond in (II) is bridged by a single S atom, with Re-Re 2.946(1), Re-S 2.375 and 2.387(4), Re-Cp(centroid) 1.95 and 1.99 Å.

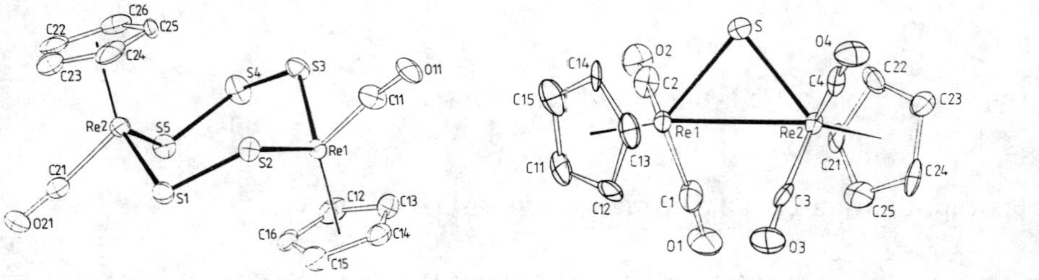

Fig. 1. Views of $C_{12}H_{10}O_2Re_2S_5$ and $C_{14}H_{10}O_4Re_2S$.

[μ-[BIS(METHYLTHIO)METHYLENE-C:C,S]]-μ-CARBONYLCARBONYL(η^5-2,4-CYCLOPENTADIEN-1-YL)-DICARBONYLCOBALT)IRON
$C_{12}H_{11}CoFeO_4S_2$

J.R. MATACHEK, R.J. ANGELICI, K.A. SCHUGART, K.J. HALLER and R.F. FENSKE, 1984. Organometallics, 3, 1038-1044.

Orthorhombic, Pbca, a = 14.252, b = 20.235, c = 10.353 Å, D_m = 1.63, Z = 8. Mo radiation, R = 0.029 for 3126 reflexions (at -60°C).

The metal-metal bond is bridged by a carbonyl group (Co-Fe 2.502, Fe-C(3) 1.952, Co-C(3) 1.899 Å) and a carbene ligand (Co-C(10) 1.916 and Fe-C(10) 1.966 Å) (Fig. 1). One S atom of the bridging $C(SMe)_2$ unit is bonded to the Co, giving a triangular Co-C-S grouping and a highly distorted bridging carbene C atom. Pertinent bond distances and angles are: Co-S(1) 2.275, S(1)-C(10) 1.789, S(2)-C(10) 1.770, S(1)-C(11) 1.810 Å, Co-C(10)-Fe 80.25, Co-S(1)-C(10) 54.70, S(1)-C(10)-Co 75.66,

S(1)-Co-C(10) 49.64, S(2)-C(10)-Fe 127.4 and S(1)-C(10)-S(2) 117.7°. Atoms C(10), S(1), S(2) and Fe are nearly coplanar, the sum of the angles around C(10) to S(1), S(2) and Fe is 356.8°.

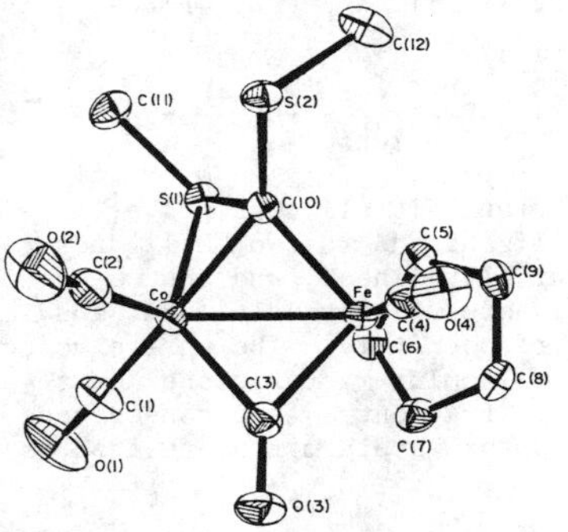

Fig. 1. Molecular structure of $C_{12}H_{11}CoFeO_4S_2$.

ACETYLFERROCENE
$C_{12}H_{12}FeO$

K. SATO, M. KATADA, H. SANO and M. KONNO, 1984. Bull. Chem. Soc. Japan, 57, 2361-2365.

Monoclinic, $P2_1/c$, a = 20.595, b = 5.7909, c = 18.840 Å, β = 116.870°, Z = 8. Mo radiation, R = 0.036 for 3194 reflexions.

Views of one molecule and the crystal packing are in Fig. 1. The C-C distances in the Cp rings are 1.392-1.442 Å, with Fe-C 2.028-2.061 Å for the two independent molecules in the cell. Weissenberg and oscillation photographs at temperatures below the melting point (86°C) are explained in terms of a model in which the molecules rotate around the central Fe atoms.

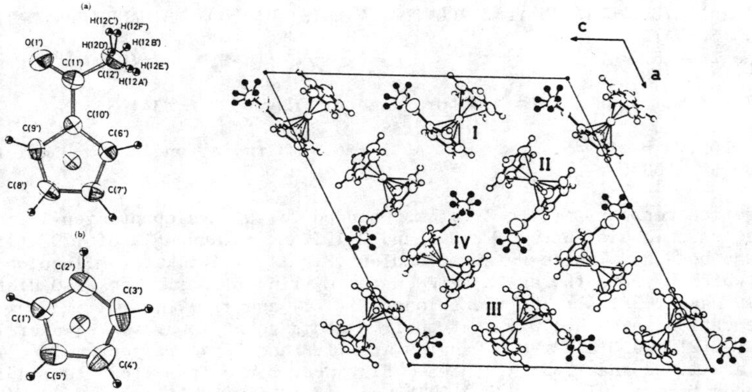

Fig. 1. The two halves of one of the $C_{12}H_{12}FeO$ molecules (acetyl H atoms disordered) and a projection of the crystal structure along b.

(η^5-PENTAMETHYLCYCLOPENTADIENYL)DICARBONYLRHODIUM

$C_{12}H_{15}O_2Rh$ $Rh(C_5(CH_3)_5)(CO)_2$

D.L. LICHTENBERGER, C.H. BLEVINS, II, and R.B. ORTEGA, 1984. Organometallics, 3, 1614-1622.

Monoclinic, $P2_1/n$, a = 7.765, b = 10.741, c = 15.267 Å, β = 98.75°, Z = 4. Mo radiation, R = 0.034 for 1610 reflexions.

The molecular configuration (Fig. 1) consists of an Rh atom bonded to one pentamethylcyclopentadienyl ligand and to two CO ligands. The two Rh-C-O groups are nearly equivalent, are essentially linear, and the OC-Rh-CO bond angle is 90°. The distances Rh-C(1) and Rh-C(2) are respectively 1.863 and 1.845 Å, and C(1)-O(1) and C(2)-O(2) 1.121 and 1.129 Å respectively. The entire molecule has approximate bilateral symmetry with the molecular mirror plane normal to the cyclopentadienyl ring and bisecting the $Rh(CO)_2$ fragment. As a consequence of the pseudo-mirror plane, C(12) and C(15) are approximately mirror-related as are atoms C(13) and C(14) (Fig. 1).

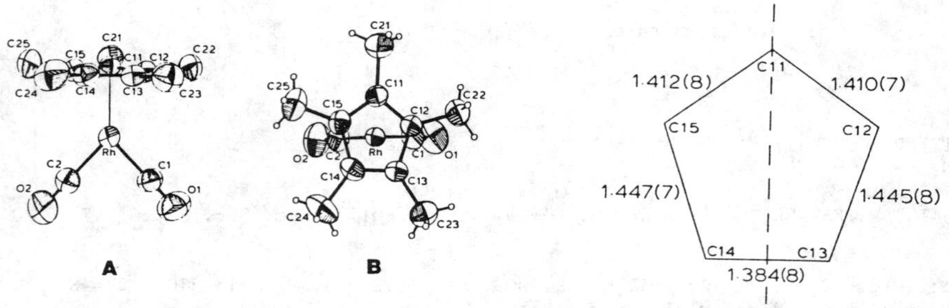

Fig. 1. Two views of the molecule $C_{12}H_{15}O_2Rh$, and bond distances in the five-membered ring. The dashed line represents the plane of pseudo-mirror symmetry.

TRIAQUA-BIS(η^5-CYCLOPENTADIENYL)ZIRCONIUM(IV) BIS(TRIFLUOROMETHANESULPHONATE) TETRA-HYDROFURAN SOLVATE

$C_{12}H_{16}F_6O_9S_2Zr,C_4H_8O$ $[(\pi-C_5H_5)_2Zr(H_2O)_3]^{2+}2CF_3SO_3^-,C_4H_8O$

U. THEWALT and W. LASSER, 1984. J. Organometal. Chem., 276, 341-347.

Hexagonal, $P6_3$, a = 21.945, c = 8.711 Å, Z = 6. Mo radiation, R = 0.054 for 2166 reflexions (at -100°C).

The cation consists of the Zr atom π-bonded to two cyclopentadienyl rings (with Zr-Cp ring centroid distances of 2.210 and 2.193 Å, and an angle of 129° between these vectors) and bonded to three water molecules (Fig. 1). The water molecules lie in the plane which bisects the angle between the C_5 ring planes. The Zr-O distances are 2.239, 2.195 and 2.261 Å. The two anions (Fig. 1) and the THF molecule are packed around the cation in such a way that their O atoms point towards the water ligands. The O(1)-Zr-O(2) and O(2)-Zr-O(3) angles are 71.8 and 73.4° respectively. The C-S distances are 1.820 and 1.836 Å, the S-O distances range from 1.421 to 1.453 Å and the C-F distances from 1.303 to 1.322 Å. The contacts between O(1) and O(22) and O(23), O(2) and O(11) and O(4), and O(3) and O(13) range from 2.59 to 2.74 Å.

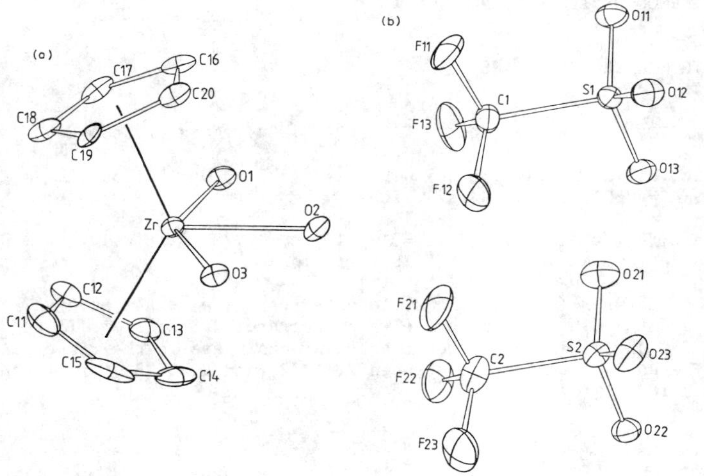

Fig. 1. The cation and anions in $C_{12}H_{16}F_6O_9S_2Zr$.

[3]FERROCENOPHANE-1,3-DIONE

$C_{13}H_{10}FeO_2$ $(C_5H_4COCH_2OC(C_5H_4))Fe$

E. GYEPES, T. GLOWIAK, Š. TOMA and J. SOLDÁNOVÁ, 1984. J. Organometal. Chem., 276, 209-214.

Monoclinic, $P2_1/c$, a = 7.775, b = 9.697, c = 14.669 Å, β = 107.01°, Z = 4. Mo radiation, R = 0.036 for 2087 reflexions.

The three-carbon bridge $-CO-CH_2-CO-$ causes the two cyclopentadienyl rings to tilt, and the dihedral angle between the rings is 9.8° (Fig. 1). The strain in the molecule is apparent by the fact that the bond angles C(11)-C(1)-C(3) (117.0), C(21)-C(2)-C(3) (116.9) and C(1)-C(3)-C(2) (103.7°) are less than normal by about 3-6°. The two exocyclic bonds C(11)-C(1) and C(21)-C(2) are also bent 8.3 and 9.7°, respectively, from the ring planes towards the Fe atom. The Fe-C(Cp) distances vary from 2.004 (C(21)) and 2.005 Å (C(11)) to 2.076 Å (C(23)) and 2.068 Å (C(13)). Other distances are: C(11)-C(1) 1.487, C(1)-C(3) 1.528, C(1)-O(1) 1.218, C(3)-C(2) 1.537, C(2)-O(2) 1.212 and C(2)-C(21) 1.479 Å.

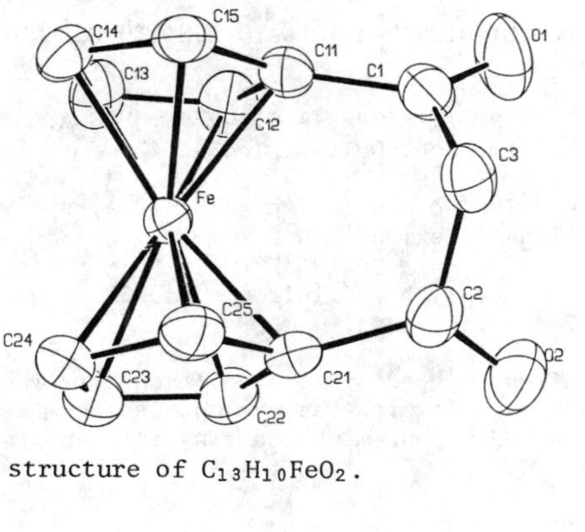

Fig. 1. Molecular structure of $C_{13}H_{10}FeO_2$.

1,3-(1,1'-RUTHENOCENEDIYL)PROPANE
$C_{13}H_{14}Ru$ (I)

1,4-(1,1'-RUTHENOCENEDIYL)BUTANE
$C_{14}H_{16}Ru$ (II)

S. OHBA, Y. SAITO, S. KAMIYAMA and A. KASAHARA, 1984. Acta Cryst., C40, 53-55.

I. Monoclinic, Pa, a = 11.448, b = 7.499, c = 6.273 Å, β = 111.30°, D_m = 1.80, Z = 2. Mo radiation, R = 0.016 for 1389 reflexions.

II. Monoclinic, $P2_1/c$, a = 11.997, b = 12.718, c = 7.725 Å, β = 107.03°, D_m = 1.66, Z = 4. Mo radiation, R = 0.033 for 2025 reflexions.

In both structures (Fig. 1) the cyclopentadienyl (Cp) rings have an eclipsed conformation and the dihedral angles between them are 14.8(2)° for (I) and 1.7(2)° for (II). The C-C-C angles in the carbon connecting chain are in the range 114.5(3) to 116.7(3)°. The Ru-C distances lie between 2.137(2) and 2.208(3) for (I) and between 2.166(3) and 2.188(3) for (II).

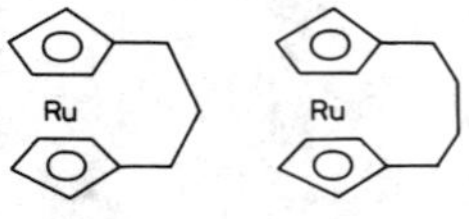

(I) (II)

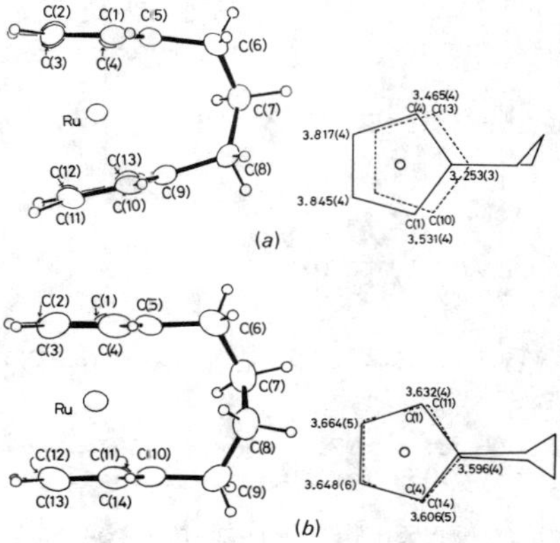

Fig. 1. The $C_{13}H_{14}Ru$ and $C_{14}H_{16}Ru$ molecules. C...C distances between rings are also shown.

DICARBONYL(η^5-CYCLOPENTADIENYL)(3,3-DIMETHYLDITHIOBUTANOATO-S,S')MOLYBDENUM
$C_{13}H_{16}MoO_2S_2$ (I)

DICARBONYL(η^5-CYCLOPENTADIENYL)(DITHIO-p-TOLUATO-S,S')TUNGSTEN
$C_{15}H_{12}O_2S_2W$ (II)

D.S. GILL, M. GREEN, K. MARSDEN, I. MOORE, A.G. ORPEN, F.G.A. STONE, I.D. WILLIAMS and P. WOODWARD, 1984. J. Chem. Soc. Dalton, 1343-1347.

I. Monoclinic, $P2_1/c$, a = 9.392, b = 6.344, c = 25.412 Å, β = 94.79°, Z = 4. Mo radiation, R = 0.033 for 2165 reflexions.

II. Monoclinic, $P2_1/c$, a = 16.001, b = 6.161, c = 14.932 Å, β = 96.32°, Z = 4. Mo radiation, R = 0.042 for 2382 reflexions.

Both molecules adopt 'four-legged piano stool' geometry in which the metal atoms can be regarded as seven-coordinate with the cyclopentadienyl ligands occupying three sites (Fig. 1). In both the MS_2C four-membered ring is planar with M-S 2.47, C-S 1.68 Å, S-M-S 68 and S-C-S 110°.

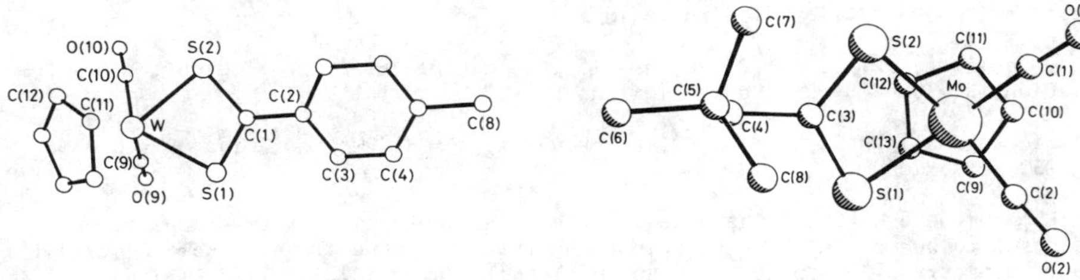

Fig. 1. The molecular structure viewed approximately parallel to ($C_{15}H_{12}O_2S_2W$) and perpendicular to ($C_{13}H_{16}MoO_2S_2$) the cyclopentadienyl plane.

DICARBONYL(η^5-CYCLOPENTADIENYL)(3-t-BUTYL-1,3,2-OXOZAPHOSPHOLIUM-P)MOLYBDENUM
$C_{13}H_{18}MoNO_3P$

L.D. HUTCHINS, E.N. DUESLER and R.T. PAINE, 1984. Organometallics, 3, 399-403.

Monoclinic, $P2_1/n$, a = 12.072, b = 9.957, c = 13.349 Å, β = 108.09°, Z = 4. Mo radiation, R = 0.035 for 2310 reflexions.

The structure consists of a pseudooctahedral molybdenum atom bonded to a (η^5-C_5H_5) ring, two terminal CO groups and the P atom of the phosphenium ion (Fig. 1). The phosphenium ion is planar, and the Mo-P distance of 2.207 Å indicates multiple bonding. The average Mo-C(carbonyl) distance is 1.948 Å and the C(1)-Mo-C(2) angle is 86.1°. The average Mo-C(ring) distance is 2.333 Å and the Mo-Cp(centroid) distance is 2.021 Å. Atoms Mo, P, N and O form a plane which is nearly perpendicular to the $Mo(CO)_2$ plane (dihedral angle 85.5°) and to the Cp plane (angle 89.3°). The P-N and P-O(3) distances are 1.660 and 1.616 Å respectively and the remaining ring bonds O(3)-C(9), C(8)-C(9) and C(8)-N are respectively 1.435, 1.495 and 1.479 Å in length.

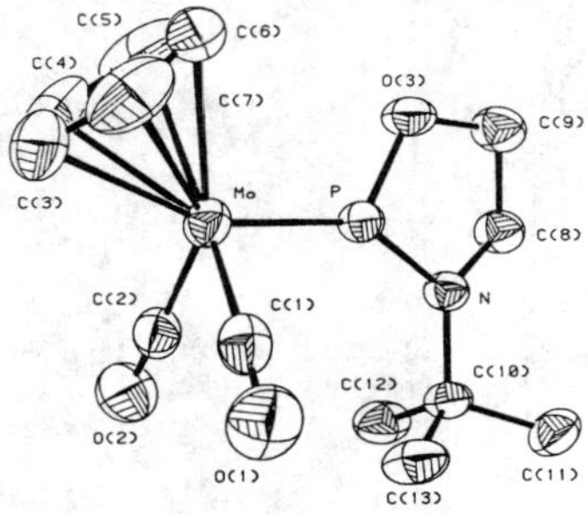

Fig. 1. Molecular structure of $C_{13}H_{18}MoNO_3P$.

BIS(η^5-CYCLOPENTADIENYL)TRIMETHYLPHOSPHINEMANGANESE(II)
$C_{13}H_{19}MnP$ (I) $(Cp)_2Mn(PMe_3)$

BIS(η^5-CYCLOPENTADIENYL)METHYLDIPHENYLPHOSPHINOMANGANESE(II)
$C_{23}H_{23}MnP$ (II) $(Cp)_2Mn(PMePh_2)$

BIS(η^5-CYCLOPENTADIENYL)-1,2-BIS(DIMETHYLPHOSPHINO)ETHANEMANGANESE(II)
$C_{16}H_{26}MnP_2$ (III) $(Cp)_2Mn(dmpe)$

C.G. HOWARD, G.S. GIROLAMI, G. WILKINSON, M. THORNTON-PETT and M.B. HURSTHOUSE, 1984. J. Am. Chem. Soc., 106, 2033-2040.

I. Monoclinic, $P2_1/c$, a = 8.243, b = 15.303, c = 11.476 Å, β = 108.04°, Z = 4. Mo radiation, R = 0.0421 for 2195 reflexions.

II. Monoclinic, $P2_1/c$, a = 9.080, b = 10.161, c = 21.769 Å, β = 100.66°, Z = 4. Mo radiation, R = 0.0400 for 2909 reflexions.

III. Tetragonal, $P\bar{4}n2$, a = 13.892, c = 9.050 Å, Z = 4. Mo radiation, R = 0.0360 for 1252 reflexions.

The compounds (Fig. 1) have pseudotrigonal geometry if the Cp groups are each considered to occupy only one coordination site. Compound III has twofold crystallographic symmetry. The Cp(cent)-Mn-Cp(cent) angles are 142.3(2), 142.1(3) and 137° (in I, II and III respectively). Some dimensions (for I, II, III in that order) are Mn-P 2.577(4), 2.613(4), 2.674(4), Mn-C 2.481(5)-2.501(6), 2.336-2.724(8), 2.492-2.742(8) Å. One of the rings in II is disordered over two sites and in I both rings are disordered. Tilt angles between Cp ring planes and the Mn-Cp(centroid) vector are 6.1 and 1.9° (for I), 0 and 10° (for II) and 7.3° (for III).

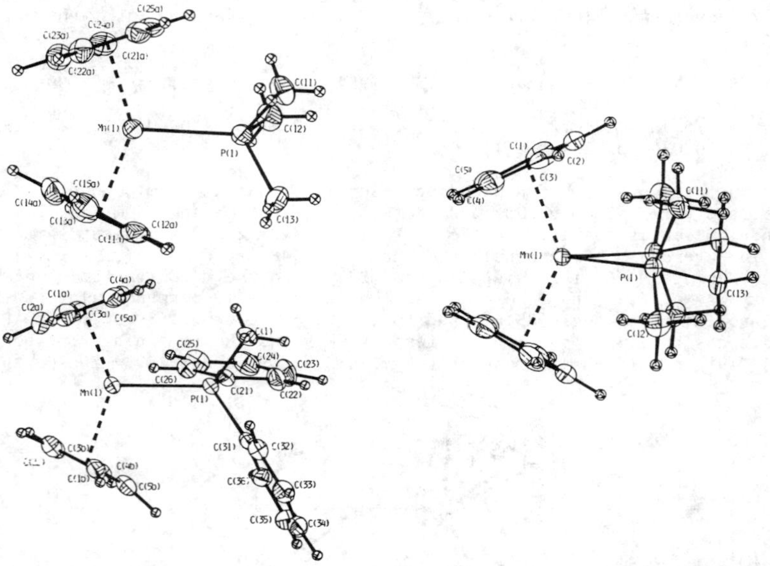

Fig. 1. Views of $(Cp)_2Mn(PMe_3)$ (I) (top left), $(Cp)_2Mn(PMePh_2)$ (II) (lower Left) and $(Cp)_2Mn(dmpe)$ (III) (right).

BIS(μ-HYDRIDO)TRIMETHYLALUMINUMBIS(η^5-CYCLOPENTADIENYL)TUNGSTEN

$C_{13}H_{21}AlW$ $(C_5H_5)_2W(\mu\text{-}H)_2Al(CH_3)_3$

J.W. BRUNO, J.C. HUFFMAN and K.G. CAULTON, 1984. J. Am. Chem. Soc., 106, 444-445.

Orthorhombic, Pmcn, a = 13.166, b = 8.981, c = 11.073 Å, Z = 4. Mo radiation, R = 0.019 (at -165°C).

The molecule (Fig. 1) has crystallographic mirror symmetry. The two hydride ligands interact with Al as well as W resulting in five- or six-coordinate Al. Important structural parameters: W-H 1.70(5) and 1.22(17), Al-H 1.96(5) and 2.08(16), W-Al 3.110(3) Å, H-W-H 60(7)°.

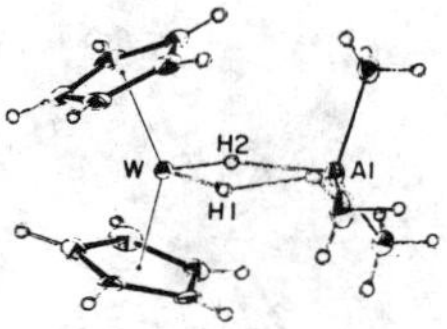

Fig. 1. The $(C_5H_5)_2W(\mu\text{-}H)_2Al(CH_3)_3$ molecule.

(η^5-CYCLOPENTADIENYL)NONACARBONYLRHODIUMDIOSMIUM
$C_{14}H_5O_9Os_2Rh$ $(C_5H_5)RhOs_2(CO)_9$

L.-Y. HSU, W.-L. HSU, D.-Y. JAN, A.G. MARSHALL and S.G. SHORE, 1984. Organometallics, 3, 591-595.

Triclinic, $P\bar{1}$, a = 9.128, b = 13.077, c = 8.224 Å, α = 106.94, β = 113.51, γ = 80.95°, Z = 2. Mo radiation, R = 0.057 for 4497 reflexions (at -80°C).

The molecule (Fig. 1) has approximate C_s symmetry and consists of an Os_2Rh triangle with four terminal CO groups bound to each Os atom, and a terminal CO and a η^5-C_5H_5 group bound to the Rh atom. The metal-metal distances are: Rh-Os(1) 2.7850 Å, Rh-Os(2) 2.7673 Å, and Os(1)-Os(2) 2.8455 Å. The average Rh-C(Cp) distance is 2.249 Å and the Rh-C(carbonyl) distance is 1.833 Å. The Os(1)-C(carbonyl) and Os(2)-C(carbonyl) distances average 1.924 and 1.934 Å respectively. The angles Os(1)-Rh-Os(2), Rh-Os(1)-Os(2) and Rh-Os(2)-Os(1) are respectively 61.65, 58.89 and 59.46°.

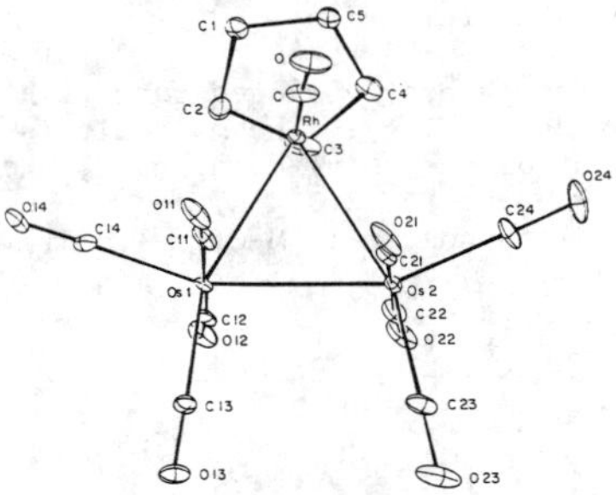

Fig. 1. Molecular structure of $C_{14}H_5O_9Os_2Rh$.

BIS(μ_2-CARBONYL)(η^5-METHYLCYCLOPENTADIENYL)(μ_3-THIO)BIS(TRICARBONYLIRON)MANGANESE
$C_{14}H_7Fe_2MnO_8S$ $Fe_2(CO)_6CH_3C_5H_4Mn(\mu_2\text{-}CO)_2(\mu_3S)$

A. WINTER, L. ZSOLNAI and G. HUTTNER, 1984. J. Organometal. Chem., 269, C29-C32.

The complex consists of a triangle of metal atoms triply-bridged by a sulphide group (Fig. 1). Both Mn-Fe bonds are also bridged by CO groups. The main bond distances of interest are: Fe(1)-Fe(2) 2.609, Fe(1)-Mn 2.660, Fe(2)-Mn 2.689, Fe(1)-S 2.175, Fe(2)-S 2.178, Mn-S 2.206 Å, Fe(1)-C(7), Fe(2)-C(8) 2.306 and 2.336 Å respectively, and Mn-C(7), Mn-C(8) 1.827 and 1.829 Å respectively. The Fe-CO(terminal) Fe-C distances range from 1.780 to 1.814 Å. The Mn atom is π-bonded to a methylcyclopentadienyl ligand, with Mn-C(Cp) distances ranging from 2.122 to 2.190 Å.

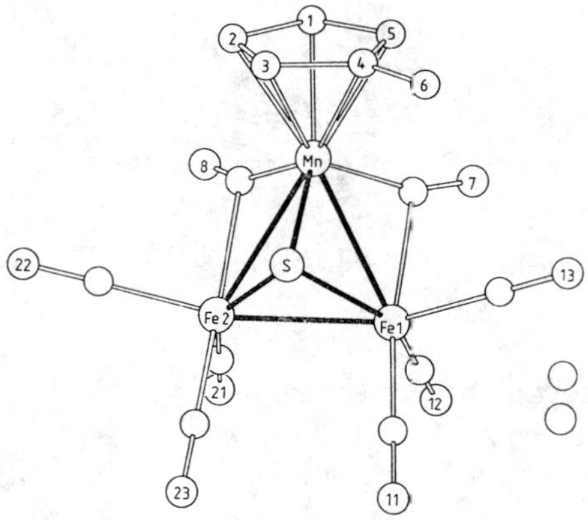

Fig. 1. Molecular structure of $C_{14}H_7Fe_2MnO_8S$.

trans-BIS(μ_2-CARBONYL)-DICARBONYL-BIS(η^5-CYCLOPENTADIENYL)-IRON-RUTHENIUM
$C_{14}H_{10}FeO_4Ru$ trans-[$FeRu(CO)_2(\mu-CO)_2(\eta-C_5H_5)_2$]

B.P. GRACEY, S.A.R. KNOX, K.A. MACPHERSON, A.G. ORPEN and S.R. STOBART, 1984. J. Organometal. Chem., 272, C45-C48.

Monoclinic, $P2_1/c$, a = 7.064, b = 12.518, c = 8.011 Å, β = 106.23°, D_m = 1.92, Z = 2. Mo radiation, R = 0.028 for 1532 reflexions.

The complex exists as the carbonyl-bridged form with a trans arrangement of the terminal CO and η-C_5H_5 ligands (Fig. 1). The molecule is disordered about a centre of inversion at the mid-point of the metal-metal bond, the length of which (2.626 Å) is approximately midway between Fe-Fe (2.534 Å) and Ru-Ru (2.735 Å) single bond distances. Other averaged distances are: M-C(1) 1.974, M-C(2) 1.788, C(1)-O(1) 1.179 and C(2)-O(2) 1.148 Å.

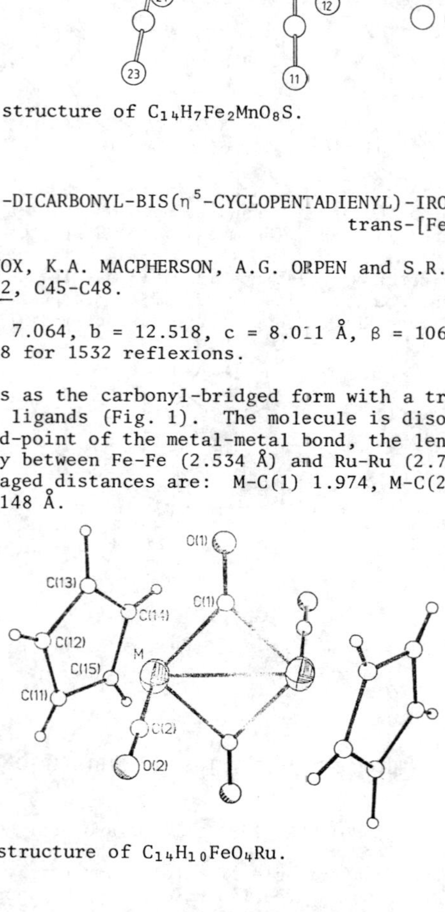

Fig. 1. Molecular structure of $C_{14}H_{10}FeO_4Ru$.

(μ_2-η^2-DIPHOSPHORUS)BIS((η^5-CYCLOPENTADIENYL)DICARBONYLMOLYBDENUM)
$C_{14}H_{10}Mo_2O_4P_2$ $P_2[Mo(CO)_2(\eta^5-C_5H_5)]_2$

O.J. SCHERER, H. SITZMANN and G. WOLMERSHAUSER, 1984. J. Organometal. Chem., 268, C9-C12.

Monoclinic, C2/c, a = 13.435, b = 7.312, c = 16.971 Å, β = 105.44°, Z = 4. Mo radiation, R = 0.021 for 1315 reflexions.

The complex has a 'tetrahedral' skeleton consisting of two Mo and two bridging P atoms (Fig. 1) and it lies on a crystallographic twofold axis. The important bond distances and angles are: P-P' 2.079, Mo-Mo' 3.022, P-Mo 2.552, P-Mo' 2.463 Å; Mo-P-Mo' 74.1, P-Mo-Mo' 51.6, P-Mo'-Mo 54.3, and P-Mo-P' 49.0°.

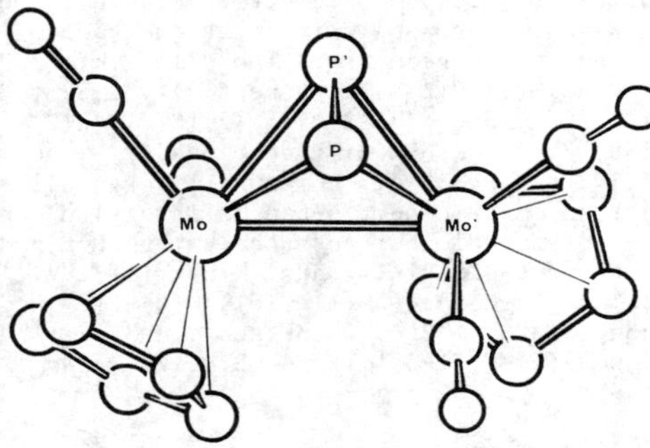

Fig. 1. Structure of $C_{14}H_{10}Mo_2O_4P_2$.

(μ-VINYLIDENE)BIS[(η[5]-CYCLOPENTADIENYL)CARBONYLCOBALT]
$C_{14}H_{12}Co_2O_2$ (μ-CCH_2)$(CpCoCO)_2$

E.N. JACOBSEN and R.G. BERGMAN, 1984. Organometallics, 3, 329-331.

Triclinic, P$\bar{1}$, a = 8.2535, b = 9.0316, c = 9.1980 Å, α = 74.427, β = 76.745, γ = 84.514°, Z = 2. [?] radiation, R = 0.023 for 2522 reflexions.

The molecule has approximate twofold rotational symmetry (Fig. 1) and the vinylic H atoms are nearly coplanar with the metal atoms (3° twist angle). The metal-metal distance is 2.500 Å. Other distances of interest are: Co(1)-C(1) 1.726, Co(2)-C(2) 1.722, Co(1)-C(3) 1.886, Co(2)-C(3) 1.882, C(3)-C(4) 1.312, Co(1)-CP(1) 1.716, and Co(2)-CP(2) 1.724 Å. The angles Co(1)-Co(2)-C(3), Co(2)-Co(1)-C(3) and Co(1)-C(3)-Co(2) are respectively 48.50, 48.37, and 83.13°.

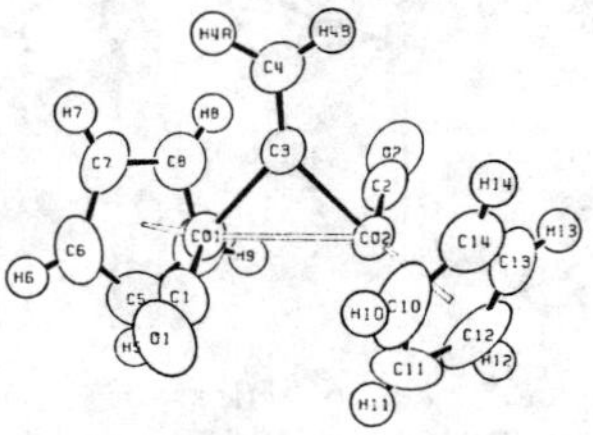

Fig. 1. Molecular structure of $C_{14}H_{12}Co_2O_2$.

BIS[($μ_2$-SULPHIDO)(η[5]-CYCLOPENTADIENYL)ETHENEDITHIOLATOTUNGSTEN]
$C_{14}H_{14}S_6W_2$ (I) $[(η^5\text{-}C_5H_5)W(S)SC_2H_2S]_2$

BIS[($μ_2$-η[4]-ETHANEDITHIOLATO)(η[5]-CYCLOPENTADIENYL)VANADIUM]
$C_{14}H_{18}S_4V_2$ (II) $[(η^5\text{-}C_5H_5)VSC_2H_4S]_2$

O.A. RAJAN, M. McKENNA, J. NOORDIK, R.C. HALTIWANGER and M.R. DUBOIS, 1984. Organometallics, 3. 831-840.

I. Monoclinic, $P2_1/c$, a = 7.992, b = 23.659, c = 14.506 Å, β = 100.62°, Z = 6. Mo radiation, R = 0.031 for 1821 reflexions.

II. Monoclinic, $P2_1/c$, a = 17.394, b = 7.647, c = 13.035 Å, β = 104.98°, Z = 4. Mo radiation, R = 0.041 for 2382 reflexions.

Compound I crystallises with 1.5 molecules per asymmetric unit. Both molecules (Fig. 1) involve similar bis(μ-sulphido) geometries with a 1,2-ethenedithiolate ligand chelated to each W ion. Each CpW unit is at the apex of an approximate square pyramid in a typical "piano stool" geometry. The W(1)...W(1') distance is 3.021 Å and W(1)-S(1), W(1)-S(2) and W(1)-S(3) are respectively 2.328, 2.387 and 2.412 Å. The W(2)...W(3) distance is 3.056 Å and W(2)-S(4), W(2)-S(5), W(2)-S(6) and W(2)-S(7) are respectively 2.350, 2.359, 2.392, 2.392 Å. Additionally, bonds W(3)-S(4) and W(3)-S(5) are respectively 2.342 and 2.338 Å. Compound II crystallises with two independent molecules per asymmetric unit sited on inversion centres. The two dithiolate ligands form a plane which bisects the V-V vector and which is parallel to the planes of the two cyclopentadienyl ligands. The V(1)...V(1') distance is 2.541 Å with V(1)-S(1), V(1)-S(2), V(1)-Cp(av), S(1)-C(6), S(2)-C(7) and C(6)-C(7) respectively 2.416, 2.434, 2.300, 1.819, 1.835 and 1.493 Å. The angles V(1)-S(2)-V(1'), S(1)-V(1)-S(2), and S(1)-V(1)-S(1') are 63.4, 76.0 and 116.6° respectively.

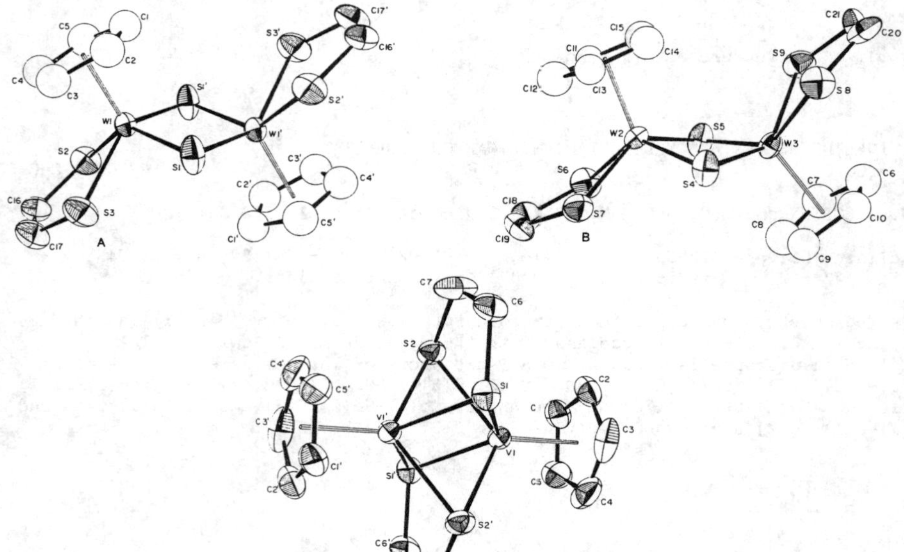

Fig. 1. Perspective drawings of the molecules of $C_{14}H_{14}S_6W_2$ (top left and right), and of one of the molecules of $C_{14}H_{18}S_4V_2$ (bottom).

BIS((η⁵-CYCLOPENTADIENYL)VINYLOXY)TITANIUM

$C_{14}H_{16}O_2Ti$ $(C_5H_5)_2Ti(OCHCH_2)_2$

M.D. CURTIS, S. THANEDAR and W.M. BUTLER, 1984. Organometallics, 3, 1855-1859.

Orthorhombic, Pbcn, a = 14.046, b = 6.099, c = 14.709 Å, Z = 4. Mo radiation, R = 0.038 for 744 reflexions.

The molecule has crystallographically-imposed C_2 symmetry and the cyclopentadienyl rings are staggered (Fig. 1). The structural parameters are all normal and are consistent with some degree of multiple bonding between the O atoms and Ti. The most interesting part of the structure is the enolate ligand which is wrapped back toward the Ti atom - a rather unusual feature, thought to be due to packing considerations in the solid state.

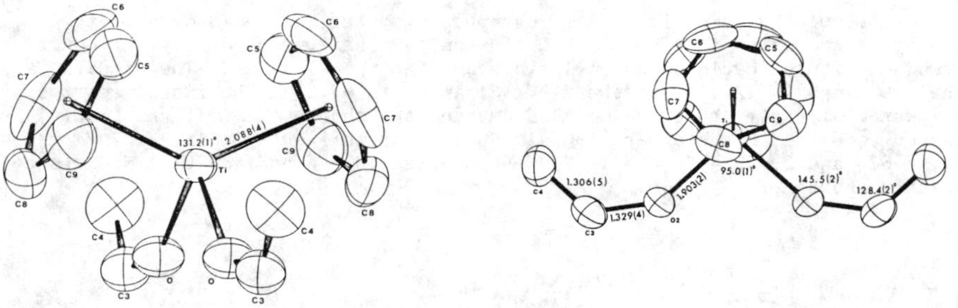

Fig. 1. Two views of $C_{14}H_{16}O_2Ti$ showing selected bond distances (Å) and angles (°).

2-ACETYL-1-(η^5-CYCLOPENTADIENYL)-1,1-DICARBONYL-3-TRIMETHYLPHOSPHINE-1-TUNGSTACYCLOPROPANE DICHLOROMETHANE SOLVATE
$C_{14}H_{19}O_3PW,CH_2Cl_2$

H.G. ALT and U. THEWALT, 1984. J. Organometal. Chem., 268, 235-245.

Monoclinic, $P2_1/n$, a = 9.747, b = 19.486, c = 10.163 Å, β = 101.20°, Z = 4. Mo radiation, R = 0.041 for 2943 reflexions (-40°C).

Of especial interest is the presence of a three-membered metallacyclopropane ring (Fig. 1). The W-C(3) and W-C(4) distances are 2.206 and 2.279 Å respectively, compared to the W-C(CO) distances of 1.938 and 1.931 Å. The distance from the centroid of the Cp ring to the metal atom is 2.009 Å. Angles within the three-membered ring are: C(3)-W-C(4) 37.6, W-C(3)-C(4) 73.9 and W-C(4)-C(3) 68.5°, and the C(3)-C(4) distance is 1.45 Å. In the trimethylphosphine group the P-C(3) distance is 1.762 Å and the $P-CH_3$ distances are 1.79, 1.81 and 1.81 Å. The C(4)-C(5) distance is 1.48 Å and the angle C(1)-W-C(2) is 77.6°.

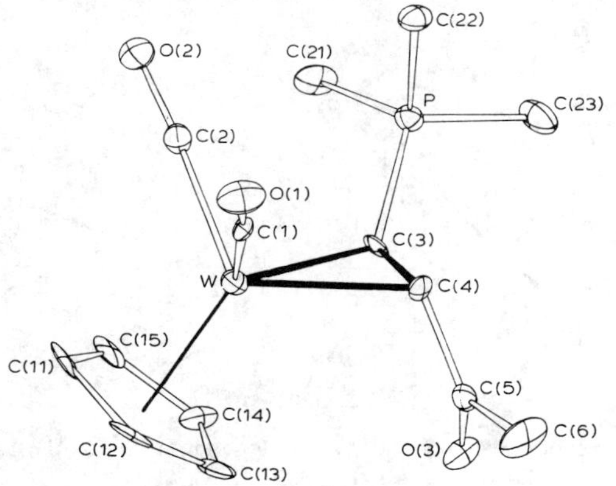

Fig. 1. Molecular structure of $C_{14}H_{19}O_3PW$.

1,1'-BIS(N,N-DIMETHYLAMINO)TITANOCENE DICHLORIDE
$C_{14}H_{20}Cl_2N_2Ti$ $[C_5H_4N(CH_3)_2]_2TiCl_2$

K.-P. STAHL, G. BOCHE and W. MASSA, 1984. J. Organometal. Chem., <u>277</u>, 113-125.

Monoclinic, C2/c, a = 13.014, b = 6.472, c = 18.285 Å, β = 98.81°, Z = 4. Mo radiation, R = 0.042 for 1175 reflexions.

The molecule has twofold crystallographic symmetry, the axis of symmetry indicated by the arrow in Fig. 1. The Ti atom is quasi-tetrahedrally surrounded by the two substituted cyclopentadienyl rings and the two Cl atoms. The dihedral angle of the two rings is 125.7° and the Cl-Ti-Cl angle is 97.5°. The rings' centres are not situated directly above the metal atom, and the shorter Ti-C(3) and Ti-C(4) distances compared to the other Ti-C distances bear this out (2.318, 2.329 Å vs. 2.564, 2.376 and 2.423 Å). The Ti-Cl distance is 2.373 Å and the C(1)-N distance is remarkably short at 1.347 Å.

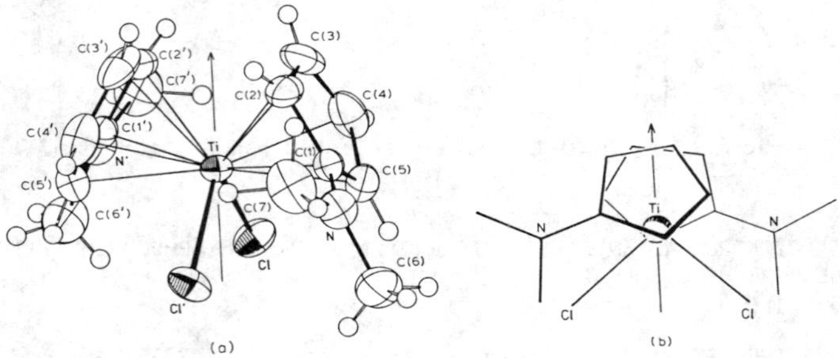

Fig. 1. Molecular structure of $C_{14}H_{20}Cl_2N_2Ti$ (left), and a schematic diagram of the projection on the $TiCl_2$ plane (right).

1,1-BIS(η^5-CYCLOPENTADIENYL)-3,3-DIMETHYL-1-TITANA-3-SILACYCLOBUTANE
$C_{14}H_{20}SiTi$ $(\eta^5-C_5H_5)_2Ti(CH_2Si(CH_3)_2CH_2)$

W.R. TIKKANEN, J.Z. LIU, J.W. EGAN, Jr., and J.L. PETERSEN, 1984. Organometallics, 3, 825-830.

Monoclinic, $P2_1/m$, a = 7.864, b = 11.637, c = 15.279 Å, β = 102.01°, Z = 4. R = 0.047 for 2074 reflexions.

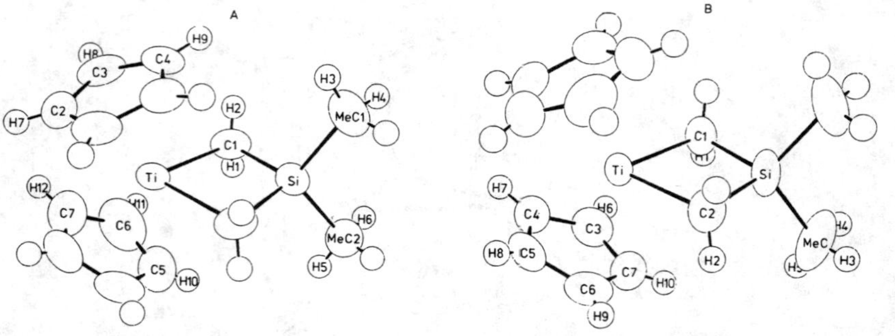

Fig. 1. Molecular structures of the two independent molecules in $C_{14}H_{20}SiTi$.

The asymmetric unit consists of two independent molecules each on a crystallographic mirror plane (Fig. 1). In one molecule the mirror plane contains Ti, Si, MeC(1) and MeC(2) atoms and C(2) and C(5) of the Cp rings whereas in the second molecule the mirror plane contains Ti, Si, C(1) and C(2). The two cyclopentadienyl rings are thus staggered in the former and eclipsed in the latter. The Ti coordination is pseudotetrahedral with Ti-C(1), Ti-C(1) and Ti-C(2) distances of 2.146, 2.169 and 2.152 Å respectively in the two molecules. The Ti-Cp(centroid) distances are: in molecule 1, 2.088 and 2.084 Å, and in molecule 2, 2.072 Å. The Cp(centroid)-Ti-Cp(centroid) angles are 132.9 and 132.5°, and the angles C(1)-Ti-C(1') and C(1)-Ti-

C(2) are respectively 84.1 and 83.7°. The Si-C(1) distance is 1.863 Å in molecule 1 and Si-C(1) and Si-C(2) are 1.865 and 1.855 Å in molecule 2, with C(1)-Si-C(1') and C(1)-Si-C(2) angles of 101.0 and 101.6° respectively.

BIS(μ_2-HYDRIDO)BIS(η^5-CYCLOPENTADIENYL)TITANIUMDICHLORO(DIETHYL ETHER)ALUMINIUM
$C_{14}H_{22}AlCl_2OTi$ $(\eta^5-C_5H_5)_2Ti(\mu-H)_2AlCl_2.O(C_2H_5)_2$

E.B. LOBKOVSKII, G.L. SOLOVEICHIK, B.M. BULYCHEV, R.G. GERR and Yu.T. STRUCHKOV, 1984. J. Organometal. Chem., 270, 45-51.

Orthorhombic, Pbca, a = 12.322, b = 15.090, c = 37.941 Å, Z = 16. Mo radiation, R = 0.071 for 1735 reflexions.

The Ti atom has distorted tetrahedral coordination involving two cyclopentadienyl moieties and two bridging H atoms (Fig. 1). The coordination polyhedron at the Al atom is a distorted trigonal bipyramid in which the axial positions are occupied by the bridging H atom and the O atom from diethyl ether. The average Ti-C(Cp) bond length is 2.33 Å and the dihedral angle between the two Cp rings is 137.2 and 137.0° (in the two independent molecules per asymmetric unit). The H-Ti-H angle is 70 and 80° in the two molecules and the Ti-H distance averages 1.80 Å. The Al-Cl(1), Al-Cl(2) and Al-O(1) distances in the two molecules are: 2.149 and 2.160 Å, 2.172 and 2.143 Å, and 1.956 and 1.979 Å. The average Al-H distance is 1.73 Å and the H(1)-Al(1)-O(1) angle (and its equivalent in molecule 2) is 158°.

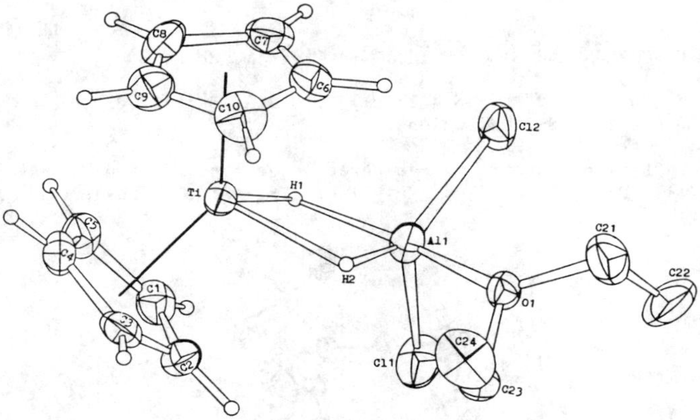

Fig. 1. Molecular structure of $C_{14}H_{22}AlCl_2OTi$.

1-CARBONYL-1-(η^5-CYCLOPENTADIENYL)-2,2-DIMETHYL-3-METHYL-1-(TRIMETHYLPHOSPHINE)-1-TUNGSTA-2-PHOSPHABICYCLO[1.1.0]BUTAN-4-ONE CHLORIDE TRIS(METHYLENE CHLORIDE) SOLVATE
$C_{14}H_{23}ClO_2P_2W,3CH_2Cl_2$

F.R. KREISSL, M. WOLFGRUBER, W.J. SIEBER and K. ACKERMANN, 1984. Organometallics, 3, 777-782.

Monoclinic, $P2_1/c$, a = 15.447, b = 10.908, c = 17.603 Å, β = 106.08°, Z = 4. Mo radiation, R = 0.046 for 3410 reflexions (at -40°C).

The structure can be regarded as a ketene complex with a P-substituted ketene ligand and a cyclopentadienylbis-(phosphine)carbonyltungsten fragment (Fig. 1). Within the W-P-C ring, the W-C(7) (2.364 Å), W-P(2) (2.396 Å) and the P(2)-C(7) (1.757 Å) bond distances are considered normal. The W-C(8) (2.126 Å) and the C(7)-C(8) (1.44 Å) bond lengths are considered rather short. The W-C-C and W-P-C planes form a dihedral angle of 118.8°. The W-C (Cp) distances range from 2.284 to 2.364 Å (average 2.314 Å) and the bonds W-P(1) and W-C(1) are respectively 2.500 and 2.019 Å in length.

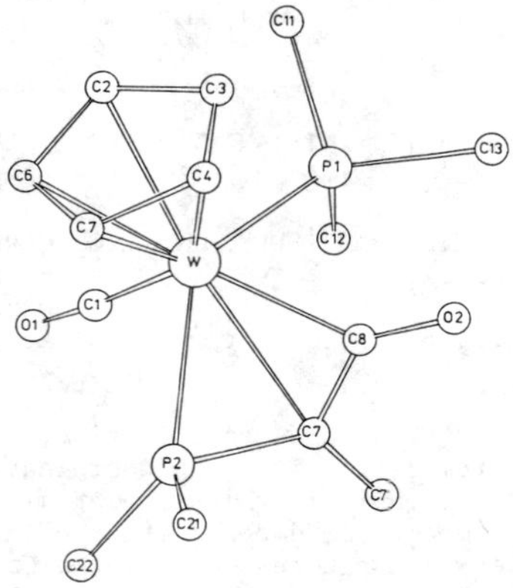

Fig. 1. Molecular structure of $C_{14}H_{23}ClO_2P_2W$.

(η[5]-CYCLOPENTADIENYL)DICARBONYL(BIS(TRIMETHYLSILYL)METHYLIDENEPHOSPHINE-P)MOLYBDENUM
$C_{14}H_{23}MoO_2PSi_2$

A.H. COWLEY, N.C. NORMAN and S. QUASHIE, 1984. J. Am. Chem. Soc., 106, 5007-5008.

Monoclinic, $P2_1/c$, a = 18.90, b = 6.432, c = 17.009 Å, β = 107.83°, Z = 4. Mo radiation, R = 0.040 for 2857 reflexions.

The Mo coordination geometry is the three-legged "piano-stool" type (Fig. 1), with Mo-P 2.174(1), P1-C1 1.649(4) Å, Mo1-P1-C1 178.3(2)°. Methylene carbon C1 is trigonal planar.

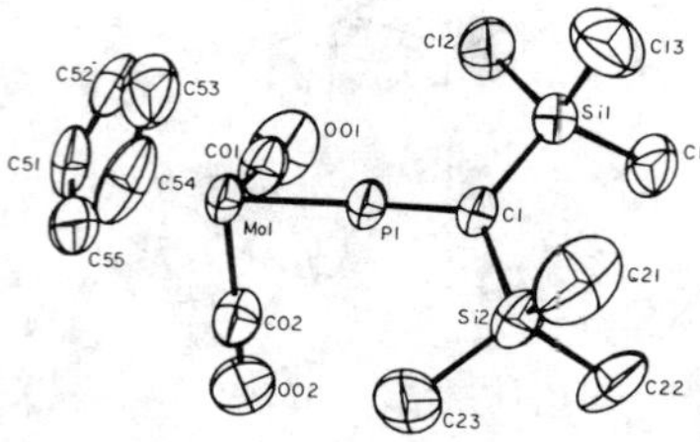

Fig. 1. A view of $C_{14}H_{23}MoO_2PSi_2$.

DICARBONYL(η[5]-CYCLOPENTADIENYL) (η[1]-CYMANTRENYL) IRON
$C_{15}H_9FeMnO_5$ (I) $(OC)_3MnC_5H_4Fe(C_5H_5)(CO)_2$

CARBONYL(η[5]-CYCLOPENTADIENYL) (η[1]-CYMANTRENYL) (TRIPHENYLPHOSPHINE) IRON
$C_{32}H_{24}FeMnO_4P$ (II) $(OC)_3MnC_5H_4Fe(C_5H_5)(CO)(P(C_6H_5)_3)$

A.S. BATSANOV and Yu.T. STRUCHKOV, 1984. J. Organometal. Chem., 266, 295-301.

I. Monoclinic, $P2_1/c$, a = 8.592, b = 13.521, c = 12.481 Å, β = 94.92°, D_m = 1.74, Z = 4. Mo radiation, R = 0.057 for 1546 reflexions.

II. Monoclinic, $P2_1/c$, a = 10.783, b = 14.294, c = 18.105 Å, β = 92.77°, D_m = 1.46, Z = 4. Mo radiation, R = 0.098 for 2053 reflexions.

In both complexes the Fe atom is coordinated by a planar η^5-C_5H_5 ligand, two two-electron donors (CO or phosphine) and a σ-bonded C(6) atom of the cymantrenyl moiety in a usual "piano-stool" arrangement (Fig. 1). The Fe-C(6) σ-bonds are 2.001 Å in I and 1.99 Å in II and are essentially equal. In the cymantrenyl moiety itself, the Mn atom is further from the C(6) atom than from the other ring carbons. The Mn-C(6) distances in I and II are 2.216 and 2.25 Å respectively, versus the average of the other four Mn-C(Cp) distances which are 2.14 Å in I and 2.15 Å in II. The differences in bond distances and angles between I and II are not significant. Conformational differences are more pronounced e.g. the dihedral angle between the planes of the Cp rings is 81.3° in I and 66° in II. The Fe...Mn distances are 3.861 Å in I and 3.852 Å in II, and the Fe-P distance in II is 2.201 Å.

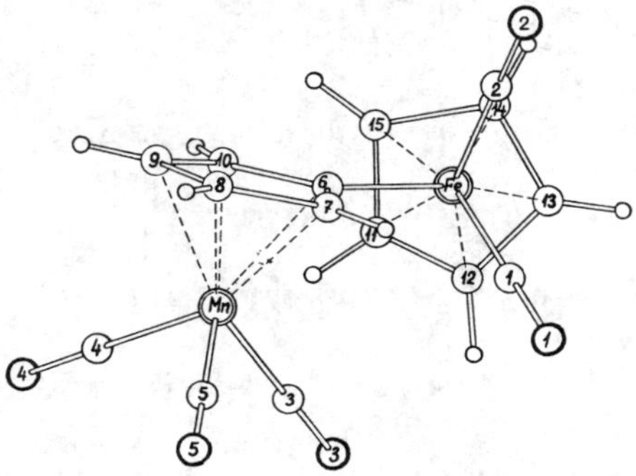

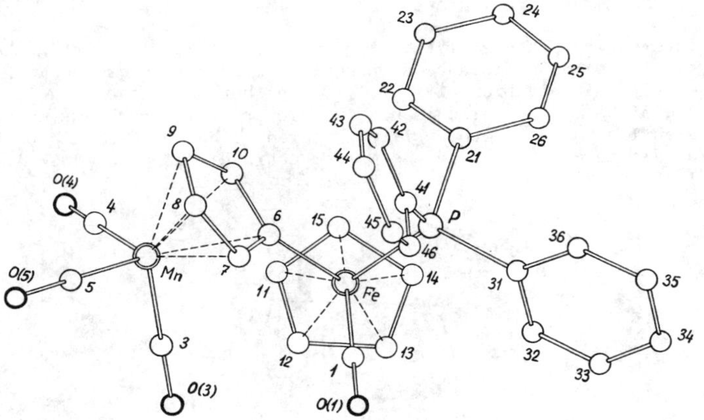

Fig. 1. Molecular structure of $C_{15}H_9FeMnO_5$ projected on the mean plane of the Cp ring C(11)...C(15) (top), and molecular structure of $C_{32}H_{24}FeMnO_4P$ (same projection) (bottom).

(η^5-CYCLOPENTADIENYL)(p-TOLYLMETHYLHYDRAZIDO(1-)-N[1])DICARBONYLRHENIUM(III) TETRA-FLUOROBORATE

$C_{15}H_{16}BF_4N_2O_2Re$ $[(\eta^5-C_5H_5)Re(CO)_2[p\text{-}NHN(CH_3)C_6H_4CH_3]][BF_4]$

C.F. BARRIENTOS-PENNA, C.F. CAMPANA, F.W.B. EINSTEIN, T. JONES, D. SUTTON and A.S. TRACEY, 1984. Inorg. Chem., 23, 363-368.

Monoclinic, $P2_1/c$, a = 10.060, b = 9.549, c = 18.193 Å, β = 93.32°, D_m = 2.02, Z = 4. Mo radiation, R = 0.043 for 1850 reflexions.

The cation (Fig. 1) has a "three-legged piano stool" geometry. The Re-N distance of 1.949(9) and the N-N distance of 1.32(1) Å are indicative of some multiple bond character in each bond. Other distances are: Re-C(C_5H_5) 2.27(1)-2.29(1), Re-CO 1.89(1) and 1.91(1) Å.

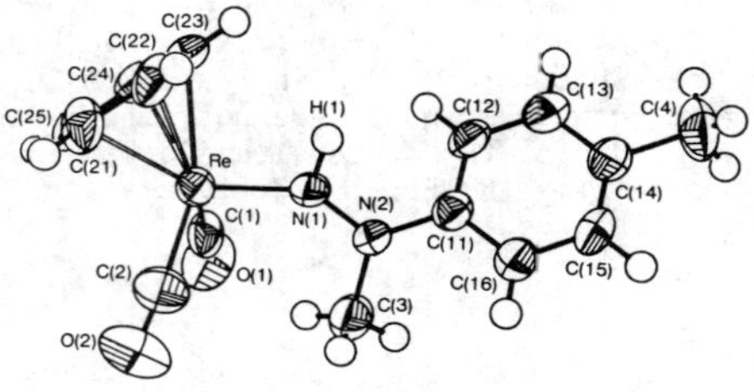

Fig. 1. The $[(\eta^5\text{-}C_5H_5)Re(CO)_2]p\text{-}NHN(Me)C_6H_4CH_3]]^+$ cation.

(η^5-CYCLOPENTADIENYL)(DIHYDRIDOBIS(η^5-CYCLOPENTADIENYL)NIOBIUM)ZINC

$C_{15}H_{17}NbZn$ $(\eta^5\text{-}C_5H_5)_2NbH_2Zn(\eta^5\text{-}C_5H_5)$

P.H.M. BUDZELAAR, K.H. DEN HAAN, J. BOERSMA, G.J.M. VAN DER KERK and A.L. SPEK, 1984. Organometallics, 3, 156-159.

Monoclinic, C2/c, a = 13.762, b = 9.679, c = 11.199 Å, β = 115.90°, Z = 4. Mo radiation, R = 0.023 for 924 reflexions.

The most interesting feature of the structure is the bonding within the NbH_2Zn fragment (Fig. 1). The Nb-Zn distance is normal at 2.5407 Å as is the Nb-H distance of 1.69 Å. The Zn...H interaction of 1.97 Å is long but nonetheless real in view of the H-Nb-Zn angle of 51°. The molecule has twofold crystallographic symmetry in which the cyclopentadienyl ring bonded to Zn is disordered. The Cp-Nb-Cp angle is 145°, the average Nb-C(Cp) distance 2.373 Å and the Nb-Cp(centroid) distance 2.056 Å. The Zn-Cp(centroid) distance is 2.00 Å.

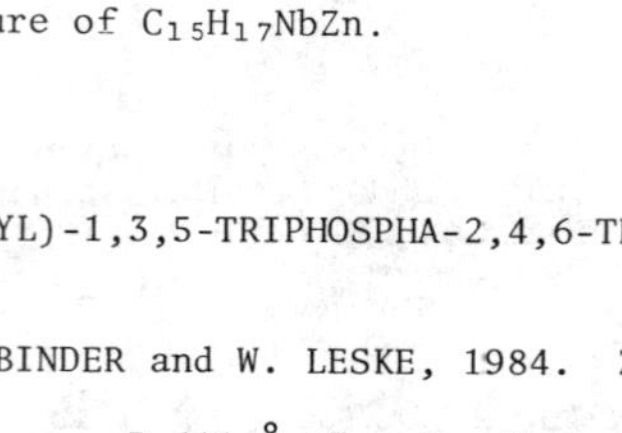

Fig. 1. Molecular structure of $C_{15}H_{17}NbZn$.

2,4,6-TRIS(η^5-CYCLOPENTADIENYL)-1,3,5-TRIPHOSPHA-2,4,6-TRINICKELACYCLOHEXANE

$C_{15}H_{21}Ni_3P_3$

B. DEPPISCH, H. SCHAFER, D. BINDER and W. LESKE, 1984. Z. Anorg. Chem., 519, 53-66.

Rhombohedral, R3, a = 16.861, c = 5.611 Å, Z = 3. Mo radiation, R = 0.041 for 843 reflexions.

The molecule (Fig. 1) has crystallographic threefold symmetry and the six-membered Ni_3P_3 ring has a chair conformation. The cyclopentadienyl rings were disordered.

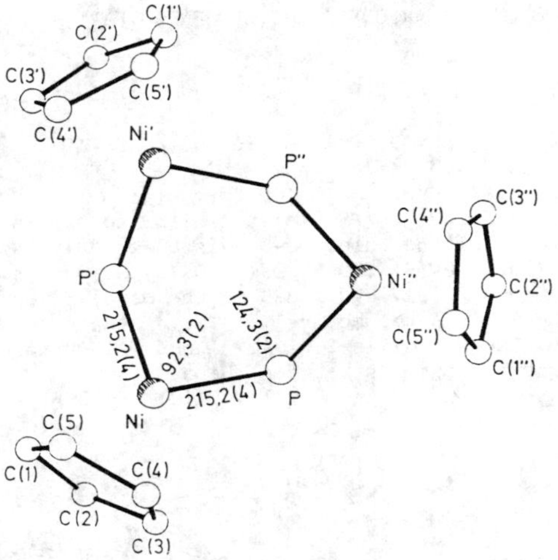

Fig. 1. The $C_{15}H_{21}Ni_3P_3$ molecule and some dimensions (pm and °).

(μ-CHLOROCHLOROTETRAMETHYLDIALUMINIUM)(η^5-2,4-CYCLOPENTADIEN-1-YL)BIS(TRIMETHYL-PHOSPHINE)RHODIUM

$C_{15}H_{35}Al_2Cl_2P_2Rh$ $CpRh(PMe_3)_2(Al_2Me_4Cl_2)$

J.M. MAYER and J.C. CALABRESE, 1984. Organometallics, 3, 1292-1298.

Monoclinic, $P2_1/n$, a = 15.696, b = 12.664, c = 12.544 Å, β = 108.32°, Z = 4. Mo radiation, R = 0.025 for 4297 reflexions (at -100°C).

The molecule (Fig. 1) has a normal "three-legged piano stool" geometry with a short, unbridged Rh-Al bond (2.458 Å). The Rh-P distances are 2.2612 and 2.2472 Å and the Rh-C(Cp) distances are 2.234, 2.306, 2.311, 2.228 and 2.244 Å. The bond angles Rh-Al(2)-Cl(2), Al(2)-Cl(2)-Al(1) and Cl(2)-Al(1)-Cl(1) are respectively 107.99, 113.60 and 100.57°. The Al(2)-Cl(2) distance is 2.419 Å, the Cl(2)-Al(1) distance 2.322 Å, and Al(1)-Cl(1) 2.182 Å. The site labelled C(1) consists of disordered C(1) and Cl(1*) with occupancies of 0.67 for C(1) and 0.33 for Cl(1*) and a separation of 0.014 Å. The angles about Rh are: P(1)-Rh-P(2) 98.24, P(1)-Rh-Al(2) 88.23 and P(2)-Rh-Al(2) 90.69°.

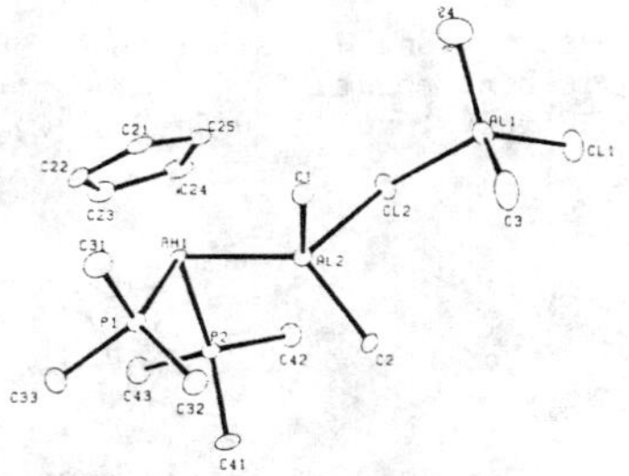

Fig. 1. Molecular structure of $C_{15}H_{35}Al_2Cl_2P_2Rh$.

(μ_3-SULPHIDO)(μ_2-HYDRIDO)UNDECACARBONYL(η^5-CYCLOPENTADIENYL)TUNGSTENTRIOSMIUM
$C_{16}H_6O_{11}Os_3SW$ $(\mu_2\text{-H})Os_3SW(\eta^5\text{-}C_5H_5)(CO)_{11}$

G. SÜSS-FINK, U. THEWALT and H.-P. KLEIN, 1984. J. Organometal. Chem., 262, 315-322.

Monoclinic, C2/c, a = 16.591, b = 9.824, c = 26.904 Å, β = 97.16°, D_m = 3.48, Z = 8. Mo radiation, R = 0.081 for 3103 reflexions.

The molecule has a pyramidal Os_3WS skeleton with, to a first approximation, a planar Os_3S base (Fig. 1). Of the three Os-Os distances only two appear to be genuine covalent bonds (Os(1)-Os(2) 2.945, Os(1)-Os(3) 2.807, Os(2)...Os(3) 3.859 Å). The W-Os distances are 3.004, 2.917 and 2.885 Å, the Os(2)-S, Os(3)-S and W-S bond lengths 2.362, 2.418 and 2.395 Å respectively. The distance between the cyclopentadienyl ring centroid and the W atom is 2.01 Å. Atoms C(10) and C(11) may be considered to act as bridging ligands between W and Os(3) and W and Os(1) respectively. The distances Os(1)...C(11) and Os(3)...C(10) are 2.59 and 2.50 Å.

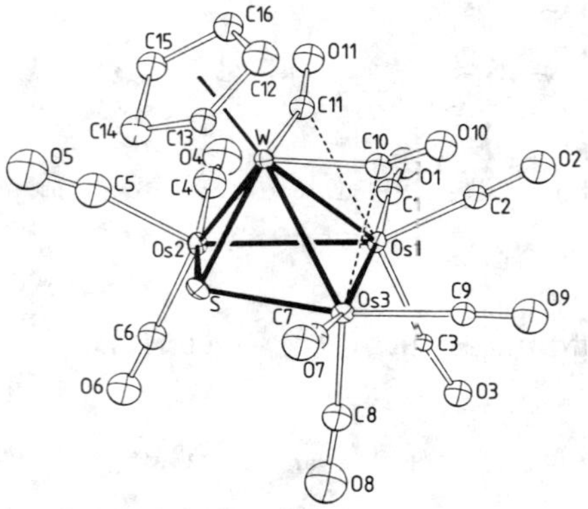

Fig. 1. Molecular structure of $C_{16}H_6O_{11}Os_3SW$.

HEXACARBONYL(η^5:η^5-FULVALENE)DITUNGSTEN
$C_{16}H_8O_6W_2$ $(\eta^5\text{:}\eta^5\text{-}C_5H_4C_5H_4)W_2(CO)_6$

H.B. ABRAHAMSON and M.J. HEEG, 1984. Inorg. Chem., 23, 2281-2286.

Triclinic, P$\bar{1}$, a = 6.708, b = 9.626, c = 12.450 Å, α = 76.16, β = 84.37, γ = 74.95°, Z = 2. Mo radiation, R = 0.049 for 2306 reflexions.

The molecule (Fig. 1) possesses a long W-W single bond and has approximate C_2 symmetry. The fulvalene ligand is bent toward the metal centres with a 16.1(5)° dihedral angle. W-W 3.347(1), W-C(fulvalene) 2.30(2)-2.35(2), W-CO 1.95(2)-1.99(2) Å.

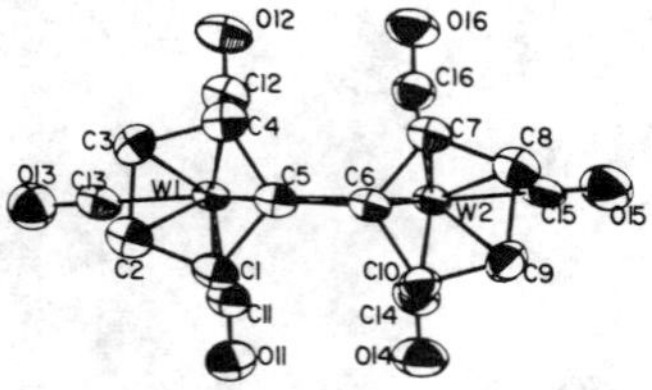

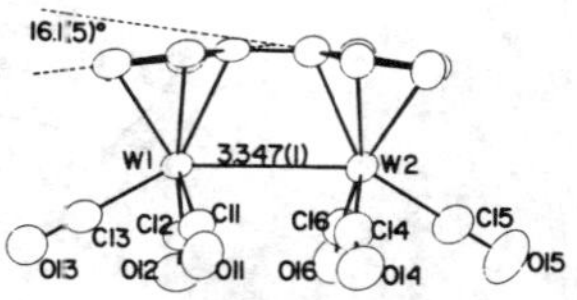

Fig. 1. Two views of $(\eta^5\text{:}\eta^5\text{-}C_5H_4C_5H_4)W_2(CO)_6$.

BIS(η^5-TRIMETHYLSILYLCYCLOPENTADIENYL)TITANIUM(IV) PENTASULPHIDE
$C_{16}H_{26}S_5Si_2Ti$ $Ti(C_8H_{13}Si)_2S_5$

N. KLOURAS, S. VOLIOTIS and G. GERMAIN, 1984. Acta Cryst., C40, 1791-1793.

Monoclinic, $P2_1/c$, a = 7.884, b = 14.018, c = 22.352 Å, β = 96.28°, Z = 4. Mo radiation, R = 0.10 for 2382 reflexions.

Ti is irregularly, tetrahedrally coordinated by the bidentate pentasulphide fragment (Ti-S 2.416(5) and 2.450(5) Å) and the two centroids (R) of the planar cyclopentadienyl rings (Ti-R 2.076 and 2.059 Å) (Fig. 1). The TiS_5 ring has a chair conformation with S-S in the range 2.040(5)-2.076(6) Å. The R-Ti-R and S_2-Ti-S_6 angles are 131.0(2) and 93.4(2)° respectively with Ti-C in the range 2.339(15)-2.429(14) and Si-C 1.841-1.895(17) Å.

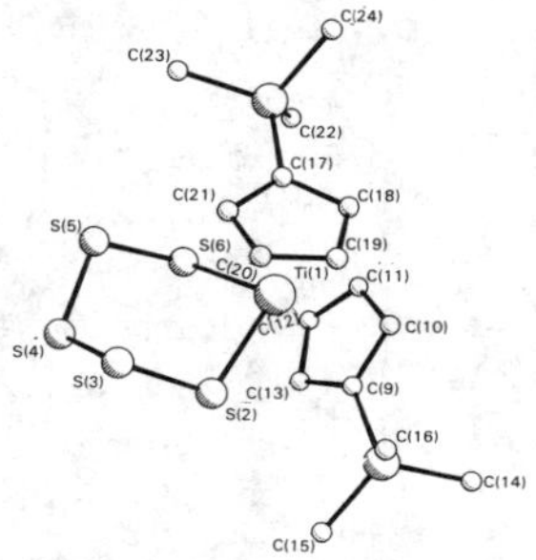

Fig. 1. The molecular structure of $Ti(C_8H_{13}Si)_2S_5$.

CYCLODICARBONYL(η^5-CYCLOPENTADIENYL)BIS(DICARBONYLCOBALT)-μ_3-SULPHIDO-μ-[TETRACARBONYL(DIMETHYLARSENIDO)IRON]MOLYBDENUM
$C_{17}H_{11}AsCo_2FeMoO_{10}S$

F. RICHTER, M. MULLER, N. GARTNER and H. VAHRENKAMP, 1984. Chem. Ber., 117, 2438-2451.

Monoclinic, $P2_1/c$, a = 15.761, b = 11.369, c = 13.640 Å, β = 93.91°, D_m = 2.07, Z = 4. Mo radiation, R = 0.041 for 3137 reflexions.

The structure of this cluster is shown in Fig. 1. Some bond lengths are: Mo-Co(1) 2.777(1), Mo-Co(2) 2.749(2), Co(1)-Co(2) 2.474(2), Co(2)-Fe 2.744(2), Co(1)-As 2.314(1), Fe-As 2.360(2), Mo-S 2.360(2), Co(1)-S 2.171(2), Co(2)-S 2.177(2), Mo-C(cp) 2.31-2.38(1) Å.

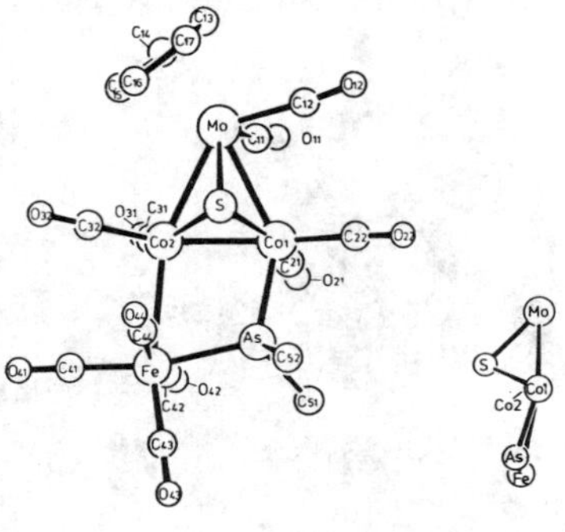

Fig. 1. A view of $C_{17}H_{11}AsCo_2FeMoO_{10}S$; also shown is a view along the Co-Co bond.

(1-5-η-6-NITROMETHYLCYCLOHEXADIENYL)(1-5-η-PENTAMETHYLCYCLOPENTADIENYL)IRIDIUM(III) TETRAFLUOROBORATE

$C_{17}H_{23}BF_4IrNO_2$

S.L. GRUNDY, A.J. SMITH, H. ADAMS and P.M. MAITLIS, 1984. J. Chem. Soc. Dalton, 1747-1754.

Monoclinic, $P2_1/c$, a = 10.913, b = 11.194, c = 15.260 Å, β = 96.99°, Z = 4. Mo radiation, R = 0.035 for 3512 reflexions.

The pentamethylcyclopentadienyl group is η^5-bonded to the Ir atom. The average Ir-C distance is 2.179 Å. Five of the six C atoms of the cyclohexadienyl ring are also η^5-bonded to Ir (Ir-C (average) 2.180 Å). The nitromethyl group is attached to the ring sp^3 C atom which is out of range of the Ir atom in the position exo to the metal (Fig. 1). The fold angle in the C_6 ring is 46°.

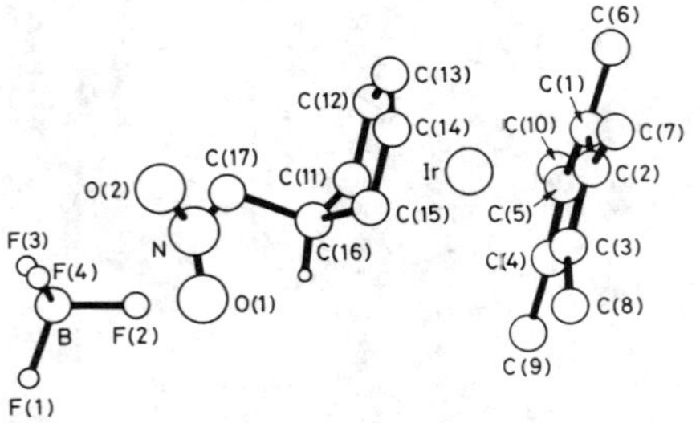

Fig. 1. $C_{17}H_{23}BF_4IrNO_2$: the molecular structure.

BIS(η^5-ISOPROPYLCYCLOPENTADIENYL)[1-MERCAPTO-1-METHANOSULPHENO(DITHIOPEROXOATE)]-TITANIUM

$C_{17}H_{24}S_4Ti$ η^5-$\{(i\text{-}C_3H_7)C_5H_4\}TiS_4CH_2$

D.M. GIOLANDO and T.B. RAUCHFUSS, 1984. Organometallics, 3, 487-489.

Monoclinic, $P2_1/c$, a = 11.584, b = 11.044, c = 15.252 Å, β = 102.95°, Z = 4. Mo radiation, R = 0.047 for 4366 reflexions.

The TiS_4CH_2 moiety assumes a boat conformation and the overall arrangement of ligands about the metal is distorted tetrahedral (Fig. 1). The angles S(1)-Ti-S(3) and Cp(B centroid)-Ti-Cp(A centroid) are respectively 91.6 and 130.1°. Other angles are: Ti-S(3)-S(4), Ti-S(1)-S(2) 109.4 and S(2)-CH_2-S(4) 118.3°. The Ti-S distances average 2.438 Å, S(1)-S(2) and S(3)-S(4) 2.055 Å, S(2)-CH_2 and S(4)-CH_2 1.818 Å, and the non-bonded distances S(1)...S(3) and S(2)...S(4) 3.489 and 3.117 Å respectively.

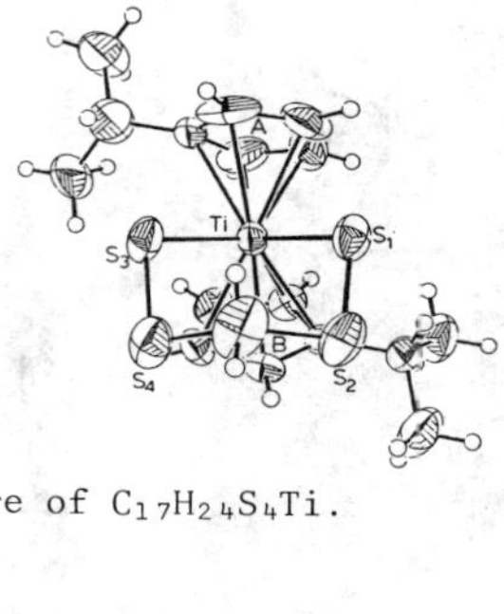

Fig. 1. Molecular structure of $C_{17}H_{24}S_4Ti$.

(μ_3-OXO)TRIS(μ_3-SULPHIDO)TRIS[(η^5-CYCLOPENTADIENYL)CHROMIUM]TRICARBONYLCOBALT PIVALIC ACID BENZENE SOLVATE

$C_{18}H_{15}CoCr_3O_4S_3,C_5H_{10}O_2,0.5(C_6H_6)$ (I) $Cp_3Cr_3S_3OCo(CO)_3,HOOCCMe_3,0.5C_6H_6$

BIS((μ_2-t-BUTYLTHIOLATO)TRIS(μ_3-SULPHIDO)BIS(η^5-CYCLOPENTADIENYL)DICHROMIUM)COBALT

$C_{28}H_{38}CoCr_4S_6$ (II) $[(C_5H_5)_2Cr_2(\mu\text{-}SCMe_3)(\mu_3\text{-}S)_2]_2Co$

I.L. EREMENKO, A.A. PASYNSKII, G.Sh. GASANOV, B. ORAZSAKHATOV, Yu.T. STRUCHKOV and V.E. SHKLOVER, 1984. J. Organometal. Chem., 275, 71-81.

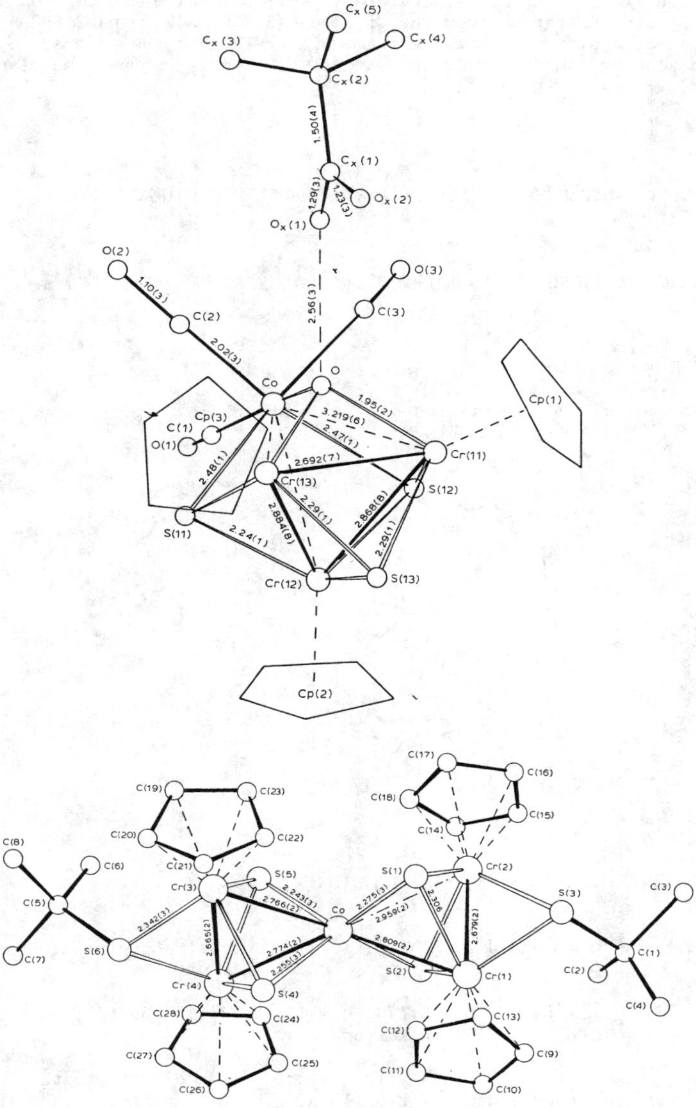

Fig. 1. The structures of $C_{18}H_{15}CoCr_3O_4S_3,C_5H_{10}O_2$ (top), and $C_{28}H_{38}CoCr_4S_6$ (bottom) showing selected bond distances (Å).

I. Triclinic, $P\bar{1}$, a = 8.961, b = 10.719, c = 16.075 Å, α = 108.72, β = 87.26, γ = 91.52°, Z = 2. Mo radiation, R = 0.091 for 1967 reflexions, (-120°C).

II. Monoclinic, $P2_1/n$, a = 11.496, b = 9.647, c = 29.240 Å, β = 91.87°, Z = 4. Mo radiation, R = 0.070 for 2689 reflexions, (-120°C).

Compound I contains a pseudotetrahedral cluster in which the base is an isosceles metal triangle Cp_3Cr_3 which is capped by the tridentate bridging S atom (Fig. 1). The short Cr-Cr bond is bridged by an O atom, and the long bonds have S bridges which in their turn are weakly bonded to the $Co(CO)_3$ group. The μ_3-O bridging atom forms a hydrogen bond with the trimethylacetic acid molecule while the Co atom is situated at weakly-bonding distances from the Cr atoms (3.216, 3.238 and 2.892 Å). Compound II (Fig. 1) has a distorted bow-tie framework Cr_4Co with the central Co atom surrounded also by four μ_3-bridging S atoms. The angle between the planes of the Cr_2Co metal triangles is 85.4°.

(μ_2-HYDRIDO)[(η^5-CYCLOPENTADIENYL)DICARBONYLMANGANESE][CARBONYLBIS(η^5-CYCLOPENTADIENYL)NIOBIUM]
$C_{18}H_{16}MnNbO_3$

W.A. HERRMANN, H. BIERSACK, B. BALBACH and M.L. ZIEGLER, 1984. Chem. Ber., 117, 95-106.

Monoclinic, $P2_1/n$, a = 9.088, b = 14.239, c = 12.801 Å, β = 95.03°, Z = 4. Mo radiation, R = 0.047 for 1526 reflexions.

The Nb and Mn atoms are linked by a hydrido bridge (Fig. 1) with Mn-Nb 3.330(2), Mn-H 1.56(8), Nb-H 1.98(8) Å, Nb-H-Mn 141(5)°.

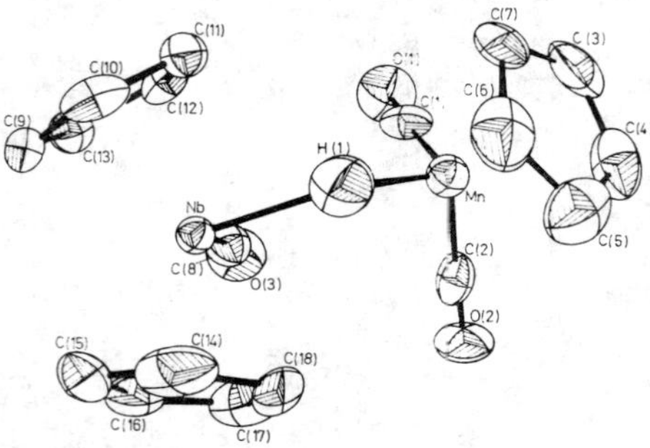

Fig. 1. A view of $C_{18}H_{16}MnNbO_3$.

BIS(η^5-CYCLOPENTADIENYL)BIS(PYRROLYL)HAFNIUM
$C_{18}H_{18}N_2Hf$ $(C_5H_5)_2Hf(NC_4H_4)_2$

R.D. ROGERS, R.V. BYNUM and J.L. ATWOOD, 1984. J. Cryst. Spectrosc. Res., 14, 21-28.

Monoclinic, $P2_1/c$, a = 9.605, b = 8.174, c = 19.487 Å, β = 90.85°, D_m = 1.91, Z = 4. Mo radiation, R = 0.068 for 2154 reflexions.

The two cyclopentadienyl rings (Fig. 1) are η^5-coordinated to the hafnium atom at an average Hf-C(η^5-) distance of 2.51(1) Å. The Hf-N-centroid (σ-pyrrolyl) angles average 166°. The Hf-N distances are 2.16(1) and 2.18(2) Å, with N-Hf-N 95.4(5)°.

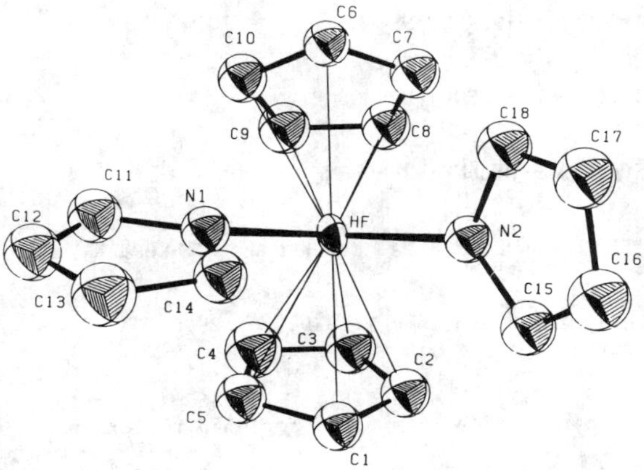

Fig. 1. A view of $C_{18}H_{18}N_2Hf$.

1,3,5-TRIS(η^5-METHYLCYCLOPENTADIENYL)-4,2,6,7-OXATRITHIA-1-ARSA-3,5-DITITANABICYCLO[3.1.1]HEPTANE

$C_{18}H_{21}AsOS_3Ti_2$ (I) $(MeCp)_3Ti_2OAsS_3$

BIS(TETRAPHENYLPHOSPHONIUM) CYCLOTETRAARSENICTETRADECASULPHURDIOXODIMOLYBDATE

$C_{48}H_{40}As_4Mo_2O_2P_2S_{14}$ (II) $(Ph_4P)_2(Mo_2O_2As_4S_{14})$

G.A. ZANK, T.B. RAUCHFUSS, S.R. WILSON and A.L. RHEINGOLD, 1984. J. Am. Chem. Soc., 106, 7621-7623.

I. Monoclinic, $P2_1/m$, a = 10.404, b = 12.932, c = 7.491 Å, β = 98.82°, Z = 2. Mo radiation, R = 0.027 for 2092 reflexions.

II. Triclinic, $P\bar{1}$, a = 13.140, b = 14.634, c = 18.589 Å, α = 112.52, β = 95.99, γ = 108.65°, Z = 2. Mo radiation, R = 0.094 for 4737 reflexions.

Molecule (I) has crystallographic mirror symmetry (Fig. 1) with As-S(1) 2.1922(10), As-S(2) 2.2621(7), Ti(1)-O 1.794(2), Ti(2)-O 1.872(2) Å, Ti-O-Ti 145.2(1)°. The anion in (II) (Fig. 1) consists of a $Mo_2O_2S_2^{2+}$ fragment strapped by an $As_4S_{12}^{4-}$ chelate. The $Mo_2O_2S_2^{2+}$ core appears to be in a relatively strain-free environment. The S-Mo-S bite angles of the outer chelates are 79.5(2) and 78.5(2)°. The As(4)-S(7) and As(1)-S(8) distances of 2.335(6) and 2.301(7) Å are elongated.

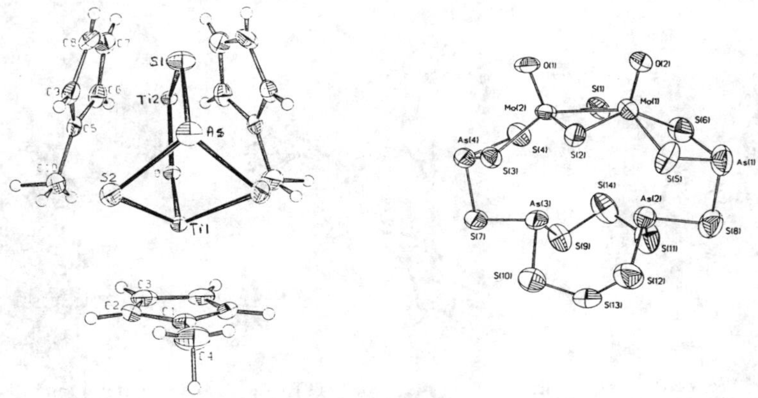

Fig. 1. Views of $(MeCp)_3Ti_2OAsS_3$ and the anion $(Mo_2O_2As_4S_{14})^{2-}$.

1,1',2,2'-BIS(TETRAMETHYLENE)FERROCENE
$C_{18}H_{22}Fe$ (I)

1,1',2,2',4,4'-TRIS(TETRAMETHYLENE)FERROCENE
$C_{22}H_{28}Fe$ (II)

1,1',2,2',4,4'-TRIS(PENTAMETHYLENE)FERROCENE
$C_{25}H_{34}Fe$ (III)

M. MILLMAN, E. FUJITA, H. DAUPLAISE, A. KVICK and R.C. KERBER, 1984. Organometallics, 3, 1170-1177.

I. Orthorhombic, Pnaa, a = 8.956, b = 11.930, c = 25.641 Å, Z = 8. Mo radiation, R = 0.100 for 1238 reflexions.

II. Triclinic, $P\bar{1}$, a = 10.132, b = 13.378, c = 13.565 Å, α = 103.45, β = 99.22, γ = 102.62°, Z = 4. Mo radiation, R = 0.053 for 6663 reflexions.

III. Monoclinic, $P2_1/c$, a = 15.274, b = 7.124, c = 17.617 Å, β = 93.27°, Z = 4. Mo radiation, R = 0.056 for 3003 reflexions.

Of interest in these structure determinations is the Fe-ring distance and also the ring-ring tilt and twist. In the tetramethylene structures (I and II) the Fe-ring distances are 1.635 Å in I and 1.629 Å in II whilst in the pentamethylene structure (III) the distance is 1.647 Å, consistent with the longer pentamethylene chain causing less strain in bridging the two rings. The angles between the ring planes in I, II and III are 6.9, 4.5 and 5.5, and 3.1° respectively again consistent with the type of bridging between the rings. In I and II the twist angles were found to be 12-14° and in III the angle is only 0.7° meaning that the rings in I and II are staggered and in III they are eclipsed (Fig. 1). Proton-proton interactions are believed to be responsible for the conformations.

(μ_2-t-BUTYLTHIOLATO)BIS(μ_3-SULPHIDO)DICARBONYLCOBALTBIS(η^5-METHYLCYCLOPENTADIENYL)DI-CHROMIUM
$C_{18}H_{23}CoCr_2O_2S_3$ (I) $(CH_3C_5H_4)_2Cr_2(SC(CH_3)_3)S_2Co(CO)_2$

(μ_2-t-BUTYLTHIOLATO)BIS(μ_3-SULPHIDO)TRICARBONYLMANGANESE-BIS(η^5-CYCLOPENTADIENYL)DI-CHROMIUM
$C_{17}H_{19}Cr_2MnO_3S_3$ (II) $(C_5H_5)_2Cr_2(\mu\text{-}SC(CH_3)_3)(\mu_3\text{-}S)_2Mn(CO)_3$

A.A. PASYNSKII, I.L. EREMENKO, B. ORAZSAKHATOV, G.Sh. GASANOV, V.M. MOVOTORTSEV, O.G. ELLERT, Z.M. SEIFULINA, V.E. SHKLOVER and Yu.T. STRUCHKOV, 1984. J. Organometal. Chem., 270, 53-64.

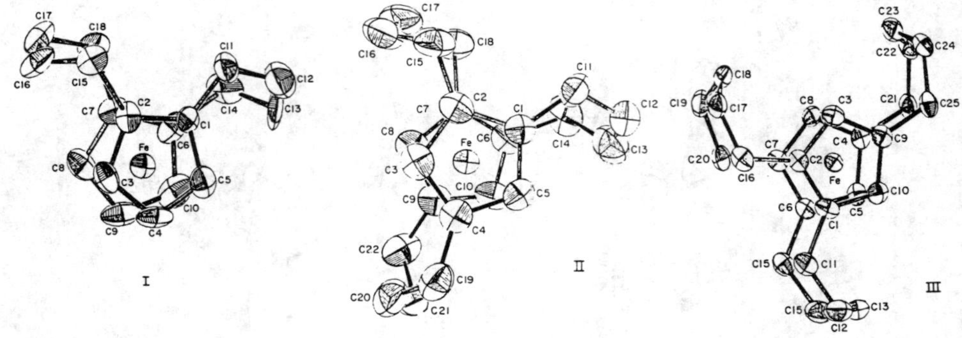

Fig. 1. Molecular structures of $C_{18}H_{22}Fe$ (left), $C_{22}H_{28}Fe$ (centre) and $C_{25}H_{34}Fe$ (right).

I. Monoclinic, $P2_1$, a = 9.437, b = 15.9460, c = 7.2540 Å, β = 109.399°, Z = 4. Mo radiation, R = 0.047 for [?] reflexions.

II. Monoclinic, C2, a = 18.709, b = 9.883, c = 11.470 Å, β = 109.65°, Z = 4. Mo radiation, R = 0.041 for 2464 reflexions (-120°C).

Compound I has a triangular Cr_2Co metal skeleton with a short Cr=Cr bond (2.590 Å) (Fig. 1). The conformation of the cyclopentadienyl ligands (average Cr-C(Cp) 2.25 Å) is staggered. Bond lengths of interest are: Cr(2)-S(2) 2.232, Co-S(1) 2.144, Cr(1)-Co 2.569, Cr(1)-S(1) 2.280, Cr(2)-S(1) 2.284 Å. Compound II also has a metal cyclic (Cr_2Mn) frame with a short, evidently partially double Cr=Cr bond (2.646 Å). The cyclopentadienyl ligands are again staggered (average Cr-C(Cp) 2.23 Å). Bond lengths in addition to those shown are: Cr(1)-S(2) 2.306, Cr(1)-S(1) 2.286, Cr(2)-S(1) 2.259 and Mn-S(1) 2.272 Å.

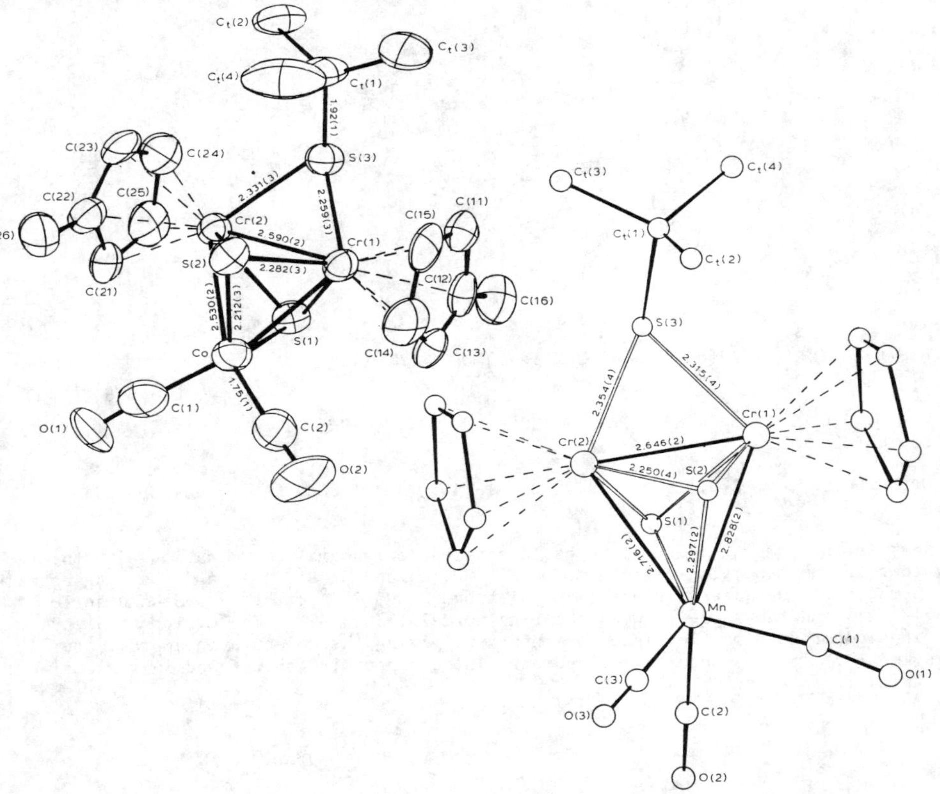

Fig. 1. Molecular structures of $C_{18}H_{23}CoCr_2O_2S_3$ (left), and $C_{17}H_{19}Cr_2MnO_3S_3$ (right) showing selected bond lengths (Å).

DICHLOROBIS(η^5-t-BUTYLCYCLOPENTADIENYL)TITANIUM(IV)

$C_{18}H_{26}Cl_2Ti$ $(\eta^5\text{-t-}C_4H_9C_5H_4)_2TiCl_2$

R.A. HOWIE, G.P. McQUILLAN and D.W. THOMPSON, 1984. J. Organometal. Chem., 268, 149-154.

Orthorhombic, $P2_12_12$, a = 12.862, b = 10.709, c = 6.602 Å, Z = 2. Mo radiation, R = 0.025 for 524 reflexions.

The overall molecular symmetry is C_2, with the twofold axis bisecting the $TiCl_2$ angle (Fig. 1). The two t-butyl groups are directed away from each other on opposite sides of the molecule. The individual Ti-C(Cp) distances are not equal, varying from 2.330 to 2.475 Å, the distortion being in a direction which increases the separation between the Ti atom and the C atom carrying the t-butyl substituent. The Ti-Cl distance and the Cl-Ti-Cl' angle are respectively 2.370 and 92.5°. The rings are partially staggered, with a relative displacement of 20° compared to 36° required for a 'perfect' staggered conformation.

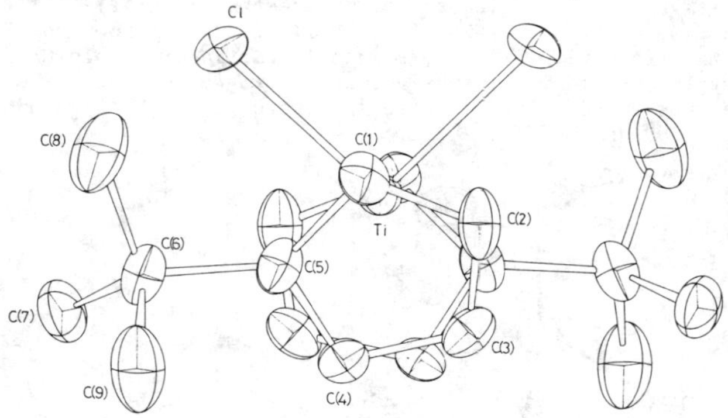

Fig. 1. Molecular structure of $C_{18}H_{26}Cl_2Ti$.

BIS(t-BUTYLTHIO)BIS(η^5-CYCLOPENTADIENYL)MOLYBDENUM(IV)

$C_{18}H_{28}MoS_2$ $[Mo(C_5H_5)_2(C_4H_9S)_2]$

M.A.A.F. DE C.T. CARRONDO, P.M. MATIAS and G.A. JEFFREY, 1984. Acta Cryst., C40, 932-934.

Orthorhombic, $P2_12_12_1$, a = 9.3424, b = 13.461, c = 14.884 Å, Z = 4. Cu radiation, R = 0.035 for 1879 reflexions.

In the complex (Fig. 1) Mo has distorted tetrahedral coordination with the two S atoms of the t-butylthio ligands (Mo-S 2.501(1) and 2.491(1) Å) and the ring normals to the cyclopentadienyl rings (mean distance 2.006(7) Å); the S-Mo-S bond angle is 71.1(1)° and the angle between the ring normals 130.4(3)°. The cyclopentadienyl rings adopt an almost eclipsed orientation, giving the molecule an approximate C_{2v} symmetry. The Mo coordination geometry is compared with that found in various similar complexes.

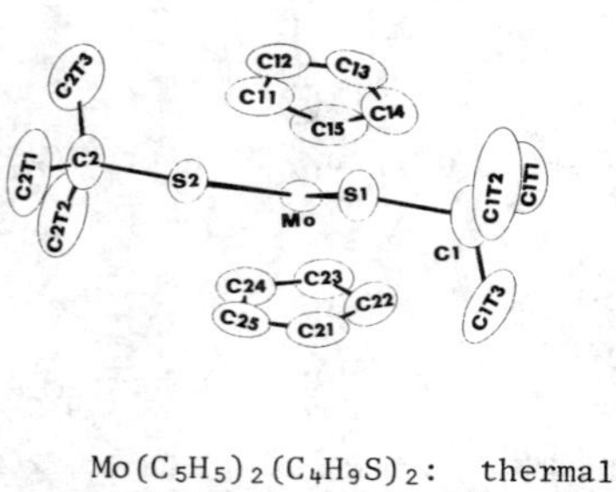

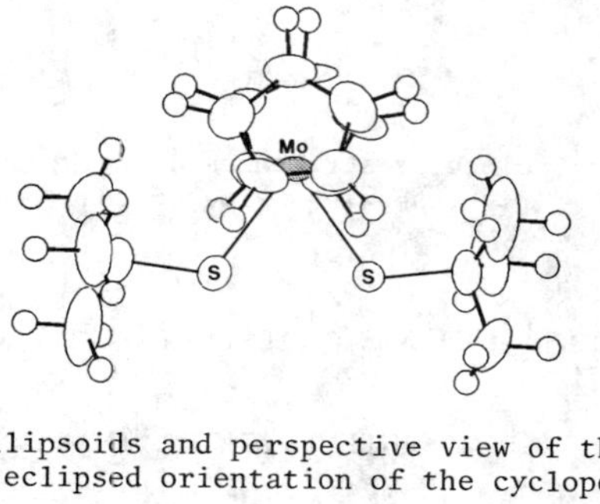

Fig. 1. $Mo(C_5H_5)_2(C_4H_9S)_2$: thermal ellipsoids and perspective view of the structure, showing the almost eclipsed orientation of the cyclopentadienyl rings.

(1,2-BIS(DIMETHYLPHOSPHINO)ETHANE-P,P')BIS(η^5-METHYLCYCLOPENTADIENYL)TITANIUM
$C_{18}H_{30}P_2Ti$

G.S. GIROLAMI, G. WILKINSON, M. THORNTON-PETT and M.B. HURSTHOUSE, 1984. J. Chem. Soc. Dalton, 2347-2350.

Monoclinic, $P2_1/n$, a = 8.073, b = 14.782, c = 15.777 Å, β = 93.34°, Z = 4. Mo radiation, R = 0.043 for 2513 reflexions.

The molecule (Fig. 1) has a classical bent structure with a centroid-Ti-centroid angle of 135.3(2)°. One of the cyclopentadienyl rings shows some disorder. However the C_5H_4Me groups are staggered with respect to each other and slightly tilted towards the dmpe ligand. The Ti-P distance is 2.533(6), Ti-C distances range from 2.310(5) to 2.441(6) Å and the P-Ti-P angle is 76.9(2)°.

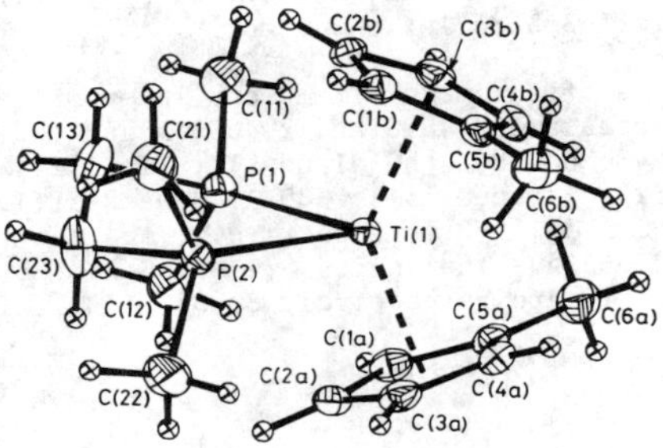

Fig. 1. $C_{18}H_{30}P_2Ti$: the molecular structure.

CARBONYLBIS(η^5-CYCLOPENTADIENYL)(μ_2-HYDRIDO)[TRICARBONYL(η^5-CYCLOPENTADIENYL)-VANADIUM]NIOBIUM
$C_{19}H_{16}NbO_4V$ (I)

CARBONYLBIS(η^5-CYCLOPENTADIENYL)(μ_2-HYDRIDO)[TRICARBONYL(η^5-CYCLOPENTADIENYL)NIOBIUM]-NIOBIUM
$C_{19}H_{16}Nb_2O_4$ (II)

W.A. HERRMANN, H. BIERSACK, B. BALBACH, P. WULKNITZ and M.L. ZIEGLER, 1984. Chem. Ber., 117, 79-94.

I. Triclinic, $P\bar{1}$, a = 11.033, b = 13.206, c = 14.763 Å, α = 68.90, β = 71.17, γ = 68.91°, Z = 4. Mo radiation, R = 0.048 for 2766 reflexions.

II. Triclinic, $P\bar{1}$, a = 11.152, b = 13.419, c = 14.840 Å, α = 67.60, β = 72.05, γ = 69.10°, Z = 4. Mo radiation, R = 0.046 for 2449 reflexions.

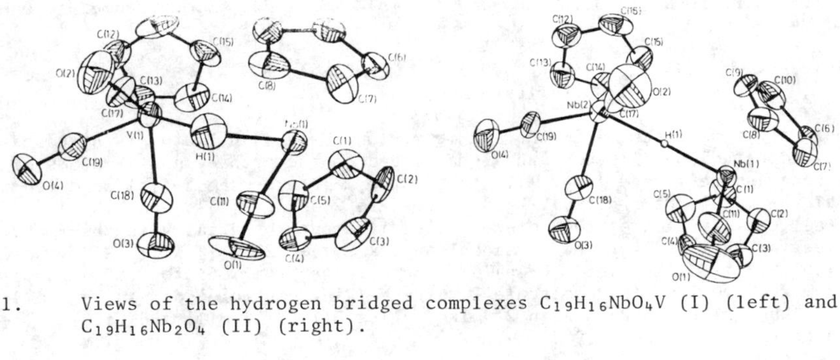

Fig. 1. Views of the hydrogen bridged complexes $C_{19}H_{16}NbO_4V$ (I) (left) and $C_{19}H_{16}Nb_2O_4$ (II) (right).

In both I and II there are two independent molecules in the asymmetric unit (Fig. 1). The hydrogen bridges are considered as three-centre two-electron functions and lead to dimensions in I of Nb-V 3.701(3) and 3.725(4), Nb-H 2.1(2) and 1.9(2), V-H 1.6(2) and 1.7(2) Å. In molecule II, Nb-Nb 3.727(2) and 3.738(3), Nb-H 1.86(15) to 1.90(24) Å.

[(4-METHOXYPHENYL)PHOSPHOTRITHIOATO(2-)-S,S']BIS[(1,2,3,4,5-η)-1-METHYL-2,4-CYCLOPENTADIEN-1-YL]TITANIUM HEMI BENZENE SOLVATE

$C_{19}H_{21}OPS_3Ti,0.5(C_6H_6)$ $(CH_3C_5H_4)_2TiS_3(CH_3OC_6H_4),0.5(C_6H_6)$

G.A. ZANK and T.B. RAUCHFUSS, 1984. Organometallics, 3, 1191-1199.

Monoclinic, $P2_1/c$, a = 7.304, b = 13.716, c = 23.995 Å, β = 111.28°, D_m = 1.35, Z = 4. Mo radiation, R = 0.035 for 3365 reflexions.

The structure consists of an $(MeCp)_2Ti$ moiety incorporated in a TiS_2P ring (Fig. 1). The Ti-S distances are 2.4898 and 2.4639 Å and the Ti...P distance is 3.099 Å. The dihedral angle between the planes Ti,S(1),S(2) and P,S(1),S(2) is 167.10° and the S(1)...S(2) distance is 3.2380 Å. The P-S(1), P-S(2) and P-S(3) distances are 2.062, 2.060 and 1.970 Å respectively, and the Ti-Cp centroid distances are 2.073 and 2.065 Å. The angles S(1)-Ti-S(2), Cp(A)-Ti-Cp(B), S(1)-P-S(2), P-S(1)-Ti and P-S(2)-Ti are respectively 81.63, 132.1, 103.52, 85.27 and 85.99°.

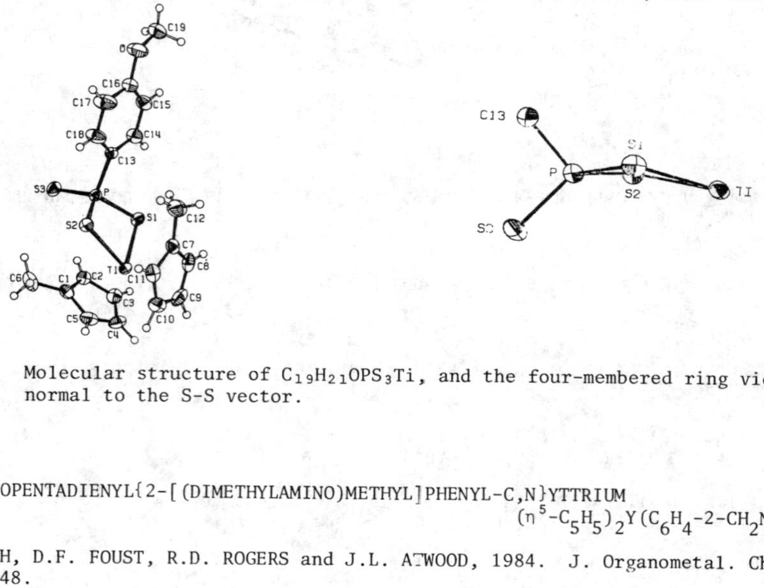

Fig. 1. Molecular structure of $C_{19}H_{21}OPS_3Ti$, and the four-membered ring viewed normal to the S-S vector.

DI-η^5-CYCLOPENTADIENYL{2-[(DIMETHYLAMINO)METHYL]PHENYL-C,N}YTTRIUM

$C_{19}H_{22}NY$ $(\eta^5-C_5H_5)_2Y(C_6H_4-2-CH_2N(CH_3)_2)$

M.D. RAUSCH, D.F. FOUST, R.D. ROGERS and J.L. ATWOOD, 1984. J. Organometal. Chem., 265, 241-248.

Monoclinic, $P2_1/c$, a = 8.310, b = 26.972, c = 15.548 Å, β = 95.88°, Z = 8. Mo radiation, R = 0.102 for 1739 reflexions.

There are two molecules in the asymmetric unit and each Y atom is η^5-coordinated to two cyclopentadienyl ligands and sigma-bonded to the ortho-C and -N atoms of the dimethylbenzylamine ligand (Fig. 1). In the two molecules the Y-C(σ) bond lengths are equal to 2.41 Å but the two Y-N distances are 2.43 Å (N(1)), and 2.54 Å (N(2)). The overall conformation of the molecules is different, especially in the N atom environment as indicated by the torsional angles C(1)-C(2)-C(7)-N(1) 32.2° and C(20)-C(21)-C(26)-N(2) -34.1°. The Y-C(Cp) distances range from 2.61 to 2.68 Å in molecule 1 and from 2.62 to 2.67 Å in molecule 2. The Y-(Ring centroid) distances are 2.35 and 2.39 Å in 1 and 2.36 and 2.38 Å in 2. The angles C(1)-Y(1)-N(1) and C(20)-Y(2)-N(2) are respectively 72.2 and 71.9°.

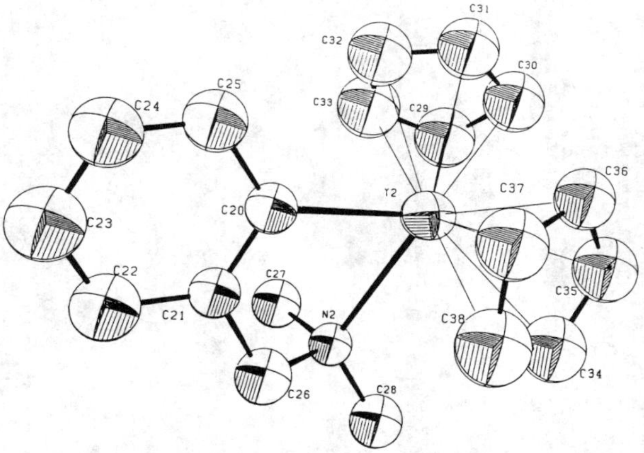

Fig. 1. Molecular structure of one of the independent molecules of $C_{19}H_{22}NY$.

2-(CHLOROBIS(FERROCENYL)PHOSPHINE-IMINO)-2,4,4,6,6-PENTACHLOROCYCLOTRIPHOSPHAZENE
$C_{20}H_{18}Cl_6Fe_2N_4P_4$ (I)

2-FERROCENYL-2-(HEPTACHLOROCYCLOTETRAPHOSPHAZEN-2-YL)-4,4,6,6-TETRACHLOROCYCLOTRIPHOSPHAZENE
$C_{10}H_9Cl_{11}FeN_7P_7$ (II)

H.R. ALLCOCK, K.D. LAVIN, G.H. RIDING and R.R. WHITTLE, 1984. Organometallics, 3, 663-669.

I. Monoclinic, $P2_1/c$, a = 11.199, b = 22.774, c = 11.504 Å, β = 104.81°, Z = 4. Mo radiation, R = 0.055 for 2737 reflexions.

II. Triclinic, $P\bar{1}$, a = 12.899, b = 14.206, c = 10.003 Å, α = 99.77, β = 105.76, γ = 112.99°, Z = 2. Mo radiation, R = 0.053 for 2562 reflexions.

Compound I consists of two ferrocene units linked via P-C covalent bonds to a pendent P atom which is outside the triphosphazene ring (Fig. 1). The unusual feature of the structure is the distortion of the angle at N(4). The wide (142.7°) angle is probably due to steric distortions imposed by the two adjacent ferrocene units. Compound II consists of a ferrocene unit linked via a P-C covalent bond to a cyclotriphosphazene ring which itself is linked geminally to a cyclotetraphosphazene ring via a P-P covalent bond. The cyclic tetrameric ring adopts a 'boat-like' conformation.

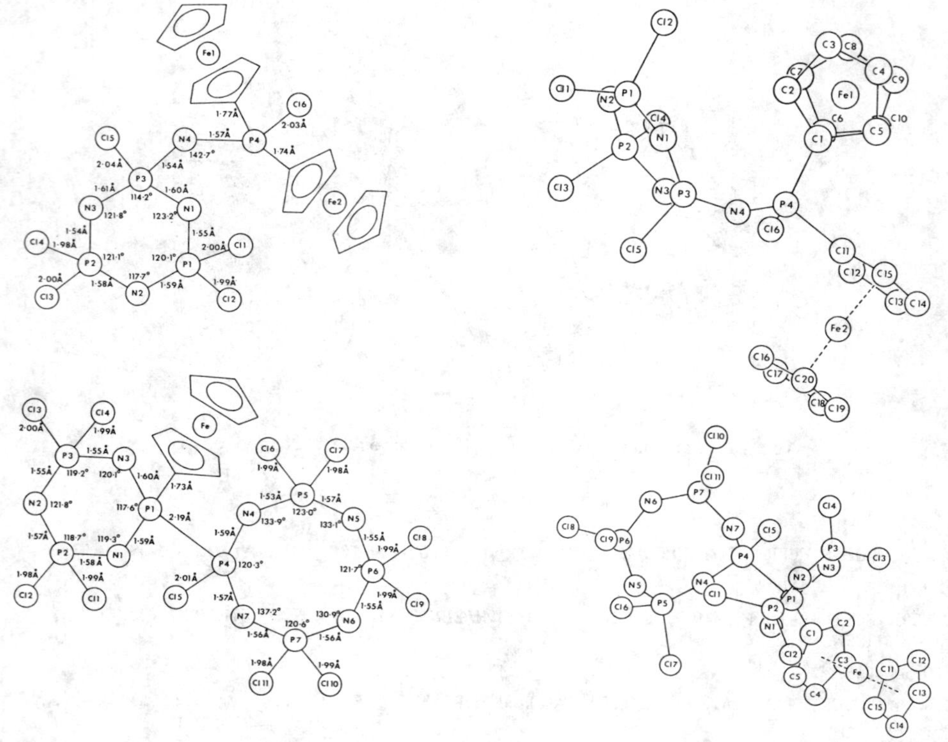

Fig. 1. Molecular structure of $C_{20}H_{18}Cl_6Fe_2N_4P_4$ showing selected bond lengths and angles (top left), and showing details of the pendant group (top right); molecular structure of $C_{10}H_9Cl_{11}FeN_7P_7$ with selected bond lengths and angles (bottom left), and showing orientation of the ferrocenyl group (bottom right).

DI-μ-CHLOROBIS(BIS-η[5]-CYCLOPENTADIENYLYTTRIUM(III))

$C_{20}H_{20}Cl_2Y_2$ $(Cp_2YCl)_2$

E.B. LOBKOVSKII, G.L. SOLOVEICHIK, B.M. BULYCHEV and A.B. EROFEEV, 1984. J. Struct. Chem., 25, 151-153 [Zh. Strukt. Khim. 25(1), 170-172].

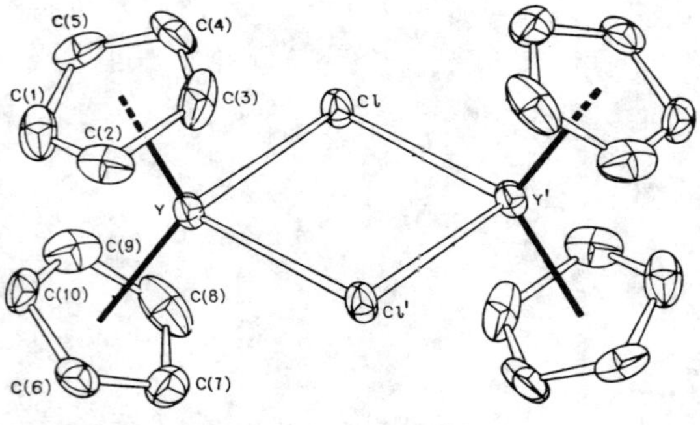

Fig. 1. The structure of the complex $(Cp_2YCl)_2$.

Monoclinic, $P2_1/c$, a = 11.141, b = 8.085, c = 12.063 Å, β = 109.89°, Z = 2. Mo radiation, R = 0.060 for 1400 reflexions.

The structure of the centrosymmetric dimer is shown in Fig. 1. Important structural parameters: Y-Cl 2.677(3) and 2.691(3), Y-C 2.56(2)-2.63(2), mean 2.60(3) Å, Y-Cl-Y' 98.58(4)°, and Cl-Y-Cl' 81.44(4)°.

DIFERRICENIUM (μ_2-OXO)BIS[TRICHLOROFERRATE(III)]
$C_{20}H_{20}Cl_6Fe_4O$ $[(C_5H_5)_2Fe]_2 \cdot (FeCl_3)_2O$

J.C.A. BOEYENS, F.B.D. KHAN and E.W. NEUSE, 1984. S. Afr. J. Chem., 37, 187-192.

Orthorhombic, Pnma, a = 14.600, b = 13.390, c = 14.232 Å, Z = 4. Mo radiation, R = 0.0901 for 1923 reflexions.

The cations and anions lie on mirror planes with the anionic Fe atom in a general position. The cations have disordered cyclopentadiene rings. In the anion, the Fe atom has tetrahedral geometry with mean Fe-Cl 2.197(3), Fe-O 1.754(2) Å, Fe-O-Fe 160.8(7), Cl-Fe-Cl 108.6(2), O-Fe-Cl 110.3(4)°.

BIS((μ_2-SULPHIDO)NIOBOCENE)
$C_{20}H_{20}Nb_2S_2$ $[(\eta^5-C_5H_5)_2Nb(\mu-S)]_2$

Yu.V. SKRIPKIN, I.L. EREMENKO, A.A. PASYNSKII, Yu.T. STRUCHKOV and V.E. SHKLOVER, 1984. J. Organometal. Chem., 267, 285-292.

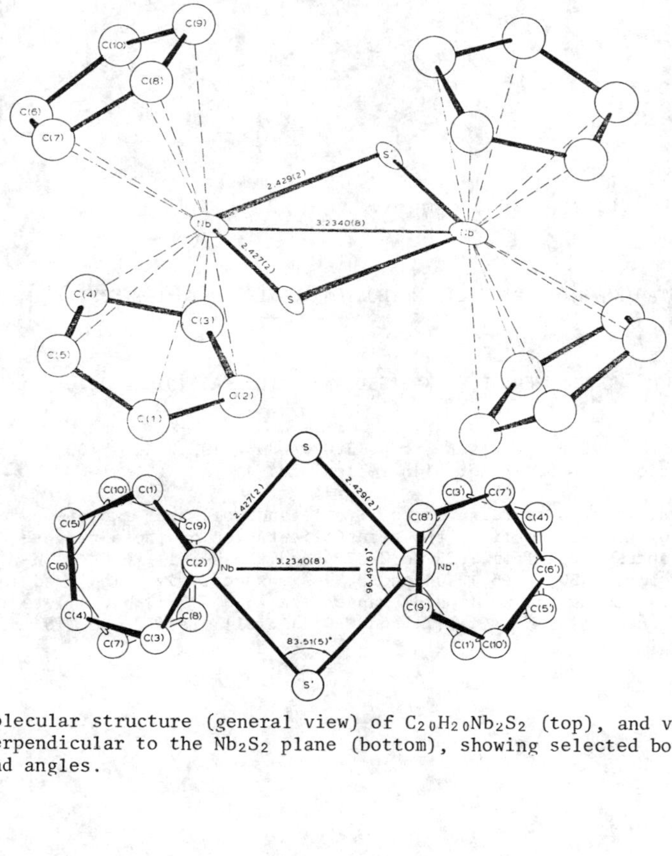

Fig. 1. Molecular structure (general view) of $C_{20}H_{20}Nb_2S_2$ (top), and viewed perpendicular to the Nb_2S_2 plane (bottom), showing selected bond distances and angles.

Monoclinic, $P2_1/n$, a = 13.260, b = 7.964, c = 8.230 Å, β = 95.41°, Z = 2. Mo radiation, R = 0.059 for 1256 reflexions (-120°C).

The complex has crystallographic inversion symmetry with a planar Nb_2S_2 ring (Fig. 1). The Nb atoms belong to niobocene fragments whose inclined rings (dihedral angle 47.3°) are in a staggered conformation. The average Nb-C(Cp) distance is 2.452 Å.

BIS(μ_2-IODO)BIS(BIS(η^5-CYCLOPENTADIENYL)HYDRIDORHENIUM)DICOPPER(I)
$C_{20}H_{22}Cu_2I_2Re_2$ $[Cp_2Re(H)CuI]_2$

V.K. BEL'SKII, V.M. ISHCHENKO, B.M. BULYCHEV and G.L. SOLOVEICHIK, 1984. Polyhedron, 3, 749-752.

Monoclinic, I2/a, a = 16.070, b = 7.788, c = 17.439 Å, β = 96.62°, Z = 4. Mo radiation, R = 0.044 for 1426 reflexions.

The molecule (Fig. 1) has twofold symmetry. The Rh-Cu bond (2.599(3) Å) is of the donor-acceptor type; dimerization occurs by the way of formation of the double bridge Cu-I_2-Cu and the direct inter-metal bond (Cu-Cu 2.552(4), Cu-I(1) 2.582(4), Cu-I(2) 2.745(3) Å, Re-Cu-Cu 140.51(14)°). The cyclopentadienyl rings form a bent sandwich with the angle between the ring centres and rhenium atom being equal to 158°. It is suggested that the Cu-Cu inter-metal bonding takes place on account of the transition of the non-bonding d-electrons of copper atoms to a high-spin state.

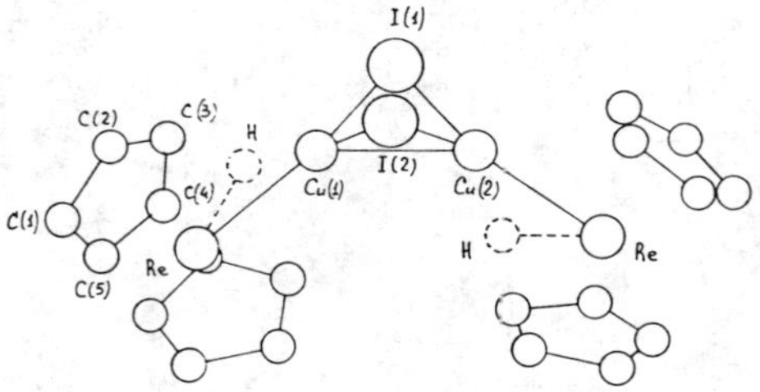

Fig. 1. The $[Cp_2Re(H)CuI]_2$ complex.

(η^5-CYCLOPENTADIENYL)DICARBONYL(α-METHOXYCARBONYL-4-METHYL-BENZYL)TRIMETHYLPHOSPHINE-TUNGSTEN
$C_{20}H_{25}O_4PW$

W.J. SIEBER, M. WOLFGRUBER, F.R. KREISSL and O. ORAMA, 1984. J. Organometal. Chem., 270, C41-C44.

Triclinic, $P\bar{1}$, a = 9.68, b = 9.81, c = 12.07 Å, α = 103.2, β = 109.4, γ = 92.4°, Z = 2. Mo radiation, R = 0.078 for 3146 reflexions.

The metal atom is coordinated to five ligands of which two are carbonyl groups. The structure analysis confirms the trans orientation of these groups (Fig. 1). The W-C(Cp) distances range from 2.28 to 2.37 Å and the W-P(1), W-C(4), W-C(5) and W-C(6) bond lengths are 2.450, 1.96, 2.03 and 2.34 Å respectively. The bond angles about the W atom (excluding the Cp ligand) range from 75.0° (P(1)-W-C(4)) to 139.4° (P(1)-W-C(6)), and the angles W-C(6)-C(7) and W-C(6)-C(21) are 112 and 115° respectively.

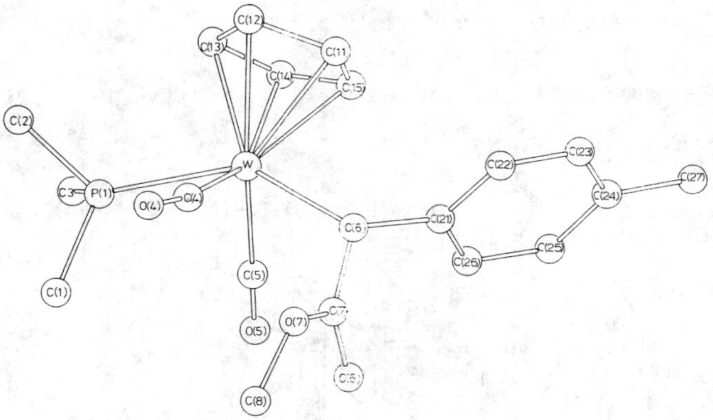

Fig. 1. Molecular structure of $C_{20}H_{25}O_4PW$.

meso-1,12-DIMETHYL-2,11-DITHIA-5,8-DIOXA[12](1,1')FERROCENOPHANE
$C_{20}H_{28}FeO_2S_2$

P.L. BELLON, F. DEMARTIN, V. SCATTURIN and B. CZECH, 1984. J. Organometal. Chem., 265, 65-71.

Triclinic, $P\bar{1}$, a = 8.879, b = 10.135, c = 12.030 Å, α = 100.14, β = 99.02, γ = 103.93°, Z = 2. Mo radiation, R = 0.030 for 2978 reflexions.

In Fig. 1 atoms C(11) and C(18) possess enantiomeric configurations, C(11) is an S centre and C(18) an R centre. The presence of another molecule, related by an inversion centre, in the unit cell confirms that the species is the meso form. The ferrocenyl moiety exhibits an almost perfect eclipsed conformation and the angle between the least-squares planes defined by the two rings is 2.3°. The Fe atom is 1.655 Å from plane C(1)...C(5) and 1.649 Å from plane C(6)...C(10). The bond distances in the bridging chain appear normal. The C-S distances range from 1.799 to 1.844 Å and the C-O distances from 1.410 to 1.427 Å. The chain has a steric arrangement of the type aag^+ aag^- g^+ aag^- a.

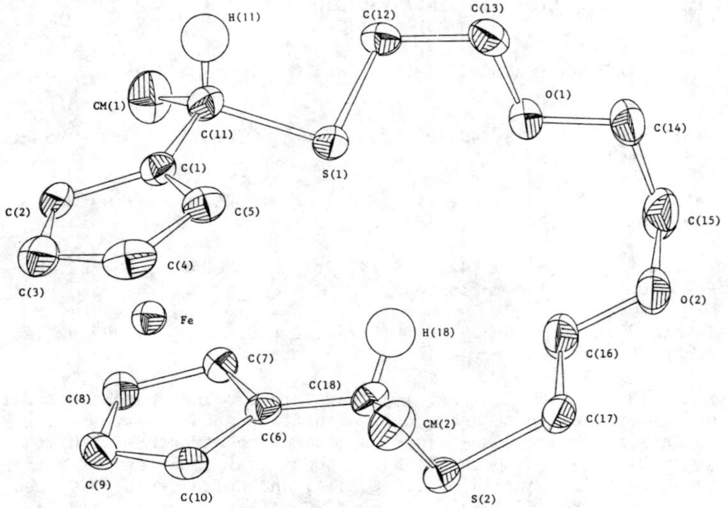

Fig. 1. Perspective view of the meso-diastereoisomer of $C_{20}H_{28}FeO_2S_2$.

BIS(η^5-PENTAMETHYLCYCLOPENTADIENYL)SAMARIUM
$C_{20}H_{30}Sm$ $(C_5Me_5)_2Sm$

W.J. EVANS, L.A. HUGHES and T.P. HANUSA, 1984. J. Am. Chem. Soc., 106, 4270-4272.

Monoclinic, $P2_1/n$, a = 9.815, b = 13.436, c = 14.163 Å, β = 94.98°, Z = 4. R = 0.035 for 2636 reflexions.

The molecule (Fig. 1) has a (ring centroid)-Sm-(ring centroid) angle of 140.1° and a mean C-Sm bond length of 2.79(1) Å.

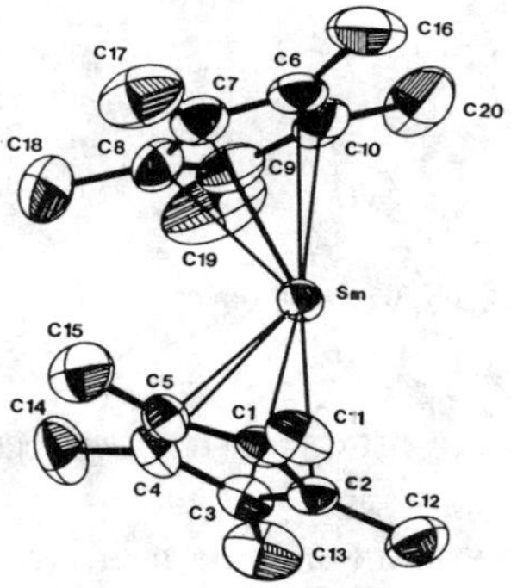

Fig. 1. A view of $(C_5Me_5)_2Sm$.

BIS(η^5-PENTAMETHYLCYCLOPENTADIENYL)VANADIUM
$C_{20}H_{30}V$ (I) $[C_5(CH_3)_5]_2V$

CARBONYLBIS(η^5-PENTAMETHYLCYCLOPENTADIENYL)VANADIUM
$C_{21}H_{30}OV$ (II) $[C_5(CH_3)_5]_2V(CO)$

CYANO(CYCLOHEXYLISOCYANIDE)BIS(η^5-PENTAMETHYLCYCLOPENTADIENYL)VANADIUM
$C_{28}H_{41}N_2V$ (III) $[C_5(CH_3)_5]_2V(C_6H_{11}NC)(CN)$

CYANOBIS(η^5-PENTAMETHYLCYCLOPENTADIENYL)VANADIUM
$C_{21}H_{30}NV$ (IV) $[C_5(CH_3)_5]_2V(CN)$

S. GAMBAROTTA, C. FLORIANI, A. CHIESI-VILLA and C. GUASTINI, 1984. Inorg. Chem., 23, 1739-1747.

I. Orthorhombic, Cmcm, a = 15.049, b = 8.660, c = 29.663 Å, Z = 8. Mo radiation, R = 0.12 for 464 reflexions.

II. Monoclinic, $P2_1/n$, a = 15.699, b = 13.353, c = 9.452 Å, β = 99.26°, Z = 4. Mo radiation, R = 0.055 for 1563 reflexions.

III. Monoclinic, $P2_1/c$, a = 10.342, b = 13.146, c = 18.956 Å, β = 92.60°, Z = 4. Mo radiation, R = 0.067 for 1163 reflexions.

IV. Monoclinic, $P2_1/n$, a = 15.895, b = 13.441, c = 9.209 Å, β = 97.92°, Z = 4. Mo radiation, R = 0.051 for 1508 reflexions.

Complexes I-IV are shown in Fig. 1. There are two independent molecules of I, one having mm symmetry while the second is disordered about a site having 2/m symmetry. I contains planar C_5Me_5 rings in a sandwich structure with mean Cp(centroid)-V 1.91(5) Å. Compound II has a bent arrangement of C_5Me_5 rings with V-CO 1.879(8) Å, mean Cp(centroid)-V 1.928(5) Å, and Cp(centroid)-V-Cp(centroid) 153.6(4)°. The structure of III shows a bent $(C_5Me_5)_2V$ unit (142.1(7)°) bonding in the equatorial plane of the CN^- and RNC ligands. The V-C and C-N bond lengths are very close for both ligands (mean V-C 2.04(2) and mean C-N 1.18(2) Å). Compound IV is isomorphous with II: V-CN 2.088(7), mean Cp(centroid)-V 1.971(7) Å, and Cp-V-Cp 151.5(3)°.

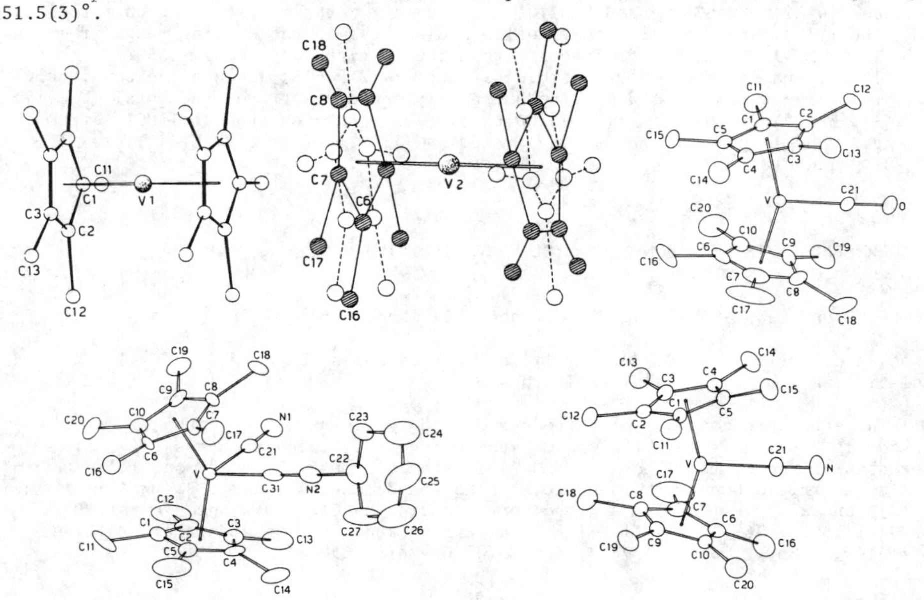

Fig. 1. The ordered (mm, upper left) and disordered (2/m. upper middle) $[C_5(CH_3)_5]_2V$ molecules, $[C_5(CH_3)_5]_2V(CO)$ (upper right), $[C_5(CH_3)_5]_2V$-$(C_6H_{11}NC)(CN)$ (lower left), and $[C_5(CH_3)_5]_2V(CN)$ (lower right).

(η^5-CYCLOPENTADIENYL)(η^3-1,2-DI-t-BUTYL-3-METHYLCYCLOPROPENYL)DICHLOROTRIMETHYLPHOSPHINETUNGSTEN

$C_{20}H_{35}Cl_2PW$

M.R. CHURCHILL, J.C. FETTINGER, L.G. McCULLOUGH and R.R. SCHROCK, 1984. J. Am. Chem. Soc., 106, 3356-3357.

Monoclinic, $P2_1/c$, a = 10.302, b = 15.314, c = 14.130 Å, β = 98.04°, Z = 4. Mo radiation, R = 0.034 for 3290 reflexions.

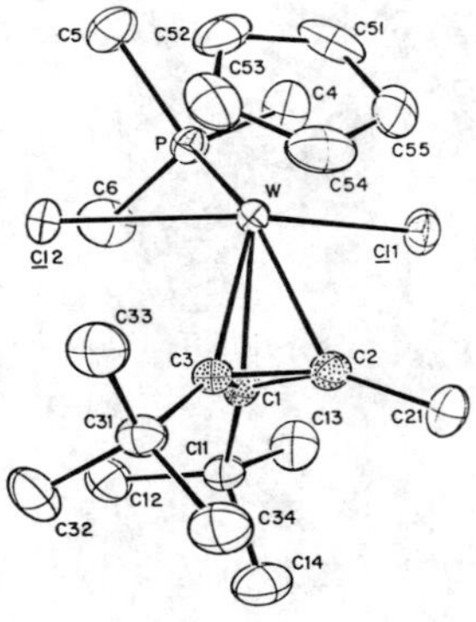

Fig. 1. A view of $C_{20}H_{35}Cl_2PW$.

The molecule (Fig. 1) is approximately a "four-legged piano stool" with the C_3 ring taking up the position trans to the PMe_3 ligand. The two tungsten-chloride distances (W-Cl(1) 2.538(2) and W-Cl(2) 2.504(2) Å) and the tungsten-phosphorus distance (W-P 2.600(2) Å) are normal. Individual W-C distances to the C_5 ring range from 2.334(8) to 2.373(9) Å, C-C distances within the C_5 ring range from 1.359(12) to 1.408(11) Å, and the W-C_5 ring centroid distance is 2.046 Å; all are unexceptional. The three carbon atoms of the C_3 ring are nearly equidistant from the metal (W-C(1) 2.139(5), W-C(2) 2.150(6), W-C(3) 2.200(6) Å). The tungsten atom is 1.991 Å from the C_3 ring. Bond lengths in the C_3 ring are C(1)-C(2) 1.466(7), C(2)-C(3) 1.416(8), C(3)-C(1) 1.473(8) Å.

(μ_3-GERMYLENE)TRIS(DICARBONYL(η^5-CYCLOPENTADIENYL)MANGANESE)

$C_{21}H_{15}GeMn_3O_6$ $[(\eta^5-C_5H_5(Mn(CO)_2))]_3Ge$

D. MELZER and E. WEISS, 1984. J. Organometal. Chem., 263, 67-73.

Monoclinic, $P2_1/n$, a = 14.344, b = 11.289, c = 14.882 Å, β = 117.4°, Z = 4. Mo radiation, R = 0.039 for 4251 reflexions.

The structure consists of three cyclopentadienyl-manganesedicarbonyl moieties linked via Mn-Ge bonds in which the Ge atom bridges all three Mn atoms (Fig. 1). The shortness of the Ge-Mn(2) bond (2.250 Å) indicates a multiple bond, one in between a double and triple bond. The average Mn-C(Cp) bond lengths are 2.138 (Mn(1)), 2.118 (Mn(2)) and 2.130 Å (Mn(3)), and the corresponding Mn-C(CO) distances are 1.790, 1.770; 1.780, 1.768; 1.754, 1.781 Å. The angles about the Ge atom are: Mn(1)-Ge-Mn(2) 140.5, Mn(1)-Ge-Mn(3) 78.4, and Mn(2)-Ge-Mn(3) 140.8°.

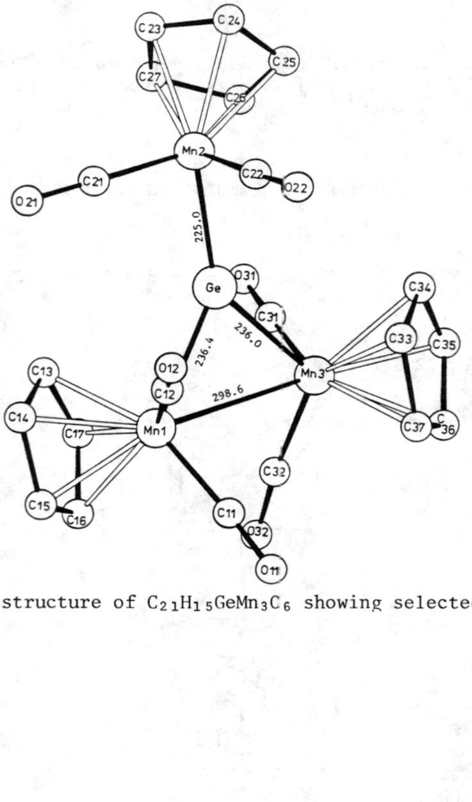

Fig. 1. Molecular structure of $C_{21}H_{15}GeMn_3C_6$ showing selected bond lengths (pm).

(μ_2-σ^1,η^2-PHENYL ISOCYANIDE)TETRACARBONYL-BIS(η^5-CYCLOPENTADIENYL)DIMOLYBDENUM
$C_{21}H_{15}Mo_2NO_4$ $[C_5H_5(CO)_2Mo]_2(\mu,\eta^2\text{-C≡NC}_6H_5)$

I. BERNAL, H. BRUNNER and J. WACHTER, 1984. J. Organometal. Chem., 277, 395-401.

Triclinic, P$\bar{1}$, a = 9.795, b = 10.219, c = 11.505 Å, α = 81.53, β = 70.83, γ = 66.91°, Z = 2. Mo radiation, R = 0.053 for 3688 reflexions.

This structure determination is of a polymorphic form of a compound whose structure has already been determined using a monoclinic form (1). Detailed comparison of bond lengths, angles, torsional angles and superposition of the two molecules show that the two determinations yield very close but not identical results, and that the largest deviations are associated with the Cp and phenyl groups which are able to respond readily to differences in packing forces (Fig. 1). Differences in bond lengths and valency angles between the two determinations range from 0.008 to 0.026 Å and 0.1 to 1.9° respectively.

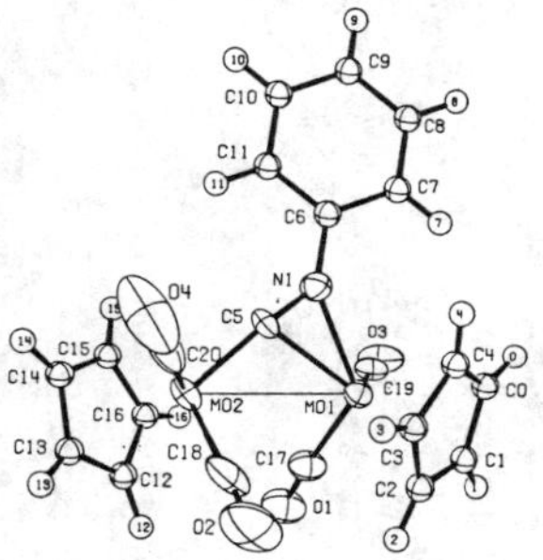

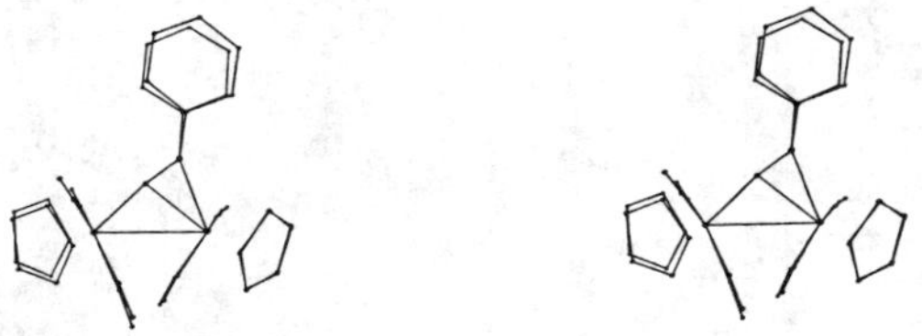

Fig. 1. A view of $C_{21}H_{15}Mo_2NO_4$, and a stereoview of the superposition of the monoclinic and triclinic structure determinations.

1. Structure Reports, 49B, 973.

N-ACETYL-3-FERROCENYL-5-p-BROMOPHENYLPYRAZOLINE-2
$C_{21}H_{19}BrFeN_2O$ (I)

N-ACETYL-3-p-BROMOPHENYL-5-FERROCENYLPYRAZOLINE-2
$C_{21}H_{19}BrFeN_2O$ (II)

V.G. ANDRIANOV, Yu.T. STRUCHKOV, V.N. POSTNOV, E.I. KLIMOVA and V.A. SAZONOVA, 1984. J. Organometal. Chem., 272, 81-89.

I. Triclinic, P$\bar{1}$, a = 10.342, b = 10.604, c = 10.637 Å, α = 65.04, β = 63.20, γ = 75.68°, Z = 2. Mo radiation, R = 0.052 for 2965 reflexions.

II. Monoclinic, P2_1/n, a = 17.042, b = 8.304, c = 13.108 Å, β = 93.01°, Z = 4. Mo radiation, R = 0.040 for 3495 reflexions.

The pyrazoline ring in both isomers (I and II) is non-planar with an envelope conformation, folding along the N(1)-C(4) vector (Fig. 1). The dihedral angle between the relevant planes is 20.4 and 21.9° respectively for I and II. The orientation of the ferrocenyl substituent is pseudo-axial. A feature is the almost ideal co-planarity of the cyclopentadienyl ferrocene ring in I (and the phenyl ring in II) and the acetyl substituent with the planar four atom fragment of the heterocycle.

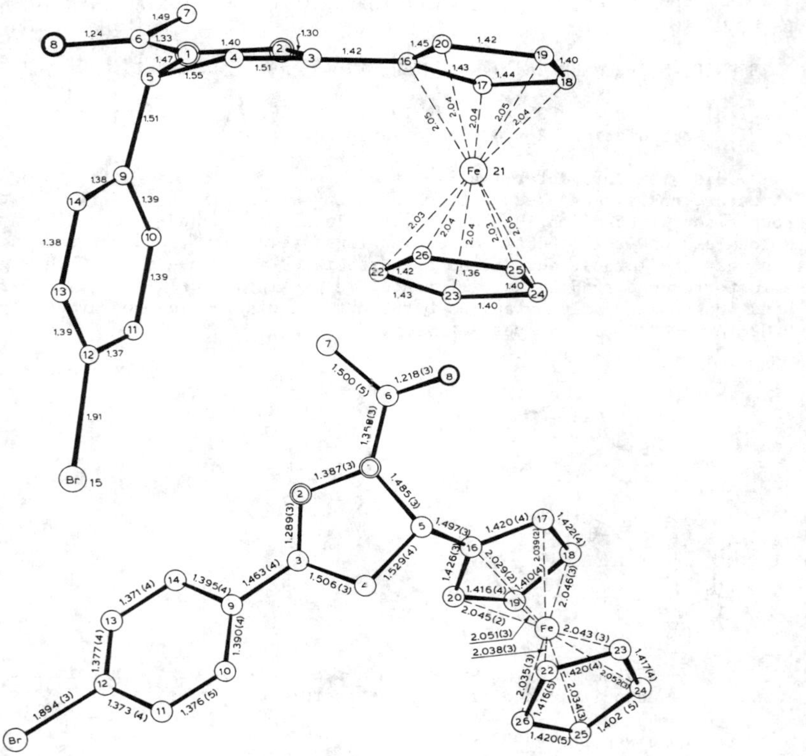

Fig. 1. Molecular structure showing $C_{21}H_{19}BrFeN_2O$ bond lengths (Å) in I (top), and II (bottom).

BIS(η^5-CYCLOPENTADIENYL)-(1,1,3,3-TETRAMETHYL-2-THIOCARBONYL-CYCLOHEXANE-C,S)-VANADIUM

$C_{21}H_{28}SV$ (I) $(C_5H_5)_2V(C_{11}H_{18}S)$

BIS(η^5-CYCLOPENTADIENYL)-(DI-t-BUTYL-THIOKETENE-C,S,)-VANADIUM

$C_{20}H_{28}SV$ (II) $(C_5H_5)_2V(C_{10}H_{18}S)$

R. DREWS, D. WORMSBÄCHER and U. BEHRENS, 1984. J. Organometal. Chem., 272, C40-C44.

I. Monoclinic, $P2_1/m$, a = 8.478, b = 12.074, c = 9.142 Å, β = 99.91°, Z = 2. Mo rqdiation, R = 0.064 for 1768 reflexions.

II. Orthorhombic, Pnma, a = 13.443, b = 12.583, c = 11.071 Å, Z = 4. Mo radiation, R = 0.066 for 1773 reflexions.

Both compounds have similar molecular geometries, and both molecules have crystallographic mirror symmetry (Fig. 1). In I of course this means disorder in the cyclohexane ring. In I, bond distances are: V-S 2.383, V-C(Cp average) 2.299, V-C(1) 2.148, C(1)-C(2) 1.346, C(2)-C(7) 1.552, C(2)-C(3) 1.554 Å. The corresponding distances in II are: 2.366, 2.299, 2.172, 1.340, C(2)-C(6) 1.572 and C(2)-C(3) 1.544 Å. The S(1)-C(1)-C(2) angles are 133.5 and 132.5° in I and II respectively, and the V-S-C(1) and V-C(1)-S angles are 60.6 and 75.1° in I and 61.8 and 73.7° in II.

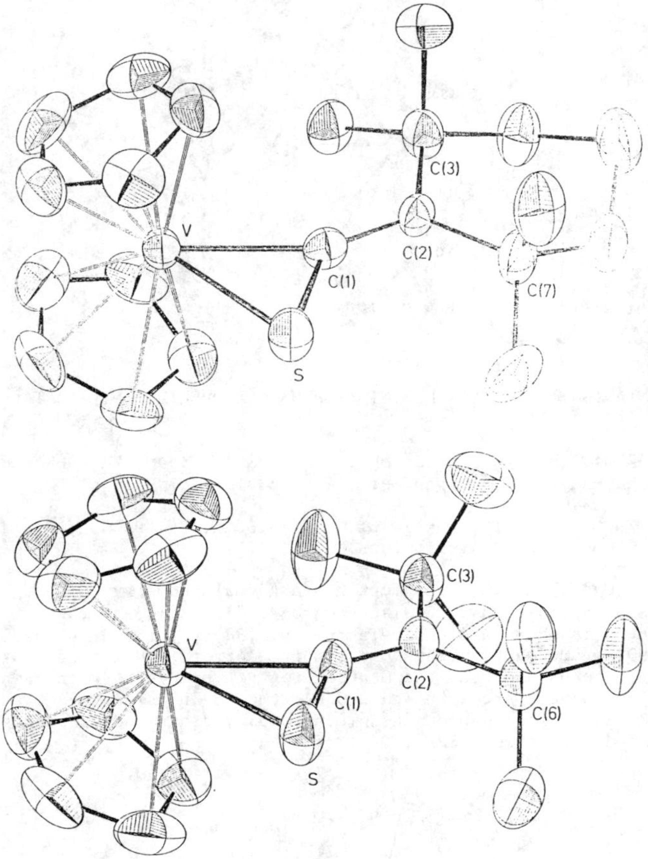

Fig. 1. Molecular structures of $C_{21}H_{28}SV$ (top), and $C_{20}H_{28}SV$ (bottom).

(μ_2-CARBONYL)(μ_2-NITROSYL)BIS[(η^5-PENTAMETHYLCYCLOPENTADIENYL)RHODIUM] HEXAFLUORO-PHOSPHATE

$C_{21}H_{30}F_6NO_2PRh_2$

S. CLAMP, N.G. CONNELLY, J.A.K. HOWARD, I. MANNERS, J.D. PAYNE and W.E. GEIGER, 1984. J. Chem. Soc. Dalton, 1659-1665.

Monoclinic, A2, a = 10.325, b = 11.760, c = 11.258 Å, β = 110.00°, Z = 2. Mo radiation, R = 0.034 for 1519 reflexions.

The X-ray determination revealed a short Rh-Rh bond (2.577(1) Å) which was symmetrically bridged by the nitrosyl (Rh-N(O) 1.95(1) Å) and carbonyl (Rh-C(O) 2.00(2) Å) ligands (Fig. 1). The PF_6^- anion is disordered.

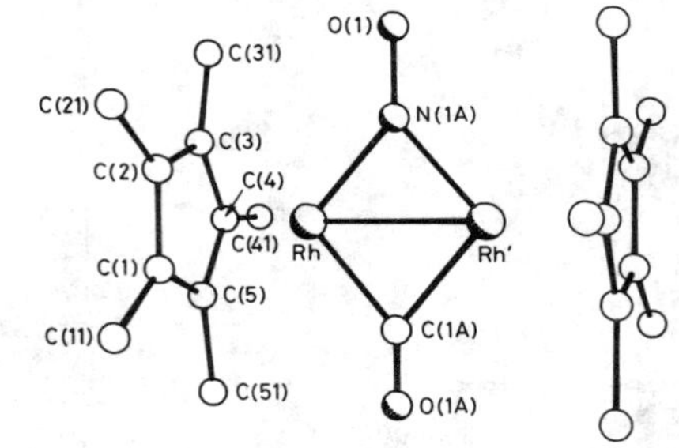

Fig. 1. $C_{21}H_{30}F_6NO_2PRh_2$: the cation.

BIS(η^5-CYCLOPENTADIENYLZINC)HYDRIDOBIS(η^5-METHYLCYCLOPENTADIENYL)TANTALUM
$C_{22}H_{25}TaZn_2$ $(CH_3C_5H_4)_2TaH(ZnC_5H_5)_2$

P.H.M. BUDZELAAR, A.A.H. VAN DER ZEIJDEN, J. BOERSMA, G.J.M. VAN DER KERK, A.L. SPEK and A.J.M. DUISENBERG, 1984. Organometallics, 3, 159-163.

Monoclinic, $P2_1/n$, a = 15.519, b = 8.461, c = 15.52 Å, β = 99.99°, Z = 4. Mo radiation, R = 0.036 for 1845 reflexions.

The Cp'_2Ta ($Cp' \equiv CH_3C_5H_4$) fragment has a normal geometry, with a bending angle of 137.5° and average Ta-C and C-C distances of 2.37 and 1.39 Å respectively (Fig. 1). The Zn-Ta-Zn angle is 62.8° and the H-Ta-Zn angle 54°. The two Ta-Zn distances (2.588 and 2.590 Å) are equal within experimental error, indicating that both bonds are normal single bonds and that the H atom does not bridge between Ta and Zn(2). The Zn-Cp(centroid) distances are 2.06 and 2.03 Å, the Ta-Zn-Cp(centroid) angles 161.4 and 164.2°, and the angles between Zn-Cp(centroid) vectors and Cp planes 82.2 and 83.0°. The Ta-H distance is 1.52 Å.

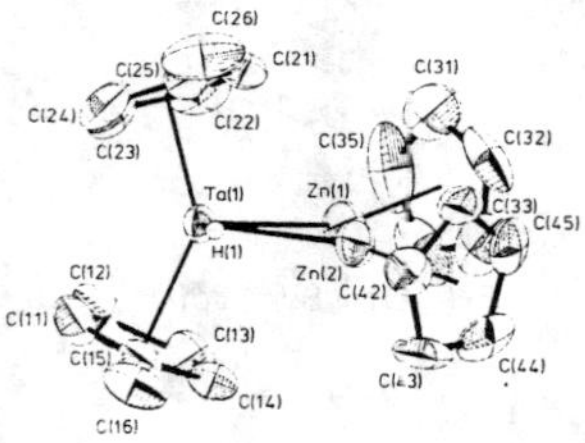

Fig. 1. Molecular structure of $C_{22}H_{25}TaZn_2$.

DIHYDRIDOBIS(TRIETHYLSILYL)(η^5-PENTAMETHYLCYCLOPENTADIENYL)RHODIUM(V)
$C_{22}H_{47}RhSi_2$

M.-J. FERNANDEZ, P.M. BAILEY, P.O. BENTZ, J.S. RICCI, T.F. KOETZLE and P.M. MAITLIS, 1984. J. Am. Chem. Soc., 106, 5458-5463.

Orthorhombic, Pbcm, a = 8.66, b = 16.75, c = 17.74 Å, Z = 4. Mo radiation, R = 0.057 for 1429 reflexions.
a = 8.671, b = 16.049, c = 17.958 Å. Neutron radiation, R = 0.061 for 4212 reflexions (at 20 K).

The molecule (Fig. 1) has crystallographic mirror symmetry and "four-legged piano stool" geometry with mean Rh-C 2.283(9) and Rh-Si 2.379(2) Å. The neutron study located the hydride (mean Rh-H 1.581(3) Å, H-Rh-H 94.8(2)°).

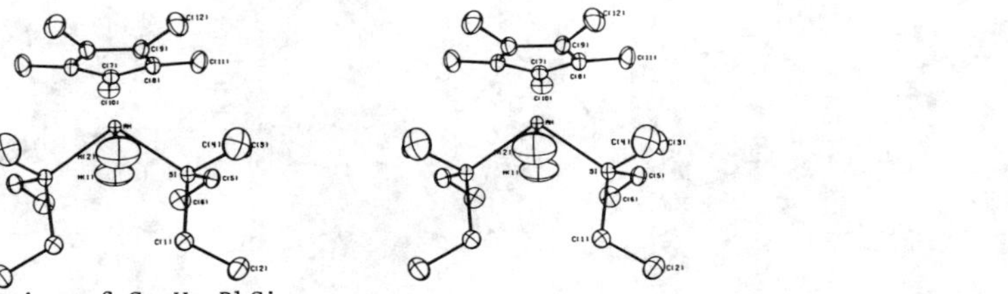

Fig. 1. A stereoview of $C_{22}H_{47}RhSi_2$.

TRICARBONYLBIS(η^5-CYCLOPENTADIENYL)(TETRAHYDRO-2-PHENYL-4H-1,3,6,2-DIOXAZAPHOSPHOCINE-P^2-DIMOLYBDENUM
$C_{23}H_{24}Mo_2NO_5P$ (I) $Cp_2Mo_2(CO)_3\{PhP(OCH_2CH_2)_2NH\}$

TRICARBONYLBIS(η^5-CYCLOPENTADIENYL)TRIMETHYLPHOSPHITEDIMOLYBDENUM
$C_{16}H_{19}Mo_2O_6P$ (II) $Cp_2Mo_2(CO)_3\{P(OMe)_3\}$

J. WACHTER, J.G. RIESS, A. MITSCHLER, 1984. Organometallics, 3, 714-722.

I. Monoclinic, $P2_1/c$, a = 14.902, b = 16.681, c = 9.631 Å, β = 104.84°, Z = 4. Cu radiation, R = 0.067 for 1625 reflexions.

II. Tetragonal, $P4_2/n$, a = 21.392, c = 8.357 Å, Z = 8. Cu radiation, R = 0.034 for 2187 reflexions.

Both compounds display short Mo-Mo distances (Fig. 1), with that in I being 2.506 Å, and that in II 2.502 Å. The Cp-Mo-Mo-Cp axes are non-linear in both compounds and the Mo-CO units are bent back over the Mo-Mo bond. The angles Mo(1)-Mo(2)-C(CO) range from 70.0 to 75.9° and the Mo(2)-Mo(1)-C(17) angles are 67.1 and 69.5° respectively in I and II. The Mo-P distances in I and II are 2.383 and 2.371 Å respectively.

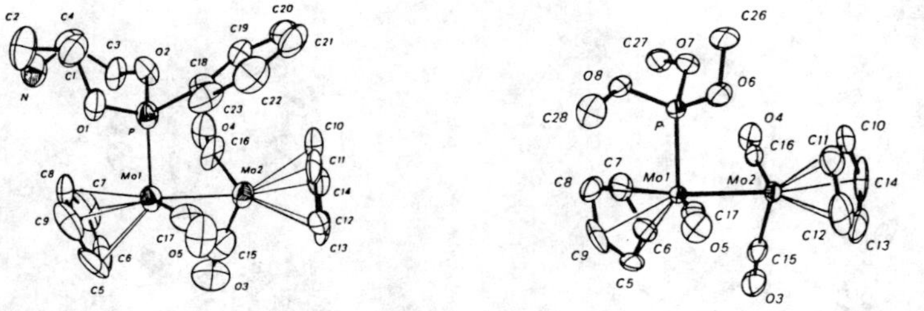

Fig. 1. Perspective diagrams of the molecules $C_{23}H_{24}Mo_2NO_5P$ (left), and of $C_{16}H_{19}Mo_2O_6P$ (right).

TRIS(μ_2-CARBONYL)-BIS[(η^5-PENTAMETHYL-CYCLOPENTADIENYL)MANGANESE]
$C_{23}H_{30}Mn_2O_3$

I. BERNAL, J.D. KORP, W.A. HERRMANN and R. SERRANO, 1984. Chem. Ber., 117, 434-444.

Monoclinic, $P2_1/c$, a = 9.723, b = 14.267, c = 16.747 Å, β = 106.1°, Z = 4. Mo radiation, R = 0.031 for 1877 reflexions.

The molecule (Fig. 1) has two C_5Me_5 rings π-bonded to a core containing two Mn atoms bridged by three CO groups. The Mn-Mn triple bond distance is 2.170(1) Å. Some other (mean) dimensions are Mn-C(CO) 1.929(5), Mn-C(C_5Me_5) 2.147(5), Mn-(ring centre) 1.776(5) Å.

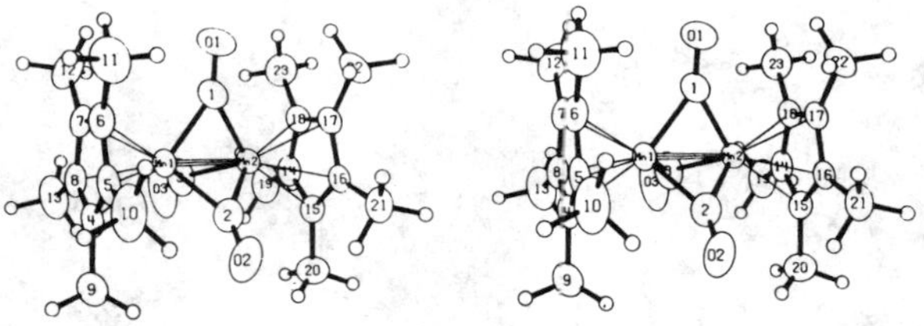

Fig. 1. A stereoview of $C_{23}H_{30}Mn_2O_3$.

(μ_2-ACETYL-C,O)(μ_2-DIPHENYLPHOSPHINO)(CARBONYL(η^5-CYCLOPENTADIENYL)IRON)TETRACARBONYL-MANGANESE

$C_{24}H_{18}FeMnO_6P$ (I)

(μ_2-DIPHENYLPHOSPHINO)(μ_2-HYDRIDO)(CARBONYL(η^5-CYCLOPENTADIENYL)IRON)TETRACARBONYL-MANGANESE

$C_{22}H_{16}FeMnO_5P$ (II)

R.P. ROSEN, J.B. HOKE, R.R. WHITTLE, G.L. GEOFFROY, J.P. HUTCHINSON and J.A. ZUBIETA, 1984. Organometallics, 3, 846-855.

I. Triclinic, $P\bar{1}$, a = 11.485, b = 12.989, c = 16.932 Å, α = 68.22, β = 87.17, γ = 88.88°, Z = 4. Mo radiation, R = 0.044 for 3353 reflexions.

II. Monoclinic, $P2_1/n$, a = 10.795, b = 13.931, c = 14.497 Å, β = 102.08°, Z = 4. Mo radiation, R = 0.057 for 2412 reflexions.

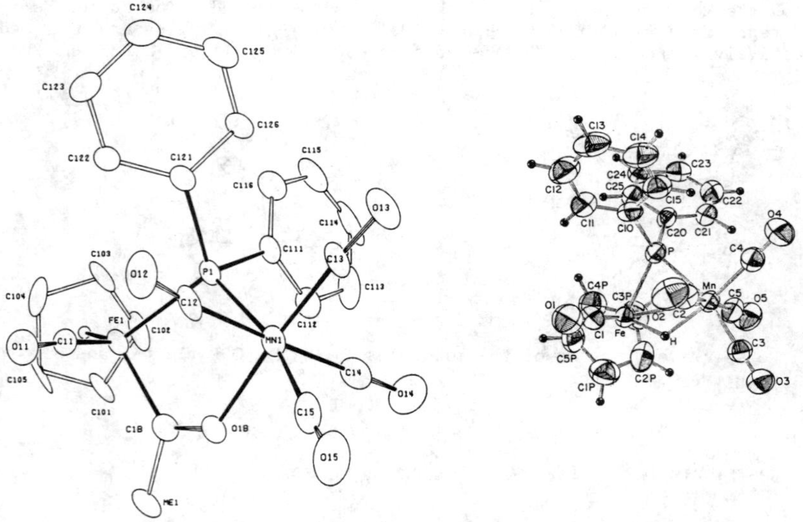

Fig. 1. Molecular structures of $C_{24}H_{18}FeMnO_6P$ (left), and $C_{22}H_{16}FeMnO_5P$ (right).

In I the two independent molecules are essentially identical. The μ-PPh_2 ligand bridges the Mn and Fe atoms which are also bridged by the μ-acetyl ligand with the C attached to Fe and the O to Mn (Fig. 1). The Mn...Fe distances are 3.718 and 3.713 Å in the two molecules. Coordination about Fe is distorted tetrahedral, and about Mn is octahedral. Mn-C(12) distance is 1.881 Å, Mn-P(1) 2.395, Mn-O(1B) 2.024, Fe-P(1) 2.249, Fe-C(1B) 1.919 Å. The acyl distance C(1B)-O(1B) is 1.262 Å. In II, the PPh_2 ligand bridges the metal atoms which are also bridged by a hydride ligand (Fe-Mn

2.806, Fe-H 1.56, Mn-H 1.72, Fe-P 2.174, Fe-C(1) 1.724, Mn-P 2.278 Å). The angles Fe-H-Mn, Fe-P-Mn, Fe-Mn-P and Mn-Fe-P are respectively 117, 78.10, 49.30 and 52.59°.

BIS(DICARBONYL(η^5-CYCLOPENTADIENYL)RUTHENIUM)BIS(η^5-CYCLOPENTADIENYL)ZIRCONIUM

$C_{24}H_{20}O_4Ru_2Zr$ $(\eta^5-C_5H_5)_2Zr[Ru(CO)_2(\eta^5-C_5H_5)]_2$

C.P. CASEY, R.F. JORDAN and A.L. RHEINGOLD, 1984. Organometallics, 3, 504-506.

Orthorhombic, Pbca, a = 15.000, b = 17.425, c = 16.957 Å, Z = 8. R = 0.040 for 3232 reflexions.

The two $Ru(CO)_2Cp$ groups are linked to the Cp_2Zr group by metal-metal bonds Zr-Ru(1) 2.938 Å and Zr-Ru(2) 2.948 Å, and are related by an approximate twofold rotational axis (Fig. 1). The carbonyl ligands C(1)-O(1) and C(4)-O(4) are almost coplanar with the $ZrRu_2$ framework whilst C(2)-O(2) and C(3)-O(3) are nearly perpendicular to the metal atom plane. Distances of interest are: Ru(1)-C(1) 1.850, Ru(2)-C(2) 1.824, Ru(1)-Cp(centroid) 1.942, Zr-Cp(centroids) 2.271 and 2.273 Å. Bond angles of interest are: Ru(1)-Zr-Ru(2) 100.5°, Cp(centroid)-Zr-Cp(centroid) 124.7°, Cp(centroid)-Ru(1)-C(2) 126.4°, and Cp(centroid)-Ru(1)-C(1) 131.7°.

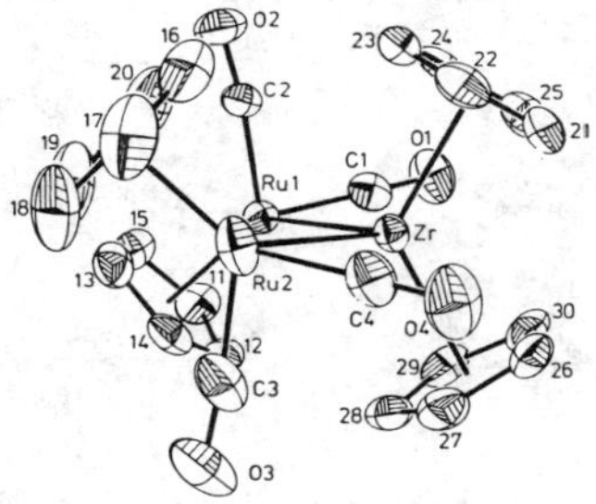

Fig. 1. Molecular structure of $C_{24}H_{20}O_4Ru_2Zr$.

(η^5-CYCLOPENTADIENYL)NITROSYL(η^2-SELENOFORMALDEHYDE)TRIPHENYLPHOSPHINERHENIUM HEXAFLUOROPHOSPHATE

$C_{24}H_{22}F_6NOP_2ReSe$ $[(C_5H_5)Re(NO)(PPh_3)(CH_2{=}Se)]^+.PF_6^-$

F.B. McCORMICK, 1984. Organometallics, 3, 1924-1927.

Monoclinic, $P2_1/c$, a = 9.689, b = 18.608, c = 15.054 Å, β = 104.98°, Z = 4. R = 0.038 for 4152 reflexions.

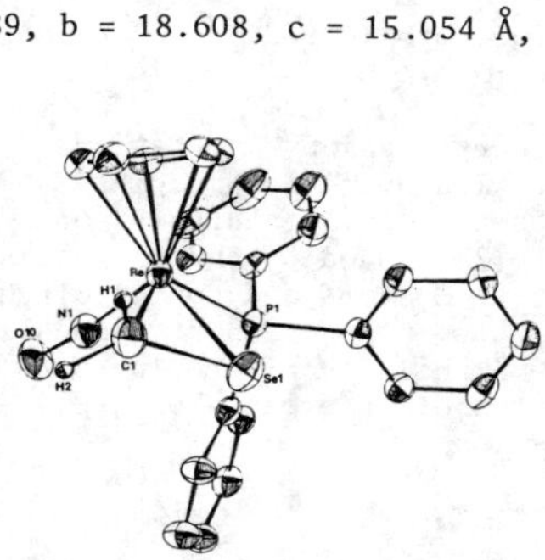

Fig. 1. Molecular structure of the cation of $C_{24}H_{22}F_6NOP_2ReSe$.

In the cation (Fig. 1), the CH_2=Se ligand is η^2-bound to the metal with the Re-C(1)=Se plane twisted 21.8° away from the Re-P(1) bond vector (toward the C_5H_5 ligand). Pertinent bond lengths and angles are: Re-C(1) 2.173, Re-Se(1) 2.522, Re-P(1) 2.447, Re-N(1) 1.765, Re-C(Cp) 2.299 (average), C(1)-H(1) 0.87, C(1)-H(2) 1.09, C(1)-Se(1) 1.879, N(1)-O(10) 1.169 Å, Re-C(1)-Se(1) 76.6, Re-Se(1)-C(1) 57.0, C(1)-Re-Se(1) 46.5, P(1)-Re-C(1) 123.8, Re-N(1)-O(10) 173.2, N(1)-Re-Se(1) 107.5, P(1)-Re-Se(1) 81.08 and N(1)-Re-P(1) 88.9°.

1,12-DIMETHYL[1.1]FERROCENOPHANIUM TRIIODIDE
$C_{24}H_{24}Fe_2I_3$

M.F. MOORE, S.R. WILSON, D.N. HENDRICKSON and U.T. MUELLER-WESTERHOFF, 1984. Inorg. Chem., 23, 2918-2920.

Monoclinic, $P2_1/c$, a = 9.458, b = 19.794, c = 13.776 Å, β = 108.23°, Z = 4. Mo radiation, R = 0.053 for 2672 reflexions.

The $C_{24}H_{24}Fe_2^+$ cation (Fig. 1) assumes a syn conformation. The two metallocene units are inequivalent. The centroid-centroid distances of 3.31(2) and 3.39(2) Å for Fe(1) and Fe(2), respectively, indicate that the former is in the Fe(II) oxidation state and the latter in the Fe(III) oxidation state.

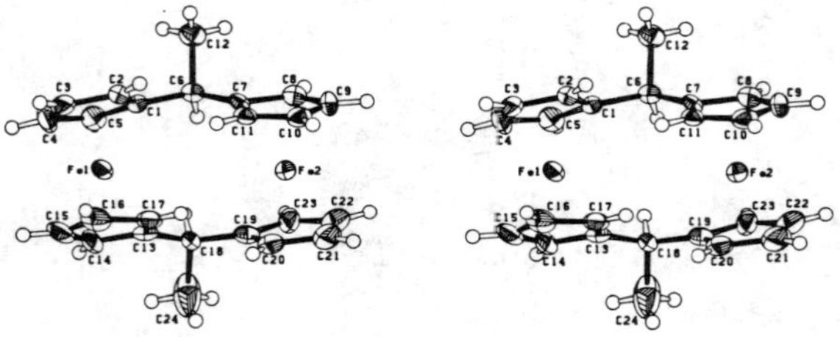

Fig. 1. Stereoview of the $C_{24}H_{24}Fe_2^+$ cation.

PENTAKIS(μ_3-SULPHIDO)TETRAKIS(η^5-CYCLOPENTADIENYLIRON) TETRACHLOROOXOTETRAHYDROFURAN-MOLYBDATE
$C_{24}H_{28}Cl_4Fe_4MoO_2S_5$ $[Fe_4S_5Cp_4]^+ . [MoOCl_4(C_4H_8O)]^-$

N. DUPRE, H.M.J. HENDRIKS, J. JORDANOV, J. GAILLARD and P. AURIC, 1984. Organometallics, 3, 800-802.

Orthorhombic, $P2_12_12_1$, a = 28.253, b = 14.972, c = 7.487 Å, Z = 4. Mo radiation, R = 0.057 for 2925 reflexions.

The complex consists of discrete anions and cations. The cation (Fig. 1) contains a distorted Fe_4S_5 core with a triply bridging S_2^{2-} ligand and with three different types of Fe atoms present: Fe(1) and Fe(4) each bonded to three S atoms and interacting with one Fe atom, Fe(2) bonded to three S atoms and interacting with two Fe atoms, and Fe(3) bonded to four S atoms and not involved in any Fe-Fe bond. The Fe-S distances divide into two sets: those "short" bonds involving Fe(1), Fe(2) and Fe(4) range from 2.186 to 2.240 Å while the four "long" Fe(3)-S bonds range from 2.271 to 2.284 Å. Other distances of interest are: Fe(1)-Fe(2) 2.723, Fe(2)-Fe(4) 2.866 and S(2)-S(4) 2.067 Å.

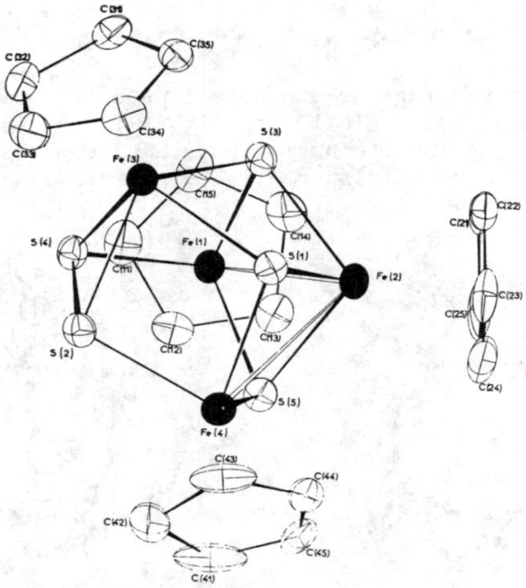

Fig. 1. The cation in $C_{24}H_{28}Cl_4Fe_4MoO_2S_5$.

(μ₂-DIACETATOLEAD)BIS[BIS(η⁵-CYCLOPENTADIENYL)HYDRIDOMOLYBDENUM]
$C_{24}H_{28}Mo_2O_4Pb$

M.M. KUBICKI, R. KERGOAT, J.-E. GUERCHAIS and P. L'HARIDON, 1984. J. Chem. Soc. Dalton, 1791-1793.

Monoclinic, P2/n, a = 11.772, b = 7.796, c = 12.951 Å, β = 96.26°, Dm = 2.17, Z = 2. Mo radiation, R = 0.039 for 1483 reflexions.

The X-ray analysis established the presence of a Mo-Pb-Mo linkage and a hydride ligand (Fig. 1). The molecules are located on a twofold axis of symmetry. The Mo-Pb bond length is 2.808(1) and the Mo-H distance is 1.7 Å. The Pb-Mo(1)-H(Mo) angle is 76.3°. The Mo-C(C_5H_5) distances range from 2.235 to 2.338(14) Å and the Pb-O(1) and Pb...O(2) distances are 2.342(8) and 3.041(9) Å respectively.

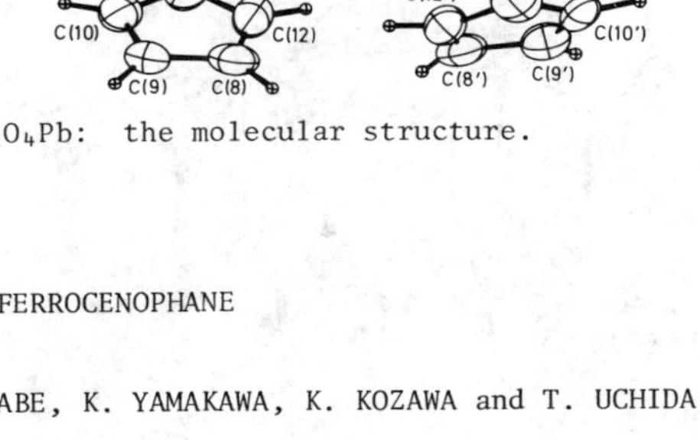

Fig. 1. $C_{24}H_{28}Mo_2O_4Pb$: the molecular structure.

[4][3][4][3](1,2,3,4)FERROCENOPHANE
$C_{24}H_{30}Fe$

M. HISATOME, J. WATANABE, K. YAMAKAWA, K. KOZAWA and T. UCHIDA, 1984. J. Organometal. Chem., 262, 365-377.

Monoclinic, $P2_1/a$, a = 17.490, b = 8.489, c = 12.157 Å, β = 101.51°, D_m = 1.41, Z = 4. Mo radiation, R = 0.069 for 2759 reflexions.

The two cyclopentadienyl rings are in an eclipsed conformation and almost parallel to each other (dihedral angle 1.1°) (Fig. 1). The bond lengths and angles in the Cp rings and methylene chains are normal but the distances between the Cp ring planes and the Fe atom are unusually short (1.589 and 1.591 Å).

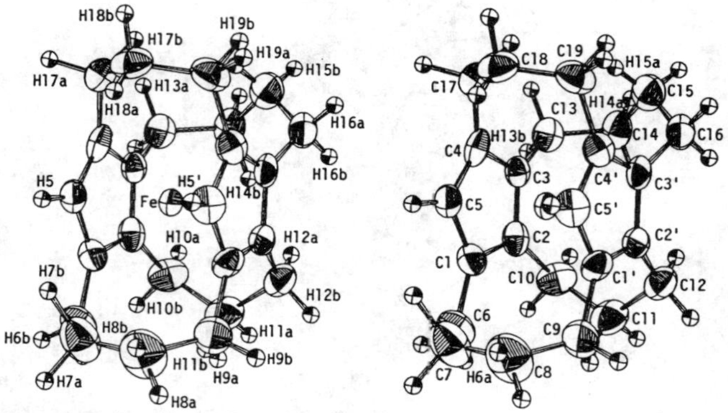

Fig. 1. Stereoscopic view of $C_{24}H_{30}Fe$.

(μ-OXO)BIS[DICARBONYL(η^5-PENTAMETHYLCYCLOPENTADIENYL)RHENIUM](Re-Re)

$C_{24}H_{30}O_5Re_2$ $(\eta^5-C_5(CH_3)_5)_2Re_2(CO)_4(\mu-O)$

W.A. HERRMANN, R. SERRANO, A. SCHÄFER, U. KÜSTHARDT, M.L. ZIEGLER and E. GUGGOLZ, 1984. J. Organometal. Chem., 272, 55-71.

Triclinic, $P\bar{1}$, a = 9.074, b = 10.402, c = 14.143 Å, α = 79.99, β = 88.42, γ = 66.18°, Z = 2. Mo radiation, R = 0.037 for 3283 reflexions.

The structure contains a core consisting of an isosceles Re_2O triangle (Fig. 1). The metal atoms have distorted square pyramidal geometry and the Re-Re distance is 2.817 Å. The two pentamethylcyclopentadienyl ligands are almost parallel to each other (interplanar angle 2°) and occupy trans-positions with respect to the central Re_2O unit. The bridging bonds Re(1)-O(5) and Re(2)-O(5) are respectively 1.973 and 1.972 Å and the Re(1)-O(5)-Re(2) angle is 91.1°. The Re-C(Cp ring) distances range from 2.25 to 2.370 Å.

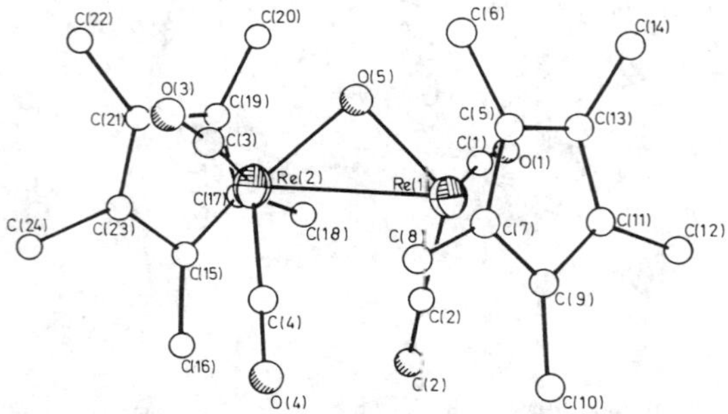

Fig. 1. Molecular structure of $C_{24}H_{30}O_5Re_2$.

PHENYLENEDISELENO-t-BUTYLZIRCONOCENE

$C_{24}H_{30}Se_2Zr$ $(\eta^5\text{-t-BuC}_5H_4)_2ZrSe_2C_6H_4$

B. GAUTHERON, G. TAINTURIER, S. POULY, F. THEOBALD, H. VIVIER and A. LAARIF, 1984. Organometallics, 3, 1495-1499.

Monoclinic, $P2_1/n$, a = 13.679, b = 22.649, c = 7.749 Å, β = 103.06°, Z = 4. Cu radiation, R = 0.052 for 4668 reflexions.

The central ring consists of a Zr atom, two Se atoms, and two aromatic C atoms (Fig. 1). The coordination polyhedron around the Zr atom (defined by the cyclopentadienyl ring centroids and the Se atoms) is a distorted tetrahedron of nearly C_{2v} symmetry. The dihedral angles between the two cyclopentadienyl rings and between the phenyl-Se-Se and Se-Zr planes are 53.8 and 56.1° respectively. The Zr atom is 1.67 Å from the phenyl group plane. The main bond distances are: Zr-Se(1) 2.670, Zr-Se(2) 2.639, Se(1)-C(31) 1.943, Se(2)-C(32) 1.929 Å. The Zr-C(Cp) distances range from 2.458 to 2.610 Å and the angles Se(1)-Zr-Se(2), Zr-Se(1)-C(31), Zr-Se(2)-C(32) are respectively 81.1, 90.3 and 91.2°. The Cp(centroid)-Zr-Cp-(centroid) angle is 129.9°.

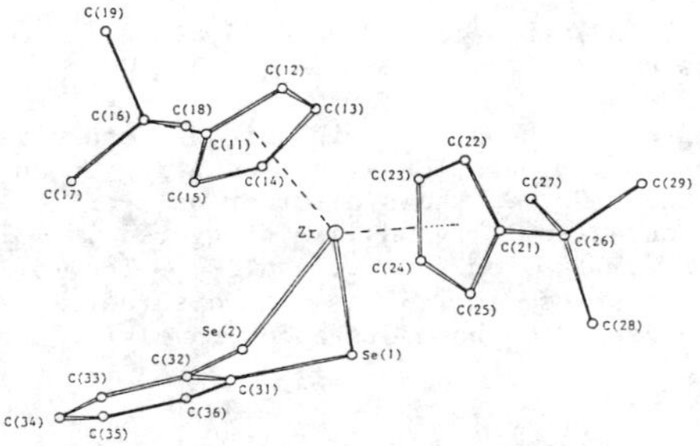

Fig. 1. Molecular structure of $C_{24}H_{30}Se_2Zr$.

BIS(μ_2-CARBONYL)(μ_2-ETHYLIDENE)BIS(η^5-PENTAMETHYLCYCLOPENTADIENYL)DICOBALT

$C_{24}H_{34}Co_2O_2$ $(\mu CHCH_3)[(\eta^5\text{-}C_5Me_5)Co(\mu\text{-CO})]_2$

W.A. HERRMANN, J.M. HUGGINS, C. BAUER, M.L. ZIEGLER and H. PFISTERER, 1984. J. Organometal. Chem., 262, 253-262.

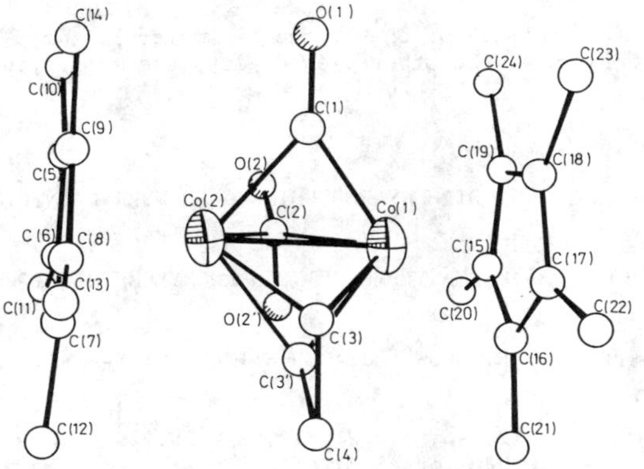

Fig. 1. Molecular structure of $C_{24}H_{34}Co_2O_2$ showing disordered bridging groups.

Triclinic, $P\bar{1}$, a = 8.922, b = 9.801, c = 13.968 Å, α = 89.78, β = 85.08, γ = 69.33°, Z = 2. Mo radiation, R = 0.057 for 2186 reflexions.

Structural analysis reveals an almost C_s-symmetrical structure with a triply-bridged metal-metal bond (Fig. 1). The bridging hydrocarbon ligand C(3)-C(4) is disordered as is the bridging carbonyl group C(2)-O(2). The 'normal' bridging CO group has Co(1)-C(1) and Co(2)-C(1) distances of 1.89 and 1.910 Å respectively, and an angle Co(1)-C(1)-Co(2) of 75.5°. The Co-Co distance is 2.327 Å and the distances from metal atoms to ring planes for Co(1) and Co(2) are 1.72 and 1.73 Å respectively. The ring planes are almost parallel, the interplanar angle being 2.2°.

AMMINEBIS[(1,2,3,4,5-η)-1,2,3,4,5-PENTAMETHYL-2,4-CYCLOPENTADIEN-1-YL](TETRAHYDRO-FURAN)YTTERBIUM

$C_{24}H_{41}NOYb$ $(C_5Me_5)_2Yb(NH_3)(OC_4H_8)$

A.L. WAYDA, J.L. DYE and R.D. ROGERS, 1984. Organometallics, 3, 1605-1610.

Orthorhombic, $Pca2_1$, a = 22.877, b = 15.489, c = 15.018 Å, Z = 8. Mo radiation, R = 0.069 for 3350 reflexions.

There are two molecules in the asymmetric unit and one molecule (No. 1) exhibits disorder between the NH_3 and THF sites resulting in an overlap of the N and O positions, while the other (No. 2) shows disorder only of the THF ring (Fig. 1). The two pentamethylcyclopentadienyl ligands are staggered in molecule No. 2 with average Yb-C distances of 2.77 Å (for both molecules). The Yb(2)-O(2) distance is 2.46 Å and the Yb(2)-N(2) distance is 2.55 Å. The disorder in molecule No. 1 precludes any minute examination of bond distances involving the NH_3 and THF ligands.

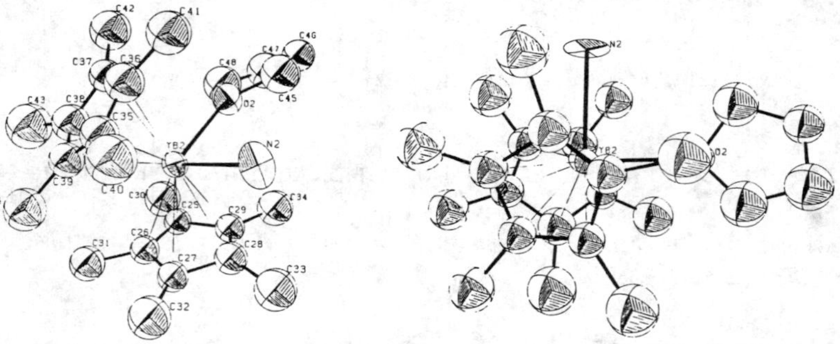

Fig. 1. Molecular structure of one of the two molecules of $C_{24}H_{41}NOYb$ in the asymmetric unit, the other view depicts the staggered C_5Me_5 rings.

BROMO(η^5-CYCLOPENTADIENYL)(η^2-DIPHENYLPHOSPHINO-o-PHENYLENEVINYL)OSMIUM

$C_{25}H_{22}BrOsP$ (I)

(η^5-CYCLOPENTADIENYL)(η^3-2-(DIPHENYLPHOSPHINO-o-PHENYLENE)DITHIOPROPIONATE)RUTHENIUM

$C_{26}H_{23}PRuS_2$ (II)

M.I. BRUCE, T.W. HAMBLEY, M.R. SNOW and A.G. SWINCER, 1984. J. Organometal. Chem., 273, 361-376.

I. Monoclinic, $P2_1/n$, a = 12.696, b = 21.719, c = 15.929 Å, β = 79.77°, D_m = 1.92, Z = 8. Mo radiation, R = 0.040 for 2867 reflexions.

II. Orthorhombic, Pbca, a = 8.921, b = 15.982, c = 32.216 Å, Z = 8. Mo radiation, R = 0.027 for 1685 reflexions.

In I the Os atom is coordinated by the cyclopentadienyl group, the Br atom and the P and C=C double bond of the chelating sp ligand (Fig. 1). The Os-C(Cp) distances average 2.223 Å, the Os-Br distance is 2.545 Å and the Os-P separation is 2.284 Å. The unusual Os-C(1) and Os-C(2) bonds are 2.249 and 2.179 Å in length and the C(1)-C(2) separation is 1.57 Å. The orientation of the vinyl group is such that the C=C bond is almost coplanar with the Os-Br vector. In II, the Ru atom is bonded to the Cp group, a P atom and, unusually, to the two S and C atoms of the dithiocarboxylate group (Fig. 1). The Ru-C(Cp) distances average 2.215 Å and the Ru-P distance is 2.288 Å. The Ru-S bond lengths are 2.418 and 2.426 Å and the Ru-C separation is 2.175 Å. The C-S distances are 1.722 and 1.728 Å and the angles S(1)-C(1)-S(2), and S(1)-Ru-S(2) are 121.3 and 76.7° respectively with other angles at C(1) being 121.4 and 116.6°, consistent with sp^2 hybridization.

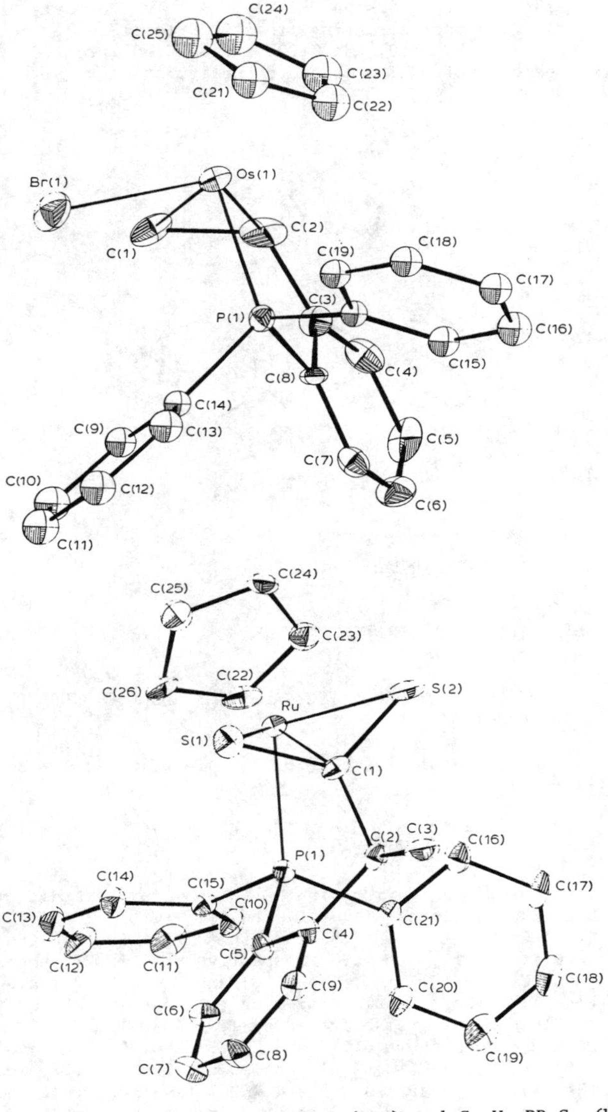

Fig. 1. Molecular structure of $C_{25}H_{22}BrOsP$ (top) and $C_{26}H_{23}PRuS_2$ (bottom).

DICARBONYLHYDRO[(1,2,3,4,5-η)-1-METHYL-2,4-CYCLOPENTADIEN-1-YL](METHYL-1-NAPHTHALENYLPHENYLSILYL)MANGANESE
$C_{25}H_{23}MnO_2Si$

F. CARRÉ, E. COLOMER, R.J.P. CORRIU and A. VIOUX, 1984. Organometallics, 3, 1272-1278.

Orthorhombic, $P2_12_12_1$, a = 8.746, b = 31.19, c = 7.814 Å, D_m = 1.33, Z = 4. Mo radiation, R = 0.075 for 721 reflexions.

The cyclopentadienyl ring shows no unusual features; the atom CP6 is coplanar with the ring and lies away from the naphthyl group. The average Mn-CP distance is 2.12 Å and the average Mn-C(CO) distance is 1.77 Å. The geometry around Si is tetrahedral (Fig. 1) with angles ranging from 104.9 to 117.0°. The Mn-Si distance is 2.461 Å, and the bonds Si-C(10), Si-C(20) and Si-C(30) are 1.86, 1.91 and 1.85 Å in length respectively. The absolute configuration at Si is S.

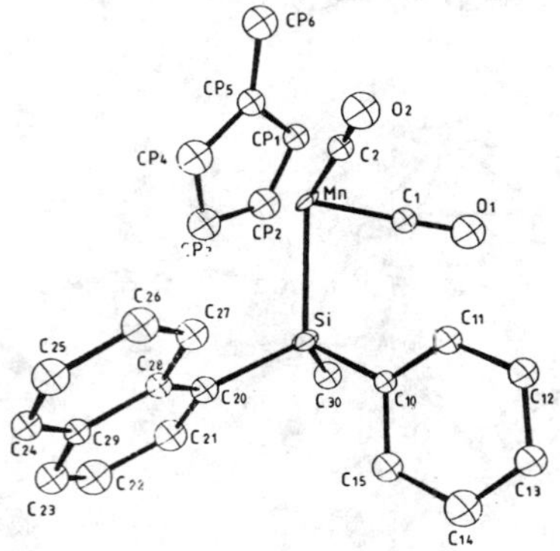

Fig. 1. Molecular structure of $C_{25}H_{23}MnO_2Si$.

2,2'-TRIMETHYLENE[1.1]FERROCENOPHANE
$C_{25}H_{24}Fe_2$ (I)

3,3'-TRIMETHYLENE[1.1]FERROCENOPHANE
$C_{25}H_{24}Fe_2$ (II)

N.J. SINGLETARY, M. HILLMAN, H. DAUPLAISE, A. KVICK and R.C. KERBER, 1984. Organometallics, 3, 1427-1434.

I. Orthorhombic, $Pna2_1$, a = 21.754, b = 20.446, c = 8.128 Å, Z = 8. Mo radiation, R = 0.088 for 1711 reflexions.

II. Monoclinic, $P2_1/c$, a = 22.667, b = 7.528, c = 10.797 Å, β = 97.34°, Z = 4. Mo radiation, R = 0.059 for 3341 reflexions.

The tilting of the pairs of rings in the two compounds is different, in I the dihedral angle between the planes C(1)-C(5) and C(1')-C(5') is 14.1° and between C(6)-C(10) and C(6')-C(10') is 4.9° (the corresponding values in the 2nd molecule in the asymmetric unit are 10.4 and 2.3° respectively). In II, the angles between the planes C(1)-C(5) and C(1')-C(5') and C(6)-C(10) and C(6')-C(10') are respectively 10.0 and 3.9° (Fig. 1). The directions of ring-tilt opening are quite different also. The 5- and -10 H atoms are separated by 1.91 Å in I and 2.08 Å in II, whilst the separation of the endo H atoms C(11) and C(12) is 2.30 Å in I and 2.68 Å in II. The C(1)-C(11)-C(6) and C(1')-C(12)-C(6') angles are 120 and 127° in I and 122.2 and 121.9° in II. The Fe...Fe separations are 4.752 Å (average) in I and 4.749 Å in II. Atom C(14) in II has disorder over two sites.

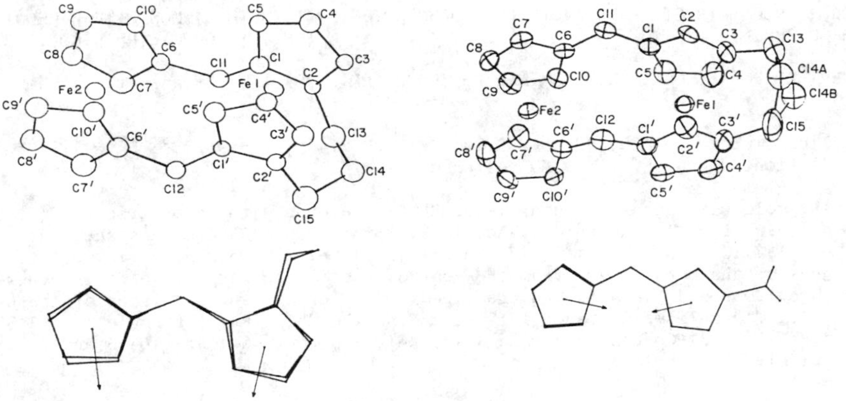

Fig. 1. Molecular structure of $C_{25}H_{24}Fe_2$ (I) (top left), II (top right), and relative twisting of the two ferrocene units and directions (arrows) of the ring-tilt openings (bottom left and right).

BIS(η^5-PENTAMETHYLCYCLOPENTADIENYL)YTTERBIUM(III) PENTACARBONYLMANGANESE(-1) TOLUENE

$C_{25}H_{30}MnO_5Yb,0.25(C_7H_8)$ $(C_5(CH_3)_5)_2YbMn(CO)_5,0.25(CH_3C_6H_5)$

J.M. BONCELLA and R.A. ANDERSEN, 1984. Inorg. Chem., 23, 432-437.

Monoclinic, C2/m, a = 18.942, b = 32.592, c = 19.029 Å, β = 109.92°, Z = 16. Mo radiation, R = 0.028 for 5487 reflexions.

The structure is composed of a polymeric chain of $[(Me_5C_5)_2Yb(\mu\text{-}OC)_3Mn(CO)_2]$ units with dimeric units $[(Me_5C_5)_2Yb(\mu\text{-}OC)_2Mn(CO)_3]$ sacked between the polymeric sheets. The structural entities are shown in Fig. 1. Mean bond lengths are: Yb-C(Cp) 2.574(13) and 2.589(15), Yb-O 2.271(2) and 2.389(27), Mn-CO_t 1.833(7) and 1.820(5), Mn-CO_b 1.748(13) and 1.791(13) Å respectively for the dimer and polymer. The Yb atoms are eight-coordinate in the dimer and nine-coordinate in the polymer.

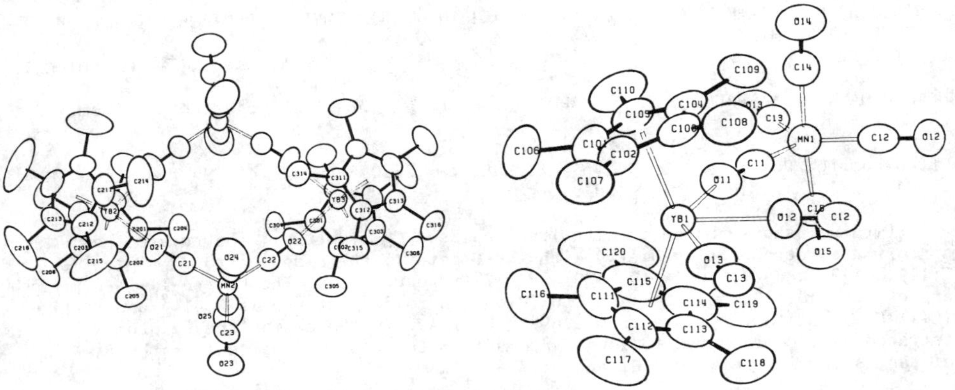

Fig. 1. The dimeric (left) and polymeric portions of $(Me_5C_5)_2YbMn(CO)_5$.

BIS(μ_2-CARBONYL)BIS(η^5-PENTAMETHYLCYCLOPENTADIENYL)(TRIMETHYL PHOSPHITE)DIRHODIUM

$C_{25}H_{39}O_5PRh_2$ $C_5(CH_3)_5[P(OCH_3)_3]Rh(\mu\text{-}CO)_2RhC_5(CH_3)_5$

H. WERNER, B. KLINGERT, R. ZOLK and P. THOMETZEK, 1984. J. Organometal. Chem., 266, 97-107.

Monoclinic, C2/c, a = 20.409, b = 9.291, c = 29.774 Å, β = 95.82°, D_m = 1.54, Z = 8. Mo radiation, R = 0.057 for 3290 reflexions.

The molecule contains a non-planar Rh_2C_2 skeleton with asymmetric Rh-C(O) bond lengths (Fig. 1). The Rh-Rh distance is 2.685 Å and the Rh-C(O) distances are: Rh(1)-C(1) 2.134, Rh(1)-C(2) 2.127, Rh(2)-C(1) 1.892 and Rh(2)-C(2) 1.906 Å. The distances of the metal atoms from the centroids of their respective five-membered rings are 1.919 and 1.933 Å, and the dihedral angle between the planes of these rings is 62.6°. The Rh(1)-P distance is 2.215 Å and the Rh(2)-Rh(1)-P angle is 103.3°. The angles Rh(1)-Rh(2)-Cp(centroid) and Rh(2)-Rh(1)-Cp(centroid) are 168.3 and 127.8° respectively.

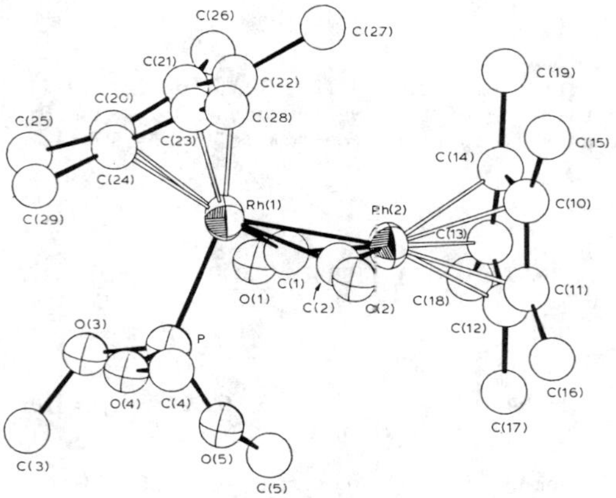

Fig. 1. Molecular structure of $C_{25}H_{39}O_5PRh_2$.

BIS(N,N,N',N'-TETRAMETHYLETHYLENEDIAMINE)LITHIUM TRIMETHYL(η^5-PENTAMETHYLCYCLOPENTADIENYL)LUTETIUM(III)

$C_{25}H_{56}LiLuN_4$ $[Li(tmed)_2]^+[C_5Me_5Lu(CH_3)_3]^-$

H. SCHUMANN, I. ALBRECHT, J. PICKARDT and E. HAHN, 1984. J. Organometal. Chem., 276, C5-C9.

Orthorhombic, $Pc2_1n$, a = 9.516, b = 13.908, c = 25.078 Å, Z = 4. Mo radiation, R = 0.084 for 1082 reflexions, (-100°C)

Figure 1 shows the anion with its idealised C_s symmetry. The Lu atom has approximate tetrahedral geometry with the C(methyl)-Lu-(Ring centroid) angles for C(11), C(12) and C(13) of 112, 120 and 117° respectively. The C(methyl)-Lu-C(methyl) angles are 110, 106 and 90°. The Lu-(Ring centroid) distance is 2.41 Å and the Lu-(C(1)-C(5)) distances range from 2.69 to 2.71 Å. In the cation the Li atom is tetrahedrally coordinated to four N atoms from the two tmed molecules; the Li-N distances varying between 2.02 and 2.27 Å.

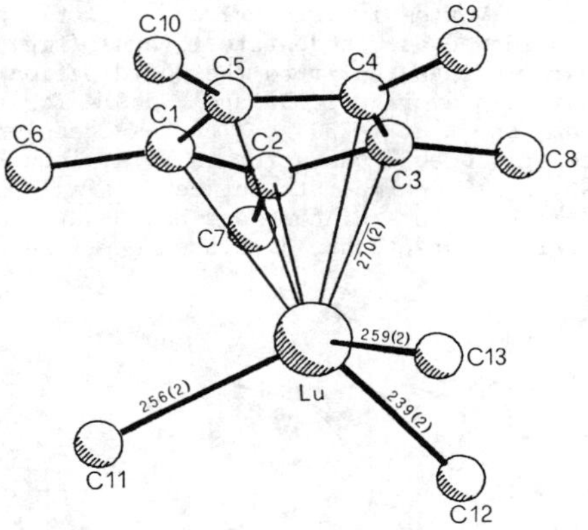

Fig. 1. Molecular structure of the anion in $C_{25}H_{56}LiLuN_4$.

(η⁵-CYCLOPENTADIENYL)DICARBONYL[C-(DIPHENYLTHIOPHOSPHINE)-N-PHENYLTHIOFORMAMIDO-N,S]-MOLYBDENUM(II)
$C_{26}H_{20}MoNO_2PS_2$ $Mo(CO)_2(\eta^5\text{-}C_5H_5)[Ph_2P(S)C(S)NPh]$

H.P.M.M. AMBROSIUS, A.W. VAN HEMERT, W.P. BOSMAN, J.H. NOORDIK and G.J.A. ARIAANS, 1984. Inorg. Chem., 23, 2678-2685.

Triclinic, $P\bar{1}$, a = 10.390, b = 14.328, c = 9.602 Å, α = 73.02, β = 108.49, γ = 112.39°, Z = 2. Mo radiation, R = 0.048 for 2356 reflexions.

The molecule (Fig. 1) has a four-legged piano stool geometry with Mo-S 2.490(2), Mo-N 2.175(5), Mo-CO 1.92(2) and 1.94(2), Mo-C(Cp) 2.27(2)-2.40(2) Å.

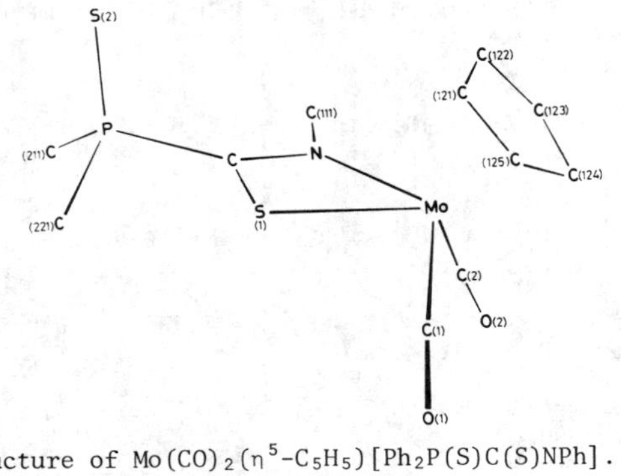

Fig. 1. The structure of $Mo(CO)_2(\eta^5\text{-}C_5H_5)[Ph_2P(S)C(S)NPh]$.

PHENYLIMIDODECAMETHYLVANADOCENE
$C_{26}H_{35}NV$ $(\pi\text{-}C_{10}H_{15})_2VNC_6H_5$

S. GAMBAROTTA, A. CHIESI-VILLA and C. GUASTINI, 1984. J. Organometal. Chem., 270, C49-C52.

Monoclinic, $P2_1/c$, a = 13.951, b = 12.199, c = 13.845 Å, β = 103.92°, Z = 4. Mo radiation, R = 0.054 for 1577 reflexions.

Coordination around the V atom is trigonal and strictly planar if the cyclopentadienyl rings are considered as monodentate ligands (Fig. 1). The phenylimido ligand is planar and nearly perpendicular to the coordination plane (dihedral angle 93.8°). The V-C(Cp) distances average 2.058 and 2.049 Å for the two ligands, and the V-N distance is rather short at 1.730 Å. The V-N-C angle is almost linear (178.2°) suggesting some triple bond character in V≡NR, in acreement with the N-C bond length (1.345 Å) which is intermediate between a single and a double bond. The Cp-V-Cp angle is 138.4°. The methyl C atoms are bent away from the ring planes to reduce methyl-methyl steric interactions, displacements ranging from 0.019 to 0.348 Å.

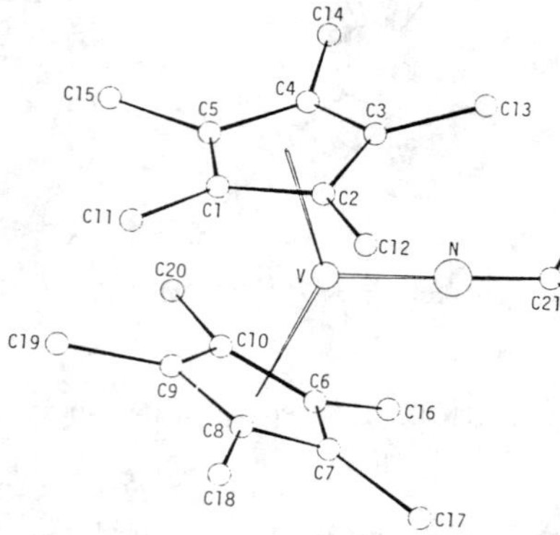

Fig. 1. Molecular structure of $C_{26}H_{35}NV$.

BIS(μ_2-METHOXO)(μ_2-3-METHYLPYRAZOLATO-N^1,N^2)BIS[(η^5-PENTAMETHYLCYCLOPENTADIENYL)-RHODIUM(III)] PERCHLORATE
$C_{26}H_{41}ClN_2O_6Rh_2$ (I)

BIS(μ_2-HYDROXO)BIS[(3,5-DIMETHYLPYRAZOLE-N^2)(η^5-PENTAMETHYLCYCLOPENTADIENYL)-RHODIUM(III)] BIS(TETRAFLUOROBORATE)
$C_{30}H_{48}B_2F_8N_4O_2Rh_2$ (II)

(μ_2-HYDROXO)BIS(μ_2-PYRAZOLATO-N^1,N^2)BIS[(η^5-PENTAMETHYLCYCLOPENTADIENYL)RHODIUM(III)] PERCHLORATE
$C_{26}H_{37}ClN_4O_5Rh$ (III)

L.A. ORO, D. CARMONA, M.P. LAMATA, M.C. APREDA, C. FOCES-FOCES, F.H. CANO and P.M. MAITLIS, 1984. J. Chem. Soc. Dalton, 1823-1829.

I. Monoclinic, C2/c, a = 22.366, b = 11.085, c = 12.252 Å, β = 107.59°, Z = 4. Cu radiation, R = 0.072 for 2297 reflexions.

II. Monoclinic, $P2_1/n$, a = 11.7094, b = 15.612, c = 10.172 Å, β = 92.44°, Z = 2. Cu radiation, R = 0.048 for 3003 reflexions.

III. Monoclinic, C2/c, a = 25.425, b = 16.343, c = 16.448 Å, β = 120.53°, Z = 8. Cu radiation, R = 0.061 for 3600 reflexions.

Each cation (Fig. 1) contains two Rh atoms each η^5-bonded to a C_5Me_5 group (Rh-C 2.151(4), 2.140(3), 2.153, 2.152(3) Å for I, II, III respectively) and bridged by two methoxo groups and a pyrazolate group in I (Rh-O 2.100(6), Rh-N 2.082(7) Å), two hydroxo groups in II (Rh-O 2.186(4), 2.172(6) Å) and by one hydroxo and two pyrazolate groups in III (Rh-O 2.110(4), 2.104(4), Rh-N 2.089(7), 2.095(6), 2.098(6), 2.088(7) Å). The Rh...Rh distances are 3.200, 3.360 and 3.488(1) Å respectively of I, II and III. Compound I exhibits internal disorder giving rise to a crystallographic twofold axis relating the two halves of the cation and passing through the methyl pyrazolato ligand. II has an internal crystallographic centre of symmetry but shows no disorder while the cation in III is crystallographically independent.

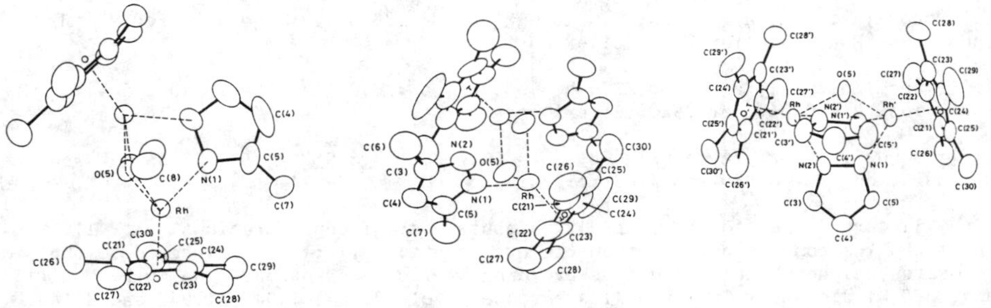

Fig. 1. The cations in $C_{26}H_{41}ClN_2O_6Rh_2$ (left), $C_{30}H_{48}B_2F_8N_4O_2Rh_2$ (centre) and $C_{26}H_{37}ClN_4O_5Rh$ (right).

(μ_2-N,N'-TETRAMETHYLETHYLENEDIAMINE-N,N')BIS(BIS(μ_2-HYDRIDO)(DIHYDRIDOALUMINIUM)BIS-(η^5-CYCLOPENTADIENYL)TITANIUM) BENZENE

$C_{26}H_{44}Al_2N_2Ti_2,C_6H_6$ $[(C_5H_5)_2TiAlH_4]_2[(CH_3)_2N(C_2H_4)N(CH_3)_2],C_6H_6$

E.B. LOBKOVSKII, G.L. SOLOVEICHIK, A.I. SISOV, B.M. BULYCHEV, A.I. GUSEV and N.I. KIRILLOVA, 1984. J. Organometal. Chem., 265, 167-173.

Triclinic, $P\bar{1}$, a = 8.406, b = 10.117, c = 11.269 Å, α = 112.01, β = 109.25, γ = 87.04°, Z = 1. Mo radiation, R = 0.056 for 834 reflexions.

The crystals consist of centrosymmetric, dumbell-shaped molecules of the metallic complex (Fig. 1), and solvent molecules of benzene. Both complex and benzene molecules have crystallographic inversion symmetry. Two moieties of $Cp_2TiH_2AlH_2$ are linked by a tetramethylethylenediamine molecule. The Al-N distance is 2.11 Å. The Al atom is bonded to a Ti atom by a double hydride bridge (Al-H(1) 1.8, Al-H(2) 1.68, Ti-H(1) 1.63, Ti-H(2) 1.60 Å) and has trigonal bipyramidal geometry (Al-H(3) 1.45, Al-H(4) 1.65 Å). The distance of the Ti atom to the mean planes through the Cp rings is 2.04 Å in each case with Ti-C distances ranging from 2.32 to 2.39 Å. The dihedral angle between the ring planes is 40.6° and the Ti...Al distance is 2.788 Å.

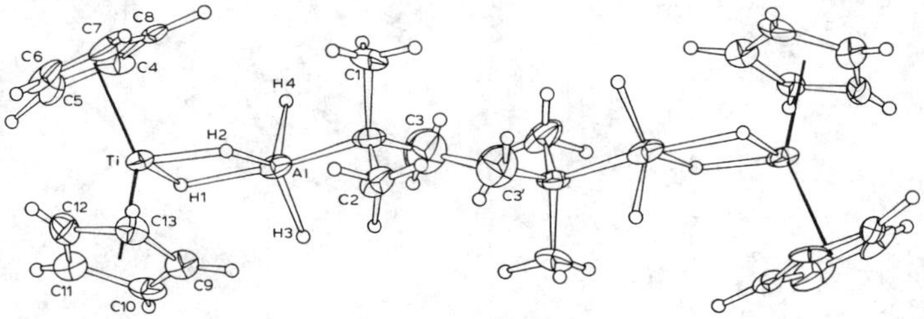

Fig. 1. Molecular structure of $C_{26}H_{44}Al_2N_2Ti_2$.

1-(η^5-CYCLOPENTADIENYL)-1-TRIPHENYLPHOSPHINE-1-COBALTACYCLOPENTADIENE
$C_{27}H_{24}CoP$ (I)

1-(η^5-METHOXYCARBONYLCYCLOPENTADIENYL)-1-TRIPHENYLPHOSPHINE-1-COBALTCYCLOPENTADIENE
$C_{29}H_{26}CoO_2P$ (II)

(μ_2-σ^3,η^2-1-(DIPHENYLPHOSPHINO-o-PHENYLENE)ETHYLIDENE-C,P)BIS((η^5-METHOXYCARBONYL-CYCLOPENTADIENYL)COBALT)
$C_{34}H_{31}Co_2O_4P$ (III)

H. YAMAZAKI and Y. WAKATSUKI, 1984. J. Organometal. Chem., 272, 251-253.

I. Monoclinic, Pa, a = 8.529, b = 16.010, c = 8.028 Å, β = 100.31°, Z = 2. Mo radiation, R = 0.034 for 2242 reflexions.

II. Monoclinic, $P2_1/a$, a = 8.327, b = 36.468, c = 8.021 Å, β = 98.75°, Z = 4. Mo radiation, R = 0.047 for 2810 reflexions.

III. Monoclinic, $P2_1/c$, a = 10.681, b = 30.722, c = 8.912 Å, β = 93.55°, Z = 4. Mo radiation, R = 0.050 for 4107 reflexions.

All three molecules have approximate piano stool configurations (Fig. 1). In II the methoxycarbonyl substituent on the cyclopentadienyl ring is directed to the side opposite the phosphine ligand thus avoiding steric interaction. The Co-P, Co-C(1) and Co-C(4) distances in I and II are respectively 2.144, 1.943, 1.947 and 2.148, 1.928 and 1.933 Å. The average Co-C(Cp) distances in I and II are 2.114 and 2.107 Å and the C(1)-C(2), C(2)-C(3) and C(3)-C(4) distances are 1.338(9), 1.447(10) and 1.351(10) Å in I, and 1.333(10), 1.442(10) and 1.341(9) Å in II. Thus these C_4 moieties are considered butadiene fragments. In III the important features are (a) a bridging carbene C bonded to a methyl group and C(32), an ortho-C in a phenyl group, and (b) the coordination of one edge of a phenyl group, C(31) and C(32), to Co(1) making a $\mu_2\eta^3$-benzyl type bond. Pertinent bond distances are: Co(1)-Co(2) 2.491, Co(1)-C(1) 1.971, Co(2)-C(1) 1.918, Co(2)-P 2.104, Co(1)-C(31) 2.087, Co(1)-C(32) 2.032, C(31)-C(32) 1.446 Å. The average Co-C(Cp) distances for Co(1) and Co(2) respectively are: 2.075 and 2.103 Å.

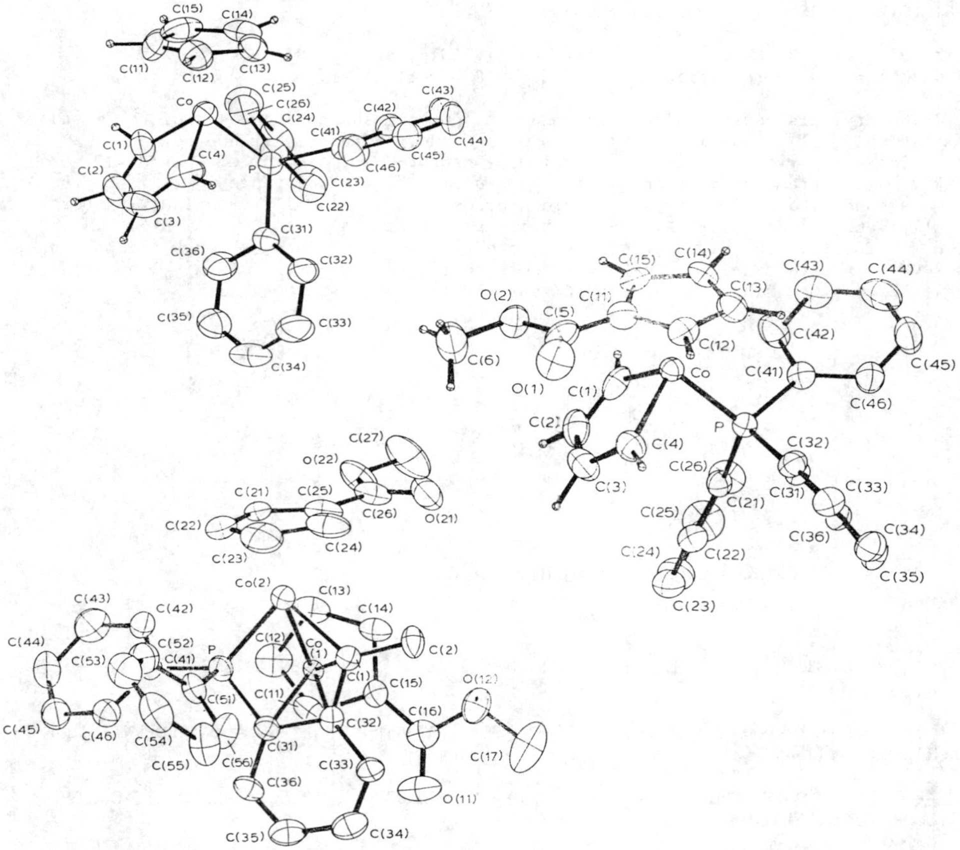

Fig. 1. Molecular structures of $C_{27}H_{24}CoP$ (top), $C_{29}H_{26}CoO_2P$ (middle), and $C_{34}H_{31}Co_2O_4P$ (bottom).

BIS[(μ_2-CARBONYL)(η^5-PENTAMETHYLCYCLOPENTADIENYL)RHODIUM]PENTACARBONYLMOLYBDENUM
$C_{27}H_{30}MoO_7Rh_2$

R.D. BARR, M. GREEN, J.A.K. HOWARD, T.B. MARDER, A.G. ORPEN and F.G.A. STONE, 1984. J. Chem. Soc. Dalton, 2757-2763.

Monoclinic, $P2_1/n$, a = 9.625, b = 17.500, c = 16.714 Å, β = 97.72°, Z = 4. Mo radiation, R = 0.026 for 4347 reflexions (at 200 K).

The core of the molecule (Fig. 1) consists of an $MoRh_2$ isosceles triangle (Mo-Rh(mean) 3.009(1), Rh-Rh 2.623(1) Å). The Rh-Rh vector is staggered with respect to the equatorial ligands of the $Mo(CO)_5$ group and is symmetrically bridged by two CO ligands (Rh-C 1.991-2.003(3) Å). The mean Rh-C(η-C_5Me_5) distance is 2.248(3) Å.

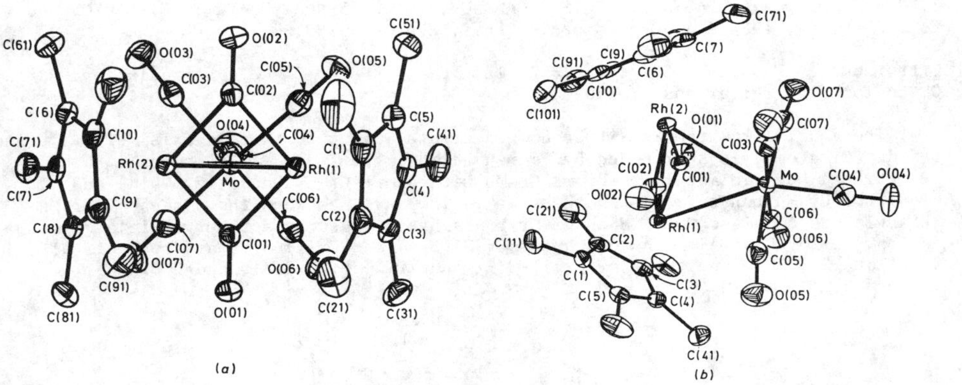

Fig. 1. Two views of $C_{27}H_{30}MoO_7Rh_2$.

1,1'-(μ_2-DIPHENYLPHOSPHINO-CARBONYL-(η^5-CYCLOPENTADIENYL)IRON-Fe,P)FERROCENE
$C_{28}H_{23}Fe_2OP$ (η^5-C_5H_4Fe(CO)Cp)Fe(η^5-C_5H_4PPh$_2$)

I.R. BUTLER and W.R. CULLEN, 1984. Organometallics, 3, 1846-1851.

Triclinic, $P\bar{1}$, a = 10.540, b = 11.507, c = 9.737 Å, α = 100.41, β = 98.49, γ = 88.95°, Z = 2. Mo radiation, R = 0.030 for 4756 reflexions.

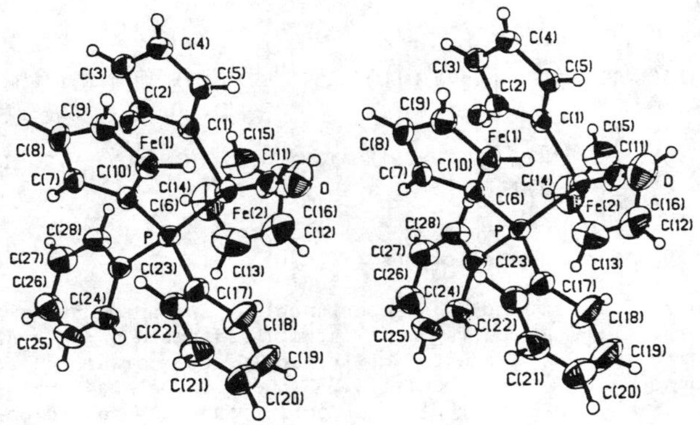

Fig. 1. Stereoview of the molecule $C_{28}H_{23}Fe_2OP$.

The molecule (Fig. 1) is the first of its kind (a [2]-ferrocenophane with heteroatoms in the bridge) to be characterized by X-ray crystallography. The rings are staggered by 10.3° with ring tilt being only 8.3° mainly due to the longer bridge P-Fe(2) distance of 2.21 Å. The dihedral angle C(1)-Fe(2)-P-C(6) is 18.32°. The Fe to Cp(centroid) distances are 1.635 (Fe(1)-Cp(1)), 1.637 (Fe(1)-Cp(2)) and 1.731 Å (Fe(2)-Cp(3)). The angles around Fe(2) are: P-Fe(2)-Cp(2) 128.8, C(1)-Fe(2)-Cp(3) 121.9, C(16)-Fe(2)-Cp(3) 123.0, and P-Fe(2)-C(1) 90.46°.

TETRAKIS(μ_3-SULPHIDO)BIS((μ_2-t-BUTYLTHIOLATO)BIS(η^5-CYCLOPENTADIENYL)DICHROMIUM(II))-MANGANESE(II)

$C_{28}H_{38}Cr_4MnS_6$ $[(C_5H_5)_2Cr_2(\mu\text{-}SCMe_3)(\mu_3\text{-}S)_2]_2Mn$

I.L. EREMENKO, A.A. PASYNSKII, G.Sh. GASANOV, B. ORAZSAKHATOV, Yu.T. STRUCHKOV and V.E. SHKLOVER, 1984. J. Organometal. Chem., 275, 183-189.

Orthorhombic, Fdd2, a = 39.715, b = 21.127, c = 7.808 Å, Z = 8. Mo radiation, R = 0.066 for 1166 reflexions, (-120°C).

The central Mn atom is situated on a crystallographic twofold axis (Fig. 1), and, besides Cr atoms, is surrounded by four μ^3-bridging S atoms. The molecule has a symmetrical bow-tie metal framework Cr_4Mn with short Cr-Cr bonds (2.690 Å) and elongated Cr-Mn bonds (2.936 and 2.947 Å). The angle between the planes of Cr_2Mn triangles is equal to 90°. Both Cr-Cr bonds are bridged by a t-butylthiolate group (average Cr-S 2.347 Å). The distances Cr(2)-S(1) and Cr(1)-S(2) are respectively 2.321 and 2.307 Å, and those of Cr(1)-S(1) and Cr(2)-S(2) 2.310 and 2.323 Å.

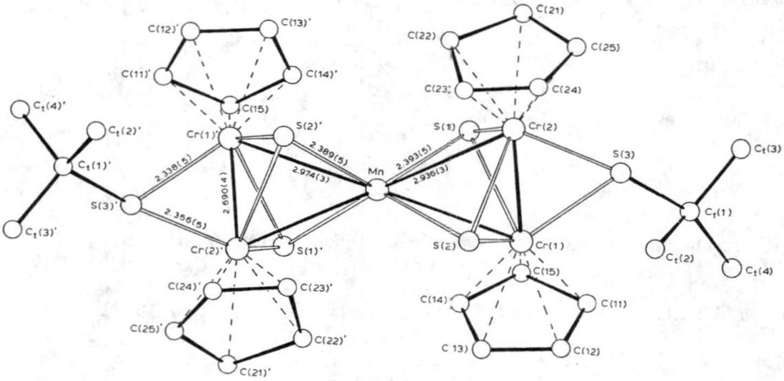

Fig. 1. View of $C_{28}H_{38}Cr_4MnS_6$ down the twofold axis.

BIS((μ_2-t-BUTYLTHIOLATO)-BIS(μ_3-SULPHIDO)-BIS(η^5-CYCLOPENTADIENYL)CHROMIUM))CHROMIUM

$C_{28}H_{38}Cr_5S_6$ $[(C_5H_5)_2Cr_2(\mu_2\text{-}SC(CH_3)_3)(\mu_3\text{-}S)_2]_2Cr$

A.A. PASYNSKII, I.L. EREMENKO, B. ORAZSAKHATOV, G.Sh. GASANOV, V.E. SHKLOVER and Yu.T. STRUCHKOV, 1984. J. Organometal. Chem., 269, 147-153.

Monoclinic, C2/c, a = 30.123, b = 7.279, c = 17.749 Å, β = 122.94°, Z = 4. Mo radiation, R = 0.064 for 2696 reflexions (-100°C).

The framework of the molecule is a pentanuclear spirane containing $[Cr^{III}]_2 \cdot Cr^{II} \cdot [Cr^{III}]_2$ ions (Fig. 1). The central Cr(1) atom is situated on a crystallographic twofold axis and the planes of the two Cr_3 triangles are normal to each other. Each metal triangle has two long Cr-Cr bonds (2.933 and 2.889 Å) and one short Cr(III)-Cr(III) bond (2.655 Å), the latter being a double bond. Above and below each Cr_3 triangle is a μ_3-sulphide bridge, and the peripheral Cr(III) atoms are bridged by a μ_2-$SC(CH_3)_3$ bridge and coordinated by a π-cyclopentadienyl ligand (average Cr-C distance 2.26 Å.

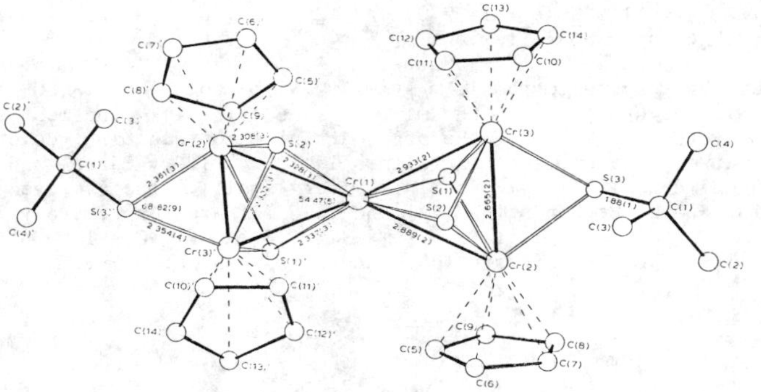

Fig. 1. Molecular structure of the pentanuclear cluster $C_{28}H_{38}Cr_5S_6$ showing selected bond distances (Å) and angles (°).

BIS((BIS(η^5-CYCLOPENTADIENYL))(μ_3-HYDRIDO)YTTRIUM-(μ_2-HYDRIDO)DIHYDRIDOALLUMINUM-(DIETHYL ETHER))

$C_{28}H_{44}Al_2O_2Y_2$ $\{[(\eta^5-C_5H_5)_2Y(\mu_3-H)][(\mu_2-H)AlH_2.OC_4H_8]\}_2$

V.K. BEL'SKII, B.M. BULYCHEV, A.B. EROFEEV and G.L. SOLOVEICHIK, 1984. J. Organometal. Chem., 268, 107-111.

Monoclinic, $P2_1/b$, a = 8.750, b = 11.043, c = 16.453 Å, γ = 95.57°, Z = 2. R = 0.041 for 1281 reflexions.

The crystals are composed of dimeric, symmetric molecules (Fig. 1) in which the Y, Al, O, H(1) and H(2) atoms are practically coplanar. This plane is the bisector of both Cp_2Y wedge-like sandwiches. The Y-Cp distance and the Cp-Y-Cp angle are respectively 2.36 Å and 127.5°. The Y(1)-Y(2), Y(1)-Al(2) and Y(1)-Al(1) distances are 3.753, 3.242 and 3.989 Å respectively, whilst other relevant bond distances are: Y(1)-H(1) 2.18, Y(1)-H(1') 2.3, Y(1)-H(2) 2.17, Al-O 1.972, Al-H(1) 2.0, Al-H(2) 1.57, Al-H(3) 1.49 and Al-H(4) 1.46 Å. The angles about atom H(1) are 117, 102 and 142°.

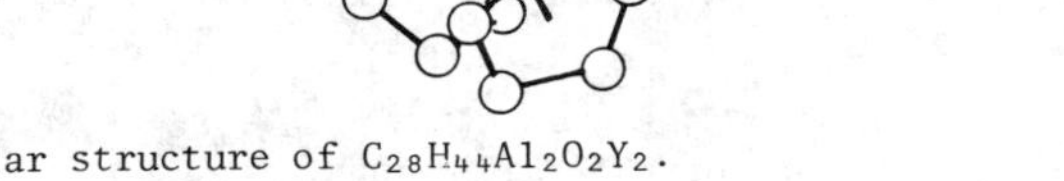

Fig. 1. Molecular structure of $C_{28}H_{44}Al_2O_2Y_2$.

(η^5,η^5-DIMETHYLBIS(TETRAMETHYLCYCLOPENTADIENYL)SILANE)BIS(TRIMETHYLSILYLMETHYL)THORIUM

$C_{28}H_{52}Si_3Th$ $(CH_3)_2Si[(CH_3)_4C_5]_2Th[CH_2Si(CH_3)_3]_2$

C.M. FENDRICK, E.A. MINTZ, L.D. SCHERTZ, T.J. MARKS and V.W. DAY, 1984. Organometallics, 3, 819-821.

Monoclinic, $P2_1/m$, a = 11.960, b = 11.270, c = 12.395 Å, β = 99.32°, Z = 2. Mo radiation, R = 0.076 for 2386 reflexions.

The molecule has crystallographic m symmetry with atoms Th, Si_C, C_1, C_2, C_{1A}, Si_A, C_{2A}, C_{1B}, Si_B and C_{2B} lying in the mirror plane (Fig. 1). The most striking feature is the severe distortion of the organoactinide coordination geometry. The dihedral angle between the Cp ring mean planes is 105.4° (very large) and the Cp(centroid)-Th-Cp'(centroid) angle is 118.4° (very small). The average Th-C(Cp) distance is 2.791 Å and the bonds Th-C(1A) and Th-C(1B) are respectively 2.54 and 2.48 Å. The Si_C - Cp_1 distance is 1.88 Å, C_{1A}-Si_A 1.81 Å, Si_B-C_{1B} 1.90 Å, and the angles Cp_1-Si-Cp'_1 and C_{1A}-Th-C_{1B} are 100.6 and 98.9° respectively.

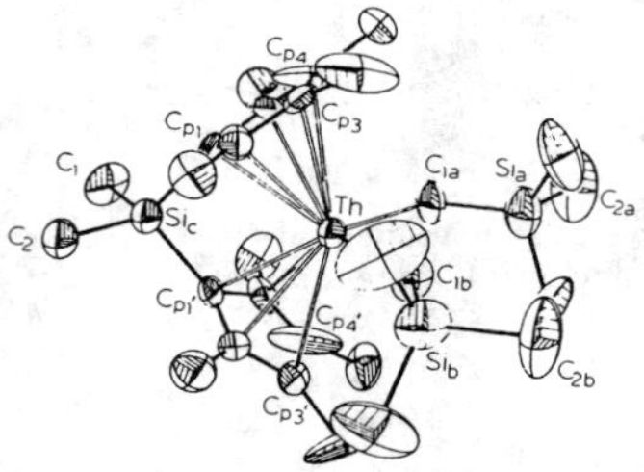

Fig. 1. Perspective drawing of the molecule $C_{28}H_{52}Si_3Th$.

(η[5]-CYCLOPENTADIENYL)(η[6]-TETRAPHENYLBORATO)IRON

$C_{29}H_{25}BFe$ $[Fe(C_5H_5)\{B(C_6H_5)_4\}]$

R.D. ROGERS AND D.C. HRNCIR, 1984. Acta Cryst., C40, 1160-1161.

Monoclinic, $P2_1/c$, a = 9.537, b = 15.432, c = 15.390 Å, β = 106.92°, Z = 4. Mo radiation, R = 0.036 for 1174 reflexions.

In the compound, which is isostructural with $(\eta^5\text{-}C_5H_5)Ru(\eta^6\text{-}C_6H_5)B(C_6H_5)_3$ (1), the average Fe-C(η^5-C_5H_5) distance is 2.05(1) and the average Fe-C[(η^6-C_6H_5)B(C_6H_5)$_3$] separation 2.09(4) Å. The Fe atom is 1.66 Å from the plane defined by the five cyclopentadienyl C atoms, and 1.537 Å from the plane defined by the C atoms of the η^6-bonded phenyl ring.

1. Structure Reports, 40B, 780.

BIS((μ_2-NEOPENTYLIDENEAMIDO)-BIS(η[5]-CYCLOPENTADIENYL)YTTRIUM)

$C_{30}H_{40}N_2Y_2$ $\{(Cp)_2Y[N\text{-}C(H)CMe_3]\}_2$

W.J. EVANS, J.H. MEADOWS, W.E. HUNTER and J.L. ATWOOD, 1984. J. Am. Chem. Soc., 106, 1291-1300.

Triclinic, $P\bar{1}$, a = 9.261, b = 10.398, c = 16.323 Å, α = 86.48, β = 88.95, γ = 70.29°, Z = 2. Mo radiation, R = 0.056 for 2532 reflexions.

The crystal lattice has two independent dimers lying on inversion centres (Fig. 1). One molecule (A) has some disorder of t-butyl groups; molecule B is well behaved. Some dimensions are; Y...Y 3.617(2), Y-N 2.314 and 2.382(9), Y-C(Cp) 2.63 to 2.67(1), N-C 1.25(2) Å.

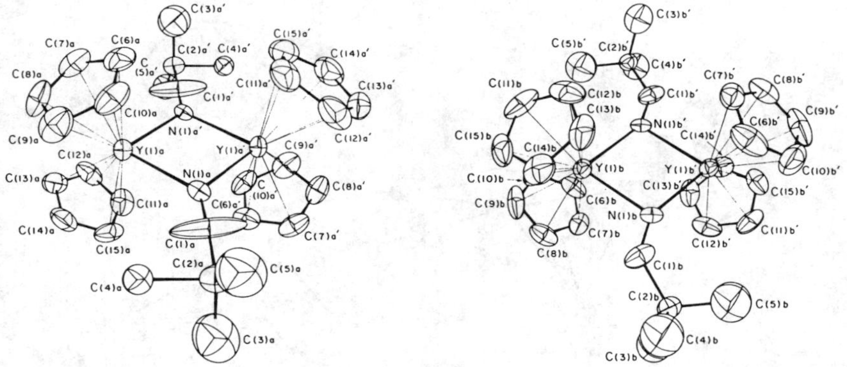

Fig. 1. Molecules A and B of $C_{30}H_{40}N_2Y_2$.

TRIS(η[5]-CYCLOPENTADIENYL)-(2-(METHYLDIPHENYLPHOSPHINO)-1-METHYLVINYLENEIMIDO-N)-URANIUM TOLUENE SOLVATE

$C_{31}H_{32}NPU,C_7H_8$

R.E. CRAMER, K. PANCHANATHESWARAN and J.W. GILJE, 1984. J. Am. Chem. Soc., 106, 1853-1854.

Triclinic, P$\bar{1}$, a = 12.789, b = 14.479, c = 10.389 Å, α = 94.39, β = 110.7, γ = 67.68°, Z = 2. R = 0.067 for 5144 reflexions.

In the molecule (Fig. 1) each of the Cp groups is disordered over two positions. Main dimensions include U-N 2.06(1), N-C 1.34(2) Å, U-N-C 163(1)°. The dimensions are consistent with the highly delocalized π system and a U-N bond order between 2 and 3.

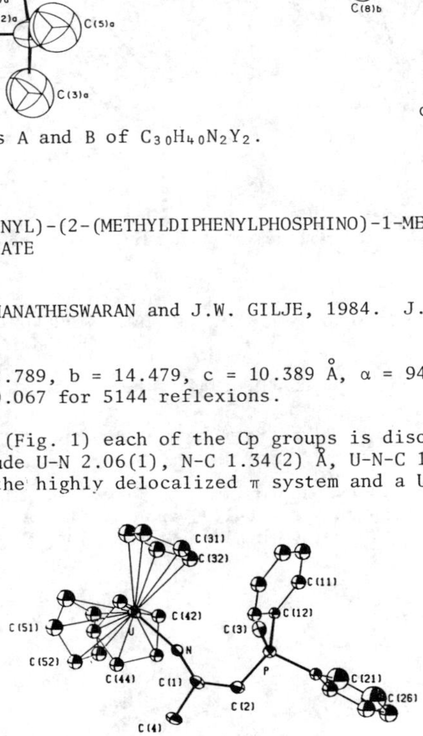

Fig. 1. A view of $C_{31}H_{32}NPU$.

1,1,1,2,2,2,3,3,3-NONACARBONYL-[μ3-(η5-CYCLOPENTADIENYL)NICKEL]-1,2,1,3-BIS(μ2-HYDRIDO)-2,3-(μ2-TRIPHENYLPHOSPHINEGOLD)TRIOSMIUM

$C_{32}H_{22}AuNiO_9Os_3P$

P. BRAUNSTEIN, J. ROSÉ, A.M. MANOTTI-LANFREDI, A. TIRIPICCHIO and E. SAPPA, 1984. J. Chem. Soc. Dalton, 1843-1848.

Triclinic, P$\bar{1}$, a = 13.923, b = 14.213, c = 9.544 Å, α = 93.81, β = 105.48, γ = 102.83°, Z = 2. Mo radiation, R = 0.055 for 2867 reflexions.

The metal cluster in the molecule (Fig. 1) consists of an $NiOs_3$ tetrahedron in which the Os(2)-Os(3) edge is bridged by the Au atom. The Os-Os distances range from 2.863(2) to 2.881(2), Ni-Os distances from 2.569(5) to 2.581(4) and Au-Os 2.747(2) to 2.775(2) Å. The Au-P distance is 2.292(8) and the Os-C (carbonyl) distances range from 1.83(3) to 1.97(4) Å.

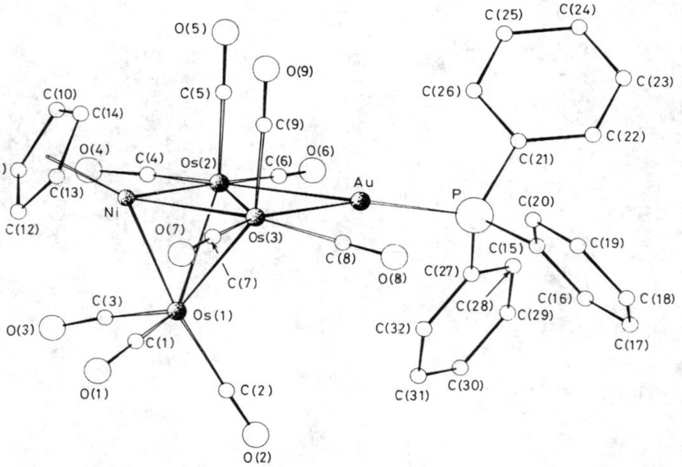

Fig. 1. $C_{32}H_{22}AuNiO_9Os_3P$: the molecular structure minus the hydride ligands.

TETRA-n-BUTYLAMMONIUM BIS[(η^5-CYCLOPENTADIENYL)TRICARBONYLCHROMIUM]GOLD(I)

$C_{32}H_{46}AuCr_2NO_6$ $[(n\text{-}C_4H_9)_4N][Au[Cr(CO)_3(\eta^5\text{-}C_5H_5)]_2]$

P. BRAUNSTEIN, U. SCHUBERT and M. BURGARD, 1984. Inorg. Chem., 23, 4057-4064.

Monoclinic, $P2_1/c$, a = 11.106, b = 16.905, c = 22.092 Å, β = 122.46°, Z = 4. Mo radiation, R = 0.103 for 1524 reflexions.

The complex anion (Fig. 1) has approximate C_2 symmetry. The Cr-Au-Cr angle is 162.2(3)°, and the Au-Cr distances are 2.641(9) and 2.635(9) Å. The coordination about the Cr atoms is of the "four-legged piano-stool" type (Cr-C(O) 1.78(5)-1.96(6), Cr-Cp(centroid) 1.88, 1.82 Å).

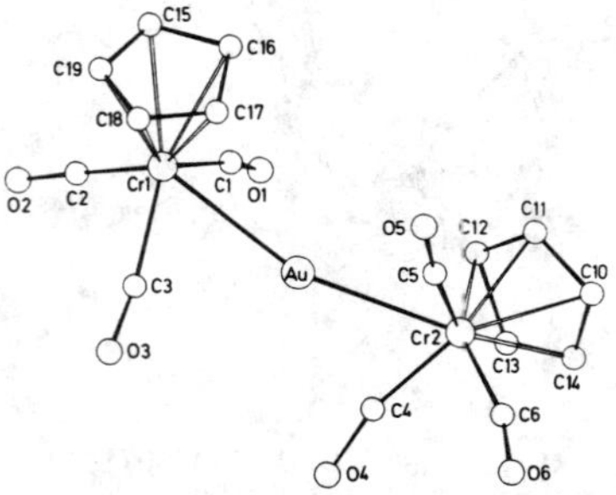

Fig. 1. The $[Au[Cr(CO)_3(\eta^5\text{-}C_5H_5)]_2]^-$ anion.

BIS((μ_2-t-BUTYLTHIOLATO)BIS(μ_3-SULPHIDO)BIS(η^5-METHYLCYCLOPENTADIENYL)DICHROMIUM)-MANGANESE

$C_{32}H_{46}Cr_4MnS_6$ (I) $[(CH_3C_5H_4)_2Cr_2(\mu\text{-}SCMe_3)(\mu_3\text{-}S)_2]_2Mn(II)$

BIS((μ_2-t-BUTYLTHIOLATO)BIS(μ_3-SULPHIDO)BIS(η^5-METHYLCYCLOPENTADIENYL)DICHROMIUM)IRON

$C_{32}H_{46}Cr_4FeS_6$ (II)

A.A. PASYNSKII, I.L. EREMENKO, G.Sh. GASANOV, Yu.T. STRUCHKOV and V.E. SHKLOVER, 1984. J. Organometal. Chem., 276, 349-362.

I. Monoclinic, Cc, a = 25.540, b = 9.208, c = 21.595 Å, β = 135.30°, Z = 4. Mo radiation, R = 0.066 for 2576 reflexions.

II. Monoclinic, Cc, a = 26.738, b = 9.195, c = 21.419 Å, β = 135.976°, Z = 4. Mo radiation, R = 0.048 for 3630 reflexions.

In I the metallospirane core Cr_4Mn is heavily distorted (Fig. 1). The two Cr_4Mn units are unsymmetrical in terms of bond lengths and the angle between the metal triangle planes is 109.60°. In II, the Cr_2Fe triangles make an angle of 113.84° with each other and in contrast to I, the Cr-Cr distances are quite similar while the Fe-Cr bond lengths are markedly different.

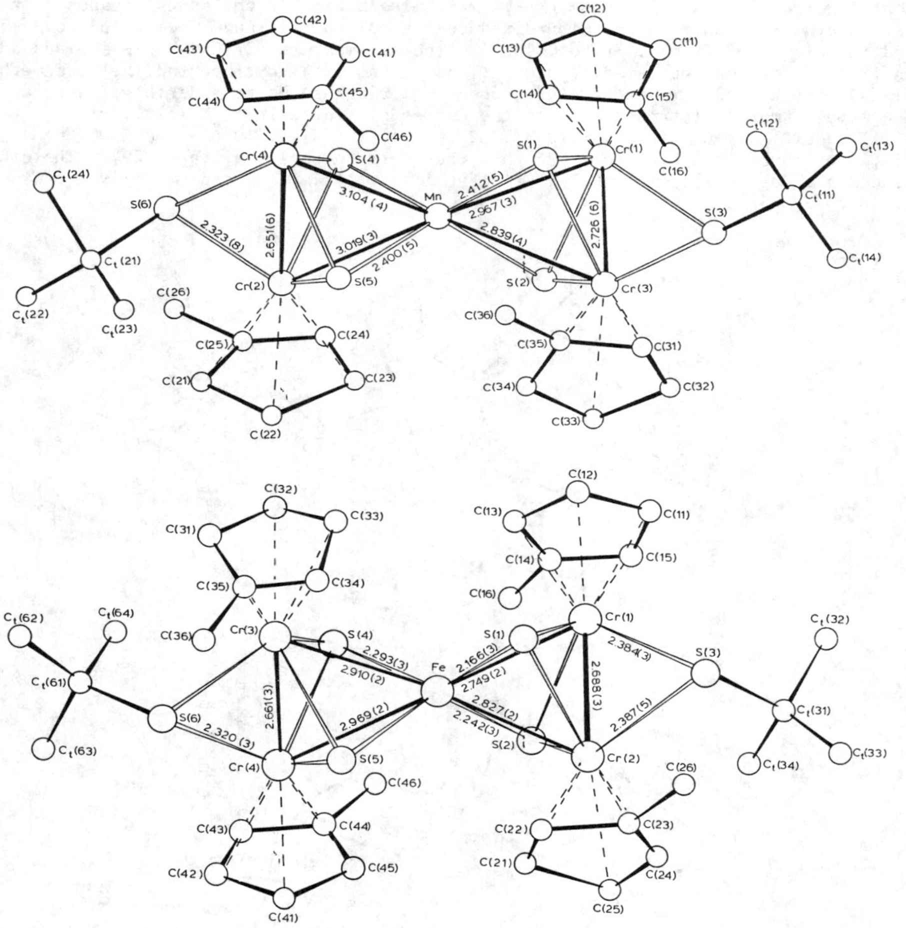

Fig. 1. Molecular structures of $C_{32}H_{46}Cr_4MnS_6$ (top), and $C_{32}H_{46}Cr_4FeS_6$ (bottom) showing selected bond lengths (Å).

TETRAKIS(μ_2-DIMETHYLPHOSPHINOMETHYL-C,P)BIS(BIS(η^5-CYCLOPENTADIENYL)ZIRCONIUM)NICKEL TETRAHYDROFURAN SOLVATE

$C_{32}H_{52}NiP_4Zr_2,C_4H_8O$

H.H. KARSCH, G. MULLER and C. KRUGER, 1984. J. Organometal. Chem., 273, 195-212.

Triclinic, $P\bar{1}$, a = 13.433, b = 18.062, c = 19.505 Å, α = 64.35, β = 76.82, γ = 71.28°, Z = 4. Mo radiation, R = 0.055 for 8826 reflexions.

The crystal structure consists of two crystallographically independent monomers and tetrahydrofuran solvent molecules per asymmetric unit (Fig. 1). Each monomer contains two bidentate $Cp_2Zr(CH_2PMe_2)_2$ molecules acting as chelating ligands to a tetrahedral Ni° centre. The Ni-P distances in molecule 1 range from 2.164 to 2.171 Å, and in molecule 2 from 2.156 to 2.186 Å. The six-membered $Zr(CH_2PMe_2)_2Ni$ rings adopt chair conformations but steric interactions of the axial cyclopentadienyl and methyl ligands, at Zr and P respectively, cause these rings to be considerably flattened. The metal...metal distances reflect the degree of flattening: Ni(1)...Zr(1) 4.367, Ni(1)...Zr(2) 4.416, Ni(2)...Zr(3) 3.962, Ni(2)...Zr(4) 4.319 Å. The Zr-C(Cp) distances range from 2.49 to 2.58 Å and the Cp-Zr-Cp angles average 129°. The other Zr-C distances (C(1), C(4), C(7), C(10), C(13), C(16), C(19) and C(22)) range from 2.270 to 2.323 Å.

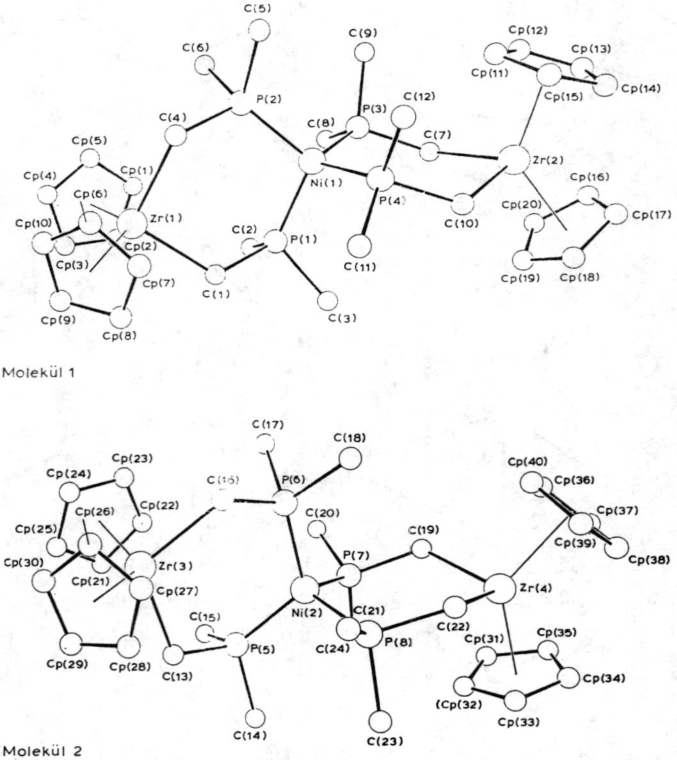

Fig. 1. Both of the independent molecules of $C_{32}H_{52}NiP_4Zr_2$ in the asymmetric unit.

BIS(BIS(η^5-CYCLOPENTADIENYL)(μ_3-HYDRIDO)(μ_2-HYDRIDO)DIHYDRIDOTRIETHYLAMINOALUMINIUM-YTTRIUM)

$C_{32}H_{58}Al_2N_2Y_2$ (I) $[(C_5H_5)_2YAlH_4N(C_2H_5)_3]_2$

TETRAKIS(η^5-CYCLOPENTADIENYL)(μ_3-HYDRIDO)TETRAKIS(μ_2-HYDRIDO)(DIETHYL ETHER)HYDRIDO-ALUMINIUMDIHYDROALUMINIUM-DI-YTTRIUM

$C_{24}H_{38}Al_2OY_2$ (II) $[(C_5H_5)_2YAlH_4]_2O(C_2H_5)_2$

V.K. BELSKY, A.B. EROFEEV, B.M. BULYCHEV and G.L. SOLOVEICHIK, 1984. J. Organometal. Chem., 265, 123-133.

I. Monoclinic, $P2_1/b$, a = 11.381, b = 13.388, c = 14.155 Å, γ = 102.01°, Z = 2. Mo radiation, R = 0.072 for 1351 reflexions.

II. Monoclinic, $P2_1/b$, a = 18.984, b = 16.981, c = 8.896 Å, γ = 81.41°, Z = 4. Mo radiation, R = 0.076 for 1479 reflexions.

Complex I consists of centrosymmetric, isolated dimer molecules (Fig. 1). The hydride atoms could not be located by X-ray structure analysis so those positions indicated are based on IR spectral data and by comparison with similar compounds (1). The dotted lines represent atomic positions of the compound in ref. (1). Relevant distances are: Y...Y' 3.700, Al...Al' 6.780, Y-Al 3.308, Y-Al' 4.108, Y-C(Cp) mean 2.64, Al-N 2.13 Å. Complex II (Fig. 1) contains two Cp_2Y wedge-shaped sandwiches and two non-equivalent Al atoms, one of which is connected to the ether molecule. Important distances are: Y(1)-Al(1) 3.546, Y(1)-Al(2) 3.207, Y(2)-Al(1) 3.583, Y(2)-Al(2) 3.196, Y-C(mean) 2.62, Al(2)-O 1.94, Y(1)...Y(2) 4.378 and Al(1)...Al(2) 4.972 Å.

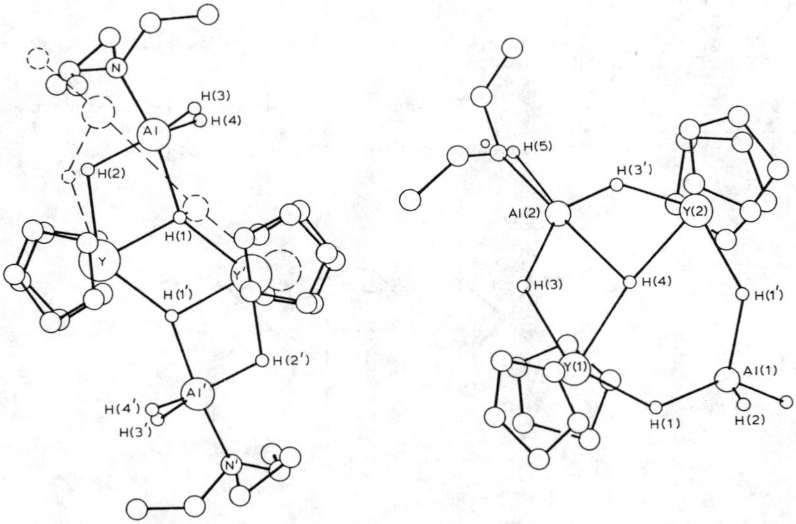

Fig. 1. Molecular structure of $C_{32}H_{58}Al_2N_2Y_2$ (left), and $C_{24}H_{38}Al_2OY_2$ (right).

1. This volume, p. 1227.

TETRAPHENYLPHOSPHONIUM TRICARBONYL(η^5-FORMYLCYCLOPENTADIENYL)CHROMATE

$C_{33}H_{25}CrO_4P$ $PPh_4^+.C_9H_5CrO_4^-$

F. EDELMANN and U. BEHRENS, 1984. Chem. Ber., 117, 3463-3472.

Monoclinic, Cc, a = 23.021, b = 7.034, c = 17.109 Å, β = 98.60°, Z = 4. Mo radiation, R = 0.043 for 3071 reflexions.

The orientation of the $Cr(CO)_3$ moiety relative to the five-membered ring is shown in Fig. 1. The Cr atom is 1.871 Å from the centroid of the ring. The C_5 ring and formyl group are 6.3° from planarity.

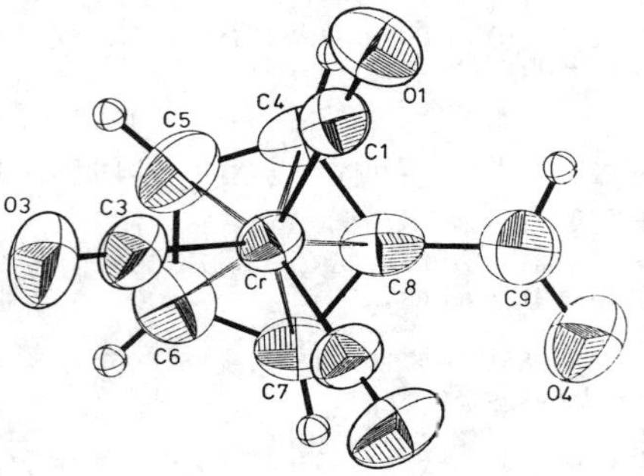

Fig. 1. A view of the $C_9H_5CrO_4^-$ anion.

DICHLORO(1,1'-BIS(DIPHENYLPHOSPHINO)FERROCENE)PALLADIUM(II) CHLOROFORM SOLVATE
$C_{34}H_{28}Cl_2FeP_2Pd,CHCl_3$ $PdCl_2[Fe(C_5H_4PPh_2)_2].CHCl_3$

T. HAYASHI, M. KONISHI, Y. KOBORI, M. KUMADA, T. HIGUCHI and K. HIROTSU, 1984. J. Am. Chem. Soc., 106, 158-163.

Monoclinic, $P2_1/n$, a = 16.872, b = 16.535, c = 12.275 Å, β = 93.85°, Z = 4. Mo radiation, R = 0.036 for 3544 reflexions.

The molecule is shown in Fig. 1. The coordination about Pd is cis-square planar with Pd-Cl 2.292(1) and 2.348(1) Å, Pd-P 2.283(1) and 2.301(1) Å, P-Pd-P 99.07(5)° and Cl-Pd-Cl 87.8(1)°. The ferrocene is in a staggered conformation.

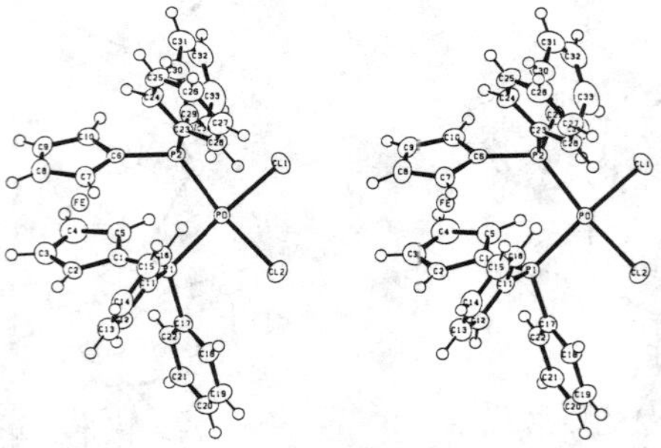

Fig. 1. Stereoview of $PdCl_2[Fe(C_5H_4PPh_2)_2]$.

(μ_2-σ^3-DIPHENYLPHOSPHINOBUTENE-C,O,P)BIS[CHLOROBIS(η^5-CYCLOPENTADIENYL)ZIRCONIUM]
$C_{34}H_{31}Cl_2OPZr_2$ $((C_5H_5)_2ZrCl)_2[O=C=CH(P(C_6H_5)_2)]$

R. CHOUKROUN, F. DAHAN and D. GERVAIS, 1984. J. Organometal. Chem., 266, C33-C36.

Monoclinic, C2/m, a = 17.002, b = 16.474, c = 11.706 Å, β = 112.72°, Z = 4. Mo radiation, R = 0.037 for 2712 reflexions.

The molecule (Fig. 1) has crystallographic m symmetry with atoms Zr(1),Zr(2),-Cl(1),Cl(2),O,C(1),C(2),P,H(C2) lying in the mirror plane. The C(1)-C(2) distance of 1.327 Å together with the coplanarity of C(1), C(2), H(C2) and P and an angle of

119.5° for C(1)-C(2)-H(C2) suggests a double-bonded C=CH(PPh_2) arrangement. The Zr(1)-O-Zr(2) angle is 162.0° with significantly different Zr-O distances (Fig. 1). Both Zr atoms are five-coordinated and the Zr-Cp(centroid) distances are 2.227 Å (Cp(1)) and 2.217 Å (Cp(2)). See also the following Report.

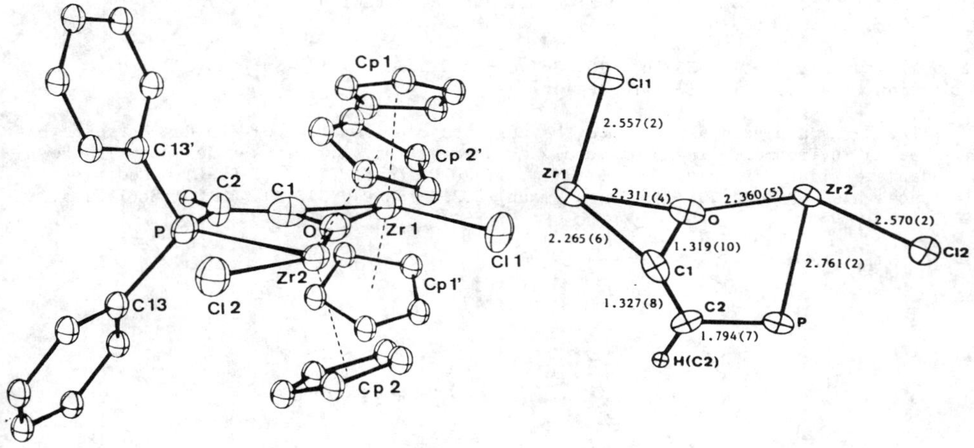

Fig. 1. Molecular structure of $C_{34}H_{31}Cl_2OPZr_2$ (left), and bond lengths (Å) within the coplanar moiety (right).

(μ_2-σ^3-DIPHENYLPHOSPHINOBUTENE-C,O,P)BIS[CHLOROBIS(η^5-CYCLOPENTADIENYL)ZIRCONIUM]
$C_{34}H_{31}Cl_2OPZr_2$

S.J. YOUNG, H. HOPE and N.E. SCHORE, 1984. Organometallics, 3, 1585-1587.

Monoclinic, C2/m, a = 16.762, b = 16.366, c = 11.488 Å, β = 112.65°, Z = 4. Mo radiation, R = 0.039 for 1347 reflexions (at 140K).

The compound consists of an oxygen-bridged dinuclear species with two distinct Cp_2ZrCl fragments linked by a P-C-C bridge (Fig. 1). The eight non-H atoms of the central framework and the H atom on C(7) lie on a crystallographic mirror plane. In the bridge, the P-C-C bond angle is 116°, the O-C-C angle 125° and the C(7)-C(8) bond distance 1.31 Å, implying a C=C double bond. Other bond distances of note are: Zr(1)-Cl(1) 2.564, Zr(1)-P 2.752, Zr(1)-O 2.303, Zr(2)-Cl(2) 2.563, Zr(2)-O 2.278, Zr(2)-C(8) 2.231, P-C(7) 1.78 and C(8)-O 1.33 Å. See also the preceding Report.

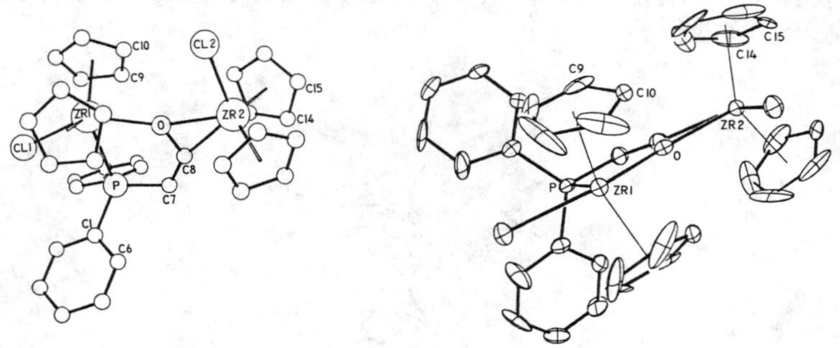

Fig. 1. Two views of the molecular structure of $C_{34}H_{31}Cl_2OPZr_2$.

(R)-CARBONYLIODO[η^5-(2-ISOPROPYL-5-METHYLCYCLOHEXAN-1-YL)CYCLOPENTADIENYL]TRIPHENYL-PHOSPHINERUTHENIUM
$C_{34}H_{38}IOPRu$

E. CESAROTTI, A. CHIESA, G.F. CIANI, A. SIRONI, R. VEFGHI and C. WHITE, 1984. J. Chem. Soc. Dalton, 653-661.

Monoclinic, $P2_1$, a = 12.833, b = 15.186, c = 7.972 Å, β = 101.50°, Z = 2. Mo radiation, R = 0.046 for 1773 reflexions.

The X-ray analysis established the absolute configuration of the Ru centre to be R. The Ru environment is approximately octahedral (Fig. 1) with angles between the monodentate ligands ranging from 92.3(1)° (I-Ru-P) to 86.2(4)° (I-Ru-C). Coordination distances are: Ru-I 2.708(1), Ru-Cp (mean) 2.234, Ru-P 2.313(4), Ru-C 1.866(13) Å.

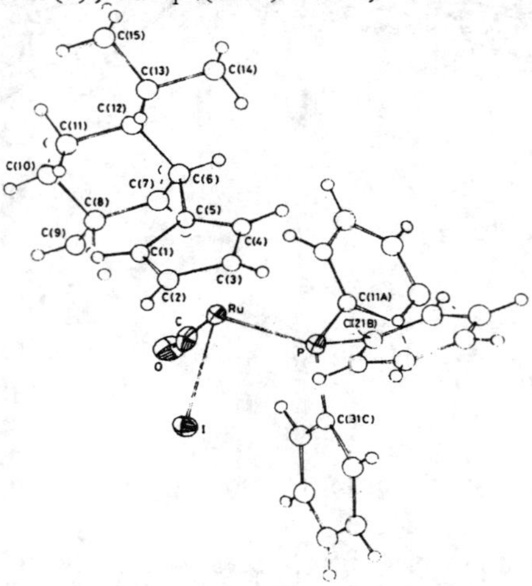

Fig. 1. $C_{34}H_{38}IOPRu$: the molecular structure.

TRIS(μ_2-HYDRIDO)[CHLOROBIS(η^5-CYCLOPENTADIENYL)ZIRCONIUM]TRIS(DIMETHYLPHENYLPHOSPHINE)-OSMIUM TOLUENE n-HEXANE SOLVATE
$C_{34}H_{46}ClOsP_3Zr,C_7H_8,C_6H_{14}$

J.W. BRUNO, J.C. HUFFMAN, M.A. GREEN and K.G. CAULTON, 1984. J. Am. Chem. Soc., 106, 8310-8312.

Triclinic, $P\bar{1}$, a = 15.111, b = 26.541, c = 10.931 Å, [angles not quoted], Z = 4. R = 0.052 for three-dimensional data (at -162°C).

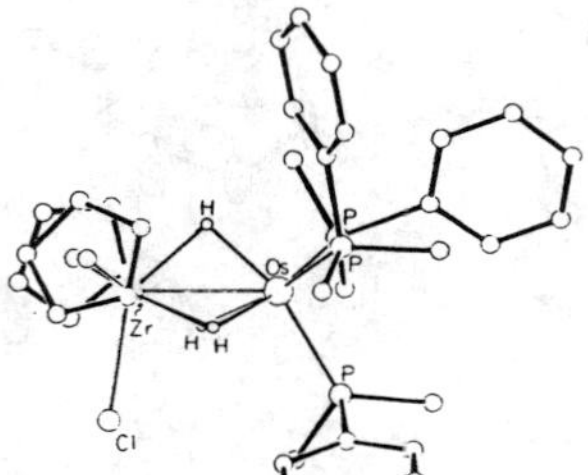

Fig. 1. A view of a $C_{34}H_{46}ClOsP_3Zr$ molecule; presumed hydride bridge locations are shown trans to Os-P bonds.

There are two independent molecules in the asymmetric unit; one is shown in Fig. 1. Main bond lengths and angles are Os-Zr 2.805(1), Os-P 2.300-2.330(3), Zr-Cl 2.521(3) Å, Zr-Os-P 117.1-123.4(1), P-Os-P 94.1-100.8(1), Cl-Zr-Os 98.6-98.9(1), Cp-Zr-Cp 121.4°.

BIS(FERROCENE ALDEHYDE BENZOYLHYDRAZONATO-N[1],O)DIOXOMOLYBDENUM(VI) DICHLOROMETHANE HEXANE SOLVATES

$C_{36}H_{30}Fe_2MoN_4O_4,0.18(C_6H_{14}),0.2(CH_2Cl_2)$

P. GOUZERH, Y. JEANNIN and V. MARECHAL, 1984. Transition Met. Chem., 9, 482-486.

Monoclinic, $P2_1/n$, a = 11.149, b = 20.463, c = 15.541 Å, β = 90.40°, D_m = 1.55, Z = 4. Mo radiation, R = 0.049 for 2886 reflexions.

The MoO_2^{2+} unit has the expected cis geometry (Fig. 1) (Mo-O 1.687 and 1.695(4) Å) and the Mo-N bonds are trans to the oxo groups (Mo-N 2.317 and 2.328(7), Mo-O(1) 1.987, Mo-O(2) 1.990(6) Å). The ferrocene moieties appear to have normal geometry.

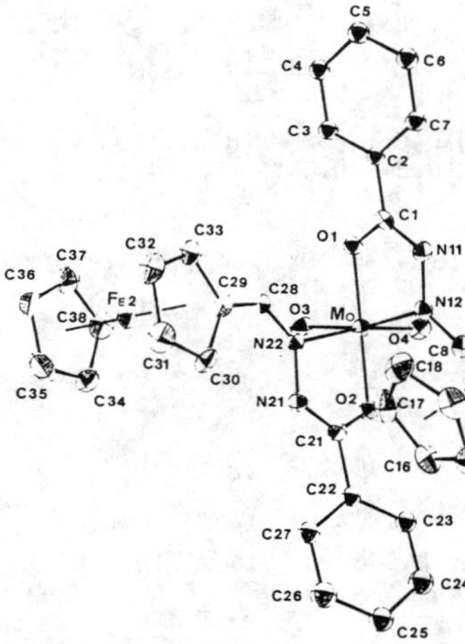

Fig. 1. A view of $C_{36}H_{30}Fe_2MoN_4O_4$.

(μ_2-BIS(DIPHENYLPHOSPHINO)METHANE) (μ_2-CARBONYL) BIS(η^5-CYCLOPENTADIENYL)DIRHODIUM(I)

$C_{36}H_{32}OP_2Rh_2$

F. FARAONE, G. BRUNO, S.L. SCHIAVO and G. BOMBIERI, 1984. J. Chem. Soc. Dalton, 533-538.

Monoclinic, C2/c, a = 23.738, b = 11.731, c = 26.883 Å, β = 123.2°, Z = 8. Mo radiation, R = 0.050 for 2809 reflexions.

The molecule (Fig. 1) consists of two Rh atoms bridged by a carbonyl group and a diphenylphosphinomethane ligand; a terminal η-C_5H_5 ligand completes the coordination sphere of each Rh atom. The Rh(1)-CO-Rh(2) and P(1)-CH_2-P(2) fragments are folded on the same side. The cyclopentadienyl ligands are in cis positions with an interplanar angle of 82.1°. Some bond distances are: Rh(1)-Rh(2) 2.683(1), Rh(1)-P(1) 2.205(2), Rh(1)-C(2) 1.978(9), Rh(2)-P(2) 2.218(2), Rh(2)-C(2) 1.971(9), Rh(1)-C (average) 2.285(10) and Rh(2)-C (average) 2.281(10) Å.

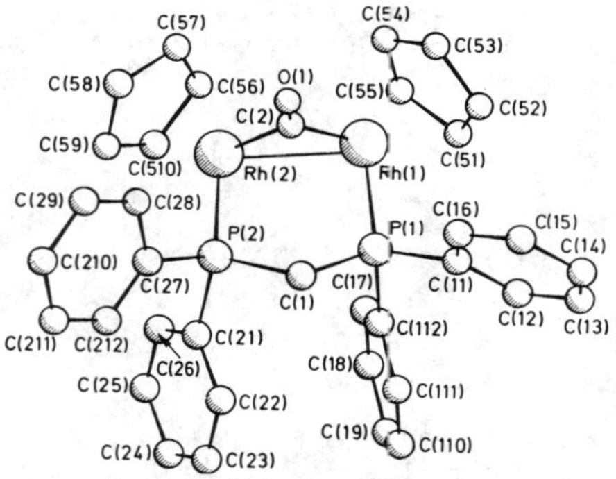

Fig. 1. $C_{36}H_{32}OP_2Rh_2$: a view of the molecule.

μ_2-CARBONYL-1,1-μ-2,4-μ'-(4-OXO-1,2-DIPHENYL-3-OXA-1,1,2,4-BUTANTETRAYL)BIS[(η^5-PENTAMETHYLCYCLOPENTADIENYL)RHODIUM]
$C_{36}H_{40}O_3Rh_2$ (I)

μ_2-CARBONYL-1,1-μ^2-2,4-μ'-(1-METHOXYOXOACETYL-2-METHOXYCARBONYL-4-OXO-3-OXA-1,1,2,4-BUTANETETRAYL)BIS[(η^5-PENTAMETHYLCYCLOPENTADIENYL)RHODIUM]
$C_{30}H_{40}O_7Rh_2$ (II)

(μ_2-CARBONYL)(μ_2-2-CARBONYL-5,5-DIMETHYL-3-OXOCYCLOHEXANE-1,2-DIYL-1-OLATO-C,C',C'',-O)BIS((η^5-PENTAMETHYLCYCLOPENTADIENYL)RHODIUM)
$C_{30}H_{40}O_4Rh_2$ (III)

W.A. HERRMANN, G.W. KRIECHBAUM, C. BAUER, B. KOUMBOURIS, H. PFISTERER, E. GUGGOLZ and M.L. ZIEGLER, 1984. J. Organometal. Chem., 262, 89-122.

I. Monoclinic, C2/c, a = 17.169, b = 16.575, c = 22.257 Å, β = 99.69°, Z = 8. Mo radiation, R = 0.043 for 3628 reflexions.

II. Monoclinic, $P2_1/n$, a = 8.622, b = 21.06, c = 16.757 Å, β = 96.47°, Z = 4. Mo radiation, R = 0.058 for 2270 reflexions.

III. Monoclinic, $P2_1/n$, a = 9.493, b = 20.495, c = 14.563 Å, β = 93.79°, Z = 4. Mo radiation, R = 0.044 for 3702 reflexions.

The Rh-Rh distances in I, II and III are respectively 2.697, 2.701 and 2.708 Å. Compounds I and II have one bridging CO group each (Rh(1)-C(5) 2.063, Rh(2)-C(5) 1.939 Å, Rh(1)-C(5)-Rh(2) 84.7° in I; Rh(1)-C(10) 1.96, Rh(2)-C(10) 2.07 Å, Rh(1)-C(10)-Rh(2) 84.0° in II), whilst III has two such groups (Fig. 1) (Rh(1)-C(1) 2.135, Rh(2)-C(1) 2.097 Å, Rh(1)-C(1)-Rh(2) 79.6°; Rh(1)-C(2) 1.948, Rh(2)-C(2) 2.060 Å, Rh(1)-C(2)-Rh(2) 85.0°). The metal-C(4) distances in I are 2.028 and 2.056 Å and the Rh(2)-C(3) distance is 2.198 Å. In II, the Rh-C(1) distances are 2.04 and 2.03 Å and the Rh(1)-C(5) and Rh(2)-C(9) bond lengths are 2.18 and 2.01 Å respectively. In III, the Rh(1)-C(3) and Rh(2)-O(3) distances are 2.245 and 2.107 Å respectively.

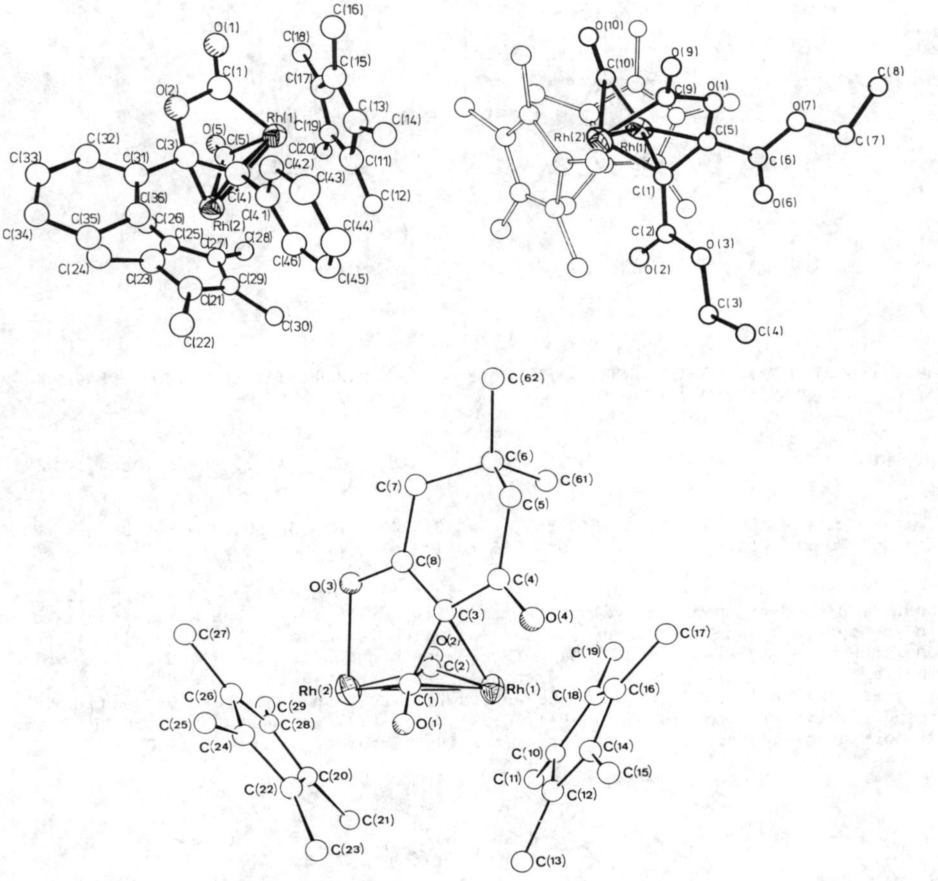

Fig. 1. Molecular structures of $C_{36}H_{40}O_3Rh_2$ (top left), $C_{30}H_{40}O_7Rh_2$ (top right), and $C_{30}H_{40}O_4Rh_2$ (bottom).

TRIS(μ_2-CARBONYL)TRICARBONYLTRIS(η^5-PENTAMETHYLCYCLOPENTADIENYL)TRIIRIDIUMTRIPLATINUM
$C_{36}H_{45}Ir_3O_6Pt_3$

M.J. FREEMAN, A.D. MILES, M. MURRAY, A.G. ORPEN and F.G.A. STONE, 1984. Polyhedron, 3, 1093-1097.

Monoclinic, P2/n, a = 41.811, b = 9.183, c = 21.892 Å, β = 90.99°, Z = 8. Mo radiation, R = 0.0903 for 6549 reflexions.

The two independent molecules in the asymmetric unit are shown in Fig. 1. The main feature is a near-planar array of metal atoms with a central triangle of platinum atoms (Pt-Pt(mean) 2.703(3) Å) each edge-bridged by an iridium atom (Ir-Pt(mean) 2.667(3) Å). The iridium atoms are each ligated by an η^5-C_5Me_5 group and by two Co ligands. Three of the latter are essentially terminally bound to Ir and lie approximately orthogonal to the Ir_3Pt_3 plane, and three bridge between Pt and Ir and lie close to the hexametal plane (mean Ir-CO(bridge) 2.06(4), mean Ir-CO(terminal) 1.80(4) Å).

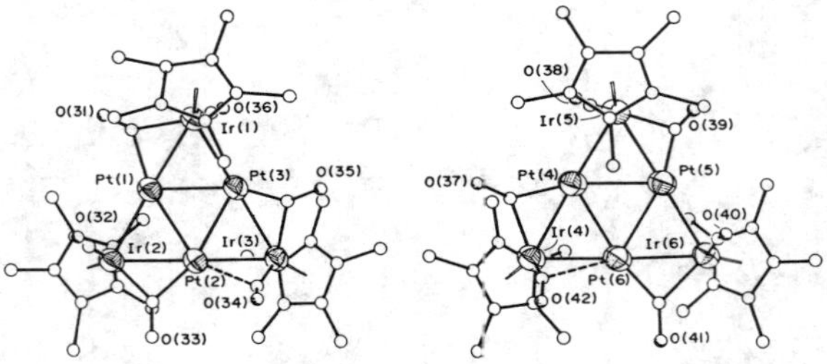

Fig. 1. The two independent $C_{36}H_{45}Ir_3O_6Pt_3$ molecules.

(1,2-BIS(DIPHENYLPHOSPHINO)NORBORNENE)CARBONYL(η^5-CYCLOPENTADIENYL)IRON HEXAFLUOROPHOSPHATE BENZENE SOLVATE

$C_{37}H_{33}F_6FeOP_3,C_6H_6$

H. BRUNNER, A.F.M. MOKHLESUR RAHMAN and I. BERNAL, 1984. Inorg. Chim. Acta, 83, L93-L96.

Triclinic, P1, a = 10.718, b = 10.859, c = 10.924 Å, α = 58.80, β = 68.19, γ = 68.90°, Z = 1. Mo radiation, R = 0.065 for 3285 reflexions.

The structure consists of an Fe atom, in a five-membered chelate ring, also bonded to a CO group and a cyclopentadienyl ligand (Fig. 1). The metal atom is almost in the centre of an octahedron, one face of which is formed by the C_5H_5 ligand and one face by P_{exo}, P_{endo}, and C(CO). The Fe-P distances are 2.231 and 2.237 Å while the Fe-C and Fe-Cp(ring centroid) distances are respectively 1.86 and 1.794 Å. The P(1)-Fe-P(2) angle is 90.6°, and the angles P(1)-Fe-C and P(2)-Fe-C 95.4 and 86.5° respectively. The torsion angle P(1)-C(30)-C(31)-P(2) is -66.94° indicating severe distortion of the norbornene skeleton (the ideal geometry requires ±120°).

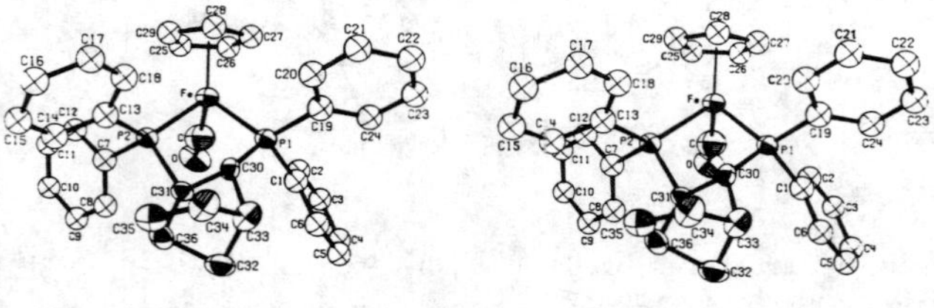

Fig. 1. Stereoscopic view of the cation in $C_{37}H_{33}F_6FeOP_3,C_6H_6$.

TETRAKIS(μ_2-CARBONYL)BIS(η^5-CYCLOPENTADIENYL)TITANIUMBIS(CARBONYL(η^5-METHYLCYCLOPENTADIENYL)MOLYBDENUM)

$C_{38}H_{34}Mo_2O_6Ti_2$ $[(C_5H_5)_2TiOCMo(CO)_2(C_5H_4CH_3)]_2$

J.S. MEROLA, K.S. CAMPO, R.A. GENTILE, M.A. MODRICK and S. ZENTZ. 1984. Organometallics, 3, 334-337.

Monoclinic, $P2_1/n$, a = 13.510, b = 10.951, c = 13.361 Å, β = 114.66°, Z = 2. Mo radiation, R = 0.068 for 1546 reflexions.

The molecule, which has crystallographic inversion symmetry, has a most unusual 12-membered ring in its core: Ti-O-C-Mo-C-O-Ti-O-C-Mo-C-O (Fig. 1). Molecular geometry information in addition to that shown is: Mo-C(1) 1.915, C(1)-O(1) 1.180 Å; O(2')-Ti-O(3) 78.72, Ti-O(2')-C(2') 141.18, Ti-O(3)-C(3) 143.38, Mo-C(2)-O(2) 177.08, Mo-C(3)-O(3) 178.48 and Mo-C(1)-O(1) 178.95°.

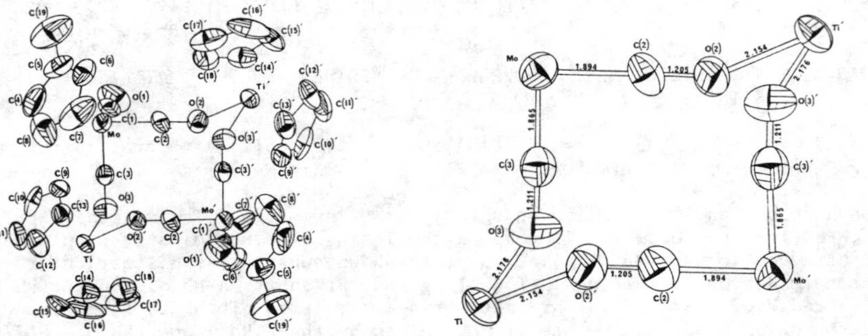

Fig. 1. Molecular structure of $C_{38}H_{34}Mo_2O_6Ti_2$ (left), and core atom portion showing bond lengths (Å) (right).

BIS(μ_2-t-BUTYLTHIOLATO)(μ_3-SULPHIDO)BIS[CHLORO(η^5-METHYLCYCLOPENTADIENYL)CHROMIUM]-TRIPHENYLPHOSPHINEPALLADIUM BENZENE SOLVATE

$C_{38}H_{47}Cl_2Cr_2PPdS_3,C_6H_6$

A.A. PASYNSKII, I.L. EREMENKO, G.Sh. GASANOV, O.G. ELLERT, V.M. NOVOTORTSEV, Yu.V. RAKITIN, T.Kh. KURBANOV, V.T. KALINNIKOV, Yu.T. STRUCHKOV and V.E. SHKLOVER, 1984. Polyhedron, 3, 775-780.

Triclinic, $P\bar{1}$, a = 11.403, b = 14.933, c = 14.131 Å, α = 99.13, β = 112.72, γ = 95.65°, Z = 2. Mo radiation, R = 0.046 for 7643 reflexions (at -120°C).

In the molecule (Fig. 1) metal atoms are separated by non-bonding distances (Cr...Cr 4.079(1), Cr...Pd 3.230(1) and 3.380(1) Å) but linked by the bridging tridentate sulphur atom (Cr-S 2.339(2) and 2.329(2), Pd-S 2.327(2) Å), and two $SCMe_3$ groups between Pd and Cr (Cr-S 2.396(2) and 2.403(2), Pd-S 2.350(2) and 2.381(2) Å, Cr-S-Pd 85.14(6) and 89.92(6)°). The Cr-Cl distance is 2.308(2) Å.

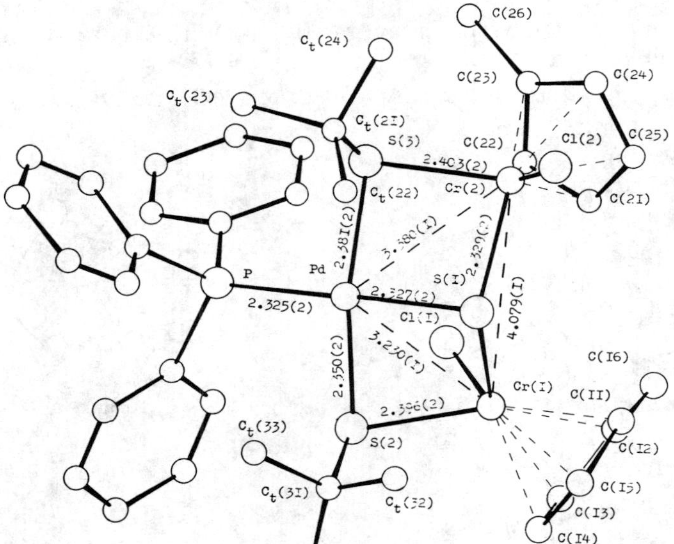

Fig. 1. A view of $C_{38}H_{47}Cl_2Cr_2PPdS_3$ with some dimensions.

TETRAKIS[(μ_3-HYDRIDO)(η^5-PENTAMETHYLCYCLOPENTADIENYL)RHODIUM] BIS(TETRAFLUOROBORATE)
$C_{40}H_{64}B_2F_8Rh_4$ $[Rh_4(\eta^5\text{-}C_5Me_5)_4H_4][BF_4]_2$

J.S. RICCI, T.F. KOETZLE, R.J. GOODFELLOW, P. ESPINET and P.M. MAITLIS, 1984. Inorg. Chem., 23, 1828-1831.

Tetragonal, $I\bar{4}$, a = 11.459, c = 15.481 Å, Z = 2. 1.0468(2) Å neutron radiation, $R(F^2)$ = 0.069 for 1371 reflexions (at 12 K).

Both anions and cations lie on sites of $\bar{4}$-S_4 symmetry. The cation (Fig. 1) is a stable, 58-electron, diamagnetic complex consisting of a Rh_4 cluster with a C_5Me_5 ligand attached to each metal. The Rh_4 core shows substantial distortion from a regular tetrahedron, resulting in four long and two short Rh-Rh distances (Rh-Rh 2.610(5) and 2.829(5) Å). The hydrides bridge the faces of the tetrarhodium cluster at a distance of 0.96 Å above the planes formed by three Rh atoms (Rh-H 1.847(6)-1.874(6) Å). The Rh atoms are 1.838 Å from the C_5 planes.

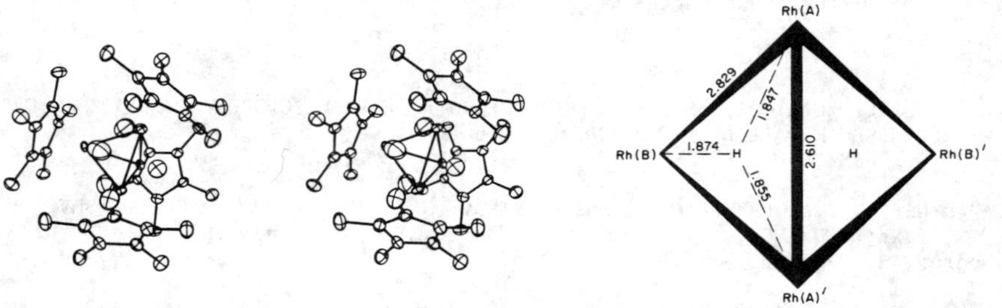

Fig. 1. Stereoview of $[Rh_4(\eta^5\text{-}C_5\text{-}Me_5)_4H_4]^{2+}$, core geometry at right.

TRICHLORO(η^5-CYCLOPENTADIENYL)BIS(TRIPHENYLPHOSPHINE OXIDE)URANIUM(IV) TETRAHYDRO-FURAN SOLVATE
$C_{41}H_{35}Cl_3O_2P_2U,C_4H_8O$ (I)

TRICHLORO(η^5-CYCLOPENTADIENYL)BIS(HEXAMETHYLPHOSPHORAMIDE)URANIUM(IV)
$C_{17}H_{41}Cl_3N_6O_2P_2U$ (II)

K.W. BAGNALL, F. BENETOLLO, G. BOMBIERI and G. DE PAOLI, 1984. J. Chem. Soc. Dalton, 67-73.

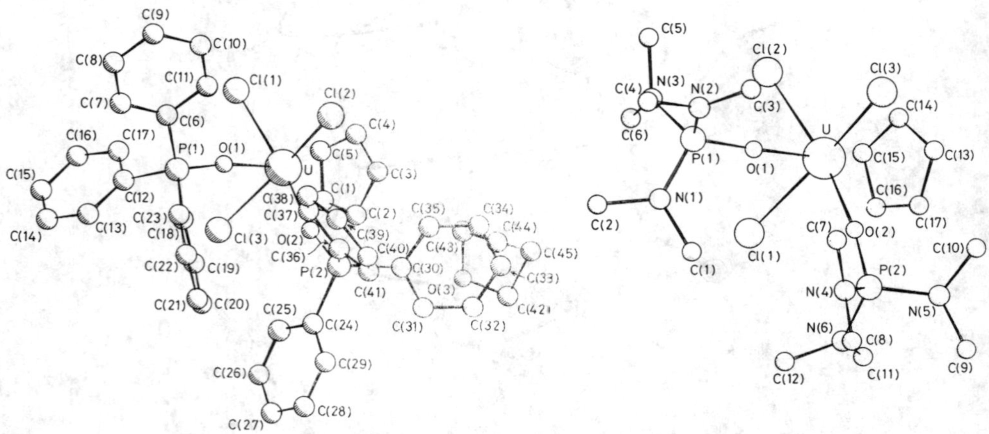

Fig. 1. The molecular structures: $C_{41}H_{35}Cl_3O_2P_2U,C_4H_8O$ (left) and $C_{17}H_{41}Cl_3N_6O_2P_2U$ (right).

I. Monoclinic, $P2_1/c$, a = 21.725, b = 11.699, c = 17.269 Å, β = 97.9°, Z = 4. Mo radiation, R = 0.054 for 3888 reflexions.

II. Monoclinic, $P2_1/n$, a = 9.942, b = 32.005, c = 9.576 Å, β = 106.3°, Z = 4. Mo radiation, R = 0.032 for 3777 reflexions.

In both I and II coordination about the U atom is distorted octahedral (Fig. 1) with the two O atoms of the PPh_3O and $P(NMe_2)_3O$ ligands respectively in cis positions. The cyclopentadienyl ligand occupies a single coordination position and is trans with respect to one oxygen ligand molecule. Three Cl atoms complete the octahedron in mer positions. Bond lengths in I include: U-Cl 2.651-2.657(4); U-O(1) 2.258(9); U-O(2) 2.312(8); U-cp centroid 2.483; P(1)-O(1) 1.53(1); P(2)-O(2) 1.51(1) Å. Corresponding distances in II are: U-Cl 2.653-2.677(3); U-O(1) 2.284(6); U-O(2) 2.301(7); U-cp centroid 2.491; P(1)-O(1) 1.504(7); P(2)-O(2) 1.497(6) Å.

BIS(TRIPHENYLPHOSPHINE)IMINIUM (η^5-METHYLCYCLOPENTADIENYL)NITROSYLCOBALT ACETONITRILE SOLVATE

$C_{42}H_{37}CoN_2OP_2,C_2H_3N$ (I)

BIS(μ_2-NITROSYL)[(η^5-CYCLOPENTADIENYL)COBALT](η^5-CYCLOPENTADIENYL)TRIMETHYLPHOSPHINE-MANGANESE

$C_{13}H_{19}CoMnN_2O_2P$ (II)

W.P. WEINER, F.J. HOLLANDER and R.G. BERGMAN, 1984. J. Am. Chem. Soc., 106, 7462-7468.

I. Triclinic, $P\bar{1}$, a = 8.795, b = 15.446, c = 15.761 Å, α = 77.73, β = 75.11, γ = 72.69°, Z = 2. Mo radiation, R = 0.031 for 3998 reflexions.

II. Triclinic, $P\bar{1}$, a = 9.741, b = 13.000, c = 13.677 Å, α = 77.55, β = 77.96, γ = 68.94°, Z = 4. Mo radiation, R = 0.026 for 3537 reflexions.

The structure of (I) consists of well separated cations, anions and acetonitrile molecules. A view of the anion is in Fig. 1. Main bond lengths are Co-N(1) 1.590(2), Co-Cp 1.710, N(1)-O 1.229(2) Å, N(1)-Co-Cp 176.4, Co-N-O 176.2°. A feature of (II) (Fig. 1) is the symmetrical placement of the bridging nitrosyl ligands with respect to the metal centres in the two independent molecules. Some dimensions are: Mn-Co 2.447 and 2.444(1), Mn-N 1.797-1.812(2), Co-N 1.796-1.804(2), Co-Cp 1.759 and 1.744(1), Mn-Cp 1.798 and 1.805(1), Mn-P 2.267 and 2.270(1), N-O 1.223-1.229(2) Å, Cp-Co-Mn 165.42 and 164.74(2)°.

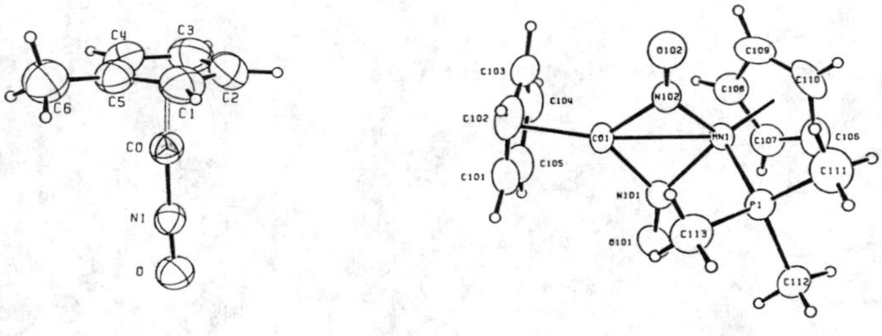

Fig. 1. A view of the anion from (I) and the structure of $C_{13}H_{19}CoMnN_2O_2P$ (II).

(μ_2-PHOSPHINO)TETRAKIS(η^5-PENTAMETHYLCYCLOPENTADIENYL)BIS(METHOXY)DIURANIUM

$C_{42}H_{67}O_2PU_2$ $[(Cp')_2UOMe]_2PH$

M.R. DUTTERA, V.W. DAY and T.J. MARKS, 1984. J. Am. Chem. Soc., 106, 2907-2912.

Monoclinic, $P2/n$, a = 13.926, b = 10.786, c = 15.282 Å, β = 107.63°, Z = 2. Mo radiation, R = 0.041 for 1677 reflexions.

The molecule has twofold crystallographic symmetry (Fig. 1) with the P atom on the twofold axis. The U-coordination is typical "pseudotetrahedral" $Cp'_2M(X)y$ type, with U-P 2.743(1), U-O 2.046(14), U-C 2.70-2.80(2) Å, U-P-U 157.7(2), U-O-C 178(1)°.

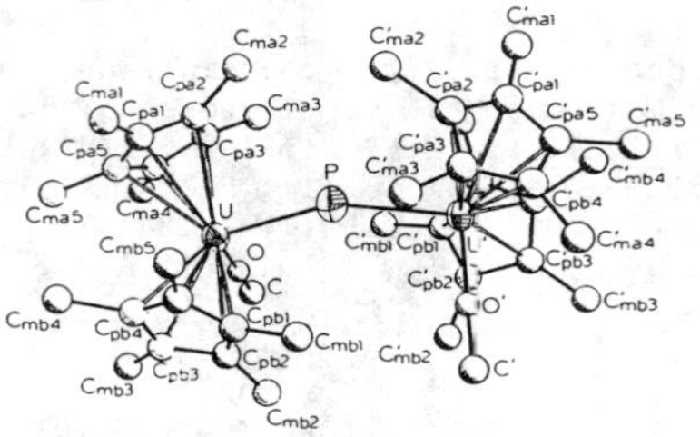

Fig. 1. $C_{42}H_{67}O_2PU_2$: a perspective view.

BIS(μ_2-CARBONYL)-2-CARBONYL-2-(η^5-CYCLOPENTADIENYL)-1,1-BIS(TRIPHENYLPHOSPHINE)-RHODIUMMOLYBDENUM
$C_{44}H_{35}MoO_3P_2Rh$ (I)

2,2,2-TRICARBONYL-2-(η^5-CYCLOPENTADIENYL)-1,1-BIS(TRIPHENYLPHOSPHINE)COPPERTUNGSTEN (isomer A)
$C_{44}H_{35}CuO_3P_2W$ (II)

2,2,2-TRICARBONYL-2-(η^5-CYCLOPENTADIENYL)-1,1-BIS(TRIPHENYLPHOSPHINE)COPPERTUNGSTEN (isomer B)
$C_{44}H_{35}CuO_3P_2W$ (III)

L. CARLTON, W.E. LINDSELL, K.J. McCULLOUGH and P.N. PRESTON, 1984. J. Chem. Soc. Dalton, 1693-1701.

I. Monoclinic, $P2_1/c$, a = 18.102, b = 10.407, c = 20.736 Å, β = 104.99°, D_m = 1.55, Z = 4. Mo radiation, R = 0.036 for 5233 reflexions.

II. Triclinic, $P\bar{1}$, a = 12.404, b = 11.756, c = 14.208 Å, α = 101.4, β = 79.5, γ = 110.3°, D_m = 1.63, Z = 2. Mo radiation, R = 0.058 for 3997 reflexions.

III. Monoclinic, $P2_1/c$, a = 17.991, b = 10.093, c = 21.608 Å, β = 105.66°, D_m = 1.63, Z = 4. Mo radiation, R = 0.034 for 4009 reflexions.

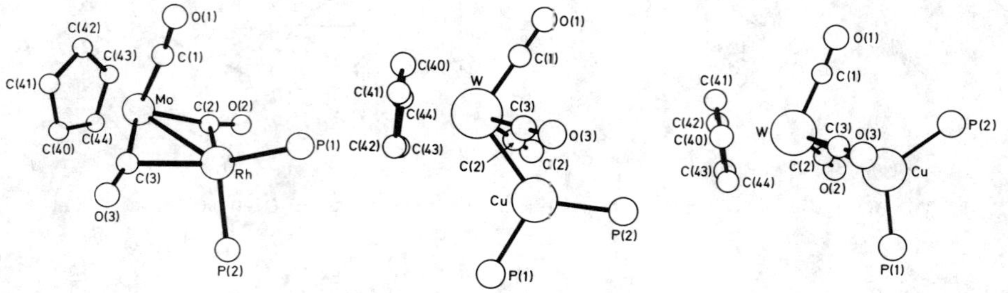

Fig. 1. The complex structures: $C_{44}H_{35}MoO_3P_2Rh$ (left), $C_{44}H_{35}CuO_3P_2W$ (isomer A) (centre) and $C_{44}H_{35}CuO_3P_2W$ (isomer B) (right).

In compound I, Fig. 1, the $Mo(CO)(\eta-C_5H_5)$ group is bonded to $Rh(PPh_3)_2$ with a short Mo-Rh separation (2.588(1)) and two asymmetric (μ-CO) ligands, with Mo-C 1.980(6) and 2.026(5) and Rh-C 2.182(6) and 2.092(5) Å. The $Mo(\mu-CO)_2Rh$ has a dihedral angle, Mo-Rh-C/Mo-Rh-C, of 161°. II and III both contain $W(CO)_3(\eta\text{-}C_5H_5)$ groups bonded to $Cu(PPh_3)_2$. Two of the CO groups bonded to the W atom show some semibridging interactions with the Cu atom especially in III. The Cu-C distances are 2.309(13) and 2.405(13) in II and 2.260(8) and 2.197(9) Å in III. In II the $Cu(PPh_3)_2$ unit is coordinated to W such that W has a 'four-legged' piano stool geometry (W-Cu 2.721(1) Å). In III the $W(\mu\text{-}CO)_2Cu$ fragment is almost planar with W-Cu 2.771(1) Å.

BIS(TETRA-n-BUTYLAMMONIUM) (η^5-PENTAMETHYLCYCLOPENTADIENYL)(μ_6-OXO)TRIS(μ_3-OXO)-NONAKIS(μ_2-OXO)HEXAOXODINIOBIUMTETRATUNGSTENRHODIUM
$C_{46}H_{87}NNbO_{19}RhW$

C.J. BESECKER, V.W. DAY, W.G. KLEMPERER and M.R. THOMPSON, 1984. J. Am. Chem. Soc., 106, 4125-4136.

Triclinic, P1, a = 11.081, b = 11.773, c = 12.444 Å, α = 115.85, β = 99.30, γ = 80.57°, Z = 1. Mo radiation, R = 0.049 for 4695 reflexions.

The crystals contain discrete disordered $[(C_5Me_5)Rh(Nb_2W_4O_{19})]^{2-}$ ions and all three diastereoisomers shown in Fig. 1 are present at equilibrium. There are six disordered niobium/tungsten sites.

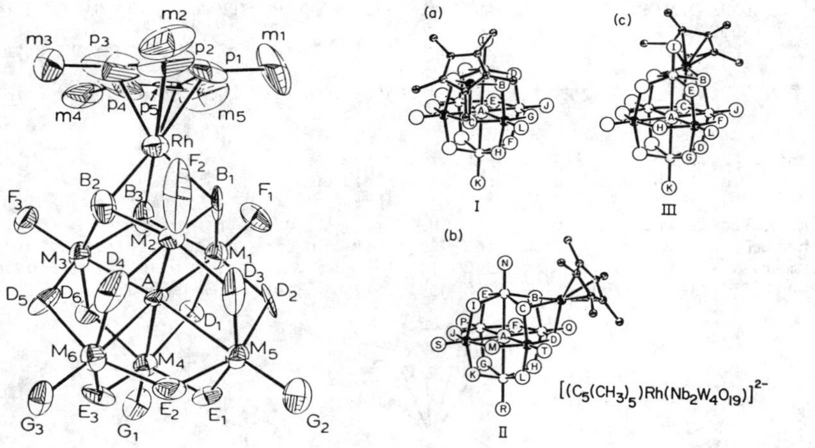

Fig. 1. A perspective drawing (left) of the non-H atoms in $[(C_5Me_5)Rh(Nb_2W_4O_{19})]^{2-}$ and the three possible diastereoisomers present in the lattice (right).

(4-PHENYL-1,2,4-TRIAZOLINE-3,5-DIONE)[BIS(TRIPHENYLPHOSPHINE)PLATINUMDISULPHIDO]-(SULPHIDO)BIS(η^5-METHYLCYCLOPENTADIENYL)DIVANADIUM
$C_{56}H_{49}N_3O_2P_2PtS_3V_2$

C.M. BOLINGER, T.B. RAUCHFUSS and S.R. WILSON, 1984. J. Am. Chem. Soc., 106, 7800-7807.

Monoclinic, $P2_1/c$, a = 22.181, b = 13.176, c = 19.303 Å, β = 111.10°, Z = 4. Mo radiation, R = 0.059 for 6934 reflexions.

The structure (Fig. 1) consists of an isosceles triangle of metal atoms capped by two μ_3-S ligands, with the V-V edge spanned by a μ-S and a $\mu\text{-}\eta^1$-PTD ligand. Main dimensions are: V-V 2.566(2) Å, V-S-V 63.5-68.1(3)°, Pt...V 3.563 and 3.651(2), V-S1 2.420(3), V-S(2) 2.443(3), V-S3 2.263(3) (indicates significant V-S π-bonding), V-N 1.990(8), N-N 1.39(1), Pt-S1 2.316(3), Pt-S2 2.327(2) Å.

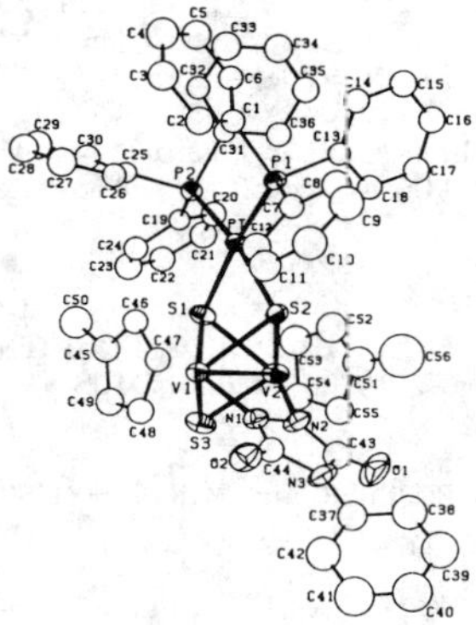

Fig. 1. A view of $C_{56}H_{49}N_3O_2P_2PtS_3V_2$.

BIS{(η^5-CYCLOPENTADIENYL)[1,3-PROPANEDIYLBIS(DIPHENYLPHOSPHINE)-P,P']NICKEL(II)} DODECAHYDRODODECABORATE

$C_{64}H_{74}B_{12}Ni_2P_4$ $[CpNi(Ph_2P(CH_2)_3PPh_2)]_2.B_{12}H_{12}$

CHEN LIQING, PAN KEZHEN, CHEN XINLAN and ZHANG LUN, 1984. J. Struct. Chem., 3, 125-128.

Monoclinic, $P2_1/n$, a = 21.214, b = 16.019, c = 21.300 Å, β = 119.71°, Dm = 1.27, Z = 4. Mo radiation, R = 0.071 for 5458 reflexions.

The molecule has a six-membered NiP_2C_3 ring in a chair conformation (Fig. 1) with the NiP_2 plane almost normal (88.5°) to the Cp ring plane (Ni-Cp(centroid) 1.733, mean Ni-P 2.146 Å, P-Ni-P 95.26°). The anion lies about an inversion centre with B-B 1.733-1.814, mean 1.779 Å in a slightly distorted icosahedron.

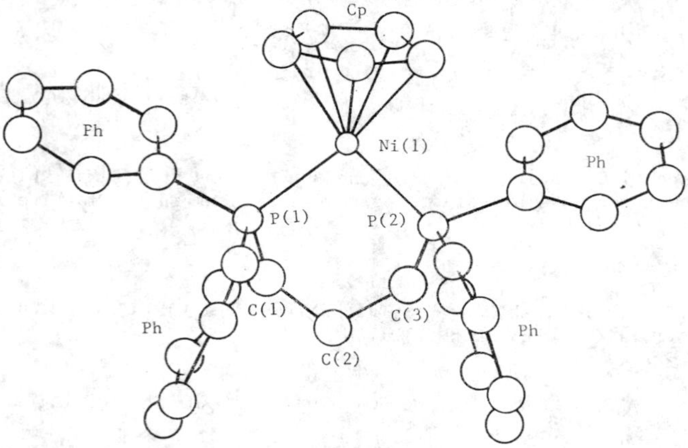

Fig. 1. A view of the cation $[CpNiPh_2P(CH_2)PPH_2]^+$.

1,2,4,2,3,4,1,3,4-TRIS(μ_3-CARBONYL)-1,2,3-TRICARBONYL-4-(η^5-CYCLOPENTADIENYL)-1,2,3-TRIS(TRIPHENYLPHOSPHINE)TRIRHODIUMMOLYBDENUM DICHLOROMETHANE SOLVATE

$C_{65}H_{50}MoO_6P_3Rh_3,CH_2Cl_2$ (I)

2,3,2,3-BIS(μ_2-CARBONYL)-3-(η^5-CYCLOPENTADIENYL)-1,3-(μ_2-DIPHENYLPHOSPHINO)(μ_2-σ^3-PHENYL)-1,2-BIS(TRIPHENYLPHOSPHINE)PLATINUMRHODIUMMOLYBDENUM
$C_{61}H_{50}MoO_2P_3PtRh$ (II)

L.J. FARRUGIA, A.D. MILES and F.G.A. STONE, 1984. J. Chem. Soc. Dalton, 2415-2422.

I. Triclinic, $P\bar{1}$, a = 14.510, b = 13.486, c = 18.935 Å, α = 89.22, β = 114.88, γ = 110.91°, Z = 2. Mo radiation, R = 0.030 for 6254 reflexions.

II. Monoclinic, $P2_1/n$, a = 11.116, b = 16.902, c = 27.792 Å, β = 93.84°, Z = 4. Mo radiation, R = 0.036 for 4988 reflexions.

In I (Fig. 1) the metal-atom core approximates to a regular tetrahedron, with Rh-Rh separations 2.710(1)-2.729(1) Å and Mo-Rh distances 2.764(1)-2.778(1) Å. The Mo atom carries the η-C_5H_5 ligand and a CO and a PPh_3 group are attached to each Rh atom. The terminal CO groups lie below the Rh_3 triangle and the three coordinated P atoms above. A CO group also triply bridges each of the $MoRh_2$ faces of the cluster. The three metal atoms in II form a triangle (Mo-P 2.958(1), Mo-Rh 2.619(1), Pt-Rh 2.662(1) Å. The Mo atom carries the η-C_5H_5 ligand, and the Pt and Rh atoms are coordinated by PPh_3 groups. The three edges of the metal triangle are bridged; the Mo-Rh edge by two CO ligands, the Mo-Pt by PPh_2 and the Rh-Pt by a C_6H_5 group. One C atom of the phenyl group is σ-bonded to Pt and together with an adjacent C atom is η^2 coordinated to the Rh atom.

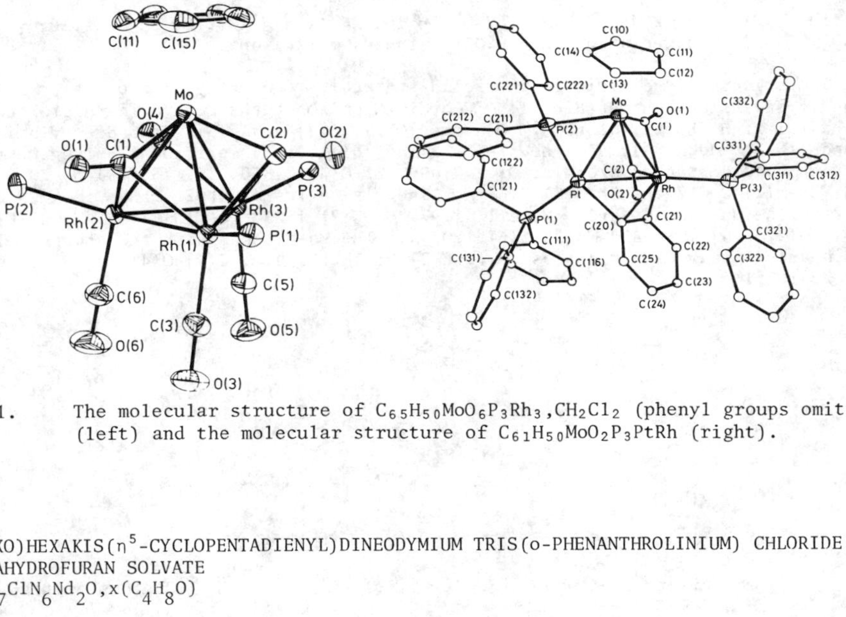

Fig. 1. The molecular structure of $C_{65}H_{50}MoO_6P_3Rh_3,CH_2Cl_2$ (phenyl groups omitted) (left) and the molecular structure of $C_{61}H_{50}MoO_2P_3PtRh$ (right).

(μ_2-OXO)HEXAKIS(η^5-CYCLOPENTADIENYL)DINEODYMIUM TRIS(o-PHENANTHROLINIUM) CHLORIDE TETRAHYDROFURAN SOLVATE
$C_{66}H_{57}ClN_6Nd_2O,x(C_4H_8O)$

JIN XIANGLIN and ZHOU JIALIE, 1984. J. Struct. Chem., 3, 235-240.

Monoclinic, $P2_1/c$, a = 20.982, b = 11.217, c = 32.954 Å, β = 104.17°, Z = 4. Cu radiation, R = 0.159 for 3381 reflexions.

Two $Nd(C_5H_5)_3$ groups are linked by a bridging oxygen (Fig. 1) with Nd-O 2.35(6) and 2.32(6) Å, Nd-O-Nd 169(2)°. Three o-phenanthrolinium groups are linked to the chloride ion by N-H...Cl hydrogen bonds (N...Cl 2.46-2.52 Å).

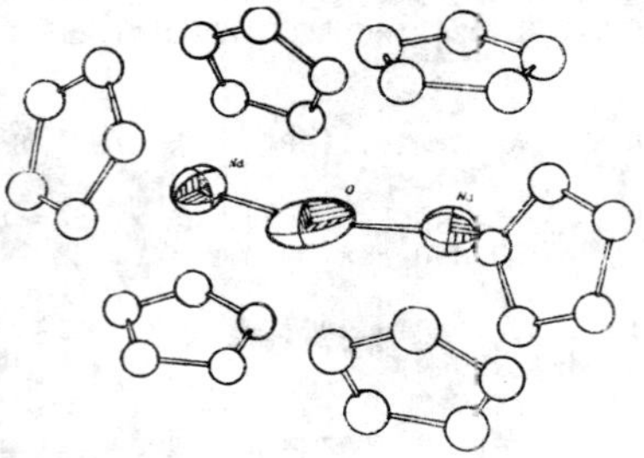

Fig. 1. The $\{[(C_5H_5)_3Nd]_2O\}^{2-}$ anion.

HEXAKIS(μ^2-CARBONYL-C,O)-BIS(CARBONYL-(η^5-CYCLOPENTADIENYL)MOLYBDENUM)BIS((η^5-CYCLOPENTADIENYL)DICARBONYLMOLYBDENUM)BIS(BIS(DIETHYLAMIDO)DIETHYLAMINOZIRCONIUM)
$C_{80}H_{82}Mo_4N_6O_{12}Zr_2$

W.J. SARTAIN and J.P. SELEGUE, 1984. Organometallics, 3, 1922-1924.

Triclinic, P$\bar{1}$, a = 12.448, b = 13.871, c = 10.724 Å, α = 104.79, β = 100.36, γ = 76.40°, Z = 1. Mo radiation, R = 0.059 for 4741 reflexions.

The molecule is dimeric in the solid state, its centre lying about a crystallographic inversion centre (Fig. 1). Coordination about the Zr atom is distorted octahedral with three isocarbonyl O atoms (O(2), O(4) and O(5)), two diethylamido ligands (N(1) and N(2)), and one diethylamine (N(3)) in a meridional arrangement. The relevant bond distances are: Zr-O(2) 2.099, Zr-O(4) 2.330, Zr-O(5) 2.146, Zr-N(1) 2.03, Zr-N(2) 2.017 and Zr-N(3) 2.52 Å. The cis-angles about Zr range from 79.1 to 101.2°. The Mo-C bonds are: Mo-C(1) 1.97, Mo-C(2) 1.83, Mo-C(3) 1.94, Mo(2)-C(4) 1.90, Mo(2)-C(5) 1.86 and Mo(2)-C(6) 1.97 Å. The Mo-C bonds with the shortest length have nearly linear C-O-Zr angles, e.g. C(2)-O(2)-Zr 174.6, C(4)-O(4)-Zr 158.0, C(5)-O(5)-Zr 163.1°.

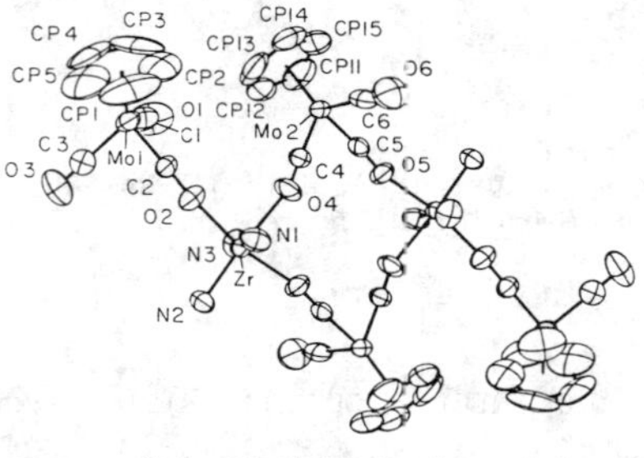

Fig. 1. Molecular structure of $C_{80}H_{82}Mo_4N_6O_{12}Zr_2$ with ethyl C atoms omitted for clarity.

DICARBONYL-(η^5-METHYLCYCLOPENTADIENYL)MOLYBDENUM(BIS(DIPHENYLPHOSPHINO)METHANE-P,P'-PLATINUM)) (μ_2-OXO)BIS(TRIOXOMOLYBDENUM) DICHLOROMETHANE
$C_{116}H_{102}Mo_4O_{11}P_8Pt_2,CH_2Cl_2$ $[C_{58}H_{51}O_2MoP_4Pt]_2^+[Mo_2O_7]^{2-},CH_2Cl_2$

P. BRAUNSTEIN, C. de MÉRIC de BELLEFON, M. LANFRANCHI and A. TIRIPICCHIO, 1984. Organometallics, 3, 1772-1774.

Monoclinic, P2$_1$/n, a = 34.841, b = 24.686, c = 12.921 Å, β = 98.41°, Z = 4. Cu radiation, R = 0.084 for 8638 reflexions.

The crystal structure contains two independent but almost equivalent cations (one of which is illustrated) and a dimolybdate anion (Fig. 1). Two different dppm ligands are bonded to the metals. The former is bridging the Mo and Pt atoms (Mo(1)-P(4) 2.417, Pt-P(3) 2.284, Pt-Mo(1) 2.912 Å averaged over two cations), and the latter chelates the Pt atom (Pt(1)-P(1) 2.373, Pt(1)-P(2) 2.334 Å). Other distances are: Pt(1)...C(1) 2.51, Mo(1)-C(1) 1.92 Å. In the anion a pseudomirror plane contains O(4), O(3) and O(7) and the two Mo atoms. The Mo-O bond distances range from 1.68 to 1.79 Å except those involving the μ-O(3) atom: Mo(3)-O(3) 1.82, Mo(4)-O(3) 1.95 Å, Mo(3)-O(3)-Mo(4) 160°.

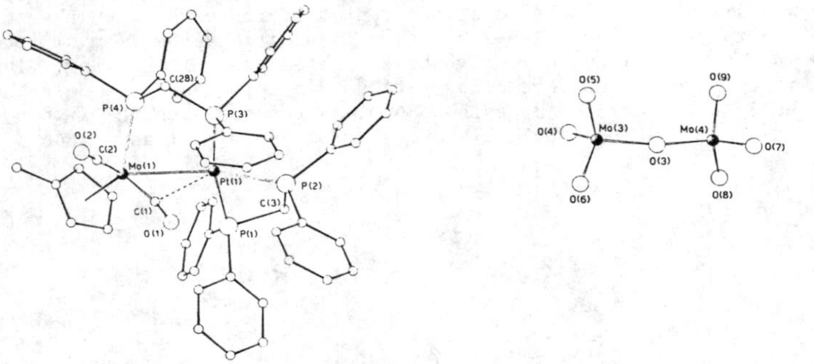

Fig. 1. View of one of the crystallographically-independent cations in $C_{116}H_{102}Mo_4O_{11}P_8Pt_2$ (left), and of the anion (right).

5-(η^6-1,3,5-TRIMETHYLBENZENE)-5-FERRADECABORANE
$C_9H_{25}B_9Fe$ (I) 5-[η^6-$C_6(CH_3)_3H_3$]FeB_9H_{13}

2-(η^6-1,3,5-TRIMETHYLBENZENE)-2-FERRA-6-OXADECABORANE
$C_9H_{22}B_8FeO$ (II) 2-[η^6-$C_6(CH_3)_3H_3$]Fe-6-OB_8H_{10}

R.P. MICCICHE, J.J. BRIGUGLIO and L.G. SNEDDON, 1984. Inorg. Chem., 23, 3992-3999.

I. Monoclinic, $P2_1/c$, a = 9.348, b = 9.842, c = 17.335 Å, β = 90.70°, Z = 4. Mo radiation, R = 0.049 for 1755 reflexions.

II. Monoclinic, $P2_1/c$, a = 12.647, b = 13.437, c = 9.281 Å, β = 104.57°, Z = 4. Mo radiation, R = 0.059 for 1449 reflexions.

Both compounds have a decaborane(14) structure (Fig. 1). In I the Fe atom occupies the cage five-position on the open face (Fe-B 2.077(5)-2.241(5), Fe-C 2.095(4)-2.144(3), Fe-H 1.55(3) Å). In compound II the Fe atom occupies the five-coordinate two-position while the oxygen is in the six-position on the open face (Fe-B 2.053(7)-2.098(7), Fe-O 1.942(3), Fe-C 2.097(5)-2.123(5) Å).

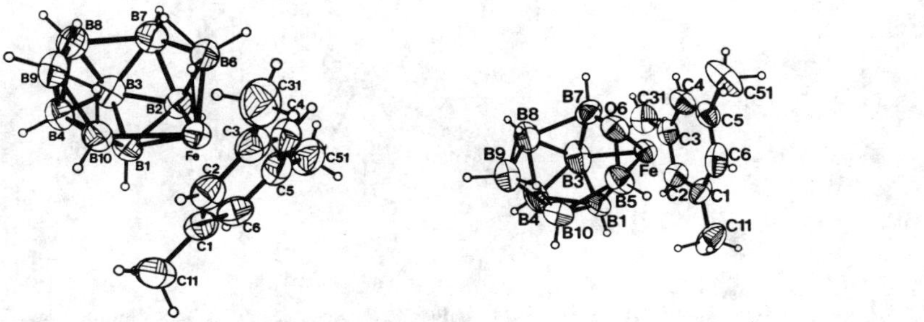

Fig. 1. The structures of 5-[η^6-$C_6(CH_3)_3H_3$]FeB_9H_{13} (left) and 2-[η^6-$C_6(CH_3)_3H_3$]Fe-6-OB_8H_{10} (right).

TRICARBONYL-4-CHROMANONECHROMIUM(0)
$C_{12}H_8CrO_5$

K.R. STEWART, S.G. LEVINE and A.T. McPHAIL, 1984. J. Organometal. Chem., 263, 45-53.

Monoclinic, $P2_1/c$, a = 6.751, b = 14.158, c = 12.236 Å, β = 104.72°, Z = 4. Cu radiation, R = 0.038 for 1526 reflexions.

The reduced symmetry of the substituted benzene ring (C-C distances range from 1.376 to 1.420 Å, mean 1.399 Å) and the slight bowing of the ring atoms away from the $Cr(CO)_3$ moiety (Fig. 1) contribute to significant differences between the individual Cr-C(benzene) distances (range 2.179-2.254 Å). In addition, the Cr atom is displaced 0.031 Å towards C(6) from lying directly above the centroid of the ring. The heterocyclic ring approximates more closely to a sofa form than to a half-chair form. The O(1)-C(2)-C(3)-C(4) torsion angle is 55.6°, and the bond lengths within the ring are as expected. The Cr-C(carbonyl) distances are 1.846, 1.851 and 1.846 Å.

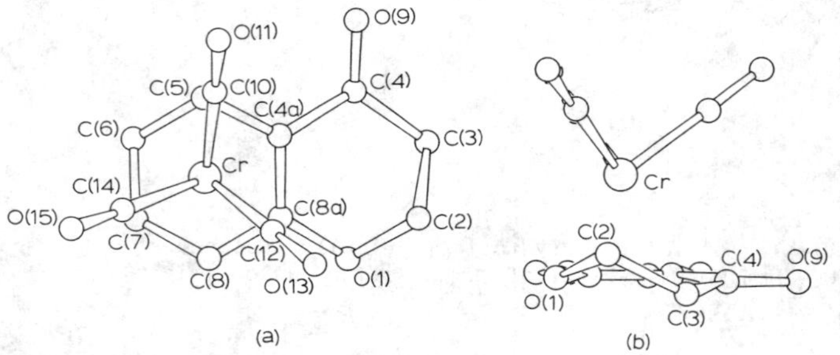

Fig. 1. Conformation of $C_{12}H_8CrO_5$ viewed (a) perpendicular to the mean plane through C(4a)-C(8a), (b) along the plane.

TRICARBONYL(η^6-1,3,5-TRIMETHYLBENZENE)RHENIUM(II) TETRABROMOALUMINIUM
$C_{12}H_{12}AlBr_4O_3Re$ (I)

TRICARBONYL(η^6-1,3,5-TRIMETHYLBENZENE)TUNGSTEN(0)
$C_{12}H_{12}O_3W$ (II)

F. CALDERAZZO, R. POLI, A. BARBATI and P.F. ZANAZZI, 1984. J. Chem. Soc. Dalton, 1059-1064.

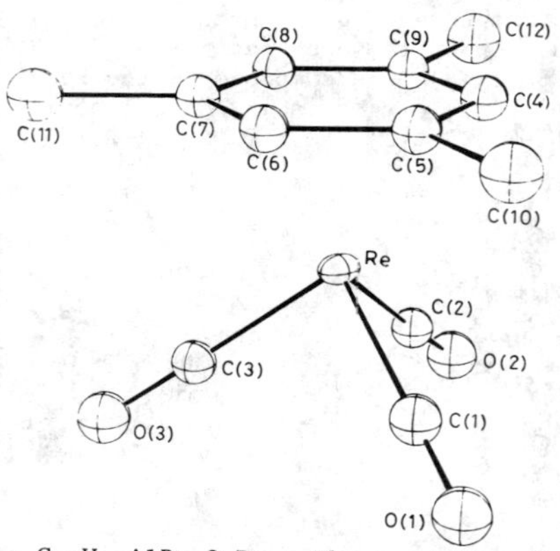

Fig. 1. The cation in $C_{12}H_{12}AlBr_4O_3Re$. The geometry of the tungsten complex is similar.

I. Monoclinic, $P2_1/a$, a = 15.687, b = 12.876, c = 9.690 Å, β = 101.95°, Z = 4. Mo radiation, R = 0.046 for 1220 reflexions.

II. Monoclinic, $P2_1/a$, a = 8.959, b = 16.979, c = 8.797 Å, β = 118.46°, Z = 4. Mo radiation, R = 0.036 for 861 reflexions.

The cation in I and the complex in II both have an eclipsed orientation of the CO groups with respect to the methyl substituents on the ring (Fig. 1). In I, a three-fold axis of symmetry of the $Re(CO)_3$ moiety coincides with the centre of the arene ring (Re-ring plane distance 1.855(2), mean Re-C 1.89, mean C-O 1.15 Å). In II the corresponding distances are 1.886(2), 1.94 and 1.16 Å respectively. One C-Me distance is longer than expected (C(7)-C(11) 1.58(3) (I), 1.62(3) Å (II)). The $AlBr_4^-$ anion in I has almost perfect tetrahedral geometry. Packing in I resembles that of sodium chloride.

η^6-BENZENEDICARBONYL-1,3-DITHIANECHROMIUM(0)
$C_{12}H_{14}CrO_2S_2$ (I)

η^6-METHYL BENZOATEDICARBONYL-1,3-DITHIANECHROMIUM(0)
$C_{14}H_{16}CrO_4S_2$ (II)

P.H. VAN ROOYEN, J.L.M. DILLEN, S. LOTZ and M. SCHINDEHUTTE, 1984. J. Organometal. Chem., 273, 61-68.

I. Monoclinic, $P2_1/n$, a = 22.014, b = 7.403, c = 8.264 Å, β = 98.64°, Z = 4. Mo radiation, R = 0.079 for 1304 reflexions.

II. Triclinic, $P\bar{1}$, a = 12.991, b = 9.278, c = 7.001 Å, α = 90.57, β = 102.06, γ = 105.76°, Z = 2. Mo radiation, R = 0.067 for 1393 reflexions.

The complexes have different conformations with the $Cr(CO)_2[S(CH_2)_3SCH_2]$ fragment in a near eclipsed orientation relative to the planar η^6-methylbenzoate ligand in II and in a staggered orientation in relation to the C atoms of the benzene ring in I (Fig. 1). The Cr-S bond distances in I and II respectively are 2.355 and 2.346 Å and both dithiane rings have chair conformations. The Cr-C(arene) distances range from 2.154 to 2.214 Å in I and from 2.171 to 2.204 Å in II.

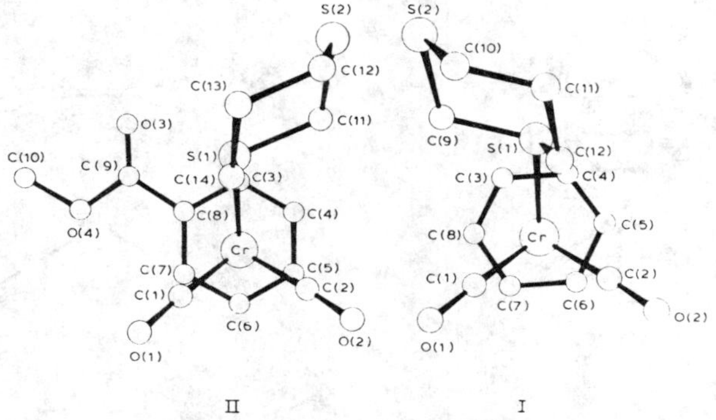

Fig. 1. Views of $C_{14}H_{16}CrO_4S_2$ (left) and $C_{12}H_{14}CrO_2S_2$ (right) along the Cr-arene bond perpendicular to the arene rings.

TRICARBONYL(2-HYDROXYMETHYL-2-PHENYLPROPANOL)CHROMIUM(III)
$C_{13}H_{14}O_5Cr$ $[Cr(CO)_3(C_{10}H_{14}O_2)]$

F. ABRAHAM, G. NOWOGROCKI, J. BROCARD, J. LEBIBI and C. BRÉMARD, 1984. Acta Cryst., C40, 1181-1183.

Monoclinic, $P2_1/c$, a = 6.579, b = 31.249, c = 8.547 Å, β = 133.04°, D_m = 1.53, Z = 4. Mo radiation, R = 0.035 for 1884 reflexions.

The molecular structure (Fig. 1) shows that the arene C atoms are nearly staggered with respect to the carbonyl ligands. The phenyl ring and one hydroxymethyl group are in eclipsed conformation. The Cr atom is pseudo-octahedrally coordinated by the carbonyl groups. Bond distances are: Cr-C(carbonyl) 1.884(4), 1.842(6) and 1.837(3), Cr-C(arene) 2.231(7), 2.232(5), 2.210(3), 2.209(3), 2.267(4) and 2.229(6) Å. Crystal packing is by hydrogen bonds and van der Waals forces.

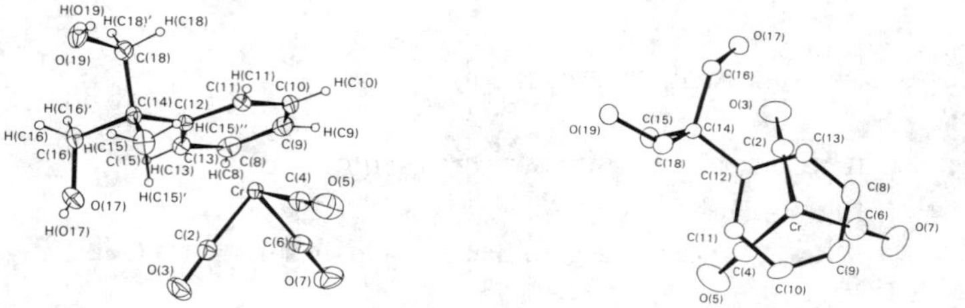

Fig. 1. $[Cr(CO)_3(C_{10}H_{14}O_2)]$: perspective view of the molecule and a view in projection on the mean plane of the benzene ring.

(BIPHENYLENE)TRICARBONYLCHROMIUM
$C_{15}H_8CrO_3$ (I) $(\eta^6-C_{12}H_8)Cr(CO)_3$

(TRIPHENYLENE)TRICARBONYLCHROMIUM
$C_{21}H_{12}CrO_3$ (II) $(\eta^6-C_{18}H_{12})Cr(CO)_3$

R.D. ROGERS, J.L. ATWOOD, T.A. ALBRIGHT, W.A. LEE and M.D. RAUSCH, 1984. Organometallics, 3, 263-270.

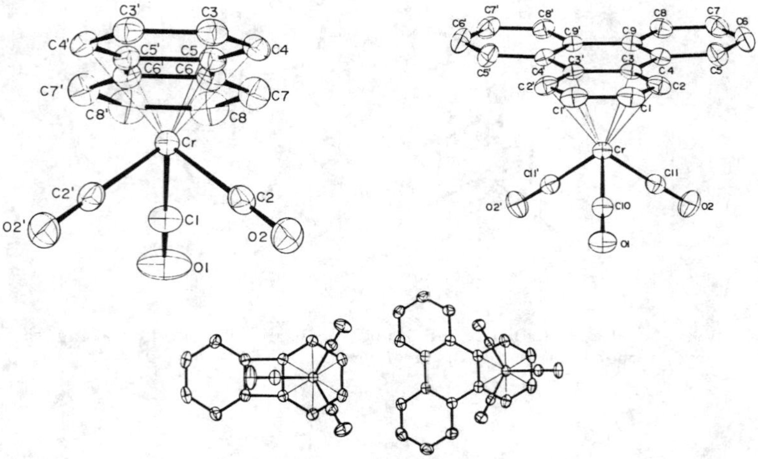

Fig. 1. Side view of molecular structures of $C_{15}H_8CrO_3$ (top left), and $C_{21}H_{12}CrO_3$ (top right), and top views of both molecules (bottom).

I. Monoclinic, $P2_1/m$, a = 7.065, b = 11.066, c = 8.030 Å, β = 98.43°, Z = 2. Mo radiation, R = 0.028 for 850 reflexions.

II. Monoclinic, C2/m, a = 12.994, b = 14.766, c = 9.393 Å, β = 121.63°, Z = 4. Mo radiation, R = 0.023 for 944 reflexions.

The biphenylene and triphenylene ligands in I and II are not much distorted from planarity - the greatest atomic deviation occurs for C(7) and C(8) which lie 0.04 and 0.08 Å respectively from the plane. Bond localization is reduced in biphenylene with C-C distances ranging from 1.365 to 1.508 Å. The Cr-C(arene) distances in I range from 2.199 to 2.249 Å with the Cr-(arene centroid) 1.721 Å in length. In II the metal-C(arene) distances range from 2.199 to 2.258 Å with the metal-centroid distance 1.720 Å. The orientation of the $Cr(CO)_3$ group with respect to the benzenoid ring is of interest (Fig. 1). In II, it adopts the exo-conformation whilst in I it adopts the endo-conformation.

(METHYL (3R,4R,6S)-1,3,4,6-TETRAHYDRO-2-OXO-3,6-EPOXY-2H-5,1-BENZOXAZOCINE-4-CARBOXYLATE)CHROMIUM TRICARBONYL
$C_{15}H_{11}CrNO_8$

K.R. STEWART, S.G. LEVINE and J. BORDNER, 1984. J. Org. Chem., 49, 4082-4084.

Monoclinic, $P2_1$, a = 6.997, b = 7.897, c = 14.357 Å, β = 90.60°, D_m = 1.59, Z = 2. Cu radiation, R = 0.043 for 895 reflexions.

The X-ray analysis shows that complexation has taken place on the face of the aromatic ring which is syn to the 3,6-epoxy bridge (Fig. 1).

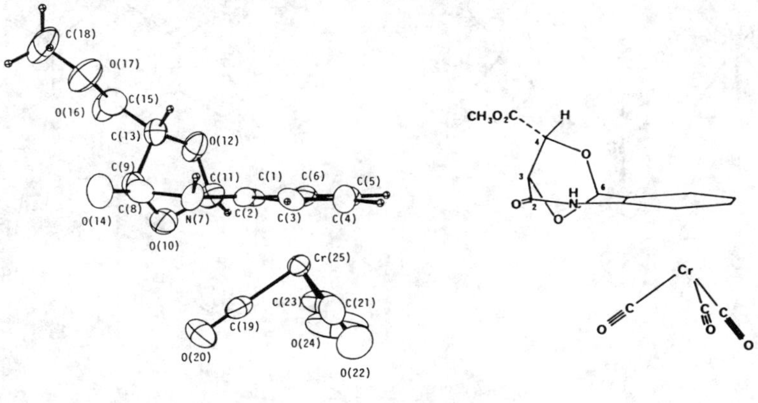

Fig. 1. A view of the $C_{15}H_{11}CrNO_8$ complex.

(η^6-9,9,10,10-TETRAMETHYL-9,10-DISILADIHYDROANTHRACENE)TRICARBONYLCHROMIUM
$C_{19}H_{20}CrO_3Si_2$ (I) $[Cr(CO)_3][C_6H_4(SiMe_2)_2C_6H_4]$

trans-(η^6,η^6-9,9,10,10-TETRAMETHYL-9,10-DISILADIHYDROANTHRACENE)BIS(TRICARBONYLCHROMIUM)
$C_{22}H_{20}Cr_2O_6Si_2$ (II) trans-$[Cr(CO)_3]_2[C_6H_4(SiMe_2)_2C_6H_4]$

cis-(η^6,η^6-9,9,10,10-TETRAMETHYL-9,10-DISILADIHYDROANTHRACENE)BIS(TRICARBONYLCHROMIUM)
$C_{22}H_{20}Cr_2O_6Si_2$ (III) cis-$[Cr(CO)_3]_2[C_6H_4(SiMe_2)_2C_6H_4]$

V.A. SHARAPOV, A.I. GUSEV, N.I. KIRILLOVA, N.V. ALEKSEEV, O.B. AFANASOVA, E.A. CHERNYSHEV and Yu.T. STRUCHKOV, 1984. Organometallics, 3, 1755-1758.

I. Monoclinic, $P2_1/c$, a = 16.941, b = 6.954, c = 17.489 Å, β = 105.93°, Z = 4. Mo radiation, R = 0.037 for 2274 reflexions.

II. Orthorhombic, $Cmc2_1$, a = 13.093, b = 13.871, c = 12.926 Å, Z = 4. Mo radiation, R = 0.026 for 1062 reflexions.

III. Orthorhombic, $Pbc2_1$, a = 8.783, b = 14.893, c = 18.226 Å, Z = 4. Mo radiation, R = 0.025 for 1607 reflexions.

Coordination of 9,9,10,10-tetramethyl-9,10-disiladihydroanthracene both by one (I) and by two (in II and III) Cr atoms causes only relatively small changes in its conformation (Fig. 1). A boat conformation of the ligand is retained wherein the central six-membered heterocycle is bent along the Si...Si line with angles of 145.2, 160.0 and 160.7° respectively for I, II and III. In all three complexes, endocyclic Si-C bonds (average 1.875, 1.879 and 1.882 Å in I, II and III respectively) are longer than exocyclic Si-CH_3 bonds (average 1.850, 1.863 and 1.854 Å respectively). The Cr-C(arene) average distances are 2.221, 2.217 and 2.216 Å in I, II and III respectively. The two $Cr(CO)_3$ groups are on opposite sides of the average ligand plane in II, whilst in III they are on the same side (cis).

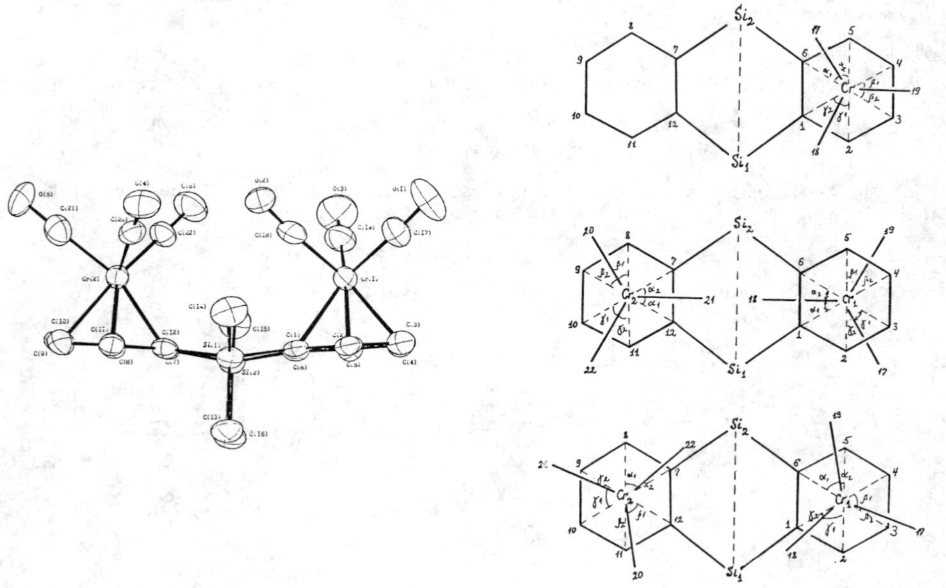

Fig. 1. Structure of $C_{22}H_{20}Cr_2O_6Si_2$ (left), and projections of the molecules in I, II and III showing relative orientation of the $Cr(CO)_3$ moieties.

8-(η^6-HEXAMETHYLBENZENE)-2,3,4,5-TETRAMETHYL-2,3,4,5-TETRACARBA-8-FERRA-1-DEBOR[C_{2v}-($1\nu^6 4\nu^4 2\nu^5 2\nu^5$)-$\Delta^{14}$-closo]NONABORANE(7)

$C_{20}H_{33}B_3Fe$ (I) 1-[η^6-$C_6(CH_3)_6$]Fe-4,5,7,8-$(CH_3)_4C_4B_3H_3$

8-(η^6-TOLUENE)-2,3,4,6-TETRAMETHYL-2,3,4,6-TETRACARBA-8-FERRA-1-DEBOR[C_{2v}-($1\nu^6 2\nu^4 4\nu^5$-$2\nu^5 2\nu^5$)-Δ^{18}-closo]UNDECABORANE(9)

$C_{15}H_{25}B_5Fe$ (II) 2-[η^6-$CH_3C_6H_5$]Fe-6,7,9,10-$(CH_3)_4C_4B_5H_5$

R.P. MICCICHE, J.J. BRIGUGLIO and L.G. SNEDDON, 1984. Organometallics, 3, 1396-1402.

I. Monoclinic, $P2_1/c$, a = 8.753, b = 10.190, c = 22.118 Å, β = 95.08°, Z = 4. Mo radiation, R = 0.060 for 2597 reflexions.

II. Triclinic, $P\bar{1}$, a = 9.368, b = 11.399, c = 8.708 Å, α = 90.84, β = 116.51, γ = 94.03°, Z = 2. Mo radiation, R = 0.074 for 760 reflexions.

Compound I has an arachno-type cage geometry derived from a bicapped-square antiprism missing two vertices, with the Fe atom occupying the five-coordinate 1-position in the cage and the four carbons occupying positions on the open face of the cage (Fig. 1). The Fe-C(7) and Fe-C(8) distances are 2.062 and 2.067 Å, and the Fe-B(3), Fe-B(2) and Fe-B(6) distances are 2.211, 2.043 and 2.196 Å respectively. The bond C(5)-C(4) is unusually short (1.367 Å) for the non-metal-bonded C atoms in the cage. Compound II is shown (Fig. 1) to be an analogue of decaborane(14) in which the Fe atom occupies the five-coordinate 2-position in the cage and the C atoms occupy the 6,7- and 9,10- positions on the open face. Bonds Fe-C(6) and Fe-C(7) are 2.00 and 2.04 Å in length, and the bonds Fe-B(5), Fe-B(1) and Fe-B(3) are respectively 2.13, 2.05 and 2.11 Å. The B-B bond lengths range from 1.79 to 1.83 Å in II, and the B(3)-B(2) and B(2)-B(6) distances in I are 1.860 and 1.848 Å respectively.

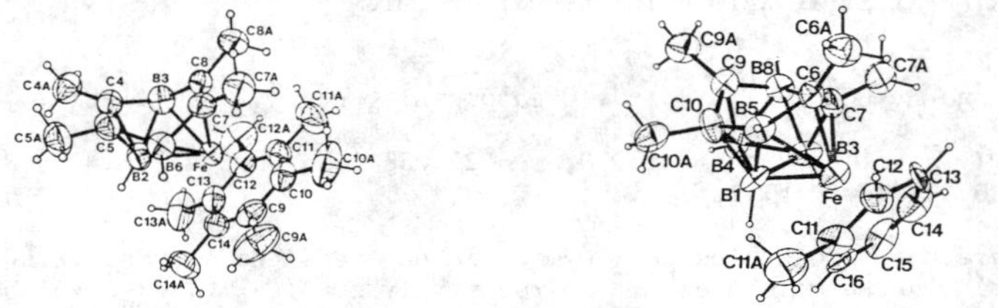

Fig. 1. Molecular structure of $C_{20}H_{33}B_3Fe$ (left) and $C_{15}H_{25}B_5Fe$ (right).

(η^6-1,4-DIPHENYLBENZENE)(η^4-TRIMETHYLTETRAFLUOROBENZOBARRELENE)IRIDIUM PERCHLORATE
$C_{23}H_{22}ClF_4IrO_4$ (I) $[Ir(Me_3TFB)(1,4-C_6H_4Me_2)]^+ClO_4^-$

(η^5-DIPHENYLIMINOCYCLOHEXADIENYL)(η^4-TETRAFLUOROBENZOBARRELENE)IRIDIUM TETRAFLUOROBORATE DICHLOROMETHANE SOLVATE
$C_{30}H_{21}BF_8IrN,CH_2Cl_2$ (II) $[Ir(TFB)(PhNPh_2)]^+BF_4^-,CH_2Cl_2$

R. USÓN, L.A. ORO, D. CARMONA, M.A. ESTERUELAS, C. FOCES-FOCES, F.H. CANO, S. GARCÍA-BLANCO and A. VÁZQUEZ DE MIGUEL, 1984. J. Organometal. Chem., 273, 111-128.

I. Orthorhombic, Pbca, a = 17.695, b = 15.807, c = 16.002 Å, Z = 8. Cu radiation, R = 0.063 for 3389 reflexions.

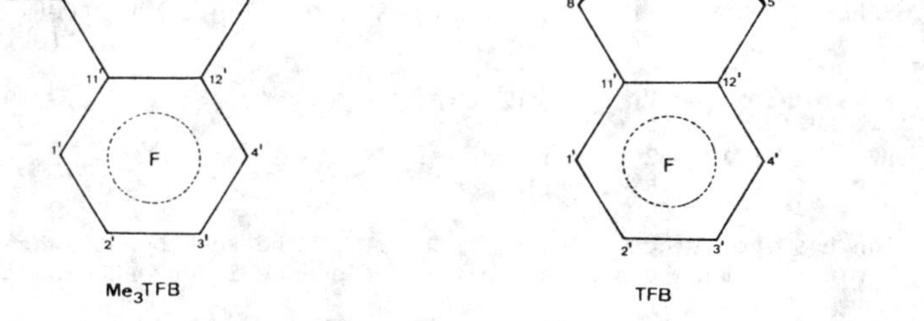

Fig. 1. Schematic structures of the cations in $C_{23}H_{22}ClF_4IrO_4$ (left), and $C_{30}H_{21}BF_8IrN$ (right).

II. Orthorhombic, $Pna2_1$, a = 9.8059, b = 20.8097, c = 14.3367 Å, Z = 4. Cu radiation, R = 0.042 for 2514 reflexions.

Both complexes show a staggered arrangement between the arene and the TFB moieties, and deviation from planarity of the coordinated arene ligands (Fig. 1). In II, the Ir-C(arene) distances are 2.29, 2.22, 2.30, 2.31, 2.29 and 2.46 Å for C(1) to C(6) respectively suggesting arene puckering towards an η^5-coordinated iminocyclohexadienyl form. Also in II, the N-C(6) distance is 1.37, suggesting double bond character. The Ir-C(6') to C(10') distances are in I: 2.165, 2.128, 2.135 and 2.166 Å, and in II: 2.12, 2.10, 2.14 and 2.15 Å. The bonds C(6')-C(7') and C(9')-C(10') are 1.43 and 1.44 Å in I and 1.40 and 1.44 Å in II.

(η^6-ETHYL TRIPHENYLSILYL KETONE)TRICARBONYLCHROMIUM
$C_{24}H_{20}CrO_4Si$

U. SCHUBERT and H. HORNIG, 1984. J. Organometal. Chem., 273, C11-C16.

Triclinic, $P\bar{1}$, a = 11.462, b = 11.930, c = 27.053 Å, α = 95.23, β = 95.88, γ = 113.94°, Z = 6. R = 0.097 for 4467 reflexions.

There are three molecules per asymmetric unit and the most striking structural characteristic in the molecules is the relatively small Si-C(40)-O(40) bond angle of 113° (Fig. 1). The Cr-C(arene) distances average 2.23 Å, and the Si-C(phenyl), Si-C(40) and C(40)-O(40) distances averaged over the three molecules are respectively 1.87, 1.93 and 1.24 Å. The angles Si-C(40)-C(41) and O(40)-C(40)-C(41) are 124 and 123° respectively.

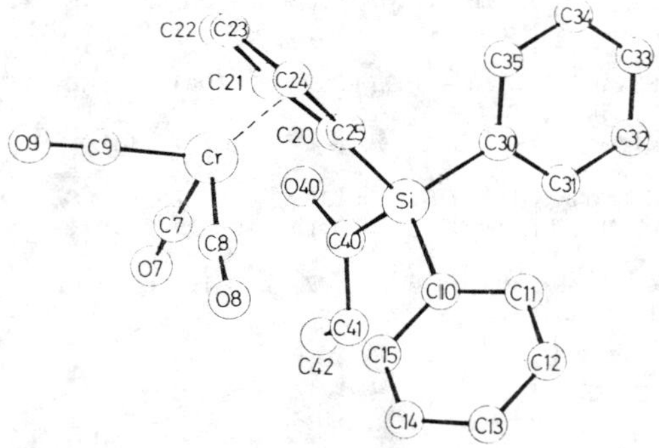

Fig. 1. Molecular structure of one of the three molecules of $C_{24}H_{20}CrO_4Si$ per asymmetric unit.

(η^6-HEXAMETHYLBENZENE)(η^5-3H-[2_2](1,4)CYCLOPHANE)RUTHENIUM(II) HYDROGEN DICHLORIDE
$C_{28}H_{36}Cl_2Ru$ $(C_{28}H_{35}Ru)^+(HCl_2)^-$

R.T. SWANN, A.W. HANSON and V. BOEKELHEIDE, 1984. J. Am. Chem. Soc., 106, 818-819.

Monoclinic, $P2_1/n$, a = 18.362, b = 14.946, c = 9.219 Å, β = 94.79°, D_m = 1.43, Z = 4. Mo radiation, R = 0.044 for 6242 reflexions.

The cation has the structure shown in Fig. 1. The Ru-C bonds range from 2.17(1)-2.24(1) Å. The proton has been added to the cyclophane in an endo fashion at C(3).

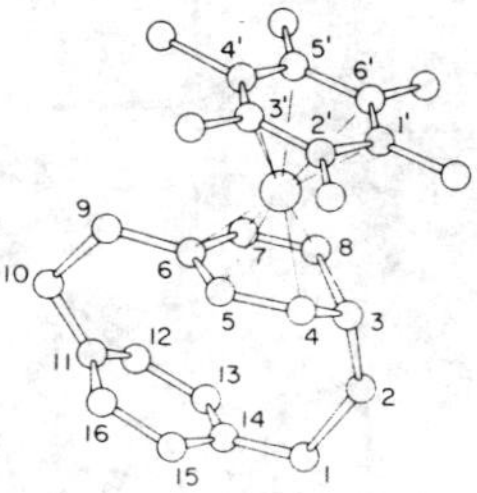

Fig. 1. The $(C_{28}H_{35}Ru)^+$ cation.

BIS[(η^5-CYCLOPENTADIENYL)(η^6-ISOPROPYLBENZENE)IRON(II)] DODECAHYDRODODECABORATE
$C_{28}H_{46}B_{12}Fe_2$

CHEN LIQING, PAN KEZHEN, ZHANG LUN and HU PEIZHI, 1984. J. Struct. Chem., 3, 5-8.

Monoclinic, C2/c, a = 18.703, b = 25.120, c = 15.190 Å, β = 94.70°, Z = 8. Mo radiation, R = 0.070 for 2931 reflexions.

In each cation (Fig. 1) the planes of the five- and six-membered rings are nearly parallel and the anion a slightly distorted icosahedron. Mean bond lengths are: B-B 1.777, Fe-C(Cp) 2.035, Fe-C(Ph) 2.075, Fe-Cp(centroid) 1.664, Fe-Ph (centroid) 1.537 Å.

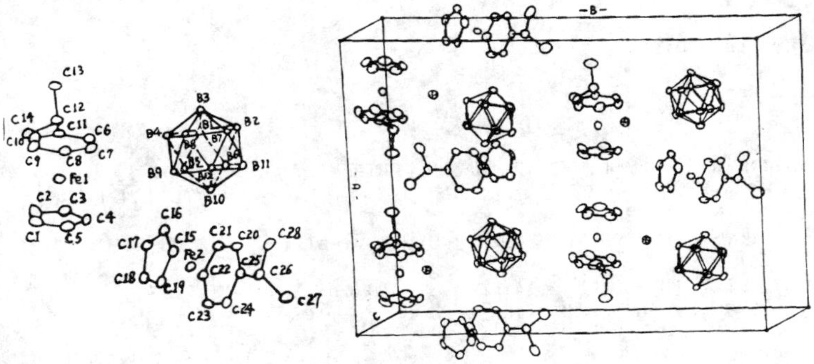

Fig. 1. The asymmetric unit of $C_{28}H_{46}B_{12}Fe_2$ and a view of the unit cell contents.

(η^6-HEXAMETHYLBENZENE)(η^6-TRIPHENYLPHOSPHINE)RUTHENIUM BIS(HEXAFLUOROPHOSPHATE)
$C_{30}H_{33}F_{12}P_3Ru$ $[C_6Me_6Ru(\eta^6-C_6H_5PPH_2)]^{2+}2PF_6^-$

H. WERNER, H. KLETZIN and Ch. BURSCHKA, 1984. J. Organometal. Chem., 276, 231-240.

Orthorhombic, $P2_1ab$, a = 11.341, b = 15.657, c = 18.110 Å, Z = 4. Mo radiation, R = 0.049 for 2591 reflexions.

The metal atom is sandwiched between the two six-membered rings in a symmetrical fashion, there being no difference in the Ru-C bond lengths (average 2.23 Å) (Fig. 1). The two rings have a conformation midway between that of eclipsed and staggered, and are almost parallel. The dihedral angle between their respective mean planes is 2.93°. The P atom of the η^5-$C_6H_5PPh_2$ ligand is bent away from the plane of the ring towards the metal by some 0.122 Å. Distances associated with the P atom are: P-C(13) 1.847, P-C(19) 1.805 and P-C(25) 1.848 Å.

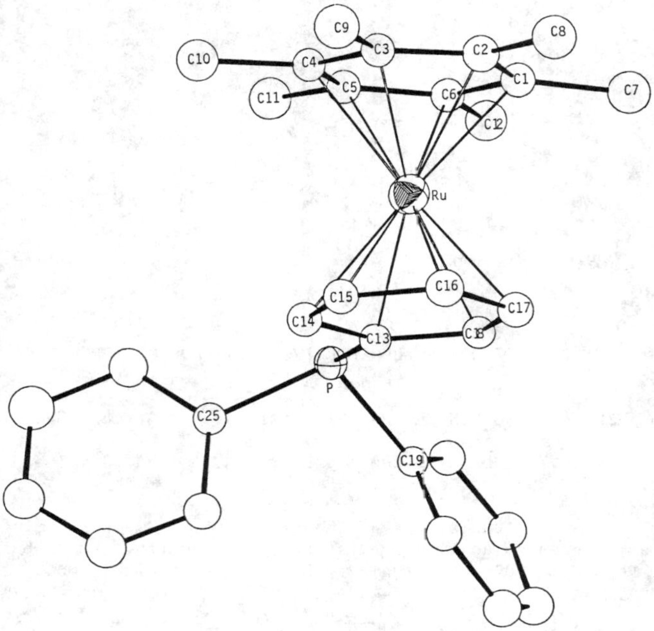

Fig. 1. Structure of $C_{30}H_{33}F_{12}P_3Ru$.

(η^6-HEXAMETHYLPHENYL(η^4-OCTAMETHYLNAPHTHALENE)RUTHENIUM

$C_{30}H_{42}Ru$ $(\eta^6-C_6(CH_3)_6)Ru(\eta^4-C_{18}H_{24})$

J.W. HULL, Jr. and W.L. GLADFELTER, 1984. Organometallics, 3, 605-613.

Monoclinic, $P2_1/n$, a = 19.768, b = 14.524, c = 19.676 Å, β = 117.13°, Z = 8. Mo radiation, R = 0.042 for 6661 reflexions.

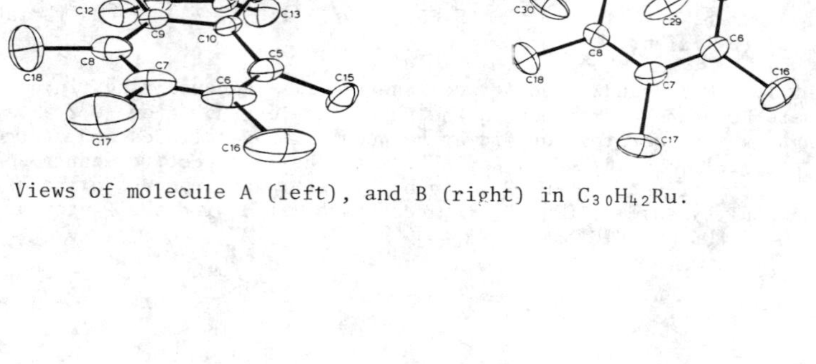

Fig. 1. Views of molecule A (left), and B (right) in $C_{30}H_{42}Ru$.

The coordination sphere about the Ru atom has an η^6-C_6Me_6 ring and an η^4-octamethylnaphthalene (OMN) ring (Fig. 1). The C_6Me_6 ring shows only slight deviations from planarity, and the average Ru-C(ring) (both independent molecules) distance is 2.25 Å. The methyl groups are pushed slightly out-of-plane towards the distal side of the ring with respect to the Ru. The metal atom is located 1.750 Å above the ring plane in A, and 1.759 Å in B. The Ru-C bonds to the OMN ring follow the pattern having the Ru-C(1), Ru-C(4) bonds longer (2.18 Å in both molecules A and B) and the Ru-C(2), Ru-C(3) bonds shorter (2.150 Å in A, 2.129 Å in B). The distance between Ru and the best plane through atoms C(1)-C(4) is 1.726 Å for A and 1.713 Å for B. The two planes comprising atoms C(1)-C(4) and C(19)-C(24) intersect with dihedral angles of 8.3° for A and 6.4° for B. Also, the C(1)-C(4) plane makes an angle of 41.5° (43.3° for B) with the C(1),C(9),C(10),C(4) plane.

BIS[(η^5-CYCLOPENTADIENYL)(η^6-TETRAHYDRONAPHTHALENE)IRON(II)] DODECAHYDRODODECABORATE
$C_{30}H_{46}B_{12}Fe_2$ $(C_5H_5FeC_{10}H_{12})_2(B_{12}H_{12})$

CHEN LIQING, PAN KEZHEN, ZHANG LUN and HU PEIZHI, 1984. J. Struct. Chem., 3, 1-4.

Orthorhombic, Pbca, a = 20.854, b = 14.629, c = 21.051 Å, Z = 8. Mo radiation, R = 0.080 for 1899 reflexions.

The cyclopentadienyl rings and the six-membered aromatic rings are individually essentially planar and are nearly parallel to each other in each cation (Fig. 1). The anion is a slightly distorted icosahedron. Mean dimensions are: Fe-C(Cp) 2.031, Fe-C(Ph) 2.085, B-B 1.776 Å. The mean Fe-Cp(centroid) and Fe-Ph(centroid distances are 1.658 and 1.530 Å respectively.

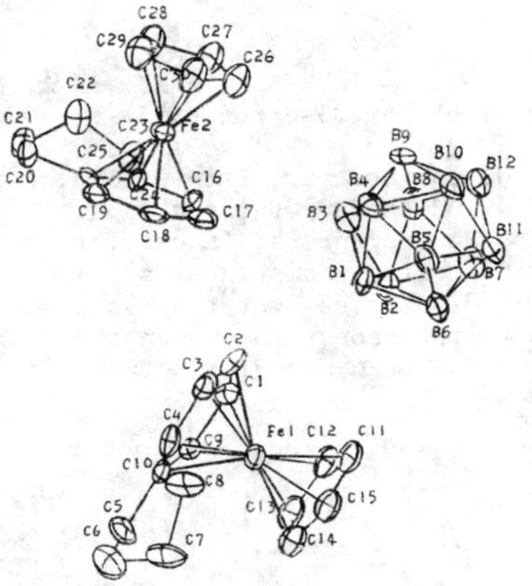

Fig. 1. The ions in the asymmetric unit of $C_{30}H_{46}B_{12}Fe_2$.

TETRAKIS(TRIMETHYLPHOSPHINE)COBALT TETRAPHENYLBORATE
$C_{36}H_{56}BCoP_4$ (I)

(η^6-PHENYL)TRIPHENYLBORATOBIS(TRIMETHYLPHOSPHINO)COBALT
$C_{30}H_{38}BCoP_2$ (II)

L.C.A. DE CARVALHO, M. DARTIGUENAVE, Y. DARTIGUENAVE and A.L. BEAUCHAMP, 1984. J. Am. Chem. Soc., 106, 6848-6849.

I. Monoclinic, $P2_1/c$, a = 16.190, b = 12.422, c = 20.90 Å, β = 111.87°, Z = 4. Mo radiation, R = 0.037 for 2167 reflexions.

II. Monoclinic, $P2_1/c$, a = 11.457, b = 13.804, c = 19.339 Å, β = 115.21°, Z = 4. Cu radiation, R = 0.051 for 2857 reflexions.

The Co atom in (I) has distorted tetrahedral geometry with normal Co-P bonds (mean 2.227 Å) and mean P-Co-P 109.0°. Molecule (II) (Fig. 1) has a phenyl of BPh_4 π-bonded to Co. The mean Co-P distance is 2.167(3) Å, the Co-centroid distance is 1.602 Å and mean Co-C 2.118 Å. The C-C distances in the π-bonded ring (mean 1.412 Å) are longer than those in the free rings (mean 1.386 Å).

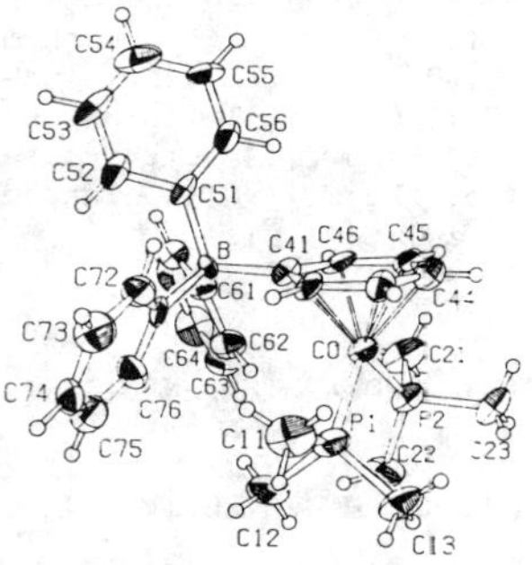

Fig. 1. $C_{30}H_{38}BCoP_2$: a perspective view.

TRI-μ-METHOXOBIS[(η-BENZENE)RUTHENIUM(II)] TETRAPHENYLBORATE

$C_{39}H_{41}BO_3Ru_2$ $[Ru_2(\eta\text{-}C_6H_6)_2(OMe)_3]^+[BPh_4]^-$

R.O. GOULD, T.A. STEPHENSON and D.A. TOCHER, 1984. J. Organometal. Chem., 263, 375-384.

Monoclinic, $P2_1/n$, a = 11.725, b = 15.573, c = 18.739 Å, β = 103.29°, D_m = 1.51, Z = 4. Mo radiation, R = 0.052 for 2546 reflexions.

The complex consists of a binuclear, tri-methoxo-bridged structure (Fig. 1). The Ru...Ru distance is 3.005 Å and the Ru-O distances lie between 2.044 and 2.080 Å, with an average of 2.060 Å. The Ru-C(arene) distances for the two rings are 2.155 and 2.158 Å while the metal-ring centroid distances are 1.643 and 1.646 Å respectively. The two rings are partially disordered over two positions, the major sites (80% occupancy) are twisted 26° with respect to one another. The Ru-O-Ru angles lie in the range 92.6 to 94.6° and the O-Ru-O angles between 71.4 and 74.9°.

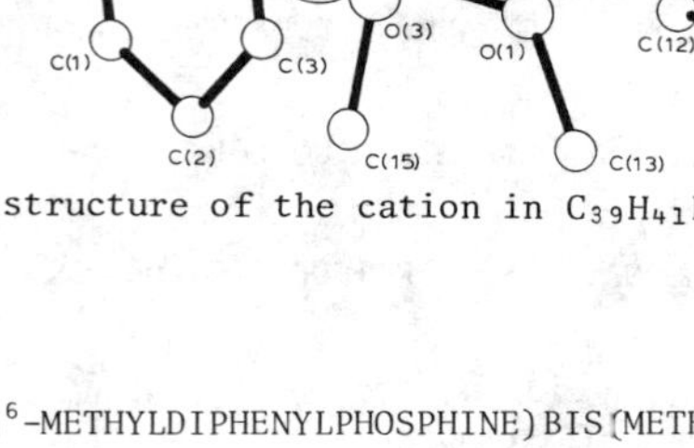

Fig. 1. Molecular structure of the cation in $C_{39}H_{41}BO_3Ru_2$.

(t-BUTYL ISOCYANIDE)(η^6-METHYLDIPHENYLPHOSPHINE)BIS(METHYLDIPHENYLPHOSPHINO)MOLYBDENUM

$C_{44}H_{48}MoNP_3$

R.L. LUCK, R.H. MORRIS and J.F. SAWYER, 1984. Organometallics, 3, 247-255.

Monoclinic, $P2_1$, a = 9.256, b = 11.497, c = 18.782 Å, β = 103.92°, Z = 2. Mo radiation, R = 0.064 for 2755 reflexions.

The geometry about the Mo atom is distorted octahedral with the σ-bonded $PMePh_2$ and $CNCMe_3$ ligands occupying one face of the octahedron and the η^6-bonded ring the other face (Fig. 1). The Mo-P distances are 2.416 and 2.420 Å and the P-Mo-P angle is 94.0° with P-Mo-C(4) angles of 88.5 and 83.5°. The Mo-C(arene) distances range from 2.24 to 2.32 Å with an average of 2.28 Å. The Mo-C(4) distance is 1.98 Å and the C(4)-N bond length 1.22 Å. The bond angles of interest in this part of the structure are C(4)-N-C(5) (150°), and Mo-C(4)-N (178°). Atom P(3) is bent away from the plane of the arene ring and from the metal atom by some 0.265 Å.

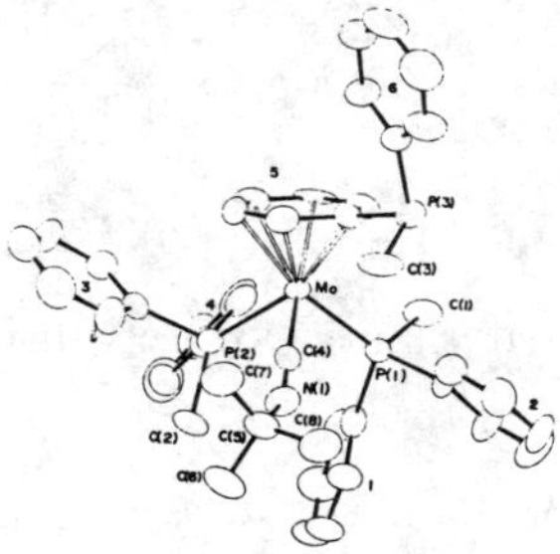

Fig. 1. Molecular structure of $C_{44}H_{48}MoNP_3$.

(μ_4-OXO)TETRAKIS((μ_2-HYDROXO)(η^6-BENZENE)RUTHENIUM) BIS(TETRAPHENYLBORATE) ACETONE SOLVATE

$C_{72}H_{68}B_2O_5Ru_4,C_6H_{12}O_2$ (I) $[Ru_4(\eta\text{-}C_6H_6)_4(\mu_2\text{-}OH)_4(\mu_4\text{-}O)](BPh_4)_2,2(CH_3)_2CO$

(μ_4-OXO)TETRAKIS((μ_2-HYDROXO)(η^6-BENZENE)OSMIUM) BIS(TETRAPHENYLBORATE) ACETONE SOLVATE

$C_{72}H_{68}B_2O_5Os_4,C_6H_{12}O_2$ (II) $[Os_4(\eta\text{-}C_6H_6)_4(\mu_2\text{-}OH)_4(\mu_4\text{-}O)](BPh_4)_2,2(CH_3)_2CO$

TRIS(μ_2-HYDROXO)BIS(η^6-1,3,5-TRIMETHYLBENZENE)-DIRUTHENIUM CHLORIDE TRIHYDRATE

$C_{18}H_{27}ClO_3Ru_2,3H_2O$ (III) $[Ru_2(\eta\text{-}1,3,5\text{-}C_6H_3Me_3)_2(OH)_3]Cl,3H_2O$

R.O. GOULD, C.L. JONES, T.A. STEPHENSON and D.A. TOCHER, 1984. J. Organometal. Chem., 264, 365-378.

I. Monoclinic, $P2_1/c$, a = 13.267, b = 12.227, c = 41.864 Å, β = 98.23°, D_m = 1.50, Z = 4. Mo radiation, R = 0.076 for 3110 reflexions.

II. Monoclinic, $P2_1/c$, a = 13.262, b = 12.226, c = 41.834 Å, β = 98.21°, Z = 4. Mo radiation, R = 0.049 for 2490 reflexions.

III. Rhombohedral, R3m, a = 14.675, c = 9.630 Å, Z = 3. Mo radiation, R = 0.047 for 398 reflexions.

Compounds I and II can be thought of as two binuclear units which share a common bridging O vertex (Fig. 1). The Ru-Ru and Os-Os distances are 3.001 and 3.008 Å, and 3.077 and 3.078 Å respectively. The metal-O(5) distances range from 2.09 to 2.16 Å (Ru) and from 2.13 to 2.16 Å (Os) with angles about O(5) of 89.8 to 124.0° (Ru) and 90.9 to 120.6° (Os). The metal-O(hydroxy) distances range from 2.047 to 2.118 Å (Ru) and from 2.055 to 2.147 Å (Os). The tetraphenylborate anions have the expected geometry with B-C distances ranging from 1.61 to 1.78 Å. The distance between the Ru atoms in III is 2.989 Å, ruling out any metal...metal bond (Fig. 1). The molecules stack in columns up the c axis and have crystallographic threefold symmetry. The metal atoms are bridged by three OH groups with Ru(2)-OH 2.11 Å and Ru(1)-OH 2.09 Å. The Ru(1)-O-Ru(2) angle is 90.9° compared with M-OH-M angles in I and II which range from 90.4 to 95.7°.

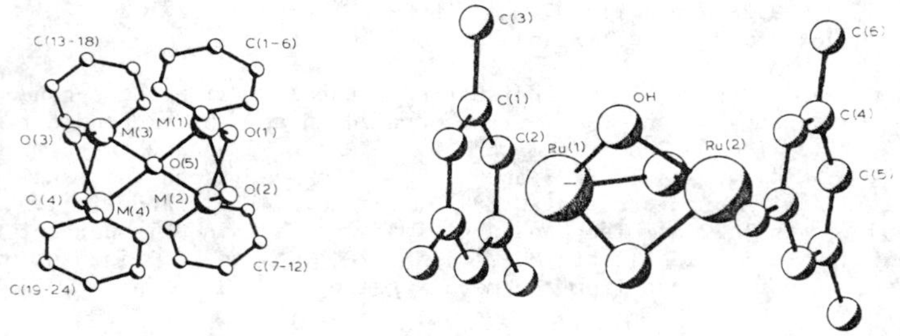

Fig. 1. View of the cations in $C_{72}H_{68}B_2O_5M_4$ (where M = Ru, Os) (left), and molecular structure of the cation in $C_{18}H_{27}ClO_3Ru_2$ (right).

BIS(1-ISOCYANOBUTANE)BIS[μ-[(η^6-PHENYL)DIPHENYLPHOSPHINE-P]]BIS(TRIPHENYLPHOSPHINE)-DIMOLYBDENUM

$C_{82}H_{78}Mo_2N_2P_4$ $[Mo(PPh_3)(PPh_3)(CN(CH_2)_3Me)]_2$

R.L. LUCK, R.H. MORRIS and J.F. SAWYER, 1984. Organometallics, <u>3</u>, 1009-1014.

Triclinic, P$\bar{1}$, a = 15.141, b = 15.300, c = 16.912 Å, α = 63.50, β = 80.14, γ = 79.37°, Z = 2. Mo radiation, R = 0.059 for 4147 reflexions.

The crystal contains discrete binuclear molecules of C_1 symmetry which are racemic with respect to the chiral Mo centres. The six-membered Mo(1)-C(321)-P(3)-Mo(2)-C(231)-P(2) ring is in a flattened chair conformation (Fig. 1) with Mo(1) and Mo(2) 0.716 and 0.529 Å respectively, out of the P(2),P(3),C(231),C(321) plane. The Mo...Mo separation is 4.804 Å. Averaged principal dimensions are Mo-P 2.465, Mo-C(ring) 2.29, Mo-C(isocyanide) 1.96, and C-N 1.18 Å. The isocyanide ligands have different C-N-C angles (C(11)-N(12)-C(13) 156, C(21)-N(22)-C(23) 163°) because they occupy axial and equatorial positions, respectively, on a Mo-C-P-Mo-C-P cyclohexane-like ring.

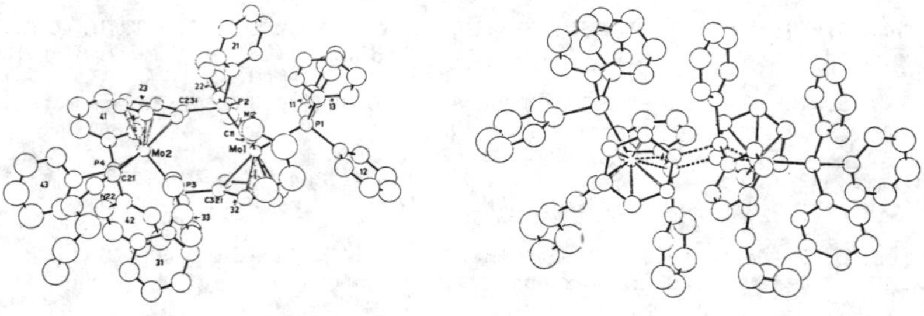

Fig. 1. Molecular structure of $C_{82}H_{78}Mo_2N_2P_4$ (left), and viewed at right angles to the η^6-bonded rings showing the chair conformation of the Mo-C-P-Mo-C-P ring (right).

(μ_3-TRIFLUOROMETHANESULPHONATO)(η^2-CYCLOHEXENE)COPPER(I)

$C_7H_{10}CuF_3O_3S$ $(C_6H_{10})CuOSO_2CF_3$

P.J.J.A. TIMMERMANS, A. MACKOR, A.L. SPEK and B. KOJIĆ-PRODIĆ, 1984. J. Organometal. Chem., <u>276</u>, 287-295.

Triclinic, P$\bar{1}$, a = 10.288, b = 10.412, c = 11.059 Å, α = 65.81, β = 81.25, γ = 70.45°, Z = 4. Cu radiation, R = 0.062 for 2511 reflexions.

The asymmetric unit contains two independent formula units which are joined to form a tetramer (through the inversion centre) (Fig. 1). Both Cu ions are four-coordinate with geometry intermediate between trigonal-pyramidal and tetrahedral. The ligands involved are three triflate groups and one π-bonding cyclohexene. Relevant metal-ligand distances are: Cu(11)-O(23) 2.048, Cu(11)-O(23A) 2.587, Cu(11)-C(11) 2.076, Cu(11)-C(12) 2.060, Cu(11)-O(13) 2.043, Cu(21)-O(12B) 2.061, Cu(21)-C(22) 2.070, Cu(21)-O(11) 2.385, Cu(21)-O(22) 2.110, and Cu(21)-C(21) 2.091 Å. The Cu ions in the tetramer are joined by O-S-O bridges of the triflate anion. The O-S distances range from 1.429 to 1.460 Å and the C=C bond lengths of the coordinated cyclohexenes are 1.33 Å (C(11)=C(12)) and 1.37 Å (C(21)=C(22)). The C(11)→C(16) ring has a symmetrical half-chair conformation whilst that of the C(21)→C(26) ring is irregular half-chair.

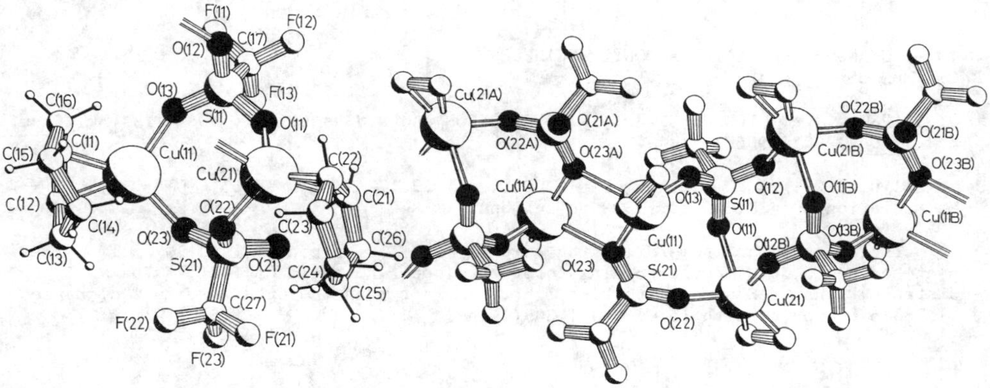

Fig. 1. The asymmetric unit of the $C_7H_{10}CuF_3O_3S$ complex (left), and part of the infinite chain running in the a direction (right). The cyclohexene ring is represented by the double bond only.

(η[4]-HEXAFLUORODEHYDROCYCLOOCTATETRAENE)BIS(TRICARBONYLCOBALT)

$C_8Co_2F_6O_6$ (I) $[Co(CO)_3]_2(C_8F_6)$

$C_{48}H_{30}Co_2F_6O_4P_2$ (II) $[Co(CO)_2(PPh_3)]_2(C_8F_6)$

S.J. DOIG, R.P. HUGHES, R.E. DAVIS, S.M. GADOL and K.D. HOLLAND, 1984. Organometallics, 3, 1921-1922.

I. Monoclinic, C2/c, a = 26.808, b = 6.953, c = 17.081 Å, β = 98.10°, Z = 8. Mo radiation, R = 0.056 for 3145 reflexions (at -110°C).

II. Monoclinic, C2/c, a = 20.670, b = 10.038, c = 21.557 Å, β = 108.13°, D_m = 1.497, Z = 4. Mo radiation, R = 0.055 for 3270 reflexions (at -110°C).

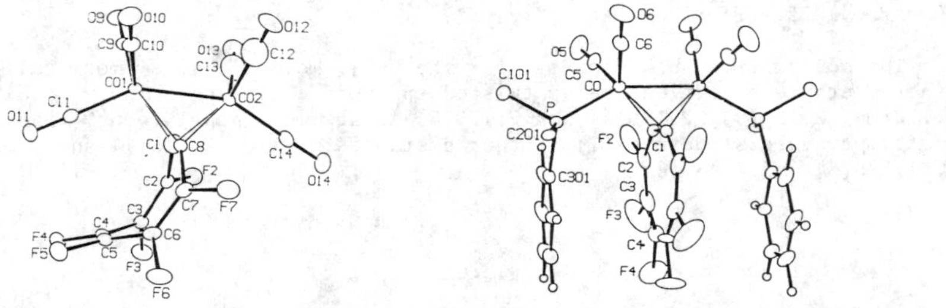

Fig. 1. Molecular structures of $C_8Co_2F_6O_6$ (left), and of $C_{48}H_{30}Co_2F_6O_4P_2$ (right).

In I, the hexafluorodehydrocyclooctatetraene is bound via its triple bond to the hexacarbonyldicobalt fragment (Fig. 1), and the non-planarity of the tub-shaped ring is indicated by the interplanar angles of 33.0° between planes C(1),C(2),C(7),C(8) and C(2),C(3),C(6),C(7) and 42.4° between planes C(2),C(3),C(6),C(7) and C(3),C(4), C(5),C(6). In II, the molecule occupies a crystallographic twofold axis and the C_8F_6 ring is planar to within 0.04 Å. It is sandwiched between phenyl rings of the two phosphine ligands, 3.2 Å from each ring. Selected bond lengths are: in I, Co(1)-Co(2) 2.473, Co(1)-C(1) 1.972, Co(1)-C(8) 1.958, Co(2)-C(1) 1.930, Co(2)-C(8) 1.936, C(1)-C(8) 1.371 Å, and in II, Co-Co' 2.467, Co - C(1) 1.940, Co-C(1') 1.969, C(1)-C(1') 1.348 Å.

(η^4-2-CYANOTHIOPHENE)TRICARBONYLMANGANESE
$C_8H_4MnNO_3S$

D.A. LESCH, J.W. RICHARDSON Jr., R.A. JACOBSON and R.J. ANGELICI, 1984. J. Am. Chem. Soc., 106, 2901-2906.

Monoclinic, $P2_1/c$, a = 12.008, b = 6.6935, c = 12.172 Å, β = 95.63°, D_m = 1.65, Z = 4. Mo radiation, R = 0.059 for 1598 reflexions.

The molecule and some dimensions are in Fig. 1. Carbon C7 is 0.597 Å from the plane of C4,C5,C6,S (corresponding to an envelope flap angle of 147°). The Mn-C distance trans to S appears to be slightly shorter than the other two Mn-C distances (1.783(6) compared with 1.803 and 1.805(6) Å).

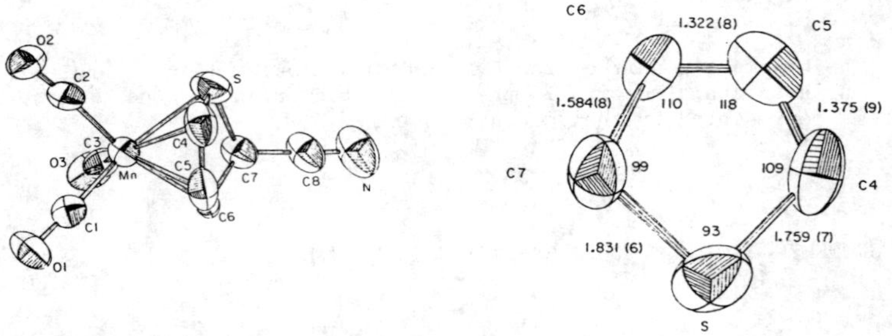

Fig. 1. A view of $C_8H_4MnNO_3S$ and some dimensions.

DICHLORO(η^4-CYCLOOCTA-1,5-DIENE)PLATINUM(II)
$C_8H_{12}Cl_2Pt$ $(C_8H_{12})PtCl_2$

A. SYED, E.D. STEVENS and S.G. CRUZ, 1984. Inorg. Chem., 23, 3673-3674.

Orthorhombic, $P2_12_12_1$, a = 6.818, b = 10.664, c = 12.423 Å, Z = 4. Mo radiation, R = 0.028 for 2032 reflexions (at 110(2) K).

The molecule is shown in Fig. 1. The C_8H_{12} molecule is symmetrically coordinated with the ethylene portions being twisted to avoid eclipsing. The Pt-C distances (2.172(6), 2.154(5), 2.176(6) and 2.178(6) Å) are more nearly equal than those of the room temperature structure (1). Other distances: Pt-Cl 2.315(1) and 2.309(1) Å, C=C 1.375(8) and 1.387(8) Å.

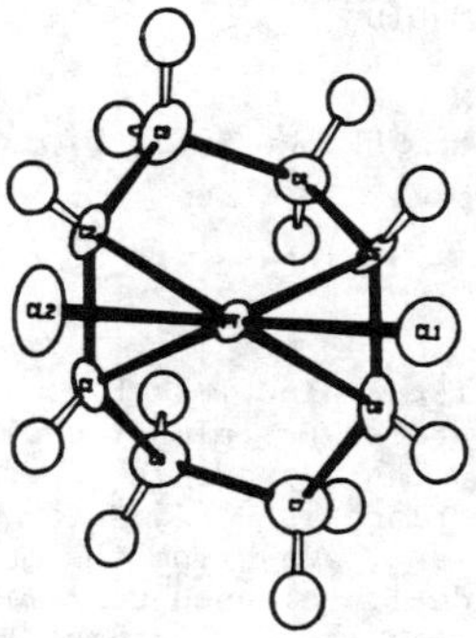

Fig. 1. The $C_8H_{12}Cl_2Pt$ molecule.

1. A.B. GOEL, S. GOEL and D. van der VEER, 1982. Inorg. Chim. Acta, 65, L205-L206.

DIBROMODICARBONYL(η[4]-NORBORNADIENE)TUNGSTEN(II)
$C_9H_8Br_2O_2W$ (I) $WBr_2(CO)_2(C_7H_8)$

DIBROMODICARBONYLBIS(TRIPHENYLPHOSPHINE)TUNGSTEN(II)
$C_{38}H_{30}Br_2O_2P_2W$ (II) $WBr_2(CO)_2[P(C_6H_5)_3]_2$

F.A. COTTON and J.H. MEADOWS, 1984. Inorg. Chem., 23, 4688-4693.

I. Monoclinic, $P2_1/n$, a = 7.506, b = 12.643, c = 12.296 Å, β = 99.60°, Z = 4. Mo radiation, R = 0.032 for 1431 reflexions.

II. Triclinic, $P\bar{1}$, a = 9.418, b = 10.021, c = 19.898 Å, α = 90.54, β = 112.80, γ = 92.02°, Z = 2. Mo radiation, R = 0.064 for 1968 reflexions.

Both structures are shown in Fig. 1. Compound I has pseudo-octahedral geometry with its CO ligands trans to each other. I has relatively short W-Br bonds (2.493(1) and 2.489(1) Å); other distances in I are: W-CO 2.05(1) and 2.03(1), W-C(C_7H_8) 2.26(1)-2.30(1) Å. Compound II has an overall geometry greatly distorted from octahedral (Br-W-Br 82.0(1)°, P-W-P 128.3(3)°, C-W-C 110(1)°). Bond lengths for II: W-Br 2.565(3) and 2.571(4), W-P 2.481(9) and 2.486(9), W-C 1.95(4) and 1.94(4) Å.

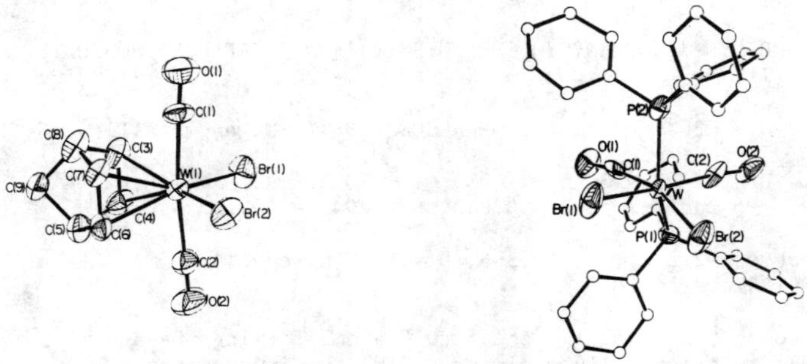

Fig. 1. The structures of $WBr_2(CO)_2(C_7H_8)$ (left) and $WBr_2(CO)_2[P(C_6H_5)_3]_2$ (right).

TETRACARBONYL(1-METHYLBORINATO)VANADIUM

$C_{10}H_8BO_4V$ $(\pi-C_5H_5BCH_3)V(CO)_4$

G.E. HERBERICH, W. BOVELETH, B. HESSNER, W. KOCH, E. RAABE and D. SCHMITZ, 1984. J. Organometal. Chem., 265, 225-235.

Orthorhombic, Pbca, a = 13.970, b = 12.760, c = 12.945 Å, Z = 8. Mo radiation, R = 0.036 for 772 reflexions.

The crystal consists of discrete molecules (Fig. 1) separated by van der Waals contact distances. The $V(CO)_4$ fragment has almost undistorted C_{4v} symmetry with virtually linear V-C-O groupings. The C atoms in the six-membered ring lie in a plane (maximum deviation from the plane is 0.015 Å for C(8)) and the B-containing moiety is tilted 6.5° from this plane, away from the metal. The $V(CO)_4$ fragment is rotated by 10.4° relative to an ideal, eclipsed conformation where two CO groups are present in the plane through the B, V and C(4) atoms (Fig. 1). The four V-C(carbonyl) distances are 1.930, 1.934, 1.924 and 1.927 Å, and the V-C (ring) distances are 2.336, 2.296, 2.265, 2.275 and 2.347 Å. The V-B distance is 2.496 Å, and other distances involving B are: B-C(5) 1.50, B-C(10) 1.59 and B-C(9) 1.50 Å.

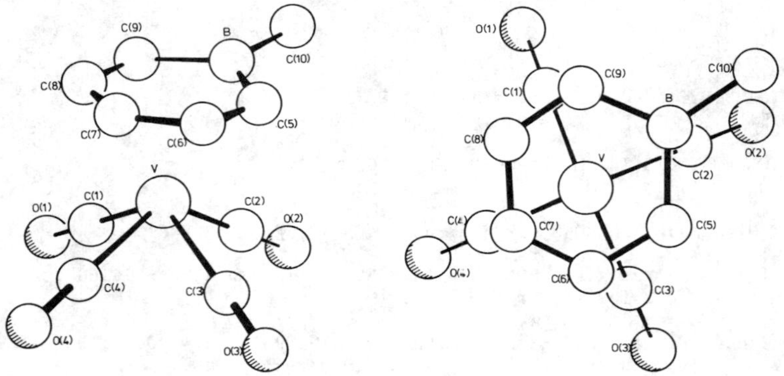

Fig. 1. Molecular structure of $C_{10}H_8BO_4V$ (left), and viewed perpendicular to the ring (right).

TRICARBONYL(2-METHOXYCYCLOHEXADIENYL)IRON HEXAFLUOROPHOSPHATE

$C_{10}H_9F_6FeO_4P$ (I) $[\eta^5-(CH_3OC_6H_6)Fe(CO)_3]^+PF_6^-$

TRICARBONYL(1-METHYL-4-METHOXYCYCLOHEXADIENYL)IRON HEXAFLUOROPHOSPHATE

$C_{11}H_{11}F_6FeO_4P$ (II) $[\eta^5-(CH_3OC_7H_8)Fe(CO)_3]^+PF_6^-$

O. EISENSTEIN, W.M. BUTLER and A.J. PEARSON, 1984. Organometallics, 3, 1150-1157.

I. Triclinic, P$\bar{1}$, a = 9.923, b = 7.059, c = 10.831 Å, α = 108.82, β = 90.74, γ = 76.52°, Z = 2. Mo radiation, R = 0.049 for 1294 reflexions.

II. Orthorhombic, $P2_12_12_1$, a = 7.556, b = 13.556, c = 14.250 Å, Z = 4. Mo radiation, R = 0.051 for 1357 reflexions.

In both compounds (Fig. 1), the five C atoms forming the dienyl part, C(4),C(5), C(7),C(8),C(9) are practically coplanar. C(6) is situated above the dienyl plane on the opposite side of the metal (the perpendicular distance from C(6) to the dienyl plane is 0.549 Å in I and 0.576 Å in II). One CO group of $Fe(CO)_3$ eclipses the methylene group in both compounds, and the angles between the planes C(4),C(5),C(7), C(8),C(9) and C(5),C(6),C(7) are 37.5° in I and 39.1° in II. In I, the distances Fe-C(5), Fe-C(7), Fe-C(4) and Fe-C(8) are respectively 2.172, 2.186, 2.190 and 2.124 Å whilst the corresponding distances in II are 2.165, 2.245, 2.172 and 2.122 Å.

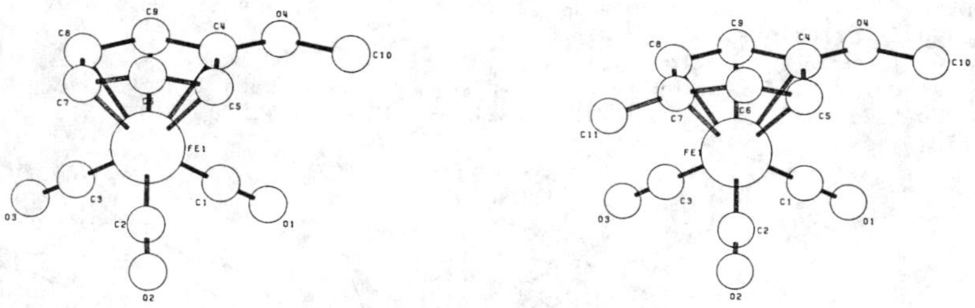

Fig. 1. Molecular structures of $C_{10}H_9F_6FeO_4P$ and $C_{11}H_{11}F_6FeO_4P$.

TRICARBONYL(η-METHOXYTROPYLIUM)CHROMIUM TETRAFLUOROBORATE

$C_{11}H_9BCrF_4O_4$ $[(C_7H_6OCH_3)Cr(CO)_3]^+BF_4^-$

U. BEHRENS, J. KOPF, K. LAL and W.E. WATTS, 1984. J. Organometal. Chem., 276, 193-198.

Triclinic, P$\bar{1}$, a = 7.403, b = 9.567, c = 10.384 Å, α = 89.10, β = 73.32, γ = 74.31°, Z = 2. Mo radiation, R = 0.045 for 2914 reflexions.

The seven-membered ring in the cation (Fig. 1) is approximately planar and is symmetrically-bonded to the $Cr(CO)_3$ group. The ring C-C bond lengths deviate by less than 0.02 Å from an average value of 1.40 Å and the internal bond angles by less than 2.5° from the value (128.6°) for a regular heptagon. The ring C-H bonds are bent by ca. 7° from the mean ring plane towards the metal atom. The Cr-C(ring) distances range from 2.217 to 2.263 Å and the Cr-C(carbonyl) distances are 1.908, 1.894 and 1.903 Å. The distances C(1)-O(4) and O(4)-C(8) are respectively 1.338 and 1.443 Å. The orientations of the tetrahedral BF_4^- anions in the crystal lattice are disordered.

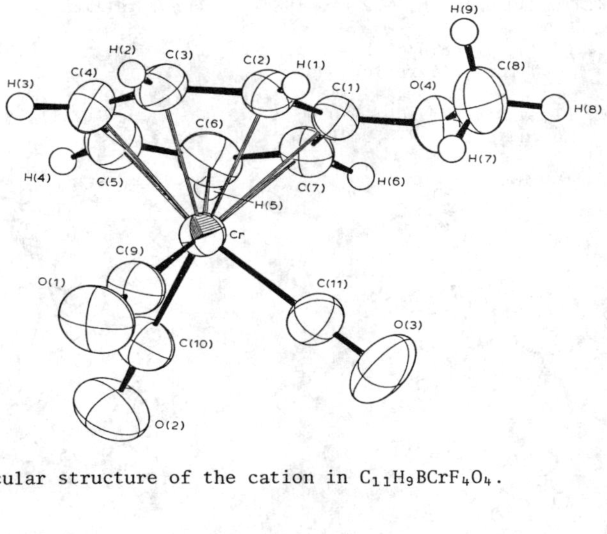

Fig. 1. Molecular structure of the cation in $C_{11}H_9BCrF_4O_4$.

(η^3-ALLYL)(η^5-1-t-BUTYL-2-METHYL-1,2-AZABOROLINYL)DICARBONYLMOLYBDENUM

$C_{13}H_{20}BMoNO_2$

G. SCHMID, U. HOHNER, D. KAMPMANN, F. SCHMIDT, D. BLASER and R. BOESE, 1984. Chem. Ber., 117, 672-681.

Orthorhombic, Pbca, a = 16.974, b = 14.418, c = 12.194 Å, Z = 8. Mo radiation, R = 0.044 for 2092 reflexions.

The molecular conformation is shown in Fig. 1. Some bond lengths are: Mo-N 2.327(3), Mo-B 2.524(6), Mo-C(Cp) 2.252-2.241(3), Mo-C(O) 1.930, 1.946(5), Mo-C(allyl) 2.234-2.352(6) Å.

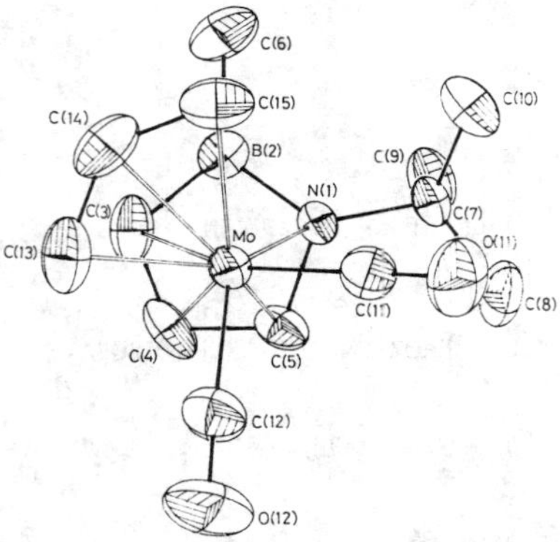

Fig. 1. A view of $C_{13}H_{20}BMoNO_2$.

CARBONYL[(1,2-η)-1,3-CYCLOOCTADIENE](η⁵-2,4-CYCLOPENTADIEN-1-YL)NITROSYLMOLYBDENUM
$C_{14}H_{17}MoNO_2$ (I) CpMo(NO)(CO)(C_8H_{12})

CARBONYL[(2,3-η)-2-CYCLOOCTEN-1-OL](η⁵-2,4-CYCLOPENTADIEN-1-YL)NITROSYLMOLYBDENUM
$C_{14}H_{19}MoNO_3$ (II) CpMo(NO)(CO)($C_8H_{13}OH$)

J.W. FALLER, K.H. CHAO and H.H. MURRAY, 1984. Organometallics, 3, 1231-1240.

I. Monoclinic, $P2_1/n$, a = 8.836, b = 7.308, c = 21.474 Å, β = 98.32°, Z = 4. Mo radiation, R = 0.020 for 1746 reflexions.

II. Triclinic, $P\bar{1}$, a = 8.477, b = 9.050, c = 9.906 Å, α = 87.11, β = 85.60, γ = 72.41°, Z = 2. Mo radiation, R = 0.019 for 2093 reflexions.

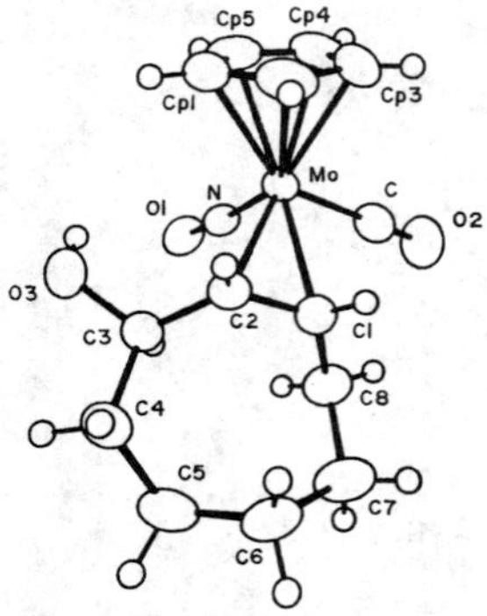

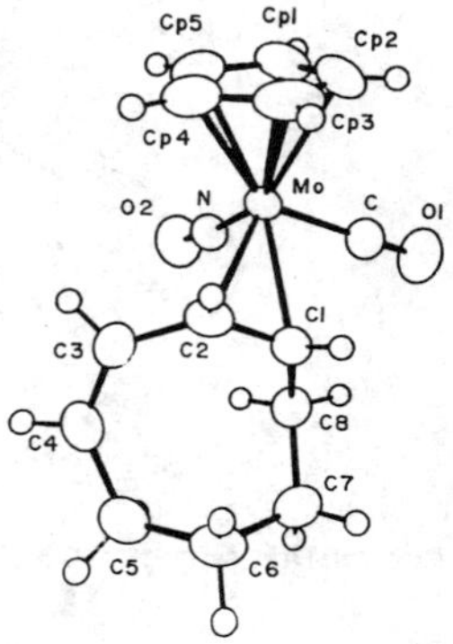

Fig. 1. Molecular structures of $C_{14}H_{17}MoNO_2$ (left), and of $C_{14}H_{19}MoNO_3$ (right).

In compound II (Fig. 1) the important features are: that the OH is cis to the nitrosyl, the OH is β to the Mo, and that the eight-membered ring is in the exo conformation. In I (Fig. 1) the structure of the cyclooctadiene complex was determined in order to establish which proton had been removed during proton abstraction, in this case it is the one at C(4) (cis to the nitrosyl). The distances C(1)-C(2) (1.396 Å), C(2)-C(3) (1.492 Å), C(3)-C(4) (1.331 Å) and C(4)-C(5) (1.498 Å) confirm this. The Mo-C(1) and Mo-C(2) distances in II are 2.344 and 2.292 Å respectively, and in I are 2.353 and 2.295 Å respectively. The Mo-N distances in I and II are 1.794 and 1.786 Å, and the Mo-C(carbonyl) distances are 1.973 and 1.997 Å respectively. The bond angles in I (C(2)-C(1)-C(8), C(1)-C(2)-C(3), C(2)-C(3)-C(4), C(3)-C(4)-C(5) and C(4)-C(5)-C(6)) are 123.3, 126.4, 130.1, 132.8 and 118.9° respectively. In II, angles C(2)-C(1)-C(8) and C(1)-C(2)-C(3) are 122.9 and 125.2° respectively, the other angles in the ring ranging from 104.8 to 116.4°.

1-(η^8-CYCLOOCTATETRAENE)-2,3-DIETHYL-1-TITANADICARBAHEPTABORANE
$C_{14}H_{22}B_4Ti$ (I) $(\eta^8-C_8H_8)Ti[(C_2H_5)_2C_2B_4H_4]$

1-(η^8-CYCLOOCTATETRAENE)-2,3-DIETHYL-1-VANADA-2,3-DICARBAHEPTABORANE
$C_{14}H_{22}B_4V$ (II) $(\eta^8-C_8H_8)V[(C_2H_5)_2C_2B_4H_4]$

1-(η^7-CYCLOHEPTATRIENE)-2,3-DIETHYL-1-CHROMADICARBAHEPTABORANE
$C_{13}H_{21}B_4Cr$ (III) $(\eta^7-C_7H_7)Cr[(C_2H_5)_2C_2B_4H_4]$

R.G. SWISHER, E. SINN and R.N. GRIMES, 1984. Organometallics, 3, 599-605.

I. Orthorhombic, $Pna2_1$, a = 14.309, b = 12.660, c = 8.476 Å, Z = 4. R = 0.049 for 1902 reflexions.

II. Orthorhombic, $Pna2_1$, a = 14.264, b = 12.595, c = 8.449 Å, Z = 4. R = 0.047 for 1271 reflexions.

III. Orthorhombic, $Pna2_1$, a = 14.334, b = 12.230, c = 8.356 Å, Z = 4. R = 0.065 for 1289 reflexions.

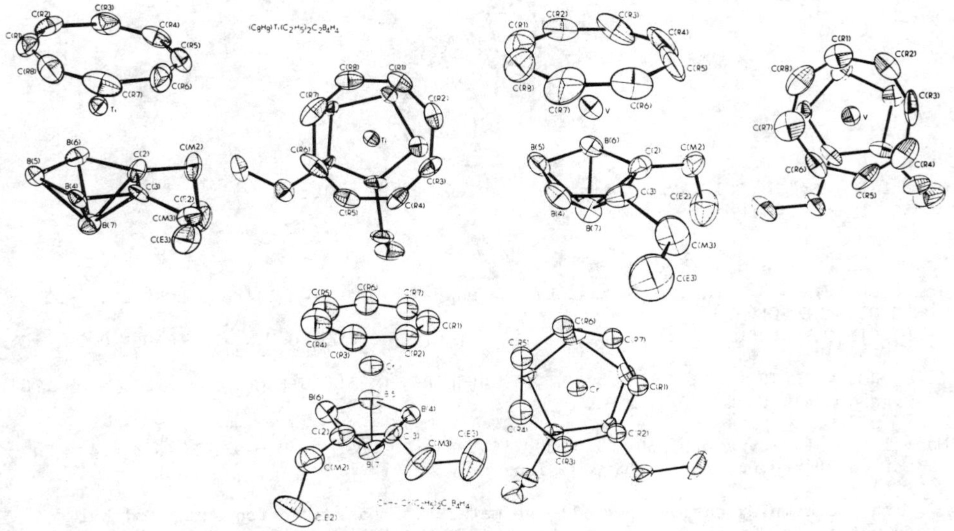

Fig. 1. Molecular structures of $C_{14}H_{22}B_4Ti$ (top left), $C_{14}H_{22}B_4V$ (top right), and $C_{13}H_{21}B_4Cr$ (bottom).

In I the COT ring is planar within experimental error and is nearly parallel to the C_2B_3 ring of the carborane ligand (dihedral angle 2.8°) (Fig. 1). The Ti atom is equidistant from the eight ring C atoms with a mean Ti-C distance of 2.290 Å. The H atoms in the ring are bent out of the ring plane towards the metal by an average of 4.7°. The Ti-carborane ring centroid distance is 1.916 Å. Distances within the carborane ring are: C(2)-C(3) 1.481, C(2)-B(6) 1.55, C(3)-B(4) 1.58, B(4)-B(5) 1.68 and B(5)-B(6) 1.66 Å. II is essentially isostructural (but not isomorphous) with I with a planar COT ring symmetrically bonded to the metal. The V atom is closer to the carborane ligand ring (1.830 Å) but the V-C_8H_8 distance is almost the same as the Ti-C_8H_8 distance. The C(2)-C(3) distance in II is longer (1.488 Å). In III, the C_7H_7 ring approximates a regular heptagon with mean C-C distance 1.426 Å (1.402 Å in I, 1.394 Å in II) and is almost planar and parallel to the C_2B_3 ring (dihedral angle 2.76°). The average Cr-C(C_7H_7) distance is 2.188 Å and the C(2)-C(3) distance is 1.492 Å.

1-CHLORO-2-(1-2-η-CYCLOHEXENE)-μ-[HYDROTRI(1-PYRAZOLYL)BORATO-N:N',N'']-DICOPPER(I)
$C_{15}H_{20}BClCu_2N_6$

J.S. THOMPSON and J.F. WHITNEY, 1984. Acta Cryst., C40, 756-759.

Triclinic, P$\bar{1}$, a = 9.951, b = 11.737, c = 9.130 Å, α = 100.51, β = 115.80, γ = 79.15°, Z = 2. Mo radiation, R = 0.026 for 3582 reflexions (at 173K).

In the complex (Fig. 1) one of the copper(I) ions is coordinated in a trigonal planar arrangement to two pyrazole N atoms, with Cu-N 1.950(1) and 1.966(1) Å, and a cyclohexene molecule with Cu-C 2.039 and 2.022(2) Å. The other copper(I) ion is coordinated linearly to a pyrazole N atom and a Cl^- ion, with Cu-N 1.904(1) and Cu-Cl 2.155(1) Å.

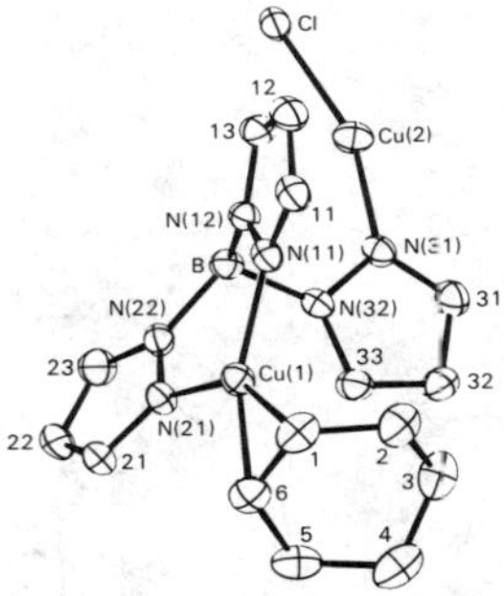

Fig. 1. $C_{15}H_{20}BClCu_2N_6$: perspective view of the molecule.

(η^4-CYCLOOCTA-1,5-DIENE)(BIS(PYRAZOLYL)METHANE-N,N')RHODIUM(I) PERCHLORATE 1,2-DICHLOROETHANE SOLVATE
$C_{15}H_{20}ClN_4O_4Rh,0.5C_2H_4Cl_2$ $[Rh(C_8H_{12})(CH_2(C_3H_3N_2)_2)]^+ClO_4^-,0.5C_2H_4Cl_2$

L.A. ORO, M. ESTEBAN, R.M. CLARAMUNT, J. ELGUERO, C. FOCES-FOCES and F.H. CANO, 1984. J. Organometal. Chem., 276, 79-97.

Monoclinic, $P2_1/c$, a = 9.5602, b = 20.7383, c = 10.0692 Å, β = 92.130°, Z = 4. Cu radiation, R = 0.061 for 2436 reflexions.

In the complex cation (Fig. 1) the metal atom has distorted square planar coordination and is bonded to two N atoms of the $CH_2(pz)_2$ group and to the olefinic bonds of the cyclooctadiene ligand. The Rh-N distances are 2.111 Å (N(2)), and 2.097 Å (N(7)) and the distances of the midpoints of bonds C(11)-C(12) and C(15)-C(16) to the metal atom are 2.015 and 2.007 Å respectively. The cis-angles about Rh range from 87.4 to 92.1°. The COD ring has a twisted boat conformation with the following bond distances: C(11)-C(12) 1.39, C(12)-C(13) 1.51, C(13)-C(14) 1.46, C(14)-C(15) 1.54, C(15)-C(16) 1.36, C(16)-C(17) 1.52, C(17)-C(18) 1.50 and C(18)-C(11) 1.51 Å. The angle between the planes of the two pyrazolyl rings is 47.1°. The solvent molecule has crystallographic inversion symmetry.

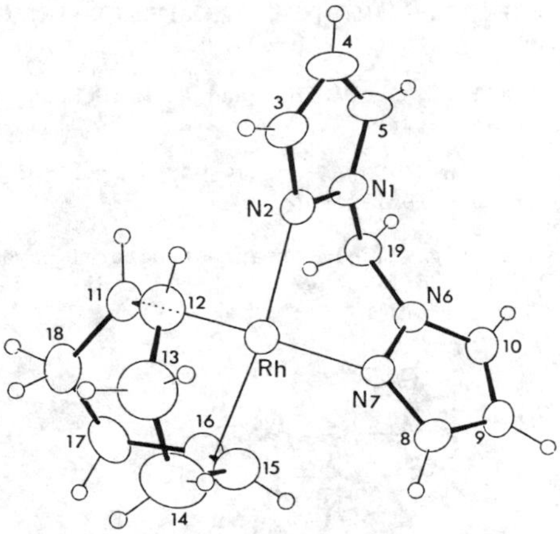

Fig. 1. View of the complex cation in $C_{15}H_{20}ClN_4O_4Rh$.

TRICHLORO[(1,2,3-η)-1-(1,1-DIMETHYLETHYL)-2,3-DIMETHYL-2-CYCLOPROPEN-1-YL](N,N,N',N'-TETRAMETHYL-1,2-ETHANEDIAMINE-N,N')TUNGSTEN

$C_{15}H_{31}Cl_3N_2W$ $W[C_3Me_2(CMe_3)][Me_2N(CH_2)_2NMe_2]Cl_3$

R.R. SCHROCK, S.F. PEDERSEN, M.R. CHURCHILL and J.W. ZILLER, 1984. Organometallics, 3, 1574-1583.

Monoclinic, $P2_1/n$, a = 10.982, b = 13.257, c = 13.957 Å, β = 105.33°, Z = 4. Mo radiation, R = 0.030 for 2285 reflexions.

The complex is approximately octahedral with a tightly and roughly symmetrically-bound η^3-cyclopropenyl ring, a mer arrangement of chloride ligands, and one TMEDA N atom more loosely bound than the other in a position trans to the η^3-C_3 ring (Fig. 1). The three W-Cl distances are similar (W-Cl(1) 2.425, W-Cl(2) 2.399, W-Cl(3) 2.433 Å) but the axial W-N(1) bond length of 2.492 Å is 0.167 Å longer than the equatorial W-N(2) bond length of 2.325 Å. The W-C distances are: W-C(1) 2.040, W-C(2) 2.115 and W-C(3) 2.133 Å resulting in a relatively short (1.920 Å) W-centroid distance, where the centroid is the centroid of the triangle formed by C(1)-C(2)-C(3). The angle N(1)-W-centroid is 173.4° and the Cl(1)-W-Cl(2) angle is 161.4°. The angle Cl(3)-W-N(2) is 154.9°. Other angles of interest are N(1)-W-N(2) 75.8 and C(1)-C(3)-C(2) 66.6°.

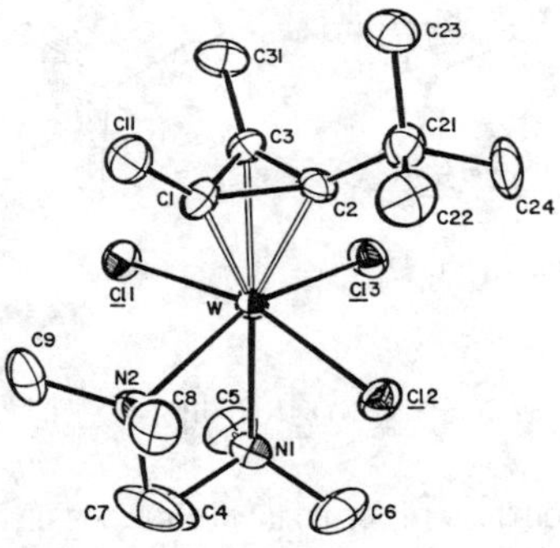

Fig. 1. Molecular structure of $C_{15}H_{31}Cl_3N_2W$.

(η^2-1-(PENTACARBONYL-TUNGSTEN)-3,4-DIMETHYL-2H-PHOSPHOLE)-PENTACARBONYL-TUNGSTEN
$C_{16}H_9O_{10}PW_2$

S. HOLAND, C. CHARRIER, F. MATHEY, J. FISCHER and A. MITSCHLER, 1984. J. Am. Chem. Soc., 106, 826-828.

Monoclinic, $P2_1/c$, a = 15.063, b = 7.306, c = 19.903 Å, β = 104.56°, Dm = 2.33, Z = 4. Cu radiation, R = 0.089 for 2074 reflexions.

The molecule is shown in Fig. 1. Important structural parameters are: W1-C21 2.46(1), W1-P 2.607(4), P-C21 1.78(1), P-W2 2.490(4) Å, W1-P-W2 125.3(1)°.

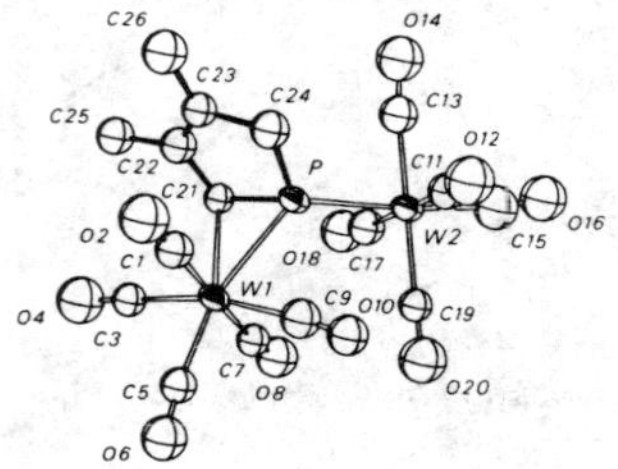

Fig. 1. The $C_{16}H_9O_{10}PW_2$ molecule.

TRICARBONYL(2,6-DI-t-BUTYLPYRIDINE)CHROMIUM(0)
$C_{16}H_{21}CrNO_3$

R.E. SCHMIDT and W. MASSA, 1984. Z. Naturforsch., B39, 213-216.

Monoclinic, $P2_1/n$, a = 6.806, b = 13.833, c = 17.630 Å, β = 96.53°, Z = 4. Mo radiation, R = 0.048 for 1672 reflexions (at 200 K).

The Cr atom is η^6 bonded to the essentially planar pyridine ring (Fig. 1) with Cr-C 2.186-2.221(5), Cr-N 2.212(4) Å. The CO ligands are eclipsed with respect to the 2, 4 and 6 positions of the pyridine ring.

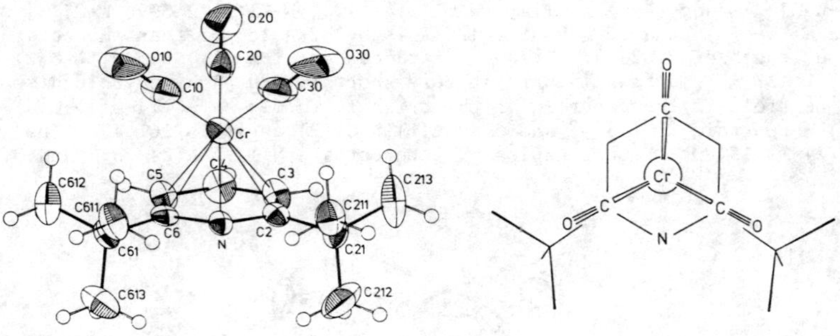

Fig. 1. Two views of $C_{16}H_{21}CrNO_3$.

BIS(η^3-1-t-BUTYL-2-METHYL-1,2-AZABOROLINYL)NICKEL
$C_{16}H_{30}B_2N_2Ni$

G. SCHMID, D. KAMPMANN, U. HOHNER, D. BLASER and R. BOESE, 1984. Chem. Ber., 117, 1052-1060.

Monoclinic, $P2_1/c$, a = 11.408, b = 10.722, c = 7.838 Å, β = 103.03°, Z = 2. Mo radiation, R = 0.061 for 462 reflexions.

The Ni atom lies on an inversion centre and is bonded to only three atoms of the ring as shown in Fig. 1, with Ni-C(3) 2.04(1), Ni-C(4) 1.97(1), Ni-C(5) 2.11(1), Ni...N(1) 2.50(1), Ni...B(2) 2.41(1), N(1)-B(2) 1.47(2) Å.

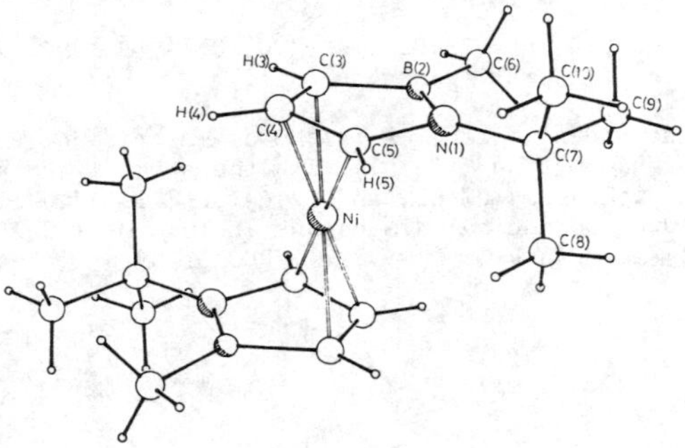

Fig. 1. The $C_{16}H_{30}B_2N_2Ni$ molecule.

BIS[1-t-BUTYL-2-METHYL-(η^5-1,2-AZABOROLINYL)]RUTHENIUM
$C_{16}H_{30}B_2N_2Ru$

G. SCHMID, O. BOLTSCH, D. BLASER and R. BOESE, 1984. Z. Naturforsch., B39, 1082-1087.

Monoclinic, $P2_1/c$, a = 8.495, b = 12.037, c = 18.045 Å, β = 96.24°, Z = 4. Mo radiation, R = 0.037 for 1773 reflexions.

The molecule (Fig. 1) has the five-membered rings in an eclipsed conformation and forming an interplanar angle of 8.6°. Main dimensions are: Ru-N 2.184 and 2.179(4), Ru-B 2.318 and 2.330(6), Ru-C 2.082-2.245(5), Ru-(ring centre) 1.829 and 1.835 Å.

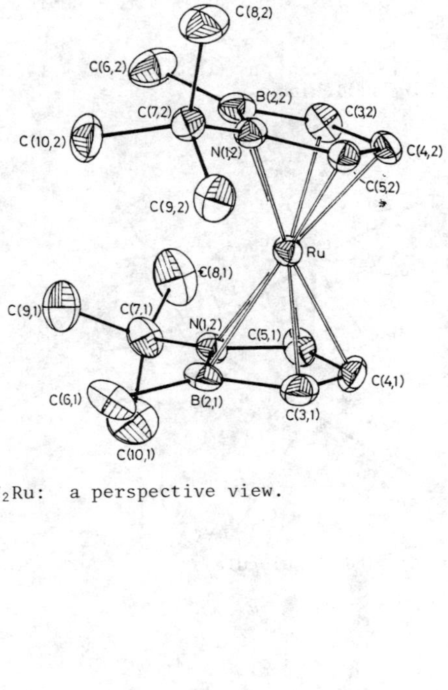

Fig. 1. $C_{16}H_{30}B_2N_2Ru$: a perspective view.

[exo-6-n-BUTYL(η^5-CYCLOHEXADIENYL)]IODOBIS(TRIMETHYLPHOSPHINE)OSMIUM(II)
$C_{16}H_{33}IOsP_2$

H. WERNER, R. WERNER and C. BURSCHKA, 1984. Chem. Ber., 117, 152-160.

Monoclinic, $P2_1/c$, a = 9.019, b = 23.220, c = 20.503 Å, β = 97.22°, Z = 8. Mo radiation, R = 0.039 for 2914 reflexions.

There are two independent molecules in the asymmetric unit. The cyclohexadienyl ring has an envelope conformation (Fig. 1) with the n-butyl group exo. The angle between the planes through atoms C7-C11 and C7,C11,C12 is 44.0° in one molecule and 42.5° in the other. Other principal dimensions include Os-I 2.757(1) and 2.763(1), Os-P 2.287(5)-2.296(4) Å, mean P-Os-P 95.2(2), P-Os-I 89.0(1) to 91.5(1)°.

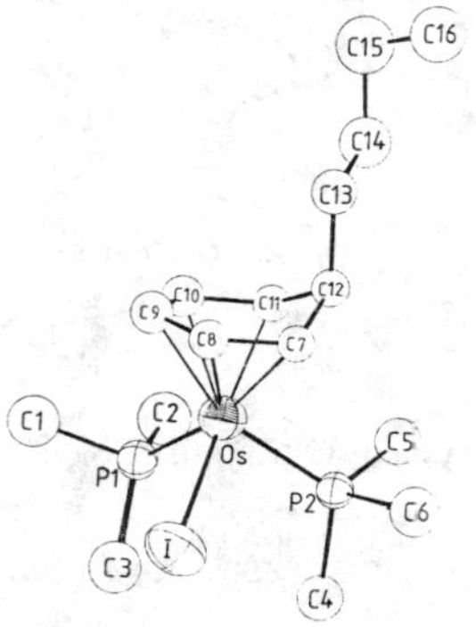

Fig. 1. The $C_{16}H_{33}IOsP_2$ molecule.

(μ_3-HYDRIDO)(μ-HYDRIDO)(η^5-CYCLOHEPTADIENYL)DECACARBONYLTRIRHENIUM
$C_{17}H_{11}O_{10}Re_3$ $Re_3H_2(CO)_{10}(\eta^5-C_7H_9)$

T. BERINGHELLI, G. CIANI, G. D'ALFONSO, P. ROMITI, A. SIRONI and M. FRENI, 1984. Inorg. Chem., 23, 2849-2852.

Monoclinic, C2/c, a = 31.507, b = 8.561, c = 16.503 Å, β = 106.60°, Z = 8. Mo radiation, R = 0.027 for 2677 reflexions.

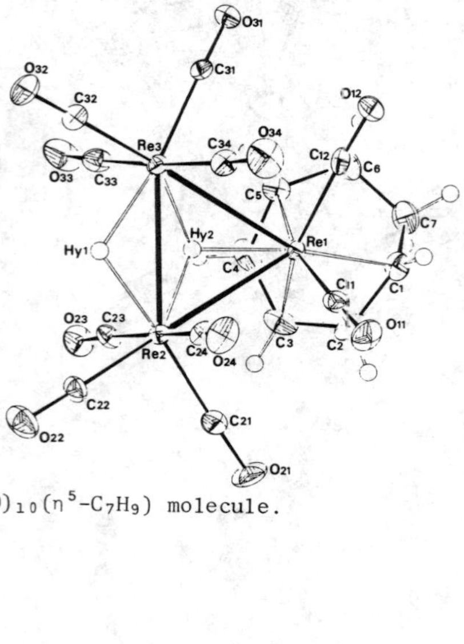

Fig. 1. The $Re_3H_2(CO)_{10}(\eta^5-C_7H_9)$ molecule.

The molecule (Fig. 1) contains a triangular arrangement of Re atoms. Two metals, Re2 and Re3, each bear four terminal CO groups, while Re1 is linked to two such ligands and to the η^5-C_7H_9 ligand. The two hydrides have been indirectly located, one double-bridging the Re2-Re3 edge and the second one in a triple-bridging position. Bond lengths: Re1-Re2 3.160(1), Re1-Re3 3.193(1), Re2-Re3 3.076(1), Re-CO 1.900(10)-2.012(10), Re1-C(η^5-C_7H_9) 2.224(10)-2.367(11) Å.

TRICARBONYL[2-5-η-(METHYL 1α-PHENYL-2,4-CYCLOHEXADIENE-1β-CARBOXYLATE)]IRON(0)
$C_{17}H_{14}FeO_5$ $[Fe(C_{14}H_{14}O_2)(CO)_3]$

B.F. ANDERSON and G.B. ROBERTSON, 1984. Acta Cryst., C40, 251-253.

Monoclinic, $P2_1/c$, a = 17.499, b = 7.546, c = 13.243 Å, β = 114.87°, Z = 4. Cu radiation, R = 0.044 for 1437 reflexions.

The carbonyl ligands are directed trans to C(1), C(4), and the mid-point of the C(2)-C(3) diene bond (Fig. 1). The Fe-C(ring) distances are in the range 2.026(7) to 2.093(5) Å.

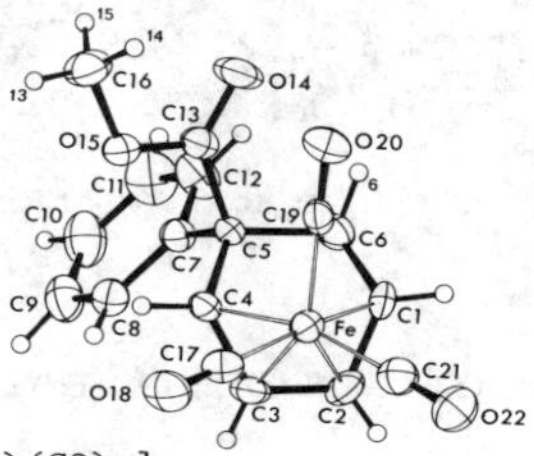

Fig. 1. A view of $[Fe(C_{14}H_{14}O_2)(CO)_3]$.

(1,5,6-η^3,2-4-η^3-DIMETHYLCYCLOOCTATETRAENE-1,8-DICARBOXYLATE)BIS(TRICARBONYLIRON)
$C_{18}H_{12}Fe_2O_{10}$

R.C. KERBER, D. GLASSER and B. LUCK, 1984. Organometallics, 3, 840-845.

Monoclinic, $P2_1/n$, a = 11.845, b = 20.845, c = 8.000 Å, β = 96.67°, D_m = 1.69, Z = 4. Mo radiation, R = 0.035 for 2879 reflexions.

The $Fe_2(CO)_6$ unit has a familiar "sawhorse" arrangement with the cyclooctatetraene ring severely bent to accommodate the bonding needs of the $Fe_2(CO)_6$ unit (Fig. 1). Fe(1) is attached to the ring via a π-allyl linkage to C(2)-C(3)-C(4) (Fe(1)-C(2) 2.153, Fe(1)-C(3) 2.047, Fe(1)-C(4) 2.101 Å) and to Fe(2) by a single bond of length 2.771 Å. Fe(2) is attached to C(1) through a 2.121 Å σ bond and to the C(5)=C(6) double bond (Fe(2)-C(5) 2.209, Fe(2)-C(6) 2.157 Å). Bond angles about C(1) are within the range 106-114° consistent with sp^3 hybridization of that C atom. Bond distances within the cyclooctatetraene ring are: C(1)-C(2) 1.486, C(2)-C(3) 1.401, C(3)-C(4) 1.404, C(4)-C(5) 1.457, C(5)-C(6) 1.405, C(6)-C(7) 1.463, C(7)-C(8) 1.321 and C(8)-C(1) 1.517 Å.

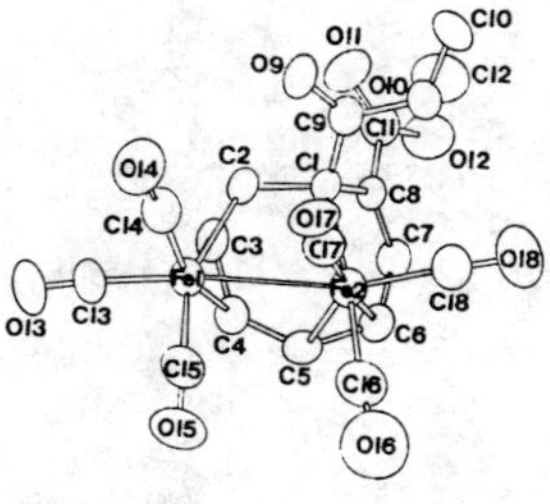

Fig. 1. Molecular structure of $C_{18}H_{12}Fe_2O_{10}$.

TRICARBONYL-(1—4-η-7-STYRYLCYCLOHEPTA-1,3,5-TRIENE)IRON
$C_{18}H_{14}FeO_3$

K. BROADLEY, N.G. CONNELLY, R.M. MILLS, M.W. WHITELEY and P. WOODWARD, 1984. J. Chem. Soc. Dalton, 683-688.

Monoclinic, C2/c, a = 25.035, b = 6.195, c = 20.938 Å, β = 102.69°, Z = 8. Mo radiation, R = 0.063 for 1977 reflexions.

The X-ray analysis established the structure of the title compound (Fig. 1). The C_7 ring adopts an envelope conformation (interplanar angle 138°) to which the $Fe(CO)_3$ fragment is η^4 bonded on the convex face. The atoms of the styryl substituent are all coplanar, and this plane lies approximately orthogonal to the mean plane of the C_7 ring. Some bond distances are: Fe-C(1) 2.111(4), Fe-C(3) 2.047(5), Fe-C(2) 2.067(4), Fe-C(4) 2.140(5), Fe-C(16) 1.785(4), Fe-C(17) 1.793(4), Fe-C(18) 1.802(4), C(7)-C(8) 1.511(5), C(8)-C(9) 1.318(6), C(9)-C(10) 1.471(6) Å.

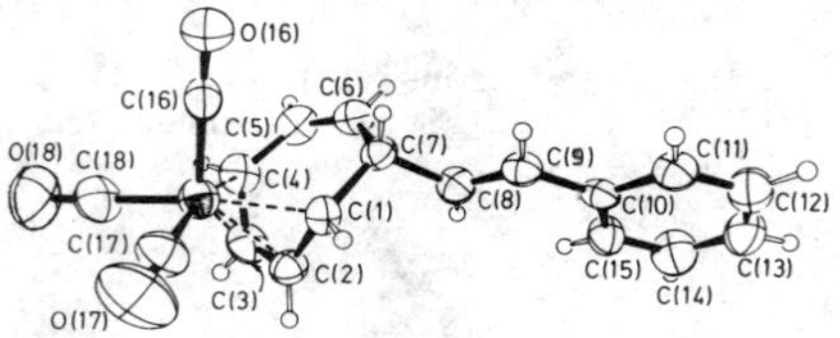

Fig. 1. $C_{18}H_{14}FeO_3$: the molecular configuration.

(μ_2-η^5,η^5-CYCLOOCTATETRAENE)BIS(η^5-CYCLOPENTADIENYL)DIRHODIUM BIS(HEXAFLUOROPHOSPHATE)
$C_{18}H_{18}F_{12}P_2Rh_2$ $[(CpRh)_2(cot)](PF_6)_2$

J. EDWIN, W.E. GEIGER and A.L. RHEINGOLD, 1984. J. Am. Chem. Soc., 106, 3052-3053.

Orthorhombic, $P2_12_12$, a = 14.073, b = 11.353, c = 6.748 Å, Z = 2. Mo radiation, R = 0.0229.

The cation has twofold crystallographic symmetry (Fig. 1) and the metal atoms are slipped to the side of the C_8 ring which has two nearly planar C_4 fragments twisted by 34.8°. The Rh-(centroid Cp) distance is 1.823 Å.

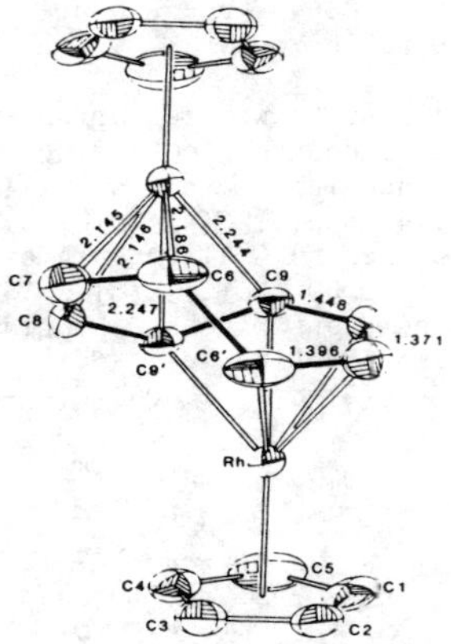

Fig. 1. A view along the twofold axis of the $[(CpRh)_2(cot)]^{2+}$ cation.

(CYCLOOCTA-1,5-DIENE)(METHYL trans-β-(PHENYLSULPHONYL)-ACRYLATE)NICKEL
$C_{18}H_{22}NiO_4S$

H.M. BÜCH, P. BINGER and C. KRÜGER, 1984. Organometallics, 3, 1504-1509.

Monoclinic, $P2_1/c$, a = 8.428, b = 17.645, c = 12.168 Å, β = 104.35°, Z = 4. Cu radiation, R = 0.062 for 2616 reflexions.

The complex is trigonal planar with the midpoints of the three coordinated C=C double bonds occupying vertices of a triangle centred at the Ni atom (Fig. 1). Two of the double bonds are perpendicular to the coordination plane, an unusual feature. The important bond distances are: Ni-C(1) 1.953, Ni-C(2) 1.966, Ni-C(11) 2.121, Ni-C(12) 2.092, Ni-C(15) 2.128 and Ni-C(16) 2.087 Å. The C(1)-C(2) distance is 1.430 Å and the C(11)-C(12) and C(15)-C(16) distances are respectively 1.342 and 1.359 Å. The non-bonded distance between Ni and S is very short at 3.074 Å. The NiC=CR and RC=CR torsional angles (δ, δ' and γ) are respectively 103.4, 100.3 and 156.3°.

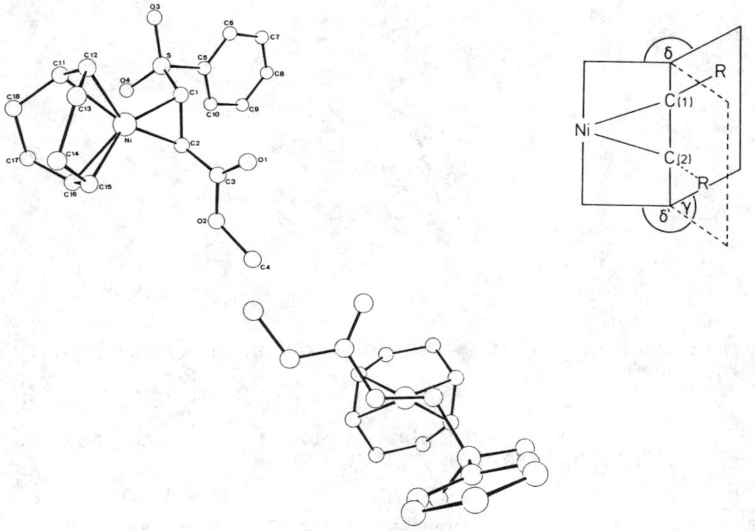

Fig. 1. Molecular structure of $C_{18}H_{22}NiO_4S$, general view (top left) and along the vector joining the midpoint of the olefinic C=C bond of the SMA ligand and the Ni atom (bottom), and illustration of the torsional angles δ, δ' and γ (top right).

TRICARBONYL[(η-2,3,4,5)-9-METHYL-9-PHENYLBICYCLO[5.2.0]NONAN-2,4-DIEN-8-ONE]IRON
$C_{19}H_{16}FeO_4$

Z. GOLDSCHMIDT, S. ANTEBI, D. COHEN and I. GOLDBERG, 1984. J. Organometal. Chem., 273, 347-359.

Monoclinic, $P2_1/n$, a = 12.701, b = 6.570, c = 21.048 Å, β = 106.46°, Z = 4. Mo radiation, R = 0.035 for 2856 reflexions.

The coordinated cycloheptadiene ring has essentially a boat conformation, with C(14) forming the prow (Fig. 1). The slightly twisted central plane defined by C(8), C(9), C(12) and C(13) and the plane containing the four coordinated C atoms C(9)-C(12) intersect with an average dihedral angle of 44°. Distances within the seven-membered ring are: C(8)-C(9) 1.507, C(9)-C(10) 1.421, C(10)-C(11) 1.391, C(11)-C(12) 1.425, C(12)-C(13) 1.504, C(13)-C(14) 1.503, C(14)-C(8) 1.556 Å. Coordination about the Fe atom is roughly square pyramidal with C(4) at the apex and the remaining carbonyls and the two butadiene double bonds defining the basal plane. The distances from Fe to C(9), C(10), C(11) and C(12) are 2.102, 2.047, 2.056 and 2.121 Å respectively. The conformation of the four-membered cyclobutanone ring is puckered with a dihedral angle of pucker 24.6°. Relevant distances and angles are: C(8)-C(17) 1.590, C(15)-C(17) 1.524, C(14)-C(15) 1.516, C(15)-O(16) 1.195 Å, C(8)-C(17)-C(15) 85.8, C(17)-C(15)-C(14) 93.0, C(15)-C(14)-C(8) 87.3 and C(14)-C(8)-C(17) 88.9°.

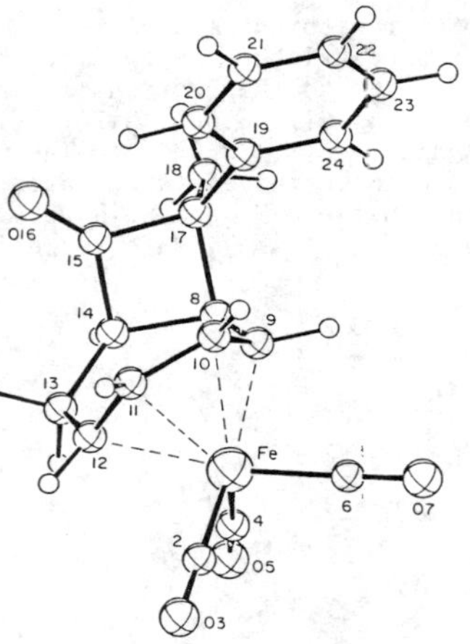

Fig. 1. View of $C_{19}H_{16}FeO_4$.

[μ-[(1,2,3,4-η:1,4-η)-2,3-BIS(DIETHYLAMINO)-1,4-DIMETHYL-1,3-BUTADIENE-1,4-DIYL]]-HEXACARBONYLDIIRON(Fe-Fe)

$C_{20}H_{26}Fe_2N_2O_6$ (I) $[Fe_2(CO)_6\{C_4(CH_3)_2[N(C_2H_5)_2]_2\}]$

BIS[μ_3-[(1,2,3,4,5-η)-2,4-BIS(DIETHYLAMINO)-3,5-DIMETHYL-2,4-CYCLOPENTADIEN-1-OLATO(2-)-N^2,O^1:O^1]]OCTACARBONYLTETRARUTHENIUM(2Ru-Ru)

$C_{38}H_{52}N_4O_{10}Ru_4$ (II) & (III) $[Ru_4(CO)_4\{C_5O(CH_3)_2[N(C_2H_5)_2]_2\}_2]$

J.-C. DARAN and Y. JEANNIN, 1984. Organometallics, 3, 1158-1163.

I. Monoclinic, $P2_1/c$, a = 16.001, b = 11.209, c = 13.550 Å, β = 103.64°, Z = 4. Mo radiation, R = 0.065 for 1947 reflexions.

(II) & (III). Monoclinic, $P2_1/n$, a = 14.135(8.059), b = 14.158(16.440), c = 10.696 (16.05) Å, β = 97.55(96.65)°, Z = 2. Mo radiation, R = 0.043(0.068) for 2403(1818) reflexions.

In I the butadiene fragment results from the tail-to-tail coupling of two 1-(diethylamino)propyne molecules whereas in II and III there is head-to-tail coupling (Fig. 1). The electronic structure of I could be described by at least two mesomeric forms: one is a ferrole π-bonded to another Fe atom and the second is one based upon the ketenimmonium ylide form of the aminoalkyne in which both Fe atoms appear identical. I has a molecular structure intermediate between the two. Fe(1) is 0.43 Å above the butadiene plane, and Fe(2)-C(1) (2.052 Å) is significantly shorter than Fe(2)-C(4) (2.110 Å). Other distances are: Fe(2)-C(2) 2.236, Fe(2)-C(3) 2.268, Fe(1)-Fe(2) 2.506, Fe(1)-C(1) 2.033, Fe(1)-C(4) 2.000, C(1)-C(2) 1.41, C(2)-C(3) 1.46, and C(3)-C(4) 1.44 Å. Compounds II and III are identical but both ethyl groups attached to N(3) are disordered in II. The whole molecule has an inversion centre which relates two dimetallic subunits containing a metal-metal bond. Ru(1) may be considered as tetrahedral with one vertex occupied by a cyclopentadiene ring and the three others occupied by two CO groups and Ru(2) (Ru(1)-Ru(2) 2.744 Å) in a piano-stool arrangement. Ru(2) is octahedral with two CO groups in cis positions, each trans to a bridging O atom; the last two trans vertices are occupied by Ru(1) and N(5)*. Both distances of interest in III are: Ru(2)-N(5)* 2.31, Ru(2)-O* 2.204, Ru(2)-O 2.211 and Ru(2)...Ru(2)* 3.492 Å.

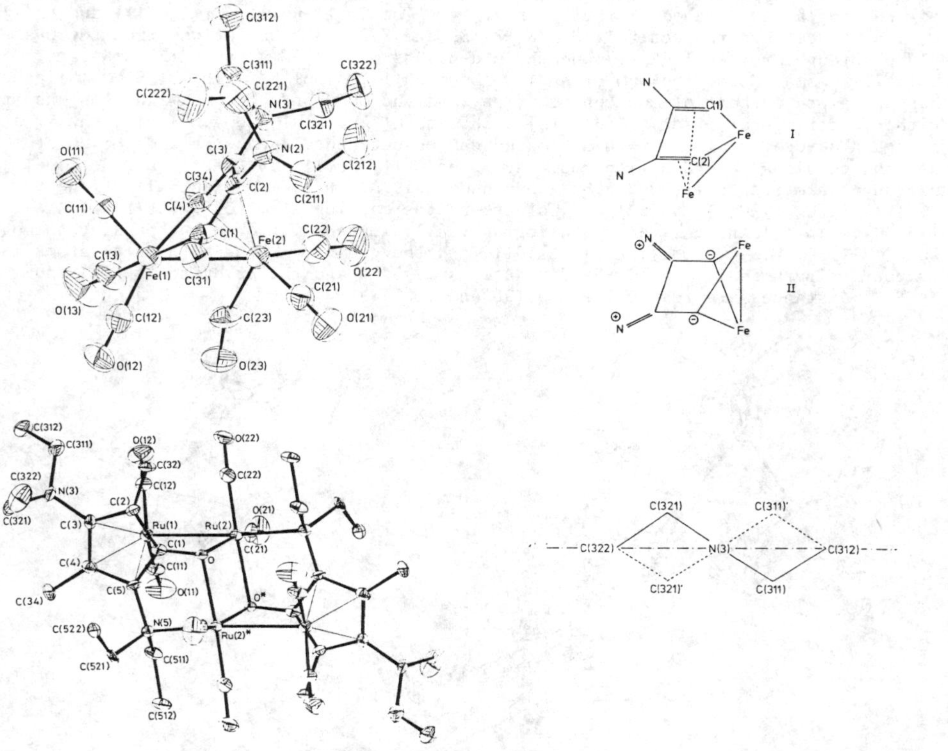

Fig. 1. Molecular structure of $C_{20}H_{26}Fe_2N_2O_6$ (top left), two possible mesomeric structures for I (top right), molecular structure of $C_{38}H_{52}N_4O_{10}Ru_4$ (bottom left), and scheme of the disordered ethyl groups in II (bottom right).

(μ_2-DI-t-BUTYLPHOSPHINO)(η^4-CYCLOOCTA-1,5-DIENE)RHODIUMPENTACARBONYLCHROMIUM

$C_{21}H_{30}CrO_5PRh$ (I) $(CO)_5Cr(\mu\text{-t-}Bu_2P)Rh(COD)$ (Cr-Rh)

(μ_2-CARBONYL)(μ_2-DI-t-BUTYLPHOSPHINO)TETRACARBONYLBIS(TRIMETHYLPHOSPHINE)CHROMIUM-COBALT

$C_{19}H_{36}CoCrO_5P_3$ (II) $(CO)_3(PMe_3)Cr(\mu\text{-t-}Bu_2P)(\mu\text{-}CO)Co(PMe_3)CO$

(μ_2-DI-t-BUTYLPHOSPHINO)PENTACARBONYLBIS(TRIMETHYLPHOSPHINE)IRONCOBALT

$C_{19}H_{36}CoFeO_5P_3$ (III) $(CO)_3(PMe_3)Fe(\mu\text{-t-}Bu_2P)Co(CO)_2(PMe_3)$

D.J. CHANDLER, R.A. JONES, A.L. STUART and T.C. WRIGHT, 1984. Organometallics, 3, 1830-1838.

I. Monoclinic, $P2_1/c$, a = 9.535, b = 28.146, c = 9.399 Å, β = 113.506°, Z = 4. Mo radiation, R = 0.049 for 2786 reflexions.

II. Monoclinic, $P2_1/n$, a = 8.527, b = 18.772, c = 16.870 Å, β = 101.59°, Z = 4. Mo radiation, R = 0.045 for 3037 reflexions.

III. Orthorhombic, Pnma, a = 11.469, b = 13.007, c = 17.509 Å, Z = 4. Mo radiation, R = 0.067 for 1060 reflexions.

In I the approximate coordination geometry of each metal atom is square planar for Rh and capped octahedral for Cr (Fig. 1). The Cr(1)-Rh(1), Cr(1)-P(1) and Rh(1)-P(1) distances are respectively 2.738, 2.389 and 2.286 Å. Deviations from the best plane through Cr(1)-P(1)-Rh(1) and the mid-points of C(3)-C(4) and C(7)-C(8) are relatively small, and the sum of angles about Rh(1) is 361.8°. The (1,5-COD) ligand has the expected boat or tub conformation, and the C=C bonds have similar lengths (to within experimental error) (C(3)-C(4) 1.385 and C(7)-C(8) 1.426 Å. In II, the Cr atom is again seven-coordinate with distorted capped octahedral geometry and the Co atom has square planar coordination geometry. The Cr(1)-Co(1), Cr(1)-P(1) and Co(1)-P(1) distances are respectively 2.505, 2.396 and 2.219 Å. P(1) is essentially trans to C(2) (P(1)-Cr(1)-C(2) 166.5°) and Co acts as the capping atom (Co(1)-Cr(1)-C(2) 113.3°). In III there is a crystallographically-imposed mirror plane passing through Co, P(2), Fe, P(3), C(301), P(1), C(101), and the geometry about both metal atoms is trigonal bipyramidal. The Co-Fe distance is 2.752 Å and the bridging Co-P(2) and Fe-P(2) distances are respectively 2.196 and 2.274 Å.

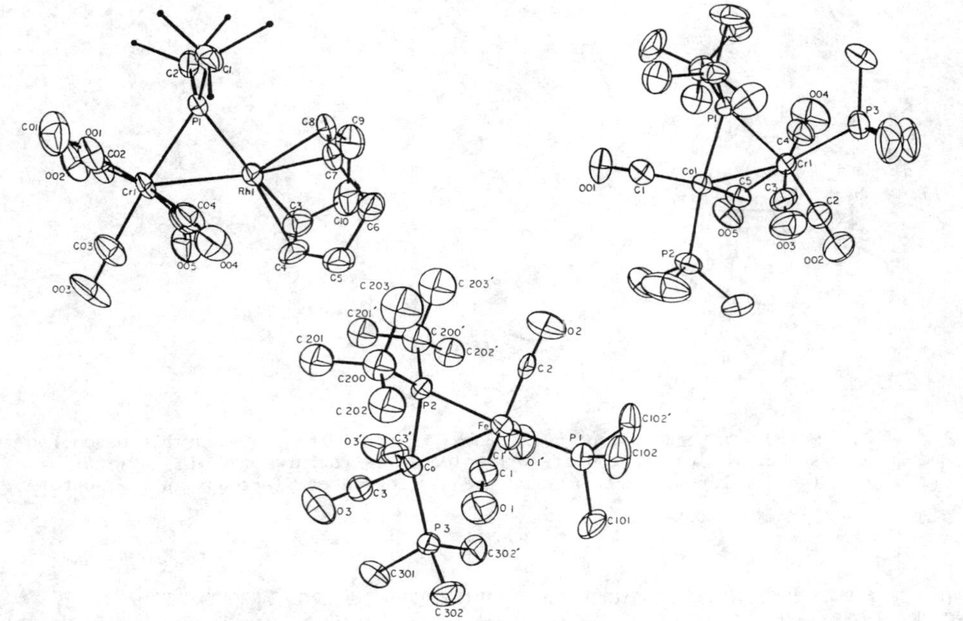

Fig. 1. General views of the molecular structure of $C_{21}H_{30}CrO_5PRh$ (top left), $C_{19}H_{36}CoCrO_5P_3$ (top right), and $C_{19}H_{36}CoFeO_5P_3$ (bottom).

(η^5-AZULENE)(η^4-TETRAFLUOROBENZOBARRELENE)RHODIUM HEXAFLUOROPHOSPHATE

$C_{22}H_{14}F_{10}PRh$ $[Rh(TFB)(C_{10}H_8)]^+PF_6^-$

L.A. ORO, M. VALDERRAMA, P. CIFUENTES, C. FOCES-FOCES and F.H. CANO, 1984. J. Organometal. Chem., 276, 67-77.

Monoclinic, $P2_1/c$, a = 8.4241, b = 16.6911, c = 15.0026 Å, β = 95.897°, Z = 4. Cu radiation, R = 0.027 for 2991 reflexions.

In the complex cation the Rh atom is coordinated only to the five-membered ring of the azulene ligand, and to the diolefinic double bonds of the TFB ligand (Fig. 1). The Rh-C (five-membered ring) distances range from 2.229-2.282 Å and the Rh-ring centroid distance is 1.898 Å. Within the azulene ligand the C(25)-C(26) and C(27)-C(28) bonds (1.371 and 1.362 Å) lie between double and aromatic bond lengths while the C(21)-C(30) and C(23)-C(24) bonds (1.441 and 1.435 Å) are longer than aromatic bonds. The shortest Rh-C (five-membered ring) distances are those which are trans to the d-olefinic double bonds. The five- and seven-membered rings show small deviations from planarity and the bond angles range from 107.1-109.5° and 127.4-130.3° respectively compared with theoretical values of 108 and 128.47° for regular planar rings. The Rh-C(6), C(7), C(9) and C(10) distances are respectively 2.159, 2.167, 2.153 and 2.155 Å, and the C(6)-C(7) and C(9)-C(10) distances are 1.397 and 1.388 Å respectively. The anion is disordered between two different orientations.

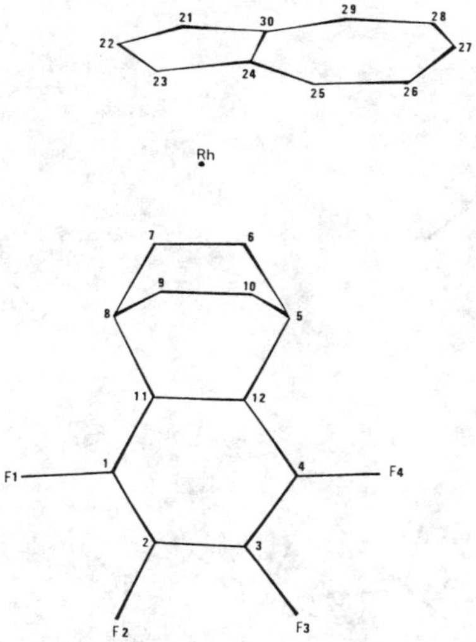

Fig. 1. Perspective view of the complex cation in $C_{22}H_{14}F_{10}PRh$.

(1,5-CYCLOOCTADIENE)CHLORO[BIS{BIS(TRIMETHYLSILYL)METHYL}CHLOROPHOSPHINE]RHODIUM(I) TOLUENE SOLVATE

$C_{22}H_{50}Cl_2PRhSi_4,C_7H_8$ (I) $[RhCl(COD)(PCl\{CH(SiMe_3)_2\}_2)],C_6H_5CH_3$

BIS[trans-{(CYCLOOCTENE)μ-CHLORO[BIS{BIS(TRIMETHYLSILYL)METHYL}PHOSPHINE]-DIRHODIUM(I)}]

$C_{44}H_{106}Cl_2P_2Rh_2Si_8$ (II)

trans-[CARBONYLCHLOROBIS[BIS{BIS(TRIMETHYLSILYL)METHYL}PHOSPHINE]RHODIUM(I)]

$C_{29}H_{78}ClOP_2RhSi_8$ (III)

B.D. MURRAY, H. HOPE, J. HVOSLEF and P.P. POWER, 1984. Organometallics, 3, 657-663.

I. Triclinic, $P\bar{1}$, a = 11.158, b = 11.930, c = 13.132 Å, α = 97.24, β = 104.02, γ = 93.82°, Z = 2. Mo radiation, R = 0.044 for 4635 reflexions (at 140K).

II. Monoclinic, $P2_1/n$, a = 11.118, b = 16.403, c = 17.782 Å, β = 104.77°, Z = 2. Mo radiation, R = 0.023 for 3443 reflexions (at 140K).

III. Monoclinic, $P2_1/n$, a = 10.620, b = 22.705, c = 20.375 Å, β = 96.00°, Z = 4. Mo radiation, R = 0.038 for 4350 reflexions (at 140K).

Molecule I has an approximately square-planar structure (Fig. 1) with Rh-Cl(1) 2.363 Å, Rh-P(1) 2.287 and Rh-C 2.124 to 2.253 Å. Other bond lengths are: P(1)-C(8) 1.838, P(1)-Cl(2) 2.090, Si(1)-C(1) 1.920, Si(2)-C(1) 1.923, C(15)-C(22) 1.401 and C(18)-C(19) 1.368 Å. The angles Cl(1)-Rh-P(1), P(1)-Rh-C(15) and Cl(1)-Rh-C(15) are respectively 85.0, 97.0 and 157.4°. Compound II exists as a trans dimer with crystallographic inversion symmetry. The Rh_2Cl_2 core is planar with Rh...Rh' 3.690, Rh-P 2.203, Rh-Cl 2.391 and 2.475, and Rh-C(1) 2.098 and Rh-C(2) 2.110 Å. Other distances are: P-C(01) 1.845, P-C(02) 1.857, Si(1)-C(01) 1.904, Si(2)-C(01) 1.911, C(1)-C(2) 1.413 Å. The angles Cl-Rh-P, Rh-Cl-Rh' and Cl-Rh-Cl' are respectively 96.0, 98.6 and 81.4°. In compound III the Rh atom is in a slightly distorted square-planar environment. The bonds Rh-Cl(1), Rh-P(2), Rh-P(1) and Rh-C(5) are respectively 2.376, 2.337, 2.345 and 1.801 Å in length. Angle C(5)-Rh-P(1) is 95.9°, Cl(1)-Rh-P(1) 85.0 and Cl(1)-Rh-P(2) 86.4°.

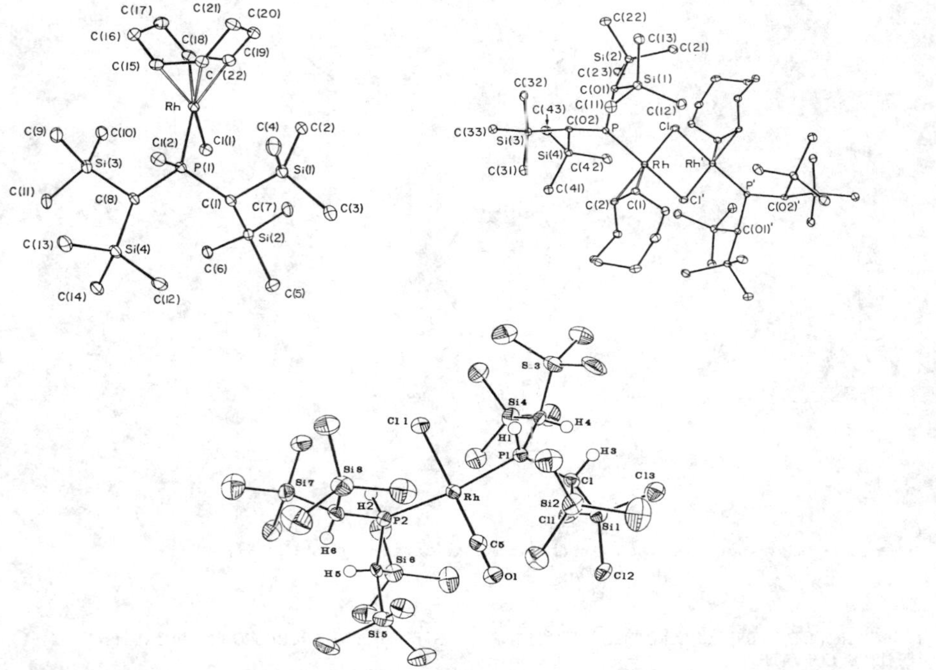

Fig. 1. Molecular structures of $C_{22}H_{50}Cl_2PRhSi_4$ (top left), $C_{44}H_{106}Cl_2P_2Rh_2Si_8$ (top right), and $C_{29}H_{78}ClOP_2RhSi_8$ (bottom).

BROMO[(1,2,3-η)-2-CYCLOOCTEN-1-YL][(1,2,3,4,5-η)-1-[5-METHYL-2-(1-METHYLETHYL)CYCLOHEXYL]-2,4-CYCLOPENTADIEN-1-YL]NITROSYL-MOLYBDENUM

$C_{23}H_{36}BrMoNO$ $(-)_D$-(S)-exo-NMCpMo(NO)(Br)(η^3-C_8H_{13})

J.W. FALLER and K.-H. CHAO, 1984. Organometallics, 3, 927-932.

Triclinic, P1, a = 6.697, b = 9.710, c = 9.908 Å, α = 108.74, β = 104.46, γ = 95.10°, Z = 1. Mo radiation, R = 0.032 for 1674 reflexions.

The molecule consists of two π-bonded moieties and two σ-bonded ligands about the Mo atom (Fig. 1). One of the π-bonded groups is neomenthylcyclopentadienyl in which the ring (C_5) atoms average 2.368 Å per bond to the metal atom. The other is an unsymmetrically-bound η^3-cyclooctenyl group. The ring appears to constrain the C(1)-C(2) and C(2)-C(3) lengths to comparable distances (1.403(9) and 1.364(9) Å) whereas the difference is usually significant in open-chain allyls. The Mo-C(1) bond length is considerably shorter than the Mo-C(3) length (2.373 and 2.544 Å respectively). The Mo-C(2) distance is 2.307 Å. Other distances of interest are: Mo-Br 2.623, Mo-N 1.764, N-O 1.204, C(3)-C(4) 1.505, C(4)-C(5) 1.558, C(5)-C(6) 1.46, C(6)-C(7) 1.55, C(7)-C(8) 1.541 and C(1)-C(8) 1.495 Å.

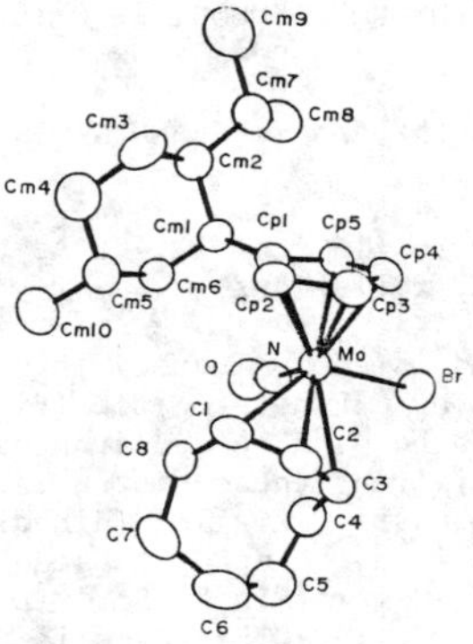

Fig. 1. Molecular structure of $C_{23}H_{36}BrMoNO$.

TRICARBONYL[(η-5,6,7,8)-2,3-DIPHENYL-4H-CYCLOHEPTA[b]FURAN]IRON

$C_{24}H_{16}FeO_4$ $(CO)_3Fe(C_7H_6C_2O(C_6H_5)_2)$

Z. GOLDSCHMIDT, S. ANTEBI and I. GOLDBERG, 1984. J. Organometal. Chem., 260, 105-113.

Monoclinic, $P2_1/c$, a = 9.653, b = 18.243, c = 11.293 Å, β = 102.54°, Z = 4. Mo radiation, R = 0.039 for 2398 reflexions.

The pentacoordinate Fe atom has square pyramidal geometry, with two CO groups (C(2)O(3) and C(6)O(7)) and the two diene double bonds defining the basal plane, and the remaining carbonyl (C(4)O(5)) forming the apex of the pyramid (Fig. 1). The Fe-C(carbonyl) distances range from 1.769 to 1.797 Å and the individual distances Fe-C(9), Fe-C(10), Fe-C(11) and Fe-C(12) are respectively 2.161, 2.059, 2.057 and 2.145 Å. The C(16)-C(24) bond is significantly shorter (1.459 Å) than its counterpart C(15)-C(18) bond (1.480 Å), reflecting the stronger conjugation between the furan and the phenyl ring at C(16) which are almost in the same plane.

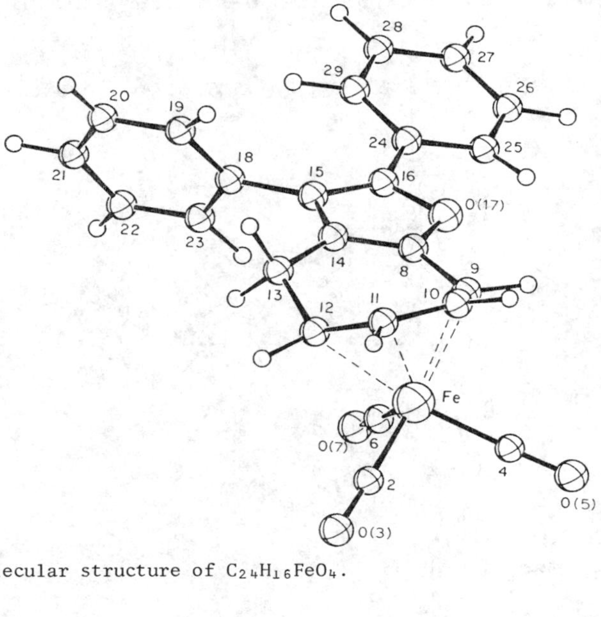

Fig. 1. Molecular structure of $C_{24}H_{16}FeO_4$.

$(\eta^5$-1,2-DIPHENYL-4,5-DIMETHYL-1-HYDROPHOSPHORIN-1-OXIDE)CYCLOPENTADIENYLIRON
$C_{24}H_{23}FeOP$

E. DESCHAMPS, F. MATHEY, C. KNOBLER and Y. JEANNIN, 1984. Organometallics, 3, 1144-1150.

Monoclinic, Cc, a = 10.428, b = 16.616, c = 12.402 Å, β = 111.13°, D_m = 1.37, Z = 4. Mo radiation, R = 0.061 for 1362 reflexions.

The cyclopentadienyl ligand is approximately in an eclipsed configuration with respect to C(2), C(3), C(4) of the C_5P heterocycle (Fig. 1). Only the five C atoms of the latter are coordinated to Fe with Fe-C distances which range from 2.05 to 2.13 Å. For the cyclopentadienyl ring these distances range from 2.06 to 2.11 Å and the two coordination planes are nearly parallel (dihedral angle ∿2.2°). The P atom is out of the C_5 plane by 0.733 Å on the opposite side of Fe, and the folding of the heterocyclic ligand around the $C\text{-}C_\alpha$ axis is 38.5°. In the heterocyclic ring the C-C distances average 1.43 Å and the C(1)-P-C(5) angle is 94.8° with C(1)-P 1.795 Å and C(5)-P 1.758 Å. The P=O and P-C(211) distances are 1.479 and 1.837 Å respectively, and there is no direct interaction between Fe and the endo phosphoryl O atom. The Fe...P distance is 2.748 Å.

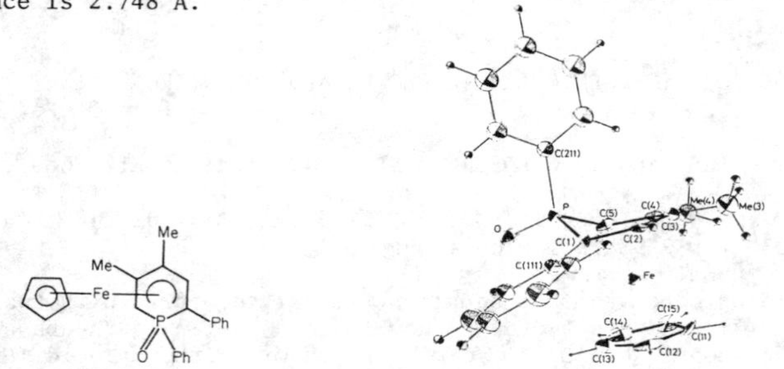

Fig. 1. Molecular structure of $C_{24}H_{23}FeOP$.

$(\mu_2$-AQUO)BIS$(\mu_2$-TRIFLUOROACETATO)BIS$[\eta^4$-CYCLOOCTA-1,5-DIENE)(TRIFLUOROACETATO)-RUTHENIUM(II)]
$C_{24}H_{26}F_{12}O_9Ru_2$ $[Ru_2(C_8H_{12})_2(OOCCF_3)_2(OOCCF_3)_2(H_2O)]$

M.O. ALBERS, D.C. LILES, E. SINGLETON and J.E. YATES, 1984. J. Organometal. Chem., 272, C62-C66.

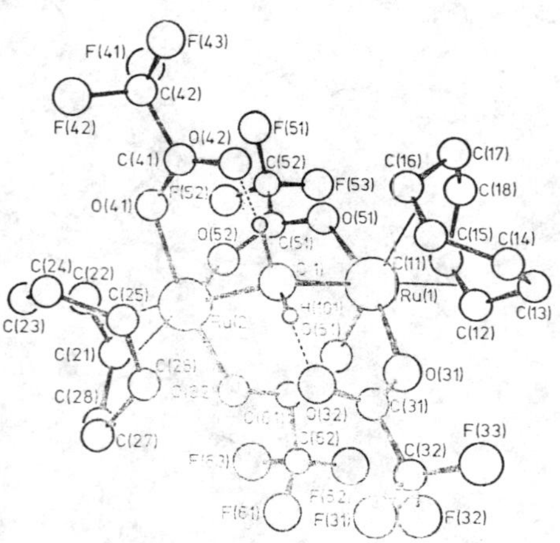

Fig. 1. The structure of $C_{24}H_{26}F_{12}O_9Ru_2$.

Triclinic, $P\bar{1}$, a = 13.731, b = 11.001, c = 10.593 Å, α = 101.04, β = 100.21, γ = 97.77°, Z = 2. Mo radiation, R = 0.055 for 3694 reflexions.

The complex is dinuclear with the two Ru atoms bridged by two trifluoroacetato groups and a water molecule (Fig. 1). A cycloocta-1,5-diene ligand and a monodentate trifluoroacetato ligand complete the coordination sphere of each Ru atom. The Ru...Ru distance is 3.733 Å. Selected bond distances are: Ru(1)-O(1) 2.160, Ru(2)-O(1) 2.168, Ru(1)-O(31) 2.094, Ru(2)-O(41) 2.104, Ru(1)-O(51) 2.097, Ru(1)-O(61) 2.131, Ru(2)-O(52) 2.124, Ru(2)-O(62) 2.078 Å. A system of strong intramolecular hydrogen bonds exists between the non-coordinated atoms of the monodentate trifluoroacetato ligands and the bridging aquo ligand with O(1)...O(32) and O(1)...O(42) distances of 2.530 and 2.495 Å respectively.

(η^4-CYCLOOCTADIENE)(2-BROMO-1-DIPHENYLPHOSPHINOPHENYL-Br,P)IRIDIUM HEXAFLUORO-ANTIMONATE

$C_{26}H_{26}BrF_6IrPSb$ $[Ir(C_8H_{12})(\eta^2-PPh_2C_6H_4Br)]^+SbF_6^-$

M.J. BURK, R.H. CRABTREE and E.M. HOLT, 1984. Organometallics, 3, 638-640.

Triclinic, $P\bar{1}$, a = 13.127, b = 10.239, c = 12.007 Å, α = 68.12, β = 77.07, γ = 104.16°, Z = 2. Mo radiation, R = 0.096 for 3213 reflexions.

The quality of the diffraction data is limited due to disorder within the anion and in atoms C(4) and C(5) of the cyclooctadiene ligand but the main features of the structure are shown in Fig. 1. The COD ligand, although partially disordered shows the vinyl groups to be trans to the chelating phosphine group. The distances Ir-Br (2.473 Å) and C-Br (1.86 Å), and angle C-Br-Ir (102.2°) all appear to be normal.

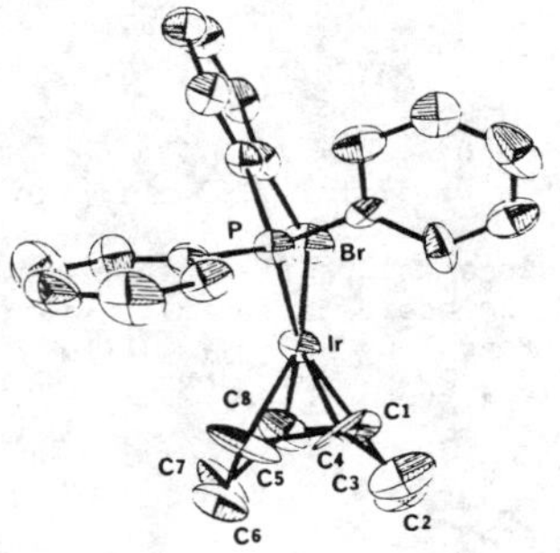

Fig. 1. Molecular structure of $C_{26}H_{26}BrF_6IrPSb$.

BIS[(μ_2-PENTAFLUOROTHIOPHENOLATO-S,S)(η^4-CYCLOOCTA-1,5-DIENE)RHODIUM(I)]

$C_{28}H_{24}F_{10}Rh_2S_2$

D. CRUZ-GARRITZ, B. RODRIGUEZ, H. TORRENS and J. LEAL, 1984. Transition Met. Chem., 9, 284-285.

Triclinic, $P\bar{1}$, a = 9.734, b = 11.831, c = 13.985 Å, α = 104.59, β = 106.58, γ = 104.22°, Z = 2. Mo radiation, R = 0.074 for 8000 reflexions.

In this molecule (Fig. 1) two essentially planar $RhSC_6H_5$(COD) groups intersect at an angle of 117°. The Rh...Rh separation is 2.955 Å; other distances are Rh-S 2.386-2.420, mean Rh-C 2.155, mean S-C 1.755 Å.

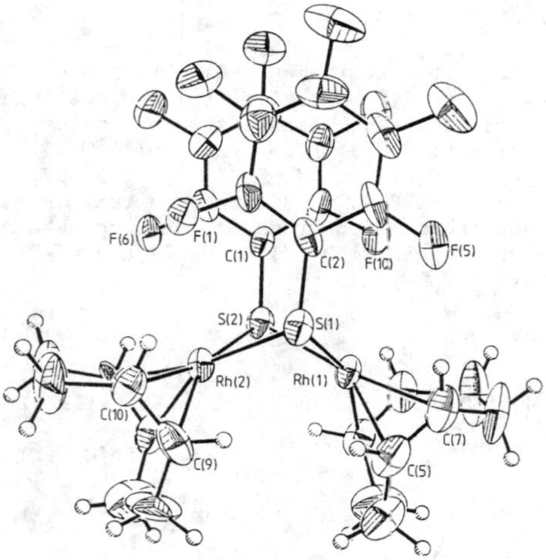

Fig. 1. The structure of $C_{28}H_{24}F_{10}Rh_2S_2$.

(4,4',5,5'-TETRACYANO-2,2'-BIIMIDAZOLATO)BIS(1,5-CYCLOOCTADIENE)(ACETONITRILE)-DIIRIDIUM(I) ACETONITRILE SOLVATE

$C_{28}H_{27}Ir_2N_9,C_2H_3N$ $Ir_2(C_8H_{12})_2(CH_3CN)(C_{10}N_8).CH_3CN$

P.G. RASMUSSEN, O.H. BAILEY and J.C. BAYÓN, 1984. Inorg. Chem., 23, 338-343.

Triclinic, $P\bar{1}$, a = 11.699, b = 14.043, c = 10.127 Å, α = 110.87, β = 91.06, γ = 71.56°, Z = 2. Mo radiation, R = 0.037 for 2863 reflexions.

The structure (Fig. 1) shows a unique tridentate bridging $C_{10}N_8^{2-}$ ligand due to the unexpected displacement of one coordination site by acetonitrile while the Ir atoms remain four-coordinate. Bond lengths: Ir-C(COD) 2.092(13)-2.147(13), Ir-N($C_{10}N_8^{2-}$) 2.080(10)-2.094(10), and Ir-N(CH_3CN) 2.044(12) Å.

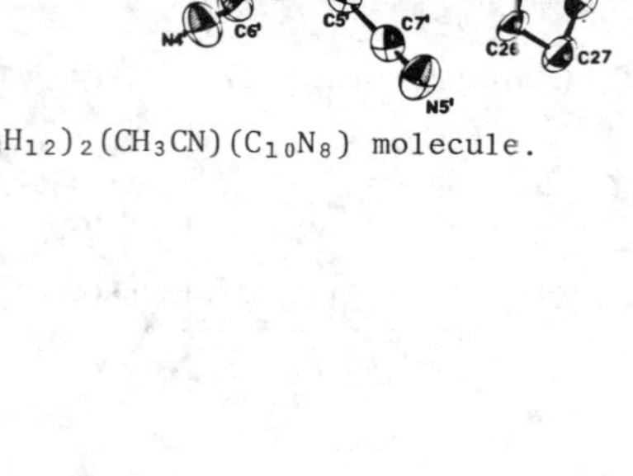

Fig. 1. The $Ir_2(C_8H_{12})_2(CH_3CN)(C_{10}N_8)$ molecule.

BIS(DICYCLOHEXYLPHOSPHINE)[(1,2,3-η)-2-METHYL-2-PROPENYL]RHODIUM
$C_{28}H_{53}P_2Rh$ (I)

[BIS(1,1-DIMETHYLETHYL)PHOSPHINE][(1,2,5,6-η)-1,5-CYCLOOCTADIENE][(1,2,3-η)-2-METHYL-2-PROPENYL]RHODIUM
$C_{20}H_{38}PRh$ (II)

[BIS(1,1-DIMETHYLETHYL)PHOSPHINE][(1,2,5,6-η)-1,5-CYCLOOCTADIENE][(1,2,3-η)-2-METHYL-2-PROPENYL]IRIDIUM
$C_{20}H_{38}IrP$ (III)

B.D. MURRAY and P.P. POWER, 1984. Organometallics, 3, 1199-1204.

I. Triclinic, $P\bar{1}$, a = 10.637, b = 12.242, c = 12.453 Å, α = 104.88, β = 94.962, γ = 111.57°, Z = 2. Mo radiation, R = 0.023 for 4458 reflexions, at 140K.

II. Triclinic, $P\bar{1}$, a = 9.130, b = 9.821, c = 12.262 Å, α = 105.13, β = 99.34, γ = 105.70°, Z = 2. Cu radiation, R = 0.068 for 3436 reflexions, at 140K.

III. Triclinic, $P\bar{1}$, a = 9.151, b = 9.778, c = 12.304 Å, α = 104.77, β = 99.34, γ = 106.05°, Z = 2. Mo radiation, R = 0.057 for 3175 reflexions, at 140K.

In compound I the geometry at Rh is approximately square planar, with the Rh, P(1), P(2), H(P1), and H(P2) atoms and the centres of the C(1)-C(2) and C(2)-C(3) bonds being almost coplanar (Fig. 1). The cyclohexyl groups are in the expected chair conformation. The average Rh-C(allyl) distance is 2.166 Å, Rh-P(1) is 2.229, P(1)-C(5) 1.847 and C(1)-C(2) 1.400 Å. Compounds II and III crystallize with similar cell dimensions but exhibit some different bond distances and angles. Bond distances of interest are: Rh-P 2.457, Ir-P 2.402 Å, the average Rh-C(COD) distance is 2.189 Å and the corresponding average Ir-C(COD) distance is 2.169 Å. The methyl group (C(12)) in II is almost coplanar with C(9), C(10) and C(11) but in III it is bent some 4° out of this plane towards the Ir atom. In I the corresponding methyl group C(4) is bent 5° out of the C(1), C(2), C(3) plane towards the Rh atom.

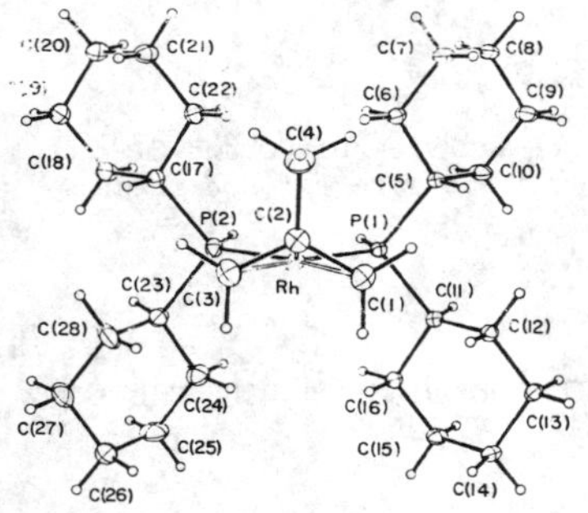

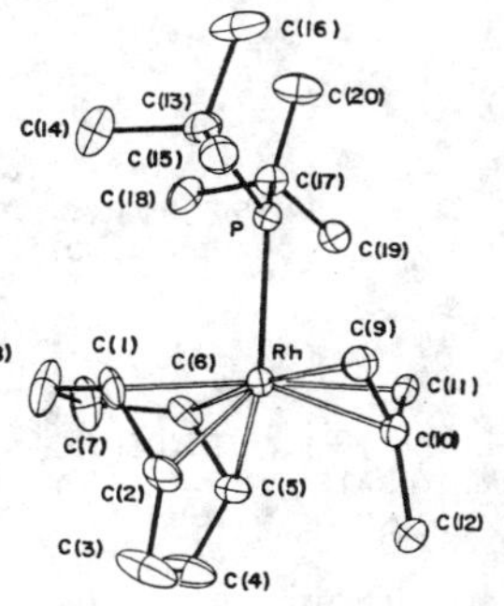

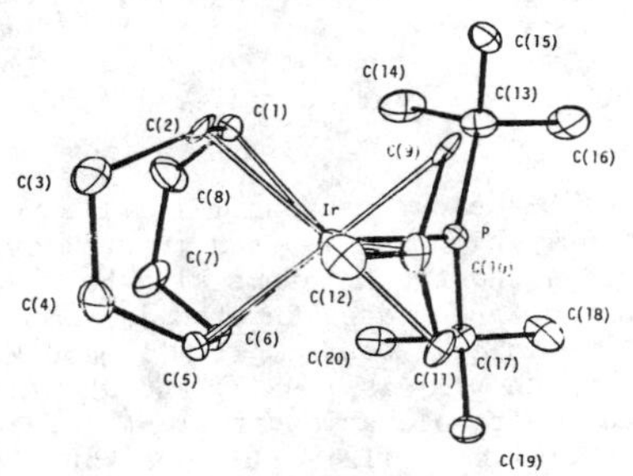

Fig. 1. Molecular structures of $C_{28}H_{53}P_2Rh$ (top left), $C_{20}H_{38}PRh$ (top right), and $C_{20}H_{38}IrP$ (bottom).

CARBONYL-CHLORO-(η^5-CYCLOPENTADIENYL)-(η^4-1-METHYL-2,3,4-TRIPHENYLCYCLOBUTADIENE)-MOLYBDENUM ACETONE SOLVATE

$C_{29}H_{23}ClMoO,C_3H_6O$ $Mo(C_5H_5)(C_4Ph_3Me)(CO)(Cl),(CH_3)_2CO$

R.P. HUGHES, J.W. REISCH and A.L. RHEINGOLD, 1984. Organometallics, 3, 1761-1763.

Triclinic, P$\bar{1}$, a = 11.041, b = 14.556, c = 16.969 Å, α = 103.29, β = 92.73, γ = 107.64°, Z = 4. Mo radiation, R = 0.052 for 4767 reflexions.

The compound crystallizes as discrete, well separated molecules in a distorted tetrahedral complex, and bears a formal resemblance to seven-coordinate "four-legged piano stool" complexes if the cyclobutadiene ligand is treated as a two-lobed, four-electron donor (Fig. 1). Distances within the cyclobutadiene ring are: C(1)-C(2) 1.44, C(2)-C(3) 1.46, C(3)-C(4) 1.46 and C(1)-C(4) 1.45 Å. The ring is planar with the endocyclic angles averaging 89.9°. The Mo-(4-membered ring centroid) distance is 2.021 Å with Mo-C(1) 2.273, Mo-C(2) 2.231, Mo-C(3) 2.219 and Mo-C(4) 2.338 Å. The distance from the centroid of the five-membered ring to the Mo atom is 1.998 Å and the angle CNT4-Mo-CNT5 (CNT = centroid) is 132.8°. There is disorder between the Cl and CO positions and also in the solvent molecule positions.

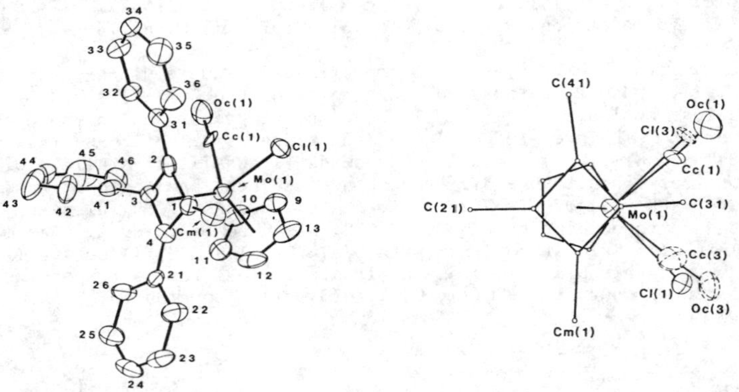

Fig. 1. Molecular structure of $C_{29}H_{23}ClMoO$ (left), and view of CO and Cl disorder (right) in which the broken line atoms were refined to an occupancy of 25%.

BIS[(μ$_2$-1,8-NAPHTHYRIDINE-N,N')(η^2,η^2-NORBORNA-2,5-DIENE)RHODIUM(I)] DIPERCHLORATE MONOHYDRATE

$C_{30}H_{28}Cl_2N_4O_8Rh_2,H_2O$

A. TIRIPICCHIO, M.T. CAMELLINI, R. USON, L.A. ORO, M.A. CIRIANO and F. VIGURI, 1984. J. Chem. Soc. Dalton, 125-131.

Orthorhombic, Pccn, a = 17.539, b = 10.252, c = 17.211 Å, Z = 4. Mo radiation, R = 0.063 for 1427 reflexions.

The structure consists of binuclear cations (Fig. 1), perchlorate anions and water molecules. In the cation, which has imposed crystallographic C_2 symmetry, two 1,8-naphthyridine ligands bridge the two Rh atoms via the N atoms (Rh-N(1) 2.12(1), Rh-N(2) 2.15(1) Å). Coordination about each Rh atom is completed by a norbornadiene molecule interacting through two olefin bonds. If M(1) and M(2) are the midpoints of C(9)-C(10) and C(12)-C(13) then Rh-M(1) is 1.98(2) and Rh-M(2) is 1.99(2) Å. The Rh atom has a slightly distorted square planar environment. The Rh-C distances range from 2.09 to 2.13 Å and C(9)-C(10) and C(12)-C(13) are both 1.45(3) Å. The naphthyridine ligands are planar and show no unusual features. The Rh...Rh distance is 2.916(3) Å. There is no hydrogen bonding in the crystal.

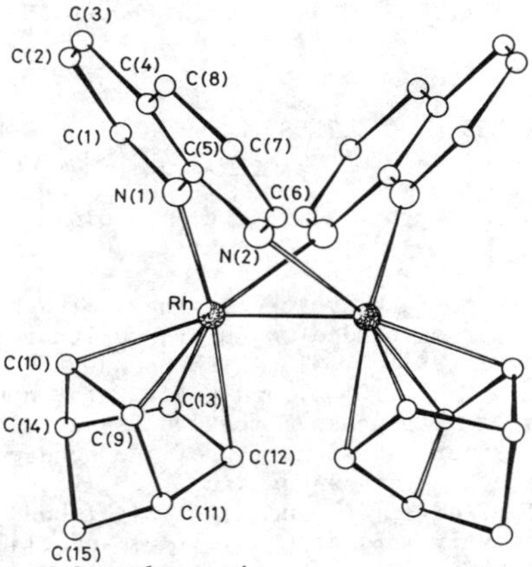

Fig. 1. $C_{30}H_{28}Cl_2N_4O_8Rh_2,H_2O$: the cation.

4a-METHYL-1,3,9-TRIPHENYL-4a-H-FLUORENE
$C_{32}H_{24}$ (I)

(η^4-4a-METHYL-1,3,9-TRIPHENYL-4a-FLUORENE)TRICARBONYLIRON
$C_{35}H_{24}FeO_3$ (II) $C_{32}H_{24}Fe(CO)_3$

M.A.A.F.C.T. CARRONDO and D. ROGERS, 1984. J. Organometal. Chem., 276, 219-230.

I. Triclinic, $P\bar{1}$, a = 12.506, b = 10.237, c = 10.121 Å, α = 108.53, β = 84.93, γ = 112.32°, Z = 2. Cu radiation, R = 0.043 for 3013 reflexions (at 10°C).

II. Monoclinic, $B2_1/c$, a = 30.283, b = 12.709, c = 14.514 Å, β = 92.73°, Z = 8. Cu radiation, R = 0.045 for 3261 reflexions.

In I, atoms C(1) to C(8) and C(13) form an indene fragment, the diene ring C(8) to C(13) approximates to half-chair configuration, and the three phenyl rings are considerably rotated about their connecting bonds (51° about C(1)-C(21), 36° about C(12)-C(31) and 36° about C(10)-C(41)), (Fig. 1). In II, a marked change has occurred in the conformation of ring C(8)-C(13) to allow Fe coordination on the side opposite to the methyl group. The complexed Fe-C distances are Fe-C(9) 2.117, Fe-C(10) 2.049, Fe-C(11) 2.067 and Fe-C(12) 2.220 Å. Bond lengths in the complexed diene are C(9)-C(10) 1.435, C(10)-C(11) 1.421, C(11)-C(12) 1.430 and C(12)-C(31) 1.505 Å compared with 1.340, 1.469, 1.352 and 1.482 Å respectively in the uncomplexed compound I. All other bonds correspond closely in length in the two compounds.

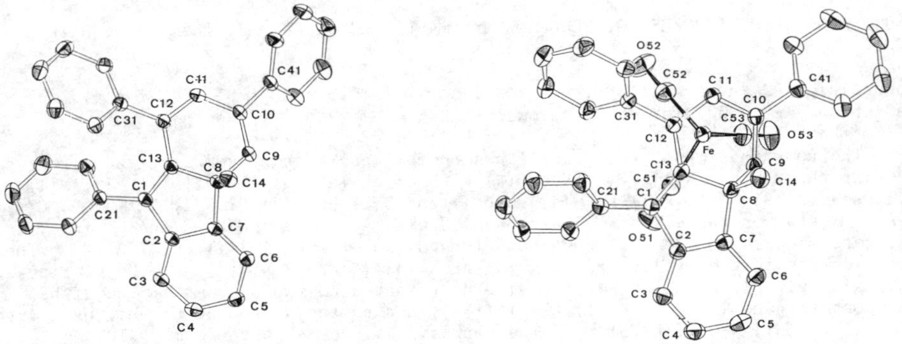

Fig. 1. Views of $C_{32}H_{24}$ (left) and $C_{35}H_{24}FeO_3$ (right).

(η^5-BICYCLO[5.1.0]OCTADIENYL)TRIS(DIMETHYLPHENYLPHOSPHINE)RUTHENIUM(II) HEXAFLUORO-PHOSPHATE
$C_{32}H_{42}F_6P_4Ru$

T.V. ASHWORTH, A.A. CHALMERS, D.C. LILES, E. MEINTJIES, H.E. OOSTHUIZEN and E. SINGLETON, 1984. J. Organometal. Chem., 276, C49-C52.

Orthorhombic, Pbca, a = 23.522, b = 14.937, c = 19.019 Å, Z = 8. Mo radiation, R = 0.043 for 3356 reflexions.

The structure consists of a bicyclo-octadienyl ligand bonded to a $Ru(PMe_2Ph)_3^+$ fragment (Fig. 1). The C atoms bonded to Ru are coplanar (within 0.03 Å) with Ru-C bond lengths varying from 2.195 to 2.315 Å, the bonds to C(41) and C(45) being ca. 0.1 Å longer than those to C(42), C(43) and C(44). The cyclopropyl ring lies in a position exo to the Ru atom. An unusual feature is the elongated (1.63 Å) C(46)-C(48) bond indicating considerable weakening of the bond common to the three- and seven-membered rings. The Ru-P distances are 2.359, 2.341 and 2.349 Å, and the C(41)-C(42), C(42)-C(43), C(43)-C(44) and C(44)-C(45) bond lengths are 1.390, 1.425, 1.415 and 1.398 Å respectively. Bonds C(45)-C(46) and C(41)-C(48) are respectively 1.482 and 1.505 Å in length.

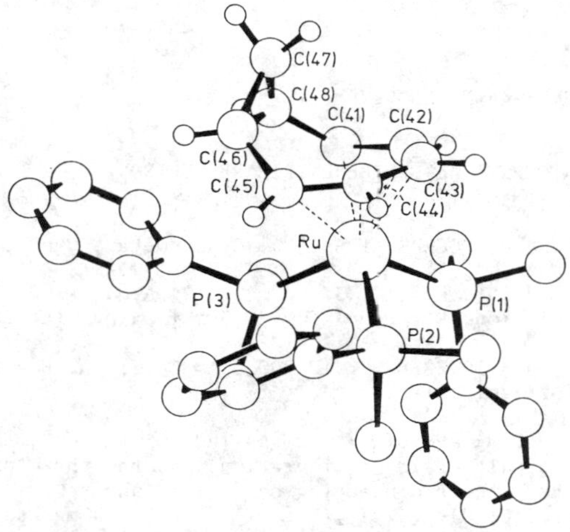

Fig. 1. Perspective view of the cation in $C_{32}H_{42}F_6P_4Ru$.

(N,N-DIETHYLDITHIOCARBAMATO-S,S')NITROSYL(η^4-TETRAPHENYLCYCLOBUTADIENE)IRON(0)
$C_{33}H_{30}FeN_2OS_2$

F.J. LALOR, L.H. BROOKES, G. FERGUSON and M. PARVEZ, 1984. J. Chem. Soc. Dalton, 245-248.

Monoclinic, $P2_1/n$, a = 11.383, b = 16.211, c = 15.529 Å, β = 93.23°, Z = 4. Mo radiation, R = 0.027 for 2998 reflexions.

The coordination at Fe (Fig. 1) has approximate mirror symmetry, the mirror plane passing through Fe, N(1), C(2) and C(4) with S-Fe-N(nitrosyl) 104.9(1), 106.9(1), N(1)-Fe-C(2) 93.7(1), S(1)-Fe-C(4) 98.6(1) and S(2)-Fe-C(4) 92.9(1)°. The Fe-C(η^4-C_4Ph_4) distances range from 2.034(2) to 2.070(2) Å, and the C-Fe-C angles involving adjacent C atoms in the cyclobutadiene ligand range from 41.2(1) to 41.8(1)°. The C-C bond lengths in the cyclobutadiene ring are 1.454(3) to 1.464(3) Å. The C_4 ring is approximately planar. The Fe atom also coordinates to two S atoms of the dithio-carbamato ligand (Fe-S 2.317(1), 2.316(1) Å) and to the N atom of the nitrosyl ligand (Fe-N 1.649(3), N-O 1.167(3) Å). The nitrosyl ligand is bent (Fe-N-O 170.4(3)°). In the crystal the discrete monomeric molecules are separated by normal van der Waals contacts.

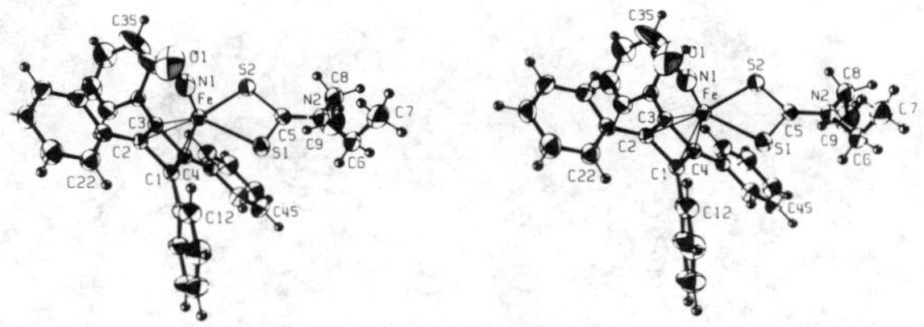

Fig. 1. $C_{33}H_{30}FeN_2OS_2$: stereoview of the molecule.

DICHLORO(η^4-1,2-DIMETHYL-3,4-DIPHENYLCYCLOBUTADIENE)TRIPHENYLPHOSPHINEPLATINUM(II)
$C_{36}H_{31}Cl_2PPt$

F. CANZIANI, C. ALLEVI, L. GARLASCHELLI, M.C. MALATESTA, A. ALBINATI and F. GANAZZOLI, 1984. J. Chem. Soc. Dalton, 2637-2643.

Monoclinic, $P2_1/c$, a = 10.900, b = 23.437, c = 13.289 Å, β = 113.5°, Z = 4. Mo radiation, R = 0.046 for 4358 reflexions.

The structure contains discrete molecules with the Pt coordination distorted tetrahedral (Fig. 1). Selected dimensions are: Pt-Cl(1) 2.425(3), Pt-Cl(2) 2.430(2), Pt-P 2.353(2), Pt-C(ring) 2.116(9)-2.214(8), mean C_4 ring C-C 1.47(1) Å.

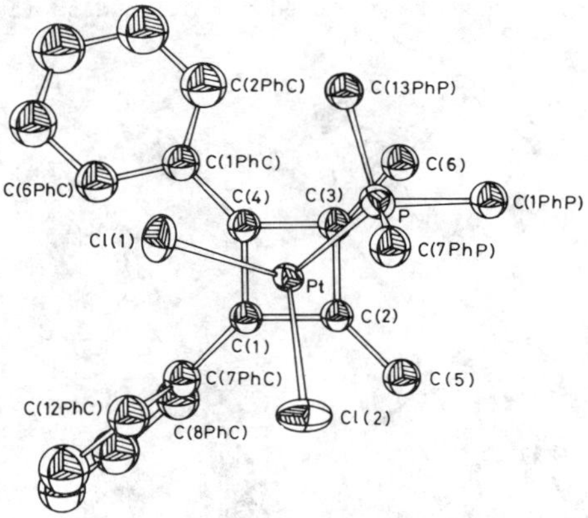

Fig. 1. A view of $C_{36}H_{31}Cl_2PPt$ along the normal to the cyclobutadiene plane; phosphine phenyl rings are omitted for clarity.

CHLOROTRIS(η^5-1,4,7-TRIMETHYLINDENYL)URANIUM
$C_{36}H_{39}ClU$

J. MEUNIER-PIRET and M. VAN MEERSSCHE, 1984. Bull. Soc. Chim. Belg., 93, 299-305.

Monoclinic, $P2_1/c$, a = 9.763, b = 16.480, c = 18.763 Å, β = 101.07°, Z = 4. Mo radiation, R = 0.0566 for 3210 reflexions.

The U atom is π-bonded to the C_5 ring of the indenyl ligands and has a distorted tetrahedral environment (Fig. 1) with U-Cl 2.601(3), mean U-C(indenyl) 2.79 Å.

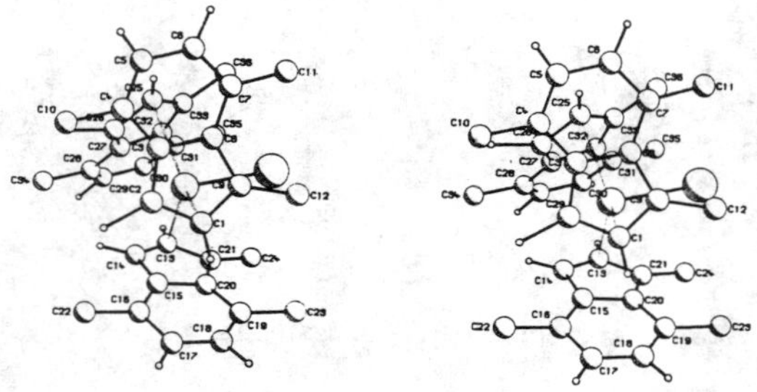

Fig. 1. A stereoview of $C_{36}H_{39}ClU$.

CYCLO-TETRAKIS((η^5-4,5,6,7-TETRAHYDROINDENYL)(μ_2-OXO)CHLOROTITANIUM)

$C_{36}H_{44}Cl_4O_4Ti_4$ $[(C_9H_{11})TiCl(O)]_4$

E. SAMUEL, R.D. ROGERS and J.L. ATWOOD, 1984. J. Cryst. Spectrosc. Res., 14, 573-579.

Tetragonal, $P\bar{4}2_1c$, a = 15.704, c = 7.776 Å, Z = 2. Mo radiation, R = 0.0032 for 554 reflexions.

The molecule (Fig. 1) lies on a crystallographic $\bar{4}$ site. Both independent Ti-O distances are equal, 1.802(5) and 1.809(5) Å, and the Ti-Cl length is 2.281(3) Å. The five Ti-C(π) distances show the expected variation associated with an indenyl moiety: 2.310(9) to 2.437(9) Å.

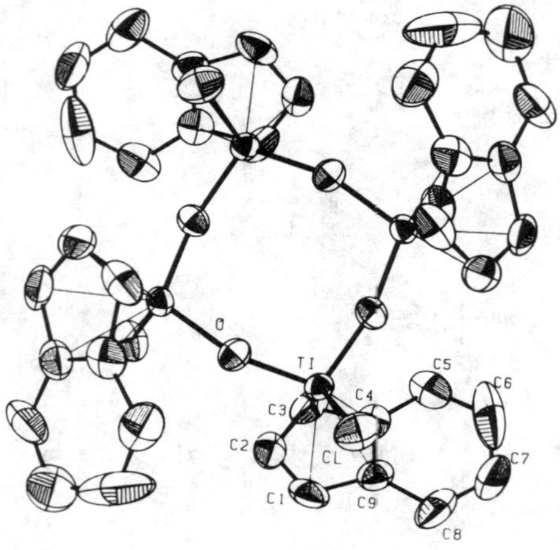

Fig. 1. A view of $C_{35}H_{44}Cl_4O_4Ti_4$.

2,2-DICARBONYL-2-η^5-CYCLOPENTADIENYL-μ-(1,2-DIPHENYL-3-p-TOLYLPROPAN-2-YL-1,3-DIYLIDENE)-1-η^5-INDENYLRHODIUMTUNGSTEN

$C_{38}H_{29}O_2RhW$

M. GREEN, J.A.K. HOWARD, S.J. PORTER, F.G.A. STONE and D.C. TYLER, 1984. J. Chem. Soc. Dalton, 2553-2559.

Monoclinic, $P2_1/n$, a = 10.709, b = 15.737, c = 18.192 Å, β = 93.75°, Dm = 1.75, Z = 4. Mo radiation, R = 0.032 for 4184 reflexions.

In the molecule (Fig. 1) a Rh-W bond (2.754(1)) is bridged by a three carbon chain (mean C-C 1.439(8) Å). Whereas all the atoms of the chain are bound to Rh (2.105(6), 2.135(6) and 2.103(5)), only the two end carbons are attached to W (2.175(5), 2.196(6) Å). The W atom carries two essentially orthogonal terminal CO groups as well as the cyclopentadienyl ring. The indenyl group is asymmetrically attached to the Rh atom, with three short Rh-C (2.185(7), 2.198(7), 2.204(7)) and two long Rh-C (2.334(7), 2.330(7) Å) separations to the C_5 ring, implying slippage to a η^3-mode of attachment.

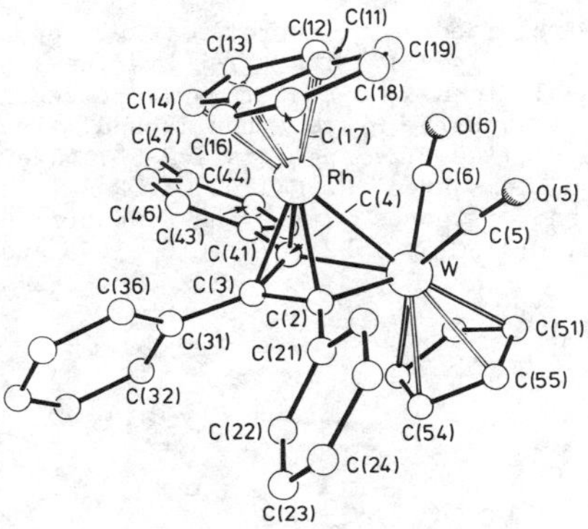

Fig. 1. $C_{38}H_{29}O_2RhW$: the molecular structure.

(BICYCLO[2.2.1]HEPTA-2,5-DIENE)[(-)-(R,R)-2-exo,3-endo-BIS(DIPHENYLPHOSPHINO)BICYCLO-[2.2.1]HEPTANE]RHODIUM(I) PERCHLORATE TETRAHYDROFURAN SOLVATE
$C_{38}H_{38}ClO_4P_2Rh,C_4H_8O$

R.E. DAVIS, B.B. MEYER, K.L. HASSETT, P.N. JURI and E.P. KYBA, 1984. Acta Cryst., C40, 21-24.

Orthorhombic, $P2_12_12_1$, a = 15.794, b = 20.541, c = 11.399 Å, Z = 4. Mo radiation, R = 0.044 for 4372 reflexions.

This structure is essentially isomorphous with that previously found for the related unsaturated bicyclo[2.2.1]hepta-2,5-diene species (1). The most pronounced effect of the clockwise torque imposed upon the rigid hydrocarbon framework upon chelation of Rh(I) by the diphosphine ligand is the distortion of the P-C-C-P torsion angle from an ideal value of 120° to 64°. Other angles are also affected; thus angles C(1)-C(6)-P(2) (129°) and C(4)-C(5)-P(1) (133°) have opened considerably from the tetrahedral value (Fig. 1). Rh-P distances are 2.323 and 2.325(1), Rh-(olefin midpoint) 2.084 and 2.102(8) Å.

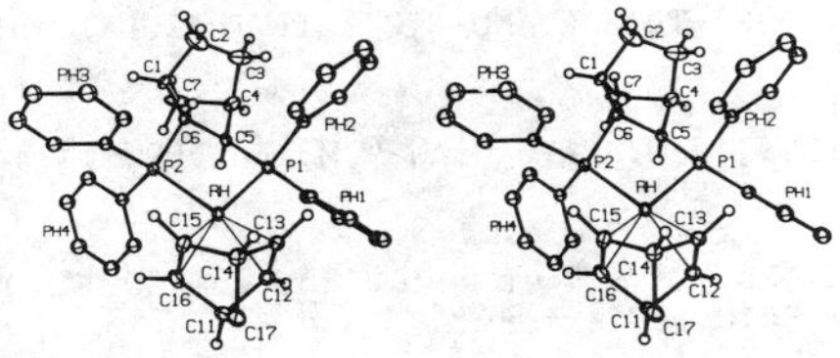

Fig. 1. A stereoview of $C_{38}H_{38}P_2Rh$; the C(17) and phenyl H atoms are omitted.

1. Structure Reports, 48B, 1201.

BIS(TRIPHENYLPHOSPHINE)(η^4-CYCLOHEXA-1,3-DIENE)TRIHYDRIDORHENIUM
$C_{42}H_{41}P_2Re$ $(Ph_3P)_2(\eta-C_6H_8)ReH_3$

D. BAUDRY, M. EPHRITIKHINE, H. FELKIN, Y. DROMZEE and Y. JEANNIN, 1984. J. Organometal. Chem., 272, 403-409.

Monoclinic, C2/c, a = 22.76, b = 10.14, c = 29.813 Å, β = 97.69°, Z = 8. Mo radiation, R = 0.064 for 1580 reflexions.

The structure may be regarded as an approximate pentagonal bipyramid with the two phosphine ligands apical (Fig. 1). The three hydride ligands were not observed directly but it is reasonable to assume that they lie between the two phosphine ligands, on the opposite side to the hydrocarbon ligand. The Re-P distances are 2.410 and 2.381 Å and the P(1)-Re-P(2) angle is 142.5°. The distances of the metal atom to the coordinating C atoms are: Re-C(1) 2.33, Re-C(2) 2.07, Re-C(3) 2.27 and Re-C(4) 2.32 Å, and the bond distances within the ring are: C(1)-C(2) 1.35, C(2)-C(3) 1.35, C(3)-C(4) 1.36, C(4)-C(5) 1.52, C(5)-C(6) 1.56 and C(6)-C(1) 1.55 Å.

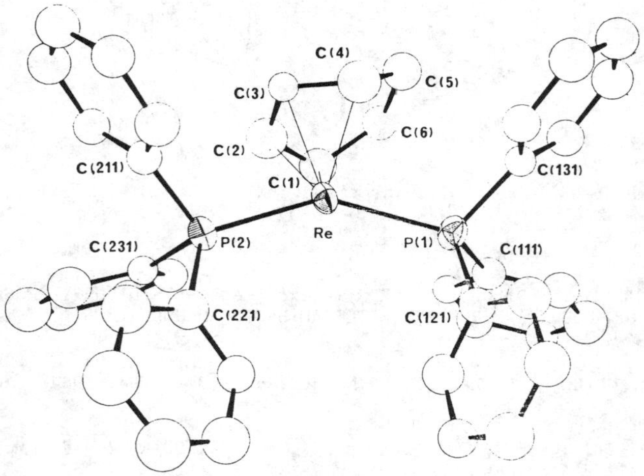

Fig. 1. Perspective view of the structure of $C_{42}H_{41}P_2Re$.

TETRAETHYLAMMONIUM (4,4',5,5'-TETRACYANO-2,2'-BIIMIDAZOLATO)(1,5-CYCLOOCTADIENE)-(TRIPHENYLPHOSPHINE)IRIDIUM(I) DICHLOROMETHANE SOLVATE
$C_{44}H_{47}IrN_9P,CH_2Cl_2$ (I) $[N(C_2H_5)_4][Ir(C_8H_{12})(P(C_6H_5)_3)(C_{10}N_8)].CH_2Cl_2$

TETRAETHYLAMMONIUM (4,4',5,5'-TETRACYANO-2,2'-BIIMIDAZOLATO)DICARBONYLIRIDIUM(I)
$C_{20}H_{20}IrN_9O_2$ (II) $[N(C_2H_5)_4][Ir(CO)_2(C_{10}N_8)]$

P.G. RASMUSSEN, O.H. BAILEY, J.C. BAYÓN and W.M. BUTLER, 1984. Inorg. Chem., 23, 343-349.

I. Orthorhombic, $P2_12_12_1$, a = 9.927, b = 11.226, c = 40.119 Å, D_m = 1.52, Z = 4. Mo radiation, R = 0.048 for 3010 reflexions.

II. Orthorhombic, Pnnm, a = 7.267, b = 20.697, c = 15.360 Å, D_m = 1.76, Z = 4. Mo radiation, R = 0.034 for 1284 reflexions.

The structures of the two complex anions are shown in Fig. 1. The Ir atom in I is five-coordinate, having an irregular trigonal bipyramidal coordination geometry with the axial sites occupied by one olefinic (C41-C42) midpoint and by one N atom of the bidentate tetracyanobiimidazole dianion (N1'). Bond lengths for I: Ir-P 2.352(5),

Ir-N_{eq} 2.322(15), Ir-N_{ax} 2.116(13), Ir-C(C=C)$_{eq}$ 2.122(17) and 2.118(15), Ir-C(C=C)$_{ax}$ 2.160(16) and 2.113(15) Å. The anions in II, which have cis-square planar geometry, pack in pairs about an inversion centre with an Ir-Ir distance of 3.183(1) Å. Bond lengths in II (which has mirror symmetry): Ir-N 2.083(6) and Ir-C 1.855(12) Å.

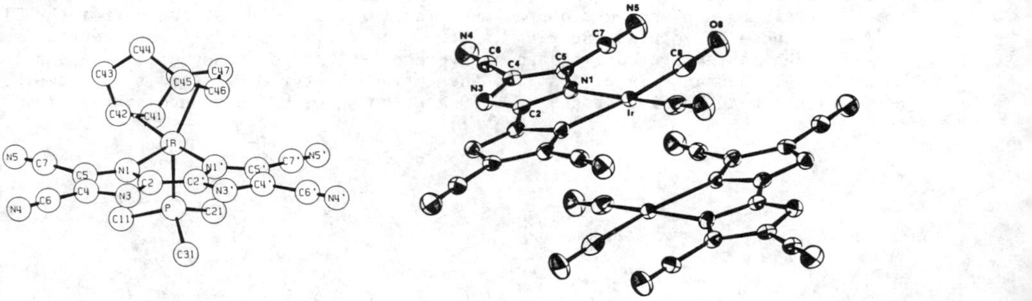

Fig. 1. The anions $[Ir(C_8H_{12})(PPh_3)(C_{10}N_8)]^-$ (left) and $[Ir(CO)_2(C_{10}N_8)]^-$ (right).

(η^3-CYCLOOCTENYL)TRIS(DIPHENYLMETHOXYPHOSPHINE)RUTHENIUM HEXAFLUOROPHOSPHATE

$C_{47}H_{52}F_6O_3P_4Ru$ $[Ru(C_8H_{13})\{P(OMe)Ph_2\}_3]^+PF_6^-$

T.V. ASHWORTH, D.C. LILES and E. SINGLETON, 1984. Organometallics, 3, 1851-1855.

Orthorhombic, $Pna2_1$, a = 20.510, b = 12.992, c = 16.954 Å, Z = 4. Mo radiation, R = 0.048 for 3257 reflexions.

The crystal structure consists of discrete cations and anions with no unusually short intermolecular distances. Coordination geometry around the Ru atom is irregular square pyramidal (Fig. 1) with P(3) occupying the axial site (Ru-P(3) 2.223 Å), P(1) and P(2) situated in basal sites cis to each other (Ru-P(1) 2.350, Ru-P(2) 2.302 Å), and the η^3-enyl portion of the cyclooctenyl ligand coordinated across the two remaining basal sites (Ru-C(41) 2.246, Ru-C(42) 2.216, Ru-C(43) 2.197 Å). An agostic Ru...H-C interaction is also present with Ru...H 2.08 Å and C-H 1.14 Å. The hydrogen atom occupies what could be the sixth coordination site of a distorted octahedron. The shortest Ru-P bond is trans to the Ru...H interaction.

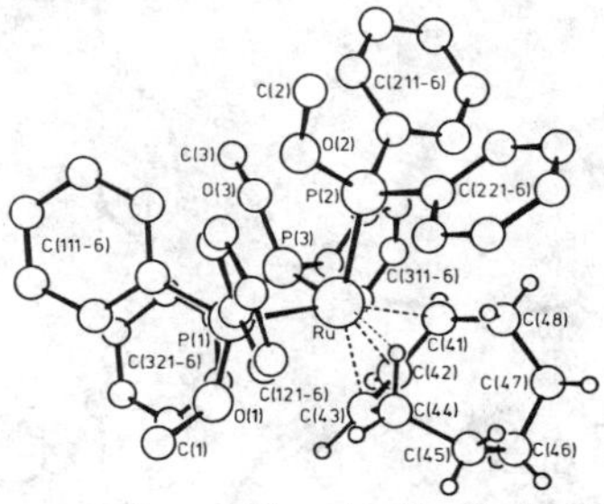

Fig. 1. Perspective view of the cation in $C_{47}H_{52}F_6O_3P_4Ru$.

TRICHLORO(η^4-TETRAPHENYLCYCLOBUTADIENE)TITANIUM TRIS(μ_2-CHLORO)HEXAKIS(TETRAHYDROFURAN)DIMAGNESIUM

$C_{52}H_{68}Cl_6Mg_2O_6Ti$ $[Mg_2Cl_3(C_4H_8O)_6]^+[(C_6H_5)_4C_4TiCl_3]^-$

M.E.E. MEIJER-VELDMAN, J.L. DE BOER, H.J. DE LIEFDE MEIJER, A.M.M. SCHREURS, J. KROON and A.L. SPEK, 1984. J. Organometal. Chem., 269, 255-265.

Monoclinic, C2, a = 17.74, b = 14.45, c = 22.03 Å, β = 92.94°, Z = 4. Mo radiation, R = 0.095 for 1703 reflexions.

The complex consists of a cation in which the Mg atoms are octahedrally surrounded by three bridging Cl atoms and three disordered THF molecules, and an anion containing a planar cyclobutadiene ring (Fig. 1). The distance from the Ti atom to the π-bonded cyclobutadiene ring is 2.0 Å and the average Ti-C bond length is 2.24 Å. The four-membered ring has an average C-C bond distance of 1.45 Å and the average Ti-Cl distance is 2.24 Å. In the cation the three Cl atoms bridge two Mg atoms with a mean Mg-Cl distance of 2.50 Å. The Mg-O(THF) distances range from 2.04 to 2.11 Å.

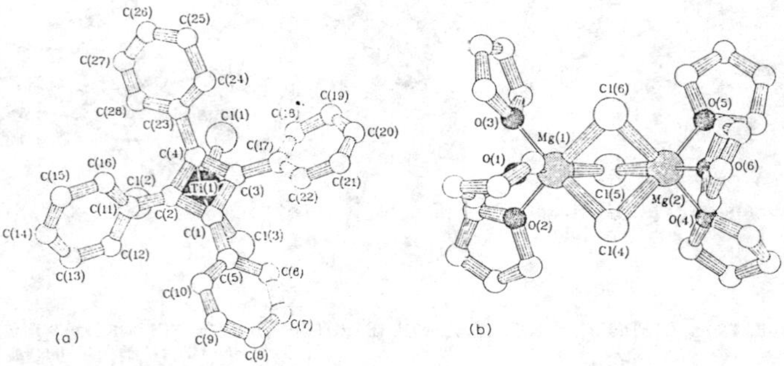

Fig. 1. Structures of the anion and cation in $C_{52}H_{68}Cl_6Mg_2O_6Ti$.

(μ2-trans-(1—4-η-CYCLOHEPTA-1,3,5-TRIEN-7-YL)(1'—4'-η-5',7'-BIS(TRIPHENYLPHOSPHONIUM)CYCLOHEPTA-1',3'-DIEN-6'-YL)ETHYLENE)BIS(TRICARBONYLIRON) BIS(HEXAFLUOROPHOSPHATE) TOLUENE SOLVATE

$C_{58}H_{46}F_{12}Fe_2O_6P_4,0.5(C_7H_8)$

N.G. CONNELLY, A.R. LUCY, R.M. MILLS, M.W. WHITELEY and P. WOODWARD, 1984. J. Chem. Soc. Dalton, 161-168.

Monoclinic, P2₁/c, a = 26.795, b = 19.699, c = 26.297 Å, β = 116.37°, Z = 8. Mo radiation, R = 0.142 for 6436 reflexions (at 220 K). [Crystal quality very poor.]

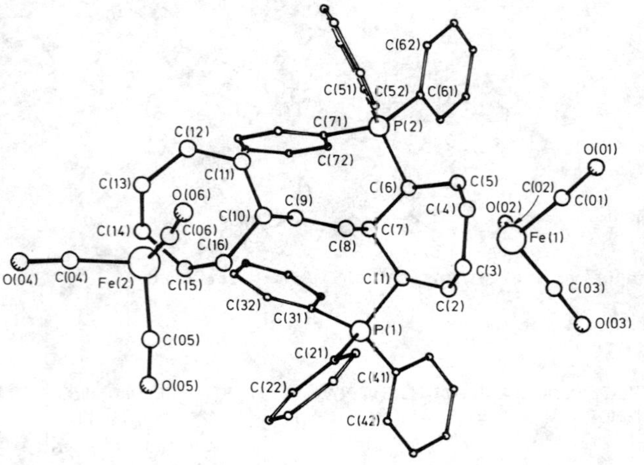

Fig. 1. $C_{58}H_{46}F_{12}Fe_2O_6P_4,0.5(C_7H_8)$: the dication.

The asymmetric unit comprises two complex dications, four PF_6^- anions and two half molecules of toluene solvate, one of which is disordered. The dication (Fig. 1) comprises two open C_7 rings joined by a trans ethylenic bridge of four coplanar carbon atoms with the two phosphine ligands bonded to the two C atoms adjacent to the bridgehead of one C_7 ring. The remaining carbon atoms of this ring are η^4-bonded to an $Fe(CO)_3$ group. In the other C_7 ring four atoms adjacent to the bridghead are η^4-bonded to another $Fe(CO)_3$ group while the other two atoms form an ethylenic moiety not bonded to the metal. The high R factor precludes any discussion of the molecular dimensions.

BIS[(μ-CHLORO)(CHLORO)(μ-3,3',4,4'-TETRAMETHYL-1,1'-DIPHOSPHAFERROCENE-P,P')(TRIPHENYLPHOSPHINE)RUTHENIUM] CHLOROFORM SOLVATE

$C_{60}H_{62}Cl_4Fe_2P_6Ru_2,2CHCl_3$ $[(PPh_3)Ru((C_6H_8P)_2Fe)(\mu\text{-}Cl)Cl]_2.2CHCl_3$

B. DESCHAMPS, F. MATHEY, J. FISCHER and J.H. NELSON, 1984. Inorg. Chem., 23, 3455-3462.

Triclinic, P$\bar{1}$, a = 12.566, b = 13.100, c = 12.202 Å, α = 110.58, β = 104.24, γ = 66.08°, Z = 1. Cu radiation, R = 0.069 for 3814 reflexions.

The centrosymmetric molecule is shown in Fig. 1. The phospholyl rings remain fully aromatic in the complex (mean C-C 1.415(4), P-C 1.744(3) Å). The Ru atoms are six coordinate with asymmetric chloride bridges (Ru-Cl 2.421(1) and 2.496(1) Å), terminal chlorides (Ru-Cl 2.409(1) Å), terminal PPh_3 (Ru-P 2.311(1) Å), and two bridging $(C_6H_8P)_2Fe$ ligands with short Ru-P bonds (mean 2.2935(4) Å). The diphosphaferrocene is in an eclipsed conformation.

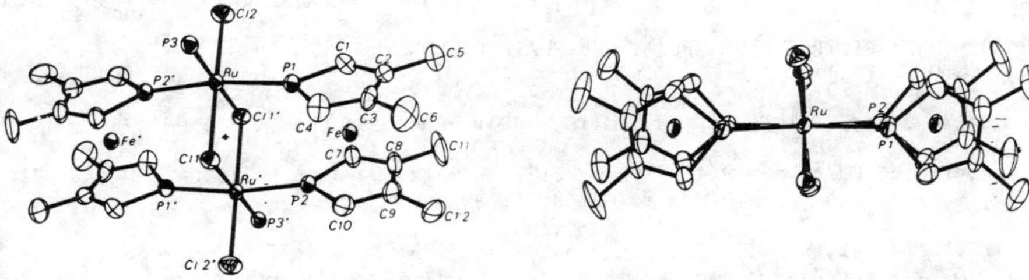

Fig. 1. Two views of $[(PPh_3)Ru((C_6H_8P)_2Fe)(\mu\text{-}Cl)Cl]_2$, phenyl rings omitted.

(μ_3-σ^2,η^3-CYCLODODECA-1,4,8-TRIENE-8,10-DIYL)TRIS(TRIPHENYLPHOSPHINEGOLD)OCTACARBONYL-TRI-RUTHENIUM

$C_{74}H_{60}Au_3O_8P_3Ru_3$ $Au_3Ru_3(C_{12}H_{15})(CO)_8(PPh_3)_3$

M.I. BRUCE, O.B. SHAWKATALY and B.K. NICHOLSON, 1984. J. Organometal. Chem., 275, 223-232.

Monoclinic, $P2_1/c$, a = 13.574, b = 40.634, c = 14.617 Å, β = 92.58°, Z = 4. Cu radiation, R = 0.078 for 3233 reflexions.

The complex contains a capped trigonal-bipyramidal Ru_3Au_3 core in which the $C_{12}H_{15}$ ligand is bonded to the Ru_3 face (Fig. 1). There are twelve metal-metal bonds: three Au-Au (2.840-2.911 Å); six Au-Ru (2.737-2.943 Å), and three Ru-Ru (2.845-2.929 Å). The Au-Ru distances separate into three short (2.737-2.775 Å) involving Ru(3), two longer ones which are the "opposite" edges of the Au(1)Au(3)Ru_3 trigonal bipyramid not involved with the capping Au(2) atom (2.882, 2.896 Å), and the equatorial edge of the same trigonal bipyramid (2.929 Å). The Au-P distances are 2.307, 2.261 and 2.215 Å and the $C_{12}H_{15}$ ligand-metal distances are: Ru(1)-C(R3) 2.14, Ru(2)-C(R1) 2.05, Ru(3)-C(R1) 2.22, Ru(3)-C(R2) 2.23, and Ru(3)-C(R3) 2.11 Å.

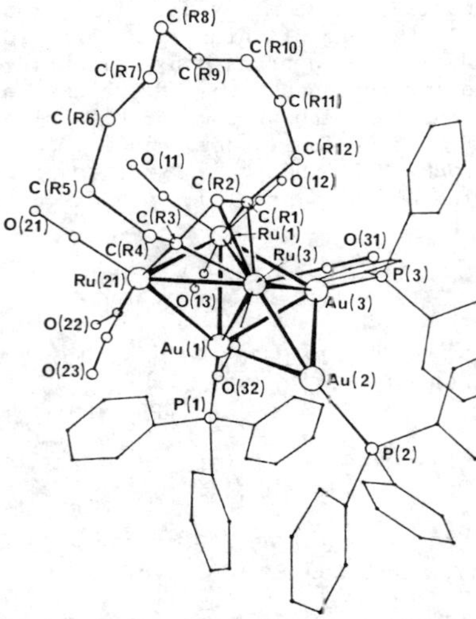

Fig. 1. View of $C_{74}H_{60}Au_3O_8P_3Ru_3$.

TETRAAQUA(ETHYLENEDIAMINE)NICKEL(II) SULPHATE DIHYDRATE
$C_2H_{20}N_2NiO_{10}S$ (I) $[Ni(en)(H_2O)_4][SO_4],(H_2O)_2$

TETRAAQUA(2,2'-BIPYRIDYL)NICKEL(II) SULPHATE DIHYDRATE
$C_{10}H_{20}N_2NiO_{10}S$ (II) $[Ni(bpy)(H_2O)_4][SO_4],(H_2O)_2$

P.C. HEALY, J.M. PATRICK and A.H. WHITE, 1984. Aust. J. Chem., 37, 921-928.

I. Monoclinic, C2/c, a = 9.459, b = 12.192, c = 12.294 Å, β = 119.54°, D_m = 1.72, Z = 4. Mo radiation, R = 0.028 for 2248 reflexions.

II. Triclinic, P$\bar{1}$, a = 11.476, b = 9.351, c = 7.793 Å, α = 77.63, β = 83.52, γ = 87.40°, D_m = 1.68, Z = 4. Mo radiation, R = 0.031 for 4323 reflexions.

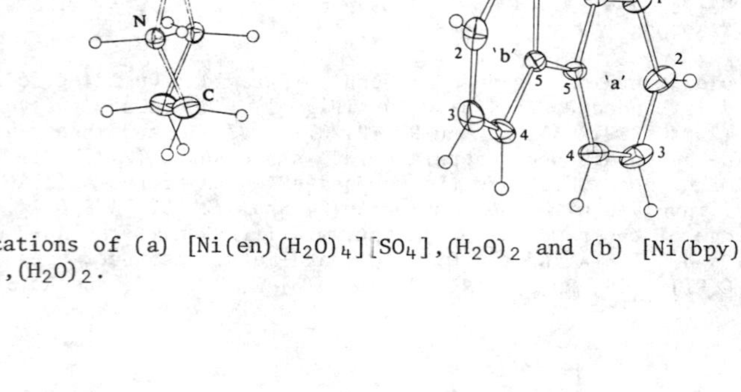

Fig. 1. The cations of (a) $[Ni(en)(H_2O)_4][SO_4],(H_2O)_2$ and (b) $[Ni(bpy)(H_2O)_4]$-$[SO_4],(H_2O)_2$.

Both I and II are examples of octahedrally coordinated Ni(II) (Fig. 1). In I, the cation, tetraaqua(ethylenediamine)nickel(II) and anion both lie about a crystallographic twofold axis. For I and II respectively, Ni-N 2.061(2), 2.063(2), Ni-O (trans to O,N respectively) 2.106(2), 2.051(2), 2.063(2), 2.087(2) Å; N-Ni-N 84.84(7), 79.37(5)°.

DICHLOROBIS(ETHYLENDIAMINE)GOLD CHLORIDE NITRITE

$C_4H_{16}AuCl_{2.5}N_{4.5}O$ $[Au(en)_2Cl_2]^+.Cl_{0.5}^-(NO_2^-)_{0.5}$

L.Kh. MINACHEVA, G.G. SADIKOV, V.G. SAKHAROVA, A.Sh. GLADKAYA and M.A. PORAI-KOSHITS, 1983. Koord. Khim. 9, 566-570 [Sov. J. Coord. Khim., 9, 329-333.]

Monoclinic, Cc, a = 16.521, b = 7.149, c = 16.325 Å, β = 143.13°, Z = 4. Mo radiation, R = 0.030 for 934 reflexions.

In the cation (Fig. 1) the Au atom has coordination 4+2. The four N atoms of two bidentate ethylenediamine groups form the equatorial plane of the Au polyhedron (Au-N 2.03-2.09(4) Å). Square-pyramidal coordination about Au is completed by two Cl^- ions (Au-Cl 3.116 Å on average). Both chelate rings have the gauche conformation. The Cl^- and NO_2^- anions randomly occupy one position. In the crystal, cations are bound in a layer by a system of H bonds of the Cl...H(N) type (2.06-2.40 Å). The outersphere Cl^- and NO_2^- anions bind the layers of hydrogen bonded cations (Cl...H(N) 2.25, 2.44, O...H(N) 1.83, 1.91 Å) to form a three dimensional network.

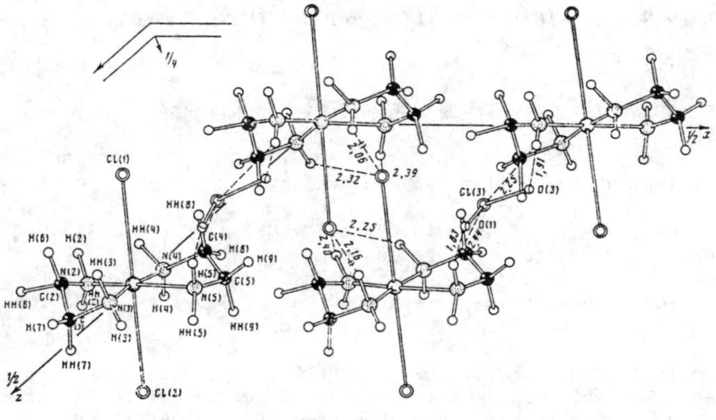

Fig. 1. $C_4H_{16}AuCl_{2.5}N_{4.5}O$: projection of the structure onto the (010) plane.

ETHYLENEDIAMMONIUM ETHYLENEDIAMINETETRAFLUOROCHROMIUM(III) CHLORIDE

$C_4H_{18}ClCrF_4N_4$ $(en\text{-}H_2)[CrF_4(en)]Cl$

C. DIAZ, A. SEGUI, J. RIBAS, X. SOLANS, M. FONT-ALTABA, A. SOLANS and J. CASABO, 1984. Transition Met. Chem., 9, 469-473.

Monoclinic, $P2_1$, a = 7.364, b = 10.510, c = 7.540 Å, β = 100.13°, Z = 2. Mo radiation, R = 0.057 for 1000 reflexions.

The Cr atom has octahedral coordination (Fig. 1) with trans Cr-F 1.893(8) and 1.885(10), Cr-F (trans to N) 1.930(4) and 1.889(4), Cr-N 2.081(6) and 2.078(7) Å, N-Cr-N 82.4(3)°. The ions are linked by N-H...F and N-H...Cl hydrogen bonds.

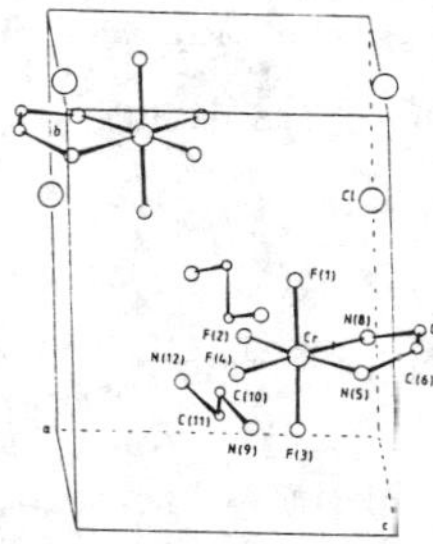

Fig. 1. A view of the (en-H_2)[CrF_4(en)]Cl crystal structure.

AQUAPENTACARBONYLRHENIUM(I) HEXAFLUOROARSENATE(V)

$C_5H_2AsF_6O_6Re$ (I) [Re(H_2O)$(CO)_5$][AsF_6]

AQUATRICARBONYL-(N,N,N',N'-TETRAMETHYLETHANE-1,2-DIAMINE)RHENIUM(I) HEXAFLUORO-ARSENATE(V)

$C_9H_{18}AsF_6N_2O_4Re$ (II) [Re(H_2O)$(CO)_3$(tmen)][AsF_6]

AQUATRICARBONYL-(N,N,N',N'-TETRAMETHYLETHANE-1,2-DIAMINE)RHENIUM(I) TETRAFLUORO-BORATE(III)

$C_9H_{18}BF_4N_2O_4Re$ (III) [Re(H_2O)$(CO)_3$(tmen)][BF_4]

E. HORN and M.R. SNOW, 1984. Aust. J. Chem., 37, 1375-1393.

I. Monoclinic, C2/c, a = 27.085, b = 7.208, c = 13.404 Å, β = 105.50°, D_m > 2.4, Z = 8. Mo radiation, R = 0.027 for 1514 reflexions.

II. Monoclinic, $P2_1$, a = 7.858, b = 15.68, c = 7.887 Å, β = 117.11°, D_m = 2.23, Z = 2. Mo radiation, R = 0.020 for 1362 reflexions.

III. Monoclinic, C2/m, a = 14.134, b = 16.010, c = 7.729 Å, β = 113.16°, D_m = 2.04, Z = 4. Mo radiation, R = 0.061 for 1374 reflexions.

The Re atom in I is octahedral (Re-O 2.206(8), Re-C 2.00(1) Å) and the AsF_6^- anion disordered. The metal in II has a distorted octahedral coordination with the carbonyl groups in a mutually cis stereochemistry (Re-O 2.268(8), Re-N 2.26(3), Re-C 1.89(3) Å, N-Re-N 80.2(4)°). III has a non-planar bidentate ligand tmen disordered as indicated in Fig. 1. The BF_4^- is also disordered (Re-O 2.24(1), Re-N 2.23(1), Re-C 1.88(2) Å, N-Re-N 81.6(7)°).

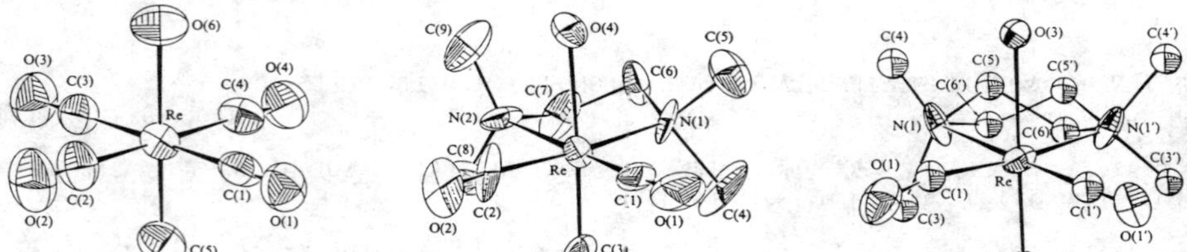

Fig. 1. The structures of [Re(H_2O)$(CO)_5$]$^+$ (left), [Re(H_2O)$(CO)_3$(tmen)]$^+$ (centre) and [Re(H_2O)$(CO)_3$(tmen)]$^+$ (right).

cis-AQUACHLOROBIS(ETHANE-1,2-DIAMINE)RUTHENIUM(III) TRIFLUOROMETHANESULPHONATE HYDRATE
$C_6H_{18}ClF_6N_4O_7RuS_2,H_2O$ $[Ru(en)_2(H_2O)Cl](CF_3O_3S)_2,H_2O$

T.W. HAMBLEY and G.A. LAWRANCE, 1984. Aust. J. Chem., 37, 435-441.

Triclinic, P$\bar{1}$, a = 11.661, b = 9.961, c = 9.130 Å, α = 78.83, β = 70.42, γ = 85.29°, D_m = 2.01, Z = 2. Mo radiation, R = 0.031 for 4656 reflexions.

The Ru in this pale yellow complex (Fig. 1) is octahedral (mean Ru-N, Ru-Cl, Ru-O distances 2.100(2), 2.323(1), 2.093(2) Å, bidentate angle N-Ru-N 82.2(1)°).

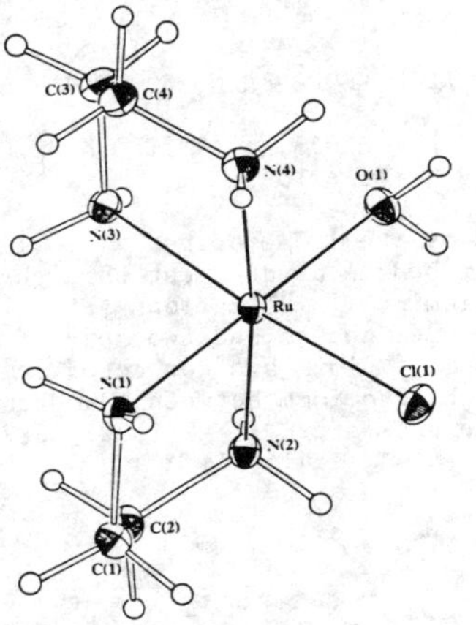

Fig. 1. The structure of cis-$[Ru(en)_2(H_2O)Cl]$.

BIS(1,3-DIAMINOPROPANE)PLATINUM(II)BIS(1,3-DIAMINOPROPANE)DIIODOPLATINUM(IV) TETRAPERCHLORATE
$(C_6H_{20}Cl_2IN_4O_8Pt)_n$

M. CANNAS, G. MARONGIU, H.J. KELLER, B. MULLER and R. MARTIN, 1984. Z. Naturforsch., B39, 197-200.

Monoclinic, $P2_1$, a = 8.74, b = 11.36, c = 8.63 Å, β = 107.6°, D_m = 2.65, Z = 2. Mo radiation, R = 0.063 for 1068 reflexions.

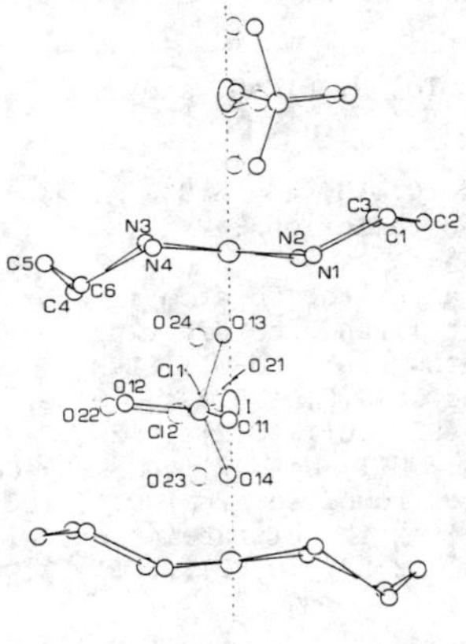

Fig. 1. A view of part of the $(C_6H_{20}Cl_2IN_4O_8Pt)_n$ structure.

The structure (Fig. 1) consists of chains formed by equivalent $Pt(tn)_2$ cations separated by b/2 and bridged by I^- ions near to the mid-point. Hydrogen bonds link the chains to the perchlorate anions. Main bond lengths are: Pt-I 2.75, 2.77(2), Pt-N 2.04-2.16(4) Å. The chelate rings are related by a pseudo inversion centre and have chair conformation. The diffraction photographs show diffuse patterns and many crystals are twinned.

BIS(1,3-PROPANEDIAMINE)MERCURY(II) SULPHATE DIHYDRATE

$C_6H_{20}HgN_4O_4S,2H_2O$ $[Hg(C_3H_{10}N_2)_2]SO_4,2H_2O$

B. KAMENAR, D. GRDENIĆ and A. HERGOLD-BRUNDIĆ, 1984. Acta Cryst., C40, 1836-1838.

Monoclinic, P2, a = 13.830, b = 6.023, c = 8.250 Å, β = 97.45°, D_m = 2.34, Z = 2. Mo radiation, R = 0.033 for 1914 reflexions.

Hg^{2+} ions are bridged by the 1,3-propanediamine ligands (which also coordinate to them in a distorted tetrahedral arrangement through the N atoms) forming polymeric chains of $[Hg(1,3\text{-propanediamine})_2]_n^{2n+}$ cations (Fig. 1). There are two crystallographically independent polymer chains and two sets of SO_4^{2-} ions since Hg and S lie on special positions. The SO_4^{2-} ions are connected in chains with the H_2O molecules through hydrogen bonds which also form between NH_2 groups and the SO_4^{2-} and H_2O oxygen atoms. Some bond lengths are Hg(1)-N(1) 2.365(7), Hg(1)-N(2') 2.244(9), Hg(2)-N(3) 2.179(10), Hg(2)-N(4') 2.416(11) Å.

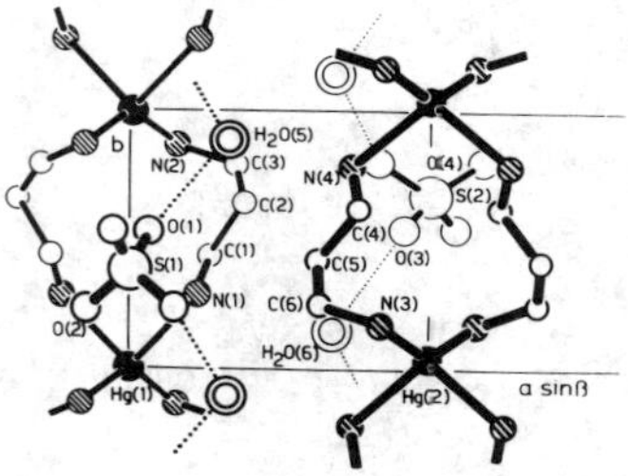

Fig. 1. The crystal structure of bis(1,3-propanediamine)mercury(II) sulphate dihydrate viewed along z.

$(-)^{CD}_{500}$-(2-AMINOETHANESELENINATO)BIS(ETHYLENEDIAMINE)COBALT(III) NITRATE

$C_6H_{22}CoN_7O_8Se$

K.-I. OKAMOTO, T. KONNO, M. NOMOTO, H. EINAGA and J. HIDAKA, 1984. Bull. Chem. Soc. Japan, 57, 1494-1498.

Monoclinic, $P2_1$, a = 9.572, b = 10.410, c = 8.838 Å, β = 115.35°, D_m = 1.90, Z = 2. Mo radiation, R = 0.035 for 2358 reflexions.

The coordination geometry about the Co atom is approximately octahedral (Fig. 1) with the 2-aminoethaneseleninate ligand coordinating to the metal atom with N and O atoms. The important bond lengths are: Co-O(1) 1.922, Co-N(4) 1.988, Co-N(1) 1.966, Co-N(2) 1.967, Co-N(3) 1.950 and Co-N(5) 1.972 Å. Other distances are: Se-O(1) 1.698, Se-O(2) 1.680, Se-C(1) 1.952, C(1)-C(2) 1.520, C(2)-N(4) 1.479 Å. The coordination angles O(1)-Co-N(4), N(1)-Co-N(2) and N(3)-Co-N(5) are respectively 95.6, 85.6 and 85.9°. The angles round Se are 100.1, 100.1 and 104.5° (the latter O(1)-Se-O(2)). Hydrogen bonds link the nitrate anions with the complex cations (O(2)...N(2') 2.802, O(2)...N(3') 2.827, O(12)...N(1) 2.948, O(21)...N(5) 2.920 Å.

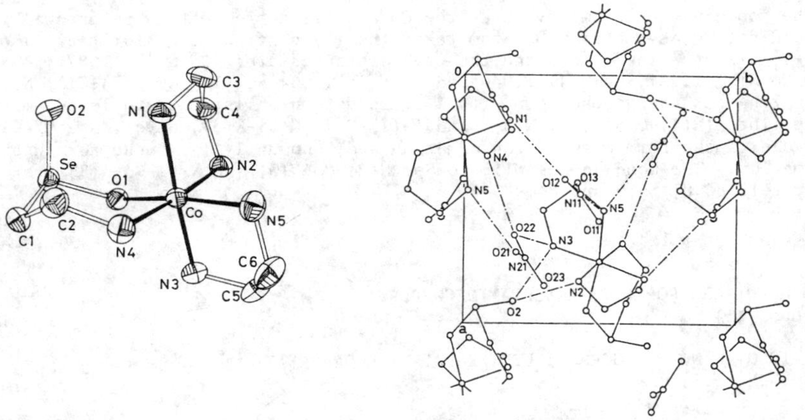

Fig. 1. Perspective drawing of the Co coordination sphere (left), and projection of the crystal structure down c showing hydrogen bonds (right).

TRIS(ETHYLENEDIAMINE-N,N')NICKEL DODECAHYDRODODECABORATE MONOHYDRATE

$C_6H_{36}B_{12}N_6Ni,H_2O$ $[Ni(en)_3]^{2+}.[B_{12}H_{12}]^{2-},H_2O$

FU ZHUJI, CAI ZHAOPING, PAN KEZHEN and ZHANG LUN, 1984. J. Struct. Chem., 3, 231-234.

Trigonal, $R\bar{3}$, a = 12.890, c = 23.955 Å, Z = 6. Mo radiation, R = 0.069 for 985 reflexions.

The $[Ni(en)_3]^{2+}$ ion has crystallographic D_3 symmetry with Ni-N 2.135(4), 2.142(4), N-C 1.505 and 1.511(5), C-C 1.512(6) Å. The icosahedral anion has threefold symmetry with mean B-B 1.786 Å.

$(+)^{CD}_{500}$-(2-(METHYLSELENO)ETHYLAMINE)BIS(ETHYLENEDIAMINE)COBALT(III) PERCHLORATE

$C_7H_{25}Cl_3CoN_5O_{12}Se$ $[Co(mseea)(en)_2]^{3+}3ClO_4^-$

T. KONNO, K.-I. OKAMOTO and J. HIDAKA, 1984. Bull. Chem. Soc. Japan, 57, 3104-3110.

Orthorhombic, $P2_12_12_1$, a = 12.220, b = 19.224, c = 8.977 Å, D_m = 1.91, Z = 4. Mo radiation, R = 0.087 for 2176 reflexions.

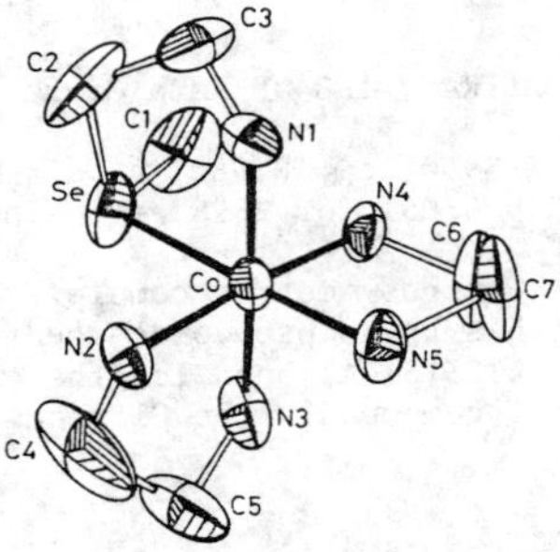

Fig. 1. Perspective drawing of the cation in $C_7H_{25}Cl_3CoN_5O_{12}Se$.

The coordination geometry about the Co atom is approximately octahedral (Fig. 1). The coordinated (asymmetric) Se atom takes the (R) configuration for the Λ isomer. The Co-Se distance is 2.386 Å and the Co-N distances are: Co-N(1) 1.987, Co-N(2) 1.983, Co-N(3) 1.921, Co-N(4) 1.940 and Co-N(5) 1.964 Å. The Se-C(2)-C(3)-N(1) chelate ring has the gauche form with λ conformation whilst the ethylenediamine chelate rings (N(4)-C(6)-C(7)-N(5)) and (N(2)-C(4)-C(5)-N(3)) have respectively a reasonable gauche form with δ conformation and, unusually, a gauche form with λ conformation. The bond angles N(1)-Co-Se, N(2)-Co-N(3) and N(4)-Co-N(5) are respectively 88.0, 85.9 and 84.9°.

BROMO-N,N'-DI-ISOPROPYLETHYLENEDIAMINECOPPER(I)
$C_8H_{20}BrCuN_2$ (I) DIEDCuBr

BIS[(μ_2-IODO)-N,N'-DI-ISOPROPYLETHYLENEDIAMINECOPPER(I)]
$C_{16}H_{40}Cu_2I_2N_4$ (II) $[DIEDCuI]_2$

D.A. HAITKO, 1984. J. Coord. Chem., 13, 119-122.

I. Monoclinic, C2/c, a = 7.650, b = 8.792, c = 19.402 Å, β = 96.50°, Z = 4. Cu radiation, R = 0.067 for 728 reflexions.

II. Monoclinic, $P2_1/c$, a = 14.286, b = 11.349, c = 15.449 Å, β = 95.46°, Z = 4. Cu radiation, R = 0.030 for 3708 reflexions.

The bromo compound (I) has twofold crystallographic symmetry (Fig. 1) (with a planar CuN_2Br moiety) with Cu-Br 2.263(2), Cu-N 2.062(6) Å, N-Cu-N 85.0 and N-Cu-Br 137.5°. Compound (II) is an iodo-bridged dimer in which each Cu atom has distorted tetrahedral coordination, with Cu(1)-I(1) 2.661(1), Cu(1)-I(2) 2.581(1), Cu(2)-I(1) 2.579(1), Cu(2)-I(2) 2.666(1), Cu...Cu 2.73, Cu-N 2.124-2.230(4) Å.

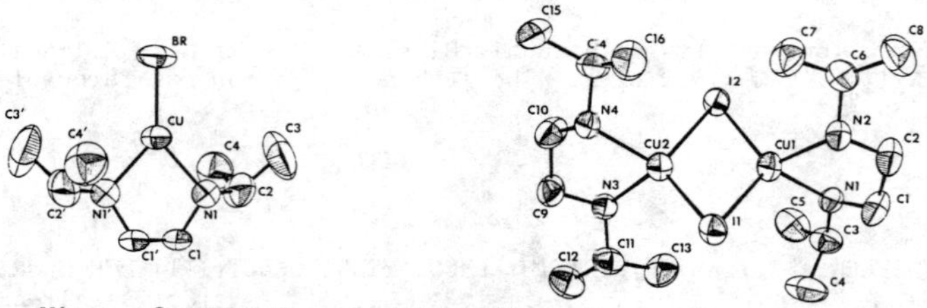

Fig. 1. Views of DIEDCuBr and $[DIEDCuI]_2$.

trans-DIAQUABIS(ETHYLENEDIAMINE)NICKEL(II) BIS(TRICHLOROACETATE)
$C_8H_{20}Cl_6N_4NiO_6$ $[Ni(C_2H_8N_2)_2(H_2O)_2](C_2Cl_3O_2)_2$

S. GARCÍA-GRANDA and F. GÓMEZ-BELTRÁN, 1984. Acta Cryst., C40, 1145-1147.

Triclinic, $P\bar{1}$, a = 7.2099, b = 8.5966, c = 8.6881 Å, α = 67.72, β = 82.74, γ = 87.08°, Z = 1. Mo radiation, R = 0.051 for 2328 reflexions.

The crystal is composed of centrosymmetric complex $[Ni(en)_2(H_2O)_2]^{2+}$ cations and $(C_2Cl_3O_2)^-$ anions (Fig. 1). Ni has trans pseudo-octahedral coordination with Ni-N 2.093(4) and 2.096(4) and Ni-O 2.123(3) Å; bond lengths and angles in both cation and anion are normal for this type of compound. The Cl atoms of the trichloroacetate are not disordered.

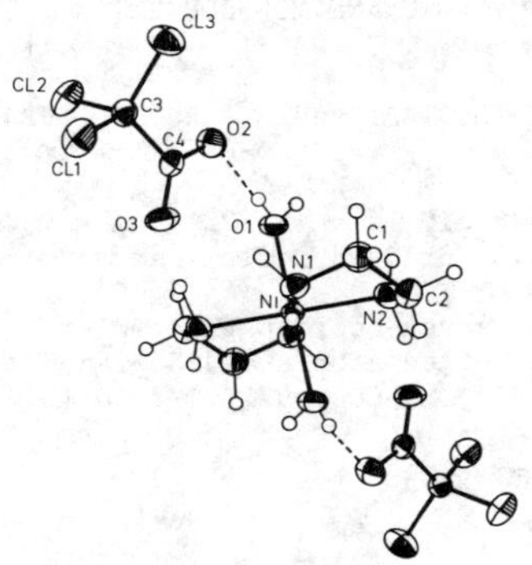

Fig. 1. $[Ni(C_2H_8N_2)_2(H_2O)_2](C_2Cl_3O_2)_2$: perspective view of the complex molecule.

AQUA[3,3'-(ETHYLENEDIIMINO)DIPROPIONAMIDE]NITRATOCOPPER(II) NITRATE MONOHYDRATE
$C_8H_{20}CuN_6O_9,H_2O$ (I) $[Cu(NO_3)(C_8H_{18}N_4O_2)(H_2O)],NO_3,H_2O$

AQUA[3,3'-(ETHYLENEDIIMINO)DIPROPIONAMIDO(2-)]COPPER(II) MONOHYDRATE
$C_8H_{18}CuN_4O_3,H_2O$ (II) $[Cu(C_8H_{16}N_4O_2)(H_2O)],H_2O$

T.-J. LEE, T.-H. LU, S.-H. LIU, C.-S. CHUNG and T.-Y. LEE, 1984. Acta Cryst., C40, 1131-1135.

I. Triclinic, $P\bar{1}$, a = 7.663, b = 10.725, c = 10.767 Å, α = 91.82, β = 108.18, γ = 90.14°, D_m = 1.653, Z = 2. Mo radiation, R = 0.068 for 2197 reflexions.

II. Monoclinic, $P2_1/c$, a = 6.795, b = 16.872, c = 12.103 Å, β = 118.02°, D_m = 1.653, Z = 4. Mo radiation, R = 0.060 for 1861 reflexions.

Compound I (Fig. 1) is a six-coordinate octahedral complex with a nitrate O atom and a water O atom in trans positions, with Cu-O 2.548(9) and 2.399(9) Å. The quadridentate ligand bcen coordinates to Cu with two tertiary N atoms and two O atoms, with Cu-N 1.983(7) and 2.002(6) and Cu-O 1.957(5) and 1.974(6) Å. In contrast, compound II exhibits fivefold tetragonal pyramidal coordination (Fig. 1), where the deprotonated ligand $(H_{-2}bcen)^{2-}$, coordinates to Cu with four N atoms and the fifth site is occupied by a water O atom, with Cu-N 1.957(5), 2.017(7), 2.040(5) and 1.954(6) and Cu-O 2.586(4) Å. Some intermolecular N...O and O...O distances in both compounds are sufficiently short to indicate hydrogen bonding.

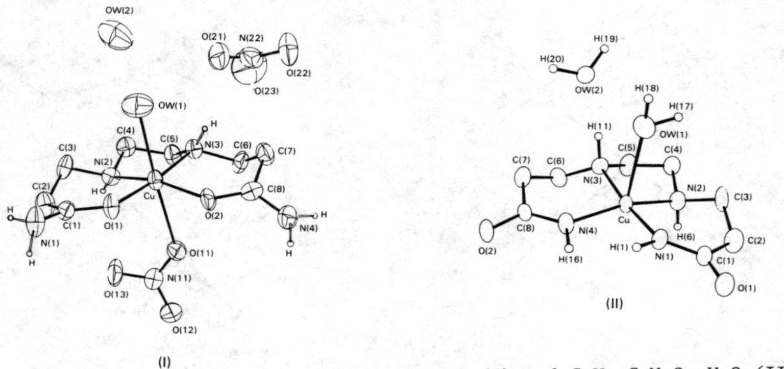

Fig. 1. Perspective views of $C_8H_{20}CuN_6O_9,H_2O$ (I) and $C_8H_{18}CuN_4O_3,H_2O$ (II).

BROMOTRICARBONYL(1,4,7-TRIAZACYCLONONANE)MOLYBDENUM(II) PERCHLORATE MONOHYDRATE

$C_9H_{15}BrClMoN_3O_4,H_2O$ $[(C_6H_{15}N_3)Mo(CO)_3Br](ClO_4).H_2O$

P. CHAUDHURI, K. WIEGHARDT, Y.-H. TSAI and C. KRÜGER, 1984. Inorg. Chem., 23, 427-432.

Triclinic, P$\bar{1}$, a = 8.375, b = 8.610, c = 13.122 Å, α = 90.68, β = 90.68, γ = 118.24°, Z = 2. Mo radiation, R = 0.043 for 2894 reflexions.

The structure consists of seven-coordinate complex cations, ClO_4^- anions, and water molecules (Fig. 1). The cation adopts a 4:3 piano-stool configuration with Mo-Br 2,676(1), Mo-N 2.240(5)-2.273(5), Mo-CO 1.972(7)-2.008(6) Å.

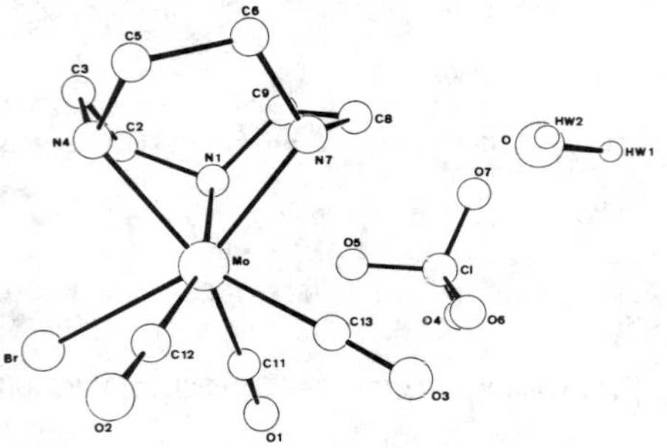

Fig. 1. Structure of the $[(C_6H_{15}N_3)Mo(CO)_3Br]^+$ cation.

FLUORO(N,N,N',N'-TETRAMETHYLETHYLENEDIAMINE)TRICARBONYLRHENIUM(I)

$C_9H_{16}FN_2O_3Re$ (I) $Re(CO)_3(tmen)F$

HYDROGEN-BIS[FLUORO(N,N,N',N'-TETRAMETHYLETHYLENEDIAMINE)TRICARBONYLRHENIUM(I)] TRIFLUOROBORATE

$C_{18}H_{34}BF_5N_4O_7Re_2$ (II) $[Re(CO)_3(tmen)F]_2H,HOBF_3$

E. HORN and M.R. SNOW, 1984. Aust. J. Chem., 37, 35-45.

I. Monoclinic, P2_1/c, a = 8.202, b = 13.115, c = 12.048 Å, β = 102.24°, D_m = 2.12, Z = 4. Mo radiation, R = 0.037 for 2876 reflexions.

II. Monoclinic, P2_1/c, a = 17.495, b = 10.772, c = 15.447 Å, β = 101.409°, D_m = 2.08, Z = 4. Cu radiation, R = 0.057 for 2904 reflexions.

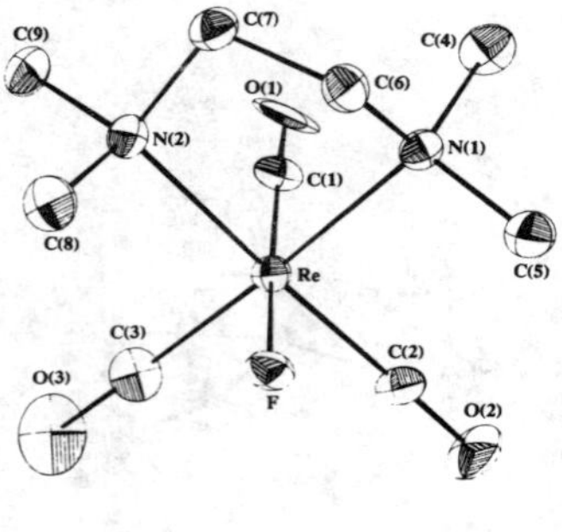

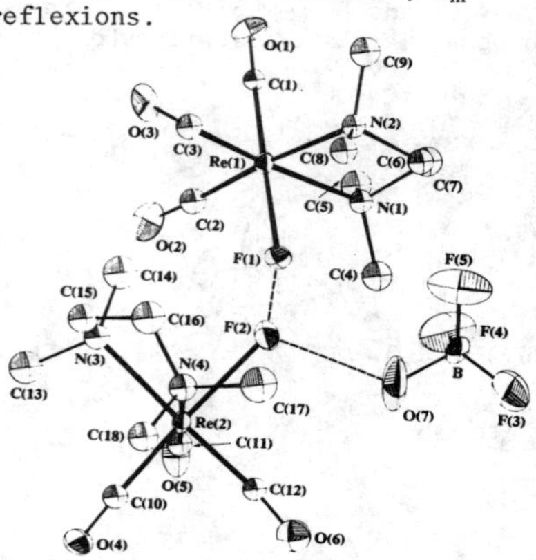

Fig. 1. The structure of $Re(CO)_3(tmen)F$ (left) and $[Re(CO)_3(tmen)F]_2H,HOBF$ (right).

In I, the rhenium atom (Fig. 1) is at the centre of an approximate octahedron with carbonyls in a fac arrangement (Re-C, mean, 1.906 Å) with the bidentate (Re-N, mean, 2.262 Å) and halide (Re-F 2.039 Å) completing the coordination. In II, there are two octahedral $Re(CO)_3$(tmen)F molecules (Fig. 1) joined by a bifluoride bridge (F...H-F 2.50 Å) and one fluorine hydrogen bonded to a fluoroborate anion (F...H-O 2.67 Å). The rhenium atoms are at the centre of a distorted octahedron with fac carbonyls (mean Re-C 1.910, 1.885; Re-N 2.245, 2.219; Re-F 2.085, 2.233 Å respectively).

BROMO(1-OXA-4,7,11-TRIAZACYCLOTRIDECANE)COPPER BROMIDE MONOHYDRATE
$C_9H_{21}Br_2CuN_3O,H_2O$ (I) $[Cu(13\text{-}aneN_3O)Br]Br,H_2O$

1,4,8,11-TETRAAZACYCLOTETRADECANEBIS(NITRATO)NICKEL
$C_{10}H_{24}N_6NiO_6$ (II) $[Ni(14\text{-}aneN_4)(NO_3)_2]$

V.J. THOM, C.C. FOX, J.C.A. BOEYENS and R.D. HANCOCK, 1984. J. Am. Chem. Soc., 106 5947-5955.

I. Monoclinic, $P2_1/c$, a = 7.753, b = 14.78, c = 13.83 Å, β = 104.4°, Z = 4. Mo radiation, R = 0.099 for 1328 reflexions.

II. Monoclinic, $P2_1/c$, a = 7.664, b = 12.87, c = 8.93 Å, β = 114.2°, Z = 2. Mo radiation, R = 0.063 for 1030 reflexions.

In (I) the cation has an approximate square-pyramidal arrangement as shown in Fig. 1. The Cu atom lies 0.47 Å above the plane of the N and O donor atoms. Molecule (II) lies about an inversion centre and has axial nitrate ions with a long Ni-O distance of 2.169(4) compared with 2.050 and 2.060(6) Å for Ni-N (ring) distances.

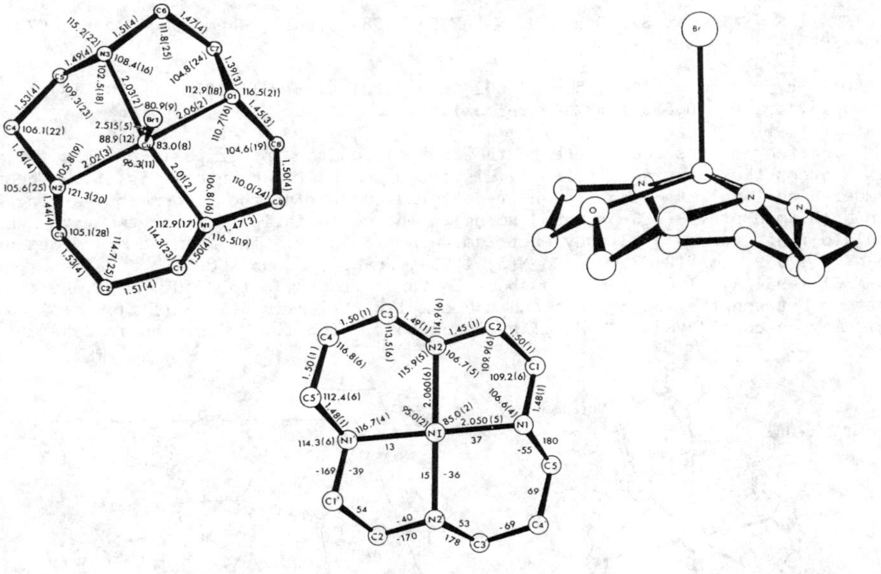

Fig. 1. Views of the $[Cu(13\text{-}aneN_3O)Br]^+$ cation and of the $Ni(14\text{-}aneN_4)$ system.

(NITRATO-O,O')(1,4,7-TRIAZACYCLODODECANE-N,N',N'')COPPER(II) NITRATE
$C_9H_{21}CuN_5O_6$ $[Cu(C_9H_{21}N_3)(NO_3)](NO_3)$

H. GAMPP, M.M. ROBERTS and S.J. LIPPARD, 1984. Inorg. Chem., 23, 2793-2798.

Orthorhombic, Pnma, a = 14.104, b = 8.882, c = 12.054 Å, Z = 4. Mo radiation, R = 0.080 for 1028 reflexions.

The structure of the cation (Fig. 1) is irregular and may be described as trigonal bipyramidal with axial positions occupied by one O atom of a bidentate NO_3^- ligand and the central N atom of the $C_9H_{21}N_3$ group. The major distortion found is the N4-Cu-N4' angle of 138.6(5)°. Bond lengths: Cu-N_{ax} 1.982(12), Cu-N_{eq} 2.011(8), Cu-O_{ax} 2.026(7), and Cu-O_{eq} 2.291(8) Å. The $-(CH_2)_5-$ bridge between N4 and N4' is disordered across a crystallographic mirror plane.

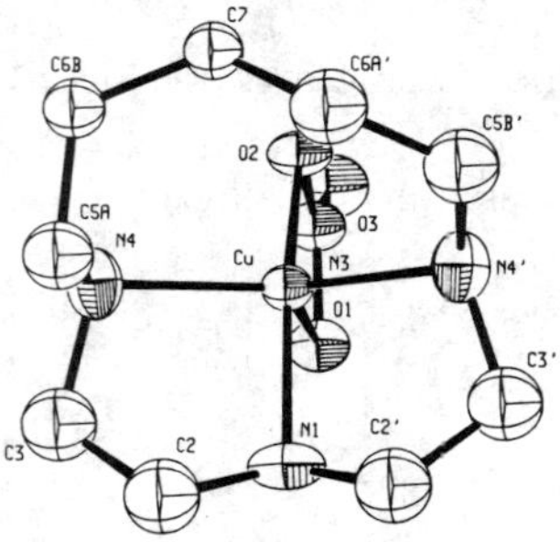

Fig. 1. The $[Cu(C_9H_{21}N_3)(NO_3)]^+$ cation.

$(-)^{CD}_{474}$-mer-{N-(2-AMINOETHYL)-1,3-PROPANEDIAMINE}(DIETHYLENETRIAMINE)COBALT(III) IODIDE MONOHYDRATE

$C_9H_{28}CoI_3N_6,H_2O$ $Co(C_9H_{28}N_6)I_3,H_2O$

M. ISHII, S. SATO, Y. SAITO and M. NAKAHARA, 1984. Bull. Chem. Soc. Japan, 57, 3094-3098.

Monoclinic, $P2_1$, a = 9.662, b = 14.602, c = 8.201 Å, β = 112.87°, D_m = 2.19, Z = 2. Mo radiation, R = 0.060 for 2441 reflexions.

The two ligands coordinate to the central Co atom in mer positions via their six N atoms thus forming a slightly distorted octahedral complex (Fig. 1). The six-membered chelate ring adopts a chair conformation whilst the five-membered ring is δ (in N-(2-aminoethyl)-1,3-propanediamine). The conformations of the two five-membered chelate rings formed by diethylenetriamine are δ and λ. The metal-N distances are: Co-N(1) 1.939, Co-N(2) 2.006, Co-N(3) 1.994, Co-N(4) 1.993, Co-N(5) 1.988 and Co-N(6) 1.984 Å. The cis-angles about Co range from 84.6 to 96.0°. The shortest contact between the iodide ions and the complex cation is 3.55 Å (I(1)...N(4)) and this may be considered an N-H...I hydrogen bond. The I...H distance is 2.55 Å.

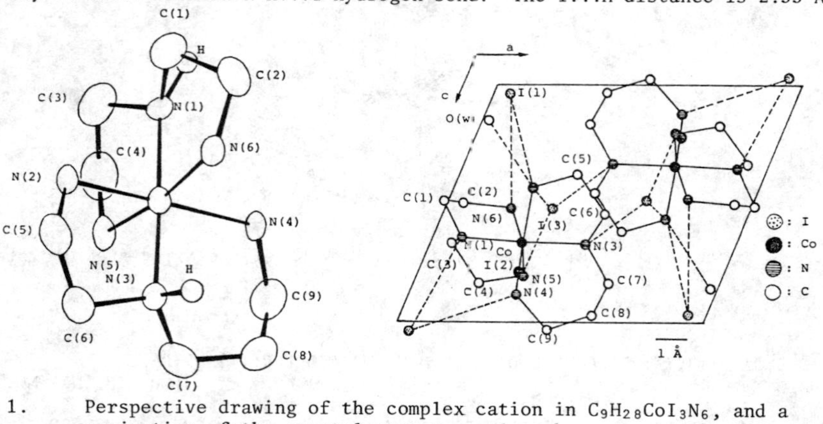

Fig. 1. Perspective drawing of the complex cation in $C_9H_{28}CoI_3N_6$, and a projection of the crystal structure along b.

SODIUM AQUA[ETHYLENEDIAMINETETRAACETATO(4-)]FERRATE(III) DIHYDRATE

$C_{10}H_{14}FeN_2NaO_9,2H_2O$ $Na[Fe(C_{10}H_{12}N_2O_8)(H_2O)],2H_2O$

J.M. LÓPEZ-ALCALÁ, M.C. PUERTA-VIZCAÍNO, F. GONZÁLEZ-VÍLCHEZ, E.N. DUESLER and R.E. TAPSCOTT, 1984. Acta Cryst., C40, 939-941.

Monoclinic, Cc, a = 8.895, b = 11.924, c = 15.043 Å, β = 100.06°, Dm = 1.73, Z = 4. Mo radiation, R = 0.027 for 2370 reflexions.

In the complex anion (Fig. 1) Fe^{III} is 7-coordinate with a hexadentate ethylenediaminetetraacetate (4-) ligand (Fe-N 2.318(2) and 2.319(2), Fe-O 1.972(2), 1.977(2), 2.096(2) and 2.096(2) Å) and a water molecule on a pseudo twofold axis (Fe-O 2.122(2) Å). The geometry is pseudo pentagonal bipyramidal. The unit-cell volume is 7% larger than that reported for this compound in an earlier and less precise study (1) and there are a number of significant differences between the two determinations in many of the bond lengths and bond angles found.

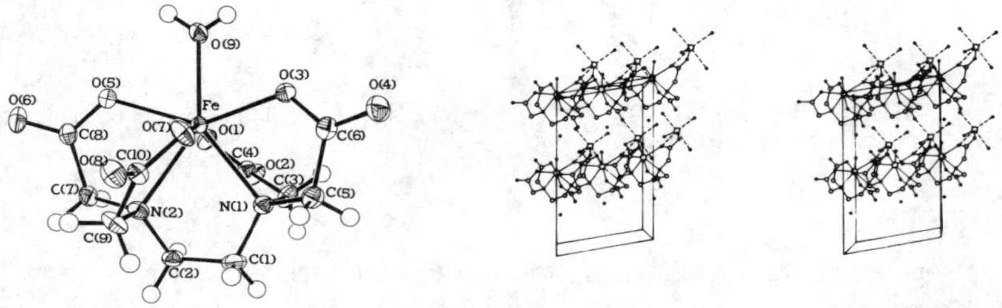

Fig. 1. $Na[Fe(C_{10}H_{12}N_2O_8)(H_2O)],2H_2O$: perspective view of the complex anion and packing diagram for the crystal structure as viewed down the b axis.

1. Structure Reports, 41B, 1033. See also the following Report.

SODIUM [AQUA(ETHYLENEDIAMINETETRAACETATO)FERRATE(III)] DIHYDRATE

$C_{10}H_{14}FeN_2NaO_9,2H_2O$ (I) $Na[Fe(C_{10}H_{12}N_2O_8)(H_2O)],2H_2O$

SILVER [AQUA(ETHYLENEDIAMINETETRAACETATO)FERRATE(III)] DIHYDRATE

$C_{10}H_{14}AgFeN_2O_9,2H_2O$ (II) $Ag[Fe(C_{10}H_{12}N_2O_8)(H_2O)],2H_2O$

POTASSIUM [AQUA(ETHYLENEDIAMINETETRAACETATO)FERRATE(III)] DIHYDRATE

$C_{10}H_{14}FeKN_2O_9,2H_2O$ (III) $K[Fe(C_{10}H_{12}N_2O_8)(H_2O)],2H_2O$

THALLIUM [AQUA(ETHYLENEDIAMINETETRAACETATO)FERRATE(III)] DIHYDRATE

$C_{10}H_{14}FeN_2O_9Tl,2H_2O$ (IV) $Tl[Fe(C_{10}H_{12}N_2O_8)(H_2O)],2H_2O$

X. SOLANS, M. FONT-ALTABA and J. GARCIA-ORICAIN, 1984. Acta Cryst., C40, 635-638.

I. Monoclinic, Cc, a = 8.896, b = 11.931, c = 15.065 Å, β = 100.15°, Z = 4. Mo radiation, R = 0.047 for 1403 reflexions.

II. Monoclinic, Cc, a = 8.929, b = 11.871, c = 15.116 Å, β = 99.85°, Z = 4. Mo radiation, R = 0.046 for 1438 reflexions.

III. Monoclinic, $P2_1/n$, a = 11.361, b = 13.396, c = 10.484 Å, β = 106.00°, Z = 4. Mo radiation, R = 0.049 for 3229 reflexions.

IV. Triclinic, $P\bar{1}$, a = 13.825, b = 8.591, c = 6.807 Å, α = 87.79, β = 97.75, γ = 104.34°, Z = 2. Mo radiation, R = 0.040 for 2350 reflexions.

In the anion (Fig. 1) Fe is surrounded by four O and two N atoms of an EDTA ligand, which acts in a sexidentate fashion, and by a water molecule; the coordination polyhedron is of roughly pentagonal bipyramidal shape. Each cation is linked to six (in I and II) or seven (in III and IV) oxygen atoms, which produces variations in the

Fe coordination polyhedron. In the crystal single layers of anions are loosely linked by the cations into electrically neutral double layers; cohesion between these layers depends largely on van der Waals interactions and weak hydrogen bonds.

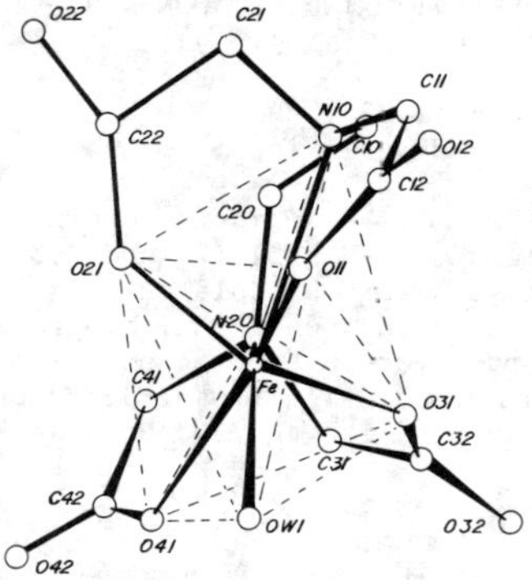

Fig. 1. M[Fe($C_{10}H_{12}N_2O_8$)(H_2O)],$2H_2O$ (M = Na, Ag, K or Tl): view of the Fe coordination polyhedron in the anion.

SODIUM TRIAQUA(ETHYLENEDIAMINETETRACETATO)SAMARATE(III) PENTAHYDRATE

$C_{10}H_{18}N_2NaO_{11}Sm,5H_2O$

D.W. ENGEL, F. TAKUSAGAWA and T.F. KOETZLE, 1984. Acta Cryst., C40, 1687-1693.

Orthorhombic, Fdd2, a = 19.416, b = 35.315, c = 12.014 Å, Z = 16. Neutron radiation, R = 0.059 for 1795 reflexions.

Bond lengths in the anion are summarised in Fig. 1. There is an extensive hydrogen-bond network incorporating two water molecules in a disordered arrangement around a twofold axis.

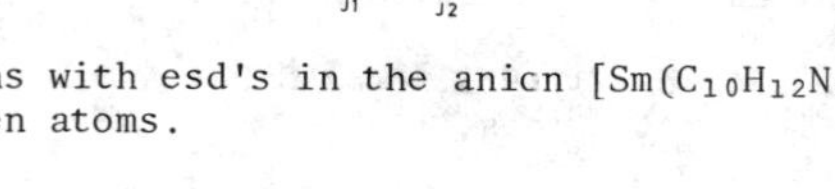

Fig. 1. Bond lengths with esd's in the anion [Sm($C_{10}H_{12}N_2O_8$)(H_2O)$_3$]$^-$. H and J are hydrogen atoms.

CHLORO(CARBONYL)(6,9-DIMETHYL-2-METHOXY-1,3-DIOXA-6,9-DIAZA-2-PHOSPHACYCLOUNDECANE-N,N'P)RHODIUM(I)

$C_{10}H_{21}ClN_2O_4PRh$ [Rh(CO)(Cl)(6,9-Me_2-2-OMe-1,3,6,9,2-$O_2N_2PC_6H_{12}$)]

C. VACCHER, A. MORTREUX, F. PETIT, J.-P. PICAVET, H. SLIWA, N.W. MURRALL and A.J. WELCH, 1984. Inorg. Chem., 23, 3613-3617.

Orthorhombic, Pbca, a = 13.5964, b = 15.367, c = 14.621 Å, D_m = 1.75, Z = 8. Mo radiation, R = 0.055 for 2108 reflexions (at 268 K).

The molecule is shown in Fig. 1. While all three donor atoms of the 11-membered macrocycle are formally associated with the rhodium centre, one Rh-N bond is extremely long, 2.570(5) Å, affording a geometry at the metal centre that is intermediate between four- and five-coordinate (planar and trigonal bipyramidal geometry). Other distances: Rh-Cl 2.423(2), Rh-P 2.1392(16), Rh-C 1.787(7), Rh-N 2.330(5) Å.

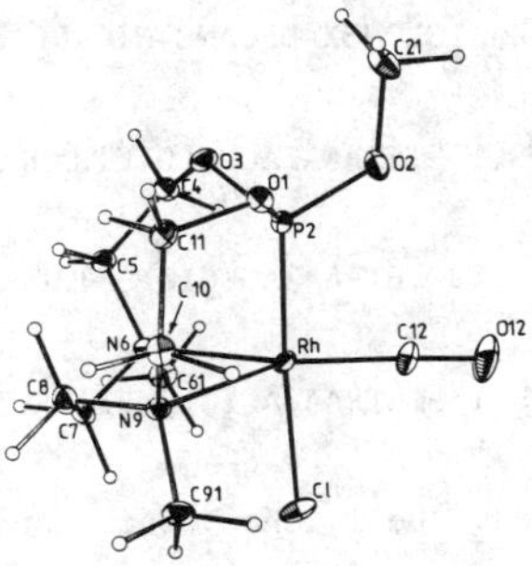

Fig. 1. The $C_{10}H_{21}ClN_2O_4PRh$ molecule.

BIS(PERCHLORATO)(1,4,7,10-TETRAAZACYCLOTETRADECANE)COPPER(II)

$C_{10}H_{24}Cl_2CuN_4O_8$ $[Cu(C_{10}H_{24}N_4)](ClO_4)_2$

A. BIANCHI, L. BOLOGNI, P. DAPPORTO, M. MICHELONI and P. PAOLETTI, 1984. Inorg. Chem., 23, 1201-1205.

Orthorhombic, $P2_12_12$, a = 22.219, b = 13.753, c = 9.097 Å, Z = 6. Mo radiation, R = 0.076 for 854 reflexions.

The asymmetric unit contains two complex molecules, one being disordered about a crystallographic C_2 axis. The Cu atoms of both molecules are in a tetragonally distorted octahedral environment where the four N atoms are in a planar arrangement (Cu-N 1.91(3)-2.04(5) Å) and the two O atoms of two ClO_4^- ions are in the apical positions (Cu-O 2.64(4)-2.70(4) Å) (see Fig. 1).

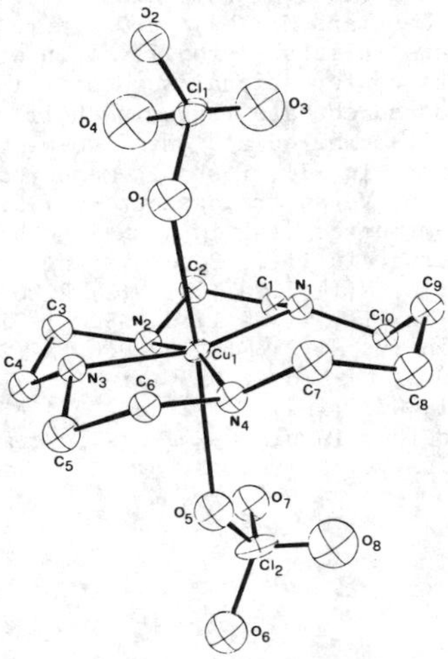

Fig. 1. The ordered $[Cu(C_{10}H_{24}N_4)](ClO_4)_2$ molecule.

trans-DICHLORO(1,4,8,11-TETRAAZACYCLOTETRADECANE)NICKEL(II)
$C_{10}H_{24}Cl_2N_4Ni$ (I) $[NiCl_2(C_{10}H_{24}N_4)]$

trans-DICHLORO(1,4,8,12-TETRAAZACYCLOPENTADECANE)NICKEL(II)
$C_{11}H_{26}Cl_2N_4Ni$ (II) $[NiCl_2(C_{11}H_{26}N_4)]$

trans-DICHLORO(1,5,9,13-TETRAAZACYCLOHEXADECANE)NICKEL(II)
$C_{12}H_{28}Cl_2N_4Ni$ (III) $[NiCl_2(C_{12}H_{28}N_4)]$

trans-BIS(ISOTHIOCYANATO)(1,4,8,11-TETRAAZACYCLOTETRADECANE)NICKEL(II)
$C_{12}H_{24}N_6NiS_2$ (IV) $[Ni(NCS)_2(C_{10}H_{24}N_4)]$

trans-BIS(ISOTHIOCYANATO)(1,4,8,12-TETRAAZACYCLOPENTADECANE)NICKEL(II)
$C_{13}H_{26}N_5NiS_2$ (V) $[Ni(NCS)_2(C_{11}H_{26}N_4)]$

trans-BIS(ISOTHIOCYANATO)(1,5,9,13-TETRAAZACYCLOHEXADECANE)NICKEL(II)
$C_{14}H_{28}N_6NiS_2$ (VI) $[Ni(NCS)_2(C_{12}H_{28}N_4)]$

T. ITO, M. KATO and H. ITO, 1984. Bull. Chem. Soc. Japan, 57, 2641-2649.

I. Monoclinic, $P2_1/n$, a = 13.607, b = 8.3685, c = 6.6207 Å, β = 103.585°, D_m = 1.495, Z = 2. Mo radiation, R = 0.027 for 5374 reflexions.

II. Monoclinic, $P2_1/a$, a = 14.504, b = 8.3019, c = 6.6771 Å, β = 100.24°, D_m = 1.448, Z = 2. Mo radiation, R = 0.059 for 1980 reflexions.

III. Triclinic, $P\bar{1}$, a = 10.169, b = 11.530, c = 9.639 Å, α = 65.17, β = 115.57, γ = 120.76°, D_m = 1.415, Z = 2. Mo radiation, R = 0.032 for 2801 reflexions.

IV. Triclinic, $P\bar{1}$, a = 14.172, b = 15.201, c = 8.394 Å, α = 79.80, β= 102.94, γ = 90.17°, D_m = 1.437, Z = 4. Mo radiation, R = 0.033 for 5374 reflexions.

V. Monoclinic, $P2_1/a$, a = 14.653, b = 14.386, c = 9.151 Å, β = 105.92°, D_m = 1.391, Z = 4. Mo radiation, R = 0.055 for 2236 reflexions.

VI. Monoclinic, $P2_1/a$. a = 14.822, b = 14.522, c = 14.522 Å, β = 106.00°, D_m = 1.407, Z = 4. Mo radiation, R = 0.039 for 2801 reflexions.

Figure 1 shows the perspective views of complexes I to VI. The geometries about Ni in all the compounds are of the trans-NiX_2N_4 type. There is disorder in compounds II and V, one linkage of the macrocyclic skeleton is shown with solid bonds and the alternative with open bonds. The conformational sequences in compounds I to VI are: in I, two six- and two five-membered chelate rings in chair and gauche conformation respectively; in II, the three six-membered and one five-membered chelate rings are in a chair-skew-chair-gauche sequence; in III, four six-membered chelate rings in a chair-chair-skew-skew sequence; in IV, an arrangement similar to that in I; in V, three six-membered and one five-membered ring in a chair-chair-skew-gauche sequence; in VI an arrangement similar to that in III. Bond distances of interest in the complexes are in I, Ni-Cl 2.510, Ni-N(1) 2.067, Ni-N(2) 2.066 Å; in II, Ni-Cl 2.497, Ni-N(1) 2.104, Ni-N(2) 2.124 Å; in III, Ni-Cl(1) 2.535, Ni-Cl(2) 2.428, Ni-N (four values from 2.142 to 2.211 Å); in IV, Ni(2)-N(21) 2.066, Ni(2)-N(22) 2.076, Ni(2)-N(23) 2.128 Å; in V, Ni-N(5) 2.099, Ni-N(6) 2.059, Ni-N(1-4) 2.102 to 2.161 Å, and in VI, Ni-N(5) 2.103, Ni-N(6) 2.051, Ni-N(1-4) 2.152 to 2.224 Å. The deviation of Ni from the plane defined by Ni and four in-plane N atoms in compounds III, V and VI is 0.072, 0.040 and 0.058 Å respectively.

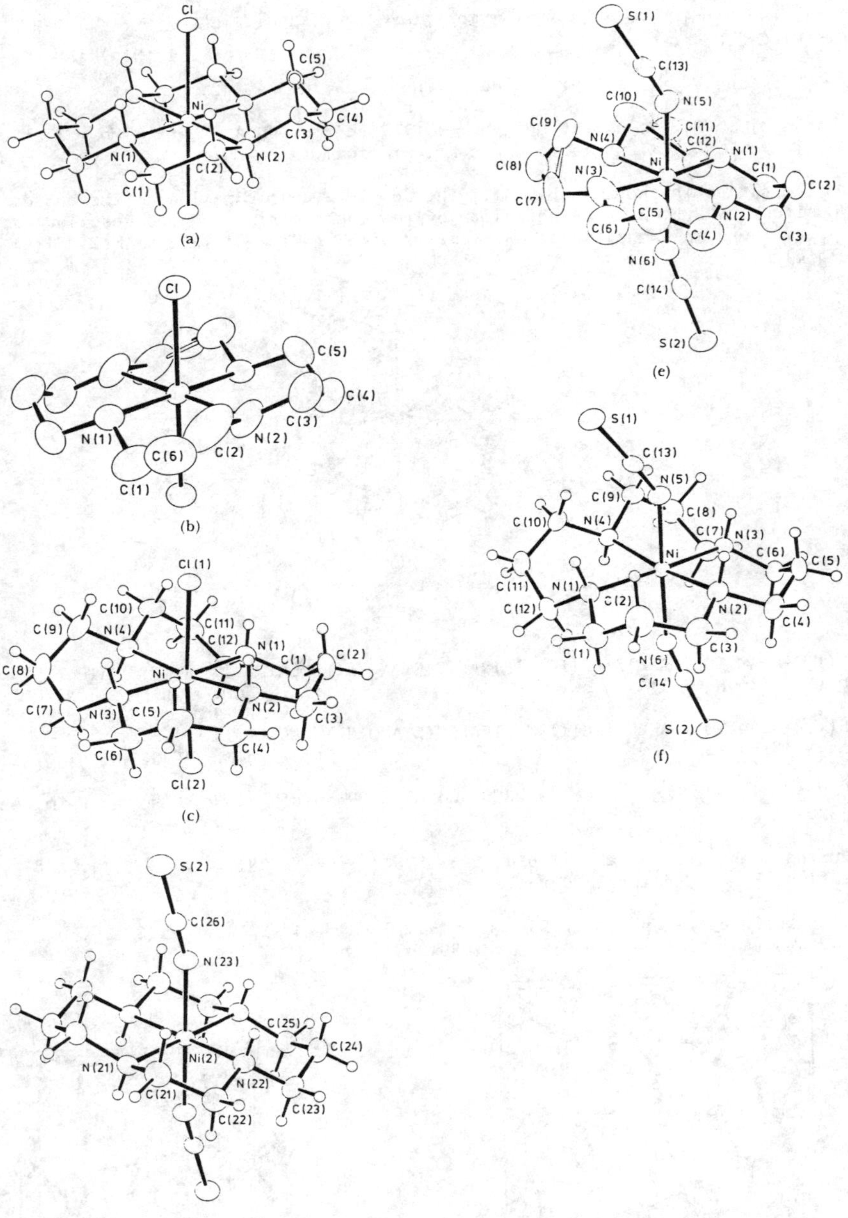

Fig. 1. Perspective views of $C_{10}H_{24}Cl_2N_4Ni$ (a), $C_{11}H_{26}Cl_2N_4Ni$ (b), $C_{12}H_{28}Cl_2N_4Ni$ (c), $C_{12}H_{24}N_6NiS_2$ (d), $C_{13}H_{26}N_6NiS_2$ (e), $C_{14}H_{28}N_6NiS_2$ (f).

(−)-cis-(RRRR)-DICHLORO(1,4,8,11-TETRAAZACYCLOTETRADECANE)CHROMIUM(III) PERCHLORATE
$C_{10}H_{24}Cl_3CrN_4O_4$ $[CrCl_2(C_{10}H_{24}N_4)](ClO_4)$

D.A. HOUSE and V. McKEE, 1984. Inorg. Chem., 23, 4237-4242.

Orthorhombic, $P2_12_12$, a = 10.097, b = 13.747, c = 6.349 Å, Z = 2. Mo radiation, R = 0.053 for 811 reflexions (absolute configuration).

The cation is shown in Fig. 1. The absolute configuration at the coordinated s-NH sites is RRRR, and the chirality of the chelate rings can be described as Λ_{en} or Δ_{tn}. The complex has exact C_2 symmetry with Cr-Cl 2.331(2), Cr-N 2.084(6) and 2.076(6) Å.

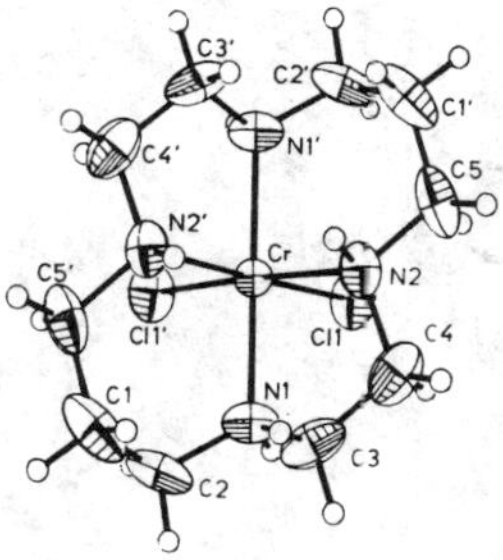

Fig. 1. The $[CrCl_2(C_{10}H_{24}N_4)]^+$ cation.

TRIS(1,2-DIAMINOETHANE)ZINC(II) TETRAKIS(CYANO)NICKELATE(II) MONOHYDRATE
$C_{10}H_{24}N_{10}NiZn,H_2O$ (I) $[Zn(NH_2CH_2CH_2NH_2)_3]^{2+}[Ni(CN)_4]^{2-},H_2O$

TRIS(1,2-DIAMINOETHANE)ZINC(II) TETRAKIS(CYANO)NICKELATE(II)
$C_{10}H_{24}N_{10}NiZn$ (II) $[Zn(NH_2CH_2CH_2NH_2)_3]^{2+}[Ni(CN)_4]^{2-}$

J. ČERNÁK, J. CHOMIČ, M. DUNAJ-JURČO and C. KAPPENSTEIN, 1984. Inorg. Chim. Acta, 85, 219-226.

I. Orthorhombic, Pbca, a = 11.680, b = 15.844, c = 19.981 Å, D_m = 1.54, Z = 8. Mo radiation, R = 0.086 for 1890 reflexions.

II. Monoclinic, $P2_1/n$, a = 7.957, b = 15.170, c = 14.932 Å, β = 96.41°, D_m = 1.49, Z = 4. Mo radiation, R = 0.058 for 1689 reflexions.

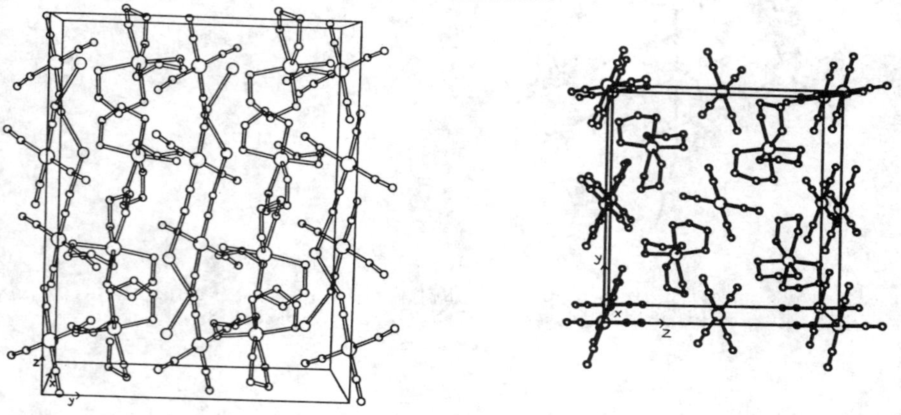

Fig. 1. Views of the unit cell of $C_{10}H_{24}N_{10}NiZn,H_2O$ (left), and $C_{10}H_{24}N_{10}NiZn$ (right).

The structure of I is built up of discrete anions, cations and water molecules (Fig. 1). In II the structure is also ionic but two crystallographically independent anions are on centres of inversion. In both compounds the Zn atoms are distorted octahedra coordinated by three en molecules in gauche conformation. The Zn-N bond lengths range from 2.20 to 2.26 Å in I, and from 2.20 to 2.24 Å in II. The Ni atoms in both complexes exhibit classical square planar coordination with Ni-C distances of 1.78-2.00 Å in I and 1.86-1.88 Å in II. Each water molecule in I forms two hydrogen bonds with the N atoms of trans cyano groups (N...O 2.89 and 2.91 Å).

AQUABROMO(1,7-DIMETHYL-1,4,7,10-TETRAAZACYCLODODECANE-N^1,N^4,N^7,N^{10})NICKEL(II) BROMIDE
$C_{10}H_{26}Br_2N_4NiO$

M. CIAMPOLINI, M. MICHELONI, N. NARDI, P. PAOLETTI, P. DAPPORTO and F. ZANOBINI, 1984. J. Chem. Soc. Dalton, 1357-1362.

Orthorhombic, $P2_12_12_1$, a = 13.478, b = 11.186, c = 10.585 Å, Z = 4. Mo radiation, R = 0.046 for 910 reflexions.

The structure (Fig. 1) consists of complex cations and bromide anions. The cations have cis-octahedral geometry with the macrocycle coordinated in a folded conformation to four sites around the central Ni atom, the other two sites being occupied by the Br atom and an O atom of coordinated water. The Ni-N (secondary) distances (2.051(12), 2.072(12)) and Ni-N (tertiary) distances (2.153(12), 2.160(12) Å) are significantly different. The Ni-Br and Ni-O distances are 2.610(12) and 2.136(9) Å respectively. The chelate rings are all asymmetric and the chirality of each N atom is the same. The Br atoms are involved in short intermolecular distances.

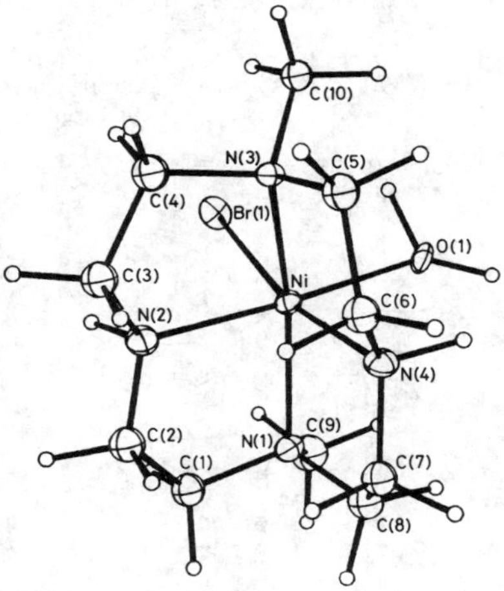

Fig. 1. $C_{10}H_{26}Br_2N_4NiO$: the complex cation.

(2-[(2-AMINOETHYL)AMINO]-2-DEOXY-L-SORBOSE)(ETHYLENEDIAMINE)NICKEL(II) DICHLORIDE HEMIMETHANOL SOLVATE
$C_{10}H_{26}Cl_2N_4NiO_5,0.5(CH_4O)$ $[Ni(en)(L\text{-}sor\text{-}en)]^{2+}2Cl^-,0.5(CH_3OH)$

T. TSUBOMURA, S. YANO, K. TORIUMI, T. ITO and S. YOSHIKAWA, 1984. Bull. Chem. Soc. Japan, 57, 1833-1838.

Monoclinic, C2, a = 17.862, b = 8.212, c = 12.644 Å, β = 90.89°, D_m = 1.55, Z = 4. Mo radiation, R = 0.032 for 2680 reflexions.

The Ni atom is octahedrally-coordinated in the complex cation (Fig. 1). The ethylenediamine ligand is bidentate, and the glycosylamine ligand of L-sor-en is tetradentate. The bonds to the metal atom are: Ni-O(1) 2.137, Ni-O(3) 2.118, Ni-N(1) 2.109, Ni-N(2) 2.068, Ni-N(3) 2.059 and Ni-N(4) 2.078 Å. The L-sorbose moiety adopts a six-membered α-$^2C_5$ chair conformation. The cis angles about the Ni atom range from 76.1 to 105.5°, and the lowest trans-angle is 162.8°. Hydrogen bonds of the type O-H...Cl and N-H...Cl probably exist between O(4)...Cl(2) (3.032 Å), O(5)...Cl(2) (2.980 Å), N(1)...Cl(1) (3.275 Å) and N(3)...Cl(1) (3.316 Å). The interatomic distance of O(3)...O(4) between two molecules is short (2.678 Å) so this indicates another possible hydrogen bond.

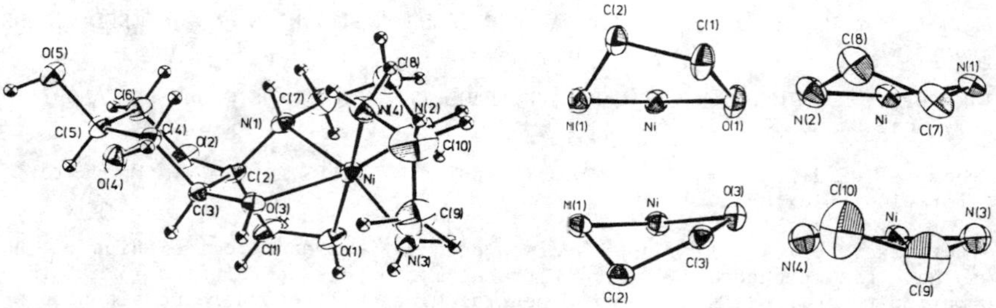

Fig. 1. Molecular structure of the cation in $C_{10}H_{26}Cl_2N_4NiO_5$ (left), and perspective drawings of the chelate rings (right).

BIS(2,2-DIMETHYLPROPANE-1,3-DIAMINE)(DINITRATO-O)COPPER(II)
$C_{10}H_{28}CuN_6O_6$

L. MENABUE, G.C. PELLACANI, L.P. BATTAGLIA, A.B. CORRADI, F. SANDROLINI, A. MOTORI, R.J. PYLKKI and R.D. WILLETT, 1984. J. Chem. Soc. Dalton, 2187-2192.

Monoclinic, $P2_1/n$, a = 5.951, b = 19.333, c = 7.457 Å, γ = 91.00°, Dm = 1.55, Z = 2. Cu radiation, R = 0.035 for 1542 reflexions.

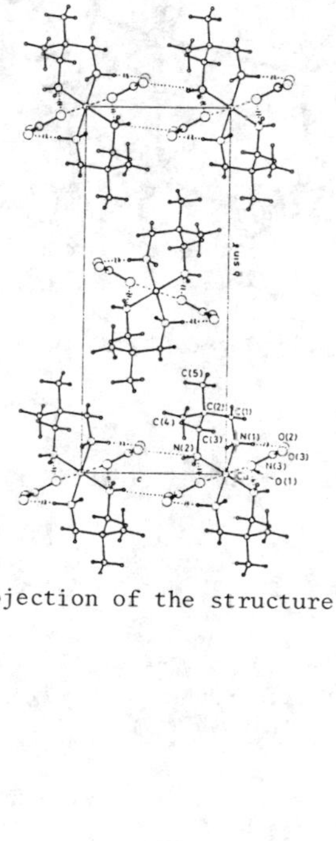

Fig. 1. $C_{10}H_{28}CuN_6O_6$: projection of the structure parallel to the [100] axis.

The Cu atom coordinates four N atoms of two ligand molecules in a square planar arrangement. Two O atoms from two centrosymmetrically related nitrate ions occupy further coordination sites forming an elongated tetragonal bipyramid (Fig. 1). An angle of 8.9° between the line through Cu...O(1) and the normal to the Cu-N(1)-N(2) equatorial plane and angles at Cu (N(1)-Cu-N(2) 92.56(2), N(1)-Cu-O(1) 97.62(2), N(2)-Cu-O(1) 82.93(2)°) indicate further distortion of the coordination polyhedron. The Cu-N distances are 2.029(4) and 2.020(3) Å. The ligands form two six-membered chelate rings with the chair conformation. The Cu...O(1) distances are 2.566(3) and the NO_3^- ion has regular trigonal planar geometry.

s-fac-[BIS(3-AMINOPROPYL)AMINE](DIETHYLENETRIAMINE)COBALT(III) DITHIONATE PERCHLORATE MONOHYDRATE

$C_{10}H_{30}ClCoN_6O_{10}S_2,H_2O$ $[Co(C_4H_{13}N_3)(C_6H_{17}N_3)]^{3+}.S_2O_6^{2-}.ClO_4^-.H_2O$

T.W. HAMBLEY and G.H. SEARLE, 1984. Acta Cryst., C40, 383-386.

Orthorhombic, $Pc2_1b$, a = 9.592, b = 13.454, c = 16.748 Å, D_m = 1.77, Z = 4. Mo radiation, R = 0.037 for 1058 reflexions.

The two tridentate ligands coordinate to Co giving an s-fac topology (Fig. 1). The two six-membered chelate rings adopt chair and flattened-chair conformations. The detailed geometry of the complex cation as calculated by molecular-mechanics analysis agrees well with the geometry observed in the crystal, with distortions of bond lengths and angles well reproduced. The Co-N distances are in the range 1.954 - 2.028(8) Å.

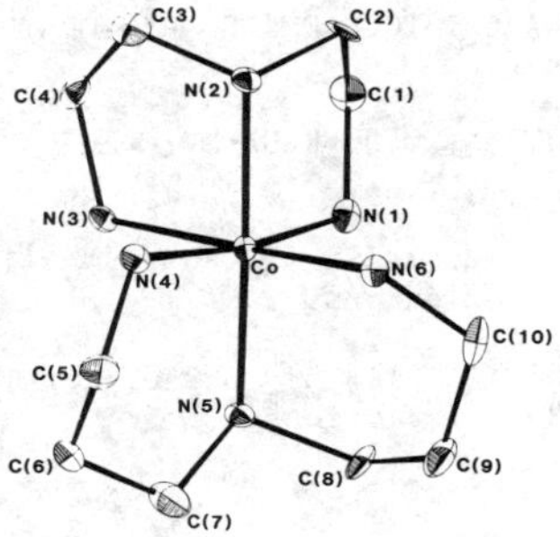

Fig. 1. A view of the $[Co(dien)(dpt)]^{3+}$ cation.

μ-fac-(DIETHYLENETRIAMINE)-[DI(3-AMINOPROPYL)AMINE]COBALT(III) PERCHLORATE HYDRATE

$C_{10}H_{30}Cl_3CoN_6O_{12},H_2O$ $[Co(dien)(dpt)](ClO_4)_3,H_2O$

T.W. HAMBLEY and G.H. SEARLE, 1984. Aust. J. Chem., 37, 249-256.

Triclinic, $P\bar{1}$, a = 8.979, b = 11.436, c = 11.463 Å, α = 91.15, β = 101.84, γ = 93.41°, D_m = 1.75, Z = 2. Mo radiation, R = 0.038 for 1988 reflexions.

The two diethylenetriamine, dien, rings adopt λ and δ conformations (N-C-C-N torsion angles +43.0(5), -43.5(5)°). The conformations of the six-membered chelate rings of di(3-aminopropyl)amine, dpt, adopt a chair and a flattened chair. Co-N distances range from 1.973 to 2.033 Å, mean 1.922(4) Å. The longest bond (Co-N(5)) is consistent with the strain of both six-membered chelate rings. N-Co-N bond angles for dien are 85.0, 84.4°, and for dpt 96.6, 88.7° (Fig. 1).

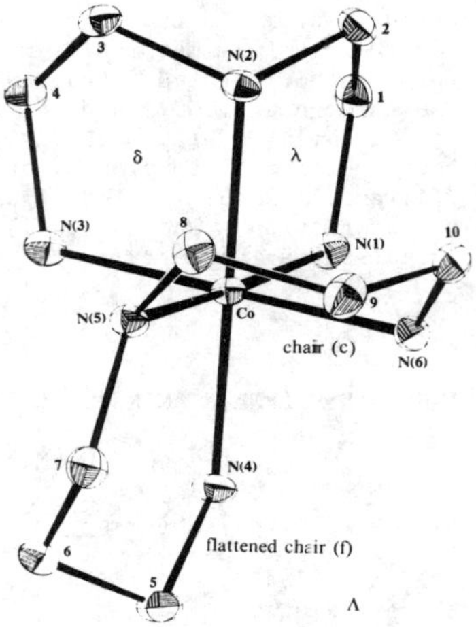

Fig. 1. The structure of $[Co(dien)(dpt)](ClO_4)_3,H_2O$.

SODIUM 1,3-PROPANEDIAMINETETRAACETATORHODATE TRIHYDRATE
$C_{11}H_{14}CrN_2NaO_8,3H_3O$ (I) Na[Cr(1,3-PDTA)],$3H_2O$

SODIUM 1,3-PROPANEDIAMINETETRAACETATOCHROMATE TRIHYDRATE
$C_{11}H_{14}N_2NaO_8Rh,3H_2O$ (II) Na[Rh(1,3-PDTA)],$3H_2O$

R. HERAK, G. SRDANOV, M.I. DJURAN, D.J. RADANOVIĆ and M. BRUVO, 1984. Inorg. Chim. Acta, <u>83</u>, 55-64.

I. Orthorhombic, $P2_12_12_1$, a = 16.514, b = 8.809, c = 11.429 Å, D_m = 1.716, Z = 4. Mo radiation, R = 0.053 for 1696 reflexions.

II. Orthorhombic, $P2_12_12_1$, a = 16.511, b = 8.836, c = 11.431 Å, D_m = 1.910, Z = 4. Mo radiation, R = 0.025 for 2928 reflexions.

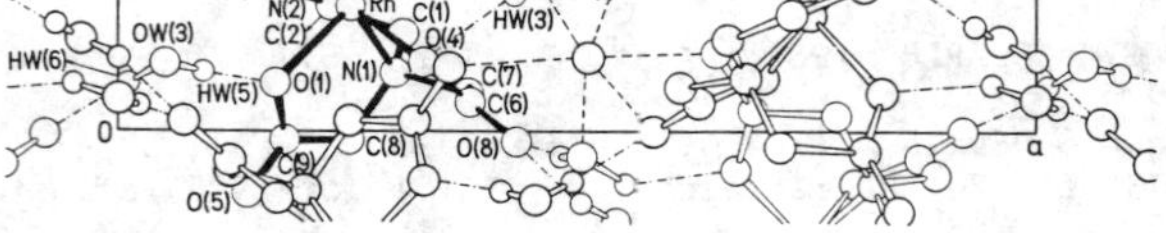

Fig. 1. Crystal structure of $C_{11}H_{14}N_2NaO_8Rh,3H_2O$ projected along the c axis. The broken lines represent the coordination sphere of Na^+ ions.

The crystal structures of the two isomorphous compounds consist of Na^+ cations, complex anions and water molecules (Fig. 1). Na^+ ions are surrounded by six O atoms; three from water molecules and three from carboxylato groups belonging to three different complex anions, in a distorted octahedral configuration. The average Na-O distance in I is 2.437 Å, and 2.449 Å in II. In the complex anion, the metal ion is coordinated octahedrally by two N and four O atoms from a sexidentate 1,3-propanediaminetetraacetate (1,3-PDTA) ligand. In both complexes the six-membered 1,3-PD chelate ring T takes a twist boat δ conformation. Equatorial glycinato G rings have an envelope λ conformation, while the two axial R rings are relatively flat. The absolute configuration of both complexes I and II is Λ. The Cr-O and Cr-N distances (average) are 1.956 and 2.065 Å respectively in I, and the corresponding distances in II are 2.024 and 2.032 Å.

trans-DICHLORO[3,3'-(1,3-PROPANEDIYLDINITRILO)BIS(2-BUTANONE)DIOXIMATO-N^1,N^2,N^3,N^4]-RHODIUM(III)
$C_{11}H_{19}Cl_2N_4O_2Rh$ (I)

trans-DICHLORO[3,3'-(1,3-PROPANEDIYLDINITRILO)BIS(2-BUTANONE)DIOXIMATO-N^1,N^2,N^3,N^4]-COBALT(III)
$C_{11}H_{19}Cl_2CoN_4O_2$ (II)

S. SIRIPAISARNPIPAT and E.O. SCHLEMPER, 1984. J. Coord. Chem., 13, 281-289.

I. Monoclinic, $P2_1/n$, a = 9.952, b = 13.751, c = 11.780 Å, β = 96.43°, D_m = 1.72, Z = 4. Mo radiation, R = 0.027 for 2435 reflexions.

II. Monoclinic, $P2_1/n$, a = 9.846, b = 13.748, c = 11.648 Å, β = 95.46°, D_m = 1.55, Z = 4. Mo radiation, R = 0.042 for 1643 reflexions.

Details of dimensions and conformation of these isomorphous complexes are shown in Fig. 1. For previous work on (II) see (1). In each case the central metal atom has distorted octahedral coordination. The presence of an intramolecular O-H...O hydrogen bond is confirmed.

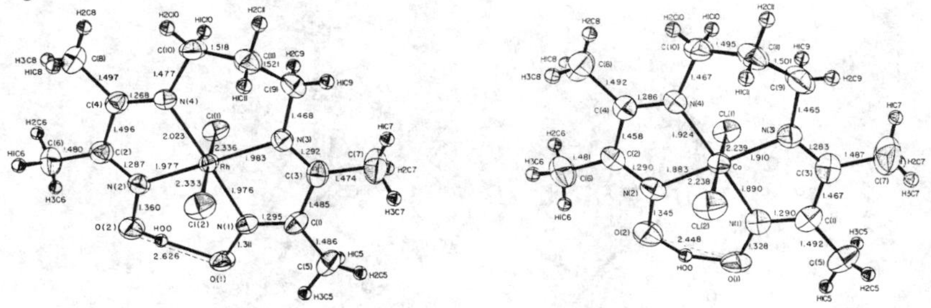

Fig. 1. Views of the isomorphous Rh and Co complexes with dimensions.

1. Structure Reports, 45B, 1162.

CARBONATO(1,7-DIMETHYL-1,4,7,10-TETRAAZACYCLODODECANE)COBALT(III) PERCHLORATE
$C_{11}H_{24}ClCoN_4O_7$ $[Co(CO_3)(C_{10}H_{24}N_4)]^+ClO_4^-$

J. GIUSTI, S. CHIMICHI, M. CIAMPOLINI, M. SABAT and D. MASI, 1984. Inorg. Chim. Acta, 88, 51-54.

Orthorhombic, Imcb, a = 18.458, b = 13.608, c = 13.436 Å, Z = 8. Mo radiation, R = 0.061 for 1256 reflexions.

The geometry of the complex cation is octahedral with the macrocyclic ligand bonded through its four N atoms (Fig. 1). The carbonato moiety is bound to the metal atom by two O atoms. The cation exhibits C_s point symmetry with the Co(III) ion, the carbonato group and the N atoms N(1) and N(2) located on the mirror plane. The Co-O(1) and Co-O(2) distances are 1.910 and 1.909 Å respectively and the Co-N(1), Co-N(2) and Co-N(3) distances 1.915, 1.925 and 2.007 Å respectively. The angle O(1)-Co-O(2) is 68.6°.

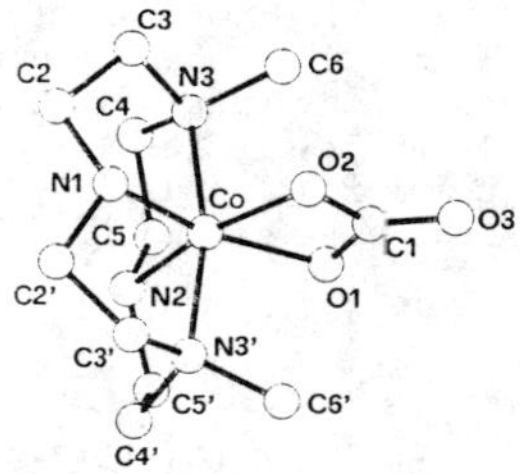

Fig. 1. Structure of the complex cation in $C_{11}H_{24}ClCoN_4O_7$.

(6-METHYL-6-(4-AMINO-2-AZABUTYL)-1,4-DIAZACYCLOHEPTANE)(ETHYLENEDIAMINE)COBALT(III) TRIPERCHLORATE

$C_{11}H_{30}Cl_3CoN_6O_{12}$ $[Co(C_9H_{22}N_4)(C_2H_8N_2)](ClO_4)_3$

K. TOMIOKA, U. SAKAGUCHI and H. YONEDA, 1984. Inorg. Chem., 23, 2863-2867.

Orthorhombic, Pbca, a = 14.059, b = 15.319, c = 20.497 Å, D_m = 1.82, Z = 8. Mo radiation, R = 0.046 for 4290 reflexions.

The Co(III) ion is surrounded by six N atoms, four from the $C_9H_{22}N_4$ ligand and two from $C_2H_8N_2$ (Fig. 1). The N-Co-N angle involving the N atoms of the 1,4-diazacycloheptane moiety is small (76.4(1)°). The Co-N bonds trans to these two N atoms have normal lengths (1.966(3) and 1.979(3) Å) while the remaining Co-N distances are elongated (1.995(3)-2.000(3) Å) as a result of steric interactions.

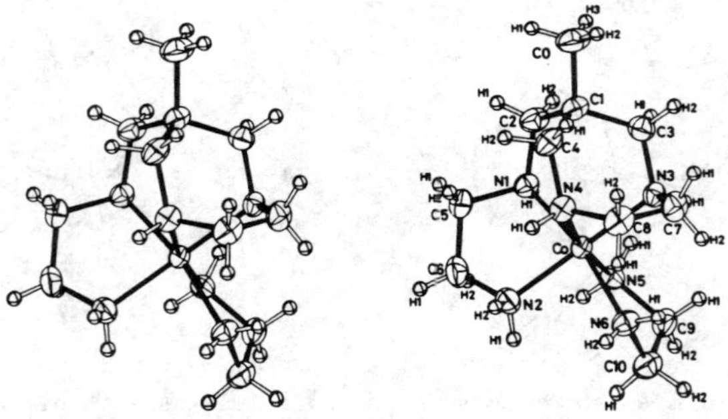

Fig. 1. A stereoview of the $[Co(C_9H_{22}N_4)(C_2H_8N_2)]^+$ cation.

1,9-DIAMINO-4-METHYL-3,7-DIAZANONANE-3,7-DIACETATO(2-)COBALT(III) NITRATE HYDRATE

$C_{12}H_{24}CoN_5O_7,H_2O$ $[Co(NH_2CH_2CH_2N(CH_2COO^-)CH_2CH_2CH(CH_3)N(CH_2COO^-)CH_2CH_2NH_2)]^+NO_3^-,H_2O$

C.A. CHANG, E.N. DUESLER and R.E. TAPSCOTT, 1984. Inorg. Chim. Acta, 85, 73-78.

Monoclinic, $P2_1/c$, a = 9.756, b = 14.912, c = 11.783 Å, β = 95.70°. Z = 4. Mo radiation, R = 0.058 for 2367 reflexions.

The crystal contains a racemic mixture of ΔΛΔ and ΛΔΛ cationic complexes (Fig. 1), NO_3^- anions, and water molecules held together by ionic interactions and hydrogen bonds. The Co atom has distorted octahedral coordination with Co-N distances of 1.944, 1.968, 2.008 and 1.934 Å for N(1), N(4), N(8) and N(11) respectively. The Co-O(14) and Co-O(18) distances are 1.895 and 1.884 Å. Around the meridian of the complex are a skewed five-membered ethylene-diamine chelate ring, a central six-membered 1,3-diaminobutane chelate ring with a skew-boat conformation, and a second skewed five-membered ring. In this order, the absolute configurations of the enantiotopic chelate ring conformations are δλδ in the ΔΛΔ isomer and λδλ in the ΛΔΛ isomer.

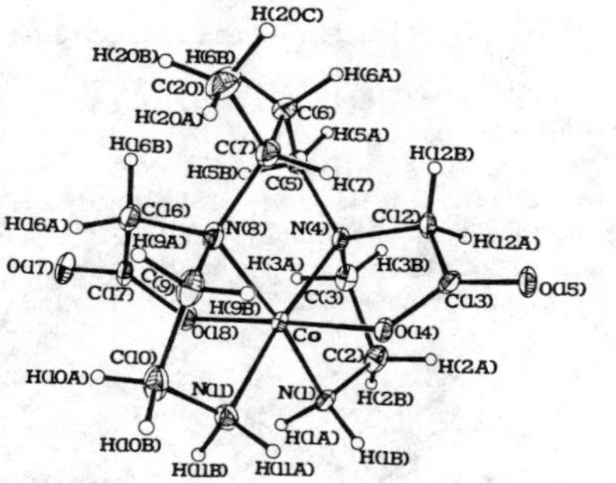

Fig. 1. Molecular structure of the cation in $C_{12}H_{24}CoN_5O_7$.

trans-BIS(ISOTHIOCYANATO)(1,4,8,11-TETRAAZACYCLOTETRADECANE)ZINC(II)

$C_{12}H_{24}N_6S_2Zn$ (I) $[Zn(NCS)_2(C_{10}H_{24}N_4)]$

trans-BIS(ISOTHIOCYANATO)(1,4,8,12-TETRAAZACYCLOPENTADECANE)ZINC(II)

$C_{13}H_{26}N_6S_2Zn$ (II) $[Zn(NCS)_2(C_{11}H_{26}N_4)]$

T. ITO, M. KATO and H. ITO, 1984. Bull. Chem. Soc. Japan, 57, 2634-2640.

I. Orthorhombic, $P2_1nb$, a = 14.502, b = 18.163, c = 6.522 Å, D_m = 1.472, Z = 4. Mo radiation, R = 0.042 for 1598 reflexions.

II. Monoclinic, $P2_1/a$, a = 14.453, b = 14.436, c = 9.213 Å, β = 104.823°, D_m = 1.419, Z = 4. Mo radiation, R = 0.052 for 2533 reflexions.

Coordination geometries about Zn(II) in I and II are of the pseudo-octahedral type with two NCS groups trans to each other (Fig. 1). The metal ion in each compound deviates from a plane defined by the four N atoms of the macrocyclic ligand by 0.179 Å on average in I, (the Zn(II) ion is disordered above and below the plane) and by 0.193 Å in II. The occupancy factors of the metal ion sites in I are 0.64 for Zn and 0.36 for Zn'. The alternating six- and five-membered chelate rings in I are in the stable chair and gauche conformations respectively. In II, three six-membered and one five-membered chelate rings adopt a chair-chair-skew-gauche conformational sequence. Main bond lengths are: (I) Zn-N(14-ane) 2.08-2.14(1), Zn-N(NCS) 2.17(1) and 2.58(1); (II) Zn-N(14-ane) 2.125(5)-2.185(5), Zn-N(NCS) 2.095(5) and 2.479(6) Å.

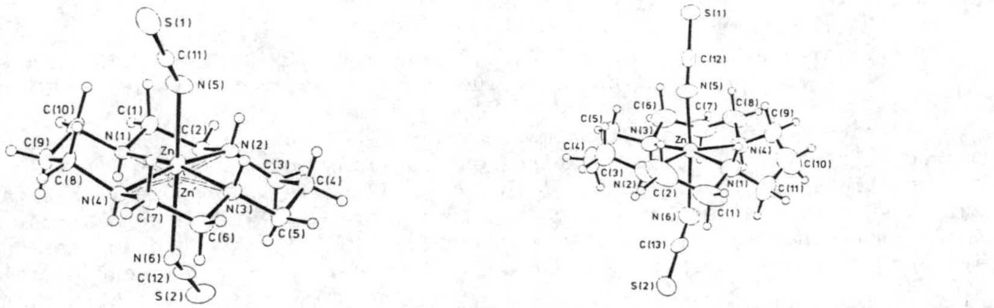

Fig. 1. Molecular structures of $C_{12}H_{24}N_6S_2Zn$ (disordered Zn with minor occupancy factor shown with open bonds) (left), and $C_{13}H_{26}N_6S_2Zn$ (right).

catena-(μ-BROMO)BIS((1R,2R)-CYCLOHEXANEDIAMINE)NICKEL BROMIDE
$C_{12}H_{28}Br_{2.77}N_4Ni$ $[Ni(C_6H_{14}N_2)_2Br]Br_{1.77}$

H. TOFTLUND and O. SIMONSEN, 1984. Inorg. Chem , 23, 4261-4265.

Orthorhombic, I222, a = 23.544, b = 7.095, c = 5.232 Å, D_m = 1.95, Z = 2. Mo radiation, R = 0.076 for 600 reflexions.

The structure contains chains of $Ni(C_6H_{14}N_2)_2$ units (Fig. 1) connected by bromide ions with the Ni-Br distances along the chain =c/2=2.616 Å. An 11.5% deficiency in the occupation of the bromide counterionic site has been observed, leaving the Ni atoms with an average oxidation state of 2.77. The Ni-N distance is 1.95(3) Å.

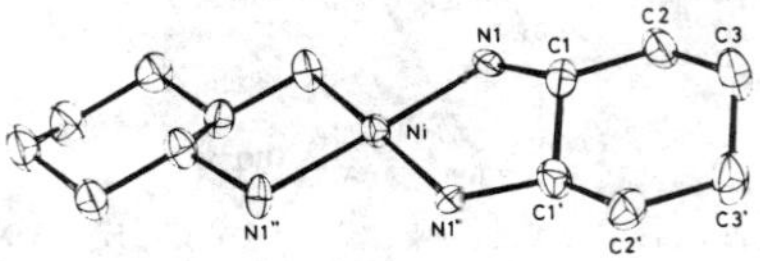

Fig. 1. The $[Ni(C_6H_{14}N_2)_2]^{+2.77}$ unit in $[Ni(C_6H_{14}N_2)_2Br]Br_{1.77}$.

AQUA(NITRATO-O,O')(1,5,9,13-TETRAAZACYCLOHEXADECANE-N[1],N[5],N[9],N[13])CADMIUM NITRATE MONOHYDRATE
$C_{12}H_{30}CdN_6O_7,H_2O$ (I)

BIS(μ_2-CHLORO)BIS[(1,5,9,13-TETRAAZACYCLOHEXADECANE-N[1],N[5],N[9],N[13])MERCURY] TETRACHLORO-MERCURATE
$(C_{24}H_{56}Cl_6Hg_3N_8)_n$ (II)

CHLORO(1,5,9,13-TETRAAZACYCLOHEXADECANE-N[1],N[5],N[9],N[13])LEAD CHLORIDE
$C_{12}H_{28}Cl_2N_4Pb$ (III)

(1,5,9,13-TETRAAZACYCLOHEXADECANE-N[1],N[5],N[9],N[13])LEAD DINITRATE
$C_{12}H_{28}N_6O_6Pb$ (IV)

N.W. ALCOCK, E.H. CURZON and P. MOORE, 1984. J. Chem. Soc. Dalton, 2813-2820.

I. Monoclinic, $P2_1/c$, a = 8.159, b = 15.092, c = 17.153 Å, β = 96.97°, Z = 4. Mo radiation, R = 0.033 for 2935 reflexions.

II. Tetragonal, P4/nmm, a = 13.4265, c = 10.973 Å, D_m = 2.11, Z = 2. Mo radiation, R = 0.040 for 1046 reflexions.

III. Orthorhombic, Fddd, a = 13.897, b = 18.631, c = 27.663 Å, Z = 16. Mo radiation, R = 0.044 for 1442 reflexions (at -80°C).

IV. Triclinic, P1, a = 8.139, b = 14.077, c = 20.729 Å, α = 92.13, β = 87.71, γ = 100.01°, Z = 5. Mo radiation, R = 0.068 for 5679 reflexions (at -80°C).

In each compound, all four macrocyclic N-H groups point to the same side of the macrocyclic plane; the other ligands occupy trans positions for I and II but are in cis positions for the larger Pb^{2+} ions in III and IV. Compound II has the tetrahedral anions with crystallographic $\bar{4}2m$ symmetry; there are two independent cations each with 4mm symmetry but one is slightly disordered. In III, the Pb atom lies on a site with symmetry 2 and the chloride ligand and a lone pair occupy adjacent positions on one side of the Pb^{2+} ion. There are 5 independent molecules in the structure of IV; one with two symmetrically bidentate nitrate groups, two with one bidentate and one unidentate, and two Pb atoms linked by two nitrate ions. Views of the various complexes are in Fig. 1. Some bond lengths (averages or ranges) are: Cd-N 2.334(6), Hg-N 2.38(3), Pb-N 2.54(1), Cd-O 2.413-2.832(8), Hg-Cl 2.64-2.83(2), Pb-Cl 2.98(1), Pb-O 2.74(3)-3.17(4) Å

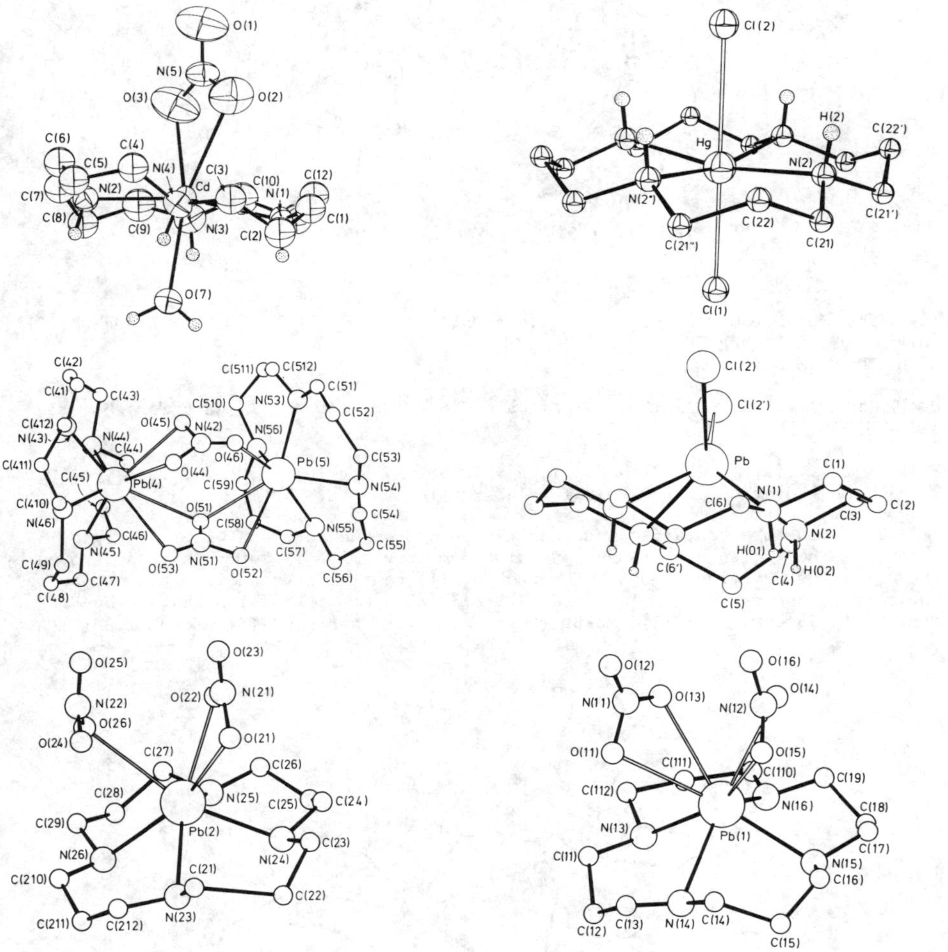

Fig. 1. Views of the cations in I, II and III and four of the Pb environments in IV (clockwise from top left).

(1,3,6,8,10,13,16,19-OCTAAZABICYCLO[6.6.6]EICOSANE)NICKEL(II) BIS(PERCHLORATE)
$C_{12}H_{30}Cl_2N_8NiO_8$ $[Ni(C_{12}H_{30}N_8)](ClO_4)_2$

M.P. SUH, W. SHIN, D. KIM and S. KIM, 1984. Inorg. Chem., 23, 618-620.

Cubic, $P2_13$, a = 12.707 Å, D_m = 1.74, Z = 4. Cu radiation, R = 0.073 for 497 reflexions.

The cation (Fig. 1) has exact C_3 symmetry and approximate D_3 symmetry. The conformation of the tris(ethylenediamine) portion is lel, and its absolute configuration is either Δλλλ or Δδδδ. The Ni atom is octahedrally coordinated by the six N atoms of the hexadentate ligand, Ni-N 2.10(1)-2.11(1) Å.

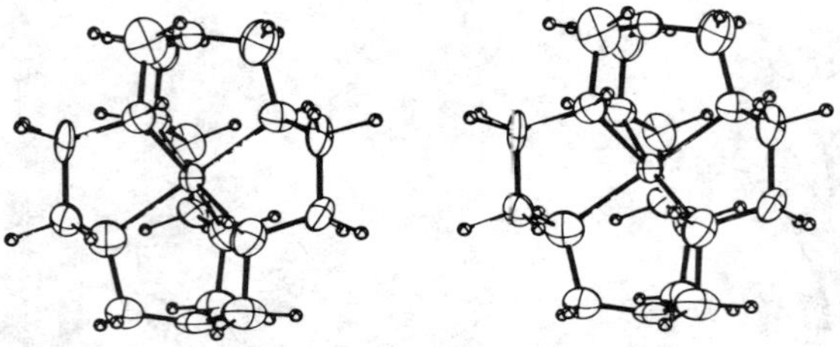

Fig. 1. Stereoview of the $[Ni(C_{12}H_{30}N_8)]^{2+}$ cation.

(μ_2-BROMO)(μ_2-HYDROXO)DIBROMO-BIS[(1,4,7-TRIAZACYCLONONANE-N^1,N^4,N^7)MOLYBDENUM(III)] DIBROMIDE TETRAHYDRATE
$C_{12}H_{31}Br_5Mo_2N_6O,4H_2O$

P. CHAUDHURI, K. WIEGHARDT, I. JIBRIL and G. HUTTNER, 1984. Z. Naturforsch., B39, 1172-1176.

Triclinic, P1, a = 8.213, b = 8.377, c = 10.081 Å, α = 103.63, β = 98.92, γ = 79.54°, Z = 1. Mo radiation, R = 0.0625 for 1427 reflexions (at -50°C).

The structure contains dinuclear cations, bromide ions and water of crystallization. The cation almost has inversion symmetry (Fig. 1) but this is destroyed by the differing bridging groups. The two Mo(III) centres have a distorted octahedral environment and a strong Mo-Mo bond (2.558(3) Å). Other bond lengths are Mo-Br (bridge) 2.531 and 2.558(5), Mo-Br (terminal) 2.547 and 2.581(7), Mo-OH 1.84(2) Å.

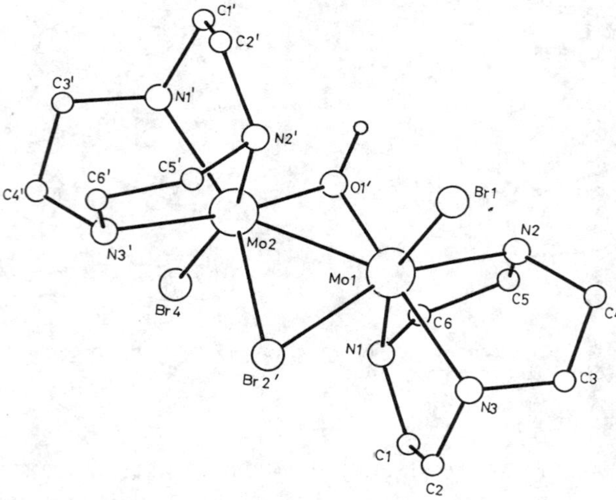

Fig. 1. A view of the dinuclear cation $C_{12}H_{31}Br_5Mo_2N_6O,4H_2O$.

BIS(μ-HYDROXO)BIS[OXO(1,4,7-TRIAZACYCLONONANE)VANADIUM(IV)] DIBROMIDE
$C_{12}H_{32}Br_2N_6O_4V_2$ $[V_2O_2(\mu\text{-}OH)_2(C_6H_{15}N_3)_2]Br_2$

K. WIEGHARDT, U. BOSSEK, K. VOLCKMAR, W. SWIRIDOFF and J. WEISS, 1984. Inorg. Chem., 23, 1387-1389.

Monoclinic, $P2_1/c$, a = 7.236, b = 14.003, c = 10.405 Å, β = 97.37°, Z = 2. Mo radiation, R = 0.043 for 1199 reflexions.

The centrosymmetric dinuclear cation is shown in Fig. 1. The geometry around each V atom is distorted octahedral: a terminal oxo group, V=O 1.603(5) Å, two μ-OH ligands, V-O 1,956(5) and 1.969(5) Å, and the tridentate cyclic amine, V-N 2.303(6) Å trans to oxo group, V-N 2.151(5) and 2.159(5) Å trans to μ-OH.

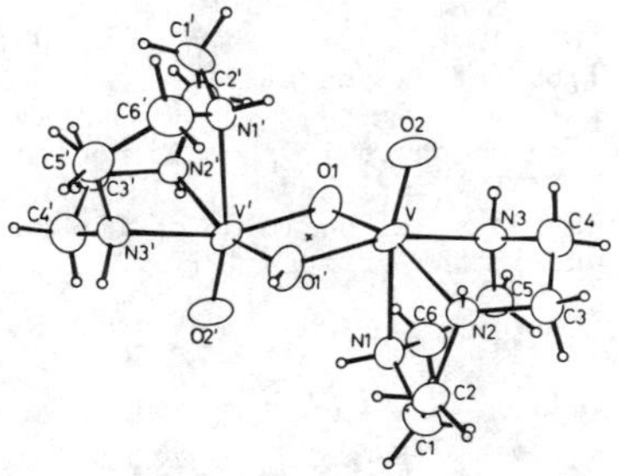

Fig. 1. The $[V_2O_2(\mu\text{-}OH)_2(C_6H_{15}N_3)]^{2+}$ cation.

BIS(μ-HYDROXO)-BIS[DICHLORO(1,4,7-TRIAZACYCLONONANE)MOLYBDENUM(III)] IODIDE
$C_{12}H_{32}Cl_2I_2Mo_2N_6O_2$ (I) $[(C_6H_{15}N_3)_2Mo_2(\mu\text{-}OH)_2Cl_2]I_2$

(μ-ACETATO)BIS(μ-HYDROXO)-BIS[(1,4,7-TRIAZACYCLONONANE)MOLYBDENUM(III)] IODIDE MONOHYDRATE
$C_{14}H_{35}I_3Mo_2N_6O_4,H_2O$ (II) $[(C_6H_{15}N_3)_2Mo_2(\mu\text{-}OH)_2(\mu\text{-}CH_3CO_2)]I_3.H_2O$

K. WIEGHARDT, M. HAHN, W. SWIRIDOFF and J. WEISS, 1984. Inorg. Chem., 23, 94-99.

I. Monoclinic, C2/c, a = 12.889, b = 14.219, c = 14.280 Å, β = 116.70°, Z = 4. Mo radiation, R = 0.057 for 1109 reflexions.

II. Monoclinic, $P2_1/a$, a = 9.833, b = 19.261, c = 14.967 Å, β = 100.00°, Z = 4. Mo radiation, R = 0.053 for 2063 reflexions.

Both cations (Fig. 1) consist of binuclear species containing formal Mo≡Mo bonds (2.501(3) and 2.471(2) Å). In I, two distorted octahedra share an edge, the cation possessing a centre of symmetry (Mo-O 2.126(9) and 2.135(11), Mo-Cl 2.438(5), Mo-N 2.197(12)-2.231(14) Å). In II, two distorted octahedra are bridged by an acetate and two OH ligands (Mo-O 2.066(10)-2.094(11), Mo-N 2.196(12)-2.234(12) Å). The Mo(μ-OH)$_2$-Mo rings are planar in both cations.

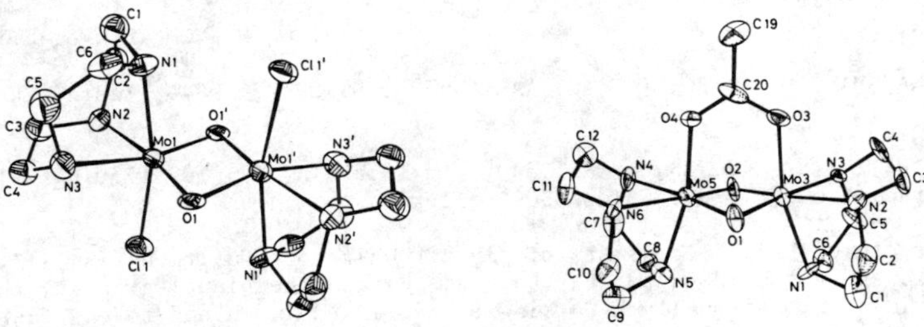

Fig. 1. The cations $[(C_6H_{15}N_3)_2Mo_2(\mu\text{-}OH)_2Cl_2]^{2+}$ (left) and $[(C_6H_{15}N_3)_2Mo_2(\mu\text{-}OH)_2(\mu\text{-}CH_3COO)]^{3+}$ (right).

BIS(N,N-DIETHYLETHYLENEDIAMINE)COPPER(II) DINITRATE
$C_{12}H_{32}CuN_6O_6$

A. WALSH and B.J. HATHAWAY, 1984. J. Chem. Soc. Dalton, 15-18.

Monoclinic, $P2_1/n$, a = 9.76, b = 12.81, c = 8.24 Å, β = 105.87°, D_m = 1.41, Z = 2. Cu radiation, R = 0.079 for 1165 reflexions.

The centrosymmetric CuN_4 chromophore is rhombic coplanar with Cu-N distances of 2.081(5) (tetriary N) and 2.011(5) Å (primary N). The conformations of the ethyl substituents effectively block the axial position of the CuN_4 chromophore and the nitrate ions exist as free ions in the crystal (Fig. 1).

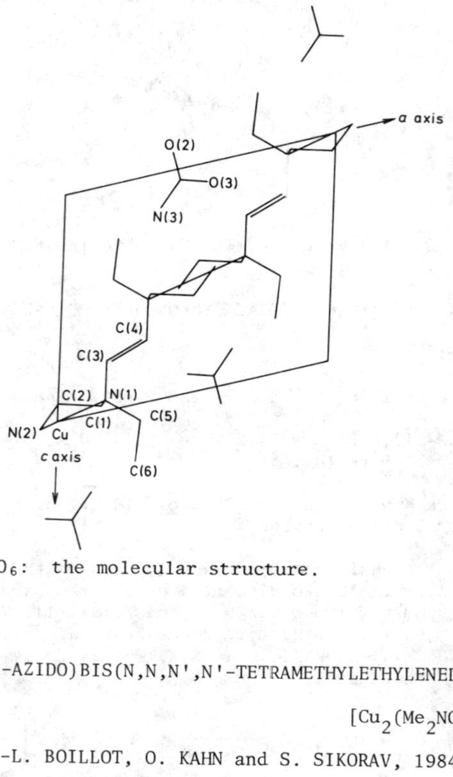

Fig. 1. $C_{12}H_{32}CuN_6O_6$: the molecular structure.

catena-BIS(μ-AZIDO)[(μ-AZIDO)BIS(N,N,N',N'-TETRAMETHYLETHYLENEDIAMINECOPPER(II))] HEXAFLUOROPHOSPHATE
$C_{12}H_{32}Cu_2F_6N_{13}P$ $[Cu_2(Me_2NCH_2CH_2NMe_2)_2(N_3)_3](PF_6)$

I. BKOUCHE-WAKSMAN, M.-L. BOILLOT, O. KAHN and S. SIKORAV, 1984. Inorg. Chem., 23, 4454-4459.

Monoclinic, $P2_1/c$, a = 12.713, b = 31.240, c = 13.259 Å, β = 102.17°, D_m = 1.60, Z = 8. Cu radiation, R = 0.075 for 3309 reflexions.

The structure (Fig. 1) consists of zigzag chains of $[Cu_2(tmen)_2(N_3)_3]^+$ cations parallel to c and noncoordinated PF_6^- anions. Within the binuclear unit, the Cu atoms are bridged symmetrically by a single N_3^- ligand in an end-to-end fashion (Cu...Cu 4.439 Å, Cu-N 2.02(1)-2.06(1) Å). The binuclear units are linked to each other by a double asymmetric end-to-end azido bridge (Cu...Cu 5.131 Å, Cu-N 1.99(1)-2.01(1) and 2.28(1)-2.34(1) Å). Each Cu has distorted square pyramidal coordination geometry with the long Cu-μ-N_3 bonds apical. Cu-N(tmen) 2.01(1)-2.08(1) Å.

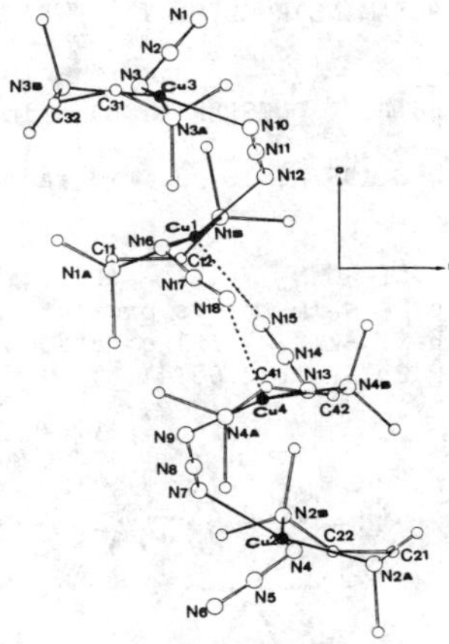

Fig. 1. The asymmetric unit for $[Cu(tmen)(N_3)_3]^+$.

μ,μ'-DIIODODI-[BIS(N,N,N',N'-TETRAMETHYLETHYLENEDIAMINE)DICOPPER(I)]
$C_{12}H_{32}Cu_2I_2N_4$ (I) $[(tmeda)CuI_2Cu(tmeda)]$

BIS-(N,N,N',N'-TETRAMETHYLETHYLENEDIAMINE)COPPER(I) DICHLOROCUPRATE(I)
$C_{12}H_{32}Cl_2Cu_2N_4$ (II) $[Cu(tmeda)_2][CuCl_2]$

L.M. ENGELHARDT, R.I. PAPASERGIO and A.H. WHITE, 1984. Aust. J. Chem., 37, 2207-2213.

I. Monoclinic, $P2_1/c$, a = 12.370, b = 12.119, c = 15.36 Å, β = 109.16°, Z = 4.
Mo radiation, R = 0.067 for 1240 reflexions.

II. Monoclinic, $P2_1/n$, a = 14.477, b = 15.063, c = 9.326 Å, β = 96.99°, Z = 4.
Mo radiation, R = 0.055 for 1195 reflexions.

I is a μ,μ'-diiodobridged dimer $(tmeda)CuI_2Cu(tmeda)$ (Fig. 1, Cu-I 2.600(5), Cu-N 2.18(2), Cu-Cu 2.566(4) Å, I-Cu-I 119.4(1), Cu-I-Cu 59.1(1)°), unlike II, an ionic complex $[Cu(tmeda)_2]^+[CuCl_2^-]$ which has the metal atoms for two different anions at crystallographically imposed centres of symmetry producing linear $[CuCl_2^-]$ (Fig. 1, cation, Cu-N 2.15(1) Å, N-Cu-N 85.6(5)°; anion, Cu-Cl 2.086(4) Å).

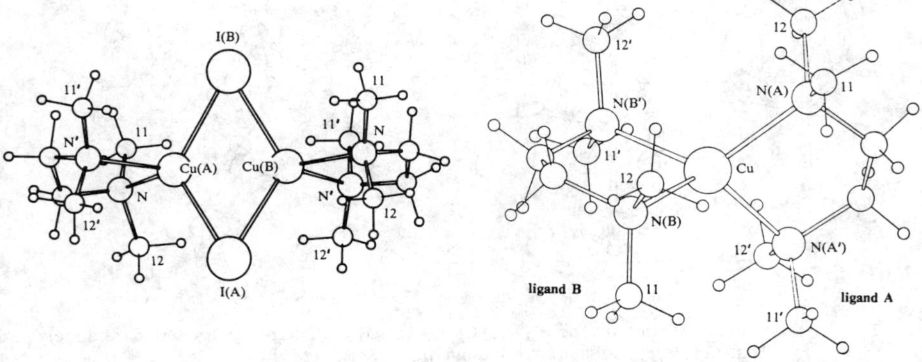

Fig. 1. The structures of (a) $[(tmeda)CuI_2Cu(tmeda)]$ and (b) $[Cu(tmeda)_2][CuCl_2]$.

lel,lel,lel-TRIS(rac-2,3-BUTANEDIAMINE)RHODIUM(III) TRIBROMIDE
$C_{12}H_{36}Br_3N_6Rh$ $[Rh(C_4H_{12}N_2)_3]Br_3$

M.F. GARGALLO, R.E. TAPSCOTT and E.N. DUESLER, 1984. Inorg. Chem., 23, 918-922.

Trigonal, P$\bar{3}$1c, a = 12.549, c = 8.195 Å, Z = 2. Mo radiation, R = 0.033 for 751 reflexions.

A racemic mixture of Δ-$[Rh(R,R-C_4H_{12}N_2)_3]^{3+}$ (Fig. 1) and Λ-$[Rh(S,S-C_4H_{12}N_2)_3]^{3+}$ cations having crystallographic D_3 symmetry is present in the structure. The Rh-N distance is 2.055(4) Å and N-Rh-N' 82.2(2)°. The cations and Br^- anions are linked by N-H...Br hydrogen bonds (Br...H 2.77(3), 2.65(4) Å).

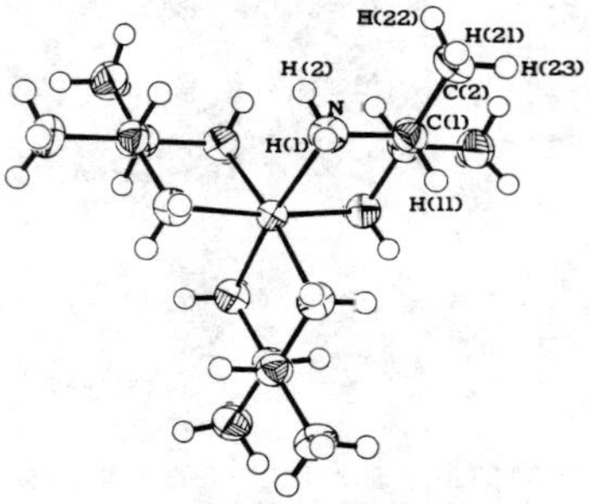

Fig. 1. The lel,lel,lel-Δ-$[Rh(R,R-C_4H_{12}N_2)_3]^{3+}$ cation.

(μ_2-SUPEROXO)BIS[(DIETHYLENETRIAMINE)(ETHYLENEDIAMINE)COBALT(III)] DIPERCHLORATE TRICHLORIDE DIHYDRATE
$C_{12}H_{42}Cl_5Co_2N_{10}O_{10},2H_2O$

D.D. DEXTER, C.N. SUTHERBY, M.W. GRIEB and R.C. BEAUMONT, 1984. Inorg. Chim. Acta, 86, 19-31.

Monoclinic, $P2_1/c$, a = 8.811, b = 11.535, c = 16.477 Å, β = 68.894°, Z = 2. Cu radiation, R = 0.085 for 1713 reflexions (at liquid N_2 temperature).

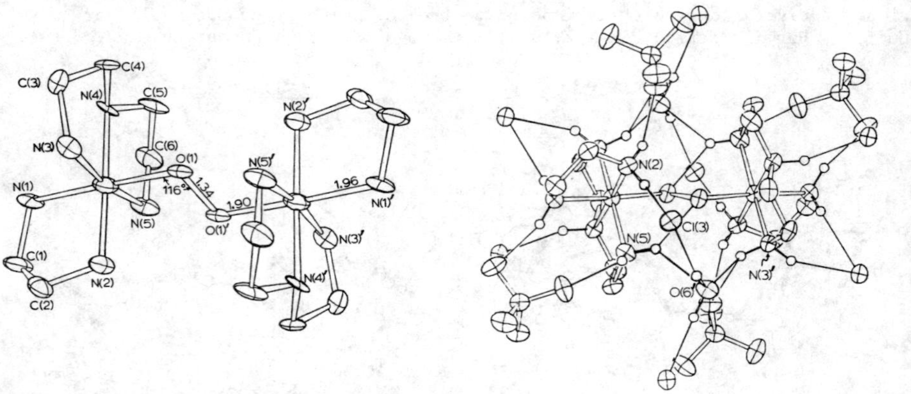

Fig. 1. Molecular structure of $C_{12}H_{42}Cl_5Co_2N_{10}O_{10}$ showing selected distances and angles (left), and hydrogen bonding scheme (right).

The molecule consists of two Co centres each having octahedral coordination, connected via a bridging dioxygen ligand which lies on a crystallographic centre of symmetry (Fig. 1). The bridging dioxygen is best considered a superoxide because of its O-O bond distance of 1.34 Å. The Co-N distances are: Co-N(1) 1.96, Co-N(2) 1.96, Co-N(3) 1.96, Co-N(4) 1.94 and Co-N(5) 1.98 Å while the angles associated with the metal atom range from 85.3 to 95.1°. There are two intramolecular hydrogen bonds N(5)...O(1') and N(5')...O(1) (2.96 Å) and intermolecular bonds between the perchlorate, chloride and water molecules with amine H atoms. The water oxygen O(6) accepts hydrogen bonds from N(3) and N(5') while donating bonds to Cl(2) and Cl(3) which in turn bridges to H(4) on N(2) (Fig. 1). The distances for N(3)...O(6), N(5')...O(6), O(6)...Cl(2), O(6)...Cl(3) and Cl(3)...N(2) are respectively 2.92, 3.08, 3.16, 3.14 and 3.26 Å.

OXO[3,3'-(1,3-PROPANEDIYLDIIMINO)BIS(3-METHYL-2-BUTANONE OXIMATO)(3-)-N,N',N'',N''']-TECHNETIUM(V)

$C_{13}H_{25}N_4O_3Tc$ $TcO(C_{13}H_{25}N_4O_2)$

C.K. FAIR, D.E. TROUTNER, E.O. SCHLEMPER, R.K. MURMANN and M.L. HOPPE, 1984. Acta Cryst., C40, 1544-1546.

Monoclinic, Pc, a = 6.950, b = 11.187, c = 11.060 Å, β = 104.13°, Z = 2. Mo radiation, R = 0.024 for 2601 reflexions.

The Tc(V) ion is approximately square pyramidally coordinated (Fig. 1). Tc is displaced 0.678 Å from the N_4 plane towards the oxo oxygen; the carbon atoms attached to N all lie in a plane parallel to N_4. Bond distances are: Tc-O 1.679(3); Tc-N (imino) 1.917(3), 1.908(3); Tc-N(oxime) 2.086(3), 2.093(4) Å. Within the intramolecular hydrogen bond O(1)...O(2) is 2.420(5) with O(2)-H 1.16(8) and O(1)...H 1.35(8) Å.

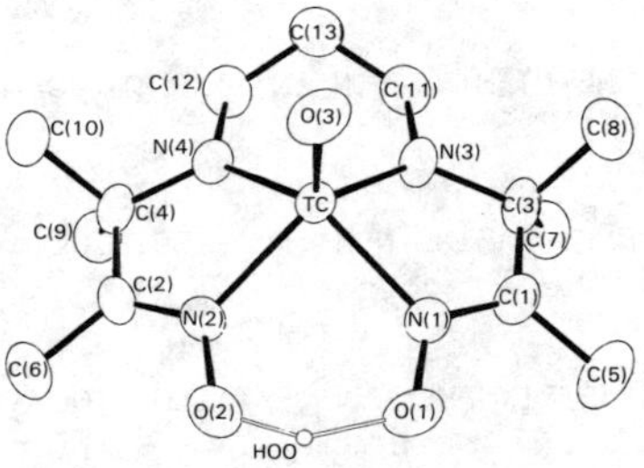

Fig. 1. The $TcO(C_{13}H_{25}N_4O_2)$ molecule.

trans-DICHLORO[HYDROGEN-3,3'-(1,3-PROPANEDIYLDIIMINO)BIS(3-METHYL-2-BUTANONE OXIMATO)]RHODIUM(III)

$C_{13}H_{27}Cl_2N_4O_2Rh$ (I)

trans-DICHLORO[HYDROGENBIS(3-AMINO-3-METHYL-2-2-BUTANONE OXIMATO)]RHODIUM(III)

$C_{10}H_{23}Cl_2N_4O_2Rh$ (II)

S. SIRIPAISARNPIPAT and E.O. SCHLEMPER, 1984. Inorg. Chem., 23, 330-334.

I. Orthorhombic, $P2_12_12_1$, a = 8.400, b = 13.149, c = 16.065 Å, D_m = 1.66, Z = 4. Mo radiation, R = 0.043 for 2579 reflexions.

II. Orthorhombic, Pnam, a = 11.108, b = 11.993, c = 11.661 Å, D_m = 1.73, Z = 4. Mo radiation, R = 0.020 for 2609 reflexions.

The two trans-dichlororhodium(III) complexes (Fig. 1) have strong O-H...O hydrogen bonds between cis oxime groups (O...O 2.474(7) and 2.459(2) Å in I and II respectively). The average Rh-ligand distances are Rh-Cl 2.336(10), Rh-N(oxime) 1.988(2), and Rh-N(amine) 2.064(8) Å. Complex II has a mirror plane containing Rh, 2Cl, and the H-bonded hydrogen atom.

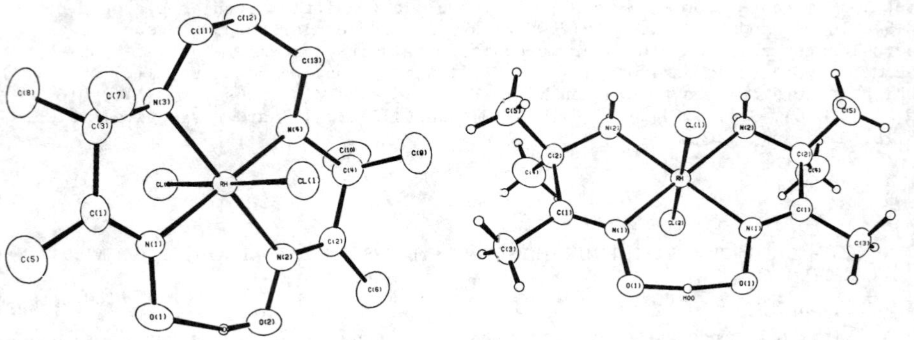

Fig. 1. The molecules $C_{13}H_{27}Cl_2N_4O_2Rh$ (left) and $C_{10}H_{23}Cl_2N_4O_2Rh$ (right).

(1,1,7,7-TETRAETHYLDIETHYLENETRIAMINE-N,N',N'')ISOTHIOCYANATOPALLADIUM HEXAFLUOROPHOSPHATE
$C_{13}H_{29}F_6N_4PPdS$ (I)

(1,1,7,7-TETRAETHYLDIETHYLENETRIAMINE-N,N',N'')ISOTHIOCYANATOPALLADIUM TETRAPHENYLBORATE ACETONE SOLVATE
$C_{37}H_{49}BN_4PdS,C_3H_6O$ (II)

(1,1,7,7-TETRAETHYLDIETHYLENETRIAMINE-N,N',N'')ISOTHIOCYANATOPALLADIUM TETRAPHENYLBORATE
$C_{37}H_{49}BN_4PdS$ (III)

(1,1,7,7-TETRAETHYLDIETHYLENETRIAMINE-N,N',N'')ISOTHIOCYANATOPALLADIUM TETRAPHENYLBORATE BROMOCHLOROMETHANE SOLVATE
$C_{37}H_{49}BN_4PdS,CH_2BrCl$ (IV)

(1,1,7,7-TETRAETHYLDIETHYLENETRIAMINE-N,N',N'')ISOTHIOCYANATOPALLADIUM TETRAPHENYLBORATE METHANOL SOLVATE
$C_{37}H_{49}BN_4PdS,CH_4O$ (V)

C.P. BROCK, J.L. HUCKABY and T.G. ATTIG, 1984. Acta Cryst., B40, 595-606.

I. Orthorhombic, $P2_12_12_1$, a = 11.824, b = 11.838, c = 15.626 Å, Z = 4. Mo radiation, R = 0.033 for 2500 reflexions.

II. Monoclinic, $P2_1/c$, a = 11.927, b = 19.250, c = 17.252 Å, β = 100.33°, Z = 4. Mo radiation, R = 0.044 for 5221 reflexions at 143 K.

III. Orthorhombic, $P2_12_12_1$, a = 9.566, b = 14.745, c = 24.419 Å, Z = 4. Mo radiation, R = 0.045 for 5144 reflexions at 150 K.

IV. Monoclinic, $P2_1/n$, a = 11.058, b = 20.833, c = 16.955 Å, β = 100.72°, Z = 4. Mo radiation, R = 0.045 for 4940 reflexions at 150 K.

V. Monoclinic, $P2_1/n$ or $P2_1$, a = 12.748, b = 9.929, c = 29.112 Å, β = 97.39°, Z = 4. Mo radiation, R = 0.050 for 1346 reflexions at 137 K.

Compounds I - IV have N-bonded thiocyanate while V is S-bonded (Fig. 1). Some packing motifs common to the five crystal structures are described.

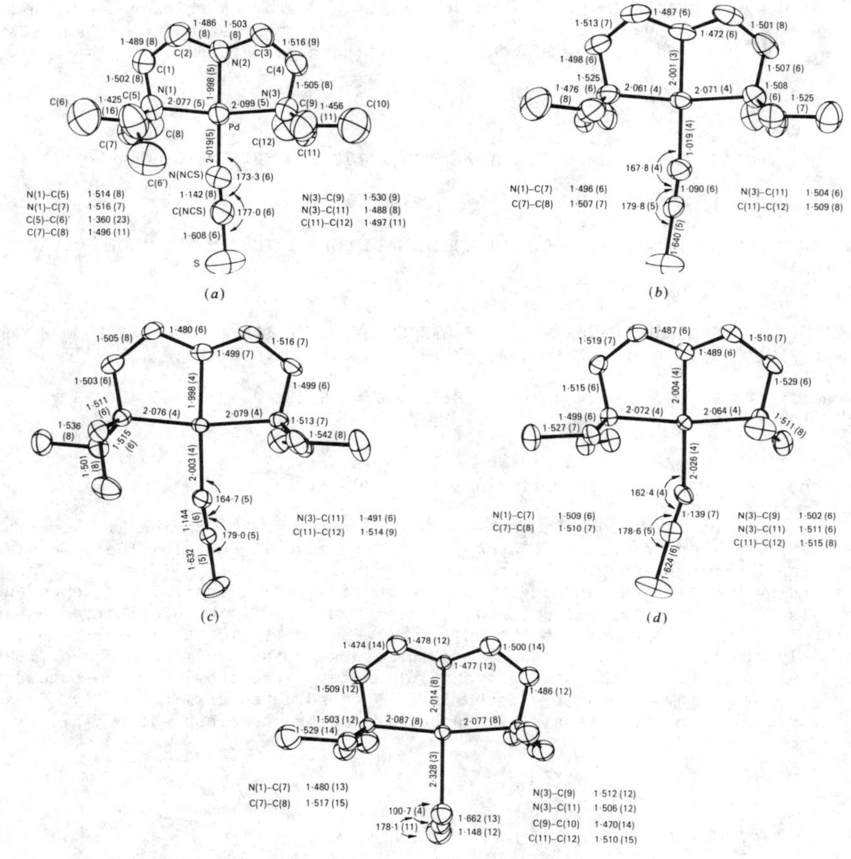

Fig. 1. Views (a) - (e) of the five NCS complexes (I) - (V).

BARIUM AQUA(DIETHYLENETRIAMINEPENTAACETATO)NEODYMIUM(III) DIHYDRATE
$C_{14}H_{21}BaN_3NdO_{11},2(H_2O)$

J.J. STEZOWSKI and J.L. HOARD, 1984. Israel J. Chem., 24, 323-334.

Triclinic, $P\bar{1}$, a = 8.845, b = 13.970, c = 10.015 Å, α = 99.2, β = 109.1, γ = 96.1°, Z = 2. R = 0.037 for 4832 reflexions.

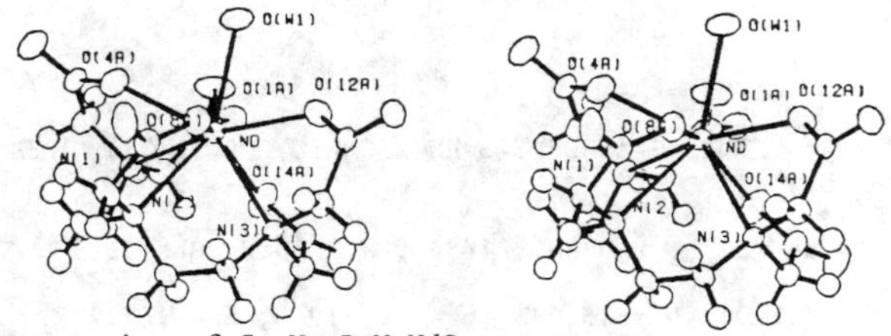

Fig. 1. A stereoview of $C_{14}H_{21}BaN_3NdO_{11}$.

The Nd coordination (Fig. 1) is described as a distorted tricapped trigonal prism in which the tridentate ends of the ligand form the trigonal prism and the central amino acid and a water molecule provide the capping atoms.

ETHYLETHYLENEDIAMINE[N-(2-AMINOETHYL)-7-METHYLSALICYLIDENEIMINATO]COBALT(III) BROMIDE
$C_{14}H_{26}BrCoN_4O$ (I)

BIS[N-(2-AMINOETHYL)-7-METHYLSALICYLIDENEIMINATO]COBALT IODIDE METHANOL SOLVATE HEMIHYDRATE
$C_{20}H_{26}CoIN_4O_2$,0.5(CH_4O),0.5(H_2O) (II)

A.I. YANOVSKII, G.G. ALEKSANDROV, Yu.T. STRUCHKOV, I.Ya. LEVITIN, R.M. BODNAR and M.E. VOL'PIN, 1983. Koord. Khim., 9, 825-834 [Sov. J. Coord. Chem., 9, 473-482].

I. Monoclinic, $P2_1/n$, a = 14.481, b = 8.320, c = 15.382 Å, β = 108.65°, Z = 4. No other details given.

II. Monoclinic, $P2_1/n$, a = 10.017, b = 31.605, c = 14.227 Å, β = 94.70°, Z = 8. Mo radiation, R = 0.040 for 5410 reflexions (at -120°C).

Compound I contains complex [Co(7-mesalen)(en)(σ-C_2H_5)]$^+$ cations in which the ethyl group is σ-bonded to the Co atom, and Br^- anions; II contains [Co(7-mesalen)$_2$]$^+$ cations and I^- anions along with water and methanol of solvation (Fig. 1). The corresponding bond lengths and angles in the two crystallographically independent cations in II have similar values. The Co atoms in I and II have distorted octahedral coordination. Co-N distances to the NH_2 of mesalen in II (1.955, 1.956, 1.964) are appreciably greater than the distances to the atoms of the aldiminato group (1.896, 1.922, 1.927 Å). Due to the strong trans influence of the ethyl group σ-bonded to the Co atom, the ethylenediamine ligand in I is coordinated asymmetrically (Co-N 2.128, 1.973 Å). In the crystals of I and II there are numerous short N-H...Br^- and N-H...I^- contacts.

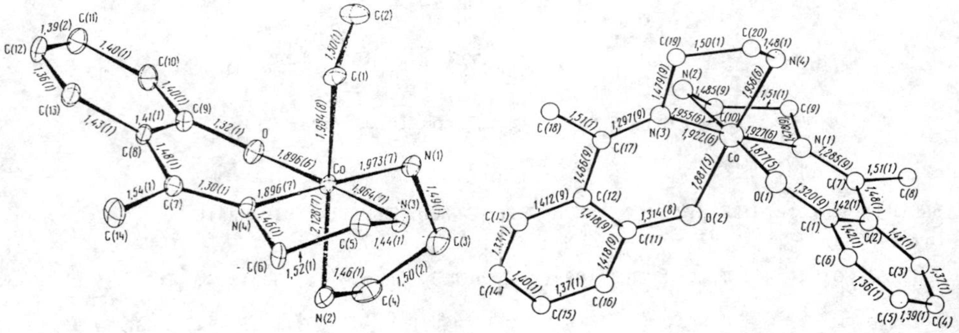

Fig. 1. The structure of the cation in $C_{14}H_{26}BrCoN_4O$ (left) and in $C_{20}H_{26}CoIN_4O_2$ (right).

BIS[(μ_2-BROMO)(7-AMINO-4-METHYL-5-AZAHEPT-3-EN-2-ONATO)COPPER(II)] HYDROBROMIDE
$C_{14}H_{26}Br_2Cu_2N_4O_2$,0.3(HBr) [CuAEBr]$_2$,0.3(HBr)

J.-P. COSTES, F. DAHAN and J.-P. LAURENT, 1984. J. Coord. Chem., 13, 355-362.

Monoclinic, $P2_1/n$, a = 6.1707, b = 13.045, c = 12.719 Å, β = 102.237°, D_m = 1.96, Z = 2. Mo radiation, R = 0.036 for 1582 reflexions.

The crystals contain centrosymmetric dimers (Fig. 1) with the geometry at the unique Cu atom described as a distorted tetragonal pyramid (the basal plane consisting of atoms N(1),N(2),O,Br with the apical site occupied by the inversion related Br atom). Some dimensions are: Cu-Br 2.4693(7), Cu-Br' 2.9561(7), Cu-N(1) 1.951(3), Cu-N(2) 1.990(3), Cu-O 1.912(3) Å, Br-Cu-Br' 98.96(2), N-Cu-N 85.0(1), Br-Cu-N(1) 156.7(1), N(2)-Cu-O 175.2°.

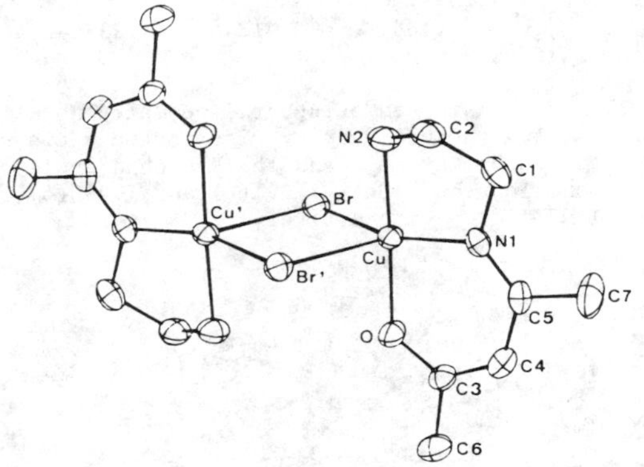

Fig. 1. A view of the centrosymmetric $[CuAEBr]_2$ dimer.

BIS(CYANODIETHYLENETRIAMINEPLATINUM(II)) TETRACYANOPLATINATE(II) MONOHYDRATE
$C_{14}H_{26}N_{12}Pt_3,H_2O$

A.I. AKHMEDOV, A.I. YANOVSKII, A.V. BABKOV and Yu.T. STRUCHKOV, 1983. Koord. Khim., 9, 1138-1142 [Sov. J. Coord. Chem., 9, 650-655].

Monoclinic, $P2_1/n$, a = 18.008, b = 9.483, c = 14.247 Å, β = 105.84°, Z = 4. Mo radiation, R = 0.078 for 3026 reflexions.

The structure is composed of two symmetrically independent $[Pt(CN)_4]^{2-}$ anions which are located at crystallographic centres of symmetry, two crystallographically independent $[Pt(dien)CN]^+$ cations and a water molecule of crystallisation. The anions are loosely stacked along the a axis with the planes of neighbouring square-planar anions almost parallel (dihedral angle 12.6°). The shortest Pt...Pt distance in the stack is however 9.004(3) Å. The shortest Pt(anion)...Pt(cation) and Pt(cation)... Pt(anion) distances are 4.216(1) and 8.557(1) Å respectively. In the cation the Pt atom has square-planar coordination with a small tetrahedral distortion. Bond lengths in the anion and cation are shown in Fig. 1. The anions and cations form hydrogen bonds with the molecules of water.

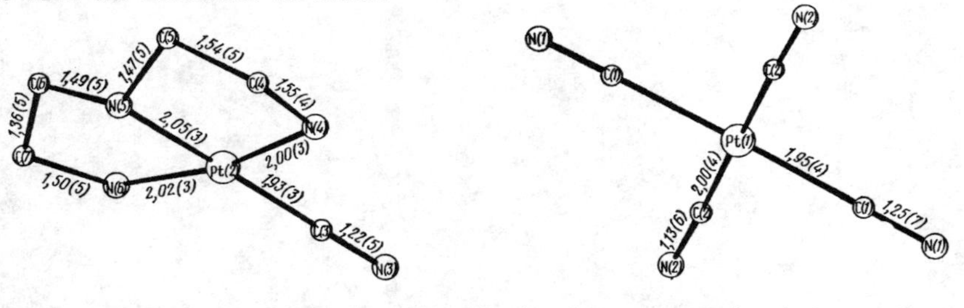

Fig. 1. $C_{14}H_{26}N_{12}Pt_3,H_2O$: bond lengths in the cation and anion.

s-cis-[N,N'-DIMETHYL-(1R,2R)-1,2-CYCLOHEXANEDIAMINE-N,N'-DIACETATO](ETHYLENEDIAMINE)-COBALT(III) PERCHLORATE

$C_{14}H_{28}ClCoN_4O_8$ $[Co(C_{12}H_{20}N_2O_4)(C_2H_8N_2)](ClO_4)$

S. YANO, T. MURATA, M. ASAMI, M. INOUE, M. YASHIRO, M. KOSAKA and S. YOSHIKAWA, 1984. Inorg. Chem., 23, 4127-4128.

Monoclinic, $P2_1$, a = 10.589, b = 11.184, c = 8.412 Å, β = 111.45°, Z = 2. Mo radiation, R = 0.074 for 2418 reflexions.

The cation (Fig. 1) has a distorted octahedral geometry about Co with the carboxylate O atoms trans to one another, giving rise to the s-cis configuration. The absolute configuration of the complex is Δ and the R,R-chxn chelate ring adopts the λ gauche conformation. Bond lengths: Co-O 1.891(6) and 1.890(6), Co-N(1-4) 1.972(7), 1.990(6), 1.995(8), and 1.973(11) Å, respectively.

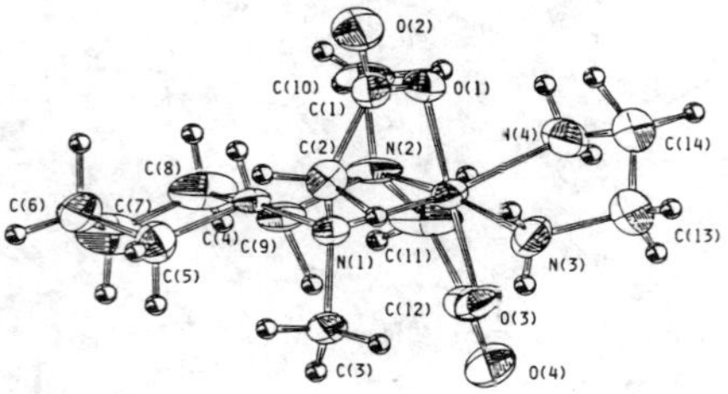

Fig. 1. The $[Co(C_{12}H_{20}N_2O_4)(C_2H_8N_2)]^+$ cation.

(1,8-DINITRO-3,6,10,13,16,19-HEXAAZABICYCLO[6.6.6]ICOSANE-N^3,N^6,N^{10},N^{16})COBALT DICHLORIDE TETRAHYDRATE (DINITROSARCOPHAGINECOBALT DICHLORIDE TETRAHYDRATE)

$C_{14}H_{29}Cl_2CoN_8O_4,4(H_2O)$

R.J. GEUE, T.W. HAMBLEY, J.M. HARROWFIELD, A.M. SARGESON and M.R. SNOW, 1984. J. Am. Chem. Soc., 106, 5478-5488.

Orthorhombic, $P2_12_12_1$, a = 10.821, b = 15.929, c = 14.259 Å, D_m = 1.579, Z = 4. Mo radiation, R = 0.048 for 2423 reflexions.

The analysis establishes the cage structure shown in Fig. 1, with Co-N(1) 2.016(7), Co-N(2) 1.980(7), Co-N(3) 1.967(7), Co-N(4) 1.975(7), Co-N(5) 1.975(7), Co-N(6) 1.946(7) Å. The absolute configuration is Λ.

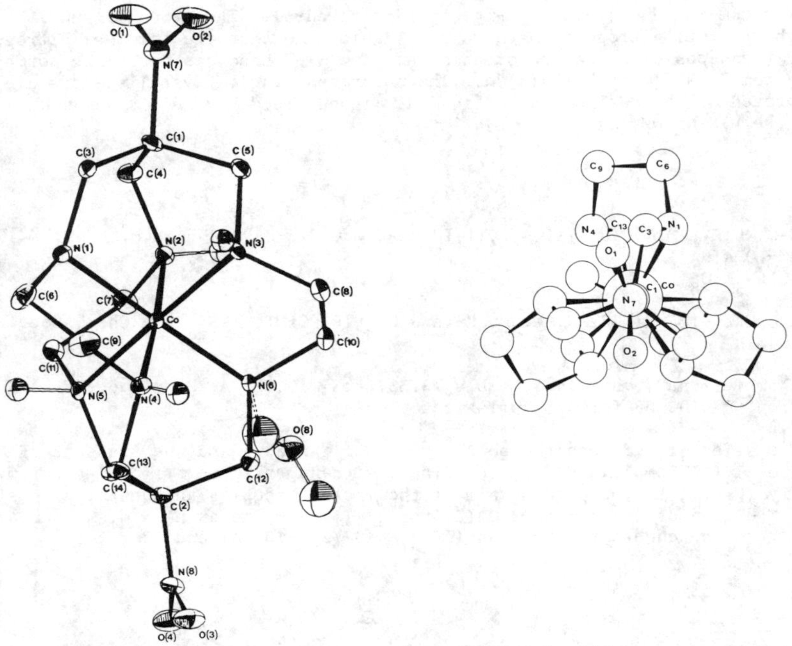

Fig. 1. Views of the complex cation in $C_{14}H_{29}Cl_2CoN_8O_4,4(H_2O)$.

1,4,7,10-TETRAAZADECANEBIS(N-METHYLIMIDAZOLE)NICKEL(II) PERCHLORATE

$C_{14}H_{30}Cl_2N_8NiO_8$ $[C_{14}H_{30}N_8Ni]^{2+}.2(ClO_4^-)$

M.R. BURKE, M.F. RICHARDSON and P.J. McCARTHY, 1984. J. Cryst. Spectrosc. Res., 14, 143-155.

Triclinic, P$\bar{1}$, a = 7.934, b = 9.770, c = 16.108 Å, α = 98.54, β = 97.16, γ = 98.10°, Z = 2. Mo radiation, R = 0.049 for 3729 reflexions.

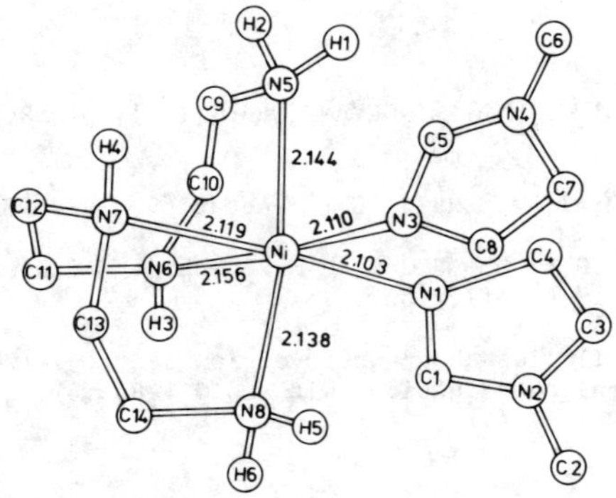

Fig. 1. A view of the $[C_{14}H_{30}N_8Ni]^{2+}$ ion.

The complexed nickel ion (Fig. 1) is octahedral. The trien ligand is bonded through its four nitrogens in the usual fashion, so that the terminal nitrogen atoms occupy trans-positions on the octahedron. The Ni-N bond distances are normal, and range from 2.103(3) to 2.156(3) Å. The polarized single-crystal spectra are interpreted on the basis of an octahedral ligand field having a tetragonal distortion along the H_2N-Ni-NH_2 axis.

[11-(2-DIMETHYLAMINOETHYL)-1,4,7,11-TETRAAZACYCLOTETRADECANE-N^1,N^2,N^3,N^4,N^5]NICKEL(II) DIPERCHLORATE
$C_{14}H_{33}Cl_2N_5NiO_8$

N.W. ALCOCK, R.G. KINGSTON, P. MOORE and C. PIERPOINT, 1984. J. Chem. Soc. Dalton, 1937-1943.

Orthorhombic, $Pna2_1$, a = 14.559, b = 15.337, c = 10.322 Å, D_m = 1.6, Z = 4. Mo radiation, R = 0.108 for 978 reflexions.

The R-factor reflects the poor quality of the data and the presence of disorder. The blue Ni(II) complex is five-coordinate with approximate trigonal-bipyramidal geometry (Fig. 1). The pendant arm of the ligand occupies an equatorial position. The Ni-N distances range from 2.04(2) to 2.17(3) Å. The axial N-Ni-N bond angle is 172(1)° and the equatorial bond angles are 107(2), 107(1) and 144(2)°.

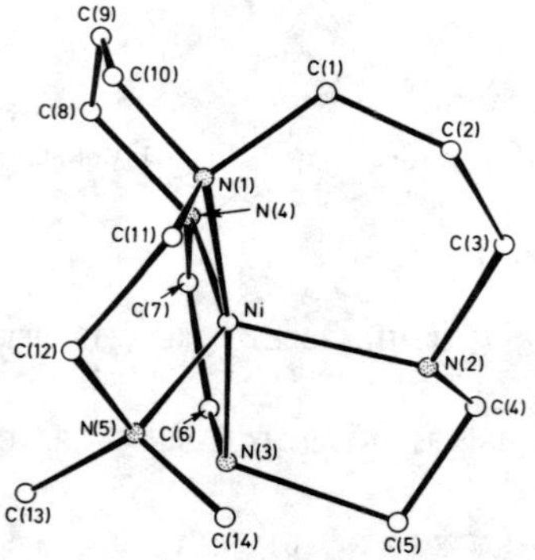

Fig. 1. $C_{14}H_{33}Cl_2N_5NiO_8$: the cation showing only one set of positions for the disordered atoms.

(μ_2-PEROXO)BIS[TRIS(2-AMINOETHYL)AMINECYANOCOBALT(III)] DIPERCHLORATE
$C_{14}H_{36}Cl_2Co_2N_{10}O_{10}$ $[C_{14}H_{36}Co_2N_{10}O_2]^{2+}.2ClO_4^-$

S. FALLAB and M. ZEHNDER, 1984. Helv. Chim. Acta, 67, 392-398.

Monoclinic, C2/c, a = 21.024, b = 8.061, c = 16.245 Å, β = 95.46°, Z = 4. Mo radiation, R = 0.08 for 1736 reflexions.

The molecule (Fig. 1) lies about an inversion centre, with Co-O-O 113.4(10)°. The Co atom has octahedral coordination, with Co-O 1.899(6), Co-N 1.949-1.998(7), Co-C 1.912(10), O-O 1.494(11) Å.

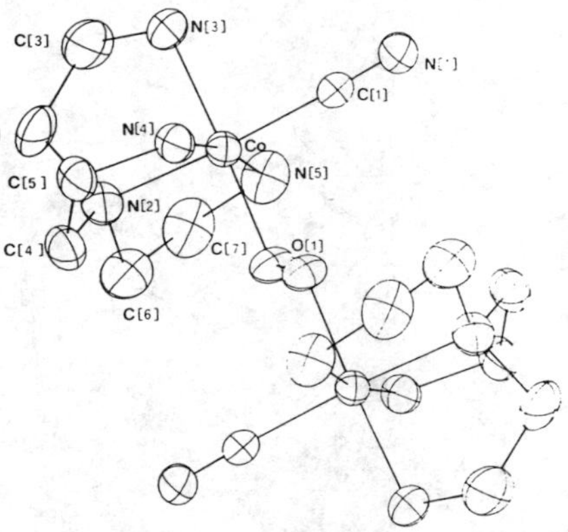

Fig. 1. A view of the $[C_{14}H_{36}Co_2N_{10}O_2]^{2+}$ cation.

(μ-OXALATO)BIS[AQUA(N,N,N',N'-TETRAMETHYLETHYLENEDIAMINE)COPPER(II)] DIPERCHLORATE HYDRATE

$C_{14}H_{36}Cl_2Cu_2N_4O_{14}$,$1.25H_2O$ (I) $[Cu_2(C_2O_4)(Me_2NCH_2CH_2NMe_2)_2(H_2O)_2](ClO_4)_2.1.25H_2O$

(μ-OXALATO)(DIETHYLENETRIAMINECOPPER(II))DIAQUA(N,N,N',N'-TETRAMETHYLETHYLENEDIAMINE)-COPPER(II) DIPERCHLORATE

$C_{12}H_{33}Cl_2Cu_2N_5O_{14}$ (II) $[Cu_2(C_2O_4)(Me_2NCH_2CH_2NMe_2)(H_2NCH_2CH_2NHCH_2CH_2NH_2)(H_2O)_2](ClO_4)_2$

(μ-OXALATO)BIS[2-METHYLIMIDAZOLE(N,N,N',N'-TETRAMETHYLETHYLENEDIAMINE)COPPER(II)] BIS(HEXAFLUOROPHOSPHATE)

$C_{22}H_{42}Cu_2F_{12}N_8O_4P_2$ (III) $[Cu_2(C_2O_4)(Me_2NCH_2CH_2NMe_2)_2(CH_3C_3H_3N_2)_2](PF_6)_2$

M. JULVE, M. VERDAGUER, A. GLEIZES, M. PHILOCHE-LEVISALLES and O. KAHN, 1984. Inorg. Chem., 23, 3808-3818.

I. Triclinic, $P\bar{1}$, a = 18.955, b = 10.019, c = 7.658 Å, α = 98.30, β = 98.37, γ = 88.19°, Z = 2. Mo radiation, R = 0.052 for 3204 reflexions.

II. Monoclinic, $P2_1/c$, a = 11.821, b = 9.093, c = 23.998 Å, β = 96.50°, Z = 4. Mo radiation, R = 0.064 for 2985 reflexions.

III. Triclinic, $P\bar{1}$, a = 8.224, b = 10.414, c = 11.754 Å, α = 94.63, β = 108.57, γ = 103.15°, Z = 1. Mo radiation, R = 0.038 for 3633 reflexions.

The binuclear cations of I-III are shown in Fig. 1 and each features a μ-oxalato ligand. In I the Cu atom is in a square pyramidal environment with the two N atoms of $Me_2NCH_2CH_2NMe_2$ and two O atoms of $C_2O_4^{2-}$ in the basal plane (Cu-N 2.005(5)-2.011(5), Cu-O 1.975(4)-1.980(4) Å) and a water molecule occupying the apical position (Cu-O 2.364(5) and 2.312(5) Å). In II both Cu atoms are in 4+2 environments with the three N atoms of $H_2NCH_2CH_2NHCH_2CH_2NH_2$ basal (Cu-N 1.975(7)-2.004(7) Å) and $C_2O_4^{2-}$ oxygen atoms basal (Cu-O 1,969(5) Å) and axial (Cu-O 2.461(5) and 2.752(5) Å at Cu(2) forming polymeric chains. At Cu(1), $Me_2NCH_2CH_2NMe_2$ nitrogen and $C_2O_4^{2-}$ oxygen atoms are basal (Cu-N 2.001(7) and 2.008(8), Cu-O 1.954(6) and 1.969(7) Å and water molecules apical (Cu-O 2.62(1) and 2.72(1) Å). The Cu atoms in the centrosymmetric cation of III have coordination intermediate between the square pyramid with only one oxygen atom of $C_2O_4^{2-}$ in the basal plane and the trigonal bipyramid (Cu-O 1.997(2) and 2.208(2), Cu-N 1.991(2)-2.056(2) Å).

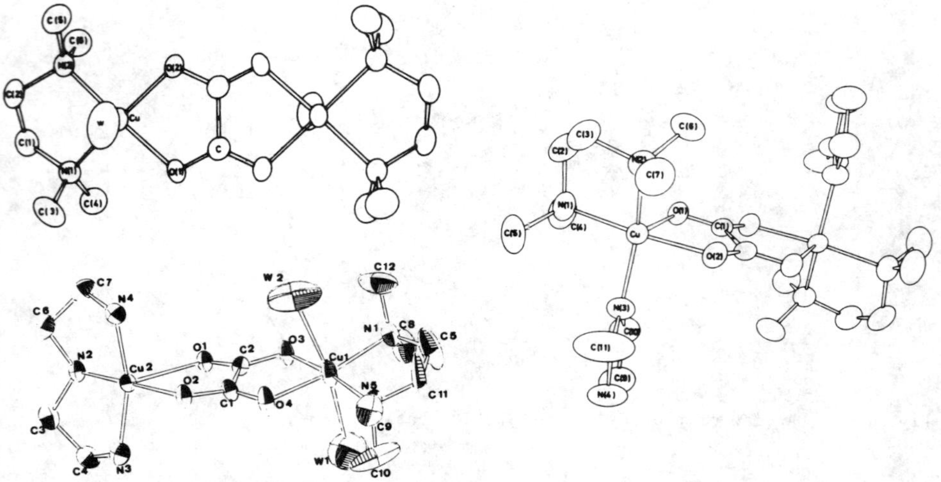

Fig. 1. The cations $[Cu_2(C_2O_4)(Me_2NCH_2CH_2NMe_2)_2(H_2O)_2]^{2+}$ (top left), $[Cu_2(C_2O_4)(Me_2NCH_2CH_2NMe_2)(H_2NCH_2CH_2NHCH_2CH_2NH_2)(H_2O)_2]^{2+}$ (bottom left), and $[Cu_2(C_2O_4)(Me_2NCH_2CH_2NMe_2)_2(CH_3C_3H_3N_2)_2]^{2+}$ (right).

[N,N'-BIS(2'-PYRIDINECARBOXAMIDO)-1,3-PROPANE]NICKEL(II) MONOHYDRATE

$C_{15}H_{14}N_4NiO_2,H_2O$ $Ni(C_7H_6N_2O)_2CH_2,H_2O$

F.S. STEPHENS and R.S. VAGG, 1984. Inorg. Chim. Acta, 90, 17-24.

Monoclinic, $P2_1/c$, a = 7.174, b = 18.590, c = 11.641 Å, β = 110.69°, Z = 4. Mo radiation, R = 0.030 for 1826 reflexions.

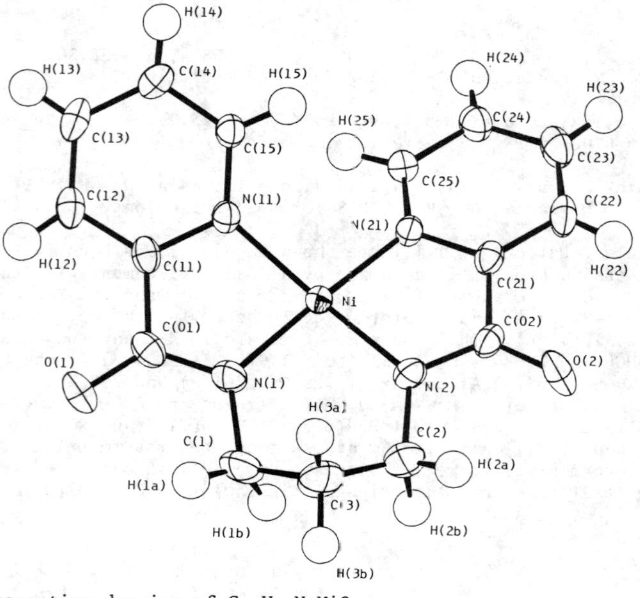

Fig. 1. Perspective drawing of $C_{15}H_{14}N_4NiO_2$.

The N_4 ligand coordinates to the Ni atom in an irregular square plane with a tetrahedral twist of 15.9° at the metal atom (Fig. 1). Main dimensions are: Ni-N(amide) 1.861 and 1.867, Ni-N(pyridine) 1.911 and 1.914 Å, N(11)-Ni-N(1) 84.0, N(11)-Ni-N(21) 98.7, N(21)-Ni-N(2) 83.8 and N(1)-Ni-N(2) 96.0°. The planes of the chelate molecules lie approximately perpendicular to [100] and the water molecules hydrogen-bond to amide O atoms (O(W)...O(1) 2.889, O(W)...O(2) 2.873 Å) to form chains parallel to (101).

α-AQUA[N,N'-BIS(2'-PYRIDINECARBOXAMIDO)-1,3-PROPANE]COPPER(II) DIHYDRATE
$C_{15}H_{16}CuN_4O_3,2H_2O$ $Cu(H_2O)(C_{15}H_{14}N_4O_2),2H_2O$

F.S. STEPHENS and R.S. VAGG, 1984. Inorg. Chim. Acta, 88, 7-14.

Monoclinic, $P2_1/c$, a = 11.719, b = 13.092, c = 12.663 Å, β = 119.56°, Z = 4. Mo radiation, R = 0.026 for 2398 reflexions.

The Cu atom is five-coordinate (Fig. 1) with the N_4-tetradentate ligand forming the base of a distorted square pyramid. The average Cu-N(amide), Cu-N(pyridine), N(amide)-Cu-N(amide) and N(pyridine)-Cu-N(pyridine) distances and angles are: 1.950, 2.043 Å; 94.5 and 100.2° respectively. The metal atom lies 0.27 Å above the N_4 plane towards the coordinated apical water molecule (Cu-O 2.236 Å). The central six-membered chelate ring adopts a skewed boat conformation. The molecules pack in sheets parallel to (10$\bar{1}$) with a hydrogen bonding network involving the water molecules and amide O atoms of the ligand (O(W1)...O(2) 2.917, O(W1)...O(W3) 2.742, O(W2)...O(1) 2.794, O(W2)...O(2) 2.795, O(W3)...O(1) 2.831, O(W3)...O(W2) 2.731 Å).

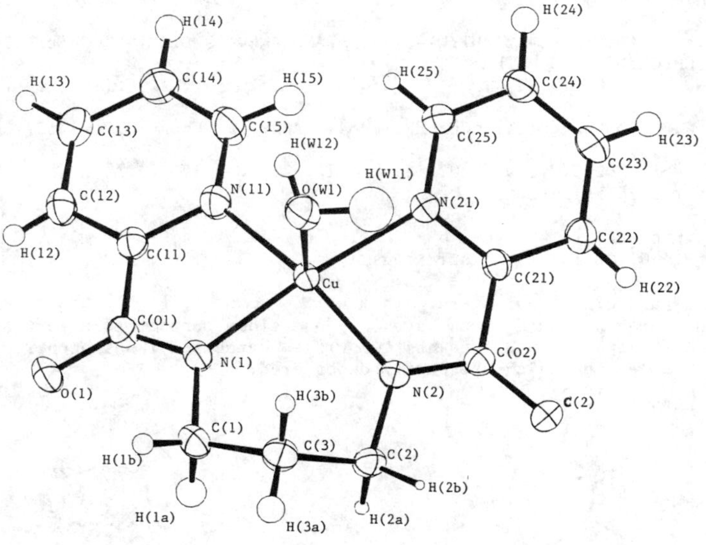

Fig. 1. Perspective drawing of $C_{15}H_{16}CuN_4O_3$.

(2,3,9,10-TETRAMETHYL-1,4,8,11-TETRAAZACYCLOTETRADECA-1,3,8,10-TETRAENE)(THIOCYANATO-N)COPPER(II) HEXAFLUOROPHOSPHATE
$C_{15}H_{24}CuF_6N_5PS$ $[Cu(NCS)(C_{14}H_{24}N_4)](PF_6)$

A.E. ELIA, E.C. LINGAFELTER and V. SCHOMAKER, 1984. Acta Cryst., C40, 1313-1315.

Orthorhombic, Pnam, a = 12.003, b = 15.365, c = 11.500 Å, D_m = 1.62, Z = 4. Mo radiation, R = 0.091 for 2424 reflexions.

Cu lies on a mirror plane and is five-coordinate, square pyramidal (Fig. 1) with the coplanar, macrocyclic N atoms forming the basal plane and the NCS^- nitrogen atom the apex. An S atom from another cation is 3.331(6) Å from Cu at the sixth coordination site. Cu is displaced from the basal plane towards NCS^- by 0.07 Å. The Cu-N distances are 1.963(8) and 1.920(8) - equatorial and 2.384(13) Å - axial. In the anion four F atoms lie in the mirror plane, the other two are disordered and lie approximately perpendicular to the mirror plane.

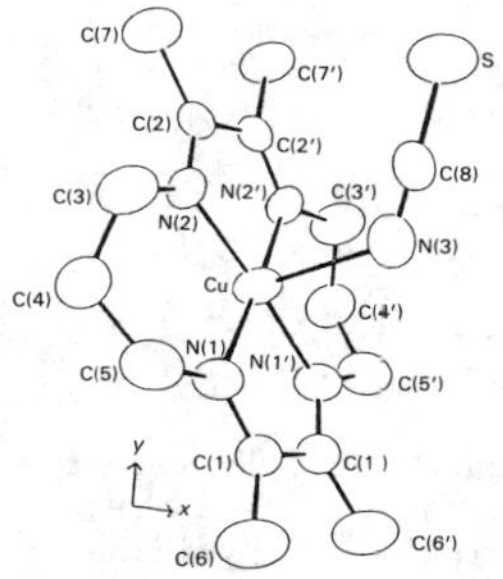

Fig. 1. The $Cu(NCS)(C_{14}H_{24}N_4)$ cation.

Λ-[{5-METHYL-5-(4-AMINO-2-AZABUTYL)-3,7-DIAZANONANE-1,9-DIAMINE}COBALT(III)][(R,R)(+)-TARTRATE] CHLORIDE HEXAHYDRATE

$C_{15}H_{34}ClCoN_6O_6,6(H_2O)$ (I) Λ-[Co(sen)(tart)]Cl,$(H_2O)_6$

Δ-[{5-METHYL-5-(4-AMINO-2-AZABUTYL)-3,7-DIAZANONANE-1,9-DIAMINE}COBALT(III)][(R,R)(+)-TARTRATE] CHLORIDE, 4.5 WATER

$C_{15}H_{34}ClCoN_6O_6,4.5(H_2O)$ (II) Δ-[Co(sen)(tart)]Cl$(H_2O)_{4.5}$

I. BERNAL, J.D. KORP and I.I. CREASER, 1984. Aust. J. Chem., 37, 2365-2369.

I. Orthorhombic, $P2_12_12_1$, a = 10.813, b = 13.626, c = 17.724 Å, Z = 4. Mo radiation, R = 0.038 for 2026 reflexions.

II. Monoclinic, $P2_1$, a = 9.469, b = 12.116, c = 12.132 Å, β = 95.94°, Z = 2. Mo radiation, R = 0.108 for 2514 reflexions.

I and II are diastereoisomers. In the I Λ isomer (Fig. 1), the amino hydrogens involved in hydrogen bonding each have only one close contact to a tartrate oxygen but in the Δ isomer each hydrogen has two close contacts. I has been reported (1). II has water molecules which are partly disordered.

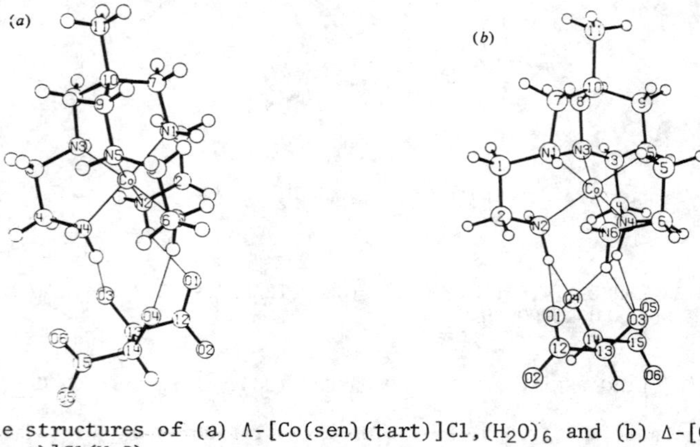

Fig. 1. The structures of (a) Λ-[Co(sen)(tart)]Cl,$(H_2O)_6$ and (b) Δ-[Co(sen)-(tart)]Cl$(H_2O)_{4.5}$.

1. Structure Reports, 50B.

CHLORO[N,N'-ETHYLENEBIS(THIOSALICYLIDENEIMINATO-N,S)]IRON(III) (α-Form)
$C_{16}H_{14}ClFeN_2S_2$ (I) [Fe(tsalen)Cl]

CHLORO[N,N'-ETHYLENEBIS(THIOSALICYLIDENEIMINATO-N,S)]IRON(III) (β-Form)
$C_{16}H_{14}ClFeN_2S_2$ (II)

CHLORO[N,N'-ETHYLENEBIS(THIOSALICYLIDENEIMINATO-N,S)]IRON(III) (γ-Form)
$C_{16}H_{14}ClFeN_2S_2$ (III)

G.D. FALLON, B.M. GATEHOUSE, P.J. MARINI, K.S. MURRAY and B.O. WEST, 1984. J. Chem. Soc. Dalton, 2733-2739.

I. Monoclinic, $P2_1/n$, a = 18.249, b = 9.994, c = 17.823 Å, β = 91.04°, D_m = 1.59, Z = 8. Mo radiation, R = 0.071 for 1464 reflexions.

II. Monoclinic, $P2_1/n$, a = 14.847, b = 7.650, c = 14.169 Å, β = 96.17°, D_m = 1.63, Z = 4. Mo radiation, R = 0.061 for 736 reflexions.

III. Orthorhombic, Pbca, a = 13.764, b = 17.788, c = 13.087 Å, D_m = 1.61, Z = 8. Mo radiation, R = 0.067 for 367 reflexions.

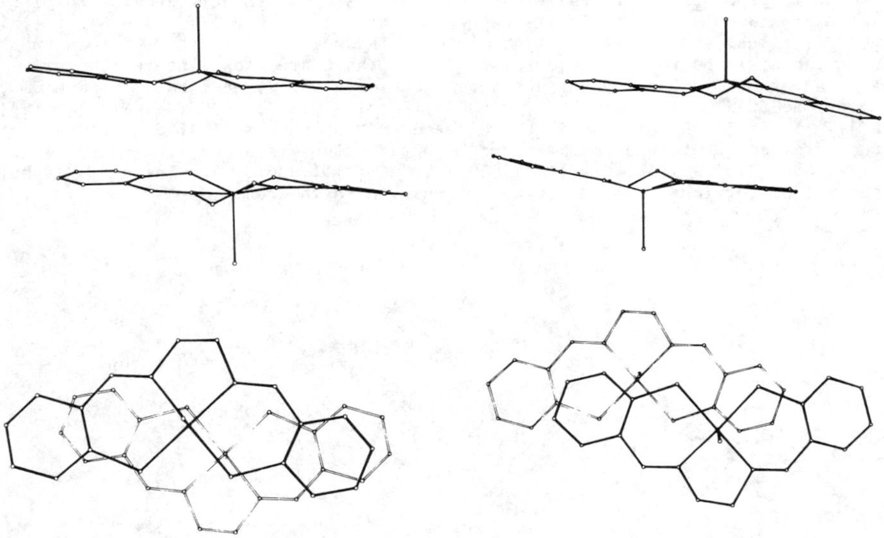

Fig. 1. Side and top views of the two α-form [Fe(tsalen)Cl] molecules (type 1 (left), type 2 (right)).

The bulk sample of [Fe(tsalen)Cl] used for crystallographic analysis was found to contain three distinct crystalline forms termed hereafter the α, β, and γ forms. The α form represented the bulk of the sample, ca. 96%, while each of the other forms made up 2% of the sample. The α form was found to contain two independent molecules in the asymmetric unit which are termed 'type 1' and 'type 2' molecules. Both these molecules have an approximate square-pyramidal stereochemistry with the Fe atoms out of the mean planes through the donor N and S atoms by 0.374 Å in 'molecule 1' and 0.402 Å in 'molecule 2'. The two thiosalicylideneimine chelate rings are virtually coplanar with the S_2N_2 plane for molecules of type 1. The ethylene bridge in this type is in the gauche position (Fig. 1) with a torsion analg of 45(2)°. Type 2 molecules have a stepped configuration (Fig. 1) with the angles of the ligand planes to the S_2N_2 plane being small, 4.5 and 12.4°. The ethylene bridge in these molecules has a torsion angle of 52(2)°. In the β- and γ-forms, five-coordinate square pyramidal geometry is maintained. The ligand chelate rings assume a stepped configuration in the β-form with the angles of the ligand planes to the S_2N_2 plane being 21.9 and 15.0°; the γ-form assumes a flat arrangement. Fe-S, Fe-Cl and Fe-N distances are in the range 2.187-2.216, 2.295-2.336 and 1.95-2.03 Å respectively.

6,11-DIMETHYL-7,10-DIAZAHEXADECA-5,11-DIENE-2,4,13,15-TETRONE
$C_{16}H_{24}N_2O_4$ (I)

(6,11-DIMETHYL-7,10-DIAZAHEXADECA-5,11-DIENE-2,4,13,15-TETRAONATO-N,N',O^4,O^{13})-COPPER(II)
$C_{16}H_{22}CuN_2O_4$ (II)

N.A. BAILEY, D.E. FENTON, S.E. GAYDA and C.A. PHILLIPS, 1984. J. Chem. Soc. Dalton, 2289-2292.

I. Orthorhombic, Pccn, a = 7.627, b = 10.158, c = 22.099 Å, D_m = 1.21, Z = 4. Mo radiation, R = 0.048 for 666 reflexions.

II. Triclinic, P$\bar{1}$, a = 10.915, b = 8.689, c = 9.593 Å, α = 104.87, β = 89.66, γ = 106.342°, D_m = 1.49, Z = 2. Mo radiation, R = 0.039 for 2492 reflexions.

Molecule I possesses crystallographically imposed C_2 symmetry and the terminal β-diketone residue is present in its keto tautomeric form (Fig. 1). The β-keto-imine fragment is planar and stabilised by an internal hydrogen bond. The C-C torsion angle of the di-iminoethane fragment is -61.1°. There are close intermolecular contacts across an inversion centre between keto oxygen O(2) and the imino N atom to a symmetry related O(2) (O(2)...N(1) 2.687(4), C(2)...O(2'') 3.000(4), N(1)...O(2'') 3.009(4) Å). In II the Cu atom is coordinated in the inner N_2O_2 site. The β-diketone chains again adopt keto tautomeric forms. Coordination distances are: Cu-O(2) 1.911(3), Cu-O(3) 1.901(3), Cu-N(1) 1.930(2), Cu-N(2) 1.926(3) Å. There is slight tetrahedral distortion (10.9°) in the coordination polyhedron.

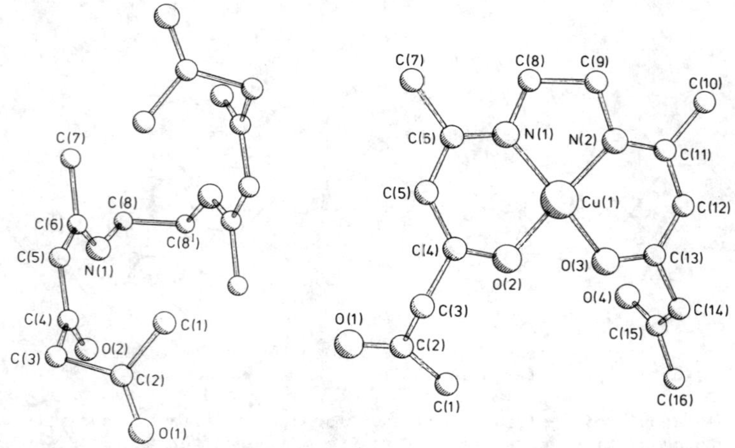

Fig. 1. The molecular structures: $C_{16}H_{24}N_2O_4$ (left) and $C_{16}H_{22}CuN_2O_4$ (right).

SODIUM AQUO(1,4,7,10-TETRAAZACYCLODODECANE-1,4,7,10-TETRAACETATO)EUROPATE(III) TETRAHYDRATE
$C_{16}H_{26}EuN_4NaO_9,4H_2O$ Na[Eu($C_{16}H_{24}N_4O_8$)(H_2O)].$4H_2O$

M.-R. SPIRLET, J. REBIZANT, J.F. DESREUX and M.-F. LONCIN, 1984. Inorg. Chem., 23, 359-363.

Triclinic, $P\bar{1}$, a = 8.710, b = 9.120, c = 15.742 Å, α = 92.87, β = 85.43, γ = 81.36°, Z = 2. Mo radiation, R = 0.032 for 3142 reflexions.

Two views of the anion are shown in Fig. 1. The Eu atom is nine-coordinate, being coordinated to four ring N atoms (Eu-N 2.519(3)-2.900(3) Å) and four carboxylate O atoms (Eu-O 2.247(3)-2.511(3) Å) of the macrocyclic ligand as well as one water molecule (Eu-O 2.480(3) Å). The coordination polyhedron is an irregular capped (by H_2O) square antiprism.

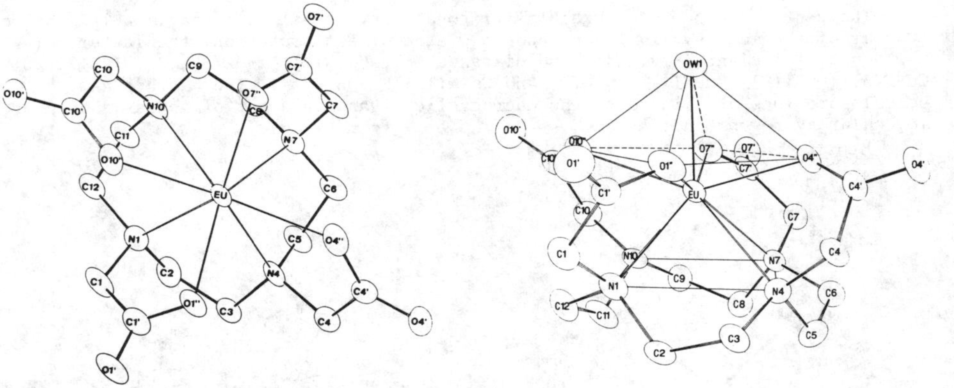

Fig. 1. The $[Eu(C_{16}H_{24}N_4O_8)(H_2O)]^-$ anion.

(5,7,7,12,14,14-HEXAMETHYL-1,4,8,11-TETRAAZACYCLOTETRADECA-4,11-DIENE)COPPER(II) PERCHLORATE (Orange isomer)

$C_{16}H_{32}Cl_2CuN_4O_8$ $[Cu(C_{16}H_{32}N_4)](ClO_4)_2$

T.J. LEE, T.H. LU, C.-S. CHUNG and T.Y. LEE, 1984. Acta Cryst., C40, 70-72.

Monoclinic, $P2_1/c$, a = 10.331, b = 10.461, c = 11.050 Å, β = 111.96°, D_m = 1.61, Z = 2. Mo radiation, R = 0.050 for 1468 reflexions.

The Cu atom lies on an inversion centre (Fig. 1) and has square-planar coordination, with Cu-N(1) 2.027(7), Cu-N(2) 1.974(5) Å, N(1)-Cu-N(2) 94.8(2)°. The five-membered rings are in gauche conformation and the six-membered rings are in twist-boat form. The orange isomer differs from the red isomer (1) in orientation of ring substituents.

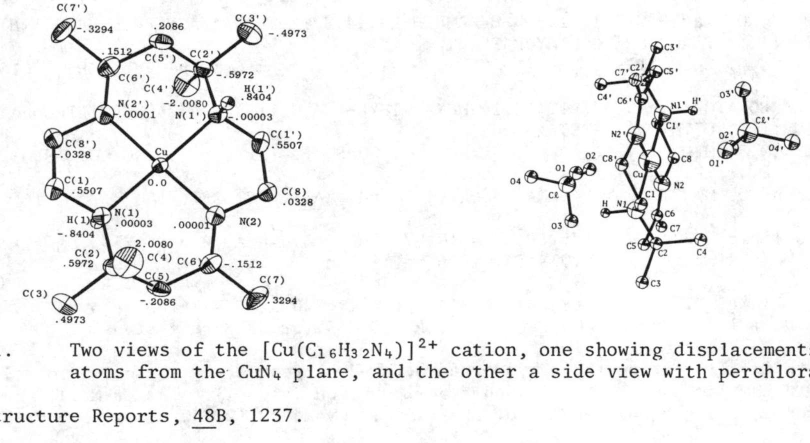

Fig. 1. Two views of the $[Cu(C_{16}H_{32}N_4)]^{2+}$ cation, one showing displacements of atoms from the CuN_4 plane, and the other a side view with perchlorate ions.

1. Structure Reports, 48B, 1237.

{[7R(S),14R(S)]-5,5,7,12,12,14-HEXAMETHYL-1,4,8,11-TETRAAZACYCLOTETRADECANE}NITRATO-COPPER(II) PERCHLORATE

$C_{16}H_{36}ClCuN_5O_7$ $[Cu(NO_3)(C_{16}H_{36}N_4)](ClO_4)$

T.-J. LEE, H.Y.J. LEE, C.-S. LEE and C.-S. CHUNG, 1984. Acta Cryst., C40, 641-644.

Monoclinic, $P2_1/c$, a = 8.769, b = 17.809, c = 14.344 Å, D_m = 1.48, Z = 4. Mo radiation, R = 0.070 for 1754 reflexions.

The central copper atom exhibits trigonal bipyramidal coordination with the four N atoms of the macrocyclic ligand and an O atom of a nitrate ion, the latter being on the equatorial plane (Fig. 1); bond distances are Cu-N(1) 1.992(9), Cu-N(2) 2.128(9), Cu-N(3) 1.971(9), Cu-N(4) 2.080(9) and Cu-O(11) 2.106(8) Å. The nitrate and perchlorate ions are linked to the macrocyclic ligand by a net of intra- and inter-molecular hydrogen bonds.

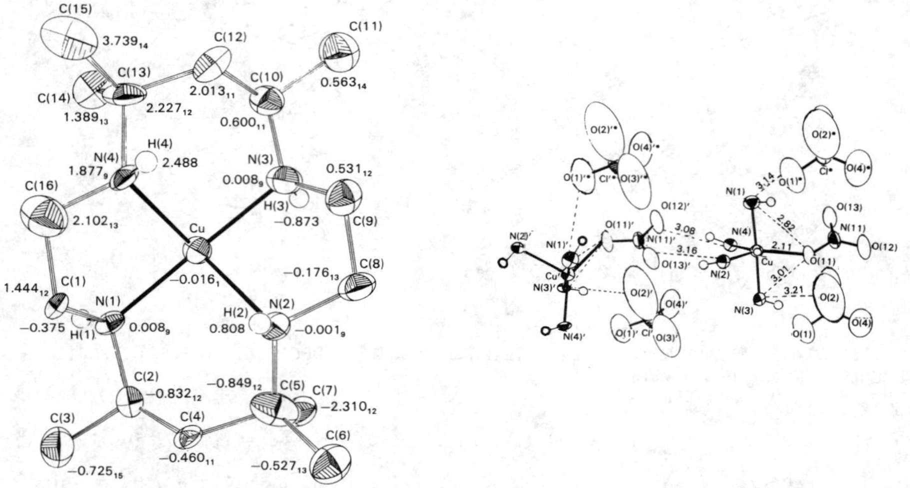

Fig. 1. $[Cu(NO_3)(C_{16}H_{36}N_4)](ClO_4)$: perspective view of the $Cu(C_{16}H_{36}N_4)$ unit showing displacements of atoms from the least-squares plane of Cu, N(1), N(2) and N(3) (Å) (left) and a view showing the coordination sphere and adjacent perchlorate ions of two neighbouring molecules with hydrogen bonds denoted by broken lines (right).

cis-DIHYDROXO-(rac-5,5,7,12,12,14-HEXAMETHYL-1,4,8,11-TETRAAZACYCLOTETRADECANE)-CHROMIUM(III) PERCHLORATE DIHYDRATE

$C_{16}H_{38}ClCrN_4O_6,2H_2O$ (I) $cis[Cr(cycb)(OH)_2]^+ClO_4^-,2H_2O$

DI-cis-CARBONATO-(rac-5,5,7,12,12,14-HEXAMETHYL-1,4,8,11-TETRAAZACYCLOTETRADECANE)-CHROMIUM(III) DITHIONATE TETRAHYDRATE

$C_{17}H_{36}CrN_4O_6S,2H_2O$ (II) $0.5\{cis-[Cr(C_{16}H_{36}N_4)O_2CO]_2{}^+S_2O_6{}^{2-},4H_2O\}$

E. BANG and O. MØNSTED, 1984. Acta Chem. Scand., A38, 281-287.

I. Monoclinic, C2/c, a = 17.260, b = 10.573, c = 13.346 Å, β = 96.75°, D_m = 1.41, Z = 4. Mo radiation, R = 0.039 for 1996 reflexions.

II. Triclinic, $P\bar{1}$, a = 8.788, b = 10.748, c = 14.240 Å, α = 74.63, β = 83.70, γ = 68.06°, D_m = 1.41, Z = 2. Mo radiation, R = 0.033 for 3338 reflexions.

Both Cr(III) cations have distorted octahedral geometry with the macrocyclic ligand in the same so-called "folded" configuration (Fig. 1). The Cr-N(1), Cr-N(8) distances in I are equal because of the twofold symmetry (2.142 Å), as are Cr-N(4) and

Cr-N(11) (2.140 Å), and Cr-O(1), Cr-O(2) (1.918 Å). In II, these distances are: 2.091, 2.106, 2.108, 2.111, 1.945 and 1.959 Å respectively. The N(4)-Cr-N(11), N(1)-Cr-N(8) and O(1)-Cr-O(2) angles are in I and II respectively 165.30, 169.70; 94.85, 99.31 and 92.15 and 67.56°. The configurations of the 5- and 6-membered rings are gauche and chair respectively and the N(4) and N(11) atoms are coordinated in a position trans to each other. Both compounds have nets of intermolecular hydrogen bonds (Fig. 1) involving the complex ions, anions and water molecules. In I the strongest hydrogen bond is O(water)...O(1) 2.748 Å and in II there are three: O(1)(water)...O(1)(dithionate) 2.820, O(1)(water)...O(2)(water) 2.808 and O(2)(water)-...O(3)(dithionate) 2.773 Å.

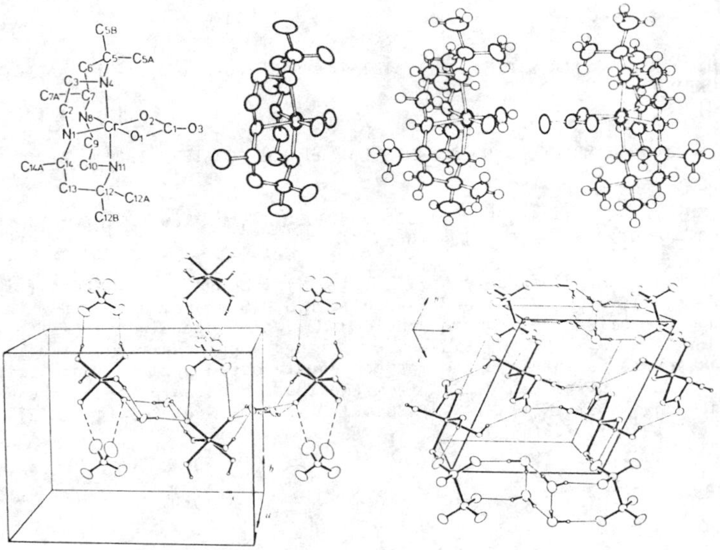

Fig. 1. Numbering scheme in the complex cations with and without H atoms in $C_{16}H_{38}ClCrN_4O_6$ and $C_{17}H_{36}CrN_4O_6S$ (top), and the unit cells of $C_{16}H_{38}ClCrN_4O_6,2H_2O$ (bottom left) and $C_{17}H_{36}CrN_4O_6S,2H_2O$ (bottom right) showing the two-dimensional net of hydrogen bonds.

trans-AQUACHLORO[(1S,4S,7S,8R,11R,14R)-5,5,7e,12,12,14e-HEXAMETHYL-1,4,8,11-TETRA-AZACYCLOTETRADECANE]CHROMIUM(III) NITRATE

$C_{16}H_{38}CrClN_6O_7$ $[CrCl(H_2O)(C_{16}H_{36}N_4)](NO_3)_2$

R. TEMPLE, D.A. HOUSE and W.T. ROBINSON, 1984. Acta Cryst., C40, 1789-1791.

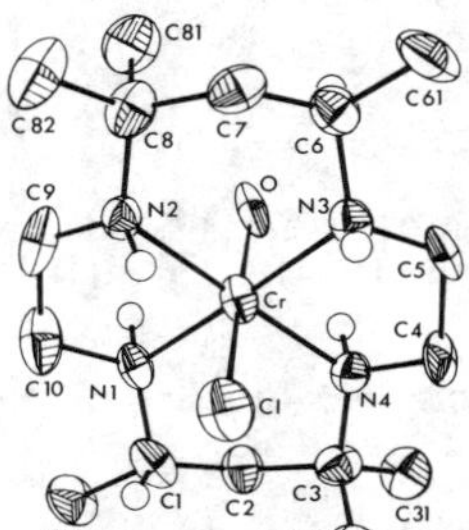

Fig. 1. The $CrCl(H_2O)(C_{16}H_{36}N_4)^{2+}$ cation.

Monoclinic, Cc, a = 14.560, b = 11.740, c = 14.772 Å, β = 110.83°, Z = 4. Mo radiation, R = 0.061 for 1894 reflexions.

The six-coordinate cations (Fig. 1) contain a CrN_4 plane with Cl^- and H_2O in trans positions; the five-membered rings have a gauche (δλ) and the six-membered rings a chair (pp) conformation. The sec-NH protons adopt the meso (RSSR) configuration and the CH_3- groups attached to C at the 7 and 14 positions of the ligand are equatorial. Bond lengths are normal with Cr-N (mean) 2.083, Cr-Cl 2.307(2) and Cr-O 2.090(6) Å.

(N,N-DIMETHYLFORMAMIDE)(R,S,R,S)-(1,4,8,11-TETRAMETHYL-1,4,8,11-TETRAAZACYCLOTETRADECANE)NICKEL(II)

$C_{17}H_{39}F_6N_5NiO_7S_2$ $[Ni(L)_2(dmf)](CF_3O_3S)_2$

S.F. LINCOLN, T.W. HAMBLEY, D.L. PISANIELLO and J.H. COATES, 1984. Aust. J. Chem., 37, 713-723.

Monoclinic, $P2_1/c$, a = 14.45, b = 14.00, c = 15.07 Å, β = 89.95°, Z = 4. Mo radiation, R = 0.072 for 2363 reflexions.

The tetraaza ligand bonds to Ni(II) (Fig. 1) through four N atoms (Ni-N 2.113(9), 2.098(9), 2.121(9), 2.691(9) Å) and assumes a R,S,R,S conformation in which the four methyl groups are orientated towards the $OCHNMe_2$ which occupies the fifth coordination side, Ni-O 1.974(6) Å. The two six-membered rings adopt chair conformations. The Ni coordination sphere is intermediate between square based pyramidal and trigonal bipyramidal. Ni is displaced 0.29 Å from the least squares plane of the N atoms. The bidentate angles are N(1)-Ni-N(2) 86.4(4), N(3)-Ni-N(4) 86.4(4), N(2)-Ni-N(3) 90.9(5), N(1)-Ni-N(4) 92.8(4)°.

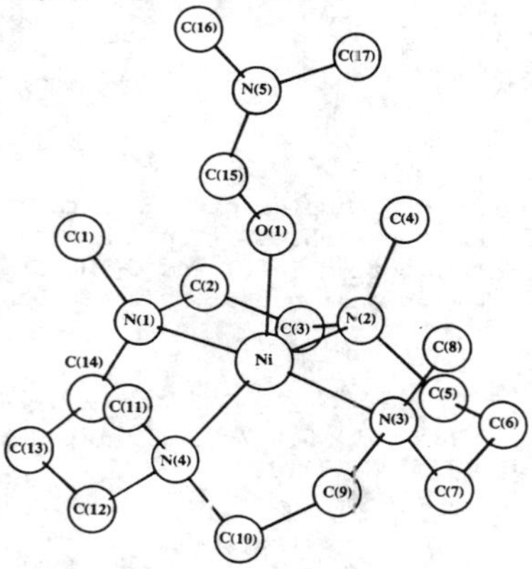

Fig. 1. The structure of $[Ni(L)_2(dmf)]^{2+}$.

AQUA(3,6,14,17,23,24-HEXA-AZATRICYCLO[17,3,1,1^{8,12}]TETRACOSA-1(23),2,6,8,10,12(24),-13,17,19,21-DECAENE)HYDROXONITRATOSAMARIUM(III) NITRATE METHANOL SOLVATE

$C_{18}H_{21}N_8O_8Sm,2(CH_4O)$

K.K. ABID, D.E. FENTON, U. CASELLATO, P.A. VIGATC and R. GRAZIANI, 1984. J. Chem. Soc. Dalton, 351-354.

Triclinic, $P\bar{1}$, a = 16.02, b = 9.64, c = 8.29 Å, α = 91.06, β = 95.26, γ = 92.18°, Z = 2. Mo radiation, R = 0.073 for 3289 reflexions.

The structure consists of discrete complex cations (Fig. 1), nitrate anions and methanol solvent molecules. Sm^{3+} is ten-coordinated being bound to the hetero atoms of the hexadentate ligand, to one chelating nitrate group and to the O atoms of the OH^- ion of one water molecule. The coordination polyhedron can best be described as an irregular antiprism capped on the square faces by N(1) and N(4). The Sm-N(pyridine) distances are 2.66(1) and 2.65(1) and the mean Sm-N(imino) 2.62 Å. The Sm-O(1,2,4,5) distances are 2.51(1), 2.52(1), 2.52(1) and 2.46(1) Å respectively.

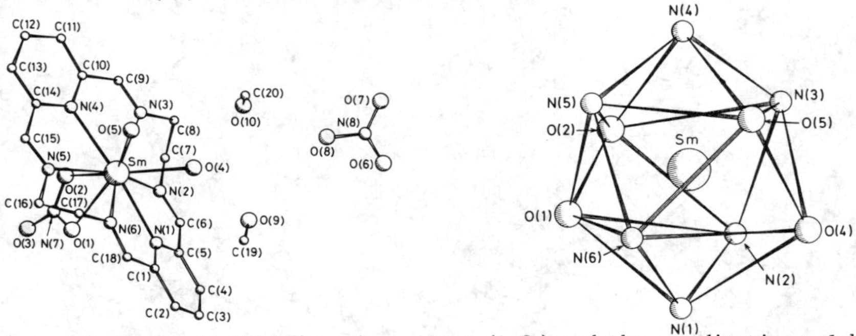

Fig. 1. $C_{18}H_{21}N_8O_8Sm,2(CH_4O)$: the complex (left) and the coordination polyhedron (right).

DICHLORO(3,4,9,10-DIBENZO-1,12-DIAZA-5,8-DIOXACYCLOTETRADECANE)NICKEL CHLOROFORM SOLVATE
$C_{18}H_{22}Cl_2N_2NiO_2,CHCl_3$ (I)

(3,4,9,10-DIBENZO-1,5,8,12-TETRA-AZACYCLOTETRADECANE)BIS(ISOTHIOCYANATO)NICKEL
$C_{20}H_{24}N_6NiS_2$ (II)

3,4,9,10-DIBENZO-1,5,8,12-TETRA-AZACYCLOTETRADECANE
$C_{18}H_{24}N_4$ (III)

K. HENRICK, L.F. LINDOY, M. McPARTLIN, P.A. TASKER and M.P. WOOD, 1984. J. Am. Chem. Soc., 106, 1641-1645.

I. Triclinic, $P\bar{1}$, a = 11.525, b = 11.307, c = 9.757 Å, α = 106.74, β = 90.78, γ = 103.22°, Z = 2. Mo radiation, R = 0.075 for 1233 reflexions.

II. Monoclinic, C2/c, a = 15.280, b = 9.765, c = 15.488 Å, β = 108.13°, Z = 4. Mo radiation, R = 0.038 for 2151 reflexions.

III. Monoclinic, $P2_1/c$, a = 11.844, b = 8.118, c = 17.788 Å, β = 100.53°, Z = 4. Mo radiation, R = 0.068 for 1049 reflexions.

The three 14-membered ring systems are shown in Fig. 1. I and II have trans-octahedral geometries; molecule II has a crystallographic twofold axis (through the Ni atom and mid-point of the ethane bridges). The conformation of the free ligand (III) is very similar to that in its complex (II).

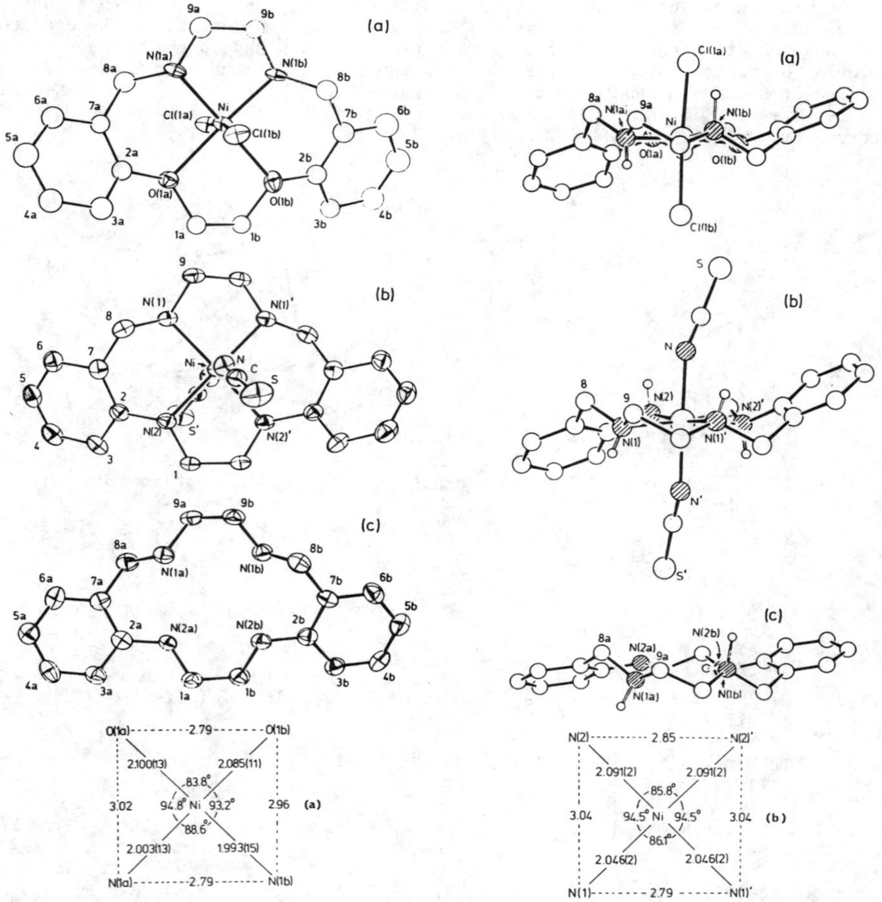

Fig. 1. Views of the 14-membered ring systems I (top), II (centre) and III (bottom). Also shown are coordination details for I (left) and II (right).

(PERCHLORATO-O)[N,N,N',N'-TETRAKIS(PYRAZOL-1-YLMETHYL)-1,2-DIAMINOETHANE]MANGANESE(II) PERCHLORATE

$C_{18}H_{24}Cl_2MnN_{10}O_8$ $[Mn(C_{18}H_{24}N_{10})(ClO_4)](ClO_4)$

F.B. HULSBERGEN, W.L. DRIESSEN, J. REEDIJK and G.C. VERSCHOOR, 1984. Inorg. Chem., 23, 3588-3592.

Monoclinic, $P2_1/c$, a = 9.522, b = 12.190, c = 22.913 Å, β = 94.201°, D_m = 1.63, Z = 4. Mo radiation, R = 0.046 for 1855 reflexions.

The cation (Fig. 1) has a coordination geometry around Mn(II) which can be described as derived from octahedral, with one axial donor atom replaced by two less strongly bonded donor atoms (bicapped), resulting in a seven-coordinate geometry. Mn is coordinated by six N atoms of the ligand and an O atom of one of the ClO_4^- ions (Mn-N four between 2.223(6) and 2.287(7), two long: 2.496(6) and 2.508(6), Mn-O 2.262(5) Å).

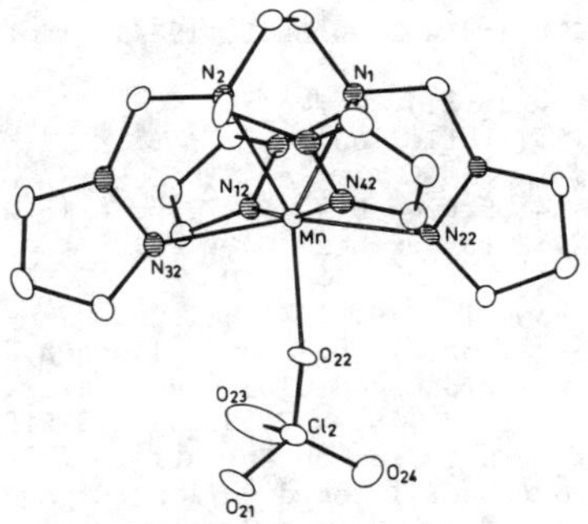

Fig. 1. The $[Mn(C_{18}H_{24}N_{10})(ClO_4)]^+$ cation.

SODIUM (1,4,8,11-TETRAAZACYCLOTETRADECANE-1,4,8,11-TETRAACETATO-$N^1,N^4,N^8,N^{11},O^1,O^4$,-$O^8,O^{11}$)TERBIUM(III) HEXAHYDRATE - SODIUM CHLORIDE (2:1)

$C_{18}H_{28}N_4NaO_8Tb,6H_2O,0.5NaCl$ $NaTb[C_{18}H_{28}N_4O_8],6H_2O,0.5NaCl$

M.-R. SPIRLET, J. REBIZANT, M.-F. LONCIN and J.F. DESREUX, 1984. Inorg. Chem., 23, 4278-4283.

Monoclinic, P2/c, a = 17.980, b = 9.081, c = 17.713 Å, β = 106.00°, Z = 4. Mo radiation, R = 0.036 for 4111 reflexions.

The Tb atom is eight-coordinated, being linked to the four N atoms and the four carboxylic O atoms of the ligand. The coordination polyhedron is a strongly distorted dodecahedron (Fig. 1), having approximate C_2 symmetry. The macrocyclic ligand exhibits a biangular {77} structure. The complex anions are associated through sodium ions and water molecules into layers that alternate with aqueous layers containing the Cl^- ions. Bond lengths: Tb-O 2.302(3)-2.330(3), Tb-N 2.575(3)-2.620(3) Å, Na-O 2.320(4)-2.533(3) Å.

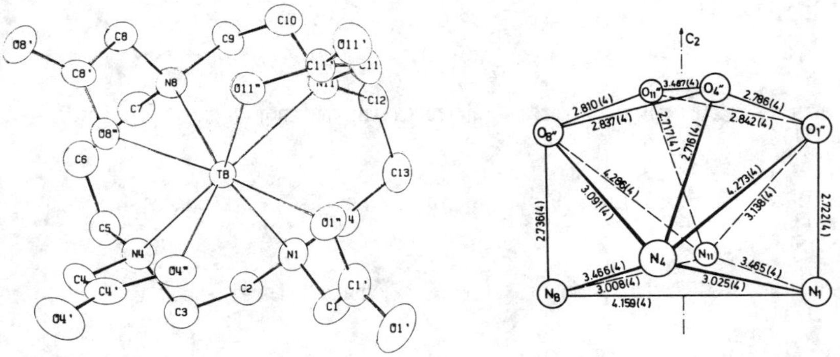

Fig. 1. The $Tb[C_{18}H_{28}N_4O_8]^-$ anion, coordination polyhedron about Tb at right.

(4,4,9,9-TETRAMETHYL-5,8-DIAZADODECANE-2,11-DIONE 2-ISOPROPYLIDENEHYDRAZONE 11-ACETYLHYDRAZONE)(PERCHLORATO)NICKEL(II) PERCHLORATE

$C_{19}H_{38}Cl_2N_6NiO_9$ (I) $[Ni(C_{19}H_{38}N_6O)(ClO_4)](ClO_4)$

(PENTANE-2,4-DIONATO)(4,4,7,7-TETRAMETHYL-5,8-DIAZADODECANE-2,11-DIONE 2,11-BIS(ISOPROPYLIDENEHYDRAZONE))NICKEL(II) PERCHLORATE

$C_{25}H_{47}ClN_6NiO_6$ (II) $[Ni(C_{20}H_{40}N_6)(C_5H_7O_2)](ClO_4)$

N.F. CURTIS, F.W.B. EINSTEIN and A.C. WILLIS, 1984. Inorg. Chem., 23, 3444-3449.

I. Orthorhombic, $Pn2_1a$, a = 26.314, b = 9.567, c = 11.122 Å, D_m = 1.47, Z = 4. Mo radiation, R = 0.037 for 1851 reflexions.

II. Orthorhombic, Pbca, a = 25.624, b = 13.476, c = 18.456 Å, D_m = 1.31, Z = 8. Mo radiation, R = 0.048 for 2412 reflexions.

The two cations are shown in Fig. 1. In I the $C_{19}H_{38}N_6O$ ligand is coordinated in a square-pyramidal arrangement, with isopropylidenehydrazone N atom in the axial site, and trans to a weakly coordinated ClO_4 oxygen atom. Bond lengths for I: Ni-O(ClO_4) 2.455(2), Ni-O 2.065(4), Ni-N 2.052(4)-2.075(5) Å. In II the tetradentate $C_{20}H_{40}N_6$ ligand is present in a folded conformation similar to that in I, with chelate $C_5H_7O_2$ completing octahedral coordination, with one O atom trans to a secondary amino N and the other trans to a diazine N atom. Bond lengths for II: Ni-O 2.013(4) and 2.025(3), Ni-N 2.100(4)-2.161(4) Å.

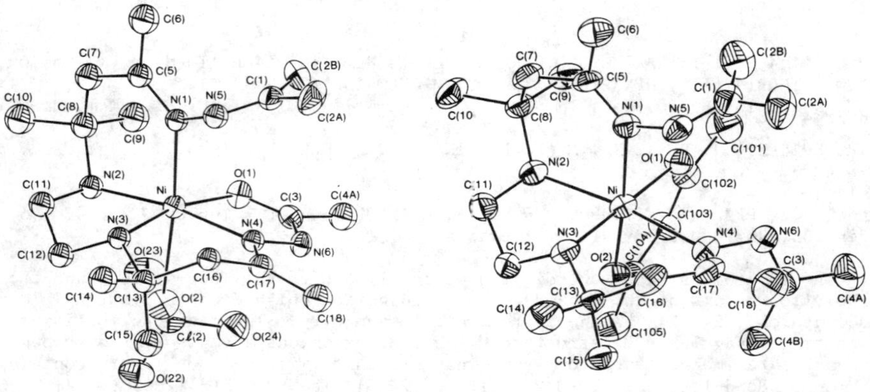

Fig. 1. The cations $[Ni(C_{19}H_{38}N_6O)(ClO_4)]^+$ (left) and $[Ni(C_{20}H_{40}N_6)(C_5H_7O_2)]^+$ (right).

[3,3'-(ETHANE-1,2-DIYLDIIMINO)BIS[1-(o-HYDROXYPHENYL)PROP-2-ENONATO]-N,N',O^1,$O^{1'}$]-NICKEL(II)
$C_{20}H_{18}N_2NiO_4$ (I)

[3,3'-(ETHANE-1,2-DIYLDIIMINO)BIS[1-(o-HYDROXYPHENYL)BUT-2-ENONATO]-N,N',O^1,$O^{1'}$]-NICKEL(II)
$C_{22}H_{22}N_2NiO_4$ (II)

[3,3'-(PROPANE-1,3-DIYLDIIMINO)BIS[1-(o-HYDROXYPHENYL)PROP-2-ENONATO]-N,N',O^1,$O^{1'}$]-NICKEL(II)
$C_{21}H_{20}CuN_2O_4$ (III)

N.A. BAILEY, S.F. DAVISON, J.R. ELLIOT, D.E. FENTON, E. GODBEHERE, S.K. HOLDROYD and C. RODRIGUEZ DE BARBARIN, 1984. J. Chem. Soc. Dalton, 1073-1081.

I. Monoclinic, C2/c, a = 23.373, b = 7.800, c = 20.819 Å, β = 110.14°, D_m = 1.55, Z = 8. Mo radiation, R = 0.134 for 603 reflexions.

II. Monoclinic, $P2_1/c$, a = 26.11, b = 7.797, c = 38.42 Å, β = 93.656°, D_m = 1.48, Z = 16. Mo radiation, R = 0.0376 for 4344 reflexions.

III. Monoclinic, a = 23.098, b = 9.525, c = 18.046 Å, β = 104.15°, D_m = 1.52, Z = 8. Mo radiation, R = 0.1063 for 891 reflexions.

In all four structures of the three compounds the metal atom is coordinated in the inner (N_2O_2) compartment of the Schiff-base 'compartmental' ligand (Fig. 1). In I one iminoketone chelate fragment is buckled giving a stepped arrangement of the two iminoketone molecules about the Ni atom. Coordination distances in I are: Ni-N(1) 1.88(4), Ni-N(2) 1.89(4), Ni-O(2) 1.87(4), Ni-O(3) 1.86(4) Å. II contains two crystallographically independent molecules in the asymmetric unit. One of these has a stepped arrangement of the two iminoketones at the Ni atom while the other has a 'butterfly' arrangement of these two groups at Ni. The Ni atom has an essentially square planar environment. Coordination distances are: Ni-O(1) 1.860, 1.859(3); Ni-O(2) 1.850, 1.849(3); Ni-N(1) 1.862, 1.846(2), Ni-N(2) 1.846, 1.851(4) Å. In III the coordination distances are: Cu(1)-N(1) 1.99(2), Cu(1)-N(2) 1.90(3), Cu(1)-O(2) 1.93(2), Cu-O(3) 1.95(2) Å. Both Ni complexes exhibit short intermolecular contacts to centrosymmetrically related molecules. In I there are two unequal contacts, one on either side of the molecule (Ni(1)...O(2) 3.40; Ni(1)...O(3) 3.69 Å). In II the contacts occur on only one side of the molecular plane (Ni(1)...N(2) 3.56; Ni(1')... O(3') 3.61 Å). There are no such contacts in the crystal structure of III.

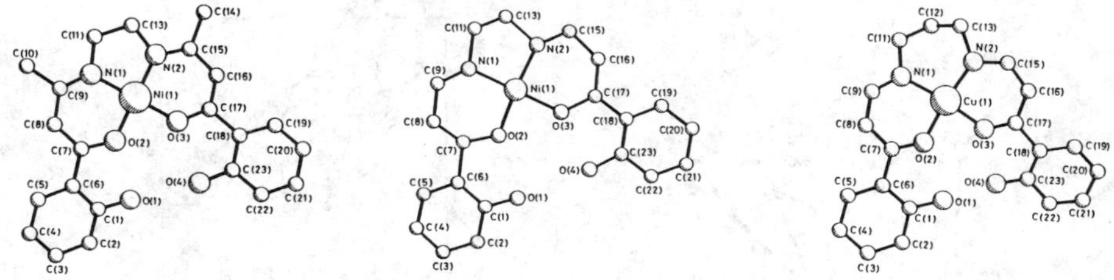

Fig. 1. The molecular structures: $C_{20}H_{18}N_2NiO_4$ (left), $C_{22}H_{22}N_2NiO_4$ (centre) and $C_{21}H_{20}CuN_2O_4$ (right).

CALCIUM BIS(μ-ETHYLENEDIAMINETETRAACETATO)TETRAKIS(μ_3-SULPHIDO)TETRAMOLYBDENUM TRIDECAHYDRATE

$C_{20}H_{24}Ca_{1.5}Mo_4N_4O_{16}S_4,13H_2O$ $Ca_{1.5}[Mo_4S_4(edta)_2].13H_2O$

T. SHIBAHARA, H. KUROYA, K. MATSUMOTO and S. OOI, 1984. J. Am. Chem. Soc., 106, 789-791.

Monoclinic, $P2_1/c$, a = 15.276, b = 12.447, c = 28.10 Å, β = 123.16°, D_m = 2.00, Z = 4. Mo radiation, R = 0.059 for 5658 reflexions.

The $[Mo_4S_4(edta)_2]^{3-}$ anion (Fig. 1) has a cubane type core of approximate S_4 symmetry. Each Mo atom has facial S_3O_2N octahedral coordination and the four Mo atoms form a nearly regular tetrahedron. Bond lengths: Mo-Mo 2.755(2)-2.880(2), Mo-S 2.351(4)-2.369(4), Mo-O 2.129(10)-2.153(9), and Mo-N 2.283(11)-2.300(11) Å.

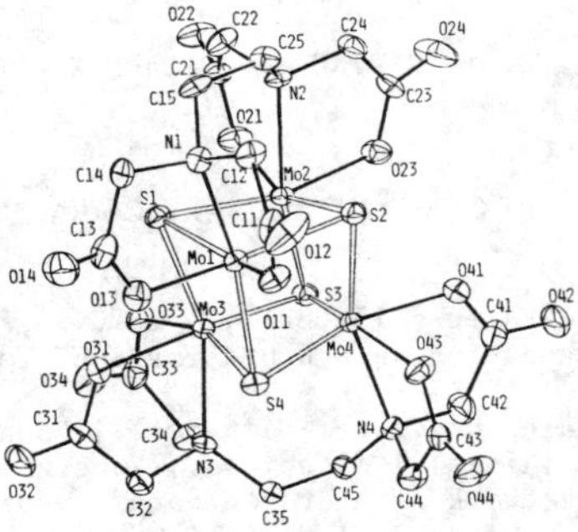

Fig. 1. The $[Mo_4S_4(edta)_2]^{3-}$ anion.

BARIUM BIS(ETHYLENEDIAMINETETRAACETATOFERRATE(II)) TETRAHYDRATE

$C_{20}H_{28}BaFe_2N_4O_{10},4H_2O$ $Ba^{2+}[(C_{10}H_{12}N_2O_4)H_2OFe]_2^{2-},4H_2O$

X. SOLANS, M. FONT-ALTABA and J. GARCIA-ORICAIN, 1984. Afinidad, 41, 572-574.

Monoclinic, C2/c, a = 30.407, b = 7.659, c = 14.487 Å, β = 108.53°, Z = 4. Mo radiation, R = 0.065 for 816 reflexions.

In the anion the Fe atom is coordinated to four oxygen and two nitrogen atoms of the EDTA ligand, and to one water molecule. The coordination polyhedron is a distorted pentagonal bipyramid (Fig. 1). Relevant distances are: Fe-N(10) 2.19, Fe-N(20) 2.35, Fe-O(11) 2.08, Fe-O(21) 1.95, Fe-O(31) 1.98, Fe-O(41) 2.07 and Fe-O(water)(W1) 2,08 Å. The Ba^{++} cation is coordinated to eleven O atoms, with Ba-O distances ranging from 2.78 to 3.07 Å.

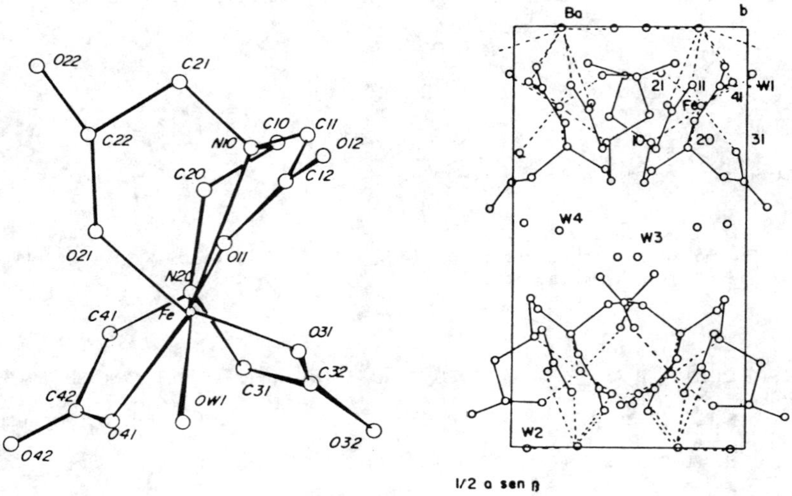

Fig. 1. The coordination polyhedron in the anion of $C_{20}H_{28}BaFe_2N_4O_{10}$ (left), and a projection of the unit cell down c (right).

TRIS(1,2-DIAMINOETHANE)NICKEL(II) BIS(PYRIDINE-2,6-DICARBOXYLATO)NICKEL(II) TETRA-HYDRATE

$C_{20}H_{30}N_8Ni_2O_8,4(H_2O)$

J.G.H. DU PREEZ, H.E. ROHWER, B.J. VAN BRECHT and M.R. CAIRA, 1984. J. Chem. Soc. Dalton, 975-980.

Triclinic, P$\bar{1}$, a = 9.395, b = 10.924, c = 15.662 Å, α = 70.95, β = 101.36, γ = 104.91°, D_m = 1.60, Z = 2. Mo radiation, R = 0.084 for 1667 reflexions.

The structure (Fig. 1) contains discrete $[Ni(pydca)_2]^{2-}$ anions of approximate D_{2d} symmetry in association with $[Ni(en)_3]^{2+}$ cations. Ions are linked by N-H...O hydrogen bonds and an extensive network of hydrogen bonds involving water molecules. The geometry at Ni(1) is distorted octahedral with Ni-O 2.09(1)-2.16(1) Å. Both pydca ligands are tridentate. The centrosymmetric cell contains cations with Λδδδ and Λλλλ configurations.

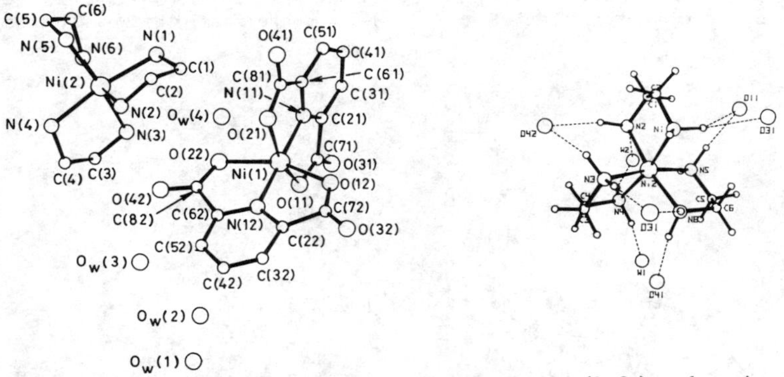

Fig. 1. $C_{20}H_{30}N_8Ni_2O_8,4(H_2O)$: the molecular structure (left) and a view of the cation $\Lambda[Ni(en)_3\delta\delta\delta]^{2+}$ showing hydrogen bonding.

catena-DI(μ-CYANO)DICYANO(N-meso-5,7,7,12,14,14-HEXAMETHYL-1,4,8,11-TETRAAZACYCLO-TETRADECA-1,11-DIENE)DINICKEL(II) HYDRATE

$C_{20}H_{32}N_8Ni_2,H_2O$ $[Ni(hmtd),Ni(CN)_4],H_2O$

G.J. GAINSFORD and N.F. CURTIS, 1984. Aust. J. Chem., 37, 1799-1816.

Monoclinic, C2/c, a = 17.803, b = 10.634, c = 14.093 Å, β = 120.54°, D_m = 1.50, Z = 4. Mo radiation, R = 0.021 for 2178 reflexions, olive green crystals.

The compound has a chain polymeric structure with trans $[Ni(CN)_4]^{2-}$ anions linking parallel stacked $[Ni(hmtd)]^{2+}$ cations (Fig. 1, cation Ni-N 2.045(1) Å, N-Ni-N 95.6(1)°; Ni-N(C) 2.207(1) Å; anion Ni-C 1.868(2) Å).

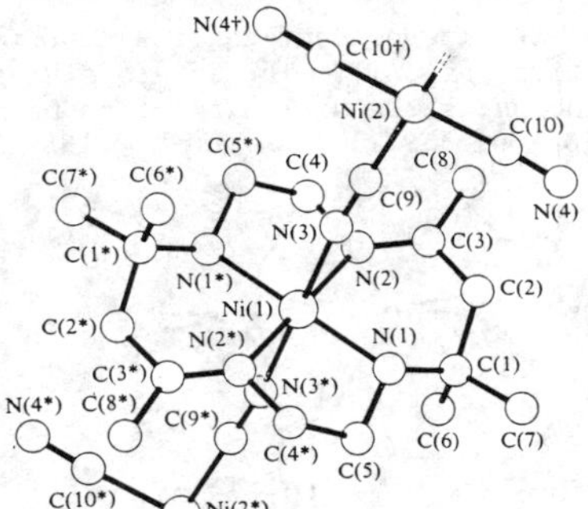

Fig. 1. The structure of $[Ni(hmtd),Ni(CN)_4]$.

TRIS(ETHYLENEDIAMINE)CADMIUM(II) BIS(BENZOHYDROXAMATE) MONOHYDRATE

$C_{20}H_{36}CdN_8O_4,H_2O$

M. BREITWIESER, S. GOTTLICHER and H. PAULUS, 1984. Z. Kristallogr., 166, 207-212.

Orthorhombic, Fdd2, a = 41.004, b = 30.272, c = 8.840 Å, Z = 16. Mo radiation, R = 0.023 for 3470 reflexions.

There are one $[Cd(en)_3]^{2+}$ ion, two benzohydroxamate anions and one water molecule in the asymmetric unit (Fig. 1). H-bonds are formed between anions and water.

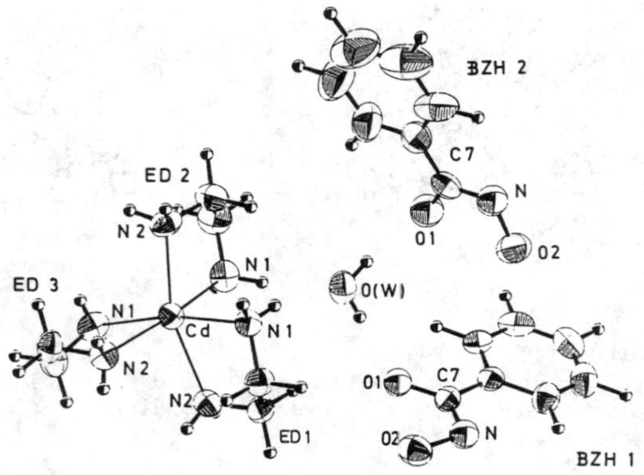

Fig. 1. The asymmetric unit of $C_{20}H_{36}CdN_8O_4,H_2O$.

BIS(μ-BROMO)BIS(N,N,N',N'-TETRAETHYLETHYLENEDIAMINE)DICOPPER(I)
$C_{20}H_{48}Br_2Cu_2N_4$ $[(C_2H_5)_2N(CH_2)_2N(C_2H_5)_2]_2Cu_2Br_2$

M.R. CHURCHILL, G. DAVIES, M.A. EL-SAYED, J.A. FOURNIER, J.P. HUTCHINSON and J.A. ZUBIETA, 1984. Inorg. Chem., 23, 783-787.

Triclinic, $P\bar{1}$, a = 7.860, b = 9.109, c = 19.773 Å, α = 101.04, β = 100.68, γ = 90.13°, Z = 2. Mo radiation, R = 0.032 for 2814 reflexions.

The structure contains two independent centrosymmetric molecules having virtually identical structures (Fig. 1). The Cu atoms have distorted tetrahedral coordination geometry, principal structural parameters being: Cu-Cu 2.606(6), Cu-(μ-Br) 2.467(12) Å, Cu-Br-Cu 63.9(1)°, and Cu-N 2.188(11) Å (mean values).

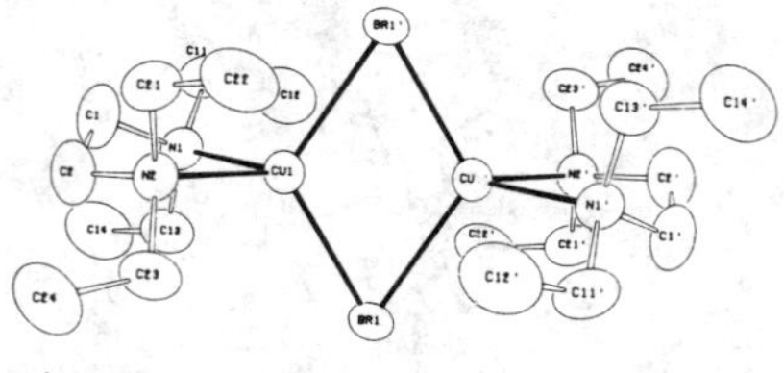

Fig. 1. The $[(C_2H_5)_2N(CH_2)_2N(C_2H_5)_2]_2Cu_2Br_2$ molecule.

1,4,8,11-TETRAMETHYL-1,4,8,11-TETRAAZADIBENZO[b,i]CYCLOTETRADECANE
$C_{22}H_{32}N_4$ (I)

PERCHLORATO(1,4,8,11-TETRAMETHYL-1,4,8,11-TETRAAZADIBENZO[b,i]CYCLOTETRADECANE)-COPPER(II) PERCHLORATE
$C_{22}H_{32}Cl_2CuN_4O_8$ (II) $[Cu(C_{22}H_{32}N_4)(OClO_3)](ClO_4)$

AZIDO(1,4,8,11-TETRAMETHYL-1,4,8,11-TETRAAZADIBENZO[b,i]CYCLOTETRADECANE)COPPER(II) PERCHLORATE
$C_{22}H_{32}ClCuN_7O_4$ (III) $[Cu(C_{22}H_{32}N_4)(N_3)](ClO_4)$

D.-D. KLAEHN, H. PAULUS, R. GREWE and H. ELIAS, 1984. Inorg. Chem., 23, 483-490.

I. Monoclinic, $P2_1/c$, a = 11.988, b = 8.122, c = 22.738 Å, β = 115.55°, Z = 4. Mo radiation, R = 0.077 for 2336 reflexions.

II. Monoclinic, $P2_1/c$, a = 15.187, b = 10.487, c = 16.688 Å, β = 100.77°, Z = 4. Mo radiation, R = 0.082 for 3225 reflexions.

III. Triclinic, $P\bar{1}$, a = 15.649, b = 10.463, c = 7.531 Å, α = 88.71, β = 100.37, γ = 91.53°, Z = 2. Mo radiation, R = 0.056 for 3440 reflexions.

Crystal packing of I and II and the structure of cation in III are shown in Fig. 1. The irregular conformation of the free ligand I becomes highly symmetric upon complexation. The Cu atoms in II and III are five-coordinate with square pyramidal coordination geometry. The Cu lies slightly above the N_4 plane formed by the macrocyclic ligand, the azide and ClO_4^- ligands being apical. All four N-methyl groups are located on one side of the N_4 coordination plane, opposite the apical ligand. Bond lengths: Cu-N(basal) 2.075(4)-2.090(4), Cu-N(N_3) 2.160(5) Å for II, Cu-N(basal) 2.058(5)-2.084(5), Cu-O 2.219(5) Å for III.

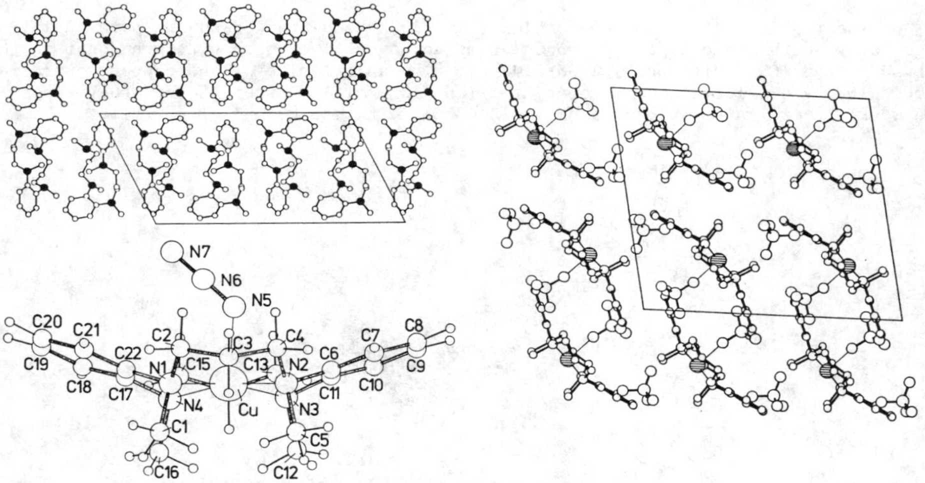

Fig. 1. The structures of $C_{22}H_{32}N_4$ (top left), $[Cu(C_{22}H_{32}N_4)(OClO_3)](ClO_4)$ (right) and the $[Cu(C_{22}H_{32}N_4)(N_3)]^+$ cation (lower left).

(NITRATO-O,O')(1,4,8,11-TETRAMETHYL-1,4,8,11-TETRAAZADIBENZO[b,i]CYCLOTETRADECANE-N^1,N^4,N^8,N^{11})NICKEL(II) NITRATE

$C_{22}H_{32}N_6NiO_6$

H. ELIAS, R. GREWE, D.-D. KLAEHN and H. PAULUS, 1984. Z. Naturforsch., B39, 903-906.

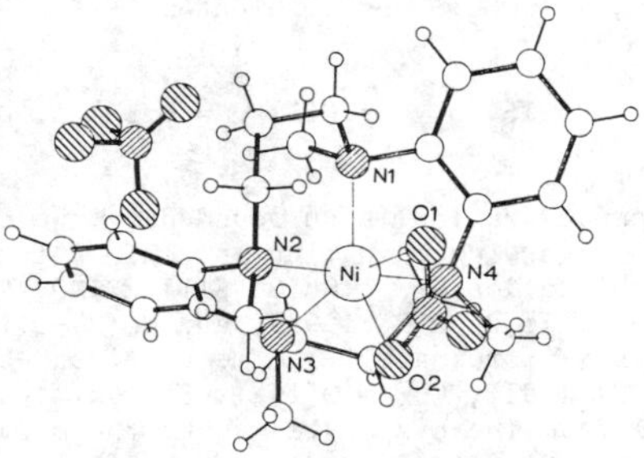

Fig. 1. A view of the coordination geometry in $C_{22}H_{32}N_6NiO_6$.

Monoclinic, $P2_1/c$, a = 8.961, b = 20.599, c = 12.986 Å, β = 91.37°, Z = 4. Mo radiation, R = 0.045 for 2364 reflexions.

The Ni atom has a distorted octahedral coordination (Fig. 1) with tetradentate N_4 macrocycle and bidentate cis nitro group. Main dimensions are Ni-N 2.114-2.136(4), Ni-O 2.136 and 2.163(4) Å, O-Ni-O 60.3(2)°. The uncoordinated nitrate is disordered.

(1,4,7,13,16,19-HEXAAZACYCLOTETRACOSANE(N^1,N^4,N^7-$Cu^1,N^{13},N^{16},N^{19}$-$Cu^2$))TRIS(ISOTHIOCYANATO)THIOCYANATODICOPPER(II)
$C_{22}H_{42}Cu_2N_{10}S_4$

A.E. MARTIN and S.J. LIPPARD, 1984. J. Am. Chem. Soc., 106, 2579-2583.

Monoclinic, $P2_1/c$, a = 14.718, b = 14.472, c = 16.569 Å, β = 113.98°, Z = 4. Mo radiation, R = 0.052 for 4074 reflexions.

The two Cu centres in the complex (Fig. 1) have distorted square-pyramidal geometry, with three ligand nitrogen atoms and one thiocyanate nitrogen donor in the basal plane (Cu-N distances, 1.93-2.07 Å). The apical site of one copper centre is occupied by a thiocyanate nitrogen atom (Cu-N 2.240(6) Å) whereas the other copper has a weakly coordinated sulphur atom (Cu-S 2.733(2) Å) in the apical position.

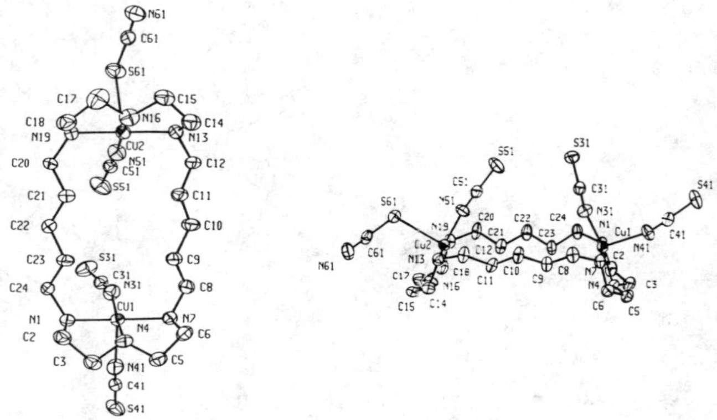

Fig. 1. Views of the $C_{22}H_{42}Cu_2N_{10}S_4$ complex.

(+)$^{CD}_{560}$-((2R,9S)-DIAMINO-3,3,8,8-TETRAMETHYL-4,7-DITHIA-1,10-DECANEDIOATO)COBALT(III) ETHYLENEDIAMINETETRAACETATECOBALTATE
$C_{22}H_{44}Co_2N_4O_{17}S_2$

M. KANESAKA, K.-I. OKAMOTO, M. NOMOTO and J. HIDAKA, 1984. Bull. Chem. Soc. Japan, 57, 3111-3116.

Orthorhombic, $P2_12_12_1$, a = 13.215, b = 37.035, c = 6.527 Å, D_m = 1.72, Z = 4. Mo radiation, R = 0.088 for 2322 reflexions.

The Co atom in the complex cation is surrounded by two cis-S, two cis-O, and two cis-N atoms, the whole ligand coordinating in sixfold fashion (Fig. 1). Bond distances of interest in the cation are: Co(2)-S(1) 2.199, Co(2)-S(2) 2.221, Co(2)-N(3) 1.944, Co(2)-N(4) 1.972, Co(2)-O(9) 1.945, and Co(2)-O(11) 1.933 Å. The Co atom in the anion is also six-coordinate. The metal-ligand atom distances are: Co(1)-N(1) 1.942, Co(1)-N(2) 1.932, Co(1)-O(1) 1.886, Co(1)-O(3) 1.943, Co(1)-O(5) 1.928, and Co(1)-O(7) 1.915 Å. The cis angles about the metal atoms in cation and anion respectively range from 81.2 to 96.0° and from 81.6 to 94.3°.

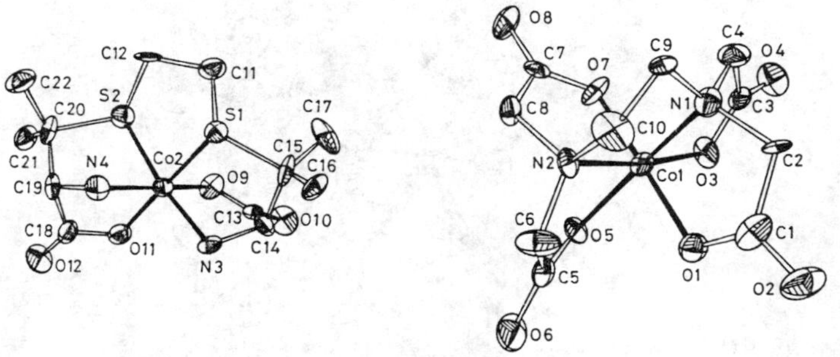

Fig. 1. Perspective drawings of the complex cation (left), and complex anion (right) in $C_{22}H_{44}Co_2N_4O_{17}S_2$.

(N,N'-BIS(2,3,5,6-TETRAFLUOROPHENYL)ETHANE-1,2-DIAMINATO)DIPYRIDINEPLATINUM(II) ACETONE SOLVATE

$C_{24}H_{16}F_8N_4Pt,C_3H_6O$

G.B. DEACON, B.M. GATEHOUSE, I.L. GRAYSON and M.C. NESBIT, 1984. Polyhedron, 3, 753-755.

Monoclinic, C2/c, a = 19.044, b = 14.006, c = 11.223 Å, β = 110.73°, D_m = 1.82, Z = 4. Mo radiation, R = 0.054 for 2806 reflexions.

The molecule has twofold crystallographic symmetry (Fig. 1), with Pt-N(1) 1.98(2), Pt-N(2) 2.08(2) Å, N(1)-Pt-N(1) 90(1), N(1)-Pt-N(2) 95(1), N(2)-Pt-N(2) 80(1)°.

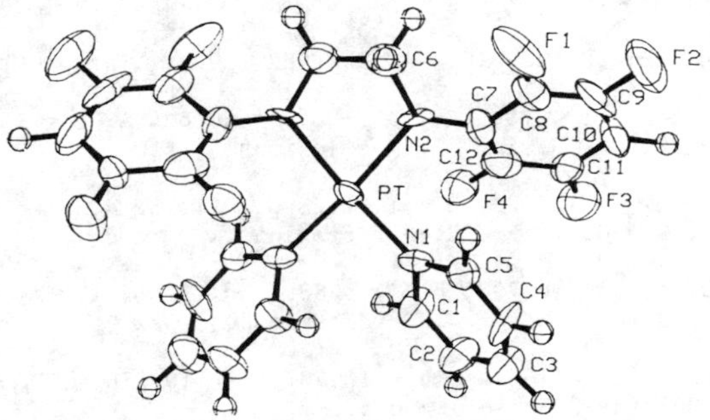

Fig. 1. A view of $C_{24}H_{16}F_8N_4Pt$.

DIOXO(N,N'-BIS(2-HYDROXY-3-t-BUTYLBENZYL)-1,2-DIAMINOETHANE-N,N',O,O')MOLYBDENUM(VI) DICHLOROMETHANE SOLVATE

$C_{24}H_{34}MoN_2O_4,CH_2Cl_2$

P. SUBRAMANIAN, J.T. SPENCE, R. ORTEGA and J.H. ENEMARK, 1984. Inorg. Chem., 23, 2564-2567.

Orthorhombic, Pbca, a = 13.728, b = 28.48, c = 27.50 Å, Z = 16. Mo radiation, R = 0.072 for 2286 reflexions.

The molecule (Fig. 1) has effective C_2 symmetry. The coordination about Mo is distorted cis-MoO_4N_2 octahedral, with the N atoms trans to the terminal oxo groups. Mean bond lengths: Mo-O_t 1.708(6), Mo-O 1.938(17), and Mo-N 2.359(6) Å.

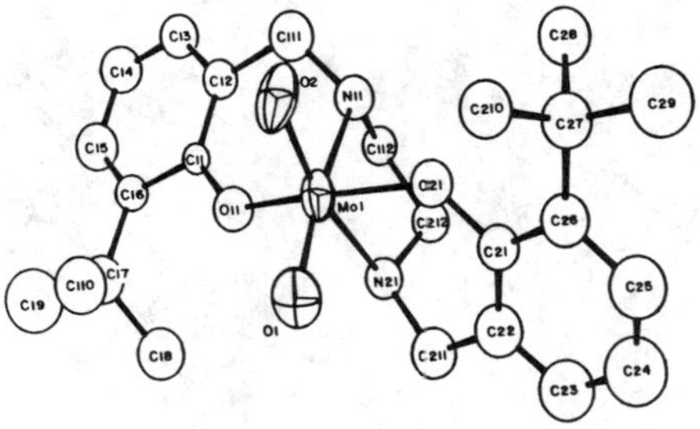

Fig. 1. The $C_{24}H_{34}MoN_2O_4$ molecule.

α-trans-[1,4,8,11-TETRAKIS(2-CYANOETHYL)-1,4,8,11-TETRAAZACYCLOTETRADECANE]BIS(ISOTHIOCYANATO)NICKEL(II)
$C_{24}H_{36}N_{10}NiS_2$ (I)

β-trans-[1,4,8,11-TETRAKIS(2-CYANOETHYL)-1,4,8,11-TETRAAZACYCLOTETRADECANE]BIS(ACETONITRILE)NICKEL(II) DIPERCHLORATE MONOHYDRATE
$C_{26}H_{42}Cl_2N_{10}NiO_8,H_2O$ (II)

β-[1,4,8,11-TETRAKIS(2-CARBAMOYLETHYL)-1,4,8,11-TETRAAZACYCLOTETRADECANE]NICKEL(II) DIPERCHLORATE NITROMETHANE SOLVATE
$C_{22}H_{44}Cl_2N_8NiO_{12},CH_3NO_2$ (III)

[1,8-BIS(2-CARBAMOYLETHYL)-4-(2-CYANOETHYL)-1,4,8,11-TETRAAZACYCLOTETRADECANE]-NICKEL(II) DIPERCHLORATE
$C_{19}H_{37}Cl_2N_7NiO_{10}$ (IV)

G.M. FREEMAN, E.K. BAREFIELD and D.G. VAN DERVEER, 1984. Inorg. Chem., 23, 3092-3103.

I. Monoclinic, $P2_1/c$, a = 10.526, b = 13.719, c = 20.625 Å, β = 107.05°, D_m = 1.37, Z = 4. Mo radiation, R = 0.057 for 3774 reflexions.

II. Monoclinic, $P2_1/n$, a = 12.726, b = 11.712, c = 11.182 Å, β = 95.40°, D_m = 1.51, Z = 2. Mo radiation, R = 0.070 for 2149 reflexions.

III. Monoclinic, $P2_1/n$, a = 16.798, b = 11.588, c = 17.882 Å, β = 91.84°, D_m = 1.53, Z = 4. Mo radiation, R = 0.066 for 4652 reflexions.

IV. Orthorhombic, $Pna2_1$, a = 16.454, b = 16.730, c = 10.338 Å, D_m = 1.51, Z = 4. Mo radiation, R = 0.067 for 2254 reflexions.

The four Ni complexes are shown in Fig. 1. Complex I has the 'trans-I' stereochemistry while II and III have the 'trans-III' stereochemistry for the macrocyclic ligand. In I-III the Ni is octahedrally coordinated, the equatorial plane being formed by the four N atoms of the macrocycle and the axial sites occupied by NCS ligands for I, CH_3CN for II, and carbamoyl O atoms for III. The complex cation in II is centrosymmetric. Complex IV also contains six-coordinate Ni with two carbamoyl O atoms coordinated in cis positions as a result of folding of the macrocycle along the 1,8-direction. Bond lengths: for I; Ni-NCS 2.034(3) and 2.091(4), Ni-N 2.154(4)-2.232(4) Å; for II; Ni-N(CH_3CN) 2.109(5), Ni-N 2.172(5) and 2.177(4) Å; for III; Ni-O 2.058(3) and 2.084(3), Ni-N 2.140(4)-2.166(4) Å; for IV; Ni-O 2.079(8) and 2.081(7), Ni-N 2.106(8)-2.174(8) Å.

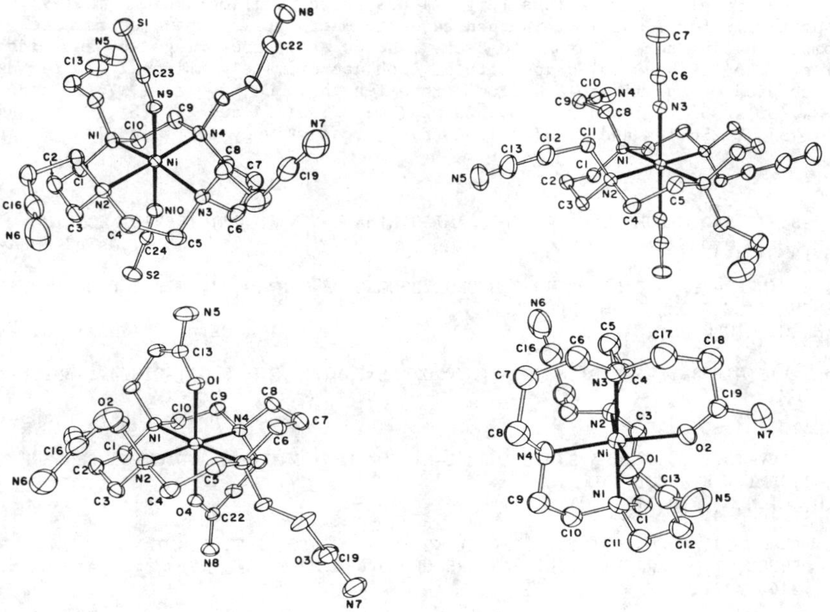

Fig. 1. The structures of $C_{24}H_{36}N_{10}NiS_2$ (upper left), $[C_{26}H_{42}N_{10}Ni]^{2+}$ (upper right), $[C_{22}H_{44}N_8NiO_4]^{2+}$ (lower left), and $[C_{19}H_{37}N_7NiO_2]^{2+}$ (lower right).

(μ-IMIDAZOLATO)BIS(IMIDAZOLE)(1,4,7,13,16,19-HEXAAZA-10,22-DIOXACYCLOTETRACOSANE)-DICOPPER(II) TRIPERCHLORATE

$C_{25}H_{49}Cl_3Cu_2N_{12}O_{14}$ (I) $[Cu_2(C_3H_3N_2)(C_3H_4N_2)_2(C_{16}H_{38}N_6O_2)](ClO_4)_3$

(μ-IMIDAZOLATO)BIS(1-METHYLIMIDAZOLE)(1,4,7,13,16,19-HEXAAZACYCLOTETRACOSANE)-DICOPPER(II) TRIPERCHLORATE

$C_{29}H_{57}Cl_3Cu_2N_{12}O_{12}$ (II) $[Cu_2(C_3H_3N_2)(C_4H_6N_2)_2(C_{18}H_{42}N_6)](ClO_4)_3$

P.K. COUGHLIN, A.E. MARTIN, J.C. DEWAN, E.-I. WANATABE, J.E. BULKOWSKI, J.-M. LEHN and S.J. LIPPARD, 1984. Inorg. Chem., 23, 1004-1009.

I. Monoclinic, C2/c, a = 15.008, b = 10.186, c = 26.062 Å, β = 100.48°, D_m = 1.68, Z = 4. Mo radiation, R = 0.074 for 1632 reflexions.

II. Orthorhombic, Pnma, a = 16.670, b = 28.112, c = 9.195 Å, D_m = 1.55, Z = 4. Mo radiation, R = 0.124 for 912 reflexions.

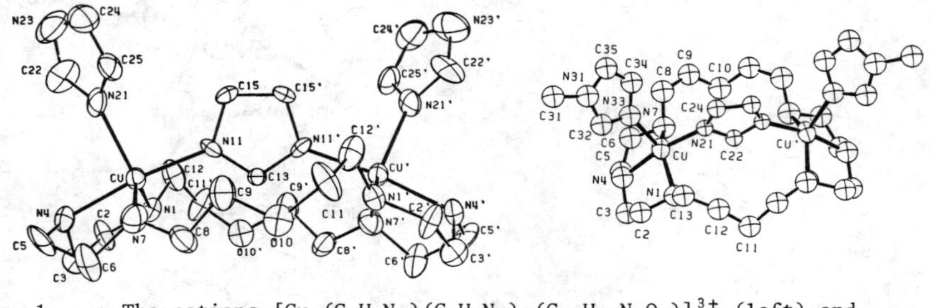

Fig. 1. The cations $[Cu_2(C_3H_3N_2)(C_3H_4N_2)_2(C_{16}H_{38}N_6O_2)]^{3+}$ (left) and $[Cu_2(C_3H_3N_2)(C_4H_6N_2)_2(C_{18}H_{42}N_6)]^{3+}$ (right).

The two binuclear cations (Fig. 1) possess crystallographic symmetry; C_2 and C_s respectively for I and II. In each case the macrocyclic ligand coordinates to two Cu atoms, donating three N atoms to each. The Cu atoms are bridged by an imidazolate anion. The fifth coordination site of each approximately trigonal bipyramidal Cu atom is occupied by an imidazole N atom. Bond lengths: $Cu\text{-}N_{ax}$(im) 1.976(9) and 1.95(2), $Cu\text{-}N_{ax}$(am) 2.021(11) and 2.03(3), $Cu\text{-}N_{eq}$(im) 2.135(11) and 2.08(3), $Cu\text{-}N_{eq}$(am) 2.118(9), 2.256(10) and 2.09(4), 2.13(3) Å, respectively for I and II.

(1,2-BIS(3,5-DICHLORO-2-HYDROXYBENZAMIDO)ETHANE-N,N',O,O')DIPYRIDINEOSMIUM
$C_{26}H_{18}Cl_4N_4O_4Os$ (I) Os(η^4-CHBA-Et)(py)$_2$

cis-BIS(N-FORMYL-3,5-DICHLORO-2-HYDROXYBENZAMIDE-N,N',O,O')-BIS(4-t-BUTYLPYRIDINE)-OSMIUM DICHLOROMETHANE SOLVATE
$C_{34}H_{32}Cl_4N_4O_6Os,CH_2Cl_2$ (II) Os(η^2-Fo-CHBA)$_2$-cis-(t-Bupy)$_2$

trans-BIS(N-FORMYL-3,5-DICHLORO-2-HYDROXYBENZAMIDE-N,N',O,O')-BIS(4-t-BUTYLPYRIDINE)-OSMIUM
$C_{34}H_{32}Cl_4N_4O_6Os$ (III) Os(η^2-Fo-CHBA)$_2$-trans-(t-Bupy)$_2$

2,2'-BIPYRIDYL(1,2-BIS(3,5-DICHLORO-2-HYDROXYBENZAMIDO)-4,5-DICHLOROBENZENE-N,N',-O,O')OSMIUM ETHANOL SOLVATE
$C_{30}H_{14}Cl_6N_4O_4Os,0.5(C_2H_6O)$ (IV) Os(η^4-CHBA-DCB)(η^2-bpy)

F.C. ANSON, J.A. CHRISTIE, T.J. COLLINS, R.J. COOTS, T.T. FURUTANI, S.L. GIPSON, J.T. KEECH, T.E. KRAFFT, B.D. SANTARSIERO and G.H. SPIES, 1984. J. Am. Chem. Soc. 106, 4460-4472.

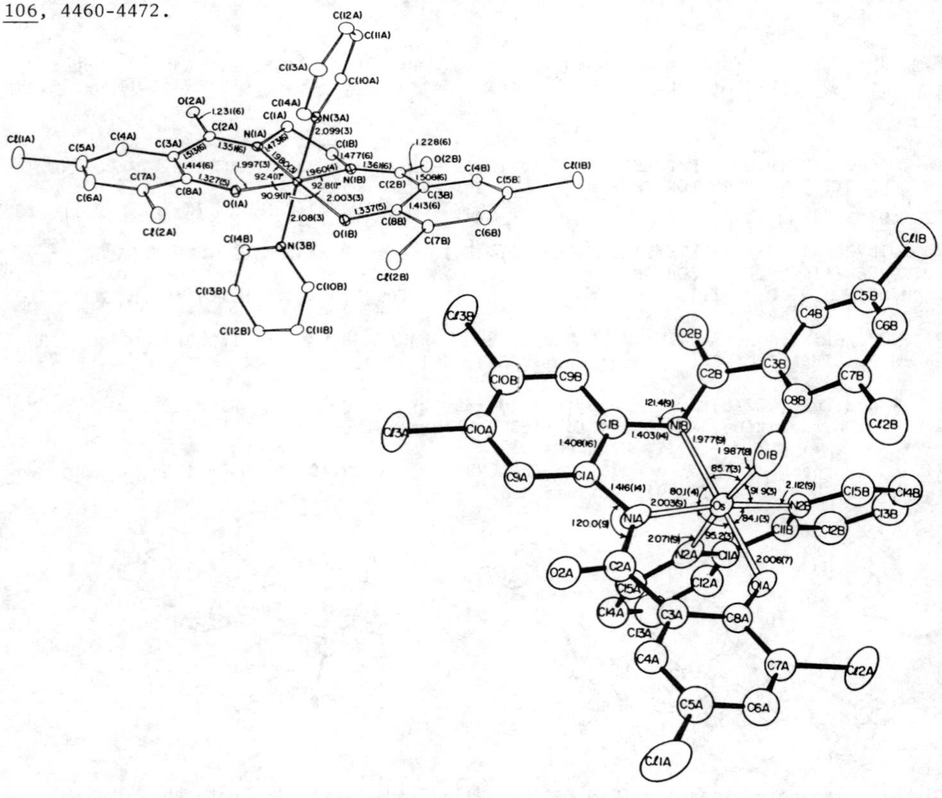

Fig. 1. Views of $C_{26}H_{18}Cl_4N_4Os$ (I) (top) and $C_{30}H_{14}Cl_6N_4O_4Os$ (IV) (bottom).

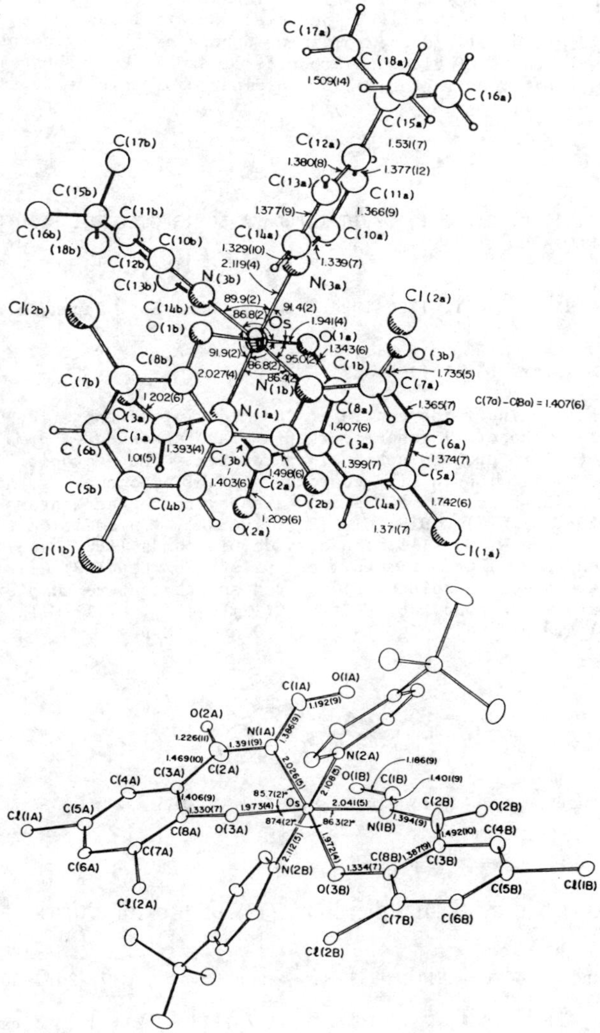

Fig. 2. Views of the cis- (top) and trans- (bottom) forms of $C_{34}H_{32}Cl_4N_4O_6Os$.

I. Triclinic, $P\bar{1}$, a = 10.083, b = 12.310, c = 12.241 Å, α = 97.38, β = 96.49, γ = 113.0°, Z = 2. Mo radiation, R = 0.029 for 5805 reflexions.

II. Monoclinic, C2/c, a = 19.907, b = 28.850, c = 9.445 Å, β = 131.79°, Z = 4. Mo radiation, R = 0.031 for 3903 reflexions.

III. Triclinic, $P\bar{1}$, a = 8.5333, b = 13.6879, c = 16.4148 Å, α = 104.35, β = 97.33, γ = 79.18°, Z = 2. Mo radiation, R = 0.043 for 6871 reflexions.

IV. Triclinic, $P\bar{1}$, a = 10.860, b = 12.633, c = 12.844 Å, α = 117.47, β = 90.42, γ = 95.90°, Z = 2. Mo radiation, R = 0.044 for 2710 reflexions.

The octahedral Os in I has axial pyridine ligands (Fig. 1) with the tetradentate anion equatorial. In IV, the bpy ligand occupies one axial and one equatorial site, with one phenolic donor axial and the remaining donors of the tetradentate anion equatorial. The structures of cleavage products II and III are shown in Fig. 2, and are the first structurally characterized N-coordinated organic imido complexes of osmium.

BIS[ACETONITRILE(11,13-DIMETHYL-1,4,7,10-TETRAAZA-10,13-CYCLOTRIDECADIEN-12-YLIDENE)-NICKEL(II)] PERCHLORATE

$C_{26}H_{46}Cl_4N_{10}Ni_2O_{16}$ $[Ni(CH_3CN)(Me_2[13]dienN_4\text{-ylidene}]_2{}^{4+}4ClO_4{}^-$

K. MOCHIZUKI, K. TORIUMI and T. ITO, 1984. Bull. Chem. Soc. Japan, 57, 881-882.

Monoclinic, $P2_1/n$, a = 13.533, b = 11.471, c = 13.300 Å, β = 93.76°, Z = 2. Mo radiation, R = 0.087 for 2857 reflexions.

The complex has an inversion centre at the midpoint of the C(12)-C(12') bond in the cation (Fig. 1). These bridgehead carbon atoms are connected by a double bond (C(12)-C(12') 1.366 Å) to produce a binuclear complex. The acetonitrile molecule is coordinated to the Ni(II) ion in five-coordinate fashion, occupying the apical site (Ni-N(S) 2.510 Å). The basal positions are taken by the other N atoms (Ni-N(1) 1.857, Ni-N(4) 1.898, Ni-N(7) 1.913, Ni-N(10) 1.865 Å). In addition to C(12)-C(12'), the bonds N(1)-C(13) and N(10)-C(11) are also double bonds (1.271 and 1.262 Å respectively in length). The four methyl groups strain the C(12)-C(12') bond so that the double bonds are almost completely localised and the entire complex adopts a chair-like structure. The bond angles C(11)-C(12)-C(12'), C(11)-C(12)-C(13) and C(13)-C(12)-C(12') are respectively 119.5, 116.0 and 118.9°.

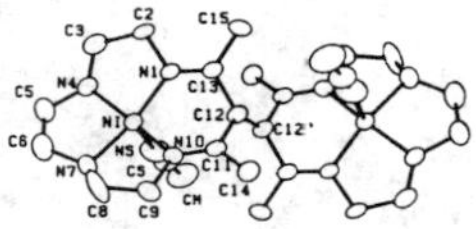

Fig. 1. Structure of the cation in $C_{26}H_{46}Cl_4N_{10}Ni_2O_{16}$.

BIS(meso-1,2-DIPHENYL-1,2-ETHANEDIAMINE)NICKEL(II) DICHLORIDE DIHYDRATE

$C_{28}H_{32}Cl_2N_4Ni,2H_2O$ $[Ni(C_{14}H_{16}N_2)_2]Cl_2.2H_2O$

S. GARCIA-GRANDA and F. GOMEZ-BELTRAN, 1984. Acta Cryst., C40, 2037-2038.

Orthorhombic, Pccn, a = 26.831, b = 15.366, c = 7.1773 Å, Z = 4. Mo radiation, R = 0.046 for 1075 reflexions.

The Ni atom lies on an inversion centre (Fig. 1), has necessarily planar coordination geometry, with Ni-N 1.922(4), 1.921(4) Å, N-Ni-N 86.4(2)°.

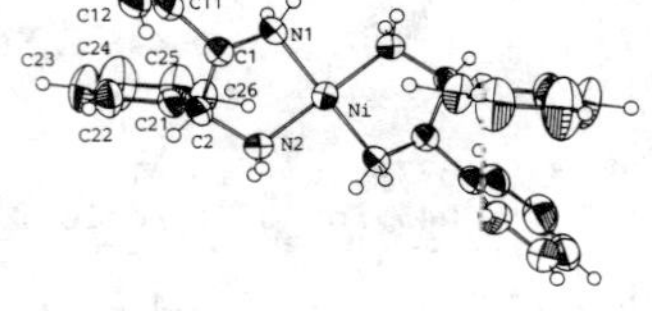

Fig. 1. A drawing of the $[Ni(C_{14}H_{16}N_2)_2]^{2+}$ cation.

BIS(μ-HYDROXO)BIS[[N,N'-BIS(2-PYRIDYLMETHYL)-1,2-ETHANE]CHROMIUM(III)] TETRA-PERCHLORATE MONOHYDRATE

$C_{28}H_{38}Cl_4Cr_2N_8O_{18},H_2O$ $[Cr(C_{14}H_{18}N_4)OH]_2(ClO_4)_4.H_2O$

M.A. HEINRICHS, D.J. HODGSON, K. MICHELSEN and E. PEDERSEN, 1984. Inorg. Chem., 23, 3174-3180.

Monoclinic, $P2_1/n$, a = 12.434, b = 21.449, c = 15.495 Å, β = 98.63°, D_m = 1.66, Z = 4. Mo radiation, R = 0.077 for 2366 reflexions.

The dinuclear cation (Fig. 1) contains one α-cis skeleton and one β-cis skeleton, the configuration at both centres being the same (ΔΔ or ΛΛ). The configurations at the secondary N atoms in the β-skeleton are the same, so the complex isolated may be described as a racemic mixture of Λ(α)-SS-Λ(β)-RR' and Δ(α)-RR-Δ(β)-SS' forms. Bond lengths: Cr-O 1.918(7)-1.976(5), Cr-N(pyr) 2.035(7)-2.101(7), Cr-N(am) 2.049(7)-2.096(7) Å.

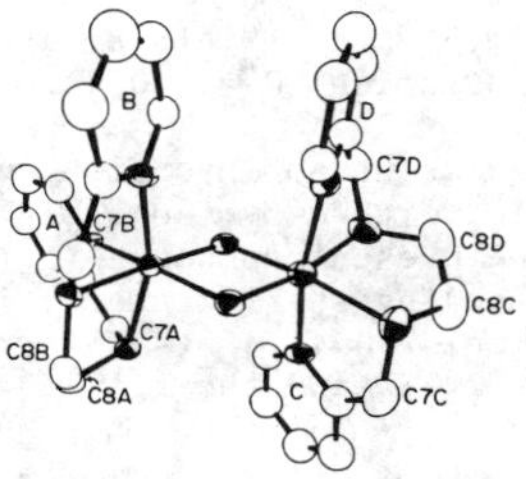

Fig. 1. The $[Cr(C_{14}H_{18}N_4)OH]_2^{4+}$ cation (ΔΔδλ conformation).

BIS(PHENYLMETHANETHIOLATO)(2,3,9,10-TETRAMETHYL-1,4,8,11-TETRAAZA-1,3,8,10-CYCLOTETRA-DECATETRAENE)IRON(III) HEXAFLUOROPHOSPHATE

$C_{28}H_{38}F_6FeN_4PS_2$ $C_{28}H_{38}FeN_4S_2^+.PF_6^-$

A.A. ARUFFO, B.D. SANTARSIERO, V. SCHOMAKER and E.C. LINGAFELTER, 1984. Acta Cryst., C40, 1693-1695.

Triclinic, $P\bar{1}$, a = 12.385, b = 12.483, c = 12.155 Å, α = 97.446, β = 111.408, γ = 63.960°, Z = 2. Mo radiation, R = 0.082 for 3622 reflexions.

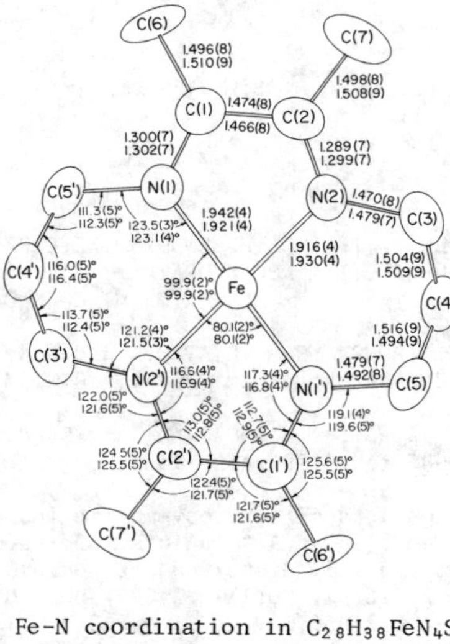

Fig. 1. A view of the Fe-N coordination in $C_{28}H_{38}FeN_4S_2^+$ and some dimensions.

Two independent cations (Fig. 1) lie on inversion centres. The average Fe-N(TIM) and Fe-S bond lengths are 1.928(3) and 2.288(1) Å. Comparison with other Fe(TIM) complexes seems to imply increased donation of electron density from the axial ligands (here sulfur) to Fe^{III}, and from Fe^{III} to the diimines of TIM. The two cations have different axial ligand conformations, one fully extended, away from the TIM plane, and the other with the phenyls lying next to the TIM plane (Fe-S-C-C torsion angles 176.7(3) and -80.2(5)°; Fe-S-CH_2 108.8(2) and 114.4(2)°].

TETRAKIS[(μ_2-α-PYRIDONE-N,O)(ETHYLENEDIAMINE-N,N')PLATINUM] PENTANITRATE MONOHYDRATE
$C_{28}H_{48}N_{17}O_{19}Pt_4,H_2O$ $[Pt_4(en)_4(C_5H_4NO)_4].(NO_3)_5,H_2O$

T.V. O'HALLORAN, M.M. ROBERTS and S.J. LIPPARD, 1984. J. Am. Chem. Soc., 106, 6427-6428.

Triclinic, P$\bar{1}$, a = 10.652, b = 13.068, c = 9.413 Å, α = 108.14, β = 96.62, γ = 68.39°, D_m = 2.485, Z = 1. Mo radiation, R = 0.030 for 2918 reflexions.

The tetranuclear cation lies about an inversion centre and is shown in Fig. 1; it is very similar to a Pt complex with α-pyridone blue (1) except for the Pt-Pt distances which are longer in the present compound (Pt(1)-Pt(2) 2.8296(5), Pt(2)-Pt(2') 2.9058(6) Å) due to non-bonding steric interactions with ethylenediamine. Other (mean) distances are Pt-N(en) 2.04(1), Pt-N(α-pyridone) 2.04(1), Pt-O 2.04(1) Å. The Pt(1)-Pt(2) twist angle ω is 24.3° and the tilt angle τ is 32.1°.

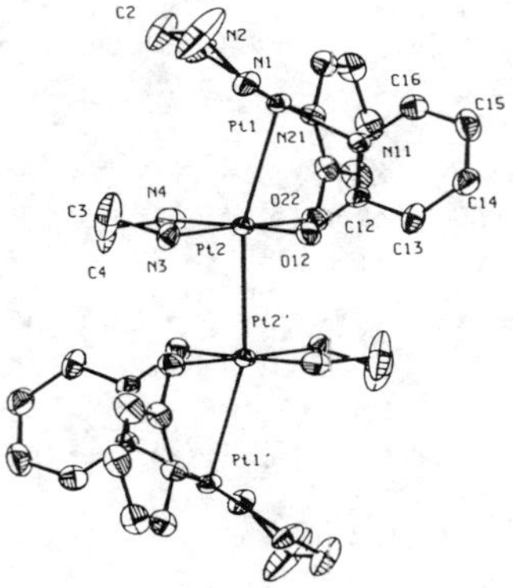

Fig. 1. A view of the $[Pt_4(en)_4(C_5H_4NO)_4]^{5+}$ cation.

1. Structure Reports, 45B, 1268.

[μ_2-BIS(ETHYLENEDIAMINE)COPPER(II)]-BIS[ETHYLENEDIAMINETETRAACETATO-COPPER(II)] DIAQUABIS(ETHYLENEDIAMINE)CUPRATE(II) DECAHYDRATE
$C_{28}H_{60}Cu_4N_{12}O_{18},10H_2O$ $(C_{24}H_{40}Cu_3N_8O_{16})^{2-}(C_4H_{20}CuN_4O_2)^{2+},10H_2O$

T.F. SYSOEVA, V.M. AGRE, V.K. TRUNOV, N.M. DYATLOVA and N.N. BARKHANOVA, 1984. J. Struct. Chem., 25, 434-439 [Zh. Strukt. Khim. 25(3), 107-113].

Triclinic, P$\bar{1}$, a = 8.88, b = 10.931, c = 14.204 Å, α = 95.4, β = 104.3, γ = 101.5°, Z = 1. Mo radiation, R = 0.043 for 2799 reflexions.

The structure (Fig. 1) consists of centrosymmetric $[Cu(en)_2(H_2O)_2]^{2+}$ cations, centrosymmetric trinuclear $[CuACu(en)_2CuA]^{2-}$ anions, and water molecules. All three independent Cu atoms have tetragonally distorted octahedral coordination geometry: Cu(1)-O 1.956, 1.962, 2.351, 2.323(4), Cu(1)-N 2.005 and 2.014(4), Cu(2)-O 2.437(4), Cu(2)-N 2.019, 2.026(4), Cu(3)-O 2.581(5), Cu(3)-N 2.017, 2.002(4) Å.

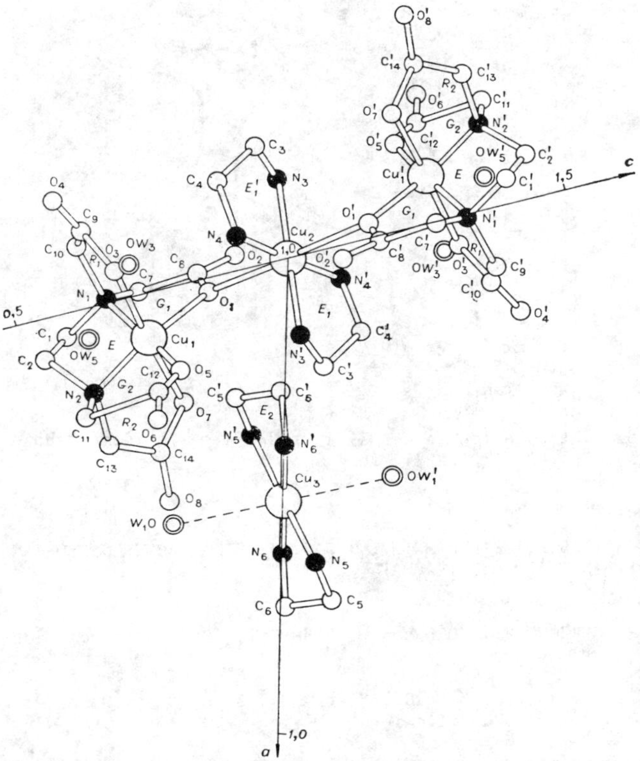

Fig. 1. The structure of [μ_2-bis(ethylenediamine)copper(II)]-bis[ethylenediamine-tetraacetato-copper(II)] diaquabis(ethylenediamine)cuprate(II) decahydrate projected onto the ac plane.

DI-μ-HYDROXOBIS[{N,N'-BIS(2-PYRIDYLMETHYL)-1,3-PROPANEDIAMINE}CHROMIUM(III)] PERCHLORATE TRIHYDRATE

$C_{30}H_{42}Cl_4Cr_2N_8O_{18},3H_2O$ $[(C_{15}H_{20}N_4)Cr(OH)]_2^{4+}4ClO_4^-,3H_2O$

H.R. FISCHER, D.J. HODGSON, K. MICHELSEN and E. PEDERSEN, 1984. Inorg. Chim. Acta, 88, 143-150.

Monoclinic, C2/c, a = 38.323, b = 14.458, c = 17.027 Å, β = 106.24°, D_m = 1.58, Z = 8. Mo radiation, R = 0.066 for 3710 reflexions.

The structure consists of binuclear cations (Fig. 1), perchlorate anions and water molecules. The geometry at each Cr centre is approximately octahedral. The relevant distances associated with the metal atoms are: Cr(1)-O(1) 1.966, Cr(1)-O(2) 1.962, Cr(2)-O(1) 1.973 and Cr(2)-O(2) 1.967 Å. The Cr-N distances range from 2.053 to 2.116 Å. The six-membered chelate rings have different conformations. The ring Cr(1), N(2)A, C(8)A, C(9)A, C(8)B and N(2)B adopts a classical chair conformation, whilst the corresponding ring at Cr(2) is a skew boat. The Cr(1)-Cr(2) distance is 3.054 Å and the angles Cr(1)-O(1)-Cr(2) and Cr(1)-O(2)-Cr(2) are 101.7 and 102.0° respectively.

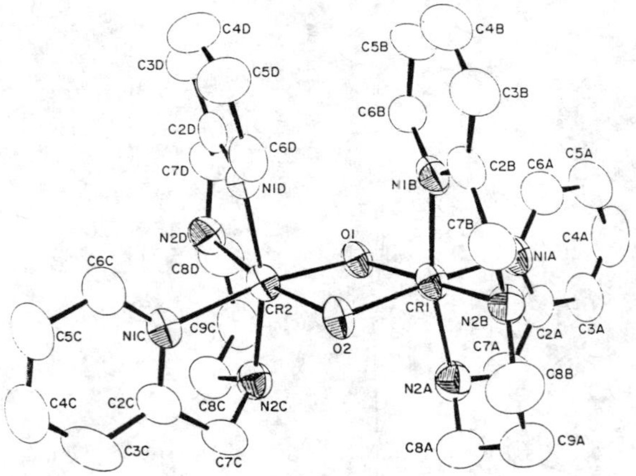

Fig. 1. View of the cation in $C_{30}H_{42}Cl_4Cr_2N_8O_{18}$. The isomer shown is the cis-{Λ(β)(RR')Λ(β)(RR')} form.

BIS[(NITRATO-O,O')(1,5,9,13-TETRAMETHYL-1,5,9,13-TETRAAZACYCLOHEXADECANE-N^1,N^5,N^9,N^{13})-CADMIUM] TETRAKIS(NITRATO-O,O')CADMATE
$C_{32}H_{72}Cd_3N_{14}O_{18}$

N.W. ALCOCK, E.H. CURZON, P. MOORE and C. PIERMONT, 1984. J. Chem. Soc. Dalton, 605-610.

Monoclinic, C2/c, a = 53.360, b = 10.248, c = 31.938 Å, β = 119.51°, Z = 12. Mo radiation, R = 0.063 for 4920 reflexions.

The cations have pseudo-trigonal-bipyramidal geometry, with bidentate nitrate groups occupying one equatorial position. They show disordered ligand conformations comprising a 'three-up and one-down' and two types of 'all four-up' arrangements of the N-methyl groups (Fig. 1). The anions are dodecahedral, with bidentate nitrate groups. Cd-N (axial) is 2.41 Å (average) and Cd-N (equatorial) is 2.37 Å, while Cd-O range from 2.27 to 2.60 Å, with the anions showing greater asymmetry.

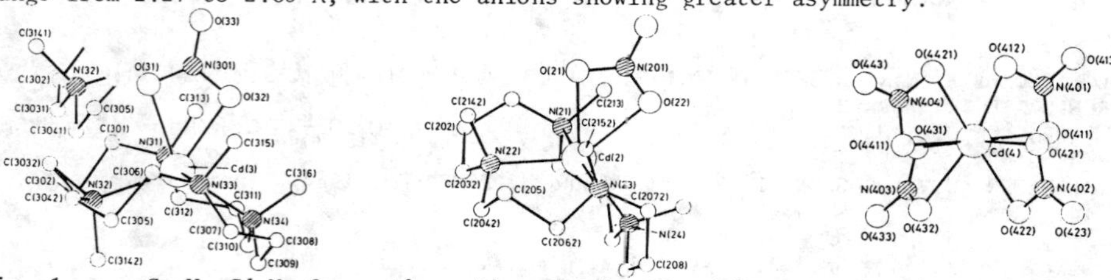

Fig. 1. $C_{32}H_{72}Cd_3N_{14}O_{18}$: the cation showing the 'three-up, one-down' (left) and the 'four-up' (centre) forms and the anion.

(μ-N,N'-BIS(2-((o-HYDROXYBENZHYDRYLIDENE)AMINO)ETHYL)PIPERAZINE-N,N',O)DIACETATODI-COPPER METHANOL SOLVATE
$C_{38}H_{40}Cu_2N_4O_6,2CH_4O$ $Cu_2(C_{34}H_{34}N_4O_2)(CH_3COO)_2.2CH_3OH$

B. CHIARI, O. PIOVESANA, T. TARANTELLI and P.F. ZANAZZI, 1984. Inorg. Chem., 23, 2542-2547.

Triclinic, P$\bar{1}$, a = 10.034, b = 14.057, c = 6.897 Å, α = 92.93, β = 97.47, γ = 89.64°, Z = 1. Mo radiation, R = 0.062 for 1204 reflexions.

The centrosymmetric binuclear molecule (Fig. 1) contains Cu atoms in distorted square planar environments with two cis N atoms (Cu-N 1.945(7) and 2.048(7) Å), one phenolic O atom (Cu-O 1.895(6) Å), and one O atom of an acetate anion (Cu-O 1.974(6) Å). The second O atom of each carboxylate ligand is at a very weakly linking distance (2.795(9) Å) to only one Cu atom.

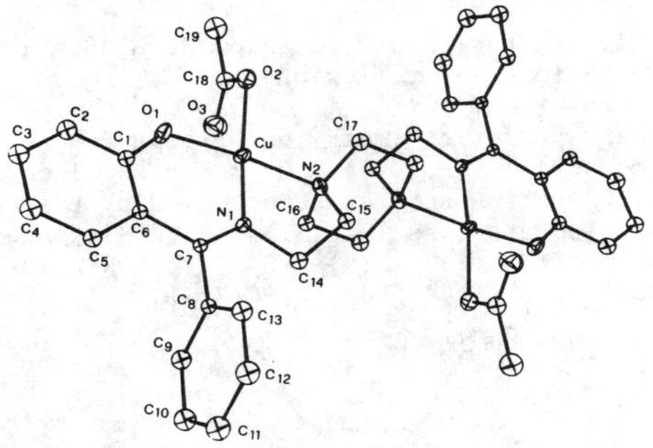

Fig. 1. The $Cu_2(C_{34}H_{34}N_4O_2)(CH_3COO)_2$ molecule.

CHLORO[(2RS,5RS,8RS,11SR)-1,4,7,10-TETRABENZYL-2,5,8,11-TETRAETHYL-1,4,7,10-TETRAAZACYCLODODECANE]COPPER(II) DICHLOROCUPRATE(I) ETHANOL SOLVATE

$C_{44}H_{60}Cl_3Cu_2N_4,3C_2H_6O$ $[CuCl(C_{44}H_{60}N_4)][CuCl_2],3C_2H_6O$

S. TSUBOYAMA, K. KOBAYASHI, T. SAKURAI and K. TSUBOYAMA, 1984. Acta Cryst., C40, 1178-1181.

Triclinic, $P\bar{1}$, a = 14.912, b = 15.054, c = 13.809 Å, α = 110.5, β = 103.7, γ = 67.68°, D_m = 1.27, Z = 2. Mo radiation, R = 0.112 for 3016 reflexions.

The crystal structure consists of complex cations, disordered anions and solvent molecules. In the cation (Fig. 1), Cu is coordinated to four N atoms and a Cl^- ion in a distorted square pyramid which is displaced approximately 41% towards the trigonal-bipyramidal form; the 12-membered ring takes the [2334] conformation. The $[CuCl_2]^-$ anion moiety is nearly linear.

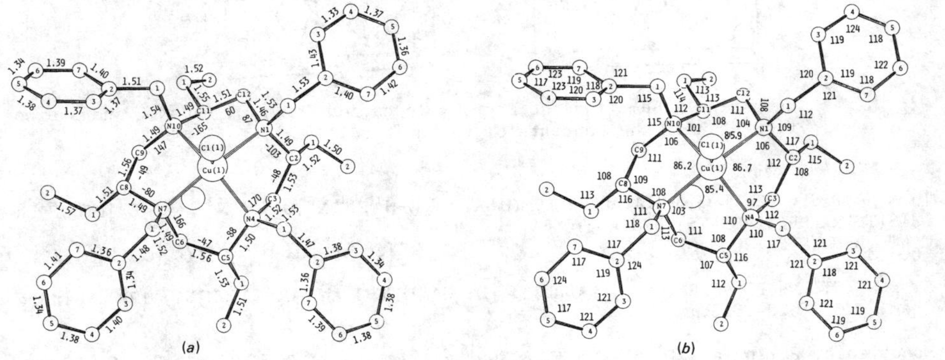

Fig. 1. $[CuCl(C_{44}H_{60}N_4)][CuCl_2],3C_2H_6O$: perspective views of the complex cation showing (a) bond lengths and torsion angles and (b) bond angles.

(2,3,9,10-TETRAMETHYL-1,4,8,11-TETRAAZA-1,3,8,10-CYCLOTETRADECATETRAENE)COPPER(II) TETRAPHENYLBORATE

$C_{62}H_{64}B_2CuN_4$ $[CuC_{14}H_{24}N_4]^{2+}[B(C_6H_5)_4^-]_2$

A.E. ELIA, E.C. LINGAFELTER and V. SCHOMAKER, 1984. Croat. Chem. Acta, 57, 653-659.

Monoclinic, $P2_1/c$, a = 11.997, b = 13.042, c = 16.391 Å, β = 102.10°, D_m = 1.36, Z = 2. Mo radiation, R = 0.055 for 1756 reflexions.

The cation, planar except for two methylene groups is one of the rare cases of planar four-coordinate copper(II) (Fig. 1). Of interest also is the fact that the apical regions of the coordination sphere are occupied by the centres of the phenyl groups from two tetraphenylborate ions (Fig. 1). Bond distances are: Cu-N(1) 1.948, Cu-N(2) 1.940, N(1)-C(1) 1.257, N(1)-C(5) 1.448, N(2)-C(2) 1.275, N(2)-C(3) 1.455, C(1)-C(2) 1.51, C(1)-C(6) 1.49 and C(2)-C(7) 1.50 Å.

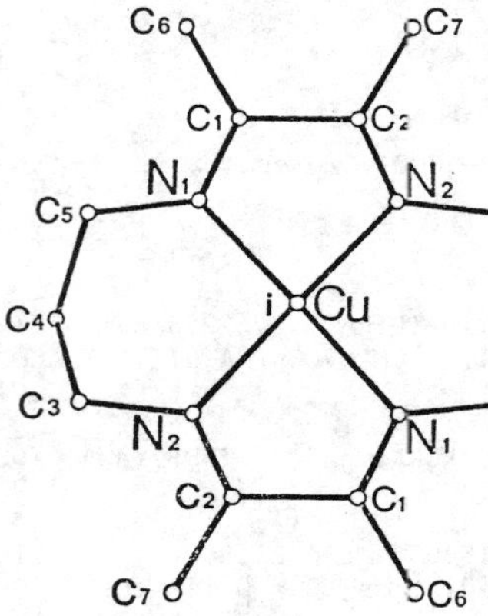

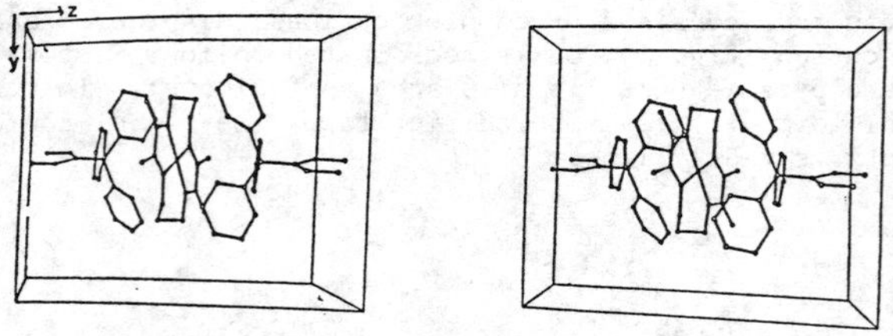

Fig. 1. Numbering scheme for the macrocycle cation in $C_{62}H_{64}B_2CuN_4$, and stereo view of half the contents of the unit cell.

[μ-N,N'-BIS(6-ETHYL-3,6-DIAZAOCTYL)OXAMIDATO(2-)-N[1],N[3],N[6],O:N[1]',N[3]',N[6]',O']DICOPPER(II) BIS(TETRAPHENYLBORATE) BIS(ACETONE)

$C_{66}H_{78}B_2Cu_2N_6O_2, 2C_3H_6O$ $[Cu_2(C_{18}H_{38}N_6O_2)][(C_6H_5)_4B]_2, 2(CH_3)_2CO$

A. BENCINI, M. DI VAIRA, A.C. FABRETTI, D. GATTESCHI and C. ZANCHINI, 1984. Inorg. Chem., 23, 1620-1623.

Monoclinic, $P2_1/n$, a = 16.639, b = 18.305, c = 11.089 Å, β = 103.2°, Z = 2. Cu radiation, R = 0.055 for 2021 reflexions.

The structure consists of binuclear cations (Fig. 1), Ph_4B^- anions, and acetone solvent molecules. Each of the Cu atoms, which are bridged by the oxamide moiety, is coordinated by one O (Cu-O 1.988(5) Å), one amidic N (Cu-N 1.899(6) Å), and two aminic N atoms (Cu-N 1.994(6) and 2.008(7) Å) which form a distorted square planar environment. At a longer distance (2.609(8) Å), approximately on the perpendicular to the coordination plane, there is an oxygen of an acetone molecule.

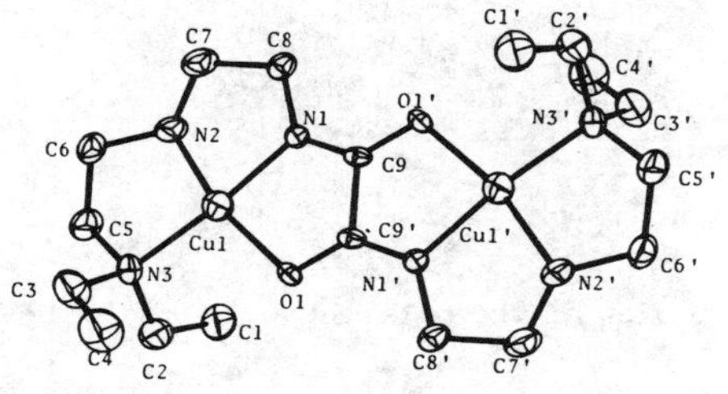

Fig. 1. The $[Cu_2(C_{18}H_{38}N_6O_2)]^{2+}$ cation.

trans-(ACETYLACETONATO)DICHLORO(1,5-CYCLOOCTADIENE)IRIDIUM(III)
$C_{13}H_{19}Cl_2IrO_2$ $[IrCl_2(C_5H_7O_2)(C_8H_{12})]$

P.A. TUCKER, 1984. Acta Cryst., C40, 620-622.

Orthorhombic, Pnam, a = 13.822, b = 17.117, c = 11.653 Å, Z = 8. Mo radiation, R = 0.030 for 2904 reflexions (at 133K).

The two independent molecules, which are separated by normal van der Waals contacts, each have crystallographic mirror symmetry, the mirror plane bisecting the Ir(acetylacetonate) ring(Fig. 1). Their molecular geometries differ only in the orientation of the acetylacetonato methyl group. Mean bond lengths are Ir-Cl 2.348, Ir-O 2.022, Ir-C 2.182 and C=C 1.369 Å. The molecular dimensions are compared with those in acetylacetonato(1,5-cyclooctadiene)iridium(I) (1).

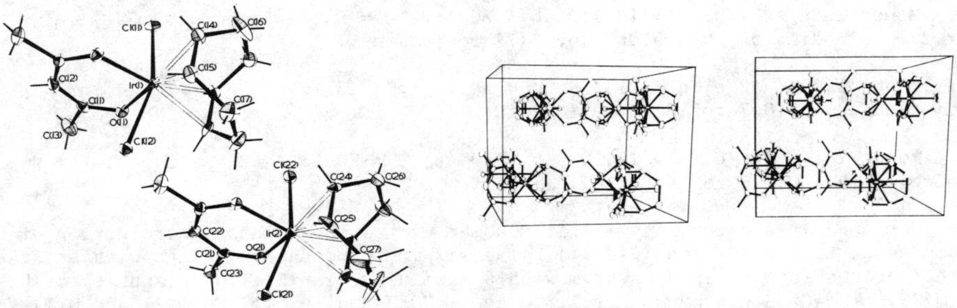

Fig. 1. $[IrCl_2(C_5H_7O_2)(C_8H_{12})]$: perspective view of the two independent molecules and a stereoscopic drawing of the molecular packing viewed roughly along c.

1. Structure Reports, 48B, 1185.

DIOXOBIS(PENTANE-2,4-DIONATO)PYRIDINEURANIUM
$C_{15}H_{19}NO_6U$

N.W. ALCOCK, D.J. FLANDERS and D. BROWN, 1984. J. Chem. Soc. Dalton, 679-681.

Orthorhombic, Fdd2, a = 29.702, b = 11.433, c = 10.593 Å, D_m = 2.03, Z = 8. Mo radiation, R = 0.032 for 832 reflexions.

The molecule exhibits pentagonal bipyramidal coordination about the U atom involving two bidentate pentane-2,4-dionate ions and a pyridine molecule (Fig. 1). Coordination distances, which are of low precision, are: U-O(1) 1.83(1), U-O(2) 2.34(1), U-O(3) 2.44(1) and U-N(1) 2.47(1) Å.

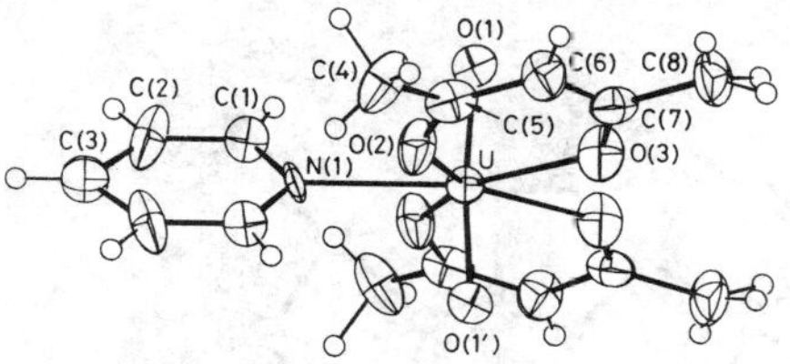

Fig. 1. $C_{15}H_{19}NO_6U$: a view of the molecule.

BIS(2,2,6,6-TETRAMETHYL-3,5-HEPTANEDIONATO)COPPER(II)
$C_{22}H_{38}CuO_4$ (I) $Cu(C_{11}H_{19}O_2)_2$

DI-μ_3-METHOXY-DI-μ-METHOXY-TETRAKIS[(2,2,6,6-TETRAMETHYL-3,5-HEPTANEDIONATO)COPPER(II)]
$C_{48}H_{88}Cu_4O_{12}$ (II) $Cu_4(OCH_3)_4.(C_{11}H_{19}O_2)_4$

μ_4-OXA-BISPYRAZOLATO-TETRAKIS[(2,2,6,6-TETRAMETHYL-3,6-HEPTANEDIONATO)COPPER(II)]
$C_{50}H_{82}Cu_4N_4O_9$ (III) $Cu_4O(C_3H_3N_2)_2.(C_{11}H_{19}O_2)_4$

HEXA-μ_3-HYDROXY-μ_6-(HYDROGENDIOXO-O,O')-NONAKIS-[(2,2,6,6-TETRAMETHYL-3,5-HEPTANE-DIONATO)COPPER(II)]
$C_{99}H_{178}Cu_9O_{26}$ (IV) $Cu_9(OHO)(OH)_6.(C_{11}H_{19}O_2)_9$

W.H. WATSON and W.W. HOLLEY, 1984. Croat. Chem. Acta, 57, 467-476.

I. Monoclinic, $P2_1/n$, a = 12.273, b = 11.040, c = 10.383 Å, β = 117.98°, D_m = 1.14, Z = 2. Mo radiation, R = 0.107 for 1930 reflexions.

II. Monoclinic, $P2_1/a$, a = 10.110, b = 30.162, c = 9.657 Å, β = 103.20°, D_m = 1.29, Z = 2. Cu radiation, R = 0.067 for 3177 reflexions.

III. Monoclinic, A2/a, a = 21.430, b = 24.70, c = 13.759 Å, β = 111.24°, Z = 4. Cu radiation, R = 0.067 for 3293 reflexions.

IV. Monoclinic, $P2_1/c$, a = 15.108, b = 29.48, c = 30.53 Å, β = 113.56°, Z = 4. Cu radiation, R = 0.125 for 6621 reflexions.

Compound I is a centrosymmetric monomer with square planar coordination of O atoms about the Cu(II) ion (Fig. 1). The six-membered chelate rings with attached methyl groups form a planar system with a maximum deviation from planarity of 0.04 Å by O(2). The Cu-O distances are 1.876 and 1.903 Å while O(1)-Cu-O(2) and O(1)-Cu-O(1') are respectively 93.2 and 86.8°. Compound II is a step tetramer (Fig. 1) with a Cu_4O_4 core which can be described in terms of two interacting methoxy-bridged dimers. In each dimer Cu(1) is square-planar with metal-plane distance of 0.05 Å. The Cu(2) ion lies 0.16 Å out of the basal plane of a distorted square pyramid of O atoms. The distances Cu(1)-O(1), Cu(1)-O(2) are 1.880 and 1.935 Å while the two Cu(1)-O(DMP) distances of 1.899 and 1.910 Å are equivalent. The Cu(2)-OCH_3 distances are statistically different with Cu(2)-O(1) 1.917 Å and Cu(2)-O(2) 1.982 Å but the Cu(2)-O(DPM) distances of 1.914 and 1.923 Å are again equal. Compound III is a tetramer with the Cu(II) ions arranged tetrahedrally about the μ_4-oxygen. A crystallographic twofold axis passes through C(24), O(1) and C(26). The coordination about each Cu(II) ion is approximately square-planar. Surprisingly, all Cu-X (X = O,N) distances are equivalent with an average value of 1.930 Å. Compound IV has in its central core nine Cu(II) ions and eight O atoms. Cu atoms A, B, C, G, H, and I form a trigonal prism with Cu atoms D, E and F lying above the rectangular faces. Atoms O(3) to O(8) are OH ions which bridge three Cu(II) ions. Each OH bridges Cu(II) ions from each end of the trigonal prism and a facial Cu(II) ion. The facial Cu(II) ions are square-planar with coordination of two O atoms from a chelating DPM and two bridging OH ions. The six Cu-O(1) and Cu-O(2) distances average 1.93 Å with a range of 1.91 to 1.95 Å. The 12 distances from five-coordinate Cu(II) ions to bridging OH ions average 2.20 Å while the six distances from the four-coordinate Cu(II) ions average 1.89 Å. The Cu-O(H)-Cu angle involving five-coordinate Cu(II) ions average

123° while the average value of the angle involving five-coordinate and a four-coordinate Cu(II) ion is 96°.

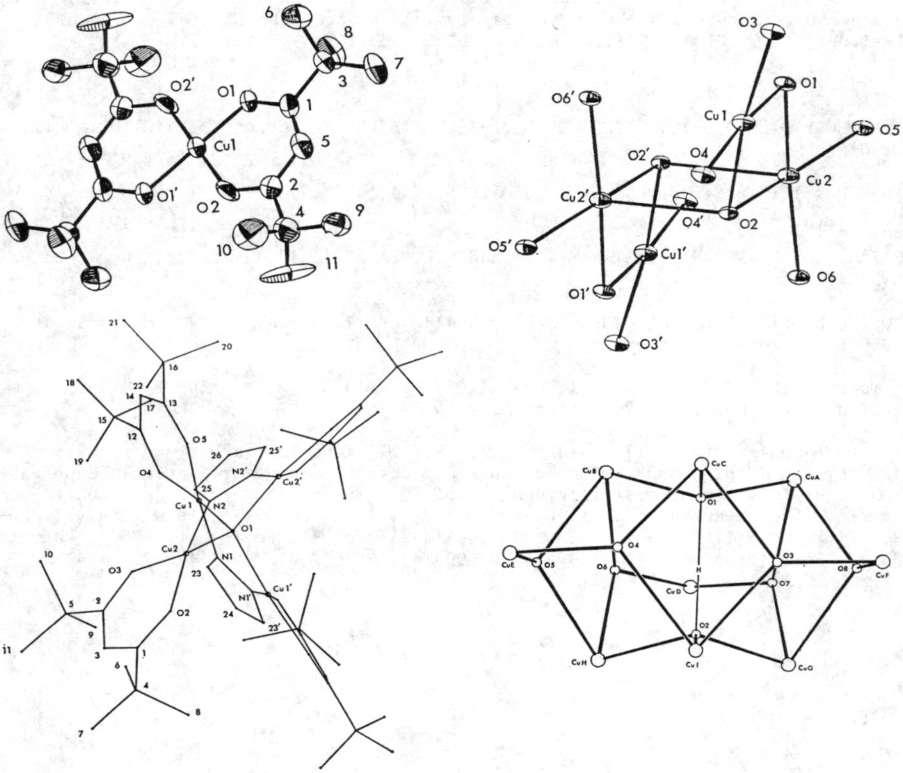

Fig. 1. Molecular structure of $C_{22}H_{38}CuO_4$ (top left), central core of the step dimer $C_{48}H_{88}Cu_4O_{12}$ (top right), molecular structure of $C_{50}H_{82}Cu_4N_4O_9$ (lower left), and the core of $C_{99}H_{178}Cu_9O_{26}$ (lower right).

BIS(BIGUANIDE)COPPER(II) BIS[BIS(ACETYLACETONATO)CHLOROCOPPER(II)]

$C_{24}H_{42}Cl_2Cu_3N_{10}O_8$ $[Cu(Hbg)_2].[Cu(acac)_2Cl]_2$

S.K. HOTA, C.R. SAHA and H. PRITZKOW, 1984. J. Coord. Chem., 13, 131-142.

Orthorhombic, Pbca, a = 15.218, b = 15.986, c = 14.565 Å, Z = 4. Mo radiation, R = 0.052 for 4267 reflexions.

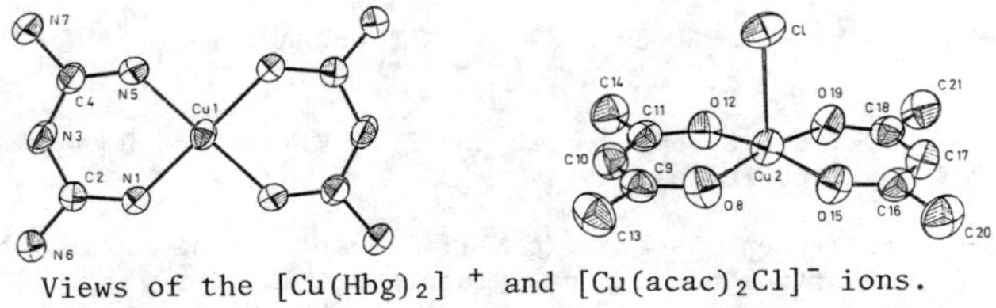

Fig. 1. Views of the $[Cu(Hbg)_2]^+$ and $[Cu(acac)_2Cl]^-$ ions.

The structure consists of square-planar $[Cu(Hbg)_2]^{2+}$ cations (with the Cu atom on an inversion centre) and square pyramidal $[Cu(acac)_2Cl]^-$ anions (Fig. 1). In the cation the Cu-N distances are 1.933 and 1.939(3) Å with chelate N-Cu-N 88.2(1)°. The Cu of the anion lies 0.194 Å from the basal plane towards the Cl atom, with Cu-O 1.918-1.933(2), Cu-Cl 2.635(1) Å, Cl-Cu-O 95.2-96.9(1)°.

BIS(μ-ETHOXY)DICHLORO[1,4-PIPERAZINEDIYLBIS(N-ETHYLENESALICYLALDIMINATO)]DIIRON(III)
$C_{26}H_{36}Cl_2Fe_2N_4O_4$ (I) $Fe_2(C_{22}H_{26}N_4O_2)(OC_2H_5)_2Cl_2$

BIS(μ-ETHOXY)BIS[BIS(2,4-PENTANEDIONATO)IRON(III)]
$C_{24}H_{38}Fe_2O_{10}$ (II) $Fe_2(C_5H_7O_2)_4(OC_2H_5)_2$

B. CHIARI, O. PIOVESANA, T. TARANTELLI and P.F. ZANAZZI, 1984. Inorg. Chem., 23, 3398-3404.

I. Orthorhombic, Pccn, a = 10.840, b = 13.414, c = 19.850 Å, Z = 4. Mo radiation, R = 0.035 for 1066 reflexions.

II. Monoclinic, $P2_1/n$, a = 9.761, b = 14.627, c = 10.485 Å, β = 105.83°, Z = 2. Mo radiation, R = 0.052 for 1526 reflexions.

Both complexes (Fig. 1) consist of two Fe(III) ions that are symmetrically bridged by two OC_2H_5 groups. The Fe atoms are octahedrally coordinated in each case. Molecules of I lie on a crystallographic twofold axis and those of II are centrosymmetric. Bond lengths: for I; Fe-(μ-O) 1.977(3) and 2.005(3), Fe-O 1.897(3), Fe-Cl 2.357(1), Fe-N(1) 2.136(4), Fe-N(2) 2.401(4) Å; for II; Fe-(μ-O) 1.979(4) and 1.985(4), Fe-O (trans to μ-O) 1.996(4), 1.998(4), Fe-O(2,4) 2.023(4), 2.027(4) Å.

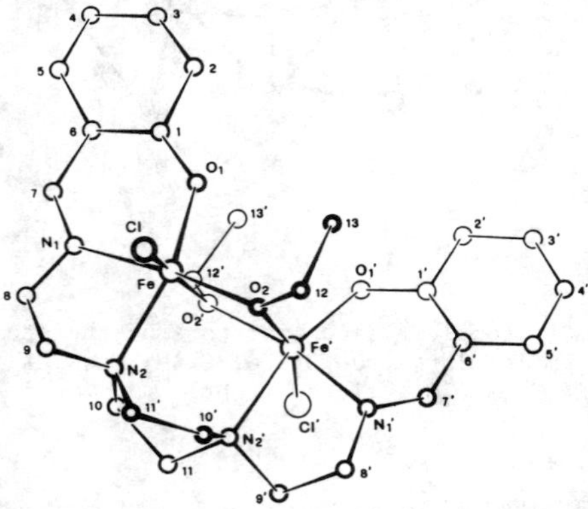

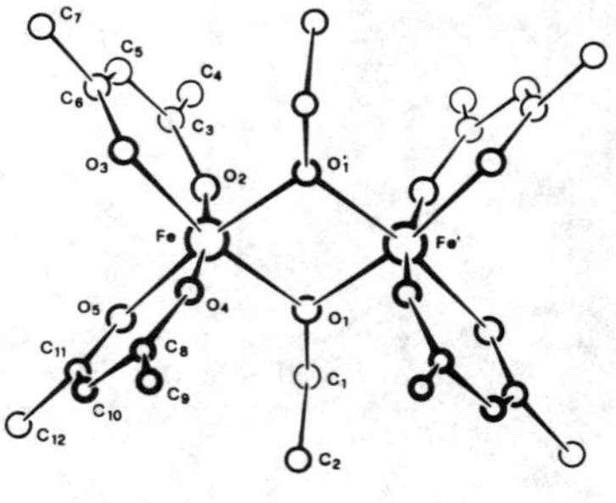

Fig. 1. The molecules $Fe_2(C_{22}H_{26}N_4O_2)(OC_2H_5)_2Cl_2$ (left) and $Fe_2(C_5H_7O_2)_4(OC_2H_5)_2$ (right).

BIS(BIS(ACETYLACETONATO)ISOPROPANOLNICKEL(II))
$C_{26}H_{44}Ni_2O_{10}$ $[Ni(OC(CH_3)CHC(CH_3)O)_2(i\text{-}C_3H_7OH)]_2$

C.J. O'CONNOR, E.D. STEVENS, C.E. PFLUGER and K.A. KLANDERMAN, 1984. Inorg. Chim. Acta, 81, 91-97.

Monoclinic, $P2_1/c$, a = 9.402, b = 8.554, c = 20.252 Å, β = 109.79°, D_m = 1.36, Z = 2. Mo radiation, R = 0.029 for 2160 reflexions.

The structure consists of centrosymmetric Ni(II) complexes dimerized through a bridge formed by one O atom of one of the acetylacetonate ligands (Fig. 1). The other acetylacetonate ligand acts as a simple chelate to the Ni ion and the sixth coordination site is occupied by the solvent molecule. The resulting geometry about the Ni ion is a distorted octahedron. The Ni...Ni' distance is 3.203 Å. Other distances are: Ni-O(1) 2.040, Ni-O(1') 2.135, Ni-O(2) 2.003, Ni-O(3) 2.014, Ni-O(4) 1.985 and Ni-O(5) 2.137 Å. The O(1)-Ni-O(1') angle is 79.85° and that of Ni-O(1)-Ni' is 100.15°. There is an intramolecular hydrogen bond between O(5) and O(3') (O...O 2.824 Å.

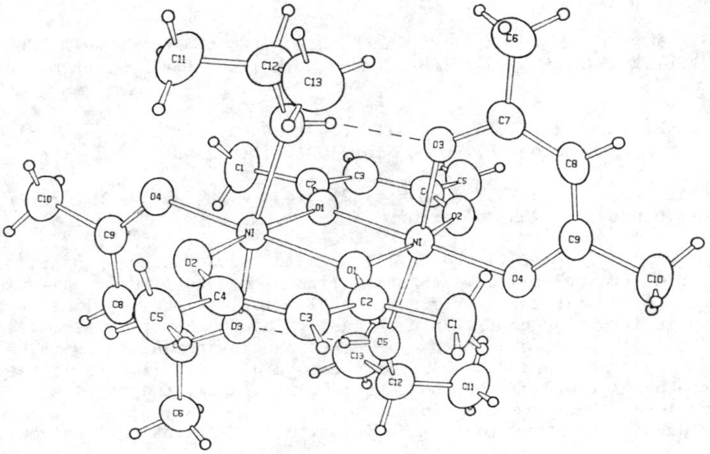

Fig. 1. Molecular structure of $C_{26}H_{44}Ni_2O_{10}$.

BIS(HEXAFLUOROACETYLACETONATO)PALLADIUM(II) 3-CYANO-2,2,5,5-TETRAMETHYL-2,5-DIHYDRO-PYRROLYL-1-OXYL (1:2 ADDUCT)

$C_{28}H_{28}F_{12}N_4O_6Pd$ $C_{10}H_2F_{12}O_4Pd,2C_9H_{13}N_2O$

L.C. PORTER and R.J. DOEDENS, 1984. Acta Cryst., C40, 1814-1817.

Monoclinic, $P2_1/n$, a = 8.932, b = 19.139, c = 10.928 Å, β = 108.42°, Z = 2. Mo radiation, R = 0.045 for 2498 reflexions.

Neutral, co-crystallised species are arranged in alternating layers perpendicular to b with only van der Waals contacts between layers (Fig. 1). The nitroxyl O atom is 4.016 Å from Pd. In the centrosymmetric $Pd(F_6acac)_2$ molecule the OCCCO group is planar within 0.002 Å with Pd displaced 0.018 Å from the plane. In the tmpCN molecule, the five-membered ring, cyano group and nitroxyl O atom are all coplanar. Pd-O bond lengths are 1.968(3) and 1.975(3) Å.

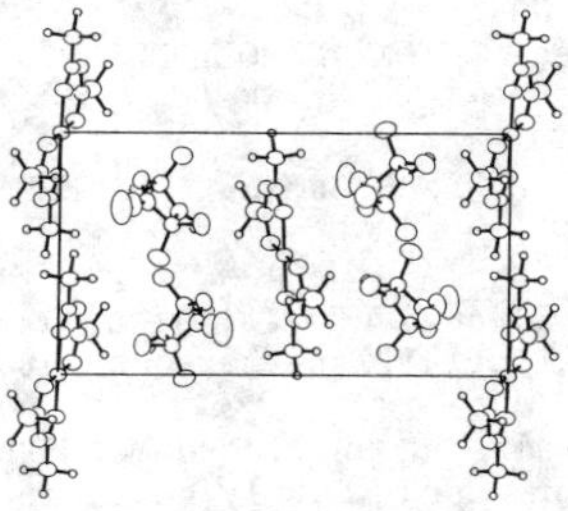

Fig. 1. The contents of one unit cell of $C_{10}H_2F_{12}O_4Pd,2C_9H_{13}N_2O$ down a with the b axis horizontal.

BIS(1,4-DIOXANE-O)BIS(p-METHOXYBENZOYL)TRIFLUOROACETONATO-O,O')COPPER(II) 1,4-DIOXANE SOLVATE
$C_{30}H_{32}CuF_6O_{10}, 2(C_4H_8O_2)$ (I)

catena-(μ_2-1,4-DIOXANE-O,O')BIS[(p-NITROBENZOYL)TRIFLUOROACETONATO-O,O']COPPER(II)
$(C_{24}H_{18}CuF_6N_2O_{10})_n$ (II)

C.J. EVENHUIS, M.A. HITCHMAN, R.G. McDONALD, D.M.L. GOODGAME, E. KWIATKOWSKI, U. DETTLAFF-WEGLIKOWSKA, C. RAKAWATCHAI and A.H. WHITE, 1984. J. Chem. Soc. Dalton, 943-952.

I. Triclinic, $P\bar{1}$, a = 12.635, b = 11.513, c = 8.094 Å, α = 70.57, β = 81.07, γ = 70.47°, Z = 1. Mo radiation, R = 0.043 for 1931 reflexions.

II. Monoclinic, $P2_1/n$, a = 15.753, b = 11.262, c = 7.587 Å, β = 91.72°, Z = 2. Mo radiation, R = 0.048 for 1405 reflexions.

In both compounds (Fig. 1) the Cu(II) atom is pseudo-square planar, coordinated by a pair of O,O'-chelating acetylacetonato ligands each unsymmetrically substituted at the β-carbon by aromatic and trifluoromethyl moieties. Cu(II) lies on a crystallographic inversion centre in each case. The CuO_4 unit is essentially planar (Cu-O 1.927(2)-1.940(4) Å) and the C_3O_2 β-diketonate system planar with Cu 0.033 Å from the mean plane in I and 0.254 Å from the mean plane in II. In I, tetragonally distorted octahedral coordination about Cu is completed by two trans monodentate dioxane molecules. There are two dioxane solvent molecules. In II, trans dioxane molecules bridge the Cu atoms to form an infinite, one-dimensional polymer.

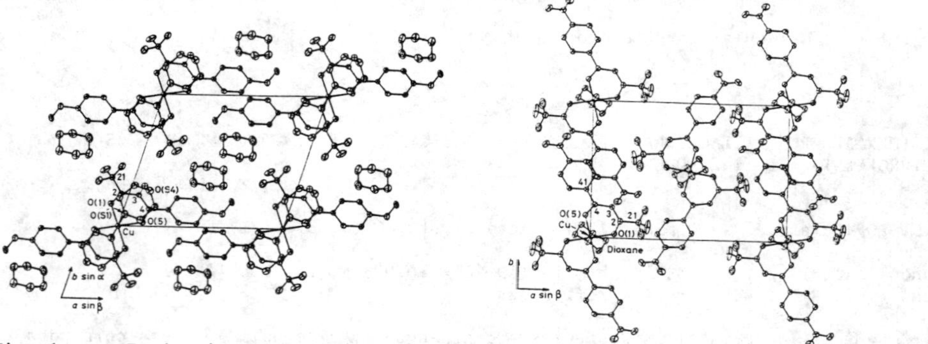

Fig. 1. Projections viewed down the c-axis: $C_{30}H_{32}CuF_6O_{10}, 2(C_4H_8O_2)$ (left) and $(C_{24}H_{18}CuF_6N_2O_{10})_n$ (right).

BIS(ACETYLACETONATO)TETRAKIS(neo-PENTOXY)DIMOLYBDENUM
$C_{30}H_{58}Mo_2O_8$ $Mo_2(OCH_2C(CH_3)_3)_4(C_5H_7O_2)_2$

M.H. CHISHOLM, K. FOLTING, J.C. HUFFMAN and A.L. RATERMANN, 1984. Inorg. Chem., 23, 613-618.

Triclinic, $P\bar{1}$, a = 14.008, b = 11.462, c = 11.497 Å, α = 87.84, β = 92.94, γ = 88.82°, Z = 2. Mo radiation, R = 0.032 for 4177 reflexions (at -143°C).

The molecule (Fig. 1) contains an unbridged Mo≡Mo bond of 2.237(1) Å. Each Mo atom is also coordinated to four approximately coplanar O atoms (Mo-O(acac) 2.064(3)-2.117(3), Mo-O(OR) 1.888(3)-1.919(3) Å) and the Mo_2O_8 moiety is staggered.

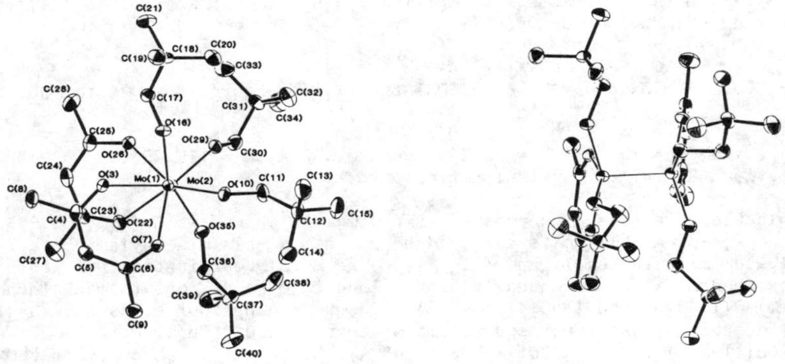

Fig. 1. Two views of $Mo_2(OCH_2C(CH_3)_3)_4(C_5H_7O_2)_2$.

BIS(1,1,1,5,5,5-HEXAFLUORO-2,4-PENTANEDIONATO)(TRI-o-TOLYLPHOSPHINE)PALLADIUM(II)
$C_{31}H_{23}F_{12}O_4PPd$ (I)

BIS(1,1,1,5,5,5-HEXAFLUORO-2,4-PENTANEDIONATO)(TRICYCLOHEXYLPHOSPHINE)PLATINUM(II)
$C_{28}H_{35}F_{12}O_4PPt$ (II)

S. OKEYA, T. MIYAMOTO, S. OOI, Y. NAKAMURA and S. KAWAGUCHI, 1984. Bull. Chem. Soc. Japan, 57, 395-404.

I. Monoclinic, $P2_1$, a = 13.998, b = 11.430, c = 10.897 Å, β = 108.48°, D_m = 1.66, Z = 2. Mo radiation, R = 0.039 for 3148 reflexions.

II. Orthorhombic, Pcab, a = 25.652, b = 20.397, c = 12.781 Å, D_m = 1.75, Z = 8. Mo radiation, R = 0.045 for 3270 reflexions.

Both complexes have highly distorted square-pyramidal structures, one of the two Hfac anions in either of them spanning the apical and basal coordination sites (Fig. 1). The M-O(11) bond intersects the basal coordination plane defined by O(12), O(21), O(22) and P at 65.3 and 66.9° respectively in I and II. The main bond lengths in I are: Pd-P 2.253, Pd-O(11) 2.796, Pd-O(12) 2.016, Pd-O(21) 2.086 and Pd-O(22) 2.000 Å. The corresponding Pt-ligand distances in II are: 2.232, 2.773, 1.988, 2.099 and 1.990 Å. The angles O(11)-Pd-O(12) and O(11)-Pt-O(12) in I and II respectively are 76.7 and 84.2°.

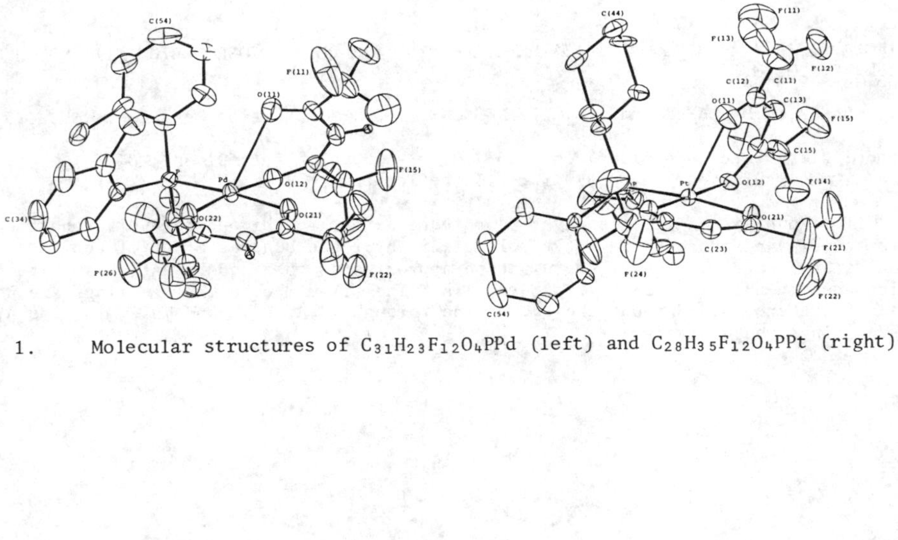

Fig. 1. Molecular structures of $C_{31}H_{23}F_{12}O_4PPd$ (left) and $C_{28}H_{35}F_{12}O_4PPt$ (right).

(μ_2-N,N'-BIS(DIMETHYLAMINOETHYL-N,N'-DIMETHYL)-1,2-DIAMINOETHANE)BIS(BIS(1,1,1,5,5,5-HEXAFLUORO-2,4-PENTANEDIONATO)COPPER(II))
$C_{32}H_{34}Cu_2F_{24}N_4O_8$

N.A. BAILEY, D.E. FENTON and M.S. LEAL GONZALEZ, 1984. Inorg. Chim. Acta, 88, 125-134.

Monoclinic, $P2_1/a$, a = 10.268, b = 25.255, c = 10.392 Å, β = 119.29°, D_m = 1.72, Z = 2. Mo radiation, R = 0.121 for 1261 reflexions.

The molecule comprises a centrosymmetric dimer in which two Cu(1,1,1,5,5,5-hexafluoropentane-2,4-dionate)$_2$ units are bridged by an N,N'-bis(dimethylaminoethyl-N,N'-dimethyl)-1,2-diaminoethane) ligand (Fig. 1). The metal coordination polyhedron is distorted octahedral with two mutually trans, long Cu-O bonds, one to each (hfpd) ligand (Cu(1)-O(1) 2.32, Cu(1)-O(4) 2.36 Å). The remaining four bonds show a slight tetrahedral distortion away from square planar coordination (Cu(1)-O(2) 2.00, Cu(1)-O(3) 1.98, Cu(1)-N(1) 2.05 and Cu(1)-N(2) 2.08 Å). The limited accuracy of the structural determination precludes detailed discussions of molecular geometry but bond lengths and bond angles within the ligands appear normal.

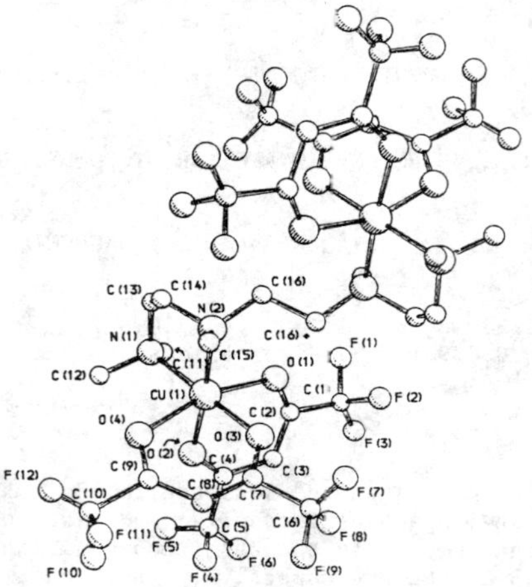

Fig. 1. Molecular structure of $C_{32}H_{34}Cu_2F_{24}N_4O_8$. Atom C(16)* is centrosymmetrically related to C(16) through the crystallographic inversion centre at (0,0,0).

DIAQUATRIS[4,4,4-TRIFLUORO-1-(3'-METHYLPHENYL)-1,3-BUTANEDIONE]ERBIUM(III) MONOHYDRATE
$C_{33}H_{28}ErF_9O_8,H_2O$

A.T. BAKER, A.M. HAMER and S.E. LIVINGSTONE, 1984. Transition Met. Chem., 9, 423-432.

Monoclinic, $P2_1/c$, a = 11.641, b = 16.479, c = 18.967 Å, β = 91.80°, D_m = 1.654, Z = 4. Mo radiation, R = 0.059 for 4136 reflexions.

The complex (Fig. 1) is eight-coordinate with three bidentate β-diketonato and two aquo ligands; the third water molecule is hydrogen bonded to one of the coordinated waters. The Er coordination approximates to a square antiprism with Er-O distances in the range 2.286(7)-2.388(7) Å. The six-membered chelate rings are not planar but are folded about the O...O line through 10.4, 27.7 and 19.2° for the A, B and C rings respectively.

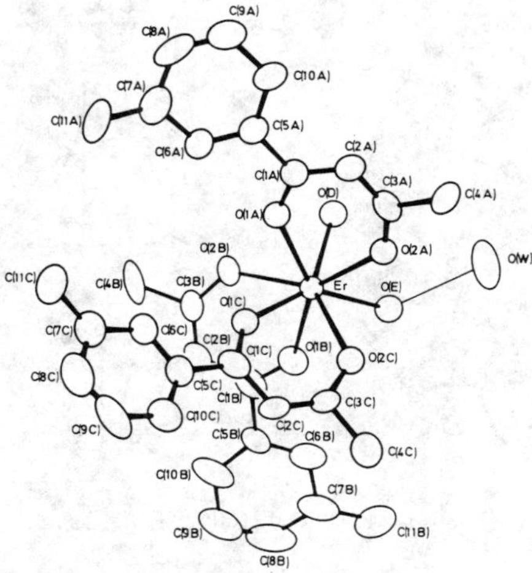

Fig. 1. A view of $C_{33}H_{28}ErF_9O_8,H_2O$.

BIS[μ-(3,3'-[1,3-PHENYLENEBIS(METHYLENE)]BIS(2,4-PENTANEDIONATO))]DICOPPER(II)
$C_{36}H_{40}Cu_2O_8$

A.W. MAVERICK and F.E. KLAVETTER, 1984. Inorg. Chem., 23, 4129-4130.

Monoclinic, $P2_1/c$, a = 12.002, b = 7.434, c = 19.431 Å, β = 99.47°, Z = 2. Mo radiation, R = 0.044 for 1262 reflexions.

The centrosymmetric binuclear molecule (Fig. 1) has approximate D_{2h} symmetry. The Cu...Cu distance is 4.908(2) Å, other geometry is normal. The "$Cu(acac)_2$" moieties are essentially planar and make an angle of 89.8° with the bridging aromatic rings.

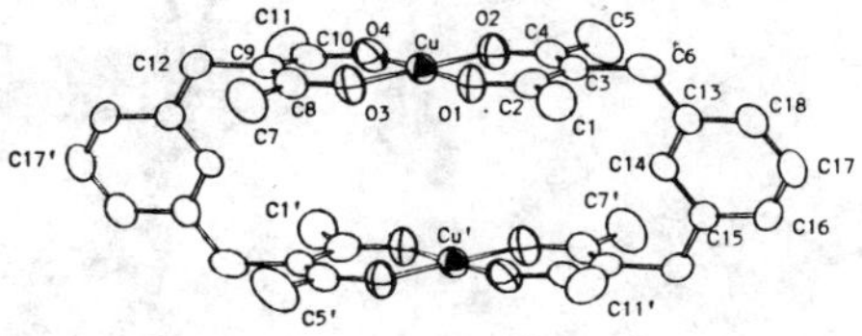

Fig. 1. The $C_{36}H_{40}Cu_2O_8$ molecule.

BIS(μ-NEOPENTOXY)BIS[(2,4-PENTANEDIONATO)BIS(NEOPENTOXY)(t-BUTYL ISOCYANIDE)-MOLYBDENUM](Mo-Mo)
$C_{40}H_{74}Mo_2N_2O_8$ $Mo_2(OC_5H_{11})_4(C_5H_7O_2)_2(CNC_4H_9)_2$

M.H. CHISHOLM, J.F. CORNING, K. FOLTING, J.C. HUFFMAN, A.L. RATERMANN, I.P. ROTHWELL and W.E. STREIB, 1984. Inorg. Chem., 23, 1037-1042.

Monoclinic, $P2_1/n$, a = 13.817, b = 12.518, c = 14.045 Å, β = 92.61°, Z = 2. Mo radiation, R = 0.066 for 2273 reflexions (at -163°C).

The molecule (Fig. 1) is an edge-shared bioctahedral dimer with a Mo-Mo bond of 2.508(2) Å, possessing a centre of symmetry. The trans CN-t-Bu and OCH_2-t-Bu ligands are disordered. Other bond lengths include: Mo-(μ-OR) 2.042(6) and 2.054(6), Mo-O (ketonate) 2.105(6) and 2.098(7), Mo-O/C 1.998(8) and 2.020(8) Å.

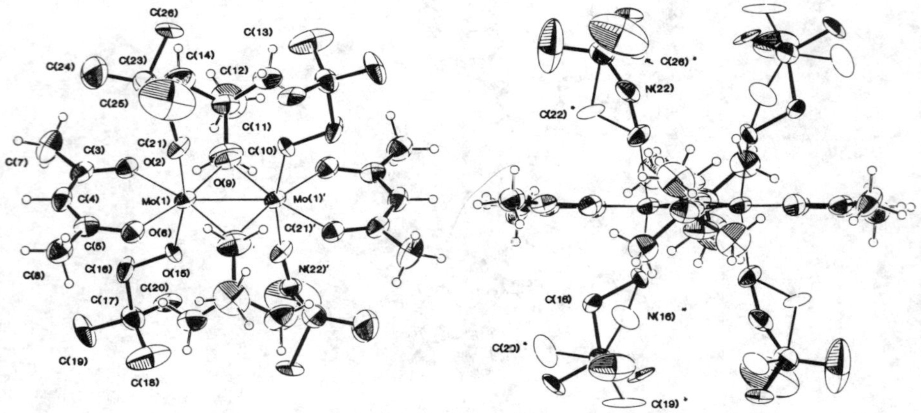

Fig. 1. Two views of $Mo_2(OC_5H_{11})_4(C_5H_7O_2)_2(CNC_4H_9)_2$.

BIS(μ-METHOXO)BIS[BIS(2,2,6,6-TETRAMETHYL-3,5-HEPTANEDIONATO)CHROMIUM(III)]

$C_{46}H_{82}Cr_2O_{10}$ $[(C_{11}H_{19}O_2)_2Cr(OCH_3)]_2$

H.R. FISCHER, D.J. HODGSON and E. PEDERSEN, 1984. Inorg. Chem., 23, 4755-4758.

Monoclinic, $P2_1/n$, a = 15.508, b = 10.862, c = 16.629 Å, β = 111.86°, D_m = 1.25, Z = 2. Mo radiation, R = 0.057 for 1838 reflexions.

The dimeric molecule (Fig. 1) is centrosymmetric with roughly octahedral coordination around the Cr(III) centres. Bond lengths: Cr-(μ-O) 1.950(3) and 1.958(3), Cr-O 1.945(3)-1.975(3) Å. The bonds trans to the μ-OMe groups are significantly shorter than the other bonds to the $C_{11}H_{19}O_2$ ligands.

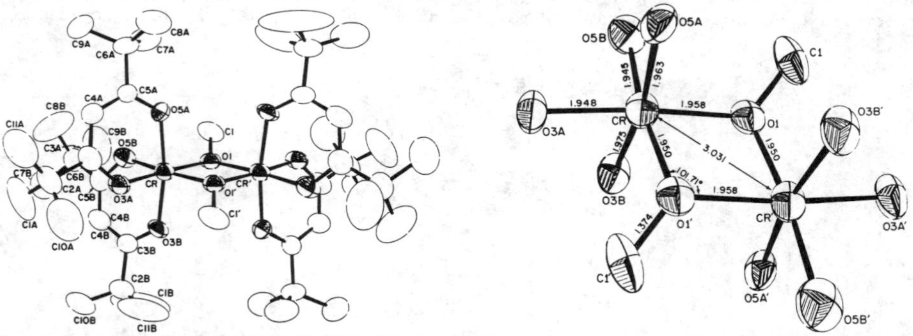

Fig. 1. The $[(C_{11}H_{19}O_2)_2Cr(OCH_3)]_2$ molecule; coordination about Cr shown at right.

DICHLOROBIS{μ_2-[o-DIPHENYLPHOSPHINO)BENZOYL]PINACOLONE-O,O'-Cu,P,Pt}COPPERPLATINUM MONOHYDRATE DICHLOROMETHANE SOLVATE

$C_{50}H_{48}Cl_2CuO_4P_2Pt,H_2O,2(CH_2Cl_2)$

D.A. WROBLESKI, C.S. DAY, B.A. GOODMAN and T.B. RAUCHFUSS, 1984. J. Am. Chem. Soc., 106, 5464-5472.

Triclinic, $P\bar{1}$, a = 11.914, b = 11.646, c = 21.125 Å, α = 102.34, β = 103.71, γ = 99.83°, Z = 2. Mo radiation, R = 0.053 for 8881 reflexions.

The molecule (Fig. 1) contains square-planar $PtCl_2P_2$ and Cu(β-diketonate)$_2$ subunits with a non-bonding Pt...Cu separation of 3.966(1), Cu-O 1.879-1.904(5), Pt-Cl 2.358 and 2.370(2), Pt-P 2.274 and 2.291(2) Å.

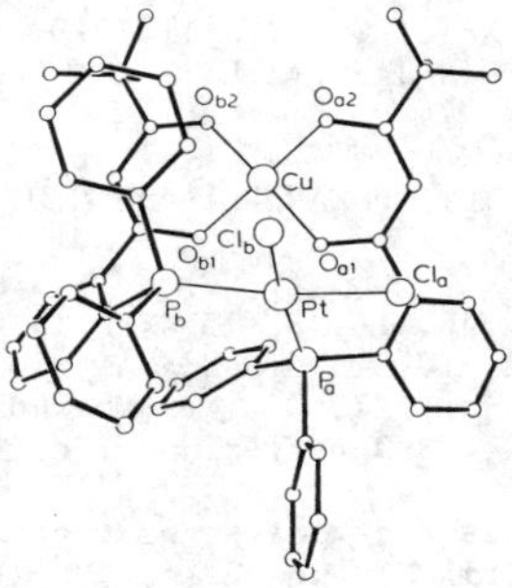

Fig. 1. A diagram of $C_{50}H_{48}Cl_2CuO_4P_2Pt$.

(1,2-BIS(DIPHENYLPHOSPHINO)ETHANE)(2,4-PENTANEDIONATO)NICKEL(II) TETRAPHENYLBORATE
$C_{55}H_{51}BNiO_2P_2$

M. BASATO, B. CORAIN, G. FAVERO and G. VALLE, 1984. Z. Kristallogr., 167, 153-157.

Monoclinic, $P2_1/n$, a = 17.290, b = 19.888, c = 13.734 Å, β = 99.3°, Z = 4. Mo radiation, R = 0.089 for 5001 reflexions.

The molecule is shown in Fig. 1. The Ni atom has slightly distorted square-planar coordination.

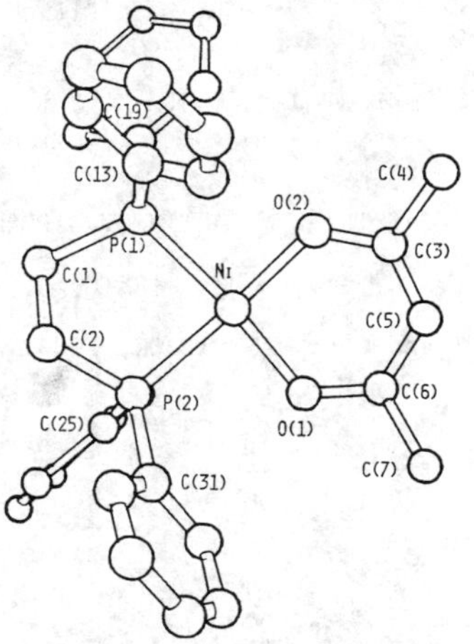

Fig. 1. A view of Ni(acac)(PPh_2-$(CH_2)_2$-PPh_2).

BIS[μ3-1,7-DIPHENYL-1,3,5,7-HEPTANETETRONATO(3-)]BIS[DIOXOPYRIDINEURANIUM(VI)]-DIPYRIDINENICKEL(II) BIS(PYRIDINE) SOLVATE
$C_{58}H_{46}N_4NiO_{12}U_2,2C_5H_5N$ (I)

BIS[μ3-1,7-DIPHENYL-1,3,5,7-HEPTANETETRONATO(3-)]BIS[DIOXOPYRIDINEURANIUM(VI)]-DIPYRIDINECOBALT(II) BIS(PYRIDINE) SOLVATE
$C_{58}H_{46}CoN_4O_{12}U_2,2C_5H_5N$ (II)

BIS[μ_3-1,7-DIPHENYL-1,3,5,7-HEPTANETETRONATO(3-)]BIS[DIOXOPYRIDINEURANIUM(VI)]-DIPYRIDINEIRON(II) BIS(PYRIDINE) SOLVATE
$C_{58}H_{46}FeN_4O_{12}U_2,2C_5H_5N$ (III)

BIS[μ_3-1,7-DIPHENYL-1,3,5,7-HEPTANETETRONATO(3-)]BIS[DIOXOPYRIDINEURANIUM(VI)]-DIPYRIDINEMANGANESE(II) BIS(PYRIDINE) SOLVATE
$C_{58}H_{46}MnN_4O_{12}U_2,2C_5H_5N$ (IV)

R.L. LINTVEDT, B.A. SCHOENFELNER, C. CECCARELLI and M.D. GLICK, 1984. Inorg. Chem., 23, 2867-2874.

I-IV. Monoclinic, $P2_1/c$, a = 13.346, 13.45, 13.475, 13.473, b = 24.011, 23.93, 23.778, 23.507, c = 10.556, 10.62, 10.718, 10.834 Å, β = 108.12, 108.3, 108.85, 109.08°, D_m = 1.733, 1.729, 1.724, 1.727, Z = 2. Mo radiation, R = 0.031, 0.039, 0.035, 0.040 for 2961, 2568, 2644, 2899 reflexions (for I-IV, respectively).

Complexes I-IV are isomorphous and consist of centrosymmetric trinuclear moieties (Fig. 1) with an octahedral Ni, Co, Fe, or Mn atom in the central position and pentagonal bipyramidal U atoms in the terminal coordination positions. Bond lengths: axial U-O(oxo) 1.734(8)-1.769(7), equatorial U-O (bridging) 2.453(5)-2.476(7), equatorial U-O (terminal) 2.247(6)-2.265(7), equatorial U-N 2.572(14)-2.584(9), Ni-O 2.032(5) and 2.029(5), Ni-N 2.175(6), Co-O 2.040(9) and 2.046(8), Co-N 2.234(10), Fe-O 2.053(7) and 2.056(6), Fe-N 2.303(8), Mn-O 2.086(6) and 2.099(6), Mn-N 2.377(9) Å.

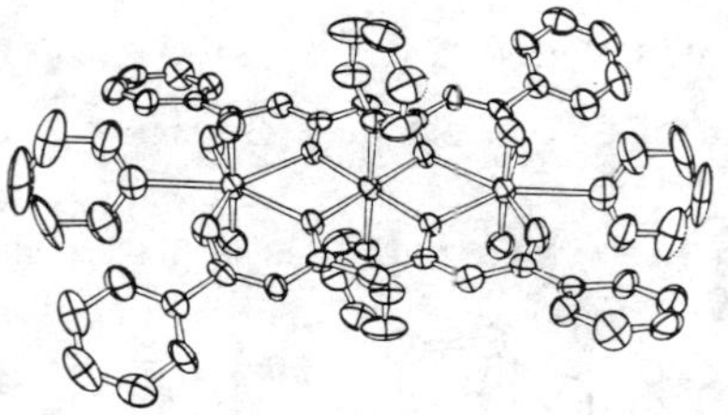

Fig. 1. The structure of $C_{58}H_{46}MN_4O_{12}U_2$ (M = Ni, Co, Fe, Mn).

1,3-DIPHENYL-1,3-PROPANDIONATOBIS(TRIPHENYLANTIMONY)DIPHENYLRHODIUM(III) DIBENZENE SOLVATE
$C_{63}H_{51}O_2RhSb_2,2C_6H_6$ $[Rh(C_{15}H_{11}O_2)(C_6H_5)_2(Sb(C_6H_5)_3)_2],2C_6H_6$

G.J. LAMPRECHT, J.G. LEIPOLDT and C.P. VAN BILJON, 1984. Inorg. Chim. Acta, 88, 55-58.

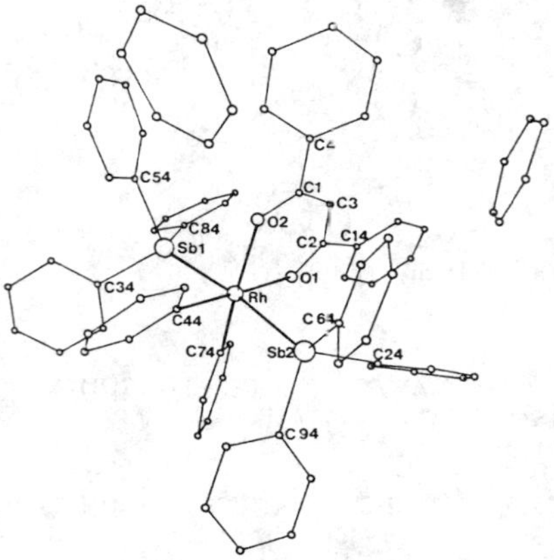

Fig. 1. Molecular configuration for $C_{63}H_{51}O_2RhSb_2,2C_6H_6$.

Triclinic, $P\bar{1}$, a = 19.083, b = 13.167, c = 13.536 Å, α = 81.81, β = 111.59, γ = 100.49°, Z = 2. Mo radiation, R = 0.079 for 6637 reflexions.

The Rh atom has a slightly-distorted octahedral coordination polyhedron (Fig. 1) in which the metal is coordinated by two phenyl groups, two O atoms of a chelate ring which are cis to each other, and two Sb atoms of the two $SbPh_3$ ligands which are in trans positions. The Rh-Sb(1) and Rh-Sb(2) distances are 2.551 and 2.588 Å respectively, the Rh-O distances 2.13 and 2.16 Å, and the Rh-C(phenyl) distances 2.04 and 2.06 Å. The chelate ring is planar within experimental error and the mean Sb-C distance is 2.145 Å.

(DIETHYL ETHER)BIS[3-(1-PHENYL-2-BENZOYLVINYL)DIBENZOYLMETHANATO-O,O']COPPER(II)
$C_{64}H_{52}CuO_7$

D.S. YUFIT, Yu.T. STRUCHKOV, N.A. DOLGOPOLOVA and L.Yu. UKHIN, 1983. Koord. Khim., 9, 1121-1125 [Sov. J. Coord. Chem., 9, 634-639].

Monoclinic, C2/c, a = 20.673, b = 8.8649, c = 29.519 Å, β = 111.74°, Z = 4. Cu radiation, R = 0.079 for 3231 reflexions.

The Cu-O(4) bond in the molecule coincides with the crystallographic twofold axis (Fig. 1). The Cu atom is coordinated to four O atoms of two bidentate diketone ligands (Cu-O(1) 1.904(3), Cu-O(2) 1.910(3) Å) as well as to the O atom of the diethyl ether moiety (Cu-O(4) 2.502(7) Å). The overall geometry of the coordination polyhedron is tetragonal pyramidal. The mean C-O and C-C distances in the chelate rings are 1.314 and 1.399 Å respectively. Intermolecular contacts are van der Waals in nature. There is considerable steric hindrance in the complex molecule primarily between neighbouring phenyl rings on the ligand.

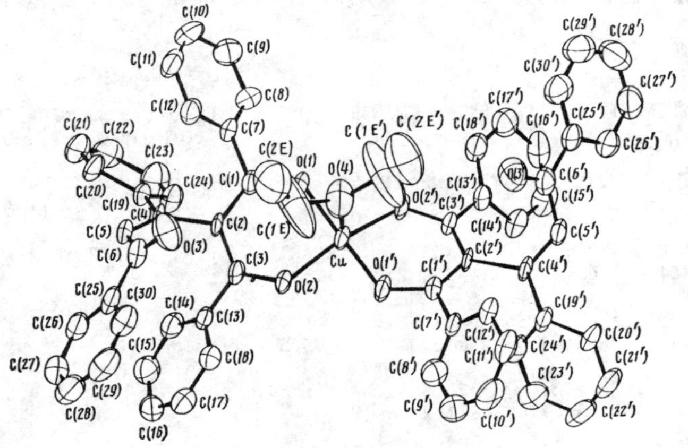

Fig. 1. $C_{64}H_{52}CuO_7$: the molecular geometry.

BIS[μ-(o-(DIPHENYLPHOSPHINO)BENZOYL)PINACOLONE-O,O',P][μ-o-(DIPHENYLPHOSPHINE OXIDE)-BENZOATO]DICOPPER METHANOL SOLVATE
$C_{69}H_{62}Cu_2O_7P_3,4CH_4O$ (I)

BIS(μ-METHOXO)BIS[(o-(DIPHENYLPHOSPHINE OXIDE)BENZOYL)PINACOLONE-O,O']BIS(METHOXO)-DICOPPER(II)
$C_{54}H_{62}Cu_2O_{10}P_2$ (II)

D.A. WROBLESKI, T.B. RAUCHFUSS, A.L. RHEINGOLD and K.A. LEWIS, 1984. Inorg. Chem., 23, 3124-3129.

I. Triclinic, $P\bar{1}$, a = 13.592, b = 14.913, c = 18.501 Å, α = 85.78, β = 86.19, γ = 72.61°, Z = 2. Mo radiation, R = 0.098 for 6295 reflexions.

II. Monoclinic, $P2_1/n$, a = 9.750, b = 16.399, c = 16.698 Å, β = 97.25°, Z = 2. Mo radiation, R = 0.105 for 2304 reflexions.

The two molecules are shown in Fig. 1. I consists of a trigonal planar Cu(I) atom (Cu-P 2.229(3) and 2.233(3), Cu-O 2.102(6) Å) linked to a square pyramidal Cu(II) (Cu-O(basal) 1.897(7)-1.961(7), Cu-O(apical) 2.358(7) Å). The most unusual feature in I is the carboxylate ligand which bridges the two Cu atoms via a single O atom. Compound II is a centrosymmetric dimer with two bridging OMe ligands (Cu-O 1.925(9) and 1.922(9) Å). The basal plane of each square pyramidal Cu(II) atom is completed by the two O atoms of the pinacolone ligand (Cu-O 1.936(9) and 1.939(8) Å). A terminal OMe ligand occupies the axial site (Cu-O 2.418(10) Å).

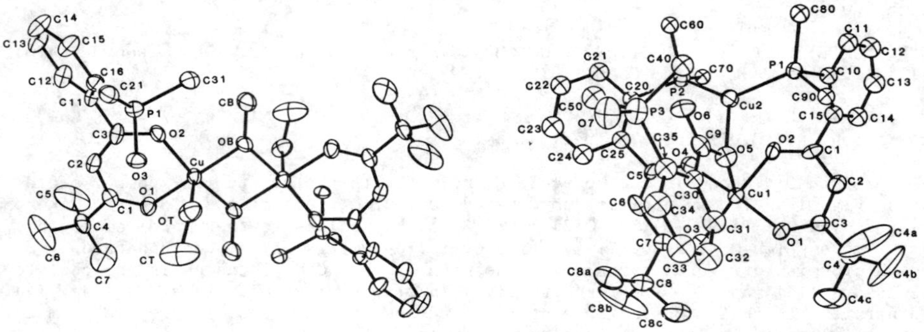

Fig. 1. The structures of $C_{54}H_{62}Cu_2O_{10}P_2$ (left) and $C_{69}H_{62}Cu_2O_7P_3$ (right).

AQUA[N,N'-ETHYLENEBIS(ACETYLACETONE IMINATO)]OXOTECHNETIUM(V) BROMIDE CHLORIDE
$C_{12}H_{20}Br_{0.75}Cl_{0.25}N_2O_4Tc$ (I) $[TcO(H_2O)(C_{12}H_{18}N_2O_2)]Br_{0.75}Cl_{0.25}$

CHLORO[N,N'-ETHYLENEBIS(SALICYLIDENEAMINATO)]OXOTECHNETIUM(V)
$C_{16}H_{14}ClN_2O_3Tc$ (II) $[TcO(C_{16}H_{14}N_2O_2)Cl]$

S. JURISSON, L.F. LINDOY, K.P. DANCEY, M. McPARTLIN, P.A. TASKER, D.K. UPPAL and E. DEUTSCH, 1984. Inorg. Chem., 23, 227-231.

I. Triclinic, $P\bar{1}$, a = 10.935, b = 9.877, c = 7.324 Å, α = 87.27, β = 102.50, γ = 97.85°, Z = 2. Mo radiation, R = 0.022 for 2370 reflexions.

II. Orthorhombic, $P2_12_12_1$, a = 26.653, b = 11.697, c = 10.529 Å, Z = 8. Mo radiation, R = 0.069 for 2105 reflexions.

Complex I was isolated as a mixed Br/Cl salt. The cation in I (Fig. 1) has distorted octahedral coordination geometry, containing a trans-oxoaquatechnetium(V) core. Important bond lengths in I: Tc=O 1.648(2), Tc-OH_2 2.282(2), Tc-O 2.020(2) and 2.012(2), Tc-N 2.001(2) and 2.003(2) Å. The two independent molecules of II (Fig. 1) each have pseudooctahedral coordination geometry and contain trans-oxochlorotechnetium(V) cores. Bond lengths in II: Tc=O 1.626(11), Tc-Cl 2.527(4), Tc-O 1.993(9) and 1.987(8), Tc-N 2.022(11) and 2.067(12) Å.

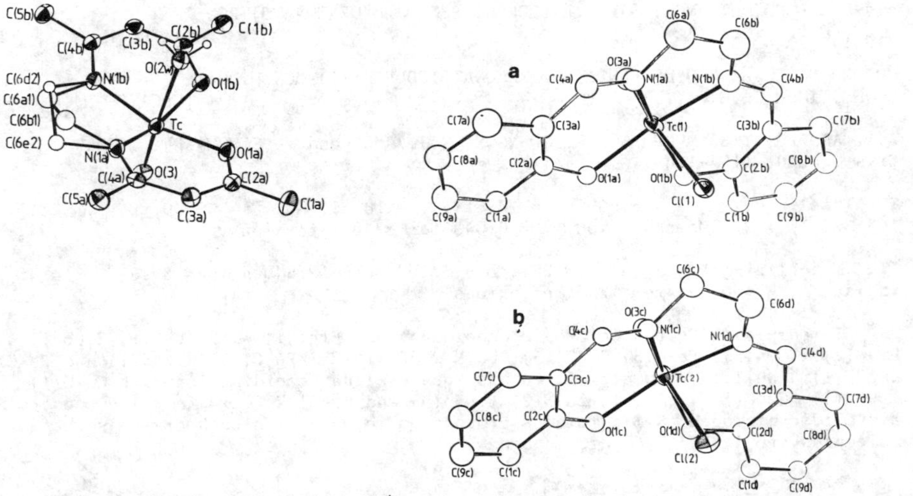

Fig. 1. The $[TcO(H_2O)(C_{12}H_{18}N_2O_2)]^+$ cation (left) and the two independent $[TcO(C_{16}H_{14}N_2O_2)Cl]$ molecules (right).

(N-{2-[2-(2-AMMONIOETHYLAMINO)ETHYLAMINO]ETHYL}SALICYLIDENEAMINATO-O,N,N',N'')-PERCHLORATOCOPPER(II) PERCHLORATE

$C_{13}H_{22}Cl_2CuN_4O_9$ $[Cu(ClO_4)(C_{13}H_{22}N_4O)],ClO_4$

E. ROTONDO, F. CUSMANO PRIOLO, G. BOMBIERI and G. BRUNO, 1984. Acta Cryst., C40, 960-962.

Triclinic, $P\bar{1}$, a = 8.787, b = 9.907, c = 12.393 Å, α = 93.05, β = 109.51, γ = 97.93°, Z = 2. Mo radiation, R = 0.047 for 2479 reflexions.

In the complex cation (Fig. 1) the coordination polyhedron around Cu is an irregular square pyramid with the protonated $[2\text{-}O^-\text{-}C_6H_4CH{=}N(CH_2)_2NH(CH_2)_2NH(CH_2)_2NH_2]$ unit acting as a tetradentate ligand through one O and three N atoms. The terminal N of the ligand is protonated and the apical position of the square pyramid is occupied by an oxygen of the perchlorate ligand. Bond distances are: Cu-N 1.931(5), 1.989(4) and 2.033(5), Cu-O 1.894(3) and Cu-O(perchlorate) 2.634(5) Å.

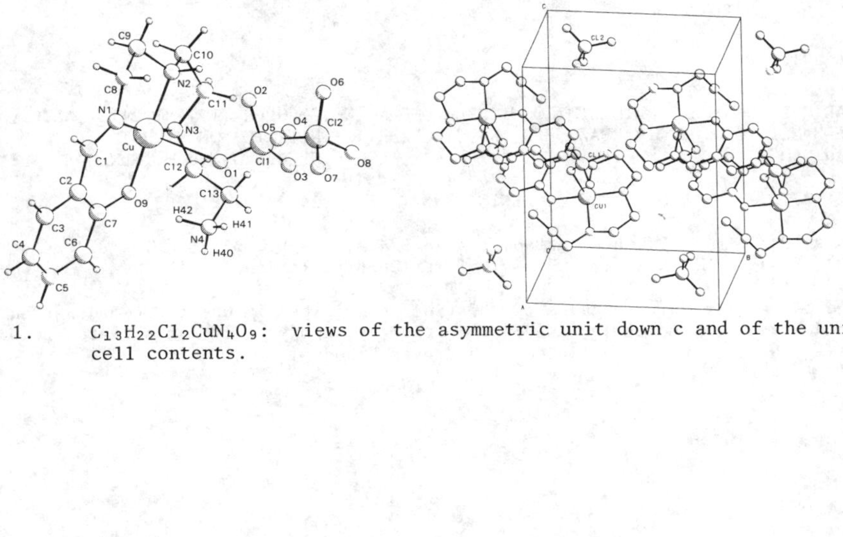

Fig. 1. $C_{13}H_{22}Cl_2CuN_4O_9$: views of the asymmetric unit down c and of the unit-cell contents.

cis-DICHLORO(METHANOL)(SALICYLALDEHYDE BENZOYLHYDRAZONATO)IRON(III)
$C_{15}H_{15}Cl_2FeN_2O_3$ (I) $[FeCl_2(C_{14}H_{11}N_2O_2)(CH_4O)]$

CHLORO(SALICYLALDEHYDE BENZOYLHYDRAZONATO)COPPER(II) MONOHYDRATE
$C_{14}H_{11}ClCuN_2O_2,H_2O$ (II) $[CuCl(C_{14}H_{11}N_2O_2)],H_2O$

A.A. ARUFFO, T.B. MURPHY, D.K. JOHNSON, N.J. ROSE and V. SCHOMAKER, 1984. Acta Cryst., C40, 1164-1169.

I. Triclinic, $P\bar{1}$, a = 6.665, b = 13.818, c = 10.122 Å, α = 108.40, β = 73.23, γ = 103.32°, Z = 2. Mo radiation, R = 0.043 for 3740 reflexions.

II. Monoclinic, $P2_1/a$, a = 16.201, b = 7.107, c = 12.540 Å, β = 89.87°, Z = 4. Mo radiation, R = 0.091 for 1250 reflexions.

In complex I Fe has octahedral coordination (Fig. 1) with Fe-O(1) 2.068(2), Fe-O(2) 1.874(2), Fe-O(3) 2.171(3), Fe-N(2) 2.119(2), Fe-Cl(1) 2.345(1) and Fe-Cl(2) 2.303(1) Å whilst in complex II Cu has square-planar coordination with Cu-O(1) 1.962(8), Cu-O(2) 1.891(8), Cu-N(2) 1.933(9) and Cu-Cl 2.211(4) Å. The uncoordinated hydrazidic nitrogen, N(1), remains protonated in both. In both crystals the molecules are linked by hydrogen bonds.

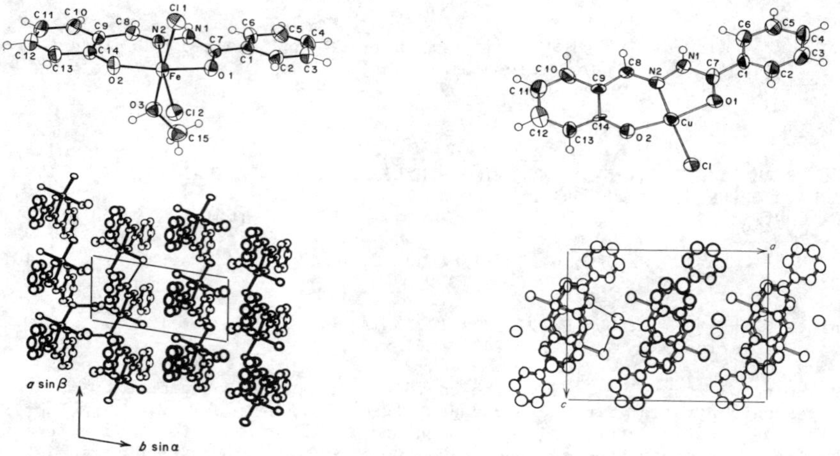

Fig. 1. Perspective views of the complex molecules and molecular packing of $[FeCl_2(C_{14}H_{11}N_2O_2)(CH_4O)]$ (I) (left) and $[CuCl(C_{14}H_{11}N_2O_2)],H_2O$ (II) (right).

AQUA(SALICYLALDEHYDE-S-METHYLTHIOSEMICARBAZONE-N^1,N^3,S)COPPER(II) NITRATO(SALICYLALDEHYDE-S-METHYLTHIOSEMICARBAZONE-N^1,N^3,S)COPPER(II) NITRATE MONOHYDRATE
$C_{18}H_{22}Cu_2N_8O_9S_2,H_2O$ $C_9H_{10}CuN_4O_4S,C_9H_{12}CuN_3O_2S^+,NO_3^-,H_2O$

D.M. PETROVIC, V.M. LEOVAC and S. LUKIC, 1984. Z. Kristallogr., 168, 31-38.

Triclinic, $P\bar{1}$, a = 9.001, b = 11.890, c = 13.924 Å, α = 90.465, β = 112.954, γ = 105.099°, Z = 2. Mo radiation, R = 0.088 for 2639 reflexions.

The asymmetric unit contains two distinct complexes; in one the coordination is completed by a water molecule, and in the other by an oxygen from a nitrate group (Fig. 1). The molecules are linked by hydrogen bonds.

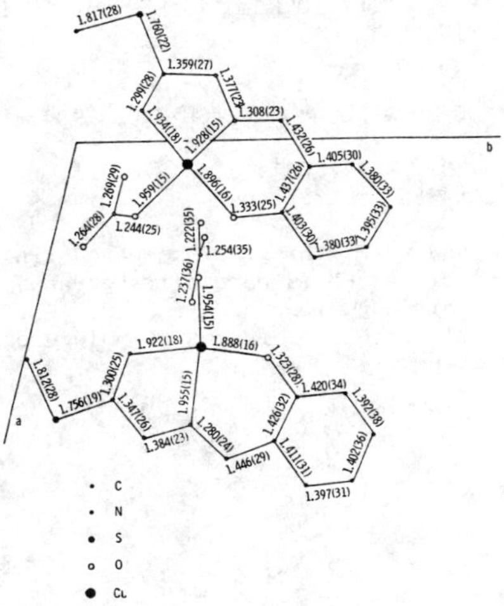

Fig. 1. Bond lengths in $C_9H_{10}CuN_4O_4S,C_9H_{12}CuN_3O_2S^+,NO_3^-$.

SALICYLIDENEANTHRANYLALDEHYDE-4-PHENYL-THIOSEMICARBAZONATOCOPPER(II) NITRATE $C_{21}H_{17}CuN_5O_4S$

V.A. NEVEROV and V.N. BIYUSHKIN, 1983. Koord. Khim., 9, 1695-1700 [Sov. J. Coord. Chem., 9, 963-968].

Monoclinic, $P2_1/b$, a = 10.337, b = 21.010, c = 9.748 Å, β = 94.73°, Z = 4. Cu radiation, R = 0.079 for 2175 reflexions.

The Cu coordination is shown in Fig. 1 with Cu-S 2.297(2), Cu-O 1.885(5), Cu-N 1.974(6), Cu-N 1.977(6) Å. The Cu atom lies 0.064 Å from the plane of the donor atoms in the direction of an S atom of an adjacent complex. The Cu...S' distance is 3.084(2) Å. The complex cations are linked to form centrosymmetric dimers, in which the S atoms act as bridges.

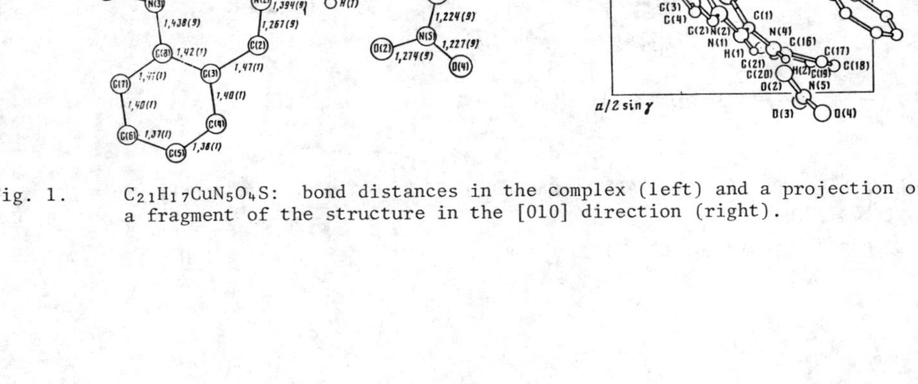

Fig. 1. $C_{21}H_{17}CuN_5O_4S$: bond distances in the complex (left) and a projection of a fragment of the structure in the [010] direction (right).

[4-METHOXYMETHYL-6-METHYL-2-(SALICYLIDENEAMINO)-3-(SALICYLIDENEAMINOMETHYL)PYRIDINATO-(2-)-N,N',O,O']COPPER HEMIHYDRATE
$C_{23}H_{21}CuN_3O_3,1.5H_2O$

N. GALESIC, R. TROJKO, Z. CIMERMAN and Z. STEFANAC, 1984. Acta Cryst., C40, 232-235.

Monoclinic, $P2_1/c$, a = 13.788, b = 7.881, c = 19.899 Å, β = 100.74°, D_m = 1.49, Z = 4. Cu radiation, R = 0.074 for 2857 reflexions.

The quadridentate ligand forms three six-membered rings (Fig. 1) with the metal atom. The coordination sphere at Cu is partially tetrahedral; the dihedral angle CuON-CuO'N' is 33.8(2)°. Main bond lengths are Cu-O(1) 1.900(5), Cu-O(1A) 1.906(5), Cu-N(1) 1.972(5), Cu-N(1A) 1.931(5) Å. The water molecules are disordered in cavities along a twofold screw axis.

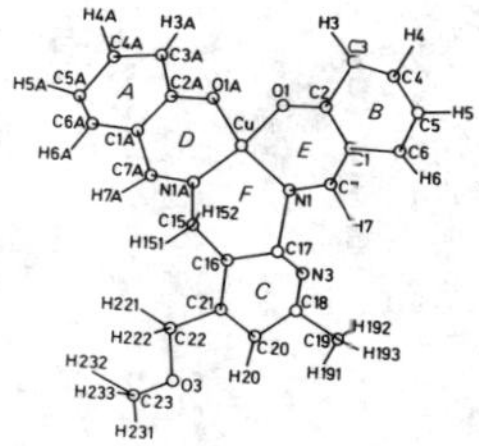

Fig. 1. A view of the $C_{23}H_{21}CuN_3O_3$ complex.

(N,N'-o-PHENYLENEBIS(SALICYLALDIMINATO))(2-METHYLIMIDAZOLE)COBALT(II)
$C_{24}H_{20}CoN_4O_2$ $[Co(C_{20}H_{14}N_2O_2)(C_4H_6N_2)]$

B.J. KENNEDY, G.D. FALLON, B.M.K.C. GATEHOUSE and K.S. MURRAY, 1984. Inorg. Chem., 23, 580-588.

Monoclinic, $P2_1/n$, a = 14.380, b = 14.355, c = 9.971 Å, β = 98.33°, D_m = 1.47, Z = 4. Mo radiation, R = 0.064 for 2960 reflexions.

The molecule (Fig. 1) displays a distorted square-pyramidal environment about Co in which one of the basal Co-N bonds is longer than the apical Co-N(2-MeImd) distance. The Co atom is displaced 0.45 Å from the basal plane. Bond lengths: Co-O1 1.973(3), Co-O2 1.977(3), Co-N1 2.032(4), Co-N2 2.127(4) and Co-N3 2.073(4) Å.

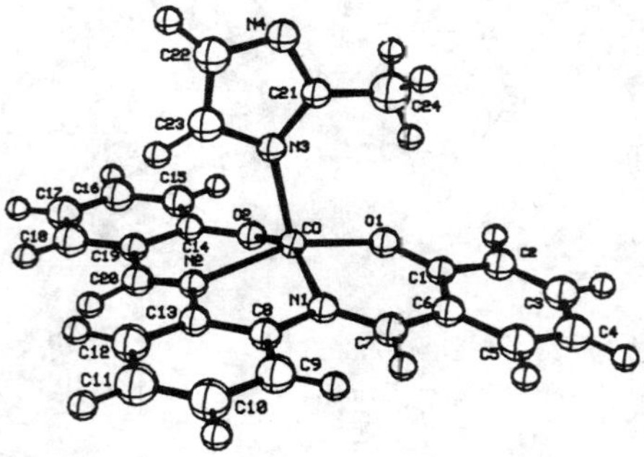

Fig. 1. The $[Co(C_{20}H_{14}N_2O_2)(C_4H_6N_2)]$ molecule.

catena-TETRAAQUA-DI-μ_3-(N-SALICYLIDENE-DL-GLUTAMATO)TRICOPPER(II) HEPTAHYDRATE
$C_{24}H_{28}Cu_3N_2O_{14},7H_2O$ $[Cu_3(C_{12}H_{10}NO_5)_2(H_2O)_4],7H_2O$

K. KORHONEN, R. HÄMÄLÄINEN and U. TURPEINEN, 1984. Acta Cryst., C40, 1175-1177.

Triclinic, $P\bar{1}$, a = 7.874, b = 8.702, c = 13.278 Å, α = 97.32, β = 104.06, γ = 97.83°, D_m = 1.68, Z = 1. Mo radiation, R = 0.037 for 3054 reflexions.

In the complex (Fig. 1) each Schiff-base ligand is hexadentate, coordinated through N and four O atoms to two Cu atoms in the trimeric unit and through one carboxyl O atom to Cu in the adjacent trimeric unit, thus forming a polymeric structure. The coordination geometries around Cu atoms are square-pyramidal (4+1) for Cu(1) and strongly tetragonally distorted octahedral (4+2) for Cu(2). Bond distances are: Cu(1)-O(1) 1.962(3), Cu(1)-O(3) 1.913(3), Cu(1)-O(4) 1.977(3), Cu(1)-N 1.928(3), Cu(1)-O(5) 2.485(4), Cu(2)-O(6) 1.944(3), Cu(2)-O(7) 2.708(5) and Cu(2')-O(2) 1.953(3) Å. The structure is composed of sixteen-membered rings in addition to the usual five- and six-membered rings.

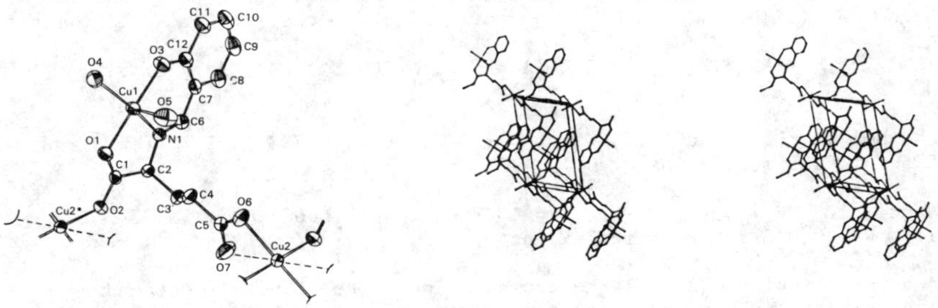

Fig. 1. $[Cu_3(C_{12}H_{10}NO_5)_2(H_2O)_4],7H_2O$: views of the asymmetric unit and molecular packing; in the latter two positions of the complex units the water molecules have been omitted.

BIS(N'-(2-PYRIDYLMETHYLENE)-N''-SALICYLOYLHYDRAZIDO-N,N',O)ZINC(II) METHANOL SOLVATE

$C_{26}H_{20}N_6O_4Zn,0.5(CH_4O)$

P. DOMIANO, C. PELIZZI, G. PREDIERI, C. VIGNALI and G. PALLA, 1984. Polyhedron, 3, 281-286.

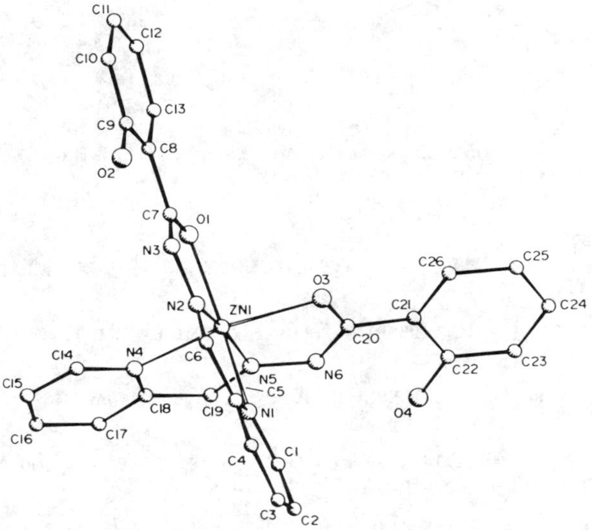

Fig. 1. A view of $C_{26}H_{20}N_6O_4Zn$.

Monoclinic, C2/c, a = 18.193, b = 24.532, c = 11.535 Å, β = 103.71°, Z = 8. Cu radiation, R = 0.072 for 1086 reflexions.

The zinc atom (Fig. 1) has a distorted octahedral environment, with Zn-O(carbonyl) 2.10, 2.16(1), Zn-N(azomethine) 2.04, 2.07(1), Zn-N(pyridine) 2.24, 2.26(1) Å. Bond angles at Zn range from 74.5(1) to 115.1(5)°.

trans-DICHLORO(N-METHYLSALICYLIDENEAMINATO)OXO(TRIPHENYLPHOSPHINE)RHENIUM(V)
$C_{26}H_{23}Cl_2NO_2PRe$ (I) trans-[Re(C_8H_8NO)Cl_2O{P(C_6H_5)$_3$}]

cis-DICHLORO(N-METHYLSALICYLIDENEAMINATO)OXO(TRIPHENYLPHOSPHINE)RHENIUM(V)
$C_{26}H_{23}Cl_2NO_2PRe$ (II) cis-[Re(C_8H_8NO)Cl_2O{P(C_6H_5)$_3$}]

V. BERTOLASI, V. FERRETTI, M. SACERDOTI and A. MARCHI, 1984. Acta Cryst., C40, 971-974.

I. Orthorhombic, Pca2_1, a = 14.802, b = 12.375, c = 13.798 Å, Z = 4. Mo radiation, R = 0.025 for 2125 reflexions.

II. Monoclinic, P2_1/n, a = 9.929, b = 15.860, c = 15.666 Å, β = 93.33°, Z = 4. Mo radiation, R = 0.045 for 2794 reflexions.

In both molecules (Fig. 1) the coordination of Re is nearly octahedral, differing only by the cis and trans arrangement of the two Cl atoms; bond distances are: Re-Cl 2.430(2) and 2.375(2) (I) and 2.397(3) and 2.355(3) (II), Re-P 2.471(2) (I) and 2.485(3) (II), Re-O 1.701(5) and 1.949(5) (I) and 1.660(8) and 1.955(6) (II) and Re-N 2.125(6) (I) and 2.150(10) Å (II). The geometry of the coordination polyhedron is compared with those of other oxorhenium(V) quasi-octahedral complexes.

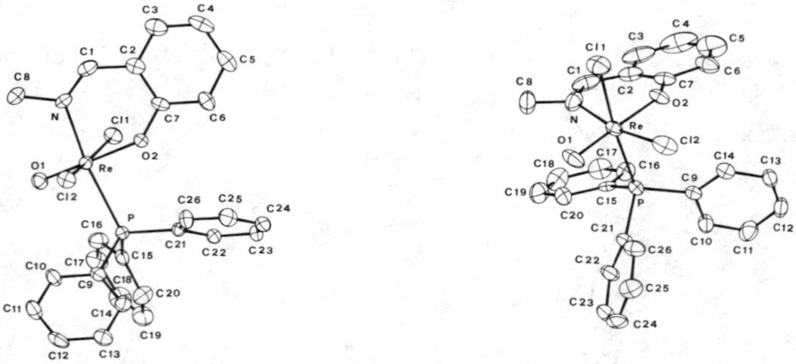

Fig. 1. Perspective views of molecules of trans- and cis-[Re(C_8H_8NO)Cl_2O-{P(C_6H_5)$_3$}] (I and II respectively).

ETHYLPYRIDINE(BIS(SALICYLIDENE)-o-PHENYLENEDIAMINE-N,N',O,O')COBALT MONOHYDRATE
$C_{27}H_{24}CoN_3O_2,H_2O$ (I)

CYANOMETHYLPYRIDINE(BIS(SALICYLIDENE)-o-PHENYLENEDIAMINE-N,N',O,O')COBALT MONOHYDRATE
$C_{27}H_{21}CoN_4O_2,H_2O$ (II)

M.F. SUMMERS, L.G. MARZILLI, N. BRESCIANI-PAHOR, and L. RANDACCIO, 1984. J. Am. Chem. Soc., 106, 4478-4485.

I. Monoclinic, P2_1/c, a = 12.140, b = 12.736, c = 15.822 Å, β = 100.2°, D_m = 1.37, Z = 4. Mo radiation, R = 0.035 for 2041 reflexions.

II. Monoclinic, P2_1/n, a = 13.964, b = 12.219, c = 14.903 Å, β = 110.6°, D_m = 1.42, Z = 4. Mo radiation, R = 0.049 for 2859 reflexions.

The most unusual structural feature in these compounds is the orientation of the py ligand of I (Fig. 1) in that it is positioned over the five-membered chelate ring made by the phenylenediamine moiety. This orientation has not been observed previously. The reason for this unusual orientation is discussed in terms of the very long Co-N(py) bond in I of 2.215(4) Å, which is the longest bond found to date in any B_{12} model. In II, this bond is shorter, 2.098(4) Å, reflecting the weaker trans influence of CH_2CN compared to C_2H_5. The py lies over the six-membered chelate rings (the normal situation). The bond lengths and bond angles at the alkyl moiety are typical. However, the saloph moieties are distinctly nonplanar with the two salicylaldiminato units forming dihedral angles of -25.4 and 17.7° in I and II, respectively. Thus the distortions are in opposite directions. This difference can also be understood in terms of the Co-N bond lengths. These bond lengths are ∿0.1 Å longer than those found with comparable axial ligand sets in cobaloximes, where the equatorial ligand system comprises two dioximato ligands.

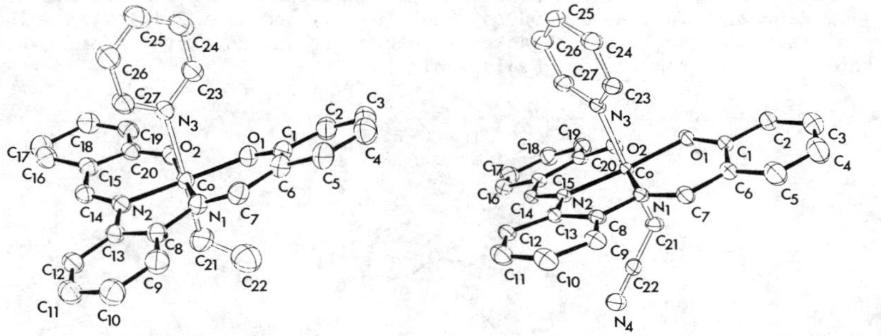

Fig. 1. Views of $C_{27}H_{24}CoN_3O_2$ (I) (left) and $C_{27}H_{21}CoN_4O_2$ (II) (right).

cis-DICHLOROBIS[DIMETHYL(PHENYL)PHOSPHINE](N-PHENYLSALICYLIDENEAMINATO)RHENIUM(III)
$C_{29}H_{32}Cl_2NOP_2Re$ $[Re(C_{13}H_{10}NO)Cl_2\{P(CH_3)_2(C_6H_5)\}_2]$

V. FERRETTI, M. SACERDOTI, V. BERTOLASI and R. ROSSI, 1984. Acta Cryst., C40, 974-976.

Monoclinic, $P2_1/c$, a = 17.411, b = 10.710, c = 20.967 Å, β = 129.58°, Z = 4. Mo radiation, R = 0.036 for 3895 reflexions.

In the molecule (Fig. 1) Re has nearly octahedral coordination with the Cl atoms cis and P atoms trans; bond distances are Re-Cl 2.383(3) and 2.468(2), Re-P 2.450(1) and 2.448(2), Re-O 1.990(7) and Re-N 2.088(7) Å. The Re^{III}-X (X = O, N, Cl, P) distances are compared with Re^{V}-X distances and the differences are interpreted in terms of Pearson's hard-soft acid-base theory.

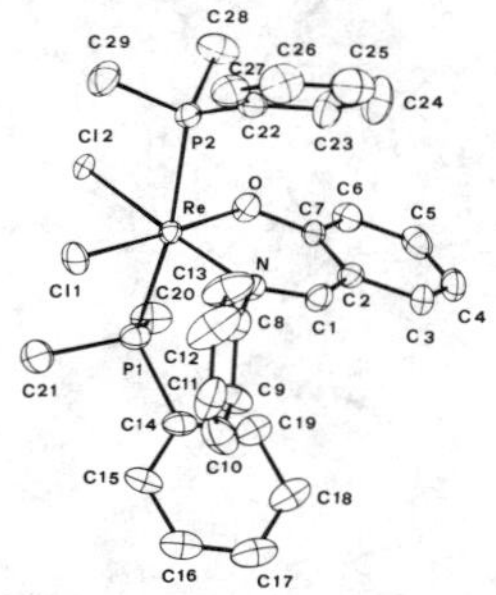

Fig. 1. $C_{29}H_{32}Cl_2NOP_2Re$: perspective view of the molecule.

trans-DIBROMOOXO(N-PHENYLSALICYLIDENEAMINATO)(TRIPHENYLPHOSPHINE)RHENIUM(V), FORM II (TRICLINIC)

$C_{31}H_{25}Br_2NO_2PRe$ $[ReBr_2(C_{13}H_{10}NO)O\{P(C_6H_5)_3\}]$

M. SACERDOTI, V. BERTOLASI, G. GILLI and A. DUATTI, 1984. Acta Cryst., C40, 968-970.

Triclinic, $P\bar{1}$, a = 8.943, b = 10.701, c = 19.917 Å, α = 120.65, β = 83.49, γ = 116.83°, Z = 2. Mo radiation, R = 0.038 for 4331 reflexions.

The molecular constitution of the title compound (Fig. 1) is identical to that of the previously studied form I (1). The molecules have slightly different geometries in the two crystals and can be considered as two different conformational isomers. The co-ordination of Re is nearly octahedral with the two O atoms in apical positions, two trans Br atoms in the equatorial plane and the chelate ligand bridging an equatorial and an axial position. Bond distances are: Re-Br 2.545(1) and 2.547(1), Re-P 2.482(3), Re-O 1.680(4) and 1.972(4) and Re-N 2.141(9) Å. The crystal consists of monomeric units packed together without contacts significantly shorter than the sum of the van der Waals radii.

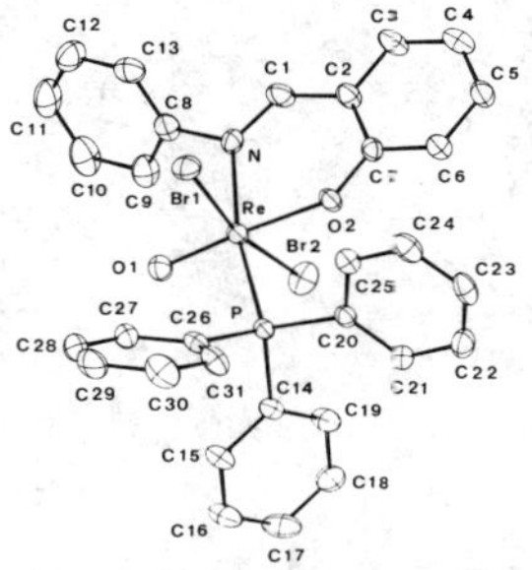

Fig. 1. $C_{31}H_{25}Br_2NO_2PRe$: perspective view of the molecule.

1. Structure Reports, 49B,1264.

DICHLORO(SALICYLALDEHYDATO-O,O')(p-TOLYLNITRENE)TRIPHENYLPHOSPHINERHENIUM

$C_{32}H_{27}Cl_2NO_2PRe$

A. MARCHI, R. ROSSI, A. DUATTI, L. MAGON, U. CASELLATO and R. GRAZIANI, 1984. Transition Met. Chem., 9, 299-302.

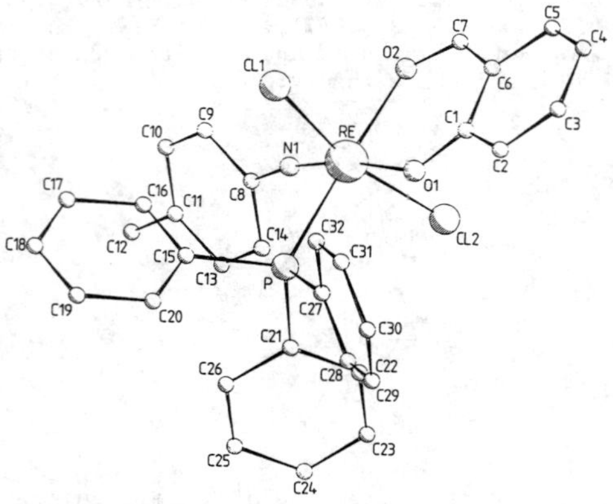

Fig. 1. A view of $C_{32}H_{27}Cl_2NO_2PRe$.

Triclinic, $P\bar{1}$, a = 11.698, b = 11.445, c = 11.725 Å, α = 74.12, β = 102.65, γ = 86.89°, Z = 2. R = 0.058 for 4400 reflexions.

The Re atom has distorted octahedral geometry with the Cl atoms trans (Fig. 1). Main bond lengths are: Re-Cl 2.381 and 2.424(3), Re-P 2.427(3), Re-O 2.01 and 2.08(1), Re-N 1.72(1) Å. Distortions from idealized geometry are attributed to the presence of a relatively bulky ligand and to multiple or partial-multiple metal-ligand bonds.

(μ-SULPHIDO)BIS[(N,N'-ETHYLENEBIS(SALICYLALDIMINATO))IRON(III)]
$C_{32}H_{28}Fe_2N_4O_4S$ $[Fe(C_{16}H_{14}N_2O_2)]_2S$

J.R. DORFMAN, J.-J. GIRERD, E.D. SIMHON, T.D.P. STACK and R.H. HOLM, 1984. Inorg. Chem., 23, 4407-4412.

Triclinic, $P\bar{1}$, a = 10.792, b = 11.746, c = 13.215 Å, α = 110.56, β = 99.23, γ = 72.27°, D_m = 1.49, Z = 2. Mo radiation, R = 0.043 for 2720 reflexions.

The molecule (Fig. 1) is binuclear with two $[Fe(salen)]^+$ units bridged by a sulphide atom. Ligands are nonplanar and adopt the umbrella conformation. The Fe atoms are displaced 0.591 and 0.625 Å from the N_2O_2 mean planes toward the μ-S atom, forming pyramidal Fe N_2O_2S units. The Fe-S-Fe bridge is unsymmetrical, with bond distances of 2.150(2) and 2.190(2) Å and Fe-S-Fe angle of 121.8(1)°. The Fe-O and Fe-N distances range 1.912(4)-1.929(4) and 2.097(4)-2.138(4) Å, respectively.

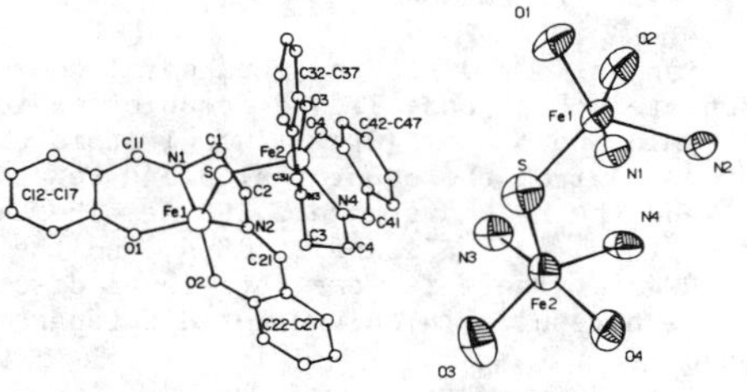

Fig. 1. The structure of $[Fe(salen)]_2S$, coordination spheres at right.

(μ_2-OXO)BIS{OXO[N,N'-PROPANE-1,3-DIYL-BIS(SALICYLIDENEIMINATO)]TECHNETIUM(V)}
$C_{34}H_{32}N_4O_7Tc_2$ (I) $[TcO(salpd)]_2O$

CHLOROOXO[N,N'-PROPANE-1,3-DIYLBIS(SALICYLIDENEIMINATO)]TECHNETIUM(V)
$C_{17}H_{16}ClN_2O_3Tc$ (II) TcOCl(salpd)

G. BANDOLI, M. NICOLINI, U. MAZZI and F. REFOSCO, 1984. J. Chem. Soc. Dalton, 2505-2511.

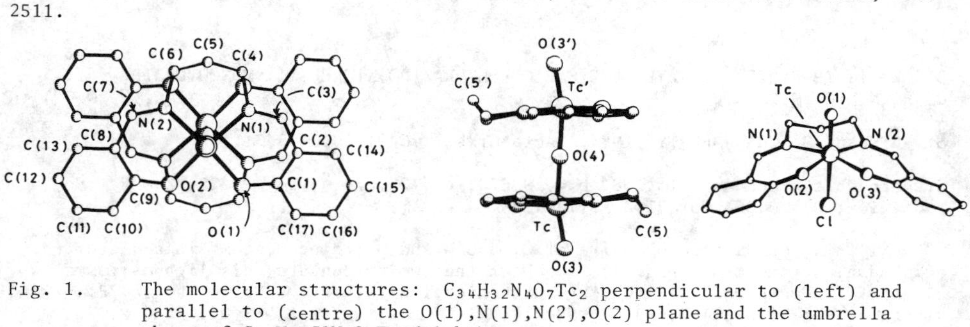

Fig. 1. The molecular structures: $C_{34}H_{32}N_4O_7Tc_2$ perpendicular to (left) and parallel to (centre) the O(1),N(1),N(2),O(2) plane and the umbrella shape of $C_{17}H_{16}ClN_2O_3Tc$ (right).

I. Monoclinic, $P2_1/c$, a = 15.041, b = 12.630, c = 16.522 Å, β = 95.35°, D_m = 1.77, Z = 4. Mo radiation, R = 0.059 for 1958 reflexions.

II. Orthorhombic, $Pn2_1a$, a = 12.010, b = 11.702, c = 11.625 Å, D_m = 1.76, Z = 4. Mo radiation, R = 0.076 for 788 reflexions.

The 'dimeric' structure of I consists of two crystallographically independent and chemically well-separated TcO(salpd) moieties bridged by an oxygen atom with the bridging Tc-O-Tc angle symmetrically imposed at 180° (Fig. 1). The O=Tc-O-Tc=O moiety is almost linear. The Tc atoms have octahedral coordination, with the equatorial plane formed by the N_2O_2 donor set. The mean Tc-N distance is 2.12 and the mean Tc-O distance 2.01 Å. The bridging Tc-O distance is 1.90(1) Å. The structure of II consists of discrete molecules of [TcOCl(salpd)] with a pseudooctahedral coordination around Tc. The salpd ligand occupies the four equatorial positions, whereas the Cl and the O atoms are trans to each other in axial positions. The two salicylaldimine groups are bent in an 'umbrella shape'. Bond distances in the inner core are: Tc=O 1.66(1), Tc-O (bridging) 1.98(1), Tc-N 2.12(2) and Tc-Cl 2.44(1) Å.

TETRAPHENYLARSONIUM OXODICHLORO(N-2-OXOPHENYLSALICYLIDENIMINATO)MOLYBDATE(V)

$C_{37}H_{29}AsCl_2MoNO_3$ $[(C_6H_5)_4As][C_{13}H_9NO_3Cl_2Mo]$

K. YAMANOUCHI, S. YAMADA and J.H. ENEMARK, 1984. Inorg. Chim. Acta, 85, 129-135.

Monoclinic, $P2_1/n$, a = 11.829, b = 16.149, c = 17.410 Å, β = 97.49°, D_m = 1.57, Z = 4. Mo radiation, R = 0.035 for 4722 reflexions.

The Mo atom in the anion is in a distorted octahedral coordination environment (Fig. 1). A planar terdentate Schiff base ligand occupies meridional positions with the N atom trans to the terminal oxo atom (O_t). Two Cl atoms are cis to the O_t atom. The Mo atom is displaced 0.33 Å from the equatorial plane towards the O_t atom. The Mo-O_t distance is 1.673 Å and the Mo-N distance 2.298 Å. The two Mo-Cl bond lengths are significantly different (30σ) at 2.371 and 2.408 Å, and the Mo-O(phenol) distances are 1.959 and 1.974 Å. There is some rotational disorder in the anion, about the Mo-O(1) vector which results in only minor discrepancies in the bond distances within the ligand.

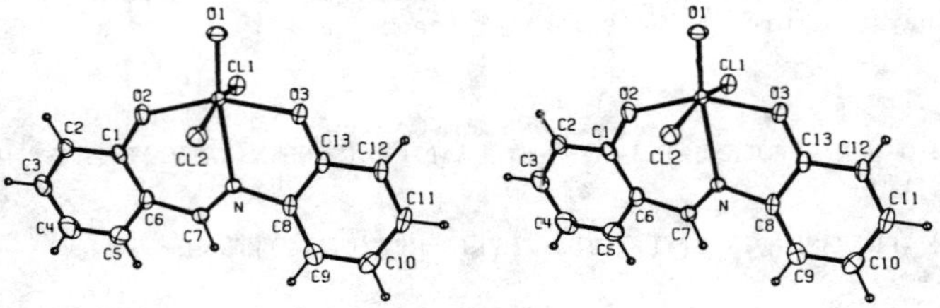

Fig. 1. Stereoview of the anion in $C_{37}H_{29}AsCl_2MoNO_3$.

BIS{2'-[α-(2-PYRIDYL)BENZYLIDENE]SALICYLOHYDRAZIDO}NICKEL(II) MONOHYDRATE

$C_{38}H_{28}N_6NiO_4,H_2O$ $Ni(C_{19}H_{14}N_3O_2)_2 \cdot H_2O$

S. SETH and S. CHAKRABORTY, 1984. Acta Cryst., C40, 1530-1533.

Orthorhombic, Pbca, a = 16.268, b = 19.670, c = 23.070 Å, D_m = 1.28, Z = 8. Mo radiation, R = 0.064 for 2103 reflexions.

Ni^{2+} is coordinated by pairs of O, azo N and pyridine N atoms in a distorted octahedral arrangement (Fig. 1). Each of the two tridentate, cis ligands forms two, planar, five-membered chelate rings with nickel with average "bite" angle 78°. Bond distances are Ni-N(py) 2.077, 2.090(11); Ni-N(azo) 1.974, 1.985(9); Ni-O 2.086, 2.090(8) Å. The two azo N atoms are trans (N-Ni-N 178.2°) with the pyridine N atoms cis (N-Ni-N 92.1°). O(2)...N(1) and O(4)...N(6) intramolecular hydrogen bonds stabilise the ligand but the water molecule is not hydrogen bonded and shows large thermal motion.

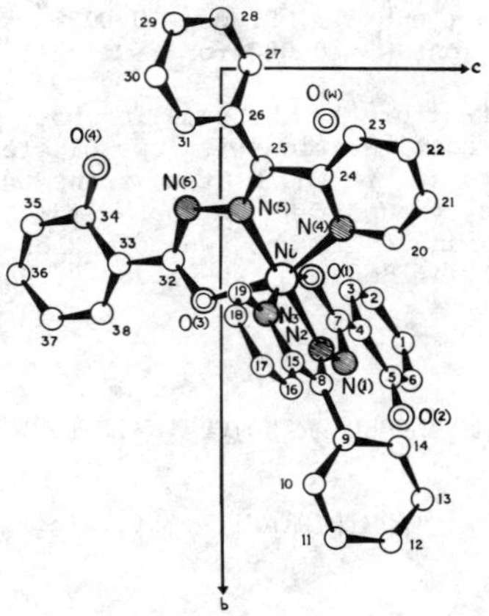

Fig. 1. The asymmetric unit of $Ni(C_{19}H_{14}N_3O_2)_2.H_2O$.

DICHLORO(N-PHENYLSALICYLALDIMINATO)(p-TOLYLIMIDO)TRIPHENYLPHOSPHINERHENIUM(V)
$C_{38}H_{32}Cl_2N_2OPRe$ (I) $ReCl_2(NC_6H_4CH_3)(PhN{=}CH{-}C_6H_4O)(PPh_3)$

DICHLORO(N-METHYLSALICYLALDIMINATO)(p-TOLYLIMIDO)TRIPHENYLPHOSPHINERHENIUM(V)
$C_{33}H_{30}Cl_2N_2OPRe$ (II) $ReCl_2(NC_6H_4CH_3)(MeN{=}CH{-}C_6H_4O)(PPh_3)$

R. ROSSI, A. MARCHI, A. DUATTI, L. MAGON, U. CASELLATO, R. GRAZIANI and G. POLIZZOTTI, 1984. Inorg. Chim. Acta, 90, 121-131.

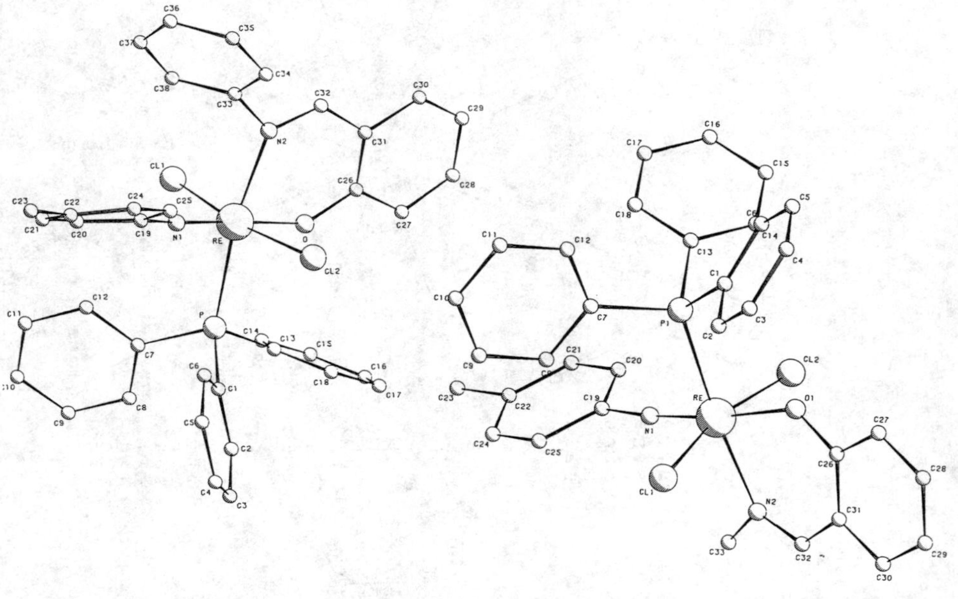

Fig. 1. Molecular structures of $C_{38}H_{32}Cl_2N_2OPRe$ (left), and $C_{33}H_{30}Cl_2N_2OPRe$ (right).

I. Monoclinic, $P2_1/c$, a = 15.340, b = 21.115, c = 10.605 Å, β = 91.36°, Z = 4. Mo radiation, R = 0.047 for 3931 reflexions.

II. Triclinic, $P\bar{1}$, a = 13.457, b = 11.795, c = 10.997 Å, α = 112.16, β = 108.17, γ = 86.36°, Z = 2. Mo radiation, R = 0.080 for 4891 reflexions.

The overall molecular structure is the same for both complexes. The Re atom is in a distorted octahedral configuration with two Cl atoms trans to one another, the arylimido group (bonded to Re via the N atom) trans to the O atom of the bidentate ligand, and the PPh_3 group trans to the N atom of the bidentate ligand (Fig. 1). Bond distances and angles in the two molecules agree well. Mean values for some bond distances involving Re are: Re-Cl 2.418, Re-P 2.44, Re-O 1.98, Re-N (arylimido) 1.73 and Re-N 2.14 Å.

BIS(N,N'-DISALICYLIDENE-1,2-PHENYLENEDIAMINO)CERIUM(IV)

$C_{40}H_{28}CeN_4O_4$

A. TERZIS, D. MENTZAFOS and H.A. TAJMIR-RIAHI, 1984. Inorg. Chim. Acta, 84, 187-193.

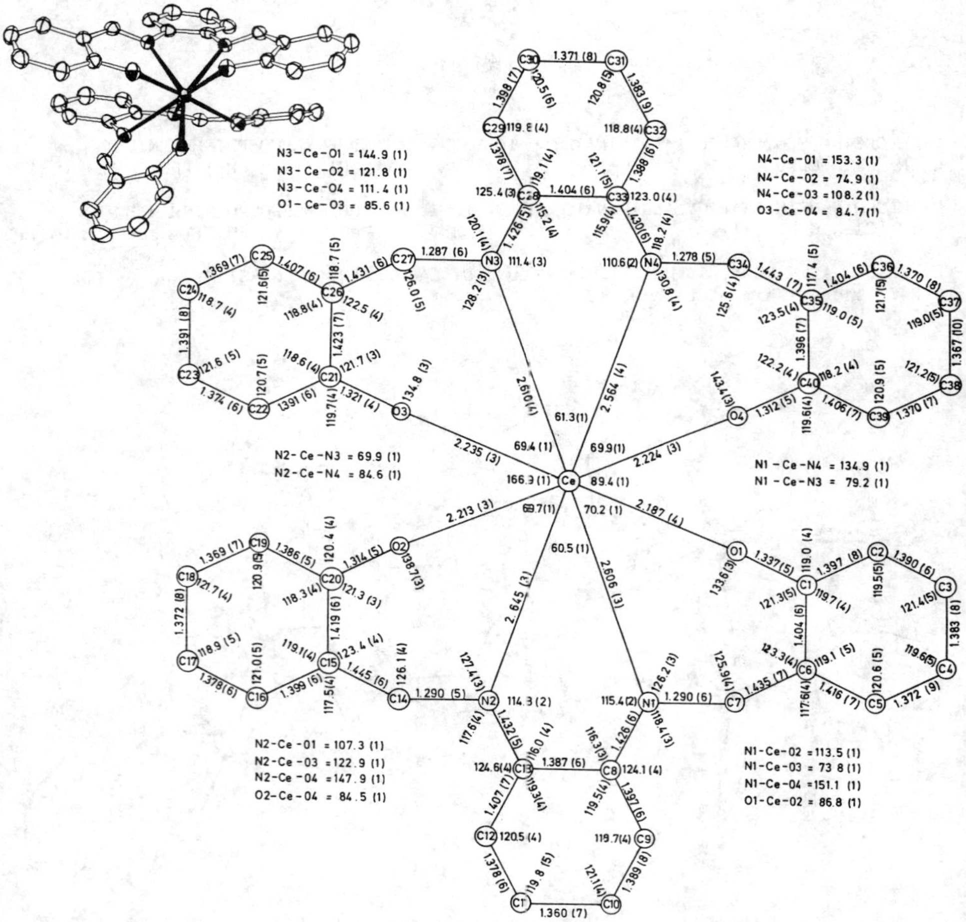

Fig. 1. ORTEP plot of $C_{40}H_{28}CeN_4O_4$ showing square antiprismatic coordination (top), and interatomic bond distances and angles (bottom).

Monoclinic, $P2_1/c$, a = 11.090, b = 16.332, c = 18.916 Å, β = 70.10°, D_m = 1.57, Z = 4. Mo radiation, R = 0.038 for 4043 reflexions.

The eight-coordinate complex belongs to a square antiprismatic species (Fig. 1) with each chelating ligand spanning sss edges. The Ce-O bonds average 2.214 Å, appreciably shorter than the Ce-N distance of 2.606 Å. The two ligands have a sandwich orientation and the trapezoidal planes intersect at 88.2°.

2-THIOIMIDAZOLIDINE SILVER(I) CHLORIDE COMPLEX
$C_6H_{12}AgClN_4S_2$ (I) $AgCl(C_3H_6N_4S)_2$

2-THIOIMIDAZOLIDINE SILVER(I) BROMIDE COMPLEX
$C_6H_{12}AgBrN_4S_2$ (II) $AgBr(C_3H_6N_4S)_2$

2-THIOIMIDAZOLIDINE
$C_3H_6N_2S$ (III)

L.P. BATTAGLIA, A.B. CORRADI and M. NARDELLI, 1984. Croat. Chem. Acta, 57, 545-563.

I. Monoclinic, $P2_1/c$, a = 11.078, b = 14.046, c = 7.749 Å, β = 103.71°, D_m = 1.96, Z = 4. Mo radiation, R = 0.056 for 1372 reflexions.

II. Monoclinic, $P2_1/n$, a = 9.050, b = 10.557, c = 13.191 Å, β = 103.38°, D_m = 2.00, Z = 4. Mo radiation, R = 0.038 for 1976 reflexions.

III. Monoclinic, $P2_1/c$, a = 5.801, b = 14.540, c = 5.774 Å, β = 101.3°, D_m = 1.45, Z = 4. Mo radiation, R = 0.024 for 506 reflexions.

Metal coordination is distorted tetrahedral in both complexes (I and II) but the structure of I is polymeric with one ethylenethiourea (etu) ligand terminal (Ag-S 2.512 Å) and one bridging through sulphur (Ag-S 2.551, 2.655 Å). In II (Fig. 1) however, the complex consists of centrosymmetric dimers with two etu ligands terminal (Ag-S 2.482, 2.489 Å) and Br acting as a bridge (Ag-Br 2.817, 2.847 Å). Relevant parameters for the uncomplexed etu (III) are: S-C(sp^2) 1.694, N-C(sp^2) 1.349 av., N-C(sp^3) av. 1.460, C(sp^3)-C(sp^3) 1.561 Å, N-C-N 108.4, C-N-C 113.2° av., N-C-C 102.5° av. Metal coordination seems to influence the S-C-N angles; that on the side of the lone pair interacting with metal is always larger (126.6° av.) than that on the other side (124.3° av.) while the thioureic moiety of the ligand maintains its planarity.

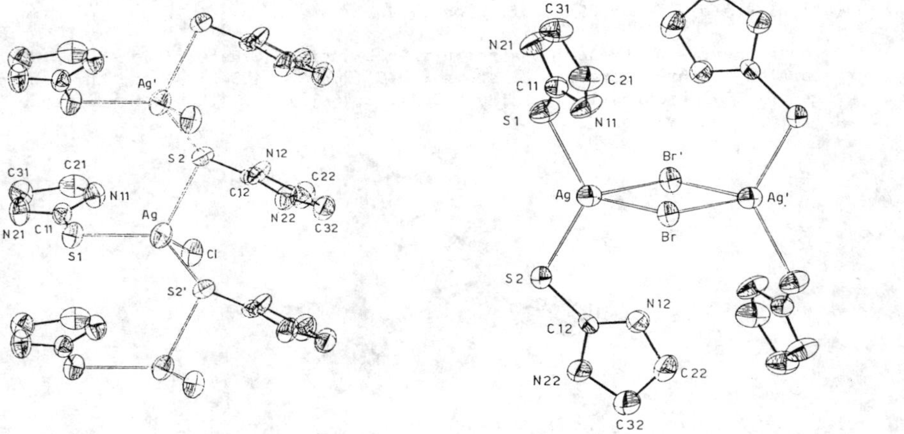

Fig. 1. Chains of coordination polyhedra in $C_6H_{12}AgClN_4S_2$ (left), and dimers in $C_6H_{12}AgBrN_4S_2$ (right).

HEXAKIS(THIOUREA-S)TECHNETIUM(III) TRICHLORIDE TETRAHYDRATE

$C_6H_{24}Cl_3N_{12}S_6Tc,4H_2O$ $[Tc(SC(NH_2)_2)_6]Cl_3.4H_2O$

M.J. ABRAMS, A. DAVISON, R. FAGGIANI, A.G. JONES and C.J.L. LOCK, 1984. Inorg. Chem., 23, 3284-3288.

Monoclinic, C2/c, a = 11.876, b = 12.048, c = 19.662 Å, β = 95.32°, D_m = 1.74, Z = 4. Mo radiation, R = 0.039 for 1753 reflexions.

The cation (Fig. 1) is roughly octahedrally coordinated by the six thiourea S atoms. The Tc-S bonds (2.431(1), 2.412(1), and 2.440(1) Å) are significantly longer than Tc(V)-S bonds, but bond lengths within the thiourea groups are normal. There is an extensive H-bonding network.

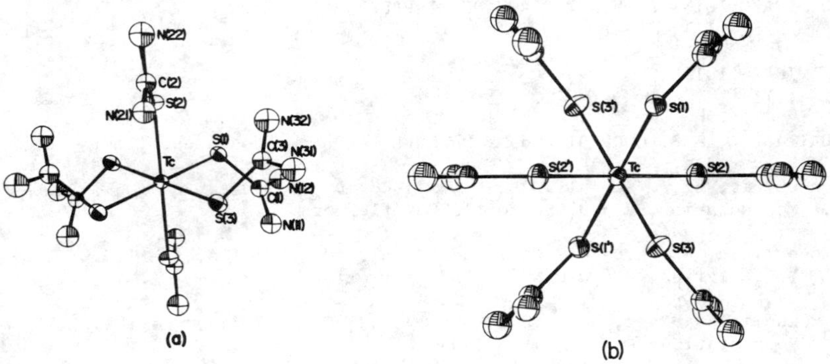

Fig. 1. Two views of the $[Tc(SC(NH_2)_2)_6]^{3+}$ cation.

HEXACHLOROTRIS(μ_2-ALLYLTHIOUREA)DIMOLYBDENUM(III) MONOHYDRATE

$C_{12}H_{24}Cl_6Mo_2N_6S_3,H_2O$

LU SHAOFANG, SHANG MAOYU, HUANG JINLING and LU JIAXI, 1984. J. Struct. Chem., 3, 21-24.

Monoclinic, $P2_1$, a = 8.484, b = 18.792, c = 8.576 Å, β = 91.35°, Z = 2. Mo radiation, R = 0.023 for 2922 reflexions.

The geometry about each Mo atom in this binuclear cluster (Fig. 1) is distorted octahedral with three terminal Cl atoms (Mo-Cl 2.409-2.453(2) Å) and three bridging S atoms of allylthiourea ligands (Mo-S 2.447-2.468(2) Å). The Mo-Mo distance is 2.446(1) Å.

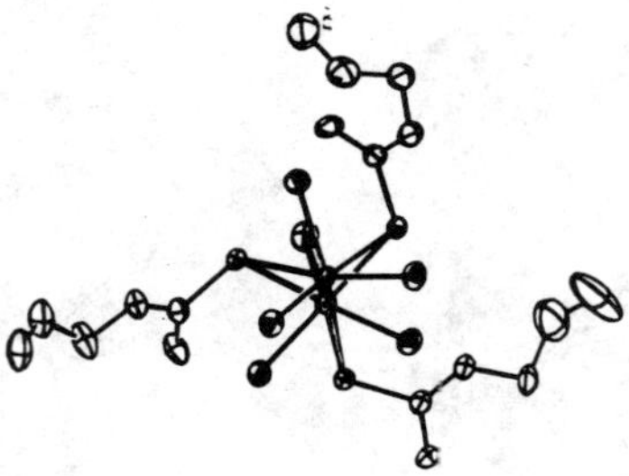

Fig. 1. A view of $C_{12}H_{24}Cl_6Mo_2N_6S_3$.

DICHLOROBIS(ACETOPHENONE THIOSEMICARBAZONE)NICKEL(II)
$C_{18}H_{22}Cl_2N_6NiS_2$

P. DAPPORTO, G. DE MUNNO and A.A.G. TOMLINSON, 1984. J. Chem. Res., (S) 40-41, (M) 0501-0503.

Monoclinic, C2/c, a = 11.389, b = 11.063, c = 18.820 Å, β = 105.30°, Z = 4. Mo radiation, R = 0.043 for 1663 reflexions.

The Ni atom lies on a crystallographic twofold axis and has distorted tetrahedral geometry (Fig. 1) with a dihedral angle of 78.43° between the SNiS and ClNiCl planes. The Cl-Ni-S angles are 114.1 and 96.5°, the difference in these values being attributed to an intramolecular N-H...Cl interaction.

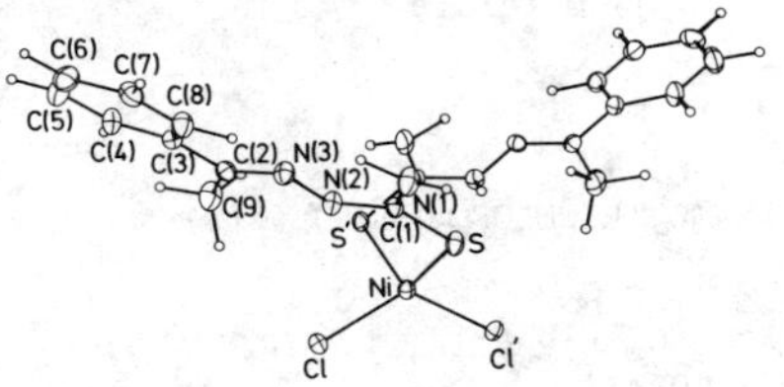

Fig. 1. A view of $C_{18}H_{22}Cl_2N_6NiS_2$.

BIS[N-(DIETHYLAMINOTHIOCARBONYL)-N'-PHENYLBENZAMIDINATO]NICKEL(II)
$C_{36}H_{40}N_6NiS_2$

J. SIELER, R. RICHTER, L. BEYER, O. LINDQVIST and L. ANDERSEN, 1984. Z. Anorg. Chem., 515, 41-50.

Orthorhombic, $P2_12_12_1$, a = 16.36, b = 16.94, c = 12.96 Å, Z = 4. Mo radiation, R = 0.055 for 1764 reflexions.

The Ni coordination is distorted towards tetrahedral (Fig. 1) with Ni-N 1.90, 1.91(1); Ni-S 2.164, 2.169(4) Å, S-Ni-S 89.0(1), N-Ni-N 93.5(4), N-Ni-S 92.2 and 93.4(3)°. Both chelate rings are considerably distorted from planarity.

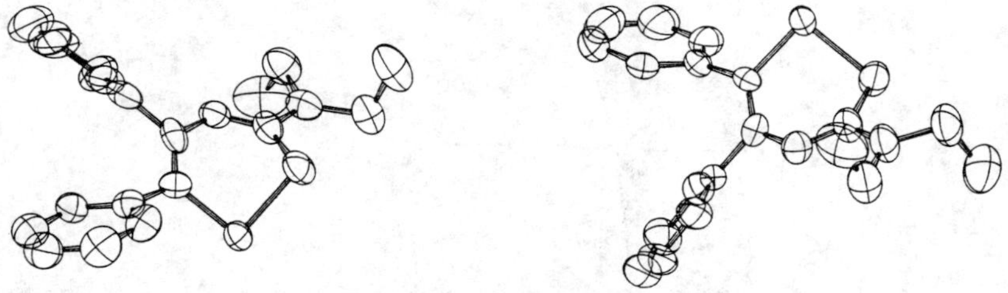

Fig. 1. Views of the two halves of $C_{36}H_{40}N_6NiS_2$.

DIOXOBIS(μ-SULPHIDO)BIS(DITHIOCARBAMATO)DIMOLYBDENUM(V) ACETONITRILE SOLVATE
$C_2H_4Mo_2N_2O_2S_6,2C_2H_3N$ $[Mo_2O_2S_2(NH_2CS_2)_2],2CH_3CN$

N.C. HOWLANDER, G.P. HAIGHT, Jr., T.W. HAMBLEY, M.R. SNOW and G.A. LAWRANCE, 1984. Inorg. Chem., 23, 1811-1815.

Orthorhombic, Pnma, a = 9.407, b = 20.877, c = 9.010 Å, Z = 4. Mo radiation, R = 0.030 for 983 reflexions.

Complex molecules having exact mirror symmetry (Fig. 1) are linked to solvent molecules via N-H...N hydrogen bonds. Each Mo atom is centred above the base of a square pyramid, with the O atoms coordinated on the same side of the bent MoS_2Mo bridge and a Mo-Mo bond of 2.820(1) Å. Other distances: Mo-(μ-S) 2.324(1) and 2.325(1), Mo-S 2.460(1) and 2.461(1), Mo-O 1.677(3) Å.

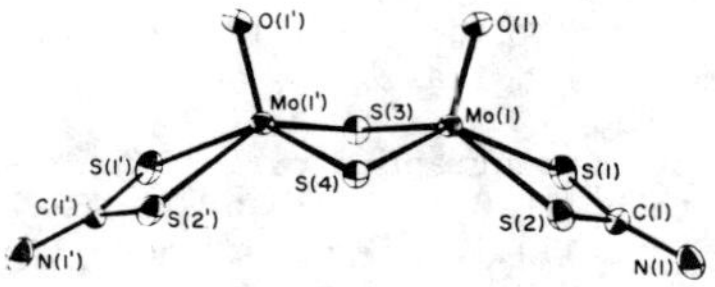

Fig. 1. The $Mo_2O_2S_2(NH_2CS_2)_2$ molecule.

mer-TRIS(O-ETHYLTHIOCARBAMATO)TRICHLORORHODIUM(III) ACETONE SOLVATE
$C_9H_{21}Cl_3N_3O_3RhS_3,C_3H_6O$

E. FORSELLINI, U. CASELLATO, G. FARAGLIA, L. VOLPONI and R. GRAZIANI, 1984. Transition Met. Chem., 9, 419-423.

Monoclinic, $P2_1$, a = 9.101, b = 15.785, c = 8.776 Å, β = 103.00°, Z = 2. Mo radiation, R = 0.040 for 2018 reflexions.

The Rh atom has distorted octahedral coordination (Fig. 1) with Rh-Cl(1) 2.354(2), Rh-Cl(2) 2.388(2), Rh-Cl(3) 2.353(2), Rh-S(1) 2.376(3), Rh-S(2) 2.369(3) and Rh-S(3) 2.332(3) Å. Bond distances in the ligands indicate partial double C-O and C-N bonds and double C-S bonds (mean 1.69(1) Å). Some N...Cl contacts (3.08 - 3.22 Å) are taken as indicating N-H...Cl hydrogen bonding.

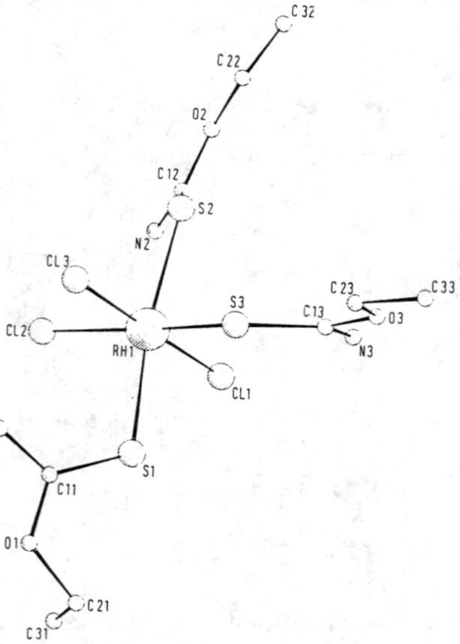

Fig. 1. A view of $C_9H_{21}Cl_3N_3O_3RhS_3$.

DICHLOROBIS(DIETHYLDITHIOCARBAMATO)THIONITROSYLTECHNETIUM(III)
$C_{10}H_{20}Cl_2N_3S_5Tc$ $Tc(S_2CNEt_2)_2Cl_2(NS)$

J. BALDAS, J. BONNYMAN, M.F. MACKAY and G.A. WILLIAMS, 1984. Aust. J. Chem., 37, 751-759.

Orthorhombic, Pcmn, crystal (1) a = 8.860, b = 15.690, c = 29.02 Å, crystal (2) a = 8.936, b = 15.681, c = 28.445 Å, Z = 8. Cu radiation, R = 0.202 for 1026 reflexions for crystal (1) with 21.9% disorder, and R = 0.078 for 2078 reflexions for crystal (2) with 4.0% disorder.

There are two independent molecules in the asymmetric unit with each molecule having Tc, Cl and S_2CNEt_2 ligands on the mirror plane. The Tc atoms have seven coordinate pentagonal, bipyramidal environments (Tc-N 1.75(2), Tc-S 2.472(5) Å, Tc-Cl 2.421(7) Å, S-Tc-S 68.4(2)°) (Fig. 1).

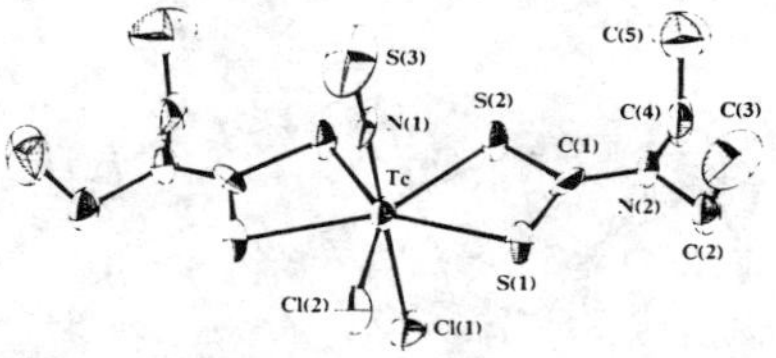

Fig. 1. The structure of $Tc(S_2CNEt_2)_2Cl_2(NS)$.

BIS(N,N-DIMETHYLTHIOCARBAMATO-S,S')(METHOXYCARBONYLDIAZENIDO-N)(METHOXYCARBONYL-HYDRAZIDO-N,N')MOLYBDENUM DIETHYL ETHER SOLVATE
$C_{10}H_{20}MoN_6O_4S_4,0.5(C_4H_{10}O)$

J.R. DILWORTH and J.A. ZUBIETA, 1984. Transition Met. Chem., 9, 39-40.

Monoclinic, C2/c, a = 19.964, b = 18.855, c = 13.127 Å, β = 114.41°, Z = 8. Mo radiation, R = 0.066 for 1585 reflexions.

The Mo atom has distorted pentagonal bipyramidal geometry (Fig. 1) with the $NNCO_2Me$ group in one axial site and dithiocarbamate S(4) in the other. The trans influence of the diazenido ligand is shown by the Mo-S4 bond length (2.597(5) Å) being much longer than the other Mo-S distances (2.482-2.519(4) Å). The diazenido ligand is bonded through N(1) (Mo-N(1) 1.71(1), N(1)-N(2) 1.33(2) Å) and the hydrazido ligand is bonded side on (Mo-N(3) 2.13(1), Mo-N(4) 2.11(1), N(3)-N(4) 1.40(2) Å).

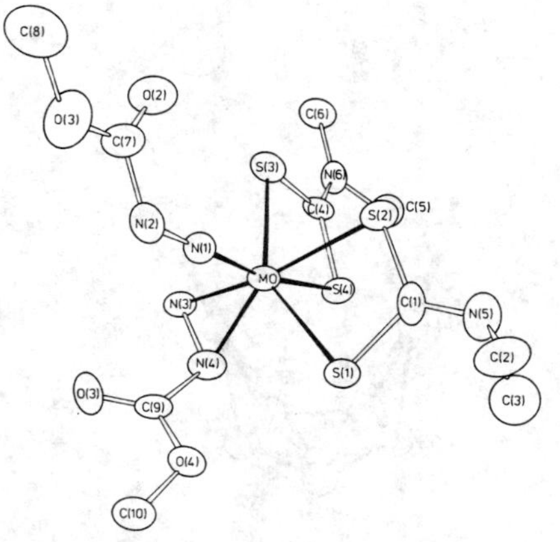

Fig. 1. A view of $C_{10}H_{20}MoN_6O_4S_4$.

BIS[N,N-BIS(2-HYDROXYETHYL)DITHIOCARBAMATO]NICKEL(II)
$C_{10}H_{20}N_2NiO_4S_4$

K. RAMALINGAM, K. RADHA, G. ARAVAMUDAN, C. MAHADEVAN, Ch. SUBRAMANYAM and M. SESHASAYEE, 1984. Acta Cryst., C40, 1838-1839.

Monoclinic, $P2_1/c$, a = 6.372, b = 11.844, c = 11.474 Å, β = 93.56°, Z = 2. Mo radiation, R = 0.046 for 1551 reflexions.

Ni lies on a symmetry centre and is planar-coordinated to four S atoms (Fig. 1) with Ni-S(1) 2.209(1) and Ni-S(2) 2.193(1) Å; the S(1)-Ni-S(2) angle is 79.25°. The NiS_2CN fragment is coplanar to within 0.05 Å. S-C distances are 1.713(4) and 1.728(4) Å. The molecules are linked by O(1)-H...O(2) and by O(2)-H...O(1) hydrogen bonds.

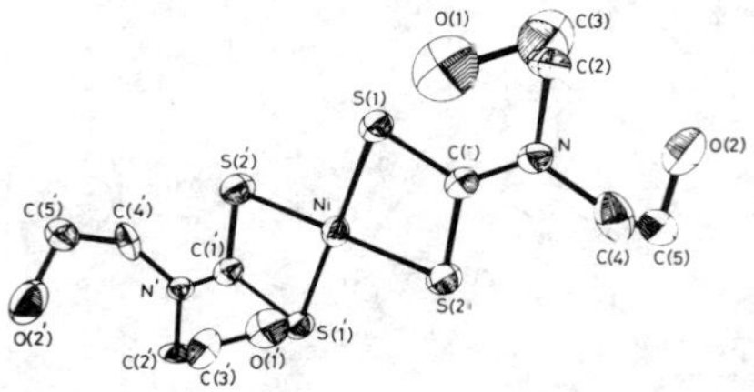

Fig. 1. The $C_{10}H_{20}N_2NiO_4S_4$ molecule.

6,9-BIS(N,N-DIETHYLDITHIOCARBAMATO)-arachno-6,9-DIAURADECABORANE
$C_{10}H_{30}Au_2B_8N_2S_4$

M.A. BECKETT, J.E. CROOK, N.N. GREENWOOD and J.D. KENNEDY, 1984. J. Chem. Soc. Dalton, 1427-1431.

Monoclinic, $P2_1/m$, a = 6.862, b = 17.751, c = 10.050 Å, β = 92.82°, Z = 2. Mo radiation, R = 0.043 for 1493 reflexions.

The metallaborane cluster has an open-faced ten vertex cluster structure (Fig. 1). The Au_2B_8 cluster and its ligands have idealised C_{2v} symmetry. The molecule itself lies on a crystallographic mirror plane. The two $Au(S_2CNEt_2)$ moieties are each bonded to an open triangle of B atoms and each Au atom is five-coordinate with Au(6)-S(1) 2.401(3), Au(6)-S(2) 2.398(3), Au(6)-B(5) 2.255(13), Au(6)-B(7) 2.213(13), Au(6)-B(2) 2.219(14) Å. The Au(6)...Au(9) distance is 3.581(1) Å. B-B distances range from 1.74(2) to 1.87(2) Å.

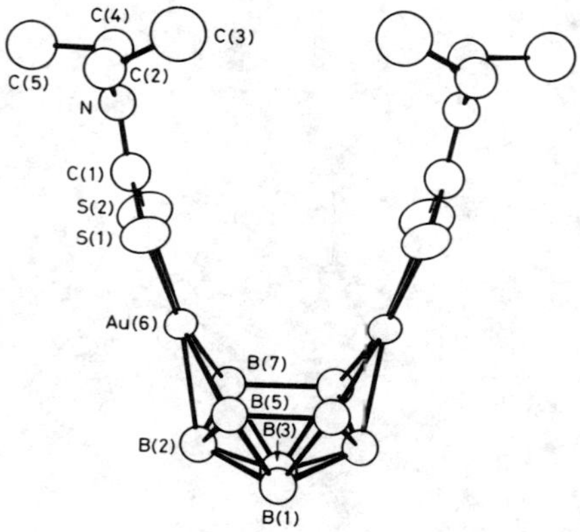

Fig. 1. $C_{10}H_{30}Au_2B_8N_2S_4$: the molecular structure viewed along the crystallographic mirror plane.

BROMOBIS(N,N-DIISOPROPYLDITHIOCARBAMATO)IRON(III) METHYLENE CHLORIDE SOLVATE
$C_{14}H_{28}BrFeN_2S_4,0.7CH_2Cl_2$ $[FeBr(S_2CN(C_3H_7)_2)_2],0.7CH_2Cl_2$

P.C. CHRISTIDIS, P.J. RENTZEPERIS, F. VAKOULIS and C.A. TSIPIS, 1984. Inorg. Chim. Acta, 83, 87-91.

Monoclinic, C2/c, a = 16.579, b = 11.774, c = 14.533 Å, β = 118.67°, D_m = 1.42, Z = 4. Mo radiation, R = 0.092 for 477 reflexions.

The molecule has twofold crystallographic symmetry, the axis of symmetry passing through Fe and Br (Fig. 1). The coordination about the Fe atom is approximately square pyramidal and the metal atom lies 0.64 Å above the mean plane of the four S atoms of the dithiocarbamate ligands. Pertinent distances are: Fe-Br 2.41, Fe-S(1) 2.28, Fe-S(2) 2.28, S(1)-C(1) 1.74, S(2)-C(1) 1.68 and C(1)-N 1.39 Å. The S(1)-Fe-S(2) and S(1)-C(1)-S(2) angles are respectively 76.0 and 110°.

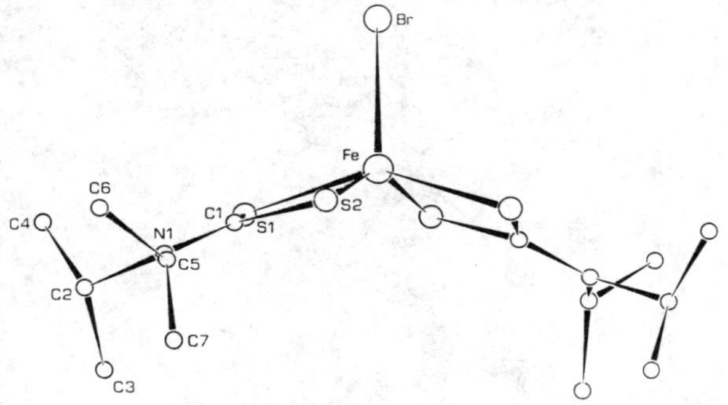

Fig. 1. Perspective view of $C_{14}H_{28}BrFeN_2S_4$.

DISULPHIDO-TRIS(N,N-DIETHYLDITHIOCARBAMATO)TANTALUM(V)
$C_{15}H_{30}N_3S_8Ta$ $[Ta(S_2)(S_2CN(C_2H_5)_2)_3]$

M.G.B. DREW, D.M. WILLIAMS and D.A. RICE, 1984. Inorg. Chim. Acta, 89, L19-L20.

Monoclinic, C2/c, a = 16.319, b = 10.887, c = 30.215 Å, β = 88.00°, D_m = 1.75, Z = 8. R = 0.099 for 1814 reflexions.

The asymmetric unit contains discrete monomers (Fig. 1). The metal atom is bonded to eight S atoms, six from the chelating dialkyldithiocarbamate residues, and two from a 'side on' coordinated S_2 group, the resulting coordination polyhedron being a distorted dodecahedron. The presence of the S_2 group with its short (2.45, 2.38 Å) Ta-S distances, small S(7)-Ta-S(8) angle (50.9°) and S-S distance of 2.07 Å is responsible for the distortion. The Ta-S distances involving the chelate ligands range from 2.51 to 2.60 Å, and the mean S-Ta-S angle and S...S distance are respectively 66.7° and 2.82 Å.

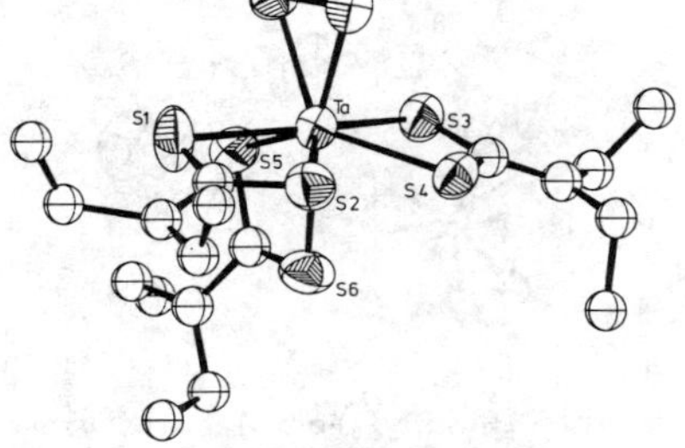

Fig. 1. Molecular structure of $C_{15}H_{30}N_3S_8Ta$.

TRIS(N,N-DIETHYLDITHIOCARBAMATO)HYDROGENSULPHIDOMOLYBDENUM(IV) TETRAHYDROFURAN SOLVATE
$C_{15}H_{31}MoN_3S_7,C_4H_8O$

N. DUPRE, H.M.J. HENDRIKS and J. JORDANOV, 1984. J. Chem. Soc. Dalton, 1463-1465.

Triclinic, $P\bar{1}$, a = 10.148, b = 14.104, c = 10.759 Å, α = 72.39, β = 89.04, γ = 97.84°, Z = 2. Mo radiation, R = 0.070 for 2774 reflexions.

The Mo atom is surrounded by seven S atoms in a distorted pentagonal arrangement (Fig. 1) (Mo-S 2.398(4) (Mo-SH) -2.524(4) Å). SH is axial and two dithiocarbamate ions occupy four of the five equatorial positions. A third dithiocarbamate ligand spans an axial and an equatorial site.

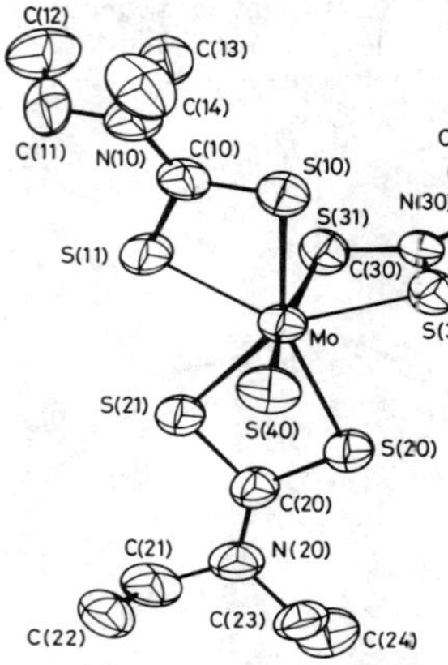

Fig. 1. $C_{15}H_{31}MoN_3S_7,C_4H_8O$: the molecular structure.

TRIS(N-METHYL-N-n-BUTYLDITHIOCARBAMATO)IRON(III)
$C_{18}H_{36}FeN_3S_6$ $[C_6H_{12}NS_2]_3Fe$

A. TERZIS, S. FILIPPAKIS, D. MENTZAFOS, V. PETROULEAS and A. MALLIARIS, 1984. Inorg. Chem., 23, 334-337.

Hexagonal, $P3_1/c$, a = 14.970, c = 14.390 Å, D_m = 1.27, Z = 4. Mo radiation, R = 0.042 for 1098 reflexions.

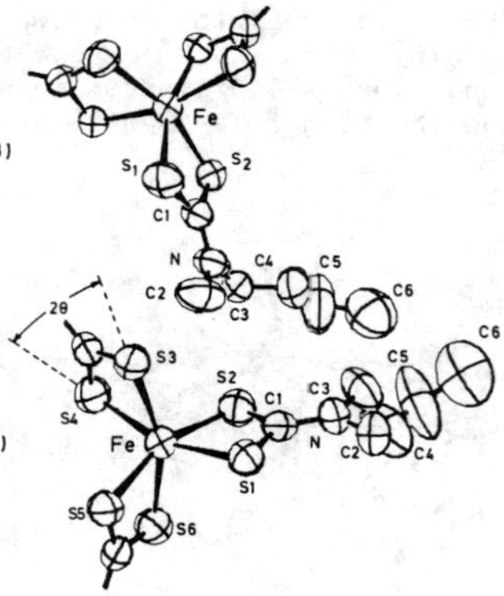

Fig. 1. The two independent $[C_6H_{12}NS_2]_3Fe$ molecules viewed along the threefold axis.

The structure consists of two independent molecules, each having exact threefold symmetry (Fig. 1). The FeS_6 cores are trigonally distorted octahedra with a mean Fe-S distance of 2.367 Å.

(N-BENZYLDITHIOCARBAMATO-S,S')BIS(TRIETHYLPHOSPHINE)PLATINUM
$C_{20}H_{37}NP_2PtS_2$ (I)

(N-BENZYLDITHIOCARBAMATO-S,S')BIS(TRIETHYLPHOSPHINE)PALLADIUM
$C_{20}H_{37}NP_2PdS_2$ (II)

R. SCHIERL, U. NAGEL and W. BECK, 1984. Z. Naturforsch., B39, 649-660.

I. Orthorhombic, $Pna2_1$, a = 14.789, b = 11.478, c = 14.915 Å, D_m = 1.60, Z = 4. Mo radiation, R = 0.063 for 3153 reflexions.

II. Orthorhombic, $Pna2_1$, a = 14.657, b = 11.417, c = 14.801 Å, D_m = 1.398, Z = 4. Mo radiation, R = 0.074 for 2176 reflexions (at -100°C).

The structures are isomorphous; a view of the Pd complex is in Fig. 1. Main dimensions are as follows (with Pt data followed by Pd data in parentheses): Pt-P 2.276 and 2.264(6) (2.305 and 2.278(4)), Pt-S 2.341 and 2.327(5) (2.329 and 2.310(5)), C-S 1.84 and 1.63(2) (1.76 and 1.71(2)) Å, P-Pt-P 100.3(2) (99.7(2)), S-Pt-S 74.7(2) (74.7(2))°.

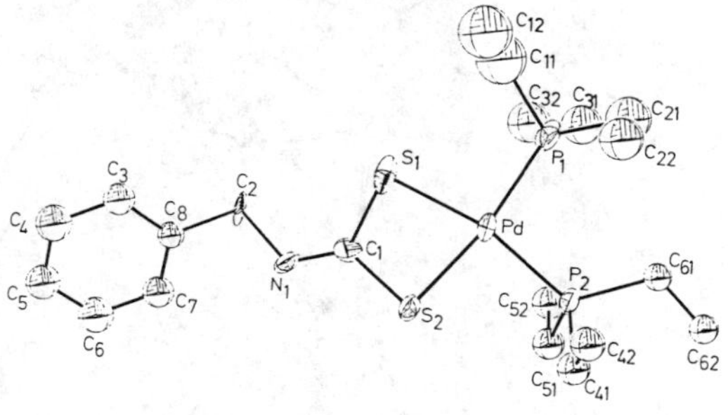

Fig. 1. A view of $C_{20}H_{37}NP_2PdS_2$.

CHLORO(DIETHYLDITHIOCARBAMATO)(TRIPHENYLPHOSPHINE)NICKEL(II)
$C_{23}H_{25}ClNNiPS_2$ $[NiCl(C_5H_{10}NS_2)\{P(C_6H_5)_3\}]$

K. RAMALINGAM, G. ARAVAMUDAN, M. SESHASAYEE and CH. SUBRAMANYAM, 1984. Acta Cryst., C40, 965-967.

Triclinic, $P\bar{1}$, a = 9.808, b = 11.690, c = 14.026 Å, α = 97.63, β = 118.35, γ = 108.81°, Z = 2. Mo radiation, R = 0.052 for 1773 reflexions.

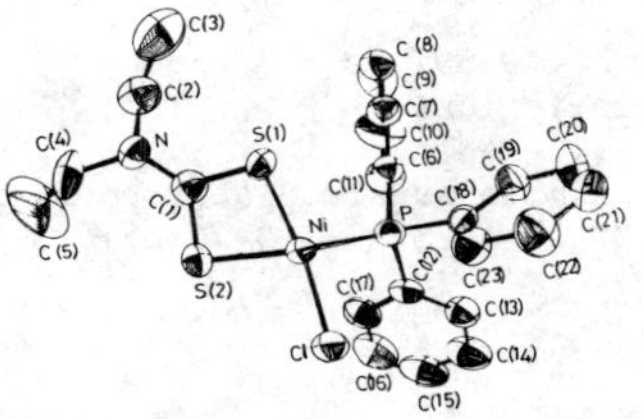

Fig. 1. $C_{23}H_{25}ClNNiPS_2$: perspective view of the molecule.

In the molecule (Fig. 1) Ni displays planar fourfold coordination with Ni-S (trans to Cl) 2.170(4), Ni-S (trans to P) 2.232(3), Ni-Cl 2.187(3), Ni-P 2.204(3) Å, S(1)-Ni-S(2) 78.4(1), S(2)-Ni-Cl 93.2(1), Cl-Ni-P 93.7(1), P-Ni-S(1) 94.7(1)°. There are no significant intermolecular interactions.

TRIS(DIETHYLDITHIOCARBAMATO)DIMETHYLPHENYLPHOSPHINETECHNETIUM(III)
$C_{23}H_{41}N_3PS_6Tc$ $Tc(S_2CNEt_2)_3(Me_2PhP)$

A.S. BATSANOV, Yu.T. STRUCHKOV, B. LORENZ and M. WAHREN, 1984. Z. Anorg. Chem., 510, 117-122.

Orthorhombic, $P2_12_12_1$, a = 8.708, b = 12.012, c = 29.626 Å, Z = 4. Mo radiation, R = 0.044 for 1992 reflexions.

The Tc atom has distorted pentagonal bipyramidal coordination (Fig. 1) with a very short Tc-P distance 2.330(3) Å. The Tc-S distances are in the range 2.430(3)-2.505(3) Å.

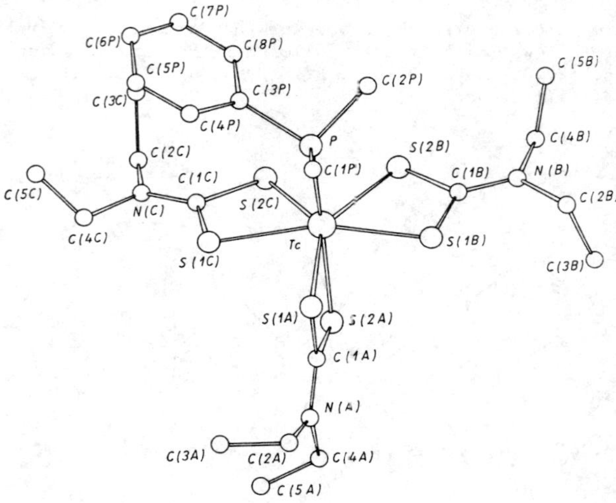

Fig. 1. The $Tc(S_2CNEt_2)_3(Me_2PhP)$ molecule.

TETRA-n-BUTYLAMMONIUM (MALEONITRILEDITHIOLATO)(DIETHYLDISELENOCARBAMATO)NICKELATE(II)
$C_{25}H_{46}N_4NiS_2Se_2$

J. STACH, R. KIRMSE, U. ABRAM, W. DIETZSCH, J.H. NOORDIK, K. SPEE and K.P. KEIJZERS, 1984. Polyhedron, 3, 433-442.

Monoclinic, C2/c, a = 19.932, b = 9.597, c = 37.580 Å, β = 101.32°, Z = 8. Mo radiation, R = 0.101 for 528 reflexions.

The nickel atom has approximate square planar coordination (Fig. 1), with Ni-S 2.08, 2.15(3); Ni-Se 2.31, 2.37(2) Å, S-Ni-S 94.0(12), Se-Ni-Se 80.6(6)°.

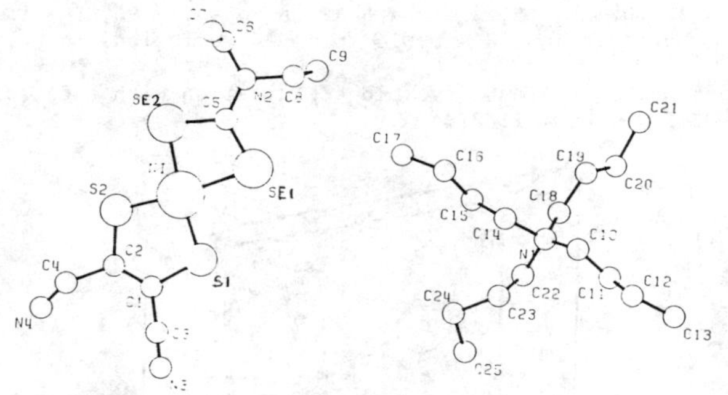

Fig. 1. The anion and cation in $C_{25}H_{46}N_4NiS_2Se_2$.

(μ-OXO)HEXAKIS(N,N-DIETHYLDITHIOCARBAMATO)DIMOLYBDENUM(IV,V) TETRAFLUOROBORATE

$C_{30}H_{60}BF_4Mo_2N_6OS_{12}$ $[Mo_2O(S_2CN(C_2H_5)_2)_6]BF_4$

J.A. BROOMHEAD, M. STEARNS and C.G. YOUNG, 1984. Inorg. Chem., 23, 729-737.

Monoclinic, C2/c, a = 13.24, b = 30.86, c = 12.42 Å, β = 97.89°, D_m = 1.563, Z = 4. Mo radiation, R = 0.035 for 3131 reflexions.

The cation (Fig. 1) contains two pentagonal-bipyramidal units linked by an almost linear (Mo-O-Mo 175.6(2)°) bridging oxo ligand having Mo-O 1.848(2) Å. The basal Mo-S distances are 2.484(3)-2.519(2) Å and the axial Mo-S distance is 2.526(2) Å. The cation has exact C_2 symmetry.

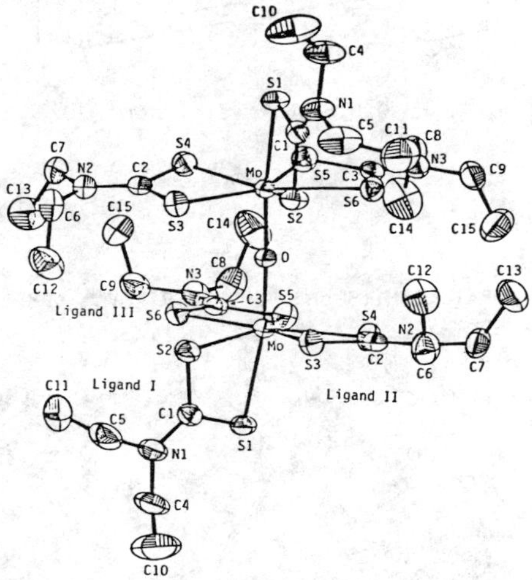

Fig. 1. The $[Mo_2O(S_2CN(C_2H_5)_2)_6]^+$ cation.

TRISODIUM DIOXOOXALATOTRIFLUOROURANIUM HEXAHYDRATE

$C_2F_3Na_3O_6U,6(H_2O)$ $3Na^+[UO_2F_3(C_2O_4)]^{3-},6H_2O$

NGUYEN QUY DAO, I. BKOUCHE-WAKSMAN, M. WALEWSKI and D. CACERES, 1984. Bull. Soc. Chim. Fr., I-129-132.

Triclinic, $P\bar{1}$, a = 12.102, b = 8.227, c = 8.193 Å, α = 117.93, β = 99.26, γ = 95.39°, D_m = 2.79, Z = 2. Mo radiation, R = 0.076 for 1944 reflexions.

The uranyl anion (Fig. 1) has fivefold coordination with U-O1 1.80, U-O2 1.82, U-F 2.24-2.30, U-O3 2.43, U-O4 2.40(4) Å.

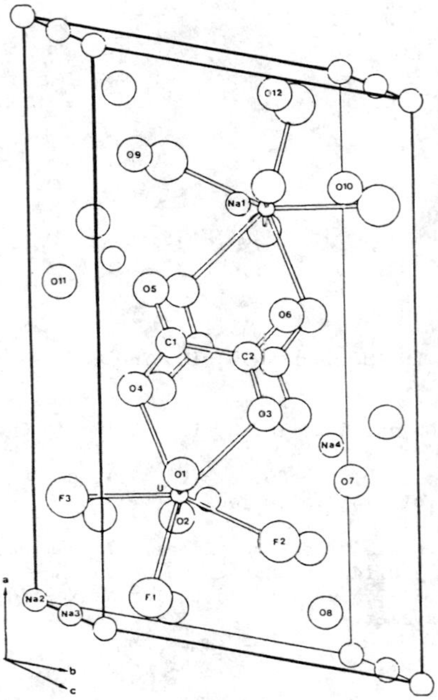

Fig. 1. A view of the $C_2F_3Na_3O_6U,6H_2O$ crystal structure.

SODIUM TRIS(FORMATO)CADMATE
$C_3H_3CdNaO_6$

A.S. ANTSYSHKINA, M.A. PORAI-KOSHITS and V.N. OSTRIKOVA, 1983. Koord. Khim., 9, 418-421 [Sov. J. Coord. Chem., 9, 244-247].

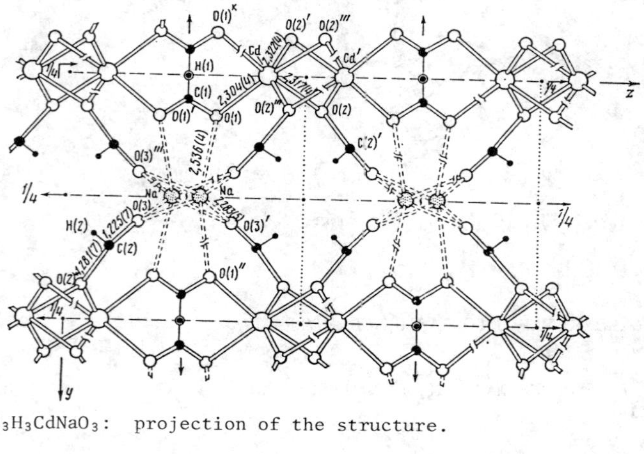

Fig. 1. $C_3H_3CdNaO_3$: projection of the structure.

Orthorhombic, Pcca, a = 6.396, b = 7.275, c = 14.120 Å, Z = 4. Mo radiation, R = 0.033 for 770 reflexions.

The tris(formato)cadmate anions form infinite layers with the Na cations located between the layers (Fig. 1). Within a layer every Cd atom is linked to four different Cd atoms by bridging formato groups. There are two types of formato bridge: single Cd-OCO-Cd bridges with an anti-anti conformation (C-O 1.258(6)) and bridges in which only one O atom of the formato group is coordinated to the Cd atoms (C-O (coord.) 1.281(7), C-O (terminal) 1.229(8) Å). The Cd-O distances range from 2.304 to 2.322(4) Å. The Cd coordination number is six and the coordination polyhedron a distorted octahedron.

TRIS(μ_3-FORMATO)YTTRIUM
$C_3H_3O_6Y$ $Y(HCOO)_3$

N.G. FURMANOVA, Z.P. RAZMANOVA, L.V. SOBOLEVA, I.A. MASLYANITSYN, G. SIEGERT, V.D. SHIGORIN and G.P. SHIPULO, 1984. Kristallografiya, 29, 476-479 [Sov. Phys. Crystallogr., 29, 285-287].

Rhombohedral, R3m, a = 6.132 Å, α = 115.54°, Z = 3. Mo radiation, R = 0.050 for 142 reflexions.

The coordination polyhedron of the Y atoms is a triply-capped trigonal prism with crystallographic symmetry 3m. Two triplets of O(2) atoms related by the c translation form a prism, while a triplet of O(1) atoms forms the caps, (Fig. 1). The O(1) atom in a formate group forms a bond in anti configuration with one Y atom, while the O(2) atom forms syn and anti bonds with two Y atoms. The formate groups in the crystal are oriented in such a way that the H atom of one group lies toward the O(1) atom of another. The C...O distance is 3.14 Å and may represent a C-H...O hydrogen bond.

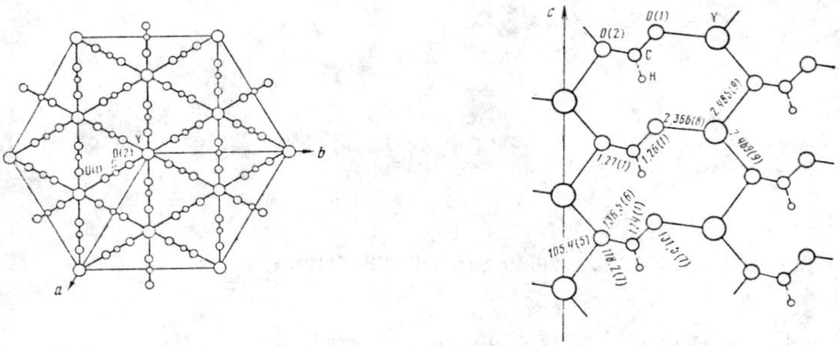

Fig. 1. Projection of the structure of $C_3H_3O_6Y$ on the ab plane (left), and a fragment of the structure in the m plane (right).

AMMONIUM TRIS(μ_2-FORMATO)CADMATE
$(C_3H_7CdNO_6)_n$ (I)

POTASSIUM TRIS(μ_2-FORMATO)CADMATE
$(C_3H_3CdKO_6)_n$ (II)

A.S. ANTSYSHKINA, M.A. PORAI-KOSHITS, V.N. OSTRIKOVA and G.G. SADIKOV, 1983. Koord. Khim., 9, 855-858 [Sov. J. Coord. Chem., 9, 489-493].

I. Orthorhombic, $Pna2_1$, a = 7.0063, b = 9.5232, c = 10.8632 Å, Z = 4. Mo radiation, R = 0.042 for 1903 reflexions.

II. Monoclinic, C2/c, a = 10.059, b = 9.171, c = 7.145 Å, β = 94.46°, Z = 4. Mo radiation, R = 0.070 for 770 reflexions.

The structural motifs of I and II are very similar. They are based on a $[Cd(HCOO)_3]^{3-}$ formatocadmate anionic skeleton whose spaces accommodate the NH_4^+ or K^+ cations (Fig. 1). The anionic skeleton is formed by bidentate bridging HCOO groups which link two different Cd atoms. The O(1)C(1)O(2) and O(5)C(3)O(6) formate groups join the Cd atoms to form polymer layers which are parallel to the xyO plane of the crystal. The layers are joined to form a three-dimensional network by the O(3)C(2)O(4) formate groups. In I all the formates have an anti-syn conformation while in II the O(3)C(2)O(4) formates have an anti-anti conformation. Each Cd atom coordinates six O atoms (Cd-O 2.282-2.322(6) Å). The K ions in II form eight bonds with K-O distances ranging from 2.73 to 2.89 Å.

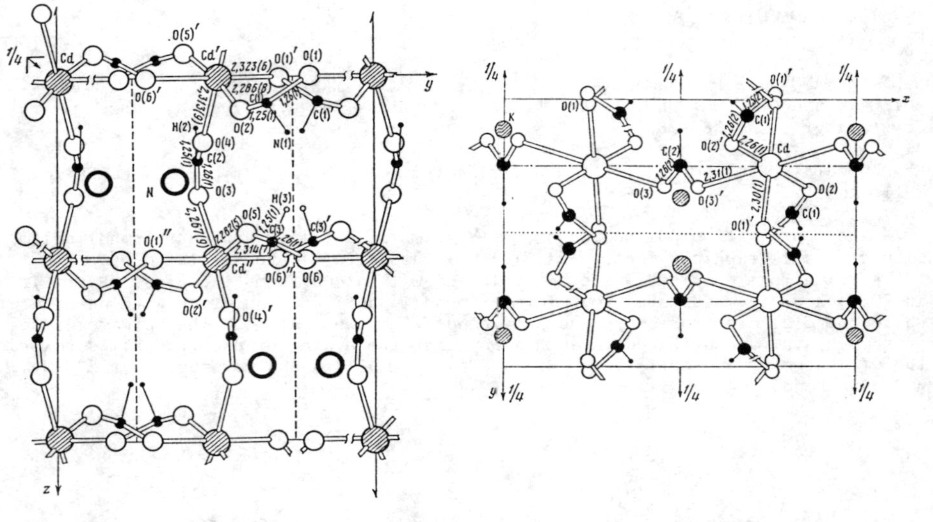

Fig. 1. Projections of the structures: $(C_3H_7CdNO_6)_n$ (left) onto the Oyz plane and $(C_3H_3CdKO_6)_n$ onto the xyO plane.

POTASSIUM (μ-CHLORO)-TETRAKIS(μ-FORMATO)DIMOLYBDENUM(II)
$C_4H_4ClKMo_2O_8$ (I) $Mo_2(O_2CH)_4.KCl$

TETRAKIS(μ-FORMATO)DIMOLYBDENUM(II) (β and γ forms)
$C_4H_4Mo_2O_8$ (II,III) $Mo_2(O_2CH)_4$

G.A. ROBBINS and D.S. MARTIN, 1984. Inorg. Chem., 23, 2086-2093.

I. Triclinic, P$\bar{1}$, a = 8.253, b = 10.684, c = 6.769 Å, α = 89.52, β = 109.73, γ = 87.03°, Z = 2. Mo radiation, R = 0.055 for 1803 reflexions.

II β-form. Monoclinic, $P2_1/c$, a = 5.485, b = 12.365, c = 19.862 Å, β = 90.24°, Z = 6. Mo radiation, R = 0.049 for 2492 reflexions.

III γ-form. Monoclinic, $P2_1/a$, a = 7.939, b = 11.193, c = 5.5271 Å, β = 110.86°, Z = 2. Mo radiation, R = 0.032 for 743 reflexions.

The structure of I consists of zigzag $Mo_2(O_2CH)_4$-Cl^- chains (Fig. 1) containing two crystallographically independent, centrosymmetric, $Mo_2(O_2CH)_4$ molecules. K^+ ions pack with eightfold coordination of five O's and three Cl^-'s. Bond lengths for I: Mo-Mo 2.109(2) and 2.102(2), Mo-Cl 2.848(2) and 2.880(2), Mo-O 2.103(5)-2.142(5), K-Cl 3.096(3)-3.708(3), K-O 2.730(6)-3.149(6) Å. The structures of II and III, like that of the α-form (1), consist of one-dimensional chains of molecules linked by intermolecular axial Mo-O interactions (Mo-O 2.643 and 2.646 for α; 2.639(5), 2.638(5), and 2.643(5) for β; 2.701(3) Å for γ-$Mo_2(O_2CH)_4$). Packing for the β and γ forms (II and III) is shown in Fig. 2. In the β-form there are two sets of molecules, a twofold set occupying positions of $\bar{1}$ symmetry, and a fourfold set occupying general positions with no symmetry while in the γ-form all molecules are of the same type having $\bar{1}$ symmetry. Bond lengths for II and III: Mo-Mo 2.093(1) and 2.0916(9), Mo-O_{eq} 2.102(5)-2.145(5) Å for the β-form, Mo-Mo 2.0910(8), Mo-O_{eq} 2.103(4)-2.130(4) Å for the γ-form.

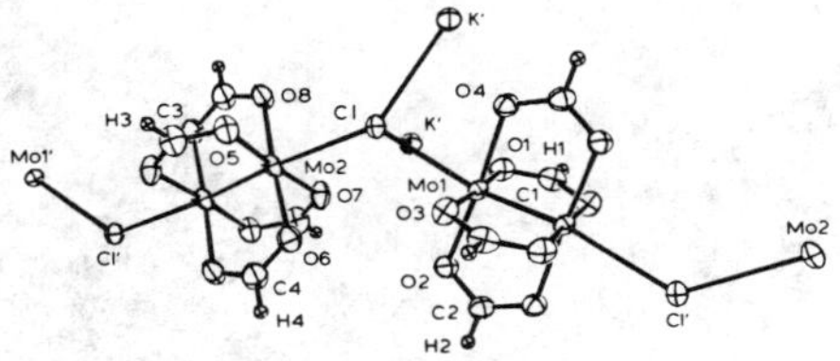

Fig. 1. The structure of $Mo_2(O_2CH)_4$.KCl.

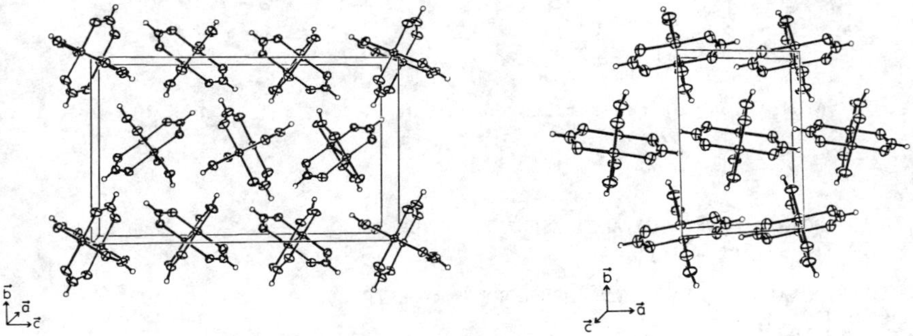

Fig. 2. The structures of β (left) and γ-$Mo_2(O_2CH)_4$ (right) viewed down the one-dimensional chain direction.

1. Structure Reports, 42B, 778.

POTASSIUM AQUABIS(μ_2-FORMATO)DIFORMATO(O,O'.O)YTTRATE
$(C_4H_6KO_9Y)_n$ (I)

PENTAPOTASSIUM OCTAKIS(FORMATO)YTTRATE
$C_8H_8K_5O_{16}Y$ (II)

A.S. ANTSYSHKINA, M.A. PORAI-KOSHITS, V.N. OSTRIKOVA and E.V. PETROVA, 1983. Koord. Khim., 9, 986-992 [Sov. J. Coord. Chem., 9, 561-567].

I. Triclinic, P1, a = 6.198, b = 6.620, c = 7.249 Å, α = 107.65, β = 113.62, γ = 98.32°, Z = 1. Mo radiation, R = 0.043 for 3659 reflexions.

II. Monoclinic, C2/c, a = 11.829, b = 11.868, c = 14.890 Å, β = 109.02°, Z = 4. Mo radiation, R = 0.058 for 2377 reflexions.

The structural units of I are polymeric layers which form as a result of the bridging function of two of the four independent HCOO groups (Fig. 1). The third HCOO group coordinates to yttrium as a monodentate ligand, and the fourth forms a four-membered chelate ring with it. The chelating group and the two bridging groups are symmetric (C-O 1.25(1) Å), and the monodentate group is asymmetric (C-O_b 1.28, C-O_t 1.22(1) Å). The coordination number of yttrium is eight and the overall geometry of the polyhedron is close to bicapped trigonal prismatic. The Y-O distances lie in the range 2.26 to 2.43 Å and the potassium ions form bonds with seven O atoms (K-O 2.64-3.0 Å). The structure of II is ionic with the potassium cations, which occupy three crystallographically inequivalent positions, arranged in layers which alternate with layers of octaformatoyttrate anions. All the HCOO groups are coordinated to the yttrium atom as monodentate ligands, forming a polyhedron in the shape of a square antiprism. The Y-O distances range from 2.35 to 2.38 Å. The K-O distances range from 2.72-3.0 Å.

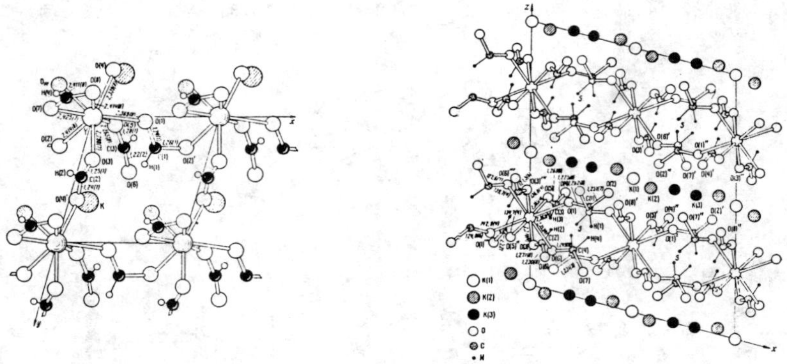

Fig. 1. Projections of the structures: $(C_4H_6KO_9Y)_n$ (left) and $C_8H_8K_5O_{16}Y$ (right).

BIS(μ_2-DITHIOOXALATO)TRIAQUAMANGANESENICKEL HYDRATE
$(C_4H_6MnNiO_7S_4)_n$,4.5n(H_2O) (I)

BIS(μ_2-DITHIOOXALATO)TRIAQUAMANGANESECOPPER HYDRATE
$(C_4H_6CuMnO_7S_4)_n$,4.5n(H_2O) (II)

BIS(μ_2-DITHIOOXALATO)TRIAQUAMANGANESEPLATINUM HYDRATE
$(C_4H_6MnO_7PtS_4)_n$,4.5n(H_2O) (III)

A. GLEIZES and M. VERDAGUER, 1984. J. Am. Chem. Soc., 106, 3727-3737.

I. Monoclinic, $P2_1/c$, a = 11.575, b = 20.654, c = 7.323 Å, β = 103.73°, D_m = 1.92, Z = 4. Mo radiation, R = 0.059 for 894 reflexions.

II. Monoclinic, $P2_1/c$, a = 11.692, b = 20.665, c = 7.360 Å, β = 103.84°, Z = 4. Mo radiation, R = 0.064 for 1149 reflexions.

III. Monoclinic, $P2_1/c$, a = 11.772, b = 20.806, c = 7.266 Å, β = 103.52°, Z = 4. Mo radiation, R = 0.066.

The compounds (and the Pd complex for which cell data are also determined) are isostructural and consist (Fig. 1) of infinite chain molecules $\{[Mn(H_2O)_3](O_2C_2S_2)\text{-}A(S_2C_2O_2)\}_\infty$ crisscrossing c-glide planes and therefore stacked along these planes. The Pt complex appears to be twinned. In all the structures the A atoms (A = Cu, Ni, Pd, Pt) have square-planar coordination (Ni-S 2.167-2.178(5), Cu-S 2.29-2.30(1), Pt-S 2.267-2.2886 Å) and the Mn atoms are heptacoordinated to O atoms in a pentagonal bipyramidal fashion with Mn-O 2.12-2.38(1) Å.

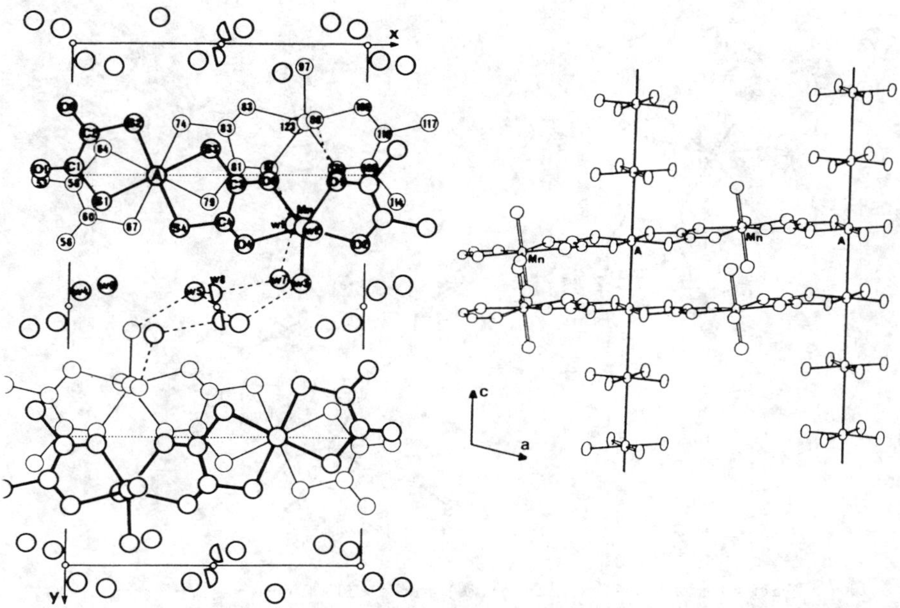

Fig. 1. Two views of the crystal structures of catena-μ-dithiooxalato structures I - III (A = Ni, Cu, Pd or Pt).

cis-DICHLOROBIS(OXAMIDE OXIME)NICKEL(II)
$C_4H_{12}Cl_2N_8NiO_4$ (I)

cis-BIS(SULPHANILATO)BIS(OXAMIDE OXIME)NICKEL(II) DIHYDRATE
$C_{16}H_{24}N_{10}NiO_{10}S_2,2(H_2O)$ (II)

catena-(μ_2-PHTHALATO)BIS(OXAMIDE OXIME)NICKEL(II) TETRAHYDRATE
$(C_{12}H_{16}N_8NiO_8)_n,4_n(H_2O)$ (III)

H. ENDRES, 1984. Z. Anorg. Chem., 513, 78-88.

I. Monoclinic, $P2_1/n$, a = 6.641, b = 14.086, c = 13.473 Å, β = 96.26°, Z = 4. Mo radiation, R = 0.031 for 4090 reflexions.

II. Monoclinic, $P2_1$, a = 6.879, b = 14.305, c = 13.390 Å, β = 103.62°, Z = 2. Mo radiation, R = 0.067 for 2693 reflexions.

III. Monoclinic, $P2_1/c$, a = 10.633, b = 11.324, c = 17.680 Å, β = 98.25°, Z = 4. Mo radiation, R = 0.110 for 3290 reflexions.

Views of the three complexes are in Fig. 1. In I and II the Ni atom has distorted octahedral coordination. III has polymeric structure with infinite chains formed about 2_1 screw axes. Some dimensions are: (I) Ni-Cl 2.466, 2.470(1), Ni-N 2.019-2.106(2) Å; (II) Ni-N 2.026-2.148(9), Ni-O 2.158 Å; (III) Ni-N 2.049-2.101(8), Ni-O 2.020, 2.046(7) Å.

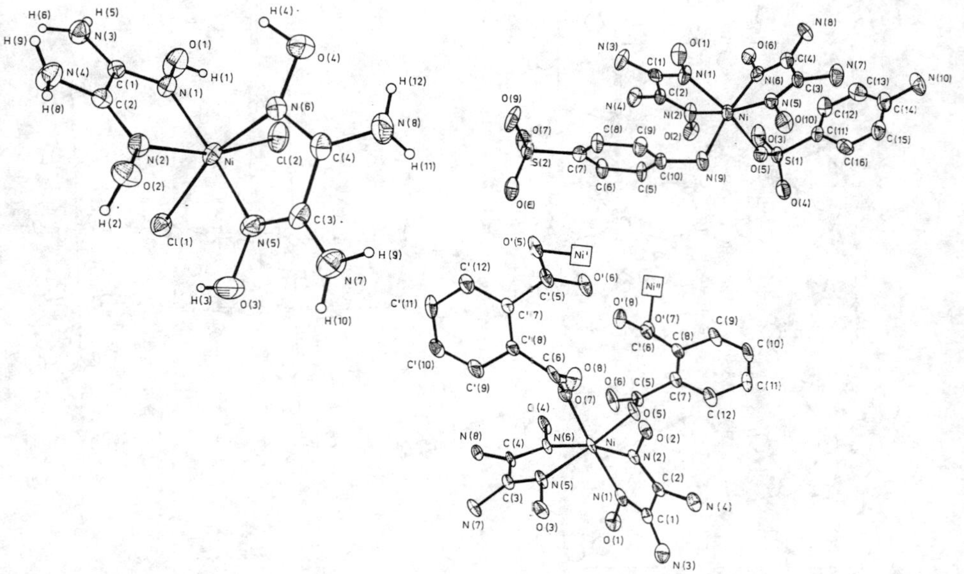

Fig. 1. Views of the bis(oxamide oxime) nickel(II) complexes (I) (top left), (II) (top right) and (III) (bottom).

SILVER(I) FUROATE
$C_5H_3AgO_3$ (I) $Ag(C_5H_3O_3)$

COPPER(II) FUROATE DIHYDRATE
$C_{10}H_6CuO_6,2H_2O$ (II) $Cu(C_5H_3O_3)_2,2H_2O$

F. CHARBONNIER, M. PETIT-RAMEL, R. FAURE and H. LOISELEUR, 1984. Rev. Chim. Miner., 21, 601-610.

I. Monoclinic, $P2_1/c$, a = 5.668, b = 17.307, c = 6.500 Å, β = 125.11, Z = 4. Mo radiation, R = 0.098 for 1261 reflexions.

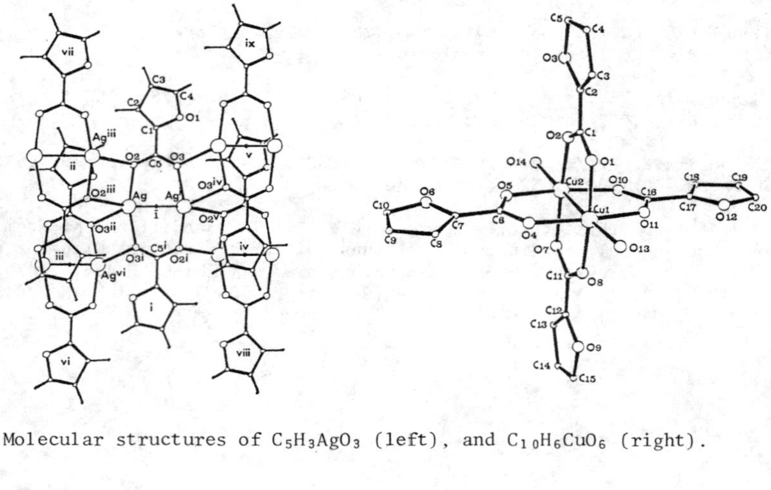

Fig. 1. Molecular structures of $C_5H_3AgO_3$ (left), and $C_{10}H_6CuO_6$ (right).

II. Monoclinic, $P2_1/c$, a = 10.119, b = 13.018, c = 19.950 Å, β = 90.69(1)°, Z = 8. Mo radiation, R = 0.062 for 3574 reflexions.

In (I) two Ag atoms and two carboxylato groups form a centrosymmetric eight-membered ring $(-Ag-O-C-O-)_2$ with an Ag-Ag bridge (2.881(2) Å) (Fig. 1). Each Ag atom is also coordinated by two carboxylic O atoms belonging to two neighbouring rings (Ag-O(3)^{i} 2.19(1), Ag-O(2) 2.24(1), Ag-O(2)^{iii} 2.66(1), and Ag-O(3)^{ii} 2.70(1) Å. The copper complex (II) consists of binuclear dimeric units with approximate $\bar{1}$ symmetry. The copper atoms are coordinated to two water molecules along the Cu-Cu vector (Cu(1)-Cu(2) 2.638(1), Cu(1)-O(13) 2.131(5), Cu(2)-O(14) 2.140(5) Å), and to four carboxylato groups (Cu(1)-O(1) 1.961(5), Cu(1)-O(8) 1.993(5), Cu(1)-O(4) 1.952(5), Cu(1)-O(11) 1.946(5), Cu(2)-O(2) 1.959(5), Cu(2)-O(5) 1.973(5), Cu(2)-O(7) 1.968(4), Cu(2)-O(10) 1.948(5) Å).

POTASSIUM catena-μ-OXALATO-BIS(OXALATO)AQUANEODYMATE(III) DIHYDRATE
$C_6H_2K_3NdO_{13},2H_2O$ (I) $K_3[Nd(C_2O_4)_3(OH_2)],2H_2O$

POTASSIUM catena-μ-OXALATO-BIS(OXALATO)AQUAGADOLINATE(III) DIHYDRATE
$C_6H_2GdK_3O_{13},2H_2O$ (II) $K_3[Gd(C_2O_4)_3(OH_2)],2H_2O$

I.A. KAHWA, F.R. FRONCZEK and J. SELBIN, 1984. Inorg. Chim. Acta, 82, 161-166.

I. Triclinic, $P\bar{1}$, a = 9.416, b = 8.502, c = 9.799 Å, α = 88.91, β = 82.17, γ = 96.28°, Z = 2. Mo radiation, R = 0.036 for 2529 reflexions.

II. Triclinic, $P\bar{1}$, a = 9.374, b = 8.474, c = 9.739 Å, α = 88.34, β = 81.62, γ = 96.58°, Z = 2. Mo radiation, R = 0.021 for 2512 reflexions.

The Nd^{3+} and Gd^{3+} cations in the isomorphous structures I and II are 9-coordinate in a distorted tricapped trigonal prismatic environment formed by four oxalates and a water molecule (Fig. 1). The oxalates coordinate to the metal ions via five-membered rings; two of these four oxalate ligands serve as bridging ligands to coordinate to neighbouring metal ions leading to an infinite chain of $[Ox(LnOx_2.H_2O)]_n$ units. The average Nd-O distance is 2.494 and that of Gd-O 2.447 Å. Compounds with Ln = Sm, Eu and Tb are isomorphous with these two structures as determined by their cell dimensions.

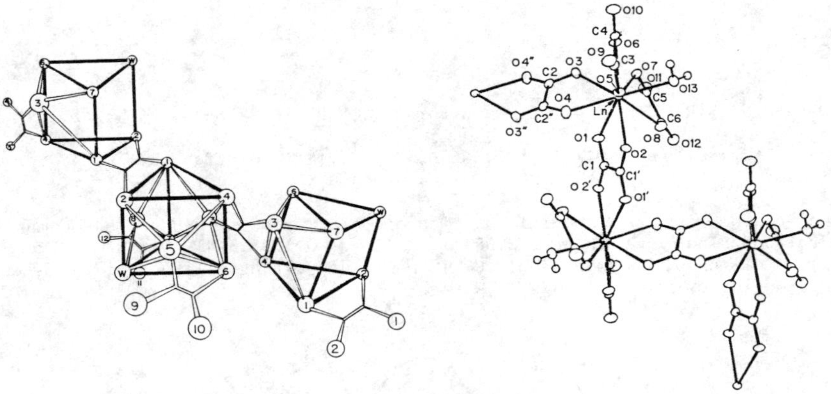

Fig. 1. A portion of the structure of $C_6H_2K_3LnO_{13},2H_2O$ showing all nine coordinated O atoms comprising the tricapped trigonal prism and portions of the two neighbouring oxalato bridged trigonal prisms (left), and a portion of the molecular structure showing atomic numbering scheme (right).

DIPOTASSIUM TRIS(HYDROXYACETATO)OXOVANADIUM(IV) MONOHYDRATE
$C_6H_8K_2O_{10}V,H_2O$ $K_2(C_6H_8O_{10}V),H_2O$

HUANG JINLING, ZHOU KANGJING and LU JIAXI, 1984. J. Struct. Chem., 3, 57-60.

Monoclinic, $P2_1/b$, a = 10.075, b = 9.857, c = 14.314 Å, γ = 115.29°, Dm = 1.96, Z = 4. Mo radiation, R = 0.060 for 1714 reflexions.

The V atom has distorted octahedral coordination (Fig. 1) with two hydroxyacetato groups as bidentate and one as a monodentate ligand. Main bond lengths are V-O(8) 1.613(4), V-O(1) 2.246(4), V-O(2) 2.039(4), V-O(4) 2.013(4), V-O(5) 2.047(4), V-O(6) 1.923(4) Å, cis O-V-O 73.6-98.9(2)°.

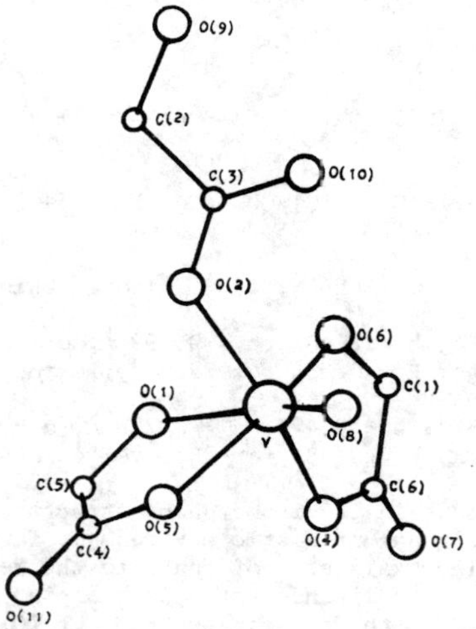

Fig. 1. A view of the $[C_6H_8O_{10}V]^{2-}$ anion.

catena-TRIS(μ_2-ACETATO)SCANDIUM(III)
$(C_6H_9O_6Sc)_n$

R. FUCHS and J. STRAHLE, 1984. Z. Naturforsch., B39, 1662-1663.

Hexagonal, $P6_3/mcm$, a = 7.911, c = 8.866 Å, Z = 2. Mo radiation, R = 0.048 for 213 reflexions.

The Sc atom lies on a special position of point symmetry D_{3d} and has compressed octahedral coordination with Sc-O 2.084(1) Å, O1-Sc-O1' 93.5(4), O1-Sc-O1" 86.5(4)° (Fig. 1). Columns of polymeric $Sc(MeCO_2)_3$ units are arranged like an ideal hexagonal close packing of cylindrical rods.

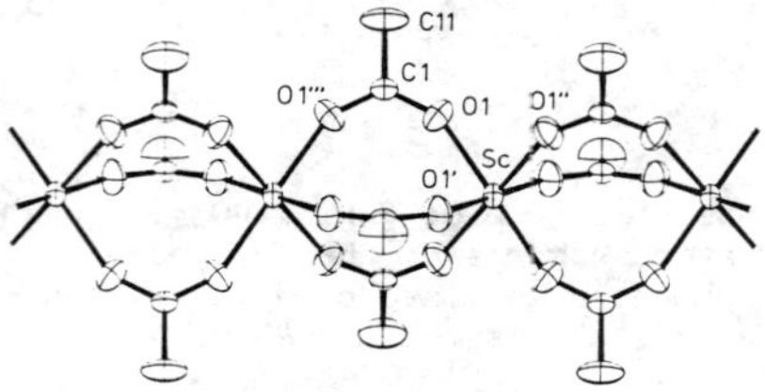

Fig. 1. A view of the polymer structure of $(C_6H_9O_6Sc)_n$.

BIS[(μ_2-β-ALANINATO-O,O')SILVER] DINITRATE
$C_6H_{14}Ag_2N_4O_{10}$ $C_6H_{14}Ag_2N_2O_4^{2+}$,2(NO_3^-)

M.E. KAMWAYA, E. PAPAVINASAM, S.G. TEOH and R.K. RAJARAM, 1984. Z. Kristallogr., 169, 51-56. Acta Cryst., C40, 1318-1320.

Monoclinic, $P2_1/n$, a = 6.656, b = 8.280, c = 12.975 Å, β = 94.90°, D_m = 2.382, Z = 2. Cu radiation, R = 0.089 for 545 reflexions.

In this centrosymmetric dimeric complex (Fig. 1) the Ag...Ag separation is 2.855 Å. The nitrate ions are bonded to alanine moieties by N-H...O hydrogen bonds.

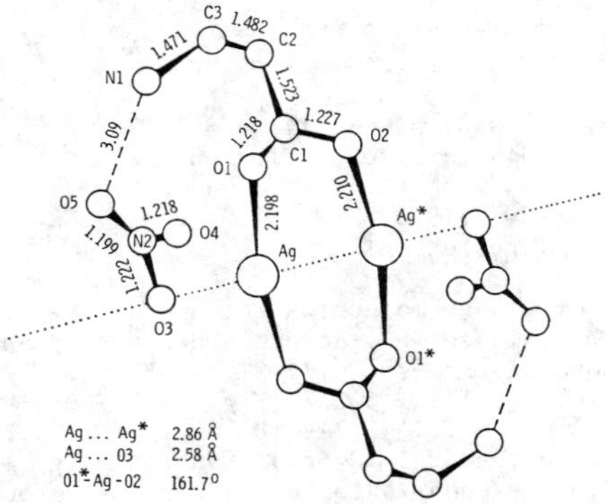

Fig. 1. $C_6H_{14}Ag_2N_4O_{10}$ with bond lengths (Å).

DIAQUANITROSYL(PYRIDINE-2,6-DICARBOXYLATO)CHROMIUM
$C_7H_7CrN_2O_7$ $Cr(C_7H_3NO_4)(NO)(H_2O)_2$

K. WIEGHARDT, U. QUILITZSCH and J. WEISS, 1984. Inorg. Chim. Acta, 89, L43-L45.

Orthorhombic, Pcnb, a = 9.245, b = 9.745, c = 11.186 Å, Z = 4. Mo radiation, R = 0.046 for 1109 reflexions.

The overall symmetry of the neutral molecule is C_2; the twofold rotation axis passes through atoms O(4), N(2), Cr, N(1), C(4) and H(4) (Fig. 1). The nitrosyl group is bound in a trans-position with respect to the pyridine N(1) atom, and the Cr-N(2) distance of 1.699 Å indicates considerable double bond character. Other important distances are: Cr-N(1) 2.012, Cr-O(2) 2.027, Cr-O(3) 2.009 and N(2)-O(4) 1.182 Å. Distances within the carboxyl group are: C(1)-O(2) 1.292, C(1)-O(1) 1.223 and C(1)-C(2) 1.501 Å. The coordination about the Cr atom is distorted octahedral, with angles ranging from 77.4° (N(1)-Cr-O(2)) to 102.6° (N(2)-Cr-O(2)).

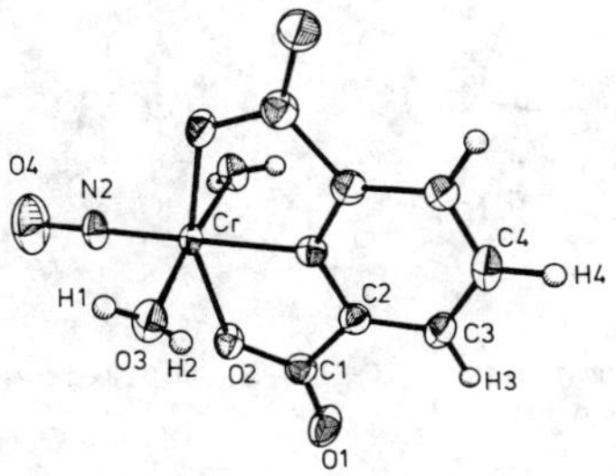

Fig. 1. Molecular structure of $C_7H_7CrN_2O_7$.

(PHENYLTHIO)ACETIC ACID
$C_8H_8O_2S$ (I) $C_6H_5SCH_2COOH$

DIAQUABIS[(PHENYLTHIO)ACETATO]ZINC(II)
$C_{16}H_{18}O_6S_2Zn$ (II) $(H_2O)_2Zn(C_6H_5SCH_2COO)_2$

catena-{AQUABIS[(PHENYLTHIO)ACETATO]CADMIUM(II)}
$C_{16}H_{16}CdO_5S_2$ (III) $[(H_2O)Cd(C_6H_5SCH_2COO)_2]_n$

T.C.W. MAK, W.-H. YIP, G. SMITH, E.J. O'REILLY and C.H.L. KENNARD, 1984. Inorg. Chim. Acta, 84, 57-64.

I. Monoclinic, $P2_1/n$, a = 7.666, b = 5.638, c = 18.939 Å, β = 99.33°, D_m = 1.38, Z = 4. Mo radiation, R = 0.038 for 1048 reflexions.

II. Monoclinic, C2/c, a = 32.048, b = 5.314, c = 10.725 Å, β = 101.20°, D_m = 1.62, Z = 4. Mo radiation, R = 0.049 for 1205 reflexions.

III. Orthorhombic, $Pca2_1$, a = 33.826, b = 5.119, c = 9.872 Å, D_m = 1.87, Z = 4. Mo radiation, R = 0.042 for 951 reflexions.

Compound I consists of planar molecules which exist as centrosymmetric hydrogen-bonded cyclic dimers. The thioacetic side chain has a synplanar-synplanar conformation with the following bond distances: C(1)-S(7) 1.770, S(7)-C(8) 1.792, C(8)-C(9) 1.495, C(9)-O(10) 1.216, C(9)-O(11) 1.302 Å (Fig. 1). The O-H...O hydrogen bonds have O...O distances of 2.660 Å. The C(1)-S(7)-C(8) angle is 103.3°. Compound II is monomeric with a distorted octahedral MO_6 coordination (Fig. 1) involving four O atoms from two symmetrical bidentate phenylthioacetate carboxyl groups (Zn-O 2.176, 2.204 Å) and two from cis-related water molecules (Zn-O 2.002 Å). Atoms C(2) to C(6) of the phenyl ring are disordered, owing to rotation about the S(7)-C(1) bond. Bonds C(9)-O(10), C(9)-O(11), C(9)-C(8), C(8)-S(7) and S(7)-C(1) are respectively 1.245, 1.256, 1.500, 1.792, and 1.747 Å. In compound III an octahedral MO_5S arrangement occurs, including a bidentate chelate (S,O) ring and a unidentate carboxylate system. Important distances are: Cd-O(10) 2.335, Cd-S(7) 2.738, Cd-O(water) 2.294 and Cd-O(11) (ligand A) 2.276 Å (Fig. 1). The second carboxyl O from each of ligands A and B acts as a bridging atom between complex centres, completing the octahedral coordination (Cd-O(10A) 2.263 and Cd-O(11B'') 2.254 Å). This results in a two-dimensional polymeric structure down c.

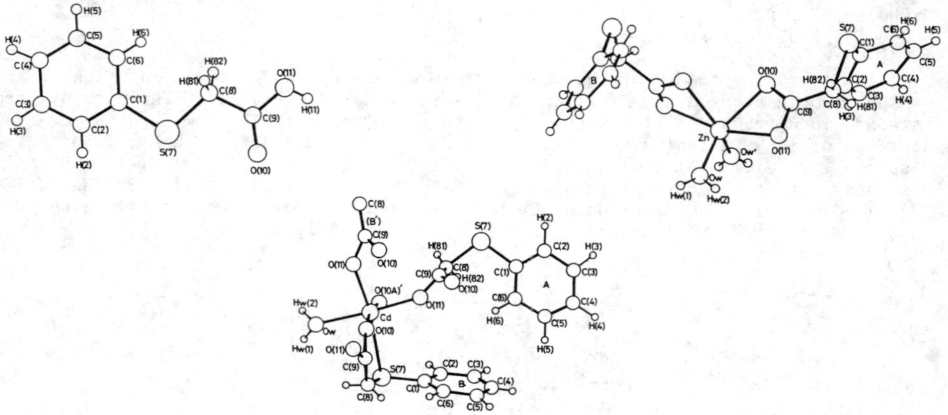

Fig. 1. Molecular conformations of $C_8H_8O_2S$ (top left), $C_{16}H_{18}O_6S_2Zn$ (top right), and $C_{16}H_{16}CdO_5S_2$ (bottom).

DILITHIUM TETRAKIS(μ-ACETATO)DICHLORODIRHODIUM(II) OCTAHYDRATE
$C_8H_{12}Cl_2Li_2O_8Rh_2,8H_2O$ $Li_2Rh_2(O_2CCH_3)_4Cl_2,8H_2O$

V.M. MISKOWSKI, W.P. SCHAEFER, B. SADEGHI, B.D. SANTARSIERO and H.B. GRAY, 1984. Inorg. Chem., 23, 1154-1162.

Tetragonal, I4/m, a = 8.8498, c = 14.5989 Å, Z = 2. Mo radiation, R = 0.023 for 1029 reflexions.

The $[Rh_2(O_2CCH_3)_4Cl_2]^{2-}$ anion resides in a position of C_{4h} site symmetry. Bond lengths: Rh-Rh 2.397(1), Rh-Cl 2.601(1), Rh-O 2.034(1) Å.

DICHLOROTETRAKIS(μ_2-ACETATO)DIOSMIUM
$C_8H_{12}Cl_2O_8Os_2$ (I) $Os_2(OOCCH_3)_4Cl_2$

DICHLOROTETRAKIS(μ_2-PROPIONATO)DIOSMIUM
$C_{12}H_{20}Cl_2O_8Os_2$ (II) $Os_2(OOCCH_2CH_3)_4Cl_2$

F.A. COTTON, A.R. CHAKRAVARTY, D.A. TOCHER and T.A. STEPHENSON, 1984. Inorg. Chim. Acta, 87, 115-119.

I. Monoclinic, $P2_1/n$, a = 6.546, b = 8.950, c = 12.533 Å, β = 90.17°, Z = 2. Mo radiation, R = 0.037 for 1540 reflexions.

II. Monoclinic, $P2_1/n$, a = 6.792, b = 10.512, c = 13.372 Å, β = 89.27°, Z = 2. Mo radiation, R = 0.028 for 775 reflexions.

The crystals of I and II are quasi-isomorphous. In each case the dinuclear molecule lies on a crystallographic centre of symmetry (Fig. 1) and the molecules have approximate D_{4h} symmetry. The coordination about each unique Os atom is essentially octahedral and the mean Os-O distances of 2.009 Å in I and 2.014 Å in II are not significantly different. The Os-Os distances are 2.314 and 2.316 Å in I and II respectively, and the Os-Cl distances 2.448 and 2.430 Å.

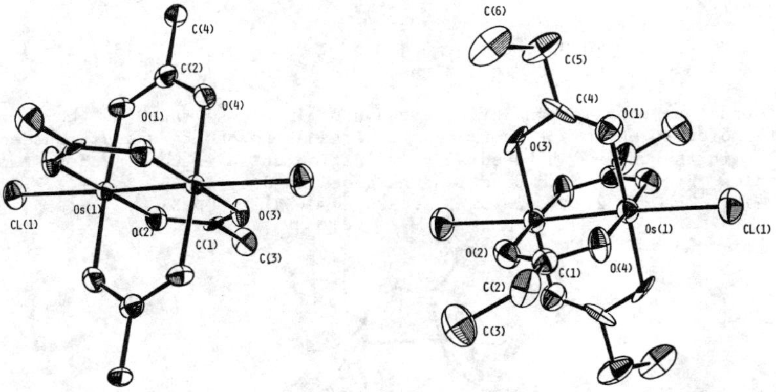

Fig. 1. Molecular structures of $C_8H_{12}Cl_2O_8Os_2$ (left), and $C_{12}H_{20}Cl_2O_8Os_2$ (right).

DIAQUABIS((R,R)-TARTRATO-O[1],O[2]:O[3],O[4])DINICKEL(II) TRIHYDRATE
$C_8H_{12}Ni_2O_{14},3H_2O$ $Ni_2((R,R)\text{-}C_4H_4O_6)_2(H_2O)_5$

L.J. BOSTELAAR, R.A.G. de GRAAFF, F.B. HULSBERGEN, J. REEDIJK and W.M.H. SACHTLER, 1984. Inorg. Chem., 23, 2294-2297.

Orthorhombic, $P2_12_12$, a = 7.805, b = 11.068, c = 8.974 Å, D_m = 2.09, Z = 2. Mo radiation, R = 0.026 for 2083 reflexions (at -162°C).

Each dimeric unit (Fig. 1) contains one tartrato dianion chelated to two Ni atoms, with the use of two hydroxyl oxygen (Ni-O 2.065(3) and 2.093(3) Å) and two carboxyl oxygen atoms (Ni-O 1.997(3) and 2.001(3) Å) as donor atoms. A second tartrato dianion chelates the same two Ni atoms in the same way, and also binds to two Ni atoms of a neighboring dimeric unit through the other carboxyl O atoms (Ni-O 2.042(3) Å) giving rise to a polymeric structure. The octahedral geometry about Ni is completed by such a carboxyl O atom of a neighboring dimer and a water molecule (disordered) (Ni-OH_2 1.979(5), 2.133(6) Å).

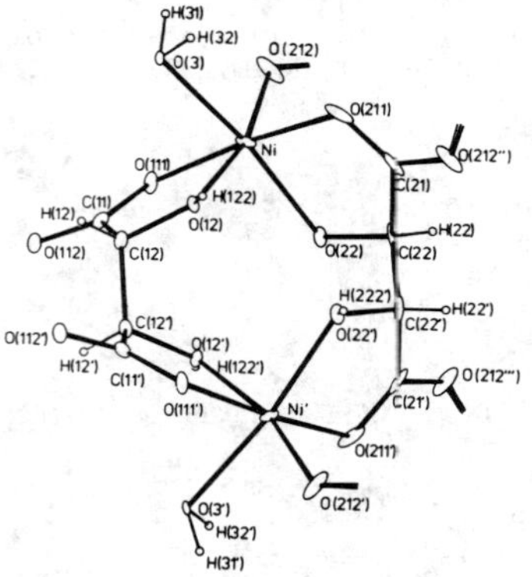

Fig. 1. The dimeric unit of $Ni_2((R,R)\text{-}C_4H_4O_6)_2(H_2O)_2$.

DIAQUA(DEMETHYLCANTHARATE)ZINC(II)
$(C_8H_{12}O_7Zn)_n$

WANG HONGHUI, ZHU NAIJUE, FU HENG, LI RONGCHANG and WANG KUI, 1984. J. Struct. Chem., 3, 87-90.

Monoclinic, $P2_1/a$, a = 10.907, b = 8.502, c = 11.083 Å, β = 107.15°, Z = 4. Mo radiation, R = 0.0406 for 1587 reflexions.

The structure contains an infinite chain of Zn atoms linked by the demethylcantharate moieties which act as tridentate ligands to one Zn and as monodentate to the next Zn in the chain. Octahedral coordination at Zn is completed by two cis H_2O molecules (Fig. 1). Main bond lengths are: Zn-O(1) 2.064, Zn-O(4) 2.143, Zn-O(5) 2.181, Zn-O(6) 2.114, Zn-O(7) 2.052, Zn-O(3a) 2.009(3) Å.

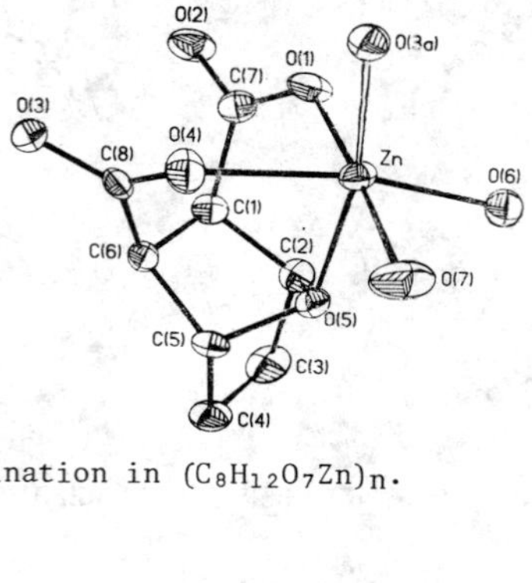

Fig. 1. The Zn coordination in $(C_8H_{12}O_7Zn)_n$.

TETRAAQUABIS(HYDROGEN MALEATO)NICKEL(II)
$C_8H_{14}NiO_{12}$ $[Ni(C_4H_3O_4)_2(H_2O)_4]$

M.P. GUPTA, H.J. GEISE and A.T.H. LENSTRA, 1984. Acta Cryst., C40, 1152-1154.

Triclinic, P$\bar{1}$, a = 7.309, b = 9.731, c = 5.170 Å, α = 105.97, β = 87.12, γ = 117.05°, D_m = 1.890, Z = 1. Mo radiation, R = 0.030 for 1336 reflexions.

In the complex molecule (Fig. 1) Ni lies on a centre of symmetry and is coordinated by six O atoms forming a slightly distorted octahedron with Ni-O(1) 2.074(1), Ni-O(W1) 2.078(1) and Ni-O(W2) 2.018(1) Å. The hydrogen maleate ion behaves as a unidentate ligand to the metal. A very short asymmetric intramolecular hydrogen bond (2.438(3) Å) connects the hydroxyl-group O atom of the non-ionized carboxyl group with an O atom of the ionized one, effectively closing a non-planar seven-membered ring. A three centre bifurcated donor hydrogen bond is observed.

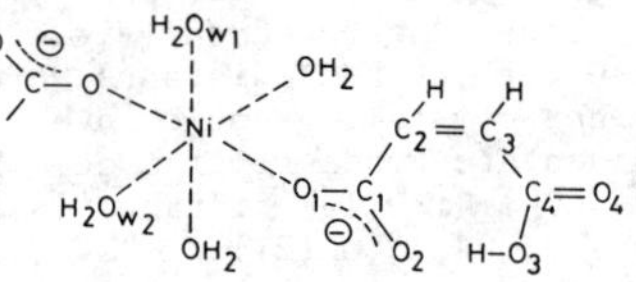

Fig. 1. $[Ni(C_4H_3O_4)_2(H_2O)_4]$: perspective view of the complex molecule and geometrical details of the three-centre bifurcated hydrogen bond; O(2) is at x,y,z and O(3) at -x,1-y,1-z.

DIBROMOTETRAKIS(μ-GLYCINE-O,O')DICHROMIUM(II) DIBROMIDE TETRAHYDRATE
$C_8H_{20}Br_4Cr_2N_4O_8,4H_2O$ (I) $Cr_2(O_2CCH_2NH_3)_4Br_4.4H_2O$

DICHLOROTETRAKIS(μ-GLYCINE-O,O')DICHROMIUM(II) DICHLORIDE TRIHYDRATE
$C_8H_{20}Cl_4Cr_2N_4O_8,3H_2O$ (II) $Cr_2(O_2CCH_2NH_3)_4Cl_4.3H_2O$

M. ARDON, A. BINO, S. COHEN and T.R. FELTHOUSE, 1984. Inorg. Chem., 23, 3450-3455.

I. Triclinic, P$\bar{1}$, a = 8.455, b = 11.340, c = 6.799 Å, α = 98.74, β = 104.38, γ = 77.68°, Z = 1. Mo radiation, R = 0.036 for 1751 reflexions.

II. Monoclinic, $P2_1/c$, a = 8.325, b = 9.307, c = 14.675 Å, β = 103.27°, Z = 2. Mo radiation, R = 0.062 for 1187 reflexions.

The two complex cations (Fig. 1) have crystallographic inversion symmetry and have the familiar $M_2(\mu\text{-}O_2CR)_4X_2$ structure. Bond lengths: for I; Cr-Cr 2.513(1), Cr-Br 2.736(1), Cr-O 2.011(5)-2.032(5) Å; for II; Cr-Cr 2.524(1), Cr-Cl 2.581(1), Cr-O 2.022(4)-2.041(4) Å. Intraligand N-H...O hydrogen bonds are present in both I and II, mean N...O 2.69 Å in I, 2.70 Å in II.

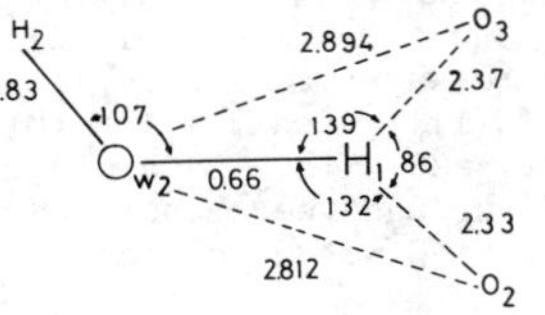

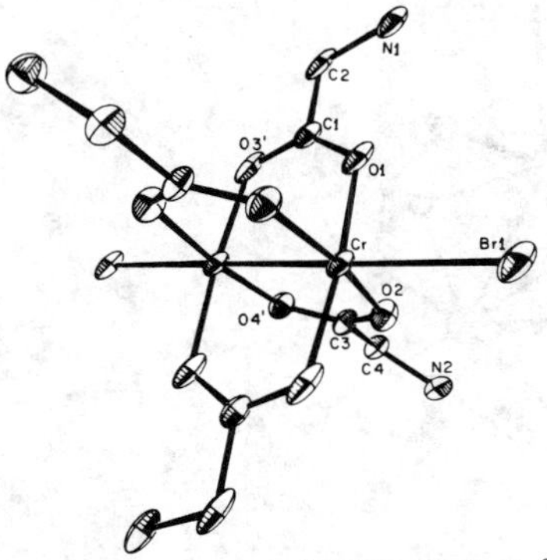

Fig. 1. The cations $Cr_2(O_2CCH_2NH_3)_4X_2^{2-}$ (X = Br, left, and X = Cl, right).

catena-[TETRAAQUA(4-CARBOXYLATOPHENOXYACETATO)MANGANESE(II)]
$C_9H_{14}MnO_9$ (I) $[Mn(OOCC_6H_4OCH_2COO)(H_2O)_4]_n$

TETRAAQUABIS(4-CARBOXYPHENOXYACETATO)NICKEL(II)
$C_{18}H_{22}NiO_{14}$ (II) $[Ni(OOCCH_2OC_6H_4COOH)_2(H_2O)_4]$

C.H.L. KENNARD, G. SMITH, E.J. O'REILLY and P.T. MANOHARAN, 1984. Inorg. Chim. Acta, 82, 35-40.

I. Orthorhombic, Pbca, a = 10.356, b = 9.086, c = 25.673 Å, D_m = 1.77, Z = 8. Mo radiation, R = 0.056 for 991 reflexions.

II. Triclinic, P$\bar{1}$, a = 4.896, b = 5.679, c = 13.403 Å, α = 93.75, β = 91.78, γ = 95.76°, D_m = 1.69, Z = 1. Mo radiation, R = 0.046 for 1160 reflexions.

Compound I has a polymeric structure involving both cis-related carboxylate groups and four water molecules about a slightly distorted octahedron (Fig. 1). The mean Mn-O(water) distance is 2.214 Å while the Mn-O(carboxylate) distances are 2.147 (O(11)), and 2.111 Å (O(411)). Compound II has crystallographic inversion symmetry and exists as a discrete monomer with unidentate trans-related acid ligands bonded through the phenoxy carboxylate O (Fig. 1). The Ni-O(10) distance is 2.058 Å and the Ni-O(water) distances are 2.026 (Ow(1)) and 2.084 Å (Ow(2)).

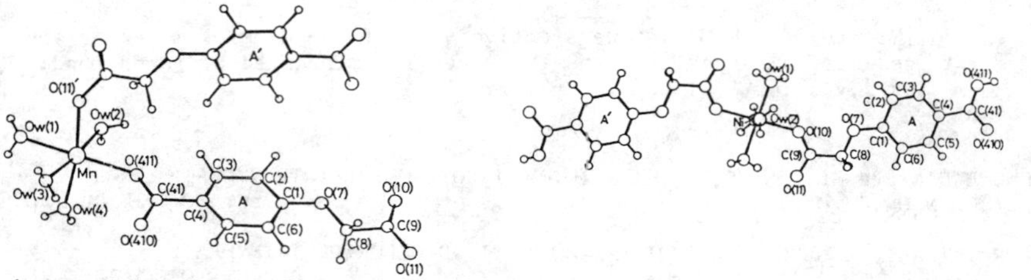

Fig. 1. Molecular configurations of $C_9H_{14}MnO_9$ (left) and (right) $C_{18}H_{22}NiO_{14}$.

3,5-DIMETHYL-1,2-DITHIOLIUM (μ_3-OXO)TRIS(μ_2-ACETATO)TRIS(μ_2-CHLORO)TRICHLOROTRIMOLYBDATE
$C_{11}H_{16}Cl_6Mo_3O_7S_2$ (I) $(C_5H_7S_2)[Mo_3(O)(Cl)_3(OAc)_3]$

3,5-DIMETHYL-1,2-DITHIOLIUM (μ_3-OXO)TRIS(μ_2-ACETATO)TRIS(μ_2-BROMO)TRICHLOROTRIMOLYBDATE
$C_{11}H_{16}Br_3Cl_3Mo_3O_7S_2$ (II)

WU DINGMING, ZHUANG HONGHUI, ZHENG JIANJI, HUANG JINLING, LU JIAXI and HUANG JIANQUAN, 1984. J. Struct. Chem., 3, 51-55.

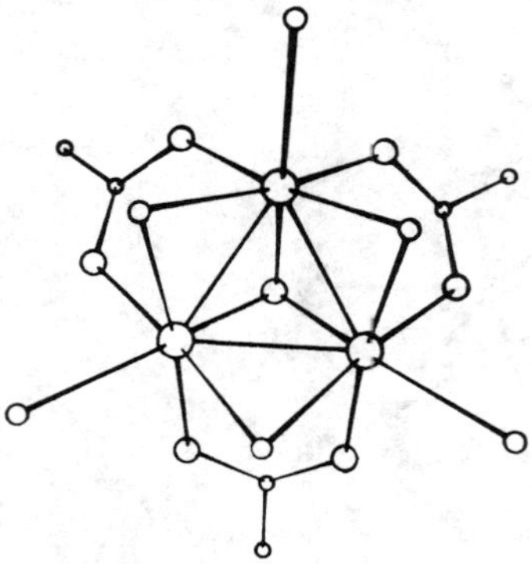

Fig. 1. The $[Mo_3(O)(Cl)_3(OAc)_3Cl_3]^-$ anion.

I. Monoclinic, $P2_1/c$, a = 10.423, b = 17.059, c = 14.062 Å, β = 97.54°, D_m = 2.17, Z = 4. Mo radiation, R = 0.061 for 2621 reflexions.

II. Monoclinic, $P2_1/c$, a = 10.517, b = 17.213, c = 14.053 Å, β = 96.72°, D_m = 2.53, Z = 4. Mo radiation, R = 0.057 for 2548 reflexions.

In these isomorphous structures the anion has three Mo atoms in an equilateral triangle (Mo-Mo 2.568-2.603(2) Å). On one side of the Mo_3 plane (Fig. 1) there is a μ_3-O atom (Mo-O 1.967-2.000(9) Å) and three acetato ligands (Mo-O 2.061-2.094(9) Å). Three terminal Cl atoms and three bridging halogens are on the other side of the Mo_3 plane (Mo-Cl(t) 2.411-2.456(3), Mo-Cl(b) 2.415-2.425(4), Mo-Br(b) 2.544-2.551(2) Å).

catena-DIAQUABIS(μ_2-ISONICOTINATO N-OXIDE)CADMIUM(II)
$C_{12}H_{12}CdN_2O_8$ (I)

HEXAAQUAZINC(II) BIS(ISONICOTINATE N-OXIDE)
$C_{12}H_{40}N_2O_{12}Zn$ (II) $[Zn(H_2O)_6]^{2+}.2(C_6H_{14}NO_3^-)$

P. KNUUTTILA, 1984. Polyhedron, 3, 303-305.

I. Triclinic, $P\bar{1}$, a = 5.696, b = 7.898, c = 11.900 Å, α = 90.86, β = 124.48, γ = 121.00°, D_m = 2.06, Z = 1. Mo radiation, R = 0.035 for 1717 reflexions.

II. Monoclinic, $P2_1/c$, a = 7.775, b = 9.753, c = 11.405 Å, β = 91.16°, D_m = 1.71, Z = 2. Mo radiation, R = 0.038 for 1315 reflexions.

In I, coordination to Cd takes place through the N-O oxygen (Cd-O 2.332(1) Å), one of the carboxyl oxygens (Cd-O 2.253(1) Å) and a water molecule (Cd-O 2.299(1) Å). The Cd atom lies on an inversion centre and an endless chain is produced by double bridges between Cd^{2+} ions (Fig. 1). Compound II is ionic, with the Zn^{2+} ion on an inversion centre and Zn-O 2.056-2.104(2) Å. The metal coordination in I and II is an almost undistorted octahedron.

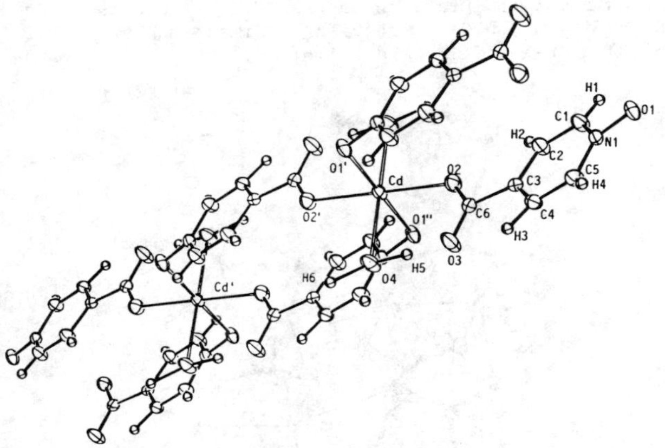

Fig. 1. A view of $C_{12}H_{12}CdN_2O_8$.

TRISODIUM BIS(NITRILOTRIACETATO)FERRATE(III) PENTAHYDRATE
$C_{12}H_{12}FeN_2Na_3O_{12},5H_2O$ $Na_3[Fe\{N(CH_2CO_2)_3\}_2],5H_2O$

W. CLEGG, A.K. POWELL and M.J. WARE, 1984. Acta Cryst., C40, 1822-1824.

Monoclinic, $P2_1/n$, a = 10.102, b = 10.859, c = 20.441 Å, β = 102.78°, Z = 4. Mo radiation, R = 0.042 for 3402 reflexions.

Fe has a distorted, pentagonal, bipyramidal coordination (Fig. 1) with the O(21)-Fe-O(15) angle 166.4(1)°. One nitrilotriacetate group is tetradentate, the other tridentate with one carboxylate group uncoordinated. Each Na^+ ion is coordinated by six O atoms from the water molecules and carboxylate groups with Na^+...O in the range 2.310(3)-2.993(4) Å. The Fe-N distances are 2.291(3) and 2.314(3) Å with Fe-O in the range 2.010(2)-2.078(2) Å.

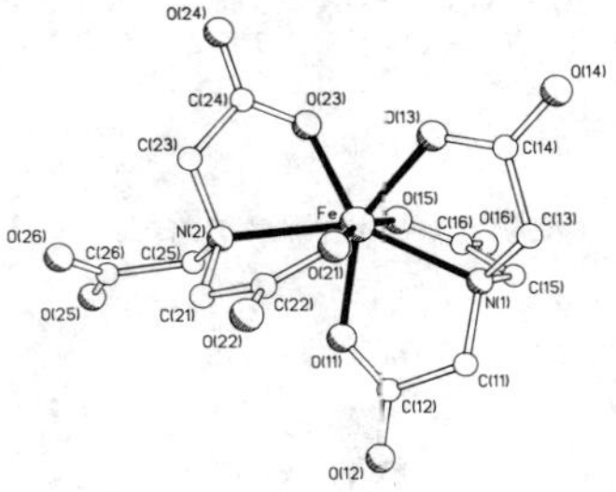

Fig. 1. The $[Fe\{N(CH_2CO_2)_3\}_2]^{3-}$ anion.

TETRAKIS(μ2-PROPIONATO-O,O')DITUNGSTEN

$C_{12}H_{20}O_8W_2$ $W_2(O_2CEt)_4$

M.H. CHISHOLM, H.T. CHIU and J.C. HUFFMAN, 1984. Polyhedron 3, 759-760.

Triclinic, $P\bar{1}$, a = 9.377, b = 8.271, c = 5.527 Å, α = 102.49, β = 84.61, γ = 89.45°, Z = 1. Mo radiation, R = 0.016 for 1464 reflexions (at -160°C).

The complex lies about an inversion centre (Fig. 1) and is connected to yield infinite chains by additional W-O interactions. Dimensions are W-W 2.189(1), W-O 2.08(2), W-O 2.665(4) Å, W-W-O 91(1) and 161.6(1)°.

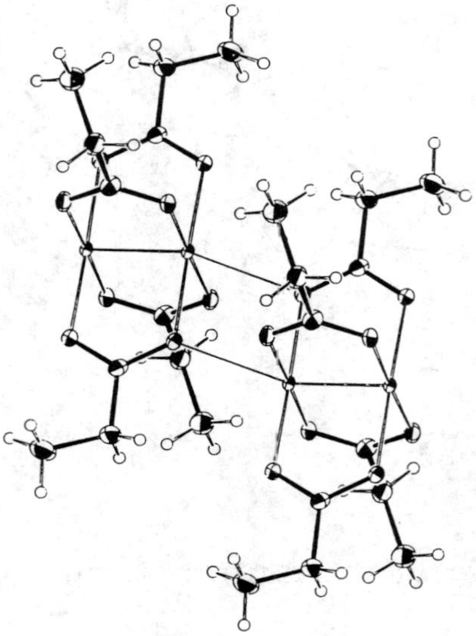

Fig. 1. Part of the $[W_2(O_2CEt)_4]_n$ chain.

DIPOTASSIUM BIS(IMINODIACETATO)NICKEL(II) TRIAQUA(IMINODIACETATO)NICKELATE(II) TRIHYDRATE

$C_{12}H_{21}K_2N_3Ni_2O_{15}.3H_2O$ $(C_8H_{10}N_2NiO_8)^{2-}(C_4H_{11}NNiO_7)2K^+.3H_2O$

V.M. AGRE, T.F. SYSOEVA, V.K. TRUNOV, N.M. DYATLOVA and A.Ya. FRIDMAN, 1984. J. Struct. Chem., 25, 297-303 [Zh. Strukt. Khim. 25(2), 141-148].

Triclinic, P$\bar{1}$, a = 9.433, b = 9.176, c = 17.464 Å, α = 91.74, β = 115.14, γ = 108.63°, Z = 2. Mo radiation, R = 0.053 for 4191 reflexions.

The structure (Fig. 1) consists of K^+ cations, $[Ni(Imda)_2]^{2-}$ anions, neutral $Ni(Imda)(H_2O)_3$ complexes, and water molecules. The Ni atom of the complex anion is octahedrally coordinated by four O and two N atoms of two tridentate Imda ligands. The nitrogen atoms are cis to one another. Bond lengths for the complex anion include: Ni-N 2.101(4) and 2.115(4), Ni-O(trans to N) 2.030(3) and 2.039(3), Ni-O(trans to O) 2.055(3) and 2.063(3) Å. The Ni atom in the neutral species is octahedrally coordinated by the ONO donor set of one Imda ligand and by three water O atoms, Ni-N 2.064(4), Ni-O 2.045(4) and 2.052(3), Ni-OH_2 2066(4)-2.077(3) Å. The tridentate Imda ligands are facially coordinated in both species. Both potassium ions are coordinated by six oxygen atoms, K-O distances ranging 2.702(5)-2.978(5) Å.

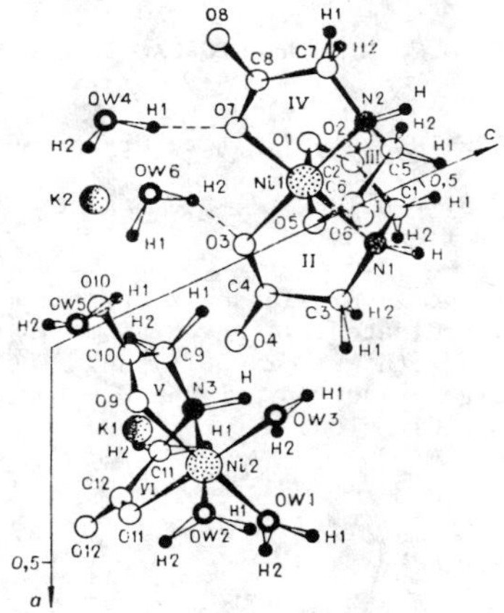

Fig. 1. The asymmetric unit of dipotassium bis(iminodiacetato)nickel(II) triaqua-(iminodiacetato)nickelate(II) trihydrate viewed perpendicular to the ac plane.

(μ_3-OXO)HEXAKIS(μ_2-ACETATO-O,O')TRIAQUAIRONDICHROMIUM TETRACHLOROFERRATE ACETONE METHANOL SOLVATE

$C_{12}H_{24}Cl_4Cr_2Fe_2O_{16},C_6H_{12}O_2,CH_4O$ $[FeCr_2(\mu_3\text{-}O)(CH_3COO)_6(H_2O)_3]^+FeCl_4^-,2(CH_3)_2CO,CH_3OH$

W. CLEGG, O.M. LAM and B.P. STRAUGHAN, 1984. Inorg. Chim. Acta, 90, L75-L76.

Monoclinic, $P2_1/c$, a = 10.164, b = 14.026, c = 27.371 Å, β = 94.29°, Z = 4. Mo radiation, R = 0.060 for 6847 reflexions.

The Cr and Fe atoms appear as a statistical mixture in the triangulo-structure of the cation, and they are represented as M (Fig. 1). Average bond lengths are: M-(μ_3-O) 1.891, M-O(acetate) 1.966 and M-O(H_2O) 2.069 Å. The solvent molecules are hydrogen-bonded to the three coordinated water molecules present in the cationic structure.

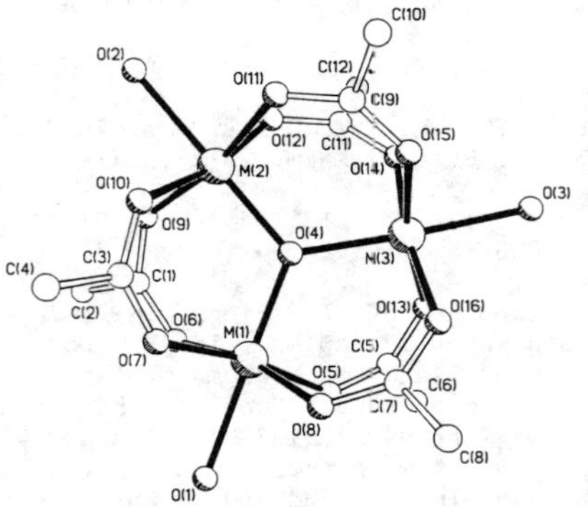

Fig. 1. Structure of the complex cation in $C_{12}H_{24}Cl_4Cr_2Fe_2O_{16}$.

BIS(2-METHYL-1,2-PROPANEDIAMINE-N,N')BIS(TRICHLOROACETATO-O)NICKEL(II)
$C_{12}H_{24}Cl_6N_4NiO_4$ $[Ni(C_2Cl_3O_2)_2(C_4H_{12}N_2)_2]$

S. GARCÍA-GRANDA and F. GÓMEZ-BELTRÁN, 1984. Acta Cryst., C40, 949-951.

Monoclinic, $P2_1/n$, a = 7.3163, b = 16.198, c = 9.943 Å, β = 95.44°, Z = 2. Mo radiation, R = 0.060 for 1481 reflexions.

In the complex molecule (Fig. 1) Ni lies on a crystallographic centre of symmetry and is octahedrally coordinated by four N atoms in a plane and two trichloroacetate ions in trans positions, with Ni-N 2.080(4) and 2.063(5), Ni-O 2.103(4) Å and N(1)-Ni-N(2) 82.5(2)°. The large anisotropy of the Cl-atom thermal parameters indicates some disorder in the $-CCl_3$ group.

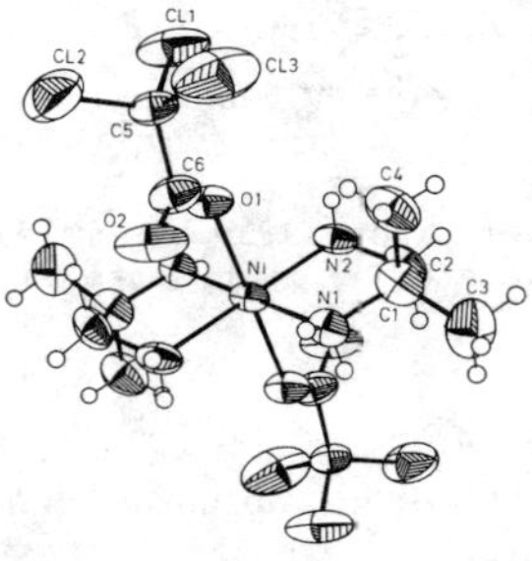

Fig. 1. Perspective view of the $C_{12}H_{24}Cl_6N_4NiO_4$ molecule.

TETRAKIS[μ-(β-ALANINE)-O,O']-DIAQUADICOPPER(II) TETRANITRATE TETRAHYDRATE
$C_{12}H_{32}Cu_2N_8O_{22},4H_2O$ $[Cu_2(H_2O)_2(C_3H_7NO_2)_4](NO_3)_4.4H_2O$

M.E. KAMAYA, E. PAPAVINASAM and S.G. TEOH, 1984. Acta Cryst., C40, 2043-2045.

Triclinic, $P\bar{1}$, a = 12.499, b = 8.717, c = 8.070 Å, α = 77.74, β = 82.44, γ = 71.48°, D_m = 1.758, Z = 1. Mo radiation, R = 0.048 for 2225 reflexions.

Four alanine groups form centrosymmetric dimers (Fig. 1) with two Cu atoms; they are nearly mutually perpendicular and are in cis configuration. The Cu-Cu separation is 2.639(1) Å. The four carboxyl O atoms and an axial water O coordinate with Cu to form a square pyramid. The Cu-O(alanyl) bonds are in the range 1.961(4)-1.981(4) Å and Cu-O(water) is 2.124(3) Å.

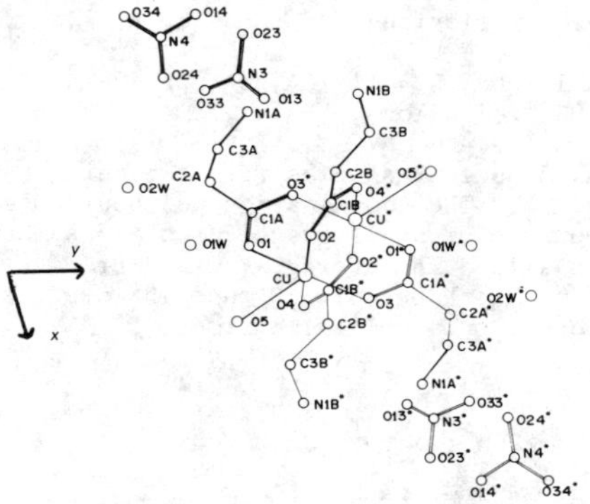

Fig. 1. A schematic drawing of the $C_{12}H_{32}Cu_2N_8O_{22},4H_2O$ structure.

AQUA(N^1-ISOPROPYL-2-METHYL-1,2-PROPANEDIAMINE)[(S)-LACTATO]COPPER(II) (S)-LACTATE

$C_{13}H_{30}CuN_2O_7$ $[(C_3H_5O_3)Cu(H_2O)C_7H_{18}N_2]^+CH_3CH(OH)COO^-$

M. AHLGRÉN, U. TURPEINEN and R. HÄMÄLÄINEN, 1984. Acta Chem. Scand., A38, 169-173.

Monoclinic, $P2_1$, a = 10.046, b = 8.972, c = 10.676 Å, β = 92.63°, Z = 2. Mo radiation, R = 0.035 for 2203 reflexions.

The structure consists of monomeric cations and (S)-lactate anions which are connected to each other by hydrogen bonds in layers perpendicular to the a axis (Fig. 1). The geometry around Cu(II) in the cation is distorted square pyramidal, the bidentate organic ligands forming the base (Cu-N(1) 2.025, Cu-N(2) 1.982, Cu-O(11) 1.952 and Cu-O(13) 1.995 Å) and the water O atom occupying the apical position (Cu-O(1) 2.284 Å). The Cu atom sits 0.211 Å above the mean plane through N(1), N(2), O(11) and O(13). The main hydrogen bonds are between the α-hydroxyl group of the lactate anion (O(23)) and the atoms O(12) and O(1) of the cation (O(23)...O(12) 2.732, O(23)...O(1) 2.773 Å) and the carboxyl O atoms of the anion (O(21), O(22)) and O(13) (O(21)...O(13) 2.499 Å) and N(2) (O(22)...N(2) 2.860 Å).

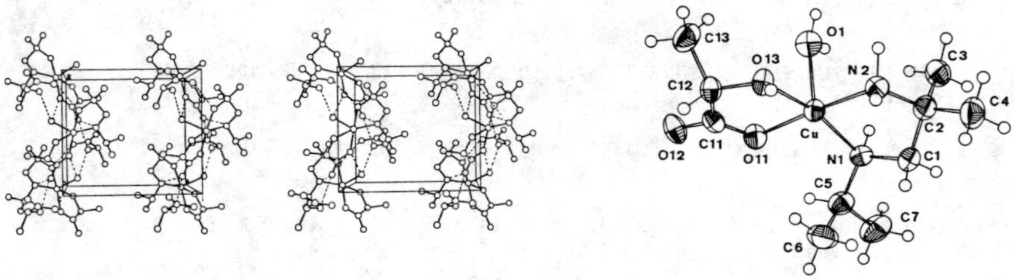

Fig. 1. Stereoview of the unit cell viewed along c (left), and the complex cation (right), in $C_{13}H_{30}CuN_2O_7$.

POTASSIUM μ-OXALATO-DI[TRIS(OXALATO)ERBATE(III)] HYDRATE

$C_{14}Er_2K_8O_{28},14H_2O$ (I)

POTASSIUM μ-OXALATO-DI[TRIS(OXALATO)YTTRATE(III)] HYDRATE

$C_{14}K_8O_{28}Y_2,14H_2O$ (II) $K_8[(C_2O_4)_3Y(C_2O_4)Y(C_2O_4)_3],14H_2O$

I.A. KAHWA, F.R. FRONCZEK and J. SELBIN, 1984. Inorg. Chim. Acta, 82, 167-172.

I. Monoclinic, C2/c, a = 16.079, b = 11.773, c = 23.371 Å, β = 91.21°, Z = 4. Mo radiation, R = 0.035 for 3266 reflexions.

II. Monoclinic, C2/c, a = 16.068, b = 11.764, c = 23.402 Å, β = 90.89°, Z = 4. Mo radiation, R = 0.037 for 2931 reflexions.

The complexes exist as dinuclear species with two metal cations linked by a single oxalate bridge (Fig. 1). The coordination polyhedron is a slightly-distorted D_{2d} dodecahedron. The dimer has crystallographic twofold symmetry with the bridging group normal to the twofold axis. The average Y-O and Er-O distances are respectively 2.344 and 2.343 Å. Structures with Ln = Tb, Dy and Yb were determined to be isomorphous by determination of unit cell dimensions.

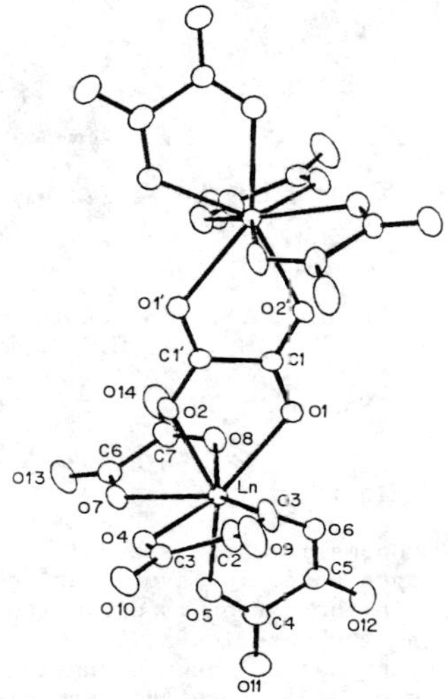

Fig. 1. Molecular structure of $C_{14}K_8Ln_2O_{28}$ viewed slightly oblique to the crystallographic twofold axis.

(μ_3-N,N'-BIS(METHOXYCARBONYL)HYDRAZIDE-N,N',O,O')DECACARBONYLTRIOSMIUM
$C_{14}H_6N_2O_{14}Os_3$ (I)

(μ_3-N,N'-BIS(METHOXYCARBONYL)HYDRAZIDE-N,N)DODECACARBONYLTRIOSMIUM
$C_{16}H_6N_2O_{16}Os_3$ (II)

F.W.B. EINSTEIN, S. NUSSBAUM, D. SUTTON and A.C. WILLIS, 1984. Organometallics, 3, 568-574.

I. Triclinic, P$\bar{1}$, a = 8.606, b = 9.412, c = 14.326 Å, α = 104.50, β = 97.90, γ = 103.29°, Z = 2. Mo radiation, R = 0.030 for 3328 reflexions.

II. Monoclinic, $P2_1/c$, a = 12.136, b = 12.298, c = 16.027 Å, β = 91.37°, Z = 4. Mo radiation, R = 0.021 for 3359 reflexions.

Compound I consists of a triangular $Os_3(CO)_{10}$ fragment with the azo ligand spanning one edge (Fig. 1). As a result the Os(1)-Os(3) edge has been lengthened to a non-bonding distance of 4.161 Å. Other distances are: Os(1)-Os(2) 2.964, Os(2)-Os(3) 2.970, Os(1)-N(5) 2.120, N(4)-N(5) 1.439, N(4)-C(3) 1.302, C(3)-O(3) 1.281, Os(1)-Os(3) 2.144, Os(3)-N(4) 2.128, N(5)-C(6) 1.338, C(6)-O(6) 1.253, Os(3)-O(6) 2.144 Å. Angle Os(1)-Os(2)-Os(3) is 89.06°. Compound II is related to I by having the extra two CO groups occupying the sites previously filled by the coordinated methoxycarbonyls in I. Selected distances are: Os(1)-Os(2) 2.967, Os(2)-Os(3) 3.008, Os(1)...Os(3) 4.198, Os(1)-N(5) 2.095, N(4)-N(5) 1.411, N(4)-C(3) 1.351, C(3)-O(3) 1.220, Os(3)-N(4) 2.092, N(5)-C(6) 1.336, C(6)-O(6) 1.226 Å. Angle Os(1)-Os(2)-Os(3) is 89.27°.

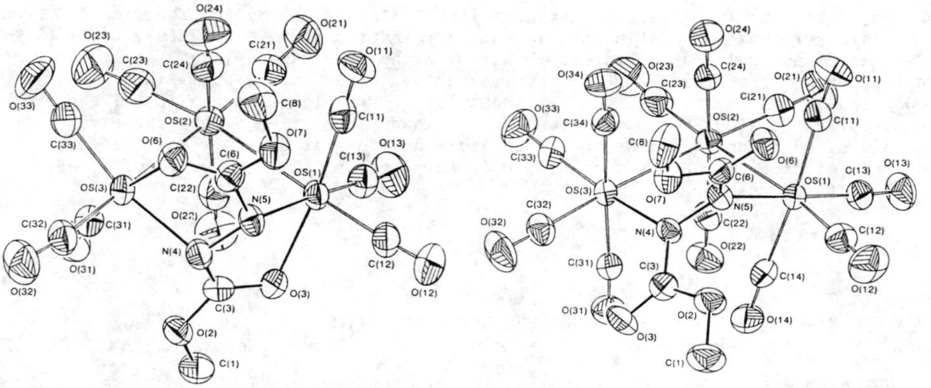

Fig. 1. Molecular structures of $C_{14}H_6N_2O_{14}Os_3$ (left), and $C_{16}H_6N_2O_{16}Os_3$ (right).

DIAQUABIS(p-NITROBENZOATO)ZINC(II)
$C_{14}H_{12}N_2O_{10}Zn$ (I)

DIAQUABIS(THIOBENZOATO)ZINC(II)
$C_{14}H_{14}O_4S_2Zn$ (II)

G.A. GUSEINOV, F.N. MUSAEV, I.R. AMIRASLANOV, B.T. USUBALIEV and Kh.S. MAMEDOV, 1983. Koord. Khim., 9, 1687-1694 [Sov. J. Coord. Chem., 9, 956-963].

I. Triclinic, $P\bar{1}$, a = 5.126, b = 12.187, c = 13.636 Å, α = 109.25, β = 100.89, γ = 90.05°, Z = 2. Mo radiation, R = 0.055 for 1988 reflexions.

II. Monoclinic, C2, a = 9.854, b = 6.794, c = 11.920 Å, β = 108.40°, Z = 2. [See ref. 1.]

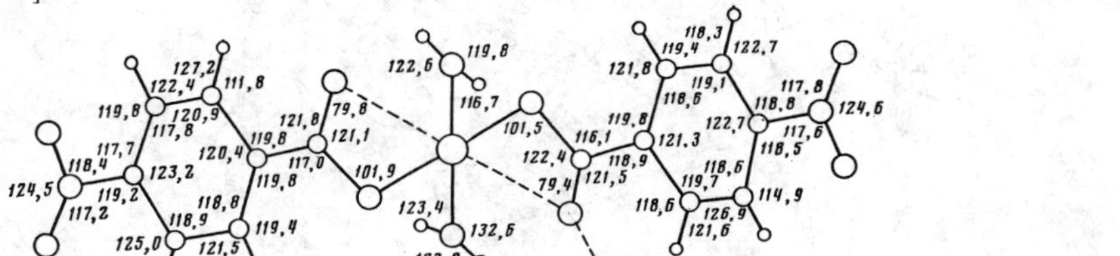

Fig. 1. $C_{14}H_8N_2O_8Zn,2(H_2O)$: bond lengths and angles.

The coordination polyhedron of the Zn atom in I is a highly distorted tetrahedron which is formed by two O atoms from the carboxyl groups of two monodentate ligands and two water molecules (Fig. 1). The Zn-O (carboxyl) distances are 2.002(5) and

2.010(5) and the Zn-O (water) distances 1.996(7) and 1.994(8) Å. The second O atom of the carboxyl does not participate in the coordination (Zn...O 2.522(6), 2.511(6) Å). Each water molecule forms hydrogen bonds with the free carboxylate O atoms of neighbouring molecules along the b axis and the carboxylate O atoms involved in coordination of the translationally identical groups along the a axis. The hydrogen bonds directed along the b axis cause significant separation of the carboxyl O atoms not involved in coordination to the Zn atoms and this is probably one reason for the absence of chelate coordination of the carboxyl group. The structure of II agrees with that reported in 1.

1. Structure Reports, 42B, 787.

HEXAAQUAMANGANESE(II) BIS(p-HYDROXYBENZOATE) DIHYDRATE
$C_{14}H_{22}MnO_{12}.2H_2O$ (I) $[Mn(H_2O)_6]^{2+}2(C_7H_5O_3)^-.2H_2O$

catena-DIAQUABIS(μ_2-p-HYDROXYBENZOATO)MANGANESE(II) MONOHYDRATE
$C_{14}H_{14}MnO_8.H_2O$ (II)

A.N. SHNULIN, G.N. NADZHAFOV and Kh.S. MAMEDOV, 1984. J. Struct. Chem., 25, 421-429 [Zh. Strukt. Khim., 25(3), 91-101].

I. Orthorhombic, Pcab, a = 7.752, b = 11.224, c = 24.12 Å, Z = 4. Cu radiation, R = 0.069 for 992 reflexions.

II. Monoclinic, A2/a, a = 7.332, b = 6.645, c = 32.69 Å, β = 94.32°, Z = 4. Cu radiation, R = 0.098 for 1058 reflexions.

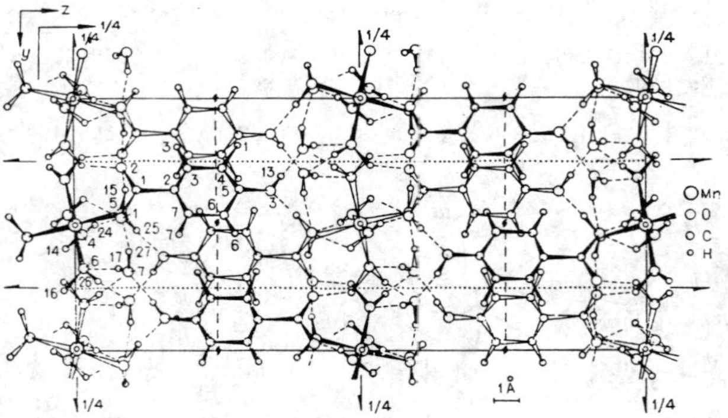

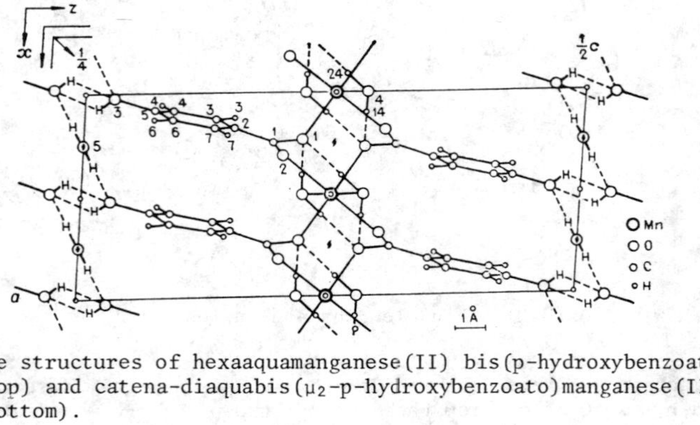

Fig. 1. The structures of hexaaquamanganese(II) bis(p-hydroxybenzoate) dihydrate (top) and catena-diaquabis(μ_2-p-hydroxybenzoato)manganese(II) monohydrate (bottom).

The two structures are shown in Fig. 1. Complex I consists of octahedral-$[Mn(H_2O)_6]^{2+}$ cations (Mn-O 2.162(5)-2.228(5) Å) and acid anions which are joined by hydrogen bonds between all the oxygen atoms in the coordination sphere, the anion, and the water molecules. In II, the octahedral environment about the central Mn ion consists of two water molecules and four oxygen atoms from bridging carboxylate groups (Mn-O 2.152(6)-2.215(7) Å). Linking of octahedra in II is facilitated by hydrogen bonds formed by the coordinated water molecules. The phenolic ends of the p-hydroxy-benzoate ligands in II combine with lattice water molecules to form a network of strong hydrogen bonds (O...O 2.73 Å). Crystals of I decompose upon heating to II.

(μ_2-ACETATO)BIS(μ_2-HYDROXO)BIS[(1,4,7-TRIAZACYCLONONANE-N,N',N'')RUTHENIUM(III)] TRIIODIDE MONOHYDRATE
$C_{14}H_{35}I_3N_6O_4Ru_2,H_2O$

K. WIEGHARDT, W. HERRMANN, M. KOPPEN, I. JIBRIL and G. HUTTNER, 1984. Z. Naturforsch., B39, 1335-1343.

Monoclinic, $P2_1/c$, a = 14.70, b = 19.00, c = 9.714 Å, β = 99.62°, Z = 4. Mo radiation, R = 0.0615 for 2519 reflexions.

The structure contains μ-hydroxo- and μ-acetato-bridged dinuclear cations, iodide ions and water molecules. The Ru-Ru bond (Fig. 1) is 2.572(3) Å, indicative of a single bond. Other dimensions include Ru-O(OH) 2.014-2.033(11), Ru-O(acetate) 2.060 and 2.074(14), Ru-N 2.041-2.096(17) Å. The iodide ions form N-H...I and O-H...I hydrogen bonds.

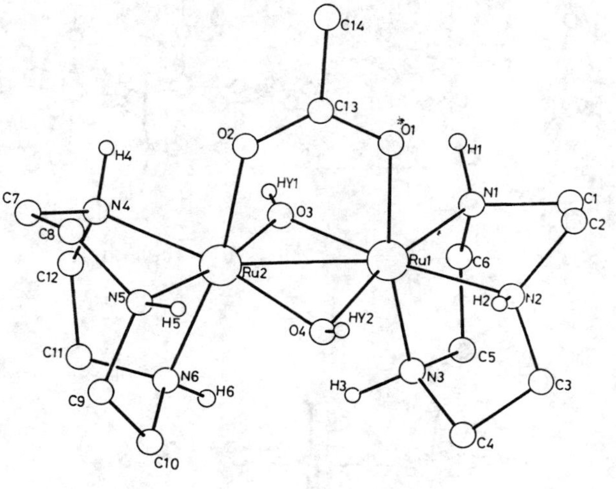

Fig. 1. A view of the cation in $C_{14}H_{35}I_3N_6O_4Ru_2$.

OCTAPOTASSIUM TETRAKIS(μ_2-OXO)OCTAKIS(OXALATO)TETRATITANATE DECAHYDRATE (Triclinic form)
$C_{16}K_8O_{36}Ti_4,10(H_2O)$ (I) $K_8[TiO(C_2O_4)_2]_4,10(H_2O)$

OCTAPOTASSIUM TETRAKIS(μ_2-OXO)OCTAKIS(OXALATO)TETRATITANATE DECAHYDRATE (Monoclinic form)
$C_{16}K_8O_{36}Ti_4,10(H_2O)$ (II)

OCTAPOTASSIUM TETRAKIS(μ_2-OXO)OCTAKIS(OXALATO)TETRATITANATE DODECAHYDRATE
$C_{16}K_8O_{36}Ti_4,12(H_2O)$ (III)

LI XIAOPIN, LIN YUJUAN and ZHOU KANGJING, 1984. J. Struct. Chem., 3, 45-50.

I. Triclinic, $P\bar{1}$, a = 9.794, b = 11.073, c = 11.761 Å, α = 68.10, β = 84.97, γ = 76.33°, Z = 1. R = 0.067 for 3644 reflexions.

II. Monoclinic, $P2_1/c$, a = 11.687, b = 14,181, c = 13.927 Å, β = 103.72°, Z = 2. R = 0.051 for 3512 reflexions.

III. Tetragonal, $P4_2/nmc$, a = 14.361, c = 11.736 Å, Z = 2. R = 0.050 for 731 reflexions.

Views of the three $[TiO(C_2O_4)_2]_4{}^{8-}$ ions are in Fig. 1. Each Ti has octahedral coordination; in I and II the anions lie about inversion centres and have similar geometries with a chair-like Ti_4O_4 ring and asymmetric Ti-O bond lengths (1.786, 1.834 in (I); 1.795, 1.837 Å in (II), with Ti-O-Ti 155.0, 146.6° in (I); 151.5, 140.2° in (II)). The O-Ti-O angles are 98.7 and 98.9 in (I) and 96.1 and 96.8 in (II). Compound III has $\bar{4}2m$ crystallographic symmetry and the four Ti atoms form a square with Ti-O 1.798 Å, Ti-O-Ti 164.4, O-Ti-O 99.3°.

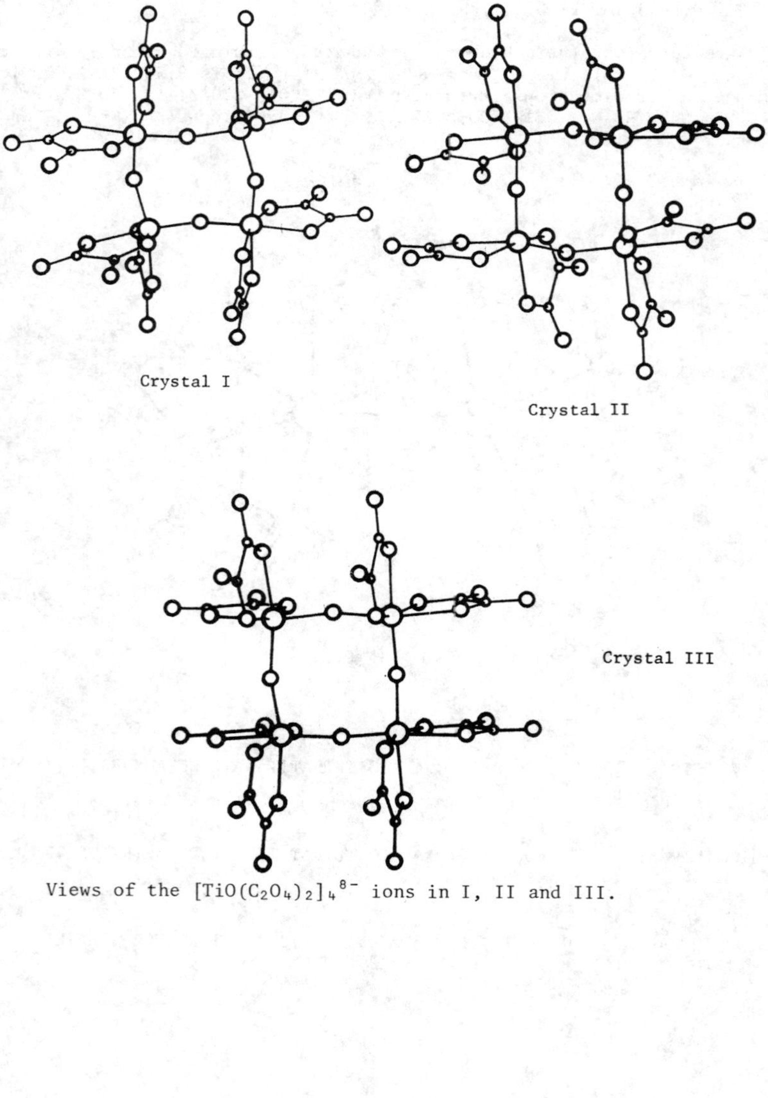

Fig. 1. Views of the $[TiO(C_2O_4)_2]_4{}^{8-}$ ions in I, II and III.

catena-{μ₄-(DIHYDROGEN MELLITATO)-BIS[BIS(OXAMIDE OXIME-N,N')NICKEL(II)]} TETRAHYDRATE
$C_{16}H_{14}N_8Ni_2O_{16},4H_2O$ $[Ni_2(C_{12}H_2O_{12})(C_2H_6N_4O_2)_2],4H_2O$

H. ENDRES and A. KNIESZNER, 1984. Acta Cryst., C40, 770-772.

Triclinic, P$\bar{1}$, a = 8.683, b = 8.854, c = 12.435 Å, α = 85.68, β = 75.32, γ = 72.24°, Z = 1. Mo radiation, R = 0.103 for 1767 reflexions.

In the complex molecule Ni is octahedrally coordinated by two bidentate oxamide oxime chelate ligands and two O atoms of different mellitic anions in cis positions (Fig. 1); bond distances are: Ni-N 2.107(12), 2.059(16), 2.154(11) and 2.082(12) and Ni-O 2.116(9) and 2.030(13) Å. Each carboxylic acid anion bridges four Ni atoms, which leads to a one-dimensional ribbon-like structure, empty spaces being filled by water molecules.

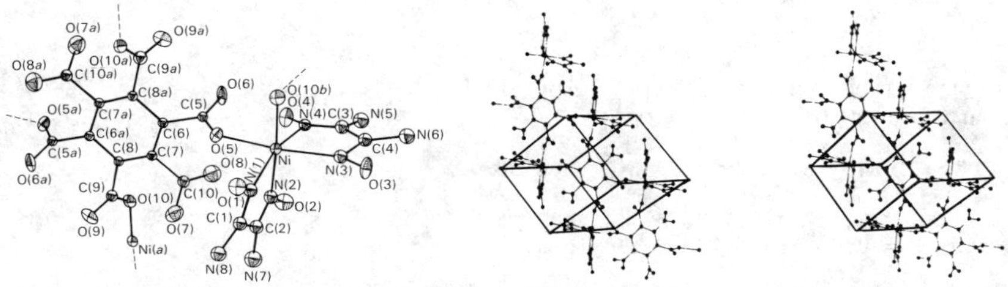

Fig. 1. $C_{16}H_{14}N_8Ni_2O_{16},4H_2O$: perspective view of a section of the crystal structure with dashed lines indicating the polymer formation and a stereoview of part of a one-dimensional ribbon extending parallel to a.

DIAQUABIS(4-FLUOROPHENOXYACETATO)COPPER(II) BIS(4-FLUOROPHENOXYACETIC ACID) DIHYDRATE
$C_{16}H_{16}CuF_2O_8,2C_8H_7FO_3,2H_2O$ (I) $[Cu(C_8H_6FO_3)_2(H_2O)_2],2C_8H_7FO_3,2H_2O$

TETRA-μ-(4-FLUOROPHENOXYACETATO-O,O')BIS[(2-AMINOPYRIDINE)COPPER(II)]
$C_{40}H_{34}Cu_2F_4N_6O_{12}$ (II) $Cu_2(C_8H_6FO_3)_4(C_4H_3N_2NH_2)_2$

E.J. O'REILLY, G. SMITH and C.H.L. KENNARD, 1984. Inorg. Chim. Acta, 90, 63-71.

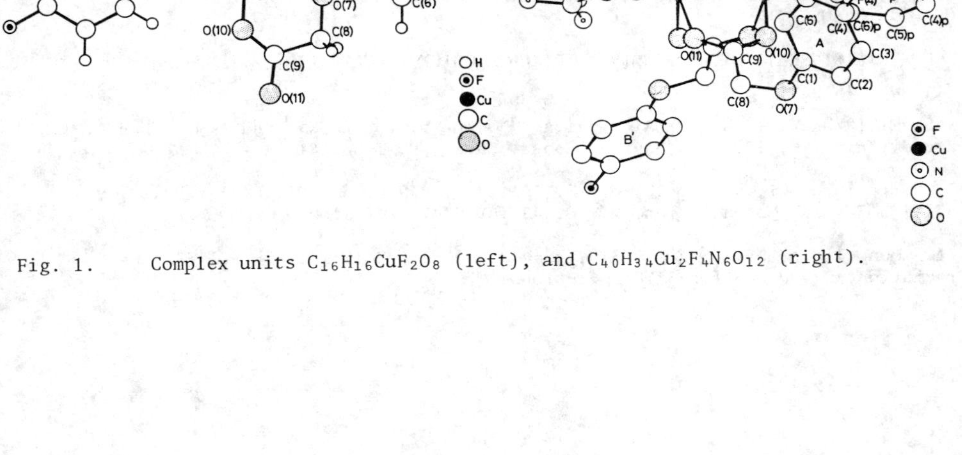

Fig. 1. Complex units $C_{16}H_{16}CuF_2O_8$ (left), and $C_{40}H_{34}Cu_2F_4N_6O_{12}$ (right).

I. Triclinic, $P\bar{1}$, a = 14.808, b = 9.832, c = 6.847 Å, α = 87.77, β = 98.41, γ = 112.33°, D_m = 1.50, Z = 1. Mo radiation, R = 0.038 for 1697 reflexions.

II. Triclinic, $P\bar{1}$, a = 12.688, b = 11.422, c = 7.951 Å, α = 78.74, β = 107.51, γ = 75.78°, D_m = 1.61, Z = 1. Mo radiation, R = 0.042 for 2683 reflexions.

The structure of I consists of centrosymmetric six-coordinate CuO_6 units (Fig. 1) which are bound by short hydrogen bonds (O(11)A...O(11)B 2.547 Å) to free acid molecules. Bond distances involving Cu are: Cu-Ow(1) 1.940, Cu-O(10)A 1.942, Cu-O(7)A 2.471 Å. Hydrogen bonds between the lattice water molecules and the complex stabilise the structure further (O(10)B...Ow(2) 2.840, Ow(1)...Ow(2) 2.656 Å. Complex II forms discrete centrosymmetric bridged dimers of the copper acetate type (Fig. 1). The Cu-O(equatorial) distances range from 1.967-1.987 (mean 1.977 Å). The axial donor is a hetero-N of the 2-aminopyridine ligand, Cu-N 2.176 Å. The Cu...Cu separation is 2.710 Å.

DIGOLD(III) STRONTIUM OCTAACETATE DIHYDRATE
$C_{16}H_{24}Au_2O_{16}Sr,2H_2O$ $Au_2Sr(CH_3COO)_8,2H_2O$

P.G. JONES, 1984. Acta Cryst., C40, 804-805.

Monoclinic, C2/c, a = 12.332, b = 11.487, c = 19.496 Å, β = 101.97°, Z = 4. Mo radiation, R = 0.065 for 2703 reflexions.

The structure consists of isolated $SrAu_2(CH_3COO)_8$ units with crystallographic symmetry 2 (Fig. 1). Coordination geometry is square planar and Archimedean anti-prismatic respectively at Au and Sr. Each acetate ion bridges Au and Sr. The C-O bond lengths indicate covalent Au-O but electrostatic Sr-O interactions; Au-O bond lengths are 1.974(8)-1.992(9) Å, Sr-O bond lengths are 2.548(8)-2.596(9) Å. The water molecules occupy "holes" in the lattice without forming hydrogen bonds.

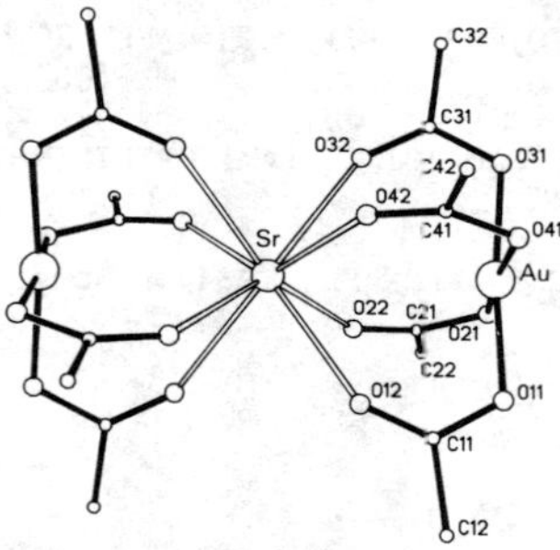

Fig. 1. $C_{16}H_{24}Au_2O_{16}Sr,2H_2O$: perspective view of the $SrAu_2(CH_3COO)_8$ unit.

BIS[(μ_2-ACETATO)ACETATO-N,N-DIMETHYLACETAMIDODIOXOURANIUM(VI)]
$C_{16}H_{30}N_2O_{14}U_2$ (I)

ACETATOTRIS(UREA)DIOXOURANIUM(VI) TRIACETATODIOXOURANATE(VI)
$C_{11}H_{24}N_6O_{15}U_2$ (II)

V.E. MISTRYUKOV, Yu.N. MIKHAILOV, I.A. YURANOV, V.V. KOLESNIK and K.M. DUNAEVA, 1983. Koord. Khim., 9, 272-277 [Sov. J. Coord. Chem., 9, 163-168.]

I. Triclinic, $P\bar{1}$, a = 8.422, b = 8.730, c = 9.279 Å, α = 93.26, β = 92.25, γ = 83.11°, Z = 1. Mo radiation, R = 0.045 for 3337 reflexions.

II. Monoclinic, $P2_1/c$, a = 7.310, b = 14.392, c = 24.36 Å, β = 93.18°, Z = 4. Mo radiation, R = 0.043 for 3138 reflexions.

The molecules of I (Fig. 1) consist of isolated dimeric units with only van der Waals contacts between these units. Each unit consists of two pentagonal bipyramids joined by two bridging acetato groups. The axial positions are occupied by O(1) and O(2) of the uranyl group (U-O 1.82, 1.81(1)) and the equatorial positions by O(3) and O(4) (2.31, 2.41(1)) of bidentate bridging acetato groups, O(5) and O(6) (2.49, 2.46(1)) of terminal acetato groups and O(7) (2.33(1) Å) of the coordinated dimethylacetamide group. II is built up of complex cations and anions which are linked by H bonds of the type N-H...O (2.86-2.98 Å). U atoms in the cation have pentagonal bipyramidal coordination with axial positions occupied by O(9), O(10) of the uranyl group and the equatorial plane is occupied by O atoms of the bidentate acetato group (2.41, 2.42(1)) and three urea molecules (2.28-2.32(1) Å). The complex anion in II has distorted hexagonal pyramidal geometry with three bidentate acetato groups in the equatorial plane (U-O 2.45-2.48(1) Å).

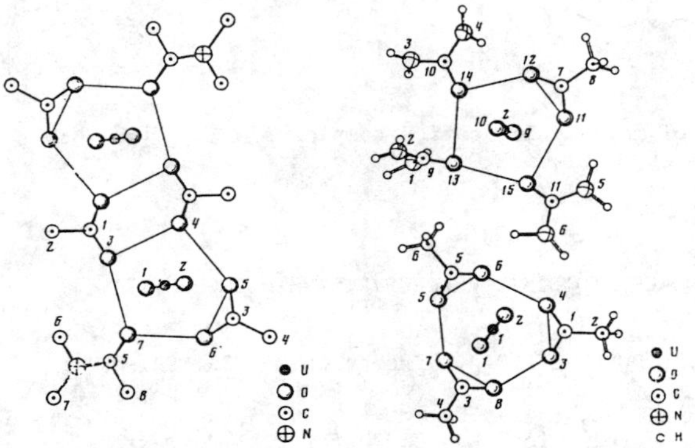

Fig. 1. The $C_{16}H_{30}N_2O_{14}U_2$ complex (left) and the cation and anion in $C_{11}H_{24}N_6O_{15}U_2$ (right).

(μ_2-1,4,7,13,16,19-HEXAAZA-10,22-DIOXACYCLOTETRACOSANE)(μ_2-HYDROXO)(μ_2-PERCHLORATO)-DICOPPER(II) DIPERCHLORATE CHLOROFORM SOLVATE
$C_{16}H_{39}Cl_3Cu_2N_6O_{15},CHCl_3$ (I)

BIS(μ_2-ACETATO)(μ_2-1,4,7,13,16,19-HEXAAZA-10,22-DIOXACYCLOTETRACOSANE)DICOPPER(II) DIPERCHLORATE
$C_{20}H_{44}Cl_2Cu_2N_6O_{14}$ (II)

P.K. COUGHLIN and S.J. LIPPARD, 1984. J. Am. Chem. Soc., 106, 2328-2336.

I. Monoclinic, $P2_1/c$, a = 14.487, b = 15.282, c = 15.573 Å, β = 91.79°, D_m = 1.754, Z = 4. Mo radiation, R = 0.058 for 4305 reflexions.

II. Monoclinic, $P2_1/n$, a = 8.507, b = 10.417, c = 17.653 Å, β = 99.18°, D_m = 1.694, Z = 2. Mo radiation, R = 0.087 for 2051 reflexions.

The (μ-hydroxo)dicopper(II) complex (I) is characterised by short Cu-O bond lengths (1.912(4) and 1.922(4) Å) and a Cu-OH-Cu angle of 143.6(2)° (Fig. 1). The Cu geometry is distorted square pyramidal with the basal plane occupied by amine nitrogen atoms (Cu-N 1.993-2.056(5) Å) and hydroxide oxygen atoms and the apical position by a weakly bonded oxygen atom of a bridging perchlorate group (Cu-O 2.540 and 2.568(8) Å). The structure of II is a linear centrosymmetric polymer propagated by acetate-bridged dicopper(II) centres and macrocycle rings (Fig. 1). The acetate groups have a syn-anti conformation and link Cu atoms through basal and apical positions of a square pyramidal structure with Cu-O(1) 1.948(6), Cu-O(2) 2.277(7) (apical), Cu-N 2.009-2.063(8) Å. The macrocycle adopts a very different conformation in (I) and (II).

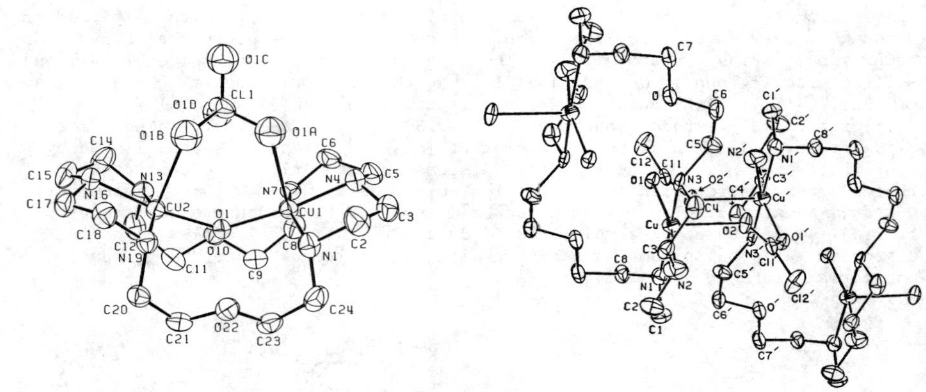

Fig. 1. Views of macrocyclic cation complexes $C_{16}H_{39}Cl_3Cu_2N_6O_{15}$ (I) (left) and in $C_{20}H_{44}Cl_2Cu_2N_6O_{14}$ (II) (right).

TETRAKIS(μ-TRIBROMOACETATO)BIS(2-CHLOROPYRIDINE)DICOPPER(II)
$C_{18}H_8Br_{12}Cl_2Cu_2N_2O_8$ $[Cu(O_2CCBr_3)_2(NC_5H_4Cl)_2]_2$

L.C. PORTER and R.J. DOEDENS, 1984. Inorg. Chem., 23, 997-999.

Triclinic, $P\bar{1}$, a = 10.452, b = 9.767, c = 10.697 Å, α = 69.87, β = 116.44, γ = 98.57°, D_m = 2.77, Z = 1. Mo radiation, R = 0.070 for 2285 reflexions.

The centrosymmetric dimer (Fig. 1) has the familiar copper acetate structure with a Cu-Cu distance of 2.766(3) Å. Other details of the geometry were deposited as supplementary material.

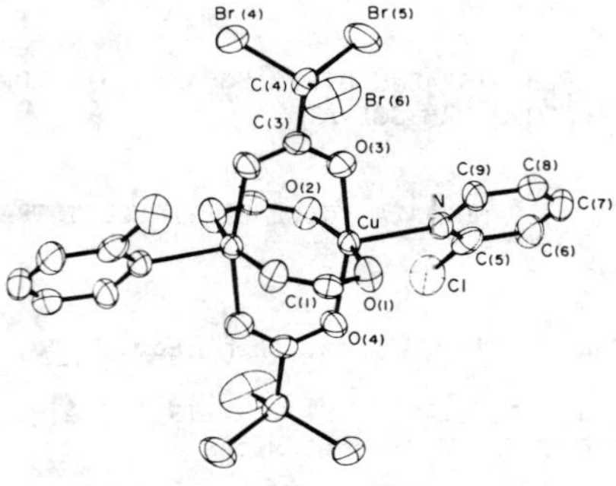

Fig. 1. The $[Cu(O_2CCBr_3)_4(NC_5H_4Cl)_2]_2$ molecule.

DIPYRIDINIUM TETRAACETATODIMOLYBDENUM(II) DIBROMIDE
$C_{18}H_{24}Br_2Mo_2N_2O_8$ (I) $(C_5H_5NH)_2^{2+}[Mo_2(OOCCH_3)_4Br_2]^{2-}$

DIPYRIDINIUM TETRAACETATODIMOLYBDENUM(II) DIIODIDE
$C_{18}H_{24}I_2Mo_2N_2O_8$ (II) $(C_5H_5NH)_2^{2+}[Mo_2(OOCCH_3)_4I_2]^{2-}$

DIPYRIDINIUM TETRAACETATODIMOLYBDENUM(II) DIIODIDE
$C_{18}H_{24}I_2Mo_2N_2O_8$ (III) $(C_5H_5NH)_2^{2+}[Mo_2(OOCCH_3)_4I_2]^{2-}$

L. GOLIĆ, I. LEBAN and P. ŠEGEDIN, 1984. Croat. Chem. Acta, 57, 565-576.

I. Tetragonal, I4/m, a = 9.746, c = 13.948 Å, D_m = 1.88, Z = 2. Mo radiation, R = 0.031 for 705 reflexions.

II. Triclinic, $P\bar{1}$, a = 10.016, b = 10.092, c = 16.325 Å, α = 74.75, β = 71.38, γ = 61.12°, D_m = 2.04, Z = 2. Mo radiation, R = 0.034 for 2888 reflexions.

III. Tetragonal, I4/mcm, a = 13.086, c = 14.712 Å, D_m = 2.21, Z = 4. Mo radiation, R = 0.028 for 655 reflexions.

All three compounds contain $[Mo_2(OOCCH_3)_4X_2]^{2-}$ anions (X = Br,I) and pyridinium cations. In II there are two crystallographically independent forms of the anion. In the anions the halide ions are coordinated axially to each of the quadruply-bonded Mo atoms (Fig. 1). The Mo-X distances are: in I, Mo-Br 2.879 Å, in II Mo-I 3.262 and 3.299 Å and in III, Mo-I 3.204 Å; these long distances indicate only weak axial coordination. The Mo-Mo distances are 2.102 Å in I, 2.103, 2.103 Å in II, and 2.102 Å in III. The Mo-O distances are: 2.114 Å in I, 2.118 and 2.116 Å in II and 2.130 Å in III, and the corresponding Mo-Mo-O angles are respectively 91.35, 91.69, 91.62 and 91.59°.

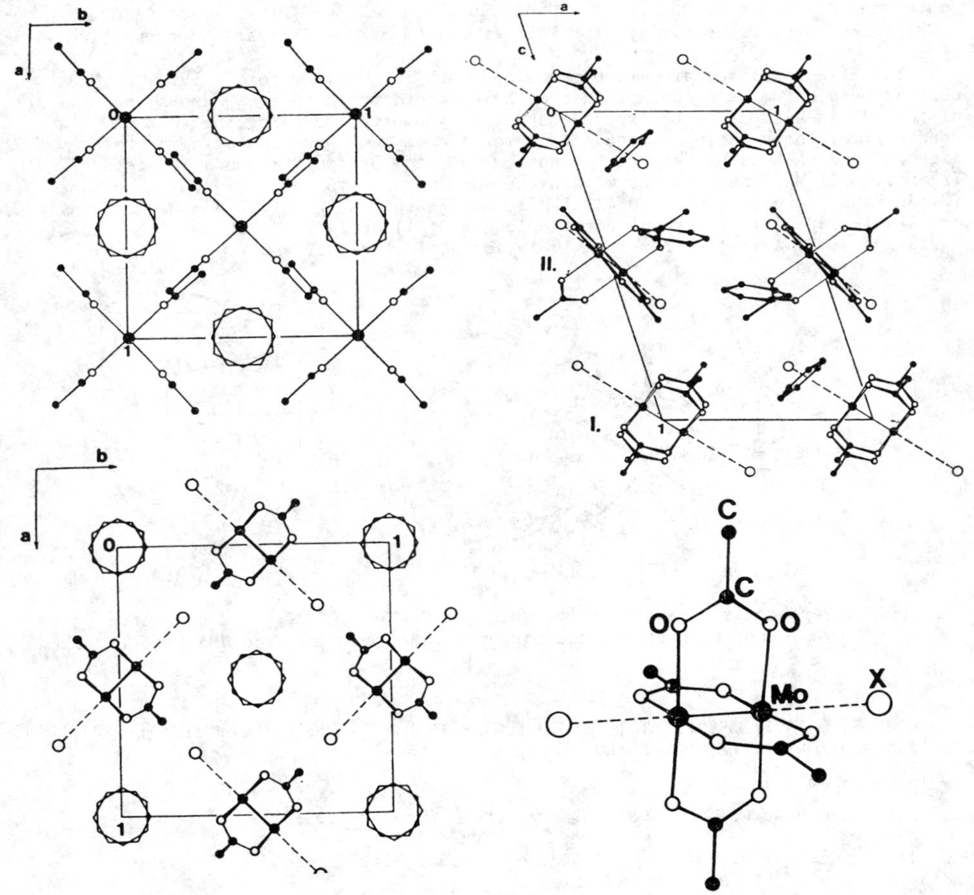

Fig. 1. Projection of the structure of $C_{18}H_{24}Br_2Mo_2N_2O_8$ along [001] (top left), projection of the structure of the triclinic modification of $C_{18}H_{24}I_2Mo_2$-N_2O_8 along [010] (top right), projection of the structure of the tetragonal modification of $C_{18}H_{24}I_2Mo_2N_2O_8$ along [001] (bottom left), and view of the cation $[Mo_2(OOCCH_3)_4X_2]^{2}$ (X = Br,I) (bottom right).

GUANIDINIUM DI(μ-ACETATO)BIS(TRIACETATOAQUALANTHANOIDATE(III)) (LANTHANOID METALS, La, Ce, Pr, Nd and Sm)

$C_{18}H_{40}La_2N_6O_{18}$ (I) $[(NH_2)_3C]_2[M_2(CH_3COO)_8(H_2O)_2]$

$C_{18}H_{40}Ce_2N_6O_{18}$ (II)

$C_{18}H_{40}N_6O_{18}Pr_2$ (III)

$C_{18}H_{40}N_6Nd_2O_{18}$ (IV)

$C_{18}H_{40}N_6O_{18}Sm_2$ (V)

H. SAWASE, Y. KOIZUMI, Y. SUZUKI, M. SHIMOI and A. OUCHI, 1984. Bull. Chem. Soc. Japan, 57, 2730-2737.

Monoclinic, $P2_1/n$, a = 18.398 (18.373, 18.364, 18.370, 18.409), b = 8.524 (8.495, 8.458, 8.432, 8.405), c = 11.003 (10.960, 10.931, 10.911, 10.830) Å, β = 106.60 (106.60, 106.65, 106.76, 106.92)°, D_m = 1.81 (1.83, 1.85, 1.89, 1.92), Z = 2. Mo radiation, R = 0.030 (0.032, 0.037, 0.033, 0.030) for 1801 (2119, 2150, 2380 and 2198) reflexions, respectively (for La, Ce, Pr, Nd and Sm derivatives respectively).

The five complexes are isomorphous and each is in dimer form (Fig. 1). The central metal atom is deca-coordinated, being coordinated to ten O atoms, and has a distorted cis-bicapped cube (tetradecahedron) geometry including the weak M-O bonding from the bridging ligand. Each metal atom has three chelating acetato ligands, and each is bridged to the next metal atom by two "tridentate" acetato ligands. All M-O bond lengths within the same metal complex are roughly the same (2.535-2.661 Å in I, 2.505-2.634 Å in II, 2.483-2.615 Å in III, 2.465-2.605 Å in IV and 2.401-2.588 Å in V) except for the long M-O(41) which is 2.860, 2.880, 2.891, 2.919 and 3.074 Å in length respectively for I, II, III, IV and V.

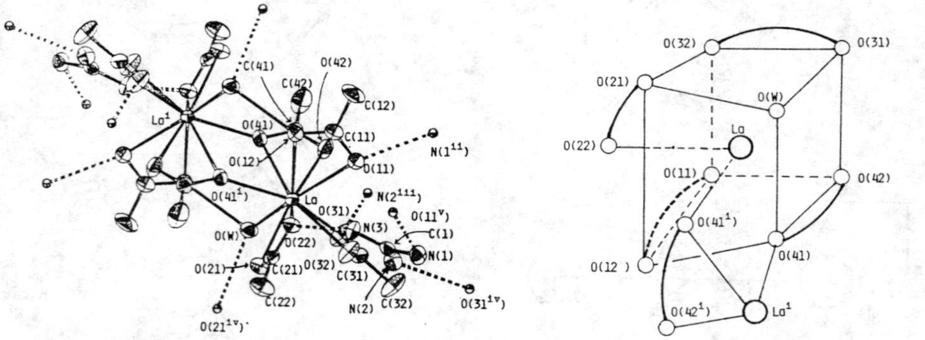

Fig. 1. Perspective drawing of the anion in $C_{18}H_{40}La_2N_6O_{18}$, and schematic representation of the cis-bicapped cube geometry around a La(III) atom.

DICHLOROBIS(μ_2-TRIFLUOROACETATO)BIS(PROPIONITRILE)DIMOLYBDENUM CHLOROTRIS(μ_2-TRIFLUOROACETATO)PROPIONITRILEDIMOLYBDENUM

$C_{19}H_{15}Cl_3F_{15}Mo_4N_3O_{10}$

P. AGASKAR and F.A. COTTON, 1984. Inorg. Chim. Acta, 83, 33-39.

Monoclinic, Cc, a = 22.324, b = 12.640, c = 17.547 Å, β = 113.41°, Z = 4. Mo radiation, R = 0.059 for 2659 reflexions.

The compound contains two different dinuclear molecules linked by three Cl...Mo bridges (2.894-3.005 Å) and one O...Mo bridge (3.053 Å) into infinite chains in which the two dinuclear units alternate (Fig. 1). The bridges occur between Mo(1) and Cl(3), Mo(2) and Cl(2), Mo(3) and Cl(1) and Mo(4) and O(2). The Mo-Mo distances in molecules a and b are respectively 2.127 and 2.134 Å. In molecule a the average distances and angles are: Mo-O 2.13, Mo-N 2.10, Mo-Cl 2.394 Å, Mo-Mo-O 90.7, Mo-Mo-N 99 and Mo-Mo-Cl 108.1°. In molecule b the distances are: Mo-O 2.10, Mo-N 2.12, Mo-Cl 2.425 Å and the angles are: Mo-Mo-O 91, Mo-Mo-N 99.1, Mo-Mo-Cl 105.4°.

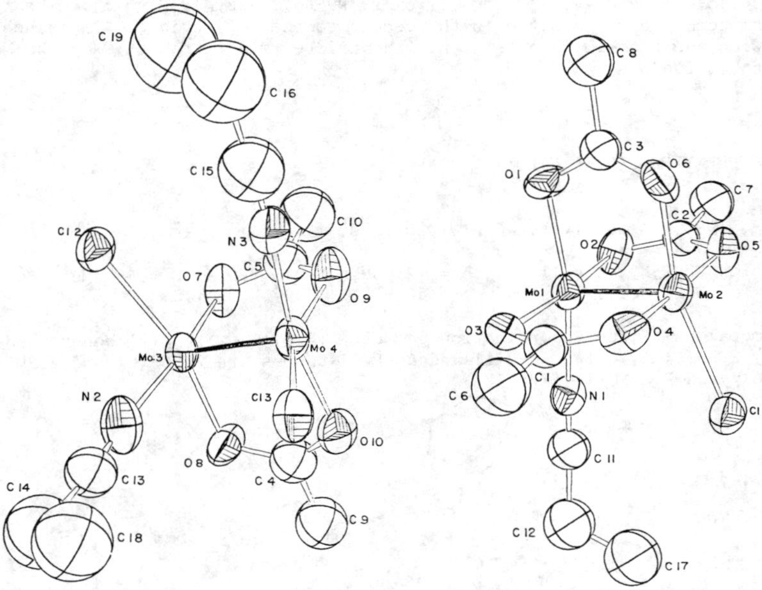

Fig. 1. Molecular structures of molecule a (left) and b (right) in $C_{19}H_{15}Cl_3F_{15}$-$Mo_4N_3O_{10}$ (F atoms have been omitted for clarity).

PYRIDINIUM μ-SALICYLATOTETRACHLORO-μ-OXO-μ-ETHOXODIOXODIMOLYBDATE(V)

$C_{19}H_{22}Cl_4Mo_2N_2O_7$ $[C_5H_5NH]_2{}^{2+}[Mo_2O_3Cl_4(C_2H_5O)(HOC_6H_4CO_2)]^{2-}$

B. KAMENAR, M. PENAVIĆ and B. MARKOVIĆ, 1984. Croat. Chem. Acta, 57, 637-643.

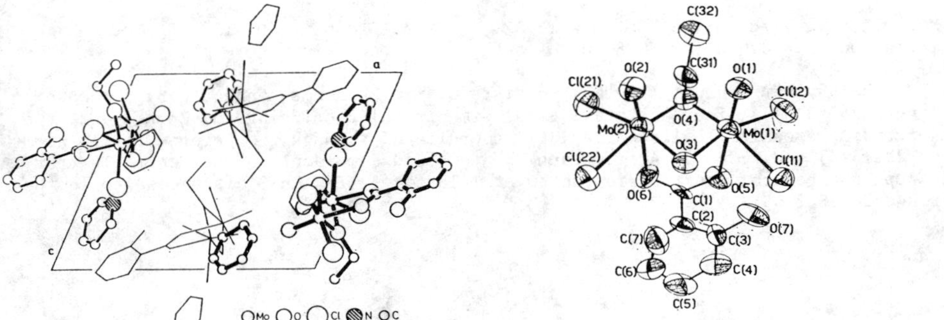

Fig. 1. Crystal structure of $C_{19}H_{22}Cl_4Mo_2N_2O_7$ along the b axis and the binuclear anion.

Monoclinic, $P2_1/n$, a = 13.643, b = 19.34, c = 11.146 Å, β = 114.07°, D_m = 1.70, Z = 4. Cu radiation, R = 0.060 for 1711 reflexions.

Each Mo atom of the binuclear anion is octahedrally coordinated, being bonded to two oxo-O atoms, one terminal (1.669 and 1.670 Å) and one bridging (1.908 and 1.916 Å), one bridging ethoxo-O atom (1.963 and 1.964 Å), one salicylato-O atom (2.295 and 2.317 Å), and two Cl atoms (from 2.399 to 2.432 Å) (Fig. 1). The terminal oxo-O atoms occupy a cis-position with respect to the MoO_2 bridge. The dihedral angle between the two MoO_2 planes within the bridge amounts to 164.9°. The Mo-Mo bond length is 2.646 Å.

ACETATO(TRIPHENYLPHOSPHINE)GOLD(I)

$C_{20}H_{18}AuO_2P$ $Au(C_2H_3O_2)(C_{18}H_{15}P)$

P.G. JONES, 1984. Acta Cryst., C40, 1320-1322.

Orthorhombic, $P2_12_12_1$, a = 11.088, b = 12.050, c = 13.839 Å, Z = 4. Mo radiation, R = 0.028 for 3055 reflexions.

The metal has linear coordination geometry (Fig. 1) with O(2) non-bonded to it (Au...O(2) 2.93(1) Å). The Au-O(1) and Au-P distances are 2.063(6) and 2.207(3) Å respectively and P-Au-O is 177.3(2)°.

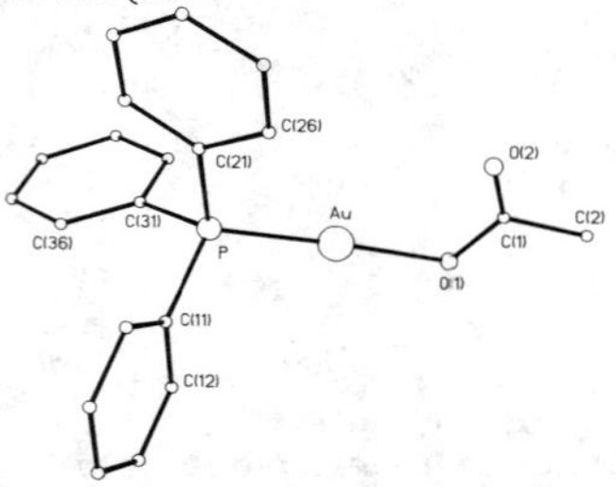

Fig. 1. The acetato(triphenylphosphine)gold(I) molecule.

TETRAKIS(ACETATOBIS(4-PYRIDINECARBONITRILE)DIRHODIUM(II)) ACETONITRILE SOLVATE

$C_{20}H_{20}N_4O_8Rh_2, C_2H_3N$ $[Rh_2(O_2CMe)_4(C_6H_4N_2)_2], CH_3CN$

F.A. COTTON and T.R. FELTHOUSE, 1984. Acta Cryst., C40, 42-45.

Monoclinic, $P2_1/c$, a = 8.385, b = 12.024, c = 26.132 Å, β = 97.15°, Z = 4. Mo radiation, R = 0.030 for 3098 reflexions.

The asymmetric unit (Fig. 1) consists of two Rh atoms connected by a bond of length 2.393(1) Å, four acetate ligands symmetrically bridging the Rh-Rh bond, two 4-pyridinecarbonitrile ligands axially coordinated through the heterocyclic N atoms at 2.244(4) and 2.243(4) Å from each Rh atom, and a molecule of acetonitrile that occupies a general position in the crystal lattice. The Rh-O distances are in the range 2.025-2.051(3) Å.

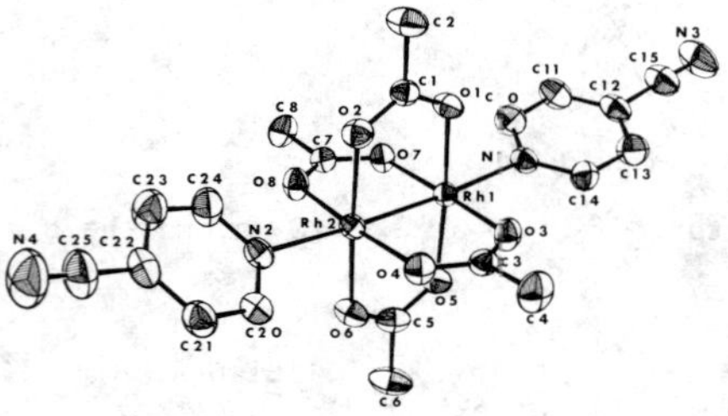

Fig. 1. A view of $C_{20}H_{20}N_4O_8Rh_2$.

TETRAKIS(μ_2-ACETATO-O,O')-BIS(NICOTINAMIDE)DICOPPER(II) DIHYDRATE

$C_{20}H_{24}Cu_2N_4O_{10}, 2H_2O$ $Cu_2(CH_3COO)_4(C_6H_6N_2O)_2 . 2H_2O$

G.V. TSINTSADZE, R.A. KIGURADZE, A.N. SHNULIN and Kh.S. MAMEDOV, 1984. J. Struct. Chem., 25, 911-916 [Zh. Strukt. Khim., 25(6), 82-87].

Triclinic, $P\bar{1}$, a = 10.974, b = 8.132, c = 7.899 Å, α = 96.55, β = 110.75, γ = 89.12°, Z = 1. Mo radiation, R = 0.045 for 2096 reflexions.

The centrosymmetric binuclear complex (Fig. 1) has a typical $Cu_2(\mu\text{-carboxylato})_4L_2$ structure with a Cu-Cu distance of 2.644(2) Å. The copper atom is displaced by 0.21 Å from the plane of the four oxygen atoms towards the axial nicotinamide N atom. Bond lengths (σ = 0.002-0.008 Å for bonds between non-hydrogen atoms) are shown in Fig. 1. Hydrogen bonds between the water molecules and ligands link the complexes in the solid state.

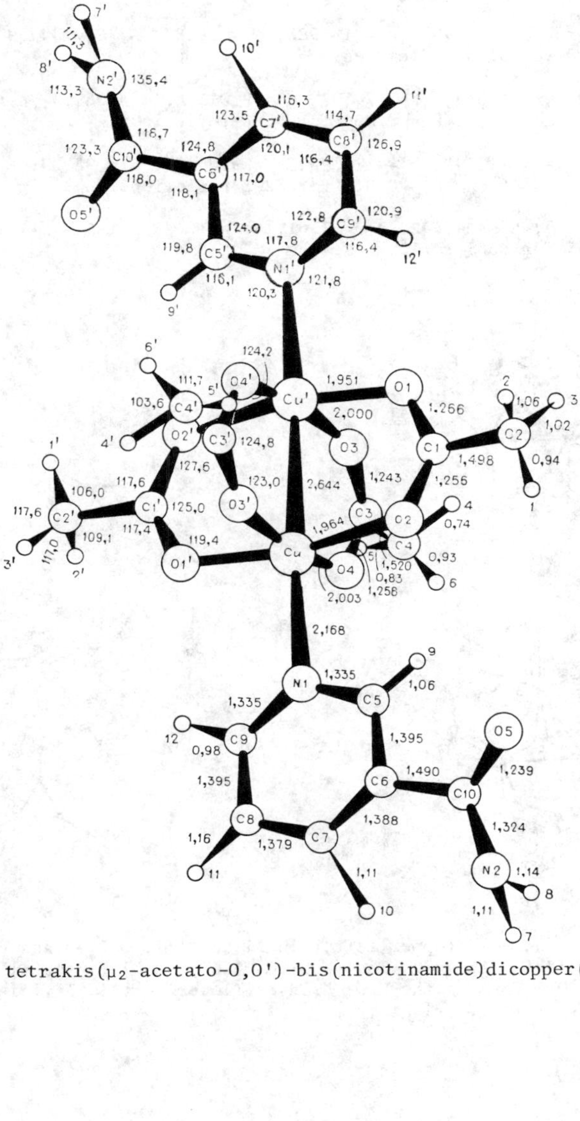

Fig. 1. The tetrakis(μ_2-acetato-O,O')-bis(nicotinamide)dicopper(II) complex.

BIS(μ_3-OXO)HEXAKIS(μ-PROPIONATO)TRIS(AQUATUNGSTEN) BIS(THIOCYANATE)
$C_{20}H_{36}N_2O_{17}S_2W_3$ (I) $[W_3O_2(C_2H_5COO)_6(H_2O)_3](NCS)_2$

BIS[BIS(μ_3-OXO)HEXAKIS(μ-PROPIONATO)DIAQUATRITUNGSTEN]TRIHYDRODIOXONIUM TRIS-(THIOCYANATE) MONOHYDRATE
$C_{39}H_{71}N_3O_{34}S_3W_3,H_2O$ (II) $\{[W_3O_2(C_2H_5COO)_6(H_2O)_2]_2(H_3O_2)\}(NCS)_3.H_2O$

BIS[BIS(μ_3-OXO)HEXAKIS(μ-PROPIONATO)DIAQUATRIMOLYBDENUM]TRIHYDRODIOXONIUM TRIS-(THIOCYANATE) MONOHYDRATE
$C_{39}H_{71}Mo_3N_3O_{34}S_3,H_2O$ (III) $\{[Mo_3O_2(C_2H_5COO)_6(H_2O)_2]_2(H_3O_2)\}(NCS)_3.H_2O$

BIS(μ_3-OXO)HEXAKIS(μ-PROPIONATO)AQUA(TRIHYDRODIOXONIUM) TRITUNGSTEN THIOCYANATE
$C_{19}H_{35}NO_{17}SW_3$ (IV) $[W_3O_2(C_2H_5COO)_6(H_2O)(H_3O_2)]NCS$

A. BINO and D. GIBSON, 1984. Inorg. Chem., 23, 109-115.

I. Monoclinic, $P2_1/c$, a = 17.953, b = 18.923, c = 10.034 Å, β = 94.59°, Z = 4. Mo radiation, R = 0.058 for 3724 reflexions.

II. Triclinic, $P\bar{1}$, a = 11.334, b = 17.393, c = 9.011 Å, α = 98.63, β = 110.48, γ = 100.09°, Z = 1. Mo radiation, R = 0.038 for 3508 reflexions.

III. Triclinic, $P\bar{1}$, a = 11.347, b = 17.399, c = 9.004 Å, α = 98.69, β = 110.57, γ = 100.15°, Z = 1. Mo radiation, R = 0.049 for 3298 reflexions.

IV. Monoclinic, $P2_1/c$, a = 14.842, b = 13.611, c = 16.432 Å, β = 111.29°, Z = 4. Mo radiation, R = 0.035 for 3385 reflexions.

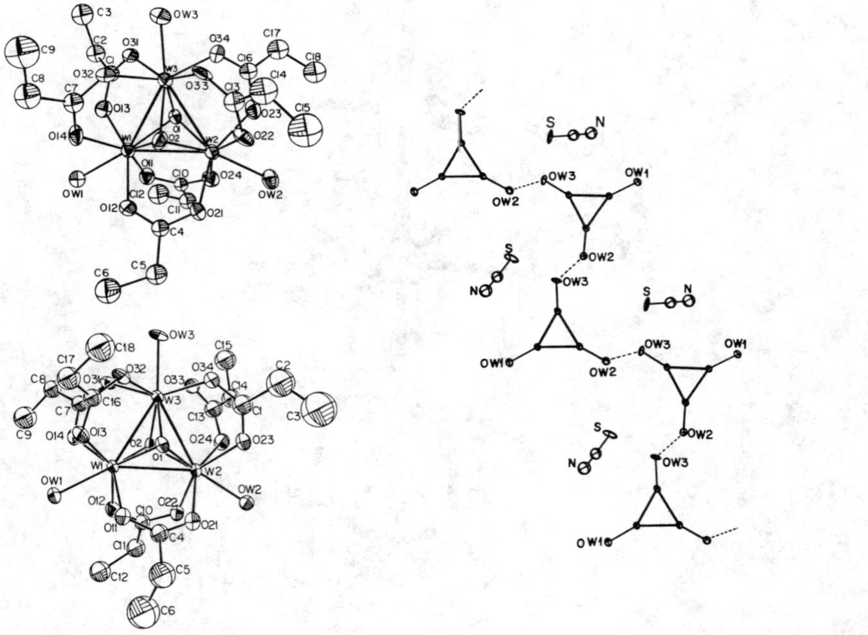

Fig. 1. Clusters in I-III (upper left), in II and III, OW2 is associated with the $H_3O_2^-$ bridging ligand. The $[W_2O_3(C_2H_5COO)_6(H_2O)(H_3O_2)_{0.5}]^+$ ion in IV, OW2 and OW3 forming the H_3O_2 bridges (lower left). Infinite chain of the polycluster in IV (right).

Compound I contains discrete $[W_3O_2(C_2H_5COO)_6(H_2O)_3]^{2+}$ units (Fig. 1) and two NCS^- anions. In the isomorphous W and Mo structures II and III two trinuclear clusters are bridged by a μ-$H_3O_2^-$ ligand and the 3+ charge of the dicluster is balanced by three NCS^- ions. Structure IV consists of infinite chains of triangular W_3 clusters (polycluster); each μ-$H_3O_2^-$ ligand bridges two adjacent clusters. The O...O separations in the $H_3O_2^-$ units in II, III, and IV are 2.46(1), 2.52(1), and 2.44(1) Å respectively. Structural parameters for I-IV: M-M 2.749(1)-2.784(1), mean M-O(H_2O) 2.10(3)-2.13(3), mean M-O(H_3O_2) 2.01(1)-2.07(1) Å.

BIS(TETRAETHYLAMMONIUM) BIS(μ-ACETATO-O,O')-TRI-μ-CHLORO-PENTACHLORO-μ_3-OXO-cyclo-TRIMOLYBDATE(3Mo-Mo)

$C_{20}H_{46}Cl_8Mo_3N_2O_5$ $2C_8H_{20}N^+,C_4H_6Cl_8Mo_3O_5^{2-}$

SHANG MAOYU, HUANG JINLING and LU JIAXI (CHAI-SI LU), 1984. Acta Cryst., C40, 761-762.

Orthorhombic, $Pna2_1$, a = 17.815, b = 16.629, c = 12.003 Å, D_m = 1.80, Z = 4. Mo radiation, R = 0.048 for 2584 reflexions.

In the mono-oxo-capped trinuclear Mo cluster anion the three Mo atoms, the μ_3-O atom O(1) and three μ-Cl atoms, Cl(1), Cl(2) and Cl(3), form an apex-deficient cubane-like configuration (Fig. 1). The three Mo atoms form a triangular configuration with average Mo-Mo bond length of 2.60(1) Å. The symmetry of the anion is C_s.

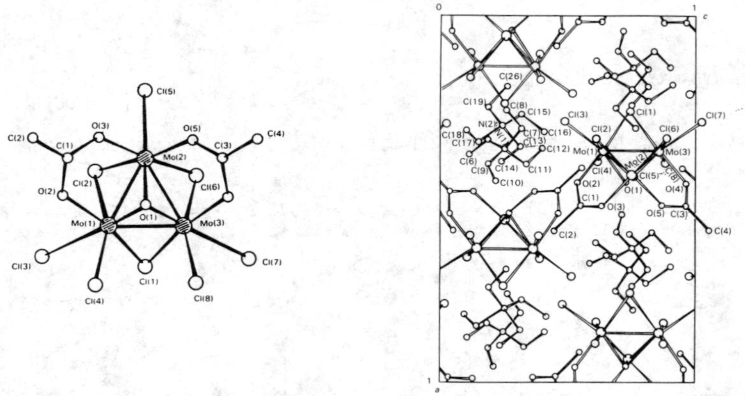

Fig. 1. $C_{20}H_{46}Cl_8Mo_3N_2O_5$: perspective views of the cluster and of the packing of the cluster in the unit cell.

TRIS(TRIFLUOROACETATO)TRIS(PYRIDINE)CHROMIUM

$C_{21}H_{15}CrF_9N_3O_6$

L.M. DIKAREVA, A.S. ANTSYSHKINA, M.A. PORAI-KOSHITS, Yu.V. ZEFIROV and V.N.OSTRIKOVA, 1983. Koord. Khim., 9, 1426-1434 [Sov. J. Coord. Chem., 9, 809-816].

Monoclinic, $P2_1/c$, a = 9.450, b = 16.464, c = 16.981 Å, β = 107.44°, Z = 4. Mo radiation, R = 0.081 for 2188 reflexions.

The structure is composed of discrete monomers (Fig. 1) in which the Cr atom has an octahedral environment and coordinates three N atoms of three pyridine moieties (Cr-N(1,2,3) 2.085(9), 2.108(8), 2.109(9)) and three O atoms of three trifluoroacetato moieties (Cr-O(1,3,5) 1.938(8), 1.951(9), 1.958(9) Å) which are monodentate.

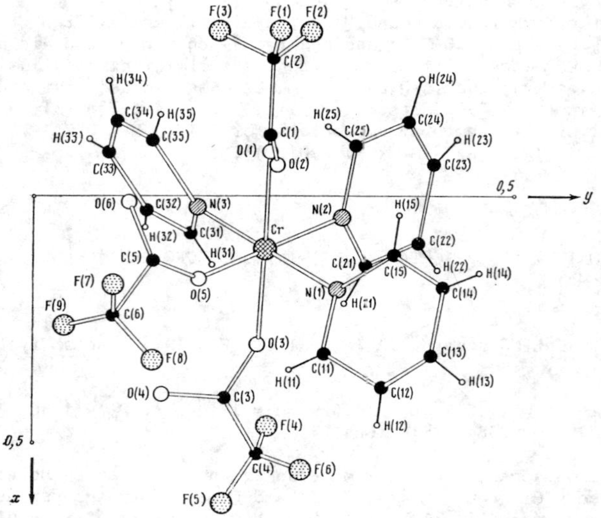

Fig. 1. $C_{21}H_{15}CrF_9N_3O_6$: the molecular structure.

AQUATRIS(m-HYDROXYBENZOATO)CERIUM(III)

$C_{21}H_{17}CeO_{10}$ $[Ce(HOC_6H_4COO)_3(H_2O)]_n$

Y. KOIZUMI, H. SAWASE, Y. SUZUKI, M. SHIMOI and A. OUCHI, 1984. Bull. Chem. Soc. Japan, 57, 1677-1678.

Monoclinic, $P2_1/c$, a = 13.401, b = 19.173, c = 8.079 Å, β = 98.94°, D_m = 1.84, Z = 4. Mo radiation, R = 0.032 for 2479 reflexions.

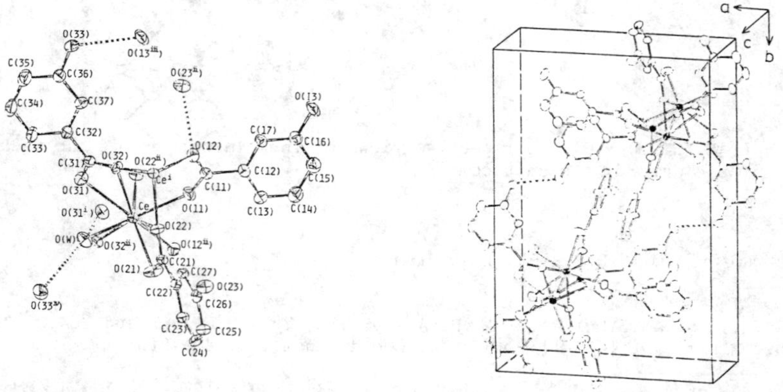

Fig. 1. Perspective drawing of the complex $C_{21}H_{17}CeO_{10}$ (left), and crystal packing diagram (right).

The central Ce(III) atom is ennea-coordinated and has deformed tri-capped trigonal-prism geometry (Fig. 1). One of the O atoms of each ligand, O(11), O(21) and O(31) takes each cap position, while O(22) or O(32) take the tips of the upper triangle of the prism. The water O atom, O(W), occupies the other tip of the triangle. All three ligands form bridges to the next adjacent Ce(III) atom, forming a linear polymer by means of the carboxylate ligands. Bonds associated with Ce are: Ce-O(11) 2.392, Ce-O(21) 2.468, Ce-O(22) 2.687, Ce-O(31) 2.535, Ce-O(32) 2.671, Ce-O(W) 2.628, Ce-O(12'') 2.489, Ce-O(22'') 2.526 and Ce-O(32'') 2.506 Å.

[TRIS(μ-ACETATO-O,O')DIZINC]-BIS(μ3-OXO)-HEXAKIS(μ-ACETATO-O,O')TRIMETHOXYTRI-MOLYBDENUM(IV) DIHYDRATE

$C_{21}H_{36}Mo_3O_{23}Zn_2,2H_2O$ $[Zn_2(O_2CCH_3)_3][Mo_3O_2(O_2CCH_3)_6(OCH_3)_3],2H_2O$

A. BIRNBAUM, F.A. COTTON, Z. DORI and M. KAPON, 1984. Inorg. Chem., 23, 1617-1619.

Orthorhombic, Pnma, a = 20.733, b = 14.042, c = 13.285 Å, D_m = 1.900, Z = 4. Mo radiation, R = 0.090 for 1780 reflexions.

The anion in this compound is a typical bicapped $[Mo_3(\mu_3\text{-}O)_2(O_2CR)_6X_3]$ cluster species with a six-electron cluster bonding configuration, but it is novel in being the first such species to have as X groups alkoxide ions (OCH_3^-). The cation consists of two Zn(II) ions bridged by three acetate ions, and the crystal structure consists of infinite zigzag chains of alternating binuclear cations and cluster anions (Fig. 1). In the anion two of the OCH_3 oxygen atoms form donor bonds to Zn (Zn-O 1.94(1) and 1.95(1) Å); these have longer Mo-O bonds, 2.01(1) and 2.06(1) Å, than the third one, 1.95(1) Å. In a corresponding way the Mo-Mo bonds differ: that between the two Mo atoms that have longer Mo-OMe bonds is shorter, 2.763(3) Å, than the other two, 2.801(3) and 2.817(3) Å.

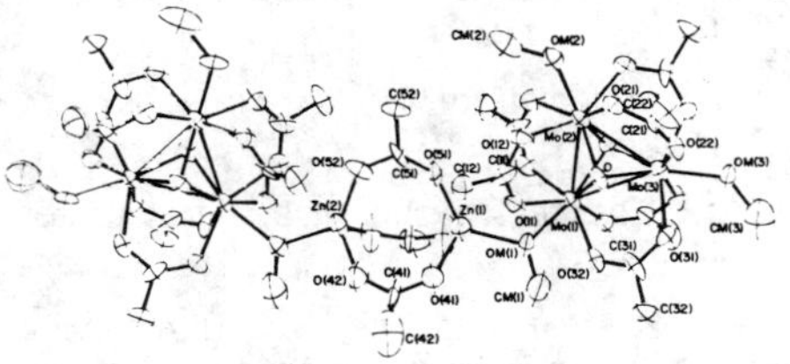

Fig. 1. $[Zn_2(O_2CCH_3)_3][Mo_3O_2(O_2CCH_3)_6(OCH_3)_3]$ structure.

(PYRIDINE-2,6-DICARBOXYLATO-N,O,O')(2,2'.6',2''-TERPYRIDYL-N,N',N'')COPPER(II) HEXAHYDRATE

$C_{22}H_{14}CuN_4O_4,6(H_2O)$ (I) [Cu(terpy)(pydca)],$6H_2O$

catena-(IMINODIACETATO-O,O')(2,2'.6',2''-TERPYRIDYL-N,N',N'')COPPER(II) ACETONE SOLVATE HYDRATE

$(C_{19}H_{16}CuN_4O_4)_n,n(C_3H_6O),n(H_2O)$ (II) $\{[Cu(terpy)ida)],C_3H_6O,H_2O\}_n$

N. BRESCIANI-PAHOR, G. NARDIN, R.P. BONOMO and E. RIZZARELLI, 1984. J. Chem. Soc. Dalton, 2625-2630.

I. Triclinic, $P\bar{1}$, a = 10.964, b = 13.209, c = 9.724 Å, α = 98.3, β = 112.9, γ = 142.2°, D_m = 1.55, Z = 2. Mo radiation, R = 0.041 for 3706 reflexions.

II. Monoclinic, $P2_1/n$, a = 23.41, b = 10.538, c = 8.372 Å, β = 91.3°, D_m = 1.64, Z = 4. Mo radiation, R = 0.044 for 1714 reflexions.

The Cu atom in (I) (Fig. 1) has distorted octahedral coordination, the terpy and pydca ligands each being tridentate, with Cu-O(1) 2.314(4), Cu-O(3) 2.419(3), Cu-N(1) 2.060(4), Cu-N(2) 1.935(4), Cu-N(3) 2.038(4) and Cu-N(4) 1.983(3) Å. E.s.r. and visible spectra indicate that the solid-state stereochemistry is preserved in solution. In (II), the Cu(terpy) units are bridged by unidentate ida ligands and the Cu atom (Fig. 1) has distorted trigonal-bipyramidal geometry with Cu-O(1) 2.175(6), Cu-O(3') 1.966(5), Cu-N(1) 2.015(6), Cu-N(2) 1.943(5) and Cu-N(3) 2.034(6) Å.

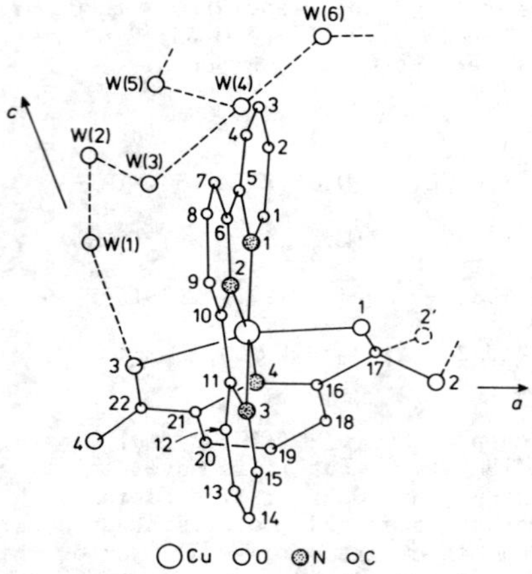

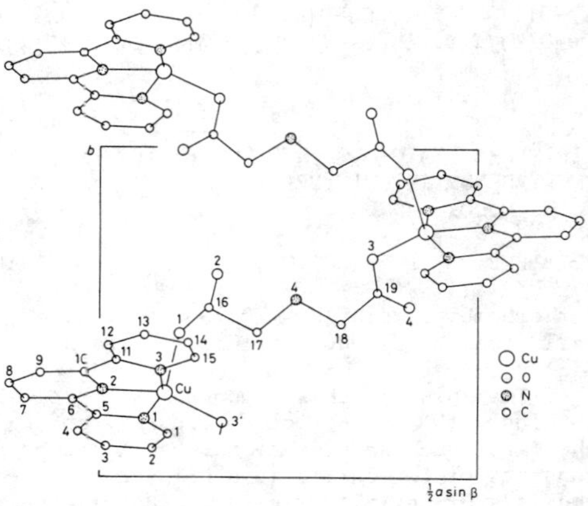

Fig. 1. Views of the [Cu(terpy)(pydca)] (left) and [Cu(terpy)(ida)] structures.

(μ_2-OXO)BIS(μ_2-ACETATO)BIS(TRIS(1-PYRAZOLYL)BORATO)DIIRON(III) ACETONITRILE SOLVATE
$C_{22}H_{26}B_2Fe_2N_{12}O_5,4(C_2H_3N)$ (I)

(μ_2-OXO)BIS(μ_2-FORMATO)BIS(TRIS(1-PYRAZOLYL)BORATO)DIIRON(III) ACETONITRILE SOLVATE
$C_{20}H_{22}B_2Fe_2N_{12}O_5,4(C_2H_3N)$ (II)

W.H. ARMSTRONG, A. SPOOL, G.C. PAPAEFTHYMIOU, R.B. FRANKEL and S.J. LIPPARD, 1984.
J. Am. Chem. Soc., 106, 3653-3667.

I. Monoclinic, $P2_1/n$, a = 13.236, b = 15.414, c = 21.697 Å, β = 107.26°, Dm = 1.319, Z = 4. Mo radiation, R = 0.040 for 5277 reflexions.

II. Orthorhombic, Pnma, a = 15.515, b = 19.764, c = 12.720 Å, Dm = 1.38, Z = 4. Mo radiation, R = 0.038 for 2597 reflexions (at -25°C).

In both structures (Fig. 1) the Fe atoms are linked by a μ-oxo and two μ-carboxylato ligands and the Fe atoms are capped by the tridentate pyrazolylborate ligands. The (μ-oxo)diiron(III) centre is bent, with average Fe-O 1.783(3) Å and Fe-O-Fe 124.6(1)°. Fe-N distances are 2.148-2.159(3) for I and 2.135 and 2.151(3) Å for II. Molecule II lies on a mirror plane (Fig. 1).

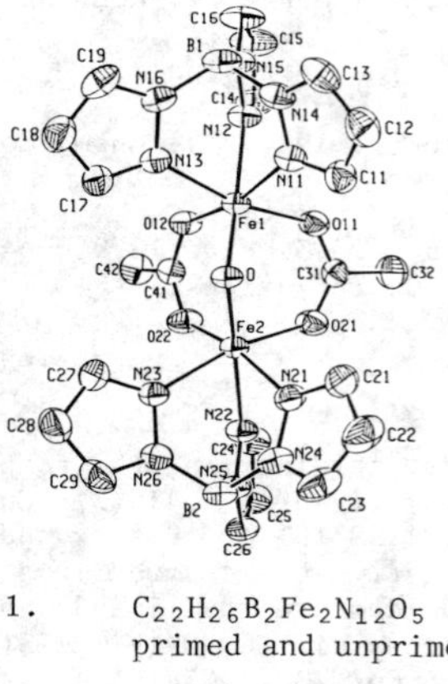

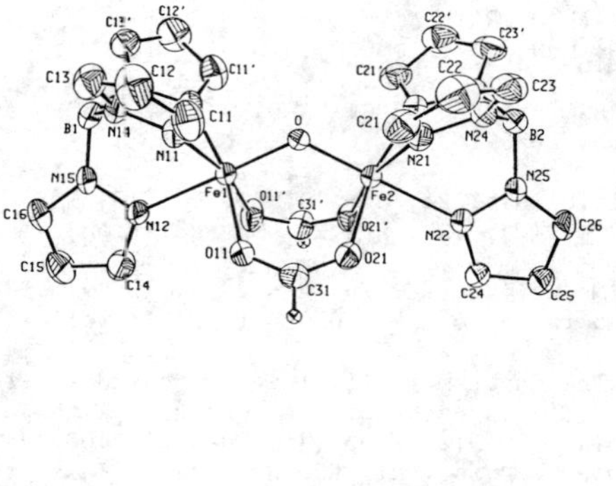

Fig. 1. $C_{22}H_{26}B_2Fe_2N_{12}O_5$ (I) (left) and $C_{20}H_{22}B_2Fe_2N_{12}O_5$ (II) (right); in II, primed and unprimed atoms are related by the crystallographic mirror plane.

(μ_2-HYDROXO)BIS(μ_2-ACETATO)BIS(HYDROTRIS(1-PYRAZOLYL)BORATO)DIIRON(III) PERCHLORATE DICHLOROMETHANE SOLVATE
$C_{22}H_{27}B_2ClFe_2N_{12}O_9,CH_2Cl_2$

W.H. ARMSTRONG and S.J. LIPPARD, 1984. J. Am. Chem. Soc., 106, 4632-4633.

Monoclinic, $P2_1/n$, a = 11.756, b = 19.925, c = 15.580 Å, β = 92.03°, D_m = 1.55, Z = 4. Mo radiation, R = 0.048 for 3658 reflexions (at 258 K).

Two Fe atoms (Fig. 1) are bridged by hydroxide and two acetate groups with two capping tridentate hydrotris(1-pyrazolyl)borate ligands. The H atom position of the hydroxo group was located on a difference Fourier map and refined. Main dimensions are Fe1...Fe2 3.439, Fe-O (bridge) 1.960 1.952(4), mean Fe-N 2.102, mean Fe-O(acetate) 1.999 Å, Fe-O-Fe 123.1(2)°.

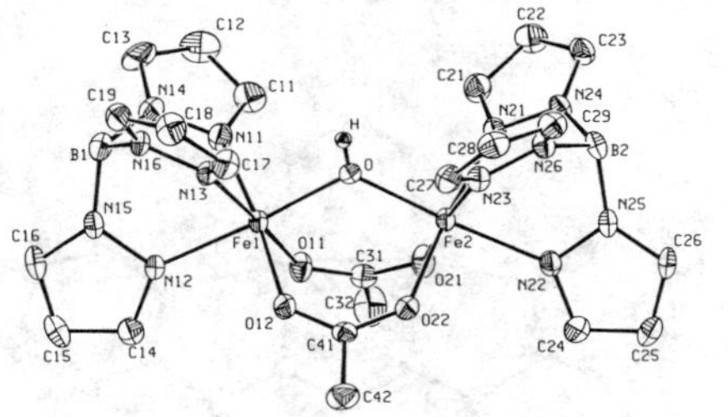

Fig. 1. Structure of $C_{22}H_{27}B_2ClFe_2N_{12}O_9$ (hydrogens omitted except for the hydroxyl H (shown as a small sphere)).

BIS[(o-METHYLBENZOATO)(1,3-PROPANEDIAMINE)]NICKEL(II)
$C_{22}H_{34}N_4NiO_4$ $[Ni(C_8H_7O_2)_2(C_3H_{10}N_2)_2]$

M. KLINGA, 1984. Acta Chem. Scand., A38, 63-66.

Orthorhombic, Pbca, a = 10.031, b = 22.540, c = 10.458 Å, D_m = 1.35, Z = 4. Mo radiation, R = 0.051 for 1454 reflexions.

The structure is mononuclear with the Ni(II) ion at the centre of symmetry (Fig. 1). Octahedral coordination about Ni is established by the bonds Ni-N(1) 2.123, Ni-N(2) 2.104 and Ni-O(1) 2.121 Å. The 1,3-propanediamine molecule displays a chair conformation and the carboxyl group in the methylbenzoato group is twisted 64.5° relative to the benzene ring plane. Hydrogen bonds form an intra- and intermolecular network. The intramolecular bond is O(2)...H(2)-N(1) (angle 139°) with O(2)...N(1) 3.10 Å. The intermolecular hydrogen bonds are: O(2)...H(1)-N(1) (0.5+x, y, -z+0.5) angle 156° and O(2)...N(1) 3.09 Å, and O(2)...H(1)-N(2) (0.5-x, -y, 0.5+z) angle 172° and O(2)...N(2) 3.11 Å.

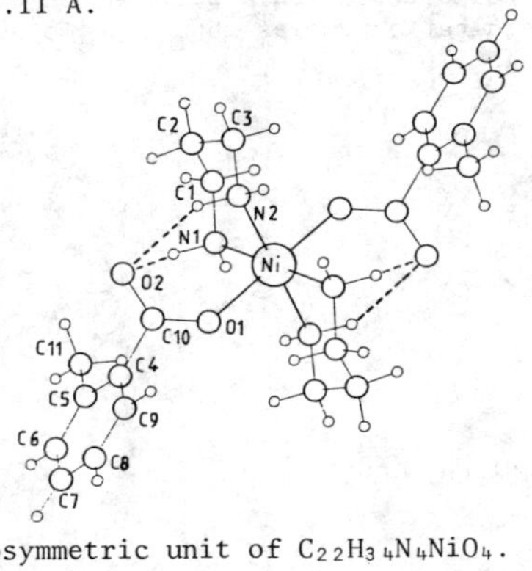

Fig. 1. One centrosymmetric unit of $C_{22}H_{34}N_4NiO_4$.

TRIS(μ-ACETATO)(2,7-BIS(2-PYRIDYL)-1,8-NAPHTHYRIDINE)DIRHODIUM(II) HEXAFLUOROPHOSPHATE

$C_{24}H_{21}F_6N_4O_6PRh_2$ $[Rh_2(O_2CCH_3)_3(C_{18}H_{12}N_4)]PF_6$

W.R. TIKKANEN, E. BINAMIRA-SORIAGA, W.C. KASKA and P.C. FORD, 1984. Inorg. Chem., 23, 141-146.

Tetragonal, $I4_1cd$, a = 28.720, c = 13.906 Å, D_m = 1.879, Z = 16. Mo radiation, R = 0.068 for 2144 reflexions (at 115 K).

The binuclear cation is shown in Fig. 1. The two Rh atoms, each having irregular octahedral geometry, are bridged by three acetate groups (Rh-O 1.993(14)-2.091(18) Å) and the naphthyridine fragment of the $C_{18}H_{12}N_4$ ligand (Rh-N 1.992(13) and 2.015(14) Å). The 2-pyridyl N atoms occupy the axial sites (Rh-N 2.202(15) and 2.189(16) Å) trans to the Rh-Rh bond (2.405(2) Å).

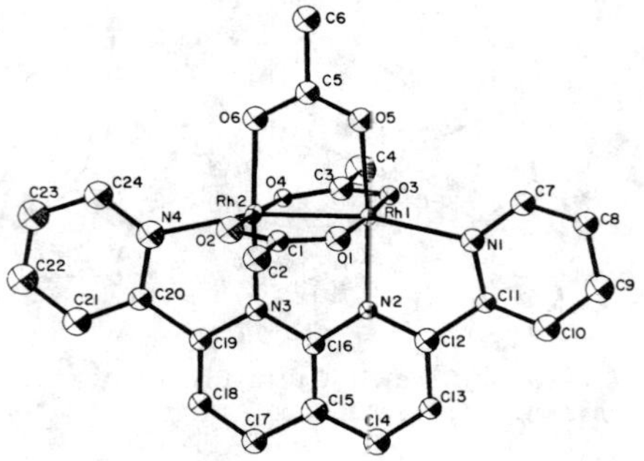

Fig. 1. The $[Rh_2(O_2CCH_3)_3(C_{18}H_{12}N_4)]^+$ cation.

TETRAKIS[μ-(3-CHLOROPROPIONATO)]-BIS(4-PICOLINE)DICOPPER(II)

$C_{24}H_{30}Cl_4Cu_2N_2O_8$ $Cu_2(3-ClC_2H_4COO)_4(NC_6H_7)_2$

K. SMOLANDER and M. MELNÍK, 1984. Acta Chem. Scand., A38, 619-623.

Triclinic, $P\bar{1}$, a = 8.475, b = 8.903, c = 10.204 Å, α = 97.27, β = 92.45, γ = 93.72°, Z = 1. Mo radiation, R = 0.038 for 3509 reflexions.

The structure consists of centrosymmetric dinuclear units in which the two Cu(II) ions are bridged by four carboxyl groups in syn-syn configuration so that the Cu-Cu distance is 2.675 Å (Fig. 1). The Cu coordination is elongated octahedral, with the carbonyl O atoms in the basal plane (Cu-O(1) 1.974, Cu-O(2') 1.973, Cu-O(3) 1.977 and Cu-O(4') 1.962 Å) and a 4-picoline molecule (Cu-N 2.170 Å) and Cu(II) ion in the terminal positions. The cis coordination angles about the metal ion range from 81.9 to 98.6° and the angle N-Cu-Cu' is 176.2°.

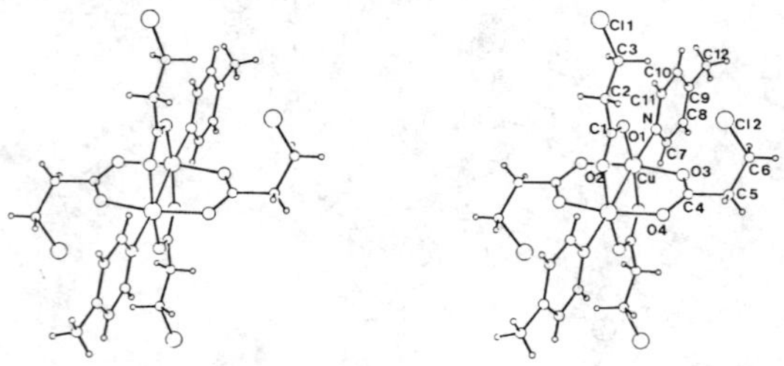

Fig. 1. Stereoview of $C_{24}H_{30}Cl_4Cu_2N_2O_8$.

ACETATOBIS(1,10-PHENANTHROLINE)COPPER(II) TETRAFLUOROBORATE
$C_{26}H_{19}BCuF_4N_4O_2$

W. FITZGERALD and B. HATHAWAY, 1984. Acta Cryst., C40, 243-245.

Triclinic, $P\bar{1}$, a = 15.087, b = 10.615, c = 8.443 Å, α = 106.43, β = 76.67, γ = 109.40°, D_m = 1.52, Z = 2. Mo radiation, R = 0.036 for 3287 reflexions.

The cation (Fig. 1) has a cis CuN_2N_2'OO' chromophore with a square-pyramidal distortion to give a (4+1+1*) stereochemistry. The acetate coordination is asymmetric (Cu-O 1.996(2) and 2.670(3) Å). Cu-N distances are 2.010, 2.062, 2.025 and 2.218(2) Å.

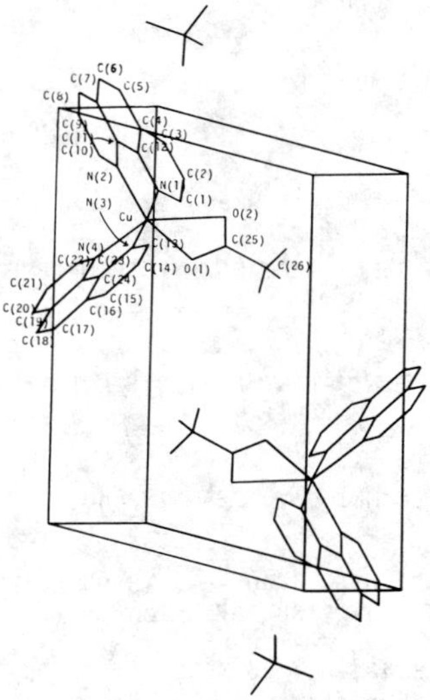

Fig. 1. The molecular structure of $[Cu(phen)_2(MeCO_2)]BF_4$.

BIS(μ_3-OXO)HEXAKIS(μ-BUTYRATO)TRIAQUATRITUNGSTEN BIS(TRIFLUOROMETHYLSULPHONATE) DIHYDRATE
$C_{26}H_{48}F_6O_{23}S_2W_3,2H_2O$ (I) $[W_3(\mu_3\text{-}O)_2(O_2CC_3H_7)_6(H_2O)_3](CF_3SO_3)_2.2H_2O$

BIS(μ_3-OXO)HEXAKIS(μ-BUTYRATO)TRIAQUATRIMOLYBDENUM BIS(PERCHLORATE) TRIHYDRATE
$C_{24}H_{48}Cl_2Mo_3O_{25},3H_2O$ (II) $[Mo_3(\mu_3\text{-}O)_2(O_2CC_3H_7)_6(H_2O)_3](ClO_4)_2.3H_2O$

F.A. COTTON, Z. DORI, D.O. MARLER and W. SCHWOTZER, 1984. Inorg. Chem., 23, 4738-4742.

I. Monoclinic, C2/c, a = 24.434, b = 23.628, c = 20.790 Å, β = 124.61°, Z = 8. Mo radiation, R = 0.053 for 3298 reflexions.

II. Cubic, $P2_13$, a = 16.272 Å, Z = 4. Cu radiation, R = 0.061 for 940 reflexions.

The two trinuclear cluster cations are shown in Fig. 1. Each cation contains an equilateral triangular M_3 cluster, capped by two μ_3-O ligands, with two μ-carboxylate ligands bridging each edge. The cation in II has exact C_3 symmetry. Bond lengths: for I; W-W 2.741(1)-2.744(1), W-(μ_3-O) 1.97(1)-2.01(1), W-O(O_2CR) 2.07(1)-2.12(1), W-OH_2 2.06(1)-2.12(1) Å; for II; Mo-Mo 2.762(2), Mo-(μ_3-O) 1.980(4) and 2.032(2), Mo-O(O_2CR) 2.078(9)-2.102(9), Mo-OH_2 2.10(1) Å. A comparison of Mo and W bond radii as a function of M-M bond order is made.

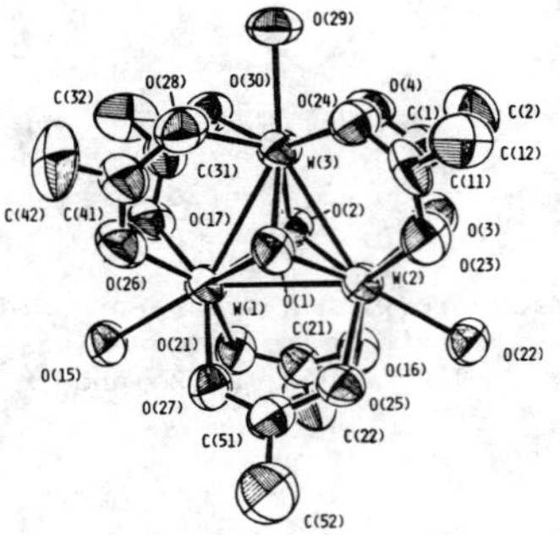

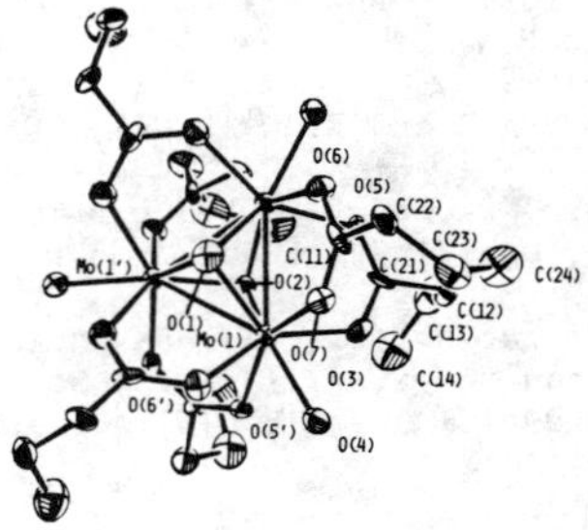

Fig. 1. The cations $[M_3(\mu_3\text{-}O)_2(O_2CC_3H_7)_6(H_2O)_3]^{2+}$ where M = W (left) and M = Mo (right).

(μ-TEREPHTHALATO)BIS[(N,N,N',N'',N''-PENTAMETHYLDIETHYLENETRIAMINE)AQUACOPPER(II)] DIPERCHLORATE

$C_{26}H_{54}Cl_2Cu_2N_6O_{14}$ (I)

(μ-TEREPHTHALATO)BIS[(N,N,N',N'',N''-PENTAETHYLDIETHYLENETRIAMINE)COPPER(II)] DIPERCHLORATE MONOHYDRATE

$C_{36}H_{70}Cl_2Cu_2N_6O_{12},H_2O$ (II)

M. VERDAGUER, J. GOUTERON, S. JEANNIN, Y. JEANNIN and O. KAHN, 1984. Inorg. Chem., 23, 4291-4296.

I. Monoclinic, $P2_1/c$, a = 8.42, b = 14.40, c = 15.99 Å, β = 100.65°, D_m = 1.52, Z = 2. Mo radiation, R = 0.049 for 2588 reflexions.

II. Orthorhombic, Pbca, a = 28.48, b = 26.363, c = 12.591 Å, D_m = 1.40, Z = 8. Cu radiation, R = 0.065 for 2760 reflexions.

The structures of I and II (Fig. 1) consist of μ-terephthalato(μ-tp)-bridged Cu(II) binuclear complexes and non-coordinated ClO_4^- anions. In I, tp bridges in a bis-monodentate fashion, and in II, in a bis-bidentate fashion. In both compounds, the environment of the metal ions is intermediate between the square-based pyramid and the trigonal-bipyramid, the latter being more pronounced in the case of II. Bond lengths: for I; Cu-O 1.931(3), Cu-OH_2 2.225(4), Cu-N 2.029(4)-2.067(4) Å; for II; Cu-O 1.972(6), 1.971(6), 2.347(6), and 2.379(6), Cu-N 1.972(8)-2.083(8) Å. Complex I is centrosymmetric.

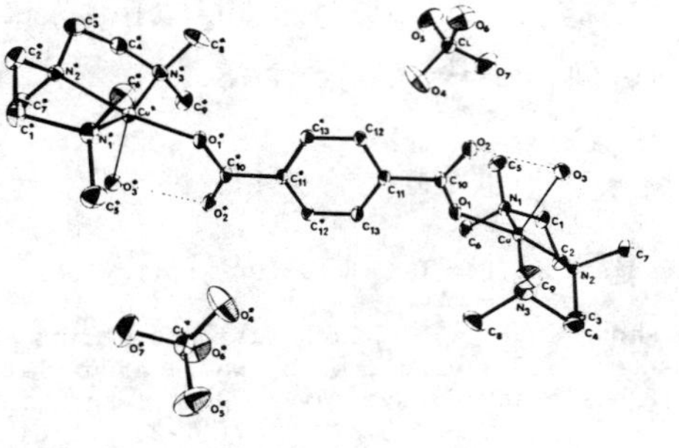

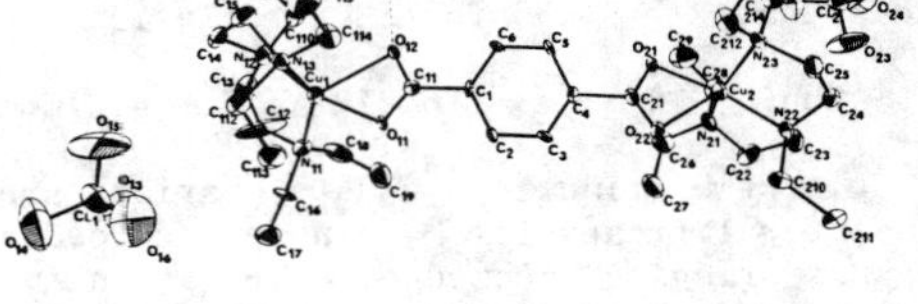

Fig. 1. The structures of $C_{26}H_{54}Cl_2Cu_2N_6O_{14}$ (left) and $C_{36}H_{70}Cl_2Cu_2N_6O_{12}.H_2O$ (right).

μ-AMIDO-μ-FORMATO-BIS[TETRAAMMINECOBALT(III)] TETRACHLORIDE TETRAHYDRATE
$CH_{27}Cl_4Co_2N_9O_2,4H_2O$ $Co_2NH_2(CHO_2)(NH_3)_8Cl_4,4H_2O$

W. CLEGG and A.G. SYKES, 1984. Acta Cryst., C40, 1820-1821.

Orthorhombic, Pnma, a = 11.661, b = 19.811, c = 9.105 Å, Z = 4. Mo radiation, R = 0.033 for 1627 reflexions.

The cation (Fig. 1) lies on a mirror plane with Co-NH_2 1.958(2), Co-O 1.912(2), Co-NH_3 (mean) 1.963(17) Å; the shortest Co-NH_3 bond is trans to Co-O while the longest is trans to Co-NH_2. The Co-N-Co angle is 127.8(2)°. Cations, anions and water molecules are linked by a three-dimensional network of hydrogen bonds.

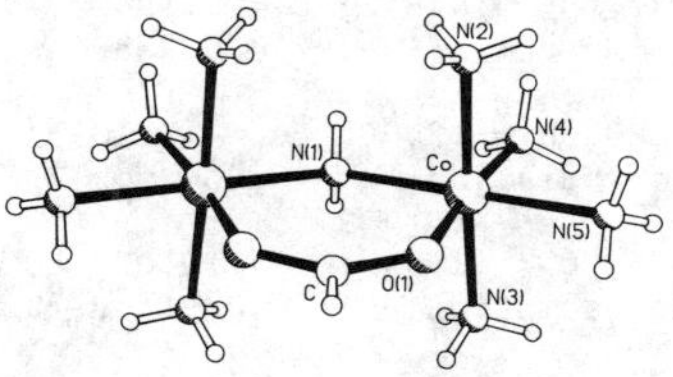

Fig. 1. The $Co_2NH_2(CHO_2)(NH_3)_8$ cation.

trans-BIS(TRICHLOROACETATO)TETRAKIS(4-CHLOROANILINE)NICKEL(II) BIS(4-CHLOROANILINE) SOLVATE
$C_{28}H_{24}Cl_{10}N_4NiO_4.2C_6H_6ClN$ $(Cl_3CCO_2)_2(4\text{-}H_2NC_6H_4Cl)_4.2(4\text{-}H_2NC_6H_4Cl)$

T.S. KUNTSEVICH, T.N. TARKHOVA and E.E. RAKOVA, 1984. J. Struct. Chem., 25, 134-137 [Zh. Strukt. Khim., 25(1), 152-156].

Monoclinic, $P2_1/b$, a = 10.905, b = 11.44, c = 19.93 Å, γ = 90.45°, Z = 2. Mo radiation, R = 0.12 for 667 reflexions.

The structure (Fig. 1) contains centrosymmetric complex molecules having trans octahedral coordination geometry with Ni-O 2.03(2) and Ni-N 2.25(4) and 2.30(4) Å and 4-chloroaniline solvate molecules. The carboxylate ligands are monodentate.

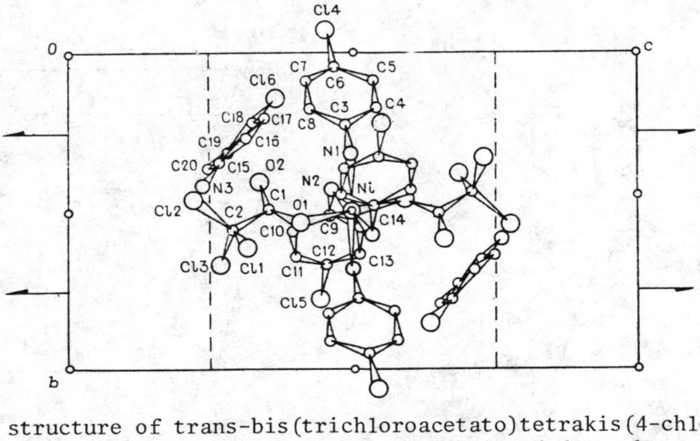

Fig. 1. The structure of trans-bis(trichloroacetato)tetrakis(4-chloroaniline)-nickel(II) bis(4-chloroaniline) solvate viewed down the a axis.

DIAQUATRIS(m-HYDROXYBENZOATO)LANTHANUM(III)-(m-HYDROXYBENZOIC ACID) DIHYDRATE POLYMER
$C_{28}H_{29}LaO_{14},2H_2O$ (I) $[(C_6H_4(OH)COO)_3La(H_2O)_2]C_6H_4(OH)COOH,2H_2O$

DIAQUATRIS(m-HYDROXYBENZOATO)NEODYMIUM(III)-(m-HYDROXYBENZOIC ACID) DIHYDRATE POLYMER
$C_{28}H_{29}NdO_{14},2H_2O$ (II) $[(C_6H_4(OH)COO)_3Nd(H_2O)_2]C_6H_4(OH)COOH,2H_2O$

DI-μ-(m-HYDROXYBENZOATO)BIS{TRIAQUABIS(m-HYDROXYBENZOATO)SAMARIUM(III)} DIHYDRATE
$C_{42}H_{42}O_{24}Sm_2,2H_2O$ (III) $[Sm(C_6H_4(OH)COO)_3(H_2O)_3]_2,2H_2O$

DI-μ-(m-HYDROXYBENZOATO)BIS{TRIAQUABIS(m-HYDROXYBENZOATO)ERBIUM(III)} DIHYDRATE
$C_{42}H_{42}Er_2O_{24},2H_2O$ (IV) $[Er(C_6H_4(OH)COO)_3(H_2O)_3]_2,2H_2O$

Y. KOIZUMI, H. SAWASE, Y. SUZUKI, T. TAKEUCHI, M. SHIMOI and A. OUCHI, 1984. Bull. Chem. Soc. Japan, 57, 1809-1817.

I. Triclinic, P$\bar{1}$, a = 12.351, b = 15.533, c = 9.250 Å, α = 109.86, β = 101.40, γ = 62.95°, D_m = 1.69, Z = 4(?). Mo radiation, R = 0.039 for 3087 reflexions.

II. Triclinic, P$\bar{1}$, a = 12.287, b = 15.542, c = 9.194 Å, α = 109.64, β = 101.46, γ = 62.83°, D_m = 1.71, Z = 4(?). Mo radiation, R = 0.038 for 3022 reflexions.

III. Triclinic, P$\bar{1}$, a = 10.580, b = 12.006, c = 10.105 Å, α = 111.14, β = 90.27, γ = 97.16°, D_m = 1.78, Z = 2. Mo radiation, R = 0.030 for 2958 reflexions.

IV. Triclinic, P$\bar{1}$, a = 10.586, b = 11.943, c = 10.027 Å, α = 111.11, β = 90.53, γ = 97.26°, D_m = 1.82, Z = 2. Mo radiation, R = 0.039 for 2925 reflexions.

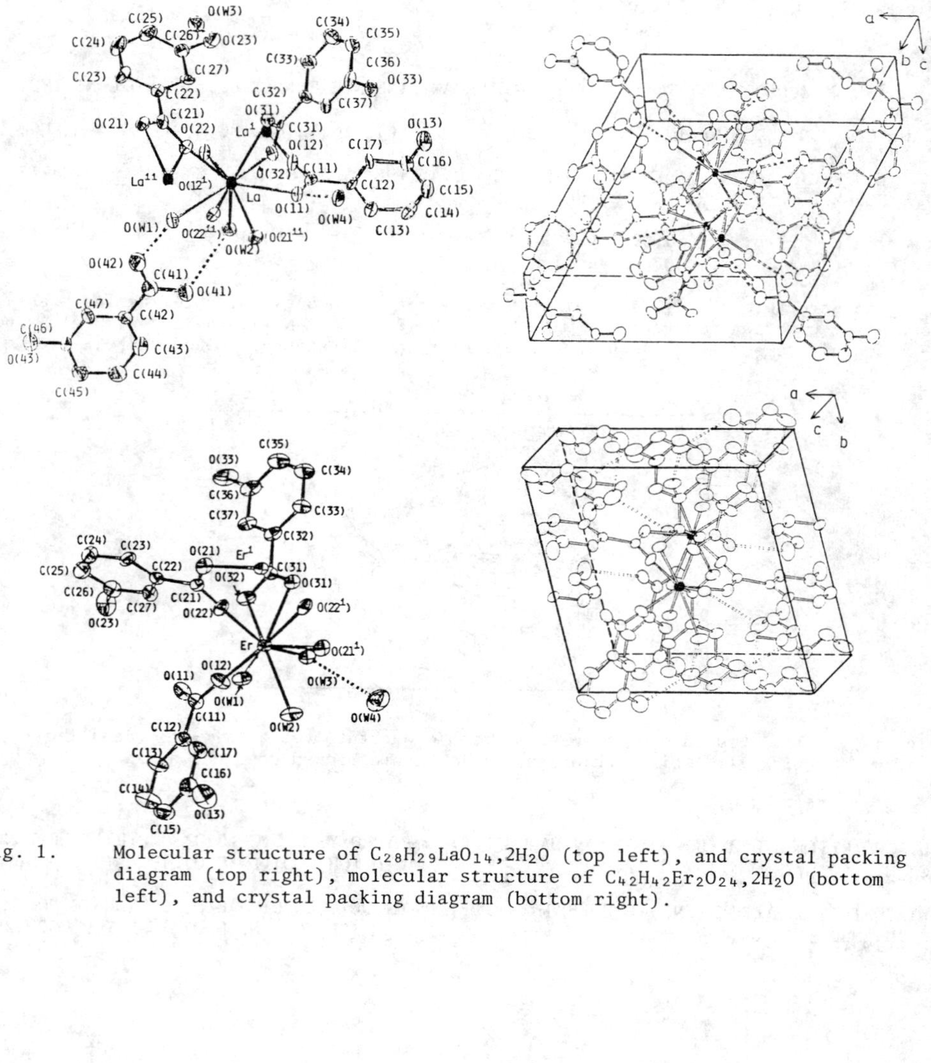

Fig. 1. Molecular structure of $C_{28}H_{29}LaO_{14},2H_2O$ (top left), and crystal packing diagram (top right), molecular structure of $C_{42}H_{42}Er_2O_{24},2H_2O$ (bottom left), and crystal packing diagram (bottom right).

Compounds I and II are isostructural and III and IV are isostructural. The central metal atoms of each structure are eight-coordinate, and each is in a square-antiprism geometry (Fig. 1). In I and II, one free acid molecule is present, and the molecule is hydrogen-bonded to the O atoms of the coordinated water molecules through its carboxyl O atoms. I and II have linear, polymeric structures, and each pair of metal atoms is bridged by a pair of the carboxylate ligands, while III and IV have a dimeric structure where metal atoms are bridged by two carboxylate ligands. The metal-oxygen distances range from 2.536-2.723 Å in I, 2.356-2.615 Å in II, 2.273-2.594 Å in III, and 2.198-2.534 Å in IV.

BIS[TETRAKIS(μ_2-ACETATO)(μ_2-4-AMINO-5-AMINOMETHYL-2-METHYLPYRIMIDINE)DIRHODIUM] HEPTAHYDRATE
$(C_{28}H_{44}N_8O_{16}Rh_4)_n, 7n(H_2O)$

K. AOKI and H. YAMAZAKI, 1984. J. Am. Chem. Soc., 106, 3691-3692.

Triclinic, P$\bar{1}$, a = 14.583, b = 10.986, c = 8.291 Å, α = 108.90, β = 93.43, γ = 95.48°, Z = 2. Mo radiation, R = 0.061 for 3533 reflexions.

The polymeric structure (Fig. 1) contains two independent dirhodium-tetraacetate cages with each involving a crystallographic inversion centre at the mid-point of the Rh-Rh bond. A feature of the complex is the metal bonding to ring nitrogen N(1) and the absence of bonding to N(3). Some bond lengths are: mean Rh-O 2.040(9), Rh(1)-N(1) 2.293(7), Rh(2)-N(5β) 2.291(9), Rh-Rh 2.404 and 2.405(1) Å.

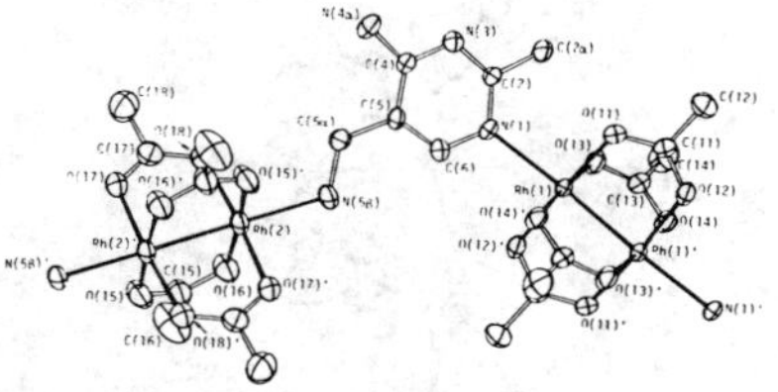

Fig. 1. A segment of the polymer $(C_{28}H_{44}N_8O_{16}Rh_4)_n$.

α-CARBONYL((DIPHENYLPHOSPHINO)ACETATO-P,O)((DIPHENYLPHOSPHINO)ACETIC ACID-P)RHODIUM(I)
$C_{29}H_{25}O_5P_2Rh$ $Rh(CO)(Ph_2PCH_2CO_2)(Ph_2PCH_2CO_2H)$

A. JEGOROV, B. KRATOCHVÍL, V. LANGER and J. PODLAHOVÁ, 1984. Inorg. Chem., 23, 4288-4291.

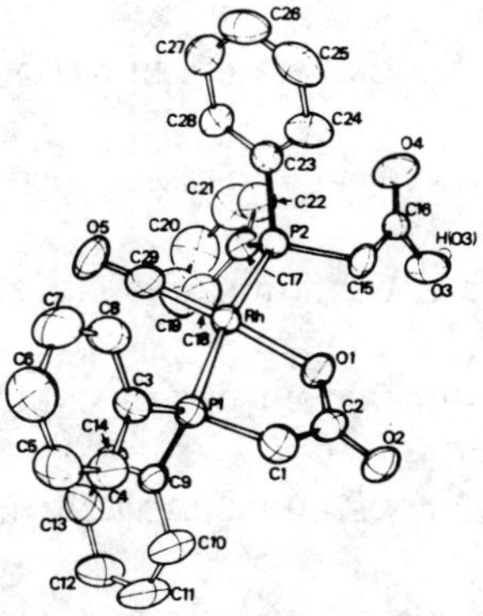

Fig. 1. The $Rh(CO)(Ph_2PCH_2CO_2)(Ph_2PCH_2CO_2H)$ molecule.

Triclinic, $P\bar{1}$, a = 9.783, b = 11.702, c = 13.120 Å, α = 76.87, β = 81.21, γ = 71.77°, D_m = 1.494, Z = 2. Mo radiation, R = 0.053 for 2367 reflexions.

The molecule (Fig. 1) has a distorted square planar arrangement around Rh with Rh-C 1.773(11), Rh-P (monodentate) 2.346(2), Rh-P (chelating) 2.302(2), Rh-O 2.064(6) Å, and P-Rh-P 170.8(1)°. There is a short Rh-H interaction of 2.80(7) Å with a phenyl ortho H atom of the $Ph_2PCH_2CO_2H$ ligand, which forces the Rh atom 0.03 Å out of the square plane. Centrosymmetric H-bonded dimers are observed.

AQUA(ETHANOL)TETRAKIS(μ-SALICYLATO-O,O')DIRHODIUM(II) ETHANOL SOLVATE MONOHYDRATE
$C_{30}H_{28}O_{14}Rh_2,C_2H_6O,H_2O$ $[Rh_2(O_2CC_6H_4OH)_4(C_2H_5OH)(H_2O)].C_2H_5OH,H_2O$

D.P. BANCROFT, F.A. COTTON and S. HAN, 1984. Inorg. Chem., 23, 2408-2411.

Monoclinic, $P2_1/n$, a = 14.830, b = 15.953, c = 15.130 Å, β = 90.90°, Z = 4. Mo radiation, R = 0.049 for 2120 reflexions.

The molecule (Fig. 1) contains a Rh-Rh bond of 2.385(2) Å, and the eight equatorial ligand sites are occupied by carboxylate O atoms, mean Rh-O 2.040(4) Å. One axial site has an ethanol molecule, Rh-O 2.30(1) Å, and the other a water molecule, Rh-O 2.30(1) Å. Intraligand O-H...O hydrogen bonds are present, O...O 2.49(2)-2.60(2) Å.

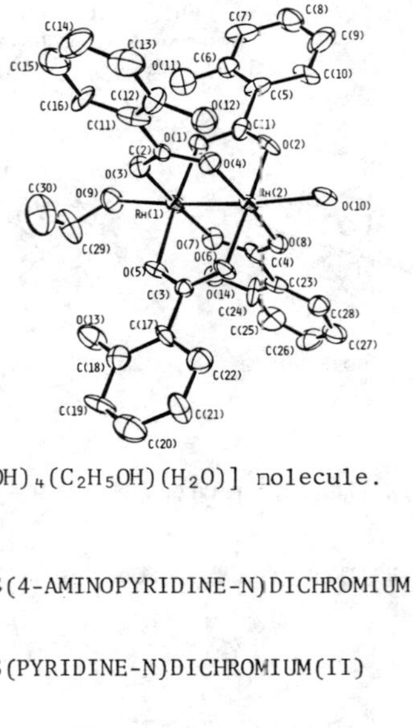

Fig. 1. The $[Rh_2(O_2CC_6H_4OH)_4(C_2H_5OH)(H_2O)]$ molecule.

TETRAKIS(μ_2-PIVALATO-O,O')BIS(4-AMINOPYRIDINE-N)DICHROMIUM(II)
$C_{30}H_{48}Cr_2N_4O_8$ (I)

TETRAKIS(μ_2-PIVALATO-O,O')BIS(PYRIDINE-N)DICHROMIUM(II)
$C_{30}H_{46}Cr_2N_2O_8$ (II)

TETRAKIS(μ_2-PIVALATO-O,O')BIS(4-CYANOPYRIDINE-N)DICHROMIUM(II) TOLUENE SOLVATE
$C_{32}H_{44}Cr_2N_4O_8,C_7H_8$ (III)

TETRAKIS(μ_2-PIVALATO-O,O')BIS(2-CYANOPYRIDINE-N)DICHROMIUM(II) TOLUENE SOLVATE
$C_{32}H_{44}Cr_2N_4O_8,C_7H_8$ (IV)

TETRAKIS(μ_2-ACETATO-O,O')BIS(4-DIMETHYLAMINOPYRIDINE-N)DICHROMIUM(II) BENZENE SOLVATE
$C_{22}H_{32}Cr_2N_4O_8,C_6H_6$ (V)

TETRAKIS(μ_2-ACETATO-O,O')BIS(4-CYANOPYRIDINE)DICHROMIUM(II)
$C_{20}H_{20}Cr_2N_4O_8$ (VI)

TETRAKIS(μ_2-FORMATO-O,O')BIS(4-DIMETHYLAMINOPYRIDINE-N)DICHROMIUM(II)
$C_{18}H_{24}Cr_2N_4O_8$ (VII)

TETRAKIS(μ_2-FORMATO-O,O')BIS(4-CYANOPYRIDINE-N)DICHROMIUM(II)
$C_{16}H_{12}Cr_2N_4O_8$ (VIII)

TRIS(μ_2-ACETATO-O,O')(μ_2-CHLOROACETATO-O,O')BIS(PYRIDINE-N)DICHROMIUM(II)
$C_{18}H_{21}ClCr_2N_2O_8$ (IX)

TETRAKIS(μ_2-CHLOROACETATO-O,O')BIS(4-CYANOPYRIDINE-N)DICHROMIUM(II) ACETIC ACID SOLVATE
$C_{20}H_{16}Cl_4Cr_2N_4O_8,C_2H_4O_2$ (X)

TETRAKIS(μ_2-DIFLUOROACETATO-O,O')BIS(DIETHYL ETHER-O)DICHROMIUM(II)
$C_{16}H_{24}Cr_2F_8O_{10}$ (XI)

TETRAKIS(μ_2-DIFLUOROACETATO-O,O')BIS(4-DIMETHYLAMINOPYRIDINE-N)DICHROMIUM(II)
$C_{22}H_{24}Cr_2F_8N_4O_8$ (XII)

TETRAKIS(μ_2-DIFLUOROACETATO-O,O')BIS(4-t-BUTYLPYRIDINE-N)DICHROMIUM(II)
$C_{26}H_{30}Cr_2F_8N_2O_8$ (XIII)

F.A. COTTON and W. WANG, 1984. Nouv. J. Chim., 8, 331-340.

I. Triclinic, P$\bar{1}$, a = 10.468, b = 11.004, c = 9.605 Å, α = 101.76, β = 95.61, γ = 117.33°, Z = 1. Mo radiation, R = 0.070 for 2814 reflexions.

II. Triclinic, P$\bar{1}$, a = 11.959, b = 16.304, c = 9.806 Å, α = 96.86, β = 103.96, γ = 97.80°, Z = 2. Mo radiation, R = 0.058 for 1670 reflexions.

III. Monoclinic, $P2_1/c$, a = 11.399, b = 22.28, c = 17.933 Å, β = 98.98°, Z = 4. Cu radiation, R = 0.089 for 2480 reflexions (at 3°C).

IV. Monoclinic, $P2_1/n$, a = 11.324, b = 10.576, c = 18.707 Å, β = 95.35°, Z = 2. Mo radiation, R = 0.077 for 2173 reflexions.

V. Triclinic, P$\bar{1}$, a = 8.298, b = 11.918, c = 8.226 Å, α = 101.78, β = 91.45, γ = 102.86°, Z = 1. Mo radiation, R = 0.080 for 1263 reflexions.

VI. Triclinic, P$\bar{1}$, a = 8.511, b = 9.736, c = 7.737 Å, α = 112.65, β = 92.01, γ = 95.74°, Z = 1. Mo radiation, R = 0.048 for 1312 reflexions.

VII. Triclinic, P$\bar{1}$, a = 7.481, b = 12.073, c = 6.842 Å, α = 100.39, β = 110.85, γ = 94.44°, Z = 1. Mo radiation, R = 0.058 for 1035 reflexions.

VIII. Orthorhombic, Pbca, a = 6.783, b = 28.496, c = 20.118 Å, Z = 8. Mo radiation, R = 0.066 for 1019 reflexions.

IX. Orthorhombic, Pbca, a = 13.106, b = 8.676, c = 19.630 Å, Z = 4. Mo radiation, R = 0.049 for 881 reflexions.

X. Tetragonal, $I4_1/amd$, a = 9.862, c = 36.84 Å, Z = 4. Mo radiation, R = 0.079 for 895 reflexions.

XI. Triclinic, P$\bar{1}$, a = 12.002, b = 12.021, c = 9.458 Å, α = 100.93, β = 106.01, γ = 86.29°, Z = 2. Mo radiation, R = 0.082 for 2637 reflexions.

XII. Triclinic, P$\bar{1}$, a = 9.996, b = 10.976, c = 7.650 Å, α = 91.67, β = 97.06, γ = 117.00°, Z = 1. Mo radiation, R = 0.063 for 1475 reflexions.

XIII. Tetragonal, $I4_1/a$, a = 20.257, c = 16.116 Å, Z = 8. Mo radiation, R = 0.084 for 811 reflexions.

The thirteen compounds provide data for a study between correlations between (a) Cr-Cr and Cr-N bond lengths, (b) each of these bond lengths with the pK_a values of the carboxylic acids, and (c) each of these bond lengths with the pK_a value of the pyridine ligand. The observed Cr-Cr bond lengths are in the range 2.315(2) to 2.514(1) Å and Cr-N bond lengths range from 2.23(2) to 2.388(4) Å. With the exception of compounds (VI) and (X) the correlations sought were found. The Cr-Cr and Cr-N bond lengths are inversely related; the Cr-Cr bond lengths vary directly with the pK_a of ligand L and inversely with pK_a of RCO_2H while the Cr-N bond lengths show the opposite correlations with pK_a values.

BIS(μ_2-BENZOATO-O,O')BIS(BENZOIC ACID)TETRACARBONYLDIRUTHENIUM
$C_{32}H_{22}O_{12}Ru_2$ (I) $[Ru(PhCO_2)(CO)_2PhCO_2H]_2$

BIS(BIS(μ_2-sec-BUTYRATO-O,O')(sec-BUTYRIC ACID)TETRACARBONYLDIRUTHENIUM)
$C_{38}H_{56}O_{20}Ru_4$ (II) $[Ru_2(C_4H_9CO_2)_2(CO)_4(C_4H_9CO_2H)]_2$

M. ROTEM, Y. SHVO, I. GOLDBERG and U. SHMUELI, 1984. Organometallics, 3, 1758-1759.

I. Monoclinic, C2/c, a = 19.598, b = 16.909, c = 10.900 Å, β = 118.41°, Z = 4. Mo radiation, R = 0.037 for 2087 reflexions.

II. Triclinic, $P\bar{1}$, a = 9.988, b = 11.006, c = 12.916 Å, α = 107.63, β = 90.42, γ = 109.91°, D_m = 1.6, Z = 1. Mo radiation, R = 0.055 for 3512 reflexions.

The unique feature of both structures is the side ligation of acid molecules (Fig. 1). Complex I has twofold symmetry and II has a centre of symmetry. The Ru coordination octahedron in I is severely distorted (Ru(1')-Ru-O(17) 162.8°) probably as a consequence of the intramolecular hydrogen bonding. Selected bond lengths are: Ru(1)-Ru(1') 2.637, Ru(1)-O(6) 2.146, Ru(1')-O(14) 2.121, Ru(1)-O(17) 2.265, O(6)-C(7) 1.265, C(7)-O(14) 1.241, O(16)...O(6) 2.605 Å. Compound II may be viewed as a dimer of two bis(μ-carboxylato)Ru-Ru cores, capped by two acid molecules. Selected bond lengths are: Ru(1)-Ru(13) 2.630, Ru(1)-O(30) 2.26, Ru(1)-O(6) 2.13, Ru(1)-O(24) 2.10, Ru(13)-O(12) 2.10, Ru(13)-O(18) 2.16, C(7)-O(6) 1.26, C(7)-O(12) 1.25, C(19)-O(18) 1.24, C(19)-O(24) 1.26, Ru(13)-O(18') 2.314, O(31)...O(6) 2.57 and Ru(13)... Ru(13') 3.544 Å.

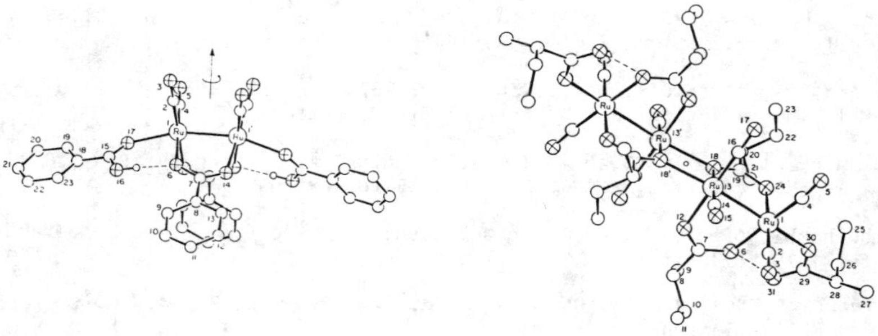

Fig. 1. Molecular structure of $C_{32}H_{22}O_{12}Ru_2$ (left), and $C_{38}H_{56}O_{20}Ru_4$ (right).

BIS(μ-BENZOATO-O,O')(DIMETHYLSULPHOXIDE)COPPER(II)
$C_{32}H_{32}Cu_2O_{10}S_2$ $[Cu(C_6H_5COO)_2(CH_3)_2SO]_2$

M. MELNÍK, M. DUNAH-JURČO and M. HANDLOVIČ, 1984. Inorg. Chim. Acta, 86, 185-190.

Monoclinic, C2/c, a = 19.030, b = 15.494, c = 23.828 Å, β = 103.73°, D_m = 1.49, Z = 4. Mo radiation, R = 0.071 for 1892 reflexions.

The compound is dimeric, with square pyramidal geometry at each Cu centre (Fig. 1). The two metal atoms are bridged by four carboxylate groups (Cu-O distances range from 1.93 to 2.00 Å) forming the equatorial ligands, whilst the apical ligands are DMSO's (Cu(1)-O(5) 2.16 Å, Cu(2)-O(10) 2.17 Å). The Cu atoms are displaced by 0.197 Å (Cu(1)) and 0.207 Å (Cu(2)) from the planes containing the four equatorial O atoms, towards the DMSO ligands. The Cu...Cu distance is 2.627 Å, and the S-O distances in the DMSO ligands are 1.47 and 1.49 Å for S(1) and S(2) respectively. The C-O distances in the benzoate ligands range from 1.22 to 1.30 Å with an average of 1.26 Å.

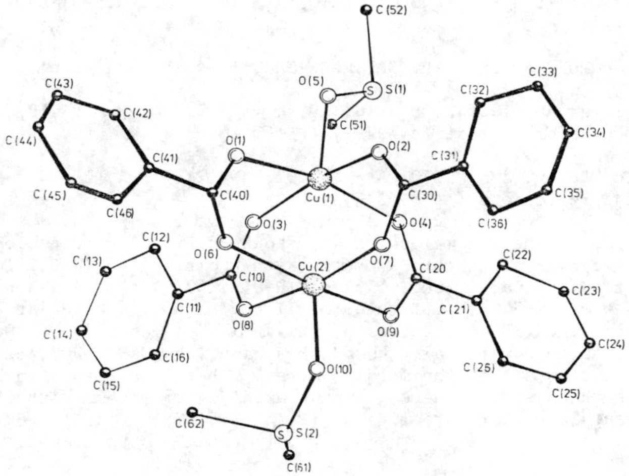

Fig. 1. Molecular structure of $C_{32}H_{32}Cu_2O_{10}S_2$.

BIS(1-AZONIA-4-AZABICYCLO[2.2.2]OCTANE) (NITRILOTRIACETATO-N,O,O',O'')(3,5-DI-t-BUTYL-CATECHOLATO-O,O')FERRATE(III) DIMETHYLFORMAMIDE SOLVATE

$C_{32}H_{52}FeN_5O_8,C_3H_7NO$ (dabcoH)$_2$[Fe(NTA)DBC],C_3H_7NO

L.S. WHITE, P.V. NILSSON, L.H. PIGNOLET and L. QUE Jr., 1984. J. Am. Chem. Soc., 106, 8312-8313.

Monoclinic, $P2_1/n$, a = 17.874, b = 9.963, c = 23.117 Å, β = 105.38°, D_m = 1.28, Z = 4. Mo radiation, R = 0.067 for 3413 reflexions.

The anion has the Fe atom with distorted octahedral coordination geometry with a bidentate catecholate and a tetradentate NTA ligand (Fig. 1), with Fe-N1 2.224(3), Fe-O2 2.034(3), Fe-O(4) 2.002(3), Fe-O6 2.039(3), Fe-O7 1.887(3), Fe-O8 1.979(3) Å. The shortest Fe-O bond is trans to Fe-N.

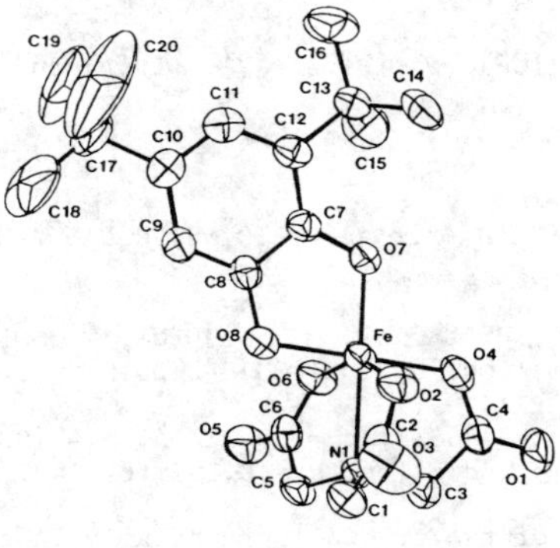

Fig. 1. The structure of the $[Fe(NTA)DBC]^{2-}$ anion.

BIS(μ-TRIFLUOROACETATO)BIS(η^1-TRIFLUOROACETATO)BIS(TRI-n-BUTYLPHOSPHINE)DITUNGSTEN
$C_{32}H_{54}F_{12}O_8P_2W_2$ (I) $W_2(CF_3COO)_4[P(C_4H_9)_3]_2$

(μ-TRIFLUOROACETATO)TRIS(η^1-TRIFLUOROACETATO)TRIS(TRIMETHYLPHOSPHINE)DITUNGSTEN
$C_{17}H_{27}F_{12}O_8P_3W_2$ (II) $W_2(CF_3COO)_4[P(CH_3)_3]$

D.J. SANTURE, J.C. HUFFMAN and A.P. SATTELBERGER, 1984. Inorg. Chem., 23, 938-945.

I. Monoclinic, I2/a, a = 19.240, b = 10.379, c = 21.522 Å, β = 93.39°, Z = 4. Mo radiation, R = 0.042 for 2514 reflexions (at -162°C).

II. Orthorhombic, $Pc2_1n$, a = 9.870, b = 15.002, c = 21.118 Å, Z = 4. Mo radiation, R = 0.036 for 1841 reflexions (at -163°C).

The structures of the two molecules are shown in Fig. 1. Molecules of I are centrosymmetric and have a C_{2h} core symmetry with two bridging (W-O 2.107(7) and 2.106(8) Å) and two terminal CF_3COO^- ligands (W-O 2.132(7) Å). The W-W distance is 2.224(1) Å and W-P 2.489(3) Å. Molecules of II have a single bridging CF_3COO^- ligand (W-O 2.101(11) and 2.064(12) Å), three terminal CF_3COO^- ligands (W-O 2.064(13)-2.124(12) Å), and three equatorially bound phosphine ligands (W-P 2.462(5)-2.504(5) Å), two of which are cis on one of the metal centres. The W-W distance in II is 2.246(1) Å. In both molecules there are weak axial interactions with the uncoordinated O atoms of monodentate carboxylate ligands (W...O 2.845(8) Å in I, 2.84(1) and 3.10(1) Å in II).

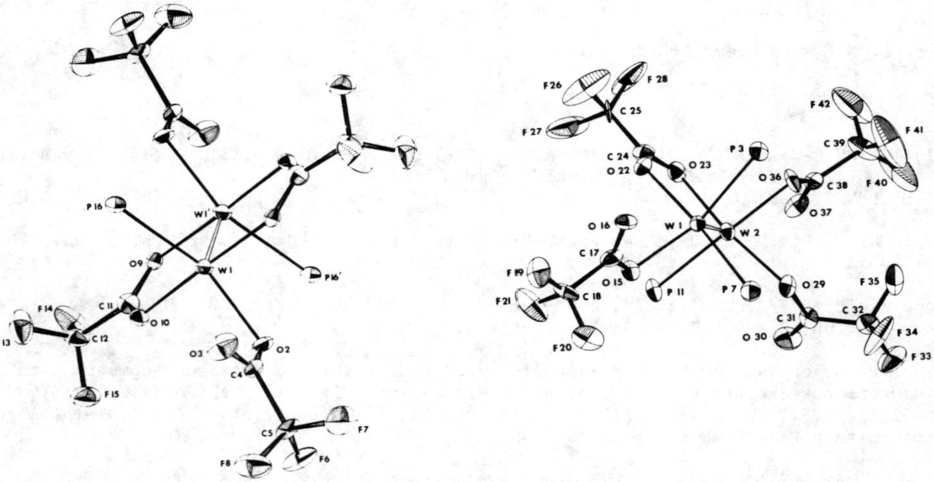

Fig. 1. The molecules $W_2(CF_3COO)_4[P(C_4H_9)_3]_2$ (left) and $W_2(CF_3COO)_4[P(CH_3)_3]_3$ (right).

BIS(N-BENZYLOXYCARBONYLGLYCINATO)(2,2'-BIPYRIDINE)PROPAN-2-OL-COPPER(II)
$C_{33}H_{36}CuN_4O_9$

L. ANTOLINI, L. MENABUE, G.C. PELLACANI, M. SALADINI, M. SOLA, L.P. BATTAGLIA and A.B. CORRADI, 1984. J. Chem. Soc. Dalton, 2319-2323.

Triclinic, $P\bar{1}$, a = 22.199, b = 15.357, c = 10.223 Å, α = 103.78, β = 83.06, γ = 88.89°, D_m = 1.36, Z = 4. Mo radiation, R = 0.095 for 3437 reflexions.

The structure consists of two crystallographically independent, but chemically equivalent, molecules linked in dimeric units by hydrogen-bonding interactions (Fig. 1). Each Cu atom has slightly distorted square-pyramidal five-coordination with the bidentate 2,2'-bipyridine ligand and two N-benzoylcarbonylglycinate ions (bglyO) in the equatorial plane, and a propan-2-ol molecule in an apical position. Coordination distances include: Cu(1)-N(1) 1.98(2), Cu(1)-N(2) 2.01(1), Cu(1)-O(1) 1.96(1), Cu(2)-N(5) 2.02(1), Cu(2)-N(6) 2.00(2), Cu(2)-O(9) 1.96(1), Cu(1)-O(5) 1.97(1), Cu(1)-O(17) 2.27(2), Cu(1)...O(6) 2.62(1), Cu(2)-O(13) 1.93(1), Cu(2)-O(18) 2.34(1) and Cu(2)... O(10) 2.84(1) Å. Packing in the crystal is a result of intermolecular stacking and hydrophobic interactions between 2,2'-bipyridine molecules and between phenyl groups.

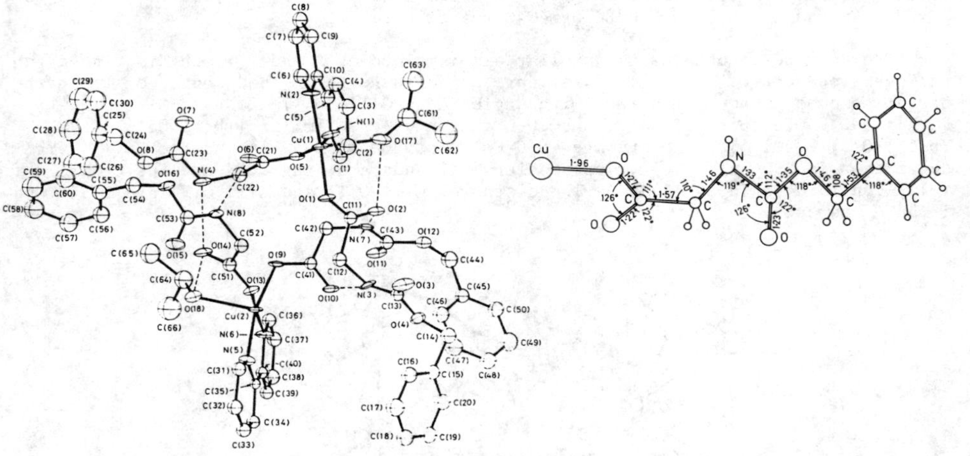

Fig. 1. $C_{33}H_{36}CuN_4O_9$: a view of the complex (left) and average dimensions of the bglyO ligands (right).

HEXAKIS(μ_2-ACETATO)(μ_3-OXO)TRIS(4-ETHYLPYRIDINE)TRIIRON 4-ETHYLPYRIDINE SOLVATE

$C_{33}H_{45}Fe_3N_3O_{13},C_7H_9N$

S.M. OH, D.N. HENDRICKSON, K.L. HASSETT and R.E. DAVIS, 1984. J. Am. Chem. Soc., 106, 7984-7985.

Monoclinic, C2/c, a = 17.060, b = 16.934, c = 16.183 Å, β = 104.49°, Z = 4. R = 0.045 for 2847 reflexions (at 163 K); a = 17.142, b = 17.004, c = 16.571 Å, β = 104.27°, Z = 4. R = 0.052 for 2809 reflexions (at room temperature).

There is no space group change in the system on cooling from room temperature to 163 K. The molecule (Fig. 1) has an oxo-centred triangular arrangement of three Fe atoms. There is a twofold crystallographic symmetry axis in the Fe_3O plane and the 4-Etpy ligand is disordered about this axis. Significantly, the Fe_3 triangle becomes more equilateral as the crystal is warmed from 163 to 298 K. At 163 K the bond distances to the central oxide ion are Fe^{II}-O 2.011(4), Fe^{III}-O 1.856(1) Å; at 298 K these values become 1.954(5) and 1.880(1) Å respectively.

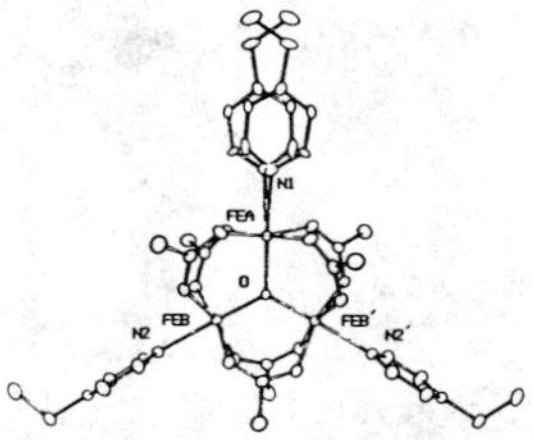

Fig. 1. A view of the $C_{33}H_{45}Fe_3N_3O_{13}$ structure at 163 K.

BIS(BIS(η-CYCLOPENTADIENYL)(PENTAFLUOROBENZOATO)YTTERBIUM)
$C_{34}H_{20}F_{10}O_4Yb_2$ $[(\eta\text{-}C_5H_5)_2YbO_2CC_6F_5]_2$

G.B. DEACON, G.D. FALLON, P.I. MacKINNON, R.H. NEWNHAM, G.N. PAIN, T.D. TUONG and D.L. WILKINSON, 1984. J. Organometal. Chem., 277, C21-C24.

Triclinic, $P\bar{1}$, a = 23.506, b = 8.417, c = 8.805 Å, α = 91.01, β = 109.71, γ = 99.79°, Z = 2. Cu radiation, R = 0.048 for 4048 reflexions.

The unit cell contains two similar dimer molecules A and B, each dimer being on an inversion centre. Molecule A is represented in Fig. 1. The bonding of the carboxylate groups is not symmetric, with angles C(1)-O(1)-Yb(1) 171° and C(1)-O(2)-Yb(1') 145°. In spite of this, the Yb-O distances are similar (Yb(1)-O(1) 2.178, Yb(1')-O(2) 2.201 Å). The average Yb-C(Cp) distance is 2.595 Å and the Yb(1)...Yb(1') separation is 5.06 Å. The cyclopentadienyl ligands are staggered and the centroid-Yb(1)-centroid angle is 131.9°. The distances C(1)-O(1) and C(1)-O(2) are 1.246 and 1.258 Å respectively.

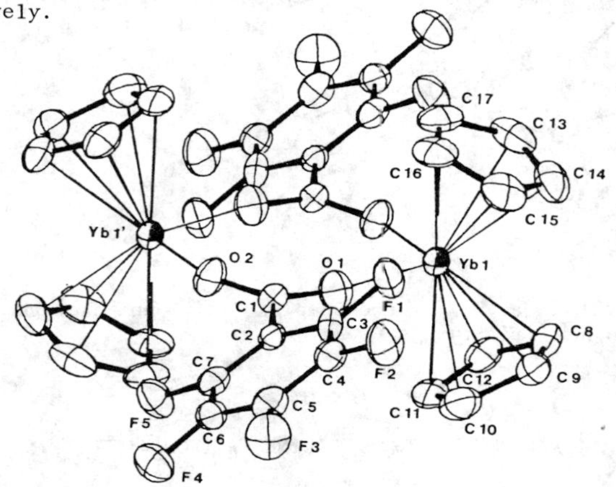

Fig. 1. Molecular structure of $C_{34}H_{20}F_{10}O_4Yb_2$ (molecule A shown).

DIAQUA(μ_2-OXALATO)BIS(2,2'-BIPYRIDYL)DICOPPER(II) OXALATO(2,2'-BIPYRIDYL)COPPER(II) DINITRATE
$C_{34}H_{28}Cu_3N_8O_{16}$ $[Cu_2(H_2O)_2(bpy)_2(C_2O_4)]^{2+}.[Cu(bpy)(C_2O_4)].2NO_3^-$

M. JULVE, J. FAUS, M. VERDAGUER and A. GLEIZES, 1984. J. Am. Chem. Soc., 106, 8306-8308.

Monoclinic, C2/c, a = 21.739, b = 10.458, c = 16.023 Å, β = 95.69°, Z = 4. R = 0.029 for 3290 reflexions.

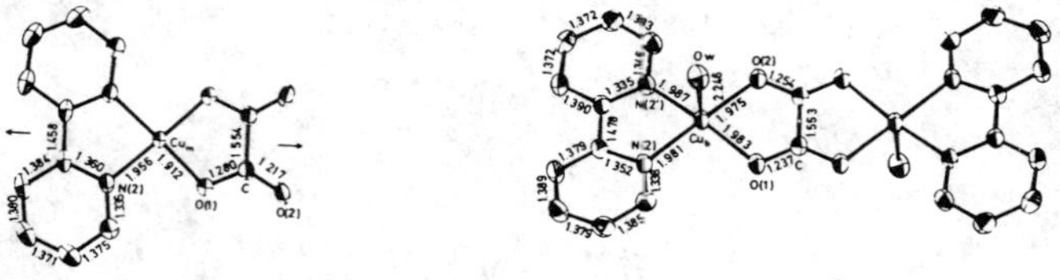

Fig. 1. The $[Cu(bpy)(C_2O_4)]$ and $[Cu_2(H_2O)_2(bpy)_2(C_2O_4)]^{2+}$ complexes.

Two different Cu complexes co-crystallize in this structure; the mononuclear complex has twofold crystallographic symmetry and the binuclear one lies about an inversion centre. In the mononuclear complex the Cu fourfold environment deviates slightly from planarity; the Cu environment in the binuclear complex is square-pyramidal, with Cu 0.155 Å above the base.

TETRAKIS(μ-BENZOATO)BIS(TETRAHYDROFURAN)DITUNGSTEN
$C_{36}H_{36}O_{10}W_2$ (I) $W_2(O_2CC_6H_5)_4(C_4H_8O)_2$

TETRAKIS(μ-4-METHOXYBENZOATO)BIS(TETRAHYDROFURAN)DITUNGSTEN BIS(TETRAHYDROFURAN) SOLVATE
$C_{40}H_{44}O_{14}W_2,2C_4H_8O$ (II) $W_2(O_2CC_6H_4\text{-}4\text{-}OCH_3)_4(C_4H_8O)_2,2C_4H_8O$

TETRAKIS(μ-2,4,6-TRIMETHYLBENZOATO)DITUNGSTEN BIS(TOLUENE) SOLVATE
$C_{40}H_{44}O_8,2C_7H_8$ (III) $W_2[O_2CC_6H_2[2,4,6\text{-}(CH_3)_3]]_4,2CH_3C_6H_5$

F.A. COTTON and W. WANG, 1984. Inorg. Chem., 23, 1604-1610.

I. Monoclinic, $P2_1/c$, a = 9.511, b = 17.716, c = 10.843 Å, β = 109.52°, Z = 2. Mo radiation, R = 0.033 for 2011 reflexions.

II. Triclinic, $P\bar{1}$, a = 11.590, b = 14.421, c = 8.278 Å, α = 100.77, β = 103.23, γ = 68.55°, Z = 1. Mo radiation, R = 0.057 for 2486 reflexions.

III. Monoclinic, $P2_1/c$, a = 8.141, b = 21.873, c = 14.244 Å, β = 102.41°, Z = 2. Mo radiation, R = 0.035 for 2790 reflexions.

Each of the three molecules resides on an inversion centre (Fig. 1). The W-W distances in I-III are 2.196(1), 2.203(1), and 2.176(1) Å, respectively. The shorter distance in III may be due to the absence of axial ligands. In I and II each phenyl group is nearly coplanar with its carboxyl group whereas in III it is nearly perpendicular. The W-O (carboxylate) distances average 2.077(4), 2.085(6), and 2.071(3) Å for I-III, respectively, and the axial W-O distances are 2.628(8) Å for I and 2.57(1) Å for II.

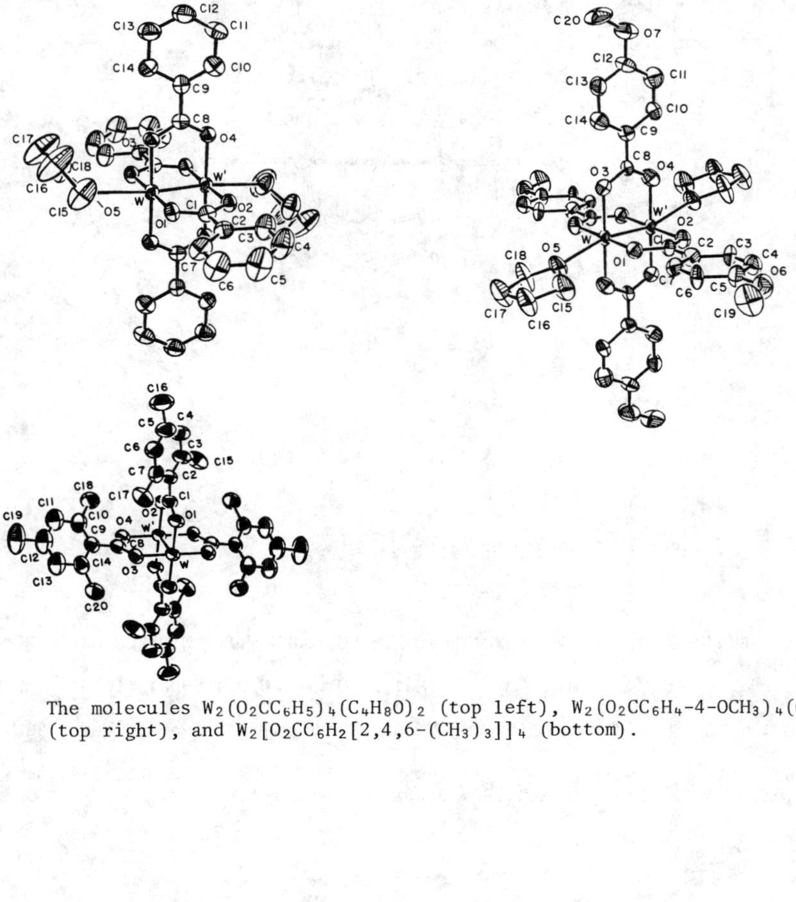

Fig. 1. The molecules $W_2(O_2CC_6H_5)_4(C_4H_8O)_2$ (top left), $W_2(O_2CC_6H_4\text{-}4\text{-}OCH_3)_4(C_4H_8O)_2$ (top right), and $W_2[O_2CC_6H_2[2,4,6\text{-}(CH_3)_3]]_4$ (bottom).

TETRAKIS(μ-ACETATO)BIS(2,4-DIAMINO-5-(3',4',5'-TRIMETHOXYBENZYL)PYRIMIDINE)-DICOPPER(II) DIBENZENE METHANOL SOLVATE

$C_{36}H_{48}Cu_2N_8O_{14},2C_6H_6,CH_4O$ $[Cu_2(OOCCH_3)_4(C_{14}H_{18}N_4O_3)_2],2C_6H_6,CH_3OH$

L. NALDINI, M.A. CABRAS, M.A. ZORODDU, F. DEMARTIN, M. MANASSERO and M. SANSONI, 1984. Inorg. Chim. Acta, 88, 45-50.

Monoclinic, C2/c, a = 24.109, b = 15.256, c = 16.532 Å, β = 116.89°, Z = 4. Mo radiation, R = 0.054 for 1931 reflexions.

Crystals of the complex consist of $[Cu_2(O_2CCH_3)_4(TMP)_2]$ molecules with benzene and methanol molecules clathrated in the molar ratio 1:2:1. The Cu complex molecules (Fig. 1) lie on crystallographic inversion centres as do the benzene molecules. The four bidentate acetate groups bridge two Cu atoms (Cu-O(4) 1.968, Cu-O(5) 1.966, Cu-O(6) 1.961, Cu-O(7) 1.992 Å), and the distorted octahedral coordination about the metal is completed by a pyrimidinic N atom N(1) and the centrosymmetrically-related Cu atom (Cu-N(1) 2.170, Cu-Cu' 2.679 Å) at the apical sites. Each metal atom is displaced from the plane of the four equatorial O atoms by 0.23 Å towards the N(1) atom and away from the other Cu atom.

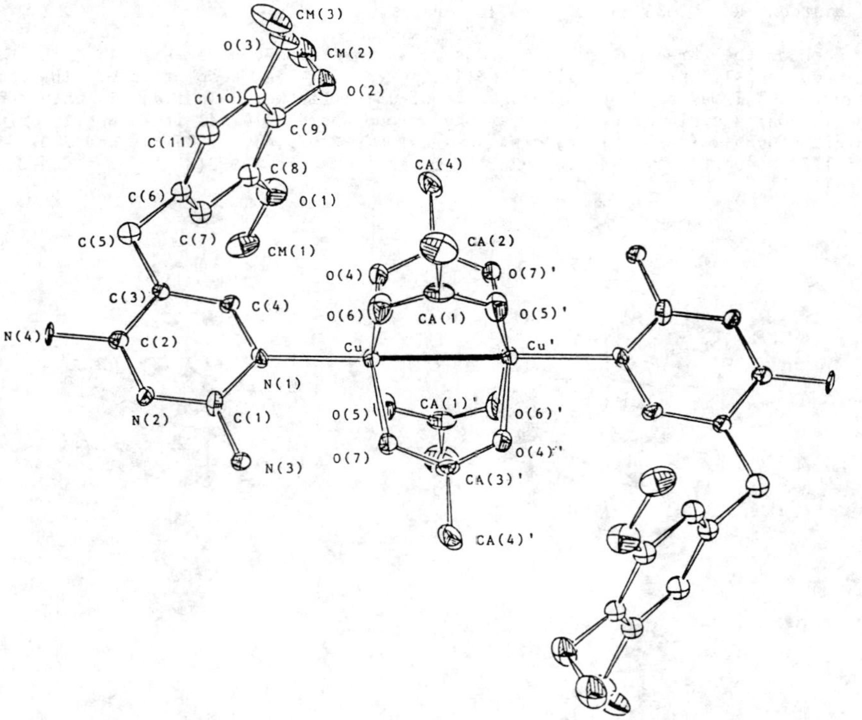

Fig. 1. Molecular structure of $C_{36}H_{48}Cu_2N_8O_{14}$.

HEXAKIS(μ-BENZOATO-O,O')BIS(μ₃-OXO)TRIAQUATRIMOLYBDENUM(IV) DIPERCHLORATE PERCHLORIC ACID ETHANOL SOLVATE

$C_{42}H_{36}Cl_2Mo_3O_{25},3C_2H_6O,HClO_4$ $[Mo_3(\mu_3\text{-}O)_2(O_2CC_6H_5)_6(H_2O)_3](ClO_4)_2.HClO_4.3C_2H_6O$

F.A. COTTON, Z. DORI, D.O. MARLER and W. SCHWOTZER, 1984. Inorg. Chem., 23, 4033-4038.

Triclinic, P$\bar{1}$, a = 13.837, b = 22.189, c = 11.394 Å, α = 92.39, β = 114.35, γ = 89.02°, D_m = 1.63, Z = 2. Cu radiation, R = 0.066 for 5416 reflexions.

The cation (Fig. 1) is a trinuclear bi-oxo-capped cluster. Important mean dimensions are: Mo-Mo 2.761(1), Mo-(μ_3-O) 1.983(6), Mo-O(benzoate) 2.081(6), and Mo-O(H_2O) 2.123(6) Å.

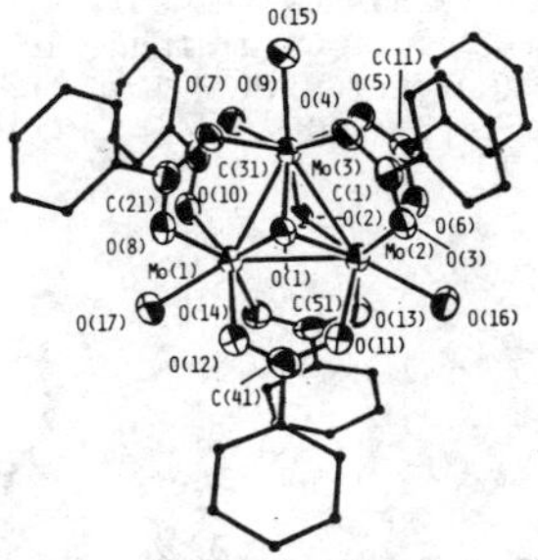

Fig. 1. The $[Mo_3(\mu_3\text{-}O)_2(O_2CC_6H_5)_6(H_2O)_3]^{2+}$ cation.

TETRA-μ-[(PHENYLTHIO)ACETATO]BIS[PYRIDINECOPPER(II)]
$C_{42}H_{38}Cu_2N_2O_8S_4$

E.J. O'REILLY, G. SMITH, C.H.L. KENNARD, T.C.W. MAK and W.-H. YIP, 1984. Inorg. Chim. Acta, 83, L63-L66.

Triclinic, $P\bar{1}$, a = 7.139, b = 10.370, c = 15.386 Å, α = 89.91, β = 107.95, γ = 103.63°, Z = 1. Mo radiation, R = 0.055 for 2636 reflexions.

The complex consists of centrosymmetric tetracarboxylate bridged dimers of the Cu(II) acetate monohydrate and Cu(II) acetate monopyridine type (Fig. 1). The four equatorial Cu-O(carboxylate) distances are 1.989 (O(10A), 1.998 (O(11A')), 1.969 (O(10B)) and 1.973 Å (O(11B')) and the axial Cu-N(pyridine) distance 2.151 Å. The Cu...Cu'distance is 2.685 Å and the O(eq.)-Cu-O(eq.) angles range from 88.1 to 91.4°. The angles C(1)-S(7)-C(8) are 105.6 and 105.1° for ligands A and B respectively.

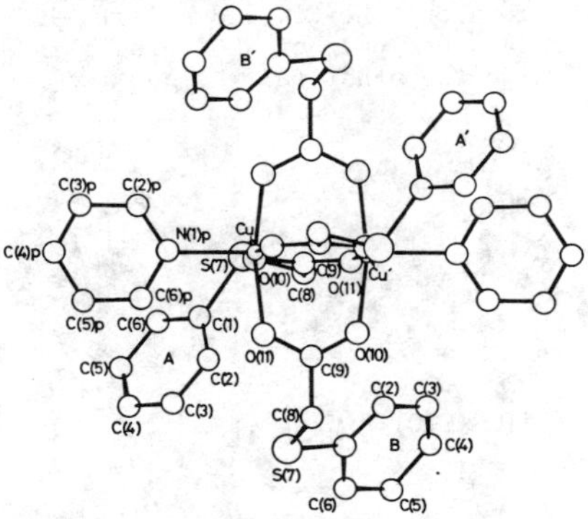

Fig. 1. Molecular structure of $C_{42}H_{38}Cu_2N_2O_8S_4$. Primed atoms are related to those unprimed by inversion symmetry.

BIS[AQUA(μ-PHENYLPROPIONATO-O,O')(1,10-PHENANTHROLINE-N,N')COPPER(II)] DINITRATE DIHYDRATE
$C_{42}H_{38}Cu_2N_6O_{12},2H_2O$ $[Cu_2(C_{12}H_8N_2)_2(C_6H_5CH_2CH_2COO)_2(H_2O)_2](NO_3)_2.2H_2O$

E. DUBLER, U.K. HÄRING, K.H. SCHELLER, P. BALTZER and H. SIGEL, 1984. Inorg. Chem., 23, 3785-3792.

Monoclinic, C2/c, a = 22.876, b = 12.284, c = 16.320 Å, β = 111.03°, D_m = 1.55, Z = 4. Mo radiation, R = 0.054 for 2607 reflexions.

The binuclear cation (Fig. 1) has C_2 symmetry with the metal atoms bridged by a pair of carboxylate ligands. The Cu...Cu distance is 3.054(1) Å. The Cu atoms have square pyramidal coordination geometry with H_2O in the apical site (Cu-O(basal) 1.952(3) and 1.956(3), Cu-N 2.014(3) and 2.004(4), Cu-OH_2 2.201(3) Å). The anions are disordered.

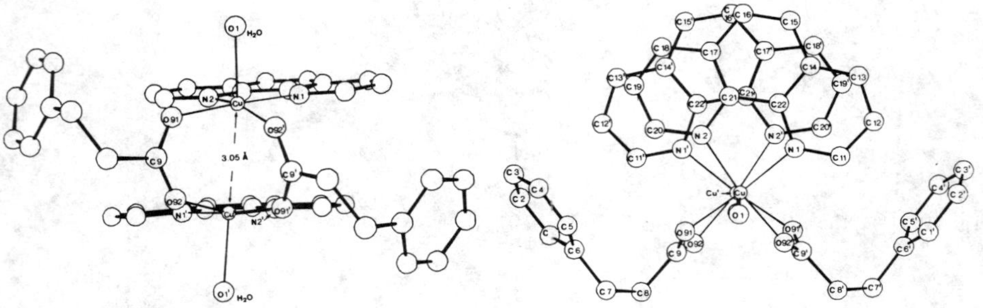

Fig. 1. Two views of the $[Cu_2(C_{12}H_8N_2)_2(C_6H_5CH_2CH_2COO)_2(H_2O)_2]^{2+}$ cation.

TETRAKIS(μ-TRIFLUOROACETATO)BIS[(TRIPHENYLPHOSPHINE)CADMIUM(II)]

$C_{44}H_{30}Cd_2F_{12}O_8P_2$ $[Cd(CF_3COO)_2P(C_6H_5)_3]_2$

T. ALLMAN, R.G. GOEL, N.K. JHA and A.L. BEAUCHAMP, 1984. Inorg. Chem., <u>23</u>, 914-918.

Monoclinic, $P2_1/c$, a = 10.011, b = 18.062, c = 15.135 Å, β = 118.34°, Z = 2. Mo radiation, R = 0.038 for 1741 reflexions.

The centrosymmetric molecule (Fig. 1) contains a $(\mu\text{-}CF_3COO)_4Cd_2$ unit with Cd-O 2.249(6)-2.298(6) Å. The apical sites are occupied by PPh_3 ligands (Cd-P 2.524(2) Å), but the Cd-P bond does not lie along the Cd...Cd axis (Cd-Cd-P 153.7(1)°). The Cd...Cd distance is 3.452(1) Å. One pair of RCOO groups is symmetrically coordinated (Cd-O-C 128.2 and 132.8°) while the other pair is not (Cd-O-C 109.0 and 154.2°).

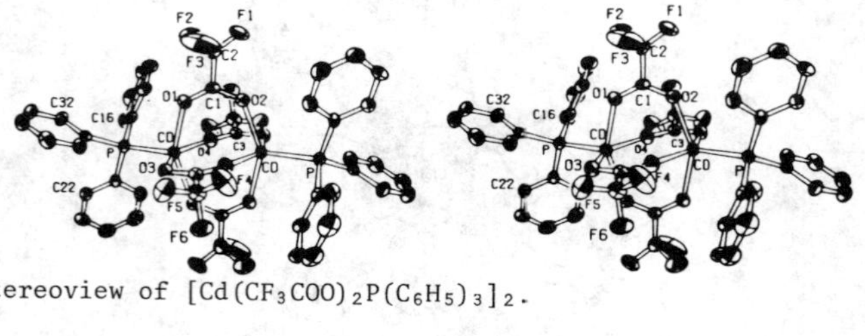

Fig. 1. Stereoview of $[Cd(CF_3COO)_2P(C_6H_5)_3]_2$.

(μ_2-ACETATO)(μ_2-N,N,N',N'-TETRAKIS(2-(1-ETHYLBENZIMIDAZOLYL))-2-HYDROXY-1,3-DIAMINOPROPANE)DICOPPER(II) DIPERCHLORATE

$C_{45}H_{52}Cl_2Cu_2N_{10}O_{11}$ (I)

(μ_2-1,3-AZIDO)-(μ_2-N,N,N',N'-TETRAKIS(2-(1-ETHYLBENZIMIDAZOLYL))-2-HYDROXY-1,3-DIAMINOPROPANE)DICOPPER(II) BIS(TETRAFLUOROBORATE)

$C_{43}H_{49}B_2Cu_2F_8N_{13}O$ (II)

V. MCKEE, M. ZVAGULIS, J.V. DAGDIGIAN, M.G. PATCH and C.A. REED, 1984. J. Am. Chem. Soc., <u>106</u>, 4765-4772.

I. Monoclinic, $P2_1/n$, a = 14.190, b = 22.707, c = 15.883 Å, β = 97.20°, Z = 4. Mo radiation, R = 0.096 for 3808 reflexions.

II. Monoclinic, C2/m, a = 19.082, b = 23.896, c = 13.230 Å, β = 116.21°, Z = 4. Mo radiation, R = 0.097 for 1631 reflexions.

Views of I and II are in Fig. 1; II has twofold crystallographic symmetry and its azide bridging mode had not been previously found. Some dimensions for I (with data for II in parentheses) are Cu...Cu 3.459(2) (3.615(3)), Cu-O 1.89-1.93(2) (1.944(8)). The Cu-N(azide) distances in II are 2.04(1) Å. The stereochemistry about Cu2 in (I) is close to trigonal bipyramidal; Cu1 is similar but more distorted. In the azide (II), the Cu stereochemistry approximates to a tetragonal pyramid. It is thought that the different coordination geometries offer a logical explanation for their different magnetic behaviour.

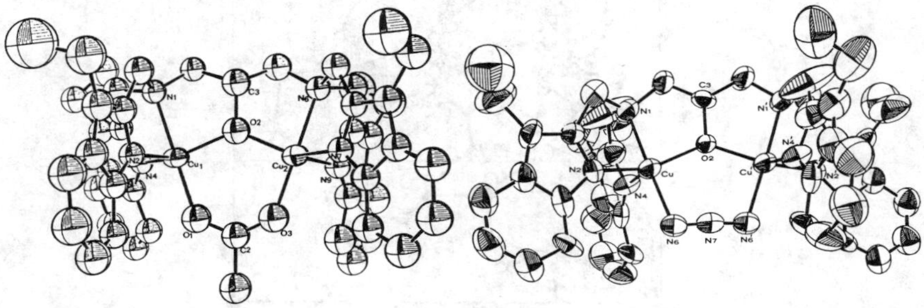

Fig. 1. Perspective views of the acetate (I) and azide (II) complexes.

HEXAKIS(ε-CAPROLACTAM)COPPER(II) TRICHLORO(ε-CAPROLACTAM)CUPRATE(II)
$C_{48}H_{88}Cl_6Cu_3N_8O_8$ $Cu_3(C_6H_{11}NO)_8Cl_6$

S.S. KUKALENKO, Yu.T. STRUCHKOV, S.I. SHESTAKOVA, A.G. TSYBULEVSKII, A.S. BATSANOV and E.B. NAZAROVA, 1983. Koord. Khim., 9, 306-311 [Sov. J. Coord. Chem., 9, 182-187].

Monoclinic, $P2_1/c$, a = 10.757, b = 19.787, c = 14.528 Å, β = 101.26°, Z = 2. Mo radiation, R = 0.051 for 3023 reflexions.

The structure consists of complex cations which lie on inversion centres and complex anions in general positions (Fig. 1). The coordination of the Cu atom in the cation is tetragonal bipyramidal (Cu-O 1.962, 1.951, 2.514(4)) and its coordination in the anion is intermediate between tetrahedral and square planar (Cu-Cl 2.237-2.242(2), Cu-O 1.954(4) Å). All the seven-membered heterocyclic rings of the caprolactam ligands have chair conformations.

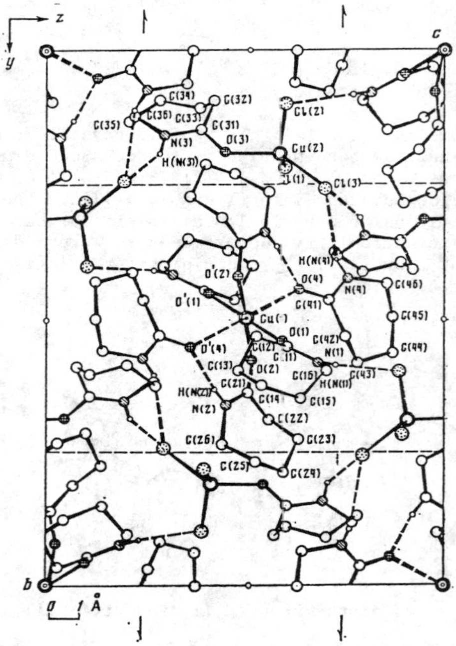

Fig. 1. $C_{48}H_{88}Cl_6Cu_3N_8O_8$: a-axis projection of the structure.

(μ_2-AQUA)BIS[μ_2-(2-CHLOROPHENOXY)ETHANOATO-O,O']BIS[(2-CHLOROPHENOXY)ETHANOATOBIS-(PYRIDINE)NICKEL(II)] MONOHYDRATE
$C_{52}H_{46}Cl_4N_4Ni_2O_{13},H_2O$

C.H.L. KENNARD, E.J. O'REILLY and G. SMITH, 1984. Polyhedron, 3, 689-693.

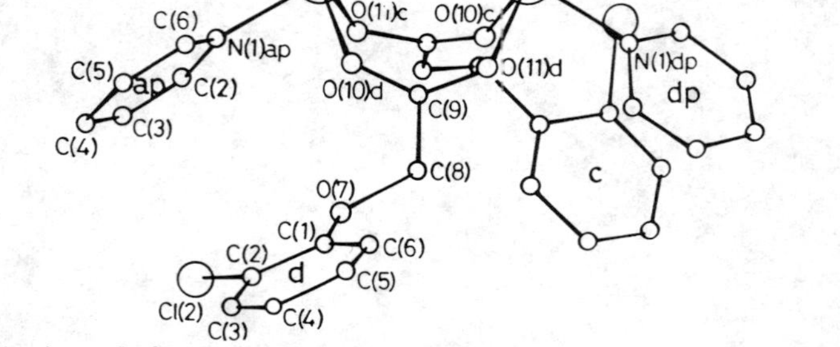

Fig. 1. A view of the $C_{52}H_{46}Cl_4N_4Ni_2O_{13}$ complex.

Triclinic, $P\bar{1}$, a = 21.740, b = 14.312, c = 9.044 Å, α = 96.69, β = 84.99, γ = 101.20°, D_m = 1.48, Z = 2. Mo radiation, R = 0.059 for 2682 reflexions.

The complex (Fig. 1) consists of two slightly distorted octahedra facially linked by two carboxylate bridges (Ni-O 2.026(9)-2.066(9) Å; mean 2.040(9) Å) from two 2-CPA ligands, and an almost symmetrical bridge from a single aqua ligand (Ni(1)-Ow(1) 2.137(8) Å; Ni(2)-Ow(1) 2.097(8) Å. Ni(1)-Ow(1)-Ni(2) 116.9(4) Å). The three remaining coordination positions about each Ni are occupied by a carboxyl oxygen of a unidentate 2-CPA ligand (Ni-O 2.058(8), 2.080(9) Å) and two pyridine nitrogens (Ni-N 2.102(14)-2.113(14) Å; mean 2.108(14) Å). The structure is stabilized by intramolecular hydrogen bonding involving the bridging water molecule (Ow(1)) and the uncomplexed oxygens of the carboxyl groups (O(10)a, 2.56 Å and O(10)b, 2.65 Å).

BIS(ACETONITRILE)TETRAKIS(μ2-2-PHENYLBENZOATO)DIRHODIUM BENZENE SOLVATE
$C_{56}H_{42}N_2O_8Rh_2,3C_6H_6$ (I) $Rh_2(O_2C(C_6H_4C_6H_5))_4(CH_3CN)_2,3C_6H_6$

TETRAKIS(μ2-1-ADAMANTYLCARBOXYLATO)BIS(METHANOL)DIRHODIUM METHANOL
$C_{46}H_{68}O_{10}Rh_2,5CH_3OH$ (II) $Rh_2[O_2C(C_{10}H_{15})]_4(CH_3OH)_2,5CH_3OH$

BIS[(μ2-ACETATO)(ACETONITRILE)(μ2-TRIPHENYLACETATO)RHODIUM] TOLUENE
$C_{48}H_{42}N_2O_8Rh_2,C_6H_5CH_3$ (III) $Rh_2(O_2CCH_3)_2(O_2CCPh_3)_2(CH_3CN)_2,C_6H_5CH_3$

F.A. COTTON and J.L. THOMPSON, 1984. Inorg. Chim. Acta, 81, 193-203.

I. Triclinic, $P\bar{1}$, a = 13.469, b = 13.507, c = 9.908 Å, α = 110.44, β = 92.78, γ = 107.92°, Z = 1. Mo radiation, R = 0.040 for 4469 reflexions.

II. Orthorhombic, Pbca, a = 10.587, b = 20.640, c = 24.693 Å, Z = 4. Mo radiation, R = 0.060 for 1442 reflexions.

III. Triclinic, $P\bar{1}$, a = 12.664, b = 21.876, c = 9.686 Å, α = 99.00, β = 105.27, γ = 75.61°, Z = 2. Mo radiation, R = 0.056 for 3094 reflexions.

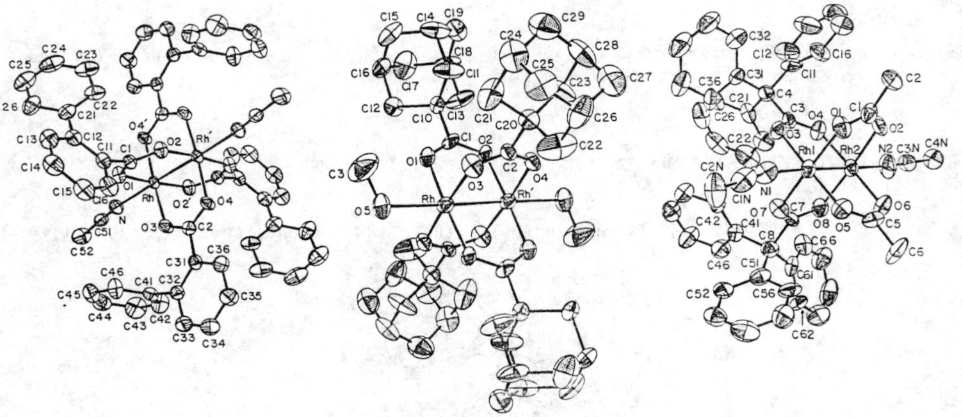

Fig. 1. Molecular structures of $C_{56}H_{42}N_2O_8Rh_2$ (left), $C_{46}H_{68}O_{10}Rh_2$ (centre), and $C_{48}H_{42}N_2O_8Rh_2$ (right).

In I the dirhodium unit is located on an inversion centre (Fig. 1), and the central portion of the molecule has approximate D_{4h} symmetry. The Rh-Rh bond length is 2.396 Å, the Rh-N distance 2.233 Å and the Rh-O distances range from 2.026 to 2.051 Å. The N-C distance in the acetonitrile molecule is 1.126 Å. In II the dirhodium unit is again located on an inversion centre (Fig. 1). The Rh-Rh distance is 2.371 Å and the Rh-O(methanol) distance is 2.296 Å. The other Rh-O distances range from 2.01 to 2.066 Å. Molecule III (Fig. 1) has no crystallographically-imposed symmetry, the carboxylate groups are bonded to the Rh atoms as before and the Rh-Rh distance is 2.388 Å. The Rh-N distances are 2.21 and 2.17 Å, the Rh-O distances range from 2.008 to 2.053 Å, and the N-C distances in the acetonitrile molecules are 1.20 and 1.13 Å.

BIS(μ_2-CARBONYL)(BIS(DIPHENYLPHOSPHINO)METHANE-P,P')(TRIFLUOROACETATOSILVER)HEXACARBONYLTRIRUTHENIUM DICHLOROMETHANE

$C_{60}H_{44}AgF_3P_4O_{10}Ru_3,0.5(CH_2Cl_2)$ $Ru_3(\mu\text{-dpm})_2(CO)_8(\mu\text{-}AgO_2CCF_3),0.5(CH_2Cl_2)$

J.A. LADD, H. HOPE and A.L. BALCH, 1984. Organometallics, 3, 1838-1846.

Orthorhombic, $Pca2_1$, a = 20.706, b = 15.693, c = 18.447 Å, Z = 4. Mo radiation, R = 0.043 for 3716 reflexions (at 140K).

The silver trifluoroacetate has added across the otherwise exposed Ru(2)-Ru(3) edge of the Ru_3 cluster (Fig. 1). The adduct contains two adjacent, nearly coplanar triangles of the four metal atoms, and the addition of Ag^+ to the Ru-Ru bond results in a lengthening of the Ru-Ru separation by 0.167 Å.

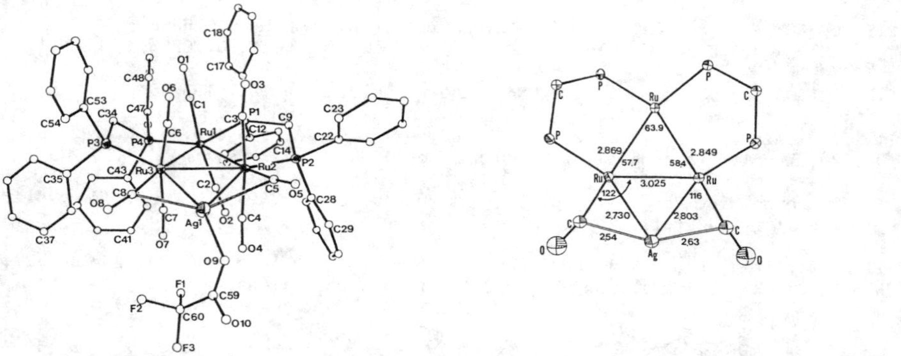

Fig. 1. The structure of $C_{60}H_{44}AgF_3P_4O_{10}Ru_3$ and selected bond distances (Å) and angles (°) in part of the molecule.

BIS(TRIFLUOROACETATO)(TRIMESITYLPHOSPHINE)MERCURY(II) DIMER

$C_{62}H_{66}F_{12}Hg_2O_8P_2$ $[HgPmes_3(O_2CCF_3)_2]_2$

E.C. ALYEA, S.A. DIAS, G. FERGUSON and P.Y. SIEW, 1984. Croat. Chem. Acta, 57, 457-465.

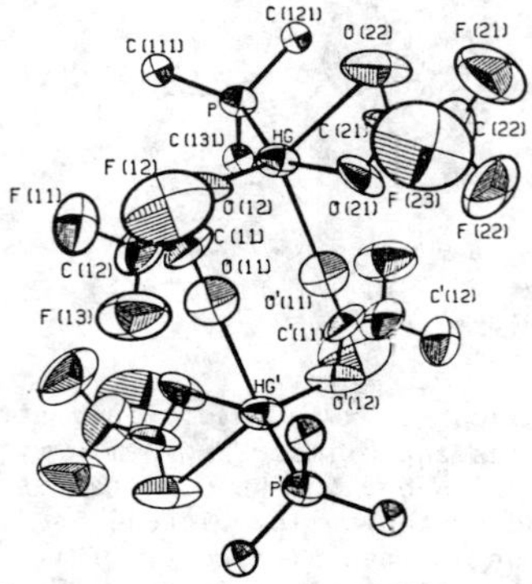

Fig. 1. A view of $C_{62}H_{66}F_{12}Hg_2O_8P_2$. All mesityl C atoms are omitted for clarity except those bonded to P.

Triclinic, $P\bar{1}$, a = 12.854, b = 12.877, c = 12.405 Å, α = 107.34, β = 118.55, γ = 63.65°, Z = 1. Mo radiation, R = 0.051 for 1656 reflexions.

The Hg coordination is characterized by three strong nearly coplanar bonds (Hg-P 2.415, Hg-O(12) 2.29, Hg-O(21) 2.18 Å) and two weaker bonds (Hg-O(22) 2.95 and Hg-O(11') 2.66 Å). The $Pmes_3$ ligand has a regular propeller conformation (Hg-P-C-C torsion angles from 45 to 49°), enlarged C-P-C angles (mean 112.6°), decreased Hg-P-C angles (mean 106.2°), and a maximum cone angle of 208°. Bridging in the solid state utilizes both O atoms of the CF_3CO_2 moiety (Fig. 1).

BIS(μ3-OXO)HEXAKIS(μ2-PIVALATO-O,O')TRIS(TETRAHYDROFURAN)TRINIOBIUM TETRAPHENYLBORATE
$C_{66}H_{98}BNb_3O_{17}$ $C_{42}H_{78}Nb_3O_{17}^{+}.Ph_4B^{-}$

F.A. COTTON, S.A. DURAJ and W.J. ROTH, 1984. J. Am. Chem. Soc., 106, 3527-3531.

Orthorhombic, Pbca, a = 27.033, b = 26.251, c = 20.609 Å, Z = 8. Cu radiation, R = 0.0518 for 5217 reflexions.

The trinuclear cation is shown in Fig. 1 and has approximate D_{3h} symmetry. Mean values of principal bond lengths are: Nb-Nb 2.842(3), Nb-O(cap) 2.035(6), Nb-O(carbox) 2.150(4), Nb-O(THF) 2.25(1) Å, Nb-O(cap)-Nb 88.5(3)°.

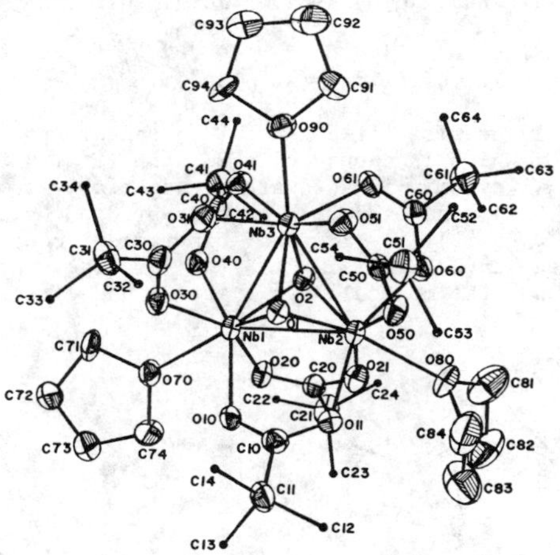

Fig. 1. A view of the $C_{42}H_{78}Nb_3O_{17}^{+}$ ion.

DICARBONYLBIS(1-(o-CARBOXYMETHOXYPHENOXY)-2-(o-HYDROXYPHENOXY)ETHANATO)BIS(TRIPHENYLPHOSPHINE)RUTHENIUM(II) DICHLOROMETHANE METHANOL SOLVATE
$C_{70}H_{60}O_{14}P_2Ru,CH_2Cl_2,CH_4O$

D.L. HUGHES and J.N. WINGFIELD, 1984. J. Chem. Soc. Dalton, 739-742.

Triclinic, $P\bar{1}$, a = 9.677, b = 23.161, c = 15.432 Å, α = 92.91, β = 91.29, γ = 102.16°, D_m = 1.36, Z = 2. Mo radiation, R = 0.094 for 4681 reflexions.

The Ru atom is octahedrally coordinated by two mutually cis carbonyl ligands, two mutually trans phosphine ligands and two mutually cis [1-(o-carboxymethoxyphenoxy)-2-(o-hydroxyphenoxy)ethanato] ligands (L) each bound to Ru by one oxygen of the carboxylate group (Fig. 1). The two L ligands diverge and there are no interactions between them. O(1) and O(31) both appear to be involved in hydrogen bonding (O(7)...O(22) 2.80, O(37)...O(41) (solvent methanol).). The carbonyl and phosphine ligands have normal dimensions and are arranged about the Ru atom in pseudo-twofold symmetry. Coordination distances are: Ru(1)-O(21) 2.084(8), Ru(1)-C(7) 2.411(4), Ru(1)-O(51) 2.092(7), Ru(1)-C(81) 1.832(13), Ru(1)-P(6) 2.430(4), Ru(1)-C(83) 1.857(12) Å.

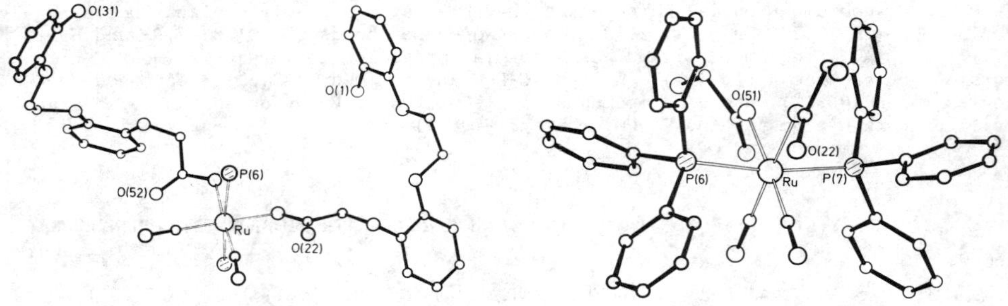

Fig. 1. $C_{70}H_{60}O_{14}P_2Ru,CH_2Cl_2,CH_4O$: two views of the molecular structure.

TETRAKIS[(μ_4-OXO)TRIS(μ-BENZOATO)(TETRAHYDROFURAN)VANADIUMZINC] TETRAHYDROFURAN

$C_{100}H_{92}O_{32}V_4Zn_4,2C_4H_8O$ $[VZnO(O_2CC_6H_5)_3(C_4H_8O)]_4.2C_4H_8O$

F.A. COTTON, S.A. DURAJ and W.J. ROTH, 1984. Inorg. Chem., 23, 4042-4045.

Monoclinic, C2/c, a = 22.780, b = 23.492, c = 20.222 Å, β = 91.45°, Z = 4. Mo radiation, R = 0.048 for 6277 reflexions (at -100°C).

The molecule (Fig. 1) consists of a central tetrahedron of V atoms (V-V 2.938(1)-2.987(1) Å), each face of which is capped by a Zn atom (V...Zn 3.417(1)-3.495(1) Å). At the centre of each ZnV_3 pyramid is an O atom, essentially equidistant from all four metal atoms (mean M-O 2.0 Å), and each Zn-V edge is bridged by a benzoate anion (mean Zn-O 2.007(6), mean V-O 2.009(3) Å). Trigonal bipyramidal coordination about Zn is completed by a THF oxygen (Zn-O 2.228(2), 2.187(2) Å). The V atoms have distorted octahedral coordination and the molecule has exact C_2 and effective T_d symmetry.

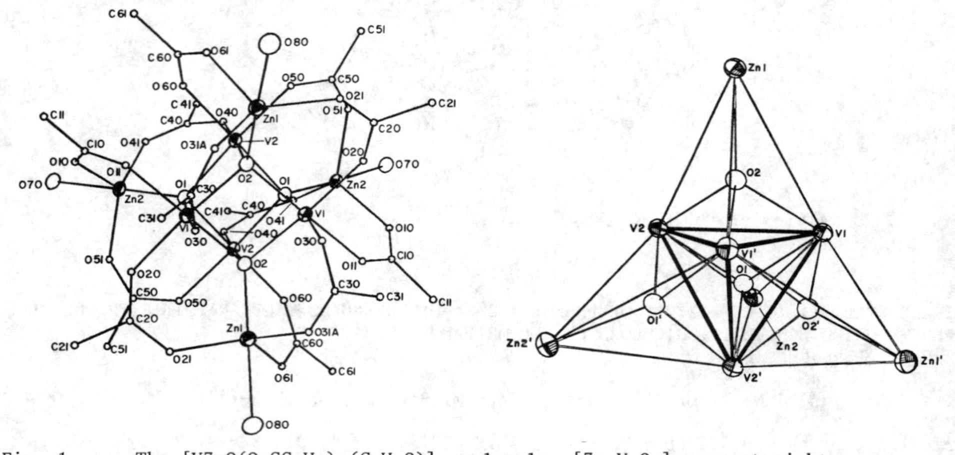

Fig. 1. The $[VZnO(O_2CC_6H_5)_3(C_4H_8O)]_4$ molecule, $[Zn_4V_4O_4]$ core at right.

DISODIUM BIS(L-CYSTEINATO-N,S)ZINCATE(II) HEXAHYDRATE

$C_6H_{10}N_2Na_2O_4S_2Zn,6(H_2O)$ (I)

BIS(O-ETHYL-L-CYSTEINATO-N,S)ZINC(II)
$C_{10}H_{20}N_2O_4S_2Zn$ (II)

DIBROMOBIS(DL-PENICILLAMINE)ZINC(II)
$C_{10}H_{22}Br_2N_2O_4S_2Zn$ (III)

DIBROMOBIS(D-PENICILLAMINE)ZINC(II)
$C_{10}H_{22}Br_2N_2O_4S_2Zn$ (IV)

P. BELL and W.S. SHELDRICK, 1984. Z. Naturforsch., B39, 1732-1737.

I. Monoclinic, C2, a = 5.206, b = 23.313, c = 15.052 Å, β = 92.53°, Z = 4. Mo radiation, R = 0.051 for 2532 reflexions.

II. Monoclinic, C2, a = 15.119, b = 5.063, c = 11.129 Å, β = 114.86°, Z = 2. Mo radiation, R = 0.019 for 1469 reflexions.

III. Monoclinic, C2/c, a = 22.071, b = 6.317, c = 13.983 Å, β = 102.24°, Z = 4. Mo radiation, R = 0.051 for 1210 reflexions.

IV. Monoclinic, C2, a = 15.423, b = 6.406, c = 11.131 Å, β = 116.36°, Z = 2. Mo radiation, R = 0.033 for 919 reflexions.

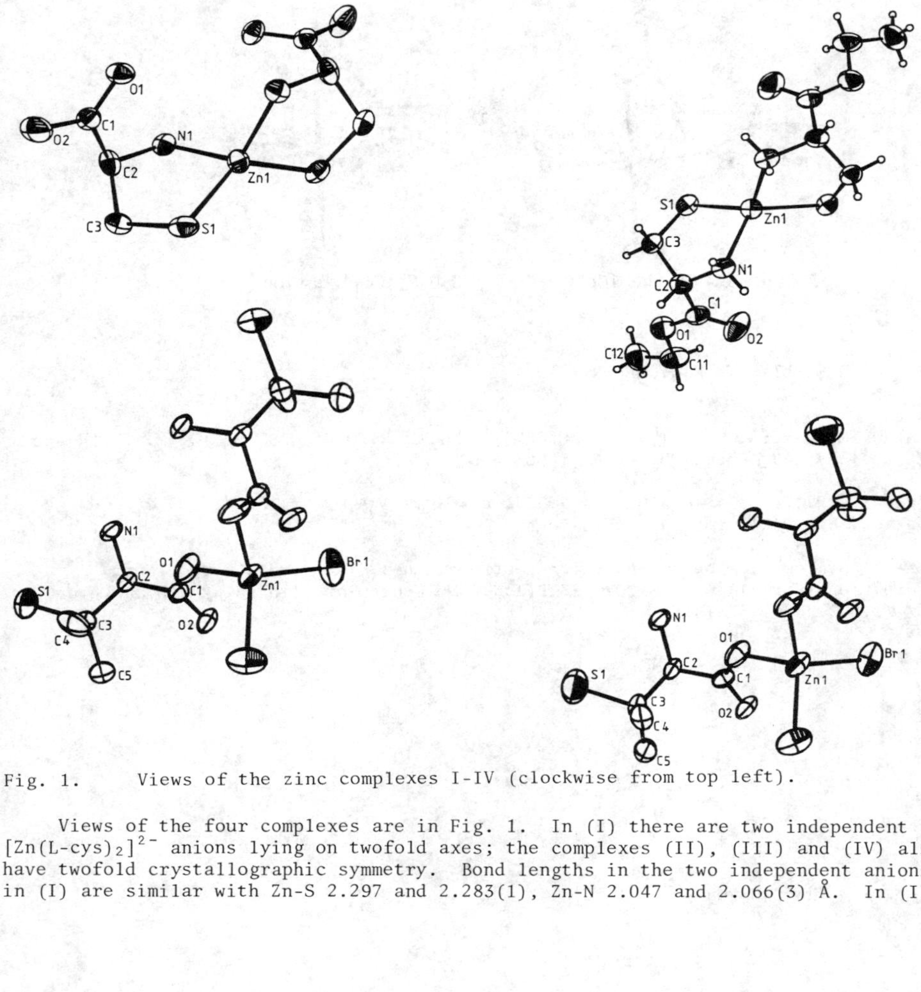

Fig. 1. Views of the zinc complexes I-IV (clockwise from top left).

Views of the four complexes are in Fig. 1. In (I) there are two independent $[Zn(L\text{-}cys)_2]^{2-}$ anions lying on twofold axes; the complexes (II), (III) and (IV) also have twofold crystallographic symmetry. Bond lengths in the two independent anions in (I) are similar with Zn-S 2.297 and 2.283(1), Zn-N 2.047 and 2.066(3) Å. In (II) the Zn-S distance is 2.273(1) and with Zn-N 2.090(1) Å. The penicillamine moieties in (III) and (IV) are present in zwitterionic form and behave as monodentate ligands with the deprotonated carboxyl group as the binding site, with Zn-Br 2.370 and 2.369(1), Zn-O 1.952 and 1.965(3) Å.

CHLORO(HISTIDINATO-N,N,O)(ETHYLENEDIAMINE)COBALT(III) CHLORIDE

$C_8H_{16}Cl_2CoN_5O_2$ $[Co(C_6H_8N_3O_2)(C_2H_8N_2)Cl]Cl$

N.R. BRODSKY, N.M. NGUYEN, N.S. ROWAN, C.B. STORM, R.J. BUTCHER and E. SINN, 1984. Inorg. Chem., 23, 891-897.

Triclinic, P$\bar{1}$, a = 7.939, b = 9.247, c = 11.021 Å, α = 118.75, β = 97.76, γ = 97.72°, Z = 2. Mo radiation, R = 0.037 for 3533 reflexions.

The structure consists of complex cations (Fig. 1) and Cl^- anions linked by an extensive network of hydrogen bonds. The octahedral coordination about Co(III) consists of a facially bound tridentate histidine (Co-O 1.921(2), Co-N(ring) 1.946(3), Co-N(amine) 1.938(3) Å), an ethylenediamine (Co-N 1.946(3) and 1.949(3) Å), and a Cl atom (Co-Cl 2.244(1) Å) bound trans to the oxygen.

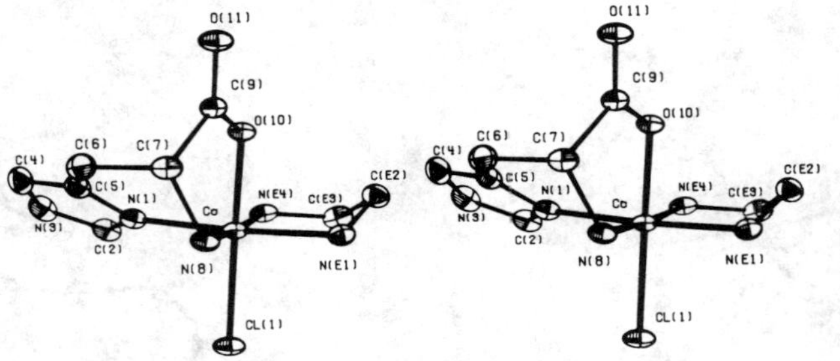

Fig. 1. Stereoview of the $[Co(C_6H_8N_3O_2)(C_2H_8N_2)Cl]^+$ cation.

BIS(α-AMINOISOBUTYRATO-N,O)COPPER(II)

$C_8H_{16}CuN_2O_4$

R. CALVO, M.A. MESA, G. OLIVA, J. ZUKERMAN-SCHPECTOR, O.R. NASCIMENTO, M. TOVAR and R. ARCE, 1984. J. Chem. Phys., 81, 4584-4591.

Monoclinic, $P2_1/c$, a = 10.470, b = 5.335, c = 10.201 Å, β = 115.6°, Z = 2. Cu radiation, R = 0.053 for 840 reflexions.

The Cu atom lies on an inversion centre and has distorted square planar coordination (Fig. 1) with N-Cu-O 83.9(1)°. Molecules are linked by N-H...O hydrogen bonds (3.002 and 3.325(4) Å).

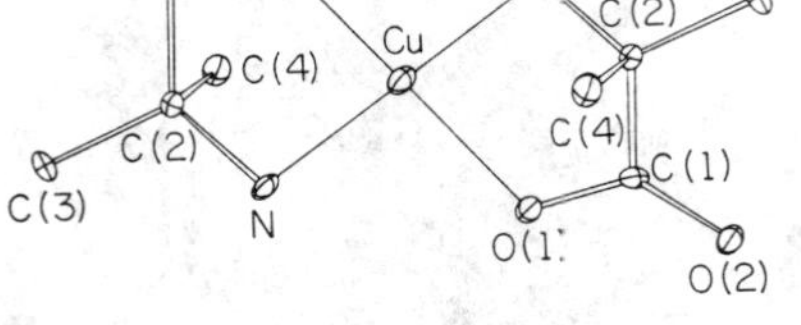

Fig. 1. The $C_8H_{16}CuN_2O_4$ molecule.

Λ-DIOXOBIS(METHYL L-CYSTEINATO)MOLYBDENUM(VI)
$C_8H_{16}MoN_2O_6S_2$ $[MoO_2(C_4H_8NO_2S)_2]$

I. BUCHANAN, M. MINELLI, M.T. ASHBY, T.J. KING, J.H. ENEMARK and C.D. GARNER, 1984. Inorg. Chem., 23, 495-500.

Orthorhombic, $P2_12_12$, a = 7.830, b = 16.202, c = 5.861 Å, Z = 2. Mo radiation, R = 0.038 for 1065 reflexions.

The molecule (Fig. 1) has crystallographic C_2 symmetry and possesses the expected cis-dioxo (Mo-O 1.714(4) Å), trans-dithiolate (Mo-S 2.414(2) Å), and cis-diamino (Mo-N 2.375(5) Å) distorted octahedral configuration. The molecules crystallize with the Λ configuration at Mo, and the L-cysteinato chelate rings have the λ conformation.

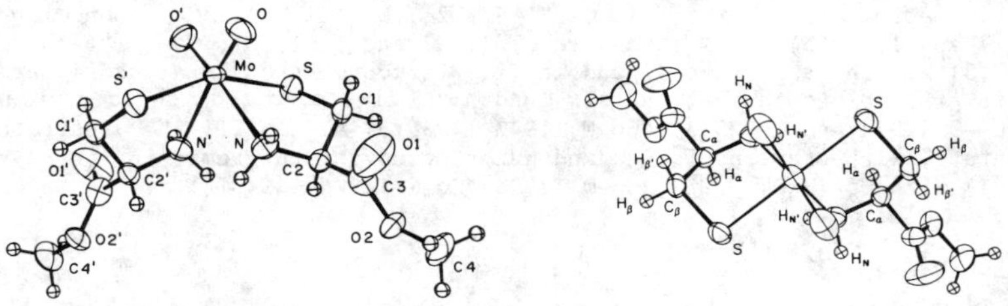

Fig. 1. Two views of $[MoO_2(C_4H_8NO_2S)_2]$.

DICHLOROBIS(GLYCINE ETHYL ESTER)PALLADIUM(II)
$C_8H_{18}Cl_2N_2O_4Pd$

U. TAUBALD, U. NAGEL and W. BECK, 1984. Chem. Ber., 117, 1003-1012.

Monoclinic, $P2_1/c$, a = 9.420, b = 12.211, c = 6.128 Å, β = 80.77°, D_m = 1.83, Z = 2. Mo radiation, R = 0.064 for 1533 reflexions.

The Pd atom lies on an inversion centre (Fig. 1), with Pd-Cl 2.299(2), Pd-N 2.057(6) Å, Cl(1)-Pd-N(1) 90.7(2)°.

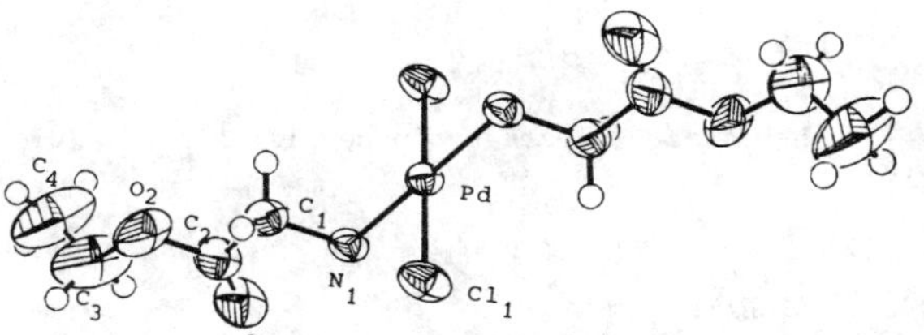

Fig. 1. A view of $C_8H_{18}Cl_2N_2O_4Pd$.

Λ-BIS(ETHYLENEDIAMINE)(MALEOATO-O,O')COBALT(III) HEXAFLUOROPHOSPHATE DIHYDRATE
$C_8H_{18}CoF_6N_4O_4P,2(H_2O)$ (I)

Δ-fac(5,5)-(R)-N-[(2-AMINOETHYL)ASPARTATE]ETHYLENEDIAMINECOBALT (+)-α-BROMOCAMPHOR-SULPHONATE TRIHYDRATE
$C_{18}H_{32}BrCoN_4O_8S,3(H_2O)$ (II)

rac-AQUA-(R)-N-[(2-AMINOETHYL)ASPARTATE]ETHYLENEDIAMINECOBALT PERCHLORATE DIHYDRATE
$C_8H_{20}ClCoN_4O_9,2(H_2O)$ (III)

A. HAMMERSHOI, A.M. SARGESON and W.L. STEFFEN, 1984. J. Am. Chem. Soc., 106, 2819-2837.

I. Monoclinic, $P2_1$, a = 11.629, b = 7.701, c = 9.379 Å, β = 97.18°, D_m = 1.822, Z = 2. Mo radiation, R = 0.045 for 4336 reflexions.

II. Triclinic, P1, a = 6.976, b = 7.242, c = 14.248 Å, α = 76.89, β = 84.61, γ = 70.86°, D_m = 1.634, Z = 1. Mo radiation, R = 0.040 for 4731 reflexions.

III. Monoclinic, $P2_1/c$, a = 7.191, b = 10.915, c = 20.723 Å, β = 91.41°, D_m = 1.808, Z = 4. Mo radiation, R = 0.045 for 4796 reflexions.

The Co atom in (I) closely approximates octahedral geometry (Fig. 1). The seven-membered ring has a pseudo-boat conformation which places olefinic carbon atoms C2 and C3 into close proximity to N2 (N2...C2 3.15, N2...C3 3.14 Å). Main bond lengths are: Co-O 1.917 and 1.905(3), Co-N (trans to O) 1.932 and 1.919(3), Co-N (trans to N) 1.951 and 1.943(3) Å. The aspartate ligand in (II) is coordinated in a quadridentate fashion to produce one seven-, one six- and two five-membered chelate rings (Fig. 1) with Co-O 1.914 and 1.925(3), Co-N 1.943-1.951(3) Å. In III the aspartate ligand is tridentate, with ethylenediamine and water occupying the remaining three sites at Co (Fig. 1) with Co-(H_2O) 1.935, Co-O 1.904, Co-N 1.926-1.948(2) Å.

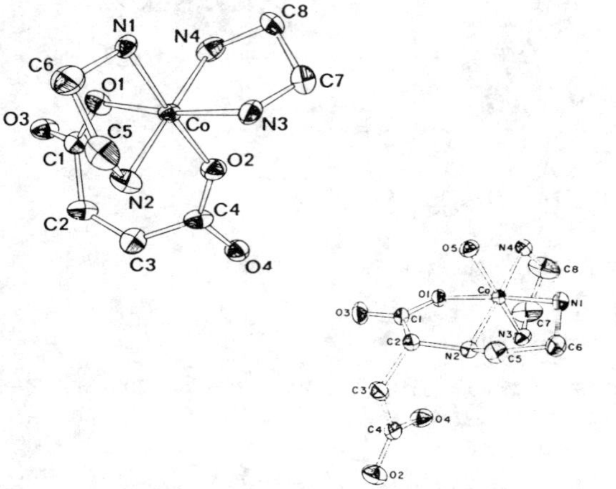

Fig. 1. Views of the maleato/aspartate Co complexes I-III (clockwise from top left).

N-BENZENESULPHONYL-DL-ALANINE
$C_9H_{11}NO_4S$ (I)

AQUA(N-BENZENESULPHONYL-DL-ALANINATO)COPPER(II) MONOHYDRATE
$(C_9H_{11}CuNO_5S)_n$,n(H_2O) (II)

S. CHAUDHURI, 1984. J. Chem. Soc. Dalton, 779-783.

I. Monoclinic, $P2_1/n$, a = 5.251, b = 16.638, c = 12.628 Å, β = 99.19°, D_m = 1.41, Z = 4. Mo radiation, R = 0.073 for 1585 reflexions.

II. Monoclinic, $P2_1/c$, a = 13.533, b = 7.908, c = 12.066 Å, β = 111.04°, D_m = 1.79, Z = 4. Cu radiation, R = 0.059 for 1965 reflexions.

Compound I (Fig. 1) forms typical carboxylic acid hydrogen-bonded dimers which are linked by N-H...O bonds. In the formation of II, the molecule reorientates about the C(carboxy)-C_α and C_α-N bonds to accommodate the Cu atom. II is polymeric and the

Cu atom has an irregular stereochemistry. The alanine moiety forms a five-membered chelate ring in the equatorial square plane by coordination through one of the carboxy oxygens and the deprotonated N, the third and fourth corners being occupied by a water oxygen and the free carboxy oxygen of a screw-related complex. One of the sulphonyl O atoms of a glide-related complex coordinates axially while the oxygen from the chelate ring of the above screw-related complex coordinates in an off-axial direction. The coordinated water molecule forms a hydrogen bond with the uncoordinated sulphonyl oxygen, forming a six-membered water-bridged chelate ring. In packing, II segregates into hydrophobic and hydrophilic regions with the water molecules forming a network of hydrogen bonds. Distances in the $CuNO_5$ chromophore of II are: Cu(1)-O(2) 1.945(2), Cu(1)-N(7) 1.968(2), Cu(1)-O(17) 1.974(2), Cu(1)-O(4') 1.966(2), Cu(1)-O(10'') 2.648(2), Cu(1)-O(2') 2.838(2) Å.

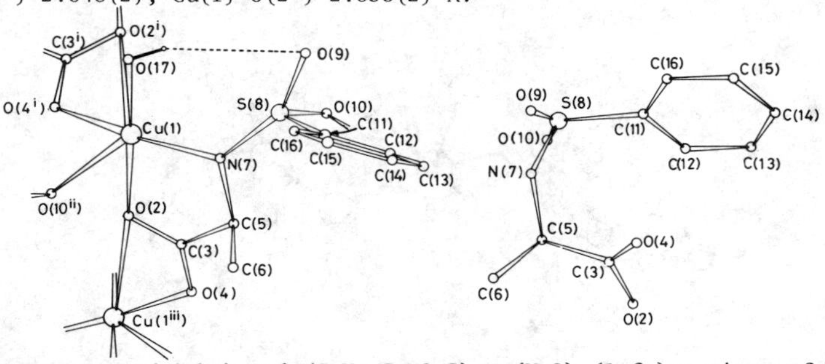

Fig. 1. $C_9H_{11}NO_4S$ (right) and $(C_9H_{11}CuNO_5S)_n,n(H_2O)$ (left): views of both molecules down the 'a' axis.

(L-ALANYL-L-HISTIDINATO)COPPER(II) HYDRATE
$C_9H_{12}CuN_4O_3,3.5(H_2O)$

Y. MAUGUEN, E. VILKAS and C. AMAR, 1984. Acta Cryst., C40, 82-85.

Orthorhombic, $P2_12_12$, a = 10.037, b = 14.477, c = 9.735 Å, Z = 4. Mo radiation, R = 0.045 for 1588 reflexions.

The Cu atom (Fig. 1) has approximate square-planar geometry and a fifth weaker interaction with a carboxyl oxygen (Cu-N(1) 1.989(6), Cu-N(2) 1.933(5), Cu-N(3) 1.944(5), Cu-O(3') 2.038(5), and Cu-O(2') 2.527(5) Å. Polymeric Cu-dipeptide-Cu-dipeptide chains are formed parallel to the crystallographic a axis and are linked by a network of hydrogen bonds involving the water molecules (one of which lies on a crystallographic twofold symmetry axis).

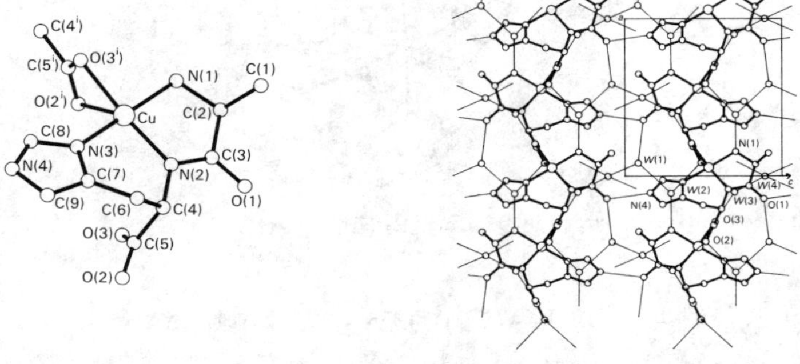

Fig. 1. $C_9H_{12}CuN_4O_3,3.5(H_2O)$: views of the complex and of the molecular packing.

AQUABIS(L-VALINATO)COPPER(II)
$C_{10}H_{22}CuN_2O_5$

ZHOU KANGJING, HUANG JINLING, LI JIANMIN, ZHANG YONGFENG and ZHOU PING, 1984. J. Struct. Chem., 3, 155-157.

Monoclinic, C2, a = 21.255, b = 9.567, c = 7.399 Å, β = 108.8°, Z = 4. Mo radiation, R = 0.089 for 1644 reflexions.

The Cu atom has square-pyramidal geometry (Fig. 1) with the apical site occupied by a water molecule (Cu-O 2.351(9) Å). The L-valine ligands are in cis-orientation with Cu-O 1.939 and 1.951(10), Cu-N 1.978 and 1.989(15) Å, O(H_2O)-Cu-O(val) 98.3 and 108.7(5)°, O(H_2O)-Cu-N 89.4 and 90.0(5)°.

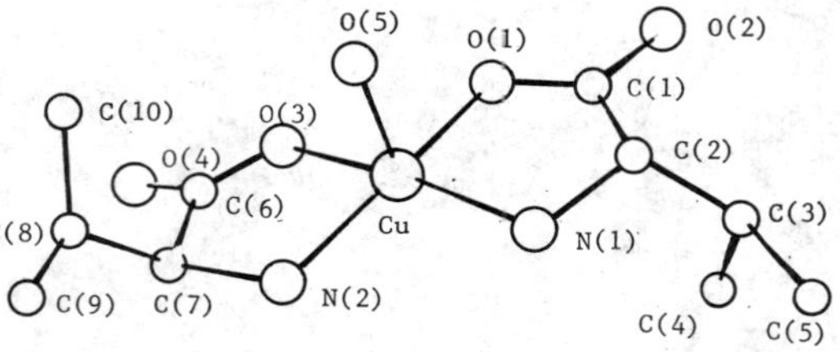

Fig. 1. A view of $C_{10}H_{22}CuN_2O_5$.

BIS(L-HISTIDINATO-O,N,N')CHROMIUM(II) NITRATE
$C_{12}H_{16}CrN_7O_7$ $Cr(C_6H_8N_3O_2)_2NO_3$

W.T. PENNINGTON, A.W. CORDES, D. KYLE and E.W. WILSON, Jr., 1984. Acta Cryst., C40, 1322-1324.

Monoclinic, $P2_1$, a = 7.404, b = 7.209, c = 15.663 Å, β = 100.68°, Z = 2. Mo radiation, R = 0.036 for 1812 reflexions.

The L-histidine anions are tridentate ligands (Fig. 1) with the planar imidazole rings trans. The Cr atom lies 0.184 and 0.299(1) Å out of the imidazole ring planes which are twisted at 44.6(1)° to each other. Cr-O distances are 1.972 and 1.970(3) Å and Cr-H are in the range 2.043-2.062(3) Å. Cations and anions are linked into zig-zag chains along a by N-H...O hydrogen bonds (N...O 2.897(5) and 2.963(6) Å).

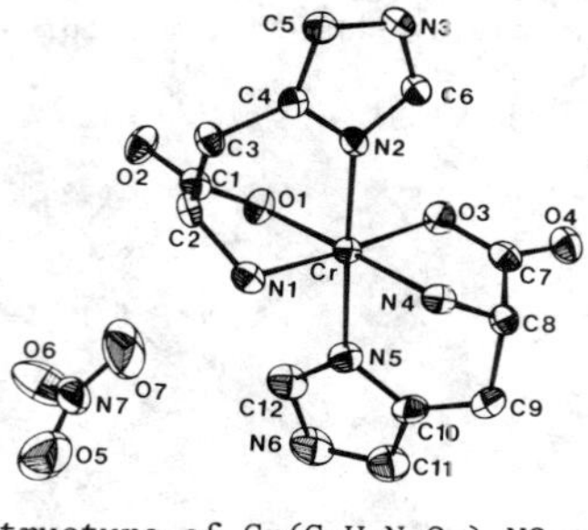

Fig. 1. The molecular structure of $Cr(C_6H_8N_3O_2)_2NO_3$.

LITHIUM BIS(β-HYDROXYETHYLIMINODIACETATO)YTTERBATE(III) TETRAHYDRATE
$C_{12}H_{18}LiN_2O_{10}Yb,4(H_2O)$

I.N. POLYAKOVA, T.N. POLYNOVA and M.A. PORAI-KOSHITS, 1983. Koord. Khim., 9, 1563-1567 [Sov. J. Coord. Chem., 9, 881-886].

Monoclinic, P2/b, a = 9.978, b = 15.148, c = 7.496 Å, γ = 117.63°, Z = 2. Mo radiation, R = 0.024 for 1755 reflexions.

The structure (Fig. 1) is built up of ytterbate complex anions and $[Li(H_2O)_4]^+$ cations. The Yb and Li atoms lie on twofold axes. The coordination polyhedron is a distorted dodecahedron. A system of hydrogen bonds (O...O 2.597-2.937 Å) and Li-O bonds (1.99(1), 1.92(1) Å links the cations and anions in layers.

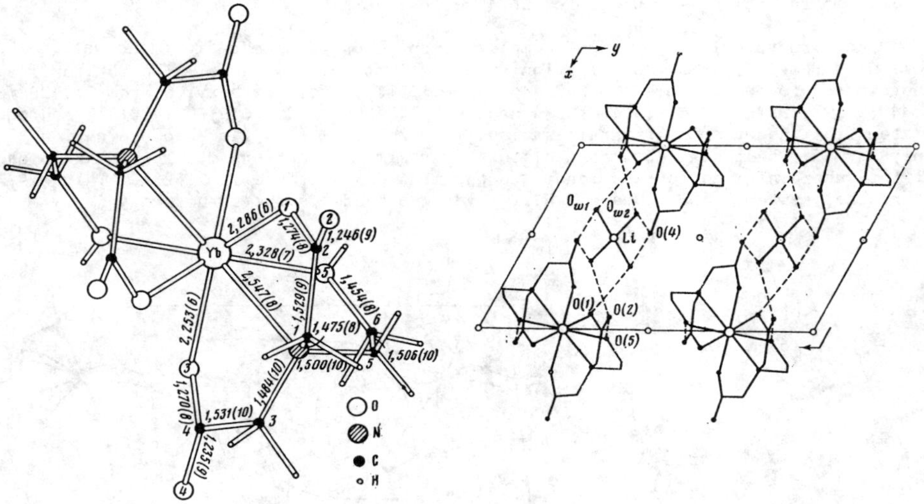

Fig. 1. $C_{12}H_{18}LiN_2O_{10}Yb,4(H_2O)$: the structure of the complex anion (left) and a projection of the crystal structure onto the xy plane (right).

BIS(O-METHYL-(S)-PENICILLAMINATO)-cis-DIOXOMOLYBDENUM(VI)
$C_{12}H_{24}MoN_2O_6S_2$

I. BUCHANAN, C.D. GARNER and W. CLEGG, 1984. J. Chem. Soc. Dalton, 1333-1342.

Monoclinic, $P2_1$, a = 11.794, b = 12.519, c = 13.040 Å, β = 106.32°, Z = 4. Mo radiation, R = 0.039 for 5502 reflexions.

The two crystallographically independent molecules in the asymmetric unit (Fig. 1) have opposite absolute configurations (Λ and Δ) for the ligand chelate rings about the metal. The penicillamine ester ligands behave as S,N-bidentate ligands with the amino groups mutually cis and each trans to an oxygen. The S atoms are mutually trans. The average Mo-O, Mo-S and Mo-N distances are 1.712(7), 2.407(5) and 2.362(2) Å respectively. The Mo atom is in an octahedral environment.

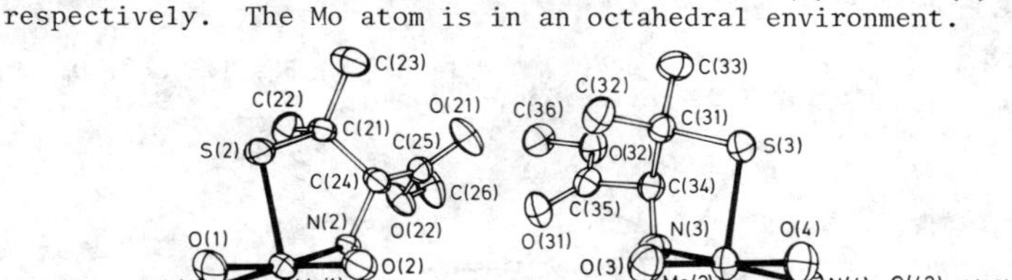

Fig. 1. $C_{12}H_{24}MoN_2O_6S_2$: the two independent molecules.

GUANIDINIUM AQUABIS(β-HYDROXYETHYLIMINODIACETATO-O,O',O'')YTTERBATE(III) DIHYDRATE
$C_{13}H_{26}N_5O_{11}Yb,2(H_2O)$

I.N. POLYAKOVA, T.A. SENINA, T.N. POLYNOVA and M.A. PORAI-KOSHITS, 1983. Koord. Khim., 9, 1131-1137 [Sov. J. Coord. Chem., 9, 644-650].

Triclinic, $P\bar{1}$, a = 8.727, b = 13.364, c = 9.686 Å, α = 105.37, β = 90.03, γ = 95.90°, Z = 2. Mo radiation, R = 0.026 for 2674 reflexions.

The structure is composed of complex anions, guanidinium cations and water molecules (Fig. 1). The coordination number of ytterbium is nine. The Yb atom is coordinated to two N atoms (Yb-N 2.646(6), 2.572(6)), four carboxyl O atoms (2.292(5), 2.349(5), 2.298(5), 2.339(5)), two hydroxyethyl O atoms (2.386(5), 2.420(5)) and one O atom from the water of crystallisation (Yb-O 2.392(5) Å). The Yb coordination polyhedron is best described as a distorted Archimedes' antiprism. The CN skeleton of the quanidinium cation is planar and the mean C-N distance is 1.32 Å. There is a branched system of OH...O and N-H...O hydrogen bonds in the structure.

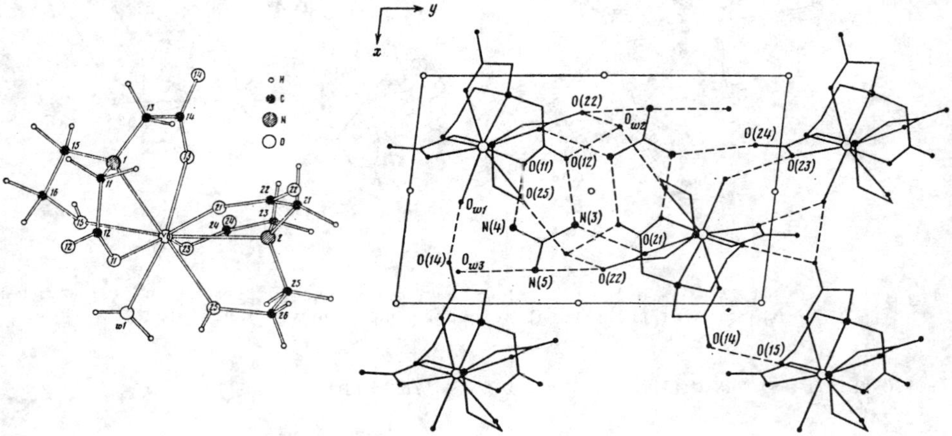

Fig. 1. $C_{13}H_{26}N_5O_{11}Yb,2(H_2O)$: the complex anion (left) and a projection of the crystal structure onto the xy plane (right).

[N,N'-(1,2-DITHIOXOETHANE-1,2-DIYL)BIS(METHYL METHIONINATO)]BIS(BROMOCOPPER(II))
$C_{14}H_{22}Br_2Cu_2N_2O_4S_4$ $Cu_2\{C_2S_2[NCH(CH_2CH_2SCH_3)COOCH_3]_2\}Br_2$

R. VEIT, J.-J. GIRERD, O. KAHN, F. ROBERT, Y. JEANNIN and NABIL EL MURR, 1984. Inorg. Chem., 23, 4448-4454.

Monoclinic, $P2_1/c$, a = 12.907, b = 6.701, c = 14.878 Å, β = 111.06°, D_m = 1.92, Z = 2. Mo radiation, R = 0.045 for 1296 reflexions.

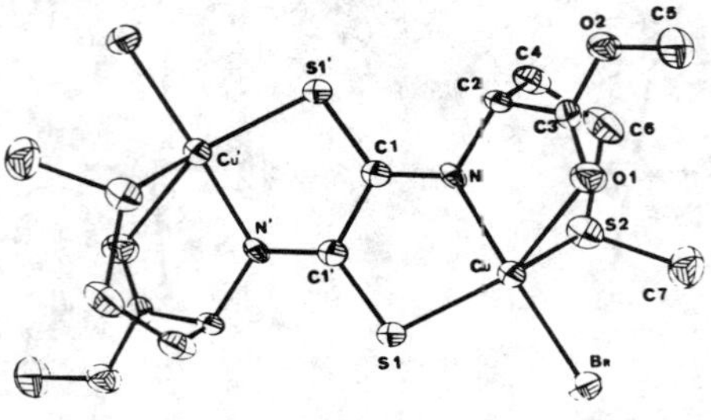

Fig. 1. The $Cu_2\{C_2S_2[NCH(CH_2CH_2SCH_3)COOCH_3]_2\}Br_2$ molecule.

The structure consists of centrosymmetric binuclear units (Fig. 1). The Cu atoms have trigonal bipyramidal coordination geometry with N and Br ligands axial. The bridging network $CuC_2S_2N_2Cu$ is almost planar with a Cu...Cu separation of 5.666(3) Å. Bond lengths: Cu-Br 2.332(1), Cu-S1 2.265(3), Cu-S2 2.405(3), Cu-O 2.524(7), Cu-N 1.950(7) Å.

(μ_3-GLYCYL-L-HISTIDYL-L-LYSINE-N,N',N'',O,O')COPPER PENTADECAHYDRATE
$C_{14}H_{22}CuN_6O_4,15H_2O$

C.M. PERKINS, N.J. ROSE, B. WEINSTEIN, R.E. STENKAMP, L.H. JENSEN and L. PICKART, 1984. Inorg. Chim. Acta, 82, 93-99.

Tetragonal, $P4_12_12$, a = 14.937, c = 25.903 Å, D_m = 1.40. Cu radiation, R = 0.16 for 1193 reflexions.

Each Cu atom has pseudo-square pyramidal coordination (Fig. 1) with Cu-N distances of 1.96, 2.06 and 1.88 Å for N(1), N(2) and ND1(2) respectively. One of the carboxyl terminal O atoms from another complex ($O^2(3)'$) fills the remaining Cu coordination site in the basal plane (Cu-$O^2(3)'$ 1.96 Å). The apical site is filled by a carboxyl terminal O atom from yet another complex (Cu-$O^2(3)''$ 2.49 Å). The carboxyl O atoms form bridges between Cu centres and thus the system is polymeric in the solid state.

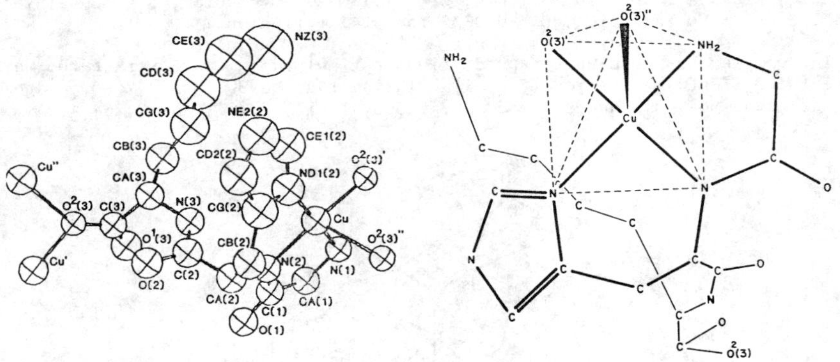

Fig. 1. ORTEP plot of $C_{14}H_{22}CuN_6O_4$ (left), and schematic diagram of Cu coordination (right).

(N-ETHYLIDENEETHYLENEDIAMINE)GLYCYLGLYCINATO-N,N',O)(NITRO-N)COBALT(III) HYDRATE
$C_{16}H_{32}Co_2N_{10}O_{10},2H_2O$ $2[CoC_8H_{16}N_5O_5.H_2O]$

B.V. PRELESNIK and R.M. HERAK, 1984. Croat. Chem. Acta, 57, 577-585.

Monoclinic, Pc, a = 7.884, b = 15.149, c = 12.206 Å, β = 120.82°, D_m = 1.790, Z = 2. Mo radiation, R = 0.057 for 2519 reflexions.

The crystal structure consists of enantiomeric pairs of complex molecules (Fig. 1) and molecules of water. There are two crystallographically independent but chemically equivalent complex molecules which differ only in conformation. The coordination around the Co atoms is distorted octahedral (Co(1)-O(1) 1.911, Co(1)-N(1) 1.881, Co(1)-N(2) 1.944, Co(1)-N(3) 1.963, Co(1)-N(4) 1.956, Co(1)-N(5) 1.932, Co(2)-O(6) 1.913, Co(2)-N(6) 1.881, Co(2)-N(7) 1.941, Co(2)-N(8) 1.950, Co(2)-N(9) 1.952 and Co(2)-N(10) 1.921 Å). The glycylglycinato ligand is tridentate and the diamine ligand is bidentate. The cis angles about the metal atoms range from 83.6 to 100.4°. The molecules of water connect molecules I and II by O-H...O hydrogen bonds in zig-zag chains running parallel to the x axis (O...O distances range from 2.729 to 2.784 Å) and N-H...O bonds join the chains in a three-dimensional network (N...O distances range from 2.869 to 3.176 Å).

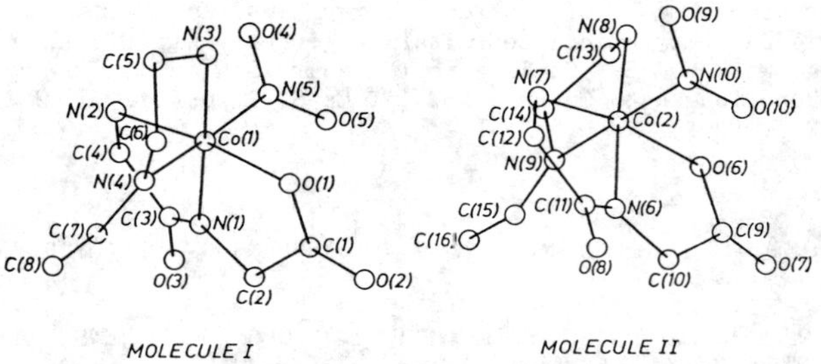

Fig. 1. Perspective drawings of the independent molecules of $C_{16}H_{32}Co_2N_{10}O_{10}$.

BARIUM TRIS(μ_2-OXO)-(μ_3-SULPHIDO)TRIS(NITRILODIACETATOACETIC ACID)TRIMOLYBDENUM DECAHYDRATE

$C_{18}H_{21}BaMo_3N_3O_{21}S,10(H_2O)$

T. SHIBAHARA, H. HATTORI and H. KUROYA, 1984. J. Am. Chem. Soc., 106, 2710-2711.

Triclinic, $P\bar{1}$, a = 15.602, b = 11.614, c = 11.150 Å, α = 100.89, β = 95.42, γ = 99.70°, Z = 2. Mo radiation, R = 0.0335 for 4245 reflexions.

The complex has a sulphur-capped trinuclear molybdenum core structure, $Mo_3O_3S^{4+}$, in the $[Mo_3O_3S(Hnta)_3]^{2-}$ anion (Fig. 1). The average Mo-Mo, Mo-S, Mo-O(bridge), Mo-O(ligand), and Mo-N distances are 2.589, 2.360, 1.917, 2.097, and 2.264 Å, respectively.

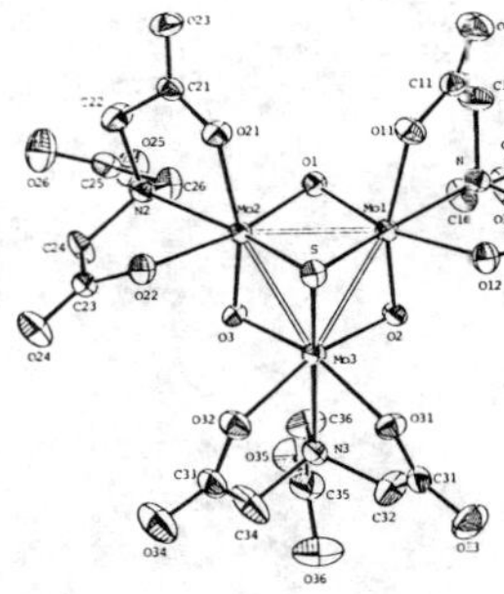

Fig. 1. A view of the $[Mo_3O_3S(Hnta)_3]^{2-}$ anion.

β_1-{S-ALANINATO[2,5-DIAZA-3-METHYL-1,6-DI(2-PYRIDYL)HEXANE]COBALT(III)} PERCHLORATE

$C_{18}H_{26}Cl_2CoN_5O_{10}$

M.W. MULQI, P.A. WILLIAMS, F.S. STEPHENS and R.S. VAGG, 1984. Inorg. Chim. Acta, 88, 183-192.

Monoclinic, $P2_1$, a = 17.204, b = 9.037, c = 17.631 Å, β = 116.85°, Z = 4. Mo radiation, R = 0.040 for 2313 reflexions.

The complex cation co-crystallizes in four distinct diastereoisomeric forms (Fig. 1), each with cis-β_1 topology in a quasi-racemic mixture. The four forms, together with their respective abundances are: Δ-β_1-S,S-endo-$[Co(S\text{-picpn})(S\text{-ala})]^{2+}$ (62%), Δ-β_1-S,S-exo-$[Co(S\text{-picpn})(S\text{-ala})]^{2+}$ (38%), Λ-β_1-R,R-endo-$[Co(R\text{-picpn})(S\text{-ala})]^{2+}$ (62%), and Λ-β_1-R,R-exo-$[Co(R\text{-picpn})(S\text{-ala})]^{2+}$ (38%). The octahedral coordination spheres are slightly distorted, with average Co-O, Co-N(amine) and Co-N(pyridine) bond lengths of 1.89, 1.94 and 1.93 Å respectively. The perchlorate groups are disordered.

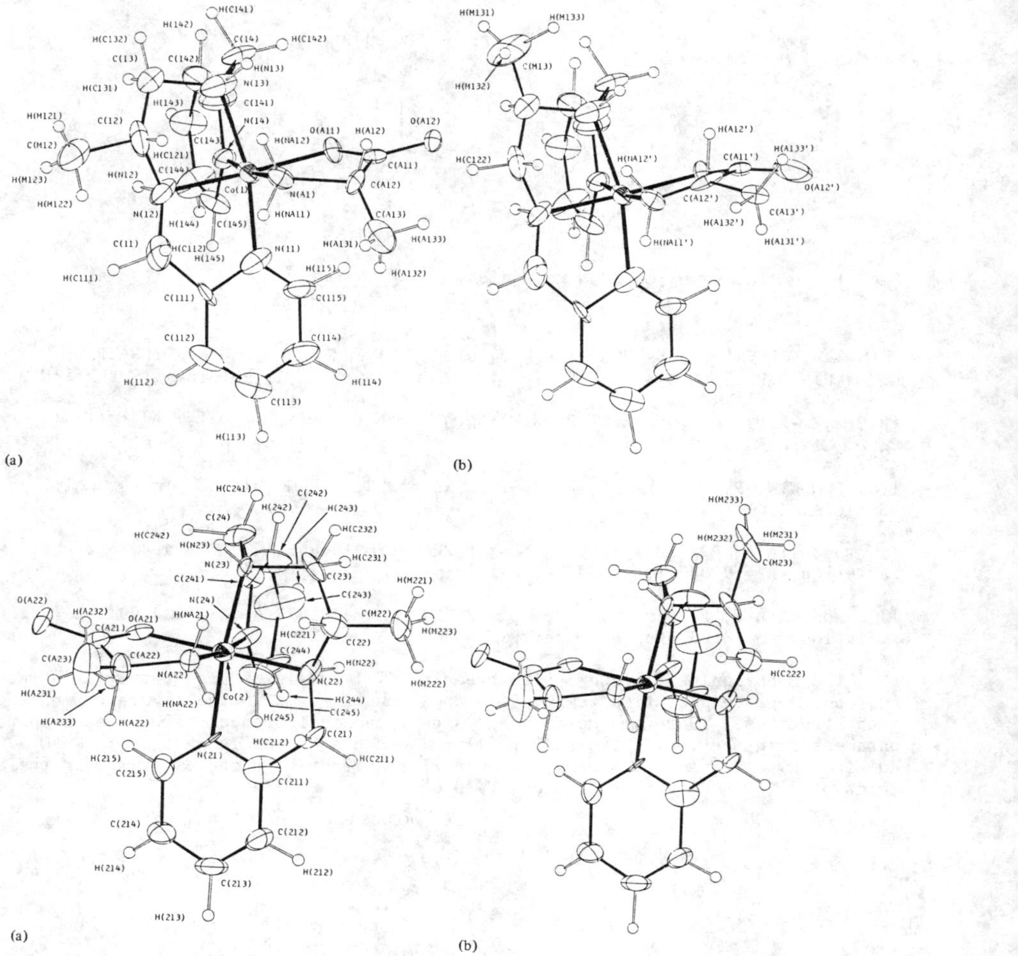

Fig. 1. Molecular structures of the cations Λ-β_1-R,R-endo-[Co(R-pipcn)(S-ala)]$^{2+}$ (top left), Λ-β_1-R,R=exo-[Co(R-picpn)(S-ala)]$^{2+}$ (top right), Δ-β_1-S,S-endo-[Co(S-picpn)(S-ala)]$^{2+}$ (bottom left), and Δ-β_1-S,S-exo-[Co(S-picpn)-(S-ala)]$^{2+}$ (Bottom right) in $C_{18}H_{26}Cl_2CoN_5O_{10}$.

(ETHANOL)(N-TOSYLGLYCINATO)(2,2'-BIPYRIDINE)COPPER(II)
$C_{21}H_{23}CuN_3O_5S$

L. ANTOLINI and L. MENABUE, 1984. Inorg. Chem., 23, 1418-1422.

Monoclinic, $P2_1/n$, a = 9.677, b = 10.460, c = 21.488 Å, β = 98.27°, D_m = 1.50, Z = 4. Cu radiation, R = 0.039 for 1705 reflexions.

The Cu atom shows tetrahedrally distorted square pyramidal N_3O_2 coordination (Fig. 1). The Tsgly dianion and the bpy molecule act as bidentate ligands in the equatorial plane (Cu-O 1.940(4), Cu-N(am) 1.946(4), Cu-N(bpy) 1.996(5) and 2.001(5) Å), and the ethanol occupies the apical position (Cu-O 2.382(4) Å).

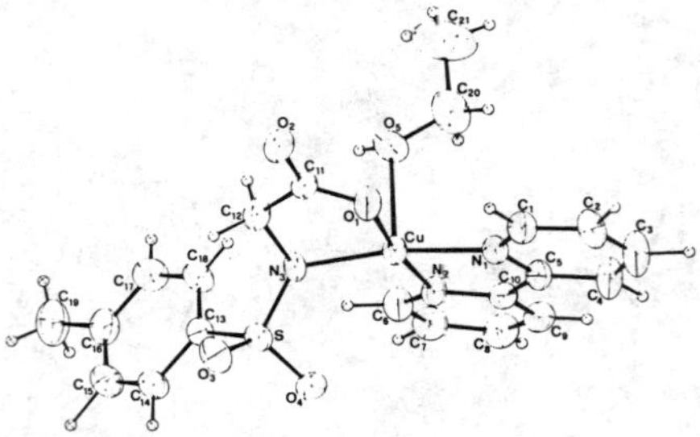

Fig. 1. The $C_{21}H_{23}CuN_3O_5S$ molecule.

[BIS(2,2'-BIPYRIDYL)(S-THREONINE-N,O)RUTHENIUM] PERCHLORATE PENTAHYDRATE
$C_{24}H_{24}ClN_5O_7Ru,5H_2O$ (I) Δ,Λ-$[Ru(bipy)_2(S\text{-}thr)]^+ClO_4^-,5H_2O$

[BIS(2,2'-BIPYRIDYL)(S-ALLOTHREONINE-N,O)RUTHENIUM] PERCHLORATE PENTAHYDRATE
$C_{24}H_{24}ClN_5O_7Ru,5H_2O$ Δ,Λ-$[Ru(bipy)_2(S\text{-}allothr)]^+ClO_4,5H_2O$

T.J. GOODWIN, P.A. WILLIAMS, F.S. STEPHENS and R.S. VAGG, 1984. Inorg. Chim. Acta, 88, 165-181.

I. Monoclinic, $P2_1$, a = 10.165, b = 28.220, c = 10.896 Å, β = 104.80°, Z = 4. Mo radiation, R = 0.042 for 3327 reflexions.

II. Monoclinic, $P2_1$, a = 10.226, b = 27.979, c = 10.875 Å, β = 104.99°, Z = 4. Mo radiation, R = 0.037 for 4456 reflexions.

The S-threonine (I) and S-allothreonine (II) complexes are grossly isostructural. A pseudo a-glide operation relates the two Δ and Λ forms in the asymmetric unit of each structure. There are no apparent significant differences between corresponding parameters in the four cations (Fig. 1). The average bond lengths are: Ru-O 2.094, Ru-N(amine) 2.128 and Ru-N(bipy) 2.050 Å with average bite angles at Ru for the amino acids of 79.7 and for bipy ligands of 79.0°.

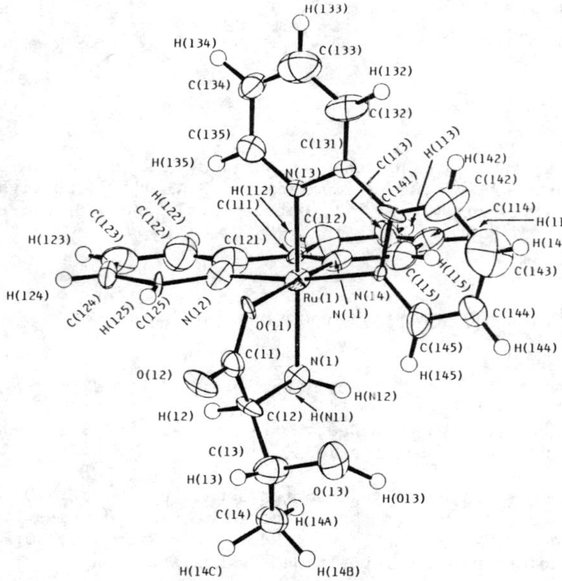

Fig. 1. Molecular structure of Δ-$[Ru(bipy)_2(S\text{-}thr)]^+$.

2,2'-BIPYRIDYLBIS(N-ACETYL-DL-LEUCINATO-O)COPPER(II)
$C_{26}H_{36}CuN_4O_6$ (I) $[Cu(C_{10}H_8N_2)(C_8H_{14}NO_3)_2]$

2,2'-BIPYRIDYLBIS(N-TOSYL-β-ALANINATO-O)COPPER(II)
$C_{30}H_{32}CuN_4O_8S_2$ (II) $[Cu(C_{10}H_8N_2)(C_{10}H_{12}NO_4S)_2]$

L. ANTOLINI, L. MENABUE, M. SALADINI, L.P. BATTAGLIA and A.B. CORRADI, 1984. Inorg. Chim. Acta, 90, 97-103.

I. Monoclinic, C2/c, a = 17.465, b = 19.740, c = 9.080 Å, β = 115.0°, D_m = 1.30, Z = 4. Cu radiation, R = 0.041 for 1311 reflexions.

II. Triclinic, P$\bar{1}$, a = 14.489, b = 14.308, c = 8.659 Å, α = 75.0, β = 74.6, γ = 66.6°, D_m = 1.47, Z = 2. Cu radiation, R = 0.038 for 3403 reflexions.

The structure of I is built up of monomeric units in which the Cu atom, which lies on the twofold axis, exhibits tetrahedrally-distorted square planar geometry (Fig. 1). The dihedral angle between planes N(1),Cu,N(1') and O(1),Cu,O(1') is 20.8°. The distances Cu-N(1) and Cu-O(1) are respectively 1.996 and 1.936 Å and the angles N(1)-Cu-N(1'), O(1)-Cu-O(1') and O(1)-Cu-N(1') are 81.0, 91.7 and 95.7° respectively. Hydrogen bonds link peptide N atoms with uncoordinated O carboxylate atoms (N(2)...O(2') 2.871 Å). The Cu atom in II has strict square planar coordination with bond angles of N(1)-Cu-N(2) 80.9, N(1)-Cu-O(1) 93.1, N(2)-Cu-O(5) 92.5, and O(1)-Cu-O(5) 93.6° and bond distances Cu-N(1) 2.005, Cu-N(2) 1.996, Cu-O(1) 1.934 and Cu-O(5) 1.944 Å (Fig. 1). There are interactions between the metal and uncoordinated carboxylate O atoms in the axial octahedral sites (Cu...O(2) 2.704 and Cu...O(6) 1.751 Å). Hydrogen bonds (N(3)...O(6') 2.857, N(4)...O(2') 2.858 Å) hold the crystal structure together.

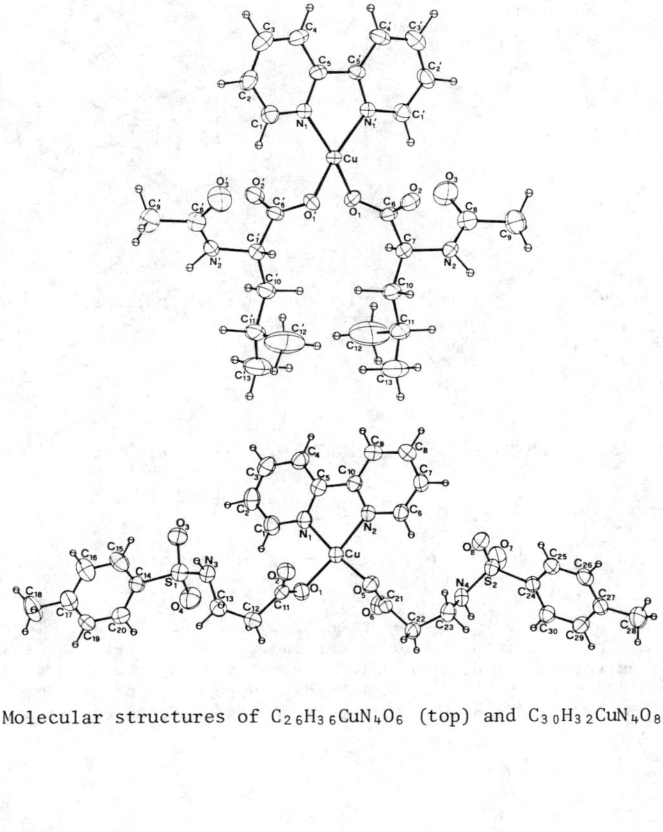

Fig. 1. Molecular structures of $C_{26}H_{36}CuN_4O_6$ (top) and $C_{30}H_{32}CuN_4O_8S_2$ (bottom).

FERRICROCIN HYDRATE
$C_{29}H_{44}FeN_9O_{13},7H_2O$

C.L. BARNES, D.L. ENG-WILMOT and D. VAN DER HELM, 1984. Acta Cryst., C40, 922-926.

Monoclinic, $P2_1$, a = 11.178, b = 17.809, c = 21.805 Å. β = 104.14°, Z = 4. Mo radiation, R = 0.0716 for 6879 reflexions (at 138K).

In the crystal structure (Fig. 1) one of the two independent molecules is conformationally very similar to ferrichrome (1) and ferrichrome A (2), but the other molecule assumes a somewhat different conformation of the cyclic hexapeptide ring. The molecules pack as A-B dimers with an approximate twofold axis relating the independent molecules. The A-B dimers are formed by nearly symmetric hydrogen bonds involving donor and recipient atoms of the 18-membered cyclic hexapeptide rings. Adjacent dimers are linked by direct intermolecular hydrogen bonds and by extensive bridging interactions with solvent molecules.

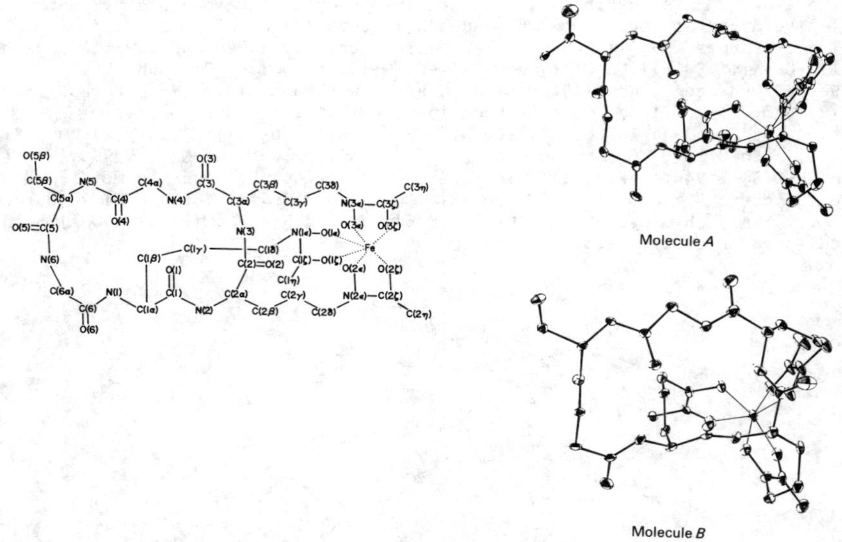

Fig. 1. $C_{29}H_{44}FeN_9O_{13},7H_2O$: perspective views of molecules A and B.

1. Structure Reports, 46B, 1158.
2. Ibid, 31B, 414.

DIAQUA-BIS[μ-(N-SALICYLIDENE-L-PHENYLALANINATO)-O,N,μ-O']-DICOPPER(II)
$C_{32}H_{30}Cu_2N_2O_8$

J.E. DAVIES, 1984. Acta Cryst., C40, 903-904.

Trigonal, $P3_121$, a = 12.724, c = 16.476 Å, Z = 3. Mo radiation, R = 0.138 for 2017 reflexions.

The structure has already been reported as monoclinic, C2 (1). The present refinement, using the same data, has been carried out in the trigonal space group $P3_121$. The two halves of the dimeric molecule are related by a diad axis which is approximately parallel to the two Cu-O(4) bonds, where O(4) is the coordinated water oxygen atom (Fig. 1). Bond lengths and angles do not differ significantly from those already reported (1).

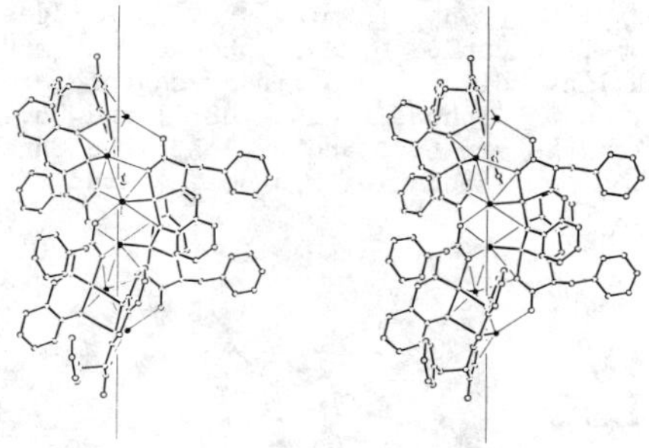

Fig. 1. $C_{32}H_{30}Cu_2N_2O_8$: stereoview of three molecules viewed perpendicular to the triad (c) axis.

1. Structure Reports, 44B, 903.

FERRIC NEUROSPORIN ACETONITRILE SOLVATE
$C_{33}H_{51}FeN_6O_{15}, 6(C_2H_3N)$

D.L. ENG-WILMOT, A. RAHMAN, J.V. MENDENHALL, S.L. GRAYSON and D. VAN DER HELM, 1984. J. Am. Chem. Soc., 106, 1285-1290.

Rhombohedral, R3, a = 19.774, c = 12.107 Å, Z = 3. Mo radiation, R = 0.058 for 1655 reflexions (at -135°C).

The molecule (Fig. 1) has crystallographic threefold symmetry with no unusual dimensions (eg. Fe-O(1) 1.963(2), Fe-O(2) 2.059(2) Å). The configuration at the asymmetric carbon atom, C(3)*, of the acyl group is R, while that in the ornithinyl backbone, C(5)*, is D. In the crystalline state and in solution, the cyclic hexadentate ligand assumes the Λ-cis configuration about the central ferric ion (vide infra).

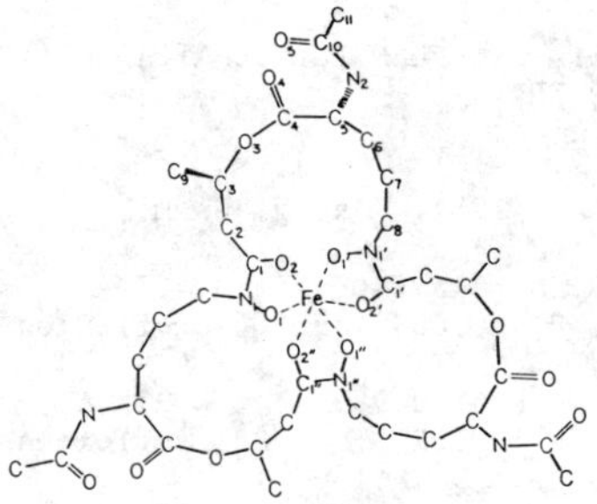

Fig. 1. The structure of ferric neurosporin.

TETRAKIS(μ_2-N-TOSYLGLYCINATO)DICOPPER(II)
$(C_{36}H_{40}Cu_2N_4O_{16}S_4)_n$

L. ANTOLINI, L. MENABUE, G.C. PELLACANI, G.B. GAVIOLI, G. GRANDI, L.P. BATTAGLIA, A.B. CORRADI and G. MARCOTRIGIANO, 1984. J. Chem. Soc. Dalton, 1687-1692

Monoclinic, $P2_1/n$, a = 24.655, b = 7.697, c = 12.378 Å, β = 87.34°, D_m = 1.48, Z = 4. Cu radiation, R = 0.052 for 1677 reflexions.

The structure (Fig. 1) consists of one-dimensional polymeric chains, extended along b, of binuclear centrosymmetric $Cu_2(tsglyO)_4$ units. The coordination polyhedron has four bidentate carboxylate groups forming syn,syn bridges between Cu atoms. The axial site of the Cu atom is occupied by a sulphonic O atom from a neighbouring tosylglycinato ion which links dimers into linear chains. Bond lengths include: Cu-O(carboxyl) 1.955(7), Cu-O(sulphonamide) 2.168(7) and Cu...Cu' 2.577(3) Å. The Cu-O-C-O-Cu (bridging) lengths are 6.41 and 6.42 Å. The Cu atom is displaced 0.171 Å from the mean plane of the four chelating O atoms towards the apical O atom.

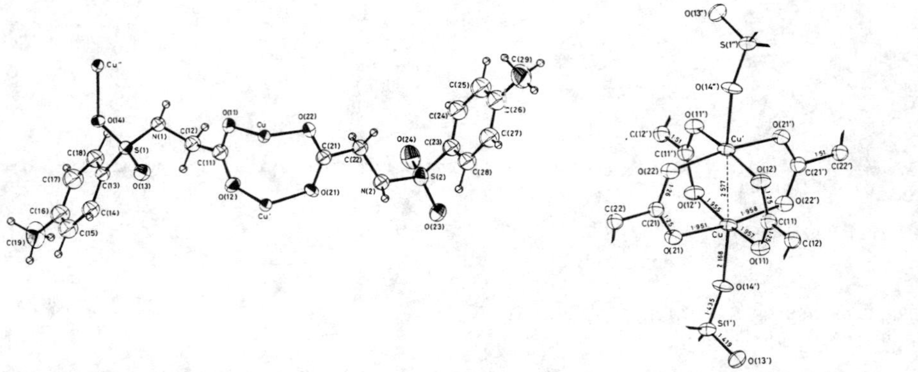

Fig. 1. $(C_{36}H_{40}Cu_2N_4O_{16}S_4)_n$: the structure (left) and a portion of the linear chain structure (right).

DIIODOBIS(μ-D-PHENYLALANINATO)BIS(μ-L-PHENYLALANINATO)DIMOLYBDENUM(II) DIIODIDE HEXAHYDRATE

$C_{36}H_{44}I_4Mo_2N_4O_8,6H_2O$ (I)

DIIODOBIS(μ-D-TYROSINATO)BIS(μ-L-TYROSINATO)DIMOLYBDENUM(II) DIIODIDE HEXAHYDRATE

$C_{36}H_{44}I_4Mo_2N_4O_{12},6H_2O$ (II)

BIS(μ-D-C-PHENYLGLYCINATO)BIS(μ-L-C-PHENYLGLYCINATO)DIMOLYBDENUM(II) TETRAKIS(p-TOLUENESULPHONATE) TETRAHYDRATE

$C_{60}H_{64}Mo_2N_4O_{20}S_4,4H_2O$ (III)

F. APFELBAUM-TIBIKA and A. BINO, 1984. Inorg. Chem., 23, 2902-2905.

I. Triclinic, P$\bar{1}$, a = 11.969, b = 13.746, c = 9.457 Å, α = 104.90, β = 107.49, γ = 66.47°, Z = 1. Mo radiation, R = 0.053 for 2661 reflexions.

II. Triclinic, P$\bar{1}$, a = 12.266, b = 12.952, c = 8.918 Å, α = 98.50, β = 92.31, γ = 63.58°, Z = 1. Mo radiation, R = 0.039 for 2822 reflexions.

III. Triclinic, P$\bar{1}$, a = 11.361, b = 16.538, c = 10.110 Å, α = 95.44, β = 110.84, γ = 75.68°, Z = 1. Mo radiation, R = 0.065 for 3765 reflexions.

The three dimeric molecules (Fig. 1) reside on inversion centres, the four chiral ligands being (necessarily) arranged around the Mo-Mo unit in the cyclic order of DDLL. Iodido ligands occupy the axial sites of the $Mo_2(\mu\text{-RCOO})_4$ units in I and II. Bond lengths are: for I; Mo-Mo 2.114(1), Mo-I 3.253(1), Mo-O 2.10(1)-2.126(7) Å; for II; Mo-Mo 2.116(1), Mo-I 3.212(1), Mo-O 2.122(4)-2.127(5) Å; and for III; Mo-Mo 2.113(1), Mo-O 2.115(5)-2.131(4) Å. In III, sulphonate O atoms occupy the equatorial positions (Mo-O 2.560(6) Å).

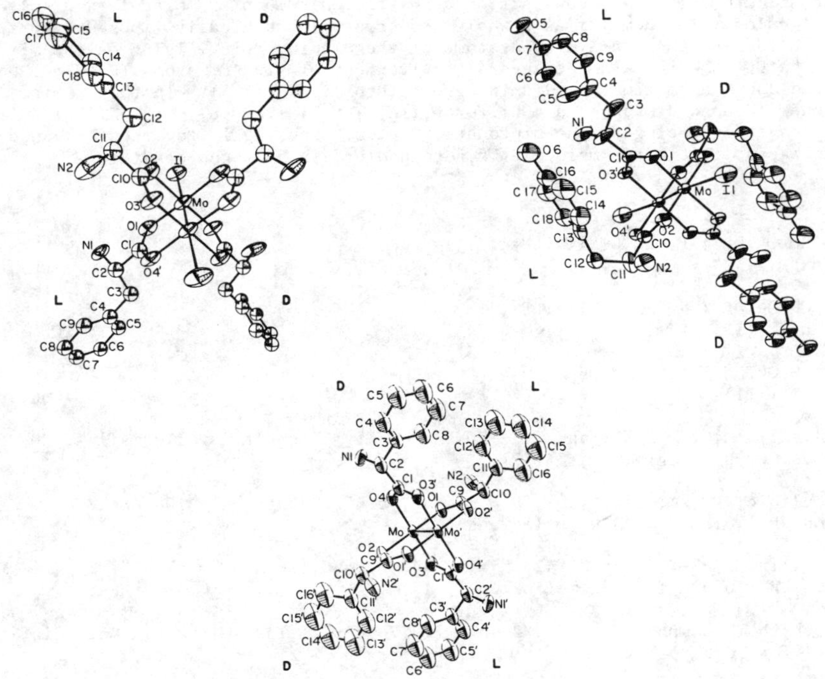

Fig. 1. The structures of $C_{36}H_{44}I_2Mo_2N_4O_8{}^{2+}$ (top left), $C_{36}H_{44}I_2Mo_2N_4O_{12}{}^{2+}$ (top right), and $C_{32}H_{36}Mo_2N_4O_8{}^{4+}$ (bottom, axial $CH_3C_6H_4SO_3$ ligands omitted).

[(2-AMINOETHYL)AMMONIUM]PENTACHLOROPLATINUM(IV) MONOHYDRATE
$C_2H_9Cl_5N_2Pt,H_2O$

F.P. FANIZZI, G. NATILE, L. MARESCA, A.M.M. LANFREDI and A. TIRIPICCHIO, 1984. J. Chem. Soc. Dalton, 1467-1470.

Orthorhombic, $Ccm2_1$, a = 15.388, b = 6.890, c = 9.882 Å, Z = 4. Mo radiation, R = 0.027 for 543 observed reflexions.

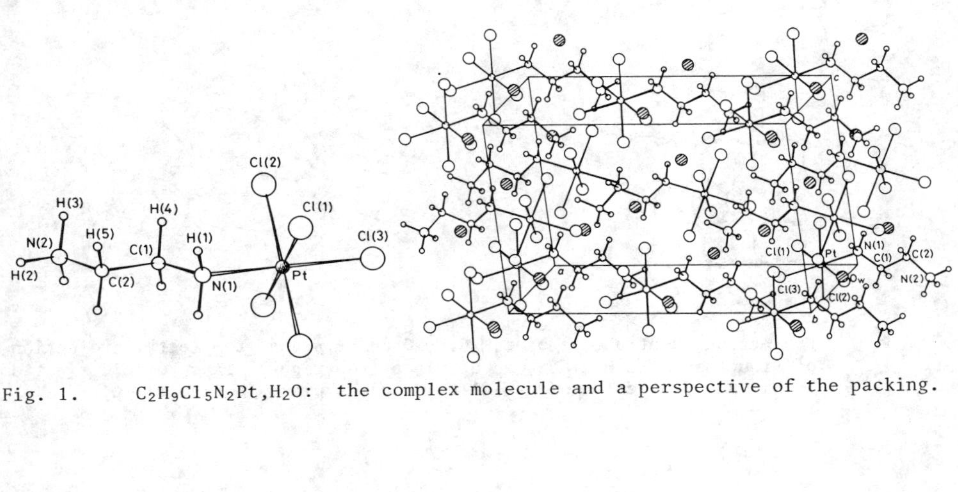

Fig. 1. $C_2H_9Cl_5N_2Pt,H_2O$: the complex molecule and a perspective of the packing.

The structure (Fig. 1) consists of a zwitterionic complex molecule and solvent water molecules. The complex as a whole has crystallographically imposed m symmetry with Pt, Cl(3) and the non-hydrogen atoms of the diamine moiety lying on the mirror plane. The Pt atom is in an octahedral environment of five Cl atoms (Pt-Cl 2.294-2.306) and one N atom from the diamine group (Pt-N 2.050(4) Å). In the diamine ligand the C-N bond distances are different (C(1)-N(1) 1.420(8), C(2)-N(2) 1.501(7) Å), the longer one being to the protonated N atom. Packing is governed by van der Waals contacts. The water molecule is not involved in hydrogen bonding.

CUPROUS CHLORIDE : FUMARODINITRILE (1:1)
$C_4H_2ClCuN_2$ (I) $CuCl.C_4H_2N_2$

CUPROUS CHLORIDE : FUMARODINITRILE (2:1)
$C_4H_2Cl_2Cu_2N_2$ (II) $2CuCl.C_4H_2N_2$

CUPROUS CHLORIDE : ACRYLAMIDE
$C_3H_5ClCuNO$ (III) $CuCl.C_3H_5ON$

P.Yu. ZAVALII, M.G. MYS'KIV and V.S. FUNDAMENSKII, 1984. Kristallografiya, 29, 60-65 [Sov. Phys. Crystallogr., 29, 34-37].

I. Monoclinic, $P2_1/m$, a = 6.398, b = 3.6945, c = 12.789 Å, γ = 100.48, Z = 2. Mo radiation, R = 0.047 for 420 reflexions.

II. Monoclinic, $P2_1/b$, a = 3.742, b = 15.546, c = 6.166 Å, γ = 97.98°, Z = 2. Mo radiation, R = 0.072 for 189 reflexions.

III. Orthorhombic, Pbca, a = 6.472, b = 7.957, c = 21.139 Å, Z = 8. Mo radiation, R = 0.041 for 1009 reflexions.

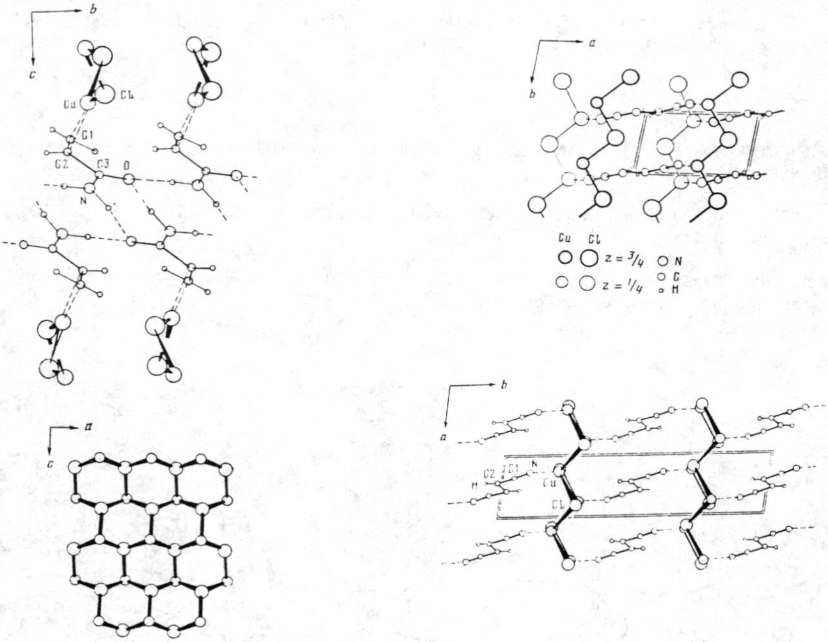

Fig. 1. Projection of structure of $C_3H_5ClCuNO$ on bc plane (top left), projection of structure of $C_4H_2ClCuN_2$ on ab plane (top right), layer of CuCl (bottom left), and projection of $C_4H_2Cl_2Cu_2N_2$ on ab plane (bottom right).

Compounds I and II are polynuclear σ complexes in which fumarodinitrile, acting as a bridging bidentate ligand, coordinates the Cu via the N atoms of both cyano groups (Fig. 1). Compound III, however, is a π-complex in which acrylamide interacts with the Cu atom via the C=C bond, and the amino group takes part only in forming a bifurcated system of hydrogen bonds N-H...O (N...O distances 2.920 and 2.945 Å). The Cu-Cl distances in I are 2.393 and 2.305 Å, in II are 2.46, 2.35 and 2.33 Å, and in III are 2.260 and 2.232 Å. The Cu-N distances in I and II are 1.981 and 1.93 Å respectively, while the Cu-C(1) and Cu-C(2) distances in III are 2.056 and 2.062 Å respectively.

cis-BIS(ACETONITRILE)DICHLOROPLATINUM(II)
$C_4H_6Cl_2N_2Pt$

F.D. ROCHON, R. MELANSON, H.E. HOWARD-LOCK, C.J.L. LOCK and G. TURNER, 1984. Can. J. Chem., 62, 860-869.

Monoclinic, $P2_1/c$, a = 6.660 (6.668), b = 12.500 (12.519), c = 10.277 (10.308) Å, β = 105.42 (105.41)°, D_m = 2.80 (2.79), Z = 4. Mo radiation, R = 0.054 (0.058) for 1578 (1526) reflexions.

The results of two independent analyses are presented. The Pt coordination (Fig. 1) is square planar with bond lengths (Pt-N 1.98(1), 1.98(1) (1.95(2), 1.99(2)) Å and Pt-Cl 2.264(4), 2.263(3) (2.270(6), 2.277(5)) Å) slightly shorter than normal. The N-C bonds are 1.10(2), 1.11(2) (1.11(3), 1.16(2)) Å. The $C-CH_3$ bonds are shorter than normal single bonds (1.48(2), 1.45(2) (1.49(3), 1.45(3)) Å) and the N-C-C framework is almost linear.

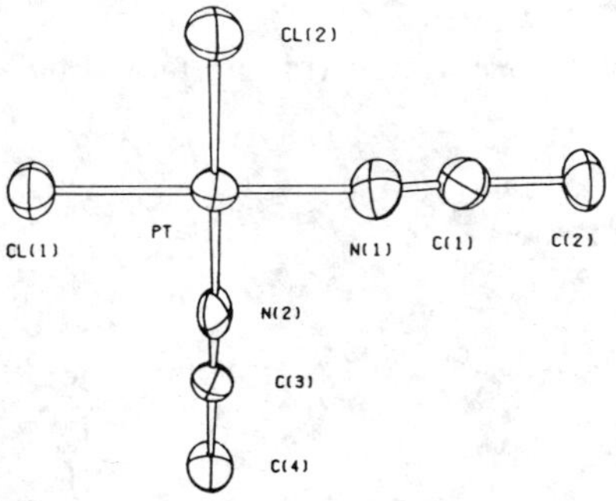

Fig. 1. The cis-$[Pt(CH_3CN)_2Cl_2]$ molecule.

DICHLOROBIS(DIMETHYLAMINE)PALLADIUM(II)
$C_4H_{14}Cl_2N_2Pd$ $Cl_2Pd(HN(CH_3)_2)_2$

G. BOMBIERI, G. BRUNO, G. CUM, R. GALLO and A. SPADARO, 1984. Inorg. Chim. Acta, 86, 121-125.

Monoclinic, $P2_1/c$, a = 7.699, b = 5.999, c = 10.454 Å, β = 98.3°, Z = 2. Mo radiation, R = 0.035 for 1059 reflexions.

The molecule lies on an inversion centre so the PdN_2Cl_2 moiety is trans and planar (Fig. 1). The Pd-N and Pd-Cl bond distances are 2.055 and 2.305 Å respectively whilst the two independent Cl-Pd-N angles are 87.7 and 92.3°. The C-N bond values of 1.481 and 1.468 Å correspond to standard single bond lengths. None of the intermolecular contact distances is less than the sum of the van der Waals radii.

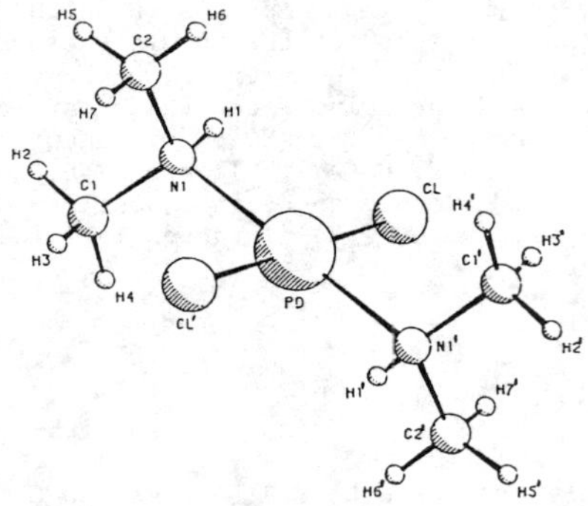

Fig. 1. Perspective view of $C_4H_{14}Cl_2N_2Pd$.

AQUABIS(OXAMIDE OXIME)COPPER(II) SULPHATE
$C_4H_{14}CuN_8O_9S$ (I) $[Cu(C_2H_6N_4O_2)_2(H_2O)]SO_4$

BIS(μ-SULPHATO)BIS[AQUA(OXAMIDE OXIME)COPPER(II)]
$C_4H_{16}Cu_2N_8O_{14}S_2$ (II) $[Cu(C_2H_6N_4O_2)(H_2O)(SO_4)]_2$

catena-(μ-SULPHATO)[DIAQUA(OXAMIDE OXIME)COPPER(II)]
$C_2H_{10}CuN_4O_8S$ (III) $[Cu(C_2H_6N_4O_2)(H_2O)_2(SO_4)]_n$

H. ENDRES, D. NÖTHE, E. ROSSATO and W.E. HATFIELD, 1984. Inorg. Chem., 23, 3467-3473.

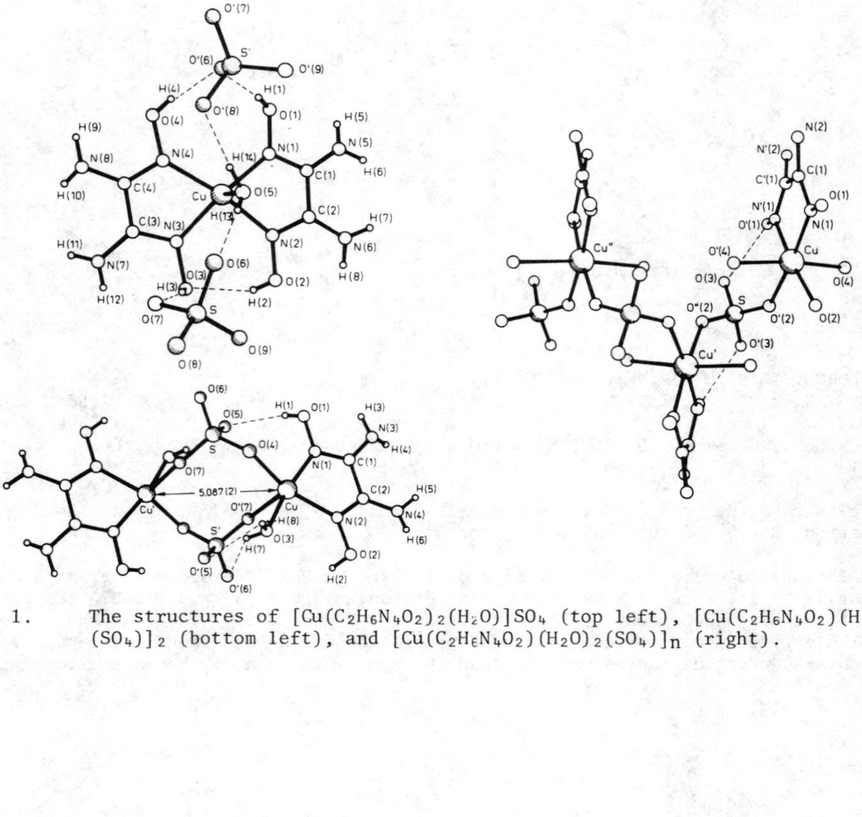

Fig. 1. The structures of $[Cu(C_2H_6N_4O_2)_2(H_2O)]SO_4$ (top left), $[Cu(C_2H_6N_4O_2)(H_2O)-(SO_4)]_2$ (bottom left), and $[Cu(C_2H_6N_4O_2)(H_2O)_2(SO_4)]_n$ (right).

I. Orthorhombic, Pbca, a = 7.829, b = 12.568, c = 28.284 Å, Z = 8. Mo radiation, R = 0.034 for 1320 reflexions.

II. Triclinic, $P\bar{1}$, a = 6.726, b = 7.982, c = 9.462 Å, α = 66.93, β = 74.37, γ = 68.81°, Z = 1. Mo radiation, R = 0.028 for 1457 reflexions.

III. Orthorhombic, Pcca, a = 7.121, b = 9.734, c = 13.770 Å, Z = 4. Mo radiation, R = 0.031 for 800 reflexions.

The structures of I-III are shown in Fig. 1. Complex I is monomeric with Cu coordinated by four oxime N atoms and a water molecule in a square pyramidal arrangement (Cu-N 1.945(4)-2.034(4), Cu-O 2.320(4) Å). Compound II exists as centrosymmetric SO_4-bridged binuclear species in which Cu has square pyramidal N_2O_3 coordination with a SO_4 oxygen atom apical (Cu-N 1.972(2) and 1.990(4) (trans to H_2O and SO_4 respectively), Cu-OH_2 1.945(2), Cu-OSO_3 1.946(3) and 2.375(2) Å). III is a polymeric SO_4-bridged structure with Cu and S atoms residing on crystallographic twofold axes. The Cu atoms have "4+1+1" distorted octahedral geometry with the basal plane consisting of two oxime N atoms (Cu-N 2.000(3) Å), and two SO_4 oxygen atoms from different sulphate groups (Cu-O 1.997(2) Å), and two water molecules axial (Cu-O 2.400(3) Å).

DIAQUABIS(2 - AMINOACETAMIDOXIME-N,N')NICKEL(II) DICHLORIDE
$C_4H_{18}Cl_2N_6NiO_4$ (I) $[Ni(C_2H_7N_3O)_2(H_2O)_2]^{2+}2Cl^-$

(2-AMINOACETAMIDOXIMATO-N,N')(2-AMINOACETAMIDOXIME-N,N')NICKEL(II) CHLORIDE SESQUIHYDRATE
$C_4H_{13}ClN_5NiO_2,1.5H_2O$ (II) $[Ni(C_2H_7N_3O)(C_2H_6N_3O)]^+Cl^-,1.5H_2O$

DIAQUABIS(2-(DIMETHYLAMINO)ACETAMIDOXIME-N,N')NICKEL(II) DICHLORIDE MONOHYDRATE
$C_8H_{26}Cl_2N_6NiO_4,H_2O$ (III) $[Ni(C_4H_{11}N_3O)_2(H_2O)_2]^{2+}2Cl^-,H_2O$

H. SAARINEN, J. KORVENRANTA, M. ORAMA and T. RAIKAS, 1984. Acta Chem. Scand., A38, 265-274

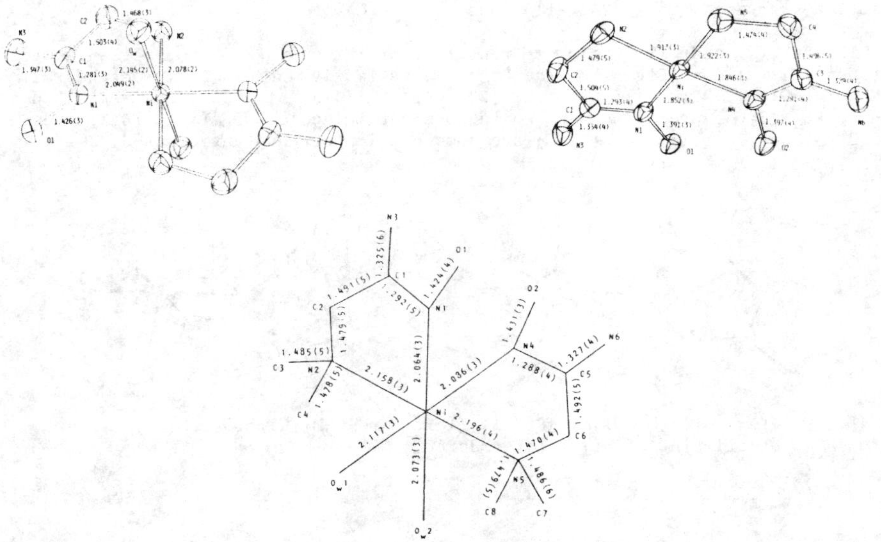

Fig. 1. The complex ion in $C_4H_{18}Cl_2N_6NiO_4$ (left), and in $C_4H_{13}ClN_6NiO_2$ (right), and schematic representation of $C_8H_{26}Cl_2N_6NiO_4$ (bottom), all showing bond lengths (Å).

I. Monoclinic, $P2_1/n$, a = 7.556, b = 10.923, c = 8.005 Å, β = 100.49°, D_m = 1.75, Z = 2. Mo radiation, R = 0.028 for 1633 reflexions.

II. Monoclinic, C2/c, a = 10.214, b = 15.634, c = 14.007 Å, β = 100.45°, D_m = 1.81, Z = 8. Mo radiation, R = 0.046 for 2164 reflexions.

III. Monoclinic, $P2_1/c$, a = 15.555, b = 8.041, c = 16.993 Å, β = 114.61°, D_m = 1.45, Z = 4. Mo radiation, R = 0.036 for 3496 reflexions.

In all three structures the bidentate ligand is bonded to the Ni atom through its amino and oxime N atoms (Fig. 1). In I the complex cation is centrosymmetric and displays a distorted trans octahedral coordination. In II the metal atom is in a square planar coordination environment in which there is a strong intramolecular hydrogen bond between the cis oxime O atoms (O(1)...O(2) 2.489 Å). O-H distances of 0.98 and 1.62 Å and an O-H-O angle of 145° suggest the presence of an asymmetric hydrogen bond. In III, the two bidentate ligands in cis-configuration and two of the water molecules are coordinated to the Ni atom to form a distorted octahedron in which the amino N's are trans to each other.

(5-CHLORO-1-METHYLURACIL-N[3])DIAMMINETRICHLOROPLATINUM(IV) DIHYDRATE
$C_5H_{10}Cl_4N_4O_2Pt,2(H_2O)$ (I)

(5,5-DICHLORO-6-HYDROXY-1-METHYL-5,6-DIHYDROURACIL-N[3])DIAMMINETRICHLOROPLATINUM(IV) DIHYDRATE
$C_5H_{11}Cl_5N_4O_3Pt,2(H_2O)$ (II)

G. MULLER, J. RIEDE, R. BEYERLE-PFNUR and B. LIPPERT, 1984. J. Am. Chem. Soc., 106, 7999-8001.

I. Monoclinic, $P2_1/c$, a = 13.468, b = 11.345, c = 18.932 Å, β = 105.54°, Z = 8. Mo radiation, R = 0.06 for 3605 reflexions (at -35°C).

II. Monoclinic, $P2_1/c$, a = 9.484, b = 11.426, c = 15.469 Å, β = 106.69°, Z = 4. Mo radiation, R = 0.056 for 2279 reflexions (at -40°C).

In both (I) and (II) the uracil ring is coordinated to N(3) to a cis-diammine-platinum(IV) residue (Fig. 1). The Pt coordination is approximately octahedral. The $Pt-NH_3$ (2.05(1)-2.08(2) Å) and Pt-Cl (2.310(6)-2.334(3) Å) distances are as expected, but the Pt-N(3) of uracil in the two independent molecules of (I) (2.14(2) and 2.09(2) Å) and in (II) (2.08(1) Å) are somewhat longer than anticipated.

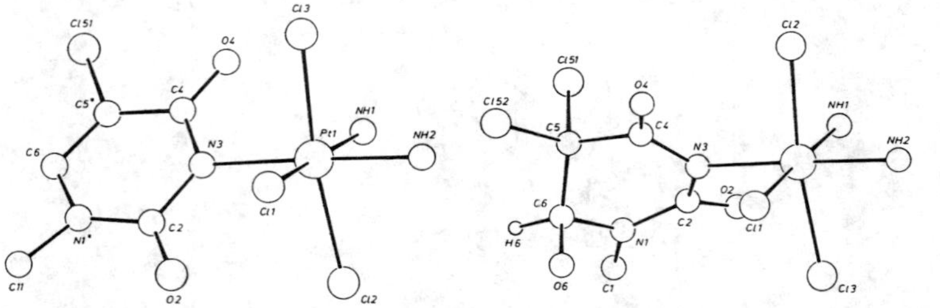

Fig. 1. Views of $C_5H_{10}Cl_4N_4O_2Pt$ (I) (left) and $C_5H_{11}Cl_5N_4O_3Pt$ (II) (right). The uracil ring of (I) is disordered.

trans-DICHLORO(DIMETHYL SULPHOXIDE)(ISOPROPYLAMINE)PLATINUM(II)
$C_5H_{15}Cl_2NOPtS$ trans-$[Pt(Me_2SO)\{(CH_3)_2CHNH_2\}Cl_2]$

R. MELANSON and F.D. ROCHON, 1984. Acta Cryst., C40, 793-795.

Triclinic, $P\bar{1}$, a = 7.789, b = 8.781, c = 8.874 Å, α = 108.23, β = 101.67, γ = 93.05°, D_m = 2.38, Z = 2. Mo radiation, R = 0.035 for 2829 reflexions.

The Pt has square-planar coordination with Pt-Cl 2.290(2) and 2.315(2), Pt-N 2.067(6) and Pt-S 2.220(2) Å (Fig. 1). The crystal is stabilised by intermolecular hydrogen bonds.

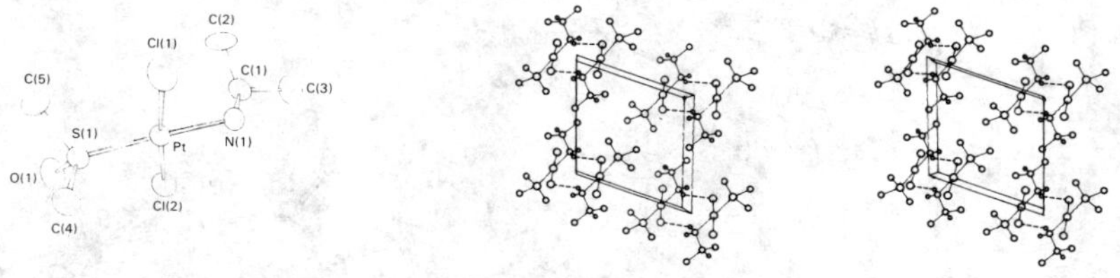

Fig. 1. $C_5H_{15}Cl_2NOPtS$: a perspective view of the molecule and a stereoview of the packing of the molecules in the crystal.

catena-BIS(μ_2-AZIDO)[BIS(μ_2-AZIDO)BIS(3-PICOLINE)DICOPPER(II)]
$(C_6H_7CuN_7)_n$ $Cu(N_3)_2(C_5H_4CH_3N)$

M.A.S. GOHER and T.C.W. MAK, 1984. Inorg. Chim. Acta, 89, 119-123.

Monoclinic, C2/c, a = 23.120, b = 13.751, c = 6.663 Å, β = 100.43°, Z = 8. Mo radiation, R = 0.063 for 979 reflexions.

The coordination geometry about the Cu atom is distorted trigonal bipyramidal, with the 3-picoline ligand occupying one of the axial positions (Fig. 1). Pertinent distances and angles are: Cu-N(1) 2.020, Cu-N(7) 1.993, Cu-N(7)^a 2.436, Cu-N(4) 1.960, Cu-N(10) 1.999 Å, N(1)-Cu-N(10) 170.5° and N(7)-N(8)-N(9) 177.2°. Two azido ligands (N(1) to N(3) and N(4) to N(6)) lie on the same crystallographic diad and bridge a pair of Cu atoms to form a planar Cu_2N_2 ring. The third azido ligand (N(7)-N(9)) asymmetrically-bridges Cu atoms related by the c glide, giving rise to a composite polymeric structure (hence the weak Cu-N(7)^a bond of 2.436 Å).

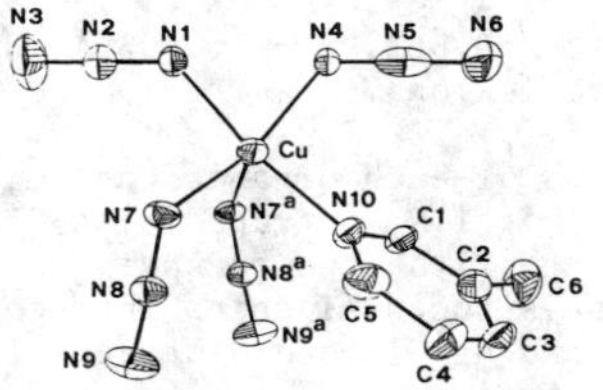

Fig. 1. Perspective view of the coordination geometry about the Cu(II) atom in $C_6H_7CuN_7$. The superscript a denotes symmetry transformation x, -y, 0.5+z.

(ADENINATO)METHYLMERCURY(II)
$C_6H_7HgN_5$ $CH_3HgC_5H_4N_5$

J.-P. CHARLAND and A.L. BEAUCHAMP, 1984. Croat. Chem. Acta, 57, 679-687.

Orthorhombic, Pbcn, a = 14.658, b = 8.407, c = 13.006 Å, D_m = 2.87, Z = 8. Cu radiation, R = 0.026 for 1132 reflexions.

The CH_3Hg group is bonded to the N(9) position of deprotonated adenine (Fig. 1), and these molecules are linked to form infinite ribbons by two pairs of complementary N-H...N bonds involving both amino protons and the lone pairs on N(1) and N(7) of each molecule. The Hg(9)-N(9) and Hg(9)-C(9) bonds are roughly equal (2.070 and 2.076 Å) and the C(9)-Hg(9)-N(9) bond is significantly non-linear (176.5°). The geometry of the adenine moieties is as expected, but high esds mask any fine details. The N...N interactions are 2.969 Å for N(1)...N(6') and 3.029 Å for N(6)...N(7').

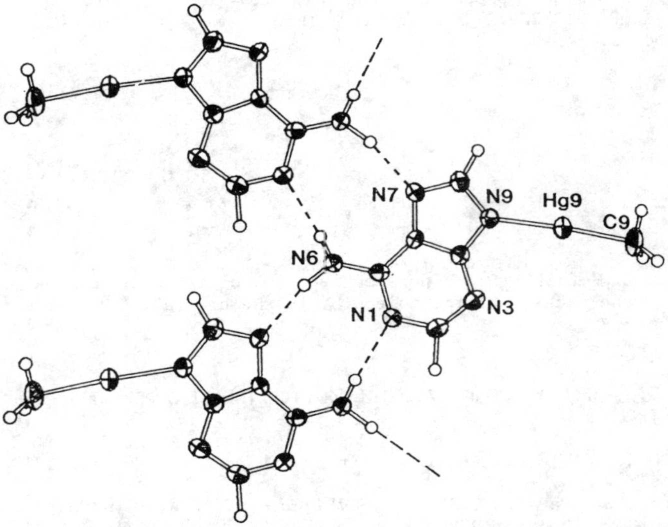

Fig. 1. Portion of the ribbon of hydrogen-bonded molecules of $C_6H_7HgN_5$.

HEXAMETHYLENETETRAMINE : SILVER(I) CHLORIDE (1:4)

$C_6H_{12}Ag_4Cl_4N_4$ $(CH_2)_6N_4.AgCl$

T.C.W. MAK, 1984. Inorg. Chim. Acta, 84, 19-23.

Orthorhombic, Pnam, a = 8.594, b = 9.567, c = 16.403 Å, D_m = 3.40, Z = 4. Mo radiation, R = 0.041 for 1824 reflexions.

All four nitrogen lone pairs of each organic cage molecule serve as ligand sites, giving rise to octahedral $AgCl_5N$, tetrahedral $AgCl_3N$, and tetragonal pyramidal $AgCl_4N$ coordination polyhedra, the first two occupying sites of crystallographic mirror symmetry (Fig. 1). These structural units interconnect to form a three-dimensional network. The Ag-N and Ag-Cl bond distances lie in the ranges 2.349 to 2.411 and 2.545-3.191 Å, respectively.

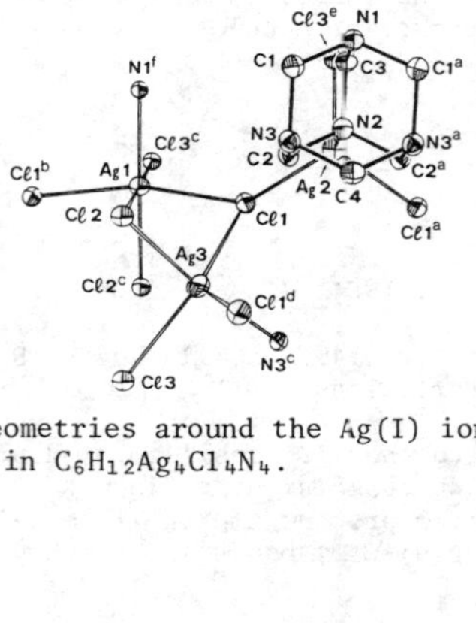

Fig. 1. Coordination geometries around the Ag(I) ions viewed approximately parallel to a, in $C_6H_{12}Ag_4Cl_4N_4$.

AQUO(ACETAMIDOACETONETHIOSEMICARBAZONATO)COPPER(II) DIMETHYLFORMAMIDE

$C_6H_{12}CuN_4O_2,C_3H_7NO$ $Cu(H_2O)C_6H_{10}N_4O,(CH_3)_2NCHO$

V.A. NEVEROV, V.N. BIYUSHKIN, L.A. NEZHEL'SKAYA and N.I. BELICHUK, 1984. Kristallografiya, 29, 731-735 [Sov. Phys. Crystallogr., 29, 432-434].

Monoclinic, $P2_1/b$, a = 7.140, b = 8.933, c = 24.214 Å, γ = 109.1°, Z = 4. Mo radiation, R = 0.051 for 2332 reflexions.

The structure consists of neutral complexes and dimethylformamide molecules (Fig. 1). The Cu atom has distorted square coordination with a water molecule (Cu-O(2) 1.952 Å) and a doubly-deprotonized molecule of acetamidoacetone thiosemicarbazone, which is bound to the metal via the S atom (Cu-S 2.256 Å) and the N atoms of the hydrazine (Cu-N(2) 1.936 Å) and amide (Cu-N(3) 1.963 Å). Other bond lengths of interest are: S-C(1) 1.748, C(1)-N(4) 1.364, C(1)-N(1) 1.314, N(1)-N(2) 1.392, N(2)-C(2) 1.286, C(2)-C(3) 1.494, C(2)-C(4) 1.491, C(4)-N(3) 1.461, N(3)-C(5) 1.328, C(5)-O(1) 1.280 and C(5)-C(6) 1.510 Å. The coordination angles around Cu range from 83.0 to 96.2°. There is one intramolecular hydrogen bond (O(1)...O(2) 2.59 Å) and three intermolecular bonds (O(2)...O(1') 2.67, N(4)...O(3) 2.98, and N(4)...O(3'') 2.91 Å).

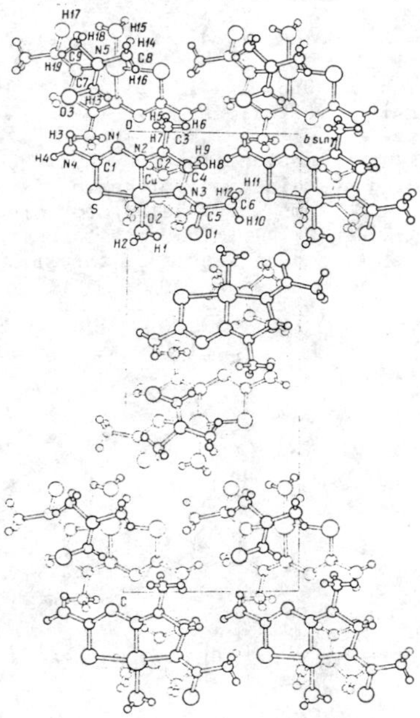

Fig. 1. Projection of the structure of $C_6H_{12}CuN_4O_2,C_3H_7NO$ along [100].

AQUA(4-AZA-1-AZONIABICYCLO[2.2.2]OCTANE)TRINITRATOCOBALT(II)

$C_6H_{15}CoN_5O_{10}$ $Co(C_6H_{13}N_2)(NO_3)_3(H_2O)$

B. VIOSSAT, S. BÉNAZETH, P. KHODADAD and N. RODIER, 1984. Acta Cryst., C40, 1340-1342.

Monoclinic, $P2_1$, a = 6.961, b = 6.982, c = 13.68 Å, γ = 95.4°, Dm = 1.85, Z = 2. Mo radiation, R = 0.039 for 1958 reflexions.

Layers of molecules parallel to (001) are linked by O-H...O, N-H...O and C-H...O hydrogen bonds. Co is trigonal bipyramidally coordinated (Fig. 1) with Co-O 2.060 to 2.111(6), Co-N 2.207(5) Å. The Co atom lies 0.033 Å from the O(2),O(5),O(8) plane.

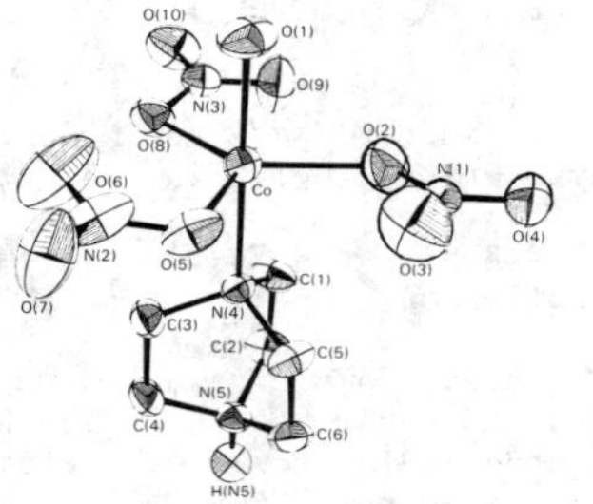

Fig. 1. The $Co(C_6H_{13}N_2)(NO_3)_3(H_2O)$ molecule.

AQUA(4-AZA-1-AZONIABICYCLO[2.2.2]OCTANE)TRINITRATOCADMIUM(II)
$C_6H_{15}CdN_5O_{10}$

B. VIOSSAT, P. KHODADAD and N. RODIER, 1984. Acta Cryst., C40, 24-28.

Monoclinic, $P2_1$, a = 7.032, b = 7.068, c = 13.791 Å, γ = 94.01°, D_m = 2.02, Z = 2. Mo radiation, R = 0.025 for 1905 reflexions.

The structure consists of molecules linked by hydrogen bonds and van der Waals interactions (Fig. 1). The Cd coordination polyhedron is a bipyramid with a hexagonal base made up of six O atoms which belong to three unsymmetrically bidentate nitro groups (Cd-O 2.315 to 2.816(8) Å). The apical positions are occupied by a water molecule (Cd-O 2.262(5) Å) and an N atom of the ligand (Cd-N 2.340(4) Å).

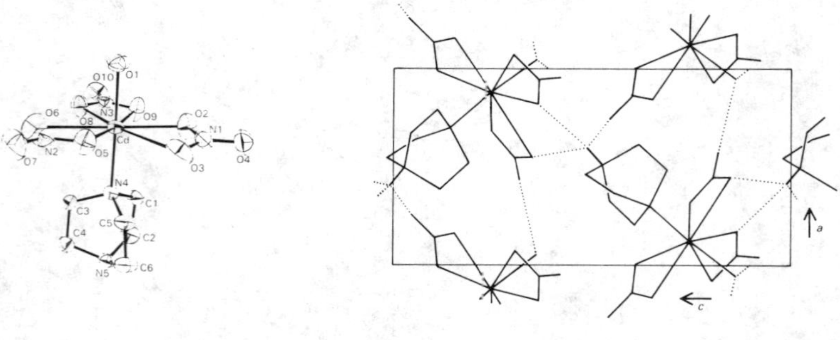

Fig. 1. The Cd coordination in $C_6H_{15}CdN_5O_{10}$ and view of the crystal structure.

TRIMETHYLAMMONIUM TRICHLOROTRIMETHYLAMINEZINC
$C_6H_{19}Cl_3N_2Zn$ (I) $[(CH_3)_3NH]^+[ZnCl_3(N(CH_3)_3]^-$

TRICHLORO-BIS(TRIMETHYLAMINE)IRON
$C_6H_{18}Cl_3FeN_2$ (II) $FeCl_3(N(CH_3)_3)_2$

K.R. MILLINGTON, S.R. WADE, G.R. WILLEY and M.G.B. DREW, 1984. Inorg. Chim. Acta, 89, 185-191.

I. Orthorhombic, Pman, a = 9.785, b = 12.438, c = 11.497 Å, D_m = 1.41, Z = 4. Mo radiation, R = 0.085 for 754 reflexions.

II. Orthorhombic, Pnma, a = 9.753, b = 10.150, c = 13.156 Å, D_m = 1.41, Z = 4. Mo radiation, R = 0.052 for 1257 reflexions.

Compound I has an ionic structure consisting of disordered cations and ordered anions both with crystallographic m symmetry (Fig. 1). The anion is tetrahedral about the Zn and N atoms with Zn-Cl bond lengths of 2.232 and 2.266 Å. The Zn-N(1) distance is 2.071 Å and the N(1)-C(12) and N(1)-C(13) distances are respectively 1.423 and 1.495 Å. Compound II consists of monomeric discrete units with crystallographic m symmetry again. This time the metal atom has trigonal bipyramidal coordination with the N atoms at the apical sites (Fe-N(1) 2.275, Fe-N(2) 2.270 Å). The Fe-Cl distances are 2.228 and 2.207 Å and the N-C distances range from 1.457 to 1.505 Å. There is little distortion from regular D_{3h} symmetry.

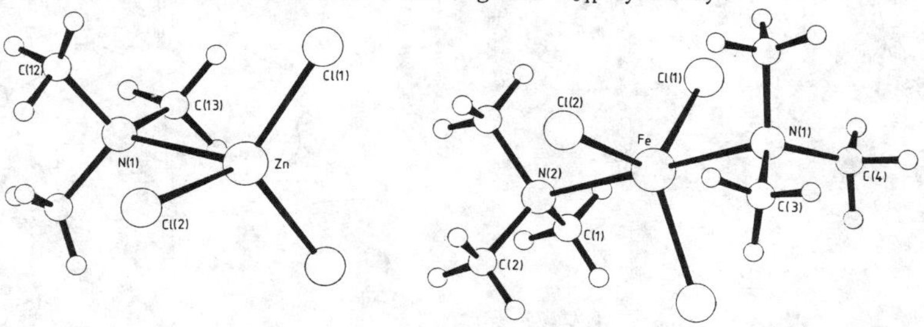

Fig. 1. Structure of the anion in $C_6H_{19}Cl_3N_2Zn$ (left), and of the molecule in $C_6H_{18}Cl_3FeN_2$ (right).

(μ_3-THIOCYANATO)-2-METHYLPYRIDINESILVER(I)
$(C_7H_7AgN_2S)_n$ (I)

(μ_3-THIOCYANATO)-3-METHYLPYRIDINESILVER(I)
$(C_7H_7AgN_2S)_n$ (II)

(μ_3-THIOCYANATO)-2,6-DIMETHYLPYRIDINESILVER(I)
$(C_8H_9AgN_2S)_n$ (III)

(μ_2-THIOCYANATO)BIS(QUINOLINE)SILVER(I)
$(C_{19}H_{14}AgN_3S)_n$ (IV)

N.K. MILLS and A.H. WHITE, 1984. J. Chem. Soc. Dalton, 229-237.

I. Orthorhombic, $Pn2_1a$, a = 18.305, b = 11.209, c = 4.180 Å, Z = 4. Mo radiation, R = 0.061 for 1139 reflexions.

II. Orthorhombic, Pbca, a = 18.055, b = 11.385, c = 8.404 Å, Z = 8. Mo radiation, R = 0.046 for 1478 reflexions.

III. Monoclinic, $P2_1/c$, a = 8.481, b = 8.102, c = 13.676 Å, β = 97.41°, Z = 4. Mo radiation, R = 0.055 for 956 reflexions.

IV. Orthorhombic, $P2_12_12_1$, a = 15.977, b = 13.544, c = 8.009 Å, Z = 4. Mo radiation, R = 0.031 for 2399 reflexions.

The structure of I (Fig. 1) which shows slight disorder due to a stacking fault, is an infinite two-dimensional polymer parallel to bc with Ag atoms pseudotetrahedrally four-coordinated by the S and N atoms of bridging thiocyanate groups (Ag-S 2.552(3), 2.637(3), Ag-N 2.27(1)) and by 2-methylpyridine (Ag-N 2.34(1) Å). Likewise II also forms an infinite two-dimensional polymer parallel to bc. The Ag-S and Ag-N (thiocyanate) distances are 2.578(2), 2.615(2) and 2.293(5) Å and the Ag-N (3-methylpyridine) 2.331(5) Å. III, however, forms a one-dimensional polymer generated by the 2_1 screw axis along b. The Ag atom is three-coordinated by S and N atoms of bridging thiocyanato groups (Ag-S 2.474(3), Ag-N 2.28(1)) and by 2,6-dimethylpyridine (Ag-N 2.275(8) Å). IV also forms a one-dimensional polymer, this time generated by the 2_1 screw axis along c. Four-coordinate pseudotetrahedral Ag atoms are linked by S and N bridging thiocyanato groups (Ag-S 2.495(1), Ag-N 2.332(5)) and quinoline N atoms (Ag-N 2.364(3), 2.361(4) Å).

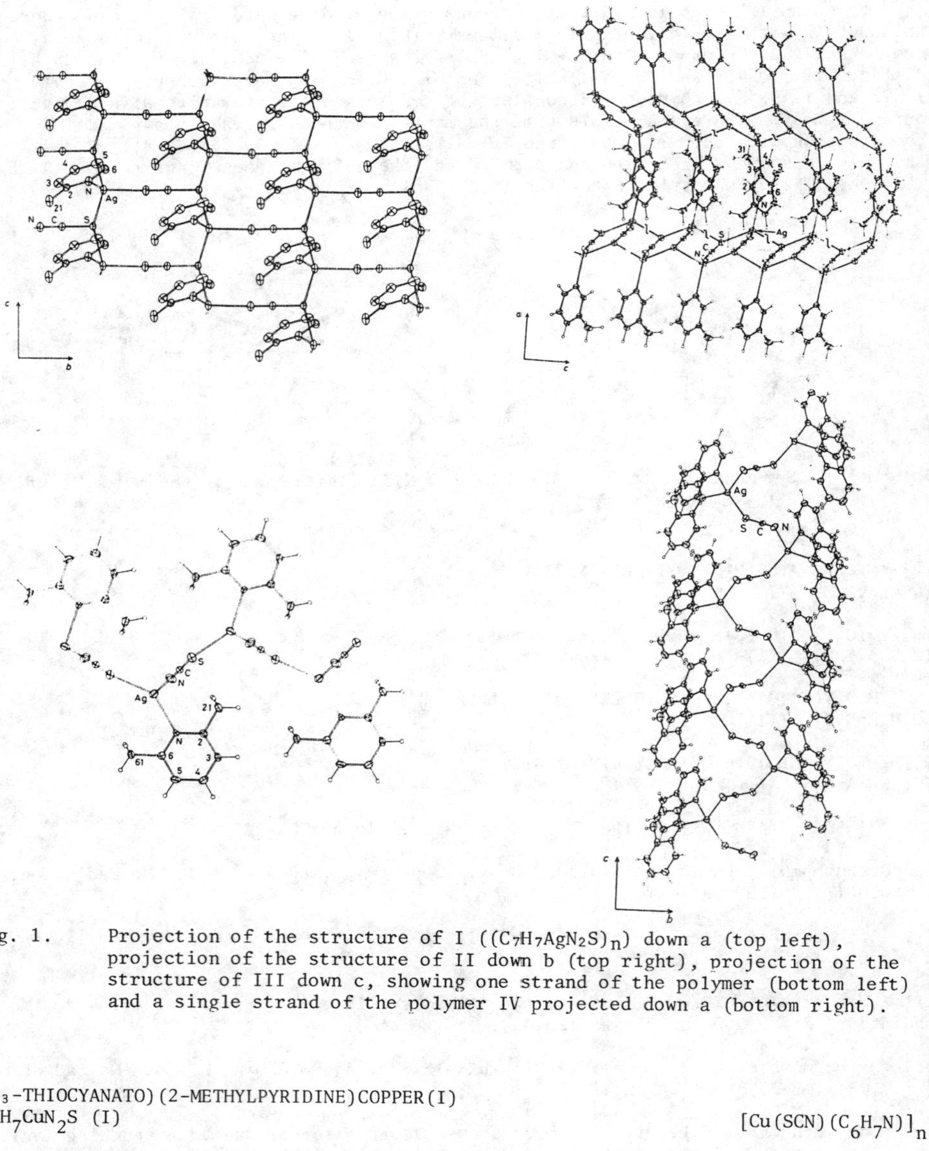

Fig. 1. Projection of the structure of I ($(C_7H_7AgN_2S)_n$) down a (top left), projection of the structure of II down b (top right), projection of the structure of III down c, showing one strand of the polymer (bottom left) and a single strand of the polymer IV projected down a (bottom right).

(μ_3-THIOCYANATO)(2-METHYLPYRIDINE)COPPER(I)
$C_7H_7CuN_2S$ (I) $[Cu(SCN)(C_6H_7N)]_n$

(μ_2-THIOCYANATO)(2,6-DIMETHYLPYRIDINE)COPPER(I)
$C_8H_9CuN_2S$ (II) $[Cu(SCN)(C_7H_9N)]_n$

(μ_2-THIOCYANATO)BIS(2-METHYLPYRIDINE)COPPER(I)
$C_{13}H_{14}CuN_3S$ (III) $[Cu(SCN)(C_6H_7N)_2]_n$

(μ_2-THIOCYANATO)BIS(3-METHYLPYRIDINE)COPPER(I)
$C_{13}H_{14}CuN_3S$ (IV) $[Cu(SCN)(C_6H_7N)_2]_n$

(μ_2-THIOCYANATO)BIS(4-METHYLPYRIDINE)COPPER(I)
$C_{13}H_{14}CuN_3S$ (V) $[Cu(SCN)(C_6H_7N)_2]_n$

(μ_2-THIOCYANATO)BIS(2,4-DIMETHYLPYRIDINE)COPPER(I)
$C_{15}H_{18}CuN_3S$ (VI) $[Cu(SCN)(C_7H_9N)_2]_n$

(μ_2-THIOCYANATO)BIS(QUINOLINE)COPPER(I)
$C_{19}H_{14}CuN_3S$ (VII) $[Cu(SCN)(C_9H_7N)_2]_n$

P.C. HEALY, C. PAKAWATCHAI, R.I. PAPASERGIO, V.A. PATRICK and A.H. WHITE, 1984.
Inorg. Chem., 23, 3769-3776.

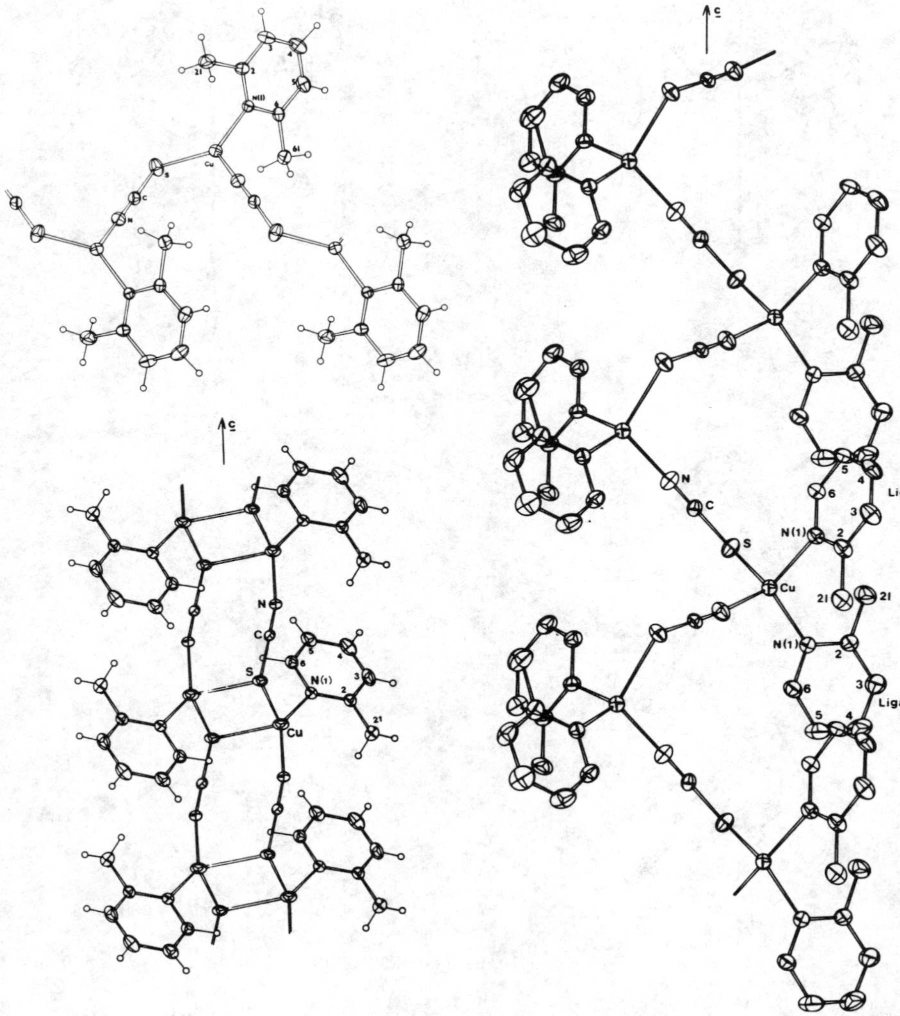

Fig. 1. The structures of $[Cu(SCN)(C_6H_7N)]_n$ (lower left), $[Cu(SCN)(C_7H_9N)]_n$ (upper left) and $[Cu(SCN)(C_6H_7N)_2]_n$ (III, right).

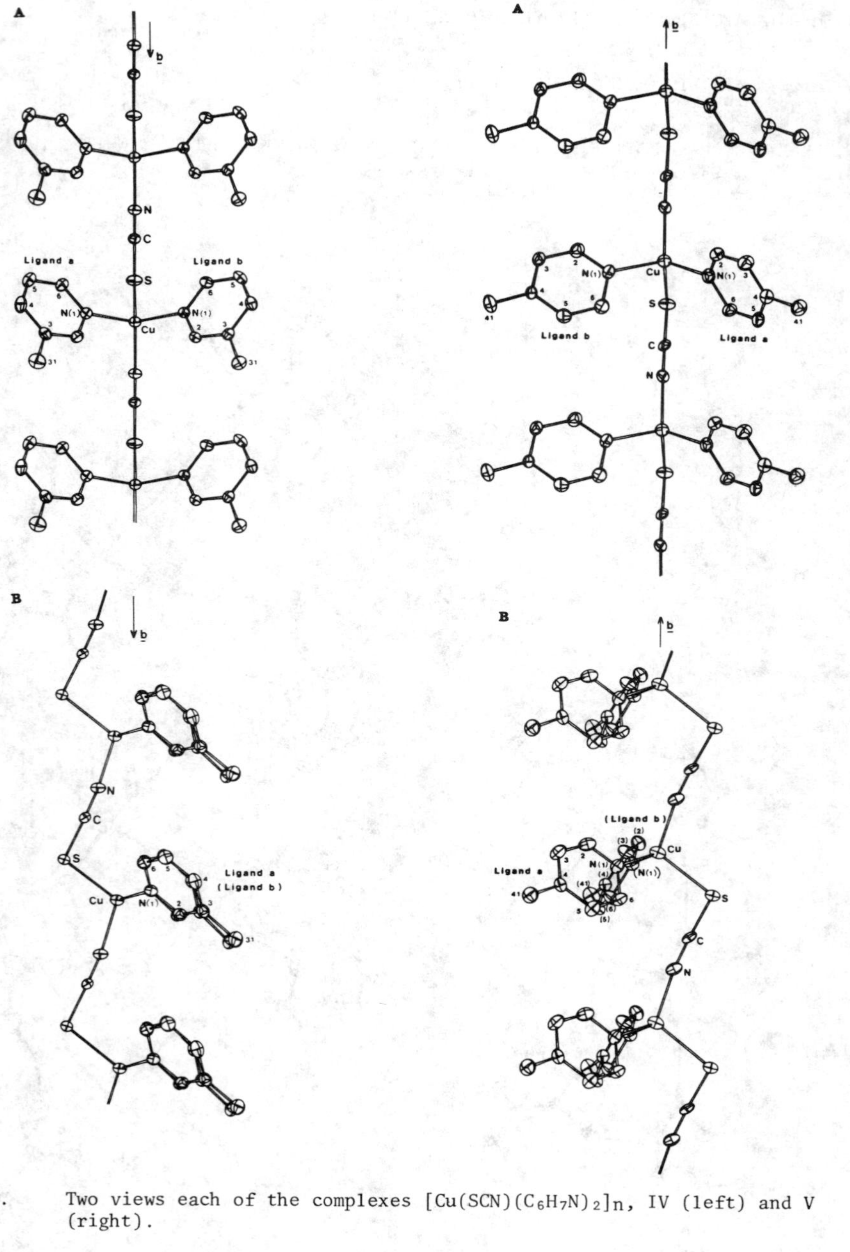

Fig. 2. Two views each of the complexes $[Cu(SCN)(C_6H_7N)_2]_n$, IV (left) and V (right).

I. Triclinic, $P\bar{1}$, a = 9.452, b = 8.077, c = 5.779 Å, α = 104.33, β = 96.64, γ = 100.83°, Z = 2. Mo radiation, R = 0.026 for 1164 reflexions.

II. Monoclinic, $P2_1/c$, a = 7.721, b = 7.753, c = 16.113 Å, β = 90.67°, Z = 4. Mo radiation, R = 0.039 for 1339 reflexions.

III. Orthorhombic, Fdd2, a = 43.68, b = 16.402, c = 8.064 Å, Z = 16. Mo radiation, R = 0.048 for 1499 reflexions (preferred polarity).

IV. Monoclinic, $P2_1/n$, a = 17.68, b = 5.874, c = 14.367 Å, β = 112.91°, Z = 4. Mo radiation, R = 0.032 for 1731 reflexions.

V. Monoclinic, $P2_1$, a = 14.59, b = 5.823, c = 8.953 Å, β = 106.44°, Z = 2. Mo radiation, R = 0.054 for 903 reflexions (preferred chirality).

VI. Monoclinic, $P2_1/c$, a = 28.93, b = 5.867, c = 20.19 Å, β = 112.70°, Z = 8. Mo radiation, R = 0.058 for 1748 reflexions.

VII. Orthorhombic, $P2_12_12_1$, a = 16.219, b = 12.838, c = 8.032 Å, Z = 4. Mo radiation, R = 0.029 for 468 reflexions (preferred chirality).

The structures of I-III are shown in Fig. 1 and those of IV and V in Fig. 2. The structure of I is a polymer consisting of pairs of CuL(SCN)CuL(SCN) strands linked by centrosymmetric Cu_2S_2 and $Cu(SCN)_2Cu$ rings. The Cu atoms are four-coordinate, Cu-S 2.263(1) and 2.806(1), Cu-NCS 1.918(2), Cu-N(pyr) 2.000(3) Å with angles at Cu 95.16(7)-136.3(1)°. The structure of II is a single-stranded polymer-CuL(SCN)CuL(SCN) in which the Cu atoms have trigonal planar coordination, Cu-S 2.232(1), Cu-NCS 1.915(4), Cu-N(pyr) 1.993(3) Å. Complexes III-VII all contain single-stranded $CuL_2(SCN)CuL_2(SCN)CuL_2$ polymers, with different conformational types. The Cu atoms in III-VII all have irregular tetrahedral coordination with Cu-S 2.290(4)-2.380(4), Cu-NCS 1.944(9)-2.00(1), and Cu-N 2.06(1)-2.121(8) Å.

(2,2',2''-(S-METHYL)TRITHIOTRIETHYLAMINE-N,S,S',S'')DIOXOMOLYBDENUM(VI) (FORM 1)
$C_7H_{15}MoNO_2S_3$ (I) $MoO_2((SCH_2CH_2)_2NCH_2CH_2SCH_3)$

(2,2',2''-(S-METHYL)TRITHIOTRIETHYLAMINE-N,S,S',S'')DIOXOMOLYBDENUM(VI) (FORM 2)
$C_7H_{15}MoNO_2S_3$ (II) $MoO_2((SCH_2CH_2)_2NCH_2CH_2SCH_3)$

(BIS(ETHYL-2-THIOLATO)(2-DIMETHYLAMINOETHYL)AMINE)DIOXOMOLYBDENUM
$C_8H_{18}MoN_2O_2S_2$ (III) $MoO_2((SCH_2CH_2)_2NCH_2CH_2N(CH_3)_2)$

J.M. BERG, K.O. HODGSON, A.E. BRUCE, J.L. CORBIN, N. PARIYADATH and E.I. STIEFEL, 1984. Inorg. Chim. Acta, 90, 25-33.

Fig. 1. View of $C_7H_{15}MoNO_2S$ (I) (left), and view of $C_8H_{18}MoN_2O_2S_2$ (III) (right).

I. Monoclinic, $P2_1/n$, a = 7.235, b = 7.717, c = 24.527 Å, β = 119.86°, Z = 4. Mo radiation, R = 0.039 for 2260 reflexions.

II. Monoclinic, $P2_1/c$, a = 14.945, b = 11.925, c = 14.878 Å, β = 114.51°, Z = 8. Mo radiation, R = 0.053 for 3615 reflexions.

III. Orthorhombic, $P2_12_12_1$, a = 7.188, b = 22.708, c = 7.746 Å, Z = 4. Mo radiation, R = 0.023 for 2186 reflexions.

The molecules of I and II are very similar (Fig. 1), having an octahedral structure with cis oxo groups, trans thiolates (cis to both oxo groups) and N and thioether S atoms trans to oxo groups. Average distances are: Mo-O 1.70, Mo-S (thiolate) 2.40, Mo-N 2.40 and M.-S(thioether) 2.79 Å. In III the coordination is also octahedral, with cis oxo groups, trans thiolates and N atoms trans to oxo (Fig. 1). Important distances are: Mo-O 1.705, 1.699, Mo-S 2.420, 2.409, and Mo-N 2.372, 2.510 Å.

BIS(DIIMINOSUCCINONITRILO)NICKEL(II)
$C_8H_4N_8Ni$ $Ni(C_4H_2N_4)_2$

S.-M. PENG, Y. WANG and C.-K. CHIANG, 1984. Acta Cryst., C40, 1541-1542.

Monoclinic, $P2_1/n$, a = 3.7046, b = 8.773, c = 15.999 Å, β = 93.34°, D_m = 1.72, Z = 2. Mo radiation, R = 0.037 for 873 reflexions.

The molecule is almost planar (the maximum deviation from the central NiN_4 plane being 0.264 Å for N(6)) with close to D_{2h} symmetry. The molecules pack in parallel along a with Ni...Ni 3.705 Å and an interplanar spacing of 3.357 Å.

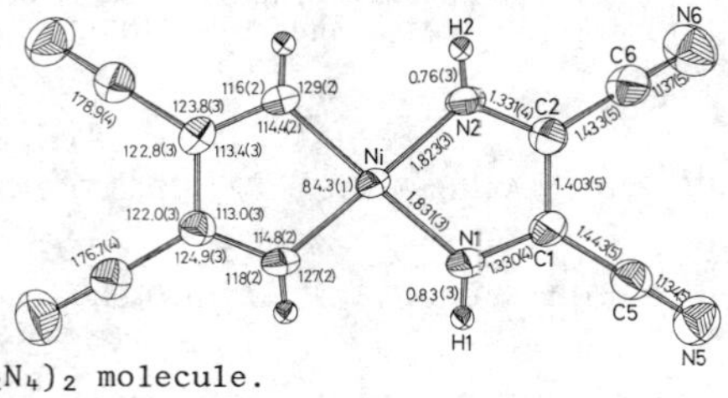

Fig. 1. The $Ni(C_4H_2N_4)_2$ molecule.

catena-POLY{μ-(2,2'-BIPYRIMIDINE-N,N':N'',N''')[(NITRATO-O,O')COPPER(II)]DI(μ-NITRATO-μ-O)[(NITRATO-O,O')COPPER(II)]}
$C_8H_6Cu_2N_8O_{12}$ $[Cu_2(NO_3)_4(C_8H_6N_4)]$

G. DE MUNNO and G. BRUNNO, 1984. Acta Cryst., C40, 2030-2032.

Monoclinic, C2/c, a = 19.314, b = 7.914, c = 15.060 Å, β = 137.4°, D_m = 2.26, Z = 8. Mo radiation, R = 0.039 for 1248 reflexions.

The compound (Fig. 1) is a polymer built up by complex dimeric units formed by two Cu atoms coordinated by a molecule of the organic ligand and four nitrate groups. Both Cu atoms are six-coordinated with a distorted octahedral geometry, being bonded by two N atoms of the organic ligand, acting as a doubly-bidentate, by two O atoms of a bidentate nitrate group and by two O atoms of two monodentate nitrate groups bridging two Cu atoms. Main bond lengths are: Cu-N1 2.011(4), Cu-N2 2.008(3), Cu-O1 1.952(4), Cu-O3 2.547(2), Cu-O4 1.965(3), Cu-O4' 2.300(2) Å.

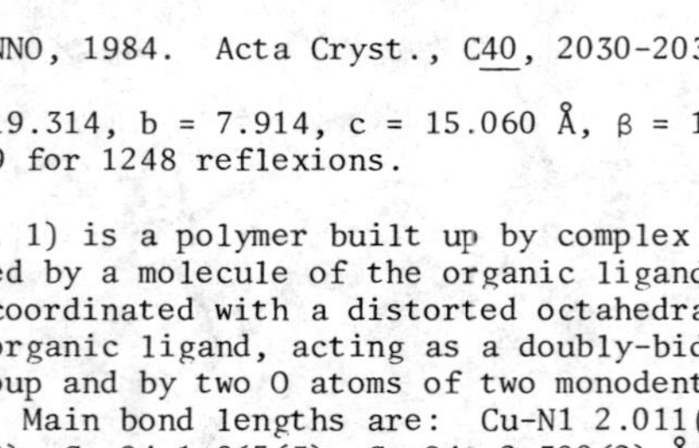

Fig. 1. A view of part of the $[Cu_2(NO_3)_4(C_8H_6N_4)]$ polymer.

SODIUM BIS(N,N'-SUCCINIMIDO)ARGENTATE(I) PENTAHYDRATE

$C_8H_8AgN_2NaO_4,5H_2O$ $Na[Ag(C_4H_4NO_2)_2],5H_2O$

W. KHAYATA, D. BAYLOCQ, F. PELLERIN and N. RODIER, 1984. Acta Cryst., C40, 765-767.

Orthorhombic, $Cmc2_1$, a = 6.856, b = 21.74, c = 20.589 Å, D_m = 1.80, Z = 8. Mo radiation, R = 0.030 for 1704 reflexions.

The crystal structure consists of $[Ag(C_4H_4NO_2)_2]^-$ and Na^+ ions and water molecules and is mainly made up of plane sheets at x = 0 and x = 0.5 (Fig. 1). The Ag atom is twofold coordinated and the Ag-N bonds, which lie between 2.06(1) and 2.08(1) Å, seem strongly covalent. Interatomic distances indicate there is an unlocated orbital in which the O-C-N-C-O chain is involved. The coordination polyhedron of the Na^+ ions is either a slightly distorted octahedron or a bipyramid with a trigonal base constituted of O atoms. Na...O distances are in the range 2.236(9) to 2.457(6) Å.

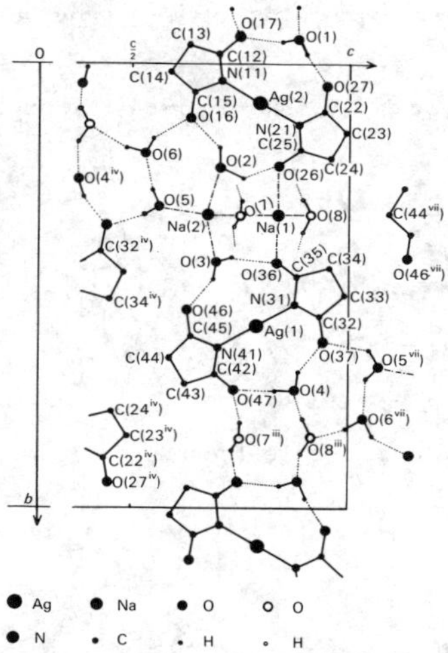

Fig. 1. $C_8H_8AgN_2NaO_4,5H_2O$: perspective view of the crystal structure.

TETRAACETONITRILESILVER(I) PERCHLORATE

$C_8H_{12}AgClN_4O_4$ $[Ag(CH_3CN)_4]^+ClO_4^-$

K. NILSSON and A. OSKARSSON, 1984. Acta Chem. Scand., A38, 79-85.

Orthorhombic, $Pn2_1a$, a = 24.500, b = 20.756, c = 8.567 Å, Z = 12. Mo radiation, R = 0.048 for 1123 reflexions (at 240K).

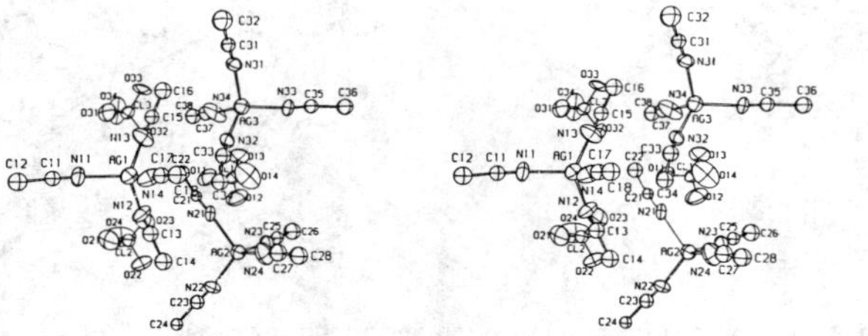

Fig. 1. Stereoscopic view of the asymmetric unit of $C_8H_{12}AgClN_4O_4$.

The structure is composed of discrete anions and cations and the asymmetric unit (Fig. 1) comprises three Ag^+ complexes and three ClO_4^- ions which alternate along the c axis. The cation and the anion have slightly distorted tetrahedral symmetry with the N-Ag-N angles and O-Cl-O angles ranging from 103.2 to 116.1° and 100 to 113° respectively. The Ag-N distances range from 2.18 to 2.33 Å, and the Cl-O distances in the perchlorate anions range from 1.34 to 1.49 Å over the three independent ionic pairs.

TETRACHLORO(2,2'-BI-2-THIAZINE)ZIRCONIUM(IV)
$C_8H_{12}Cl_4N_2S_2Zr$ (I) $ZrCl_4(C_8H_{12}N_2S_2)$

TETRACHLOROBIS(2,2'-BI-2-THIAZOLINE)ZIRCONIUM(IV) ACETONITRILE SOLVATE
$C_{12}H_{16}Cl_4N_4S_4Zr,2C_2H_3N$ (II) $ZrCl_4(C_6H_8N_2S_2)_2.2CH_3CN$

BIS(2,2'-BI-2-THIAZOLINE)TETRAKIS(ISOTHIOCYANATO)ZIRCONIUM(IV)
$C_{16}H_{16}N_8S_8Zr$ (III) $Zr(NCS)_4(C_6H_8N_2S_2)_2$

R.A. JOHNSON, R.B. VON DREELE and T.M. BROWN, 1984. Inorg. Chem., 23, 4302-4309.

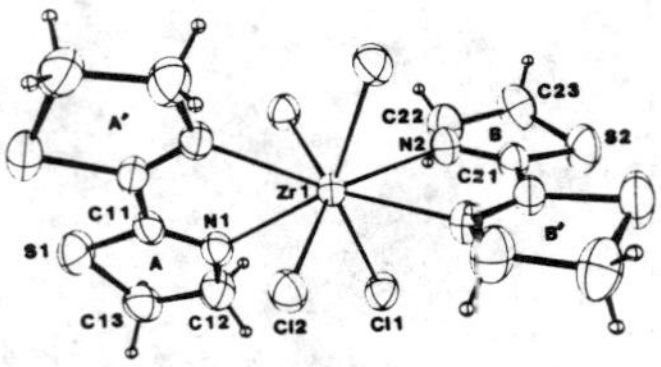

Fig. 1. The $ZrCl_4(C_6H_8N_2S_2)_2$ molecule (compound II, crystallographic C_2 symmetry).

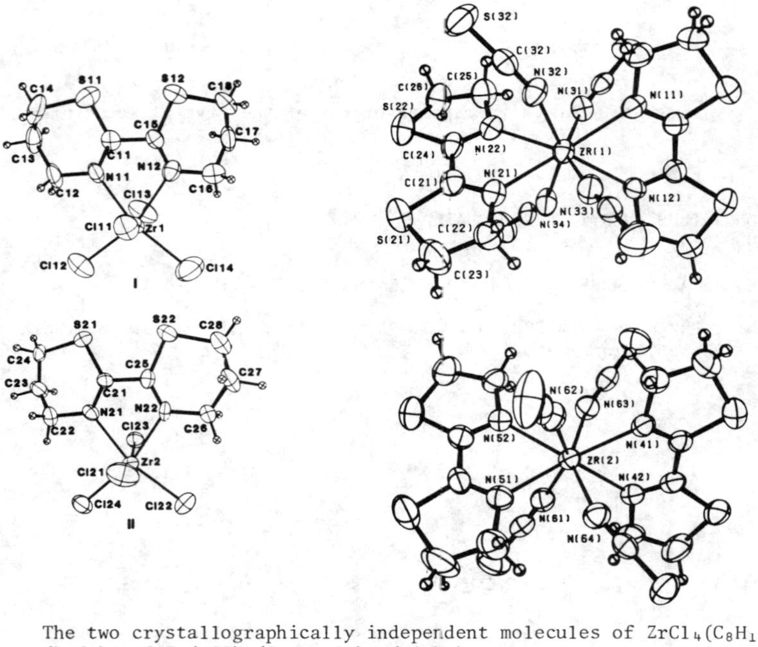

Fig. 2. The two crystallographically independent molecules of $ZrCl_4(C_8H_{12}N_2S_2)$ (left) and $Zr(NCS)_4(C_6H_8N_2S_2)_2$ (right).

I. Monoclinic, $P2_1/n$, a = 18.478, b = 10.930, c = 16.656 Å, β = 109.76°, D_m = 1.80, Z = 8. Mo radiation, R = 0.069 for 3072 reflexions.

II. Monoclinic, C2/c, a = 14.199, b = 14.933, c = 13.825 Å, β = 112.34°, D_m = 1.62, Z = 4. Mo radiation, R = 0.053 for 3534 reflexions.

III. Monoclinic, $P2_1/c$, a = 15.161, b = 21.627, c = 16.350 Å, β = 96.99°, D_m = 1.68, Z = 8. Mo radiation, R = 0.053 for 6456 reflexions.

Comparison of the three structures (Figs. 1 and 2) reveals a great deal about the factors that influence the shape of the coordination polyhedron. The two bithiazoline complexes, II and III, have geometries intermediate between an ideal dodecahedron and an ideal square antiprism. The bithiazine ligand is too large to permit the formation of an eight-coordinate $ZrCl_4(L\text{-}L)_2$ complex; thus, the six-coordinate compound I results. The average Zr-N bond lengths follow the crowding effects with 2.300(10) Å for I, 2.463(10) Å for II, and 2.388(15) Å for III. Other bond lengths are Zr-Cl 2.370(3)-2.431(3) Å for I, Zr-Cl 2.497(1) and 2.484(1) Å for II, Zr-NCS 2.156(5)-2.187(5) Å for III.

TETRAKIS(ACETONITRILE)TETRACHLOROURANIUM(IV)

$C_8H_{12}Cl_4N_4U$ $[UCl_4(C_2H_3N)_4]$

F.A. COTTON, D.O. MARLER and W. SCHWOTZER, 1984. Acta Cryst., C40, 1186-1188.

Monoclinic, C2/c, a = 14.709, b = 8.459, c = 13.938 Å, β = 91.79°, Z = 4. Mo radiation, R = 0.040 for 1051 reflexions.

The molecules are discrete monomeric units with eight ligands coordinated in an almost ideal dodecahedral arrangement (Fig. 1). The four nitrogen atoms form an elongated tetrahedron and the four chlorine atoms a flattened tetrahedron. Important bond distances are: U-Cl(1) 2.627(4), U-Cl(2) 2.620(4), U-N(1) 2.601(13) and U-N(2) 2.577(14) Å.

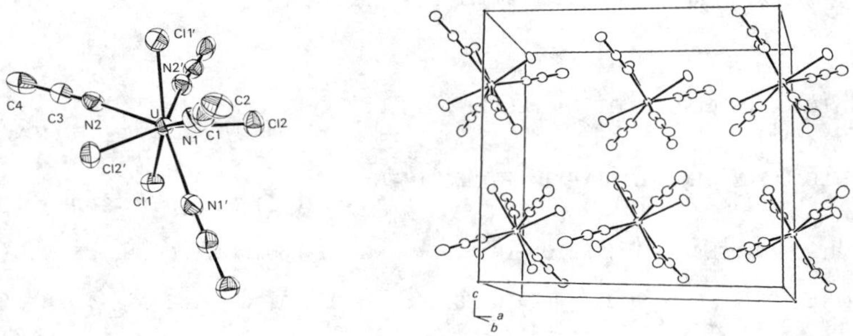

Fig. 1. $UCl_4(C_2H_3N)_4$: perspective view of the molecule and packing diagram of the unit cell.

BIS(3-OXIMATOPIPERAZINE 2-OXIME)NICKEL(II) DIHYDRATE

$C_8H_{14}N_8NiO_4,2H_2O$

X. SOLANS, M. FONT ALTABA, J.L. BRIANSÓ and A. POVEDA, 1984. Afinidad, 41, 440-441.

Orthorhombic, $P2_12_12$, a = 13.207, b = 12.425, c = 4.353 Å, Z = 2. Mo radiation, R = 0.115

The molecule has crystallographic twofold symmetry in which the Ni atom has square-planar coordination (Fig. 1). The five-membered ring (Ni, N(2), C(3), C(8), N(9)) has an envelope conformation with the metal atom deviating from the plane of the remaining atoms. The six-membered ring has a distorted chair conformation and some atoms suffer from positional disorder. The low accuracy in atomic positions results in low accuracy in molecular geometry: Ni-N(2) 1.87, Ni-N(9) 1.85, O(1)-N(2) 1.36, O(10)-N(9) 1.37, N(2)-C(3) 1.29, N(9)-C(8) 1.29, C(3)-C(8) 1.50 Å, N(2)-Ni-N(9) 74.6°.

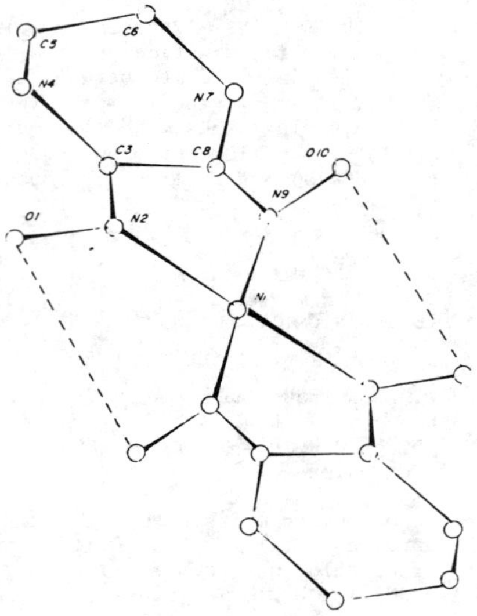

Fig. 1. Molecular structure diagram of $C_8H_{14}N_8NiO_4$.

BIS((DIMETHYLGALLIUM)GLYOXIMATO)NICKELATE
$C_8H_{16}Ga_2N_4NiO_4$ (I) $((CH_3)_2Ga)_2[Ni(ONCHCHNO)_2]$

BIS((DIMETHYLGALLIUM)DIMETHYLGLYOXIMATO)PLATINATE
$C_{12}H_{24}Ga_2N_4O_4Pt$ (II) $((CH_3)_2Ga)_2[Pt(ONC(CH_3)C(CH_3)NO)_2]$

U. KOHLER, H.-D. HAUSEN and J. WEIDLEIN, 1984. J. Organometal. Chem., 272, 337-350.

I. Monoclinic, C2/c, a = 20.251, b = 9.181, c = 8.458 Å, β = 102.2°, Z = 4. Mo radiation, R = 0.065 for 1483 reflexions.

II. Triclinic, $P\bar{1}$, a = 6.185, b = 7.517, c = 10.307 Å, α = 92.9, β = 102.9, γ = 90.1°, Z = 1. Mo radiation, R = 0.044 for 2521 reflexions.

Both compounds have non-planar structures (Fig. 1). (I) has crystallographic twofold symmetry and (II) has inversion symmetry. (I) has an overall 'tub' or 'boat' conformation whilst II has a "chair" conformation in which the Ni...Ga and Pt...Ga distances are 3.563 and 3.509 Å respectively. The Ni-N(1), Ni-N(2), Pt-N(1) and Pt-N(2) distances are 1.906, 1.909, 1.99 and 2.031 Å respectively. The Ga-O distances in I are 1.933 and 1.918 Å, and in II are 1.93 and 1.92 Å. The N-Ni-N and N-Pt-N angles are 82.2 and 98.0° in I, and 77.2 and 102.8° in II, and the O-Ga-O angles in I and II are 93.2 and 97.5° respectively.

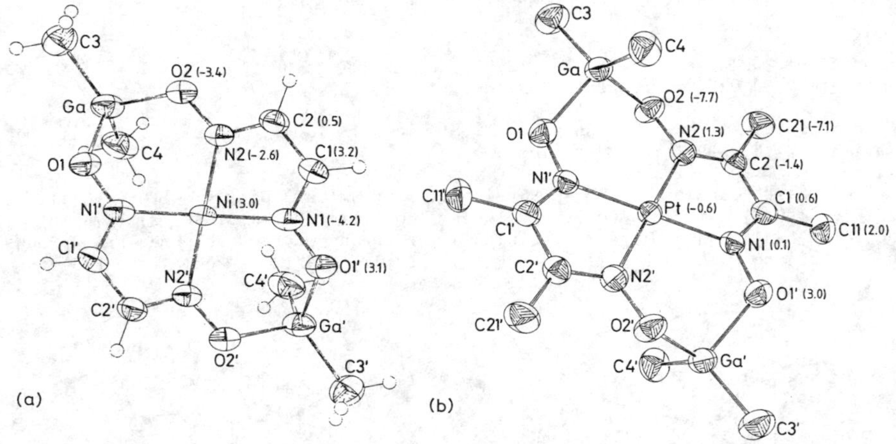

Fig. 1. Molecular structures of $C_8H_{16}Ga_2N_4NiO_4$ (left), and $C_{12}H_{24}Ga_2N_4O_4Pt$ (right). The distance in pm of atoms from the best plane through metal, N(1),N(2),C(1),C(2) is shown in parentheses.

BIS(N^1-ACETIMIDOYLACETAMIDINE-N^1,N^3)NICKEL(II) CHLORIDE TRIHYDRATE

$C_8H_{17.5}Cl_{1.5}N_6Ni,3H_2O$ $C_8H_{17.5}N_6Ni^{1.5+},Cl_{1.5}^-,3H_2O$

R. NORRESTAM, 1984. Acta Cryst., C40, 955-957.

Triclinic, $P\bar{1}$, a = 7.378, b = 8.862, c = 12.813 Å, α = 72.99, β = 79.96, γ = 77.27°, Z = 2. Cu radiation, R = 0.034 for 1385 reflexions (at 117K).

In the complex molecule (Fig. 1) Ni is coordinated by an almost ideal square-planar arrangement of four imino N atoms, with Ni-N 1.849(4), 1.861(4), 1.858(4) and 1.873(4) Å and N-Ni-N angles in the range 89.8(2)-90.3(2)°. The complex molecules are stacked along the a direction, with chlorine atoms and water molecules occupying the space between the molecular stacks.

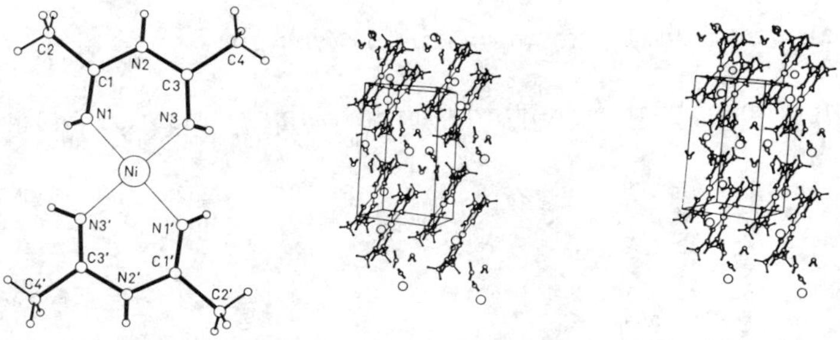

Fig. 1. $C_8H_{17.5}Cl_{1.5}N_6Ni,3H_2O$: molecular structure and packing diagram (c is vertical, a horizontal and b out of the plane of the paper).

AZIDODICARBONYLBIS(TRIMETHYLPHOSPHINE)COBALT

$C_8H_{18}CoN_3O_2P_2$ (I) $Co(N_3)(CO)_2(PMe_3)_2$

AZIDODIMETHYLTRIS(TRIMETHYLPHOSPHINE)COBALT(III)

$C_{11}H_{33}CoN_3P_3$ (II) $CoMe_2(N_3)(PMe_3)_3$

K.W. CHIU, G. WILKINSON, M. THORNTON-PETT and M.B. HURSTHOUSE, 1984. Polyhedron, 3, 79-85.

I. Monoclinic, $P2_1$, a = 6.656, b = 12.038, c = 9.421 Å, β = 103.59°, Z = 2. Mo radiation, R = 0.0256 for 1522 reflexions.

II. Monoclinic, $P2_1/c$, a = 9.056, b = 15.815, c = 13.156 Å, β = 91.0°, Z = 4. Mo radiation, R = 0.0399 for 3524 reflexions.

The cobalt coordination in I is trigonal bipyramidal (Fig. 1) with Co-N(1) 2.029(5), Co-P(1) 2.192(3), Co-P(2) 2.184(3), Co-C(1) 1.749(5), Co-C(2) 1.739(6) Å. Molecule II has octahedral geometry, with a mer arrangement of the three PMe_3 groups. Bond lengths are Co-N(1) 2.071(4), Co-P(1) 2.222(3), Co-P(2) 2.266(3), Co-P(3) 2.232(3), Co-C(1) 2.031(5), Co-C(2) 2.027(5) Å.

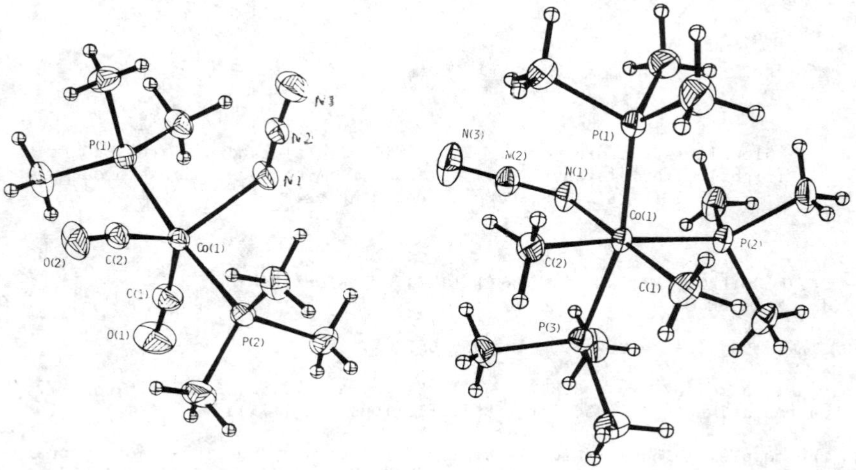

Fig. 1. Views of $Co(N_3)(CO)_2(PMe_3)_2$ (I) (left) and $CoMe_2(N_3)PMe_3)_3$ (II) (right).

BIS(2-METHYL-2-AMINOPROPANE-1-THIOLATO-N,S)DIOXOMOLYBDENUM
$C_8H_{20}MoN_2O_2S_2$ (I) $MoO_2[NH_2C(CH_3)_2CH_2S]_2$

(N,N-BIS(2-METHYLPROPANE-2-THIOLATO)ETHYLENEDIAMINE-N,N',S,S')DIOXOMOLYBDENUM
$C_{10}H_{22}MoN_2O_2S_2$ (II) $MoO_2[SC(CH_3)_2CH_2NHCH_2CH_2NHCH_2C(CH_3)_2S]$

J.M. BERG, D. SPIRA, K. WO, B. McCORD, R. LYE, M.S. CO, J. BELMONT, C. BARNES, K. KOSYDAR, S. RAYBUCK, K.O. HODGSON, A.E. BRUCE, J.L. CORBIN and E.I. STIEFEL, 1984. Inorg. Chim. Acta, 90, 35-39.

I. Orthorhombic, Pbca, a = 11.234, b = 11.822, c = 20.179 Å, Z = 8. Mo radiation, R = 0.038 for 1682 reflexions.

II. Monoclinic, $P2_1/c$, a = 10.798, b = 6.911, c = 20.333 Å, β = 95.20°, Z = 4. Mo radiation, R = 0.032 for 2025 reflexions.

Compound I has octahedral coordination (Fig. 1) with cis oxo groups (Mo-O 1.705, 1.705 Å), trans thiolate donors cis to the oxc groups (Mo-S 2.416, 2.402 Å) and N donors trans to oxo (Mo-N 2.325, 2.385 Å). The structure of II is very similar (Fig. 1), small differences in the geometries of the two compounds being due to the constraints of the extra chelate ring in the complex with the tetradentate ligand. Distances in II are: Mo-O 1.714, 1.710 Å, Mo-S 2.415, 2.404 Å, and Mo-N 2.316, 2.362 Å.

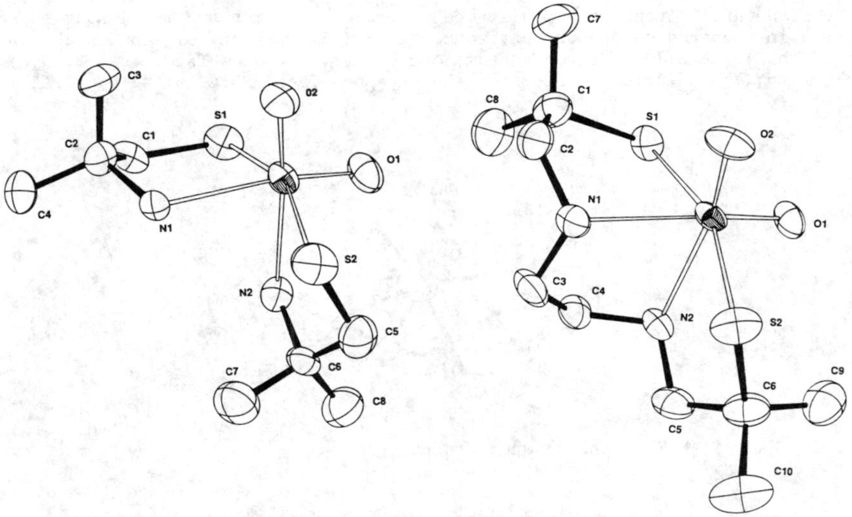

Fig. 1. Molecular structure of $C_8H_{20}MoN_2O_2S_2$ (left), and $C_{10}H_{22}MoN_2O_2S_2$ (right).

trans-DICHLOROBIS(2-AMINO-2-METHYL-1-PROPANOL)PLATINUM
$C_8H_{22}Cl_2N_2O_2Pt$

F.D. ROCHON, P.C. KONG and R. MELANSON, 1984. Can. J. Chem., 62, 2657-2660.

Monoclinic, $P2_1/c$, a = 6.282, b = 20.901, c = 10.769 Å, β = 92.64°, D_m = 2.10, Z = 4. Mo radiation, R = 0.049 for 2372 reflexions.

The Pt atom (Fig. 1) has square-planar coordination, with Pt-Cl 2.306(2), 2.309(3), Pt-N 2.060(8) and 2.071(7) Å. The terminal hydroxyl groups are disordered. Molecules are linked in the crystal by an extensive hydrogen bond network.

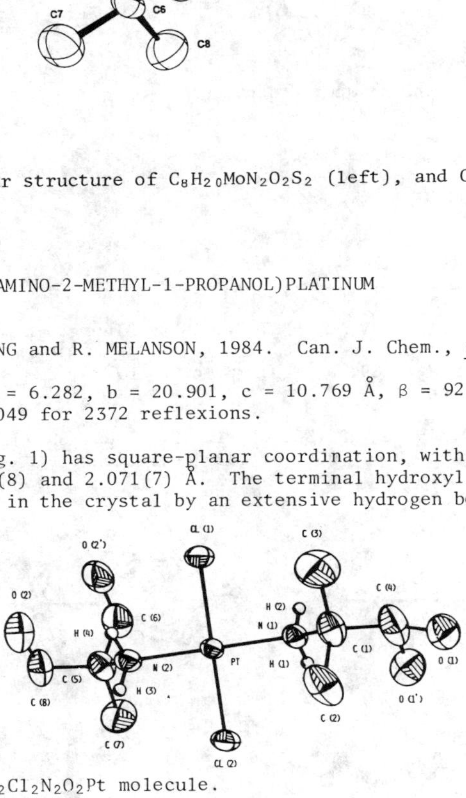

Fig. 1. The $C_8H_{22}Cl_2N_2O_2Pt$ molecule.

TETRACHLOROTRIS(TRIMETHYLAMINE)THORIUM(IV)
$C_9H_{27}Cl_4N_3Th$

M.G.B. DREW and G.R. WILLEY, 1984. J. Chem. Soc. Dalton, 727-729.

Monoclinic, $P2_1/a$, a = 15.336, b = 8.320, c = 15.263 Å, β = 101.0°, D_m = 1.91, Z = 4. Mo radiation, R = 0.085 for 1557 reflexions.

The structure is composed of discrete molecules in which the metal atom is seven coordinate in a capped octahedral environment (Fig. 1) with one Cl atom in the capping position (Th-Cl 2.678(10)), three N atoms occupying the capped face (Th-N 2.79(3), 2.83(3), 2.66(4)) and three Cl atoms in the uncapped face (Th-Cl 2.642(9), 2.637(12), 2.685(11) Å).

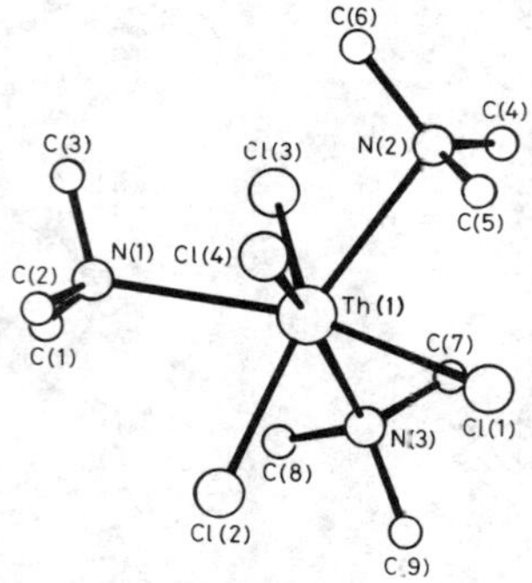

Fig. 1. $C_9H_{27}Cl_4N_3Th$: the molecular structure.

PENTACARBONYLPYRIDINECHROMIUM
$C_{10}H_5CrNO_5$ (I) $Cr(CO)_5NC_5H_5$

TETRACARBONYLBIS(PYRIDINE)CHROMIUM
$C_{14}H_{10}CrN_2O_4$ (II) $Cr(CO)_4(NC_5H_5)_2$

W. RIES, I. BERNAL, M. QUAST and T.A. ALBRIGHT, 1984. Inorg. Chim. Acta, 83, 5-15.

I. Orthorhombic, Pbam, a = 15.289, b = 19.276, c = 7.677 Å, Z = 8. Mo radiation, R = 0.026 for 1085 reflexions.

II. Triclinic, $P\bar{1}$, a = 7.365, b = 8.136, c = 13.491 Å, α = 89.49, β = 88.89, γ = 63.09°, Z = 2. Mo radiation, R = 0.036 for 1520 reflexions.

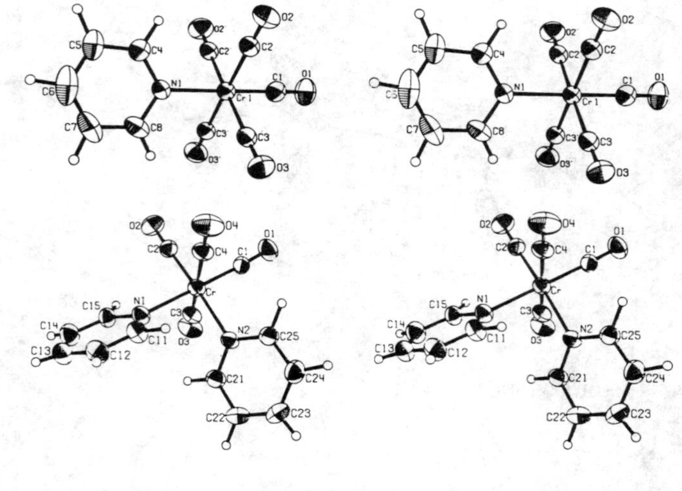

Fig. 1. Stereoscopic view of one independent molecule of $C_{10}H_5CrNO_5$ (top), and stereoscopic view of $C_{14}H_{10}CrN_2O_4$ (bottom).

Crystals of I contain two independent half molecules per asymmetric unit with imposed crystallographic m symmetry for each (Fig. 1). The coordination about Cr is almost ideal octahedral and bond distances and angles in the independent molecules differ by negligible amounts. The average Cr-N distance is 2.157 Å and the average Cr-C distance 1.884 Å. In II (Fig. 1) the coordination about the metal is a slightly distorted octahedron with angles involving the metal ranging from 85.4 to 94.3°. The Cr-N and Cr-C distances average 2.165 and 1.865 Å respectively. In both compounds the pyridine rings are planar.

(2,2'-BIPYRIDINE)DICHLOROZINC(II)
$C_{10}H_8Cl_2N_2Zn$ $Zn(C_{10}H_8N_2)Cl_2$

M.A. KHAN and D.G. TUCK, 1984. Acta Cryst., C40, 60-62.

Triclinic, $P\bar{1}$, a = 8.922, b = 9.054, c = 7.587 Å, α = 93.94, β = 94.61, γ = 111.30°, D_m = 1.70, Z = 2. Mo radiation, R = 0.0274 for 1643 reflexions.

The Zn atom has distorted tetrahedral geometry with Zn-Cl 2.210 and 2.198(1), Zn-N 2.064 and 2.053(2) Å, Cl-Zn-Cl 117.1(1), N-Zn-N 80.3(1) and Cl-Zn-N 104.2(1) to 119.6(1)°. The twist angle between the two individually planar rings of the bipyridine moiety is 4.3(5)°.

OXODIPEROXOBIPYRIDYLMOLYBDENUM(VI)
$C_{10}H_8MoN_2O_5$

E.O. SCHLEMPER, G.N. SCHRAUZER and L.A. HUGHES, 1984. Polyhedron, 3, 377-380.

Monoclinic, $P2_1/n$, a = 6.261, b = 12.726, c = 13.752 Å, β = 91.84°, Z = 4. Mo radiation, R = 0.034 for 1125 reflexions.

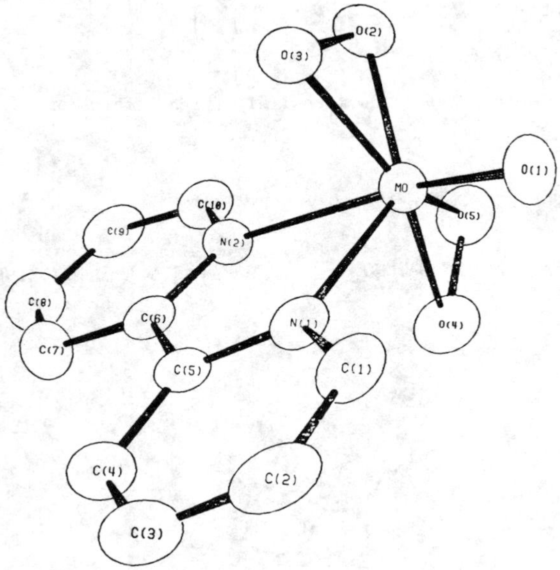

Fig. 1. A view of $C_{10}H_8MoN_2O_5$.

The molybdenum coordination is distorted trigonal bipyramidal (Fig. 1) with Mo-N(1) 2.199(5), Mo-N(2) 2.312(5), Mo-O(3) 1.948(4), Mo-O(4) 1.953(4), Mo-O(2) 1.912(5), Mo-O(4) 1.908(4) Å. The axial Mo-O(1) distance is 1.692(4) Å. The O-O distances are 1.459(6) and 1.465(5) Å.

(μ_2-DIBROMOGOLD)(μ_2-DIPYRIDINEGOLD)
$(C_{10}H_{10}Au_2Br_2N_2)_n$ (I),(II)

W. CONZELMANN, W. HILLER, J. STRAHLE and G.M. SHELDRICK, 1984. Z. Anorg. Chem., 512. 169-176.

I. Triclinic, P$\bar{1}$, a = 7.915, b = 9.356, c = 10.051 Å, α = 85.78, β = 102.07, γ = 109.31°, Z = 2. Mo radiation, R = 0.049 for 1702 reflexions (at 170 K).

II. Monoclinic, C2/c, a = 12.259, b = 15.222, c = 14.597 Å, β = 97.82°, Z = 8. Mo radiation, R = 0.071 for 1067 reflexions (at 170 K).

Form I is obtained from absolute ethanol, form II from toluene. I is built up from Au_4 zig-zag chains $AuBr_2$-$Au(py)_2$, $Au(py)_2$-$AuBr_2$ (Fig. 1) with pairs of Au atoms related by inversion centres. In II there are infinite chains with sequence $AuBr_2$-$Au(py)_2$ with Au1 and Au3 (Fig. 1) lying in crystallographic twofold axes. The Au...Au separations are 3.246(1) and 3.496(2) in I and 3.302(2) and 3.563(2) Å in II. The Au-N distances are 1.95-2.05(3) Å and Au-Br 2.385-2.390(3) Å.

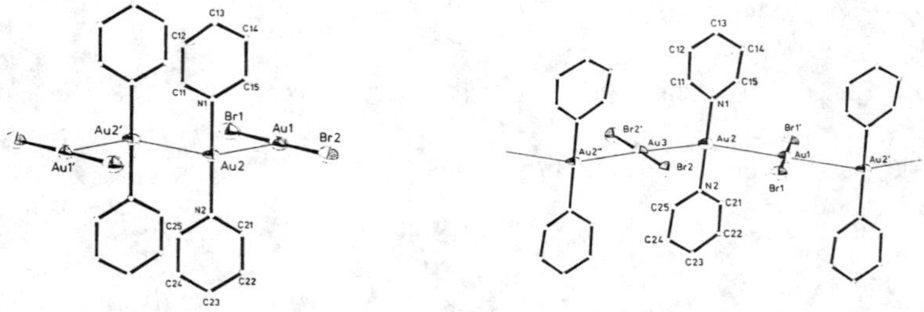

Fig. 1. The triclinic (left) and monoclinic (right) forms of $(C_{10}H_{10}Au_2Br_2N_2)_n$.

(2-METHOXYETHYL)(THEOPHYLLINATO-N^7)MERCURY(II)
$C_{10}H_{14}HgN_4O_3$ $Hg(C_3H_7O)(C_7H_7N_4O_2)$

K. CALDWELL, G.B. DEACON, B.M. GATEHOUSE, S.C. LEE and A.J. CANTY, 1984. Acta Cryst., C40, 1533-1536.

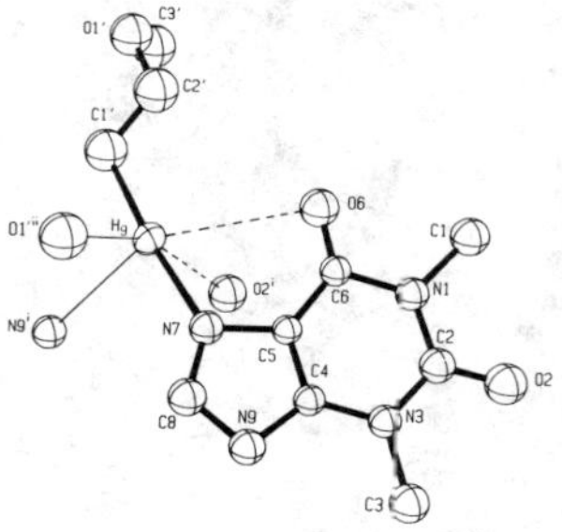

Fig. 1. The $Hg(C_3H_7O)(C_7H_7N_4O_2)$ structure. (Dotted lines indicate short, non-bonded contact distances.)

Monoclinic, $P2_1/c$, a = 8.305, b = 9.349, c = 16.370 Å, β = 97.37°, Z = 4. Mo radiation, R = 0.057 for 1461 reflexions.

The structure consists of molecules linked into a three dimensional network by weak (2.89 and 2.91(1) Å) Hg...O(methoxyethyl) and Hg...N(9) theophylline interactions (Fig. 1). The four-coordinate mercury atom is bonded to the deprotonated N(7) with Hg-N(7) 2.08(1), Hg-C 2.13(2) Å, C(1')-Hg-N(7) 173.6°. Hg is 0.548(1) Å from the theophylline mean plane.

DICHLOROBIS(3,5-DIMETHYLPYRAZOLE-N²)ZINC(II)
$C_{10}H_{16}Cl_2N_4Zn$ $ZnCl_2(C_5H_8N_2)_2$

E. BOUWMAN, W.L. DRIESSEN, R.A.G. DE GRAAFF and J. REEDIJK, 1984. Acta Cryst., C40, 1562-1563.

Monoclinic, C2/c, a = 15.013, b = 8.300, c = 23.988 Å, β = 95.85°, Z = 8. Mo radiation, R = 0.033 for 2430 reflexions.

The zinc ion has distorted tetrahedral coordination (Fig. 1) with Cl-Zn-Cl 119.2(1)°, Zn-Cl 2.216 and 2.224(1), Zn-N 2.017 and 2.019(2) Å. The dimethylpyrazole ligands are planar within ±0.03 Å with their least-squares planes intersecting at 84.9(1)°. Molecules are linked by weak N-H...Cl hydrogen bonds (N...Cl 3.333 and 3.470(3) Å).

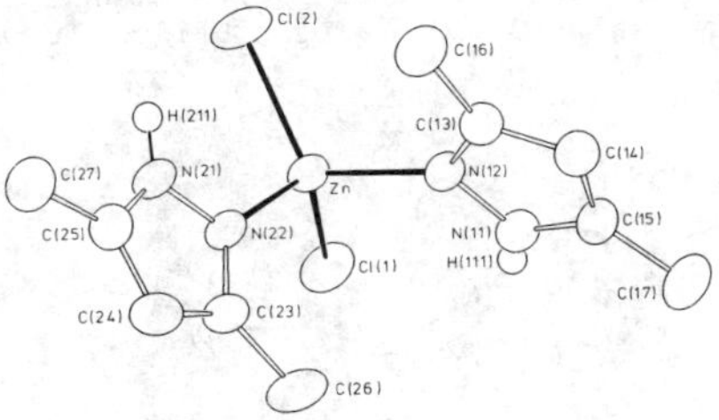

Fig. 1. The $ZnCl_2(C_5H_8N_2)_2$ molecule.

DI-μ-AZIDO-BIS[AZIDO(2-AMINOPYRIDINE)AQUO]DICOPPER(II)
$C_{10}H_{16}Cu_2N_{16}O_2$ $[Cu(C_5H_4NNH_2)(N_3)_2(H_2O)]_2$

M.A.S. GOHER and T.C.W. MAK, 1984. Inorg. Chim. Acta, 85, 117-122.

Triclinic, $P\bar{1}$, a = 7.142, b = 7.812, c = 9.727 Å, α = 96.52, β = 92.52, γ = 113.47°, D_m = 1.77, Z = 1. Mo radiation, R = 0.030 for 1960 reflexions.

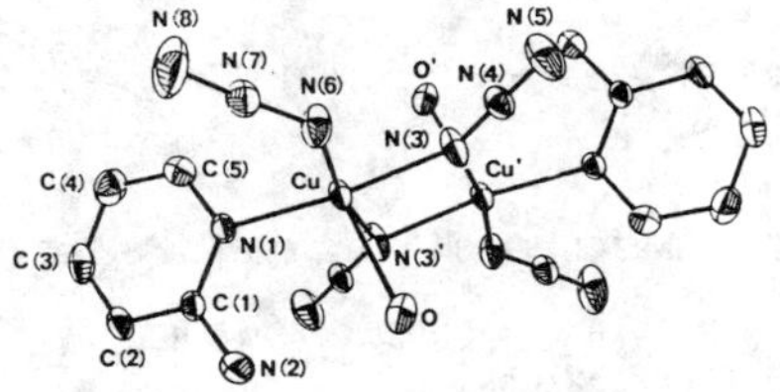

Fig. 1. Molecular structure of $C_{10}H_{16}Cu_2N_{16}O_2$.

The dimeric molecule, which has crystallographic inversion symmetry, contains both terminal and μ(1)-bridging azido groups (Fig. 1). Each Cu atom is further coordinated by the ring N atom of a 2-aminopyridine ligand and a water molecule to give a distorted square pyramid with the metal displaced 0.17 Å above the N_4 basal plane towards the apical aquo ligand. Bond distances involving Cu are: Cu-N(1) 2.001, Cu-N(3) 2.018, Cu-N(3') 2.003, Cu-N(6) 1.962, Cu-O 2.317 Å. The angles Cu-N(6)-N(7), and Cu-N(3)-N(4) are respectively 126.5 and 127.2° whilst that of Cu-N(3)-Cu' is 103.1°. Each water molecule forms an intramolecular O...H-N(amine) acceptor hydrogen bond (O...N(2) 2.852 Å), and is linked by two O-H...N (terminal azide) intermolecular donor hydrogen bonds (O...N(8") 2.901, O...N(6''') 2.878 Å) to adjacent dimeric complexes to yield a layer structure parallel to (001).

trans-CHLORO(DIMETHYLAMINE-N)BIS[DIMETHYLGLYOXIMATO(1-)-N,N']COBALT(III)
$C_{10}H_{21}ClCoN_5O_4$ $[CoCl(C_4H_7N_2O_2)_2\{NH(CH_3)_2\}]$

A.T.H. LENSTRA, H.J. GEISE and S.K. TYRLIK, 1984. Acta Cryst., C40, 749-751.

Orthorhombic, $P2_1nb$, a = 9.136, b = 12.650, c = 13.981 Å, Z = 4. Mo radiation, R = 0.042 for 1120 reflexions.

In the complex molecule (Fig. 1) Co is coordinated by four N atoms of two dimethylglyoxime monoanions, one N atom of the dimethylamine and a chlorine in a slightly distorted octahedron, with Co-N 1.939(5), 1.855(5), 1.866(5), 1.893(6) and 1.994(5) Å respectively and Co-Cl 2.248(2) Å. Comparison of the geometry of the title compound with other chloro-cobaloximes carrying trans-related N-donor ligands shows that the Co-Cl distance is rather insensitive to changes in the nature of the N-donor ligand. There are two short O...O hydrogen bridges of 2.45(1) and 2.42(1) Å.

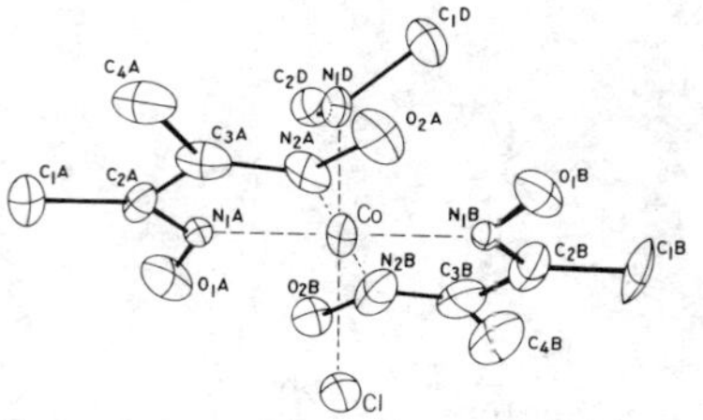

Fig. 1. $C_{10}H_{21}ClCoN_5O_4$: perspective view of the molecule.

1,1,1,2,2,2,3,3,3-NONACARBONYL-1,2,1,3-BIS(μ_2-HYDRIDO)(μ_3-2',2',2'-TRIFLUOROETHYLIMIDO)-triangulo-TRIOSMIUM
$C_{11}H_4F_3NO_9Os_3$ (I)

1,1,2,2,2,3,3,3-OCTACARBONYL-1,2,1,3,2,3-TRIS(μ_2-HYDRIDO)-1-HYDRIDO(μ_3-2',2',2'-TRIFLUOROETHYLIMIDO)-triangulo-TRIOSMIUM
$C_{10}H_6F_3NO_8Os_3$ (II)

1,1,1,1,2,2,2,3,3,3-DECACARBONYL-2,3-(μ_2-HYDRIDO)-2,3-(μ_2-2',2',2'-TRIFLUOROETHYLAMIDO)-triangulo-TRIOSMIUM
$C_{12}H_4F_3NO_{10}Os_3$ (III)

Z. DAWOODI, M.J. MAYS and K. HENRICK, 1984. J. Chem. Soc. Dalton, 433-440.

I. Monoclinic, $P2_1/c$, a = 9.383, b = 12.388, c = 16.171 Å, β = 97.79°, Z = 4. Mo radiation, R = 0.051 for 3091 reflexions.

II. Monoclinic, C2/c, a = 17.387, b = 8.137, c = 26.142 Å, β = 99.15°, Z = 8. Mo radiation, R = 0.046 for 1944 reflexions.

III. Monoclinic, C2/c, a = 19.983, b = 15.562, c = 14.089 Å, β = 118.80°, Z = 8. Mo radiation, R = 0.042 for 2248 reflexions.

I and II are similar (Fig. 1) consisting of a triangle of Os atoms with all CO ligands terminal in each case. In I the NCH_2CF_3 ligand is triply bridging with Os-N distances of 2.056(12), 2.101(13) and 2.119(13) Å. Os(1)-Os(2) and Os(2)-Os(3) are bridged by H atoms and are 0.14 Å longer (2.863(1), 2.856(1)) than Os(1)-Os(3) (2.717(1) Å). The N-C(1) distance is 1.44(2) Å. In II the hydride atoms were not located. However, the equivalence of the Os-Os bonds (2.862(1), 2.870(1) and 2.861(1) Å) indicates that they are located above each Os-Os bond. The NCH_2CF_3 ligand is again triply-bridging with Os-N 2.067(16), 2.050(16), 2.072(15) Å. The terminal H atom is disordered between the sites occupied by C(22) and C(23) in I leading to half occupancy of these sites by CO groups. The disorder is accompanied by a slight disorder in the CF_3 group. The C(1)-N distance is 1.48(2) Å. In III there is also a triangle of Os atoms, this time with ten terminal CO groups (Os-C 1.86-1.95(2) Å) and the organic ligand attached to the cluster via the N atom which bridges Os(1)-Os(3) (Os(1)-N 2.166(14), Os(3)-N 2.150(15) Å). The N-C(1) distance is 1.46(3) Å.

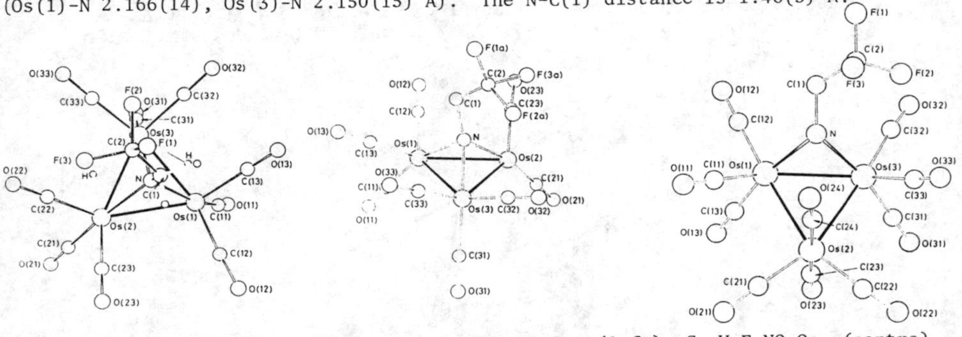

Fig. 1. The molecular structures: $C_{11}H_4F_3NO_9Os_3$ (left), $C_{10}H_6F_3NO_8Os_3$ (centre) and $C_{12}H_4F_3NO_{10}Os_3$ (right).

(E)-PYRIDINE-2-CARBALDEHYDE PYRIDIN-2'-YLHYDRAZONE HYDRATE

$C_{11}H_{10}N_4, 2.5(H_2O)$ (I) papyH, $(H_2O)_{2.5}$

CHLORO[(E)-PYRIDINE-2-CARBALDEHYDE PYRIDIN-2'-YLHYDRAZONATO]PALLADIUM(II)

$C_{11}H_9ClN_4Pd$ (II) [Pd(papy)Cl]

A.T. CASEY, B.F. HOSKINS and I.P. TRAVERSO, 1984. Aust. J. Chem., 37, 739-749.

I. Monoclinic, $P2_1/c$, a = 17.762, b = 3.8717, c = 18.503 Å, β = 107.99°, D_m = 1.30, Z = 4. Mo radiation, R = 0.061 for 1295 reflexions. Pale yellow crystals.

II. Triclinic, $P\bar{1}$, a = 6.9721, b = 9.0739, c = 9.1304 Å, α = 88.116, β = 88.778, γ = 81.392°, D_m = 1.97, Z = 2. Cu radiation, R = 0.028 for 2224 reflexions. Dark red crystals.

I is planar (Fig. 1) which suggests extensive π-electron delocalisation over the whole molecule with N(2)-N(3) being short (1.351 Å). In the Pd complex, the deprotonated ligand coordinates as a tridentate (Pd-N 1.995(3) Å, N-Pd-N 80.5(1)°) with Cl completing the square plane (Pd-Cl 2.3102(9) Å) (Fig. 1).

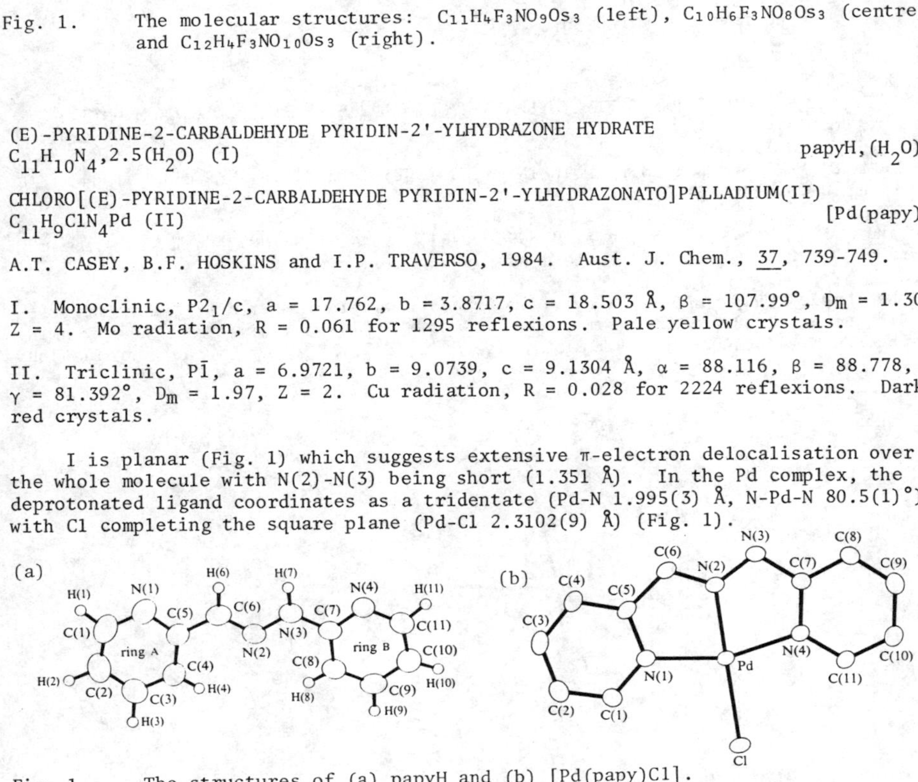

Fig. 1. The structures of (a) papyH and (b) [Pd(papy)Cl].

1,3,5,7-TETRAAZAADAMANTANEPENTACARBONYLMOLYBDENUM(0)
$C_{11}H_{12}MoN_4O_5$ $(CH_2)_6N_4.Mo(CO)_5$

K.Y. HUI, P.C. CHAN and T.C.W. MAK, 1984. Inorg. Chim. Acta, 84, 25-29.

Orthorhombic, Pbca, a = 12.800, b = 13.377, c = 17.524 Å, D_m = 1.67, Z = 8. Mo radiation, R = 0.039 for 2491 reflexions.

The structure can be described as an axially-distorted octahedron with an apical N lone pair of the organic ligand coordinated to the metal atom (Fig. 1). The Mo-C (carbonyl) bond lengths range from 1.960 Å (C(7)) to 2.050 Å with the equatorial bond lengths significantly longer than the axial. The Mo-N(1) distance is 2.337 Å and the Mo-C-O angles range from 172.5 to 178.8°. The bond lengths within the organic ligand show the expected variation generated by metal coordination: N(1)-C(1) 1.500, N(1)-C(2) 1.503, N(1)-C(3) 1.510 Å, compared to the remaining N-C bond distances which range from 1.450 to 1.473 Å.

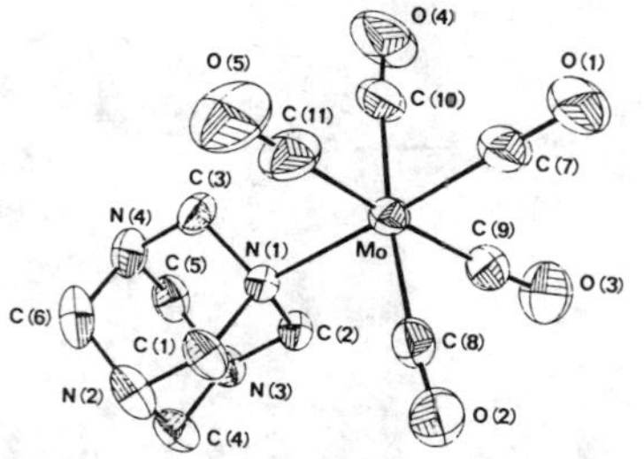

Fig. 1. Perspective view of $C_{11}H_{12}MoN_4O_5$.

DIBROMO[5-(1,2,5-DITHIAZEPAN-5-YLMETHYLENE)-4-METHYL-2-ETHYLIMIDAZOLE]CADMIUM(II)
$C_{11}H_{19}Br_2CdN_3S_2$ $[Cd(C_{11}H_{19}N_3S_2)Br_2]$

G.C. BAUMANN, M.F. SUMMERS, J.P. HUTCHINSON, J. ZUBIETA and L.G. MARZILLI, 1984. Inorg. Chem., 23, 3104-3108.

Monoclinic, $P2_1/n$, a = 12.988, b = 10.899, c = 13.148 Å, β = 108.74°, Z = 4. Mo radiation, R = 0.0235 for 1474 reflexions.

The complex (Fig. 1) is a highly distorted trigonal bipyramid with 2Br (Cd-Br 2.542(1) and 2.581(1) Å), 2N (from imidazole, Cd-N 2.270(4) Å, and the tertiary amine function, Cd-N 2.446(4) Å), and a disulphide S (Cd-S 2.938(2) Å) in the coordination sphere.

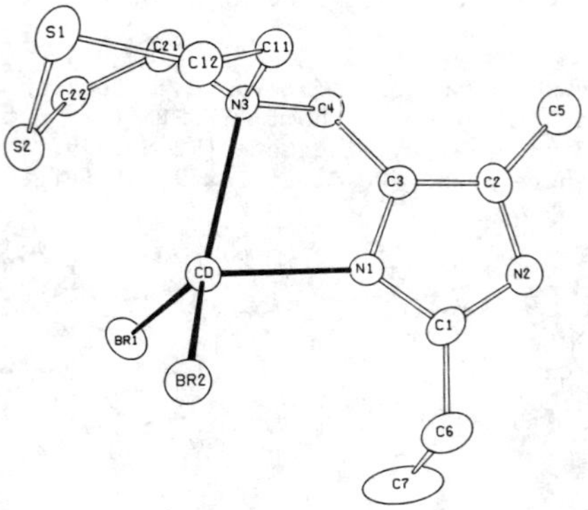

Fig. 1. The $[Cd(C_{11}H_{19}N_3S_2)Br_2]$ molecule.

catena-μ-DICHLOROBIS(BENZOTRIAZOLE)CADMIUM(II)
$C_{12}H_{10}CdCl_2N_6$ (I) $[CdCl_2(C_6H_5N_3)_2]$

catena-μ-DICHLOROBIS(BENZOTRIAZOLE)MANGANESE(II)
$C_{12}H_{10}Cl_2MnN_6$ (II) $[MnCl_2(C_6H_5N_3)_2]$

I. SØTOFTE and K. NIELSEN, 1984. Acta Chem. Scand., A38, 257-260.

I. Monoclinic, C2/c, a = 23.125, b = 3.818, c = 16.680 Å, β = 100.61°, Z = 4. Cu radiation, R = 0.053 for 1198 reflexions.

II. Monoclinic, C2/c, a = 23.441, b = 3.728, c = 16.480 Å, β = 98.50°, Z = 4. Cu radiation, R = 0.043 for 1002 reflexions.

Crystals of compounds I and II are isomorphous. The metal atoms in each compound are on twofold axes and are linked in the y direction by double chloride bridges, forming infinite $-Cd-Cl_2-Cd-$ and $-Mn-Cl_2-Mn-$ ribbons (Fig. 1). M-Cl-M angles are 92.9 and 93.5° for I and II respectively whilst Cl-M-Cl angles are 87.0 and 86.5°. The Cd-Cl distances are 2.639 and 2.631 Å and the Mn-Cl distances are 2.559 and 2.560 Å. The Cd-N and Mn-N distances are respectively 2.401 and 2.305 Å. The BTAH ligands in both complexes are nearly planar, and the plane of the ligand is almost perpendicular to the plane of the metal and Cl atoms, dihedral angles being 88.2 and 89.2° for I and II respectively.

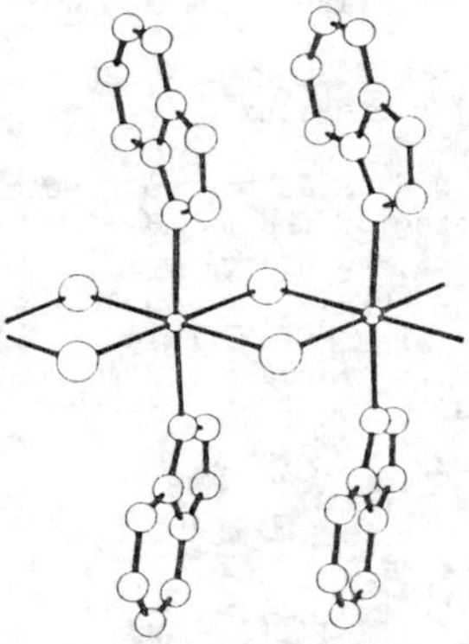

Fig. 1. Coordination about the metal atoms in $C_{12}H_{10}CdCl_2N_6$ and $C_{12}H_{10}Cl_2MnN_6$.

catena-μ-AQUA-BIS(μ-BENZOTRIAZOLATO)COPPER(II)
$C_{12}H_{10}CuN_6O$ $[Cu(H_2O)(C_6H_4N_3)_2]$

I. SØTOFTE and K. NIELSEN, 1984. Acta Chem. Scand., A38, 253-263.

Monoclinic, I2/c, a = 21.533, b = 6.862, c = 17.059 Å, β = 94.44°, Z = 8. Cu radiation, R = 0.079 for 828 reflexions.

The complex is polymeric, four Cu atoms with y ≈ 0.25 being linked to the remaining Cu atoms in the y direction (0.5-x, 0.5+y, -z) by two bridging benzotriazolato (BTA) groups (Fig. 1). The four bridging N atoms are nearly coplanar with deviations from the best plane being less than 0.1 Å. The coordination polyhedron around Cu is a distorted tetragonal bipyramid, with apices of the elongated axis occupied by water molecules. The polyhedra are linked by the sharing of these apices. The Cu-O-Cu' angle is 88.1° and the important distances are: Cu-N1(1) 2.00, Cu-N1(2) 2.01, Cu-N2(1') 2.02, Cu-N2(2'') 2.02, Cu-O 2.48 Å. The N-Cu-N cis angles range from 84.9 to 98.1°. The Cu...Cu distance is 3.447 Å.

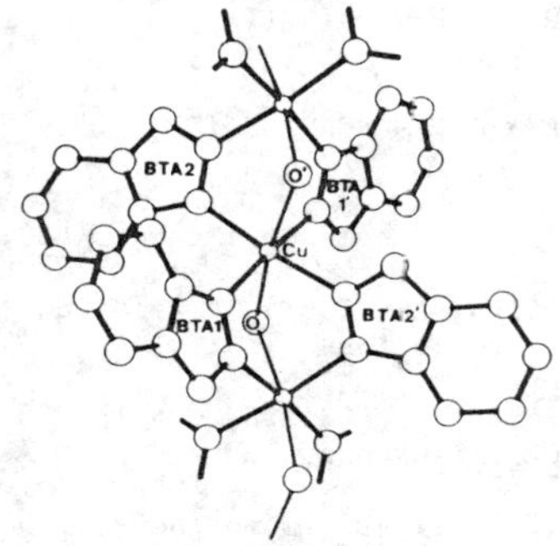

Fig. 1. Coordination around the Cu atoms in $C_{12}H_{10}CuN_6O$.

AQUADIFLUOROOXO(1,10-PHENANTHROLINO)VANADIUM(IV)
$C_{12}H_{10}F_2N_2O_2V$ $[VOF_2(H_2O)C_{12}H_8N_2]$

A. DEMŠAR and P. BUKOVEC, 1984. J. Fluor. Chem., 24, 369-375.

Monoclinic, $P2_1/n$, a = 8.456, b = 7.252, c = 19.250 Å, β = 100.28°, D_m = 1.68, Z = 4. Mo radiation, R = 0.068 for 2007 reflexions.

Vanadium is octahedrally coordinated by two F atoms, two N atoms of the phenanthroline molecule, the O atom of the vanadyl group and an O atom of a water molecule (Fig. 1). Bond distances are as follows: V-O 1.623, V-OH_2 2.012, V-F 1.896 and 1.921, V-N 2.134 and 2.306 Å. The longer V-N distance is in a trans position to the terminal O atom. The N-V-N chelate angle is 73.66°, and the F-V-F angle is 159.8°. Two short O-H...F hydrogen bonds (2.537 and 2.592 Å) bind two molecules related by the screw axis.

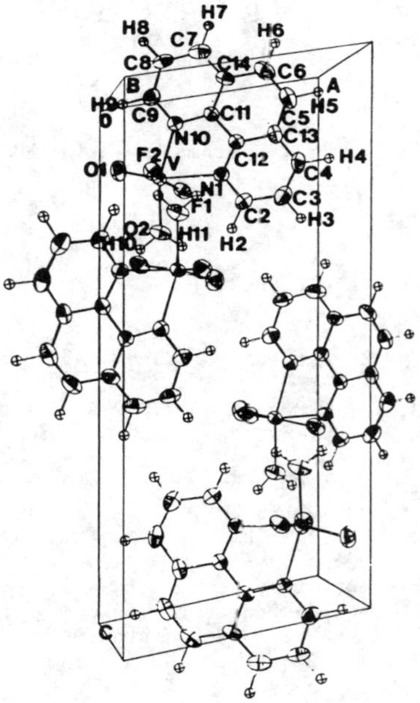

Fig. 1. View of the unit cell and numbering scheme in $C_{12}H_{10}F_2N_2O_2V$.

BIS(9-METHYLHYPOXANTHINE)SILVER(I) NITRATE DIHYDRATE

$C_{12}H_{12}AgN_9O_5,2H_2O$ $[Ag(C_6H_6N_4O)_2](NO_3),2H_2O$

K. AOKI and W. SAENGER, 1984. Acta Cryst., C40, 772-775.

Triclinic, P$\bar{1}$, a = 9.985, b = 14.655, c = 6.606 Å, α = 107.26, β = 95.92, γ = 96.71°, Z = 2. Cu radiation, R = 0.066 for 2532 reflexions.

In the complex, the Ag^+ ion is linearly bonded to the two 9-methylhypoxanthine ligands through the N(7) sites (Fig. 1); the Ag-N distances are 2.147(7) and 2.154(7) Å. There is also an Ag...O contact of 2.736(8) Å. The structure of the metal-bridged base-base pair is in the syn configuration with respect to the N(7)-Ag-N(7) bond and stabilized by a water bridge between the carbonyl O atoms O(6).

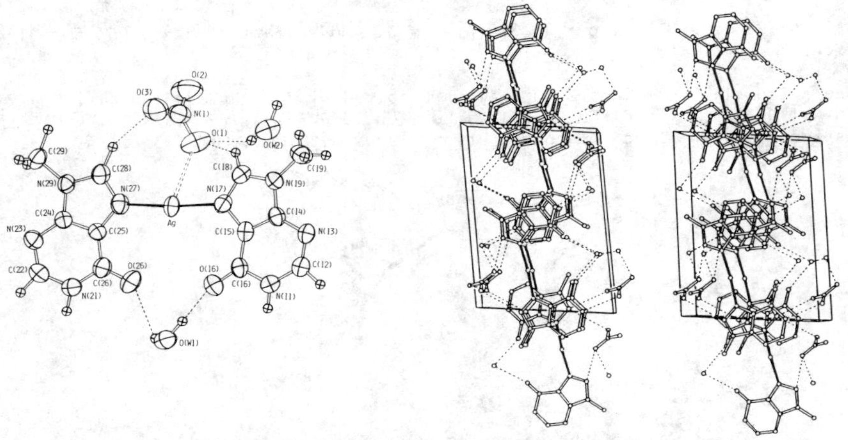

Fig. 1. $C_{12}H_{12}AgN_9O_5,2H_2O$: perspective view of the asymmetric unit and a stereoview of the crystal packing viewed down the c axis, broken lines indicate hydrogen bonds.

DIAQUA(1,10-PHENANTHROLINE)COPPER(II) SULPHATE

$C_{12}H_{12}CuN_2O_6S$ $[Cu(H_2O)_2(phen)][SO_4]$

P.C. HEALY, J.M. PATRICK and A.H. WHITE, 1984. Aust. J. Chem., 37, 1111-1115.

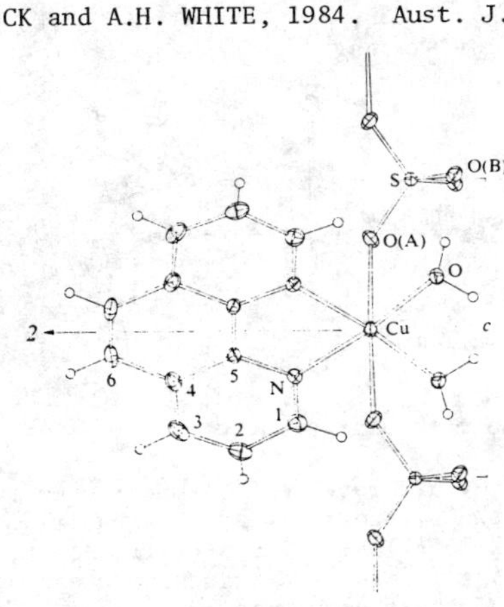

Fig. 1. The structure of $[Cu(H_2O)_2(phen)][SO_4]$.

Monoclinic, C2/c, a = 14.883, b = 13.843, c = 7.019 Å, β = 108.60°, D_m = 1.80, Z = 4. Mo radiation, R = 0.033 for 1445 reflexions.

The pseudo-square planar Cu environments (Cu-N, O: 2.009(2), 1.970(2) Å, N-Cu-N 82.0(1)°) are bridged by sulphate groups in the axial octahedral positions to give a linear polymeric array (Cu-O 2.468(3) Å, Fig. 1).

(μ_2-CHLORO)BIS[BIS(IMIDAZOLE)COPPER(I)] CHLORIDE
$C_{12}H_{16}Cl_2Cu_2N_8$

W. CLEGG, S.R. ACOTT and C.D. GARNER, 1984. J. Chem. Soc. Dalton, 2581-2583.

Trigonal, $P3_221$, a = 8.546, c = 21.370 Å, Z = 3. Mo radiation, R = 0.054 for 2124 reflexions.

The trigonally coordinated Cu atoms are joined by a Cl bridge (Cu-Cl 2.564(2), mean Cu-N 1.904(5) Å, Cu-Cl-Cu 95.5(1)°, Cu...Cu 3.797(3) Å) (Fig. 1). The Cl^- anion lies between the two hinged Cu coordination planes with Cu...Cl 3.134(2) Å.

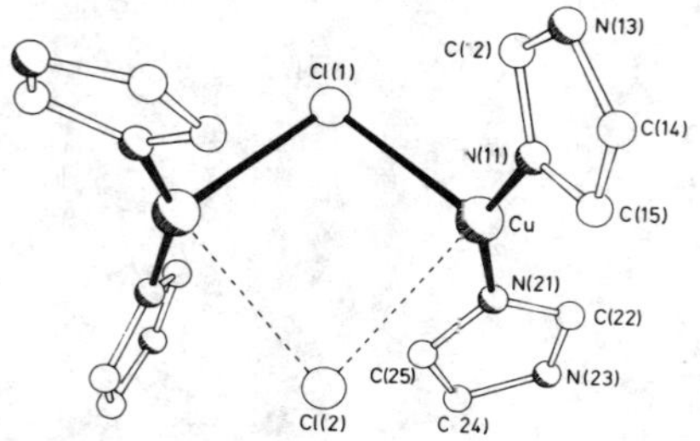

Fig. 1. $C_{12}H_{16}Cl_2Cu_2N_8$: the molecular structure.

DICHLOROBIS(1-VINYL-3-METHYLPYRAZOLE)PALLADIUM
$C_{12}H_{16}Cl_2N_4Pd$

V.I. SOKOL, M.A. PORAI-KOSHITS, T.O. VARDOSANIDZE, V.N. VORAPAEV, E.S. DOMNINA and G.G. SKVORTSOVA, 1983. Koord. Khim., 9, 1539-1543 [Sov. J. Coord. Chem., 9, 870-874].

Monoclinic, $P2_1/n$, a = 7.654, b = 9.238, c = 11.800 Å, β = 106.7°, Z = 2. Mo radiation, R = 0.044 for 1919 reflexions.

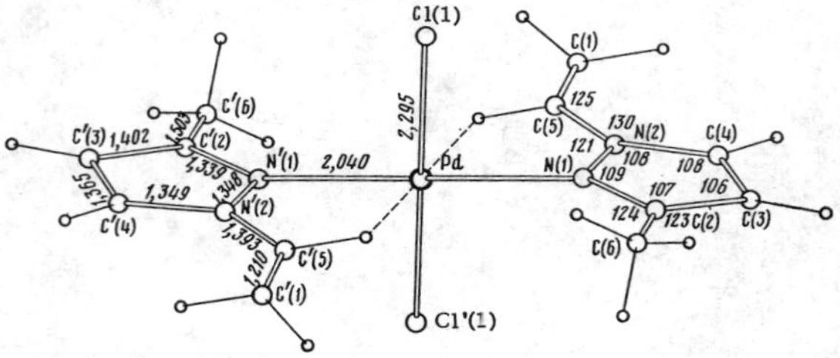

Fig. 1. $C_{12}H_{16}Cl_2N_4Pd$: the molecular structure.

The structure is composed of discrete centrosymmetric molecules (Fig. 1) in which the Pd atom coordinates two pyrazole ligands via their N atoms (Pd-N 2.040(3)) and two chlorine atoms (Pd-Cl 2.295(1) Å). The overall geometry of the molecule is square-planar. The pyrazole ring and the vinyl group are essentially planar and the angle between the mean planes of these groups is 10°. The olefinic moiety of the vinyl group does not participate in bonding to the Pd atom. The mean plane of the pyrazole ligand is perpendicular to the Cl-Pd-Cl grouping. There is a close approach (2.743 Å) of one of the H atoms of the vinyl group to the Pd atom. This H atom occupies one vertex of a distorted octahedron.

DICHLOROTHIAMINECOPPER(I)
$C_{12}H_{17}Cl_2CuN_4OS$ $Cu(C_{12}H_{17}N_4OS)Cl_2$

R.E. CRAMER, R.B. MAYNARD and R.S. EVANGELISTA, 1984. J. Am. Chem. Soc., 106, 111-116.

Triclinic, $P\bar{1}$, a = 9.203, b = 14.866, c = 5.985 Å, α = 99.56, β = 94.59, γ = 104.23°, D_m = 1.699, Z = 2. Mo radiation, R = 0.035 for 3227 reflexions.

The Cu(I) ion is coordinated to two Cl^- ions and the N(1)' atom of the pyrimidine ring of the thiamine molecule in a distorted trigonal planar arrangement (Fig. 1, Cu-Cl 2.2002(6) and 2.3366(6), Cu-N 1.993(2) Å). The torsion angles ϕ_T -10.1° and ϕ_P -83.8° correspond to the frequently observed F conformation. A hydrogen-bonding network and base-stacking interactions between parallel pyrimidine planes are the predominant intermolecular forces in the crystal.

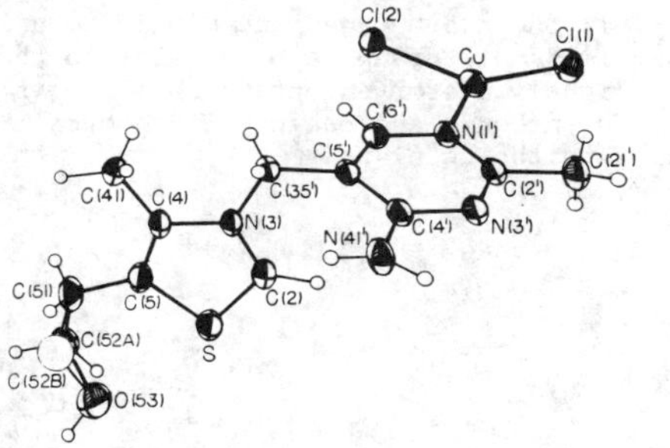

Fig. 1. The $Cu(C_{12}H_{17}N_4OS)Cl_2$ molecule.

DIAQUATETRAKIS(IMIDAZOLE)COPPER(II) DIFLUORIDE
$C_{12}H_{20}CuF_2N_8O_2$

W. VREUGDENHIL, P.J.M.W.L. BIRKER, R.W.M. ten HOEDT, G.C. VERSCHOOR and J. REEDIJK, 1984. J. Chem. Soc. Dalton, 429-432.

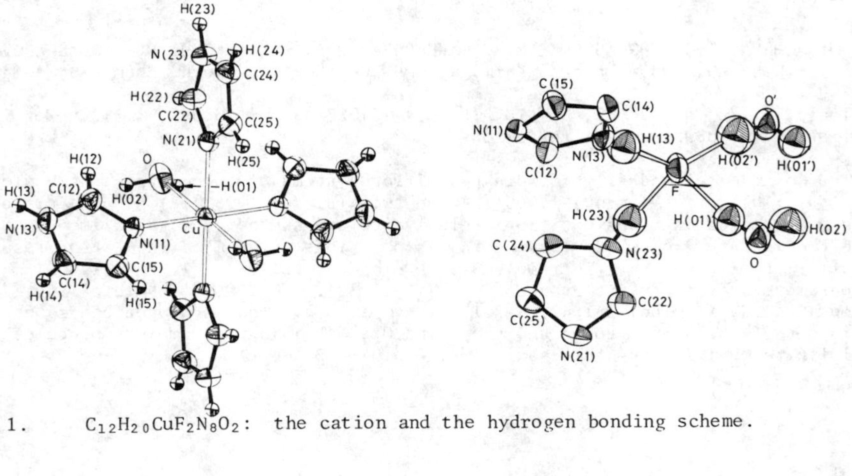

Fig. 1. $C_{12}H_{20}CuF_2N_8O_2$: the cation and the hydrogen bonding scheme.

Monoclinic, C2/c, a = 12.708, b = 10.229, c = 13.915 Å, β = 106.63°, D_m = 1.57, Z = 4. Mo radiation, R = 0.027 for 1468 reflexions.

The F atoms in the complex are strongly hydrogen-bonded to two imidazole N-H groups and two water molecules (Fig. 1) such that each F is in a tetrahedral environment (F-N 2.661(2), 2.647(2); F-O 2.717(2), 2.695(2) Å). Infinite chains of O-Cu-O(F_2)OCuO units are formed. Coordination around Cu is elongated tetragonal with four imidazole groups as equatorial ligands and water molecules as axial ligands.

trans-BIS(2,3-BUTANEDIONE DIOXIMATO-N,N')(IMIDAZOLE-N^3)METHYLCOBALT(III)

$C_{12}H_{21}CoN_6O_4$

V. PATTABHI, M. NETHAJI, E.J. GABE, F.L. LEE and Y. LE PAGE, 1984. Acta Cryst., C40, 1155-1156.

Monoclinic, $P2_1/n$, a = 8.075, b = 13.969, c = 14.892 Å, β = 97.98°, Z = 4. Mo radiation, R = 0.043 for 2765 reflexions.

In the molecule (Fig. 1) the two butanedione dioximato units and the Co atom lie in a plane with the planar imidazole group perpendicular to the butanedione plane. The Co coordination is a slightly distorted octahedron. The structure has two O-H...O intramolecular hydrogen bonds and chains of hydrogen-bonded molecules are formed almost perpendicular to the bc plane.

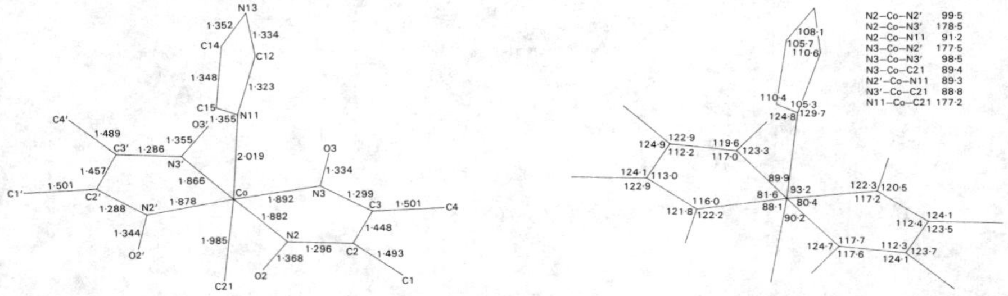

Fig. 1. $C_{12}H_{21}CoN_6O_4$: bond lengths and angles in the complex molecule.

DINITROBIS(2-METHYL-2-AMINO-3-PENTANONE OXIME)COBALT(III)

$C_{12}H_{27}CoN_6O_6$

L.Kh. MINACHEVA, T.S. KHODASHOVA, V.A. SAKHAROVA, M.A. PORAI-KOSHITS, L.P. SHAROVA and A.I. STETSENKO, 1983. Koord. Khim. 9, 699-704 [Sov. J. Coord. Chem., 9, 401-406].

Monoclinic, $P2_1/c$, a = 9.603, b = 16.316, c = 12.257 Å, β = 93.98°, Z = 2. Mo radiation, R = 0.089 for 1715 reflexions.

The Co atom octahedrally coordinates nitrogen atoms of three types: two N(oxime) (Co-N 1.910(10)), two N(amino) 1.960(10) and two N(nitro) 1.958(12) Å. The nitro groups are located in trans positions relative to one another and the oxime groupings in the two chelate groups are located in cis positions (Fig. 1). The latter results in the formation of a short intramolecular hydrogen bond (O...H 1.1 and 1.4 Å). The mean parameters of the nitro groups are: N-O 1.23 Å, ONO 119.9°; chelate ring, N(oxime)-C 1.29, N(amino)-C 1.42, C-C 1.56, N-O 1.35 Å. Weak hydrogen bonds between the N atoms of the amino groups and the oxime O atoms join the complex molecules to form infinite chains along the z-axis (N...O 3.01-3.2 Å).

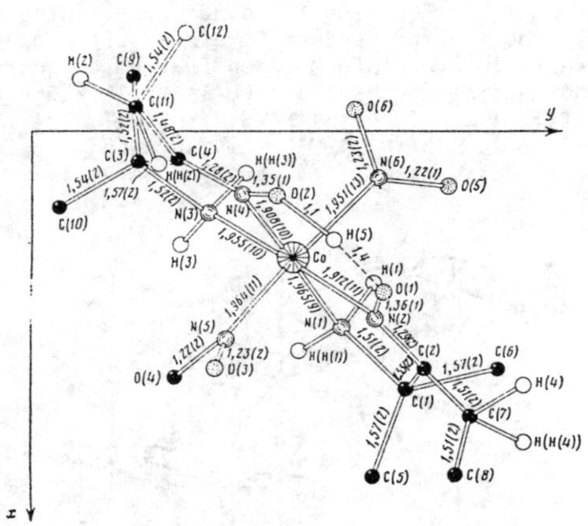

Fig. 1. $C_{12}H_{27}CoN_6O_6$: the molecular geometry.

[2,2-DI(2-PYRIDYL)-1,3-DIOXOLANE-N,N']DICHLOROPALLADIUM(II)

$C_{13}H_{12}Cl_2N_2O_2Pd$ $PdCl_2(C_{13}H_{12}N_2O_2)$

G.R. NEWKOME, V.K. GUPTA, K.J. THERIOT, J.C. EWING, S.P. WICELINSKI, W.R. HUIE, F.R. FRONCZEK and S.F. WATKINS, 1874. Acta Cryst., C40, 1352-1355.

Monoclinic, $P2_1/c$, a = 9.767, b = 19.224, c = 15.351 Å, β = 90.34°, Z = 8. Mo radiation, R = 0.031 for 4502 reflexions.

In both independent molecules Pd is square planar coordinated with slight tetrahedral distortion (Fig. 1). The chlorine atom displacements from the PdNN plane are 0.03 and -0.038 Å in molecule A, 0.00 and 0.030 Å in B; the N-Pd-N angles are 86.8(2)° in both molecules. Both acetal rings have different envelope conformations. Some bond lengths are Pd-N (mean) 2.032(2), Pd-Cl (mean) 2.290(1) in A, 2.285(1) and 2.297(1) Å in B. The dihedral angles between the pyridine rings are 64.5(3) and 63.4(3) Å. There is a short intramolecular Pd...O contact in both molecules (2.987(4) and 2.935(3) Å).

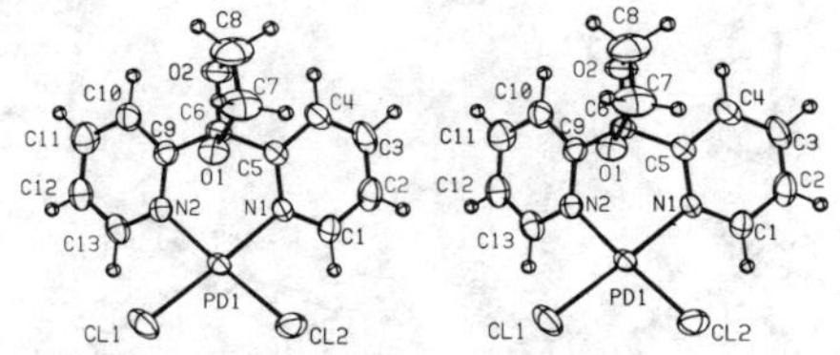

Fig. 1. Stereoview of molecule A in the $PdCl_2(C_{13}H_{12}N_2O_2)$ molecule.

(μ_2-BROMO)(μ_2-HYDROXO)(μ_2-3,6-BIS(2'-PYRIDYL)PYRIDAZINE-N(1),N(2),N(3),N(4))DIBROMO-COPPER

$C_{14}H_{11}Br_3Cu_2N_4O$ $[Cu_2(C_{14}H_{10}N_4)(OH)Br_3]_n$

P. DAPPORTO, G. DE MUNNO, A. SEGA and C. MEALLI, 1984. Inorg. Chim. Acta, 83, 171-176.

Monoclinic, Pn, a = 13.691, b = 6.245, c = 10.298 Å, β = 103.92°, Z = 2. Mo radiation, R = 0.066 for 655 reflexions.

The structure consists of binuclear units joined by bridging Br atoms to form a polymeric array (Fig. 1). The two independent Cu atoms of the binuclear moiety are five-coordinate with geometry intermediate between that of a square pyramid and a trigonal bipyramid. Important distances are: Cu(1)-Br(1) 2.666, Cu(1)-Br(2) 2.378, Cu(2)-Br(3) 2.814, Cu(2)-Br(1') 2.515, Cu(1)-N (2.03, 2.01), Cu(2)-N (2.05, 1.99), Cu(1)-O 1.92 and Cu(2)-O 1.86 Å. The N(2)-N(3) distance and the Cu(1)-O-Cu(2) angle are 1.25 Å and 128° respectively. The Cu(1)...Cu(2) separation is 3.398 Å.

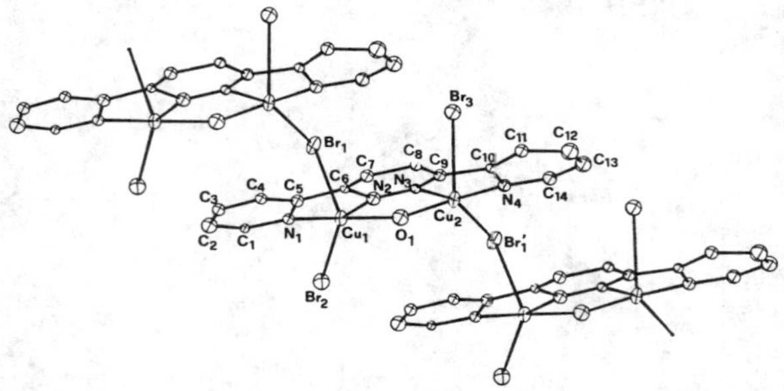

Fig. 1. Perspective view of the polymeric $C_{14}H_{11}Br_3Cu_2N_4O$ molecule.

catena-DICHLORO(DIPHENYLETHANEDIONE DIOXIME)COPPER(II)

$C_{14}H_{12}Cl_2CuN_2O_2$ $[Cu(C_{14}H_{12}N_2O_2)Cl_2]_n$

M. MÉGNAMISI-BÉLOMBÉ, P. SINGH, D.E. BOLSTER and W.E. HATFIELD, 1984. Inorg. Chem., 23, 2578-2582.

Orthorhombic, Pbcn, a = 13.382, b = 19.925, c = 6.901 Å, Z = 4. Mo radiation, R = 0.049 for 796 reflexions.

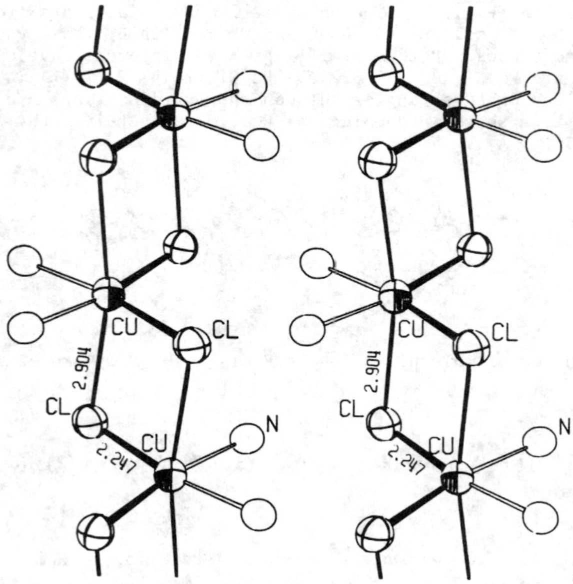

Fig. 1. Stereoview of the doubly bridged Cu_2Cl_2 polymeric chain in $[Cu(C_{14}H_{12}N_2O_2)Cl_2]n$.

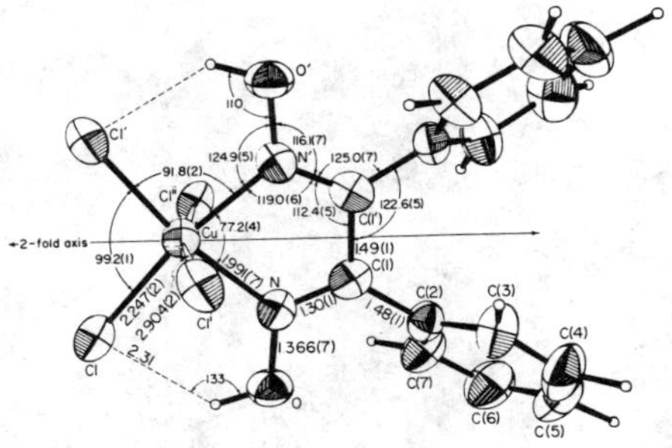

Fig. 2. Geometry of the repeating unit of $[Cu(C_{14}H_{12}N_2O_2)Cl_2]_n$.

The structure may be regarded as a chain of stacked planar $Cu(C_{14}H_{12}N_2O_2)Cl_2$ units, with Cu sites equidistantly arranged in zigzag infinite chains along c with Cu...Cu 3.649(1) Å (Fig. 1). Monomeric units, having exact C_2 symmetry, are linked to each other by two chloro bridges (Cu-Cl-Cu 89.3°). Cu has distorted octahedral 4+2 coordination with two cis Cl and two cis N atoms forming the basal plane (Cu-Cl 2.247(2), Cu-N 1.991(7) Å), and two Cl atoms of adjacent monomer units occupying the axial sites (Cu-Cl 2.904(2) Å (Fig. 2).

AQUA-μ-CHLORO-TRICHLORO-μ-[3,6-DI(2-PYRIDYL)PYRIDAZINE]DICOPPER(II)

$C_{14}H_{12}Cl_4Cu_2N_4O$ $[Cu_2Cl_4(H_2O)(C_{14}H_{10}N_4)]$

G. DE MUNNO and G. BRUNNO, 1984. Acta Cryst., C40, 2022-2024.

Monoclinic, $P2_1/c$, a = 9.809, b = 9.212, c = 19.316 Å, β = 100.4°, D_m = 2.02, Z = 4. Mo radiation, R = 0.048 for 1227 reflexions.

The complex is shown in Fig. 1. Both Cu atoms are five-coordinated, the apical sites being occupied by a Cl atom and a water molecule. Main bond lengths are: Cu-Cl 2.195-2.500(3), Cu(2)-O(1) 2.313(7), Cu-N 1.976-2.038(8) Å. For another form of this compound, see (1).

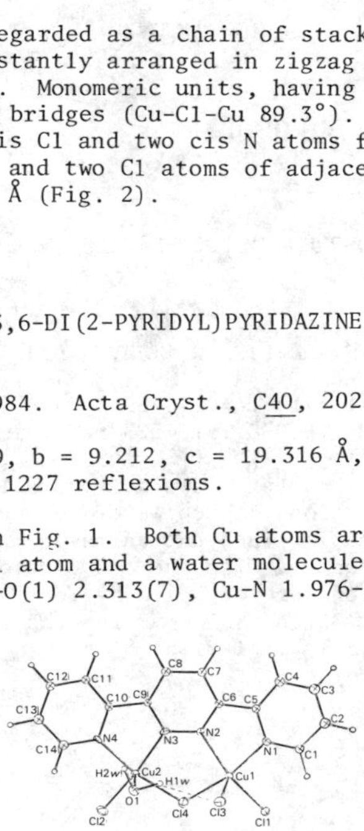

Fig. 1. A view of $C_{14}H_{12}Cl_4Cu_2N_4O$.

1. G. DE MUNNO, G. DENTI and P. DAPPORTO, 1983. Inorg. Chim. Acta, 74, 199-203.

AQUADICHLORO[2-(3,5-DIMETHYLPYRAZOL-1-YL)QUINOLINE]COBALT(II)

$C_{14}H_{15}Cl_2CoN_3O$ $Co(L)Cl_2(H_2O)$

A.T. BAKER, N.J. FERGUSON and H.A. GOODWIN, 1984. Aust. J. Chem., 37, 2421-2429.

Triclinic, $P\bar{1}$, a = 7.435, b = 9.701, c = 11.845 Å, α = 71.02, β = 77.75, γ = 69.86°, D_m = 1.632, Z = 2. Mo radiation, R = 0.023 for 2310 reflexions.

Cobalt(II) has a trigonal bipyramidal coordination and is monomeric (Fig. 1, Co-N 2.116(2), Co-Cl 2.3029(6), Co-O 2.141 Å, N-Co-N 75.67(6)°).

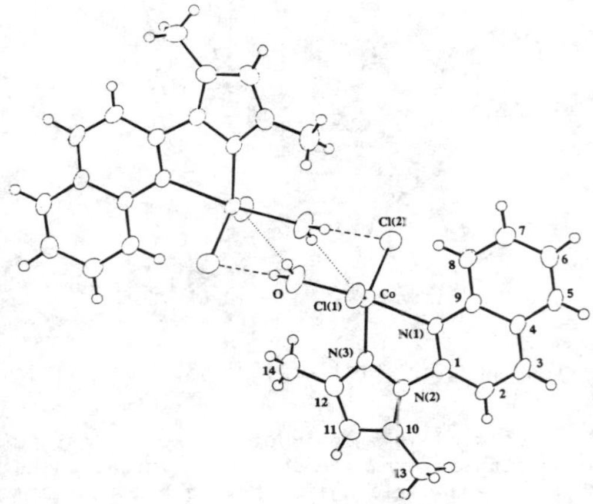

Fig. 1. The structure of $Co(L)Cl_2(H_2O)$.

[2,9-BIS(N-METHYLHYDRAZINO)-1,10-PHENANTHROLINE]DICHLORORHODIUM(III) TETRAFLUOROBORATE SESQUIHYDRATE
$C_{14}H_{16}BCl_2F_4N_6Rh,1.5H_2O$

C.W.G. ANSELL, E. EGERT, J. LEWIS and P.R. RAITHBY, 1984. Acta Cryst., C40, 359-362.

Triclinic, P$\bar{1}$, a = 12.200, b = 13.171, c = 14.259 Å, α = 110.31, β = 93.90, γ = 102.60°, Z = 4. Mo radiation, R = 0.058 for 4925 reflexions.

The asymmetric unit (Fig. 1) contains two structurally similar mononuclear Rh^{3+} cations and the associated BF_4^- anions; three water molecules are involved in hydrogen bonding with both the cations and the anions. In each cation the Rh atom adopts a distorted octahedral geometry with the two Cl atoms in trans axial positions while the four equatorial positions are occupied by four N atoms of the planar hydrazino-phenanthroline ligand. Principal bond lengths are: Rh-Cl 2.312-2.343(3), Rh-N(phen) 1.923-1.945(7), Rh-N(H_2) 2.131-2.154(8) Å.

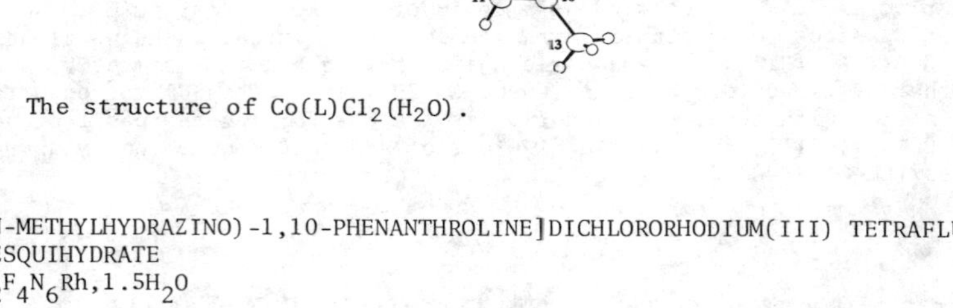

Fig. 1. The structures of the two complex ions in $C_{14}H_{16}BCl_2F_4N_6Rh,1.5H_2O$.

TETRAKIS(2-AMINOTHIAZOLE)-trans-BIS-(N-THIOCYANATO)COBALT(II)
$C_{14}H_{16}CoN_{10}S_6$ $Co(C_3H_4N_2S)_4(NCS)_2$

E.S. RAPER, J.R. CREIGHTON, R.E. OUGHTRED and I.W. NOWELL, 1984. Inorg. Chim. Acta, 87, 19-24.

Monoclinic, $P2_1/n$, a = 10.640, b = 13.070, c = 16.942 Å, β = 91.78°, D_m = 1.60, Z = 4. Mo radiation, R = 0.044 for 3273 reflexions.

The complex is monomeric with a distorted CoN_6 octahedron and Co-N distances of 2.165, 2.213, 2.173 and 2.176 Å (average 2.182 Å) to the hetero N atoms of 2-aminothiazole (Fig. 1). The Co-N distances to the thiocyanate ligands are 2.119 and 2.106 Å and the angles about Co range from 86.3 to 94.2°. The S-C and C-N distances in the thiocyanate ligands are respectively 1.626 and 1.164 Å, and 1.625 and 1.149 Å. The S-C-N angles are 177.9 and 178.6°. In the aminothiazole rings the C-S distances range from 1.702 to 1.742 Å. These ligands are essentially planar but are significantly twisted out of the CoN_4 plane.

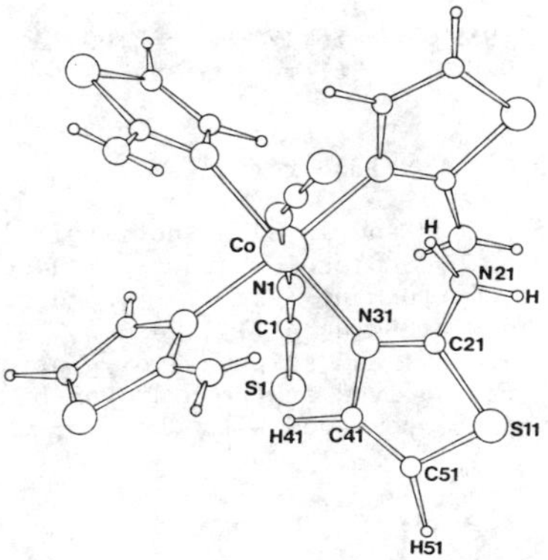

Fig. 1. Perspective view of $C_{14}H_{16}CoN_{10}S_6$.

BIS(SACCHARINATO-N)TETRAAQUAZINC(II) DIHYDRATE
$C_{14}H_{16}N_2O_{10}S_2Zn,2H_2O$ (I) $[Zn(C_7H_4NO_3S)_2(H_2O)_4].2H_2O$

BIS(SACCHARINATO-N)TETRAAQUACHROMIUM(II) DIHYDRATE
$C_{14}H_{16}CrN_2O_{10}S_2,2H_2O$ (II) $[Cr(C_7H_4NO_3S)_2(H_2O)_4].2H_2O$

F.A. COTTON, G.E. LEWIS, C.A. MURILLO, W. SCHWOTZER and G. VALLE, 1984. Inorg. Chem., 23, 4038-4041.

I. Monoclinic, $P2_1/c$, a = 7.934, b = 16.118, c = 7.691 Å, β = 99.90°, Z = 2. Mo radiation, R = 0.028 for 1814 reflexions (at 5°C).

II. Monoclinic, $P2_1/c$, a = 8.042, b = 16.032, c = 7.804 Å, β = 100.90°, Z = 2. Mo radiation, R = 0.030 for 1317 reflexions (at 5°C).

The structures are isomorphous and contain centrosymmetric complex molecules and water molecules hydrogen bonded to one another (Fig. 1). The metal atoms are octahedrally coordinated by four H_2O molecules and two $(C_7H_4NO_3S)^-$ anions. The Cr complex shows a Jahn-Teller distortion. Bond lengths: Zn-N 2.201(2), Zn-O(1W) 2.058(2), Zn-O(2W) 2.159(2) Å, Cr-N 2.196(3), Cr-O(1W) 2.048(3), Cr-O(2W) 2.396(4) Å.

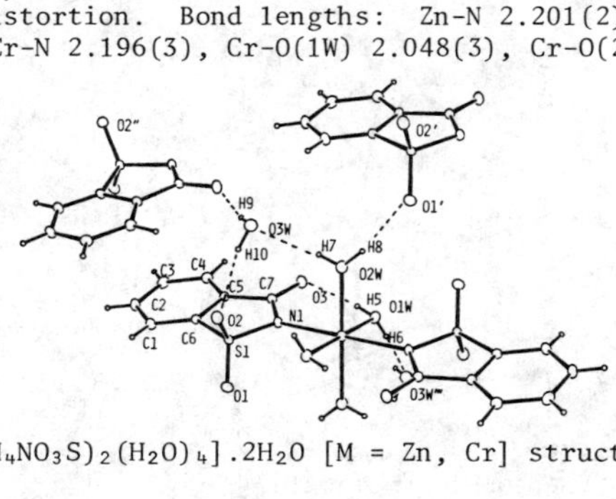

Fig. 1. The $[M(C_7H_4NO_3S)_2(H_2O)_4].2H_2O$ [M = Zn, Cr] structure.

TETRAAQUABIS(SACCHARINATO-N)ZINC(II) DIHYDRATE
$C_{14}H_{16}N_2O_{10}S_2Zn,2H_2O$ (I) $[Zn(C_7H_4NO_3S)_2(H_2O)_4],2H_2O$

TETRAAQUABIS(SACCHARINATO-N)CADMIUM(II) DIHYDRATE
$C_{14}H_{16}CdN_2O_{10}S_2,2H_2O$ (II) $[Cd(C_7H_4NO_3S)_2(H_2O)_4],2H_2O$

S.Z. HAIDER, K.M.A. MALIK, S. DAS and M.B. HURSTHOUSE, 1984. Acta Cryst., C40, 1147-1150.

I. Monoclinic, $P2_1/c$, a = 7.939, b = 16.120, c = 7.691 Å, β = 99.87°, D_m = 1.83, Z = 2. Mo radiation, R = 0.028 for 1581 reflexions.

II. Monoclinic, $P2_1/c$, a = 8.036, b = 16.145, c = 7.870 Å, β = 100.24°, D_m = 1.92, Z = 2. Mo radiation, R = 0.025 for 1850 reflexions.

The complexes are isomorphous and in each the metal lies on a centre of symmetry and is bonded to four H_2O molecules and two monodentate $C_7H_4NO_3S$ ligands in a distorted trans octahedral environment (Fig. 1). Bond distances are: Zn-O 2.059(4) and 2.159(4), and Zn-N 2.200(4) Å (I); Cd-O 2.279(4) and 2.342(4), and Cd-N 2.323(4) Å (II). The carbonyl and sulphonyl oxygens of the $C_7H_4NO_3S$ ligand are not bound to the metal but are involved in the hydrogen-bonding network. The structures of these complexes are compared with those of other, similar complexes.

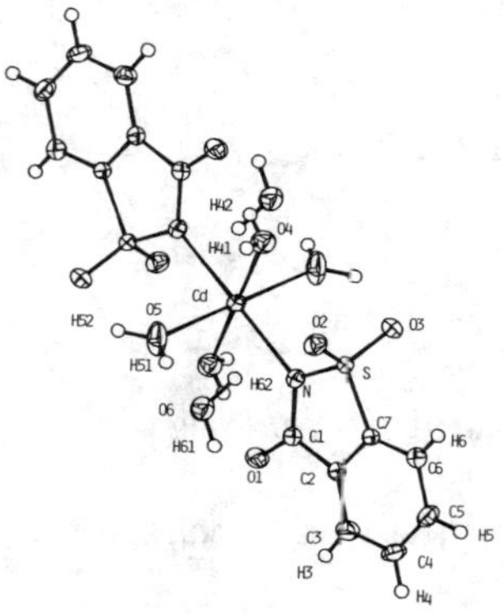

Fig. 1. Perspective view of the $[Cd(C_7H_4NO_3S)_2(H_2O)_4],2H_2O$ complex molecule.

[N,N-BIS(3,5-DIMETHYLPYRAZOL-1-YLMETHYL)AMINOETHANE](NITRATO-O)(NITRATO-O,O')NICKEL(II)
$C_{14}H_{23}N_7NiO_6$

J.W.F.M. SCHOONHOVEN, W.L. DRIESSEN, J. REEDIJK and G.C. VERSCHOOR, 1984. J. Chem. Soc. Dalton, 1053-1058.

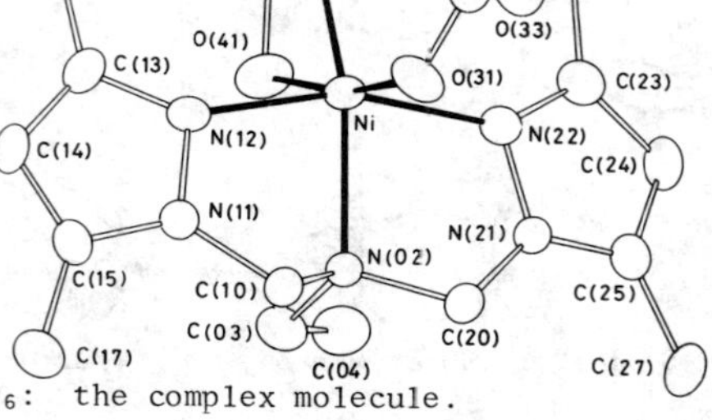

Fig. 1. $C_{14}H_{23}N_7NiO_6$: the complex molecule.

Monoclinic, $P2_1/c$, a = 8.511, b = 14.103, c = 15.865 Å, β = 92.75°, Z = 4. Mo radiation, R = 0.041 for 1690 reflexions.

The Ni atom has a NiN_3O_3(mer)chromophore in a distorted octahedral geometry (Fig. 1). One nitrate group is monodentate and the other bidentate with Ni-O 2.052(5) and 2.157(5) and 2.145(5) Å respectively. Ni-N distances are 2.061(5) and 2.034(5) Å for the pyrazole nitrogens and 2.128(5) Å for the amine nitrogen. Both pyrazole rings are planar and the methyl groups are coplanar with the respective pyrazole rings. The dihedral angle between the planes of the pyrazole rings is 17.6(2)°. Packing is via normal van der Waals contacts.

TRIS[μ-3,5-BIS(TRIFLUOROMETHYL)PYRAZOLATO-N,N']TRIGOLD(I)
$C_{15}H_3Au_3F_{18}N_6$ $[Au(C_3HN_2(CF_3)_2)]_3$

B. BOVIO, F. BONATI and G. BANDITELLI, 1984. Inorg. Chim. Acta, 87, 25-33.

Triclinic, $P\bar{1}$, a = 11.904, b = 12.171, c = 9.346 Å, α = 98.63, β = 81.94, γ = 113.47°, Z = 2. Mo radiation, R = 0.066 for 2360 reflexions.

The molecule contains a nine-membered ring which is rather irregular and non-planar (Fig. 1). Each Au atom is two-coordinated in nearly linear fashion with an average Au-N distance of 1.93 Å and an average N-Au-N angle of 178.7°. These Au-N distances are very short and the N-N, C-N and C-C average distances within the pyrazole rings are longer than usual, suggesting a π-interaction between the Au atoms and the pyrazolato rings. Distances between metal atoms are: 3.355, 3.344 and 3.346 Å, and distances between metals in different trimers are at least 3.998 Å, ruling out intermolecular interaction.

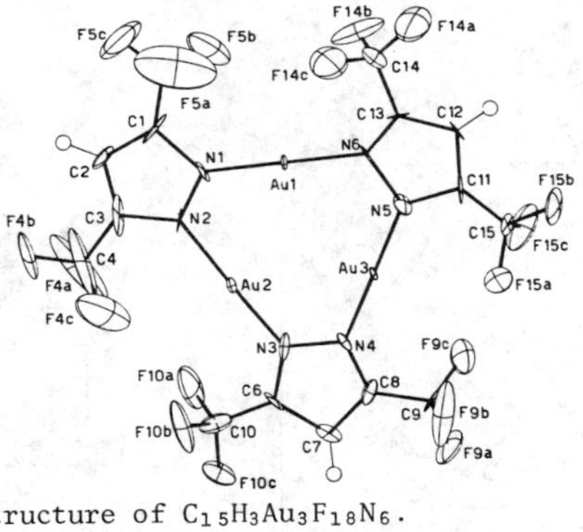

Fig. 1. Molecular structure of $C_{15}H_3Au_3F_{18}N_6$.

NONACARBONYLDI-μ-HYDRIDO-μ_3-PHENYLIMIDO-triangulo-TRIRUTHENIUM
$C_{15}H_7NO_9Ru$

S. BHADURI, K.S. GOPALKRISHNAN, W. CLEGG, P.G. JONES, G.M. SHELDRICK and D. STALKE, 1984. J. Chem. Soc. Dalton, 1765-1767.

Monoclinic, $P2_1/c$, a = 11.697, b = 9.265, c = 18.589 Å, β = 100.18°, Z = 4. Mo radiation, R = 0.039 for 3019 reflexions.

The X-ray analysis established the nature of the complex (Fig. 1). The nine CO groups are all terminal with three bonded to each metal atom. There are two long (2.805(1)) and one short (2.680(1) Å) Ru-Ru bonds. The former are H-bridged and the latter unbridged. The Ru-H distances lie in the range 1.71(7) to 1.81(7) Å. The two Ru_2H planes subtend angles of 130 and 127° with the Ru_3 plane. The N atom lies 1.315 Å above the Ru_3 plane (Ru-N 2.068(8) Å). The NPh ligand triply bridges the Ru atoms.

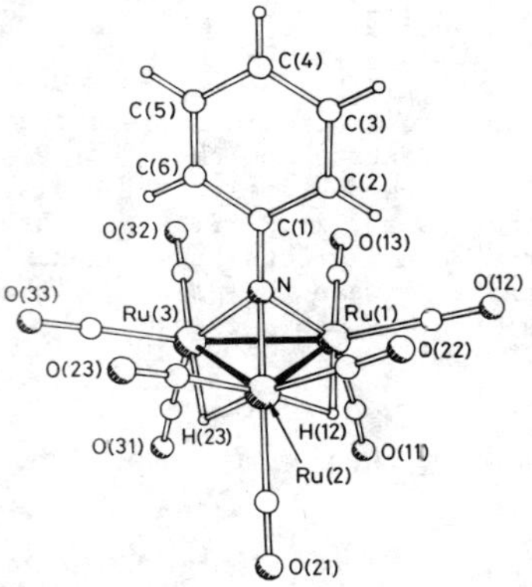

Fig. 1. $C_{15}H_7NO_9Ru$: molecular structure.

SODIUM OXO-(1,10-PHENANTHROLINE-N,N')TRICYANOMOLYBDENUM(IV) BIS(PHENANTHROLINE) METHANOL SOLVATE MONOHYDRATE

$C_{15}H_8MoN_5NaO,2C_{12}H_8N_2,CH_4O,H_2O$ $Na^+[MoO(CN)_3(C_{12}H_8N_2)]^-,2C_{12}H_8N_2,CH_3OH,H_2O$

S.S. BASSON, J.G. LEIPOLDT and I.M. POTGIETER, 1984. Inorg. Chim. Acta, 90, 57-62.

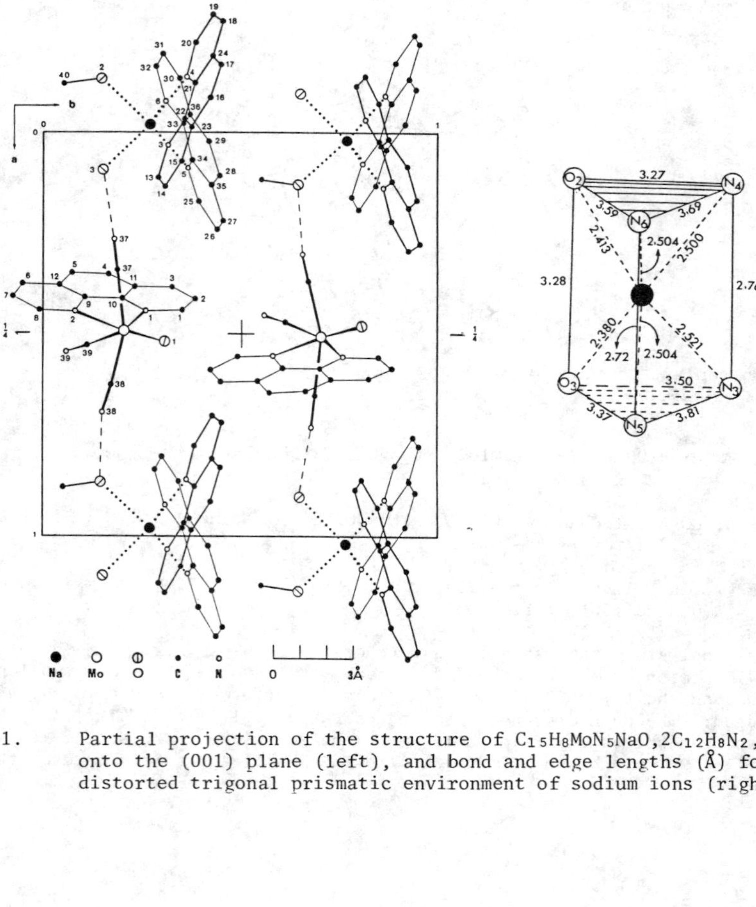

Fig. 1. Partial projection of the structure of $C_{15}H_8MoN_5NaO,2C_{12}H_8N_2,CH_4O,H_2O$ onto the (001) plane (left), and bond and edge lengths (Å) for the distorted trigonal prismatic environment of sodium ions (right).

Monoclinic, $P2_1/c$, a = 15.090, b = 14.584, c = 18.126 Å, β = 111.55°, D_m = 1.54, Z = 4. Mo radiation, R = 0.075 for 3655 reflexions.

The anion is a distorted octahedron with the O and a CN ligand trans-orientated to the bidentate ligand (Fig. 1). The two Mo-N bonds (2.173 and 2.363 Å) differ significantly with the longer one trans to the strong Mo=O 1.669 Å bond. The CN ligands have two normal Mo-C(av.) 2.15 Å bond distances and a significantly shorter Mo-C (2.09 Å) bond trans to the shorter Mo-N bond. The distorted six atom environment of each Na ion originates from one methanol, one water and two phenanthroline molecules.

TRIS(NITRATO-O,O')(2,2'.6,2''-TERPYRIDYL-N,N',N'')SCANDIUM(III)
$C_{15}H_{11}N_6O_9Sc$ (I)

[1,2-BIS(PYRIDINE-α-CARBALDIMINO)ETHANE-N,N',N'',N''']BIS(μ_2-HYDROXO)BIS(NITRATO-O,O')-DISCANDIUM(III) DINITRATE ACETONITRILE SOLVATE
$C_{28}H_{30}N_{12}O_{14}Sc_2$,2($C_3H_3N$) (II)

A.M. ARIF, M.B. HURSTHOUSE, M. THORNTON-PETT and WENXIANG ZHU, 1984. J. Chem. Soc. Dalton, 2449-2454.

I. Monoclinic, $P2_1/c$, a = 8.982, b = 16.486, c = 13.272 Å, β = 108.40°, D_m = 1.69, Z = 4. Mo radiation, R = 0.043 for 3272 reflexions.

II. Monoclinic, $P2_1/n$, a = 17.658, b = 9.937, c = 11.827 Å, β = 92.96°, Z = 2. Mo radiation, R = 0.045 for 3523 reflexions.

Compound I consists of discrete monomers ($Sc(NO_3)_3$terpy) in which the Sc atom is nine-coordinate and in an irregular coordination polyhedron (Fig. 1). The Sc atom lies in the plane of the terpyridyl ligand and the six O atoms from the three nitrate groups are arranged in the remainder of the coordination sphere such that ligand-ligand repulsions are minimised. One of the nitrate groups is unsymmetrically coordinated. Excluding Sc-O(32) (2.482) the mean Sc-O distance is 2.263 and the mean Sc-N distance 2.302 Å. II is a centrosymmetric dimeric complex with two bridging OH groups. The Sc atom is eight-coordinate in this case. Each 1,2-bis(pyridine-α-carbaldimino)ethane is approximately coplanar with the Sc atom while the two hydroxy and two nitrate O atoms which are coordinated to one particular Sc atom form an approximately coplanar system which is perpendicular to the plane of the tetramine. The result is a distorted dodecahedral coordination polyhedron. The Sc-N distances range from 2.341 to 2.403, Sc-O(nitrate) 2.333, 2.295 and Sc-OH 2.053 and 2.089 Å. As in I, the ligands are arranged to minimise ligand-ligand repulsion. The Sc-(μ-OH)$_2$-Sc system has idealised local symmetry D_{2h}, and the Sc-Sc distance is 3.347 Å. The OH...OH separation is 2.439 Å.

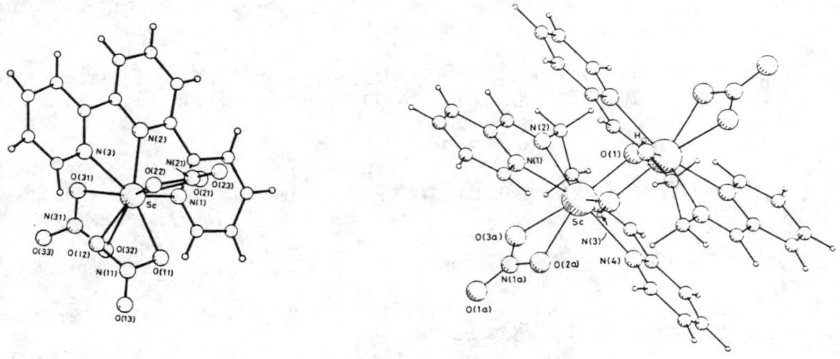

Fig. 1. The molecular structures: $C_{15}H_{11}N_6O_9Sc$ (left) and $C_{28}H_{30}N_{12}O_{14}Sc_2$,-2($C_3H_3N$) (right).

(μ_2-CYANO)(2,9-DIMETHYL-1,10-PHENANTHROLINE-N,N')COPPER
$(C_{15}H_{12}CuN_3)_n$ (I)

(μ_2-THIOCYANATO)(2,9-DIMETHYL-1,10-PHENANTHROLINE-N,N')COPPER
$(C_{15}H_{12}CuN_3S)_n$ (II)

G.O. MORPURGO, G. DESSY and V. FARES, 1984. J. Chem. Soc. Dalton, 785-791.

I. Monoclinic, $P2_1/n$, a = 9.839, b = 7.788, c = 16.559 Å, β = 92.33°, D_m = 1.55, Z = 4. Cu radiation, R = 0.056 for 942 reflexions.

II. Orthorhombic, Pbca, a = 16.283, b = 9.485, c = 18.115 Å, D_m = 1.56, Z = 8. Mo radiation, R = 0.045 for 1279 reflexions.

The structure of I consists of infinite non-linked zig-zag chains of Cu(CN). Each Cu atom has four neighbours in a distorted tetrahedral arrangement from two symmetry-related bridging CN^- groups and by two N atoms from the dimethylphenanthroline (dmphen) ligand (Cu-N(1) 2.01(1), Cu-N(2) 2.121(9), Cu-N(3) 2.131(9) and Cu-C(1^I) 1.89(1) Å). The result is an infinite one-dimensional polymeric structure along the [010] direction. Each polymer sequence is 'fitted' with its centrosymmetrically related one such that stacks of aromatic systems are formed with alternate dmphen interplanar distances of 3.26 and 3.90 Å (Fig. 1) each dmphen forming an angle of 23° with the (010) plane. II is related to I in that NCS^- groups act as a bridge between two Cu atoms and the dmphen molecules tend to couple by facing each other. The Cu atom has a tetrahedral environment of two N atoms from one dmphen ligand (Cu-N(2) 2.078(6), Cu-N(3) 2.076(6)) and S and N from two different symmetry related NCS^- groups (Cu-S 2.323(2), Cu-N 1.943(2) Å) to form a CuN_3S unit. II also has a one-dimensional polymeric structure along the [010] direction.

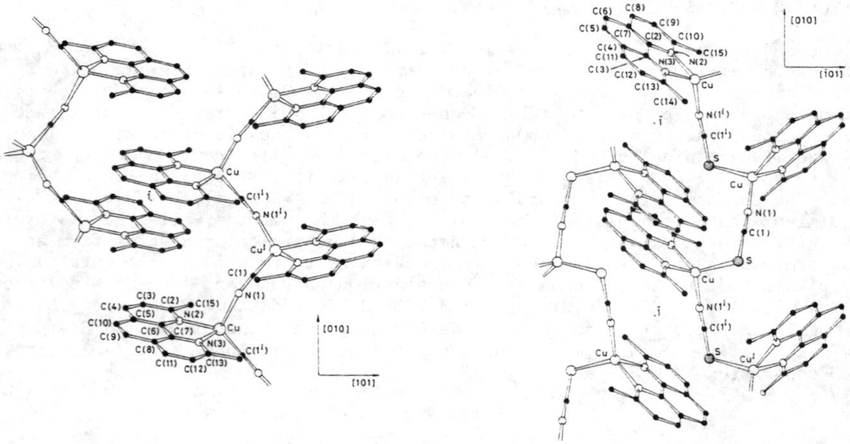

Fig. 1. Projection of the structure of $(C_{15}H_{12}CuN_3)_n$ on the ($\bar{1}$01) plane (left) and of the $(C_{15}H_{12}CuN_3S)_n$ structure on the ($\bar{1}0\bar{1}$) plane (right).

AQUANITRITO(2,2',6',2''-TERPYRIDYL)COPPER NITRITE MONOHYDRATE
$C_{15}H_{13}CuN_5O_5,H_2O$ $[Cuterpy(ONO)OH_2]NO_2,H_2O$

R. ALLMANN, S. KREMER and D. KUCHARZCYK, 1984. Inorg. Chim. Acta, 85, L19-L21.

Triclinic, $P\bar{1}$, a = 13.508, b = 8.761, c = 8.177 Å, α = 110.07, β = 86.06, γ = 105.31°, Z = 2. Mo radiation, R = 0.035 for 2304 reflexions.

The metal ion has almost planar coordination to the three N atoms of the terpyridyl ligand (Cu-N 2.04, 1.94 and 2.05 Å) and one nitrite ligand (Cu-O(6) 1.97 Å) (Fig. 1). Square pyramidal coordination is completed by addition of the coordinated water molecule in an apical position (Cu-O(1) 2.24 Å). The Cu ion lies 0.11 Å out of the best plane through atoms N(1), N(2), N(3) and O(6). The angle Cu-O(6)-N(5) is 112.1° and the angles O(5)-N(5)-O(6) and O(3)-N(4)-O(4) (bound and free nitrites respectively) are 115.6 and 107.6° respectively. All four water protons are engaged in hydrogen bond formation (O(1)...O(2) 2.76, O(1)...O(4) 2.74, O(2)...O(3') 2.89 and O(2)...O(4') 2.90 Å).

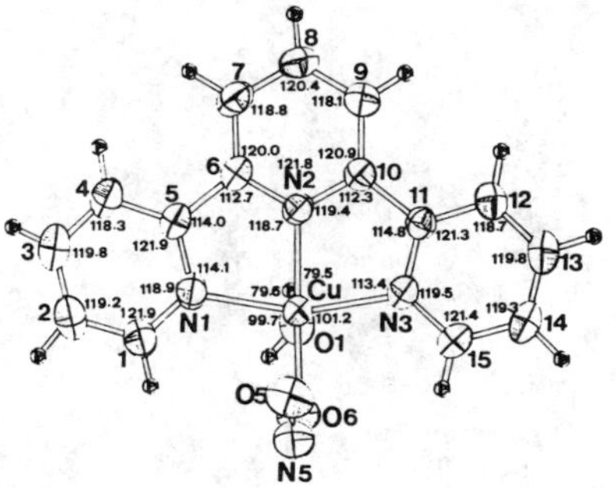

Fig. 1. Molecular geometry of the cation in $C_{15}H_{13}CuN_5O_5$.

TRIBROMOTRIS(PYRIDINE)THALLIUM(III)
$C_{15}H_{15}Br_3N_3Tl$

S.E. JEFFS, R.W.H. SMALL and I.J. WORRALL, 1984. Acta Cryst., C40, 1827-1829.

Orthorhombic, $P2_12_12_1$, a = 13.39, b = 15.60, c = 8.840 Å, D_m = 2.56, Z = 4. Mo radiation, R = 0.050 for 2064 reflexions.

In the neutral $TlBr_3(C_5H_5N)_3$ molecules (Fig. 1), Tl has a distorted octahedral coordination with mer geometry; trans bonds show deviations from linearity within the range 7-10°. Some bond lengths are Tl-Br(1) 2.618(2), Tl-Br(2) 2.626(2), Tl-Br(3) 2.670(2), Tl-N 2.42-2.51(1) Å.

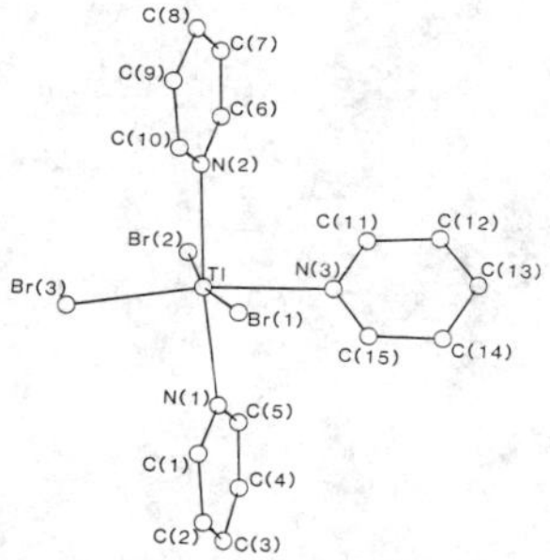

Fig. 1. The molecular structure of $TlBr_3(C_5H_5N)_3$.

mer-TRICHLOROTRIS(PYRIDINE)RHODIUM(III)
$C_{15}H_{15}Cl_3N_3Rh$ $RhCl_3(C_5H_5N)_3$

K.R. ACHARYA, S.S. TAVALE and T.N. GURU ROW, 1984. Acta Cryst., C40, 1327-1328.

Monoclinic, $P2_1/n$, a = 9.008, b = 12.457, c = 15.445 Å, β = 90.87°, D_m = 1.63, Z = 4. Mo radiation, R = 0.052 for 1171 reflexions.

Rhodium has an almost regular octahedral coordination (Fig. 1) with the three, planar pyridine rings perpendicular to each other. Mean bond lengths are Rh-Cl 2.334(5) and Rh-N 2.057(13) Å.

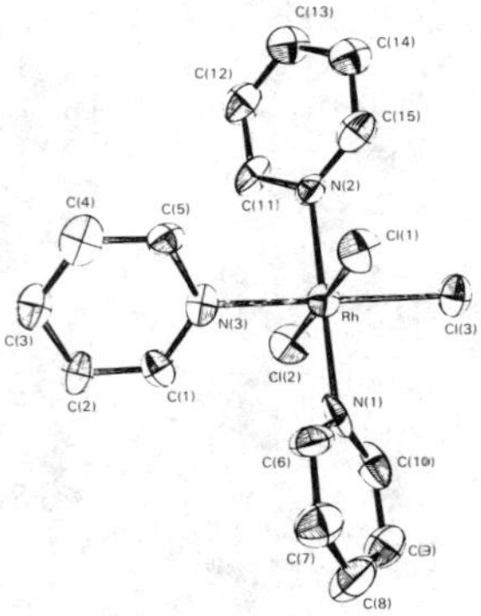

Fig. 1. The structure of $RhCl_3(C_5H_5N)_3$.

μ-OXO-1,1,1,2,2,2-HEXAOXO-1,1,2-TRIS(PYRIDINE)DIRHENIUM(VII)
$C_{15}H_{15}N_3O_7Re_2$ $[Re_2O_7(C_5H_5N)_3]$

J.W. JOHNSON, J.F. BRODY, G.B. ANSELL and S. ZENTZ, 1984. Acta Cryst., C40, 2024-2026.

Monoclinic, $P2_1/c$, a = 9.068, b = 12.280, c = 17.494 Å, β = 95.02°, Z = 4. Mo radiation, R = 0.026 for 1764 reflexions.

The complex (Fig. 1) contains two inequivalent rhenium atoms bridged by a single oxygen atom, Re(1)-O 1.806(7), Re(2)-O 2.056(7) Å. Re(1) is five-coordinate in a trigonal bipyramidal geometry with three oxo ligands (Re(1)-O 1.695-1.706(9) Å), the bridging oxygen atom, and a pyridine nitrogen atom in its coordination sphere. Re(2) is six-coordinate in a distorted octahedral geometry with a fac-trioxo group (Re(2)-O 1.710-1.724(9) Å), the bridging oxygen atom, and two pyridine nitrogen atoms.

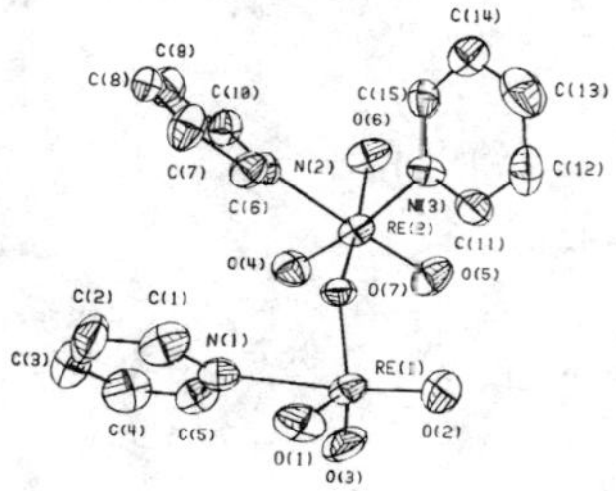

Fig. 1. A view of $[Re_2O_7py_3]$.

[3-(HYDROXYAMINO)-3-METHYL-2-BUTANONE-OXIMATO-N,N'(2-)][3-(HYDROXYAMINO)-3-METHYL-2-BUTANONE-OXIMATO-N,N']PYRIDINECOBALT(III)
$C_{15}H_{24}CoN_5O_4$

Yu.A. SIMONOV, V.E. ZAVODNIK. L.P. SHAROVA, A.A. DVORKIN, S.A. D'YACHENKO and A.I. STETSENKO, 1984. J. Struct. Chem., 25, 430-433 [Zh. Strukt. Khim., 25(3), 102-106].

Triclinic, $P\bar{1}$, a = 10.855, b = 9.567, c = 9.463 Å, α = 70.61, β = 94.23, γ = 94.33°, Z = 2. Mo radiation, R = 0.025 for 2026 reflexions.

The structure (Fig. 1) is molecular. The Co atom is coordinated by five nitrogen atoms in a distorted square planar geometry with the pyridine nitrogen atom apical. The Co atom is displaced 0.272 Å from the basal plane toward the pyridine N atom. The basal Co-N distances differ: Co-N(O) 1.795 Å and Co-N(OH) 1.952 Å. There is a single intramolecular O-H...O hydrogen bond (O...O 2.417 Å).

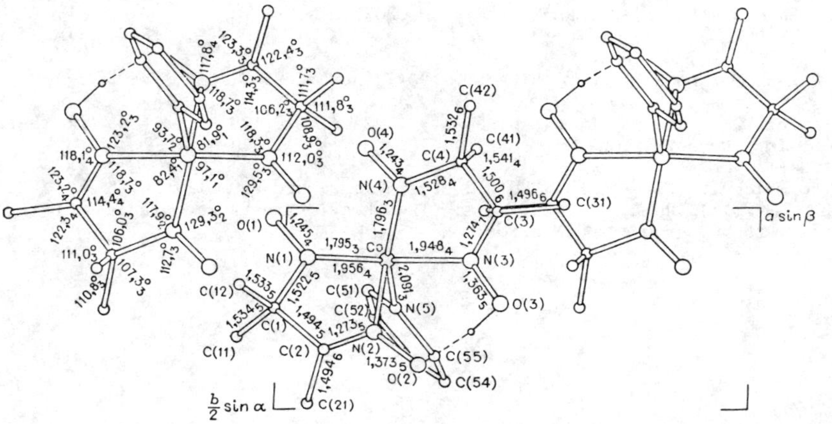

Fig. 1. The structure of [3-(hydroxyamino)-3-methyl-2-butanone-oximato-N,N'(2-)]-[3-(hydroxyamino)-3-methyl-2-butanone-oximato-N,N']pyridinecobalt(III) projected on the ab plane.

(μ-HYDRIDO)(μ-η^2-PHENYLDIAZENIDO)DECACARBONYLTRIOSMIUM

$C_{16}H_6N_2O_{10}Os_3$ $(\mu\text{-}H)Os_3(CO)_{10}(N{=}NC_6H_5)$

D.E. SAMKOFF, J.R. SHAPLEY, M.R. CHURCHILL and H.J. WASSERMAN, 1984. Inorg. Chem., 23, 397-402.

Orthorhombic, $P2_12_12_1$, a = 9.482, b = 12.542, c = 17.588 Å, Z = 4. Mo radiation, R = 0.077 for 2114 reflexions.

The molecule is shown in Fig. 1. The (μ-η^2-N=NPh) ligand spans the Os(2)-Os(3) linkage, with Os(3)-N(1) 2.14(2), Os(2)-N(2) 2.11(3), and N(1)-N(2) 1.20(4) Å. The (μ-H,μ-N=NPh) dibridged Os(2)-Os(3) distance is 2.895(2) Å as compared to Os(1)-Os(2) 2.868(2) and Os(1)-Os(3) 2.862(2) Å. The Os-CO distances range from 1.80(3) to 2.07(4) Å.

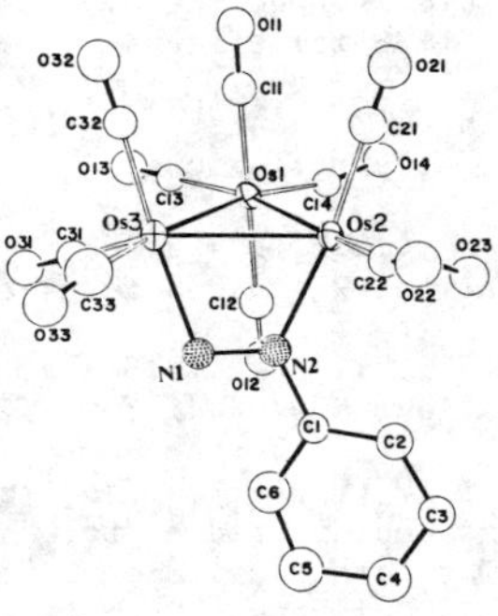

Fig. 1. The $(\mu\text{-}H)Os_3(CO)_{10}(\mu\text{-}\eta^2\text{-}N{=}NC_6H_5)$ molecule.

ab-BIS(BENZONITRILE)-df-DICARBONYL-ce-DICHLORORUTHENIUM
$C_{16}H_{10}Cl_2N_2O_2Ru$ $[RuCl_2(CO)_2(C_6H_5CN)_2]$

J.-C. DARAN, Y. JEANNIN and C. RIGAULT, 1984. Acta Cryst., C40, 249-251.

Monoclinic, C2/c, a = 18.18, b = 9.95, c = 12.22 Å, β = 129.4°, Z = 4. Mo radiation, R = 0.066 for 998 reflexions.

The molecule (Fig. 1) has twofold crystallographic symmetry, with Ru-Cl 2.380(4), Ru-C(2) 1.877(13), Ru-N(1) 2.119(12) Å. The phenyl rings are coplanar with the plane defined by the CO groups and N atoms.

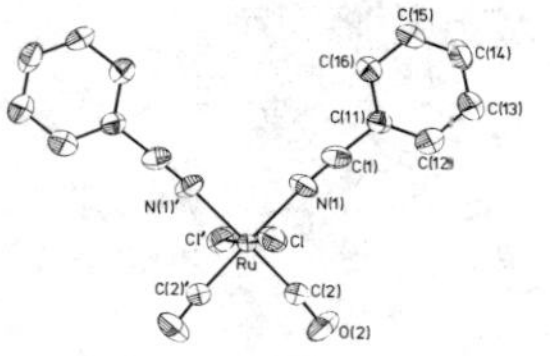

Fig. 1. A view of $[RuCl_2(CO)_{10}(C_6H_5CN)_2]$.

cis-CARBONYLDICHLORO-(2,2':6',2''-TERPYRIDYL)RUTHENIUM(II)
$C_{16}H_{11}Cl_2N_3ORu$ (I) $[Ru(tpy)(CO)Cl_2]$

trans-CARBONYLDICHLORO-(2,2':6',2''-TERPYRIDYL)RUTHENIUM(II)
$C_{16}H_{11}Cl_2N_3ORu$ (II) $[Ru(tpy)(CO)Cl_2]$

red-DIBROMODICARBONYL-(2,2':6',2''-TERPYRIDYL)RUTHENIUM(II)
$C_{17}H_{11}Br_2N_3O_2Ru$ (III) $[Ru(tpy)(CO)_2Br_2)]$

yellow-DIBROMODICARBONYL-(2,2':6',2''-TERPYRIDYL)RUTHENIUM(II)
$C_{17}H_{11}Br_2N_3O_2Ru$ (IV) $[Ru(tpy)(CO)_2Br_2]$

G.B. DEACON, J.M. PATRICK, B.W. SKELTON, N.C. THOMAS and A.H. WHITE, 1984. Aust. J. Chem., 37, 929-945.

I. Monoclinic, $P2_1/n$, a = 14.112, b = 14.330, c = 8.085 Å, β = 101.21°, D_m = 1.79, Z = 4. Mo radiation, R = 0.043 for 1738 reflexions.

II. Monoclinic, $P2_1/n$, a = 14.40, b = 13.196, c = 8.755 Å, β = 103.01°, D_m = 1.79, Z = 4. Mo radiation, R = 0.030 for 2146 reflexions.

III. Triclinic, $P\bar{1}$, a = 15.668, b = 14.272, c = 8.468 Å, α = 80.29, β = 77.96, γ = 88.68°, D_m = 1.98, Z = 4. Mo radiation, R = 0.038 for 4686 reflexions.

IV. Triclinic, $P\bar{1}$, a = 13.837, b = 12.744, c = 10.832 Å, α = 106.22, β = 96.11, γ = 91.37°, D_m = 2.00, Z = 4. Mo radiation, R = 0.047 for 3578 reflexions.

Ru is octahedral in compounds I-IV (Fig. 1). The two polymorphs (III, red, and IV, yellow) have octahedral Ru atoms coordinated to tpy which behaves as a bidentate ligand with trans Br and cis CO. The colour difference is likely to be found in interligand rather than intramolecular effects. Both III and IV have two molecules in the asymmetric unit.

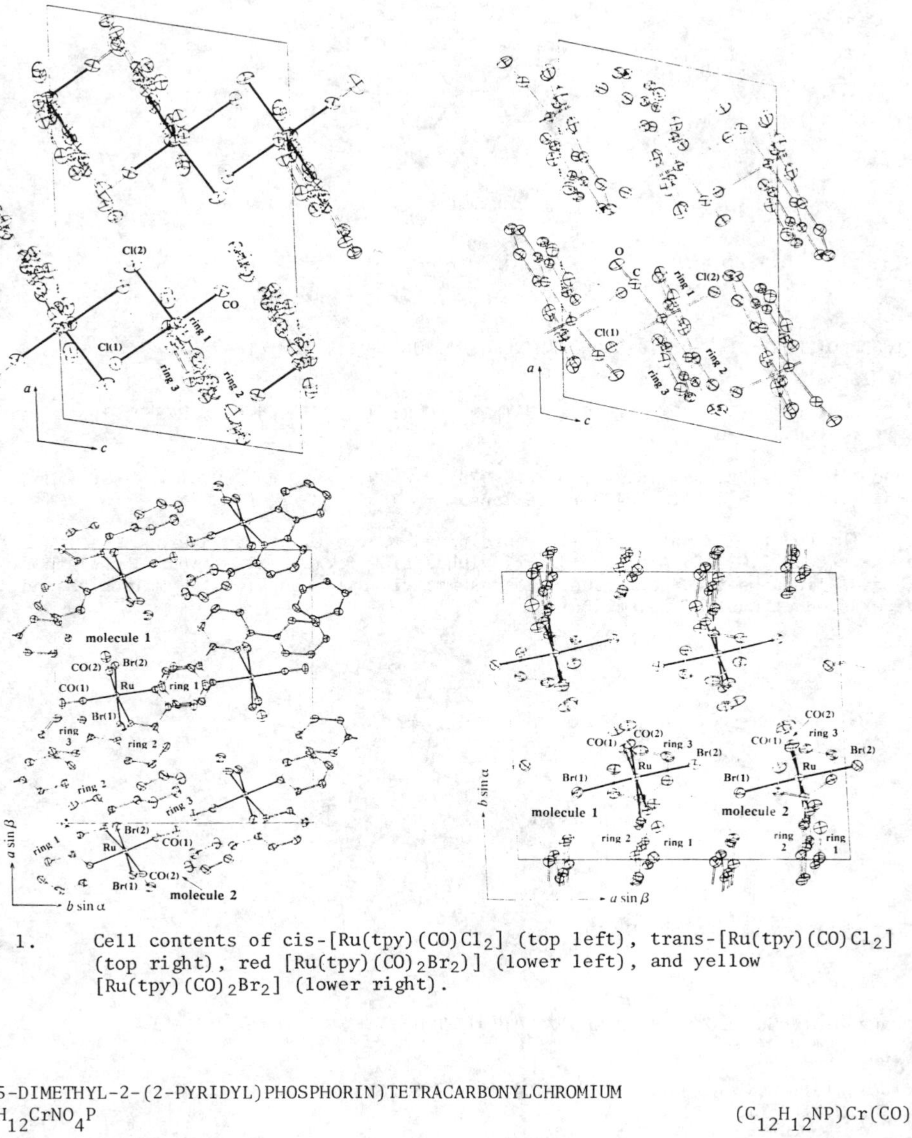

Fig. 1. Cell contents of cis-[Ru(tpy)(CO)Cl_2] (top left), trans-[Ru(tpy)(CO)Cl_2] (top right), red [Ru(tpy)$(CO)_2Br_2$)] (lower left), and yellow [Ru(tpy)$(CO)_2Br_2$] (lower right).

(4,5-DIMETHYL-2-(2-PYRIDYL)PHOSPHORIN)TETRACARBONYLCHROMIUM

$C_{16}H_{12}CrNO_4P$ $(C_{12}H_{12}NP)Cr(CO)_4$

A. BRÈQUE, C.C. SANTINI, F. MATHEY, J. FISCHER and A. MITSCHLER, 1984. Inorg. Chem., 23, 3463-3467.

Monoclinic, $P2_1/n$, a = 12.955, b = 18.943, c = 6.635 Å, β = 91.83°, D_m = 1.47, Z = 4. Cu radiation, R = 0.075 for 1640 reflexions.

The Cr atom has octahedral coordination geometry (Fig. 1) with Cr-P 2.280(1), Cr-N 2.193(4), Cr-CO 1.928(7) and 1.889(7) (trans to CO), 1.825(9) (trans to N), 1.843(5) Å (trans to P). The intracyclic O-C bonds are especially short (mean 1.798(4) Å).

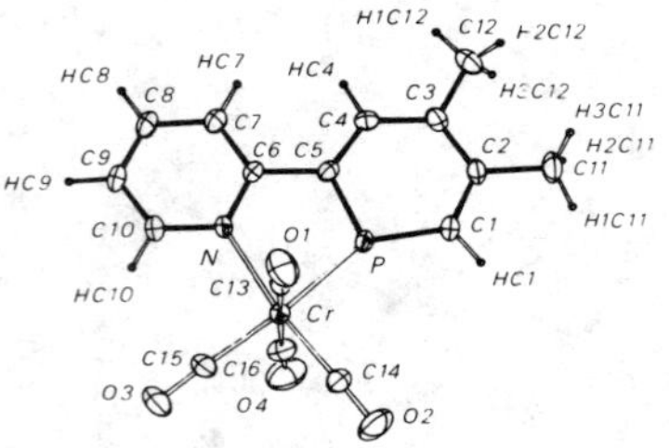

Fig. 1. The $(C_{12}H_{12}NP)Cr(CO)_4$ molecule.

DICHLORO[PHENYLDI(2-PYRIDYL)PHOSPHINE-N,N']COBALT(II) HEMIETHANOLATE
$C_{16}H_{13}Cl_2CoN_2P,0.5(C_2H_6O)$

M.G. EHRLICH, F.R. FRONCZEK, S.F. WATKINS, G.R. NEWKOME and D.C. HAGER, 1984. Acta Cryst., C40, 78-80.

Monoclinic, $P2_1/c$, a = 16.249, b = 12.078, c = 19.642 Å, β = 99.21°, Z = 8. Mo radiation, R = 0.029 for 2425 reflexions.

The Co atom is tetrahedrally coordinated to two Cl atoms (average Co-Cl 2.221(1), average Co-N 2.031(2) Å). The dipyridylphosphine forms a six-membered chelate ring (Fig. 1) in a boat conformation. The asymmetric unit contains two molecules and a disordered ethanol of solvation.

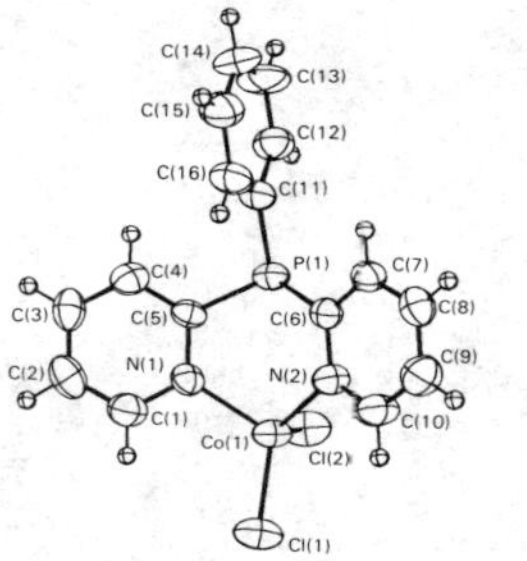

Fig. 1. A view of $C_{16}H_{13}Cl_2CoN_2P$.

trans-BIS[(PHENYLAZO)ACETALDOXIMATO]PLATINUM(II)
$C_{16}H_{16}N_6O_2Pt$ (I)

trans-BIS[(PHENYLAZO)ACETALDOXIMATO]PALLADIUM(II)
$C_{16}H_{16}N_6O_2Pd$ (II)

D. BANDYOPADHYAY, P. BANDYOPADHYAY, A. CHAKRAVORTY, F.A. COTTON and L.R. FALVELLO, 1984. Inorg. Chem., 23, 1785-1787.

I. Monoclinic, I2/a, a = 20.033, b = 3.956, c = 20.578 Å, β = 91.45°, Z = 4. Mo radiation, R = 0.017 for 930 reflexions.

II. Monoclinic, C2/c, a = 20.679, b = 7.759, c = 11.405 Å, β = 114.76°, Z = 4. Mo radiation, R = 0.025 for 1272 reflexions.

The two complexes are shown in Fig. 1. The Pt complex I is centrosymmetric with the metal atom and the atoms of the two chelate rings being planar to within 0.08 Å. Bond lengths: Pt-N 2.020(6) and 2.015(6) Å. Molecules of II have exact C_2 symmetry, each chelate ring being planar to within 0.06 Å. There is a large deviation of the PdN_4 coordination sphere from planarity toward the tetrahedral configuration. The dihedral angle between the two planar chelate rings is 25.6°. Bond lengths: Pd-N 2.033(3) and 2.027(3) Å.

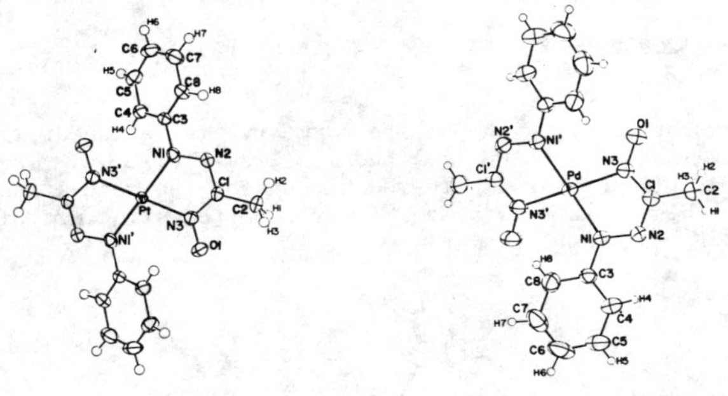

Fig. 1. The $C_{16}H_{16}N_6O_2M$ molecules: M=Pt (left) and Pd (right).

AQUA[μ_2-1,4-BIS(1'-METHYL-2'-IMIDAZOLYL)PHTHALAZINE-N^1,N^3,N^4,N^6](μ_2-HYDROXO)TRICHLORO-DICOPPER(II) MONOHYDRATE

$C_{16}H_{17}Cl_3Cu_2N_6O_2,H_2O$

L.K. THOMPSON, F.W. HARTSTOCK, P. ROBICHAUD and A.W. HANSON, 1984. Can. J. Chem., 62, 2755-2762.

Monoclinic, $P2_1/n$, a = 10.063, b = 20.967, c = 9.919 Å, β = 96.48°, Z = 4. Cu radiation, R = 0.034 for 2442 reflexions.

The copper coordination geometry (Fig. 1) is square pyramidal and the two copper(II) centres are bridged by two groups: N_2 (phthalazine) and hydroxide. A large metal...metal separation (3.425 Å) and Cu-O-Cu bridge angle (126.2°) are considered to be a dominant factor responsible for the very large antiferromagnetic exchange present in this system (-2J = 800 cm^{-1}). A comparison of -2J and the Cu-O-Cu bridge angle shows that for a related series of complexes with the same basic binuclear centre and a $d_{x^2-y^2}$ copper ion ground state an approximately linear relationship exists. Some bond lengths are Cu(1)-Cl(1) 2.235(1), Cu(1)-O(1) 1.921(3), Cu(1)-O(2) 2.286(4), Cu(1)-N(1) 2.002(4), Cu(1)-N(3) 2.039(4), Cu(2)-Cl(2) 2.247(1), Cu(2)-O(1) 1.919(3), Cu(2)-Cl(3) 2.679(2), Cu(2)-N(6) 2.003(4), Cu(2)-N(4) 2.049(3) Å.

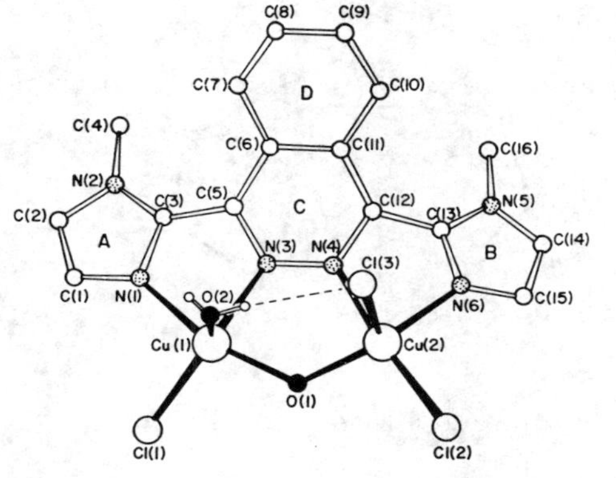

Fig. 1. Molecular structure of $C_{16}H_{17}Cl_3Cu_2N_6O_2$.

TETRAKIS(N-METHYLIMIDAZOLE-N')COPPER(I) PERCHLORATE
$C_{16}H_{24}ClCuN_8O_4$ $[Cu(C_4H_6N_2)_4],[ClO_4]$

W. CLEGG, S.R. ACOTT and C.D. GARNER, 1984. Acta Cryst., C40, 768-769.

Tetragonal, I$\bar{4}$, a = 12.952, c = 6.617 Å, Z = 2. Mo radiation, R = 0.044 for 1347 reflexions.

In the complex cation Cu, with S_4 site symmetry, is tetrahedrally coordinated, with Cu-N 2.054(2) Å and N-Cu-N 108.1(1) and 112.2(2)° (Fig. 1). The perchlorate anion is disordered. The cation is ordered except for the methyl substituent on the imidazole ligand. This appears to be the first reported structure of a simple monodentate-imidazole Cu^I complex.

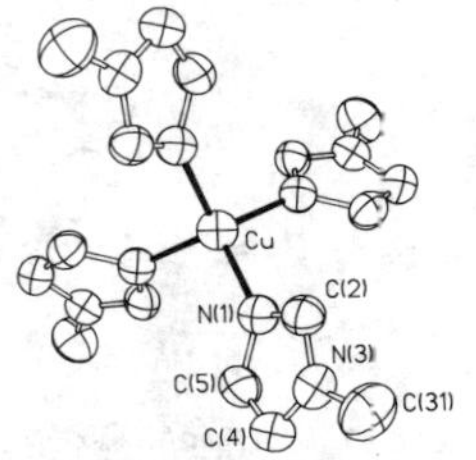

Fig. 1. $C_{16}H_{24}ClCuN_8O_4$: perspective view of the $[Cu(C_4H_6N_2)_4]^+$ cation.

TETRAKIS(DIETHYLAMIDO)TETRACOPPER(I)
$C_{16}H_{40}Cu_4N_4$ $[CuN(C_2H_5)_2]_4$

H. HOPE and P.P. POWER, 1984. Inorg. Chem., 23, 936-937.

Tetragonal, $I4_1/a$, a = 16.012, c = 8.473 Å, D_m = 1.66, Z = 4. Mo radiation, R = 0.032 for 1033 reflexions.

The tetrameric molecule (Fig. 1) has crystallographic $\bar{4}$ symmetry and contains an almost perfect square plane of two-coordinate Cu atoms bridged by diethylamido groups. The Cu-Cu distance is 2.664(2) Å, and the average Cu-N bond length is 1.904(3) Å. The N-Cu-N angle is 175.4(1)°, and the Cu-Cu-Cu angle is 89.9(1)°.

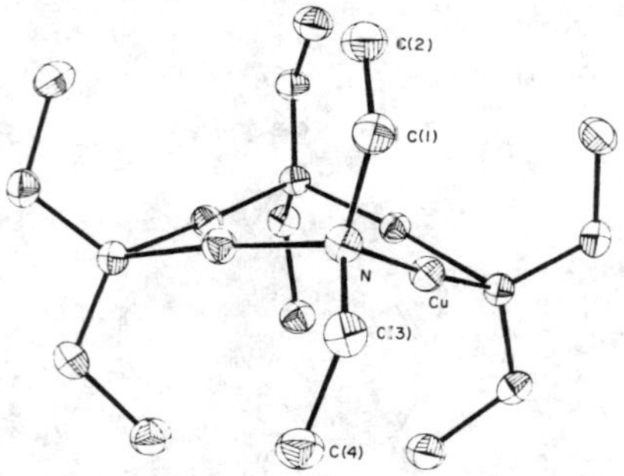

Fig. 1. The $[CuN(C_2H_5)_2]_4$ molecule.

BIS[BIS(TRIMETHYLSILYL)AMIDO]CHLORO-t-BUTYLIMIDOTANTALUM(V)
$C_{16}H_{45}ClN_3Si_4Ta$ (I)

BIS[BIS(TRIMETHYLSILYL)AMIDO]BIS(μ_2-BROMO)DIBROMOBIS(TRIMETHYLSILYLIMIDO)DITANTALUM(V)
$C_{18}H_{54}Br_4N_4Si_6Ta_2$ (II)

D.C. BRADLEY, M.B. HURSTHOUSE, K.M.A. MALIK, A.J. NIELSON and G.B.C. VURU, 1984. J. Chem. Soc. Dalton, 1069-1072.

I. Monoclinic, $P2_1/n$, a = 12.343, b = 19.089, c = 12.135 Å, β = 94.98°, Z = 4. Mo radiation, R = 0.041 for 3826 reflexions.

II. Triclinic, $P\bar{1}$, a = 9.254, b = 9.340, c = 12.287 Å, α = 87.57, β = 84.04, γ = 102.30°, Z = 1. Mo radiation, R = 0.053 for 2658 reflexions.

The molecule of I (Fig. 1) has a mononuclear pseudotetrahedral configuration with the t-butyl-imidotantalum moiety linear (C-N-Ta 165.8(6)°). The Ta-N(3) bond distance of 1.763(6) Å is indicative of triple bond character. Other bond distances are Ta-N(1,2) 1.988(5), 2.029(5), Ta-Cl 2.322(2) and Si-N 1.755(5)-1.774(6) Å. Compound II exists in the crystal as a centrosymmetric bromine bridged dimer. The Ta-Br(1) and Ta-Br(1') distances are 2.713(1) and 2.655(1) Å respectively. The trimethylsilyl-imidotantalum moiety is nearly linear (Si-N-Ta 166.1°) with Ta=N 1.761(9) and Si-N 1.733(9) Å. The Ta-N(2) distance is 1.965(9) Å. The Ta atoms are five-coordinate with distorted square based pyramidal geometry with the trimethylsilyl group axial.

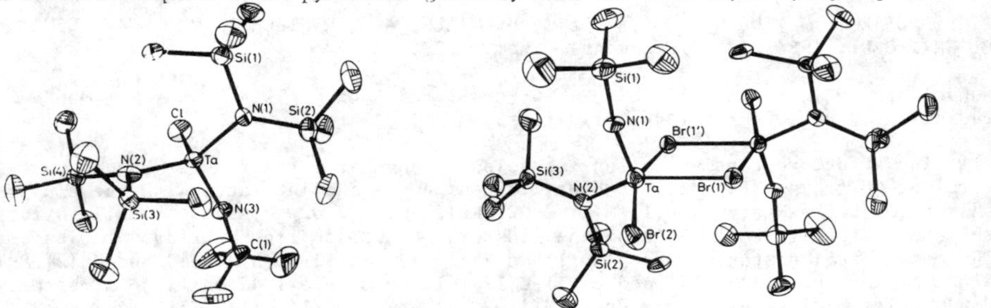

Fig. 1. The molecular structures: $C_{16}H_{45}ClN_3Si_4Ta$ (left) and $C_{18}H_{54}Br_4N_4Si_6Ta_2$ (right).

2,3-BIS(ACETONITRILE)-1,1,1,1,2,2,2,3,3,3-DECACARBONYL-2,3-(μ_2-HYDRIDO)TRIOSMIUM TRICARBONYLTRICHLOROOSMATE

$C_{17}H_7Cl_3N_2O_{13}Os_4$ $[Os_3(H)(CO)_{10}(MeCN)_2]^+.[Os(CO)_3Cl_3]^-$

C.E. ANSON, E.J. DITZEL, M. FAJARDO, H.D. HOLDEN, B.F.G. JOHNSON, J. LEWIS, J. PUGA and P.R. RAITHBY, 1984. J. Chem. Soc. Dalton, 2723-2727.

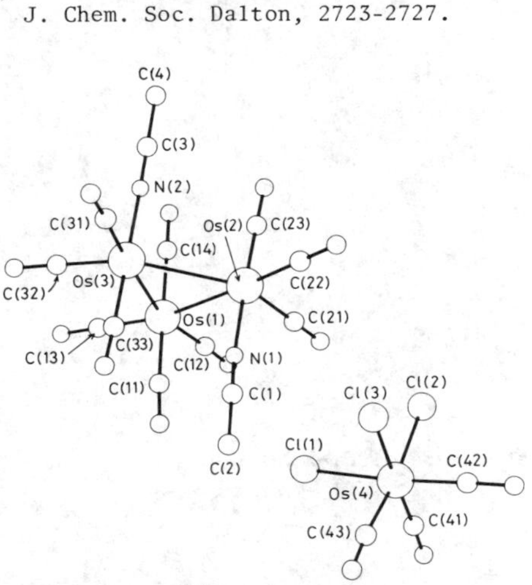

Fig. 1. The structure of $[Os_3(H)(CO)_{10}(MeCN)_2]^+.[Os(CO)_3Cl_3]^-$.

Monoclinic, $P2_1/n$, a = 9.825, b = 21.751, c = 13.715 Å, β = 90.72°, Z = 4. Mo radiation, R = 0.070 for 5167 reflexions.

In the cation (Fig. 1) the Os atoms lie at the vertices of an isosceles triangle with the two MeCN ligands occupying trans-axial sites on two of the Os atoms. The Os-Os distance (Os(2)-Os(3)) between the two atoms bonded to MeCN is relatively long (3.002(2) Å) compared with the other Os-Os bond lengths (2.886 and 2.882(2) Å). The presence of a hydride ligand bridging Os(2) and Os(3) is inferred from analysis of the ligand distribution in the cluster. The anion adopts the fac configuration (Fig. 1) with close to ideal octahedral geometry at Os(4) and mean Os-Cl 2.39(2), Os-C(carbonyl) 1.95(5) Å.

(μ_2-BENZYLIDENEAMINO)(μ_2-HYDRIDO)DECACARBONYLTRIRUTHENIUM

$C_{17}H_7NO_{10}Ru_3$ $(H)Ru_3(CO)_{10}(N{=}CHC_6H_5)$

P.M. LAUSAROT, M. TURINI, G.A. VAGLIO, M. VALLE, A. TIRIPICCHIO, M.T. CAMELLINI and P. GARIBOLDI, 1984. J. Organometal. Chem., 273, 239-246.

Monoclinic, $P2_1$, a = 13.127, b = 9.296, c = 9.624 Å, β = 111.41°, Z = 2. Mo radiation, R = 0.076 for 1446 reflexions.

The structure consists of an isosceles triangular metal cluster bonded to ten terminal CO groups (Fig. 1). A benzylideneamino ligand interacts with the cluster only through the N atom (Ru(1)-N(1) 2.02, Ru(2)-N(1) 2.02 Å), symmetrically bridging the shortest Ru(1)-Ru(2) edge of the cluster. A hydride ligand bridges the same edge. The metal-metal distances are: Ru(1)-Ru(2) 2.772, Ru(1)-Ru(3) 2.819 and Ru(2)-Ru(3) 2.822 Å, and the N(1)-C(11) and C(11)-C(12) distances are 1.42 and 1.46 Å respectively. Bond angles of interest are: Ru(1)-N(1)-Ru(2) 86.6, Ru(1)-N(1)-C(11) 127, Ru(2)-N(1)-C(11) 146 and N(1)-C(11)-C(12) 123°.

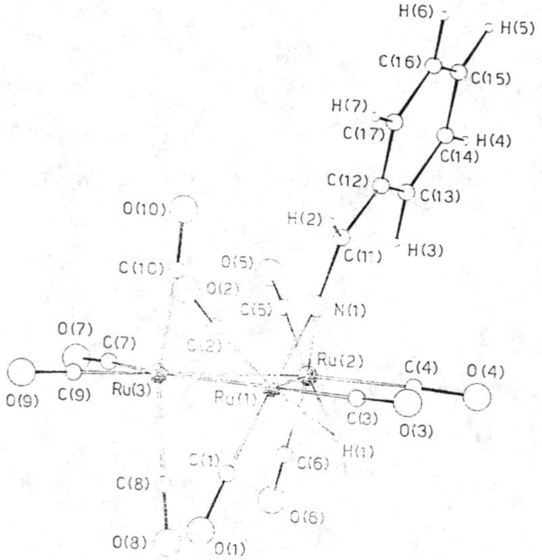

Fig. 1. Molecular structure of $C_{17}H_7NO_{10}Ru_3$.

TRICARBONYL-μ-CHLOROCHLORO(TETRAHYDRO-4,4,8,8-TETRAMETHYL-2-PHENYL-4H-1,3,6,2-DIOXAZAPHOSPHOCINE-N^6,P^2)DIRHODIUM
$C_{17}H_{22}Cl_2NO_5PRh_2$

J. WACHTER, F. JEANNEAUX, G. LE BORGNE and J.G. RIESS, 1984. Organometallics, 3, 1034-1038.

Triclinic, $P\bar{1}$, a = 10.18, b = 12.77, c = 10.04 Å, α = 104.71, β = 100.31, γ = 70.55°, Z = 2. Mo radiation, R = 0.032 for 2805 reflexions.

The structure is best described as two square-planar arrangements of ligands, each centred around one of the metal atoms, these two moieties being linked by a single bridging chlorine atom Cl(1) in such a way that the CO groups C(1) and C(2) are cis-oriented to each other (Fig. 1). The two square planes form a dihedral angle of 86.2°, and the angle Rh(1)-Cl(1)-Rh(2) is 89.36°. Important bond distances are: Rh(1)-Cl(1) 2.372, Rh(1)-Cl(2) 2.352, Rh(1)-C(1) 1.838, Rh(1)-C(3) 1.829, Rh(2)-Cl(1) 2.454, Rh(2)-P 2.156, Rh(2)-N 2.164 and Rh(2)-C(2) 1.842 Å. The Rh(1)...Rh(2) distance is 3.393 Å. There appears to be a hydrogen bond of the type Cl...H-N. The intramolecular distance for N...Cl(2) is 3.269 Å which is at the upper limit of the range of values usually accepted for such hydrogen bonds (3.16-3.26 Å).

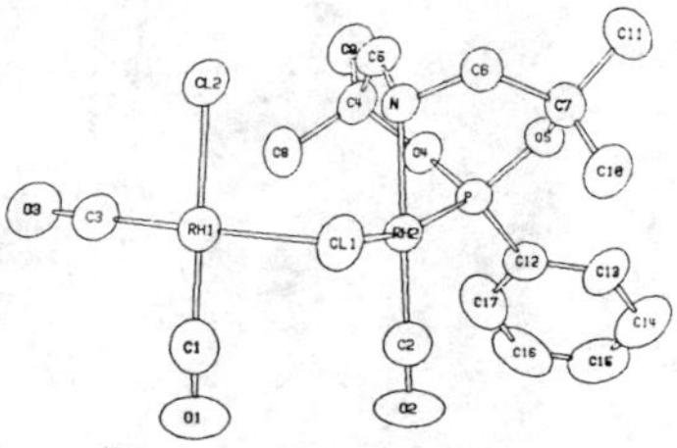

Fig. 1. View of the molecule $C_{17}H_{22}Cl_2NO_5PRh_2$.

(1-CYANOETHYL)BIS(DIMETHYLGLYOXIMATO)(4-METHYLPYRIDINE)COBALT(II)
$C_{17}H_{25}CoN_6O_4$

A. UCHIDA, Y. OHASHI, Y. SASADA, Y. OHGO and S. BABA, 1984. Acta Cryst., B40, 473-478.

Monoclinic, $P2_1$, a = 8.875, b = 28.525, c = 8.695 Å, β = 115.59°, D_m = 1.47, Z = 4. Mo radiation, R = 0.063 for 3255 reflexions.

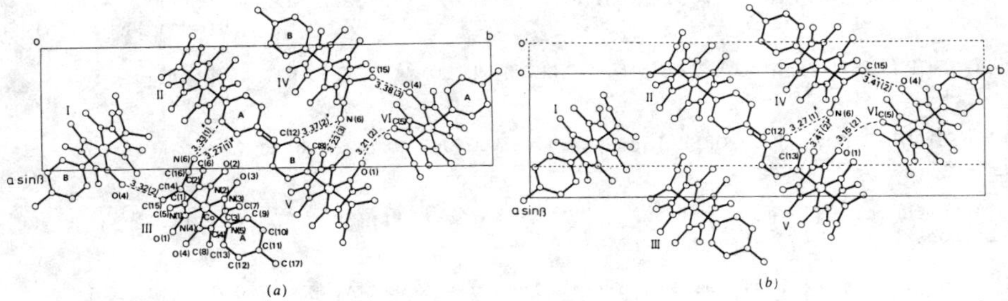

Fig. 1. The crystal structure of $C_{17}H_{25}CoN_6O_4$ viewed along the c-axis; (a) initial stage, (b) final stage.

The two crystallographically independent molecules are related by a pseudo inversion centre. On irradiation with X-rays, the unit-cell dimensions gradually change, converging to a = 8.691(10), b = 28.05(2), c = 8.864(7) Å, β = 114.49(7)° without degradation of crystallinity. The space group is transformed to $P2_1/a$. R = 0.120 for 2452 observed reflections. One of the two cyanoethyl groups in the crystallographically independent molecules changes its configuration from R to S. The volume of the cavity for the reactive cyanoethyl group (12.61 $Å^3$) is greater than that for the non-reactive group (11.05 $Å^3$). Details of the initial and final stages of the structure are in Fig. 1.

{BIS[2-(1-METHYL-2-THIOIMIDAZOLE)ETHYL]AMINE-N,N',N''}(1,2-DIMETHYLIMIDAZOLE)-COPPER(II) BIS(HEXAFLUOROPHOSPHATE) ACETONE SOLVATE
$C_{17}H_{27}CuF_{12}N_7P_2S_2,0.5(C_3H_6O)$

K.D. KARLIN, R.W. CRUSE, J.W. MCKOWN, J.C. HAYES, P.L. DAHLSTROM and J. ZUBIETA, 1984. Transition Met. Chem., 9, 404-405.

Monoclinic, $P2_1/c$, a = 14.764, b = 14.630, c = 15.087 Å, β = 113.45°, Z = 4. Mo radiation, R = 0.068 for 1152 reflexions.

The coordination around the Cu(II) ion is essentially square (Fig. 1) with Cu-N(1) 2.045(16), Cu-N(2) 2.023(14), Cu-N(3) 1.965(15), Cu-N(6) 1.987(17) Å, cis N-Cu-N 89.1-91.4(6)°.

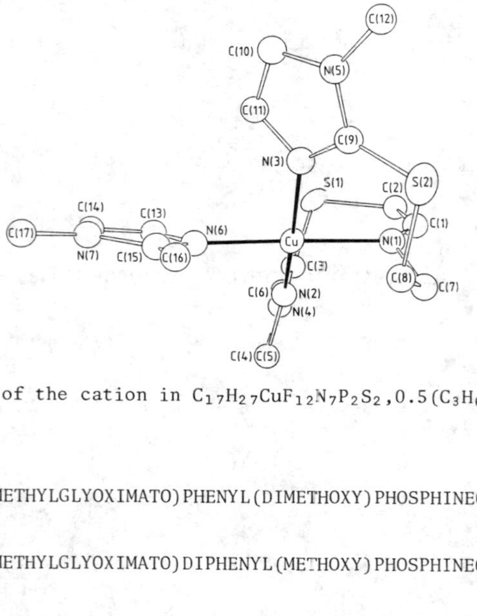

Fig. 1. A view of the cation in $C_{17}H_{27}CuF_{12}N_7P_2S_2,0.5(C_3H_6O)$.

trans-METHYLBIS(DIMETHYLGLYOXIMATO)PHENYL(DIMETHOXY)PHOSPHINECOBALT(III)
$C_{17}H_{28}CoN_4O_6P$ (I) $(P(OMe)_2Ph)Co(DH)_2CH_3$

trans-METHYLBIS(DIMETHYLGLYOXIMATO)DIPHENYL(METHOXY)PHOSPHINECOBALT(III)
$C_{22}H_{30}CoN_4O_5P$ (II) $(P(OMe)Ph_2)Co(DH)_2CH_3$

N. BRESCIANI-PAHOR, J.D. ORBELL and L. RANDACCIO, 1984. Croat. Chem. Acta 57, 433-441.

I. Orthorhombic, Pbca, a = 14.652, b = 26.45, c = 11.330 Å, D_m = 1.43, Z = 8. Mo radiation, R = 0.033 for 2108 reflexions.

II. Monoclinic, $P2_1$, a = 9.423, b = 15.184, c = 8.948 Å, β = 100.50°, D_m = 1.36, Z = 2. Mo radiation, R = 0.034 for 2063 reflexions.

In both compounds the Co atom has distorted octahedral geometry (Fig. 1), the two DH units occupying equatorial positions and Me and the P ligand axial ones. The four N-donors are coplanar to within ±0.004 Å in (I) and ±0.005 Å in (II) with the Co atom displaced 0.06 and 0.09 Å respectively above this plane. The Co-C and Co-P bond lengths are 2.013 and 2.287 Å in (I), and 2.019 and 2.352 Å in (II). The Co-N distances range from 1.875-1.885 Å in (I) and from 1.859 to 1.900 Å in (II), and the P-Co-C(Me) bond angle is 178.4 and 1.75.1° respectively in (I) and (II).

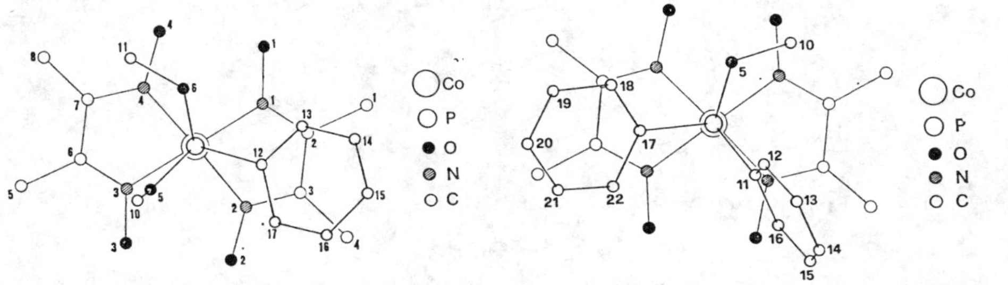

Fig. 1. Numbering schemes for $C_{17}H_{28}CoN_4O_6P$ (left), and $C_{22}H_{30}CoN_4O_5P$ (right). The molecules are viewed down the P-Co-C(Me) bond.

mer-TRICHLOROTRIS(ISONICOTINONITRILE)IRON(III)
$C_{18}H_{12}Cl_3FeN_6$ $[FeCl_3(C_6H_4N_2)_3]$

J.-C. DARAN, Y. JEANNIN and L. MARTIN, 1984. Acta Cryst., C40, 406-409.

Monoclinic, $P2_1/c$, a = 10.152, b = 12.216, c = 16.906 Å, β = 97.4°, D_m = 1.53, Z = 4. Mo radiation, R = 0.045 for 2324 reflexions.

The analysis establishes that the ligand geometry around Fe is meridional (Fig. 1), with Fe-Cl 2.262(2) (trans to N), 2.326 and 2.330(3) Å, Fe-N 2.280(4) trans to Cl), 2.202 and 2.223(4) Å.

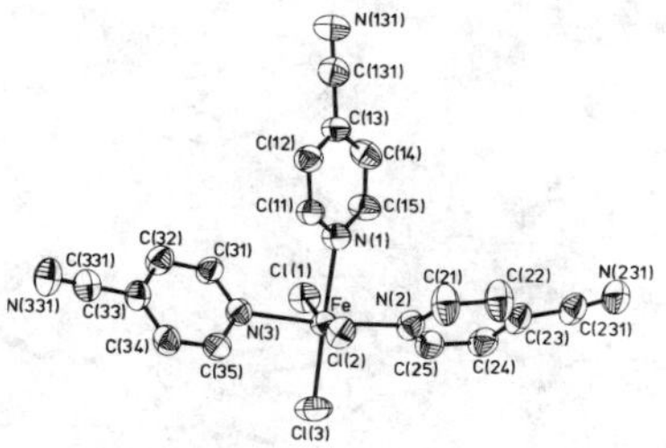

Fig. 1. A view of $[FeCl_3(C_6H_4N_2)_3]$.

CHLOROTRIS(3-METHYLPYRIDINE)COPPER(I)
$C_{18}H_{21}ClCuN_3$

E.W. AINSCOUGH, A.G. BINGHAM, A.M. BRODIE and K.L. BROWN, 1984. J. Chem. Soc. Dalton, 989-991.

Rhombohedral, R3, a = 14.384, c = 7.451 Å, Z = 3. Cu radiation, R = 0.048 for 765 reflexions.

The complex consists of monomeric units in which the Cu atom has tetrahedral coordination geometry (Fig. 1). The Cu-N(1) bond distance is 2.021(5) and the Cu-Cl bond is 2.458(2) Å. The monomeric units are stacked in such a manner that infinite linear Cu-Cl...Cu chains are formed, the intermolecular Cu...Cl distance being 4.994(2) and the intermolecular Cu...Cu distance 7.451(2) Å.

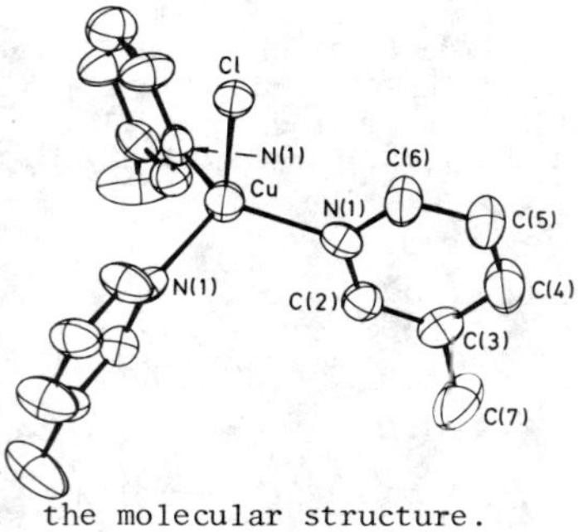

Fig. 1. $C_{18}H_{21}ClCuN_3$: the molecular structure.

BIS[2-(n-PROPYLMERCAPTOMETHYL)PYRIDINE]COPPER(II) DIPERCHLORATE
$C_{18}H_{26}Cl_2CuN_2O_8S_2$ $[Cu(Pr^nSCH_2py)_2]^{2+}[ClO_4^-]_2$

N. AOI, G.-E. MATSUBAYASHI, T. TANAKA and K. NAKATSU, 1984. Inorg. Chim. Acta, 85, 123-127.

Triclinic, P$\bar{1}$, a = 8.955, b = 8.794, c = 8.517 Å, α = 104.37, β = 109.723, γ = 78.86°, Z = 1. Mo radiation, R = 0.044 for 2444 reflexions.

The complex consists of a Cu atom centrosymmetrically coordinated to two pyridyl N atoms (Cu-N 2.015 Å), two thioether S atoms (Cu-S 2.366 Å) and two O atoms of perchlorate ions (Cu-O 2.501 Å) in a distorted octahedral fashion (Fig. 1). The Cu-N distance is longer than usual but the Cu-O distance is shorter than normal, consistent with the long Cl-O(1) distance of 1.449 Å compared to the other three Cl-O distances of 1.403, 1.409 and 1.411 Å. The angles about the metal atom range from 83.8 to 96.2°, and the angles Cu-O(1)-Cl, Cu-S-C(6) and Cu-S-C(7) are respectively 147.2, 93.6 and 104.7°.

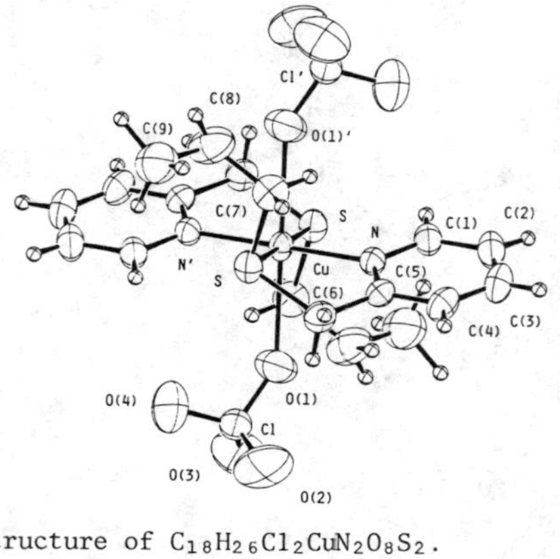

Fig. 1. Molecular structure of $C_{18}H_{26}Cl_2CuN_2O_8S_2$.

2,2'-DIAMINOBIPHENYL-R,R-trans-1,2-DIAMINOCYCLOHEXANEPLATINUM(II) CHLORIDE TRIHYDRATE
$C_{18}H_{26}Cl_2N_4Pt,3H_2O$ $[(C_6H_{10}(NH_2)_2)Pt(C_{12}H_8(NH_2)_2)]^{2+}2Cl^-,3H_2O$

C.A. RADLOWSKI, C.F. LIU and M.-J. JUN, 1984. Inorg. Chim. Acta, 86, 101-106.

Orthorhombic, $P2_12_12_1$, a = 6.219, b = 17.633, c = 21.523 Å, D_m = 1.74, Z = 4. Mo radiation, R = 0.059 for 3279 reflexions.

The metal atom has square planar stereochemistry (Fig. 1) with N(1)-Pt-N(2) and N(3)-Pt-N(4) angles of 83.1 and 88.5°. The conformation of the diaminobiphenyl rings are unambiguously found to be λ. Cyclohexanediamine was also found to be in the λ conformation. The Pt-N distances are 2.035, 2.023, 2.063 and 2.073 Å respectively for N(1), N(2), N(3) and N(4). The Pt-N(1)-N(2) plane is at an angle of 1.8° to the Pt-N(3)-N(4) plane so the N atoms of the opposite rings do not eclipse each other exactly. The Pt-N-C angles are 109.2, 109.8, 117.2 and 114.3° for N(1), N(2), N(3) and N(4) respectively.

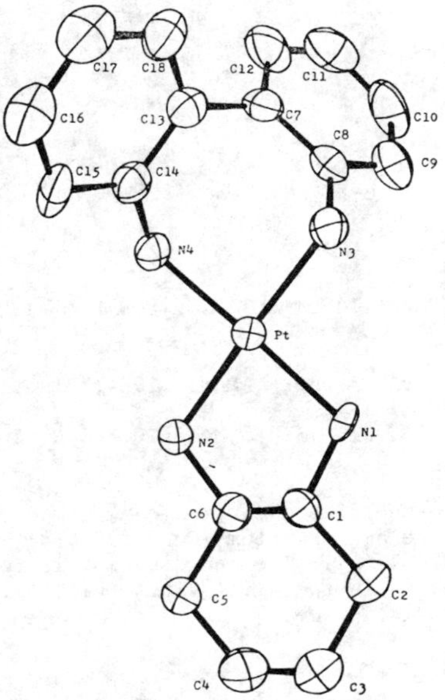

Fig. 1. Molecular conformation of $C_{18}H_{26}Cl_2N_4Pt$.

(N,N-DIETHYLNICOTINAMIDE)DIACETYL-BIS(HYDRAZONATO)-S-METHYLCARBODITHIOATEZINC(II)
$C_{18}H_{26}N_5OS_4Zn$ (I)

DI-μ-(N,N-DIETHYLNICOTINAMIDE-O,N)TETRACHLORODIZINC(II)
$C_{10}H_{14}Cl_2N_2OZn$ (II)

G. DAVIES, A. EL-TOUKHY, K.D. ONAN and M. VEIDIS, 1984. Inorg. Chim. Acta, 84, 41-50.

I. Triclinic, $P\bar{1}$, a = 10.258, b = 10.708, c = 12.548 Å, α = 107.69, β = 99.70, γ = 106.49°, Z = 2. Cu radiation, R = 0.038 for 3642 reflexions.

II. Triclinic, $P\bar{1}$, a = 9.427, b = 14.106, c = 8.589 Å, α = 120.87, β = 116.17, γ = 106.76°, Z = 2. Cu radiation, R = 0.062 for 1851 reflexions.

Compound I consists of isolated units of cis-N_2S_2 coordination in the equatorial plane and an axial DENC ligand (Fig. 1). The N_2S_2 plane is inclined at 83.7° to the plane of the pyridine ring, and the Zn atom is displaced 0.46 Å out of the N_2S_2 plane towards the pyridine N atom. The Zn-S and Zn-N equatorial distances are 2.360, 2.362 and 2.132, 2.120 Å respectively, with Zn-N(15) 2.139 Å. Compound II consists of discrete centrosymmetric dimers in which the Zn atoms are tetrahedrally coordinated (Fig. 1). The coordination angles about Zn are distorted, ranging from 100.8 to 120.1° and the bond distances are: Zn-O(13) 2.010, Zn-Cl(1) 2.199, Zn-Cl(2) 2.203 and Zn-N(1') 2.073 Å.

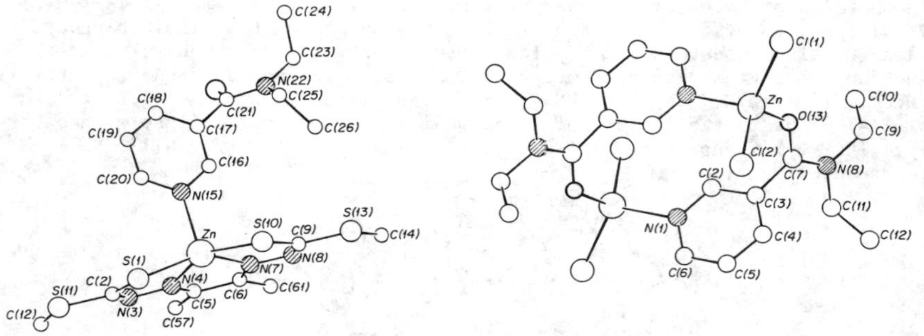

Fig. 1. Perspective views of $C_{18}H_{26}N_5OS_4Zn$ (left), and $C_{10}H_{14}Cl_2N_2OZn$ (right).

AQUO[TRIS[(3,5-DIMETHYL-1-PYRAZOLYL)METHYL]AMINE]COBALT(II) PERCHLORATE

$C_{18}H_{29}CoCl_2N_7O_9$ $[Co(H_2O)(N(CH_2C_5H_7N_2)_3)](ClO_4)_2$

C. BENELLI, I. BERTINI, M. DI VAIRA and F. MANI, 1984. Inorg. Chem., 23, 1422-1425.

Orthorhombic, $Pca2_1$, a = 18.456, b = 13.480, c = 10.888 Å, Z = 4. Mo radiation, R = 0.061 for 1820 reflexions.

The Co atom is in a five-coordinate environment formed by three pyrazole N atoms (Co-N 2.005(8)-2.060(9) Å), the amine N atom (Co-N 2.293(8) Å), and by the O atom of the water molecule (Co-O 2.024(7) Å). The coordination is approximately trigonal bipyramidal with amine-N and OH_2 in the axial sites (Fig. 1). The metal is displaced 0.48 toward O from the equatorial plane.

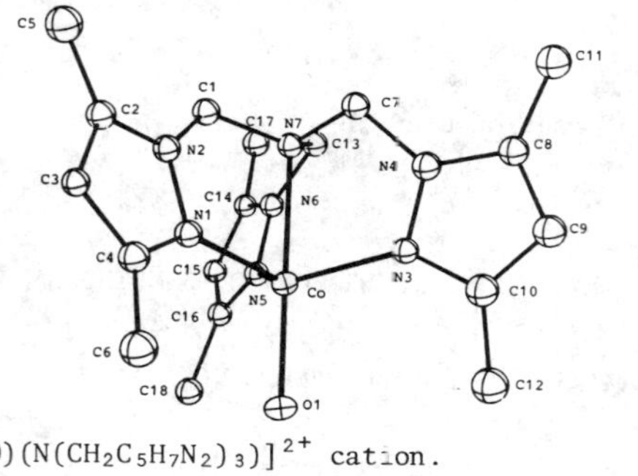

Fig. 1. The $[Co(H_2O)(N(CH_2C_5H_7N_2)_3)]^{2+}$ cation.

DIAQUOBIS(1,1'-DIMETHYL-2,2'-DIIMIDAZOLYLSULPHIDE)COPPER(II) METHYLSULPHATE

$C_{18}H_{30}CuN_8O_{10}S_4$

D.A. BALDWIN, J.C.A. BOEYENS, R.G. COPPERTHWAITE, J.H.N. LOUBSER and A.J. MARKWELL, 1984. J. Cryst. Spectrosc. Res., 14, 157-167.

Monoclinic, $P2_1/n$, a = 8.460, b = 8.503, c = 20.341 Å, β = 98.41°, D_m = 1.61, Z = 2. Mo radiation, R = 0.0973 for 2570 reflexions.

The Cu atom lies on an inversion centre (Fig. 1) and has tetragonally distorted octahedral geometry; four N atoms occupy the equatorial plane and two H_2O molecules occupy the axial positions. The average Cu-N and Cu-O distances are 2.005(6) and 2.47(1) Å, respectively. The measured spectral parameters (visible, EPR, and photoelectron spectroscopy) correlate well with the molecular structure.

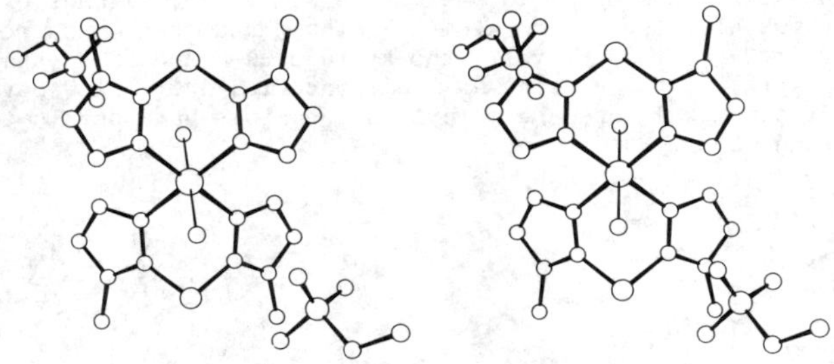

Fig. 1. A stereoview of $C_{18}H_{30}CuN_8O_{10}S_4$.

(μ_2-η^2,η^2-HYDRAZIDO)(μ_2-η^1,η^1-HYDRAZINE)HEXAMETHYLBIS(PHENYLIMIDO)DITUNGSTEN TOLUENE SOLVATE
$C_{18}H_{34}N_6W_2$,0.5(C_7H_8)

L. BLUM, I.D. WILLIAMS and R.R. SCHROCK, 1984. J. Am. Chem. Soc., 106, 8316-8317.

Orthorhombic, Pbca, a = 23.957, b = 15.329, c = 14.051 Å, Z = 8. Mo radiation, R = 0.043 for 2500 reflexions (at 250 K).

The molecule (Fig. 1) has the hitherto unknown μ-η^2,η^2-$NHNH^{2-}$ ligand as well as a μ-η^1,η^1-NH_2NH_2 link. The metal centres are chemically equivalent and the molecule has non-crystallographic C_{2v} symmetry. No W-W bond is present (W...W 3.731(1) Å) and the dihedral angle between the W(1)N_2 and W(2)N_2 planes is large (~135°) consistent with complex containing W(6+) and $NHNH^{2-}$. Some bond lengths are in Fig. 1; other values are W(1)-N(3) 2.106(12), W(1)-N(4) 2.115(11), W(2)-N(3) 2.157(11), W(2)-N(4) 2.170(12), N(3)-N(4) 1.391(15), N(1)-N(2) 1.434(14), W(1)-N(1) 2.320(11), W(2)-N(2) 2.353(11) Å.

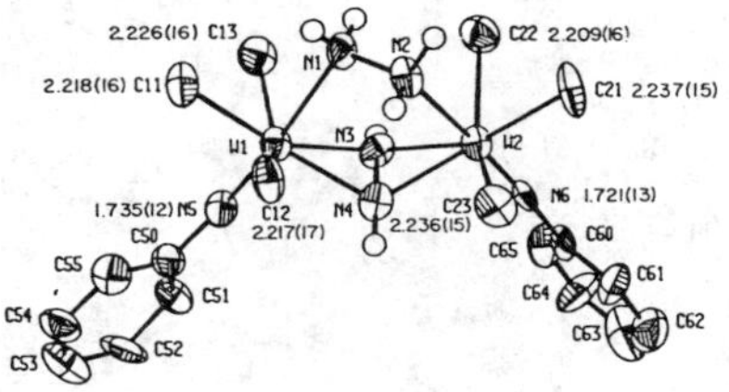

Fig. 1. A view of $C_{18}H_{34}N_6W_2$.

SODIUM TRIS[BIS(TRIMETHYLSILYL)AMIDO]EUROPATE(II)
$C_{18}H_{54}EuN_3NaSi_6$ (I) $NaEu(N(SiMe_3)_2)_3$

SODIUM TRIS[BIS(TRIMETHYLSILYL)AMIDO]YTTERBATE(II)
$C_{18}H_{54}N_3NaSi_6Yb$ (II) $NaYb(N(SiMe_3)_2)_3$

T.D. TILLEY, R.A. ANDERSEN and A. ZALKIN, 1984. Inorg. Chem., 23, 2271-2276.

I. Monoclinic, $P2_1/n$, a = 17.586, b = 19.170, c = 21.808 Å, β = 107.90°, Z = 8. Mo radiation, R = 0.031 for 6751 reflexions.

II. Orthorhombic, Pbca, a = 19.334, b = 18.534, c = 20.020 Å, Z = 8. Mo radiation, R = 0.056 for 2546 reflexions.

In the Eu complex I there are two crystallographically distinct, but chemically equivalent, molecules. The molecular structures of both complexes are similar (see

Fig. 1). The lanthanide atom bonds to one of the silylamide ligands exclusively and shares the two others with the sodium atom. The Eu-N and Yb-N (unshared) distances are 2.46(1) and 2.38(2) Å, respectively; the average Eu-N and Yb-N (shared) distances are 2.55(2) and 2.46(2) Å; the average Na-N distance is 2.46(1) Å. Several noteworthy close contacts between metal ions and methyl carbon atoms have been noted, i.e. Na-C 2.71, Eu-C 2.97, and Yb-C 2.86 Å.

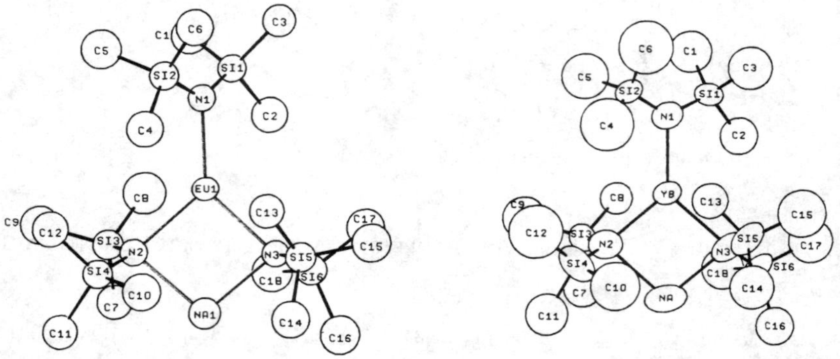

Fig. 1. One of the two $NaEu(N(SiMe_3)_2)_3$ molecules (left) and the $NaYb(N(SiMe_3)_2)_3$ molecule (right).

CYANO(DIHYDRODIBENZO[b,i][1,4,8,11]TETRAAZACYCLOTETRADECINATO)COBALT(III)
$C_{19}H_{14}CoN_5$

S. DEGER, M. HANACK, W. HILLER and J. STRAHLE, 1984. Justus Liebigs Ann. Chem., 1791-1797.

Orthorhombic, $P2_12_12_1$, a = 6.986, b = 14.662, c = 15.537 Å, Z = 4. Mo radiation, R = 0.074 for 1128 reflexions.

The Co atom (Fig. 1) is 0.196(2) Å from the plane of the four N atoms to which it is bonded, with Co-N1 1.916(8), Co-N2 1.86(1), Co-N3 1.914(9), Co-N(4) 1.852(9), Co-C19 1.82(1), C19-N5 1.16(2) Å.

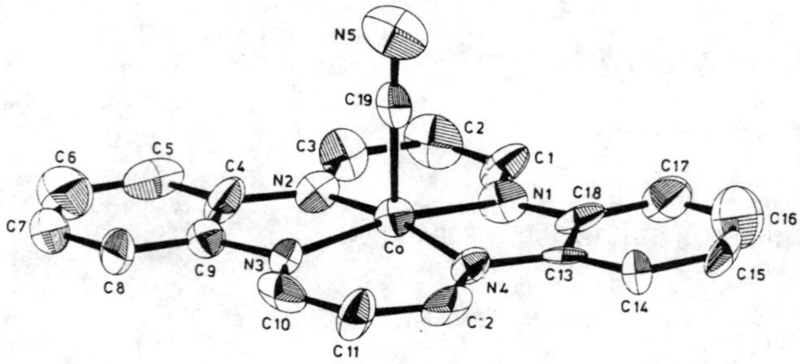

Fig. 1. A view of $C_{19}H_{14}CoN_5$.

TRICHLORO[7-CHLORO-1-(CYCLOPROPYLMETHYL)-1,3-DIHYDRO-5-PHENYL-2H-1,4-BENZODIAZEPIN-2-ONE]GOLD(III)
$C_{19}H_{17}AuCl_4N_2O$

G. MINGHETTI, M.L. GANADU, C. FODDAI, M.A. CINELLU, F. CARIATI, F. DEMARTIN and M. MANASSERO, 1984. Inorg. Chim. Acta, 86, 93-100.

Monoclinic, $P2_1/n$, a = 12.542, b = 9.269, c = 19.574 Å, β = 102.67°, Z = 4. Mo radiation, R = 0.056 for 1380 reflexions.

The monomeric molecules (Fig. 1) are separated by normal van der Waals contacts in the crystal. The Au(III) ion is bonded to three Cl atoms and to the organic ligand via the N(4) atom in a square planar fashion. The metal lies 0.026 Å above the best plane through atoms Au, Cl(1), Cl(2), Cl(3) and N(4). The three Au-Cl bond lengths are 2.272, 2.260 and 2.278 Å and the Au-N(4) distance is 2.030 Å. The seven-membered diazepine ring displays a boat-type conformation and the geometry around N(1) is trigonal planar. The C(5)-N(4) distance of 1.230 Å indicates a localized double bond.

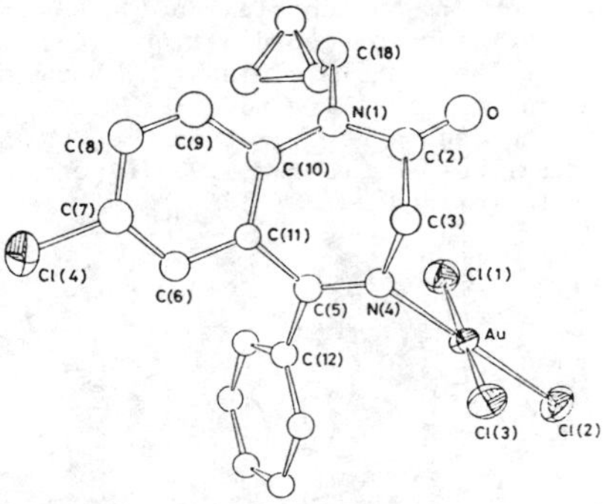

Fig. 1. Molecular structure of $C_{19}H_{17}AuCl_4N_2O$.

BIS(μ_2-1,8-NAPHTHYRIDINE-2(1H)-ONE-N,N')TETRACARBONYLDIRHODIUM(I) HEMIHYDRATE
$C_{20}H_{10}N_4O_6Rh_2,0.5H_2O$ $Rh_2(\mu-C_8H_5N_2O)_2(CO)_4,0.5H_2O$

A.M.M. LANFREDI, A. TIRIPICCHIO, R. USÓN, L.A. ORO, M.A. CIRIANO and B.E. VILLAROYA, 1984. Inorg. Chim. Acta, 88, L9-L10.

Monoclinic, C2/c, a = 33.910, b = 8.859, c = 16.279 Å, β = 113.65°, Z = 8. Cu radiation, R = 0.076 for 3281 reflexions.

The complex is binuclear with two Onapy ligands bridging the metal atoms through the N atoms (Rh(1)-N(2) 2.12, Rh(1)-N(3) 2.098, Rh(2)-N(1) 2.14, Rh(2)-N(4) 2.11 Å), the square planar coordination around each Rh atom being completed by two terminal CO groups (Rh(1)-C(1) 1.85, Rh(1)-C(2) 1.90, Rh(2)-C(3) 1.83, Rh(2)-C(4) 1.88 Å). The Onapy ligands are roughly planar and the dihedral angle between their mean planes is 81.5°. The Rh-Rh distance, at 2.880 Å, is considered short and indicative of a metal-metal interaction. There are no hydrogen bonds in the crystal structure.

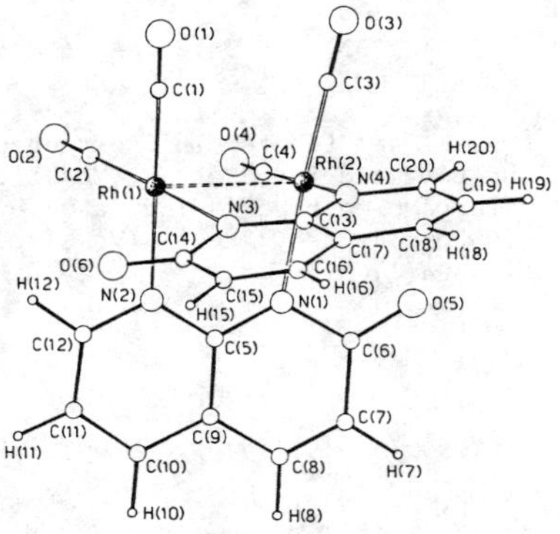

Fig. 1. View of the complex in $C_{20}H_{10}N_4O_6Rh_2$.

BIS(2,2'-BIPYRIDYL)CHLOROCOPPER(II) HEXAFLUOROPHOSPHATE MONOHYDRATE
$C_{20}H_{16}ClCuF_6N_4P,H_2O$

S. TYAGI, B. HATHAWAY, S. KREMER, H. STRATEMEIER and D. REINEN, 1984. J. Chem. Soc. Dalton, 2087-2091.

Monoclinic, $P2_1/n$, a = 21.403, b = 12.235, c = 8.599 Å, β = 92.68°, D_m = 1.63, Z = 4. Mo radiation, R = 0.052 for 2094 reflexions.

The structure consists of a complex cation, a PF_6^- anion and a water molecule (Fig. 1). The CuN_4Cl chromophore in the complex cation has regular trigonal-bipyramidal geometry with no evidence of coordination of either the anion or the water molecule. The individual pyridine rings are planar with twists of 9.07 and 4.17° with respect to one another in each bipyridyl ligand. The mean planes of each bipyridyl ligand make an angle of 127° to each other. Within the CuN_4Cl chromophore the Cu-Cl distance is 2.344(2), the axial Cu-N distances 1.996(6) and 2.005(6) and the equatorial Cu-N distances 2.105(6) and 2.108(6) Å.

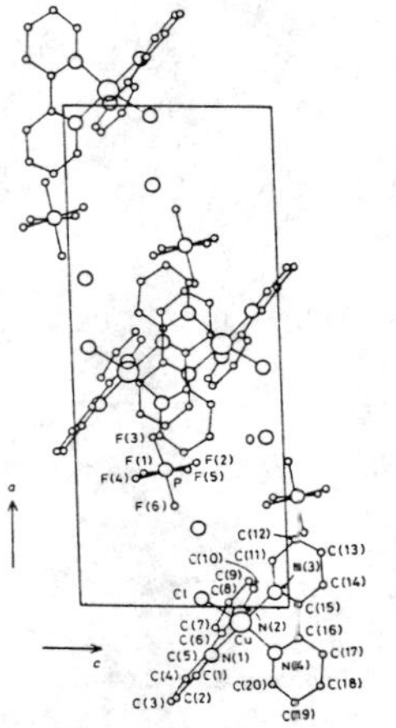

Fig. 1. $C_{20}H_{16}ClCuF_6N_4P,H_2O$: the molecular structure and crystal packing.

BIS(μ_2-OXO)DIOXODICHLOROBIS(2,2'-BIPYRIDYL)MOLYBDENUM(V) NITROMETHANE SOLVATE
$C_{20}H_{16}Cl_2Mo_2N_4O_4,CH_3NO_2$ $Mo_2O_4Cl_2(bipy)_2,CH_3NO_2$

J. BECK, W. HILLER, E. SCHWEDA and J. STRAHLE, 1984. Z. Naturforsch., B39, 1110-1113.

Monoclinic, $P2_1/n$, a = 9.541, b = 19.253, c = 14.232 Å, β = 102.48°, Z = 4. Cu radiation, R = 0.062 for 1576 reflexions.

The binuclear complex (Fig. 1) has two terminal oxygen atoms in a cis-arrangement (Mo-O 1.607 and 1.651(7) Å) and two bridging O atoms (Mo-O 1.915-1.957(6) Å) forming a non-planar Mo_2O_2 ring. A strong Mo-Mo bond is indicated by the short distance of 2.565(1) Å. Mo coordination is completed by the Cl (Mo-Cl 2.447 and 2.492(3) Å) and bipyridyl ligands (Mo-N 2.266-2.304(7) Å.

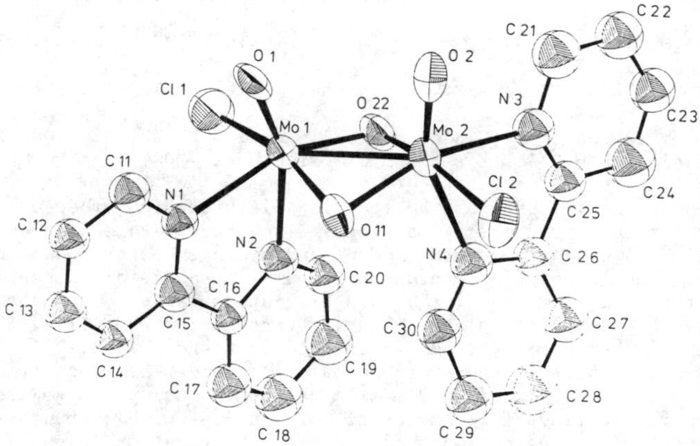

Fig. 1. A view of $Mo_2O_4Cl_2(bipy)_2$.

BIS(2,2'-BIPYRIDYL)COPPER(II) BIS(HEXAFLUOROPHOSPHATE)
$C_{20}H_{16}CuF_{12}N_4P_2$

J. FOLEY, S. TYAGI and B.J. HATHAWAY, 1984. J. Chem. Soc. Dalton, 1-5.

Tetragonal, $I4_1/acd$, a = 16.228, c = 18.954 Å, D_m = 1.75, Z = 8. Mo radiation, R = 0.048 for 319 reflexions.

The structure (Fig. 1) is composed of complex cations and PF_6^- anions. The cation contains a unique compressed tetrahedral CuN_4 chromophore (dihedral angle 44.6°). The Cu atom occupies a special position of 222 symmetry. The Cu-N distances (1.985 Å) are equivalent and the N(1)-Cu-N(1''') angle is 152.5°. The bipyridyl ligands bond symmetrically and the pyridine rings are planar. The PF_6^- anions have regular octahedral coordination and the P atom lies on a special twofold position. The P-F distance is 1.570 Å and the F-P-F angle 88.3°. There is slight evidence of disorder involving the F atoms. The PF_6^- anions occupy positions relative to the CuN_4 chromophore such that four long Cu...F distances are present (3.3 Å).

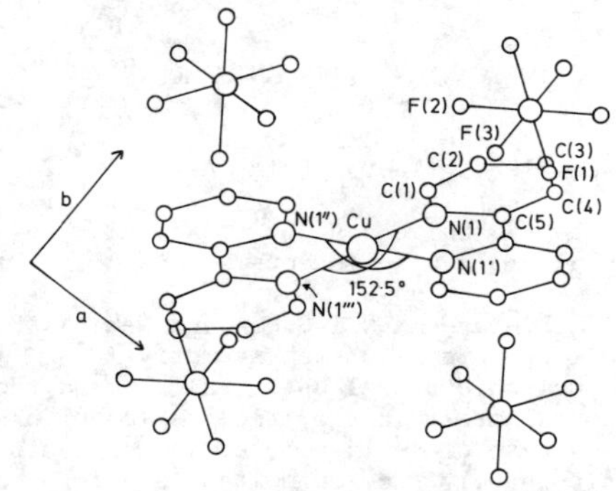

Fig. 1. $C_{20}H_{16}CuF_{12}N_4P_2$: the molecular structure viewed down the c axis.

2,2'-BIPYRIDINIUM(1+) (2,2'-BIPYRIDINE)OXODIPERCXOVANADATE(1-) HYDROGEN PEROXIDE HYDRATE

$C_{20}H_{17}N_4O_5V,3.4H_2O_2,1.6H_2O$ $(Hbipy)^+[VO(O_2)_2bipy]^-,3.4H_2O_2,1.6H_2O$

R. STOMBERG and H. SZENTIVANYI, 1984. Acta Chem. Scand., A38, 121-128.

Triclinic, $P\bar{1}$, a = 14.236, b = 10.848, c = 8.316 Å, α = 95.22, β = 104.98, γ = 93.71°, Z = 2. Mo radiation, R = 0.052 for 3336 reflexions (at -100°C).

The compound crystallises from the reaction mixture which produced a complex whose structure has been previously described (1). The crystals are composed of 2,2'-bipyridinium(1+) cations, complex anions and hydrogen peroxide and water of crystallization. The geometry of the anion (Fig. 1) is a distorted pentagonal bipyramid with the vanadyl O atom (O(5)) and N(2) at the apices, and the peroxo groups and N(1) forming the pentagonal plane. Bond distances are: V-N(1) 2.137, V-N(2) 2.284, V=O(5) 1.619, V-O(peroxo) from 1.862 to 1.942 Å and O(1)-O(2) 1.473, O(3)-O(4) 1.474 Å. The O-O distances in the hydrogen peroxide molecules range from 1.458 to 1.475 Å. The V atom is displaced 0.28 Å from the pentagonal plane towards the O(5) atom. The two pyridine rings form angles of 7.1 and 5.4° with each other in the cation and ligand respectively. The cation adopts a cis configuration (Fig. 1).

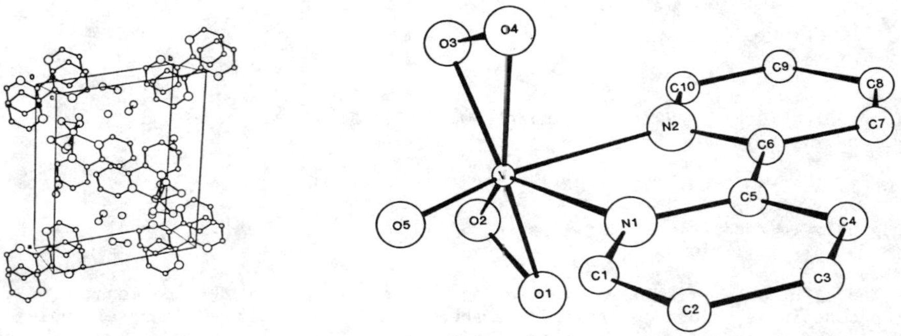

Fig. 1. Views of the unit cell, and the structure of the anion in $C_{20}H_{17}N_4O_5V$,-$3.4H_2O_2,1.6H_2O$.

1. This volume, p1607.

BIS(DIFLUOROBORONMETHYLGLYOXIMATO)NICKEL(II) ANTHRACENE

$C_{20}H_{18}B_2F_4N_4NiO_4$ $(\pi-C_{14}H_{10})Ni(C_6H_8B_2F_4N_4O_4)$

F.S. STEPHENS and R.S. VAGG, 1984. Inorg. Chim. Acta, 89, 203-208

Triclinic, $P\bar{1}$, a = 7.448, b = 7.780, c = 9.366 Å, α = 91.4, β = 92.5, γ = 92.1°, Z = 1. Mo radiation, R = 0.075 for 1250 reflexions.

The planar macrocycle (average Ni-N 1.850 Å and N-Ni-N 82.4°) and the anthracene molecule each have crystallographic inversion symmetry (Fig. 1). The planes of the macrocycle and anthracene molecule are almost parallel, and their stacking is such that the molecules lie approximately in the $(21\bar{1})$ series of planes with an interplanar spacing of 3.17 Å. These close intermolecular contacts indicate π-π interactions, giving rise to alternately stacked molecules aligned along a. Other bond lengths of interest are: N(1)-O(1) 1.383, N(2)-O(2') 1.348, O(1)-B 1.500, O(2)-B 1.499, B-F(1) 1.38, B-F(2) 1.38, N-C(1) 1.302, and N(2)-C(2) 1.303 Å.

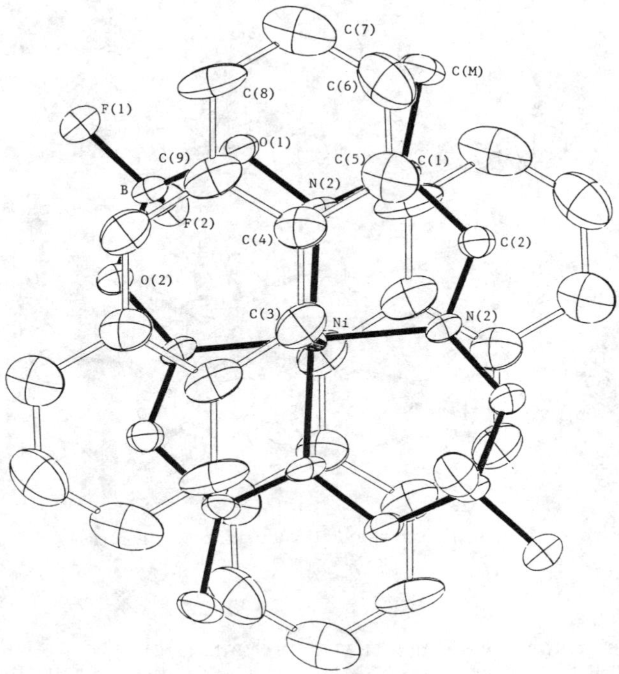

Fig. 1. Orientation of the anthracene molecule relative to the macrocycle in $C_{20}H_{18}B_2F_4N_4NiO_4$.

PHENAZINE BIS(DIFLUOROBORONDIMETHYLGLYOXIMATO)NICKEL(II)

$C_{20}H_{20}B_2F_4N_6NiO_4$ $Ni(C_4H_6BF_2N_2O_2)_2{:}C_{12}H_8N_2$

F.S. STEPHENS and R.S. VAGG, 1984. Inorg. Chim. Acta, 89, 47-52.

Monoclinic, $P2_1/c$, a = 9.130, b = 8.031, c = 15.492 Å, β = 91.20°, Z = 2. Cu radiation, R = 0.095 for 1571 reflexions.

The planar macrocycle (average Ni-N distance 1.850 Å, and N-Ni-N angle 82.3°) and phenazine molecules each have crystallographic inversion symmetry (Fig. 1), with non-bonded contacts between the phenazine N atoms and the metal atom of 3.38 Å. Within the difluoroborondimethylglyoximato ligand the N-O distances are 1.382, 1.343 Å; the N-C distances 1.289 and 1.291 Å; the O-B distances 1.491 and 1.483 Å and the B-F distances 1.373 and 1.358 Å. π-π Interactions give rise to two sets of alternatively stacked chains of molecules directed along b. The molecules lie on the $(11\bar{2})$ and $(\bar{1}22)$ series of planes which intersect at 67.7° and have separations of 3.32 Å.

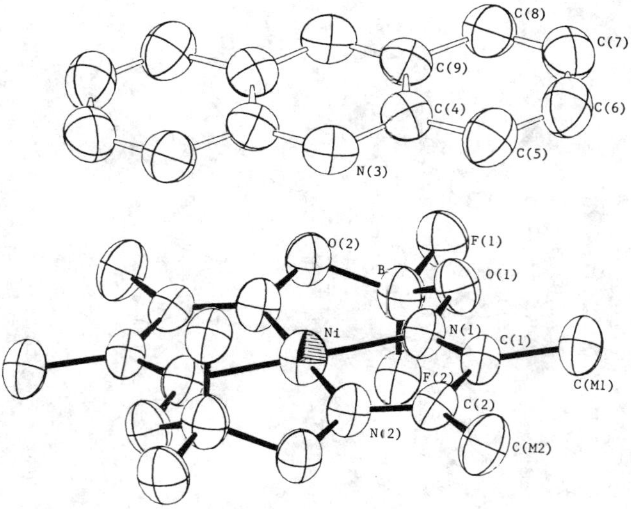

Fig. 1. Perspective drawing of the 1:1 adduct in $C_{20}H_{20}B_2F_4N_6NiO_4$.

trans-DIIODOTETRAKIS(PYRIDINE)CADMIUM(II) - PYRIDINE(1/2)

$C_{20}H_{20}CdI_2N_4,2C_5H_5N$ $[CdI_2(C_5H_5N)_4].2C_5H_5N$

M. ITO, T. SHIBATA and Y. SAITO, 1984. Acta Cryst., C40, 2041-2043.

Orthorhombic, Ccca, a = 11.620, b = 15.847, c = 17.245 Å, D_m = 1.7, Z = 4. Mo radiation, R = 0.058 for 1589 reflexions.

The complex has 222 symmetry and approximate 422 symmetry (Fig. 1). Cd is octahedrally coordinated with the N atoms of four pyridine molecules and two I^- ions in trans positions at average distances of 2.411(7) and 2.9693(7) Å, respectively. The crystal structure consists of layers perpendicular to c, built up of trans-$CdI_2(py)_4$ units linked together by van der Waals interactions between the ligating pyridine(py) molecules. The guest pyridine molecules are enclathrated in the cavities formed between the layers.

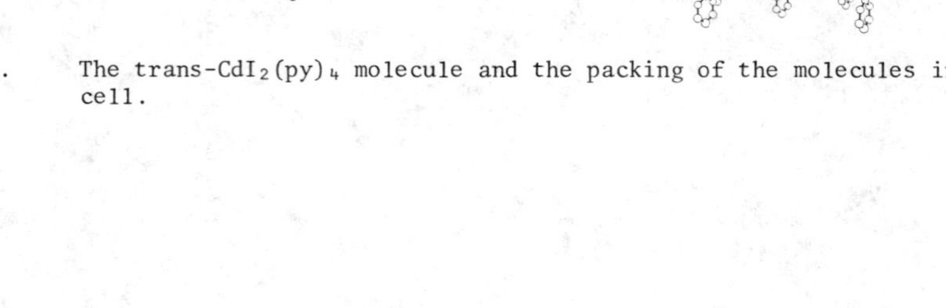

Fig. 1. The trans-$CdI_2(py)_4$ molecule and the packing of the molecules in the unit cell.

trans-DICHLOROTETRAKIS(PYRIDINE)IRIDIUM(III) CHLORIDE HEXAHYDRATE
$C_{20}H_{20}Cl_3IrN_4,6H_2O$ $[Ir(py)_4Cl_2]Cl,6H_2O$

R.D. GILLARD, S.H. MITCHELL, P.A. WILLIAMS and R.S. VAGG, 1984. J. Coord Chem., 13, 325-330.

Monoclinic, $P2_1/c$, a = 12.595, b = 10.498, c = 21.243 Å, β = 98.26°, Z = 4. Mo radiation, R = 0.034 for 3998 reflexions.

The cation has trans-octahedral geometry with Ir-Cl 2.348 and 2.351(1), Ir-N 2.053(16), 2.082(23) (mean 2.06 Å). The pyridine ligands are arranged in a four-bladed propellor orientation with pitch angles between 42.3 and 51.2° (Fig. 1). The six water molecules form a hydrogen-bonded sheet with Cl^- ions in holes in this sheet.

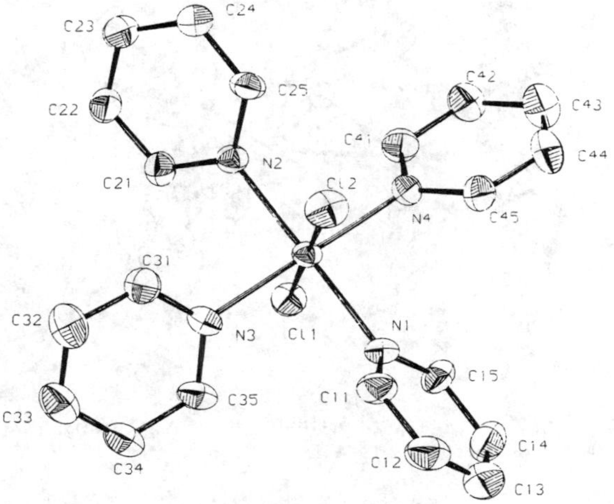

Fig. 1. A view of the $[Ir(py)_4Cl_2]^+$.

μ,μ'-DIIODO-[DI-(BIS-PYRIDINE)DICOPPER(I)]
$C_{20}H_{20}Cu_2I_2N_4$ $[py_2CuI_2Cupy_2]$

J.C. DYASON, L.M. ENGELHARDT, P.C. HEALY and A.H. WHITE, 1984. Aust. J. Chem., 37, 2201-2205.

Orthorhombic, $C222_1$, a = 14.578, b = 10.448, c = 31.559 Å, Z = 8. Mo radiation, R = 0.076 for 1355 reflexions. A twinned crystal was used in analysis.

This molecule is a μ,μ'-diiodobridged dimer with the metal atoms tetrahedrally coordinated (Fig. 1, Cu-I 2.639(4), Cu-N 2.06(3), Cu-Cu 2.699(5) Å, I-Cu-I 118.5(1), Cu-I-Cu 61.5(1)°).

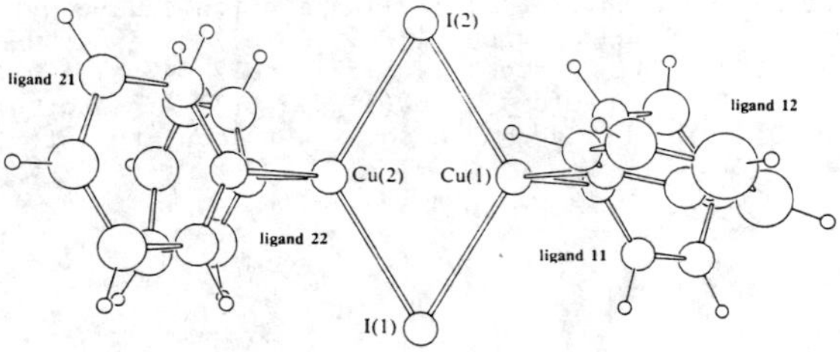

Fig. 1. The structure of $[py_2CuI_2Cupy_2]$.

BIS(μ_2-OXO)BIS(μ_2-SULPHIDO)TETRAOXOTETRAKIS(PYRIDINE)TRIMOLYBDENUM
$C_{20}H_{20}Mo_3N_4O_6S_2$

JIN XIANGLIN, BAI BO and TANG YOUQI, 1984. J. Struct. Chem., 3, 139-141.

Monoclinic, C2/c, a = 23.969, b = 13.315, c = 21.213 Å, β = 128.73°, D_m = 1.91, Z = 8. Mo radiation, R = 0.051 for 2984 reflexions.

The molecule is shown in Fig. 1 and has Mo(1)-Mo(2) 2.832(2), Mo-S 2.326-2.341(3), Mo(1)-O(3) 2.094, Mo(3)-O(3) 1.799, Mo(2)-O(2) 2.091, Mo(3)-O(2) 1.790(4), Mo-N 2.288-2.307(7), Mo-O(t) 1.677-1.701(12) Å. There is no bonding between Mo(3) and the Mo(1) or Mo(2) atoms (Mo...Mo 3.379 and 3.387 Å).

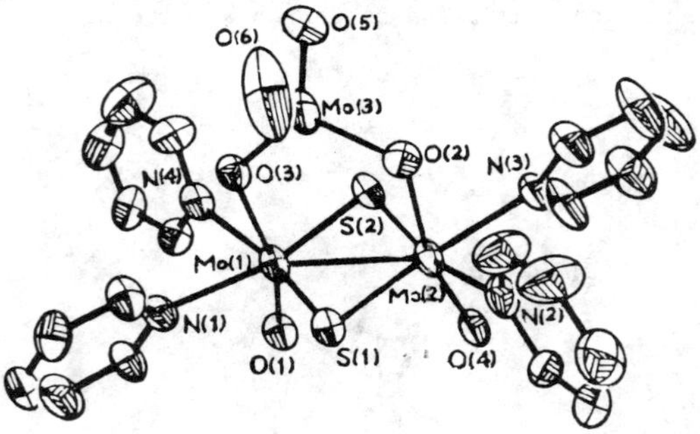

Fig. 1. The $C_{20}H_{20}Mo_3N_4O_6S_2$ structure.

(μ-HYDROXO)(μ-NITRATO-O,O')(μ-1,4-BIS((4-METHYL-2-PYRIDYL)AMINO)PHTHALAZINE-N^1,N^3,-N^4,N^6)DIAQUA(NITRATO-O)DICOPPER(II) NITRATE
$C_{20}H_{23}Cu_2N_9O_{12}$ (I)

(μ_3-SULPHATO-O,O',O'')(μ-HYDROXO)(μ-1,4-BIS(2-PYRIDYLAMINO)PHTHALAZINE-N^1,N^3,N^4,N^6)-CHLORODICOPPER(II) DIHYDRATE
$C_{18}H_{15}ClCu_2N_6O_5S,2H_2O$ (II)

L.K. THOMPSON, A.W. HANSON and B.S. RAMASWAMY, 1984. Inorg. Chem., 23, 2459-2465.

I. Monoclinic, $P2_1/a$, a = 27.078, b = 13.4451, c = 7.3744 Å, β = 105.575°, Z = 4. Cu radiation, R = 0.039 for 3710 reflexions.

II. Monoclinic, $P2_1/a$, a = 17.2850, b = 9.5248, c = 14.5520 Å, β = 117.198°, Z = 4. Cu radiation, R = 0.030 for 2566 reflexions.

The two complexes are shown in Fig. 1. Both Cu atoms in I are distorted six-coordinate, with bridging NO_3 and OH groups, Cu...Cu 3.138(1) Å, and a Cu-O-Cu bridge angle of 115.3(1)°, Cu-(μ-O) 1.856(2) and 1.859(2), Cu-(μ-O_3N) 2.428(3) and 2.330(3), Cu-OH_2 2.007(3) and 2.620(3), Cu-O_3N 2.035(2) and 2.738(2), Cu-N(pyr) 1.952(3) and 1.964(3), Cu-N 2.022(3) and 2.042(3) Å. In II distorted square pyramidal Cu atoms are found, with bridging O_4S and OH groups, Cu...Cu 3.211(1) Å, Cu-O-Cu 115.5(1)°, Cu-N(ax) 1.963(3) and 1.973(3), Cu-N(eq) 2.039(3) and 2.047(3), Cu-(μ-O) 1.900(3) and 1.892(3), Cu-Cl 2.301(1), Cu-O(eq) 2.060(3)-2.346(3) Å. The sulphate acts both as a bidentate intramolecular bridge and also as a tridentate intermolecular bridge (Fig. 2).

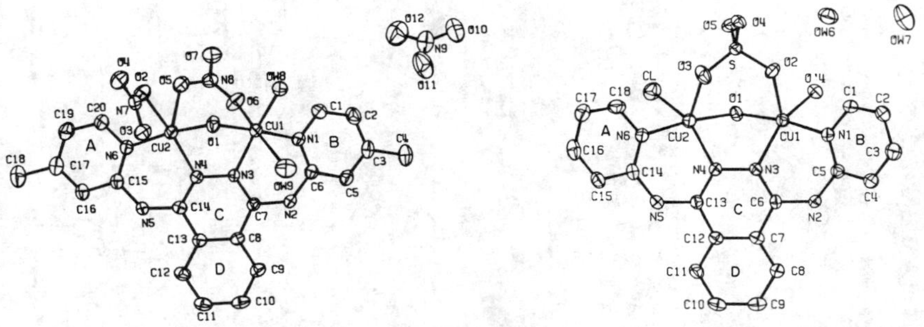

Fig. 1. The $C_{20}H_{23}Cu_2N_9O_{12}$ (left) and $C_{18}H_{15}ClCu_2N_6O_5S,2H_2O$ (right) structures.

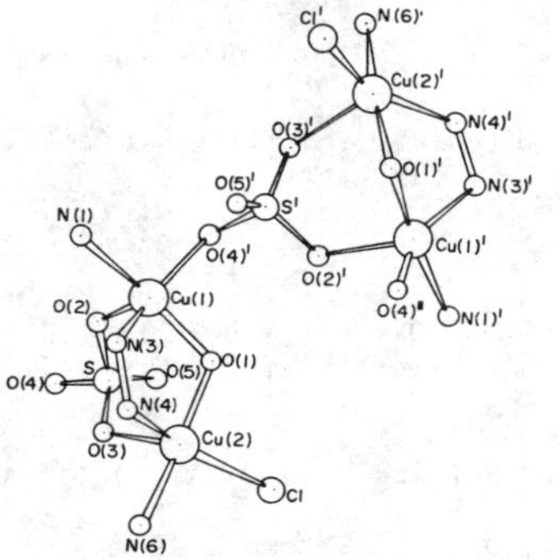

Fig. 2. The polymeric sulphate-bridged chains in $C_{18}H_{15}ClCu_2N_6O_5S,2H_2O$.

6,19-DIOXA-27,28-DITHIA-3,9,16,22-TETRAAZATRICYCLO[22.2.1.1^{11,14}]OCTACOSA-2,9,11,13,-15,22,24,26-OCTAENE

$C_{20}H_{24}N_4O_2S_2$ (I)

(6,9,22,25-TETRAOXA-33,34-DITHIA-3,12,19,28-TETRAAZATRICYCLO[28.2.1.1^{14,17}]TETRATRIACONTA-2,12,14,16,18,28,30,32-OCTAENE-N^3,N^{12},N^{19},N^{28})DISILVER(I) DIPERCHLORATE DIHYDRATE

$C_{24}H_{32}Ag_2Cl_2N_4O_{12}S_2,2(H_2O)$ (II)

N.A. BAILEY, M.M. EDDY, D.E. FENTON, S. MOSS, A. MUKHOPADHYAY and G. JONES, 1984. J. Chem. Soc. Dalton, 2281-2288.

I. Orthorhombic, $P2_12_12_1$, a = 25.820, b = 14.275, c = 5.938 Å, D_m = 1.26, Z = 4. Mo radiation, R = 0.044 for 1629 reflexions.

II. Monoclinic, C2/c, a = 27.941, b = 8.612, c = 15.359 Å, β = 102.92°, D_m = 1.78, Z = 4. Mo radiation, R = 0.122 for 855 reflexions.

The macrocycle in I has approximate C_s symmetry through two S atoms and perpendicular to the thiophene rings which are planar and although not parallel (24.7°) are aligned and approximately eclipsed with minimum separation of 3.49 Å. Bond lengths and angles are unexceptional. The disilver cation of II (Fig. 1) possesses crystallographically imposed C_2 symmetry and has a twisted loop configuration for the macrocycle. The two symmetry related Ag atoms are accomodated at opposite ends of the macrocycle each being coordinated most strongly by a pair of imine N atoms (Ag-N 2.27(4), 2.17(4) Å) and a water molecule (Ag-O 2.72(7) Å). The coordinated water molecule forms H bonds to two disordered perchlorate ions. The thiophene fragments are planar and torsion angles unexceptional.

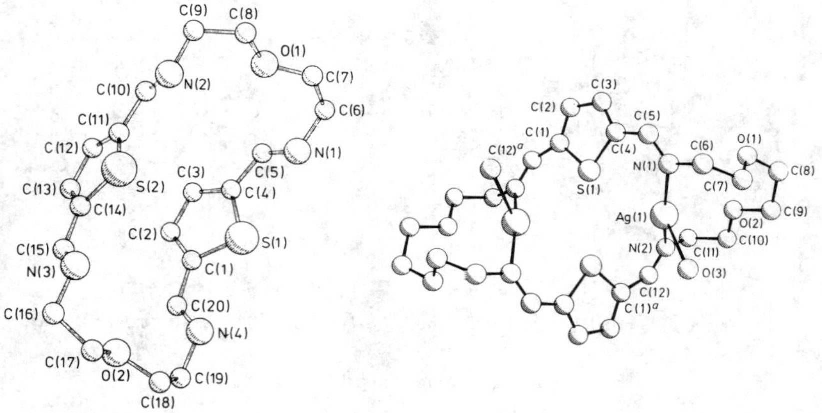

Fig. 1. Structure of the $C_{20}H_{24}N_4O_2S_2$ molecule (left) and the cation in $C_{24}H_{32}Ag_2Cl_2N_4O_{12}S_2,2(H_2O)$ (right).

BIS(DIMETHYLGLYOXIMATO)((S)-1-METHOXYCARBONYLETHYL)-((R)-α-METHYLBENZYLAMINE)-COBALT(III) (Triclinic and monoclinic forms)
$C_{20}H_{32}CoN_5O_6$

T. KURIHARA, A. UCHIDA, Y. OHASHI and Y. SASADA, 1984. Acta Cryst., B40, 478-483.

I. Triclinic, P1, a = 8.637, b = 8.955, c = 8.C94 Å, α = 105.71, β = 100.18, γ = 96.23°, Z = 1. Mo radiation, R = 0.052 for 2646 reflexions.

II. Monoclinic, $P2_1$, a = 8.017, b = 16.715, c = 8.972 Å, β = 101.88°, Z = 2. Mo radiation, R = 0.046 for 2697 reflexions.

The two different forms result from a conformational change around the Co-C and Co-N bonds; the triclinic form is shown in Fig. 1. The Co-C(17) distance is 2.080(7) in I and 2.086(5) Å in (II). The Co-N(5) (trans to C(17)) distances are 2.075(5) in (I) and 2.074(4) Å in (II). The Co-N (other) distances are in the range 1.847-1.903(5) Å. The torsion angles C(19)-C(17)-Co-N(1) and N(4)-Co-N(5)-C(9) are -51.7(5) and 22.7(5)° for the triclinic form, 5.4(4) and -30.6(4)° for the monoclinic form.

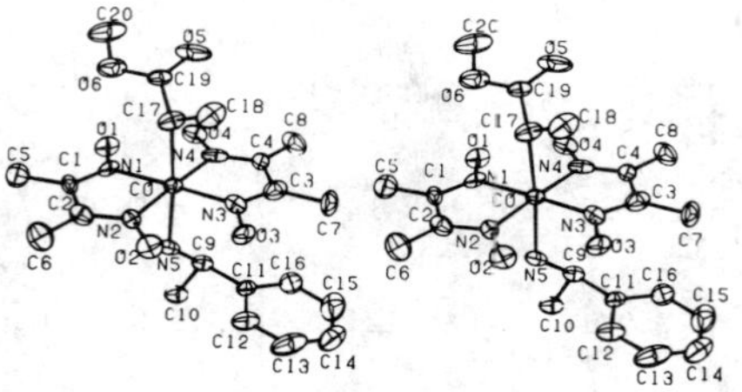

Fig. 1. A stereoview of the triclinic form of $C_{20}H_{32}CoN_5O_6$.

BIS(μ_2-BROMO)DIBROMOBIS(N,N'-DI-t-BUTYLDIAZABUTADIENE)DINICKEL
$C_{20}H_{40}Br_4N_4Ni_2$ (I)

DIBROMO(N,N'-DI-t-BUTYLDIAZABUTADIENE)NICKEL
$C_{10}H_{20}Br_2N_2Ni$ (II)

G.B. JAMESON, H.R. OSWALD and H.R. BEER, 1984. J. Am. Chem. Soc., 106, 1669-1675.

I. Monoclinic, C2/c, a = 20.429, b = 7.156, c = 20.504 Å, β = 98.50°, D_m = 1.72, Z = 4. Mo radiation, R = 0.039 for 2560 reflexions.

II. Monoclinic, $P2_1/n$, a = 7.125, b = 19.717, c = 10.396 Å, β = 90.91°, D_m = 1.66, Z = 4. Mo radiation, R = 0.055 for 1545 reflexions (at -150°C).

Molecule I is transformed to II upon heating. I (Fig. 1) is a distorted trigonal-bipyramidal centrosymmetric dimer: Ni-Br(terminal 2.457(1) Å, Ni-Br(bridging) 2.497(1) and 2.583 Å, Ni-N 2.042(4) and 2.039(4) Å, Br(terminal)-Ni-Br(bridging, long) 165.79(3)° (defines the pseudotrigonal axis), Br(bridging)-Ni-Br(bridging) 83.11(3)°, N-Ni-N 80.8(2)°. The monomer has tetrahedral D_{2h} symmetry: Ni-Br 2.333(2) and 2.343(2) Å, Ni-N 1.996(7) and 2.002(8) Å, Br-Ni-Br 126.78(6)°, N-Ni-N 82.5(3)°. The reaction mechanism involves cleavage of the long Ni-Br bond and concerted movement of $NiBr_2$(dab) monomers with concomitant rearrangement to a tetrahedral D_{2d} system such that centrosymmetrically related Ni centres, formerly 3.806 Å separated, become separated by 9.893 Å and related by a 2_1 screw axis.

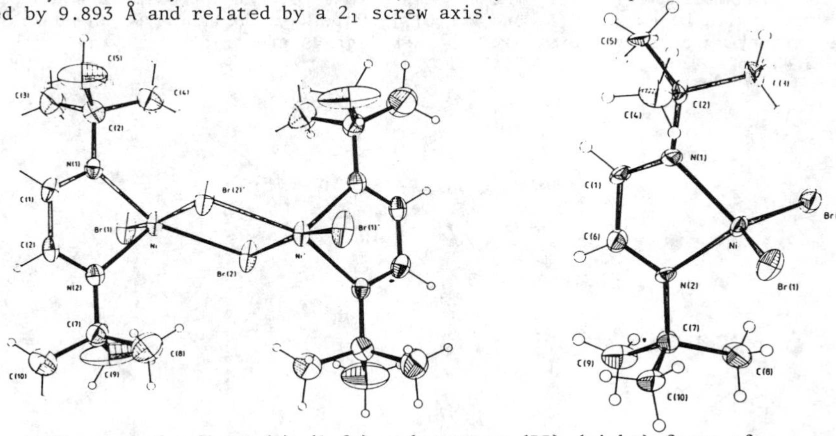

Fig. 1. Views of the dimer (I) (left) and monomer (II) (right) forms of $C_{10}H_{20}Br_2N_2Ni$.

BIS(μ_3-PHENYLIMIDO)TRIS(TRICARBONYLIRON)(2Fe-Fe)

$C_{21}H_{10}Fe_3N_2O_9$ $[Fe_3(C_6H_5N)_2(CO)_9]$

W. CLEGG, G.M. SHELDRICK, D. STALKE, S. BHADURI and H.K. KHWAJA, 1984. Acta Cryst., C40, 2045-2047.

Monoclinic, $P2_1/n$, a = 11.282, b = 13.658, c = 15.379 Å, β = 99.81°, Z = 4. Mo radiation, R = 0.029 for 3495 reflexions.

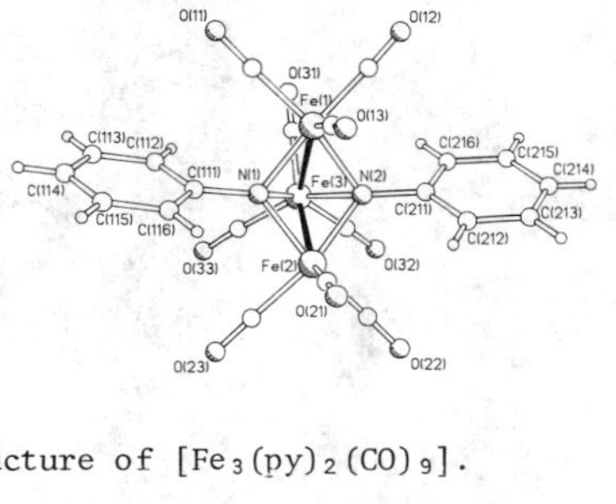

Fig. 1. Molecular structure of $[Fe_3(py)_2(CO)_9]$.

The three Fe atoms (Fig. 1) form an approximately isosceles triangle, with Fe-Fe 2.484(1) and 2.421(1) Å, Fe...Fe 3.044(1) Å, Fe-Fe-Fe 76.7(1)°. All carbonyl groups are terminal. The NPh ligands are triply bridging, with Fe-N 1.926(2) to 1.969(2) Å, N...N 2.301(6) Å.

BIS-μ_3-PHENYLIMIDO-TRIS(TRICARBONYLRUTHENIUM)
$C_{21}H_{10}N_2O_9Ru_3$ $[Ru_3(C_6H_5N)_2(CO)_9]$

W. CLEGG, G.M. SHELDRICK, D. STALKE, S. BHADURI and K.S. GOPALKRISHNAN, 1984. Acta Cryst., C40, 927-929

Orthorhombic, $Pna2_1$, a = 19.730, b = 9.667, c = 13.043 Å, Z = 4. Mo radiation, R = 0.025 for 2164 reflexions.

In the molecule (Fig. 1) the three Ru atoms form an approximately isosceles triangle, with Ru-Ru 2.710(1) and 2.626(1), Ru...Ru 3.289(1) Å, Ru-Ru-Ru 76.1(1)°. All carbonyl groups are terminal. The NPh ligands are triply bridging, with Ru-N 2.052(4) to 2.129(6) Å, N...N 2.405(11) Å.

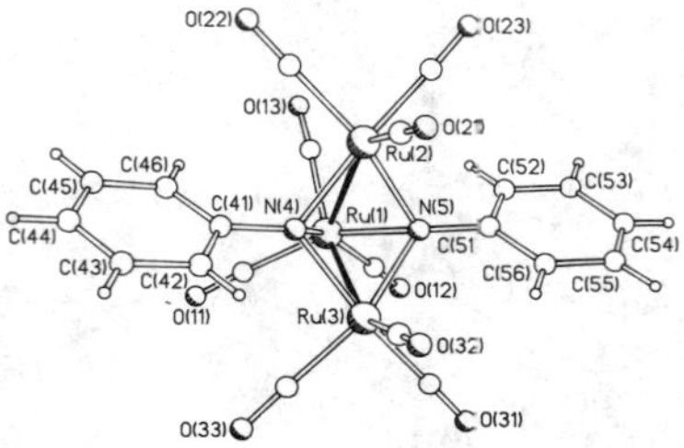

Fig. 1. Molecular structure of $Ru_3(C_6H_5N)_2(CO)_9$.

[(R)-1-CYANOETHYL][DIETHYL(PHENYL)PHOSPHINE]BIS(DIMETHYLGLYOXIMATO)COBALT(III)
$C_{21}H_{33}CoN_5O_4P$ (I)

[(R)-1-CYANOETHYL]BIS(DIMETHYLGLYOXIMATO)(ETHYLDIPHENYLPHOSPHINE)COBALT(III)
$C_{25}H_{33}CoN_5O_4P$ (II)

Y. TOMOTAKE, A. UCHIDA, Y. OHASHI, Y. SASADA, Y. OHGO and S. BABA, 1984. Acta Cryst., C40, 1684-1687.

I. Orthorhombic, $P2_12_12_1$, a = 14.340, b = 14.451, c = 11.820 Å, D_m = 1.38, Z = 4. Mo radiation, R = 0.038 for 3020 reflexions.

II. Monoclinic, $P2_1$, a = 8.966, b = 17.567, c = 8.906 Å, β = 109.98°, D_m = 1.41, Z = 2. Mo radiation, R = 0.032 for 3045 reflexions.

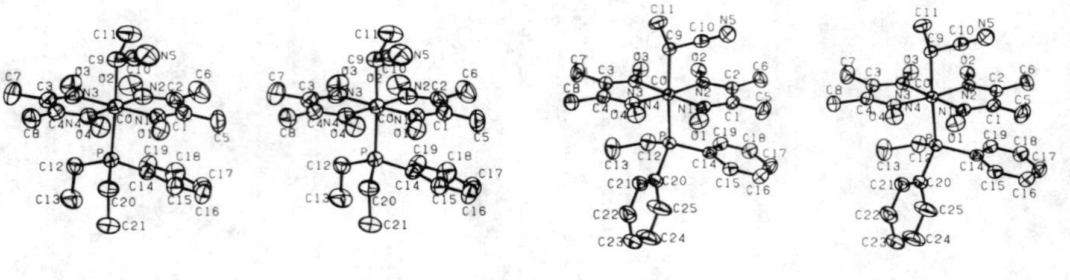

Fig. 1. Stereoviews of $C_{21}H_{33}CoN_5O_4P$ (I) (left), and of $C_{25}H_{33}CoN_5O_4P$ (II) (right).

The structures (Fig. 1) are not racemized by X-rays. Principal dimensions are (I) Co-P 2.313(1), Co-N 1.878-1.883(3), Co-C 2.074(4); II Co-P 2.370(1), Co-N 1.880-1.893(3), Co-C 2.098(3) Å.

TETRACARBONYL(6-p-STYRYL-2,2'-BIPYRIDYL)TUNGSTEN(O)
$C_{22}H_{15}N_2O_4W$ $(CO)_4W(C_{18}H_{15}N_2)$

C. LONG, J.G. VOS, R.A. HOWIE and J.M. KELLY, 1984. J. Organometal. Chem., 272, 385-390.

Monoclinic, $P2_1/n$, a = 13.735, b = 8.839, c = 17.105 Å, β = 109.06°, Z = 4. Mo radiation, R = 0.055 for 1734 reflexions.

The monomeric molecule contains a tungsten atom bound to four terminal carbonyl groups and to one 6-p-steryl-2,2'-bipyridyl ligand via its two N atoms (Fig. 1). The W-C distances are 1.99, 1.99, 1.94 and 1.98 Å. The two W-N distances are significantly different, W-N(1) is 2.29(1) Å compared with the W-N(2) distance of 2.23(1) Å. This difference is assumed to be due to the presence of the styryl group α to one coordinating N, (N(1)). The metal cis coordination angles range from 72.4° (N(1)-W-N(2)) to 96.0° (N(1)-W-C(2)). The two pyridine rings are practically co-planar, while the angle between the bipyridyl and styryl ring systems is 111.2°.

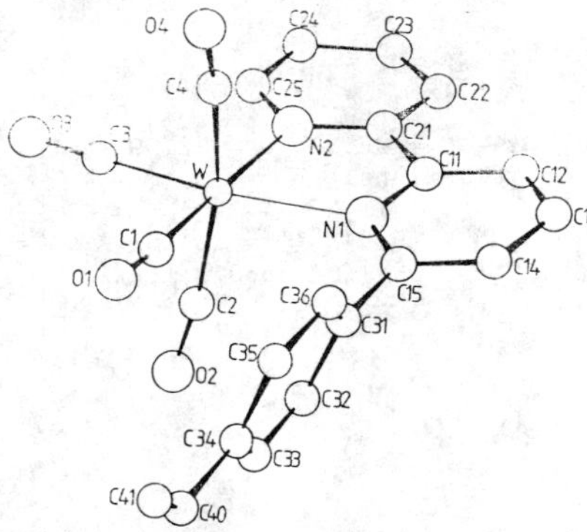

Fig. 1. Molecular structure of $C_{22}H_{15}N_2O_4W$.

cis-DICARBONYLBIS(2,2'-BIPYRIDINE)MOLYBDENUM(O)
$C_{22}H_{16}MoN_4O_2$ $Mo(CO)_2(C_{10}H_8N_2)_2$

M.H. CHISHOLM. J.A. CONNOR, J.C. HUFFMAN, E.M. KOBER and C. OVERTON, 1984. Inorg. Chem., 23, 2298-2303.

Monoclinic, $P2_1/a$, a = 26.830, b = 9.467, c = 8.756 Å, β = 119.91°, Z = 4. Mo radiation, R = 0.0475 for 1994 reflexions (at -162°C).

The molecule (Fig. 1) has cis-octahedral coordination geometry (Mo-N 2.239(5) and 2.220(5) Å trans to CO, 2.155(6) and 2.143(6) Å trans to N, Mo-CO 1.922(7) and 1.935(7) Å) and approximate C_2 symmetry. The bipyridine ligand appears to be slightly distorted as a result of M(d)→bpy(π*) bonding.

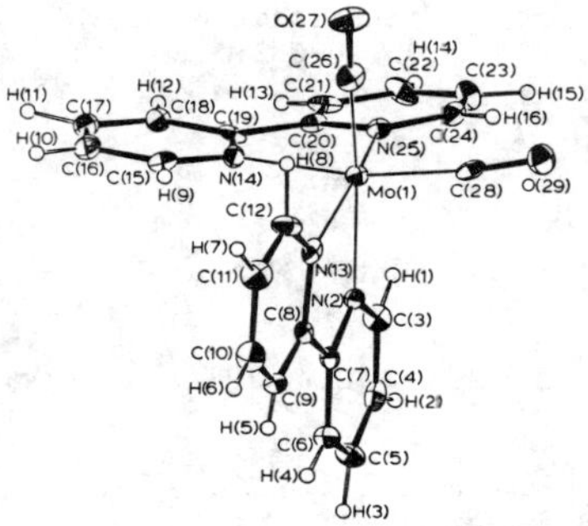

Fig. 1. The $Mo(CO)_2(C_{10}H_8N_2)_2$ molecule.

DICHLOROBIS(2-PHENYLAZOPYRIDINE)RUTHENIUM(II) (C_2 ISOMER)
$C_{22}H_{18}Cl_2N_6Ru$ (I) $[RuCl_2(C_{11}H_9N_3)_2]$

DICHLOROBIS(2-PHENYLAZOPYRIDINE)RUTHENIUM(II) (C_1 ISOMER)
$C_{22}H_{18}Cl_2N_6Ru$ (II) $[RuCl_2(C_{11}H_9N_3)_2]$

A. SEAL and S. RAY, 1984. Acta Cryst., C40, 929-932.

I. Monoclinic, $P2_1/c$, a = 8.421, b = 22.88, c = 12.99 Å, β = 116.3°, Z = 4. Mo radiation, R = 0.051 for 3257 reflexions.

II. Triclinic, $P\bar{1}$, a = 13.371, b = 15.345, c = 13.101 Å, α = 114.75, β = 93.75, γ = 67.61°, Z = 4. Mo radiation, R = 0.054 for 3660 reflexions.

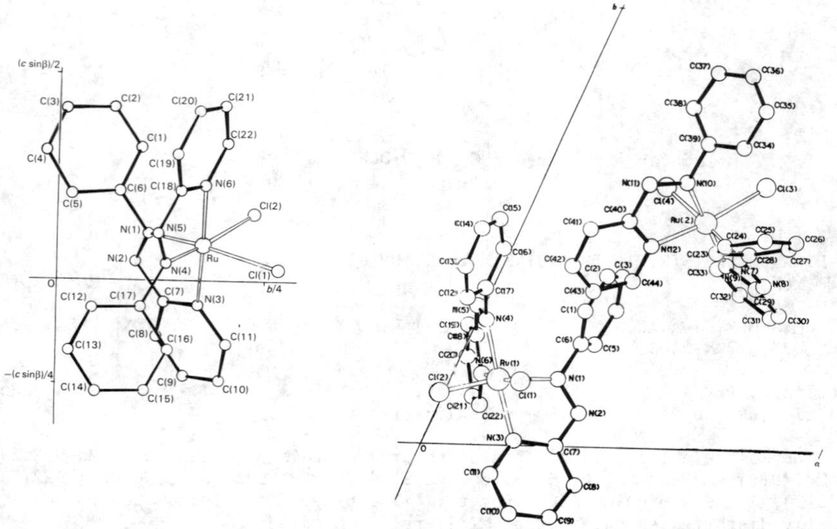

Fig. 1. $RuCl_2(C_{11}H_9N_3)_2$: asymmetric units of the structures of the C_2 isomer (I) projected down the a axis and cf the C_1 isomer (II) projected on the ab plane.

In both compounds (Fig. 1) Ru has distorted octahedral coordination. With reference to the pairs of Cl, N(pyridine) and N(azo) atoms coordinating with Ru, the C_2 isomer (I) has cis-trans-cis configuration and the C_1 isomer (II) cis-cis-cis. In both structures the Ru-N(azo) distances, 1.977(4) and 1.984(4) in I and 1.960(9) and 2.003(9) (molecule 1) and 1.958(9) and 1.96(1) (molecule 2) in II, are relatively shorter than the Ru-N(pyridine) distances, 2.045(4) and 2.051(4) in I and 2.059(9) and 2.020(9) (molecule 1) and 2.069(9) and 2.028(9) Å (molecule 2) in II. This indicates stronger bonding in the former case, caused by the considerable π-backbonding power of the azo function. The Ru-Cl distances, in the range 2.379(4) to 2.410(3) Å, are normal.

BIS(ISOTHIOCYANATO)TETRAKIS(PYRIDINE)NICKEL(II)
$C_{22}H_{20}N_6NiS_2$

F. VALACH, P. SIVÝ and B. KOREŇ, 1984. Acta Cryst., C40, 957-959.

Monoclinic, C2/c, a = 12.434, b = 12.944, c = 16.461 Å, β = 118.78°, D_m = 1.39, Z = 4. Mo radiation, R = 0.063 for 1300 reflexions.

In the complex molecule (Fig. 1) Ni^{II} lies on a crystallographic centre of symmetry and is pseudo-octahedrally trans-coordinated by two N atoms from monodentate NCS groups, Ni-N 2.062(8), and by four N atoms from the pyridine rings, Ni-N 2.169(31) and 2.151(8) Å. The N-C-S bond angle is 180(1)° and the distances N-C and C-S are 1.15(1) and 1.63(2) Å. The pyridine rings are almost planar. For earlier work, see 1.

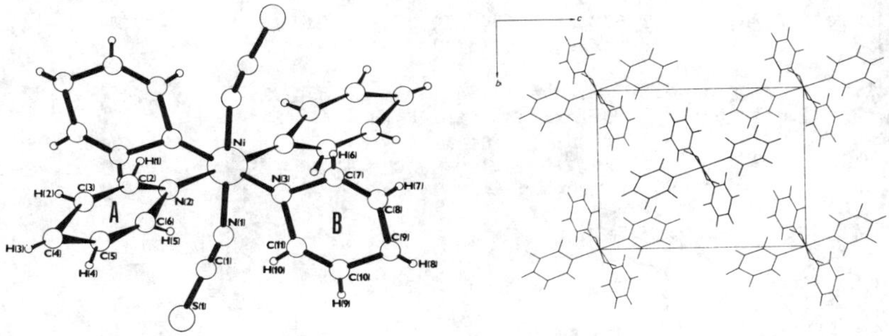

Fig. 1. $C_{22}H_{20}N_6NiS_2$: view of the molecule and projection of the crystal structure on (100).

1. Structure Reports, 22, 784.

trans-BIS(DIMETHYLPHENYLPHOSPHINE)BIS(PYRAZOLE)PLATINUM
$C_{22}H_{30}N_4P_2Pt$ $[Pt(C_3H_4N_2)_2\{P(CH_3)_2(C_6H_5)\}_2]$

J.D. OLIVER, D.F. MULLICA, D.A. GROSSIE, W.O. MILLIGAN and H.O. PERKINS, 1984. Acta Cryst., C40, 746-748.

Monoclinic, C2/c, a = 21.509, b = 6.290, c = 17.934 Å, β = 102.54°, D_m = 1.70, Z = 4. Mo radiation, R = 0.028 for 1416 reflexions.

In the complex molecule Pt lies on a crystallographic inversion centre with a trans arrangement of the dimethylphenylphosphine and pyrazole ligands (Fig. 1). The plane of the pyrazole ring forms an angle of 73.8° with the square-planar coordination plane defined by the Pt, P and coordinated N atoms of the pyrazole ligand. The coordinated pyrazole ligand is flat and virtually isometric. Some bond distances and angles in the complex are: Pt-P 2.305(3), Pt-N 2.027(7), mean C-C(phenyl) 1.382(14) Å and P-Pt-N 91.7(2)°.

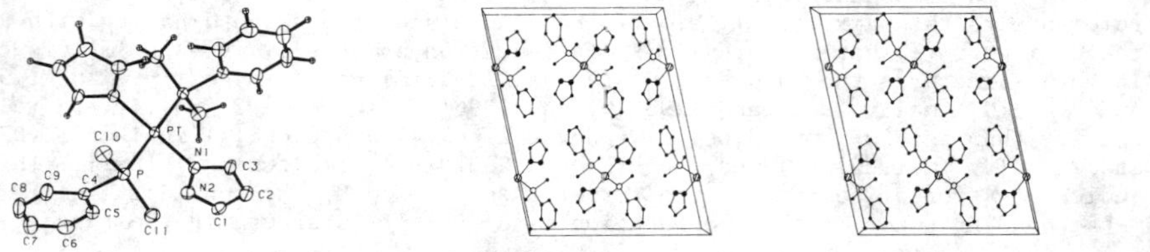

Fig. 1. $C_{22}H_{30}N_4P_2Pt$: perspective view of the molecule and stereoscopic packing diagram viewed along the b axis.

BIS[μ-BIS(TRIMETHYLSILYL)AMIDO][BIS(TRIMETHYLSILYL)AMIDOMANGANESE(II)][(TETRAHYDRO-FURAN)LITHIUM]

$C_{22}H_{62}LiMnN_3OSi_6$ (I) $[Mn(N(SiMe_3)_2)_3Li(THF)]$

BIS[μ-BIS(TRIMETHYLSILYL)AMIDO]BIS[BIS(TRIMETHYLSILYL)AMIDOMANGANESE(II)]

$C_{24}H_{72}Mn_2N_4Si_8$ (II) $[Mn_2(N(SiMe_3)_2)_4]$

BIS[μ-BIS(TRIMETHYLSILYL)AMIDO]BIS[BIS(TRIMETHYLSILYL)AMIDOCOBALT(II)]

$C_{24}H_{72}Co_2N_4Si_8$ (III) $[Co_2(N(SiMe_3)_2)_4]$

B.D. MURRAY and P.P. POWER, 1984. Inorg. Chem., 23, 4584-4588.

I. Monoclinic, $P2_1/n$, a = 11.678, b = 19.362, c = 17.020 Å, β = 108.77°, Z = 4. Mo radiation, R = 0.031 for 3628 reflexions (at 140 K).

II. Monoclinic, C2/c, a = 17.997, b = 14.942, c = 18.636 Å, β = 121.24°, Z = 4. Mo radiation, R = 0.031 for 2672 reflexions (at 140 K).

III. Monoclinic, C2/c, a = 17.907, b = 14.644, c = 18.633 Å, β = 120.47°, Z = 4. Mo radiation, R = 0.037 for 1969 reflexions (at 140 K).

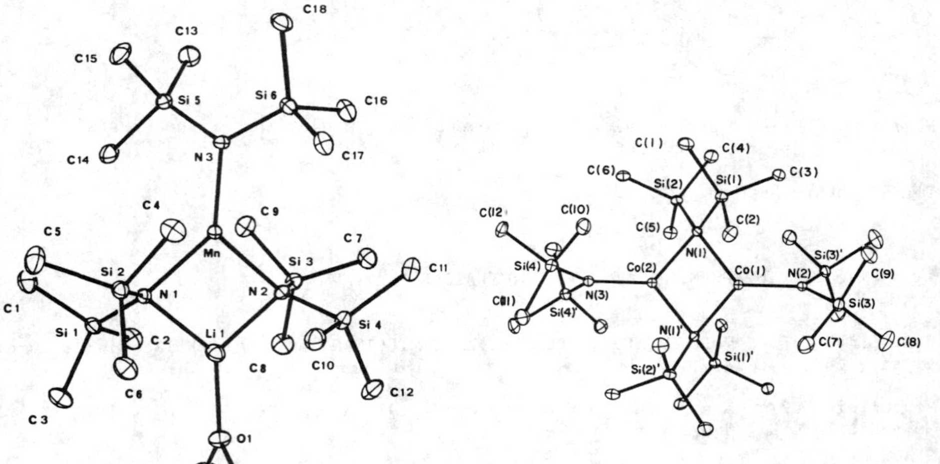

Fig. 1. The structures of $[Mn(N(SiMe_3)_2)_3Li(THF)]$ (left) and $[M_2(N(SiMe_3)_2)_4]$ (M = Mn, Co) (right).

The molecular structures of I, II, and III have the amido group acting both as a bridging and as a terminal ligand. I is the first homoleptic tris(silylamide) of Mn. Both II and III are dimeric in the solid state and are isomorphous. The metal and lithium atoms in I-III have distorted trigonal planar coordination geometry. Bond lengths: for I; Mn-(μ-N) 2.143(3) and 2.143(2), Mn-N 2.023(3), Li-N 2.101(6) and 2.104(6), Li-O 1.939(6), Mn...Li 2.718(6) Å; for II; Mn-(μ-N) 2.174(3) and 2.170(3), Mn-N 1.999(3) and 1.997(3), Mn...Mn 2.811(1) Å; for III; Co-(μ-N) 2.062(4) (x2), Co-N 1.922(5) and 1.910(5), Co...Co 2.583(1) Å. II and III possess exact C_2 symmetry.

{2,6-BIS[1-((1-CARBOXYMETHYL)-2-IMIDAZOL-4-YLETHYL)IMINO)ETHYL]PYRIDINE}ZINC(II) DIPERCHLORATE

$C_{23}H_{27}Cl_2N_7O_{12}Zn$ $[Zn(C_{23}H_{27}N_7O_4)](ClO_4)_2$

L. CASELLA, M.E. SILVER and J.A. IBERS, 1984. Inorg. Chem., 23, 1409-1418.

Orthorhombic, $P2_12_12_1$, a = 28.872, b = 9.567, c = 10.344 Å, D_m = 1.65, Z = 4. Mo radiation, R = 0.041 for 4848 reflexions (at -152°C).

The Zn atom is bonded to five N atoms and displays a coordination geometry intermediate between trigonal bipyramidal and square planar (Fig. 1). Both of the six-membered amino acid chelate rings adopt the λ configuration.

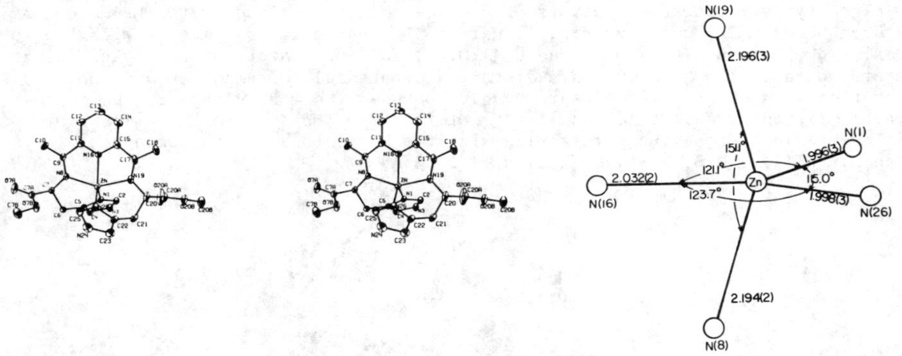

Fig. 1. Stereoview of the $[Zn(C_{23}H_{27}N_7O_4)]^{2+}$ cation; inner coordination geometry at right.

TETRAETHYLAMMONIUM HYDROTRIS(3-METHYLPYRAZOLYL)BORATOTRICARBONYLMOLYBDENUM(0)

$C_{23}H_{36}BMoN_7O_3$ $[NEt_4]^+[Mo(CO)_3HB(C_4H_5N_2)_3]^-$

T.J. DESMOND, F.J. LALOR, G. FERGUSON and M. PARVEZ, 1984. J. Organometal. Chem., 277, 91-97.

Monoclinic, $P2_1$, a = 9.092, b = 16.265, c = 9.871 Å, β = 109.43°, Z = 2. Mo radiation, R = 0.029 for 1813 reflexions.

In the anion the Mo atom has octahedral geometry with the three CO groups trans to the three N atoms of the tridentate ligand (Fig. 1). The principal dimensions are: Mo-N 2.281-2.290, mean 2.285 Å, and Mo-C 1.916-1.934, mean 1.928 Å. The dimensions of the tridentate ligand are unexceptional with mean bond lengths B-N 1.532, N-N 1.363, N-C 1.405, C_{sp^2}-C_{sp^2} 1.374 and C_{sp^2}-C_{sp^2} 1.474 Å. The cation $[NEt_4]^+$ has a fully staggered conformation with mean N-C and C-C bond lengths 1.504 and 1.544 Å.

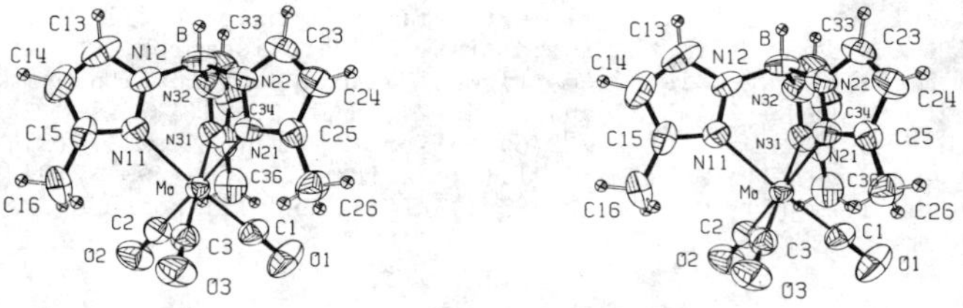

Fig. 1. Stereoview of the anion in $C_{23}H_{36}BMoN_7O_3$.

TRIS(μ_2-BENZOTRIAZOLE-N,N')HEXACARBONYLDIMOLYBDENUM TETRAHYDROFURAN SOLVATE

$C_{24}H_{15}Mo_2N_9O_6,5(C_4H_8O)$ $(C_6H_4N_3H)_3Mo_2(CO)_6,5(C_4H_8O)$

G.G. ALEKSANDROV, V.N. BABIN, A.P. KHARCHEVNIKOV, Yu.T. STRUCHKOV and N.S. KOCHETKOVA, 1984. J. Organometal. Chem., 266, 109-116.

Orthorhombic, $P2_1mn$, a = 10.006, b = 17.312, c = 16.140 Å, Z = 2. Mo radiation, R = 0.107 for 1890 reflexions (-120°C).

The crystal structure consists of complex molecules linked with THF molecules via hydrogen bonds. The complex (Fig. 1) has crystallographic C_s symmetry with the Mo atoms, two CO groups (C(2)-O(2) and C(4)-O(4)) and one benzotriazole ligand in the symmetry plane. The Mo atoms have distorted octahedral coordination and the Mo...Mo separation is 4.089 Å. The dihedral angles between the mean planes of the benzotriazole ligands are 117, 117 and 126°. Of the five THF solvate molecules two participate in hydrogen bond formation with the NH protons of the benzotriazole ligands: OT(1)...N(3) 2.56, OT(2)...N(6) 2.68 Å.

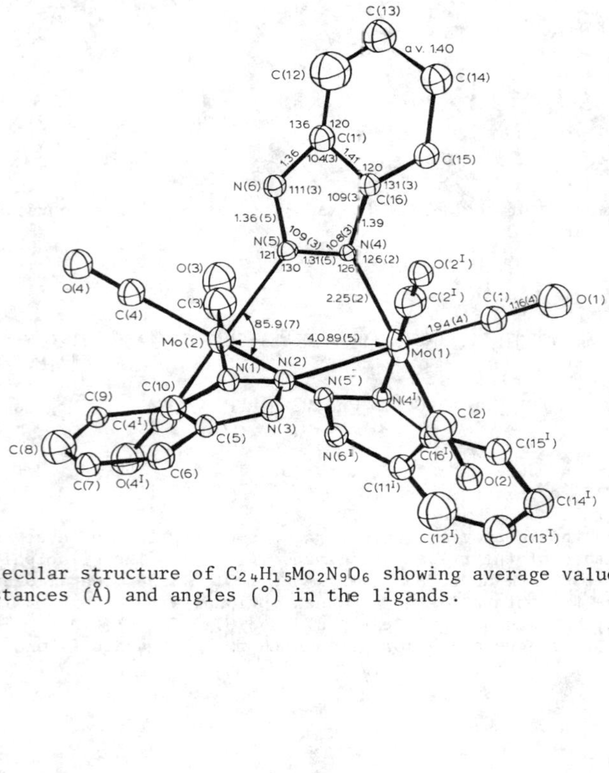

Fig. 1. Molecular structure of $C_{24}H_{15}Mo_2N_9O_6$ showing average values for bond distances (Å) and angles (°) in the ligands.

TRIS[(4-PYRIDIN-2-YL)THIAZOLATO]NICKEL(II) PERCHLORATE TRIHYDRATE
$C_{24}H_{18}Cl_2N_6NiO_8S_3,3(H_2O)$ $[Ni(L)_3][ClO_4]_2(H_2O)_3$

A.T. BAKER, H.A. GOODWIN and A.D. RAE, 1984. Aust. J. Chem., 37, 2431-2440.

Monoclinic, $P2_1/c$, a = 9.404, b = 29.169, c = 17.900 Å, β = 137.76°, D_m = 1.60, Z = 4. Mo radiation, R = 0.087 for 2307 reflexions.

The Ni atom has distorted octahedral coordination (Fig. 1). Ligand orientational disorder was observed, with ligands interchanging about a pseudo twofold axis (Ni-N 2.073(7) Å, N-Ni-N 79.4(3)°).

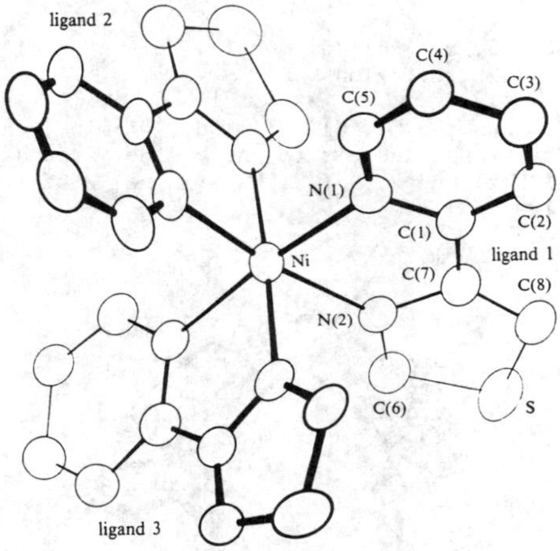

Fig. 1. The structure of $[Ni(L)_3]^{2+}$.

TETRAKIS(μ_2-ISOSELENOCYANATO)MERCURYTETRAKIS(PYRIDINE)STRONTIUM
$(C_{24}H_{20}HgN_8Se_4Sr)_n$

K. BRODERSEN, M. CYGAN and H.-U. HUMMEL, 1984. Z. Naturforsch., B39, 582-585.

Tetragonal, $P\bar{4}n2$, a = 9.552, c = 17.610 Å, Z = 2. Ag radiation, R = 0.043 for 1050 reflexions.

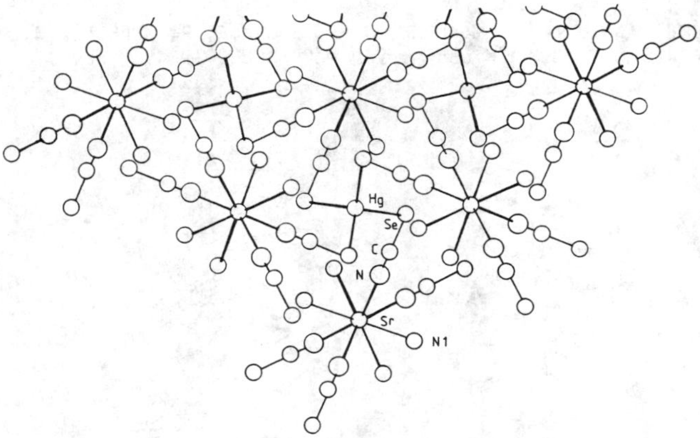

Fig. 1. A view of part of the $(C_{24}H_{20}HgN_8Se_4Sr)_n$ structure.

The structure contains nearly tetrahedral $Hg(SeCN)_4$ units (Fig. 1) with Hg-Se 2.641(2) Å and Se-Hg-Se 104.92 and 111.80(5)°. The N atoms of the pyridines and SeCN bond to the Sr atom to produce Archimedean antiprismatic coordination (Sr-N(SeCN) 2.667(12), Sr-N(py) 2.762(10) Å).

cis-DICHLOROBIS[2-(m-TOLYLAZO)PYRIDINE]OSMIUM(II)
$C_{24}H_{22}Cl_2N_6Os$ $OsCl_2(C_{12}H_{11}N_3)_2$

B.K. GHOSH, A. MUKHOPADHYAY, S. GOSWAMI, S. RAY and A. CHAKRAVORTY, 1984. Inorg. Chem., 23, 4633-4639.

Monoclinic, $P2_1/c$, a = 8.035, b = 24.176, c = 13.416 Å, β = 114.34°, Z = 4. Mo radiation, R = 0.026 for 3326 reflexions.

The complex (Fig. 1) has the Cl, Cl, N(imine), N(imine), and N(azo), N(azo) pairs, respectively, cis, trans, and cis to one another around octahedral Os(II). Bond lengths: Os-Cl 2.399(2) and 2.393(2), Os-N(imine) 2.075(6) and 2.055(6), Os-N(azo) 1.972(6) and 1.975(6) Å.

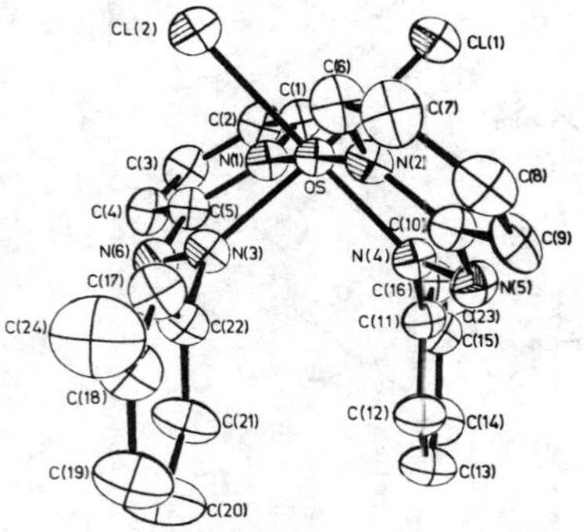

Fig. 1. The $OsCl_2(C_{12}H_{11}N_3)_2$ molecule.

AQUA(5,5',3'',5'''-TETRAMETHYL-2,2'.6',2''.6'',2'''-QUATERPYRIDINE-N^1,N^2,N^3,N^4)-COPPER(II) DIPERCHLORATE
$C_{24}H_{24}Cl_2CuN_4O_9$ $[Cu(H_2O)(C_{24}H_{24}N_4)](ClO_4)_2$

J.-P. GISSELBRECHT, M. GROSS, J.-M. LEHN, J.-P. SAUVAGE, R. ZIESSEL, C. PICCINNI-LEOPARDI, J.M. ARRIETA G. GERMAIN and M. VAN MEERSSCHE, 1984. Nouv. J. Chim., 8, 661-667.

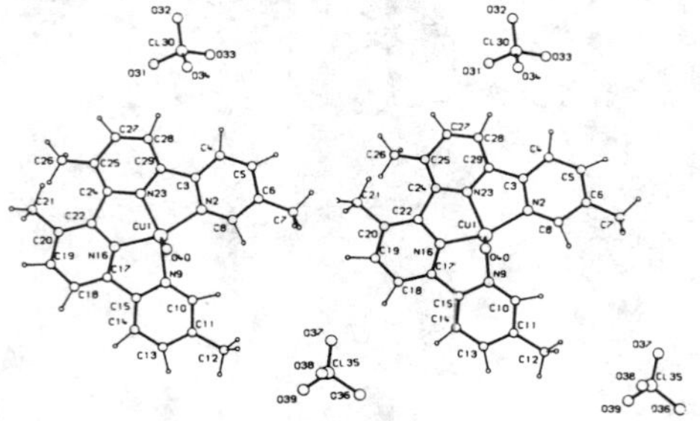

Fig. 1. A stereoview of $C_{24}H_{24}Cl_2CuN_4O_9$.

Orthorhombic, Pbca, a = 24.671, b = 13.938, c = 15.376 Å, Z = 8. Mo radiation, R = 0.057 for 2374 reflexions.

The Cu atom (Fig. 1) has distorted square pyramidal coordination geometry with four N atoms at the base and an apical water molecule. Cu-N bond lengths involving the N atoms of terminal rings are longer (2.024 and 2.026(5) Å) than those for inner N atoms (1.937 and 1.947(6) Å). The Cu-O distance is 2.307(6) Å with the Cu atom 0.14 Å above the N_4 plane in the direction of the water molecule. Each individual pyridine ring is planar, but dihedral angles between ring planes are from -26 to +16°.

BIS[μ_2-3,5-BIS(PYRIDIN-2-YL)-4-AMINO-1,2,4-TRIAZOLE-N',N^1,N^2,N'')BIS(AQUACHLORO-NICKEL(II)] DICHLORIDE TETRAHYDRATE

$C_{24}H_{24}Cl_4N_{12}Ni_2O_2,4(H_2O)$ $[Ni_2(abpt)_2Cl_2(H_2O)_2]Cl_2.4H_2O$

F.S. KEIJ, R.A.G. DE GRAAFF, J.G. HAASNOOT and J. REEDIJK, 1984. J. Chem. Soc. Dalton, 2093-2097.

Triclinic, P$\bar{1}$, a = 6.840, b = 10.199, c = 12.480 Å, α = 78.62, β = 76.47, γ = 84.17°, D_m = 1.67, Z = 1. Mo radiation, R = 0.022 for 4015 reflexions.

The structure consists of complex cations, two Cl^- anions and four lattice water molecules (Fig. 1). There is an inversion centre at the centre of the cation. The $[Ni_2(abpt)_2]^{4+}$ system is nearly planar. The Ni atoms are coordinated by four N atoms with N-Ni-N angles of 93.00(4), 76.74(4), 113.67(4) and 76.47(4)° and Ni-N distances of 2.013(1), 2.029(1), 2.155(1) and 2.164(1) Å. The axial ligands are a water molecule and a chloride ion. Coordination about the metal centres is rhombically distorted octahedral.

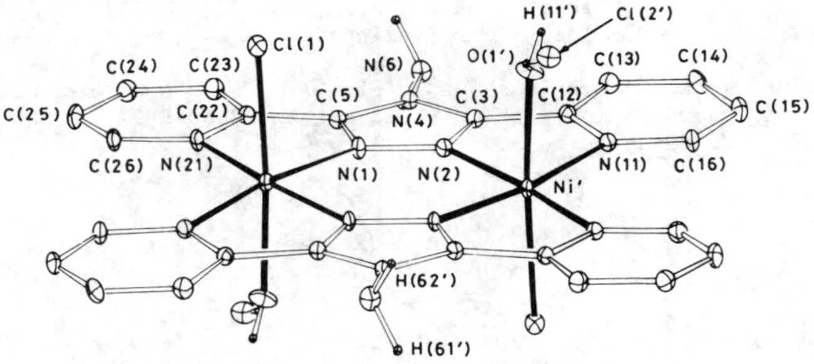

Fig. 1. $C_{24}H_{24}Cl_4N_{12}Ni_2O_2,4(H_2O)$: projection of the structure.

(N^5,N^5-BIS(2-BENZIMIDAZOLYLMETHYL)BENZYLAMINE-N^1,N^3,N^5)(NITRATO-O,O')METHANOLCOPPER(II) NITRATE

$C_{24}H_{25}CuN_7O_7$

K. TAKAHASHI, Y. NISHIDA and S. KIDA, 1984. Polyhedron, 3, 113-116.

Monoclinic, $P2_1/n$, a = 10.409, b = 16.237, c = 15.828 Å, β = 103.82°, D_m = 1.50, Z = 4. Mo radiation, R = 0.057 for 2760 reflexions.

The Cu atom has octahedral coordination (Fig. 1) elongated along the O1A-Cu-O2B axis, with Cu-N1 1.943(5), Cu-N3 1.940(6), Cu-N(5) 2.109(6), Cu-O1A 2.266(6), Cu-O1B 1.998(4) and Cu-O2B 2.637(5) Å. In the crystal, pairs of molecules are arranged about an inversion centre forming a loose dimeric unit.

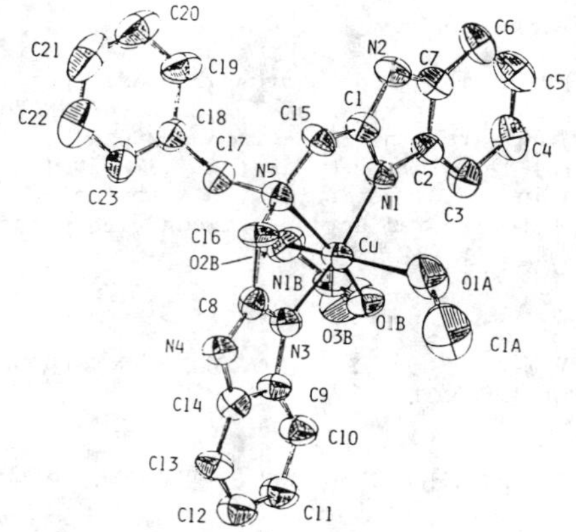

Fig. 1. The $C_{24}H_{25}CuN_7O_7$ molecule.

DI-μ-BROMO-BIS[BIS(2-METHYLPYRIDINE)COPPER(I)]
$C_{24}H_{28}Br_2Cu_2N$ $[CuBr(C_6H_7N)_2]_2$

V. SCHRAMM, A. PIERRE and W. HILLER, 1984. Acta Cryst., C40, 1840-1841.

Triclinic, P$\bar{1}$, a = 9.041, b = 9.182, c = 9.593 Å, α = 94.37, β = 117.78, γ = 108.59°, Z = 1. Mo radiation, R = 0.036 for 1456 reflexions.

The centrosymmetric structure is shown in Fig. 1. The copper atoms have a distorted tetrahedral coordination with Cu-Br 2.5831 and 2.6045(7), Cu-N 2.022 and 2.043(5) Å. The Cu...Cu distance is 3.3491(6) Å.

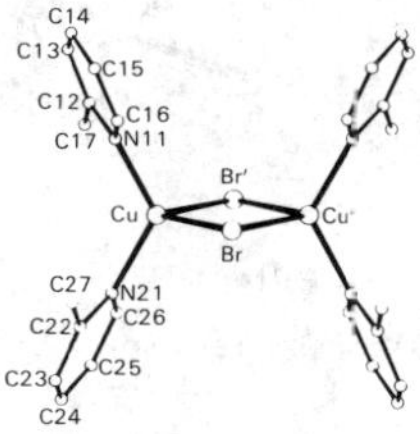

Fig. 1. The $[CuBr(C_6H_7N)_2]_2$ structure.

μ-OXO-BIS-fac-[TRIAQUA-1,10-PHENANTHROLINEIRON(III)] TETRAKIS(NITRATE) MONOHYDRATE
$C_{24}H_{28}Fe_2N_8O_{19},H_2O$ $[Fe_2O(phen)_2(H_2O)_6](NO_3)_4(H_2O)$

P.C. HEALY, J.M. PATRICK and A.H. WHITE, 1984. Aust. J. Chem., 37, 1405-1410.

Monoclinic, $P2_1/c$, a = 18.202, b = 9.771, c = 19.459 Å, β = 94.46°, D_m = 1.60, Z = 4. Mo radiation, R = 0.059 for 3467 reflexions.

The geometry of the metal atoms in the $[(OH_2)_3(phen)FeOFe(phen)(OH_2)_3]$ cation is octahedral (Fig. 1) and typical for high-spin Fe(III) (Fe-Fe 3.506(2), Fe-O (bridge) 1.765(4), 1.784(4), Fe-O(H_2O) 2.068(5), Fe-N 2.155(6) Å, Fe-O-Fe 162.0(3), N-Fe-N 76.8(2)°). The three water molecules are fac within an iron coordination sphere. Two of the four nitrate anions are disordered.

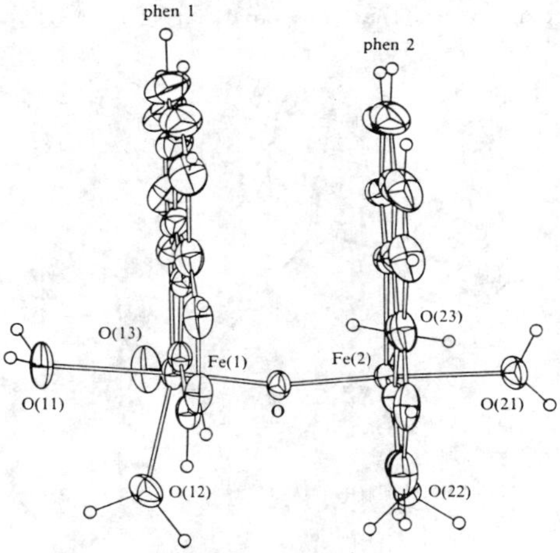

Fig. 1. The structure of $[Fe_2O(phen)_2(H_2O)_6]^{4+}$.

DIOXOTETRAKIS(4-METHYLPYRIDINE)RHENIUM(V) PERRHENATE

$C_{24}H_{28}N_4O_6Re_2$ $[ReO_2(C_6H_7N)_4][ReO_4]$

J.W. JOHNSON, J.F. BRODY, G.B. ANSELL and S. ZENTZ, 1984. Inorg. Chem., 23, 2415-2418.

Triclinic, $P\bar{1}$, a = 10.180, b = 10.695, c = 13.261 Å, α = 103.17, β = 99.64, γ = 96.47°, Z = 2. Mo radiation, R = 0.036 for 948 reflexions.

The structure contists of two crystallographically independent centrosymmetric trans-$[ReO_2(C_6H_7N)_4]^+$ cations (Re-O 1.75(2) and 1.74(2), Re-N 2.11(2)-2.16(2) Å, see Fig. 1) and tetrahedral ReO_4^- anions (Re-O 1.59(3)-1.67(3) Å).

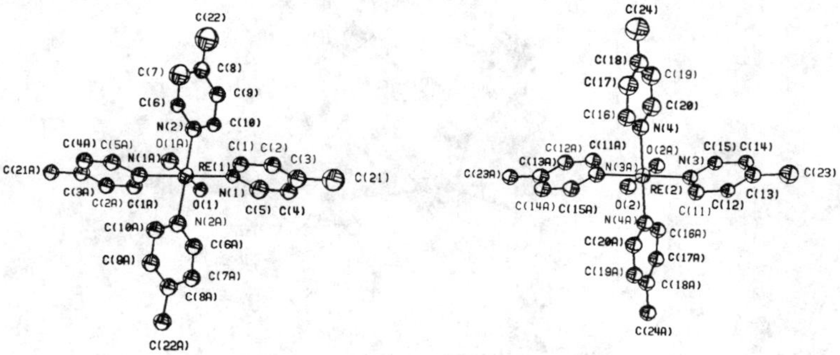

Fig. 1. The two independent $[ReO_2(C_6H_7N)_4]^+$ cations.

1-(TRICYCLOHEXYLPHOSPHINEGOLD)-2-ISOPROPYLIMIDAZOLE BENZENE SOLVATE

$C_{24}H_{42}AuN_2P,C_6H_6$

B. BOVIO, F. BONATI, A. BURINI and B.R. PIETRONI, 1984. Z. Naturforsch., B39, 1747-1754.

Monoclinic, $P2_1/c$, a = 17.373, b = 14.833, c = 11.796 Å, β = 99.17°, Z = 4. Mo radiation, R = 0.039 for 2804 reflexions.

The Au atom is shown to be linearly coordinated (Fig. 1) with Au-N 2.019(5), Au-P 2.255(2) Å, N-Au-P 177.0(1)°.

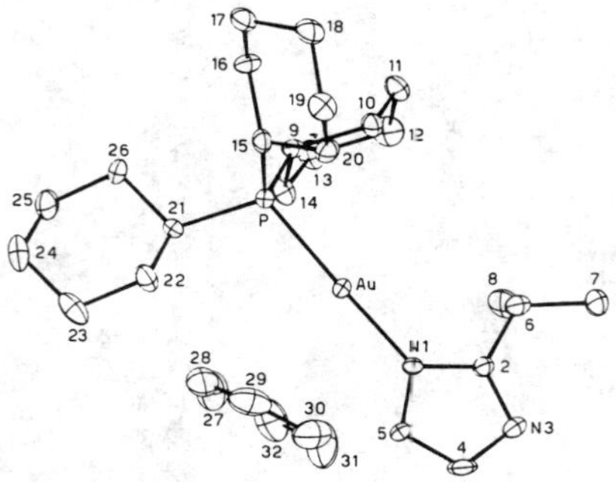

Fig. 1. A view of $C_{24}H_{42}AuN_2P,C_6H_6$.

TETRAKIS(CYCLOHEXYLAMINE)COPPER(II) DINITRATE
$C_{24}H_{52}CuN_6O_6$ $[Cu(C_6H_{11}NH_2)_4](NO_3)_2$

J. GOUTERON, S. JEANNIN, Y. JEANNIN, J. LIVAGE and C. SANCHEZ, 1984. Inorg. Chem., 23, 3387-3393.

Orthorhombic, $P2_12_12_1$, a = 13.68, b = 16.21, c = 13.55 Å, D_m = 1.3, Z = 4. Mo radiation, R = 0.076 for 1387 reflexions.

The cation is shown in Fig. 1. The Cu atom has flattened tetrahedral coordination geometry with Cu-N 1.97(1)-2.05(1) Å and N-Cu-N 92.7(7)-152.7(3) °. The cation has near D_{2d} symmetry.

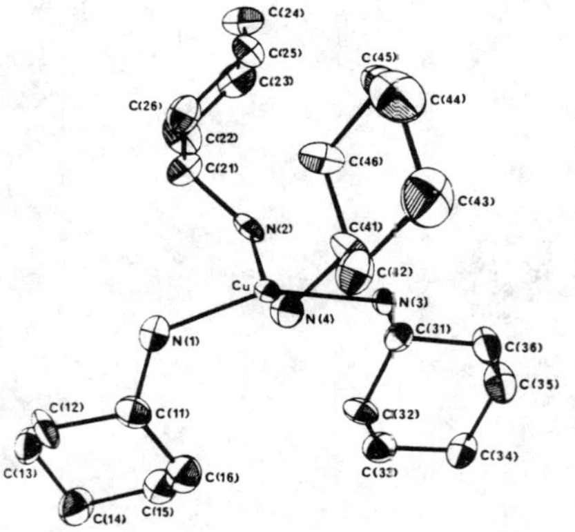

Fig. 1. The $[Cu(C_6H_{11}NH_2)_4]^{2+}$ cation.

PENTACARBONYL(CYANOMETHYLENETRIPHENYLPHOSPHORANE)CHROMIUM(0)
$C_{25}H_{16}CrNO_5P$

U. KUNZE, H. JAWAD, W. CONZELMANN and J. STRAHLE, 1984. Z. Anorg. Chem., 516, 93-101.

Triclinic, $P\bar{1}$, a = 11.265, b = 11.536, c = 9.514 Å, α = 103.47, β = 102.04, γ = 84.00°, Z = 2. Cu radiation, R = 0.044 for 2427 reflexions.

Main dimensions in the molecule (Fig. 1) are Cr-N 2.062(6), N-C6 1.153(7), C6-C7 1.372(8), C7-P 1.709(5) Å, Cr-N-C6 168.1(5), N-C6-C7 178.0(6), C6-C7-P 120.1(4)°.

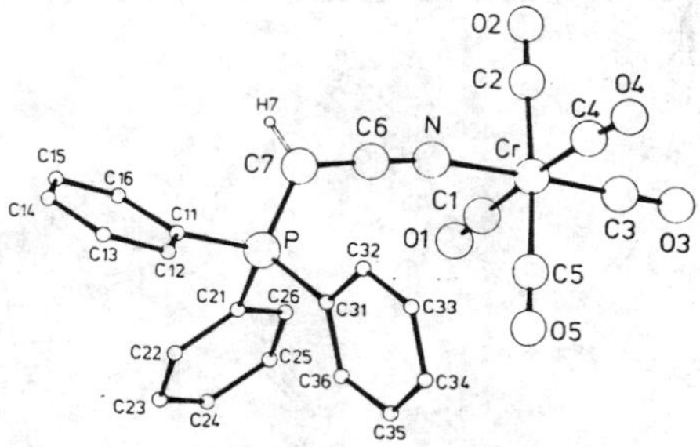

Fig. 1. A view of $C_{25}H_{16}CrNO_5P$.

BIS[2-PYRIDIN-2-YLAMINO)-4-(PYRIDIN-2-YL)THIAZOLE]IRON(II) BIS(TETRAFLUOROBORATE) TRIHYDRATE

$C_{26}H_{20}B_2F_8FeN_8S_2,3(H_2O)$ $[Fe(paptH)_2[BF_4]_2,(H_2O)_3$

A.T. BAKER, H.A. GOODWIN and A.D. RAE, 1984. Aust. J. Chem., 37, 443-447.

Monoclinic, $P2_1/c$, a = 8.968, b = 9.038, c = 41.15 Å, β = 94.81°, D_m = 1.577, Z = 4. Cu radiation, R = 0.0826 for 2549 reflexions.

The Fe(II) coordination sphere is a distorted octahedron (Fig. 1) (mean Fe-N 2.131 Å, mean N-Fe-N angle 77.4°). Both ligands and anions are orientationally disordered with the waters positional disordered. The tridentate ligands disorder around a pseudo twofold axis through N(3)-S with occupancy for two orientations at 0.700(4), 0.300(4) for ligand A, and 0.742(5), 0.258(5) for ligand B.

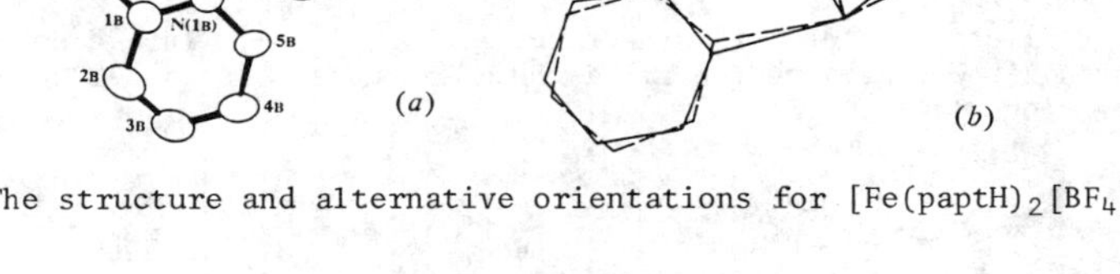

Fig. 1. The structure and alternative orientations for $[Fe(paptH)_2[BF_4]_2,(H_2O)_3$.

TETRAKIS(PYRIDINIUM) (μ_2-OXO)(μ_2-SULPHATO)BIS[OXO(TRIISOTHIOCYANATO)MOLYBDATE(V)]

$C_{26}H_{24}Mo_2N_{10}O_7S_7$

YANG RONGSHENG and YU XIUFEN, 1984. J. Struct. Chem., 3, 101-104.

Triclinic, P1, a = 11.060, b = 8.526, c = 12.059 Å, α = 95.40, β = 114.08, γ = 97.52°, D_m = 1.640, Z = 1. Mo radiation, R = 0.069 for 2939 reflexions.

Both Mo atoms have distorted octahedral geometry (Fig. 1) with Mo-O(3) 1.860 and 1.854, Mo-O(sulphate) 2.228 and 2.236, Mo-O(t) 1.647 and 1.661, Mo-N 2.02-2.12 Å.

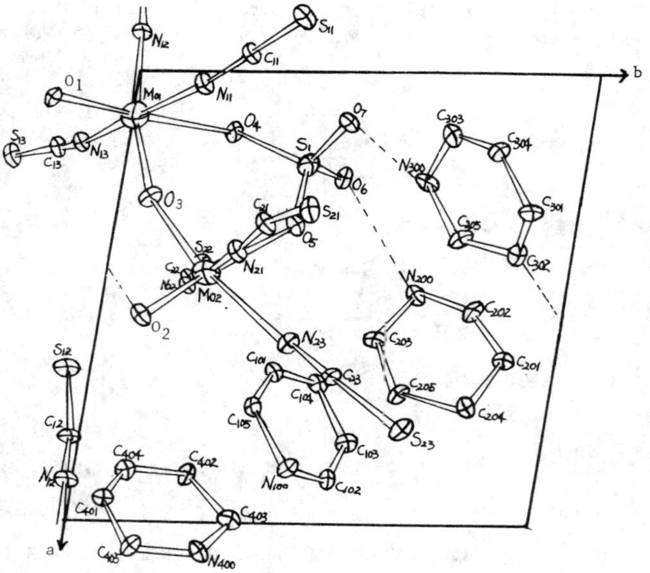

Fig. 1. The unit cell for $C_{26}H_{24}Mo_2N_{10}O_7S_7$.

1-BENZHYDRYL-3,3-DIMETHYL-4-PHENYLAZETIDINE-2-DIMETHYLIMINIUM CHLORIDE TRIHYDRATE
$C_{26}H_{29}ClN_2,3(H_2O)$ (I)

DIBROMOBIS(2-DIMETHYLAMINO-3,3-DIMETHYL-4-PHENYL-1-AZETINE-N)ZINC
$C_{26}H_{36}Br_2N_4Zn$ (II)

F. STIERLI, R. PREWO, J.H. BIERI and H. HEIMGARTNER, 1984. Helv. Chim. Acta, 67, 927-933.

I. Monoclinic, $P2_1/c$, a = 9.170, b = 17.630, c = 16.523 Å, β = 107.50°, Z = 4. Mo radiation, R = 0.117 for 1880 reflexions.

II. Monoclinic, $P2_1/c$, a = 8.772, b = 18.528, c = 18.909 Å, β = 111.71°, Z = 4. Mo radiation, R = 0.089 for 4757 reflexions.

Molecular dimensions for both structures are in Fig. 1. In I atom C5 is 0.14 Å from the plane of the four-membered ring; the Zn atom is 0.56 and 0.22 Å from the corresponding ring planes in II.

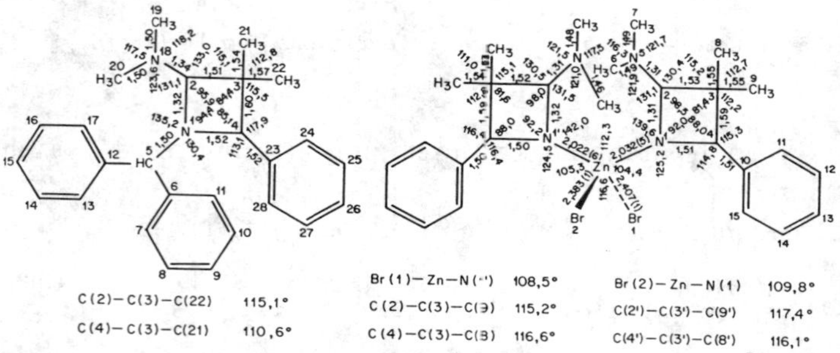

Fig. 1. Molecular dimensions for $C_{26}H_{29}ClN_2$ (I) (left) and $C_{26}H_{36}Br_2N_4Zn$ (II) (right).

AZIDOBIS(DIMETHYLGLYOXIMATO)(TRIPHENYLPHOSPHINE)COBALT(III)
$C_{26}H_{29}CoN_7O_4P$ (I)

[4,5-BIS(METHOXYCARBONYL)-1,2,3-TRIAZOLATO]BIS(DIMETHYLGLYOXIMATO)(TRIPHENYLPHOSPHINE)-COBALT(III)
$C_{32}H_{35}CoN_7O_8P$ (II)

J.H. NELSON, N.E. TAKACH, N. BRESCIANI-PAHOR, L. RANDACCIO and E. ZANGRANDO, 1984. Acta Cryst., C40, 742-746.

I. Orthorhombic, $Pca2_1$, a = 15.156, b = 11.120, c = 16.557 Å, Dm = 1.42, Z = 4. Mo radiation, R = 0.025 for 1988 reflexions.

II. Triclinic, $P\bar{1}$, a = 10.261, b = 18.190, c = 10.425 Å, α = 76.01, β = 88.15, γ = 110.19°, Dm = 1.41, Z = 2. Mo radiation, R = 0.037 for 4469 reflexions.

In both complexes the cobalt atom is octahedrally distorted with the dimethylglyoxime monoanions at the equatorial positions (Fig. 1). The N-Co-P fragment is characterised by Co-P, Co-N bond lengths of 2.311(1) and 2.014(4) Å respectively and an N-Co-P angle of $17\bar{5}.4(1)°$ in the azido derivative (I). The corresponding data for the triazolate derivative (II) are 2.333(1), 1.953(3) Å and 178.0(1)° respectively. The trans influence of the axial ligands is compared with that of other ligands.

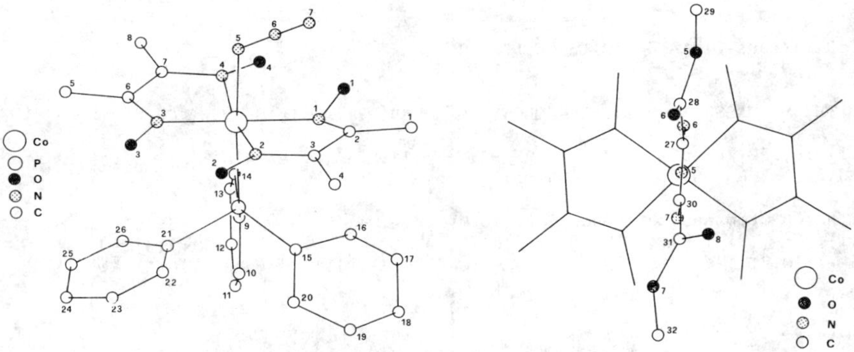

Fig. 1. Perspective views of molecules of $C_{26}H_{29}CoN_7O_4P$ (I) and $C_{32}H_{35}CoN_7O_8P$ (II).

TETRAKIS(DIMETHYLAMIDO)BIS[TRIS(TRIMETHYLSILYL)SILYL]DIMOLYBDENUM
$C_{26}H_{78}Mo_2N_4Si_8$ (I) $Mo_2[Si(SiMe_3)_3]_2(NMe_2)_4$

TETRAKIS(DIMETHYLAMIDO)BIS[TRIS(TRIMETHYLTIN)TIN]DIMOLYBDENUM TOLUENE
$C_{26}H_{78}Mo_2N_4Sn_8,2C_7H_8$ (II) $Mo_2[Sn(SnMe_3)_3]_2(NMe_2)_4.2CH_3C_6H_5$

M.H. CHISHOLM, H.-T. CHIU, K. FOLTING and J.C. HUFFMAN, 1984. Inorg. Chem., 23, 4097-4102.

I. Triclinic, $P\bar{1}$, a = 9.400, b = 9.949, c = 13.304 Å, α = 83.24, β = 97.81, γ = 69.80°, Z = 1. Mo radiation, R = 0.026 for 2768 reflexions (at -165°C).

II. Orthorhombic, $Pbn2_1$, a = 14.600, b = 31.737, c = 13.545 Å, Z = 4. Mo radiation, R = 0.046 for 4105 reflexions (at -162°C).

These compounds (Fig. 1) have a branched 10 metal atom chain with M'-M' and Mo-M' single bonds and a central Mo≡Mo triple bond. There is considerable steric crowding at the Mo centres. The compounds adopt the anti-ethane-like geometry. Important mean geometrical parameters: for I; Mo-Mo 2.216(1), Mo-Si 2.670(2), Mo-N 1.95(1), Si-Si 2.38(1) Å, Mo-Mo-Si 106.9(1), Mo-Mo-N 105(1)°; for II; Mo-Mo 2.201(2), Mo-Sn 2.77(1), Sn-Sn 2.77(1), Mo-N 1.95(2) Å, Mo-Mo-Sn 100.7(2), Mo-Mo-N 105(2)°.

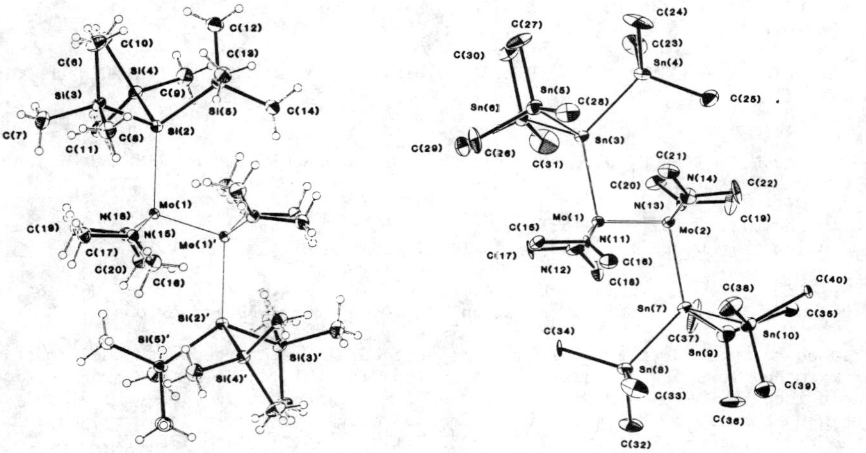

Fig. 1. The $Mo_2[M'(M'Me_3)_3]_2(NMe_2)_4$ molecules (M' = Si, left, and M' = Sn, right).

BROMOTRIS(trans-CINNAMONITRILE)COPPER(I)
$C_{27}H_{21}BrCuN_3$

M. BOLTE, M. MASSAUX and J. ZAREMBOWITCH, 1984. Transition Met. Chem., 9, 461-465.

Monoclinic, $P2_1/c$, a = 7.77, b = 25.92, c = 11.97 Å, β = 94.5°, D_m = 1.49, Z = 4. Mo radiation, R = 0.067 for 1685 reflexions.

The Cu atom has tetrahedral geometry (Fig. 1) with Cu-Br 2.431(5), Cu-N(1) 1.95(1), Cu-N(2) 1.97(1), Cu-N(3) 2.01(1) Å, Br-Cu-N 107.8 - 111.7(3), N-Cu-N 108.5 - 110.2(7)°. The C≡N lengths were 1.09(3), 1.13(3) and 1.15(3) Å.

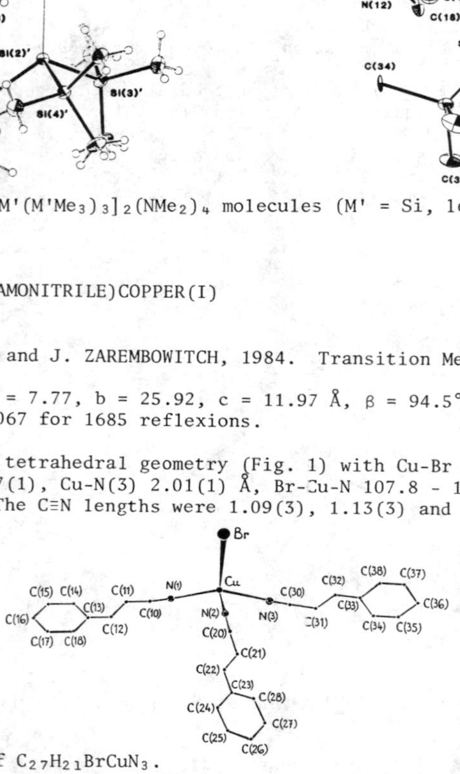

Fig. 1. A view of $C_{27}H_{21}BrCuN_3$.

(2,2'-BIPYRIDINE)CHLORO(TRIPHENYLPHOSPHINE)COPPER(I) MONOHYDRATE
$C_{28}H_{23}ClCuN_2P,H_2O$

B.E. GREEN, C.H.L. KENNARD, G. SMITH, B.D. JAMES and A.H. WHITE, 1984. Acta Cryst., C40, 426-428.

Triclinic, $P\bar{1}$, a = 16.967, b = 9.426, c = 8.712 Å, α = 65.50, β = 88.23, γ = 89.53°, D_m = 1.39, Z = 2. Mo radiation, R = 0.049 for 2371 reflexions.

The Cu atom (Fig. 1) is tetrahedrally coordinated to 2,2'-bipyridine (Cu-N 2.076(5), 2.087(6) Å; N-Cu-N 79.2(2)°) Cl (Cu-Cl 2.330(2) Å) and triphenylphosphine (Cu-P 2.180(2) Å), and is distorted (N-Cu-Cl 102.9(2), 104.5(2)°; N-Cu-P 122.8(2), 123.2(1)°; and Cl-Cu-P 117.1(1)°). The water mclecule is hydrogen bonded to Cl (Cl...O 3.25(1), 3.30(1) Å).

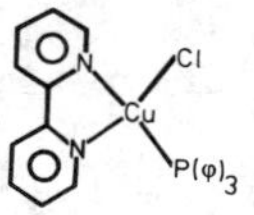

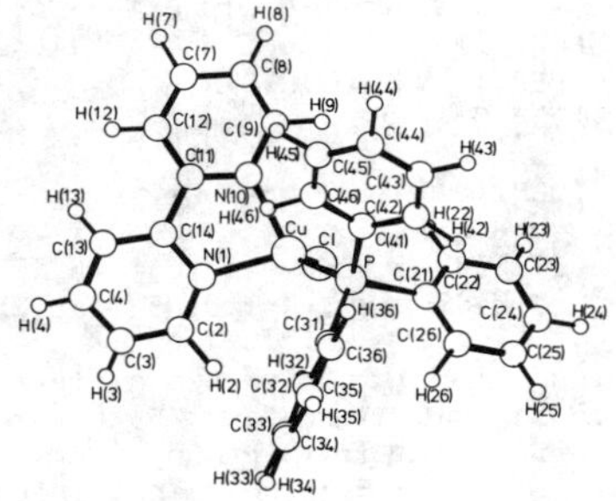

Fig. 1. A view of $C_{28}H_{23}ClCuN_2P$.

BIS(2,9-DIMETHYL-1,10-PHENANTHROLINE)COPPER(I) BROMIDE HYDRATE
$C_{28}H_{24}BrN_4,H_2O$ (I) $[Cu(dmp)_2]Br,H_2O$

BIS(4,4',6,6'-TETRAMETHYL-2,2'-BIPYRIDINE)COPPER(I) CHLORIDE DIHYDRATE
$C_{28}H_{32}ClCuN_4,2(H_2O)$ (II) $[Cu(tmbp)_2]Cl,(H_2O)_2$

BIS(2,9-DIMETHYL-1,10-PHENANTHROLINE)COPPER(I) NITRATE DIHYDRATE
$C_{28}H_{24}CuN_5O_3,2(H_2O)$ (III) $[Cu(dmp)_2]NO_3,(H_2O)_2$

J.F. DOBSON, B.E. GREEN, P.C. HEALY, C.H.L. KENNARD, C. PAKAWATCHAI and A.H. WHITE, 1984. Aust. J. Chem., 37, 649-659.

I. Monoclinic, $P2_1/c$, a = 14.067, b = 17.687, c = 10.337 Å, β = 91.04°, D_m = 1.50, Z = 4. Mo radiation, R = 0.038 for 1855 reflexions.

II. Triclinic, $P\bar{1}$, a = 15.305, b = 13.164, c = 7.344 Å, α = 82.03, β = 93.83, γ = 108.11°, Z = 2. Mo radiation, R = 0.051 for 1293 reflexions.

III. Triclinic, $P\bar{1}$, a = 7.619, b = 11.967, c = 15.534 Å, α = 71.26, β = 82.13, γ = 86.86°, Z = 2. Mo radiation, R = 0.043 for 2335 reflexions.

Each cation has a distorted tetrahedral stereochemistry (I: Cu-N 2.044(5) Å, N-Cu-N 82.7(2)°; II: Cu-N 2.040(5) Å, N-Cu-N 81.3(2)°; III: (redetermination of 1) Cu-N 2.040(4) Å, N-Cu-N 82.9(2)°.

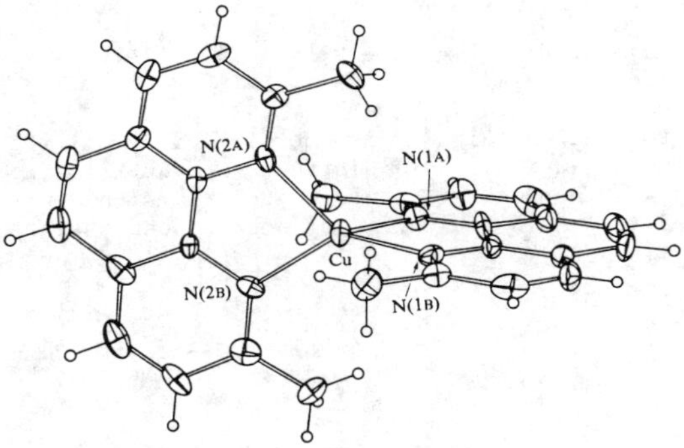

Fig. 1. The structure of $[Cu(dmp)_2]^2$.

1. R. HAMALAINEN, U. TURPEINEN, M. AHLGREN and T. RAIKAS, 1978. Finn. Chem. Lett. 199.

AQUADICHLORO-μ-[3,6-DI(2-PYRIDYL)PYRIDAZINE-N,N',N'',N''']-μ-HYDROXO-NITRATO-O-DICOPPER(II) DICHLORO-μ[3,6-DI(2-PYRIDYL)PYRIDAZINE-N,N',N'',N''']-μ-HYDROXO-μ-NITRATO-O,O'-DICOPPER(II) SESQUIHYDRATE
$C_{28}H_{24}Cl_4Cu_4N_{10}O_9,1.5H_2O$

G. DE MUNNO and G. DENTI, 1984. Acta Cryst., C40, 616-618.

Triclinic, $P\bar{1}$, a = 14.500, b = 13.049, c = 10.874 Å, α = 84.8, β = 110.1, γ = 110.5°, D_m = 1.96, Z = 2. Mo radiation, R = 0.050 for 2114 reflexions.

The asymmetric unit consists of two different neutral dinuclear complexes (Fig. 1) and of 1.5 solvent molecules. Both complexes contain the tetradentate organic ligand, a bridging hydroxo group and two terminal chlorine ligands. In one complex the copper atoms are five-coordinate, in a distorted square-pyramidal geometry, the two apical sites being occupied by the oxygen atoms of a monodentate nitrato group and of a water molecule, respectively. In the second complex the copper atoms are bridged by a bidentate nitrato group. Such complexes form a dimeric structure in which one of the copper atoms achieves a distorted octahedral geometry, because of a long interaction with the μ-hydroxo oxygen of the second monomeric unit.

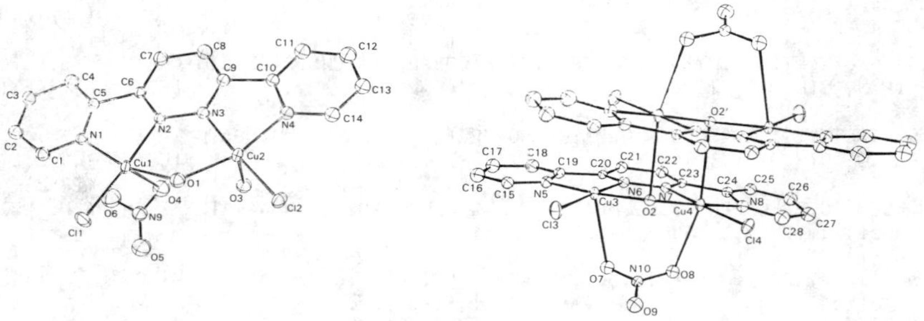

Fig. 1. $C_{28}H_{24}Cl_4Cu_4N_{10}O_9,1.5H_2O$: perspective views of the two different complex molecules in the asymmetric unit.

(BIPYRIDINE)(ISOPROPYLIDENEAMIDO)(TERPYRIDINE)RUTHENIUM TRIPERCHLORATE MONOHYDRATE
$C_{28}H_{25}Cl_3N_6O_{12}Ru,H_2O$ $[Ru(C_{15}H_{11}N_3)(C_{10}H_8N_2)(NCMe_2)](ClO_4)_3,H_2O$

P.A. ADCOCK, F.R. KEENE, R.S. SMYTHE and M.R. SNOW, 1984. Inorg. Chem., 23, 2336-2343.

Orthorhombic, $Pna2_1$, a = 22.034, b = 10.689, c = 14.251 Å, Z = 4. Mo radiation, R = 0.036 for 2005 reflexions.

The Ru atom of the cation (Fig. 1) is coordinated by six N atoms in a distorted octahedral geometry. The Ru-N(pyridyl) distances are Ru-N(1,2,5,6) 2.055(6)-2.076(7) and Ru-N(4) 1.937(7) Å. The Ru-$NCMe_2$ linkage is linear (Ru-N-C 175.2(10)°) and the Ru-N distance of 1.831(10) Å is indicative of double bond character. The N-CMe_2 distance is 1.259(17) Å.

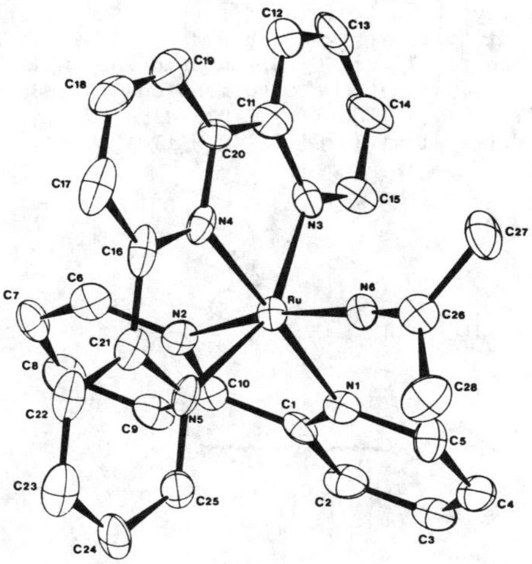

Fig. 1. The $[Ru(C_{15}H_{11}N_3)(C_{10}H_8N_2)(NCMe_2)]^{3+}$ cation.

(2,2'-BIPYRIDINE)[BIS(2,6-DIACETYLPYRIDINE)DIHYDRAZINE-N[1],N[2],N[4],N[5]]IRON(II) DIPERCHLORATE DIHYDRATE

$C_{28}H_{26}Cl_2FeN_8O_8,2H_2O$ $[Fe(C_{10}H_8N_2)(C_{18}H_{18}N_6)](ClO_4)_2.2H_2O$

G.A. BREWER and E. SINN, 1984. Inorg. Chem., 23, 2532-2537.

Monoclinic, $P2_1/n$, a = 8.943, b = 24.332, c = 15.954 Å, β = 100.77°, Z = 4. Mo radiation, R = 0.080 for 1512 reflexions.

The cation (Fig. 1) has Fe in a distorted octahedral environment, coordinated by two N atoms of the bipyridine ligand (Fe-N 1.96(1) and 2.01(1) Å) and by two pyridinic N atoms (Fe-N 1.92(1) (x2)) and two hydrazine N atoms (Fe-N 1.96(1) and 1.95(1) Å) of the macrocyclic ligand. The macrocyclic ligand is clearly bent.

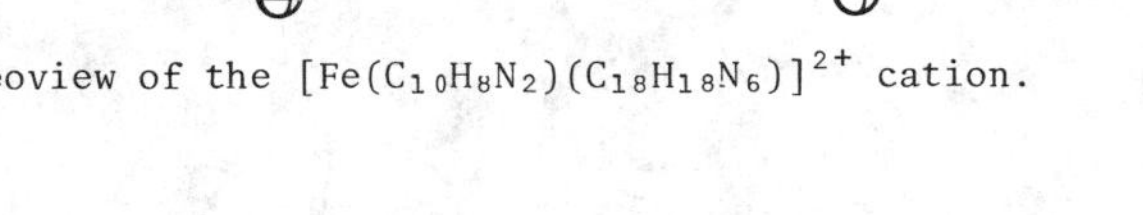

Fig. 1. Stereoview of the $[Fe(C_{10}H_8N_2)(C_{18}H_{18}N_6)]^{2+}$ cation.

(μ_2-CHLORO)DICHLOROBIS[2-(2-PYRIDYL)ETHYLAMINE-N,N',N'']DICOPPER HEXAFLUOROPHOSPHATE

$C_{28}H_{34}Cl_3Cu_2F_6N_6P$ $[Cu_2(py2)_2Cl_3]^+.PF_6^-$

K.D. KARLIN, J.W. MCKOWN, J.C. HAYES, J.P. HUTCHINSON and J. ZUBIETA, 1984. Transition Met. Chem., 9, 405-406.

Monoclinic, C2/c, a = 16.974, b = 17.144, c = 11.987 Å, β = 108.41°, Z = 4. Mo radiation, R = 0.048 for 2484 reflexions.

The cation (Fig. 1) lies on a twofold axis (passing through Cl(2)) and the Cu coordination geometry is distorted trigonal bipyramidal with N(2) and N(3) axial. Unusual features are the nearly linear Cu-Cl-Cu bridging angle (178.0(1)°) and the long Cu-Cl(2) bond length (2.549 Å). Other dimensions include Cu-Cl(1) 2.279(1), Cu-N(1) 2.066(5), Cu-N(2) 2.029(4), Cu-N(3) 2.022(4) Å, N(2)-Cu-N(3) 176.6(1), Cl(1)-Cu-Cl(2) 117.8, N(1)-Cu-Cl(2) 91.4(1), Cl(1)-Cu-N(1) 150.4(1)°.

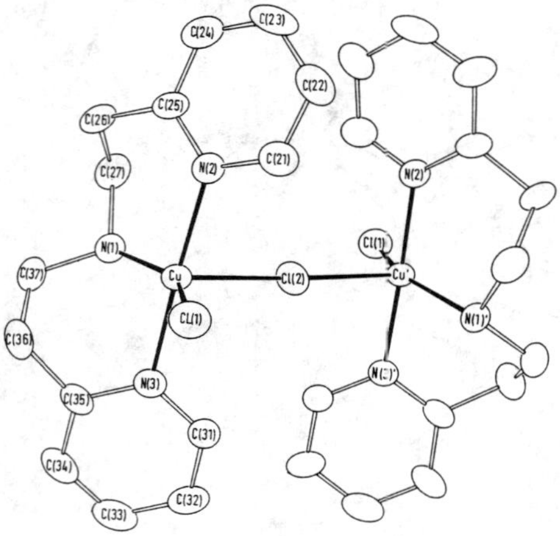

Fig. 1. A view of the $[Cu_2(py2)_2Cl_3]^+$ cation.

trans-DICHLOROBIS(11-PHENYL-9,10,11-TRIAZABICYCLC[6.3.0]UNDEC-9-ENE-N[9])PALLADIUM(II) CHLOROFORM SOLVATE

$C_{28}H_{38}Cl_2N_6Pd,2(CHCl_3)$

F. PORTA, M. PIZZOTTI, G. LA MONICA, L.A. FINESSI, S. CENINI, P.L. BELLON and F. DEMARTIN, 1984. J. Chem. Soc. Dalton, 2409-2414.

Triclinic, $P\bar{1}$, a = 9.258, b = 10.702, c = 11.156 Å, α = 113.94, β = 99.60, γ = 104.87°, Z = 1. Mo radiation, R = 0.039 for 2854 reflexions.

The Pd atom lies on a crystallographic inversion centre and thus has a strictly planar trans conformation (Fig. 1). The two chloroform molecules are hydrogen bonded to the coordinated Cl atoms. The mean plane of the inner coordination sphere is perpendicular to the 1,2,3-triazapentane ring. The Pd-Cl(1) and Pd-N(1) distances are 2.298(1) and 2.001(2) Å respectively. The N-substituted triazapentane exhibits pronounced delocalisation of the π-system with N(1)-N(2), N(2)-N(3) and N(3)-C(11) distances of 1.266(3), 1.331(3) and 1.412(3) Å. The cyclo-octane moiety shows the effect of strain with an average C-C-C angle of 116.5° and an average C-C distance of 1.52 Å. The C(1)-C(8) junction has a synperiplanar conformation.

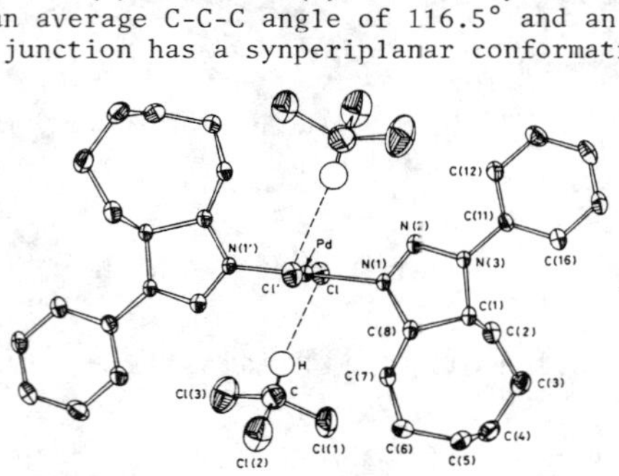

Fig. 1. $C_{28}H_{38}Cl_2N_6Pd,2(CHCl_3)$: the molecular structure.

BIS(ETHANOL)(6,6''-(2,3,6,7-TETRAAZAOCTA-3,5-DIENE-2,7-DIYL)-4'-PHENYL-2,2'.6',2''-TERPYRIDINE)NICKEL BIS(TETRAFLUOROBORATE)
$C_{29}H_{33}B_2F_8N_7NiO_2$ (I)

6,6'-(3,6-DIAZA-2,7-DIAZONIAOCTA-3,5-DIENE-2,7-DIYL)-4'-PHENYL-2,2'.6',2''-TERPYRIDINE BIS(HEXAFLUOROPHOSPHATE)
$C_{25}H_{23}F_{12}N_7P_2$ (II)

E.C. CONSTABLE, J. LEWIS, M.C. LIPTROT and P.R. RAITHBY, 1984. J. Chem. Soc. Dalton, 2177-2185.

I. Monoclinic, C2/c, a = 14.271, b = 17.365, c = 12.364 Å, β = 98.03°, Z = 4. Cu radiation, R = 0.068 for 1944 reflexions.

II. Monoclinic, $P2_1/c$, a = 8.199, b = 12.592, c = 27.399 Å, β = 92.32°, Z = 4. Cu radiation, R = 0.119 for 2011 reflexions.

In I the cation lies on a crystallographic twofold axis which passes through N(1), N(4), C(10), C(11) and C(14) (Fig. 1). The X-ray analysis established the presence of a seven-coordinate pentagonal-bipyramidal Ni(II) centre. The N_5 donor set of atoms which define the vertices of an equatorial plane are coplanar and the Ni atom is located in the cavity of the macrocycle (Ni(1)-N(1) 2.324(3), Ni(1)-N(3) 2.104(3), Ni(1)-N(4) 2.015(4) Å). The ethanol groups occupy axial sites (Ni-O(1) 2.084(4) Å). The BF_4^- anions are not coordinated. The phenyl ring is twisted 14.9° with respect to the plane of the central pyridine ring whilst the pyridines are inclined at an angle of 3°. In II, one hexafluorophosphate moiety is disordered. Here the fifteen atoms that make up the macrocycle are essentially coplanar and the phenyl ring makes an angle of 5.7° with the plane of the pyridine ring to which it is attached. Hole-size calculations based on the circle described by the five donor atoms show no significant differences between the two structures - 2.10 Å for the Ni complex and 2.12 Å for the free macrocycle.

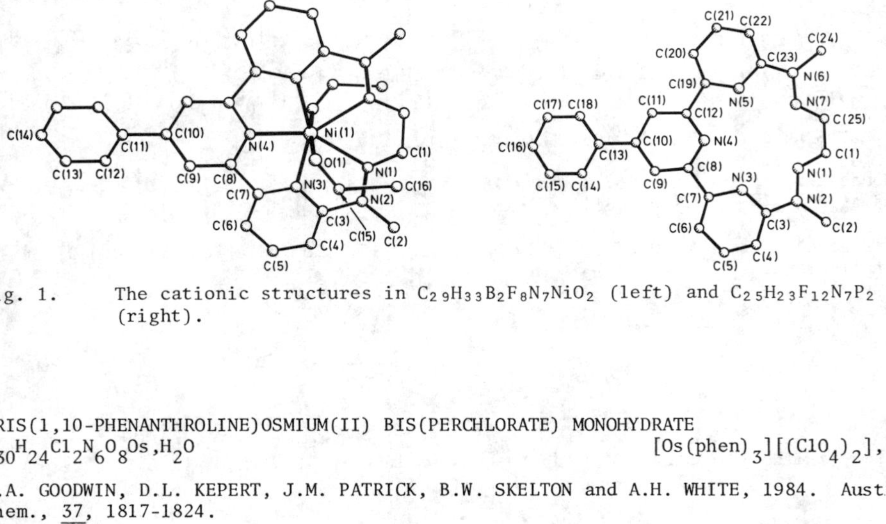

Fig. 1. The cationic structures in $C_{29}H_{33}B_2F_8N_7NiO_2$ (left) and $C_{25}H_{23}F_{12}N_7P_2$ (right).

TRIS(1,10-PHENANTHROLINE)OSMIUM(II) BIS(PERCHLORATE) MONOHYDRATE
$C_{30}H_{24}Cl_2N_6O_8Os,H_2O$ $[Os(phen)_3][(ClO_4)_2],(H_2O)$

H.A. GOODWIN, D.L. KEPERT, J.M. PATRICK, B.W. SKELTON and A.H. WHITE, 1984. Aust. J. Chem., 37, 1817-1824.

Monoclinic, C2/c, a = 35.94, b = 16.020, c = 12.370 Å, β = 102.38°, Z = 8. Mo radiation, R = 0.043 for 3477 reflexions.

The osmium atom is octahedral with approximate D_3 point symmetry, and is chelated to three phen groups (Fig. 1, Os-N 2.074(9) Å, N-Os-N 78.9(3)°).

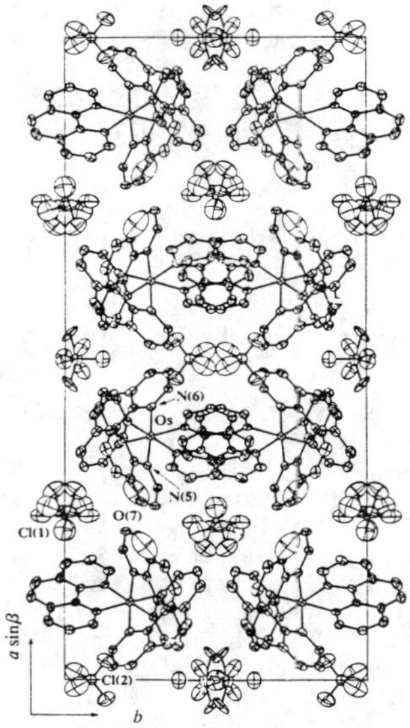

Fig. 1. The packing of $[Os(phen)_3][(ClO_4)_2],(H_2O)$.

BIS(2,2'-BIPYRIDINE-N,N')(NH$^+$-2,2'-BIPYRIDINIUM-3-YL-N')IRIDIUM(III) TRISPERCHLORATE WATER

$C_{30}H_{24}Cl_3IrN_6O_{12},H_2O$ (I) $[Ir(C_{10}H_8N_2)_3](ClO_4)_3,H_2O$

TRIS(2,2'-BIPYRIDINE-N,N')IRIDIUM(III) TRISPERCHLORATE WATER

$C_{30}H_{24}Cl_3IrN_6O_{12},2.333H_2O$ (II) $[Ir(C_{10}H_8N_2)_3](ClO_4)_3,2.333H_2O$

A.C. HAZELL and R.G. HAZELL, 1984. Acta Cryst., C40, 806-811.

I. Monoclinic, $P2_1/c$, a = 16.700, b = 16.279, c = 13.600 Å, β = 108.86°, Z = 4. Mo radiation, R = 0.031 for 2531 reflexions.

II. Trigonal, R3c, a = 30.370, c = 20.922 Å, Z = 18. Mo radiation, R = 0.040 for 2870 reflexions.

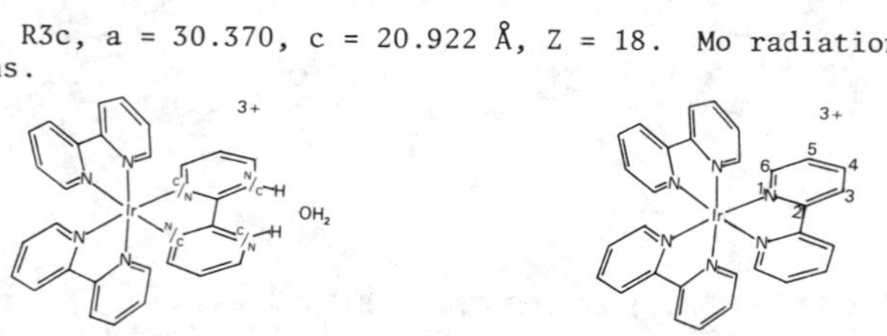

Fig. 1. Molecular structures of the cations in $C_{30}H_{24}Cl_3IrN_6O_{12},H_2O$ (I) (left) and $C_{30}H_{24}Cl_3IrN_6O_{12},2.333H_2O$ (II) (right).

In complex I iridium is bonded to two bipyridine groups via the nitrogen atoms, the bipyridiniumyl group is bonded via one nitrogen and one carbon atom (Fig. 1). The non-coordinating nitrogen atom is hydrogen bonded to a water molecule which is in the plane of the ligand. The carbon-bonded bipyridiniumyl ligand is disordered. The mean Ir-N distance is 2.082(5) Å for N trans to C, and 2.042(4) Å for the others; the value for the long bonds is the average of a normal Ir-N bond and of one trans to C. In complex II iridium is coordinated via the nitrogen atoms to all three bipyridine groups, with mean Ir-N distance 2.021(6) Å. In both complexes iridium has distorted octahedral coordination. The crystal structure of complex I has been reported elsewhere (1).

1. W.A. WICKRAMASINGHE, P.H. BIRD and N. SERPONE, 1981. J. Chem. Soc. Chem. Commun., p. 1284.

TRIS(2,2'-BIPYRIDINE)NICKEL(III) TRIPERCHLORATE ACETONITRILE DICHLOROMETHANE
$C_{30}H_{24}Cl_3N_6NiO_{12},2C_2H_3N,0.5CH_2Cl_2$

D.J. SZALDA, D.H. MACARTNEY and N. SUTIN, 1984. Inorg. Chem., 23, 3473-3479.

Triclinic, $P\bar{1}$, a = 11.193, b = 17.634, c = 10.494 Å, α = 95.41, β = 107.02, γ = 95.28°, D_m = 1.57, Z = 2. Cu radiation, R = 0.096 for 5087 reflexions (at 160 K).

The cation (Fig. 1) has tetragonally distorted octahedral coordination about the Ni(III) atom. The six Ni-N bonds form three sets with two pairs of longer equatorial bonds, 2.022(6) and 2.000(5) Å, and a pair of shorter axial bonds, 1.924(6) Å.

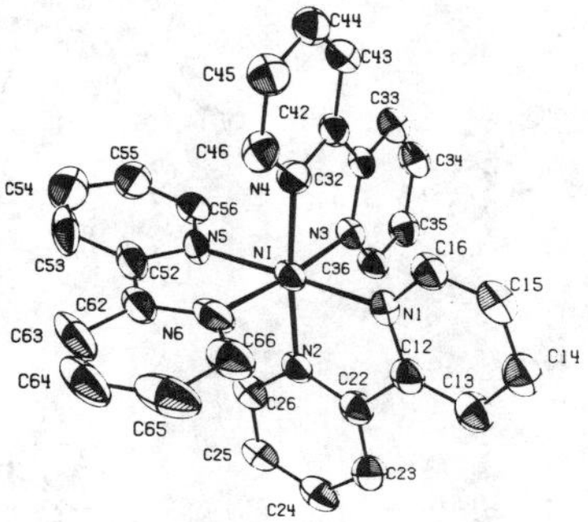

Fig. 1. The $[Ni(C_{10}H_8N_2)_3]^{3+}$ cations.

α-(1,10-PHENANTHROLINE)TETRAHYDROBORATO(TRIPHENYLPHOSPHINE)COPPER(I)
$C_{30}H_{27}BCuN_2P$ (I) $(C_{12}H_8N_2)Cu(P(C_6H_5)_3)BH_4$

β-(1,10-PHENANTHROLINE)TETRAHYDROBORATO(TRIPHENYLPHOSPHINE)COPPER(I)
$C_{30}H_{27}BCuN_2P$ (II) $(C_{12}H_8N_2)Cu(P(C_6H_5)_3)BH_4$

β-(1,10-PHENANTHROLINE)TETRADEUTEROBORATO(TRIPHENYLPHOSPHINE)COPPER(I)
$C_{30}H_{23}BCuD_4N_2P$ (III) $(C_{12}H_8N_2)Cu(P(C_6H_5)_3)BD_4$

(2,9-DIMETHYL-4,7-DIPHENYL-1,10-PHENANTHROLINE)TETRAHYDROBORATOCOPPER(I)
$C_{26}H_{24}BCuN_2$ (IV)

B.E. GREEN, C.H.L. KENNARD, G. SMITH, M.M. ELCOMBE, F.H. MOORE, B.D. JAMES and A.H. WHITE, 1984. Inorg. Chim. Acta, 83, 177-189.

I. Triclinic, $P\bar{1}$, a = 16.247, b = 9.750, c = 9.322 Å, α = 62.92, β = 84.77, γ = 84.34°, D_m = 1.31, Z = 2. Mo radiation, R = 0.08 for 2929 reflexions.

II. Triclinic, $P\bar{1}$, a = 13.675, b = 10.115, c = 9.700 Å, α = 95.22, β = 96.22, γ = 101.02°, D_m = 1.32, Z = 2. Mo radiation, R = 0.06 for 2713 reflexions.

III. Triclinic, $P\bar{1}$, a = 13.703, b = 10.096, c = 9.74 Å, α = 95.23, β = 96.51, γ = 101.04°, D_m = 1.33, Z = 2. Neutron radiation (λ = 1.238 Å), R = 0.12 for 1320 reflexions.

IV. Monoclinic, Cc, a = 14.522, b = 20.07, c = 7.718 Å, β = 106.17°, D_m = 1.33, Z = 4. Mo radiation, R = 0.06 for 1359 reflexions.

The structure of the α-form (I) revealed atomic parameters only marginally improved on those determined previously (1). I is yellow while II (the β-form) forms red crystals and their structures, while basically similar, have subtle differences (Fig. 1). The neutron refinement of III unequivocally located the BD_4^- group linked to the metal atom. The structural differences between I, II and III are: (a) the Cu-P distances of 2.205, 2.227 and 2.26 Å, (b) the bond angles N(1)-Cu-P 117.0, 109.7 and 109.6°, N(10)-Cu-P 109.6, 108.4 and 109.0°. In both α and β isomers the BH_4^- moiety behaves as a bidentate ligand bonding through three-centred hydrogen bonds to the metal atom, with Cu-D distances of 2.03 and 1.63 Å in III. In IV the metal is four-coordinate unlike compounds I-III, with Cu-N distances of 2.03 and 2.06 Å and Cu-H (from BH_4^-) distances of 1.50, 1.50 Å. The N-Cu-N angle is 82.6°.

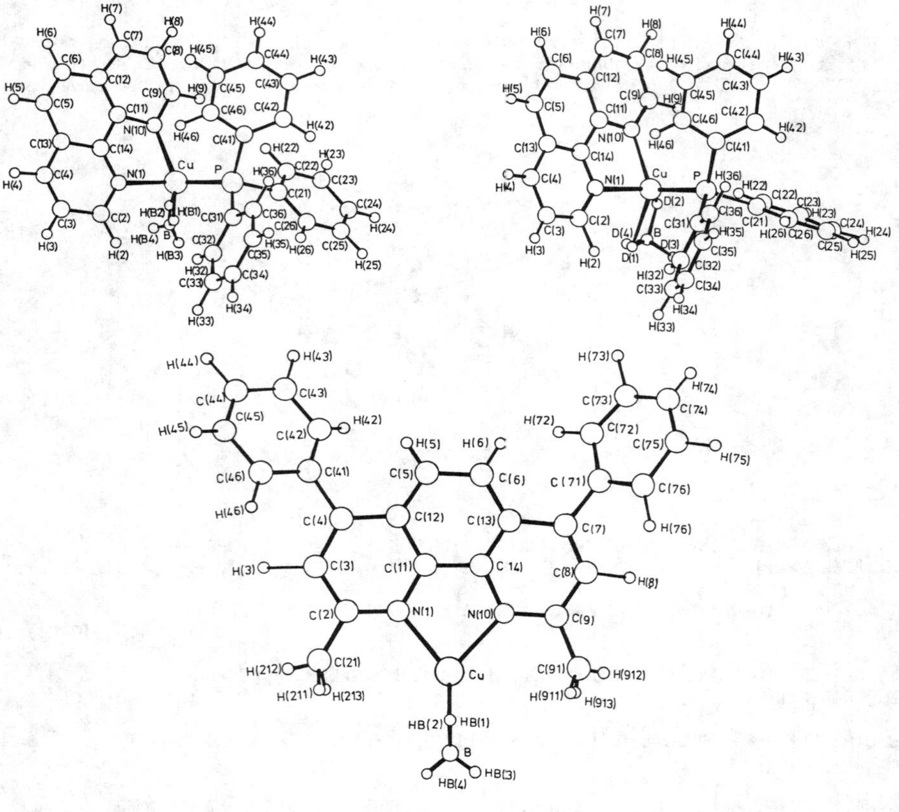

Fig. 1. Molecular conformations of α-$C_{30}H_{27}BCuN_2P$ (top left), β-$C_{30}H_{23}BCuD_4N_2P$ (top right), and $C_{26}H_{24}BCuN_2$ (bottom).

1. Structure Reports, 48B, 1455.

BIS(ISOTHIOCYANATO)TETRAKIS(4-VINYLPYRIDINE)NICKEL (α-Phase)
$C_{30}H_{28}N_6NiS_2$ $Ni(NCS)_2(4\text{-}Vipy)_4$

J. LIPKOWSKI, K. SUWINSKA, J. HATT, A. ZIELENKIEWICZ and W. ZIELENKIEWICZ, 1984. J. Inclusion Phenomena, 2, 317-325.

Orthorhombic, Pbca, a = 11.27, b = 16.86, c = 32.06 Å, Z = 8. Mo radiation, R = 0.056 for 950 reflexions.

The Ni atom has trans-octahedral coordination and the four equatorial pyridine rings adopt a propeller conformation (tilted 28, 38, 38 and 52° from a planar conformation).

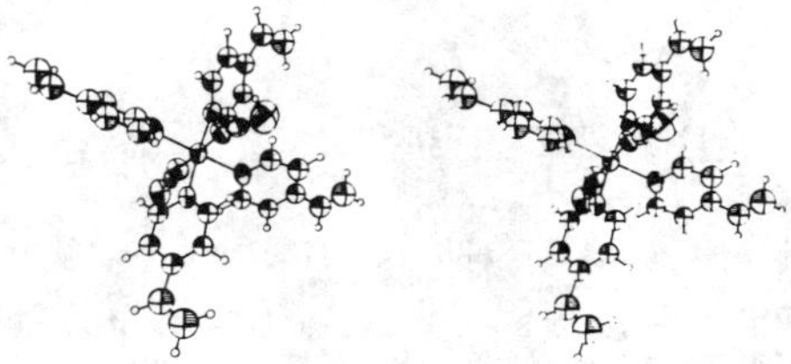

Fig. 1. A stereoview of $Ni(NCS)_2(4\text{-}Vipy)_4$.

BIS(μ_2-SULPHATO)HEXAKIS(PYRIDINE)DIMOLYBDENUM(II) PYRIDINE SOLVATE
$C_{30}H_{30}Mo_2N_6O_8S_2,2(C_5H_5N)$

F.A. COTTON and A.H. REID Jr., 1984. Nouv. J. Chim., 8, 203-206.

Monoclinic, $P2_1/n$, a = 9.945, b = 15.893, c = 13.547 Å, β = 94.98°, Z = 2. Mo radiation, R = 0.0547 for 1352 reflexions.

The molecule (Fig. 1) lies about an inversion centre, but the effective symmetry is D_{2h}. Main dimensions are: Mo-Mo 2.134(2), Mo-O 2.108 and 2.116(6), Mo-N (trans to N) 2.221 and 2.236(8), Mo-N (trans to Mo) 2.712(8), S-O(Mo) 1.511 and 1.520(6), S-O (terminal) 1.420 and 1.448(8) Å. An unusual feature is that the five-membered Mo_2O_2S ring is virtually planar.

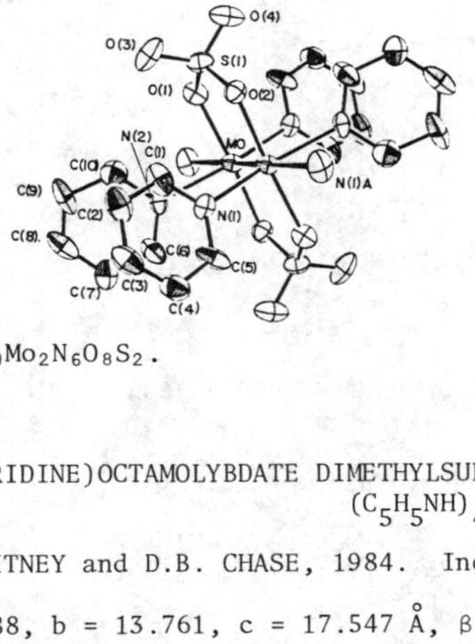

Fig. 1. A view of $C_{30}H_{30}Mo_2N_6O_8S_2$.

TETRAKIS(PYRIDINIUM) BIS(PYRIDINE)OCTAMOLYBDATE DIMETHYLSULPHOXIDE
$C_{30}H_{34}Mo_8N_6O_{26},2C_2H_6OS$ $(C_5H_5NH)_4[(C_5H_5N)_2Mo_8O_{26}].2(CH_3)_2SO$

E.M. McCARRON, III, J.F. WHITNEY and D.B. CHASE, 1984. Inorg. Chem., 23, 3275-3280.

Monoclinic, $P2_1/n$, a = 11.388, b = 13.761, c = 17.547 Å, β = 100.70°, Z = 2. Mo radiation, R = 0.038 for 4937 reflexions.

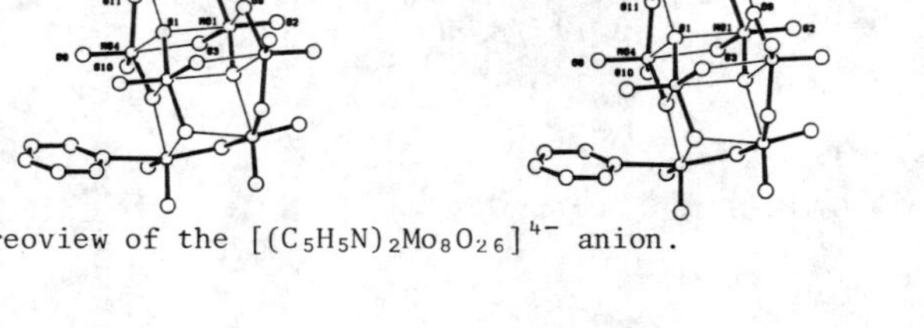

Fig. 1. Stereoview of the $[(C_5H_5N)_2Mo_8O_{26}]^{4-}$ anion.

The anion (Fig. 1) is built up of two $C_5H_5NMoO_5$ and six MoO_6 edge-sharing octahedra, all of which exhibit the typical 2+2+2 octahedral distortion. Bond lengths: Mo-N 2.279(5), Mo=O 1.701(4)-1.721(4), Mo-O 1.746(4)-2.406(3) Å.

TETRABENZYLAMINEPALLADIUM(II) ACETATE CHLORIDE
$C_{30}H_{39}ClN_4O_2Pd$ $[Pd(NH_2CH_2Ph)_4][Cl][O_2CMe]$

T.C. JONES, A.J. NIELSON and C.E. RICKARD, 1984. Aust. J. Chem., 37, 2179-2192.

Triclinic, P$\bar{1}$, a = 10.242, b = 11.378, c = 14.392 Å, α = 109.50, β = 94.77, γ = 100.01°, Z = 2. Mo radiation, R = 0.084 for 2321 reflexions.

Pd has a square planar coordination (Fig. 1). Hydrogen bonds from the NH proton hold the cation to chlorine and the oxygen atoms of the anions. No ortho aromatic ring hydrogen is close to a metal centre (Pd-N 2.028(10) Å).

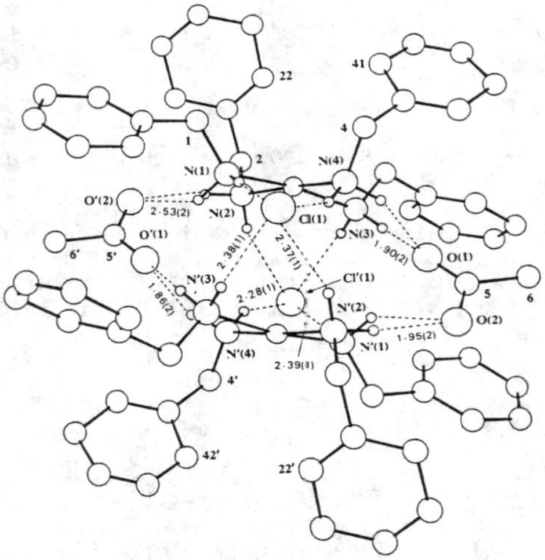

Fig. 1. The structure of $[Pd(NH_2CH_2Ph)_4][Cl][O_2CMe]$.

μ-OXO-BIS{[HYDROTRIS(3,5-DIMETHYL-1-PYRAZOLYL)BORATO]DIOXOMOLYBDENUM(VI)}
$C_{30}H_{44}B_2Mo_2N_{12}O_5$ $Mo_2(C_{15}H_{22}BN_6)_2O_5$

K.M. BARNHART and J.H. ENEMARK, 1984. Acta Cryst., C40, 1362-1364.

Monoclinic, C2/c, a = 28.253, b = 9.550, c = 15.761 Å, β = 115.78°, D_m = 1.51, Z = 4. Mo radiation, R = 0.028 for 2576 reflexions.

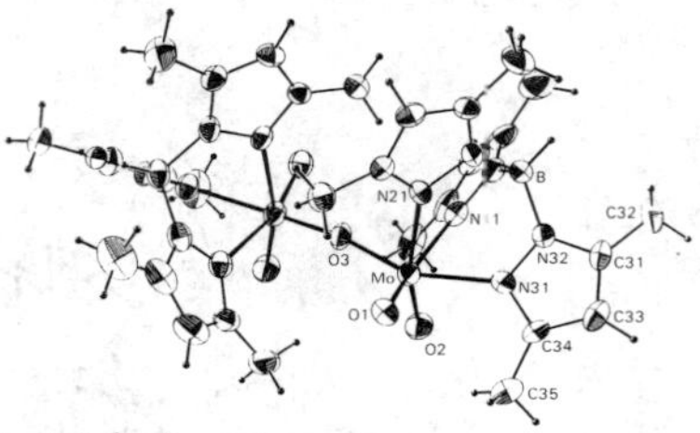

Fig. 1. The $Mo_2(C_{15}H_{22}BN_6)_2O_5$ molecule.

The molecule contains an $O_2MoOMoO_2$ core with the bridging O atom on a twofold axis. The tridentate ligand is coordinated through N atoms (Fig. 1). The Mo-N distance (2.208(2) Å) trans to the bridging oxygen is significantly shorter than those trans to the terminal oxygens (2.318(2) and 2.314(2) Å). Further dimensions are Mo-O (non-bridging) 1.701(2) and 1.696(2), (bridging) 1.889(1) Å, Mo-O-Mo 167.1(2)°. Terminal and bridging oxo groups have fac stereochemistry.

BIS(TRIAQUATRIS(4-ETHYL-1,2,4-TRIAZOLE-N^1)IRON(II)-N^2,$N^{2'}$,$N^{2''}$)IRON(II) HEXAKIS(TRIFLUOROMETHANESULPHONATE)

$C_{30}H_{54}F_{18}Fe_3N_{18}O_{24}S_6$ $[Fe_3(C_4H_7N_3)_6(H_2O)_6](CF_3SO_3)_6$

G. VOS, R.A.G. de GRAAFF, J.G. HAASNOOT, A.M. van der KRAAN, P. de VAAL and J. REEDIJK, 1984. Inorg. Chem., 23, 2905-2910.

Trigonal, P$\bar{3}$1c, a = 14.488, c = 19.858 Å, Z = 2. Mo radiation, R = 0.046 for 922 reflexions at 300 K. a = 14.265, c = 19.632 Å, R = 0.063 for 1112 reflexions at 105 K.

The structure consists of disordered $CF_3SO_3^-$ anions and $[Fe_3(C_4H_7N_3)_6(H_2O)_6]^{6+}$ cations (Fig. 1) having exact $\bar{3}$m symmetry. The two pairs of metal centres are each linked by three 1,2-bicoordinating triazole ligands. The terminal coordination sites are occupied by water molecules. A spin crossover involving the central Fe(II) occurs upon cooling between 230 and 190 K. Fe-OH_2 distances at 300 K are 2.156(4) Å, whereas the Fe-N distances are 2.157(4) Å (terminal Fe) and 2.174(4) Å (central Fe). The Fe-L distances of the terminal Fe hardly change (0.01-0.02 Å) upon cooling. However, the observed decrease in the Fe-N bond lengths (~0.14 Å) upon cooling for the central Fe(II) ion reflects a clear strengthening of these bonds in the low-spin state.

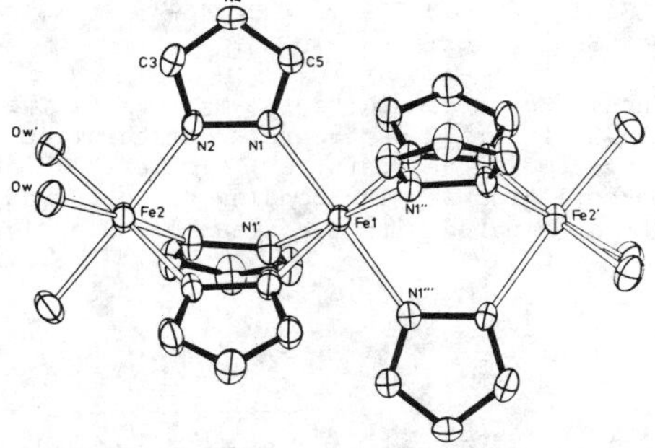

Fig. 1. The $[Fe_3(C_4H_7N_3)_6(H_2O)_6]^{6+}$ cation (ethyl groups omitted).

BIS(BIPYRIDYL)-4,5-DIAZAFLUORENERUTHENIUM(II) BIS(HEXAFLUOROPHOSPHATE)

$C_{31}H_{24}F_{12}N_6P_2Ru$

L.J. HENDERSON Jr., F.R. FRONCZEK and W.R. CHERRY, 1984. J. Am. Chem. Soc., 106, 5876-5879.

Trigonal, P3c1, a = 11.002, c = 16.350, Z = 2. Mo radiation, R = 0.063 for 480 reflexions.

The molecule (Fig. 1) has crystallographic D_3 symmetry requiring the ligands to be disordered. The Ru-N distance is 2.074(6) Å and the N-Ru-N bite angle is 80.1(4)°.

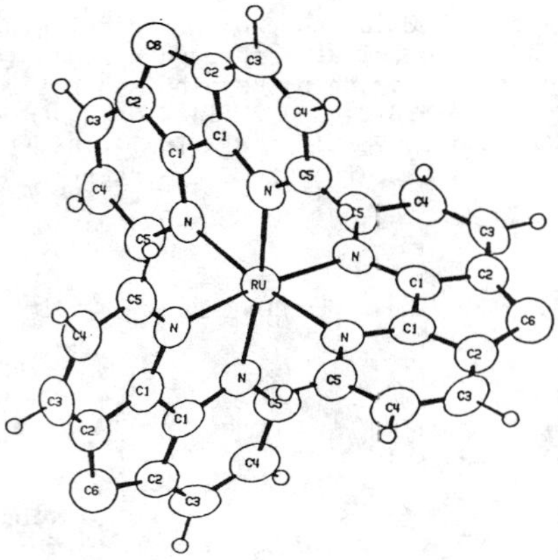

Fig. 1. A view of the disordered cation in $C_{31}H_{24}F_{12}N_6P_2Ru$.

DIAQUABIS(4-METHYL-N-8-QUINOLYLBENZENESULPHONAMIDO-N,N')COBALT(II) ACETONE (1/1)

$C_{32}H_{30}CoN_4O_6S_2,C_3H_6O$ $[Co(C_{16}H_{13}N_2O_2S)_2(H_2O)_2],C_3H_6O$

J.M. CASTRESANA, M.P. ELIZALDE, J.M. ARRIETA, G. GERMAIN and J.-P. DECLERCQ, 1984. Acta Cryst., C40, 763-765.

Monoclinic, $P2_1/c$, a = 13.555, b = 10.519, c = 24.469 Å, β = 101.07°, Z = 4. Mo radiation, R = 0.048 for 2833 reflexions.

In the complex molecule Co is bonded to the N atoms of the chelating agent and to two water molecules in a distorted octahedral configuration (Fig. 1). Bond distances and angles in the two ligands are very similar. The isoquinoline groups are planar, as are the phenyl groups. The bonding to Co causes changes in geometry around the amido N(11) atom compared with the free molecule (1).

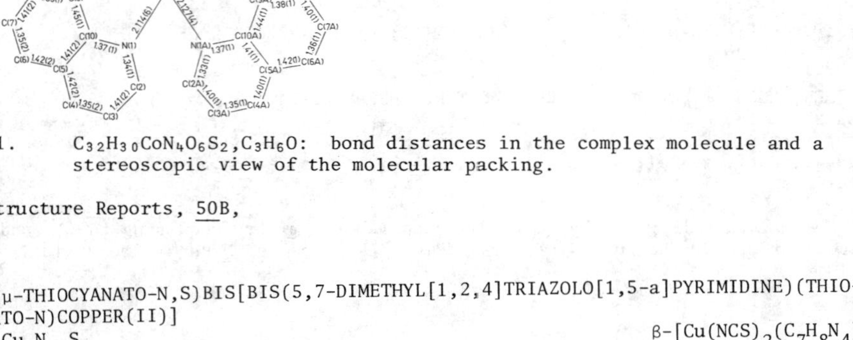

Fig. 1. $C_{32}H_{30}CoN_4O_6S_2,C_3H_6O$: bond distances in the complex molecule and a stereoscopic view of the molecular packing.

1. Structure Reports, 50B,

β-BIS(μ-THIOCYANATO-N,S)BIS[BIS(5,7-DIMETHYL[1,2,4]TRIAZOLO[1,5-a]PYRIMIDINE)(THIOCYANATO-N)COPPER(II)]

$C_{32}H_{32}Cu_2N_{20}S_4$ β-$[Cu(NCS)_2(C_7H_8N_4)_2]_2$

J.G. HAASNOOT, W.L. DRIESSEN and J. REEDIJK, 1984. Inorg. Chem., 23, 2803-2807.

Monoclinic, $P2_1/n$, a = 10.928, b = 14.709, c = 12.804 Å, β = 95.31°, D_m = 1.535, Z = 2. Mo radiation, R = 0.025 for 2492 reflexions.

The two Cu atoms in the centrosymmetric binuclear complex (Fig. 1) are linked by a pair of NCS bridges. The Cu atoms have distorted square pyramidal coordination geometry with the N atom of the μ-NCS ligand in the apical site. Bond lengths: Cu-S 2.4125(7), Cu-NCS_t 1.937(2), Cu-NCS_b 2.205(2), Cu-N (trans to S) 2.058(2), Cu-N (trans to N) 1.995(2) Å.

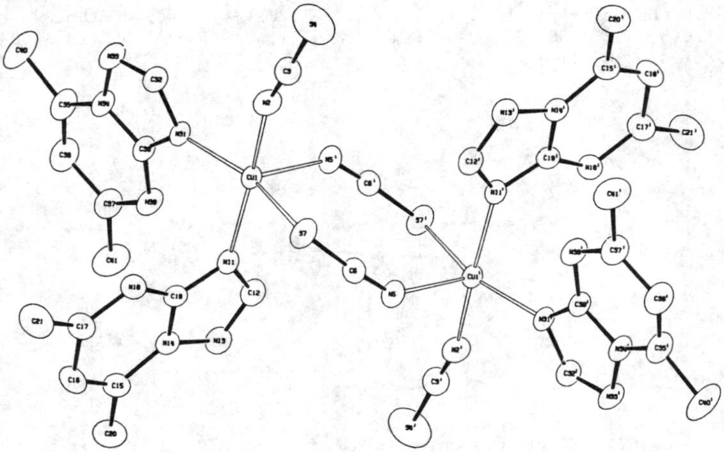

Fig. 1. The β-$[Cu(NCS)_2(C_7H_8N_4)_2]_2$ molecule.

TETRAPHENYLPHOSPHONIUM DIOXO[4-(1,3-THIAZOL-2-YLAZO)RESORCINOLATO]VANADATE(V) MONOHYDRATE

$C_{33}H_{25}N_3O_4PSV,H_2O$ $[P(Ph)_4][VO_2(C_9H_5N_3O_2S)],H_2O$

N. GALESIC and M. SIROKI, 1984. Acta Cryst., C40, 378-381.

Triclinic, $P\bar{1}$, a = 11.363, b = 14.932, c = 9.273 Å, α = 96.30, β = 107.10, γ = 91.65°, D_m = 1.46, Z = 2. Cu radiation, R = 0.036 for 4328 reflexions.

The 4-(1,3-thiazol-2-ylazo)resorcinolato anion acts as a terdentate ligand complexing V through the phenolic O atom, the azo N adjacent to the resorcinol ring, and the thiazolyl N atom, thus forming two five-membered chelate rings (Fig. 1). Two oxo O atoms complete the pentagonal V coordination sphere with an O-V-O angle of 109.1(1)° and V-O bond lengths of 1.613(2) and 1.622(3) Å. A resonance effect through the azo group is evident from the lengthened azo N-N bond of 1.343(3) Å and shortened neighbouring N-C bonds of 1.318(4) and 1.355(4) Å.

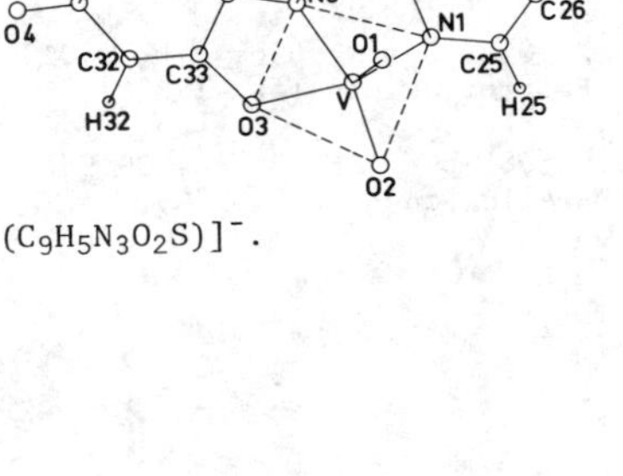

Fig. 1. A view of $[VO_2(C_9H_5N_3O_2S)]^-$.

TRIS(μ-2-ANILINOPYRIDINATO)TRICHLORODIOSMIUM TOLUENE SOLVATE
$C_{33}H_{27}Cl_3N_6Os_2,0.25(C_7H_8)$

A.R. CHAKRAVARTY, F.A. COTTON and D.A. TOCHER, 1984. Inorg. Chem., 23, 4693-4697.

Triclinic, P$\bar{1}$, a = 9.269, b = 12.023, c = 16.191 Å, α = 90.84, β = 100.73, γ = 97.01°, Z = 2. Mo radiation, R = 0.035 for 2693 reflexions (at 5°C).

The compound (Fig. 1) is unusual in that each Os atom is in a different environment. The dinuclear Os core is bridged by three $PhNpy^-$ ligands, with the fourth equatorial position on each metal occupied by Cl^-. One Os atom has an axial Cl^- ligand. The Os-Os triple bond is long at 2.392(1) Å, other distances are: Os-Cl(eq) 2.326(5) and 2.354(5), Os-Cl(ax) 2.449(5), Os-N(py) 2.058(14)-2.099(14), Os-N(an) 1.986(14)-2.041(14) Å.

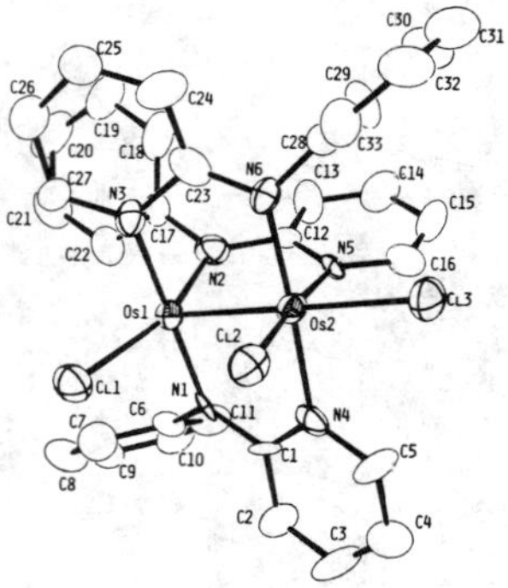

Fig. 1. The $Os_2Cl_3(PhNpy)_3$ molecule.

(μ_2-t-BUTOXY)(μ_2-3-PHENYL-3-NITRILOPROP-2-EN-1-YLIDENE-C,N,N)PENTA-t-BUTOXYDITUNGSTEN
$C_{33}H_{61}NO_6W_2$ (I)

BIS(μ_2-ISOPROPOXY)(μ_2-3-PHENYL-3-NITRILOPROP-2-EN-1-YL-C,N,N)PENTAISOPROPOXYDITUNGSTEN
$C_{30}H_{57}NO_7W_2$ (II)

BIS(μ_2-NEOPENTOXY)(μ_2-1,2,3,4-TETRAMETHYLBUTA-1,3-DIENE-1-IMINO-4-NITRILO-N,N',N')-TETRANEOPENTOXYPYRIDYLDITUNGSTEN
$C_{43}H_{83}N_3O_6W_2$ (III)

BIS(μ_2-ISOPROPOXY)(μ_2-1,4-DIMETHYLBUTA-1,3-DIENE-1,4-DIIMINO-N,N'N')PENTAISOPROPOXY-DITUNGSTEN
$C_{27}H_{58}N_2O_7W_2$ (IV)

M.H. CHISHOLM, D.M. HOFFMAN and J.C. HUFFMAN, 1984. J. Am. Chem. Soc., 106, 6815-6826.

I. Triclinic, P$\bar{1}$, a = 19.237, b = 10.619, c = 10.180 Å, α = 111.97, β = 97.26, γ = 73.00°, Z = 2. Mo radiation, R = 0.023 for 4832 reflexions (at -158°C).

II. Triclinic, P$\bar{1}$, a = 18.346, b = 11.579, c = 10.180 Å, α = 107.08, β = 61.54, γ = 98.19°, Z = 2. Mo radiation, R = 0.033 for 4135 reflexions (at -152°C).

III. Monoclinic, $P2_1/a$, a = 26.006, b = 17.056, c = 12.015 Å, β = 110.29°, Z = 4. Mo radiation, R = 0.052 for 4666 reflexions (at -168°C).

IV. Monoclinic, $P2_1/c$, a = 10.898, b = 16.366, c = 20.942 Å, β = 111.16°, Z = 4. Mo radiation, R = 0.041 for 3759 reflexions (at -165°C).

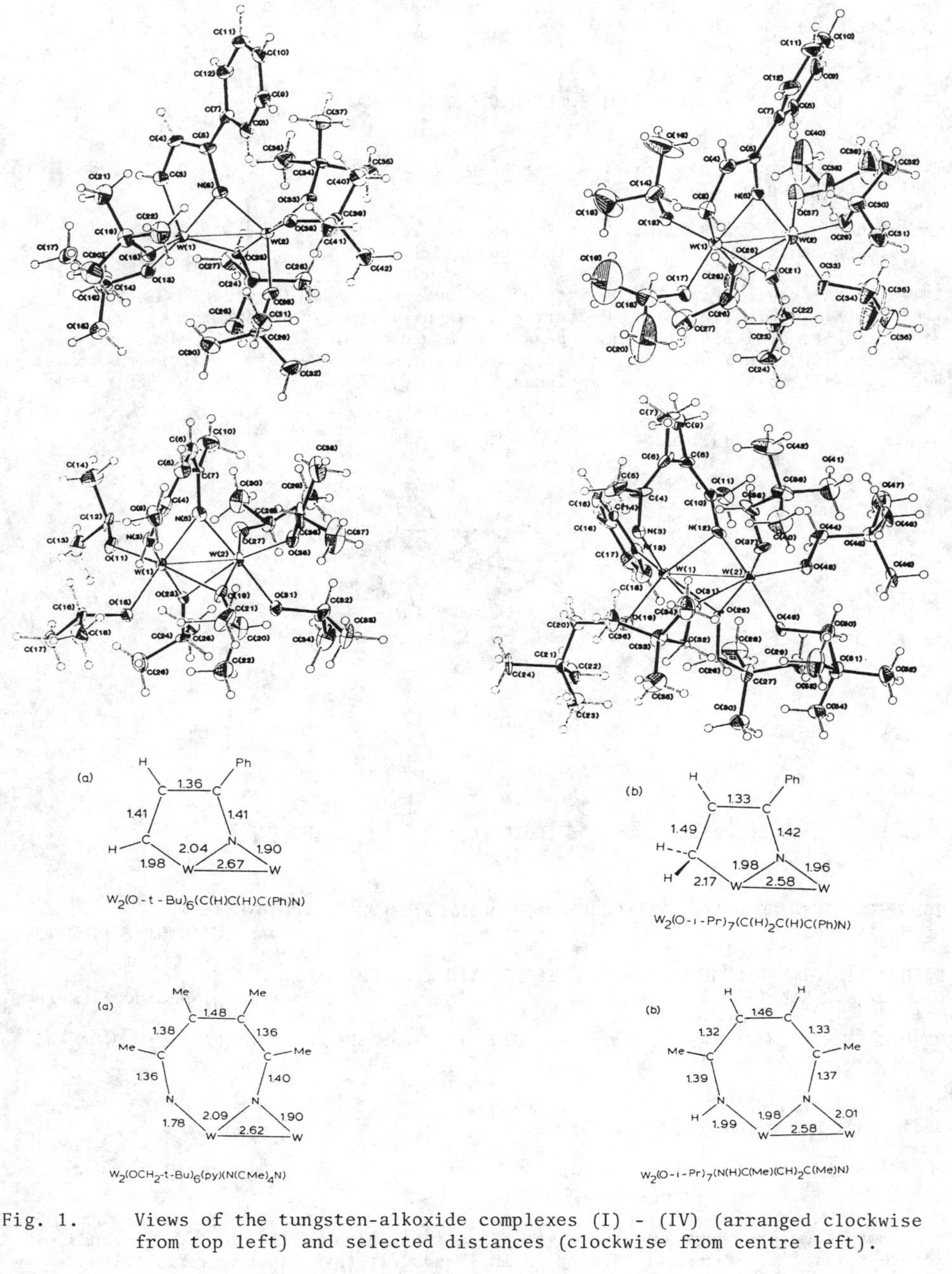

Fig. 1. Views of the tungsten-alkoxide complexes (I) - (IV) (arranged clockwise from top left) and selected distances (clockwise from centre left).

Views of (I) - (IV) are in Fig. 1. In (I) each W atom is in a distorted trigonal-bipyramidal environment with W=C 1.980(6), W-N 2.041(5), W=N 1.903(5), W-W 2.674(1) Å. Compound II has each W atom in a distorted octahedral environment; some dimensions are: W-C 2.174(8), W-N 1.980(6) and 1.962(7), W-W 2.585(1) Å. Complex III has a seven-membered metallacycle (W-N 1.78(1), W-N(μ) 1.90(1) and 2.09(1) Å). Each W atom has distorted octahedral geometry; W-W is 2.617(1) Å. Molecule IV also has a seven-membered metallacycle, with W-N(amido) 1.986(8), W-N(μ-imido) 1.983 and 2.007(8) Å. Each W atom has distorted octahedral geometry and W-W is 2.576(1) Å.

TRIS(2,2'-BIPYRIDYL)IRON(II) NITROPRUSSIDE TETRAHYDRATE
$C_{35}H_{24}Fe_2N_{12}O,4H_2O$ $[Fe(C_{10}H_8N_2)_3][Fe(CN)_5NO],4H_2O$

M.E. GARCÍA POSSE, M.A. JURI, P.J. AYMONINO, O.E. PIRO, H.A. NEGRI and E.E. CASTELLANO, 1984. Inorg. Chem., 23, 948-952.

Monoclinic, $P2_1/c$, a = 11.067, b = 14.576, c = 23.076 Å, β = 90.91°, D_m = 1.44, Z = 4. Mo radiation, R = 0.045 for 2228 reflexions.

The structure (Fig. 1) consists of complex ions and H_2O molecules. The $[Fe(C_{10}H_8N_2)_3]^{2+}$ cation has a distorted octahedral coordination geometry and a mean Fe-N distance of 1.956(8) Å and a mean bite angle of 81.6(3)°. The nitroprusside anion has the usual umbrella-like distorted octahedral geometry with mean parameters: Fe-C 1.94(1), Fe-N 1.650(7) Å, and mean N-Fe-C(eq) 95°.

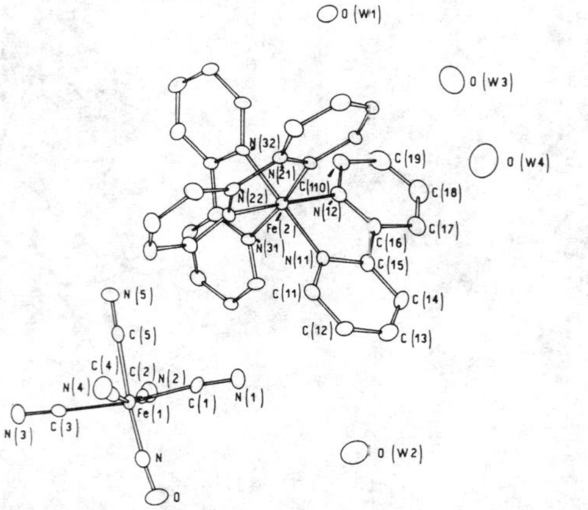

Fig. 1. The asymmetric unit of $[Fe(C_{10}H_8N_2)_3][Fe(CN)_5NO],4H_2O$.

IODOTETRAKIS(DIMETHYLPHENYLPHOSPHINO)TRIMETHYLSILYLDIAZENIDOTUNGSTEN
$C_{35}H_{53}IN_2P_4SiW$ (I) trans-$[WI(NNSiMe_3)(PMe_2Ph)_4]$

DIIODOTRIS(DIMETHYLPHENYLPHOSPHINE)TRIMETHYLSILYLHYDRAZIDOTUNGSTEN
$C_{27}H_{43}I_2N_2P_3SiW$ (II) mer-$[WI_2(NNHSiMe_3)(PMe_2Ph)_3]$

M. HIDAI, K. KOMORI, T. KODAMA, D.-M. JIN, T. TAKAHASHI, S. SUGIURA, Y. UCHIDA and Y. MIZOBE, 1984. J. Organometal. Chem., 272, 155-167.

I. Monoclinic, $P2_1/n$, a = 18.376, b = 18.723, c = 11.727 Å, β = 96.08°, Z = 4. Mo radiation, R = 0.10 for 6482 reflexions.

II. Monoclinic, $P2_1/n$, a = 17.418, b = 18.031, c = 11.370 Å, β = 91.61°, Z = 4. Mo radiation, R = 0.12 for 6239 reflexions.

Both complexes have octahedral coordination spheres (Fig. 1). The ligands of interest (viz. trimethylsilyldiazenido in I and trimethylsilylhydrazido II) have W-N-N angles of 168 and 171° respectively. Bond distances are: in I, W-N(1) 1.82, N(1)-N(2) 1.24, N(2)-Si 1.68, W-I 2.917, W-P 2.498, 2.507, 2.497 and 2.529 Å; in II W-N(1) 1.78, N(1)-N(2) 1.33, N(2)-Si 1.75, W-I(1) 2.860, W-I(2) 2.896, W-P 2.500, 2.464 and 2.511 Å. The difference in Si-N bond lengths between compounds I and II indicates contributions from canonical structures having predominantly double and single bond orders respectively.

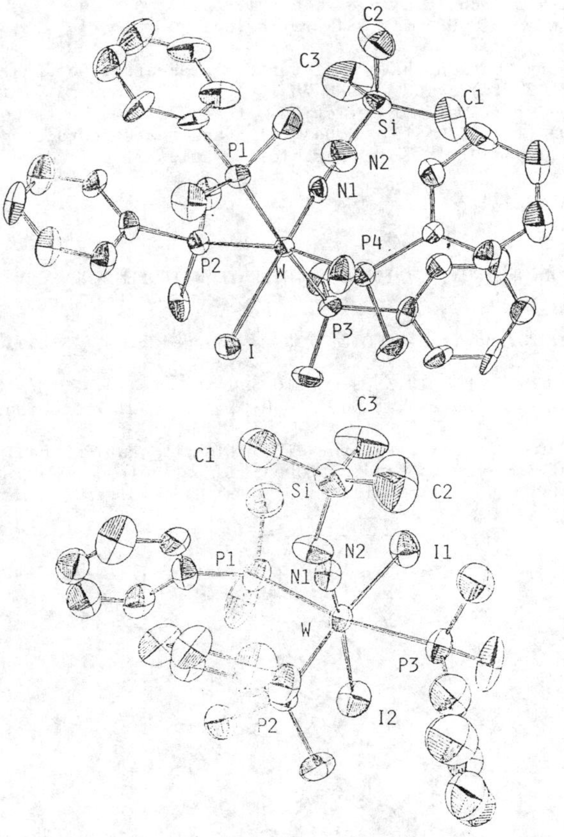

Fig. 1. Molecular structures of $C_{35}H_{53}IN_2P_4SiW$ (top), and $C_{27}H_{43}I_2N_2P_3SiW$ (bottom).

TRIS(1,10-PHENANTHROLINE)COBALT(II) DIPERCHLORATE MONOHYDRATE

$C_{36}H_{24}Cl_2CoN_6O_8,H_2O$ $Co(C_{12}H_8N_2)_3(ClO_4)_2.H_2O$

D. BOYS, C. ESCOBAR and O. WITTKE, 1984. Acta Cryst., C40, 1359-1362.

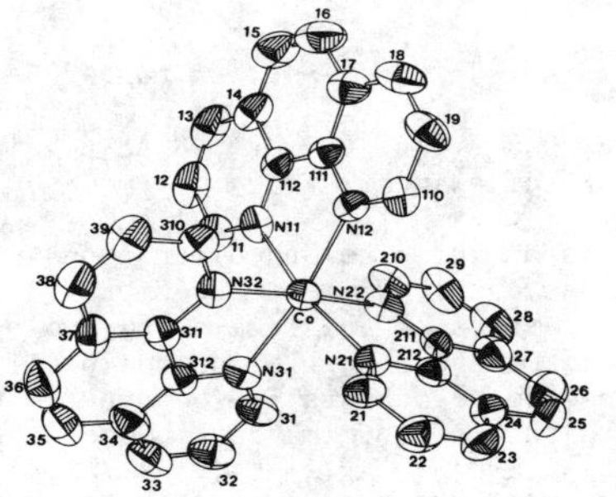

Fig. 1. The $Co(C_{12}H_8N_2)_3$ cation.

Monoclinic, C2/c, a = 36.366, b = 15.854, c = 12.341 Å, β = 102.83°, D_m = 1.547, Z = 8. Mo radiation, R = 0.086 for 3849 reflexions.

The discrete metal cations have a distorted octahedral coordination (Fig. 1) with average Co-N 2.127(14) Å and N-Co-N 78.1(4)°. The compound is isostructural with the Cu analogue (1). Maximum deviations of atoms from the three phenanthroline ring planes are 0.031(11), 0.035(9), 0.064(10) Å and the dihedral angles between these planes are 92.4, 97.2 and 97.2°. One perchlorate anion is disordered.

1. Structure Reports, 39B, 757.

TRIS(1,10-PHENANTHROLINE)IRON(II) IODIDE DIHYDRATE NITROBENZENE SOLVATE
$C_{36}H_{24}FeI_2N_6,C_6H_5NO_2,2H_2O$ $[Fe(C_{12}H_8N_2)_3]I_2.2H_2O.C_6H_5NO_2$

T. FUJIWARA, E. IWAMOTO and Y. YAMAMOTO, 1984. Inorg. Chem., 23, 115-117.

Triclinic, P$\bar{1}$, a = 16.917, b = 12.350, c = 10.590 Å, α = 114.47, β = 80.96, γ = 97.55°, D_m = 1.65, Z = 2. Mo radiation, R = 0.054 for 5301 reflexions (at -40°C).

The structure (Fig. 1) consists of $[Fe(C_{12}H_8N_2)_3]^{2+}$ cations having normal geometry, I^- ions, H_2O molecules, and nitrobenzene molecules. The nitrobenzene solvent molecules were found to be disordered over three sites.

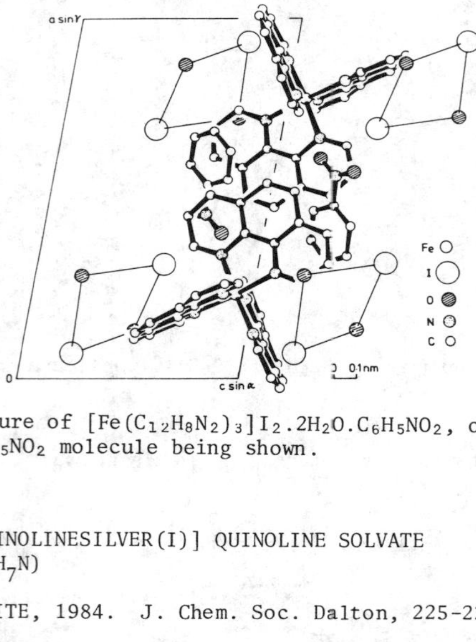

Fig. 1. The structure of $[Fe(C_{12}H_8N_2)_3]I_2.2H_2O.C_6H_5NO_2$, one of the three locations of the $C_6H_5NO_2$ molecule being shown.

TETRAKIS[(μ_3-CHLORO)QUINOLINESILVER(I)] QUINOLINE SOLVATE
$(C_{36}H_{28}Ag_4Cl_4N_4)_n,n(C_9H_7N)$

N.K. MILLS and A.H. WHITE, 1984. J. Chem. Soc. Dalton, 225-228.

Monoclinic, C2/c, a = 30.508, b = 7.216, c = 22.698 Å, β = 122.11°, Z = 4. Mo radiation, R = 0.042 for 1258 reflexions.

The structure (Fig. 1) is an infinite one-dimensional polymer about the twofold axes 0,y 1/4 etc. It comprises Ag_5Cl_4 face-centred motifs linked in a linear array through Ag atoms at opposite corners of the face. The two vacant coordination sites at the other corners each accommodate a pair of quinoline ligands. The Ag atom at the centre of the motif is pseudo-tetrahedrally coordinated to the four Cl atoms such that the Ag_5Cl_4 unit adopts a saddle conformation. Some distances include: Ag(2)...Ag(3) 3.130(2), Ag(1)...Ag(3) 3.590(3), Ag(2)-Cl(1) 2.725(4), Cl(1)...Cl(2) 4.311(5), Ag(3)...Cl(2) 2.603(4), Ag(1)-Cl(2) 2.630(4), Ag(2)-N(11) 2.27(1) Å. The solvent quinoline molecules are located so that the centres of the centre bonds lie at 1/4, 3/4, 1/2 and the molecules are centrosymmetric i.e. they are disordered. The bromo derivative is isomorphous with the chloro derivative.

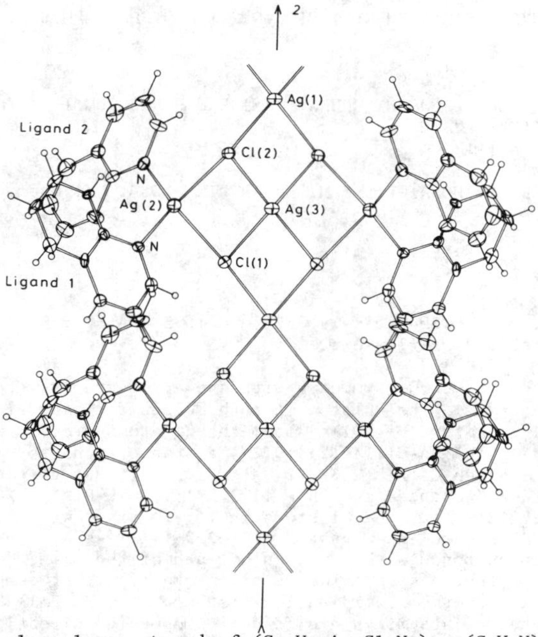

Fig. 1. A single polymer strand of $(C_{36}H_{28}Ag_4Cl_4N_4)_n,n(C_9H_7N)$.

BIS(μ_2-IODO)BIS(BIS(QUINOLINE)COPPER(I)) ACETONITRILE SOLVATE
$C_{36}H_{28}Cu_2I_2N_4,C_2H_3N$

W. HILLER, 1984. Z. Naturforsch., B39, 861-863.

Monoclinic, C2/c, a = 19.740, b = 7.644, c = 25.217 Å, β = 112.64°, Z = 4. Mo radiation, R = 0.027 for 1521 reflexions (at 220 K).

The complex lies about an inversion centre and has a planar Cu_2I_2 moiety. Tetrahedral coordination at Cu is distorted by the bulky ligands (N-Cu-N 118.5°, I-Cu-I' 108.0°.

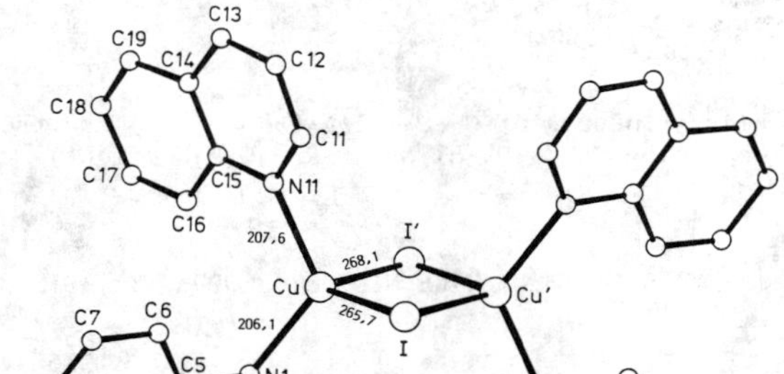

Fig. 1. A view of $[(quinoline)_2CuI]_2$.

cis-DICHLORO[(o-DIPHENYLPHOSPHINOPHENYL)AMIDO-N,P][(o-DIPHENYLPHOSPHINOPHENYL)AMINE-N,P]RHODIUM(III)
$C_{36}H_{31}Cl_2N_2P_2Rh$ (I)

trans-DICHLOROBIS[(o-DIPHENYLPHOSPHINOPHENYL)AMINE-N,P]RHODIUM(III) TOLUENE p-SULPHONATE ACETONITRILE SOLVATE
$C_{43}H_{39}Cl_2N_2O_3P_2RhS,C_2H_3N$ (II)

G.J. ORGAN, M.K. COOPER, K. HENRICK and M. McPARTLIN, 1984. J. Chem. Soc. Dalton, 2377-2382.

I. Monoclinic, $P2_1/n$, a = 17.153, b = 18.667, c = 9.891 Å, β = 91.77°, Z = 4. Mo radiation, R = 0.049 for 3501 reflexions.

II. Monoclinic, I2/c, a = 18.394, b = 14.380, c = 16.378 Å, β = 96.67°, Z = 4. Mo radiation, R = 0.064 for 2741 reflexions.

The cation in II has exact C_2 symmetry with the anion and solvate molecule disordered about the C_2 axis. The Rh atom in both molecules I and II exhibits approximate octahedral geometry (Fig. 1). In I the Cl atoms are cis with a Cl(1)-Rh-Cl(2) angle of 88.7(1)°. The Rh-Cl(1) (2.419(2)) distance trans to P(1) is significantly shorter than that of Rh-Cl(2) (2.384(2) Å) which is trans to the amido N donor N(2). Other coordination distances are Rh-P(1) 2.271(2), Rh-P(2) 2.262(2), Rh-N(1) 2.143(5) and Rh-N(2) 2.018(5) Å. The Cl atoms in II are trans with equivalent Rh-Cl distances of 2.342(2) Å and a Cl-Rh-Cl angle of 169.9(1)°. As a result the two Cl atoms form strong hydrogen bonds with a proton of each of the amino donor groups (Cl(1)...H(1b) 2.51 Å). There is no evidence of interactions between the Cl ligands and the phenyl groups. The other coordination distances are Rh-P(1) 2.295(2) and Rh-N(1) 2.145(8) Å. The chelate rings are folded in opposite directions about the donor atoms.

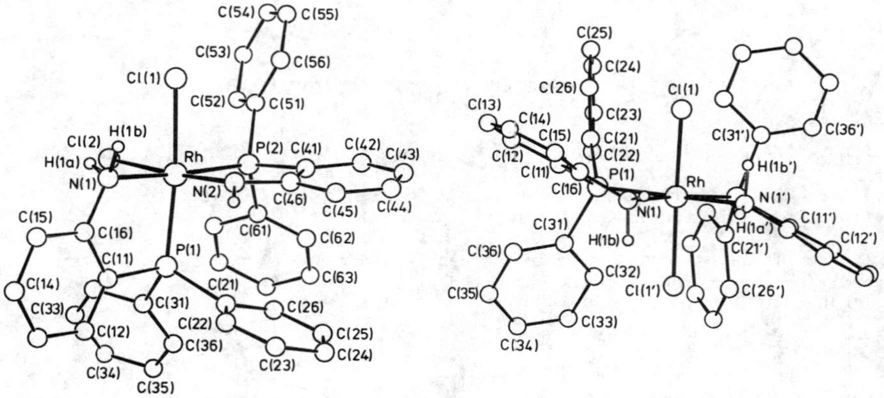

Fig. 1. The molecular structure of $C_{36}H_{31}Cl_2N_2P_2Rh$ (left) and the structure of the complex cation in $C_{43}H_{39}Cl_2N_2O_3P_2RhS,C_2H_3N$ (right).

m-XYLYL-α,α'-BIS[((2-(2-PYRIDYL)ETHYL)AMINE-N,N',N'')COPPER(I)] BIS(HEXAFLUOROPHOSPHATE)
$C_{36}H_{40}Cu_2F_{12}N_6P_2$ (I) $(C_{36}H_{40}Cu_2N_6)^{2+}.2(PF_6^-)$

(μ_2-HYDROXO)((μ_2-2,6-BIS(2-(2-PYRIDYL)ETHYL)AMINEMETHYL)PHENOXO-N,N,N,O,O)DICOPPER(I) BIS(HEXAFLUOROPHOSPHATE)
$C_{36}H_{40}Cu_2F_{12}N_6O_2P_2$ (II) $(C_{36}H_{40}Cu_2N_6O_2)^{2+}.2(PF_6^-)$

K.D. KARLIN, J.C. HAYES, Y. GULTNEH, R.W. CRUSE, J.W. MCKOWN, J.P. HUTCHINSON and J. ZUBIETA, 1984. J. Am. Chem. Soc., 106, 2121-2128.

I. Triclinic, $P\bar{1}$, a = 11.264, b = 11.448, c = 15.722 Å, α = 95.72, β = 102.05, γ = 94.77°, Z = 2. Mo radiation, R = 0.0696 for 2761 reflexions.

II. Monoclinic, $P2_1/c$, a = 18.221, b = 13.323, c = 18.643 Å, β = 102.39°, Z = 4. Mo radiation, R = 0.0969 for 2711 reflexions.

In I (Fig. 1) each Cu(I) atom is three-coordinate (Cu-N(py) 1.904-1.937(9), Cu-N(tertiary amine) 2.196 and 2.121(8) Å). Cu(1) is 0.147 Å from the N(1)-N(3) plane and Cu(2) is 0.228 Å from the N(4)-N(6) plane. In II each Cu is pentacoordinate (Fig. 1). The geometry around each copper is essentially undistorted from a tetragonal pyramid with N(3) and N(6) pyridines occupying axial positions on opposite sides of the Cu_2O_2 plane. The copper atoms Cu(1) and Cu(2) lie 0.31 and 0.26 Å, respectively, out of their corresponding basal planes towards the apical N_{py} atom. In each case, the $Cu-N_{py-axial}$ vector forms an angle of ca. 87° with its least-squares basal plane. The Cu_2O_2 unit is essentially planar, with the maximum deviation from the least-squares plane being 0.012 Å. The Cu(I-I)...Cu(II) separation is 3.082(3) Å. Other dimensions are in the expected ranges (Cu-O 1.938-1.979(10) Å, Cu-N(amino) 2.03(1) Å).

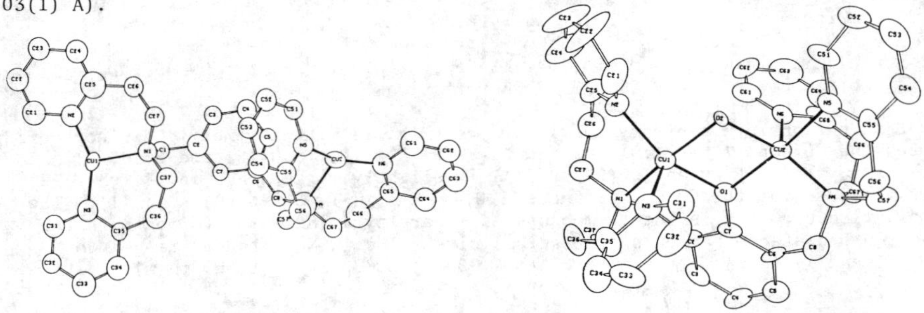

Fig. 1. Views of binuclear complexes $(C_{36}H_{40}Cu_2N_6)^{2+}$ (I) (left) and $(C_{36}H_{40}Cu_2N_6O_2)^{2+}$ (II) (right).

BIS(μ-AZIDO)TETRAKIS(4-t-BUTYLPYRIDINE)DICOPPER(II) PERCHLORATE

$C_{36}H_{52}Cl_2Cu_2N_{10}O_8$ $[Cu_2(N_3)_2(C_9H_{13}N)_4](ClO_4)_2$

S. SIKORAV, I. BKOUCHE-WAKSMAN and O. KAHN, 1984. Inorg. Chem., 23, 490-495.

Monoclinic, $P2_1/c$, a = 12.819, b = 13.76, c = 13.65 Å, β = 100.00°, D_m = 1.30, Z = 2. Mo radiation, R = 0.065 for 2172 reflexions.

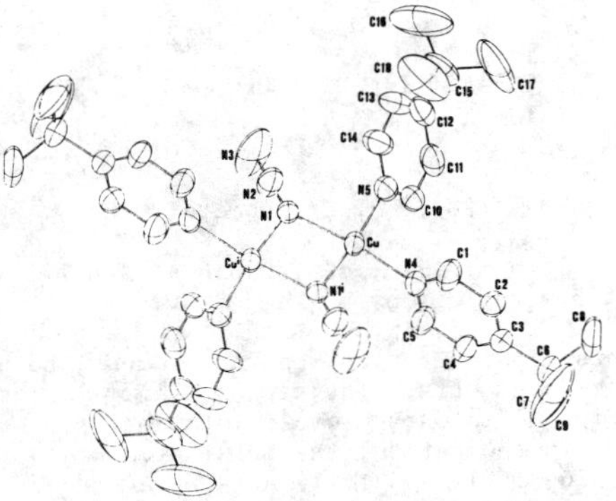

Fig. 1. The $[Cu_2(N_3)_2(C_9H_{13}N)_4]^{2+}$ cation.

The structure consists of discrete, centrosymmetric, binuclear cations (Fig. 1) and ClO_4^- anions. The Cu atoms have square-planar coordination geometry and are bridged by azido groups in an end-on fashion. The Cu...Cu distance is 3.045(3) Å and the Cu-N-Cu angle is 100.5(6)°. Bond lengths: Cu-N(pyr) 1.99(1) and 2.00(1) Å, Cu-N(N_3) 1.99(1) and 1.97(1), N-N(Cu) 1.18(2), N-N 1.11(3) Å.

BIS(μ-FLUORO)BIS[TRIS(3,4,5-TRIMETHYLPYRAZOLE-N^2)COPPER(II)] BIS(TETRAFLUOROBORATE)
$C_{36}H_{60}B_2Cu_2F_{10}N_{12}$ (I) $[CU_2F_2(C_6H_{10}N_2)_6](BF_4)_2$

BIS(μ-FLUORO)BIS[(5-METHYLPYRAZOLE-N^2)BIS(3,5-DIMETHYLPYRAZOLE-N^2)COPPER(II)] BIS(TETRAFLUOROBORATE)
$C_{28}H_{44}B_2Cu_2F_{10}N_{12}$ (II) $[Cu_2F_2(C_4H_6N_2)_2(C_5H_8N_2)_4](BF_4)_2$

F.J. RIETMEIJER, R.A.G. de GRAAFF and J. REEDIJK, 1984. Inorg. Chem., 23, 151-156.

I. Monoclinic, $P2_1/n$, a = 11.514, b = 16.439, c = 13.159 Å, β = 101.82°, D_m = 1.34, Z = 2. Mo radiation, R = 0.033 for 2553 reflexions.

II. Monoclinic, $P2_1/c$, a = 9.954, b = 11.862, c = 17.679 Å, β = 99.45°, D_m = 1.43, Z = 2. Mo radiation, R = 0.035 for 2258 reflexions.

Both cations are centrosymmetric dimers (Fig. 1) with asymmetric difluoro bridges between the Cu ions (Cu-F 1.911(2) and 2.183(2) Å in I, 1.901(2) and 2.195(2) Å in II). In I, the Cu-F-Cu angle (94.59(7)°) is slightly larger than in II (93.73(8)°), as is the case for the Cu-Cu distances (3.0141(8) Å in I, 2.9962(9) Å in II). The coordination geometry in both compounds is square pyramidal, the F^- ion with the longer Cu-F distance occupying the apical site. The Cu-N distances range from 1.965(3) and 1.964(3) Å (trans to F) to 1.984(4)-2.009(3) Å (trans to N).

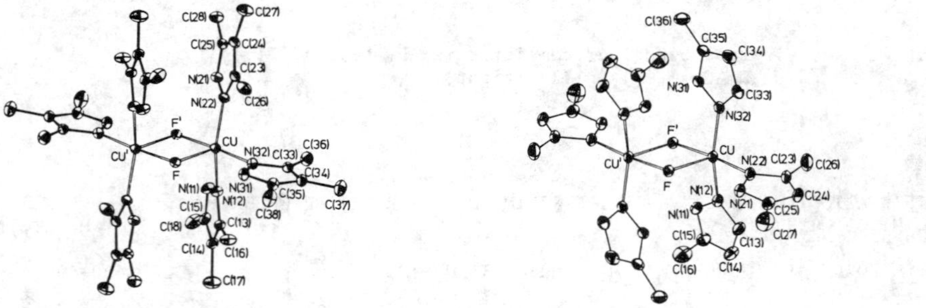

Fig. 1. The complex cations $[Cu_2F_2(C_6H_{10}N_2)_6]^{2+}$ (left) and $[Cu_2F_2(C_4H_6N_2)_2(C_5H_8N_2)_4]^{2+}$ (right).

BIS(TRIS(1-PYRAZOLYL)BORATE)-COBALT(III) BIS(CHLCRO-BIS(TETRACARBONYL-COBALT)-TIN)-TRICARBONYL-COBALT
$C_{37}H_{20}B_2Cl_2Co_6N_{12}O_{19}Sn_2$ $Co(HB(C_3H_3N_2)_3)_2^+(Sn_2Co_5Cl_2(CO)_{19})^-$

O.J. CURNOW and B.K. NICHOLSON, 1984. J. Organometal. Chem., 267, 257-263.

Monoclinic, C2/c, a = 27.325, b = 9.516, c = 23.316 Å, β = 113.51°, Z = 4. Mo radiation, R = 0.043 for 2839 reflexions.

The structure consists of discrete cations and anions, both of which have crystallographically-imposed symmetry. The cation (Fig. 1) lies on an inversion centre while the anion (Fig. 1) is on a twofold rotation axis coincident with the Co(1)-C(12)-O(12) vector. In the cation, the metal atom has octahedral coordination, with Co-N distances of 1.926, 1.911 and 1.918 Å. The average B-N and N-N distances are 1.537 and 1.365 Å respectively. The anion consists of two $Sn(Cl)(Co(CO)_4)_2$ groups bonded trans to one another about the planar $Co(CO)_3$ unit. Each $Co(CO)_4$ unit has normal trigonal bipyramidal geometry. The important distances are: Co(1)-Sn(1) 2.508, Co(2)-Sn(1) 2.668, Co(3)-Sn(1) 2.656, Sn(1)-Cl(1) 2.418.

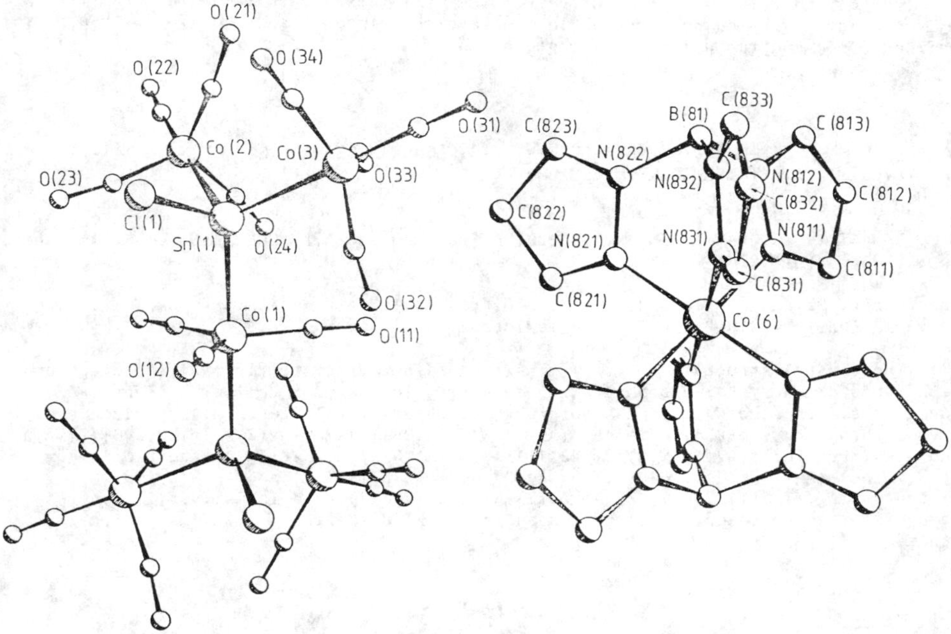

Fig. 1. Molecular structure of the anion (left) and cation (right) in $C_{37}H_{20}B_2Cl_2$-$Co_6N_{12}O_{19}Sn_2$.

trans-DIOXOTETRAKIS(4-t-BUTYLPYRIDINE)TECHNETIUM(V) TRIFLUOROMETHYLSULPHONATE MONO-HYDRATE

$C_{37}H_{52}F_3N_4O_5STc,H_2O$ trans-$[O_2(C_4H_9C_5H_4N)_4Tc]F_3CSO_3.H_2O$

M.E. KASTNER, P.H. FACKLER, M.J. CLARKE and E. DEUTSCH, 1984. Inorg. Chem., 23, 4683-4688.

Triclinic, $P\bar{1}$, a = 13.876, b = 15.429, c = 11.822 Å, α = 111.83, β = 108.77, γ = 65.24°, D_m = 1.33, Z = 2. Cu radiation, R = 0.074 for 3682 reflexions.

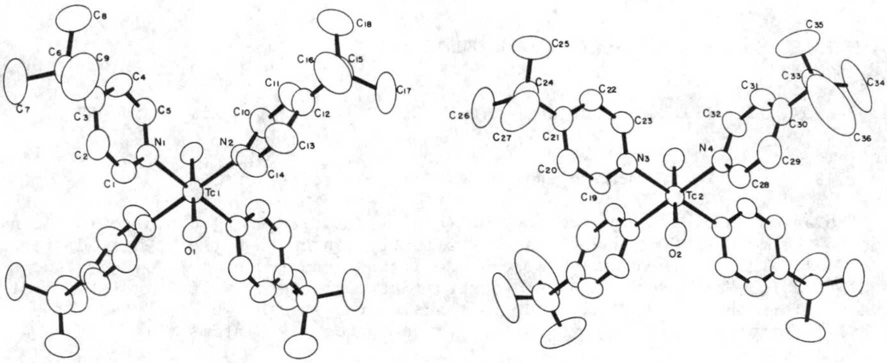

Fig. 1. The two independent trans-$[O_2(C_4H_9C_5H_4N)_4Tc]^+$ cations.

The structure contains two independent centrosymmetric cations (Fig. 1). The coordination about Tc is nominally octahedral with average Tc=O and Tc-N bond lengths of 1.743(6) and 2.15(2) Å, respectively. The pyridines are twisted away from the O=Tc=O axis by varying degrees and form an average dihedral angle of 74° with a plane defined by Tc and two adjacent nitrogen atoms.

BIS[μ_2-(N,N'-BIS(2-PYRIDINYLMETHYLENE)-1,2-CYCLOHEXANEDIAMINE-$N^1,N^{1'},N^2,N^{2'}$)]-DISILVER(II) BIS(TRIFLUOROMETHYLSULPHONATE)
$C_{38}H_{40}Ag_2F_6N_8O_6S_2$

G.C. VAN STEIN, G. VAN KOTEN, K. VRIEZE, C. BREVARD and A.L. SPEK, 1984. J. Am. Chem. Soc., 106, 4486-4492.

Monoclinic, $P2_1/c$, a = 10.412, b = 33.310, c = 13.197 Å, β = 105.38°, D_m = 1.65, Z = 4. Mo radiation, R = 0.067 for 3092 reflexions.

The crystal structure (Fig. 1) consists of four discrete $[Ag_2(N_4)_2]^{2+}$ dications, with twofold noncrystallographic axial symmetry, in the unit cell with eight $O_3SCF_3^-$ monoanions. The two N_4 ligands coordinate to the silver(I) centres as dibidentates with alternating short Ag-N(imine) (Ag(1)-N(2) 2.251(11) Å; Ag(2)-N(6) 2.240(12) Å) and Ag-N(pyridine) (Ag(1)-N(8) 2.242(11) Å; Ag(2)-N(4) 2.252(13) Å) and two longer Ag-N(imine) (Ag(1)-N(7) 2.440(11) Å; Ag(2)-N(3) 2.430(11) Å) and Ag-N(pyridine) distances (Ag(1)-N(1) 2.455(12) Å; Ag(2)-N(5) 2.429(11) Å). The silver(I) centres have distorted tetrahedral geometries and are at a distance of 3.254(2) Å from each other.

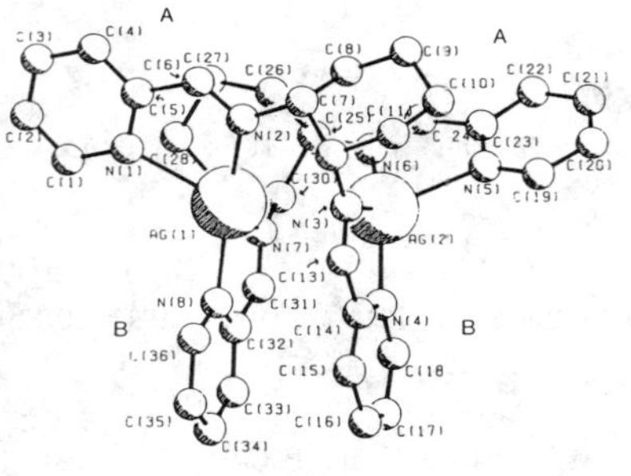

Fig. 1. A view of the $[Ag_2(N_4)_2]^{2+}$ unit.

SODIUM TRIS(CYANO)PHENANTHROLINEOXOMOLYBDATE(IV)
$C_{39}H_{24}MoN_9NaO$ $Na^+[MoO(CN)_3(C_{12}H_8N_2)]^-.2C_{12}H_8N_2$

S.S. BASSON, J.G. LEIPOLDT and I.M. POTGIETER, 1984. Inorg. Chim. Acta, 87, 71-78.

Triclinic, $P\bar{1}$, a = 13.354, b = 14.380, c = 9.294 Å, α = 93.53, β = 97.05, γ = 104.93°, D_m = 1.48, Z = 2. Mo radiation, R = 0.069 for 3353 reflexions.

The anion is a distorted octahedron (Fig. 1) The Mo-N(2) distance (2.363 Å) trans to the Mo=O bond (1.659 Å) is significantly longer than the Mo-N(1) distance of 2.174 Å. The mer-arrangement of cyanide ligands has two normal Mo-C distances (Mo-C(37) 2.16, Mo-C(38) 2.18 Å) and a significantly shorter Mo-C(39) distance of 2.092 Å. This shorter Mo-C distance is trans to the shorter Mo-N(1) bond. The Na^+ ion has a distorted trigonal prismatic environment of six N atoms (Fig. 1).

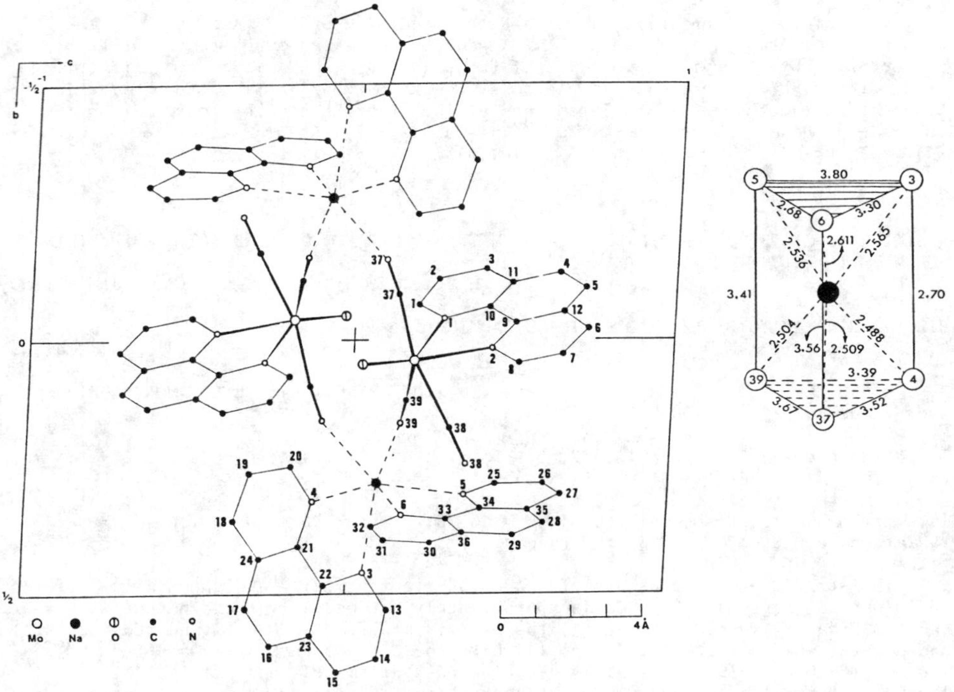

Fig. 1. Projection of the structure of $C_{39}H_{24}MoN_9NaO$ onto the (100) plane (left), and the environment of the sodium ions (right).

DICHLOROBIS[μ-2-(2-PYRIDYL)-1,8-NAPHTHYRIDINE-N^1,N^2,N^3][2-(2-PYRIDYL)-1,8-NAPHTHYRIDINE-N^1,N^2]DIRHODIUM(II) BIS(HEXAFLUOROPHOSPHATE) ACETONITRILE SOLVATE

$C_{39}H_{27}Cl_2F_{12}N_9P_2Rh_2,C_2H_3N$ $[Rh_2(C_{13}H_9N_3)_3Cl_2][PF_6]_2.CH_3CN$

A.T. BAKER, W.R. TIKKANEN, W.C. KASKA and P.C. FORD, 1984. Inorg. Chem., 23, 3254-3256.

Triclinic, P$\bar{1}$, a = 11.003, b = 12.534, c = 16.685 Å, α = 73.49, β = 87.17, γ = 89.12°, D_m = 1.841, Z = 2. Mo radiation, R = 0.044 for 4603 reflexions.

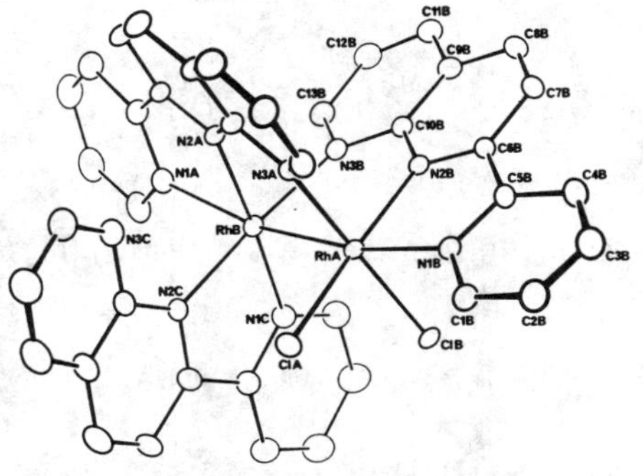

Fig. 1. The $[Rh_2(C_{13}H_9N_3)_3Cl_2]^{2+}$ cation.

The complex cation (Fig. 1) has a Rh-Rh bond of 2.5668(7) Å bridged by the naphthyridine fragments of two $C_{13}H_9N_3$ ligands. The sites trans to the Rh-Rh bond are occupied by the pyridyl moieties of the bridging $C_{13}H_9N_3$ ligands. Hexacoordination of the Rh(II) centres is achieved at one Rh by coordinating the bipyridyl fragment of the third $C_{13}H_9N_3$ ligand and at the other Rh atom by coordinating two chlorides. Bond lengths: Rh-Cl 2.343(2) and 2.363(2), Rh-N(axial) 2.160(5) and 2.190(5), Rh-N(equatorial) 2.014(5)-2.101(5) Å.

BIS(NITRATO-O)BIS(2,2'-BIPYRIDINE)CADMIUM(II) - (NITRATO-O)AQUABIS(2,2'-BIPYRIDINE)-CADMIUM(II) NITRATE (1:1)

$C_{40}H_{34}Cd_2N_{12}O_{13}$ (I) $(C_{10}H_8N_2)_2Cd(NO_3)_2,[(C_{10}H_8N_2)_2(H_2O)Cd(NO_3)](NO_3)$

BIS(ISOTHIOCYANATO)BIS(2,2'-BIPYRIDINE)CADMIUM(II)

$C_{22}H_{16}CdN_6S_2$ (II) $(C_{10}H_8N_2)_2Cd(NCS)_2$

P.F. RODESILER, R.W. TURNER, N.G. CHARLES, E.A.H. GRIFFITH and E.L. AMMA, 1984. Inorg, Chem., 23, 999-1004.

I. Orthorhombic, Pcab, a = 14.372, b = 35.134, c = 17.174 Å, D_m = 1.93, Z = 8. Mo radiation, R = 0.066 for 3636 reflexions.

II. Orthorhombic, Pcab, a = 16.274, b = 16.75, c = 16.136 Å, D_m = 1.62, Z = 8. Mo radiation, R = 0.032 for 2331 reflexions.

The structure of I consists of two crystallographically and chemically different Cd^{2+} species (Fig. 1). Each Cd is approximately octahedrally coordinated, with four N and two O donor ligands. For Cd(1) the O atoms are from a NO_3^- ligand (Cd-O 2.43(1) Å) and a water molecule (Cd-O 2.246(9) Å) while for Cd(2) both O atoms are from NO_3^- ligands (Cd-O 2.435(9) and 2.41(1) Å). The Cd-N distances in I are 2.33(1)-2.43(1) Å. The Cd atom is also approximately octahedral in II, being bound to six N atoms (Cd-N(pyr) 2.374(4)-2.398(4), Cd-N(NCS) 2.239(4) and 2.260(5) Å). The NCS ligands are mutually cis (Fig. 1).

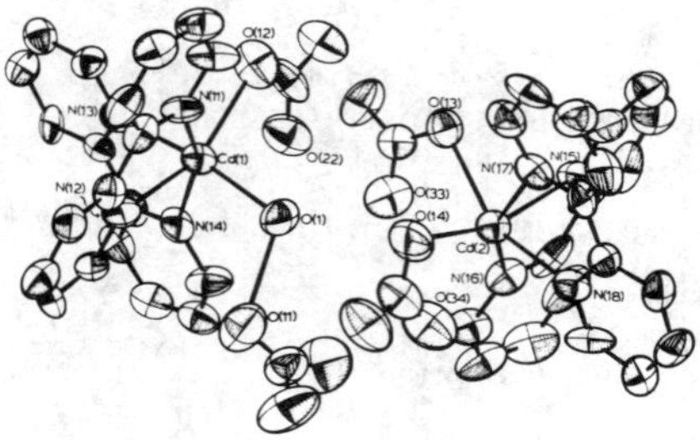

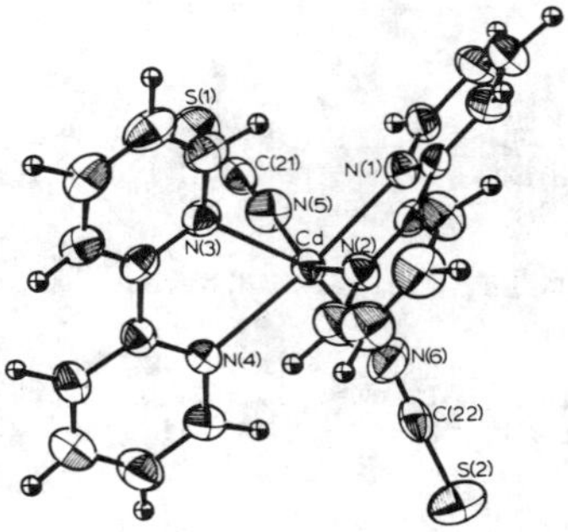

Fig. 1. The isolated $(C_{10}H_8N_2)_2Cd(NO_3)_2$ and $[(C_{10}H_8N_2)_2(H_2O)Cd(NO_3)](NO_3)$ molecules (left) and the $(C_{10}H_8N_2)_2Cd(NCS)_2$ molecule (right).

mer-BIS[2,2'-{1,2-PHENYLENEBIS(IMINOMETHYLENE)}-BIS(PHENOL)ATO(3-)]MOLYBDENUM(VI) METHANOL

$C_{40}H_{34}MoN_4O_4,CH_4O$

S.F. GHELLER, T.W. HAMBLEY, M.R. SNOW, K.S. MURRAY and A.G. WEDD, 1984. Aust. J. Chem., 37, 911-919.

Monoclinic, $P2_1/c$, a = 9.963, b = 17.244, c = 20.703 Å, β = 93.32°, D_m = 1.40, Z = 4. Mo radiation, R = 0.037 for 1687 reflexions, deep red-brown prism.

mer-$[Mo(HL)_2]$,MeOH (Fig. 1) features the tridentate ligand HL^{3-}, where two deprotonated amine nitrogens and one deprotonated phenolic oxygen are the ligand atoms, and the other phenolic oxygen is protonated and not bound to the metal (Mo-O 1.974(5), Mo-N 2.023(6) Å, N-Mo-N 77.3(3), O-Mo-N 84.9(3)°). The stereochemistry about the Mo is close to octahedral.

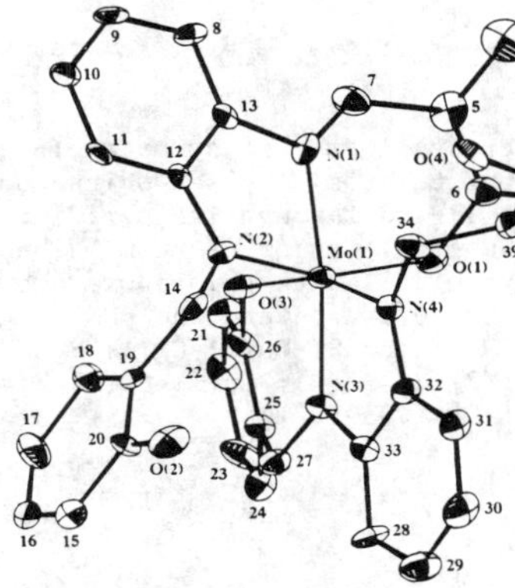

Fig. 1. The structure of mer-$[Mo(HL)_2]$,MeOH.

2,2'-BIPYRIDINIUM(1+) μ-HYDROGEN-BIS[(2,2'-BIPYRIDINE)OXODIPEROXOVANADATE](1) HYDROGEN PEROXIDE WATER

$C_{40}H_{36}N_8O_{10}V_2,0.5H_2O_2,5.5H_2O$ $(Hbiby)^+[H\{VO(O_2)_2bipy\}_2],0.5H_2O_2,5.5H_2O$

H. SZENTIVANYI and R. STOMBERG, 1984. Acta Chem. Scand., A38, 101-107.

Triclinic, $P\bar{1}$, a = 9.454, b = 9.103, c = 12.076 Å, α = 111.32, β = 100.68, γ = 104.05°, Z = 1. Mo radiation, R = 0.040 for 2583 reflexions (at -100°C).

Two centrosymmetrically-related $VO(O_2)_2$bipy entities are held together by a very strong, asymmetric hydrogen bond, the O(1)...O(1') distance being 2.456 Å (Fig. 1). The hydrogen atoms H(01') and H(01) have occupancy factors of 0.5 each on either side of the inversion centre. The V atoms have a pentagonal bipyramidal arrangement of ligands; the peroxo O atoms and one N atom (N(1)) form the pentagonal plane while the vanadyl O atom and N(2) occupy the apical positions. Bond distances are: V=O(5) 1.612, V-O(peroxo) from 1.869 to 1.995 Å, V-N(1) 2.154, V-N(2) 2.290, O(1)-O(2) 1.471 and O(3)-O(4) 1.461 Å. In each bipyramid the V atom is displaced 0.31 Å from the pentagonal plane towards O(5). The $Hbipy^+$ ions are disordered in the crystals, they have two sites (occupancy factor 0.5) about 0.6 Å from each other.

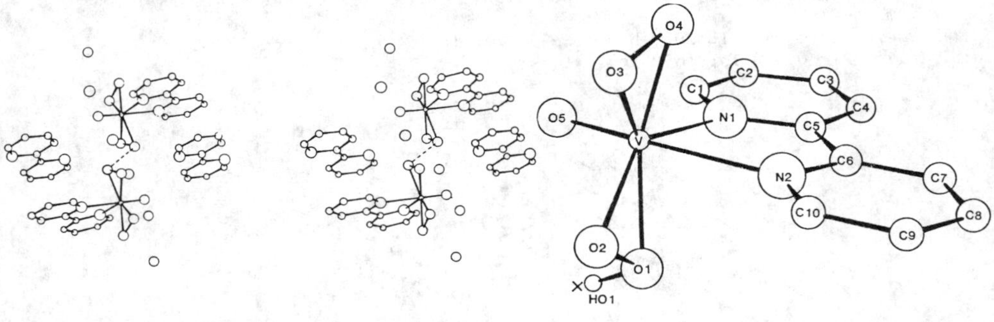

Fig. 1. Views of the unit cell contents, and the asymmetric part of the anion in $C_{40}H_{36}N_8O_{10}V_2,0.5H_2O_2,5.5H_2O$.

(μ_4-OXO)HEXAKIS(μ_2-CHLORO)TETRAKIS(N,N-DIETHYLNICOTINAMIDE-N'-COPPER(II))
$C_{40}H_{56}Cl_6Cu_4N_8O_5$ $(DENC)_4Cu_4Cl_6O$

A. EL-TOUKHY, GUANG-ZUAN CAI, G. DAVIES, T.R. GILBERT, K.D. ONAN and M. VEIDIS, 1984. J. Am. Chem. Soc., 106, 4596-4605.

Tetragonal, $I4_1$, a = 15.213, c = 22.864 Å, Z = 4. Cu radiation, R = 0.069 for 2461 reflexions.

The complex has twofold crystallographic symmetry (Fig. 1) and consists of a μ_4-oxo group coordinated to four Cu atoms each of which coordinates a monodentate N,N-diethylnicotinamide (DENC) ligand through its pyridine nitrogen atom. Each pair of copper atoms is bridged by a single chlorine atom. The coordination about each copper centre is trigonal bipyramidal, with chlorine atoms in the equatorial plane and the copper atom displaced ca. 0.2 Å toward the DENC ligand. Main bond lengths are: Cu(1)-O(1) 1.930, Cu(2)-O(1) 1.875, Cu(1)-Cl(1) 2.504(3), Cu(2)-Cl(1)' 2.352(4), Cu(1)-Cl(2) 2.336(4), Cu(2)-Cl(2) 2.458(4), Cu(1)-Cl(4) 2.433, Cu(2)-Cl(3) 2.385, Cu(1)-N(1) 1.975(10), Cu(2)-N(13) 1.930(12) Å.

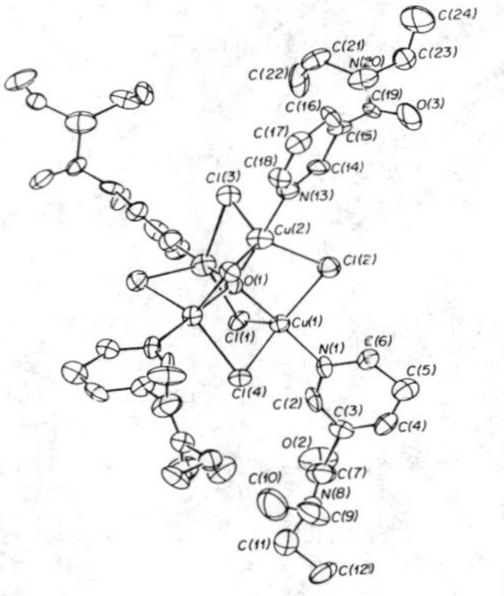

Fig. 1. View of $(DENC)_4Cu_4Cl_6O$.

TRIS(1,2-DIMETHOXYETHANE)LITHIUM μ-CHLORO-μ-OXO-BIS[CHLORO(PENTAMETHYLCYCLOPENTADIENYL)(1-PYRROLYL)ZIRCONATE(IV)] DIMETHOXYETHANE SOLVATE
$C_{40}H_{68}Cl_3LiN_2O_7Zr_2,C_4H_{10}O_2$ $[Li(C_4H_{10}O_2)_3][Zr_2Cl_3O(C_4H_4N)_2(C_{10}H_{15})_2],C_4H_{10}O_2$

J.L. ATWOOD, R.D. ROGERS and R.V. BYNUM, 1984. Acta Cryst., C40, 1812-1814.

Monoclinic, C2/c, a = 12.203, b = 30.853, c = 15.194 Å, β = 106.82°, Z = 4. Mo radiation, R = 0.082 for 1850 reflexions.

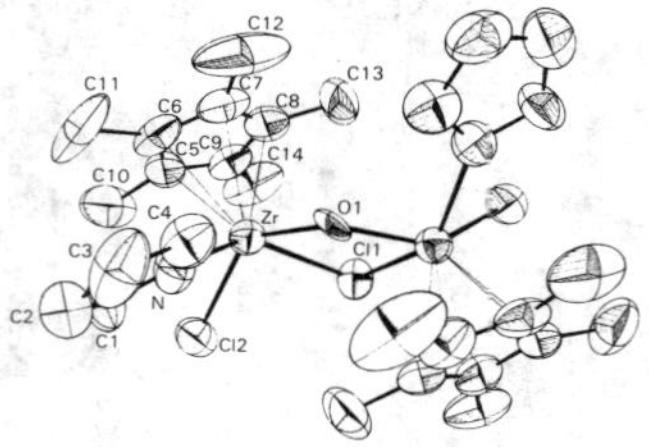

Fig. 1. The $[Zr_2Cl_3O(C_4H_4N)_2(C_{10}H_{15})_2]^-$ anion.

The cation and anion (Fig. 1) both lie on twofold axes. The pyrrolyl ligand is bonded in a head-on fashion with the Zr-N-centroid(pyrrolyl) angle 174°. Bridge angles are 132.5(8) for Zr-O-Zr and 82.0(2)° for Zr-Cl-Zr. Some bond lengths are Zr-O 1.910(6), Zr-Cl (bridge) 2.665(4) and Zr-Cl (terminal) 2.472(4) Å. The uncoordinated molecule of solvent lies highly disordered about a twofold axis.

CARBONYL-3,5-BIS(TRIFLUOROMETHYL)PYRAZOLEBIS(TRIPHENYLPHOSPHINE)IRIDIUM(I)
$C_{42}H_{31}F_6IrN_2OP_2$

A.L. BANDINI, G. BANDITELLI, F. BONATI, G. MINGHETTI, F. DEMARTIN and M. MANASSERO, 1984. J. Organometal. Chem., 269, 91-105.

Monoclinic, $P2_1/n$, a = 21.106, b = 19.700, c = 9.437 Å, β = 94.34°, Z = 4. Mo radiation, R = 0.036 for 4474 reflexions.

The Ir atom is in an essentially square planar environment (Fig. 1). The monodentate pyrazolato ligand is planar and approximately normal to the metal coordination plane (dihedral angle 85.8°). Main bond lengths are: Ir-P 2.328, 2.336, Ir-C 1.811, Ir-N 2.079, C-O 1.155, N-N 1.360, N(1)-C(2) 1.347, N(2)-C(4) 1.340, C-F 1.288-1.344 Å

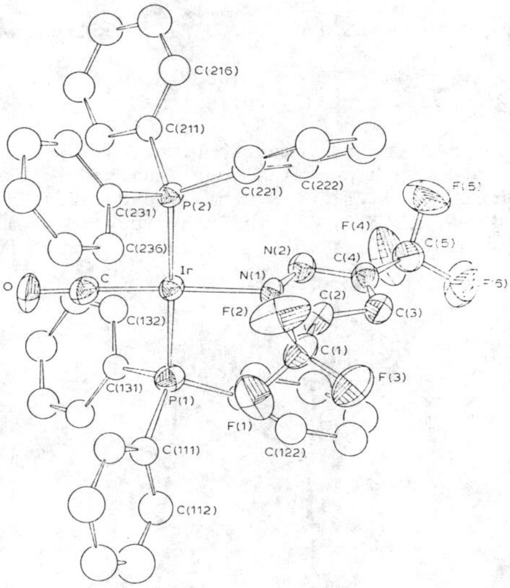

Fig. 1. Molecular structure of $C_{42}H_{31}F_6IrN_2OP_2$.

TRICHLORO(PHENYLIMIDO)BIS(TRIPHENYLPHOSPHINE)RHENIUM(V) DICHLOROMETHANE (1/1)
$C_{42}H_{35}Cl_3NP_2Re,CH_2Cl_2$ $ReCl_3(C_6H_5N)\{P(C_6H_5)_3\}_2,CH_2Cl_2$

E. FORSELLINI, U. CASELLATO, R. GRAZIANI, M.C. CARLETTI and L. MAGON, 1984. Acta Cryst., C40, 1795-1797.

Triclinic, $P\bar{1}$, a = 12.296, b = 16.961, c = 11.022 Å, α = 74.00, β = 99.72, γ = 108.46°, D_m = 1.61, Z = 2. Mo radiation, R = 0.045 for 5812 reflexions.

The slightly distorted octahedral coordination about Re is shown in Fig. 1. The atoms Re, N, Cl(1), Cl(2), Cl(3) are coplanar to within ±0.01 Å; the phenylimido phenyl ring is tilted by 5.2(3)° with respect to the mean equatorial plane. Some bond lengths are, Re-P 2.503 and 2.490(2), Re-Cl 2.415, 2.411, 2.402(2), Re-N 1.726(6) Å. The Re-N-C angle is 172.6(6)°.

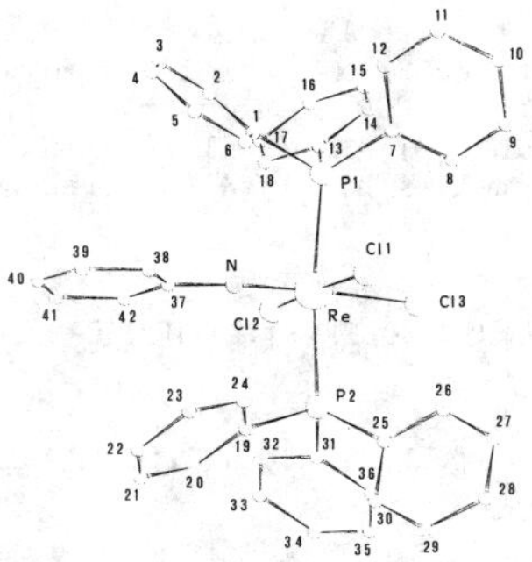

Fig. 1. The molecular structure of $ReCl_3(C_6H_5N)\{P(C_6H_5)_3\}_2$.

DI(μ-HYDROXO)BIS[(BENZYLBIS(2-(2'-PYRIDYL)ETHYL)AMINE)COPPER(II)] BIS(HEXAFLUOROPHOSPHATE) DIHYDRATE

$C_{42}H_{48}Cu_2F_{12}N_6O_2P_2,2H_2O$ $[Cu(C_{21}H_{23}N_3)(OH)]_2(PF_6)_2,2H_2O$

K.D. KARLIN, Y. GULTNEH, J.C. HAYES and J. ZUBIETA, 1984. Inorg. Chem., 23, 519-521.

Triclinic, $P\bar{1}$, a = 9.426, b = 11.415, c = 11.586 Å, α = 74.78, β = 88.72, γ = 74.14°, Z = 1. Mo radiation, R = 0.059 for 2270 reflexions.

The cation (Fig. 1) is a centrosymmetric binuclear species possessing a parallel-planar structure. The two square-pyramidal coordination spheres are joined by axial- and basal-edged OH groups. Two pyridyl N atoms (Cu-N 2.021(7) and 2.000(7) Å), the tertiary amino N atom (Cu-N 2.063(5) Å), and a hydroxyl O atom (Cu-O 1.931(3) Å) form the basal group. The axial Cu-O distance is 2.298(4) Å and the Cu...Cu distance is 3.271 Å.

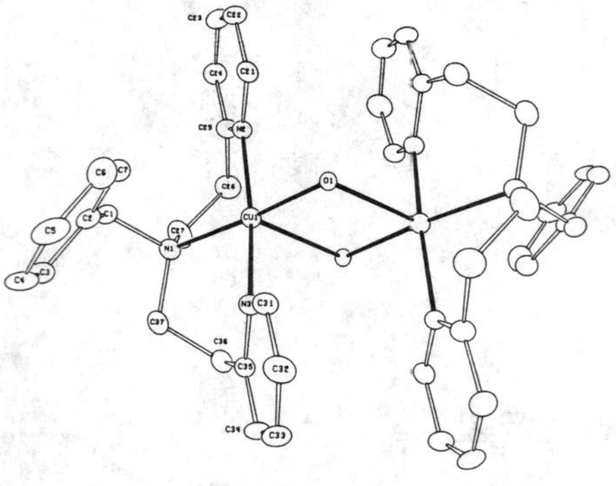

Fig. 1. The $[Cu(C_{21}H_{23}N_3)(OH)]_2^{2+}$ cation.

BIS[(μ-THIOCYANATO-N)BIS(μ-3,5-DIETHYL-1,2,4-TRIAZOLE-N¹,N²)BIS(THIOCYANATO-N)(3,5-DIETHYL-1,2,4-TRIAZOLE-N')NICKEL(II)N,N¹,N¹']NICKEL(II) DIHYDRATE

$C_{42}H_{66}N_{24}Ni_3S_6,2H_2O$ $Ni_3(C_6H_{11}N_3)_6(NCS)_6,2H_2O$

G.A. van ALBADA, R.A.G. de GRAAFF, J.G. HAASNOOT and J. REEDIJK, 1984. Inorg. Chem., 23, 1404-1408.

Orthorhombic, Pbca, a = 14.447, b = 23.047, c = 18.613 Å, D_m = 1.40, Z = 4. Mo radiation, R = 0.058 for 1804 reflexions.

All three Ni atoms in the centrosymmetric trinuclear complex (Fig. 1) have NiN_6 chromophases having octahedral geometry with Ni-N 2.02(1)-2.18(1) Å. The usual structural features are the N-bonding bridging NCS groups between the Ni(II) atoms and the 1,2-bridging triazole ligands (two between each pair of Ni atoms).

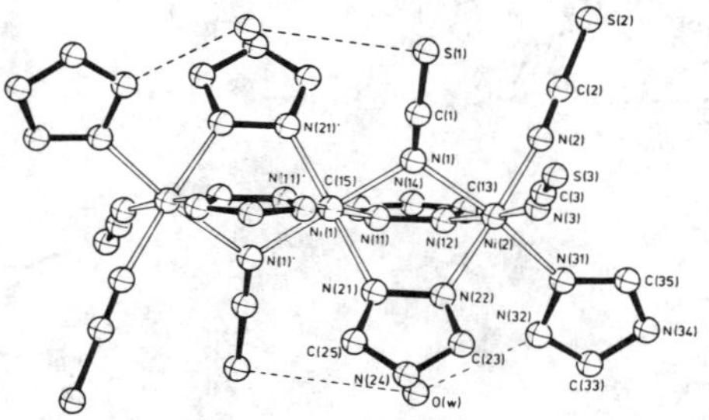

Fig. 1. The $Ni_3(C_6H_{11}N_3)_6(NCS)_6,2H_2O$ structure, ethyl groups omitted.

DICARBONYLBIS(μ-PYRAZOLYL-N,N')BIS(TRIPHENYLPHOSPHINE)DIIRIDIUM(I)
$C_{44}H_{36}Ir_2N_4O_2P_2$ (I) $[Ir(PPh_3)(CO)(\mu\text{-}C_3H_3N_2)]_2$

DICARBONYLBIS(μ-PYRAZOLYL-N,N')BIS(TRIPHENYLPHOSPHINE)DIIRIDIUM(I) n-HEXANE SOLVATE
$C_{44}H_{36}Ir_2N_4O_2P_2,C_6H_{14}$ (II) $[Ir(PPh_3)(CO)(\mu\text{-}C_3H_3N_2)]_2.C_6H_{14}$

DICARBONYLDICHLOROBIS(μ-4-CHLOROPYRAZOLYL-N,N')BIS(TRIPHENYLPHOSPHINE)DIIRIDIUM(II) DICHLOROMETHANE SOLVATE
$C_{44}H_{34}Cl_4Ir_2N_4O_2P_2,2CH_2Cl_2$ (III) $[Ir(PPh_3)(CO)(Cl)(\mu\text{-}C_3H_2ClN_2)]_2.2CH_2Cl_2$

J.L. ATWOOD, K.A. BEVERIDGE, G.W. BUSHNELL, K.R. DIXON, D.T. EADIE, S.R. STOBART and M.J. ZAWOROTKO, 1984. Inorg. Chem., 23, 4050-4057.

I. Triclinic, $P\bar{1}$, a = 9.992, b = 10.402, c = 20.278 Å, α = 90.01, β = 90.10, γ = 105.13°, Z = 2. Mo radiation, R = 0.046 for 1481 reflexions.

II. Monoclinic, $P2_1/c$, a = 15.528, b = 9.713, c = 30.544 Å, β = 93.98°, Z = 4. Mo radiation, R = 0.055 for 2611 reflexions.

III. Monoclinic, $P2_1/n$, a = 15.851, b = 28.617, c = 11.092 Å, β = 95.82°, Z = 4. Mo radiation, R = 0.050 for 3003 reflexions.

The complex molecules in all three structures are shown in Fig. 1. The pyrazolyl-bridged dinuclear complex in I and II have the terminal PPh_3 and CO ligands mutually trans across the Ir_2 axis with formally nonbonded Ir...Ir separations of 3.163(2) Å in I and 3.290(1) in II. The Ir atoms in I and II have square planar coordination geometry with Ir-P 2.24(1)-2.262(7), Ir-N 2.02(4)-2.10(2), and Ir-C(O) 1.67(3)-1.89(5) Å. Complex III is formed by oxidative of addition of Cl_2 to I. The trans configuration is retained in III. Axial substitution at each Ir with Ir-Cl 2.443(6) Å has been accompanied by formation of a metal-metal bond (Ir-Ir 2.737(1) Å) and electrophilic substitution by Cl of the four-position in each bridging pyrazolyl ring. Other distances for III are: Ir-P 2.316(6) (x2), Ir-N 2.04(2)-2.10(2), Ir-C(O) 1.81(2) and 1.84(2) Å

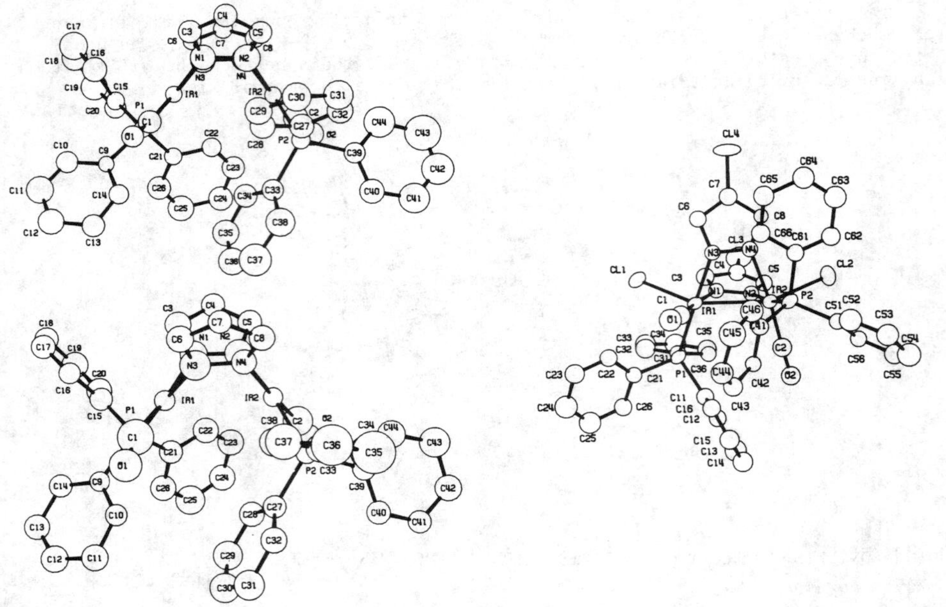

Fig. 1. The $[Ir(PPh_3)(CO)(\mu-C_3H_3N_2)]_2$ complexes in I and II (upper and lower left, respectively) and $[Ir(PPh_3)(CO)(Cl)(\mu-C_3H_2ClN_2)]_2$ (right).

TETRAKIS(2-ANILINOPYRIDINATO)DIMOLYBDENUM(II)

$C_{44}H_{40}Mo_2N_8$ (I) $Mo_2(C_6H_5NC_5H_5N)_4$

TETRAKIS(2-ANILINOPYRIDINATO)DITUNGSTEN(II) TETRAHYDROFURAN SOLVATE

$C_{44}H_{40}N_8W_2,0.67(C_4H_8O)$ (II) $W_2(C_6H_5NC_5H_5N)_4.0.66C_4H_8O$

A.R. CHAKRAVARTY, F.A. COTTON and E.S. SHAMSHOUM, 1984. Inorg. Chem., 23, 4216-4221.

I. Monoclinic, $P2_1/c$, a = 19.867, b = 9.763, c = 20.082 Å, β = 100.4°, Z = 4. Mo radiation, R = 0.060 for 2976 reflexions.

II. Triclinic, $P\bar{1}$, a = 10.535, b = 19.194, c = 18.598 Å, α = 107.4, β = 99.7, γ = 103.6°, Z = 3. Cu radiation, R = 0.067 for 4861 reflexions (at 5°C).

Molecules of each compound are shown in Fig. 1. The structures of I and II contain two and three independent centrosymmetric molecules with mean Mo-Mo and W-W quadruple bond lengths of 2.071(2) and 2.165(2) Å. The average M-N(py) and M-Na bond lengths are 2.156(8) and 2.141(8) Å for I, 2.144(14) and 2.123(8) Å, respectively, for II.

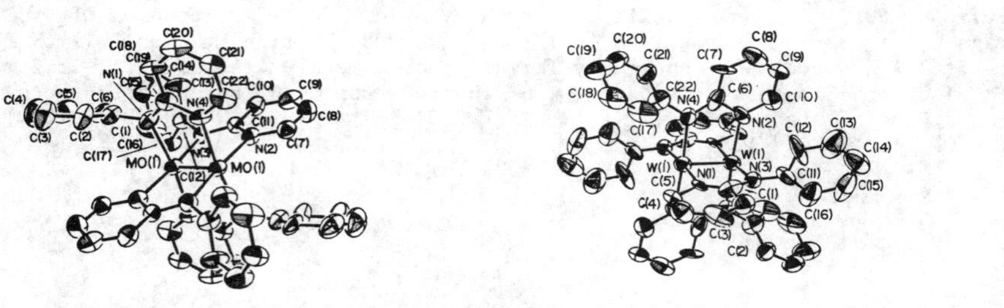

Fig. 1. Molecules of $M_2(PhNpy)_4$, M = Mo (left) and W (right).

BIS[(4,5-BIS[2-(2-PYRIDYL)ETHYLIMINOMETHYL]IMIDAZOLATE)(IMIDAZOLATE)DICOPPER] NITRATE TRIHYDRATE

$C_{44}H_{50}Cu_4N_{20}O_{15}$ (I) $[Cu_2(bpim)(im)]_2(NO_3)_4.3H_2O$

BIS[2-[2-(2-PYRIDYL)ETHYLIMINOMETHYL]PYRIDINECOPPER] IMIDAZOLATE NITRATE HYDRATE

$C_{29}H_{34}Cu_2N_{11}O_{11.5}$ (II) $[Cu(pip)]_2(im)(NO_3)_3.2.5H_2O$

BIS[N,N,N',N'',N''-PENTAMETHYLDIETHYLENETRIAMINECOPPER] 2-METHYLIMIDAZOLATE PERCHLORATE

$C_{22}H_{51}Cl_3Cu_2N_8O_{12}$ (III) $[Cu(pmdt)]_2(2\text{-}Meim)(ClO_4)_3$

BIS[N,N,N',N'',N''-PENTAMETHYLDIETHYLENETRIAMINECOPPER]BENZIMIDAZOLATE PERCHLORATE HYDRATE

$C_{25}H_{53}Cl_3Cu_2N_8O_{13}$ (IV) $[Cu(pmdt)]_2(Bzim)(ClO_4)_3.H_2O$

G. KOLKS and S.J. LIPPARD, 1984. Acta Cryst., C40, 261-271.

I. Tetragonal, $I4_1/a$, a = 27.204, c = 14.706 Å, D_m = 1.650, Z = 8. Mo radiation, R = 0.059 for 2816 reflexions.

II. Monoclinic, $P2_1/c$, a = 17.475, b = 17.809, c = 23.565 Å, β = 105.85°, D_m = 1.581, Z = 8. Mo radiation, R = 0.105 for 3181 reflexions.

III. Orthorhombic, Imm2, a = 14.226, b = 16.214, c = 7.970 Å, D_m = 1.544, Z = 2. Mo radiation, R = 0.048 for 900 reflexions.

IV. Orthorhombic, Iba2, a = 17.443, b = 37.510, c = 12.468 Å, D_m = 1.48, Z = 8. Mo radiation, R = 0.091 for 2048 reflexions.

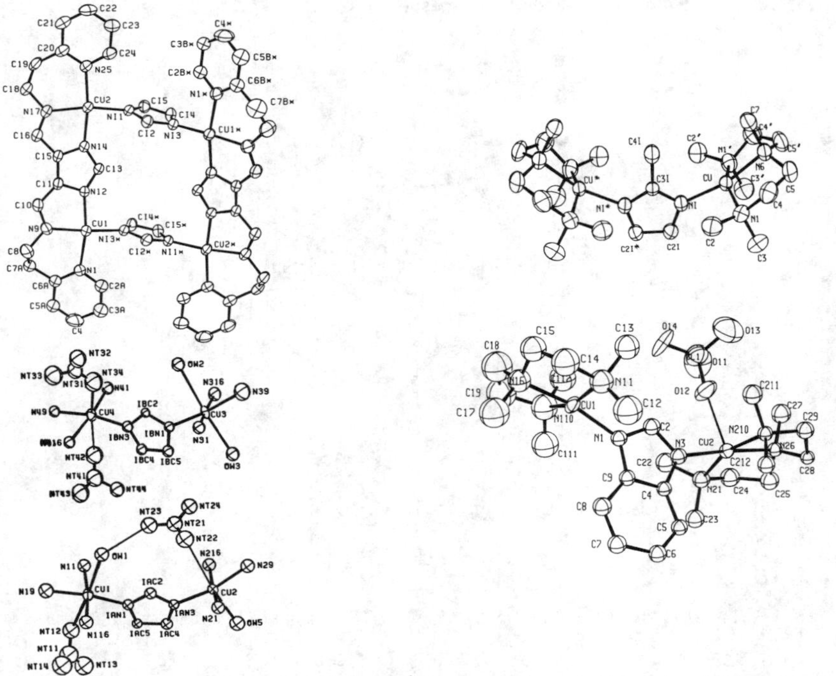

Fig. 1. Views of the $[Cu_2(bpim)(im)]_2{}^{4+}$ cation (top left), the (im) bridge and Cu coordination of the four Cu atoms in (II) (lower left), the $[Cu(pmdt)]_2$-$(2\text{-}Meim)^{3+}$ cluster of (III) (top right), and the $[Cu(pmdt)]_2(Bzim)$-$(ClO_4)^{2+}$ cation in (IV) (lower right).

In I (Fig. 1) the tetranuclear cation consists of two Cu_2bpim^{3+} units joined by two bridging im groups and has a crystallographically required twofold axis. Three N atoms of the bpim ligand (Cu-N distances 1.99-2.09 Å) and one N of the im ion (Cu-N 1.967(5) Å) form the principal copper coordination plane. Two geometrically distinct bridging im groups, one parallel (bpim ligand) and one perpendicular (im ligand) to the copper coordination plane, occur in the complex. Axial sites are occupied by water and nitrate O atoms. In II (Fig. 1) each Cu atom is bonded to three N atoms of the pip ligand (Cu-N distances 2.02-2.09 Å) and to one N atom of the bridging, perpendicular-type im ring (av. Cu-N 1.978(12) Å), all in an approximate plane. Water and nitrate-group O atoms are coordinated at the axial sites. Both III and IV (Fig. 1) have bridged binuclear structures with features similar to those of II. The cation of III has crystallographic mm symmetry.

BIS(TETRABUTYLAMMONIUM) DICYANO(TETRACYANOBIIMIDAZOLE)PLATINUM

$C_{44}H_{72}N_{12}Pt$ $[(C_4H_9)_4N^+]_2[Pt(CN)_2(C_{10}N_8)]^{2-}$

P.G. RASMUSSEN, O.H. BAILEY and J.C. BAYÓN, 1984. Inorg. Chim. Acta, 86, 107-111.

Triclinic, P$\bar{1}$, a = 16.664, b = 12.422, c = 13.842 Å, α = 112.47, β = 111.01, γ = 89.40°, Z = 2. Mo radiation, R = 0.038 for 5359 reflexions.

The complex anion (Fig. 1) is approximately planar with only the nitriles of N(4) and N(5) and the C(12) cyanide deviating slightly from the mean plane. The coordination around the Pt atom is distorted square planar (the N(1)-Pt-N(1') angle is 78.8°). The distances Pt-C(11), Pt-C(12), Pt-N(1) and Pt-N(1') are 1.915, 1.918, 2.077 and 2.049 Å respectively and the angles N(1)-Pt-C(11), N(1')-Pt-C(12) and C(11)-Pt-C(12) are 97.0, 96.3 and 87.7° respectively.

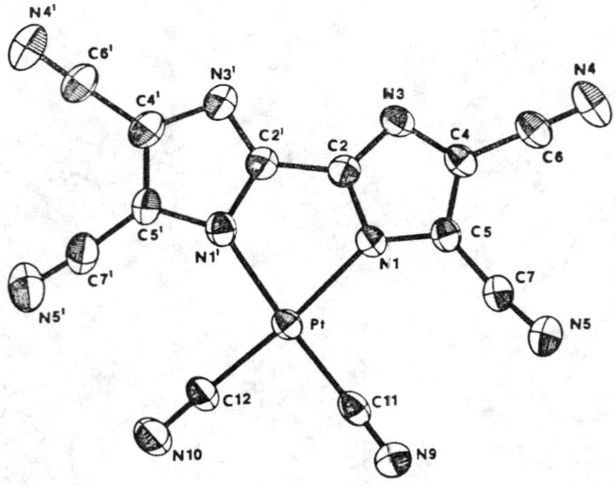

Fig. 1. Structure of the complex anion in $C_{44}H_{72}N_{12}Pt$.

TRIS[(TRIS(1-PYRAZOLYLETHYL)AMINE-N,N',N'',N''')BROMOCOPPER] TRIBROMOCOPPER BROMIDE TRIHYDRATE

$C_{45}H_{63}Br_7Cu_4N_{21},3H_2O$ $[CuBr(C_{15}H_{21}N_7)^+]_3.[CuBr_3^{2-}].Br^-,3H_2O$

A. BENCINI and F. MANI, 1984. Inorg. Chim. Acta, 87, L9-L13.

Triclinic, P$\bar{1}$, a = 22.430, b = 15.488, c = 9.738 Å, α = 100.0, β = 88.33, γ = 106.4°, Z = 2. Mo radiation, R = 0.044 for 4135 reflexions.

The compound consists of three non-equivalent five-coordinate cations, a $CuBr_3^{2-}$ anion, a Br^- ion and water molecules. Cations I and II have distorted square-pyramidal geometry about the Cu atom whilst in III the geometry is intermediate between that of a trigonal bipyramid and that of a square pyramid (Fig. 1). The Cu-Br distances in I, II and III respectively are 2.474, 2.484 and 2.470 Å and the Cu-N distances range from 1.95 to 2.27 Å in I, 1.96 to 2.23 Å in II, and 1.96 to 2.16 Å in III. The longest Cu-N bonds in I and II correspond to the apical bonds (N(1) and N(8)). The $CuBr_3^{2-}$ anion has almost trigonal planar coordination with Cu-Br distances of 2.394, 2.388 and 2.374 Å and Br-Cu-Br angles of 116.1, 122.9 and 121.0°.

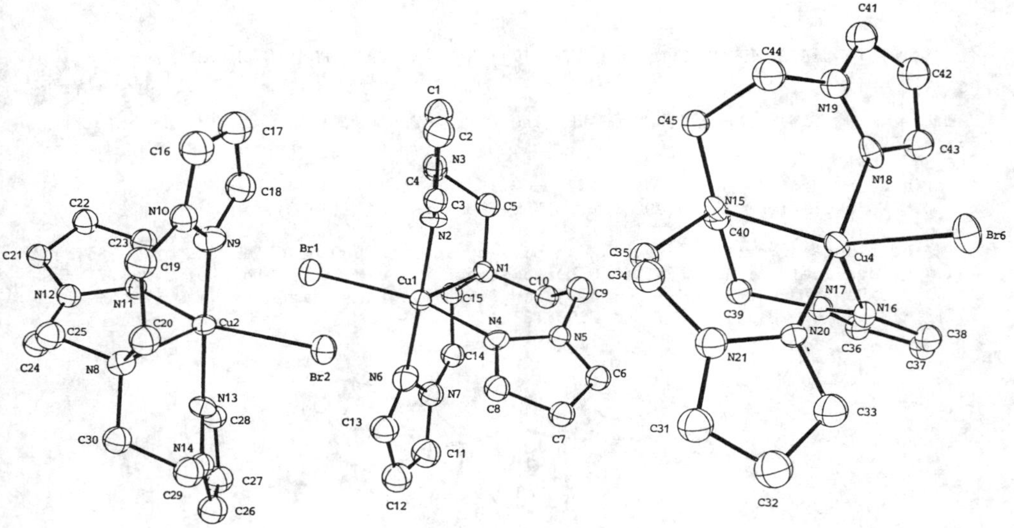

Fig. 1. Cations I, II and III in $C_{45}H_{63}Br_7Cu_4N_{21},3H_2O$.

BIS[(μ-2-PYRIDYLAMIDE-N,N')(μ-2-PYRIDYLAMIDE-N,N)(2-PYRIDYLAMIDE-N)(DIMETHYLPHENYL-PHOSPHINE)RUTHENIUM]

$C_{46}H_{52}N_{12}P_2Ru_2$ $Ru_2(C_5NH_4NH)_6(P(CH_3)_2C_6H_5)_2$

A.R. CHAKRAVARTY, F.A. COTTON and D.A. TOCHER, 1984. Inorg. Chem., 23, 4030-4033.

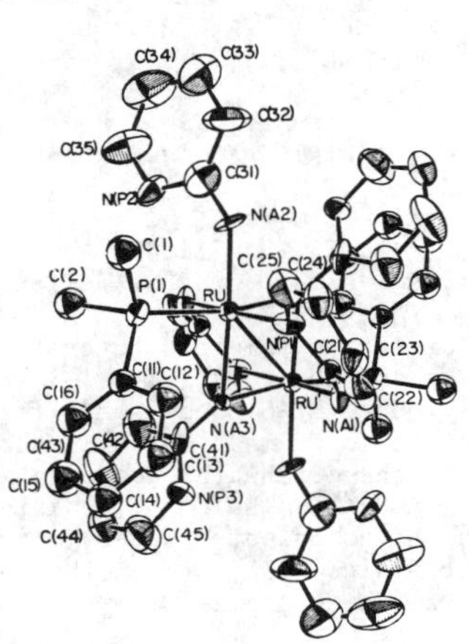

Fig. 1. The $Ru_2(C_5NH_4NH)_6(P(CH_3)_2C_6H_5)_2$ molecule.

Monoclinic, $P2_1/c$, a = 11.159, b = 10.067, c = 20.193 Å, β = 82.97°, Z = 2. Mo radiation, R = 0.054 for 1315 reflexions.

The molecule (Fig. 1) consists of two Ru atoms that have distorted-octahedral coordination united by a shared edge and a Ru-Ru bond. The anionic ligand, $(C_5NH_4NH)^-$, exhibits three modes of coordination: (1) with the two Ru atoms bridged through N(py) and N(am), (2) with the amine N as a bridging (μ,η^1) atom, and (3) with the amine N atom as a terminal ligand atom. The Ru-Ru distance is 2.573(2) Å, Ru-P 2.388(4), and Ru-N 2.05-2.13 Å. The molecule is centrosymmetric.

(μ-OXO)BIS[AQUABIS(PHENANTHROLINE)IRON(III)] TETRANITRATE PENTAHYDRATE

$C_{48}H_{36}Fe_2N_{12}O_{15}.5H_2O$ $[Fe_2O(C_{12}H_8N_2)_4(H_2O)_2](NO_3)_4.5H_2O$

J.E. PLOWMAN, T.M. LOEHR, C.K. SCHAUER and O.P. ANDERSON, 1984. Inorg. Chem., 23, 3553-3559.

Triclinic, $P\bar{1}$, a = 13.828, b = 15.122, c = 15.890 Å, α = 84.44, β = 64.99, γ = 68.46°, Z = 2. Mo radiation, R = 0.090 for 5613 reflexions (at -130°C).

The binuclear cation (Fig. 1) exhibits typical Fe-μ-O bond lengths of 1.785(2) Å and a bridge angle of 155.1(4)°. Other distances: Fe-N (trans to μ-O) 2.263(7) and 2.253(7), Fe-N (trans to OH_2) 2.154(9) and 2.198(7), Fe-N (trans to N) 2.126(6)-2.145(8), and Fe-OH_2 2.027(9) and 2.021(5) Å.

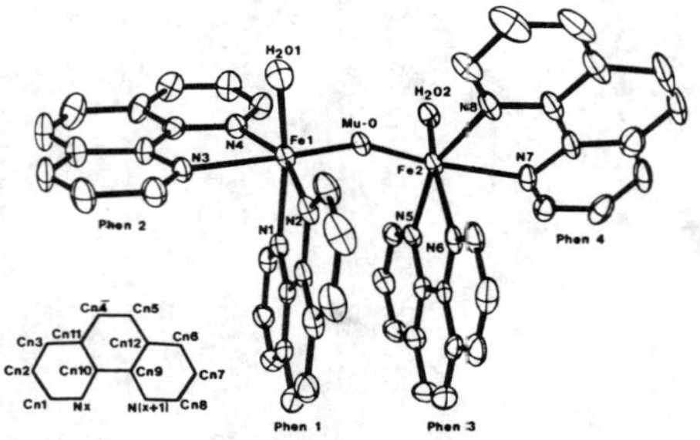

Fig. 1. The $[Fe_2O(C_{12}H_8N_2)_4(H_2O)_2]^{4+}$ cation.

BIS(μ-TETRAKIS(THIOCYANATO-S)MERCURY(II)-N,N')DIAQUAHEXAKIS(5,7-DIMETHYL[1,2,4]-TRIAZOLO[1,5a]PYRIMIDINE-N³)DICOBALT(II)

$C_{50}H_{52}Co_2Hg_2N_{32}O_2S_8$ $[CoHg(dmtp)_3(SCN)_4(H_2O)]_2$

M.B. CINGI, A.M.M. LANFREDI, A. TIRIPICCHIO, J.G. HAASNOOT and J. REEDIJK, 1984. Inorg. Chim. Acta, 86, 137-143.

Triclinic, $P\bar{1}$, a = 11.481, b = 13.975, c = 11.479 Å, α = 101.08, β = 90.11, γ = 77.24°, Z = 1. Mo radiation, R = 0.038 for 2392 reflexions.

The structure consists of centrosymmetric heterotetranuclear complexes formed by thiocyanate groups bridging Hg and Co atoms (Fig. 1). Hg is tetrahedrally coordinated by S atoms from thiocyanate groups (Hg-S bonds range from 2.457 to 2.601 Å), two of which bridge two centrosymmetric Co atoms. Three N atoms from DMTP ligands and one O from a water molecule complete the octahedral coordination of Co (Co-N bonds: 2.11, 2.14 and 2.18 Å; Co-O bond: 2.08 Å). The Co-N(1) distance is 2.08 Å. Intermolecular hydrogen bonds occur between the coordinated water molecule and N(3) of a thiocyanate group, and a non-bridging pyrimidine N(33) atom. The relevant distances are: O...N(3) 2.774 and O...N(33) 2.722 Å with angles N(3)-H(2)...O 161° and N(33)-H(1)...O 165°.

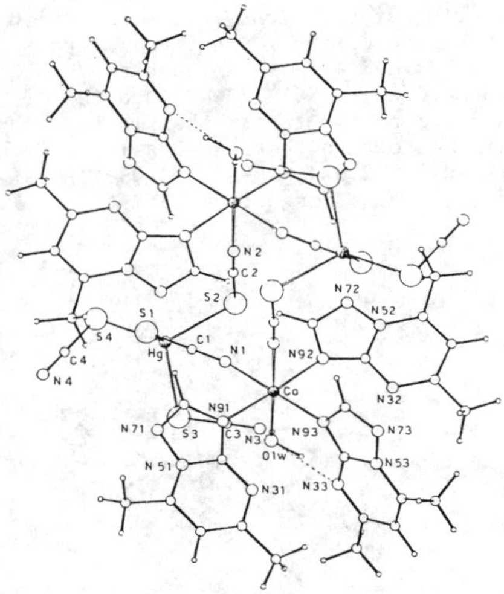

Fig. 1. Perspective view of the heterotetranuclear complex $C_{50}H_{52}Co_2Hg_2N_{32}O_2S_8$.

(3,11-DIBENZYL-14,2-DIMETHYL-2,12-DIPHENYL-3,11,15,19,22,26-HEXAAZATRICYCLO-[11.7.7.1^{5,9}]OCTACOSA-1,5,7,9(28),12,14,19,21,26-NONAENE-κ^{4}N)NICKEL(II) HEXAFLUOROPHOSPHATE ETHANOL SOLVATE

$C_{50}H_{52}F_{12}N_6NiP_2.C_2H_6O$

B. KORYBUT-DASZKIEWICZ, M. KOJIMA, J.H. CAMERON, N. HERRON, M.Y. CHAVAN, A.J. JIRCITANO, B.K. COLTRAIN, G.L. NEER, N.W. ALCOCK and D.H. BUSCH, 1984. Inorg. Chem., 23, 903-914.

Monoclinic, $P2_1/c$, a = 10.126, b = 17.629, c = 30.666 Å, β = 100.87°, D_m = 1.38, Z = 4. Mo radiation, R = 0.083 for 3363 reflexions.

The structure of the cation (Fig. 1) illustrates the demanding steric requirements placed upon the lacuna size and shape by the peripheral substituents of the macrocycle. The Ni atom has square-planar coordination geometry with Ni-N 1.88(2)-1.89(2) Å.

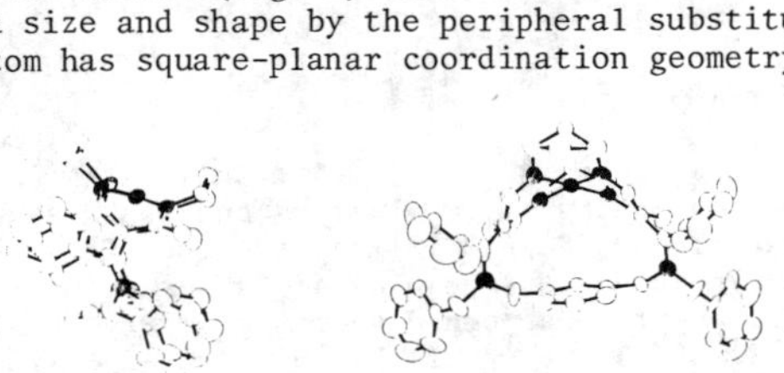

Fig. 1. Two views of the $(C_{50}H_{52}NiN_6)^{2+}$ cation.

BIS(ACETONITRILE)BIS(2-ISOCYANO-2-METHYLPROPANE)BIS(TRIPHENYLPHOSPHINE)RHENIUM TETRAFLUOROBORATE DICHLOROMETHANE SOLVATE

$C_{50}H_{54}B_2F_8N_4P_2Re,0.5CH_2Cl_2$ $[Re(CNCMe_3)_2(NCCH_3)_2(PPh_3)_2]^{2+}2BF_4^-,0.5(CH_2Cl_2)$

J.D. ALLISON, P.E. FANWICK and R.A. WALTON, 1984. Organometallics, 3, 1515-1520.

Monoclinic, C2/c, a = 44.729, b = 12.170, c = 21.018 Å, β = 106.15°, Z = 8. Mo radiation, R = 0.047 for 6036 reflexions.

The complex consists of discrete anions and cations with the solvent molecules located on the twofold axis. The cation is in the all-trans form (Fig. 1). The six-coordinate Re atom is quite crowded - this is indicated by the bent Re-C-N angles (Re-C(10)-N(10) 171.5° and Re-C(20)-N(20) 174.4°) and the C(10)-Re-C(20) angle of 170.4°. The three phenyl rings on the two triphenylphosphines are in nearly an eclipsed configuration with respect to the P-Re-P axis. The bonds associated with Re are: Re-C(10) 2.023, Re-C(20) 2.034, Re-N(1) 2.058, Re-N(2) 2.055, Re-P(1) 2.454 and Re-P(2) 2.445 Å. The cis angles range from 83.7 to 95.1°.

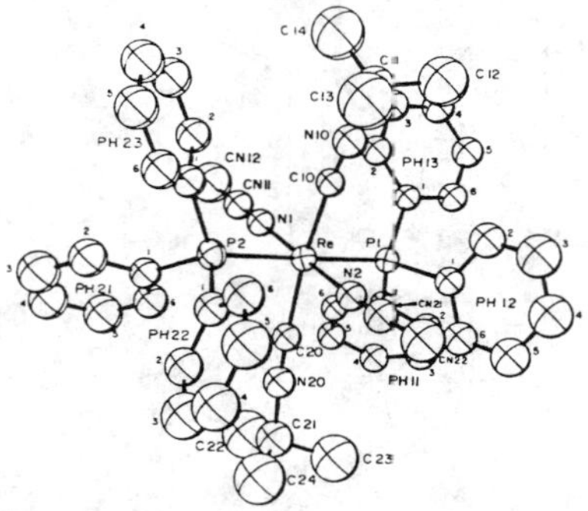

Fig. 1. Perspective view of the cation in $C_{50}H_{54}B_2F_8N_4P_2Re$.

BIS(TRIPHENYLPHOSPHINE)[2,7-BIS(2-PYRIDYL)-1,8-NAPHTHYRIDINE]COPPER(I) HEXAFLUORO-PHOSPHATE

$C_{54}H_{42}CuF_6N_4P_3$ (I) $[Cu(C_{18}H_{12}N_4)(PPh_3)_2]PF_6$

DICHLORO(μ-CHLORO)(μ-HYDROXO)[μ-2,7-BIS(2-PYRIDYL)-1,8-NAPHTHYRIDINE]DICOPPER(II)

$C_{18}H_{13}Cl_3Cu_2N_4O$ (II) $[Cu_2(C_{18}H_{12}N_4)(\mu\text{-Cl})(\mu\text{-OH})Cl_2]$

W.R. TIKKANEN, C. KRÜGER, K.D. BOMBEN, W.L. JOLLY, W.C. KASKA and P.C. FORD, 1984. Inorg. Chem., 23, 3633-3638.

I. Monoclinic, $P2_1/n$, a = 14.855, b = 21.914, c = 15.111 Å, β = 98.690°, Z = 4. Mo radiation, R = 0.057 for 4961 reflexions.

II. Monoclinic, $P2_1/m$, a = 7.694, b = 16.434, c = 7.921 Å, β = 107.974°, Z = 2. Mo radiation, R = 0.036 for 1851 reflexions.

The two structures are shown in Fig. 1. The Cu(I) atom in I is in a distorted tetrahedral environment consisting of two P atoms and two N atoms from a bipyridyl fragment of the $C_{18}H_{12}N_4$ ligand (Cu-P 2.285(1) and 2.287(2), Cu-N(1) 2.125(4), Cu-N(21) 2.287(2) Å). The binuclear complex II has exact mirror symmetry with Cu(II) atoms having square pyramidal coordination geometry (μ-Cl ligand axial). Bond lengths for II: Cu-Cl 2.288(1), Cu-(μ-Cl) 2.555(1), Cu-O 1.946(1), Cu-N(1) 2.000(3), Cu-N(2) 2.083(3) Å.

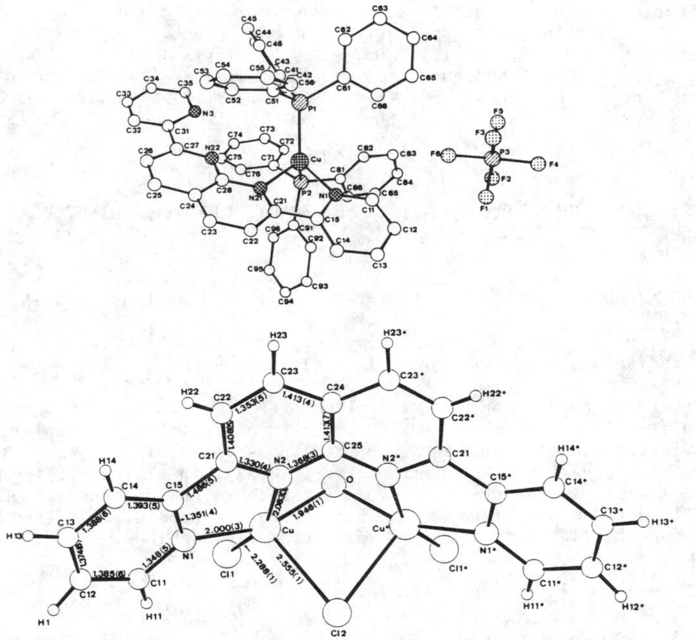

Fig. 1. The structures of $[Cu(C_{18}H_{12}N_4)(PPh_3)_2]PF_6$ (top) and $[Cu_2(C_{18}H_{12}N_4)$-$(\mu\text{-Cl})(\mu\text{-OH})Cl_2]$ (bottom).

(μ_2-4,4',5,5'-TETRACYANO-2,2'-BIIMIDAZOLE-N,N',N'',N''')DIPYRIDYLBIS(TRIPHENYLPHOSPHINE)DICOPPER(I) PYRIDINE SOLVATE

$C_{56}H_{40}Cu_2N_{10}O_6P_2, C_5H_5N$ $Cu_2[P(OC_6H_5)_3]_2(C_5H_5N)_2(C_{10}N_8), C_5H_5N$

P.G. RASMUSSEN, J.E. ANDERSON and J.C. BAYÓN, 1984. Inorg. Chim. Acta, 87, 159-164.

Triclinic, $P\bar{1}$, a = 10.456, b = 15.060, c = 10.438 Å, α = 94.89, β = 115.93, γ = 72.76°, D_m = 1.30, Z = 1. Mo radiation, R = 0.035 for 2743 reflexions.

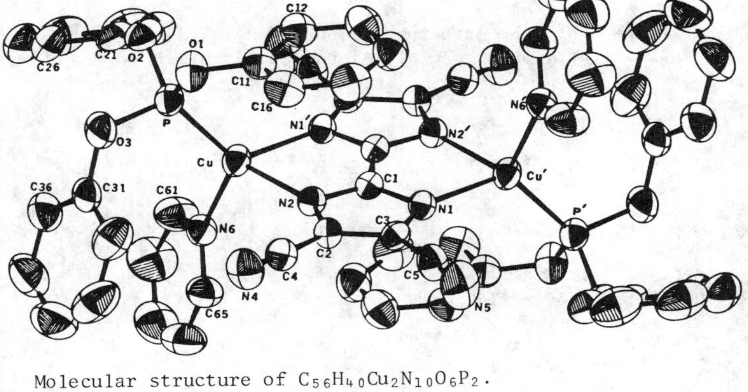

Fig. 1. Molecular structure of $C_{56}H_{40}Cu_2N_{10}O_6P_2$.

The dimeric molecule (Fig. 1) has an inversion centre and each Cu atom is tetrahedrally coordinated, with angles ranging from 101-126°. The coordination distances are: Cu-P 2.148, Cu-N(6) 2.036, Cu-N(1) 2.185 and Cu-N(2) 2.154 Å. The P-O distances are 1.605, 1.602 and 1.599 Å. The $TCBIIM^{2-}$ ion and the Cu atom are coplanar, the largest deviation from the idealized plane of these atoms being 0.05 Å. The Cu...Cu separation is 5.37 Å and the C(1)-C(1') distance is 1.428 Å.

trans-BIS(1,2-BIS(DIPHENYLPHOSPHINO)ETHANE)BROMO(2-DIAZO-1,3-DIMETHYL-1,3-DIAZACYCLOPENTANE)TUNGSTEN HEXAFLUOROPHOSPHATE
$C_{57}H_{58}BrF_6N_4P_5W$

H.M. COLQUHOUN and D.J. WILLIAMS, 1984. J. Chem. Soc. Dalton, 1675-1679.

Triclinic, P$\bar{1}$, a = 12.546, b = 14.332, c = 17.174 Å, α = 74.07, β = 76.79, γ = 87.05°, Z = 2. Mo radiation, R = 0.047 for 5502 reflexions.

The organonitrogen ligand in the molecule (Fig. 1) adopts a planar diaminodiazomethane conformation. There is severe steric strain within the complex, notably in the 17° deviation from linearity at N(1), the non-coplanarity of W, N(1), N(2) and C(53), in the wide variation of bond angles at C(53), in the partial pyramidalisation of N(3) and finally in the skewed nature of the coordination sphere about tungsten. All these distortions appear to arise from severe non-bonded repulsions between the N-methyl group of C(56) and one of the chelating diphosphines.

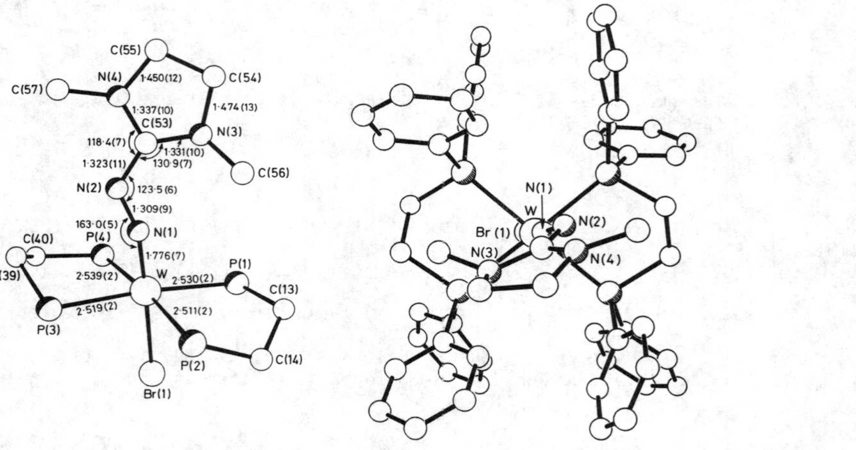

Fig. 1. $C_{57}H_{58}BrF_6N_4P_5W$: the structure (minus phenyl groups) showing bond lengths and angles (left) and the distortions and pyramidalisation of N(3).

BIS(2-(N-ANILINO)PYRIDINE-N,N')BIS(TRIPHENYLPHOSPHINERUTHENIUM (TWO FORMS)
$C_{58}H_{48}N_4P_2Ru$

A.R. CHAKRAVARTY, F.A. COTTON and E.S. SHAMSHOUM, 1984. Inorg. Chim. Acta, 86, 5-11.

I. Monoclinic, $P2_1/c$, a = 13.852, b = 18.872, c = 18.206 Å, β = 84.32°, Z = 4. Cu radiation, R = 0.053 for 5069 reflexions.

II. Monoclinic, $P2_1/c$, a = 11.166, b = 20.404, c = 20.716 Å, β = 97.14°, Z = 4. Mo radiation, R = 0.055 for 3152 reflexions.

In the orange C_1-isomer (I) the coordinating atoms N-py, N-Ph and PPh_3 are all in cis-arrangement whilst in II, the yellow C_2-isomer, the py-rings are trans to each other (Fig. 1). The complexes are six-coordinate with four-membered chelate rings; the N-Ru-N angle in the chelate rings is ca. 62° in each case. The Ru-P bond lengths lie in the range 2.286 to 2.337 Å and the average Ru-N(py) and Ru-N(Ph) bond lengths are 2.083 and 2.171 Å respectively.

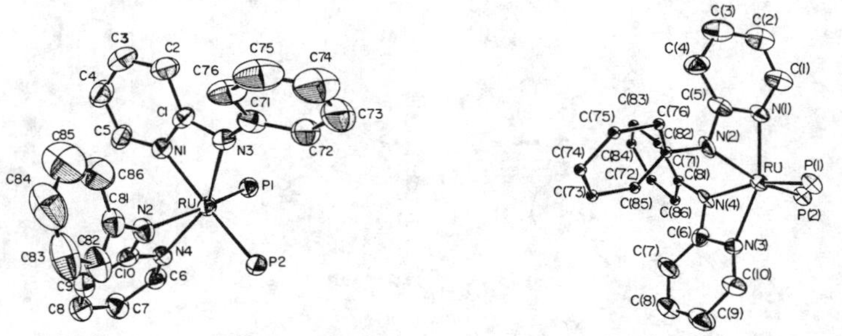

Fig. 1. Molecular structures of the orange C_1-isomer (left), and the yellow C_2-isomer of $C_{58}H_{48}N_4P_2Ru$ (right) excluding phenyl rings.

BIS(1,5-DIAZACYCLOOCTANE)NICKEL(II) BIS(TETRAPHENYLBORATE) DIMETHYLSULPHOXIDE SOLVATE
$C_{60}H_{68}B_2N_4Ni,2(C_2H_6OS)$

J.C.A. BOEYENS and C.C. FOX, 1984. S. Afr. J. Chem., 37, 1-4.

Triclinic, P1. a = 18.29, b = 10.802, c = 15.198 Å, α = 90.65, β = 100.80, γ = 91.16°, Z = 2. Mo radiation, R = 0.13 for 6815 reflexions.

There are two independent molecules of the complex and the structure would not refine well. The two independent cations have their 8-membered rings in a chair-boat form (Fig. 1).

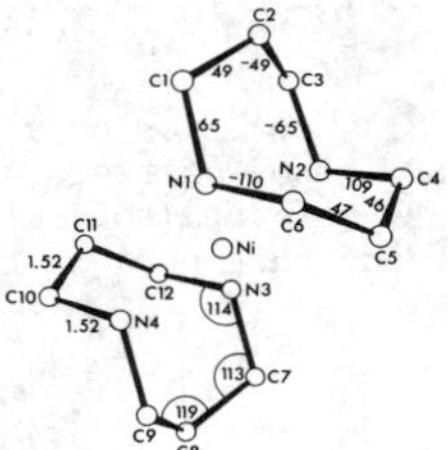

Fig. 1. A bis(1,5-diazacyclooctane)nickel(II) cation with mean torsion angles, bond lengths and bond angles.

TETRAKIS(μ_2-CHLORO)DICHLOROHEXAKIS(N,N-DIETHYLNICOTINAMIDE-N')TRICOPPER
$C_{60}H_{84}Cl_6Cu_3N_{12}O_6$ $[(C_5H_4NC(O)N(C_2H_5)_2)_2CuCl_2]_3$

K.D. ONAN, M. VEIDIS, G. DAVIES, M.A. EL-SAYED and A. EL-TOUKHY, 1984. Inorg. Chim. Acta, 81, 7-13.

Monoclinic, $P2_1/c$, a = 16.618, b = 8.253, c = 28.194 Å, β = 117.3°, D_m = 1.38, Z = 2. Cu radiation, R = 0.059 for 3692 reflexions.

The trimeric unit (Fig. 1) utilizes a crystallographic centre of symmetry. Each molecule consists of two five-coordinate and one six-coordinate copper centre. The μ,μ-dichloro bridging system is asymmetric with a Cu...Cu separation of 3.686 Å. The Cu-N distances range from 2.004 to 2.027 Å and the Cu-Cl distances from 2.264 to 2.982 Å. The Cl(1)-Cu(1)-Cl(2), Cl(1)-Cu(2)-Cl(2) and Cl(1)-Cu(1)-Cl(1') angles are respectively 87.8, 92.56, 92.2°.

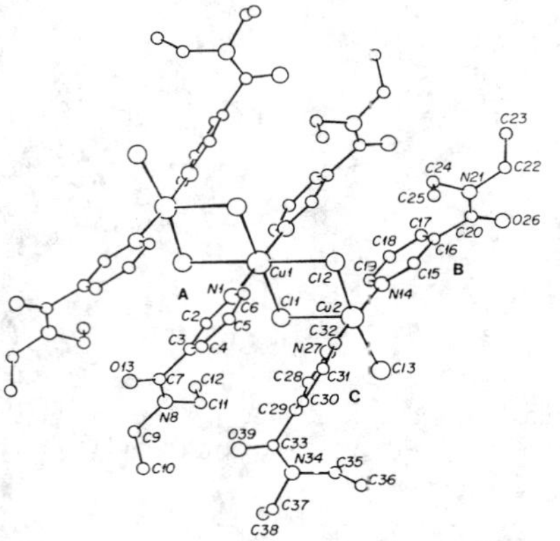

Fig. 1. Molecular structure of $C_{60}H_{84}Cl_6Cu_3N_{12}O_6$.

BIS(5,6-DIHYDROPYRIMIDINO[5,4-c]CARBAZOLE-N^3)TETRAKIS(METHANOL)COBALT(II) 5,6-DIHYDROPYRIMIDINO[5,4-c]CARBAZOLE DINITRATE METHANOL

$C_{62}H_{68}CoN_{14}O_{12}$ $2C_{14}H_{11}N_3,2CH_4O,[Co(C_{14}H_{11}N_3)_2(CH_4O)_4](NO_3)_2$

B. VIOSSAT, NGUYEN-HUY DUNG, J.-C. LANCELOT and M. ROBBA, 1984. Acta Cryst., C40, 935-938.

Triclinic, $P\bar{1}$, a = 14.43, b = 10.70, c = 13.966 Å, α = 111.94, β = 92.77, γ = 124.9°, D_m = 1.42, Z = 1. Mo radiation, R = 0.053 for 2929 reflexions.

In the complex cation (Fig. 1) Co^{II} lies on a centre of symmetry and has octahedral coordination with two N atoms of two 5,6-dihydropyrimidino[5,4-c]carbazole molecules and four O atoms of four methanol molecules: Co-N 2.158(3), Co-O 2.077(5) and 2.121(3) Å. Many hydrogen bonds take part in the cohesion of the structure of the title compound.

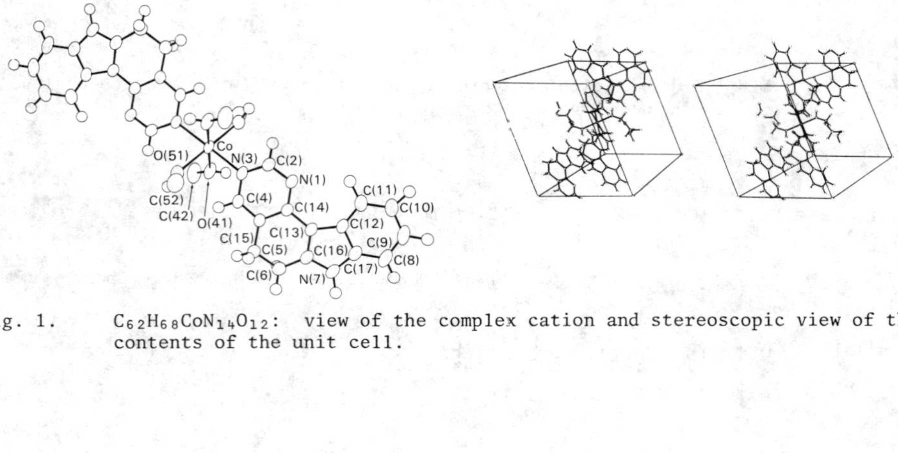

Fig. 1. $C_{62}H_{68}CoN_{14}O_{12}$: view of the complex cation and stereoscopic view of the contents of the unit cell.

(TRIS(2-DIPHENYLPHOSPHINOETHYL)AMINE-N,P,P',P'')-THIOCARBONYLCOBALT TETRAPHENYLBORATE
$C_{67}H_{62}BCoNP_3S$ $[(np_3)Co(CS)]^+BPh_4^-$

G. BIANCHINI, D. MASI, C. MEALLI, A. MELI, M. SABAT and G. SCAPACCI, 1984. J. Organometal. Chem., 273, 91-103.

Monoclinic, $P2_1/a$, a = 18.749, b = 34.494, c = 18.229 Å, β = 90.73°, Z = 8. Mo radiation, R = 0.097 for 1914 reflexions.

The crystal structure consists of two independent formula units with discrete cations and anions (Fig. 1). Both cations exhibit trigonal-bipyramidal coordination about the Co atoms, with the np_3 tripod ligands chelated through the three P atoms and the N atoms bonded in the apical position. The thiocarbonyl groups located trans to the N atoms complete the coordination sphere. Disorder in the cations prevents great accuracy in molecular geometry, but those bond distances of interest are: Co-P (2.15 to 2.23 Å), Co-N 2.03 Å in both molecules, Co-C 1.50 Å in both molecules, and C-S 1.66 and 1.76 Å in molecules I and II respectively. The Co-C-S angles in the two molecules are 176 and 174°.

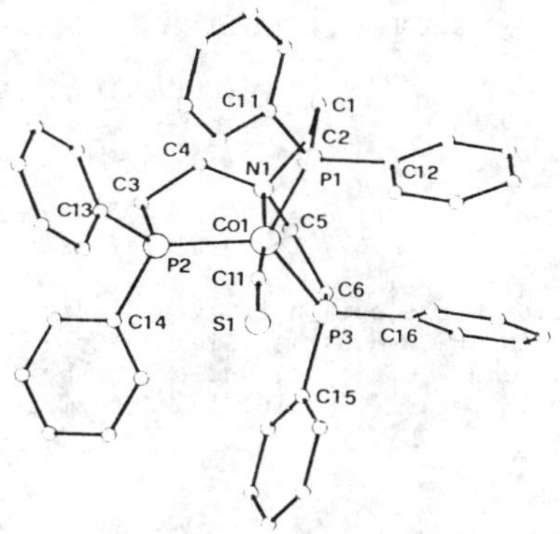

Fig. 1. Perspective view of the complex cation in one of the two independent molecules of $C_{67}H_{62}BCoNP_3S$.

{2,2'-BIS[[o-(DIPHENYLPHOSPHINO)BENZYLIDENE]AMINO]-6,6'-DIMETHYLBIPHENYL}RHODIUM(I) TETRAPHENYLBORATE METHANOL SOLVATE
$C_{76}H_{62}BN_2P_2Rh,CH_4O$ (I) $[Rh(C_{52}H_{42}N_2P_2)][B(C_6H_5)_4].CH_3OH$

{2,2'-BIS[[o-(DIPHENYLPHOSPHINO)BENZYLIDENE]AMINO]-6,6'-DIMETHYLBIPHENYL}IRIDIUM(I) TETRAPHENYLBORATE METHANOL SOLVATE
$C_{76}H_{62}BIrN_2P_2,CH_4O$ (II) $[Ir(C_{52}H_{42}N_2P_2)][B(C_6H_5)_4].CH_3OH$

T.L. MARXEN, B.J. JOHNSON, P.V. NILSSON and L.H. PIGNOLET, 1984. Inorg. Chem., 23, 4663-4670.

I. Triclinic, $P\bar{1}$, a = 13.473, b = 13.506, c = 17.424 Å, α = 100.20, β = 98.17, γ = 91.41°, Z = 2. Mo radiation, R = 0.034 for 6168 reflexions.

II. Triclinic, $P\bar{1}$, a = 13.48, b = 13.551, c = 17.438 Å, α = 100.28, β = 98.00, γ = 91.38°, Z = 2. Mo radiation, R = 0.055 for 4477 reflexions.

The two essentially isostructural compounds have approximately square-planar P_2N_2 coordination core geometries (Fig. 1) with average Rh-P and Ir-P distances of 2.222(1) and 2.244(3) Å and with average Rh-N and Ir-N distances of 2.106(3) and 2.09(1) Å, respectively. The PPNN ligand is tetradentate in both complexes, and the twisted substituted biphenyl group imposes an asymmetry on the entire complex.

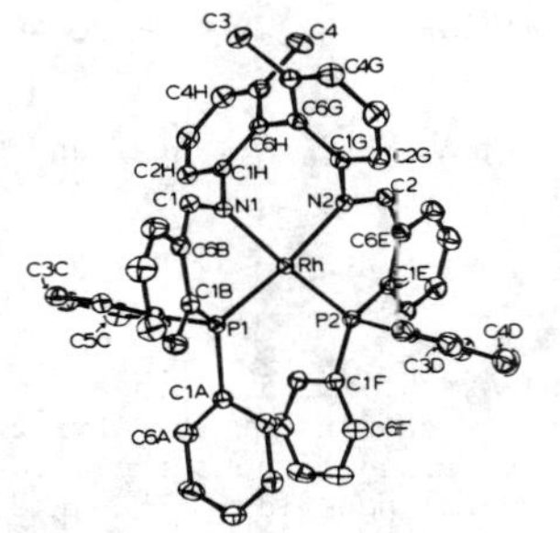

Fig. 1. The structure of the $[Rh(C_{52}H_{42}N_2P_2)]^+$ cation.

μ-OXO-BIS[THIOCYANATO(TRIS(2-PYRAZOL-1-YLETHYL)AMINE)CHROMIUM(III)] BIS(TETRAPHENYL-BORATE)

$C_{80}H_{82}B_2Cr_2N_{16}OS_2$ $[\{Cr(NCS)(N(CH_2CH_2N_2C_3H_3)_3)\}_2O][B(C_6H_5)_4]_2$

M. DI VAIRA and F. MANI, 1984. Inorg. Chem., 23, 409-412.

Triclinic, $P\bar{1}$, a = 19.098, b = 14.688, c = 13.826 Å, α = 105.83, β = 86.85, γ = 101.78°, Z = 2. Mo radiation, R = 0.077 for 3361 reflexions.

Each of the metal atoms in the cation is in a distorted octahedral environment formed by four N atoms of a $N(CH_2CH_2N_2C_3H_3)_3$ ligand (Cr-N 2.07(1)-2.14(1) Å), by the N atom of NCS (Cr-N 2.00(1) and 2.02(1) Å), and by the bridging O atom (Fig. 1). The Cr-O-Cr bridge is almost linear (176.5(6)°) and symmetrical with Cr-O 1.82(1) and 1.81(1) Å.

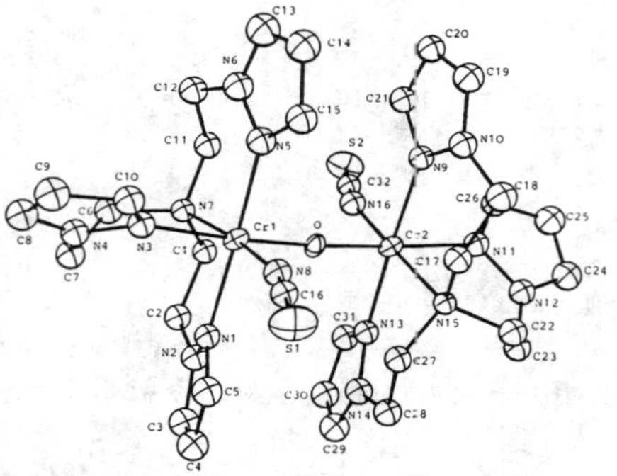

Fig. 1. The $[\{Cr(NCS)(N(CH_2CH_2N_2C_3H_3)_3)\}_2O]^{2+}$ cation.

TETRAAMMINE(METHYLENEDIPHOSPHATO)COBALT(III) HYDROCHLORIDE

$CH_{16}ClCoN_4O_6P_2$ $[Co(NH_3)_4H_2P_2O_6CH_2]^+Cl^-$

T.P. HAROMY, W.B. KNIGHT, D. DUNAWAY-MARIANO and M. SUNDARALINGAM, 1984. Inorg. Chem., 23, 2412-2415.

Monoclinic, $P2_1/c$, a = 9.259, b = 8.586, c = 14.448 Å, β = 126.23°, Z = 4. Cu radiation, R = 0.045 for 2044 reflexions.

The structure is shown in Fig. 1 along with selected geometrical parameters. The six-membered chelate ring is in a half-chair conformation.

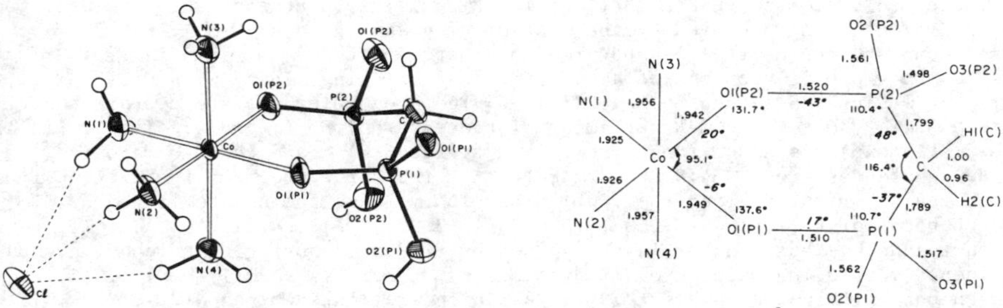

Fig. 1. The structure of $[Co(NH_3)_4H_2P_2O_6CH_2]^+Cl^-$, (σ 0.004 Å and 0.3°).

DIOXO(μ_2-SELENONO)BIS(UREA)URANIUM(VI)
$(C_2H_8N_4O_7SeU)_n$ (I)

DIOXO(μ_2-MALONAMIDE-O,O')(μ_3-SELENONO)URANIUM
$(C_3H_6N_2O_7SeU)_n$ (II)

V.E. MISTRYUKOV and Yu.N. MIKHAILOV, 1983. Koord. Khim., 9, 97-102 [Sov. J. Coord. Chem., 9, 58-63].

I. Monoclinic, $P2_1/m$, a = 9.427, b = 13.855, c = 6.556 Å, β = 105.35°, Z = 2. Mo radiation, R = 0.036 for 1573 reflexions.

II. Monoclinic, $P2_1/n$, a = 8.935, b = 10.698, c = 10.037 Å, β = 90.37°, Z = 4. Mo radiation, R = 0.036 for 1756 reflexions.

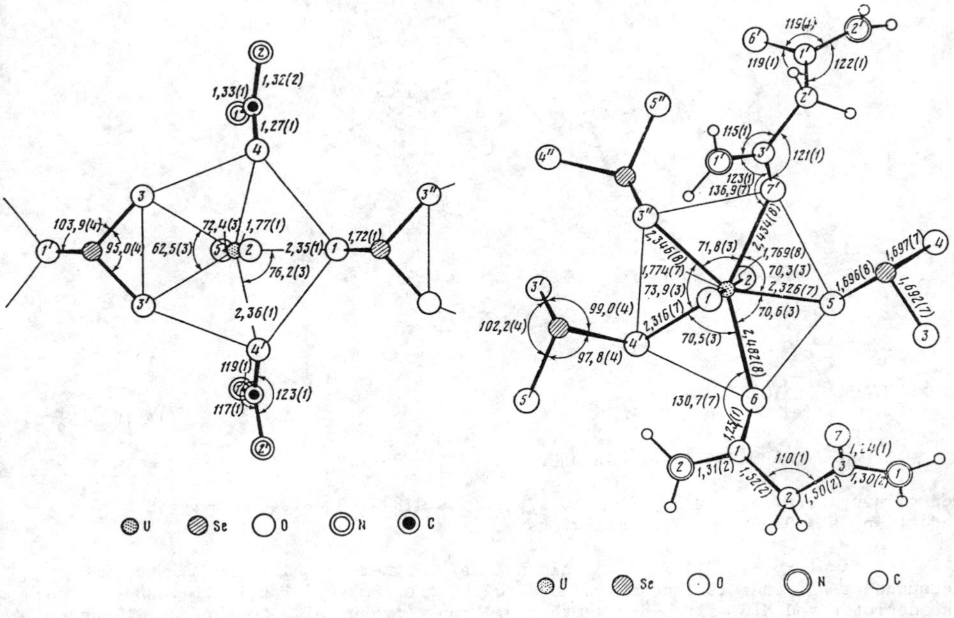

Fig. 1. The coordination polyhedra showing bond lengths and angles: $(C_2H_8N_4O_7SeU)_n$ (left) and $(C_3H_6N_2O_7SeU)_n$ (right).

The coordination polyhedra in I and II (Fig. 1) are pentagonal bipyramidal, the axial positions being occupied by the O atoms of uranyl groups (U-O 1.76-1.79 Å). The selenono and the neutral ligand are coordinated in the equatorial plane. In I the selenono group is coordinated as a tridentate bridging ligand with U-O(1)b 2.35(1), U-O(3)c 2.45(1) and Se-O 1.72(1) Å. The urea molecules are coordinated via O with U-O(4) 2.36 Å. The C-N carbamide distances are 1.33(1) and 1.34(1) Å. In II the U-O(selenono) distances range from 2.316 to 2.346(8) Å and the U-O(malonic diamide) distances are 2.434(8) and 2.482(8) Å. The structure of I is built up from polymer chains oriented along the (001) direction. These chains are linked by hydrogen bonds of the type N-H...O (O(2)...N(1) 2.98, O(3)...N(2) 2.82 Å). II, on the other hand, is built up of layers joined by tridentate bridging selenono groups and bidentate bridging malonic diamide molecules. There is an N(1)-H(2)...O(3) hydrogen bond within the layers (2.91(1) Å) which are linked by an N(2)-H(2)...O(5) hydrogen bond 2.83(1) Å.

DIBROMOBIS(DIMETHYLSULPHOXIDE)DINITROSYLMOLYBDENUM

$C_4H_{12}Br_2MoN_2O_4S_2$ $[MoBr_2(NO)_2(DMSO)_2]$

C. SCHUMACHER, F. WELLER and K. DEHNICKE, 1984. Z. Anorg. Chem., 508, 79-85.

Monoclinic, $P2_1/c$, a = 12.358, b = 8.922, c = 13.045 Å, β = 95.21°, Z = 4. Mo radiation, R = 0.038 for 1552 reflexions.

The Mo atom is octahedrally coordinated (Fig. 1) with Mo-Br 2.587, 2.597(1), Mo-N 1.804, 1.807(6), Mo-O 2.127, 2.140(4) Å.

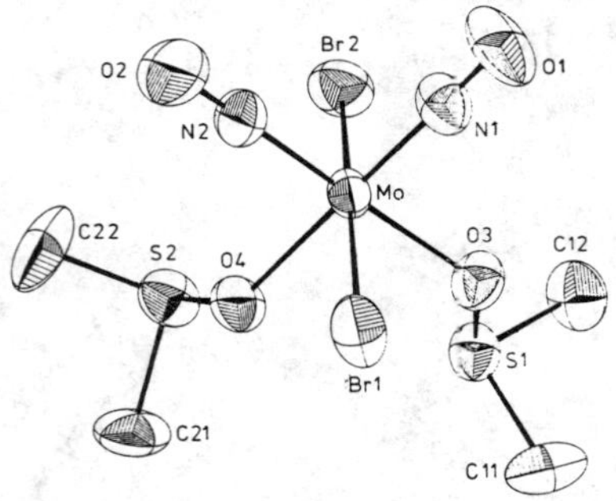

Fig. 1. A view of $[MoBr_2(NO)_2(DMSO)_2]$.

(1-METHYLURACILATO)SILVER(I)

$C_5H_5AgN_2O_2$ $[Ag(C_5H_5N_2O_2)]$

K. AOKI and W. SAENGER, 1984. Acta Cryst., C40, 775-778.

Monoclinic, C2/c, a = 12.537, b = 10.278, c = 9.918 Å, β = 104.84°, Z = 8. Mo radiation, R = 0.099 for 479 reflexions.

The complex (Fig. 1) is a one-dimensional polymer in which one of the two independent Ag^+ ions is linearly bonded to the two 1-methyluracilate ligands through the deprotonated N(3) sites and another Ag^+ ion tetrahedrally to the four ligands through the two O(2) and two O(4) sites; an N(3)-metal-bonded base-base pair, which involves an inversion centre at the metal site, is arranged in anti configuration with respect to the N-Ag-N bond and is further cross-linked by two additional metal ions with the formation of eight-membered chelate rings.

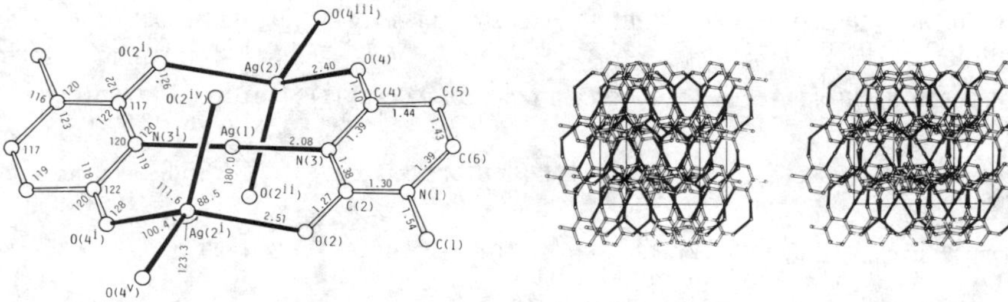

Fig. 1. $C_5H_5AgN_2O_2$: a perspective view of the basic dimeric unit with bond lengths and angles, showing the anti configuration of the unit with respect to the N-Ag-N bond and also showing the formation of the eight-membered chelate rings (left); and a stereoview of the crystal packing viewed down the a axis, with the c axis horizontal and the b axis vertical (right).

SODIUM ZINC NITRILOTRIACETATE MONOHYDRATE
$C_6H_6NNaO_6Zn,H_2O$ (I)

DISODIUM HYDROGEN NITRILOTRIACETATE
$C_6H_7NNa_2O_6$ (II)

J.D. OLIVER, B.L. BARNETT and L.C. STRICKLAND, 1984. Acta Cryst., B40, 377-381.

I. Orthorhombic, $P2_12_12_1$, a = 7.869, b = 9.783, c = 12.292 Å, Z = 4. Mo radiation, R = 0.052 for 859 reflexions.

II. Cubic, $P2_13$, a = 9.417, Z = 4. Cu radiation, R = 0.034 for 208 reflexions.

In (I) (Fig. 1) the nitrilotriacetate ion is coordinated to the Zn^{2+} ion in a tetradentate fashion. The remaining two octahedral sites about the Zn^{2+} ion are occupied by O atoms of different nitrilotriacetate ions. A comparison of the structure with other published structures identifies a common geometry, termed the 'butterfly' geometry, for transition-metal complexes with nitrilotriacetate ligands. The Na^+ ions bridge the nitrilotriacetatozinc ions thereby holding the structure intact. The water molecule is coordinated to the five-coordinate Na^+ ion. The structure of (II) (Fig. 1) is the first observation of the zwitterionic form of the nitrilotriacetate ion as a tridentate chelating ion. The $[C_6H_7NO_6Na]^-$ ions are linked together by a complex network of bridging Na^+ ions. The differences in the coordination of the nitrilotriacetate ligand to the metal ions in these structures illustrate the versatile, multidentate chelating capabilities of the ligand.

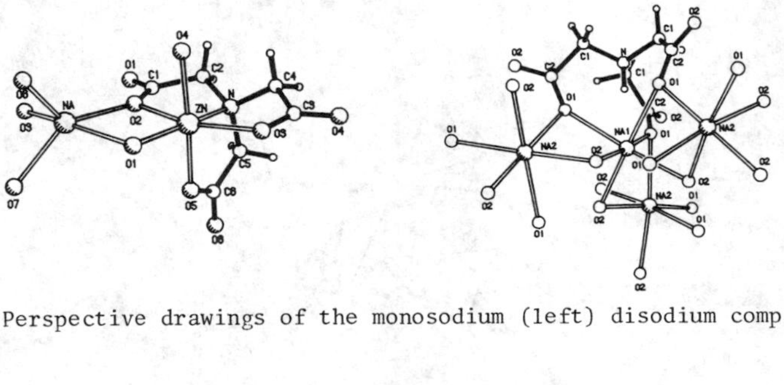

Fig. 1. Perspective drawings of the monosodium (left) disodium complexes.

DI-μ-CHLOROBIS[CHLORO-(cis-1,2-CYCLOHEXANEDIOL)COPPER(II)] DICYCLOHEXANEDIOL
$C_6H_{12}Cl_2CuO_2,C_6H_{12}O_2$ (I) $[CuCl_2(C_6H_{12}O_2)],C_6H_{12}O_2$

DI-μ-CHLOROBIS[CHLORO-(trans-1,2-CYCLOHEXANEDIOL)COPPER(II)] DICYCLOHEXANDIOL
$C_6H_{12}Cl_2CuO_2,C_6H_{12}O_2$ (II) $[CuCl_2(C_6H_{12}O_2)],C_6H_{12}O_2$

L. HILTUNEN, M. LESKELÄ, L. NIINISTÖ and R. SILLANPÄÄ, 1984. Acta Chem. Scand., A38, 201-209.

I. Monoclinic, $P2_1/n$, a = 12.542, b = 8.830, c = 14.265 Å, β = 90.24°, Z = 4. Mo radiation, R = 0.042 for 3096 reflexions.

II. Monoclinic, $P2_1/c$, a = 10.282, b = 14.675, c = 10.379 Å, β = 96.79°, Z = 4. Mo radiation, R = 0.047 for 3022 reflexions.

In both structures there is a 4+1 coordination around the metal atom (Fig. 1). In I, atoms Cl(1), Cl(2), O(1) and O(2) are situated on a plane 2.241, 2.261, 2.005 and 2.010 Å from Cu, respectively, while Cl(2') is 2.76 Å above Cu and the plane. In the trans form (II) the corresponding distances are 2.252, 2.261, 2.016, 2.002 and 2.650 Å. The $CuCl_2$(chd) units are joined to form dimers via chloro bridges and only one cyclohexanediol (chd) molecule is coordinated to the Cu(II) ion. The two uncoordinated diol molecules per dimer are held in the structure by hydrogen bonding (O(4)...O(1) 2.644, O(4)...Cl(1) 3.133 Å).

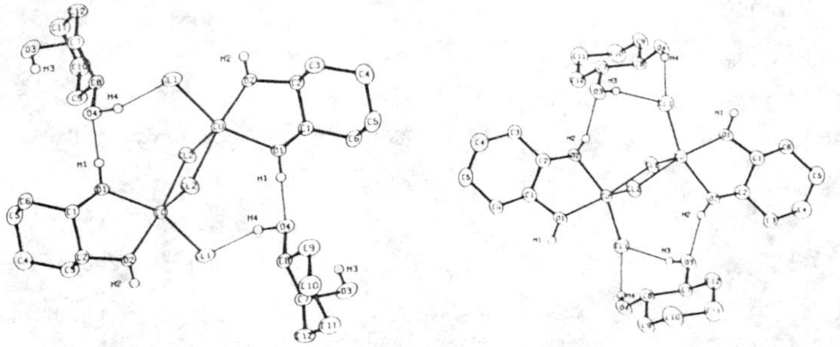

Fig. 1. Perspective views of $C_6H_{12}Cl_2CuO_2,C_6H_{12}O_2$ (cis) (left) and trans (right).

TRINITRATO-(O,O')(TRIETHYLENE GLYCOL)EUROPIUM(III)
$C_6H_{14}EuN_3O_{13}$ $Eu(NO_3)_3(C_6H_{14}O_4)$

E. FORSELLINI, U. CASTELLATO, G. TOMAT, R. GRAZIANI and P. DI BERNARDO, 1984. Acta Cryst., C40, 795-797.

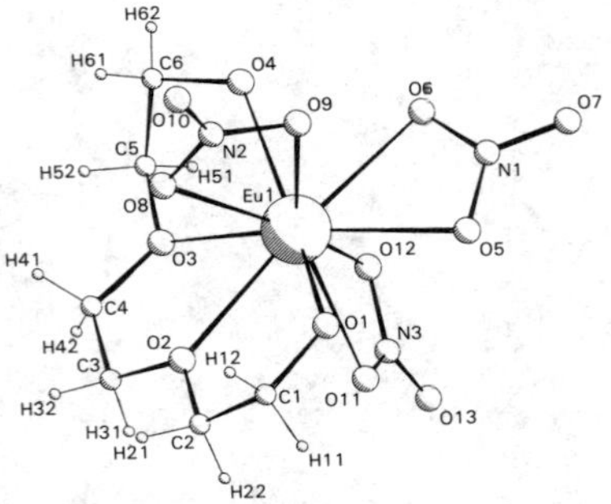

Fig. 1. Perspective view of the $Eu(NO_3)_3(C_6H_{14}O_4)$ molecule.

Monoclinic, $P2_1/c$, a = 12.342, b = 9.305, c = 14.201 Å, β = 110.88°, Dm = 2.12, Z = 4. Mo radiation, R = 0.0316 for 2286 reflexions.

In the complex, Eu is ten-coordinated, being directly bonded to the oxygen atoms of the glycol and of three bidentate nitrates (Fig. 1). The glycol molecule forms a ring-like structure around the metal together with a nitrate ion; two other nitrates are axial above and below the ring plane. Eu-O bond distances are in the range 2.433(4)-2.508(4) Å for the glycol and 2.488(5)-2.522(5) Å for the nitrato oxygen atoms.

DISODIUM OCTAKIS(μ_3-CHLORO)HEXAKIS(METHOXYMOLYBDATE) METHANOL SOLVATE
$C_6H_{18}Cl_8Mo_6Na_2,10(CH_4O)$ (I)

DISODIUM OCTAKIS(μ_3-METHOXO)HEXAKIS(METHOXYMOLYBDATE) METHANOL SOLVATE
$C_{14}H_{42}Mo_6Na_2O_{14},10(CH_4O)$ (II)

M.H. CHISHOLM, J.A. HEPPERT and J.C. HUFFMAN, 1984. Polyhedron, 3, 475-478.

I. Triclinic, $P\bar{1}$, a = 12.218, b = 10.603, c = 10.673 Å, α = 59.46, β = 76.80, γ = 84.70°, Z = 1. Mo radiation, R = 0.026 for 2915 reflexions (at -162°C)

II. Monoclinic, $P2_1/n$, a = 12.199, b = 11.500, c = 18.689 Å, β = 88.47°, Z = 2. Mo radiation, R = 0.045 for 2687 reflexions (at -160°C).

Compound I consists of $[Mo_6(\mu_3\text{-}Cl)_8(OMe)_6]^{2-}$ anions (Fig. 1) within a lattice of hydrogen bonded MeOH molecules and with each Na^+ ion being surrounded by five MeOH molecules (Mo-Mo 2.607(4), Mo-O(terminal) 2.051(29), Mo-Cl(μ_3) 2.490(6) Å. In II (Fig. 1) there are $[Mo_6(\mu_3\text{-}OMe)_8(OMe_6]^{2-}$ ions (Fig. 1) within a lattice of extensively hydrogen bonded methanol and solvated Na^+ ions (Mo-Mo 2.536(6), Mo-O(terminal) 2.130(38), Mo-O(μ_3) 2.173(13) Å).

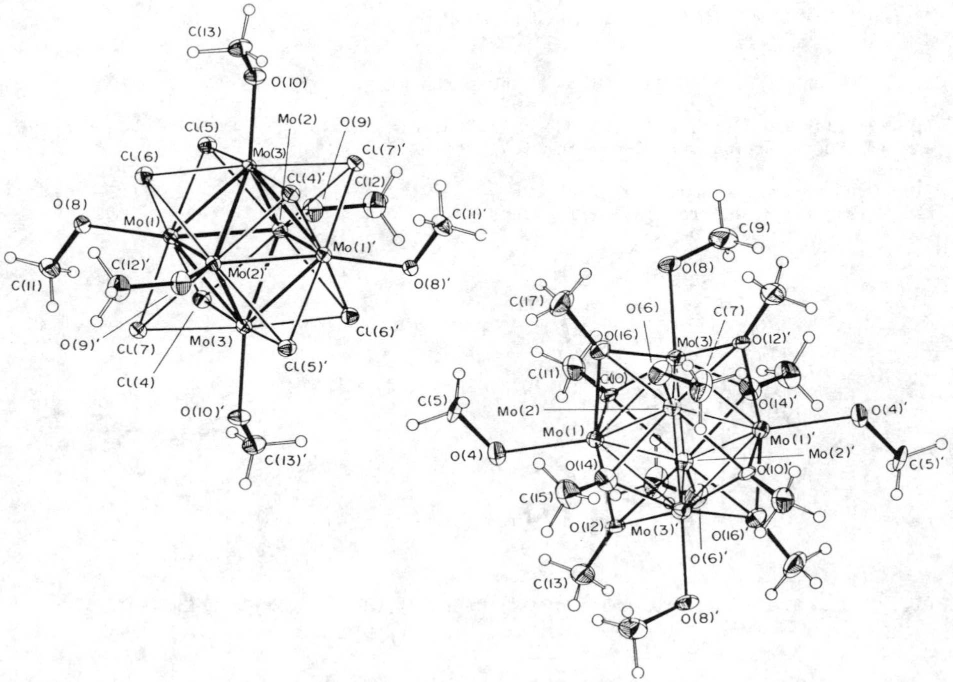

Fig. 1. The $[Mo_6(\mu_3\text{-}Cl)_8(OMe)_6]^{2-}$ (top) and $[Mo_6(\mu_3\text{-}OMe)_8(OMe)_6]^{2-}$ (bottom) ions.

catena-AQUA-DI-μ-CHLORO-CHLORO(2,6-DIMETHYL-4-PYRONE)LANTHANUM(III)
$C_7H_{10}Cl_3LaO_3$ $LaCl_3(C_7H_8O_2)(H_2O)$

C.B. CASTELLANI and V. TAZZOLI, 1984. Acta Cryst., C40, 1834-1836.

Monoclinic, $P2_1/m$, a = 7.255, b = 9.152, c = 9.781 Å, β = 103.82°, Z = 2. Mo radiation, R = 0.079 for 1734 reflexions.

The La^{3+} ion has a pentagonal bipyramidal coordination (Fig. 1) with La-O($C_7H_8O_2$) 2.33(2), La-O(H_2O) 2.50(2), La-Cl(eq) 2.90, 2.92(1), La-Cl(ax) 2.72(1) Å. The coordination polyhedra form infinite polymeric chains with $2_1/m$ symmetry by sharing Cl atoms.

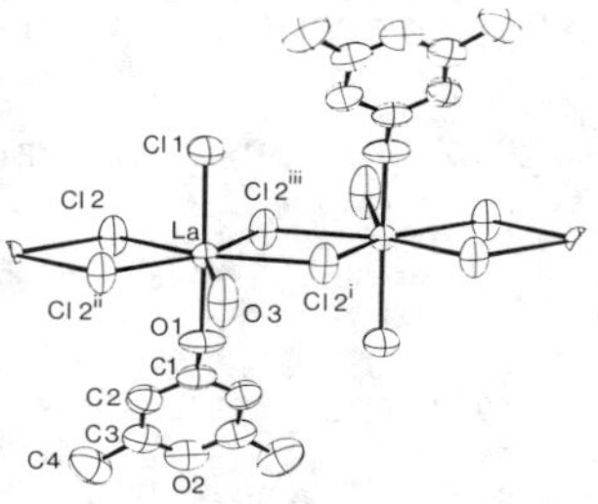

Fig. 1. The $LaCl_3(C_7H_8O_2)(H_2O)$ complex.

(μ_3-SUCCINIMIDE-N,O,O')(μ-SUCCINIMIDE-N,O)AQUADISILVER(I)
$C_8H_{10}Ag_2N_2O_5$ (I) $Ag_2(C_4H_4NO_2)_2(H_2O)$

BIS(SUCCINIMIDE-N)SILVER(I) TETRAAQUALITHIUM
$C_8H_{16}AgLiN_2O_8$ (II) $Li(H_2O)_4[Ag(C_4H_4NO_2)_2]$

J. PERRON and A.L. BEAUCHAMP, 1984. Inorg. Chem., 23, 2853-2859.

I. Monoclinic, $P2_1/c$, a = 8.031, b = 13.306, c = 10.345 Å, β = 97.16°, D_m = 2.59, Z = 4. Mo radiation, R = 0.038 for 1671 reflexions.

II. Orthorhombic, Pnma, a = 12.091, b = 8.934, c = 12.981 Å, D_m = 1.82, Z = 4. Mo radiation, R = 0.026 for 1351 reflexions.

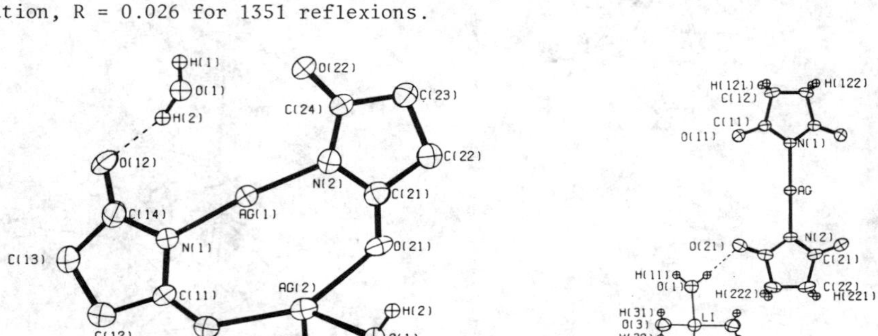

Fig. 1. The asymmetric units of $Ag_2(C_4H_4NO_2)_2(H_2O)$ (left) and $Li(H_2O)_4[Ag-C_4H_4NO_2)_2]$ (right).

Complex I (Fig. 1) has a polymeric structure which can be regarded as the Ag^+ salt of the $[Ag(C_4H_4NO_2)_2]^-$ ion, in which Ag is coordinated to two deprotonated N atoms of two succinimide anions (Ag-N 2.079(5) and 2.067(5) Å, N-Ag-N 169.0(2)°). The remaining Ag atoms are coordinated by two carbonyl O atoms of one anion, one carbonyl O atom of another anion, and a water molecule, in a roughly tetrahedral pattern (Ag-O 2.357(5)-2.448(5), Ag-OH_2 2.367(6) Å). The structure of II consists of linear, C_s symmetry, $[Ag(C_4H_4NO_2)_2]^-$ ions (Ag-N 2.088(2), 2.087(2) Å, N-Ag-N 178.48(8)°) which are H-bonded to tetrahedral $Li(OH_2)_4^+$ cations (Li-O 1.922(4)-1.981(6) Å) (Fig. 1).

BIS(μ_2-CHLORO)BIS[CHLOROBIS(2-PYRROLIDONO)COPPER(II)]
$C_8H_{14}Cl_2CuN_2O_2$

S.S. KUKALENKO, Yu. T. STRUCHKOV, S.I. SHESTAKOVA, A.G. TSYBULEVSKII, A.S. BATSANOV and E.B. NAZAROVA, 1983. Koord. Khim., 9, 312-318 [Sov. J. Coord. Chem., 9, 187-193].

Triclinic, P$\bar{1}$, a = 7.278, b = 8.862, c = 9.973 Å, α = 108.69, β = 103.38, γ = 90.72°, Z = 2. Mo radiation, R = 0.035 for 1654 reflexions.

The molecules in the crystal are joined to form dimers (Fig. 1) in which the Cu atoms have trigonal-bipyramidal coordination with one terminal Cl atom, two asymmetric bridging Cl atoms and two O atoms of the pyrrolidone ligand (Cu-Cl 2.238(1), Cu-Cl 2.320, 2.670(1), Cu-O 1.962, 1.979(3) Å). In one pyrrolidone ligand the five-membered ring has an envelope conformation while in the other this ring is essentially planar. The dimers are linked by hydrogen bonds to form chains.

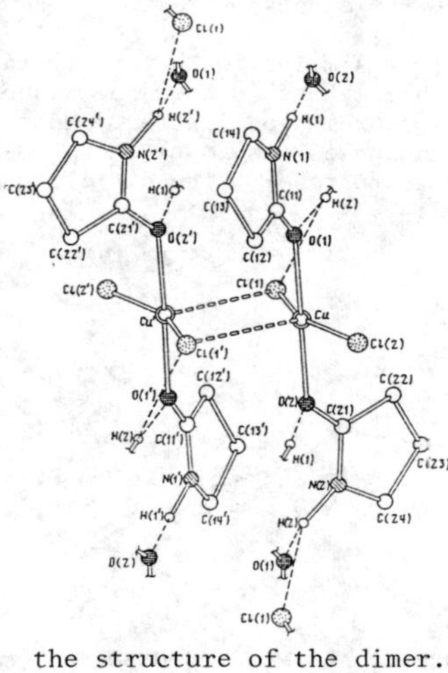

Fig. 1. $C_8H_{14}Cl_2CuN_2O_2$: the structure of the dimer.

BIS[2-AMINO-2-HYDROXYMETHYL-1,3-PROPANEDIOLATO(1-)-O,N]COPPER(II) SODIUM PERCHLORATE MONOHYDRATE
$C_8H_{20}ClCuN_2NaO_{10},H_2O$ $[Cu(C_4H_{10}NO_3)_2]NaClO_4,H_2O$

G.J.M. IVARSSON, 1984. Acta Cryst., C40, 67-69.

Monoclinic, C2/c, a = 14.2872, b = 10.9454, c = 11.1921 Å, β = 96.290°, D_m = 1.70, Z = 4. Mo radiation, R = 0.044 for 2826 reflexions.

The copper and sodium atoms lie on independent inversion centres and the chlorine of the perchlorate lies on a twofold axis. The structure consists of uncharged $Cu[NH_2C(CH_2O)(CH_2OH)_2]_2$ complexes (Fig. 1), Na^+ and ClO_4^- ions and water molecules. The Cu-coordination is square-planar (Cu-N 1.999(2), Cu-O 1.908(2) Å). The Na^+ ion is surrounded octahedrally by oxygens from the Cu-bonded ligand, the perchlorate ion and the water molecule (Na-O 2.334(2), 2.516(3) and 2.305(3) Å respectively).

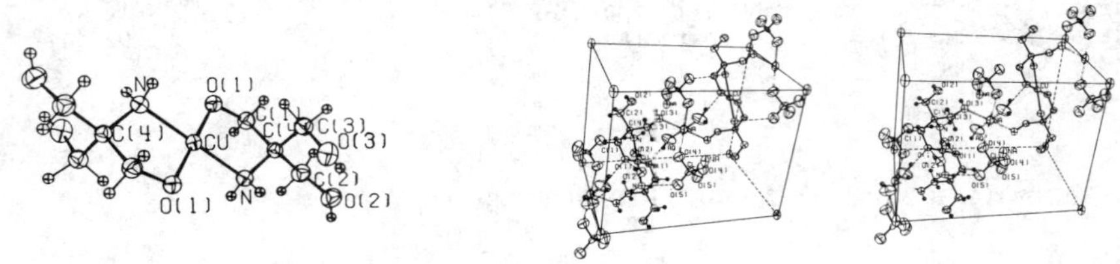

Fig. 1. A view of the $[Cu(C_4H_{10}NO_3)_2]$ complex and a stereoview of the crystal structure.

MOLYBDENUM TRIOXIDE - DIMETHYL SULPHOXIDE (FORM 2)

$C_8H_{24}Mo_3O_{10}S_4$ $Mo_3O_9.4(CH_3)_2SO$

V. McKEE and C.J. WILKINS, 1984. Acta Cryst., C40, 1678-1681.

Orthorhombic, $P2_12_12$, a = 10.446, b = 13.565, c = 7.900 Å, D_m = 2.20, Z = 2. Mo radiation, R = 0.0338 for 1508 reflexions.

This form of the material is similar to that reported previously (1) in that it contains a chain (Fig. 1) in which there is the same sequence $Mo_{tet}Mo_{oct}Mo_{oct}$ of tetrahedrally and octahedrally coordinated centres. However, the asymmetric unit is smaller and the bridging O bond angles are lower (Mo_{tet}-O-Mo_{oct} 134.6(2), Mo_{oct}-O-Mo_{oct} 145.1(4)° in the present structure). The differing chain-packing arrangements in the two forms are discussed.

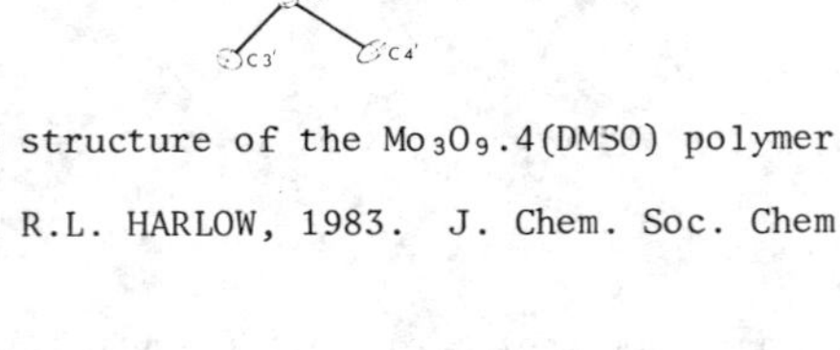

Fig. 1. The chain structure of the $Mo_3O_9.4$(DMSO) polymer.

1. E.M. McCARRON and R.L. HARLOW, 1983. J. Chem. Soc. Chem. Commun., 90-91.

DI-μ-AQUA-TETRAKIS[μ-(METHANOL)-(METHANOL)SODIUM] OCTAMOLYBDATE

$C_8H_{36}Mo_8Na_4O_{36}$ $[Na_4(CH_4O)_8(H_2O)_2][Mo_8O_{26}]$

E.M. McCARRON III and R.L. HARLOW, 1984. Acta Cryst., C40, 1140-1141.

Monoclinic, C2/m, a = 13.838, b = 17.324, c = 9.087 Å, β = 118.47°, Z = 2. Mo radiation, R = 0.027 for 2017 reflexions (at 173K).

The structure (Fig. 1) comprises apparently unique $[Na_4(\mu\text{-}CH_3OH)_4(CH_3OH)_4(\mu\text{-}H_2O)_2]^{4+}$ tetrameric cations and $Mo_8O_{26}{}^{4-}$ anions in the well characterized β conformation with extensive interactions between the two. Both the cation and the anion have crystallographic 2/m symmetry with the ion centres located at 0.5,0,0 and 0,0,0 respectively. Mo-O distances are in the range 1.695(2) to 2.431(2) Å and Na-O distances 2.333(3) to 2.421(2) Å.

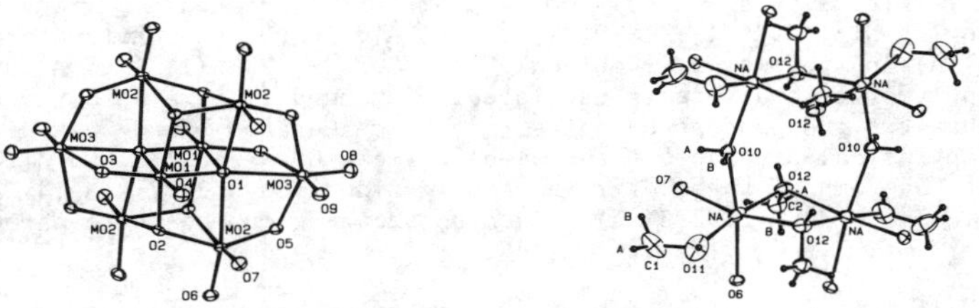

Fig. 1. $[Na_4(CH_4O)_8(H_2O)_2][Mo_8O_{26}]$: projections of the $Mo_8O_{26}{}^{4-}$ anion and $[Na_4(CH_3OH)_8(H_2O)_2]^{4+}$ cation, each of which has 2/m symmetry.

2'-DEOXYCYTIDINE-5'-MONOPHOSPHATOCADMIUM(II)
$(C_9H_{16}CdN_3O_9P)_n$ (I)

2'-DEOXYCYTIDINE-5'-MONOPHOSPHATOCADMIUM(II) TRIHYDRATE
$(C_{18}H_{29}Cd_2N_6O_{16}P_2)_n,3n(H_2O)$ (II)

K. AOKI and W. SAENGER, 1984. J. Inorg. Biochem., 20, 225-245.

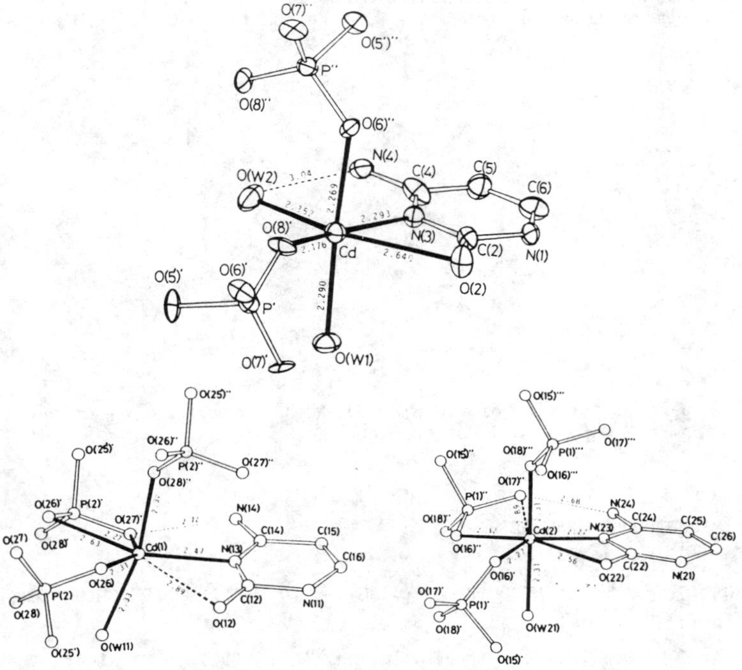

Fig. 1. Views of the Cd coordination in $C_9H_{16}CdN_3O_9P$ (top) and at the two independent moieties in $C_{18}H_{29}Cd_2N_6O_{16}P_2,3H_2O$ (bottom).

I. Orthorhombic, $P2_12_12_1$, a = 8.422, b = 24.248, c = 7.292 Å, Z = 4. Cu radiation, R = 0.052 for 1273 reflexions.

II. Monoclinic, C2, a = 30.809, b = 5.369, c = 25.126 Å, β = 127.61°, Z = 4. Cu radiation, R = 0.106 for 2287 reflexions.

Complex I (Fig. 1) has the Cd atom in a distorted octahedral arrangement to the cytosine base (Cd-N(3) 2.29(1), Cd-O(2) 2.64(1) Å), to two phosphate O atoms (Cd-O 2.27, 2.18(1) Å) and to two water molecules (Cd-C 2.29, 2.35(1) Å). A three-dimensional polymeric structure is developed. Compound II is also a three-dimensional polymer and has two independent Cd units, with each Cd atom bound in a distorted pentagonal bipyramidal arrangement to the base (mean Cd-N(3) 2.35, Cd-O(2) 2.73 Å), to four phosphate O atoms of three different nucleotides (Cd-O 2.27-2.89(4) Å) and to a water molecule (mean Cd-O 2.32(3) Å). Each phosphate group bonds to three Cd atoms.

(6-HYDROXY-4,8-DIAZAUNDECANEDIAMIDE)COPPER(II) PERCHLORATE

$C_9H_{20}Cl_2CuN_4O_{11}$ $[Cu(C_9H_{20}N_4O_3)](ClO_4)_2$

T.-J. LEE, C.-Y. HONG, S.-H. LIU, C.-S. CHUNG and T.-Y. LEE, 1984. Acta Cryst., C40, 1673-1675.

Monoclinic, $P2_1/c$, a = 13.997, b = 10.656, c = 14.254 Å, β = [?]°, D_m = 1.81, Z = 4. Mo radiation, R = 0.076 for 2195 reflexions.

The coordination geometry around the Cu atom (Fig. 1) is a tetragonally distorted octahedron comprising two O atoms from the two amide groups and two N atoms from the two amine groups in an equatorial plane, and the two O atoms of the perchlorate ions above and below this plane. Main bond lengths are Cu-O(1) 1.966(7), Cu-O(2) 1.970(7), Cu-N(2) 1.998(10), Cu-N(3) 2.008, Cu-O(11) 2.448(11), Cu-O(21) 2.641(11) Å. The two perchlorate ions are situated on opposite sides of the equatorial plane; they form intra- and intermolecular hydrogen bonds with the multidentate ligand and neighbouring molecules.

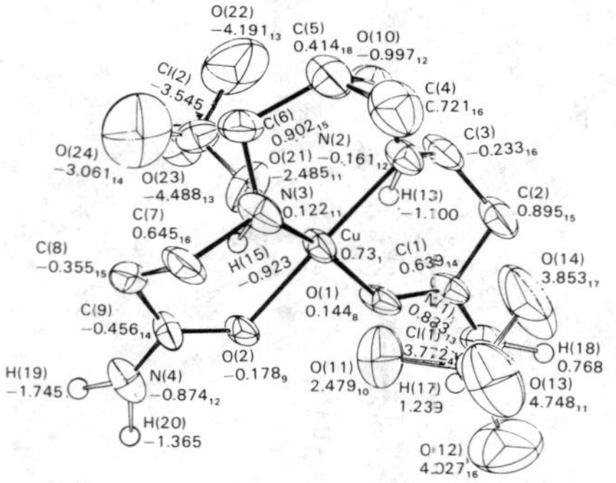

Fig. 1. A perspective view of the $C_9H_{20}Cl_2CuN_4O_{11}$ complex, showing deviations from the major coordination plane.

NONAMETHYLIMIDODIPHOSPHORAMIDE

$C_9H_{27}N_5O_2P_2$ (I)

DIOXOETHANOLBIS(NONAMETHYLIMIDODIPHOSPHORAMIDE-O,O')URANIUM DIPERCHLORATE

$C_{20}H_{60}Cl_2N_{10}O_{15}P_4U$ (II)

K. BOKOLO, A. COURTOIS, J.-J. DELPUECH, E. ELKAIM, J. PROTAS, D. RINALDI, L. RODEHUSER and P. RUBINI, 1984. J. Am. Chem. Soc., 106, 6333-6338.

I. Monoclinic, $P2_1/c$, a = 6.866, b = 13.612, c = 17.818 Å, β = 101.65°, Z = 4. Ag radiation, R = 0.04 for 2367 reflexions.

II. Orthorhombic, $P2_1mn$, a = 8.407, b = 15.464, c = 18.365 Å, Z = 2. Mo radiation, R = 0.062 for 2104 reflexions.

Views of the free ligand (NIPA) and its complex are in Fig. 1. (I) in the solid is in a trans-conformation with P(1),P(2),N(1),C(1) coplanar. The complex (II) has mirror symmetry; a UO_2^{2+} unit (U-O 1.73, 1.75(1) Å) is perpendicular to a pentagon of O atoms (two each from two NIPA ligands and one from an ethanol molecule) with U-O(1) 2.36(1), U-O(2) 2.37(1) and U-O(3) 2.40(1) Å.

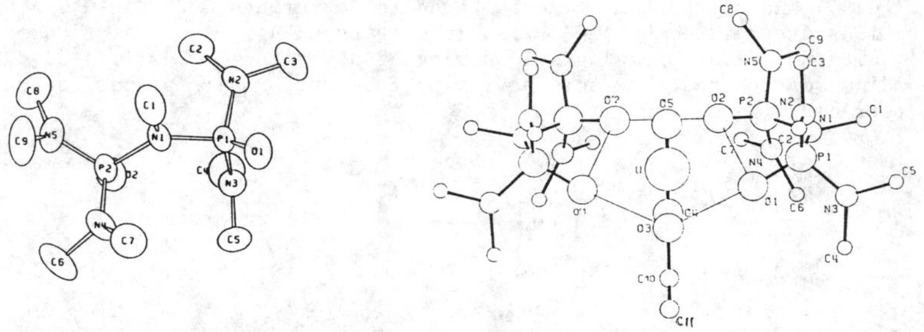

Fig. 1. A view of the NIPA ligand (I) and its metal complex (II).

BIS(GLUTARIMIDE-N,O,O')DISILVER
$C_{10}H_{12}AgN_2O_4$

J. PERRON and A.L. BEAUCHAMP, 1984. Can. J. Chem., 62, 1287-1291.

Monoclinic, C2/c, a = 13.497, b = 9.503, c = 9.600 Å, β = 92.30°, D_m = 2.32, Z = 4. Mo radiation, R = 0.027 for 1086 reflexions.

In this polymeric system (Fig. 1) one Ag atom (Ag(1)) lies on an inversion centre and the other is on a twofold axis. Ag(1) has linear coordination and Ag(2) is tetrahedrally coordinated.

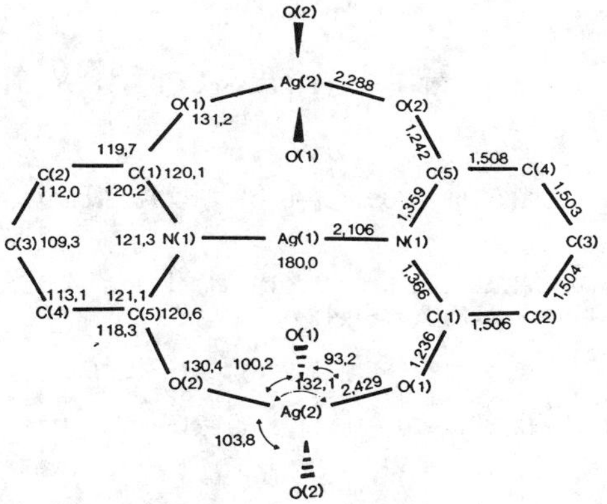

Fig. 1. Part of the structural motif for the $C_{10}H_{12}AgN_2O_4$ polymer.

BIS(μ_3-1-METHYLURACILATO)BIS(cis-DIAMMINEPLATINUM(II))DINITRATODISILVER DINITRATE DIHYDRATE
$C_{10}H_{22}Ag_2N_{12}O_{16}Pt_2,2H_2O$

U. THEWALT, D. NEUGEBAUER and B. LIPPERT, 1984. Inorg. Chem., 23, 1713-1718.

Monoclinic, C2/c, a = 13.810, b = 16.279, c = 11.871 Å, β = 95.58°, D_m = 3.00, Z = 4. Mo radiation, R = 0.051 for 1797 reflexions (at -40°C).

The structure is shown in Fig. 1. The ligands are arranged in head-tail fashion. Each 1-MeU ligand bridges two cis-$(NH_3)_2Pt$ through N(3) and O(4) (Pt-N 2.026(11), Pt-O 2.037(10) Å) and at the same time is bound to Ag through O(2) (Ag-O 2.386(11) Å). Adjacent cations are related by a C_2 axis, leading to an Ag...Ag separation of 3.954(3) Å, with NO_3^- ions bridging neighboring Ag atoms (Ag-O 2.421(13)-2.744(12) Å). The Pt coordination spheres show some deviation from pure square planar toward a distorted tetrahedral geometry.

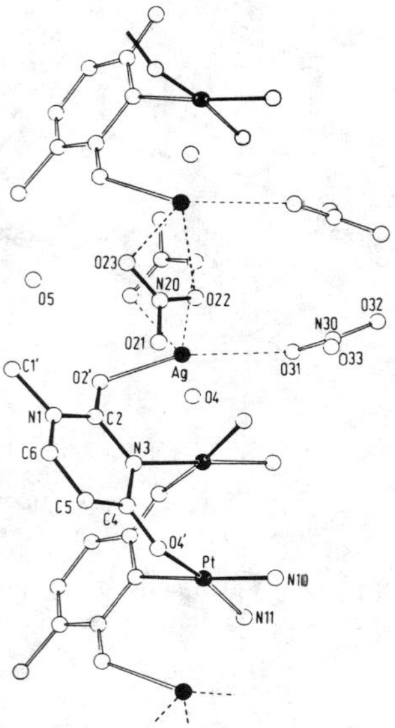

Fig. 1. Section of the cis-$[(NH_3)_2Pt(1\text{-}MeU)Ag]_2(NO_3)_4,2H_2O$ structure.

CHLOROPENTAKIS(DIMETHYLSULPHOXIDE)IRON (μ_2-OXO)BIS(TRICHLOROIRON)
$C_{10}H_{30}Cl_7Fe_3O_6S_5$ $[FeCl(DMSO)_5]^{2+}[Fe_2Cl_6O]^{2-}$

V.I. PONOMAREV, L.D. ARUTYUNYAN and L.O. ATOVMYAN, 1984. Kristallografiya, 29. 910-922 [Sov. Phys. Crystallogr., 29, 538-544].

Monoclinic, $P2_1/b$, a = 12.962, (12.861, 12.708), b = 16.839, (16.673, 16.335), c = 15.712, (15.614, 15.470 Å), γ = 81.47, (81.27, 79.97°), Z = 4. Mo radiation, R = 0.058 (0.064, 0.051) for 2422, (2670, 3531) reflexions (at 343, 300 and 103K respectively).

At 103K the crystal structure is ordered and is built up of octahedral cations and binuclear anions. The DMSO molecules are bound to the metal atom (Fig. 1) by their O atoms (Fe(1)-O(18) 2.066, Fe(1)-O(19) 2.032, Fe(1)-O(20) 2.003 Å). The Fe(1)-Cl(10) distance is 2.291 Å. O-S and S-CH_3 bond lengths range from 1.530 to 1.557 Å and from 1.779 to 1.794 Å respectively. In the binuclear anion the tetrahedrally-coordinated Fe atoms are linked by a bridging O atom. The Fe-O-Fe angle is 146.5° and the Fe-Cl distances are 2.214 to 2.231 Å while the Fe-O distances are 1.774 and

1.778 Å. At 300K disorder in one of the DMSO ligands of the cation appears (Fig. 1), and at 343K there is a tendency to disorder in the two adjacent DMSO groups lying in the plane of the octahedron perpendicular to the Fe-Cl bond.

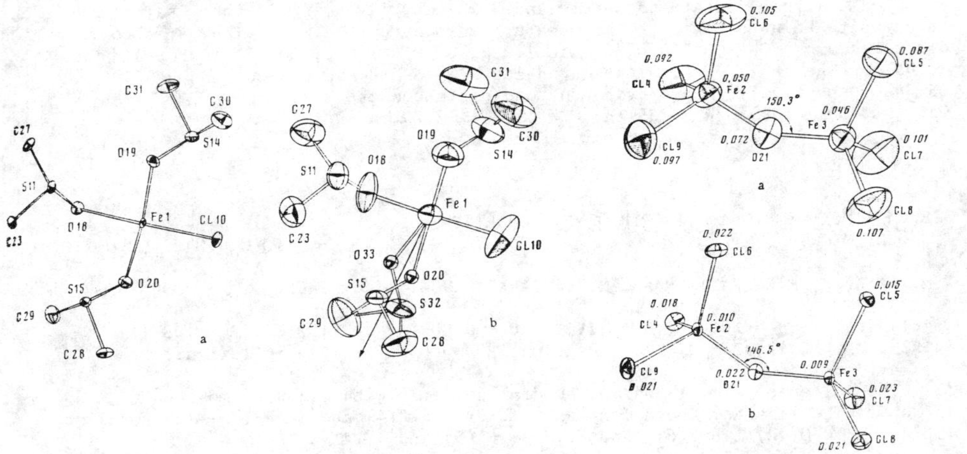

Fig. 1. Equatorial fragment of octahedral cation at 103K (a) and 300K (b) (left), the anion at 300K (a) and 103K (b) (right).

catena-μ-(CARBONATO-O,O':O'')-2,2'-DIPYRIDYLAMINECOPPER(II) TRIHYDRATE

$C_{11}H_9CuN_3O_3,3H_2O$ $[Cu(C_{10}H_9N_3)CO_3],3H_2O$

J. SLETTEN, 1984. Acta Chem. Scand., A38, 491-496.

Monoclinic, $P2_1/c$, a = 11.243, b = 7.144, c = 17.343 Å, β = 101.15°, D_m = 1.70, Z = 4. Mo radiation, R = 0.023 for 2162 reflexions.

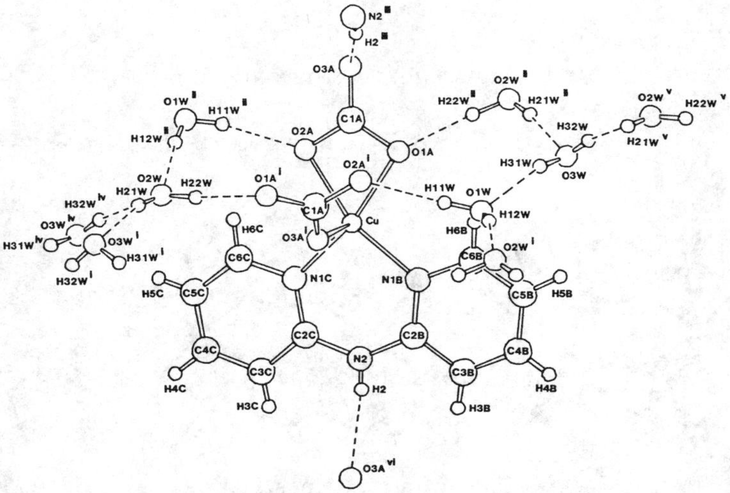

Fig. 1. The molecular unit and its environment in $C_{11}H_9CuN_3O_3,3H_2O$.

The structure consists of polymeric chains in which the carbonate ions serve as bridges between metal ions. The carbonate is bidentately coordinated to one Cu(II) ion and monodentately to another. The Cu coordination is slightly distorted square pyramidal, with bidentate carbonate and 2,2'-dipyridylamine in the basal plane, the apical site being occupied by monodentate carbonate (Fig. 1). The metal coordination distances are: Cu-N1B 1.970, Cu-N1C 1.975, Cu-O1A 1.980, Cu-O2A 1.986, and Cu-O3A' 2.295 Å. The chelate coordination angles N1C-Cu-N1B and O1A-Cu-O2A are 93.41 and 65.84° respectively. Some hydrogen bond distances are $N(2)^{iii}$...O(3A) 2.798, O(1A)...$O2W^{ii}$ 2.790, O1W...O3W 2.733 and O3W...$O2W^{ii}$ 2.752 Å.

DIBROMOETHOXOOXOBIS(4-NITROPYRIDINE)TECHNETIUM(V)

$C_{12}H_{13}Br_2N_4O_6Tc$ $[O(C_2H_5O)Br_2(C_5H_4N_2O_2)_2Tc]$

P.H. FACKLER, M.E. KASTNER and M.J. CLARKE, 1984. Inorg. Chem., 23, 3968-3972.

Triclinic, $P\bar{1}$, a = 10.820, b = 11.146, c = 9.006 Å, α = 102.82, β = 108.11, γ = 64.53°, D_m = 2.14, Z = 2. Cu radiation, R = 0.043 for 1514 reflexions.

The molecule (Fig. 1) has octahedral geometry with trans-bromo and trans-4-nitropyridine ligands. Bond lengths: Tc-Br 2.570(4) and 2.538(4), Tc-O(oxo) 1.684(6), Tc-O(OR) 1.855(6), and Tc-N 2.147(8) and 2.141(8) Å.

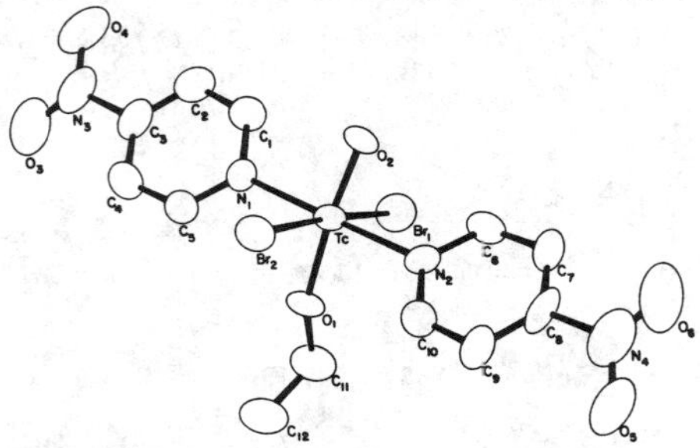

Fig. 1. The $[O(C_2H_5O)Br_2(C_5H_4N_2O_2)_2Tc]$ molecule.

BIS(μ_2-1,3-DIOXANE-O,O')-1,3-DIOXANE(HEXAFLUOROARSENIC-F)SILVER

$(C_{12}H_{24}AgAsF_6O_6)_n$

P.G. JONES, H.W. ROESKY, J. LIEBERMANN and G.M. SHELDRICK, 1984. Z. Naturforsch., B39, 1729-1731.

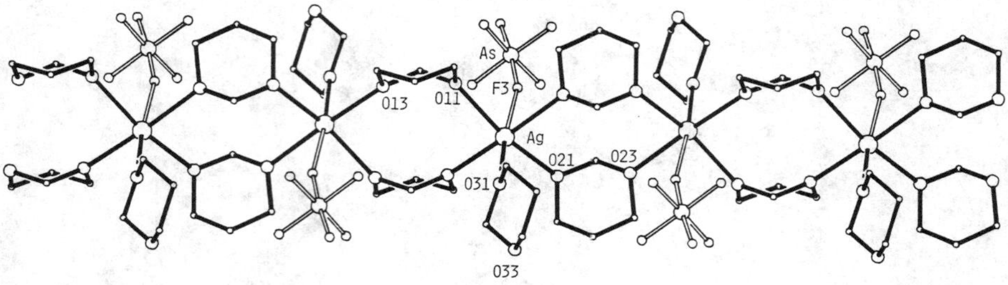

Fig. 1. The polymer structure $(C_{12}H_{24}AgAsF_6O_6)_n$.

Monoclinic, $P2_1/n$, a = 9.399, b = 18.140, c = 11.476 Å, β = 92.98°, Z = 4. Mo radiation, R = 0.049 for 2825 reflexions.

The compound is a chain polymer with two bridging and one terminal dioxane ligands (Fig. 1). Ag-O bond lengths are in the range 2.422-2.489(6) Å and a weak Ag...F contact (2.697(7) Å) completes a distorted octahedral coordination at silver.

DICHLORO-18-CROWN-6-MERCURY
$C_{12}H_{24}Cl_2HgO_6$ (I)

DICHLORO-18-CROWN-6-CADMIUM
$C_{12}H_{24}CdCl_2O_6$ (II)

C.R. PAIGE and M.F. RICHARDSON, 1984. Can. J. Chem., 62, 332-335.

I. Rhombohedral, $R\bar{3}$, a = 7.746 Å, α = 95.26°, D_m = 1.87, Z = 1. Mo radiation, R = 0.026 for 543 reflexions.

II. Rhombohedral, $R\bar{3}$, a = 7.712 Å, α = 95.57°, D_m = 1.61, Z = 1. Mo radiation, R = 0.022 for 534 reflexions.

In these isomorphous structures (Fig. 1) the metal coordination is hexagonal bipyramidal, with the chlorine atoms axial and the 18-crown-6 equatorial. The complexes are required to have $\bar{3}$ symmetry but have almost $\bar{3}m$ symmetry. Principal bond lengths are Hg-Cl 2.314(1), Hg-O 2.825(4), Cd-Cl 2.3645(5), Cd-O 2.752(1) Å.

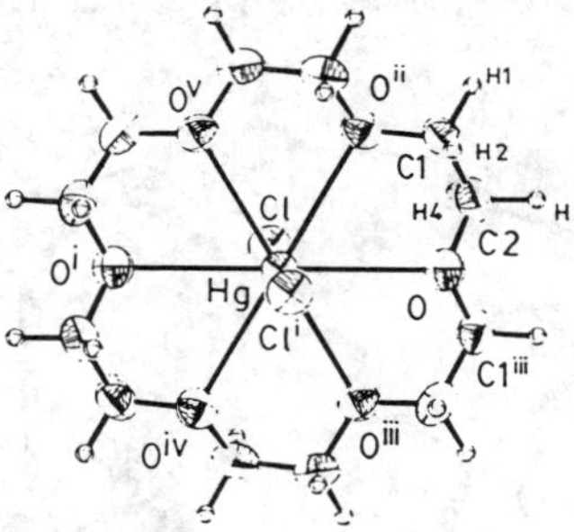

Fig. 1. A molecule of $HgCl_2$.18-crown-6.

BIS(1,5-DITHIACYCLOOCTAN-3-OL-O,S,S')NICKEL(II) DIPERCHLORATE DIMETHANOL
$C_{12}H_{24}Cl_2NiO_{10}S_4, 2CH_4O$ (I) $[Ni(C_6H_{12}OS_2)_2](ClO_4)_2, 2CH_3OH$

BIS(1,5-DITHIACYCLOOCTAN-3-OL-O,S,S')IRON(II) DIPERCHLORATE
$C_{12}H_{24}Cl_2FeO_{10}S_4$ (I) $[Fe(C_6H_{12}OS_2)_2](ClO_4)_2$

M.M. OLMSTEAD, W.K. MUSKER and R.M. KESSLER, 1984. Acta Cryst., C40, 1172-1175.

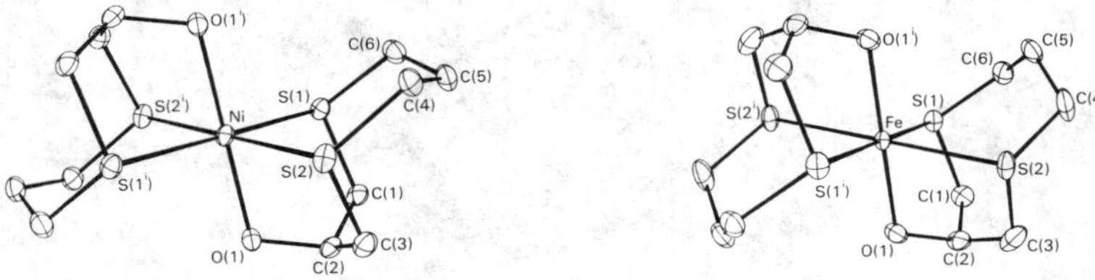

Fig. 1. Perspective views of the complex cations in $[Ni(C_6H_{12}OS_2)_2](ClO_4)_2, 2CH_3OH$ (I) and $[Fe(C_6H_{12}OS_2)_2](ClO_4)_2$ (II)

I. Monoclinic, $P2_1/c$, a = 10.734, b = 11.853, c = 10.392 Å, β = 103.62°, Z = 2. Mo radiation, R = 0.037 for 1960 reflexions (at 140K).

II. Monoclinic, $P2_1/c$, a = 7.990, b = 8.573, c = 15.681 Å, β = 95.50°, Z = 2. Mo radiation, R = 0.046 for 1601 reflexions (at 140K).

The nickel complex (I) is centrosymmetric with Ni-S 2.398(1) and 2.396(1), and Ni-O 2.046(3) Å (Fig. 1). A molecule of methanol present in the lattice is hydrogen-bonded to the hydroxyl group of the ligand and to an oxygen atom of an uncoordinated perchlorate iron. The iron complex (II) (Fig. 1) resembles the nickel complex except that the thioether-metal bonds are much longer: Fe-S 2.565(1) and 2.514(1), and Fe-O 2.062(3) Å.

BIS(1-OXA-4,7-DIAZACYCLONONANE-O(1),N(4),N(7))NICKEL DIPERCHLORATE

$C_{12}H_{28}Cl_2N_4NiO_{10}$ $C_{12}H_{28}N_4O_2Ni.2(ClO_4)$

J.C.A. BOEYENS, R.D. HANCOCK and V.J. THOM, 1984. J. Cryst. Spectrosc. Res., 14, 261-268.

Monoclinic, Cm, a = 16.392, b = 8.792, c = 7.304 Å, β = 99.70°, D_m = 1.65, Z = 2. Mo radiation, R = 0.089 for 831 reflexions.

The cation is disordered about the mirror plane in space group Cm; the disorder was successfully resolved and a view of one of the two components is shown in Fig. 1 along with a summary of the dimensions. The Ni-coordination geometry is octahedral and the cation has a pseudo threefold axis.

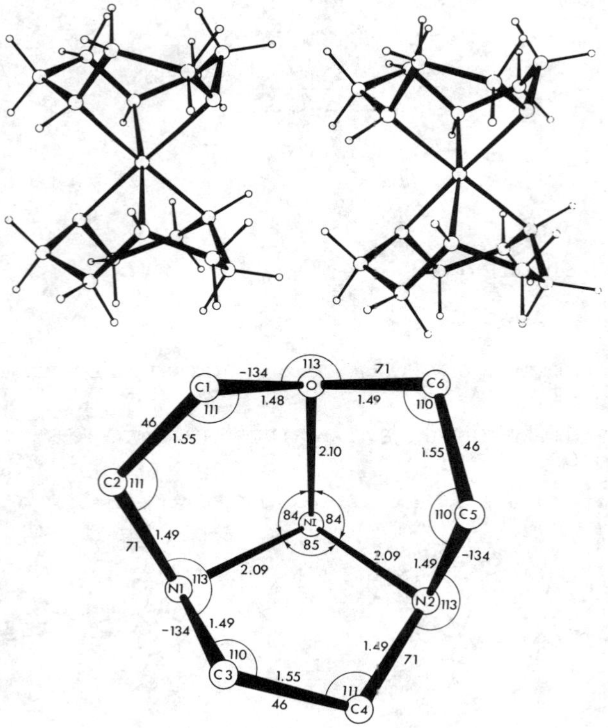

Fig. 1. A stereoview of the $(C_{12}H_{28}N_4O_2Ni)^{2+}$ cation and some dimensions.

BIS[(μ_2-CHLORO)CHLORO(PINACOL-O,O')COPPER(II)]
$C_{12}H_{28}Cl_4Cu_2O_4$ (I) $[CuCl_2(HOC(CH_3)_2C(CH_3)_2OH)]_2$

DICHLORO(PINACOL-O,O')COPPER(II) PINACOL
$C_6H_{14}Cl_2CuO_2,C_6H_{14}O_2$ (II) $[CuCl_2(HOC(CH_3)_2C(CH_3)_2OH)],C_6H_{14}O_2$

R. SILLANPÄÄ, 1984. Inorg. Chim. Acta, 82, 75-80.

I. Orthorhombic, Pbcn, a = 20.710, b = 11.401, c = 17.469 Å, D_m = 1.62, Z = 8. Mo radiation, R = 0.037 for 2359 reflexions.

II. Monoclinic, C2/c, a = 18.878, b = 12.115, c = 7.541 Å, β = 104.94°, D_m = 1.46, Z = 4. Mo radiation, R = 0.045 for 2070 reflexions.

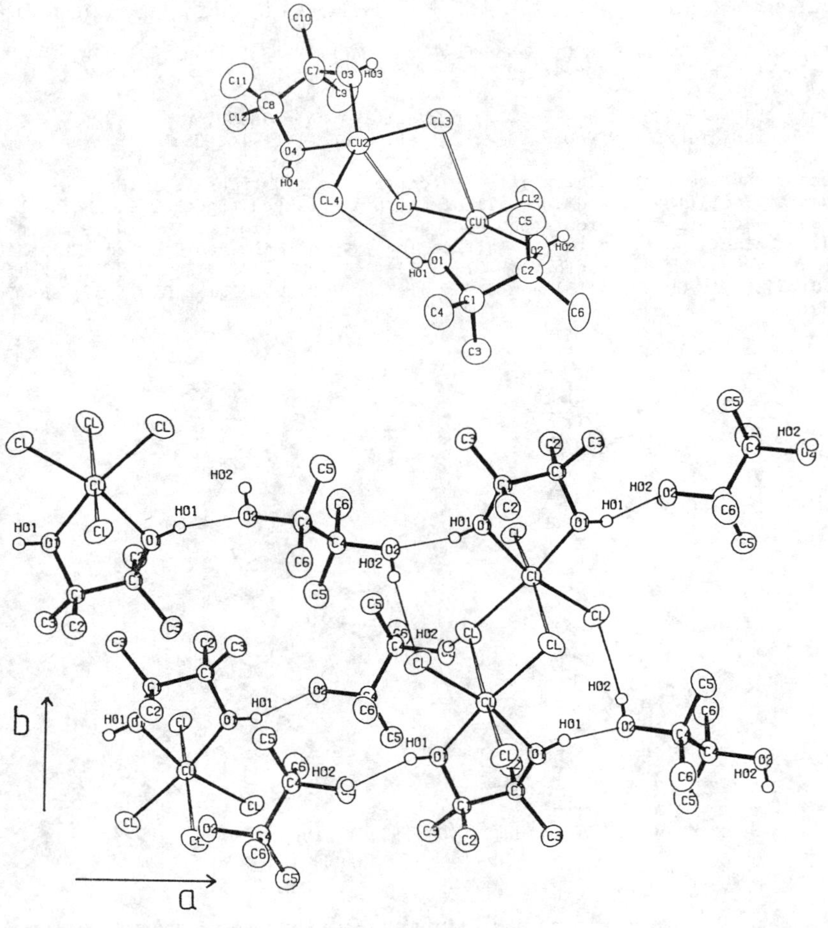

Fig. 1. Molecular structure of $C_{12}H_{28}Cl_4Cu_2O_4$ (top), and molecular packing in $C_6H_{14}Cl_2CuO_2,C_6H_{14}O_2$ (bottom).

In I the dimeric molecules consist of two five-coordinate Cu(II) ions with two different CuO_2Cl_2Cl' chromophores (Fig. 1). The symmetry round one Cu atom (Cu(1)) can be described as a distorted square based pyramid and that round Cu(2) as a distorted trigonal bipyramid. Selected distances are: Cu(1)-Cl(3) 3.080, Cu(1)-Cl(1) 2.218, Cu(1)-Cl(2) 2.201, Cu(1)-O(1) 1.946, Cu(1)-O(2) 1.974, Cu(2)-Cl(3) 2.231,

Cu(2)-Cl(4) 2.258, Cu(2)-Cl(1) 2.664, Cu(2)-O(3) 1.972 and Cu(2)-O(4) 1.976 Å. One O-H...Cl intramolecular hydrogen bond exists (O...Cl 3.168 Å) and three such intermolecular bonds, (O...Cl 3.143, 3.082 and 3.159 Å). The structure of II is a polymer (Fig. 1). The Cu atom is surrounded by two O atoms of one pinacol molecule (Cu-O 1.991 Å) and by two Cl^- ions (Cu-Cl 2.223 Å). Two extra Cl^- ions from adjacent $CuCl_2(H_2pin)$ units are at 3.223 Å. The second pinacol molecule is uncoordinated but has an anti conformation and connects $CuCl_2(H_2pin)$ units together by hydrogen bonds (O(1)...O(2) and O(2)...Cl' are 2.727 and 3.236 Å respectively).

trans-DICHLORODIAQUABIS(TETRAHYDROFURAN)VANADIUM(III) TRICHLORO(TETRAHYDROFURAN)-ZINC(II) TETRAHYDROFURAN SOLVATE

$C_{12}H_{28}Cl_5O_5VZn,C_4H_8O$ $[VCl_2(C_4H_8O)_2(H_2O)_2][ZnCl_3(C_4H_8O)].C_4H_8O$

K. FOLTING, J.C. HUFFMAN, R.L. BANSEMER, K.G. CAULTON, J.L. MARTIN and P.D. SMITH, 1984. Inorg. Chem., 23, 4589-4592.

Triclinic, P$\bar{1}$, a = 8.330, b = 20.222, c = 8.161 Å, α = 98.50, β = 97.82, γ = 99.85°, Z = 2. Mo radiation, R = 0.032 for 2874 reflexions (at -162°C).

The structure (Fig. 1) consists of centrosymmetric trans-$VCl_2(THF)_2(H_2O)_2^+$, tetrahedral $ZnCl_3(THF)^-$, and THF, all of which are linked together in chains by hydrogen bonding from the four aquo hydrogens of the cation to the three choro ligands on zinc as well as to the lattice THF oxygen. All coordinated THF's are planar at oxygen. Bond lengths: V-Cl 2.3292(14) and 2.3301(13), V-O(THF) 2.0135(26) and 2.0125(26), V-OH_2 2.034(3) and 2.035(3), Zn-Cl 2.2265(15)-2.2507(16), and Zn-O 2.025(3) Å.

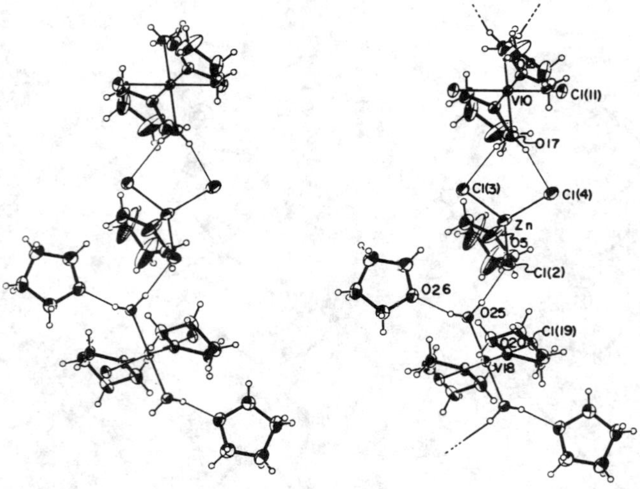

Fig. 1. Stereoview of $[VCl_2(THF)_2(H_2O)_2][ZnCl_3(THF)].THF$.

[N,N'-BIS(5-AMINO-3-HYDROXYPENTYL)MALONAMIDE-N,N',N'',N''',O,O']COPPER(II) TRIHYDRATE

$C_{13}H_{24}Cu_2N_4O_4,3(H_2O)$ (I)

BIS(μ_2-AQUA)({[N,N'-BIS(2-HYDROXYETHYL)MALONAMIDE-N,N',O,O']COPPER(II)}-O,O')TETRAAQUA-BARIUM] TETRAHYDRATE

$C_{14}H_{40}Ba_2Cu_2N_4O_{18},4(H_2O)$ (II)

T. TOKI, M. MIKURIYA, H. ŌKAWA, I. MURASE and S. KIDA, 1984. Bull. Chem. Soc. Japan, 57, 2098-2105.

I. Monoclinic, $P2_1/n$, a = 22.61, b = 8.890, c = 9.496 Å, β = 93.00°, D_m = 1.66, Z = 4. Mo radiation, R = 0.067 for 2596 reflexions.

II. Triclinic, $P\bar{1}$, a = 9.642, b = 11.374, c = 8.761 Å, α = 96.93, β = 112.27, γ = 102.83°, D_m = 2.02, Z = 2. Mo radiation, R = 0.080 for 3184 reflexions.

Compound I consists of an alkoxo-oxygen bridged binuclear molecule in which the organic ligand functions as a sexadentate $N_2O_2N_2$ donor set (Fig. 1). The N_2O_2 donor sets are planar with Cu(1) and Cu(2) 0.081 and 0.018 Å respectively from the relevant N_2O_2 plane. The dihedral angle between the planes is 7.4°. The Cu-O bond lengths (1.910-1.940 Å) and the Cu-N(amine) bond lengths (1.978 and 1.990 Å) are normal for in-plane coordination. The Cu-N(amide) distances are 1.930 and 1.940 Å. The water molecules participate in intermolecular hydrogen bonds (OW(1)...O(3) 2.728, OW(2)... O(2) 2.739, OW(2)...N(4) 2.429 Å) and do not coordinate to the metal atoms. In II, two water molecules lie between two Ba atoms to form a Cu-Ba-Ba-Cu tetranuclear entity (Fig. 1) about a crystallographic inversion centre. Each Cu atom has square planar coordination geometry with Cu-O bond lengths of 1.918 and 1.955 Å and Cu-N bond lengths of 1.943 and 1.923 Å. The six-membered Cu,N(1),C(3),C(4),C(5),N(2) chelate ring is planar and the two five-membered Cu,O(1),C(1),C(2),N(1) and Cu,O(4),C(7),C(6),N(2) chelate rings assume a gauche conformation. The Ba-O distances vary from 2.741 to 2.960 Å and the coordination about the Ba atom is a distorted square antiprism. The Ba-O(1) and Ba-O(4) lengths are substantially shorter than the Ba-O (water) distances.

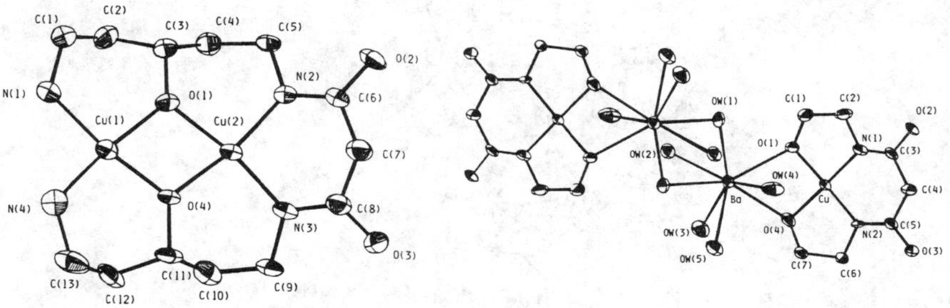

Fig. 1. Perspective views of $C_{13}H_{24}Cu_2N_4O_4,3H_2O$ (left) and $C_{14}H_{40}Ba_2Cu_2N_4O_{18}$ (right).

BIS(2,6-DIMETHYL-4H-PYRAN-4-ONE)DINITRATOZINC
$C_{14}H_{16}N_2O_{10}Zn$ $[Zn(NO_3)_2(C_7H_8O_2)_2]$

C.J. BROWN and J.F.P. LEWIS, 1984. Acta Cryst., C40, 368-370.

Monoclinic, A2/a, a = 16.786, b = 12.922, c = 9.089 Å, β = 106.80°, D_m = 1.536, Z = 4. Cu radiation, R = 0.043 for 1500 reflexions.

The Zn atom (Fig. 1) lies on a twofold axis and has distorted tetrahedral coordination (Zn-O(1) 1.979(2), Zn-O(2) 2.037(2) Å, O-Zn-O 87.0-118.8(1)°).

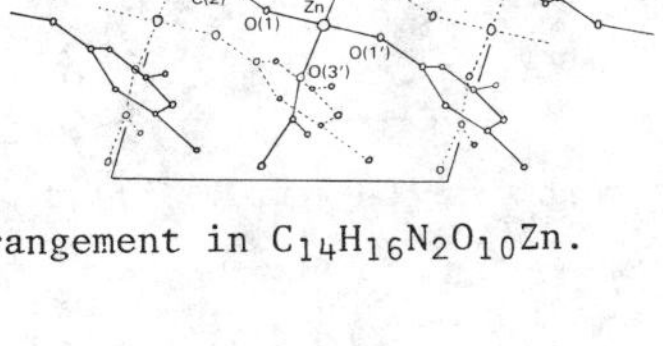

Fig. 1. The packing arrangement in $C_{14}H_{16}N_2O_{10}Zn$.

TETRAKIS(μ_2-CHLORO)(trans-1,2-CYCLOHEXANEDIOL)BIS(TETRAHYDROFURAN)DICOPPER
$C_{14}H_{28}Cl_4Cu_2O_4$ $[CuCl_2(C_6H_{10}(OH)_2)(\mu\text{-}Cl)_2Cu(C_4H_8O)_2]$

R. SILLANPÄÄ, T. NORTIA and L. HILTUNEN, 1984. Inorg. Chim. Acta, 83, 111-114.

Monoclinic, C2/c, a = 16.561, b = 13.217, c = 13.290 Å, β = 134.05°, D_m = 1.68, Z = 4. Mo radiation, R = 0.047 for 1873 reflexions.

The structure consists of alternating $CuCl_2$(trans-cyclohexanediol) and $CuCl_2(THF)_2$ units which are joined by two chloride bridges (Cu(1)-Cl(1) 2.88, Cu(1)-Cl(2) 2.726, Cu(2)-Cl(1) 2.350, Cu(2)-Cl(2) 2.240 Å) and one O-H...O(THF) hydrogen bond (O(1)...O(2) 2.681 Å) (Fig. 1). Both Cu(II) ions are six-coordinate. Cu(1) lies on a twofold axis of symmetry and Cu(2) on a centre of symmetry. The Cu(1)-O(1) distance is 1.996 Å and the Cu(2)-O(2) distance is 2.719 Å. The diol molecule has a chair conformation and the THF molecule exhibits an envelope conformation with C(4) as the 'flap' atom.

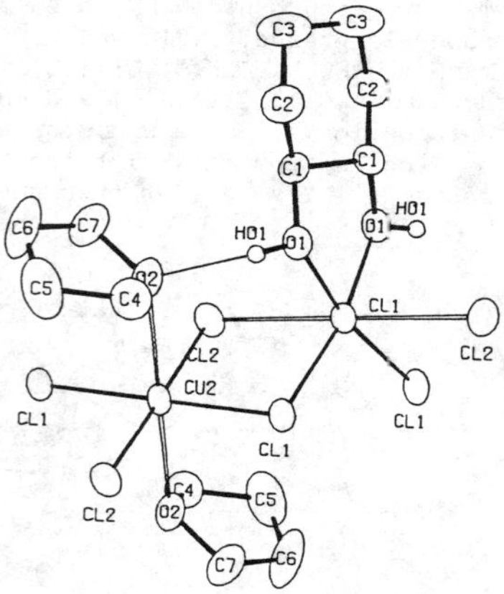

Fig. 1. Molecular structure of $C_{14}H_{28}Cl_4Cu_2O_4$.

AQUA(N,N'-TRIMETHYLENE-DI-2-PYRIDINECARBOXAMIDE)COPPER(II) DIHYDRATE (α FORM)
$C_{15}H_{16}CuN_4O_3,2(H_2O)$ (I)

BIS(CHLORO(N,N'-TRIMETHYLENE-DI-2-PYRIDINECARBOXAMIDE)COPPER(II)) SULPHATE OCTAHYDRATE
$C_{30}H_{32}Cl_2Cu_2N_8O_8S,8(H_2O)$ (II)

BIS(SULPHATO(N,N'-TRIMETHYLENE-DI-2-PYRIDINECARBOXAMIDE)COPPER(II)) TRIDECAHYDRATE
$C_{30}H_{32}Cu_2N_8O_{12}S_2,13(H_2O)$ (III)

S. TSUBOYAMA, T. SAKURAI, K. KOBAYASHI, N. AZUMA, Y. KAJIKAWA and K. ISHIZU, 1984. Acta Cryst., B40, 466-473.

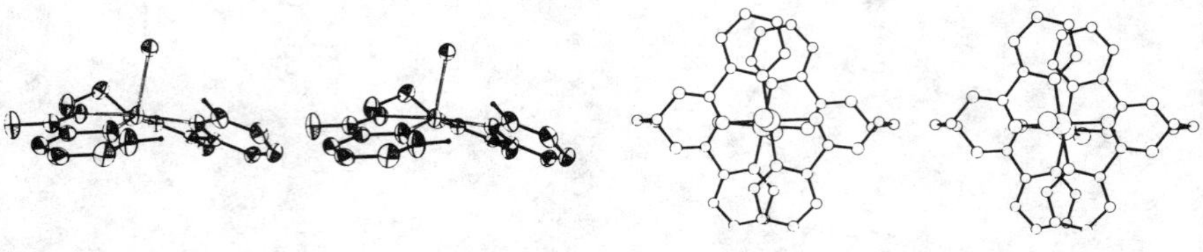

Fig. 1. Stereoviews of the monomeric Cu complex (I) and of the dimeric complex (II).

I. Monoclinic, $P2_1/c$, a = 11.730, b = 13.097, c = 12.668 Å, β = 119.59°, D_m = 1.59, Z = 4. Mo radiation, R = 0.046 for 2279 reflexions.

II. Monoclinic, $P2_1/c$, a = 14.679, b = 13.070, c = 23.958 Å, β = 111.56°, D_m = 1.53, Z = 4. Mo radiation, R = 0.061 for 5415 reflexions.

III. Orthorhombic, $C222_1$, a = 15.42, b = 24.92, c = 25.73 Å, D_m = 1.52, Z = 8. Mo radiation, R = 0.20 for 910 reflexions.

In the monomeric complex (I) (Fig. 1) prepared in an alkaline solution, the Cu^{II} ion is surrounded by four N atoms of the two deprotonated amide and two pyridine moieties, and a water molecule at the apex, yielding a distorted square-pyramidal structure, with average Cu-N(amide) 1.950, Cu-N(pyridine) 2.048, Cu-O 2.369(1) Å. Both complex cations in the dimeric complexes (II) and (III) obtained in acidic solution have approximate D_2 symmetry, the Cu^{II} ions being surrounded by the two N in pyridine rings and two amide O atoms in each half of two picolinoyl groups in an approximate plane, and two Cl^- ions and SO_4^{2-} ions occupying axial positions respectively. Some bond lengths for II are: Cu-N(pyridine) 1.969, Cu-O 1.981, Cu-Cl 2.498 Å. There are two independent molecules of III in the asymmetric unit; the structure is not well defined but is similar to II, the main differences between II and III being in the mode of packing.

BIS(METHYL-N,N-DIETHYLCARBAMYLMETHYLENEPHOSPHONATO)DYSPROSIUM THIOCYANATE
$C_{15}H_{30}DyN_3O_8P_2S$ $Dy[O_2P(OCH_3)CH_2C(O)N(C_2H_5)_2]_2(NCS)$

S.M. BOWEN, E.N. DUESLER and R.T. PAINE, 1984. Inorg. Chim. Acta, 84, 221-227.

Monoclinic, $P2_1/c$, a = 13.282, b = 19.168, c = 9.648 Å, β = 90.09°, D_m = 1.78, Z = 4. Mo radiation, R = 0.047 for 3727 reflexions.

The Dy atom is seven coordinate (a pentagonal bipyramid) and bonded in a bidentate fashion to two anionic phosphonate ligands through the carbonyl O atoms and one of two phosphonate O atoms (Fig. 1). In addition each Dy atom is coordinated to two phosphonate O atoms from two neighbouring complexes and to the N atom of a thiocyanate ion. This yields a two-dimensional polymeric structure. Some important bond distances include Dy-NCS 2.433 Å, Dy-O(carbonyl) (average) 2.39 Å, Dy-O(equat. phosphoryl) (average) 2.303 Å, Dy-O(axial phosphoryl) (average) 2.25 Å, P-O (phosphoryl) (average) 1.493 Å and C-O(carbonyl) (average) 1.25 Å.

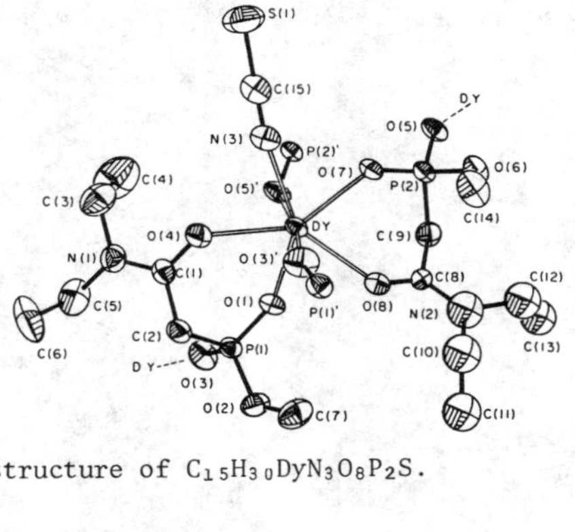

Fig. 1. Molecular structure of $C_{15}H_{30}DyN_3O_8P_2S$.

BIS(3-HYDROXYPHENYLHYDRAZIDE-N,O)DIISOTHIOCYANATOCOBALT(II) MONOHYDRATE
$C_{16}H_{16}CoN_6O_4S_2,H_2O$

I. MATSICHEK and L.A. ASLANOV, 1983. Koord. Khim., 9, 970-976 [Sov. J. Coord. Chem., 9, 555-561].

Triclinic, $P\bar{1}$, a = 7.706, b = 8.318, c = 9.576 Å, α = 101.47, β = 107.23, γ = 108.01°, Z = 1. Mo radiation, R = 0.031 for 2408 reflexions.

The structure (Fig. 1) is composed of discrete complex molecules linked by a highly developed system of hydrogen bonds involving water molecules and van der Waals interactions. The Co atom is located at a centre of symmetry and is surrounded octahedrally by the N and O atoms of two chelating hydrazide ligands (Co-N 2.179(2), Co-O 2.092(1) Å, N-Co-O 76.40(6)°) and the N atoms of two thiocyanate groups (Co-N 2.103(2) Å, Co-N-C 170.9(2)°, N-C 1.152(3), C-S 1.645(2) Å, NSC 178.8(2)°). The five-membered chelate ring takes an envelope form. In the hydrazide ligand the phenyl ring forms a dihedral angle of 14° with the plane of the -CCON- carbamide grouping.

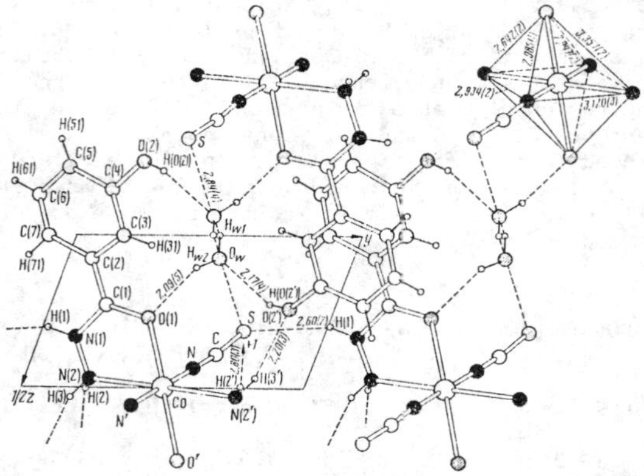

Fig. 1. $C_{16}H_{16}CoN_6O_4S_2,H_2O$: projection of the structure onto the (Oyz) plane.

(o-CATECHOLATO-O,O')(η^5-PENTAMETHYLCYCLOPENTADIENYL)COBALT
$C_{16}H_{19}CoO_2$ $[\eta^5-C_5(CH_3)_5]Co(OOC_6H_4)$

E.J. MILLER, A.L. RHEINGOLD and T.B. BRILL, 1984. J. Organometal. Chem., 273, 377-383.

Orthorhombic, Pnma, a = 12.942, b = 12.902, c = 8.543 Å, Z = 4. Mo radiation, R = 0.038 for 1416 reflexions.

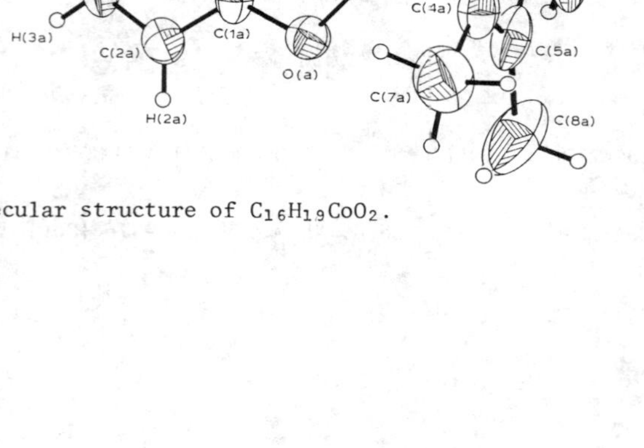

Fig. 1. Molecular structure of $C_{16}H_{19}CoO_2$.

The molecule has crystallographic m symmetry, the mirror plane bisecting the phenylene ring and the C(4)-C(4a) bond of the Cp ring (Fig. 1). The metallacycle ring is planar and is coplanar with the phenylene ring (dihedral angle 1.6°). The O-Co-O angle is 86.9° and the Co-O distance 1.837 Å. The cyclopentadienyl ring is planar and forms a dihedral angle of 91.7° with the metallacycle fragment. The Cp ring contains two short (1.412 Å) and three longer (average 1.430 Å) bonds, consistent with a slight preference for diolefin bonding. The $O_2C_6H_4$ fragment is best described as a catecholate with a C-O bond distance of 1.338 Å.

DINITRATOBIS(PYRIDOXINE)COPPER(II)
$C_{16}H_{22}CuN_4O_{12}$ (I)

(μ_2-PYRIDOXINE)PYRIDOXINECOPPER(II) NITRATE HYDRATE
$(C_{16}H_{21}CuN_3O_9)_n$,2.5n(H_2O) (II)

V.Kh. SABIROV, A.S. BATSANOV, Yu.T. STRUCHKOV and M.A. AZIZOV, 1983. Koord. Khim., 9, 1701-1707 [Sov. J. Coord. Chem., 9, 969-975].

I. Triclinic, P$\bar{1}$, a = 10.483, b = 6.928, c = 8.198 Å, α = 73.92, β = 67.62, γ = 97.81°, Z = 1. Cu radiation, R = 0.061 for 1663 reflexions.

II. Monoclinic, $P2_1/n$, a = 14.839, b = 10.977, c = 13.028 Å, β = 99.78°, Z = 4. Cu radiation, R = 0.061 for 2035 reflexions.

The structure of I (Fig. 1) is built up of discrete centrosymmetric molecules in which the coordination polyhedron of Cu is an extended tetragonal bipyramid and the monodentate NO_3 groups are axial. The structure of II comprises ...Cu^{2+}(PN)(μ-PN-H)$^-$ ... cationic polymer chains (PN = pyridoxine). The coordination about Cu is trigonal bipyramidal. The PN and PN-H ligands chelate the Cu atoms by means of the deprotonated phenol group and the adjacent CH_2OH group.

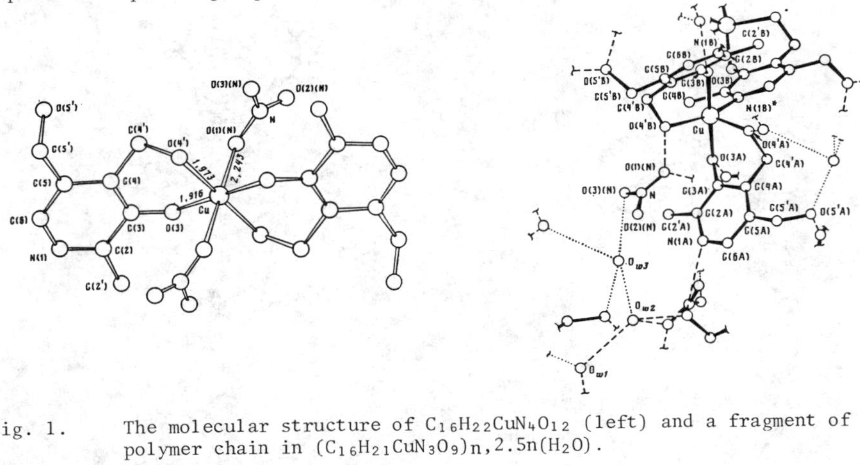

Fig. 1. The molecular structure of $C_{16}H_{22}CuN_4O_{12}$ (left) and a fragment of the polymer chain in $(C_{16}H_{21}CuN_3O_9)_n$,2.5n(H_2O).

DIOXOBIS(ISOPROPOXY)(2,2'-BIPYRIDINE)MOLYBDENUM(VI) HEMITOLUENE SOLVATE
$C_{16}H_{22}MoN_2O_4$,0.5C_7H_8 (I) $MoO_2(OC_3H_7)_2(C_{10}H_8N_2)$,0.5$CH_3C_6H_5$

(μ_3-OXO)(μ_3-ISOPROPOXY)TRIS(μ-ISOPROPOXY)HEXAKIS(ISOPROPOXY)TRIMOLYBDENUM
$C_{30}H_{70}Mo_3O_{11}$ (II) $Mo_3O(OC_3H_7)_{10}$

(μ_3-OXO)(μ_3-NEOPENTOXY)TRIS(μ-NEOPENTOXY)HEXAKIS(NEOPENTOXY)TRIMOLYBDENUM DICHLORO-METHANE SOLVATE
$C_{50}H_{110}Mo_3O_{11}$,0.33(CH_2Cl_2) (III) $Mo_3O(OC_5H_{11})_{10}$,0.33CH_2Cl_2

TETRAKIS(μ-OXO)HEXAKIS(μ-ISOPROPOXY)HEXAKIS(ISOPROPOXY)HEXAKIS(OXOMOLYBDENUM)
$C_{36}H_{84}Mo_6O_{22}$ (IV) $Mo_6O_{10}(OC_3H_7)_{12}$

M.H. CHISHOLM, K. FOLTING, J.C. HUFFMAN and C.C. KIRKPATRICK, 1984. Inorg. Chem., 23, 1021-1037.

I. Triclinic, $P\bar{1}$, a = 13.907, b = 8.413, c = 19.999 Å, α = 111.02, β = 71.37, γ = 88.98°, Z = 4. Mo radiation, R = 0.067 for 4021 reflexions (at -161°C).

II. Triclinic, $P\bar{1}$, a = 21.274, b = 21.808, c = 10.207 Å, α = 98.69, β = 92.92, γ = 118.03°, Z = 4. Mo radiation, R = 0.070 for 7711 reflexions (at -162°C).

III. Orthorhombic, Pbcn, a = 35.557, b = 18.969, c = 19.342 Å, Z = 8. Mo radiation, R = 0.075 for 2830 reflexions (at -161°C).

IV. Triclinic, $P\bar{1}$, a = 13.082, b = 11.478, c = 9.760 Å, α = 106.40, β = 91.85, γ = 99.81°, Z = 1. Mo radiation, R = 0.027 for 4388 reflexions (at -162°C).

The structures of I-IV are shown in Fig. 1. The two virtually identical, independent, molecules of I have approximate C_{2v} symmetry with cis oxo ligands and trans (O-i-Pr) ligands. Bond lengths in I include: Mo=O 1.71, Mo-O 1.93, and Mo-N 2.34 Å (mean values). Molecules of II and III both have the same basic Mo_3O_{11} skeleton. Important bond lengths for II and III are: Mo-Mo 2.523(6)-2.539(2), Mo-(μ_3-O) 2.026(30)-2.068(8), Mo-(μ_3-OR) 2.153(32)-2.238(33), Mo-(μ-OR) 1.965(28)-2.054(7), Mo-OR 1.842(30)-1.912(7) (trans to μ_3-OR), 1.930(25)-1.960(26) Å (trans to μ_3-O). In the centrosymmetric hexanuclear molecule of IV there are a pair of Mo-Mo bonds (Mo(1)-Mo(2) 2.585(1) Å). The terminal Mo(3) atoms have a +6 charge while the other four are +5. The four central Mo atoms form a chain of fused square-pyramidal units while the outer Mo atoms have approximate octahedral geometry. Bond lengths in IV are: Mo=O 1.671(2)-1.691(2), Mo-(μ-O) 1.921(2)-1.939(2), Mo-(μ-OR) 2.044(2)-2.197(2), and Mo-OR 1.806(2)-1.919(2) Å. O(12) may be semibridging (Mo(2)...O(12) 2.88(1) Å.

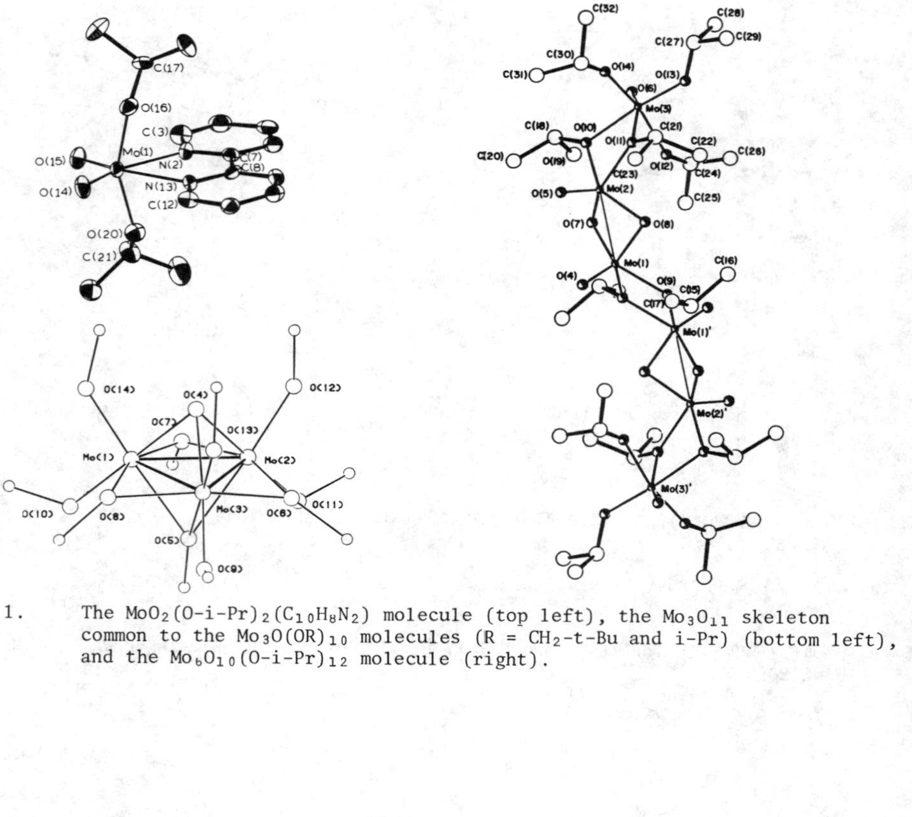

Fig. 1. The $MoO_2(O\text{-}i\text{-}Pr)_2(C_{10}H_8N_2)$ molecule (top left), the Mo_3O_{11} skeleton common to the $Mo_3O(OR)_{10}$ molecules (R = CH_2-t-Bu and i-Pr) (bottom left), and the $Mo_6O_{10}(O\text{-}i\text{-}Pr)_{12}$ molecule (right).

1,4,10,13-TETRAOXA-7,16-DIAZACYCLOOCTADECANE-7,16-DIACETATOCOPPER(II) COPPER(II) DICHLORIDE METHANOL SOLVATE

$C_{16}H_{28}Cl_2Cu_2N_2O_8,CH_4O$ $[Cu(C_{16}H_{28}N_2O_8)],CuCl_2,CH_3OH$

P. GLUZIŃSKI, J.W. KRAJEWSKI, Z. URBAŃCZYK-LIPKOWSKA, G.D. ANDREETTI and G. BOCELLI, 1984. Acta Cryst., C40, 778-781.

Monoclinic, $P2_1/n$, a = 16.711, b = 9.671, c = 15.829 Å, β = 111.00°, Z = 4. Cu radiation, R = 0.078 for 1877 reflexions.

In the complex, both Cu atoms have coordination spheres of irregular octahedral shapes (Fig. 1). The first (inside the ligand cavity) is coordinated symmetrically to pairs of N, ether O, and carbonyl O atoms, with Cu-N 2.056(8) and 2.065(8), Cu-O(ether) 2.781(8) and 2.701(8) and Cu-O(carbonyl) 1.899(5) and 1.911(6) Å. The second (of $CuCl_2$) is coordinated to two Cl and four carbonyl O atoms of neighbouring ligands, with Cu-Cl 2.214(3) and 2.202(4), Cu-O 1.983(6), 2.725(5), 1.977(6) and 2.759(6) Å, thus joining the molecules into chains parallel to [101]. Additional complexation by $CuCl_2$ results in an inequality in N configurations and loss of symmetry in the ligand. The structure is compared with that of 1,4,10,13-tetraoxa-7,16-diazacyclooctadecane-7,16-diacetatocopper(II) itself (1).

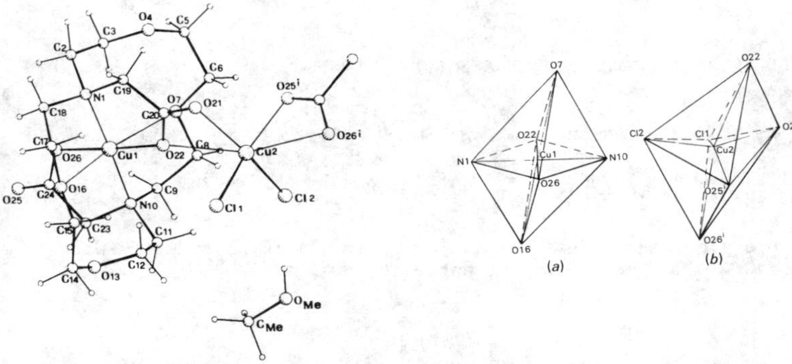

Fig. 1. $C_{16}H_{28}Cl_2Cu_2N_2O_8,CH_3OH$: projection of the structure on the ab plane and shapes of coordination polyhedra for (a) Cu(1), and (b) Cu(2).

1. Structure Reports, 49B, 368, 1314.

BIS(μ_2-ETHOXO)BIS(μ_2-PINACOLATO-O,O')TETRACHLORODITUNGSTEN(IV)

$C_{16}H_{34}Cl_4O_6W_2$ (I)

BIS(μ_2-ETHOXO)BIS(μ_2-3,4-DIMETHYLHEXANE-3,4-DIOLATO-O,O')TETRACHLORODITUNGSTEN(IV)

$C_{20}H_{42}Cl_4O_6W_2$ (II)

L.B. ANDERSON, F.A. COTTON, D. DEMARCO, L.R. FALVELLO, S.M. TETRICK and R.A. WALTON, 1984. J. Am. Chem. Soc., 106, 4743-4749.

I. Monoclinic, $P2_1/n$, a = 8.681, b = 15.743, c = 9.235 Å, β = 90.98°, Z = 2. Mo radiation, R = 0.0375 for 1753 reflexions.

II. Monoclinic, $P2_1/n$, a = 10.660, b = 12.311, c = 11.268 Å, β = 102.23°, Z = 2. Mo radiation, R = 0.0364 for 1603 reflexions.

Both molecules (Fig. 1) have crystallographic inversion symmetry and similar molecular dimensions. Details for I (with data for II in parentheses) are: W-W 2.701(1) (2.691(1)), W1-O1 2.062(5) (2.048(6)), W1-O1' 2.036(5) (2.039(7)), W1-O2 1.827(5) (1.822(7)), W1-O3' 1.815(5) (1.812(7)), W-Cl1 2.359(2) (2.365(3)), W-Cl2 2.368(2) (2.355(3)) Å. For molecule II there is some disorder involving the β-carbon of the ethyl group. The conformation of the six-membered $W_2O_2C_2$ ring is unusual (Fig. 1) and indicates strain.

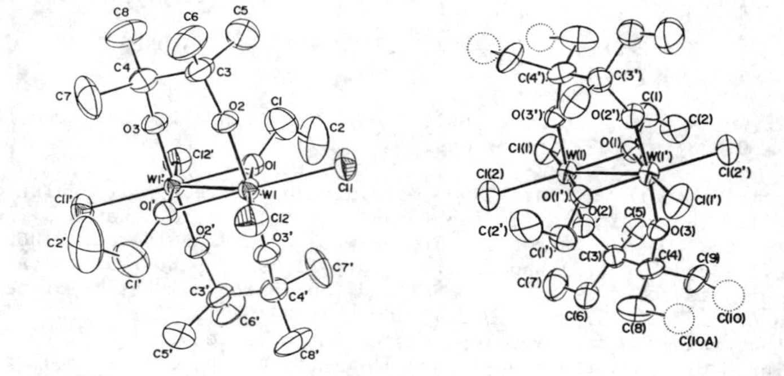

Fig. 1. Views of $C_{16}H_{34}Cl_4O_6W_2$ (I) (left) and $C_{20}H_{42}Cl_4O_6W_2$ (II) (right).

trans-DIIODOTETRAKIS(t-BUTOXY)TUNGSTEN(VI)

$C_{16}H_{36}I_2O_4W$ $WI_2(OCMe_3)_4$

F.A. COTTON, W. SCHWOTZER and E.S. SHAMSHOUM, 1984. Inorg. Chem., 23, 4111-4113.

Tetragonal, I4/m, a = 10.443, c = 10.483 Å, Z = 2. Mo radiation, R = 0.027 for 776 reflexions (at -104°C).

The molecule (Fig. 1) has crystallographic 4/m (C_{4h}) symmetry with W-I 2.8293(8) and W-O 1.840(4) Å. The W-O-C bond angle is 160.5(4)°.

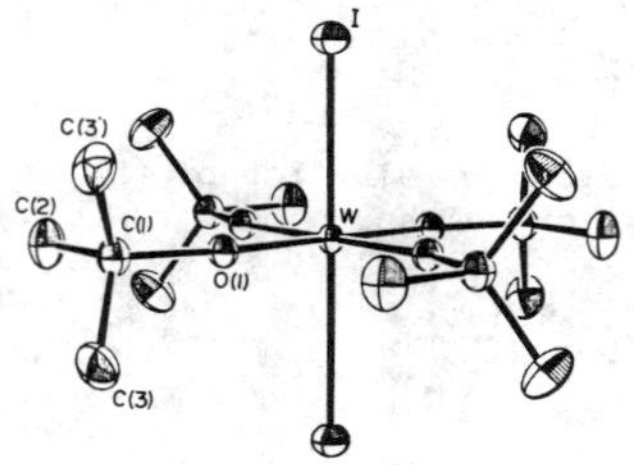

Fig. 1. The $WI_2(OCMe_3)_4$ molecule.

TETRAKIS[μ-(ACETONE OXIMATO)-μ-O,N]-TETRAKIS(METHYLZINC)

$C_{16}H_{36}N_4O_4Zn_4$ $Zn_4(CH_3)_4(C_3H_6NO)_4$

N.A. BELL, H.M.M. SHEARER and C.B. SPENCER, 1984. Acta Cryst., C40, 613-616.

Monoclinic, $P2_1/c$, a = 9.258, b = 13.58, c = 24.10 Å, β = 105.4°, D_m = 1.40, Z = 4. Cu radiation, R = 0.101 for 3081 reflexions (photographic data).

The molecules exist in the crystal as distinct tetramers (Fig. 1). The four Zn atoms lie at the corners of a slightly distorted tetrahedron and above each face is an acetoximate group in which the O atom is attached to two Zn atoms and the N to a third. A cage structure results which has four five-membered rings and two adjacent six-membered rings but no four-membered rings so that the four Zn atoms are in three different environments. Some bond distances are: Zn-C 1.940(18)-1.958(17), Zn-O 2.024(8)-2.069(8) and Zn-N 2.047(10)-2.098(10) Å.

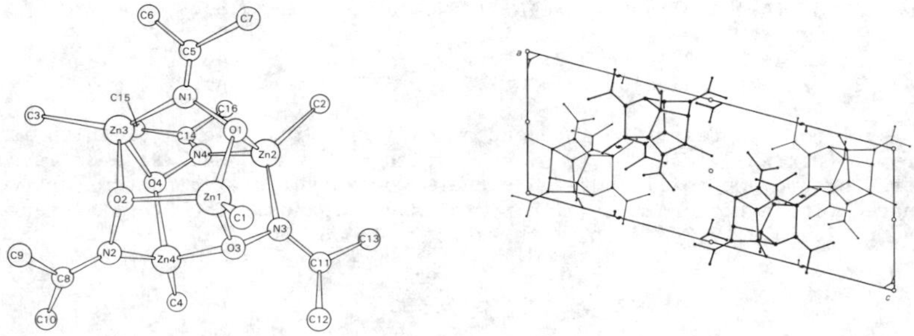

Fig. 1. $Zn_4(CH_3)_4(C_3H_6NO)_4$: perspective view of the molecule and projection of the crystal structure on [010].

TETRAKIS[(μ_2-TRIS(HYDROXYMETHYL)METHYLAMINE-N,O)CHLOROCOPPER(II)]
$C_{16}H_{40}Cl_4Cu_4N_4O_{12}$ (I)

BIS(TRIS(HYDROXYMETHYL)METHYLAMINE-N,O)COPPER(II) BROMIDE
$C_8H_{12}BrCuN_2O_6$ (II)

D. MASI, C. MEALLI, M. SABAT, A. SABATINI, A. VACCA and F. ZANOBINI, 1984. Helv. Chim. Acta, 67, 1818-1826.

I. Triclinic, $P\bar{1}$, a = 9.182, b = 9.120, c = 8.817 Å, α = 88.95, β = 87.01, γ = 84.13°, Z = 1. Mo radiation, R = 0.041 for 1429 reflexions.

II. Monoclinic, $P2_1/c$, a = 11.384, b = 10.049, c = 12.149 Å, β = 95.83°, Z = 4. Mo radiation, R = 0.062 for 955 reflexions.

Compound I lies about an inversion centre (Fig. 1); the compound consists of two dinuclear units held together by weak Cu-O interactions. The central core has a step like geometry with Cu1...Cu2 2.841(1), Cu2-Cu2' 3.233(2), Cu-O 1.939-2.020(6) and 2.407(6) (linking the dinuclear units), Cu-Cl 2.260 and 2.253(2), Cu-N 1.985 and 1.984(7) Å. There is a 42° dihedral angle between the Cu-O-O planes in the dinuclear unit. In II, the Cu coordination is square-planar, with Cu-O1 1.943(11), Cu-O2 2.018(11), Cu-N1 1.993(13), Cu-N2 1.995(13) Å. Molecules are linked to form dimers by O-H...O hydrogen bonds about an inversion centre O...O 2.50(2) Å.

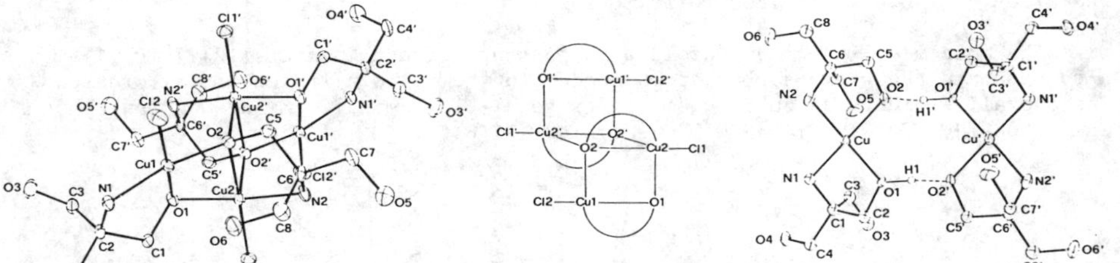

Fig. 1. The tetrameric $C_{16}H_{40}Cl_4Cu_4N_4O_{12}$ and hydrogen-bonded $C_8H_{12}BrCuN_2O_6$ structures.

TETRAKIS(μ_2-α-PYRROLIDONE-N,O)OCTAAMMINETETRAPLATINUM NITRATE TRIHYDRATE
$C_{16}H_{48}N_{12}O_4Pt_4.5.48(NO_3),3H_2O$

K. MATSUMOTO, H. TAKAHASHI and K. FUWA, 1984. J. Am. Chem. Soc., 106, 2049-2054.

Triclinic, P1, a = 12.962, b = 17.931, c = 9.778 Å, α = 99.84, β = 82.45, γ = 108.61°, D_m = 2.64, Z = 2. Mo radiation, R = 0.118 for 3373 reflexions.

The structure contains two independent cations (Fig. 1) each of which is actually a mixture of $[Pt_4(NH_3)_8(C_4H_6NO)_4]^{n+}$ ions, with n = 5 or 6. Each Pt atom is cis-coordinated by two ammine ligands and either two exocyclic oxygen atoms or two deprotonated ring nitrogen atoms of α-pyrrolidcne ligands. The Pt-Pt distances are in the range 2.724(8)-2.764(8) Å, with Pt-Pt-Pt 166.6-170.9(3)°.

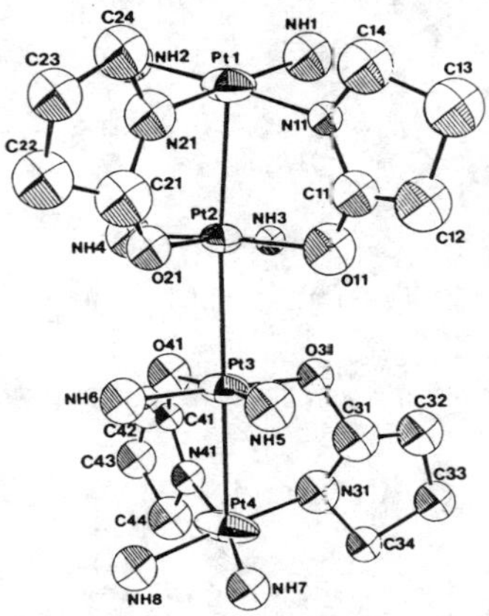

Fig. 1. A view of one of the $[Pt_4(NH_3)_8(C_4H_6NO)_4]$ cations.

[5,5'-(6-METHYL-2,6,10-TRIAZAUNDECA-1,10-DIENE-1,11-DIYL)BIS(URACIL)-O,O',N,N',N'']-COPPER(II) ETHANOL SOLVATE

$C_{17}H_{21}CuN_7O_4,C_2H_6O$

I. SASAKI, M.N. DUFOUR, A. GAUDEMER, A. CHIARONI, C. RICHE, D. PARKER-DECROUEZ and P. BOUCLY, 1984. Nouv. J. Chim., 8, 237-242.

Monoclinic, C2/c, a = 19.597, b = 8.109, c = 17.628 Å, β = 115.87°, Z = 4. Cu radiation, R = 0.079 for 1300 reflexions.

The molecule lies on a twofold axis passing through Cu and N(12) (Fig. 1); as a consequence the three carbon atoms bonded to the amino nitrogen are disordered. The overall geometry at Cu is close to trigonal-bipyramidal. Main bond lengths are Cu-O 2.089(4), Cu-N(8) 1.938(6), Cu-N(12) 2.151(8) Å.

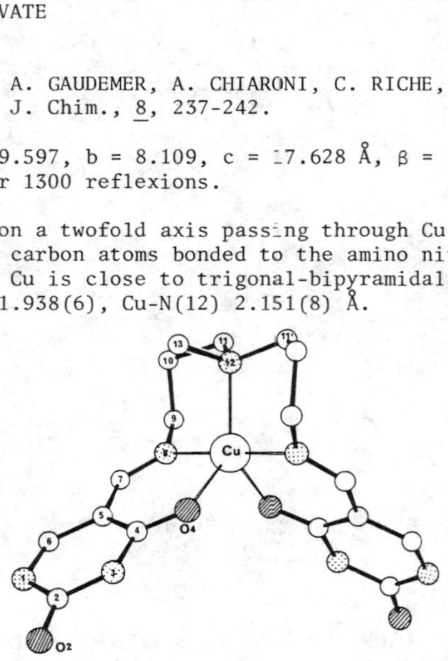

Fig. 1. A view of $C_{17}H_{21}CuN_7O_4$.

{1,2-BIS[2-(5',5',5'-TRIFLUORO-4'-(TRIFLUOROMETHYL)-4'-OLATO-2'-PENTYL)IMINOETHOXY]-ETHANE}COBALT(II)
$C_{18}H_{22}CoF_{12}N_2O_4$

E. KONEFAL, S.J. LOEB, D.W. STEPHAN and C.J. WILLIS, 1984. Inorg. Chem., 23, 538-545.

Triclinic, $P\bar{1}$, a = 10.144, b = 15.740, c = 15.579 Å, α = 102.56, β = 92.03, γ = 96.58°, Z = 4. Mo radiation, R = 0.050 for 4470 reflexions.

One of the two independent, but virtually identical, molecules is shown in Fig. 1. The Co atom is coordinated by four O and two N atoms of the hexadentate ligand in a distorted octahedral geometry. Two alkoxy O atoms in the ligand (Co-O 1.933(4)-1.961(4) Å) make an O-Co-O angle of 122°; two imino N atoms (Co-N 2.127(5)-2.141(4) Å) are trans collinear about Co, while the two ethereal O atoms (Co-O 2.317(4)-2.341(5) Å) in the chain make an unusually small (69-70°) bite angle at Co.

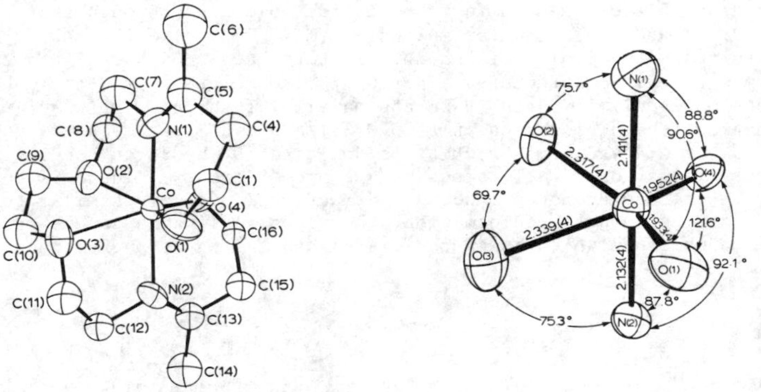

Fig. 1. One of the $C_{18}H_{22}CoF_{12}N_2O_4$ molecules, coordination group at right.

DICHLOROBIS(1-ETHYL-3-p-TOLYLTRIAZENE 1-OXIDATO)RUTHENIUM(IV)
$C_{18}H_{24}Cl_2N_6O_2Ru$ $RuCl_2[CH_3C_6H_4NNN(C_2H_5)O]_2$

S. BHATTACHARYA, A. CHAKRAVORTY, F.A. COTTON, R. MUKHERJEE and W. SCHWOTZER, 1984. Inorg. Chem., 23, 1709-1713.

Monoclinic, $P2_1/n$, a = 9.824, b = 17.059, c = 6.564 Å, β = 99.57°, Z = 2. Cu radiation, R = 0.055 for 1139 reflexions (at 5°C).

The coordination sphere is $RuCl_2N_2O_2$, with the metal atom located at the centre of symmetry (Fig. 1). The chelate dimensions are normal, and the phenyl ring is only slightly (10°) tilted from the chelate-ring plane. Bond lengths: Ru-Cl 2.367(1), Ru-O 1.974(4), Ru-N 2.007(4), O-N 1.311(6), N(1)-N(2) 1.288(6), N(2)-N(3) 1.321(6) Å.

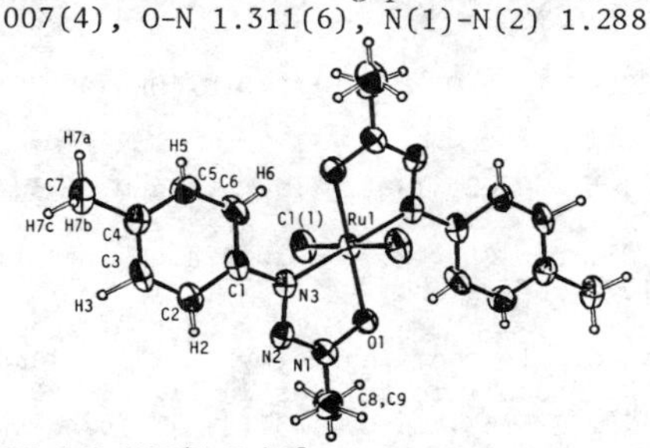

Fig. 1. The $RuCl_2[CH_3C_6H_4NNN(C_2H_5)O]_2$ molecule.

(μ_4-URIDINE-5'-MONOPHOSPHATE)DIAQUACADMIUM(II) DIHYDRATE
$(C_{18}H_{30}Cd_2N_4O_{22}P_2)_n,4n(H_2O)$ (I)

BIS(μ_3-2'-DEOXYURIDINE-5'-MONOPHOSPHATE)HEXAAQUADICADMIUM(II)
$(C_{18}H_{34}Cd_2N_4O_{22}P_2)_n$ (II)

K. AOKI and W. SAENGER, 1984. J. Chem. Soc. Dalton, 1401-1409.

I. Monoclinic, C2, a = 28.835, b = 14.241, c = 4.928 Å, β = 91.37°, Z = 4. Cu radiation, R = 0.079 for 1236 reflexions.

II. Triclinic, P1, a = 18.278, b = 7.572, c = 5.356 Å, α = 89.76, β = 98.22, γ = 93.74°, Z = 1. Mo radiation, R = 0.060 for 2530 reflexions.

Compound I has a one dimensional polymeric structure (Fig. 1). Pairs of crystallographically independent Cd atoms (Cd(1)...Cd(2) 3.249(3) Å) which lie on a twofold axis along b are bound to four nucleotide molecules through their phosphate groups (Cd-O 2.29) and to four water molecules (Cd-O 2.29 Å) with each Cd in a slightly distorted octahedral environment. In turn each phosphate links two dinuclear units to form a cage-like structure. The sugar-base moieties project to both sides of the one dimensional polymer chain in an antiparallel fashion. The uracil base has the anti-conformation, the ribose ring an endo-C(3')/exo-C(2') pucker and a gauche-gauche conformation about C(4')-C(5'). Packing is dominated by hydrogen bonding between two waters of crystallisation, uracil bases and ribose OH groups. There is no base-base stacking nor inter-base hydrogen bonding. Structure II is very similar to that of I. Here the base is in the anti-conformation, the glycosyl link is gauche-gauche (except for one molecule which is gauche-trans) and the sugar pucker is endo-C(1'), exo-C(4').

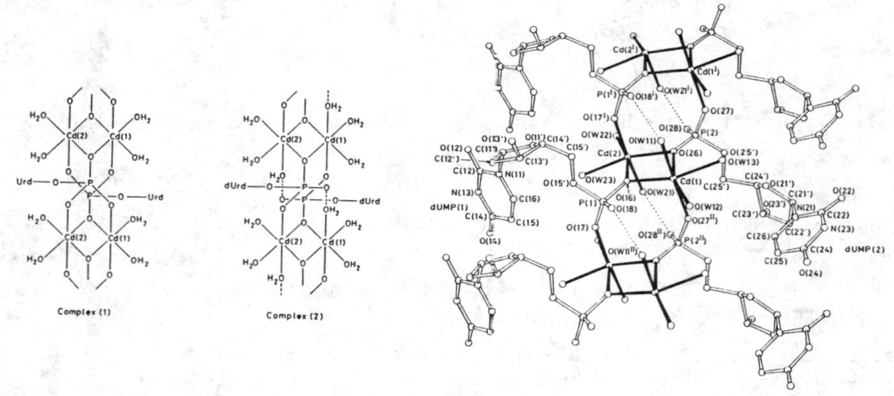

Fig. 1. The polymeric structures: $(C_{18}H_{30}Cd_2N_4O_{22}P_2)_n,4n(H_2O)$ (left) and $(C_{18}H_{34}Cd_2N_4O_{22}P_2)_n$.

[BIS(4,5-DIISOPROPYLIMIDAZOL-2-YL)PHOSPHINIC ACID]DICHLOROZINC HYDRATE
$C_{18}H_{31}Cl_2N_4O_2PZn.H_2O$

R.G. BALL, R.S. BROWN and J.L. COCHO, 1984. Inorg. Chem., 23, 2315-2318.

Monoclinic, $P2_1/c$, a = 9.805, b = 26.202, c = 10.924 Å, β = 111.79°, Z = 4. Mo radiation, R = 0.066 for 1674 reflexions.

The Zn^{2+} ion is tetrahedrally coordinated by two Cl^-'s and by N(3) of an imidazole and by O(1) of the phosphinic acid as in Fig. 1. Bond lengths: Zn-Cl 2.243(2) and 2.160(3), Zn-N 2.025(6), Zn-O 2.037(5) Å. Bond angles at Zn vary from 87.3(2) to 121.4(2)°.

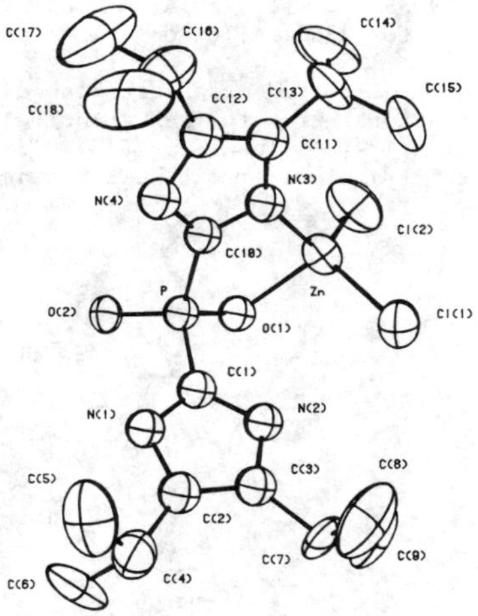

Fig. 1. The $C_{18}H_{31}Cl_2N_4O_2PZn$ molecule.

[2,2'-(1,4,10,13-TETRAOXA-7,16-DIAZA-7,16-CYCLOOCTADECYLENE)DIPROPIONATO]COPPER(II)
$C_{18}H_{32}CuN_2O_8$ $Cu(C_{18}H_{32}N_2O_8)$

J.W. KRAJEWSKI, P. GLUZIŃSKI, Z. URBAŃCZYK-LIPKOWSKA and M. DOBLER, 1984. Acta Cryst., C40, 1135-1137.

Monoclinic, $P2_1/c$, a = 7.032, b = 15.533, c = 9.836, β = 113.46°, Z = 2. Mo radiation, R = 0.034 for 1431 reflexions.

The molecule is centrosymmetric with Cu at the centre of symmetry (Fig. 1). Cu is coordinated to N(1), O(13) and O(7) and to their symmetry-related equivalents, forming an irregular octahedron with one elongated and skewed (by 11.58(7)°) diagonal. Interatomic distances are Cu-N(1) 2.050(3), Cu-O(13) 1.924(3) and Cu-O(7) 2.615(3) Å.

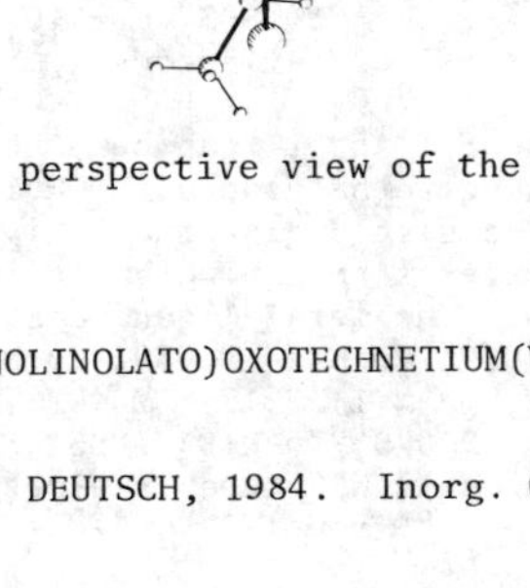

Fig. 1. $Cu(C_{18}H_{32}N_2O_8)$: perspective view of the molecule.

cis-CHLOROBIS(2-METHYL-8-QUINOLINOLATO)OXOTECHNETIUM(V)
$C_{20}H_{16}ClN_2O_3Tc$ $TcO(CH_3C_9H_5NO)_2Cl$

B.E. WILCOX, M.J. HEEG and E. DEUTSCH, 1984. Inorg. Chem., 23, 2962-2967.

Triclinic, $P\bar{1}$, a = 7.693, b = 9.337, c = 12.739 Å, α = 86.52, β = 85.99, γ = 84.12°, Z = 2. Mo radiation, R = 0.035 for 2882 reflexions.

The molecule (Fig. 1) has a distorted octahedral coordination geometry, with the Cl and oxo ligands cis. A structural trans effect is noted. Bond lengths: Tc-Cl 2.360(1), Tc=O 1.649(3), Tc-N1 2.179(3), Tc-N2 2.215(3), Tc-O2 1.947(3), and Tc-O3 1.994(3) Å. There is a short intramolecular oxo...Me contact (C11...O1 2.836(5) Å).

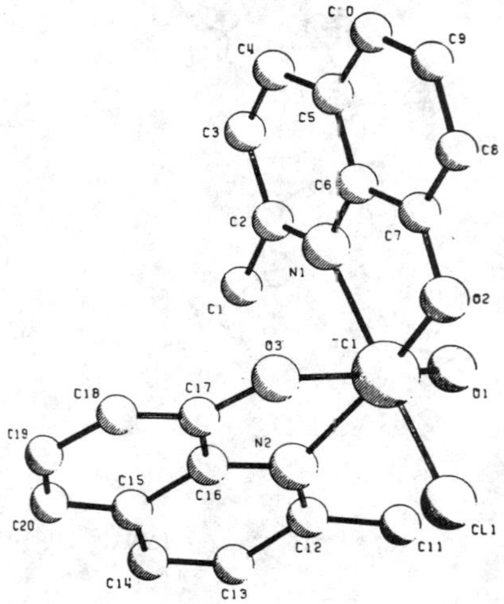

Fig. 1. The $TcO(CH_3C_9H_5NO)_2Cl$ molecule.

2-[BIS-(2-BENZIMIDAZOLYLMETHYL)AMINO]ETHANOLCOBALT(II)
$C_{20}H_{19}CoN_7OS_2$ (I)

2-[BIS-(2-BENZIMIDAZOLYLMETHYL)AMINO]ETHANOLMANGANESE(II)
$C_{20}H_{19}MnN_7OS_2$ (II)

2-[BIS-(2-BENZIMIDAZOLYLMETHYL)AMINO]ETHANOLZINC(II)
$C_{20}H_{19}N_7OS_2Zn$ (III)

K. TAKAHASHI, Y. NISHIDA and S. KIDA, 1984. Bull. Chem. Soc. Japan, 57, 2628-2633.

I. Orthorhombic, $P2_12_12_1$, a = 12.881, b = 17.970, c = 9.076 Å, D_m = 1.57, Z = 4. Mo radiation, R = 0.044 for 2007 reflexions.

II. Orthorhombic, $P2_12_12_1$, a = 18.179, b = 13.073, c = 9.097 Å, D_m = 1.49, Z = 4. Mo radiation, R = 0.054 for 2583 reflexions.

III. Monoclinic, $P2_1/a$, a = 15.421, b = 14.803, c = 10.589 Å, β = 115.81°, D_m = 1.56, Z = 4. Mo radiation, R = 0.079 for 2462 reflexions.

The coordination geometries of I and II can be described as distorted octahedral (Fig. 1). In III however, the alocholic OH is not coordinated so the Zn^{2+} ion is five-coordinate with a highly-distorted trigonal bipyramidal configuration. In all three complexes the thiocyanate ions coordinate in cis configurations via their N atoms. The metal-ligand distances are: in I, Co-N(A) 2.086, Co-N(B) 2.040, Co-N(1) 2.089, Co-N(3) 2.089, Co-N(5) 2.302, Co-O 2.262 Å; in II, Mn-N(A) 2.140, Mn-N(B) 2.107, Mn-N(1) 2.175, Mn-N(3) 2.183, Mn-N(5) 2.517, Mn-O 2.400 Å; and in III, Zn-N(A) 2.045, Zn-N(B) 1.951, Zn-N(1) 2.008, Zn-N(3) 2.090 and Zn-N(5) 2.652 Å. Each metal-NCS bond is almost linear, the M-N-C angle deviations being in the range 6 - 28° from 180°.

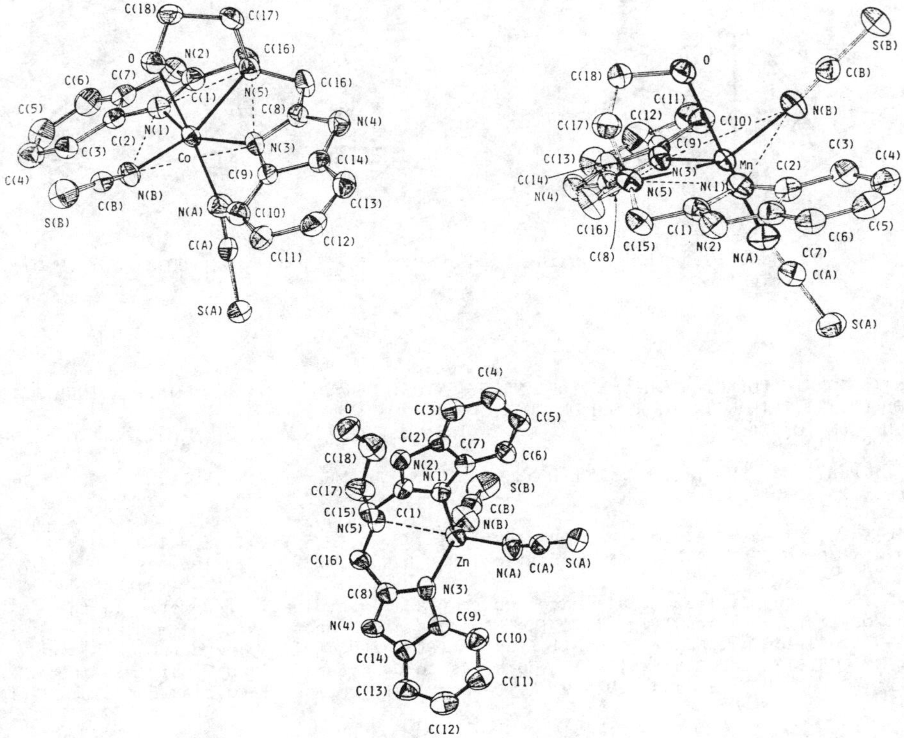

Fig. 1. Molecular structure of $C_{20}H_{19}CoN_7OS_2$ (top left), $C_{20}H_{19}MnN_7OS_2$ (top right), and $C_{20}H_{19}N_7OS_2Zn$ (bottom).

DICYCLOHEXYL(18-CROWN-6)DIOXOURANIUM DIPERCHLORATE
$C_{20}H_{36}Cl_2O_{16}U$ (I) $(C_{20}H_{36}O_8U)^{2+}.2(ClO_4^-)$

TRIS[DICYCLOHEXYL(18-CROWN-6)] BIS[(μ_2-HYDROXO)DIOXOTRIAQUAURANIUM] DIPERCHLORATE ACETONITRILE SOLVATE
$C_{60}H_{122}Cl_2O_{38}U_2,C_2H_3N$ (II) $3[C_{20}H_{36}O_6].[(UO_2)_2(OH)_2(H_2O)_6].(ClO_4)_2,CH_3CN$

A. NAVAZA, F. VILLAIN and P. CHARPIN, 1984. Polyhedron, 3, 143-149.

I. Monoclinic, $P2_1/c$, a = 12.947, b = 12.850, c = 17.539 Å, β = 91.4°, Z = 4. Mo radiation, R = 0.079 for 1195 reflexions.

II. Triclinic, $P\bar{1}$, a = 9.693, b = 13.836, c = 17.888 Å, α = 70.90, β = 75.83, γ = 74.02°, Z = 1. Cu radiation, R = 0.09 for 3162 reflexions.

In I (Fig. 1) the uranyl moiety is coordinated to six additional oxygen atoms; the uranyl O-U-O system is almost linear (176.7(1.7)°), with U-O 1.77(4) Å. U-O(crown) distances are in the range 2.43-2.68(4) Å. In II, the uranyl groups are not associated with the crown ether, but instead form a centrosymmetric dimeric ion $[(UO_2)_2(OH)_2(H_2O)_6]^{2-}$, with two pentagonal bipyramids sharing an edge. The U...U separation is 3.827(8) Å and U-O(axial) 1.70, 1.73(2), U-O(equatorial) 2.27-2.47(2) Å, uranyl O-U-O 174.6°.

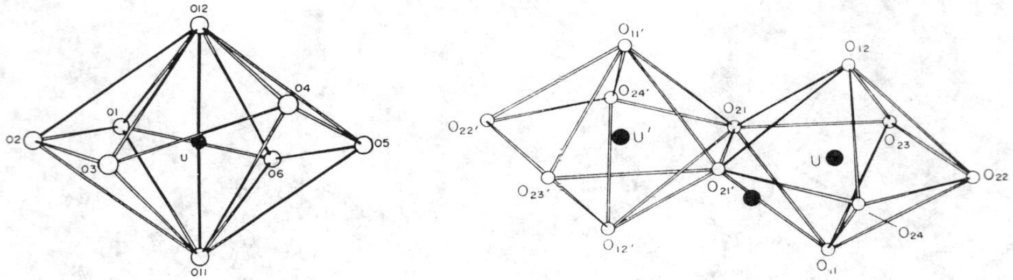

Fig. 1. The UO_2 coordination in $(C_{20}H_{36}O_8U)^{2+}$ (left) and in $[(UO_2)_2(OH)_2(H_2O)_6]^{2-}$ (right).

BIS[(μ-1-METHYLURACILATO-N^3,O^4)(μ-1-METHYLCYTOSINE-N^3,O^2)-cis-DIAMMINEPLATINUM(II)]-COPPER(II) TETRANITRATE HEXAHYDRATE

$C_{20}H_{36}CuN_{18}O_{18}Pt_2,6H_2O$ $\{[Pt(NH_3)_2(C_5H_5N_2O_2)(C_5H_7N_3O)]_2Cu\}(NO_3)_4 \cdot 6H_2O$

B. LIPPERT, U. THEWALT, H. SCHÖLLHORN, D.M.L. GOODGAME and R.W. ROLLINS, 1984. Inorg. Chem., 23, 2807-2813.

Triclinic, $P\bar{1}$, a = 11.522, b = 10.924, c = 10.736 Å, α = 91.51, β = 109.08, γ = 114.43°, D_m = 2.01, Z = 1. Mo radiation, R = 0.051 for 2603 reflexions.

The three metal atoms all have square planar coordination geometry in the centrosymmetric cation (Fig. 1). The Pt atoms are bound to N3 of 1-methyluracil (Pt-N 2.045(13) Å), N3 of 1-methylcytosine (Pt-N 2.036(11) Å), and two NH_3 ligands (Pt-N 2.081(12) and 2.068(14) Å). Cu binds to O4 of uracil and O2 of cytosine in pairs (Cu-O 1.931(12) and 1.988(9) Å, respectively). The Pt-Cu distance is 2.681(1) Å.

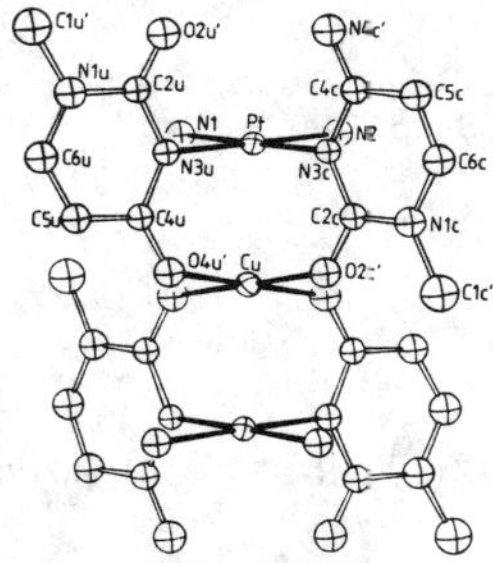

Fig. 1. The $\{[Pt(NH_3)_2(C_5H_5N_2O_2)(C_5H_7N_3O)]_2Cu\}^{4+}$ cation.

trans-DICHLOROTETRAKIS(TETRAHYDROFURAN)TITANIUM(III) TRICHLORO(TETRAHYDROFURAN)ZINC(II)

$C_{20}H_{40}Cl_5O_5TiZn$ $[TiCl_2(OC_4H_8)_4][ZnCl_3(OC_4H_8)]$

K. FOLTING, J.C. HUFFMAN, R.L. BANSEMER and K.G. CAULTON, 1984. Inorg. Chem., 23, 3289-3292.

Orthorhombic, $P2_12_12_1$, a = 21.970, b = 9.065, c = 14.134 Å, Z = 4. Mo radiation, R = 0.053 for 2040 reflexions (at -158°C).

The structure consists of trans-$[TiCl_2(OC_4H_8)_4]^+$ cations having octahedral geometry (Ti-Cl 2.361(3), 2.359(3), Ti-O 2.064(8)-2.098(7) Å) and $[ZnCl_3(OC_4H_8)]^-$ anions having quasi-tetrahedral geometry (Zn-O 2.121(8), Zn-Cl 2.239(3)-2.242(3) Å) (see Fig. 1). Conformational features and Ti-O distances are consistent with THF being capable of donating to octahedral Ti(III) with both lone pairs.

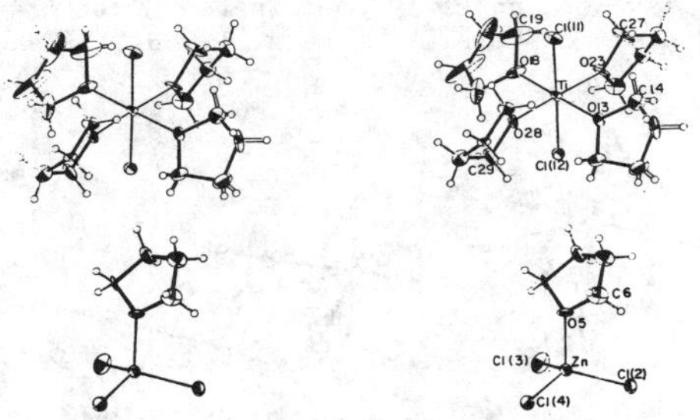

Fig. 1. Stereoviews of $[TiCl_2(OC_4H_8)_4]^+$ (top) and $[ZnCl_3(OC_4H_8)]^-$ (bottom).

TETRAKIS(μ_2-1-METHYLURACIL-N,O)OCTAAMMINETETRAPLATINUM PENTANITRATE PENTAHYDRATE
$C_{20}H_{44}N_{21}O_{23}Pt_4,5H_2O$ $[Pt(NH_3)_8(C_5H_5N_2O_2)_4].(NO_3)_5,5H_2O$

P.K. MASCHARAK, I.D. WILLIAMS and S.J. LIPPARD, 1984. J. Am. Chem. Soc., 106, 6428-6430.

Triclinic, $P\bar{1}$, a = 10.123, b = 13.084, c = 19.508 Å, α = 92.28, β = 101.09, γ = 107.49°, D_m = 2.435, Z = 2. Mo radiation, R = 0.038 for 2168 reflexions.

The tetranuclear cation (Fig. 1) has two cis-diammineplatinum units bridged by two 1-methyluracilate anions which have deprotonated N3 and exocyclic O4 as the donor atoms. Main dimensions are: outer Pt-Pt 2.793 and 2.810(1), central Pt-Pt 2.865(1) Å, Pt-Pt-Pt 164.8 and 165.0(1)°. The twist angle about Pt2-Pt3 is 22.3° and the tilt angle between outer Pt coordination planes is 26.7°.

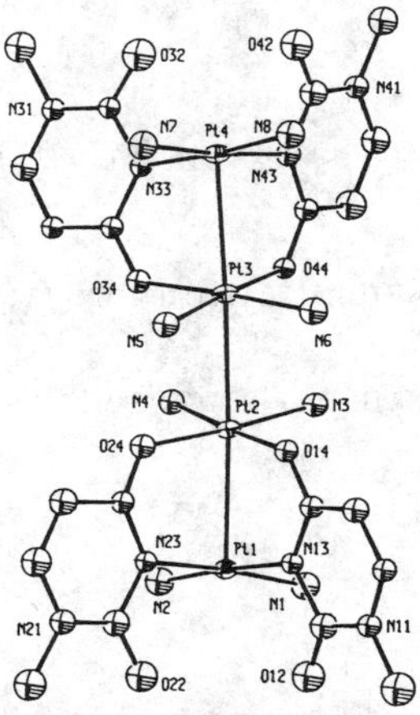

Fig. 1. A view of $[Pt_4(NH_3)_8(C_5H_5N_2O_2)_4].(NO_3)_5,5H_2O$.

HEXAKIS(UREA)CHROMIUM(III) HEXACYANOCHROMATE(III) BIS(DIMETHYLSULPHOXIDE) BIS(ETHANOL)
$C_{20}H_{48}Cr_2N_{18}O_{10}S_2$ $[Cr(urea)_6][Cr(CN)_6],(dmso)_2,(EtOH)_2$

B.N. FIGGIS, E.S. KUCHARSKI, J.M. PATRICK and A.H. WHITE, 1984. Aust. J. Chem., 37, 265-271.

Monoclinic, $P2_1/n$, a = 17.449, b = 10.315, c = 11.726 Å, β = 102.51°, D_m = 1.38, Z = 2. Mo radiation, R = 0.039 for 4203 reflexions.

Both ions (Fig. 1) have octahedrally coordinated Cr(III) atoms at centres of symmetry. Metal-ligand distances are Cr-O, mean, 1.968(2) Å and Cr-C, mean, 2.078(2) Å.

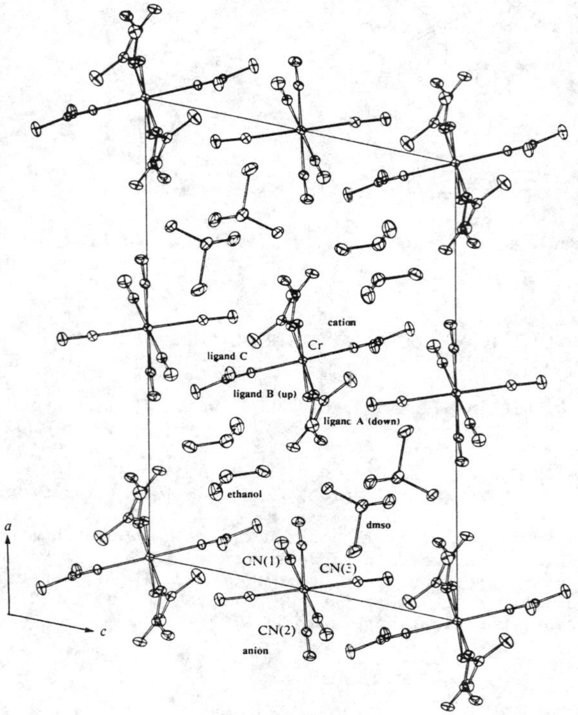

Fig. 1. Unit cell contents of $[Cr(urea)_6][Cr(CN)_6],(dmso)_2,(EtOH)_2$.

TRIS(2,6-DIMETHYL-4H-PYRAN-4-ONE)DINITRATOCADMIUM(II)

$C_{21}H_{24}CdN_2O_{12}$ $[Cd(NO_3)_2(C_7H_8O_2)_3]$

A. BANERJEE, C.J. BROWN, P.C. JAIN and P. GAUTAM, 1984. Acta Cryst., C40, 1161-1164.

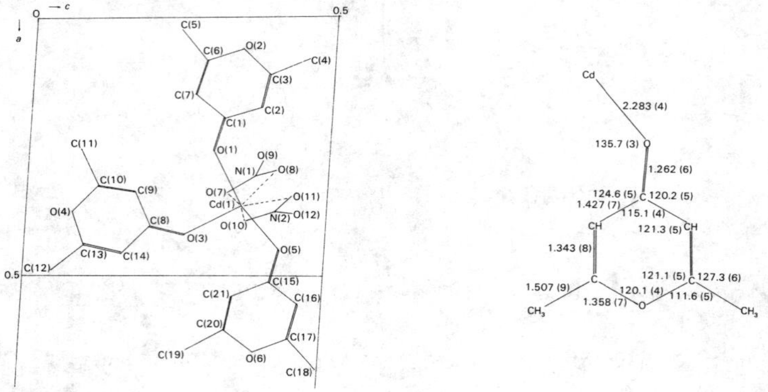

Fig. 1. $[Cd(NO_3)_2(C_7H_8O_2)_3]$: projection of part of the unit cell on (010) showing one of the molecules and averaged bond lengths and angles in the 2,6-dimethyl-4-pyrone ring.

Monoclinic, $P2_1/n$, a = 19.372, b = 11.345, c = 23.023 Å, β = 93.77°, D_m = 1.60, Z = 8. Mo radiation, R = 0.061 for 7325 reflexions.

The structure comprises two discrete non-equivalent molecules in each of which Cd is seven-coordinated, to three O atoms of the dimethyl-4-pyrone ligands, and bidentately to four O atoms of two nitrato groups (Fig. 1). The pyrone rings are planar and there are no strong intermolecular interactions.

cis-TRICHLOROTRIS(2,6-DIMETHYL-4-PYRONE)GADOLINIUM(III)
$C_{21}H_{24}Cl_3GdO_6$

C.B. CASTELLANI and V. TAZZOLI, 1984. Acta Cryst., C40, 1832-1834.

Monoclinic, $P2_1/c$, a = 10.612, b = 14.759, c = 18.028 Å, β = 120.07°, Z = 4. Mo radiation, R = 0.046 for 5657 reflexions.

The Gd^{3+} ion is coordinated by three dimethyl-4-pyrone ligands through the carbonyl groups and by three Cl^- ions (Fig. 1) forming a slightly distorted octahedron with Gd-O 2.328, 2.328, 2.355(5) and Gd-Cl 2.622, 2.634, 2.645(3) Å. The carbonyl bond lengths are longer than are found in uncoordinated 2,6-dimethyl-4-pyrone molecules (1.247(9)-1.262(8) Å).

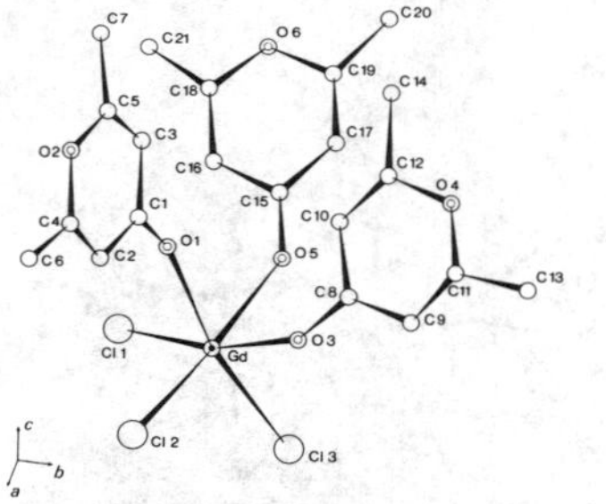

Fig. 1. The cis-trichlorotris(2,6-dimethyl-4-pyrone)gadolinium(III) molecule.

BIS(4-CHLORO-1-BENZOQUINONE-2-OXIMATO)(2,2'-BIPYRIDINE)COPPER(II)
$C_{22}H_{14}Cl_2CuN_4O_4$

C.B. CASTELLANI, G. GATTI and R. MILLINI, 1984. Inorg. Chem., 23, 4004-4008.

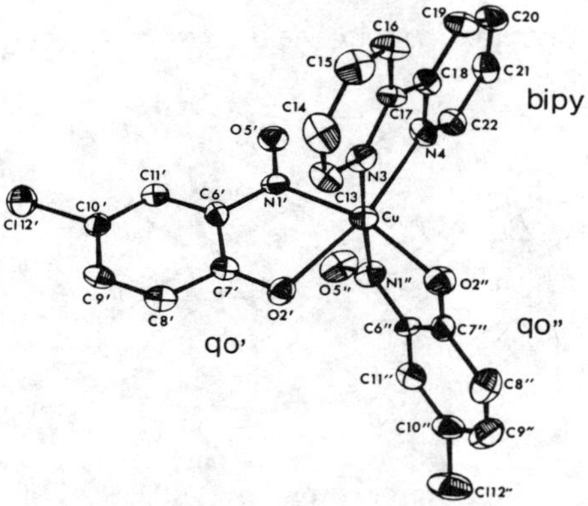

Fig. 1. One of the two independent $C_{22}H_{14}Cl_2CuN_4O_4$ molecules.

Triclinic, $P\bar{1}$, a = 15.499, b = 14.260, c = 11.087 Å, α = 71.26, β = 75.25, γ = 81.07°, Z = 4. Mo radiation, R_W = 0.034 for 3860 reflexions.

The complex (Fig. 1) has Cu atoms in a tetragonally distorted cis,cis-octahedral coordination geometry. Bond lengths: Cu-N1' 2.021(5), 2.113(7); Cu-N1'' 1.999(6), 1.988(5); Cu-O2' 2.320(4), 2.262(4)' Cu-O2'' 2.028(4), 2.106(5); Cu-N3 2.055(6), 2.012(5); Cu-N4 2.181(5), 2.203(5) Å.

TETRAKIS(μ-6-FLUORO-2-OXOPYRIDINATO)(DIMETHYLSULPHOXIDE-S)DIRHODIUM(II)

$C_{22}H_{18}F_4N_4O_5Rh_2S$ $Rh_2(C_5H_3NOF)_4((CH_3)_2SO)$

F.A. COTTON, S. HAN and W. WONG, 1984. Inorg. Chem., 23, 4762-4765.

Monoclinic, $P2_1/n$, a = 17.031, b = 9.177, c = 17.993 Å, β = 117.15°, Z = 4. Mo radiation, R = 0.049 for 2149 reflexions.

The molecule is shown in Fig. 1. All four bridging C_5H_3NOF ligands are aligned in the same direction so that the fluorine atom screens the axial position of Rh(2). A Me_2SO ligand is axially coordinated to the dirhodium unit through S at the other end (Rh-S 2.332(3) Å). The Rh-Rh distance is 2.410(1) Å and the mean Rh-O and Rh-N distances are 2.023(4) and 2.045(5) Å, respectively.

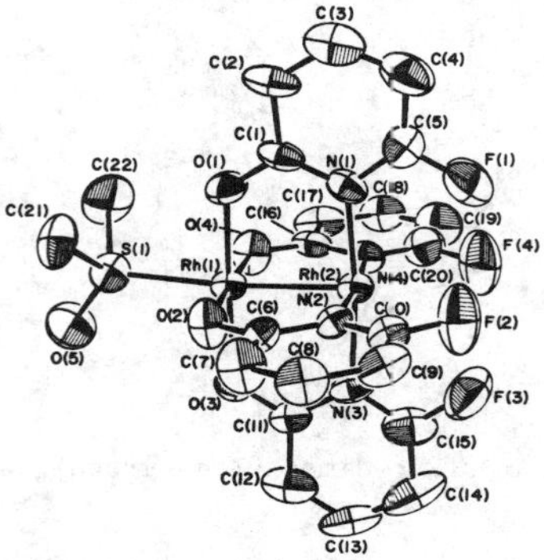

Fig. 1. The $Rh_2(C_5H_3NOF)_4((CH_3)_2SO)$ molecule.

N,N'-(1,2-PHENYLENE)BIS(PYRIDOXAL-HYDROCHLORIDE-IMINATO)COPPER(II) HYDRATE

$C_{22}H_{22}Cl_2CuN_4O_4$,3.5(H_2O)

F. NEPVEU, F. DAHAN, R. HARAN, P. CASSOUX and J.-J. BONNET, 1984. J. Cryst. Spectrosc. Res., 14, 129-142.

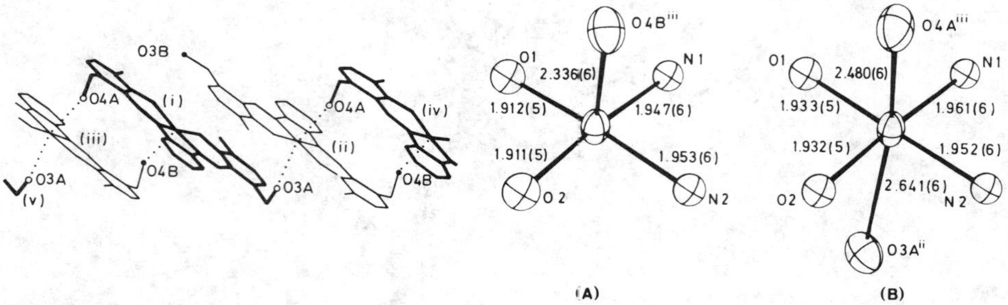

Fig. 1. Packing of A and B conformers in $C_{22}H_{22}Cl_2CuN_4O_4$ and details of Cu environment.

Triclinic, $P\bar{1}$, a = 12.028, b = 20.378, c = 11.387 Å, α = 113.10, β = 84.13, γ = 95.99°, D_m = 1.57, Z = 4. Mo radiation, R = 0.054 for 3904 reflexions.

The asymmetric unit contains two independent molecules (A and B) of the complex; the copper atom (Fig. 1) has a square pyramidal environment in A and an octahedral one in B. Intermolecular Cu-O (hydroxymethyl groups) contacts observed in the solid state are partially present in aqueous solutions. In the crystal there is a dense hydrogen bond network involving the ions and water molecules.

trans-BIS(BENZO[1,3]THIAZOLE)BIS(N,N-DIMETHYLFORMAMIDE)BIS(N-THIOCYANATO)COBALT(II)
$C_{22}H_{24}CoN_6O_2S_4$ $Co(C_7H_5NS)_2(NCS)_2((CH_3)_2NCHO)_2$

E.S. RAPER, R.E. OUGHTRED, J.R. CREIGHTON and I.W. NOWELL, 1984. Inorg. Chim. Acta, 87, 201-205.

Monoclinic, $P2_1/c$, a = 9.138, b = 10.301, c = 14.426 Å, β = 91.40°, D_m = 1.42, Z = 2. Mo radiation, R = 0.037 for 2051 reflexions.

The cobalt atoms lie on centres of symmetry and are octahedrally coordinated to pairs of trans related BZT, DMF and thiocyanate groups (Fig. 1). Both the BZT and DMF molecules are essentially planar and are twisted out of their respective equatorial planes by 42.3 and 56.3°. In the coordination sphere the Co-O(1), Co-N(2) and Co-N(3) (BZT) distances are 2.118, 2.082 and 2.185 Å respectively. The angles about Co range from 89.8 to 90.0°. Within the thiocyanate group, N(2)-C(13) is 1.147 Å and C(13)-S(2) is 1.633 Å, and the angles N(2)-C(13)-S(2) and Co-N(2)-C(13) are 179.4 and 155.7° respectively. The angle Co-O(1)-C(10) is 126.2°.

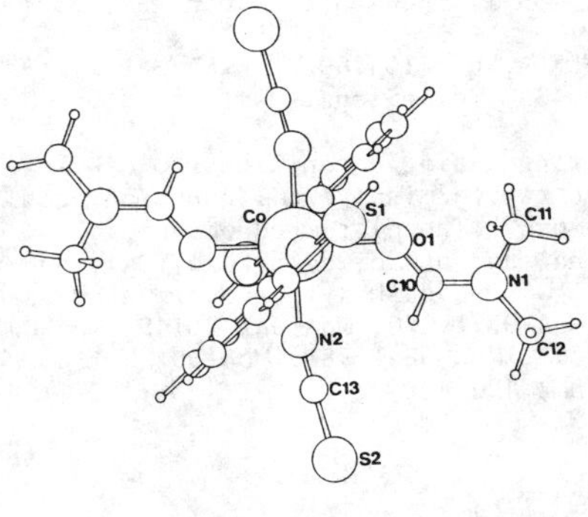

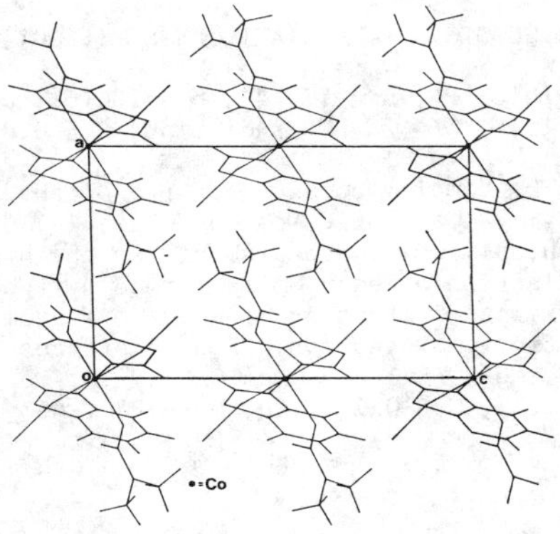

Fig. 1. Molecular structure of $C_{22}H_{24}CoN_6O_2S_4$ (left), and unit cell contents viewed down b (right).

1,4,8,11-TETRATHIACYCLOTETRADECANEDIPICRATOMERCURY(II) HEMINITROMETHANE
$C_{22}H_{24}HgN_6O_{14}S_4, 0.5CH_3NO_2$ $[Hg(C_{10}H_{20}S_4).(C_6H_2N_3O_7)_2], 0.5CH_3NO_2$

M. HERCEG, B. MATKOVIĆ, D. SEVDIĆ, D. MATKOVIĆ-ČALOGOVIĆ and A. NAGL, 1984. Croat. Chem. Acta, 57, 609-619.

Triclinic, $P\bar{1}$, a = 12.794, b = 13.108, c = 10.090 Å, α = 92.85, β = 107.67, γ = 94.54°, D_m = 1.986, Z = 2. Mo radiation, R = 0.030 for 7321 reflexions.

The structure consists of discrete, well-separated molecules. Coordination of the Hg atom is a distorted octahedron with four S atoms (Hg-S distances are 2.458, 2.519, 2.629 and 3.050 Å respectively for S(1), S(3), S(4) and S(2)) from the macrocycle and two O atoms (Hg-O distances are 2.531 and 2.558 Å for O(1) and O(2) respectively) from the two picrates. The macrocycle is folded so that three S atoms with short Hg-S bonds are in an equatorial and the fourth in an apical position.

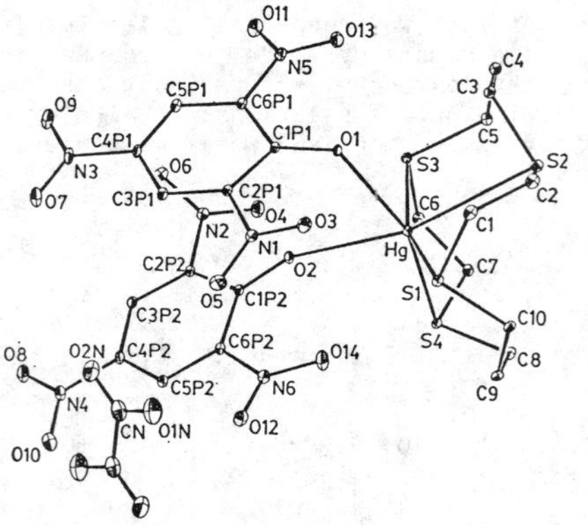

Fig. 1. Perspective drawing of the molecule $C_{22}H_{24}HgN_6O_{14}S_4$.

PYRIDINIUM μ-OXO-BIS[NITRILOTRIACETATODIOXOMOLYBDATE(VI)]

$C_{22}H_{26}Mo_2N_4O_{17}$ $(pyH)_2^+[Mo_2O_5(ntaH)_2]_2^{2-}$

K. MATSUMOTO, Y. MARUTANI and S.-I. OOI, 1984. Bull. Chem. Soc. Japan, 57, 2671-2672.

Triclinic, $P\bar{1}$, a = 17.244, b = 12.161, c = 7.558 Å, α = 104.99, β = 83.36, γ = 74.88°, D_m = 1.84, Z = 2. Mo radiation, R = 0.031 for 4338 reflexions.

The complex consists of two octahedra sharing a single non-linear oxo bridge (Fig. 1). The angle MoA-OB-MoB is 167.0°. The two terminal oxo ligands on each Mo are mutually cis, and also cis to the bridging oxo. The distorted octahedral coordination of each Mo is formed by two terminal and one bridging oxo (O_b), and two carboxylato O atoms and one amino N atom of ntaH. One carboxyl group in ntaH is free from coordination. Bond distances are: MoA-OB 1.870, MoA-OA1 2.145, MoA-OA3 2.071, MoA-OA8 1.695, MoA-OA7 1.695, MoA-NA 2.413, MoB-OB 1.881, MoB-OB1 2.156, MoB-OB3 2.075, MoB-OB7 1.700, MoB-OB8 1.683, MoB-NB 2.426 Å.

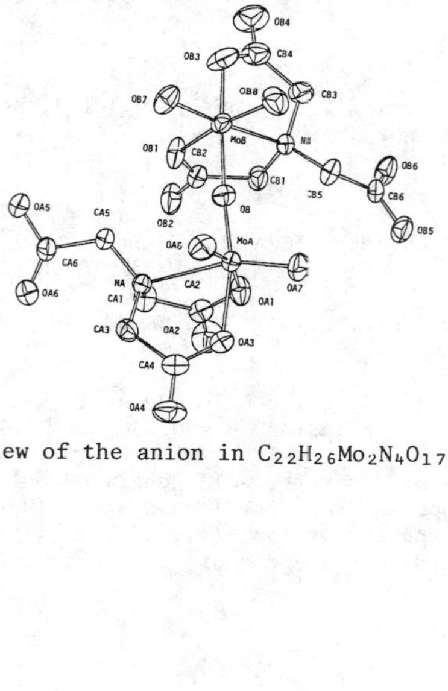

Fig. 1. Perspective view of the anion in $C_{22}H_{26}Mo_2N_4O_{17}$.

(8-HYDROXYQUINOLINATO)[2-(4'-METHOXYPHENYLAZO-N^2)-5-METHOXYPHENYL]PALLADIUM(II)
$C_{23}H_{19}N_3O_3Pd$

A.M.M. LANFREDI, F. UGOZZOLI, M. GHEDINI and S. LOCOCCIA, 1984. Inorg. Chim. Acta, 86, 165-168.

Triclinic, P$\bar{1}$, a = 10.225, b = 11.092, c = 9.365 Å, α = 105.65, β = 94.08, γ = 105.69°, Z = 2. Mo radiation, R = 0.033 for 3395 reflexions.

The Pd atom is in a square planar arrangement involving N and C atoms from the 2-(4'-methoxyphenylazo)-5-methoxyphenyl ligand (Pd-N(2) 2.012, Pd-C(10) 1.939 Å) and O and N atoms from the 8-hydroxyquinolate ligand (Pd-O(1) 2.018, and Pd-N(1) 2.118 Å) (Fig. 1). The four coordinated atoms are strictly coplanar with the Pd atom 0.051 Å from this plane. Angles about Pd are: N(1)-Pd-N(2) 107.7, N(1)-Pd-O(1) 80.9, N(2)-Pd-C(10) 78.3 and O(1)-Pd-C(10) 93.0°. The complexes are associated in pairs in columnar stacks parallel to [100]. The stacking separation in each diadic unit is 3.304 Å with metal-metal distance of 3.424 Å.

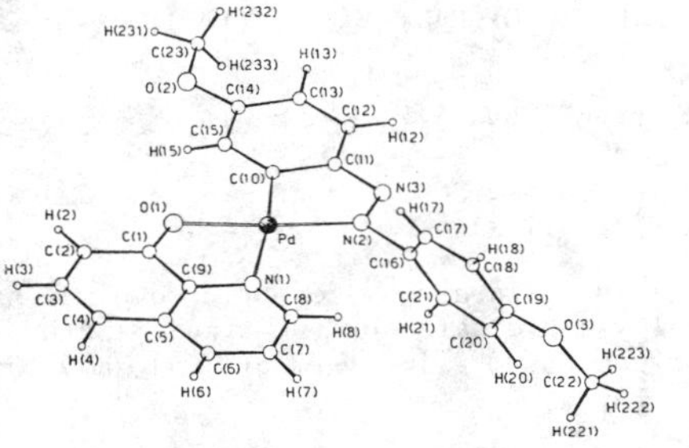

Fig. 1. Molecular structure of $C_{23}H_{19}N_3O_3Pd$.

(μ_2-ETHOXO)(μ_2-ISOTHIOCYANATO)BIS[(2,5-DIFORMYLIMINOFURAN)-N,N'-(μ_2-TRIMETHYLENE)]-DIISOTHIOCYANATODICOBALT(II)
$C_{23}H_{25}Co_2N_7O_3S_3$ (I)

(μ_2-AZIDO)(μ_2-METHOXO)DIAZIDOBIS[(2,5-DIFORMYLIMINOFURAN)-N,N'-(μ_2-TRIMETHYLENE)]-DICOBALT(II)
$C_{19}H_{23}Co_2N_{13}O_3$ (II)

M.G.B. DREW, F.S. ESHO, A. LAVERY and S.M. NELSON, 1984. J. Chem. Soc. Dalton, 545-556.

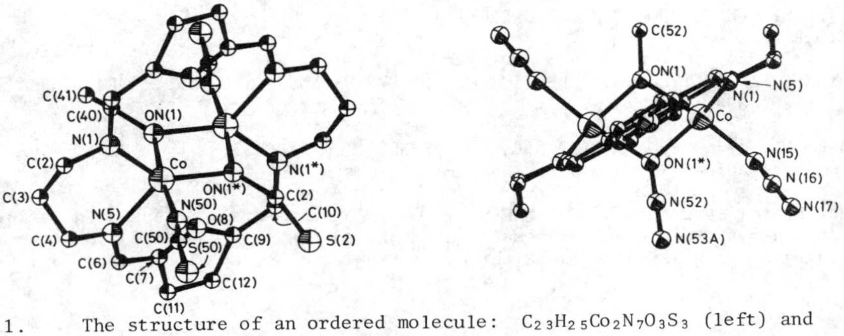

Fig. 1. The structure of an ordered molecule: $C_{23}H_{25}Co_2N_7O_3S_3$ (left) and $C_{19}H_{23}Co_2N_{13}O_3$ (right).

I. Monoclinic, $P2_1/c$, a = 9.03, b = 10.73, c = 14.64 Å, β = 99.0°, D_m = 1.57, Z = 2. Mo radiation, R = 0.099 for 692 reflexions.

II. Monoclinic, $P2_1/n$, a = 8.381, b = 10.09, c = 14.185 Å, β = 96.1°, D_m = 1.67, Z = 2. Mo radiation, R = 0.115 for 700 reflexions.

In both structures the Co atoms are in a distorted trigonal-bipyramidal environment, being bonded to two N atoms of the macrocycle (Co-N 2.09(2), 2.15(2) in I; 2.07(2), 2.12(2) Å in II) and a disorder-d N/O bridge atom (Co-N/O 1.97(2) in I, 2.01(2) Å in II) in the equatorial plane. The axial sites are occupied by a N atom of a terminal anion (Co-NCS^- 2.03(2), Co-N_3^- 2.04(2) Å) and the other disordered N/O bridge atom (Co-N/O 2.09(2) in I, 2.14(2) Å in II). The Co...Co distance is 3.119(6) Å in I and 3.195(6) Å in II. Fig. 1 shows one possible orientation of an ordered molecule of each complex.

(μ-OXO)BIS[DICHLOROTRIS(TETRAHYDROFURAN)VANADIUM(III)]

$C_{24}H_{48}Cl_4O_7V_2$ $[(C_4H_8O)_3Cl_2VOVCl_2(C_4H_8O)_3]$

P. CHANDRASEKHAR and P.H. BIRD, 1984. Inorg. Chem., 23, 3677-3679.

Monoclinic, C2/c, a = 19.181, b = 9.200, c = 18.935 Å, β = 103.09°, Z = 4. Mo radiation, R = 0.097 for 937 reflexions.

The molecule (Fig. 1) is a linear, oxygen-bridged dimer having exact C_2 symmetry. The V atoms are octahedrally coordinated by two trans Cl atoms (V-Cl 2.390(6) and 2.384(6) Å), three THF oxygens (V-O 2.16(1)-2.26(1) Å), and the μ-O atom (V-O_B 1.769(5) Å).

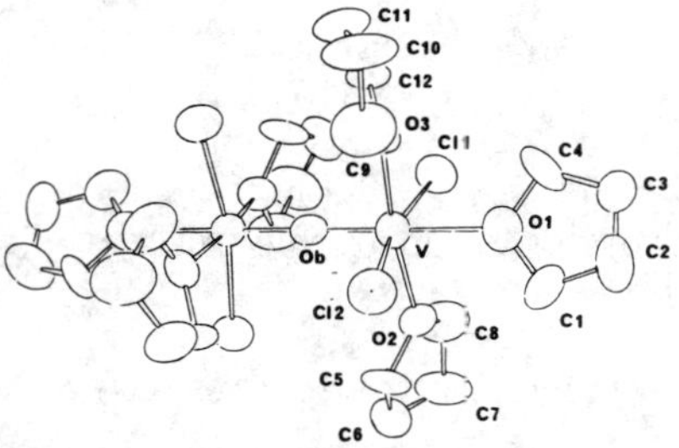

Fig. 1. The $[(C_4H_8O)_3Cl_2VOVCl_2(C_4H_8O)_3]$ molecule.

catena-(μ_2-DI-n-HEXYLPHOSPHINATO)COPPER(II)

$C_{24}H_{52}CuO_4P_2$ $Cu[(n-C_6H_{13})_2PO_2]_2$

J.S. HAYNES, K.W. OLIVER, S.J. RETTIG, R.C. THOMPSON and J. TROTTER, 1984. Can. J. Chem., 62, 891-898.

Triclinic, $P\bar{1}$, a = 9.800, b = 12.336, c = 13.352 Å, α = 88.53, β = 74.02. γ = 82.33°, Z = 2. Mo radiation, R = 0.045 for 3095 reflexions.

The crystal structure of $Cu[(n-C_6H_{13})_2PO_2]_2$ consists of well-separated infinite chains of centrosymmetric spiro-fused eight-membered rings propagating along the crystallographic a axis (Fig. 1), with Cu-O 1.896-1.932(3), P-O 1.513-1.521(3), P-C 1.779-1.807(5) Å.

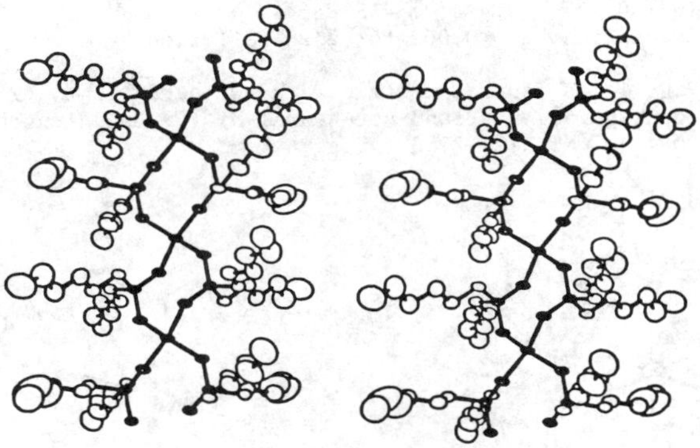

Fig. 1. A stereoview of the $Cu[(n\text{-}C_6H_{13})_2PO_2]_2$ polymeric structure.

BIS(2-AMINOBENZOPHENONE)DICHLOROCOPPER(II)

$C_{26}H_{22}Cl_2CuN_2O_2$ $CuCl_2(C_{13}H_{11}NO)_2$

I.M. VEZZOSI and L. ANTOLINI, 1984. Inorg. Chim. Acta, 85, 155-159.

Triclinic, $P\bar{1}$, a = 13.397, b = 10.752, c = 9.205 Å, α = 72.26, β = 91.58, γ = 106.86°, D_m = 1.48, Z = 2. Cu radiation, R = 0.034 for 2581 reflexions.

The structure consists of discrete monomers showing distorted trigonal bipyramidal coordination geometry about the Cu^{2+} ion (Fig. 1). The amino N atoms are axial ligands, with the equatorial positions occupied by two Cl atoms and a carbonyl O from one aminobenzophenone molecule acting as a bidentate ligand. The amount of distortion may be gauged from the angles Cl(1)-Cu-Cl(2) (141.7°) and N(2)-Cu-O(2) (79.3°). There is a strong intramolecular hydrogen bond with an N(1)...O(1) separation of 2.772 Å, and two intermolecular hydrogen bonds (N-H...Cl interactions) with N...Cl distances of 3.463 and 3.340 Å, (N(1)...Cl(1') and N(2)...Cl(2'') respectively).

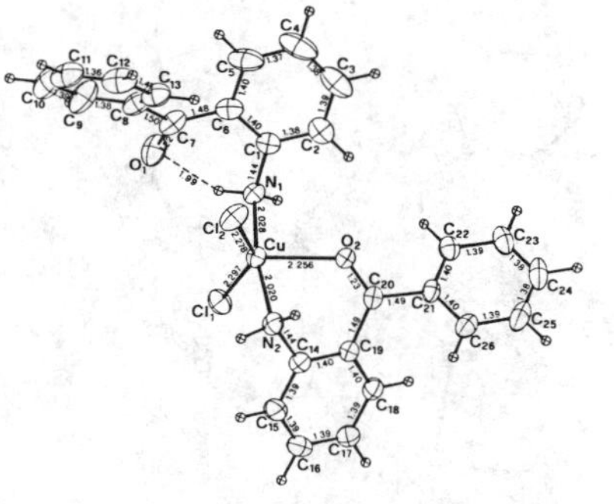

Fig. 1. Molecular structure of $C_{26}H_{22}Cl_2CuN_2O_2$ showing selected bond distances (Å).

BIS[5,5,5-TRIFLUORO-4-(TRIFLUOROMETHYL)-2-(2-(2-PYRIDYL)ETHYL)IMINO-4-PENTANOLATO-O,N,N']NICKEL(II)

$C_{26}H_{26}F_{12}N_4NiO_2$ $Ni[OC(CF_3)_2C(CH_3)NCH_2CH_2(C_5H_4N)]_2$

S.J. LOEB, D.W. STEPHAN and C.J. WILLIS, 1984. Inorg. Chem., 23, 1509-1512.

Triclinic, $P\bar{1}$, a = 10.180, b = 10.444, c = 8.246 Å, α = 110.13, β = 105.00, γ = 106.69°, Z = 1. Mo radiation, R = 0.062 for 2323 reflexions.

The molecule (Fig. 1) is centrosymmetric with two equivalent, tridentate, ligands in a facial configuration around octahedral Ni^{2+}. Bond lengths: Ni-O 2.033(3), Ni-N(1) 2.084(4), Ni-N(2) 2.209(4) Å.

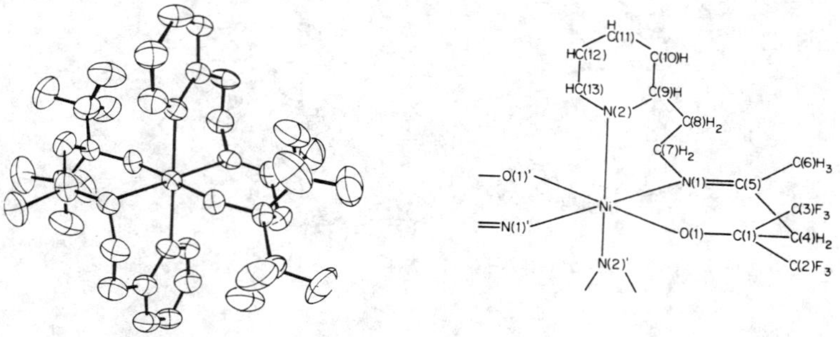

Fig. 1. The $Ni[OC(CF_3)_2C(CH_3)NCH_2CH_2(C_5H_4N)]_2$ molecule.

METHANOLBIS(N-PHENYLBENZOYLHYDROXAMATO)DIOXOURANIUM(VI)

$C_{27}H_{24}N_2O_7U$ $(CH_3OH)UO_2(C_6H_5N(O)C(O)C_6H_5)_2$

U. CASELLATO, P.A. VIGATO, S. TAMBURINI, R. GRAZIANI and M. VIDALI, 1984. Inorg. Chim. Acta, 81, 47-54.

Triclinic, $P\bar{1}$, a = 14.24, b = 10.60, c = 9.52 Å, α = 94.03, β = 105.80, γ = 107.92°, Z = 2. Mo radiation, R = 0.046 for 4026 reflexions.

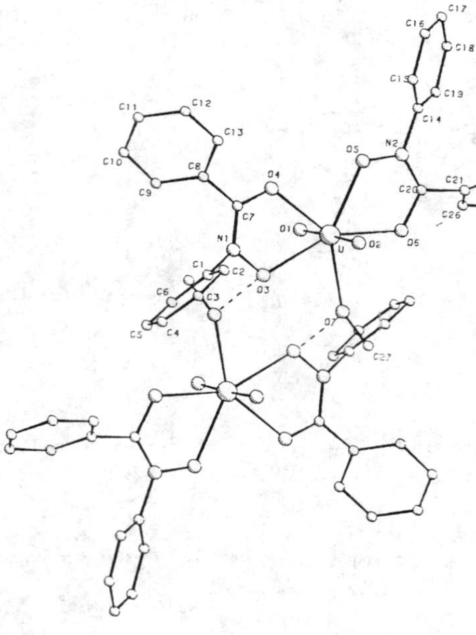

Fig. 1. Molecular structure of $C_{27}H_{24}N_2O_7U$.

Two bidentate hydroxamato ligands and one methanol molecule are equatorially bonded to the linear uranyl group in pentagonal bipyramidal geometry (Fig. 1). Distances in the coordination polyhedron are: U-O(1) 1.75, U-O(2) 1.75, U-O(3) 2.42, U-O(4) 2.40, U-O(5) 2.38, U-O(6) 2.35 and U-O(7) (methanol) 2.37 Å. The O(1)-U-O(2) angle is 177.3°. The phenyl rings are inclined by 50° to the equatorial (pentagonal) plane.

DICHLOROTETRAKIS(BENZAMIDATO)DIOSMIUM(III)

$C_{28}H_{24}Cl_2N_4O_4Os_2$ $Os_2Cl_2(C_6H_5CONH)_4$

A.R. CHAKRAVARTY, F.A. COTTON and D.A. TOCHER, 1984. Inorg. Chim. Acta, 89, L15-L16.

Triclinic, P$\bar{1}$, a = 10.840, b = 14.381, c = 10.682 Å, α = 110.91, β = 105.80, γ = 68.10°, Z = 2. Mo radiation, R = 0.089 for 2019 reflexions.

There are two independent 'half molecules' per asymmetric unit, and their molecular geometries are almost identical. Each molecule has inversion symmetry with each Os atom in an octahedral environment (Fig. 1). The molecule can exist in four possible isomeric forms and in the absence of substituents on the N atom it is difficult to assign atom sites. The R factor is high presumably because of disorder in the molecular framework. The Os-Os and Os-Cl distances are 2.371, 2.364 and 2.504 and 2.466 Å respectively. The Os-O and Os-N distances range from 1.88 to 2.03 Å and from 1.94 to 2.04 Å, and the Os-Os-Cl angles are 175.1 and 172.4°. The Os-Os-O and Os-Os-N angles lie in the ranges 88.7-93.5° and 86.5-89.2° respectively. The average O-C-N angle is 121°.

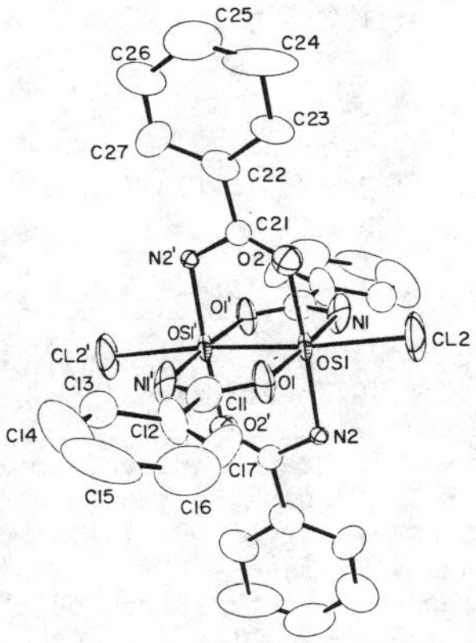

Fig. 1. Molecular structure of one of the independent molecules of $C_{28}H_{24}Cl_2$-$N_4O_4Os_2$.

BIS(2,6-DI-t-BUTYLPHENOXY)TRICHLOROTANTALUM(V)

$C_{28}H_{42}Cl_3O_2Ta$ $Ta(OC_6H_3(t-C_4H_9)_2]_2Cl_3$

L.R. CHAMBERLAIN, I.P. ROTHWELL and J.C. HUFFMAN, 1984. Inorg. Chem., 23, 2575-2578.

Monoclinic, C2/c, a = 17.477, b = 11.909, c = 28.319 Å, β = 97.71°, Z = 8. Mo radiation, R = 0.032 for 3288 reflexions (at -160°C).

The molecular structure (Fig. 1) shows a distorted square pyramidal arrangement about the Ta atom with one OAr ligand occupying the apex and the other a basal coordination site. A t-Bu group of the basal OAr ligand occupies the "open" site trans to the apical OAr, there being a Ta...H distance of 2.67 Å. Bond lengths: Ta-O(apical) 1.836(4), Ta-O(basal) 1.872(5), Ta-Cl 2.339(2), 2.335(2), and 2.358(2) Å.

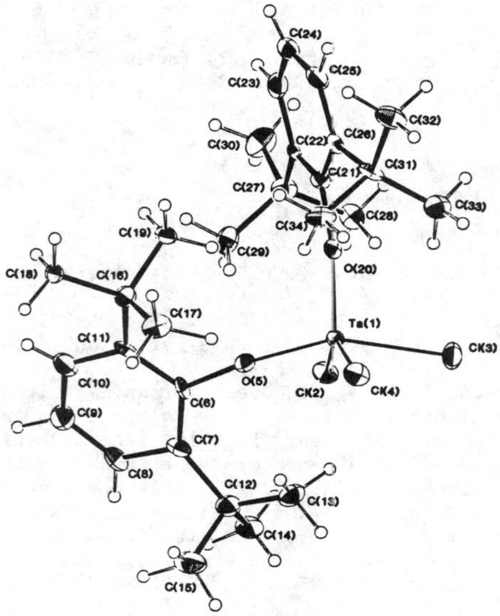

Fig. 1. The $Ta[OC_6H_3(C_4H_9)_2]_2Cl_3$ molecule.

BIS(μ-METHOXO)-BIS[PIMELYL BIS(N-ISOPROPYLHYDROXAMATO-O,O')]DIIRON(III)
$C_{28}H_{54}Fe_2N_4O_{10}$

S.J. BARCLAY, P.E. RILEY and K.N. RAYMOND, 1984. Inorg. Chem., 23, 2005-2010.

Monoclinic, $P2_1/n$, a = 11.453, b = 14.986, c = 11.948 Å, β = 119.046°, D_m = 1.33, Z = 2. Mo radiation, R = 0.027 for 3023 reflexions.

The complex (Fig. 1) has crystallographic C_i and approximate C_{2h} symmetry and is composed of dihydroximate ligands that span the two Fe atoms, in addition to the μ-OCH_3 bridges. The Fe-O distances average 2.002(4) (alkoxide), 2.003(2) (N-O), and 2.038(8) Å (C-O). The O-Fe-O angles average 77.72(4) (intra-ring) and 75.42(5)° (alkoxide). The distortion of the FeO_6 centres from octahedral symmetry results in three trans O-Fe-O angles 163.3, 165.6, and 158.3°.

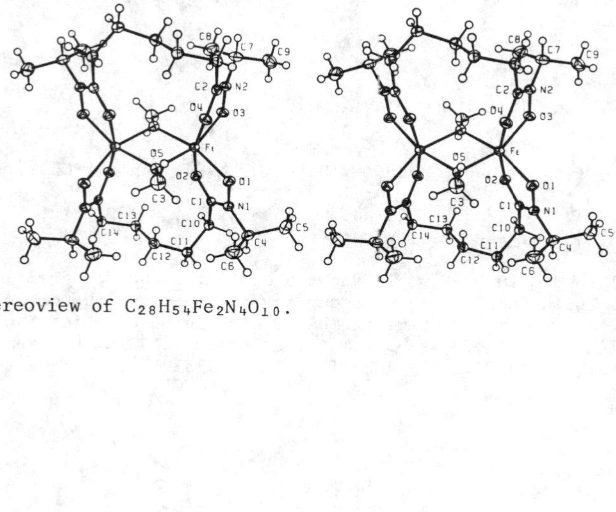

Fig. 1. Stereoview of $C_{28}H_{54}Fe_2N_4O_{10}$.

TRIS(μ_2-CHLORO)HEXAKIS(TETRAHYDROFURAN)DIMAGNESIUM(II) PENTACHLORO(TETRAHYDROFURAN)-TITANIUM(IV)

$C_{28}H_{56}Cl_8Mg_2O_7Ti$ $[C_{24}H_{48}Cl_3Mg_2O_6][C_4H_8Cl_5OTi]$

P. SOBOTA, J. UTKO and T. LIS, 1984. J. Chem. Soc. Dalton, 2077-2089.

Monoclinic, $P2_1/c$, a = 17.16, b = 12.845, c = 20.82 Å, β = 112.80°, D_m = 1.38, Z = 4. Mo radiation, R = 0.044 for 2424 reflexions.

The structure (Fig. 1) consists of complex cations and anions. In the cation the two Mg atoms are bridged by three Cl atoms (average Mg-Cl 2.502(4) Å). The Mg...Mg distance is 3.169(4) Å. The six-coordinate Mg atoms are also coordinated to three thf molecules (Mg-O 2.041(6)-2.102(5) Å). In the complex anion the Ti atom is octahedrally coordinated by five Cl atoms and one thf molecule such that the axial Ti-Cl distance (2.249(3) Å) is shorter than the equatorial Ti-Cl distances (Ti-Cl 2.296(3)-2.318(3) Å). Ti-O(7) is 2.158(5) Å. The average Cl(4)-Ti-Cl(5,6,7,8) distance is 94.2(7)°.

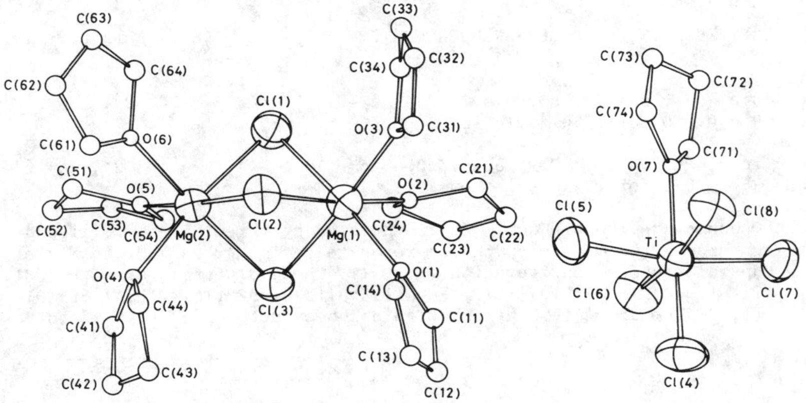

Fig. 1. $C_{28}H_{56}Cl_8Mg_2O_7Ti$: the complex cation (left) and anion (right).

(μ-HYDROXY)[2,6-BIS[[BIS[2-(1-PYRAZOLYL)ETHYL]AMINO]METHYL]-p-CRESOLATO]DICOPPER(II) BIS(TETRAFLUOROBORATE)

$C_{29}H_{38}B_2Cu_2F_8N_{10}O_2$

T.N. SORRELL, D.L. JAMESON and C.J. O'CONNOR, 1984. Inorg. Chem., 23, 190-195.

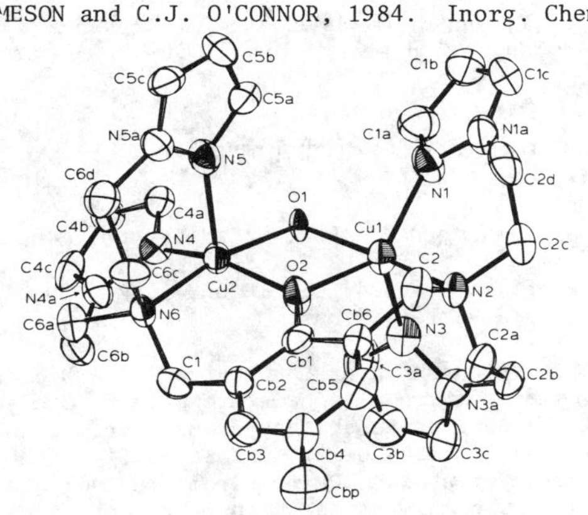

Fig. 1. The $[Cu_2(C_{29}H_{37}N_{10}O)(OH)]^{2+}$ cation.

Monoclinic, $P2_1/c$, a = 22.32, b = 9.551, c = 16.715 Å, β = 95.76°, D_m = 1.61, Z = 4. Mo radiation, R = 0.073 for 2988 reflexions.

The cation is shown in Fig. 1. The two Cu atoms, each bonded to two bridging O atoms and three terminal N atoms, have different coordination geometries. Cu1 is trigonal bipyramidal, O1 and N2 axial, Cu1-O1 1.942(6), Cu1-O2 2.002(8), Cu1-N1 1.987(11), Cu1-N2 2.032(9), and Cu1-N3 2.100(10) Å. Cu2 has tetragonal pyramidal geometry, N5 apical, Cu2-O1 1.941(6), Cu2-O2 1.927(8), Cu2-N4 1.959(10), Cu2-N5 2.148(11), and Cu2-N6 2.049(10) Å.

TETRAPHENYLPHOSPHONIUM TRIS(1-NITROSOETHANAL-N,O'-OXIMATO)FERRATE(II)
$C_{30}H_{29}FeN_6O_6P$ (I) $[P(C_6H_5)_4][Fe(C_2H_3N_2O_2)_3]$

TETRAPHENYLARSONIUM TRIS(1-NITROSOETHANAL-N,O'-OXIMATO)FERRATE(II)
$C_{30}H_{29}AsFeN_6O_6$ (II) $[As(C_6H_5)_4][Fe(C_2H_3N_2O_2)_3]$

P. GOUZERH, Y. JEANNIN, E. MILER-SRENGER and F. VALENTINI, 1982. Acta Cryst., C40, 797-801.

I. Monoclinic, $P2_1$, a = 11.663, b = 12.974, c = 10.272 Å, β = 102.9°, Z = 2. Mo radiation, R = 0.070 for 1330 reflexions.

II. Monoclinic, $P2_1$, a = 11.719, b = 13.038, c = 10.342 Å, β = 103.2°, Z = 2. Mo radiation, R = 0.056 for 2041 reflexions.

The compounds are isomorphous. Fe is chelated by three bidentate ligands coordinating through the O atom of the deprotonated oxime group and the N atom of the nitroso group in a facial configuration (Fig. 1). Bond distances are: Fe-O 1.927(11)-1.963(17) (I), 1.904(7)-1.927(9) (II), Fe-N 1.840(12)-1.922(16) (I), 1.855(9)-1.884(9) Å (II). The absolute configuration has been determined to be Λ in both I and II.

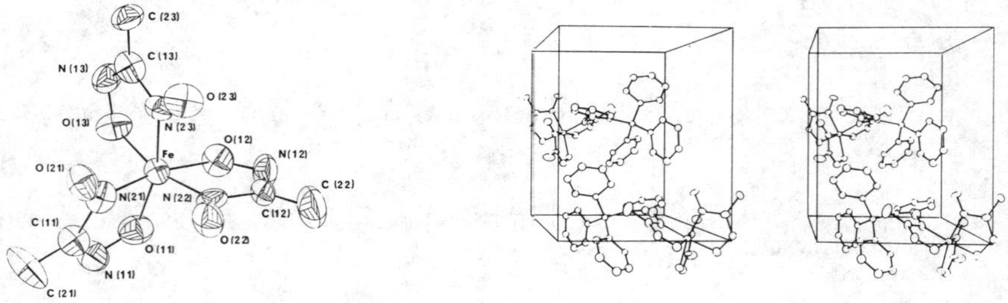

Fig. 1. $C_{30}H_{29}FeN_6O_6P$ (I) and $C_{30}H_{29}AsFeN_6O_6$ (II): perspective view of the $Fe(C_2H_3N_2O_2)_3^-$ anion and stereoview of the molecular packing.

HEXAKIS(PYRIDINE N-OXIDE)COPPER(II) NITRATE DIHYDRATE
$C_{30}H_{30}CuN_8O_{12},2H_2O$ $[Cu(C_5H_5NO)_6](NO_3)_2,2H_2O$

J.S. WOOD, C.P. KEIJZERS and R.O. DAY, 1984. Acta Cryst., C40, 404-406.

Monoclinic, $P2_1/c$, a = 9.556, b = 9.924, c = 18.558 Å, β = 100.43°, D_m = 1.50, Z = 2. Cu radiation, R = 0.042 for 2798 reflexions.

The cation (Fig. 1) lies on an inversion centre and exhibits a statically distorted Jahn-Teller geometry at room temperature with Cu-O distances of 1.954(2), 1.970(2) and 2.478(2) Å in contrast to the anhydrous complex, which has previously been shown to undergo a dynamic Jahn-Teller distortion (1). The cations pack in a pseudo-ferrodistortive arrangement and EPR measurements indicate no exchange between the two magnetically distinct complexes in the unit cell.

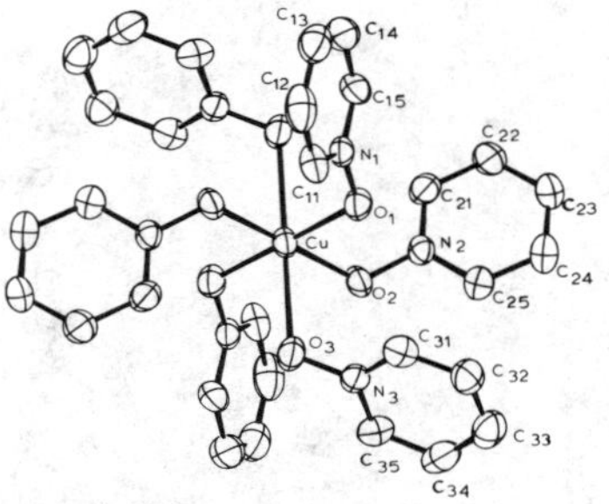

Fig. 1. A perspective view of the $[Cu(C_5H_5NO)_6]^{2+}$ complex ion.

1. Structure Reports, 48B, 1502.

BIS(TRIETHYLAMMONIUM) TRIS(CATECHOLATO(2-))TITANIUM(IV)
$C_{30}H_{44}N_2O_6Ti$ (I) $[(C_2H_5)_3NH]_2(Ti(C_6H_4O_2)_3]$

TETRAPOTASSIUM DI(μ-OXO)BIS[BIS(CATECHOLATO(2-))TITANIUM(IV)] NONAHYDRATE
$C_{24}H_{16}K_4O_{10}Ti_2,9H_2O$ (II) $K_4[TiO(C_6H_4O_2)_2]_2,9H_2O$

BIS(TRIETHYLAMMONIUM) BIS[BIS(3,5-DI-t-BUTYLCATECHOLATO(2-))(3,5-DI-t-BUTYLCATECHOLATO(1-))TITANIUM(IV)] BIS(CHLOROFORM) SOLVATE
$C_{96}H_{154}N_2O_{12}Ti_2,2CHCl_3$ (III) $[(C_2H_5)_3NH]_2[Ti(C_{14}H_{20}O_2)_2(C_{14}H_{21}O_2)]_2,2CHCl_3$

B.A. BORGIAS, S.R. COOPER, Y.B. KOH and K.N. RAYMOND, 1984. Inorg. Chem., 23, 1009-1016.

I. Monoclinic, Ia, a = 9.431, b = 32.634, c = 10.055 Å, β = 100.07°, D_m = 1.25, Z = 4. Mo radiation, R = 0.028 for 3602 reflexions.

II. Monoclinic, I2/a, a = 20.602, b = 6.237, c = 28.099 Å, β = 102.37°, Z = 4. Mo radiation, R = 0.034 for 3371 reflexions.

III. Triclinic, P$\bar{1}$, a = 14.369, b = 14.617, c = 15.330 Å, α = 62.12, β = 81.85, γ = 70.35°, Z = 1. Mo radiation, R = 0.062 for 6722 reflexions.

The complex anions in I-III are shown in Fig. 1. That in I is pseudooctahedral with an average Ti-O bond length of 1.966(27) Å and a mean trigonal twist angle of 43.5°. Strong N-H...O hydrogen bonding with the cations in I perturbs the geometry significantly from the ideal D_3 symmetry. The bis-(μ-oxo) dimer in II has exact C_i and approximate C_{2h} symmetry. The Ti_2O_2 unit has an internal O-Ti-O angle of 82.0(1)° and Ti-O 1.859(1) and 1.876(1) Å, Ti...Ti 2.819(1) Å. The Ti-O(cat) bonds trans to the oxide anions are 0.061 Å longer (2.051(19) vs. 1.990(3) Å) than the cis bonds. In contrast, the binuclear anion in III contains both mono- and dianionic catecholate ligands, and has exact C_i symmetry. The monoanion is monodentate, but one of the bidentate dianion O atoms bridges the two metal atoms. The average Ti-O bridge bonds are longer (2.060(19) vs. 1.887(2) Å) than the bonds trans to them. The monodentate Ti-O bond length is 1.936(2) Å.

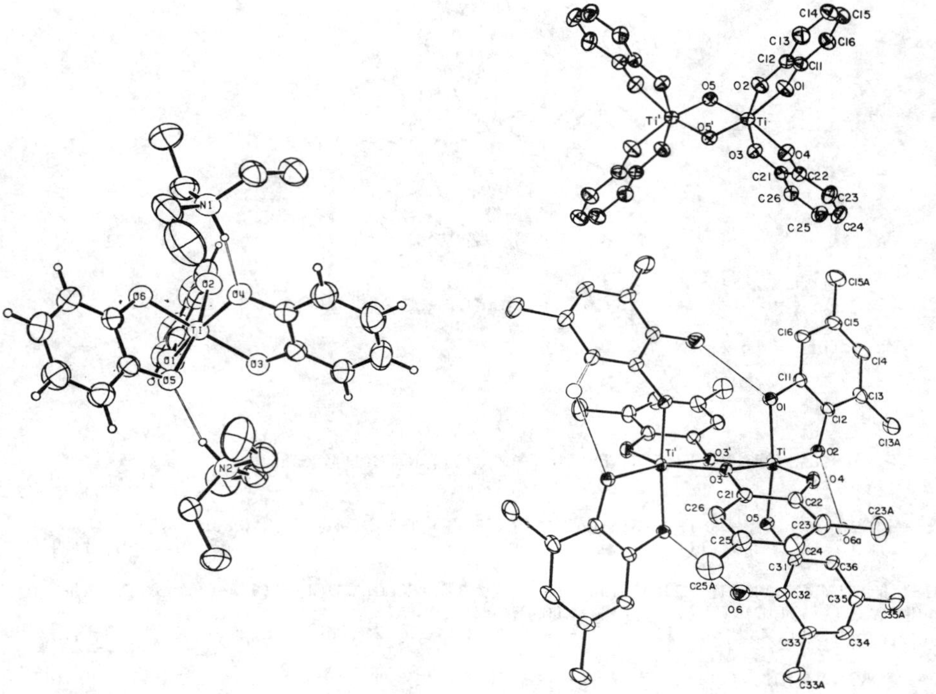

Fig. 1. The $[Et_3NH]_2[Ti(C_6H_4O_2)_3]$ structure (left), the $[TiO(C_6H_4O_2)_2]_2{}^{4-}$ anion (upper right), and the $[Ti(C_{14}H_{20}O_2)_2(C_{14}H_{21}O_2)]_2{}^{2-}$ anion (lower right).

(5,5-DIMETHYL-1,2,8,9-TETRAPHENYL-3,4,6,7-TETRAAZANONA-1,3,6,8-TETRAENE-1,9-DIOLATO-O,O',N,N')NICKEL(II)

$C_{31}H_{26}N_4NiO_2$

F. ABRAHAM, J.M. CAPON, G. NOWOGROCKI, S. SUEUR and C. BRÉMARD, 1984. Acta Cryst., C40, 1355-1357.

Monoclinic, $P2_1/c$, a = 12.63, b = 13.00, c = 19.44 Å, β = 121.47°, D_m = 1.31, Z = 4. Mo radiation, R = 0.042 for 1947 reflexions.

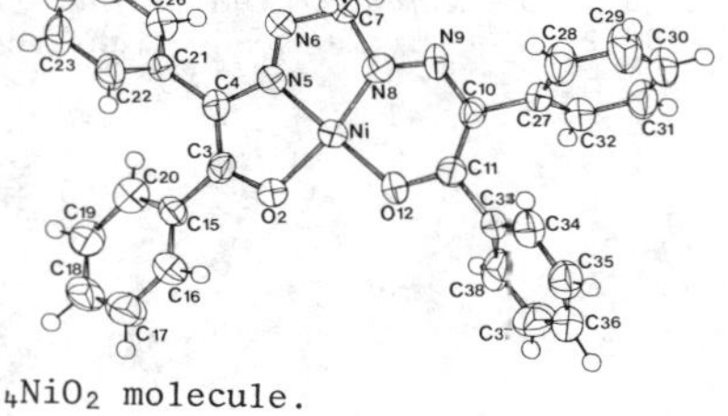

Fig. 1. The $C_{31}H_{26}N_4NiO_2$ molecule.

The macrocycle is tetradentate (Fig. 1) with a cis N_2O_2 donor in an approximately square-planar geometry; Ni is displaced 0.024 Å from the N_2O_2 mean plane. The atoms of the five-membered monoaza and triaza chelate rings and the six-membered diaza chelate ring are approximately coplanar (maximum displacement is 0.103(8) Å for C(10) from the mean plane). The phenyl rings are inclined at 34.6, 59.0, 62.5, 51.6(4)° to the mean plane of the macrocycle.

{N,N'-BIS[1-(PYRROL-2-YL-CARBONYL)PROP-1-EN-2-YL]-1,3-BIS(2-AMINOPHENOXY)PROPANE-N,N',O,O'}COPPER(II)
$C_{31}H_{30}CuN_4O_4$ (I)

(ACETATO-O,O')[N,N'-BIS(PYRROLE-2-CARBALDEHYDE)-3,6-DIOXA-1,8-DIAMINE-N,N',N'',O]-COPPER(II) HEMIHYDRATE
$C_{18}H_{24}CuN_4O_4,0.5(H_2O)$ (II)

N.A. BAILEY, A. BARRASS, D.E. FENTON, M.S.L. GONZALEZ, R. MOODY and C.O.R. de BARBARIN, 1984. J. Chem. Soc. Dalton, 2741-2746.

I. Monoclinic, $P2_1/c$, a = 12.069, b = 20.909, c = 11.958 Å, β = 113.95°, D_m = 1.43, Z = 4. Mo radiation, R = 0.0549 for 1901 reflexions.

II. Triclinic, $P\bar{1}$, a = 15.873, b = 9.693, c = 13.104 Å, α = 93.08, β = 93.38, γ = 95.29°, D_m = 1.48, Z = 4. Mo radiation, R = 0.0363 for 4800 reflexions.

The Cu atom in (I) has square-planar coordination geometry (Fig. 1) with a significant tetrahedral distortion (mean Cu-O 1.910(6), mean Cu-N 2.015(7) Å, O-Cu-O 156.8(3), N-Cu-N 162.8(3)°). There are two independent molecules in (II) with identical conformations and very similar geometry; the Cu coordination is tetragonally distorted octahedral (with mean Cu-N(1) 1.957(3), Cu-N(2) 2.020(3), Cu-N(3) 2.015(3), Cu-O(3) 1.974(3) Å). The axial Cu-O distances are in the range 2.527-2.765(3) Å.

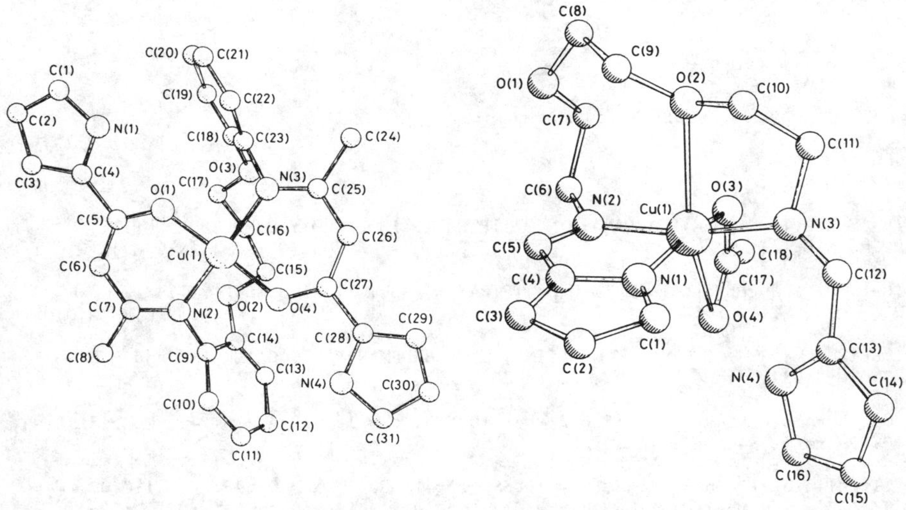

Fig. 1. Views of $C_{31}H_{30}CuN_4O_4$ (I) (left) and one of the $C_{18}H_{24}CuN_4O_4$ molecules (II) (right).

TETRAKIS[μ_3-METHOXO(2,4-DINITROPHENOLATO)(METHANOL)COBALT(II)]
$C_{32}H_{40}Co_4N_8O_{28}$ $[Co(OCH_3)\{2,4-(NO_2)_2C_6H_3O\}CH_3OH)]_4$

Yu.A. SIMONOV, V.K. BEL'SKII, G.S. MATUZENKO and N.V. GÉRBÉLÉU, 1984. Kristallografiya, 29, 82-85 [Sov. Phys. Crystallogr., 29, 47-49].

Tetragonal, $I4_1/a$, a = 12.921, c = 27.233 Å, D_m = 1.75, Z = 4. Mo radiation, R = 0.026 for 1321 reflexions.

The crystal structure consists of cubane complexes (Fig. 1) linked by van der Waals interaction. The intrinsic symmetry of the tetrahedron is S_4. In the tetramer the Co-Co distances are 3.168 Å (x2) and 3.079 Å (x4) and the Co atoms are linked by μ_3-bridging methoxo groups; the Co-O distances are 2.042, 2.094 and 2.100 Å. The coordination polyhedron of Co is formed by three symmetry-related O atoms of the methoxo groups, the O atoms of the methyl alcohol (2.108 Å), and two O atoms of 2,4-dinitrophenol (2.001 and 2.133 Å). The latter acts as a chelate ligand and is linked to the Co via the phenol O and the O of the nitro group in position 2.

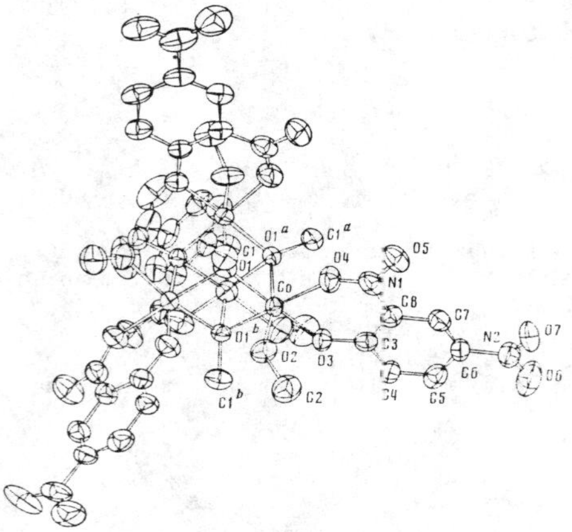

Fig. 1. Structure of the cubane complex $C_{32}H_{40}Co_4N_8O_{28}$.

BIS(μ-ISOPROPYL N-PHENYLCARBAMATO-N,O')TETRAKIS(ISOPROPOXY)DIMOLYBDENUM(III)
$C_{32}H_{52}Mo_2N_2O_8$ (I) $Mo_2(O\text{-}i\text{-}Pr)_4[N(C_6H_5)C(O)O\text{-}i\text{-}Pr]_2$

BIS(μ-t-BUTYL N-PHENYLCARBAMATO-N,O')TETRAKIS(t-BUTOXY)DITUNGSTEN(III)
$C_{38}H_{64}N_2O_8W_2$ (II) $W_2(OCMe_3)_4[N(C_6H_5)C(O)OCMe_3]_2$

M.H. CHISHOLM, F.A. COTTON, K. FOLTING, J.C. HUFFMAN, A.L. RATERMANN and E.S. SHAMSHOUM, 1984. Inorg. Chem., 23, 4423-4427.

I. Orthorhombic, Pnab, a = 18.652, b = 10.808, c = 18.912 Å, Z = 4. Mo radiation, R = 0.099 for 1784 reflexions (at -55°C).

II. Triclinic, $P\bar{1}$, a = 10.48, b = 11.38, c = 11.80 Å, α = 96.96, β = 110.02, γ = 106.93°, Z = 1. Mo radiation, R = 0.037 for 3012 reflexions.

The two molecules are shown in Fig. 1. Molecule I has a C_2 axis bisecting the Mo-Mo bond, and consists of two cis bridging bidentate ligands and four isopropoxy groups. Bond lengths for I: Mo-Mo 2.221(5), Mo-O 2.18(2), Mo-O-i-Pr 1.91(2) and 1.88(2), Mo-N 2.13(3) Å. Molecule II is centrosymmetric and consists of two trans bridging bidentate ligands and four tert-butoxy ligands. Bond lengths for II: W-W 2.290(1), W-O 2.139(5), W-O-t-Bu 1.895(5) and 1.942(5), W-N 2.100(6) Å.

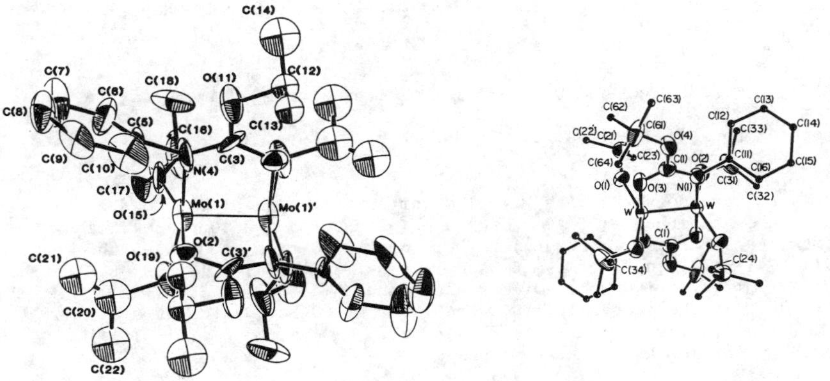

Fig. 1. The structures of $Mo_2(O\text{-}i\text{-}Pr)_4[N(C_6H_5)C(O)O\text{-}i\text{-}Pr]_2$ (left) and $W_2(OCMe_3)_4[N(C_6H_5)C(O)OCMe_3]_2$ (right).

BIS(BENZAMIDE)TRIS(O,O'-DIISOPROPYL DITHIOPHOSPHATO)NEODYMIUM(III)

$C_{32}H_{56}N_2NdO_8P_3S_6$ $[Nd\{(C_3H_7O)_2PS_2\}_3(C_6H_5CONH_2)_2]$

R. NAKAMURA, K. NAGAI, M. SHIMOI and A. OUCHI, 1984. Bull. Chem. Soc. Japan, 57, 2919-2924.

Monoclinic, C2/c, a = 28.887, b = 28.913, c = 24.814 Å, β = 106.72°, D_m = 1.36, Z = 16. Mo radiation, R = 0.066 for 7761 reflexions.

The sixteen Nd(III) ions in one unit cell fall into three categories - Nd(1) is in a general position while Nd(2) and also Nd(3) and P(21) as well as P(31) atoms sit on twofold axes. The general features of the three types of complex molecules are almost the same and are represented by Fig. 1. Each complex molecule is isolated and no bridgings are found between them. The central Nd(III) ion of the complex is octa-coordinated and has dodecahedral geometry (Fig. 1). Three O,O'-diisopropyl dithiophosphato ligands, chelating by both S atoms, span the two "m" edges of a trapezium as well as one "a" edge of the other trapezium, and two O atoms of both the benzamide ligates at both tips of the base edge of the latter. The Nd(1)-S distances range from 2.896 to 2.985 Å and the Nd(1)-O(11) and Nd(1)-O(12) bond lengths are 2.366 and 2.364 Å respectively. The Nd(2)-S and Nd(3)-S distances range from 2.895 to 2.984 and 2.898 to 2.956 Å. respectively, and the Nd(2)-O(21) and Nd(3)-O(31) bonds are 2.380 and 2.381 Å in length respectively.

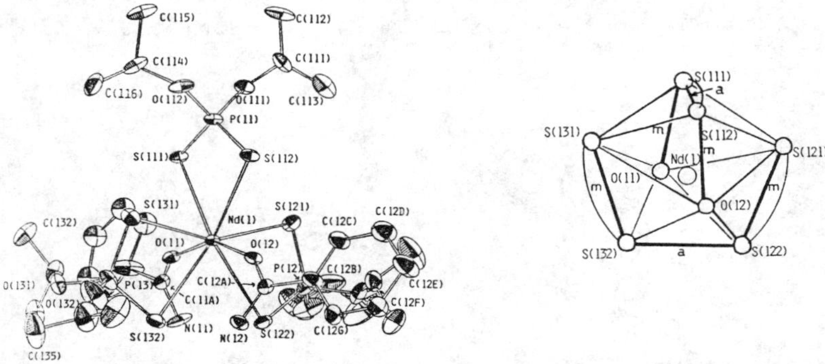

Fig. 1. Perspective drawing of $C_{32}H_{56}N_2NdO_8P_3S_6$, and the dodecahedron geometry around Nd(1).

(μ_2-OXO)(μ_2-1-n-PROPYL-2α-HYDROXYBENZYLBENZIMIDAZOLE-N,O,O)(1-n-PROPYL-2α-HYDROXYBENZYLBENZIMIDAZOLE-N,O)TETRAOXODIMOLYBDENUM

$C_{34}H_{34}Mo_2N_4O_7$ $Mo_2O_5(C_{17}H_{17}N_2O)_2$

B. PIGGOTT, R.N. SHEPPARD and D.J. WILLIAMS, 1984. Inorg. Chim. Acta, 86, L65-L66.

Orthorhombic, Pbca, a = 16.854, b = 18.453, c = 21.720 Å, Z = 8. Cu radiation, R = 0.032 for 3418 reflexions.

The structure consists of a Mo(VI) dimer in which one of the Mo atoms is six-coordinate and the other is five-coordinate (octahedral and square pyramidal arrangements respectively) (Fig. 1). There are two O-bridges between the metal atoms, the Mo(1)-O(2)-Mo(2) angle is 116.6° and that of Mo(1)-O(1)-Mo(2) is 95.8°. The O(1) bridge is asymmetric with Mo(1)-O(1) 2.026 Å and Mo(2)-O(1) 2.358 Å whilst Mo(1)-O(2) and Mo(2)-O(2) are respectively 1.872 and 1.959 Å. Other distances are: Mo(1)-O(3) 1.698, Mo(1)-O(4) 1.691, Mo(1)-N(1) 2.141, Mo(2)-O(5) 1.683, Mo(2)-O(6) 1.699, Mo(2)-O(7) 1.928, Mo(2)-N(20) 2.273 Å.

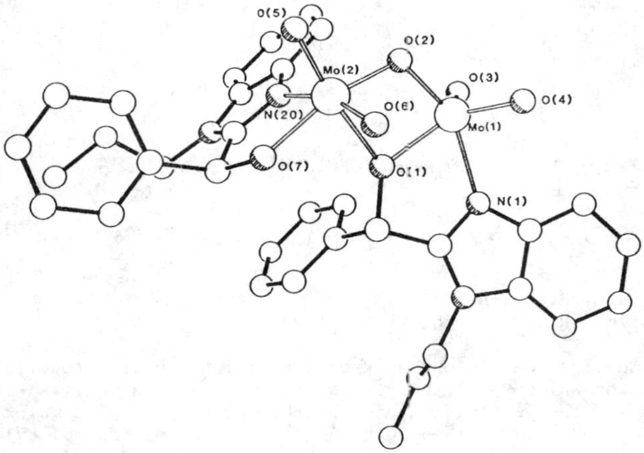

Fig. 1. Molecular structure of $C_{34}H_{34}Mo_2N_4O_7$.

BIS(μ_2-CHLORO)[(DIETHYL ETHER-O,O')LITHIUM]CHLOROBIS(TRI-t-BUTYLMETHOXY)ZIRCONIUM

$C_{34}H_{74}Cl_3LiO_4Zr$ $\{((CH_3)_3C)_3CO\}_2ZrCl_3.Li(O(C_2H_5)_2)_2$

T.V. LUBBEN, P.T. WOLCZANSKI and G.D. VAN DUYNE, 1984. Organometallics, 3, 977-983.

Monoclinic, $P2_1/c$, a = 12.946, b = 14.026, c = 26.257 Å, β = 121.02°, Z = 4. Mo radiation, R = 0.079 for 4346 reflexions (at -75°C).

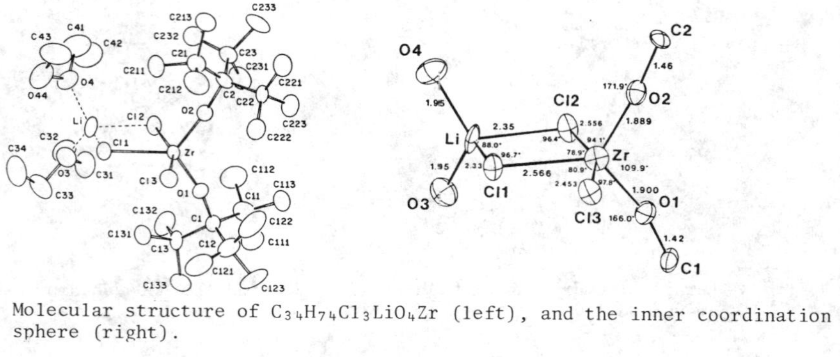

Fig. 1. Molecular structure of $C_{34}H_{74}Cl_3LiO_4Zr$ (left), and the inner coordination sphere (right).

The overall coordination sphere is pseudo-trigonal bipyramidal; the equatorial plane comprising Cl(1), O(1) and O(2) with Cl(2) and Cl(3) axially disposed about Zr (Fig. 1). With the exception of a rather acute (88°) Cl(1)-Li-Cl(2) angle, the geometry about Li is near tetrahedral. The Zr-O bonds are short (average 1.895 Å) and the corresponding Zr-O-C angles are nearly linear (average 169°) indicating the possibility of strong pπ-dπ donation. In the tritox ligand the six central bonds (C-C) average 1.63 Å and this, together with a flattening of the tri-tert-butylmethyl groups (O-C-C angles average 105°; $C\text{-}C_{1,2}\text{-}C$ angles average 113°) shows dramatically the distortion in the bulky ligand.

TETRA-n-BUTYLAMMONIUM (μ_2-BENZAMIDO-N,O)TRIS(μ_2-HYDRIDO)DECACARBONYLTRIRHENIUM
$C_{33}H_{45}N_2O_{11}Re_3$ $[(C_4H_9)_4N]^+[Re_3(\mu\text{-}H)_3(CO)_{10}(\mu\text{-}\eta^2\text{-}OC(C_6H_5)NH)]^-$

T. BERINGHELLI, E. CESAROTTI, G. CIANI, G. D'ALFONSO and A. SIRONI, 1984. J. Organometal. Chem., 268, C18-C22.

Monoclinic, Cc, a = 16.979, b = 16.496, c = 15.124 Å, β = 108.30°, Z = 4. Mo radiation, R = 0.033 for 2366 reflexions.

The anion (Fig. 1) contains an isosceles triangle of Re atoms, with two longer hydrogen-bridged edges (Re(1)-Re(3) 3.289, Re(2)-Re(3) 3.274 Å) and one shorter edge doubly-bridged by a hydride ligand and by the diaxial chelate benzamide ligand (Re(1)-Re(2) 3.153 Å). The hydride atoms, not directly located, are assumed to be almost coplanar with the Re_3 plane on the basis of the disposition of the other ligands. The Re(1)-N, Re(2)-O, N-C and C-O distances are 2.14, 2.13, 1.34 and 1.29 Å respectively, and the Re(1)-N-C, Re(2)-O-C and N-C-O angles are 128, 130, and 124° respectively. The dihedral angle between the phenyl ring plane and the N,C,O,C(1) plane is 30.6°.

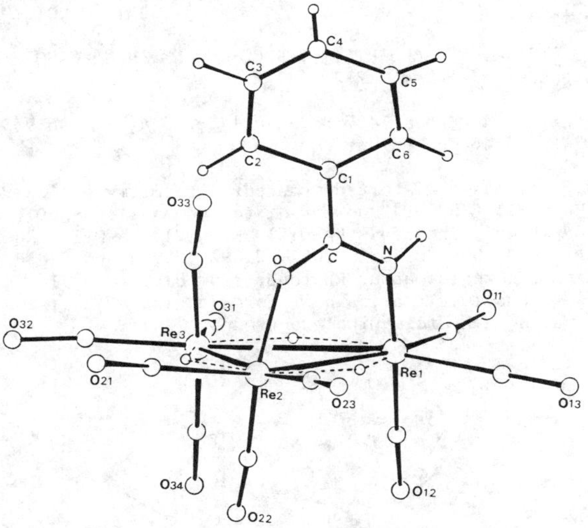

Fig. 1. A view of the anion in $C_{33}H_{45}N_2O_{11}Re_3$.

HEXAKIS(NEOPENTOXY)DIBROMO(PYRIDINE)DIMOLYBDENUM(Mo-Mo)
$C_{35}H_{71}Br_2Mo_2NO_6$ $Mo_2(OCH_2\text{-}t\text{-}Bu)_6Br_2(C_5H_5N)$

M.H. CHISHOLM, J.C. HUFFMAN, A.L. RATERMANN and C.A. SMITH, 1984. Inorg. Chem., 23, 1596-1599.

Triclinic, $P\bar{1}$, a = 19.137, b = 11.215, c = 11.303 Å, α = 113.53, β = 77.57, γ = 95.42°, Z = 2. Mo radiation, R = 0.050 for 2273 reflexions (at -164°C).

The molecule (Fig. 1) has a highly distorted confacial bioctahedral structure with one bridging bromide (Mo-Br 2.736(1) and 2.720(1) Å), two bridging alkoxy ligands (Mo-O 2.020(5)-2.118(5) Å, and Mo-Mo 2.534(1) Å. Other bond lengths: Mo-O (terminal) 1.872(5)-1.905(5), Mo-N(pyr) 2.257(6), and Mo-Br (terminal) 2.584(1) Å.

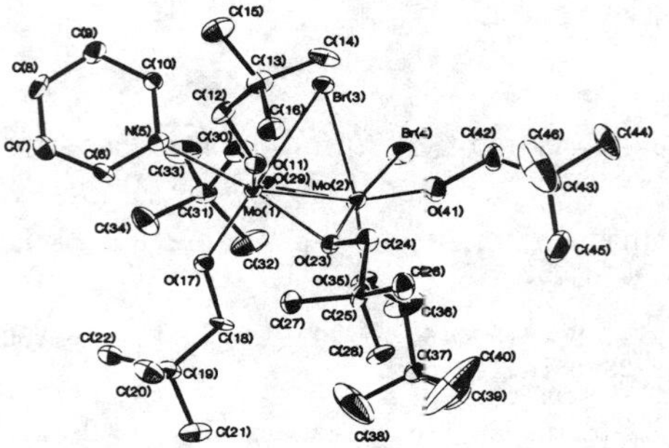

Fig. 1. The $Mo_2(OCH_2\text{-t-Bu})_6Br_2(C_5H_5N)$ molecule.

(2-DIPHENYLPHOSPHINO-3,4,5,6-TETRAFLUOROPHENYL-C,P)BIS(8-HYDROXYQUINOLINATE-N,O)-RHODIUM DICHLOROMETHANE SOLVATE

$C_{36}H_{22}F_4N_2O_2PRh,CH_2Cl_2$ $[Rh(C_9H_6NO)_2(PPh_2C_6F_4)],CH_2Cl_2$

F.L. BARCELÓ, J.C. BESTEIRO, P. LAHUERTA, C. FOCES-FOCES, F.H. CANO and M. MARTÍNEZ-RIPOLL, 1984. J. Organometal. Chem., 270, 343-351.

Monoclinic, $P2_1/n$, a = 15.8494, b = 14.7188, c = 14.6675 Å, β = 96.933°, Z = 4. Cu radiation, R = 0.067 for 5346 reflexions.

The complex (Fig. 1) has a distorted octahedral geometry with a four atom metallocycle-ring (Rh-C(31)-C(32)-P) showing distorted angles of 69.8 and 84.8° at the Rh and P atoms respectively. The bonds Rh-O(2), Rh-N(2), Rh-O(1), Rh-N(1), Rh-P and Rh-C(31) are respectively 2.039, 2.089, 2.034, 2.093, 2.274 and 2.038 Å, and the remaining cis coordination angles about Rh range from 81.0 to 99.1°. The four atom metallocycle ring is virtually planar, and it forms angles of 81.1 and 88.9° with the other two coordinating 8-hydroxyquinolate groups.

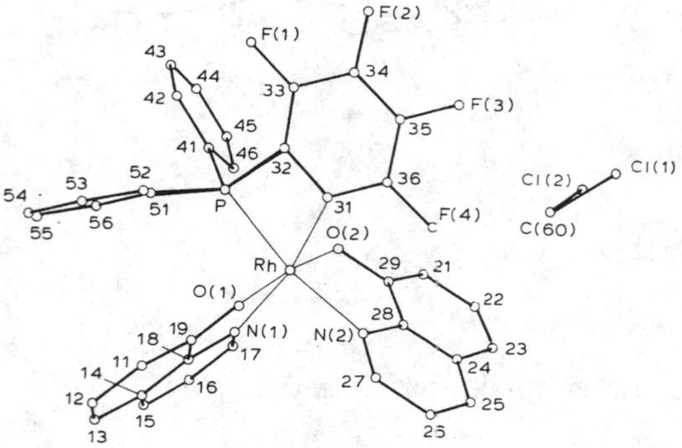

Fig. 1. A view of the complex $C_{36}H_{22}F_4N_2O_2PRh,CH_2Cl_2$.

(μ-OXO)BIS[OXOBIS(8-QUINOLINATO)VANADIUM(V)]
$C_{36}H_{24}N_4O_7V_2$ $[VO(C_9H_6NO)_2]_2O$

S. YAMADA, C. KATAYAMA, J. TANAKA and M. TANAKA, 1984. Inorg. Chem., 23, 253-255.

Monoclinic, C2/c, a = 25.034, b = 9.674, c = 14.003 Å, β = 112.03°, D_m = 1.53, Z = 4. Cu radiation, R = 0.075 for 1707 reflexions.

The dinuclear oxo-bridged molecule (Fig. 1) has exact C_2 symmetry. Each V atom has distorted octahedral coordination geometry with the N atoms coordinated trans to the bridging and terminal oxo ligands. The V-O-V angle is 173.4(4)° and bond lengths involving V are: V=O 1.587(6), V-(μ-O) 1.779(2), V-O 1.893(6) and 1.925(6), V-N 2.310(6) and 2.199(7) Å.

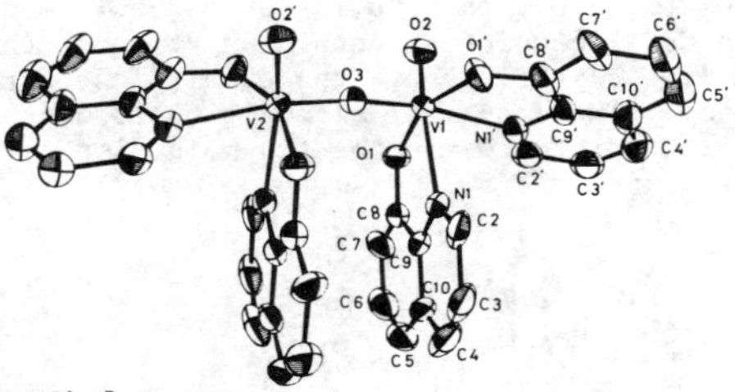

Fig. 1. The $[VO(C_9H_6NO)_2]_2O$ molecule.

TRIBROMOBIS(TRIPHENYLPHOSPHINE OXIDE)THALLIUM(III)
$C_{36}H_{30}Br_3O_2P_2Tl$ $TlBr_3(OPPh_3)_2$

S.E. JEFFS, R.W.H. SMALL and I.J. WORRALL, 1984. Acta Cryst., C40, 381-383.

Monoclinic, $P2_1/c$, a = 14.85, b = 14.60, c = 17.37 Å, β = 107.7°, D_m = 1.712, Z = 4. Mo radiation, R = 0.0542 for 2670 reflexions.

The thallium atom has distorted trigonal-bipyramidal geometry (Fig. 1), with Tl-Br 2.497-2.515(2), Tl-O 2.38, 2.39(2) Å.

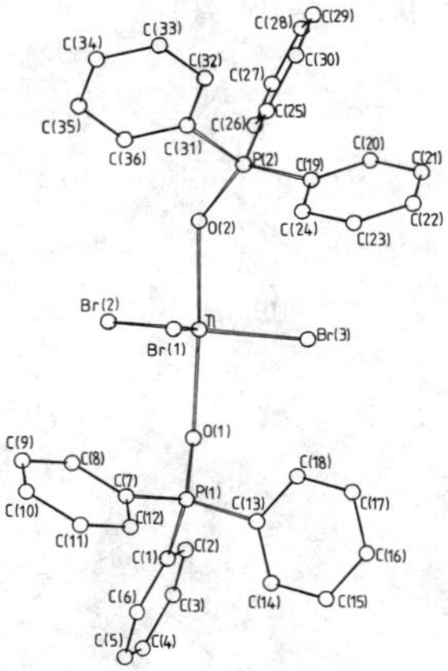

Fig. 1. The structure of $TlBr_3(OPPh_3)_2$.

(μ_2-2,6-BIS(N,N-(BIS(2-PYRIDYL)ETHYL)AMINOMETHYL)PHENOXO-N,N',N'',O)DICOPPER(I) HEXAFLUOROPHOSPHATE
$C_{36}H_{39}Cu_2F_6N_6OP$

K.D. KARLIN, R.W. CRUSE, Y. GULTNEH, J.C. HAYES and J. ZUBIETA, 1984. J. Am. Chem. Soc., 106, 3372-3374.

Monoclinic, $P2_1/n$, a = 13.861, b = 31.482, c = 16.956 Å, β = 98.20°, Z = 8. Mo radiation, R = 0.059 for 2927 reflexions.

There are two crystallographically independent molecules in each asymmetric unit; each molecule consists of two four-coordinate Cu(I) ions ligated to the amino nitrogen and two pyridine donors of the py2 tridentate group. The fourth donor is the bridging phenoxo oxygen. The geometry about each Cu(I) ion is best described as pyramidal, similar to that found in Cu(I) complexes containing "tripod" tetradentate ligands (Fig. 1). Each Cu atom lies slightly above the basal plane formed by the two pyridine donors and the oxygen atom. The longer axial position is occupied by the tertiary amino nitrogen. The Cu...Cu distances in the two molecules are 3.619 and 3.715 Å with corresponding Cu-O-Cu angles of 128.0(4) and 130.0(4)°.

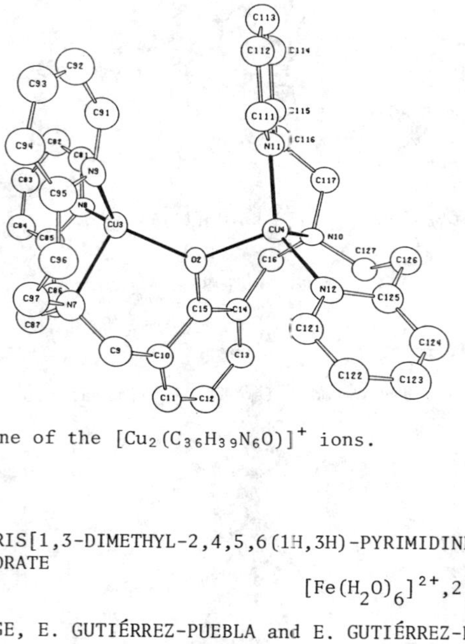

Fig. 1. A view of one of the $[Cu_2(C_{36}H_{39}N_6O)]^+$ ions.

HEXAAQUAIRON(II) BIS{TRIS[1,3-DIMETHYL-2,4,5,6(1H,3H)-PYRIMIDINETETRONE 5-OXIMATO]-FERRATE(II)} DODECAHYDRATE
$C_{36}H_{48}Fe_3N_{18}O_{30},12H_2O$ $[Fe(H_2O)_6]^{2+},2[Fe(C_6H_6N_3O_4)_3]^-,12H_2O$

C. RUIZ-VALERO, A. MONGE, E. GUTIÉRREZ-PUEBLA and E. GUTIÉRREZ-RIOS, 1984. Acta Cryst., C40, 811-812.

Hexagonal, $R\bar{3}$, a = 16.329, c = 21.275 Å, Z = 3. Mo radiation, R = 0.049 for 2078 reflexions.

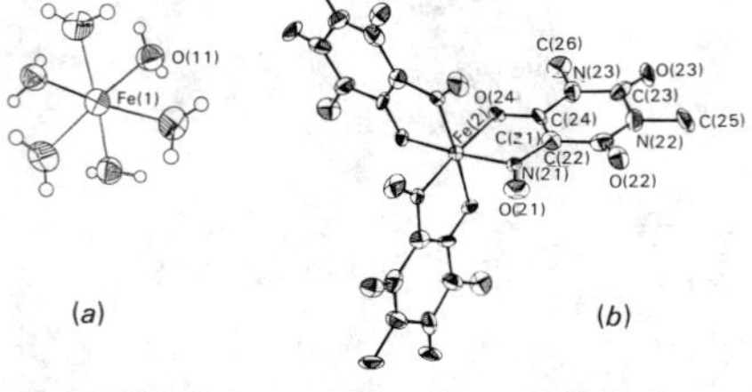

Fig. 1. $C_{36}H_{48}Fe_3N_{18}O_{30},12H_2O$: coordination polyhedron of (a) the $[Fe(H_2O)_6]^{2+}$ cation, and (b) the $[Fe(C_6H_6N_3O_4)_3]^-$ anion.

In the crystal structure the two Fe atoms are located in special positions along the threefold axis, and exhibit octahedral coordination; bond distances are (in the cation) Fe-O 2.136(5) and (in the anion) Fe-N 1.881(5) and Fe-O 1.990(5) Å (Fig. 1). An array of cations, each lying between two anions, is repeated along the threefold axis. Coordinated and uncoordinated water molecules contribute to structural packing through hydrogen bonds.

1,2-BIS(DIMETHYLAMIDO)TETRAKIS(2,6-DIMETHYLPHENOLATO)DIMOLYBDENUM
$C_{36}H_{48}Mo_2N_2O_4$ (I) 1,2-$Mo_2(N(CH_3)_2)_2(OC_8H_9)_4$

BIS(μ-DIMETHYLCARBAMATO)TETRAKIS(2,6-DIMETHYLPHENOLATO)DIMOLYBDENUM
$C_{38}H_{48}Mo_2N_2O_8$ (II) $Mo_2(\mu\text{-}O_2CN(CH_3)_2)_2(OC_8H_9)_4$

T.W. COFFINDAFFER, I.P. ROTHWELL and J.C. HUFFMAN, 1984. Inorg. Chem., 23, 1433-1436.

I. Monoclinic, $P2_1/n$, a = 16.618, b = 10.208, c = 10.369 Å, β = 95.25°, Z = 2. Mo radiation, R = 0.030 for 1865 reflexions (at -168°C).

II. Orthorhombic, Pbna, a = 8.421, b = 20.648, c = 21.665 Å, Z = 4. Mo radiation, R = 0.048 for 1775 reflexions (at -168°C).

Centrosymmetric molecules of I contain a Mo-Mo bond of 2.227(1) Å and a very short Mo-NMe_2 distance of 1.908(4) Å. One of the aryl groups lies over the Mo-Mo bond, while the other lies away from it. The former has a short Mo-O distance (1.926(3) Å) with a large Mo-O-Ar angle (148.7°) while the other has a longer Mo-O bond (1.961(3) Å) and a smaller Mo-O-Ar angle (115.3°). Molecules of II possess exact C_2 symmetry and contain two four-coordinate Mo atoms bridged by two $Mo_2NCO_2^-$ groups which are mutually cis, giving rise to a nearly eclipsed Mo_2O_8 skeleton. The Mo-OAr distances (1.941(5) and 1.953(5) Å) are shorter than those to the carbamate oxygens (2.106(5) and 2.043(5) Å). The Mo-Mo distance in II is 2.220(1) Å.

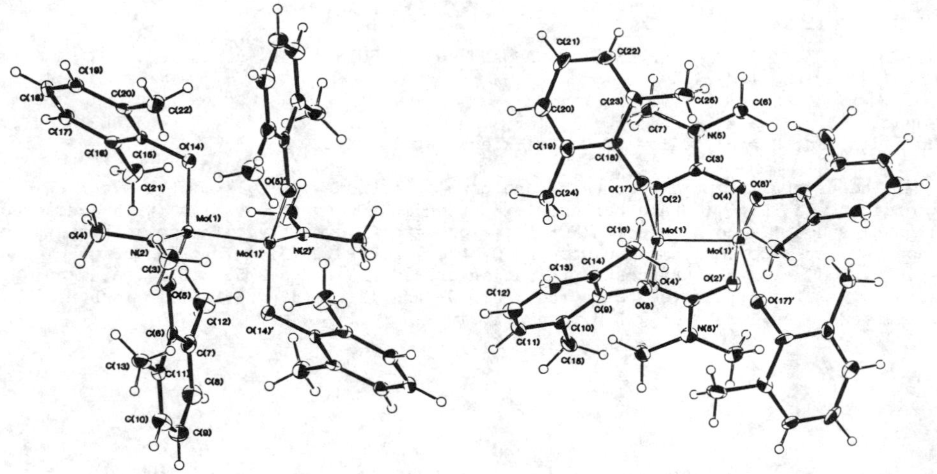

Fig. 1. The 1,2-$Mo_2(N(CH_3)_2)_2(OC_8H_9)_4$ (left) and $Mo_2(\mu\text{-}O_2CN(CH_3)_2)_2(OC_8H_9)_4$ (right) molecules).

BIS(μ_2-t-BUTYLOXO)(μ_2-OXO)BIS(PHENYLIMINO)TETRAKIS(t-BUTYLOXY)DITUNGSTEN
$C_{36}H_{64}N_2O_7W_2$

F.A. COTTON and E.S. SHAMSHOUM, 1984. J. Am. Chem. Soc., 106, 3222-3225.

Monoclinic, C2/c, a = 27.589, b = 18.075, c = 18.393 Å, β = 116.0°, Z = 8. Mo radiation, R = 0.0384 for 5925 reflexions (at -104°C).

The molecule (Fig. 1) consists of face sharing octahedra with two t-butoxide units and an oxygen atom as bridging groups. The W atoms have formal oxidation numbers of VI and no W-W bond exists. Principal dimenaions are: W-O(oxo) 1.937(4), W-O(t-butyloxo) 2.078, 2.092, 2.264, 2.269(4), W-N 1.737 and 1.738(5) Å, mean W-N-C 164.7(3)°.

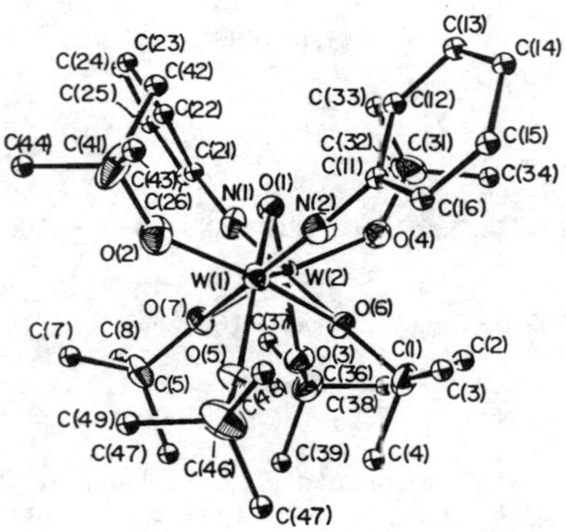

Fig. 1. A view of $C_{36}H_{64}N_2O_7W_2$.

TRIS[DINITRATO(PENTAETHYLENEGYLCOL DIMETHYL ETHER)LANTHANUM(III)] HEXANITRATOLANTHANATE(III)

$C_{36}H_{78}La_4N_{12}O_{54}$

J.-C.G. BUNZLI, J.-M. PFEFFERLE, B. AMMANN, G. CHAPUIS and F.-J. ZUNIGA, 1984. Helv. Chim. Acta, 67, 1121-1127.

Monoclinic, I2, a = 12.234, b = 11.105, c = 26.613 Å, β = 90.65°, D_m = 1.865, Z = 2. Mo radiation, R = 0.068 for 2770 reflexions (at 150 K).

The crystal structure contains one complex cation with C_2 symmetry with ten-coordinate La (Fig. 1), another complex cation in a general location (with a disordered ligand) and a 12-coordinate $[La(NO_3)_6]^{3-}$ ion with C_2 symmetry. The polyether systems adopt a ring-like conformation with La-O(ether) 2.54-2.79(3) Å for the cation with C_2 symmetry. The La-O distances in the anion are in the range 2.56-2.67(3) Å, mean 2.64(4) Å. The effective atomic radii of 10- and 12-coordinate La(III) ions are estimated to be 1.28 and 1.33 Å respectively.

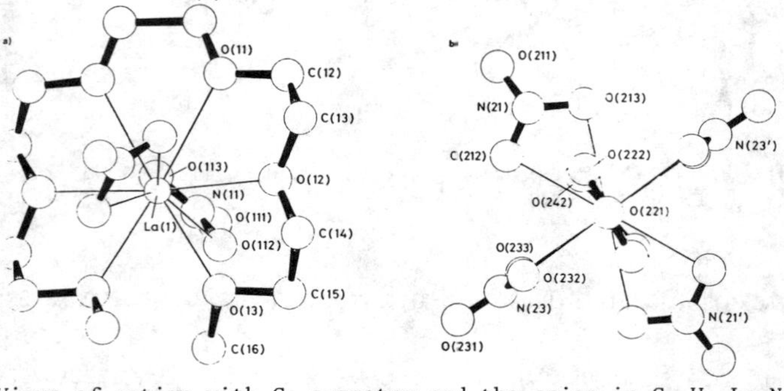

Fig. 1. Views of cation with C_2 symmetry and the anion in $C_{36}H_{78}La_4N_{12}O_{54}$.

POTASSIUM TRIS(μ-t-BUTOXO)HEXAKIS(t-BUTOXY)DIURANIUM(IV) HEXANE SOLVATE
$C_{36}H_{81}KO_9U_2,C_6H_{14}$ (I) $KU_2(O\text{-}t\text{-}C_4H_9)_9.C_6H_{14}$

TRIS(μ-t-BUTOXO)HEXAKIS(t-BUTOXY)DIURANIUM(V,IV)
$C_{36}H_{81}O_9U_2$ (II) $U_2(O\text{-}t\text{-}C_4H_9)_9$

BIS(μ-ISOPROPOXO)OCTAKIS(ISOPROPOXY)DIURANIUM(V)
$C_{30}H_{70}O_{10}U_2$ (III) $U_2(O\text{-}i\text{-}C_3H_7)_{10}$

F.A. COTTON, D.O. MARLER and W. SCHWOTZER, 1984. Inorg. Chem., 23, 4211-4215.

I. Monoclinic, $P2_1/c$, a = 10.713, b = 25.990, c = 19.480 Å, β = 91.13°, Z = 4. Mo radiation, R = 0.065 for 2527 reflexions.

II. Orthorhombic, Pbcm, a = 13.749, b = 19.977, c = 16.923 Å, Z = 4. Mo radiation, R = 0.044 for 2307 reflexions (at -105°C).

III. Triclinic, $P\bar{1}$, a = 10.974, b = 12.226, c = 10.002 Å, α = 111.56, β = 110.09, γ = 67.87°, Z = 1. Mo radiation, R = 0.048 for 2719 reflexions.

The three complexes are shown in Fig. 1. I and II each contain confacial bioctahedra with U...U distances of 3.631(2) and 3.549(1) Å, respectively. Molecules of III consist of edge-sharing octahedra with a U...U distance of 3.789(1) Å. The terminal alkoxide groups in all three compounds display nearly linear U-O-C arrangements, with U-O-C angles between 160 and 176°. The bridging and terminal U-O distances are: 2.35(2)-2.44(2) and 2.10(2)-2.14(2) Å for I, 2.457(9)-2.502(7) and 2.106(7)-2.107(10) Å for the U(IV) atom in II, 2.251(7)-2.271(9) and 2.062(10)-2.073(7) Å for the U(V) atom in II, 2.28(1)-2.29(1) and 2.02(1)-2.05(1) Å for III. Compound I is a tight ion pair with the K^+ ion cradled by two μ-O and two terminal O atoms and mean distances of 2.75(2) and 2.93(2) Å respectively.

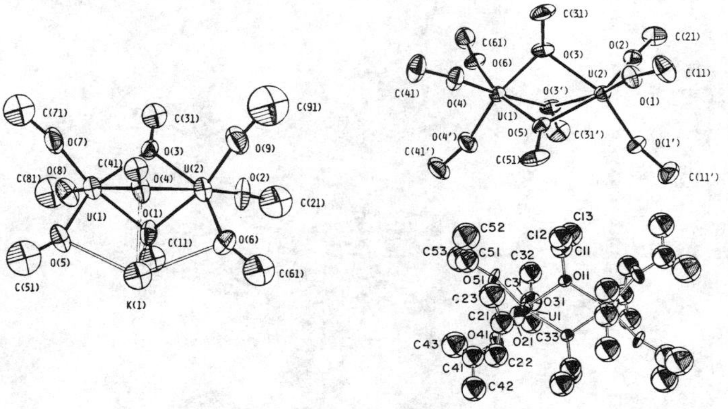

Fig. 1. $KU_2(O\text{-}t\text{-}C_4H_9)_9$ ion pair (left), and the $U_2(O\text{-}t\text{-}C_4H_9)_9$ (upper right) and $U_2(O\text{-}i\text{-}C_3H_7)_{10}$ (lower right) molecules. Central portions only shown.

TETRA-μ-NITRITO(O,N)-TETRAKIS[NITRO(TRI-n-PROPYLPHOSPHINE)PALLADIUM(II)]
$C_{36}H_{84}N_8O_{16}P_4Pd_4$ $[Pd_4(NO_2)_4(ONO)_4\{P(C_3H_7)_3\}_4]$

G. BOMBIERI, G. BRUNO, M. CUSUMANO and G. GUGLIELMO, 1984. Acta Cryst., C40, 409-412.

Monoclinic, $P2_1/n$, a = 12.870, b = 27.642, c = 8.565 Å, β = 101.0°, Z = 2. Mo radiation, R = 0.052 for 1997 reflexions.

The tetrameric molecule (Fig. 1) has an approximate square of Pd atoms linked by nitro-nitrito bridges. The Pd coordination is square-planar, with Pd-P 2.247, 2.248(4), Pd-N 2.00-2.04(1), Pd-O 2.14, 2.16(1) Å.

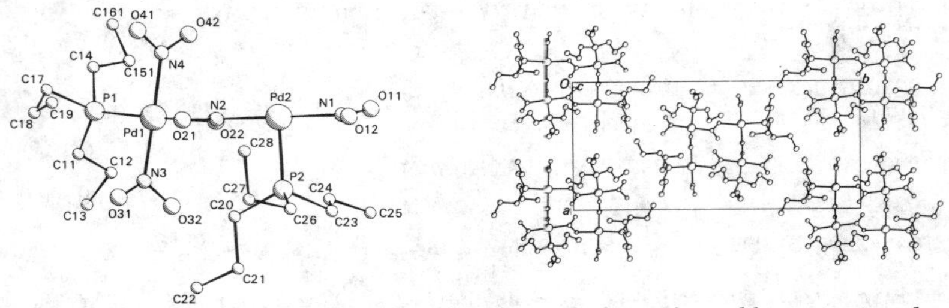

Fig. 1. The asymmetric unit and projection of the unit cell contents along c for $C_{36}H_{84}N_8O_{16}P_4Pd_4$.

trans-[η^4-1,2-BIS(3,5-DICHLORO-2-HYDROXYBENZAMIDO)-4,5-DICHLOROBENZENE]BIS(4-t-BUTYLPYRIDINE)COBALT(IV)
$C_{38}H_{32}Cl_6CoN_4O_4$

F.C. ANSON, T.J. COLLINS, R.J. COOTS, S.L. GIPSON and T.G. RICHMOND, 1984. J. Am. Chem. Soc., 106, 5037-5038.

Monoclinic, $P2_1/c$, a = 11.73, b = 24.22, c = 13.90 Å, β = 100.8°, Z = 4. Mo radiation, R = 0.075 for 2004 reflexions.

The structure was determined to be as shown in Fig. 1.

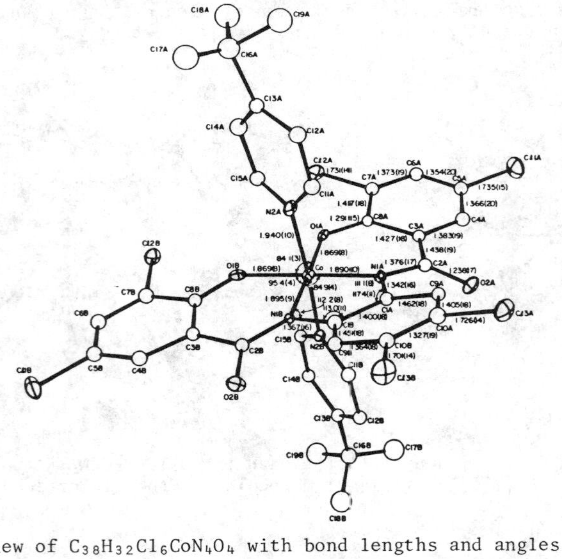

Fig. 1. A view of $C_{38}H_{32}Cl_6CoN_4O_4$ with bond lengths and angles.

BIS(TETRAETHYLAMMONIUM) BIS[(μ_2-SULPHIDO)(O,O'-BIPHENOLATO-O,O')IRON]
$C_{40}H_{56}Fe_2N_2O_4S_2$ (I)

BIS(TETRAETHYLAMMONIUM) BIS(μ_2-SULPHIDO)TETRAKIS(PYRROLATO-N)DIIRON
$C_{32}H_{56}Fe_2N_6S_2$ (II)

D. COUCOUVANIS, A. SALIFOGLOU, M.G. KANATZIDIS, A. SIMOPOULOS and V. PAPAEFTHYMIOU, 1984. J. Am. Chem. Soc., 106, 6081-6082.

I. Monoclinic, C2/c, a = 11.500, b = 13.541, c = 26.612 Å, β = 92.25°, D_m = 1.30, Z = 4. Mo radiation, R = 0.032 for 2108 reflexions.

II. Monoclinic, $P2_1/n$, a = 9.689, b = 16.362, c = 11.910 Å, β = 97.74°, D_m = 1.32, Z = 2. Mo radiation, R = 0.069 for 1197 reflexions.

The dianions of (I) and (II) are located on crystallographic inversion centres (Fig. 1) and the Fe_2S_2 moieties are similar. Details of geometry for (I) are: mean Fe-S 2.215, Fe-O(1) 1.895(2), Fe-O(2) 1.892(2) Å, S-Fe-S 104.9(1), O-Fe-O 96.1(1)°; for (II) mean Fe-S 2.18, Fe-N(1) 1.96(1), Fe-N(2) 2.09(3) Å, S-Fe-S 104.3(4), N-Fe-N 110.6(6)°.

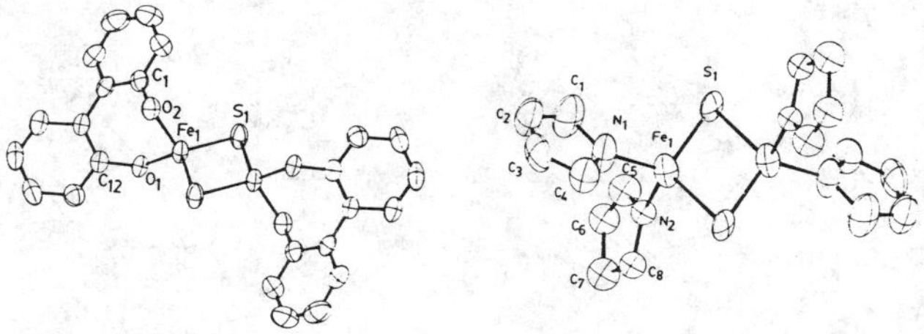

Fig. 1. Views of the anions in $C_{40}H_{56}Fe_2N_2O_4S_2$ and $C_{32}H_{56}Fe_2N_6S_2$.

BIS(2,4,6-TRI-t-BUTYLPHENOXY-O)BIS(DIMETHYLAMINO)TITANIUM(IV)
$C_{40}H_{70}N_2O_2Ti$

R.A. JONES, J.G. HEFNER and T.C. WRIGHT, 1984. Polyhedron, 3, 1121-1124.

Monoclinic, $P2_1/n$, a = 11.641, b = 24.237, c = 14.375 Å, β = 93.606°, Z = 4. Mo radiation, R = 0.0816 for 7689 reflexions.

The Ti atom has distorted tetrahedral geometry (Fig. 1), with Ti-O1 1.813(4), Ti-O2 1.860(3), Ti-N1 1.897(5), Ti-N2 1.864(4) Å, O-Ti-O 115.2(2), N-Ti-N 98.6(2), Ti-O-C 152.0(3) and 171.0(3)°. Both N atoms have planar geometry. The dimensions are interpreted as indicating the presence of substantial O→Ti and N→Ti π-bonding.

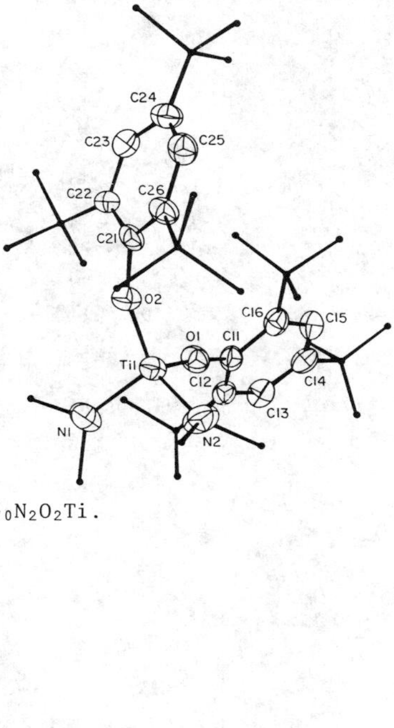

Fig. 1. A view of $C_{40}H_{70}N_2O_2Ti$.

DIPOTASSIUM TRIS(3,5-DI-t-BUTYLCATECHOLATO)MANGANATE(IV) HEXAKIS(ACETONITRILE) SOLVATE
$C_{42}H_{20}K_2MnO_6,6C_2H_3N$ $K_2[Mn(C_{14}H_{20}O_2)_3],6CH_3CN$

J.R. HARTMAN, B.M. FOXMAN and S.R. COOPER, 1984. Inorg. Chem., 23, 1381-1387.

Trigonal, R$\bar{3}$, a = 14.760, c = 50.752 Å, Z = 6. Mo radiation, R = 0.069 for 1838 reflexions.

The $[Mn(C_{14}H_{20}O_2)_3]^{2-}$ anion has crystallographic threefold symmetry with short Mn-O bond lengths and no evidence of Jahn-Teller distortion (Fig. 1).

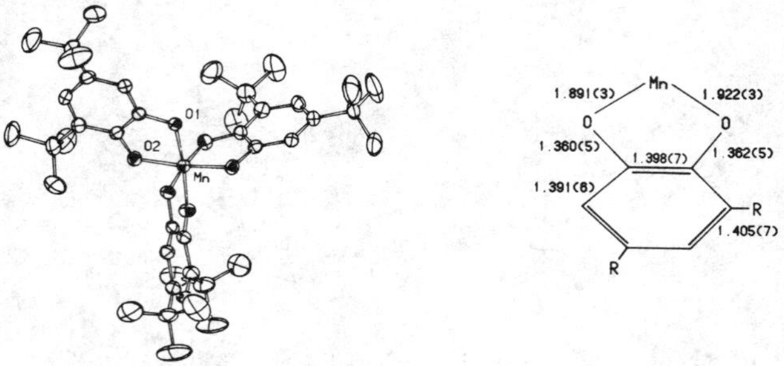

Fig. 1. The $[Mn(C_{14}H_{20}O_2)_3]^{2-}$ anion, bond lengths at right.

CUPFERRONATOBIS(TRIPHENYLPHOSPHINE)COPPER(I)
$C_{42}H_{35}CuN_2O_2P_2$

J. CHARALAMBOUS, L.I.B. HAINES, N.J. HARRIS, K. HENRICK and F.B. TAYLOR, 1984. J. Chem. Res., (S) 220-221, (M) 2101-2162.

Triclinic, P$\bar{1}$, a = 10.097, b = 12.955, c = 14.740 Å, α = 88.384, β = 104.438, γ = 103.332°, Z = 2. Mo radiation, R = 0.042 for 4642 reflexions.

The structure is shown in Fig. 1. The Cu atom has tetrahedral geometry, with Cu-O(1) 2.102, Cu-O(2) 2.078, Cu-P(1) 2.240, Cu-P(2) 2.283 Å, P-Cu-P 125.6, O-Cu-O 75.1, O-Cu-P 104.2-115.0°.

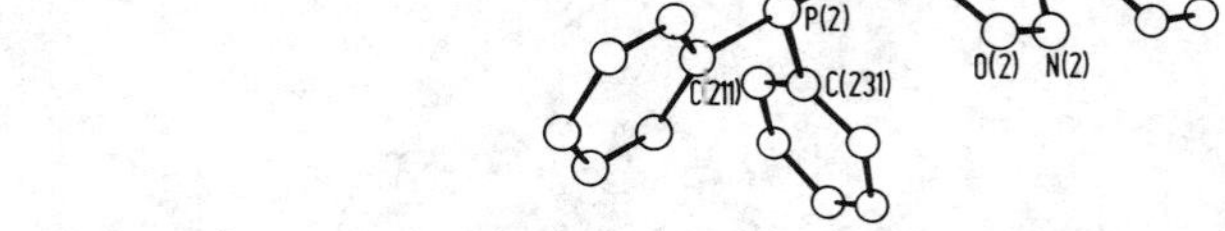

Fig. 1. A view of cupferronatobis(triphenylphosphine)copper(I).

DIHYDRIDOBIS(ACETONE)BIS(TRIPHENYLPHOSPHINE)IRIDIUM(III) TETRAFLUOROBORATE
$C_{42}H_{44}BF_4IrO_2P_2$ $[IrH_2((CH_3)_2CO)_2(P(C_6H_5)_3)_2]BF_4$

R.H. CRABTREE, G.G. HLATKY, C.P. PARNELL, B.E. SEGMÜLLER and R.J. URIARTE, 1984. Inorg. Chem., 23, 354-358.

Monoclinic, $P2_1/c$, a = 10.565, b = 25.315, c = 15.460 Å, β = 95.74°, Z = 4. Mo radiation, R = 0.056 for 2235 reflexions.

The complex cation (Fig. 1) has octahedral coordination geometry with trans phosphine ligands (I...P 2.313(2) and 2.321(2) Å). The trans influence of the H^- ligands on the O-bound acetone groups is pronounced: Ir-O 2.235(5) and 2.220(5) Å, 10% longer than expected.

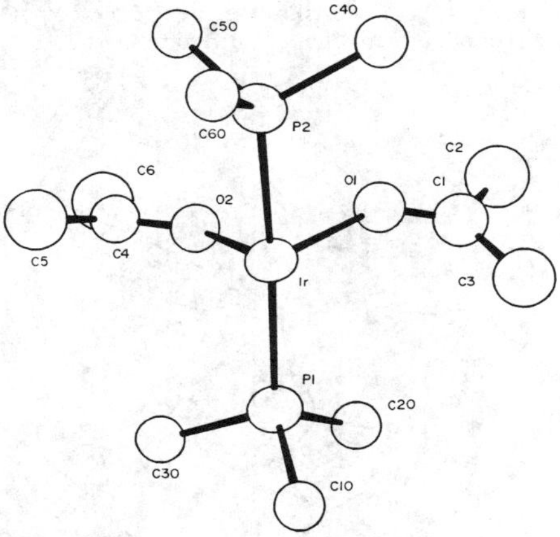

Fig. 1. The $[IrH_2(Me_2CO)_2(PPh_3)_2]^+$ cation, H^- ligands not shown.

DIFLUORO-(μ_2-1,1,10,10-TETRAKIS(1-METHYLBENZIMIDAZOL-2-YL-METHYL)-1,10-DIAZA-4,7-DIOXADECANE)DICOPPER(II) BIS(TETRAFLUOROBORATE) DIHYDRATE
$C_{42}H_{48}B_2Cu_2F_{10}N_{10}O_2,2(H_2O)$ $[Cu_2(megtb)F_2](BF_4)_2,2H_2O$

J. VAN RIJN and J. REEDIJK, 1984. Rec. J.R. Neth. Chem. Soc., 103, 78-79.

Orthorhombic, Pccn, a = 12.159, b = 16.909, c = 25.654 Å, Z = 4. R = 0.068 for 1685 reflexions.

The molecule has crystallographic twofold symmetry and the Cu coordination is close to square pyramidal (Fig. 1) with Cu-N(7) 2.107(8), Cu-N(11) 1.945(9), Cu-N(21) 1.955(9), Cu-F(2) 1.861(6) and Cu-O(4A) 2.35(2) Å.

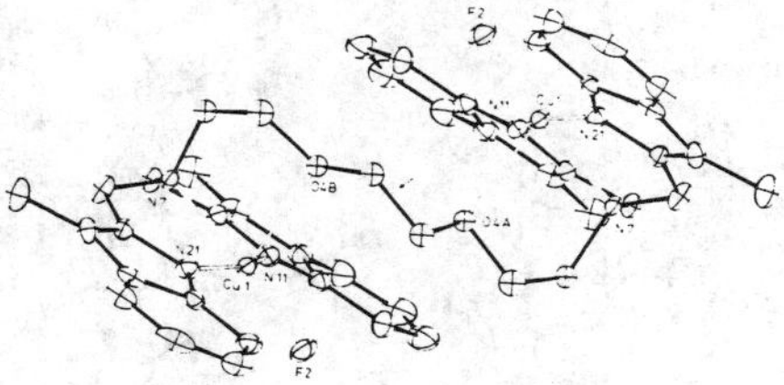

Fig. 1. A view of the $[Cu_2(megtb)F_2]^{2+}$ cation.

LANTHANUM TRIS(CYCLOPENTADIENYLMOLYBDENOTRICARBONYLATE) PENTA(TETRAHYDROFURANATE) TETRAHYDROFURAN SOLVATE

$C_{44}H_{55}LaMo_3O_{14},C_4H_8O$ $(C_4H_8O)_5La[C_5H_5Mo(CO)_3]_3,C_4H_8O$

A.A. PASYNSKII, I.L. EREMENKO, G.Z. SULEIMANOV, Yu.A. NURIEV, I.P. BELETSKAYA, V.E. SHKLOVER and Yu.T. STRUCHKOV, 1984. J. Organometal. Chem., 266, 45-52.

Triclinic, P$\bar{1}$. a = 11.316, b = 11.388, c = 20.757 Å, α = 98.42, β = 101.65, γ = 92.53°, Z = 2. Mo radiation, R = 0.048 for 6904 reflexions (-120°C).

The La atom is bound to three equivalent CpMo(CO)$_3$ fragments via almost linear CO bridges Mo-C-O-La (Mo-C-O 177.9°) and the coordination (Fig. 1) is complete with the binding of five THF molecules (La-O av. 2.537 Å). The coordination of La with the terminal CO group (La-O av. 2.444 Å) leads to elongation of the C-O bond from an average value of 1.16 to 1.21 Å, and to shortening of the Mo-C bond from an average value of 1.943 to 1.874 Å compared to the unbridged CO.

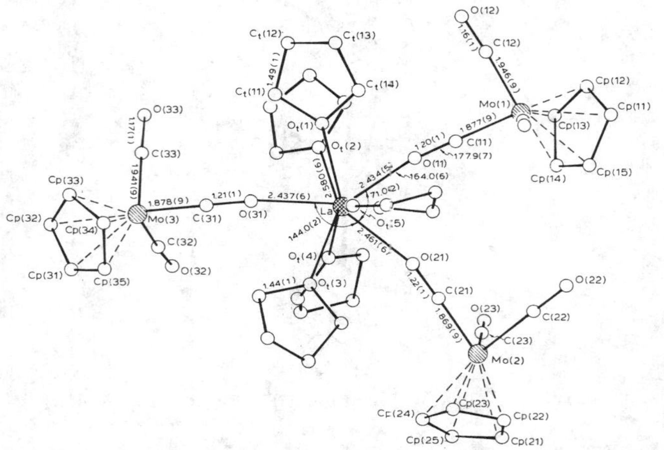

Fig. 1. Molecular structure of $C_{44}H_{55}LaMo_3O_{14}$ showing main bond lengths (Å) and angles (°).

trans-(N,N'-ETHYLENEBIS(ACETYLACETONE IMINATO))BIS(TRIPHENYLPHOSPHINE)TECHNETIUM(III) HEXAFLUOROPHOSPHATE

$C_{48}H_{48}F_6N_2O_2P_3Tc$ $[Tc((acac)_2en)(PPh_3)_2]PF_6$

S.S. JURISSON, K. DANCEY, M. McPARTLIN, P.A. TASKER and E. DEUTSCH, 1984. Inorg. Chem., 23, 4743-4749.

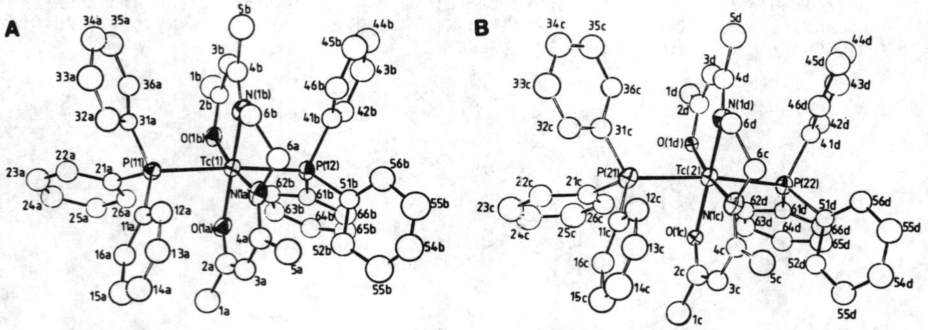

Fig. 1. The two independent $[Tc((acac)_2en)(PPh_3)_2]^+$ cations.

Triclinic, $P\bar{1}$, a = 19.037, b = 16.339, c = 15.454 Å, α = 95.37, β = 96.47, γ = 91.32°, Z = 4. Mo radiation, R = 0.109 for 7899 reflexions.

In each of the two crystallographically independent cations (Fig. 1) the Tc(III) centre is approximately octahedrally coordinated with the two N and two O donor atoms of the tetradentate Schiff base ligand in the equatorial plane and the two phosphine P atoms in the trans apical positions. Bond lengths: Tc-P 2.496(3)-2.524(4), Tc-O 2.007(10)-2.038(9), and Tc-N 2.041(11)-2.084(12) Å.

BIS(μ_2-N,N'-DIBENZYLTARTRAMIDE)TETRAISOPROPOXYDITITANIUM TOLUENE SOLVATE
$C_{48}H_{64}N_4O_{12}Ti_2,C_7H_8$ (I)

BIS(μ_2-DIETHYL-TARTRATO-O,O')DIETHOXYBIS(DIPHENYLHYDROXAMATO-O,O')DITITANIUM
$C_{46}H_{54}N_2O_{18}Ti_2$ (II)

I.D. WILLIAMS, S.F. PEDERSEN, K.B. SHARPLESS and S.J. LIPPARD, 1984. J. Am. Chem. Soc., 106, 6430-6431.

I. Orthorhombic, $C222_1$, a = 18.139, b = 16.240, c = 19.210 Å, Z = 4. Mo radiation, R = 0.071 for 1823 reflexions (at 225 K).

II. Monoclinic, $P2_1$, a = 12.488, b = 19.215, c = 11.838 Å, β = 114.79°, Z = 2. Mo radiation, R = 0.067 for 3179 reflexions (at 250 K).

The structures of these binuclear titanium tartrate catalysts are shown in Fig. 1. Compound (I) has twofold symmetry, with one diolate O atom of each tartramide bridging two Ti atoms to produce six-coordinate pseudoctahedral coordination. The Ti...Ti separation is 3.348(3) Å and the central planar Ti_2O_2 core has Ti-O2 2.160(9) and Ti-O2' 1.973(9) Å. The longer distance is attributed to ring strain and the influence of a good π-donor isopropoxide ligand (Ti-O6 1.805(9) Å) in the trans position. Compound II also has a binuclear structure; it has virtual but not crystallographic twofold symmetry.

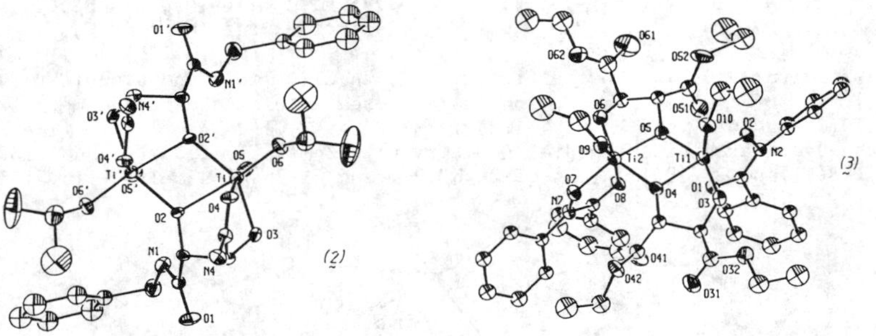

Fig. 1. Views of the binuclear Ti tartrates: (I) (left) and (II).

TETRA(TETRAETHYLAMMONIUM) TETRAKIS[(μ_3-METHOXO)TRICARBONYLCHROMATE(0)]
$C_{48}H_{92}Cr_4N_4O_{16}$ $[Et_4N]_4{}^+[Cr_4(CO)_{12}(\mu_3\text{-}OCH_3)_4]^{4-}$

T.J. McNEESE, M.B. COHEN and B.M. FOXMAN, 1984. Organometallics, 3, 552-556.

Orthorhombic, Fddd, a = 12.766, b = 23.694, c = 38.601 Å, D_m = 1.33, Z = 8. Mo radiation, R = 0.086 for 1017 reflexions.

The anion consists of a distorted cubic Cr_4O_4 core with triply-bridging μ_3-methoxide groups and $Cr(CO)_3$ units (Fig. 1). Each Cr atom is six-coordinate and each μ_3-O atom is four-coordinate. The anion has 222 (D_2) symmetry. The non-bonding Cr...Cr distances are 3.362, 3.301 and 3.254 Å, the Cr-O(4) distance is 2.114 Å, the Cr-O(4)-Cr angles are 102.7, 100.8 and 105.2°, and the O(4)-Cr-O(4) angles are 75.5, 76.6 and 73.8°. The C(4)-O(4) distance is 1.42 Å and the Cr-C(1), Cr-C(2), Cr-C(3) distances 1.78, 1.77 and 1.78 Å respectively. Two symmetry-independent cations are present, each with crystallographic 2 (C_2) symmetry. The α-C atoms of the ethyl groups of one of the cations is disordered.

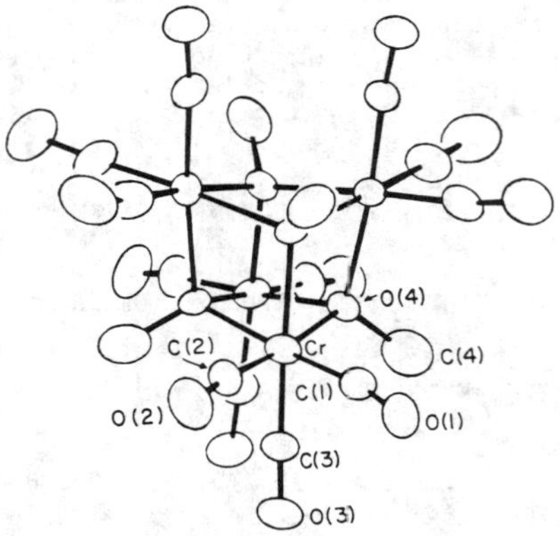

Fig. 1. Structure of the anion in $C_{48}H_{92}Cr_4N_4O_{16}$.

BIS[TRIS(μ-CHLORO)HEXAKIS(TETRAHYDROFURAN-O)DIVANADIUM(II)] BIS(μ-CHLORO)TETRACHLORO-DIZINC(II)

$C_{48}H_{96}Cl_{12}O_{12}V_4Zn_2$ $[V_2Cl_3(C_4H_8O)_6]_2[Zn_2Cl_6]$

R.J. BOUMA, J.H. TEUBEN, W.R. BEUKEMA, R.L. BANSEMER, J.C. HUFFMAN and K.G. CAULTON, 1984. Inorg. Chem., 23, 2715-2718.

Triclinic, P$\bar{1}$, a = 11.444, b = 15.241, c = 10.771 Å, α = 109.09, β = 98.61, γ = 75.66°, Z = 1. Mo radiation, R = 0.027 for 4143 reflexions (at -160°C).

The structure consists of centrosymmetric $Zn_2Cl_6^{2-}$ ions and binuclear $[V_2(\mu\text{-}Cl)_3(C_4H_8O)_6]^+$ ions (Fig. 1). The cation has face-shared bioctahedral geometry with V-Cl 2.470(1)-2.490(1) and V-O 2.131(2)-2.158(2) Å, V-Cl-V 73.6-74.0°. The $Zn_2Cl_6^{2-}$ anion has edge-shared bitetrahedral geometry with Zn-(μ-Cl) 2.351(1) and 2.363(1), Zn-Cl 2.218(1) and 2.215(1) Å, Zn-Cl-Zn 87.4°, and Cl-Zn-Cl 92.61(3)-116.51(3)°.

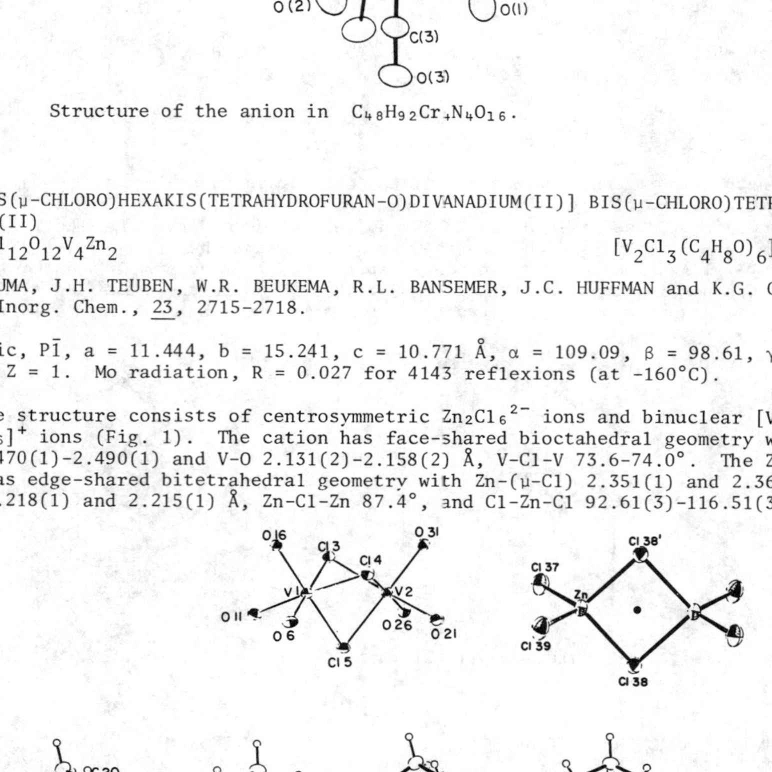

Fig. 1. Inner coordination of the $[V_2Cl_3(C_4H_8O)_6]^+$ cation and the $Zn_2Cl_6^{2-}$ anion (top); stereoview of $[V_2Cl_3(C_4H_8O)_6]^+$ (bottom).

CARBONYL(o-NITROPHENOLATO-O)(o-NITROSOPHENOLATO-N,O)BIS(TRIPHENYLPHOSPHINE)-RUTHENIUM(II) BENZENE SOLVATE

$C_{49}H_{38}N_2O_6P_2Ru,0.5(C_6H_6)$

M. PIZZOTTI, C. GROTTI and F. DEMARTIN, 1984. J. Chem. Soc. Dalton, 735-737.

Triclinic, $P\bar{1}$, a = 10.525, b = 10.818, c = 20.915 Å, α = 86.53, β = 83.53, γ = 72.21°, Z = 2. Mo radiation, R = 0.053 for 4876 reflexions.

The Ru atom is in a distorted octahedral environment with the two triphenylphosphine ligands trans to each other in an almost eclipsed conformation (Fig. 1). The o-nitrosophenolate ligand is coordinated in a chelating fashion through nitroso-N and phenolic O atoms. The o-nitrophenolate acts as a monodentate ligand via the phenolic O atom. The nitro group is not coplanar with the phenyl ring of the nitrophenolate anion (dihedral angle 43,4°). Bond distances include: Ru-P(1) 2.419(1), Ru-P(2) 2.412(1), Ru-O(1) 2.129(2), Ru-O(3) 2.080(2), Ru-N(2) 1.961(3), Ru-C 1.844(4) Å.

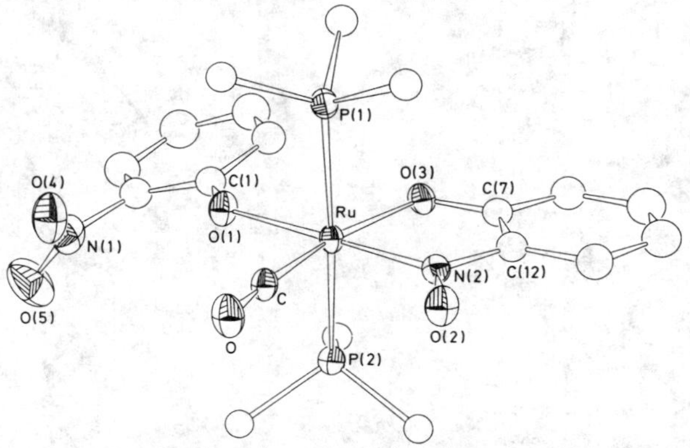

Fig. 1. $C_{49}H_{38}N_2O_6P_2Ru,0.5(C_6H_6)$: the molecular structure.

TRIS(μ-ISOPROPOXO)TETRAKIS(ISOPROPOXY)BIS(α-PHENYLBENZALDEHYDEHYDRAZONO)(PYRIDINE)-DIMOLYBDENUM

$C_{49}H_{67}Mo_2N_5O_6$ (I) $Mo_2(O\text{-}i\text{-}C_3H_7)_6(N_2C(C_6H_5)_2)_2(C_5H_5N)$

BIS(μ-α-p-TOLYL-4-METHYLBENZALDEHYDEHYDRAZONO)HEXAKIS(t-BUTOXY)DITUNGSTEN n-HEXANE SOLVATE

$C_{54}H_{82}N_4O_6W_2,C_6H_{14}$ (II) $W_2(O\text{-}t\text{-}C_4H_9)_6(N_2C(CH_3C_6H_5)_2)_2,n\text{-}C_6H_{14}$

TETRAKIS(t-BUTOXY)(α-PHENYLBENZALDEHYDEHYDRAZONO)MOLYBDENUM

$C_{29}H_{46}MoN_2O_4$ (III) $Mo(O\text{-}t\text{-}C_4H_9)_4(N_2C(C_6H_5)_2)$

M.H. CHISHOLM, K. FOLTING, J.C. HUFFMAN and A.L. RATERMANN, 1984. Inorg. Chem., 23, 2303-2311.

I. Monoclinic, $P2_1$, a = 12.512, b = 14.795, c = 14.099 Å, β = 109.51°, Z = 2. Mo radiation, R = 0.051 for 3143 reflexions (at -162°C).

II. Monoclinic, C2/c, a = 23.672, b = 12.214, c = 21.221 Å, β = 97.24°, Z = 4. Mo radiation, R = 0.053 for 2228 reflexions (at -165°C).

III. Triclinic, $P\bar{1}$, a = 17.652, b = 12.366, c = 9.060 Å, α = 76.95, β = 87.92, γ = 124.92°, Z = 2. Mo radiation, R = 0.033 for 2565 reflexions (at -164°C).

The structures of I-III are shown in Fig. 1. The structure of I has three O-i-Pr ligands bridging a Mo-Mo single bond (Mo-Mo 2.662(1), Mo-O 2.040(7)-2.239(7) Å), with a terminal $NNCPH_2^{2-}$ ligand coordinated to each Mo atom (Mo-N 1.758(9) and 1.781(8) Å). One Mo atom is also coordinated by two terminal O-i-Pr ligands (Mo-O 1.962(7) and

1.966(7) Å) and the other by a terminal O-i-Pr and pyridine (Mo-O 1.989(8), Mo-N 2.229(9) Å). The N-N distances in I are 1.297(11) and 1.319(12) Å. Molecules of II have exact C_2 symmetry with a central $[O_3W\text{-}\mu\text{-}N]_2$ moiety based on the fusing of two trigonal-bipyramidal units sharing a common equatorial-axial edge formed by the agency of a pair of $NNCAr_2^{2-}$ ligands. The W-W and N-N distances of 2.675(1) and 1.410(16) Å are indicative of single bonds. Other distances in II: $W\text{-}O_{eq}$ 1.874(11) and 1.858(11), $W\text{-}O_{ax}$ 1.944(9), $W\text{-}N_{eq}$ 1.942(11), and $W\text{-}N_{ax}$ 1.991(11) Å. The molecule III has a trigonal-bipyramidal MoO_4N moiety with a short Mo-N (1.797(3) Å) and long Mo-O (1.944(3) Å) bond in the axial positions. The equatorial Mo-O bonds average 1.889(3) Å.

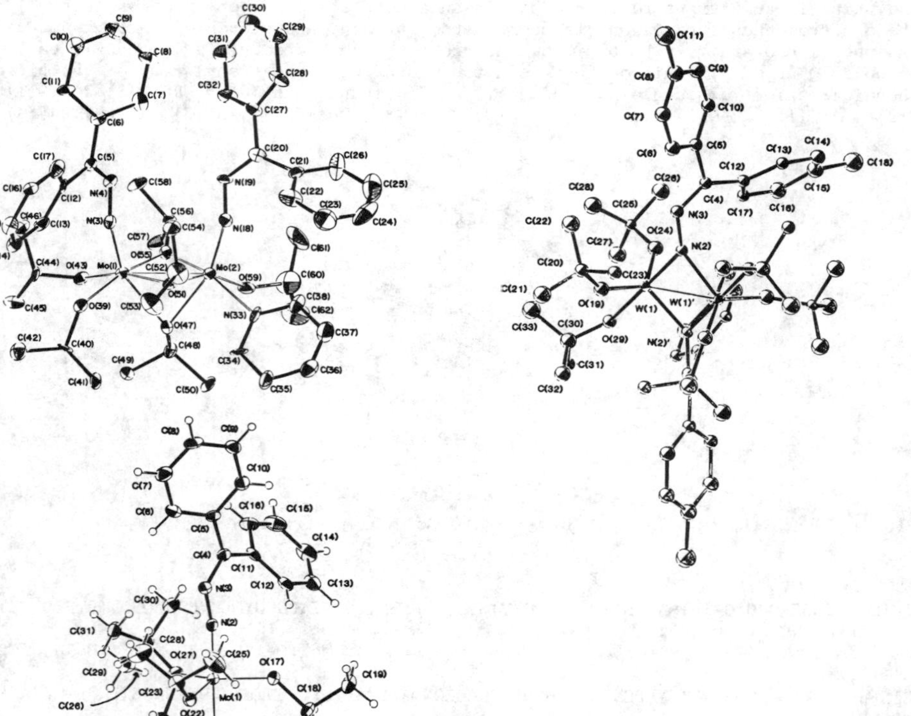

Fig. 1. The molecules $Mo_2(O\text{-}i\text{-}Pr)_6(N_2CPh_2)_2(C_5H_5N)$ (top left), $W_2(O\text{-}t\text{-}Bu)_6(N_2C(CH_3C_6H_5)_2)_2$ (right), and $Mo(O\text{-}t\text{-}Bu)_4(N_2CPh_2)$ (bottom left).

BIS[μ-BIS(DIPHENYLPHOSPHINO)METHANE]CARBONYL(TETRACHLOROCATECHOLATO-O,O')DIRHODIUM

$C_{57}H_{44}Cl_4O_3P_4Rh_2$ $Rh_2(Ph_2PCH_2PPh_2)_2(1,2\text{-}O_2C_6Cl_4)(CO)$

J.A. LADD, M.M. OLMSTEAD and A.L. BALCH, 1984. Inorg. Chem., 23, 2318-2323.

Triclinic, $P\bar{1}$, a = 11.126, b = 11.759, c = 20.065 Å, α = 91.82, β = 101.41, γ = 92.23°, D_m = 1.62, Z = 2. Mo radiation, R = 0.054 for 5018 reflexions (at 140 K).

The structure (Fig. 1) consists of two distinct Rh atoms: one five-coordinate and square pyramidal (with bonds to two O, two P, and one axial Rh) and the other four-coordinate and nearly planar (with bonds to two P, one C, and 1 Rh atoms). The two μ-dpm ligands are arranged in an unusual fashion so that they are trans to one another at one Rh atom and cis at the other.

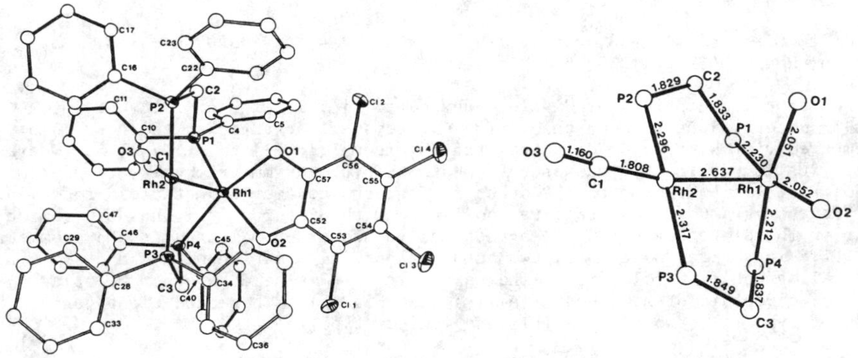

Fig. 1. The $Rh_2(Ph_2PCH_2PPh_2)_2(1,2\text{-}O_2C_6Cl_4)(CO)$ molecule, inner coordination distances (σ = 0.001-0.007 Å for Rh-X) at right.

BIS{3',5,5'-TRIMETHYL-6-[(4-CARBOXYLATO-5-METHYLAMINO-1,3-BENZOXAZOLE)METHYL]-6'-[1-(IMIDAZOL-2-YL-CARBONYL)ETHYL]-2,2'-SPIROBI(TETRAHYDROPYRAN)-N,O,O'}IRON ETHANOL SOLVATE

$C_{58}H_{74}FeN_6O_{12},2(C_2H_6O)$

E. BAKER, E.N. MASLEN, K.J. WATSON and A.H. WHITE, 1984. J. Am. Chem. Soc., 106, 2860-2864.

Orthorhombic, $P2_12_12_1$, a = 24.174, b = 17.642, c = 14.947 Å, Z = 4. Mo radiation, R = 0.10 for 2256 reflexions.

Coordination of the ligands to the ferrous ion is approximately octahedral through a carboxyl oxygen, the nitrogen of the benzoxazole system, and the carbonyl oxygen in such a manner that the whole complex has pseudo twofold symmetry (Fig. 1). The Fe-O distances are 2.01-2.19, Fe-N 2.22 Å. For details of the ligand and calcium complex see (1).

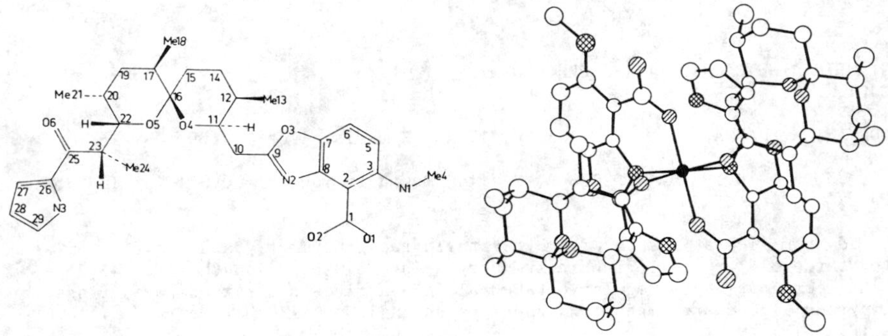

Fig. 1. $C_{58}H_{74}FeN_6O_{12}$: the ligand and the Fe complex.

1. Structure Reports, 40B, 481; 42B, 394.

BIS(μ_4-N,N-DIETHYLCARBAMATO)TETRAKIS(μ_3-N,N-DIETHYLCARBAMATO)HEXAKIS(μ_2-N,N-DIETHYLCARBAMATO)HEXACOBALT
$C_{60}H_{120}Co_6N_{12}O_{24}$

D.B. DELL'AMICO, F. CALDERAZZO, B. GIOVANNITTI and G. PELIZZI, 1984. J. Chem. Soc. Dalton, 647-653.

Monoclinic, C2/c, a = 18.428, b = 19.189, c = 24.654 Å, β = 95.90°, Z = 4. Mo radiation, R = 0.063 for 1258 reflexions.

The structure consists of discrete hexamers (Fig. 1) with no relevant intermolecular interactions between them. The Co atoms are arranged at the corners of a six-membered ring with Co-Co distances ranging from 3.161(5) to 3.197(4) Å. There are three crystallographically independent Co atoms in the asymmetric unit. The primed atoms of Fig. 1 are related to the other half of the hexamer by a twofold rotation axis. Four of the Co atoms are five-coordinate with distorted trigonal bipyramidal geometry. There are seven crystallographically independent ligands in the asymmetric unit. Five in general positions, two occupy a special position with N and sp^2 C atoms on crystallographic twofold axes. The hexameric unit is held together by bridging ligands of three different types, (a)-(c) (see Fig. 1). There are eight carbamato-ligands of type (a), two of type (b), and four of type (c).

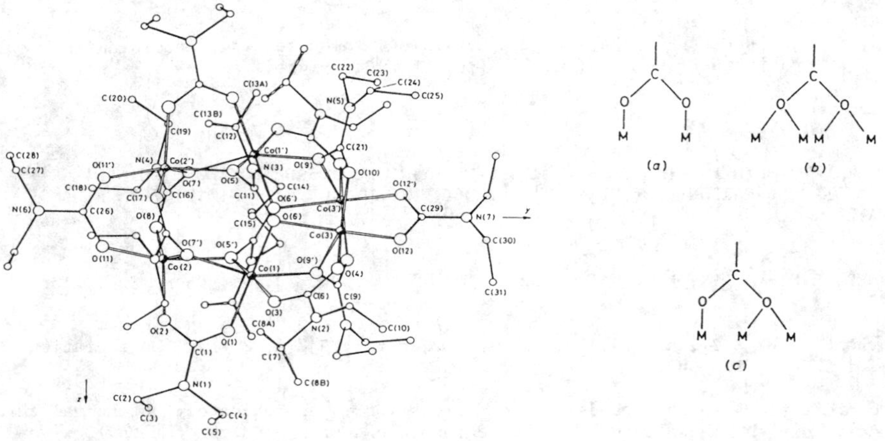

Fig. 1. $C_{60}H_{120}Co_6N_{12}O_{24}$: the hexameric unit and the three types of carbamato-ligand.

HEXAKIS(ANTIPYRINE)CADMIUM(II) DIPERCHLORATE
$C_{66}H_{72}CdCl_2N_{12}O_{14}$

C. MAHADEVAN, A. RADHA and M. SESHASAYEE, 1984. Z. Kristallogr., 169, 159-163.

Trigonal, P$\bar{3}$, a = 14.196, c = 9.558 Å, Z = 1. Mo radiation, R = 0.071 for 1097 reflexions.

The Cd atom lies at a site with $\bar{3}$ crystallographic symmetry and the Cl atom at a site with threefold crystallographic symmetry. The Cd-O bond length is 2.243(7) Å and the Cd coordination is regular octahedral. Both five- and six-membered rings in the antipyrene ligand are planar and there is an angle of 113° between the plane normals (Fig. 1).

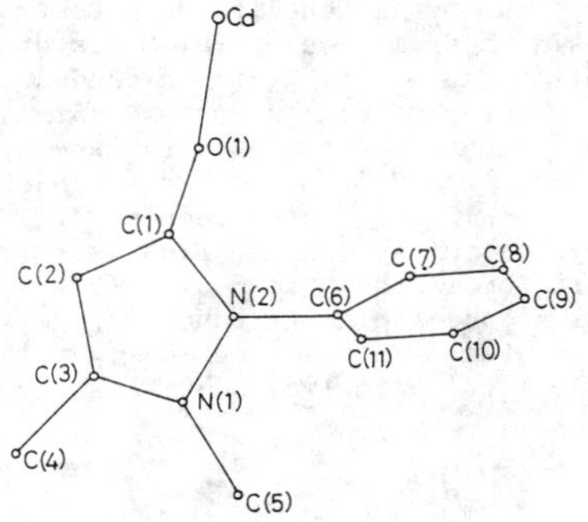

Fig. 1. The Cd(antipyrene) asymmetric unit.

TETRAKIS(TRIPHENYLARSINE OXIDE) HEXAAQUANICKEL(II) DIBROMIDE DIHYDRATE
$C_{72}H_{72}As_4Br_2NiO_{10},2H_2O$ (I) $\{Ni(H_2O)_6.4[(C_6H_5)_3AsO]\}^{2+}2Br^-,2H_2O$

BROMO-TETRAKIS(TRIPHENYLARSINE OXIDE)NICKEL(II) BROMIDE HYDRATE TOLUENE SOLVATE
$C_{72}H_{60}As_4Br_2NiO_4,1.5C_7H_8,H_2O$ (II) $\{NiBr[(C_6H_5)_3AsO]_4\}^+Br^-,1.5C_6H_5CH_3,H_2O$

G. OLIVA, E.E. CASTELLANO, J. ZUKERMAN-SCHPECTOR and A.C. MASSABNI, 1984. Inorg. Chim. Acta, 89, 9-17.

I. Monoclinic, $P2_1/c$, a = 13.731, b = 16.267, c = 17.647 Å, β = 112.04°, Z = 2. Cu radiation, R = 0.039 for 3203 reflexions.

II. Triclinic, $P\bar{1}$, a = 12.647, b = 13.953, c = 22.853 Å, α = 90.91, β = 96.70, γ = 111.16°, Z = 2. Mo radiation, R = 0.087 for 4293 reflexions.

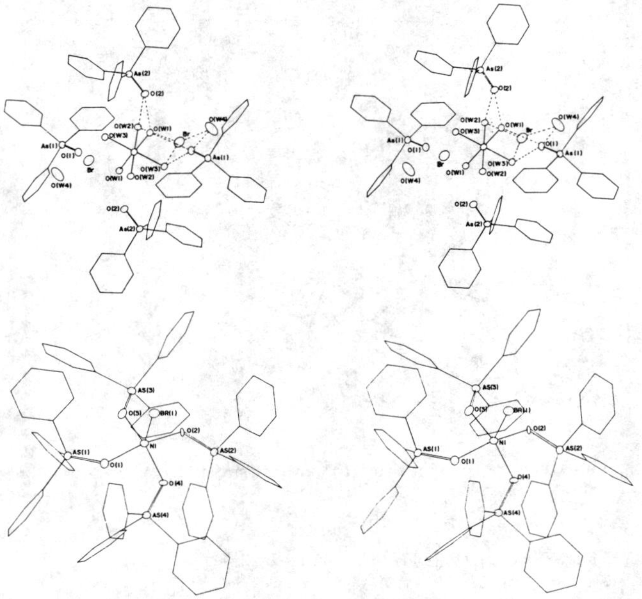

Fig. 1. Stereoview of the green complex (I) $C_{72}H_{72}As_4Br_2NiO_{10},2H_2O$ (top) and stereoview of the orange complex (II), $C_{72}H_{60}As_4BrNiO_4$ (bottom).

In the green complex (I), the metal ion lies on a centre of symmetry and is octahedrally coordinated to six water molecules (Ni-O distances 2.053, 2.080 and 2.089 Å) (Fig. 1). The water molecules are hydrogen-bonded to the O atoms of the Ph_3AsO ligands and to the free water molecules and Br^- anions, forming a second coordination sphere in an approximate octahedral arrangement. The O...O(water) hydrogen bonding distances range from 2.665 to 2.784 Å, and the Br...O(water) distances from 3.227 to 3.297 Å. In the orange complex (II) the cation is penta-coordinated, with the four O atoms of the Ph_3AsO ligands forming the base of a tetragonal pyramid and one Br atom in the apical site (Fig. 1). The Ni-O distances are 2.01, 1.96, 2.02 and 2.01 Å, and the Ni-Br bond length is 2.506 Å. The remaining Br^- anion is hydrogen-bonded to two crystallographically-related water molecules (O...Br 3.33 and 3.34 Å). The metal atom lies 0.361 Å above the plane of the four O atoms of the tetragonal pyramid.

BIS(TRI-ISOBUTYL PHOSPHONATO-O)TRIS(4-BENZOYL-3-METHYL-1-PHENYL-5-PYRAZOLONATO-O,O')-EUROPIUM

$C_{75}H_{93}EuN_6O_{14}P_2$ $Eu(C_{17}H_{13}N_2O_2)_3.2(C_{12}H_{27}PO_4)$

É.T. KARASEVA, V.E. KARASEV, A.A. UDOVENKO and N.I. SIGULA, 1983. Koord. Khim., 9, 692-698 [Sov. J. Coord. Chem., 9, 394-400].

Monoclinic, Bb, a = 24.424, b = 24.729, c = 14.377 Å, γ = 116.33°, Z = 4. Mo radiation, R = 0.09 for 3176 reflexions.

The complexes are mononuclear with the Eu atom in a distorted tetragonal anti-prismatic environment (Fig. 1). The vertices of the coordination polyhedron are occupied by six O atoms in three phenylmethylbenzoylpyrazolonate ligands and two O atoms of two triisobutyl phosphate groups. The Eu-O distances range from 2.29 to 2.51 Å. The molecular complexes are held together by van der Waals interactions.

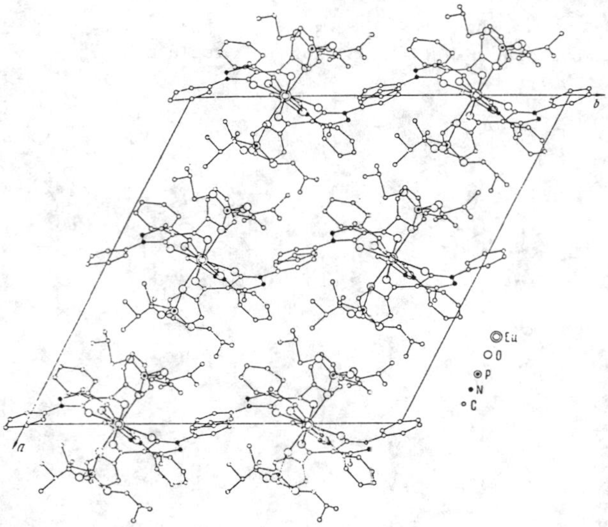

Fig. 1. $C_{75}H_{93}EuN_6O_{14}P_2$: projection of the structure onto the ab plane.

HEXADECAPHENYLOCTASILOXYSPIRO[9.9]TITANATE(IV)

$C_{96}H_{80}O_{10}Si_8Ti$

M.B. HURSTHOUSE and M.A. HOSSAIN, 1984. Polyhedron, 3, 95-97.

Monoclinic, $P2_1/n$, a = 13.829, b = 27.594, c = 23.514 Å, β = 94.39°, Z = 4. Cu radiation, R = 0.066 for 10652 reflexions.

In the spiro molecule (Fig. 1), the metal atom has a geometry very close to tetrahedral, with O-Ti-O angles of 107.9-111.0(2)° and very short Ti-O bonds of length 1.777-1.791(5) Å. The two TiO_5Si_4 rings have different, ill-defined conformations; the Si-O bond lengths and Si-O-Si angles are similar to those in $(SiO)_n$ rings with Si-O 1.608(5)-1.637(5) Å. The Si-C(phenyl) distances are in the range 1.850-1.868(6) Å.

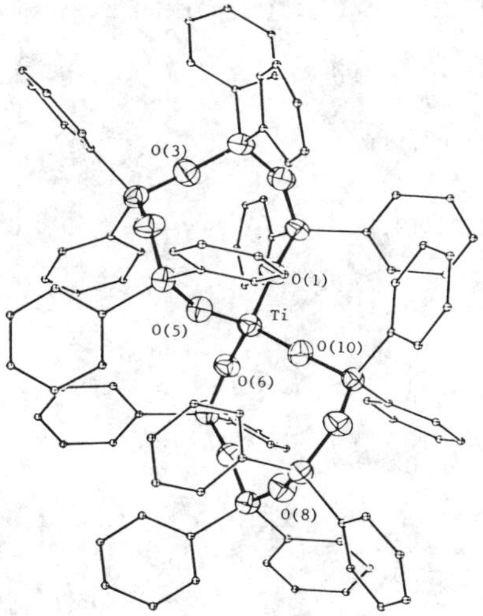

Fig. 1. A view of $C_{96}H_{80}O_{10}Si_8Ti$.

BIS(μ_2-3,5-DI-t-BUTYLBENZENE-1,2-DIOLATO-O,O,O')TETRAKIS(μ_3-3,5-DI-t-BUTYLBENZENE-1,2-DIOLATO-O,O,O')BIS(3,5-DI-t-BUTYLBENZENE-1,2-DIOLATO-O,O')TETRAMANGANESE
$C_{112}H_{160}Mn_4O_{16}$ (I) $Mn_4(3,5\text{-DBSQ})_8$

BIS(3,5-DI-t-BUTYLBENZENE-1,2-DIOLATO-O,O')BIS(PYRIDINE)MANGANESE PYRIDINE SOLVATE
$C_{38}H_{50}MnN_2O_4,2(C_5H_5N)$ $Mn(3,5\text{-DBCat})_2(py)_2$

M.W. LYNCH, D.N. HENDRICKSON, B.J. FITZGERALD and C.G. PIERPONT, 1984. J. Am. Chem. Soc., 106, 2041-2049.

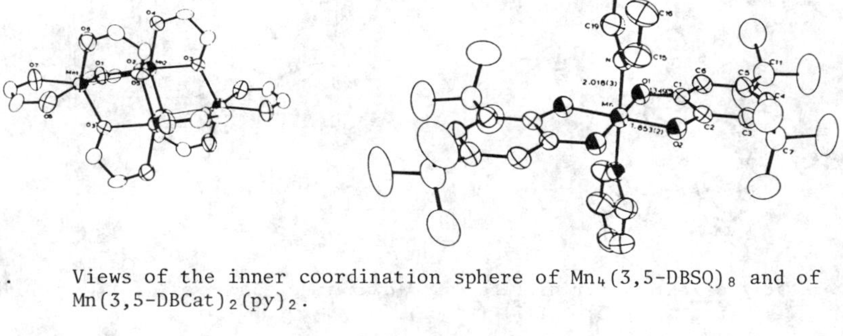

Fig. 1. Views of the inner coordination sphere of $Mn_4(3,5\text{-DBSQ})_8$ and of $Mn(3,5\text{-DBCat})_2(py)_2$.

I. Triclinic, $P\bar{1}$, a = 14.822, b = 14.803, c = 16.428 Å, α = 118.19, β = 110.63, γ = 92.86°, D_m = 1.14, Z = 1. Mo radiation, R = 0.059 for 2756 reflexions.

II. Monoclinic, $P2_1/c$, a = 9.032, b = 18.862, c = 13.296 Å, β = 96.59°, D_m = 1.21, Z = 2. Mo radiation, R = 0.049 for 2429 reflexions.

The tetrameric structure I lies about an inversion centre; in II the Mn atom lies on an inversion centre (Fig. 1). In I the two independent Mn ions have distorted octahedral coordination with Mn-O(av) 2.138(6) Å, (except Mn-O(5) 2.526(7) Å, O(5) bridges two Mn ions). Dimensions for II (Fig. 1) are as expected.

SILVER NITRATE - 1,4-OXATHIANE (1:1)
$C_4H_8AgNO_4S$ (I)

SILVER NITRATE - 1,4-OXATHIANE (2:1)
$C_4H_8Ag_2N_2O_7S$ (II)

SILVER NITRATE - 1,4-OXATHIANE (6:1)
$C_4H_8Ag_6N_6O_{19}S$ (III)

J.C. BARNES and J.D. PATON, 1984. Acta Cryst., C40, 72-78.

I. Monoclinic, $P2_1/c$, a = 6.12, b = 7.39, c = 18.13 Å, β = 97.0°, Z = 4. Cu radiation, R = 0.090 for 886 reflexions.

II. Monoclinic, $P2_1/c$, a = 12.065, b = 6.259, c = 13.527 Å, β = 82.42°, Z = 4. Mo radiation, R = 0.060 for 1821 reflexions.

III. Triclinic, $P\bar{1}$, a = 10.7347, b = 13.1367, c = 7.2488 Å, α = 91.556, β = 92.668, γ = 89.141°, Z = 2. Mo radiation, R = 0.057 for 5308 reflexions.

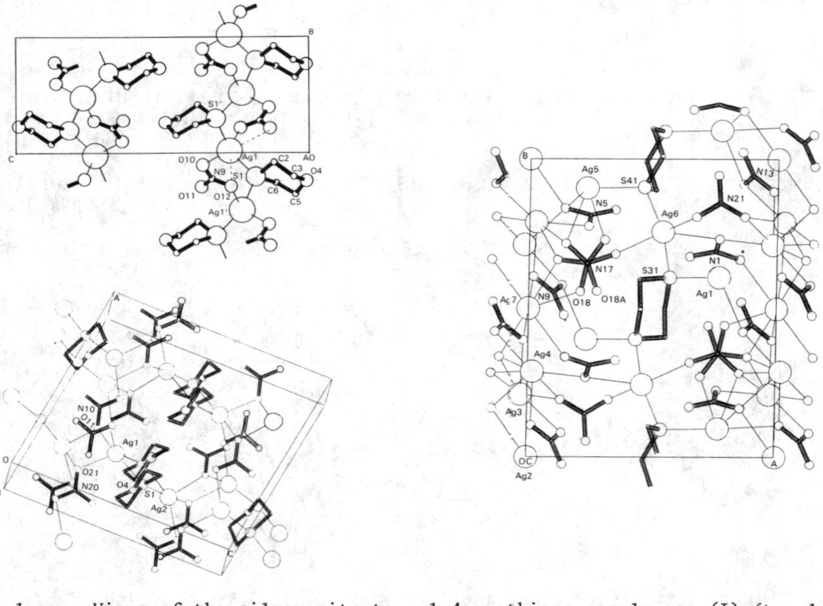

Fig. 1. Views of the silver nitrate - 1,4-oxathiane complexes, (I) (top left), (II) (bottom left), (III) (right).

The complexes (I), (II) and (III) (Fig. 1) are polymeric lattice complexes in which Ag atoms are coordinated by the S atoms of the 1,4-oxathiane at about 2.5 Å and by nitrate groups with Ag-O in the range 2.4 to 3.1 Å. In (II) and (III) the O atom of 1,4-oxathiane is also coordinated to Ag. The coordination numbers at Ag vary from 6 to 10 with irregular geometries. There are no strongly bonded polymeric fragments common throughout the series.

cis-DICHLORO[1,2-BIS(METHYLTHIO)1,1-DIFLUOROETHANE]PLATINUM(II)
$C_4H_8Cl_2F_2PtS_2$ $[PtCl_2(CH_3SCF_2(CH_2SCH_3))]$

O. CANO, J. LEAL, P. QUINTANA and H. TORRENS, 1984. Inorg. Chim. Acta, 89, L9-L10.

Monoclinic, $P2_1/n$, a = 8.953, b = 13.628, c = 9.105 Å, β = 114.95°, Z = 4. Mo radiation, R = 0.032 for 8500 reflexions.

The molecule has the expected cis-square planar coordination at the Pt atom (Fig. 1) and the methyl substituents of the S atoms are mutually syn. Valency angles around the Pt atom are all within 2° of the ideal values of 90 and 180°. The Pt-Cl bond lengths are 2.299 and 2.306 Å and the Pt-S distances are 2.240 and 2.270 Å, the shorter bond being adjacent to the S-CF_2 group. Both S atoms adopt quasi-tetrahedral coordinations and the four S-C bond lengths agree to within experimental error, with a mean of 1.797 Å. The Pt-S(2)-C(4) and S(2)-C(3)-C(2) angles are respectively 106.6 and 115.2°.

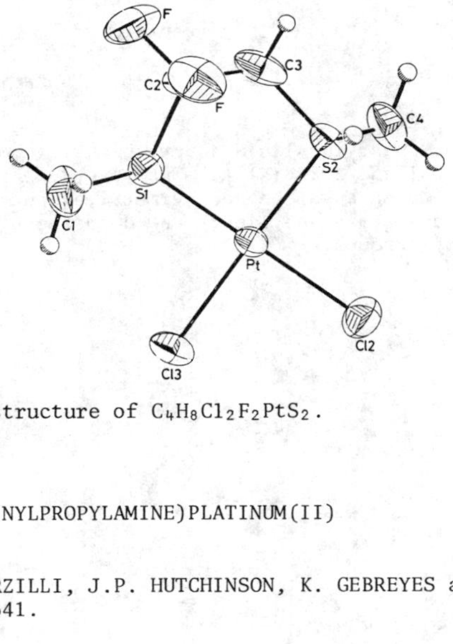

Fig. 1. Molecular structure of $C_4H_8Cl_2F_2PtS_2$.

DICHLORO(3-METHYLSULPHINYLPROPYLAMINE)PLATINUM(II)
$C_4H_{11}Cl_2NOPtS$ $[PtCl_2(C_4H_{11}NOS)]$

G.R. CHAMPION, L.G. MARZILLI, J.P. HUTCHINSON, K. GEBREYES and J. ZUBIETA, 1984. Acta Cryst., C40, 639-641.

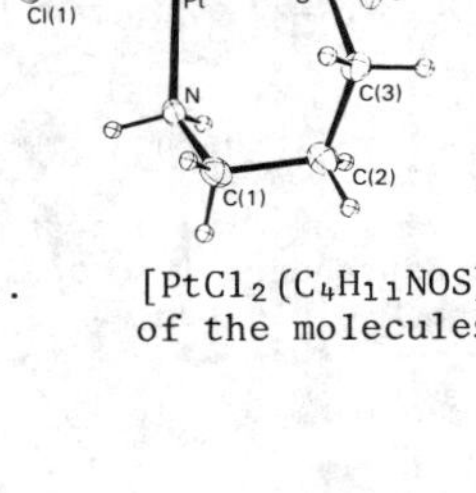

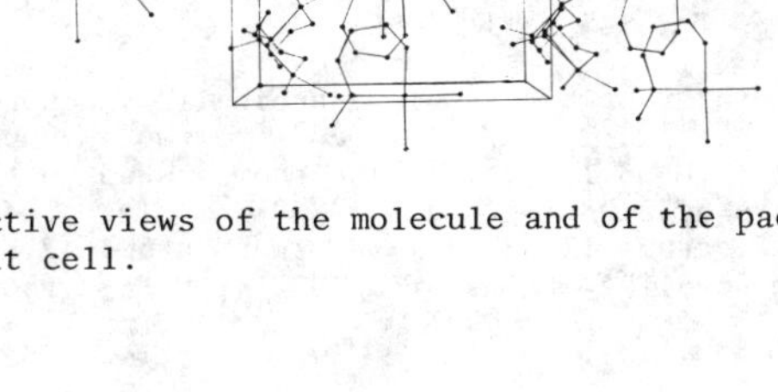

Fig. 1. $[PtCl_2(C_4H_{11}NOS)]$: perspective views of the molecule and of the packing of the molecules in the unit cell.

Monoclinic, $P2_1/c$, a = 13.159, b = 9.372, c = 15.468 Å, β = 100.99°, Z = 8. Mo radiation, R = 0.031 for 2122 reflexions.

The two independent molecules are identical and display a square-planar arrangement of donor atoms about Pt (Fig. 1). Pt-N distances are 2.062(7) and 2.063(9) Å for N and N'. Pt-S distances are 2.200(2) and 2.204(3) Å for S and S'. Pt-Cl distances range from 2.307(3) to 2.317(3) Å.

CHLORO(N-ETHYL-1,3-IMIDAZOLIDINE-2-THIONE)GOLD(I)
$C_5H_{10}AuClN_2S$ [(EtImt)AuCl]

M.S. HUSSAIN and A.A. ISAB, 1984. Transition Met. Chem., 9, 398-401.

Triclinic, $P\bar{1}$, a = 7.213, b = 7.857, c = 9.056 Å, α = 108.54, β = 100.06, γ = 107.27°, D_m = 2.69, Z = 2. Mo radiation, R = 0.073 for 2157 reflexions.

The structure contains monomeric units with linear coordination around the Au(I) atom with Au-Cl 2.26(1), Au-S 2.25(1) Å, S-Au-Cl 174.2(3)°.

cis-[3,4-BIS(TRIFLUOROMETHYL)-2,5-DITHIAHEXANE]DICHLOROPLATINUM(II)
$C_6H_8Cl_2F_6PtS_2$

W.N. HUNTER, K.W. MUIR and D.W.A. SHARP, 1984. Acta Cryst., C40, 37-39.

Tetragonal, $P4_2/nbc$, a = 14.081, c = 27.121 Å, Z = 16. Mo radiation, R = 0.018 for 1869 reflexions.

The Pt atom (Fig. 1) has near ideal cis-square planar geometry, with Pt-Cl 2.316 and 2.324(2), Pt-S 2.231 and 2.244(2) Å. The five-membered ring adopts a conformation intermediate between envelope and symmetrical puckered. The sulphur methyl substituents are mutually syn and lie on the opposite sides of the chelate ring from the trifluoromethyl groups.

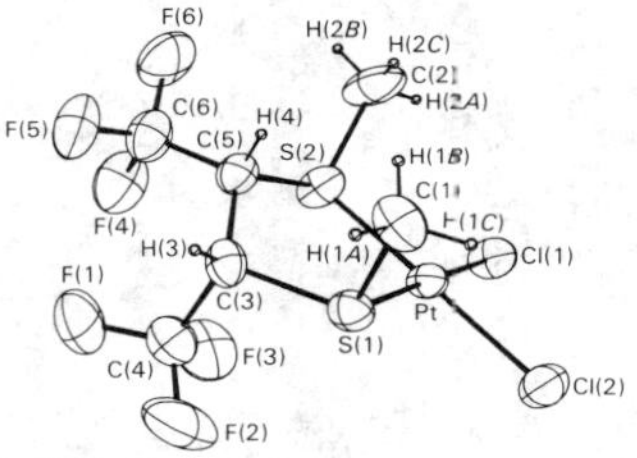

Fig. 1. A view of $C_6H_8Cl_2F_6PtS_2$.

BIS(μ_2-ISOPROPYLTHIO)DICHLORODIMERCURY(II)
$(C_6H_{14}Cl_2Hg_2S_2)n$

P. BISCARINI, E. FORESTI and G. PRADELLA, 1984. J. Chem. Soc. Dalton, 953-957.

Monoclinic, C2, a = 21.424, b = 4.668, c = 6.734 Å, β = 90.45°, Z = 2. Mo radiation, R = 0.056 for 585 reflexions.

There are two crystallographically independent mercury atoms in each asymmetric unit (Fig. 1). One of them bonds to two sulphur atoms at 2.378(6) Å in a linear fashion (S-Hg-S 178.2°). Also coordinated to this Hg atom are four Cl atoms at distances of 2.985(8) and 3.342(8) Å. The other Hg atom is tetrahedrally coordinated to two Cl atoms (2.534(8)) and two S atoms (2.537(6) Å). The whole structure is built up from parallel chains of $Hg(SPr^i)_2.HgCl_2$ groups.

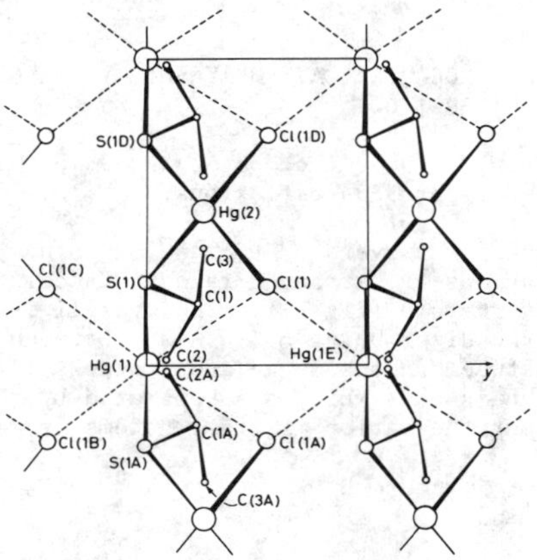

Fig. 1. $(C_6H_{14}Cl_2Hg_2S_2)_n$: projection of the (100) plane.

DITHIODIIRON HEPTACARBONYL

$C_7Fe_2O_7S_2$ $Fe_2(CO)_6(S_2CO)$

N.S. NAMETKIN, B.I. KOLOBKOV, V.D. TYURIN, A.N. MURATOV, A.I. NEKHAEV, M. MAVLONOV, A.Ya. SIDERIDU, G.G. ALEKSANDROV, A.V. LEBEDEV, M.T. TASHEV and H.B. DUSTOV, 1984. J. Organometal. Chem., 276, 393-397.

Orthorhombic, Pnam, a = 14.131, b = 6.764, c = 12.900 Å, Z = 4. Mo radiation, R = 0.029 for 720 reflexions.

The complex has crystallographic m symmetry (Fig. 1), the mirror plane containing atoms of the S_2CO ligand which pass normally to the Fe-Fe bond. The bond angles about the metal atoms range from 56.73° (Fe'-Fe-S(1)) to 159.6° (S(1)-Fe-C(2)), and the angles Fe'-S(1)-Fe and Fe'-S(2)-Fe are respectively 66.55 and 66.44°.

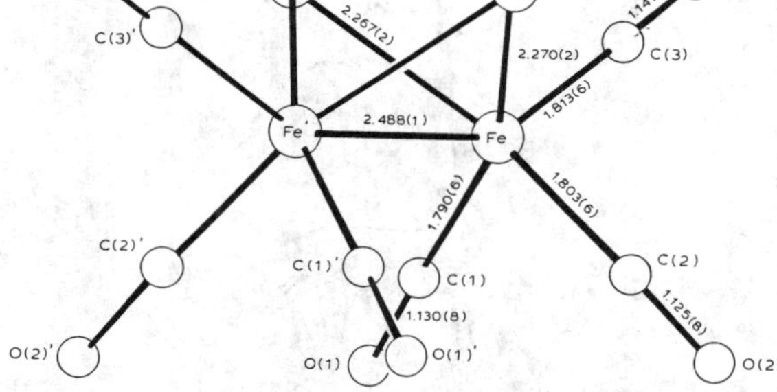

Fig. 1. Molecular structure of $C_7Fe_2O_7S_2$ showing bond lengths (Å) and angles (°).

RUBIDIUM BIS[(Z)-1,2-DICYANOETHYLENE-1,2-DITHIOLATO]PLATINATE DIHYDRATE
$C_8N_4PtRbS_4,2(H_2O)$

M.M. AHMAD, D.J. TURNER, A.E. UNDERHILL, A. KOBAYASHI, Y. SASAKI and H. KOBAYASHI, 1984. J. Chem. Soc. Dalton, 1759-1762.

Triclinic, $P\bar{1}$, a = 14.659, b = 7.179, c = 6.734 Å, α = 98.37, β = 97.39, γ = 92.55°, Z = 2. Mo radiation, R = 0.045 for 3305 reflexions.

The anion is almost planar with average Pt-S and S-C bond lengths of 2.260 and 1.697 Å respectively. The anions are arranged face to face in an eclipsed conformation to form a columnar stacked structure (Fig. 1). The overall structure is dimeric, the interplanar spacing within the dimer being 3.356 Å and between dimers, 3.512 Å. There are short S...S distances between anions in adjacent stacks (3.655 Å). The columns of anions form two-dimensional sheets which are separated by the cations and water molecules. Coordination about the cation is from O atoms of water and the cyano groups of the anions.

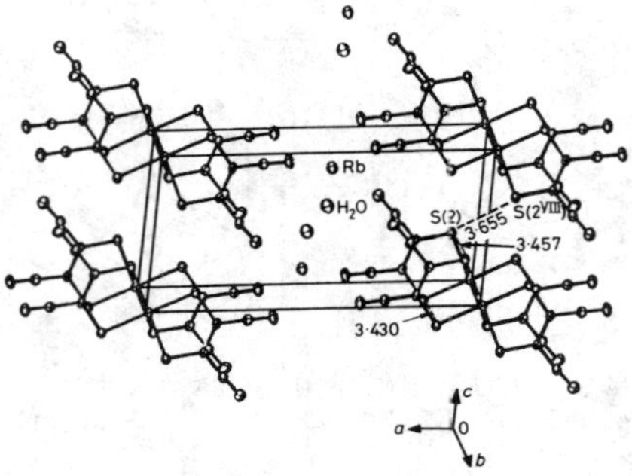

Fig. 1. $C_8N_4PtRbS_4,H_2O$: the crystal structure.

LITHIUM HYDROXONIUM BIS(1,2-DICYANO-1,2-ETHYLENEDITHIOLATO-S,S')PLATINUM HYDRATE
$C_8H_{0.9}Li_{0.7}N_4O_{0.3}PtS_4,1.7(H_2O)$ $0.7(Li^+).0.3(H_3O^+).C_8N_4PtS_4^-,1.7(H_2O)$

A. KOBAYASHI, T. MORI, Y. SASAKI, H. KOBAYASHI, M.M. AHMAD and A.E. UNDERHILL, 1984. Bull. Chem. Soc. Japan, 57, 3262-3268.

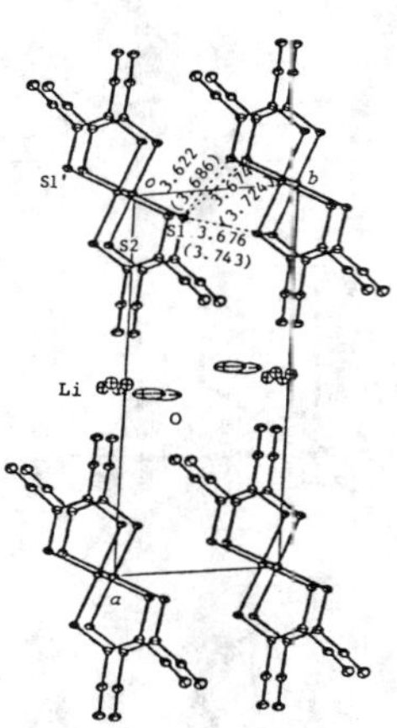

Fig. 1. The c* axis projection of the average structure of α-LiPt(mnt).

Triclinic, $P\bar{1}$, a = 15.596 (15.564), b = 6.410 (6.320), c = 3.639 (3.597) Å, α = 100.52 (99.35), β = 90.75 (90.54), γ = 96.28 (96.25)°, Z = 1. Mo radiation, R_W = 0.065 (0.068) for 1686 (2015) reflexions (at room temperature and in parentheses at 153 K).

At both temperatures anions stack face-to-face along the c-axis (Fig. 1). The compound undergoes a metal-insulator transition at 220 K. The Pt-S distances are 2.271(2) at room temperature and 2.267(3) Å at 153 K.

BIS(TETRAHYDROTHIOPHENE)SILVER(I) TETRAFLUOROBORATE
$C_8H_{16}AgBF_4S_2$ $[Ag(C_4H_8S)_2]^+BF_4^-$

B. NORÉN and Å. OSKARSSON, 1984. Acta Chem. Scand., A38, 479-484.

Orthorhombic, $P2_12_12_1$, a = 7.497, b = 11.432, c = 15.249 Å, D_m = 1.77, Z = 4. Mo radiation, R = 0.031 for 1368 reflexions (at 200K).

The coordination geometry around Ag is close to trigonal planar (Fig. 1), with angles of S(1)-Ag-S(2) 118.78, S(1)-Ag-S(2') 124.69 and S(2)-Ag-S(2') 116.26°. There are two crystallographically different ligands in the structure: S(1) of ligand 1 is bound only to one Ag(I) ion while the atom S(2) of ligand 2 bridges two metal ions in such a way as to form infinite ...S-Ag-S-Ag... chains along a. These chains are packed with the BF_4^- anions situated in voids in the vicinity of the metal ions. The closest contacts B...Ag in adjacent chains are 3.73 and 6.69 Å indicating that the BF_4^- ions have little bridging effects between chains. The Ag-S distances are: Ag-S(1) 2.445, Ag-S(2) 2.555 and Ag-S(2') 2.519 Å.

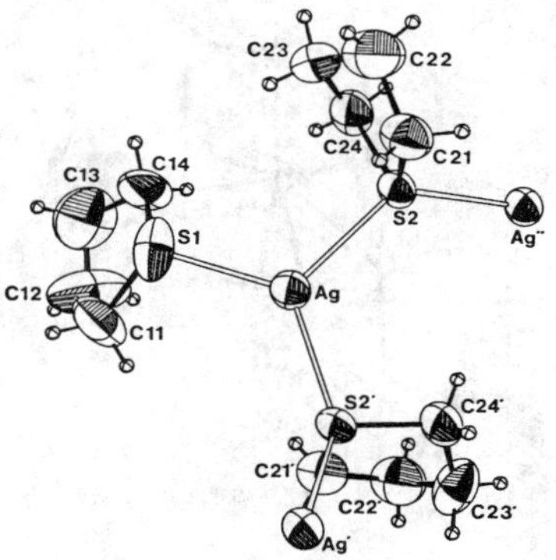

Fig. 1. Coordination around Ag viewed perpendicular to the best plane through Ag, S(1), S(2) and S(2') in $C_8H_{16}AgBF_4S_2$.

DIBROMOTETRAKIS(DIMETHYL SULPHOXIDE)RUTHENIUM(II)
$C_8H_{24}Br_2O_4RuS_4$ $RuBr_2[(CH_3)_2SO]_4$

J.D. OLIVER and D.P. RILEY, 1984. Inorg. Chem., 23, 156-158.

Tetragonal, I4/m, a = 9.181, c = 11.121°, Z = 2. Mo radiation, R = 0.018 for 403 reflexions.

The molecule (Fig. 1) has crystallographic 4/m (D_{4h}) symmetry with the Ru, S and O atoms lying on the mirror plane. The Me_2SO ligands are all S-bonded. Principal metrical details are: Ru-Br 2.540(1), Ru-S 2.360(1), S-O 1.484(3), S-C 1.789(3) Å, Ru-S-O 112.5(1), Ru-S-C 116.0(1), C-S-O 105.7(1), and C-S-C' 99.6(2)°.

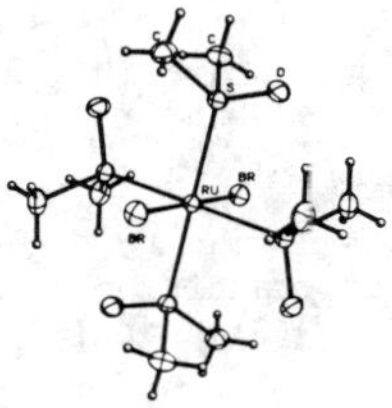

Fig. 1. The $RuBr_2[(CH_3)_2SO]_4$ molecule.

BIS[DICHLOROBIS(DIMETHYL SULPHIDE)(μ_2-DISELENIDO)NIOBIUM(IV)]
$C_8H_{24}Cl_4Nb_2S_4Se_4$

M.G.B. DREW, D.A. RICE and D.M. WILLIAMS, 1984. J. Chem. Soc. Dalton, 1087-1090.

Monoclinic, $P2_1/n$, a = 8.366, b = 13.121, c = 11.204 Å, β = 99.53°, D_m = 2.35, Z = 2. Mo radiation, R = 0.068 for 1855 reflexions.

The structure contains discrete centrosymmetric molecules (Fig. 1) in which Nb(IV) is bonded to two chlorine atoms (Nb-Cl(1) 2.527(5), Nb-Cl(2) 2.510(5)), four Se atoms (Nb-Se 2.638, 2.630, 2.640, 2.625(2)) and two S atoms from the dimethylsulphides (Nb-S(1,2) 2.739(4) Å). The Nb,S(1),S(2) atoms and midpoint of Se_2 are coplanar with the Se_2 group perpendicular to this plane. The midpoints of Se_2, Cl and two S atoms provide an octahedral environment for the metal. The Nb...Nb distance is 2.962(3) Å.

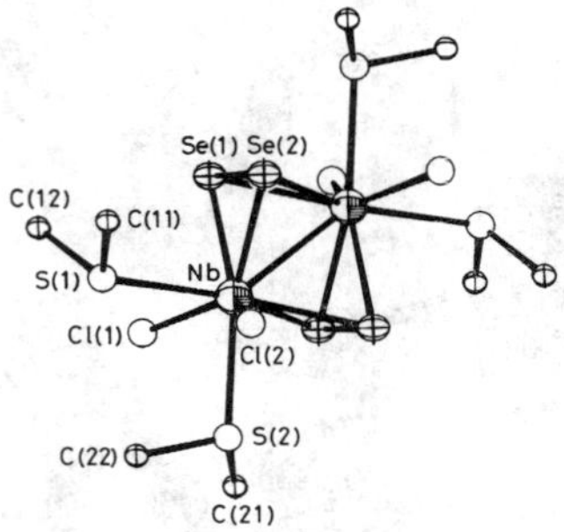

Fig. 1. $C_8H_{24}Cl_4Nb_2S_4Se_4$: the molecular structure.

TRICALCIUM BIS[BIS(DITHIOOXALATO-S,S')NICKEL] OXALATE TETRADECAHYDRATE
$C_{10}Ca_3Ni_2O_{12}S_8$,14(H_2O) (I)

CALCIUM BIS(DITHIOOXALATO-S,S')NICKEL HEXAHYDRATE
$C_4CaNiO_4S_4$,6(H_2O) (II)

A. GLEIZES, F. MAURY and J. GALY, 1984. Nouv. J. Chim., 8, 521-529.

I. Orthorhombic, Ccca, a = 22.459, b = 22.727, c = 7.206 Å, D_m = 1.95, Z = 4. Mo radiation, R = [?] for 562 reflexions.

II. Monoclinic, $P2_1/c$, a = 11.788, b = 18.975, c = 7.215 Å, β = 106.69°, Z = 4. Mo radiation, R = [?] for 864 reflexions.

Both structures (Fig. 1) are built up of zigzag chains of nickel atoms and of either dihydrated (I) or tetrahydrated (II) calcium atoms with dithiooxalate as the bridging ligand. These chains are stacked in layers and in II these layers are weakly linked by a few water molecules. In I, the chains are firmly linked by oxalate ions at the calcium atoms. These anions also link up with a second series of tetrahydrated calcium atoms to produce a structure with tunnels in which water molecules reside. In (I) Ca(1) lies on a twofold axis and has pseudo pentagonal bipyramidal geometry; Ca(2) has 222 symmetry and its coordination is rhombic prismatic.

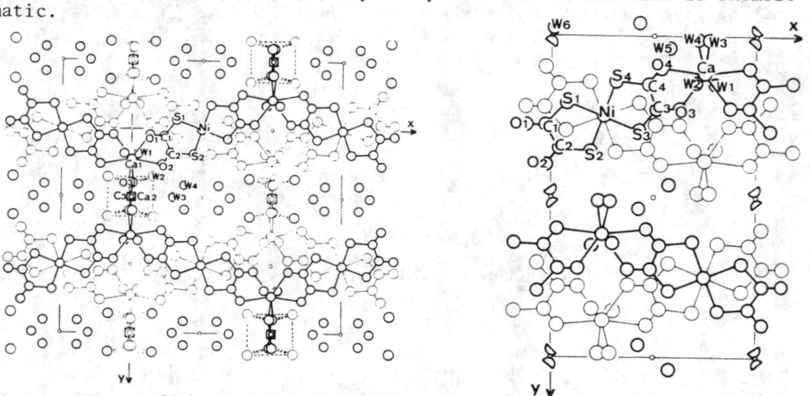

Fig. 1. Views of the calcium thiooxalate structures (I) (left) and (II) (right).

(η^5-CYCLOPENTADIENYL)[1,2-DIMERCAPTO-3-(MERCAPTODITHIO)CYCLOPENTENE-S,S',S'',S''']-TITANIUM
$C_{10}H_{10}S_5Ti$

D.M. GIOLANDO and T.B. RAUCHFUSS, 1984. J. Am. Chem. Soc., 106, 6455-6456.

Monoclinic, $P2_1/c$, a = 18.60, b = 7.508, c = 18.66 Å, β = 92.18°, Z = 8. Mo radiation. R = 0.054 for 2197 reflexions.

The two independent molecules in the asymmetric unit have similar geometry; one is shown in Fig. 1. A CpTi moiety is coordinated to four S atoms which define a plane (Ti is 0.92 Å above this plane). Ti-S distances quoted are 2.355-2.537(3) Å. The dithiolene is coordinated in an unusual manner as the Ti...C(1) (2.421(8) Å) and Ti...C(2) (2.483(8) Å) distances are only slightly longer than the mean Ti...C(Cp) distance (2.328 Å). Other dimensions are: S3-S4 2.025(3), S4-S5 2.061(3) Å, S3-S4-S5 112.8(2)°.

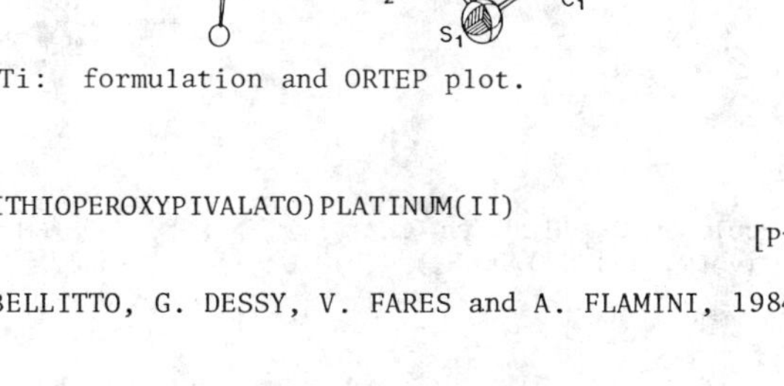

Fig. 1. $C_{10}H_{10}S_5Ti$: formulation and ORTEP plot.

(DITHIOPIVALATO)(TRITHIOPEROXYPIVALATO)PLATINUM(II)
$C_{10}H_{18}PtS_5$ $[Pt(C_5H_9S_2)(C_5H_9S_3)]$

E. AGOSTINELLI, C. BELLITTO, G. DESSY, V. FARES and A. FLAMINI, 1984. Acta Cryst., C40, 619-620.

Monoclinic, C2/c, a = 18.785, b = 6.204, c = 27.372 Å, β = 95.35°, D_m = 2.04, Z = 8. Mo radiation, R = 0.038 for 1919 reflexions.

In the molecule (Fig. 1) Pt is coordinated, in a tetrahedrally distorted square-planar arrangement, by four S atoms belonging to two different, four- and five-membered chelate rings, both lying on the same plane, with Pt-S(1) 2.322(4), Pt-S(2) 2.323(3), Pt-S(3) 2.238(4), Pt-S(4) 2.248(3) and S(3)-S(5) 2.037(5) Å. The S-Pt-S angles within the two rings are 73.0(1) and 93.6(1)° respectively.

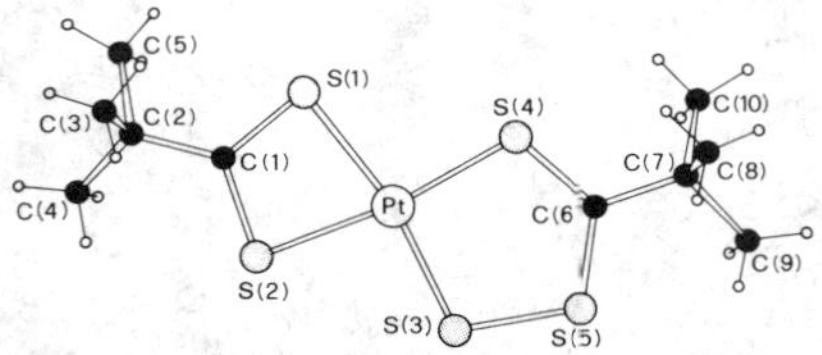

Fig. 1. A perspective view of the $[Pt(C_5H_9S_2)(C_5H_9S_3)]$ molecule.

BIS[2-METHYL-1-(METHYLAMINO)-2-PROPANETHIOLATO-N,S)DIOXOMOLYBDENUM(VI)
$C_{10}H_{24}MoN_2O_2S_2$ (I) $MoO_2[CH_3NHCH_2C(CH_3)_2S]_2$

BIS[2,3-DIMETHYL-3-(METHYLAMINO)-2-BUTANETHIOLATO-N,S)DIOXOMOLYBDENUM(VI)
$C_{14}H_{32}MoN_2O_2S_2$ (II) $MoO_2[CH_3NHC(CH_3)_2C(CH_3)_2S]_2$

BIS[1-(DIMETHYLAMINO)-2-METHYL-2-PROPANETHIOLATO-N,S)DIOXOMOLYBDENUM(VI)
$C_{12}H_{28}MoN_2O_2S_2$ (III) $MoO_2[(CH_3)_2NCH_2C(CH_3)_2S]_2$

J.M. BERG, D.J. SPIRA, K.O. HODGSON, A.E. BRUCE, K.F. MILLER, J.L. CORBIN and E.I. STIEFEL, 1984. Inorg. Chem., 23, 3412-3418.

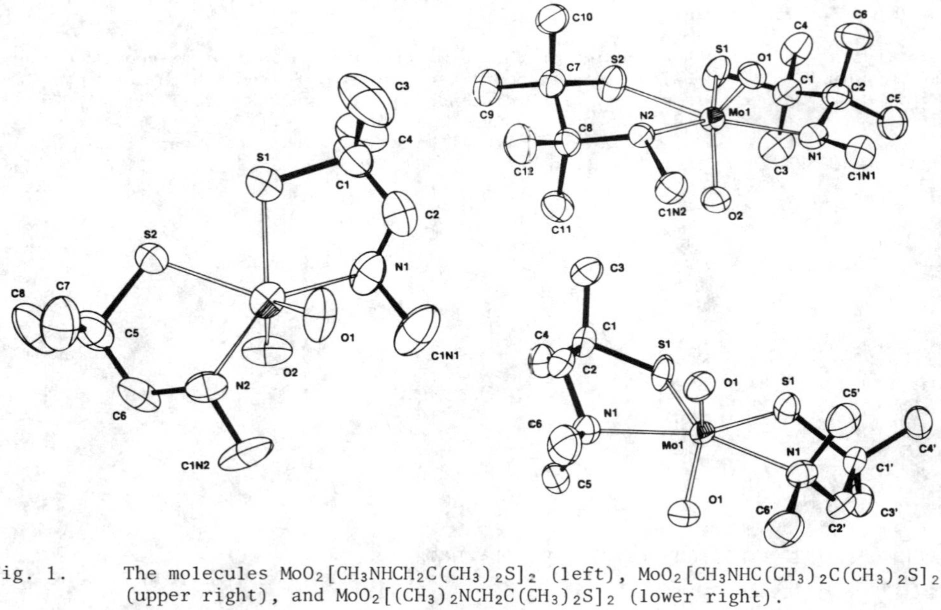

Fig. 1. The molecules $MoO_2[CH_3NHCH_2C(CH_3)_2S]_2$ (left), $MoO_2[CH_3NHC(CH_3)_2C(CH_3)_2S]_2$ (upper right), and $MoO_2[(CH_3)_2NCH_2C(CH_3)_2S]_2$ (lower right).

I. Monoclinic, $P2_1/c$, a = 10.688, b = 11.923, c = 14.032 Å, β = 106.65°, Z = 4. Mo radiation, R = 0.049 for 1516 reflexions.

II. Monoclinic, $P2_1/c$, a = 10.771, b = 11.281, c = 16.021 Å, β = 90.43°, Z = 4. Mo radiation, R = 0.028 for 4774 reflexions.

III. Orthorhombic, Fdd2, a = 18.441, b = 22.998, c = 8.006 Å, Z = 8. Mo radiation, R = 0.031 for 1377 reflexions.

The six-coordinate structures (Fig. 1) are remarkably similar with an unusual nonoctahedral geometry described as a skew trapezoidal bipyramid. An approximate N_2S_2 plane contains the Mo atom with equivalent O atoms above and below the plane. Mean bond lengths for I-III: Mo-O 1.72, Mo-N 2.30, Mo-S 2.42 Å. An additional unusual feature of these molecules is the short S...S contact of 2.73-2.82 Å with an S-Mo-S angle of 68.8-71.0° attributable to partial disulphide bond formation.

$(-)_{589}$-[2-(METHYLTHIO)PROPIONATO-S,O][TRIS(2-AMINOETHYL)AMINE]COBALT(III) DIPERCHLORATE

$C_{10}H_{25}Cl_2CoN_4O_{10}S$ $[Co(C_4H_7O_2S)(C_6H_{18}N_4)](ClO_4)_2$

S. OHBA and Y. SAITO, 1984. Acta Cryst., C40, 398-401.

Orthorhombic, $P2_12_12_1$, a = 11.346, b = 15.554, c = 11.044 Å, D_m = 1.78, Z = 4. Mo radiation, R = 0.042 for 2310 reflexions.

The Co atom has octahedral coordination (Fig. 1) with the S atom trans to the tertiary N of the tren ligand. Principal dimensions include Co-S 2.239(1), Co-O(1) 1.902(3), Co-N 1.946-1.962(5) Å.

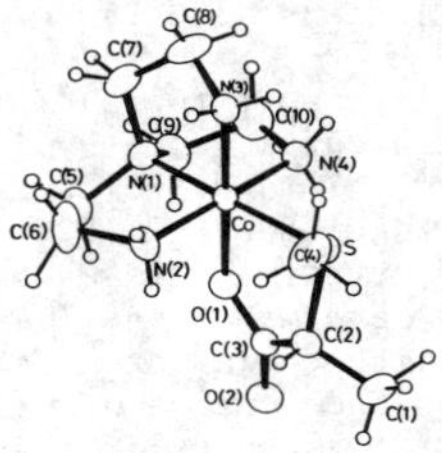

Fig. 1. The structure and absolute configuration of the complex cation $[Co(C_4H_7O_2S)(C_6H_{18}N_4)]^{2+}$.

(3,3-BIS(ETHYLTHIO)PROP-2-ENETHIAL)PENTACARBONYLCHROMIUM(0)

$C_{12}H_{12}CrO_5S_3$

G.J. KRUGER, L. LINFORD and H.G. RAUBENHEIMER, 1984. J. Chem. Soc. Dalton, 2337-2340.

Orthorhombic, Pbcm, a = 11.530, b = 18.745, c = 7.889 Å, D_m = 1.48, Z = 4. Mo radiation, R = 0.038 for 1148 reflexions.

The Cr atom is in an octahedral environment with angles varying between 89.3 and 94.0° (Fig. 1). The trans Cr-CO bond (1.850(5) Å) is significantly shorter than the four cis Cr-CO bonds (average 1.900 Å). There is extensive delocalisation in the thio ligand which is reflected in the bond lengths (C(4)-C(5) 1.401(5), C(5)=C(6) 1.377, S(1)=C(4) 1.642(4), S(2)-C(6) 1.729(4) and S(3)-C(6) 1.741(4) Å). The Cr-S distance is 2.383(1) Å.

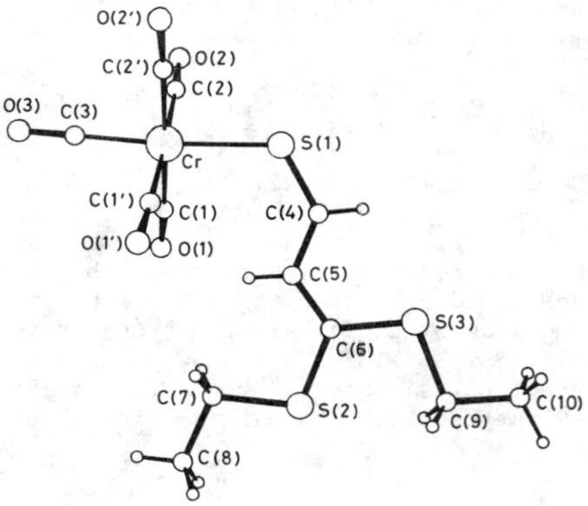

Fig. 1. $C_{12}H_{12}CrO_5S_3$: a perspective view of the molecule perpendicular to the mirror plane.

(3,6-DITHIAOCTANEDIOATO-S,S'-(3,6-DITHIAOCTANEDIOIC ACID-S,S'))COPPER(I)
$C_{12}H_{19}CuO_8S_4$

J. HELDER, P.J.M.W.L. BIRKER, G.C. VERSCHOOR and J. REEDIJK, 1984. Inorg. Chim. Acta, 85, 169-173.

Monoclinic, P2/n, a = 11.625, b = 7.664, c = 9.874 Å, β = 95.16°, D_m = 1.80, Z = 2. Mo radiation, R = 0.026 for 2993 reflexions (173 K).

The Cu(I) ion lies on a crystallographic twofold axis and is coordinated by four thioether S atoms in a distorted tetrahedral geometry (Fig. 1). The bond angles about the metal atom range from 92.62 to 127.77° and the Cu-S bond distances are 2.3321 and 2.2920 Å. The S-C distances range from 1.808 to 1.826 Å. Each molecule is hydrogen-bonded to four neighbouring molecules by six hydrogen bonds, a fact not reported in a previous, room temperature structure determination (1). Bonds between O(2) and O(4) of adjacent molecules result in infinite chains of molecules in the b direction (O...O 2.609 Å), and strong hydrogen bonds between pairs of O(1) atoms in adjacent molecules related by an a-translation have O...O distances of 2.441 Å.

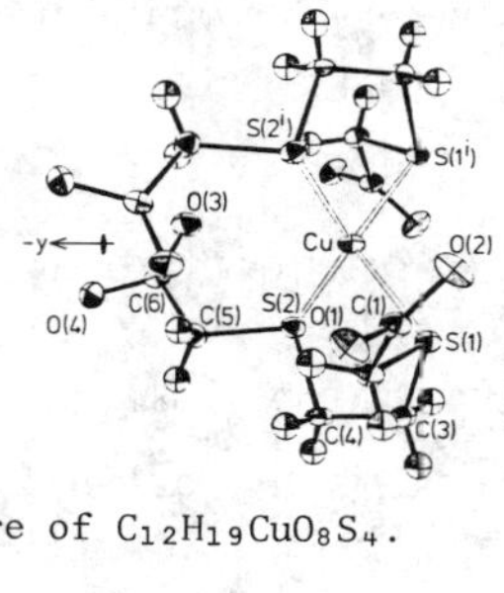

Fig. 1. Molecular structure of $C_{12}H_{19}CuO_8S_4$.

1. Structure Reports, 39B, 818.

[1,2-BIS(t-BUTYLTHIO)ETHYLENE-S,S']DICARBONYLRHODIUM(I) PERCHLORATE
$C_{12}H_{20}ClO_6RhS_2$

A. RUIZ, C. CLAVER, J.C. RODRIGUEZ, M. AGUILO, X. SOLANS and M. FONT-ALTABA, 1984. J. Chem. Soc. Dalton, 2665-2669.

Orthorhombic, $Ccm2_1$, a = 11.129, b = 15.877, c = 10.742 Å, Z = 4. Mo radiation, R = 0.035 for 734 reflexions.

The cation (Fig. 1) lies on a crystallographic mirror plane (passing through Rh and the mid-point of the C(1)-C(1)' bond). The Rh atom has square-planar coordination (Rh-S 2.364(3), Rh-C(3) 1.86(1) Å, C(3)-Rh-S 92.3(5)°) and the five-membered ring is planar.

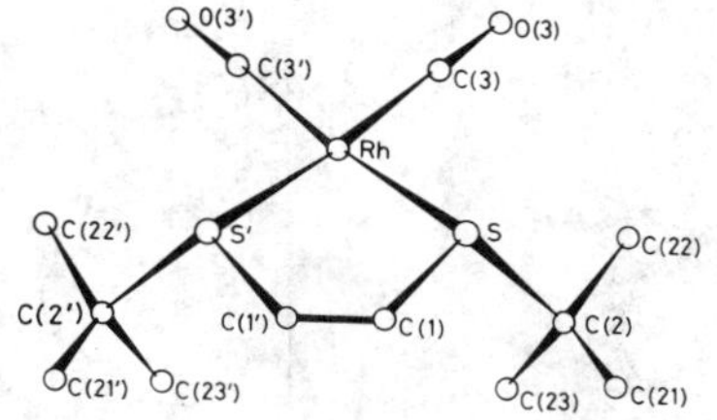

Fig. 1. A view of the cation in $C_{12}H_{20}ClO_6RhS_2$.

BIS(3-HYDROXY-1,5-DITHIACYCLOOCTANE)COPPER(II) DIPERCHLORATE
$C_{12}H_{24}Cl_2CuO_{10}S_4$ $[Cu(C_6H_{12}OS_2)_2][ClO_4]_2$

W.K. MUSKER, M.M. OLMSTEAD and R.M. KESSLER, 1984. Inorg. Chem., 23, 1764-1768.

Monoclinic, $P2_1/c$, a = 6.811, b = 9.225, c = 16.575 Å, β = 90.54°, Z = 2. Mo radiation, R = 0.027 for 2034 reflexions (at 140 K).

The centrosymmetric cation (Fig. 1) has irregular fac-octahedral coordination geometry with Cu-S 2.333(1), 2.335(1), Cu-O 2.350(1) Å, S-Cu-S 91.0(1), and S-Cu-O 79.7(1) and 81.6(1)°.

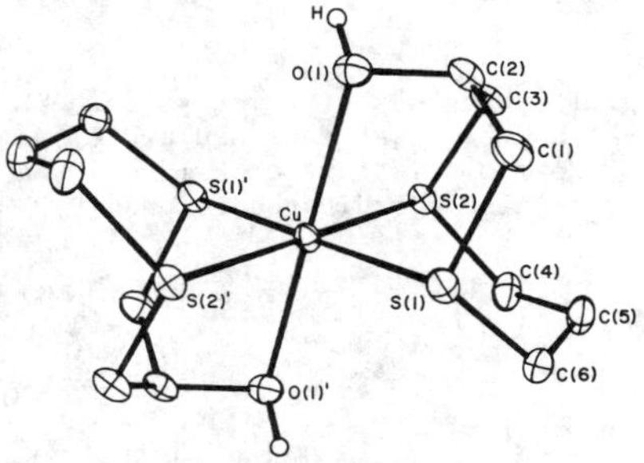

Fig. 1. The $[Cu(C_6H_{12}OS_2)_2]^{2+}$ cation.

μ_3-THIO-μ_3-TRIS(DISULPHIDO)CHLORATO-cyclo-TRIS[(DIETHYL DIOXODITHIOPHOSPHATO-S,S')-MOLYBDENUM](3Mo-Mo)
$C_{12}H_{30}ClMo_3O_6P_3S_{13}$

SHANG MAOYU, HUANG JINLING and LU JIAXI(CHAI-SI LU), 1984. Acta Cryst., C40, 759-761.

Triclinic, $P\bar{1}$, a = 11.570, b = 13.093, c = 14.399 Å, α = 102.62, β = 102.15, γ = 111.87°, Z = 2. Mo radiation, R = 0.055 for 4113 reflexions.

In the thio-capped trinuclear Mo cluster some of the skeleton atoms are in a distorted pseudocubane configuration with three S atoms capped by a Cl atom (Fig. 1). The three Mo atoms form an equilateral-triangle configuration with average Mo-Mo bond length of 2.725(3) Å. The S-Cl distance is 2.90(3) Å and the bond is covalent in character.

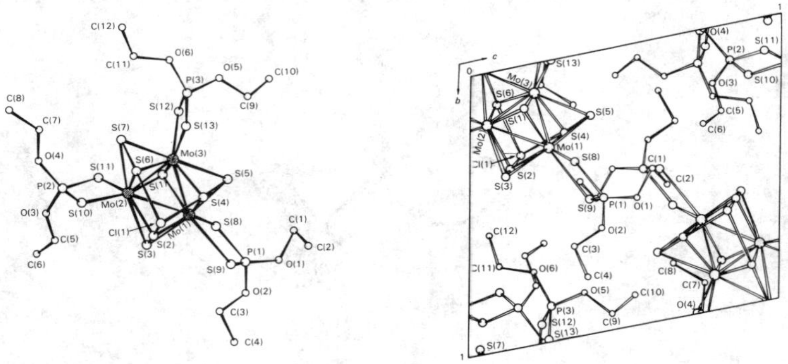

Fig. 1. $C_{12}H_{30}ClMo_3O_6P_3S_{13}$: perspective views of the cluster and of the packing of the cluster in the unit cell.

BIS(μ_3-SULPHIDO)TRIS[(μ_2-CHLORO)(O,O'-DIETHYLDITHIOPHOSPHATO-S,S')MOLYBDENUM]

$C_{12}H_{30}Cl_3Mo_3O_6P_3S_8$ $[Mo_3(S)_2(Cl)_3(S_2P(OEt)_2)_3]$

LU SHAOFANG, LIN YUHUI, HUANG JINLING, LU JIAXI and HUANG JIANQUAN, 1984. J. Struct. Chem., 3, 147-149.

Hexagonal, $P6_3/m$, a = 15.492, c = 8.530 Å, Z = 2. Mo radiation, R = 0.042 for 611 reflexions.

In this cluster (Fig. 1) which has $\bar{6}$ symmetry, three Mo atoms form an equilateral triangle (Mo-Mo 2.606(1) Å) and two μ_3-S caps are above and below the triangle (Mo-S 2.381(4) Å). Each pair of Mo atoms is further bridged by a Cl atom (mean Mo-Cl 2.457(3) Å). Each Mo atom is also coordinated to one $[S_2P(OEt)_2]^{2-}$ ligand (Mo-S 2.492(3) Å).

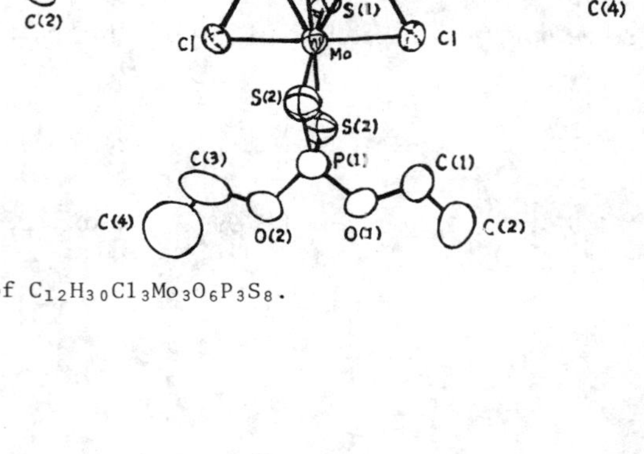

Fig. 1. A view of $C_{12}H_{30}Cl_3Mo_3O_6P_3S_8$.

(μ_3-OXO)TRIS[(μ_2-DISULPHIDO)(DIETHYLDITHIOPHOSPHATO-S,S')MOLYBDENUM] IODIDE
$C_{12}H_{30}IMo_3O_7P_3S_{12}$

SHANG MAOYU, HUANG JINLING and LU JIAXI, 1984. J. Struct. Chem., 3, 17-20.

Monoclinic, $P2_1/n$, a = 16.570, b = 12.370, c = 20.007 Å, β = 99.97°, Z = 4. Mo radiation, R = 0.047 for 4387 reflexions.

In the cluster (Fig. 1) three Mo atoms form an equilateral triangle (mean Mo-Mo 2.626(4) Å) which is capped by an O atom (mean Mo-O 2.036(2) Å). Each Mo pair is further bridged by a S_2^{2-} moiety which has one S atom essentially on the Mo_3 plane and the other S is on the side of the plane opposite the μ_3-O atom. Each Mo is also coordinated to two S atoms from $[S_2P(OEt)_2]^-$ ligands. An iodine atom is located above the S_3 face with a mean S...I distance 3.20 Å (less than the van der Waals radii sum).

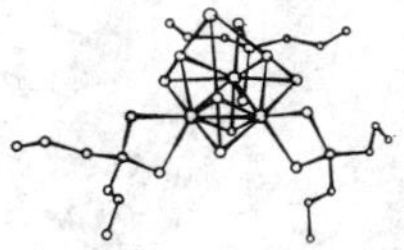

Fig. 1. A view of the cluster $C_{12}H_{30}IMo_3O_7P_3S_{12}$.

TETRAKIS(μ_2-3-AMINOPROPANE-1-THIOLATO-N,S,S)TRINICKEL(II) DICHLORIDE
$C_{12}H_{32}Cl_2N_4Ni_3S_4$ $[Ni_3(MPA)_4]Cl_2$

H. BARRERA, J. SUADES, M.C. PERUCAUD and J.L. BRIANSO, 1984. Polyhedron, 3, 839-843.

Monoclinic, $P2_1/c$, a = 8.136, b = 10.666, c = 13.150(6) Å, β = 103.32°, Z = 2. Cu radiation, R = 0.100 for 1909 reflexions.

In the centrosymmetric molecule (Fig. 1) atom Ni(1) lies on an inversion centre, with Ni-S 2.182-2.202(5) Å. The analysis is of low accuracy and no attempt was made to locate H atoms.

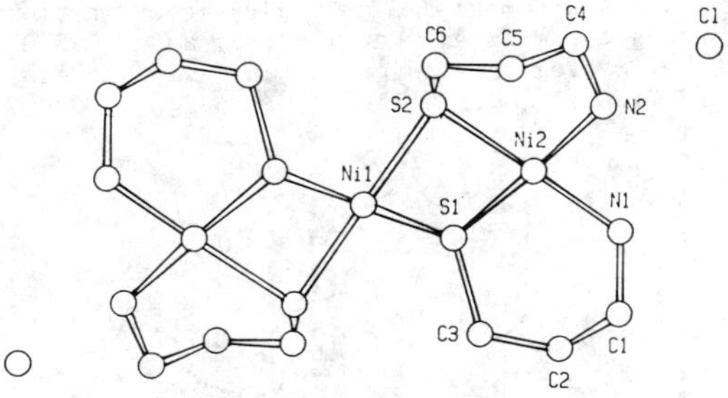

Fig. 1. The structure of $[Ni_3(MPA)_4]Cl_2$.

AQUABIS(PERCHLORATO-O)[2-(METHYLTHIO-o-PHENYLENE-IMINOMETHYL)PYRIDYL-N,N',S]COPPER(II) MONOHYDRATE
$C_{13}H_{14}Cl_2CuN_2O_9S,H_2O$

A.W. ADDISON, T.N. RAO and E. SINN, 1984. Inorg. Chem., 23, 1957-1967.

Monoclinic, $P2_1/c$, a = 9.212, b = 14.439, c = 15.117 Å, β = 95.60°, Z = 4. Mo radiation, R = 0.063 for 3195 reflexions.

The structure (Fig. 1) contains elongated pseudooctahedral Cu(II) ions, with the slightly distorted equator occupied by an H_2O (Cu-O 1.931(2) Å) and the N and S atoms of the tridentate ligand (Cu-S 2.320(1), Cu-N(1) 2.000(3), Cu-N(8) 1.943(2) Å). Perchlorates are weakly bound on the axial positions (Cu-O 2.490(3) and 2.560(4) Å).

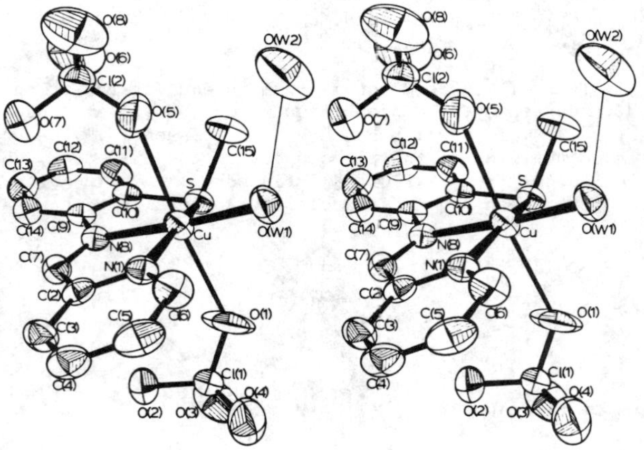

Fig. 1. Stereoview of $C_{13}H_{14}Cl_2CuN_2O_9S,H_2O$.

[1,7-BIS(5-METHYLIMIDAZOL-4-YL)-2,6-DITHIAHEPTANE]DICHLOROCOPPER(II)
$C_{13}H_{20}Cl_2CuN_4S_2$ $CuCl_2(C_{13}H_{20}N_4S_2)$

J. van RIJN, W.L. DRIESSEN, J. REEDIJK and J.-M. LEHN, 1984. Inorg. Chem., 23, 3584-3588.

Orthorhombic, Pbca, a = 12.927, b = 18.933, c = 14.469 Å, D_m = 1.62, Z = 8. Mo radiation, R = 0.022 for 1641 reflexions.

The complex (Fig. 1) has a compressed octahedral coordination geometry, with the two imidazole N atoms along the main axis at 1.949(2) and 1.955(2) Å and both Cl^- ions and the two thioether S atoms at longer distances in the equatorial plane, (Cu-Cl 2.395(1), 2.457(1); Cu-S 2.886(1), 2.970(1) Å).

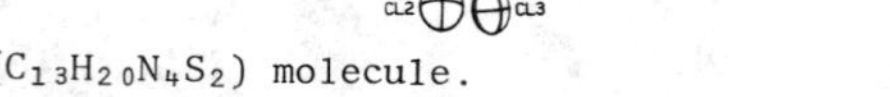

Fig. 1. The $CuCl_2(C_{13}H_{20}N_4S_2)$ molecule.

BROMODIBENZYLSULPHIDEGOLD(I)
$C_{14}H_{14}AuBrS$ (I)

CHLORODIBENZYLSULPHIDEGOLD(I)
$C_{14}H_{14}AuClS$ (II)

J. STRAHLE, W. HILLER and W. CONZELMANN, 1984. Z. Naturforsch., B39, 538-541.

I. Tetragonal, $P4_1$, a = 10.567, c = 12.863 Å, Z = 4. Mo radiation, R = 0.023 for 632 reflexions.

II. Tetragonal, $P4_1$, a = 10.462, c = 12.885 Å, Z = 4. Mo radiation, R = 0.51 for 1004 reflexions.

The structures are isomorphous and a view of (I) is in Fig. 1. The Au atoms have linear coordination with Au-Br 2.387(2), Au-Cl 2.222(9), Au-S 2.275(4) (I), 2.290(9) Å (II), Br-Au-S 178.7(1), Cl-Au-S 172.3(3)°. The S atoms are sp^3 hybridised.

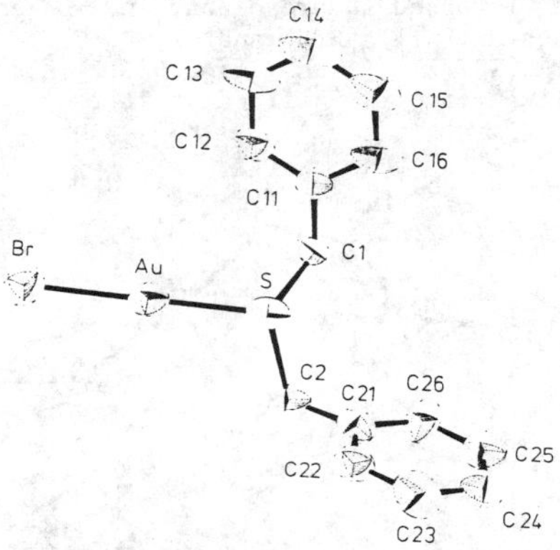

Fig. 1. A view of $C_{14}H_{14}AuBrS$.

(1,2-BIS(PHENYLTHIO)ETHANE-S,S')TRICHLOROTHIOTANTALUM(V)

$C_{14}H_{14}Cl_3S_3Ta$

M.G.B. DREW, D.A. RICE and D.M. WILLIAMS, 1984. J. Chem. Soc. Dalton, 845-848.

Monoclinic, $P2_1/c$, a = 15.538, b = 7.352, c = 16.619 Å, β = 110.88°, D_m = 2.20, Z = 4. R = 0.062 for 2003 reflexions.

The metal atom (Fig. 1) is in a distorted octahedral environment with a terminal Ta=S bond (2.204(5)), three Ta-Cl interactions (2.322(5), 2.366(5), 2.241(6) Å) and two bonds to the S atoms of the ligand. The bond that is trans to the terminal sulphur atom is longer (2.863(4)) than the bond that is cis (2.705(5) Å). The S=Ta-S(cis) angle is 86.5(2)°.

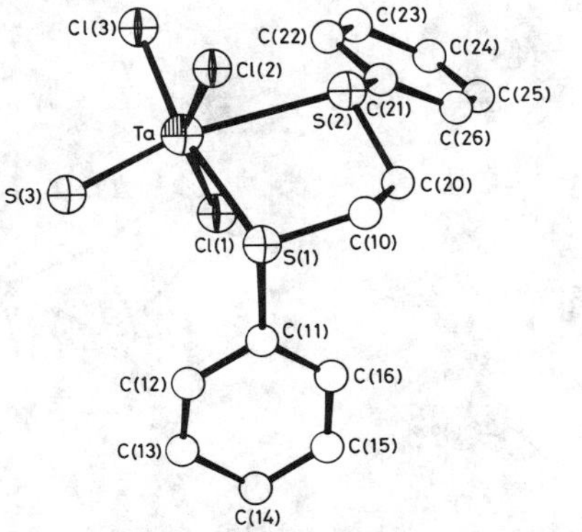

Fig. 1. $C_{14}H_{14}Cl_3S_3Ta$: the molecular structure.

(3-ALLYLPENTANE-2,4-DITHIONATO)CHLORO(TRIETHYLPHOSPHINE)NICKEL(II)

$C_{14}H_{26}ClNiPS_2$ Ni(asacsac)(PEt_3)Cl

K.J. CAVELL, D.G. HAY, A.F. MASTERS and G.A. WILLIAMS, 1984. Aust. J. Chem., 37, 273-279.

Monoclinic, $P2_1/c$, a = 10.078, b = 8.202, c = 22.984 Å, β = 97.57°, Z = 4. Cu radiation, R = 0.049 for 1888 reflexions.

The metal atom is in a distorted square planar environment (Fig. 1) with the dihedral angle between planes Ni, S(1), S(2) and Ni, P, Cl of 8.3° (Ni-P 2.234(2), Ni-Cl 2.208(2), Ni-S (trans P) 2.127(2), (trans Cl) 2.083(2) Å and S-Ni-S 96.8(1)°). The plane containing the allyl group is inclined at 71.2° to the dithionato ligand.

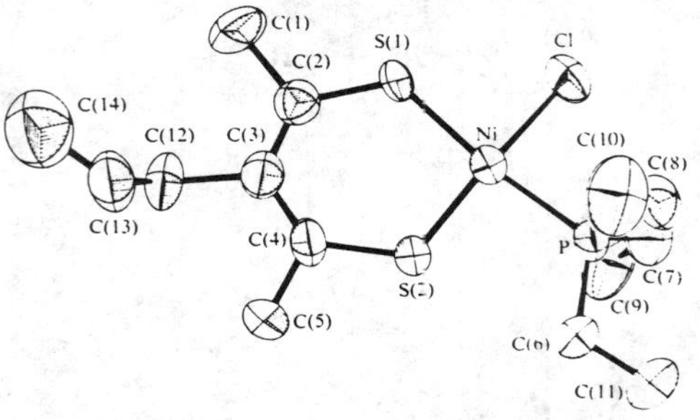

Fig. 1. The structure of Ni(asacsac)(PEt_3)Cl.

BIS(3-METHOXY-1,5-DITHIACYCLOOCTANE-S,S')COPPER(I) PERCHLORATE METHANOL SOLVATE

$C_{14}H_{28}ClCuO_6S_4,CH_4O$ (I) $[Cu(C_7H_{14}OS_2)_2][ClO_4].CH_3OH$

BIS(3-METHOXY-1,5-DITHIACYCLOOCTANE-O,S,S')COPPER(II) BIS(PERCHLORATE)

$C_{14}H_{28}Cl_2CuO_{10}S_4$ (II) $[Cu(C_7H_{14}OS_2)_2][ClO_4]_2$

W.K. MUSKER, M.M. OLMSTEAD and R.M. KESSLER, 1984. Inorg. Chem., 23, 3266-3269.

I. Orthorhombic, $Pna2_1$, a = 14.987, b = 19.483, c = 7.563 Å, D_m = 1.65, Z = 4. Mo radiation, R = 0.052 for 2339 reflexions (at 140 K).

II. Triclinic, $P\bar{1}$, a = 6.699, b = 9.391, c = 9.741 Å, α = 67.73, β = 79.03, γ = 87.02°, D_m = 1.79, Z = 1. Mo radiation, R = 0.025 for 1714 reflexions (at 140 K).

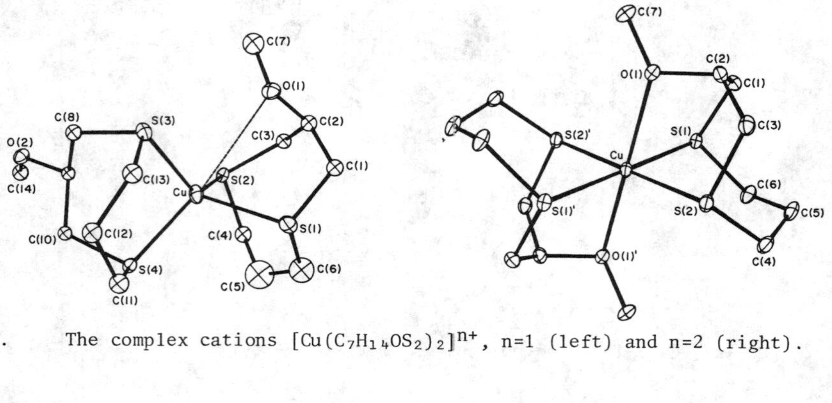

Fig. 1. The complex cations $[Cu(C_7H_{14}OS_2)_2]^{n+}$, n=1 (left) and n=2 (right).

The two complex cations are shown in Fig. 1. The Cu(I) atom in I has distorted tetrahedral CuS_4 coordination with Cu-S 2.294(3)-2.351(2) Å, in addition there is a Cu...O interaction of 3.069(5) Å which has a stereochemical effect. The Cu(II) atom in II lies at a centre of symmetry and has distorted octahedral geometry. The four S atoms form the equatorial plane (Cu-S 2.346(1) and 2.340(1) Å), with the two O atoms (Cu-O 2.369(2) Å) tilted 10.5° off a vertical to the CuS_4 plane.

(o-BENZENEDITHIOLATO-S,S')TRIS(TRIMETHYLPHOSPHINE)IRON
$C_{15}H_{31}FeP_3S_2$ (I) $[Fe(S_2C_6H_4)(PMe_3)_3]$

BIS(TETRAPHENYLARSONIUM) BIS(o-BENZENEDITHIOLATO-S,S')IRON
$C_{60}H_{48}As_2FeS_4$ (II) $[AsPh_4]_2{}^+[Fe(S_2C_6H_4)_2]^{2-}$

D. SELLMANN, U. KLEINE-KLEFFMANN, L. ZAPF, G. HUTTNER and L. ZSOLNAI, 1984. J. Organometal. Chem., 263, 321-331.

I. Monoclinic, $P2_1/c$, a = 10.62, b = 11.13, c = 17.59 Å, β = 96.92°, Z = 4. Mo radiation, R = 0.056 for 2355 reflexions for data at 253 K.

II. Monoclinic, $P2_1/c$, a = 11.172, b = 14.929, c = 19.50 Å, β = 129.83°, Z = 2. Mo radiation, R = 0.049 for 2437 reflexions.

Complex I consists of pentacoordinate Fe in a distorted square pyramidal arrangement of ligands (Fig. 1). The Fe-S distances are 2.201 and 2.208 Å, and the basal Fe-P distances 2.223 and 2.223 Å with the apical Fe-P(4) distance being 2.145 Å. The angles Fe-S(1)-C(26) and Fe-S(2)-C(21) are 106.9 and 107.2° respectively and the S(1)-Fe-S(2) angle is 88.2°. Complex II consists of discrete tetrahedral $AsPh_4^+$ cations and mononuclear $[Fe(S_2C_6H_4)_2]^{2-}$ anions. In the anion, which has crystallographic inversion symmetry, the Fe atom is four-coordinate with Fe-S distances of 2.233 and 2.219 Å. The S-Fe-S angles are 89.6 and 90.4°, and the Fe-S-C values are 105.3 and 106.0°. The As-C distances in the cation of II range from 1.912 to 1.922 Å and the angles about As from 106.0 to 112.9°.

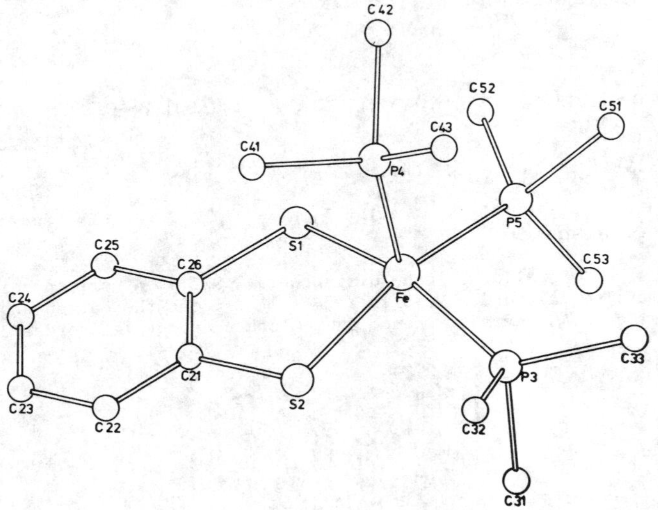

Fig. 1. Molecular structure of $C_{15}H_{31}FeP_3S_2$.

DIIODOTETRAKIS(μ-PYRIMIDINE-2-THIONATO-N1,S2)DIPLATINUM(III)
$C_{16}H_{12}I_2N_8Pt_2S_4$

D.M.L. GOODGAME, R.W. ROLLINS and A.C. SKAPSKI, 1984. Inorg. Chim. Acta, 83, L11-L12.

Monoclinic, $P2_1/c$, a = 15.874, b = 9.729, c = 15.983 Å, β = 98.02°, Z = 4. Cu radiation, R = 0.043 for 2044 reflexions.

The complex is dimeric (Fig. 1) with four bridging pyrimidine-2-thionates which form a cis-planar array of two S and two N atoms about Pt, with an I atom occupying each terminal position of the dimer. The Pt-Pt distance is 2.554 Å and the Pt-I distances are 2.768 and 2.779 Å. The PtN_2S_2 planes are not fully eclipsed, there is a twist of ca. 26° about the Pt-Pt axis. The mean Pt-N and Pt-S distances are 2.109 and 2.294 Å respectively, and the mean Pt-Pt-I angle is 173.2°.

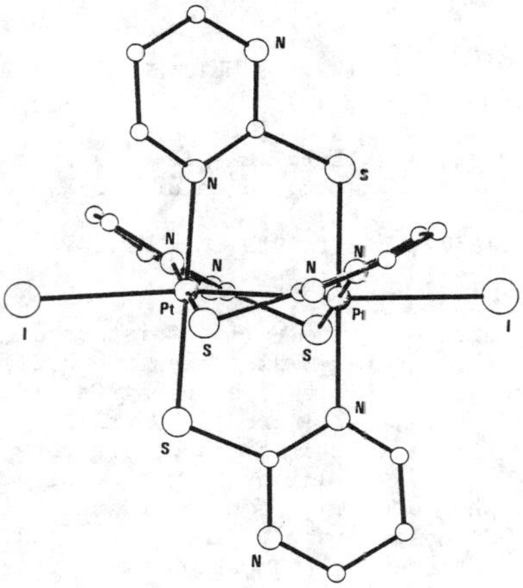

Fig. 1. Molecular structure of $C_{16}H_{12}I_2N_8Pt_2S_4$.

BIS(μ-ETHYLTHIO-S)-BIS[CARBONYL(η^5-CYCLOPENTADIENYL)IRON](Fe-Fe) CHLORATE
$C_{16}H_{20}ClFe_2O_5S_2$

$C_{16}H_{20}Fe_2O_2S_2^+ . ClO_3^-$

R.B. ENGLISH, 1984. Acta Cryst., C40, 1567-1570.

Tetragonal, $P4_32_12$, a = 9.180, c = 24.156 Å, D_m = 1.70, Z = 4. Mo radiation, R = 0.078 for 862 reflexions.

In the cation (Fig. 1) the Fe-Fe distance is 2.957(4) Å corresponding to a one-electron metal-metal bond. Fe-S is 2.240(4) Å. Both the cation and the chlorate anion lie on sites of C_2 symmetry hence the chlorate anion is disordered.

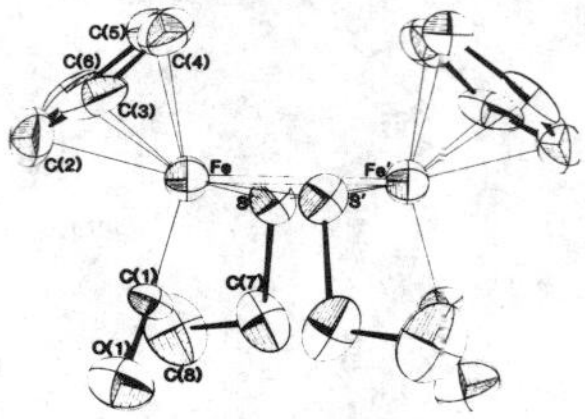

Fig. 1. The structure of the $C_{16}H_{20}Fe_2O_2S_2^+$ cation.

BIS(μ_2-CHLORO)BIS(DIMETHYLSULPHIDE)BIS(PHENYLIMIDO)TETRACHLORODINIOBIUM
$C_{16}H_{22}Cl_6N_2Nb_2S_2$

F.A. COTTON, S.A. DURAJ and W.J. ROTH, 1984. J. Am. Chem. Soc., 106, 4749-4751.

Monoclinic, C2/c, a = 20.736, b = 7.671, c = 16.780 Å, β = 96.86°, Z = 4. Mo radiation, R = 0.0489 for 1190 reflexions.

The molecule (Fig. 1) has crystallographic inversion symmetry, with Nb-Cl1 2.762(2), Nb-Cl1' 2.462(2), Nb-Cl2 2.344(3), Nb-Cl(3) 2.325(3), Nb-S 2.654(3), Nb=N 1.733(7) Å and Nb=N-C 175.1(7)°. The Nb...Nb separation is 4.096(2) Å.

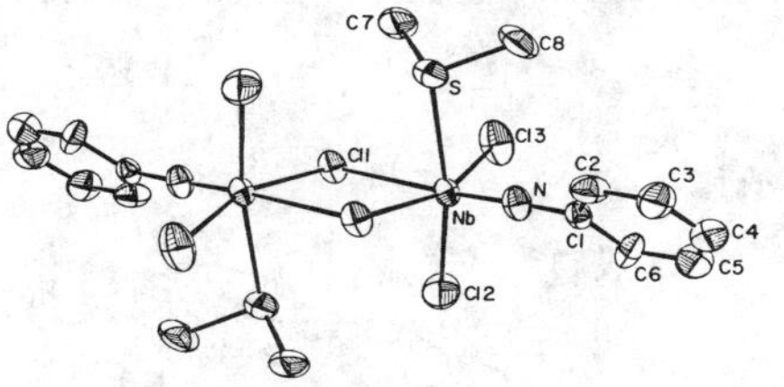

Fig. 1. A drawing of $C_{16}H_{22}Cl_6N_2Nb_2S_2$.

TETRAKIS(1-METHYLIMIDAZOLINE-2(3H)-THIONE)DIBROMONICKEL(II)
$C_{16}H_{24}Br_2N_8NiS_4$ $[Ni(C_4H_6N_2S)_4]^{2+}2Br^-$

R.E. OUGHTRED, E.S. RAPER and I.W. NOWELL, 1984. Inorg. Chim. Acta, 84, L5-L8.

Orthorhombic, Pbca, a = 18.629, b = 13.918, c = 20.367 Å, D_m = 1.66, Z = 8. Mo radiation, R = 0.050 for 2250 reflexions.

The crystal structure consists of distorted square planar cations (Fig. 1) and bromide anions. The Ni-S distances are 2.207, 2.215, 2.212 and 2.217 Å and the S-Ni-S angles 91.9, 87.7, 91.2 and 88.7°. The Ni-S-C(2) angles range from 109.8 to 113.2°, and the average S(2)-C(2) distance is 1.724 Å. The Br^- anions are effectively removed from the coordination sphere, their function being to link neighbouring complex cations by means of N-H...Br...H-N contacts. The four Br...N distances range from 3.218 to 3.260 Å.

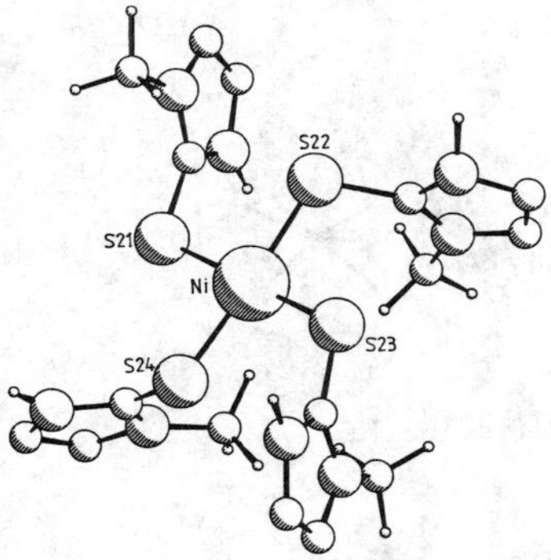

Fig. 1. Perspective view of the complex cation in $C_{16}H_{24}Br_2N_8NiS_4$.

BIS(TETRAMETHYLAMMONIUM) BIS(MALEONITRILEDITHIOLATO)CUPRATE(II)
$C_{16}H_{24}CuN_6S_4$

C. MAHADEVAN and M. SESHASAYEE, 1984. J. Cryst. Spectrosc. Res., 14, 215-226.

Monoclinic, $P2_1/c$, a = 13.230, b = 11.518, c = 16.269 Å, β = 95.07°, Z = 4. Mo radiation, R = 0.059 for 1907 reflexions.

The anions (Fig. 1) have approximately 222 (D_2) symmetry, with an angle of 41.14(1)° between the normals to planes of the ligands and with Cu-S 2.252 to 2.267(4) Å. The amino atoms of the cations have an approximate tetrahedral configuration. The anions are well separated from each other, the closest Cu...Cu distance being 7.811(2) Å.

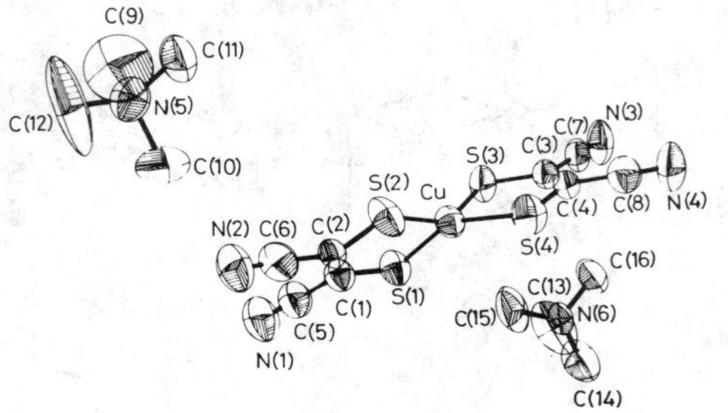

Fig. 1. A view of $C_{16}H_{24}CuN_6S_4$.

BIS(μ-DISELENIDO-μ-Se,μ-Se')-BIS[DICHLOROBIS(TETRAHYDROTHIOPHENE)NIOBIUM(IV)](Nb-Nb)
$C_{16}H_{32}Cl_4Nb_2S_4Se_4$ $Nb_2Cl_4Se_4(C_4H_8S)_4$

M.G.B. DREW, D.A. RICE and D.M. WILLIAMS, 1984. Acta Cryst., C40, 1547-1549.

Triclinic, $P\bar{1}$, a = 8.678, b = 8.293, c = 11.103 Å, α = 102.9, β = 101.6, γ = 103.5°, D_m = 2.20, Z = 1. Mo radiation, R = 0.101 for 1384 reflexions.

In the discrete, centrosymmetric dimers (Fig. 1) the two Nb atoms are bridged by two diselenide bridges with Nb-Se (mean) 2.636(4), Se-Se 2.260(5), Nb...Nb 2.972(5) Å. The coordination around each Nb atom is completed by two terminal chlorine atoms (Nb-Cl 2.490(7) and 2.481(8) Å) and by two sulphur atoms from the thiophene ligands (Nb-S 2.717(7) and 2.745(7) Å).

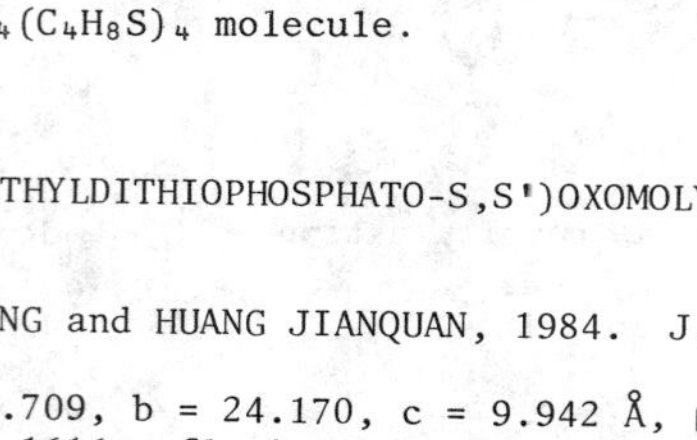

Fig. 1. The $Nb_2Cl_4Se_4(C_4H_8S)_4$ molecule.

(μ_2-OXO)BIS[BIS(O,O'-DIETHYLDITHIOPHOSPHATO-S,S')OXOMOLYBDENUM]
$C_{16}H_{40}Mo_2O_{11}P_4S_8$

WU DINGMING, HUANG JINLING and HUANG JIANQUAN, 1984. J. Struct. Chem., 3, 109-111.

Monoclinic, $P2_1/n$, a = 8.709, b = 24.170, c = 9.942 Å, β = 112.35°, Z = 2. Mo radiation, R = 0.064 for 1616 reflexions.

The unit cell packing is shown in Fig. 1. The molecule lies about an inversion centre (linear Mo-O-Mo) with Mo-O(b) 1.860(1), Mo-O(5) 1.652(8), Mo-S(1) 2.501(4), Mo-S(2) (trans to O(5)) 2.801(4), Mo-S(3) 2.457(4), Mo-S(4) 2.535(3) Å. For a solvated form see (1).

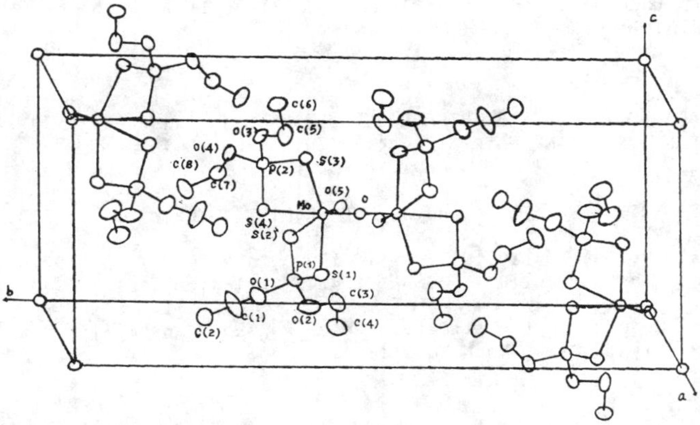

Fig. 1. The unit cell contents of $C_{16}H_{40}Mo_2O_{11}P_4S_8$.

1. Structure Reports, 34B, 598.

(μ_3-SULPHIDO)(μ_2-SULPHIDO)(μ_2-O,O'-DIETHYLTHIOPHOSPHATO-S,S')TRIS(O,O'-DIETHYLDITHIO-PHOSPHATO-S,S')DIOXOTRIMOLYBDENUM

$C_{16}H_{40}Mo_3O_{11}P_4S_{10}$

LU SHAOFANG, HUANG JIANQUAN, HUANG JINLING and LU JIAXI, 1984. J. Struct. Chem., 3, 151-154.

Monoclinic, $P2_1/c$, a = 12.748, b = 14.320, c = 23.118 Å, β = 101.07°, Z = 4. Mo radiation, R = 0.039 for 4160 reflexions.

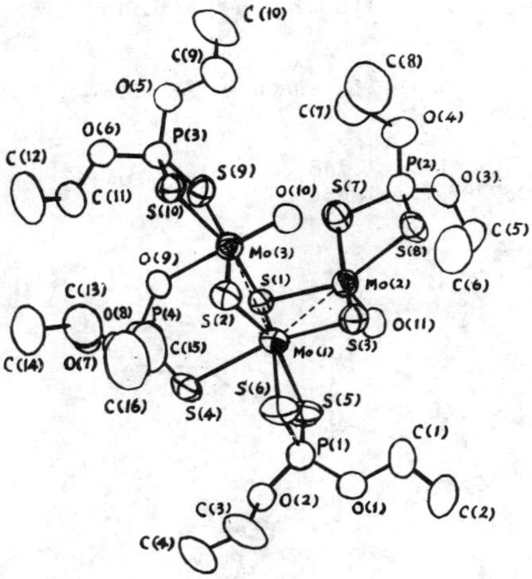

Fig. 1. A view of $C_{16}H_{40}Mo_3O_{11}P_4S_{10}$.

The compound (Fig. 1) is a rather irregular trinuclear molybdenum cluster with Mo(1)-Mo(2) 2.839(1), Mo(1)-Mo(3) 2.808(1) and Mo(2)...Mo(3) 3.337(1) Å. The three Mo atoms are linked by a triply bridging S atom (Mo-S 2.367-2.419(2) Å). Mo(1) has distorted octahedral coordination with six S atoms. Mo(2) has square-pyramidal coordination with four Mo-S and one Mo=O bond (Mo(2)-O(11) 1.651(6) Å). Mo(3) has octahedral coordination (four S and two O atoms).

1,1,1,6,6,6-HEXACARBONYL-5,5,8,8-TETRAMETHYL-2-PHENYL-4,7λ^3,9λ^3-TRITHIA-2λ^4-AZA-5λ^4,8λ^4-DIPHOSPHA-1,6-DIMANGANATRICYCLO[4.2.1.0^{3,9}]NONA-2,7-DIENE
$C_{17}H_{17}Mn_2NO_6P_2S_3$

E. LINDNER, C.-P. KRIEG, W. HILLER and D. HUBNER, 1984. Chem. Ber., 117, 489-501.

Triclinic, P$\bar{1}$, a = 9.775, b = 10.195, c = 14.190 Å, α = 104.87, β = 92.63, γ = 116.76°, Z = 2. Mo radiation, R = 0.033 for 4543 reflexions.

The analysis determines the structure to be as shown in Fig. 1, with Mn-S 2.333-2.398(5), Mn-P 2.291-2.350(5), Mn-N 2.09(1) Å.

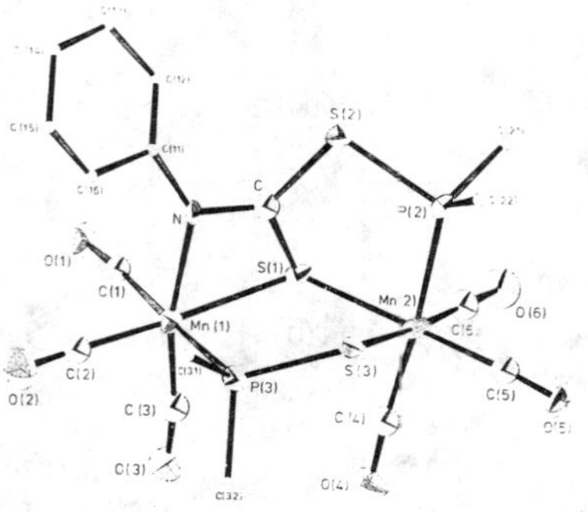

Fig. 1. A view of $C_{17}H_{17}Mn_2NO_6P_2S_3$.

[DISULPHIDOTHIONITRATO(1-)](TRIPHENYLPHOSPHINE SULPHIDE)COPPER(I)
$C_{18}H_{15}CuNPS_4$ $Cu(NS_3)(C_{18}H_{15}PS)$

H.W. ROESKY, K.K. PANDEY, M. NOLTEMEYER and G.M. SHELDRICK, 1984. Acta Cryst., C40, 1555-1556.

Monoclinic, P2_1/c, a = 10.523, b = 21.748, c = 8.880 Å, β = 98.74°, Z = 4. Mo radiation, R = 0.046 for 2677 reflexions.

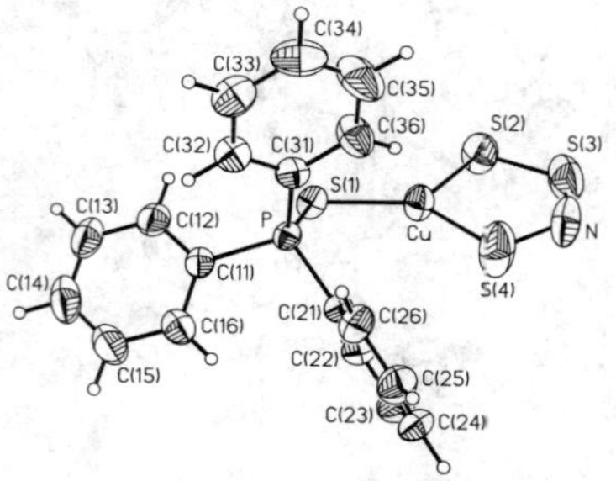

Fig. 1. The $Cu(NS_3)(SPPh_3)$ complex.

Cu is trigonally coordinated by Ph_3PS and the two terminal S atoms of the bidentate $[SSNS]^-$ ligand (Fig. 1) with Cu-S(1) 2.214(2), Cu-S(2) 2.200(3), Cu-S(4) 2.191(3), N-S(3) 1.489(6), N-S(4) 1.655(5) Å.

(1,13-DIAZA-4,10,16,22-TETRATHIACYCLOTETRACOSANE)BIS(DICHLOROCOPPER(II)) DIHYDRATE
$C_{18}H_{38}Cl_4Cu_2N_2S_4 \cdot 2H_2O$ (I) $[(CuCl_2)_2(C_{18}H_{38}N_2S_4)] \cdot 2H_2O$

BIS(μ-AZIDO)(1,13-DIAZA-4,10,16,22-TETRATHIACYCLOTETRACOSANE)BIS(AZIDOCOPPER(II))
$C_{18}H_{38}Cu_2N_{14}S_4$ (II) $[Cu_2(N_3)_4(C_{18}H_{38}N_2S_4)]$

Y. ANGUS, R. LOUIS, J.-P. GISSELBRECHT and R. WEISS, 1984. J. Am. Chem. Soc., 106, 93-102.

I. Monoclinic, $P2_1/c$, a = 7.943, b = 22.481, c = 17.538 Å, β = 100.97°, D_m = 1.56, Z = 4. Cu radiation, R = 0.048 for 3150 reflexions.

II. Monoclinic, C2/m, a = 10.126, b = 13.246, c = 11.156 Å, β = 93.47°, D_m = 1.58, Z = 2. Cu radiation, R = 0.038 for 954 reflexions.

In both structures (Fig. 1), the two Cu atoms are located inside the molecular cavity. In I the two (inequivalent) Cu^{2+} ions are pentacoordinate $CuNS_2Cl_2$ centres with square pyramidal geometry (Cu-Cl (basal) 2.260(1), 2.240(1), Cu-Cl (apical) 2.453(1), 2.499(1), Cu-S 2.236(1)-2.435(1), Cu-N 2.033(5), 2.015(4) Å). The Cu...Cu distance is 7.228(1) Å and the macrocycle is in a boat-type conformation. Complex II has exact C_{2h} symmetry with the two Cu^{2+} centres bridged by two oxide ligands end to end. The coordination geometry is elongated octahedral with four equatorial N atoms and two axial S atoms (Cu-S 2.919(1), Cu-N 2.072(3), Cu-N_3 2.013(3) and 1.994(3) (bridging), 1.987(3) Å (terminal)). The Cu...Cu distance is 5.145(1) Å and the macrocycle is in a chair-type conformation.

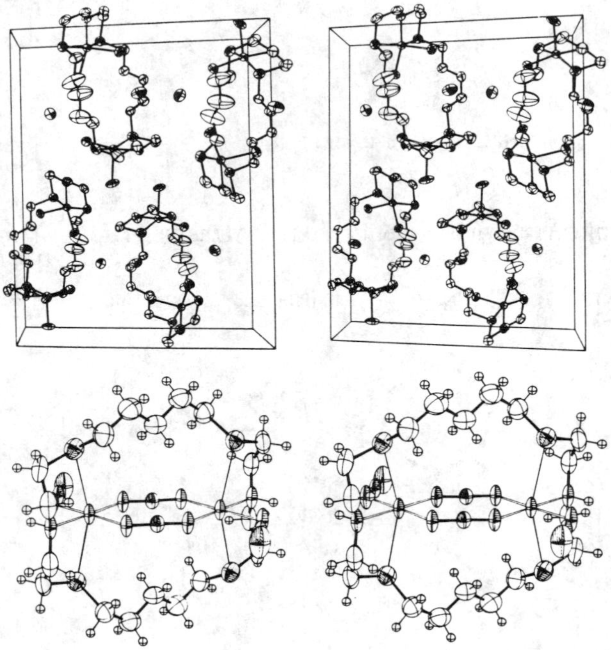

Fig. 1. Stereoviews of the unit cell of $[(CuCl_2)_2(C_{18}H_{38}N_2S_4)] \cdot 2H_2O$ (top) and the complex $[Cu_2(N_3)_4(C_{18}H_{38}N_2S_4)]$ (bottom).

PYRIDINIUM BIS(PYRIDINE-2,6-DICARBOTHIOATO)FERRATE(III) DIHYDRATE
$C_{19}H_{12}FeN_3O_4S_4,2(H_2O)$

U. HILDEBRAND, J. LEX, K. TARAZ, S. WINKLER, W. OCKELS and H. BUDZIKIEWICZ, 1984. Z. Naturforsch., B39, 1607-1613.

Triclinic, P$\bar{1}$, a = 7.184, b = 10.195, c = 16.269 Å, α = 76.10, β = 77.69, γ = 71.45°, Z = 2. R = 0.048 for 4136 reflexions.

The blue pigment obtained from Pseudomonas putida on addition of Fe^{++} is shown in Fig. 1. The Fe atom has octahedral coordination with Fe-S 2.267-2.299(1), Fe-N 1.939 and 1.948(2), S-C 1.715-1.722(3) Å, S(1)-Fe-S(2) 171.9, S(3)-Fe-S(4) 171.9, N(1)-Fe-N(2) 177.1(1)°.

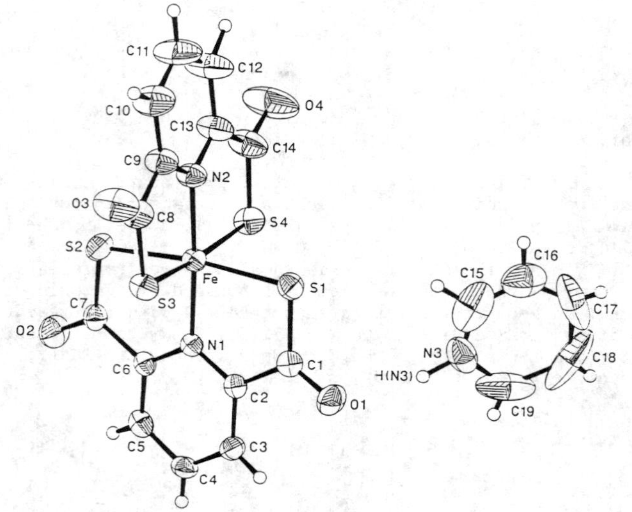

Fig. 1. The anion and cation in $C_{19}H_{12}FeN_3O_4S_4$.

cis-DICHLORO(N-THIONITROSODIMETHYLAMINE)(TRIPHENYLARSINE)PALLADIUM(II)
$C_{20}H_{21}AsCl_2N_2PdS$ $Pd(SNN(CH_3)_2)(As(C_6H_5)_3)Cl_2$

G. TRESOLDI, G. BRUNO, P. PIRAINO, G. FARAONE and G. BOMBIERI, 1984. J. Organometal. Chem., 265, 311-322.

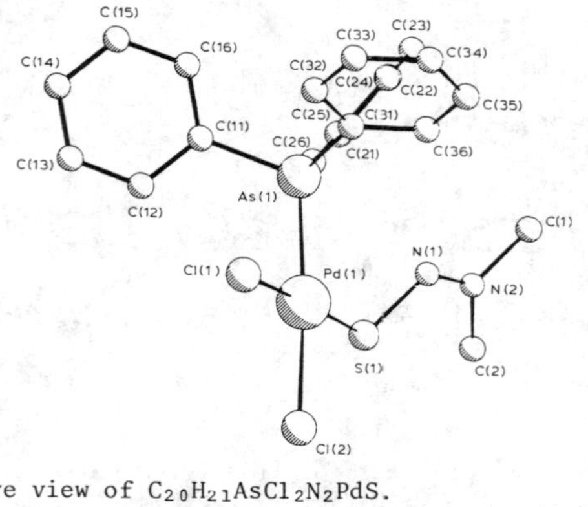

Fig. 1. Perspective view of $C_{20}H_{21}AsCl_2N_2PdS$.

Triclinic, $P\bar{1}$, a = 9.407, b = 10.540, c = 12.265 Å, α = 68.1, β = 78.3, γ = 86.3°, Z = 2. Mo radiation, R = 0.029 for 2664 reflexions.

The Pd coordination geometry (Fig. 1) is distorted square planar. The S-Pd-As angle is considerably greater than 90° (98.9°) presumably because of steric hindrance between As and the Me_2NNS ligand. The remaining angles are Cl(1)-Pd-As 84.2, Cl(1)-Pd-Cl(2) 92.0 and Cl(2)-Pd-S 84.9°. The bonds involving Pd are: Pd-As 2.362, Pd-Cl(1) 2.313, Pd-Cl(2) 2.359 and Pd-S 2.249 Å. The S-N(1) and N(1)-N(2) bond distances are respectively 1.639 and 1.272 Å, and the angles Pd-S-N(1) and S-N(1)-N(2) are 107.2 and 118.0° respectively. The Me_2NNS ligand is almost planar, being inclined at 37.8° to the coordination plane.

TETRAETHYLAMMONIUM BIS[1,2-DI(CARBOMETHOXY)ETHANE-1,2-DITHIOLATO]OXOTECHNETATE(V)
$C_{20}H_{36}NO_9S_4Tc$

G. BANDOLI, M. NICOLINI, U. MAZZI, H. SPIES and R. MUNZE, 1984. Transition Met. Chem., 9, 127-129.

Monoclinic, $P2_1/c$, a = 12.722, b = 13.820, c = 18.213 Å, β = 107.13°, Z = 4. Mo radiation, R = 0.065 for 3378 reflexions.

In the molecule (Fig. 1) the Tc atom is 0.78 Å from the base of square pyramid of S atoms with the oxo-oxygen atom at the apex (Tc-S 2.306-2.322(2), Tc-O 1.672(6) Å, O-Tc-S 107.6-112.1(2)°).

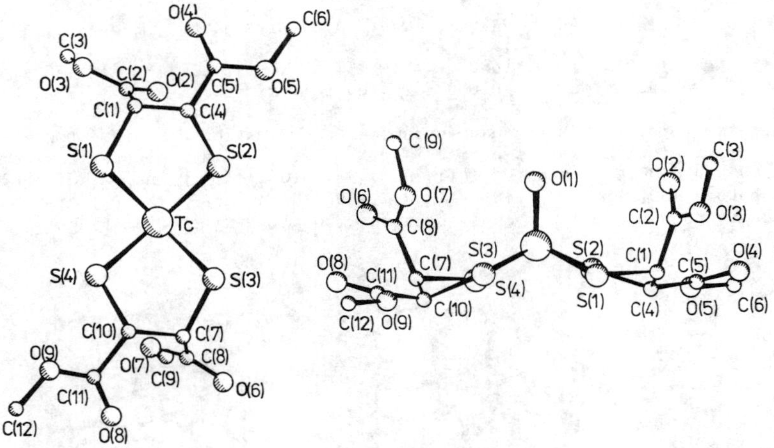

Fig. 1. Two views of the anion in $C_{20}H_{36}NO_9S_4Tc$.

AQUA(1,7-BIS(N-METHYLBENZIMIDAZOL-2'-YL)-2,6-DITHIAHEPTANE)COPPER(II) DIPERCHLORATE
$C_{21}H_{26}Cl_2CuN_4O_9S$

A.W. ADDISON, T.N. RAO, J. REEDIJK, J. VAN RIJN and G.C. VERSCHOOR, 1984. J. Chem. Soc. Dalton, 1349-1356.

Monoclinic, $P2_1/n$, a = 18.459, b = 10.362, c = 16.365 Å, β = 117.14°, D_m = 1.61, Z = 4. Mo radiation, R = 0.047 for 3342 reflexions.

The compound consists of complex cations (Fig. 1) and perchlorate anions. Coordination about the Cu atom is intermediate between trigonal bipyramidal and square pyramidal with Cu-N 1.950(4), 1.997(4), Cu-O(water) 2.225(4), Cu-S 2.328(1), 2.337(1) Å. The perchlorate anion shows some evidence of disorder.

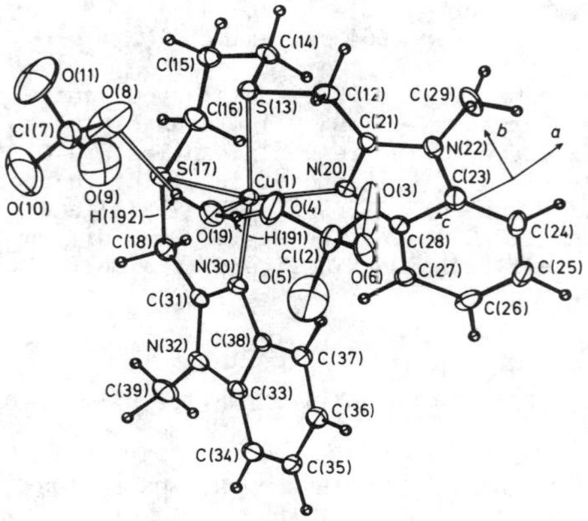

Fig. 1. $C_{21}H_{26}Cl_2CuN_4O_9S$: the complex cation.

ACETATO(3,3,7,7,11,11,15,15-OCTAMETHYL-1,9-DITHIA-5,13-DIAZACYCLOHEXADECANE)SILVER(I) DIHYDRATE

$C_{22}H_{45}AgN_2O_2S_2,2H_2O$

G. FERGUSON, R. McCRINDLE and M. PARVEZ, 1984. Acta Cryst., C40, 354-356.

Triclinic, $P\bar{1}$, a = 10.818, b = 12.662, c = 10.457 Å, α = 101.11, β = 103.23, γ = 89.77°, Z = 2. Mo radiation, R = 0.026 for 4164 reflexions.

The Ag atom (Fig. 1) has a square-pyramidal coordination distorted towards trigonal bipyramidal with axial N atoms (N-Ag-N 179.3(1)°) and equatorial acetato O and S atoms (S-Ag-O 98.0(1) and 101.4(1), S-Ag-S 160.6(1)°). Principal dimensions are: Ag-S both 2.589(1), Ag-N 2.481(2) and 2.430(2), and Ag-O 2.686(2) Å; the macrocyclic ligand has the t,t,t configuration with all six-membered rings in twist-boat conformations.

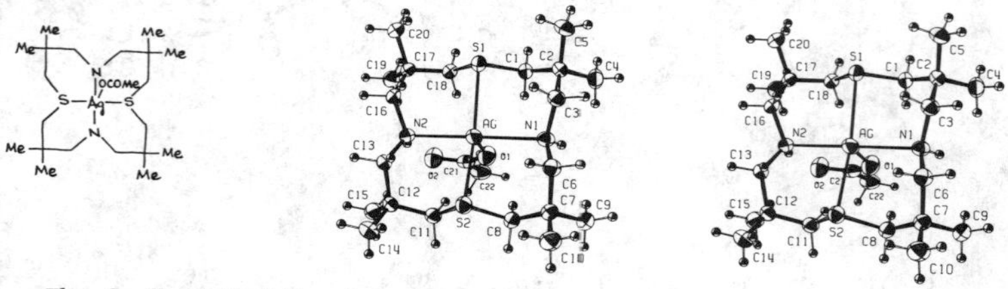

Fig. 1. The $C_{22}H_{45}AgN_2O_2S_2$ structure.

BIS(TETRAETHYLAMMONIUM) (μ_3-SULPHIDO)TRIS(μ_2-SULPHIDO)TRIS(1,2-ETHANEDITHIOLATO)TRI-MOLYBDENUM

$C_{22}H_{52}N_2Mo_3S_{10}$ $2(Et_4N^+).C_6H_{12}Mo_3S_{10}{}^{2-}$

T.R. HALBERT, K. MCGAULEY, W.-H. PAN, R.S. CZERNUSZEWICZ and E.I. STIEFEL, 1984. J. Am. Chem. Soc., 106, 1849-1851.

Monoclinic, $P2_1/n$, a = 13.694, b = 13.654, c = 20.013 Å, β = 101.7°, Z = 4. Mo radiation, R = 0.041 for 4607 reflexions.

The crystal contains ordered $Mo_3S_4(SCH_2CH_2S)_3^{2-}$ dianions as illustrated in Fig. 1, along with ordered and partially disordered tetraethylammonium cations. The $Mo_3S_4^{2+}$ core has an equilateral triangle of Mo(IV) atoms bound by Mo-Mo bonds of 2.77 to 2.79 Å, a single capping μ^3-sulphido, and three bridging μ^2-sulphido ligands, Mo-S1(av) 2.35 Å.

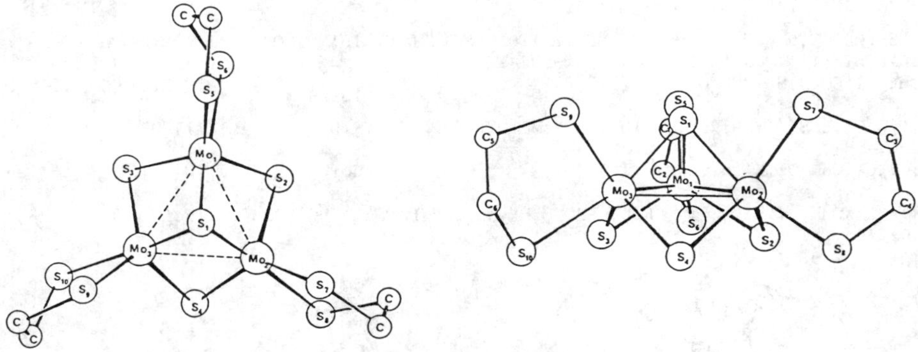

Fig. 1. Two views of the $C_6H_{12}Mo_3S_{10}^{2-}$ anion.

TETRAETHYLAMMONIUM HEXAKIS(METHANETHIOLATO)DIMERCURATE(II)

$C_{22}H_{58}Hg_2N_2S_6$ $[(Et)_4N]_2[Hg_2(SMe)_6]$

G.A. BOWMAKER, I.G. DANCE, B.C. DOBSON and D.A. ROGERS, 1984. Aust. J. Chem., 37, 1607-1618.

Triclinic, P$\bar{1}$, a = 8.656, b = 9.130, c = 12.368 Å, α = 102.16, β = 105.51, γ = 105.29°, Z = 1. Mo radiation, R = 0.034 for 2593 reflexions.

The structure has centrosymmetric dinuclear anions $[(MeS)_2Hg(\mu\text{-}SMe)_2\text{-}Hg(SMe)_2]^{2-}$ with distorted tetrahedral geometry about the Hg atoms (Fig. 1, Hg-S_μ 2.668(2), Hg-S_t 2.456(2), S_μ-Hg-S_μ 94.27(5), Hg-S_μ-Hg' 85.73(5)°) with the methyl substituents on the bridging S_μ adopting an anti configuration.

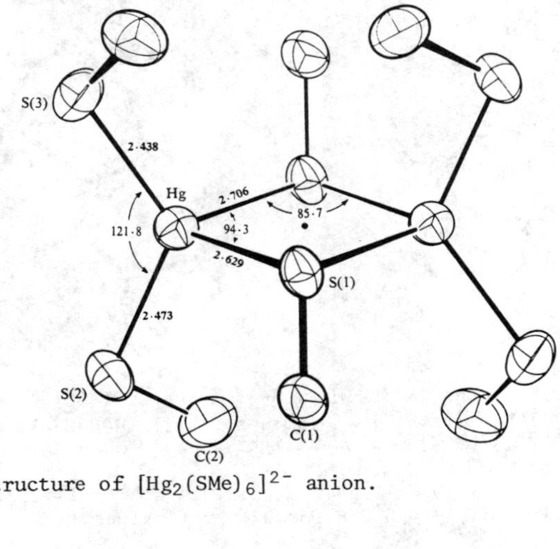

Fig. 1. The structure of $[Hg_2(SMe)_6]^{2-}$ anion.

[6,6'-BIS(((2-PYRIDYLMETHYL)THIO)METHYL)-2,2'-BIPYRIDINE]NICKEL(II) DICHLORIDE ETHANOL SOLVATE DIHYDRATE

$C_{24}H_{22}Cl_2N_4NiS_2,C_2H_6O,2H_2O$ (I) $[Ni(C_{24}H_{22}N_4S_2)]Cl_2,C_2H_5OH,2H_2O$

DIAQUA[2,6-BIS(((2-PYRIDYLMETHYL)OXY)METHYL)PYRIDINE]COBALT(II) DICHLORIDE

$C_{19}H_{23}Cl_2CoN_3O_4$ (II) $[Co(C_{19}H_{19}N_3O_2)(H_2O)_2]Cl_2$

BIS{CHLORO[2,6-BIS(((2-PYRIDYLMETHYL)THIO)METHYL)PYRIDINE]COBALT(II)} TETRACHLOROCOBALTATE(II) METHANOL SOLVATE DIHYDRATE

$C_{38}H_{38}Cl_6Co_3N_6S_4,2CH_4O,2H_2O$ (III) $[Co(C_{19}H_{19}N_3S_2)Cl]_2[CoCl_4],2H_2O,2CH_3OH$

AQUA[6,6'-BIS(((2-PYRIDYLMETHYL)OXY)METHYL-2,2'-BIPYRIDINE]COBALT(II) DICHLORIDE (SOLVATE)

$C_{24}H_{24}Cl_2CoN_4O_3$,X(solvent) (IV) $[Co(C_{24}H_{22}N_4O_2)(H_2O)]Cl_2$,X(solvent)

G.R. NEWKOME, V.K. GUPTA, F.R. FRONCZEK and S. PAPPALARDO, 1984. Inorg. Chem., 23, 2400-2408.

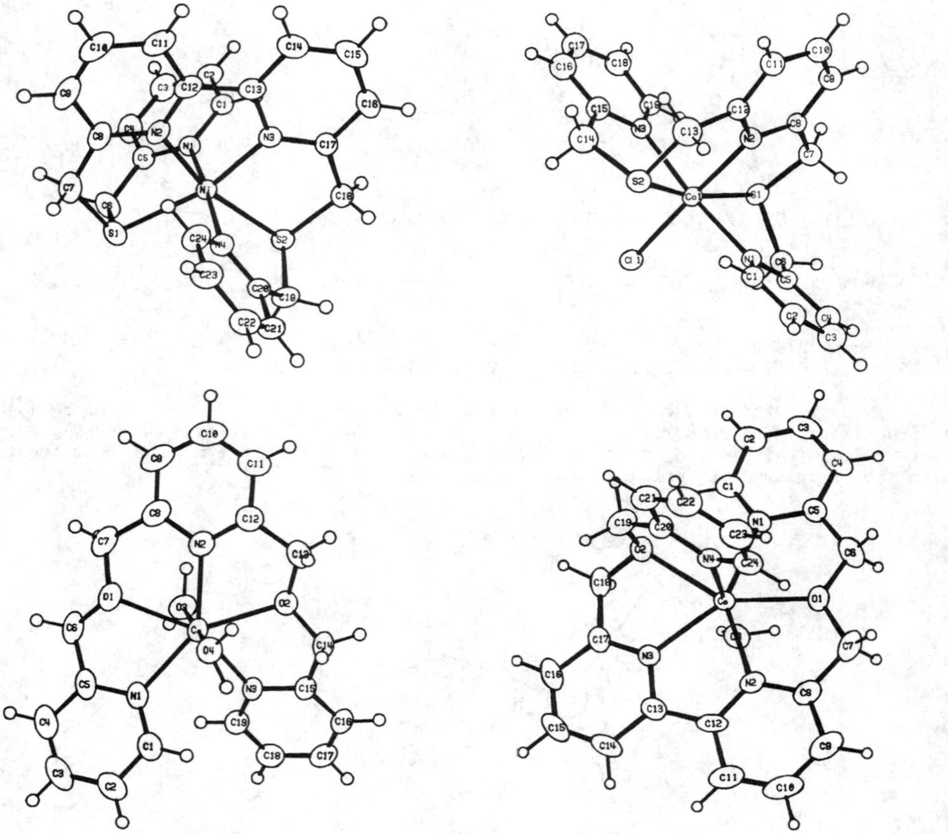

Fig. 1. The complex cations $[Ni(C_{24}H_{22}N_4S_2)]^{2+}$ (upper left), $[Co(C_{19}H_{19}N_3O_2)(H_2O)_2]^{2+}$ (lower left), $[Co(C_{19}H_{19}N_3S_2)Cl]^{+}$ (upper right), and $[Co(C_{24}H_{22}N_4O_2)(H_2O)]^{2+}$ (lower right).

I. Triclinic, P$\bar{1}$, a = 11.588, b = 11.644, c = 12.421 Å, α = 114.00, β = 99.78, γ = 100.98°, Z = 2. Mo radiation, R = 0.038 for 3719 reflexions.

II. Monoclinic, $P2_1/c$, a = 13.099, b = 8.052, c = 20.952 Å, β = 106.92°, Z = 4. Mo radiation, R = 0.035 for 2374 reflexions.

III. Monoclinic, $P2_1/c$, a = 30.780, b = 9.278, c = 17.653 Å, β = 90.80°, Z = 4. Mo radiation, R = 0.049 for 1752 reflexions.

IV. Triclinic, $P\bar{1}$, a = 9.155, b = 12.879, c = 15.072 Å, α = 111.22, β = 93.89, γ = 102.30°, Z = 2. Mo radiation, R = 0.051 for 2873 reflexions (unresolved disorder of unknown quantity of unknown solvent).

The complex cations of I-IV are shown in Fig. 1. In the two thioether complexes, I and III, the metal atoms are in octahedral environments. In I all coordination sites are occupied by the cis-S_2N_4 donor set of the hexadentate ligand, Ni-S 2.398(1) and 2.392(1), Ni-N 2.116(2) and 2.103(2) (trans to N) and 2.030(2) and 2.021(2) Å (trans to S). The cobalt atom in III is coordinated to the trans-S_2N_3 donor set of the pentadentate ligand and to a chloride ligand, Co-S 2.445(2)-2.462(2), Co-Cl 2.344(2) and 2.349(2), Co-N 2.151(6)-2.182(6) (trans to N) and 2.133(6) and 2.103(6) Å (trans to Cl). The Co atoms in the two diether complexes (II and IV) have pentagonal bipyramidal coordination geometry. In II the five equatorial positions are occupied by the ligand heteroatoms, while two water molecules reside at the axial positions, Co-O 2.327(2) and 2.280(2), Co-N 2.208(2)-2.226(2), Co-OH_2 2.112(2) and 2.123(2) Å. Complex IV, with a bipyridine ligand, retains the pentagonal bipyramidal arrangement except that one axial H_2O is replaced by the terminal pyridine group.

BIS(N,N'-DICYCLOPENTYLDITHIOOXAMIDE-S,S')PALLADIUM(II)

$C_{24}H_{40}N_4PdS_4$ Pd(DcPDTO)$_2$

H.C. HOFMANS, H.O. DESSEYN, R.A. DOMMISSE, P. VAN NUFFEL and A.T.H. LENSTRA, 1984. Transition Met. Chem., 9, 213-217.

Orthorhombic, Pbca, a = 12.795, b = 20.790, c = 10.107 Å, Z = 4. Mo radiation, R = 0.062 for 1931 reflexions (at 150 K).

The Pd atom lies on an inversion centre and has square planar geometry with Pd-S 2.295 and 2.303(2) Å, S-Pd-S 89.74(6)°, C-S 1.709 and 1.718(8), C(1)-C(2) 1.497(10) Å.

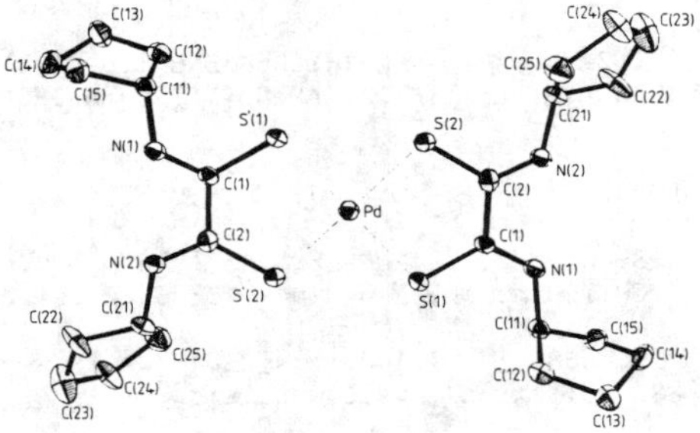

Fig. 1. The structure of Pd(DcPDTO)$_2$.

BIS(TETRAETHYLAMMONIUM) BIS[DIMERCAPTOMALEODINITRILATO(2-)-S,S']NICKELATE(II)

$C_{24}H_{40}N_6NiS_4$ $[N(C_2H_5)_4]_2[Ni(C_4N_2S_2)_2]$

C. MAHADEVAN, M. SESHASAYEE, A. RADHA and P.T. MANOHARAN, 1984. Acta Cryst., C40, 2032-2034.

Triclinic, $P\bar{1}$, a = 7.5608, b = 8.7017, c = 12.626 Å, α = 93.73, β = 104.27, γ = 75.257°, D_m = 1.286, Z = 1. Mo radiation, R = 0.058 for 1550 reflexions.

The Ni atom of the anion lies on an inversion centre and has a (necessarily) planar coordination geometry slightly distorted from square, with Ni-S 2.171 and 2.177(2) Å, S-Ni-S 92.3(1)°.

BIS(μ_2-ETHANETHIOLATO)(μ_2-SULPHIDO)BIS[(DIMETHYLPHENYLPHOSPHINE)(ETHANETHIOLATO)OXO-TUNGSTEN(V)]
$C_{24}H_{42}O_2P_2S_5W_2$

D.C. POVEY and R.L. RICHARDS, 1984. J. Chem. Soc. Dalton, 2585-2587.

Triclinic, $P\bar{1}$, a = 9.104, b = 10.806, c = 17.642 Å, α = 92.47, β = 94.14, γ = 105.43°, Z = 2. Mo radiation, R = 0.049 for 3439 reflexions.

The complex (Fig. 1) is best described as a confacial bioctahedron, with the two SEt ligands and the S ligand occupying the bridging positions. The oxo ligands are cis. The W-W distance is 2.801(1) Å and the angle subtended at the bridging sulphide, S(4), is 73.5(1)°. Coordination distances are: W(1)-O(1) 1.674(9), W(1)-P(1) 2.604(5), W(1)-S(1) 2.404(5), W(1)-S(3) 2.488(4), W(1)-S(4) 2.338(4), W(1)-S(5) 2.647(4), W(2)-O(2) 1.677(9), W(2)-S(2) 2.400(5), W(2)-S(3) 2.465(4), W(2)-S(4) 2.343(5), W(2)-S(5) 2.675(4) and W(2)-P(2) 2.586(4) Å.

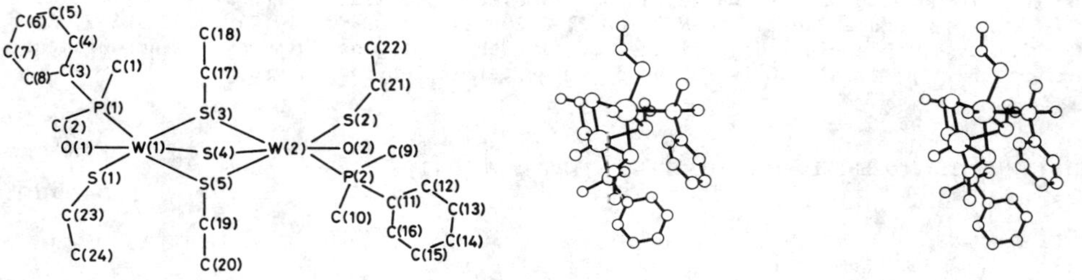

Fig. 1. $C_{24}H_{42}O_2P_2S_5W_2$: the atom numbering and a stereoview of the complex molecule.

(μ-ACETATO-O,O')[(μ-S)-α,α'-{(2-MERCAPTO-4-METHYLBENZENE-1,3-DIYL)-DI(METHYLIDYNE)DI-(AZINO)}BIS(CYCLOHEXANEMETHANOL)ATO(3-)]DIPALLADIUM(II)
$C_{25}H_{32}N_4O_4Pd_2S$ (I) [Pd_2(L)(OAc)]

(μ-ACETATO-O,O')[(μ-S)-α,α'-{(2-MERCAPTO-4-METHYLBENZENE-1,3-DIYL)-DI(METHYLIDYNE)DI-(AZINO)}BIS(CYCLOHEXANEMETHANOL)ATO(3-)]DIPALLADIUM(II) CHLOROFORMATE
$C_{27}H_{32}Cl_6N_4O_4Pd_2S$ (II) [Pd_2(L)(OAc)($CHCl_3$)$_2$]

B.F. HOSKINS and I.A.S. McDONALD, 1984. Aust. J. Chem., 37, 725-737.

I. Triclinic, $P\bar{1}$, a = 15.513, b = 22.238, c = 8.0743 Å, α = 100.81, β = 99.54, γ = 77.23°, D_m = 1.745, Z = 4. Cu radiation, R = 0.044 for 5084 reflexions.

II. Monoclinic, $P2_1/c$, a = 14.668, b = 9.448, c = 25.483 Å, β = 95.39°, Z = 4. Cu radiation, R = 0.063 for 4883 reflexions, yellow prisms.

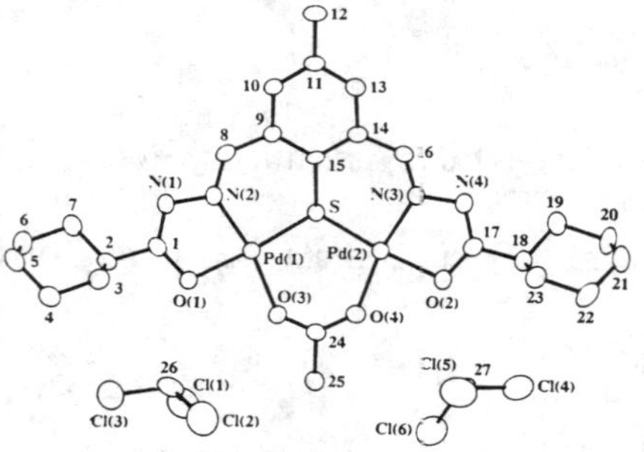

Fig. 1. The structure of [Pd_2(L)(OAc)($CHCl_3$)$_2$].

In I, there are two discrete binuclear molecules in the asymmetric unit with a pronounced buckling of the ligand involving a distinct V-shape (ignoring the cyclohexyl substituents) so that the acetate and toluene groups are aligned on opposite sides of the V (Pd-O 2.016(6), Pd-S 2.214(2), Pd-N 1.939(7), Pd-Pd 3.667(1) Å). In the solvated structure, (II), the dimeric unit is near planar (ignoring the cyclohexyl substituents) with the acetate and toluene groups symmetrically disposed with respect to the rest of the dimer. The molecules pack in columns down the b axis with centrosymmetrically related Pd atoms 3.52 Å away from each other (Pd-O 2.025(6), Pd-S 2.218(2), Pd-N 1.942(7) Å, (Fig. 1)).

(μ-THIO)(μ-METHYLTHIO)(μ-TRIFLUOROACETATO)-BIS[(DIETHYLDITHIOPHOSPHATO-S,S')(p-TOLYLIMIDO)MOLYBDENUM]

$C_{25}H_{37}F_3Mo_2N_2O_6P_2S_6$ $[Mo(NC_6H_4CH_3)(S_2P(OC_2H_5)_2)]_2(\mu\text{-}S)(\mu\text{-}SCH_3)(\mu\text{-}O_2CCF_3)$

M.E. NOBLE, K. FOLTING, J.C. HUFFMAN and R.A.D. WENTWORTH, 1984. Inorg. Chem., 23, 631-634.

Monoclinic, A2/a, a = 33.891, b = 16.169, c = 13.902 Å, β = 90.52°, Z = 8. Mo radiation, R = 0.043 for 5400 reflexions (at -165°C).

The molecule is shown in Fig. 1 and possesses an edge-shared bioctahedral structure. Important bond lengths: Mo-Mo 2.844(1), Mo-(μ-S) 2.349(2) and 2.353(2), Mo-(μ-SMe) 2.437(2) and 2.437(1), Mo-S 2.493(2)-2.548(2), Mo-O 2.275(4) and 2.272(4), Mo-N 1.722(5) and 1.729(5), S-C 1.817(6) Å.

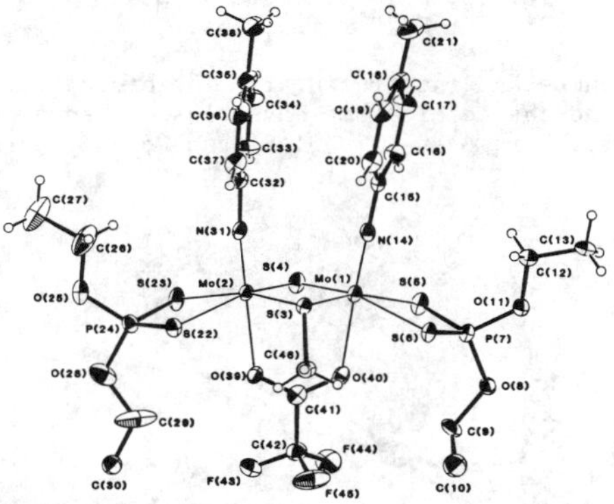

Fig. 1. The $[Mo(NC_6H_4CH_3)(S_2P(OC_2H_5)_2)]_2(\mu\text{-}S)(\mu\text{-}SCH_3)(\mu\text{-}O_2CCF_3)$ molecule.

BIS(N,N-DIETHYL-2-BENZOYLTHIOACETAMIDATO)COPPER(II)

$C_{26}H_{32}CuN_2O_2S_2$

R. RICHTER, J. SIELER, E. LUDWIG, E. UHLEMANN and L. GOLIC, 1984. Z. Anorg. Chem., 513, 114-122.

Triclinic, P$\bar{1}$, a = 10.709, b = 12.150, c = 15.651 Å, α = 90.13, β = 103.73, γ = 98.30°, D_m = 1.35, Z = 3. Mo radiation, R = 0.055 for 4151 reflexions.

There are two independent molecules in the structure, one in a general position with a nearly planar cis-arrangement and one lying on an inversion centre with an exactly trans-planar arrangement (Fig. 1). In the cis-molecule principle bond lengths are: Cu-S 2.236, 2.244(2); Cu-O 1.884, 1.912(4) Å. In the trans molecule Cu-S is 2.254(1) and Cu-O 1.910(4) Å.

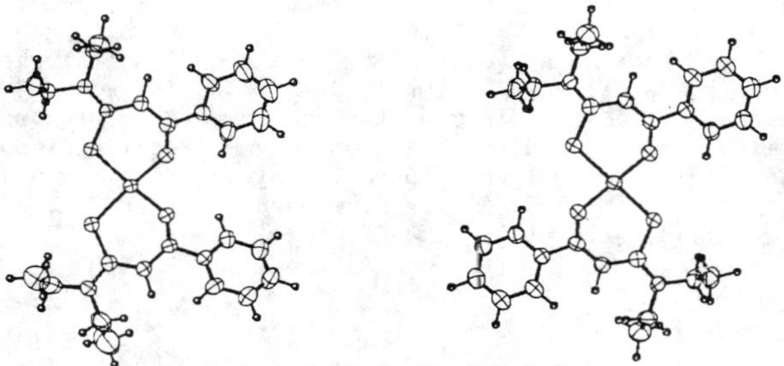

Fig. 1. The cis- and trans- forms of $C_{26}H_{32}CuN_2O_2S_2$.

BIS(DICYCLOHEXYLPHOSPHINODITHIOFORMATO-S,P)PLATINUM(II)
$C_{26}H_{44}P_2S_4Pt$ (I) $Pt[ptf]_2$

BIS(DICYCLOHEXYLPHOSPHINODITHIOFORMATO-S,P)PALLADIUM(II)
$C_{26}H_{44}P_2S_4Pd$ (II) $Pd[ptf]_2$

D. DAKTERNIEKS, B.F. HOSKINS and E.R.T. TIEKINK, 1984. Aust. J. Chem., 37, 197-203.

I. Monoclinic, $P2_1/c$, a = 14.732, b = 15.118, c = 16.000 Å, β = 120.90°, D_m = 1.59, Z = 4. Mo radiation, R = 0.058 for 3454 reflexions.

II. Monoclinic, $P2_1/c$, a = 14.721, b = 15.142, c = 15.986 Å, β = 120.88°, D_m = 1.40, Z = 4. Cu radiation, R = 0.038 for 5025 reflexions.

I and II are isomorphous and isostructural. The bidentate ligands have a cis configuration (Fig. 1), and the metal atoms have close to square planar coordination. Mean dimensions are Pt-P 2.274(4), Pt-S 2.331(5) and Pd-P 2.287(1) and Pd-S 2.336(1) Å respectively.

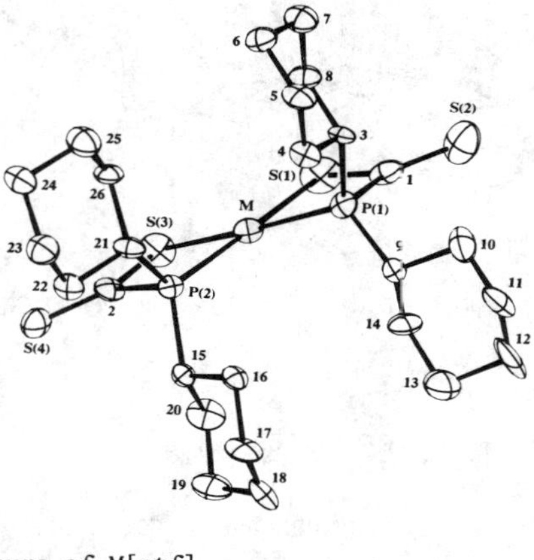

Fig. 1. The structure of $M[ptf]_2$.

BIS(TETRAETHYLAMMONIUM) anti-BIS(μ-ETHANETHIOLATO)-BIS[BIS(ETHANETHIOLATO)IRON(II)]
$C_{28}H_{70}Fe_2N_2S_6$ (I) $[N(C_2H_5)_4]_2[Fe_2(SC_2H_5)_6]$

BIS(TETRAETHYLAMMONIUM) anti-TRIS(o-XYLENE-α,α'-DITHIOLATO)DIIRON(II) BIS(ACETONITRILE) SOLVATE
$C_{40}H_{64}Fe_2N_2S_6,2C_2H_3N$ (II) $[N(C_2H_5)_4]_2[Fe_2((SCH_2)_2C_6H_4)_3].2CH_3CN$

BIS(TETRA-n-BUTYLAMMONIUM) syn-BIS(μ-ETHANETHIOLATO)BIS[BIS(ETHANETHIOLATO)COBALT(II)]
$C_{44}H_{86}Co_2N_2S_6$ (III) $[N(n\text{-}C_4H_9)_4]_2[Co_2(SC_2H_5)_6]$

BIS(TETRAETHYLAMMONIUM) anti-BIS(μ-ETHANETHIOLATO)BIS[BIS(ETHANETHIOLATO)COBALT(II)]
$C_{28}H_{70}Co_2N_2S_6$ (IV) $[N(C_2H_5)_4]_2[Co_2(SC_2H_5)_6]$

BIS(TETRAMETHYLAMMONIUM) HEXAKIS(μ-ETHANETHIOLATO)TETRAKIS[(ETHANETHIOLATO)IRON(II)]
$C_{28}H_{74}Fe_4N_2S_{10}$ (V) $[N(CH_3)_4]_2[Fe_4(SC_2H_5)_{10}]$

K.S. HAGEN and R.H. HOLM, 1984. Inorg. Chem., 23, 418-427.

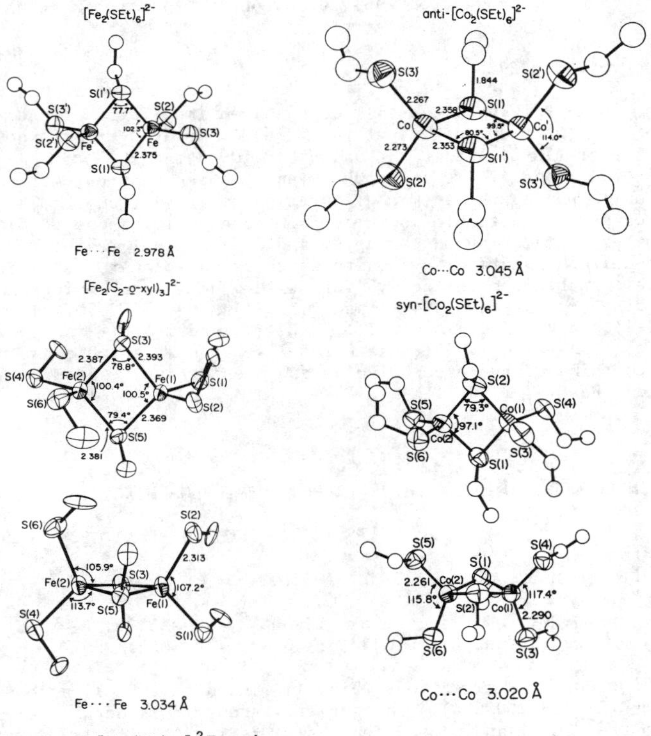

Fig. 1. The four $[M_2(SR)_6]^{2-}$ anions.

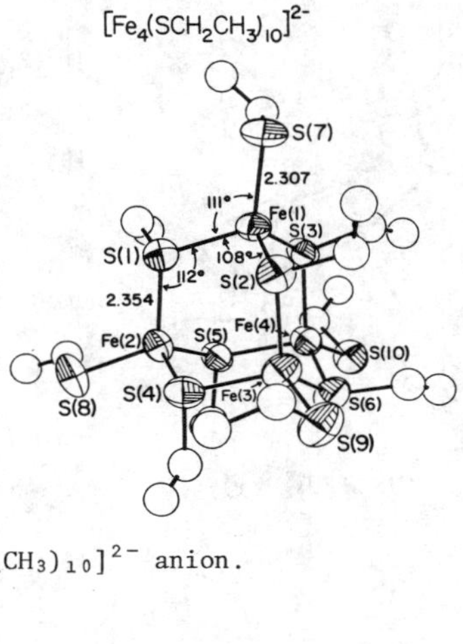

Fig. 2. The $[Fe_4(SCH_2CH_3)_{10}]^{2-}$ anion.

I. Monoclinic, $P2_1/n$, a = 9.815, b = 14.588, c = 14.791 Å, β = 100.97°, D_m = 1.18, Z = 2. Mo radiation, R = 0.043 for 1980 reflexions.

II. Monoclinic, $P2_1/n$, a = 10.62, b = 25.06, c = 18.927 Å, β = 101.94°, Z = 4. Mo radiation, R = 0.066 for 2042 reflexions.

III. Orthorhombic, Pbca, a = 16.672, b = 21.017, c = 33.849 Å, D_m = 1.29, Z = 8. Mo radiation, R = 0.062 for 3067 reflexions.

IV. Monoclinic, $P2_1/n$, a = 9.829, b = 14.504, c = 14.772 Å, β = 100.90°, D_m = 1.20, Z = 2. Mo radiation, R = 0.047 for 1961 reflexions. (Isomorphous with I).

V. Orthorhombic, Pbca, a = 18.879, b = 24.981, c = 21.599 Å, D_m = 1.29, Z = 8. Mo radiation, R = 0.046 for 2928 reflexions.

The anions in I-IV (Fig. 1) are all edge-shared imperfect tetrahedral dimers in which the S-M-S and the M-S-M angles of the $M_2(\mu\text{-}S)_2$ bridges are respectively smaller (97-102°) and larger (78-81°) than the values of 109.5 and 70.5° for a perfect tetrahedral dimer. I and IV are isomorphous and contain centrosymmetric anions with planar bridge units having the anti configuration of ethyl groups. The anion II exists in one of several possible isomeric forms and contains one terminal chelate ring and two others furnishing one bridging and one terminal S atom. This connectivity requires an anti bridge unit, which is non-planar. When crystallized III, the anion is stabilized in the syn form with a non-planar bridge unit. Mean $M\text{-}S_t$ and $M\text{-}S_b$ distances are: 2.304(1) and 2.375(1) Å for I, 2.307(14) and 2.383(10) Å for II, 2.273(14) and 2.363(12) Å for III, 2.271(4) and 2.355(2) Å for IV. The anion in V (Fig. 2) contains a $Fe_4(\mu\text{-}S)_6$ adamantine-like cage in which each Fe atom is tetrahedrally coordinated by three bridging and one terminal S atom (mean $Fe\text{-}S_{(b)}$ 2.354(9) and mean $Fe\text{-}S_{(t)}$ 2.307(11) Å).

CARBONYL[(DIPHENYLPHOSPHINO)ETHANETHIOLATO][(DIPHENYLPHOSPHINO)ETHANETHIOL]HYDRIDO-IRIDIUM CHLORIDE

$C_{29}H_{30}ClIrOP_2S_2$ $[IrH(SCH_2CH_2PPh_2)(HSCH_2CH_2PPh_2)(CO)]^+Cl^-$

D.W. STEPHAN, 1984. Inorg. Chem., 23, 2207-2210.

Triclinic, $P\bar{1}$, a = 10.261, b = 11.269, c = 14.506 Å, α = 85.20, β = 73.02, γ = 111.69°, Z = 2. Mo radiation, R = 0.024 for 4393 reflexions.

The structure consists of cations (Fig. 1) linked to the Cl^- anions via S-H...Cl hydrogen bonding (Cl...H 2.056 Å). The geometry about Ir is octahedral with the H^- ligand trans to the RSH ligand. Bond lengths: Ir-S1 2.411(1), Ir-S2 2.462(1), Ir-P1 2.327(1), Ir-P2 2.348(1), Ir-C 1.885(5), S-H 1.354(10) Å.

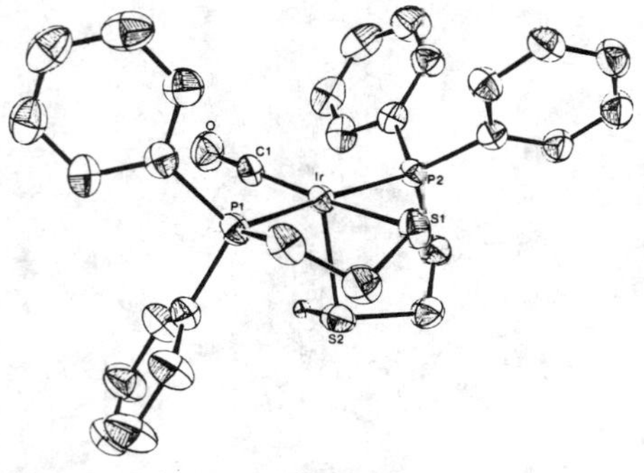

Fig. 1. The $[IrH(SCH_2CH_2PPh_2)(HSCH_2CH_2PPh_2)(CO)]^+$ cation.

[(μ-S)-α,α'-{(2-MERCAPTO-4-METHYLBENZENE-1,3-DIYL)DI(METHYLIDYNE)-DI(AZINO)}BIS-(CYCLOHEXANEMETHANOL)ATO(3-)]-(μ-1-PROPYLIMIDAZOLATO-$C^{(4)}$,$N^{(3)}$)DIPALLADIUM(II) ETHANOL
$C_{29}H_{38}N_6O_2Pd_2S,0.75(C_2H_6O)$

B.F. HOSKINS and I.A.S. McDONALD, 1984. Aust. J. Chem., 37, 1395-1403.

Triclinic, P$\bar{1}$, a = 7.078, b = 13.615, c = 17.565 Å, α = 78.52, β = 80.21, γ = 83.83°, D_m = 1.592, Z = 2. Mo radiation, R = 0.054 for 4133 reflexions.

The quinquedentate ligand embraces two Pd atoms. Both metal atoms (Fig. 1) have similar distorted square planar coordination (Pd...Pd 3.605 Å), the fourth positions being taken by either C(4) or N(3) from a propylimidazolato group (Pd-O 2.038(6), Pd-N 1.983(7), Pd-C 2.009(7), Pd-S 2.200(2) Å, N-Pd-O 80.5(3)°).

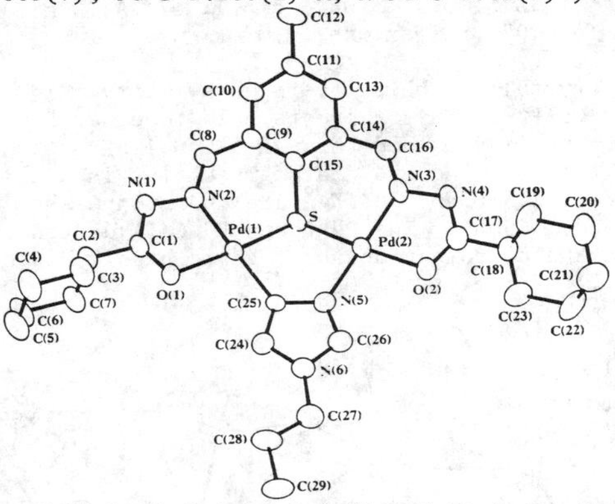

Fig. 1. The structure of $C_{29}H_{38}N_6O_2Pd_2S,0.75(C_2H_6O)$ (note that N(3) and C(4) atoms are labelled N(5) and C(25) respectively).

TETRAKIS(μ_3-SULPHIDO)(DICHLOROHEXAPYRIDINETETRACOPPER)TUNGSTEN
$C_{30}H_{30}Cl_2Cu_4N_6S_4W$ $WCu_4S_4Cl_2(py)_6$

ZHOU GONGDU, GUO YAN, ZHANG SHIWEI and TANG YOUQI, 1984. J. Struct. Chem., 3, 41-44.

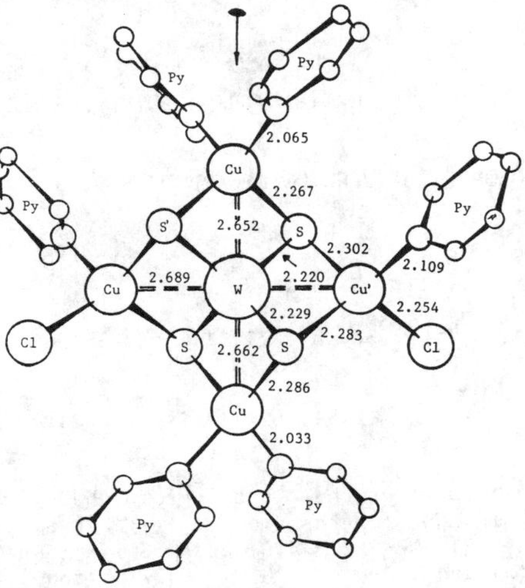

Fig. 1. A view of $WCu_4S_4Cl_2(py)_6$.

Monoclinic, C2/c, a = 14.883, b = 12.580, c = 20.318 Å, β = 100.50°, D_m = 1.98, Z = 4. Mo radiation, R = 0.069 for three-dimensional data.

The molecule has twofold crystallographic symmetry (Fig. 1). The central W atom links with four Cu atoms in square and with four S atoms in a tetrahedral arrangement. Main bond lengths are in the figure.

BIS(μ_2-σ^1,η^2-CARBON DISULPHIDE)BIS((DI-t-BUTYLPHENYLPHOSPHINE)PALLADIUM(0))
$C_{30}H_{46}P_2Pd_2S_4$ $Pd_2(CS_2)_2(Pt\text{-}Bu_2Ph)_2$

D.H. FARRAR, R.R. GUKATHASAN and K. WON, 1984. J. Organometal. Chem., 275, 263-271.

Monoclinic, $P2_1/c$, a = 10.373, b = 13.684, c = 12.173 Å, β = 94.04°, Z = 2. Mo radiation, R = 0.032 for 2071 reflexions.

The complex has crystallographic inversion symmetry (Fig. 1) with two centrosymmetrically-related $Pd(Pt\text{-}Bu_2Ph)$ fragments held together by two doubly bridging CS_2 moieties. Each CS_2 group is π-bonded through a C=S linkage to one metal atom and σ-bonded to the other through the exo-S atom resulting in a six-membered ring. The two Pd-S bond distances, 2.316 and 2.313 Å are indistinguishable as are the two C-S bond lengths, 1.643 and 1.650 Å. The Pd-P and Pd-C(1) distances are respectively 2.408 and 1.980 Å and the coordination about Pd is highly-distorted square planar.

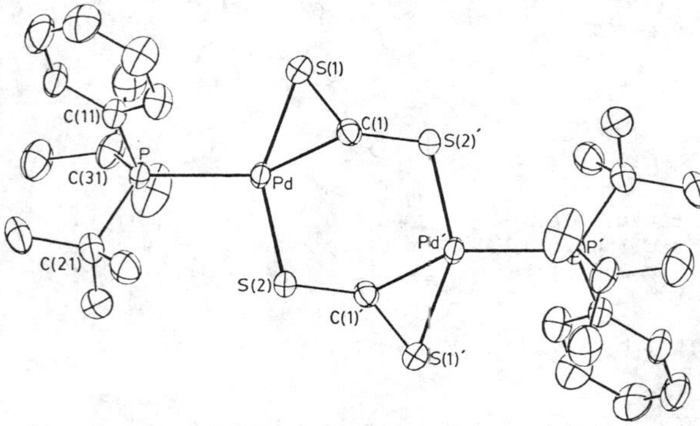

Fig. 1. Molecular structure of $C_{30}H_{46}P_2Pd_2S_4$.

(μ_2-t-BUTYLTHIO)DICARBONYL(μ_2-CHLORO)BIS(TRI-t-BUTYLPHOSPHINE)DIRHODIUM
$C_{30}H_{63}ClO_2P_2Rh_2S$ (I)

(μ_2-CHLORO)(μ_2-(p-CHLOROPHENYL)SULPHIDO)DICARBONYLBIS(TRI-t-BUTYLPHOSPHINE)DIRHODIUM
$C_{32}H_{58}Cl_2O_2P_2Rh_2Si$ (II)

H. SCHUMANN, G. CIELUSEK, S. JURGIS, E. HAHN, J. PICKARDT, J. BLUM, Y. SASSON and A. ZORAN, 1984. Chem. Ber., 117, 2825-2838.

I. Orthorhombic, Pbca, a = 27.356, b = 17.711, c = 15.347 Å, Z = 8. Mo radiation, R = 0.068 for 1212 reflexions.

II. Orthorhombic, $P2_12_12_1$, a = 12.51, b = 14.54, c = 21.70 Å, Z = 4. Mo radiation, R = 0.077 for 2538 reflexions.

The X-ray analyses show that the t-butyl groups have a cis-configuration in both complexes (Fig. 1). Main bond lengths are for (I): Rh-S 2.351(8) and 2.375(8), Rh-Cl 2.404(8) and 2.419(8), Rh-P 2.374(9) and 2.391(8), Rh-Cl 1.70(4) and 1.77(4) Å; for (II): Rh-S 2.368(5) and 2.390(6), Rh-Cl 2.427(5) and 2.462(6), Rh-P 2.372(6) and 2.376(6), Rh-C 1.79(3) and 1.81(3) Å.

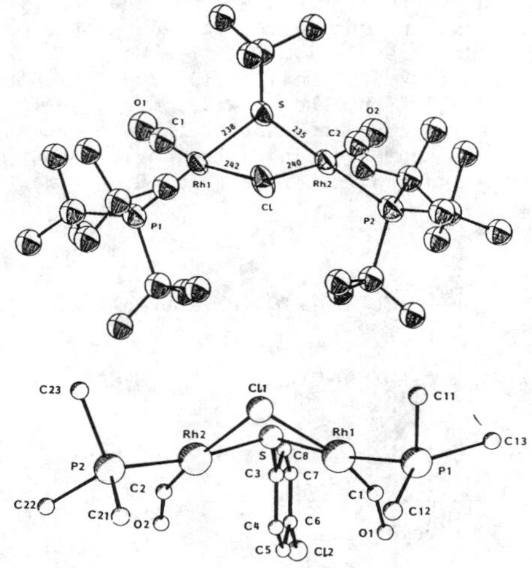

Fig. 1. Views of the dirhodium complexes (I) (top) and (II) (bottom).

(μ_2-p-PHENYLDIIMIDO)BIS(BIS(O,O'-DIETHYLDITHIOPHOSPHATO-S,S')(O,O'-DIETHYLDITHIOPHOSPHATO-S)MOLYBDENUM

$C_{30}H_{64}Mo_2N_2O_{12}P_6S_{12}$ $[Mo(S_2P(OC_2H_5)_2)_3]_2(\mu\text{-p-}NC_6H_4N)$

K.L. WALL, K. FOLTING, J.C. HUFFMAN and R.A.D. WENTWORTH, 1984. Inorg. Chim. Acta. 86, L25-L27.

Triclinic, P$\bar{1}$, a = 15.279, b = 17.019, c = 11.640 Å, α = 90.17, β = 95.51, γ = 74.65°, Z = 2. Mo radiation, R = 0.063 for 5772 reflexions (-165°C).

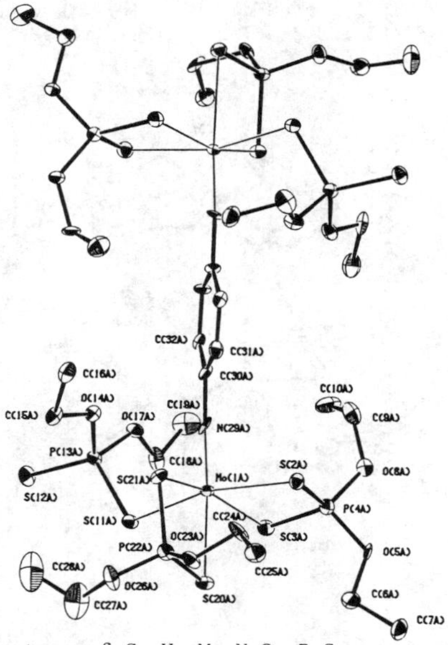

Fig. 1. Molecular structure of $C_{30}H_{64}Mo_2N_2O_{12}P_6S_{12}$.

The metal atoms have distorted octahedral coordination symmetry, being coordinated to S atoms from two bidentate, and one monodentate dithiophosphate ligands, and to the N atom of the p-phenyldiimido ligand (Fig. 1). The Mo-S distances are: Mo(1)-S(2) 2.544, 2.533; Mo(1)-S(3) 2.486, 2.486; Mo(1)-S(11) 2.460, 2.466, Mo(1)-S(20) 2.630, 2.662; Mo(1)-S(21) 2.471, 2.483 Å and the Mo-N(29) distance is 1.729, 1.723 Å for each half of the molecule related to one another through the approximate centre of symmetry in the centre of the aromatic ring. The angles about the metal atoms range from 77.5 to 94.8° at the A end and from 77.8 to 95.5° at the B end of the molecule.

TRIETHYLAMMONIUM HEXAKIS(μ_2-t-BUTYLTHIOLATO)PENTACUPRATE(I)
$C_{30}H_{70}Cu_5NS_6$ (I)

TETRAETHYLAMMONIUM HEXAKIS(μ_2-t-BUTYLTHIOLATO)PENTACUPRATE(I)
$C_{32}H_{74}Cu_5NS_6$ (II)

G.A. BOWMAKER, G.R. CLARK, J.K. SEADON and I.G. DANCE, 1984. Polyhedron, 3, 535-544.

I. Monoclinic, $P2_1/c$, a = 10.519, b = 21.457, c = 20.065 Å, β = 95.11°, Z = 4. Cu radiation, R = 0.072 for 3093 reflexions.

II. Monoclinic, C2/c, a = 45.500, b = 11.805, c = 20.168 Å, β = 117.81°, Z = 8. Cu radiation, R = 0.078 for 2953 reflexions.

Both compounds contain the molecular cage $[Cu_5(\mu\text{-}SBu^t)_6]^-$, (Fig. 1), in which two copper atoms are three-coordinate (Cu(trig)), three copper atoms are two-coordinate (Cu(dig),and all thiolate ligands are doubly-bridging. The polyhedral stereochemistry of the core is trigonal bipyramido-Cu_5-trigonal antiprismo-S_6. The complete $[Cu_5(\mu\text{-}SBu^t)_6]^-$ cage in the Et_4N^+ compound closely approaches D_3 symmetry, but in the Et_3NH^+ compound one SBu^t ligand is inverted at the sulphur bridge, causing angular distortions in the cage. Two structural features, the antiprismatic twist of the S_6 polyhedron and the bending of Cu(dig) towards the cage centroid (S-Cu(dig)-S 171(1)°), provide evidence for weak Cu...Cu attractive interactions within the cage. Infrared data are discussed. Cu-S distances are in the range 2.152-2.178(4) Å for I and 2.13-2.20 for II. For I, the Cu(dig)-S and Cu(trig)-S distances are in the range 2.152-2.178(4) and 2.251-2.371(4) Å respectively. The corresponding ranges for II are 2.13-2.20 and 2.27-2.30 Å. In I the Cu(dig)...Cu(dig) and Cu(trig)-Cu(trig) distances are in the range 3.202-3.554(3) and 2.667-2.874(3) Å respectively. The corresponding data for II are 3.17-3.27 and 2.69-2.75 Å.

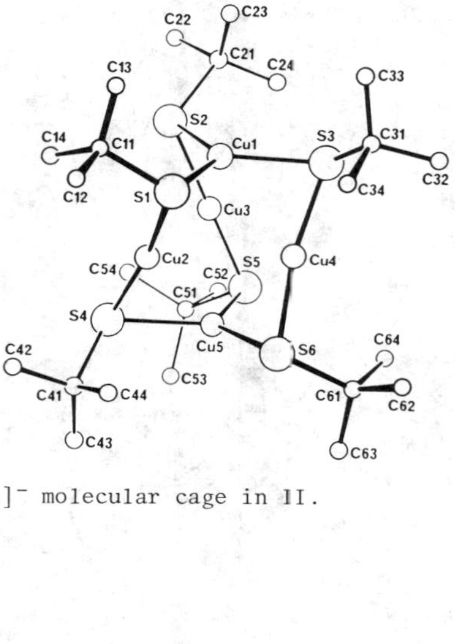

Fig. 1. The $[Cu_5(SBu)_6]^-$ molecular cage in II.

catena-BIS(μ_2-TETRAHYDROTHIOPHENESILVER)BIS(BIS(PERFLUOROPHENYL)GOLD)
$(C_{32}H_{16}Ag_2Au_2F_{20}S_2)_n$ (I)

catena-BIS(μ_2-η^2-BENZENESILVER)BIS(BIS(PERFLUOROPHENYL)GOLD)
$(C_{36}H_{12}Ag_2Au_2F_{20})_n$ (II)

R. USON, A. LAGUNA, M. LAGUNA, B.R. MANZANO, P.G. JONES and G.M. SHELDRICK, 1984.
J. Chem. Soc. Dalton, 285-292.

I. Orthorhombic, Pccn, a = 11.185, b = 22.475, c = 14.802 Å, Z = 4. Mo radiation, R = 0.041 for 2005 reflexions.

II. Monoclinic, C2/c, a = 24.231, b = 7.570, c = 22.613 Å, β = 117.49°, Z = 4. Mo radiation, R = 0.035 for 3008 reflexions.

The structure of I (Fig. 1) comprises $(AuAg)_2$ rings (Au-Ag 2.726, 2.718 Å) linked by short Au...Au contacts (2.889 Å) to form infinite metal atom chains. The Ag atoms are also coordinated to the S atoms of the tetrahydrothiophene moieties (Ag-S 2.454(4) Å). II contains the same type of metal atom chain (Au...Au 3.013; Ag-Ag 2.702, 2.792 Å) with the benzene ring coordinated by one edge to silver. In both molecules the Au atoms lie on crystallographic twofold axes.

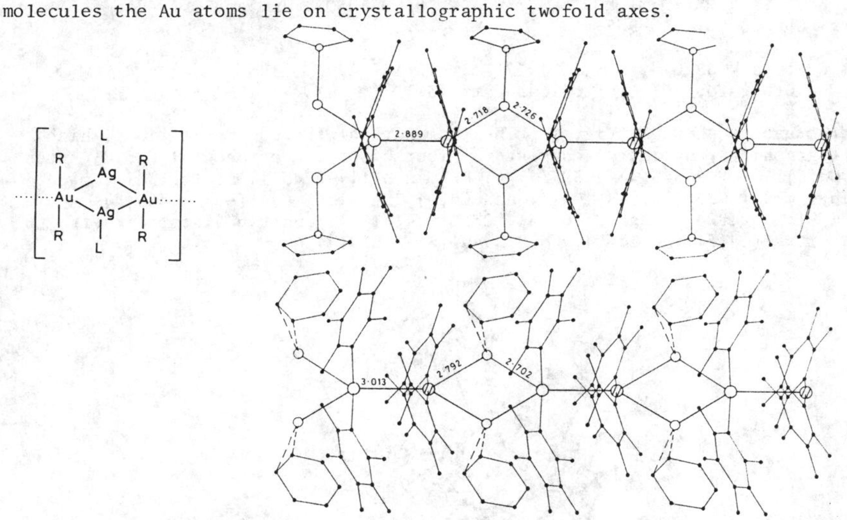

Fig. 1. The polymeric structures: $(C_{32}H_{16}Ag_2Au_2F_{20}S_2)_n$ (top) and $(C_{36}H_{12}Ag_2Au_2F_{20})_n$ (bottom. The molecular skeleton is inset.

cis-BIS(2,2'-BIPYRIDINE)BIS(BENZENETHIOLATO)NICKEL(I) DIDEUTERIOHYDRATE
$C_{32}H_{26}N_4NiS_2,D_2O$ $[Ni(C_{10}H_8N_2)_2(C_6H_5S)_2],D_2O$

K. OSAKADA, T. YAMAMOTO, A. YAMAMOTO, A. TAKENAKA and Y. SASADA, 1984. Acta Cryst., C40, 85-87.

Monoclinic, $P2_1/n$, a = 15.999, b = 12.421, c = 15.343 Å, β = 106.61°, D_m = 1.361, Z = 4. Cu radiation, R = 0.077 for 3973 reflexions.

The nickel coordination (Fig. 1) is distorted octahedral, with Ni-S 2.444 and 2.445(2), Ni-N 2.086(4) to 2.104(4) Å. The complex has approximate twofold symmetry. The D_2O molecule is hydrogen bonded to S2 with S...O 3.220(8) Å.

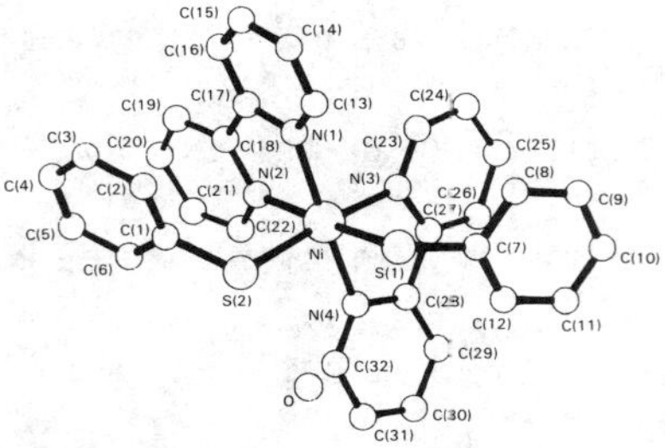

Fig. 1. A view of the $C_{32}H_{26}N_4NiS_2,D_2O$ complex.

HEXADECAKIS(2-ETHANOLTHIOLATO)DECACADMIUM(II) TETRAKIS(PERCHLORATE) OCTAHYDRATE

$C_{32}H_{80}Cd_{10}Cl_4O_{32}S_{16},8H_2O$ $[Cd_{10}(SCH_2CH_2OH)_{16}](ClO_4)_4,8H_2O$

S. LACELLE, W.C. STEVENS, D.M. KURTZ, Jr., J.W. RICHARDSON, Jr. and R.A. JACOBSON, 1984. Inorg. Chem., 23, 930-935.

Monoclinic, C2/c, a = 32.074, b = 13.1417, c = 25.162 Å, β = 126.12°, Z = 4. Mo radiation, R = 0.072 for 4911 reflexions (at 143K).

The structure is shown in Fig. 1. The framework of the complex cation (which has crystallographic C_2 symmetry) consists of four distorted octahedral CdS_3O_3 sites (Cd(1) and Cd(3), Cd-S 2.549(4)-2.595(4) Å, Cd-O 2.37(1)-2.49(1) Å, four distorted trigonal bipyramidal CdS_4O sites (Cd(4) and Cd(6), O-axial, Cd-S(eq) 2.485(4)-2.582(4), Cd-S(ax) 2.824(4) and 2.767(4), Cd-O 2.41(1) Å),and two distorted tetrahedral CdS_4 sites (Cd(2) and Cd(5), Cd-S 2.491(5)-2.550(4) Å).

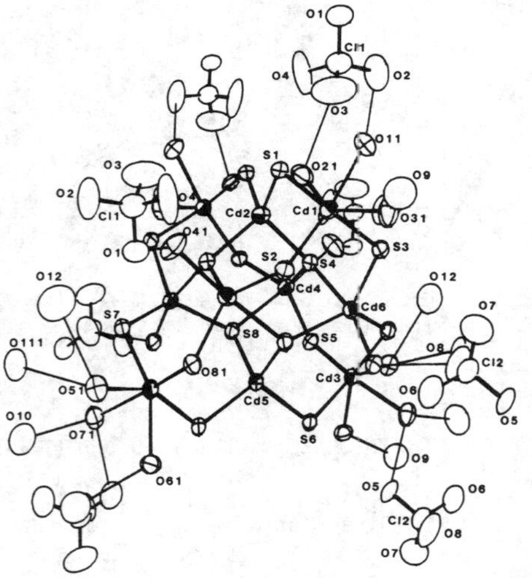

Fig. 1. The structure of $[Cd_{10}(SCH_2CH_2OH)_{16}](ClO_4)_4,8H_2O$.

BIS(N-METHYLPHENAZINIUM) BIS(DICYANOETHENEDITHIOLATO)CUPRATE(II)

$C_{34}H_{22}CuN_8S_4$ $(C_{13}H_{11}N_2)_2[(C_4N_2S_2)_2Cu]$

P. KUPPUSAMY, B.L. RAMAKRISHNA and P.T. MANOHARAN, 1984. Inorg. Chem., 23, 3886-3892.

Monoclinic, $P2_1/n$, a = 11.417, b = 8.126, c = 17.674 Å, β = 92.11°, Z = 2. Mo radiation, R = 0.048 for 1944 reflexions.

The compound forms a kind of mixed stack with a donor-acceptor sequence DAD-DAD along the a axis (Fig. 1). The centrosymmetric anions have square-planar coordination about Cu with Cu-S 2.282(2) and 2.269(2).

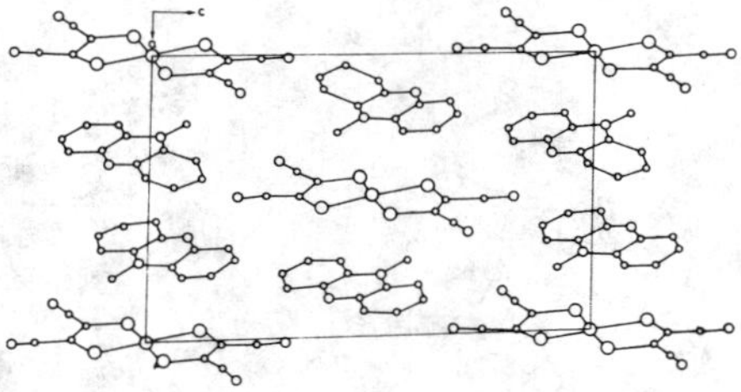

Fig. 1. Packing arrangement of $(C_{13}H_{11}N_2)_2[(C_4N_2S_2)_2Cu]$.

BIS(TETRAETHYLAMMONIUM) TRIS(PHENYLTHIOLATO)COPPER(I)
$C_{34}H_{55}CuN_2S_3$ $[N(C_2H_5)_4]_2[Cu(SC_6H_5)_3]$

C.D. GARNER, J.R. NICHOLSON and W. CLEGG, 1984. Inorg. Chem., 23, 2148-2150.

Monoclinic, $P2_1/c$, a = 9.116, b = 13.107, c = 30.677 Å, β = 95.39°, Z = 4. Mo radiation, R = 0.074 for 4075 reflexions.

The anion (Fig. 1) has Cu(I) coordinated by a trigonal planar array of three S atoms. Cu-S 2.239(2)-2.258(2), mean 2.250(9), mean S-C 1.758(6) Å, S-Cu-S 118.2(1)-121.3(1)°. The Cu atom is displaced 0.05 Å from the S_3 plane.

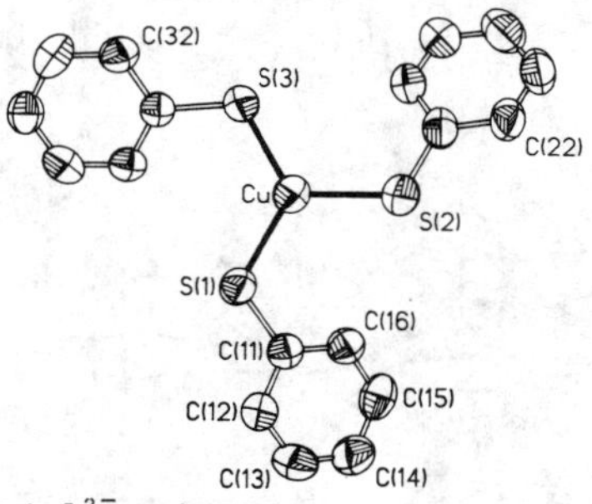

Fig. 1. The $[Cu(SC_6H_5)_3]^{2-}$ anion.

DIMETHYLSULPHOXIDEDIOXO[2,6-BIS(2',2'-DIPHENYL-2'-ETHANOLATO)PYRIDINE-N,O,O]MOLYBDENUM
$C_{35}H_{33}MoNO_5S$ (I)

DIOXO[2,6-BIS(2',2'-DIPHENYLETHYL-2'-THIOLATO)PYRIDINE-N,S,S']MOLYBDENUM
$C_{33}H_{27}MoNO_2S_2$ (II)

J.M. BERG and R.H. HOLM, 1984. J. Am. Chem. Soc., 106, 3035-3036.

I. Monoclinic, $P2_1/c$, a = 8.738, b = 17.799, c = 19.927 Å, β = 93.77°, Z = 4. Mo radiation, R = 0.037 for 2742 reflexions.

II. Orthorhombic, $P2_12_12_1$, a = 8.910, b = 16.350, c = 19.193 Å, Z = 4. Mo radiation, R = 0.036 for 1877 reflexions.

Molecule I (Fig. 1) has distorted octahedral stereochemistry with cis oxo and trans alkoxide ligands (Mo-O1 1.702(4), Mo-O2 1.708(4), Mo-O3 1.899(3), Mo-O4 1.904(3), Mo-O5 2.382(3), Mo-N1 2.417(4) Å). The Mo atom in II is five-coordinate and the MoO_2NS_2 unit has distorted trigonal bipyramidal geometry with trans S atoms and a pseudo twofold axis coincident with the Mo-N bond (mean Mo-O 1.694(6), mean Mo-S 2.416(2), Mo-N 2.244(7) Å).

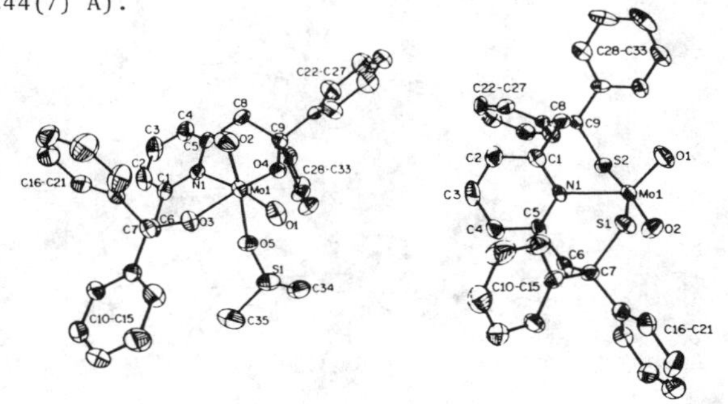

Fig. 1. Views of $C_{35}H_{33}MoNO_5S$ and $C_{33}H_{27}MoNO_2S_2$.

trans-BIS[o-(DIPHENYLPHOSPHINO)THIOANISOLE-P,S] PALLADIUM(II)
$C_{36}H_{28}P_2PdS_2$ $Pd(o\text{-}Ph_2PC_6H_4S)_2$

A. BENEFIEL, D.M. ROUNDHILL, W.C. FULTZ and A.L. RHEINGOLD, 1984. Inorg. Chem., 23, 3316-3324.

Monoclinic, $P2_1/c$, a = 9.735, b = 12.865, c = 12.732 Å, β = 110.41°, Z = 2. Mo radiation, R = 0.038 for 1330 reflexions.

The centrosymmetric molecule (Fig. 1) has trans square planar coordination geometry with Pd-S 2.308(2) and Pd-P 2.291(1) Å.

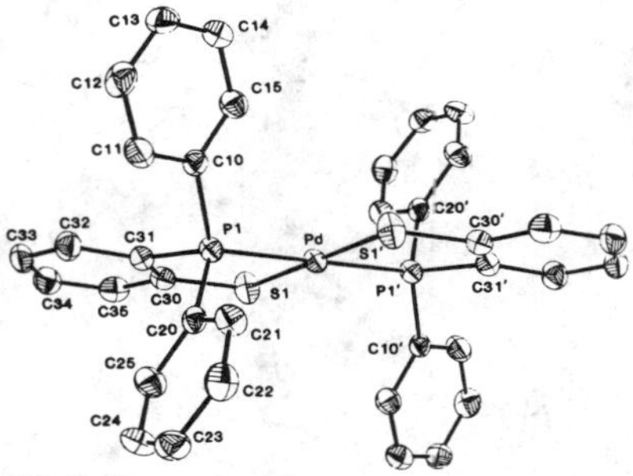

Fig. 1. The $Pd(o\text{-}Ph_2PC_6H_4S)_2$ molecule.

BIS(N-(DIETHYLAMINOTHIOCARBONYL)-N'-PHENYLBENZAMIDINATO)COPPER(II)
$C_{36}H_{40}CuN_6S_2$

R. RICHTER, L. BEYER, V.G. ANDRIANOV and Yu.T. STRUCHKOV, 1984. Z. Anorg. Chem., 513, 123-132.

Orthorhombic, Pbca, a = 27.792, b = 19.197, c = 12.325 Å, D_m = 1.34, Z = 8. Mo radiation, R = 0.06 for 1589 reflexions (at 153 K).

The Cu atom has distorted tetrahedral coordination (Fig. 1) with Cu-S 2.223(4) and 2.235(4), Cu-N 1.96(1) and 1.97(1) Å, S-Cu-S 90.7(2), N-Cu-N 98.7(5)°.

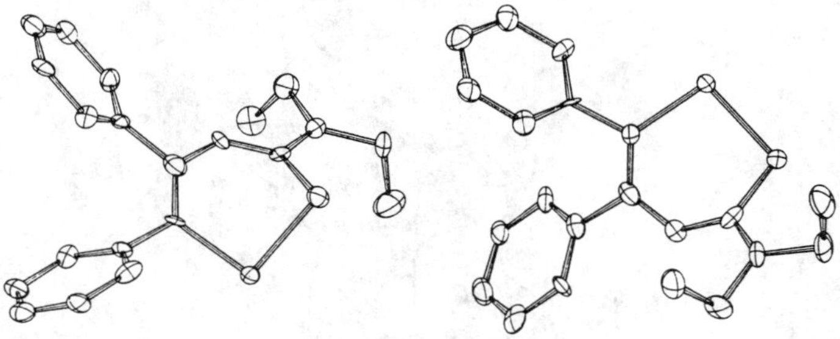

Fig. 1. Views of the two halves of $C_{36}H_{40}CuN_6S_2$.

(4,7,13,16-TETRAPHENYL-1,10-DITHIA-4,7,13,16-TETRAPHOSPHACYCLO-OCTADIENE-S^1,S^{10},P^4,-P^7,P^{13},P^{16})NICKEL DIBROMIDE DIHYDRATE
$C_{36}H_{44}Br_2NiP_4S_2$,2(H_2O)

M. CIAMPOLINI, N. NARDI, P. DAPPORTO, P. INNOCENTI and F. ZANOBINI, 1984. J. Chem. Soc. Dalton, 575-579.

Triclinic, $P\bar{1}$, a = 9.645, b = 10.159, c = 10.869 Å, α = 68.07, β = 78.09, γ = 84.69°, Z = 1. Mo radiation, R = 0.059 for 1395 reflexions.

The structure comprises complex cations, Br^- anions and lattice water molecules. The Ni atom lies on a centre of symmetry and is surrounded by six donor atoms from the ligand (Fig. 1). The four P atoms of the macrocycle lie in a plane with the two independent Ni-P distances normal (2.225(2), 2.197(2) Å). The two S atoms lie approximately in the apical positions of an elongated octahedron, the $P(CH_2)_2S(CH_2)_2P$ moieties adopting a facial arrangement. The Ni-S distances are 2.942(2) Å.

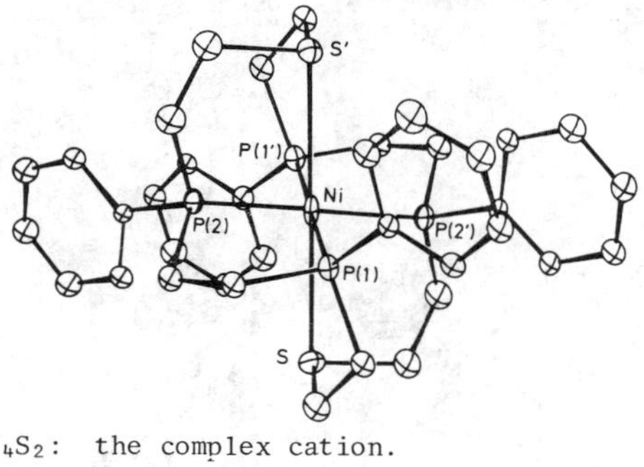

Fig. 1. $C_{36}H_{44}Br_2NiP_4S_2$: the complex cation.

[(4RS,7RS,13RS,16RS)TETRAPHENYL-1,10-DITHIA-4,7,13,16-TETRAPHOSPHACYCLO-OCTADECANE-P^4,P^7,P^{13},P^{16},S^1)NICKEL(II) DIBROMIDE DIHYDRATE
$C_{36}H_{44}Br_2NiP_4S_2$,5.5(H_2O)

M. CIAMPOLINI, N. NARDI, P. DAPPORTO and F. ZANOBINI, 1984. J. Chem. Soc. Dalton, 995-998.

Monoclinic, $P2_1/n$, a = 17.375, b = 15.047, c = 16.122 Å, β = 90.33°, Z = 4. Mo radiation, R = 0.066 for 1881 reflexions.

The Ni atom is five-coordinated, being linked to four P atoms (2.216(6)-2.250(6)) and an S atom (Ni-S 2.350(5) Å) of the macrocyclic ligand. The overall geometry can be described as distorted trigonal-bipyramidal.

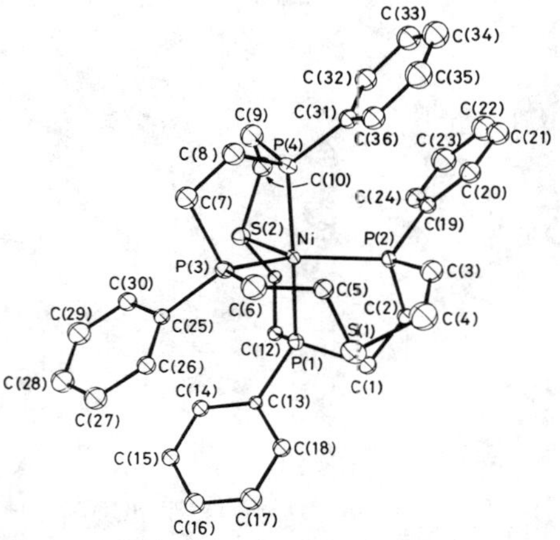

Fig. 1. $C_{36}H_{44}Br_2NiP_4S_2$,5.5(H_2O): the complex cation.

POTASSIUM 2,2,2-CRYPT TRIS(DITHIOBENZOATO)NICKEL(II)
$C_{39}H_{51}KN_2NiO_6$

J.P. FACKLER, Jr., R. DEL NIERA P., C. CAMPANA and B. TRZCINSKA-BANCROFT, 1984. J. Am. Chem. Soc., 106, 7883-7886.

Monoclinic, $P2_1/n$, a = 15.319, b = 18.277, c = 16.450 Å, β = 106.00°, D_m = 1.45, Z = 4. Mo radiation, R = 0.0962 for 3752 reflexions.

The anion is shown to have a distorted NiS_6 octahedron with mean Ni-S 2.42(2) Å. The 2,2,2-crypt portion of the structure is disordered.

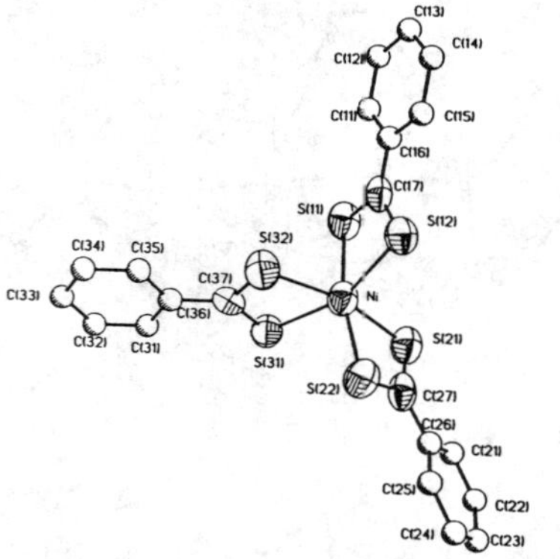

Fig. 1. A view of the tris(dithiobenzoato)nickel(II) anion.

TRIS(BENZAMIDE)BIS(O,O'-DIISOPROPYL DITHIOPHOSPHATO)TERBIUM(III) O,O'-DIISOPROPYL DITHIOPHOSPHATE
$C_{39}H_{63}N_3O_9P_3S_6Tb$ (I)

TRIS(BENZAMIDE)BIS(O,O'-DIISOPROPYL DITHIOPHOSPHATO)ERBIUM(III) O,O'-DIISOPROPYL DITHIOPHOSPHATE
$C_{39}H_{63}ErN_3O_9P_3S_6$ (II)

K. NAGAI, R. NAKAMURA, M. SHIMOI and A. OUCHI, 1984. Bull. Chem. Soc. Japan, 57, 2725-2729.

Monoclinic, $P2_1/n$, a = 25.076 (25.036), b = 15.446 (15.228), c = 14.579 (14.805) Å, β = 100.77 (99.50)°, D_m = 1.41 (1.41), Z = 4 (4). Mo radiation, R = 0.061 (0.055) for 6538 (5220) reflexions. [Er parameters in parentheses.]

Compounds I and II are isomorphous and their crystal structures consist of discrete complex cations and dithiophosphate anions (Fig. 1). The central metal atom of the cation is hepta-coordinate, and it has pentagonal bipyramidal coordination geometry in which four S atoms of two dithiophosphate ions and one O atom of a benzamide ligand assume equatorial positions, and two benzamide O atoms take axial positions. The Metal-S distances are: Tb-S(1) 2.806, Tb-S(2) 2.816, Tb-S(3) 2.845, Tb-S(4) 2.850, Er-S(1) 2.781, Er-S(2) 2.790, Er-S(3) 2.818, Er-S(4) 2.814 Å. Other bonds of interest are: Tb-O(100) 2.236, Tb-O(200) 2.267, Tb-O(300) 2.296, Er-O(100) 2.198, Er-O(200) 2.232, Er-O(300) 2.264 Å. The angles O(100)-M-O(200) are 179.7° in both compounds. The angles S(1)-M-S(2), S(1)-M-O(300), S(3)-M-S(4) are (in I and II respectively) 70.9, 71.6; 73.5, 73.5; and 70.5, 70.9°.

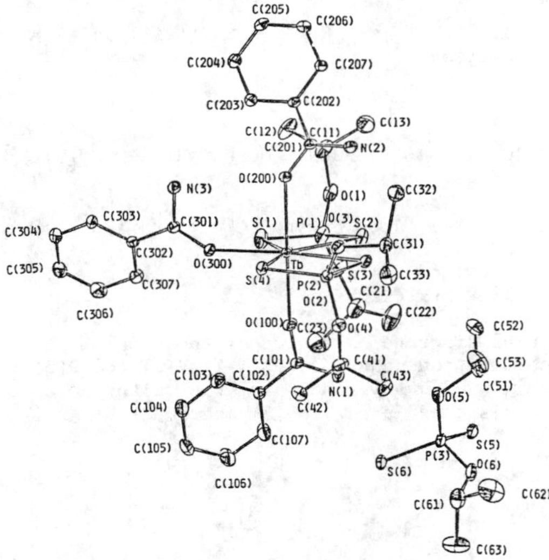

Fig. 1. Perspective view of the Tb(III) complex $C_{39}H_{63}N_3O_9P_3S_6Tb$.

TRIS(TETRAETHYLAMMONIUM) TETRAKIS[(μ_3-SULPHIDO)(t-BUTYLTHIOLATO)IRON] ACETONITRILE SOLVATE

$C_{40}H_{96}Fe_4N_3S_8,C_2H_3N$ (I) $[(C_2H_5)_4N]_3[Fe_4S_4(SC_4H_9)_4].CH_3CN$

TRIS(TETRAMETHYLAMMONIUM) TETRAKIS[(μ_3-SULPHIDO)(ETHYLTHIOLATO)IRON]

$C_{20}H_{56}Fe_4N_3S_8$ (II) $[(CH_3)_4N]_3[Fe_4S_4(SC_2H_5)_4]$

K.S. HAGEN, A.D. WATSON and R.H. HOLM, 1984. Inorg. Chem., 23, 2984-2990.

I. Monoclinic, $P2_1/c$, a = 23.14, b = 11.905, c = 24.81 Å, β = 116.16°, Z = 4. Mo radiation, R = 0.049 for 4854 reflexions.

II. Tetragonal, $P\bar{4}2_1c$, a = 12.858, c = 11.280 Å, Z = 2. Mo radiation, R = 0.026 for 1265 reflexions.

The two complex anions are shown in Fig. 1. Both clusters have similar $[Fe_4S_4]^+$ core geometries. In I the core has idealized D_{2d} symmetry with four short (mean 2.297(6) Å) and eight long (mean 2.330(7) Å) Fe-S distances roughly parallel and perpendicular, respectively, to the pseudo-$\bar{4}$ axis. II has imposed S_4 symmetry with small deviations from D_{2d} symmetry; four short (2.293(1) Å) and eight long (mean 2.324(5) Å) Fe-S distances are approximately parallel and perpendicular to the crystallographic $\bar{4}$ axis. The mean Fe-SR distance in I is 2.300(4) Å and the Fe-SR distance in II is 2.315(2) Å.

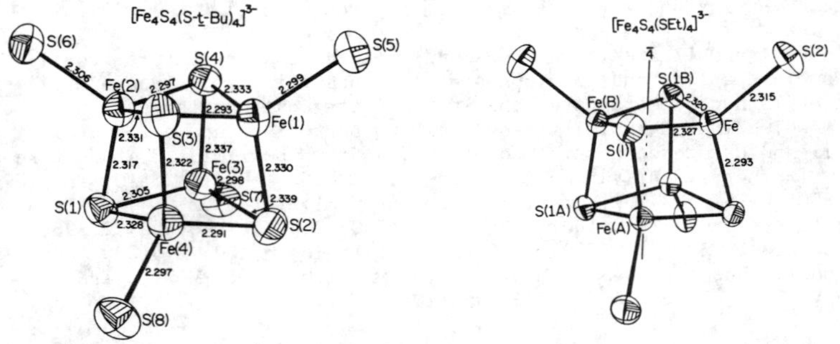

Fig. 1. The $[Fe_4S_4(SR)_4]^{3-}$ anions: R = t-C_4H_9 (left) and R = C_2H_5 (right). R groups are omitted.

cis-DIHYDRO(2-PYRIDINETHIOLATO)BIS(TRIPHENYLPHOSPHINE)IRIDIUM(III)

$C_{41}H_{36}IrNP_2S$ $IrH_2(C_5H_4NS)[P(C_6H_5)_3]_2$

P. MURA and S.D. ROBINSON, 1984. Acta Cryst., C40, 1798-1799.

Orthorhombic, Pbca, a = 9.229, b = 24.239, c = 31.965 Å, D_m = 1.54, Z = 8. Mo radiation, R = 0.058 for 1512 reflexions.

The structure contains discrete molecules of the complex (Fig. 1) with highly distorted octahedral coordination about Ir (P(1)-Ir-P(2) 168.9(3) and S-Ir-N 64.7(8)°). The triphenylphosphine ligands are axial with the two hydrido H atoms in cis positions. The Ir-P distances are 2.288(7) and 2.281(8) Å with Ir-S 2.497(9) and Ir-N 2.15(2) Å.

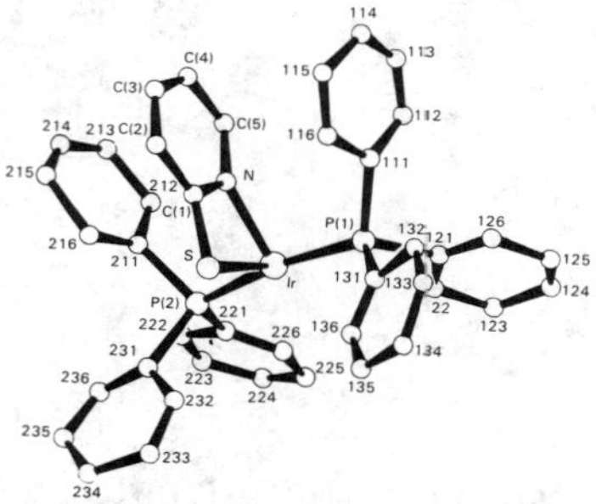

Fig. 1. The $IrH_2(C_5H_4NS)(PPh_3)_2$ molecule.

(η^3-THIADIPHOSPHIRENE)(1,1,1-TRIS(DIPHENYLPHOSPHINOMETHYL)ETHANE)COBALT(I) TETRAFLUOROBORATE BENZENE SOLVATE

$C_{41}H_{39}BCoF_4P_5S,C_6H_6$ (I)

(η^3-SELENADIPHOSPHIRENE)(1,1,1-TRIS(DIPHENYLPHOSPHINOMETHYL)ETHANE)COBALT(I) TETRAFLUOROBORATE BENZENE SOLVATE

$C_{41}H_{39}BCoF_4P_5Se,C_6H_6$ (II)

M. DI VAIRA, M. PERUZZINI and P. STOPPIONI, 1984. J. Chem. Soc. Dalton, 359-364.

I. Monoclinic, $P2_1/n$, a = 17.193, b = 20.164, c = 13.136 Å, β = 103.35°, Z = 4. Mo radiation, R = 0.061 for 3520 reflexions.

II. Monoclinic, $P2_1/n$, a = 17.049, b = 20.119, c = 13.188 Å, β = 102.74°, Z = 4. Mo radiation, R = 0.069 for 1512 reflexions.

Structure of these isomorphous compounds comprise complex cations, BF_4^- anions and interposed benzene solvate molecules. The Co atom is in a six-coordinate environment in both cations (Fig. 1). The P_2S triangular unit in I is disordered about the pseudo threefold axis of the cation, the P(4-6) sites having 0.67 P and 0.33 S occupancies. The mean interatomic distance in this ring is 2.119(10) while the mean Co-P_2S distance is 2.270(9) Å. The other Co-P distances average 2.211(9) Å. The P_2Se group in II is affected by similar disorder although one of the three possible orientations of the group is more populated than the other two (P(4) 0.94 P, 0.06 Se; P(5) 0.83 P, 0.17 Se; Se 0.23 P, 0.77 Se). Distances within the P_2Se group range from 2.144(8) to 2.247(6) and the distances from this group to the metal range from 2.272(6) to 2.403(4) Å. The anions and solvent molecules are not unusual.

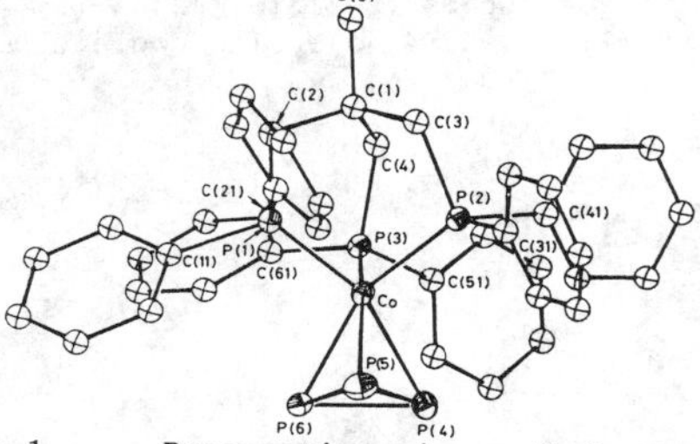

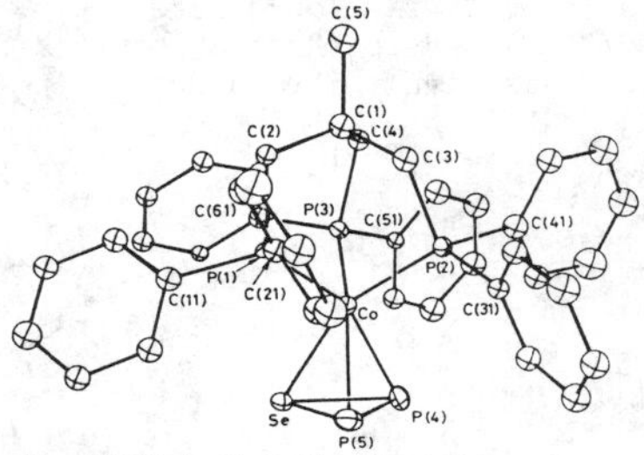

Fig. 1. Perspective views of the cations: $C_{41}H_{39}BCoF_4P_5S,C_6H_6$ (left) and $C_{41}H_{39}BCoF_4P_5Se,C_6H_6$ (right).

BIS(DINITROGEN)BIS(METHYLDIPHENYLPHOSPHINO)(1-DIPHENYLPHOSPHINO-2-METHYLTHIOETHANE-P,S)MOLYBDENUM

$C_{41}H_{43}MoN_4P_3S$

R.H. MORRIS, J.M. RESSNER, J.F. SAWYER and M. SHIRALIAN, 1984. J. Am. Chem. Soc., 106, 3683-3684.

Orthorhombic, Pbca, a = 18.48, b = 20.49, c = 20.67 Å, Z = 8. Mo radiation, R = 0.095 for 1209 reflexions.

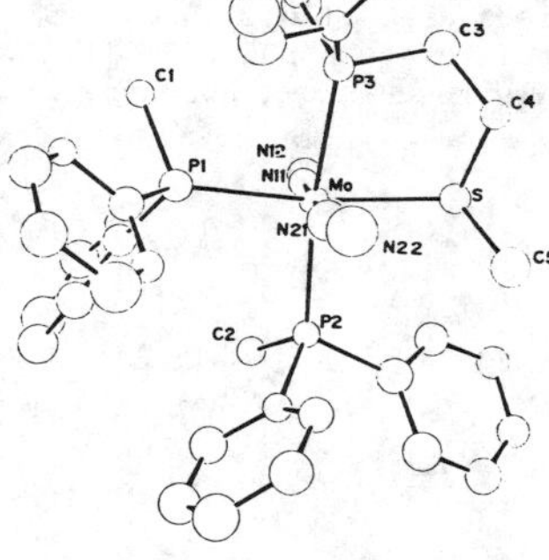

Fig. 1. The molecular structure of $C_{41}H_{43}MoN_4P_3S$.

The Mo atom has octahedral coordination with Mo-S 2.483(8), Mo-P1 2.443(9), Mo-P2 2.500(8), Mo-P3 2.457(9), Mo-N11 1.98(2), Mo-N21 2.00(3), N-N 1.04(3), 1.10(3) Å, Mo-N-N 168(3), 174(3)°.

{2,16-DIMETHYL-3,15,21-TRIAZA-6,9,12-TRITHIABICYCLO[15.3.1]HENICOSA-1(21),2,15,17,19-PENTAENE}SILVER(I) TETRAPHENYLBORATE

$C_{41}H_{45}AgBN_3S_3$ $[Ag(C_{17}H_{25}N_3S_3)][BPh_4]$

M.G.B. DREW, D.A. RICE and S. BIN SILONG, 1984. Acta Cryst., C40, 2014-2016.

Triclinic, P$\bar{1}$, a = 13.863, b = 10.965, c = 14.990 Å, α = 109.0, β = 99.2, γ = 65.0°, D_m = 1.35, Z = 2. Mo radiation, R = 0.063 for 3966 reflexions.

The structure consists of discrete complex cations (Fig. 1) and BPh_4^- anions. The Ag atom is six-coordinate, being bonded to three N atoms (2.613(9), 2.422(9), 2.399(7) Å) and to three S atoms (2.670(3), 2.655(3), 2.949(4) Å). The coordination geometry is highly irregular.

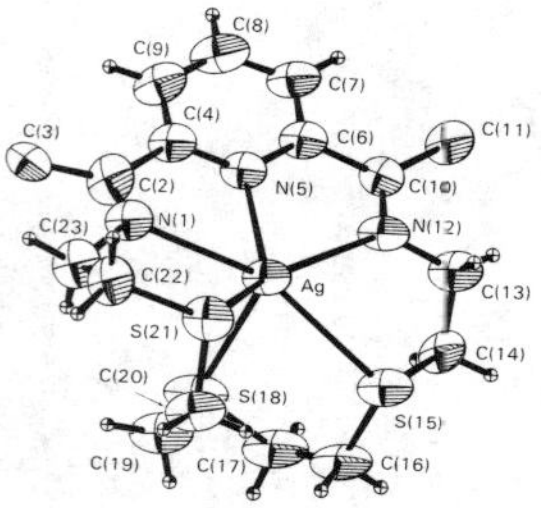

Fig. 1. The $[Ag(C_{17}H_{25}N_3S_3)]^+$ structure.

HEXAKIS(μ_2-PHENYLTHIOLATO)HEXACARBONYLTRIIRON ACETONE SOLVATE

$C_{42}H_{30}Fe_3O_6S_6,2(C_3H_6O)$

U. BERGER and J. STRAHLE, 1984. Z. Anorg. Chem., 516, 19-29.

Triclinic, P$\bar{1}$, a = 11.953, b = 13.88, c = 15.94 Å, α = 89.51, β = 98.36, γ = 99.12°, Z = 2. Mo radiation, R = 0.046 for 4014 reflexions (at -80°C).

The crystal lattice contains two unrelated complexes each lying on independent inversion centres; one complex is shown in Fig. 1. The symmetry requires that in each case the three Fe atoms are linearly arranged and there is approximate D_{3d} symmetry. The Fe...Fe separations are 3.170(1) and 3.195(4) Å with Fe-S in the range 2.473(1)-2.519(1) for the central Fe atoms and 2.335(1)-2.354(1) for the terminal Fe atoms. The mean Fe-C and C-O distances are 1.800 and 1.135 Å respectively.

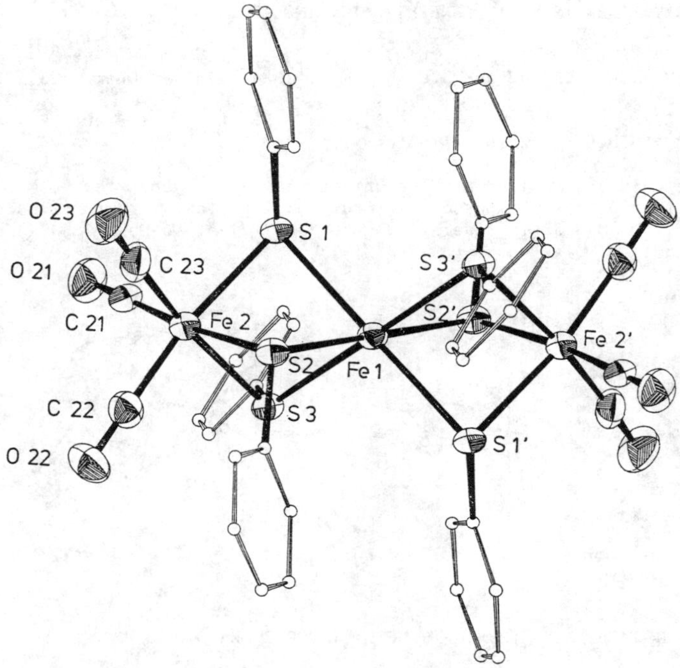

Fig. 1. A view of one of the $C_{42}H_{30}Fe_3O_6S_6$ molecules.

CHLORO-2-THIOPYRROLE-1,2-DICARBOXIMIDEBIS(TRIPHENYLPHOSPHINE)COPPER(I)
$C_{42}H_{34}ClCuN_2OP_2S$

K.K. PANDEY, M. NOLTEMEYER, G.M. SHELDRICK and R. SAHEB, 1984. Z. Naturforsch., B39, 586-589.

Monoclinic, C2/c, a = 38.580, b = 10.138, c = 29.821 Å, β = 139.76°, Z = 8. Mo radiation, R = 0.063 for 2155 reflexions.

The Cu atom has tetrahedral coordination (Fig. 1) with mean Cu-P 2.294(5), Cu-Cl 2.330(3), Cu-S 2.386(5) Å, Cl-Cu-P 112.1(2), P-Cu-P 124.3(1), P-Cu-S 102.6 and 106.2(2), Cl-Su-S 108.6(1)°.

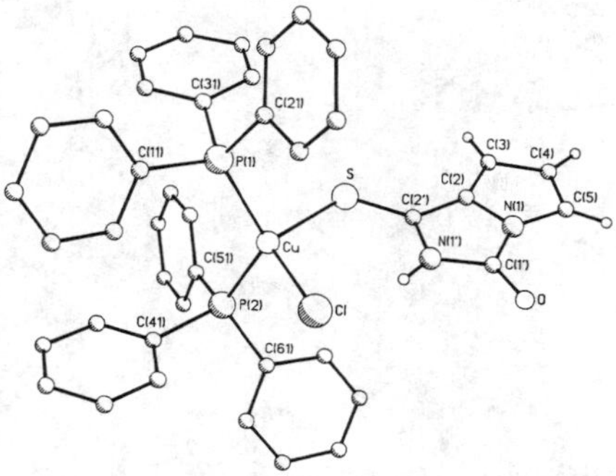

Fig. 1. A view of $C_{42}H_{34}ClCuN_2OP_2S$.

BIS(ISOPROPOXY)TETRAKIS(MESITYLENETHIOLATO)DIMOLYBDENUM

$C_{42}H_{58}Mo_2O_2S_4$ $Mo_2(OC_3H_7)_2(SC_6H_2(CH_3)_3)_4$

M.H. CHISHOLM, J.F. CORNING and J.C. HUFFMAN, 1984. Inorg. Chem., 23, 754-757.

Monoclinic, C2/c, a = 21.778, b = 8.600, c = 24.587 Å, β = 108.90°, Z = 4. Mo radiation, R = 0.026 for 2516 reflexions (at -163°C).

The molecule (Fig. 1) adopts the anti conformation about the Mo≡Mo triple bond with crystallographically imposed C_i symmetry. Important structural parameters: Mo-M 2.230(1), Mo-S 2.32(1) (average), Mo-O 1.878(2) Å, Mo-Mo-S 95.4(1) and 99.8(1), Mo-Mo-O 106.8(1)°.

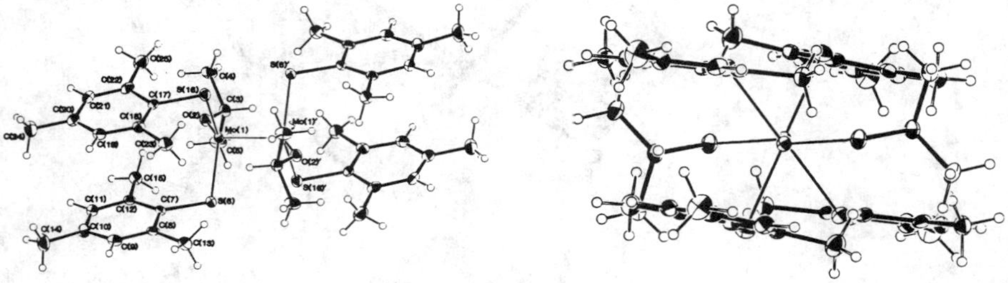

Fig. 1. Two views of $Mo_2(OC_3H_7)_2(SC_6H_2(CH_3)_3)_4$.

TRIMETHYLAMMONIUM TETRAKIS(2,4,6-TRIMETHYLTHIOPHENOLATO)OXORHENATE

$C_{42}H_{60}NOReS_4$ (I) $[HN(C_2H_5)_3]^+[ReO(SC_6H_2(CH_3)_3)_4]^-$

(2,6-DIISOPROPYLTHIOPHENOL)TRIPHENYLPHOSPHONIUM TETRAKIS(2,4,6-TRIMETHYLTHIOPHENOLATO)-OXORHENATE ETHANOL SOLVATE

$C_{78}H_{100}OPReS_5,1.5C_2H_6O$ (II) $[(C_6H_5)_3PSC_6H_3(C_3H_7)_2]^+[ReO(SC_6H_3(C_3H_7)_2)_4]^-,1.5C_2H_5OH$

P.J. BLOWER, J.R. DILWORTH, J. HUTCHINSON, T. NICHOLSON and J. ZUBIETA, 1984. Inorg. Chim. Acta, 90, L27-L30.

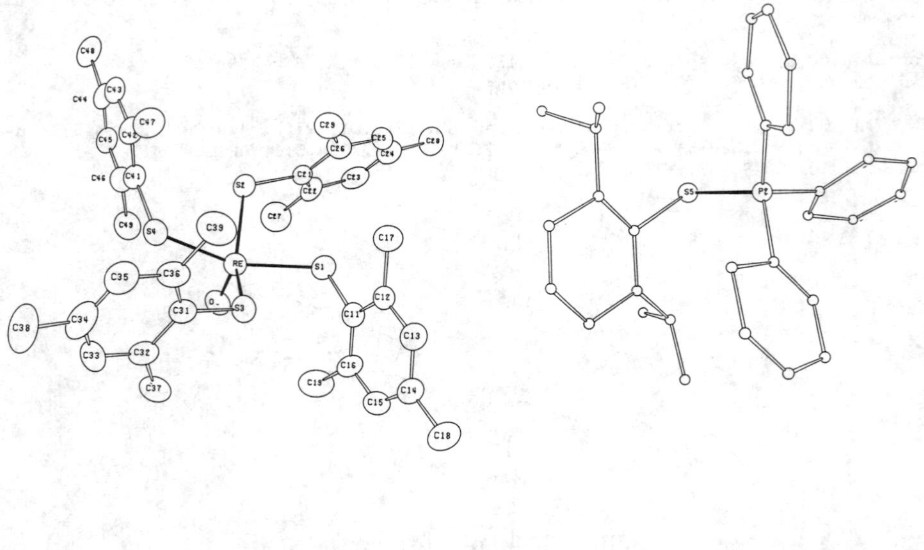

Fig. 1. Structures of the anion in $C_{42}H_{60}NOReS_4$ and the cation in $C_{78}H_{100}OPReS_5$.

I. Monoclinic, $P2_1/n$, a = 12.988, b = 16.387, c = 21.336 Å, β = 102.78°, Z = 4. R = 0.056 for 3407 reflexions.

II. Triclinic, $P\bar{1}$, a = 14.295, b = 14.620, c = 19.689 Å, α = 86.97, β = 77.15, γ = 84.60°, Z = 2. R = 0.074 for 1813 reflexions.

In I the geometry about the Re atom is distorted significantly from square pyramidal coordination with the metal atom displaced 0.71 Å from the best plane through the four S atoms, towards the apical O atom (Fig. 1). The Re-S distances are: 2.339, 2.319, 2.322, and 2.356 Å with the Re-O bond length 1.65 Å. In II the anion contains a distorted square pyramidal $ReOS_4$ arrangement (Fig. 1) once again. The Re-O group is disordered about the plane generated by the four S atoms, resulting in low bond length accuracy (Re-O 1.70, Re-S 2.33, 2.36, 2.34 and 2.38 Å). The cation is unusual (Fig. 1) in II, and the P-S bond length of 2.11 Å suggests that a structure with the positive charge localized on the P atom ($R_3\overset{+}{P}$-SAr) is the major contributing resonance structure.

TRIS(TETRAETHYLAMMONIUM) TRIS[(μ_2-PHENYLTHIOLATO)BROMOCHLOROIRON]
$C_{42}H_{75}Br_3Cl_3Fe_3N_3S_3$ $(Et_4N)_3[(FeBrClSPh)_3]$

LIU PING, HUANG LIANGREN, WANG LINGLING, KANG BEISHENG and LU JIAXI, 1984. J. Struct. Chem., 3, 135-137.

Monoclinic, C2/c, a = 24.569, b = 13.504, c = 18.348 Å, β = 110.78°, D_m = 1.41, Z = 4. Mo radiation, R = 0.070 for 2315 reflexions.

The anion (Fig. 1) has twofold crystallographic symmetry (the twofold axis passes through Fe(1) and S(2)) and the halogen atoms are mutually disordered. The Fe_3S_3 ring is planar (Fe-S 2.338-2.349(1) Å, Fe-S-Fe 138.6 and 141.4(1)°). The three phenyl rings are coplanar with the Fe_3S_3 ring.

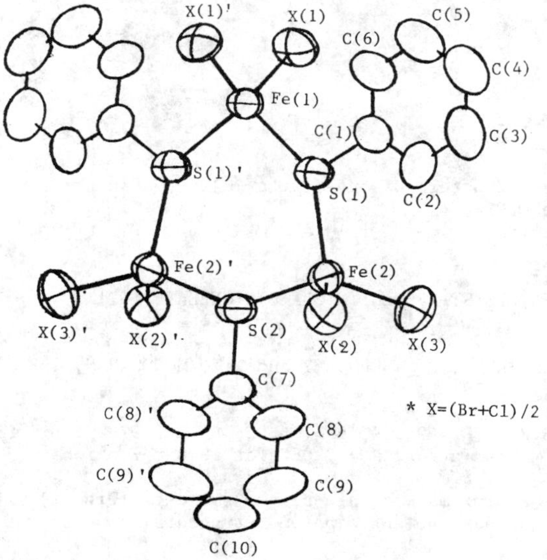

Fig. 1. The $[(FeBrClSPh)_3]^{3-}$ anion (X = (Br+Cl)/2).

TETRACARBONYL(μ_2-(2,2,6,6-TETRAMETHYLCYCLOHEXYLIDENEMETHANEDITHIOLATO-S,S))(2,2,6,6-TETRAMETHYLCYCLOHEXYLIDENEMETHYLENE)-TRIPHENYLPHOSPHINEDIIRON
$C_{44}H_{51}Fe_2O_4PS_2$

H. UMLAND and U. BEHRENS, 1984. J. Organometal. Chem., 273, C39-C42.

Monoclinic, Pn, a = 9.304, b = 16.789, c = 27.329 Å, β = 91.32°, Z = 4. Mo radiation, R = 0.046 for 4107 reflexions.

The two molecules in the asymmetric unit have molecular geometries which are almost identical. The main structural element is the cage-like arrangement of atoms Fe(1), Fe(2), S(1), S(2) and C(01) (Fig. 1). The C(01)-C(02) distance is 1.34 Å, indicating a double bond. The C(12)-C(13) distance is 1.30 Å for the vinylidene C=C double bond and the Fe(2)-C(12) bond length, at 1.79 Å, lies within the range for Fe-C(carbonyl) bond lengths. Other bond distances are: Fe(1)-Fe(2) 2.521, Fe(1)-P 2.239, and S-C(01) 1.82 Å. The angle Fe(2)-C(12)-C(13) is 177°.

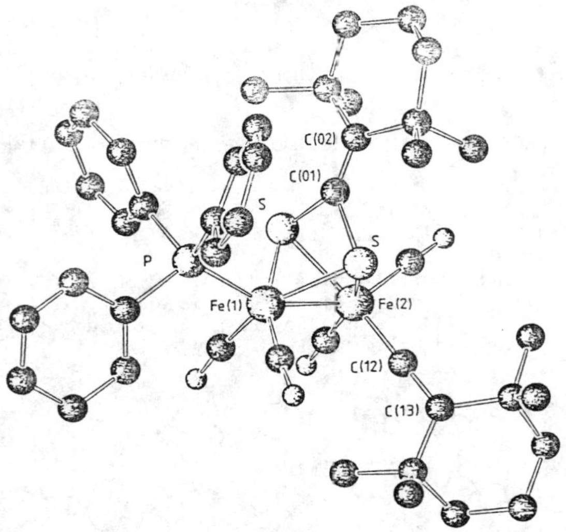

Fig. 1. Molecular structure of one of the molecules in the asymmetric unit of $C_{44}H_{51}Fe_2O_4PS_2$.

TETRA-n-BUTYLAMMONIUM BIS(STILBENEDITHIOLATO)NICKELATE(III)
$C_{44}H_{56}NNiS_4$

C. MAHADEVAN, M. SESHASAYEE, P. KUPPUSAMY and P.T. MONOHARAN, 1984. J. Cryst. Spectrosc. Res., 14, 179-191.

Triclinic, P$\bar{1}$, a = 8.874, b = 9.549, c = 26.025 Å, α = 97.22, β = 95.06, γ = 98.75°, D_m = 1.214, Z = 2. Mo radiation, R = 0.062 for 2910 reflexions.

The Ni atom has approximate square-planar geometry (Fig. 1) with mean Ni-S 2.137(2) Å. The cation has the anticipated tetrahedral geometry at N.

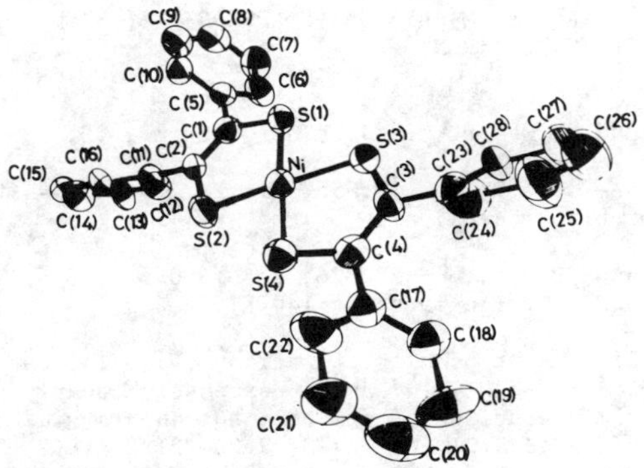

Fig. 1. A view of the anion in $C_{44}H_{56}NNiS_4$.

TETRA(n-PROPYL)AMMONIUM TRIS(2,3,5,6-TETRAMETHYLBENZENETHIOLATO)(ACETONITRILE)-COBALT(II)
$C_{44}H_{70}CoN_2S_3$ (I) $[N(C_3H_7)_4][Co(SC_{10}H_{13})_3(CH_3CN)]$

TETRA(n-PROPYL)AMMONIUM BIS(2,3,5,6-TETRAMETHYLBENZENETHIOLATO)COPPER(I)
$C_{32}H_{54}CuNS_2$ (II) $[N(C_3H_7)_4][Cu(SC_{10}H_{13})_2]$

S.A. KOCH, R. FIKAR, M. MILLAR and T. O'SULLIVAN, 1984. Inorg. Chem., 23, 121-122.

I. Monoclinic, $P2_1/n$, a = 19.714, b = 11.863, c = 19.910 Å, β = 95.59°, Z = 4. Mo radiation, R = 0.054 for 3524 reflexions.

II. Monoclinic, C2/c, a = 10.763, b = 15.035, c = 21.176 Å, β = 98.45°, Z = 4. Mo radiation, R = 0.060 for 1140 reflexions.

The two anions are shown in Fig. 1. The Co(II) ion in I is tetrahedrally coordinated by three thiolato S atoms (Co-S 2.267(1)-2.296(1) Å) and the N atom of an acetonitrile (Co-N 2.036(4) Å). In II the anion has exact C_2 symmetry with a Cu-S distance of 2.137(2) Å and a S-Cu-S angle of 178.6(1)° in this two-coordinate Cu(I) species.

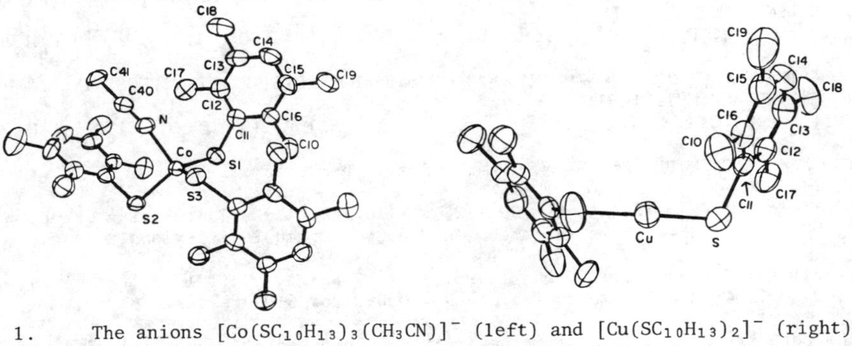

Fig. 1. The anions $[Co(SC_{10}H_{13})_3(CH_3CN)]^-$ (left) and $[Cu(SC_{10}H_{13})_2]^-$ (right).

TETRAKIS(TETRAETHYLAMMONIUM) (μ_4-SULPHIDO)BIS(μ_3-SULPHIDO)HEXAKIS(μ_2-SULPHIDO)BIS-(BENZYLTHIOLATO)HEXAIRON MONOHYDRATE
$C_{46}H_{94}Fe_6N_4S_{11},H_2O$ (I) $[(C_2H_5)_4N]_4[Fe_6S_9(SCH_2C_6H_5)_2],H_2O$

DISODIUM HEXAKIS(TETRAETHYLAMMONIUM) BIS[(μ_4-SULPHIDO)BIS(μ_3-SULPHIDO)HEXAKIS(μ_2-SULPHIDO)BIS(METHYLTHIOLATO)HEXAIRON]

$C_{52}H_{132}Fe_{12}N_6Na_2S_{22}$ (II) $[(C_2H_5)_4N]_6[(Fe_6S_9(SCH_3)_2)_2Na_2]$

H. STRASDEIT, B. KREBS and G. HENKEL, 1984. Inorg. Chem., 23, 1816-1825.

I. Monoclinic, $P2_1/c$, a = 16.339, b = 19.535, c = 21.729 Å, β = 106.56°, Z = 4. Mo radiation, R = 0.060 for 6156 reflexions.

II. Monoclinic, C2/m, a = 12.813, b = 28.626, c = 16.056 Å, β = 129.06°, Z = 2. Mo radiation, R = 0.085 for 2355 reflexions (at 140 K).

Crystals of I and II consist of discrete anions (Fig. 1), cations, and, in I, H_2O molecules. The $[Fe_6S_9(SCH_2C_6H_5)_2]^{4-}$ anion possesses a nearly planar arrangement of Fe atoms resulting from a coupling of six FeS_4 tetrahedra via common edges and corners, Fe-(μ_4-S) 2.330(2)-2.339(2), Fe-(μ_3-S) 2.244(2)-2.318(2), Fe-(μ_2-S) 2.210(3)-2.263(2), Fe-SR 2.282(3) and 2.284(3), Fe-Fe 2.692(2)-2.802(2) Å. The novel $[(Fe_6S_9(SCH_3)_2)_2Na_2]^{6-}$ anion has exact C_{2h} symmetry and may be described in terms of two hexanuclear iron-sulphide-thiolate units connected by two Na^+ cations or, alternatively, in terms of four $[Fe_3NaS_4(SCH_3)_6]^{5-}$ units condensed via adjacent Na and S corners. Bond lengths for II: Fe-Fe 2.707(4)-2.767(3), Fe-(μ_4-S) 2.345(3) and 2.348(4), Fe-(μ_3-S) 2.239(5)-2.301(4), Fe-(μ_2-S) 2.205(4)-2.259(6), Fe-SR 2.273(6), Na-S 2.883(7)-3.106(8) Å.

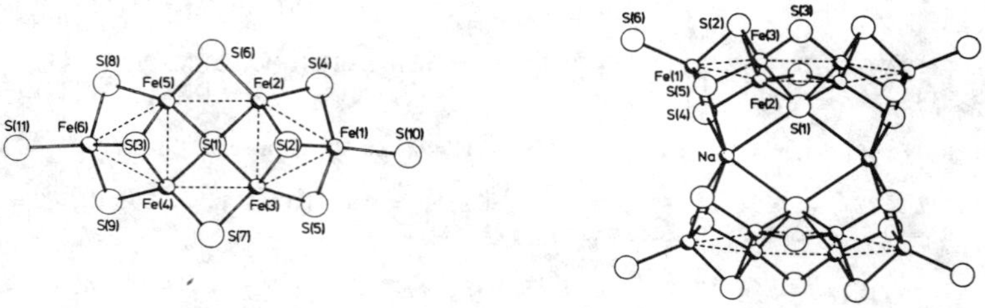

Fig. 1. The $[Fe_6S_9(SCH_2C_6H_5)_2]^{4-}$ anion (left) and the $[(Fe_6S_9(SCH_3)_2)_2Na_2]^{6-}$ anion (right).

CHLORO(TETRACHLOROTETRATHIONAPHTHALENATO-S,S)CARBONYLBIS(TRIPHENYLPHOSPHINE)IRIDIUM-(III) BENZENE ACETONITRILE

$C_{47}H_{30}Cl_5IrOP_2S_4,C_6H_6,C_2H_3N$ (I) $[(C_6H_5)_3P]_2(CO)ClIr(C_{10}Cl_4S_4).C_6H_6.CH_3CN$

HYDRIDO(TETRACHLOROTETRATHIONAPHTHALENATO-S,S)CARBONYLBIS(TRIPHENYLPHOSPHINE)IRIDIUM-(III) BENZENE ACETONITRILE

$C_{47}H_{31}Cl_4IrOP_2S_4,C_6H_6,C_2H_3N$ (II) $[(C_6H_5)_3P]_2(CO)HIr(C_{10}Cl_4S_4).C_6H_6.CH_3CN$

B.K. TEO and P.A. SNYDER-ROBINSON, 1984. Inorg. Chem., 23, 32-39.

I. Triclinic, $P\bar{1}$, a = 12.434, b = 20.312, c = 10.888 Å, α = 104.05, β = 97.98, γ = 75.48°, D_m = 1.67, Z = 2. Mo radiation, R = 0.064 for 4002 reflexions.

II. Triclinic, $P\bar{1}$, a = 12.102, b = 19.911, c = 10.853 Å, α = 101.26, β = 96.79, γ = 80.41°, D_m = 1.64, Z = 2. Mo radiation, R = 0.068 for 5109 reflexions.

For both structures (Fig. 1), the Ir atom is octahedrally coordinated by two PPh_3 ligands (in a trans configuration, Ir-P 2.421(4) and 2.417(4) Å for I, 2.354(4) and 2.336(4) Å for II), two S atoms from the $C_{10}Cl_4S_4$ ligands (Ir-S 2.374(4) and 2.346(4) Å for I, 2.382(4) and 2.426(4) Å for II), one CO ligand (Ir-C 1.860(15) and 1.895(13) Å for I and II), and one Cl^- (Ir-Cl 2.438(4) Å) or H^- ligand. The $IrS_2(CO)X$ plane is not coplanar with the plane of the tetrathiolene ligand.

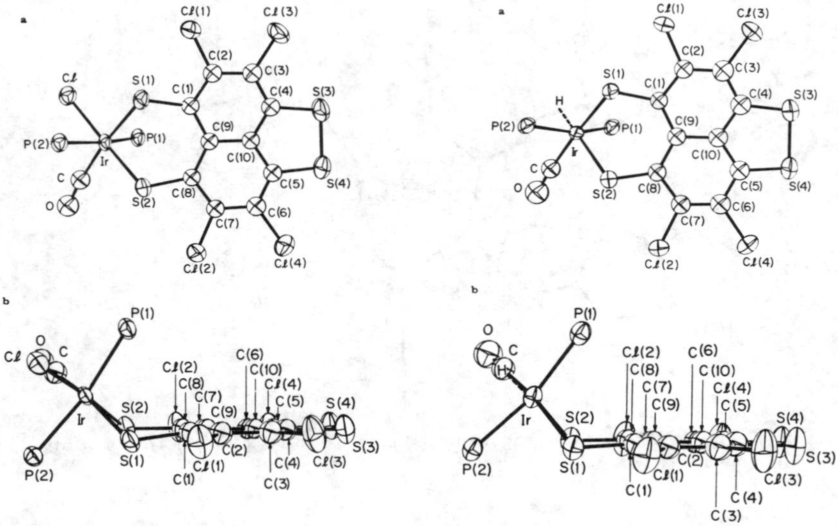

Fig. 1. The $[(C_6H_5)_3P]_2(CO)XIr(C_{10}Cl_4S_4)$ molecules: X = Cl (left) and X = H (right).

DI-μ-(DIPHENYL DISULPHIDE-SS')BIS[(DIPHENYL DISULPHIDE-S)SILVER(I)] BIS(HEXAFLUORO-ARSENATE)
$C_{48}H_{40}Ag_2As_2F_{12}S_8$ (I)

DI-μ-(DIPHENYL DISELENIDE-SeSe')BIS[(DIPHENYL DISELENIDE-Se)SILVER(I)] BIS(HEXAFLUORO-ARSENATE)
$C_{48}H_{40}Ag_2As_2F_{12}Se_8$ (II)

H.W. ROESKY, T. GRIES, P.G. JONES, K.-L. WEBER and G.M. SHELDRICK, 1984. J. Chem. Soc. Dalton, 1781-1784.

I. Triclinic, $P\bar{1}$, a = 10.902, b = 10.976, c = 13.308 Å, α = 112.82, β = 92.43, γ = 107.69°, Z = 1. Mo radiation, R = 0.041 for 5139 reflexions.

II. Triclinic, $P\bar{1}$, a = 10.676, b = 11.253, c = 13.625 Å, α = 66.40, β = 89.67, γ = 71.40°, Z = 1. Mo radiation, R = 0.044 for 4439 reflexions.

Compounds I and II are isostructural although this is not immediately apparent from the cell constants for the reduced cells. The structures contain a six-membered centrosymmetric ring in which two Ag atoms are connected by two bridging Ph_2S_2 or Ph_2Se_2 ligands (Fig. 1) (Ag-S(2) 2.503(2), S(2)-S(1) 2.065(2), Ag-S(1') 2.638(2); Ag-Se(2) 2.580(2), Se(1)-Se(2) 2.344(2), Ag-Se(1') 2.711(2) Å). The six-membered rings adopt a chair conformation. The X_2 unit of one independent Ph_2X_2 ligand bridges the Ag atoms (X = S, Se) while the other ligand coordinates in an exocyclic manner to Ag via only one X atom (Ag-S(3) 2.534(2), S(3)-S(4) 2.080(3); Ag-Se(3) 2.616(2), Se(3)-Se(4) 2.360(2) Å). The AsF_6^- ion is disordered in both compounds. The occupancies for the major and minor component are 0.8 and 0.2 in I and 0.5 for both in II. The F atoms of AsF_6^- form weak contacts with the Ag atoms such that the coordination at silver is distorted pyramidal in each case.

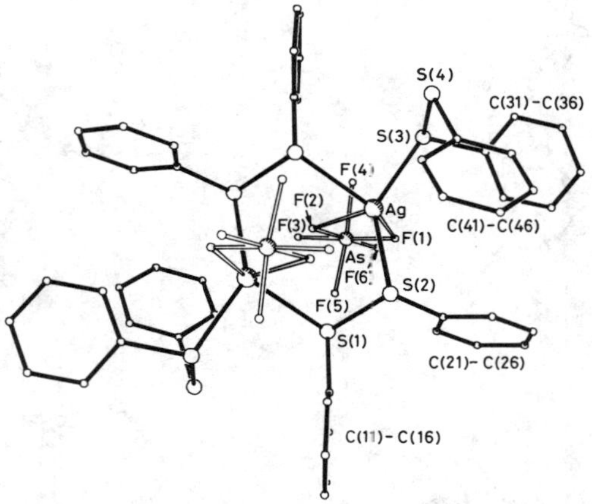

Fig. 1. $C_{48}H_{40}Ag_2As_2F_{12}S_8$: the centrosymmetric unit showing the weak Ag-F contacts.

cyclo-TETRAKIS[BIS(μ-N-METHYLPIPERIDINE-4-THIOLATO)NICKEL(II)]
$C_{48}H_{96}N_8Ni_4S_8$ $[(SC_5H_9N(CH_3))_2Ni]_4$

W. GAETE, J. ROS, X. SOLANS, M. FONT-ALTABA and J.L. BRIANSÓ, 1984. Inorg. Chem., 23, 39-43.

Monoclinic, P2/n, a = 24.345, b = 10.638, c = 12.323 Å, β = 92.34°, D_m = 1.33, Z = 2. Mo radiation, R = 0.073 for 2433 reflexions.

The complex (Fig. 1) consists of a cyclic tetranuclear array of Ni atoms, bridged by eight S atoms. The approximately square planar NiS_4 units form the faces of a skew parallelogram with S atoms at the corners. Axial and equatorial ligands alternate around the ring. Mean geometry: Ni-Ni 2.665(5), Ni-S 2.208(4), S...S 2.868(6) Å, Ni-S-Ni 74.3(1)°.

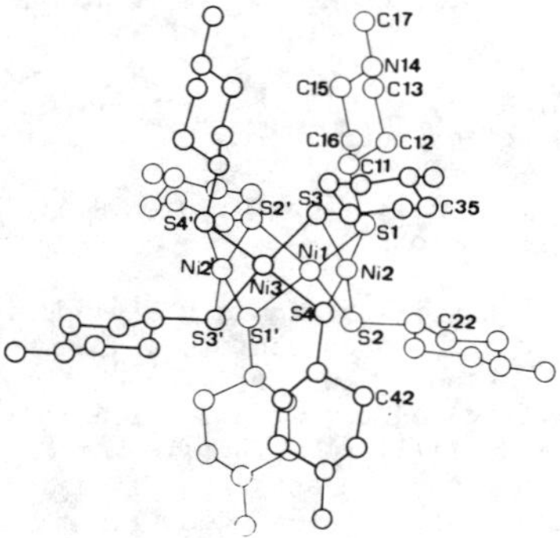

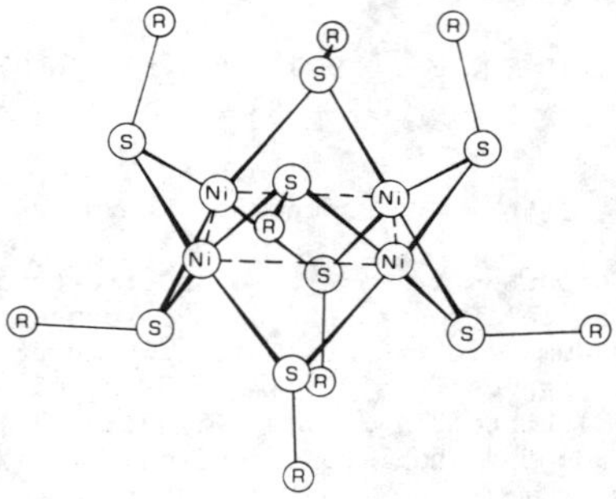

Fig. 1. The $[(SC_5H_9N(CH_3)_2)_2Ni]_4$ molecule, Ni-S skeleton shown at right.

TRIS(TETRAETHYLAMMONIUM) TRIS(μ_2-METHOXY)BIS[TETRAKIS(μ_3-SULPHIDO)TRIS(t-BUTYLTHIOLATOIRON)MOLYBDENUM]
$C_{51}H_{123}Fe_6Mo_2N_3O_3S_{14}$ $(Et_4N)_3[Fe_6Mo_2S_8(S\text{-}t\text{-}Bu)_6(OMe)_3]$

HUANG LIANGREN and LIN SHANHUO, 1984. J. Struct. Chem., 3, 25-27.

Hexagonal, $P6_3/mmc$, a = 18.191, c = 16.264 Å, D_m = 1.37, Z = 2. R = 0.069 for 536 reflexions.

The anion (Fig. 1) has crystallographic C_{3h} symmetry and the central framework comprises two Fe_3MoS_4 cubane-like clusters linked by three μ_2-methoxy groups. Principal bond lengths are: Mo-Fe 2.740(3), Mo-S(1) 2.342(3), Mo-O 2.172(14), Fe-S 2.228-2.267(3) Å. The Mo...Mo separation is 3.151(5) Å.

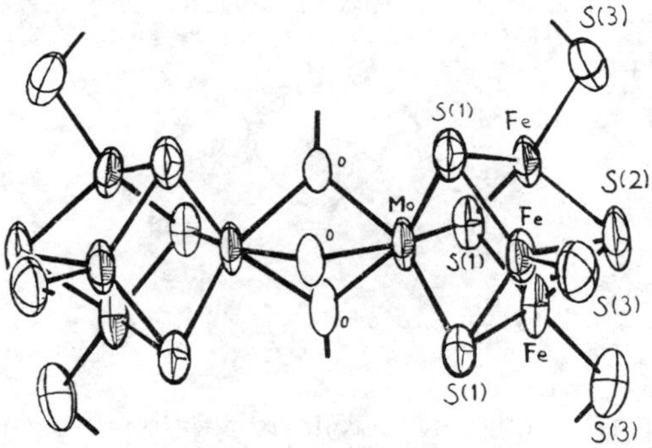

Fig. 1. The core of the $[Fe_6Mo_2S_8(S\text{-}t\text{-}Bu)_6(OMe)_3]^{3-}$.

DICHLOROBIS[4-(DI-p-TOLYLPHOSPHINO)DIBENZOTHIOPHENE-S,P]RUTHENIUM ACETONE SOLVATE
$C_{52}H_{42}Cl_2P_2RuS_2,2(C_3H_6O)$

S.M. BUCKNOR, M. DRAGANJAC, T.B. RAUCHFUSS, C.J. RUFFING, W.C. FULTZ and A.L. RHEINGOLD, 1984. J. Am. Chem. Soc., 106, 5379-5380.

Monoclinic, $P2_1/n$, a = 14.164, b = 14.478, c = 25.663 Å, β = 91.28°, D_m = 1.372, Z = 4. Mo radiation, R = 0.1019 for 3203 reflexions.

The Ru atom (Fig. 1) has distorted octahedral coordination, with Ru-S(1) 2.343(5), Ru-S(2) 2.402(5), Ru-P(1) 2.336(5), Ru-P(2) 2.314(5), Ru-Cl(1) 2.414(4), Ru-Cl(2) 2.459(5) Å. The sulphur atoms have pyramidal geometry.

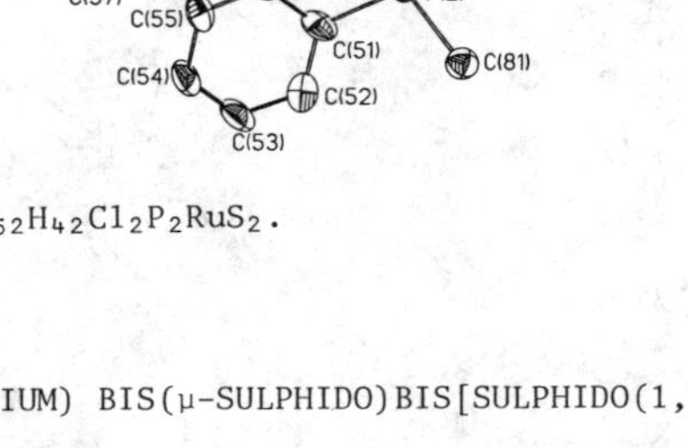

Fig. 1. A view of $C_{52}H_{42}Cl_2P_2RuS_2$.

BIS(TETRAPHENYLPHOSPHONIUM) BIS(μ-SULPHIDO)BIS[SULPHIDO(1,2-ETHANEDITHIOLATO)-TUNGSTEN(VI)]
$C_{52}H_{48}P_2S_8W_2$ $[P(C_6H_5)_4]_2[W_2S_4(C_2H_4S_2)_2]$

W.-H. PAN, T. CHANDLER, J.H. ENEMARK and E.I. STIEFEL, 1984. Inorg. Chem., 23, 4265-4269.

Monoclinic, $P2_1/n$, a = 16.126, b = 25.03, c = 13.021 Å, β = 101.73°, D_m = 1.757, Z = 4. Mo radiation, R = 0.039 for 7414 reflexions.

The dinuclear complex anion (Fig. 1) contains a syn-$W_2S_4^{2+}$ core. Important distances include: W-W 2.862(1), W-(μ-S) 2.319(2)-2.336(2), W=S 2.143(2) and 2.144(2), W-S 2.391(2)-2.415(2) Å. The W atoms are each 0.72 Å above the basal plane in each distorted square-pyramidal unit.

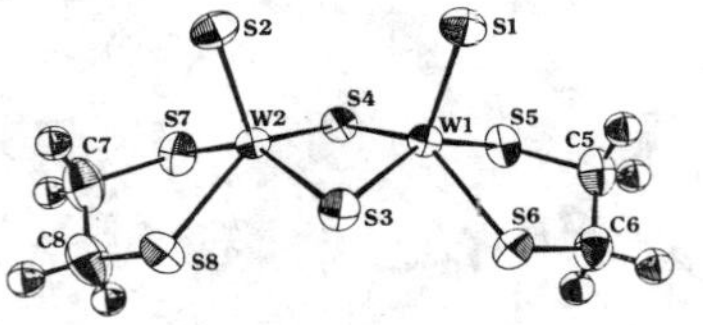

Fig. 1. The $[W_2S_4(C_2H_4S_2)_2]^{2-}$ anion.

BIS(μ-N-METHYLDIPHENYLPHOSPHINOTHIOFORMAMIDATO-P,N)BIS(μ-N-METHYLDIPHENYLPHOSPHINO-THIOFORMAMIDATO-S,N)DIMOLYBDENUM(II) BIS(DICHLOROMETHANE) SOLVATE

$C_{56}H_{52}Mo_2N_4P_4S_4,2CH_2Cl_2$ (I) $Mo_2[(C_6H_5)_2PC(S)NCH_3]_4,2CH_2Cl_2$

TETRAKIS(μ-N-METHYLDIPHENYLPHOSPHINOTHIOFORMAMIDATO-S,N)DIMOLYBDENUM(II) HEMI(DICHLOROMETHANE)SOLVATE

$C_{56}H_{52}Mo_2N_4P_4S_4,0.5(CH_2Cl_2)$ (II) $Mo_2[(C_6H_5)_2PC(S)NCH_3]_4,0.5CH_2Cl_2$

H.P.M.M. AMBROSIUS, F.A. COTTON, L.R. FALVELLO, H.T.J.M. HINTZEN, T.J. MELTON, W. SCHWOTZER, M. TOBIAS and J.G.M. van der LINDEN, 1984. Inorg. Chem., 23, 1611-1616.

I. Monoclinic, $P2_1/c$, a = 13.163, b = 14.502, c = 16.719 Å, β = 105.38°, Z = 2. Mo radiation, R = 0.077 for 2430 reflexions.

II. Triclinic, $P\bar{1}$, a = 15.513, b = 16.150, c = 14.909 Å, α = 94.68, β = 118.53, γ = 63.93°, Z = 2. Mo radiation, R = 0.058 for 3898 reflexions.

The two isomeric molecules are shown in Fig. 1. In the centrosymmetric molecule I two ligands (trans to each other) are N,S-coordinated while the other two are N,P-coordinated. Bond lengths for I: Mo-Mo 2.104(2), Mo-S 2.439(3), Mo-P 2.571(3), Mo-N(1) 2.117(8), Mo-N(2) 2.182(8) Å. In compound II all four ligands are coordinated through N and S such that like atoms on each Mo atom are trans to each other. Bond lengths for II: Mo-Mo 2.083(1), mean Mo-S 2.466(8), mean Mo-N 2.124(5) Å.

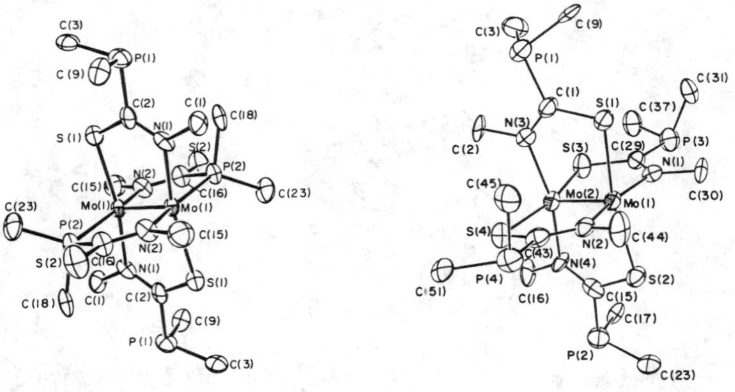

Fig. 1. The $Mo_2[Ph_2PC(S)NMe\text{-}P,N]_2[Ph_2PC(S)NMe\text{-}S,N]_2$ (left) and $Mo_2[Ph_2PC(S)NMe\text{-}S,N]_4$ (right) molecules.

BIS(TETRAPHENYLPHOSPHONIUM) BIS(THIOPHENOLATO)DICHLOROTETRAKIS(μ_3-SULPHIDO)TETRAIRON
$C_{60}H_{50}Cl_2Fe_4P_2S_6$ (I) $2(Ph_4P^+).C_{12}H_{10}Cl_2Fe_4S_6{}^{2-}$

BIS(TETRAPHENYLPHOSPHONIUM) BIS(PHENOLATO)DICHLOROTETRAKIS(μ_3-SULPHIDO)TETRAIRON
$C_{60}H_{50}Cl_2Fe_4O_2P_2S_4$ (II) $2(Ph_4P^+).C_{12}H_{10}Cl_2Fe_4O_2S_4{}^{2-}$

BIS(TETRAPHENYLPHOSPHONIUM) BIS(THIOPHENOLATO)BIS(p-METHYLPHENOLATO)TETRAKIS(μ_3-SULPHIDO)TETRAIRON
$C_{74}H_{64}Fe_4O_2P_2S_6$ (III) $2(Ph_4P^+).C_{26}H_{24}Fe_4O_2S_6{}^{2-}$

M.G. KANATZIDIS, N.C. BAENZIGER, D. COUCOUVANIS, A. SIMOPOULOS and A. KOSTIKAS, 1984. J. Am. Chem. Soc., 106, 4500-4511.

I. Orthorhombic, Pbcn, a = 13.082, b = 21.182, c = 21.376 Å, D_m = 1.49, Z = 4. Mo radiation, R = 0.062 for 2203 reflexions.

II. Orthorhombic, Pbcn, a = 13.060, b = 20.891, c = 21.423 Å, D_m = 1.46, Z = 4. Mo radiation, R = 0.061 for 2149 reflexions.

III. Orthorhombic, $P2_12_12$, a = 24.595, b = 12.588, c = 10.877 Å, D_m = 1.43, Z = 2. Mo radiation, R = 0.0658 for 3231 reflexions.

Compounds I and II are isomorphous. The anions in I and II (Fig. 1) are located on crystallographically imposed twofold axes. The mean Fe-S bond lengths in I and II are 2.278(5) and 2.285(5) Å, respectively. The Fe_4S_4 cores in I and II represent the first examples of such cores without an overall compressed D_{2d} geometry. The anion in III also is located on a crystallographic twofold axis. The Fe_4S_4 unit can be described as slightly compressed along an axis that is perpendicular to the crystallographic twofold. The terminal Fe-SPh bond lengths in I and III are 2.261(3) and 2.289(3) Å, respectively. The Fe-OAr bond lengths in II and III are found at 2.057(9) and 1.996(9) Å, respectively.

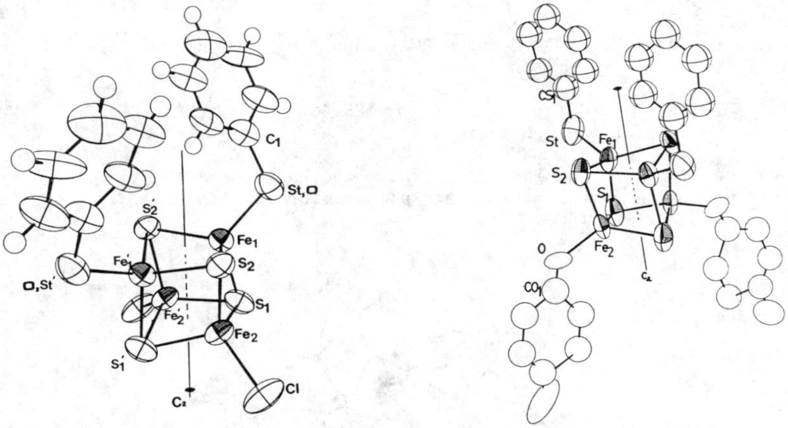

Fig. 1. Views of the Fe_4S_4 complexes I, II (left), III (right).

BIS[DICARBONYL(μ-DIPHENYLPHOSPHINE-N-METHYLCARBOTHIOAMIDO-N,S,P)(DIPHENYLPHOSPHINE-N-METHYLCARBOTHIOAMIDO-S,P)MOLYBDENUM] DICHLOROMETHANE SOLVATE
$C_{60}H_{52}Mo_2N_4O_4P_4S_4,CH_2Cl_2$ $[Mo(CO)_2[Ph_2PC(S)NMe][\mu\text{-}Ph_2PC(S)NMe]]_2.CH_2Cl_2$

H.P.M.M. AMBROSIUS, J. WILLEMSE, J.A. CRAS, W.P. BOSMAN and J.H. NOORDIK, 1984. Inorg. Chem., 23, 2672-2678.

Monoclinic, C2/c, a = 18.958, b = 19.542, c = 18.879 Å, β = 90.84°, Z = 4. Cu radiation, R = 0.090 for 2717 reflexions. (Experimental details from preliminary account (1).)

The binuclear molecule (Fig. 1) has exact C_2 symmetry. One pair of ligands bridge the two seven-coordinate Mo atoms and the other pair coordinates in a bidentate manner through P and S. Bond lengths: Mo-S(1) 2.512(4), Mo-S(2) 2.523(4), Mo-P(1) 2.453(5), Mo-P(2) 2.607(4), Mo-N(2) 2.25(2), Mo-C(3) 1.92(2), and Mo-C(4) 1.95(2) Å. The geometry at Mo is best described as a 4:3 piano-stool configuration.

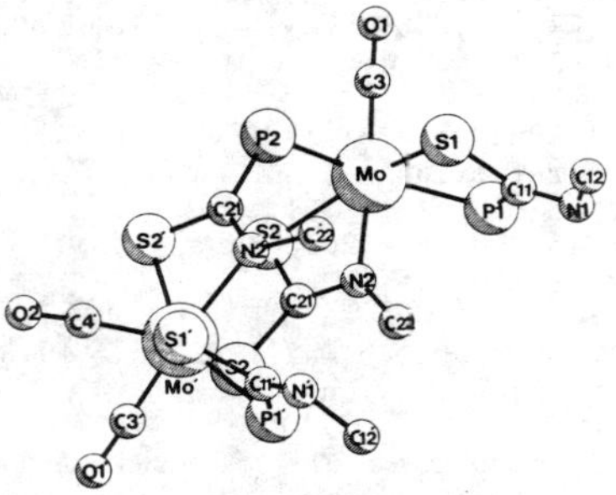

Fig. 1. The $[Mo(CO)_2[Ph_2PC(S)NMe][\mu-Ph_2PC(S)NMe]]_2$ molecule.

1. Structure Reports, 46B, 1304.

CYCLOHEXAKIS[BIS(μ_2-N,N-DIMETHYLMERCAPTOPROPYLAMINE)NICKEL] DODECAPERCHLORATE DIHYDRATE

$C_{60}H_{156}Cl_{12}N_{12}Ni_6O_{48}S_{12},2(H_2O)$ $[Ni_6(MPDMAH)_{12}]^{12+}.12(ClO_4^-),2H_2O$

H. BARRERA, J.C. BAYON, J. SUADES, G. GERMAIN and J.P. DECLERCQ, 1984. Polyhedron, 3, 969-975.

Monoclinic, C2/c, a = 36.842, b = 16.077, c = 25.403 Å, β = 118.36°, Z = 1. Mo radiation, R = 0.104 for 3387 reflexions.

The cation lies about an inversion centre and has a planar hexanuclear array of nickel atoms each with approximate square-planar coordination (Fig. 1). All Ni-S bonds are in the range 2.212-2.177(8) Å.

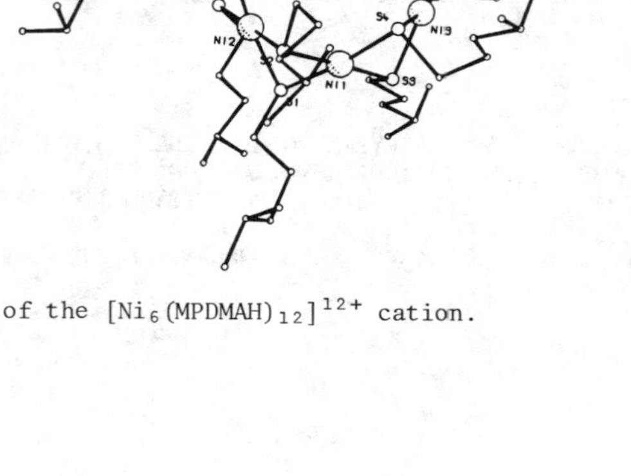

Fig. 1. A view of the $[Ni_6(MPDMAH)_{12}]^{12+}$ cation.

BIS(TETRAPHENYLPHOSPHONIUM) μ-TETRAKIS(BIS(METHOXYCARBONYL)ETHYLENE-1,2-DITHIOLATO-S,S')DIFERRATE(III) BIS(N,N-DIMETHYLFORMAMIDE) SOLVATE

$C_{72}H_{64}Fe_2O_{16}P_2S_8,2C_3H_7NO$ $[(C_6H_5)_4P]_2\{Fe_2[S_2C_2(COOCH_3)_2]_2\}.2C_3H_7NO$

M.G. KANATZIDIS and D. COUCOUVANIS, 1984. Inorg. Chem., 23, 403-409.

Triclinic, P$\bar{1}$, a = 12.343, b = 13.710, c = 14.787 Å, α = 63.79, β = 68.23, γ = 85.32°, D_m = 1.43, Z = 1. Mo radiation, R = 0.052 for 2886 reflexions.

The Fe(III) atoms in the centrosymmetric anion have tetragonal-pyramidal coordination geometry with one basal S atom (S(3)) serving as the axial ligand for the second Fe atom (Fig. 1). The Fe-S distances for the terminal ligand are both 2.230(2) Å. The other two basal Fe-S(3) and Fe-S(4) bonds of the bridging ligand are 2.256(2) and 2.243(2) Å respectively. The axial bridging Fe-S(3)' bond length is 2.466(2) Å.

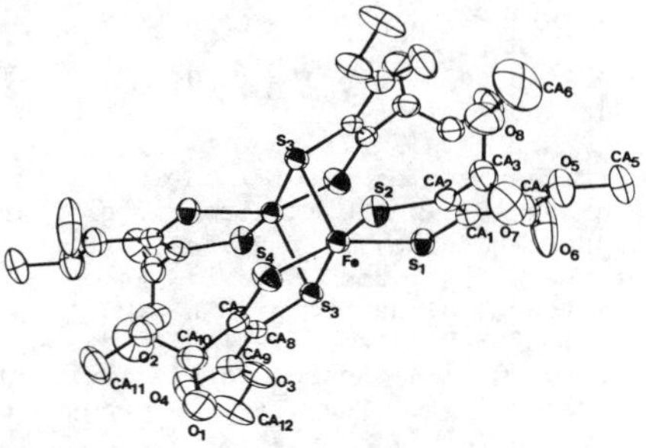

Fig. 1. The $\{Fe_2[S_2C_2(COOCH_3)_2]_2\}^{2-}$ anion.

(μ_2-METHYLSULPHIDO)(μ_2-SULPHIDO)TETRAKIS(TRIPHENYLPHOSPHINE)DIPLATINUM(II) HEXAFLUOROPHOSPHATE METHANOL SOLVATE

$C_{73}H_{63}F_6P_5Pt_2S_2,CH_4O$ (I) $[Pt_2(S)(SMe)(PPh_3)_4]PF_6,MeOH$

cis-BIS(μ_2-METHYLSULPHIDO)DINITROBIS(TRIPHENYLPHOSPHINE)DIPLATINUM(II)

$C_{38}H_{36}N_2O_4P_2Pt_2S_2$ (II) $[Pt_2(SMe)_2(NO_2)_2(PPh_3)_2]$

C.E. BRIANT, C.J. GARDNER, T.S.A. HOR, N.D. HOWELLS and D.M.P. MINGOS, 1984. J. Chem. Soc. Dalton, 2645-2651.

I. Monoclinic, $P2_1/c$, a = 22.835, b = 13.094, c = 24.357 Å, β = 110.99°, Z = 4. Mo radiation, R = 0.056 for 3551 reflexions.

II. Triclinic, P$\bar{1}$, a = 10.476, b = 12.781, c = 16.814 Å, α = 111.27, β = 93.79, γ = 76.06°, Z = 2. Mo radiation, R = 0.062 for 3138 reflexions.

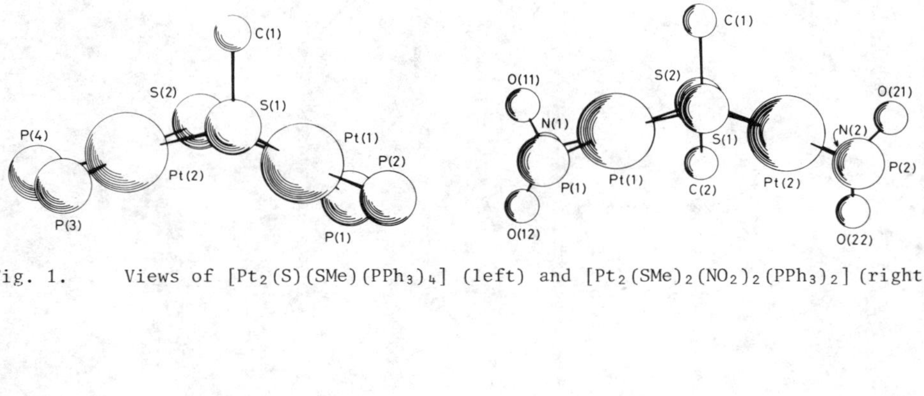

Fig. 1. Views of $[Pt_2(S)(SMe)(PPh_3)_4]$ (left) and $[Pt_2(SMe)_2(NO_2)_2(PPh_3)_2]$ (right).

Compound (I) is shown to have a hinged square-planar geometry (Fig. 1) with a 138° hinge angle and Pt(1)-S(1) 2.374(6), Pt(1)-S(2) 2.311(6), Pt(1)-P(1) 2.269(5), Pt-P(2) 2.298(5), Pt(2)-S(1) 2.351(6), Pt(2)-S(2) 2.329(6), Pt(2)-P(3) 2.286(5), Pt(2)-P(4) 2.268(5) Å. Compound II has similar geometry to (I), with a hinge angle of 141° and Pt(1)-S(1) 2.298(6), Pt(1)-S(2) 2.356(6), Pt(1)-P(1) 2.265(6), Pt(1)-N(1) 2.04(2), Pt(2)-S(1) 2.291(6), Pt(2)-S(2) 2.351(6), Pt(2)-P(2) 2.267(6), Pt(2)-N(2) 2.03(2) Å.

BIS(cis-1,2-BIS(DIPHENYLPHOSPHINO)ETHYLENE)DISULPHURDINITRIDEIRIDIUM(III) TETRAPHENYLBORATE

$C_{76}H_{64}BIrN_2P_4S_2,0.5C_3H_6O$ $[(dppen)_2Ir(S_2N_2)]^+BPh_4^-,0.5(CH_3)_2CO$

F. CECCONI, C.A. GHILARDI, S. MIDOLLINI, S. MONETI and A. ORLANDINI, 1984. J. Organometal. Chem., 275, C22-C24.

Monoclinic, $P2_1/n$, a = 26.458, b = 23.688, c = 12.045 Å, β = 102.36°, Z = 4. Mo radiation, R = 0.066 for 4186 reflexions.

The molecular structure consists of complex cations (Fig. 1) and BPh_4^- anions with acetone molecules interspersed in the lattice. The Ir atom displays distorted octahedral geometry, being surrounded by four P atoms of the two dppen molecules and by one S and one N atom of the S_2N_2 ligand. The S_2N_2 group chelates the metal to form a five-membered ring which is planar to within 0.03 Å. Unfortunately the group is disordered and the Ir-S(1) (2.379 Å) and Ir-N(1) (2.289 Å) bond lengths differ by 0.09 Å (instead of the expected 0.25 Å) hence distances within the group are not accurate enough to warrant discussion. The Ir-P distances are 2.347, 2.370, 2.341 and 2.367 Å for P(1),(2),(3) and (4) respectively.

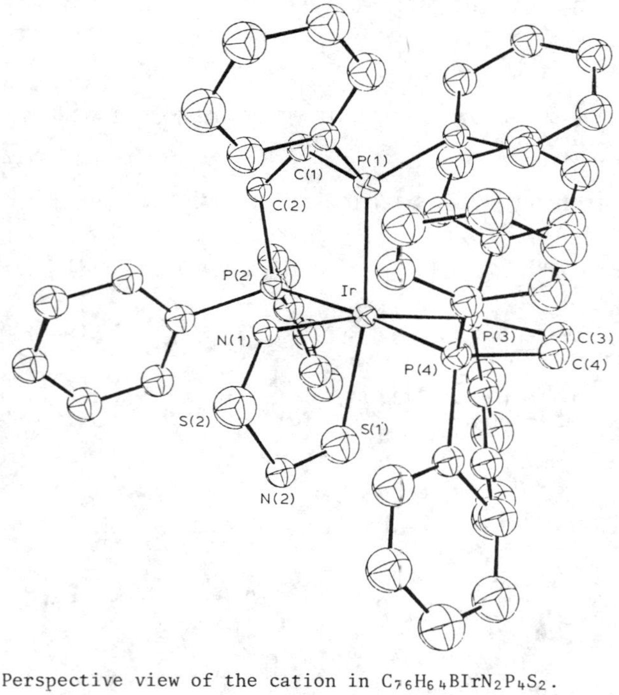

Fig. 1. Perspective view of the cation in $C_{76}H_{64}BIrN_2P_4S_2$.

TRIS(TETRAETHYLAMMONIUM) TRIS(μ_2-PHENYLTHIOLATO)[TETRAKIS(μ_3-SULPHIDO)TRIS(PHENYLTHIOLATOIRON)][TRIS(μ_3-SULPHIDO)(μ_2-SULPHIDO)TRIS(PHENYLTHIOLATOIRON)]DIMOLYBDENUM
$C_{78}H_{105}Fe_6Mo_2N_3S_{17}$ $(Et_4N)_3[Mo_2Fe_6S_8(SPh)_9]$

CAI JINHUA and CHENG CHANGNENG, 1984. J. Struct. Chem., 3, 33-35.

Monoclinic, $P2_1$, a = 19.026, b = 17.350, c = 16.035 Å, β = 104.82°, Z = 2. Mo radiation, R = 0.076 for 4666 reflexions.

The anion has approximate C_3 symmetry and consists of two $MoFe_3S_4(SPh)_3$ cubane-like clusters which are bridged by three μ_2-SPh groups (Fig. 1). Principal mean bond lengths are: Fe-S 2.249(4), Mo-S 2.337(3), Fe-S(t) 2.259(4), Fe-S(b) 2.588(3) Å. The Mo...Mo separation is 3.728(1) Å.

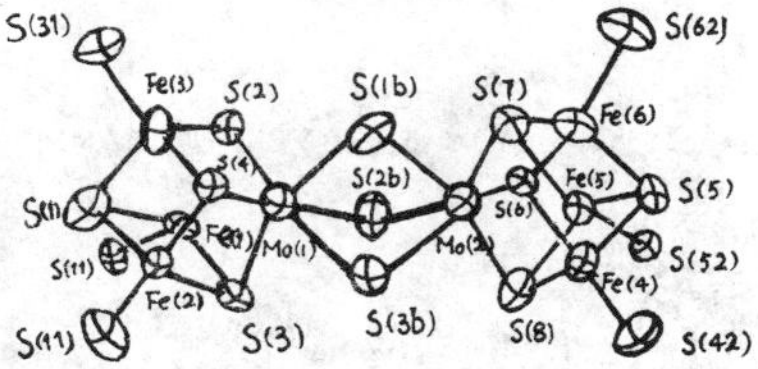

Fig. 1. The core of the $[Mo_2Fe_6S_8(SPh)_9]^{3-}$ cluster.

TETRAKIS(TETRAETHYLAMMONIUM) BIS((μ_2-p-CHLOROPHENYLTHIOLATO)TETRAKIS(μ_3-SULPHIDO)-(CATECHOLATOMOLYBDENUM)BIS(p-CHLOROPHENYLTHIOLATO)TRIIRON)
$C_{80}H_{112}Cl_6Fe_6Mo_2N_4O_4S_{14}$ (I)

TRIS(TETRAETHYLAMMONIUM) TETRAKIS(μ_3-SULPHIDO)TRIS(ETHYLTHIOLATOIRON)(CYANOTETRACHLOROCATECHOLATOMOLYBDENUM)
$C_{37}H_{75}Cl_4Fe_3MoN_4O_2S_7$ (II)

TRIS(TETRAETHYLAMMONIUM) TETRAKIS(μ_3-SULPHIDO)(AZIDOTETRACHLOROCATECHOLATOMOLYBDENUM)-TRIS(p-CHLOROPHENYLTHIOLATOIRON) ACETONITRILE SOLVATE
$C_{48}H_{72}Cl_7Fe_3MoN_6O_2S_7,C_2H_3N$ (III)

R.E. PALERMO, R. SINGH, J.K. BASHKIN and R.H. HOLM, 1984. J. Am. Chem. Soc., 106, 2600-2612.

I. Monoclinic, $P2_1/n$, a = 18.25, b = 22.08, c = 26.21 Å, β = 99.43°, Dm = 1.53, Z = 4. Mo radiation, R = 0.087 for 3364 reflexions.

II. Orthorhombic, Pnma, a = 22.793, b = 19.274, c = 12.456 Å, Z = 4. Mo radiation, R = 0.074 for 2035 reflexions.

III. Monoclinic, $P2_1/n$, a = 19.158, b = 18.56, c = 20.15 Å, β = 105.96°, Z = 4. Mo radiation, R = 0.076 for 5062 reflexions.

The anion of I (Fig. 1) consists of two $MoFe_3S_4(S\text{-}p\text{-}C_6H_4Cl)_2(cat)$ subclusters joined by two Mo-S(R)-Fe bridges whose elongated bonds contribute to the ready cleavage. The anion of II is disordered about a crystallographic mirror plane. The anion of III is not disordered. Dimensions of II and III (Fig. 1) are similar.

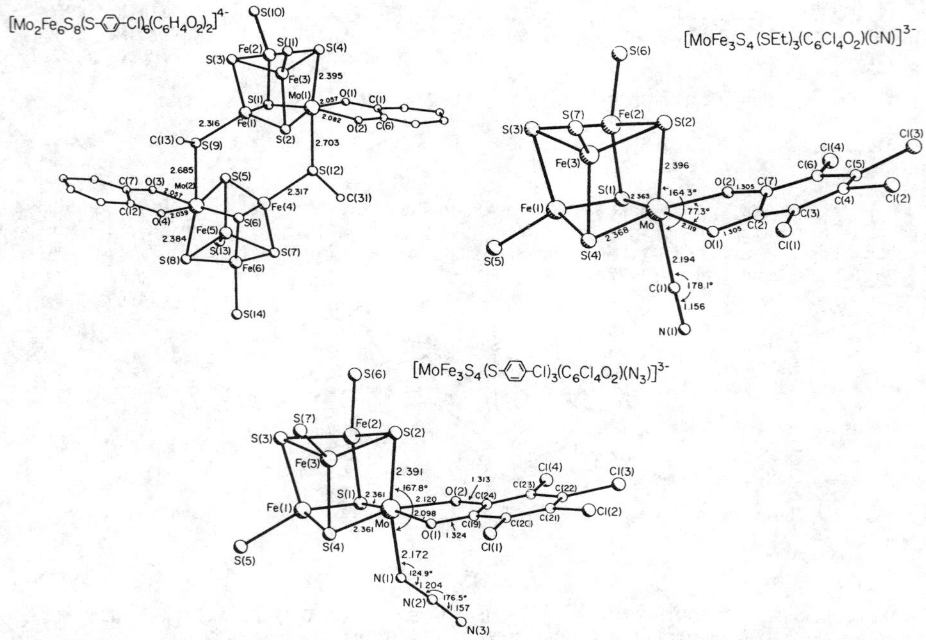

Fig. 1. Structures of the anions in the $MoFe_3S_4$ cubane-type structure (I) (top left), (II) (top right), (III) (bottom).

DIBROMO(5,8,16,19-TETRAPHENYL-1,12-DITHIA-5,8,16,19-TETRAPHOSPHACYCLODOCOSANE-S,S',-P1,P2,P1',P2')DINICKEL(II) BIS(TETRAPHENYLBORATE) METHANOL SOLVATE
$C_{88}H_{92}B_2Br_2Ni_2P_4S_2,CH_4O$

M. CIAMPOLINI, N. NARDI, P.L. ORIOLI, S. MANGANI and F. ZANOBINI, 1984. J. Chem. Soc. Dalton, 2265-2270.

Orthorhombic, Pbca, a = 17.791, b = 20.993, c = 22.758 Å, Z = 4. Mo radiation, R = 0.092 for 917 reflexions.

The cation (Fig. 1) contains two Ni atoms in a slightly distorted square-planar environment related by a crystallographic centre of symmetry. Each Ni atom is coordinated by two phosphorus and one sulphur atoms of the macrocycle and by one bromide ion (Ni-P(1) 2.19(2), Ni-P(2) 2.20(1), Ni-S 2.24(1) and Ni-Br 2.314(7) Å). The Ni...Ni separation is 5.13 Å. The macrocycle has the 5RS, 8RS, 16SR, 19SR meso centrosymmetric configuration.

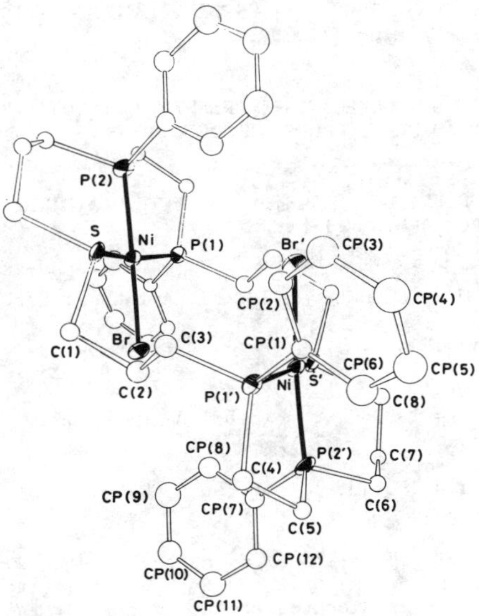

Fig. 1. $C_{88}H_{92}B_2Br_2Ni_2P_4S_2,CH_4O$: the complex cation.

TETRAKIS(TETRAETHYLAMMONIUM) BIS{TRIS(μ_2-PHENYLTHIOLATO)TETRAKIS(μ_3-SULPHIDO)TRIS-[(PHENYLTHIOLATO)IRON]MOLYBDENUM}IRON

$C_{104}H_{140}Fe_7Mo_2N_4S_{20}$ $(Et_4N)_4[Mo_2Fe_7S_8(SPh)_{12}]$

CAI JINHUA and KANG BEISHENG, 1984. J. Struct. Chem., 3, 143-145.

Triclinic, P$\bar{1}$, a = 12.775, b = 13.076, c = 20.576 Å, α = 80.00, β = 81.39, γ = 61.51°, Z = 1. Mo radiation, R = 0.077 for 4031 reflexions.

The anion (Fig. 1) lies about an inversion centre and has a double cubane configuration with $MoFe_3S_4$ moieties which are linked through their Mo atoms by a $(PhS)_3Fe(SPh)_3$ bridging subunit. Atom Fe(4) is on the inversion centre. Some bond lengths are: Mo-S(1) 2.354(4), Mo-S(2) 2.352(4), Mo-S(3) 2.356(4), Mo-S(8) 2.585(4), Mo-S(9) 2.563(4), Mo-S(10) 2.552(4) Å; Fe-S distances are in the range 2.257(5)-2.276(5) in the cubane moiety and 2.523-2.546(4) Å with the S-bridges.

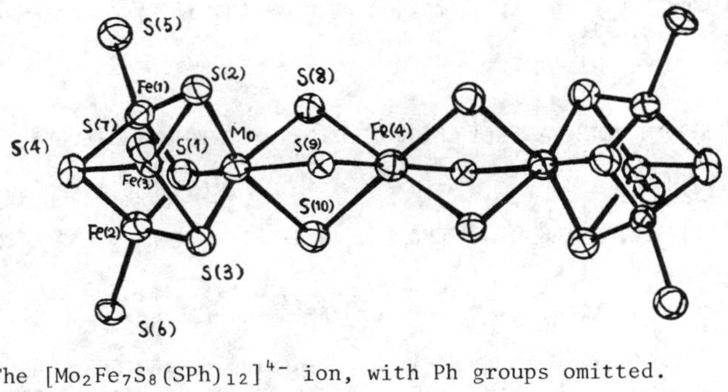

Fig. 1. The $[Mo_2Fe_7S_8(SPh)_{12}]^{4-}$ ion, with Ph groups omitted.

TETRAKIS(TETRAMETHYLAMMONIUM) DODECAKIS(μ_2-PHENYLTHIOLATO)TETRAKIS(PHENYLTHIOLATO)-TETRAKIS(μ_3-SULPHIDO)DECACADMIUM (ISOMER I)
$C_{112}H_{128}Cd_{10}N_4S_{20}$ (I)

TETRAKIS(TETRAMETHYLAMMONIUM) DODECAKIS(μ_2-PHENYLTHIOLATO)TETRAKIS(PHENYLTHIOLATO)-TETRAKIS(μ_3-SULPHIDO)DECAZINC (ISOMORPH II OF COMPOUNDS)
$C_{112}H_{128}N_4S_{20}Zn_{10}$ (II)

TETRAKIS(TETRAMETHYLAMMONIUM) DODECAKIS(μ_2-PHENYLTHIOLATO)TETRAKIS(PHENYLTHIOLATO)-TETRAKIS(μ_3-SULPHIDO)DECACADMIUM (ISOMER II)
$C_{112}H_{128}Cd_{10}N_4S_{20}$ (III)

I.G. DANCE, A. CHOY and M.L. SCUDDER, 1984. J. Am. Chem. Soc., 106, 6285-6295.

I. Tetragonal, $I\bar{4}$, a = 20.946, c = 14.779 Å, Z = 2. Cu radiation, R = 0.033 for 2971 reflexions.

II. Tetragonal, $P\bar{4}2_1c$, a = 19.783, c = 16.871 Å, D_m = 1.41, Z = 2. Mo radiation, R = 0.044 for 1140 reflexions.

III. Tetragonal, $I\bar{4}2m$, a = 20.140, c = 16.896 Å, D_m = 1.593, Z = 2. Mo radiation, R = 0.042 for 916 reflexions.

The complexes (Fig. 1) are molecular with the same $(\mu_3\text{-S})_4M_{10}S_{16}$ core structure which is a super tetrahedral fragment in the cubic (sphalerite) lattice. Three-coordinate chalcogenide ions are located at the centres of the hexagonal faces of the $(\mu\text{-S})_{12}$ truncated tetrahedron. 186 molecular configurational isomers are possible. The three analyses have found two isomers with $\bar{4}$ symmetry ((I) and (II)). Form (III) is a mirror disordered form of (II). Small distortions of the core geometry from idealized $\bar{4}3m$ symmetry can be traced to weak repulsions between phenyl substituents on the surface of the molecular anions.

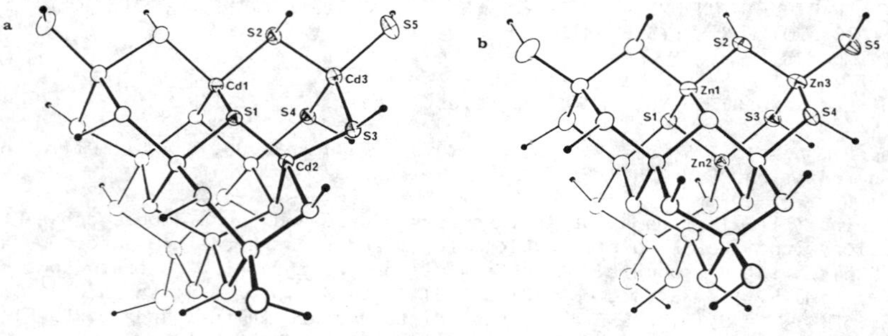

Fig. 1. Views of the $[S_4Cd_{10}(SPh)_{16}]^{4-}$ (left) and $[S_4Zn_{10}(SPh)_{16}]^{4-}$ (right) structures; only the first carbon of each phenyl ring is shown. The $\bar{4}$ axis is vertical.

PENTAKIS(TRIPHENYLPHOSPHINE)HEXAKIS(4-CHLOROBENZENETHIOLATO)HEXASILVER(I) BIS(TOLUENE) SOLVATE
$C_{126}H_{99}Ag_6Cl_6P_5S_6,2C_7H_8$ $Ag_6(SC_6H_4Cl)_6(PPh_3)_5,(CH_3C_6H_5)_2$

I.G. DANCE, L.J. FITZPATRICK and M.L. SCUDDER, 1984. Inorg. Chem., 23, 2276-2281.

Triclinic, $P\bar{1}$, a = 15.024, b = 18.488, c = 26.506 Å, α = 82.14, β = 86.19, γ = 66.45°, D_m = 1.50, Z = 2. Mo radiation, R = 0.060 for 7571 reflexions.

The $Ag_6(SC_6H_4Cl)_6(PPh_3)_5$ molecule (Fig. 1) contains an $Ag_5(SR)_6$ central cage, with PPh_3 ligands terminal at three of the Ag atoms and an $Ag(PPh_3)_2$ appendage inserted between two bridging thiolates. The $Ag_5(SR)_6$ cage, when symmetrized, contains a basal unit comprised of an irregular hexagon of alternating Ag^{per}_{bas} and S_{bas} atoms, centered on an Ag_{bas}^{cen} atom with trigonal planar $(S_{bas})_3$ coordination (Ag-S 2.480(4)-2.673(4) Å). A trigonal planar podal unit, $Ag_{pod}(S_{pod})_3$ (Ag-S 2.407(5)-2.689(4) Å), is parallel to the basal plane and is attached to it by three S_{pod}-Ag_{bas}^{per} bonds (Ag-S 2.503(4)-2.677(4) Å). Tetrahedral coordination of each Ag_{bas}^{per} atom is completed by a terminal PPh_3 ligand (Ag-P 2.405(4)-2.459(4) Å). The additional $Ag(PPh_3)_2$ moiety bridges two S_{pod} atoms (Ag-P 2.479(4), 2.529(4), Ag-S 2.628(4), 2.684(4) Å). The Ag_{bas}^{per}-S_{bas} distances are 2.593(4)-2.838(4) Å and Ag3-S5 3.322(5) Å.

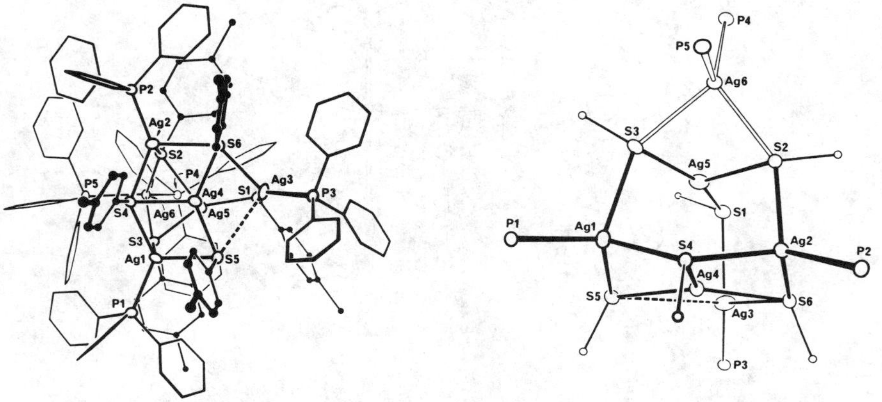

Fig. 1. The $Ag_6(SC_6H_4Cl)_6(PPh_3)_5$ molecule, central core at right.

(μ-TETRATHIOOXALATO)BIS[1,1,1-TRIS(DIPHENYLPHOSPHINOMETHYL)ETHANE-P,P',P'')RHODIUM] BIS(TETRAPHENYLBORATE) DICHLOROMETHANE SOLVATE

$C_{132}H_{118}B_2P_6Rh_2S_4,CH_2Cl_2$ $\{[Rh(C_{39}H_{41}P_3)]_2(C_2S_4)\}[(C_6H_5)_4B]_2.CH_2Cl_2$

C. BIANCHINI, C. MEALLI, A. MELI and M. SABAT, 1984. Inorg. Chem., 23, 4125-4127.

Monoclinic, $P2_1/a$, a = 15.496, b = 28.736, c = 14.216 Å, β = 106.02°, Z = 2. Mo radiation, R = 0.095 for 3195 reflexions.

The centrosymmetric binuclear cation (Fig. 1) has approximate C_{2h} symmetry and consists of two (triphos)Rh units bridged by a $C_2S_4^{2-}$ ligand. The coordination geometry is roughly square pyramidal with Rh-S 2.310(6) and 2.295(5), Rh-P 2.245(5), 2.309(5), and 2.334(6) Å.

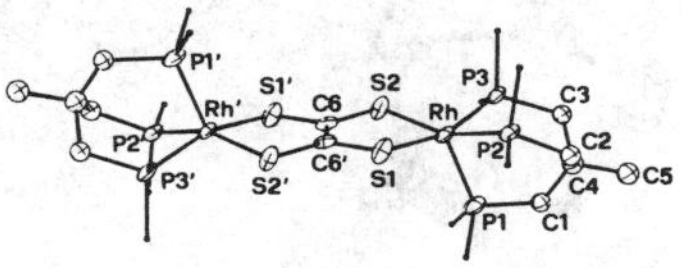

Fig. 1. The $\{[Rh(C_{39}H_{41}P_3)]_2(C_2S_4)\}^{2+}$ cation.

(μ-CHLORO)(μ-DIFLUOROPHOSPHIDO)BIS[μ-(METHYLAMINO)BIS(DIFLUOROPHOSPHINE)](METHYLAMINO-DIFLUOROPHOSPHINE)TRICARBONYLDIMOLYBDENUM(Mo-Mo)

$C_6H_{10}ClF_{12}Mo_2N_3O_3P_6$ $(CO)_2Mo[\mu-CH_3N(PF_2)_2]_2(\mu-PF_2)(\mu-Cl)Mo(CO)(PF_2NHCH_3)$

R.B. KING, M. SHIMURA and G.M. BROWN, 1984. Inorg. Chem., 23, 1398-1404.

Monoclinic, $P2_1/c$, a = 17.546, b = 9.725, c = 14.558 Å, β = 107.20°, Z = 4. Mo radiation, R = 0.029 for 4441 reflexions.

The binuclear complex is shown in Fig. 1. The coordination about each Mo atom is slightly distorted octahedral, if the Mo-Mo bond (2.9750(3) Å) is excluded. Bond lengths: Mo-(μ-Cl) 2.5280(8) and 2.5311(8), Mo-P(μ-PF_2) 2.3283(8) and 2.3297(8), Mo-P(1-4) 2.3630(8)-2.3757(8), Mo-P(6) 2.3500(9), Mo-CO 1.986(3)-2.075(3) Å.

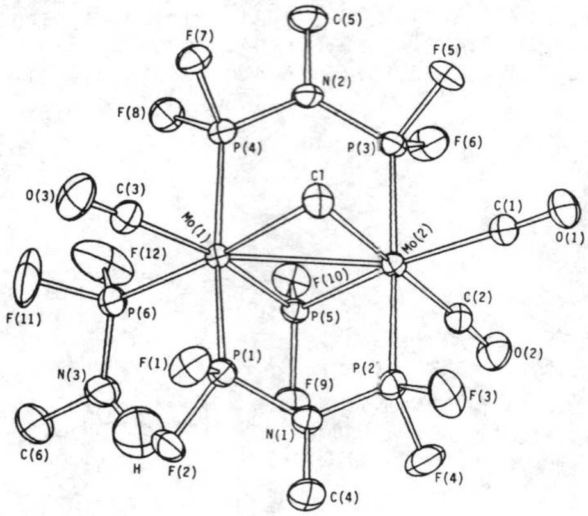

Fig. 1. The $C_6H_{10}ClF_{12}Mo_2N_3O_3P_6$ molecule.

TRIBROMO(TRIETHYLPHOSPHINE)GOLD(III)
$C_6H_{15}AuBr_3P$

D.S. EGGLESTON, D.F. CHODOSH, D.T. HILL and G.R. GIRARD, 1984. Acta Cryst., C40, 1357-1359.

Monoclinic, $P2_1/n$, a = 7.871, b = 14.171, c = 11.734 Å, β = 94.41°, D_m = 2.80, Z = 4. Mo radiation, R = 0.032 for 1470 reflexions.

The structure contains discrete monomers (Fig. 1) with Au square planar coordinated. The atoms Au,Br(1),Br(2),Br(3),P are planar within 0.043(1) Å, The Au-Br bond trans to P is lengthened compared to the other two such bonds.

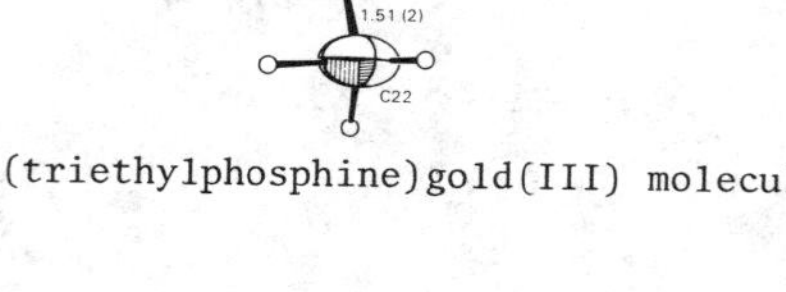

Fig. 1. The tribromo(triethylphosphine)gold(III) molecule. (Distances in Å.)

BIS(DIMETHYL HYDROGEN PHOSPHITE-P)BIS(DIMETHYL PHOSPHITO-P)PLATINUM(II)
$C_8H_{26}O_{12}P_4Pt$ $Pt(P(OH)(OCH_3)_2]_2[P(O)(OCH_3)_2]_2$

J.M. SOLAR, R.D. ROGERS and W.R. MASON, 1984. Inorg. Chem., 23, 373-377.

Monoclinic, $P2_1/c$, a = 7.147, b = 15.204, c = 10.680 Å, β = 116.53°, Z = 2. Mo radiation, R = 0.022 for 793 reflexions.

The centrosymmetric square planar molecule is shown in Fig. 1. The two unique P ligands are equivalent, being bridged by a symmetric O...H...O hydrogen bond (O(1)...O(4)' 2.38(1) Å). The Pt-P distances are 2.314(3) and 2.310(3) Å.

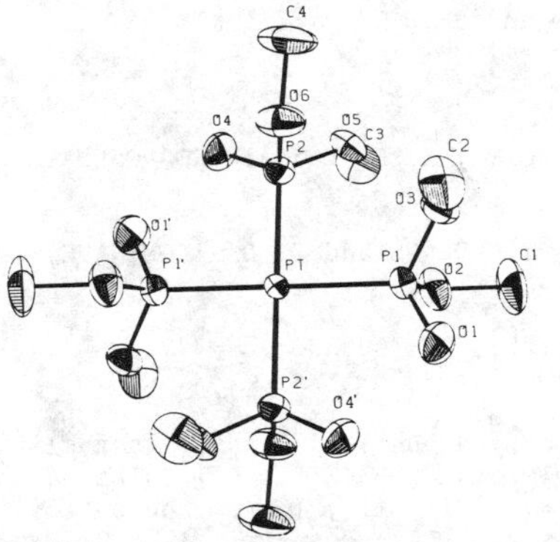

Fig. 1. The $Pt[P(OH)(OCH_3)_2]_2[P(O)(OCH_3)_2]_2$ molecule.

TRIS(TRIMETHYLPHOSPHINE)TETRACHLOROTUNGSTEN
$C_9H_{27}Cl_4P_3W$ (I) $WCl_4(PMe_3)_3$

(TRIMETHYLPHOSPHINE)BIS(ACETYLACETONATO)OXOMOLYBDENUM
$C_{13}H_{23}MoO_5P$ (II) $MoO(acac)_2PMe_3$

R.D. ROGERS, E. CARMONA, A. GALINDO, J.L. ATWOOD and L.G. CANADA, 1984. J. Organometal. Chem., 277, 403-415.

I. Monoclinic, $P2_1/c$, a = 14.835, b = 11.422, c = 11.568 Å, β = 91.92°, Z = 4. Mo radiation, R = 0.040 for 1613 reflexions.

II. Monoclinic, $P2_1/c$, a = 14.155, b = 9.984, c = 12.629 Å, β = 98.80°, Z = 4. Mo radiation, R = 0.026 for 1332 reflexions.

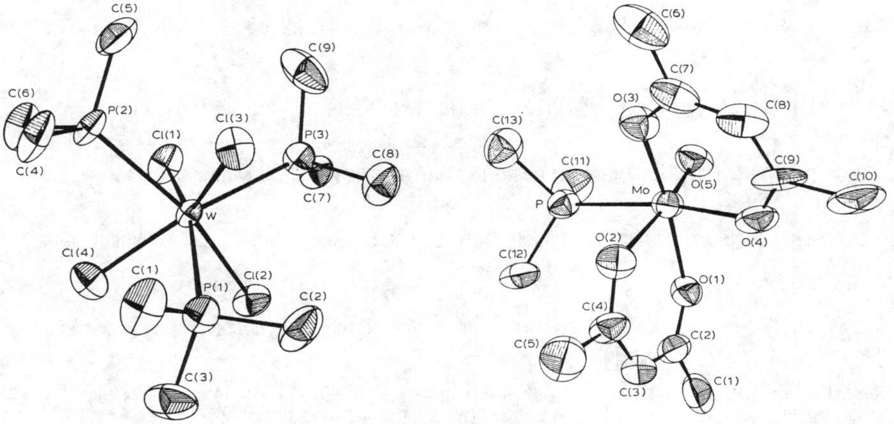

Fig. 1. Molecular structures of $C_9H_{27}Cl_4P_3W$ (left), and $C_{13}H_{23}MoO_5P$ (right).

In I the W atom is seven-coordinate, the ligands describing an approximate capped octahedron with a Cl atom capping the face defined by the three P atoms (Fig. 1). The unique W-Cl(3) distance is 2.417 Å, while the remaining W-Cl and W-P bond lengths are 2.450, 2.470, 2.416 (Cl(1), Cl(2), Cl(4)) and 2.552, 2.553 and

2.548 Å (P(1), P(2), P(3)). In II the Mo coordination is pseudo-octahedral with the oxo and phosphine ligands in a cis configuration (Fig. 1). The Mo-O(oxo) separation is 1.676 Å and the Mo-O(1), O(2), O(3), O(4) and P distances are: 2.053, 2.168, 2.109, 2.132 and 2.468 Å respectively. The chelate angles O(1)-Mo-O(2) and O(3)-Mo-O(4) are respectively 81.9 and 88.6°.

NITROSYLTRIS(TRIMETHYLPHOSPHINE)NICKEL HEXAFLUOROPHOSPHATE
$C_9H_{27}F_6NNiOP_4$ $[Ni(NO)(P(CH_3)_3)_3]^+PF_6^-$

G. ELBAZE, F. DAHAN, M. DARTIGUENAVE and Y. DARTIGUENAVE, 1984. Inorg. Chim. Acta, 87, 91-97.

Orthorhombic, Pnma, a = 16.253, b = 10.536, c = 12.228 Å, Z = 4. Mo radiation, R = 0.036 for 1085 reflexions.

The coordination geometry around the Ni atom is a slightly-distorted tetrahedron with an average P-Ni-P angle of 105.63° (Fig. 1). The molecule has crystallographic m symmetry, with atoms Ni,N,O,P(1),C(1),P(3) of the cation lying on the mirror plane. The Ni-P(1) and Ni-P(2) distances are respectively 2.229 and 2.239 Å, the Ni-N distance is 1.645 Å, and the N-O bond length 1.139 Å. The nitrosyl group is coordinated in a nearly linear fashion with an Ni-N-O angle of 175.4°. The bending in this group occurs in the C(1)-P(1)-Ni-N-O plane (the mirror plane), towards atom P(1).

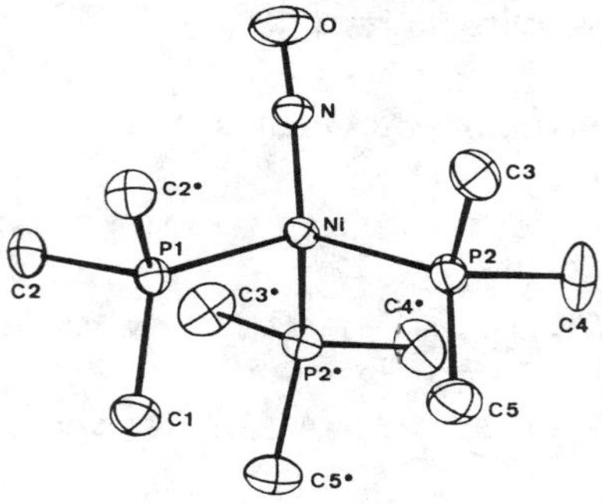

Fig. 1. Molecular structure of the cation in $C_9H_{27}F_6NNiOP_4$.

mer-HYDRIDO(TETRAHYDROBORATO-H,H')TRIS(TRIMETHYLPHOSPHINE)RUTHENIUM(II)
$C_9H_{32}BP_3Ru$ (I)

TRIS(μ_2-CHLORO)BIS[TRIS(TRIMETHYLPHOSPHINE)RUTHENIUM(II)] TETRAFLUOROBORATE
$C_{18}H_{54}BCl_3F_4P_6Ru$ (II)

BIS(cis-METHYLTETRAKIS(TRIMETHYLPHOSPHINE)RUTHENIUM)MERCURY(II) TETRAHYDROFURAN SOLVATE
$C_{26}H_{78}HgP_8Ru_2,C_4H_8O$ (III)

J.A. STATLER, G. WILKINSON, M. THORNTON-PETT and M.B. HURSTHOUSE, 1984. J. Chem. Soc. Dalton, 1731-1738.

I. Triclinic, P$\bar{1}$, a = 9.603, b = 12.566, c = 8.572 Å, α = 91.11, β = 117.43, γ = 75.30°, Z = 2. Mo radiation, R = 0.023 for 2910 reflexions.

II. Monoclinic, P2/n, a = 15.224, b = 9.706, c = 13.135 Å, β = 111.90°, Z = 2. Mo radiation, R = 0.048 for 2585 reflexions.

III. Triclinic, P$\bar{1}$, a = 14.232, b = 9.326, c = 18.728 Å, α = 90.34, β = 89.08, γ = 104.21°, Z = 2. Mo radiation, R = 0.029 for 4564 reflexions.

In I the Ru atom has an octahedral environment with the borohydride acting as a bidentate ligand (Fig. 1). The two Ru-H borohydride bonds (1.81(4), 1.85(4)) are trans to the hydride (Ru-H 1.49(4)) and one phosphine (Ru-P 2.259(3) Å). The two other phosphines are trans to one another (Ru-P 2.304(3), 2.305(3) Å). The chloro-bridged complex (II) has confacial bioctahedral geometry with Ru-P and Ru-Cl bond lengths ranging from 2.248(4) to 2.256(4) and 2.463(4) to 2.488(4) Å, respectively. In III the two Ru atoms are bridged by an Hg atom (Ru-Hg 2.798, 2.800(4) Å, Ru-Hg-Ru 170.7(2)°). The Ru-C distances are 2.21(1), 2.23(1) and the Ru-P distances range from 2.309(4) to 2.344(4) Å.

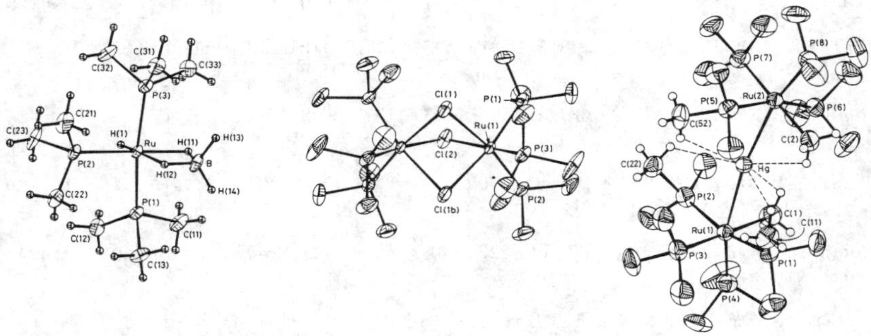

Fig. 1. The structure of $C_9H_{32}BP_3Ru$ (left), the cation in $C_{18}H_{54}BCl_3F_4P_6Ru$ (centre) and the structure of $C_{26}H_{78}HgP_8Ru_2$ (right).

HEXACARBONYL[μ-(1,2,3,4-TETRAMETHYL-1,4-TETRAPHOSPHINEDIYL-$P^1,P^4:P^1,P^4$)]DIIRON(Fe-Fe)
$C_{10}H_{12}Fe_2O_6P_4$ (I) $Fe_2(CO)_6(PCH_3)_4$

HEXACARBONYL[μ-(1,2,3,4-TETRAPHENYL-1,4-TETRAPHOSPHINEDIYL)]DIIRON(Fe-Fe)
$C_{30}H_{20}Fe_2O_6P_4$ (II) $Fe_2(CO)_6(PC_6H_5)_4$

A.L. RHEINGOLD and M.E. FOUNTAIN, 1984. Organometallics, 3, 1417-1421

I. Monoclinic, $P2_1/c$, a = 7.906, b = 17.139, c = 13.849 Å, β = 101.41°, Z = 4. Mo radiation, R = 0.064 for 1876 reflexions.

II. Monoclinic, $P2_1/c$, a = 16.988, b = 11.492, c = 16.336 Å, β = 101.09°, Z = 4. Mo radiation, R = 0.036 for 3919 reflexions.

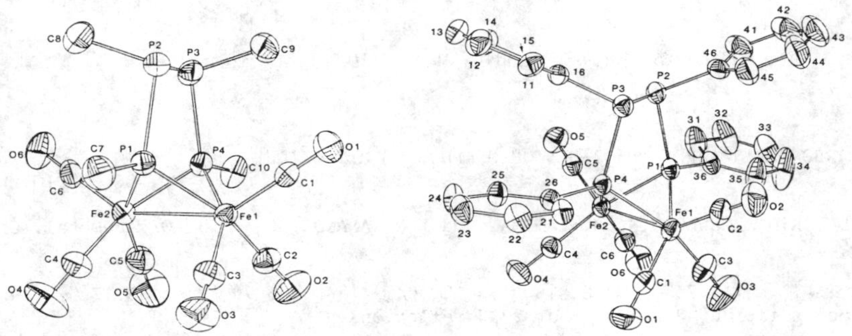

Fig. 1. Molecular structure of $C_{10}H_{12}Fe_2O_6P_4$ (left), and of $C_{30}H_{20}Fe_2O_6P_4$ (right).

Both compounds I and II possess similar distorted tetragonal-pyramidal metal atom coordination geometries joined along a basal edge formed by the mutually-bridging termini of tetraphosphorus atom catenates (Fig. 1). Both possess single Fe-Fe bonds (2.634 in I and 2.633 Å in II), and the only significant structural accommodation for the greater steric demands of phenyl vs. methyl substitution is the P(1)-P(2)-

P(3)-P(4) dihedral angle which is -10.5° in I and -23.6° in II. The average Fe-P distance is 2.219 Å in I and 2.220 Å in II and the average P-P bond distances are 2.198 Å in I and 2.208 Å in II. The individual P-P bond distances are insensitive to whether the bond is internal or terminal in the P_4 chain. The P-P-P bond angles are 96.5 and 96.3° in I and 94.6 and 94.1° in II.

TETRAKIS(METHYLTRIHYDROBORATO)(1,2-BIS(DIMETHYLPHOSPHINO)ETHANE)URANIUM(IV)
$C_{10}H_{40}B_4P_2U$ (I) $(CH_3BH_3)_4U[(CH_3)_2PCH_2CH_2P(CH_3)_2]$

TRIS(METHYLTRIHYDROBORATO)BIS(1,2-BIS(DIMETHYLPHOSPHINO)ETHANE)URANIUM(III)
$C_{15}H_{50}B_3P_4U$ (II) $(CH_3BH_3)_3U[(CH_3)_2PCH_2CH_2P(CH_3)_2]_2$

J. BRENNAN, R. SHINOMOTO, A. ZALKIN and N. EDELSTEIN, 1984. Inorg. Chem., 23, 4143-4146.

I. Monoclinic, $P2_1/c$, a = 9.439, b = 14.023, c = 16.596 Å, β = 97.48°, Z = 4. Mo radiation, R = 0.024 for 2220 reflexions.

II. Tetragonal, $P4_32_12$, a = 11.297, c = 23.030 Å, Z = 4. Mo radiation, R = 0.024 for 1395 reflexions.

The two molecules are shown in Fig. 1. The U(N) atom in I is coordinated to four $MeBH_3$ groups through tridentate H-bridges and to the two P atoms of the phosphine ligand, the UB_4P_2 arrangement being cis-octahedral. The U(IV) is 14-coordinate, mean U-P 3.02(1), mean U-B 2.57(1) Å. The U(III) atom in II is coordinated to three tridentate $MeBH_3$ ligands and by two diphos ligands, the UB_3P_4 arrangement being pentagonal bipyramidal with B atoms axial. The mean U-P distance is 3.13(6) Å, U-B(ax) 2.62(1) Å, U-B(eq) 2.66(1) Å. Complex II has C_2 symmetry.

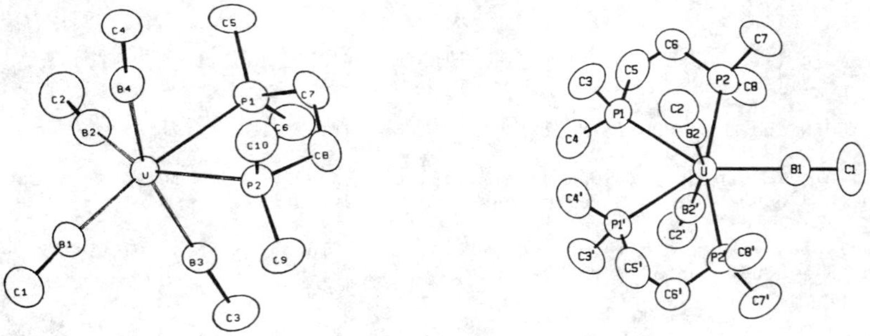

Fig. 1. The $C_{10}H_{40}B_4P_2U$ (left) and $C_{15}H_{50}B_3P_4U$ (right) molecules.

2-PENTACARBONYLMANGANESE-1,3,2-BENZODIOXAPHOSPHOLE 2-OXIDE
$C_{11}H_4MnO_8P$ $\overline{OC_6H_4OP}(O)Mn(CO)_5$

S.S.C. CHU, R.D. ROSENSTEIN, M. LATTMAN and B.N. ANAND, 1984. J. Organometal. Chem., 265, 45-52.

Triclinic, $P\bar{1}$, a = 6.753, b = 8.818, c = 12.004 Å, α = 106.79, β = 105.37, γ = 98.82°, Z = 2. Mo radiation, R = 0.073 for 2293 reflexions.

The Mn and P atoms have distorted octahedral and tetrahedral configurations, respectively (Fig. 1). The P=O bond length is normal at 1.458 Å as is the Mn-P bond length (2.307 Å). The Mn-C distances are 1.864, 1.874, 1.860, 1.861 and 1.876 Å. The phenyl ring is essentially planar but the dioxaphosphole five-membered ring is slightly puckered. The dihedral angle between the rings is 174.6°. The P=O bond has an almost eclipsed conformation with one of the equatorial carbonyls, leading to pseudo mirror symmetry.

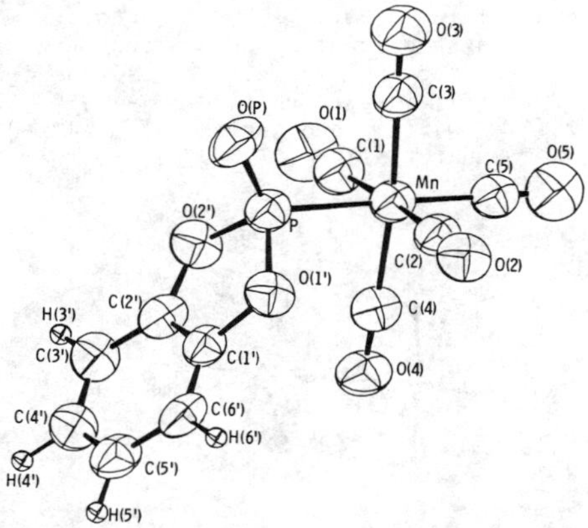

Fig. 1. Molecular structure of $C_{11}H_4MnO_8P$.

(μ_2-CARBONYL)BIS(μ_2-BIS(DIMETHYLPHOSPHINO)METHANE)DICHLORODIPALLADIUM
$C_{11}H_{28}Cl_2OP_4Pd_2$

M.L. KULLBERG and C.P. KUBIAK, 1984. Organometallics, 3, 632-634.

Orthorhombic, $P2_12_12_1$, a = 15.511, b = 11.107, c = 12.346 Å, D_m = 1.83, Z = 4. Mo radiation, R = 0.048 for 1784 reflexions.

The molecule possesses an "A-frame" structure with two Pd centres having approximately square planar geometries and sharing a common carbonyl ligand (Fig. 1). The Pd...Pd separation is 3.169 Å, suggesting a non-bonding metal-metal interaction. The bridging CO is slightly asymmetric with bond distances Pd(1)-C(1) 1.988 Å and Pd(2)-C(1) 1.948 Å, and bond angles Pd(1)-C(1)-O(1) 122.8, Pd(2)-C(1)-O(1) 130.0 and Pd(1)-C(1)-Pd(2) 107.23°. The C(1)-O(1) bond distance is 1.180 Å.

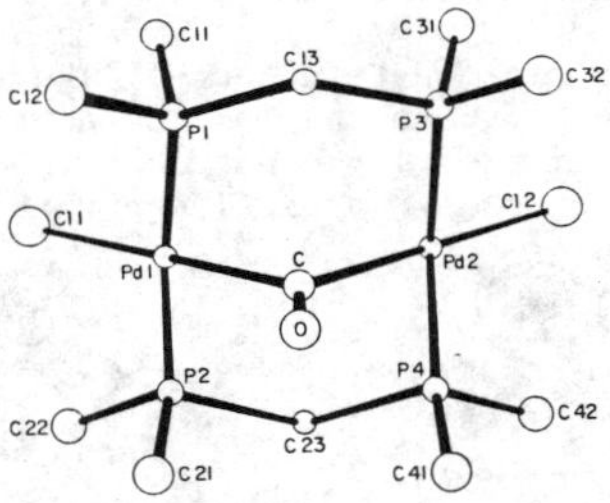

Fig. 1. Molecular structure of $C_{11}H_{28}Cl_2OP_4Pd_2$.

NONACARBONYLTRIMETHYLPHOSPHINEOSMIUMTUNGSTEN
$C_{12}H_9O_9OsPW$

F.W.B. EINSTEIN, T. JONES, R.K. POMEROY and P. RUSHMAN, 1984. J. Am. Chem. Soc., 106, 2707-2708.

Monoclinic, $P2_1/n$, a = 8.832, b = 16.241, c = 12.988 Å, β = 101.99°, Z = 4. Mo radiation, R = 0.019 for 2189 reflexions.

The 18-electron compound, $Os(CO)_4(PMe_3)$, acts as a donor ligand to the $W(CO)_5$ unit via an unbridged osmium-tungsten donor-acceptor bond (Fig. 1). The strong donor ligand trimethylphosphine is trans to this bond. The Os-W bond length is 3.0756(5) Å; other dimensions are Os-P 2.359(2), Os-C range 1.941(7)-1.954(8), W-C(1) 1.964(8), W-C(equatorial) range 2.021(7)-2.045(8), and angles (deg) W-Os-P 171.31(4); W-Os-C range 80.2(2)-89.1(2); Os-W-C(equatorial) 84.7(2)-93.9(2).

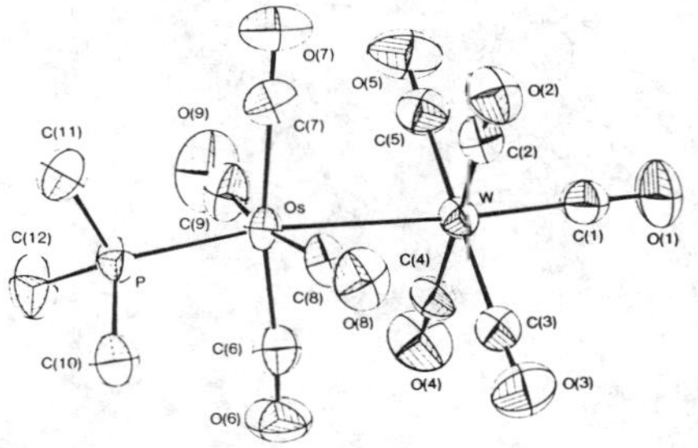

Fig. 1. A view of $(Me_3P)(OC)_4OsW(CO)_5$.

(η^3-NONAMETHYLCYCLOTETRAPHOSPHAZANIUM-P,P,P)TRICARBONYLMOLYBDENUM(0) IODIDE ACETONITRILE SOLVATE

$C_{12}H_{27}IMoN_4O_3P_4,C_2H_3N$

K.D. GALLICANO, N.L. PADDOCK, S.J. RETTIG and J. TROTTER, 1984. Can. J. Chem., 62, 1869-1873.

Orthorhombic, Pcab, a = 13.1938, b = 13.7847, c = 29.281 Å, Z = 8. Mo radiation, R = 0.030 for 2388 reflexions.

The structure consists of discrete cations (Fig. 1), anions, and solvent molecules. The complex cation has approximate mirror symmetry in the solid state, the Mo atom being coordinated to three phosphazane P atoms and three carbonyl ligands in a distorted octahedral arrangement. Important bond lengths (corrected for libration) are: Mo-P 2.483(2), 2.439(2), and 2.474(2), mean Mo-C 1.984(3), mean P-N 1.710(11) for coordinated P and 1.669(5) Å for quaternized P. The stronger bonding of the cationic ligand to the metal relative to that of the neutral species $(CH_3NPCH_3)_4$-$Mo(CO)_3$, and the strengthening of the ring bonds in both molecules relative to $(CH_3NPCH_3)_4$, are discussed in terms of d-orbital contraction.

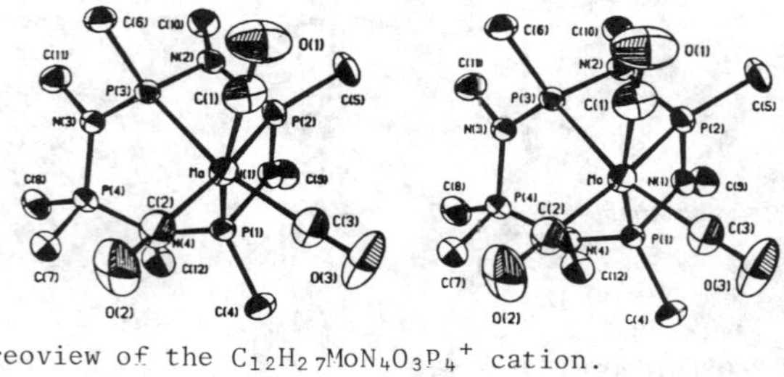

Fig. 1. A stereoview of the $C_{12}H_{27}MoN_4O_3P_4^+$ cation.

DIIODOBIS(TRIETHYLPHOSPHINE)MANGANESE(II)

$C_{12}H_{30}I_2MnP_2$ $MnI_2[P(C_2H_5)_3]_2$

N. HEBENDANZ, F.H. KÖHLER and G. MÜLLER, 1984. Inorg. Chem., 23, 3043-3044.

Orthorhombic, Pccn, a = 17.326, b = 17.302, c = 14.702 Å, D_m = 1.643, Z = 8. Mo radiation, R = 0.050 for 1866 reflexions.

The structure consists of two independent molecules (Fig. 1), each having exact C_2 symmetry and distorted tetrahedral coordination geometry. Structural parameters: Mn-I 2.662(2) and 2.670(2), Mn-P 2.528(4) and 2.539(4) Å, bond angles at Mn 105.1(1) to 120.1(1)°.

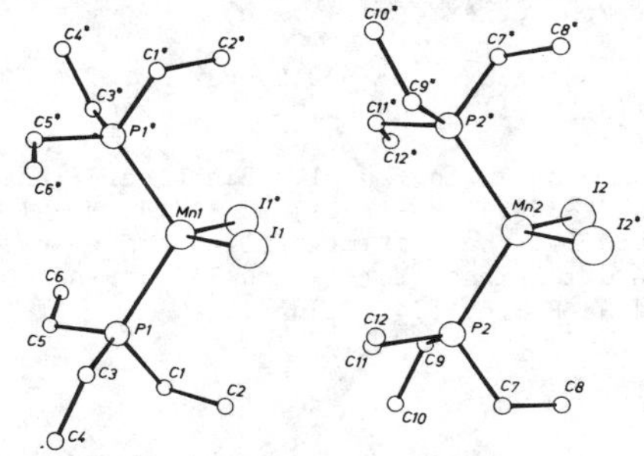

Fig. 1. The two independent $MnI_2[P(C_2H_5)_3]_2$ molecules.

TETRAKIS(μ-CHLORO)TETRACHLOROTETRAKIS(TRIMETHYLPHOSPHINE)DINIOBIUM(IV)
$C_{12}H_{36}Cl_8Nb_2P_4$ (I) $Nb_2Cl_8[P(CH_3)_3]_4$

trans-TETRACHLOROBIS(TRIETHYLPHOSPHINE)NIOBIUM(IV)
$C_{12}H_{30}Cl_4NbP_2$ (II) $NbCl_4[P(C_2H_5)_3]_2$

trans-TETRACHLOROBIS(ETHYLDIPHENYLPHOSPHINE)NIOBIUM(IV)
$C_{28}H_{30}Cl_4NbP_2$ (III) $NbCl_4[P(C_2H_5)(C_6H_5)_2]_2$

F.A. COTTON, S.A. DURAJ and W.J. ROTH, 1984. Inorg. Chem., 23, 3592-3596.

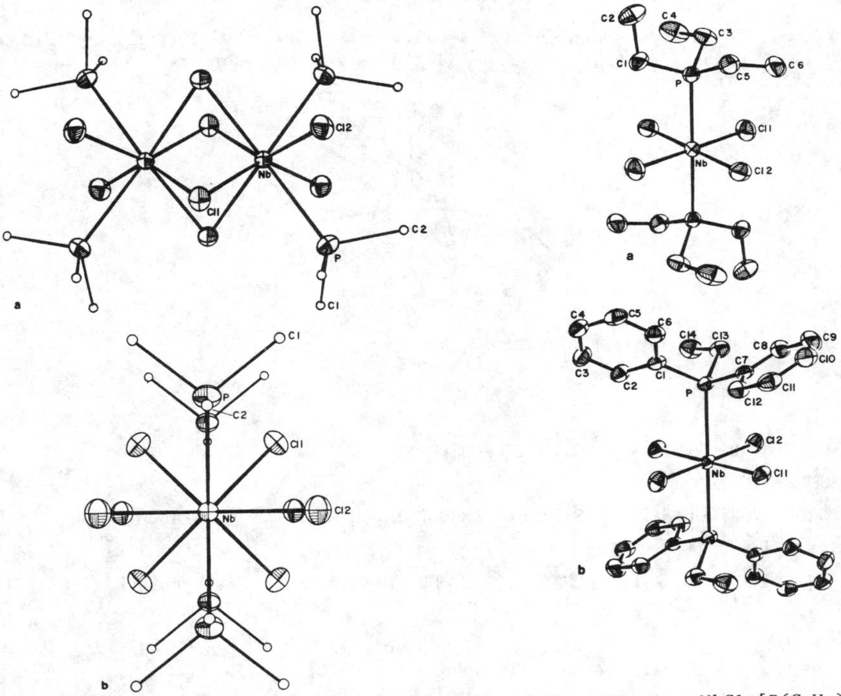

Fig. 1. Two views of $Nb_2Cl_8[P(CH_3)_3]_4$ (left), and the structure $NbCl_4[P(C_2H_5)_3]_2$ (top right) and $NbCl_4[P(C_2H_5)(C_6H_5)_2]_2$ (bottom right).

I. Cubic, Im3, a = 16.372 Å, Z = 6. Mo radiation, R = 0.022 for 557 reflexions.

II. Monoclinic, $P2_1/n$, a = 7.224, b = 12.131, c = 12.017 Å, β = 90.74°, Z = 2. Mo radiation, R = 0.0465 for 702 reflexions.

III. Monoclinic, $P2_1/n$, a = 9.851, b = 16.010, c = 9.367 Å, β = 98.67°, Z = 2. Cu radiation, R = 0.035 for 1819 reflexions.

The three molecules are shown in Fig. 1. Binuclear I has crystallographic mmm (D_{2h}) symmetry with Nb-Nb 2.833(1), Nb-(μ-Cl) 2.541(1), Nb-Cl 2.502(1), and Nb-P 2.671(1) Å. In both II and III there is crystallographic inversion symmetry imposed on the molecule, and mean dimensions (for II and III in that order) are Nb-Cl 2.366(9), 2.367(34) Å and Nb-P 2.664(3), 2.706(1) Å.

trans-DIOXOTETRAKIS(TRIMETHYLPHOSPHINE)RHENIUM(V) PERRHENATE
$C_{12}H_{36}O_6P_4Re_2$ (I) $[ReO_2(PMe_3)_4]ReO_4$

trans-DIOXOTETRAKIS(TRIMETHYLPHOSPHINE)RHENIUM(V) TETRAPHENYLBORATE
$C_{36}H_{56}BO_2P_4Re$ (II) $[ReO_2(PMe_3)_4]BPh_4$

P.G. EDWARDS, A.C. SKAPSKI, A.M.Z. SLAWIN and G. WILKINSON, 1984. Polyhedron, 3, 1083-1085.

I. Monoclinic, Pc, a = 9.941, b = 11.092, c = 11.263 Å, β = 90.55°, Z = 2. Cu radiation, R = 0.098 for 972 reflexions.

II. Monoclinic, I2/a, a = 19.219, b = 19.619, c = 21.376 Å, β = 97.19°, Z = 8. Cu radiation, R = 0.061 for 3908 reflexions.

The cation in I is shown in Fig. 1, with mean Re=O 1.79(2), mean Re-P 2.485(8) Å, O-Re-O 177°, trans P-Re-P 162(1)°. In II the cation is disordered, with three orientations for the O=Re=O axis.

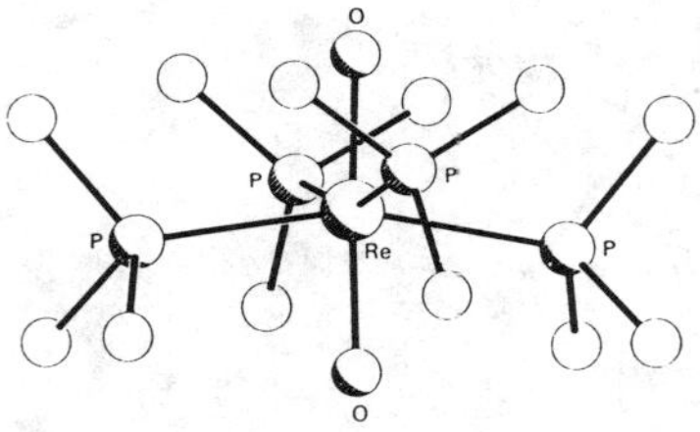

Fig. 1. The $[ReO_2(PMe_3)_4]^+$ cation.

DICHLORODIHYDRIDOTETRAKIS(TRIMETHYLPHOSPHINE)TANTALUM(IV)
$C_{12}H_{38}Cl_2P_4Ta$ (I) $TaCl_2H_2[P(CH_3)_3]_4$

DICHLORODIHYDRIDOBIS(1,2-BIS(DIMETHYLPHOSPHINO)ETHANE-P,P')TANTALUM(IV)
$C_{12}H_{34}Cl_2P_4Ta$ (II) $TaCl_2H_2[(CH_3)_2PCH_2CH_2P(CH_3)_2]_2$

M.L. LUETKENS, Jr., W.L. ELCESSER, J.C. HUFFMAN and A.P. SATTELBERGER, 1984. Inorg. Chem., 23, 1718-1726.

I. Monoclinic, Cc, a = 15.127, b = 12.005, c = 12.140 Å, β = 92.49°, Z = 4. Mo radiation, R = 0.029 for 2418 reflexions (at -160°C).

II. Orthorhombic, $P2_12_12_1$, a = 9.781, b = 15.842, c = 13.832 Å, Z = 4. Mo radiation, R = 0.036 for 2000 reflexions (at -160°C).

The two molecules are shown in Fig. 1. Compound I adopts a distorted dodecahedral geometry while II is more appropriately described as a distorted square antiprismatic complex. Bond lengths: for I; Ta-Cl 2.542(5) and 2.563(5), Ta-P 2.559(5)-2.633(6), Ta-H 1.94(11) and 1.85(11) Å; for II; Ta-Cl 2.546(3) and 2.557(3), Ta-P 2.512(3)-2.577(3), Ta-H 1.94(10) and 1.71(10) Å.

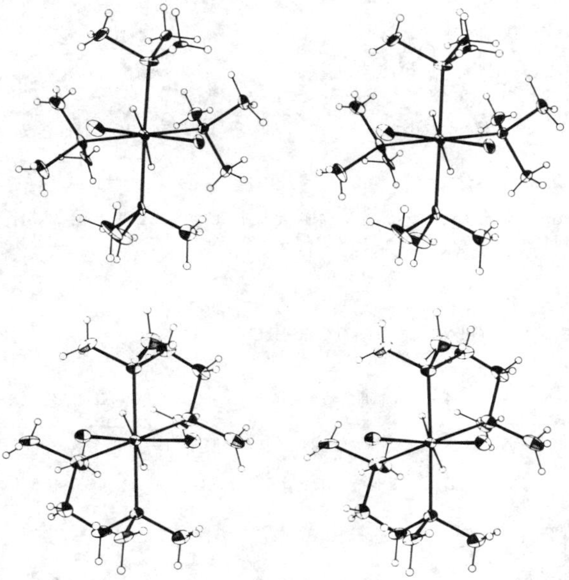

Fig. 1. Stereoviews of $TaCl_2H_2[P(CH_3)_3]_4$ (top) and $TaCl_2H_2[(CH_3)_2PCH_2CH_2P(CH_3)_2]_2$ (bottom).

trans-CARBONYLDICHLORO(DICHLOROPHOSPHIDO)BIS(TRIETHYLPHOSPHINE)IRIDIUM(III)
$C_{13}H_{30}Cl_4IrOP_3$ (I)

trans-CARBONYLDICHLORO(DICHLOROPHOSPHINE OXIDE)BIS(TRIETHYLPHOSPHINE)IRIDIUM(III)
$C_{13}H_{30}Cl_4IrO_2P_3$ (II)

E.A.V. EBSWORTH, R.O. GOULD, N.T. McMANUS, N.J. PILKINGTON and D.W.H. RANKIN, 1984. J. Chem. Soc. Dalton, 2561-2567.

I. Orthorhombic, $P2_12_12_1$, a = 9.720, b = 17.975, c = 25.881 Å, Z = 8. Mo radiation, R = 0.039 for 2460 reflexions.

II. Orthorhombic, $P2_12_12_1$, a = 9.784, b = 17.918, c = 25.980 Å, Z = 8. Mo radiation, R = 0.044 for 3342 reflexions.

I and II are isomorphous and contain two molecules per asymmetric unit. The first molecule lies near 0.75, 0.75, 0.25 and symmetry related positions in the unit cell and forms an approximately face centred array of Ir atoms. It is well-ordered in both structures (Fig. 1). The other molecule also gives an approximate face-centred array of Ir atoms near 0.75, 0.5, 0.5. It is significantly disordered. The two molecules do not differ significantly in either compound. Allowing for disorder the main differences are minor changes in the conformations of the ethyl groups. Bond lengths for I, II include: Ir-P 2.389(6), 2.418(6) and 2.399(6), 2.422(5); Ir-P' 2.324(7), 2.290(5); Ir-Cl(P) 2.431(6), 2.412(6); Ir-Cl(C) 2.372(5), 2.368(5); P'-Cl 2.09(1), 2.051(8) and 2.134(9), 2.061(8); and in II, P'-O 1.503(14) Å.

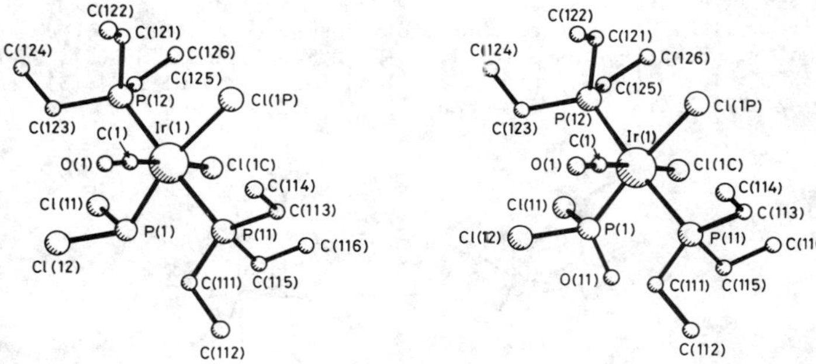

Fig. 1. The ordered molecules in $C_{13}H_{30}Cl_4IrOP_3$ (left) and $C_{13}H_{30}Cl_4IrO_2P_3$ (right).

(μ_2-t-BUTYLARSINIDENE)BIS(PENTACARBONYLCHROMIUM)
$C_{14}H_9AsCr_2O_{10}$ (I) $[Cr(CO)_5]_2(\mu\text{-t-}C_4H_9As)$

1,2,3,4-TETRAKIS(t-BUTYL)-1,4-DICHLORO-5,5,5,5-TETRACARBONYL-1,2,3,4-TETRA-ARSA-5-CHROMACYCLOPENTANE
$C_{20}H_{36}As_4Cl_2CrO_4$ (II) $(CO)_4\overline{CrAs\text{-t-}C_4H_9Cl(As\text{-t-Bu})_2As}\text{-t-BuCl}$

R.A. JONES and B.R. WHITTLESEY, 1984. Organometallics, 3, 469-475.

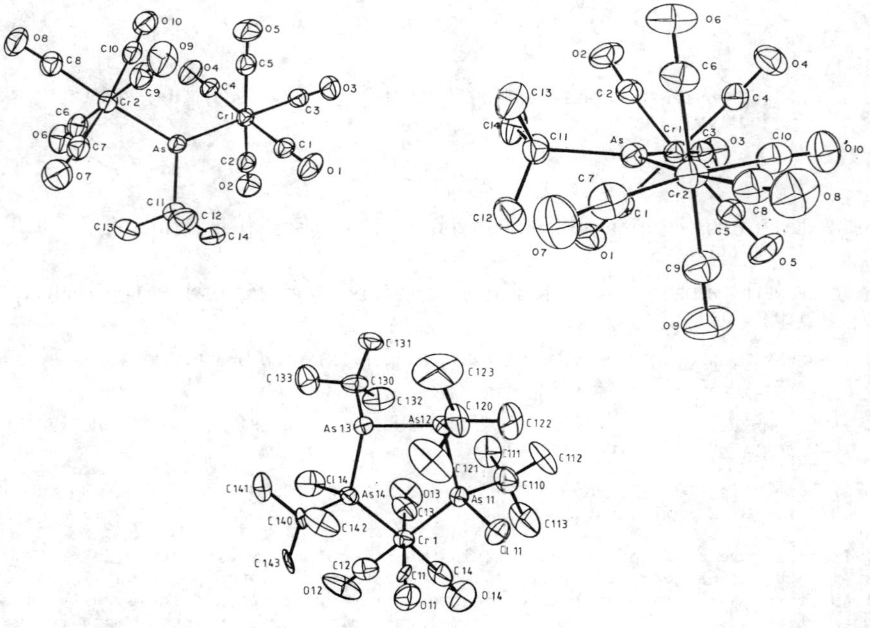

Fig. 1. Two views of the molecule $C_{14}H_9AsCr_2O_{10}$ (top left, right), and a general view of the molecule $C_{20}H_{36}As_4Cl_2CrO_4$.

I. Triclinic, $P\bar{1}$, a = 9.379, b = 9.483, c = 11.778 Å, α = 87.86, β = 70.67, γ = 77.77°, Z = 2. Mo radiation, R = 0.047 for 2804 reflexions.

II. Monoclinic, $P2_1/n$, a = 11.030, b = 35.190, c = 15.303 Å, β = 97.82°, Z = 8. Mo radiation, R = 0.054 for 4100 reflexions.

In I, coordination about the As atom is virtually planar (sum of angles about As is 359.84°), and the distances for As-Cr(1), As-Cr(2) and As-C(11) are 2.387, 2.403 and 2.027 Å respectively (Fig. 1). The equatorial CO units of Cr(1) are twisted relative to the Cr_2As plane (Cr(1)-C(2)-O(2) is inclined 46° relative to the Cr(1)-Cr(2)-As-C(11) plane while C(7)-O(7) and C(10)-O(10) are more closely aligned to this plane (8°). The structural parameters for the two independent molecules of II are very similar, and each molecule has an idealized (not crystallographic) twofold axis through Cr and bisecting the bond As(12)-As(13). The $CrAs_4$ ring is virtually planar and its As-As distances are: As(11)-As(12) 2.478, As(12)-As(13) 2.427, and As(13)-As(14) 2.479 Å. The Cr-As distances are 2.418 and 2.422 Å.

PENTACARBONYL(TRIISOPROPYLCYCLOTRIPHOSPHAN)TUNGSTEN(0)
$C_{14}H_{21}O_5P_3W$ $W(CO)_5(C_9H_{21}P_3)$

K.-F. TEBBE, 1984. Acta Cryst., C40, 1552-1555.

Monoclinic, $P2_1/n$, a = 12.535, b = 11.645, c = 14.631 Å, β = 100.75°, Z = 4. Mo radiation, R = 0.045 for 2996 reflexions.

The three-membered ring (Fig. 1) is a unidentate, two-electron donor with W-P 2.535(2) Å. Further bond lengths are P(1)-P(2) 2.181(4), P(2)-P(3) 2.207(4), P(1)-P(3) 2.188(3), W-C(5) 1.97(1), W-C (equatorial) (mean) 2.04(1) Å.

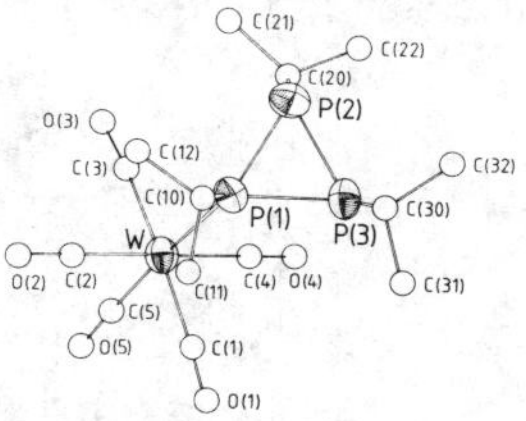

Fig. 1. The $W(CO)_5(C_9H_{21}P_3)$ molecule.

trans-BIS[1,2-BIS(DIMETHYLPHOSPHINO)ETHANE](HYDROXY)OXOTECHNETIUM(V) BIS(TRIFLUOROMETHANESULPHONATE)
$C_{14}H_{33}F_6O_8P_4S_2Tc$ (I) trans-$[Tc(Me_2PCH_2CH_2PMe_2)_2(OH)(O)](F_3CSO_3)_2$

trans-DICHLOROBIS[1,2-BIS(DIMETHYLPHOSPHINO)ETHANE]TECHNETIUM(III) TRIFLUOROMETHANESULPHONATE
$C_{13}H_{32}Cl_2F_3O_3P_4STc$ (II) trans-$[Tc(Me_2PCH_2CH_2PMe_2)_2Cl_2](F_3CSO_3)$

J.-L. VANDERHAYDEN, A.R. KETRING, K. LIBSON, M.J. HEEG, L. ROECKER, P. MOTZ, R. WHITTLE, R.C. ELDER and E. DEUTSCH, 1984. Inorg. Chem., 23, 3184-3191.

I. Monoclinic, $P2_1/c$, a = 8.052, b = 11.527, c = 16.070 Å, β = 101.96°, D_m = 1.66, Z = 2. Mo radiation, R = 0.051 for 2364 reflexions.

II. Monoclinic, $P2_1/c$, a = 8.076, b = 24.401, c = 13.435 Å, β = 96.61°, D_m = 1.56, Z = 4. Mo radiation, R = 0.056 for 3331 reflexions.

The Tc(V) atom in I is approximately octahedrally coordinated with four equatorial P atoms at an average distance of 2.477(5) Å. The O and OH ligands are disordered about a centre of symmetry, the mean Tc-O distance being 1.795(3) Å. The Tc(III) centre in II is trans octahedral with four equatorial P atoms at an average distance of 2.436(5) Å and two trans Cl atoms at 2.324(4) Å. Both cations are shown in Fig. 1.

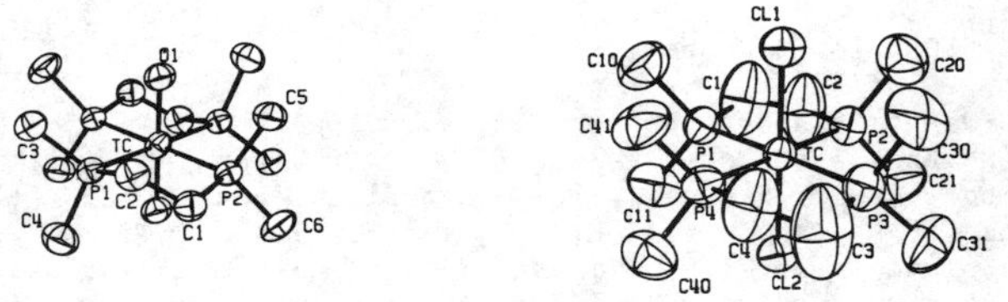

Fig. 1. The cations trans-$[Tc(Me_2PCH_2CH_2PMe_2)_2(OH)(O)]^{2+}$ (left) and trans-$[Tc(Me_2PCH_2CH_2PMe_2)_2Cl_2]^+$ (right).

IODO(METHYLENE)TETRAKIS(TRIMETHYLPHOSPHINE)TUNGSTEN TRIFLUOROMETHANESULPHONATE

$C_{14}H_{38}F_3IO_3P_4SW$ $[WI(CH_2)\{P(CH_3)_3\}_4][CF_3SO_3]$

A.J. SCHULTZ, J.M. WILLIAMS, R.R. SCHROCK and S.J. HOLMES, 1984. Acta Cryst., C40, 590-592.

Rhombohedral, R$\bar{3}$c, a = 16.730, c = 50.380 Å, Z = 18. Mo radiation, R = 0.064 for 2041 reflexions.

In the molecule (Fig. 1) tungsten has pseudo-octahedral coordination with the methylene and iodo ligands in a trans configuration. The W-C(1) distance of 1.83(2) Å indicates a W-C bond order of between 2 and 3. Other bond distances are W-I 2.881(2) and W-P 2.519(4) and 2.507(4) Å.

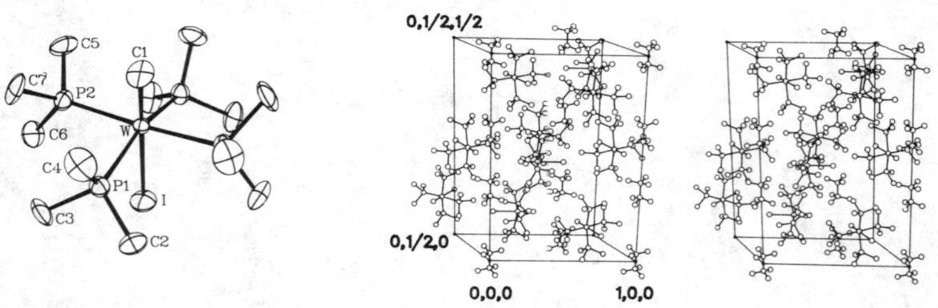

Fig. 1. $[WI(CH_2)\{P(CH_3)_3\}_4][CF_3SO_3]$: perspective view of the $[WI(CH_2)\{P(CH_3)_3\}_4]^+$ cation (left) and a stereoscopic view of the crystal packing in a portion of the unit cell (right).

BIS[(2-AMINOETHYL)DIMETHYLPHOSPHINE][1,2-BIS(DIMETHYLPHOSPHINOETHANE]COBALT(III) TRIBROMIDE DIHYDRATE

$C_{14}H_{40}Br_3CoN_2P_2,2H_2O$ $[Co\{NH_2CH_2CH_2P(CH_3)_2\}_2\{(CH_3)_2PCH_2CH_2P(CH_3)_2\}]^{3+}Br_3{}^{-}2H_2O$

T. OHISHI , S. OHBA, K. KASHIWABARA, Y. SAITO and J. FUJITA, 1984. Bull. Chem. Soc. Japan, 57, 877-878.

Monoclinic, $P2_1/a$, a = 16.380, b = 19.845, c = 9.538 Å, β = 122.34°, D_m = 1.76, Z = 4. Mo radiation, R = 0.071 for 3299 reflexions.

The complex ion is the trans (P,N) isomer, where one of the N atoms of (edmp) occupies the position trans to the P atom of the other (edmp) ligand (Fig. 1). The five-membered chelate ring formed by the ligand assumes an envelope conformation. The main bond distances are: Co-P(1) 2.294, Co-P(2) 2.245, Co-P(3) 2.273, Co-P(4) 2.295, Co-N(1) 2.047 and Co-N(2) 2.058 Å with P-C distances ranging from 1.801 to 1.843 Å and N(1)-C(2) 1.503 and N(2)-C(4) 1.476 Å. The cis angles about the metal range from 82.7 to 97.1°.

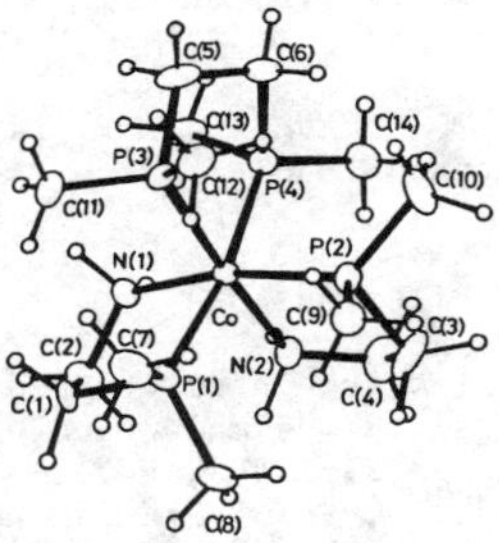

Fig. 1. Structure of the complex cation in $C_{14}H_{40}Br_3CoN_2P_2$.

DIHYDRIDOPENTAKIS(TRIMETHYLPHOSPHINE)MOLYBDENUM
$C_{15}H_{47}MoP_5$ (I)

(FORMATO-O,O')HYDRIDOTETRAKIS(TRIMETHYLPHOSPHINE)MOLYBDENUM(II)
$C_{13}H_{38}MoO_2P_4$ (II)

D. LYONS, G. WILKINSON, M. THORNTON-PETT and M.B. HURSTHOUSE, 1984. J. Chem. Soc. Dalton, 695-700.

I. Triclinic, $P\bar{1}$, a = 18.020, b = 16.767, c = 9.650 Å, α = 105.29, β = 93.25, γ = 110.97°, Z = 4. Mo radiation, R = 0.025 for 5268 reflexions.

II. Orthorhombic, Cmcm, a = 13.096, b = 12.455, c = 13.845 Å, Z = 4. Mo radiation, R = 0.027 for 922 reflexions.

The dihydride (I) contains two crystallographically independent molecules in the unit cell. The structures of these are identical within experimental error and have pentagonal bipyramidal coordination geometry (Fig. 1) with the hydride atoms in the equatorial plane along with the P atoms of three phosphines. The H atoms are almost trans to two of the phosphines (Mo-P 2.460(3)-2.478(3) Å). The unique equatorial Mo-P length is 2.403(3), 2.407(3) Å whilst the axial Mo-P lengths are 2.424(3)-2.428(3) Å. Mo-H distances range from 1.67(3)-1.76(4) Å. Molecules of the formate (II) are also pentagonal bipyramidal with mm (C_{2v}) molecular symmetry. The hydride and symmetrically bidentate formate groups are mutually trans. The Mo-P(axial) distance is 2.426(3), Mo-P(equatorial) 2.369(3), Mo-H 1.68(3) and Mo-O 2.318(4) Å.

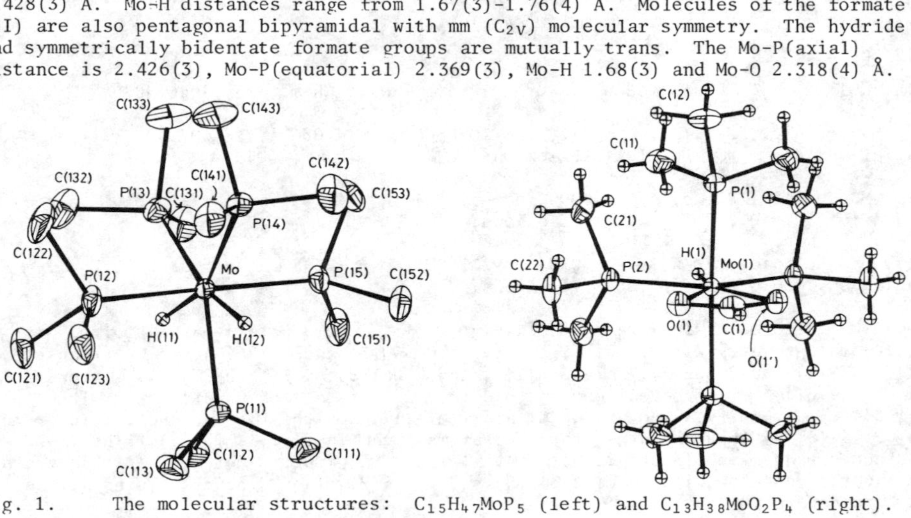

Fig. 1. The molecular structures: $C_{15}H_{47}MoP_5$ (left) and $C_{13}H_{38}MoO_2P_4$ (right).

(μ_3-SULPHIDO)(μ_3-p-TOLYPHOSPHIDO)TRIS(TRICARBONYLIRON)(2Fe-Fe)
$C_{16}H_7Fe_3O_9PS$ $SFe_3(CO)_9(PC_6H_4CH_3)$

J.P. FACKLER, Jr., A.M. MAZANY, D. SEYFERTH, H.P. WITHERS, Jr., T.G. WOOD and C.F. CAMPANA, 1984. Inorg. Chim. Acta, 82, 31-33.

Monoclinic, $P2_1/n$, a = 8.558, b = 9.022, c = 27.506 Å, β = 95.40°, Z = 4. Mo radiation, R = 0.033 for 4659 reflexions.

The complex consists of three iron atoms capped with a triply-bridging S atom and a μ_3-p-$CH_3C_6H_4P$ group (Fig. 1). The main bond distances are: Fe(1)-Fe(3) 2.636, Fe(2)-Fe(3) 2.650, Fe(1)-Fe(2) 3.495, Fe(1)-S 2.272, Fe(2)-S 2.266, Fe(3)-S 2.250, Fe(1)-P 2.177, Fe(2)-P 2.179, Fe(3)-P 2.244 and S...P 2.719 Å.

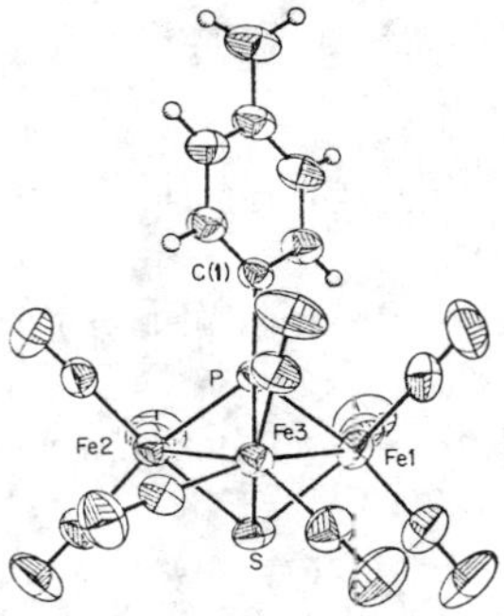

Fig. 1. Molecular structure of $C_{16}H_7Fe_3O_9PS$.

cis-BIS(DIMETHYLPHENYLPHOSPHINE)TETRACHLOROTANTALUM(IV)
$C_{16}H_{22}Cl_4P_2Ta$ (I) $TaCl_4(PMe_2Ph)_2$

TRIS(TRIMETHYLPHOSPHINE)TETRACHLOROTANTALUM(IV)
$C_9H_{27}Cl_4P_3Ta$ (II) $TaCl_4(PMe_3)_3$

trans-BIS(TRIETHYLPHOSPHINE)TETRACHLOROTANTALUM(IV)
$C_{12}H_{30}Cl_4P_2Ta$ (III) $TaCl_4(PEt_3)_2$

F.A. COTTON, S.A. DURAJ and W.J. ROTH, 1984. Inorg. Chem., 23, 4046-4050.

I. Monoclinic, $P2_1/c$, a = 15.466, b = 10.080, c = 14.086 Å, β = 97.32°, Z = 4. Mo radiation, R = 0.031 for 1914 reflexions.

II. Monoclinic, $P2_1/c$, a = 15.000, b = 11.714, c = 11.606 Å, β = 91.56°, Z = 4. Mo radiation, R = 0.037 for 2425 reflexions.

III. Monoclinic, $P2_1/n$, a = 8.032, b = 12.851, c = 10.357 Å, β = 98.55°, Z = 2. Mo radiation, R = 0.030 for 1266 reflexions.

The three molecules are shown in Fig. 1. Six-coordinate I has distorted cis-octahedral geometry with Ta-Cl (trans to P) 2.360(3) and 2.386(3), Ta-Cl (trans to Cl) 2.319(3) and 2.322(3), Ta-P 2.641(3) and 2.651(3) Å, P-Ta-P 97.7(1)°, trans Cl-Ta-Cl 159.6(1)°. The seven-coordinate molecule II is a capped octahedron with virtual C_{3v} symmetry. The unique Cl atom (Cl(1)) caps the P_3 face and is closer to Ta (2.417(3) Å) than the other three Cl atoms (mean Ta-Cl 2.447(8) Å). The Ta-P distances range 2.633(3) to 2.650(3) Å. The centrosymmetric molecule III has trans-octahedral geometry with Ta-Cl 2.347(2) and 2.373(2), Ta-P 2.636(2) Å, and all angles at Ta close to 90°.

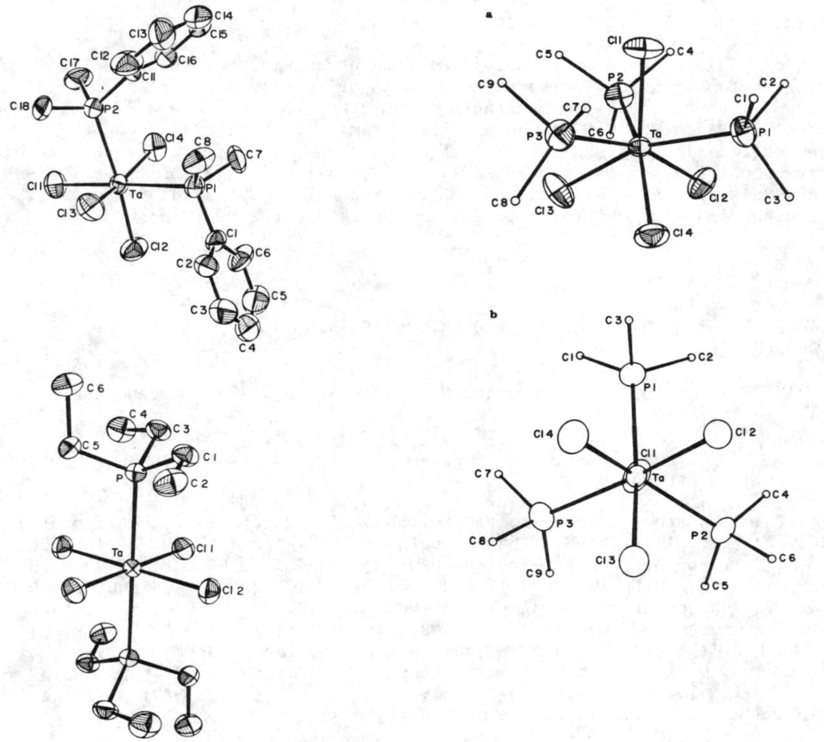

Fig. 1. The structures of $TaCl_4(PMe_2Ph)_2$ (top left), $TaCl_4(PMe_3)_3$ (right), and $TaCl_4(PEt_3)_2$ (bottom left).

BIS(μ_2-BROMO)HEXACARBONYL(μ_2-1,2-DI-t-BUTYL-3,4-DIMETHYL-1,2-DIAZA-3,4-DIPHOSPHETIDINE-P,P')DIRHENIUM

$C_{16}H_{24}Br_2N_2O_6P_2Re_2$ $Re_2(\mu\text{-}Br)_2(CO)_6(\mu\text{-}cis\text{-}CH_3\overline{P(NCMe_3)PMeNCMe_3})$

O.J. SCHERER, R. ANSELMANN and W.S. SHELDRICK, 1984. J. Organometal. Chem., 263, C26-C29.

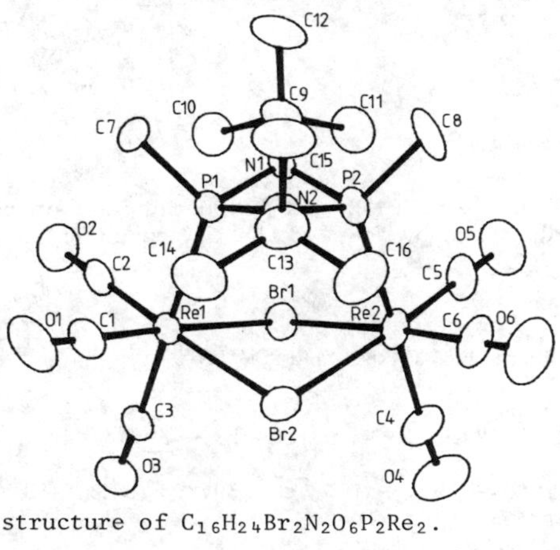

Fig. 1. Molecular structure of $C_{16}H_{24}Br_2N_2O_6P_2Re_2$.

Monoclinic, $P2_1/n$, a = 14.137, b = 18.147, c = 10.403 Å, β = 100.93°, Z = 4. Mo radiation, R = 0.037 for 3629 reflexions.

The complex consists of a "double decker" arrangement of atoms, a four-membered P_2N_2 ring on top of a four-membered Re_2Br_2 ring (Fig. 1). Each Re atom is six-coordinate, with Re-Br distances ranging from 2.662 to 2.676 Å and Re-P distances of 2.431 and 2.439 Å. The P-N distances range from 1.684 to 1.694 Å, and the normals to planes P(1),N(1),N(2) and P(2),N(1),N(2) make an angle of 32.7° with each other. The corresponding angle for planes Re(1),Br(1),Er(2) and Re(2),Br(1),Br(2) is 35.7°. The Re-Br-Re angles are 95.47 and 95.53° and the Re(1)...Re(2) distance is 3.952 Å. The corresponding P-N-P angles are 92.9 and 93.2° and the P...P distance is 2.449 Å.

PENTACARBONYLBIS(2-HYDROXYPHENOXY)PHOSPHINYLMANGANESE
$C_{17}H_{10}MnO_{10}P$

M. LATTMAN, B.N. ANAND, S.S.C. CHU and R.D. ROSENSTEIN, 1984. Organometallics, 3, 670-674.

Tetragonal, $I4_1/a$, a = 25.145, c = 12.025 Å, Z = 16. Cu radiation, R = 0.046 for 1512 reflexions (at 165K).

The P and Mn atoms have distorted tetrahedral and octahedral configurations respectively (Fig. 1). The Mn-P bond length is 2.297 Å and the Mn-C distances range from 1.836 to 1.869 Å. The cis-angles about Mn range from 85.6 to 93.2°. The distance for bond P-O(P) is 1.487 Å, that for P-O'(P) 1.606 Å, and that for P-O''(P) 1.617 Å. The angles around P range from 100.9 to 117.4°. Other bond distances and angles of interest are: C(6')-O'(P) 1.398, C(6'')-O''(P) 1.405 Å, P-O'(P)-C(6') 124.7, and P-O''(P)-C(6'') 123.6°. There are two hydrogen bonds, one intramolecular and one intermolecular, in the asymmetric unit. The latter leads to a chain arrangement of molecules related by the 4_1 axis. The O(1'')...O(P) distance is 2.831 Å, and the O(P)...O(1') (intermolecular) distance 2.650 Å.

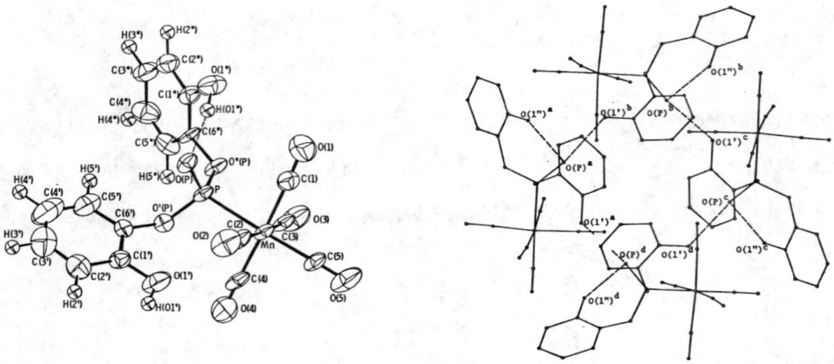

Fig. 1. Molecular structure of $C_{17}H_{10}MnO_{10}P$ (left), and intra- and intermolecular hydrogen bonds shown by four of the molecules (right).

TRI-t-BUTYLCYCLOTRIPHOSPHANEPENTACARBONYLCHROMIUM(0)
$C_{17}H_{27}CrO_5P_3$ $[(t\text{-}BuP)_3Cr(CO)_5]$

K.-F. TEBBE and M. FEHER, 1984. Z. Naturforsch., B39, 37-42.

Monoclinic, $P2_1/c$, a = 16.649, b = 9.165, c = 30.898 Å, β = 91.45°, Z = 8. Mo radiation, R = 0.048 for 3657 reflexions.

There are two independent molecules in the asymmetric unit; they have nearly identical geometry. A view of one molecule is in Fig. 1. Mean bond lengths are: cis-P-P 2.221(2), trans-P-P 2.190(4), P-Cr 2.443(8), P-C 1.895(8), Cr-C(ax) 1.847(8), Cr-C(eq) 1.896(8) Å.

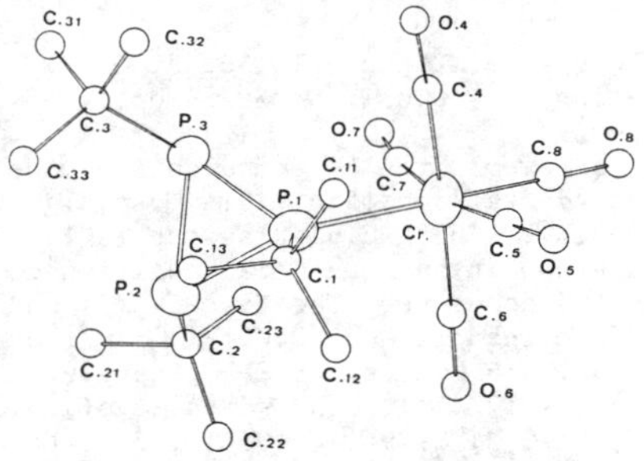

Fig. 1. A view of one molecule of $C_{17}H_{27}CrO_5P_3$.

(μ_2-CARBONYL)BIS(μ_2-N-ETHYLTETRAMETHOXYDIPHOSPHAZENE-P,P')TETRACARBONYLDIRUTHENIUM
$C_{17}H_{34}N_2O_{13}P_4Ru_2$

G. DE LEEUW, J.S. FIELD, R.J. HAINES, B. McCULLOCH, E. MEINTJIES, C. MONBERG, G.M. OLIVIER, P. RAMDIAL, C.N. SAMPSON, B. SIGWARTH, N.D. STEEN and K.G. MOODLEY, 1984. J. Organometal. Chem., 275, 99-111.

Monoclinic, $P2_1/n$, a = 14.060, b = 18.088, c = 12.305 Å, β = 90.96°, Z = 4. Mo radiation, R = 0.058 for 2384 reflexions.

The two metal atoms, separated by 2.801 Å, are bridged by two diphosphazane ligands and a carbonyl group (Fig. 1). The five carbonyl groups are almost coplanar and the Ru→P vectors are essentially orthogonal to this plane. The Ru-P distances are 2.291, 2.290, 2.302 and 2.295 Å, and the P-N distances are 1.66, 1.69, 1.67 and 1.66 Å. The Ru(1)-C(5) and Ru(2)-C(5) distances are 2.06 and 2.08 Å respectively and the Ru(1)-C(5)-Ru(2) angle is 85.1°. The P-O distances range from 1.60 to 1.63 Å.

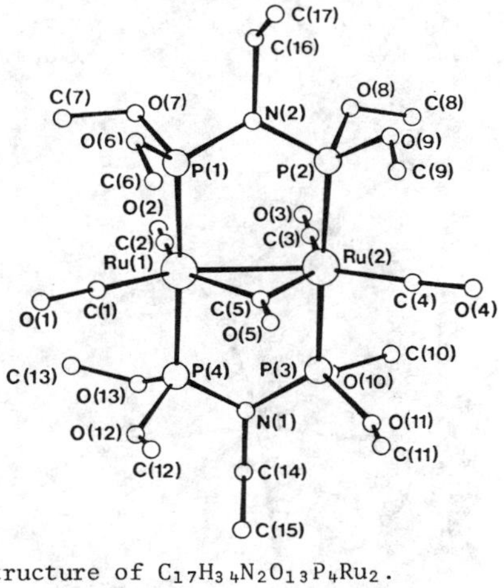

Fig. 1. Molecular structure of $C_{17}H_{34}N_2O_{13}P_4Ru_2$.

(BIS(TRIMETHYLSILYL)AMINOBIS(TRIMETHYLSILYL)METHYLENEPHOSPHINE-P)TETRACARBONYLIRON
$C_{17}H_{36}FeNO_4PSi_4$ $(Me_3Si)_2C=P[Fe(CO)_4]N(SiMe_3)_2$

R.H. NEILSON, R.J. THOMA, I. VICKOVIC and W.H. WATSON, 1984. Organometallics, 3, 1132-1133.

Triclinic, $P\bar{1}$, a = 10.567, b = 15.129, c = 10.562 Å, α = 97.44, β = 115.50, γ = 106.08°, Z = 2. Cu radiation, R = 0.051 for 3036 reflexions.

The structure analysis reveals the η^1-coordination mode of the (methylene)-phosphine and some unexpected features (Fig. 1). Most significant is the severe twist about the P=C(5) bond of 30.3°. The P=C(5) distance is 1.657 Å. Also, the phosphine ligand has unusual coordination in an equatorial position of the slightly distorted trigonal-bipyramidal geometry about Fe. The Fe, P, C(2) and C(4) atoms are coplanar with a maximum deviation from the plane of 0.001 Å. In the trigonal plane, the P-Fe-C angles are 124.1 and 125.5° while the C-Fe-C angle is 110.4°. The C-Fe-C angle for the axial ligands is 173.1°. Other details of molecular geometry are: P-Fe 2.208, Fe-C(1) 1.795, Fe-C(2) 1.794, Fe-C(3) 1.787, Fe-C(4) 1.768, P-N 1.681 and N-Si (1.788, 1.790 Å).

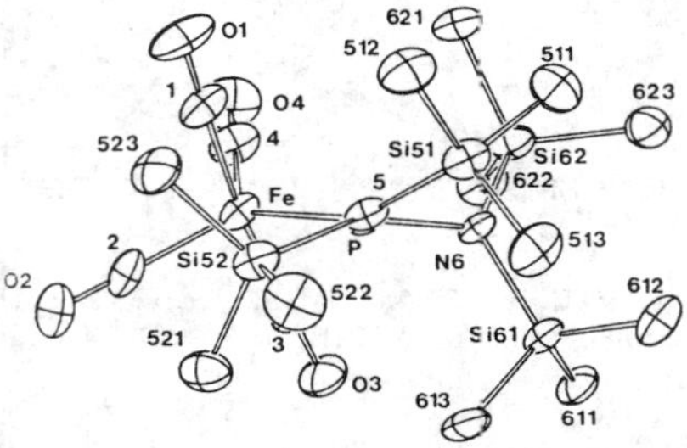

Fig. 1. Molecular structure of $C_{17}H_{36}FeNO_4PSi_4$.

TRIBROMODICARBONYLBIS(DIMETHYLPHENYLPHOSPHINE)RHENIUM(III)
$C_{18}H_{22}Br_3O_2P_2Re$

M.G.B. DREW, B.J. BRISDON and A.M. WATTS, 1984. Polyhedron, 3, 1059-1063.

Triclinic, $P\bar{1}$, a = 9.575, b = 14.705, c = 9.629 Å, α = 77.71, β = 109.06, γ = 115.02°, D_m = 2.48, Z = 2. Mo radiation, R = 0.07 for 2991 reflexions.

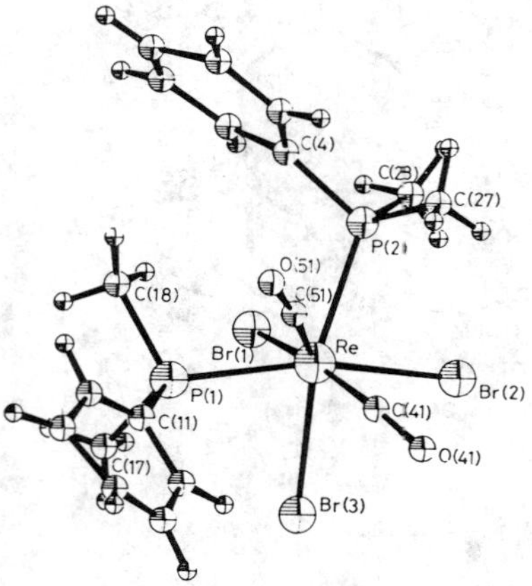

Fig. 1. A view of $C_{18}H_{22}Br_3O_2P_2Re$.

The metal atom (Fig. 1) has a capped octahedral environment with a carbonyl in the unique capping position (1.87(3) Å), two phosphines (2.456(13), 2.496(11) Å) and one carbonyl 1.85(5) Å in the capped face, and three bromine atoms in the uncapped face (2.589(5), 2.623(6), 2.622(5) Å).

DIBROMOBIS[TRIS(2-CYANOETHYL)PHOSPHINE]MERCURY(II) ACETONE SOLVATE
$C_{18}H_{24}Br_2HgN_6P_2,C_3H_6O$ (I)

catena-(μ_2-CHLORO)CHLOROTRIS(2-CYANOETHYL)PHOSPHINEMERCURY(II)
$(C_9H_{12}Cl_2HgN_3P)_n$ (II)

N.A. BELL, M. GOLDSTEIN, L.A. MARCH and I.W. NOWELL, 1984. J. Chem. Soc. Dalton, 1621-1624.

I. Monoclinic, $P2_1/n$, a = 9.174, b = 15.795, c = 20.741 Å, β = 93.01°, D_m = 1.84, Z = 4. Mo radiation, R = 0.043 for 2511 reflexions.

II. Orthorhombic, $P2_12_12_1$, a = 20.872, b = 7.854, c = 8.519 Å, D_m = 2.23, Z = 4. Mo radiation, R = 0.053 for 1252 reflexions.

Compound I (Fig. 1) is monomeric while II has a polymeric structure with almost linear Cl-Hg-$P(CH_2CH_2CN)_3$ units linked by chlorine atoms to give single chains. The geometry about Hg in the two complexes indicates tris(2-cyanoethyl)phosphine to be a strong σ-donor: I, Hg-Br(1) 2.750(2), Hg-Br(2) 2.703(2), Hg-P(1) 2.441(3), Hg-P(2) 2.440(3), Hg...O 3.91(2); II, Hg-Cl(1) 2.780(7), Hg-Cl(1') 2.714(7), Hg-Cl(2) 2.331(7), Hg-P 2.393(5) Å. In I the P-Hg-P angle is 151.3(1)° and the O atom of the solvent acetone molecule occupies an equatorial position of a trigonal bipyramidal polyhedron about Hg. The geometry about Hg in II is distorted octahedral.

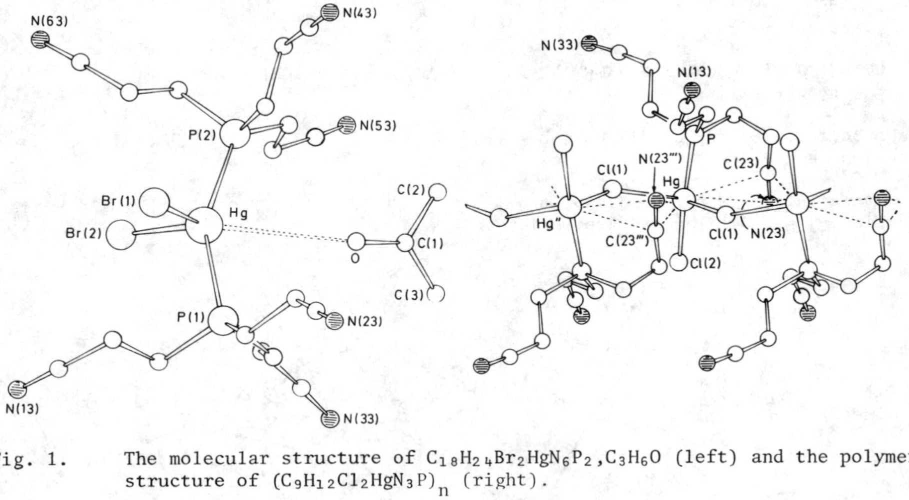

Fig. 1. The molecular structure of $C_{18}H_{24}Br_2HgN_6P_2,C_3H_6O$ (left) and the polymeric structure of $(C_9H_{12}Cl_2HgN_3P)_n$ (right).

DICHLOROBIS(t-BUTYLMERCAPTO)[(1,2-BIS(DIMETHYLARSINO)-o-PHENYLENE-As,As']TITANIUM
$C_{18}H_{34}As_2Cl_2S_2Ti$

R.A. JONES, S.T. SCHWAB and B.R. WHITTLESEY, 1984. Polyhedron, 3, 505-507.

Trigonal, $P3_121$, a = 16.892, c = 8.492 Å, Z = 3. Mo radiation, R = 0.0462 for 948 reflexions.

The coordination geometry about the Ti(IV) atom (Fig. 1) is a distorted octahedron. The ligands are arranged so that the two chloride and two t-BuS ligands occupy roughly trans and cis positions, respectively. There is a crystallographically imposed twofold axis lying in the S-S'-Ti-As-As' plane and which bisects the S-Ti-S' and As-Ti-As' angles. The sulphur, titanium and arsenic atoms are all virtually coplanar while the angles subtended at titanium reflect the notable distortions from an idealized octahedral geometry (As-Ti-As' 75.71(7)°, As-Ti-Cl and As-Ti-S angles 82.97° (average) and 85.86(8)°, respectively). The two trans chloride ligands are markedly bent back from the side of the molecule occupied by the two t-BuS ligands (Cl-Ti-Cl 162.2(2)°). The Ti-S distance is 2.332(3) Å.

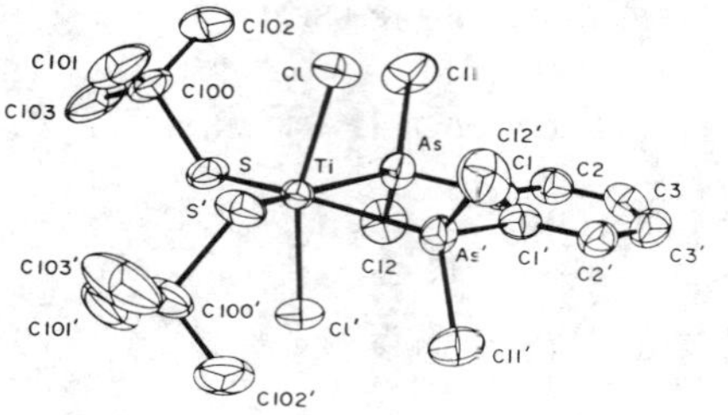

Fig. 1. A general view of $C_{18}H_{34}As_2Cl_2S_2Ti$.

(μ_2-DI-t-BUTYLPHOSPHINO)(TETRACARBONYLTRIMETHYLPHOSPHINECHROMIUM)CHLOROTRIMETHYLPHOSPHINENICKEL
$C_{18}H_{36}ClCrNiO_4P_3$ (I)

(μ_2-DI-t-BUTYLPHOSPHINO)(TETRACARBONYLTRIMETHYLPHOSPHINECHROMIUM)CARBONYLTRIMETHYLPHOSPHINERHODIUM
$C_{19}H_{36}CrO_5P_3Rh$ (II)

R.A. JONES, J.G. LASCH, N.C. NORMAN, A.L. STUART, T.C. WRIGHT and B.R. WHITTLESEY, 1984. Organometallics, 3, 114-119.

I. Monoclinic, $P2_1/n$, a = 17.682, b = 11.323, c = 13.231 Å, β = 92.582, Z = 4. Mo radiation, R = 0.054 for 2714 reflexions.

II. Monoclinic, $P2_1/n$, a = 8.459, b = 18.867, c = 17.113 Å, β = 101.42°, Z = 4. Mo radiation, R = 0.041 for 2806 reflexions.

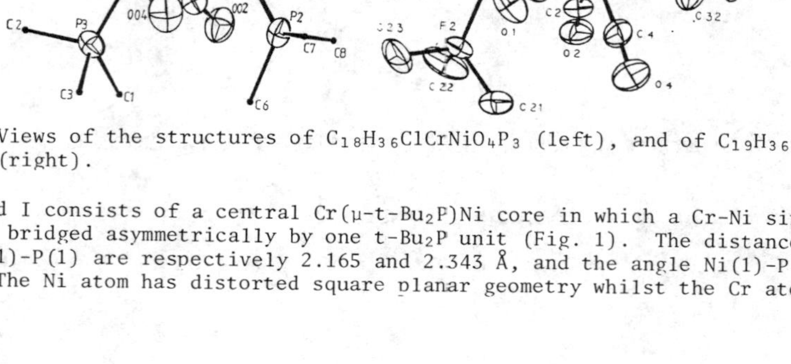

Fig. 1. Views of the structures of $C_{18}H_{36}ClCrNiO_4P_3$ (left), and of $C_{19}H_{36}CrO_5P_3Rh$ (right).

Compound I consists of a central Cr(μ-t-Bu_2P)Ni core in which a Cr-Ni single bond (2.571 Å) is bridged asymmetrically by one t-Bu_2P unit (Fig. 1). The distances Ni(1)-P(1) and Cr(1)-P(1) are respectively 2.165 and 2.343 Å, and the angle Ni(1)-P(1)-Cr(1) is 69.43°. The Ni atom has distorted square planar geometry whilst the Cr atom is seven-coordinate including the metal-metal bond. In compound II the geometry of the central Cr-P(1)-Rh core also reveals considerable distortion for the μ-t-Bu_2P group, with the angle Cr-P(1)-Rh 68.94° and the bonds Cr-P(1) and Rh-P(1) respectively 2.449 and 2.296 Å in length. The main difference is: in I the Cr-PMe_3 group is trans to the μ-t-Bu_2P group and in II it is virtually cis (P(3)-Cr-P(1) 102.82°).

BIS(TRI-n-PROPYLPHOSPHINE)MERCURY-DI(μ-IODO)DIIODOCADMIUM
$C_{18}H_{42}CdHgI_4P_2$ $I_2Cd(\mu-I)_2Hg(P(C_3H_7)_3)_2$

N.A. BELL, M. GOLDSTEIN, L.A. MARCH and I.W. NOWELL, 1984. Inorg. Chim. Acta, 83, 75-78.

Monoclinic, $P2_1/c$, a = 10.694, b = 13.794, c = 22.415 Å, β = 96.25°, D_m = 2.29, Z = 4. Mo radiation, R = 0.048 for 3053 reflexions.

The [CdI_2Hg] ring is almost planar and each metal atom is four-coordinate (Fig. 1). Within the ring the distances are: Hg-I(1) 3.077, Hg-I(2) 3.056, Cd-I(1) 2.866, Cd-I(2) 2.882 Å, and the angles are: Cd-I(1)-Hg 87.2, Cd-I(2)-Hg 87.3, I(1)-Cd-I(2) 96.3 and I(1)-Hg-I(2) 88.6°. The Hg coordination polyhedron is highly irregular with a large P-Hg-P angle of 152.8°. The Hg-P distances are 2.424 and 2.410 Å and the Cd-I(terminal) distances are both 2.705 Å. The I(3)-Cd-I(4) angle is 118.6°.

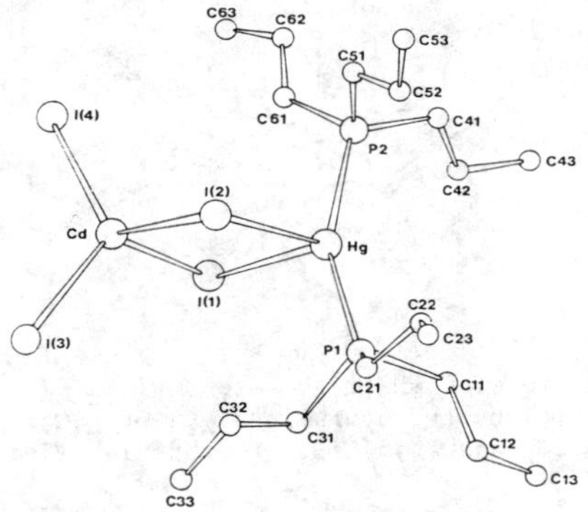

Fig. 1. Molecular structure of $C_{18}H_{42}CdHgI_4P_2$.

TRIS[BIS-1,2-(DIMETHYLPHOSPHINO)ETHANE]CHROMIUM
$C_{18}H_{48}CrP_6$ (I) $Cr(Me_2PCH_2CH_2PMe_2)_3$

TRIS[BIS-1,2-(DIMETHYLPHOSPHINO)ETHANE]MOLYBDENUM
$C_{18}H_{48}MoP_6$ (II) $Mo(Me_2PCH_2CH_2PMe_2)_3$

TRIS[BIS-1,2-(DIMETHYLPHOSPHINO)ETHANE]TUNGSTEN
$C_{18}H_{48}P_6W$ (III) $W(Me_2PCH_2CH_2PMe_2)_3$

TRIS[BIS-1,2-(DIMETHYLPHOSPHINO)ETHANE]VANADIUM
$C_{18}H_{48}P_6V$ (IV) $V(Me_2PCH_2CH_2PMe_2)_3$

TRIS[BIS-1,2-(DIMETHYLPHOSPHINO)ETHANE]NIOBIUM
$C_{18}H_{48}NbP_6$ (V) $Nb(Me_2PCH_2CH_2PMe_2)_3$

TRIS[BIS-1,2-(DIMETHYLPHOSPHINO)ETHANE]TANTALUM
$C_{18}H_{48}P_6Ta$ (VI) $Ta(Me_2PCH_2CH_2PMe_2)_3$

F.G.N. CLOKE, P.J. FYNE, V.C. GIBSON, M.L.H. GREEN, M.J. LEDOUX, R.N. PERUTZ, A. DIX, A. GOURDON and K. PROUT, 1984. J. Organometal. Chem., 277, 61-73.

I. Cubic, a = 10.922 Å, Z = 2. Mo radiation, R = 0.045 for 164 reflexions.

II. Cubic, a = 11.058 Å, Z = 2. Mo radiation, R = 0.047 for 176 reflexions.

III. Cubic, a = 11.040 Å, Z = 2. Mo radiation, R = 0.037 for 141 reflexions.

IV. Cubic, a = 11.041 Å, Z = 2. Mo radiation, R = 0.052 for 117 reflexions.

V. Cubic, a = 11.194 Å, Z = 2. Mo radiation, R = 0.066 for 217 reflexions.

VI. Cubic, a = 11.164 Å, Z = 2. Mo radiation, R = 0.036 for 147 reflexions.

The isomorphous crystals (space group Im3m) have isolated molecules with an essentially octahedral distribution of the six P atoms about the metal centre (Fig. 1). The metal atom has point symmetry 32 and each molecule occupies one of the eight possible sites with symmetry 32 so that the average of the electron distribution has m3m symmetry. The P atoms are located near mirror planes and the structure in Fig. 1 is one of several equally good and equally well-refined disordered models. The metal-P distances for compounds I to VI respectively are: 2.313, 2.421, 2.414, 2.412, 2.526 and 2.500 Å, and the metal-P-C angles range from 110 to 125°.

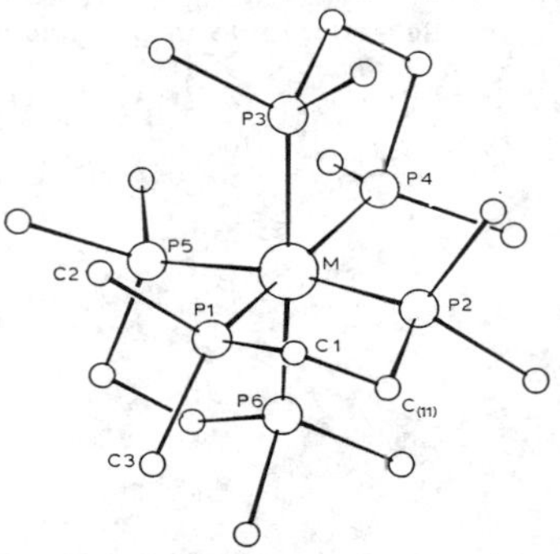

Fig. 1. Molecular structure of $C_{18}H_{48}MP_6$. Atoms P(2), P(3), P(4), P(5) and P(6) are related to P(1) by the operations y,x,0; $\bar{y}$,0,x; x,0,y; 0,$\bar{x}$,$\bar{y}$ and 0,$\bar{y}$,x respectively and atom C(11) to that labelled C(1) by the operation y,x,$\bar{z}$.

BIS(μ_3-SULPHIDO)(DIMETHYLPHENYLPHOSPHINETRICARBONYLTUNGSTEN)NONACARBONYLTRIOSMIUM
$C_{20}H_{11}O_{12}Os_3PS_2W$ (I)

BIS(μ_3-SULPHIDO)[BIS(DIMETHYLPHENYLPHOSPHINE)DICARBONYLTUNGSTEN]DECACARBONYLTRIOSMIUM
$C_{28}H_{22}O_{12}Os_3P_2S_2W$ (II)

BIS(μ_3-SULPHIDO)[BIS(DIMETHYLPHENYLPHOSPHINE)DICARBONYLTUNGSTEN]NONACARBONYLTRIOSMIUM
$C_{27}H_{22}O_{11}Os_3P_2S_2W$ (III)

(μ_4-SULPHIDO)(μ_3-SULPHIDO)[(μ_2-CARBONYL)BIS(DIMETHYLPHENYLPHOSPHINE)TETRACARBONYLDI-TUNGSTEN]NONACARBONYLTRIOSMIUM
$C_{30}H_{22}O_{14}Os_3P_2S_2W_2$ (IV)

R.D. ADAMS, I.T. HORVATH and P. MATHUR, 1984. J. Am. Chem. Soc., 106, 6296-6305.

I. Triclinic, P$\bar{1}$, a = 9.229, b = 11.785, c = 28.559 Å, α = 87.02, β = 83.82, γ = 66.82°, Z = 4. Mo radiation, R = 0.033 for 6034 reflexions.

II. Monoclinic, $P2_1/n$, a = 19.917, b = 9.560, c = 21.392 Å, β = 113.74°, Z = 4. Mo radiation, R = 0.046 for 3135 reflexions.

III. Orthorhombic, $P2_12_12_1$, a = 13.943, b = 15.674, c = 16.174 Å, Z = 4. Mo radiation, R = 0.048 for 2332 reflexions.

IV. Monoclinic, $P2_1/c$, a = 11.666, b = 21.539, c = 15.934 Å, β = 92.65°, Z = 4. Mo radiation, R = 0.057 for 3659 reflexions.

The compounds (I)-(IV) are shown in Fig 1. Compound (I) has a cluster of four metal atoms in the form of a butterfly tetrahedron with W at a wing tip. Triply bridging sulphido ligands bridge the open triangular faces. The two independent molecules of (I) in the crystal lattice have very similar geometry. One Os-W bond is longer than the other (eg. Os(2)-W(1) 3.031(1), Os(6)-W(2) 3.068(1) Å compared with Os(3)-W(1) 2.969(1), Os(5)-W(2) 2.976(1) Å). The Os-Os bonding is irregular (Os-Os 2.899(1)-3.060(1) Å). Molecule (II) contains a planar cluster of one W and three Os atoms; it has only three metal-metal bonds (Os(1)-Os(2) 2.895(1), Os(2)-Os(3) 2.887(1), Os(3)-W 3.044(1) Å) and two triply bridging sulphido ligands symmetrically disposed about the M_4 plane. Complex (III) is isoelectronic with (I) and is described as a butterfly tetrahedron with one bond missing. Main bond lengths are Os(1)-Os(3) 2.829(1), Os(2)-Os(3) 2.923(1), Os(2)-W 2.814(1), Os(3)-W 3.016(1) Å. The separations Os(1)...Os(2) (3.813(1) Å) and Os(1)...W (3.666(1) Å) are too long to be bonding. Compound (IV) has a structure analogous with that of (III) but with an additional tungsten tricarbonyl phosphine unit bridging one of the W-S bonds. (IV) is electronically unsaturated and has two unusually short adjacent metal-metal bonds (Os(2)-W(2) 2.703(1), W(1)-W(2) 2.711(1) Å).

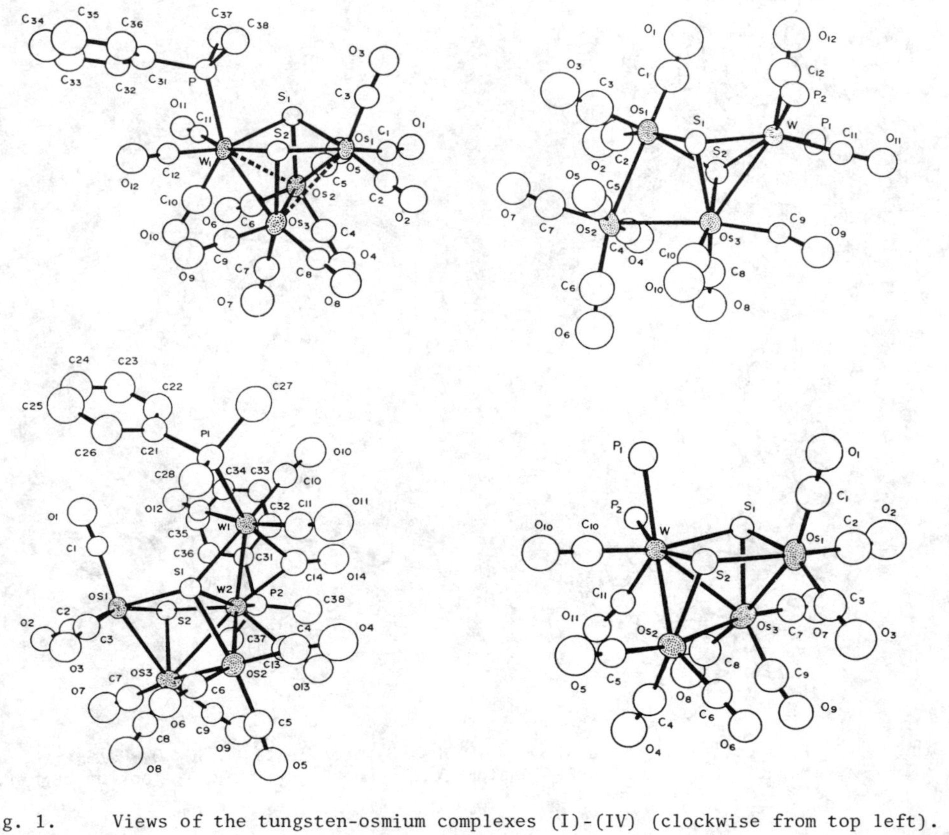

Fig. 1. Views of the tungsten-osmium complexes (I)-(IV) (clockwise from top left).

DODECACARBONYL(DIMETHYLPHENYLPHOSPHINE)DI-μ_3-THIOXOTETRAOSMIUM(3Os-Os)
$C_{20}H_{11}O_{12}Os_4PS_2$ (I) $Os_4(CO)_{12}(PMe_2Ph)(\mu_3\text{-}S)_2$

UNDECACARBONYL(DIMETHYLPHENYLPHOSPHINE)DI-μ_3-THIOXOTETRAOSMIUM(5Os-Os)
$C_{19}H_{11}O_{11}Os_4PS_2$ (II) $Os_4(CO)_{11}(PMe_2Ph)(\mu_3\text{-}S)_2$

DODECACARBONYL(2-ISOCYANO-2-METHYLPROPANE)DI-μ_3-THIOXOTETRAOSMIUM(3Os-Os)
$C_{17}H_9NO_{12}Os_4S_2,0.5C_6H_{14}$ (III) $Os_4(CO)_{12}(CN\text{-}t\text{-}Bu)(\mu_3\text{-}S)_2,0.5C_6H_{14}$

UNDECACARBONYL(2-ISOCYANO-2-METHYLPROPANE)DI-μ_3-THIOXOTETRAOSMIUM(5Os-Os)
$C_{16}H_9NO_{11}Os_4S_2$ (IV) $Os_4(CO)_{11}(CN\text{-}t\text{-}Bu)(\mu_3\text{-}S)_2$

R.D. ADAMS, I.T. HORVATH and K. NATARAJAN, 1984. Organometallics, 3, 1540-1549.

I. Monoclinic, $P2_1/a$, a = 17.941, b = 8.957, c = 17.949 Å, β = 99.75°, Z = 4. Mo radiation, R = 0.041 for 3542 reflexions.

II. Triclinic, $P\bar{1}$, a = 8.994, b = 10.057, c = 15.048 Å, α = 99.62, β = 92.22, γ = 92.55°, Z = 2. Mo radiation, R = 0.040 for 3110 reflexions.

III. Monoclinic, $P2_1/n$, a = 15.113, b = 10.798, c = 18.659 Å, β = 102.33°, Z = 4. Mo radiation, R = 0.047 for 1939 reflexions.

IV. Orthorhombic, Pbca, a = 23.82, b = 11.52, c = 18.497 Å, Z = 8. Mo radiation, R = 0.042 for 1815 reflexions.

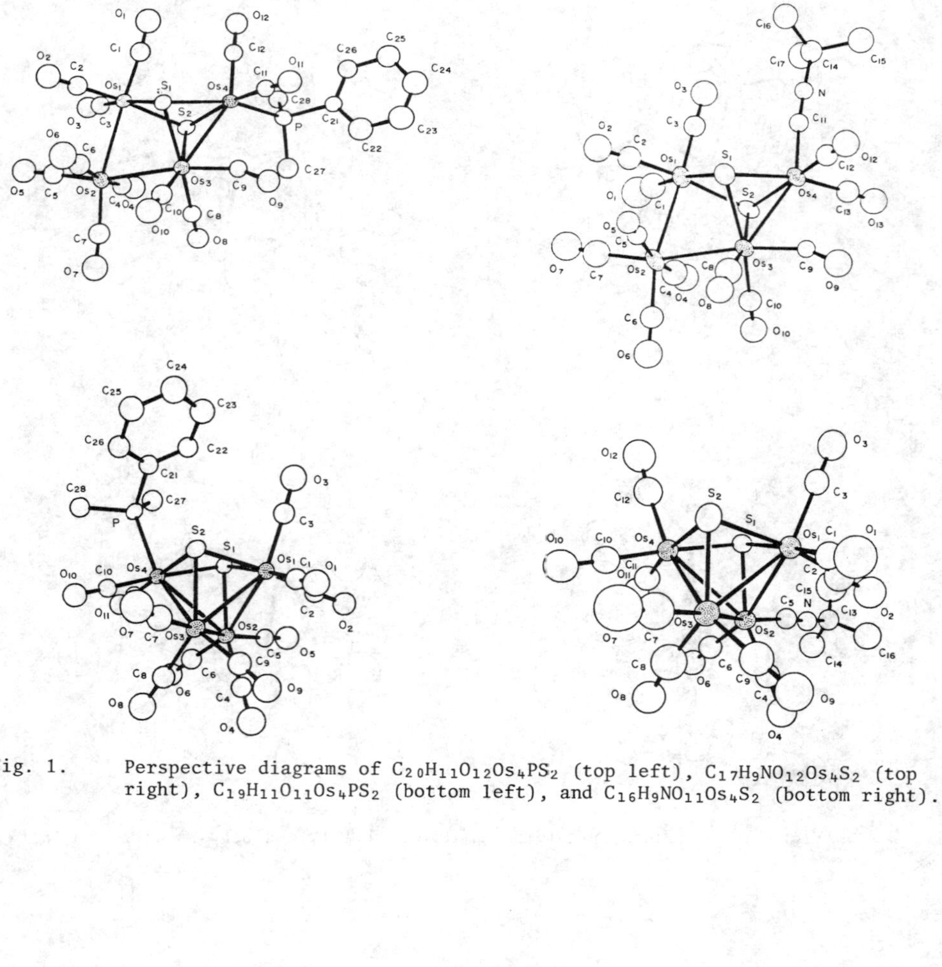

Fig. 1. Perspective diagrams of $C_{20}H_{11}O_{12}Os_4PS_2$ (top left), $C_{17}H_9NO_{12}Os_4S_2$ (top right), $C_{19}H_{11}O_{11}Os_4PS_2$ (bottom left), and $C_{16}H_9NO_{11}Os_4S_2$ (bottom right).

Both compounds I and III contain open-planar clusters of four Os atoms with three metal-metal bonds. Triply-bridging sulphido ligands are symmetrically disposed on each side of the Os_4 plane (Fig. 1). The Os(1)-Os(2), Os(2)-Os(3) and Os(3)-Os(4) distances in I are 2.915, 2.880 and 2.801 Å, and in III are 2.898, 2.867 and 2.878 Å respectively. The Os-S distances range from 2.419 to 2.465 Å in I and from 2.411 to 2.482 Å in III. The Os(1)-Os(2)-Os(3) and Os(2)-Os(3)-Os(4) angles are respectively 69.75 and 129.18° in I and 70.28 and 129.14° in III. The Os(4)-P distance in I is 2.349 Å and the angles about S(1) are 85.24, 97.5 and 69.95°, and about S(2) are 84.95, 98.3 and 70.39°. In III the corresponding angles are: 84.2, 96.9, 69.0 and 85.4, 98.3 and 70.2°. The Os(4)-C(11) distance is 1.92 Å. Compounds II and IV contain butterfly tetrahedral clusters of four metal atoms with triply-bridging sulphido ligands on the two open triangular faces. Two of the five metal-metal bonds in each compound are elongated but in II they are nearly equal in length (Os(1)-Os(3) 2.987, Os(2)-Os(4) 3.013 Å), while in IV they are significantly different (Os(1)-Os(3) 2.965, Os(2)-Os(4) 3.115 Å). The other Os-Os distances are 2.925, 2.921 and 2.893 Å in II, and 2.923, 2.922 and 2.909 Å in IV. The Os-S distances in II range from 2.376 to 2.431 Å, and in IV from 2.381 to 2.443 Å.

BIS(2,3-BUTANEDIONE DIOXIMATO)CHLORO(DIPHENYLHYDROXOARSINE)COBALT(III)
$C_{20}H_{25}AsClCoN_4O_5$

M. PIZZEY, L. MIHICHUK, R.J. BARTON and B.E. ROBERTSON, 1984. Can. J. Chem., 62, 285-288.

Orthorhombic, $Pna2_1$, a = 15.334, b = 11.173, c = 13.745 Å, D_m = 1.61, Z = 4. Mo radiation, R = 0.071 for 2345 reflexions.

The Co atom (Fig. 1) is bonded equatorially to two dmgH ligands (average Co-N 1.89(1) Å) with a Cl atom above this plane and the As atom of the $(C_6H_5)_2As(OH)$ group below it. The As-O bond length is 1.781(7) Å, corresponding to a single bond and is consistent with the observed H atom on the oxygen atom.

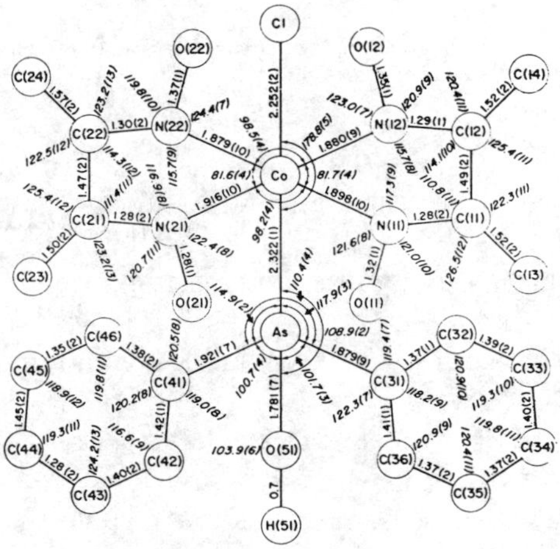

Fig. 1. Molecular dimensions for $C_{20}H_{25}AsClCoN_4O_5$.

BIS(o-PHENYLENE-BIS(DIMETHYLPHOSPHINE))NICKEL(II) BIS(TRIIODIDE) DIIODINE
$C_{20}H_{32}I_{10}NiP_4$

L.R. GRAY, S.J. HIGGINS, W. LEVASON and M. WEBSTER, 1984. J. Chem. Soc. Dalton, 1433-1439.

Triclinic, $P\bar{1}$, a = 9.672, b = 12.369, c = 9.574 Å, α = 106.55, β = 107.70, γ = 99.48°, D_m = 2.77, Z = 1. Mo radiation, R = 0.032 for 2914 reflexions.

The arrangement of ligands, four P atoms from two chelating diphosphines, about Ni is square-planar (Ni-P 2.210, 2.200(2) Å) (Fig. 1). The NiP_4 unit is essentially planar and the benzene ring backbones make dihedral angles of 14.4° with this plane. The closest approach of an iodine atom to Ni is 3.491(1) Å (Ni...I(3)) which implies there is negligible bonding to the Ni atom. There are also linear I_3^- anions (I-I 2.910, 2.932(1)) and diiodine molecules (I-I 2.728(1) Å). Weak secondary interactions between the iodine atoms of I_3^- and I_2 (3.442(1), 3.576(1) Å) lead to the formation of bifurcated chains (Fig. 1). The polyiodide unit can also be described in terms of I_5^- units.

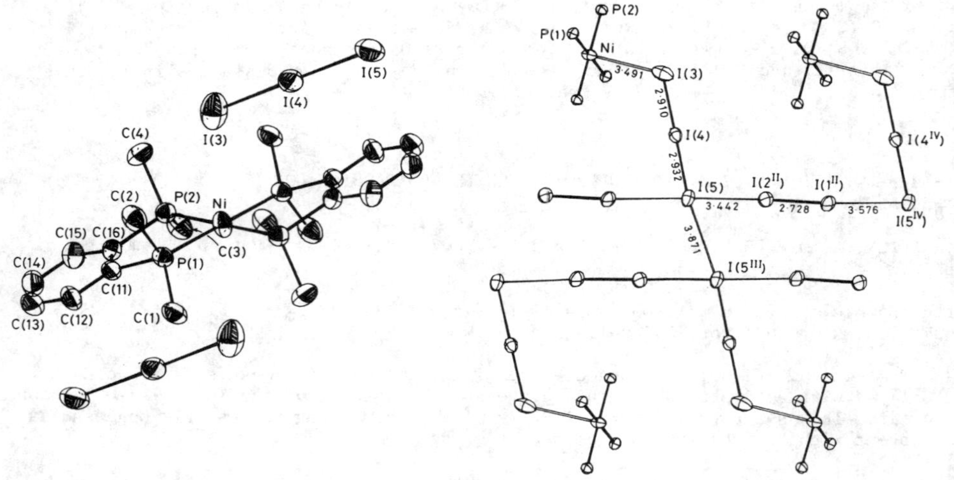

Fig. 1. $C_{20}H_{32}I_{10}NiP_4$: the molecular structure showing the nearest I_3 groups (left) and iodide links (right).

BIS[(μ_2-BROMO)(μ_2-TETRAETHYL PHOSPHITE-P,P')DICARBONYLMANGANESE(I)]

$C_{20}H_{40}Br_2Mn_2O_{14}P_4$ (I) $Mn_2(\mu\text{-}Br)_2(CO)_4\{(C_2H_5O)_2POP(OC_2H_5)_2\}_2$

(μ_2-BROMO)(μ_2-HYDRIDO)BIS((μ_2-TETRAETHYL DIPHOSPHITE-P,P')DICARBONYLMANGANESE(I))

$C_{20}H_{41}BrMn_2O_{14}P_4$ (II) $Mn_2(\mu\text{-}Br)(\mu\text{-}H)(CO)_4\{(C_2H_5O)_2POP(OC_2H_5)_2\}_2$

J. GIMENO, V. RIERA, M.A. RUIZ, A.M.M. LANFREDI and A. TIRIPICCHIO, 1984. J. Organometal. Chem., 268, C13-C17.

I. Orthorhombic, Pbca, a = 15.378, b = 16.327, c = 14.453 Å, Z = 4. Mo radiation, R = 0.065 for 1164 reflexions.

II. Orthorhombic, Pbca, a = 33.066, b = 14.435, c = 14.806 Å, Z = 8. Mo radiation, R = 0.080 for 2406 reflexions.

Complex I has crystallographic inversion symmetry with a non-bonding Mn...Mn distance of 3.517 Å (Fig. 1), whereas in II, with the presence of the hydride bridge, the Mn-Mn distance is much shorter (2.936 Å) and is consistent with a single metal-metal bond. Other distances and angles of note are: in I, Mn-Br 2.525, Mn'-Br 2.522, Mn-P(1) 2.241, Mn-P(2) 2.238 Å; Mn-Br-Mn' 88.4, P(1)-Mn-P(2) 167.6°, and in II, Mn(1)-Br 2.477, Mn(2)-Br 2.484, Mn(1)-P(1) 2.211, Mn(1)-P(2) 2.209, Mn(2)-P(3) 2.212, Mn(2)-P(4) 2.216 Å; Mn(1)-Br-Mn(2) 72.6, P(1)-Mn(1)-P(2) 176.1, P(3)-Mn(2)-P(4) 178.9°.

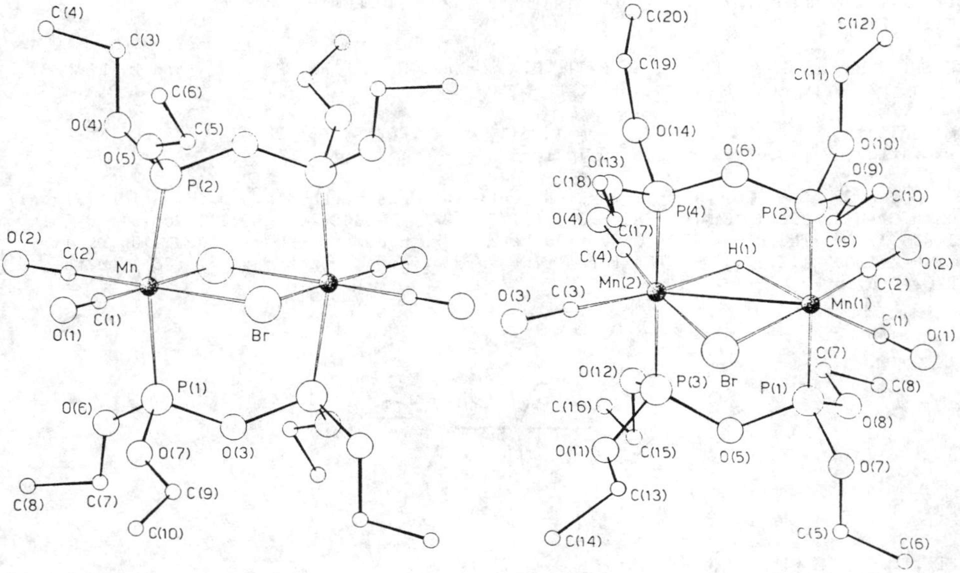

Fig. 1. View of the complexes $C_{20}H_{40}Br_2Mn_2O_{14}P_4$ (left), and $C_{20}H_{41}BrMn_2O_{14}P_4$ (right).

(σ^2-DIHYDROGEN)TRICARBONYLBIS(TRIISOPROPYLPHOSPHINE)TUNGSTEN
$C_{21}H_{44}O_3P_2W$ $W(CO)_3(P\text{-}i\text{-}Pr_3)_2(\eta^2\text{-}H_2)$

G.J. KUBAS, R.R. RYAN, B.I. SWANSON, P.J. VERGAMINI and H.J. WASSERMAN, 1984. J. Am. Chem. Soc., 106, 451-452.

Orthorhombic, Pcab, a = 15.749, b = 24.219, c = 13.333 Å, Z = 8. Mo radiation, R = 0.043 for 2118 reflexions (at -100°C). η^2-H_2 positions confirmed by neutron diffraction study.

The geometry about W (Fig. 1) is that of a regular octahedron with "cis" interligand angles ranging from 88.0(4) to 92.0(6)°. The dihydrogen ligand is symmetrically coordinated in an η^2 mode with mean W-H 1.95(23) (X-ray) and 1.75 Å (neutron, ΔF). The H-H distance is 0.75(16) (X-ray), 0.84 Å (neutron, ΔF).

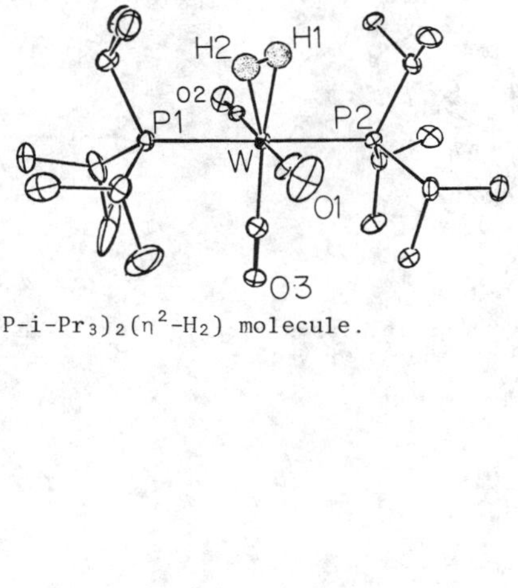

Fig. 1. The $W(CO)_3(P\text{-}i\text{-}Pr_3)_2(\eta^2\text{-}H_2)$ molecule.

trans-TETRACARBONYLBIS[BIS(2-FURYL)METHYLPHOSPHINE]CHROMIUM
$C_{22}H_{18}CrO_8P_2$

E. LINDNER, H. RAULEDER, C. SCHEYTT, H.A. MAYER, W. HILLER, R. FAWZI and P. WEGNER, 1984. Z. Naturforsch., B39, 632-642.

Monoclinic, $P2_1/c$, a = 10.048, b = 14.311, c = 8.262 Å, β = 98.87°, Z = 2. Mo radiation, R = 0.031 for 2515 reflexions.

The Cr atom lies on an inversion centre and has octahedral coordination (Fig. 1), with Cr-P 2.309(1), Cr-C 1.882 and 1.885(3), P-C11 1.808(2), P-C21 1.809(3), P-C31 1.825(3) Å. The P atom has distorted tetrahedral geometry (C-P-C 100.1-100.9(1), Cr-P-C 116.9-117.5(1)°). The furan rings are oriented almost normal (88°) to the P-Cr-P axis.

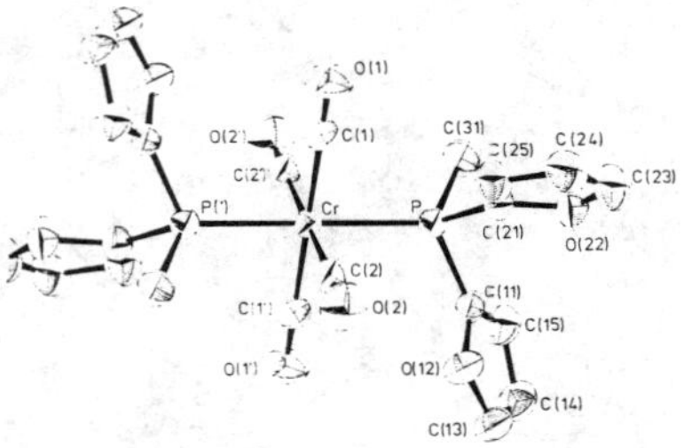

Fig. 1. A view of $C_{22}H_{18}CrO_8P_2$.

2,2,2,2,2-PENTACARBONYL(μ_2-HYDRIDO)-1-TRIPHENYLPHOSPHINEGOLDCHROMIUM
$C_{23}H_{16}AuCrO_5P$

M. GREEN, A.G. ORPEN, I.D. SALTER and F.G.A. STONE, 1984. J. Chem. Soc. Dalton, 2497-2503.

Triclinic, $P\bar{1}$, a = 9.310 (9.244), b = 12.105 (12.019), c = 11.967 (11.786) Å, α = 112.90 (112.67), β = 95.56 (95.90), γ = 109.59 (109.59)°, Z = 4. Mo radiation, R = (0.047) for (3401) reflexions. (Figures in parenthese refer to 210 K).

The molecule (Fig. 1) has a bent Au(μ-H)Cr geometry with Au-Cr 2.770(2), Au-H 1.72(11) and Cr-H 1.64(12) Å. The Au-H-Cr angle is 111(5)°. The $Cr(CO)_5$ group is staggered relative to the Au(μ-H)Cr bridge, with the hydrido-ligand lying transoid to the axial carbonyl ligand and the phosphine group. The Au-P distance is 2.290(3) and the Cr-C distances range from 1.860(12) to 1.914(12) Å. The P-Au-Cr and C(1)-Cr-Au moieties are non-linear (angles 155.4(1) and 161.2(2)°).

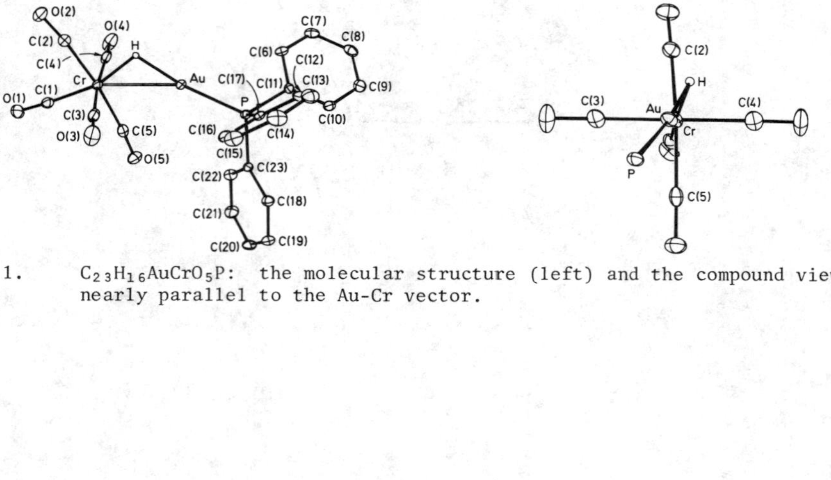

Fig. 1. $C_{23}H_{16}AuCrO_5P$: the molecular structure (left) and the compound viewed nearly parallel to the Au-Cr vector.

1,1,1,2,2,2,3,3,3-NONACARBONYL-1,2-(μ_2-DIMETHYLPHENYLPHOSPHINE)GOLD-1,3-(μ_2-HYDRIDO)-(μ_3-PHENYLPHOSPHINO)TRIRUTHENIUM
$C_{23}H_{17}AuO_9P_2Ru_3$ (I)

1,2-(μ_2-CARBONYL)-1,2,2,3,3,3,4,4,4,5,5,5-DODECACARBONYL-1,2,3,4-(μ_4-PHENYLPHOSPHIDO)-1-TRIETHYLPHOSPHINEDIRHODIUMTRIRUTHENIUM
$C_{25}H_{20}O_{13}P_2Rh_2Ru_3$ (II)

M.J. MAYS, P.R. RAITHBY, P.L. TAYLOR and K. HENRICK, 1984. J. Chem. Soc. Dalton, 959-967.

I. Monoclinic, $P2_1/c$, a = 8.805, b = 20.195, c = 16.683 Å, β = 100.62°, Z = 4. Mo radiation, R = 0.067 for 1398 reflexions.

II. Monoclinic, C2/c, a = 20.257, b = 9.679, c = 34.726 Å, β = 90.24°, Z = 8. Mo radiation, R = 0.034 for 4708 reflexions.

Complex I (Fig. 1) exists as discrete units separated by normal van der Waals distances. Three Ru atoms define an irregular triangle, the longest edge of which is bridged by the Au(1) atom of the $Au(PMe_2Ph)$ group (Au(1)-Ru 2.749(5), 2.763(5) Å). The Ru_3 triangle is asymmetrically capped, on the opposite side to the $Au(PMe_2Ph)$ unit by the P(1) atom of the PPh group. P(1) lies 1.54 Å from the Ru_3 plane. The Ru-Ru distances range from 2.869 to 3.002(6) Å. The hydride ligand, although not located, is assumed to bridge Ru(1)-Ru(3). The nine carbonyl groups are linear and terminal (Ru-C 1.91(4), C-O 1.15(4) Å). In II the five metal atoms define a square based pyramid with the μ_4-P atom of PPh 1.23 Å below the square base (Rh-P(1) 2.374, Ru-P(1) 2.347 Å). The two Rh atoms occupy adjacent positions in the basal plane with Rh-Rh 2.775(1) Å and are bridged by a carbonyl group. The other twelve CO groups are terminal and linear. The PEt_3 group occupies an axial site on Rh(1). The Ru-Ru distances range from 2.837(1) to 2.874(1), Rh-Ru (capping) are 2.758(1) and 2.783(1) and Rh-Ru (basal) are 2.930(1) and 2.862(1) Å.

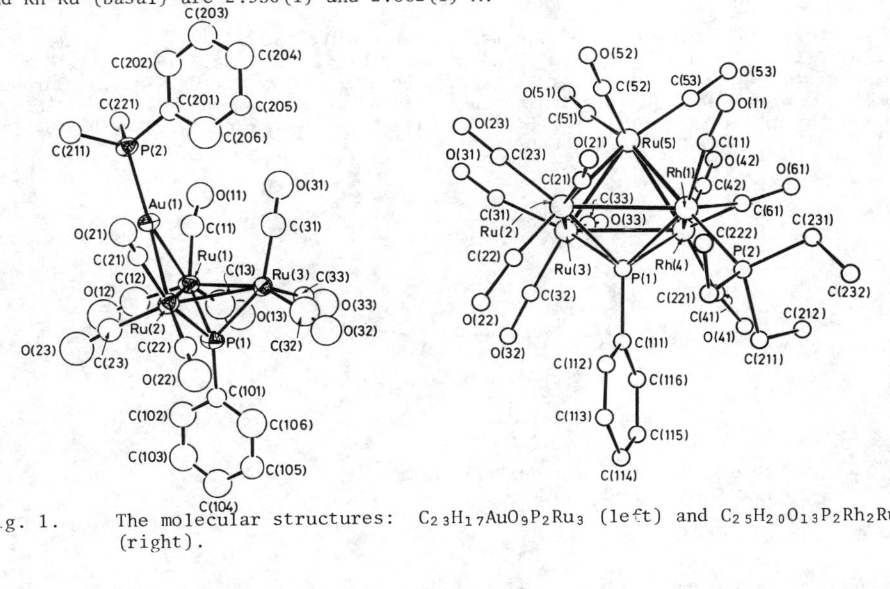

Fig. 1. The molecular structures: $C_{23}H_{17}AuO_9P_2Ru_3$ (left) and $C_{25}H_{20}O_{13}P_2Rh_2Ru_3$ (right).

(3,4-BIS(METHOXYCARBONYL)-1,5-DIMETHYL-7-PHENYL-7-PHOSPHATRICYCLO[3.2.0.0^{4,6}]HEPT-2-ENE)PENTACARBONYLCHROMIUM
$C_{23}H_{19}CrO_9P$

A. MARINETTI, F. MATHEY, J. FISCHER and A. MITSCHLER, 1984. Nouv. J. Chim., 8, 453-457.

Triclinic, P$\bar{1}$, a = 10.760, b = 11.408, c = 10.417 Å, α = 98.74, β = 105.28, γ = 85.38°, D_m = 1.40, Z = 2. Cu radiation, R = 0.050 for 2597 reflexions.

The molecule (Fig. 1) has a strained four-membered phosphetane ring with a C-P-C angle of 77.0° and a puckering of 26.3°, P-C(1) 1.805(3), P-C(5) 1.950(3), C(1)-C(6) 1.524(4), C(5)-C(6) 1.560(4) Å. The P-C(5) distance is very long but there appears to be no obvious explanation for it. The Cr atom has octahedral coordination with Cr-P 2.360(1), Cr-C 1.863-1.888(4) Å.

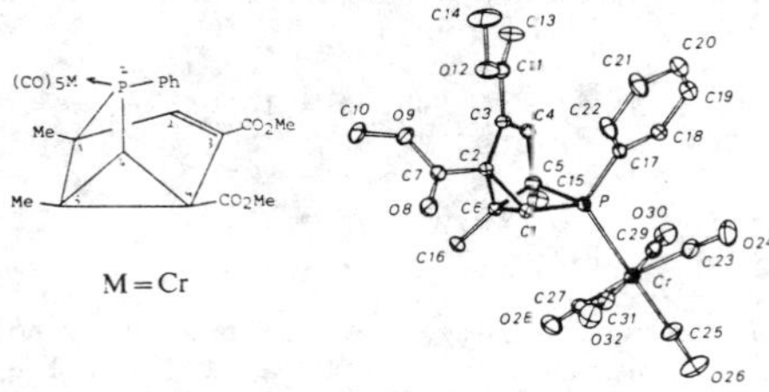

Fig. 1. $C_{23}H_{19}CrO_9P$.

TRIETHYLAMMONIUM PENTACARBONYL(DIPHENYLPHOSPHINITO-P)TUNGSTEN(0)
$C_{23}H_{26}NO_6PW$ $[NH(C_2H_5)_3][W(C_{12}H_{10}OP)(CO)_5]$

W. HILLER, C. ZEIHER and I.-P. LORENZ, 1984. Acta Cryst., C40, 1806-1808.

Triclinic, P$\bar{1}$, a = 10.795, b = 10.828, c = 12.131 Å, α = 114.95, β = 95.85, γ = 94.99°, Z = 2. Mo radiation, R = 0.023 for 4441 reflexions.

The tungsten atom has a nearly octahedral coordination (Fig. 1) with angles P-W-C(5) 171.1(2), P-W-C(1),C(4) 84.7 and 83.7, P-W-C(2),C(3) 94.3 and 98.1(1)°. P has a distorted tetrahedral coordination. There is a very strong N-H...O hydrogen bond (N...O 2.603(5) Å) linking the phosphinite and triethylammonium groups. Some bond lengths are W-P 2.554(1), W-C(5) 1.991(5), W-C(1-4) in range 2.024-2.051(5), P-O 1.521(3), P-C 1.837(4) and 1.860(4) Å.

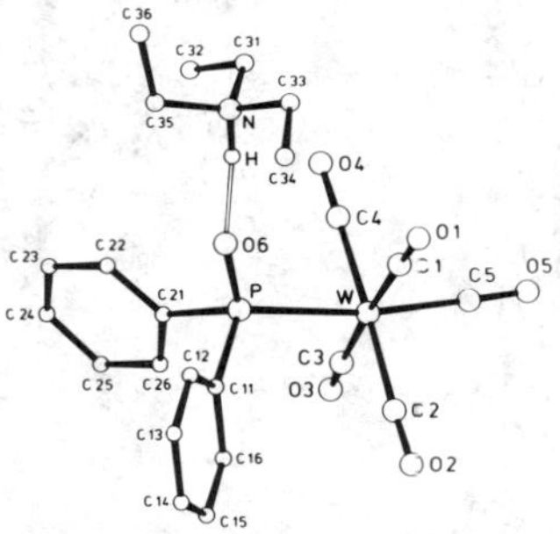

Fig. 1. The molecular structure of triethylammonium pentacarbonyl(diphenylphosphinito-P)tungsten(0).

DICHLORO-2,2'-BIS[1-PHENYL-3,4-DIMETHYLPHOSPHOL-3-ENE]NICKEL(II)
$C_{24}H_{28}Cl_2NiP_2$

F. MERCIER, F. MATHEY, J. FISCHER and J.H. NELSON, 1984. J. Am. Chem. Soc., 106, 425-426.

Monoclinic, C2/c, a = 14.127, b = 9.611, c = 17.777 Å, β = 106.81°, D_m = 1.43, Z = 4. Cu radiation, R = 0.041 for 1486 reflexions.

The molecule is shown in Fig. 1. The Ni atom deviates only slightly from square planarity with the dihedral angle between $NiCl_2$ and NiP_2 planes being 13.6°. Bond lengths: Ni-Cl 2.201, Ni-P 2.126 Å.

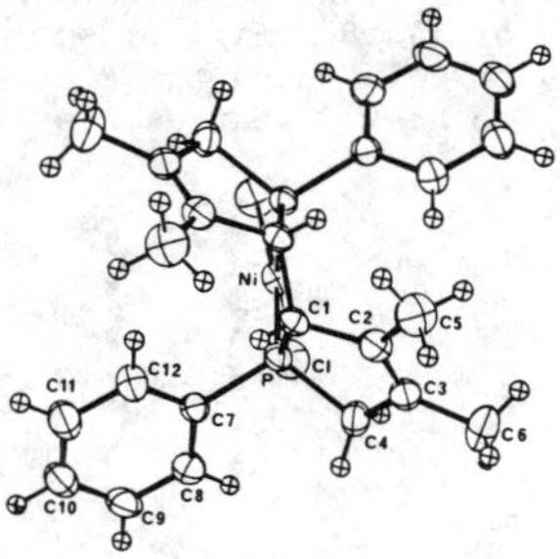

Fig. 1. The $C_{24}H_{28}Cl_2NiP_2$ molecule.

mer-DICHLORO(TRICHLOROBORYLNITRENE)TRIS(DIMETHYLPHENYLPHOSPHINE)RHENIUM(V)
$C_{24}H_{33}BCl_5NP_3Re$

R. DANTONA, E. SCHWEDA and J. STRAHLE, 1984. Z. Naturforsch., B39, 733-745.

Orthorhombic, $Pn2_1a$, a = 20.699, b = 8.870, c = 16.673 Å, Z = 4. Cu radiation, R = 0.066 for 2381 reflexions.

The Re atom has octahedral coordination (Fig. 1) with an $NBCl_3$ group trans to a chlorine. Main dimensions are: Re-Cl1 2.394(3), Re-Cl2 2.439(3), Re-P1 2.488(2), Re-P2 2.473(4), Re-P3 2.482(2), Re-N 1.728(7), N-B 1.53(1), B-Cl 1.82-1.88(1) Å, Re-N-B 176.5(6), Cl1-Re-Cl2 85.9(1), P1-Re-P2 98.2(1), P2-Re-P3 97.2(1)°.

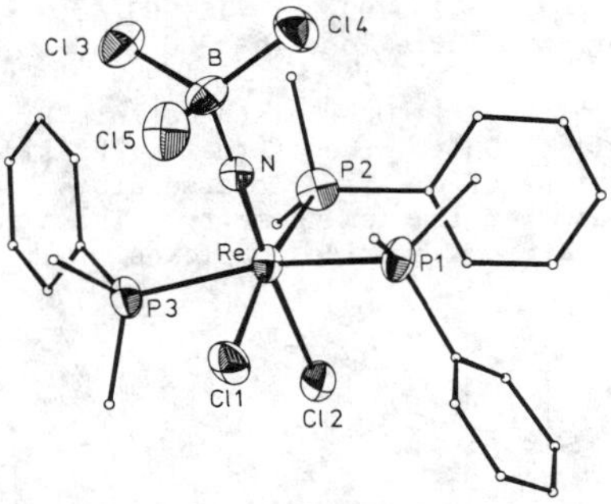

Fig. 1. $C_{24}H_{33}BCl_5NP_3Re$: a view of the borylnitrene.

PENTAHYDRIDOTRIS(DIMETHYLPHENYLPHOSPHINE)RHENIUM
$C_{24}H_{38}P_3Re$ (I) $H_5Re(P(CH_3)_2C_6H_5)_3$

HEPTAHYDRIDOBIS(DIMETHYLPHENYLPHOSPHINE)RHENIUM
$C_{16}H_{29}P_2Re$ (II) $H_7Re(P(CH_3)_2C_6H_5)_2$

R.G. TELLER, W.E. CARROLL and R. BAU, 1984. Inorg. Chim. Acta, 87, 121-127..

I. Monoclinic, $P2_1/c$, a = 6.876, b = 19.493, c = 19.646 Å, β = 103.26°, D_m = 1.58, Z = 4. Mo radiation, R = 0.067 for 2067 reflexions.

II. Monoclinic, $P2_1/n$, a = 19.083, b = 6.337, c = 15.234 Å, β = 93.72°, D_m = 1.70, Z = 4. Mo radiation, R = 0.050 for 1672 reflexions.

Although the hydride ligands could not be located, the overall structure of the complexes could be deduced from knowledge of the Re/P core of the molecules. In I the ReP_3 skeleton has distorted pyramidal geometry consistent with a dodecahedral structure for the H_5ReP_3 core (Fig. 1). The Re-P distances are: 2.340, 2.365 and 2.365 Å. In II, the P-Re-P angle of 146.8° (Fig. 1) suggests a tricapped trigonal prismatic geometry for the H_7ReP_2 core in which the P atoms are placed in opposing axial and equatorial positions. The Re-P distances are 2.396 and 2.395 Å.

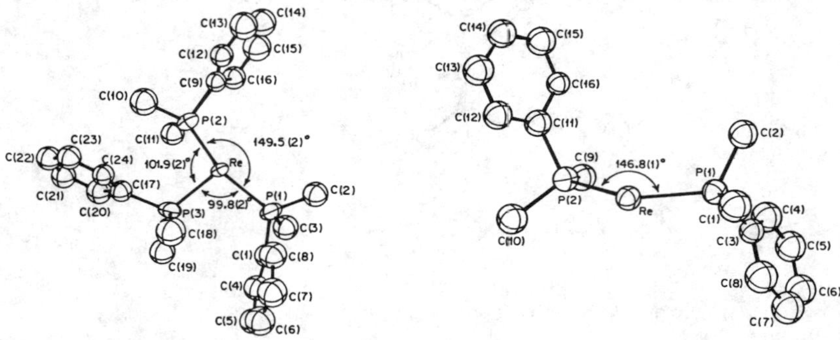

Fig. 1. Molecular structure of $C_{24}H_{38}P_3Re$ (left), and $C_{16}H_{29}P_2Re$ (right).

BIS[DIISOPROPYL(PHENYL)PHOSPHINE]HEXAHYDRIDOOSMIUM

$C_{24}H_{44}OsP_2$ $[Os(C_{12}H_{19}P)_2H_6]$

P.W. FROST, J.A.K. HOWARD and J.L. SPENCER, 1984. Acta Cryst., C40, 946-949.

Monoclinic, $P2_1/n$, a = 17.196, b = 13.799, c = 10.969 Å, β = 98.31°, D_m = 1.48, Z = 4. Mo radiation, R = 0.034 for 5696 reflexions (at 200K).

In the complex molecule (Fig. 1) Os-P distances are 2.340(1) and 2.347(1) Å and the P-Os-P angle is 156.2(1)°. Only three of the six hydride ligands were located in the analysis. The positions of the five located atoms bound to Os are consistent with triangular dodecahedral coordination geometry. The structure is compared with those of other eight-coordinate polyhydride complexes.

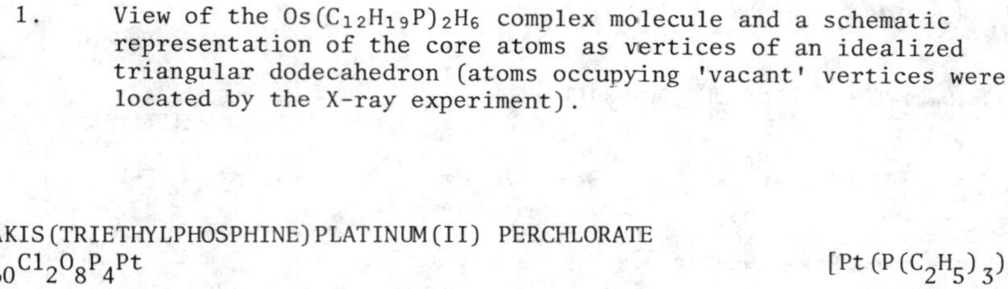

Fig. 1. View of the $Os(C_{12}H_{19}P)_2H_6$ complex molecule and a schematic representation of the core atoms as vertices of an idealized triangular dodecahedron (atoms occupying 'vacant' vertices were not located by the X-ray experiment).

TETRAKIS(TRIETHYLPHOSPHINE)PLATINUM(II) PERCHLORATE

$C_{24}H_{60}Cl_2O_8P_4Pt$ $[Pt(P(C_2H_5)_3)_4]^{2+}2ClO_4^-$

J. KOZELKA, H.-P. LÜTHI, E. DUBLER and R.W. KUNZ, 1984. Inorg. Chim. Acta, 86, 155-163.

Orthorhombic, Pbca, a = 26.839, b = 13.762, c = 19.568 Å, D_m = 1.58, Z = 8. Mo radiation, R = 0.053 for 1610 reflexions.

The expected square planar PtP_4 coordination exhibits a strong tetrahedral distortion (Fig. 1). The four Pt-P bond distances are: 2.329, 2.331, 2.351 and 2.332 Å while the P-Pt-P (trans) angles are 150.3 and 150.9°, instead of 180° for square planar coordination. The conformations of the methylene C atoms are eclipsed, with C(15),P(1),Pt,P(3),C(35) and C(25),P(2),Pt,P(4),C(45) lying in two vertical planes perpendicular to each other. The P-C distances range from 1.83 to 1.93 Å.

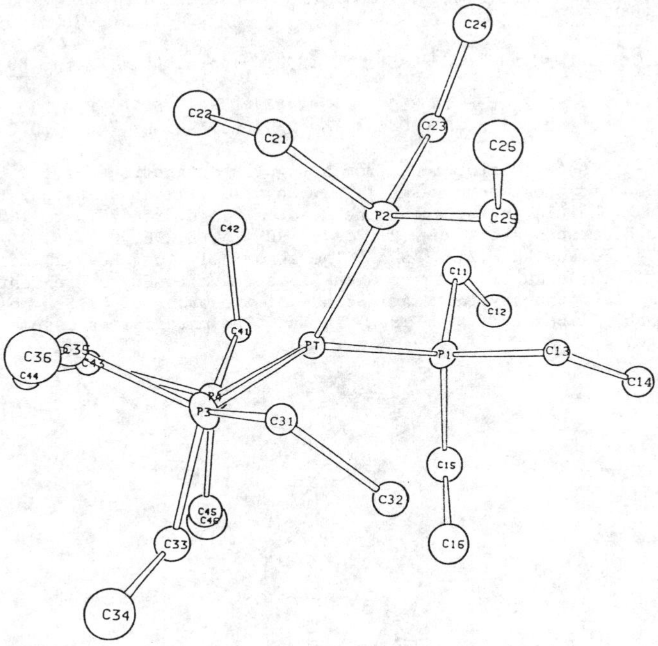

Fig. 1. Molecular structure of the cation in $C_{24}H_{60}Cl_2O_8P_4Pt$.

DIBROMO(BIS(DIPHENYLPHOSPHINO)METHANE-P,P')PALLADIUM(II)
$C_{25}H_{22}Br_2P_2Pd$

K. PETERS, E.-M. PETERS, H.G. VON SCHNERING and H.-P. ABICHT, 1984. Z. Kristallogr., 168, 149-152.

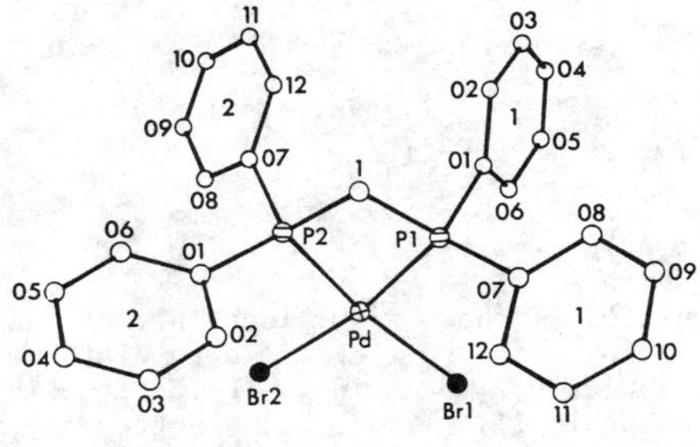

Fig. 1. Perspective drawing of $C_{25}H_{22}Br_2P_2Pd$.

Triclinic, $P\bar{1}$, a = 13.973, b = 19.740, c = 9.505 Å, α = 97.77, β = 94.34, γ = 71.19°, Z = 4. Mo radiation, R = 0.08 for three-dimensional data.

There are two independent molecules in the asymmetric unit and in each the Pd atom has square-planar coordination (Fig. 1) with Pd-P 2.225-2.242(4), Pd-Br 2.474-2.488(2) Å, P-Pd-P 73.6 and 73.7(1), Br-Pd-Br 94.5 and 95.4(1)°.

cis,mer-CARBONYLDIHYDROTRIS(DIMETHYLPHENYLPHOSPHINE)RUTHENIUM
$C_{25}H_{35}OP_3Ru$ cis,mer-$RuH_2(CO)(P(CH_3)_2C_6H_5)_3$

J.W. BRUNO, J.C. HUFFMAN and K.G. CAULTON, 1984. Inorg. Chim. Acta, 89, 167-173.

Triclinic, $P\bar{1}$, a = 12.752, b = 11.759, c = 9.487 Å, α = 75.96, β = 107.13, γ = 102.25°, Z = 2. Mo radiation, R = 0.054 for 2894 reflexions (at -160°C).

The structure shows distortions from a regular octahedron (Fig. 1). This is assumed to be due to the two very small (e.g. hydride) ligands in the presence of larger ligands. The cis angles about the Ru atom range from 77.0 to 100.81°. Important bond distances are: Ru-P(2) 2.346, Ru-P(3) 2.295, Ru-P(4) 2.288, Ru-C(29) 1.902, Ru-H(1) 1.59 and Ru-H(2) 2.39 Å. The abnormally long Ru-H(2) distance is quite reasonable for a Ru-Cl distance when Cl is trans to CO, consequently the authors suggest that the crystal examined had of the order of 1/18th mol fraction of $RuHCl(CO)(PMe_2Ph)_3$ impurity ($Z_{Cl} = 18Z_H$) in it. The material is assumed, therefore, to be ∿94° pure.

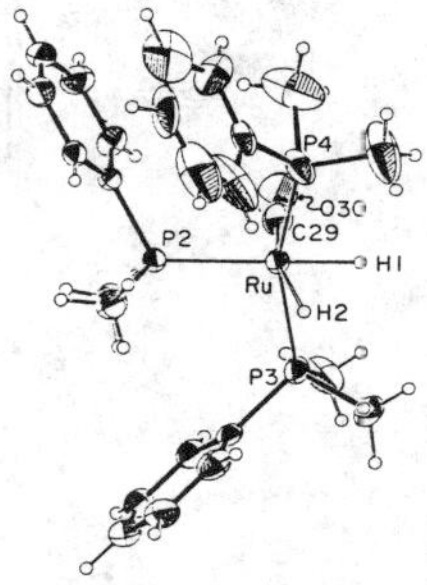

Fig. 1. Molecular structure of $C_{25}H_{35}OP_3Ru$.

HEXACARBONYL[μ-[1,2-PHENYLENEBIS[METHYLENE(PHENYLPHOSPHINIDENE)]P,P':P,P']]DIIRON(Fe-Fe) DICHLOROMETHANE SOLVATE
$C_{26}H_{18}Fe_2O_6P_2,CH_2Cl_2$ (I) $[(\mu\text{-}o\text{-}C_6H_4(CH_2)_2)(PhP)_2]Fe_2(CO)_6,CH_2Cl_2$

HEXACARBONYL[μ-[(METHYLPHENYLPHOSPHINO)METHYLENE]-1,2-PHENYLENEMETHYLENE(PHENYLPHOSPHINIDENE)]DIIRON(Fe-Fe)
$C_{27}H_{20}Fe_2O_6P_2$ (II)

J.P. FACKLER, Jr., and A.M. MAZANY, 1984. Organometallics, 3, 1121-1123.

I. Monoclinic, $P2_1/c$, a = 11.996, b = 13.592, c = 16.891 Å, β = 109.80°, Z = 4. Mo radiation, R = 0.069 for 2553 reflexions.

II. Monoclinic, $P2_1/n$, a = 11.670, b = 14.581, c = 16.674 Å, β = 109.47°, Z = 4. Mo radiation, R = 0.047 for 4158 reflexions.

The molecular structure of II shows significant structural changes from that of its precursor I (Fig. 1) although the Fe(1)-Fe(2) bond distance increases by only 0.056 Å upon the substitution of P(2) by C in the Fe(2) coordination sphere (Fe(1)-Fe(2) 2.628 Å in I and 2.684 Å in II). Other bond lengths of interest (in I and II respectively) are: P(2)-C(38) 1.876 and 1.802 Å, P(1)-C(37) 1.880 and 1.866 Å,

Fe(1)-P(1) 2.214 and 2.222 Å, Fe(1)-P(2) 2.225 and 2.235 Å, Fe(2)-P(1) 2.230 and 2.190 Å. Fe(2)-C(38) in II is 2.139 Å in length, and Fe(2)-P(2) in I is 2.221 Å.

Fig. 1. Molecular structure of $C_{26}H_{18}Fe_2O_6P_2$ (left), and of $C_{27}H_{20}Fe_2O_6P_2$ (right) with phenyl rings indicated by a single atom.

(μ_3-SULPHIDO)BIS(DIMETHYLPHENYLPHOSPHINE)DECACARBONYLPLATINUMTRIOSMIUM
$C_{26}H_{22}O_{10}Os_3P_2PtS$ (I) $PtOs_3(CO)_{10}(PMe_2Ph)_2(\mu_3\text{-}S)$

(μ_3-SULPHIDO)TRIS(DIMETHYLPHENYLPHOSPHINE)NONACARBONYLPLATINUMTRIOSMIUM
$C_{33}H_{33}O_9Os_3P_3PtS$ (II) $PtOs_3(CO)_9(PMe_2Ph)_3(\mu_3\text{-}S)$

(μ_3-SULPHIDO)TRIS(DIMETHYLPHENYLPHOSPHINE)OCTACARBONYLPLATINUMTRIOSMIUM
$C_{32}H_{33}O_8Os_3P_3PtS$ (III) $PtOs_3(CO)_8(PMe_2Ph)_3(\mu_3\text{-}S)$

(μ_3-SULPHIDO)(μ-2-DIMETHYLPHOSPHINOPHENYL-C,P')(HYDRIDO)BIS(DIMETHYLPHENYLPHOSPHINE)-HEPTACARBONYLPLATINUMTRIOSMIUM
$C_{31}H_{33}O_7Os_3P_3PtS$ (IV) $HPtOs_3(CO)_7(PMe_2Ph)_2(PMe_2C_6H_4)(\mu_3\text{-}S)$

R.D. ADAMS and T.S.A. HOR, 1984. Inorg. Chem., 23, 4723-4732.

I. Triclinic, $P\bar{1}$, a = 8.558, b = 13.814, c = 15.214 Å, α = 109.15, β = 93.05, γ = 95.75°, Z = 2. Mo radiation, R = 0.060 for 3512 reflexions.

II. Triclinic, $P\bar{1}$, a = 10.404, b = 13.746, c = 29.332 Å, α = 85.23, β = 88.43, γ = 74.25°, Z = 4. Mo radiation, R = 0.037 for 5488 reflexions.

III. Triclinic, $P\bar{1}$, a = 9.761, b = 11.926, c = 16.412 Å, α = 78.57, β = 89.16, γ = 87.98°, Z = 2. Mo radiation, R = 0.032 for 4223 reflexions.

IV. Triclinic, $P\bar{1}$, a = 10.432, b = 12.727, c = 14.448 Å, α = 98.46, β = 93.82, γ = 91.82°, Z = 2. Mo radiation, R = 0.037 for 3676 reflexions.

The structures of I-IV are shown in Fig. 1. I and II are approximately planar clusters of one Pt and three Os atoms. They have five metal-metal bonds and a triply bridging sulphido ligand on the $PtOs_2$ triangular face. III consists of a tetrahedral cluster of one Pt and three Os atoms with a μ_3-S ligand on the Os_3 face. IV consists of a tetrahedral $PtOs_3$ cluster and has a μ_3-S ligand on the Os_3 face. The phenyl ring of one of the PMe_2Ph ligands is orthometallated on a metal atom adjacent to the primary coordination site. Bond lengths: for I; Os-Os 2.867(1)-2.870(1), Os-Pt 2.768(1) and 2.836(1), Os-S 2.416(4) and 2.414(4), Pt-S 2.337(4), Pt-P 2.275(4) and 2.262(4); for II; Os-Os 2.833(1)-2.907(1), Os-Pt 2.782(1)-2.835(1), Os-S 2.392(4)-2.406(4), Pt-S 2.339(4) and 2.354(4), Os-P 2.332(4) and 2.335(5), Pt-P 2.249(5)-2.279(5); for III; Os-Os 2.770(1)-3.064(1), Os-Pt 2.740(1)-2.853(1), Os-S 2.315(2)-2.406(2), Os-P 2.317(2), Pt-P 2.285(2) and 2.292(2); for IV; Os-Os 2.837(1)-2.972(1), Os-Pt 2.684(1)-2.816(1), Os-S 2.317(3)-2.440(3), Os-P 2.324(3) and 2.332(3), Pt-P 2.289(3), Os-C(aryl) 2.099(10) Å.

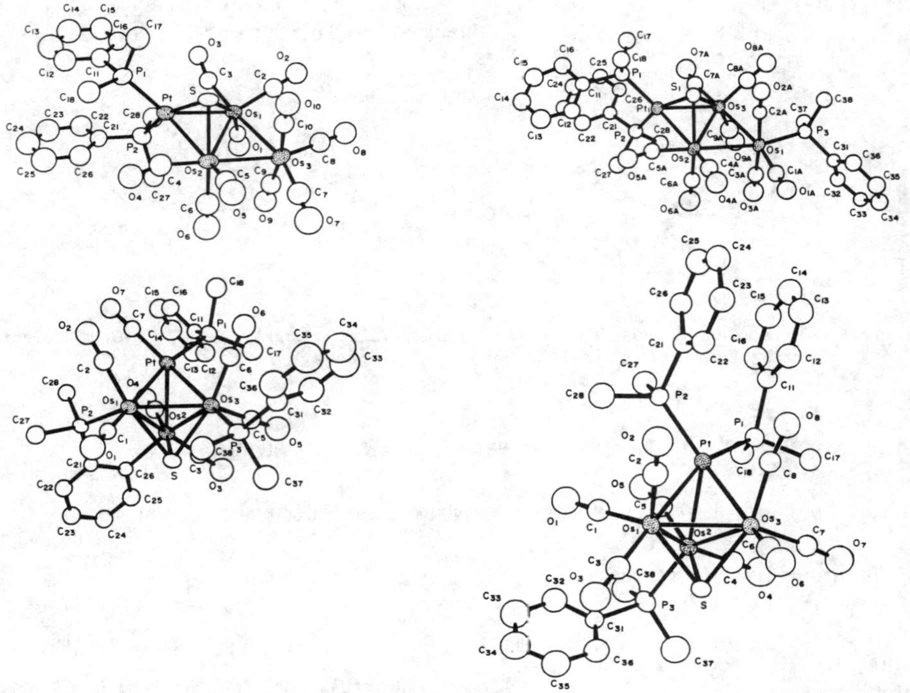

Fig. 1. The structures of $PtOs_3(CO)_{10}(PMe_2Ph)_2(\mu_3\text{-}S)$ (top left), $PtOs_3(CO)_9$-$(PMe_2Ph)_3(\mu_3\text{-}S)$ (top right, one of two independent molecules), $PtOs_3(CO)_8(PMe_2Ph)_3(\mu_3\text{-}S)$ (lower right), and $HPtOs_3(CO)_7(PMe_2Ph)_2$-$(PMe_2C_6H_4)(\mu_3\text{-}S)$ (lower left).

(1,2-BIS(DIPHENYLPHOSPHINO)ETHANE)TRIBROMONICKEL(III) TOLUENE SOLVATE
$C_{26}H_{24}Br_3NiP_2,C_7H_8$

L.R. GRAY, S.J. HIGGINS, W. LEVASON and M. WEBSTER, 1984. J. Chem. Soc. Dalton, 459-467.

Monoclinic, $P2_1/n$, a = 15.567, b = 14.627, c = 15.151 Å, β = 113.38°, D_m = 1.61, Z = 4. Mo radiation, R = 0.056 for 1583 reflexions.

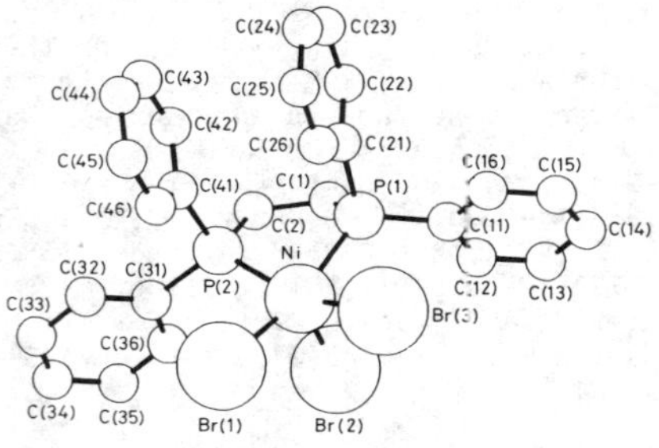

Fig. 1. $C_{26}H_{24}Br_3NiP_2,C_7H_8$: a discrete molecule of the compound.

The structure consists of discrete molecules of the compound (Fig. 1) which has approximate C_s symmetry and packing reveals that holes in the lattice are filled by toluene solvent molecules. The Ni atom has a distorted square pyramidal environment with an apical Br atom. The basal plane is distorted by the chelating diphosphine ligand (P(1)-Ni-P(2) 82.6(2)°). The Br-Ni-Br angle involving basal Br atoms is 95.3° while those involving basal and axial Br atoms are 101 and 105°. The Ni atom lies 0.34 Å above the P_2Br_2 plane. Coordination distances are: Ni-Br (basal) 2.346(3), 2.363(3); Ni-Br (apical) 2.417(3); Ni-P(1) 2.223(5), Ni-P(2) 2.236(5) Å.

DICHLORO{1,3-BIS(DIPHENYLPHOSPHINO)PROPANE}DIGOLD(I)
$C_{27}H_{26}Au_2Cl_2P_2$ $ClAuP(C_6H_5)_2CH_2CH_2CH_2P(C_6H_5)_2AuCl$

M.K. COOPER, L.E. MITCHELL, K. HENRICK, M. McPARTLIN and A. SCOTT, 1984. Inorg. Chim. Acta, 84, L9-L10.

Orthorhombic, Pbcn, a = 19.507, b = 14.404, c = 19.827 Å, Z = 8. Mo radiation, R = 0.062 for 2341 reflexions.

The two Au atoms each have approximately linear coordination (Fig. 1) with P-Au-Cl angles of 173.6 and 175.7°. Chemically equivalent bonds in the two limbs of the compound are equivalent within experimental error (Au(1)-Cl(1) 2.305, Au(2)-Cl(2) 2.296, Au(1)-P(1) 2.231, Au(2)-P(2) 2.237, P(1)-C(1) 1.80 and P(2)-C(3) 1.80 Å. The P-C(phenyl) bond lengths range from 1.81 to 1.82 Å and the C-C chain distances are C(1)-C(2) 1.55 and C(2)-C(3) 1.55 Å. The Au(1)...Au(2) distance between glide related molecules is 3.316 Å.

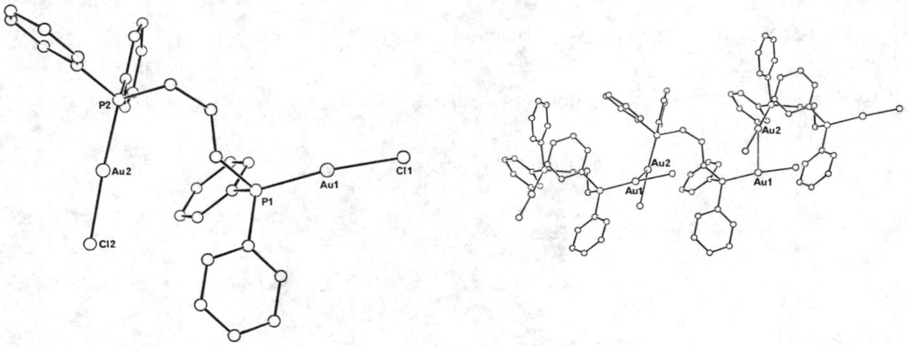

Fig. 1. Molecular structure of $C_{27}H_{26}Au_2Cl_2P_2$ (left), and the polymeric chain structure showing b-glide related metal-metal contacts (right).

(μ-CHLORO)(μ-TRIPHENYLPHOSPHINEGOLD)DECACARBONYLTRIRUTHENIUM
$C_{28}H_{15}AuClO_{10}PRu_3$ $Ru_3(\mu\text{-}AuP(C_6H_5)_3)(\mu\text{-}Cl)(CO)_{10}$

G. LAVIGNE, F. PAPAGEORGIOU and J.-J. BONNET, 1984. Inorg. Chem., 23, 609-613.

Monoclinic, $P2_1/n$, a = 12.697, b = 16.748, c = 16.078 Å, β = 113.0°, Z = 4. Mo radiation, R = 0.022 for 3249 reflexions (at -155°C).

The complex (Fig. 1) has a butterfly metal framework with the Au atom occupying one of the wing tips. Important bond lengths: Ru-Ru (unsupported) 2.8197(7) and 2.8125(8), (μ-Cl)(μ-Au)Ru-Ru 2.8742(6), Ru-Au 2.7523(6) and 2.7549(6), Au-P 2.307(1), Ru-Cl 2.430(1) and 2.439(1), Ru-C 1.871(7)-1.958(7) Å.

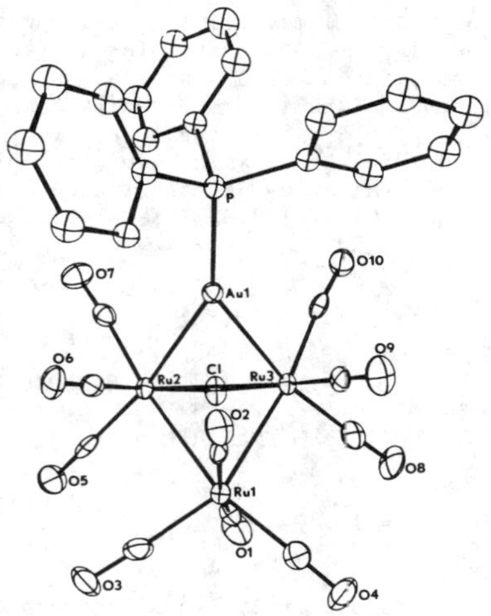

Fig. 1. The $Ru_3(\mu\text{-}AuP(C_6H_5)_3)(\mu\text{-}Cl)(CO)_{10}$ molecule.

BIS(μ_3-SULPHIDO)(TRIPHENYLPHOSPHINE)DECACARBONYLPLATINUMTRIOSMIUM
$C_{28}H_{15}O_{10}Os_3PPtS_2$ (I) $PtOs_3(CO)_{10}(PPh_3)(\mu_3\text{-}S)_2$

BIS(μ_3-SULPHIDO)BIS(TRIPHENYLPHOSPHINE)NONACARBONYLPLATINUMTRIOSMIUM
$C_{45}H_{30}O_9Os_3P_2PtS_2$ (II) $PtOs_3(CO)_9(PPh_3)_2(\mu_3\text{-}S)_2$

R.D. ADAMS, T.S.A. HOR and I.T. HORVATH, 1984. Inorg. Chem., 23, 4733-4738.

I. Triclinic, P$\bar{1}$, a = 8.994, b = 11.312, c = 17.058 Å, α = 76.20, β = 85.06, γ = 78.41°, Z = 2. Mo radiation, R = 0.032 for 3754 reflexions.

II. Monoclinic, $P2_1/n$, a = 9.492, b = 31.220, c = 15.331 Å, β = 99.34°, Z = 4. Mo radiation, R = 0.038 for 3561 reflexions.

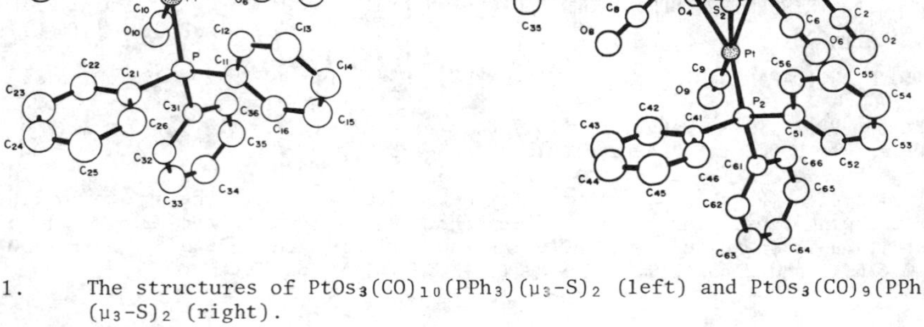

Fig. 1. The structures of $PtOs_3(CO)_{10}(PPh_3)(\mu_3\text{-}S)_2$ (left) and $PtOs_3(CO)_9(PPh_3)_2$-(μ_3-S)$_2$ (right).

Compound II is simply a PPh_3 derivative of I. Both compounds (Fig. 1) consist of triangular $PtOs_2$ clusters, with the third Os atom linked to one of the Os atoms in the triangular group via a metal-metal bond and the two μ_3-S ligands. Bond lengths: for I; Os-Os 2.826(1) and 2.990(1), Os-Pt 2.858(1) and 2.905(1), Os-S 2.399(2)-2.471(2), Pt-S 2.305(2), Pt-P 2.330(2) Å; for II; Os-Os 2.819(1) and 2.998(1), Os-Pt 2.878(1) and 2.904(1), Os-S 2.403(4)-2.484(4), Pt-S 2.304(4), Os-P 2.342(4), Pt-P 2.330(4) Å.

2,2,2,3,3,3-HEXACARBONYL-1,2.1,3.2,3-TRIS(μ_2-HYDRIDO)-2,3-(μ_2-TETRACARBONYLOSMIUM)-1-TRIPHENYLPHOSPHINECOPPERDIOSMIUM

$C_{28}H_{18}CuO_{10}Os_3P$ (I)

1,1,2,2,2,3,3,3,4,4,4-UNDECACARBONYL-1,2.2,3.3,4-TRIS(μ_2-HYDRIDO)-1-TRIPHENYLPHOSPHINEIRIDIUMTRIOSMIUM

$C_{29}H_{18}IrO_{11}Os_3P$ (II)

B.F.G. JOHNSON, J. LEWIS, P.R. RAITHBY, S.N. AZMAN, B. SYED-MUSTAFFA, M.J. TAYLOR, K.H. WHITMIRE and W. CLEGG, 1984. J. Chem. Soc. Dalton, 2111-2118.

I. Monoclinic, $P2_1/c$, a = 12.801, b = 13.539, c = 18.650 Å, β = 105.20°, Z = 4. Mo radiation, R = 0.067 for 2401 reflexions.

II. Monoclinic, $P2_1/c$, a = 8.810(2), b = 16.170, c = 23.420 Å, β = 99.27°, Z = 4. Mo radiation, R = 0.068 for 3145 reflexions.

In I the three Os atoms lie at the vertices of an isosceles triangle, the long edge (3.026(3) Å) of which is bridged by the Cu atom of the Cu(PPh_3) groups to give a 'butterfly' metal framework (Fig. 1). The dihedral angle between the Os(1)Os(2)Os(3) and Os(1)Os(2)Cu planes is 107.8°. Each of the Os atoms bonded to Cu is also bonded to three CO groups while the third Os atom is bonded to four CO groups. The hydrides were not located but evidence suggests that they bridge the long Os-Os bond and the two Os-Cu bonds. In II the Ir atom caps the Os_3 triangle to form a distorted tetrahedral metal framework. The Ir-Os bonds are 2.759(3), 2.789(2) and 2.917(3) Å. The distribution of the eleven terminal CO ligands and the phosphine group, which is coordinated to the Ir atom, indicates that the three hydrides bridge the two longer Os-Os bonds, and the longest Os-Ir bond.

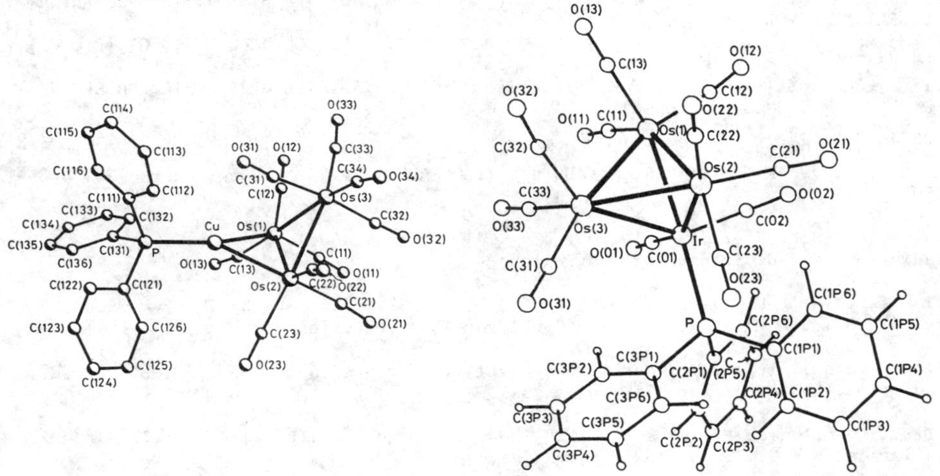

Fig. 1. The molecular structures: $C_{28}H_{18}CuO_{10}Os_3P$ (left) and $C_{29}H_{18}IrO_{11}Os_3P$ (right).

cis-TETRACARBONYLHYDROGENBIS(DIPHENYLPHOSPHINITO-P)MANGANESE(I)
$C_{28}H_{21}MnO_6P_2$ $(CO)_4Mn[(P(C_6H_5)_2O)_2H]$

R. FAWZI, W. HILLER, I.-P. LORENZ, J. MOHYLA and C. ZEIHER, 1984. J. Organometal. Chem., 262, C43-C45.

Orthorhombic, $P2_12_12_1$, a = 9.76, b = 13.72, c = 19.46 Å, Z = 4. R = 0.035 for 3331 reflexions.

The Mn atom is octahedrally-coordinated to four carbonyl and two phosphinite ligands (in a cis configuration) (Fig. 1). The P=O bonds are elongated at 1.55 Å and the two O atoms are linked by an unsymmetrical hydrogen bridge (O(5)-H 1.35, O(6)-H 1.06, O(5)...O(6) 2.41 Å, O(5)-H-O(6) 177°). The Mn-P(1) and Mn-P(2) distances are respectively 2.355 and 2.342 Å and the Mn-C distances range from 1.819 to 1.863 Å. The six-membered ring Mn-P(1)-O(5)-H-O(6)-P(2) is almost planar with a maximum deviation of 0.12 Å (P(2)) from the best plane through the atoms.

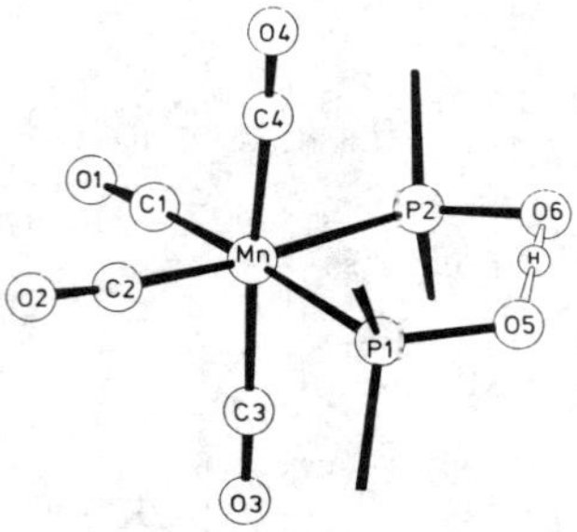

Fig. 1. Molecular structure of $C_{28}H_{21}MnO_6P_2$ with phenyl groups omitted.

BIS(μ_3-CARBONYL)TETRAKIS(μ-CARBONYL)BIS(η^5-CYCLOPENTADIENYL)BIS(TRIETHYLPHOSPHINE)-DIPLATINUMDICHROMIUM
$C_{28}H_{40}Cr_2O_6P_2Pt_2$ (I) $Pt_2Cr_2(\eta^5-C_5H_5)_2(CO)_6[P(C_2H_5)_3]_2$

BIS(μ_3-CARBONYL)TETRAKIS(μ-CARBONYL)BIS(η^5-CYCLOPENTADIENYL)BIS(TRIETHYLPHOSPHINE)-DIPLATINUMDIMOLYBDENUM
$C_{28}H_{40}Mo_2O_6P_2Pt_2$ (II) $Pt_2Mo_2(\eta^5-C_5H_5)_2(CO)_6[P(C_2H_5)_3]_2$

BIS(μ_3-CARBONYL)TETRAKIS(μ-CARBONYL)BIS(η^5-CYCLOPENTADIENYL)BIS(TRIETHYLPHOSPHINE)-DIPLATINUMDITUNGSTEN (monoclinic A and B forms)
$C_{28}H_{40}O_2P_2Pt_2W_2$ (III) $Pt_2W_2(\eta^5-C_5H_5)_2(CO)_6[P(C_2H_5)_3]_2$

R. BENDER, P. BRAUNSTEIN, J.-M. JUD and Y. DUSAUSOY, 1984. Inorg. Chem., 23, 4489-4502.

I. Monoclinic, $P2_1/c$, a = 10.765, b = 9.430, c = 17.450 Å, β = 115.37°, Z = 2. Mo radiation, R = 0.032 for 2321 reflexions.

II. Triclinic, $P\bar{1}$, a = 10.026, b = 11.155, c = 15.126 Å, α = 85.17, β = 75.44, γ = 84.33°, Z = 2. Mo radiation, R = 0.051 for 4574 reflexions.

IIIA. Monoclinic, $P2_1/n$, a = 8.768, b = 14.147, c = 13.580 Å, β = 77.96°, Z = 2. Mo radiation, R = 0.033 for 2365 reflexions.

IIIB. Monoclinic, $P2_1/n$, a = 11.920, b = 12.930, c = 12.166 Å, β = 61.72°, Z = 2. Mo radiation, R = 0.085 for 1921 reflexions.

The $Pt_2M_2Cp_2(\mu_3-CO)_2(\mu-CO)_4(PEt_3)_2$ (M = Cr, Mo, W) complexes are shown in Fig. 1. All possess a centre of symmetry; in the Mo structure there are two slightly different molecules and there are two monoclinic crystal forms of the W complex. The structures are characterized by a planar, triangulated parallelogram framework for the metallic core. The Pt-Pt distances are short, ranging from 2.612(1) Å in I to 2.677(1) Å in II. The Pt-M distances have values of 2.748(1) and 2.709(1) Å for M=Cr and range from 2.777(2) to 2.846(1) Å for M=Mo and from 2.775(1) to 2.836(1) Å for M=W. A shorter Pt-M distance is found where the contribution of the μ-CO ligands to this bond is higher. In the 58-electron clusters, the 18-electron $[CpM(CO_3)]^-$ fragments bridge the L-Pt(I)-Pt(I)-L unit in a very original way: a three-legged piano-stool structure with the two Pt atoms located within the $M(CO)_3$ cone. Each PEt_3 ligand is coordinated to a Pt atom with Pt'-Pt-P 169.7(1)-177.7(1)° and an average Pt-P distance of 2.285 Å. The planes of the η^5-Cp ligands are parallel to one another and make dihedral angles of 75.6-86.9° with the metallic plane. The C(1)-O(1) and C(3)-O(3) ligands are semi-bridging the M-Pt edges and C(2)-O(2) is semi triply bridging the MPtPt face.

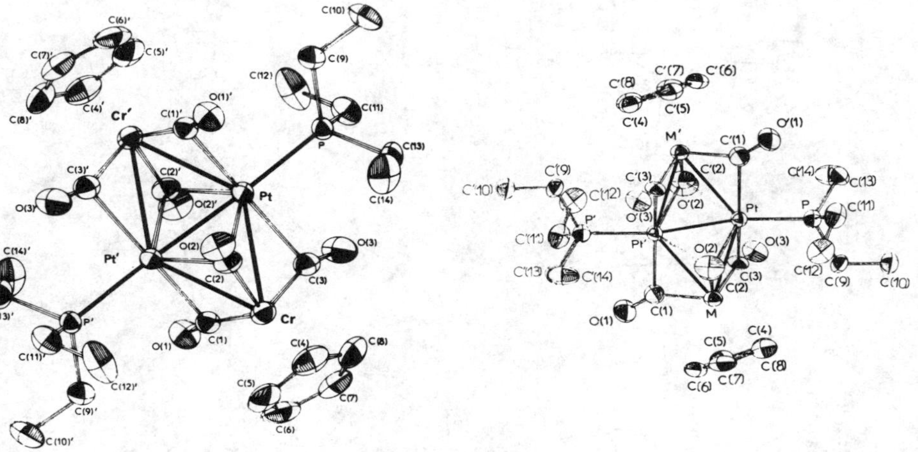

Fig. 1. The structures of $Pt_2M_2Cp_2(\mu_3\text{-}CO)_2(\mu\text{-}CO)_4(PEt_3)_2$; M = Cr (left), Mo and W (right).

UNDECACARBONYLTRIPHENYLPHOSPHINERUTHENIUMDIIRON
$C_{29}H_{15}Fe_2O_{11}PRu$ (I) $Fe_2Ru(CO)_{11}P(C_6H_5)_3$

UNDECACARBONYLTRIMETHOXYPHOSPHINERUTHENIUMDIIRON
$C_{14}H_9Fe_2O_{14}PRu$ (II) $Fe_2Ru(CO)_{11}P(OCH_3)_3$

UNDECACARBONYLTRIPHENYLPHOSPHINEDIRUTHENIUMIRON
$C_{29}H_{15}FeO_{11}PRu_2$ (III) $FeRu_2(CO)_{11}P(C_6H_5)_3$

DECACARBONYLBIS(TRIPHENYLPHOSPHINE)DIRUTHENIUMIRON
$C_{46}H_{30}FeO_{10}P_2Ru_2$ (IV) $FeRu_2(CO)_{10}(P(C_6H_5)_3)_2$

T. VENÄLÄINEN and T. PAKKANEN, 1984. J. Organometal. Chem., 266, 269-283.

I. Triclinic, $P\bar{1}$, a = 9.203, b = 11.903, c = 15.117 Å, α = 81.55, β = 87.29, γ = 66.72°, Z = 2. Mo radiation, R = 0.033 for 3936 reflexions.

II. Orthorhombic, $Pna2_1$, a = 17.220, b = 14.573, c = 8.741 Å, Z = 4. Mo radiation, R = 0.034 for 2016 reflexions.

III. Monoclinic, $P2_1/n$, a = 11.424, b = 16.033, c = 16.642 Å, β = 93.35°, Z = 4. Mo radiation, R = 0.037 for 3975 reflexions.

IV. Orthorhombic, Pccm, a = 14.854, b = 17.180, c = 16.786 Å, Z = 4. Mo radiation, R = 0.054 for 3780 reflexions.

Each of the clusters in I-IV consists of a metal triangle surrounded by ligands (Fig. 1). The range of the Fe-Ru bond distances in the monosubstituted structures I, II and III is 2.757-2.796 Å, with an average of 2.776 Å. The Ru-P distances are practically equal in the PPh_3 derivatives I, III and IV where the average is 2.371 Å. For the $P(OMe)_3$ derivative, II, the corresponding distance is 2.282 Å reflecting the smaller size of the ligand. Disubstitution changes the cluster geometry drastically. In IV the Ru-Ru bond expands 0.03 Å to 2.854 Å while the Fe-Ru bond contracts to 2.664 Å (compared to 2.788 and 2.772 Å in III). The phosphines are oriented along the Ru-Ru axis and the short Fe-Ru bonds are unsymmetrically bridged by carbonyls.

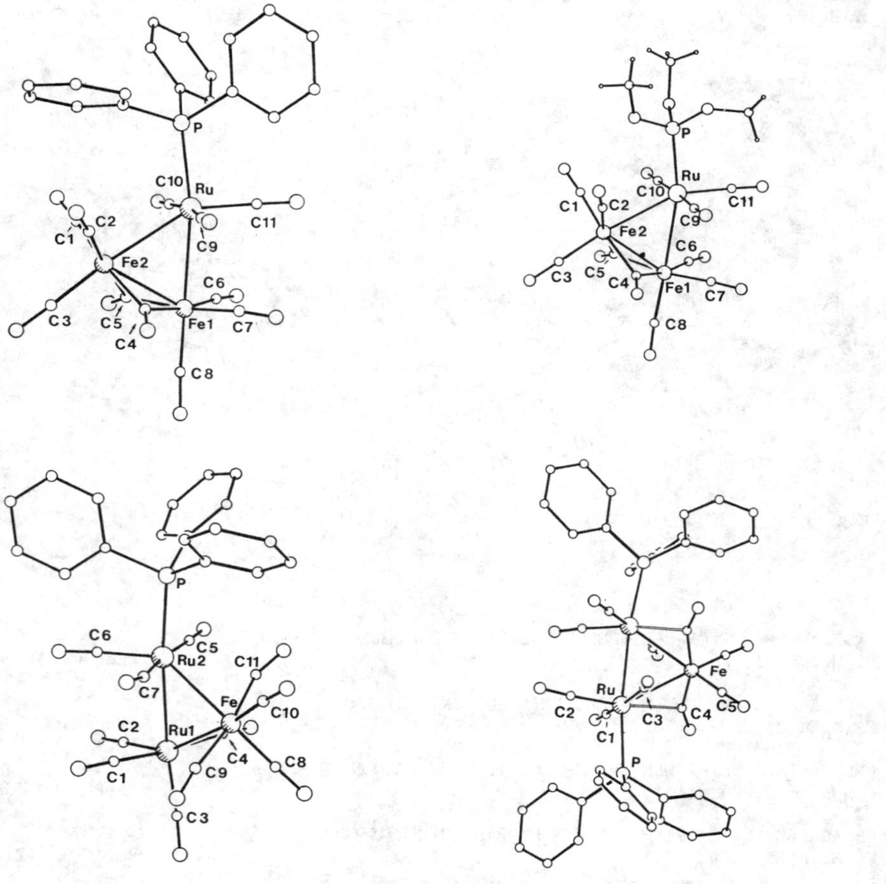

Fig. 1. Molecular structures of $C_{29}H_{15}Fe_2O_{11}PRu$ (top left), $C_{14}H_9Fe_2O_{14}PRu$ (top right), $C_{29}H_{15}FeO_{11}PRu_2$ (bottom left), and $C_{46}H_{30}FeO_{10}P_2Ru_2$ (bottom right).

BIS(μ_2-BROMO)(μ_2-TETRAPHENYLDISTIBANE)BIS(TRICARBONYLRHENIUM((I))
$C_{30}H_{20}Br_2O_6Re_2Sb_2$

I. BERNAL, J.D. KORP, F. CALDERAZZO, R. POLI and D. VITALI, 1984. J. Chem. Soc. Dalton, 1945-1950.

Monoclinic, $P2_1/n$, a = 16.585, b = 22.036, c = 19.764 Å, β = 109.34°, Z = 8. Mo radiation, R = 0.040 for 2197 reflexions.

The molecule (Fig. 1) consists of two pseudo-octahedral rhenium(I) centres joined by bromide and Sb_2Ph_4 bridges. Average distances include Re-Br 2.663(4), Sb...Br 3.640(3), Re...Re 3.970(2) and Sb-Sb 2.826(4) Å. The folding angle, the angle between the normals to the Re-Br-Br planes, is 10.7°. There are two molecules in the asymmetric unit.

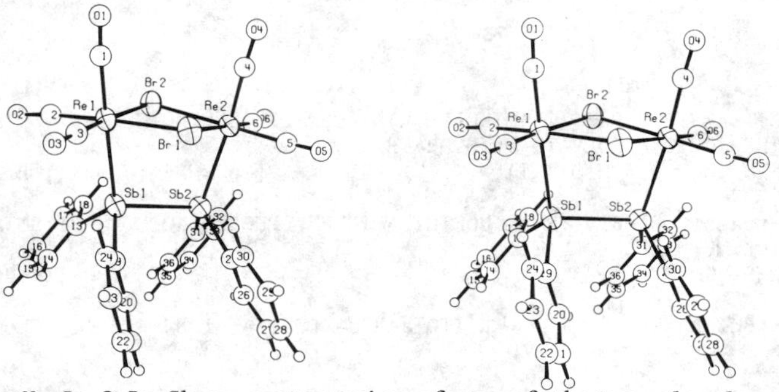

Fig. 1. $C_{30}H_{20}Br_2O_6Re_2Sb_2$: a stereoview of one of the two molecules in the asymmetric unit.

BROMO(cis-3,4-DIHYDRO-2,3,4,5-TETRAPHENYL-2H-1,2,3-DIAZAPHOSPHOLE-P)TETRACARBONYL-RHENIUM

$C_{30}H_{21}BrN_2O_4PRe$

G. BACCOLINI, L. BUSETTO, L. CONTESSA, V. ALBANO and F. DEMARTIN, 1984. Phosphorus and Sulfur, 19, 319-325.

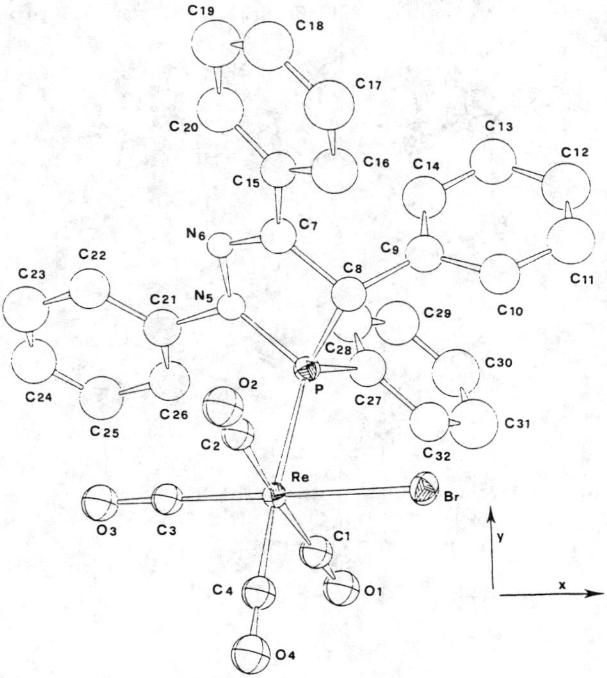

Fig. 1. A view of $C_{30}H_{21}BrN_2O_4PRe$.

Monoclinic, $P2_1/n$, a = 15.671, b = 10.343, c = 18.927 Å, β = 111.51°, Z = 4. Mo radiation, R = 0.046 for 1575 reflexions.

The analysis shows the molecule to have cis-geometry (Fig. 1). The Re coordination sphere has idealized C_s symmetry.

trans(NCS,NCS)-cis(P,P)-BIS((2-AMINOETHYL)DIPHENYLPHOSPHINO-N,P)-DIISOTHIOCYANATO COBALT BROMIDE ACETONE SOLVATE TRIHYDRATE

$C_{30}H_{32}BrCoN_4P_2S_2,C_3H_6O,3(H_2O)$ (I) $[Co(NCS)_2(edpp)_2]^+Br^-,3H_2O,(CH_3)_2Co$

cis(NCS,NCS)-trans(P,P)-BIS((2-AMINOETHYL)DIPHENYLPHOSPHINO-N,P)-DIISOTHIOCYANATO COBALT BROMIDE METHANOL SOLVATE

$C_{30}H_{32}BrCoN_4P_2S_2,CH_4O$ (II) $[Co(NCS)_2(edpp)_2]^+Br^-,CH_3OH$

M. ATOH, K. KASHIWABARA, H. ITO, T. ITO and J. FUJITA, 1984. Bull. Chem. Soc. Japan, 57, 3139-3146.

I. Monoclinic, $P2_1/a$, a = 23.094, b = 17.513, c = 9.755 Å, β = 105.23°, D_m = 1.43, Z = 4. Mo radiation, R = 0.044 for 5342 reflexions (at 168K).

II. Triclinic, $P\bar{1}$, a = 13.158, b = 14.138, c = 9.723 Å, α = 104.99, β = 97.41, γ = 103.6°, D_m = 1.46, Z = 2. Mo radiation, R = 0.056 for 6790 reflexions (at 173K).

The structural differences between I and II are clear (Fig. 1), with the trans-thiocyanate groups and cis P,P-ligands in I and the reverse in II. The Co-P(1) distances in I and II respectively are 2.272 and 2.263 Å, the Co-P(2) distances 2.268 and 2.275 Å, the Co-N(1) distances 1.883 and 1.984 Å, the Co-N(2) distances 1.881 and 1.935 Å, the Co-N(3) distances 2.016 and 1.982 Å and the Co-N(4) distances 2.003 and 1.976 Å. The octahedral cis-angles about Co in both complexes range from 85.2-103.3° in I (the 103.3° angle in P(1)-Co-P(2)) and from 85.3 to 96.5° in II.

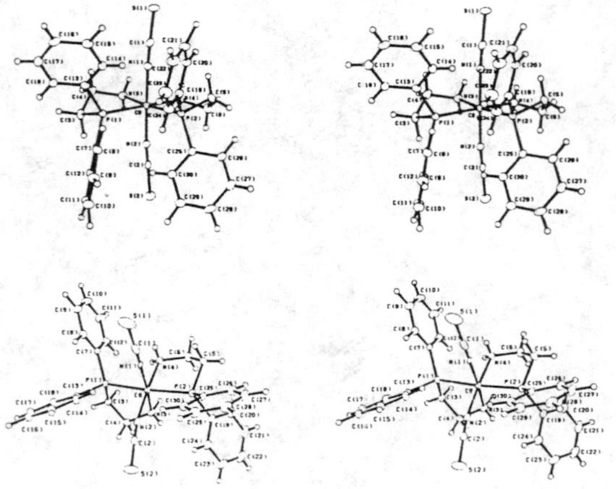

Fig. 1. Stereoviews of the cations in the trans-NCS (top) and cis-NCS (bottom) isomers of $[Co(NCS)_2(edpp)_2]^+$.

fac-TRIS(ACETONITRILE)TRIS(DIMETHYLPHENYLPHOSPHINE)OSMIUM BIS(HEXAFLUOROPHOSPHATE) DICHLOROMETHANE SOLVATE

$C_{30}H_{42}F_{12}N_3OsP_5,CH_2Cl_2$ $Os(NCMe)_3(PMe_2Ph)_3{}^{2+}.2PF_6{}^-,CH_2Cl_2$

J.W. BRUNO, J.C. HUFFMAN and K.G. CAULTON, 1984. J. Am. Chem. Soc., 106, 1663-1669.

Triclinic, $P\bar{1}$, a = 19.653, b = 11.407, c = 10.201 Å, α = 67.03, β = 95.69, γ = 94.96°, Z = 2. Mo radiation, R = 0.0506 for 5023 reflexions.

In the molecule (Fig. 1) the three Os-N distances are identical to within 2σ (2.081-2.097(8) Å), while the three Os-P distances differ by less than 5σ (2.318-2.334(2) Å). The P-Os-P angles increase from 3 to 8° above 90° while N-Os-N angles decrease to 5 to 7° below 90°, but all cis N-Os-P angles are within 1° degree of 90°.

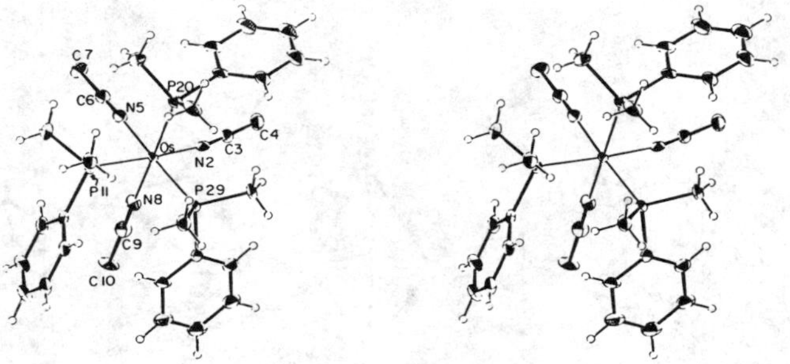

Fig. 1. A stereoview of $Os(NCMe)_3(PMe_2Ph)_3{}^{2+}$.

TRIS(μ_2-CARBONYL)(μ_3-TRIPHENYLPHOSPHINEGOLD)CARBONYLCOBALTTRIS(TRICARBONYLRUTHENIUM)

$C_{31}H_{15}AuCoO_{13}PRu_3$ (I)

(μ_2-CARBONYL)(μ_2-HYDRIDO)(μ_3-BIS(TRIPHENYLPHOSPHINE)DIGOLD)DICARBONYLCOBALTTRIS(TRICARBONYLRUTHENIUM) DICHLOROMETHANE SOLVATE

$C_{48}H_{31}Au_2CoO_{12}P_2Ru_3,0.5(CH_2Cl_2)$ (II)

M.I. BRUCE and B.K. NICHOLSON, 1984. Organometallics, 3, 101-108.

I. Triclinic, $P\bar{1}$, a = 9.345, b = 14.217, c = 14.721 Å, α = 114.83, β = 93.16, γ = 92.44°, D_m = 2.20, Z = 2. Mo radiation, R = 0.037 for 3332 reflexions.

II. Monoclinic, C2/c, a = 35.173, b = 13.548, c = 23.193 Å, β = 102.27°, D_m = 2.02, Z = 8. Mo radiation, R = 0.046 for 2874 reflexions.

The structures of compounds I and II are compared with that of a third, similar compound details of which have already been published (1), (Fig. 1). In compound II the H ligand was not located experimentally but it is assumed to occupy a bridging position along the Co-Ru(2) bond.

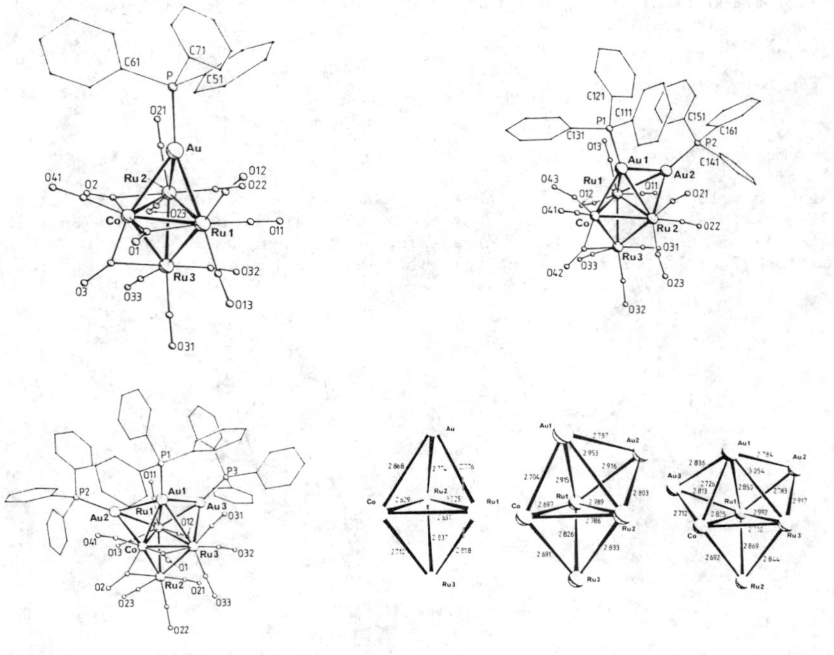

Fig. 1. Perspective views of the molecules of $C_{31}H_{15}AuCoO_{13}PRu_3$ (top left), $C_{48}H_{31}Au_2CoO_{12}P_2Ru_3$ (top right), $C_{66}H_{45}Au_3CoO_{12}P_3Ru_3$ (bottom left), and diagrams of the metal cores of the clusters showing metal-metal distances (Å) (bottom right).

1. M.I. BRUCE and B.K. NICHOLSON, 1982. J. Chem. Soc., Chem. Commun., 1141.

BIS(μ_2-DIPHENYLPHOSPHINO)HEPTACARBONYLDIMANGANESE

$C_{31}H_{20}Mn_2O_7P_2$ $Mn_2(\mu\text{-}PPh_2)_2(\mu\text{-}CO)(CO)_6$

T. KAWAMURA, T. SOWA, T. YONEZAWA, T. YAMABE, H. MASUDA, T. TAGA and K. MACHIDA, 1984. J. Organometal. Chem., 276, C10-C12.

Monoclinic, $P2_1$, a = 12.491, b = 10.066, c = 12.461 Å, β = 109.28°, Z = 2. Cu radiation, R = 0.055 for 2149 reflexions.

The dinuclear structure has a metal-metal bond of length 2.812 Å, a semi-bridging carbonyl group, and two phosphido bridges (Fig. 1). In the semi-bridging CO moiety, Mn(2)-C(7) is 1.91 and Mn(1)-C(7) 2.53 Å and the angles Mn(2)-C(7)-O(7) 163 and Mn(1)-C(7)-O(7) 119.6°. The metal-P distances are: Mn(1)-P(1) 2.214, Mn(2)-P(1) 2.467, Mn(1)-P(2) 2.176 and Mn(2)-P(2) 2.436 Å, and the angles Mn(1)-P(1)-Mn(2) and Mn(1)-P(2)-Mn(2) are 73.6 and 74.9° respectively.

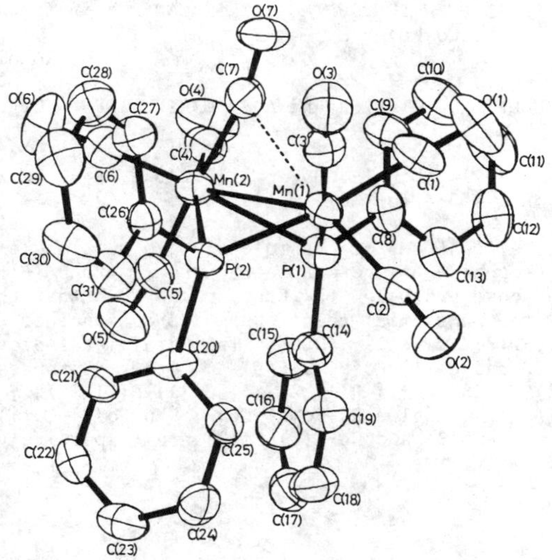

Fig. 1. Molecular structure of $C_{31}H_{20}Mn_2O_7P_2$.

(μ_2-HYDRIDO)(μ_2-HYDROXO)(μ_2-BIS(DIPHENYLPHOSPHINO)METHANE)HEXACARBONYLDIRHENIUM
$C_{31}H_{24}O_7P_2Re_2$ (μ-H)(μ-OH)$Re_2(CO)_6$(dppm)

K.-W. LEE, W.T. PENNINGTON, A.W. CORDES and T.L. BROWN, 1984. Organometallics, 3, 404-413.

Triclinic, $P\bar{1}$, a = 10.904, b = 11.099, c = 14.646 Å, α = 84.69, β = 83.18, γ = 60.69°, Z = 2. Mo radiation, R = 0.031 for 5547 reflexions.

The molecular structure consists of two $Re(CO)_3$ units bridged by hydrido, hydroxo, and dppm ligands (Fig. 1). The metal-metal distance is 3.030 Å, the Re-H distances of the H bridge are 1.84 and 2.00 Å, and the Re-O distances of the hydroxo bridge are 2.177 and 2.160 Å. The average Re-CO bond length is 1.921 Å and the Re-P distances are 2.453 and 2.463 Å. The Re(1)-O(1)-Re(2) angle is 88.6°.

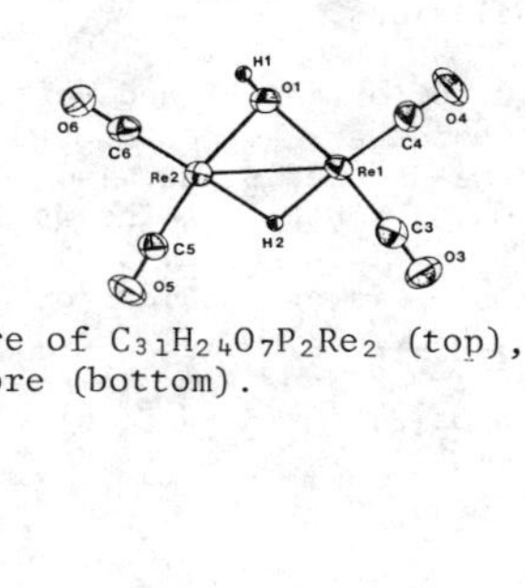

Fig. 1. Molecular structure of $C_{31}H_{24}O_7P_2Re_2$ (top), and the structure of the $Re_2(CO)_4(H)(OH)$ core (bottom).

μ-CARBONYLTRICARBONYL[μ-(DIPHENYLPHOSPHINO)][TETRACARBONYL[μ-(DIPHENYLPHOSPHINO)]-DICOBALT]RUTHENIUM(Co-Co)(2Co-Ru)
$C_{32}H_{20}Co_2O_8P_2Ru$ $RuCo_2(\mu\text{-}PPh_2)_2(\mu\text{-}CO)(CO)_7$

R. REGRAGUI, P.H. DIXNEUF, N.J. TAYLOR and A.J. CARTY, 1984. Organometallics, 3, 1020-1025.

Monoclinic, $P2_1/n$, a = 12.714, b = 15.600, c = 17.246 Å, β = 108.33°, D_m = 1.66, Z = 4. Mo radiation, R = 0.034 for 3610 reflexions.

The gross structural features of the molecule are a triangle of two Co atoms and one Ru atom with two phosphido bridges and a μ-carbonyl group (Fig. 1). The two phosphido bridges lie above and below the $RuCo_2$ plane such that the atoms P(1) and P(2) bear an approximate trans axial relationship to one another with respect to Co(1). The P(1)-Co(1)-P(2) angle is 140.5°. The metal-metal distances are: Ru-Co(1) 2.7073, Ru-Co(2) 2.7182 and Co(1)-Co(2) 2.5497 Å. Other distances of interest are: Ru-P(1) 2.318, Ru-C(8) 2.325, Co(1)-P(1) 2.183, Co(1)-P(2) 2.151, Co(2)-C(8) 1.818 and Co(2)-P(2) 2.213 Å. The angles Ru-Co(1)-Co(2), Ru-Co(2)-Co(1), Co(1)-Ru-Co(2) Ru-P(1)-Co(1), Co(1)-P(2)-Co(2) and Ru-C(8)-Co(2) are respectively 62.2, 61.8, 56.1, 73.9, 71.5 and 81.0°.

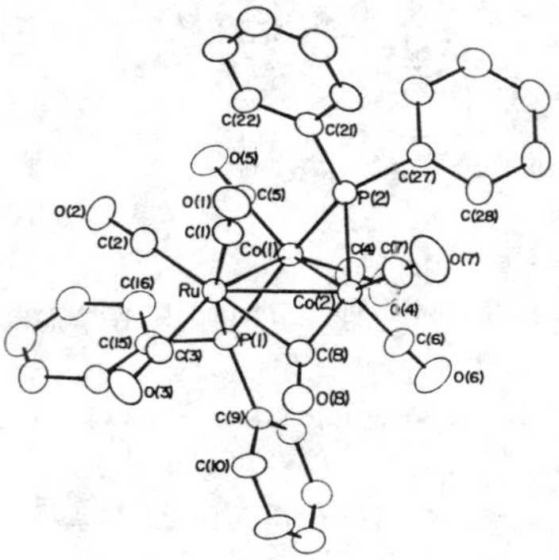

Fig. 1. Perspective view of the molecular structure of $C_{32}H_{20}Co_2O_8P_2Ru$.

(μ_3-SULPHIDO)OCTACARBONYLTRIS(DIMETHYLPHENYLPHOSPHINE)PLATINUMTRIOSMIUM
$C_{32}H_{33}O_8Os_3P_3PtS$ (I) $PtOs_3(CO)_8(PMe_2Ph)_3(S)$

(μ_3-SULPHIDO)UNDECACARBONYLBIS(DIMETHYLPHENYLPHOSPHINE)TUNGSTENTRIOSMIUM
$C_{27}H_{22}O_{11}Os_3P_2SW$ (II) $Os_3W(CO)_{11}(PMe_2Ph)_2(S)$

R.D. ADAMS, T.S.A. HOR and P. MATHUR, 1984. Organometallics, 3, 634-636.

I. Triclinic, $P\bar{1}$, a = 9.761, b = 11.926, c = 16.412 Å, α = 78.57, β = 89.16, γ = 87.98°, Z = 2. Mo radiation, R = 0.032 for 4223 reflexions.

II. Monoclinic, $P2_1/n$, a = 14.631, b = 12.654, c = 19.354 Å, β = 106.32°, Z = 4. Mo radiation, R = 0.055 for 3251 reflexions.

Compound I consists of a cluster of one Pt and three Os atoms in a distorted tetrahedron (Fig. 1). The Os-Pt bonds are asymmetric (Os(1)-Pt 2.853, Os(2)-Pt 2.740, Os(3)-Pt 2.766 Å) and the Os-Os bonding is also quite irregular (Os(1)-Os(2) 2.816, Os(1)-Os(3) 2.770, Os(2)-Os(3) 3.064 Å). The Os-S distances in the triply-bridging sulphido moiety are Os(1)-S 2.406, Os(2)-S 2.315 and Os(3)-S 2.350 Å. In II, the structure of the metallic cluster is again a distorted tetrahedron with distances Os(1)-Os(2) 2.831, Os(1)-Os(3) 2,798, Os(2)-Os(3) 2.870, Os(1)-W 2.947, Os(2)-W 2.193, Os(3)-W 2.827, Os(1)-S 2.311, Os(2)-S 2.396, W-S 2.37 Å. Carbonyl C(10)-O(10) could be weakly semi-bridging from W to Os(3) (Os(3)-W-C(10) 64.6, W-C(10)-O(10) 168.3°).

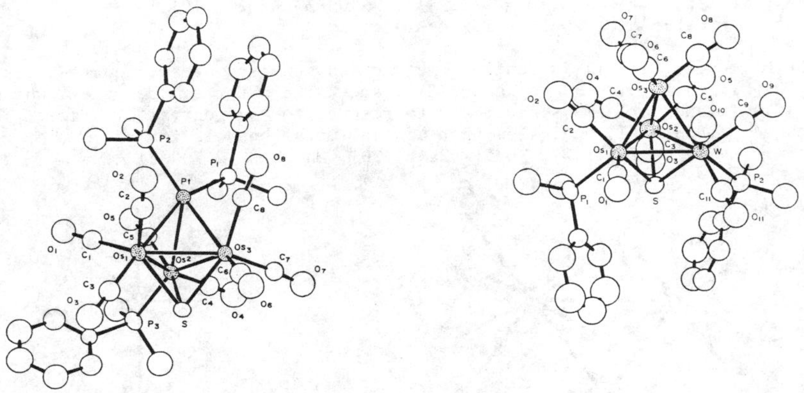

Fig. 1. Molecular structures of $C_{32}H_{33}O_8Os_3P_3PtS$ (left), and $C_{27}H_{22}O_{11}Os_3P_2SW$ (right).

(TETRAHYDROBORATO)BIS(DIPHENYL-n-BUTYLPHOSPHINE)COPPER(I)

$C_{32}H_{42}BCuP_2$ $Cu(BH_4)[P(C_6H_5)_2C_4H_9]_2$

E.B. LOBKOVSKII, V.D. MAKHAEV and A.P. BORISOV, 1984. J. Struct. Chem., 25, 496-498 [Zh. Strukt. Khim., 25(3), 172-175].

Monoclinic, $P2_1/n$, a = 12.400, b = 14.672, c = 18.115 Å, γ = 105.31°, Z = 4. Mo radiation, R = 0.062 for 1521 reflexions.

The molecule (Fig. 1) contains a bidentate BH_4^- ligand. The coordination about Cu is distorted tetrahedral with Cu-P 2.246(2) and 2.253(3), Cu-H 1.60(10) and 1.80(10) Å. One of the n-butyl side chains was found to be randomly disordered.

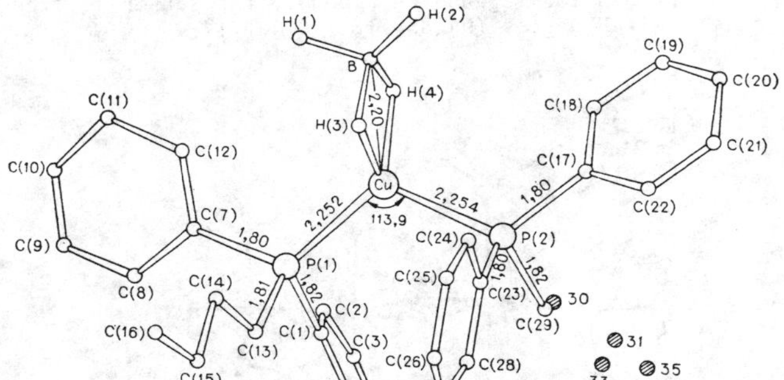

Fig. 1. The (tetrahydroborato)bis(diphenyl-n-butylphosphine)copper(I) molecule.

TETRAKIS(μ-CHLORO)-BIS[DICHLOROBIS(DIMETHYL(PHENYL)PHOSPHINE)NIOBIUM(IV)] BENZENE SOLVATE

$C_{32}H_{44}Cl_8Nb_2P_4,C_6H_6$ $[NbCl_2(P(CH_3)_2C_6H_5)_2]_2(\mu\text{-}Cl)_4,C_6H_6$

F.A. COTTON and W.J. ROTH, 1984. Inorg. Chem., 23, 945-947.

Monoclinic, $P2_1/n$, a = 8.545, b = 14.277, c = 19.634 Å, β = 101.33°, Z = 2. Mo radiation, R = 0.0385 for 2968 reflexions.

The molecule (Fig. 1) is centrosymmetric and consists of two square-antiprismatic $Nb(PMe_2Ph)_2Cl_6$ units sharing a square Cl_4 face. The outer square faces, which have a trans P_2Cl_2 ligand set, are staggered with respect to the central Cl_4 face. The Nb-Nb distance, 2.838(1) Å, indicates that a metal-metal bond is formed. Other mean bond lengths are: Nb-Cl_b 2.541(6), Nb-Cl_t 2.484(10), and Nb-P 2.700(2) Å.

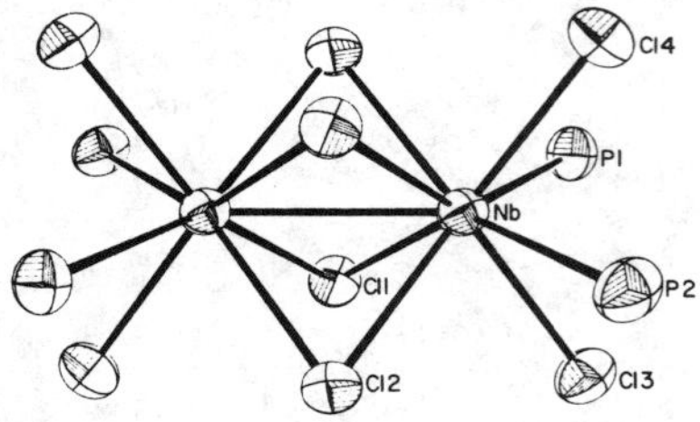

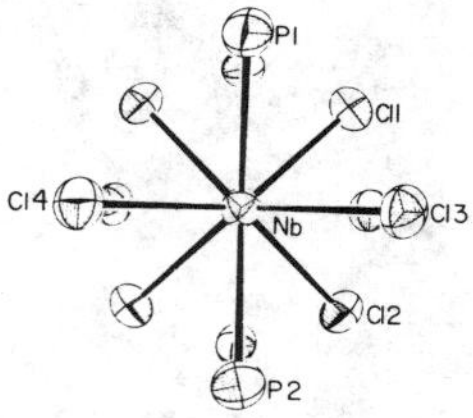

Fig. 1. Two views of the central portion of $[NbCl_2(P(CH_3)_2C_6H_5)_2]_2(\mu\text{-}Cl)_4$.

DI(μ-H)-[DIHYDROGENBIS(DIMETHYLPHENYLPHOSPHINE)IRIDIUM(IV)]

$C_{32}H_{50}Ir_2P_4$ $(\mu\text{-}H)_2[IrH_3(PMe_2Ph)_2]_2$

G.B. ROBERTSON and P.A. TUCKER, 1984. Aust. J. Chem., 37, 257-263.

Orthorhombic, Pbca, a = 11.433, b = 14.768, c = 20.616 Å, Z = 4. Cu radiation, R = 0.037 for 1897 reflexions.

A molecule (Fig. 1) consists of $(\mu\text{-}H)_2$ bridged $[IrH_3(PMe_2Ph)_2]_2$ dimers at a centre of symmetry (Ir...Ir 2.739(1) Å). The metal ion is approximately octahedral with two cis-phosphine ligands, two terminal and two bridging hydrogens in the coordination. The trans effect influences Ir-P bonds, mean 2.273(3) Å. The low steric requirements of the hydrido ligands is probably the reason for the enlarged P-Ir-P (98.33(9)°) and Ir-Ir-P angles (94.85(7), 140.43(7)°).

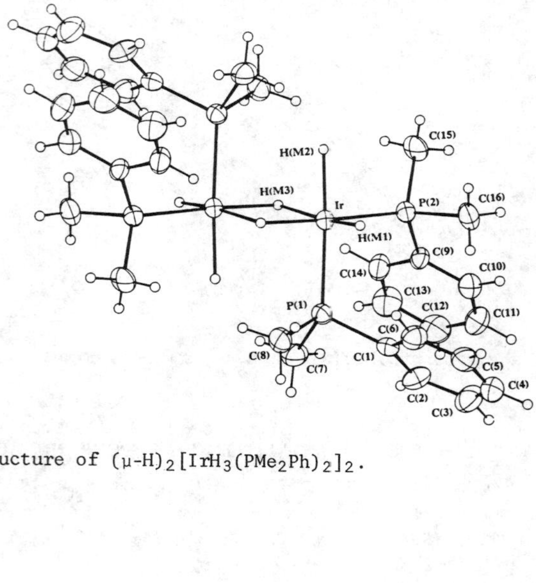

Fig. 1. The structure of $(\mu\text{-}H)_2[IrH_3(PMe_2Ph)_2]_2$.

BIS(TETRA-n-BUTYLAMMONIUM) BIS(μ-BROMO)OCTABROMODIOSMIUM(IV)
$C_{32}H_{72}Br_{10}N_2Os_2$ (I) $[N(n-C_4H_9)_4]_2[Os_2Br_{10}]$

TRIBROMO(ACETONITRILE)BIS(TRIPHENYLPHOSPHINE)OSMIUM(III)
$C_{38}H_{33}Br_3NOsP_2$ (II) $OsBr_3(P(C_6H_5)_3)_2(CH_3CN)$

F.A. COTTON, S.A. DURAJ, C.C. HINCKLEY, M. MATUSZ and W.J. ROTH, 1984. Inorg. Chem., 23, 3080-3083.

I. Monoclinic, $P2_1/n$, a = 12.186, b = 16.528, c = 12.908 Å, β = 103.02°, Z = 2. Mo radiation, R = 0.037 for 923 reflexions.

II. Monoclinic, $P2_1/n$, a = 10.210, b = 15.308, c = 23.122 Å, β = 91.15°, Z = 4. Mo radiation, R = 0.0385 for 3386 reflexions.

Compound I contains centrosymmetric edge-sharing bioctahedral $[Br_4Os(\mu\text{-}Br)_2OsBr_4]^{2-}$ anions having approximate D_{2h} symmetry (Fig. 1). The Os...Os distance is 3.788(3) Å, Os-(μ-Br)$_{avg}$ 2.544(4), Os-Br (terminal)$_{avg}$ 2.454(5) Å, and Os-Br-Os 96.3(1)°. Compound II consists of approximately octahedral molecules (Fig. 1) in which the Br ligands are meridional and the PPh_3 ligands are mutually trans. Bond lengths: Os-Br 2.481(1), 2.499(1) (trans to Br), 2.499(1) (trans to CH_3CN), Os-P 2.418(2), 2.419(2), and Os-N 2.024(10) Å.

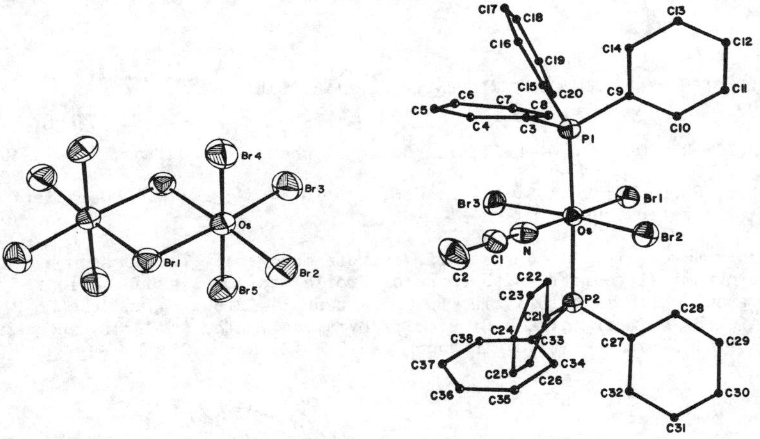

Fig. 1. The $[Os_2Br_{10}]^{2-}$ anion (left) and the $OsBr_3(PPh_3)_2(CH_3CN)$ molecule (right).

(μ_2-TETRAPHENYLDIPHOSPHOXANE-P,P')(TETRACARBONYLIRON)PENTACARBONYLMOLYBDENUM
$C_{33}H_{20}FeMoO_{10}P_2$ $Mo(CO)_5PPh_2OPPh_2Fe(CO)_4$

E.H. WONG, F.C. BRADLEY, L. PRASAD and E.J. GABE, 1984. J. Organometal. Chem., 263, 167-177.

Triclinic, $P\bar{1}$, a = 10.533, b = 10.993, c = 16.439 Å, α = 97.86, β = 92.34, γ = 116.20°, Z = 2. Mo radiation, R = 0.037 for 3830 reflexions.

The molecule contains a tetraphenyldiphosphoxane bridging a six-coordinate Mo and a five-coordinate Fe (Fig. 1). The Mo atom has distorted octahedral geometry with angles ranging from 84.3 to 94.6°, and the Fe atom has a slightly distorted trigonal bipyramidal geometry with P in the expected axial position. Distances involving Mo are: Mo-P(1) 2.483, Mo-C(5) 2.026, Mo-C(equator.) (average) 2.03 Å, and those involving Fe are: Fe-P(2) 2.209, Fe-C(9) 1.798, Fe-C(equator.) (average) 1.77 Å. The bridging P-O-P angle is 146.4° and the associated distances are P(1)-O(10) 1.641 and P(2)-O(10) 1.630 Å.

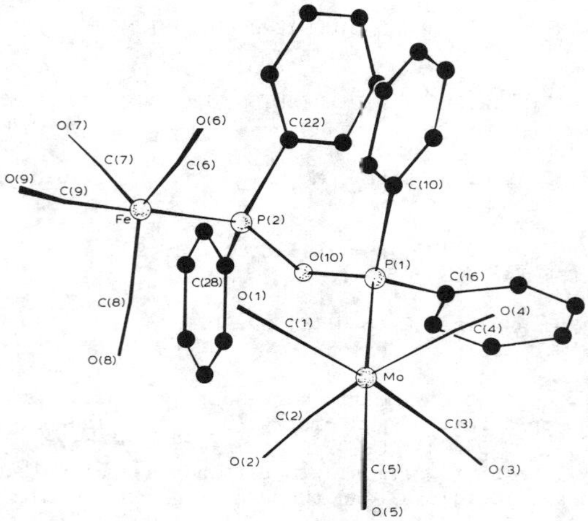

Fig. 1. Molecular structure of $C_{33}H_{20}FeMoO_{10}P_2$.

TRICARBONYL(DIPHENYLBISMUTHIDO)TRIPHENYLPHOSPHINECOBALT(I)
$C_{33}H_{25}BiCoO_3P$

F. CALDERAZZO, R. POLI and G. PELIZZI, 1984. J. Chem. Soc. Dalton, 2535-2540.

Orthorhombic, Pbca, a = 26.105, b = 21.066, c = 10.845 Å, Z = 8. Mo radiation, R = 0.052 for 1176 reflexions.

The structure (Fig. 1) consists of discrete monomers. The Co atom has distorted trigonal bipyramidal geometry with three equatorial carbonyl groups oriented towards the $BiPh_2$ group which is axial. The Bi-Co distance is 2.692(5) and the Co-P distance 2.210(9) Å. The geometry at the Bi atom is pyramidal. The Co-CO distances range from 1.75(3) to 1.84(3) and the Bi-C distances are 2.25(2) and 2.28(2) Å.

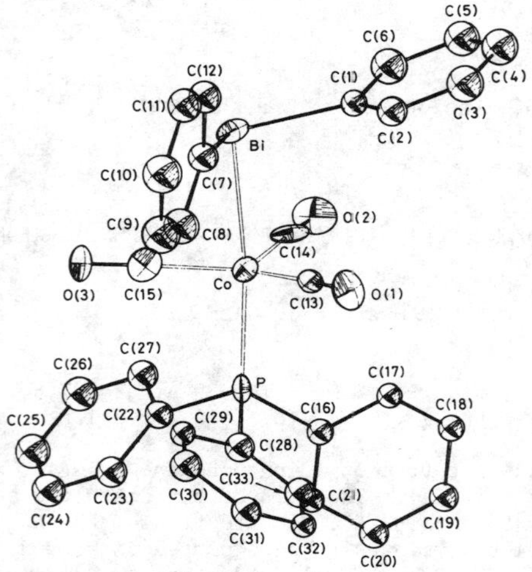

Fig. 1. $C_{33}H_{25}BiCoO_3P$: the molecular structure.

cis-DICHLOROBIS[(2-METHYLCYCLOPROPYLMETHOXY)DIPHENYLPHOSPHINE]PLATINUM(II) CHLOROFORM SOLVATE

$C_{34}H_{38}Cl_2O_2P_2Pt,CHCl_3$ $PtCl_2(C_{17}H_{19}OP)_2.CHCl_3$

T.J. BARTCZAK, W.J. YOUNGS and J.A. IBERS, 1984. Acta Cryst., C40, 1564-1567.

Monoclinic, $P2_1/c$, a = 11.628, b = 18.274, c = 17.266 Å, β = 93.63°, Z = 4. Mo radiation, R = 0.053 for 4156 reflexions (at 118K).

The Pt^{II} ion has tetrahedrally distorted, square-planar coordination (Fig. 1) with Pt-Cl 2.358(2) and 2.369(2) and Pt-P 2.222(2) and 2.224(3) Å. Deviations from the best, weighted Cl_2PtP_2 plane are Cl(1) 0.143(2), Cl(2) -0.131(2), P(1) 0.119(2), P(2) -0.153(2), Pt -0.025(2) Å; the trans P-Pt-Cl angles are 171.03(8) and 172.55(8)°. There is disorder in the methylcyclopropane rings. The closest complex...solvent distance is 3.3 Å for Cl(3)...C(16).

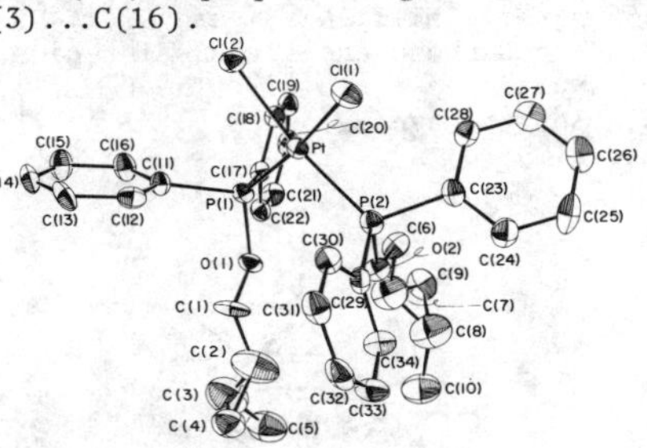

Fig. 1. The $PtCl_2(C_{17}H_{19}OP)_2$ complex.

[η^2-BIS(N,O-DIPHENYLPHOSPHINOMETHYLDIMETHYLSILYL)ACRYLOYLAMIDO-P,P']NICKEL(II)

$C_{34}H_{39}NNiO_2P_2Si_2$

M.D. FRYZUK and P.A. MACNEIL, 1984. J. Am. Chem. Soc., 106, 6993-6999.

Monoclinic, $P2_1/n$, a = 16.240, b = 11.534, c = 19.076 Å, β = 105.39°, Z = 4. Mo radiation, R = 0.050 for 1733 reflexions.

The structure is shown to be as given in Fig. 1, with Ni-P 2.214 and 2.244(3), Ni-C(3) 1.731(15), Ni-C(9) 2.08(2), Ni-C(10) 2.083(13) Å. The geometry at Ni is only slightly distorted from tetrahedral with P-Ni-P 106.5, P-Ni-C(3) 105.9 and 106.2°.

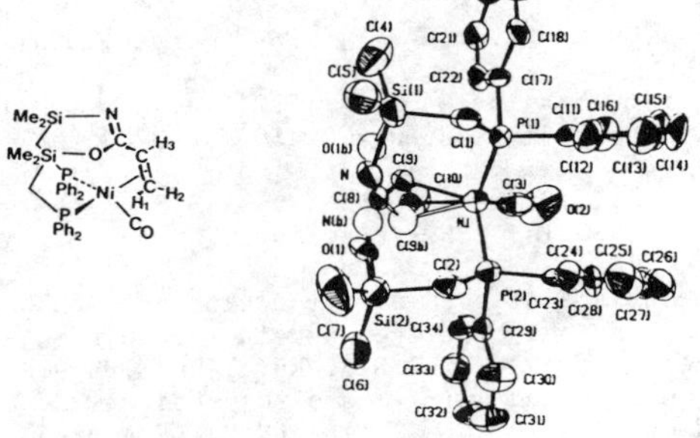

Fig. 1. A view of $C_{34}H_{39}NNiO_2P_2Si_2$.

BIS(μ_3-CARBONYL)BIS(μ_4-σ^2,η^4-BIS(BIS(TRIMETHYLSILYL)METHYL)DIPHOSPHENE)CHLOROTETRACARBONYLPENTANICKEL
$C_{34}H_{76}ClNi_5O_6P_4Si_8$

M.M. OLMSTEAD and P.P. POWER, 1984. J. Am. Chem. Soc., 106, 1495-1496.

Monoclinic, $P2_1/c$, a = 19.853, b = 13.686, c = 23.298 Å, β = 113.13°, Z = 4. Mo radiation, R = 0.07 for 5484 reflexions (at 140 K).

The molecular geometry (Fig. 1) involves four of the five nickel atoms in a butterfly arrangement. These four nickels are bound to terminal carbonyls, and two additional carbonyls triply bridge the underside of each butterfly wing. The two diphosphene ligands bridge two wing tips and hinge metal centres to the fifth nickel atom, Ni(1), which is also bonded to chlorine. The Ni-Ni distances within the butterfly average ca. 2.55 Å while Ni(1)-Ni(3) and Ni(1)-Ni(5) distances are somewhat longer, averaging 2.69 Å. The diphosphene ligands interact most closely with Ni(3) and Ni(5) (Ni-P distances average ca. 2.18 Å). The P-Ni(1), P-Ni(2), and P-Ni(3) distances are almost 0.2 Å longer, averaging ca. 2.37 Å. The P(1)-P(2) and P(3)-P(4) distances are 2.085(4) and 2.098(4) Å.

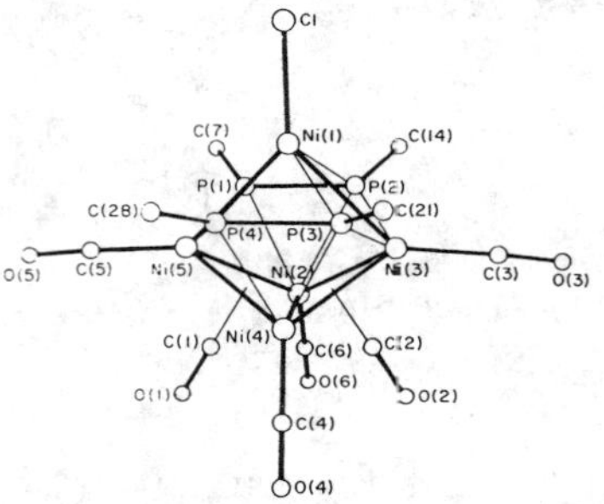

Fig. 1. A view of $C_{34}H_{76}ClNi_5O_6P_4Si_8$; the $SiMe_3$ groups are not shown.

(μ_3-1,2-DIPHENYLVINYLENE)(μ_2-HYDRIDO)(μ_2-DIPHENYLPHOSPHINO)NONACARBONYLTRIRUTHENIUM DIPHENYLACETYLENE SOLVATE
$C_{35}H_{21}O_9PRu_3,0.5C_{14}H_{10}$ (I)
(μ_5-σ^3,η^2,η^2-PHENYLACETYLENE)(μ_3-σ^2,η^2-1,4-DIPHENYLBUT-1-EN-3-YNE-1,2-DIYL)(μ_2-DIPHENYLPHOSPHINO)DODECACARBONYLPENTARUTHENIUM
$C_{48}H_{25}O_{12}PRu_5$ (II)

S.A. MACLAUGHLIN, N.J. TAYLOR and A.J. CARTY, 1984. Organometallics, 3, 392-399.

I. Triclinic, $P\bar{1}$, a = 9.616, b = 14.194, c = 15.218 Å, α = 102.17, β = 102.31, γ = 97.91°, D_m = 1.72, Z = 2. Mo radiation, R = 0.030 for 2744 reflexions.

II. Monoclinic, $P2_1/n$, a = 11.195, b = 25.638, c = 16.384 Å, β = 92.79°, D_m = 1.85, Z = 4. Mo radiation, R = 0.030 for 4257 reflexions.

The structure of I consists of a 50-electron cluster with an open triangular framework, μ-hydrido and μ-phosphido bridges and face-capping μ_3-acetylene (Fig. 1). Metal-metal distances are: Ru(1)-Ru(2) 2.813, Ru(1)-Ru(3) 2.908, Ru(2)...Ru(3) 3.838 Å. The distances Ru(3)-P and Ru(2)-P are respectively 2.387 and 2.385 Å and those of Ru(2)-C(11) and Ru(3)-C(10) are 2.118 and 2.168 Å respectively. The presence of half a molecule of diphenylacetylene of crystallization permits direct internal comparison of a -C≡C- bond lengthening on coordination. C(10)-C(11) is 1.415 Å whereas C(36)-C(36') in PhC≡CPh is 1.253 Å. In II, an acetylene caps an open triangular face as a four electron ligand. The Ru-Ru distances range from 2.7660 to 2.9200 Å with Ru(1)...Ru(2) as a non-bonded contact of 3.6280 Å. Other distances of interest are C(17)≡C(18) 1.187, C(13)≡C(14) 1.340, Ru(4)-P 2.275, Ru(5)-P 2.272 Å.

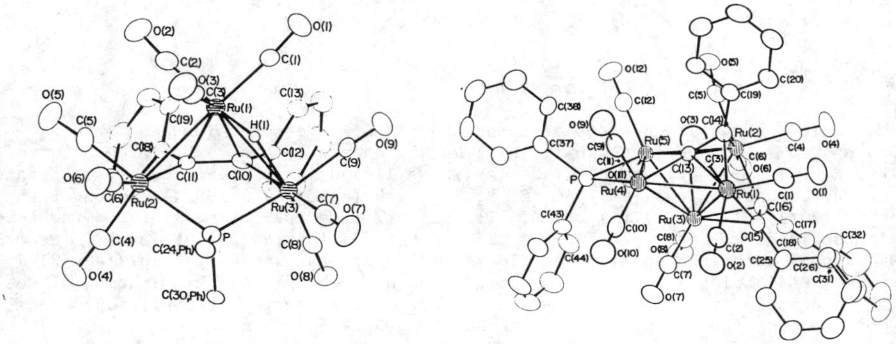

Fig. 1. Molecular structures of $C_{35}H_{21}O_9PRu_3$ (left), and of $C_{48}H_{25}O_{12}PRu_5$.

(BIS(DIPHENYLPHOSPHINO)METHANE)DECACARBONYLTRIRUTHENIUM

$C_{35}H_{22}O_{10}P_2Ru_3$ $Ru_3(CO)_{10}[(C_6H_5)_2PCH_2P(C_6H_5)_2]$

A.W. COLEMAN, D.F. JONES, P.H. DIXNEUF, C. BRISSON, J.-J. BONNET and G. LAVIGNE, 1984. Inorg. Chem., 23, 952-956.

Monoclinic, $P2_1/c$, a = 13.122, b = 12.040, c = 23.658 Å, β = 103.44°, Z = 4. Mo radiation, R = 0.036 for 4346 reflexions.

The molecule is shown in Fig. 1. The $Ph_2PCH_2PPh_2$ group that bridges a Ru-Ru bond occupies equatorial positions. Important bond lengths: Ru(1)-Ru(2) 2.834(1), Ru(1)-Ru(3) 2.841(1), Ru(2)-Ru(3) 2.860(1), Ru(1)-P(1) 2.322(2), Ru(2)-P(2) 2.334(2) Å. The shortest Ru-Ru bond is supported by the chelating phosphorus ligand. The Ru-P bonds are not coplanar, P-Ru-Ru-P 19.1(1)°. Of particular interest is the bending of every axial CO toward one Ru-Ru bond.

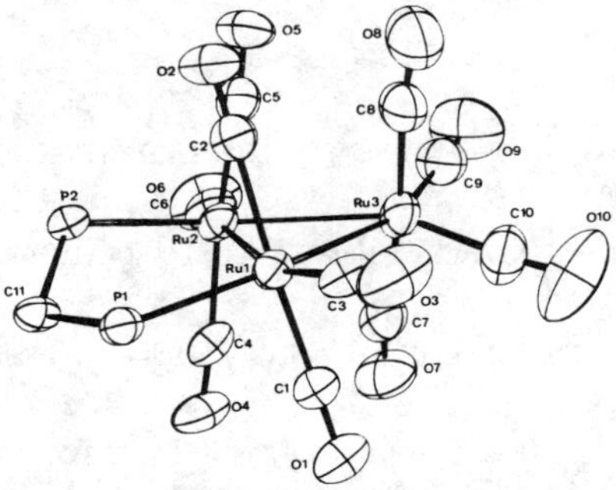

Fig. 1. The $Ru_3(CO)_{10}[(C_6H_5)_2PCH_2P(C_6H_5)_3]$ molecule, phenyl groups omitted.

(μ_4-PHENYLPHOSPHINO)(μ_3-ISOPENTYLIDYNE)(μ_2-DIPHENYLPHOSPHINO)DODECACARBONYLPENTA-RUTHENIUM

$C_{35}H_{24}O_{12}P_2Ru_5$ (I)

BIS(μ_4-PHENYLPHOSPHINO)(μ_3-PHENYLPHOSPHINO)(μ_3-HYDRIDO)(μ_2-DIPHENYLPHOSPHINO)DECA-CARBONYLPENTARUTHENIUM BENZENE SOLVATE

$C_{40}H_{26}O_{10}P_4Ru_5$,$0.5(C_6H_6)$ (II)

K. KWEK, N.J. TAYLOR and A.J. CARTY, 1984. J. Am. Chem. Soc., 106, 4636-4638.

I. Monoclinic, $P2_1/c$, a = 11.724, b = 16.752, c = 20.617 Å, β = 94.05°, D_m = 1.96, Z = 4. Mo radiation, R = 0.023 for 4309 reflexions.

II. Monoclinic, $P2_1/n$, a = 9.909, b = 23.004, c = 21.124 Å, β = 102.07°, D_m = 1.87, Z = 4. Mo radiation, R = 0.033 for 4491 reflexions.

Molecule I (Fig. 1) consists of an octahedron of five ruthenium atoms and a phosphorus atom with one face (Ru(1)-Ru(3)-Ru(4)) capped by the carbon atom C(13) of an alkylidyne (CH_2-i-Pr) group. The alkylidyne ligand is unsymmetrically coordinated, with two Ru-C bond lengths (Ru(1)-C(13) 2.174(5), Ru(4)-C(13) 2.199(4) Å) longer than the third (Ru(1)-C(13) 1.917(5) Å). In II (Fig. 1) five ruthenium atoms and the three phenylphosphinidenes make up a face-capped pentagonal-bipyriamidal skeleton. Two of the PPh groups occupy equatorial positions with the third capping the Ru(1)-Ru(2)-Ru(3) face; P(4) bridges the Ru(4)-Ru(5) edge as a μ-PPh_2 group. Examination of Ru-Ru-C(O) bond angles and the coordination surface of the cluster strongly suggested a μ_3-bonding mode for the hydride on the Ru(1)-Ru(3)-Ru(5) face.

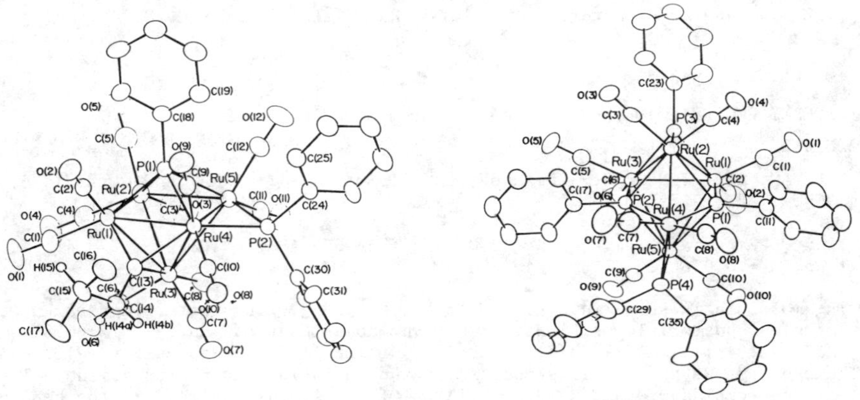

Fig. 1. Views of $C_{35}H_{24}O_{12}P_2Ru_5$ (I) (left) and $C_{40}H_{26}O_{10}P_4Ru_5$ (II) (right).

TETRACARBONYL(β-DIPHENYLPHOSPHINO-α-DIPHENYLPHOSPHINOXYSTYRENE-P,P')TUNGSTEN
$C_{36}H_{26}O_5P_2W$ (I)

TETRACARBONYL(α-DIPHENYLPHOSPHINYLACETOPHENONE-P)(HYDROXYDIPHENYLPHOSPHINO-P)TUNGSTEN
$C_{36}H_{28}O_6P_2W$ (II)

S. AL-JIBORI, M. HALL, A.T. HUTTON and B.L. SHAW, 1984. J. Chem. Soc. Dalton, 863-867.

I. Monoclinic, $P2_1/c$, a = 17.013, b = 10.454, c = 19.723 Å, β = 114.23°, Z = 4. Mo radiation, R = 0.032 for 3634 reflexions.

II. Monoclinic, $P2_1/c$, a = 18.629, b = 9.931, c = 18.182 Å, β = 94.58°, Z = 4. Mo radiation, R = 0.068 for 4027 reflexions.

Compound I (Fig. 1) contains a tungsten atom coordinated octahedrally to four terminal carbonyl groups and to the P(1) and P(2) atoms of the chelating ligand (W-C 1.990(8)-2.043(7), W-P(1) 2.508(2), W-P(2) 2.470(2) Å. In II, the photohydrolysis product of I, the six-membered chelate ring has been cleaved and the fifth and sixth coordination sites of W are now occupied by the P atoms of two monodentate ligands (W-P(1) 2.520(4), W-P(2) 2.496(4), W-C 1.86(2)-2.00(2) Å). The P(2)-O(5)...O(6)-C(6) distance (2.58(5) Å) is consistent with the presence of a strong chelating hydrogen bond. The P(2)-O(5)...O(6) and O(5)...O(6)-C(6) bond angles are 112(3) and 136°.

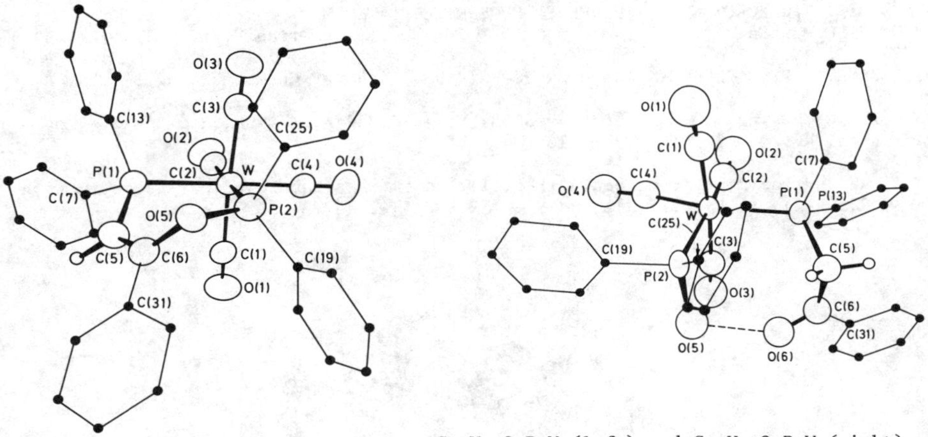

Fig. 1. The molecular structures: $C_{36}H_{26}O_5P_2W$ (left) and $C_{36}H_{28}O_6P_2W$ (right).

TRICHLORO(THIONITROSYL)BIS(TRIPHENYLPHOSPHINE)OSMIUM(II)
$C_{36}H_{30}Cl_3NOsP_2S$

H.W. ROESKY, K.K. PANDEY, W. CLEGG, M. NOLTEMEYER and G.M. SHELDRICK, 1984. J. Chem. Soc. Dalton, 719-721.

Monoclinic, I2/a, a = 16.159, b = 9.437, c = 22.506 Å, β = 102.94°, Z = 4. Mo radiation, R = 0.042 for 2076 reflexions. [The ruthenium complex is isostructural with the osmium complex (Monoclinic, I2/a, a = 16.135, b = 9.418, c = 22.493 Å, β = 102.94°, Z = 4. Mo radiation) but the crystals were too small for data collection.]

The Os atom is octahedrally coordinated (Fig. 1) with the phosphines cis to the NS group; a crystallographic twofold axis along the Cl-Os-N-S bonds imposes linearity at N (Os-N 1.779(9), N-S 1.503(10), Os-P 2.459(2), Os-Cl (cis) 2.387(3), Os-Cl (trans) 2.399(3) Å).

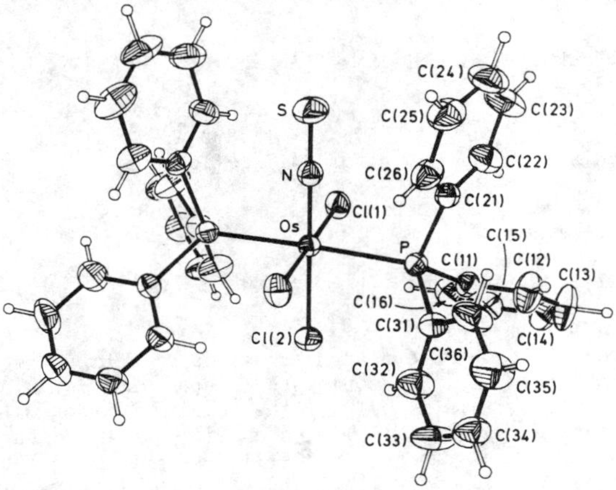

Fig. 1. $C_{36}H_{30}Cl_3NOsP_2S$: the molecular structure.

trans-BIS(TRICHLOROTIN)BIS(TRIPHENYL PHOSPHITE)PLATINUM
$C_{36}H_{30}Cl_6O_6P_2PtSn_2$ trans-$[Pt(SnCl_3)_2(P(OPh)_3)_2]$

A. ALBINATI, P.S. PREGOSIN and H. RÜEGGER, 1984. Inorg. Chem., 23, 3223-3229.

Monoclinic, $P2_1/n$, a = 11.177, b = 13.181, c = 14.489 Å, β = 98.52°, Z = 2. Mo radiation, R = 0.063 for 1907 reflexions.

The centrosymmetric molecule (Fig. 1) has a trans square planar geometry with Pt-Sn 2.599(2), Pt-P 2.272(3), and Sn-Cl 2.332(3)-2.347(6) Å.

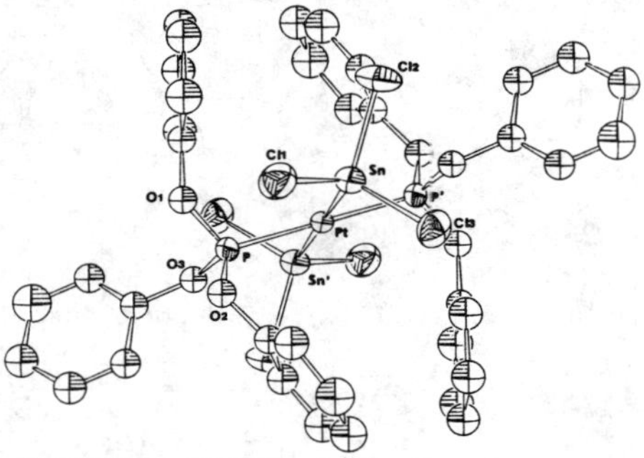

Fig. 1. The trans-$[Pt(SnCl_3)_2(P(OPh)_3)_2]$ molecule.

TETRAETHYLAMMONIUM TETRACARBONYLBIS(DIPHENYLPHOSPHINOUS ACID)MOLYBDATE DICHLOROMETHANE SOLVATE
$C_{36}H_{41}MoNO_6P_2,0.37CH_2Cl_2$ (I) $(NEt_4)^+[Mo(CO)_4(PPh_2O)_2H]^-,0.37CH_2Cl_2$

TETRACARBONYL(DIPHENYLPHOSPHINOUS ACID)(2-DIMETHYLAMINOETHYLDIPHENYLPHOSPHINITE)-MOLYBDENUM HYDRATE
$C_{32}H_{31}MoNO_6P_2,H_2O$ (II)

3,3,3,3-TETRACARBONYL-2,2,4,4-TETRAPHENYL-1-THIA-2,4-DIPHOSPHA-3-MOLYBDACYCLOBUTANE
$C_{28}H_{20}MoO_4P_2S$ (III) $Mo(CO)_4(PPh_2)_2S$

F.A. COTTON, L.R. FALVELLO, M. TOMAS, G.M. GRAY and C.S. KRAIHANZEL, 1984. Inorg. Chim. Acta, 82, 129-139.

I. Triclinic, $P\bar{1}$, a = 12.456, b = 13.757, c = 11.892 Å, α = 95.51, β = 108.30, γ = 91.91°, Z = 2. Mo radiation, R = 0.054 for 4417 reflexions.

II. Orthorhombic, Pbca, a = 20.901, b = 16.032, c = 19.686 Å, Z = 8. Mo radiation, R = 0.041 for 2032 reflexions.

III. Orthorhombic, $P2_12_12_1$, a = 11.04, b = 14.244, c = 17.233 Å, Z = 4. Mo radiation, R = 0.038 for 1843 reflexions.

The octahedral coordination sphere at Mo consists of four CO groups and two cis oriented P donor ligands (Fig. 1). The unlocated H atom in I forms a strong H bridge between O(5) and O(6) (O...O 2.415 Å). In II the acidic proton on O(6) forms a strong hydrogen bond to the tertiary amine (N(1)) with an N(1)...O(6) distance of 2.77 Å. Complex III has essentially C_{2v} symmetry; the P(1)-S-P(2) angle is 86.9°. The P-Mo-P angles are 88.06, 86.36 and 72.07° for I, II and III respectively. The average Mo-C distance for all the compounds is 2.010 Å while the Mo-P distances range from 2.487 to 2.563 Å, and average 2.51 Å.

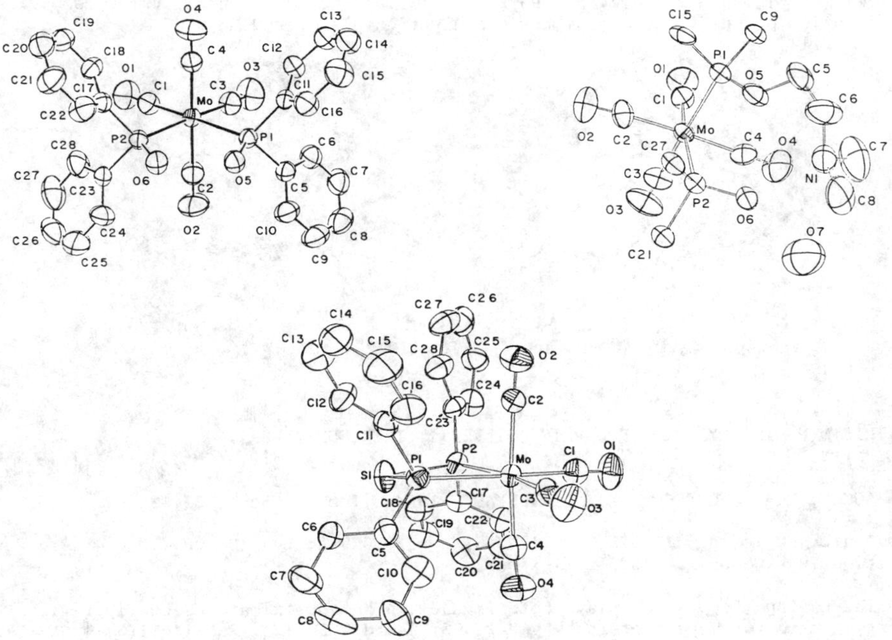

Fig. 1. Molecular structures of $C_{36}H_{41}MoNO_6P_2$ (top left), $C_{32}H_{31}MoNO_7P_2,H_2O$ (top right), and $C_{28}H_{20}MoO_4P_2S$ (below).

(2,11-BIS(DI-t-BUTYLPHOSPHINOMETHYL)BENZO[c]PHENANTHRENE-P,P')BROMOSILVER DICHLOROMETHANE SOLVATE
$C_{36}H_{50}AgBrP_2,CH_2Cl_2$ (I)

(2,11-BIS(DI-t-BUTYLPHOSPHINOMETHYL)BENZO[c]PHENANTHRENE-P,P')CHLOROSILVER DICHLOROMETHANE SOLVATE
$C_{36}H_{50}AgClP_2,CH_2Cl_2$ (II)

(2,11-BIS(DI-t-BUTYLPHOSPHINOMETHYL)BENZO[c]PHENANTHRENE-P,P')PERCHLORATOSILVER(I) ACETONE SOLVATE
$C_{36}H_{50}AgClO_4P_2,C_3H_6O$ (III)

M. CAMALLI, F. CARUSO, S. CHALOUPKA, P.N. KAPOOR, P.S. PREGOSIN and L.M. VENANZI, 1984. Helv. Chim. Acta, 67, 1603-1611.

I. Monoclinic, $P2_1/n$, a = 11.174, b = 24.323, c = 14.178 Å, β = 102.69°, Z = 4. Mo radiation, R = 0.075 for 2361 reflexions.

II. Monoclinic, $P2_1/n$, a = 11.163, b = 24.382, c = 14.094 Å, β = 76.88°, Z = 4. Mo radiation, R = 0.068 for 4347 reflexions.

III. Orthorhombic, Pbca, a = 16.015, b = 20.050, c = 25.002 Å, Z = 8. Mo radiation, R = 0.063 for 3233 reflexions.

Compounds I and II are isomorphous; I and III are shown in Fig. 1. The Ag-coordination is distorted trigonal planar; for I Ag-P1 2.463(4), Ag-P2 2.433(5), Ag-Br 2.681(2) Å. The corresponding data for II and III are 2.457(2), 2.427(2),

2.569(2) (to Cl); and 2.394(2), 2.393(2), 2.913(13) (to O4). In all three complexes the deviation of the Ag atom from the coordination plane is 0.06-0.07 Å.

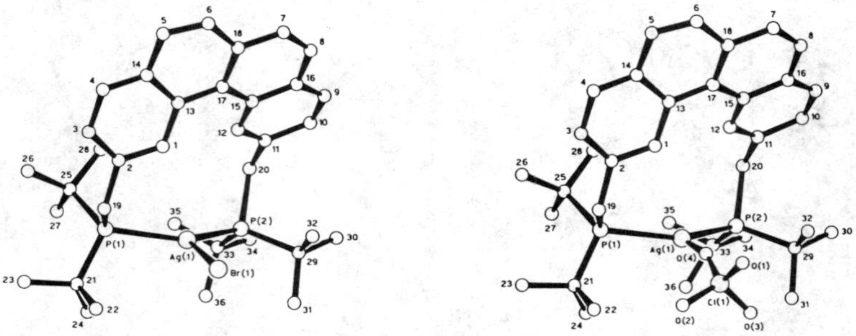

Fig. 1. Views of $C_{36}H_{50}AgBrP_2$ and $C_{36}H_{50}AgClO_4P_2$.

BIS[TRIS(N,N-DIMETHYLAMINO)PHOSPHINE]SILVER(I) TETRAPHENYLBORATE
$C_{36}H_{56}AgBN_6O_2$ $\{Ag[P(N(CH_3)_2)_3]_2\}^+[(C_6H_5)_4B]^-$

S.M. SOCOL, R.A. JACOBSON and J.G. VERKADE, 1984. Inorg. Chem., 23, 88-94.

Monoclinic, $P2_1/c$, a = 11.975, b = 17.325, c = 20.079 Å, β = 107.08°, Z = 4. Mo radiation, R = 0.055 for 4523 reflexions.

The cation (Fig. 1) displays both ligands in an approximate C_s configuration. Important structural parameters: Ag-P 2.395(2) and 2.393(2), P-N 1.651(6)-1.683(7), mean B-C 1.645 Å, P-Ag-P 166.9(1)°.

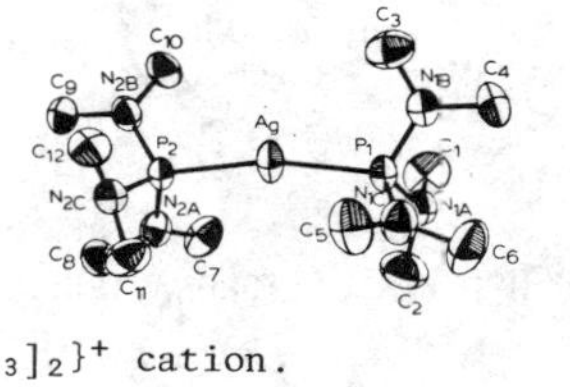

Fig. 1. The $\{Ag[P(N)CH_3)_2)_3]_2\}^+$ cation.

HEXAHYDRIDOTRIS(PHENYLDIISOPROPYLPHOSPHINE)TUNGSTEN(VI)
$C_{36}H_{63}P_3W$ $WH_6(PPh\text{-}i\text{-}Pr_2)_3$

D.A. GREGSON, S.A. MASON, J.A.K. HOWARD, J.L. SPENCER and D.G. TURNER, 1984. Inorg. Chem., 23, 4103-4107.

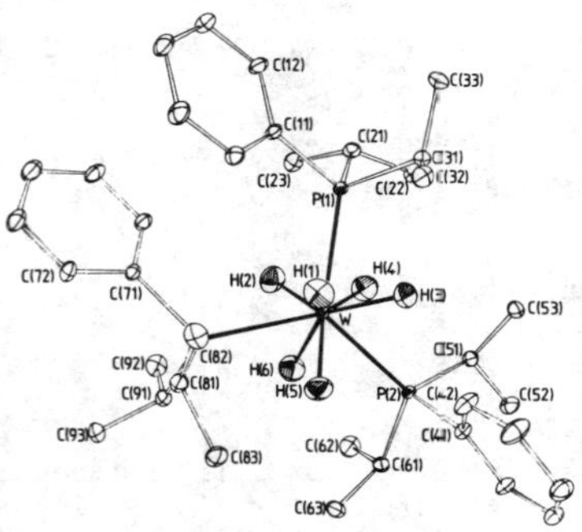

Fig. 1. The $WH_6(PPh\text{-}i\text{-}Pr_2)_3$ molecule.

Triclinic, $P\bar{1}$, a = 11.490, b = 10.610, c = 16.324 Å, α = 104.76, β = 108.04, γ = 89.92°, Z = 2. Mo radiation, R = 0.033 for 7583 reflexions (at 200 K); a = 11.427, b = 10.554, c = 16.249 Å, α = 104.59, β = 108.16, γ = 89.88°. 1.262 Å neutron radiation, R = 0.049 for 7068 reflexions (at 100 K).

The molecule (Fig. 1) shows approximate C_{2v} symmetry with an asymmetric disposition of the phosphine ligands around the W atom. The nine-coordinate W atom has a tricapped trigonal prismatic geometry with phosphine ligands occupying a pair of eclipsed prism corners and one of the equatorial sites. Bond lengths (100 K): W-P 2.424(2), 2.519(2), 2.524(2), W-H 1.718(3)-1.745(3) Å.

trans-CARBONYL(SULPHINYLNITRILO)BIS(TRIPHENYLPHOSPHINE)RHODIUM(I)
$C_{37}H_{30}NO_2P_2RhS$

H.W. ROESKY, K.K. PANDEY, B. KREBS and M. DARTMANN, 1984. J. Chem. Soc. Dalton, 2271-2273.

Triclinic, $P\bar{1}$, a = 9.248, b = 9.502, c = 10.567 Å, α = 71.52, β = 89.71, γ = 69.80°, D_m = 1.43, Z = 1. Mo radiation, R = 0.047 for 3377 reflexions.

In this square-planar molecular complex (Fig. 1) the carbonyl group (Rh-C 1.843(6)) is in the trans position relative to the NSO ligand (Rh-N 1.989(9), N-S 1.48(1), S-O 1.51(1) Å, N-S-O 122.5(5)°). The Rh-P bond lengths to the trans coordinated phosphine ligands are 2.330(1) Å. The arrangement of molecules in the crystal is disordered, with a superposition of two half-occupied CO and NSO ligand sites in the electron density map, simulating the crystallographic inversion centre at the Rh site.

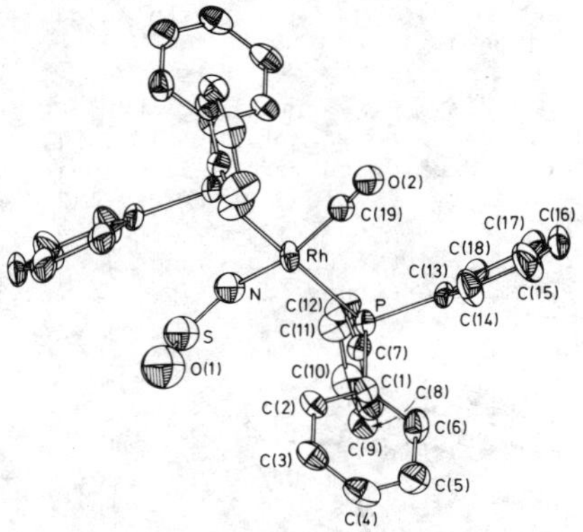

Fig. 1. $C_{37}H_{30}NO_2P_2RhS$: the molecular structure.

DIIODOMETHYLBIS(TRIPHENYLPHOSPHINE)RHODIUM(III)
$C_{37}H_{33}I_2P_2Rh$ $CH_3RhI_2(PPh_3)_2$

A.R. SIEDLE, R.A. NEWMARK and L.H. PIGNOLET, 1984. Organometallics, 3, 855-859.

Monoclinic, C2/c, a = 23.982, b = 9.828, c = 16.063 Å, β = 114.73°, Z = 4. R = 0.022 for 2618 reflexions.

The coordination geometry of Rh(III) is flattened square pyramidal with a C_2 axis through Rh and the methyl C atom (Fig. 1). The metal is displaced 0.25 Å towards the apical methyl group from the least-squares plane containing two trans I atoms and two trans P atoms which comprise the base of the pyramid. Some angles within the coordination core are: I-Rh-I' 164.47, I-Rh-P 89.73, I-Rh-P' 89.33, I-Rh-C 97.77, P-Rh-P' 173.04 and P-Rh-C 93.48°.

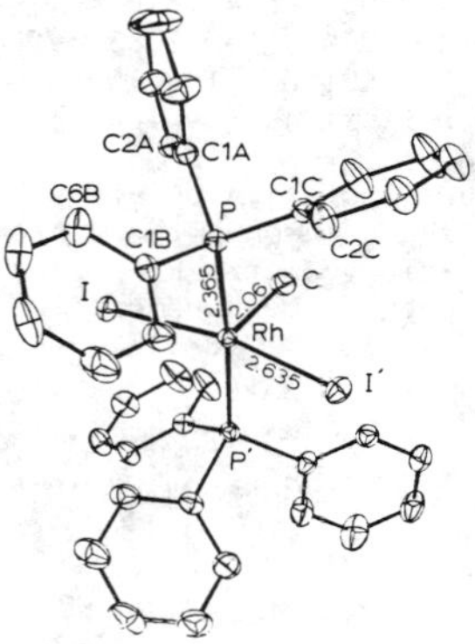

Fig. 1. Molecular structure of $C_{37}H_{33}I_2P_2Rh$.

PENTAKIS(μ_2-CARBONYL)TETRAKIS(DIMETHYLPHENYLPHOSPHINEPLATINUM)

$C_{37}H_{44}O_5P_4Pt_4$ $[Pt_4(CO)_5(P(CH_3)_2C_6H_5)_4]$

A. MOOR, P.S. PREGOSIN, L.M. VENANZI and A.J. WELCH, 1984. Inorg. Chim. Acta, 85, 103-110.

Monoclinic, $P2_1/n$, a = 19.698, b = 10.944, c = 21.360 Å, β = 112.43°, Z = 4. Mo radiation, R = 0.085 for 4751 reflexions.

The molecule has no imposed symmetry but the central $Pt_4(CO)_5P_4$ core has approximate C_{2v} symmetry (Fig. 1). The Pt_4 unit is a highly distorted, edge-opened (Pt(1)...Pt(4) 3.3347 Å) tetrahedron, with five edge-bridging CO and four terminal phosphine ligands. The Pt-Pt distances are: Pt(1)-Pt(2) 2.728, Pt(1)-Pt(3) 2.706, Pt(2)-Pt(3) 2.731, Pt(2)-Pt(4) 2.754, and Pt(3)-Pt(4) 2.754 Å. The Pt-P distances range from 2.233 to 2.271 Å and the Pt-C(CO) distances from 1.968 to 2.176 Å. The angles involving the metal atoms alone range from 59.44 to 60.34°.

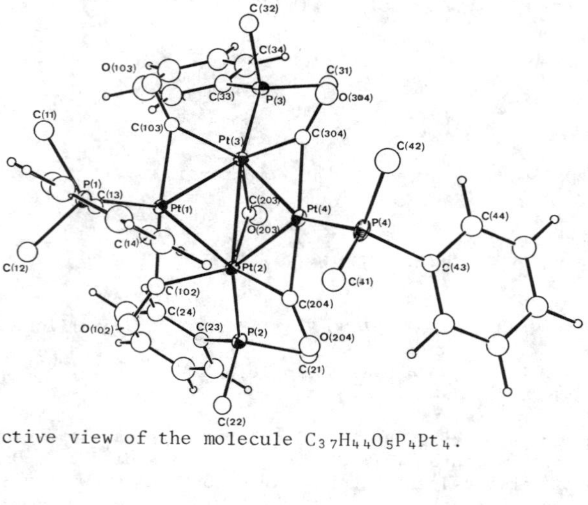

Fig. 1. Perspective view of the molecule $C_{37}H_{44}O_5P_4Pt_4$.

(μ_4-SULPHIDO)TRIS(μ_3-SULPHIDO)BIS(μ_2-DIPHENYLPHOSPHINO)TETRADECACARBONYLHEXAOSMIUM

$C_{38}H_{20}O_{14}Os_6P_2S_4$ $Os_6(CO)_{14}(\mu\text{-}PPh_2)_2(\mu_3\text{-}S)(\mu_4\text{-}S)$

R.D. ADAMS, I.T. HORVÁTH and B.E. SEGMÜLLER, 1984. J. Organometal. Chem., 262, 243-252.

Triclinic, $P\bar{1}$, a = 10.427, b = 13.552, c = 17.919 Å, α = 84.87, β = 75.41, γ = 78.43°, Z = 2. Mo radiation, R = 0.042 for 3223 reflexions.

The molecule consists of two sulphido bridged open triosmium clusters which are linked by a bridging sulphido ligand and a bridging diphenylphosphino ligand (Fig. 1). The metal-metal bonds are non-equivalent (Os(1)-Os(2) 2.907 and Os(2)-Os(3) 2.773 Å) and (Os(4)-Os(5) 2.960 and Os(5)-Os(6) 3.020 Å) in their respective clusters. The Os-S distances vary from 2.392 to 2.530 Å and the Os-P distances from 2.285 to 2.370 Å.

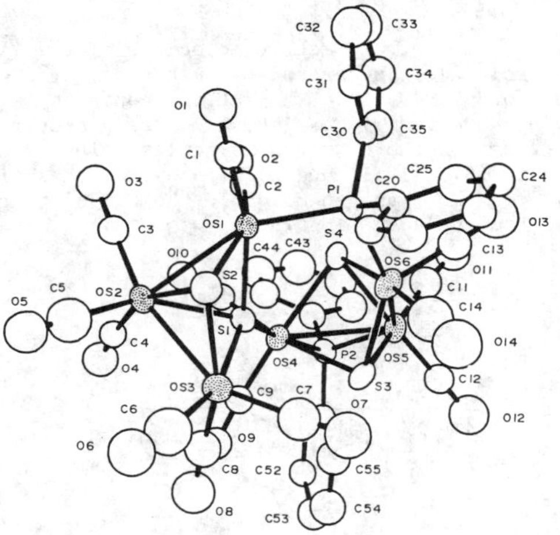

Fig. 1. Molecular structure of $C_{38}H_{20}O_{14}Os_6P_2S_4$.

CHLORODICARBONYL(DISULPHURDINITRIDE)BIS(TRIPHENYLPHOSPHINE)RUTHENIUM(II) TETRACHLOROALUMINATE

$C_{38}H_{30}AlCl_5N_2O_2P_2RuS_2$ $[(P(C_6H_5)_3)_2(CO)_2(S_2N_2)RuCl]^+[AlCl_4]^-$

H.W. ROESKY, J. ANHAUS and W.S. SHELDRICK, 1984. Inorg. Chem., 23, 75-79.

Monoclinic, Cc, a = 10.265, b = 23.185, c = 18.361 Å, β = 106.06°, Z = 4. Mo radiation, R = 0.041 for 3292 reflexions.

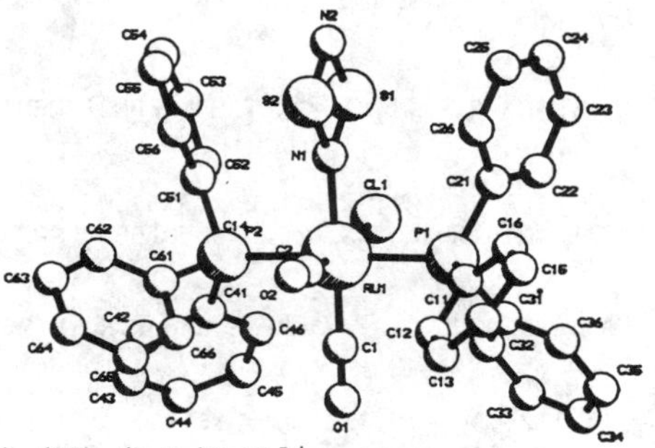

Fig. 1. The $[(PPh_3)_2(CO)_2(S_2N_2)RuCl]^+$ cation.

The cation (Fig. 1) features a monodentate S_2N_2 ligand (Ru-N 2.082(6) Å) and an intramolecular S...Cl distance of 2.918(4) Å. The geometry at Ru is octahedral, Ru-P 2.432(2) and 2.446(2) Å, Ru-Cl 2.404(2) Å, Ru-C(1) 1.926(8) Å, Ru-C(2) 1.876(9) Å, S-N 1.612(8)-1.684(7) Å, S-S 2.361(4) Å.

trans-BIS[(CYANO)TRIPHENYLPHOSPHINE]NICKEL CYANOGEN
$C_{38}H_{30}N_2NiP_2,C_2N_2$ $[Ni(CN)_2(P(C_6H_5)_3)_2].C_2N_2$

B. CORAIN, M. BASATO, G. FAVERO, P. ROSANO and G. VALLE, 1984. Inorg. Chim. Acta, 85, L27-L30.

Triclinic, P$\bar{1}$, a = 11.318, b = 9.587, c = 9.137 Å, α = 99.9, β = 107.2, γ = 108.8°, Z = 1. Mo radiation, R = 0.039 for 2677 reflexions.

The structure consists of centrosymmetric molecules of the Ni complex and centrosymmetric cyanogen molecules, separated by distances large enough to indicate a lack of intermolecular bonds (Fig. 1). The lattice contains fairly wide channels defined by two phenyl rings and two Ni-C≡N 'sticks', the diameter of the channels being ~8 Å which allows room for the cyanogen molecules. The Ni-C, C≡N and Ni-P distances are respectively 1.873, 1.155 and 2.225 Å with angles of 92.5° (C(19)-Ni-P) and 87.4° (C(19)-Ni-P'). In the cyanogen molecule the C-C distance is 1.381 Å, the C-N distance 1.127 Å and the angle C'-C-N 179.7°.

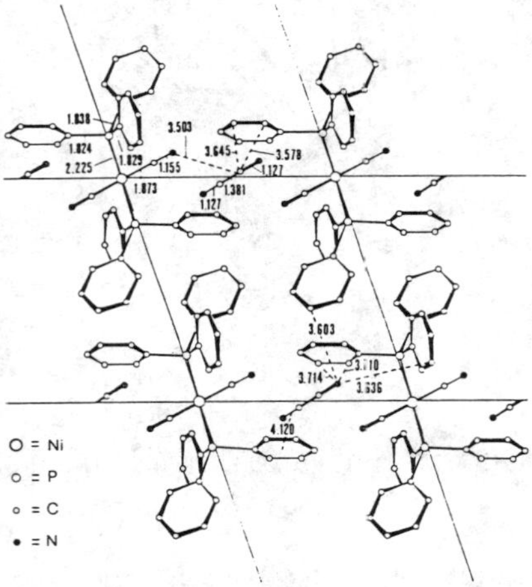

Fig. 1. View of the unit cell of $C_{38}H_{30}N_2NiP_2,C_2N_2$ along the 001 direction, with selected intra- and intermolecular distances.

[1,2-BIS(DIPHENYLPHOSPHINE)ETHANE-P,P'][(E)-η^2-DIPHENYLPHOSPHENE]PALLADIUM(0)
$C_{38}H_{34}P_4Pd$ (I)

1-(1,2-BIS(DIPHENYLPHOSPHINE)ETHANE-P,P')-2,2,2,2,2,3,3,3,3-DECACARBONYL-[(E)-μ_3-η^2-DIPHENYLDIPHOSPHENE]PALLADIUM(0)DITUNGSTEN(0) HEXADEUTEROBENZENE SOLVATE
$C_{48}H_{34}O_{10}P_4PdW_2,C_6D_6$ (II)

J. CHATT, P.B. HITCHCOCK, A. PIDCOCK. C.P. WARRENS and K.R. DIXON, 1984. J. Chem. Soc. Dalton, 2237-2244.

I. Monoclinic, C2/c, a = 11.112, b = 18.826, c = 16.356 Å, β = 99.59°, Z = 4. Mo radiation, R = 0.050 for 1351 reflexions.

II. Monoclinic, C2/c, a = 16.718, b = 18.805, c = 34.576 Å, β = 95.76°, Z = 8. Mo radiation, R = 0.099 for 4404 reflexions.

Compound II contained a molecule of C_6D_6 of crystallisation as it crystallised in an NMR tube after some NMR measurements. Both complexes (Fig. 1) reveal nearly identical and planar PdP_4 moieties comprising the donor atoms of 1,2-bis(diphenylphosphanyl)ethane and η^2-[(E)-diphenylphosphene] ligands (I: Pd-P(1) 2.366(2), Pd-P(2) 2.304(2); II: Pd-P(1) 2.375(4), Pd-P(2) 2.355(4), Pd-P(3) 2.305(4), Pd-P(4) 2.307(5) Å). In II the two diphosphene lone pairs coordinate to $W(CO)_5$ groups and the W-P bonds are bent 32° away from the Pd atom. The diphosphene P-P bond in II (2.186(6)) is 0.06 Å longer than that in I (2.121(4) Å), which is similar to that in other diphosphene complexes and is approximately midway between the P-P single and double bond lengths.

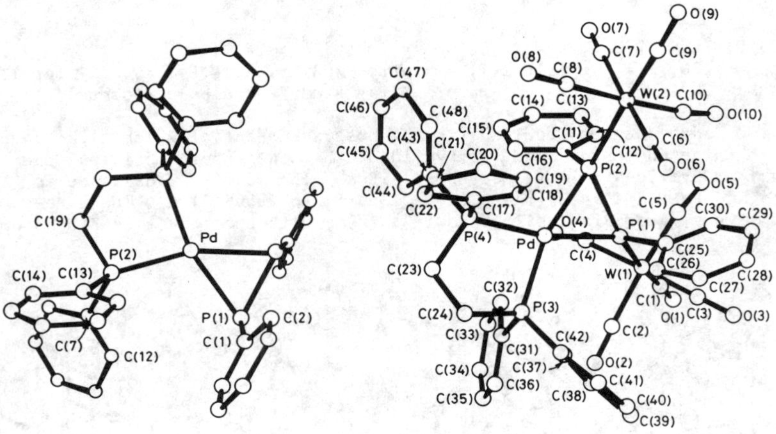

Fig. 1. The molecular structures: $C_{38}H_{34}P_4Pd$ (left) and $C_{48}H_{34}O_{10}P_4PdW_2,C_6D_6$ (right).

TRICARBONYLBIS(TRIPHENYLPHOSPHINE)RUTHENIUM(0) TETRAHYDROFURAN(2/1)

$C_{39}H_{30}O_3P_2Ru,0.5C_4H_8O$ $[Ru(CO)_3\{P(C_6H_5)_3\}_2],0.5C_4H_8O$

F. DAHAN, S. SABO and B. CHAUDRET, 1984. Acta Cryst., C40, 786-788.

Monoclinic, $P2_1/c$, a = 17.845, b = 12.315, c = 18.351 Å, β = 104.55°, Z = 4. Mo radiation, R = 0.045 for 2894 reflexions.

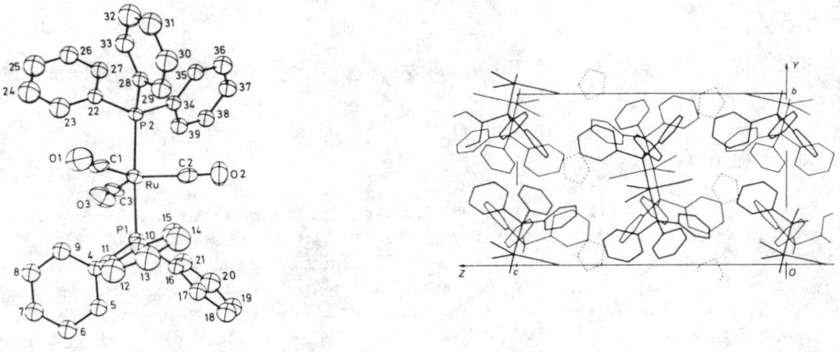

Fig. 1. $C_{39}H_{30}O_3P_2Ru,0.5C_4H_8O$: a perspective view of the complex molecule and a view of the unit-cell contents along the x axis.

In the complex molecule Ru is fivefold coordinated, the three carbonyls being in the equatorial plane and the P atoms of the triphenylphosphines at the apices of a rather regular trigonal bipyramid, the preferred configuration for complexes in low oxidation states (Fig. 1). Bond distances are: Ru-C 1.925(8), 1.880(10) and 1.895(10), Ru-P 2.332(2) and 2.330(2) Å. The structure can be considered as complex units with tetrahydrofuran molecules stabilising the crystal packing.

PENTAHYDRIDOTRIS(METHYLDIPHENYLPHOSPHINE)RHENIUM(V)
$C_{39}H_{44}P_3Re$ $ReH_5[PCH_3(C_6H_5)_2]_3$

T.J. EMGE, T.F. KOETZLE, J.W. BRUNO and K.G. CAULTON, 1984. Inorg. Chem., 23, 4012-2017.

Triclinic, P$\bar{1}$, a = 17.935, b = 20.066, c = 10.403 Å, α = 91.72, β = 100.64, γ = 69.48°, D_m = 1.50, Z = 4. 1.1611(1) Å neutron radiation, R(F^2) = 0.091 for 11374 reflexions at 20 K. (A room temperature X-ray study was also performed.)

The two independent molecules (Fig. 1) possess essentially identical dodecahedral cores, with three phosphines and one hydride in B (5 neighbor) positions and the remaining four hydrides in A (4 neighbor) sites. The mean Re-H distance is 1.688(5) Å and Re-P distances range from 2.359(3) to 2.398(3) Å. The P-Re-P angles are of two types, averaging 101(1) and 147.3(8)°, with the longer Re-P bond opposite the wider angle.

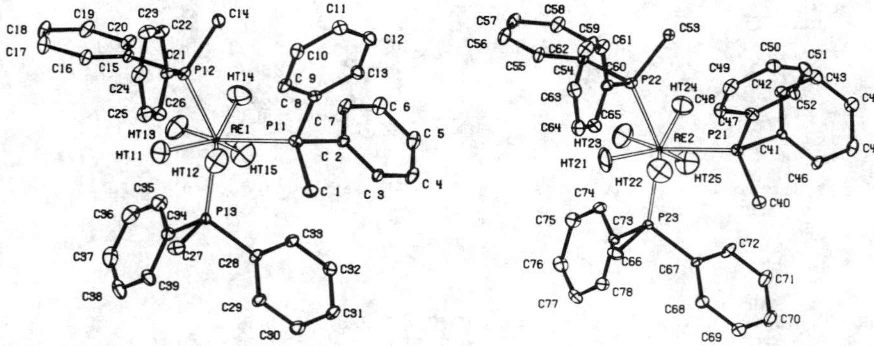

Fig. 1. The two independent molecules of $ReH_5[P(CH_3)(C_6H_5)_2]_3$.

TETRACARBONYLBIS(TRIPHENYLPHOSPHINE)CYCLODIGOLDOSMIUM
$C_{40}H_{30}Au_2O_4OsP_2$ $[Os(CO)_4\{AuP(C_6H_5)_3\}_2]$

B.F.G. JOHNSON, J. LEWIS, P.R. RAITHBY and A. SANDERS, 1984. J. Organometal. Chem., 260, C29-C32.

Orthorhombic, $Pc2_1n$, a = 13.039, b = 15.851, c = 18.499 Å, Z = 4. Mo radiation, R = 0.047 for 2923 reflexions.

The three metal atoms lie at the vertices of an approximately isosceles triangle in which the long edge (2.929 Å) is defined by the Au-Au vector (Fig. 1). The Au-Os distances are 2.667 and 2.646 Å and the Os-C distances range from 1.91 to 1.96 Å. Angles Au(1)-Os-Au(2), Os(1)-Au(1)-Au(2) and Os(1)-Au(2)-Au(1) are respectively 66.9, 56.2 and 56.9°. The Au(1)-P(1) and Au(2)-P(2) distances are 2.226 and 2.324 Å respectively and the angles P(1)-Au(1)-Os(1) and P(2)-Au(2)-Os(1) are 174.1 and 168.2°.

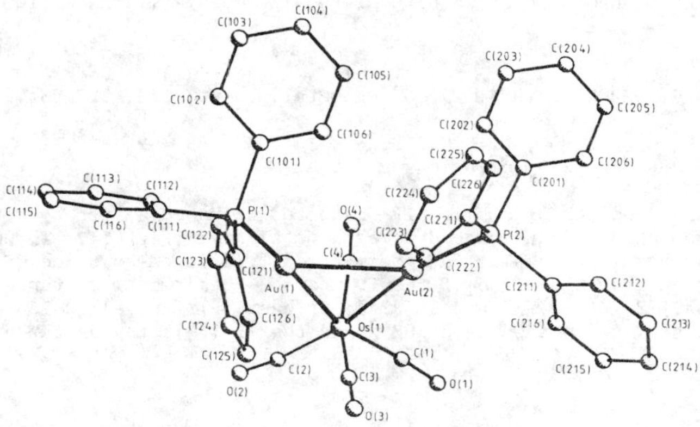

Fig. 1. Molecular structure of $C_{40}H_{30}Au_2O_4OsP_2$.

trans,trans-ACETONITRILEDIISOTHIOCYANATONITRIDOBIS(TRIPHENYLPHOSPHINE)TECHNETIUM(V) ACETONITRILE SOLVATE
$C_{40}H_{33}N_4P_2S_2Tc,0.5(C_2H_3N)$

J. BALDAS, J. BONNYMAN and G.A. WILLIAMS, 1984. J. Chem. Soc. Dalton, 833-837.

Monoclinic, $P2_1/c$, a = 9.296, b = 18.614, c = 23.307 Å, β = 109.63°, Z = 4. Cu radiation, R = 0.045 for 5498 reflexions.

The Tc atom (Fig. 1) has an octahedral environment with Tc-N-C angles of 177.1(4) and 166.7(4)° for the NCS ligands and 168.6(4)° for the CH_3CN ligand. The Tc≡N(4) bond distance is 1.629(4) Å and the Tc-N(3) bond distance is 2.491(4) Å, the latter indicating the strong trans influence of the nitrido group. The other coordination distances are: Tc-P(1) 2.494(1), Tc-P(2) 2.524(1), Tc-N(1) 2.045(4), Tc-N(2) 2.068(3) Å. The P-Tc-P angle is 174.0(1)°.

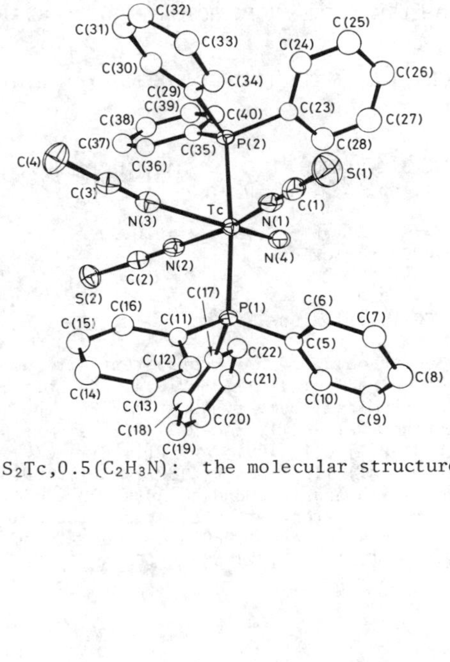

Fig. 1. $C_{40}H_{33}N_4P_2S_2Tc,0.5(C_2H_3N)$: the molecular structure.

BIS(μ_2-CARBONYL)HEXACARBONYLTETRAKIS(DIMETHYLPHENYLPHOSPHINE)TETRAIRIDIUM BENZENE SOLVATE

$C_{40}H_{44}Ir_4O_8P_4,2C_6H_6$ $Ir_4(CO)_8(PMe_2Ph)_4,2C_6H_6$

A.J. BLAKE and A.G. OSBORNE, 1984. J. Organometal. Chem., 260, 227-234.

Monoclinic, $P2_1/c$, a = 15.907, b = 10.215, c = 34.653 Å, β = 94.29°, Z = 4. Mo radiation, R = 0.095 for 3250 reflexions.

The structure consists of a slightly distorted tetrahedral arrangement of Ir atoms with one phosphine ligand bonded to each metal atom (Fig. 1). Three Ir atoms form an irregular basal triangular face symmetrically bridged by three CO groups which are displaced from the plane of the Ir atoms. Metal-metal distances are: Ir(basal)-Ir(basal) 2.726, 2.737 and 2.782 Å; Ir(basal)-Ir(apical) 2.734, 2.730 and 2.727 Å. The mean Ir-P distance is 2.308 Å and the Ir-Ir-Ir angles range from 59.19 to 61.23°.

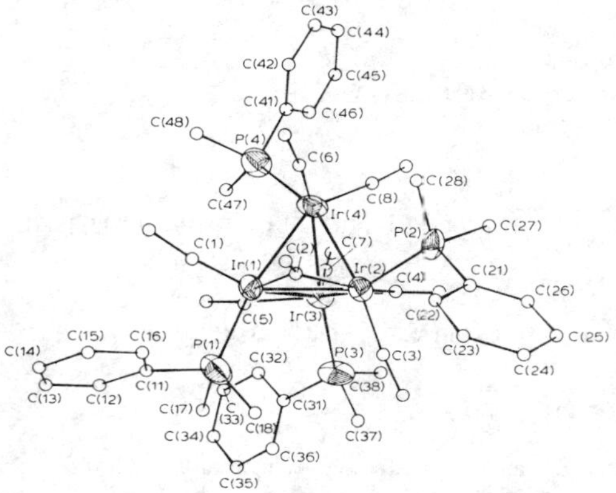

Fig. 1. Molecular structure of $C_{40}H_{44}Ir_4O_8P_4$.

[BIS[2,4,6-TRIS(1,1-DIMETHYLETHYL)PHENYL]DIPHOSPHENE-P]TETRACARBONYLIRON

$C_{40}H_{58}FeO_4P_2$ (I) $(Bu_3C_6H_2)_2P_2Fe(CO)_4$

[[BIS(TRIMETHYLSILYL)METHYL][2,4,6-TRIS(1,1-DIMETHYLETHYL)PHENYL]DIARSENE-As[1]]PENTACARBONYLCHROMIUM

$C_{30}H_{48}As_2CrO_5Si_2$ (II) $Bu_3C_6H_2As{=}AsCH(SiMe_3)_2Cr(CO)_5$

A.H. COWLEY, J.E. KILDUFF, J.G. LASCH, N.C. NORMAN, M. PAKULSKI, F. ANDO and T.C. WRIGHT, 1984. Organometallics, 3, 1044-1050.

I. Monoclinic, $P2_1/c$, a = 21.251, b = 9.793, c = 20.99 Å, β = 108.73°, Z = 4. Mo radiation, R = 0.058 for 4204 reflexions.

II. Monoclinic, $P2_1/n$, a = 10.710, b = 29.770, c = 11.985 Å, β = 98.02°, Z = 4. Mo radiation, R = 0.046 for 3891 reflexions.

In I, the P-P bond length of 2.050 Å is consistent with double bond character (Fig. 1). Atom P(1) bonds to Fe (Fe(1)-P(1) 2.215 Å) and forms one of the equatorial ligands in the trigonal bipyramidal geometry about the Fe, the remaining four ligands being carbonyls. The distances P(1)-C(1) and P(2)-C(7) are respectively 1.893 and 1.859 Å. and the angles P(2)-P(1)-C(1) 109.3 and P(1)-P(2)-C(7) 108.4°, with Fe(1)-P(1)-P(2) 135.52, Fe(1)-P(1)-C(1) 115.1 and P(2)-P(1)-C(1) 109.3°. In II, the Cr atom has octahedral coordination and its bond to As(2) is 2.454 Å in length. The As-As distance of 2.246 Å is consistent with retention of the double bond. The As(2)-C(22) and As(1)-C(1) distances are 1.989 and 1.964 Å respectively and the bond angles at As are As(1)-As(2)-C(22) (103.9°) and As(2)-As(1)-C(1) (104.0°). Atoms

C(1), As(1), As(2), C(22) and Cr(1) are coplanar with a mean deviation from the plane of 0.04 Å.

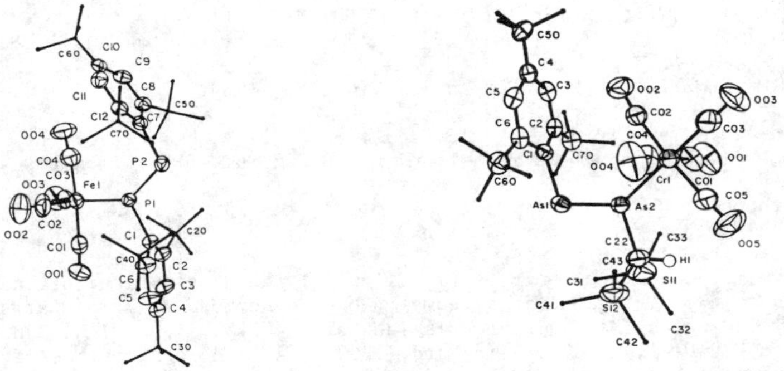

Fig. 1. Perspective views of $C_{40}H_{58}FeO_4P_2$ (left), and $C_{30}H_{48}As_2CrO_5Si_2$ (right).

(CYCLO-DIARSENIC-SULPHUR-As,As',S)(1,1,1-TRIS(DIPHENYLPHOSPHINOMETHYL)ETHANE)COBALT TETRAFLUOROBORATE BENZENE SOLVATE

$C_{41}H_{39}As_2BCoF_4P_3S,C_6H_6$

M. DI VAIRA, P. INNOCENTI, S. MONETI, M. PERUZZINI and P. STOPPIONI, 1984. Inorg. Chim. Acta, 83, 161-165.

Monoclinic, $P2_1/n$, a = 16.987, b = 20.608, c = 13.098 Å, β = 104.83°, Z = 4. Mo radiation, R = 0.058 for 3170 reflexions.

The metal atom is in a distorted six-coordinate environment formed by the triphos P atoms and by the atoms of the heterocyclic As_2S unit in a staggered arrangement (Fig. 1). The site occupancies in the As_2S ring are: As(1) (0.15 S, 0.85 As), As(2) (0.15S, 0.85 As), S (0.70 S, 0.30 As), the S atom residing with highest probability in the site forming the shortest contact distances with the other atoms in the cation. The As(1)-As(2) distance is 2.339 Å while the As(1)-S and As(2)-S distances are respectively 2.264 and 2.278 Å. The metal-P distances are respectively 2.193, 2.223 and 2.227 Å and the distances Co-As(1), Co-As(2) and Co-S are 2.389, 2.393 and 2.325 Å respectively.

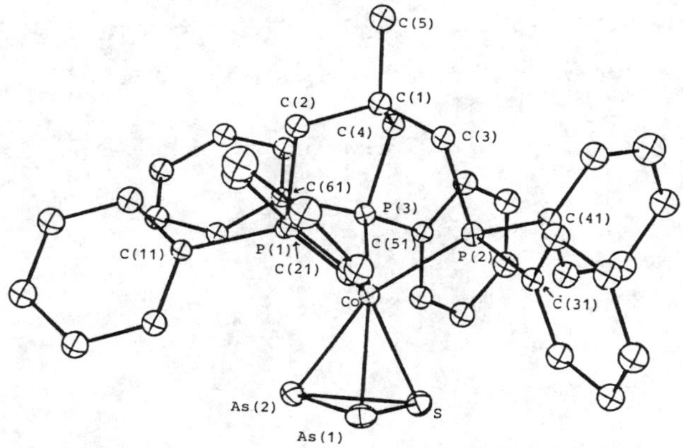

Fig. 1. Perspective view of the cation in $C_{41}H_{39}As_2BCoF_4P_3S,C_6H_6$.

DICYANOTRIS[(HYDROXYMETHYL)DIPHENYLPHOSPHINE]NICKEL(II) - 0.5 BENZENE
$C_{41}H_{39}N_2NiO_3P_3,0.5C_6H_6$ (I) $[Ni(CN)_2[P(CH_2OH)(C_6H_5)_2]_3],0.5C_6H_6$

trans-DICYANOBIS[(HYDROXYMETHYL)DIPHENYLPHOSPHINE]NICKEL(II)
$C_{28}H_{26}N_2NiO_2P_2$ (II) trans-$[Ni(CN)_2[P(CH_2OH)(C_6H_5)_2]_2]$

H. HOPE, M.M. OLMSTEAD, P.P. POWER and M. VIGGIANO, 1984. Inorg. Chem., <u>23</u>, 326-330.

I. Triclinic, $P\bar{1}$, a = 12.323, b = 12.844, c = 15.507 Å, α = 111.38, β = 90.39, γ = 119.85°, D_m = 1.35, Z = 2. Mo radiation, R = 0.057 for 2855 reflexions (at 140K).

II. Orthorhombic, Pbca, a = 10.258, b = 16.173, c = 15.055 Å, D_m = 1.41, Z = 4. Mo radiation, R = 0.038 for 1765 reflexions (at 140K).

The two molecules are shown in Fig. 1. In I the Ni atom is coordinated to three P atoms and to two CN ligands in a geometry best described as irregular square pyramidal. The apical Ni-P distance is long at 2.400(3) Å, basal distances are Ni-P 2.229(4) and 2.246(4), Ni-C 1.859(6) and 1.869(7) Å. The centrosymmetric trans-square planar complex II has Ni-P 2.210(1) and Ni-C 1.858(3) Å. There are no Ni-O interactions in either structure.

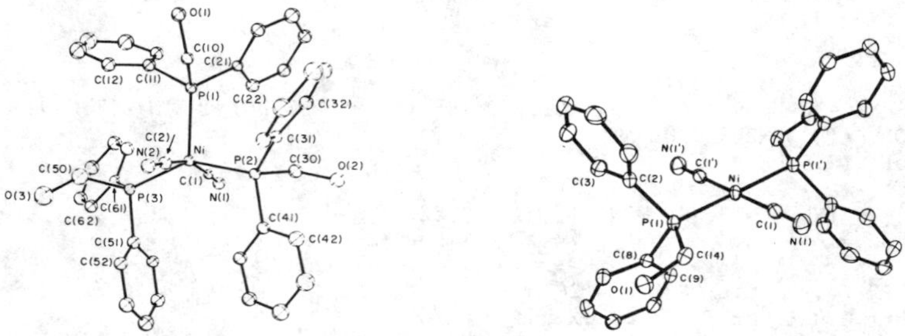

Fig. 1. The $[Ni(CN)_2[P(CH_2OH)(C_6H_5)_2]_3]$ (left) and trans-$[Ni(CN)_2[P(CH_2OH)-(C_6H_5)_2]_2]$ (right) molecules.

(CARBONATO-O,O')[1,1-BIS((DIPHENYLPHOSPHINO)METHYL)-1-(DIPHENYLPHOSPHINEOXIDEMETHYL)-ETHANE-P,P']NICKEL(II) BENZENE SOLVATE HEMIHYDRATE
$C_{42}H_{39}NiO_4P_3,C_6H_6,0.5(H_2O)$ $[OPPh_2CH_2C(CH_3)(CH_2PPh_2)_2]Ni(O_2CO),C_6H_6,0.5H_2O$

C. BIANCHINI, C. MEALLI, A. MELI and M. SABAT, 1984. Inorg. Chem., <u>23</u>, 2731-2732.

Monoclinic, C2, a = 24.928, b = 15.258, c = 11.833 Å, β = 90.91°, Z = 4. Mo radiation, R = 0.078 for 1171 reflexions.

The square-planar complex (Fig. 1) represents the first example of a Ni complex with a bidentate CO_3^{2-} ligand. Ni-P 2.143(8) and 2.153(7), Ni-O 1.95(2) and 1.91(2) Å, O-Ni-O 69.8(8)°.

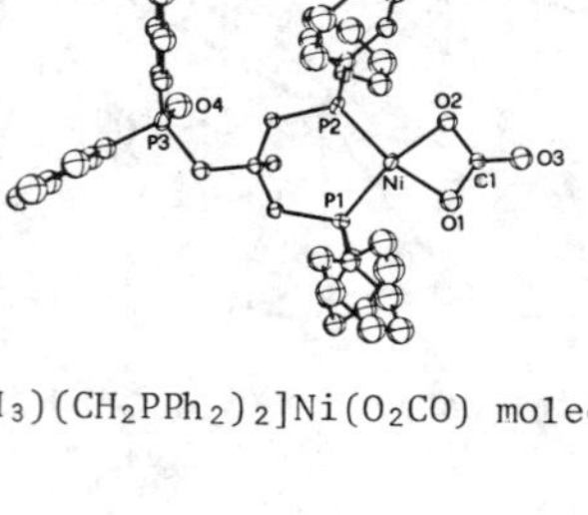

Fig. 1. The $[OPPh_2CH_2C(CH_3)(CH_2PPh_2)_2]Ni(O_2CO)$ molecule.

mer-TRICHLORO[TRIS(2-DIPHENYLPHOSPHINOETHYL)PHOSPHINE]CHROMIUM(III)
$C_{42}H_{42}Cl_3CrP_4$

L.R. GRAY, A.L. HALE, W. LEVASON, F.P. McCULLOUGH and M. WEBSTER, 1984. J. Chem. Soc. Dalton, 47-53.

Monoclinic, $P2_1/a$, a = 14.787, b = 21.617, c = 14.474 Å, β = 113.43°, D_m = 1.32, Z = 4. Mo radiation, R = 0.075 for 2101 reflexions.

The structure consists of discrete molecules (Fig. 1) in which the Cr atom is coordinated to three P and three Cl atoms in a distorted octahedral arrangement. The relationship of the three Cl and three P atoms to each other is mer. There are two long Cr-P distances (Cr-P(2) 2.466(5), Cr-P(3) 2.489(6)) and one normal Cr-P distance (2.399(4) Å). The Cr-Cl distances are normal (Cr-Cl (trans Cl) 2.320(5), 2.292(6), Cr-Cl (trans P) 2.306(4) Å). The angles at Cr differ by up to 9.4° from the ideal value of 90°. Non-bonded distances in the CrP_3Cl_3 octahedron vary from 3.09 to 3.66 Å.

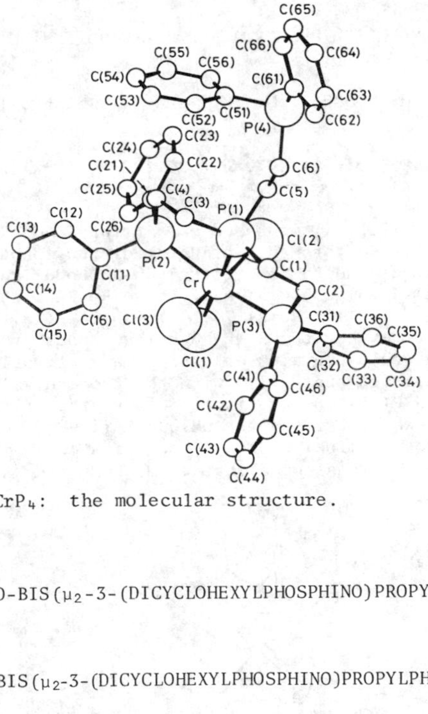

Fig. 1. $C_{42}H_{42}Cl_3CrP_4$: the molecular structure.

(R_n,S_n)-trans-DICHLORO-BIS(μ_2-3-(DICYCLOHEXYLPHOSPHINO)PROPYLPHENYLPHOSPHINO)-DIPLATINUM
$C_{42}H_{66}Cl_2P_4Pt_2$ (I)

(S_n,S_n)-cis-DICHLORO-BIS(μ_2-3-(DICYCLOHEXYLPHOSPHINO)PROPYLPHENYLPHOSPHINO)DIPLATINUM
$C_{42}H_{66}Cl_2P_4Pt_2$ (II)

R. GLASER, D.J. KOUNTZ, R.D. WAID, J.C. GALLUCCI and D.W. MEEK, 1984. J. Am. Chem. Soc., 106, 6324-6333.

I. Monoclinic, $P2_1/n$, a = 11.1177, b = 12.466, c = 16.371 Å, β = 99.751°, D_m = 1.71, Z = 2. Mo radiation, R = 0.026 for 4365 reflexions (at 148 K).

II. Orthorhombic, $P2_12_12_1$, a = 15.269, b = 17.119, c = 16.605 Å, D_m = 1.76, Z = 4. Mo radiation, R = 0.027 for 4667 reflexions (at 148 K).

Views of (I) and (II) are in Fig. 1. In (I), the meso-complex, the molecule lies about an inversion centre and the Pt_2P_2 core is planar, with Pt-P 2.2470(9) and 2.3119(10) Å. Complex (II) is chiral and its absolute configuration was determined. The core Pt_2P_2 ring is folded (dihedral angle 19.1(2)°) with Pt-P 2.243(2)-2.330(2) Å.

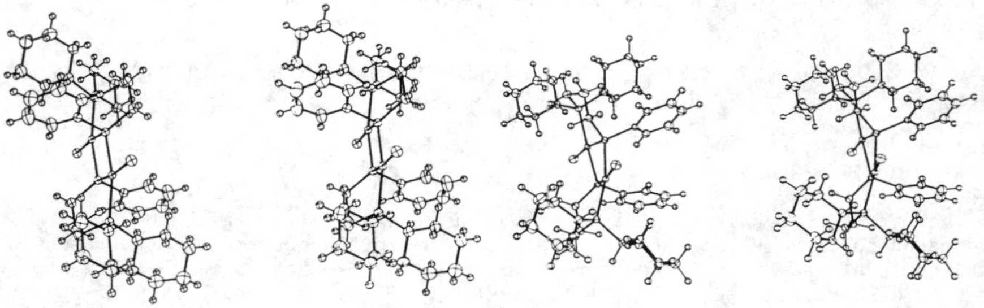

Fig. 1. Stereoview of the meso (left) and chiral (right) forms of $C_{42}H_{66}Cl_2P_4Pt_2$.

(μ_2-HYDRIDO)TRIS(μ_2-DIPHENYLPHOSPHINO)HEPTACARBONYLTRIRUTHENIUM
$C_{43}H_{31}O_7P_3Ru_3$

J.S. FIELD, R.J. HAINES, M.H. MOORE, D.N. SMIT and L.M. STEER, 1984. S. Afr. J. Chem., 37, 138-139.

Monoclinic, $P2_1/c$, a = 21.580, b = 11.947, c = 20.102 Å, β = 112.8°, Z = 4. Mo radiation, R = 0.102 for 6941 reflexions.

The molecular geometry (Fig. 1) is based on a triangle of Ru atoms with each pair of these atoms being linked through a bridging diphenylphosphido group and separated by a distance corresponding to a formal Ru-Ru bond (Ru(1)-Ru(2) 2.908(5), Ru(1)-Ru(3) 2.780(5), Ru(2)-Ru(3) 2.976(5) Å). These Ru-Ru distances, together with the Ru-P-Ru bond angles (Ru(1)-P(12)-Ru(2) 74.1(1), Ru(1)-P(13)-Ru(3) 70.6(1), Ru(2)-P(23)-Ru(3) 78.8(1)°) indicate that the Ru-P-Ru bridge bonds are of the closed and not the open type. The molecule contains an approximate mirror plane through the midpoint of the Ru(1)-Ru(2) edge and Ru(3), with the two equivalent bridging phosphorus atoms (P(13) and P(23)) being situated above the triruthenium plane and the third P atom, (P(12)), which bridges the unique Ru(1)-Ru(2) edge, lying below this plane. Ru(3) has three terminal carbonyls coordinated to it, whereas Ru(1) and Ru(2) contain two each; it is assumed that an H atom bridges these atoms.

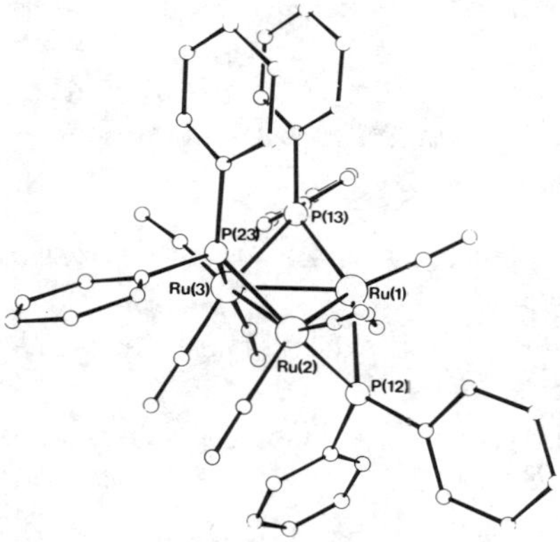

Fig. 1. A view of $C_{43}H_{31}O_7P_3Ru_3$.

(μ-CHLORO)(μ-HYDRIDO)[1,2-BIS(DIPHENYLPHOSPHINO)ETHANE-P,P'-RHODIUM]CHLOROTRIS(TRIETHYLPHOSPHINE)IRIDIUM TETRAFLUOROBORATE
$C_{44}H_{70}BCl_2F_4IrP_5Rh$ (I) [(diphos)Rh(μ-H)(μ-Cl)IrCl(PEt_3)$_3$]BF_4

(μ-CHLORO)(μ-HYDRIDO)[1,2-BIS(DIPHENYLPHOSPHINO)ETHANE-P,P'-RHODIUM]HYDRIDOTRIS(TRIETHYLPHOSPHINE)IRIDIUM TETRAFLUOROBORATE

$C_{44}H_{71}BClF_4IrP_5Rh$ (II) $[(diphos)Rh(\mu-H)(\mu-Cl)IrH(PEt_3)_3]BF_4$

H. LEHNER, D. MATT, A. TONGI, R. THOUVENOT, L.M. VENANZI and A. ALBINATI, 1984. Inorg. Chem., 23, 4254-4261.

I. Monoclinic, $P2_1/c$, a = 11.786, b = 19.673, c = 22.571 Å, β = 89.88°, Z = 4. Mo radiation, R = 0.054 for 5750 reflexions.

II. Monoclinic, $P2_1$, a = 11.102, b = 13.582, c = 16.694 Å, β = 85.06°, Z = 2. Mo radiation, R = 0.062 for 4553 reflexions.

The two binuclear cations, in which the metal atoms are bridged by H and Cl ligands, are shown in Fig. 1. In both the coordination about Rh is distorted square planar and that about Ir is distorted octahedral. In I the Rh-Ir distance is 2.903(1) Å, and the μ-Cl atom is symmetrically bonded to the metals (Rh-Cl 2.386(3), Ir-Cl 2.381(3) Å). Other distances in I are: Ir-Cl(2) 2.366(3), Ir-P(3) 2.390(3), Ir-P(4) 2.384(3), Ir-P(5) 2.367(3), Ir-H 1.67(8), Rh-H 1.85(8), Rh-P(1) 2.216(3), Rh-P(2) 2.199(3) Å. In II there is asymmetry in the Cl bridge (Rh-Cl 2.394(5), Ir-Cl 2.510(5) Å), and the Rh-Ir separation is 2.969(2) Å. Other distances in II are: Ir-P(3) 2.369(6), Ir-P(4) 2.362(6), Ir-P(5) 2.245(7), and Rh-P 2.215(6) and 2.213(6) Å.

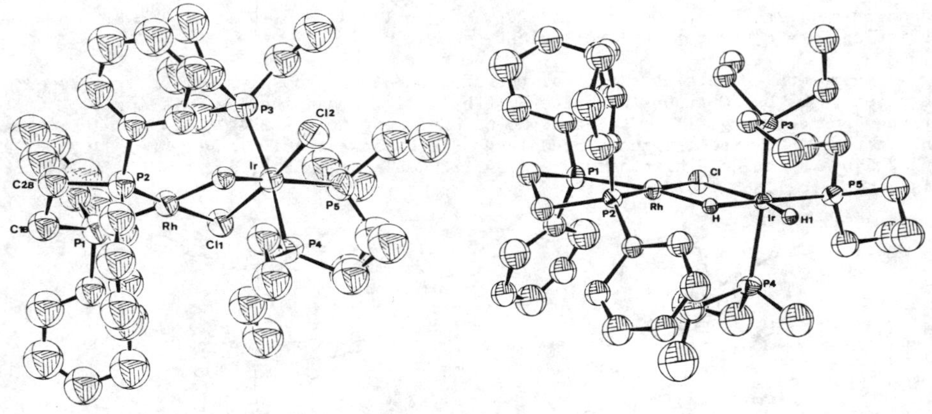

Fig. 1. The cations $[(diphos)Rh(\mu-H)(\mu-Cl)IrX(PEt_3)_3]^+$, X = Cl (left) and H (right).

μ_2-HYDRIDODECACARBONYLBIS(TRIPHENYLPHOSPHINE)IRONTRICOBALT DICHLOROMETHANE SOLVATE

$C_{46}H_{31}Co_3FeO_{10}P_2,CH_2Cl_2$ $HFeCo_3(CO)_{10}(PPh_3)_2,CH_2Cl_2$

E. IISKOLA and T. PAKKANEN, 1984. Acta Chem. Scand., A38, 731-734.

Triclinic, $P\bar{1}$, a = 11.528, b = 13.763, c = 16.736 Å, α = 81.36, β = 78.04, γ = 65.00°, Z = 2. Mo radiation, R = 0.030 for 4962 reflexions.

The metal 'framework' is tetrahedral, containing three Co atoms bound to an apical Fe atom (Fig. 1). Triphenylphosphine ligands are coordinated to Co atoms axially with respect to the Co_3 plane. Seven of the ten CO groups are terminal, three of them bridge the Co atoms, and the H atom forms a triple bridge under the Co_3 plane. Ligand substitution lengthens the Co(1)-Co(2) distance to 2.515 Å while the Co(1)-Co(3) and Co(2)-Co(3) distances are 2.498 and 2.487 Å respectively. The Co-Fe distances are 2.577, 2.577 and 2.562 Å and the Co-H distances are 1.72, 1.63 and 1.72 for Co(1), (2) and (3) respectively. The Co-P distances are 2.246 and 2.253 Å and the bond angles associated with the bridging H atom range from 94 to 96°.

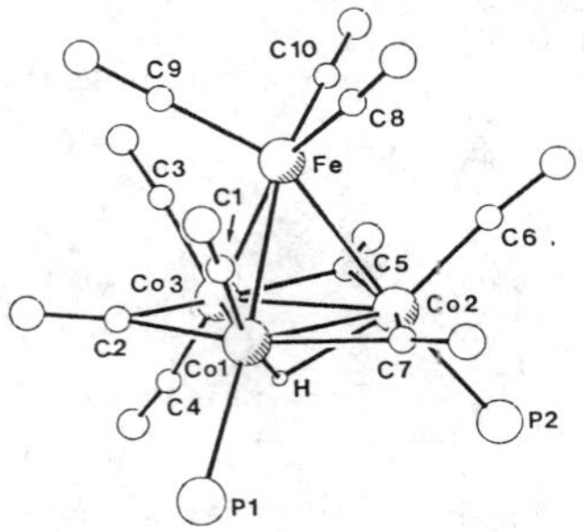

Fig. 1. Molecular structure of $C_{46}H_{31}Co_3FeO_{10}P_2$, omitting phenyl rings.

TRI-μ-CARBONYLHEXACARBONYL[μ_3-[METHYLIDYNETRIS[DIPHENYLPHOSPHINE]-P:P'P'']]TETRACOBALT
$C_{46}H_{31}Co_4O_9P_3$ $Co_4(CO)_9(HC(PPh_2)_3)$

D.J. DARENSBOURG, D.J. ZALEWSKI and T. DELORD, 1984. Organometallics, 3, 1210-1217.

Monoclinic, I2/a, a = 21.943, b = 17.289, c = 26.137 Å, β = 105.14°, Z = 8. Mo radiation, R = 0.052 for 3001 reflexions.

The molecule contains a central tetrahedron of Co atoms symmetrically bridged on three edges by CO groups with the $HC(PPh_2)_3$ ligand capping the CO-bridged triangular face (Fig. 1). The basal Co-Co bond lengths are 2.447, 2.456 and 2.467 Å whilst the Co-Co (basal-apical) bond lengths are considerably longer at 2.546, 2.529 and 2.546 Å. The Co-P distances are 2.207, 2.208 and 2.203 Å, and the P-C(100) distances are 1.88, 1.873 and 1.91 Å.

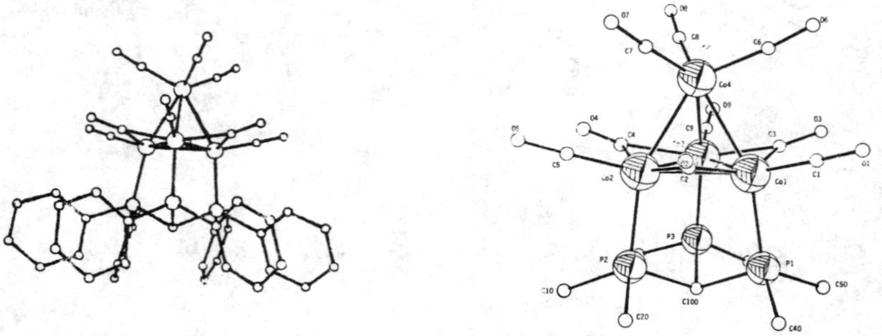

Fig. 1. A view of the $C_{46}H_{31}Co_4O_9P_3$ molecule, and the labelled molecule excluding phenyl rings (right).

TETRACARBONYLBIS[μ-(DIPHENYLPHOSPHINO)][(TRIPHENYLPHOSPHINE)PLATINUM]TUNGSTEN(Pt-W)
$C_{46}H_{35}O_4P_2PtW$ (I) $(CO)_4W(\mu\text{-}PPh_2)_2Pt(PPh_3)$

TETRACARBONYL[[(2,3-η-)-DIMETHYL-2-BUTYNEDIOATE]PLATINUM]BIS[μ-(DIPHENYLPHOSPHINO)]-TUNGSTEN
$C_{34}H_{26}O_8P_2PtW$ (II) $(CO)_4W(\mu\text{-}PPh_2)_2Pt(CH_3O_2CC{\equiv}CCO_2CH_3)$

E.D. MORRISON, A.D. HARLEY, M.A. MARCELLI, G.L. GEOFFROY, A.L. RHEINGOLD and W.C. FULTZ, 1984. Organometallics, 3, 1407-1413.

I. Monoclinic, $P2_1/c$, a = 11.672, b = 18.726, c = 19.869 Å, β = 96.08°, Z = 4. Mo radiation, R = 0.051 for 4491 reflexions.

II. Monoclinic, $P2_1/n$, a = 11.438, b = 11.095, c = 17.764 Å, β = 95.55°, Z = 4. Mo radiation, R = 0.0431 for 4098 reflexions.

In compound I the W and Pt atoms are bridged by two μ-PPh_2 ligands and joined by a metal-metal bond (2.764 Å). The W atom is further coordinated by four CO's and the Pt by a PPh_3 ligand (Fig. 1). The metal-P distances are: Pt-P(1) 2.234, Pt-P(2) 2.249, Pt-P(3) 2.266, W-P(2) 2.509, W-P(3) 2.507 Å. The trigonal angles about Pt are: P(1)-Pt-P(2) 120.2, P(1)-Pt-P(3) 121.9 and P(2)-Pt-P(3) 117.8° and the cis angles about W range from 84.0 to 92.1°. The structure of II is similar to that of I with the alkyne coordinated to Pt such that the alkyne ligand lies nearly in the plane (7.5° dihedral angle) defined by Pt and the two bridging P atoms (Fig. 1). The Pt-W distance is 2.795 Å, and the bonds Pt-C(1) and Pt-C(2) are 2.034 and 2.035 Å respectively. The C(1)-C(2), C(1)-C(3) and C(2)-C(4) distances are respectively 1.31, 1.45 and 1.44 Å.

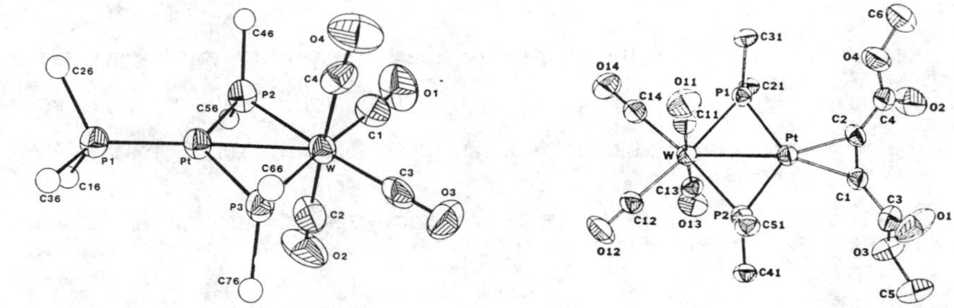

Fig. 1. Perspective drawings of $C_{46}H_{35}O_4P_3PtW$ (left), and $C_{34}H_{26}O_8P_2PtW$ (right), excluding the phenyl rings.

[μ$_3$-TRIS(DIPHENYLPHOSPHINO)METHANE-P,P',P'']TETRAKIS(μ$_2$-HYDRIDO)NONACARBONYLTETRA-RUTHENIUM DICHLOROMETHANE SOLVATE

$C_{46}H_{35}O_9P_3Ru_4,CH_2Cl_2$ $H_4Ru_4(CO)_9$(tripod),CH_2Cl_2

A.A. BAHSOUN, J.A. OSBORN, J.-P. KINTZINGER, P.H. BIRD and U. SIRIWARDANE, 1984. Nouv. J. Chim., 8, 125-132.

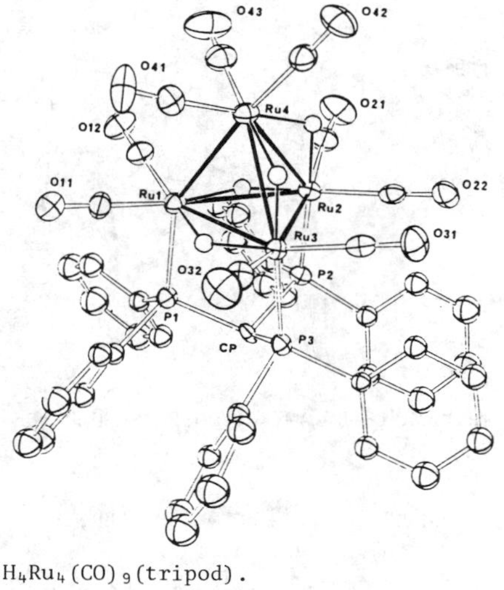

Fig. 1. A view of $H_4Ru_4(CO)_9$(tripod).

Monoclinic, $P2_1/n$, a = 17.046, b = 16.347, c = 17.870 Å, β = 87.02°, D_m = 1.76, Z = 4. Mo radiation, R = 0.043 for 2785 reflexions.

The cluster is shown in Fig. 1. The presence of the hydride ligands (which were located) produces several distortions in the geometry; Ru-Ru distances bridged by H average 2.95 Å while those not bridged average 2.80 Å, and the group of three CO ligands bound to apical Ru(4) are tilted away from hydrides H(24) and H(34) (the plane O(41),O(42),O(43) makes an angle of 15° with the Ru(1),Ru(2),Ru(3) plane). Ru-P distances are in the range 2.316(3)-2.342(3) Å.

BIS(μ₂-DIPHENYLPHOSPHINO)(CARBONYLHYDRIDOTRIPHENYLPHOSPHINEIRIDIUM)TETRACARBONYL-TUNGSTEN
$C_{47}H_{36}IrO_5P_3W$ (I) $WIrH(\mu\text{-}PPh_2)_2(CO)_5(PPh_3)$

BIS(μ₂-DIPHENYLPHOSPHINO)(CARBONYLHYDRIDOTRIPHENYLPHOSPHINEIRIDIUM)TRICARBONYL-(α-METHOXYBENZYLIDENE)TUNGSTEN
$C_{54}H_{44}IrO_5P_3W$ (II) $WIrH(\mu\text{-}PPh_2)_2(CO)_4(PPh_3)\{C(OCH_3)Ph\}$

M.J. BREEN, P.M. SHULMAN, G.L. GEOFFROY, A.L. RHEINGOLD and W.C. FULTZ, 1984. Organometallics, 3, 782-793.

I. Monoclinic, $P2_1/n$, a = 11.416, b = 19.129, c = 20.673 Å, β = 100.78°, Z = 4. Mo radiation, R = 0.033 for 5433 reflexions.

II. Monoclinic, $P2_1/n$, a = 18.773, b = 20.690, c = 12.761 Å, β = 95.56°, Z = 4. Mo radiation, R = 0.050 for 3408 reflexions.

In I the W and Ir atoms are separated by 2.876 Å and bridged by two μ-PPh_2 ligands (W-P(1) 2.528, W-P(2) 2.534, Ir-P(1) 2.302, Ir-P(2) 2.308 Å) (Fig. 1). The Ir is further coordinated by H (located and refined, Ir-H 1.60(8) Å), PPh_3 (Ir-P(3) 2.273 Å) and one CO (Ir-C(5) 1.922 Å), and the W atom is ligated by four additional CO's (W-C from 2.004 to 2.037 Å). The angles Ir-P(1)-W and Ir-P(2)-W are respectively 72.7 and 72.9°. In II, the carbene is terminally bound to W (W-C(2) 2.02 Å) which is further ligated by three CO's (1.90, 2.00, 2.02 Å). The W-Ir bond (2.858 Å) is bridged again by two μ-PPh_2 ligands (W-P(1) 2.507, W-P(2) 2,496, Ir-P(1) 2.286, Ir-P(2) 2.328 Å), and the Ir centre is further coordinated by the hydride (not located), PPh_3 (Ir-P(3) 2.283 Å) and one CO (Ir-C(5) 1.83 Å). The angles Ir-P(1)-W and Ir-P(2)-W are respectively 73.0 and 72.6°.

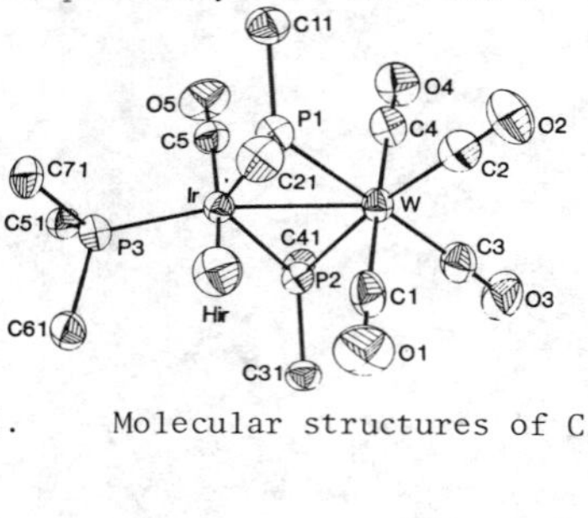

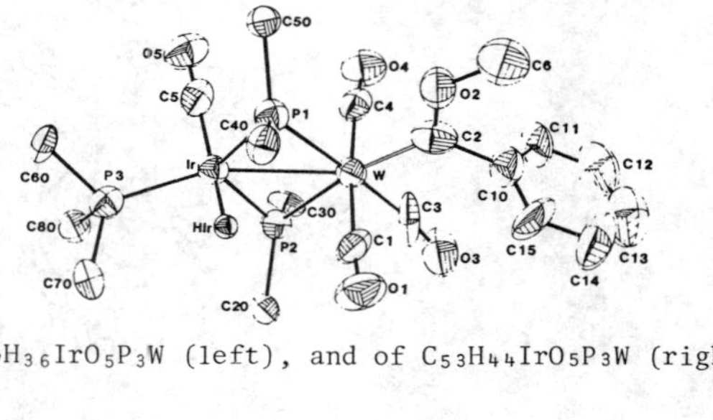

Fig. 1. Molecular structures of $C_{47}H_{36}IrO_5P_3W$ (left), and of $C_{53}H_{44}IrO_5P_3W$ (right).

(t-BUTYLIMINO(N-t-BUTYL-N-TRIMETHYLSILYLAMINO)PHOSPHANESULPHIDE-P,S)BIS(TRIPHENYL-PHOSPHINE)PLATINUM BENZENE SOLVATE
$C_{47}H_{57}N_2P_3PtSSi,1.5(C_6H_6)$

O.J. SCHERER, H. JUNGMANN, C. KRUGER and G. WOLMERSHAUSER, 1984. Chem. Ber., 117, 2382-2392.

Triclinic, $P\bar{1}$, a = 10.990, b = 13.719, c = 22.383 Å, α = 72.53, β = 78.05, γ = 69.73°, Z = 2. Mo radiation, R = 0.0434 for 8793 reflexions.

A view of the complex is shown in Fig. 1. There is a weak N→Si donor-acceptor bond (N2...Si 2.629(6) Å). Other bond lengths are: Pt-S 2.378(2), Pt-P1 2.285(2), Pt-P2 2.252(2), Pt-P3 2.386(2), P1-S 2.073, P1-N1 1.682(6), P1-N2 1.566(6), Si-N1 1.802(6) Å. The interplanar angle between the three-membered ring and the Pt,P2,P3 plane is 8.1°.

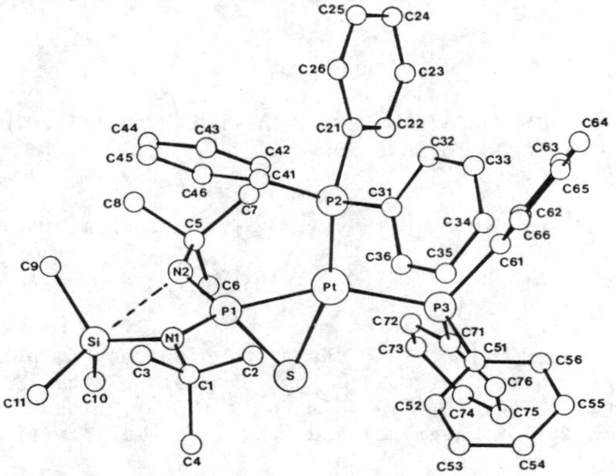

Fig. 1. A view of $C_{47}H_{57}N_2P_3PtSSi$.

BIS[(μ_2-BROMO)[HYDROGENBIS(DIPHENYLPHOSPHINITO)]NITROSYLRHODIUM] $C_{48}H_{42}Br_2N_2O_6P_4Rh_2$

R.B. ENGLISH and M.M. DE V. STEYN, 1984. S. Afr. J. Chem., 37, 177-181.

Triclinic, P$\bar{1}$, a = 16.197, b = 14.290, c = 12.808 Å, α = 116.6, β = 71.4, γ = 114.0°, D_m = 1.7, Z = 2. Cu radiation, R = 0.067 for 2357 reflexions.

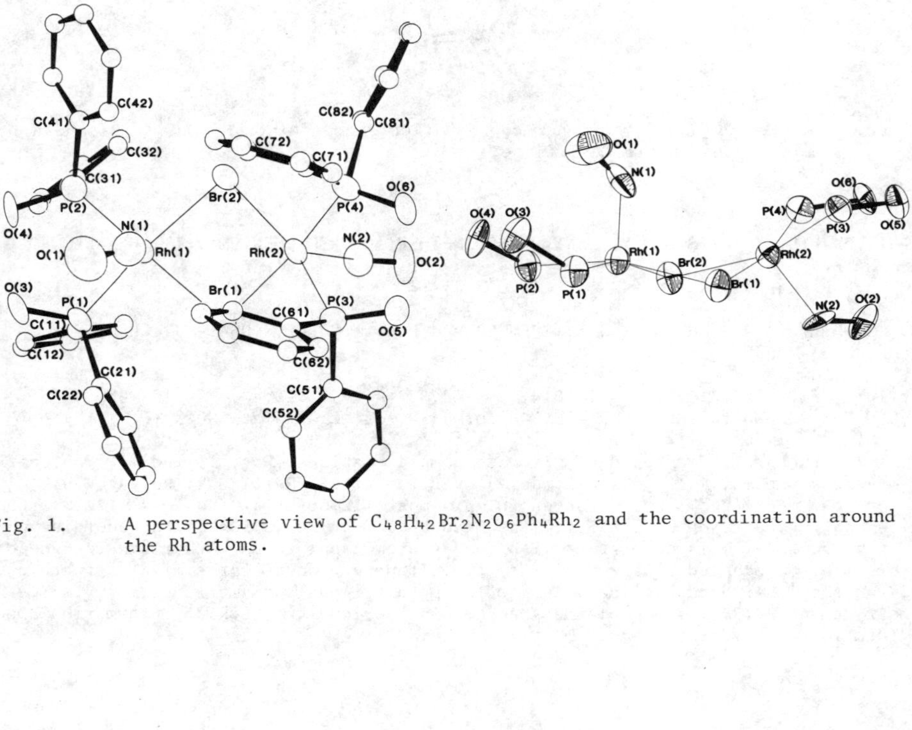

Fig. 1. A perspective view of $C_{48}H_{42}Br_2N_2O_6Ph_4Rh_2$ and the coordination around the Rh atoms.

The complex (Fig. 1) consists of two square pyramidal moieties sharing a common Br ligand, and with bent apical nitrosyl groups. Main dimensions are: Rh(1)-Br(1) 2.572(3), Rh(1)-Br(2) 2.642(3), Rh(2)-Br(1) 2.566(3), Rh(2)-Br(2) 2.626(3), Rh-N 1.88 and 1.99(2), Rh-P 2.272-2.284(6), N-O 1.10 and 1.18(2) Å, Rh-N-O 128 and 130(2)°. The phosphine O atoms are linked by short O-H...O hydrogen bonds (O...O 2.40 and 2.41(2) Å).

TETRAKIS(μ_2-HYDRIDO)[DIHYDROHEXAKIS(DIMETHYLPHENYLPHOSPHINO)DIIRIDIUM]COPPER HEXAFLUOROPHOSPHATE TETRAHYDROFURAN TOLUENE SOLVATE

$C_{48}H_{72}CuF_6Ir_2P_7,0.5(C_4H_8O),0.5(C_7H_8)$

L.F. RHODES, J.C. HUFFMAN and K.G. CAULTON, 1984. J. Am. Chem. Soc., 106, 6874-6876.

Monoclinic, $P2_1/c$, a = 17.833, b = 20.876, c = 18.630 Å, β = 105.36°, Z = 4. R_w = 0.087 for three-dimensional data (at -162°C).

The complex has a linear trimetallic chain in which the phosphines adopt an eclipsed rotational conformation (Fig. 1). Some dimensions are: Cu-Ir(1) 2.502(4), Cu-Ir(3) 2.529(4), Ir-P 2.277-2.319(10) Å, Ir-Cu-Ir 176.0°. The P-Ir-Cu angles fall into two groups, two small (105° average) and four large (124° average). The hydride ligands could not be located by the X-ray study.

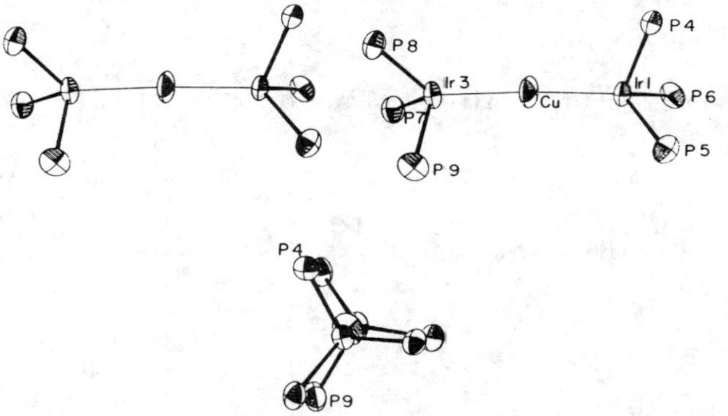

Fig. 1. A stereoview of the inner coordination sphere of $C_{48}H_{72}CuF_6Ir_2P_7$ and a view along the Ir-Cu-Ir line.

(μ_2-SILVER)BIS[BIS(DIISOPROPYLPHENYLPHOSPHINE)HEPTAHYDRIDORHENIUM(VII)] HEXAFLUOROPHOSPHATE

$C_{48}H_{90}AgF_6P_5Re_2$

N.G. CONNELLY, J.A.K. HOWARD, J.L. SPENCER and P.K. WOODLEY, 1984. J. Chem. Soc. Dalton, 2003-2009.

Triclinic, $P\bar{1}$, a = 13.188, b = 16.014, c = 14.827 Å, α = 112.47, β = 101.03, γ = 90.65°, Z = 2. Mo radiation, R = 0.048 for 4452 reflexions (at 200 K).

The structure is composed of discrete complex cations (Fig. 1) and PF_6^- anions. One of the Re atoms, Re(2), is disordered over two sites (Re(2) 60%; Re(2A) 40%). However, the metal atoms are arranged in an almost linear array (Re-Ag-Re 172°) with a mean Re-Ag separation of 2.873 Å. The mean P-Re-P angle is 134°. The coordination geometry about Re(1) is consistent with the Ag atom capping the other square face of the original monocapped square antiprism type structure found for similar structures. Re(2) on the other hand has idealised tetracapped trigonal prismatic geometry. The geometry about Re(2A) is unexplained due to disorder and the lack of information about the locations of hydride atoms.

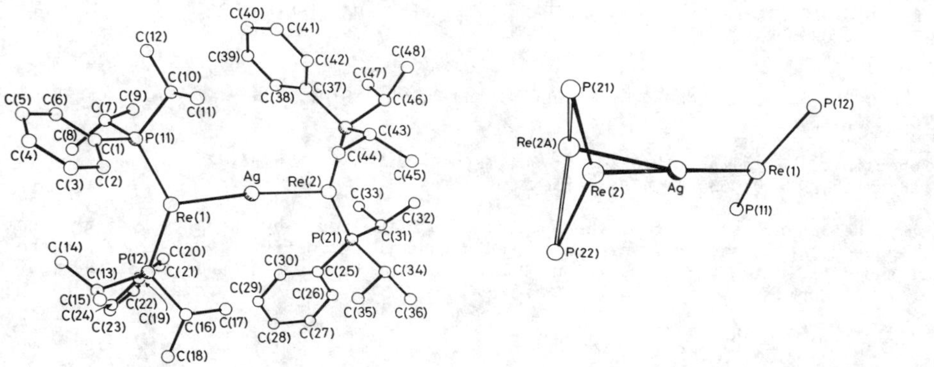

Fig. 1. $C_{48}H_{90}AgF_6P_5Re_2$: the structure of the cation and the heavy atom core.

(μ_3-BENZYLIDYNE)TRIS(μ_2-HYDRIDO)HEPTACARBONYLBIS(TRIPHENYLARSINE)TRIRUTHENIUM
$C_{50}H_{38}As_2O_7Ru_3$ $(\mu\text{-H})_3Ru_3(\mu_3\text{-CPh})(CO)_7(AsPh_3)_2$

Z.A. RAHMAN, L.R. BEANAN, L.M. BAVARO, S.P. MODI, J.B. KEISTER and M.R. CHURCHILL, 1984. J. Organometal. Chem., 263, 75-92.

Monoclinic, C2/c, a = 36.746, b = 12.743, c = 22.802 Å, β = 108.60°, Z = 8. Mo radiation, R = 0.054 for 2523 reflexions.

The molecule consists of a triangular arrangement of Ru atoms (Ru(1)-Ru(2) 2.864, Ru(2)-Ru(3) 2.863, Ru(1)-Ru(3) 2.848 Å) capped by a triply-bridging benzylidyne moiety with Ru(1)-C(1) 2.079, Ru(2)-C(1) 2.066 and Ru(3)-C(1) 2.116 Å (Fig. 1). There are three bridging hydride ligands, each spanning two Ru atoms. The molecule has six radial terminal CO ligands with Ru-C distances ranging from 1.828 to 1.935 Å. The longer Ru-C distance involving the single axial CO (Ru(3)-C(3B) 2.014 Å) illustrates the trans effect of the μ_3-CPH ligand. The Ru(1)-As(1) and Ru(2)-As(2) distances are 2.516 and 2.536 Å respectively.

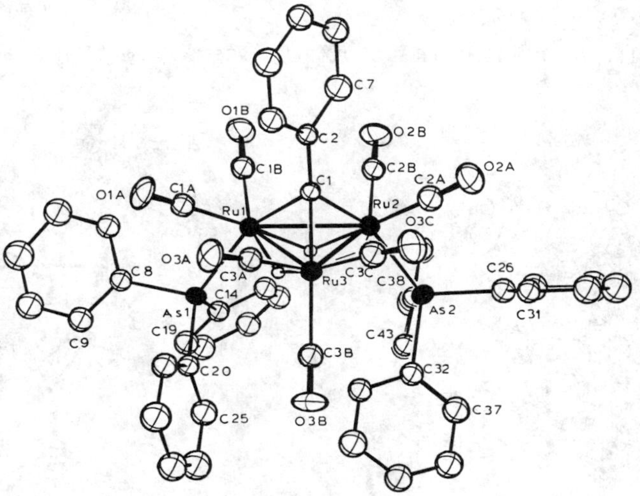

Fig. 1. Molecular structure of $C_{50}H_{38}As_2O_7Ru_3$.

{1,4-BIS[BIS(DIPHENYLPHOSPHINO)METHANE-P]TETRAPHOSPHABUTADIENE-P^1,P^2,P^3,P^4}COBALT TETRAFLUOROBORATE

$C_{50}H_{44}BCoF_4P_8$ $C_{50}H_{44}CoP_8^+.BF_4^-$

F. CECCONI, C.A. GHILARDI, S. MIDOLLINI and A. ORLANDINI, 1984. J. Am. Chem. Soc., 106, 3667-3668.

Monoclinic, $P2_1/n$, a = 19.071, b = 26.712, c = 9.709 Å, β = 93.92°, Z = 4. Mo radiation, R = 0.063 for 4703 reflexions.

The ligand results from the opening of a P_4 molecule by the attack of two dppm ligands. The Co atom has very distorted octahedral geometry (Fig. 1) in which the ligand coordinates through all four P atoms of the P_4 fragment and two of the four P atoms of dppm.

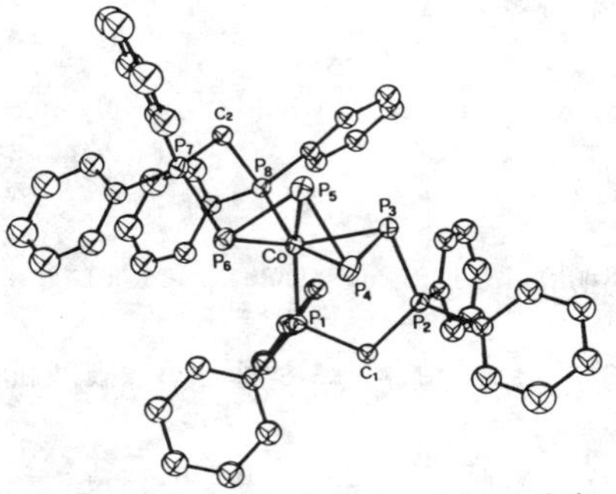

Co–P1	2.203 (2)	P2–P3	2.183 (3)
Co–P3	2.281 (2)	P3–P4	2.173 (3)
Co–P4	2.285 (2)	P4–P5	2.197 (3)
Co–P5	2.305 (2)	P5–P6	2.171 (3)
Co–P6	2.281 (2)	P6–P7	2.196 (3)
Co–P8	2.196 (2)		

Fig. 1. A view of $C_{50}H_{44}CoP_8^+$ cation with selected bond lengths.

BIS[μ-BIS(DIPHENYLPHOSPHINO)METHANE-P,P']TETRABROMODIMOLYBDENUM BIS(TETRAHYDROFURAN) SOLVATE

$C_{50}H_{44}Br_4Mo_2P_4,2C_4H_8O$ (I) $Mo_2Br_4(Ph_2PCH_2PPh_2)_2.2C_4H_8O$

BIS[μ-TRIS(DIPHENYLPHOSPHINO)METHANE-P,P']TETRACHLORODIMOLYBDENUM BIS(DICHLOROMETHANE) SOLVATE

$C_{74}H_{62}Cl_4Mo_2P_6,2CH_2Cl_2$ (II) $Mo_2Cl_4[HC(PPh_2)_3]_2.2CH_2Cl_2$

F.L. CAMPBELL III, F.A. COTTON and G.L. POWELL, 1984. Inorg. Chem., 23, 4222-4226.

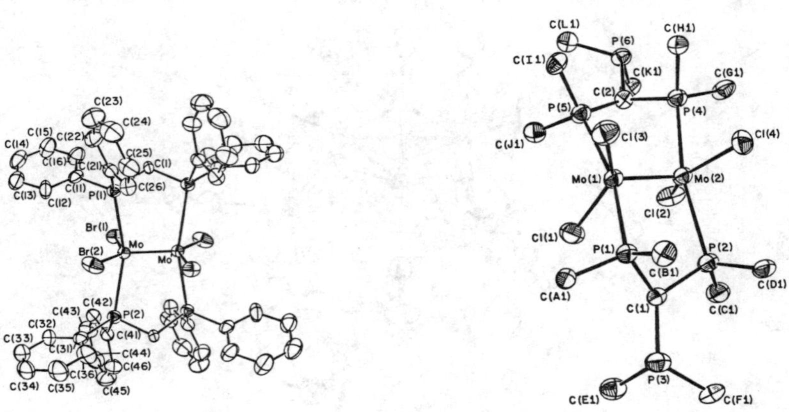

Fig. 1. The $Mo_2Br_4(Ph_2PCH_2PPh_2)_2$ (left) and $Mo_2Cl_4[HC(PPh_2)_3]_2$ (right) molecules.

I. Monoclinic, $P2_1/n$, a = 11.653, b = 17.339, c = 14.409 Å, β = 98.07°, Z = 2. Mo radiation, R = 0.047 for 3107 reflexions.

II. Triclinic, $P\bar{1}$, a = 15.078, b = 22.941, c = 12.380 Å, α = 104.55, β = 107.87, γ = 92.44°, Z = 2. Mo radiation, R = 0.0635 for 6712 reflexions.

The two dinuclear molecules are shown in Fig. 1. In I the molecule is centrosymmetric with a Mo-Mo bond length of 2.138(1) Å and an eclipsed rotational conformation. In II, only two P atoms of each $HC(PPh_2)_3$ ligand are coordinated. The Mo-Mo distance in II, with a partially staggered conformation (20(3)°), is 2.148(1) Å. Other bond lengths: for I; Mo-Br 2.537(1) and 2.532(1), Mo-P 2.549(2) and 2.674(2) Å; for II; Mo-Cl 2.378(2)-2.412(2), Mo-P 2.544(2)-2.662(2) Å. A relationship between Mo-Mo bond length and cos(2χ) (χ = twist away from an eclipsed conformation) has been established for $Mo_2X_4(LL)_2$ compounds.

cis-DICHLOROBIS(BIS(DIPHENYLPHOSPHINO)METHANE-P,P')OSMIUM CHLOROFORM SOLVATE
$C_{50}H_{44}Cl_2OsP_4,CHCl_3$ (I) cis-$OsCl_2((C_6H_5)_2PCH_2P(C_6H_5)_2)_2,CHCl_3$

cis-DICHLOROBIS(BIS(DIPHENYLPHOSPHINO)METHANE-P,P')RUTHENIUM CHLOROFORM SOLVATE
$C_{50}H_{44}Cl_2P_4Ru,CHCl_3$ (II) cis-$RuCl_2((C_6H_5)_2PCH_2P(C_6H_5)_2)_2,CHCl_3$

trans-DICHLOROBIS(BIS(DIPHENYLPHOSPHINO)METHANE-P,P')RUTHENIUM
$C_{50}H_{44}Cl_2P_4Ru$ (III) trans-$RuCl_2((C_6H_5)_2PCH_2P(C_6H_5)_2)_2$

A.R. CHAKRAVARTY, F.A. COTTON and W. SCHWOTZER, 1984. Inorg. Chim. Acta, 84, 179-185.

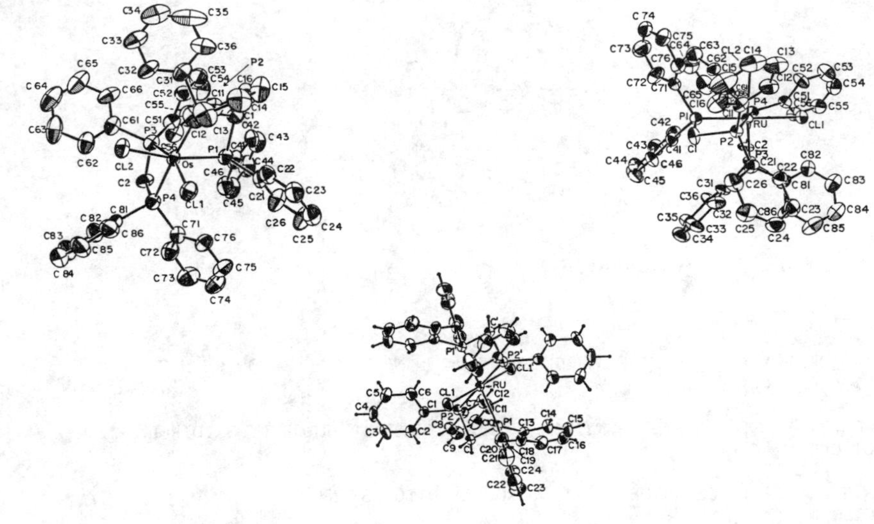

Fig. 1. Molecular structures of $C_{50}H_{44}Cl_2OsP_4$ (I) (top left), $C_{50}H_{44}Cl_2P_4Ru$ (II) (top right), and $C_{50}H_{44}Cl_2P_4Ru$ (III) (bottom).

I. Monoclinic, $P2_1/n$, a = 13.415, b = 22.859, c = 16.693 Å, β = 105.77°, Z = 4. Mo radiation, R = 0.044 for 5482 reflexions.

II. Monoclinic, $P2_1/n$, a = 13.442, b = 22.833, c = 16.750 Å, β = 105.53°, Z = 4. Mo radiation, R = 0.065 for 4403 reflexions.

III. Monoclinic, $P2_1/n$, a = 11.368, b = 10.656, c = 18.832 Å, β = 103.90°, Z = 2. Mo radiation, R = 0.028 for 2023 reflexions.

The structures of I and II are isotypic but they differ from that of the trans isomer III (Fig. 1). In I and II respectively the M-Cl(1), M-Cl(2), M-P(1), M-P(2), M-P(3) and M-P(4) distances are: 2.448, 2.440; 2.455, 2.451; 2.306, 2.303; 2.338, 2.335; 2.297, 2.318; 2.339, 2.338 Å. The Cl-M-Cl angles in I and II are 83.08 and 84.1° and the chelate ring in both compounds is heavily distorted as evidenced by the P-M-P angle (72.0° in II, 71.6° in I). Compound III has a centre of inversion and the P-M-P angle is 71.4°. The Ru-Cl(1), Ru-P(1) and Ru-P(2) distances are respectively 2.426, 2.340 and 2.367 Å. P-C distances in I range from 1.853 to 1.875 Å and in II from 1.859 to 1.884 Å. In III the unique P-C distances are 1.840 and 1.858 Å.

HEXACHLORO(μ-OXO)BIS[μ-BIS(DIPHENYLPHOSPHINO)METHANE]DIOSMIUM(IV) BIS(CHLOROFORM)
$C_{50}H_{44}Cl_6OOs_2P_4,2CHCl_3$ $Os_2(\mu\text{-}O)[\mu\text{-}CH_2(P(C_6H_5)_2)_2]_2Cl_6.2CHCl_3$

A.R. CHAKRAVARTY, F.A. COTTON and W. SCHWOTZER, 1984. Inorg. Chem., 23, 99-103.

Monoclinic, $P2_1/c$, a = 11.138, b = 24.461, c = 12.769 Å, β = 105.82°, Z = 4. Mo radiation, R = 0.044 for 3027 reflexions.

The centrosymmetric complex molecule (Fig. 1) has a linear Os-O-Os central unit (Os-O 1.792(1) Å), bridged by two $(PPh_2)_2CH_2$ ligands (Os-P 2.452(3) and 2.443(4) Å). The Os-Cl distances 2.307(3) (trans to O), 2.387(3) and 2.360(3) Å. Coordination geometry is approximately octahedral. The P-C-P angle is 124.6(7)°.

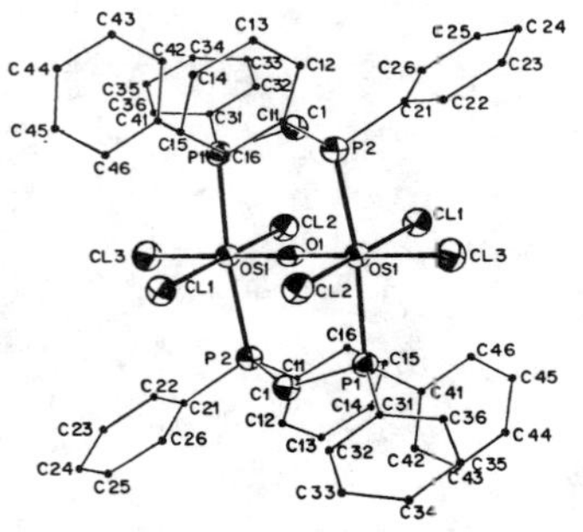

Fig. 1. The $Os_2(\mu\text{-}O)[\mu\text{-}CH_2(PPh_2)_2]_3Cl_6$ molecule.

BIS(μ_2-CHLORO)BIS(μ_2-BIS(DIPHENYLPHOSPHINO)METHANE)TETRACHLORODIRHENIUM
$C_{50}H_{44}Cl_6P_4Re_2$ (I)

BIS(μ_2-CHLORO)BIS[μ_2-BIS(DIPHENYLPHOSPHINO)METHANE]TRICHLOROETHOXYDIRHENIUM
$C_{52}H_{49}Cl_5OP_4Re_2$ (II)

BIS(μ_2-ISOPROPOXO)DICHLOROHEXAKIS(ISOPROPOXY)DITUNGSTEN
$C_{24}H_{56}Cl_2O_8W_2$ (III)

T.J. BARDER, F.A. COTTON, D. LEWIS, W. SCHWOTZER, S.M. TETRICK and R.A. WALTON, 1984. J. Am. Chem. Soc., 106, 2882-2891.

I. Monoclinic, C2/c, a = 23.083, b = 10.866, c = 23.253 Å, β = 124.25°, Z = 4. Mo radiation, R = 0.036 for 2810 reflexions.

II. Orthorhombic, $Pn2_1a$, a = 24.099, b = 18.255, c = 12.565 Å, Z = 4. Mo radiation, R = 0.053 for 3273 reflexions.

III. Monoclinic, C2/m, a = 17.706, b = 12.034, c = 9.727 Å, β = 123.74°, Z = 2. Mo radiation, R = 0.026 for 1145 reflexions.

Molecules I and II (Fig. 1) have Re-Re double bonds. In I, the dimeric units lie about crystallographic inversion centres, with Re-Re 2.616(1), Re-Cl (bridge) 2.391(2), Re-Cl (terminal) 2.388(5) Å. The Re_2P_2C rings have half-chair conformation. In II, one of the terminal chloride ligands of I has been replaced by an ethoxide group. The main bond lengths are Re-Re 2.667(1), Re-Cl 2.321-2.461(5), Re-O 2.085(14) Å. The two five-membered Re_2P_2C rings are in half-chair conformations; the two C atoms are syn. Although there are no crystallographic restrictions imposed, the virtual symmetry is C_m. Molecule III has 2/m symmetry (Fig. 1) with W-W 2.784(1), W-Cl 2.449(3), W-O(1) 2.013(7), W-O(1') 2.076(7), W-O(2) 1.913(8), W-O(3) 1.837(6) Å.

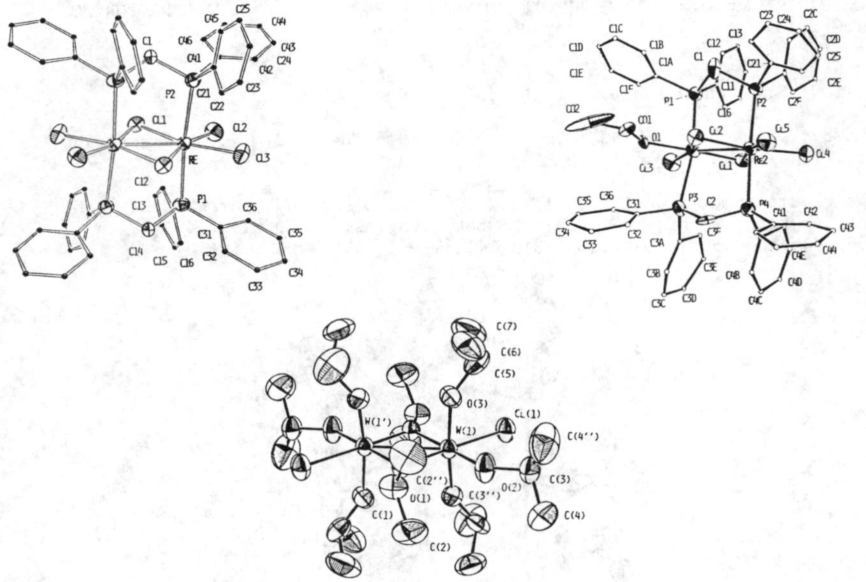

Fig. 1. Views of $C_{50}H_{44}Cl_6P_4Re_2$ (I), $C_{52}H_{49}Cl_5OP_4Re_2$ (II), and $C_{24}H_{56}Cl_2O_8W_2$ (III) (clockwise from top left).

BIS[TETRACARBONYL({DIPHENYL[METHYL(1-PHENYLETHYL)AMINO]}PHOSPHINO)MANGANESE]
$C_{50}H_{44}Mn_2N_2O_8P_2$

J. MEUNIER-PIRET, M. VAN MEERSSCHE, M. GIELEN and I. VANDEN EYNDE, 1984. Bull. Soc. Chim. Belg., 93, 307-311.

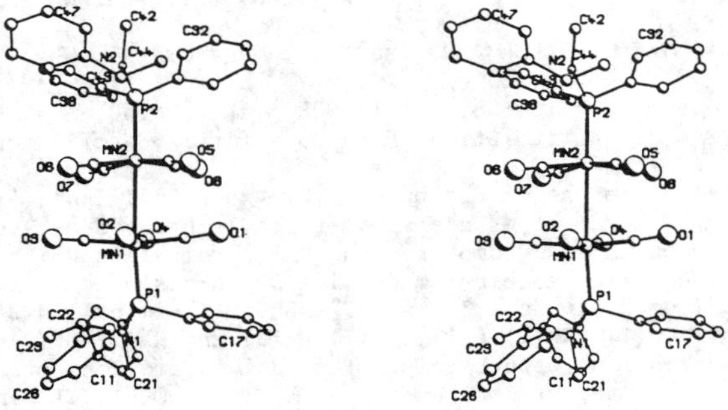

Fig. 1. A stereoview of $C_{50}H_{44}Mn_2N_2O_8P_2$.

Monoclinic, $P2_1$, a = 11.713, b = 19.552, c = 12.072 Å, β = 103.82°, Z = 2. Mo radiation, R = 0.093 for 1789 reflexions.

The Mn atoms (Fig. 1) have distorted octahedral coordination, with carbonyl groups essentially normal to the P-Mn-Mn-P line. Main dimensions are: Mn-Mn 2.957(5), Mn-P 2.247(6) and 2.279(6), mean Mn-C 1.83 Å, P-Mn-Mn 175.6(6) and 174.2(5)°.

BIS(μ-CHLORO)BIS[μ-BIS(DIPHENYLPHOSPHINO)METHANE-P,P']BIS(η^2-TETRAHYDROBORATO)DI-VANADIUM(II)

$C_{50}H_{52}B_2Cl_2P_4V_2$ $[V(\mu\text{-}Cl)(\mu\text{-}Ph_2PCH_2PPh_2)BH_4]_2$

F.A. COTTON, S.A. DURAJ and W.J. ROTH, 1984. Inorg. Chem., 23, 4113-4115.

Orthorhombic, Pbca, a = 13.743, b = 22.555, c = 16.184 Å, Z = 4. Mo radiation, R = 0.060 for 1142 reflexions.

The centrosymmetric molecule (Fig. 1) has edge-sharing bioctahedral geometry with a V...V separation of 3.112(3) Å. Bond lengths: V-(μ-Cl) 2.426(3) and 2.422(3), V-P 2.570(4) and 2.568(4), V...B 2.314(14), V-H 1.78(8) and 1.69(10) Å.

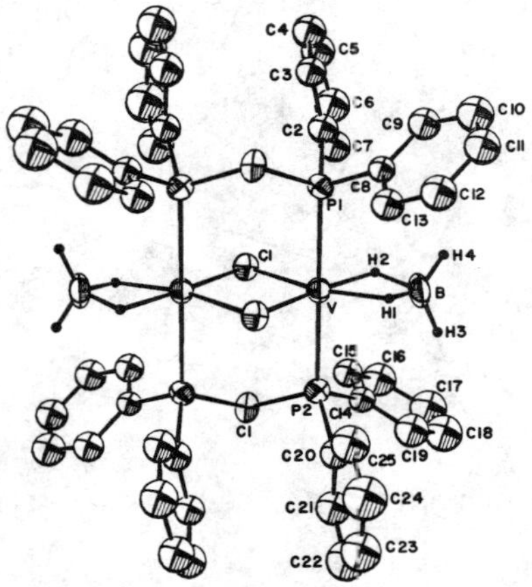

Fig. 1. The $[V(\mu\text{-}Cl)(\mu\text{-}Ph_2PCH_2PPh_2)BH_4]_2$ molecule.

μ-CARBONYL-μ-CHLORO-DI[μ-BIS(DIPHENYLPHOSPHINO)AMINE-P,P']-BIS[CARBONYLRHODIUM(I)] CHLORIDE METHANOL SOLVATE

$C_{51}H_{42}Cl_2N_2O_3P_4Rh_2,CH_4O$ $[Rh_2(\mu CO)(\mu Cl)(\mu Ph_2PNHPPh_2)_2(CO)_2]Cl,CH_3OH$

G. LIEHR, G. SZUCSÁNYI and J. ELLERMANN, 1984. J. Organometal. Chem., 265, 95-106.

Triclinic, $P\bar{1}$, a = 16.806, b = 13.806, c = 11.905 Å, α = 96.30, β = 96.00, γ = 110.75°, D_m = 1.53, Z = 2. Ag radiation, R = 0.064 for 3258 reflexions.

The two Rh atoms of the cation are bridged by two PNP ligands (Fig. 1), and in the equatorial plane the metal atoms are asymmetrically bridged by a Cl and CO ligand. A terminal CO group on each metal completes the Rh coordination which can be described as a distorted trigonal bipyramid. The Rh-Rh distance is 2.737 Å and the Rh-P distances range from 2.316 to 2.328 Å. The Rh-CO (terminal) distances are both 1.85 Å whilst the Rh(1)-C(1), Rh(2)-C(1), Rh(1)-Cl(2) and Rh(2)-Cl(2) distances are 2.12, 2.04, 2.536 and 2.591 Å respectively. The cations are connected in infinite chains along the b axis of the crystal by strong N-H...O, O-H...Cl^- and N-H...Cl^- hydrogen bonds. The distances N(1)...O(1M), O(1M)...$Cl(1)^-$ and N(2)...$Cl(1)^-$ are respectively 2.822, 3.078 and 3.129 Å.

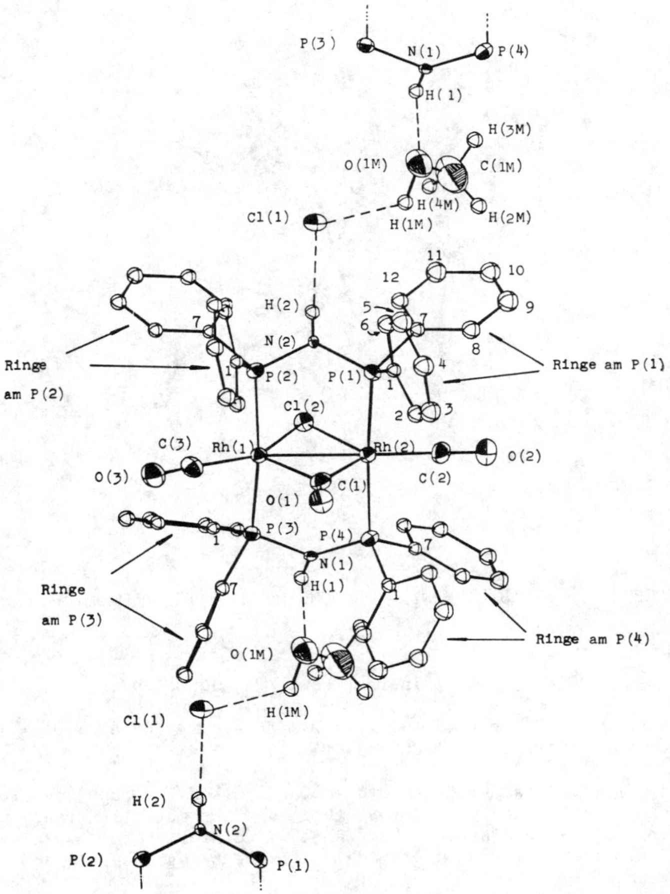

Fig. 1. Molecular structure of $C_{51}H_{42}Cl_2N_2O_3P_4Rh_2$ showing hydrogen bonds from the cation to methanol solvent (M) and chloride anion.

trans-CHLORO(CARBONYL)RHODIUM(I)-BIS[μ-(DIPHENYLARSINO)(DIPHENYLPHOSPHINO)METHANE]-cis-DICHLOROPLATINUM(II) DICHLOROMETHANE SOLVATE

$C_{51}H_{44}As_2Cl_3OP_2PtRh.CH_2Cl_2$ trans-Rh(CO)Cl[μ-$(C_6H_5)_2AsCH_2P(C_6H_5)_2]_2$-cis-$PtCl_2$,$CH_2Cl_2$

R.R. GUIMERANS, F.E. WOOD and A.L. BALCH, 1984. Inorg. Chem., 23, 1307-1308.

Monoclinic, $P2_1/c$, a = 18.760, b = 11.610, c = 22.954 Å, β = 92.81°, Z = 4. Mo radiation, R = 0.038 for 6415 reflexions (at 140K).

The molecule (Fig. 1) displays approximate mirror symmetry and consists of trans-$Rh(CO)ClAs_2$ and cis-$PtCl_2P_2$ coordination groups in a face-to-face orientation with Rh...Pt 3.043(1) Å. Bond lengths: Pt-P 2.237(2), 2.238(2), Pt-Cl 2.345(2), 2.356(2), Rh-As 2.417(1), 2.400(1), Rh-Cl 2.363(2), Rh-CO 1.818(8) Å.

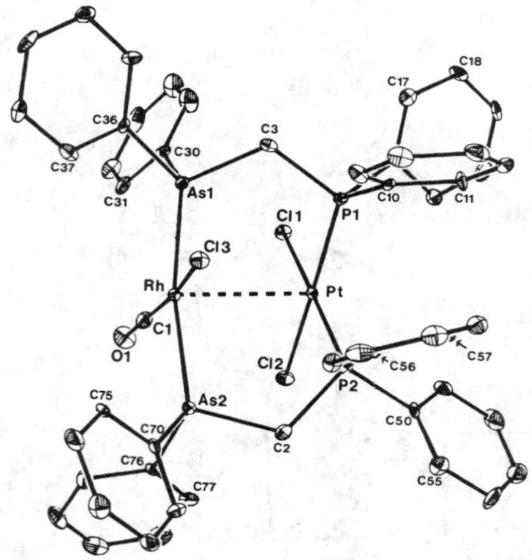

Fig. 1. The Rh(CO)Cl[μ-$Ph_2AsCH_2PPh_2$]$_2$$PtCl_2$ molecule.

(μ-HYDRIDO)(μ-CARBONYL)BIS[BIS(DIPHENYLPHOSPHINO)METHANE]TRICHLORODIRHODIUM MONOHYDRATE

$C_{51}H_{45}Cl_3OP_4Rh_2,H_2O$ $[Rh_2Cl_3(\mu\text{-}H)(\mu\text{-}CO)((C_6H_5)_2PCH_2P(C_6H_5)_2)_2],H_2O$

B.R. SUTHERLAND and M. COWIE, 1984. Inorg. Chem., 23, 1290-1297.

Monoclinic, I2/a, a = 25.021, b = 14.697, c = 27.486 Å, β = 107.57°, D_m = 1.54, Z = 8. Mo radiation, R = 0.052 for 5535 reflexions.

The molecule (Fig. 1) has bridging hydride and carbonyl groups, two terminal chloro ligands on one metal and one on the other, and two mutually trans, bridging $Ph_2PCH_2PPh_2$ groups. The coordination about Rh(1) is distorted octahedral and that about Rh(2) is distorted tetragonal pyramidal with the bridging CO ligand occupying the apical position.

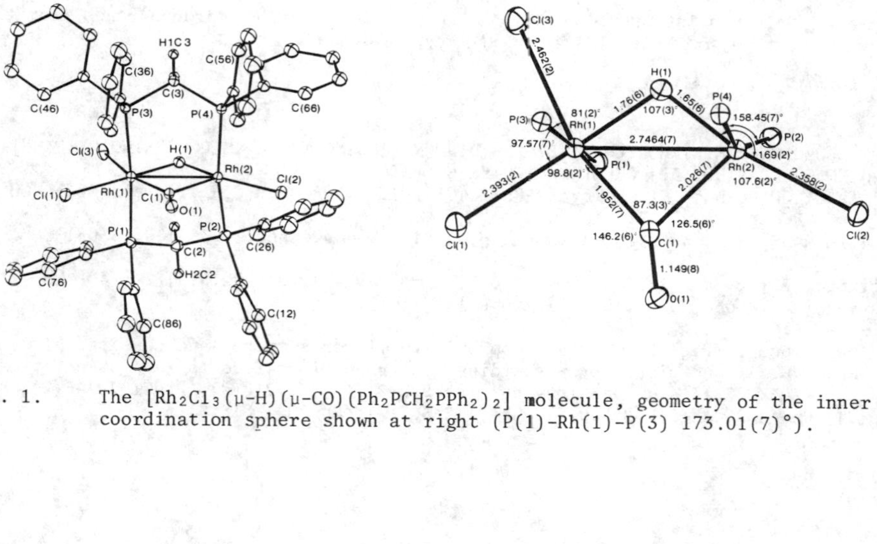

Fig. 1. The $[Rh_2Cl_3(\mu\text{-}H)(\mu\text{-}CO)(Ph_2PCH_2PPh_2)_2]$ molecule, geometry of the inner coordination sphere shown at right (P(1)-Rh(1)-P(3) 173.01(7)°).

OCTACARBONYL-BIS(μ-{IODO[TETRACARBONYL(TRIPHENYLPHOSPHINE)RHENIO]STANNIO(IV)})-DIRHENIUM

$C_{52}H_{30}I_2O_{16}P_2Re_4Sn_2$ $Re_2(CO)_8\{Sn(I)Re(CO)_4[P(C_6H_5)_3]\}_2$

H. PREUT, H.-J. HAUPT and U. FLORKE, 1984. Acta Cryst., C40, 600-602.

Monoclinic, $P2_1/c$, a = 9.096, b = 29.271, c = 11.291 Å, β = 93.73°, Z = 2. Mo radiation, R = 0.034 for 2978 reflexions.

The central fragment of the molecule (Fig. 1) consists of a planar Re_2Sn_2 rhombus with Re-Sn 2.776(1) and 2.769(1) Å and Re-Re 3.176(1) Å. The $Re(C_{18}H_{15}P)(CO)_4$ ligand pairs bonded to Sn are in the trans position with respect to the ring plane; Re-Sn is 2.714(1) Å. Intermolecular distances are normal.

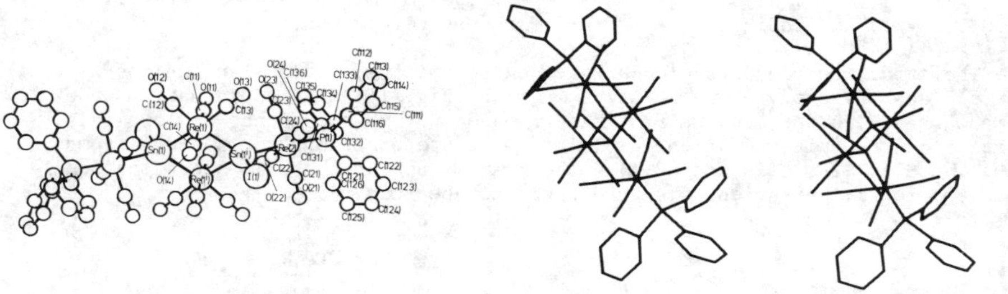

Fig. 1. $Re_2(CO)_8\{Sn(I)Re(CO)_4[P(C_6H_5)_3]\}_2$: perspective and stereoscopic views of the molecule.

BIS((μ2-HYDRIDO)(μ2-o-PHENYLENEDIPHENYLPHOSPHINO-C,P)(μ4-SULPHIDO)(μ3-SULPHIDO)OCTADECACARBONYLTRIOSMIUMPLATINUM) CHLOROFORM SOLVATE

$C_{52}H_{30}O_{16}Os_6P_2Pt_2S_4,2(CHCl_3)$ $[HPtOs_3(CO)_8(\mu_4\text{-}S)(\mu_3\text{-}S)(PPh_2C_6H_4)]_2,2(CHCl_3)$

R.D. ADAMS and T.S.A. HOR, 1984. Organometallics, 3, 1915-1916.

Monoclinic, $P2_1/n$, a = 14.308, b = 21.753, c = 21.760 Å, β = 96.20°, Z = 4. Mo radiation, R = 0.053 for 4874 reflexions.

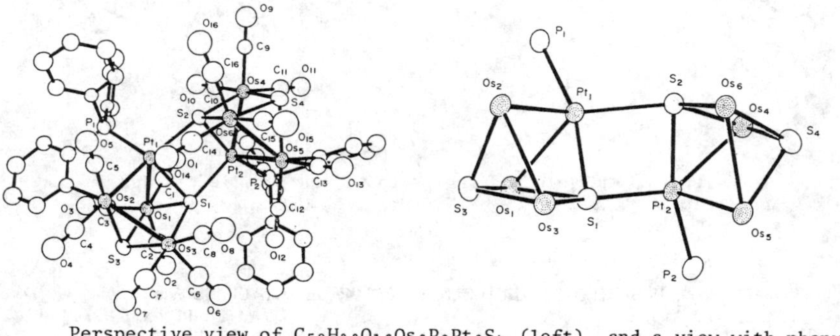

Fig. 1. Perspective view of $C_{52}H_{30}O_{16}Os_6P_2Pt_2S_4$ (left), and a view with phenyl rings and carbonyl ligands omitted (right).

The complex consists of two trigonal-prismatic $PtOs_3S_2$ clusters linked on an edge by quadruply-bridging sulphido ligands (S(1) and S(2)) which serve as six-electron donors and show inverted tetrahedral geometries (Fig. 1). The four-membered Pt(1)-S(1)-Pt(2)-S(2) ring is slightly puckered, with the dihedral angle between the two PtS_2 planes being 42°. The Pt(1)...Pt(2) distance at 3.313 Å is indicative of a non-bonding interaction. Within the trigonal prisms there are three metal-metal bonds Pt(1)-Os(1) 2.680, Pt(1)-Os(2) 2.744, and Os(2)-Os(3) 2.956 Å. Other bond distances of interest are: Pt(1)-S(2) 2.583, Pt(1)-S(1) 2.365, Pt(2)-S(1) 2.536, Pt(2)-S(2) 2.372 Å. Formation of the four-membered Pt_2S_2 ring produces considerable distortion in the bridging sulphido ligand geometry: Pt(2)-S(1)-Os(3) 145.3, Pt(1)-S(1)-Pt(2) and Pt(1)-S(2)-Pt(2) 85.02 and 83.81° respectively.

BIS[o-(DIPHENYLPHOSPHINO)PHENYLTELLURIUM]PLATINUM TETRAKIS(THIOCYANATO)PLATINATE DIMETHYLFORMAMIDE SOLVATE

$C_{52}H_{38}N_4P_2Pt_2S_4Te_2,C_6H_{14}N_2O_2$ $[Pt\{PhTe(o\text{-}PPh_2C_6H_4)\}_2]^{2+}[Pt(SCN)_4]^{2-},2(CH_3)_2NCHO$

H.J. GYSLING and H.R. LUSS, 1984. Organometallics, 3, 596-598.

Triclinic, $P\bar{1}$, a = 11.175, b = 12.438, c = 11.003 Å, α = 99.954, β = 97.51, γ = 80.30°, Z = 1. Mo radiation, R = 0.038 for 6996 reflexions.

The structure consists of discrete centrosymmetric cations and anions with strict square planar symmetry about Pt atoms in both (Fig. 1). The coordination about Te is tetrahedral, with the lone pair of electrons occupying one of the sp^3 orbitals. There is disorder in the anion involving alternate positions for S and C atoms (labelled S1A, S2A, C25A and C26A). The N atoms appear to be in about the same positions for either orientation and could not be resolved.

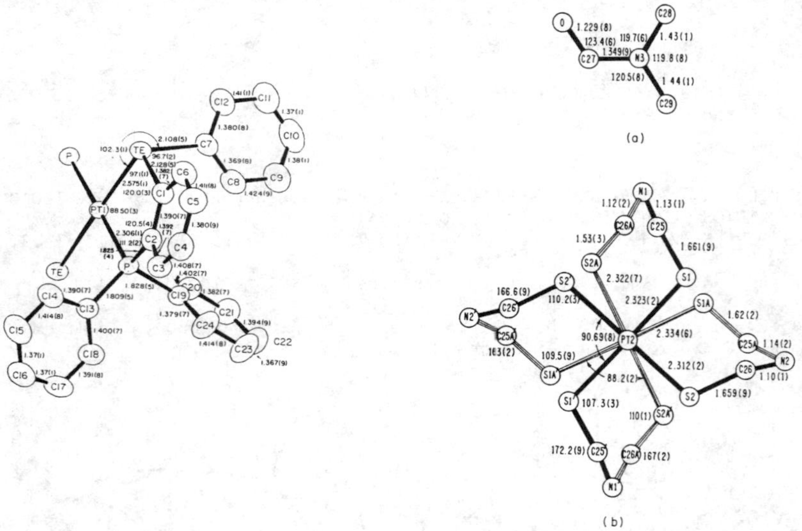

Fig. 1. Bond lengths and selected angles in the cation (left), and in the DMF molecule and anion of $C_{52}H_{38}N_4P_2Pt_2S_4Te_2,C_6H_{14}N_2O_2$ (right).

BIS[1,2-BIS(DIPHENYLPHOSPHINO)ETHANE]GOLD(I) CHLORIDE DIHYDRATE

$C_{52}H_{48}AuClP_4,2H_2O$ $[Au(Ph_2PCH_2CH_2PPh_2)_2]^+Cl^-,2H_2O$

P.A. BATES and J.M. WATERS, 1984. Inorg. Chim. Acta, 81, 151-156.

Monoclinic, $P2_1/n$, a = 10.057, b = 21.704, c = 21.724 Å, β = 94.06°, Z = 4. Mo radiation, R = 0.038 for 4263 reflexions.

The crystal structure consists of the packing of monomeric units of the Au(I) complex cation, chloride anions and water molecules. The coordination of each Au atom approximates to ideal tetrahedral geometry (Fig. 1), the dihedral angles between the planes defined by Au,P(1),P(2) and Au,P(3),P(4) being 89.7°. The Au-P distances can be grouped thus: Au-P(1) 2.412, Au-P(3) 2.402, Au-P(2) 2.384, Au-P(4) 2.391 Å and the 'bite' angles P(1)-Au-P(2) and P(3)-Au-P(4) are respectively 87.1 and 86.4°. There are weak hydrogen bonds between the Cl^- ions and the water molecules (Cl...O(1) 3.17, Cl...O(2) 3.19 Å).

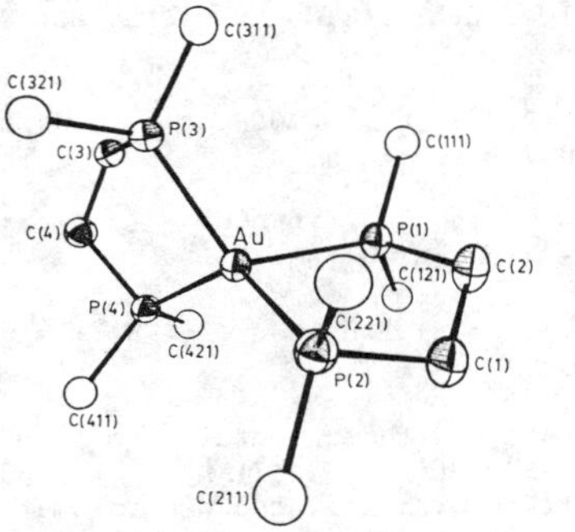

Fig. 1. Molecular structure of the cation (minus phenyl rings) in $C_{52}H_{48}AuClP_4,2H_2O$.

BIS(1,2-BIS(DIPHENYLPHOSPHINO)ETHANE)GOLD(I) HEXAFLUOROANTIMONATE ACETONE SOLVATE
$C_{52}H_{48}AuF_6P_4Sb,C_3H_6O$

S.J. BERNERS-PRICE, M.A. MAZID and P.J. SADLER, 1984. J. Chem. Soc. Dalton, 969-974.

Monoclinic, $P2_1/n$, a = 18.685, b = 19.414, c = 15.096 Å, β = 91.69°, Z = 4. Cu radiation, R = 0.062 for 4811 reflexions.

The Au(I) atom (Fig. 1) has a flattened coordination geometry and is chelated by two bis(phosphine) ligands, Au-P 2.389(3), 2.416(3) Å. Tetrahedral coordination is prevented by the small bite of the diphenylphosphinoethane ligand. The solvent molecule and the SbF_6^- moieties have normal geometries.

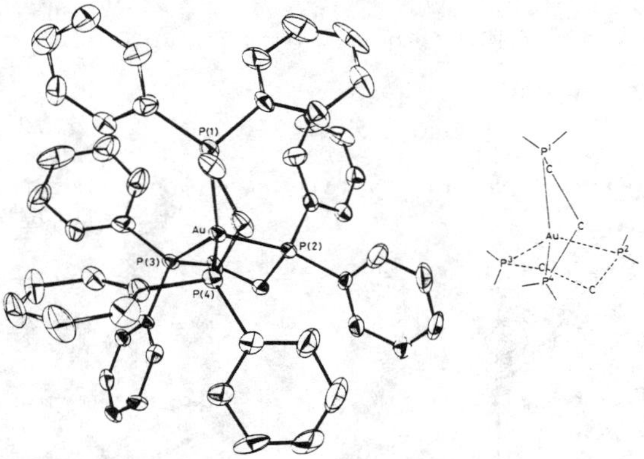

Fig. 1. $C_{52}H_{48}AuF_6P_4Sb,C_3H_6O$: the geometry of the cation.

BIS[(ETHANE-1,2-DIYL)-BIS(DIPHENYLPHOSPHINE)]PALLADIUM(II) DICHLORIDE HEXAHYDRATE
$C_{52}H_{48}Cl_2P_4Pd,6(H_2O)$ (I) $[Pd(diphos)_2]Cl_2,6H_2O$

BIS[(ETHANE-1,2-DIYL)-BIS(DIPHENYLPHOSPHINE)]PLATINUM(II) DICHLORIDE HEXAHYDRATE
$C_{52}H_{48}Cl_2P_4Pt,6(H_2O)$ (II) $[Pt(diphos)_2]Cl_2,6H_2O$

DICHLORO[(ETHANE-1,2-DIYL)BIS(DIPHENYLPHOSPHINE)]PLATINUM(II) DICHLOROMETHANE
$C_{26}H_{24}Cl_2P_2Pt,CH_2Cl_2$ (III) $[Pt(diphos)Cl_2],CH_2Cl_2$

L.M. ENGELHARDT, J.M. PATRICK, C.L. RASTON, P. TWISS and A.H. WHITE, 1984. Aust. J. Chem., 37, 2193-2200.

I. Monoclinic, $P2_1/c$, a = 12.779, b = 18.084, c = 12.942 Å, β = 119.29, Z = 2. Mo radiation, R = 0.043 for 2057 reflexions.

II. Monoclinic, $P2_1/c$, a = 12.789, b = 18.099, c = 12.965 Å, β = 119.39°, Z = 2. Mo radiation, R = 0.036 for 3176 reflexions.

III. Orthorhombic, $P2_12_12_1$, a = 21.360, b = 13.499, c = 9.551 Å, Z = 4. Mo radiation, R = 0.036 for 3468 reflexions.

I and II are isostructural, with the metal atom at a centre of symmetry (Fig. 1, Pd-P 2.345(2) Å, P-Pd-P 81.65°, Pt-P 2.334(2) Å, P-Pt-P 82.04°). II was previously reported (1). III crystallises with solvated dichloromethane molecules. The metal atom has its square plane completed with two chlorines (Fig. 1, Pt-P 2.230, 2.224, Pt-Cl 2.355 Å, P-Pt-P 86.32°).

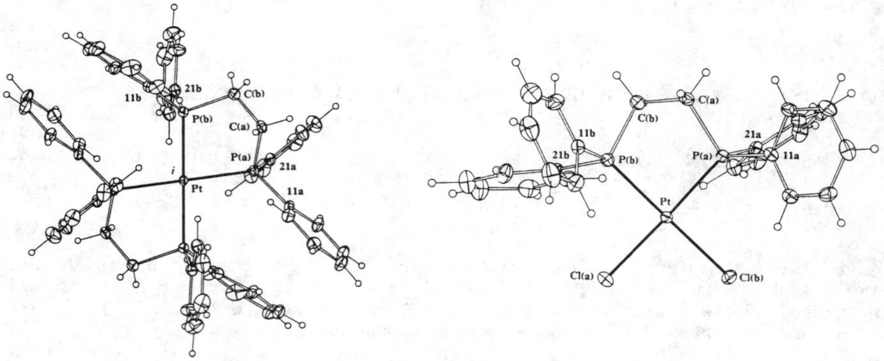

Fig. 1. The structure of $[Pt(diphos)_2]Cl_2$ and $[Pt(diphos)Cl_2]$.

1. Structure Reports, 43B, 1053.

BIS[μ-1,2-BIS(DIPHENYLPHOSPHINO)ETHANE-P,P']TETRACHLORODIMOLYBDENUM (β isomer)
$C_{52}H_{48}Cl_4Mo_2P_4$ $Mo_2Cl_4[(C_6H_5)_2PCH_2CH_2P(C_6H_5)_2]_2$

P.A. AGASKAR and F.A. COTTON, 1984. Inorg. Chem., 23, 3383-3387.

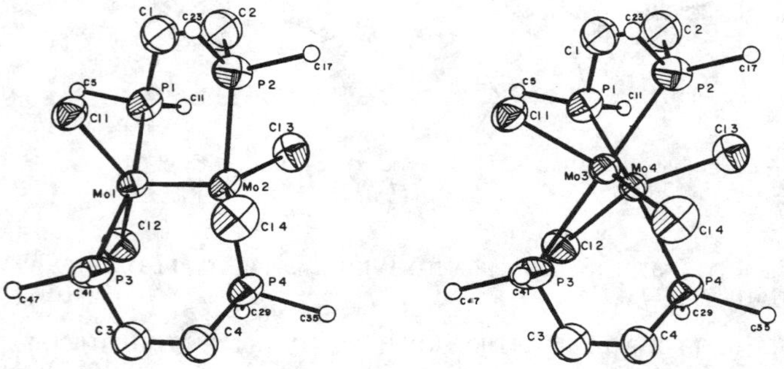

Fig. 1. The inner core of $Mo_2Cl_4[(C_6H_5)_2PCH_2CH_2P(C_6H_5)_2]_2$: dominant (left) and minor (right) orientations of the Mo atoms.

Monoclinic, $P2_1/n$, a = 23.103, b = 13.258, c = 16.792 Å, β = 107.19°, Z = 4. Mo radiation, R = 0.073 for 1861 reflexions.

The two dppe ligands bridge the Mo≡Mo unit. The midpoints of the Mo-Mo quadruple bonds reside in general positions. At each position there are two orientations, a principal one (87%) and a minor one (13%), both of which are shown in Fig. 1. In the principal orientation the helical molecule consists of two fused rings with chair conformations and a mean angle of twist about the Mo-Mo axis of 30.5°. In the minor orientation, the fused rings have twist conformations and the sense of the helix about the Mo-Mo bond is opposite to the one in the principal molecule with an angle of 27.7°. For the principal molecule: Mo-Mo 2.183(3), mean Mo-P 2.579(8), and mean Mo-Cl 2.385(7) Å.

TETRAKIS(METHYLDIPHENYLPHOSPHINE)DIHYDRIDOIRON
$C_{52}H_{54}FeP_4$ $FeH_2(PMePh_2)_4$

J.C. HUFFMAN, N.J. WELLS and K.G. CAULTON, 1984. J. Cryst. Spectrosc. Res., 14, 581-587.

Monoclinic, $P2_1/n$, a = 11.684, b = 37.793, c = 9.765 Å, β = 90.81°, Z = 4. Mo radiation, R = 0.063 for 2585 reflexions (at -162°C).

The crystal structure of $FeH_2(PMePh_2)_4$ shows this to exist as the cis isomer (Fig. 1) in the solid state with major distortions of the FeP_4 skeleton: cis P-Fe-P angles 98.1(1)-107.2(1)° and trans P-Fe-P angle 143.3(1)°. All hydrogen atoms were refined, and the mean Fe-H separation is 1.59 Å with H-Fe-H 82(4)°. Long Fe-P distances, 2.187(3)-2.251(3) Å, emphasize the crowding in this molecule.

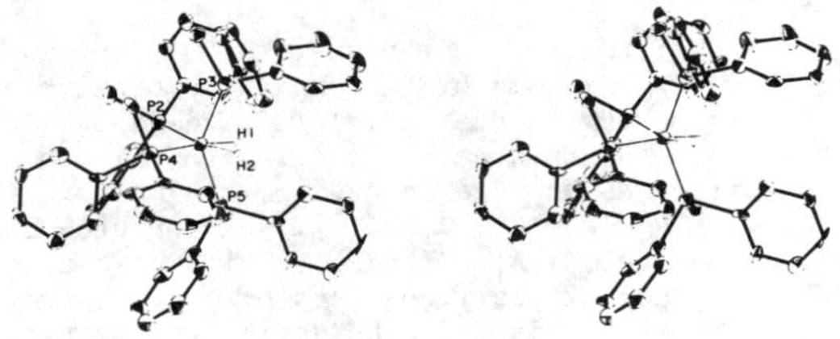

Fig. 1. A stereoview of $C_{52}H_{54}FeP_4$.

trans,trans-BIS(μ_2-LITHIO)BIS(n-BUTYLBIS(PHENYLETHYNYL)TRIETHYLPHOSPHINEPLATINUM)
$C_{52}H_{68}Li_2P_2Pt_2$

A. SEBALD, B. WRACKMEYER, C.R. THEOCHARIS and W. JONES, 1984. J. Chem. Soc. Dalton, 747-756.

Monoclinic, $P2_1/n$, a = 14.061, b = 18.408, c = 19.737 Å, β = 92.18°, Z = 4. Mo radiation, R = 0.069 for 2398 reflexions.

The four Pt-Li distances in the molecule (Fig. 1) are the same (2.81(6) Å), the lengths suggesting that this is a weak interaction. Close contact between the Li atoms and the acetylenic C atoms directly bonded to the Pt atoms (C-Li 2.07(5)-2.24(7) Å) is indicated. The Pt...Pt and Li...Li distances are 4.350 and 3.610 Å respectively. Pt(1), Pt(2), Li(1), Li(2) and the four acetylenic groups form a planar moiety. The triethylphosphine groups are trans to one another with respect to the Pt-Li plane and the two acetylenic C atoms of each phenylacetylide group are coplanar with the phenyl rings indicating delocalisation throughout. The butyl groups are disordered. Each Pt atom has a square planar environment.

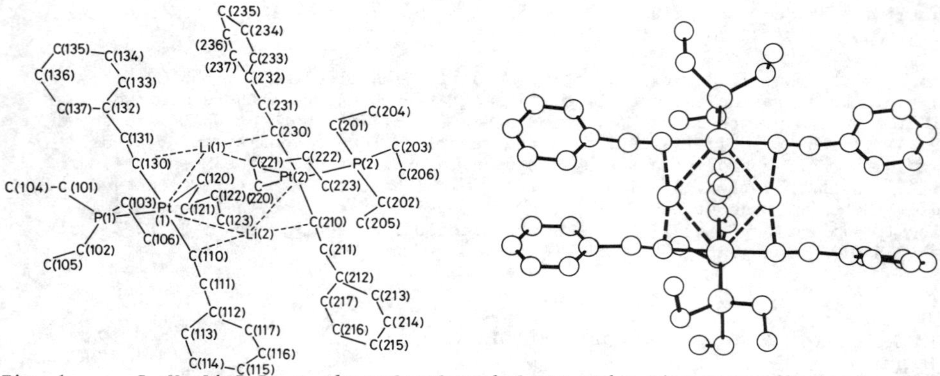

Fig. 1. $C_{52}H_{68}Li_2P_2Pt_2$: the molecular skeleton and a view perpendicular to the mean plane.

BIS(μ_2-BIS(DIPHENYLPHOSPHINO)METHANE-P,P')-1-BROMO-2,2,2-TRICARBONYLPALLADIUMMANGANESE
$C_{53}H_{44}BrMnO_3P_4Pd$

B.F. HOSKINS, R.J. STEEN and T.W. TURNEY, 1984. J. Chem. Soc. Dalton, 1831-1836.

Monoclinic, Cc, a = 18.317, b = 13.412, c = 21.013 Å, β = 111.75°, D_m = 1.51, Z = 4. Mo radiation, R = 0.057 for 7286 reflexions.

The molecule (Fig. 1) consists of a near-planar $Mn(CO)_3PdBr$ entity oriented perpendicular to the plane of the P atoms of the diphenylphosphinomethane framework. Pd has a square-planar environment with Pd-Mn 2.810(2), Pd-P(1) 2.299(4), Pd-P(3) 2.290(4) and Pd-Br 2.547(2) Å. The Mn atom has a distorted six-coordinate geometry which is probably best described as an edge-capped trigonal bipyramid. The Mn atom is coordinated to two P atoms (Mn-P 2.254(3), 2.276(3)) and to three C atoms of carbonyl groups (Mn-C 1.96(1), 1.63(2), 1.78(2) Å). The P...P bite distances in the molecule are 3.006(5) and 2.990(5) Å. The unit cell consists of four discrete molecules of complex. The crystal is disordered such that the asymmetric unit is composed on average of two enantiomers in the ratio 3:1.

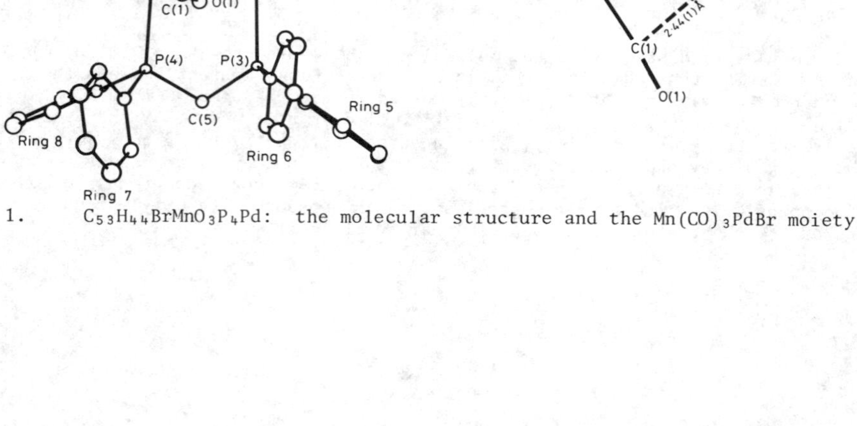

Fig. 1. $C_{53}H_{44}BrMnO_3P_4Pd$: the molecular structure and the $Mn(CO)_3PdBr$ moiety.

(μ-CARBONYL)BIS[μ-BIS(DIPHENYLPHOSPHINO)METHANE]DICARBONYLDICHLORODIIRIDIUM TRIS-(TETRAHYDROFURAN) SOLVATE

$C_{53}H_{44}Cl_2Ir_2O_3P_4,3C_4H_8O$ $[Ir_2Cl_2(CO)_2(\mu\text{-}CO)(Ph_2PCH_2PPh_2)_2],3C_4H_8O$

B.R. SUTHERLAND and M. COWIE, 1984. Inorg. Chem., 23, 2324-2331.

Monoclinic, $P2_1/c$, a = 22.890, b = 13.152, c = 22.129 Å, β = 114.39°, Z = 4. Mo radiation, R = 0.043 for 5303 reflexions (at -40°C).

The terminal CO and Cl ligands are disordered, there being two isomers present. One isomer has all CO groups mutually cis (25% abundance) and the other has the μ-CO group trans to a terminal CO on one metal and to a Cl ligand on the other (Fig. 1). Bond lengths: Ir-Ir 2.779(1), Ir-P 2.321(3)-2.341(3), Ir-Cl 2.489(7)-2.557(9), Ir-CO(bridging) 2.04(2) Å.

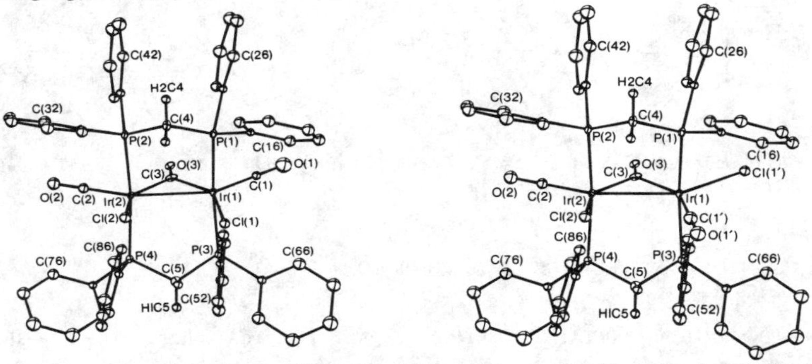

Fig. 1. The 25% (left) and 75% (right) isomers of $[Ir_2Cl_2(CO)_2(\mu\text{-}CO)(Ph_2PCH_2\text{-}PPh_2)_2]$.

HEXACARBONYLTRIS(μ_2-DIPHENYLPHOSPHINO)DIPHENYLPHOSPHINETRIRHODIUM DICHLOROMETHANE SOLVATE

$C_{54}H_{41}O_6P_4Rh_3,CH_2Cl_2$ (I)

HEPTACARBONYLTRIS(μ_2-DIPHENYLPHOSPHINO)TRIRHODIUM

$C_{43}H_{30}O_7P_3Rh_3$ (II)

PENTACARBONYLTRIS(μ_2-DIPHENYLPHOSPHINO)TRIRHODIUM

$C_{41}H_{30}O_5P_3Rh_3$ (III)

R.J. HAINES, N.D.C.T. STEEN and R.B. ENGLISH, 1984. J. Chem. Soc. Dalton, 515-525.

I. Monoclinic, $P2_1/c$, a = 12.515, b = 11.843, c = 37.102 Å, β = 94.6°, Z = 4. Cu radiation, R = 0.111 for 5587 reflexions.

II. Monoclinic, $P2_1/c$, a = 18.16, b = 17.935, c = 13.119 Å, β = 103.5°, Z = 4. Mo radiation, R = 0.134 for 1023 reflexions.

III. Monoclinic, $P2_1/n$, a = 14.003, b = 17.471, c = 16.523 Å, β = 103.76°, Z = 4. Mo radiation, R = 0.067 for 5346 reflexions.

The geometries of I and II are very similar (Fig. 1) with three rhodium atoms in each being linked to each other via bridging PPh_2 ligands and together with three P atoms occupy alternate vertices of a six-membered ring. Two of the Rh atoms have two terminal CO groups coordinated to them and adopt a geometry between square planar and tetrahedral. The third has three terminal CO groups as well as the P atoms of two PPh_2 groups and adopts trigonal bipyramidal geometry. Both Rh_3P_3 rings exhibit approximate threefold symmetry. The Rh(1)...Rh(2) distances are 3.124(11), 3.118(2), Rh(1)-Rh(3) 3.233(12), 3.246(2) and Rh(2)-Rh(3) 3.084(11), 3.130(2) Å for I, II respectively. Compound III consists of a triangular skeleton of Rh atoms with each edge bridged by a diphenylphosphido ligand such that the Rh-P framework adopts a half-chair geometry. The P atoms of the PPh_2 groups coordinated to the unique Rh atom Rh(2) are essentially coplanar with the metal atom geometry planar while the third, coordinated to Rh(1) and Rh(3) is almost orthogonal to it, the dihedral angle between the P(3)Rh(1)Rh(3) and Rh(1)Rh(3)Rh(2) being 97.2°. The Rh(1)-Rh(2), Rh(1)-Rh(3) and Rh(2)-Rh(3) distances are 2.793(1), 2.698(1) and 2.806(1) Å respectively. Rh(1) and Rh(3) are further coordinated to two terminal CO groups while Rh(2) is coordinated only to one. Rh(2) has an essentially planar configuration.

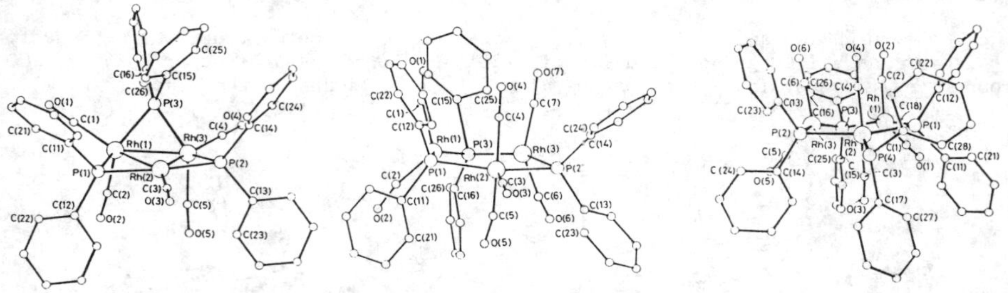

Fig. 1. The molecular structures of $C_{54}H_{41}O_6P_4Rh_3,CH_2Cl_2$ (right), $C_{43}H_{30}O_7P_3Rh_3$ (centre), $C_{41}H_{30}O_5P_3Rh_3$ (left).

CHLORO-TRIS(TRIPHENYLANTIMONY)COPPER(I) CHLOROFORM SOLVATE
$C_{54}H_{45}ClCuSb_3,CHCl_3$

A.L. RHEINGOLD and M.E. FOUNTAIN, 1984. J. Cryst. Spectrosc. Res., 14, 549-557.

Triclinic, $P\bar{1}$, a = 18.827, b = 14.279, c = 14.399 Å, α = 84.43, β = 87.39, γ = 75.18°, Z = 2. Mo radiation, R = 0.064 for 4689 reflexions.

The coordination about Cu(I) is approximately tetrahedral (Fig. 1). The Cu-Cl distance 2.235(5) Å is short and the three Cu-Sb distances are long (2.548-2.564(2) Å, average 2.554(5) Å). The co-crystallized solvent molecules are disordered.

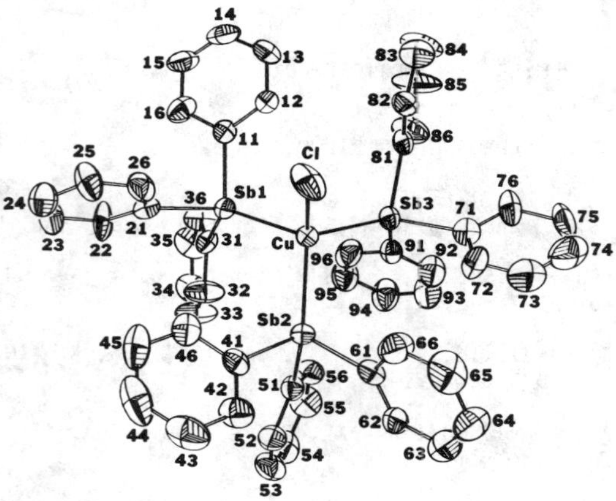

Fig. 1. A view of $(Ph_3Sb)_3CuCl$.

trans-DIISOTHIOCYANATOBIS[1,2-BIS(DIPHENYLPHOSPHINO)ETHANE-P,P']TECHNETIUM(II)
$C_{54}H_{48}N_2P_4S_2Tc$ trans-$Tc(Ph_2PCH_2CH_2PPh_2)_2(NCS)_2$

G. BANDOLI, U. MAZZI, A. ICHIMURA, K. LIBSON, W.R. HEINEMAN and E. DEUTSCH, 1984. Inorg. Chem., 23, 2898-2901.

Triclinic, $P\bar{1}$, a = 21.384, b = 12.878, c = 9.549 Å, α = 71.51, β = 81.94, γ = 83.38°, D_m = 1.35, Z = 2. Mo radiation, R = 0.069 for 3114 reflexions.

In this complex (Fig. 1) the Tc(II) centre is approximately octahedrally coordinated with four equatorial P atoms at an average distance of 2.44(1) Å and two trans axial N-bonded NCS^- ligands at 2.04(2) Å.

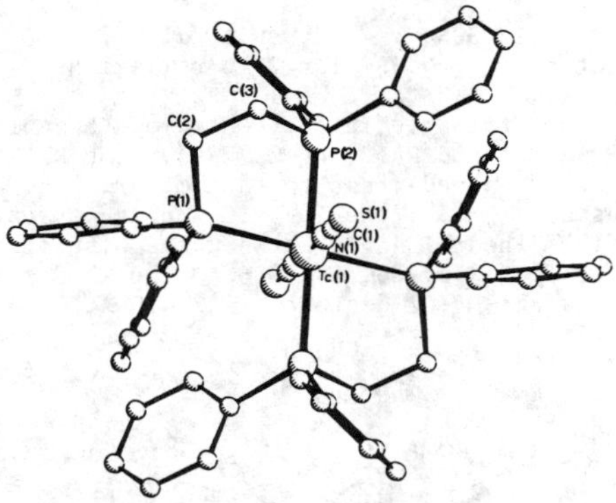

Fig. 1. The trans-$Tc(Ph_2PCH_2CH_2PPh_2)_2(NCS)_2$ molecule.

(μ_2-CHLORO)-BIS(μ_2-BIS(DIPHENYLPHOSPHINO)METHANE)CARBONYLCHLORO(N,N-DIMETHYLTHIOCARBOXAMIDO)DIRHODIUM TETRAFLUOROBORATE METHYLENE CHLORIDE SOLVATE
$C_{54}H_{50}BCl_2F_4NOP_4Rh_2S,1.833CH_2Cl_2$

J.A.E. GIBSON and M. COWIE, 1984. Organometallics, 3, 722-731.

Monoclinic, $P2_1/n$, a = 12.340, b = 32.888, c = 15.374 Å, β = 105.935°, Z = 4. Cu radiation, R = 0.070 for 4122 reflexions.

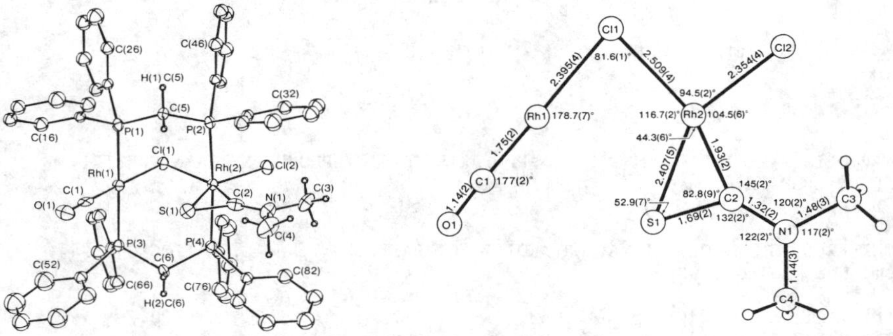

Fig. 1. Perspective view of the cation in $C_{54}H_{50}BCl_2F_4NOP_4Rh_2S$ (left), and bond lengths and angles in the inner coordination sphere of the cation (right).

The cation (Fig. 1) consists of two Rh centres bridged by two mutually trans DPM groups with the Cl, CO and $SCNMe_2$ ligands bound in the equatorial plane which is approximately perpendicular to the Rh-P vectors. Rh(1) has an essentially square-planar environment whereas Rh(2) has a distorted octahedral geometry. The Rh-P distances range from 2.319 to 2.369 Å, the longer ones associated with Rh(2) presumably because of the steric crowding about Rh(2). The η^2-dimethylthiocarboxamido ligand is essentially planar and its geometry is consistent with significant partial double bond character in the S(1)-C(2) and C(2)-N(1) bonds. The BF_4^- anion is severely disordered with B-F distances ranging from 1.10 to 1.56 Å and F-B-F angles from 90 to 133°.

(μ_2-CHLOROMERCURY)(μ_2-HYDRIDO)BIS(μ_2-N-ETHYL(P,P'-TETRAPHENOXY)DIPHOSPHAZENE-P,P')-DICARBONYLDIRHODIUM
$C_{54}H_{51}ClHgN_2O_{10}P_4Rh_2$

J.S. FIELD, R.J. HAINES, E. MEINTJIES, B. SIGWARTH and P.H. VAN ROOYEN, 1984.
J. Organometal. Chem., 268, C43-C47.

Triclinic, P$\bar{1}$, a = 13.562, b = 12.303, c = 18.990 Å, α = 97.5, β = 76.9, γ = 115.8°, D_m = 1.72, Z = 2. Mo radiation, R = 0.057 for 7576 reflexions.

The Rh and Hg atoms form a triangle in the centre of the molecule (Fig. 1) with Rh(1)-Rh(2) 2.762, Rh(1)-Hg 2.711 and Rh(2)-Hg 2.778 Å. The Rh-Rh bond is bridged by the two diphosphazene ligands disposed trans to each other. The Rh(1)-Rh(2)-C(2) and Rh(2)-Rh(1)-C(1) angles are 173.3°, the CO groups being essentially coplanar with Cl, Hg and the Rh atoms. The dihedral angle between the best plane through these atoms and that for the $Rh_2P_4N_2(CO)_2$ system is 89.6°. The Rh-P distances are 2.253, 2.254, 2.242 and 2.257 Å, and the Hg-Cl distance is 2.470 Å.

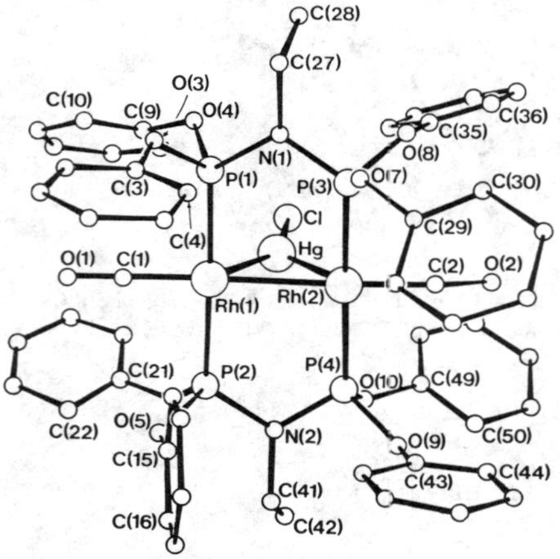

Fig. 1. Molecular structure of $C_{54}H_{51}ClHgN_2O_{10}P_4Rh_2$.

1-MERCAPTOBIS(μ_3-SULPHIDO)-1,2,2,3,3-PENTAKIS(TRIETHYLPHOSPHINE)TRINICKEL(II) TETRAPHENYLBORATE
$C_{54}H_{96}BNiP_5S_3$

C.A. GHILARDI, S. MIDOLLINI, A. ORLANDINI, C. BATTISTONI and G. MATTOGNO, 1984.
J. Chem. Soc. Dalton, 939-942.

Triclinic, P$\bar{1}$, a = 17.461, b = 17.048, c = 11.112 Å, α = 99.21, β = 90.76, γ = 102.57°, Z = 2. Mo radiation, R = 0.070 for 2990 reflexions.

The structure consists of complex cations (Fig. 1) and BPh_4^- anions. The inner core of the cation is a triangle of Ni atoms capped above and below by two triply-bridging sulphur ligands which complete a trigonal bipyramid. Ni(1) and Ni(2) are also coordinated by two triethylphosphine groups, Ni(3) by a phosphine and a mercapto group. Ni...Ni distances range from 2.878 to 3.027(3), Ni-S distances from 2.168 to 2.201(5), and Ni-P distances from 2.206 to 2.223(5) Å.

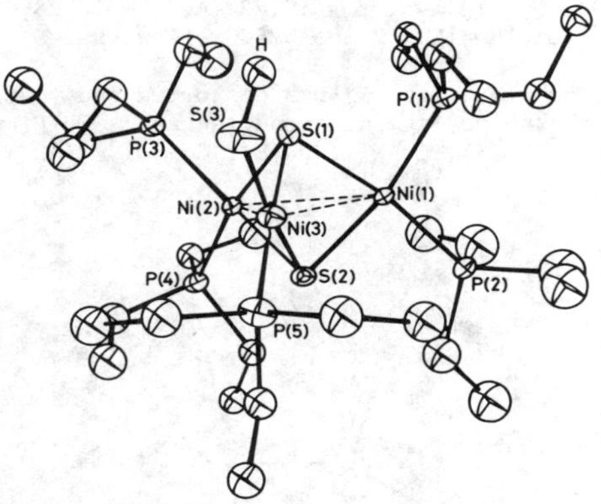

Fig. 1. $C_{54}H_{96}BNiP_5S_3$: the cation.

THIOCYANATOTRIS(TRIPHENYLPHOSPHINE)GOLD(I) MONOHYDRATE

$C_{55}H_{45}AuNP_3S,H_2O$ $(P(C_6H_5)_3)_3AuSCN,H_2O$

J.A. MUIR, M.M. MUIR, S. ARIAS, P.G. JONES and G.M. SHELDRICK, 1984. Inorg. Chim. Acta, 81, 169-174.

Orthorhombic, $P2_12_12_1$, a = 13.95, b = 30.44, c = 11.09 Å, Z = 4. Cu radiation, R = 0.085 for 2030 reflexions.

Coordination about the Au atom is fourfold (Fig. 1). The Au-phosphine geometry is considerably distorted from trigonal planar with the Au atom 0.41 Å out of the plane of the three P atoms. The Au-S bond length is 2.93 Å whilst the Au-P distances are 2.38, 2.39 and 2.41 Å. The angles about the Au atom vary from 90.2° (S-Au-P(1)) to 118.6° (P(1)-Au-P(3)). The thiocyanate ligand is almost linear (S-C-N 174.0°) and the Au-S-C angle is 99.0°. The S-C and C-N distances are 1.70 and 1.20 Å respectively and there is an interaction between the N atom and the O atom of the water molecule (N...O 2.69 Å).

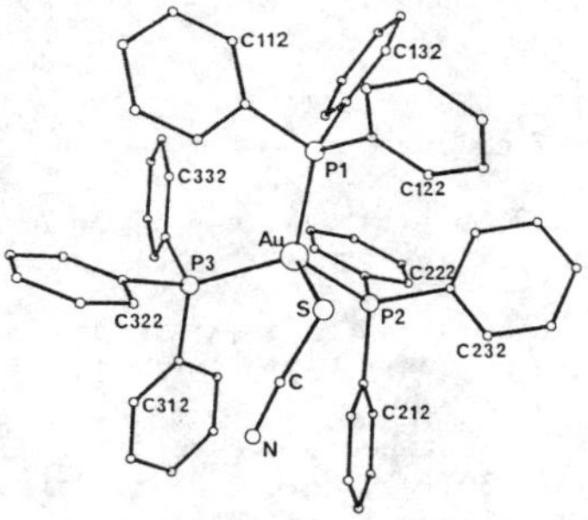

Fig. 1. Molecular structure of $C_{55}H_{45}AuNP_3S$.

HYDRIDO(CYANOTRIHYDROBORATO)TRIS(TRIPHENYLPHOSPHINE)COBALT(II)

$C_{55}H_{49}BCoNP_3$ $CoH(BH_3CN)(PPh_3)_3$

R.J. BARTON, D.G. HOLAH, H. SHENGZHI, A.N. HUGHES, S.I. KHAN and B.E. ROBERTSON, 1984. Inorg. Chem., 23, 2391-2395.

Triclinic, $P\bar{1}$, a = 13.074, b = 10.346, c = 19.722 Å, α = 75.28, β = 78.75, γ = 65.42°, Z = 2. Mo radiation, R = 0.067 for 3252 reflexions.

The five-coordinate complex (Fig. 1) has distorted square pyramidal coordination geometry with an apical phosphine ligand. Bond lengths: Co-P(apical) 2.291(3), Co-P(basal) 2.245(2) and 2.234(3), Co-N 1.904(8), Co-H 1.42 Å.

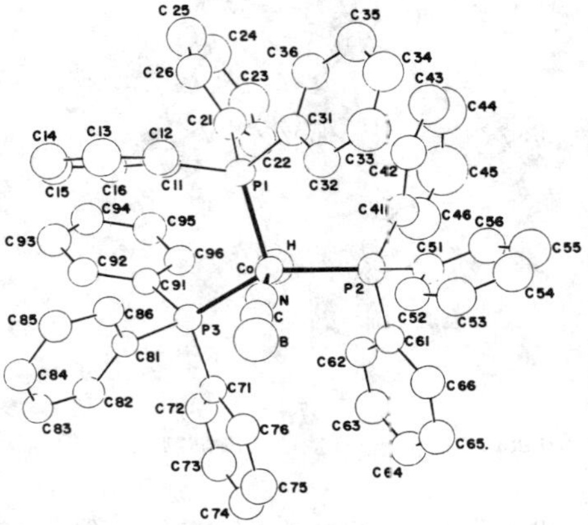

Fig. 1. The $CoH(BH_3CN)(PPh_3)_3$ molecule.

BIS(TRIPHENYLPHOSPHINE)IMINIUM (μ_2-DIPHENYLPHOSPHINO)BIS(TETRACARBONYLMANGANESE)
$C_{56}H_{40}MnNO_8P_3$ (I)

(μ_2-DIMETHYLPHENYLPHOSPHINEGOLD)(μ_2-DIPHENYLPHOSPHINO)BIS(TETRACARBONYLMANGANESE)
$C_{28}H_{21}AuMn_2O_8P_2$ (II)

(μ_2-DIPHENYLPHOSPHINO)(μ-HYDRIDO)BIS(TETRACARBONYLMANGANESE)
$C_{20}H_{11}Mn_2O_8P$ (III)

J.A. IGGO, M.J. MAYS, P.R. RAITHBY and K. HENRICK, 1984. J. Chem. Soc. Dalton, 633-641.

I. Monoclinic, $P2_1/c$, a = 10.172, b = 30.441, c = 17.076 Å, β = 100.88°, Z = 4. Mo radiation, R = 0.067 for 4005 reflexions.

II. Monoclinic, $P2_1/c$, a = 13.360, b = 14.645, c = 17.36 Å, β = 115.20°, Z = 4. Mo radiation, R = 0.079 for 1843 reflexions.

III. Triclinic, $C\bar{1}$, a = 8.157, b = 19.246, c = 14.510 Å, α = 89.10, β = 106.31, γ = 90.07°, Z = 4. Mo radiation, R = 0.049 for 2710 reflexions.

Each Mn atom in I (Fig. 1) has octahedral geometry and is coordinated to four CO groups (Mn-C (CO trans to PPh_2) 1.830(10), Mn-C (CO pseudotrans to PPh_2) 1.805(10), Mn-C (axial CO) 1.772(10), C-O (average) 1.155(13) Å), the other Mn atom (Mn-Mn 2.867(2)) and P of the bridging phosphino ligand (Mn-P 2.260(3) Å). In II the Mn centres are approximately octahedrally coordinated by the Au atom, P of the phosphino bridge and four CO ligands (if the second Mn centre is ignored). The Mn-P(1)-Mn-Au unit is nearly planar. Coordination distances are: Mn-Mn 3.066(8), Au(1)-Mn(1) 2.696(6), Au(1)-Mn(2) 2.681(6), Au(1)-P(2) 2.293(7), Mn(1)-P(1) 2.297(9) and Mn(2)-P(1) 2.279(10) Å. The average Au-C distance (CO trans to PPh_2) is 2.68(3) and the average Mn-CO distance is 1.81(3) Å. The average Mn-Mn-C(carbonyl) angle is 125.6(10) and the bite angle for the phosphino ligand is 84.1(3)°. The structure of III is similar to that reported in (1). The molecule possesses approximate C_{2h} symmetry with Mn-Mn 2.951(1), Mn-P 2.290(2) Å, Mn(1)-P-Mn(2) 80.3(1), Mn(1)-Mn(2)-P (average) 49.9(1)°, Mn-CO (average)1.837(7) Å, C(13)-Mn-C(14) 96.6(3) and P-Mn(1)-C(13) 96.2(2)°.

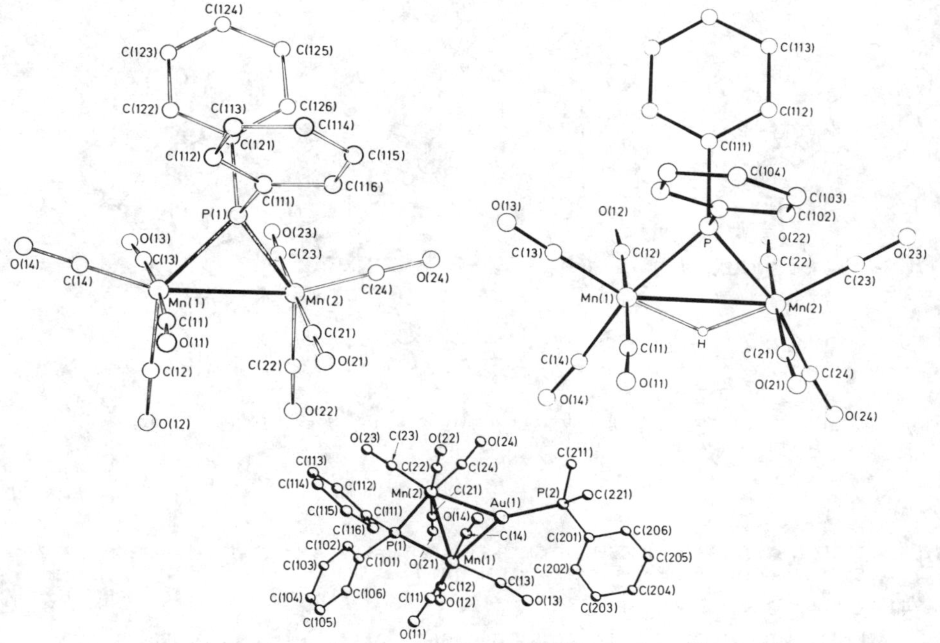

Fig. 1. The molecular structures: $C_{56}H_{40}MnNO_8P_3$ (top left), $C_{28}H_{21}AuMn_2O_8P_2$ (bottom), $C_{20}H_{11}Mn_2O_8P$ (top right).

1. Structure Reports, 32B, 466.

TETRAPHENYLPHOSPHONIUM (μ_2-HYDRIDO)-(μ_2-BIS(DIPHENYLPHOSPHINO)METHANE)OCTACARBONYL-DIMOLYBDATE

$C_{57}H_{43}Mo_2O_8P_3$ (I) $Ph_4P^+.C_{33}H_{23}Mo_2O_8P_2^-$

TETRAETHYLAMMONIUM (μ_2-HYDRIDO)-(μ_2-1,4-BIS(DIPHENYLPHOSPHINO)BUTANE)OCTACARBONYL-DIMOLYBDATE

$C_{44}H_{49}Mo_2NO_8P_2$ (II) $Et_4N^+.C_{36}H_{29}Mo_2O_8P_2^-$

M.Y. DARENSBOURG, R. EL MEHDAWI, T.J. DELORD, F.R. FRONCZEK and S.F. WATKINS, 1984. J. Am. Chem. Soc., 106, 2583-2591.

I. Monoclinic, $P2_1/c$, a = 13.886, b = 18.545, c = 20.156 Å, β = 91.03°, Z = 4. Mo radiation, R = 0.0413 for 4402 reflexions.

II. Triclinic, P1, a = 8.760, b = 13.940, c = 19.686 Å, α = 105.75, β = 93.59, γ = 97.38°, Z = 2. Mo radiation, R = 0.0471 for 4720 reflexions.

Anions I and II (Fig. 1) are bridged by both H^- and the bidentate ligands and show overall bent, staggered configurations with P_1-Mo_1-Mo_2-P_2 torsion angles of 21° for I and 42° for II. Similar framework bends (the intersection angle of $(OC)_{ax}$-Mo vectors) of ca. 160° and similar Mo_1-H-Mo_2 angles of ca. 131° are observed for I and II. The Mo...Mo separation in I is 3.4028(8) Å and in II it is 3.4995(8) Å. The distance between the P-donor atoms in I is 3.19 Å and in II it is 5.08 Å. In II the hydride bridge is directed into the area underneath the ligand bridge (the $(CH_2)_4$ unit), whereas in I the plane of the hydride bridge is roughly perpendicular to the P-CH_2-P bridge.

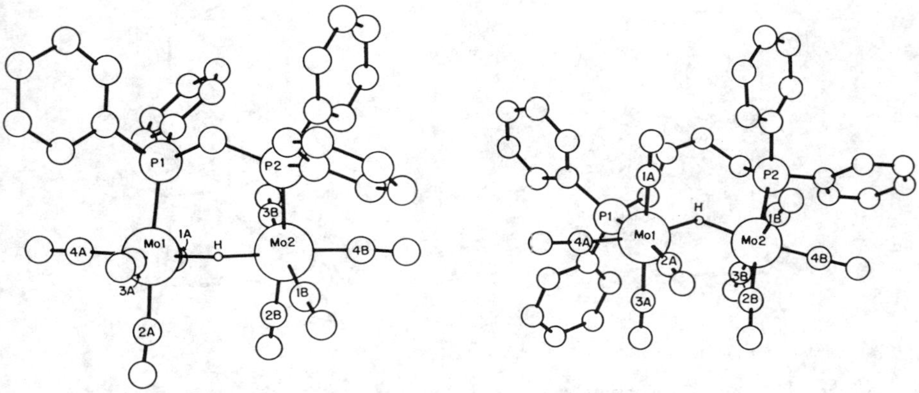

Fig. 1. Views of the hydrogen bridged anions $C_{33}H_{23}Mo_2O_8P_2^-$ (left) and $C_{36}H_{29}Mo_2O_8P_2^-$ (right).

1,2;3,4-BIS{μ-[BIS(DIPHENYLPHOSPHINO)METHANE]-P,P'}-1,2;2,4;4,1-TRI-μ-CARBONYL-1,2,-3,3,4-PENTACARBONYL-tetrahedro-TETRAIRIDIUM(0)

$C_{58}H_{44}Ir_4O_8P_4$ $Ir_4(CO)_8[(C_6H_5)_2PCH_2P(C_6H_5)_2]_2$

M.M. HARDING, B.S. NICHOLLS and A.K. SMITH, 1984. Acta Cryst., C40, 790-793.

Monoclinic, I2/a, a = 28.833, b = 17.264, c = 23.074 Å, β = 97.00°, D_m = 2.2, Z = 8. Mo radiation, R = 0.042 for 4779 reflexions.

In the complex, the Ir atoms form a nearly regular tetrahedron with Ir-Ir bonds in the range 2.692-2.746(1) Å (Fig. 1). Each bis(diphenylphosphino)methane ligand bridges two Ir atoms, with mean Ir-P 2.297(5) Å; the $Ir_4[(C_6H_5)_2PCH_2P(C_6H_5)_2]_2$ portion has approximate, and non-crystallographic, symmetry $\bar{4}(S_4)$, but the carbonyl groups, three bridging and five terminal, do not conform to this.

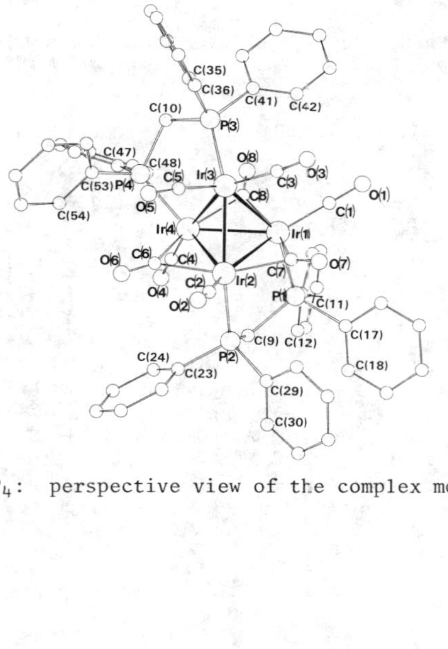

Fig. 1. $C_{58}H_{44}Ir_4O_8P_4$: perspective view of the complex molecule.

cis,cis-BIS(μ-2,6-BIS(DIPHENYLPHOSPHINO)PYRIDINE-P,P')TETRACHLORODIPLATINUM(II) DICHLOROMETHANE SOLVATE

$C_{58}H_{46}Cl_4N_2P_4Pt_2,6CH_2Cl_2$ (I) $Pt_2[\mu-(Ph_2P)_2py]_2Cl_4.6CH_2Cl_2$

trans,trans-BIS(μ-2,6-BIS(DIPHENYLPHOSPHINO)PYRIDINE-P,P')TETRAIODODIPLATINUM(II) DICHLOROMETHANE SOLVATE

$C_{58}H_{46}I_4N_2P_4Pt_2,2CH_2Cl_2$ (II) $Pt_2[\mu-(Ph_2P)_2py]_2I_4.2CH_2Cl_2$

F.E. WOOD, J. HVOSLEF, H. HOPE and A.L. BALCH, 1984. Inorg. Chem., 23, 4309-4315.

I. Monoclinic, $P2_1/c$, a = 16.17, b = 14.60, c = 16.36 Å, β = 110.9°, D_m = 1.9, Z = 2. Mo radiation, R = 0.066 for 3005 reflexions (at 140 K).

II. Monoclinic, C2/c, a = 24.093, b = 21.544, c = 12.287 Å, β = 95.08°, D_m = 2.0, Z = 4. Mo radiation, R = 0.056 for 3102 reflexions (at 140 K).

The two dinuclear complexes (Fig. 1) contain central 12-membered rings. Centrosymmetric I has cis square planar coordination about Pt with Pt-Cl 2.356(7) and 2.363(7), Pt-P 2.252(7) and 2.278(7) Å, and a separation of 3.01 Å between the parallel central pyridine rings. Compound II is a face-to-face dimer having trans square planar coordination geometry with Pt-I 2.602(1) and 2.607(1), Pt-P 2.307(4) and 2.306(5) Å.

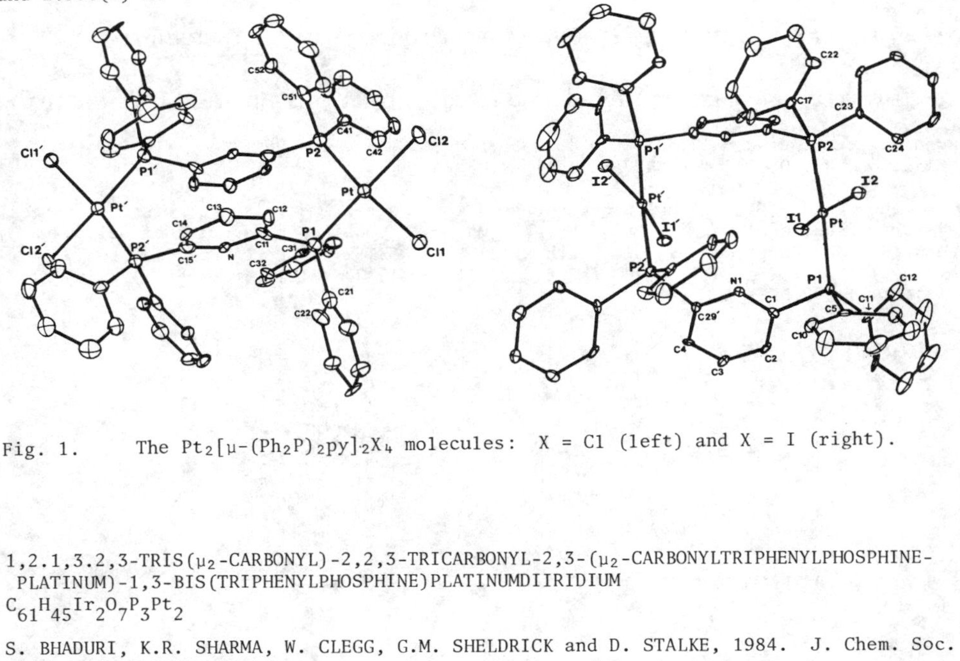

Fig. 1. The $Pt_2[\mu-(Ph_2P)_2py]_2X_4$ molecules: X = Cl (left) and X = I (right).

1,2.1,3.2,3-TRIS(μ_2-CARBONYL)-2,2,3-TRICARBONYL-2,3-(μ_2-CARBONYLTRIPHENYLPHOSPHINE-PLATINUM)-1,3-BIS(TRIPHENYLPHOSPHINE)PLATINUMDIIRIDIUM

$C_{61}H_{45}Ir_2O_7P_3Pt_2$

S. BHADURI, K.R. SHARMA, W. CLEGG, G.M. SHELDRICK and D. STALKE, 1984. J. Chem. Soc. Dalton, 2851-2853.

Triclinic, $P\bar{1}$, a = 12.728, b = 14.824, c = 15.704 Å, α = 98.23, β = 95.49, γ = 103.93°, Z = 2. Mo radiation, R = 0.050 for 5402 reflexions.

The four metal atoms adopt a "butterfly" configuration (Fig. 1) with Ir and Pt atoms at "hinge" and "wing-tip" respectively, with Pt...Pt 2.976(1), Ir-Ir 2.741(1), Pt-Ir 2.675-2.714(1), Pt(1)-P(1) 2.300(4), Pt(2)-P(2) 2.264(4) Å.

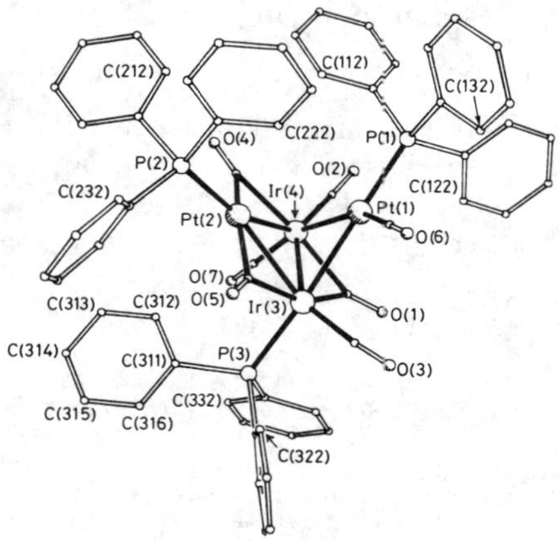

Fig. 1. A view of $C_{61}H_{45}Ir_2O_7P_3Pt_2$.

BIS(μ_3-SULPHIDO)HEXAKIS(DIMETHYLPHENYLPHOSPHINE)TRIPLATINUM BIS(TETRAETHYLBORATE)
$C_{64}H_{106}B_2P_6Pt_3S_2$ $(C_{48}H_{66}P_6Pt_3S_2)^{2+}.2(Et_4B^+)$

G.W. BUSHNELL, K.R. DIXON, R. ONO and A. PIDCOCK, 1984. Can. J. Chem., 62, 696-702.

Monoclinic, C2/c, a = 15.447, b = 18.033, c = 26.505 Å, β = 96.73°, Z = 4. Mo radiation, R = 0.0545 for 3427 reflexions.

The cation (Fig. 1) has twofold crystallographic symmetry and consists of three distorted square-planar cis-$PtS_2(PMe_2Ph)_2$ moieties combined by sharing of sulphur atoms to produce a central Pt_3S_2 unit, with Pt(2)-Pt(1) 3.182(1), Pt(1)-Pt(1') 3.108(1), S(1)-Pt(1) 2.372(3), S(1)-Pt(2) 2.358(3), P(1)-Pt(1) 2.278(3), Pt(2)-Pt(1) 2.264(3), P(3)-Pt(2) 2.265(3) Å.

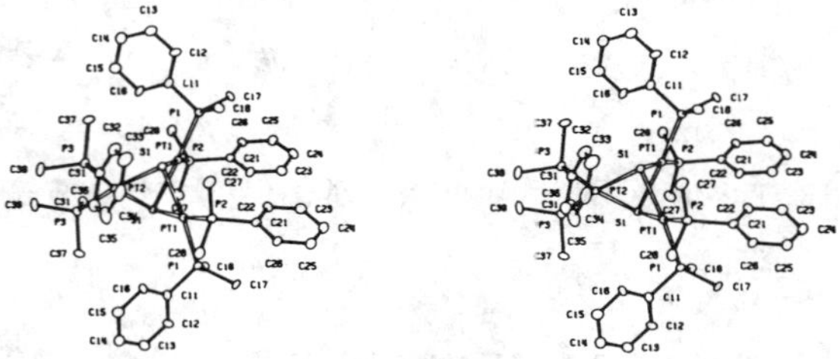

Fig. 1. A stereoview of the $[Pt_3S_2(PMe_2Ph)_6]^{2+}$ cation.

(μ_3-HYDRIDO)TRIS(TRIPHENYLPHOSPHINEGOLD)TETRAKIS(TRICARBONYLRUTHENIUM)
$C_{66}H_{46}Au_3O_{12}P_3Ru_4$

J.A.K. HOWARD, I.D. SLATER and F.G.A. STONE, 1984. Polyhedron, 3, 567-573.

Monoclinic, $P2_1/n$, a = 22.31, b = 18.54, c = 18.76 Å, β = 113.33°, Z = 4. Mo radiation, R = 0.055 for 4322 reflexions.

The metal atom core (Fig. 1) can be regarded as a trigonal bipyramid [Au(3), Ru(1), Ru(2), Ru(3), Ru(4)] with two faces capped by Au(1) and Au(2). The hydrido ligand, the presence of which was revealed by the ^1H NMR data, was not located in the X-ray diffraction study. Some bond lengths are Au(1)-Au(3) 2.822(2), Au(2)-Au(3) 2.838(2), Au-Ru 2.825-3.011(2), Ru-Ru 2.801-3.001(3), Au-P 2.276-2.289(9) Å.

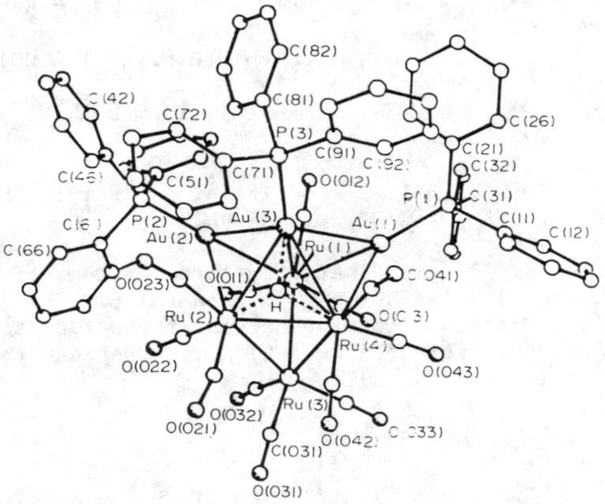

Fig. 1. The molecular structure of $[Au_3Ru_4(\mu_3\text{-}H)(CO)_{12}(PPh_3)_3]$.

BROMO(1,1,4,7,10,10-HEXAPHENYL-1,4,7,10-TETRAPHOSPHADECANE-P^1,P^4,P^7,P^{10})IRON(II) TETRAPHENYLBORATE DICHLOROMETHANE

$C_{66}H_{62}BBrFeP_4,CH_2Cl_2$ $[Fe(C_{42}H_{42}P_4)Br][(C_6H_5)_4B].CH_2Cl_2$

M. BACCI, C.A. GHILARDI and A. ORLANDINI, 1984. Inorg. Chem., 23, 2798-2802.

Triclinic, $P\bar{1}$, a = 13.549, b = 19.058, c = 12.475 Å, α = 72.0, β = 82.4, γ = 81.2°, Z = 2. Mo radiation, R = 0.079 for 3500 reflexions (at 150 K).

The structure of the anion at 150 K (Fig. 1) shows smaller deviations from C_{2v} symmetry than does the room temperature structure (1). The Fe(II) is in a singlet spin state at 150 K and in an intermediate (singlet-triplet) state at 298 K. The coordination about Fe is distorted trigonal bipyramidal with Fe-P_{ax} 2.247(4) and 2.235(4), Fe-P_{eq} 2.141(4) and 2.186(4), Fe-Br_{eq} 2.338(3) Å.

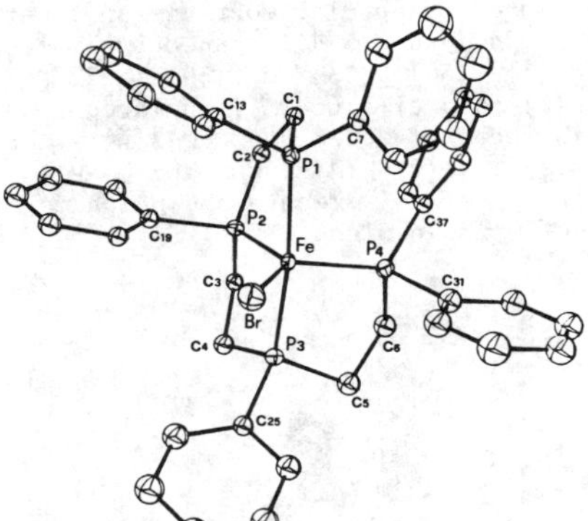

Fig. 1. The $[Fe(C_{42}H_{42}P_4)Br]^+$ cation at 150 K.

1. Structure Reports, 40B, 1067.

[TRIS(2-(DIPHENYLPHOSPHINO)ETHYL)AMMONIUM]CARBONYLNICKEL(0) TETRAPHENYLBORATE TETRAHYDROFURAN SOLVATE

$C_{67}H_{62}BNNiOP_3,0.5C_4H_8O$ $\{[HN(CH_2CH_2P(C_6H_5)_2)_3]Ni(CO)\}[B(C_6H_5)_4],0.5C_4H_8O$

F. CECCONI, C.A. GHILARDI, P. INNOCENTI, C. MEALLI, S. MIDOLLINI and A. ORLANDINI, 1984. Inorg. Chem., 23, 922-929.

Triclinic, P$\bar{1}$, a = 18.917, b = 16.341, c = 10.151 Å, α = 94.31, β = 94.27, γ = 96.94°, Z = 2. Mo radiation, R = 0.073 for 4102 reflexions.

The Ni atom is tetrahedrally coordinated by three P atoms of the $HN(CH_2CH_2PPh_2)_3$ ligand (Ni-P 2.220(3)-2.223(3) Å) and by a CO ligand (Ni-C 1.737(12) Å) (Fig. 1). The N-H distance, 1.15(9) Å, indicates a bonding between these two atoms, but the Ni-H distance of 1.95(9) Å does not exclude some interaction between the hydrogen and the metal atom. The anion geometry is normal.

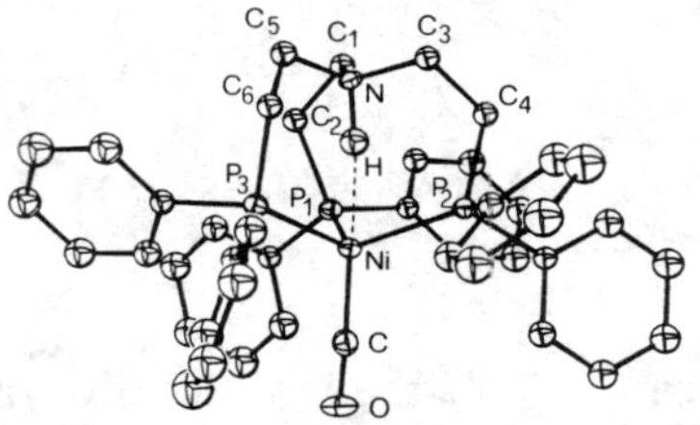

Fig. 1. The $\{[HN(CH_2CH_2PPh_2)_3]Ni(CO)\}^+$ cation.

BIS(μ2-CARBONYL)(μ2-DIPHENYLPHOSPHINO)CARBONYLTRIS(TRIPHENYLPHOSPHINE)COBALTDIPLATINUM

$C_{69}H_{55}CoO_3P_4Pt_2$ $CoPt_2(PPh_2)(CO)_2(CO)(PPh_3)_3$

R. BENDER, P. BRAUNSTEIN, B. METZ and P. LEMOINE, 1984. Organometallics, 3, 381-384.

Triclinic, P$\bar{1}$, a = 14.807, b = 15.051, c = 14.047 Å, α = 93.14, β = 108.75, γ = 94.55°, Dm = 1.67, Z = 2. Mo radiation, R = 0.058 for 4717 reflexions.

The metal atoms form a closed triangular $CoPt_2$ arrangement (Fig. 1). The Pt(1)-Co and Pt(2)-Co distances of 2.540 and 2.574 Å respectively are bridged by an asymmetric carbonyl group (Pt(1)-C(30) 2.13, Co-C(30) 1.82, Pt(2)-C(20) 2.10 and Co-C(20) 1.87 Å). The Pt(1)-Pt(2) bond of 2.664 Å is bridged by a μ-PPh_2 ligand, with Pt(1)-P(2) and Pt(2)-P(2) distances of 2.258 and 2.252 Å respectively. Other distances are: Pt(1)-P(3) 2.245, Pt(2)-P(4) 2.265 and Co-P(1) 2.163 Å. The three metal atoms and the four P atoms define a mean plane which makes a dihedral angle of 98.4° with the Co-P(1)-P(2)-C(10) mean plane.

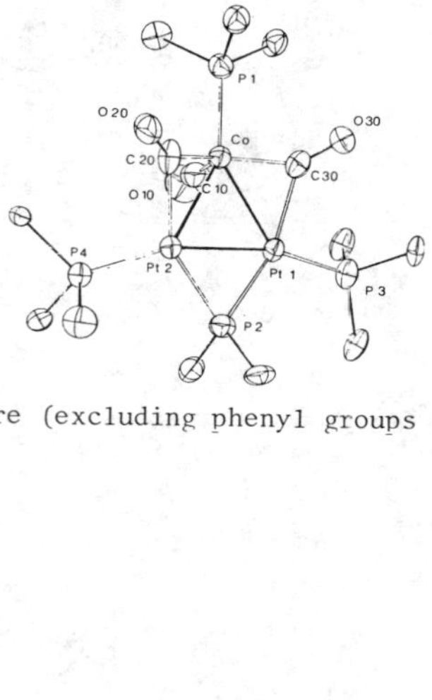

Fig. 1. Molecular structure (excluding phenyl groups on P atoms) of $C_{69}H_{55}CoO_3P_4Pt_2$.

BIS{2-[(DIPHENOXYPHOSPHINO)OXY]PHENYL-C,P}BIS(TRIPHENYL PHOSPHITE-P)RUTHENIUM
$C_{72}H_{58}O_{12}P_4Ru$ $[(C_6H_5O)_3P]_2[(C_6H_5O)_2\dot{P}(OC_6H_4)]_2Ru$

M.F. GARBAUSKAS, J.S. KASPER and L.N. LEWIS, 1984. J. Organometal. Chem., 276, 241-248.

Monoclinic, $P2_1/c$, a = 15.39, b = 15.33, c = 27.16 Å, β = 94.32°, Z = 4. Mo radiation, R = 0.078 for 4991 reflexions.

The molecule comprises a ruthenium atom bound by two conventional triphenylphosphite ligands (P(1) and P(4)) and two triphenylphosphite ligands (P(2) and P(3)) in which orthometallation has occurred, thus forming two metal carbon bonds (Fig. 1). The complex has distorted octahedral metal coordination with angles P(3)-Ru-C(38) 77.5, P(2)-Ru-C(20) 79.2 and P(2)-Ru-P(4) 105.9°. The Ru-P distances are: 2.312, 2.245, 2.264 and 2.288 Å. The main feature of the structure is the presence of two ortho-ring C to metal linkages. These linkages form planar five-membered rings in which the angles Ru-P(2)-O(4) and Ru-P(3)-O(7) are respectively 108.2 and 108.8°.

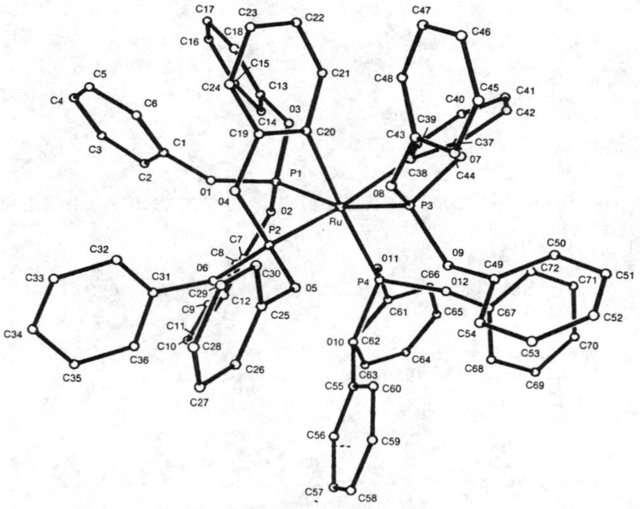

Fig. 1. Molecular structure of $C_{72}H_{58}O_{12}P_4Ru$.

TETRAKIS[(μ_3-BROMO)TRIPHENYLPHOSPHINECOPPER(I)]
$C_{72}H_{60}Br_4Cu_4P_4$ $[PPh_3CuBr]_4$

P.F. BARRON, J.C. DYASON, L.M. ENGELHARDT, P.C. HEALY and A.H. WHITE, 1984. Inorg. Chem., 23, 3766-3769.

Orthorhombic, Pbcn, a = 17.662, b = 20.748, c = 18.351 Å, D_m = 1.61, Z = 4. Mo radiation, R = 0.044 for 3708 reflexions.

The tetramer (Fig. 1) has a cubane structure and has crystallographic $\bar{1}$ symmetry. Cu(I) atoms have distorted tetrahedral coordination geometry with Cu-Br 2.491(2)-2.617(2), Cu-P 2.209(3) and 2.206(3) Å.

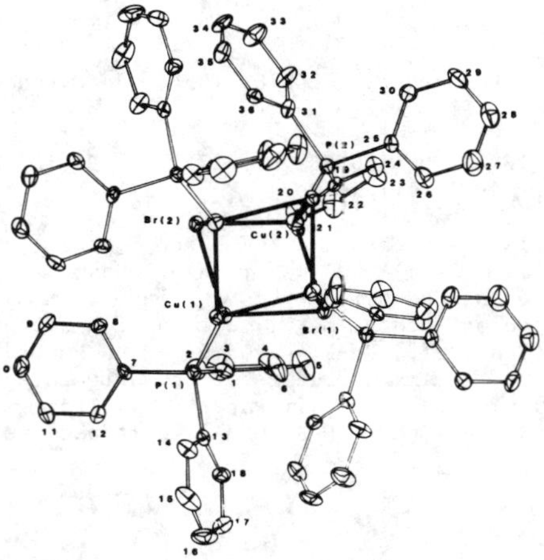

Fig. 1. The structure of $[PPh_3CuBr]_4$.

TETRAPHENYLARSONIUM TRIS(μ_2-CHLORO)DICHLOROBIS[HYDROGENBIS(DIPHENYLPHOSPHINITO)]-DIRHODIUM(III) DICHLOROMETHANE SOLVATE

$C_{72}H_{62}AsCl_5O_4P_4Rh_2,3(CH_2Cl_2)$

J.A.S. DUNCAN, T.A. STEPHENSON, M.D. WALKINSHAW, D. HEDDEN and D.M. ROUNDHILL, 1984. J. Chem. Soc. Dalton, 801-807.

Triclinic, P$\bar{1}$, a = 12.157, b = 16.191, c = 20.523 Å, α = 96.02, β = 99.88, γ = 96.10°, Z = 2. Mo radiation, R = 0.063 for 4380 reflexions.

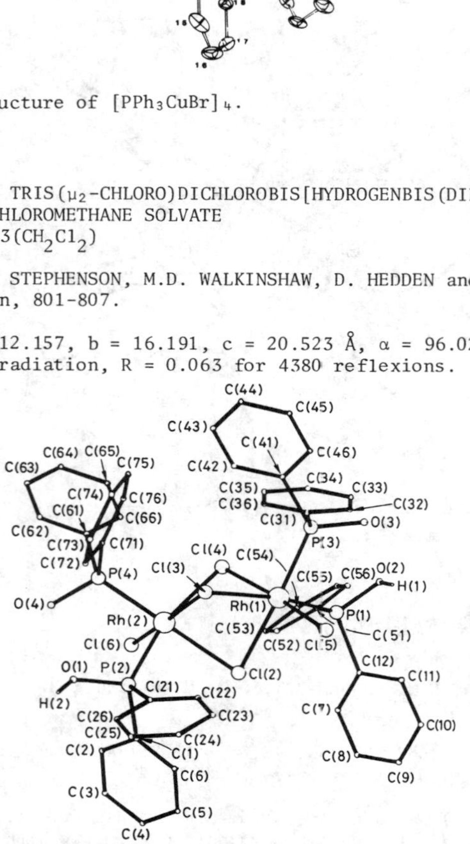

Fig. 1. $C_{72}H_{62}AsCl_5O_4P_4Rh_2,3(CH_2Cl_2)$: the structure of the anion.

The complex anion (Fig. 1) has an approximate non-crystallographic C_2 axis through Cl(2) and the bisector of the Rh...Rh vector (Fig. 1). The average Rh-Cl (bridging) bond length is 2.520 when trans to a phosphine group and 2.351 Å when trans to Cl. The average Rh-Cl (terminal) distance is 2.312 Å. The O(2)...O(3) and O(1)...O(4) distances are 2.368 and 2.353 Å respectively and are unexpectedly short. There are no symmetrical hydrogen bonds in the crystal. The P(1)-O(2)-H(1)...O(3) unit can be described as an asymmetric intramolecular hydrogen bond with O(2)-H(1) 0.97, H(1)...O(3) 2.1(2) Å, P(1)-O(2)-H(1) 108(12) and O(2)-H(1)...O(3) 98(13)° while there is no significant intramolecular hydrogen bonding in the P(2)-O(1)-H(2)...O(4) unit (O(1)-H(2) 0.81, H(2)...O(4) 2.85(16) Å, P(2)-O(1)-H(2) 122(12), O(1)-H(2)... O(4) 45(10)°).

μ-TETRAHYDROBORATO-BIS[BIS(TRIPHENYLPHOSPHINE)COPPER(I)] PERCHLORATE

$C_{72}H_{64}BClCu_2O_4P_4$ $[(Cu(P(C_6H_5)_3)_2)_2BH_4]^+[ClO_4]^-$

B.E. GREEN, C.H.L. KENNARD, G. SMITH, B.D. JAMES, P.C. HEALY and A.H. WHITE, 1984. Inorg. Chim. Acta, 81, 147-150.

Orthorhombic, Pnnb, a = 23.17, b = 18.90, c = 14.94 Å, Dm = 1.31, Z = 4. Mo radiation, R = 0.045 for 2203 reflexions.

Two tetrahedral Cu(I) atoms are linked by a tetrahedral BH_4^- group through bidentate hydrogen bridges (Fig. 1). The Cu and B atoms lie on the crystallographic twofold axis with Cu...B separations of 2.23 and 2.20 Å. The Cu(1)-H(1) and Cu(2)-H(2) distances are respectively 1.810 and 1.864 Å and the angles about the Cu and B atoms are: P(1)-Cu(1)-P(1') 119.67, P(2)-Cu(2)-P(2') 124.52, H(1)-Cu(1)-H(3) 61, H(2)-Cu(2)-H(4) 64, H(1)-B-H(3) 108 and H(2)-B-H(4) 116°. The Cu-P distances are: Cu(1)-P(1) 2.256 and Cu(2)-P(2) 2.260 Å.

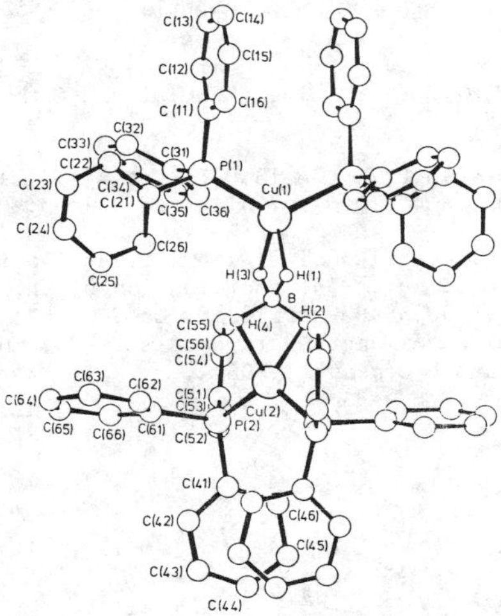

Fig. 1. Molecular structure of the cation in $C_{72}H_{64}BClCu_2O_4P_4$.

DI(μ-HYDROGENSULPHIDO)-BIS[CHLORO(HYDRIDO)BIS(TRIPHENYLPHOSPHINE)RHODIUM(III)] BIS-(DICHLOROMETHANE)

$C_{72}H_{64}Cl_2P_2S_2Rh_2,2CH_2Cl_2$ (I) $[RhCl(H)(\mu\text{-}SH)(P(C_6H_5)_3)_2]_2.2CH_2Cl_2$

CHLORO(HYDRIDO)(HYDROGENSULPHIDO)CARBONYLBIS(TRIPHENYLPHOSPHINE)IRIDIUM(III)

$C_{37}H_{32}ClOP_2SIr$ (II) $IrCl(H)(SH)CO[P(C_6H_5)_3]_2$

A.M. MUETING, P. BOYLE and L.H. PIGNOLET, 1984. Inorg. Chem., 23, 44-48.

I. Monoclinic, $P2_1/a$, a = 23.830, b = 15.531, c = 9.788 Å, β = 99.88°, Z = 2. Mo radiation, R = 0.039 for 5125 reflexions.

II. Monoclinic, C2/c, a = 23.664, b = 9.685, c = 15.782 Å, β = 115.33°, Z = 4. Mo radiation, R = 0.022 for 4077 reflexions.

The two molecules are shown in Fig. 1. Complex I is a symmetrically SH^--bridged binuclear species with a planar $Rh_2(\mu\text{-}S)_2$ grouping. The coordination geometry is octahedral with cis PPh_3 ligands and trans Cl-Rh-H. Bond lengths for I: Rh-P1 2.319(1), Rh-P2 2.338(1), Rh-Cl 2.519(1), Rh-S 2.383(1) and 2.385(1), Rh-H 1.59(6) Å. II has exact C_2 symmetry, octahedral geometry with trans PPh_3 ligands and trans H-Ir-Cl and HS-Ir-CO arrangements. Bond lengths for II: Ir-P 2.359(2), Ir-Cl 2.492(3), Ir-S 2.336(4), Ir-C 1.94(2), Ir-H 1.47(8) Å. The CO and S ligands are mutually disordered.

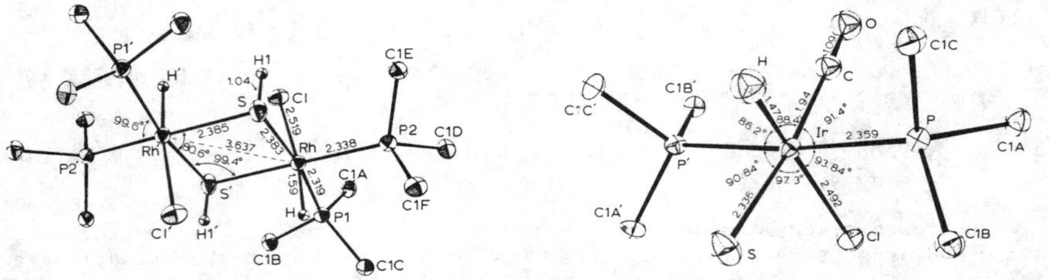

Fig. 1. The $[RhCl(H)(\mu\text{-}SH)(PPh_3)_2]_2$ (left) and $IrCl(H)(SH)CO(PPh_3)_2$ (right) molecules.

μ-DITHIOCYANATOTETRAKIS(TRIPHENYLPHOSPHINE)DICOPPER(I)

$C_{74}H_{60}Cu_2N_2P_4S_2$ $[CuNCS(PPh_3)_2]_2$

N.K. HOMSY, M. NOLTEMEYER, H.W. ROESKY, H.-G. SCHMIDT and G.M. SHELDRICK, 1984. Inorg. Chim. Acta, 90, L59-L60.

Triclinic, $P\bar{1}$, a = 10.276, b = 13.077, c = 13.041 Å, α = 114.71, β = 92.99, γ = 100.77°, Z = 1. Mo radiation, R = 0.060 for 3140 reflexions.

The structure consists of dimeric units with a crystallographic inversion centre (Fig. 1) in the middle of each dimer. The coordination of the Cu(I) ion is distorted tetrahedral with angles N-Cu-S 98.5, P-Cu-P 122.5, Cu-S-C 102.6, S-C-N 179.4 and C-N-Cu 157.1°. The principal bond lengths are: Cu-N 1.983, Cu-S 2.408, N-C 1.115, C-S 1.617, Cu-P(mean) 2.294 and P-C(mean) 1.808 Å.

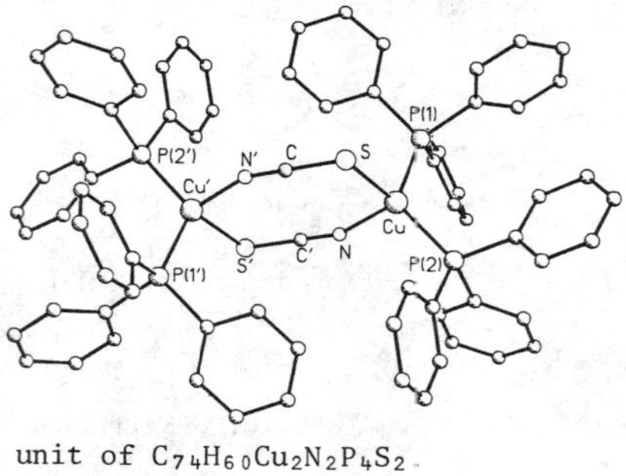

Fig. 1. The dimeric unit of $C_{74}H_{60}Cu_2N_2P_4S_2$

(μ_2-BIS(DIPHENYLPHOSPHINO)METHANE)BIS(BIS(DIPHENYLPHOSPHINO)METHANEPLATINUM) BIS(HEXAFLUOROPHOSPHATE)

$C_{75}H_{66}F_{12}P_8Pt_2$ $\{Pt_2[(C_6H_5)_2PCH_2P(C_6H_5)_2]_3\}^{2+}2PF_6^-$

S.I. AL-RESAYES, P.B. HITCHCOCK and J.F. NIXON, 1984. J. Organometal. Chem., 267, C13-C15.

Monoclinic, $P2_1/n$, a = 15.865, b = 17.499, c = 31.701 Å, β = 94.65°, Z = 4. [Mo?] radiation, R = 0.085 for 4711 reflexions.

The crystal structure consists of discrete cations and anions in a 1:2 ratio. The cation (Fig. 1) has a Pt-Pt bond length of 2.646 Å with one bridging dppm ligand and two terminal chelating dppm groups. The coordination about each Pt atom is essentially square planar with a maximum deviation of 0.08 Å for Pt(1) from the Pt(1),P(1),P(3),P(5),Pt(2) plane and 0.18 Å for Pt(2) from the Pt(2),P(2),P(4),P(6),-Pt(1) plane. The Pt-P bond lengths range from 2.295 (P(6)) to 2.328 Å (P(2) and P(4)).

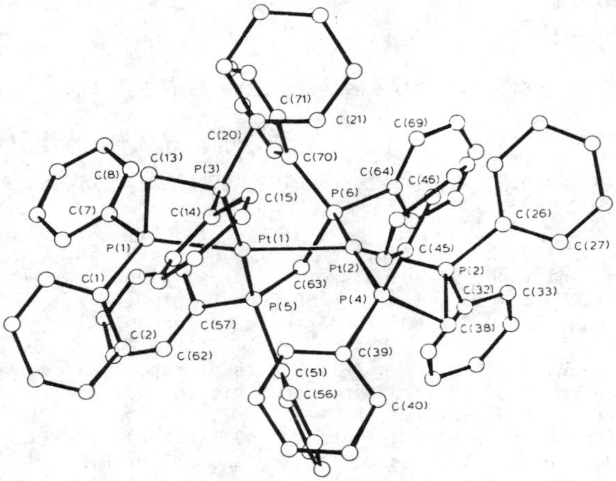

Fig. 1. Molecular structure of the cation in $C_{75}H_{66}F_{12}P_8Pt_2$.

TRIS(μ_2-CARBONYL)TETRAKIS(TRICYCLOHEXYLPHOSPHINE)TRIPLATINUMGOLD HEXAFLUOROPHOSPHATE

$C_{75}H_{132}AuF_6O_3P_5Pt_3$ $[Pt_3Au(CO)_3(P(C_6H_{11})_3)_4]^+PF_6^-$

C.E. BRIANT, R.W.M. WARDLE and D.M.P. MINGOS, 1984. J. Organometal. Chem., 267, C49-C51.

Monoclinic, C2/c, a = 19.149, b = 18.567, c = 50.968 Å, β = 95.96°, Z = 8. Cu radiation, R = 0.085 for 2028 reflexions.

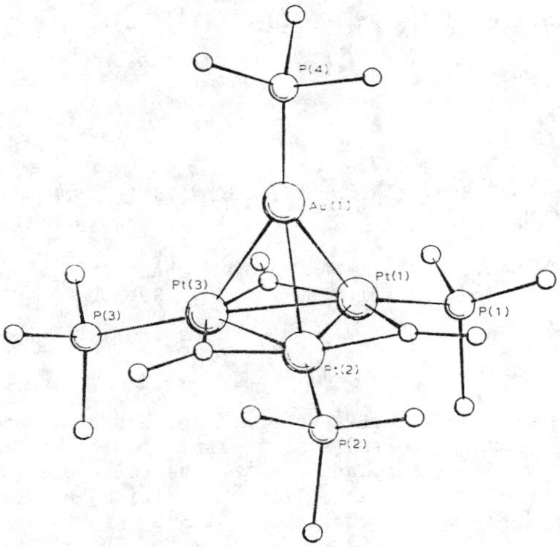

Fig. 1. Structure of the cation in $C_{75}H_{132}AuF_6O_3P_5Pt_3$ with the cyclohexyl rings omitted except for the ipso carbon atoms.

The cation (Fig. 1) has a tetrahedral Pt-Au cluster with Pt-Pt bond lengths significantly shorter (Pt(1)-Pt(2) 2.705, Pt(2)-Pt(3) 2.678, Pt(1)-Pt(3) 2.704 Å) than the Pt-Au bonds (Pt(1)-Au(1) 2.768, Pt(2)-Au(1) 2.757, Pt(3)-Au(1) 2.750 Å). Other bond lengths of interest are: Pt(1)-P(1) 2.24, Pt(2)-P(2) 2.29, Pt(3)-P(3) 2.29 and Au(1)-P(1) 2.27 Å. The PF_6^- anions are disordered about a position of twofold symmetry and a centre of symmetry in the unit cell.

BIS(μ-HYDRIDO)HYDRIDOBIS(1,2-BIS(DIPHENYLPHOSPHINO)ETHANE)DIPLATINUM(II) TETRAPHENYLBORATE

$C_{76}H_{71}BP_4Pt_2$ $[H_3Pt_2((C_6H_5)_2PCH_2CH_2P(C_6H_5)_2)_2]^+[B(C_6H_5)_4]^-$

M.Y. CHIANG, R. BAU, G. MINGHETTI, A.L. BANDINI, G. BANDITELLI and T.F. KOETZLE, 1984. Inorg. Chem., 23, 122-124.

Triclinic, P$\bar{1}$, a = 15.153, b = 11.422, c = 19.291 Å, α = 98.50, β = 100.47, γ = 97.57°, Z = 2. Mo radiation, R = 0.052 for 6222 reflexions, at room temperature; a = 15.109, b = 11.297, c = 19.111 Å, α = 98.766, β = 100.484, γ = 97.469°, 1.6113(13) Å neutrons, R = 0.043 for 12843 reflexions at ~20K.

The basic geometry of the cation (Fig. 1) is as reported for the X-ray structure of the monoclinic form (1). Pt(1) has trigonal bipyramidal geometry with Pt-(μ-H) 1.860(2) and 2.049(2), Pt-H 1.610(2), Pt-P(eq) 2.231(1), and Pt-P(ax) 2.312(1) Å. Pt(2) is square planar with Pt-(μ-H) 1.691(2) and 1.656(2), Pt-P 2.278(1) and 2.259(1) Å. The Pt...Pt distance is 2.711(1) Å and the Pt-H-Pt angles are 99.39(1) and 93.41(10)°.

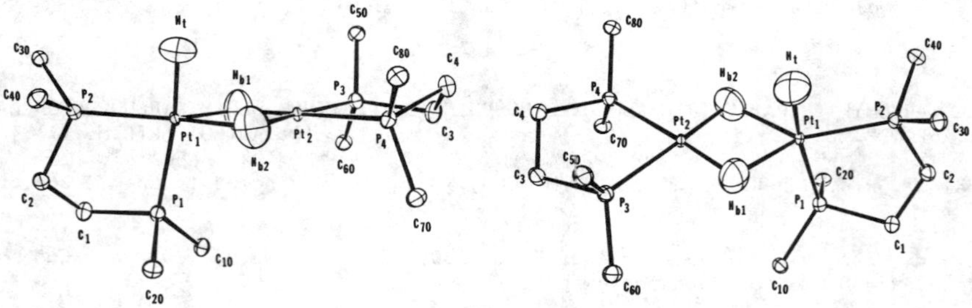

Fig. 1. Two views of the $[H_3Pt_2(Ph_2PCH_2CH_2PPh_2)_2]^+$ cation.

1. Structure Reports, 50B,

BIS[TRIS(TRIFLUOROMETHYL)GERMYLBIS(TRIPHENYLPHOSPHINE)PLATINUM]MERCURY TOLUENE

$C_{78}H_{60}F_{18}Ge_2HgP_4Pt_2,2(C_7H_8)$ $[(CF_3)_3GePt(P(C_6H_5)_3)_2]_2Hg,C_6H_5CH_3$

M.N. BOCHKAREV, N.L. ERMOLAEV, L.N. ZAKHAROV, Yu.N. SAF'YANOV, G.A. RAZUVAEV and Yu.T. STRUCHKOV, 1984. J. Organometal. Chem., 270, 289-300.

Monoclinic, C2, a = 22.025, b = 12.356, c = 28.593 Å, β = 90.58°, Z = 4. Mo radiation, R = 0.040 for 3246 reflexions.

The molecule contains a Ge-Pt-Hg-Pt-Ge chain and has approximate C_2 symmetry (Fig. 1). Both Pt atoms have distorted square-planar coordination, the coordination about Hg is significantly non-linear, and the tetrahedral geometry about the Ge atoms is distorted (Pt-Ge-C angles in the range 111 to 130° with an average of 122°). The Ge(1)-Pt(1)...Pt(2)-Ge(2) torsion angle is 101.5°.

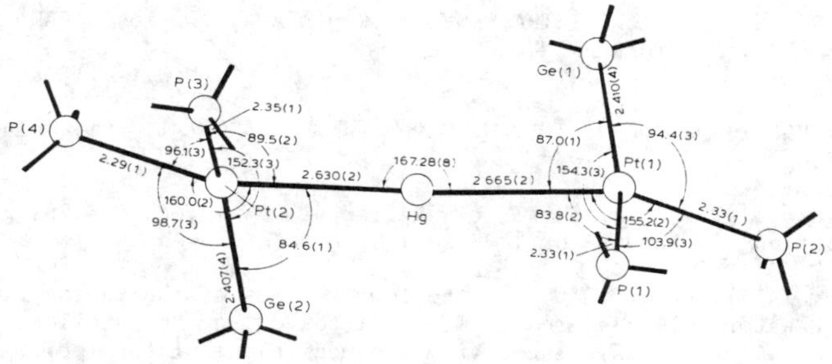

Fig. 1. The polymetallic chain in the molecule of $C_{78}H_{60}F_{18}Ge_2HgP_4Pt_2$, showing bond distances (Å) and angles (°).

BIS(BIS(DIPHENYLPHOSPHINOMETHYL)PHENYLPHOSPHINE)CARBONYLTETRAIODOTRIRHODIUM TETRA-PHENYLBORATE DEUTERIOCHLOROFORM METHYLENE CHLORIDE
$C_{89}H_{78}BI_4OP_6Rh_3,2.2CDCl_3,1.4CH_2Cl_2$

A.L. BALCH, R.R. GUIMERANS and M.M. OLMSTEAD, 1984. J. Organometal. Chem., 268, C38-C42.

Monoclinic, $P2_1/n$, a = 11.517, b = 23.301, c = 36.815 Å, β = 97.48°, Z = 4. R = 0.076 for 8877 reflexions (at 140 K).

The crystal structure consists of discrete cations, anions and solvent molecules. The cation has a near planar centre consisting of three Rh atoms and three I atoms with terminal CO and I moieties (Fig. 1). This 'planar' unit is bridged by two bis(diphenylphosphinomethyl)phenylphosphine ligands. The terminal iodo and CO ligands (I(1) and C(94)O(1)) have 80/20 disorder. The molecular dimensions and parameters of the 'planar' unit are shown in Fig. 1.

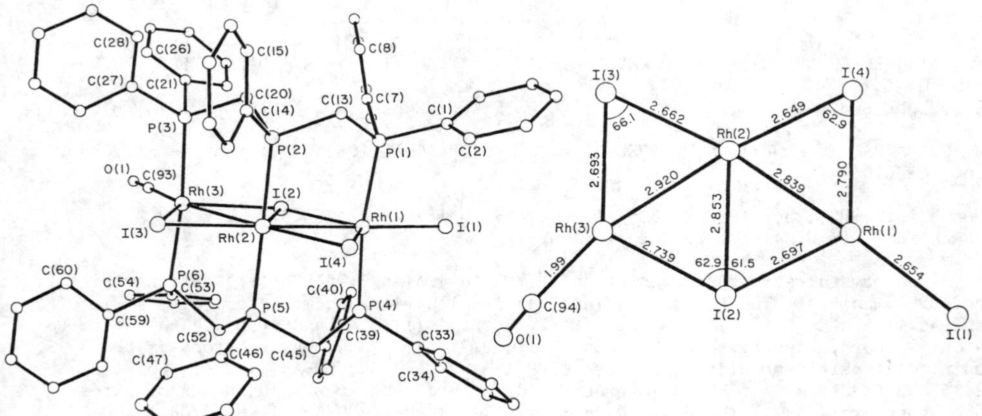

Fig. 1. Perspective drawing of the cation in $C_{89}H_{78}BI_4OP_6Rh_3$, and of the planar section showing selected distances (Å) and angles (°).

1,10-DIPROPYL-4RS,7SR,13SR,16RS-TETRAPHENYL-1,10-DIAZA-4,7,13,16-TETRAPHOSPHACYCLO-OCTADECANEAQUACOBALT(II) TETRAPHENYLBORATE

$C_{90}H_{100}B_2CoN_2OP_4$ $[Co(H_2O)(\gamma-[18]aneP_4N_2)][BPh_4]_2$

S. MANGANI, P. ORIOLI, M. CIAMPOLINI, N. NARDI and F. ZANOBINI, 1984. Inorg. Chim. Acta, 85, 65-72.

Triclinic, $P\bar{1}$, a = 13.252, b = 13.943, c = 24.316 Å, α = 70.66, β = 75.22, γ = 69.23°, Z = 2. Mo radiation, R = 0.069 for 1403 reflexions.

The Co atom is five-coordinated by the four P atoms of the macrocycle and the oxygen atom of a water molecule, resulting in distorted square pyramidal geometry (Fig. 1). The metal atom is displaced from the mean plane of the four P atoms by 0.28 Å towards the water molecule. The Co-P distances are 2.259, 2.276, 2.288 and 2.304 Å, and the Co-O(water) distance is 2.27 Å. The two N atoms of the ligand are not coordinated to the metal atom, and lie on the 'top' side of the P_4 plane at distances of 4.036 and 4.170 Å from the Co. The macrocycle has a boat-shaped configuration.

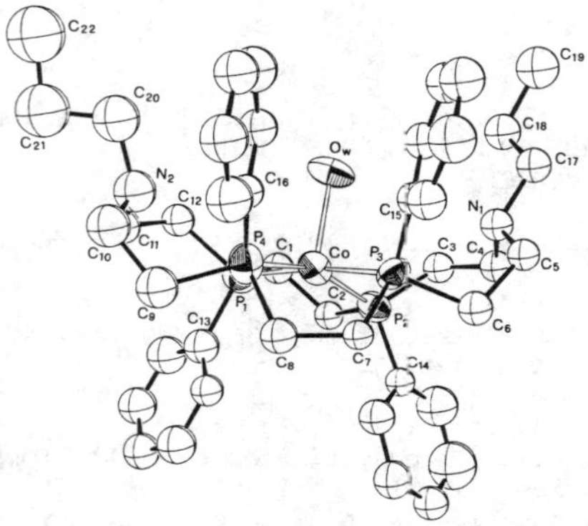

Fig. 1. Molecular structure of the complex cation in $C_{90}H_{100}B_2CoN_2OP_4$.

(μ_2-DIARSENICSULPHUR)BIS(TRIPHENYLPHOSPHINE)PLATINUM(1,1,1-TRIS(DIPHENYLPHOSPHINO-METHYL)ETHANE)COBALT TETRAPHENYLBORATE

$C_{101}H_{89}As_2BCoP_5PtS$ $[\{(C_6H_5)_2PCH_2\}_3CCH_3Co(As_2S)Pt(P(C_6H_5)_3)_2]^+B(C_6H_5)_4^-$

M. DI VAIRA, S. MONETI, M. PERUZZINI and P. STOPPIONI, 1984. J. Organometal. Chem., 266, C8-C10.

Monoclinic, $P2_1$, a = 19.298, b = 21.900, c = 10.187 Å, β = 98.37°, Z = 2. Mo radiation, R = 0.044 for 2952 reflexions.

The crystal structure consists of $[(triphos)Co(As_2S)Pt(PPh_3)_2]^+$ cations (Fig. 1) and BPh_4^- anions. The Pt atom is in a nearly planar arrangement formed by P(7), P(8) and As(1) and S (Pt-P(7) 2.33, Pt-P(8) 2.35, Pt-As(1) 2.47 and Pt-S 2.43 Å). The Co atom is coordinated by three P atoms (P(1), P(2), P(3)) and by the atoms of the As_2S unit. The pertinent distances are: Co-P(1-3) 2.21-2.26, Co-As(1) 2.45, Co-As(2) 2.36 and Co-S 2.42 Å. The As_2S group is disordered, and the occupancies of sites in the group are: As(1) (0.56 As, 0.44 S), As(2) (1.00 As, 0.00 S), S (0.44 As, 0.56 S). Bond angles of interest are: As(1)-Co-S 77.2, As(1)-Co-As(2) 58.6, As(2)-Co-S 58.8, As(1)-As(2)-S 80.4, As(1)-Pt-S 76.5 and P(4)-Pt-P(5) 101.6°.

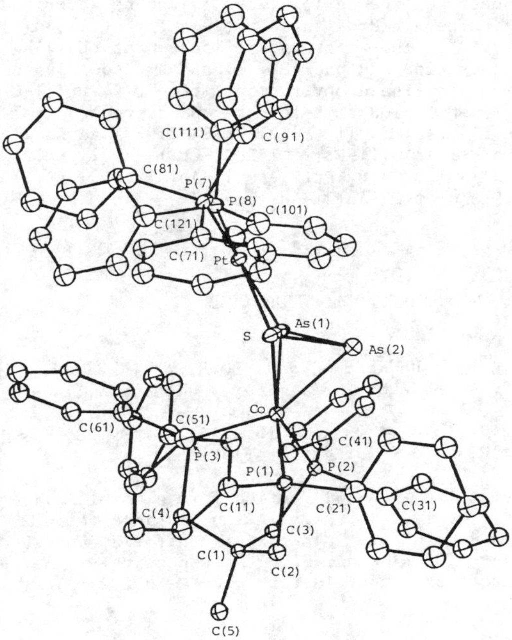

Fig. 1. Molecular structure of the cation in $C_{101}H_{89}As_2BCoP_5PtS$.

TETRAKIS(2,6-DIMETHYLPHENYL ISOCYANIDE)TETRAKIS(TRIPHENYLPHOSPHINE)DIPLATINUMDIGOLD BIS(HEXAFLUOROPHOSPHATE)

$C_{108}H_{96}Au_2F_{12}N_4P_6Pt_2$ $[Au_2Pt_2(P(C_6H_5)_3)_4(CNC_6H_3(CH_3)_2)_4]^{2+}2PF_6^-$

C.E. BRIANT, D.I. GILMOUR and D.M.P. MINGOS, 1984. J. Organometal. Chem., 267, C52-C55.

Triclinic, P$\bar{1}$, a = 13.882, b = 15.460, c = 27.578 Å, α = 77.99, β = 88.95, γ = 67.42°, Z = 2. Mo radiation, R = 0.051 for 3930 reflexions.

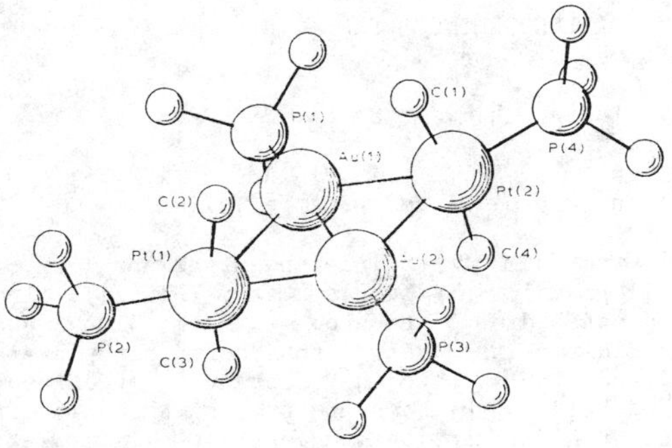

Fig. 1. Molecular structure of the cation in $C_{108}H_{96}Au_2F_{12}N_4P_6Pt_2$.

The skeletal geometry of the cation (Fig. 1) can be described in terms of a 'flattened butterfly'. The dihedral angle between the Pt(1),Au(1),Au(2) and Pt(2),Au(1),Au(2) planes is 168°. The Pt atoms are located at the wing tip positions of the butterfly. Two isocyanide and one PPh_3 ligand are coordinated to each Pt atom and define a distorted T-shaped ligand environment, e.g. P(2)-Pt(1)-C(2) 94.6, P(2)-Pt(1)-C(3) 97.0 and C(2)-Pt(1)-C(3) 168°. Relevant bond distances are: Au(1)-Au(2) 2.590, Au(1)-Pt(1) 2.717, Au(1)-Pt(2) 2.711, Au(2)-Pt(2) 2.922 and Au(2)-Pt(1) 3.026 Å. Angles of interest are Pt(1)-Au(1)-Pt(2) 135.0, Pt(1)-Au(2)-Pt(2) 114.9, Au(1)-Pt(1)-Au(2) 53.3, Au(1)-Pt(2)-Au(2) 54.6, Pt(1)-Au(1)-Au(2) 69.5, Pt(1)-Au(2)-Au(1) 57.2, Pt(2)-Au(1)-Au(2) 66.9 and Pt(2)-Au(2)-Au(1) 58.3°.

HEPTAKIS(TRIPHENYLPHOSPHINE)HEPTAGOLD HYDROXIDE

$C_{126}H_{106}Au_7OP_7$ $[Au_7(P(C_6H_5)_3)_7](OH)$

J.W.A. van der VELDEN, P.T. BEURSKENS, J.J. BOUR, W.P. BOSMAN, J.H. NOORDIK, M. KOLENBRANDER and J.A.K.M. BUSKES, 1984. Inorg. Chem., 23, 146-151.

Triclinic, P$\bar{1}$, a = 34.94, b = 44.25, c = 15.45 Å, α = 99.98, β = 102.66, γ = 88.11°, Z = 8. Mo radiation, R = 0.099 for 4787 reflexions.

The Au atoms of the $[Au_7(PPh_3)_7]^+$ cation form a novel, slightly distorted, pentagonal bipyramid (Fig. 1). The axial Au-Au distance (mean 2.584(1) Å) is the shortest ever found in Au clusters. The axial-equatorial and equatorial-equatorial Au-Au distances average 2.82(8) and 2.95(6) Å respectively. All P-Au vectors point approximately to the centre of the cluster, mean P-Au 2.27(9) Å.

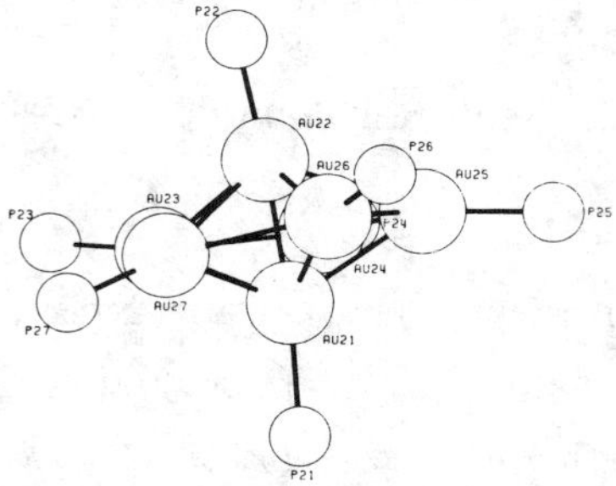

Fig. 1. The Au-P skeleton for one of the four independent $[Au_7(PPh_3)_7]^+$ cations.

HEXAKIS[(μ_2-HYDRIDO)TRI-p-TOLYLPHOSPHINECOPPER]

$C_{126}H_{132}Cu_6P_6$ $H_6Cu_6[P(p\text{-}C_6H_4CH_3)_3]_6$

D.M. HO and R. BAU, 1984. Inorg. Chim. Acta, 84, 213-220.

Triclinic, P$\bar{1}$, a = 19.785, b = 23.057, c = 15.810 Å, α = 102.53, β = 112.39, γ = 83.96°, Z = 2. Mo radiation, R = 0.091 for 6837 reflexions.

The complex exists in the form of two independent hexanuclear clusters, each cluster possessing crystallographic inversion symmetry (Fig. 1). The central core of the cluster is a distorted octahedron of Cu atoms with six 'long' and six 'short' Cu-Cu distances. In the two clusters the 'long' and 'short' bonds average, respectively, 2.68 and 2.52 Å, and 2.63 and 2.55 Å. The Cu-P distances range from 2.214 to 2.242 Å. It is thought that six edge-bridging H atoms are present though their presence will have to be confirmed by neutron diffraction studies.

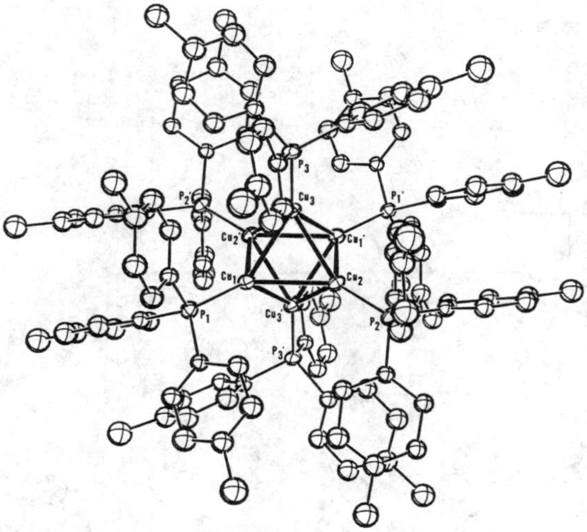

Fig. 1. Molecular structure of cluster 1 in $C_{126}H_{132}Cu_6P_6$.

POTASSIUM METHYLPHOSPHONODIOXOVANADATE MONOHYDRATE
CH_3KO_5PV,H_2O

R. MATTES, A. PREUSS and K.-L. RICHTER, 1984. Z. Naturforsch., B39, 1331-1334.

Trigonal, $P3_121$, a = 11.112, c = 10.040 Å, Z = 6. Mo radiation, R = 0.021 for 901 reflexions (at -140°C).

The structure (Fig. 1) consists of open arrangements of corner sharing V-O polyhedra and CH_3-P-O tetrahedra. Polyhedral chains form helices along the screw axes and the vanadium polyhedron is highly distorted trigonal bipyramidal with two terminal O atoms. Main dimensions are V-O 1.593-2.007(2), P-O 1.514-1.528(2), P-C 1.772(4) Å. The K^+ ion has ten O atoms in the range 2.713-3.474(3) Å.

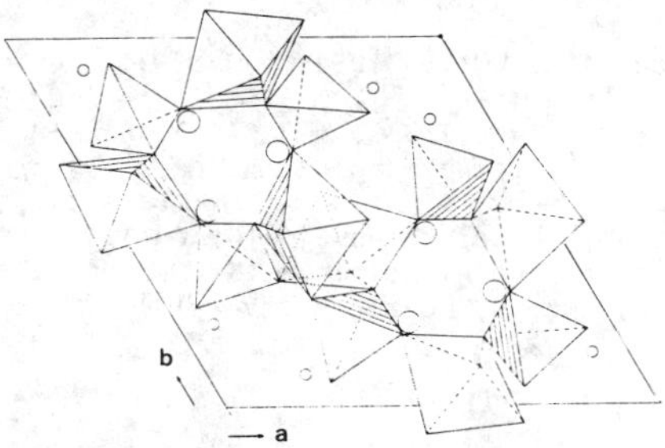

Fig. 1. A projection of the CH_3KO_5PV,H_2O structure; large circles are K^+ ions, small circles are water molecules.

ETHYLENEDIAMMONIUM HEXASULPHIDE
$C_2H_{10}N_2S_6$

P. BOTTCHER, H. BUCHKREMER-HERMANNS and J. BARON, 1984. Z. Naturforsch., B39, 416-420.

Monoclinic, Cc, a = 15.52, b = 15.64, c = 9.67 Å, β = 117.4°, Z = 8. Ag radiation, R = 0.043 for 2530 reflexions.

The two independent S_6^{2-} chains have unbranched zig-zag conformations (Fig. 1), with S-S 2.060(1)-2.077(2) Å, S-S-S 106.7-109.8°, S-S-S-S torsion angles ±58.9 to ±82.0°. The ethylenediammonium cations are maximally extended.

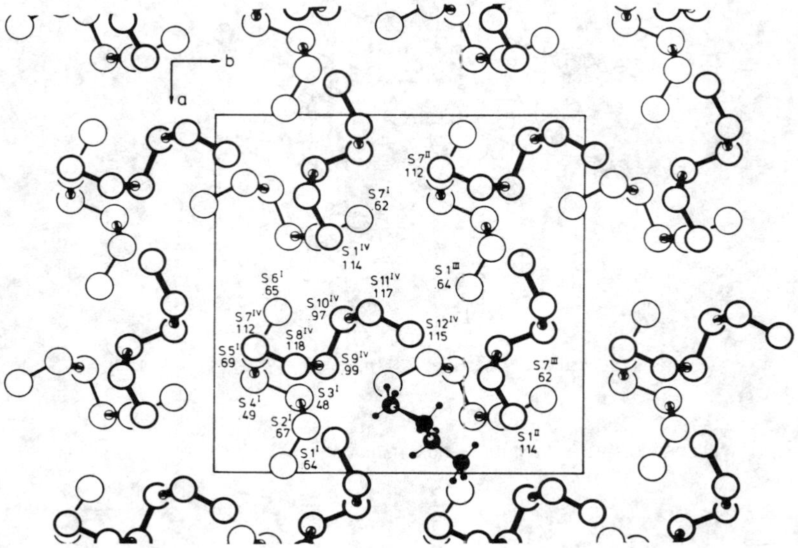

Fig. 1. A projection on the a-b plane of the ethylenediammonium hexasulphide structure.

ETHYLENEDIAMMONIUM HEXAAQUANICKEL(II) SULPHATE

$C_2H_{22}N_2NiO_{14}S_2$ $[enH_2][Ni(H_2O)_6][SO_4]_2$

P.C. HEALY, J.M. PATRICK and A.H. WHITE, 1984. Aust. J. Chem., 37, 1105-1109.

Monoclinic, $P2_1/n$, a = 13.148, b = 7.119, c = 8.245 Å, β = 100.97°, D_m = 1.8, Z = 2. Mo radiation, R = 0.038 for 2158 reflexions.

This Tutton salt (Fig. 1) contains $[enH_2]^{2+}[Ni(OH_2)_6]^{2+}(SO_4)_2{}^{2-}$ ions with all water molecules coordinating to the nickel atom (Ni-O 2.064(2) Å). Both cations are at special positions (0.5,0,0 and 0,0,0 respectively).

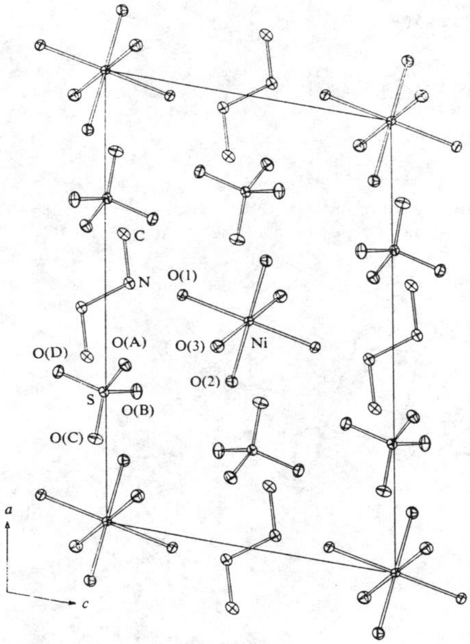

Fig. 1. The structure of $[enH_2][Ni(H_2O)_6][SO_4]_2$.

POTASSIUM TETRAKIS(ISOTHIOCYANATO)COBALT(II) MONOHYDRATE NITROMETHANE SOLVATE

$C_4CoK_2N_4S_4,H_2O,2CH_3NO_2$ $K_2[Co(NCS)_4],H_2O,2CH_3NO_2$

J.S. WOOD and R.K. McMULLAN, 1984. Acta Cryst., C40, 1803-1806.

Monoclinic, I2, a = 11.098, b = 5.227, c = 16.496 Å, β = 100.06°, D_m = 1.78, Z = 2. Neutron radiation (λ = 1.0470 Å), R (F^2) = 0.038 for 2011 reflexions (at 120K).

Chains of approximately tetrahedral $Co(NCS)_4^{2-}$ anions parallel to b are linked by O-H...S hydrogen bonds to water molecules (Fig. 1); the O...S distance is 3.377(2) Å. K^+ ions are linked by nitromethane and water molecules to give double chains parallel to b. Each K^+ is coordinated by four O atoms (K^+...O 2.839(2)-3.110(2) Å) and by three S atoms (mean K^+...S 3.349(2) Å). Mean bond lengths are Co-N 1.964(2), N-C 1.173(1), C-S 1.627(2) Å.

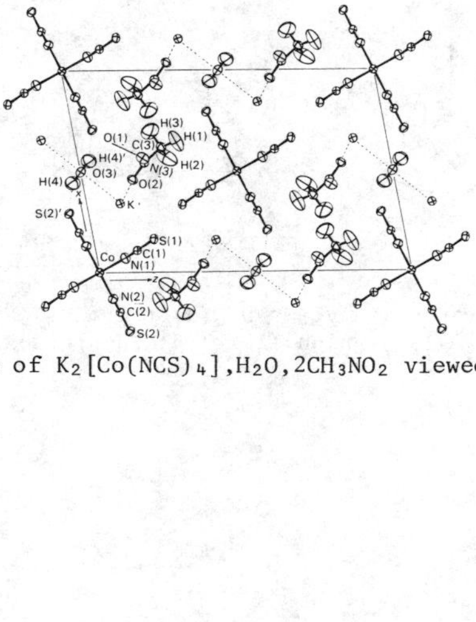

Fig. 1. The structure of $K_2[Co(NCS)_4],H_2O,2CH_3NO_2$ viewed along b.

ALANINIUM METHYL ESTER TETRACHLOROAURATE(III)
$C_4H_{10}AuCl_4NO_2$

E. AMBACH, M.M. SINGH, U. NAGEL and W. BECK, 1984. Z. Naturforsch., B39, 1129-1134.

Monoclinic, $P2_1$, a = 6.919, b = 16.481, c = 10.588 Å, β = 101.42°, D_m = 2.41, Z = 4. Mo radiation, R = 0.0505 for 3884 reflexions.

The crystal structure contains discrete $[AuCl_4]^-$ and $[H\text{-}alaOMe]^+$ ions. The $[AuCl_4]^-$ ions have square-planar geometry (mean Au-Cl 2.282(5) Å). The cation dimensions are unexceptional.

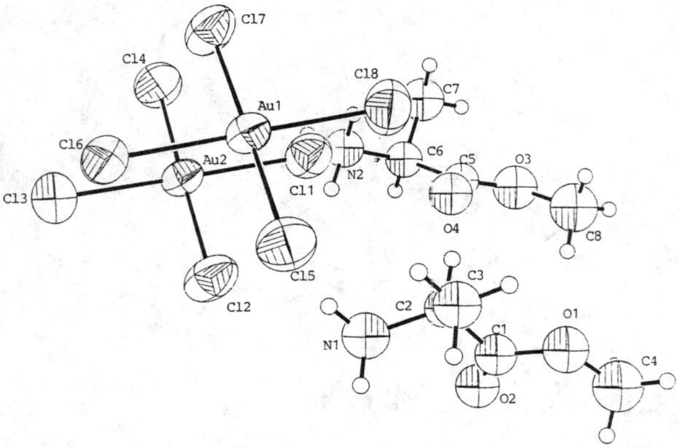

Fig. 1. The $[AuCl_4]^-$ and $[H\text{-}alaOMe]^+$ ions.

TETRAMETHYLAMMONIUM DIIODOARGENTATE(I)
$C_4H_{12}AgI_2N$ $[N(CH_3)_4][AgI_2]$

K. PETERS, H.G. VON SCHNERING, W. OTT and H.-M. SEIDENSPINNER, 1984. Acta Cryst., C40, 789-790.

Orthorhombic, Immm, a = 9.228, b = 15.353, c = 6.899 Å, Z = 4. Mo radiation, R = 0.082 for 610 reflexions.

In the crystal structure AgI_4 tetrahedra share edges only and form infinite chains parallel to the c axis (Fig. 1). The mean Ag-I bond distance is 2.781 Å.

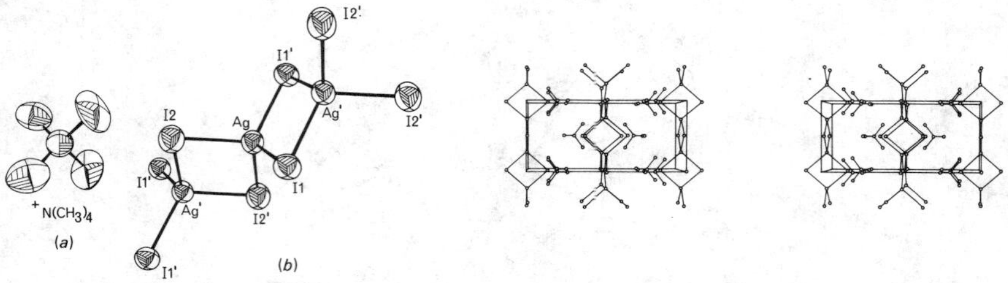

Fig. 1. $C_4H_{12}AgI_2N$: perspective views of (a) the $(CH_3)_4N^+$ cation and (b) a section of the $[AgI_2]^-$ anion (left) and a stereoscopic view of the unit cell (right).

TETRAMETHYLAMMONIUM TRICHLOROCHROMATE(II)

$C_4H_{12}Cl_3CrN$ $[(CH_3)_4N]CrCl_3$

C. BELLITTO, G. DESSY, V. FARES, D. FIORANI and S. VITICOLI, 1984. J. Phys. Chem. Solids, 45, 1129-1134.

Hexagonal, $P6_3/m$, a = 9.129, c = 6.512 Å, Z = 2. Mo radiation, R = 0.052 for 442 reflexions.

The crystal structure consists of infinite linear chains (along the z axis), of octahedral Cr atoms bridged by three halogen atoms (Fig. 1). The Cr(1)...Cr(2) distance is 3.256 Å. The Cl atom positions can be (dynamically) selected in order to describe the $CrCl_3^-$ chains as a sequence of alternating elongated and compressed octahedra where the Cr-Cl long distance is 2.685 Å and the short one 2.425 Å. The Cl-Cr-Cl angles range from 78.6 to 101.4°, and the Cl...Cl distances for Cl(1)... Cl(2m) and Cl(1)...Cl(3'm) are respectively 3.425 and 3.802 Å. The tetramethyl-ammonium ion is required to be disordered with the N atom at (0.33,0.66,0.25) and the methyl groups reflected through the mirror plane at Z = 0.25. The N-C bonds range from 1.44 to 1.57 Å, and the angles C(1)-N-C(2) and C(2)-N-C(2') are respectively 103 and 114°.

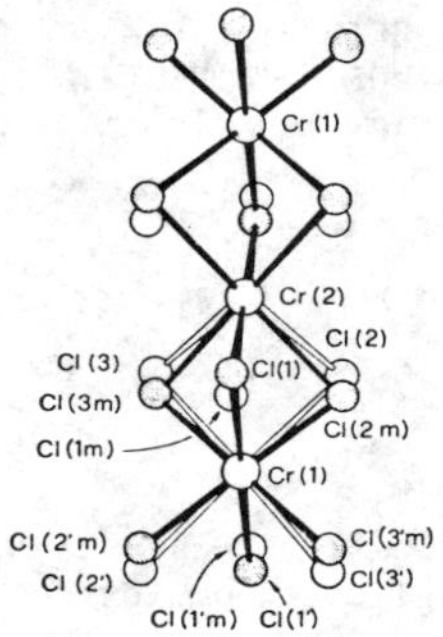

Fig. 1. Part of the unit cell of $C_4H_{12}Cl_3CrN$ showing the $CrCl_3^-$ anionic chain.

DIETHYLENETRIAMMONIUM HEXACHLOROANTIMONATE(III)

$C_4H_{16}Cl_6N_3Sb$ $[H_3NCH_2CH_2NH_2CH_2CH_2NH_3]^{3+}SbCl_6^{3-}$

I.M. VEZZOSI, L.P. BATTAGLIA and A.B. CORRADI, 1984. Inorg. Chim. Acta, 89, 151-155.

Triclinic, $P\bar{1}$, a = 10.016, b = 10.076, c = 8.123 Å, α = 67.96, β = 105.88, γ = 100.03°, D_m = 2.00, Z = 2. Mo radiation, R = 0.037 for 2288 reflexions.

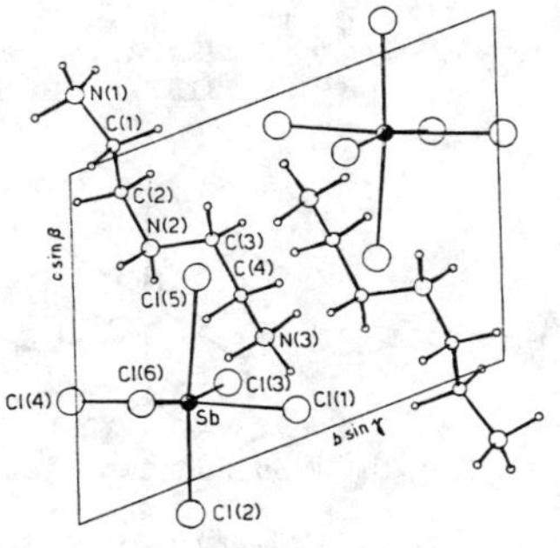

Fig. 1. Projection of the structure of $C_4H_{16}Cl_6N_3Sb$ along [100].

The structure consists of monomeric $SbCl_6^{3-}$ anions and triprotonated cations (Fig. 1). The anion has a strongly distorted octahedral geometry, presenting three short (2.495, 2.521 and 2.415 Å for Cl(1), Cl(2), Cl(3) respectively) and three long (2.836, 2.846 and 3.114 Å for Cl(4), Cl(5), Cl(6) respectively) Sb-Cl bonds. The presence of multiple hydrogen bonds, involving mainly the cation and the three long-bonded Cl atoms (N(2)...Cl(5) 3.193, N(3)...Cl(6) 3.151, N(1)...Cl(4) 3.222, N(1)...Cl(5) 3.326, N(1)...Cl(6) 3.217, N(2)...Cl(4) 3.103, N(3)...Cl(5) 3.219, N(3)...Cl(1) 3.200 and N(3)...Cl(2) 3.418 Å) is considered to be responsible for the octahedral distortion. Bond distances and angles in the aliphatic amine are normal.

TELLURIC ACID - GLYCINE (1:2) HYDRATE

$C_4H_{16}N_2O_{10}Te,2H_2O$ $Te(OH)_6.2(C_2H_5NO_2),2H_2O$

D.T. QUI, J. VICAT and A. DURIF, 1984. Acta Cryst., C40, 181-184.

Monoclinic, $P2_1/n$, a = 7.924, b = 12.154, c = 7.188 Å, β = 110.67°, Z = 2. Mo radiation, R = 0.020 for 1942 reflexions (at 120 K).

The structure consists of slightly distorted $Te(OH)_6$ octahedra (with the Te atom at an inversion centre and Te-OH 1.905 to 1.929(2) Å, average 1.920(2) Å) and glycine species in the zwitterionic form (Fig. 1). The molecules are linked by Te-OH...O-C hydrogen bonds. See also (1).

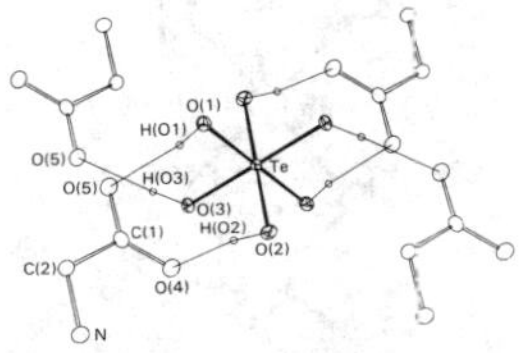

Fig. 1. A view of the $Te(OH)_6$ glycine complex.

1. Structure Reports, 50B.

HEXAAQUAVANADIUM(III) TETRAKIS(TRIFLUOROMETHANESULPHONATE) OXONIUM MONOHYDRATE

$C_4H_{17}F_{12}O_{20}S_4V$ $[V(H_2O)_6]^{3+}.H_5O_2^+.4(CF_3SO_3^-)$

F.A. COTTON, C.K. FAIR, G.E. LEWIS, G.N. MOTT, F.K. ROSS, A.J. SCHULTZ and J.M. WILLIAMS, 1984. J. Am. Chem. Soc., 106, 5319-5323.

I. Triclinic, $P\bar{1}$, a = 8.950, b = 9.112, c = 8.928 Å, α = 103.38, β = 93.75, γ = 70.19°, Z = 1. Mo radiation, R = 0.046 for 2605 reflexions.

II. Triclinic, $P\bar{1}$, a = 8.942, b = 8.902, c = 8.774 Å, α = 103.4, β = 95.2, γ = 69.9°, Z = 1. Neutron radiation, R = 0.067 for 2137 reflexions (at 20 K).

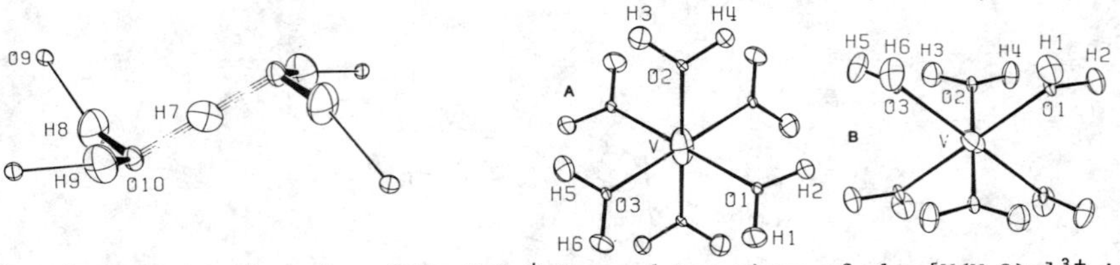

Fig. 1. A perspective view of the $H_5O_2^+$ ion and two views of the $[V(H_2O)_6]^{3+}$ ion.

The structure consists of centrosymmetric $[V(H_2O)_6]^{3+}$ and $H_5O_2^+$ cations (Fig. 1) and $CF_3SO_3^-$ anions. All hydrogen atoms participate in hydrogen bonds. The $H_5O_2^+$ ion appears to be truly centrosymmetric with an O...O distance of 2.430(3) Å at 20 K; the configuration at each oxygen atom is highly pyramidal. The $[V(H_2O)_6]^{3+}$ cation has nearly flat water molecules (i.e., sums of bond angles around the oxygen atoms are 357.7-359.7°), and these are arranged to give the hexaaquo ion essentially D_{3d} symmetry of the sort in which the six hydrogen atoms of three mutually cis water molecules are essentially coplanar. Possible reasons for the adoption of this particular highly symmetric arrangement are discussed.

BIS(1H⁺,5H⁺-(S)-METHYLISOTHIOCARBONOHYDRAZIDIUM) BIS-μ₂-CHLOROOCTACHLORODIBISMUTHATE-(III) TETRAHYDRATE

$C_4H_{20}Bi_2Cl_{10}N_8S_2,4H_2O$ (I) $(C_2H_{10}N_4S^{2+})_2(Bi_2Cl_{10})^{4-},4H_2O$

TRIS(1H⁺-(S)-METHYLISOTHIOCARBONOHYDRAZIDIUM) HEXACHLOROBISMUTHATE(III) PENTACHLORO-IODOBISMUTHATE(III)

$C_6H_{27}BiCl_{5.67}I_{0.33}N_{12}S_3$ (II) $(C_2H_9N_4S^{+})_3(BiCl_{5.67}I_{0.33})^{3-}$

F. BIGOLI, M. LANFRANCHI and M.A. PELLINGHELLI, 1984. Inorg. Chim. Acta, <u>90</u>, 215-220.

I. Triclinic, $P\bar{1}$, a = 10.621, b = 9.989, c = 7.439 Å, α = 88.31, β = 84.51, γ = 68.88°, Z = 1. Mo radiation, R = 0.043 for 2229 reflexions.

II. Triclinic, $P\bar{1}$, a = 14.109, b = 12.209, c = 8.206 Å, α = 103.54, β = 104.95, γ = 81.96°, Z = 2. Mo radiation, R = 0.041 for 3637 reflexions.

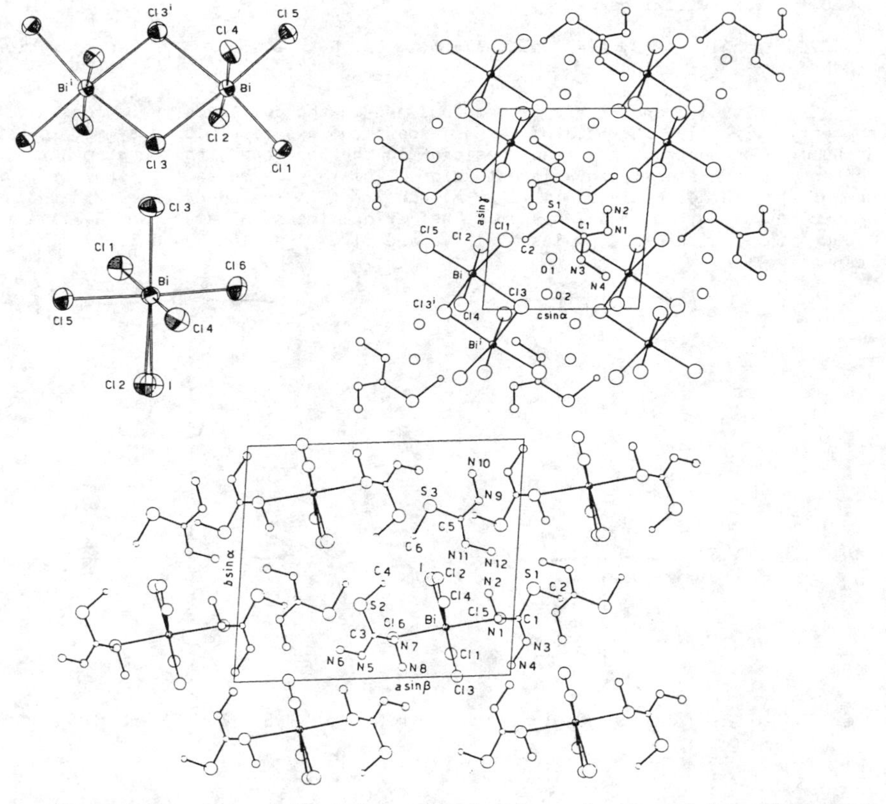

Fig. 1. Perspective views of the anions (top left), projection of $C_4H_{20}Bi_2Cl_{10}N_8S_2$ along b (top right), and projection of $C_6H_{27}BiCl_{5.67}I_{0.33}N_{12}S_3$ along c (bottom).

The structure of I consists of diprotonated organic cations, water molecules and dinuclear $Bi_2Cl_{10}^{4-}$ anions held together by N-H...Cl, N-H...O and O-H...Cl hydrogen bonds (Fig. 1). The anion consists of two edge-sharing octahedra in which three pairs of bonds of similar length are observed (Bi-Cl(average) 2.602. 2.712 and 2.855 Å), corresponding to Cl atoms Cl(1) and Cl(5), Cl(2) and Cl(4), and Cl(3) and Cl(3'). Compound II consists of monoprotonated cations and $BiCl_{5.67}I_{0.33}^{3-}$ anions held together by a three-dimensional lattice of hydrogen bonds. Each Bi atom is octahedrally coordinated to six Cl atoms, one of which is statistically substituted by an I atom (Fig. 1). The Bi-Cl distances range from 2.681 to 2.779 Å and the Bi-I distance is 2.899 Å.

TETRAMETHYLAMMONIUM DICHLOROCYANOSELENATE(II)
$C_5H_{12}Cl_2N_2Se$ (I) $(CH_3)_4N^+.NCSeCl_2^-$

TETRAMETHYLAMMONIUM DIBROMOCYANOSELENATE(II)
$C_5H_{12}Br_2N_2Se$ (II) $(CH_3)_4N^+.NCSeBr_2^-$

S. BJØRNEVÅG, P.U. GAHRE, S. HAUGE and O. VIKANE, 1984. Acta Chem. Scand., A38, 175-181.

I. Tetragonal, $P\bar{4}2_1m$, a = 9.229, c = 6.023 Å, D_m = 1.60, Z = 2. Mo radiation, R = 0.038 for 518 reflexions.

II. Tetragonal, $P\bar{4}2_1m$, a = 9.453, c = 6.131 Å, D_m = 2.05, Z = 2. Mo radiation, R = 0.035 for 371 reflexions.

Crystals of I and II are isomorphous and in the complexes the anions have mm symmetry (Fig. 1). The Se atom has a T-shape coordination with a symmetric halogen-Se-halogen sequence. In a trans position to the Se-C bond, the Se atom has a close contact to an N atom in an adjacent anion. This Se...N contact is shorter than the sum of the van der Waals radii (3.30 Å) and if it is included, the coordination around Se becomes distorted square planar. The cation dimensions are: N-C 1.511 Å in I and 1.500 Å in II with C-N-C angles of 107.1 and 110.7° and 109.0 and 109.7° in I and II respectively.

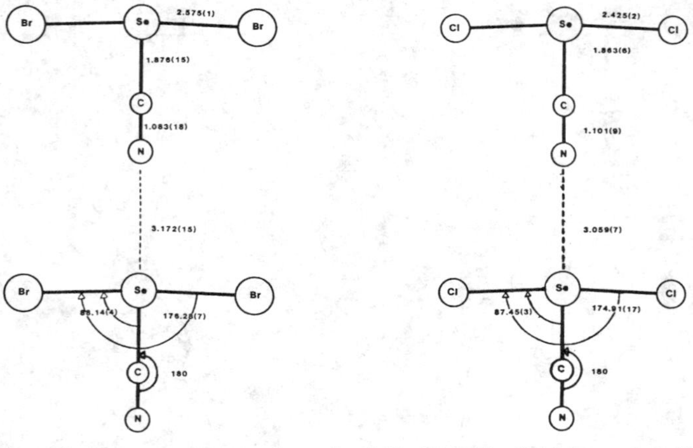

Fig. 1. The dihalocyanoselenate ions in $C_5H_{12}Cl_2N_2Se$ (left), and $C_5H_{12}Br_2N_2Se$ (right).

N-METHYLPYRIDINIUM TRIIODODICOPPER(I)
$C_6H_8Cu_2NI_3$ (I)

METHYLTRIPHENYLPHOSPHONIUM TETRAIODOTRICOPPER(I)
$C_{19}H_{18}Cu_3I_4P$ (II)

TETRAPROPYLAMMONIUM TETRAIODOTRICOPPER(I)
$C_{12}H_{28}Cu_3NI_4$ (III)

H. HARTL and F. MAHDJOUR-HASSAN-ABADI, 1984. Z. Naturforsch., B39, 149-156.

I. Orthorhombic, Pnma, a = 18.180, b = 6.288, c = 11.191 Å, Z = 4. Mo radiation, R = 0.047 for 1021 reflexions.

II. Monoclinic, $P2_1/c$, a = 12.601, b = 25.187, c = 8.014 Å, β = 107.32°, Z = 4. Mo radiation, R = 0.042 for 3931 reflexions.

III. Orthorhombic, Pnna, a = 17.942, b = 11.395, c = 11.058 Å, Z = 4. Mo radiation, R = 0.050 for 1707 reflexions.

These compounds contain chains of edge (I,II) or face (III) sharing CuI_4 tetrahedra as shown in Fig. 1. In (III) face sharing gives rise to very short Cu...Cu distances of 2.461(5) and 2.503(5) Å. The shortest Cu...Cu distances in (I) and (II) are 2.947(4) and 2.848(2) Å respectively. Cu-I distances are in the range 2.504(5)-3.293(4) Å.

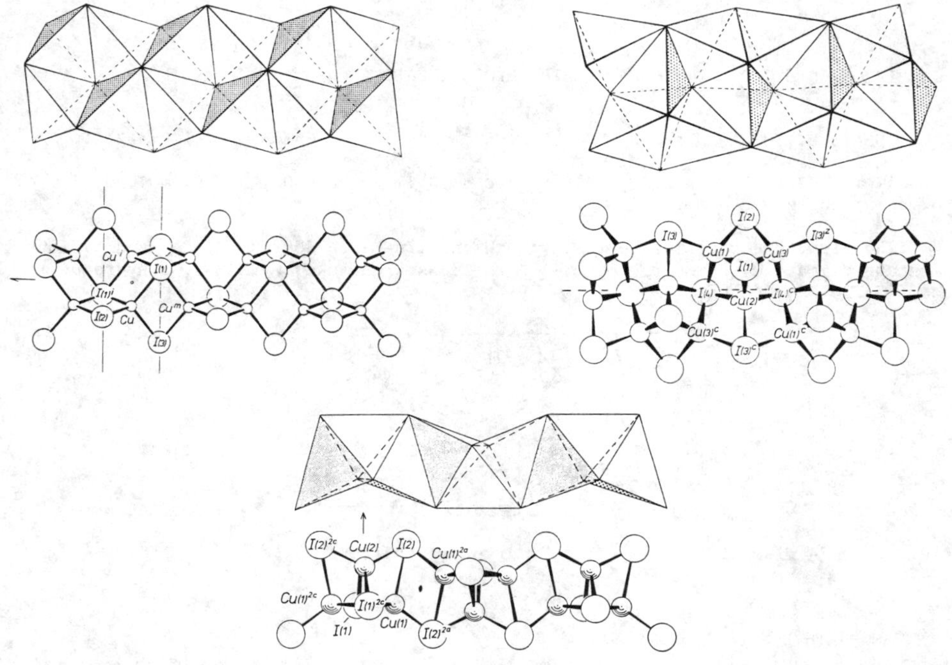

Fig. 1. Views of the CuI_4 tetrahedra: (I) (top left), (II) (top right) and (III) (lower).

3,4,5-TRIS(METHYLTHIO)-1,2-DITHIOLIUM catena-μ-IODO-μ₃-IODO-(μ-IODO-DICUPRATE(I))
$C_6H_9Cu_2I_3S_5$ $[S_2C_3(SCH_3)_3]^+[Cu_2I_3]^-$

M. ASPLUND and S. JAGNER, 1984. Acta Chem. Scand., A38, 129-134.

Monoclinic, $P2_1/c$, a = 10.856, b = 18.992, c = 8.630 Å, β = 106.74°, Z = 4. Mo radiation, R = 0.060 for 2932 reflexions.

The anion consists of an infinite I-bridged chain along c, I(1) and I(2) each bridging two Cu(I) atoms while I(3) bridges three Cu(I) atoms (Fig. 1.). One copper atom, (Cu(1)), is coordinated in distorted trigonal planar fashion whilst Cu(2) is surrounded by an approximately tetrahedral arrangement of iodine atoms. The Cu-I bond lengths are: Cu(1)-I(1) 2.573, Cu(1)-I(2) 2.523, Cu(1)-I(3) 2.730, Cu(2)-I(1) 2.644, Cu(2)-I(2') 2.708, Cu(2)-I(3) 2.781, Cu(2)-I(3') 2.597 Å, and the Cu...Cu separations are Cu(1)...Cu(2) 2.551 and Cu(1)...Cu(2'') 2.657 Å. The cations lie between the $[Cu_2I_3]^-$ chains with their molecular planes approximately perpendicular to b and the dithio groups oriented towards the chains. There are no short Cu...S contacts.

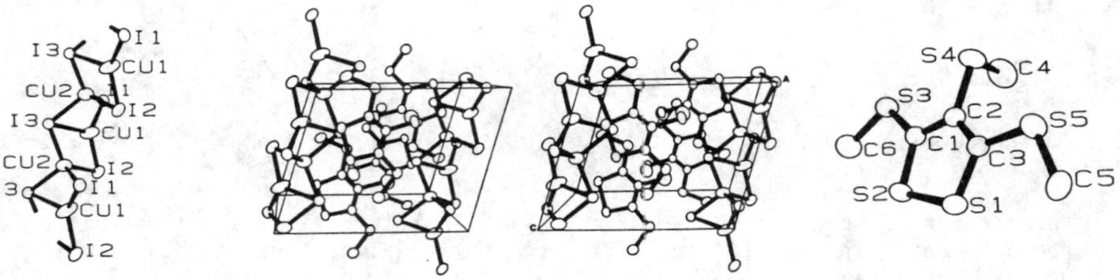

Fig. 1. Part of the chain-like anion (left), stereoscopic view of the unit cell (centre), and the cation (right) in $C_5H_9Cu_2I_3S_5$.

1,4-DIAZONIABICYCLO[2.2.2]OCTANE TETRACHLOROZINCATE
$C_6H_{14}Cl_4N_2Zn$

B. VIOSSAT, P. KHODADAD and N. RODIER, 1984. Bull. Soc. Chim. Fr., I-214-216.

Monoclinic, $P2_1/c$, a = 9.572, b = 6.767, c = 18.310 Å, β = 92.74°, D_m = 1.76, Z = 4. Mo radiation, R = 0.030 for 1716 reflexions.

The structure contains tetrahedral anions (Zn-Cl 2.253-2.276(1) Å) and cations with approximate C_3 symmetry (Fig. 1). The ions are linked by N-H...Cl hydrogen bonds 3.220-3.422(3) Å.

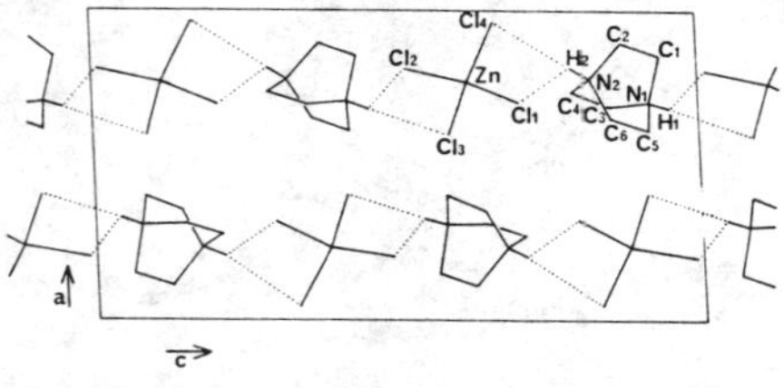

Fig. 1. A view of the $C_6H_{14}Cl_4N_2Zn$ crystal structure.

LANTHANUM ETHYL SULPHATE ENNEAHYDRATE
$C_6H_{15}LaO_{12}S_3,9H_2O$ (I) $La(C_2H_5SO_4)_3,9H_2O$

CERIUM ETHYL SULPHATE ENNEAHYDRATE
$C_6H_{15}CeO_{12}S_3,9H_2O$ (II) $Ce(C_2H_5SO_4)_3,9H_2O$

PRASEODYMIUM ETHYL SULPHATE ENNEAHYDRATE
$C_6H_{15}O_{12}PrS_3,9H_2O$ (III) $Pr(C_2H_5SO_4)_3,9H_2O$

NEODYMIUM ETHYL SULPHATE ENNEAHYDRATE
$C_6H_{15}NdO_{12}S_3,9H_2O$ (IV) $Nd(C_2H_5SO_4)_3,9H_2O$

SAMARIUM ETHYL SULPHATE ENNEAHYDRATE
$C_6H_{15}O_{12}S_3Sm,9H_2O$ (V) $Sm(C_2H_5SO_4)_3,9H_2O$

EUROPIUM ETHYL SULPHATE ENNEAHYDRATE
$C_6H_{15}EuO_{12}S_3,9H_2O$ (VI) $Eu(C_2H_5SO_4)_3,9H_2O$

GADOLINIUM ETHYL SULPHATE ENNEAHYDRATE
$C_6H_{15}GdO_{12}S_3,9H_2O$ (VII) $Gd(C_2H_5SO_4)_3,9H_2O$

TERBIUM ETHYL SULPHATE ENNEAHYDRATE
$C_6H_{15}O_{12}S_3Tb,9H_2O$ (VIII) $Tb(C_2H_5SO_4)_3,9H_2O$

DYSPROSIUM ETHYL SULPHATE ENNEAHYDRATE
$C_6H_{15}DyO_{12}S_3,9H_2O$ (IX) $Dy(C_2H_5SO_4)_3,9H_2O$

HOLMIUM ETHYL SULPHATE ENNEAHYDRATE
$C_6H_{15}HoO_{12}S_3,9H_2O$ (X) $Ho(C_2H_5SO_4)_3,9H_2O$

ERBIUM ETHYL SULPHATE ENNEAHYDRATE
$C_6H_{15}ErO_{12}S_3,9H_2O$ (XI) $Er(C_2H_5SO_4)_3,9H_2O$

THULIUM ETHYL SULPHATE ENNEAHYDRATE
$C_6H_{15}O_{12}S_3Tm,9H_2O$ (XII) $Tm(C_2H_5SO_4)_3,9H_2O$

YTTERBIUM ETHYL SULPHATE ENNEAHYDRATE
$C_6H_{15}O_{12}S_3Yb,9H_2O$ (XIII) $Yb(C_2H_5SO_4)_3,9H_2O$

LUTETIUM ETHYL SULPHATE ENNEAHYDRATE
$C_6H_{15}LuO_{12}S_3,9H_2O$ (XIV) $Lu(C_2H_5SO_4)_3,9H_2O$

R.E. GERKIN and W.J. REPPART, 1984. Acta Cryst., C40, 781-786.

Hexagonal, $P6_3/m$, Z = 2. Mo radiation (all at 171K).
I. a = 14.045, c = 7.0996 Å. R = 0.020 for 1560 reflexions.
II. a = 14.010, c = 7.0860 Å. R = 0.018 for 1521 reflexions.
III. a = 13.984, c = 7.0754 Å. R = 0.024 for 1486 reflexions.
IV. a = 13.963, c = 7.0644 Å. R = 0.017 for 1511 reflexions.
V. a = 13.930, c = 7.0468 Å. R = 0.018 for 1459 reflexions.
VI. a = 13.910, c = 7.0350 Å. R = 0.019 for 1506 reflexions.
VII. a = 13.905, c = 7.0249 Å. R = 0.019 for 1490 reflexions.
VIII. a = 13.885, c = 7.0187 Å. R = 0.021 for 1481 reflexions.
IX. a = 13.862, c = 7.0122 Å. R = 0.018 for 1484 reflexions.
X. a = 13.858, c = 6.9972 Å. R = 0.022 for 1507 reflexions.
XI. a = 13.850, c = 6.9897 Å. R = 0.020 for 1441 reflexions.
XII. a = 13.845, c = 6.9841 Å. R = 0.028 for 1444 reflexions.
XIII. a = 13.834, c = 6.9769 Å. R = 0.018 for 1439 reflexions.
XIV. a = 13.834, c = 6.9612 Å. R = 0.020 for 1438 reflexions.

In the crystal structures the lanthanide ions are coordinated by nine water molecules, with the coordination polyhedron having $\bar{6}$ symmetry (Fig. 1). The metal-oxygen distances decrease with increasing atomic number, and correlate well with the ionic radius of the metal. The ethyl sulphate ions are hydrogen-bonded to the water molecules coordinating the metal ion.

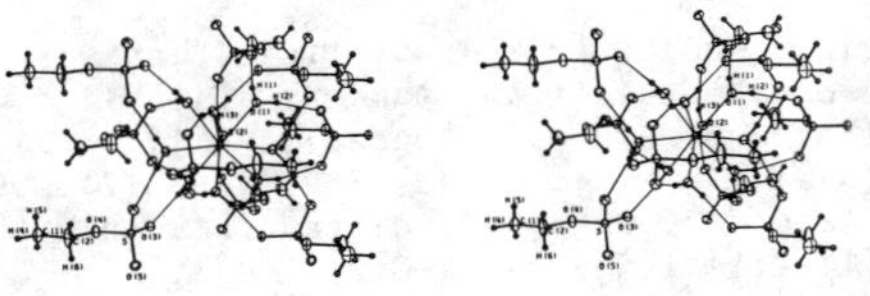

	M–O(1)	M–O(2)
La	2·517 (1)	2·616 (3)
Ce	2·491 (1)	2·600 (3)
Pr	2·470 (2)	2·583 (3)
Nd	2·457 (1)	2·570 (2)
Sm	2·430 (1)	2·550 (4)
Eu	2·415 (1)	2·542 (3)
Gd	2·401 (1)	2·536 (3)
Tb	2·382 (2)	2·527 (3)
Dy	2·371 (1)	2·517 (3)
Ho	2·362 (2)	2·511 (4)
Er	2·352 (2)	2·503 (3)
Tm	2·340 (2)	2·504 (5)
Yb	2·324 (2)	2·503 (3)
Lu	2·318 (2)	2·497 (3)

Fig. 1. $C_6H_{15}GdO_{12}S_3,9H_2O$: stereoview of the environment of the gadolinium ion, and M-O distances.

TRIETHANOLAMMONIUM TRICHLOROANTIMONY(III) CHLORIDE
$C_6H_{16}Cl_4NO_3Sb$ $[N(CH_2CH_2OH)_3]^+.Cl^-.SbCl_3$

D. BRODALLA and A. LIPKA, 1984. Z. Naturforsch., B39, 1167-1171.

Orthorhombic, $Pna2_1$, a = 18.437, b = 10.521, c = 7.197 Å, D_m = 2.01, Z = 4. Mo radiation, R = 0.055 for 1611 reflexions.

The $SbCl_3$ molecules and $[(C_2H_4OH)_3NH]^+$ cations are linked by oxygen bridges (Sb-O 2.743 Å) and these units are further cross linked by the chloride ions as shown in Fig. 1. The Sb-Cl distances in the $SbCl_3$ molecule are in the range 2.392-2.453 Å and the distances to bridging chloride ions are much longer (2.902 and 3.321 Å).

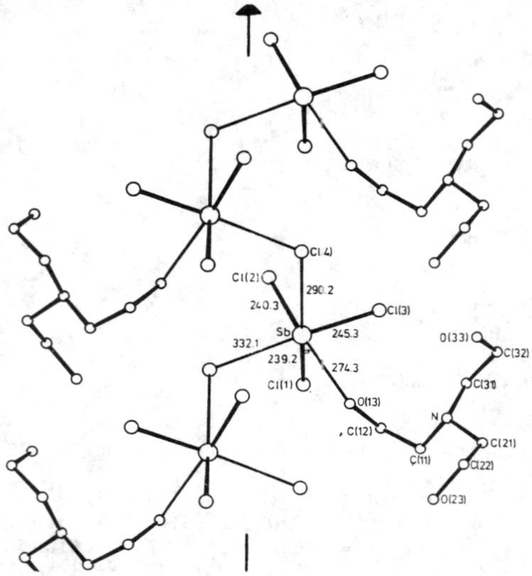

Fig. 1. The $[N(CH_2CH_2OH)_3]^+.Cl^-.SbCl_3$ structure.

N,N,N',N'-TETRAMETHYLETHYLENEDIAMMONIUM TETRACHLOROPLATINATE(II)
$C_6H_{18}Cl_4N_2Pt$ (I)

N,N,N',N'-TETRAMETHYLETHYLENEDIAMMONIUM HEXACHLCROPLATINATE(IV)
$C_6H_{18}Cl_6N_2Pt$ (II)

C. BISI-CASTELLANI, A.-M. MANOTTI-LANFREDI, A. TIRIPICCHIO, L. MARESCA and G. NATILE, 1984. Inorg. Chim. Acta, 90, 155-159.

I. Triclinic, $P\bar{1}$, a = 7.344, b = 8.345, c = 6.216 Å, α = 84.53, β = 109.22, γ = 69.43°, Z = 1. Mo radiation, R = 0.027 for 1302 reflexions.

II. Monoclinic, $P2_1/a$, a = 14.409, b = 12.736, c = 8.601 Å, β = 99.58°, Z = 4. Mo radiation, R = 0.039 for 1559 reflexions.

Compound I consists of (H_2tmen^{2+}) cations and square planar $PtCl_4^{2-}$ anions whilst II consists of the same cations and octahedral $PtCl_6^{2-}$ anions (Fig. 1). Both structures have their cations and anions linked by hydrogen bonds (I: N(1)...Cl(1) 3.190 Å; II: N(1)...Cl(5) 3.48, N(1)...Cl(6) 3.37, N(2)...Cl(2) 3.26 and N(2)...Cl(3) 3.30 Å). The Pt-Cl distances in I are 2.304 and 2.303 Å, and distances within the cation (which is centrosymmetric) are: N(1)-C(1) 1.506, N(1)-C(2) 1.493, N(1)-C(3) 1.512 and C(1)-C(1') 1.509 Å. In II, the Pt-Cl distances range from 2.304 to 2.316 Å, and the C-N distances within the cation from 1.49 to 1.53 Å with C(1)-C(2) 1.53 Å. The conformation of the N-C-C-N system in the cations is trans in both compounds, with dihedral angles of 180° exactly and 166° in I and II respectively.

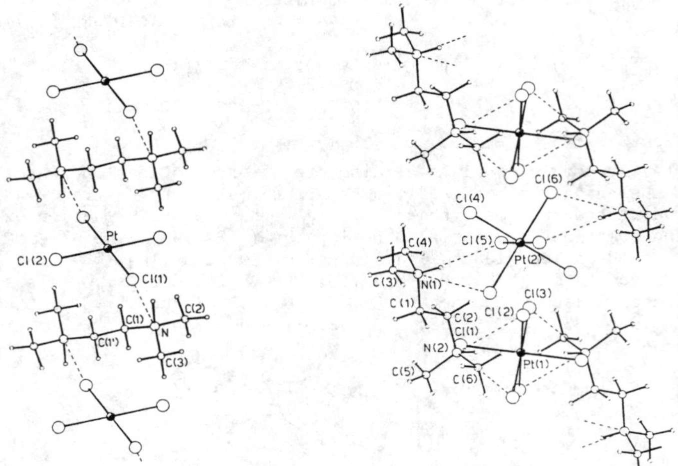

Fig. 1. Chains of cations and anions in $C_6H_{18}Cl_4N_2Pt$ (left), and $C_6H_{18}Cl_6N_2Pt$ (right).

TETRAMETHYLAMMONIUM TETRABROMOCUPRATE(II)
$C_8H_{24}Br_4CuN_2$ (I) $[NMe_4]_2CuBr_4$

TETRAMETHYLAMMONIUM TETRAZINCATE(II)
$C_8H_{24}Br_4N_2Zn$ (II)

P. TROUELAN, J. LEFEBVRE and P. DEROLLEZ, 1984. Acta Cryst., C40, 386-389.

I. Orthorhombic, Pnma, a = 12.600, b = 9.326, c = 15.825 Å, D_m = 1.92, Z = 4. Mo radiation, R = 0.060 for 585 reflexions.

II. Orthorhombic, Pnma, a = 12.681, b = 9.239, c = 16.025 Å, D_m = 1.91, Z = 4. Mo radiation, R = 0.059 for 898 reflexions.

In these isomorphous structures, the space group requires the $[MBr_4]^{2-}$ and the two independent NMe_4 tetrahedra to lie on mirror planes. The mean bond length is 2.426 Å for Cu-Br and 2.447 Å for Zn-Br. The Br-M-Br angles are in the range 102.0-126.6(2) for Cu and 108.6-112.2(1) for Zn.

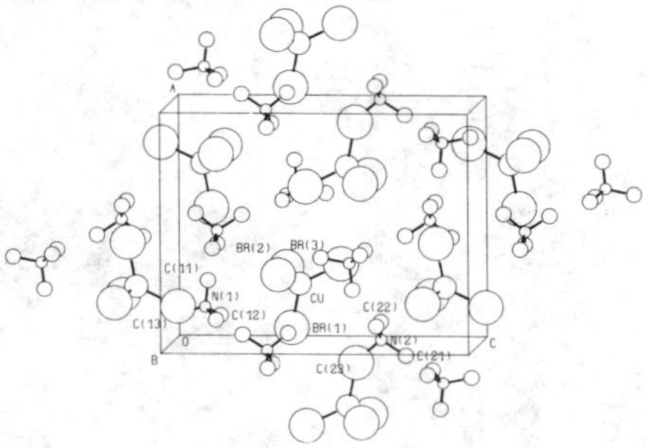

Fig. 1. A view of the $[NMe_4]_2CuBr_4$ crystal structure.

BIS(PIPERAZINIUM) OCTACHLORODIMOLYBDATE TETRAHYDRATE

$C_8H_{24}Cl_8Mo_2N_4,4H_2O$ $(C_4H_{12}N_2{}^{2+})_2(Mo_2Cl_8)^{4-},4H_2O$

I. LEBAN and P. ŠEGEDIN, 1984. Inorg. Chim. Acta, 85, 181-183.

Orthorhombic, Pbca, a = 15.154, b = 13.170, c = 12.208 Å, D_m = 1.96, Z = 4. Mo radiation, R = 0.050.

The crystal structure consists of discrete anions, cations and water molecules (Fig. 1). In the anion the Mo-Cl distances are normal, ranging from 2.445 to 2.463 Å but the Mo-Mo distance is short, being 2.129 Å. The piperazine cation exists in a usual chair conformation with normal bond lengths (C-C 1.52, 1.49 Å; C-N 1.44, 1.49, 1.51, 1.48 Å), bond angles (108 to 114°), and torsional angles (|54|-|60°|). A network of hydrogen bonds (N-H...Cl, N-H...O, O-H...Cl and O-H...O exists between the ions and the water molecules. The N...Cl, N...O, O...Cl and O...O distances are respectively 3.19, 3.23; 2.83, 3.02; 3.20, 3.23; and 2.84 Å.

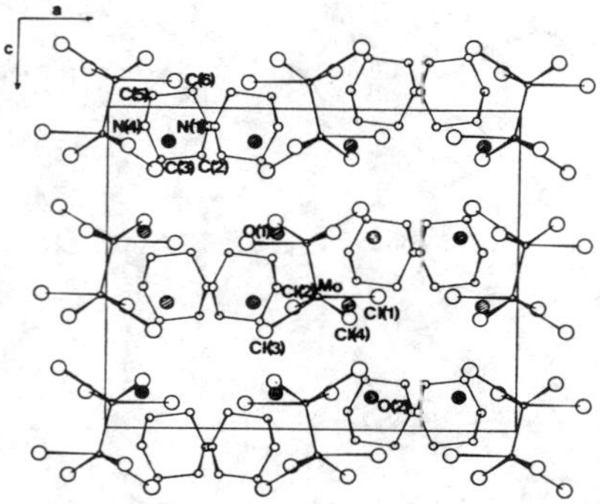

Fig. 1. A view of the unit cell of $C_8H_{24}Cl_8Mo_2N_4,4H_2O$ along [010].

TETRAKIS(DIMETHYLAMMONIUM) HEXAMOLYBDATE DIHYDRATE

$C_8H_{32}Mo_6N_4O_{20},2H_2O$ $[(CH_3)_2NH_2]_4Mo_6O_{20}.2H_2O$

H. TORAYA, F. MARUMO and I. YAMASE, 1984. Acta Cryst., B40, 145-150.

Monoclinic, $P2_1$, a = 19.800, b = 10.294, c = 7.605 Å, β = 90.68°, Z = 2. Cu radiation, R = 0.15 for 944 reflexions.

The structure (Fig. 1) is composed of infinite chains of polymolybdate anions $[Mo_6O_{20}^{4-}]_\infty$, running along the c axis, dimethylammonium cations and water molecules, connecting the chains laterally. $[Mo_6O_{20}^{4-}]_\infty$ is a new type of isopolymolybdate anion.

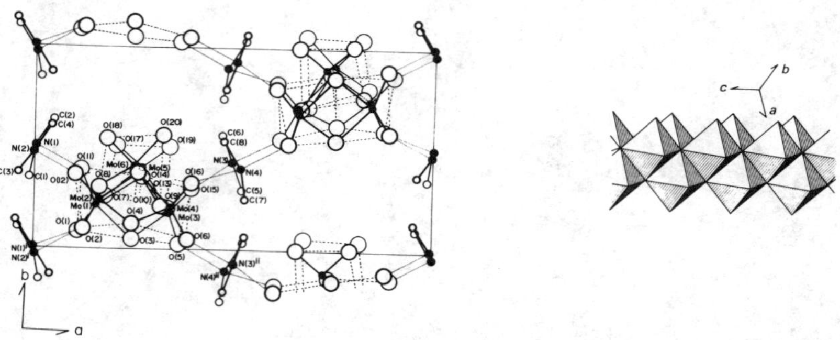

Fig. 1. The crystal structure viewed down c* and a schematic drawing of the $[Mo_6O_{20}^{4-}]_\infty$ chain.

TETRAKIS(DIMETHYLAMMONIUM) β-OCTAMOLYBDATE BIS(N,N-DIMETHYLFORMAMIDE)

$C_8H_{32}Mo_8N_4O_{26}.2C_3H_7NO$ $[NH_2(CH_3)_2]_4[Mo_8O_{26}].2C_3H_7NO$

A.J. WILSON, V. McKEE, B.R. PENFOLD and C.J. WILKINS, 1984. Acta Cryst., C40, 2027-2030.

Monoclinic, C2/c, a = 18.118, b = 14.568, c = 16.288 Å, β = 101.003°, D_m = 2.36, Z = 4. Mo radiation, R = 0.046 for 2222 reflexions.

The octamolybdate anions, $Mo_8O_{26}^{4-}$ (Fig. 1), are centrosymmetric and possess the previously established β-structure. Mean Mo-O bond lengths are: terminal O, 1.698(8); single-bridging O, 1.909(8); triple bridging O, 2.091(8); five-bridging O, 2.319(8) Å. The dimethylformamide solvate molecules are H-bonded (N...O 2.73(4) Å) to one of the two crystallographically independent cations and the nitrogen atoms of both sets of cations form a number of close contacts with the anion terminal oxygen atoms. There is also a useful discussion on relationships among octamolybdate anions.

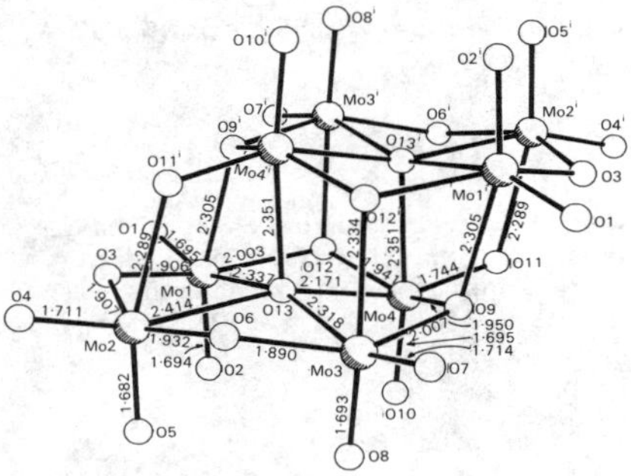

Fig. 1. The β-$Mo_8O_{26}^{4-}$ anion with bond lengths.

BIS(PYRIDINIUM) DIAQUADIBROMOCHROMATE DIBROMIDE

$C_{10}H_{16}Br_4CrN_2O_2$

M.F.C. LADD, L.F. LARKWORTHY, G.A. LEONARD, D.C. POVEY and S.S. TANDON, 1984. J. Chem. Soc. Dalton, 2351-2352.

Triclinic, P$\bar{1}$, a = 8.560, b = 12.745, c = 4.239 Å, α = 91.46, β = 90.81, γ = 74.15°, Z = 1. Mo radiation, R = 0.065 for 1168 reflexions.

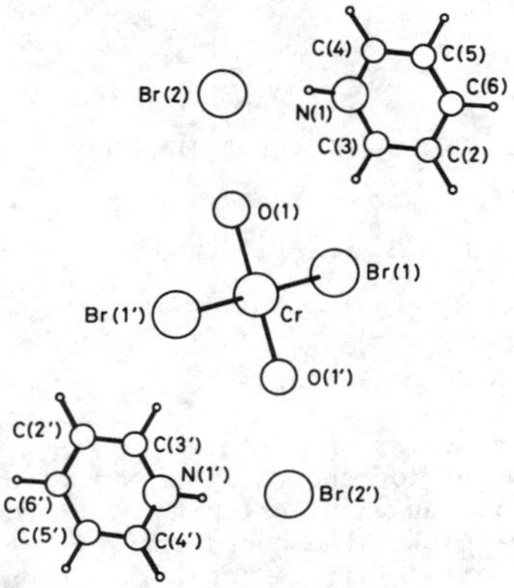

Fig. 1. $C_{10}H_{16}Br_4CrN_2O_2$: the molecular structure.

The Cr atom lies on a centre of symmetry and the $CrBr_2(OH)_2$ units are consequently strictly trans-planar (Fig. 1). The Cr-Br(1) and Cr-OH_2 distances are 2.579(1) and 2.038(2) Å respectively. The Cr-Br(2) distances are almost 5.0 Å. A Br(2)-H(1) distance of 2.34 Å is indicative of a strong hydrogen bond between the pyridinium cation and Br(2).

BENZYLTRIMETHYLAMMONIUM PENTAFLUOROSILICATE
$C_{10}H_{16}F_5NSi$ (I) $[C_6H_5CH_2N(CH_3)_3]^+[SiF_5]^-$

TETRAMETHYLAMMONIUM DIPHENYLTRIFLUOROSILICATE
$C_{16}H_{22}F_3NSi$ (II) $[(CH_3)_4N]^+[(C_6H_5)_2SiF_3]^-$

D. SCHOMBURG and R. KREBS, 1984. Inorg. Chem., 23, 1378-1381.

I. Orthorhombic, Pbcm, a = 12.077, b = 7.855, c = 13.297 Å, Z = 4. Cu radiation, R = 0.039 for 836 reflexions (at 170K).

II. Trigonal, $P3_221$, a = 8.8046, c = 18.4224 Å, Z = 3. Mo radiation, R = 0.041 for 1393 reflexions.

Both structures consist of alkylammonium cations having normal geometry and pentacoordinate anions having trigonal bipyramidal geometry (Fig. 1). The phenyl groups in II occupy equatorial positions. Both anions display exact C_2 symmetry. Bond lengths: for I; Si-F_{eq} 1.602(2) and 1.579(2), Si-F_{ax} 1.646(1) Å; for II; Si-F_{eq} 1.648(2), Si-F_{ax} 1.688(1), Si-C 1.893(2) Å.

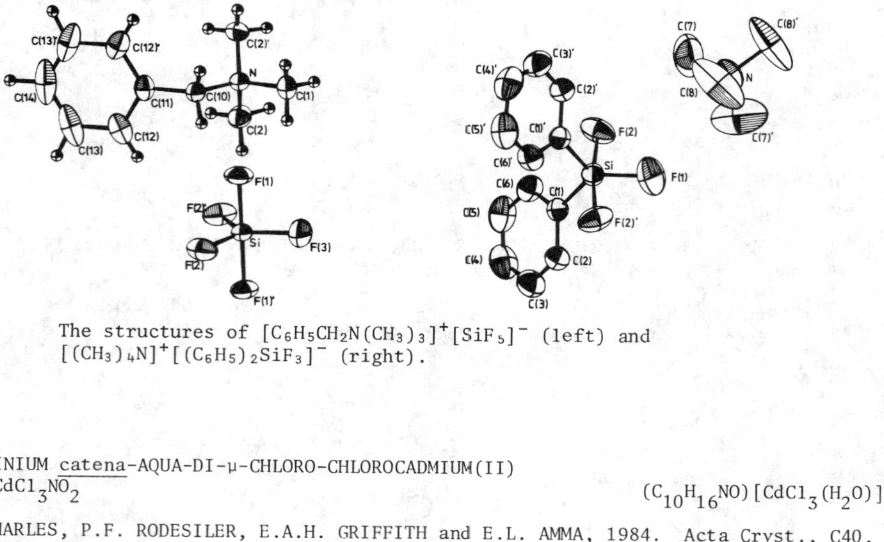

Fig. 1. The structures of $[C_6H_5CH_2N(CH_3)_3]^+[SiF_5]^-$ (left) and $[(CH_3)_4N]^+[(C_6H_5)_2SiF_3]^-$ (right).

EPHEDRINIUM catena-AQUA-DI-μ-CHLORO-CHLOROCADMIUM(II)
$C_{10}H_{18}CdCl_3NO_2$ $(C_{10}H_{16}NO)[CdCl_3(H_2O)]$

N.G. CHARLES, P.F. RODESILER, E.A.H. GRIFFITH and E.L. AMMA, 1984. Acta Cryst., C40, 1676-1678.

Monoclinic, $P2_1/c$, a = 13.686, b = 7.548, c = 14.551 Å, β = 104.16°, D_m = 1.80, Z = 4. Mo radiation, R = 0.040 for 3500 reflexions.

The structure (Fig. 1) contains ephedrinium ions and a linear chain of the type $[CdCl_3(H_2O)]_n^{n-}$. The Cd atoms are octahedrally coordinated with the four equatorial halogens forming the chain with an axial chlorine atom and an axial water molecule. The bond distances and angles within the ephedrinium ion have the expected values. The Cd-Cl distances average 2.619 Å with a total variation of 0.03 Å.

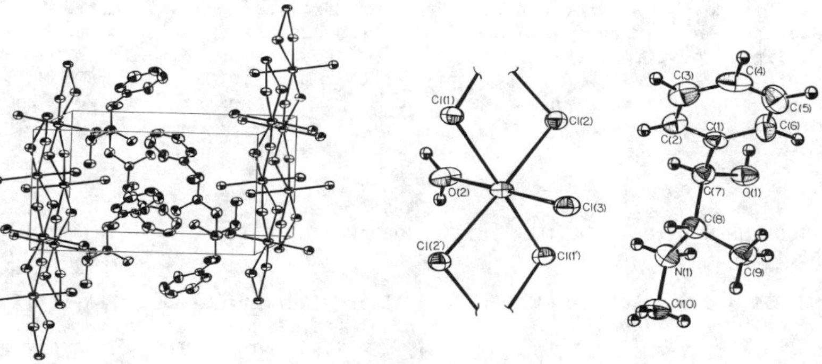

Fig. 1. ORTEP plots of the unit-cell contents for $C_{10}H_{18}CdCl_3NO_2$ and individual views of the ions.

TETRAMETHYLAMMONIUM BIS(μ_2-CYANO)TETRAAQUAMANGANESE(II)TETRACYANOMANGANESE(III) TETRAHYDRATE

$C_{10}H_{20}Mn_2N_7O_4,4H_2O$ (I) $[NMe_4][Mn(H_2O)_4Mn(CN)_6],4H_2O$

TETRAMETHYLAMMONIUM BIS(μ_2-CYANO)TETRAAQUAMANGANESE(II)TETRACYANOIRON(III) TETRAHYDRATE

$C_{10}H_{20}FeMnN_7O_4,4(H_2O)$ (II) $[NMe_4][Mn(H_2O)_4Fe(CN)_6],4H_2O$

H. HENKEL and D. BABEL, 1984. Z. Naturforsch., B39, 880-886.

I. Tetragonal, P4/n, a = 10.669, c = 10.641 Å, Z = 2. Mo radiation, R = 0.0360 for 1874 reflexions.

II. Tetragonal, P4/n, a = 10.670, c = 10.568 Å, Z = 2. Mo radiation, R = 0.0341 for 1362 reflexions.

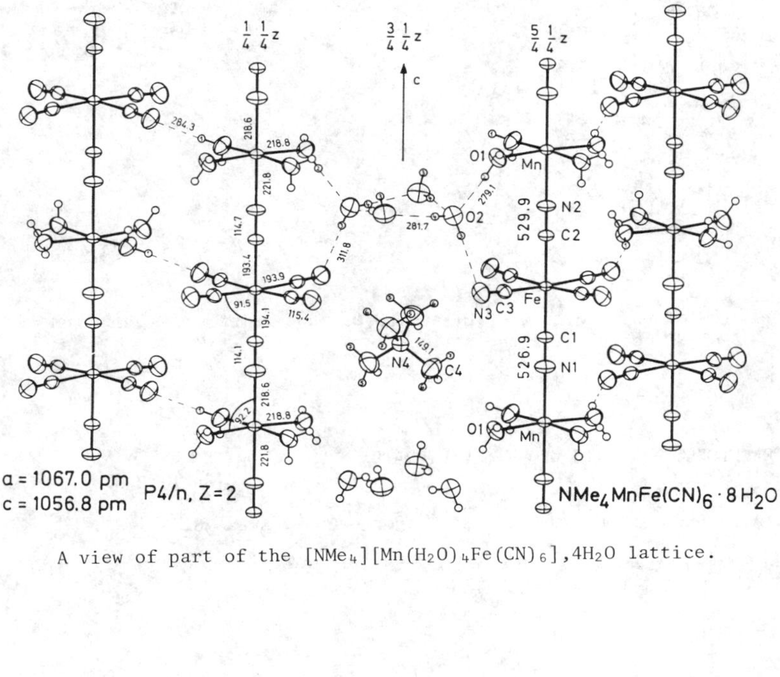

Fig. 1. A view of part of the $[NMe_4][Mn(H_2O)_4Fe(CN)_6],4H_2O$ lattice.

These isomorphous structures are built up from slightly distorted $[M(CN)_6]^{3-}$ octahedra and square-planar $[Mn(H_2O)_4]^{2+}$ cations, stacked one upon the other in nearly eclipsed positions to produce cyano-bridged chains (Fig. 1). The mean bond lengths are: Fe-C 1.938, Mn-C 1.993, C-N 1.147, Mn-N 2.202 and 2.182, Mn-O 2.188 and 2.185 Å. The NMe_4 cations and water molecules occupy positions between the cyano-bridged chains.

BIS(α-PICOLINE N-OXIDE)HYDRONIUM TETRACHLOROAURATE(III)

$C_{12}H_{15}AuCl_4N_2O_2$ $[C_{12}H_{15}N_2O_2]^+[AuCl_4]^-$

M.S. HUSSAIN and S.A.A. AL-HAMOUD, 1984. Inorg. Chim. Acta, 82, 111-117.

Monoclinic, $P2_1/n$, a = 22.376, b = 9.874, c = 7.957 Å, β = 94.98°, Dm = 2.13, Z = 4. Mo radiation, R = 0.039 for 2623 reflexions.

The structure consists of discrete $AuCl_4^-$ and $C_{12}H_{15}N_2O_2^+$ ions (Fig. 1). The dimeric cation has a short hydrogen bond with an O...O distance of 2.393 Å (O(1)-H... O(9)). The dihedral angle between the two planar moieties in the cation is 44.1° and the distances N(2)-O(1) and N(10)-O(9) are respectively 1.341 and 1.348 Å. The anion is planar with an average Au-Cl distance of 2.268 Å and Cl-Au-Cl angles ranging from 88.8 to 91.0°.

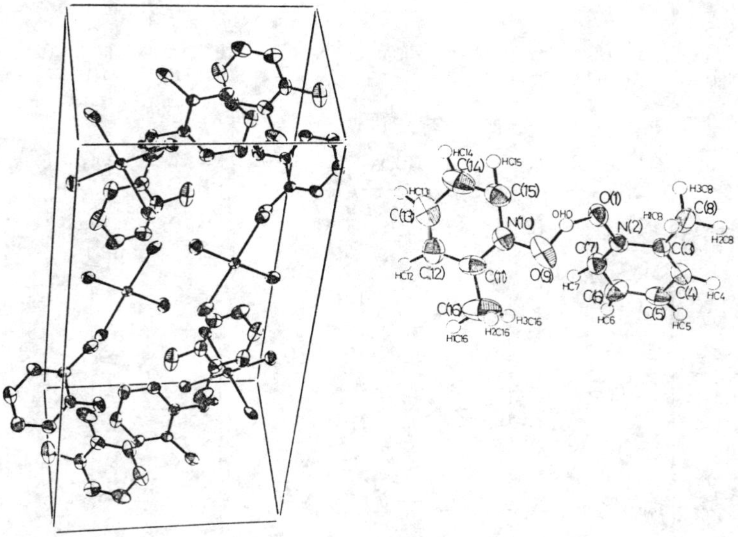

Fig. 1. View of unit cell of $C_{12}H_{15}AuCl_4N_2O_2$, and relative orientation of 2-picNO moieties in the cation.

DIAQUATETRACHLOROTIN(IV) 18-CROWN-6 DIHYDRATE

$C_{12}H_{28}Cl_4O_8Sn,2H_2O$ $[SnCl_4(H_2O)_2]$18-crown-6,$2H_2O$

G. VALLE, A. CASSOL and U. RUSSO, 1984. Inorg. Chim. Acta, 82, 81-84.

Monoclinic, $P2_1/a$, a = 15.753, b = 15.072, c = 12.209 Å, β = 97.77°, Z = 4. Cu radiation, R = 0.13 for 1394 reflexions.

The Sn atom is octahedrally coordinated to four Cl atoms (2.37, 2.37, 2.40 and 2.38 Å) and two water molecules in cis positions (Sn-O 2.16 and 2.10 Å), and this moiety, two water molecules of hydration and two half crown-ethers each with inversion symmetry are connected by a complex network of hydrogen bonds (Fig. 1). O...O contacts are: OH(1)...O(1) 2.68, OH(3)...O(1) 3.01, OH(3)...O(2) 2.83, OH(3)...O(3) 2.89, OH(2)...OH(4) 2.48, OH(2)...O(5) 2.76, OH(4)...O(4) 2.93, OH(4)...O(5) 3.04 and OH(4)...O(6) 2.86 Å.

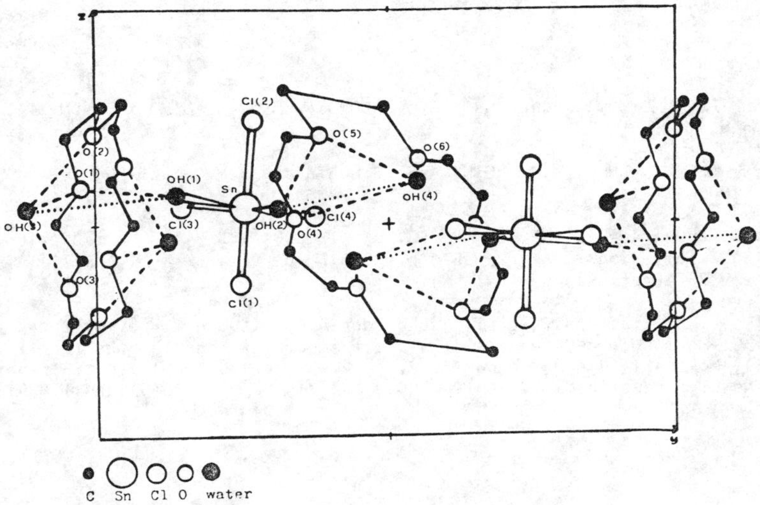

Fig. 1. Perspective view of $C_{12}H_{28}Cl_4O_8Sn,2H_2O$.

DIAQUATETRACHLOROTIN(IV) 18-CROWN-6 CHLOROFORM SOLVATE DIHYDRATE
$C_{12}H_{28}Cl_4O_8Sn,CHCl_3,2(H_2O)$

P.A. CUSACK, B.N. PATEL, P.J. SMITH, D.W. ALLEN and I.W. NOWELL, 1984. J. Chem. Soc. Dalton, 1239-1243.

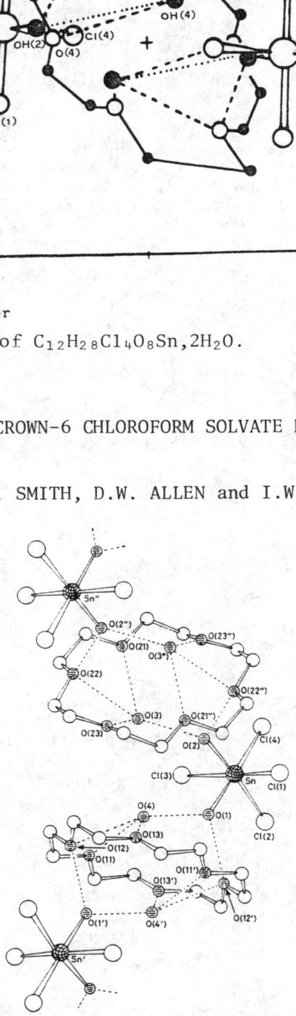

Fig. 1. $C_{12}H_{28}Cl_4O_8Sn,CHCl_3,2(H_2O)$: part of the crystal structure in which dotted lines indicate O...O contacts.

Monoclinic, P2 /n, a = 10.315, b = 13.630, c = 20.649 Å, β = 94.68°, D_m = 1.66, Z = 4. Mo radiation, R = 0.062 for 3142 reflexions.

The water molecules with the octahedral $Sn(OH_2)_2Cl_4$ units (Fig. 1) are cis to one another and are involved in an extensive hydrogen bonding scheme. The latter links together the uncoordinated water molecules, 18-crown-6, and $Sn(OH_2)_2Cl_4$ octahedra to give hydrogen-bonded chains which run parallel to the b-axis. The two crown ether molecules adopt very similar conformations.

TRIS(TETRAMETHYLAMMONIUM) HEXACHLOROBIS(μ-CHLORO)(μ-HYDRIDO)DIMOLYBDENUM(III)

$C_{12}H_{37}Cl_8Mo_2N_3$ $[N(CH_3)_4]_3[Mo_2Cl_8H]$

F.A. COTTON, P.C.W. LEUNG, W.J. ROTH, A.J. SCHULTZ and J.M. WILLIAMS, 1984. J. Am. Chem. Soc. 106, 117-120.

Orthorhombic, Pnma, a = 24.483, b = 9.655, c = 11.820 Å, Z = 4. Mo radiation, R = 0.032 for 2048 reflexions. a = 24.357, b = 9.613, c = 11.589 Å, 1.1611(1) Å neutrons, R = 0.037 for 2521 reflexions (at 18 K).

The $[Mo_2Cl_8H]^{3-}$ anions (Fig. 1) are bisected longitudinally by mirror planes containing Cl-Mo-(μ-H)-Mo-Cl chains of atoms. Important bond lengths (at 18 K): Mo-Mo 2.357(3), Mo-(μ-Cl) 2.480(2) and 2.492(2), Mo-Cl (trans to H) 2.498(3) and 2.481(3), Mo-Cl (trans to Cl) 2.409(2) and 2.393(2), Mo-H 1.816(5) and 1.829(5) Å.

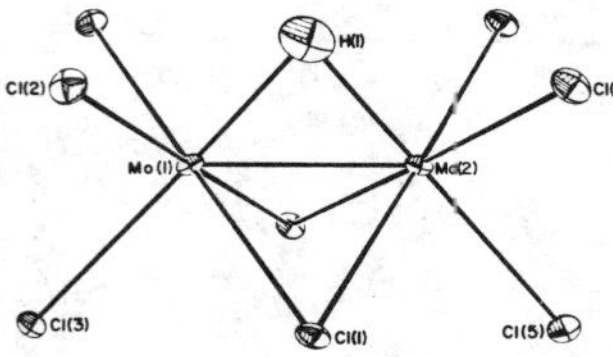

Fig. 1. The $[Mo_2Cl_8H]^{3-}$ anion (18 K neutron study).

BIS(ETHYLENEDIAMINE)BARIUM TRISELENOANTIMONATE

$C_{12}H_{48}Ba_3Sb_2Se_3$ $[Ba(NH_2CH_2CH_2NH_2)_2{}^{2+}]\ [SbSe_3{}^{3-}]_2$

K.-T. KÖNIG, B. EISENMANN and H. SCHÄFER, 1984. Rev. Chim. Miner., 21, 640-647.

Rhombohedral, R$\bar{3}$c, a = 16.144, c = 27.523 Å, Z = 6. Mo radiation, R = 0.049 for 643 reflexions.

Three Ba^{2+} ions, each coordinated to two en-molecules and two $SbSe_3{}^{3-}$ anions form neutral sphere-like clusters (Fig. 1). These are arranged in a cubic close packing sequence ABCA'B'C'. The anions lie on threefold symmetry axes with Sb-Se 2.548(2) Å, Se-Sb-Se 101.7(3)°, and Se...Se 3.952(3) Å (Fig. 1). Relevant distances within the coordination sphere of the Ba^{2+} ion are: Ba...Se 3.358(2), Ba-N(1) 2.94(2), Ba-N(4) 2.90(2), Ba...C(3) 3.74(3), Ba...C(4) 3.73(2) Å. The N(1)-Ba-N(4) angle is 62.0(6)°.

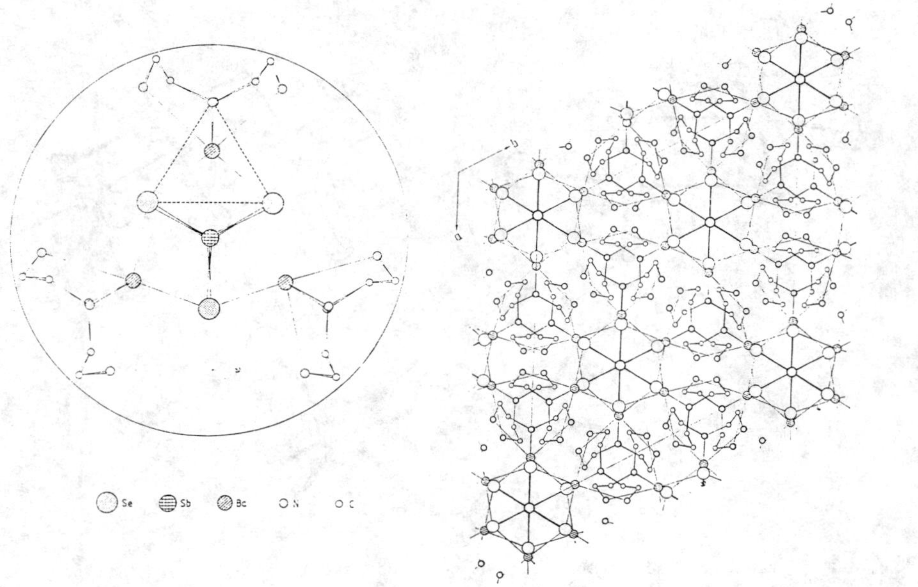

Fig. 1. View of metal ion coordination in $C_{12}H_{48}Ba_3Sb_2Se_3$ and the projection of the unit cell down the c-direction.

BIS(N-METHYLANILINIUM) BIS[(μ_2-BROMO)TRIBROMOARSENATE(III)]
$C_{14}H_{20}As_2Br_8N_2$ (I)

PYRIDINIUM TETRABROMOARSENATE(III)
$C_5H_6AsBr_4N$ (II)

TRIS(PYRIDINIUM) NONABROMODIARSENATE(III) ACETONITRILE SOLVATE
$C_{15}H_{18}As_2Br_9N_3$,1.5(C_2H_3N) (III)

J. KAUB and W.S. SHELDRICK, 1984. Z. Naturforsch., B39, 1257-1261.

I. Monoclinic, $P2_1/c$, a = 5.938, b = 12.774, c = 18.034 Å, β = 91.22°, Z = 2. Mo radiation, R = 0.054 for 1656 reflexions.

II. Triclinic, $P\bar{1}$, a = 9.303, b = 9.667, c = 7.306 Å, α = 115.29, β = 109.91, γ = 72.84°, Z = 2. Mo radiation, R = 0.053 for 1421 reflexions.

III. Triclinic, $P\bar{1}$, a = 20.309, b = 9.658, c = 9.753 Å, α = 120.31, β = 100.16, γ = 84.99°, Z = 2. Mo radiation, R = 0.141 for 2100 reflexions.

The compounds are shown to contain discrete centrosymmetric $As_2Br_8^{2-}$ ions (I), infinite $(AsBr_4^-)_n$ chains (II) and discrete $As_2Br_9^{2-}$ ions (III) (Fig. 1). Main dimensions are: for (I) As-Br (bridge) 3.152 and 3.203(1), As-Br (terminal) 2.347 (axial), 2.372 and 2.387(1) Å; for (II) As-Br(1) 3.130, As-Br(1') 2.688, As-Br(3) 2.690, As-Br(3') 3.129, As-Br(2) 2.393, As-Br(4) 2.395 Å; for (III) As-Br (bridge) 2.86-3.01(2), As-Br (terminal) 2.42-2.49(2) Å.

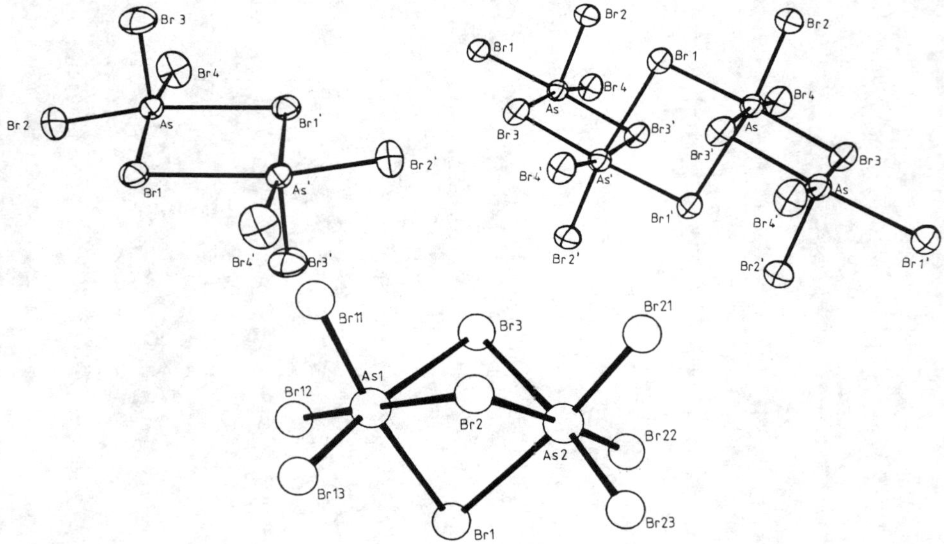

Fig. 1. Views of the $As_2Br_8^{2-}$, $(AsBr_4^-)_n$ and $As_2Br_9^{2-}$ ions.

BIS(N-METHYLANILINIUM) BIS[(μ_2-CHLORO)TRICHLOROARSENATE(III)]
$C_{14}H_{20}As_2Cl_8N_2$ (I)

TRIS(PYRIDINIUM) TRIS(μ_3-CHLORO)HEXACHLORODIARSENATE(III) ACETONITRILE SOLVATE
$C_{15}H_{18}As_9Cl_9N_3,C_2H_3N$ (II)

BIS(PYRIDINIUM) BIS(μ_2-CHLORO)(μ_2-OXO)TETRACHLORODIARSENATE(III)
$C_{10}H_{12}AsCl_6N_2O$ (III)

J. KAUB and W.S. SHELDRICK, 1984. Z. Naturforsch., B39, 1252-1256.

I. Monoclinic, $P2_1/c$, a = 5.679, b = 12.517, c = 17.466 Å, β = 90.44°, Z = 2. Mo radiation, R = 0.030 for 1758 reflexions.

II. Monoclinic, $P2_1/m$, a = 9.710, b = 21.966, c = 14.396 Å, β = 90.90°, Z = 4. Mo radiation, R = 0.085 for 1834 reflexions.

III. Triclinic, $P\bar{1}$, 9.310, b = 9.602, c = 11.534 Å, α = 112.25, β = 98.04, γ = 97.42°, Z = 2. Mo radiation, R = 0.045 for 3434 reflexions.

The compounds are shown to contain discrete $As_2Cl_8^{2-}$ (I), $As_2Cl_9^{3-}$ (II) and $As_2Cl_6O^{2-}$ (III) ions (Fig. 1). Main bond lengths in (I) are As-Cl (bridge) 3.035 and 3.063(1), As-Cl (terminal) 2.197-2.210(1) Å. In (II) there are two independent anions each with a crystallographic mirror plane containing the bridging chlorine atoms, with As-Cl (bridge) 2.829-3.080(4), As-Cl (terminal) 2.214-2.306(4) Å. In the anion of III, the principal dimensions are As-O 1.765 and 1.777(1), As-Cl (bridge) 2.836-2.965(1), As-Cl (terminal) 2.242-2.268(1) Å, As-O-As 119.3(1), As-Cl-As 62.5 and 64.9(1)°.

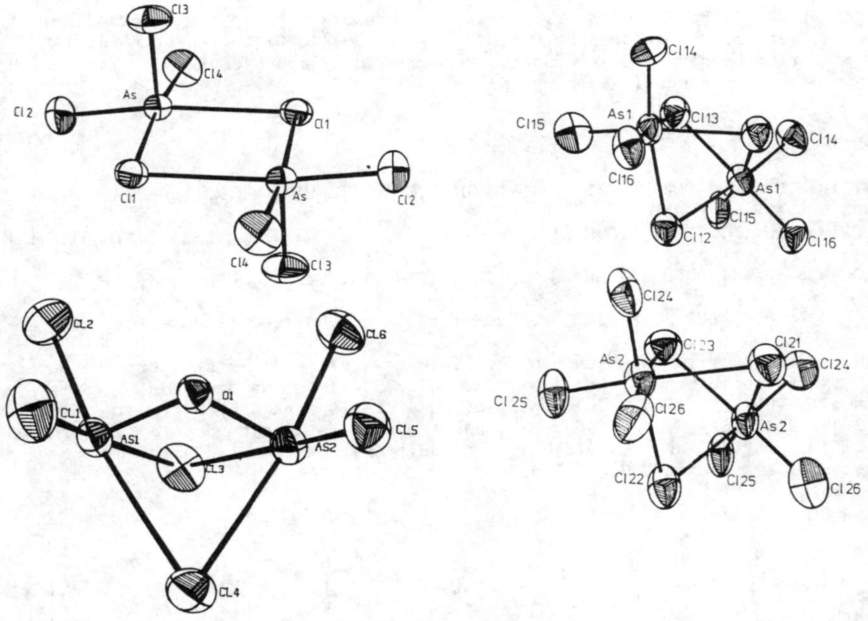

Fig. 1. Views of the $As_2Cl_8^{2-}$, $As_2Cl_9^{3-}$ and $As_2Cl_6O^{2-}$ ions.

BIS(TETRAMETHYLAMMONIUM) SODIUM HEXACYANOFERRATE MONOHYDRATE
$C_{14}H_{24}FeN_8Na,H_2O$ $[NMe_4]_2.Na.[Fe(CN)_6],H_2O$

M. WITZEL and D. BABEL, 1984. Z. Naturforsch., B39, 201-206.

Orthorhombic, Pnma, a = 18.639, b = 8.617, c = 13.172 Å, Z = 4. Mo radiation, R = 0.041 for 2515 reflexions.

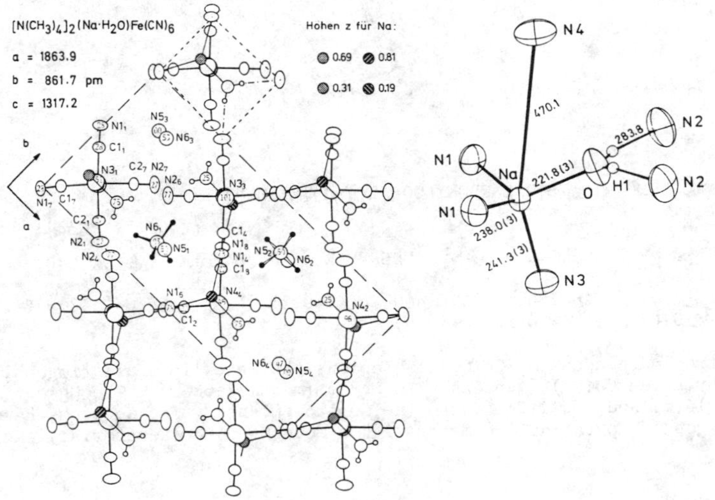

Fig. 1. [001] projection of $C_{14}H_{24}FeN_8Na,H_2O$ and details of Na coordination.

The structure has an elpasolite-related framework of slightly deformed $Fe(CN)_6^{3-}$ octahedra, with mean distances Fe-C 1.940, C-N 1.139 Å. Distorted octahedral N_6 cavities formed by six $Fe(CN)_6^{3-}$ units are occupied by monohydrated Na^+ ions which have distorted tetrahedral geometry (Na-N 2.380-2.413(3), Na-O 2.218(3) Å).

TRIS(PYRIDINIUM) (μ_3-SULPHIDO)TRIS(μ_2-DISULPHIDO)HEPTACHLOROTRIMOLYBDENUM HEMIHYDRATE ETHANOL SOLVATE

$C_{15}H_{18}Cl_7Mo_3N_3S_7$,0.5($H_2O$),2($C_2H_6O$) $(C_5H_6N)_3(Mo_3Cl_7S_7)$,0.5(H_2O),2(C_2H_6O)

LU SHAOFANG, SHANG MAOYU and HUANG JINLING, 1984. J. Struct. Chem., 3, 9-12.

Cubic, Pa3, a = 19.605 Å, Z = 8. Mo radiation, R = 0.041 for 1120 reflexions.

In the anion (Fig. 1) which has threefold crystallographic symmetry the Mo atoms form an exact equilateral triangle with Mo-Mo 2.750(1) Å. On one side there is a μ_3-S atom (Mo-S 2.356(1) Å). Each pair of Mo atoms is linked by a bridging S_2 ligand (Mo-S 2.384-2.493(3), S-S 2.029(4) Å). Each Mo atom is also bonded to two terminal Cl atoms (Mo-Cl 2.446 and 2.499(3) Å). An additional Cl atom is positioned above an S_3 face (Fig. 1) so that S-Cl is 2.969(3) Å.

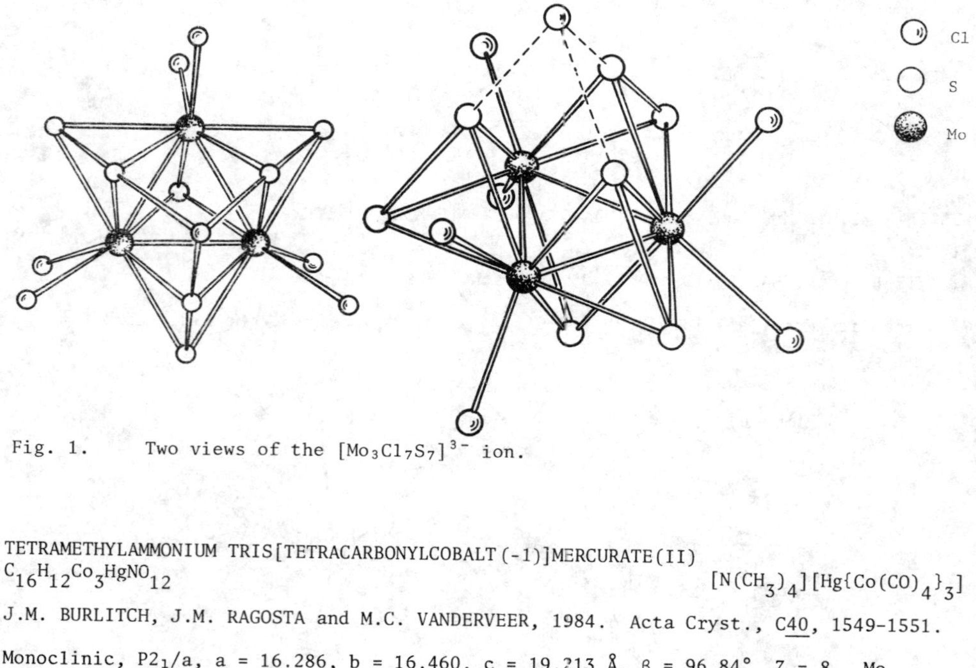

Fig. 1. Two views of the $[Mo_3Cl_7S_7]^{3-}$ ion.

TETRAMETHYLAMMONIUM TRIS[TETRACARBONYLCOBALT(-1)]MERCURATE(II)

$C_{16}H_{12}Co_3HgNO_{12}$ $[N(CH_3)_4][Hg\{Co(CO)_4\}_3]$

J.M. BURLITCH, J.M. RAGOSTA and M.C. VANDERVEER, 1984. Acta Cryst., C40, 1549-1551.

Monoclinic, $P2_1/a$, a = 16.286, b = 16.460, c = 19.213 Å, β = 96.84°, Z = 8. Mo radiation, R = 0.048 for 4470 reflexions.

In the two independent $Hg(Co(CO)_4)_3^-$ anions the Hg cations are trigonal planar coordinated by three $Co(CO)_4^-$ anions with Hg-Co (mean) 2.686 Å. The Hg cations are displaced 0.016(2) and 0.024(2) Å from the Co atom planes. The Co atoms have distorted trigonal bipyramidal geometry with a mean Hg-Co-C (equatorial) angle of 80.9°.

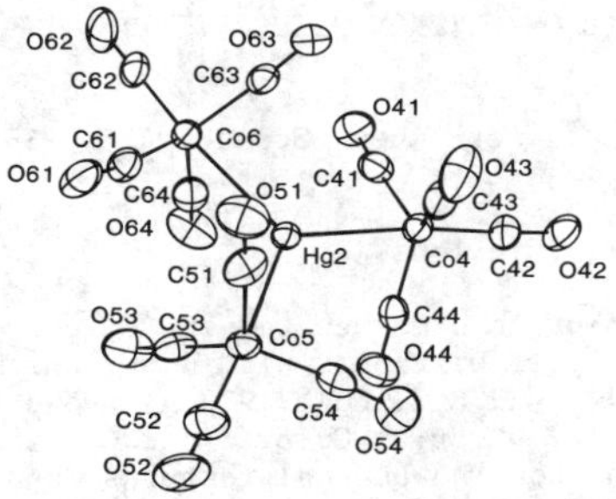

Fig. 1. One of the $Hg[Co(CO)_4]_3^-$ anions with atom numbering.

TETRA(n-BUTYL)AMMONIUM TETRAIODOALUMINATE

$C_{16}H_{36}AlI_4N$ $[NBu_4][AlI_4]$

R.D. ROGERS, J.C. BAKER and J.L. ATWOOD, 1984. J. Cryst. Spectrosc. Res., 14, 333-339.

Monoclinic, $P2_1/n$, a = 11.661, b = 20.828, c = 11.674 Å, β = 98.39°, Z = 4. Mo radiation, R = 0.051 for 1587 reflexions.

The structure consists of discreet $[NBu^n_4]^+$ and $[AlI_4]^-$ ions. The $[AlI_4]^-$ anion is approximately tetrahedral and exhibits an average Al-I bond length of 2.52(1) Å (range 2.506(8) to 2.532(7) Å). The closest cation-anion, I...C, contact is >4 Å. The compound was isolated from the reaction of I_2 with $[NBu^n_4][Al_2Me_6I]$.

TETRABUTYLAMMONIUM BROMOCHLOROCUPRATE(I)

$C_{16}H_{36}BrClCuN$ $N(C_4H_9)_4^+[ClCuBr]^-$

M. ASPLUND, S. JAGNER and M. NILSSON, 1984. Acta Chem. Scand., A38, 57-62.

Monoclinic, C2/c, a = 13.051, b = 9.942, c = 15.859 Å, β = 92.72°, Z = 4. Mo radiation, R = 0.038 for 1423 reflexions (at 168K).

The cation has crystallographic twofold symmetry whilst the anion is disordered and sited on an inversion centre (Fig. 1). The Cu atom is on the centre of symmetry and the Br and Cl atoms occupy their sites statistically with occupancy factors of 0.5. The apparent Cu-Cl distance agrees with that determined in tetrabutylammonium dichlorocuprate, 2.107 Å (1) but the Cu-Br distance appears to be slightly shorter than that in the dibromocuprate, 2.226 Å (1). The cation (Fig. 1) has the usual staggered conformation and the bond distances are normal (N-C(1) 1.520, N-C(5) 1.530, C-C distances range from 1.510 to 1.532 Å).

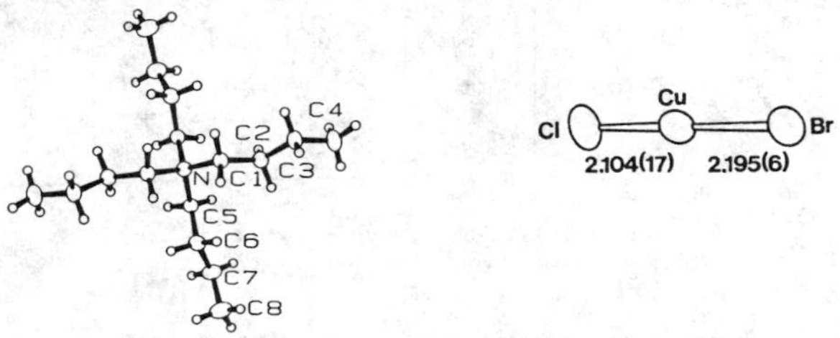

Fig. 1. The cation (left) and anion (right) in $C_{16}H_{36}BrClCuN$.

1. Structure Reports, 50B.

BIS(TETRAETHYLAMMONIUM) DI-μ-BROMO-DIBROMODICUPRATE(I)

$C_{16}H_{40}Br_4Cu_2N_2$ $2[N(C_2H_5)_4]^+[Cu_2Br_4]^{2-}$

M. ASPLUND and S. JAGNER, 1984. Acta Chem. Scand., A38, 135-139.

Monoclinic, $P2_1/c$, a = 8.375, b = 13.817, c = 11.086 Å, β = 97.03°, Z = 2. Mo radiation, R = 0.060 for 1328 reflexions.

The anion is a centrosymmetric dimer containing Cu coordinated in distorted trigonal planar fashion (Fig. 1). Distances within the anion are: Cu-Br(1) 2.319, Cu-Br(2) 2.454, Cu-Br(2') 2.441 and Cu...Cu' 2.937 Å. Bond angles are: Br(1)-Cu-Br(2) 125.7, Br(1)-Cu-Br(2') 127.9, Br(2)-Cu-Br(2') 106.3 and Cu-Br(2)-Cu' 73.7°. The tetraethylammonium cation shows no unusual features, and the closest contact between anion and cation is Cu...C(3), 3.56 Å. The N-C distances range from 1.53 to 1.54 Å and the C-C distances from 1.51 to 1.56 Å. The angles about N range from 108 to 112°.

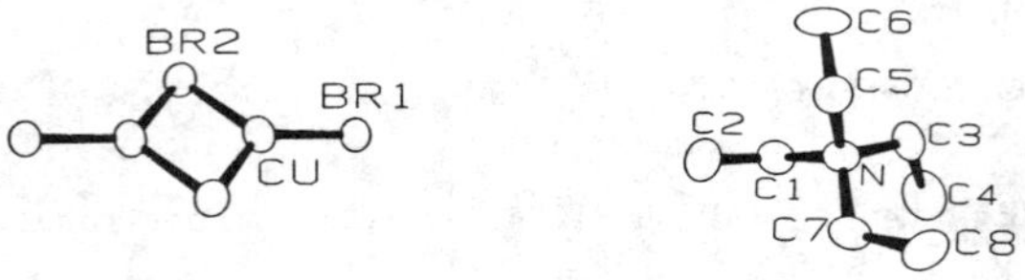

Fig. 1. The anion (left) and cation (right) in $C_{16}H_{40}Br_4Cu_2N_2$.

BIS(TETRAETHYLAMMONIUM) (μ_3-CHLORO)(μ_3-OXO)TRIS(μ_2-CHLORO)HEXACHLOROTRIMOLYBDENUM ETHANOL SOLVATE

$C_{16}H_{40}Cl_{10}Mo_3N_2O,C_2H_6O$ $(Et_4N)_2(Mo_3Cl_{10}O),C_2H_6O$

SHANG MAOYU, HUANG JINLING and LU JIAXI, 1984. J. Struct. Chem., 3, 13-16.

Monoclinic, $P2_1/c$, a = 11.379, b = 20.766, c = 14.487 Å, β = 91.49°, Dm = 1.81, Z = 4. Mo radiation, R = 0.040 for 3261 reflexions.

In the anion (Fig. 1), three Mo atoms form an isosceles triangle (Mo-Mo 2.601, 2.608, 2.642(1) Å) with a μ_3-O and a μ_3-Cl atom above and below the Mo_3 triangle with Mo-O 2.082, 2.088, 2.068(5), Mo-Cl 2.484, 2.491, 2.514(2) Å. Each Mo pair is bridged by a Cl atom (Mo-Cl 2.436-2.447(2) Å) to form a planar Mo_3Cl_3 unit. Coordination at each Mo atom is completed by two terminal Cl atoms (Mo-Cl 2.350-2.405(2) Å).

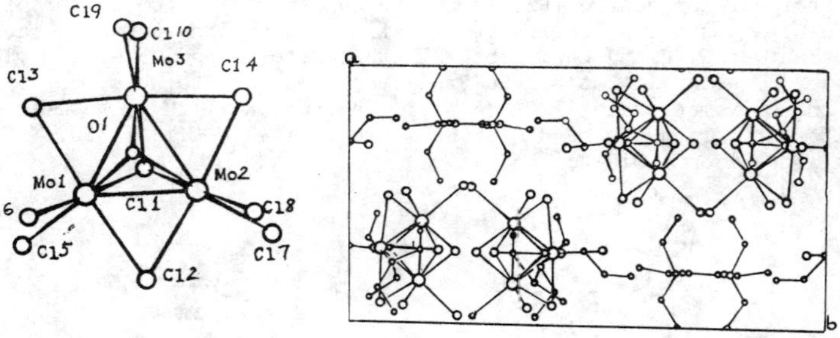

Fig. 1. The $[Mo_3Cl_{10}O]^-$ ion and a view of the crystal structure.

TETRAETHYLAMMONIUM DI-μ-FLUORO-BIS[AQUADIFLUOROOXOVANADATE(IV)]

$C_{16}H_{44}F_6N_2O_4V_2$ $[N(C_2H_5)_4]_2^{2+}[V_2O_2F_6(H_2O)_2]^{2-}$

A. DEMSAR and P. BUKOVEC, 1984. Croat. Chem. Acta, 57, 673-677.

Monoclinic, $P2_1/n$, a = 7.088, b = 13.166, c = 13.624 Å, β = 97.58°, D_m = 1.42, Z = 2. Mo radiation, R = 0.058 for 1523 reflexions.

The structure contains discrete dinuclear anions with crystallographic inversion symmetry (Fig. 1). These dimeric units are linked to form chains by hydrogen bonds (O-H...F 2.591 and 2.679 Å). The geometry about vanadium is distorted octahedral with V-F distances from 1.920 to 2.209 Å, V-O 1.594, V-OH_2 2.083, V...V 3.328 Å, and the angle V-F-V 106.0°.

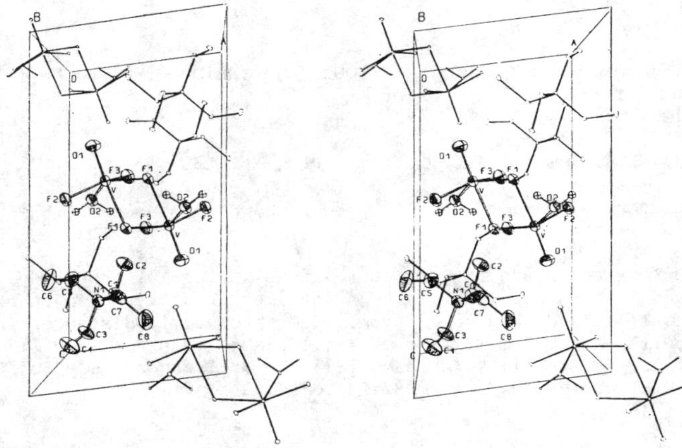

Fig. 1. Stereoscopic view of the unit cell of $C_{16}H_{44}F_6N_2O_4V_2$.

BIS(TETRAETHYLAMMONIUM) DICHLORO-BIS(μ_2-HYDROGENPHOSPHATE)BIS(μ_2-DIHYDROGENPHOSPHATE)-DIPLATINATE MONOHYDRATE

$C_{16}H_{46}Cl_2N_2O_{16}P_4Pt_2,H_2O$ $[(C_2H_5)_4N^+]_2[Pt_2(H_2PO_4)_2(HPO_4)_2Cl_2]^{2-},H_2O$

D.P. BANCROFT, F.A. COTTON, L.R. FALVELLO, S. HAN and W. SCHWOTZER, 1984. Inorg. Chim. Acta, 87, 147-153.

Triclinic, $P\bar{1}$, a = 10.136, b = 11.660, c = 8.392 Å, α = 111.30, β = 88.04, γ = 109.47°, Z = 1. Mo radiation, R = 0.047 for 2146 reflexions.

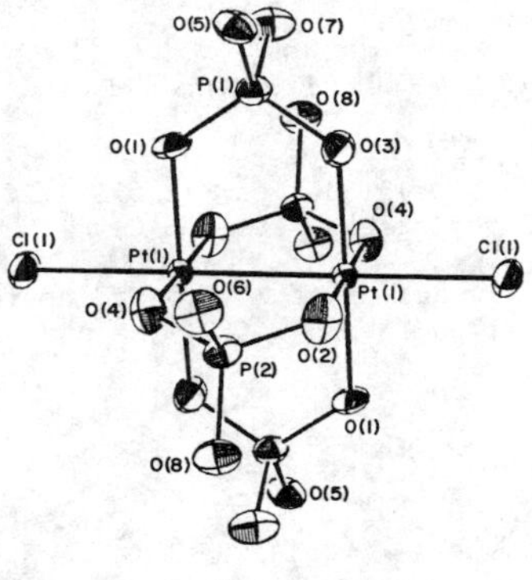

Fig. 1. The molecular structure of the anion in $C_{16}H_{46}Cl_2N_2O_{16}P_4Pt_2$.

The anion has crystallographic inversion symmetry (Fig. 1) with a Pt-Pt bond length of 2.529 Å. The Cl atoms are coordinated axially with a Pt-Cl distance of 2.448 Å and an angle Pt-Pt-Cl of 179.7°. The presence of $H_2PO_4^-$ and HPO_4^{2-} groups is established by reference to the P-O bond lengths. In the group containing P(2), the bonds P(2)-O(6) and P(2)-O(8) are respectively 1.564 and 1.489 Å, suggesting P-OH and P=O, constituting an HPO_4^{2-} group. In the group containing P(1), the bonds P(1)-O(5) and P(1)-O(7) are 1.532 and 1.560 Å respectively which are not significantly different, and both are close to standard P-OH distances, suggesting an $H_2PO_4^-$ group. The water molecule forms two hydrogen bonds, one with O(8) and one with Cl(1) (O...O 2.57, O...Cl 3.10 Å respectively).

BIS(TETRAETHYLAMMONIUM) HEXAIODOBIS(μ_2-IODO)(μ_2-HYDRIDO)DIMOLYBDENUM(II) OXONIUM DIHYDRATE

$C_{16}H_{48}I_8Mo_2N_2O_3$ $(H_7O_3)[(C_2H_5)_4N]_2[Mo_2I_8H]$

A. BINO and S. LUSKI, 1984. Inorg. Chim. Acta, 86, L35-L36.

Orthorhombic, Pnma, a = 32.683, b = 11.384, c = 10.868 Å, Z = 4. R = 0.044 for 1755 reflexions.

The dinuclear anion has the structure shown in Fig. 1, the bridging H atom was not located and is thus not shown. The anion lies on a crystallographic mirror plane which contains I(1), I(2) and the H atom. The interatomic distances in the anion are: Mo-Mo' 2.408, I(1)-Mo 2.809, I(2)-Mo 2.818, I(3)-Mo 2.869, I(4)-Mo 2.759, I(5)-Mo 2.758 Å. The angles Mo-I(1)-Mo' and Mo-I(2)-Mo' are respectively 50.76 and 50.58° whilst the I-Mo-I angles range from 87.34 to 94.99°.

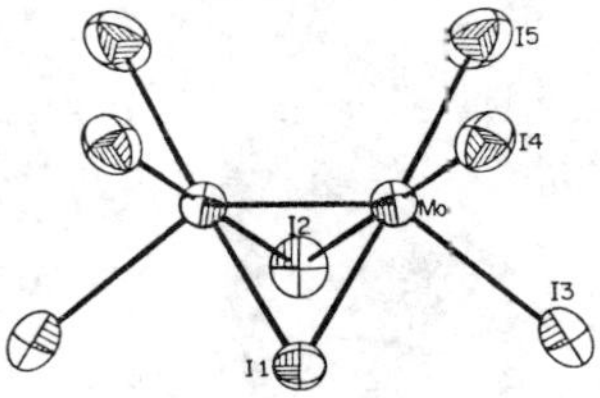

Fig. 1. Structure of the anion in $C_{16}H_{48}I_8Mo_2N_2O_3$.

TETRAKIS(TETRAMETHYLAMMONIUM) DITINHEXATELLURIDE

$C_{16}H_{48}N_4Sn_2Te_6$ $[(CH_3)_4N]_4Sn_2Te_6$

J.C. HUFFMAN, J.P. HAUSHALTER, A.M. UMARJI, G.K. SHENOY and R.C. HAUSHALTER, 1984. Inorg. Chem., 23, 2312-2315.

Monoclinic, $P2_1/c$, a = 11.229, b = 13.590, c = 12.671 Å, β = 116.96°, Z = 2. Mo radiation, R = 0.030 for 2643 reflexions (at -168°C).

The $Sn_2Te_6^{4-}$ anion (Fig. 1) has exact $\bar{1}$ (C_i) and approximate D_{2h} symmetry. The terminal Sn-Te distances are 2.700(2) and 2.689(1) Å and bridging Sn-Te distances are 2.794(1) and 2.804(1) Å.

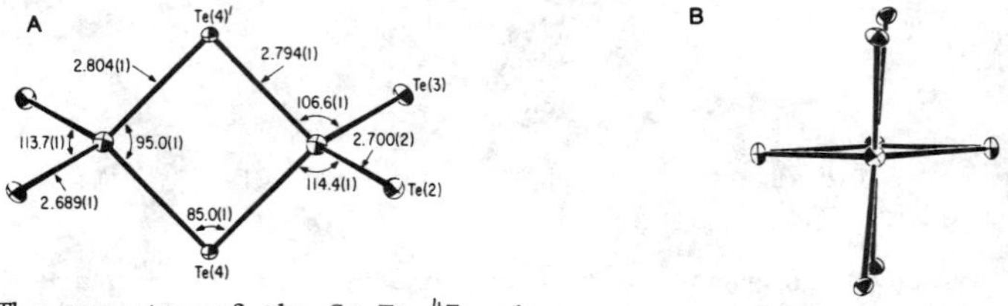

Fig. 1. The geometry of the $Sn_2Te_6^{4-}$ anion.

TRIS[2-(1',3'-DITHIOL-2'-YLIDENE)-1,3-DITHIOLE] HEXACHLOROSTANNATE(IV)
$C_{18}H_{12}Cl_6S_{12}Sn$

K. KONDO, G.-E. MATSUBAYASHI, T. TANAKA, H. YOSHIOKA and K. NAKATSU, 1984. J. Chem. Soc. Dalton, 379-384.

Tetragonal, P4/mbm, a = 11.807, c = 11.861 Å, D_m = 1.90, Z = 2. Mo radiation, R = 0.040 for 876 reflexions.

The X-ray analysis revealed a layer structure consisting of tetrathiafulvalene (ttf) trimers and $SnCl_6^{2-}$ anions. The octahedral $SnCl_6^{2-}$ anions are located at (0,0,0) and (0.5,0.5,0) and have Sn-Cl distances of 2.437(2), and 2.439(2) Å. The ttf molecules are arranged so that their planes are perpendicular to the c plane. The ttf molecules constitute a trimeric unit with the parallel planes separated by the same distance and adjacent trimers perpendicular to each other. Interplanar spacing in the trimer is 3.490 Å. The closest S...S distance between adjacent ttf trimers perpendicular to each other is 3.705 Å. The closest S...Cl distance is 3.367(2) Å.

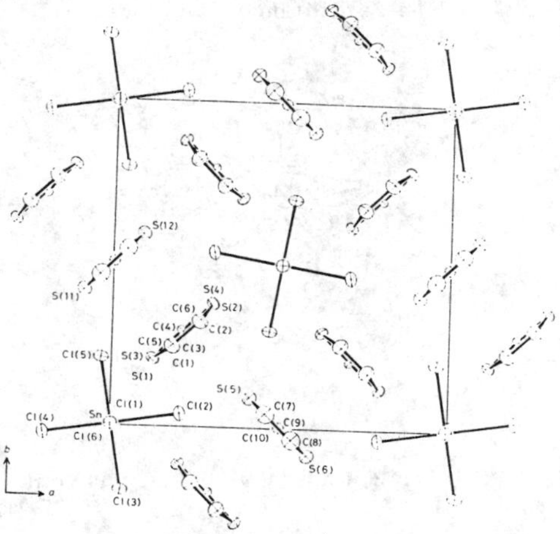

Fig. 1. $C_{18}H_{12}Cl_6S_{12}Sn$: the crystal structure projected along c.

CHLOROTRIPHENYLPHOSPHONIUM 5,5,5,5-TETRACHLORO-1,4-DIAZA-2,3-DITHIA-5-RHENACYCLO-PENTANE
$C_{18}H_{15}Cl_5N_2PReS_2$ $[ClPPh_3][ReCl_4(N_2S_2)]$

W. HILLER, J. MOHYLA, J. STRAHLE, H.G. HAUCK and K. DEHNICKE, 1984. Z. Anorg. Chem., 514, 72-78.

Monoclinic, $P2_1/n$, a = 15.865, b = 10.260, c = 16.235 Å, β = 112.69°, Z = 4. Mo radiation, R = 0.038 for 2861 reflexions (at 233 K).

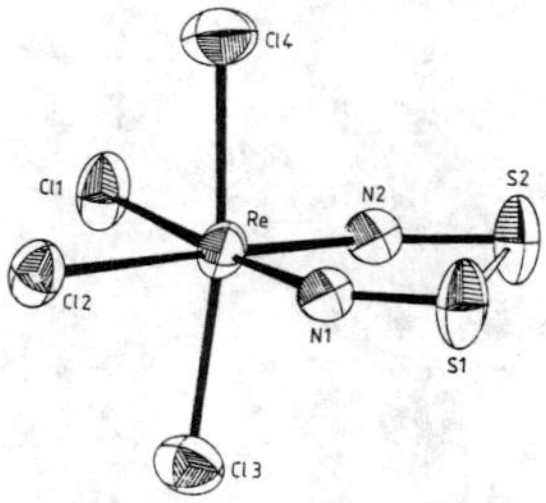

Fig. 1. A view of the $[ReCl_4(N_2S_2)]^-$ ion.

The Re atom has octahedral coordination (Fig. 1) with Re-Cl 2.374-2.394(2), Re-N 1.828 and 1.795(5), N-S 1.511 and 1.528(5) and S-S 2.530(2) Å. The Re-N and N-S bond lengths are in range of double bonds; the S-S distance is very long.

TETRAETHYLAMMONIUM (μ_3-BISMUTHIDO)NONACARBONYL(μ_3-CARBONYL)-triangulo-TRIFERRATE(1-)
$C_{18}H_{20}BiFe_3NO_{10}$ $[(C_2H_5)_4N][Fe_3Bi(CO)_9(\mu_3\text{-}CO)]$

K.H. WHITMIRE, C.B. LAGRONE, M.R. CHURCHILL, J.C. FETTINGER and L.V. BIONDI, 1984. Inorg. Chem., 23, 4227-4232.

Monoclinic, $P2_1/n$, a = 14.659, b = 12.126, c = 14.773 Å, β = 109.10°, Z = 4. Mo radiation, R = 0.054 for 3446 reflexions.

The $[Fe_3Bi(CO)_9(\mu_3\text{-}CO)]^-$ anion (Fig. 1) has approximate C_3 symmetry. Fe-Bi distances average 2.650(2) Å and Fe-Fe distances average 2.642(7) Å. The Et_4N^+ cation is disordered.

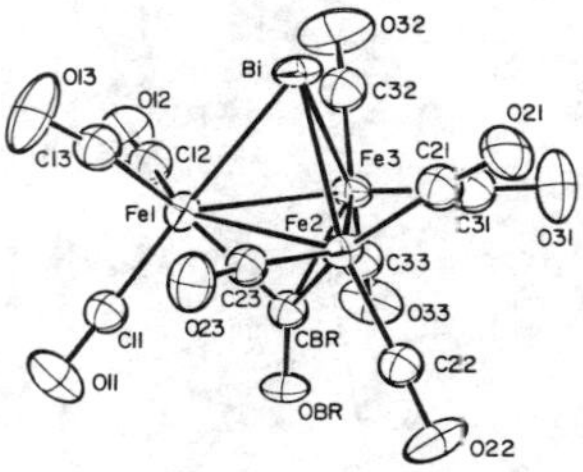

Fig. 1. The $[Fe_3Bi(CO)_9(\mu_3\text{-}CO)]^-$ anion.

TRILITHIUM BIS(HEXAMETHYLENETETRAMINE)HEXACYANOIRON(III) PENTAHYDRATE
$C_{18}H_{24}FeLi_3N_{14}, 5H_2O$ $Li_3[Fe(CN)_6].2C_6H_{12}N_4, 5H_2O$

J. PICKARDT, J. KAHLER, N. RAUTENBERG and E. RIEDEL, 1984. Z. Naturforsch., B39, 1162-1166.

Orthorhombic, I2mm, a = 9.094, b = 10.461, c = 14.558 Å, Z = 2. Mo radiation, R = 0.045 for 1158 reflexions.

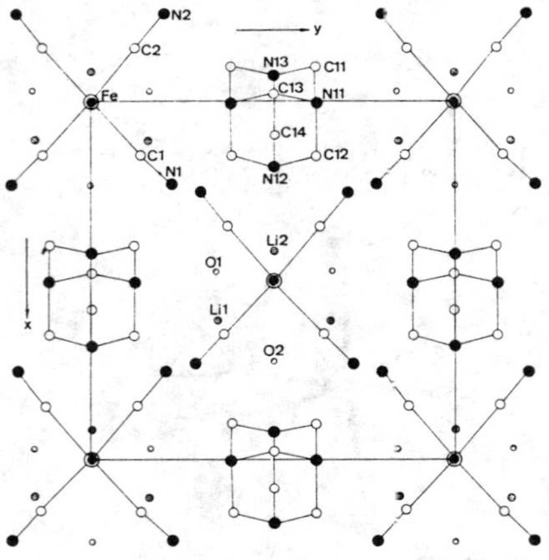

Fig. 1. The unit cell of $Li_3[Fe(CN)_6].2C_6H_{12}N_4, 5H_2O$.

The structure can be regarded as the packing of $[Fe(CN)_6]^{3-}$ octahedra and ethylenetetramine molecules, with the Li ions coordinated to N atoms of both groups and to water molecules (Fig. 1). Main dimensions are: Fe-C 1.908-1.968(5), Li-O 1.917-1.973, Li-N 1.99-2.18(2). The HMTA molecule lies on a mirror plane, the Te atom on a site with symmetry and the Li atoms on sites with symmetry.

TRIS(N-t-BUTYLACETAMIDE(1+)) OCTABROMOPENTASILVER(I)
$C_{18}H_{42}Ag_5Br_8N_3O_3$ $(C_6H_{14}NO)_3[Ag_5Br_8]$

J.D. KILDEA and A.H. WHITE, 1984. Inorg. Chem., 23, 3825-3827.

Monoclinic, C2/c, a = 28.84, b = 6.350, c = 36.48 Å, β = 108.25°, Z = 8. Mo radiation, R = 0.053 for 1706 reflexions.

The structure consists of $(C_6H_{14}NO)^+$ cations and a $(Ag_5Br_8)_n{}^{3n-}$ polyanion (Fig. 1). The Ag atoms have pseudotetrahedral coordination with Ag-Br 2.667(6)-2.745(4) Å.

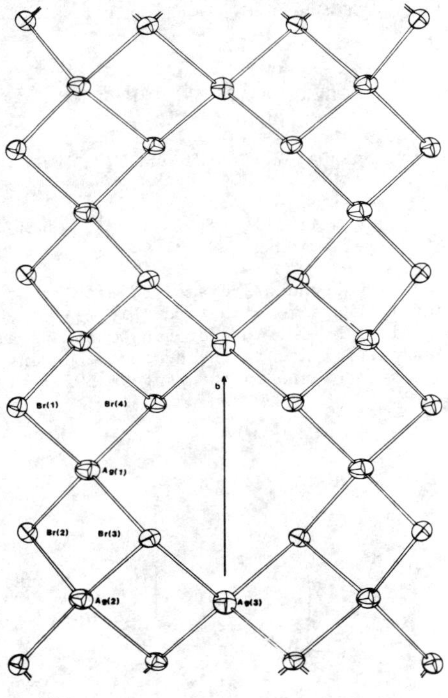

Fig. 1. The $(Ag_5Br_8)_n{}^{3n-}$ polyanion.

TRIS(CYCLOHEXYLAMMONIUM) CHLORIDE TETRACHLOROCOBALTATE(II)
$C_{18}H_{42}Cl_5CoN_3$ $(C_6H_{14}N)_3Cl(CoCl_4)$

U. GEISER, R.D. WILLETT and R.M. GAURA, 1984. Acta Cryst., C40, 1346-1349.

Monoclinic, $P2_1/c$, a = 15.365, b = 10.180, c = 22.042 Å, β = 127.84°, Z = 4. Mo radiation, R = 0.070 for 1376 reflexions.

Slightly distorted $CoCl_4^{2-}$ tetrahedra (Co-Cl 2.253-2.300(4) Å, Cl-Co-Cl 104.5-118.7(1)°) are hydrogen bonded through cyclohexylammonium groups into chains along c and via two amine nitrogens and an uncoordinated chlorine along b (Fig. 1). All cyclohexyl rings are in the chair conformation.

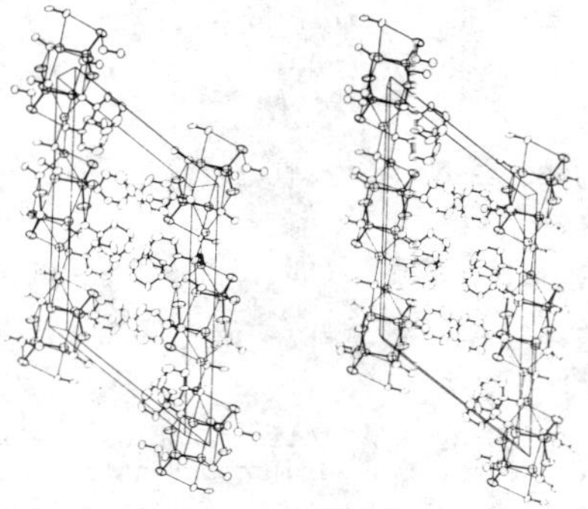

Fig. 1. Stereoview along b of the $(C_6H_{14}N)_3Cl(CoCl_4)$ structure.

OXONIUM TRIS(TRIETHYLAMMONIUM) OCTAMOLYBDATE(4-) DIHYDRATE
$C_{18}H_{51}Mo_8N_3O_{27},2H_2O$ $(Et_3N)_3(H_3O)[Mo_8O_{26}],2H_2O$

P.K. BHARADWAJ, Y. OHASHI, Y. SASADA, Y. SASAKI and T. YAMASE, 1984. Acta Cryst., C40, 48-50.

Monoclinic, $P2_1/a$, a = 21.271, b = 11.837, c = 20.189 Å, β = 117.92°, Z = 4. Mo radiation, R = 0.043 for 9441 reflexions.

Two crystallographically independent octamolybdates lie at independent inversion centres and have approximately the same structure (Fig. 1) as β-$[Mo_8O_{26}]^{4-}$ (1). The H_3O^+ cation links the two anions by OH...O hydrogen bonds and all triethylammonium cations and water molecules participate in the hydrogen bonding network. In the anions, the terminal Mo=O distances are in the range 1.692-1.717(5) Å.

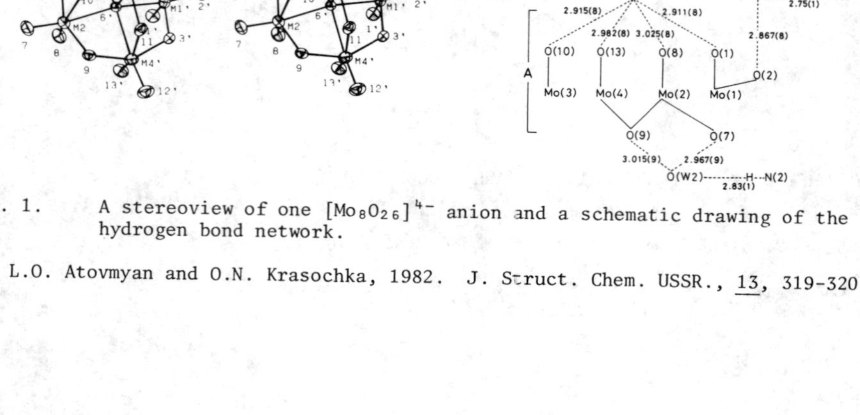

Fig. 1. A stereoview of one $[Mo_8O_{26}]^{4-}$ anion and a schematic drawing of the hydrogen bond network.

1. L.O. Atovmyan and O.N. Krasochka, 1982. J. Struct. Chem. USSR., 13, 319-320.

METHYLTRIPHENYLPHOSPHONIUM DICHLORODITHIOPHOSPHATE
$C_{19}H_{18}Cl_2P_2S_2$ (I) $[PPh_3Me][PS_2Cl_2]$

TETRAPHENYLPHOSPHONIUM DIBROMODITHIOPHOSPHATE
$C_{24}H_{20}Br_2P_2S_2$ (II) $[PPh_4][PS_2Br_2]$

U. MULLER and A.T. MOHAMMED, 1984. Z. Anorg. Chem., 514, 164-170.

I. Orthorhombic, $P2_12_12_1$, a = 10.89, b = 13.34, c = 14.76 Å, Z = 4. Mo radiation, R = 0.046 for 1116 reflexions.

II. Tetragonal, $I\bar{4}$, a = 13.01, c = 7.21 Å, Z = 2. Mo radiation, R = 0.065 for 357 reflexions.

Compound I is isotypic with $[PPh_3Me][VO_2Cl_2]$ (1) with P-S 1.948, 1.975(4); P-Cl 2.008, 2.031(4) Å, S-P-S 114.4(2), Cl-P-Cl 103.5, S-P-Cl 108.1-111.9(2)°. II is isotypic with $[AsPh_4][FeCl_4]$; the $[PS_2Br_2]^-$ ion occupies a site with $\bar{4}$ symmetry so that the S and Br atoms are disordered.

1. Structure Reports, 46B, 1327.

METHYLTRIPHENYLPHOSPHONIUM PERMANGANATE (α and β forms)
$C_{19}H_{18}MnO_4P$ (I and II)

n-HEPTYLTRIPHENYLPHOSPHONIUM PERMANGANATE
$C_{25}H_{30}MnO_4P$ (III)

H. KARAMAN, R.J. BARTON, B.E. ROBERTSON and D.G. LEE, 1984. J. Org. Chem., 49, 4509-4518.

I. α-form, Orthorhombic, $Pbc2_1$, a = 12.626, b = 14.320, c = 19.806 Å, Z = 8. Mo radiation, R = 0.066 for 1732 reflexions.

II. β-form, Monoclinic, $P2_1/c$, a = 10.637, b = 13.424, c = 15.036 Å, β = 119.44°, Z = 4. Mo radiation, R = 0.080 for 2138 reflexions.

III Orthorhombic, $P2_12_12_1$, a = 13.122, b = 10.801, c = 17.154 Å, Z = 8. Mo radiation, R = 0.10 for 1530 reflexions.

There are two independent formula units of I in the asymmetric unit. The mean Mn-O distances are 1.50(2), 1.49(2), 1.54(1) and 1.47(2) Å for I (two values), II and III respectively. The P-C(phenyl) distances are in the range 1.77-1.82(2) Å. In each of the alkyltriphenylphosphonium ions, one phenyl ring is approximately parallel to the P-C(alkyl) bond, another is inclined by 13-31° and the third is rotated 66-77°.

METHYLTRIPHENYLPHOSPHONIUM ACETONITRILETETRACHLOROOXONIOBATE
$C_{21}H_{21}Cl_4NNbOP$ $[PPh_3Me][NbOCl_4(MeCN)]$

W. HILLER, J. STRAHLE, H. PRINZ and K. DEHNICKE, 1984. Z. Naturforsch., B39, 107-110.

Triclinic, $P\bar{1}$, a = 11.735, b = 11.301, c = 9.468 Å, α = 97.70, β = 93.57, γ = 78.62°, Z = 2. R = 0.025 for 3688 reflexions (at 210 K).

The anion has distorted octahedral geometry (Fig. 1) with the acetonitrile N atom trans to the Nb-O bond. Main dimensions are: Nb-O 1.688(2), Nb-N 2.448(3), Nb-Cl 2.375(1)-2.432(1) Å, O-Nb-Cl 96.0-99.2(1), N-Nb-Cl 81.2-82.9(1)°.

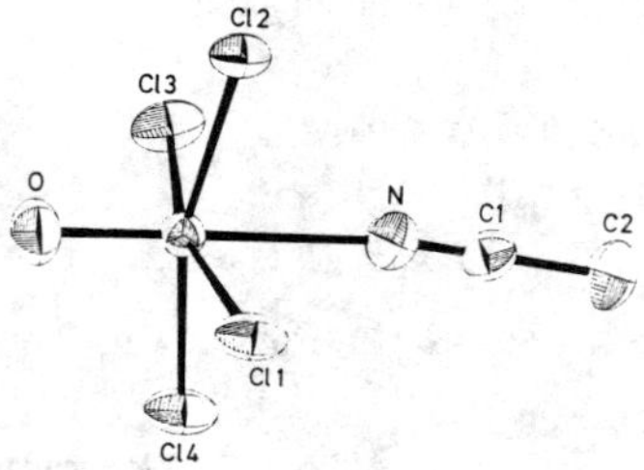

Fig. 1. A view of $[NbOCl_4(CH_3CN)]^-$.

TRIMETHYLBENZYLAMMONIUM (μ_3-HYDRIDO)TETRAKIS(μ-CARBONYL)OCTACARBONYL-tetrahedro-NICKELTRIIRON

$C_{22}H_{17}Fe_3NNiO_{12}$ $[N(CH_3)_3CH_2C_6H_5]^+[Fe_3Ni(CO)_{12}H]^-$

A. CERIOTTI, P. CHINI, A. FUMAGALLI, T.F. KOETZLE, G. LONGONI and F. TAKUSAGAWA, 1984. Inorg. Chem., 23, 1363-1368.

Triclinic, $P\bar{1}$, a = 7.416, b = 13.849, c = 14.108 Å, α = 103.13, β = 103.15, γ = 99.23°, Z = 2. Mo radiation, R = 0.030 for 3860 reflexions.

The complex anion contains a tetrahedron of metal atoms surrounded by eight terminal and four edge-bridging carbonyl groups (Fig. 1). The hydride ligand has been located over the centre of a Fe_2Ni face at a distance of 0.60(3) Å from the trimetal plane. Bond lengths: Ni-Fe 2.4712(8)-2.5252(7), Fe-Fe (bridged) 2.570(2), Fe-Fe (unbridged) 2.720(1) and 2.7139(9), M-C (terminal) 1.763(3)-1.791(3), M-C (bridge) 1.869(3)-2.232(2) (Ni) and 1.838(2)-2.002(3) (Fe), Ni-H 1.55(3), Fe-H 1.56(2) and 1.61(3) Å.

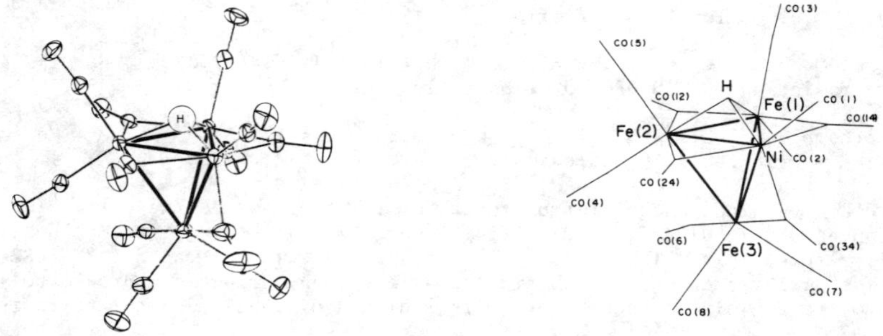

Fig. 1. Perspective and schematic views of the $[Fe_3Ni(CO)_{12}H]^-$ anion.

TETRAPHENYLARSONIUM TETRAAZIDOAURATE(III)

$C_{24}H_{20}AsAuN_{12}$ $Ph_4As^+.Au(N_3)_4^-$

W. BECK and H. NOTH, 1984. Chem. Ber., 117, 419-422.

Tetragonal, P4/n, a = 13.222, c = 7.679 Å, Z = 1. Mo radiation, R = 0.0956 for 1084 reflexions.

The As atom lies at a site with $\bar{4}$ symmetry and the gold atom at a site with fourfold symmetry (Fig. 1). Main dimensions are: Au-N1 2.028(16), As-C 1.898(10) Å, N-Au-N 89.3(7), C-As-C 111.6(3)°.

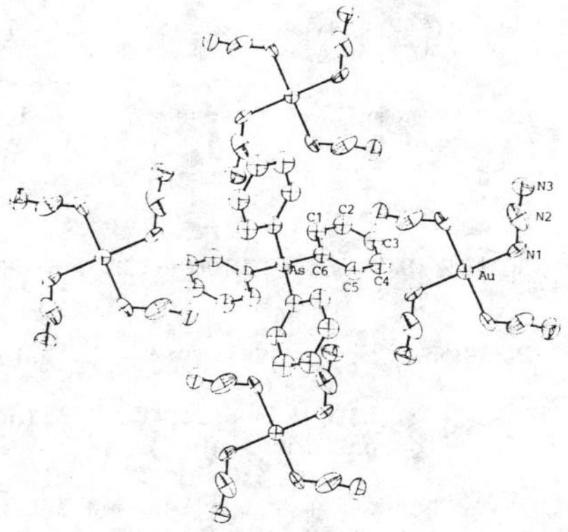

Fig. 1. A view of $Ph_4As^+.Au(N_3)_4^-$ along the c-direction.

TETRAPHENYLARSONIUM TETRACHLORONITRIDOTECHNETATE(VI)

$C_{24}H_{20}AsCl_4NTc$ $[AsPh_4][TeNCl_4]$

J. BALDAS, J.F. BOAS, J. BONNYMAN and G.A. WILLIAMS, 1984. J. Chem. Soc. Dalton, 2395-2400.

Tetragonal, P4/n, a = 12.707, c = 7.793 Å, Z = 2. Cu radiation, R = 0.031 for 1079 reflexions.

The $[TcNCl_4]^-$ anion (Fig. 1) in the compound possesses C_{4v} symmetry with Cl-Tc-Cl bond angles of 86.95(1) and 153.33(6)° and N≡Tc-Cl bond angles of 103.34(3)°. The Tc≡N bond distance is 1.581(5) Å and the Tc-Cl distances are 2.3220(9) Å. The cation has the expected geometry.

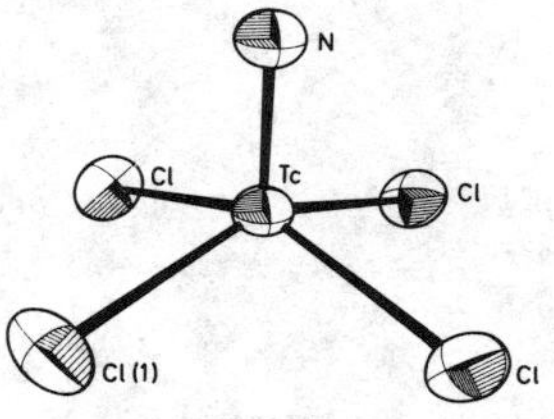

Fig. 1. $C_{24}H_{20}AsCl_4NTc$: the anion.

TETRAPHENYLARSONIUM 1,1,1,1-TETRACHLORO-1-TUNGSTA-2,4,6-TRIAZA-3,5-DITHIACYCLOHEXA-1,3,5-TRIENE

$C_{24}H_{20}AsCl_4N_3S_2W$ $[Ph_4As][WCl_4(N_3S_2)]$

U. KYNAST, E. CONRADI, U. MULLER and K. DEHNICKE, 1984. Z. Naturforsch., B39, 1680 1685.

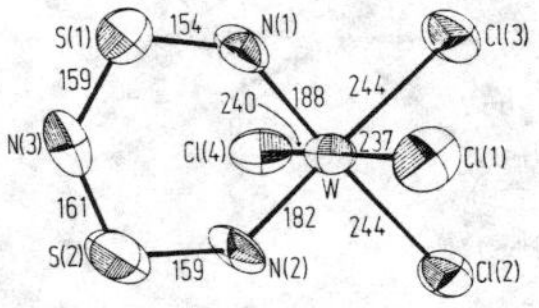

Fig. 1. A view of the $[WCl_4(N_3S_2)]^-$ anion with bond lengths (pm).

Monoclinic, $P2_1/c$, a = 13.524, b = 9.684, c = 23.938 Å, β = 115.05°, Z = 4. Mo radiation, R = 0.045 for 1845 reflexions.

In the anion (Fig. 1) the W atoms have octahedral coordination; the WN_3S_2 ring is planar. Bond lengths are in the figure.

TETRAPHENYLPHOSPHONIUM CHLOROTETRAOXOOSMIUM DICHLOROMETHANE SOLVATE

$C_{24}H_{20}ClO_4OsP,CH_2Cl_2$ $PPh_4[OsO_4Cl],CH_2Cl_2$

R. WEBER, K. DEHNICKE, U. MULLER and D. FENSKE, 1984. Z. Anorg. Chem., 516, 214-222.

Monoclinic, P2/b, a = 17.537, b = 21.839, c = 6.919 Å, γ = 106.72°, Z = 4. Mo radiation, R = 0.032 for 4212 reflexions (at -110°C).

The structure consists of PPh_4 cations (two independent PPh_4 cations lying on twofold axes) and OsO_4Cl^- anions with five-coordinate Os in a trigonal bipyramidal geometry (Fig. 1). The anion can be regarded as an OsO_4 tetrahedron, monocapped by a chlorine ion. Each chloride is linked with two CH_2Cl_2 molecules by hydrogen bridges. The As-Cl bond length (2.76 Å) is very long.

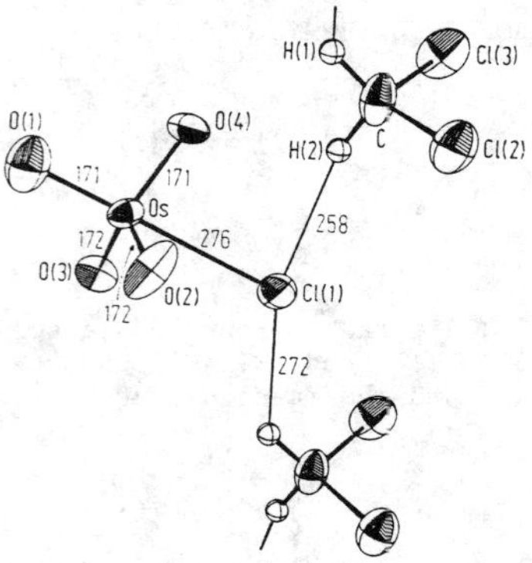

Fig. 1. A view of the $[OsO_4Cl]^-$ anion, hydrogen bonded to two CH_2Cl_2 molecules, with distances in pm.

TETRAPHENYLPHOSPHONIUM TETRACHLOROTHIONIOBIUM(V)

$C_{24}H_{20}Cl_4NbPS$ $[Ph_4P].[NbSCl_4]$

U. MULLER and P. KLINGELHOFER, 1984. Z. Anorg. Chem., 510, 109-116.

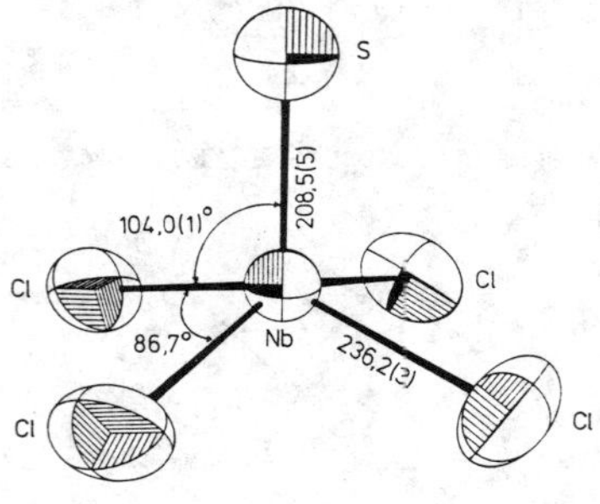

Fig. 1. The $NbSCl_4$ anion.

Tetragonal, P4/n, a = 13.03, c = 7.60 Å, Z = 2. Mo radiation, R = 0.066 for 1017 reflexions.

The cation has $\bar{4}$ and the anion 4-fold crystallographic symmetry (Fig. 1). The Nb=S bond length is 2.085(5) Å.

TETRAPHENYLPHOSPHONIUM OXOTETRACHLOROTUNGSTATE
$C_{24}H_{20}Cl_4OPW$ (I) $[PPh_4][WOCl_4]$

BIS(TETRAPHENYLPHOSPHONIUM) OXOPENTACHLOROTUNGSTATE DICHLOROMETHANE SOLVATE
$C_{48}H_{40}Cl_5OPW,2(CH_2Cl_2)$ (II) $[PPh_4]_2[WOCl_5],2(CH_2Cl_2)$

BIS(TETRAPHENYLPHOSPHONIUM) OXOPENTACHLORORHENATE DICHLOROMETHANE SOLVATE
$C_{48}H_{40}Cl_5OPRe,2(CH_2Cl_2)$ (III) $[PPh_4]_2[ReOCl_5],2(CH_2Cl_2)$

D. FENSKE, K. STAHL, E. HEY and K. DEHNICKE, 1984. Z. Naturforsch., B39, 850-854.

I. Tetragonal, P4/n, a = 12.694, c = 7.596 Å, Z = 2. Mo radiation, R = 0.054 for 1416 reflexions (at -110°C).

II. Triclinic, P$\bar{1}$, a = 10.111, b = 11.134, c = 12.402 Å, α = 70.29, β = 79.40, γ = 80.89°, Z = 1. Mo radiation, R = 0.087 for 3781 reflexions (at -110°C).

III. Triclinic, P$\bar{1}$, a = 10.057, b = 11.090, c = 12.392 Å, α = 70.52, β = 79.59, γ = 80.90°, Z = 1. Mo radiation, R = 0.050 for 5436 reflexions (at -100°C).

In (I) the $WOCl_4$ ion has fourfold symmetry and forms a tetragonal pyramid with axial oxygen and W-O 1.676(7), W-Cl 1.379(1) Å, O-W-Cl 105.3(2), Cl-W-Cl 86.0(2)°. Compounds (II) and (III) are isomorphous. The metal atoms lie on inversion centres and the $[MOCl_5]^-$ ions are disordered.

TETRAPHENYLPHOSPHONIUM HEXACHLOROOSMATE(V)
$C_{24}H_{20}Cl_6OsP$ (I) $[P(C_6H_5)_4]OsCl_6$

BIS(TETRAPHENYLPHOSPHONIUM) HEXACHLOROOSMATE(IV) N,N-DIMETHYLFORMAMIDE SOLVATE
$C_{48}H_{40}Cl_6OsP_2,C_3H_7NO$ $[P(C_6H_5)_4]_2OsCl_6.(CH_3)_2NC(O)H$

E.E. KIM, K. ERIKS and R. MAGNUSON, 1984. Inorg. Chem., 23, 393-397.

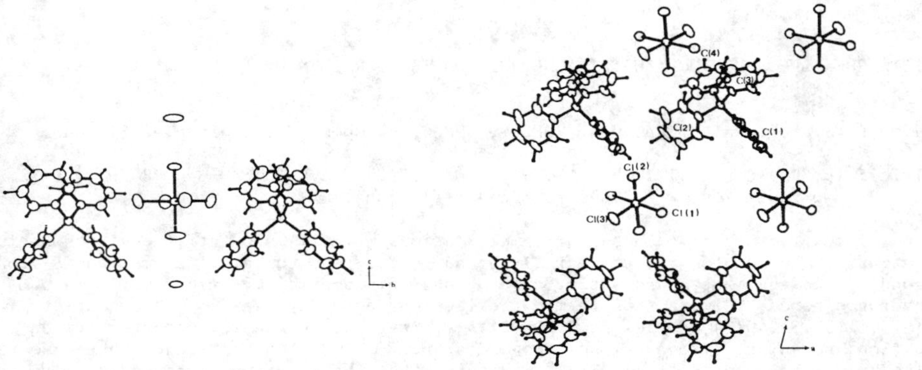

Fig. 1. The structures of $(Ph_4P)OsCl_6$ (left) and $(Ph_4P)_2OsCl_6$ (right).

I. Tetragonal, P4/n, a = 12.957, c = 7.709 Å, D_m = 1.9, Z = 2. Mo radiation, R = 0.023 for 766 reflexions.

II. Triclinic, P$\bar{1}$, a = 10.028, b = 11.228, c = 11.996 Å, α = 106.58, β = 79.37, γ = 101.24°, D_m = 1.52, Z = 1. Cu radiation, R = 0.053 for 2537 reflexions.

Both structures contain octahedral $OsCl_6$ anions and tetrahedral Ph_4P^+ cations (Fig. 1). In I the mean Os-Cl distance is 2.284 Å, and the $OsCl_6^-$ octahedron is slightly distorted so that the Os atom lies 0.026 Å below the plane of the equatorial Cl atoms. There is a short Cl...Cl contact of 3.184 Å along the z axis. In II, the $OsCl_6^{2-}$ anion is nearly regular, with a mean Os-Cl distance of 2.332 Å.

TETRAPHENYLPHOSPHONIUM AQUATETRACHLOROTHIONITROSYLOSMATE(II)
$C_{24}H_{22}Cl_4NOOsPS$ $[PPh_4]_2[(H_2O)Os(NS)Cl_4]$

K.K. PANDEY, H.W. ROESKY, M. NOLTEMEYER and G.M. SHELDRICK, 1984. Z. Naturforsch., B39, 590-593.

Monoclinic, $P2_1/c$, a = 11.033, b = 19.598, c = 12.543 Å, β = 96.51°, Z = 4. Mo radiation, R = 0.029 for 3177 reflexions.

In the anion (Fig. 1), the Os atom is octahedrally coordinated with H_2O trans to the NS group. Main dimensions are Os-O 2.165(4), Os-N 1.731(4), Os-Cl 2.345-2.374(2), N-S 1.514(5) Å, Os-N-S 174.9(3)°. The anions form centrosymmetric dimeric units in the crystal via four O-H...Cl hydrogen bonds (O...Cl 3.295(6) and 3.167(6) Å).

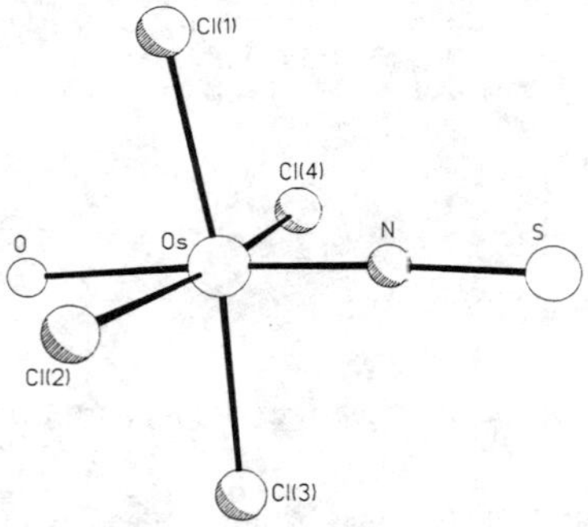

Fig. 1. The $[(H_2O)Os(NS)Cl_4]^{2-}$ anion.

TETRAPHENYLPHOSPHONIUM AQUATETRACHLOROTHIONITROSYLRUTHENIUM
$C_{24}H_{22}Cl_4NOPRuS$

J.W. BATS, K.K. PANDEY and H.W. ROESKY, 1984. J. Chem. Soc. Dalton, 2081-2083.

Triclinic, P$\bar{1}$, a = 10.173, b = 11.756, c = 12.793 Å, α = 66.24, β = 78.89, γ = 72.58°, Z = 2. Mo radiation, R = 0.040 for 2644 reflexions.

The structure is composed of complex anions (Fig. 1) and tetraphenylphosphonium cations. The Ru atom is octahedrally bound by four equatorial Cl atoms, an NS group bonding via the N atom and water. Two anions form centrosymmetric dimers via O-H...Cl hydrogen bonds. Dimers thus formed are interconnected by van der Waals contacts (S...Cl(2) 3.240(2), S...Cl(3) 3.300(2) Å). There are no short contacts between cation and anion. The axial Ru-O bond distance (2.112(3) Å) is longer than expected. The Ru-N-S bond angle of 170.9(3)° indicates that the NS group is bonded nearly linearly to the metal atom.

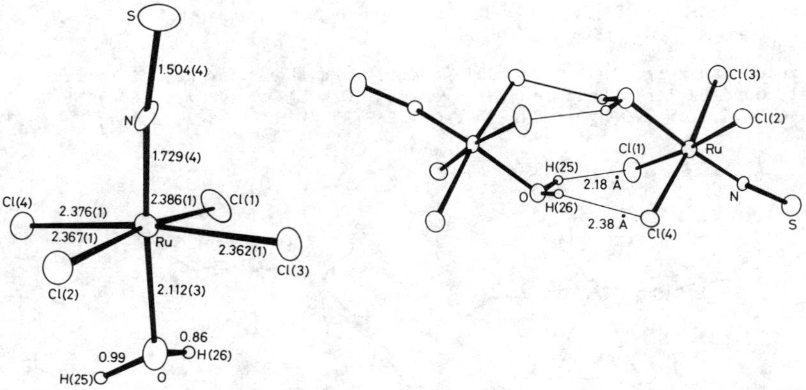

Fig. 1. $C_{24}H_{22}Cl_4NOPRuS$: the anion with bond distances (left) and the dimer formed from two anions by hydrogen bonding (right).

TETRAPHENYLPHOSPHONIUM AQUAOXOTETRACHLORONIOBIUM

$C_{24}H_{22}Cl_4NbO_2P$ (I) $PPh_4[NbOCl_4(OH_2)]$

BIS(TETRAPHENYLPHOSPHONIUM) DICHLOROPHOSPHATOOXOTETRACHLORONIOBIUM DICHLOROMETHANE SOLVATE

$C_{48}H_{40}Cl_6NbO_3P_3,2(CH_2Cl_2)$ (II) $(PPh_4)_2[NbOCl_4(O_2PCl_2)]$

P. KLINGELHOFER and U. MULLER, 1984. Z. Anorg. Chem., 516, 85-92.

I. Tetragonal, P4/n, a = 13.09, c = 7.34 Å, Z = 2. Mo radiation, R = 0.046 for 681 reflexions.

II. Triclinic, P$\bar{1}$, a = 11.72, b = 11.87, c = 21.05 Å, α = 88.40, β = 83.20, γ = 71.28°, Z = 2. Mo radiation, R = 0.059 for 2502 reflexions.

Molecule I is isomorphous with the previously reported $AsPh_4[MoOCl_4(OH_2)]$ (1). Both I and II (Fig. 1) contain $NbOCl_4$ tetragonal pyramidal moieties to which either a H_2O or $PO_2Cl_2^-$ is attached. Dimensions for II are in Fig. 1; for I, Nb=O 1.74(1), Nb-Cl 2.381(2), Nb-OH_2 2.49(1) Å.

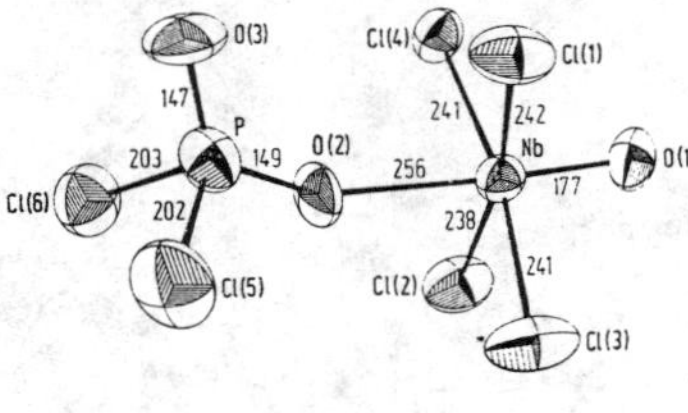

Fig. 1. A view of $[NbOCl_4(O_2PCl_2)]^{2-}$ with bond lengths (pm).

1. Structure Reports, 43B, 894.

TETRAKIS(TETRAMETHYLAMMONIUM) TETRATHIOCYANATOOXOVANADATE TETRATHIOCYANATOAQUA-OXOVANADATE

$C_{24}H_{50}N_{12}O_3S_8V_2$ $(Me_4N)_4[VO(NCS)_4][VO(H_2O)(NCS)_4]$

ZHOU KANGJING, XU YUANZHI and HUANG JINLING, 1984. J. Struct. Chem., 3, 61-64.

Orthorhombic, Pcab, a = 23.252, b = 13.015, c = 30.621 Å, D_m = 1.32, Z = 8. Mo radiation, R = 0.087 for 1903 reflexions.

One anion has tetragonal pyramidal geometry and the other has distorted octahedral coordination. Bond lengths are for $[VO(NCS)_4]^{2-}$ V-O 1.548(6), V-N 2.021-2.046(8) Å; for $[VO(H_2O)(NCS_4]^{2-}$ V-O 1.557(5), V-N 2.008-2.038(8), V-O(H_2O) 2.317(5) Å.

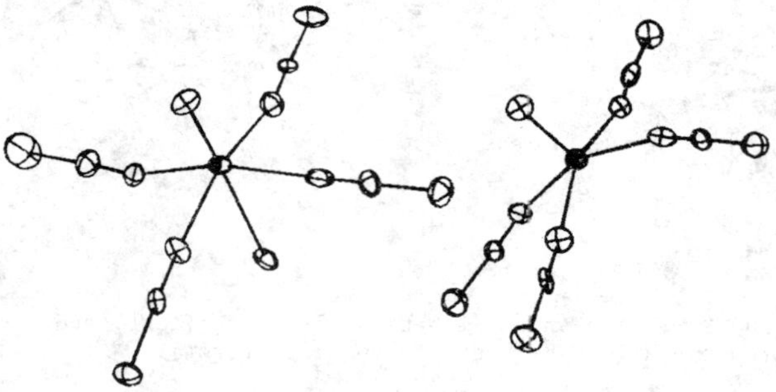

Fig. 1. Views of the two anions $[VO(H_2O)(NCS)_4]^{2-}$ and $[VO(NCS)_4]^{2-}$.

BIS(TETRAPROPYLAMMONIUM) HEXA-μ-BROMO-tetrahedro-TETRACUPRATE(I)
$C_{24}H_{56}Br_6Cu_4N_2$ $[N(C_3H_7)_4{}^+]_2[Cu_4Br_6{}^{2-}]$

M. ASPLUND and S. JAGNER, 1984. Acta Chem. Scand., A38, 725-730.

Tetragonal, $P4_2/n$, a = 15.708, c = 15.709 Å, Z = 4. Mo radiation, R = 0.062 for 1168 reflexions.

The anion is an aggregate composed of an octahedron of bromide ligands containing a tetrahedron of trigonal planar coordinated Cu(I) atoms (Fig. 1). The orientation of the Cu(I) tetrahedron is disordered, there being two equivalent Cu(I) tetrahedra related by the inversion centre of the Br octahedron. The Cu(I) atoms have occupancies of 0.5 in four general eightfold sites. Cu-Br distances range from 2.373-2.422 Å and Cu...Cu from 2.718-2.750 Å. Two of the three crystallographically-independent cations have $\bar{4}$ symmetry while the third has a disordered orientation with respect to a twofold axis.

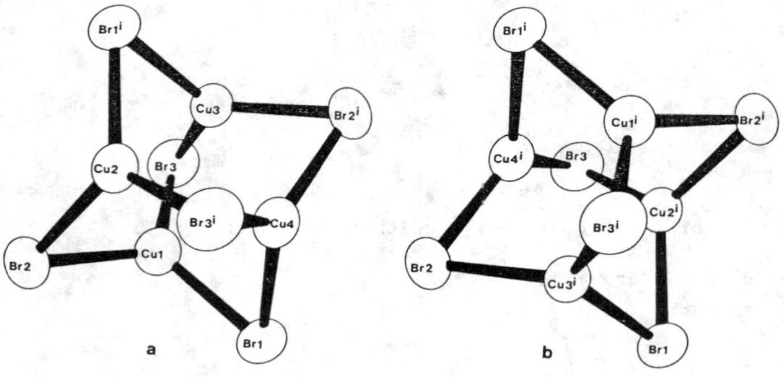

Fig. 1. The anion in $C_{24}H_{56}Br_6Cu_4N_2$ showing the two orientations of the Cu(I) tetrahedron. The superscript i denotes an atom in -x, -y, 1-z.

BIS(TETRAPROPYLAMMONIUM) DI-μ-IODO-DIIODODICUPRATE(I)
$C_{24}H_{56}Cu_2I_4N_2$ $[N(C_3H_7)_4]_2^+Cu_2I_4^{2-}$

M. ASPLUND and S. JAGNER, 1984. Acta Chem. Scand., A38, 411-414.

Monoclinic, $P2_1/n$, a = 9.017, b = 12.683, c = 16.140 Å, β = 93.15°, Z = 2. Mo radiation, R = 0.044 for 2899 reflexions.

The anion is a centrosymmetric dimer (Fig. 1) in which the configuration of ligands about Cu(I) is approximately trigonal planar. The metal atom is situated some 0.013 Å from the plane defined by the three ligand atoms (I(1), I(2) and I(2')). The Cu...Cu' distance is 2.698 Å, and other distances are Cu-I(1) 2.499, Cu-I(2) 2.571, and Cu-I(2') 2.582 Å. Important bond angles are: I(1)-Cu-I(2) 122.98, I(1)-Cu-I(2') 120.15, I(2)-Cu-I(2') 116.86 and Cu-I(2)-Cu' 63.14°. There are no close contacts between anion and cations in the unit cell.

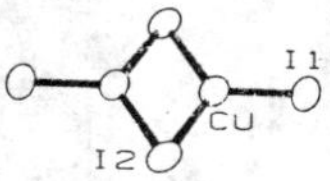

Fig. 1. The anion in $C_{24}H_{56}Cu_2I_4N_2$.

TRIS(TETRAETHYLAMMONIUM) catena-μ-CHLORO-ennea-μ-CHLORO-HEPTACUPRATE(I)
$C_{24}H_{60}Cl_{10}Cu_7N_3$ $[N(C_2H_5)_4^+]_3[Cu_7Cl_{10}]^{3-}$

M. ASPLUND and S. JAGNER, 1984. Acta Chem. Scand., A38, 807-811.

Monoclinic, C2/c, a = 19.052, b = 16.296, c = 13.782 Å, β = 98.95°, Z = 4. Mo radiation, R = 0.063 for 2743 reflexions.

The $[Cu_7Cl_{10}]^{3-}$ anion is an infinite Cl-bridged chain, the repeat unit of which contains six three-coordinate Cu(I) atoms and one two-coordinate Cu(I) atom (Fig. 1). For the latter, Cu-Cl distances (symmetry-related) are 2.165 Å while for the three-coordinate Cu(I) atoms the Cu-Cl distances are: Cu(2)-Cl(1) 2.311, Cu(2)-Cl(2) 2.206, Cu(2)-Cl(3) 2.298, Cu(3)-Cl(2) 2.261, Cu(3)-Cl(4) 2.253, Cu(3)-Cl(5) 2.346, Cu(4)-Cl(3) 2.245, Cu(4)-Cl(4) 2.258 and Cu(4)-Cl(5') 2.291 Å. The closest Cu...Cu distances are 2.890 to 3.447 Å, and the ligand configurations in the three-coordinate units are approximately trigonal planar.

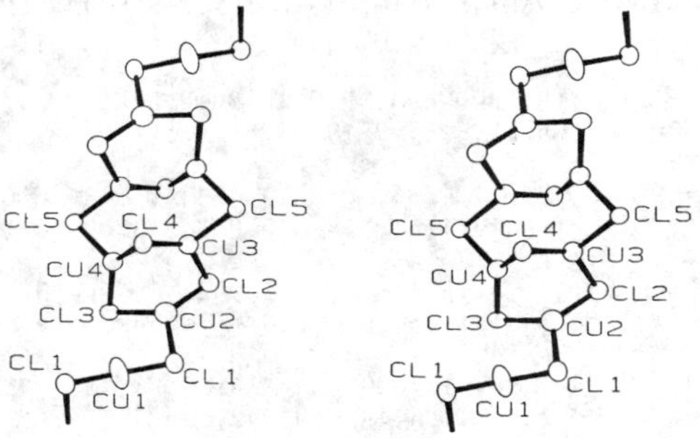

Fig. 1. Stereoview of part of the infinite chain anion in $C_{24}H_{60}Cl_{10}Cu_7N_3$.

HEXAKIS(TETRAMETHYLAMMONIUM) DIHYDROGENDODECATUNGSTATE ENNEAHYDRATE
$C_{24}H_{74}N_6O_{40}W_{12},9H_2O$ $[Me_4N]_6[H_2W_{12}O_{40}],9H_2O$

M. ASAMI, H. ICHIDA and Y. SASAKI, 1984. Acta Cryst., C40, 35-37.

Monoclinic, $P2_1$, a = 13.122, b = 21.457, c = 13.223 Å, β = 91.71°, D_m = 3.11, Z = 2. Mo radiation, R = 0.085 for 6585 reflexions.

The metadodecatungstate anion (Fig. 1) consists of four tetrahedrally corner-shared W_3O_{13} units made by edge sharing WO_6, and is therefore confirmed to be the Keggin type. The molecule has T_d symmetry and its framework is regarded as a regular cubo-octahedron. The average W-W distance is 3.33(1) Å within the same W_3O_{13} unit and 3.63(2) Å otherwise. The W-O bond lengths, formally classified into four types, are 1.71(4), 1.91(5), 1.96(9) and 2.22(9) Å. The O atoms forming the tetrahedral cavity are 2.95(6) Å distant from one another, suggesting that the cavity is capable of containing two non-acidic H atoms.

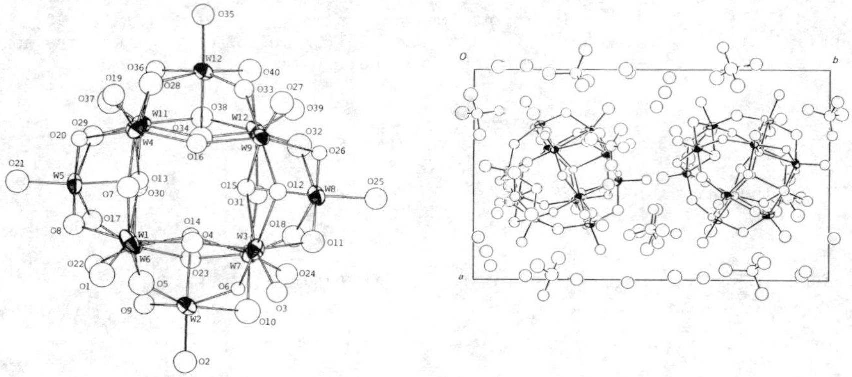

Fig. 1. The structure of the $[H_2W_{12}O_{40}]^{6-}$ anion and the crystal structure projected along c*.

BIS(TETRA-n-PROPYLAMMONIUM) CYANO(TETRATHIOMOLYBDENIO-S,S')CUPRATE(I)
$C_{25}H_{56}CuMoN_3S_4$ (I) $[(C_3H_7)_4N]_2[(CN)CuS_2MoS_2]$

BIS(TETRAPHENYLARSONIUM) (μ-TETRATHIOMOLYBDENIO-S,S':S'',S''')BIS(CYANOCUPRATE(I)) MONOHYDRATE
$C_{50}H_{40}As_2Cu_2MoN_2S_4,H_2O$ (II) $[(C_6H_5)_4As]_2[(CN)CuS_2MoS_2Cu(CN)],H_2O$

BIS(TETRA-n-PROPYLAMMONIUM) CYANO(TETRATHIOTUNGSTENIO-S,S')ARGENTATE(I)
$C_{25}H_{56}AgN_3S_4W$ (III) $[(C_3H_7)_4N]_2[(CN)AgS_2WS_2]$

S.F. GHELLER, T.W. HAMBLEY, J.R. RODGERS, R.C.T. BROWNLEE, M.J. O'CONNOR, M.R. SNOW and A.G. WEDD, 1984. Inorg. Chem., 23, 2519-2528

I. Triclinic, $P\bar{1}$, a = 9.712, b = 12.686, c = 14.474 Å, α = 76.07, β = 83.98, γ = 81.19°, D_m = 1.33, Z = 2. Mo radiation, R = 0.040 for 3709 reflexions.

II. Triclinic, $P\bar{1}$, a = 9.561, b = 15.158, c = 18.027 Å, α = 97.45, β = 101.30, γ = 91.81°, D_m = 1.56, Z = 2. Mo radiation, R = 0.048 for 4936 reflexions.

III. Triclinic, $P\bar{1}$, a = 9.546, b = 13.009, c = 14.692 Å, α = 75.62, β = 84.93, γ = 80.57°, D_m = 1.56, Z = 2. Mo radiation, R = 0.042 for 3957 reflexions.

The three complex anions are shown in Fig. 1. In I and III, binuclear dianions $[(CN)M'S_2MS_2]^{2-}$, of C_{2v} point symmetry contain a tetrahedral MS_4 group acting as a bidentate ligand to an essentially linear M'CN unit. The trinuclear dianions in II have close to D_{2d} symmetry, the tetrahedral MoS_4 unit acting as a bidentate ligand to each of the two CuCN units. Bond lengths: for I; Mo-(μ-S) 2.221(1) and 2.230(2), Mo-S 2.149(1) and 2.151(2), Cu-S 2.217(2) and 2.197(1), Cu-C 1.880(6) Å; for II; Mo-S 2.198(2)-2.219(2), Cu-S 2.216(2)-2.222(2), Cu-C 1.888(9) and 1.884(9) Å; for III; W-(μ-S) 2.229(2) and 2.225(2), W-S 2.153(3) and 2.158(3), Ag-S 2.481(3) and 2.476(2), Ag-C 2.093(11) Å. The coordination about Cu and Ag atoms is planar and the M-M' distances are 2.624(1) Å in I, 2.635(1) and 2.634(1) Å in II, and 2.890(1) Å in III.

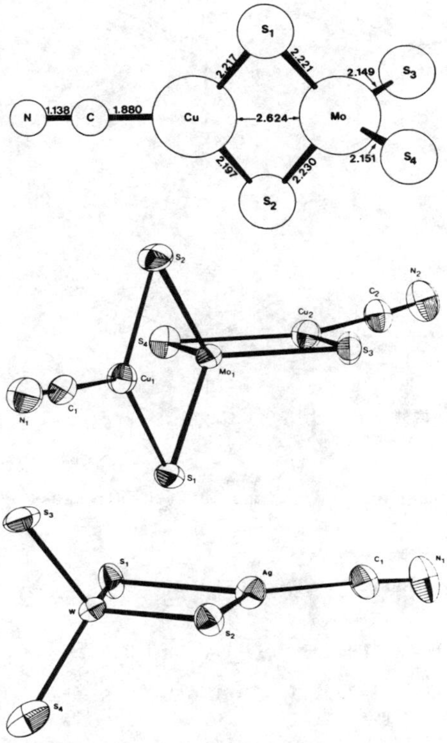

Fig. 1. The anions $[(CN)CuS_2MoS_2]^{2-}$ (top), $[(CN)CuS_2MoS_2Cu(CN)]^{2-}$ (middle), and $[(CN)AgS_2WS_2]^{2-}$ (bottom).

BIS(3-METHYLPYRIDINIUM) BIS(3-ETHYLPYRIDINIUM) β-OCTAMOLYBDATE

$C_{26}H_{36}Mo_8N_4O_{26}$ $[C_6H_8N^+]_2[C_7H_{10}N^+]_2[Mo_8O_{26}]^{4-}$

P. ROMÁN, M.E. GONZÁLEZ-AGUADO, C. ESTEBAN-CALDERÓN and M. MARTÍNEZ-RIPOLL, 1984. Inorg. Chim. Acta, 90, 115-120.

Monoclinic, $P2_1/n$, a = 13.858, b = 11.018, c = 14.005 Å, β = 91.35°, D_m = 2.46, Z = 2. Mo radiation, R = 0.064 for 5090 reflexions.

The crystal structure contains discrete $[\beta\text{-}Mo_8O_{26}]^{4-}$ centrosymmetric polyanions linked to the cations (Fig. 1) through hydrogen bonds of the type N-H...O and C-H...O, (N...O 3.09, 2.87, 2.95 Å; C...O 3.33, 3.39, 3.26, 3.33 Å). There are three classes of Mo-O bond lengths in the anion: short (1.70 Å mean, range 1.666-1.749 Å), medium (1.94 Å mean, range 1.889-1.991 Å) and long (2.34 Å mean, range 2.156-1.444 Å). The cation rings are planar, C(17), C(27) and C(28) being 0.11, 0.05 and -0.46 Å respectively out of the planes defined by the ring atoms.

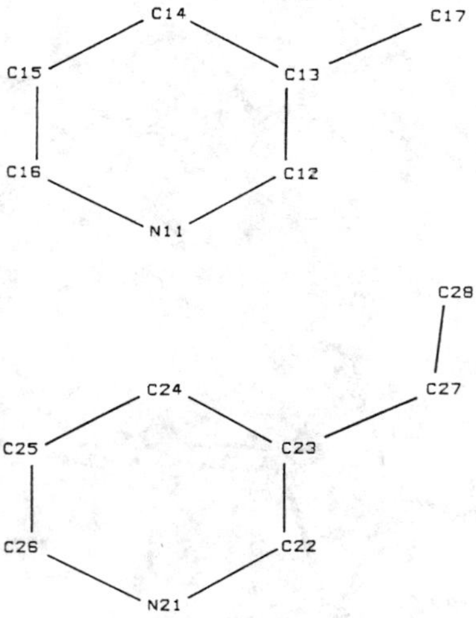

Fig. 1. Schematic drawing of the cations in $C_{26}H_{36}Mo_8N_4O_{26}$.

TETRAPHENYLPHOSPHONIUM PENTACARBONYLGERMYLCHROMIUM
$C_{29}H_{23}CrGeO_5P$ $[PPh_4].[Cr(CO)_5GeH_3]$

D. MELZER and E. WEISS, 1984. Chem. Ber., 117, 2464-2468.

Monoclinic, C2/c, a = 22.301, b = 6.989, c = 18.002 Å, β = not given, Z = 4. Mo radiation, R = 0.064 for 1974 reflexions.

In the cation (Fig. 1) the P atom lies on a twofold axis and in the anion the Cr atom is on an inversion centre, with the Ge atom disordered with a CO group.

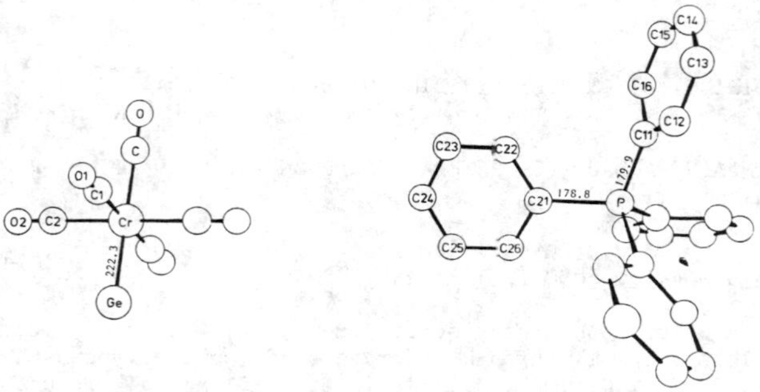

Fig. 1. The ions in $[PPh_4].[Cr(CO)_5GeH_3]$.

TETRAPHENYLPHOSPHONIUM BIS(μ3-SULPHIDO)CYCLOTRIS(DICARBONYLRHODIUM(I))
$C_{30}H_{20}O_6PRh_3S_2$ (I)

TETRAMETHYLAMMONIUM BIS(μ3-SELENIDO)CYCLOTRIS(DICARBONYLRHODIUM(I))
$C_{10}H_{12}NO_6Rh_3Se_2$ (II)

D. GALLI, L. GARLASCHELLI, G. CIANI, A. FUMAGALLI, S. MARTINENGO and A. SIRONI, 1984. J. Chem. Soc. Dalton, 55-61.

I. Orthorhombic, $Pbc2_1$, a = 13.529, b = 16.965, c = 28.365 Å, D_m = 1.75, Z = 8. Mo radiation, R = 0.042 for 3438 reflexions.

II. Orthorhombic, $Cmc2_1$, a = 14.029, b = 10.889, c = 12.748 Å, D_m = 2.40, Z = 4. Mo radiation, R = 0.039 for 1032 reflexions.

In I, there are two crystallographically independent cations and anions in the asymmetric unit while in II both anions and cations lie on mirror planes. The $[PPh_4]^+$ and NHe_4^+ cations are both normal and the complex anions have similar geometries (Fig. 1). In the $[Rh_3(\mu_3\text{-}Se)_2(CO)_6]^-$anion a crystallographic mirror plane passes through Rh(1), Se(1) and Se(2). The triangular Rh_3 moiety is capped on each side by two triply bridging Se atoms. Each Rh atom displays square planar geometry with respect to two cis Se atoms and two terminally bonded CO groups. The mean bonding parameters for I, II are: Rh...Rh 3.05(2), 3.13(2); Rh-S(Se) 2.351(2), 2.458(1); Rh-C 1.829(6), 1.858(4); C-O 1.155(5), 1.135(6) Å; Rh-S(Se)-Rh 81.0(4), 79.2(5); C-Rh-C 92.4(5), 96.7(7); S(Se)-Rh-S(Se) 82.8(1), 85.1(2); S(Se)-Rh-C(cis) 92.4(3), 89.1(4); S(Se)-Rh-C(trans) 174.7(4), 174.1(4)°. The Rh...Rh distances mean that the two anions can be considered as trinuclear rhodium(I) aggregates rather than triangular clusters.

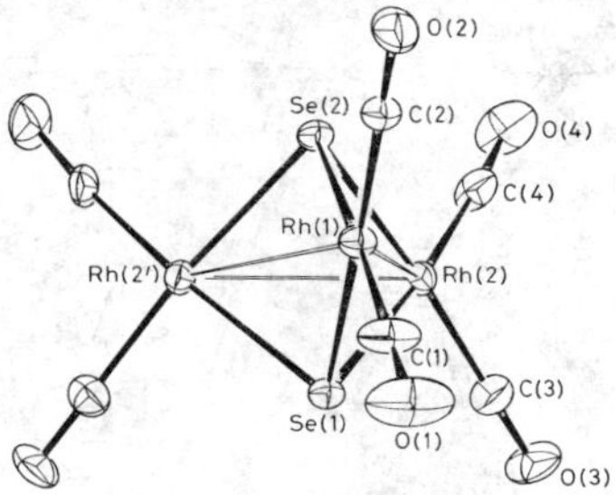

Fig. 1. $C_{10}H_{12}NO_6Rh_3Se_2$: a view of the anion.

TRIS(2,2'-BIPYRIDINE)IRON(II) (μ_2-OXO)BIS[TRICHLOROFERRATE(III)]

$C_{30}H_{24}Cl_6Fe_3N_6O$ $[Fe(bpy)_3][Fe_2OCl_6]$

H. WEISS and J. STRAHLE, 1984. Z. Naturforsch., B39, 1453-1455.

Monoclinic, $P2_1/c$, a = 9.641, b = 26.307, c = 13.789 Å, β = 99.74°, Z = 4. Mo radiation, R = 0.025 for 2491 reflexions (at -40°C).

The anion has a bent oxo-bridge (Fig. 1) with mean Fe-O 1.765, mean Fe-Cl 2.222 Å and Fe-O-Fe 148.1°. The cation has octahedral geometry with Fe-N 1.956-1.977 Å.

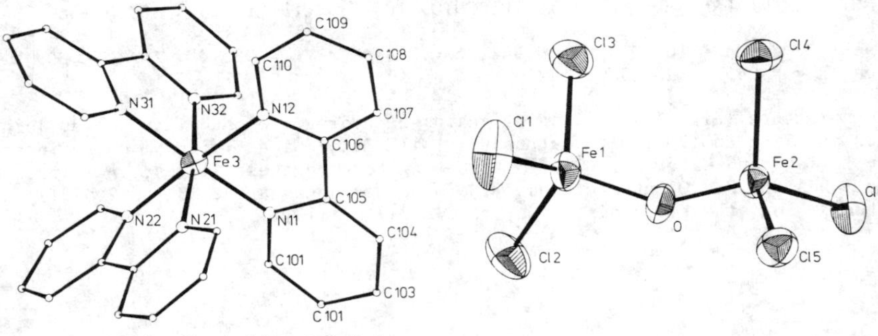

Fig. 1. The $[Fe(bpy)_3][Fe_2OCl_6]$ ions.

TRIS(2,2'-BIPYRIDINIUM) NITRATE HEXANITRATOTHORATE(IV)

$C_{30}H_{27}N_{13}O_{21}Th$ $[(bpyH)_3NO_3]^{2+}.[Th(NO_3)_6]^{2-}$

M.A. KHAN, N. KUMAR and D.G. TUCK, 1984. Can. J. Chem., 62, 850-855.

Triclinic, P$\bar{1}$, a = 9.300, b = 14.161, c = 17.037 Å, α = 108.79, β = 95.77, γ = 102.88°, Z = 2. Mo radiation, R = 0.029 for 5783 reflexions.

The structure of the $[Th(NO_3)_6]^{2-}$ anion (Fig. 1) is described as a nonsymmetric distorted icosahedron in which symmetric bidentate nitrato ligands occupy all the available sites. Three $bpyH^-$ cations are hydrogen-bonded to a nitrate ion to form the unusual $[(bpyH)_3NO_3]^{2+}$ cation. The Th-O bond lengths are in the range 2.540(4)-2.602(3) Å, average 2.567 Å. The bipyridinium cations have the two N atoms in a syn conformation.

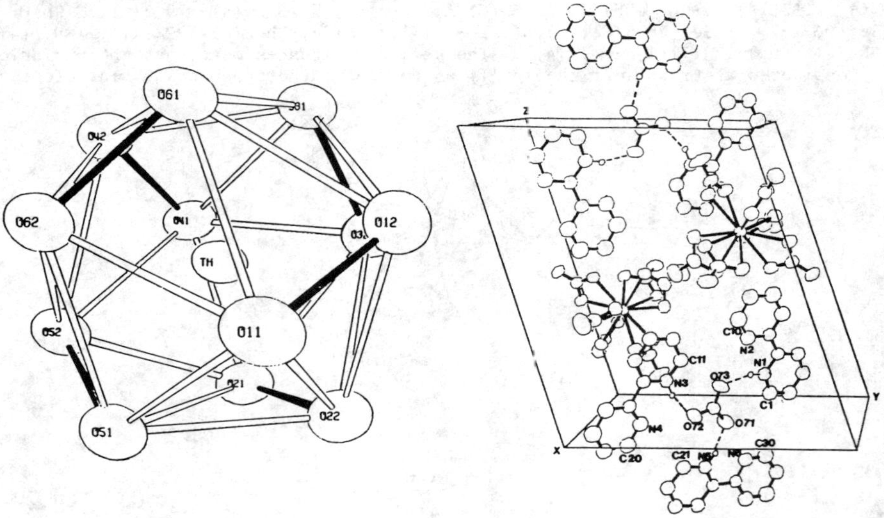

Fig. 1. The ThO_{12} polyhedron and the unit-cell packing of $C_{30}H_{27}N_{13}O_{21}Th$.

HEXAAQUANICKEL(II) DIPOTASSIUM TETRAHYDROGEN TETRA-o-PHTHALATE TETRAHYDRATE

$C_{32}H_{32}K_2NiO_{22},4H_2O$ $K_2[Ni(H_2O)_6][C_8H_5O_4]_4,4H_2O$

M.B. CINGI, A.M.M. LANFREDI and A TIRIPICCHIO, 1984. Acta Cryst., C40, 56-58.

Monoclinic, $P2_1/c$, a = 10.408, b = 6.864, c = 29.447 Å, β = 98.10°, Z = 2. Cu radiation, R = 0.045 for 3185 reflexions.

The structure (Fig. 1) consists of chains of hydrogen o-phthalate anions joined in slices parallel to (001) with octahedral $[Ni(H_2O)_6]^{2+}$ ions (with Ni on an inversion centre), K^+ ions and water molecules interposing between these slices, and linked by hydrogen bonds and K...O interactions. The Ni-O distances are 2.035 to 2.072(3) Å. Around the K^+ ion, the K...O distances are 2.679(7)-3.113(4) Å.

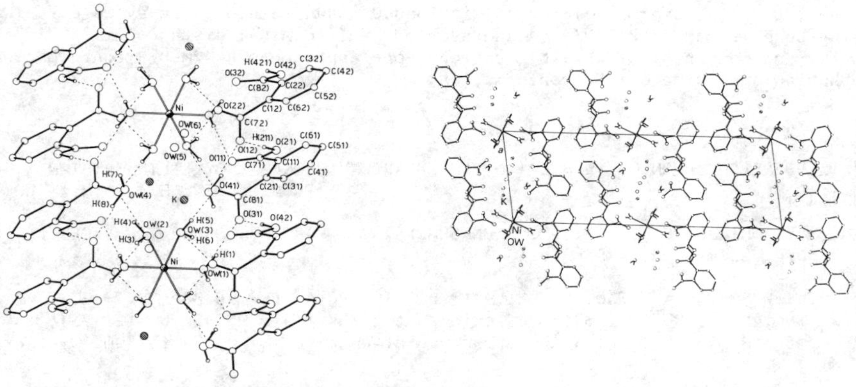

Fig. 1. Views of the $C_{32}H_{32}K_2NiO_{22},4H_2O$ structure.

BIS(TETRA-n-BUTYLAMMONIUM) DECACHLORODIOSMATE(IV)
$C_{32}H_{72}Cl_{10}N_2Os_2$ (I) $[n\text{-}Bu_4N]_2[Os_2Cl_{10}]$

TETRAETHYLAMMONIUM HEXACHLOROOSMATE(V)
$C_8H_{20}Cl_6NOs$ (II) $[Et_4N][OsCl_6]$

B. KREBS, G. HENKEL, M. DARTMANN, W. PREETZ and M. BRUNS, 1984. Z. Naturforsch., B39, 843-849.

I. Orthorhombic, Pbca, a = 15.446, b = 15.965, c = 19.431 Å, Z = 4. Mo radiation, R = 0.044 for 3121 reflexions.

II. Monoclinic, $P2_1$, a = 7.020, b = 10.163, c = 11.609 Å, β = 103.91°, Z = 2. Mo radiation, R = 0.027 for 1794 reflexions (at -133°C).

The $[Os_2Cl_{10}]^{2-}$ ion in (I) lies about an inversion centre. The Os atom has octahedral coordination with Os-Cl (bridge) 2.410 and 2.412(2), Os-Cl (terminal) 2.296-2.308(2) Å. The Os...Os separation is 3.626(1) Å. The Os atom in (II) also has octahedral geometry, with Os-Cl 2.295-2.308(3) Å, mean value 2.303 Å.

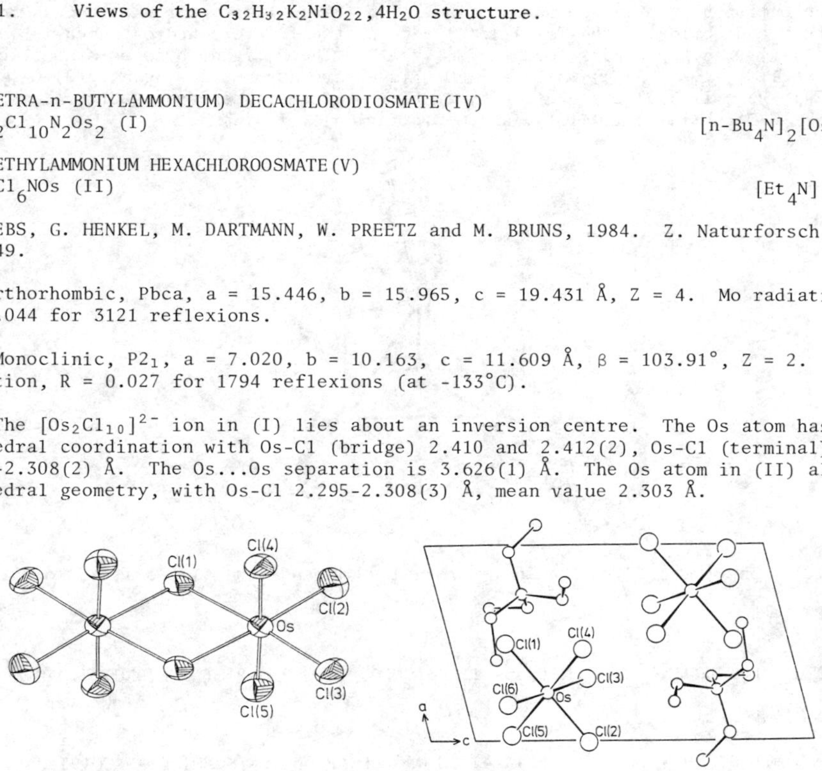

Fig. 1. A view of the $[Os_2Cl_{10}]^{2-}$ anion and of the unit cell for $[Et_4N][OsCl_6]$.

TETRA(n-BUTYL)AMMONIUM TETRAIODO-μ,-μ'-DIIODODIPLATINATE(II)
$C_{32}H_{72}I_6N_2Pt_2$ $[(C_4H_9)_4N]_2[Pt_2I_6]$

R.D. ROGERS, H. ISCI and W.R. MASON, 1984. J. Cryst. Spectrosc. Res., 14, 383-392.

Monoclinic, $P2_1/c$, a = 14.457, b = 14.036, c = 23.696 Å, β = 101.02°, Z = 4. Mo radiation, R = 0.049 for 1786 reflexions.

The $[Pt_2I_6]^{2-}$ anion is planar to within 0.07 Å and exhibits an average Pt-I bridging bond length of 2.559(4) Å and average Pt-I terminal distance of 2.571(7) Å. Of the two independent $[(n\text{-}C_4H_9)_4N]^+$ cations, one approaches a $\bar{4}2m\text{-}D_{2d}$ conformation, the other an approximate $\bar{4}\text{-}S_4$ conformation.

BENZYLTRIPHENYLPHOSPHONIUM (μ-HYDRIDO)BIS(μ-CARBONYL)NONACARBONYLTETRAIRIDIUM
$C_{36}H_{23}Ir_4O_{11}P$ $[P(CH_2C_6H_5)(C_6H_5)_3]^+[HIr_4(CO)_{11}]^-$

R. BAU, M.Y. CHIANG, C.-Y. WEI, L. GARLASCHELLI, S. MARTINENGO and T.F. KOETZLE, 1984. Inorg. Chem., 23, 4758-4762.

Monoclinic, $P2_1/c$, a = 9.330, b = 21.971, c = 19.190 Å, β = 96.86°, D_m = 2.37, Z = 4. Mo radiation, R = 0.072 for 3714 reflexions at 298 K. a = 9.243, b = 21.731, c = 18.997 Å, β = 97.12°. 1.16173 Å neutron radiation, $R(F^2)$ = 0.074 for 9627 reflexions at 80 K.

The cation has normal geometry. The anion (Fig. 1) has approximate C_s symmetry with the H atom bridging the Ir(3)-Ir(4) edge of the Ir_4 tetrahedron which is longer (2.800(1) Å) than the others (mean 2.723(17) Å). The H ligand bridges symmetrically (Ir-H 1.821(3), 1.847(3) Å) while the CO bridges are asymmetric (mean Ir(2)-C 2.167(24), Mean Ir(3/4)-C 2.020(22) Å). Other mean distances are Ir-C(terminal) 1.894(6), C-O(terminal) 1.141(1), and C-O(bridging) 1.179(2) Å.

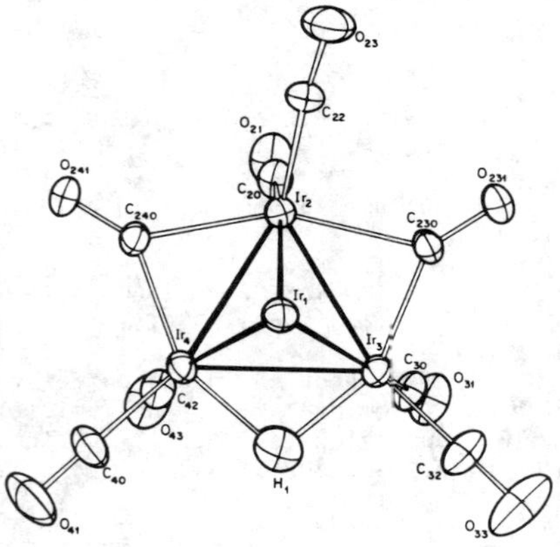

Fig. 1. The $[HIr_4(CO)_{11}]^-$ anion, terminal CO ligands on Ir(1) omitted for clarity.

BIS(TRIPHENYLPHOSPHINE)IMINIUM trans-1,2-DICHLORO-1,3,2,3-BIS(μ_2-HYDRO)TETRAHYDRO-TRIBORATE
$C_{36}H_{36}B_3Cl_2NP_2$ (I)

BIS(TRIPHENYLPHOSPHINE)IMINIUM trans-2-CHLORO-HEXAHYDRO-1-ISOTHIOCYANATOTRIBORATE DICHLOROMETHANE SOLVATE
$C_{37}H_{36}B_3ClN_2P_2S,CH_2Cl_2$ (II)

M. ARUNCHAIYA, J.H. MORRIS, S.J. ANDREWS, D.A. WELCH and A.J. WELCH, 1984. J. Chem. Soc. Dalton, 2525-2533.

I. Monoclinic, $P2_1/c$, a = 10.955, b = 12.840, c = 24.236 Å, β = 92.298°, Z = 4. Mo radiation, R = 0.073 for 3267 reflexions.

II. Triclinic, $P\bar{1}$, a = 11.285, b = 14.400, c = 15.708 Å, α = 120.68, β = 109.76, γ = 65.971°, Z = 2. Mo radiation, R = 0.192 for 5706 reflexions (at 185 K).

In the $[B_3H_6Cl_2]^-$ anion of I the Cl atoms are 1,2-substituents with a mean B-Cl distance of 1.851(9) Å. The two μ-H atoms asymmetrically bridge (Fig. 1) the B(1)-B(3) (1.75(2)) and B(2)-B(3) (1.76(2) Å) bonds. The B(1)-B(2) distance is 1.80(1) Å. The ion has overall C_2 symmetry about the axis from B(3) to the midpoint of B(1)-B(2). The $[N(PPh_3)_2]^+$ cation has effective C_2 symmetry. Serious disorder in the NCS^- group in II resulting in the very high R factor precludes discussion of that compound.

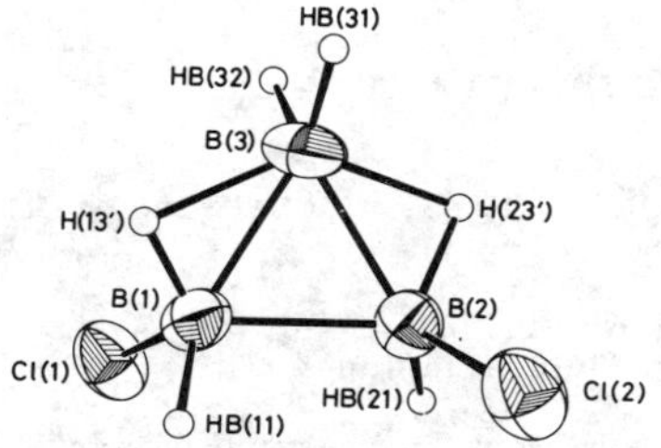

Fig. 1. $C_{36}H_{36}B_3Cl_2NP_2$: a view of the $[B_3H_6Cl_2]^-$ anion.

BIS(TRIPHENYLPHOSPHINE)IMINIUM 1-ISOTHIOCYANATOTRIBORANE(7)
$C_{37}H_{37}B_3N_2P_2S$ (I) $[((C_6H_5)_3P)_2N]^+[B_3H_7NCS]^-$

BIS(TRIPHENYLPHOSPHINE)IMINIUM 1-ISOSELENOCYANATOTRIBORANE(7)
$C_{37}H_{37}B_3N_2P_2Se$ (II) $[((C_6H_5)_3P)_2N]^+[B_3H_7NCSe]^-$

S.J. ANDREWS and A.J. WELCH, 1984. Inorg. Chim. Acta, 88, 153-160.

I. Triclinic, $P\bar{1}$, a = 10.065 (10.021), b = 11.064 (10.903), c = 17.572 (17.482) Å, α = 109.23 (109.06), β = 101.37 (101.41), λ = 77.02 (76.48)°, Z = 2. Mo radiation, R = 0.056 (0.047) for 4479 (5515) reflexions for data at 291 K and (in parentheses) at 185 K.

II. Triclinic, $P\bar{1}$, a = 10.163 (10.129), b = 11.044 (10.887), c = 17.605 (17.488) Å, α = 109.13 (108.84), β = 101.09 (101.29), γ = 76.29 (75.76)°, Z = 2. Mo radiation, R = 0.062 (0.049) for 4038 (4888) reflexions, for data at 291 K and (in parentheses) at 185 K.

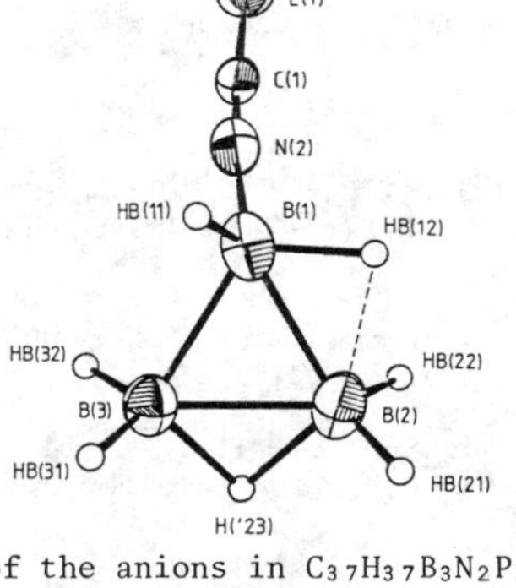

Fig. 1. Perspective view of the anions in $C_{37}H_{37}B_3N_2P_2S$ and $C_{37}H_{37}B_3N_2P_2Se$ using low temperature data. E is S in I and Se in II.

Analysis of room and low temperature data suggests different stereochemical arrangements of the borane H atoms; one edge- and one face-bridging H is suggested from the room temperature data sets, whilst the low temperature data reveal only one (edge-) bridging H atom for each compound. The anions in I and II are similar (Fig. 1) with a boron triangle which is markedly isosceles. The average B-B distances are B(1)-B(2) 1.790, B(1)-B(3) 1.800 and B(2)-B(3) 1.762 Å. All terminal B-H bonds are of expected length, and there is, in addition, a weak, semi-bridging interaction between B(2) and HB(12), 1.66 Å in I and 1.89 Å in II. Bond B(3)-HB(11) is 2.26 Å in I and 2.28 Å in II. The BNCS and BNCSe fragments are nearly linear and have normal interatomic distances. There are no close contacts between ion pairs.

BIS(METHYLTRIPHENYLPHOSPHONIUM) TRIIODOCUPRATE(I)
$C_{38}H_{36}CuI_3P_2$

G.A. BOWMAKER, G.R. CLARK, D.A. ROGERS, A. CAMUS and N. MARSICH, 1984. J. Chem. Soc. Dalton, 37-45.

Monoclinic, $P2_1$, a = 9.580, b = 12.476, c = 16.410 Å, β = 102.71°, D_m = 1.73, Z = 2. Mo radiation, R = 0.034 for 2261 reflexions.

The crystals are ionic and contain the mononuclear anion $[CuI_3]^{2-}$ and $[PMePh_3]^+$ cations (Fig. 1). The cations have normal geometry with an average P-C distance of 1.797 Å, average C-P-C angle of 109.5°, and planar phenyl rings with the P atom lying approximately in the plane of each ring. They do not however adopt the usual 'propeller blade' orientation. Instead, the plane of one ring in each cation lies perpendicular to the Me-P line, one is parallel and the third adopts an intermediate orientation. The $[CuI_3]^{2-}$ ion has planar geometry with each I atom coordinated to a central trigonally coordinated Cu. The ion does not possess strict threefold symmetry. Packing of ions into the unit cell is such that there are far more cation-cation than cation-anion contacts.

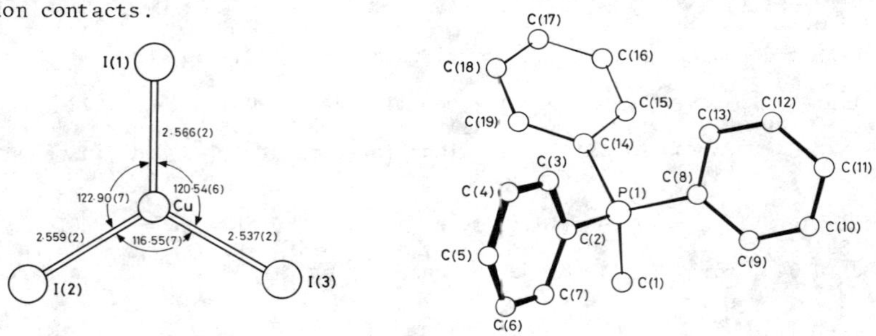

Fig. 1. $C_{38}H_{36}CuI_3P_2$: the anion and cation.

METHYLTRIPHENYLPHOSPHONIUM 1,1,1,2,2,2,3,3,4,4-DECACARBONYL-3,4-[μ_2-(μ_2-HYDRIDO)BIS-(TRICARBONYLOSMIUM)]-1,2,3.1,2,4-BIS(μ_3-TRICARBONYLOSMIUM)-tetrahedro-TETRAOSMIUM
$C_{41}H_{19}O_{22}Os_8P$

B.F.G. JOHNSON, J. LEWIS, W.J.H. NELSON, M.D. VARGAS, D. BRAGA, K. HENRICK and M. McPARTLIN, 1984. J. Chem. Soc. Dalton, 2151-2161.

Triclinic, $P\bar{1}$, a = 16.775, b = 13.509, c = 11.267 Å, α = 89.88, β = 103.97, γ = 96.99°, Z = 2. Mo radiation, R = 0.036 for 5615 reflexions.

The Os_8 core in the $[Os_8H(CO)_{22}]^-$ anion consists of fused tetrahedra (Fig. 1). Three of the tetrahedra are linked by sharing faces to give a bicapped tetrahedral unit. The fourth shares an edge with the basic Os_6 unit forming a very short Os(3)-Os(4) bond (2.600(1) Å). The position of the hydride ligand was inferred from the characteristic displacement of the close-packed carbonyl ligands produced by the steric requirements of the 'invisible' H atom. The CO ligands are apparently pushed back from the Os(7)-Os(8) bond (mean cis C-Os-Os angle 106.9°) (Fig. 1, right).

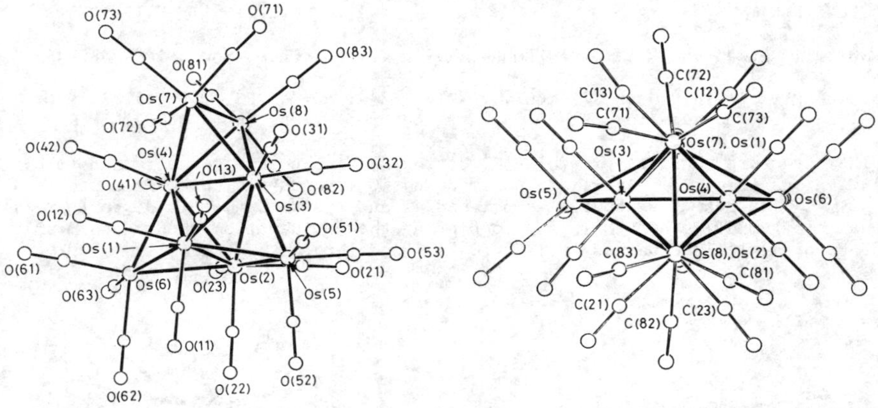

Fig. 1. $C_{41}H_{19}O_{22}Os_8P$: the anion (left) and a view down the virtual C_2 axis to show displacement of the CO ligands caused by H ligand bridging Os(7)-Os(8).

BIS(TRIPHENYLPHOSPHINE)IMINIUM AZIDOPENTACARBONYLTUNGSTEN
$C_{41}H_{30}N_4O_5P_2W$ $\{(C_6H_5)_3P\}_2N^+[W(CO_5)N_3]^-$

D. SELLMANN, W. WEBER, G. LIEHR and H.P. BECK, 1984. J. Organometal. Chem., 269, 155-170.

Triclinic, $P\bar{1}$, a = 12.435, b = 16.790, c = 9.850 Å, α = 103.88, β = 77.87, γ = 106.0°, D_m = 1.5, Z = 2. Ag radiation, R = 0.037 for 3795 reflexions.

The complex consists of discrete anions and cations. The tungsten atom in the anion is coordinated pseudo-octahedrally by five CO ligands and the azide group (Fig. 1). Bond lengths and angles indicate a covalent σ-W-N bond, the azide group exerting a strong trans-effect. The bond distances in the N_3 ligand are remarkably similar (1.185 and 1.160 Å) despite the covalent $W-N_3$ bonding mode.

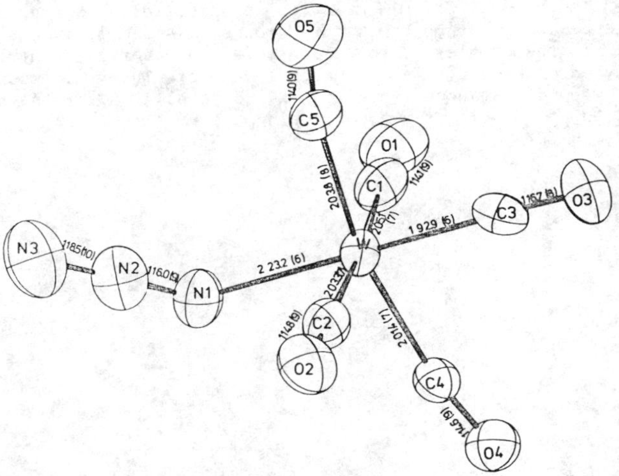

Fig. 1. Structure of the anion in $C_{41}H_{30}N_4O_5P_2W$ showing bond lengths (pm).

BIS(TRIPHENYLPHOSPHINE)IMINIUM (μ-HYDRIDO)NONACARBONYLIRONTUNGSTEN
$C_{45}H_{31}FeNO_9P_2W$ $[((C_6H_5)_3P)_2N][(\mu\text{-}H)[Fe(CO)_4W(CO)_5]]$

J. ARNDT, T. DELORD and M.Y. DARENSBOURG, 1984. J. Am. Chem. Soc., 106, 456-457.

Triclinic, $P\bar{1}$, a = 14.099, b = 11.601, c = 14.148 Å, α = 95.53, β = 118.23, γ = 92.39°, Z = 2. Mo radiation, R = 0.059 for 3635 reflexions (at -103°C).

The structure consists of $(Ph_3P)_2N^+$ cations having normal geometry and $(\mu\text{-}H)Fe\text{-}W(CO)_9^-$ anions (Fig. 1). There are no bridging CO ligands and the W-Fe distance is 2.989(2) Å. The geometry about W is octahedral and that about Fe is distorted trigonal bipyramidal, the μ-H ligand completing the coordination about each metal. The C7-Fe-C8 angle is expanded to 147.6(8)°. Bond lengths are normal.

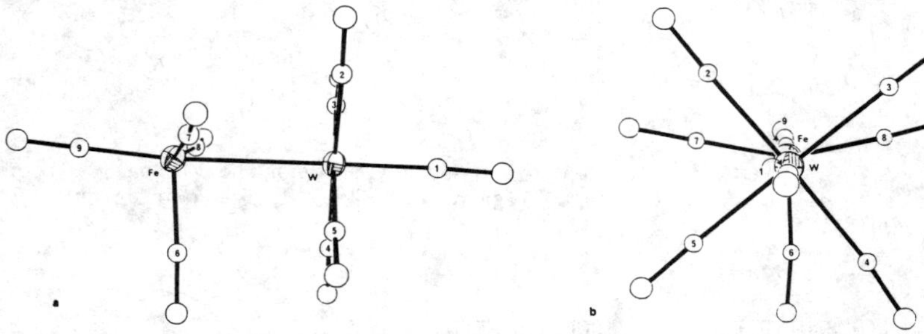

Fig. 1. Two views of the $(\mu\text{-}H)FeW(CO)_9^-$ anion.

TRIS(1,10-PHENANTHROLINE)POTASSIUM μ_2-HYDRIDOBIS[PENTACARBONYLCHROMIUM(0)]
$C_{46}H_{25}Cr_2KN_6O_{10}$ $[K(phen)_3]^+[Cr_2(CO)_{10}(H)]^-$

G. BOMBIERI, G. BRUNO, M.D. GRILLONE and G. POLIZZOTTI, 1984. J. Organometal. Chem., 273, 69-80.

Triclinic, $P\bar{1}$, a = 11.884, b = 13.968, c = 15.612 Å, α = 112.6, β = 99.7, γ = 106.2°, Z = 2. Mo radiation, R = 0.066 for 3692 reflexions.

The complex anion adopts a bent Cr-H-Cr configuration with the equatorial CO groups of the two independent $Cr(CO)_5$ moieties in an eclipsed configuration (Fig. 1). The Cr...Cr distance is 3.394 Å, the two Cr-H bonds 1.74 and 1.71 Å, and the Cr-H-Cr bond angle 159°. In the complex cation, located on an inversion centre, the coordination polyhedron around K is unexpectedly cubic. The K...K' separation is 4.15 Å and the K-N bond distances range from 2.835 to 3.117 Å for the non-bridging N atoms and from 2.907 to 3.342 Å for the bridging N's.

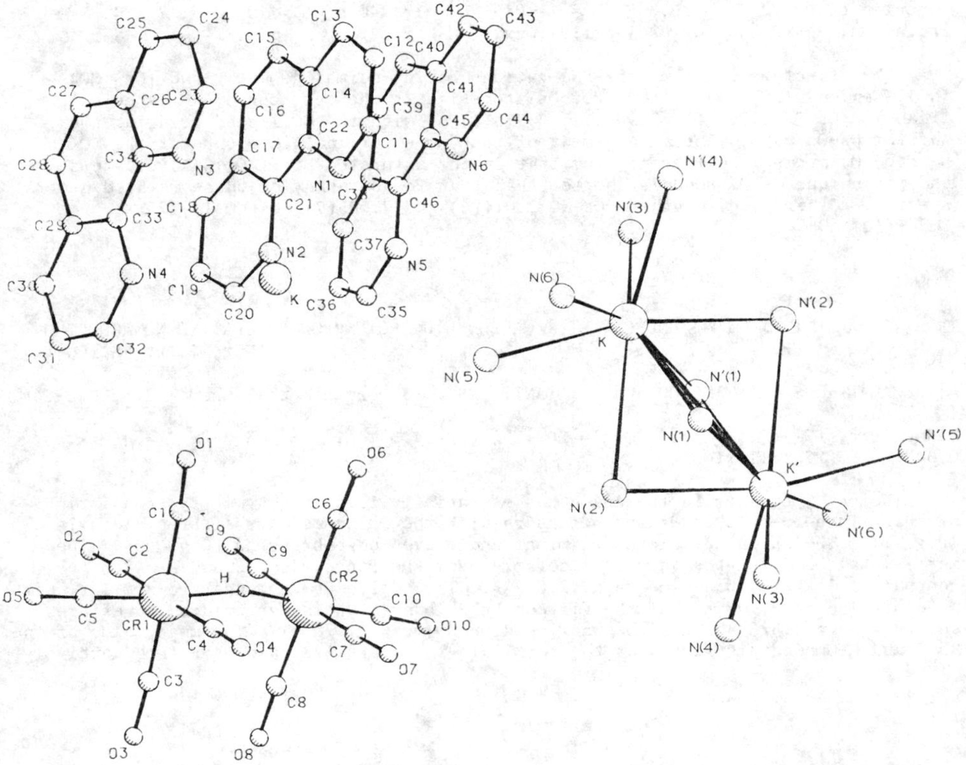

Fig. 1. View down a of $C_{46}H_{25}Cr_2KN_6O_{10}$ (left), and the K coordination polyhedron (right).

BIS(TRIPHENYLPHOSPHINE)IMINIUM DECACARBONYLNITROSYLTRIOSMATE(3Os-Os)

$C_{46}H_{30}N_2O_{11}Os_3P_2$ $[(P(C_6H_5)_3)_2N]^+[Os_3(CO)_{10}(NO)]^-$

B.F.G. JOHNSON, J. LEWIS, J.M. MACE, P.R. RAITHBY, R.E. STEVENS and W.L. GLADFELTER, 1984. Inorg. Chem., 23, 1600-1603.

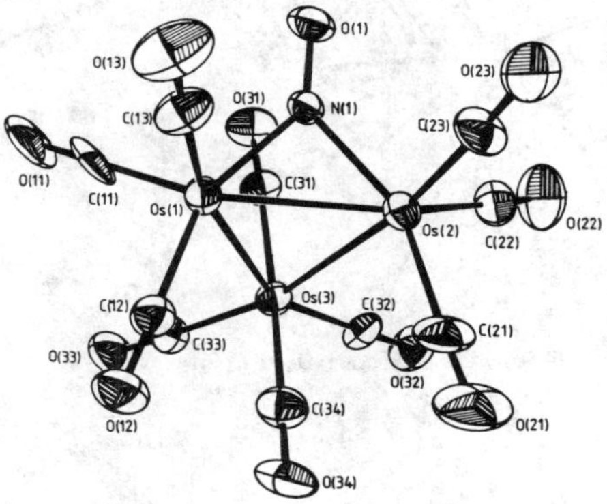

Fig. 1. The $[Os_3(CO)_{10}(\mu\text{-}NO)]^-$ anion.

Monoclinic, $P2_1/c$, a = 16.848, b = 17.015, c = 17.740 Å, β = 107.25°, Z = 4. Mo radiation, R = 0.045 for 6261 reflexions.

The structure consists of PPN^+ cations having normal geometry and $[Os_3(CO)_{10}(NO)]^-$ anions (Fig. 1). The three Os atoms lie at the vertices of an isoceles triangle (Os-Os 2.751(1), 2.906(1), 2.915(1) Å) with the NO ligand bridging the shorter Os-Os edge. The arrangement of the three terminal CO groups at the NO bridged Os atoms is different from that in the saturated $HOs_3(CO)_{10}(X)$ clusters, and the unique Os-Os bond is shorter than that found in any of these related neutral species. Other bond lengths: Os-NO 2.014(6) and 2.005(7), Os-CO 1.912(12)-1.963(10) Å.

TRIS(TETRAETHYLAMMONIUM) UNDECAKIS(μ_2-CARBONYL)UNDECACARBONYLDIPLATINUMNONARHODIUM

$C_{46}H_{60}N_3O_{22}Pt_2Rh_9$ $[NEt_4{}^+]_3[Pt_2Rh_9(CO)_{22}]^{3-}$

A. FUMAGALLI, S. MARTINENGO and G. CIANI, 1984. J. Organometal. Chem., 273, C46-C48.

Monoclinic, $P2_1/c$, a = 13.587, b = 20.175, c = 24.351 Å, β = 103.48°, Z = 4. R = 0.047 for 2322 reflexions.

The metal cluster in the anion has essentially ideal D_{3h} symmetry, consisting of three face-to-face condensed octahedra with the Pt atoms on the threefold axis (Fig. 1). The anion has eleven terminal and eleven edge-bridging CO groups. The metal-metal bond lengths vary considerably over the range 2.654-2.997 Å, with the central Pt(1)-Pt(2) bond length at 2.812 Å. The twelve Pt-Rh bonds are similar, with a mean value of 2.707 Å. Of the fifteen Rh-Rh bonds, the seven bridged edges are shorter (mean 2.822 Å) than the unbridged ones (mean 2.878 Å). The mean value of the Rh-C and C-O bond distances for the terminal groups are 1.82 and 1.15 Å respectively.

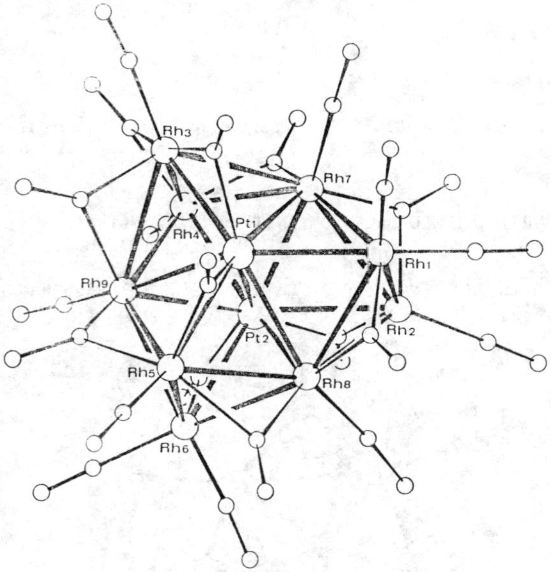

Fig. 1. A view of the anion in $C_{46}H_{60}N_3O_{22}Pt_2Rh_9$.

BIS(TRIPHENYLPHOSPHINE)IMINIUM TRIS(μ_2-CARBONYL)(NITROSYLIRON)NONACARBONYLTRIRUTHENIUM
$C_{48}H_{30}FeN_2O_{13}P_2Ru_3$ (I) PPN.$[FeRu_3(CO)_{12}(NO)]$

TETRAETHYLAMMONIUM (μ_4-NITRIDO)DODECACARBONYLIRONTRIRUTHENIUM
$C_{20}H_{20}FeN_2O_{12}Ru_3$ (II) $Et_4N.[FeRu_3N(CO)_{12}]$

D.E. FJARE and W.L. GLADFELTER, 1984. J. Am. Chem. Soc., 106, 4799-4810.

I. Monoclinic, $P2_1/c$, a = 19.690, b = 16.262, c = 15.911 Å, β = 107.24°, Z = 4. Mo radiation, R = 0.046 for 7125 reflexions.

II. Triclinic, $P\bar{1}$, a = 12.464, b = 12.697, c = 9.437 Å, α = 94.40, β = 100.17, γ = 107.75°, Z = 2. Mo radiation, R = 0.022 for 4358 reflexions.

In (I) (Fig. 1) there is a linear nitrosyl bound to Fe (Fe-N 1.643(4) Å) and the molecule has approximate threefold symmetry. In (II) there is a butterfly framework of metal atoms and the Fe and Ru atoms are mutually disordered. The nitrido N atom is coordinated to all four metals (N-M1 2.067(2), N-M(2) 2.057(2), N-M3 1.872(2), N-M4 1.846(2) Å).

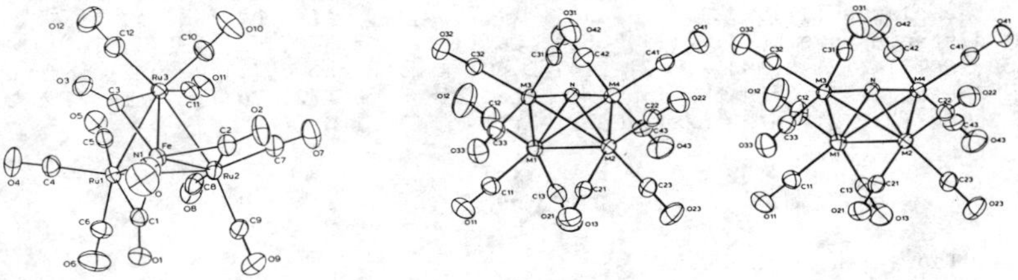

Fig. 1. A perspective view of $[FeRu_3(CO)_{12}(NO)]$ and a stereoview of $[Fe_3Ru_3N(CO)_{12}]^-$.

TETRAPHENYLARSONIUM HEXACHLOROTECHNETATE(IV)
$C_{48}H_{40}As_2Cl_6Tc$ $[As(C_6H_5)_4{}^+]_2[TcCl_6{}^{2-}]$

J. BALDAS, J. BONNYMAN, D.L. SAMUELS and G.A. WILLIAMS, 1984. Acta Cryst., C40, 1343-1346.

Triclinic, $P\bar{1}$, a = 10.111, b = 12.165, c = 10.263 Å, α = 93.86, β = 114.51, γ = 99.08°, Z = 1. Cu radiation, R = 0.025 for 3590 reflexions.

The centrosymmetric $TcCl_6{}^{2-}$ anion shows small deviations from ideal octahedral geometry with Tc-Cl bond lengths 2.3441, 2.3635 and 2.3649(7) Å and Cl-Tc-Cl angles of 88.63, 89.95 and 90.26(3)°. These deviations arise from close contacts (C...Cl 3.5, H...Cl 2.8 Å) with the bulky $As(C_6H_5)_4{}^+$ cations.

TETRAPHENYLARSONIUM HEXACHLOROTELLURATE(IV)
$C_{48}H_{40}As_2Cl_6Te$ $[Ph_4As^+]_2.TeCl_6{}^{2-}$

J. MILNE, E. PHILIPPOT and M. MAURIN, 1984. Rev. Chim. Miner., 21, 749-757.

Triclinic, $P\bar{1}$, a = 10.181, b = 10.348, c = 12.237 Å, α = 93.26, β = 99.41, γ = 115.06°, D_m = 1.60, Z = 1. Mo radiation, R = 0.041 for 4211 reflexions.

The anion has distorted octahedral symmetry and possesses a centre of inversion with two short Te-Cl bonds (2.518(3) Å) and two pairs of long Te-Cl bonds (2.544(3) and 2.548(3) Å), Cl-Te-Cl angles range from 88.27(9) to 91.73(9)°. The different

Te-Cl distances are caused by the repulsive interactions between Cl and the phenyl groups of the cation which has approximate tetrahedral symmetry (As-C(1) 1.919(8), As-C(7) 1.913(4), As-C(13) 1.904(8), As-C(19) 1.923(9) Å; C-As-C angles range from 106.3(4) to 113.6(4)°). The phenyl groups are rotated about the As-C axis in such a way as to reduce interactions between H atoms attached to C atoms ortho to As on different phenyl groups.

BIS(TETRAPHENYLARSONIUM) DI-μ-IODODIIODODICUPRATE(I)

$C_{48}H_{40}As_2Cu_2I_4$ $[As(C_6H_5)_4]_2^+Cu_2I_4^{2-}$

M. ASPLUND and S. JAGNER, 1984. Acta Chem. Scand., A38, 297-301.

Monoclinic, $P2_1/c$, a = 12.040, b = 18.046, c = 22.683 Å, β = 102.73°, Z = 4. Mo radiation, R = 0.058 for 4345 reflexions.

The anion in the complex is a dimer, and is neither centrosymmetric nor planar (Fig. 1). Each Cu atom has approximate trigonal planar coordination with Cu(1) displaced 0.04 Å from the plane through I(1), I(2) and I(3), and Cu(2) 0.07 Å from that through I(1), I(2) and I(4). The anion is folded about the I(1)...I(2) vector such that the planes I(1),I(2),I(3) and I(1),I(2),I(4) are inclined at 147°. The important distances are Cu(1)-I(1) 2.610, Cu(1)-I(2) 2.578, Cu(1)-I(3) 2.490, Cu(2)-I(1) 2.609, Cu(2)-I(2) 2.584, Cu(2)-I(4) 2.491 Å with the Cu(1)...Cu(2) interaction 2.663 Å. The angles within the four-membered ring are I(2)-Cu-I(1) 114.4, I(2)-Cu(2)-I(1) 114.2, Cu(1)-I(2)-Cu(2) 62.1, Cu(1)-I(1)-Cu(2) 61.4°.

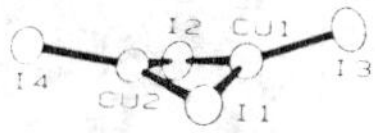

Fig. 1. The anion in $C_{48}H_{40}As_2Cu_2I_4$.

BIS(TETRAPHENYLARSONIUM) BIS[TRIAZIDO(μ_2-TRIAZADITHIA-N,N,N')VANADATE(VI)] DICHLOROMETHANE SOLVATE

$C_{48}H_{40}As_2N_{24}S_4V_2,CH_2Cl_2$ $[Ph_4As]_2[V(N_3)_3(N_3S_2)]_2,CH_2Cl_2$

J. HANICH, M. KRESTEL, U. MULLER, K. DEHNICKE and D. REHDER, 1984. Z. Naturforsch., B39, 1686-1695.

Triclinic, $P\bar{1}$, a = 10.87, b = 13.17, c = 13.50 Å, α = 58.8, β = 85.9, γ = 68.0°, Z = 1. Mo radiation, R = 0.058 for 1496 reflexions.

The anion lies about an inversion centre (Fig. 1). The N_3S_2 ligands are bonded to V atoms with short V=N bonds (1.89 and 1.72(1) Å) forming planar VN_3S_2 rings. Dimers are formed through V-N donor-accepted interactions (V-N 2.24(1) Å) involving one N atom from each VN_3S_2 ring. Vanadium coordination of 6 is completed by azido groups, with V-N 2.00-2.04 Å.

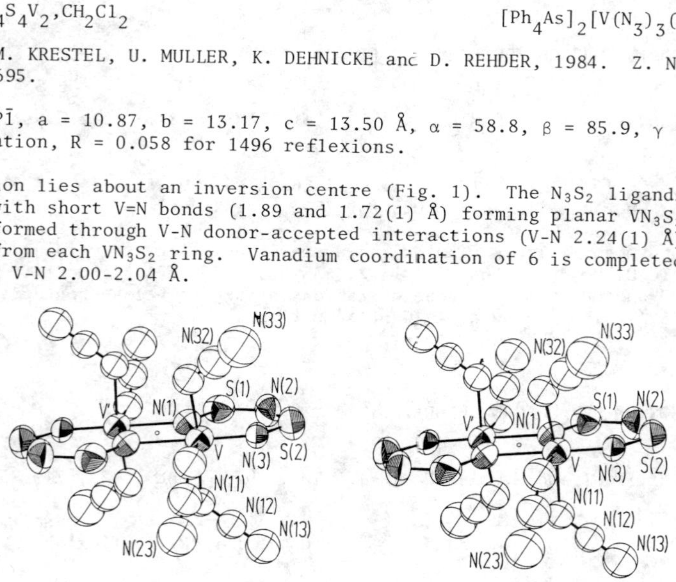

Fig. 1. A stereoview of the $[V(N_3)_3(N_3S_2)]_2^{2-}$ ion.

BIS(TETRAPHENYLPHOSPHONIUM) TETRABROMOSELENATE
$C_{48}H_{40}Br_4P_2Se$ (I) $[PPh_4]_2[SeBr_4]$

BIS(TETRAPHENYLPHOSPHONIUM) BIS(μ_2-BROMO)TETRABROMODISELENATE
$C_{48}H_{40}Br_6P_2Se_2$ (II) $[PPh_4]_2[Se_2Br_6]$

B. KREBS, A. SCHAFFER and S. POHL, 1984. Z. Naturforsch., B39, 1633-1641.

I. Triclinic, $P\bar{1}$, a = 14.190, b = 15.125, c = 10.383 Å, α = 92.69, β = 100.89, γ = 92.04°, D_m = 1.63, Z = 2. Mo radiation, R = 0.057 for 5190 reflexions.

II. Triclinic, $P\bar{1}$, a = 10.933, b = 11.154, c = 10.344 Å, α = 79.26, β = 72.63, γ = 81.35°, D_m = 1.81, Z = 1. Mo radiation, R = 0.034 for 4215 reflexions (at -120°C).

There are two independent $SeBr_4^{2-}$ ions in I, each lying on an inversion centre; consequently each ion is strictly planar. Dimensions are in Fig. 1. In compound II the $Se_2Br_6^{2-}$ ion lies about an inversion centre. The deviations of the Se and Br atoms from the best plane of the Se_2Br_6 unit is 0.01-0.03 Å. Dimensions are in Fig. 1; the Se-Br-Se angle is 94.12(2)°.

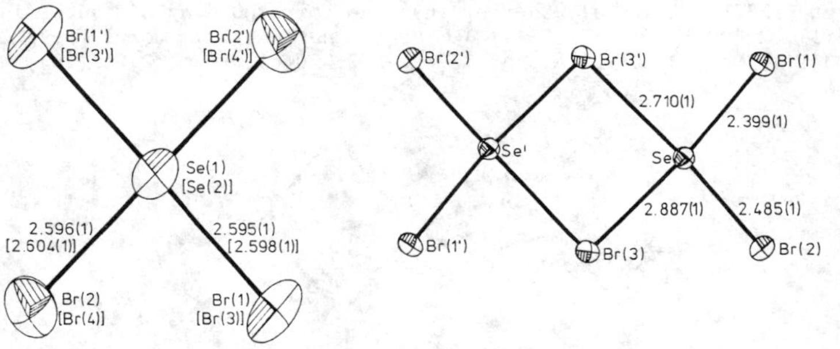

Fig. 1. Bond lengths (pm) and views of the $SeBr_4^{2-}$ and Se Br_6^{2-} ions.

BIS(TETRAPHENYLPHOSPHONIUM) (DICHLORODITHIOFERRATO-S,S')DISULPHIDOOXOMOLYBDATE(2-) METHYL CYANIDE SOLVATE
$C_{48}H_{40}Cl_2FeMoOP_2S_4,C_2H_3N$ $[P(C_6H_5)_4]_2,[FeCl_2\{MoO(S_2)\}(S)_2],C_2H_3N$

W. CLEGG, C.D. GARNER and P.A. FLETCHER, 1984. Acta Cryst., C40, 754-756.

Triclinic, $P\bar{1}$, a = 10.436, b = 12.486, c = 20.277 Å, α = 85.04, β = 75.12, γ = 81.38°, Z = 2. Mo radiation, R = 0.066 for 6036 reflexions.

In the anion the $FeCl_2$ and $(S_2)Mo(O)$ units are linked by two μ_2-S bridges with Fe...Mo 2.746(1) Å (Fig. 1). There is a folded central FeS_2Mo ring. There are no particularly short contacts between the various ions and solvent molecules in the crystal structure. The structure of the same salt as dimethylformamide solvate has recently been reported (1).

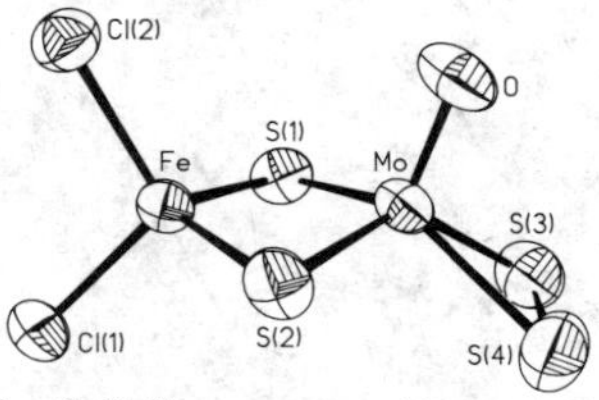

Fig. 1. $C_{48}H_{40}Cl_2FeMoOP_2S_4,C_2H_3N$: perspective view of the anion.

1. A. MÜLLER, S. SARKAR, H. BÖGGE, R. JOSTES, A. TRAUTWEIN and V. LAVER, 1983. Angew. Chem. Inst. Ed. Engl., 22, 561.

BIS(TETRAPHENYLPHOSPHONIUM) TETRAKIS(μ-THIO)[BIS(DICHLOROFERRATE(II))MOLYBDATE(VI)]
$C_{48}H_{40}Cl_4Fe_2MoP_2S_4$ (I) $[(C_6H_5)_4P]_2[Cl_2FeS_2MoS_2FeCl_2]$

BIS(TETRAPHENYLPHOSPHONIUM) TETRAKIS(μ-THIO)[BIS(DICHLOROFERRATE(II))TUNGSTATE(VI)]
$C_{48}H_{40}Cl_4Fe_2P_2S_4W$ (II) $[(C_6H_5)_4P]_2[Cl_2FeS_2WS_2FeCl_2]$

D. COUCOUVANIS, E.D. SIMHON, P. STREMPLE, M. RYAN, D. SWENSON, N.C. BAENZIGER, A. SIMOPOULOS, V. PAPAEFTHYMIOU, A. KOSTIKAS and V. PETROULEAS, 1984. Inorg. Chem., 23, 741-749.

I. Triclinic, $P\bar{1}$, a = 10.197, b = 13.054, c = 20.164 Å, α = 100.46. β = 100.61, γ = 100.92°, D_m = 1.50, Z = 2. Mo radiation, R = 0.058 for 5125 reflexions.

II. Triclinic, $P\bar{1}$, a = 10.215, b = 13.071, c = 20.187 Å, α = 100.54, β = 100.58, γ = 100.97°, D_m = 1.61, Z = 2. Mo radiation, R = 0.050 for 9711 reflexions.

The isomorphous structures consist of Ph_4P^+ cations having normal geometry and trinuclear complex anions (Fig. 1) consisting of tetrahedral MS_4^{2-} units bridging two $FeCl_2$ groups. The Fe-Mo distances in I at 2.771(2) and 2.781(2) Å and the Fe-Fe distance at 5.552(2) Å are all shorter than the corresponding distances in II, which are 2.795(1), 2.808(1), and 5.603(1) Å. Mean bond lengths: for I, Fe-S 2.296(4), Fe-Cl 2.234(1), Mo-S 2.206(2) Å, for II, Fe-S 2.320(6), Fe-Cl 2.233(10), W-S 2.209(5) Å.

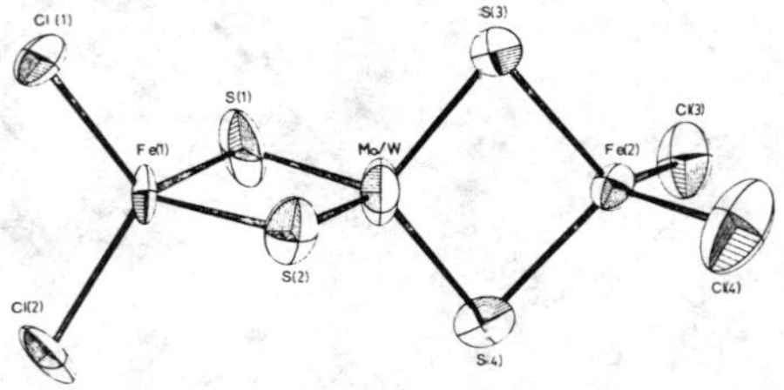

Fig. 1. The $[Cl_2FeS_2MS_2FeCl_2]^{2-}$ anion (M = Mo, W).

BIS(TETRAPHENYLPHOSPHONIUM) HEXACHLOROHEXAKIS(μ_3-SULPHIDO)HEXAFERRATE
$C_{48}H_{40}Cl_6Fe_6P_2S_6$ $(Ph_4P)_2(Fe_6S_6Cl_6)$

D. COUCOUVANIS, M.G. KANATZIDIS, W.R. DUNHAM and W.R. HAGEN, 1984. J. Am. Chem. Soc., 106, 7998-7999.

Triclinic, $P\bar{1}$, a = 13.454, b = 13.528, c = 19.211 Å, α = 74.31, β = 70.27, γ = 61.51°, D_m = 1.62, Z = 2. Mo radiation, R = 0.06 for 5467 reflexions.

The anion has a hexagonal prismatic Fe_6S_6 core system (Fig. 1) with Fe-S 2.254-2.268, Fe-Cl 2.189 Å. For details of a similar anion see (1).

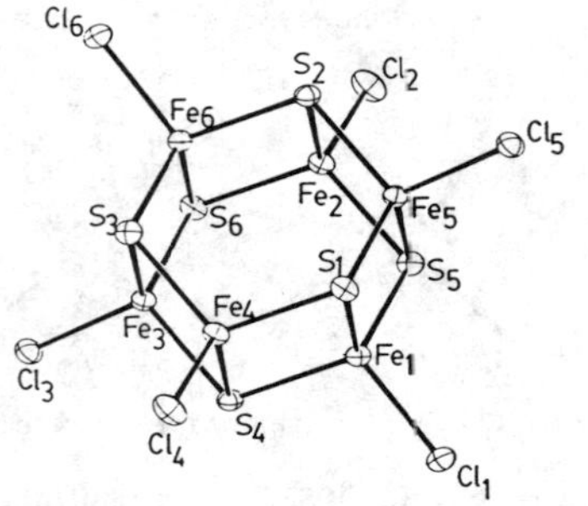

Fig. 1. A view of the $[Fe_6S_6Cl_6]^{2-}$ ion.

1. M.G. Kanatzidis, W.R. Dunham, W.R. Hagen and D. Coucouvanis, 1984. J. Chem. Soc. Chem. Commun., 356-358.

BIS(TETRAPHENYLPHOSPHONIUM) BIS(μ_2-CHLORO)OCTACHLORODIMOLYBDENUM DICHLOROMETHANE SOLVATE

$C_{48}H_{40}Cl_{10}Mo_2P_2,CH_2Cl_2$ $2[Ph_4P].[Mo_2Cl_{10}],CH_2Cl_2$

E. HEY, F. WELLER and K. DEHNICKE, 1984. Z. Anorg. Chem., 508, 86-92.

Triclinic, P$\bar{1}$, a = 11.223, b = 11.367, c = 12.838 Å, α = 63.40, β = 69.08, γ = 86.47°, Z = 2. Mo radiation, R = 0.045 for 2959 reflexions.

The $Mo_2Cl_{10}^{2-}$ anion lies about an inversion centre (Fig. 1) with Mo-Cl (bridge) 2.501 and 2.522(1), Mo-Cl (terminal) 2.307-2.326(2) Å. The Mo...Mo separation is 3.797 Å.

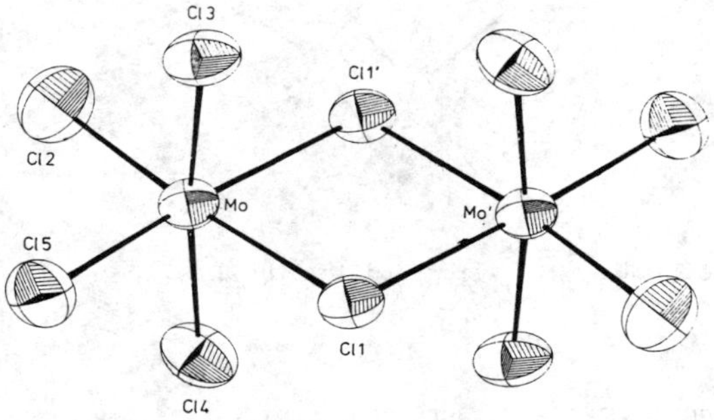

Fig. 1. A view of $[Mo_2Cl_{10}]^{2-}$.

BIS(TETRAPHENYLPHOSPHONIUM) (μ_4-PENTASULPHIDO)TRIS(μ_3-TETRASULPHIDO)HEXACOPPER

$C_{48}H_{40}Cu_6P_2S_{17}$ $2(C_{24}H_{20}P^+),Cu_6S_{17}$

A. MULLER, M. ROMER, H. BOGGE, E. KRICKEMEYER and D. BERGMANN, 1984. Z. Anorg. Chem., 511, 84-88.

Monoclinic, C2/c, a = 20.270, b = 11.605, c = 26.331 Å, β = 109.68°, Z = 4. Mo radiation, R = 0.063 for 3562 reflexions.

The structure of the $[Cu_6S_{17}]^{2-}$ ion is shown in Fig. 1. The anion has twofold crystallographic symmetry.

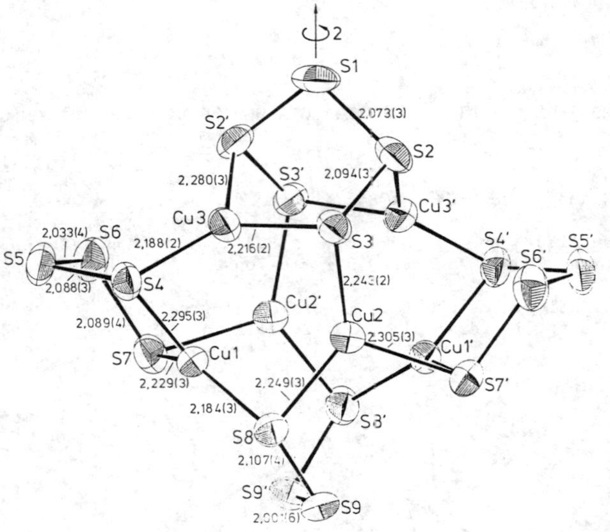

Fig. 1. A view of $[Cu_6S_{17}]^{2-}$ with bond lengths (Å).

BIS(TETRAPHENYLPHOSPHONIUM) BIS((μ_2-SELENIDO)PENTASELENIDOIRON)
$C_{48}H_{40}Fe_2P_2Se_{12}$ $((C_6H_5)_4P^+)_2[Fe_2Se_2(Se_5)_2]^{2-}$

H. STRASDEIT, B. KREBS and G. HENKEL, 1984. Inorg. Chim. Acta, 89, L11-L13.

Monoclinic, C2/c, a = 21.451, b = 16.076, c = 16.244 Å, β = 103.63°, Z = 4. Mo radiation, R = 0.072 for 3617 reflexions (at -133°C).

The anion possesses crystallographic inversion symmetry (Fig. 1) which means the central $Fe_2(\mu_2\text{-}Se)_2$ moiety is exactly planar. The Fe...Fe distance in this moiety is 2.787 Å and the angles Fe-Se(1)-Fe' and Se(1)-Fe-Se(1') are 73.73 and 106.27°. The Fe atoms are tetrahedrally-coordinated to the two bridging Se atoms (Fe-Se(1) 2.329, Fe-Se(1') 2.317 Å) and the Se atoms Se(2) (2.428 Å) and Se(6) (2.421 Å) of the chelate ligand, so the anion can be described as a spiro-tricyclic system composed of a four-membered cycle and two six-membered $FeSe_5$ chelate rings. The $FeSe_5$ rings exist in the chair conformation and the average Se-Se distance is 2.337 Å with an average of 102.4° for the Se-Se-Se angles.

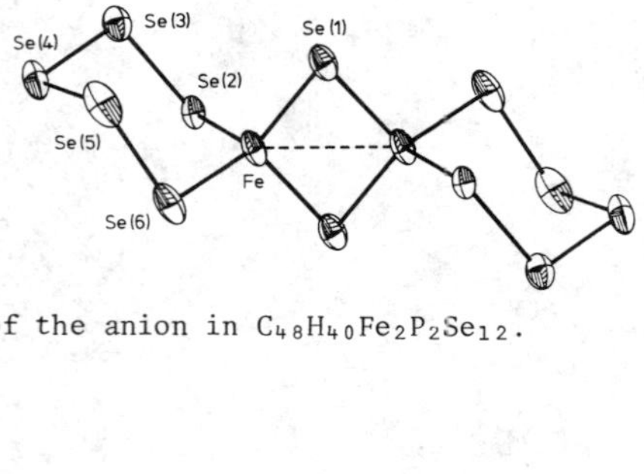

Fig. 1. Structure of the anion in $C_{48}H_{40}Fe_2P_2Se_{12}$.

BIS(TETRAPHENYLPHOSPHONIUM) BIS(μ-SULPHIDO)BIS(η^2-DISULPHIDO)DISULPHIDODIMOLYBDENUM
$C_{48}H_{40}Mo_2P_2S_8$ $[(C_6H_5)_4P]_2[Mo_2S_8]$

W.-H. PAN, M.A. HARMER, T.R. HALBERT and E.I. STIEFEL, 1984. J. Am. Chem. Soc., 106, 459-460.

Triclinic, P$\bar{1}$, a = 13.277, b = 18.933, c = 10.959 Å, α = 100.28, β = 110.39, γ = 104.23°, Z = 2. Mo radiation, R = 0.043 for 8126 reflexions.

The $Mo_2S_8^{2-}$ anion (Fig. 1) has a $Mo_2(S)_2(\mu\text{-}S)_2(S_2)_2^{2-}$ structure with distorted square-pyramidal geometry at Mo. Structural parameters: Mo-Mo 2.821(1), Mo=S 2.108(1) and 2.128(1), Mo-S (S_2) 2.375(2)-2.405(2), Mo-(μ-S) 2.306(1)-2.314(1), S-S 2.051(2) and 2.072(3) Å, Mo-S-Mo 75.3(1) and 75.2(1)°.

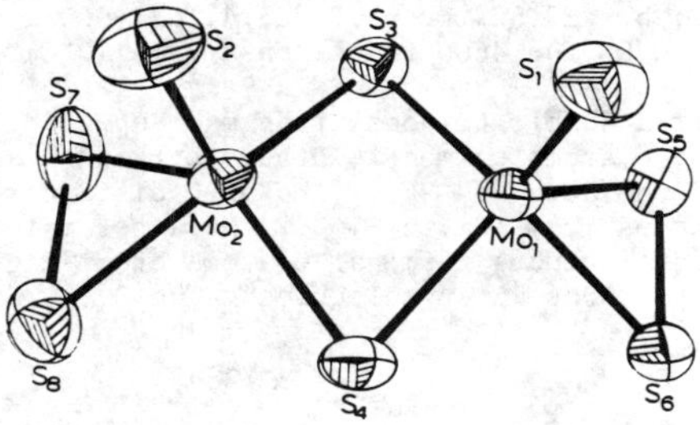

Fig. 1. The $Mo_2(S)_2(\mu\text{-}S)_2(S_2)_2^{2-}$ anion.

BIS(TETRAPHENYLPHOSPHONIUM) TETRATELLURIDE METHANOL SOLVATE
$C_{48}H_{40}P_2Te_4,2(CH_4O)$

J.C. HUFFMAN and R.C. HAUSHALTER, 1984. Z. Anorg. Chem., 518, 203-209.

Monoclinic, C2/c, a = 23.296, b = 14.720, c = 14.619 Å, β = 111.34°, Z = 4. Mo radiation, R = 0.0282 for 2587 reflexions (at -169°C).

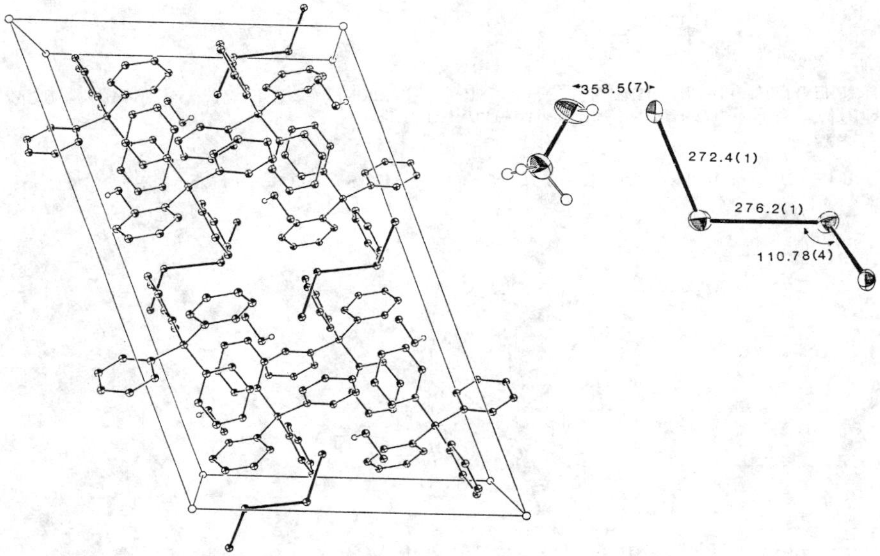

Fig. 1. The unit cell contents of $(PPh_4)_2Te_4,2(CH_3OH)$ and some dimensions.

The structure (Fig. 1) consists of isolated PPh_4^+ cations and methanol solvated Te_4^{2-} anions. The Te_4^{2-} anion lies about a crystallographic twofold axis with Te-Te 2.762(1) (central) and 2.724(1) Å (terminal), Te-Te-Te 110.78(4)° and torsion angle Te-Te-Te-Te 99.85(12)°. The O...Te distance is 3.585(7) Å.

HEXAKISTETRAETHYLAMMONIUM HEXA(OXOSILASESQUIOXANE) HYDRATE

$C_{48}H_{120}N_6O_{15}Si_6,36.5H_2O$ $6[N(C_2H_5)_4^+]Si_6O_{15}^{6-},36.5H_2O$

Yu.I. SMOLIN, Yu.F. SHEPELEV, A.S. ERSHOV, D. HOEBBEL and W. WIEKER, 1984. Kristallografiya, 29, 712-721 [Sov. Phys. Crystallogr., 29, 421-416].

Triclinic, P1, a = 13.907, b = 22.524, c = 17.235 Å, α = 96.12, β = 99.89, γ = 94.66°, Z = 2. Mo radiation, R = 0.08 for 4600 reflexions (at -110°C).

The silicate anion is a double trigonal ring made up of Si-O tetrahedra (Fig. 1). Each tetrahedron has three O atoms common to other tetrahedra and one "free" O atom, so that the Si:O ratio in the anion is 2:5. The Si-O distances range from 1.55 to 1.67 Å and the Si-O-Si angles from 128 to 146°. The water molecules form a three-dimensional framework which includes terminal O atoms of Si-O groups each of which is linked by hydrogen bonds to three water molecules. The O...O distances range from 2.46 to 3.40 Å.

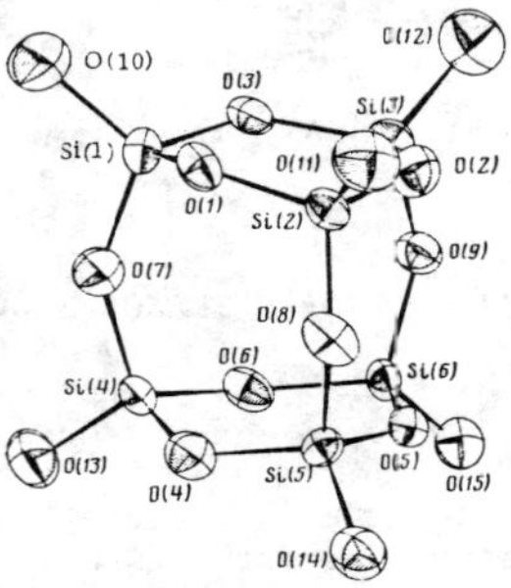

Fig. 1. The $[Si_6O_{15}]^{6-}$ anion.

BIS(TRIPHENYLPHOSPHINE)IMINIUM 1,2;1,3-BIS(μ_2-CARBONYL)-1,1,2,2,2,3,3,3,4,4,4-UNDECA-CARBONYL-2,4-(μ_2-HYDRIDO)-tetrahedro-IRONTRIOSMATE

$C_{49}H_{31}FeNO_{13}Os_3P_2$

D.T. EADIE, H.D. HOLDEN, B.F.G. JOHNSON and J. LEWIS, 1984. J. Chem. Soc. Dalton, 301-304.

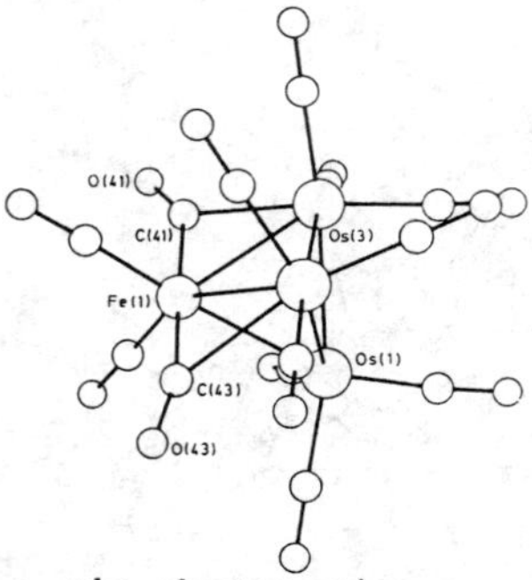

Fig. 1. $C_{49}H_{31}FeNO_{13}Os_3P_2$: the cluster anion.

Triclinic, $P\bar{1}$, a = 9.184, b = 15.223, c = 19.152 Å, α = 107.75, β = 102.81, γ = 93.67°, Z = 2. Mo radiation, R = 0.049 for 7013 reflexions.

In the mixed-metal cluster anion (Fig. 1) the four metal atoms define a tetrahedron with two carbonyl ligands semi-bridging between the iron atom and two of the osmium atoms, and the hydride lying along an Os-Os edge. The cation has the usual bent geometry with an angle P-N-P of 146.3(7)°.

BIS(TRIPHENYLPHOSPHINE)IMINIUM (μ_2-CARBONYL)(μ_2-HYDRIDO)NONACARBONYLTRIOSMIUMTETRACARBONYLTUNGSTEN

$C_{50}H_{31}NO_{14}Os_3P_2W$ $[(Ph_3P)_2N]^+[Os_3WH(CO)_{14}]^-$

M. FAJARDO, M.P. GÓMEZ-SAL, B.F.G. JOHNSON, J. LEWIS and P.R. RAITHBY, 1984. J. Organometal. Chem., 272, C57-C61.

Monoclinic, $P2_1/n$, a = 18.560, b = 15.923, c = 18.502 Å, β = 112.07°, Z = 4. Mo radiation, R = 0.057 for 6504 reflexions.

In the anion the metal atoms adopt a closo-tetrahedral geometry with the W atom coordinated to four terminal and one bridging carbonyl groups (Fig. 1). The metal-metal bond distances are: W(1)-Os(1) 2.932, W(1)-Os(2) 2.973, W(1)-Os(3) 2.970, Os(1)-Os(2) 2.779, Os(1)-Os(3) 2.790 and Os(2)-Os(3) 2.903 Å. The hydride was not located directly in the analysis but it is assumed that it bridges the long Os(2)-Os(3) edge. The CO groups adjacent to this edge bend away from it with an average cis Os-Os-C angle of 115° compared to a value of 94° for the other two Os-Os edges.

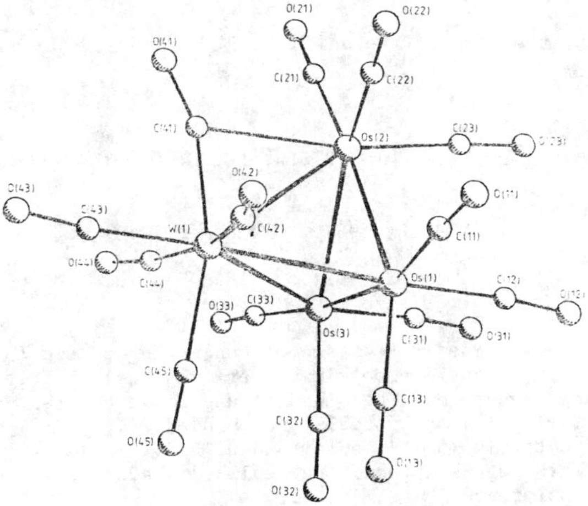

Fig. 1. The structure of the anion in $C_{50}H_{31}NO_{14}Os_3P_2W$.

BIS(TRIPHENYLPHOSPHINE)IMINIUM (BIS(HYDRIDOTRIOSMIUMDECACARBONYL))SILVER

$C_{56}H_{32}AgNO_{20}Os_6P_2$ $[((C_6H_5)_3P)_2N]^+[\{Os_3H(CO)_{10}\}_2Ag]^-$

M. FAJARDO, M.P. GÓMEZ-SAL, H.D. HOLDEN, B.F.G. JOHNSON, J. LEWIS, R.C.S. McQUEEN and P.R. RAITHBY, 1984. J. Organometal. Chem., 267, C25-C28.

Monoclinic, C2/c, a = 27.851, b = 18.218, c = 26.630 Å, β = 112.10°, Z = 8. Mo radiation, R = 0.061 for 5632 reflexions.

There are two independent anions, each lying on a centre of symmetry (Fig. 1). The Ag atoms are bonded to symmetry-related Os_3 isosceles triangles, bridging the short Os-Os bonds. Bond lengths in both anions are: Ag(1)-Os(1) 2.865, 2.874;

Ag(1)-Os(2) 2.852, 2.865; Os(1)-Os(2) 2.676, 2.666; Os(1)-Os(3) 2.830, 2.827 and Os(2)-Os(3) 2.827, 2.839 Å. The bond angle Os(1)-Ag(1)-Os(2) is 55.8° in one anion and 55.4° in the other, and the dihedral angles between the planes Ag(1),Os(1),Os(2) and Os(1),Os(2),Os(3) are 114.8 and 115.4° in each anion. Although the hydride ligands were not located directly, the CO groups adjacent to the short Os(1)-Os(2) edge bend away from it with an average cis-Os-Os-C(carbonyl) angle of 132°. This is consistent with the location of the hydride ligands along these edges.

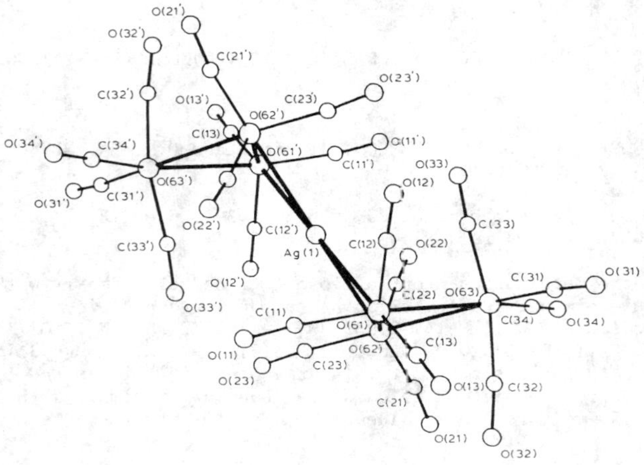

Fig. 1. Molecular structure of the anion in $C_{56}H_{32}AgNO_{20}Os_6P_2$.

BIS(TETRAPHENYLARSONIUM) (μ_4-DISULPHIDO)BIS(μ_2-SULPHIDO)TETRAKIS(TRICARBONYLIRON)

$C_{60}H_{40}As_2Fe_4O_{12}S_4$ $(Ph_4As^+)_2[Fe_4S_4(CO)_{12}]^-$

K.S. BOSE, E. SINN and B.A. AVERILL, 1984. Organometallics, 3, 1126-1128.

Monoclinic, $P2_1/n$, a = 10.662, b = 18.397, c = 16.245 Å, β = 107.66°, Z = 2. Mo radiation, R = 0.032 for 3523 reflexions.

The anion contains a "linear" Fe_4S_4 core, consisting of two $(\mu\text{-}S)_2Fe_2(CO)_6$ units linked by an S-S bond in a centrosymmetric butterfly structure (Fig. 1), rather than the anticipated cubane structure. The dihedral angle between the FeS_2 planes is 90.91°. The bonded S(1)-S(1') and formally non-bonded S(1)...S(2) distances (2.164 and 2.773 Å) are respectively ~0.1 Å longer and 0.10-0.15 Å shorter than expected for such interactions. This suggests that a delocalized bonding description involving all four S atoms may be approriate.

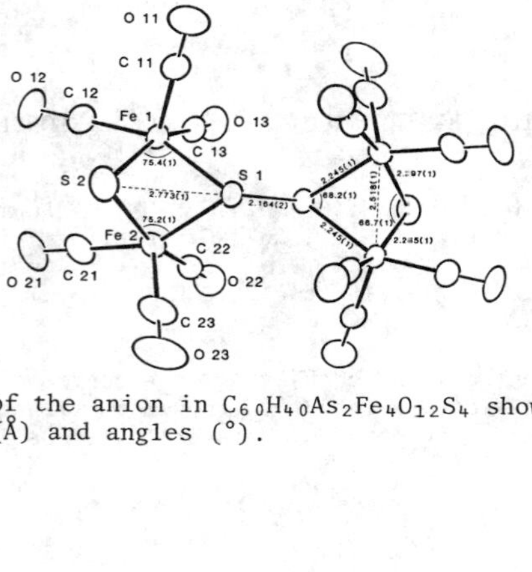

Fig. 1. Structure of the anion in $C_{60}H_{40}As_2Fe_4O_{12}S_4$ showing selected bond distances (Å) and angles (°).

BIS[BIS(TRIPHENYLPHOSPHINE)IMINIUM] TETRACYANOCOBALT DIMETHYLFORMAMIDE SOLVATE
$C_{76}H_{60}CoN_6P_2,4(C_3H_7NO)$ $2[(Ph_3P)_2N].[Co(CN)_4],4(C_3H_7NO)$

S.J. CARTER, B.M. FOXMAN and L.S. STUHL, 1984. J. Am. Chem. Soc., 106, 4265-4266.

Monoclinic, $P2_1/c$, a = 11.188, b = 12.423, c = 29.142 Å, β = 98.31°, D_m = 1.26, Z = 2. Mo radiation, R = 0.084 for 2025 reflexions.

The anion (Fig. 1) consists of crystallographically centrosymmetric square planar moiety associated with a pair of disordered DMF molecules. The Co-C distances are 1.869(15) and 1.875(14) Å.

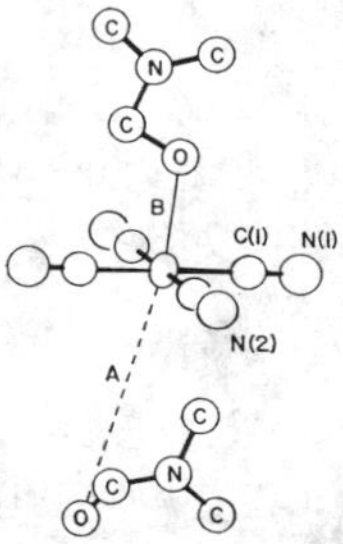

Fig. 1. The $Co(CN)_4^{2-}$ ion and its pair of associated DMF molecules. Distances A and B are 5.89 and 2.64 Å respectively.

BIS[BIS(TRIPHENYLPHOSPHINE)IMINIUM] BIS(μ-CARBONYL)DODECACARBONYLTRIIRONCHROMIUM
$C_{86}H_{60}CrFe_3N_2O_{14}P_4$ $[(Ph_3P)_2N]_2[Fe_3Cr(CO)_{14}]$

C.P. HORWITZ, E.M. HOLT and D.F. SHRIVER, 1984. Inorg. Chem., 23, 2491-2499.

Triclinic, $P\bar{1}$, a = 13.428, b = 17.734, c = 18.254 Å, α = 95.46, β = 105.74, γ = 69.82°, Z = 2. Mo radiation, R = 0.085 for 5074 reflexions.

The $[Fe_3Cr(CO)_{14}]^{2-}$ anion is shown in Fig. 1. The metal framework is pseudo-tetrahedral with Fe-Cr distances averaging 2.633(25) Å and Fe-Fe distances averaging 2.656(3) Å. The polygon formed by the carbonyl carbon atoms may be described as a hexacapped rectangular antiprism.

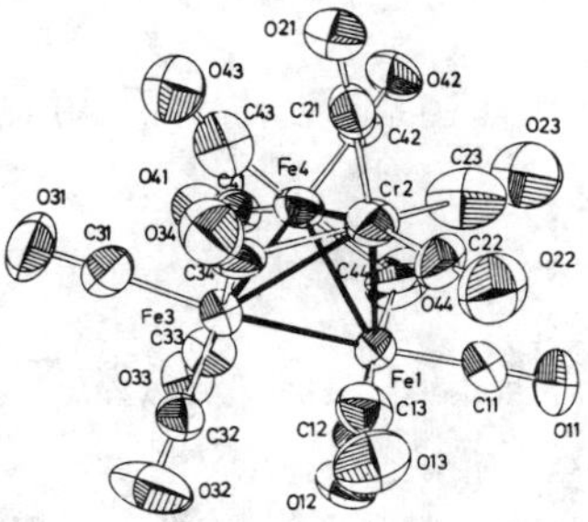

Fig. 1. The $[Fe_3Cr(CO)_{14}]^{2-}$ anion.

BIS(BIS(TRIPHENYLPHOSPHINE)IMINIUM) HEXAKIS(μ_2-CARBONYL)NONACARBONYLRUTHENIUMTETRA-RHODIUM
$C_{87}H_{60}N_2O_{15}P_4Rh_4Ru$ $2[(C_6H_5)_3P]_2N^+[RuRh_4(CO)_9(\mu-CO)_6]^{2-}$

A. FUMAGALLI and G. CIANI, 1984. J. Organometal. Chem., 272, 91-103.

Triclinic, $P\bar{1}$, a = 11.459, b = 14.171, c = 26.083 Å, α = 84.32, β = 83.46, γ = 82.81°, Z = 2. Mo radiation, R = 0.059 for 2849 reflexions.

The anion contains a trigonal bipyramidal metal atom cluster, elongated in the direction of the ideal threefold axis (Fig. 1). The Ru atom is in the apical position and the mean values of the $Rh_{eq.}$-$Rh_{eq.}$, $Rh_{eq.}$-$Rh_{ap.}$ and $Rh_{eq.}$-Ru bond lengths are 2.71, 3.03 and 2.99 Å respectively. There are nine terminal and six edge-bridging CO groups. A curious feature is that the two Ru-Rh carbonyl bridged edges are longer than the unbridged one (3.00 vs. 2.96 Å), while the Rh(1)-Rh(4) bridged edge is, as expected, shorter than the two Rh(2,3)-Rh(4) unbridged edges (2.93 vs. 3.08 Å.

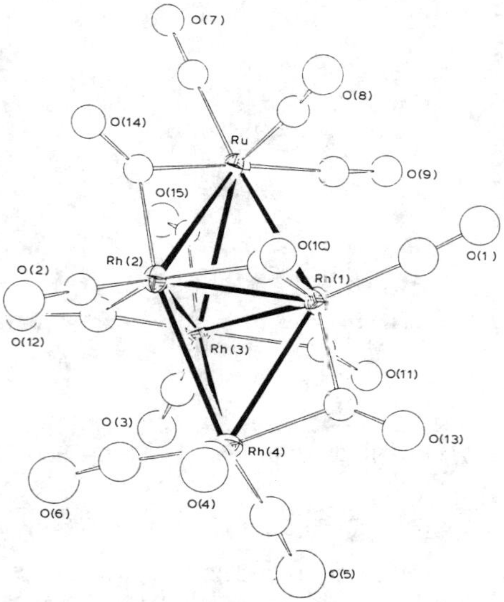

Fig. 1. Structure of the dianion in $C_{87}H_{60}N_2O_{15}P_4Rh_4Ru$.

ADDITIONAL PAPERS

The following Reports were prepared too late for inclusion in the main text.

p-TOLUQUINONE-MONOGUANYLHYDRAZONE
$C_8H_{10}N_4O$

G. RECK, R.-G. KRETSCHMER and W. SCHULZE, 1984. Crystal Res. & Technol., 19, 213-217.

Orthorhombic, $Pna2_1$, a = 7.589(3), b = 20.117(6), c = 5.699(2) Å, Z = 4. Mo radiation, R = 0.05 for 584 reflexions.

The bond distances and angles from this refinement indicate wide delocalization of π-electrons within the molecule (Fig. 1). This explains the coplanarity between the guanyl-hydrazone group and the quinone ring (largest deviation from the ring plane is 0.09 Å). There is an unusual hydrogen bridge system in the crystal structures. The carbonyl atom O(1) is a proton acceptor of three symmetry independent intermolecular bridges, namely N(11')-H(112')...O(1) with N...O 2.910 Å; N(11'')-H(111'')...O(1) with N...O 2.966 Å, and N(10')-H(101')...O(1) with N...O 2.974 Å. Atoms H(112') and H(111'') deviate 0.25 and 0.54 Å respectively from the plane O(1),C(1),C(2),C(6) whereas the distance of H(101') from the plane is -1.62 Å.

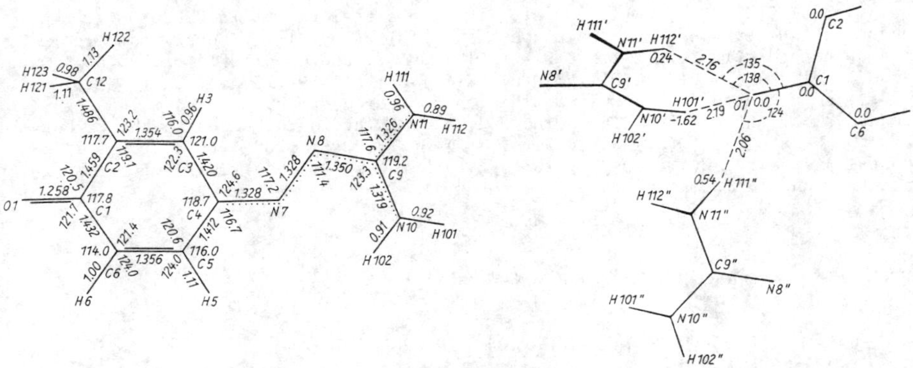

Fig. 1. Bond distances and angles in $C_8H_{10}N_4O$ (esds for non-H atoms range from 0.005 to 0.007 Å and from 0.4 to 0.5°) and the hydrogen bonds in the crystal structure.

PROCAINE

$C_{13}H_{20}N_2O_2$ $H_2NC_6H_4C(O)OCH_2CH_2N(C_2H_5)_2$

B.K. SINHA, E. SUBRAMANIAN, V. PATTABHI and J. TROTTER, 1984. Indian J. Pure & Appl. Phys., 22, 249-251.

Monoclinic, $P2_1/c$, a = 8.204, b = 14.370, c = 11.524 Å, β = 98.17°, D_m = 1.17, Z = 4. Mo radiation, R = 0.061 for 1357 reflexions.

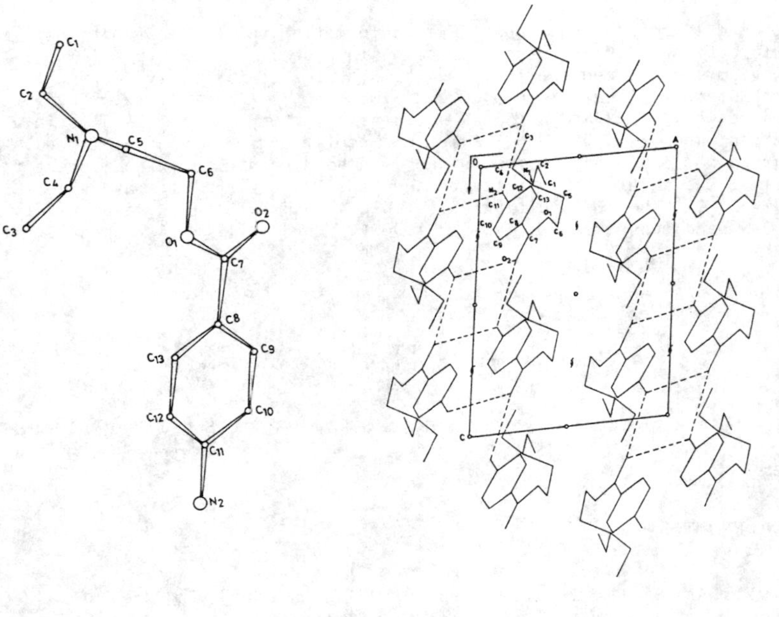

Fig. 1. Perspective view of the molecule (left), and projection of the structure down the b axis showing hydrogen bonds in broken lines.

The p-aminophenyl group is planar to within 0.004(4) Å and the ester group consisting of C(8), C(7), O(1) and C(6) is also planar to within 0.020(3) Å; the dihedral angle between these two planes is 1.5(3)° (Fig. 1). The six-membered ring has a quinonoid structure, with C(9)-C(10) 1.361(7) and C(12)-C(13) 1.358(6) Å, and C(8)-C(13) 1.404(6), C(8)-C(9) 1.390(6), C(10)-C(11) 1.396(7), and C(11)-C(12) 1.407(6) Å in length. The crystal exhibits a layered structure with molecules extended along the b axis. The amino N(2) atom is hydrogen-bonded to the carbonyl O(2) atom with N(2)...O(2') 3.110(5) Å, N(2)-H(2)...O(2') 160(4)°; N(2)...O(2'') 3.065(5) Å and N(2)-H(1)...O(2'') 163(4)°. The C(7)-O(1)-C(6)-C(5)-N(1) group is in a trans-gauche conformation.

7-BENZYL-3-THIA-7-AZABICYCLO[3.3.1]NONAN-9-ONE
$C_{14}H_{17}NOS$ (I)

7-BENZYL-3-THIA-7-AZABICYCLO[3.3.1]NONANE HYDROPERCHLORATE
$C_{14}H_{21}ClNO_4S$ (II)

7-BENZYL-9-PHENYL-3-THIA-7-AZABICYCLO[3.3.1]NONAN-9-OL HYDROPERCHLORATE
$C_{20}H_{24}ClNO_5S$

B.R. BAILEY III, K.D. BERLIN, E.M. HOLT, B.J. SCHERLAG, R. LAZZARA, J. BRACHMANN, D. VAN DER HELM, D.R. POWELL, N.S. PANTALEO and P.C. RUENITZ, 1984. J. Med. Chem., 27, 758-767.

I. Triclinic, P$\bar{1}$, a = 11.105(7), b = 12.033(6), c = 9.899(4) Å, α = 104.64(4), β = 85.40(5), γ = 92.39(4)°, Z = 4. Mo radiation, R = 0.076 for 3317 reflexions.

II. Orthorhombic, $P2_12_12_1$, a = 6.604(3), b = 14.742(9), c = 16.177(5) Å, Z = 4. Mo radiation, R = 0.037 for 1764 reflexions.

III. Orthorhombic, $Pna2_1$, a = 10.046(3), b = 14.858(6), c = 13.084(5) Å, Z = 4. Mo radiation, R = 0.033 for 1621 reflexions.

Compound I is in a chair-boat (CB) conformation, with the S-containing ring in a boat conformation and the N-containing ring in a chair conformation (Fig. 1). The fused ring portion of the two independent molecules is virtually identical. The hydroperchlorate (II) exists in a CC conformation. An intramolecular hydrogen bond is formed between the N and S (S...N 3.038(4) Å). Compound III is also in a CC conformation, and a similar N-H...S intramolecular hydrogen bond is formed. A phenyl and OH group are bonded at C(9) with the C-Ph bond equatorial with respect to the thiane ring, and axial with respect to the piperidine ring. The average C-S bond lengths in I, II, and III are 1.811(5), 1.826(4) and 1.821(4) Å respectively, and average C-N bond lengths are 1.463(4), 1.509(4), and 1.511(5) Å respectively.

(I)
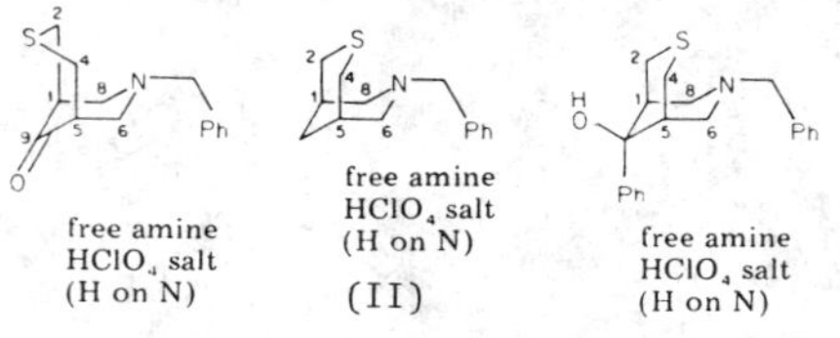

(II) (III)

Fig. 1. Schematic structures of $C_{14}H_{17}NOS$, $C_{14}H_{21}ClNO_4S$ and $C_{20}H_{24}ClNO_5S$.

BROMHEXINE HYDROCHLORIDE
$C_{14}H_{21}Br_2ClN_2$

C.H. KOO, Y.J. JUNG and S.W. LEE, 1984. Arch. Pharm. Res., 7, 115-120.

Orthorhombic, $Pca2_1$, a = 14.598(2), b = 12.461(3), c = 9.186(1) Å, D_m = 1.62, Z = 4. Cu radiation, R = 0.082 for 967 reflexions.

The cyclohexane ring has a normal chair conformation with C-C bond lengths ranging from 1.47 to 1.57 Å (esd 0.04-0.06 Å) and angles from 109 to 114° (esd 2.3-3.1°) (Fig. 1). The benzene ring is planar to within 0.02 Å. Atoms C(7) and Br(1) lie to within 0.06 Å of this plane but Br(2) and N(1) are displaced some 0.16 and 0.12 Å respectively from the plane. Bond lengths N(2)-C(7), N(2)-C(8) and N(2)-C(9) are respectively 1.48, 1.56 and 1.51 Å (esd 0.03-0.04) and the atoms around N(2) are in a distorted tetrahedral configuration. Other bond distances are: Br(1)-C(3) 1.87(3), Br(2)-C(5) 1.91(3) and N(1)-C(2) 1.37(4) Å. The molecules are linked by hydrogen bonds N(1)-H(1)...Cl(-1/2+x,1-y,z) with N...Cl 3.257 Å and N-H...Cl 156°. Two intra-molecular hydrogen bonds exist: N(1)-H(2)...Br(1) with N...Br 3.122 Å and N-$\hat{H}$...Br 128°, and N(2)-H(3)...Cl, with N...Cl 3.138 Å and N-$\hat{H}$...Cl 167°.

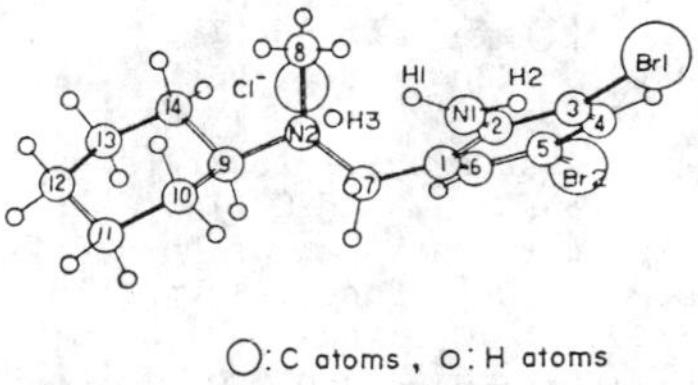

Fig. 1. Atomic numbering scheme for $C_{14}H_{21}Br_2ClN_2$.

η^5-CYCLOPENTADIENYL(CARBONYL)IRON-μ-CARBONYL-μ-THIOCARBONYL-η^5-CYCLOPENTADIENYLIRON-ISOPROPYLISONITRILE

$C_{17}H_{17}Fe_2NO_2S$ $(C_5H_5)_2Fe_2(CO)_2(CS)(CNC_3H_7)$

J.A.S. HOWELL, P. MATHUR, R. KUMAR, A.R. MANNING and F.S. STEPHENS, 1984. J. Organometal. Chem., 262, 227-241.

Orthorhombic, $P2_12_12_1$, a = 9.220, b = 10.126, c = 18.953 Å, Z = 4. Mo radiation, R = 0.071 for 904 reflexions.

The structure has the expected Fe(CO)(CS)Fe bridging system, with disorder between the bridging CO and CS groups, and with the cyclopentadienyl rings in a cis configuration (Fig. 1). Bond distances of note are: Fe...Fe 2.494, Fe(1)-C(BS) 1.90, Fe(1)-C(BO) 1.84, Fe(1)-C(T1) 1.81, Fe(1)-Cp(1) 1.74, Fe(2)-C(BS) 1.96, Fe(2)-C(BO) 1.90, Fe(2)-C(T2) 1.77, Fe(2)-Cp(2) 1.75, C(T1)-N 1.18, C(BS)-S 1.50 and C(BO)-O(B) 1.51 Å

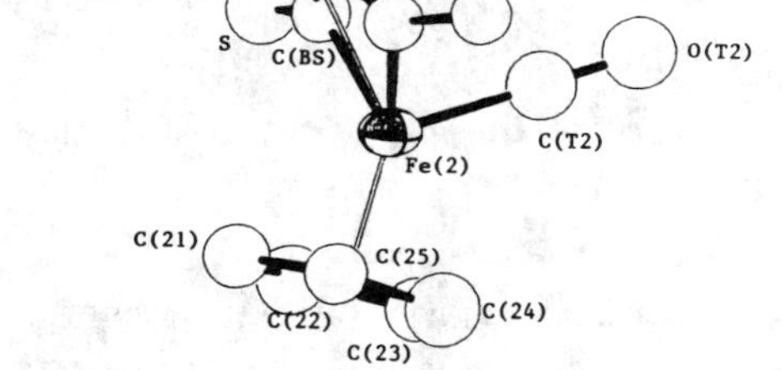

Fig. 1. Molecular structure of $C_{17}H_{17}Fe_2NO_2S$.

(RS)-1,2-DIPHENYLETHYL METHACRYLATE
$C_{18}H_{18}O_2$ (I)

(RS)-α-tert-BUTYLBENZYL METHACRYLATE
$C_{15}H_{20}O_2$ (II)

H. KAGEYAMA, K. MIKI, N. KASAI, Y. OKAMOTO, E. YASHIMA, K. HATADA and H. YUKI, 1984. Makromol. Chem., 185, 913-932.

I. Monoclinic, C2/c, a = 33.341(3), b = 5.916(1), c = 15.911(2) Å, β = 106.42(1)°, Z = 8. Cu radiation, R = 0.113 for 1100 reflexions.

II. Monoclinic, $P2_1/c$, a = 6.028(1), b = 31.009(3), c = 15.207(1) Å, β = 96.780(8)°, Z = 8. Cu radiation, R = 0.081 for 2206 reflexions.

Molecule I has an antiperiplanar-synperiplanar (ap-sp) structure (Fig. 1). The conformation of the C(2)=C(3) (1.33(3) Å) and C(1)=O(2) (1.19(2) Å) double bonds about the C(1)-C(2) bond is antiperiplanar, and that of the C(1)=O(2) double bond and the O(1)-C(5) ester bond about the C(1)-O(1) bond is synperiplanar. The methacryl residue formed by the C(1), C(2), C(3), C(4), O(1) and O(2) atoms is approximately planar. One of the two crystallographically-independent molecules (molecule 1) also has an ap-sp structure in which the ap conformation of the C(2)=C(3) (1.359(10) Å) and the C(1)=O(2) (1.193(7) Å) double bonds is observed about the C(1)-C(2) bond. Molecule 2 also has an ap-sp structure, and molecules 1 and 2 have differences only in the $CH_2=C(CH_3)$- moiety. Distance C(2')-C(4') (1.416(10) Å) in molecule 2 is not significantly shorter than the C(2')-C(3') distance (1.431(10) Å) but features around C(3') and C(4') suggest that an sp-sp structure coexists with the ap-ap structure. Bond distances of interest in I are: O(1)-C(1) 1.33(3), O(1)-C(5) 1.48(2), C(1)-C(2) 1.49(2), C(2)-C(4) 1.46(2), C(5)-C(6) 1.57(2), and C(5)-C(13) 1.51(2) Å. Corresponding (averaged) distances in II are: O(1)-C(1) 1.330, O(1)-C(5) 1.450, O(2)-C(1) 1.199, C(1)-C(2) 1.475, C(2)-C(3) 1.395, C(2)-C(4) 1.446, C(5)-C(6) 1.516, C(5)-C(12) 1.544 Å.

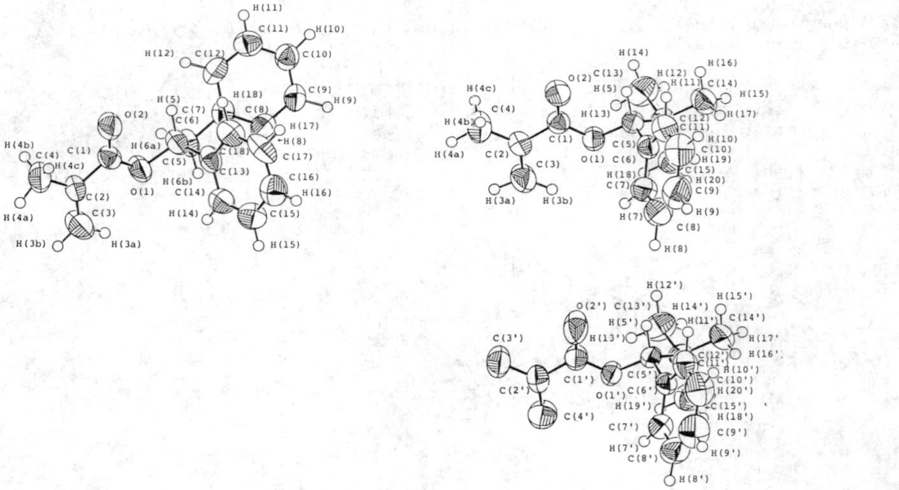

Fig. 1. Perspective views of $C_{18}H_{18}O_2$ (I) (left), and of the two $C_{15}H_{20}O_2$ (II) molecules.

trans-N-(CARBOXYMETHYL)-4-(β-1-NAPHTHYLVINYL)PYRIDINIUM BROMIDE
$C_{19}H_{16}BrNO_2$ (I)

cis-N-(2-AMINOETHYL)-4-(β-1-NAPHTHYLVINYL)-3-METHYLPYRIDINIUM BROMIDE HYDROBROMIDE
$C_{20}H_{22}Br_2N_2$ (II)

A.Y. CHWEH, J.F. De BERNARDIS, J.F. SIUDA, N.G. RONDAN, J.E. ABOLA and D.J. ABRAHAM, 1984. J. Med. Chem., 27, 825-830.

I. Monoclinic, $P2_1$, a = 7.961, b = 5.067, c = 20.452 Å, β = 95.0°, D_m = 1.458, Z = 2. Cu radiation, R = 0.038 for 1885 reflexions.

II. Monoclinic, $P2_1/c$, a = 17.732, b = 7.607, c = 15.694 Å, β = 113.6°, Z = 4. Cu radiation, R = 0.097 for 2855 reflexions.

One of the striking differences between the molecular structures of I and II is in their conformations (Fig. 1). In I, the pyridine ring and the vinyl group are coplanar with the naphthyl group twisted out of the plane by 12.6°. Since molecule II has a cis-configuration, steric interactions cause significant differences in planarity. The pyridine ring is rotated by 42.4° out of the plane of the double bond C(9)-C(10), and the naphthyl group is rotated out of this plane by 36.8°. The naphthyl and pyridinium rings are tilted at an angle of 62.5° to each other. Bond lengths of interest are: in I, C(5)-C(8) 1.427(9), C(8)-C(9) 1.331(11), C(9)-C(10) 1.442(10) Å, and in II: C(5)-C(9) 1.475(11), C(9)-C(10) 1.339(12), and C(10)-C(11) 1.493(12) Å. In I also, the proximity of the Br atom to O(1) indicates the presence of an O(1)-H...Br hydrogen bond with O(1)...Br 3.11(6) Å.

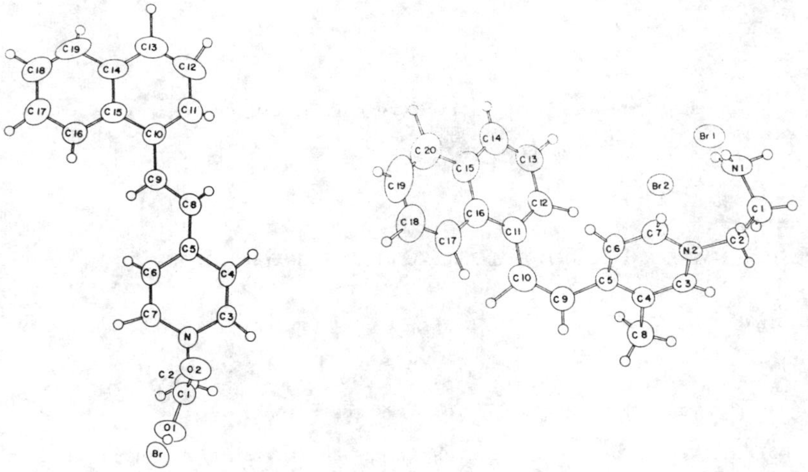

Fig. 1. Molecular structures of $C_{19}H_{16}BrNO_2$ (left), and $C_{20}H_{22}Br_2N_2$ (right).

BIS(DEOXYCHOLIC ACID) QUADRICYCLANE CLATHRATE
$2(C_{24}H_{40}O_4),C_7H_8$

V.M. COIRO, E. GIGLIO, F. MAZZA and N.V. PAVEL, 1984. J. Inclusion Phenomena, 1, 329-337.

Orthorhombic, $P2_12_12_1$, a = 27.150, b = 13.359, c = 14.161 Å, Z = 4. Mo radiation, R = 0.086 for 2079 reflexions.

The crystal packing (Fig. 1) is very similar to that found in the norbornadiene-deoxycholic acid complex (1) and is characterized by an assembly of antiparallel pleated bilayers, formed by deoxycholic acid molecules held together by O-H...O hydrogen bonds. Quadricyclane occupies approximately the same position as norbornadiene and has van der Waals interactions with methyl groups of deoxycholic acid.

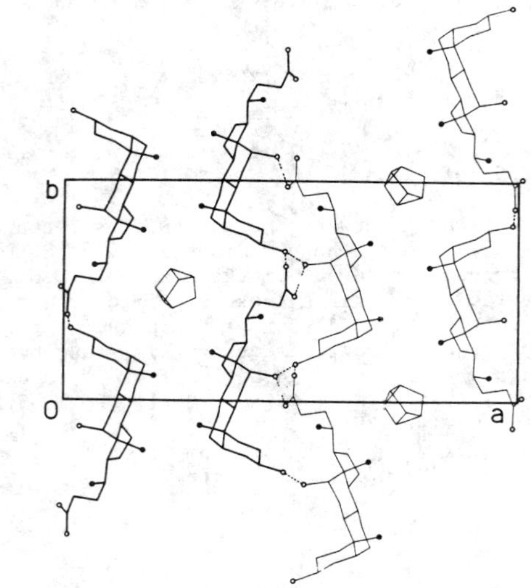

Fig. 1. A view of the bis(deoxycholic acid) quadricyclane chathrate.

1. Structure Reports, 48B, 769.

5,8-DIHYDRO-5,8-METHANO-1,4-NAPHTHALENEDIYL BIS(4-ETHOXYBENZOATE)

$C_{29}H_{26}O_6$ $(C_2H_5O\text{-}C_6H_4\text{-}COO)_2C_{11}H_8$

U. BAUMEISTER, H. HARTUNG and M. JASKÓLSKI, 1984. Crystal Res. & Technol., 19, 681-690.

Monoclinic, $P2_1/n$, a = 8.555(1), b = 15.280(2), c = 19.093(3) Å, β = 94.17(1)°, D_m = 1.21, Z = 4. Cu radiation, R = 0.062 for 2402 reflexions.

The molecule is almost fully stretched, and apart from the central benzonorbornadiene moiety, nearly centrosymmetric (Fig. 1). The norbornene skeleton, within the central part, is disordered and exists in two different conformations. The least-squares planes of the complete ethoxybenzoate groups make angles of 101.4 and 93.2° with the aromatic ring of the naphthalene moiety, and 12.4° with each other. All bond lengths and angles within the two ethoxybenzoate are in agreement with expected values, and the values within conformer A of the norbornene skeleton agree with those reported for related compounds. The molecular packing is characterized by two different lattice regions. The molecules in one region are parallel but their long axes are inclined by 49.5° to the molecules of the other region.

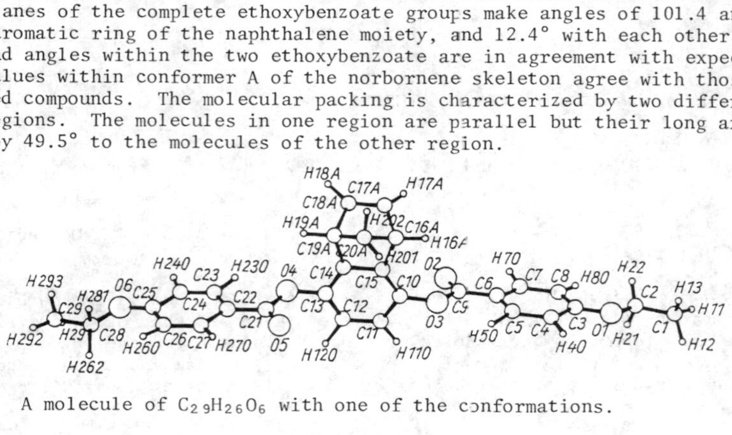

Fig. 1. A molecule of $C_{29}H_{26}O_6$ with one of the conformations.

(RS)-1,2,2,2-TETRAPHENYLETHYL METHACRYLATE BENZENE SOLVATE
$C_{30}H_{26}O_2,0.5C_6H_6$ (I)

(1S,2R)-(+)-1-BENZYL-3-DIMETHYLAMINO-2-METHYL-1-PHENYLPROPYL METHACRYLATE
$C_{23}H_{29}NO_2$ (II)

H. KAGEYAMA, Y. YANAGISAWA, N. KASAI, Y. OKAMOTO, E. YASHIMA, T. TAKEDA and K. HATADA, 1984. Makromol. Chem., 185, 933-955.

I. Triclinic, $P\bar{1}$, a = 9.271(2), b = 33.965(4), c = 9.167(2) Å, α = 90.26(2), β = 118.80(2), γ = 88.03(2)°, D_m = 1.22, Z = 4. Cu radiation, R = 0.054 for 5229 reflexions.

II. Monoclinic, $P2_1$, a = 12.482(1), b = 9.184(1), c = 9.246(1) Å, β = 93.49(1)°, D_m = 1.11, Z = 2. Cu radiation, R = 0.057 for 1387 reflexions.

Both of the crystallographically-independent molecules of (I) have the same structure (Fig. 1). Each molecule has a synperiplanar (sp) conformation with respect to the C=C and C=O double bonds about the C(1)-C(2) bond, and also an sp conformation with respect to the C=O and C-O bonds about the C(1)-O(1) bond. Molecule (II) has an antiperiplanar (ap) and an sp conformation about the C(1)-C(2) and C(1)-O(1) bonds respectively (Fig. 1). Average bond lengths of interest in (I) are: O(1)-C(1) 1.354, O(1)-C(5) 1.455, O(2)-C(1) 1.203, C(1)-C(2) 1.482, C(2)-C(3) 1.375, C(2)-C(4) 1.434, C(5)-C(6) 1.520, and C(5)-C(12) 1.579 Å. The equivalent bond lengths in (II) are: O(1)-C(1) 1.345(5), O(1)-C(5) 1.466(5), O(2)-C(1) 1.202(6), C(1)-C(2) 1.503(7), C(2)-C(3) 1.387(10), C(2)-C(4) 1.417(9), C(5)-C(6) 1.528(6), C(5)-C(12) 1.562(6), and C(5)-C(19) 1.563(6) Å.

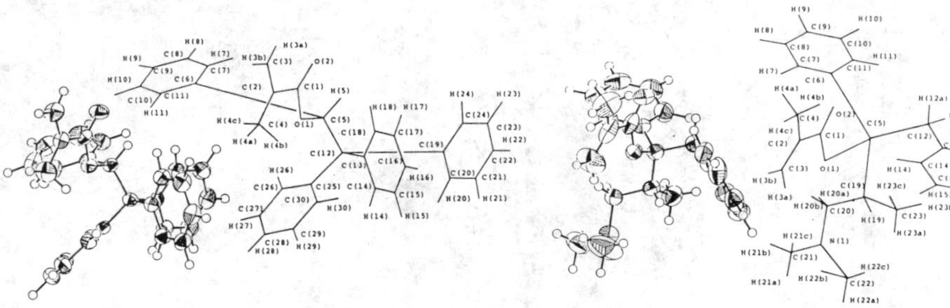

Fig. 1. Views of one of the independent molecules of $C_{30}H_{26}O_2$ (left) and of $C_{23}H_{29}NO_2$ (right).

6,6'-BIS(2-HYDROXY-5-METHYLBENZYL)-4,4'-DIMETHYL-2,2'-METHYLENEDIPHENOL
$C_{31}H_{32}O_4$ (I)

6-(2-HYDROXY-3,5-DIMETHYLBENZYL)-6'-(2-HYDROXY-5-METHYLBENZYL)-4,4'-DIMETHYL-2,2'-METHYLENEDIPHENOL TOLUENE SOLVATE
$C_{32}H_{34}O_4,0.5C_7H_8$ (II)

6-(3-t-BUTYL-2-HYDROXY-5-METHYLBENZYL)-6'-(2-HYDROXY-5-METHYLBENZYL)-4,4'-DIMETHYL-2,2'-METHYLENEDIPHENOL
$C_{35}H_{40}O_4$ (III)

E. PAULUS and V. BÖHMER, 1984. Makromol. Chem., 185, 1921-1935.

I. Triclinic, $P\bar{1}$, a = 9.931(1), b = 11.015(1), c = 12.922(2) Å, α = 85.37(1), β = 72.16(1), γ = 69.73(1)°, D_m = 1.23, Z = 2. Mo radiation, R = 0.103 for 4530 reflexions.

II. Triclinic, $P\bar{1}$, a = 9.734(1), b = 12.250(2), c = 13.757(3) Å, α = 75.73(1), β = 86.03(2), γ = 67.54(1)°, D_m = 1.19, Z = 2. Mo radiation, R = 0.149 for 4778 reflexions.

III. Triclinic, $P\bar{1}$, a = 9.290(1), b = 13.136(1), c = 14.712(1) Å, α = 63.53(1), β = 73.23(1), γ = 82.83(1)°, D_m = 1.13, Z = 2. Mo radiation, R = 0.118 for 4203 reflexions.

The conformation of the phenolic units is trans/trans(anti/anti) in I and cis/trans(syn/anti) in II and III (Fig. 1). As a result, in the crystal lattice, II and III form cyclic dimers via intermolecular head-to-tail hydrogen bonds, while the molecules of I are arranged in infinite chains by head-to-head and tail-to-tail hydrogen bonds (Fig. 1). Intramolecular hydrogen bonds between all OH groups of adjacent phenolic units are found for all molecules. They are directed to the opposite end of the molecule by the ortho-methyl or ortho-tert-butyl group in II and III, while no special orientation is found in I.

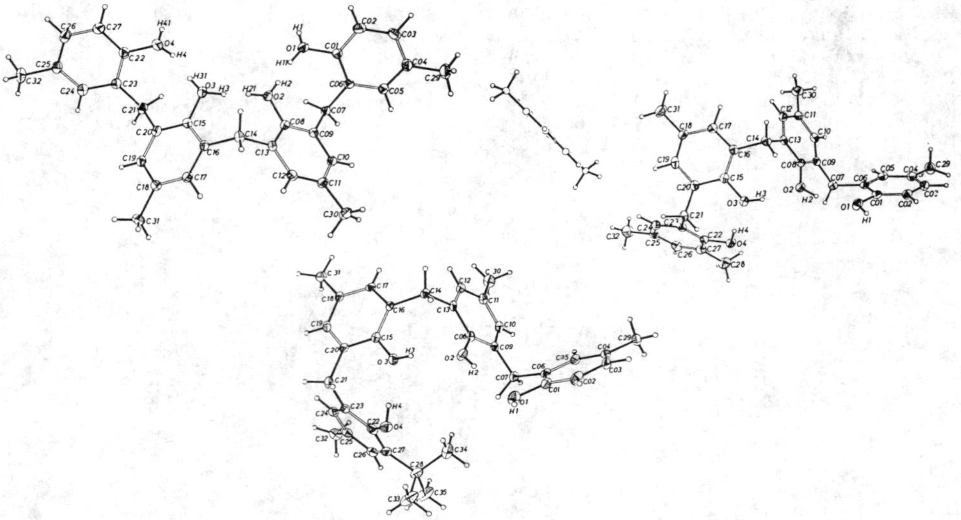

Fig. 1. Molecular structures of $C_{31}H_{32}O_4$ (I), $C_{32}H_{34}O_4$,$0.5C_7H_8$ (II), and $C_{35}H_{40}O_4$ (III) (clockwise from top left).

SUBJECT INDEX

This index contains the names of the substances printed at the heads of the reports. Leading Greek and numerical prefixes, and prefixes such as cis, trans, etc. are disregarded in fixing the letter under which the entry is listed, but such prefixes within a name are included in fixing the exact alphabetical position.

FORMULA INDEX

The entries are listed primarily according to the number of carbon atoms in the formula; within groups with the same number of carbon atoms the order is that of increasing number of hydrogen atoms, and other elements then follow alphabetically.

AUTHOR INDEX

Names beginning with a separated prefix are listed before single-word names beginning with the same letters; accents are omitted.